Tables for the calculation of friction in internal flows

Tables for the calculation of friction in internal flows

HR Wallingford and D.I.H. Barr

Thomas Telford, London

Published by Thomas Telford Services Ltd, Thomas Telford House,
1 Heron Quay, London E14 4JD, UK

Distributors for Thomas Telford books are
USA: American Society of Civil Engineers, Publications Sales
Department, 345 East 47th Street, New York, NY 10017-2398
Japan: Maruzen Co Ltd, Book Department, 3-10 Nihonbashi
2-chome, Chuo-ku, Tokyo 103
Australia: DA Information Services, 648 Whitehorse Road, Mitcham,
Victoria 3132

First published 1995

A catalogue record for this book is available from the British Library

ISBN: 0 7277 2046 5

© HR Wallingford and D. I. H. Barr, 1995

All rights, including translation, reserved. Except for fair copying, no
part of this publication may be reproduced, stored in a retrieval
system or transmitted in any form or by any means, electronic,
mechanical, photocopying or otherwise, without the prior written
permission of the Publications Manager, Publications Division,
Thomas Telford Services Ltd, Thomas Telford House, 1 Heron Quay,
London E14 4JD, UK

While all reasonable efforts have been made to ensure the accuracy
of the information given in these Tables, no warranty, express or
implied, is given by the publishers or by the authors

Set in Helvetica by D. I. H. Barr
Printed and bound in Great Britain by Redwood Books,
Trowbridge, Wiltshire

Helvetica™ is a trademark of Linotype AG and its subsidiaries in the UK and other countries

Preface

These tables are part of an ongoing effort to facilitate the use of the Colebrook-White equation. Specifically, this equation deals with uniform turbulent flow of incompressible fluid in circular tubes or pipes. It derives from the individual logarithmic equations for smooth and for rough turbulent flow. The composite equation can predict the resistance characteristics of flow in commercial pipes and conduits to a degree which is not matched by other approaches. The following sequence of events was motivated largely by these circumstances.

Two strands of development derived from the publication of P. Ackers' *Universal resistance diagram* in 1958 as part of the First edition of the Wallingford Charts. First there was the incremental value approach which was initiated by providing the individual charts for specific, progressive, values of roughness size. With each chart dealing with the flow of water at a medial temperature, there was the opportunity to provide for fully dimensional working of the basic design problems of uniform flow. The incremental approach was carried further in the Wallingford Tables, first published in 1963. These provided values of velocity and discharge for table points specified by a grid of values of diameter and gradient which both varied by logical increments.

The second strand of development centred on the reworking of the *Universal resistance diagram* as the *Generalised friction loss chart*. This was published in 1966 in the Engineering Sciences Data Unit's Item No 66027. Then the circumstances of concern required full cover of variation in viscosity. Although the development retained the advantages of explicit working of solutions which were inherent in the *Universal resistance diagram*, the solution routes, perforce, continued to be carried through non-dimensional groups of variables.

The possibility of an amalgamation of the two strands arose with the publication in 1994 of the Sixth edition of the Wallingford Tables in two volumes. Volume II contains an alternative route for table-aided solution of the Colebrook-White equation which is compressed in comparison with the established mode of full display of individual solutions. In Volume II the concern is still with the flow of water of medial temperature but the potential is apparent to cover a very wide range of viscosities, still with a logical cover of roughness sizes.

On the basis of this potential, the present volume of tables has been designed to meet a perceived need for a more generalised companion volume to the Wallingford Tables. Inevitably, to maintain a feasible overall size, there have been minor compromises in cover.

The authors acknowledge the contribution of Ronald Baron, Computer Officer, Department of Civil Engineering, University of Strathclyde, to the production of the various forms of table.

Users of these Tables are invited to provide comments or corrections, particularly on any additional conduit shapes which are in common use. The authors are grateful for various comments which have been received already. These have influenced the content of the 6th edition of the Wallingford Tables and that of this volume.

<u>Contents (continued)</u>

<u>**Contents (continued)**</u>

Contents (continued)

Tables E (continued)

<u>**Contents (continued)**</u>

Page

Tables F
Non-circular ducts - Tables of properties of unit sections

This volume of tables deals with the frictional resistance to the single-phase flow of incompressible Newtonian fluids in smooth and rough tubes and ducts. The cover extends to a wide range of transportation and process circumstances, within the foregoing limits. The volume is independent in use but is related to the Wallingford Tables[1], in particular to the 6th Edition of these Tables, as follows.

The Wallingford Tables

Since 1963, the Wallingford Tables[1] have provided a design aid for those concerned with the flow of water through pipes. The Wallingford Tables display directly the mean velocities and discharges which result from combinations of pipe diameters, piezometric gradients and roughness sizes. Throughout, the kinematic viscosity, v, used in those Tables is that for water at 15°C, i.e. $v = 1.141 \times 10^{-6} \, \mathrm{m^2 s^{-1}}$. The values of velocity and discharge are calculated from the Colebrook-White[2] equation with the incremental values of diameter chosen to match standard pipe diameters.

In 1994, the 6th edition of the Wallingford Tables[1] was published in two volumes. Volume II adopted an alternative tabular format for the solution of the Colebrook-White[2] equation. This alternative system requires significantly less space for a given coverage and so made feasible the preparation of these Tables covering a wide range of viscosities.

Scope of this volume of tables

The alternative system involves an array of solutions, for mean velocity and discharge, obtained from a simple resistance equation (Tables D). Then the ratios of these solutions to solutions of the Colebrook-White equation are given in Tables E, which cover a wide range of combinations of values of kinematic viscosity and surface roughness size. Where laminar flow solutions are applicable, the ratios derive from the Poiseuille[3] (or Hagen-Poiseuille) equation.

The range of diameters covered is from 5 to 2500 mm. There are considerably more table diameters in Tables D than would arise simply from a range of standard diameters. This gives greater continuity in increments of the flow variables which, in turn, aids solution of the inverse problems of determination of diameter or gradient. Also, the incremental system of friction gradients has been modified from that adopted in the Wallingford Tables, firstly, by reducing some of the larger increments and, secondly, by covering a greater range. For piezometric gradient, i.e. head loss per unit length of pipe in terms of the fluid flowing, the overall range is from 0.0001 to 30. This corresponds to a range of kinematic pressure gradient from 0.0098 to 294.2 $\mathrm{ms^{-2}}$.

These tables are oriented towards covering a wide range of viscosities. Tables E treat 39 incremental values of kinematic viscosity, from $0.125 \times 10^{-6} \, \mathrm{m^2 s^{-1}}$ to $2.00 \times 10^{-3} \, \mathrm{m^2 s^{-1}}$. Figure 1 overleaf shows these values and the scheme of combinations with roughness sizes, k_s, ranging from 0.0015 to 30 mm. As the value of viscosity is increased, the smaller roughness sizes are progressively omitted, with the effect of these considered to be sufficiently

Figure 1 : Combinations of kinematic viscosity, ν, and roughness size, k_s, in Tables E

R. size, k_s (mm) → Kin. visc., ν (m²s⁻¹) ↓	0.0015	0.003	0.006	0.015	0.030	0.060	0.150	0.30	0.60	1.50	3.0	6.0	15.0	30.0
2.00×10^{-3}	◇	◇	◇	◆	■	■	○	○	○	○	○	○	○	○
1.50×10^{-3}	◇	◇	◇	◆	■	■	○	○	○	○	○	○	○	○
1.00×10^{-3}	◇	◇	◇	◆	■	■	○	○	○	○	○	○	○	○
800×10^{-6}	◇	◇	◇	◆	■	■	○	○	○	○	○	○	○	○
600×10^{-6}	◇	◇	◇	◆	■	■	○	○	○	○	○	○	○	○
400×10^{-6}	◇	◇	◇	◆	■	■	○	○	○	○	○	○	○	○
300×10^{-6}	◇	◇	◇	◆	■	■	○	○	○	○	○	○	○	○
200×10^{-6}	◇	◇	◇	◆	■	■	○	○	○	○	○	○	○	○
150×10^{-6}	◇	◇	◇	◆	■	■	○	○	○	○	○	○	○	○
100×10^{-6}	◇	◇	◆	◆	■	■	■	■	○	○	○	○	○	○
80.0×10^{-6}	◇	◇	◆	◆	■	■	■	■	○	○	○	○	○	○
60.0×10^{-6}	◇	◇	◆	◆	■	■	■	■	○	○	○	○	○	○
50.0×10^{-6}	◇	◇	◆	◆	■	■	■	■	○	○	○	○	○	○
40.0×10^{-6}	◇	◇	◆	◆	■	■	■	■	○	○	○	○	○	○
30.0×10^{-6}	◇	◇	◆	◆	■	■	■	■	○	○	○	○	○	○
25.0×10^{-6}	◇	◇	◆	◆	■	■	■	■	○	○	○	○	○	○
20.0×10^{-6}	◇	◆	◆	◆	■	■	■	■	○	○	○	○	○	○
15.0×10^{-6}	◇	◆	◆	◆	■	■	■	■	○	○	○	○	○	○
12.5×10^{-6}	◇	◆	◆	◆	■	■	■	■	○	○	○	○	○	○
10.0×10^{-6}	◇	◆	◆	◆	■	■	■	■	■	■	○	○	○	○
8.00×10^{-6}	◇	◆	◆	◆	■	■	■	■	■	■	○	○	○	○
6.00×10^{-6}	◇	◆	◆	◆	■	■	■	■	■	■	○	○	○	○
5.00×10^{-6}	◇	◆	◆	◆	■	■	■	■	■	■	○	○	○	○
4.00×10^{-6}	◇	◆	◆	◆	■	■	■	■	■	■	○	○	○	○
3.00×10^{-6}	◆	◆	◆	◆	■	■	■	■	■	■	○	○	○	○
2.50×10^{-6}	◆	◆	◆	◆	◆	◆	◆	◆	■	■	○	○	○	○
2.00×10^{-6}	◆	◆	◆	◆	◆	◆	◆	◆	■	■	○	○	○	○
1.50×10^{-6}	◆	◆	◆	◆	◆	◆	◆	◆	■	■	■	■	○	○
1.25×10^{-6}	◆	◆	◆	◆	◆	◆	◆	◆	◆	◆	■	■	■	■
1.00×10^{-6}	◆	◆	◆	◆	◆	◆	◆	◆	◆	◆	■	■	■	■
0.80×10^{-6}	◆	◆	◆	◆	◆	◆	◆	◆	■	■	■	■	○	○
0.60×10^{-6}	◆	◆	◆	◆	◆	◆	◆	◆	■	■	■	■	○	○
0.50×10^{-6}	◆	◆	◆	◆	◆	◆	◆	◆	■	■	○	○	○	○
0.40×10^{-6}	◆	◆	◆	◆	◆	◆	◆	◆	■	■	○	○	○	○
0.30×10^{-6}	◆	◆	◆	◆	◆	◆	■	■	■	■	○	○	○	○
0.25×10^{-6}	◆	◆	◆	◆	■	■	■	■	○	○	○	○	○	○
0.20×10^{-6}	◆	◆	◆	◆	■	■	■	■	○	○	○	○	○	○
0.15×10^{-6}	◆	◆	◆	◆	■	■	■	■	○	○	○	○	○	○
0.125×10^{-6}	◆	◆	◆	◆	■	■	■	■	○	○	○	○	○	○
Kin. visc., ν (m²s⁻¹) R. size, k_s (mm)	0.0015	0.003	0.006	0.015	0.030	0.060	0.150	0.30	0.60	1.50	3.0	6.0	15.0	30.0

◆ For these combinations, the Tables E cover the full range of diameters from 6 mm to 2500 mm.

◇ These combinations are covered implicitly for the full range of diameters because results are virtually independent of roughness size up to the first such size available (i.e. indicated ◆) for the particular kinematic viscosity.

■ For these combinations, the Tables E cover the range of diameters from 125 mm to 2500 mm.

○ These potential combinations are not treated in Tables E.

approximated by the smallest size then tabulated. At higher values of viscosity, the combination with rougher surfaces is considered less likely to be required, and the corresponding roughness sizes are omitted also.

There are provided auxiliary tables (Tables F) to aid assessments of flows in non-circular conduits.

This volume of tables is independent of but fully compatible with both volumes of the 6th Edition of the Wallingford Tables[1]. Thus the tables here have the same identification letters, and Tables D have a colour coding for the units of discharge which matches that of the 6th Edition.

Review of hydraulic resistance

The Colebrook-White equation

In 1939, Colebrook[2] published his equation for turbulent flow in circular pipes flowing full. It derives from the smooth and rough turbulent logarithmic resistance laws for circular tubes. These had been evaluated experimentally by Nikuradse[4,5], after work by Prandtl and von Karman. The individual equations describing smooth and rough turbulent flow can be written as:

$$\frac{1}{\sqrt{\lambda}} = 2 \log \left\{ \frac{\boldsymbol{R}\sqrt{\lambda}}{2\cdot 51} \right\} \qquad \text{(Smooth turbulent flow)} \qquad (1)$$

$$\frac{1}{\sqrt{\lambda}} = 2 \log \left\{ \frac{3\cdot 71 D}{k_s} \right\} \qquad \text{(Rough turbulent flow)} \qquad (2)$$

The following equation provides a transition between the individual laws. It is known as the Colebrook-White equation to indicate the influence of C. M. White, Colebrook's collaborator and former research supervisor.

$$\frac{1}{\sqrt{\lambda}} = -2 \log \left\{ \frac{k_s}{3\cdot 71 D} + \frac{2\cdot 51}{\boldsymbol{R}\sqrt{\lambda}} \right\} \qquad (3)$$

Here :

$$\text{Friction factor, } \lambda = 2(Sg)D/V^2 = 1\cdot 2337\,(Sg)D^5/Q^2 \qquad (4)$$

$$\text{Reynolds number, } \boldsymbol{R} = VD/\nu = 4Q/\pi\nu D \qquad (5)$$

$$\text{Mean velocity, } V = Q/(\pi/4)D^2 \qquad (6)$$

where D, k_s, Q, (Sg) and ν are diameter, roughness size, discharge, product of piezometric gradient and acceleration due to gravity and kinematic viscosity respectively, with the latter given by dynamic viscosity, μ, divided by mass density, ρ. Also:

$$\text{Piezometric gradient, } S = h/l = \Delta p/\rho g l \qquad (7)$$

where h is the head difference which acts to overcome uniform flow friction over a length l, while Δp is the corresponding pressure difference. Then Eq. (4) may also be written as

$$\lambda = 2(\Delta p/\rho l)D/V^2 = 1\cdot 2337(\Delta p/\rho l)D^5/Q^2 \qquad (4a)$$

If Eqs (4), (5) and (6) are substituted in Eq. (3), one obtains

$$\frac{V}{\sqrt{(2SgD)}} = \frac{0 \cdot 9003\,Q}{\sqrt{(Sg)}\,D^{2 \cdot 5}} = -2\log\left\{\frac{k_s}{3 \cdot 71D} + \frac{1 \cdot 775\,v}{\sqrt{(Sg)}\,D^{1 \cdot 5}}\right\} \quad (8)$$

Following from Eqs (4) and (4a), substitution of $\Delta p/\rho l$ for Sg throughout Eq. (8) gives

$$\frac{V}{\sqrt{(2\Delta pD/\rho l)}}$$

$$= \frac{0 \cdot 9003\,Q}{\sqrt{(\Delta pD^5/\rho l)}} = -2\log\left\{\frac{k_s}{3 \cdot 71D} + \frac{1 \cdot 775\,v}{\sqrt{(\Delta pD^3/\rho l)}}\right\} \quad (8a)$$

Other form of the friction factor and of the Colebrook-White equation

Frequently the symbol f is used instead of λ or to denote $\lambda/4$. The former practice usually remains associated with the designation Darcy or Darcy-Weisbach friction factor. The latter is often designated the Fanning friction factor, $f_F = 2(\tau/\rho)/V^2$ or $f_F = \tau/(\rho V^2/2)$, where τ is the boundary shear stress and $\rho V^2/2$ is the dynamic pressure. In the mode of Eqs (4) and (4a), the latter definition corresponds to

$$\text{Friction factor, } f_F = 2(Sg)R/V^2 \quad = 2(Sg)D/4V^2 \quad (9)$$

$$= 2(\Delta p/\rho l)R/V^2 = 2(\Delta p/\rho l)D/4V^2 \quad (9a)$$

where R is hydraulic mean depth, or hydraulic radius, equal to $D/4$. Then Eq. (3) becomes

$$\frac{1}{\sqrt{f_F}} = -4\log\left\{\frac{k_s}{3 \cdot 71D} + \frac{1 \cdot 255}{R\sqrt{f_F}}\right\} \quad (10)$$

For the present purposes, we will adopt the form of the equations using λ (or f) rather than f_F and using D rather than R.

Eqs (8) and (8a) demonstrate that the Colebrook-White equation may be solved explicitly for mean velocity or for discharge. The main purpose of these tables is to provide accurate solutions both for the direct and the inverse problems.

Throughout both Tables D and Tables E, the corresponding values of kinematic pressure gradient $\Delta p/\rho l$ (numerically equal in value to Sg or $(h/l)g$) are given as an alternative to the values of piezometric gradient S.

The linear measure of surface roughness, k_s

In his experiments, Nikuradse[5] used pipes which were roughened by a uniformly graded sand applied to the semi-dry lacquered surface and then re-lacquered. The original diameter of those grains has provided a standard of comparison against which other surfaces may be evaluated. Thus the roughness of another surface is given as the size of sand which, when applied as a uniform coating, would give the same resistance under rough turbulent flow. Typical values for equivalent sand roughness size, k_s, are available for most of the surfaces likely to be found in practice, based on hydraulic testing. Sources of such data include Colebrook[2], Rouse[6], King[7], Lamont[8], Perkins and Gardiner[9] and ESDU International[10]. Many of these values are listed in Appendix 1.

The Colebrook-White equation is most appropriate where the roughness consists of separate protuberances of random height and spacing. This is termed commercial roughness. Some classes of pipe, not having this form of roughness, do not follow the same resistance function in the transition zone between smooth-turbulence and rough-turbulence.

The basis of Tables D and Tables E

Tables D generated by a Manning type equation

For SI (metre-second) units, the Manning[11] equation is

$$V = (1/n)\, R^{2/3}\, S^{1/2} \tag{11}$$

where n is the Manning coefficient and R is hydraulic mean depth (or hydraulic radius). The hydraulic mean depth as a unifying concept is more often associated with free surface flow. Its significance does not, however, differ in essential from that of the hydraulic or equivalent diameter, i.e. the diameter of the equivalent pipe D_{ep}. Both measures specify the ratio of area to wetted perimeter of a flow cross-section. Hydraulic mean depth is specifically this, while the diameter of the equivalent pipe is that of the circular section which has the same relation between area and wetted perimeter as has the cross-section involved. Consider a circular flow section of diameter D. Then $R = A/P = (\pi D^2/4)/\pi D = D/4$, where A and P are cross-sectional area and wetted perimeter respectively. Because D_{ep} must equal D, $D_{ep} = 4R$ and an alternative form of Eq. (11) is

$$mV = 39{\cdot}685 (D_{ep})^{2/3}\, S^{1/2} \tag{12}$$

where D_{ep} is the diameter of the equivalent pipe and m is defined as $100n$. Then with $Q = V \times (\pi/4)D^2$,

$$mQ = 31{\cdot}169 (D_{ep})^{8/3}\, S^{1/2} \tag{12a}$$

Tables D are generated by Eqs (12) and (12a) which apply only with metre-second units. The use of m rather than n aids greatly the tabular presentations and their use, this applying to both Tables D and the linked Tables E. However, Eqs (12) and (12a) continue to correspond fully with the basic Manning equation for rough turbulent flow (Eq. (11)) when m is taken as 100 times the Manning coefficient n. Thus Tables D do give Manning equation solutions directly, provided the user is confident with the value of m to be applied. For approximate evaluation of m for solutions for rough turbulent flows in terms of k_s as roughness size, the following relationship is adapted from that established by Williamson[12],

$$m_W = (k_s)^{1/6}/0{\cdot}832 \quad \text{(roughness size } k_s \text{ in mm)} \tag{13}$$

Values of m_W obtained in this way are shown in Table 1 overleaf.

Tables E generated by the Colebrook-White equation in conjunction with the Manning type equation

For solutions in the turbulent region of flow, Tables E list values of m_C obtained from

$$m_C = mV/V_C \tag{14}$$

where mV is given by Eq. (12) and the Colebrook-White velocity V_C is given by Eq. (8).

TABLE 1 : Values of m_W from Manning-Williamson equation

k_s (mm)	m_W	k_s (mm)	m_W	k_s (mm)	m_W	k_s (mm)	m_W
0·060	0·752	0·600	1·104	6·000	1·621	60·00	2·379
0·150	0·876	1·500	1·286	15·00	1·888	150·0	2·772
0·300	0·984	3·000	1·444	30·00	2·120	300·0	3·111

Note :- Values of m_W are given by Eq. (13) i.e. $M_W = (k_s)^{1/6}/0.832$ (roughness size, k_s, in mm). These values apply specifically to rough turbulent flows in circular pipes, or in conceptual circular cross-sections relating to equivalent pipes. Also note the presence of values of m_W, for smaller values of k_s, which may be less than can correspond to real flow conditions, as delineated by the values of m_C for smooth turbulent flow which are summarised in Table 2 below.

TABLE 2 : Values of m_C for smooth turbulent flow

Kinematic viscosity, ν, $0.30 \times 10^{-6}\,\mathrm{m^2 s^{-1}}$

Gradient	$(\Delta p/\rho l)$	Pipe diameters in mm		
S	$\mathrm{ms^{-2}}$	10	100	1000
0·00010	(0·00098)	-	0·934	0·939
0·00100	(0·00981)	0·850	0·810	0·850
0·01000	(0·09807)	0·751	0·715	0·777
0·10000	(0·98067)	0·636	0·640	0·715
1·00000	(9·80665)	0·552	0·579	0·662
10·0000	(98·0665)	0·487	0·529	0·616

Kinematic viscosity, ν, $1.00 \times 10^{-6}\,\mathrm{m^2 s^{-1}}$

Gradient	$(\Delta p/\rho l)$	Pipe diameters in mm		
S	$\mathrm{ms^{-2}}$	10	100	1000
0·00010	(0·00098)	-	1·111	1·055
0·00100	(0·00981)	-	0·940	0·944
0·01000	(0·09807)	-	0·815	0·854
0·10000	(0·98067)	0·757	0·719	0·719
1·00000	(9·80665)	0·641	0·643	0·717
10·0000	(98·0665)	0·555	0·582	0·664

Kinematic viscosity, ν, $3.00 \times 10^{-6}\,\mathrm{m^2 s^{-1}}$

Gradient	$(\Delta p/\rho l)$	Pipe diameters in mm		
S	$\mathrm{ms^{-2}}$	10	100	1000
0·00010	(0·00098)	-	-	1·189
0·00100	(0·00981)	-	1·102	1·049
0·01000	(0·09807)	-	0·934	0·939
0·10000	(0·98067)	-	0·810	0·850
1·00000	(9·80665)	0·751	0·715	0·777
10·0000	(98·0665)	0·636	0·640	0·715

Kinematic viscosity, ν, $10.0 \times 10^{-6}\,\mathrm{m^2 s^{-1}}$

Gradient	$(\Delta p/\rho l)$	Pipe diameters in mm		
S	$\mathrm{ms^{-2}}$	10	100	1000
0·00010	(0·00098)	-	-	1·380
0·00100	(0·00981)	-	-	1·196
0·01000	(0·09807)	-	1·111	1·055
0·10000	(0·98067)	-	0·940	0·944
1·00000	(9·80665)	-	0·815	0·854
10·0000	(98·0665)	0·757	0·719	0·780

Kinematic viscosity, ν, $30.0 \times 10^{-6}\,\mathrm{m^2 s^{-1}}$

Gradient	$(\Delta p/\rho l)$	Pipe diameters in mm		
S	$\mathrm{ms^{-2}}$	10	100	1000
0·00010	(0·00098)	-	-	1·618
0·00100	(0·00981)	-	-	1·370
0·01000	(0·09807)	-	-	1·189
0·10000	(0·98067)	-	1·102	1·049
1·00000	(9·80665)	-	0·937	0·939
10·0000	(98·0665)	-	0·810	0·850

Kinematic viscosity, ν, $100 \times 10^{-6}\,\mathrm{m^2 s^{-1}}$

Gradient	$(\Delta p/\rho l)$	Pipe diameters in mm		
S	$\mathrm{ms^{-2}}$	10	100	1000
0·00010	(0·00098)	-	-	-
0·00100	(0·00981)	-	-	1·631
0·01000	(0·09807)	-	-	1·380
0·10000	(0·98067)	-	-	1·196
1·00000	(9·80665)	-	1·111	1·055
10·0000	(98·0665)	-	0·940	0·944

Kinematic viscosity, ν, $300 \times 10^{-6}\,\mathrm{m^2 s^{-1}}$

Gradient	$(\Delta p/\rho l)$	Pipe diameters in mm		
S	$\mathrm{ms^{-2}}$	10	100	1000
0·01000	(0·09807)	-	-	1·618
0·10000	(0·98067)	-	-	1·370
1·00000	(9·80665)	-	-	1·189
10·0000	(98·0665)	-	1·102	1·049

Kinematic viscosity, ν, $1.00 \times 10^{-3}\,\mathrm{m^2 s^{-1}}$

Gradient	$(\Delta p/\rho l)$	Pipe diameters in mm		
S	$\mathrm{ms^{-2}}$	10	100	1000
0·01000	(0·09807)	-	-	-
0·10000	(0·98067)	-	-	1·631
1·00000	(9·80665)	-	-	1·380
10·0000	(98·0665)	-	-	1·196

Note :- Values of m_C derive from the smooth turbulent law; i.e. Eq. (1), in conjunction with Eq. (12).

It has long been realised that if n, and hence m, are treated as pure coefficients, Eq. (11), and hence Eq. (12), are not dimensionally homogeneous. There have been various suggestions to resolve this issue by allocating dimensions to n or to the constant 1 in Eq. (11) or to both. However, in using these Tables to obtain solutions in SI units, the sole function of m_C, as obtained from Tables E, is as a divisor of mV or of mQ, as obtained from Tables D. Thus it is convenient to ignore the issue of the units of m in these Tables and illustrative solutions and to treat m as non-dimensional, with mV and mQ then treated as having the dimensions and units of velocity and of discharge respectively.

Parallel use of the Poiseuille equation in Tables E

Where laminar flow applies in Tables E, m_C is replaced by m_P obtained from

$$m_P = mV/V_P \tag{15}$$

where the laminar flow mean velocity V_P is given by the Poiseiulle equation, i.e. $\lambda = 64/R$, arranged as

$$V_P = SgD^2/32\nu \qquad \text{or} \quad V_P = \Delta pD^2/32\rho l\nu \tag{16}$$

$$\text{hence} \qquad Q_P = \pi SgD^4/128\nu \qquad \text{or} \quad Q_P = \pi\Delta pD^4/128\rho l\nu \tag{16a}$$

In the second forms of Eqs (16) and (16a), l is the length of tube over which the pressure drop Δp occurs.

Proportioning exponents arising from the Manning equation

If, in using Tables D, interpolation is required in reaching mV or mQ from D and S, or *vice versa*, the full proportioning equations are given as Eqs (17)-(19). Given values, table values and required values are indicated by subscripts G, T and R respectively.

$$mQ_R = mQ_T \left[\frac{D_G}{D_T} \right]^{2\cdot667} \left[\frac{S_G}{S_T} \right]^{1/2\cdot00} \tag{17}$$

For mV_R, mQ_T is replaced by mV_T and the exponent 2·667 by 0·667

$$S_R = S_T \left[\frac{mQ_G}{mQ_T} \right]^{2\cdot00} \left[\frac{D_T}{D_G} \right]^{5\cdot333} \tag{18}$$

$$D_R = D_T \left[\frac{mQ_G}{mQ_T} \right]^{1/2\cdot667} \left[\frac{S_T}{S_G} \right]^{1/5\cdot333} \tag{19}$$

Throughout, values of S can be replaced consistently by corresponding values of $\Delta p/\rho l$.

Arrangement and use of Tables D and Tables E

Each of the ten Tables D treats a set of 13 diameters, ranging from 5 mm to 2500 mm. For each set of 13 diameters there is a sequence of gradients, each stated first as a piezometric gradient, S, and then as the corresponding kinetic pressure gradient, $\Delta p/\rho l$. The ranges of the sequences of gradients are varied to cover the likely applications. At each table point, as specified by a diameter and gradient combination, there are values of mV and mQ, derived from Eqs (12) and (12a).

In isolation, Tables D present an array of potential solutions; actual solutions where m happens to have the value of one. Further, departures of the value of m from one are small for a wide range of typical conditions of turbulent flow in smooth and moderately rough pipes. The user should neither overestimate nor underestimate the immediate convenience of the values in the array. On the one hand, one cannot determine the nature of the flow without recourse to Tables E. On the other hand, users will rapidly gain a feeling for approximate values of m in turbulent flows, aided by the information provided in Tables 1 and 2 (page 6) and in Appendix 2. Then approximate solutions can be obtained directly from Tables D.

Fig. 1 (page 2) shows the scheme of arrangement of Tables E, which by a mix of full and reduced tables aims to avoid repetitions and to concentrate on the more likely areas of analysis and design. Each Table E gives values of m_C, and m_P as appropriate, for one combination of values of kinematic viscosity, v, and roughness size, k_s. The full page examples of Tables E cover the whole spectrum of Tables D. Throughout Tables E, m_C varies slowly with variations of diameter and gradient, allowing increments between table points to be much greater than in Tables D. Specifically, the values of m_C are calculated to solve the Colebrook-White equation for Reynolds numbers above 2400, and the values of m_P are to solve the Poiseuille equation for Reynolds numbers below 2000; i.e. using Eqs (14) and (15) respectively. The tables show how the location of the laminar-turbulent transition changes with changing value of kinematic viscosity.

With laminar flow, roughness sizes do not affect the values of m_P in Tables E. For comparable increments of diameter and gradient, values of m_P vary much more rapidly than do values of m_C. Also, pipe size standards for small pipes are most often based on external diameters and wall thicknesses. Internal diameters for individual cases must then be measured or calculated and the number of possible internal diameters is too large for all to be included in Tables D. From Eq. (16), mean velocity in laminar flow varies with the square of the diameter and directly with the gradient measure. Hence discharge varies with the fourth power of the diameter and directly with the gradient, as in Eq. (16a). With this combination of circumstances it is convenient to have available Appendix 3. This gives the multiplying factor, F_P, which quantifies variation of discharge with change in diameter from standard values. Then one can obtain accurate laminar flow solutions by first calculating for a nearby table point in Tables E and then adjusting to the given diameter. This approach is illustrated in the following section.

The main sequencing of Tables E is by increasing value of kinematic viscosity, v, with the resulting sub-groups ordered by increasing value of roughness size, k_s.

Assessments for circular section tubes and pipes

Preliminary
Most design situations concern the relation between diameter, D, gradient, S, and discharge, Q. For turbulent flows, Tables D and Tables E, in combination, provide solutions of the Colebrook-White equation.

Use of the Tables D and E to find velocity, *V*, and/or discharge, *Q*, in turbulent flow

> *(i) Find the mean velocity and the discharge of an oil of kinematic viscosity $15 \cdot 0 \times 10^{-6}\, m^2 s^{-1}$ in a 600 mm diameter pipe, with roughness size 0·060 mm, under a piezometric gradient of 0·0020? (Numerical solution of the Colebrook-White equation gives $1 \cdot 0311\, ms^{-1}$ and $0 \cdot 2915\, m^3 s^{-1}$ respectively)*

From Table D5, read the values of mV and mQ for the stated combination of diameter and gradient, i.e. $1 \cdot 263\, \text{ms}^{-1}$ and 356·97 litres/sec. Table E213, for kinematic viscosity $v = 15 \cdot 0 \times 10^{-6}\, \text{m}^2\text{s}^{-1}$ and roughness size $k_s = 0 \cdot 060$ mm, gives, for a diameter of 600 mm and a gradient of 0·0020, the value of m_C as 1·224. This gives as solutions for mean velocity and discharge $1 \cdot 263/1 \cdot 224 = 1 \cdot 032\, \text{ms}^{-1}$ and $356 \cdot 97/1 \cdot 224 = 291 \cdot 64$ litres/sec, or $0 \cdot 2916\, \text{m}^3\text{s}^{-1}$, respectively.

The foregoing shows the inherent accuracy of the method but is atypical in that the given diameter and gradient are table values in both Tables D and Tables E. Typically, this will be true for turbulent flow conditions with Tables D, because of the closeness of the incremental values, but not for Tables E.

> *(ii) What is the discharge of hot water of kinematic viscosity $0 \cdot 4 \times 10^{-6}\, m^2 s^{-1}$ in a 110 mm diameter pipe, with roughness size 0·003 mm, under a piezometric gradient of 0·0725? (Numerical solution of the Colebrook-White equation gives 33·952 litres/sec).*

From Table D3, mQ, for the stated combination of diameter and gradient, is 23·313 litres/sec. From Table E44, for kinematic viscosity $v = 0 \cdot 40 \times 10^{-6}\, \text{m}^2\text{s}^{-1}$ and roughness size $k_s = 0 \cdot 003$ mm, estimate the value of m_C by visual interpolation as 0·688. Then the estimate of the discharge is $23 \cdot 313/0 \cdot 688 = 33 \cdot 885$ litres/sec. Using linear interpolation between the four adjoining table points in Table E44 gives $m_C = 0 \cdot 6866$ and hence an estimate of 33·954 litres/sec.

Solution for discharge, *Q*, in laminar flow

> *(iii) Estimate the discharge of an oil of kinematic viscosity $8 \cdot 0 \times 10^{-6}\, m^2 s^{-1}$ in a 4·2 mm diameter tube, under a piezometric gradient of 0·54. (Numerical solution of the Poiseuille equation gives 5·055 millilitres/sec.)*

The closest table point in Tables E is that for 6 mm diameter and 0·40 piezometric gradient. Table E184 for kinematic viscosity $8 \cdot 0 \times 10^{-6}\, \text{m}^2\text{s}^{-1}$ shows $m_P = 1 \cdot 502$ at this point and Table D1 shows $mQ = 23 \cdot 432$ millilitres/sec. Thus the discharge Q_T for this table point is $23 \cdot 432/1 \cdot 502 = 15 \cdot 601$ millilitres/sec. For laminar flow

$$Q_R \;=\; Q_T \times [S_G/S_T] \times F_P \qquad (20)$$

where the values of S can be replaced consistently by corresponding values of $\Delta p/\rho l$.

From Appendix 3 the value of F_P is 0·240. Hence the discharge for the specified conditions is

$$15 \cdot 601 \times [0 \cdot 54/0 \cdot 40] \times 0 \cdot 240 = 5 \cdot 0547 \text{ millilitres/sec.}$$

Preliminary estimates of discharge from Tables D

When calculating discharge using Tables E to support Tables D, solutions are obtained immediately which are well within the normal requirements for final accuracy. Tables E are individually specific to a combination of values of kinematic viscosity, v, and roughness size, k_s. Preliminary solutions for turbulent conditions may be obtained using Tables D only, for turbulent conditions, if one uses a value of m derived from past experience or from Appendix 2.

Solution for gradient in turbulent flow

If one wishes to determine the slope of the piezometric gradient then one has to use an iterative procedure to determine the appropriate value of m. In many cases $m = 1$ is an appropriate initial value. Alternatively, Appendix 2 may be used to guide the user in the selection of an appropriate initial value for m.

In the following examples the initial value of m is derived from that Table E which is involved in the the Colebrook-White solution.

> *(iv) Find the kinematic pressure gradient which is required to overcome pipe friction loss when 200 litres/sec of fluid of kinematic viscosity $0.125 \times 10^{-6}\, m^2 s^{-1}$ is delivered through a 225 mm diameter pipe with roughness size estimated as 0.003 mm. (Iterative numerical solution of the Colebrook-White equation gives $0.5222\, m^2 s^{-1}$)*

In the column for diameter 225 mm in Table D4, use an approximate value for mQ of 200 litres/sec (i.e. simply the required flow) to locate the general position of the solution in the column and to estimate the corresponding kinematic pressure gradient as approximately $1.15\, m^2 s^{-1}$. However, by applying this rough estimate of gradient and the stipulated diameter of 225m to Table E2, for kinematic viscosity $v = 0.125 \times 10^{-6}\, m^2 s^{-1}$ and roughness size 0.003 mm, it is made clear that the m_C value is approximately 0.665. Thus, with the product of this latter estimate and the discharge of 200 litres/sec giving $mQ \approx 133.0$, it is seen to be better (inspecting the column in Table D4) to use $0.510\, m^2 s^{-1}$ in entering Table E2 to interpolate m_C as 0.673. This gives $mQ = 134.6$. From Table D4 the kinematic pressure gradient interpolates as $0.5215\, m^2 s^{-1}$.

It is worth noting that even the immediate use of the rough estimate of $1.15\, m^2 s^{-1}$ in Table E2, i.e. accepting the m_C value as 0.665 and thus the mQ value as 133.0, leads to a solution which is just 2.3% less than the exact solution.

Turning to a case where m is clearly greater than one:

> *(v) Find the piezometric gradient which is required to overcome pipe friction loss when 380 litres/sec of kerosene of kinematic viscosity $2.0 \times 10^{-6}\, m^2 s^{-1}$ is delivered through a 450 mm diameter pipe of roughness size 1.50 mm. (Iterative numerical solution of the Colebrook-White equation gives 0.01758)*

Assuming flow to be turbulent, inspection of Table E136 shows that m_C will not be less than 1.285. Then the product of 380 and 1.285 is 488.3 and in Table D5 this gives $S \approx 0.0173$. Interpolating for m_C in Table E136 gives the value of m_C as 1.293 and hence the value of mQ as 491·3. Returning to Table D5, this gives S as 0·01757; 0·06% less than the exact solution.

Again note that even the initial approximation of $S=0.0173$ is only 1·6% less than the exact solution.

Solution for gradient in laminar flow

> *(vi) Estimate the kinematic pressure gradient necessary to pass 8·0 millilitres/sec of oil of kinematic viscosity $15.0 \times 10^{-6}\,m^2s^{-1}$ through a 17·0mm diameter tube. (Numerical solution of the Poiseuille equation gives $0.05853\,m^2s^{-1}$)*

In Tables E209-E211, for kinematic viscosity $\nu = 15.0 \times 10^{-6}\,m^2s^{-1}$, there are columns for diameter of 15mm. Thus the general location of the solution can be established by applying m_P values from this column to the corresponding mQ values from Table D1. Here table gradient $0.0981\,m^2s^{-1}$ is seen to be a suitable base because the corresponding values for mQ of 42·654 millilitres/sec and m_P of 5·251 give Q_T as 8·123 millilitres/sec for 15mm diameter. In Appendix 3, for the discharge adjustment for change of diameter from 15mm to 17mm, F_P is read as 1·650.

For laminar flow

$$S_R \;=\; \left\{\, S_T \times [Q_G/Q_T] \,\right\} / F_P \qquad (20a)$$

where values of S can be replaced consistently by corresponding values of $\Delta p/\rho l$.

Then, $\quad \left\{\, 0.0981 \times [8.0/8.123] \,\right\} / 1.650 \;=\; 0.05855\,m^2s^{-1}$

Note that with this route to a laminar flow solution, the result is not actually influenced by the table point chosen as base, but it is best practice to establish the zone of the solution as early as possible.

Solution for diameter in turbulent flow

> *(vii) Find the size of pipe to deliver $1.00\,m^3s^{-1}$ of oil of kinematic viscosity $150 \times 10^{-6}\,m^2s^{-1}$ under a piezometric gradient of 0·0075, assuming a roughness size of 0·060mm. (Iterative solution of the Colebrook-White equation gives 804·6mm)*

Assuming turbulent flow, Table E267 indicates $m_C \approx 1.40$. Then one seeks in Tables D a value of mQ of approximately $1.40\,m^3s^{-1}$ or approximately 1400 litres/sec, on the 0·0075 gradient line. This is found in Table D6, where 1367·9 and 1488·7 litres/sec appear in the columns for 775 and 800mm diameter respectively. Returning to Table E267, m_C is interpolated as 1·513, and this gives the adjusted value of mQ as 1513 litres/sec. Inspection of Table D6 shows that 800mm is the standard size closest to the solution. To obtain the actual Colebrook-White solution, substitute in Eq. (19):

$$D_R = 800 \times [1513/1488.7]^{1/2.667} = 804.9\,mm$$

Solution for diameter in laminar flow

> *(viii) Estimate the diameter of tube necessary to convey 30·0 millilitres/sec of oil of kinematic viscosity $30.0 \times 10^{-6}\,m^2s^{-1}$ under a piezometric gradient of 0·240. (Numerical solution of the Poiseuille equation gives 11·172mm)*

Inspection of Table E229 or E230 and Table D1 on the 0·20 piezometric gradient line suggests 10 mm as giving a suitable base point. For this point there are read a mQ value of 64·699 millilitres/sec and a m_P value of 4·032 in Tables D1 and E229 respectively. These give Q_T as 16·046 millilitres/sec. Proportioning directly (in laminar flow) for the greater gradient of 0·24 gives $16·046 \times [0·24/0·20] = 19·255$ millilitres/sec. With discharge proportional to the fourth power of the diameter in laminar flow (see Eq. (16a)), the required diameter is $10 \times [30/19·255]^{1/4} = 11·172$ mm.

Other routes towards solutions

For the above examples, the emphasis has been on obtaining solutions which compare in accuracy with numerical solutions. In turbulent flow, users may prefer to avoid the inverse solutions for gradient or diameter. Provided that the same indications are used for suitable location within the tables, as have been explained in the inverse solution examples, it may be found suitable to approach a solution by trial and error assessments of discharge, and then making comparisons with the given discharge. This is made easy by the gradual nature of the changes in value of m_C.

With laminar flow, m_P varies more rapidly. Here, as an alternative to the methods already detailed, preliminary estimates can be obtained using the full range of values in Tables D in conjunction with interpolation in Tables E.

Individual Tables E are for a given combination of values of kinematic viscosity and roughness size. Then these have a diagrammatic aspect, with the location of the laminar-turbulent transition shown against variation in diameter and gradient. Users must assess the circumstances that apply to the case in hand where solutions are within or are very close to this transition.

Interpolation between entries in Tables D

Interpolation should be used where intermediate values between the entries in Tables D are required. Linear interpolation may be sufficient in many cases, because the increments in diameter are small throughout Tables D, otherwise, Eqs (17) to (19) provide the exponents, as have been used where necessary in the above examples.

Turbulent flow with a non-tabulated roughness size, k_s

Consider example (v) as already treated where 380 litres/sec of kerosene of kinematic viscosity $2·0 \times 10^{-6}\,\mathrm{m^2 s^{-1}}$ was to be delivered through a 450 mm diameter pipe of roughness size 1·50 mm. The solution for S of 0·01757 in Table D5 was obtained with m_C interpolated in Table E136 as 1·293 and hence mQ as 491·3 litres/sec for use in Table D5.

Suppose an assessment is then required with the roughness size, k_s, decreased to 1·0 mm, which is not a tabulated value in these tables. Taking the nearest Tables E point as diameter 400 mm and piezometric gradient 0·020, the change in m_C values for a change in roughness size from 1·50 mm to 0·60 mm (Tables E136 and E135) is from 1·290 to 1·144, i.e. $-0·146$. Then proportioning this change simply by the increments of change in roughness sizes; $-(0·5/0·9) \times 0·146 = -0·081$; gives the change to apply to the original m_C value of 1·293. This gives the adjusted m_C value as 1·212 for the intermediate roughness size of 1·0 mm. For use in Table D5, the adjusted value for mQ is then $1·212\times380=460·6$ litres/sec giving 0·01546 as the adjusted solution for S. This is within 2% of the Colebrook-White numerical solution of 0·01577.

Where the roughness size is greater than the largest tabulated for a particular kinematic viscosity, Eq. (13) and Table 1 give approximate estimates for the value of m. If, in the immediately above example, the roughness size is increased to 15 mm, the approximation for m is 1·888. Then $1·888\times380=717·4$ litres/sec, giving in Table D5 solution for S of approximately 0·0375. This approximate solution is 3% less than the Colebrook-White numerical solution of 0·0387. In this case, use of the first equation of Fig. 3 gives $S = 0·0386$. The proximity of the latter result to the Colebrook-White solution demonstrates that, with the increase to 15 mm roughness size, the flow conditions are close to rough turbulent. However, both these approximate results indicate that care must be taken where solutions are required for combinations of kinematic viscosity and roughness sizes which it has not been feasible to include in Tables E. The equations given in Figs 2 and 3 overleaf may be used in such cases, and also in cases of intermediate roughness sizes, as an alternative to the above procedures.

Turbulent flows with intermediate values of kinematic viscosity, ν
The intention for these Tables is to provide a good coverage of the variation in kinematic viscosity by adopting small increments. If this is not sufficient, the same procedure can be followed as for intermediate roughness sizes. Using example (ii), consider the value of kinematic viscosity to be increased from $0·4\times10^{-6}\,m^2s^{-1}$ to $0·45\times10^{-6}\,m^2s^{-1}$, the roughness size remaining as 0·003 mm. Tables E44 and E54 for roughness size 0·003 mm and kinematic viscosities $0·40\times10^{-6}\,m^2s^{-1}$ and $0·50\times10^{-6}\,m^2s^{-1}$ respectively, show change of the value of m_C from 0·689 to 0·702 at the corresponding table points for a diameter of 100 mm and a gradient $S = 0·060$. Again, the proportioning is by the increments in values of viscosity. Thus the addition to be applied to the interpolated m_C value of 0·687 is $0·013\times(5/10) = 0·065$, giving 0·6935 as the adjusted value of m_C. The value of mQ is unchanged, hence $23·233/0·6935 = 33·501$ litres/sec, which is within 0·5% of the numerically obtained Colebrook-White solution of 33·657 litres/sec.

Laminar flows with intermediate values of kinematic viscosity ν
From Eq. (16a), discharge varies inversely with change in kinematic viscosity, and solutions can be so adjusted from those for tabulated viscosities.

Multiplying factors, on tabulated values of mQ, for standard but non-tabulated diameters in turbulent flows
Estimates may be required for discharges in a pipe of a standard diameter which is not included in Tables D. A list of multiplying factors is provided in Appendix 4. Values of mQ so obtained may then be operated on by the appropriate value of m_C from Tables E.

(ix) Estimate the discharge of a fluid of kinematic viscosity $1·50\times10^{-6}\,m^2s^{-1}$ in a 762 mm pipe of roughness size 0·060 mm, under a gradient, S, of 0·00120. (The Colebrook-White numerical solution is 517·4 litres/sec)

Figure 2 : Colebrook-White equation and direct solution approximations in terms of the individual variables

Solution for Q (or V) (i.e. Colebrook-White equation)

$$\frac{V}{\sqrt{(2SgD)}} = \frac{0{\cdot}9003\,Q}{\sqrt{(Sg)}\,D^{2{\cdot}5}} = -2\log\left\{\frac{k_s}{3{\cdot}71D} + \frac{1{\cdot}775\,\nu}{\sqrt{(Sg)}\,D^{1{\cdot}5}}\right\} \qquad (8)$$

Solution for S (Q as input variable for flow) (Barr [34] approximation)

$$\frac{0{\cdot}9003\,Q}{\sqrt{(Sg)}\,D^{2{\cdot}5}} = -1{\cdot}9\log\left\{\left(\frac{k_s}{3{\cdot}71D}\right)^{1{\cdot}053} + \left(\frac{4{\cdot}932\,\nu\,D}{Q}\right)^{0{\cdot}937}\right\}$$

Solution for D (Q as input variable for flow) (Pham [35] approximation)

$$\frac{0{\cdot}9003\,Q}{\sqrt{(Sg)}\,D^{2{\cdot}5}} = -1{\cdot}8844\log\left\{\frac{0{\cdot}365(Sg)^{0{\cdot}2}\,k_s}{Q^{0{\cdot}4}} + \frac{3{\cdot}55\,\nu}{Q^{0{\cdot}6}\,(Sg)^{0{\cdot}2}}\right\}$$

where D is diameter of circular pipe flowing full
           V is mean velocity
           Q is discharge
           S is piezometric gradient, but see below for alternative of $\Delta p/\rho l$ for Sg
           g is acceleration due to gravity, but see as for S
           ν is kinematic viscosity
           k_s is equivalent sand roughness size

In the above equations, Sg can be substituted by $\Delta p/\rho l$

where Δp is pressure difference acting to overcome friction in pipe length l
           ρ is mass density of the fluid flowing

The foregoing variables are in any coherent system of units such as pure SI (kg-m-s units)

Figure 3 : Logarithmic rough turbulent equations for kg-m-s units

For SI units (D and k_s in m, Q in m^3s^{-1}, $\Delta p/\rho l$ in ms^{-2}) the following apply

Solution either for Q, or for S or $\Delta p/\rho l$

$$\frac{Q}{\sqrt{S}D^{2{\cdot}5}} = 6{\cdot}957\log\left\{\frac{3{\cdot}71D}{k_s}\right\} \quad \text{or} \quad \frac{Q}{\sqrt{(\Delta p/\rho l)}D^{2{\cdot}5}} = 2{\cdot}221\log\left\{\frac{3{\cdot}71D}{k_s}\right\}$$

Solution for D (Q given as flow variable)

$$\frac{Q}{\sqrt{S}D^{2{\cdot}5}} = 6{\cdot}555\log\left\{\frac{1{\cdot}736\,Q^{0{\cdot}4}}{S^{0{\cdot}2}\,k_s}\right\} \quad \text{or} \quad \frac{Q}{\sqrt{(\Delta p/\rho l)}D^{2{\cdot}5}} = 2{\cdot}093\log\left\{\frac{2{\cdot}741\,Q^{0{\cdot}4}}{(\Delta p/\rho l)^{0{\cdot}2}\,k_s}\right\}$$

From Table D6 read mQ as 501·34 litres/sec for 750 mm diameter and gradient 0·00120. From Appendix 4, determine the discharge factor for the adjustment to 762 mm from 750 mm diameter as 1·043. Hence, mQ is adjusted to 522·9 litres/sec. From Table E120, for kinematic viscosity $1·50 \times 10^{-6}\,\mathrm{m}^2\mathrm{s}^{-1}$ and roughness size k_s of 0·060 mm, interpolate between 600 mm and 800 mm diameter and 0·0010 and 0·00150 gradient to obtain m_C as 1·012. The estimate of discharge is then 522·9/1·012 = 516·7 litres/sec.

Checks on mean velocity and Reynolds number

Tables D show values of mV where V is the mean velocity corresponding to the discharge Q at the same table point. The same value of m applies as is used to operate on mQ, after corresponding interpolation where appropriate. Then V is given in ms^{-1}. The Reynolds number is DV/v where the variables D, V and v are in consistent units. Diameter D should be in metres, together with an appropriate value of kinematic viscosity, v, in $\mathrm{m}^2\mathrm{s}^{-1}$. Where an estimated value of m is used and the solution may be for flow near the laminar-turbulent transition, it is essential to check the value of the Reynolds number.

Where Tables E are used to obtain a value of the Colebrook-White equation based m_C or the Poiseuille equation based m_P, warning is given also of closeness to the laminar-turbulent transition. If very close to this transition, the Reynolds number should be evaluated for guidance. As already stated, the entries of m_C stop below a Reynolds number of 2400, and the entries of m_P are then taken to the Reynolds number limit of 2000.

Other sources of resistance

When designing a pipe system on the basis of its full-bore capacity, allowances must be made for head losses which will occur at bends, and at appurtenances involving changes in cross-section. Hydraulic handbooks, manuals and other technical publications give head-loss coefficients, ε_1, ε_2, ... for use in the expressions $h_1 = \varepsilon_1 V^2/2g$ etc., with h_1, h_2, ... contributing cumulatively to the total head difference Σh necessary to drive the system. Here $V^2/2g$ is the kinetic head based simply on the mean velocity.

Where a pipe leads from a body of fluid, there is some loss of head at the entry, which will depend upon the sharpness of the arris. Furthermore, the kinetic energy of flow $\alpha(V^2/2g)$, where α is the Coriolis coefficient, is generally not recovered at an exit where the pipe leads into a body of the fluid. These factors may well prove important if the pipe is relatively short. The head loss with a sharp-edged re-entrant inlet is approximately $V^2/2g$. With a flush headwall the loss coefficient drops to about 0·4, whilst a rounding of as little as a sixth or seventh of the pipe diameter will almost eliminate the entrance loss, but not the need to allow in the flow assessment for the kinetic head. Indicative values of head loss coefficient are given in list (b) of Appendix 5. Much further information on losses at features and bends is given by Miller[13], by Fried and Idelchik[14], and in the General Electric Data Book for Fluid Flow[15].

Calculating with additional pressure or head losses present

A common practice is to take into account losses over and above those for uniform flow as an equivalent additional length of pipe. From the original definition of the Darcy-Weisbach friction factor λ (Eqs (4) and (4a)), the value of its reciprocal, i.e. $1/\lambda$, is the number of diameters to give the length of pipe in which the losses are equivalent to the kinetic head $V^2/2g$, or the dynamic pressure $\rho V^2/2$, on the basis of Eqs (4) and (4a) respectively. With the Fanning friction factor f_F (Eqs (9) and (9a)), $1/4f_F$ is substituted for $1/\lambda$.

Consequently, if an additional loss is expressed as a head loss $\varepsilon\,(V^2/2g)$ or as a pressure drop $\varepsilon\,(\rho V^2/2)$, ε/λ is the number of pipe diameters to give the equivalent length for the feature. This is constant in rough turbulent flow but otherwise varies with Reynolds number. For kilogram-metre-second units, $1/\lambda\ (= 1/4f_F)$ is given by $0{\cdot}051(V^2/SD)$ or by $0{\cdot}5[V^2/(\Delta p/\rho l)D]$. Then the length in metres per unit ε is $0{\cdot}051(V^2/S)$ or $0{\cdot}5[V^2/(\Delta p/\rho l)]$. This allows estimation of the required total head or pressure difference for a selected discharge through the pipe system as follows. The additional length $\Sigma\,\varepsilon\times[0{\cdot}051(V^2/S)]$, or $\Sigma\,\varepsilon\times\{0{\cdot}5[V^2/(\Delta p/\rho l)]\}$, is added to the actual pipe length before multiplying by the already calculated piezometric gradient, S, or kinetic pressure gradient, $\Delta p/\rho l$, respectively, for the simple pipe flow, to obtain the required total head or the required pressure difference. However, it follows that information provided simply in terms of equivalent length of pipe, whether supplied directly or by values of coefficient ε, may not sufficiently allow for variation in Reynolds number in smooth pipe flows.

Alternatively, we can adopt values of the coefficient

$$\varepsilon_f = SL\big/(V^2/2g) \tag{21}$$

$$\text{or}\qquad \varepsilon_f = \Delta p\big/(\rho V^2/2) \tag{21a}$$

where L is the pipe length and Δp is the pressure drop due to pipe friction over this length, so that ε_f allows for the pipe flow resistance element in the following total head (22) and total active pressure difference (22a) equations.

$$\Sigma\,h = (\,\varepsilon_1 + \varepsilon_2 \ldots + \varepsilon_f\,)(V^2/2g) \tag{22}$$

$$\Sigma\,\Delta p = (\,\varepsilon_1 + \varepsilon_2 \ldots + \varepsilon_f\,)(\rho V^2/2) \tag{22a}$$

This approach is preferable where discharge must be estimated by iteration, especially in smooth pipe systems. However, indicative values of the allowances for bends and appurtenances with either system are given in Appendix 5.

Solutions for successive trial values of mean velocity corresponding to the assessed pipe friction gradients S_f or $(\Delta p/\rho l)_f$, i.e. $V = mV/m_C$, can be obtained using Tables D to give values of mV and Tables E to give values of m_C. In the tabular solution for the following example, the kinematic pressure gradient alternative is used. Then column two is for the above step and column three for application of Eq. (21a). Column four shows the trial summations of total head loss coefficient and column five shows the check values of velocity obtained from Eq. (22a) rearranged, using these summations. In the illustrative solution, the starting assumption for the pipe friction element of the kinematic pressure gradient $(\Delta p/\rho l)_f$ is the available pressure difference, and then the gradients are adjusted by the square of the velocity balance quotients. As can be seen, convergence is normally rapid.

(x) A 200 mm pipe, roughness size 0·03 mm, is to convey a liquid which has a density, ρ, of 1200 kgm^{-3} and a dynamic viscosity, μ, of 0·003 Pa·sec (0·003 kgm^{-1}s^{-1}) over a route of 250 m length. For this route, exit, valve and bend effects are assessed to result in a combined additional loss coefficient ε_c of 10·0, and an overall pressure difference of 235 kPa (235000 kgm^{-1}s^{-2}) is available. Estimate the maximum discharge obtainable (Numerical solution using the Colebrook-White equation gives 0·1137 m^{3}s^{-1} with a pipe friction kinematic pressure gradient, $(\Delta p/\rho l)_f$, of 0·5214 ms^{-2}).

The kinematic viscosity, $\nu = \mu/\rho$, is 0·003/1200 $= 2·5 \times 10^{-6}$ m^{2}s^{-1} and, with roughness size 0·03 mm, Table E141 is applicable, in conjunction with Table D4 for the 200 mm diameter. The overall kinematic pressure difference, $\Sigma \Delta p/\rho$, is 235000/1200 = 195·83 m^{2}s^{-2} and, to start the iteration, the first trial kinematic pressure gradient, $(\Delta p/\rho l)_f$, then is 195·83/250 = 0·7833 ms^{-2}.

$$\begin{array}{ccccc}
(\Delta p/\rho l)_f & V & \varepsilon_f = \dfrac{2L(\Delta p/\rho l)}{V^2} & \Sigma \varepsilon & V = \sqrt{\left\{\dfrac{2\times(\Sigma \Delta p/\rho)}{\Sigma \varepsilon}\right\}} \\
\text{(trial)} & \text{(by Tables)} & & (\varepsilon_c + \varepsilon_f) & \text{(check)} \\
 & \text{(i.e. } mV/m_C) & & &
\end{array}$$

$(\Delta p/\rho l)_f$	V	ε_f	$\Sigma \varepsilon$	V
0·7833	4·499	19·35	29·35	3·653
	(3·838/0·853)			

[adjust gradient - 0·7833 $\times$ (3·653 / 4·499)^2 = 0·5164]

0·5164	3·595	19·98	29·98	3·614
	(3·113/0·866)			

[adjust gradient - 0·5164 $\times$ (3·614 / 3·595)^2 = 0·5219]

0·5219	3·606	20·07	30·07	3·609
	(3·130/0·868)			

The assessed mean velocity of 3·609 ms^{-1} gives a discharge of 0·1134 m^{3}s^{-1} under a kinematic pressure gradient of 0·5219 ms^{-2}.

For assessment using total head and piezometric gradient etc. (i.e. based on Eqs (4), (21) and (22)) the corresponding table headings are as follows:

$$\begin{array}{ccccc}
S_f & V & \varepsilon_f = \dfrac{(2gL)S}{V^2} & \Sigma \varepsilon & V = \sqrt{\left\{\dfrac{2g\times \Sigma h}{\Sigma \varepsilon}\right\}} \\
\text{(trial)} & \text{(by Tables)} & & (\varepsilon_c + \varepsilon_f) & \text{(check)} \\
 & \text{(i.e. } mV/m_C) & & &
\end{array}$$

Non-circular cross-sections of flow

Hydraulic equivalence and the hydraulic diameter

The method adopted here follows from the work of Chezy[16], Johnson[17], Ackers[18], ESDU International[10], Ward-Smith[19], Jones[20] and Obot[21, 22], adapted as necessary. One is concerned with a steady flow in a non-circular section and the corresponding design variables. First one considers an equivalent circular cross-section for which the mean velocity can be assumed to be the same as in the non-circular section. The kinematic viscosity, ν, and, for turbulent flow, the roughness size, k_s, remain constant. Then the resistance dependent element of the assessment process is carried out in terms of this equivalent circular pipe, which has the same hydraulic diameter as the non-circular section. The final step is to relate back from the variable so determined to the corresponding variable of the non-circular section.

Fig. 4 outlines the procedures for the three standard design problems.

Chezy's concept of hydraulic mean depth (hydraulic radius)
According to Chezy[16], the hydraulic mean depth is A/P, which is one quarter of the diameter of the equivalent pipe, D_{ep}, i.e. the hydraulic diameter. This has been demonstrated already in the sub-section 'Tables D generated from a Manning type equation'. Following Johnson[17] and Ackers[18], one can make the following comparison between a non-circular flow section and its equivalent pipe. Since it is intended that

$$V \;=\; V_{ep}$$

$$Q/A \;=\; Q_{ep}\,/\,[(\pi/4)(4A/P)^2]$$

Thus

$$Q_{ep} \;=\; Q \times 4\pi A/P^2 \tag{23}$$

Here A and P are cross-sectional flow area and wetted perimeter of the non-circular flow section, V and Q represent mean velocity and discharge and the subscript ep indicates the equivalent pipe in the sense of providing the same mean velocity, though not, of course, the same conveyance. The discharge conversion factor $J = 4\pi A/P^2$ must apply also to the relation between flow areas, because discharge is the mean velocity times the flow area. Clearly J can be expressed alternatively either as $4\pi R/P$, or as $\pi D_{ep}/P$.

'Unit size' measures for properties of conduits
For any geometrically similar series of conduit shapes, a 'unit size' can be defined in terms of a key dimension. Then, a multiplying factor M specifies the size of an actual example of the shape within the geometrically similar series. Also for each shape there are 'unit' case values of wetted perimeter, P_u, and of cross-sectional area of flow, A_u.

Hence for an actual example, MP_u and $M^2 A_u$ are the sizes of the wetted perimeter and of the area, respectively. It follows that there is a 'unit' case value of hydraulic mean depth (hydraulic radius), i.e. $R_u = A_u/P_u$ and hence a 'unit' case value of diameter of equivalent pipe, or 'unit' hydraulic diameter, i.e. $D_{ep(u)} = 4R_u$.

In these Tables, the sizes of $D_{ep(u)}$ and P_u are given in metres, with those of unit cross-sectional areas in square metres. Thus, with values of $D_{ep(u)}$ made available in Tables F, the diameter of the equivalent pipe, i.e. the hydraulic diameter, D_{ep} is given by

$$D_{ep} = MD_{ep(u)} \tag{24}$$

The need for the gradient factor, C, as an additional control
Often with turbulent flows and nearly always with laminar flows, the procedures so far outlined are insufficient to deal accurately with the changes in flow conditions as between a non-circular flow section and its equivalent circular pipe. This is because of the presence of residual shape effect affecting the assessment of mean velocity as obtained from the basic procedure just outlined. A residual shape effect is any difference in mean velocity between that in a non-circular conduit and that in the circular tube with the same hydraulic mean

Figure 4 : Solution routes for internal flow in non-circular cross-sections

The solution routes for the three main design problems are shown below. The problems are of finding one variable from (i) discharge (Q), (ii) friction gradient (S_f or $(\Delta p/\rho l)_f$) or (iii) size (i.e. factor M), given the other two variables together with values of kinematic viscosity, ν , and, for turbulent flow, roughness size, k_s. For these solutions, the values of equivalent (hydraulic) diameter for the unit case ($D_{ep(u)}$) and of the equivalent discharge factor (J) are given in the appropriate Table F. Also listed in Tables F are values of the gradient factor C for both laminar and turbulent flows. The scale factor M defines the difference is size between the case in hand and the unit case illustrated in the appropriate Table F.

Then   (i) *Find Q* :

$$M \times D_{ep(u)} = D_{ep}$$

$$S_f \times C = S_{ep}$$

$$\nu, k_s \quad \boxed{\text{R.A.}} \quad \rightarrow \quad Q_{ep} = Q \times J$$

Hence Q

(ii) *Find S_f* :

$$M \times D_{ep(u)} = D_{ep}$$

$$Q \times J = Q_{ep}$$

$$\nu, k_s \quad \boxed{\text{R.A.}} \quad \rightarrow \quad S_{ep} = S_f \times C$$

Hence S_f

(iii) *Find size (i.e. factor M)* :

$$Q \times J = Q_{ep}$$

$$S_f \times C = S_{ep}$$

$$\nu, k_s \quad \boxed{\text{R.A.}} \quad \rightarrow \quad D_{ep} = M \times D_{ep(u)}$$

Hence M

Notes : The central element of each solution route is an assessment of resistance to flow in a circular tube. These elements are indicated by the block containing 'R.A.'. Such assessment may be aided by Tables, as is emphasised here, or by Charts or may be accomplished by solution of the appropriate equation.

The measure of friction gradient which is featured in the illustrations is piezometric gradient $S_f = h_f/l$, where h_f is the friction head loss, in terms of the fluid flowing, over length l. The kinematic pressure gradient, $\Delta p/\rho l$, may be substituted consistently for S. Tables D and E are arranged to allow directly for the differences in value between these alternative measures.

depth or hydraulic radius, R, or the same hydraulic diameter, $D_{ep} = 4R$; with the friction gradient, kinematic viscosity and, in turbulent flow, roughness remaining the same.

ESDU International[10] and Ward-Smith[19] collated much of the information available[23-33] and presented it for non-dimensional working. Later, Obot[21,22] introduced the 'frictional law of corresponding states'. This provides the most complete degree of non-dimensional correlation yet achieved between the hydraulic behaviour of different conduit shapes. Obot dealt with the values of critical Reynolds numbers for the laminar-turbulent transition in non-circular conduits. These values are obtained by observation, with the hydraulic diameters of the non-circular conduits under test used in evaluating the Reynolds number. Obot showed that to then obtain correspondence in friction factor values, the Reynolds number values for critical conditions must be adjusted to be the same as that for the critical condition in circular conduits.

For dimensional working, it is convenient to adjust for the residual shape effect by modifying the friction gradient between that applying to the non-circular flow section under examination, S_f or $(\Delta p/\rho l)_f$, and that applied to the corresponding equivalent pipe, as follows:

$$S_{ep} = C \times S_f \tag{25}$$

or
$$(\Delta p/\rho l)_{ep} = C \times (\Delta p/\rho l)_f \tag{25a}$$

where C is the gradient factor which, for a given conduit geometry, varies with different states of flow. Much of the experimental work has been done near the laminar-turbulent transition and with smooth surfaced conduits. In these Tables the values for laminar and for turbulent flow are indicated by C_L and C_T, respectively. The factor C is the reciprocal of the factor adopted by ESDU International[10] and by Ward-Smith[19] for non-dimensional working, while C_L corresponds directly with Obot's[21] ratio of values of critical Reynolds numbers.

Then in using the Tables, Eqs (23), (24) and (25) or (25a) provide the essential relationships between a given non-circular flow and the corresponding equivalent pipe flow. The application of these equations is shown in Fig. 4.

It is emphasised that for refined assessments, those with critical dependence on the value of C which is adopted, the source literature for evaluation of C should be consulted, either as referenced here[10, 15, 19-33] or as traced forward from citations of these works, etc. There are some indications that values of C_T which have been obtained for smooth turbulent flows at low values of Reynolds number may not differ widely from those appropriate to other conditions of turbulent flow, although Obot[21] has sought to demonstrate that C_T is always close to unity for rough turbulent flow. For rough turbulent, internal, flows in non-circular conduits, the effect of uncertainty in the evaluation of C_T may be less than that concerning other variables.

Tables of properties of unit sections (Tables F)
These tables define the geometry of the sections treated and give the values needed to follow the procedures illustrated in Fig. 4. Confirmation of choice of C_L or C_T, for laminar or turbulent flow respectively, must depend on the findings from Tables E as a solution proceeds.

Finding the discharge in a rectangular conduit (turbulent flow)
The steps for the 'unit size' method are as follows.

1) Determine the multiplying factor M between the 'unit' size, shown in Table F1 for rectangular conduits, and the actual size.
2) Then determine, for the given shape, values of $D_{ep(u)}$ and J.
3) Determine the diameter of the equivalent circular pipe, the hydraulic diameter, D_{ep}, using Eq. (24) (i.e. $D_{ep} = MD_{ep(u)}$).
4) From Table F determine a value of gradient factor C; C_L or C_T as appears probable. Depending on gradient measure in use, find $S_{ep} = C \times S_f$ or $(\Delta p/\rho l)_{ep} = C \times (\Delta p/\rho l)_f$ (Eq. (25) or Eq. (25a)).
5) Use the Tables D and Tables E to determine the discharge for pipe with diameter D_{ep}, gradient S_{ep} or $(\Delta p/\rho l)_{ep}$, and kinematic viscosity v and, for turbulent flow, roughness size k_s i.e. find Q_{ep}. During this process, confirm that correct assessment was made whether flow is laminar or turbulent.
6) Determine the predicted discharge in the rectangular conduit, i.e. $Q = Q_{ep}/J$ (Eq. (23)).

(xi) Estimate the discharge of a liquid of kinematic viscosity $1 \cdot 50 \times 10^{-6}\,m^2 s^{-1}$ in a 270 mm by 54 mm rectangular conduit, with surface roughness size 0.015 mm, under a kinematic pressure gradient of $0 \cdot 20\,ms^{-2}$ (Numerical solution using the Colebrook-White equation gives 19·346 litres/sec).

1) From the diagram at the head of Table F1 the value of M is 0·27.
2) The ratio a/b is 0·20 and, from Table F1, $D_{ep(u)}$ and J are 0·3333 m and 0·4363 respectively.
3) Then the diameter of the equivalent circular pipe, D_{ep}, is $0 \cdot 27 \times 0 \cdot 3333 = 0 \cdot 0900$ m (90·00 mm).
4) Table F1 shows the value of C_T for a/b of 0·20 as 0·96, giving $(\Delta p/\rho l)_{ep}$ as $0 \cdot 192\,ms^2$.
4) From Table D3, mQ, for 90·00 mm diameter and $0 \cdot 192\,ms^2$ kinematic pressure gradient, is 7·094 litres/sec, and from Table E118, for k_s of 0·015 mm, m_C is 0·840, from the turbulent flow zone of the table, giving $Q_{ep} = 7 \cdot 094/0 \cdot 840$, i.e. 8·445 litres/sec.
5) The estimated discharge in the conduit is Q_{ep}/J, i.e. $8 \cdot 445/0 \cdot 4363 = 19 \cdot 356$ litres/sec. Note that this agreement provides verification of the 'unit size', tabular, method and but not of the choice of value of C_T, which is applied to both tabular and numerical routes.

Finding the pressure gradient in a fully eccentric annulus (laminar flow)

(xii) A tube of 12 mm external diameter lies inside a tube of 20 mm internal diameter forming a fully eccentric annulus, the surface roughness of both tubes being 0·006 mm. Estimate the pressure gradient necessary for a discharge in the annulus of 15 millilitres/sec of a liquid of density $900\,kgm^{-3}$ and kinematic viscosity $80 \cdot 0 \times 10^{-6}\,m^2 s^{-1}$ (Numerical solution using the Poiseuille equation gives 1·7216 kPa/metre).

For the geometrical parameters, Table F4 applies and at $D_2/D_1 = 0 \cdot 60$ the values of M, $D_{ep(u)}$ and J are 0·020, 0·400 and 0·2500, respectively. Thus D_{ep} is $0 \cdot 020 \times 0 \cdot 400 = 0 \cdot 0080$ m or 8·0 mm (Eq. (24)) and Q_{ep} is $15 \times 0 \cdot 2500 = 3 \cdot 75$ millilitres/sec (Eq. (23)).

Scanning the 8 mm columns in Table D1 and in Table E253 for kinematic viscosity $80{\cdot}0\times10^{-6}\,m^2s^{-1}$, at the kinematic pressure gradient value of $2{\cdot}942\,ms^{-2}$ there are values for mQ of $43{\cdot}704$ millilitres/sec and for m_P of $11{\cdot}82$, respectively, giving Q_T as $3{\cdot}697$ millilitres/sec.

With discharge proportional to gradient in laminar flow, the required value of $(\Delta p/\rho l)_{ep}$ is $2{\cdot}942\times(3{\cdot}75/3{\cdot}697) = 2{\cdot}984\,ms^{-2}$.

From Table F5, the value of C_L is $1{\cdot}56$, hence $(\Delta p/\rho l)_f$ is $2{\cdot}984/1{\cdot}56 = 1{\cdot}913\,ms^{-2}$ (Eq. (25a)). With a fluid density of $900\,kgm^{-3}$, this corresponds with a required pressure gradient of $1{\cdot}7217\,kPa/metre$.

Finding the size of a concentric annulus (turbulent flow)

(xiii)   A concentric annulus, $D_2/D_1 = 0{\cdot}40$, is to carry 30 litres/sec of liquid of kinematic viscosity $0{\cdot}60\times10^{-6}\,m^2s^{-1}$ under a piezometric gradient of $0{\cdot}052$. Taking the roughness size of the annular surfaces as $0{\cdot}015\,mm$, estimate the size required (Numerical solution using the Colebrook-White equation gives $142{\cdot}7\,mm$ diameter for the outer tube).

From Table F4, the values of $D_{ep(u)}$, J and C_T are $0{\cdot}6000\,m$ ($600\,mm$), $0{\cdot}4286$ and $0{\cdot}91$ respectively. Applying Eqs (23) and (25) respectively, Q_{ep} is $30\times0{\cdot}4286 = 12{\cdot}858$ litres/sec, and S_{ep} is $0{\cdot}052\times0{\cdot}91 = 0{\cdot}04732$. At this piezometric gradient in Table E66, m_C is estimated as $0{\cdot}74$ and thus mQ is estimated as $9{\cdot}515$ litres/sec. In Table D3, these gradient and mQ values combine at a diameter of approximately $85\,mm$. Then in Table E66, m_C is adjusted to $0{\cdot}751$ and thus mQ is adjusted to $9{\cdot}656$ litres/sec. Interpolation in Table D3 gives diameter D_{ep} as $85{\cdot}60\,mm$ and Eq. (24) $(D_{ep} = MD_{ep(u)})$ gives M as $85{\cdot}60/600 = 0{\cdot}1427$. Thus the required diameter is $142{\cdot}7\,mm$.

Review

Illustrations are to a greater accuracy than is physically significant. This has been done in order that the user can appreciate the accuracy that can be achieved using the Tables. The approach adopted, however, also means that appropriate initial estimates can be made as readily as possible.

These Tables can be used independently. Clearly, this is their main function. There is the possibility, however, of using them in conjunction with Volume II of the 6th Edition of the Wallingford Tables, to deal with different values of kinematic viscosity in the many conduit and channel shapes treated in that volume.

Where fluid properties vary along a pipe or conduit, because of temperature change or reduction in pressure, the user must work with elements of length so that the fluid properties can be assumed constant in each element.

Appendix 7 gives the relationships between some commonly found units of discharge and the SI units of discharge used in these Tables.

References

1. HR WALLINGFORD and BARR, D.I.H. *Tables for the hydraulic design of pipes, sewers and channels*, 6th edition in two volumes. Thomas Telford, London, 1994. (Note: the earlier editions of the Wallingford Tables were published as single volumes by H.M.S.O. in 1963, 1969 and 1977, in the first two cases under the authorship of P.Ackers and with the title *Tables for the hydraulic design of storm-drains, sewers and pipe-lines*, and in the third case with the title *Tables for the hydraulic design of pipes* and with the Hydraulics Research Station as author. Subsequent editions were published by Hydraulics Research Ltd. in 1983, and by Thomas Telford Ltd. in 1990, in both cases with the title *Tables for the hydraulic design of pipes and sewers* and with Hydraulics Research Ltd as author.)

2. COLEBROOK, C.F. Turbulent flow in pipes, with particular reference to the transition region between the smooth and the rough pipe laws. *J. Instn. Civ. Engrs*, 1939, Vol. 11, pp 133-156.

3. ROUSE, H. and INCE, S. *History of hydraulics.* Dover, New York, 1963.

4. NIKURADSE, J. Gesetzmäßigkeit der turbulenten Strömung in glatten Rohren. *Forsch. Arb. Ing.-Wes.* No.356 (1932).

5. NIKURADSE, J. Strömungsgesetze in rauhen Rohren. *Forsch. Arb. Ing.-Wes.* No.361 (1933).

6. ROUSE, H. Evaluation of boundary roughness. *Proc. 2nd Hydraulics Conf.*, Univ. of Iowa, 1943, Bulletin 27.

7. KING, H.W. *Handbook of Hydraulics.* McGraw-Hill, New York, 1954, Section 6.

8. LAMONT, P.A. A review of pipe friction data and formulae, with a proposed set of exponential formulae based on the theory of roughness. *Proc.Instn Civ.Engrs*, Part 3, 1954, 3, p 248.

9. PERKINS, J.A. and GARDINER, I.M. *The effect of sewage slime on the hydraulic roughness of pipes.* Report IT 218, Hydraulics Research Station, Wallingford, 1982.

10. ESDU International, Item No 66027. *Friction losses for fully developed flow in straight pipes.* 1966 and amended 1971 and 1994.

11. MANNING, R. On the flow of water in open channels and pipes. *Proc. Instn Civ. Engrs*, Ire., 1891, Vol. 20, p161; 1895, Vol. 24, p 179.

12. WILLIAMSON, J. The laws of flow in rough pipes, Strickler, Manning, Nikuradse and drag-velocity. *La Houille Blanche*, 1951, Vol. 6, No. 5, pp 738-57.

13. MILLER, D.S. *Internal flow systems - 2nd Edition.* BHRA (Information Services), Bedford, 1990.

14. FRIED, E. and IDELCHIK, I.E. *Flow resistance - A design guide for engineers.* Hemisphere Publishing, New York, 1989.

15. WHITE, F.M. (Gen. Ed.) *General Electric - Fluid flow data book.* General Electric, Schenectady, New York, 1943 to 1983 and Genium Publishing, New York, 1984 onwards.

16. MOURET, G. Antoine Chezy, histoire d'une formule d'hydraulique. Annales des Ponts et Chaussees, 1921-II. (See Reference 3).

17. JOHNSON, S.P. *A survey of flow calculation methods.* Pre-printed programme for June 19-21, 1934, meeting of the Amer. Soc. Mech. Engrs., University of California, Berkeley.

18. ACKERS, P. *Resistance of fluids flowing in channels and pipes.* Hydraulics Research Paper No. 1, H.M.S.O., London, 1958.

19. WARD-SMITH, A.J. *Internal fluid flow - The fluid dynamics of flow in pipes and ducts.* Clarendon Press, Oxford, 1980.

20. JONES, O.C.JNR An improvement in the calculation of turbulent friction in rectangular ducts. *Trans. Am. Soc. Mech. Engrs, Journal of Fluids Engineering*, 1976, Vol. 98, pp. 173-181.

21. OBOT, N.T. Determination of incompressible flow friction in smooth circular and noncircular passages: a generalised approach including validation of the nearly century old hydraulic diameter concept. *Trans. Am. Soc. Mech. Engrs, Journal of Fluids Engineering*, 1988, Vol. 110, pp. 431-440.

22. OBOT, N.T. The frictional law of corresponding states: its origin and applications. *Trans. Instn Chem. Engrs*, 1993, Vol. 71, Part A, pp. 3-10.

23. KNUDSEN, J.G. and KATZ, D.L. *Fluid dynamics and heat transfer.* McGraw-Hill, New York, 1958.

24. CARLSON, L.W. and IRVINE, T.P. Fully developed pressure drop in triangular shaped ducts. *Trans. Am. Soc. Mech. Engrs,* 1961, Series C, Vol. 83, pp. 82-88.

25. HARTNETT, H.P., KOH, J.C.Y.and McCOMAS, S.T. A comparison of predicted and measured friction factors for turbulent flow through rectangular ducts. *Trans. Am. Soc. Mech. Engrs,* 1962, Series C, Vol. 84, No. 1, pp. 82-88.

26. PARKINSON, G.V. and DENTON, J.D. *Laminar flow through an eccentric annular pipe.* Aeronautical Research Council, 1962, ARC 24 326

27. DODGE, N.A. Friction losses in annular flow. *Am. Soc. Mech. Engrs,* 1963, Paper No. 63-WA-11.

28. GUNN, D.J. and DARLING, C.W.W. Fluid flow and energy losses in non-circular conduits. *Trans. Instn Chem. Engrs,* 1963, Vol. 41, No. 4, pp. 163-173.

29. BRIGHTON, J.A. and JONES, J.B. Fully developed turbulent flow in annuli. *Trans. Am. Soc. Mech. Engrs,* 1964, Series D, Vol. 86, No. 4, pp. 835-844.

30. JONSSON, V.K. and SPARROW, E.M. Results of laminar flow analysis and turbulent flow experiments for eccentric annular ducts. *A.I.Ch.E.Jl.*, 1965, Vol. 11, No. 6, pp. 1143-1145.

31. SNYDER, W.T. and GOLDSTEIN, G.A. An analysis of fully developed laminar flow in an eccentric annulus. *A.I.Ch.E.Jl.*, 1965, Vol. 11, No. 3, pp. 462-467.

32. SPARROW, E.M. and HAJI-SHEIKH, A. Laminar heat transfer and pressure drop in isosceles triangular, right triangular and circular sector ducts. *Trans. Am. Soc. Mech. Engrs,* 1965, Series C, Vol. 87, No. 3, pp. 426-427.

33. JONSSON, V.K. and SPARROW, E.M. Experiments on turbulent flow phenomena in eccentric annular ducts. *Jl.Fluid Mech.*, 1966, Vol. 25, Part 1, pp. 65-86.

34. BARR, D.I.H. Explicit Colebrook-White solutions. *Civil Engineering,* Sept. 1986, pp. 19-31.

35. PHAM, Q.T. Explicit equations for the solution of turbulent pipe-flow problems. *Trans. Instn Chem. Engrs,* 1979, Vol. 57, pp. 281-283.

a	Length of smaller sides of rectangle.
A	Area of cross-section of pipe or conduit.
b	Length of larger sides of rectangle.
C	Gradient factor (see also subscripts L and T).
D	Diameter. Values are generally stated in millimetres, but in metres with respect to unit sizes. In formulae and for area calculations, values may require to be entered in metres.
f	Friction factor; see also λ and subscript F.
F	Discharge factor for effect of change of diameter.
g	Acceleration due to gravity, $9\cdot80665\,\mathrm{ms^{-2}}$ for standard gravity in SI.
h	Head loss over stipulated length of pipe or conduit, equal to $\Delta p/\rho g$.
J	Discharge conversion factor, $4\pi A/P^2$ where A and P relate to a particular flow circumstance.
l	Length over which friction head loss or pressure loss is assessed.
k_s	Nikuradse equivalent sand roughness size; the linear measure of roughness size. Stated in millimetres.
$\log$	Common logarithm (base 10).
M	Multiplying factor on unit measure of conduit size to specify size of given example.
m	$100 \times n$ as defined below; adopted to simplify conjointly Tables D, for values of mV and mQ, and Tables E, for values of m_C.
n	Manning coefficient.
Δp	Pressure difference, especially that to overcome friction with gradient $\Delta p/l$.
P	Length of wetted perimeter of conduit cross-section.
Q	Discharge (volume per unit time). Stated in cubic metres per second (m^3s^{-1}), litres/sec or millilitres/sec, the latter being $1/1000$ and $1/1\,000\,000$ cubic metres per second, respectively.
R	Hydraulic mean depth (hydraulic radius) of flow section, A/P, i.e. $(\pi/4)D^2/\pi D$ or $(1/4)D$ for circular pipe. One quarter of the hydraulic diameter.
R	Reynolds number, VD/ν or $4Q/\pi\nu D$
S	Piezometric gradient; head loss per unit length of pipe or conduit, h/l, equal to $\Delta p/\rho g l$.
V	Mean velocity of flow through cross-section.

α	Coriolis coefficient; $\alpha V^2/2g$ is kinetic energy head; angle defining isosceles triangle or sector of circle.
ε	Head or pressure loss coefficient, normally applied to $V^2/2g$ or to $\rho V^2/2$.
λ	Darcy-Weisbach friction factor, $2(Sg)D/V^2$.
μ	Dynamic viscosity, i.e. coefficient of viscosity.
ν	Kinematic viscosity, i.e. dynamic viscosity divided by mass density.
ρ	Mass density.
τ	Boundary shear stress in pipe or conduit flow.

Subscripts

C	Indicating derivation from solution of the Colebrook-White equation in respect of friction coefficients m or n.
ep	Relating to equivalent pipe - i.e. conceptual circular pipe with same hydraulic mean depth (hydraulic radius) and same mean velocity as non-circular flow. Indicating the hydraulic diameter when applied to diameter D.
ep(u)	Relating to equivalent pipe for unit case.
f	Indicating overall friction gradient.
F	Indicating Fanning friction factor, $2(\tau/\rho)/V^2$ or $2(\Delta p/\rho l)\,D/4V^2$
G	Given value between table values in Colebrook-White (or Manning) solutions.
L	Indicates factor applying to laminar flow.
P	Indicating derivation from solution of the Poiseuille equation in respect of friction coefficients m or n.
R	Required value between table values in Colebrook-White (or Manning) solutions
T	Table value in Colebrook-White (or Manning) solutions; also indicates a factor applying to turbulent flow.
u, (u)	Relating to unit section.
W	Indicating derivation from Manning-Williamson equation in relation to use of friction coefficients m or n.

APPENDIX 1 : Recommended roughness values

Classification	Suitable values of k_s (mm)			Classification	Suitable values of k_s (mm)		
	Good	Normal	Poor		Good	Normal	Poor
Smooth materials inc. non-ferrous metals				**Clayware**			
				Glazed or unglazed pipe:			
Good quality drawn pipes of aluminium, brass,copper, lead etc, and non-metallic pipes of Alkathene, glass,				With sleeve joints	0·03	0·06	0·15
				With spigot and socket joints and "O" ring seals			
				-- dia < 0·150 m	--	0·03	--
perspex etc	0·0015	0·003	--	-- dia > 0·150 m	--	0·06	--
Asbestos-cement	0·015	0·03	0·15	**uPVC**			
Pitch fibre	0·003	0·006	0·03	With chemically cemented joints	--	0·03	--
Glass fibre	--	0·06	--	With spigot and socket joints, "O" ring seals at			
Metal pipes				6 to 9 metre intervals	--	0·06	--
Spun bitumen or concrete lined	0·015	0·03	0·06	**Brickwork**			
Wrought iron (new)	0·03	0·06	0·15	Glazed	0·6	1·5	3·0
Rusty wrought iron	0·15	0·6	3·0	Well pointed	1·5	3·0	6·0
Uncoated steel (new)	0·015	0·03	0·06	Old, in need of pointing	--	15	30
Uncoated steel (scaled)	1·5	6·0	15·0				
Coated steel	0·03	0·06	0·15	**Unlined rock tunnels**			
Galvanised iron, coated cast iron	0·06	0·15	0·3	Granite and other homogeneous rocks	60	150	300
Uncoated cast iron	0·15	0·3	0·6	Diagonally bedded slates	--	300	600
Tate relined pipes	0·15	0·3	0·6	(values to be used with *design* diameter)			
Old tuberculated water mains as follows:							
Slight degree of attack	0·6	1·5	3·0				
Moderate attack	1·5	3·0	6·0				
Appreciable attack	6·0	15	30				
Severe attack	15	30	60				
(Good: up to 20 years use; Normal: 40 to 50 years use; Poor: 80 to 100 years use)							
Wood							
Wood stave pipes, planed plank conduits	0·15	0·6	1·5				
Concrete							
Very smooth pipes & joints	0·03	--	--				
Precast concrete pipes with "O" ring joints	0·06	0·15	0·6				
Spun precast concrete pipes with "O" ring joints	0·06	0·15	0·3				
Monolithic construction against steel forms	0·3	0·6	1·5				
Monolithic construction against rough forms	0·6	1·5	--				

The lists of values in Appendices 1 and 2 should not be taken as absolving the engineer of the responsibility for checking the actual surface roughness achieved in particular projects by precise, in context, hydraulic tests whenever possible. Where such direct evidence is available from comparable projects it should clearly take precedence over the values quoted here.

APPENDIX 2 : Indicative values of coefficient m_C for turbulent flow

These indicative values are conditioned by size range and gradient range. It is emphasised that the values are medial for the zones detailed. The individual tables illustrate the patterns of change of value of m_C with increasing value of kinematic viscosity, roughness size, gradient and diameter. These values, with appropriate interpolation, can be applied directly to Tables D. The solutions so obtained for turbulent flows are rough approximations to those obtainable with full involvement of Tables E by the user, the latter being accurate solutions of the Colebrook-White equation. Where there is seen to be propinquity to conditions for laminar flow, the flow state must be determined and Tables E used to give solution for laminar or turbulent flow as applicable.

It is reiterated that the lists of values in Appendices 1 and 2 should not be taken as absolving the engineer of the responsibility for checking the actual surface roughness achieved in particular projects by precise, in context, physical tests whenever possible. Where such direct evidence is available from comparable projects it should clearly take precedence over the values quoted here.

Medial values of m_C for gradients S and $(\Delta p/\rho l)$ ms^{-2}

Kinematic viscosity, ν	$c.\,0{\cdot}0001$ $c.\,(0{\cdot}001)$	$c.\,0{\cdot}0010$ $c.\,(0{\cdot}010)$	$c.\,0{\cdot}0100$ $c.\,(0{\cdot}100)$	$c.\,0{\cdot}100$ $c.\,(1{\cdot}00)$	$c.\,1{\cdot}00$ $c.\,(10)$	$c.\,10{\cdot}0$ $c.\,(100)$

* Smooth pipes - drawn non-ferrous metal, plastic, pitch fibre,
and the like: k_s = **0·0015** mm and **0·003** mm: <u>to 15 mm diameter</u>

Kinematic viscosity, ν	$c.\,0{\cdot}0001$	$c.\,0{\cdot}0010$	$c.\,0{\cdot}0100$	$c.\,0{\cdot}100$	$c.\,1{\cdot}00$	$c.\,10{\cdot}0$
$c.\,0{\cdot}30\times10^{-6}$ m^{2}s^{-1}	-	-	0·8	0·65	0·6	0·55
$c.\,1{\cdot}00\times10^{-6}$ m^{2}s^{-1}	-	-	-	0·8	0·65	0·6
$c.\,3{\cdot}00\times10^{-6}$ m^{2}s^{-1}	-	-	-	-	0·8	0·65
$c.\,10{\cdot}0\times10^{-6}$ m^{2}s^{-1}	-	-	-	-	-	0·8

Ditto; <u>15 mm to 40 mm diameter</u>

Kinematic viscosity, ν	$c.\,0{\cdot}0001$	$c.\,0{\cdot}0010$	$c.\,0{\cdot}0100$	$c.\,0{\cdot}100$	$c.\,1{\cdot}00$	$c.\,10{\cdot}0$
$c.\,0{\cdot}30\times10^{-6}$ m^{2}s^{-1}	-	0·85	0·75	0·65	0·6	0·55
$c.\,1{\cdot}00\times10^{-6}$ m^{2}s^{-1}	-	-	0·9	0·75	0·65	0·6
$c.\,3{\cdot}00\times10^{-6}$ m^{2}s^{-1}	-	-	-	0·85	0·75	0·65
$c.\,10{\cdot}0\times10^{-6}$ m^{2}s^{-1}	-	-	-	-	0·9	0·75
$c.\,30{\cdot}0\times10^{-6}$ m^{2}s^{-1}	-	-	-	-	-	0·85

Ditto; <u>40 mm to 100 mm diameter</u>

Kinematic viscosity, ν	$c.\,0{\cdot}0001$	$c.\,0{\cdot}0010$	$c.\,0{\cdot}0100$	$c.\,0{\cdot}100$	$c.\,1{\cdot}00$	$c.\,10{\cdot}0$
$c.\,0{\cdot}30\times10^{-6}$ m^{2}s^{-1}	1·0	0·85	0·75	0·65	0·6	0·6
$c.\,1{\cdot}00\times10^{-6}$ m^{2}s^{-1}	1·2	1·0	0·85	0·75	0·65	0·6
$c.\,3{\cdot}00\times10^{-6}$ m^{2}s^{-1}	-	1·2	1·0	0·85	0·75	0·65
$c.\,10{\cdot}0\times10^{-6}$ m^{2}s^{-1}	-	-	1·2	1·0	0·85	0·75
$c.\,30{\cdot}0\times10^{-6}$ m^{2}s^{-1}	-	-	-	1·2	1·0	0·85
$c.\,100\times10^{-6}$ m^{2}s^{-1}	-	-	-	-	1·2	1·0
$c.\,300\times10^{-6}$ m^{2}s^{-1}	-	-	-	-	-	1·2

Ditto; <u>100 mm to 400 mm diameter</u>

Kinematic viscosity, ν	$c.\,0{\cdot}0001$	$c.\,0{\cdot}0010$	$c.\,0{\cdot}0100$	$c.\,0{\cdot}100$	$c.\,1{\cdot}00$	$c.\,10{\cdot}0$
$c.\,0{\cdot}30\times10^{-6}$ m^{2}s^{-1}	0·95	0·85	0·75	0·7	0·65	0·65
$c.\,1{\cdot}00\times10^{-6}$ m^{2}s^{-1}	1·1	0·95	0·85	0·75	0·7	0·65
$c.\,3{\cdot}00\times10^{-6}$ m^{2}s^{-1}	1·3	1·1	0·95	0·85	0·75	0·7
$c.\,10{\cdot}0\times10^{-6}$ m^{2}s^{-1}	-	1·3	1·1	0·95	0·85	0·75
$c.\,30{\cdot}0\times10^{-6}$ m^{2}s^{-1}	-	-	1·3	1·1	0·95	0·85
$c.\,100\times10^{-6}$ m^{2}s^{-1}	-	-	-	1·3	1·1	0·95
$c.\,300\times10^{-6}$ m^{2}s^{-1}	-	-	-	-	1·3	1·1
$c.\,1{\cdot}00\times10^{-3}$ m^{2}s^{-1}	-	-	-	-	-	1·3

Ditto; <u>400 mm to 1000 mm diameter</u>

Kinematic viscosity, ν	$c.\,0{\cdot}0001$	$c.\,0{\cdot}0010$	$c.\,0{\cdot}0100$	$c.\,0{\cdot}100$	$c.\,1{\cdot}00$	$c.\,10{\cdot}0$
$c.\,0{\cdot}30\times10^{-6}$ m^{2}s^{-1}	0·95	0·85	0·8	0·75	0·7	0·7
$c.\,1{\cdot}00\times10^{-6}$ m^{2}s^{-1}	1·05	0·95	0·85	0·8	0·75	0·7
$c.\,3{\cdot}00\times10^{-6}$ m^{2}s^{-1}	1·2	1·05	0·95	0·85	0·8	0·75
$c.\,10{\cdot}0\times10^{-6}$ m^{2}s^{-1}	1·4	1·2	1·05	0·95	0·85	0·8
$c.\,30{\cdot}0\times10^{-6}$ m^{2}s^{-1}	1·7	1·4	1·2	1·05	0·95	0·85
$c.\,100\times10^{-6}$ m^{2}s^{-1}	-	1·7	1·4	1·2	1·05	0·95
$c.\,300\times10^{-6}$ m^{2}s^{-1}	-	-	1·7	1·4	1·2	1·05
$c.\,1{\cdot}00\times10^{-3}$ m^{2}s^{-1}	-	-	-	1·7	1·4	1·2

* Good asbestos cement; good uncoated steel; good lined
metal pipes; etc: k_s = **0·015** mm: <u>40 mm to 100 mm diameter</u>

Kinematic viscosity, ν	$c.\,0{\cdot}0001$	$c.\,0{\cdot}0010$	$c.\,0{\cdot}0100$	$c.\,0{\cdot}100$	$c.\,1{\cdot}00$	$c.\,10{\cdot}0$
$c.\,0{\cdot}30\times10^{-6}$ m^{2}s^{-1}	1·0	0·85	0·75	0·7	0·7	0·7
$c.\,1{\cdot}00\times10^{-6}$ m^{2}s^{-1}	1·2	1·0	0·85	0·75	0·7	0·7
$c.\,3{\cdot}00\times10^{-6}$ m^{2}s^{-1}	-	1·2	1·0	0·85	0·75	0·7
$c.\,10{\cdot}0\times10^{-6}$ m^{2}s^{-1}	-	-	1·2	1·0	0·85	0·75
$c.\,30{\cdot}0\times10^{-6}$ m^{2}s^{-1}	-	-	-	1·2	1·0	0·85
$c.\,100\times10^{-6}$ m^{2}s^{-1}	-	-	-	-	1·2	1·0
$c.\,300\times10^{-6}$ m^{2}s^{-1}	-	-	-	-	-	1·2

APPENDIX 2 : Indicative values of coefficient m_C for turbulent flow (continued)

	Medial values of m_C for gradients S and $(\Delta p/\rho l)$ ms^{-2}					
	$c.0{\cdot}0001$ $c.(0{\cdot}001)$	$c.0{\cdot}0010$ $c.(0{\cdot}010)$	$c.0{\cdot}0100$ $c.(0{\cdot}100)$	$c.0{\cdot}100$ $c.(1{\cdot}00)$	$c.1{\cdot}00$ $c.(10)$	$c.10{\cdot}0$ $c.(100)$

*** Good asbestos cement; good uncoated steel; good lined metal pipes; etc: $k_s = \mathbf{0{\cdot}015}$ mm: <u>100 mm to 400 mm diameter</u>**

Kinematic viscosity, ν,						
$c.0{\cdot}30 \times 10^{-6}$ m^{2}s^{-1}	0·95	0·85	0·8	0·75	0·75	0·75
$c.1{\cdot}00 \times 10^{-6}$ m^{2}s^{-1}	1·1	0·95	0·85	0·8	0·75	0·75
$c.3{\cdot}00 \times 10^{-6}$ m^{2}s^{-1}	1·3	1·1	0·95	0·85	0·8	0·75
$c.10{\cdot}0 \times 10^{-6}$ m^{2}s^{-1}	-	1·3	1·1	0·95	0·85	0·8
$c.30{\cdot}0 \times 10^{-6}$ m^{2}s^{-1}	-	-	1·3	1·1	0·95	0·85
$c.100 \times 10^{-6}$ m^{2}s^{-1}	-	-	-	1·3	1·1	0·95
$c.300 \times 10^{-6}$ m^{2}s^{-1}	-	-	-	-	1·3	1·1
$c.1{\cdot}00 \times 10^{-3}$ m^{2}s^{-1}	-	-	-	-	-	1·3

Ditto; <u>400 mm to 1000 mm diameter</u>

Kinematic viscosity, ν,						
$c.0{\cdot}30 \times 10^{-6}$ m^{2}s^{-1}	0·95	0·9	0·85	0·85	0·8	0·8
$c.1{\cdot}00 \times 10^{-6}$ m^{2}s^{-1}	1·1	0·95	0·9	0·85	0·85	0·8
$c.3{\cdot}00 \times 10^{-6}$ m^{2}s^{-1}	1·2	1·1	0·95	0·9	0·85	0·85
$c.10{\cdot}0 \times 10^{-6}$ m^{2}s^{-1}	1·45	1·2	1·1	0·95	0·9	0·85
$c.30{\cdot}0 \times 10^{-6}$ m^{2}s^{-1}	1·7	1·4	1·2	1·05	0·95	0·9
$c.100 \times 10^{-6}$ m^{2}s^{-1}	-	1·7	1·45	1·2	1·1	0·95
$c.300 \times 10^{-6}$ m^{2}s^{-1}	-	-	1·7	1·4	1·2	1·05
$c.1{\cdot}00 \times 10^{-3}$ m^{2}s^{-1}	-	-	-	1·7	1·45	1·2

Ditto; <u>1000 mm to 2500 mm diameter</u>

Kinematic viscosity, ν,						
$c.0{\cdot}30 \times 10^{-6}$ m^{2}s^{-1}	1·0	0·95	0·9	0·9	0·9	0·9
$c.1{\cdot}00 \times 10^{-6}$ m^{2}s^{-1}	1·1	1·0	0·95	0·9	0·9	0·9
$c.3{\cdot}00 \times 10^{-6}$ m^{2}s^{-1}	1·2	1·1	1·0	0·95	0·9	0·9
$c.10{\cdot}0 \times 10^{-6}$ m^{2}s^{-1}	1·4	1·2	1·1	1·0	0·95	0·9
$c.30{\cdot}0 \times 10^{-6}$ m^{2}s^{-1}	1·6	1·4	1·2	1·1	1·0	0·95
$c.100 \times 10^{-6}$ m^{2}s^{-1}	1·9	1·6	1·4	1·2	1·1	1·0
$c.300 \times 10^{-6}$ m^{2}s^{-1}	-	1·9	1·6	1·4	1·2	1·1
$c.1{\cdot}00 \times 10^{-3}$ m^{2}s^{-1}	-	-	1·9	1·6	1·4	1·2

*** Good wrought iron & coated steel; normal asb. cement; normal uncoated steel; spun lined metal pipes; smooth jointed uPVC: $k_s = \mathbf{0{\cdot}030}$ mm: <u>40 mm to 100 mm diameter</u>**

Kinematic viscosity, ν,						
$c.0{\cdot}30 \times 10^{-6}$ m^{2}s^{-1}	1·0	0·85	0·8	0·75	0·75	0·75
$c.1{\cdot}00 \times 10^{-6}$ m^{2}s^{-1}	1·2	1·0	0·85	0·8	0·75	0·75
$c.3{\cdot}00 \times 10^{-6}$ m^{2}s^{-1}	-	1·2	1·0	0·85	0·8	0·75
$c.10{\cdot}0 \times 10^{-6}$ m^{2}s^{-1}	-	-	1·2	1·0	0·85	0·8
$c.30{\cdot}0 \times 10^{-6}$ m^{2}s^{-1}	-	-	-	1·2	1·0	0·85
$c.100 \times 10^{-6}$ m^{2}s^{-1}	-	-	-	-	1·2	1·0

Ditto; <u>100 mm to 400 mm diameter</u>

Kinematic viscosity, ν,						
$c.0{\cdot}30 \times 10^{-6}$ m^{2}s^{-1}	0·95	0·9	0·85	0·8	0·8	0·8
$c.1{\cdot}00 \times 10^{-6}$ m^{2}s^{-1}	1·1	0·95	0·9	0·85	0·8	0·8
$c.3{\cdot}00 \times 10^{-6}$ m^{2}s^{-1}	1·3	1·1	0·95	0·9	0·85	0·8
$c.10{\cdot}0 \times 10^{-6}$ m^{2}s^{-1}	-	1·3	1·1	0·95	0·9	0·85
$c.30{\cdot}0 \times 10^{-6}$ m^{2}s^{-1}	-	-	1·3	1·1	0·95	0·9
$c.100 \times 10^{-6}$ m^{2}s^{-1}	-	-	-	1·3	1·1	0·95
$c.300 \times 10^{-6}$ m^{2}s^{-1}	-	-	-	-	1·3	1·1

Ditto; <u>400 mm to 1000 mm diameter</u>

Kinematic viscosity, ν,						
$c.0{\cdot}30 \times 10^{-6}$ m^{2}s^{-1}	1·0	0·9	0·9	0·85	0·85	0·85
$c.1{\cdot}00 \times 10^{-6}$ m^{2}s^{-1}	1·1	1·0	0·9	0·9	0·85	0·85
$c.3{\cdot}00 \times 10^{-6}$ m^{2}s^{-1}	1·2	1·1	1·0	0·9	0·9	0·85
$c.10{\cdot}0 \times 10^{-6}$ m^{2}s^{-1}	1·45	1·25	1·1	1·0	0·9	0·9
$c.30{\cdot}0 \times 10^{-6}$ m^{2}s^{-1}	1·7	1·4	1·2	1·1	1·0	0·9
$c.100 \times 10^{-6}$ m^{2}s^{-1}	-	1·7	1·45	1·25	1·1	1·0
$c.300 \times 10^{-6}$ m^{2}s^{-1}	-	-	1·7	1·4	1·2	1·1
$c.1{\cdot}00 \times 10^{-3}$ m^{2}s^{-1}	-	-	-	1·7	1·45	1·25

Ditto; <u>1000 mm to 2500 mm diameter</u>

Kinematic viscosity, ν,						
$c.0{\cdot}30 \times 10^{-6}$ m^{2}s^{-1}	1·0	0·95	0·95	0·95	0·95	0·95
$c.1{\cdot}00 \times 10^{-6}$ m^{2}s^{-1}	1·1	1·0	0·95	0·95	0·95	0·95
$c.3{\cdot}00 \times 10^{-6}$ m^{2}s^{-1}	1·2	1·1	1·0	0·95	0·95	0·95
$c.10{\cdot}0 \times 10^{-6}$ m^{2}s^{-1}	1·4	1·2	1·1	1·0	0·95	0·95
$c.30{\cdot}0 \times 10^{-6}$ m^{2}s^{-1}	1·6	1·4	1·2	1·1	1·0	0·95
$c.100 \times 10^{-6}$ m^{2}s^{-1}	1·9	1·6	1·4	1·2	1·1	1·0
$c.300 \times 10^{-6}$ m^{2}s^{-1}	-	1·9	1·6	1·4	1·2	1·1
$c.1{\cdot}00 \times 10^{-3}$ m^{2}s^{-1}	-	-	1·9	1·6	1·4	1·2

APPENDIX 2 : Indicative values of coefficient m_c for turbulent flow (continued)

Medial values of m_c for gradients S and $(\Delta p/\rho l)$ ms^{-2}

* Good galv. iron & concrete pipes; normal W.I. & coated steel; normal clayware; poor uncoated steel; glass fibre; S. & S. jointed uPVC: $k_s = \mathbf{0.060}$ mm: <u>100 mm to 400 mm diameter</u>

Kinematic viscosity, ν,	$c.0.0001$ $c.(0.001)$	$c.0.0010$ $c.(0.010)$	$c.0.0100$ $c.(0.100)$	$c.0.100$ $c.(1.00)$	$c.1.00$ $c.(10)$	$c.10.0$ $c.(100)$
$c.0.30\times10^{-6}$ m^{2}s^{-1}	1.0	0.9	0.85	0.85	0.85	0.85
$c.1.00\times10^{-6}$ m^{2}s^{-1}	1.1	1.0	0.9	0.85	0.85	0.85
$c.3.00\times10^{-6}$ m^{2}s^{-1}	1.3	1.1	1.0	0.9	0.85	0.85
$c.10.0\times10^{-6}$ m^{2}s^{-1}	-	1.3	1.1	1.0	0.9	0.85
$c.30.0\times10^{-6}$ m^{2}s^{-1}	-	-	1.3	1.1	1.0	0.9
$c.100\times10^{-6}$ m^{2}s^{-1}	-	-	-	1.3	1.1	1.0
$c.300\times10^{-6}$ m^{2}s^{-1}	-	-	-	-	1.3	1.1
$c.1.00\times10^{-3}$ m^{2}s^{-1}	-	-	-	-	-	1.3

Ditto; <u>400 mm to 1000 mm diameter</u>

Kinematic viscosity, ν,	$c.0.0001$	$c.0.0010$	$c.0.0100$	$c.0.100$	$c.1.00$	$c.10.0$
$c.0.30\times10^{-6}$ m^{2}s^{-1}	1.0	0.95	0.95	0.9	0.9	0.9
$c.1.00\times10^{-6}$ m^{2}s^{-1}	1.1	1.0	0.95	0.95	0.9	0.9
$c.3.00\times10^{-6}$ m^{2}s^{-1}	1.25	1.1	1.0	0.95	0.95	0.9
$c.10.0\times10^{-6}$ m^{2}s^{-1}	1.45	1.25	1.1	1.0	0.95	0.95
$c.30.0\times10^{-6}$ m^{2}s^{-1}	1.75	1.45	1.25	1.1	1.0	0.95
$c.100\times10^{-6}$ m^{2}s^{-1}	-	1.75	1.45	1.25	1.1	1.0
$c.300\times10^{-6}$ m^{2}s^{-1}	-	-	1.7	1.45	1.25	1.1
$c.1.00\times10^{-3}$ m^{2}s^{-1}	-	-	-	1.75	1.45	1.25

Ditto; <u>1000 mm to 2500 mm diameter</u>

Kinematic viscosity, ν,	$c.0.0001$	$c.0.0010$	$c.0.0100$	$c.0.100$	$c.1.00$	$c.10.0$
$c.0.30\times10^{-6}$ m^{2}s^{-1}	1.05	1.0	1.0	1.0	1.0	1.0
$c.1.00\times10^{-6}$ m^{2}s^{-1}	1.1	1.05	1.0	1.0	1.0	1.0
$c.3.00\times10^{-6}$ m^{2}s^{-1}	1.25	1.1	1.05	1.0	1.0	1.0
$c.10.0\times10^{-6}$ m^{2}s^{-1}	1.4	1.25	1.1	1.05	1.0	1.0
$c.30.0\times10^{-6}$ m^{2}s^{-1}	1.6	1.4	1.25	1.1	1.05	1.0
$c.100\times10^{-6}$ m^{2}s^{-1}	1.95	1.6	1.4	1.25	1.1	1.05
$c.300\times10^{-6}$ m^{2}s^{-1}	-	1.9	1.6	1.4	1.25	1.1
$c.1.00\times10^{-3}$ m^{2}s^{-1}	-	-	1.95	1.6	1.4	1.25

* Good uncoated cast iron; normal G.I. & concrete pipes; poor W.I. & coated steel; etc: $k_s = \mathbf{0.150}$ mm: <u>100 mm to 400 mm diameter</u>

Kinematic viscosity, ν,	$c.0.0001$	$c.0.0010$	$c.0.0100$	$c.0.100$	$c.1.00$	$c.10.0$
$c.0.30\times10^{-6}$ m^{2}s^{-1}	1.0	1.0	0.95	0.95	0.95	0.95
$c.1.00\times10^{-6}$ m^{2}s^{-1}	1.15	1.05	1.0	0.95	0.95	0.95
$c.3.00\times10^{-6}$ m^{2}s^{-1}	1.3	1.15	1.05	1.0	0.95	0.95
$c.10.0\times10^{-6}$ m^{2}s^{-1}	-	1.3	1.15	1.05	1.0	0.95
$c.30.0\times10^{-6}$ m^{2}s^{-1}	-	-	1.3	1.15	1.05	1.0
$c.100\times10^{-6}$ m^{2}s^{-1}	-	-	-	1.3	1.15	1.05
$c.300\times10^{-6}$ m^{2}s^{-1}	-	-	-	-	1.3	1.15

Ditto; <u>400 mm to 1000 mm diameter</u>

Kinematic viscosity, ν,	$c.0.0001$	$c.0.0010$	$c.0.0100$	$c.0.100$	$c.1.00$	$c.10.0$
$c.0.30\times10^{-6}$ m^{2}s^{-1}	1.05	1.0	1.0	1.0	1.0	1.0
$c.1.00\times10^{-6}$ m^{2}s^{-1}	1.15	1.05	1.0	1.0	1.0	1.0
$c.3.00\times10^{-6}$ m^{2}s^{-1}	1.25	1.1	1.05	1.0	1.0	1.0
$c.10.0\times10^{-6}$ m^{2}s^{-1}	1.45	1.25	1.15	1.05	1.0	1.0
$c.30.0\times10^{-6}$ m^{2}s^{-1}	1.7	1.45	1.25	1.1	1.05	1.0
$c.100\times10^{-6}$ m^{2}s^{-1}	-	1.75	1.45	1.25	1.15	1.05
$c.300\times10^{-6}$ m^{2}s^{-1}	-	-	1.7	1.45	1.25	1.1
$c.1.00\times10^{-3}$ m^{2}s^{-1}	-	-	-	1.75	1.45	1.25

Ditto; <u>1000 mm to 2500 mm diameter</u>

Kinematic viscosity, ν,	$c.0.0001$	$c.0.0010$	$c.0.0100$	$c.0.100$	$c.1.00$	$c.10.0$
$c.0.30\times10^{-6}$ m^{2}s^{-1}	1.1	1.1	1.1	1.05	1.05	1.05
$c.1.00\times10^{-6}$ m^{2}s^{-1}	1.15	1.1	1.1	1.1	1.05	1.05
$c.3.00\times10^{-6}$ m^{2}s^{-1}	1.25	1.15	1.1	1.1	1.1	1.05
$c.10.0\times10^{-6}$ m^{2}s^{-1}	1.4	1.25	1.15	1.1	1.1	1.1
$c.30.0\times10^{-6}$ m^{2}s^{-1}	1.6	1.4	1.25	1.15	1.1	1.1
$c.100\times10^{-6}$ m^{2}s^{-1}	1.95	1.6	1.4	1.25	1.15	1.1
$c.300\times10^{-6}$ m^{2}s^{-1}	-	1.9	1.6	1.4	1.25	1.15
$c.1.00\times10^{-3}$ m^{2}s^{-1}	-	-	1.95	1.6	1.4	1.25

* Good concrete against steel forms; normal uncoated C.I.; poor G.I. & concrete pipes;; etc: $k_s = \mathbf{0.30}$ mm: <u>100 mm to 400 mm diameter</u>

Kinematic viscosity, ν,	$c.0.0001$	$c.0.0010$	$c.0.0100$	$c.0.100$	$c.1.00$	$c.10.0$
$c.0.30\times10^{-6}$ m^{2}s^{-1}	1.1	1.05	1.05	1.05	1.05	1.0
$c.1.00\times10^{-6}$ m^{2}s^{-1}	1.15	1.1	1.05	1.05	1.05	1.05
$c.3.00\times10^{-6}$ m^{2}s^{-1}	1.3	1.15	1.1	1.05	1.05	1.05
$c.10.0\times10^{-6}$ m^{2}s^{-1}	-	1.35	1.15	1.1	1.05	1.05
$c.30.0\times10^{-6}$ m^{2}s^{-1}	-	-	1.3	1.15	1.1	1.05
$c.100\times10^{-6}$ m^{2}s^{-1}	-	-	-	1.35	1.15	1.1
$c.300\times10^{-6}$ m^{2}s^{-1}	-	-	-	-	1.3	1.15

APPENDIX 2 : Indicative values of coefficient m_c for turbulent flow (continued)

Medial values of m_c for gradients S and $(\Delta p/\rho l)$ ms^{-2}

* Good concrete against steel forms; normal uncoated C.I.; poor G.I. & concrete pipes;; etc: $k_s = \mathbf{0 \cdot 30}$ mm: <u>400 mm to 1000 mm diameter</u>

Kinematic viscosity, ν,	$c.\,0.0001$ / $c.\,(0.001)$	$c.\,0.0010$ / $c.\,(0.010)$	$c.\,0.0100$ / $c.\,(0.100)$	$c.\,0.100$ / $c.\,(1.00)$	$c.\,1.00$ / $c.\,(10)$	$c.\,10.0$ / $c.\,(100)$
$c.\,0{\cdot}30 \times 10^{-6}\ \mathrm{m^2 s^{-1}}$	1·1	1·1	1·1	1·1	1·1	1·1
$c.\,1{\cdot}00 \times 10^{-6}\ \mathrm{m^2 s^{-1}}$	1·15	1·1	1·1	1·1	1·1	1·1
$c.\,3{\cdot}00 \times 10^{-6}\ \mathrm{m^2 s^{-1}}$	1·25	1·15	1·1	1·1	1·1	1·1
$c.\,10{\cdot}0 \times 10^{-6}\ \mathrm{m^2 s^{-1}}$	1·45	1·25	1·15	1·1	1·1	1·1
$c.\,30{\cdot}0 \times 10^{-6}\ \mathrm{m^2 s^{-1}}$	1·7	1·45	1·25	1·15	1·1	1·1
$c.\,100 \times 10^{-6}\ \mathrm{m^2 s^{-1}}$	-	1·75	1·45	1·25	1·15	1·1
$c.\,300 \times 10^{-6}\ \mathrm{m^2 s^{-1}}$	-	-	1·7	1·45	1·25	1·15
$c.\,1{\cdot}00 \times 10^{-3}\ \mathrm{m^2 s^{-1}}$	-	-	-	1·75	1·45	1·25

Ditto; <u>1000 mm to 2500 mm diameter</u>

Kinematic viscosity, ν,	$c.\,0.0001$	$c.\,0.0010$	$c.\,0.0100$	$c.\,0.100$	$c.\,1.00$	$c.\,10.0$
$c.\,0{\cdot}30 \times 10^{-6}\ \mathrm{m^2 s^{-1}}$	1·15	1·15	1·15	1·15	1·15	1·15
$c.\,1{\cdot}00 \times 10^{-6}\ \mathrm{m^2 s^{-1}}$	1·2	1·15	1·15	1·15	1·15	1·15
$c.\,3{\cdot}00 \times 10^{-6}\ \mathrm{m^2 s^{-1}}$	1·25	1·2	1·15	1·15	1·15	1·15
$c.\,10{\cdot}0 \times 10^{-6}\ \mathrm{m^2 s^{-1}}$	1·4	1·3	1·2	1·15	1·15	1·15
$c.\,30{\cdot}0 \times 10^{-6}\ \mathrm{m^2 s^{-1}}$	1·6	1·4	1·25	1·2	1·15	1·15
$c.\,100 \times 10^{-6}\ \mathrm{m^2 s^{-1}}$	1·95	1·65	1·4	1·3	1·2	1·15
$c.\,300 \times 10^{-6}\ \mathrm{m^2 s^{-1}}$	-	1·9	1·6	1·4	1·25	1·2
$c.\,1{\cdot}00 \times 10^{-3}\ \mathrm{m^2 s^{-1}}$	-	-	1·95	1·65	1·4	1·3

* Good glazed brickwork; normal rusty W.I.; poor uncoated C.I.; very poor concrete pipes; etc: $k_s = \mathbf{0 \cdot 60}$ mm: <u>100 mm to 400 mm diameter</u>

Kinematic viscosity, ν,	$c.\,0.0001$	$c.\,0.0010$	$c.\,0.0100$	$c.\,0.100$	$c.\,1.00$	$c.\,10.0$
$c.\,0{\cdot}30 \times 10^{-6}\ \mathrm{m^2 s^{-1}}$	1·15	1·15	1·15	1·15	1·15	1·15
$c.\,1{\cdot}00 \times 10^{-6}\ \mathrm{m^2 s^{-1}}$	1·25	1·15	1·15	1·15	1·15	1·15
$c.\,3{\cdot}00 \times 10^{-6}\ \mathrm{m^2 s^{-1}}$	1·35	1·2	1·15	1·15	1·15	1·15
$c.\,10{\cdot}0 \times 10^{-6}\ \mathrm{m^2 s^{-1}}$	-	1·35	1·25	1·15	1·15	1·15
$c.\,30{\cdot}0 \times 10^{-6}\ \mathrm{m^2 s^{-1}}$	-	-	1·35	1·2	1·15	1·15
$c.\,100 \times 10^{-6}\ \mathrm{m^2 s^{-1}}$	-	-	-	1·35	1·25	1·15
$c.\,300 \times 10^{-6}\ \mathrm{m^2 s^{-1}}$	-	-	-	-	1·35	1·2
$c.\,1{\cdot}00 \times 10^{-3}\ \mathrm{m^2 s^{-1}}$	-	-	-	-	-	1·35

Ditto; <u>400 mm to 1000 mm diameter</u>

Kinematic viscosity, ν,	$c.\,0.0001$	$c.\,0.0010$	$c.\,0.0100$	$c.\,0.100$	$c.\,1.00$	$c.\,10.0$
$c.\,0{\cdot}30 \times 10^{-6}\ \mathrm{m^2 s^{-1}}$	1·2	1·2	1·2	1·2	1·2	1·2
$c.\,1{\cdot}00 \times 10^{-6}\ \mathrm{m^2 s^{-1}}$	1·25	1·2	1·2	1·2	1·2	1·2
$c.\,3{\cdot}00 \times 10^{-6}\ \mathrm{m^2 s^{-1}}$	1·3	1·25	1·2	1·2	1·2	1·2
$c.\,10{\cdot}0 \times 10^{-6}\ \mathrm{m^2 s^{-1}}$	1·45	1·3	1·25	1·2	1·2	1·2
$c.\,30{\cdot}0 \times 10^{-6}\ \mathrm{m^2 s^{-1}}$	1·75	1·45	1·3	1·25	1·2	1·2
$c.\,100 \times 10^{-6}\ \mathrm{m^2 s^{-1}}$	-	1·75	1·45	1·3	1·25	1·2
$c.\,300 \times 10^{-6}\ \mathrm{m^2 s^{-1}}$	-	-	1·75	1·45	1·3	1·25
$c.\,1{\cdot}00 \times 10^{-3}\ \mathrm{m^2 s^{-1}}$	-	-	-	1·75	1·45	1·3

Ditto; <u>1000 mm to 2500 mm diameter</u>

Kinematic viscosity, ν,	$c.\,0.0001$	$c.\,0.0010$	$c.\,0.0100$	$c.\,0.100$	$c.\,1.00$	$c.\,10.0$
$c.\,0{\cdot}30 \times 10^{-6}\ \mathrm{m^2 s^{-1}}$	1·25	1·25	1·25	1·25	1·25	1·25
$c.\,1{\cdot}00 \times 10^{-6}\ \mathrm{m^2 s^{-1}}$	1·25	1·25	1·25	1·25	1·25	1·25
$c.\,3{\cdot}00 \times 10^{-6}\ \mathrm{m^2 s^{-1}}$	1·3	1·25	1·25	1·25	1·25	1·25
$c.\,10{\cdot}0 \times 10^{-6}\ \mathrm{m^2 s^{-1}}$	1·45	1·3	1·25	1·25	1·25	1·25
$c.\,30{\cdot}0 \times 10^{-6}\ \mathrm{m^2 s^{-1}}$	1·65	1·45	1·3	1·25	1·25	1·25
$c.\,100 \times 10^{-6}\ \mathrm{m^2 s^{-1}}$	1·95	1·65	1·45	1·3	1·25	1·25
$c.\,300 \times 10^{-6}\ \mathrm{m^2 s^{-1}}$	-	1·95	1·65	1·45	1·3	1·25
$c.\,1{\cdot}00 \times 10^{-3}\ \mathrm{m^2 s^{-1}}$	-	-	1·95	1·65	1·45	1·3

* Good ordinary brickwork; normal glazed brickwork; poor concrete against steel forms; etc: $k_s = \mathbf{1 \cdot 50}$ mm: <u>400 mm to 1000 mm diameter</u>

Kinematic viscosity, ν,	$c.\,0.0001$	$c.\,0.0010$	$c.\,0.0100$	$c.\,0.100$	$c.\,1.00$	$c.\,10.0$
$c.\,0{\cdot}30 \times 10^{-6}\ \mathrm{m^2 s^{-1}}$	1·35	1·3	1·3	1·3	1·3	1·3
$c.\,1{\cdot}00 \times 10^{-6}\ \mathrm{m^2 s^{-1}}$	1·35	1·35	1·3	1·3	1·3	1·3
$c.\,3{\cdot}00 \times 10^{-6}\ \mathrm{m^2 s^{-1}}$	1·4	1·35	1·35	1·3	1·3	1·3
$c.\,10{\cdot}0 \times 10^{-6}\ \mathrm{m^2 s^{-1}}$	1·55	1·4	1·35	1·35	1·3	1·3
$c.\,30{\cdot}0 \times 10^{-6}\ \mathrm{m^2 s^{-1}}$	1·75	1·55	1·4	1·35	1·35	1·3
$c.\,100 \times 10^{-6}\ \mathrm{m^2 s^{-1}}$	-	1·8	1·55	1·4	1·35	1·35
$c.\,300 \times 10^{-6}\ \mathrm{m^2 s^{-1}}$	-	-	1·75	1·55	1·4	1·35
$c.\,1{\cdot}00 \times 10^{-3}\ \mathrm{m^2 s^{-1}}$	-	-	-	1·8	1·55	1·4

Ditto; <u>1000 mm to 2500 mm diameter</u>

Kinematic viscosity, ν,	$c.\,0.0001$	$c.\,0.0010$	$c.\,0.0100$	$c.\,0.100$	$c.\,1.00$	$c.\,10.0$
$c.\,0{\cdot}30 \times 10^{-6}\ \mathrm{m^2 s^{-1}}$	1·35	1·35	1·35	1·35	1·35	1·35
$c.\,1{\cdot}00 \times 10^{-6}\ \mathrm{m^2 s^{-1}}$	1·4	1·35	1·35	1·35	1·35	1·35
$c.\,3{\cdot}00 \times 10^{-6}\ \mathrm{m^2 s^{-1}}$	1·4	1·4	1·35	1·35	1·35	1·35
$c.\,10{\cdot}0 \times 10^{-6}\ \mathrm{m^2 s^{-1}}$	1·5	1·4	1·4	1·35	1·35	1·35
$c.\,30{\cdot}0 \times 10^{-6}\ \mathrm{m^2 s^{-1}}$	1·65	1·5	1·4	1·4	1·35	1·35
$c.\,100 \times 10^{-6}\ \mathrm{m^2 s^{-1}}$	1·95	1·65	1·5	1·4	1·4	1·35
$c.\,300 \times 10^{-6}\ \mathrm{m^2 s^{-1}}$	-	1·95	1·65	1·5	1·4	1·4

APPENDIX 3: Multiplying factor F_P for discharge (laminar flow)

D_G (mm)	D_T (mm)	Fact.	Notes
4.0	5	0.410	
4.0	6	0.198	
4.2	5	0.498	
4.2	6	0.240	
4.4	5	0.600	
4.4	6	0.289	
4.6	5	0.716	
4.6	6	0.345	
4.763	5	0.823	0.1875in
4.763	6	0.397	0.1875in
4.8	5	0.849	
4.8	6	0.410	
5.0	6	0.482	
5.2	5	1.170	
5.2	6	0.564	
5.4	5	1.360	
5.4	6	0.656	
5.6	5	1.574	
5.6	6	0.759	
5.8	6	0.873	
6.2	6	1.140	
6.35	6	1.255	0.25in
6.4	6	1.295	
6.6	6	1.464	
6.8	6	1.650	
6.8	7	0.891	
7.0	6	1.853	
7.0	8	0.586	
7.2	7	1.119	
7.2	8	0.656	
7.4	7	1.249	
7.4	8	0.732	
7.6	7	1.390	
7.6	8	0.815	
7.8	8	0.904	
7.938	8	0.969	0.3125in
8.2	8	1.104	
8.4	8	1.216	
8.6	8	1.335	
8.8	8	1.464	
9.0	8	1.602	
9.2	9	1.092	
9.2	10	0.716	
9.4	9	1.190	
9.4	10	0.781	
9.525	10	0.823	0.375in
9.6	10	0.849	
9.8	10	0.922	
10.2	10	1.082	
10.4	10	1.170	
10.6	10	1.262	
10.8	10	1.360	
10.8	11	0.929	
11.113	10	1.525	0.4375in
11.113	11	1.042	0.4375in

D_G (mm)	D_T (mm)	Fact.	Notes
11.2	10	1.574	
11.2	11	1.075	
11.4	10	1.689	
11.4	11	1.154	
11.6	11	1.237	
11.6	12.5	0.742	
11.8	12	0.935	
11.8	12.5	0.794	
12.5	12	1.177	
12.7	12	1.255	0.5in
12.7	12.5	1.066	0.5in
13.5	12.5	1.360	
13.5	13	1.623	
14.288	14	1.085	0.5625in
14.288	15	0.823	0.5625in
14.5	14	1.151	
14.5	15	0.873	
15.5	15	1.140	
15.875	15	1.255	0.625in
15.875	16	0.969	0.625in
16.5	16	1.131	
17.0	15	1.650	
17.0	16	1.274	
17.463	15	1.837	0.6875in
17.463	17	1.113	0.6875in
18.0	20	0.656	
18.5	18	1.116	
18.5	20	0.732	
19.0	20	0.815	
19.05	20	0.823	0.75in
19.5	20	0.904	
20.5	20	1.104	
20.638	20	1.134	0.8125in
21.0	20	1.216	
21.5	20	1.335	
22.225	20	1.525	0.875in
22.225	22	1.042	0.875in
23.0	22	1.195	
23.0	25	0.716	
23.813	25	0.823	0.9375in
25.4	25	1.066	1.0in
26.0	25	1.170	
27.0	28	0.865	
27.0	30	0.656	
28.575	28	1.085	1.125in
28.575	30	0.823	1.125in
29.0	30	0.873	
31.75	30	1.255	1.25in
35.0	30	1.853	
38.10	40	0.823	1.5in
44.45	40	1.525	1.75in
44.45	45	0.952	1.75in
50.8	50	1.066	2.0in
63.5	60	1.255	2.5in
67.06	60	1.560	2.64in

D_G (mm)	D_T (mm)	Fact.	Notes
76.2	80	0.823	3.0in
88.9	80	1.525	3.5in
101.6	100	1.066	4.0in
107.95	100	1.358	4.25in
107.95	110	0.928	4.25in
127.0	125	1.066	5.0in
133.35	125	1.295	5.25in
133.35	135	0.952	5.25in
152.4	150	1.066	6.0in
159.0	150	1.263	6.26in
177.8	175	1.066	7.0in
177.8	200	0.625	7.0in
203.2	200	1.066	8.0in
219.2	200	1.443	8.63in
228.6	200	1.707	9.0in
228.6	225	1.066	9.0in
254.0	250	1.066	10.0in
267	250	1.300	10.51in
279.4	275	1.066	11.0in
279.4	300	0.752	11.0in
304.8	300	1.066	12.0in
323.85	300	1.358	12.75in
355.6	350	1.066	14.0in
355.6	400	0.625	14.0in
368	400	0.716	14.49in
381.0	375	1.066	15.0in
381.0	400	0.823	15.0in
406.4	400	1.066	16.0in
419.1	400	1.205	16.5in
440	400	1.464	
457.2	450	1.066	18.0in
457.2	500	0.699	18.0in
508.0	500	1.066	20.0in
515	500	1.126	
533.4	500	1.295	21.0in
558.8	600	0.752	22.0in
560	600	0.759	
609.6	600	1.066	24.0in
635.0	600	1.255	25.0in
660.4	600	1.468	26.0in
685.8	600	1.707	27.0in
685.8	675	1.066	27.0in
711.2	600	1.974	28.0in
762.0	800	0.823	30.0in
812.8	800	1.066	32.0in
838.2	800	1.205	33.0in
914.4	800	1.707	36.0in
914.4	900	1.066	36.0in
990.6	1000	0.963	39.0in
1016.0	1000	1.066	40.0in
1066.8	1000	1.295	42.0in
1117.6	1000	1.560	44.0in
1143.0	1000	1.707	45.0in
1219.2	1250	0.905	48.0in
1295.4	1250	1.153	51.0in

To estimate the discharge for a diameter D_G, obtain the discharge for a diameter D_T, for same gradient from the appropriate tables and multiply by the factor F_P given above. The factor derives from the ratio of diameters to the exponent four. Values of D_T are chosen to concentrate on solutions using table values from Tables E.

APPENDIX 4: Multiplying factor F_T to apply to value of mQ (turbulent flow)

D_G (mm)	D_T (mm)	Fact.	Notes	D_G (mm)	D_T (mm)	Fact.	Notes	D_G (mm)	D_T (mm)	Fact.	Notes
4·0	5	0·552		21·5	22	0·941		990·6	1000	0·975	39·0in
4·2	5	0·628		22·225	22	1·028	0·875in	1016	1000	1·043	40·0in
4·4	5	0·711		23·0	22	1·126		1066·8	1050	1·043	42·0in
4·6	5	0·801		23·813	24	0·979	0·9375in	1070	1075	0·988	
4·763	5	0·878	0·1875in	25·4	25	1·043	1·0in	1117·6	1125	0·983	44·0in
4·8	5	0·897		26·0	25	1·110		1143	1150	0·984	45·0in
5·2	5	1·110		27·0	28	0·908		1207	1200	1·016	
5·4	5	1·228		28·575	28	1·056	1·125in	1219·2	1225	0·987	48·0in*
5·6	5	1·353		29·0	28	1·098		1264	1275	0·977	
5·8	6	0·914		31·75	30	1·163	1·25in	1295·4	1300	0·991	51·0in
6·2	6	1·091		38·10	35	1·254	1·5in	1320·8	1300	1·043	52·0in
6·35	6	1·163	0·25in	44·45	45	0·968	1·75in	1333·5	1350	0·968	52·5in
6·4	6	1·188		50·8	50	1·043	2·0in	1371·6	1350	1·043	54·0in
6·6	6	1·289		63·5	65	0·940	2·5in	1380	1375	1·010	*
6·8	7	0·926		67·06	65	1·087	2·64in	1422·4	1425	0·995	56·0in
7·2	7	1·078		76·2	75	1·043	3·0in	1447·8	1450	0·996	57·0in
7·4	7	1·160		88·9	90	0·968	3·5in	1511·3	1500	1·020	59·5in
7·6	7	1·245		101·6	100	1·043	4·0in	1520	1525	0·991	
7·8	8	0·935		108·0	110	0·952	4·25in	1524	1525	0·998	60·0in
7·938	8	0·979	0·3125in	127·0	125	1·043	5·0in	1530	1525	1·009	*
8·2	8	1·068		133·4	135	0·968	5·25in	1625·6	1625	1·001	64·0in
8·4	8	1·139		152·4	150	1·043	6·0in	1638·3	1650	0·981	64·5in
8·6	8	1·213		159·0	160	0·983	6·26in	1660	1650	1·016	*
8·8	9	0·942		177·8	175	1·043	7·0in	1676·4	1675	1·002	66·0in
9·2	9	1·060		203·2	200	1·043	8·0in	1680	1675	1·008	
9·4	9	1·123		219·2	225	0·933	8·63in	1727·2	1725	1·003	68·0in
9·525	9	1·163	0·375in	228·6	225	1·043	9·0in	1752·6	1750	1·004	69·0in
9·6	9	1·188		254·0	250	1·043	10·0in	1803·4	1800	1·005	71·0in
9·8	10	0·948		267	275	0·924	10·51in	1828·8	1825	1·006	72·0in*
10·2	10	1·054		279·4	275	1·043	11·0in	1866·9	1875	0·989	73·5in
10·4	10	1·110		304·8	300	1·043	12·0in	1879·6	1875	1·007	74·0in
10·6	10	1·168		323·9	315	1·077	12·75in	1905	1900	1·007	75·0in
10·8	11	0·952		355·6	350	1·043	14·0in	1910	1900	1·014	*
11·113	11	1·028	0·4375in	368	375	0·951	14·49in	1930·4	1950	0·973	76·0in
11·2	11	1·049		381·0	375	1·043	15·0in	1943·1	1950	0·991	76·5in
11·4	11	1·100		406·4	400	1·043	16·0in	1981·2	2000	0·975	78·0in
11·6	11	1·152		419·1	425	0·963	16·5in	1990	2000	0·987	*
11·8	12	0·956		440	450	0·942		2133·6	2150	0·980	84·0in
12·5	12	1·115		457·2	450	1·043	18·0in	2140	2150	0·988	*
12·7	12	1·163	0·5in	508	500	1·043	20·0in	2280	2300	0·977	*
13·5	13	1·106		515	500	1·082		2286·0	2300	0·984	90·0in
14·288	14	1·056	0·5625in	533·4	525	1·043	21·0in	2438·4	2450	0·987	96·0in*
14·5	14	1·098		558·8	550	1·043	22·0in	2590·8	2500	1·100	102·0in
15·5	15	1·091		560	550	1·049		2600	2500	1·110	
15·875	16	0·979	0·625in	609·6	600	1·043	24·0in	2700	2500	1·228	
16·5	16	1·086		630	625	1·021		2743·2	2500	1·281	108·0in
17·463	17	1·074	0·6875in	635·0	625	1·043	25·0in	2800	2500	1·353	
17·5	17	1·080		660·4	650	1·043	26·0in	2895·6	2500	1·480	114·0in
18·5	18	1·076		685·8	675	1·043	27·0in	3048·0	2500	1·696	120·0in*
19·0	18	1·155		711·2	700	1·043	28·0in	3200·4	2500	1·932	126·0in
19·05	18	1·163	0·75in	730	725	1·018		3352·8	2500	2·187	132·0in*
19·5	20	0·935		762	750	1·043	30·0in	3400	2500	2·270	
20·5	20	1·068		812·8	800	1·043	32·0in	3505·2	2500	2·463	138·0in
20·638	20	1·087	0·8125in	838·2	825	1·043	33·0in	3657·6	2500	2·754	144·0in
21·0	20	1·139		914·4	900	1·043	36·0in	4000	2500	3·502	

* indicates tunnel size; one-pass or grano. lined

To obtain the value of mQ for a non-tabulated diameter D_G, obtain the value for a diameter D_T, for same gradient, from the appropriate Table D and multiply by the factor F_T given above. The factor derives from the the ratio of diameters to the exponent 2·667.

APPENDIX 5 : Properties of fluids

List (b) illustrates variations in kinematic viscosity. Where the user of the Tables does not have direct assessment of density and/or viscosity of the fluid being considered, possible sources of that information include ref. 15 and the following. In List (b) items based on data from ref. 15 are indicated by * and 'glycol' is shown by 'g.'.

- ESDU International, Item No 66024. *Approximate data on the viscosity of some common liquids.* 1966.
- ESDU International, Item No 66024. *A guide to the viscosity of liquid petroleum products.* 1971.
- KAYE, G.W.C. and LABY, T.H. *Tables of physical and chemical constants and some mathematical functions, 15th edition.* Longmans, London, 1986 (now prepared under the direction of an Editorial Committee).
- VISWANATH, D.S. and NATARAJAN, G. *Data book on the viscosity of liquids.* Hemisphere Publishing, N.Y., 1989.
- TOULOUKIAN, Y.S.,ed. *TPRC Series - Thermophysical properties of matter.* Plenum Publishing, New York (c. 14 volumes from 1970 onwards)

(a) Temperatures, T, of fresh water in $°C$ at the standard values of kinematic viscosity, ν, in m^2s^{-1}.

ν	T	ν	T	ν	T	ν	T
$0{\cdot}30 \times 10^{-6}$	$98{\cdot}6$	$0{\cdot}50 \times 10^{-6}$	$56{\cdot}7$	$0{\cdot}80 \times 10^{-6}$	$30{\cdot}2$	$1{\cdot}25 \times 10^{-6}$	$11{\cdot}6$
$0{\cdot}40 \times 10^{-6}$	$72{\cdot}9$	$0{\cdot}60 \times 10^{-6}$	$45{\cdot}4$	$1{\cdot}00 \times 10^{-6}$	$20{\cdot}3$	$1{\cdot}50 \times 10^{-6}$	$5{\cdot}4$

(b) Values of density, ρ, in kgm^{-3} and kinematic viscosity, ν, in m^2s^{-1}, for various fluids

	ρ	ν		ρ	ν
Water (0°C)	999·9	$1{\cdot}792 \times 10^{-6}$	Lubricating oil (20°C)*	871·0	$14{\cdot}99 \times 10^{-3}$
Water(5°C)	1·000	$1{\cdot}519 \times 10^{-6}$	Lubricating oil (100°C)*	820·0	$2{\cdot}439 \times 10^{-6}$
Water (10°C)	999·6	$1{\cdot}308 \times 10^{-6}$	Mercury (20°C)*	13579	$0{\cdot}114 \times 10^{-6}$
Water (15°C)	999·1	$1{\cdot}141 \times 10^{-6}$	Mercury (20°C)*	13025	$0{\cdot}076 \times 10^{-6}$
Water (20°C)	998·2	$1{\cdot}007 \times 10^{-6}$	Octane C_8-H_{18} (7°C)*	712·7	$0{\cdot}902 \times 10^{-6}$
Water (30°C)	995·7	$0{\cdot}804 \times 10^{-6}$	Octane C_8-H_{18} (27°C)*	696·6	$0{\cdot}718 \times 10^{-6}$
Water (40°C)	992·9	$0{\cdot}661 \times 10^{-6}$	Octane C_8-H_{18} (47°C)*	680·3	$0{\cdot}588 \times 10^{-6}$
Water (50°C)	988·8	$0{\cdot}556 \times 10^{-6}$	Octane C_8-H_{18} (67°C)*	663·6	$0{\cdot}423 \times 10^{-6}$
Water (60°C)	983·2	$0{\cdot}477 \times 10^{-6}$	Pentane C_5-H_{12} (7°C)*	638·9	$0{\cdot}409 \times 10^{-6}$
Water (80°C)	974·5	$0{\cdot}367 \times 10^{-6}$	Pentane C_5-H_{12} (27°C)*	619·3	$0{\cdot}355 \times 10^{-6}$
Water (100°C)	958·4	$0{\cdot}296 \times 10^{-6}$	Propylene g. 100% (10°C)*	1042	$107{\cdot}5 \times 10^{-6}$
Salt water (0°C)	c 1027	$c\,1{\cdot}923 \times 10^{-6}$	Propylene g. 50% (10°C)*	1047	$8{\cdot}787 \times 10^{-6}$
Salt water (20°C)	c 1025	$c\,1{\cdot}085 \times 10^{-6}$	Propylene g. 30% (10°C)*	1028	$3{\cdot}891 \times 10^{-6}$
Salt water (40°C)	c 1019	$c\,0{\cdot}881 \times 10^{-6}$	Sulphuric acid H_2-SO_4 (20°C)*	1834	$13{\cdot}85 \times 10^{-6}$
Acetone C_3-H_6-O (7°C)*	805·2	$0{\cdot}457 \times 10^{-6}$	Tetraethylene g. 100% (6·55°C)*	1084	$9{\cdot}041 \times 10^{-6}$
Acetone C_3-H_6-O (27°C)*	782·3	$0{\cdot}396 \times 10^{-6}$	Toluene C_7-H_8 (7°C)*	879·0	$0{\cdot}794 \times 10^{-6}$
Acetone C_3-H_6-O (47°C)*	759·7	$0{\cdot}329 \times 10^{-6}$	Toluene C_7-H_8 (27°C)*	860·4	$0{\cdot}628 \times 10^{-6}$
Acetone C_3-H_6-O (67°C)*	736·2	$0{\cdot}287 \times 10^{-6}$	Toluene C_7-H_8 (47°C)*	841·7	$0{\cdot}513 \times 10^{-6}$
Ammonia (−40°C)*	691·7	$0{\cdot}406 \times 10^{-6}$	Toluene C_7-H_8 (67°C)*	822·6	$0{\cdot}432 \times 10^{-6}$
Ammonia (0°C)*	640·1	$0{\cdot}377 \times 10^{-6}$	Transformer oil (−40°C)*	916·0	$4{\cdot}221 \times 10^{-3}$
Ammonia (10°C)*	626·1	$0{\cdot}368 \times 10^{-6}$	Transformer oil (10°C)*	885·0	$37{\cdot}80 \times 10^{-6}$
Ammonia (20°C)*	611·7	$0{\cdot}359 \times 10^{-6}$	Triethylene g. 100% (10°C)*	1133	$82{\cdot}09 \times 10^{-6}$
Ammonia (30°C)*	596·4	$0{\cdot}349 \times 10^{-6}$	Triethylene g. 50% (10°C)*	1083	$8{\cdot}957 \times 10^{-6}$
Ammonia (40°C)*	581·0	$0{\cdot}340 \times 10^{-6}$	Triethylene g. 30% (10°C)*	1049	$3{\cdot}622 \times 10^{-6}$
Ammonia (50°C)*	564·3	$0{\cdot}330 \times 10^{-6}$	Turpentine (20°C)*	855	$1{\cdot}739 \times 10^{-6}$
Benzene C_6-H_6 (7°C)*	893·0	$0{\cdot}907 \times 10^{-6}$	Turpentine (26°C)	868	$1{\cdot}58 \times 10^{-6}$
Benzene C_6-H_6 (27°C)*	872·0	$0{\cdot}674 \times 10^{-6}$	Jet fuel (JR4)(15·6°C)	773	$1{\cdot}125 \times 10^{-6}$
Benzene C_6-H_6 (47°C)*	850·0	$0{\cdot}533 \times 10^{-6}$	Paraffin oil (20°C)	810	$2{\cdot}375 \times 10^{-6}$
Benzene C_6-H_6 (67°C)*	828·0	$0{\cdot}441 \times 10^{-6}$	Kerosene (26°C)	820	$2{\cdot}00 \times 10^{-6}$
Diethylene g. 100% (10°C)*	1126	$63{\cdot}94 \times 10^{-6}$	Ethyl alcohol	785	$1{\cdot}4 \times 10^{-6}$
Diethylene g. 50% (10°C)*	1078	$7{\cdot}050 \times 10^{-6}$	Methyl alcohol	787	$0{\cdot}71 \times 10^{-6}$
Diethylene g. 30% (10°C)*	1046	$3{\cdot}270 \times 10^{-6}$	Propyl alcohol	800	$2{\cdot}40 \times 10^{-6}$
Ethylene g. 100% (10°C)*	1119	$29{\cdot}94 \times 10^{-6}$	Ether (26°C)	714	$0{\cdot}31 \times 10^{-6}$
Ethylene g. 50% (10°C)*	1070	$5{\cdot}140 \times 10^{-6}$	Linseed oil (26°C)	929	$35{\cdot}6 \times 10^{-6}$
Ethylene g. 30% (10°C)*	1041	$2{\cdot}882 \times 10^{-6}$	Crude oil (−10°C)	925	$2{\cdot}0 \times 10^{-3}$

APPENDIX 5 : Properties of fluids (continued)

(b) Values of density, ρ, in kgm^{-3} and kinematic viscosity, ν, in m^{2}s^{-1}, for various fluids (continued)

	ρ	ν		ρ	ν
Crude oil (20°C)	855	74×10^{-6}	Glycerin (20°C)	1258	$1 \cdot 188 \times 10^{-6}$
Petrol (0°C)	c716	$0 \cdot 8 \times 10^{-6}$	SAE 10 oil (20°C)	918	$89 \cdot 3 \times 10^{-6}$
Petrol (20°C)	c716	$0 \cdot 59 \times 10^{-6}$	SAE 30 oil (20°C)	918	479×10^{-6}
Petrol (60°C)	c716	$0 \cdot 4 \times 10^{-6}$	Air (Atmos. 20°C)	1·205	$14 \cdot 9 \times 10^{-6}$
Fuel oil (20°C)	940	$1 \cdot 20 \times 10^{-3}$	Hydrogen (Atmos. 20°C)	0·0839	107×10^{-6}

APPENDIX 6 : Allowances for additional head losses in turbulent flow

As explained in the section *Other sources of resistance*, the most basic method of allowing for additional head losses at bends and at appurtenances involving change in cross-section is to add pipe length. If the adjustment rule is not conditioned by the uniform flow conditions obtaining, this approach becomes more and more fallible the greater is any change in cross-sectional area of flow, temporary or permanent, that is involved. Given below (a) are indicative values for smooth pipes in terms of number of pipe diameters to be added to the pipe length. Where two diameters (D and d) are involved, the additional length is given in terms of the smaller diameter (d) and is to be added to the length of the smaller pipe. List (a) is based on material given in the annual Reference Handbook published by *Water Engineering and Management*.

Also given below (b) are indicative values, for smooth pipe systems, of the additional loss coefficient ε (Eqs (22) or (22a)). Where two diameters (D and d) are involved, the velocity to be used is that in the smaller diameter (d). List (b) is based mainly on material compiled in *Mark's Standard handbook for mechanical engineers*, Editors E.A.Avallone and T.Baumuster, III, McGraw-Hill, New York, 1987.

The values for given items in two lists ((a) and (b)) for smooth pipes may not be fully compatible; where additional losses are a significant proportion of the total, references 13-15 can provide extension of list (b) as appropriate for individual requirements.

(a) Indicative numbers of pipe diameters to be added to flow length to allow for additional losses

Abrupt 90° elbow (mitre)	67	$d/D=3/4$ tee, through flow	25	Borda, re-entrant, entrance	30
Standard 90° bend fitting	32	$d/D=1/2$ tee, through flow	32	Ordinary abrupt entrance	17
Medium sweep 90° bend	25	Sudden enlarge't, $d/D=3/4$	7	Globe valve, fully open	330
Long sweep 90° bend	20	Sudden enlarge't, $d/D=1/2$	18	Angle valve, fully open	165
Standard 45° bend fitting	15	Sudden enlarge't, $d/D=1/4$	32	Gate valve, fully open	7
180° close return bend	72	Sudden contract'n, $d/D=3/4$	7	Gate valve, 1/4 closed	40
Standard tee, 90° turn	67	Sudden contract'n, $d/D=1/2$	12	Gate valve, 1/2 closed	200
Standard tee, through flow	20	Sudden contract'n, $d/D=1/4$	15	Gate valve, 3/4 closed	800

(b) Indicative values of additional loss coefficient ε , for summation in Eqs (22) or (22a)

Standard 90° bend fitting	0·75	Gradual contraction	0·05	'Exit loss', without venturi	1·0
Long sweep 90° bend	0·45	Sudden cont'n, $d/D=3/4$	0·26	Globe valve, fully open	6·4
Standard 45° bend fitting	0·35	Sudden cont'n, $d/D=1/2$	0·36	Globe valve, 1/2 closed	9·5
Standard tee, 90° turn	1·5	Sudden cont'n, $d/D=1/4$	0·45	Gate valve, fully open	0·2
Standard tee, through flow	0·4	Borda, re-entrant, entrance	1·0	Gate valve, 1/4 closed	0·9
Sudden enlarge't, $d/D=3/4$	0·2	Sharp edged entry	0·4	Gate valve, 1/2 closed	4·5
Sudden enlarge't, $d/D=1/2$	0·6	Rounded entrance,	0·1	Gate valve, 3/4 closed	24
Sudden enlarge't, $d/D=1/4$	0·9	Well rounded entrance	0·05		

APPENDIX 7 : Relationships between units of discharge

1 cubic foot per second	= 28·3168 litres/sec	1 U.S. gallon per second	= 3·78541 litres/sec
1 cubic foot per minute	= 0·47195 litres/sec	1 U.S. gallon per minute	= 0·06309 litres/sec
1 Imperial gallon per second	= 4·54609 litres/sec	10^6 U.S. gallon per day	= 43·8126 litres/sec
1 Imperial gallon per minute	= 0·07577 litres/sec	10^3 barrels per hour	= 44·1631 litres/sec
10^6 Imperial gallon per day	= 52·6168 litres/sec	10^3 barrels per day	= 1·84014 litres/sec
10^6 litres per day	= 11·5741 litres/sec		

m = Manning $n \times 100$
S = 0·00300 to 0·00950

i.e. kin. pr. grad., $\Delta p/\rho l$ =
(0·02942) to (0·09316) ms^{-2}

Full bore flow; Tables E give m_C or m_P for Colebrook-White or for laminar solutions resp., to divide mV and/or mQ as follows:

mV to give velocities in ms^{-1}
mQ to give discharges in millilitres/sec

Gradient — **(Equivalent) Pipe diameters in mm**

S, $(\Delta p/\rho l)$	5	6	7	8	9	10	11	12	13	14	15	16	18
0·00300	0·0636	0·0718	0·0795	0·0869	0·0940	0·1009	0·1075	0·1139	0·1202	0·1263	0·1322	0·1380	0·1493
(0·02942)	1·2479	2·0293	3·0611	4·3704	5·9831	7·9240	10·217	12·885	15·951	19·436	23·363	27·750	37·990
0·00320	0·0656	0·0741	0·0821	0·0898	0·0971	0·1042	0·1110	0·1177	0·1241	0·1304	0·1365	0·1425	0·1542
(0·03138)	1·2889	2·0959	3·1614	4·5137	6·1793	8·1838	10·552	13·308	16·474	20·074	24·129	28·660	39·236
0·00340	0·0677	0·0764	0·0847	0·0926	0·1001	0·1074	0·1145	0·1213	0·1279	0·1344	0·1407	0·1469	0·1589
(0·03334)	1·3285	2·1604	3·2587	4·6526	6·3694	8·4357	10·877	13·717	16·981	20·692	24·871	29·542	40·443
0·00360	0·0696	0·0786	0·0871	0·0952	0·1030	0·1105	0·1178	0·1248	0·1316	0·1383	0·1448	0·1512	0·1635
(0·03530)	1·3671	2·2230	3·3532	4·7875	6·5541	8·6803	11·192	14·115	17·474	21·292	25·592	30·399	41·616
0·00380	0·0715	0·0808	0·0895	0·0979	0·1058	0·1135	0·1210	0·1282	0·1353	0·1421	0·1488	0·1553	0·1680
(0·03727)	1·4045	2·2839	3·4451	4·9187	6·7337	8·9181	11·499	14·502	17·952	21·875	26·294	31·232	42·756
0·00400	0·0734	0·0829	0·0918	0·1004	0·1086	0·1165	0·1241	0·1316	0·1388	0·1458	0·1527	0·1594	0·1724
(0·03923)	1·4410	2·3432	3·5346	5·0464	6·9086	9·1498	11·798	14·879	18·419	22·443	26·977	32·043	43·867
0·00420	0·0752	0·0849	0·0941	0·1029	0·1113	0·1194	0·1272	0·1348	0·1422	0·1494	0·1564	0·1633	0·1766
(0·04119)	1·4766	2·4011	3·6219	5·1711	7·0792	9·3758	12·089	15·246	18·874	22·998	27·643	32·834	44·950
0·00440	0·0770	0·0869	0·0963	0·1053	0·1139	0·1222	0·1302	0·1380	0·1455	0·1529	0·1601	0·1671	0·1808
(0·04315)	1·5113	2·4576	3·7071	5·2928	7·2458	9·5964	12·373	15·605	19·318	23·539	28·293	33·607	46·008
0·00460	0·0787	0·0889	0·0985	0·1077	0·1165	0·1249	0·1331	0·1411	0·1488	0·1563	0·1637	0·1709	0·1849
(0·04511)	1·5453	2·5128	3·7904	5·4117	7·4087	9·8121	12·651	15·956	19·752	24·068	28·929	34·362	47·042
0·00480	0·0804	0·0908	0·1006	0·1100	0·1190	0·1276	0·1360	0·1441	0·1520	0·1597	0·1672	0·1746	0·1888
(0·04707)	1·5785	2·5669	3·8720	5·5281	7·5680	10·023	12·924	16·299	20·177	24·585	29·552	35·101	48·054
0·00500	0·0821	0·0927	0·1027	0·1122	0·1214	0·1303	0·1388	0·1471	0·1551	0·1630	0·1707	0·1782	0·1927
(0·04903)	1·6111	2·6198	3·9518	5·6421	7·7241	10·230	13·190	16·635	20·593	25·092	30·161	35·825	49·045
0·00525	0·0841	0·0949	0·1052	0·1150	0·1244	0·1335	0·1422	0·1507	0·1590	0·1670	0·1749	0·1826	0·1975
(0·05148)	1·6509	2·6845	4·0494	5·7814	7·9148	10·482	13·516	17·046	21·101	25·712	30·906	36·710	50·256
0·00550	0·0861	0·0972	0·1077	0·1177	0·1273	0·1366	0·1456	0·1543	0·1627	0·1710	0·1790	0·1869	0·2021
(0·05394)	1·6897	2·7477	4·1447	5·9175	8·1011	10·729	13·834	17·447	21·598	26·317	31·633	37·574	51·439
0·00575	0·0880	0·0994	0·1101	0·1204	0·1302	0·1397	0·1488	0·1577	0·1664	0·1748	0·1830	0·1911	0·2067
(0·05639)	1·7277	2·8094	4·2378	6·0505	8·2832	10·970	14·145	17·839	22·083	26·909	32·344	38·418	52·595
0·00600	0·0899	0·1015	0·1125	0·1230	0·1330	0·1427	0·1520	0·1611	0·1700	0·1786	0·1870	0·1952	0·2111
(0·05884)	1·7649	2·8699	4·3290	6·1806	8·4613	11·206	14·449	18·223	22·558	27·487	33·040	39·244	53·726
0·00625	0·0917	0·1036	0·1148	0·1255	0·1357	0·1456	0·1552	0·1644	0·1735	0·1822	0·1908	0·1992	0·2155
(0·06129)	1·8013	2·9290	4·4183	6·3081	8·6358	11·437	14·747	18·598	23·024	28·054	33·721	40·054	54·834
0·00650	0·0936	0·1056	0·1171	0·1280	0·1384	0·1485	0·1583	0·1677	0·1769	0·1859	0·1946	0·2032	0·2198
(0·06374)	1·8369	2·9870	4·5058	6·4330	8·8068	11·664	15·039	18·967	23·479	28·610	34·389	40·847	55·920
0·00675	0·0953	0·1077	0·1193	0·1304	0·1411	0·1513	0·1613	0·1709	0·1803	0·1894	0·1983	0·2070	0·2239
(0·06619)	1·8719	3·0440	4·5916	6·5555	8·9746	11·886	15·326	19·328	23·927	29·155	35·044	41·625	56·985
0·00700	0·0971	0·1096	0·1215	0·1328	0·1437	0·1541	0·1642	0·1740	0·1836	0·1929	0·2019	0·2108	0·2280
(0·06865)	1·9063	3·0998	4·6758	6·6758	9·1393	12·104	15·607	19·683	24·366	29·690	35·687	42·389	58·031
0·00725	0·0988	0·1116	0·1236	0·1352	0·1462	0·1568	0·1671	0·1771	0·1868	0·1963	0·2055	0·2146	0·2321
(0·07110)	1·9400	3·1547	4·7586	6·7940	9·3010	12·318	15·883	20·031	24·797	30·215	36·319	43·139	59·058
0·00750	0·1005	0·1135	0·1258	0·1375	0·1487	0·1595	0·1700	0·1801	0·1900	0·1996	0·2090	0·2182	0·2361
(0·07355)	1·9732	3·2086	4·8400	6·9101	9·4600	12·529	16·155	20·373	25·221	30·732	36·939	43·877	60·068
0·00800	0·1038	0·1172	0·1299	0·1420	0·1536	0·1648	0·1756	0·1860	0·1962	0·2062	0·2159	0·2254	0·2438
(0·07845)	2·0379	3·3138	4·9987	7·1368	9·7703	12·940	16·684	21·042	26·048	31·740	38·151	45·316	62·037
0·00850	0·1070	0·1208	0·1339	0·1464	0·1583	0·1698	0·1810	0·1918	0·2023	0·2125	0·2225	0·2323	0·2513
(0·08336)	2·1006	3·4158	5·1525	7·3564	10·071	13·338	17·198	21·689	26·850	32·716	39·325	46·710	63·947
0·00900	0·1101	0·1243	0·1378	0·1506	0·1629	0·1747	0·1862	0·1973	0·2082	0·2187	0·2290	0·2391	0·2586
(0·08826)	2·1615	3·5149	5·3019	7·5697	10·363	13·725	17·696	22·318	27·628	33·665	40·465	48·064	65·801
0·00950	0·1131	0·1277	0·1415	0·1547	0·1674	0·1795	0·1913	0·2027	0·2139	0·2247	0·2353	0·2456	0·2657
(0·09316)	2·2207	3·6112	5·4472	7·7771	10·647	14·101	18·181	22·929	28·385	34·587	41·574	49·381	67·604
S, $(\Delta p/\rho l)$	5	6	7	8	9	10	11	12	13	14	15	16	18

Gradient — **(Equivalent) Pipe diameters in mm**

S = 0·00300 to 0·00950;

$\Delta p/\rho l$ = (0·02942) to (0·09316) ms^{-2}

m = Manning $n \times 100$
S = 0·0100 to 0·0290

Full bore flow; Tables E give m_C or m_P for Colebrook-White or for laminar solutions resp., to divide mV and/or mQ as follows:

i.e. kin. pr. grad., $\Delta p/\rho l$ = (0·0981) to (0·2844) ms^{-2}

mV to give velocities in ms^{-1}
mQ to give discharges in millilitres/sec

Gradient $S, (\Delta p/\rho l)$	(Equivalent) Pipe diameters in mm												
	5	6	7	8	9	10	11	12	13	14	15	16	18
0·0100	0·116	0·131	0·145	0·159	0·172	0·184	0·196	0·208	0·219	0·231	0·241	0·252	0·273
(0·0981)	2·2784	3·7050	5·5887	7·9791	10·924	14·467	18·654	23·525	29·123	35·486	42·654	50·664	69·360
0·0105	0·119	0·134	0·149	0·163	0·176	0·189	0·201	0·213	0·225	0·236	0·247	0·258	0·279
(0·1030)	2·3347	3·7965	5·7267	8·1762	11·193	14·824	19·114	24·106	29·842	36·362	43·707	51·916	71·073
0·0110	0·122	0·137	0·152	0·166	0·180	0·193	0·206	0·218	0·230	0·242	0·253	0·264	0·286
(0·1079)	2·3896	3·8858	5·8615	8·3686	11·457	15·173	19·564	24·673	30·544	37·218	44·736	53·137	72·745
0·0115	0·124	0·141	0·156	0·170	0·184	0·198	0·210	0·223	0·235	0·247	0·259	0·270	0·292
(0·1128)	2·4433	3·9731	5·9932	8·5567	11·714	15·514	20·004	25·228	31·231	38·055	45·741	54·331	74·380
0·0120	0·127	0·144	0·159	0·174	0·188	0·202	0·215	0·228	0·240	0·253	0·264	0·276	0·299
(0·1177)	2·4959	4·0586	6·1221	8·7407	11·966	15·848	20·434	25·771	31·902	38·873	46·725	55·500	75·980
0·0125	0·130	0·147	0·162	0·177	0·192	0·206	0·219	0·233	0·245	0·258	0·270	0·282	0·305
(0·1226)	2·5474	4·1423	6·2484	8·9209	12·213	16·175	20·855	26·302	32·560	39·675	47·689	56·644	77·547
0·0130	0·132	0·149	0·166	0·181	0·196	0·210	0·224	0·237	0·250	0·263	0·275	0·287	0·311
(0·1275)	2·5978	4·2243	6·3721	9·0976	12·455	16·495	21·268	26·823	33·205	40·460	48·633	57·766	79·083
0·0135	0·135	0·152	0·169	0·184	0·200	0·214	0·228	0·242	0·255	0·268	0·280	0·293	0·317
(0·1324)	2·6473	4·3048	6·4935	9·2709	12·692	16·809	21·674	27·334	33·837	41·231	49·559	58·867	80·589
0·0140	0·137	0·155	0·172	0·188	0·203	0·218	0·232	0·246	0·260	0·273	0·286	0·298	0·323
(0·1373)	2·6959	4·3838	6·6126	9·4410	12·925	17·118	22·071	27·835	34·458	41·988	50·469	59·947	82·068
0·0145	0·140	0·158	0·175	0·191	0·207	0·222	0·236	0·250	0·264	0·278	0·291	0·303	0·328
(0·1422)	2·7436	4·4614	6·7297	9·6081	13·154	17·421	22·462	28·328	35·068	42·731	51·362	61·008	83·521
0·0150	0·142	0·160	0·178	0·194	0·210	0·226	0·240	0·255	0·269	0·282	0·296	0·309	0·334
(0·1471)	2·7905	4·5377	6·8447	9·7724	13·379	17·719	22·846	28·812	35·668	43·461	52·240	62·051	84·948
0·0160	0·147	0·166	0·184	0·201	0·217	0·233	0·248	0·263	0·278	0·292	0·305	0·319	0·345
(0·1569)	2·8820	4·6865	7·0692	10·093	13·817	18·300	23·595	29·757	36·838	44·887	53·953	64·086	87·734
0·0170	0·151	0·171	0·189	0·207	0·224	0·240	0·256	0·271	0·286	0·301	0·315	0·329	0·355
(0·1667)	2·9707	4·8307	7·2868	10·404	14·243	18·863	24·321	30·673	37·971	46·268	55·614	66·058	90·434
0·0180	0·156	0·176	0·195	0·213	0·230	0·247	0·263	0·279	0·294	0·309	0·324	0·338	0·366
(0·1765)	3·0568	4·9708	7·4980	10·705	14·655	19·410	25·026	31·562	39·072	47·609	57·226	67·973	93·056
0·0190	0·160	0·181	0·200	0·219	0·237	0·254	0·271	0·287	0·302	0·318	0·333	0·347	0·376
(0·1863)	3·1406	5·1070	7·7035	10·998	15·057	19·942	25·712	32·427	40·143	48·914	58·794	69·836	95·606
0·0200	0·164	0·185	0·205	0·224	0·243	0·261	0·278	0·294	0·310	0·326	0·341	0·356	0·385
(0·1961)	3·2222	5·2396	7·9036	11·284	15·448	20·460	26·380	33·270	41·186	50·185	60·322	71·650	98·090
0·0210	0·168	0·190	0·210	0·230	0·249	0·267	0·284	0·301	0·318	0·334	0·350	0·365	0·395
(0·2059)	3·3018	5·3690	8·0988	11·563	15·830	20·965	27·032	34·091	42·203	51·424	61·811	73·420	100·51
0·0220	0·172	0·194	0·215	0·235	0·255	0·273	0·291	0·309	0·325	0·342	0·358	0·374	0·404
(0·2157)	3·3795	5·4954	8·2894	11·835	16·202	21·458	27·668	34·893	43·196	52·634	63·266	75·147	102·88
0·0230	0·176	0·199	0·220	0·241	0·260	0·279	0·298	0·315	0·333	0·350	0·366	0·382	0·413
(0·2256)	3·4554	5·6189	8·4757	12·101	16·566	21·940	28·290	35·678	44·167	53·817	64·688	76·836	105·19
0·0240	0·180	0·203	0·225	0·246	0·266	0·285	0·304	0·322	0·340	0·357	0·374	0·390	0·422
(0·2354)	3·5297	5·7397	8·6580	12·361	16·923	22·412	28·898	36·445	45·117	54·975	66·079	78·489	107·45
0·0250	0·183	0·207	0·230	0·251	0·271	0·291	0·310	0·329	0·347	0·364	0·382	0·398	0·431
(0·2452)	3·6025	5·8581	8·8365	12·616	17·272	22·875	29·494	37·197	46·047	56·108	67·442	80·107	109·67
0·0260	0·187	0·211	0·234	0·256	0·277	0·297	0·317	0·335	0·354	0·372	0·389	0·406	0·440
(0·2550)	3·6739	5·9741	9·0115	12·866	17·614	23·328	30·078	37·933	46·959	57·219	68·777	81·694	111·84
0·0270	0·191	0·215	0·239	0·261	0·282	0·303	0·323	0·342	0·361	0·379	0·397	0·414	0·448
(0·2648)	3·7438	6·0879	9·1832	13·111	17·949	23·772	30·651	38·656	47·853	58·309	70·088	83·250	113·97
0·0280	0·194	0·219	0·243	0·266	0·287	0·308	0·328	0·348	0·367	0·386	0·404	0·422	0·456
(0·2746)	3·8125	6·1996	9·3517	13·352	18·279	24·208	31·213	39·365	48·732	59·379	71·374	84·778	116·06
0·0290	0·198	0·223	0·247	0·270	0·292	0·314	0·334	0·354	0·374	0·393	0·411	0·429	0·464
(0·2844)	3·8800	6·3094	9·5172	13·588	18·602	24·637	31·766	40·062	49·594	60·431	72·637	86·278	118·12
$S, (\Delta p/\rho l)$	5	6	7	8	9	10	11	12	13	14	15	16	18

Gradient (Equivalent) Pipe diameters in mm

S = 0·0100 to 0·0290; $\Delta p/\rho l$ = (0·0981) to (0·2844) ms^{-2}

m = Manning $n \times 100$
S = 0·0300 to 0·0950

i.e. kin. pr. grad., $\Delta p/\rho l$ = (0·2942) to (0·9316) ms^{-2}

Full bore flow; Tables E give m_C or m_P for Colebrook-White or for laminar solutions resp., to divide mV and/or mQ as follows:

mV to give velocities in ms^{-1}
mQ to give discharges in millilitres/sec

Gradient (Equivalent) Pipe diameters in mm

S, ($\Delta p/\rho l$)	5	6	7	8	9	10	11	12	13	14	15	16	18
0·0300	0·201	0·227	0·252	0·275	0·297	0·319	0·340	0·360	0·380	0·399	0·418	0·436	0·472
(0·2942)	3·9464	6·4172	9·6799	13·820	18·920	25·058	32·309	40·747	50·442	61·464	73·879	87·753	120·14
0·0320	0·208	0·234	0·260	0·284	0·307	0·330	0·351	0·372	0·392	0·412	0·432	0·451	0·488
(0·3138)	4·0758	6·6277	9·9974	14·274	19·541	25·880	33·369	42·083	52·096	63·479	76·302	90·631	124·07
0·0340	0·214	0·242	0·268	0·293	0·317	0·340	0·362	0·384	0·405	0·425	0·445	0·465	0·503
(0·3334)	4·2012	6·8316	10·305	14·713	20·142	26·676	34·396	43·378	53·700	65·433	78·650	93·420	127·89
0·0360	0·220	0·249	0·276	0·301	0·326	0·349	0·372	0·395	0·416	0·437	0·458	0·478	0·517
(0·3530)	4·3230	7·0297	10·604	15·139	20·726	27·449	35·393	44·636	55·256	67·330	80·930	96·129	131·60
0·0380	0·226	0·255	0·283	0·309	0·335	0·359	0·383	0·405	0·428	0·449	0·471	0·491	0·531
(0·3727)	4·4415	7·2223	10·894	15·554	21·294	28·202	36·363	45·859	56·771	69·175	83·148	98·763	135·21
0·0400	0·232	0·262	0·290	0·317	0·343	0·368	0·393	0·416	0·439	0·461	0·483	0·504	0·545
(0·3923)	4·5569	7·4100	11·177	15·958	21·847	28·934	37·307	47·050	58·245	70·972	85·308	101·33	138·72
0·0420	0·238	0·269	0·298	0·325	0·352	0·378	0·402	0·426	0·450	0·472	0·495	0·516	0·559
(0·4119)	4·6694	7·5929	11·453	16·352	22·387	29·649	38·229	48·212	59·684	72·725	87·415	103·83	142·15
0·0440	0·243	0·275	0·305	0·333	0·360	0·386	0·412	0·436	0·460	0·484	0·506	0·529	0·572
(0·4315)	4·7793	7·7716	11·723	16·737	22·913	30·347	39·128	49·347	61·088	74·436	89·472	106·27	145·49
0·0460	0·249	0·281	0·311	0·340	0·368	0·395	0·421	0·446	0·471	0·494	0·518	0·540	0·585
(0·4511)	4·8867	7·9463	11·986	17·113	23·428	31·029	40·008	50·456	62·461	76·109	91·483	108·66	148·76
0·0480	0·254	0·287	0·318	0·348	0·376	0·404	0·430	0·456	0·481	0·505	0·529	0·552	0·597
(0·4707)	4·9918	8·1172	12·244	17·481	23·932	31·696	40·868	51·541	63·805	77·746	93·450	111·00	151·96
0·0500	0·259	0·293	0·325	0·355	0·384	0·412	0·439	0·465	0·491	0·515	0·540	0·563	0·609
(0·4903)	5·0947	8·2846	12·497	17·842	24·426	32·349	41·711	52·604	65·120	79·349	95·377	113·29	155·09
0·0525	0·266	0·300	0·333	0·364	0·393	0·422	0·450	0·477	0·503	0·528	0·553	0·577	0·625
(0·5148)	5·2205	8·4892	12·805	18·282	25·029	33·148	42·741	53·903	66·728	81·309	97·732	116·09	158·92
0·0550	0·272	0·307	0·341	0·372	0·403	0·432	0·460	0·488	0·515	0·541	0·566	0·591	0·639
(0·5394)	5·3434	8·6889	13·107	18·713	25·618	33·928	43·747	55·171	68·299	83·222	100·03	118·82	162·66
0·0575	0·278	0·314	0·348	0·381	0·412	0·442	0·471	0·499	0·526	0·553	0·579	0·604	0·654
(0·5639)	5·4635	8·8842	13·401	19·133	26·194	34·691	44·730	56·411	69·834	85·092	102·28	121·49	166·32
0·0600	0·284	0·321	0·356	0·389	0·421	0·451	0·481	0·510	0·537	0·565	0·591	0·617	0·668
(0·5884)	5·5810	9·0753	13·689	19·545	26·757	35·437	45·692	57·625	71·336	86·923	104·48	124·10	169·90
0·0625	0·290	0·328	0·363	0·397	0·429	0·461	0·491	0·520	0·549	0·576	0·603	0·630	0·681
(0·6129)	5·6961	9·2624	13·972	19·948	27·309	36·168	46·634	58·813	72·807	88·715	106·63	126·66	173·40
0·0650	0·296	0·334	0·370	0·405	0·438	0·470	0·500	0·530	0·559	0·588	0·615	0·642	0·695
(0·6374)	5·8089	9·4459	14·248	20·343	27·850	36·884	47·558	59·978	74·249	90·472	108·75	129·17	176·83
0·0675	0·301	0·340	0·377	0·412	0·446	0·479	0·510	0·540	0·570	0·599	0·627	0·655	0·708
(0·6619)	5·9195	9·6258	14·520	20·730	28·380	37·587	48·464	61·120	75·663	92·195	110·82	131·63	180·20
0·0700	0·307	0·347	0·384	0·420	0·454	0·487	0·519	0·550	0·581	0·610	0·639	0·667	0·721
(0·6865)	6·0282	9·8025	14·786	21·111	28·901	38·276	49·353	62·242	77·051	93·887	112·85	134·05	183·51
0·0725	0·312	0·353	0·391	0·427	0·462	0·496	0·529	0·560	0·591	0·621	0·650	0·678	0·734
(0·7110)	6·1349	9·9760	15·048	21·484	29·412	38·954	50·226	63·343	78·415	95·549	114·85	136·42	186·76
0·0750	0·318	0·359	0·398	0·435	0·470	0·504	0·538	0·570	0·601	0·631	0·661	0·690	0·746
(0·7355)	6·2397	10·147	15·305	21·852	29·915	39·620	51·085	64·426	79·756	97·182	116·81	138·75	189·95
0·0800	0·328	0·371	0·411	0·449	0·486	0·521	0·555	0·588	0·621	0·652	0·683	0·713	0·771
(0·7845)	6·4444	10·479	15·807	22·568	30·896	40·919	52·760	66·539	82·371	100·37	120·64	143·30	196·18
0·0850	0·338	0·382	0·423	0·463	0·501	0·537	0·572	0·606	0·640	0·672	0·704	0·735	0·795
(0·8336)	6·6427	10·802	16·294	23·263	31·847	42·179	54·384	68·587	84·906	103·46	124·36	147·71	202·22
0·0900	0·348	0·393	0·436	0·476	0·515	0·553	0·589	0·624	0·658	0·692	0·724	0·756	0·818
(0·8826)	6·8353	11·115	16·766	23·937	32·771	43·401	55·961	70·575	87·368	106·46	127·96	151·99	208·08
0·0950	0·358	0·404	0·448	0·489	0·529	0·568	0·605	0·641	0·676	0·711	0·744	0·777	0·840
(0·9316)	7·0226	11·420	17·226	24·593	33·669	44·591	57·494	72·509	89·762	109·38	131·47	156·16	213·78
S, ($\Delta p/\rho l$)	5	6	7	8	9	10	11	12	13	14	15	16	18

Gradient (Equivalent) Pipe diameters in mm

S = 0·0300 to 0·0950; $\Delta p/\rho l$ = (0·2942) to (0·9316) ms^{-2}

m = Manning $n \times 100$
S = 0·1000 to 0·2900

Full bore flow; Tables E give m_C or m_P for Colebrook-White or for laminar solutions resp., to divide mV and/or mQ as follows:

i.e. kin. pr. grad., $\Delta p/\rho l$ = (0·9807) to (2·8439) ms^{-2}

mV to give velocities in ms^{-1}
mQ to give discharges in millilitres/sec

Gradient (Equivalent) Pipe diameters in mm

S, ($\Delta p/\rho l$)	5	6	7	8	9	10	11	12	13	14	15	16	18
0·1000	0·367	0·414	0·459	0·502	0·543	0·582	0·621	0·658	0·694	0·729	0·763	0·797	0·862
(0·9807)	7·2050	11·716	17·673	25·232	34·543	45·749	58·988	74·393	92·094	112·22	134·88	160·21	219·34
0·1050	0·376	0·425	0·471	0·514	0·556	0·597	0·636	0·674	0·711	0·747	0·782	0·817	0·883
(1·0297)	7·3830	12·006	18·109	25·855	35·396	46·879	60·445	76·230	94·368	114·99	138·21	164·17	224·75
0·1100	0·385	0·435	0·482	0·526	0·569	0·611	0·651	0·690	0·728	0·765	0·801	0·836	0·904
(1·0787)	7·5567	12·288	18·536	26·464	36·229	47·982	61·867	78·024	96·589	117·69	141·47	168·03	230·04
0·1150	0·394	0·444	0·492	0·538	0·582	0·625	0·666	0·705	0·744	0·782	0·819	0·855	0·924
(1·1278)	7·7265	12·564	18·952	27·059	37·043	49·060	63·257	79·778	98·760	120·34	144·65	171·81	235·21
0·1200	0·402	0·454	0·503	0·550	0·595	0·638	0·680	0·721	0·760	0·799	0·836	0·873	0·944
(1·1768)	7·8927	12·834	19·360	27·641	37·840	50·116	64·618	81·494	100·88	122·93	147·76	175·51	240·27
0·1250	0·410	0·463	0·513	0·561	0·607	0·651	0·694	0·735	0·776	0·815	0·853	0·891	0·964
(1·2258)	8·0555	13·099	19·759	28·210	38·620	51·149	65·950	83·174	102·96	125·46	150·80	179·13	245·22
0·1300	0·418	0·472	0·524	0·572	0·619	0·664	0·708	0·750	0·791	0·831	0·870	0·909	0·983
(1·2749)	8·2150	13·358	20·150	28·769	39·385	52·162	67·257	84·821	105·00	127·95	153·79	182·67	250·08
0·1350	0·426	0·481	0·534	0·583	0·631	0·677	0·721	0·764	0·806	0·847	0·887	0·926	1·001
(1·3239)	8·3715	13·613	20·534	29·317	40·136	53·156	68·538	86·437	107·00	130·38	156·72	186·15	254·84
0·1400	0·434	0·490	0·543	0·594	0·642	0·689	0·734	0·778	0·821	0·863	0·903	0·943	1·020
(1·3729)	8·5251	13·863	20·911	29·855	40·872	54·131	69·795	88·023	108·97	132·78	159·60	189·57	259·52
0·1450	0·442	0·499	0·553	0·604	0·654	0·701	0·747	0·792	0·835	0·878	0·919	0·960	1·038
(1·4220)	8·6760	14·108	21·281	30·384	41·596	55·089	71·031	89·581	110·90	135·13	162·42	192·92	264·12
0·1500	0·449	0·508	0·562	0·615	0·665	0·713	0·760	0·806	0·850	0·893	0·935	0·976	1·056
(1·4710)	8·8243	14·349	21·645	30·903	42·307	56·031	72·245	91·113	112·79	137·44	165·20	196·22	268·63
0·1600	0·464	0·524	0·581	0·635	0·687	0·737	0·785	0·832	0·878	0·922	0·965	1·008	1·090
(1·5691)	9·1137	14·820	22·355	31·917	43·694	57·869	74·614	94·101	116·49	141·94	170·62	202·66	277·44
0·1700	0·478	0·540	0·599	0·655	0·708	0·759	0·809	0·858	0·905	0·950	0·995	1·039	1·124
(1·6671)	9·3942	15·276	23·043	32·899	45·039	59·650	76·911	96·997	120·08	146·31	175·87	208·89	285·98
0·1800	0·492	0·556	0·616	0·673	0·728	0·782	0·833	0·883	0·931	0·978	1·024	1·069	1·156
(1·7652)	9·6666	15·719	23·711	33·853	46·345	61·379	79·141	99·809	123·56	150·55	180·97	214·95	294·27
0·1900	0·506	0·571	0·633	0·692	0·748	0·803	0·856	0·907	0·956	1·005	1·052	1·098	1·188
(1·8633)	9·9314	16·150	24·361	34·780	47·615	63·061	81·309	102·54	126·94	154·68	185·92	220·84	302·33
0·2000	0·519	0·586	0·649	0·710	0·768	0·824	0·878	0·930	0·981	1·031	1·079	1·127	1·219
(1·9613)	10·189	16·569	24·993	35·684	48·851	64·699	83·421	105·21	130·24	158·70	190·75	226·58	310·19
0·2100	0·532	0·600	0·665	0·727	0·787	0·844	0·899	0·953	1·005	1·056	1·106	1·155	1·249
(2·0594)	10·441	16·978	25·611	36·565	50·058	66·297	85·482	107·81	133·46	162·62	195·46	232·17	317·85
0·2200	0·544	0·615	0·681	0·745	0·805	0·864	0·921	0·976	1·029	1·081	1·132	1·182	1·278
(2·1575)	10·687	17·378	26·213	37·425	51·236	67·857	87·493	110·34	136·60	166·44	200·06	237·64	325·33
0·2300	0·557	0·628	0·696	0·761	0·823	0·883	0·941	0·998	1·052	1·106	1·158	1·208	1·307
(2·2555)	10·927	17·768	26·802	38·267	52·387	69·382	89·460	112·82	139·67	170·18	204·56	242·98	332·64
0·2400	0·568	0·642	0·711	0·778	0·841	0·902	0·962	1·019	1·075	1·129	1·182	1·234	1·335
(2·3536)	11·162	18·151	27·379	39·090	53·514	70·874	91·384	115·25	142·67	173·85	208·96	248·20	339·79
0·2500	0·580	0·655	0·726	0·794	0·859	0·921	0·981	1·040	1·097	1·153	1·207	1·260	1·363
(2·4517)	11·392	18·525	27·943	39·896	54·618	72·336	93·268	117·63	145·61	177·43	213·27	253·32	346·80
0·2600	0·592	0·668	0·740	0·809	0·876	0·939	1·001	1·061	1·119	1·175	1·231	1·285	1·390
(2·5497)	11·618	18·892	28·497	40·686	55·699	73·768	95·115	119·96	148·50	180·94	217·49	258·34	353·67
0·2700	0·603	0·681	0·755	0·825	0·892	0·957	1·020	1·081	1·140	1·198	1·254	1·309	1·416
(2·6478)	11·839	19·252	29·040	41·461	56·760	75·173	96·927	122·24	151·33	184·39	221·64	263·26	360·41
0·2800	0·614	0·693	0·768	0·840	0·909	0·975	1·039	1·101	1·161	1·220	1·277	1·333	1·442
(2·7459)	12·056	19·605	29·573	42·222	57·802	76·553	98·706	124·48	154·10	187·77	225·70	268·09	367·02
0·2900	0·625	0·706	0·782	0·855	0·925	0·992	1·057	1·120	1·182	1·241	1·300	1·357	1·468
(2·8439)	12·270	19·952	30·096	42·969	58·825	77·908	100·45	126·69	156·83	191·10	229·70	272·84	373·52
S, ($\Delta p/\rho l$)	5	6	7	8	9	10	11	12	13	14	15	16	18

Gradient (Equivalent) Pipe diameters in mm

S = 0·1000 to 0·2900;

$\Delta p/\rho l$ = (0·9807) to (2·8439) ms^{-2}

m = Manning $n \times 100$
S = 0·3000 to 0·9500

i.e. kin. pr. grad., $\Delta p/\rho l$ = (2·9420) to (9·3163) ms^{-2}

Full bore flow; Tables E give m_C or m_P for Colebrook-White or for laminar solutions resp., to divide mV and/or mQ as follows:

mV to give velocities in ms^{-1}
mQ to give discharges in millilitres/sec

Gradient S, ($\Delta p/\rho l$)	(Equivalent) Pipe diameters in mm												
	5	6	7	8	9	10	11	12	13	14	15	16	18
0·3000 (2·9420)	0·636	0·718	0·795	0·869	0·940	1·009	1·075	1·139	1·202	1·263	1·322	1·380	1·493
	12·479	20·293	30·611	43·704	59·831	79·240	102·17	128·85	159·51	194·36	233·63	277·50	379·90
0·3200 (3·1381)	0·656	0·741	0·821	0·898	0·971	1·042	1·110	1·177	1·241	1·304	1·365	1·425	1·542
	12·889	20·959	31·614	45·137	61·793	81·838	105·52	133·08	164·74	200·74	241·29	286·60	392·36
0·3400 (3·3343)	0·677	0·764	0·847	0·926	1·001	1·074	1·145	1·213	1·279	1·344	1·407	1·469	1·589
	13·285	21·604	32·587	46·526	63·694	84·357	108·77	137·17	169·81	206·92	248·71	295·42	404·43
0·3600 (3·5304)	0·696	0·786	0·871	0·952	1·030	1·105	1·178	1·248	1·316	1·383	1·448	1·512	1·635
	13·671	22·230	33·532	47·875	65·541	86·803	111·92	141·15	174·74	212·92	255·92	303·99	416·16
0·3800 (3·7265)	0·715	0·808	0·895	0·979	1·058	1·135	1·210	1·282	1·353	1·421	1·488	1·553	1·680
	14·045	22·839	34·451	49·187	67·337	89·181	114·99	145·02	179·52	218·75	262·94	312·32	427·56
0·4000 (3·9227)	0·734	0·829	0·918	1·004	1·086	1·165	1·241	1·316	1·388	1·458	1·527	1·594	1·724
	14·410	23·432	35·346	50·464	69·086	91·498	117·98	148·79	184·19	224·43	269·77	320·43	438·67
0·4200 (4·1188)	0·752	0·849	0·941	1·029	1·113	1·194	1·272	1·348	1·422	1·494	1·564	1·633	1·766
	14·766	24·011	36·219	51·711	70·792	93·758	120·89	152·46	188·74	229·98	276·43	328·34	449·50
0·4400 (4·3149)	0·770	0·869	0·963	1·053	1·139	1·222	1·302	1·380	1·455	1·529	1·601	1·671	1·808
	15·113	24·576	37·071	52·928	72·458	95·964	123·73	156·05	193·18	235·39	282·93	336·07	460·08
0·4600 (4·5111)	0·787	0·889	0·985	1·077	1·165	1·249	1·331	1·411	1·488	1·563	1·637	1·709	1·849
	15·453	25·128	37·904	54·117	74·087	98·121	126·51	159·56	197·52	240·68	289·29	343·62	470·42
0·4800 (4·7072)	0·804	0·908	1·006	1·100	1·190	1·276	1·360	1·441	1·520	1·597	1·672	1·746	1·888
	15·785	25·669	38·720	55·281	75·680	100·23	129·24	162·99	201·77	245·85	295·52	351·01	480·54
0·5000 (4·9033)	0·821	0·927	1·027	1·122	1·214	1·303	1·388	1·471	1·551	1·630	1·707	1·782	1·927
	16·111	26·198	39·518	56·421	77·241	102·30	131·90	166·35	205·93	250·92	301·61	358·25	490·45
0·5250 (5·1485)	0·841	0·949	1·052	1·150	1·244	1·335	1·422	1·507	1·590	1·670	1·749	1·826	1·975
	16·509	26·845	40·494	57·814	79·148	104·82	135·16	170·46	211·01	257·12	309·06	367·10	502·56
0·5500 (5·3937)	0·861	0·972	1·077	1·177	1·273	1·366	1·456	1·543	1·627	1·710	1·790	1·869	2·021
	16·897	27·477	41·447	59·175	81·011	107·29	138·34	174·47	215·98	263·17	316·33	375·74	514·39
0·5750 (5·6388)	0·880	0·994	1·101	1·204	1·302	1·397	1·488	1·577	1·664	1·748	1·830	1·911	2·067
	17·277	28·094	42·378	60·505	82·832	109·70	141·45	178·39	220·83	269·09	323·44	384·18	525·95
0·6000 (5·8840)	0·899	1·015	1·125	1·230	1·330	1·427	1·520	1·611	1·700	1·786	1·870	1·952	2·111
	17·649	28·699	43·290	61·806	84·613	112·06	144·49	182·23	225·58	274·87	330·40	392·44	537·26
0·6250 (6·1292)	0·917	1·036	1·148	1·255	1·357	1·456	1·552	1·644	1·735	1·822	1·908	1·992	2·155
	18·013	29·290	44·183	63·081	86·358	114·37	147·47	185·98	230·24	280·54	337·21	400·54	548·34
0·6500 (6·3743)	0·936	1·056	1·171	1·280	1·384	1·485	1·583	1·677	1·769	1·859	1·946	2·032	2·198
	18·369	29·870	45·058	64·330	88·068	116·64	150·39	189·67	234·79	286·10	343·89	408·47	559·20
0·6750 (6·6195)	0·953	1·077	1·193	1·304	1·411	1·513	1·613	1·709	1·803	1·894	1·983	2·070	2·239
	18·719	30·440	45·916	65·555	89·746	118·86	153·26	193·28	239·27	291·55	350·44	416·25	569·85
0·7000 (6·8647)	0·971	1·096	1·215	1·328	1·437	1·541	1·642	1·740	1·836	1·929	2·019	2·108	2·280
	19·063	30·998	46·758	66·758	91·393	121·04	156·07	196·83	243·66	296·90	356·87	423·89	580·31
0·7250 (7·1098)	0·988	1·116	1·236	1·352	1·462	1·568	1·671	1·771	1·868	1·963	2·055	2·146	2·321
	19·400	31·547	47·586	67·940	93·010	123·18	158·83	200·31	247·97	302·15	363·19	431·39	590·58
0·7500 (7·3550)	1·005	1·135	1·258	1·375	1·487	1·595	1·700	1·801	1·900	1·996	2·090	2·182	2·361
	19·732	32·086	48·400	69·101	94·600	125·29	161·55	203·73	252·21	307·32	369·39	438·77	600·68
0·8000 (7·8453)	1·038	1·172	1·299	1·420	1·536	1·648	1·756	1·860	1·962	2·062	2·159	2·254	2·438
	20·379	33·138	49·987	71·368	97·703	129·40	166·84	210·42	260·48	317·40	381·51	453·16	620·37
0·8500 (8·3357)	1·070	1·208	1·339	1·464	1·583	1·698	1·810	1·918	2·023	2·125	2·225	2·323	2·513
	21·006	34·158	51·525	73·564	100·71	133·38	171·98	216·89	268·50	327·16	393·25	467·10	639·47
0·9000 (8·8260)	1·101	1·243	1·378	1·506	1·629	1·747	1·862	1·973	2·082	2·187	2·290	2·391	2·586
	21·615	35·149	53·019	75·697	103·63	137·25	176·96	223·18	276·28	336·65	404·65	480·64	658·01
0·9500 (9·3163)	1·131	1·277	1·415	1·547	1·674	1·795	1·913	2·027	2·139	2·247	2·353	2·456	2·657
	22·207	36·112	54·472	77·771	106·47	141·01	181·81	229·29	283·85	345·87	415·74	493·81	676·04

S, ($\Delta p/\rho l$)	5	6	7	8	9	10	11	12	13	14	15	16	18

Gradient (Equivalent) **Pipe diameters in mm**

S = 0·3000 to 0·9500; $\Delta p/\rho l$ = (2·9420) to (9·3163) ms^{-2}

m = Manning $n \times 100$
S = 1·0000 to 2·9000

i.e. kin. pr. grad., $\Delta p/\rho l$ = (9·8067) to (28·439) ms^{-2}

Full bore flow; Tables E give m_C or m_P for Colebrook-White or for laminar solutions resp., to divide mV and/or mQ as follows:

mV to give velocities in ms^{-1}
mQ to give discharges in millilitres/sec

Gradient **(Equivalent) Pipe diameters in mm**

S, $(\Delta p/\rho l)$	5	6	7	8	9	10	11	12	13	14	15	16	18
1·0000	1·160	1·310	1·452	1·587	1·717	1·842	1·963	2·080	2·194	2·305	2·414	2·520	2·726
(9·8067)	22·784	37·050	55·887	79·791	109·24	144·67	186·54	235·25	291·23	354·86	426·54	506·64	693·60
1·0500	1·189	1·343	1·488	1·627	1·759	1·888	2·011	2·131	2·248	2·362	2·473	2·582	2·793
(10·297)	23·347	37·965	57·267	81·762	111·93	148·24	191·14	241·06	298·42	363·62	437·07	519·16	710·73
1·1000	1·217	1·374	1·523	1·665	1·801	1·932	2·059	2·182	2·301	2·418	2·532	2·643	2·859
(10·787)	23·896	38·858	58·615	83·686	114·57	151·73	195·64	246·73	305·44	372·18	447·36	531·37	727·45
1·1500	1·244	1·405	1·557	1·702	1·841	1·975	2·105	2·231	2·353	2·472	2·588	2·702	2·923
(11·278)	24·433	39·731	59·932	85·567	117·14	155·14	200·04	252·28	312·31	380·55	457·41	543·31	743·80
1·2000	1·271	1·435	1·591	1·739	1·881	2·018	2·150	2·279	2·404	2·525	2·644	2·760	2·986
(11·768)	24·959	40·586	61·221	87·407	119·66	158·48	204·34	257·71	319·02	388·73	467·25	555·00	759·80
1·2500	1·297	1·465	1·624	1·775	1·920	2·059	2·195	2·326	2·453	2·577	2·699	2·817	3·047
(12·258)	25·474	41·423	62·484	89·209	122·13	161·75	208·55	263·02	325·60	396·75	476·89	566·44	775·47
1·3000	1·323	1·494	1·656	1·810	1·958	2·100	2·238	2·372	2·502	2·628	2·752	2·873	3·108
(12·749)	25·978	42·243	63·721	90·976	124·55	164·95	212·68	268·23	332·05	404·60	486·33	577·66	790·83
1·3500	1·348	1·523	1·687	1·844	1·995	2·140	2·281	2·417	2·549	2·678	2·804	2·928	3·167
(13·239)	26·473	43·048	64·935	92·709	126·92	168·09	216·74	273·34	338·37	412·31	495·59	588·67	805·89
1·4000	1·373	1·550	1·718	1·878	2·032	2·180	2·322	2·461	2·596	2·728	2·856	2·982	3·225
(13·729)	26·959	43·838	66·126	94·410	129·25	171·18	220·71	278·35	344·58	419·88	504·69	599·47	820·68
1·4500	1·397	1·578	1·749	1·911	2·068	2·218	2·364	2·505	2·642	2·776	2·907	3·034	3·282
(14·220)	27·436	44·614	67·297	96·081	131·54	174·21	224·62	283·28	350·68	427·31	513·62	610·08	835·21
1·5000	1·421	1·605	1·779	1·944	2·103	2·256	2·404	2·548	2·687	2·823	2·956	3·086	3·338
(14·710)	27·905	45·377	68·447	97·724	133·79	177·19	228·46	288·12	356·68	434·61	522·40	620·51	849·48
1·6000	1·468	1·658	1·837	2·008	2·172	2·330	2·483	2·631	2·775	2·916	3·053	3·187	3·448
(15·691)	28·820	46·865	70·692	100·93	138·17	183·00	235·95	297·57	368·38	448·87	539·53	640·86	877·34
1·7000	1·513	1·709	1·893	2·070	2·239	2·402	2·559	2·712	2·861	3·006	3·147	3·285	3·554
(16·671)	29·707	48·307	72·868	104·04	142·43	188·63	243·21	306·73	379·71	462·68	556·14	660·58	904·34
1·8000	1·557	1·758	1·948	2·130	2·304	2·471	2·633	2·791	2·944	3·093	3·238	3·381	3·657
(17·652)	30·568	49·708	74·980	107·05	146·55	194·10	250·26	315·62	390·72	476·09	572·26	679·73	930·56
1·9000	1·599	1·806	2·002	2·188	2·367	2·539	2·706	2·867	3·024	3·178	3·327	3·473	3·757
(18·633)	31·406	51·070	77·035	109·98	150·57	199·42	257·12	324·27	401·43	489·14	587·94	698·36	956·06
2·0000	1·641	1·853	2·054	2·245	2·428	2·605	2·776	2·942	3·103	3·260	3·414	3·564	3·855
(19·613)	32·222	52·396	79·036	112·84	154·48	204·60	263·80	332·70	411·86	501·85	603·22	716·50	980·90
2·1000	1·682	1·899	2·104	2·300	2·488	2·669	2·844	3·014	3·180	3·341	3·498	3·652	3·950
(20·594)	33·018	53·690	80·988	115·63	158·30	209·65	270·32	340·91	422·03	514·24	618·11	734·20	1005·1
2·2000	1·721	1·944	2·154	2·354	2·547	2·732	2·911	3·085	3·254	3·419	3·580	3·738	4·043
(21·575)	33·795	54·954	82·894	118·35	162·02	214·58	276·68	348·93	431·96	526·34	632·66	751·47	1028·8
2·3000	1·760	1·987	2·202	2·407	2·604	2·794	2·977	3·155	3·328	3·496	3·661	3·822	4·134
(22·555)	34·554	56·189	84·757	121·01	165·66	219·40	282·90	356·78	441·67	538·17	646·88	768·36	1051·9
2·4000	1·798	2·030	2·250	2·459	2·660	2·854	3·041	3·222	3·399	3·571	3·739	3·904	4·223
(23·536)	35·297	57·397	86·580	123·61	169·23	224·12	288·98	364·45	451·17	549·75	660·79	784·89	1074·5
2·5000	1·835	2·072	2·296	2·510	2·715	2·912	3·104	3·289	3·469	3·645	3·816	3·984	4·310
(24·517)	36·025	58·581	88·365	126·16	172·72	228·75	294·94	371·97	460·47	561·08	674·42	801·07	1096·7
2·6000	1·871	2·113	2·342	2·560	2·769	2·970	3·165	3·354	3·538	3·717	3·892	4·063	4·395
(25·497)	36·739	59·741	90·115	128·66	176·14	233·28	300·78	379·33	469·59	572·19	687·77	816·94	1118·4
2·7000	1·907	2·153	2·386	2·608	2·821	3·027	3·225	3·418	3·605	3·788	3·966	4·141	4·479
(26·478)	37·438	60·879	91·832	131·11	179·49	237·72	306·51	386·56	478·53	583·09	700·88	832·50	1139·7
2·8000	1·942	2·193	2·430	2·656	2·873	3·082	3·284	3·481	3·671	3·857	4·039	4·216	4·561
(27·459)	38·125	61·996	93·517	133·52	182·79	242·08	312·13	393·65	487·32	593·79	713·74	847·78	1160·6
2·9000	1·976	2·231	2·473	2·703	2·924	3·137	3·343	3·542	3·736	3·926	4·110	4·291	4·642
(28·439)	38·800	63·094	95·172	135·88	186·02	246·37	317·66	400·62	495·94	604·31	726·37	862·78	1181·2
S, $(\Delta p/\rho l)$	5	6	7	8	9	10	11	12	13	14	15	16	18

Gradient **(Equivalent) Pipe diameters in mm**

S = 1·0000 to 2·9000;

$\Delta p/\rho l$ = (9·8067) to (28·439) ms^{-2}

m = Manning $n \times 100$
S = 3·0000 to 9·5000

i.e. kin. pr. grad., $\Delta p/\rho l$ = (29·420) to (93·163) ms^{-2}

Full bore flow; Tables E give m_C or m_P for Colebrook-White or for laminar solutions resp., to divide mV and/or mQ as follows:

mV to give velocities in ms^{-1}
mQ to give discharges in millilitres/sec

Gradient (Equivalent) Pipe diameters in mm

S, ($\Delta p/\rho l$)	5	6	7	8	9	10	11	12	13	14	15	16	18
3·0000	2·010	2·270	2·515	2·749	2·974	3·190	3·400	3·603	3·800	3·993	4·181	4·364	4·721
(29·420)	39·464	64·172	96·799	138·20	189·20	250·58	323·09	407·47	504·42	614·64	738·79	877·53	1201·4
3·2000	2·076	2·344	2·598	2·840	3·072	3·295	3·511	3·721	3·925	4·124	4·318	4·508	4·876
(31·381)	40·758	66·277	99·974	142·74	195·41	258·80	333·69	420·83	520·96	634·79	763·02	906·31	1240·7
3·4000	2·140	2·416	2·678	2·927	3·166	3·397	3·619	3·835	4·046	4·251	4·451	4·646	5·026
(33·343)	42·012	68·316	103·05	147·13	201·42	266·76	343·96	433·78	537·00	654·33	786·50	934·20	1278·9
3·6000	2·202	2·486	2·755	3·012	3·258	3·495	3·724	3·947	4·163	4·374	4·580	4·781	5·172
(35·304)	43·230	70·297	106·04	151·39	207·26	274·49	353·93	446·36	552·56	673·30	809·30	961·29	1316·0
3·8000	2·262	2·554	2·831	3·094	3·347	3·591	3·826	4·055	4·277	4·494	4·705	4·912	5·313
(37·265)	44·415	72·223	108·94	155·54	212·94	282·02	363·63	458·59	567·71	691·75	831·48	987·63	1352·1
4·0000	2·321	2·621	2·904	3·175	3·434	3·684	3·926	4·160	4·388	4·610	4·827	5·040	5·451
(39·227)	45·569	74·100	111·77	159·58	218·47	289·34	373·07	470·50	582·45	709·72	853·08	1013·3	1387·2
4·2000	2·378	2·685	2·976	3·253	3·519	3·775	4·023	4·263	4·497	4·724	4·947	5·164	5·586
(41·188)	46·694	75·929	114·53	163·52	223·87	296·49	382·29	482·12	596·84	727·25	874·15	1038·3	1421·5
4·4000	2·434	2·749	3·046	3·330	3·602	3·864	4·117	4·363	4·602	4·835	5·063	5·286	5·717
(43·149)	47·793	77·716	117·23	167·37	229·13	303·47	391·28	493·47	610·88	744·36	894·72	1062·7	1454·9
4·6000	2·489	2·810	3·115	3·405	3·683	3·951	4·210	4·461	4·706	4·944	5·177	5·404	5·846
(45·111)	48·867	79·463	119·86	171·13	234·28	310·29	400·08	504·56	624·61	761·09	914·83	1086·6	1487·6
4·8000	2·542	2·871	3·182	3·478	3·762	4·036	4·300	4·557	4·807	5·050	5·288	5·521	5·972
(47·072)	49·918	81·172	122·44	174·81	239·32	316·96	408·68	515·41	638·05	777·46	934·50	1110·0	1519·6
5·0000	2·595	2·930	3·247	3·550	3·839	4·119	4·389	4·651	4·906	5·155	5·397	5·635	6·095
(49·033)	50·947	82·846	124·97	178·42	244·26	323·49	417·11	526·04	651·20	793·49	953·77	1132·9	1550·9
5·2500	2·659	3·002	3·327	3·637	3·934	4·221	4·497	4·766	5·027	5·282	5·531	5·774	6·245
(51·485)	52·205	84·892	128·05	182·82	250·29	331·48	427·41	539·03	667·28	813·09	977·32	1160·9	1589·2
5·5000	2·721	3·073	3·406	3·723	4·027	4·320	4·603	4·878	5·146	5·406	5·661	5·910	6·392
(53·937)	53·434	86·889	131·07	187·13	256·18	339·28	437·47	551·71	682·99	832·22	1000·3	1188·2	1626·6
5·7500	2·783	3·142	3·482	3·806	4·117	4·417	4·707	4·988	5·261	5·528	5·788	6·042	6·536
(56·388)	54·635	88·842	134·01	191·33	261·94	346·91	447·30	564·11	698·34	850·92	1022·8	1214·9	1663·2
6·0000	2·842	3·210	3·557	3·888	4·206	4·512	4·808	5·095	5·374	5·647	5·912	6·172	6·677
(58·840)	55·810	90·753	136·89	195·45	267·57	354·37	456·92	576·25	713·36	869·23	1044·8	1241·0	1699·0
6·2500	2·901	3·276	3·630	3·968	4·293	4·605	4·907	5·200	5·485	5·763	6·034	6·300	6·814
(61·292)	56·961	92·624	139·72	199·48	273·09	361·68	466·34	588·13	728·07	887·15	1066·3	1266·6	1734·0
6·5000	2·958	3·341	3·702	4·047	4·378	4·696	5·004	5·303	5·594	5·877	6·154	6·424	6·949
(63·743)	58·089	94·459	142·48	203·43	278·50	368·84	475·58	599·78	742·49	904·72	1087·5	1291·7	1768·3
6·7500	3·015	3·404	3·773	4·124	4·461	4·786	5·100	5·404	5·700	5·989	6·271	6·547	7·082
(66·195)	59·195	96·258	145·20	207·30	283·80	375·87	484·64	611·20	756·63	921·95	1108·2	1316·3	1802·0
7·0000	3·070	3·467	3·842	4·200	4·543	4·874	5·193	5·503	5·805	6·099	6·386	6·667	7·211
(68·647)	60·282	98·025	147·86	211·11	289·01	382·76	493·53	622·42	770·51	938·87	1128·5	1340·5	1835·1
7·2500	3·124	3·528	3·910	4·274	4·623	4·960	5·285	5·601	5·908	6·207	6·499	6·785	7·339
(71·098)	61·349	99·760	150·48	214·84	294·12	389·54	502·26	633·43	784·15	955·49	1148·5	1364·2	1867·6
7·5000	3·178	3·589	3·977	4·347	4·702	5·045	5·375	5·697	6·009	6·313	6·610	6·901	7·465
(73·550)	62·397	101·47	153·05	218·52	299·15	396·20	510·85	644·26	797·56	971·82	1168·1	1387·5	1899·5
8·0000	3·282	3·706	4·107	4·490	4·857	5·210	5·552	5·883	6·206	6·520	6·827	7·127	7·709
(78·453)	64·444	104·79	158·07	225·68	308·96	409·19	527·60	665·39	823·71	1003·7	1206·4	1433·0	1961·8
8·5000	3·383	3·820	4·234	4·628	5·006	5·370	5·723	6·064	6·397	6·721	7·037	7·347	7·947
(83·357)	66·427	108·02	162·94	232·63	318·47	421·79	543·84	685·87	849·06	1034·6	1243·6	1477·1	2022·2
9·0000	3·481	3·931	4·357	4·762	5·151	5·526	5·889	6·240	6·582	6·916	7·241	7·560	8·177
(88·260)	68·353	111·15	167·66	239·37	327·71	434·01	559·61	705·75	873·68	1064·6	1279·6	1519·9	2080·8
9·5000	3·577	4·039	4·476	4·893	5·292	5·677	6·050	6·411	6·763	7·105	7·440	7·767	8·401
(93·163)	70·226	114·20	172·26	245·93	336·69	445·91	574·94	725·09	897·62	1093·8	1314·7	1561·6	2137·8

S, ($\Delta p/\rho l$)	5	6	7	8	9	10	11	12	13	14	15	16	18

Gradient (Equivalent) Pipe diameters in mm

S = 3·0000 to 9·5000;

$\Delta p/\rho l$ = (29·420) to (93·163) ms^{-2}

m = Manning $n \times 100$
S = 10·000 to 30·000

Full bore flow; Tables E give m_C or m_P for Colebrook-White or for laminar solutions resp., to divide mV and/or mQ as follows:

i.e. kin. pr. grad., $\Delta p/\rho l$ = (98·067) to (294·20) ms^{-2}

mV to give velocities in ms^{-1}
mQ to give discharges in millilitres/sec

Gradient — (Equivalent) Pipe diameters in mm

S, $(\Delta p/\rho l)$	5	6	7	8	9	10	11	12	13	14	15	16	18
10·000	3·669	4·144	4·592	5·020	5·430	5·825	6·207	6·578	6·938	7·290	7·633	7·968	8·619
(98·067)	72·050	117·16	176·73	252·32	345·43	457·49	589·88	743·93	920·94	1122·2	1348·8	1602·1	2193·4
10·500	3·760	4·246	4·706	5·144	5·564	5·969	6·360	6·740	7·110	7·470	7·821	8·165	8·832
(102·97)	73·830	120·06	181·09	258·55	353·96	468·79	604·45	762·30	943·68	1149·9	1382·1	1641·7	2247·5
11·000	3·849	4·346	4·816	5·265	5·695	6·109	6·510	6·899	7·277	7·646	8·005	8·357	9·040
(107·87)	75·567	122·88	185·36	264·64	362·29	479·82	618·67	780·24	965·89	1176·9	1414·7	1680·3	2300·4
11·500	3·935	4·444	4·925	5·383	5·823	6·247	6·656	7·054	7·441	7·817	8·185	8·545	9·243
(112·78)	77·265	125·64	189·52	270·59	370·43	490·60	632·57	797·78	987·60	1203·4	1446·5	1718·1	2352·1
12·000	4·020	4·539	5·031	5·499	5·948	6·381	6·800	7·206	7·601	7·986	8·361	8·729	9·442
(117·68)	78·927	128·34	193·60	276·41	378·40	501·16	646·18	814·94	1008·8	1229·3	1477·6	1755·1	2402·7
12·500	4·103	4·633	5·134	5·612	6·071	6·513	6·940	7·354	7·757	8·150	8·534	8·909	9·637
(122·58)	80·555	130·99	197·59	282·10	386·20	511·49	659·50	831·74	1029·6	1254·6	1508·0	1791·3	2452·2
13·000	4·184	4·725	5·236	5·723	6·191	6·641	7·077	7·500	7·911	8·312	8·703	9·085	9·828
(127·49)	82·150	133·58	201·50	287·69	393·85	521·62	672·57	848·21	1050·0	1279·5	1537·9	1826·7	2500·8
13·500	4·264	4·815	5·336	5·832	6·309	6·768	7·212	7·643	8·062	8·470	8·869	9·258	10·01
(132·39)	83·715	136·13	205·34	293·17	401·36	531·56	685·38	864·37	1070·0	1303·8	1567·2	1861·5	2548·4
14·000	4·342	4·903	5·434	5·940	6·425	6·892	7·344	7·783	8·210	8·625	9·031	9·428	10·20
(137·29)	85·251	138·63	209·11	298·55	408·72	541·31	697·95	880·23	1089·7	1327·8	1596·0	1895·7	2595·2
14·500	4·419	4·990	5·530	6·045	6·538	7·014	7·474	7·921	8·355	8·778	9·191	9·595	10·38
(142·20)	86·760	141·08	212·81	303·84	415·96	550·89	710·31	895·81	1109·0	1351·3	1624·2	1929·2	2641·2
15·000	4·494	5·075	5·624	6·148	6·650	7·134	7·602	8·056	8·498	8·928	9·348	9·759	10·56
(147·10)	88·243	143·49	216·45	309·03	423·07	560·31	722·45	911·13	1127·9	1374·4	1652·0	1962·2	2686·3
16·000	4·642	5·241	5·809	6·350	6·868	7·368	7·851	8·320	8·776	9·221	9·655	10·08	10·90
(156·91)	91·137	148·20	223·55	319·17	436·94	578·69	746·14	941·01	1164·9	1419·4	1706·2	2026·6	2774·4
17·000	4·784	5·403	5·988	6·545	7·080	7·595	8·093	8·576	9·046	9·505	9·952	10·39	11·24
(166·71)	93·942	152·76	230·43	328·99	450·39	596·50	769·11	969·97	1200·8	1463·1	1758·7	2088·9	2859·8
18·000	4·923	5·559	6·161	6·735	7·285	7·815	8·328	8·825	9·309	9·780	10·24	10·69	11·56
(176·52)	96·666	157·19	237·11	338·53	463·45	613·79	791·41	998·09	1235·6	1505·5	1809·7	2149·5	2942·7
19·000	5·058	5·712	6·330	6·919	7·485	8·029	8·556	9·067	9·564	10·05	10·52	10·98	11·88
(186·33)	99·314	161·50	243·61	347·80	476·15	630·61	813·09	1025·4	1269·4	1546·8	1859·2	2208·4	3023·3
20·000	5·189	5·860	6·494	7·099	7·679	8·238	8·778	9·302	9·812	10·31	10·79	11·27	12·19
(196·13)	101·89	165·69	249·93	356·84	488·51	646·99	834·21	1052·1	1302·4	1587·0	1907·5	2265·8	3101·9
21·000	5·318	6·005	6·655	7·274	7·869	8·441	8·995	9·532	10·05	10·56	11·06	11·55	12·49
(205·94)	104·41	169·78	256·11	365·65	500·58	662·97	854·82	1078·1	1334·6	1626·2	1954·6	2321·7	3178·5
22·000	5·443	6·146	6·811	7·446	8·054	8·640	9·207	9·756	10·29	10·81	11·32	11·82	12·78
(215·75)	106·87	173·78	262·13	374·25	512·36	678·57	874·93	1103·4	1366·0	1664·4	2000·6	2376·4	3253·3
23·000	5·565	6·284	6·964	7·613	8·235	8·834	9·414	9·976	10·52	11·06	11·58	12·08	13·07
(225·55)	109·27	177·68	268·02	382·67	523·87	693·82	894·60	1128·2	1396·7	1701·8	2045·6	2429·8	3326·4
24·000	5·685	6·419	7·114	7·777	8·412	9·024	9·616	10·19	10·75	11·29	11·82	12·34	13·35
(235·36)	111·62	181·51	273·79	390·90	535·14	708·74	913·84	1152·5	1426·7	1738·5	2089·6	2482·0	3397·9
25·000	5·802	6·552	7·261	7·937	8·585	9·210	9·814	10·40	10·97	11·53	12·07	12·60	13·63
(245·17)	113·92	185·25	279·43	398·96	546·18	723·36	932·68	1176·3	1456·1	1774·3	2132·7	2533·2	3468·0
26·000	5·917	6·682	7·405	8·094	8·755	9·392	10·01	10·61	11·19	11·75	12·31	12·85	13·90
(254·97)	116·18	188·92	284·97	406·86	556·99	737·68	951·15	1199·6	1485·0	1809·4	2174·9	2583·4	3536·7
27·000	6·030	6·809	7·546	8·248	8·922	9·571	10·20	10·81	11·40	11·98	12·54	13·09	14·16
(264·78)	118·39	192·52	290·40	414·61	567·60	751·73	969·27	1222·4	1513·3	1843·9	2216·4	2632·6	3604·1
28·000	6·140	6·934	7·684	8·400	9·086	9·747	10·39	11·01	11·61	12·20	12·77	13·33	14·42
(274·59)	120·56	196·05	295·73	422·22	578·02	765·53	987·06	1244·8	1541·0	1877·7	2257·0	2680·9	3670·2
29·000	6·249	7·057	7·820	8·548	9·247	9·920	10·57	11·20	11·82	12·41	13·00	13·57	14·68
(284·39)	122·70	199·52	300·96	429·69	588·25	779·08	1004·5	1266·9	1568·3	1911·0	2297·0	2728·4	3735·2
30·000	6·356	7·177	7·954	8·695	9·405	10·09	10·75	11·39	12·02	12·63	13·22	13·80	14·93
(294·20)	124·79	202·93	306·11	437·04	598·31	792·40	1021·7	1288·5	1595·1	1943·6	2336·3	2775·0	3799·0
S, $(\Delta p/\rho l)$	5	6	7	8	9	10	11	12	13	14	15	16	18

Gradient — (Equivalent) Pipe diameters in mm

S = 10·000 to 30·000;

$\Delta p/\rho l$ = (98·067) to (294·20) ms^{-2}

m = Manning $n \times 100$
S = 0.00100 to 0.00290

i.e. kin. pr. grad., $\Delta p/\rho l$ = (0.00981) to (0.02844) ms^{-2}

Full bore flow; Tables E give m_C or m_P for Colebrook-White or for laminar solutions resp., to divide mV and/or mQ as follows:

mV to give velocities in ms^{-1}
mQ to give discharges in millilitres/sec

Gradient — (Equivalent) **Pipe diameters in mm**

S, $(\Delta p/\rho l)$	18	20	22	25	28	30	35	40	45	50	55	60	65
0.00100	0.0862	0.0925	0.0985	0.1073	0.1157	0.1212	0.1343	0.1468	0.1588	0.1703	0.1815	0.1923	0.2029
(0.00981)	21.934	29.049	37.455	52.669	71.253	85.646	129.19	184.45	252.51	334.43	431.20	543.82	673.21
0.00105	0.0883	0.0947	0.1010	0.1099	0.1186	0.1242	0.1376	0.1504	0.1627	0.1745	0.1860	0.1971	0.2079
(0.01030)	22.475	29.766	38.380	53.970	73.013	87.761	132.38	189.00	258.75	342.69	441.85	557.25	689.84
0.00110	0.0904	0.0970	0.1033	0.1125	0.1214	0.1271	0.1408	0.1539	0.1665	0.1786	0.1904	0.2017	0.2128
(0.01079)	23.004	30.467	39.283	55.240	74.731	89.826	135.50	193.45	264.84	350.75	452.25	570.36	706.07
0.00115	0.0924	0.0992	0.1057	0.1151	0.1241	0.1299	0.1440	0.1574	0.1703	0.1827	0.1946	0.2063	0.2176
(0.01128)	23.521	31.151	40.166	56.481	76.410	91.845	138.54	197.80	270.79	358.63	462.41	583.18	721.94
0.00120	0.0944	0.1013	0.1079	0.1175	0.1268	0.1327	0.1471	0.1608	0.1739	0.1866	0.1988	0.2107	0.2222
(0.01177)	24.027	31.821	41.030	57.696	78.054	93.820	141.52	202.05	276.61	366.35	472.36	595.72	737.47
0.00125	0.0964	0.1034	0.1102	0.1200	0.1294	0.1355	0.1501	0.1641	0.1775	0.1904	0.2029	0.2150	0.2268
(0.01226)	24.522	32.478	41.876	58.886	79.663	95.755	144.44	206.22	282.32	373.90	482.10	608.01	752.67
0.00130	0.0983	0.1054	0.1123	0.1223	0.1319	0.1381	0.1531	0.1674	0.1810	0.1942	0.2069	0.2193	0.2313
(0.01275)	25.008	33.121	42.705	60.052	81.241	97.651	147.30	210.30	287.91	381.31	491.65	620.05	767.58
0.00135	0.1001	0.1074	0.1145	0.1247	0.1345	0.1408	0.1560	0.1705	0.1845	0.1979	0.2109	0.2235	0.2357
(0.01324)	25.484	33.752	43.519	61.196	82.789	99.511	150.11	214.31	293.39	388.57	501.01	631.86	782.20
0.00140	0.1020	0.1094	0.1166	0.1270	0.1369	0.1434	0.1589	0.1737	0.1879	0.2015	0.2147	0.2276	0.2400
(0.01373)	25.952	34.371	44.317	62.319	84.308	101.34	152.86	218.24	298.78	395.70	510.21	643.45	796.55
0.00145	0.1038	0.1113	0.1186	0.1292	0.1393	0.1459	0.1617	0.1767	0.1912	0.2051	0.2186	0.2316	0.2443
(0.01422)	26.412	34.979	45.102	63.422	85.800	103.13	155.57	222.11	304.07	402.70	519.24	654.84	810.65
0.00150	0.1056	0.1132	0.1207	0.1314	0.1417	0.1484	0.1645	0.1798	0.1945	0.2086	0.2223	0.2356	0.2485
(0.01471)	26.863	35.577	45.873	64.506	87.267	104.89	158.23	225.90	309.26	409.59	528.12	666.04	824.51
0.00160	0.1090	0.1170	0.1246	0.1357	0.1464	0.1533	0.1698	0.1857	0.2008	0.2154	0.2296	0.2433	0.2566
(0.01569)	27.744	36.744	47.377	66.622	90.129	108.33	163.41	233.31	319.41	423.02	545.43	687.88	851.55
0.00170	0.1124	0.1206	0.1285	0.1399	0.1509	0.1580	0.1751	0.1914	0.2070	0.2221	0.2366	0.2508	0.2645
(0.01667)	28.598	37.875	48.835	68.672	92.903	111.67	168.44	240.49	329.24	436.04	562.22	709.05	877.76
0.00180	0.1156	0.1241	0.1322	0.1440	0.1553	0.1626	0.1802	0.1969	0.2130	0.2285	0.2435	0.2580	0.2722
(0.01765)	29.427	38.973	50.251	70.663	95.596	114.91	173.33	247.46	338.78	448.68	578.52	729.61	903.21
0.00190	0.1188	0.1275	0.1358	0.1479	0.1595	0.1670	0.1851	0.2023	0.2188	0.2348	0.2502	0.2651	0.2796
(0.01863)	30.233	40.041	51.628	72.599	98.216	118.05	178.08	254.25	348.06	460.98	594.37	749.60	927.96
0.00200	0.1219	0.1308	0.1393	0.1517	0.1636	0.1714	0.1899	0.2076	0.2245	0.2409	0.2567	0.2720	0.2869
(0.01961)	31.019	41.081	52.969	74.485	100.77	121.12	182.70	260.85	357.11	472.95	609.81	769.07	952.06
0.00210	0.1249	0.1340	0.1428	0.1555	0.1677	0.1756	0.1946	0.2127	0.2301	0.2468	0.2630	0.2787	0.2940
(0.02059)	31.785	42.096	54.277	76.325	103.26	124.11	187.21	267.29	365.93	484.63	624.87	788.07	975.58
0.00220	0.1278	0.1371	0.1461	0.1591	0.1716	0.1797	0.1992	0.2177	0.2355	0.2526	0.2692	0.2853	0.3009
(0.02157)	32.533	43.086	55.555	78.121	105.69	127.03	191.62	273.58	374.54	496.04	639.58	806.61	998.53
0.00230	0.1307	0.1402	0.1494	0.1627	0.1755	0.1838	0.2036	0.2226	0.2408	0.2583	0.2753	0.2917	0.3077
(0.02256)	33.264	44.055	56.803	79.877	108.06	129.89	195.93	279.73	382.95	507.19	653.95	824.74	1021.0
0.00240	0.1335	0.1432	0.1526	0.1662	0.1793	0.1877	0.2080	0.2274	0.2460	0.2639	0.2812	0.2980	0.3143
(0.02354)	33.979	45.002	58.025	81.595	110.38	132.68	200.14	285.75	391.19	518.09	668.02	842.48	1042.9
0.00250	0.1363	0.1462	0.1558	0.1697	0.1830	0.1916	0.2123	0.2321	0.2510	0.2693	0.2870	0.3041	0.3208
(0.02452)	34.680	45.930	59.222	83.277	112.66	135.42	204.27	291.64	399.26	528.78	681.79	859.85	1064.4
0.00260	0.1390	0.1491	0.1589	0.1730	0.1866	0.1954	0.2165	0.2367	0.2560	0.2746	0.2927	0.3101	0.3271
(0.02550)	35.367	46.840	60.394	84.926	114.89	138.10	208.31	297.41	407.16	539.25	695.30	876.88	1085.5
0.00270	0.1416	0.1519	0.1619	0.1763	0.1901	0.1991	0.2206	0.2412	0.2609	0.2799	0.2982	0.3160	0.3334
(0.02648)	36.041	47.732	61.545	86.544	117.08	140.73	212.28	303.08	414.92	549.52	708.54	893.58	1106.2
0.00280	0.1442	0.1547	0.1649	0.1795	0.1936	0.2027	0.2247	0.2456	0.2657	0.2850	0.3037	0.3218	0.3395
(0.02746)	36.702	48.608	62.674	88.132	119.23	143.31	216.18	308.64	422.53	559.60	721.54	909.98	1126.5
0.00290	0.1468	0.1575	0.1678	0.1827	0.1971	0.2063	0.2287	0.2500	0.2704	0.2900	0.3091	0.3275	0.3455
(0.02844)	37.352	49.468	63.784	89.692	121.34	145.85	220.00	314.10	430.01	569.51	734.31	926.09	1146.4
S, $(\Delta p/\rho l)$	18	20	22	25	28	30	35	40	45	50	55	60	65

Gradient — (Equivalent) **Pipe diameters in mm**

S = 0.00100 to 0.00290; $\Delta p/\rho l$ = (0.00981) to (0.02844) ms^{-2}

m = Manning $n \times 100$
S = 0·00300 to 0·00950

i.e. kin. pr. grad., $\Delta p/\rho l$ = (0·02942) to (0·09316) ms^{-2}

Full bore flow; Tables E give m_C or m_P for Colebrook-White or for laminar solutions resp., to divide mV and/or mQ as follows:

mV to give velocities in ms^{-1}
mQ to give discharges in millilitres/sec

Gradient   (Equivalent) Pipe diameters in mm

S, $(\Delta p/\rho l)$	18	20	22	25	28	30	35	40	45	50	55	60	65
0·00300	0·149	0·160	0·171	0·186	0·200	0·210	0·233	0·254	0·275	0·295	0·314	0·333	0·351
(0·02942)	37·990	50·314	64·874	91·226	123·41	148·34	223·76	319·47	437·36	579·25	746·87	941·92	1166·0
0·00320	0·154	0·165	0·176	0·192	0·207	0·217	0·240	0·263	0·284	0·305	0·325	0·344	0·363
(0·03138)	39·236	51·964	67·002	94·217	127·46	153·21	231·10	329·95	451·71	598·24	771·36	972·81	1204·3
0·00340	0·159	0·170	0·182	0·198	0·213	0·223	0·248	0·271	0·293	0·314	0·335	0·355	0·374
(0·03334)	40·443	53·563	69·064	97·117	131·38	157·92	238·22	340·11	465·61	616·65	795·10	1002·7	1241·3
0·00360	0·164	0·175	0·187	0·204	0·220	0·230	0·255	0·278	0·301	0·323	0·344	0·365	0·385
(0·03530)	41·616	55·116	71·066	99·933	135·19	162·50	245·12	349·97	479·11	634·53	818·15	1031·8	1277·3
0·00380	0·168	0·180	0·192	0·209	0·226	0·236	0·262	0·286	0·309	0·332	0·354	0·375	0·395
(0·03727)	42·756	56·627	73·013	102·67	138·90	166·95	251·84	359·56	492·24	651·92	840·57	1060·1	1312·3
0·00400	0·172	0·185	0·197	0·215	0·231	0·242	0·269	0·294	0·318	0·341	0·363	0·385	0·406
(0·03923)	43·867	58·098	74·910	105·34	142·51	171·29	258·38	368·90	505·02	668·86	862·41	1087·6	1346·4
0·00420	0·177	0·189	0·202	0·220	0·237	0·248	0·275	0·301	0·325	0·349	0·372	0·394	0·416
(0·04119)	44·950	59·532	76·760	107·94	146·03	175·52	264·76	378·01	517·50	685·37	883·71	1114·5	1379·7
0·00440	0·181	0·194	0·207	0·225	0·243	0·254	0·282	0·308	0·333	0·357	0·381	0·403	0·426
(0·04315)	46·008	60·933	78·566	110·48	149·46	179·65	270·99	386·90	529·67	701·50	904·50	1140·7	1412·1
0·00460	0·185	0·198	0·211	0·230	0·248	0·260	0·288	0·315	0·341	0·365	0·389	0·413	0·435
(0·04511)	47·042	62·303	80·332	112·96	152·82	183·69	277·08	395·60	541·58	717·27	924·83	1166·4	1443·9
0·00480	0·189	0·203	0·216	0·235	0·254	0·265	0·294	0·322	0·348	0·373	0·398	0·421	0·444
(0·04707)	48·054	63·643	82·060	115·39	156·11	187·64	283·04	404·11	553·23	732·69	944·72	1191·4	1474·9
0·00500	0·193	0·207	0·220	0·240	0·259	0·271	0·300	0·328	0·355	0·381	0·406	0·430	0·454
(0·04903)	49·045	64·955	83·752	117·77	159·33	191·51	288·88	412·44	564·63	747·80	964·20	1216·0	1505·3
0·00525	0·197	0·212	0·226	0·246	0·265	0·278	0·308	0·336	0·364	0·390	0·416	0·441	0·465
(0·05148)	50·256	66·559	85·820	120·68	163·26	196·24	296·01	422·63	578·58	766·27	988·01	1246·0	1542·5
0·00550	0·202	0·217	0·231	0·252	0·271	0·284	0·315	0·344	0·372	0·399	0·426	0·451	0·476
(0·05394)	51·439	68·126	87·840	123·52	167·10	200·86	302·98	432·57	592·19	784·30	1011·3	1275·4	1578·8
0·00575	0·207	0·222	0·236	0·257	0·277	0·291	0·322	0·352	0·381	0·408	0·435	0·461	0·486
(0·05639)	52·595	69·657	89·814	126·30	170·86	205·37	309·79	442·29	605·50	801·93	1034·0	1304·0	1614·3
0·00600	0·211	0·226	0·241	0·263	0·283	0·297	0·329	0·360	0·389	0·417	0·445	0·471	0·497
(0·05884)	53·726	71·155	91·746	129·01	174·53	209·79	316·45	451·81	618·53	819·18	1056·2	1332·1	1649·0
0·00625	0·215	0·231	0·246	0·268	0·289	0·303	0·336	0·367	0·397	0·426	0·454	0·481	0·507
(0·06129)	54·834	72·622	93·637	131·67	178·13	214·11	322·98	461·12	631·28	836·07	1078·0	1359·5	1683·0
0·00650	0·220	0·236	0·251	0·274	0·295	0·309	0·342	0·374	0·405	0·434	0·463	0·490	0·517
(0·06374)	55·920	74·060	95·492	134·28	181·66	218·35	329·37	470·25	643·78	852·63	1099·4	1386·5	1716·4
0·00675	0·224	0·240	0·256	0·279	0·301	0·315	0·349	0·381	0·412	0·443	0·472	0·500	0·527
(0·06619)	56·985	75·471	97·311	136·84	185·12	222·51	335·65	479·21	656·05	868·87	1120·3	1412·9	1749·1
0·00700	0·228	0·245	0·261	0·284	0·306	0·321	0·355	0·388	0·420	0·451	0·480	0·509	0·537
(0·06865)	58·031	76·856	99·097	139·35	188·52	226·60	341·81	488·01	668·08	884·81	1140·9	1438·8	1781·1
0·00725	0·232	0·249	0·265	0·289	0·312	0·326	0·362	0·395	0·428	0·459	0·489	0·518	0·546
(0·07110)	59·058	78·216	100·85	141·82	191·85	230·61	347·86	496·64	679·91	900·47	1161·1	1464·3	1812·7
0·00750	0·236	0·253	0·270	0·294	0·317	0·332	0·368	0·402	0·435	0·466	0·497	0·527	0·556
(0·07355)	60·068	79·554	102·57	144·24	195·13	234·55	353·80	505·13	691·53	915·87	1180·9	1489·3	1843·7
0·00800	0·244	0·262	0·279	0·303	0·327	0·343	0·380	0·415	0·449	0·482	0·513	0·544	0·574
(0·07845)	62·037	82·163	105·94	148·97	201·53	242·24	365·41	521·70	714·21	945·90	1219·6	1538·1	1904·1
0·00850	0·251	0·270	0·287	0·313	0·337	0·353	0·391	0·428	0·463	0·497	0·529	0·561	0·591
(0·08336)	63·947	84·691	109·20	153·56	207·74	249·70	376·65	537·76	736·19	975·02	1257·2	1585·5	1962·7
0·00900	0·259	0·277	0·296	0·322	0·347	0·363	0·403	0·440	0·476	0·511	0·544	0·577	0·609
(0·08826)	65·801	87·147	112·36	158·01	213·76	256·94	387·57	553·35	757·54	1003·3	1293·6	1631·4	2019·6
0·00950	0·266	0·285	0·304	0·331	0·357	0·373	0·414	0·452	0·489	0·525	0·559	0·593	0·625
(0·09316)	67·604	89·535	115·44	162·34	219·62	263·98	398·19	568·51	778·30	1030·8	1329·1	1676·2	2075·0
S, $(\Delta p/\rho l)$	18	20	22	25	28	30	35	40	45	50	55	60	65

Gradient   (Equivalent) Pipe diameters in mm

S = 0·00300 to 0·00950;

$\Delta p/\rho l$ = (0·02942) to (0·09316) ms^{-2}

m = Manning $n \times 100$
S = 0·0100 to 0·0290

i.e. kin. pr. grad., $\Delta p/\rho l$ =
(0·0981) to (0·2844) ms^{-2}

Full bore flow; Tables E give m_C or m_P for Colebrook-White or for laminar solutions resp., to divide mV and/or mQ as follows:

mV to give velocities in ms^{-1}
mQ to give discharges in millilitres/sec

Gradient S, ($\Delta p/\rho l$)	18	20	22	25	28	30	35	40	45	50	55	60	65
0·0100 (0·0981)	0·273 / 69·360	0·292 / 91·861	0·312 / 118·44	0·339 / 166·55	0·366 / 225·32	0·383 / 270·84	0·425 / 408·54	0·464 / 583·28	0·502 / 798·51	0·539 / 1057·6	0·574 / 1363·6	0·608 / 1719·7	0·642 / 2128·9
0·0105 (0·1030)	0·279 / 71·073	0·300 / 94·129	0·319 / 121·37	0·348 / 170·67	0·375 / 230·89	0·393 / 277·52	0·435 / 418·63	0·476 / 597·68	0·514 / 818·23	0·552 / 1083·7	0·588 / 1397·3	0·623 / 1762·2	0·657 / 2181·5
0·0110 (0·1079)	0·286 / 72·745	0·307 / 96·344	0·327 / 124·22	0·356 / 174·68	0·384 / 236·32	0·402 / 284·05	0·445 / 428·48	0·487 / 611·75	0·527 / 837·49	0·565 / 1109·2	0·602 / 1430·1	0·638 / 1803·6	0·673 / 2232·8
0·0115 (0·1128)	0·292 / 74·380	0·314 / 98·509	0·334 / 127·02	0·364 / 178·61	0·392 / 241·63	0·411 / 290·44	0·455 / 438·11	0·498 / 625·50	0·538 / 856·31	0·578 / 1134·1	0·615 / 1462·3	0·652 / 1844·2	0·688 / 2283·0
0·0120 (0·1177)	0·299 / 75·980	0·320 / 100·63	0·341 / 129·75	0·372 / 182·45	0·401 / 246·83	0·420 / 296·69	0·465 / 447·53	0·508 / 638·95	0·550 / 874·73	0·590 / 1158·5	0·629 / 1493·7	0·666 / 1883·8	0·703 / 2332·1
0·0125 (0·1226)	0·305 / 77·547	0·327 / 102·70	0·348 / 132·42	0·379 / 186·21	0·409 / 251·92	0·428 / 302·80	0·475 / 456·76	0·519 / 652·12	0·561 / 892·77	0·602 / 1182·4	0·642 / 1524·5	0·680 / 1922·7	0·717 / 2380·2
0·0130 (0·1275)	0·311 / 79·083	0·333 / 104·74	0·355 / 135·05	0·387 / 189·90	0·417 / 256·91	0·437 / 308·80	0·484 / 465·80	0·529 / 665·04	0·572 / 910·45	0·614 / 1205·8	0·654 / 1554·7	0·693 / 1960·8	0·731 / 2427·3
0·0135 (0·1324)	0·317 / 80·589	0·340 / 106·73	0·362 / 137·62	0·394 / 193·52	0·425 / 261·80	0·445 / 314·68	0·493 / 474·68	0·539 / 677·71	0·583 / 927·79	0·626 / 1228·8	0·667 / 1584·3	0·707 / 1998·1	0·745 / 2473·5
0·0140 (0·1373)	0·323 / 82·068	0·346 / 108·69	0·369 / 140·14	0·401 / 197·07	0·433 / 266·60	0·453 / 320·46	0·502 / 483·39	0·549 / 690·14	0·594 / 944·81	0·637 / 1251·3	0·679 / 1613·4	0·720 / 2034·8	0·759 / 2518·9
0·0145 (0·1422)	0·328 / 83·521	0·352 / 110·61	0·375 / 142·62	0·409 / 200·56	0·441 / 271·32	0·461 / 326·13	0·511 / 491·94	0·559 / 702·36	0·605 / 961·54	0·649 / 1273·5	0·691 / 1642·0	0·732 / 2070·8	0·773 / 2563·5
0·0150 (0·1471)	0·334 / 84·948	0·358 / 112·51	0·382 / 145·06	0·416 / 203·99	0·448 / 275·96	0·469 / 331·70	0·520 / 500·35	0·568 / 714·37	0·615 / 977·98	0·660 / 1295·2	0·703 / 1670·0	0·745 / 2106·2	0·786 / 2607·3
0·0160 (0·1569)	0·345 / 87·734	0·370 / 116·20	0·394 / 149·82	0·429 / 210·68	0·463 / 285·01	0·485 / 342·58	0·537 / 516·76	0·587 / 737·80	0·635 / 1010·0	0·681 / 1337·7	0·726 / 1724·8	0·769 / 2175·3	0·812 / 2692·8
0·0170 (0·1667)	0·355 / 90·434	0·381 / 119·77	0·406 / 154·43	0·442 / 217·16	0·477 / 293·78	0·500 / 353·13	0·554 / 532·67	0·605 / 760·50	0·655 / 1041·1	0·702 / 1378·9	0·748 / 1777·9	0·793 / 2242·2	0·836 / 2775·7
0·0180 (0·1765)	0·366 / 93·056	0·392 / 123·24	0·418 / 158·91	0·455 / 223·46	0·491 / 302·30	0·514 / 363·36	0·570 / 548·11	0·623 / 782·55	0·674 / 1071·3	0·723 / 1418·9	0·770 / 1829·4	0·816 / 2307·2	0·861 / 2856·2
0·0190 (0·1863)	0·376 / 95·606	0·403 / 126·62	0·429 / 163·26	0·468 / 229·58	0·504 / 310·59	0·528 / 373·32	0·585 / 563·13	0·640 / 803·99	0·692 / 1100·7	0·742 / 1457·7	0·791 / 1879·6	0·838 / 2370·4	0·884 / 2934·5
0·0200 (0·1961)	0·385 / 98·090	0·414 / 129·91	0·441 / 167·50	0·480 / 235·54	0·518 / 318·65	0·542 / 383·02	0·601 / 577·76	0·656 / 824·88	0·710 / 1129·3	0·762 / 1495·6	0·812 / 1928·4	0·860 / 2432·0	0·907 / 3010·7
0·0210 (0·2059)	0·395 / 100·51	0·424 / 133·12	0·452 / 171·64	0·492 / 241·36	0·530 / 326·52	0·555 / 392·48	0·615 / 592·03	0·673 / 845·25	0·728 / 1157·2	0·781 / 1532·5	0·832 / 1976·0	0·881 / 2492·1	0·930 / 3085·0
0·0220 (0·2157)	0·404 / 102·88	0·434 / 136·25	0·462 / 175·68	0·503 / 247·04	0·543 / 334·21	0·568 / 401·71	0·630 / 605·96	0·688 / 865·14	0·745 / 1184·4	0·799 / 1568·6	0·851 / 2022·5	0·902 / 2550·7	0·952 / 3157·6
0·0230 (0·2256)	0·413 / 105·19	0·443 / 139·31	0·473 / 179·63	0·515 / 252·59	0·555 / 341·72	0·581 / 410·74	0·644 / 619·58	0·704 / 884·59	0·761 / 1211·0	0·817 / 1603·9	0·870 / 2068·0	0·922 / 2608·1	0·973 / 3228·6
0·0240 (0·2354)	0·422 / 107·45	0·453 / 142·31	0·483 / 183·49	0·526 / 258·02	0·567 / 349·07	0·594 / 419·58	0·658 / 632·90	0·719 / 903·61	0·778 / 1237·1	0·834 / 1638·4	0·889 / 2112·5	0·942 / 2664·1	0·994 / 3298·0
0·0250 (0·2452)	0·431 / 109·67	0·462 / 145·24	0·493 / 187·27	0·536 / 263·35	0·579 / 356·27	0·606 / 428·23	0·671 / 645·95	0·734 / 922·24	0·794 / 1262·6	0·852 / 1672·1	0·907 / 2156·0	0·962 / 2719·1	1·014 / 3366·1
0·0260 (0·2550)	0·440 / 111·84	0·471 / 148·12	0·502 / 190·98	0·547 / 268·56	0·590 / 363·32	0·618 / 436·71	0·685 / 658·74	0·748 / 940·51	0·810 / 1287·6	0·868 / 1705·3	0·925 / 2198·7	0·981 / 2772·9	1·034 / 3432·7
0·0270 (0·2648)	0·448 / 113·97	0·480 / 150·94	0·512 / 194·62	0·558 / 273·68	0·601 / 370·24	0·630 / 445·03	0·698 / 671·29	0·763 / 958·42	0·825 / 1312·1	0·885 / 1737·7	0·943 / 2240·6	0·999 / 2825·8	1·054 / 3498·1
0·0280 (0·2746)	0·456 / 116·06	0·489 / 153·71	0·521 / 198·19	0·568 / 278·70	0·612 / 377·04	0·641 / 453·19	0·711 / 683·61	0·777 / 976·01	0·840 / 1336·2	0·901 / 1769·6	0·960 / 2281·7	1·018 / 2877·6	1·074 / 3562·3
0·0290 (0·2844)	0·464 / 118·12	0·498 / 156·43	0·531 / 201·70	0·578 / 283·63	0·623 / 383·71	0·652 / 461·22	0·723 / 695·71	0·790 / 993·29	0·855 / 1359·8	0·917 / 1800·9	0·977 / 2322·1	1·036 / 2928·5	1·093 / 3625·4
S, ($\Delta p/\rho l$) Gradient	18	20	22	25	28	30	35	40	45	50	55	60	65

(Equivalent) Pipe diameters in mm

S = 0·0100 to 0·0290;

$\Delta p/\rho l$ = (0·0981) to (0·2844) ms^{-2}

m = Manning $n \times 100$
S = 0·0300 to 0·0950

i.e. kin. pr. grad., $\Delta p/\rho l =$
(0·2942) to (0·9316) ms^{-2}

Full bore flow; Tables E give m_C or m_P for Colebrook-White or for laminar solutions resp., to divide mV and/or mQ as follows:

mV to give velocities in ms^{-1}
mQ to give discharges in millilitres/sec

Gradient   (Equivalent) Pipe diameters in mm

S, ($\Delta p/\rho l$)	18	20	22	25	28	30	35	40	45	50	55	60	65
0·0300	0·472	0·506	0·540	0·588	0·634	0·664	0·735	0·804	0·870	0·933	0·994	1·053	1·111
(0·2942)	120·14	159·11	205·15	288·48	390·27	469·10	707·61	1010·3	1383·1	1831·7	2361·8	2978·6	3687·3
0·0320	0·488	0·523	0·557	0·607	0·655	0·685	0·760	0·830	0·898	0·963	1·027	1·088	1·148
(0·3138)	124·07	164·33	211·88	297·94	403·07	484·49	730·81	1043·4	1428·4	1891·8	2439·3	3076·3	3808·3
0·0340	0·503	0·539	0·575	0·626	0·675	0·707	0·783	0·856	0·926	0·993	1·058	1·122	1·183
(0·3334)	127·89	169·38	218·40	307·11	415·47	499·40	753·30	1075·5	1472·4	1950·0	2514·3	3171·0	3925·5
0·0360	0·517	0·555	0·591	0·644	0·694	0·727	0·806	0·881	0·953	1·022	1·089	1·154	1·217
(0·3530)	131·60	174·29	224·73	316·01	427·52	513·87	775·14	1106·7	1515·1	2006·6	2587·2	3262·9	4039·3
0·0380	0·531	0·570	0·607	0·661	0·713	0·747	0·828	0·905	0·979	1·050	1·119	1·186	1·251
(0·3727)	135·21	179·07	230·89	324·67	439·23	527·96	796·38	1137·0	1556·6	2061·6	2658·1	3352·3	4150·0
0·0400	0·545	0·585	0·623	0·679	0·732	0·766	0·849	0·928	1·004	1·077	1·148	1·216	1·283
(0·3923)	138·72	183·72	236·89	333·11	450·64	541·67	817·07	1166·6	1597·0	2115·1	2727·2	3439·4	4257·8
0·0420	0·559	0·599	0·639	0·695	0·750	0·785	0·870	0·951	1·029	1·104	1·176	1·246	1·315
(0·4119)	142·15	188·26	242·74	341·33	461·77	555·05	837·25	1195·4	1636·5	2167·3	2794·5	3524·3	4362·9
0·0440	0·572	0·613	0·654	0·712	0·768	0·804	0·891	0·974	1·053	1·130	1·204	1·276	1·346
(0·4315)	145·49	192·69	248·45	349·37	472·64	568·11	856·95	1223·5	1675·0	2218·3	2860·3	3607·3	4465·6
0·0460	0·585	0·627	0·668	0·728	0·785	0·822	0·911	0·996	1·077	1·155	1·231	1·304	1·376
(0·4511)	148·76	197·02	254·03	357·22	483·26	580·88	876·21	1251·0	1712·6	2268·2	2924·6	3688·3	4565·9
0·0480	0·597	0·641	0·683	0·743	0·802	0·839	0·930	1·017	1·100	1·180	1·257	1·333	1·406
(0·4707)	151·96	201·26	259·50	364·90	493·66	593·37	895·06	1277·9	1749·5	2317·0	2987·5	3767·7	4664·1
0·0500	0·609	0·654	0·697	0·759	0·818	0·857	0·949	1·038	1·123	1·204	1·283	1·360	1·435
(0·4903)	155·09	205·41	264·85	372·43	503·84	605·61	913·51	1304·2	1785·5	2364·8	3049·1	3845·4	4760·3
0·0525	0·625	0·670	0·714	0·777	0·838	0·878	0·973	1·064	1·150	1·234	1·315	1·394	1·470
(0·5148)	158·92	210·48	271·39	381·62	516·28	620·56	936·07	1336·5	1829·6	2423·2	3124·4	3940·3	4877·9
0·0550	0·639	0·686	0·731	0·796	0·858	0·899	0·996	1·089	1·177	1·263	1·346	1·426	1·505
(0·5394)	162·66	215·43	277·77	390·60	528·43	635·17	958·10	1367·9	1872·7	2480·2	3197·9	4033·1	4992·7
0·0575	0·654	0·701	0·747	0·814	0·877	0·919	1·018	1·113	1·204	1·292	1·376	1·458	1·538
(0·5639)	166·32	220·27	284·02	399·38	540·30	649·44	979·64	1398·7	1914·8	2535·9	3269·8	4123·7	5104·9
0·0600	0·668	0·716	0·763	0·831	0·896	0·939	1·040	1·137	1·230	1·319	1·406	1·490	1·571
(0·5884)	169·90	225·01	290·13	407·97	551·92	663·41	1000·7	1428·7	1956·0	2590·5	3340·1	4212·4	5214·7
0·0625	0·681	0·731	0·779	0·848	0·915	0·958	1·062	1·160	1·255	1·347	1·435	1·521	1·604
(0·6129)	173·40	229·65	296·11	416·39	563·31	677·09	1021·3	1458·2	1996·3	2643·9	3409·0	4299·2	5322·2
0·0650	0·695	0·745	0·794	0·865	0·933	0·977	1·083	1·183	1·280	1·373	1·463	1·551	1·636
(0·6374)	176·83	234·20	301·97	424·63	574·46	690·50	1041·6	1487·1	2035·8	2696·2	3476·5	4384·4	5427·6
0·0675	0·708	0·760	0·810	0·882	0·951	0·995	1·103	1·206	1·304	1·399	1·491	1·580	1·667
(0·6619)	180·20	238·66	307·72	432·72	585·40	703·65	1061·4	1515·4	2074·6	2747·6	3542·7	4467·9	5531·0
0·0700	0·721	0·774	0·824	0·898	0·968	1·014	1·123	1·228	1·328	1·425	1·519	1·609	1·697
(0·6865)	183·51	243·04	313·37	440·66	596·15	716·56	1080·9	1543·2	2112·7	2798·0	3607·7	4549·9	5632·5
0·0725	0·734	0·787	0·839	0·914	0·985	1·032	1·143	1·250	1·352	1·450	1·545	1·638	1·727
(0·7110)	186·76	247·34	318·92	448·46	606·70	729·25	1100·0	1570·5	2150·1	2847·6	3671·6	4630·4	5732·2
0·0750	0·746	0·801	0·853	0·929	1·002	1·049	1·163	1·271	1·375	1·475	1·572	1·666	1·757
(0·7355)	189·95	251·57	324·37	456·13	617·07	741·71	1118·8	1597·4	2186·8	2896·2	3734·3	4709·6	5830·2
0·0800	0·771	0·827	0·881	0·960	1·035	1·084	1·201	1·313	1·420	1·523	1·623	1·720	1·815
(0·7845)	196·18	259·82	335·01	471·09	637·31	766·04	1155·5	1649·8	2258·5	2991·2	3856·8	4864·0	6021·4
0·0850	0·795	0·852	0·908	0·989	1·067	1·117	1·238	1·353	1·464	1·570	1·673	1·773	1·870
(0·8336)	202·22	267·82	345·32	485·59	656·92	789·61	1191·1	1700·5	2328·0	3083·3	3975·5	5013·7	6206·7
0·0900	0·818	0·877	0·935	1·018	1·098	1·149	1·274	1·392	1·506	1·616	1·722	1·825	1·925
(0·8826)	208·08	275·58	355·33	499·66	675·97	812·51	1225·6	1749·8	2395·5	3172·7	4090·8	5159·1	6386·6
0·0950	0·840	0·901	0·960	1·046	1·128	1·181	1·309	1·431	1·547	1·660	1·769	1·875	1·977
(0·9316)	213·78	283·13	365·07	513·35	694·49	834·77	1259·2	1797·8	2461·2	3259·6	4202·9	5300·5	6561·7
S, ($\Delta p/\rho l$)	18	20	22	25	28	30	35	40	45	50	55	60	65

Gradient   (Equivalent) Pipe diameters in mm

S = 0·0300 to 0·0950;

$\Delta p/\rho l$ = (0·2942) to (0·9316) ms^{-2}

m = Manning $n \times 100$
S = 0·1000 to 0·2900

i.e. kin. pr. grad., $\Delta p/\rho l$ =
(0·9807) to (2·8439) ms^{-2}

Full bore flow; Tables E give m_C or m_P for Colebrook-White or for laminar solutions resp., to divide mV and/or mQ as follows:

mV to give velocities in ms^{-1}
mQ to give discharges in millilitres/sec

Gradient — **(Equivalent) Pipe diameters in mm**

S, ($\Delta p/\rho l$)	18	20	22	25	28	30	35	40	45	50	55	60	65
0·1000	0·862	0·925	0·985	1·073	1·157	1·212	1·343	1·468	1·588	1·703	1·815	1·923	2·029
(0·9807)	219·34	290·49	374·55	526·69	712·53	856·46	1291·9	1844·5	2525·1	3344·3	4312·0	5438·2	6732·1
0·1050	0·883	0·947	1·010	1·099	1·186	1·242	1·376	1·504	1·627	1·745	1·860	1·971	2·079
(1·0297)	224·75	297·66	383·80	539·70	730·13	877·61	1323·8	1890·0	2587·5	3426·9	4418·5	5572·5	6898·4
0·1100	0·904	0·970	1·033	1·125	1·214	1·271	1·408	1·539	1·665	1·786	1·904	2·017	2·128
(1·0787)	230·04	304·67	392·83	552·40	747·31	898·26	1355·0	1934·5	2648·4	3507·5	4522·5	5703·6	7060·7
0·1150	0·924	0·992	1·057	1·151	1·241	1·299	1·440	1·574	1·703	1·827	1·946	2·063	2·176
(1·1278)	235·21	311·51	401·66	564·81	764·10	918·45	1385·4	1978·0	2707·9	3586·3	4624·1	5831·8	7219·4
0·1200	0·944	1·013	1·079	1·175	1·268	1·327	1·471	1·608	1·739	1·866	1·988	2·107	2·222
(1·1768)	240·27	318·21	410·30	576·96	780·54	938·20	1415·2	2020·5	2766·1	3663·5	4723·6	5957·2	7374·7
0·1250	0·964	1·034	1·102	1·200	1·294	1·355	1·501	1·641	1·775	1·904	2·029	2·150	2·268
(1·2258)	245·22	324·78	418·76	588·86	796·63	957·55	1444·4	2062·2	2823·2	3739·0	4821·0	6080·1	7526·7
0·1300	0·983	1·054	1·123	1·223	1·319	1·381	1·531	1·674	1·810	1·942	2·069	2·193	2·313
(1·2749)	250·08	331·21	427·05	600·52	812·41	976·51	1473·0	2103·0	2879·1	3813·1	4916·5	6200·5	7675·8
0·1350	1·001	1·074	1·145	1·247	1·345	1·408	1·560	1·705	1·845	1·979	2·109	2·235	2·357
(1·3239)	254·84	337·52	435·19	611·96	827·89	995·11	1501·1	2143·1	2933·9	3885·7	5010·1	6318·6	7822·0
0·1400	1·020	1·094	1·166	1·270	1·369	1·434	1·589	1·737	1·879	2·015	2·147	2·276	2·400
(1·3729)	259·52	343·71	443·17	623·19	843·08	1013·4	1528·6	2182·4	2987·8	3957·0	5102·1	6434·5	7965·5
0·1450	1·038	1·113	1·186	1·292	1·393	1·459	1·617	1·767	1·912	2·051	2·186	2·316	2·443
(1·4220)	264·12	349·79	451·02	634·22	858·00	1031·3	1555·7	2221·1	3040·7	4027·0	5192·4	6548·4	8106·5
0·1500	1·056	1·132	1·207	1·314	1·417	1·484	1·645	1·798	1·945	2·086	2·223	2·356	2·485
(1·4710)	268·63	355·77	458·73	645·06	872·67	1048·9	1582·3	2259·0	3092·6	4095·9	5281·2	6660·4	8245·1
0·1600	1·090	1·170	1·246	1·357	1·464	1·533	1·698	1·857	2·008	2·154	2·296	2·433	2·566
(1·5691)	277·44	367·44	473·77	666·22	901·29	1083·3	1634·1	2333·1	3194·1	4230·2	5454·3	6878·8	8515·5
0·1700	1·124	1·206	1·285	1·399	1·509	1·580	1·751	1·914	2·070	2·221	2·366	2·508	2·645
(1·6671)	285·98	378·75	488·35	686·72	929·03	1116·7	1684·4	2404·9	3292·4	4360·4	5622·2	7090·5	8777·6
0·1800	1·156	1·241	1·322	1·440	1·553	1·626	1·802	1·969	2·130	2·285	2·435	2·580	2·722
(1·7652)	294·27	389·73	502·51	706·63	955·96	1149·1	1733·3	2474·6	3387·8	4486·8	5785·2	7296·1	9032·1
0·1900	1·188	1·275	1·358	1·479	1·595	1·670	1·851	2·023	2·188	2·348	2·502	2·651	2·796
(1·8633)	302·33	400·41	516·28	725·99	982·16	1180·5	1780·8	2542·5	3480·6	4609·8	5943·7	7496·0	9279·6
0·2000	1·219	1·308	1·393	1·517	1·636	1·714	1·899	2·076	2·245	2·409	2·567	2·720	2·869
(1·9613)	310·19	410·81	529·69	744·85	1007·7	1211·2	1827·0	2608·5	3571·1	4729·5	6098·1	7690·7	9520·6
0·2100	1·249	1·340	1·428	1·555	1·677	1·756	1·946	2·127	2·301	2·468	2·630	2·787	2·940
(2·0594)	317·85	420·96	542·77	763·25	1032·6	1241·1	1872·1	2672·9	3659·3	4846·3	6248·7	7880·7	9755·8
0·2200	1·278	1·371	1·461	1·591	1·716	1·797	1·992	2·177	2·355	2·526	2·692	2·853	3·009
(2·1575)	325·33	430·86	555·55	781·21	1056·9	1270·3	1916·2	2735·8	3745·4	4960·4	6395·8	8066·1	9985·3
0·2300	1·307	1·402	1·494	1·627	1·755	1·838	2·036	2·226	2·408	2·583	2·753	2·917	3·077
(2·2555)	332·64	440·55	568·03	798·77	1080·6	1298·9	1959·3	2797·3	3829·5	5071·9	6539·5	8247·4	10210
0·2400	1·335	1·432	1·526	1·662	1·793	1·877	2·080	2·274	2·460	2·639	2·812	2·980	3·143
(2·3536)	339·79	450·02	580·25	815·95	1103·8	1326·8	2001·4	2857·5	3911·9	5180·9	6680·2	8424·8	10429
0·2500	1·363	1·462	1·558	1·697	1·830	1·916	2·123	2·321	2·510	2·693	2·870	3·041	3·208
(2·4517)	346·80	459·30	592·22	832·77	1126·6	1354·2	2042·7	2916·4	3992·6	5287·8	6817·9	8598·5	10644
0·2600	1·390	1·491	1·589	1·730	1·866	1·954	2·165	2·367	2·560	2·746	2·927	3·101	3·271
(2·5497)	353·67	468·40	603·94	849·26	1148·9	1381·0	2083·1	2974·1	4071·6	5392·5	6953·0	8768·8	10855
0·2700	1·416	1·519	1·619	1·763	1·901	1·991	2·206	2·412	2·609	2·799	2·982	3·160	3·334
(2·6478)	360·41	477·32	615·45	865·44	1170·8	1407·3	2122·8	3030·8	4149·2	5495·2	7085·4	8935·8	11062
0·2800	1·442	1·547	1·649	1·795	1·936	2·027	2·247	2·456	2·657	2·850	3·037	3·218	3·395
(2·7459)	367·02	486·08	626·74	881·32	1192·3	1433·1	2161·8	3086·4	4225·3	5596·0	7215·4	9099·8	11265
0·2900	1·468	1·575	1·678	1·827	1·971	2·063	2·287	2·500	2·704	2·900	3·091	3·275	3·455
(2·8439)	373·52	494·68	637·84	896·92	1213·4	1458·5	2200·0	3141·0	4300·1	5695·1	7343·1	9260·9	11464
S, ($\Delta p/\rho l$)	18	20	22	25	28	30	35	40	45	50	55	60	65

Gradient — **(Equivalent) Pipe diameters in mm**

S = 0·1000 to 0·2900;

$\Delta p/\rho l$ = (0·9807) to (2·8439) ms^{-2}

m = Manning $n \times 100$
S = 0·3000 to 0·9500

Full bore flow; Tables E give m_C or m_P for Colebrook-White or for laminar solutions resp., to divide mV and/or mQ as follows:

i.e. kin. pr. grad., $\Delta p/\rho l$ = (2·9420) to (9·3163) ms^{-2}

mV to give velocities in ms^{-1}
mQ to give discharges in millilitres/sec

Gradient — (Equivalent) Pipe diameters in mm

S, ($\Delta p/\rho l$)	18	20	22	25	28	30	35	40	45	50	55	60	65
0·3000	1·493	1·602	1·707	1·858	2·004	2·099	2·326	2·542	2·750	2·950	3·144	3·331	3·514
(2·9420)	379·90	503·14	648·74	912·26	1234·1	1483·4	2237·6	3194·7	4373·6	5792·5	7468·7	9419·2	11660
0·3200	1·542	1·654	1·763	1·919	2·070	2·167	2·402	2·626	2·840	3·047	3·247	3·441	3·629
(3·1381)	392·36	519·64	670·02	942·17	1274·6	1532·1	2311·0	3299·5	4517·1	5982·4	7713·6	9728·1	12043
0·3400	1·589	1·705	1·817	1·978	2·134	2·234	2·476	2·706	2·928	3·141	3·347	3·547	3·741
(3·3343)	404·43	535·63	690·64	971·17	1313·8	1579·2	2382·2	3401·1	4656·1	6166·5	7951·0	10027	12413
0·3600	1·635	1·754	1·870	2·036	2·196	2·299	2·548	2·785	3·012	3·232	3·444	3·649	3·849
(3·5304)	416·16	551·16	710·66	999·33	1351·9	1625·0	2451·2	3499·7	4791·1	6345·3	8181·5	10318	12773
0·3800	1·680	1·802	1·921	2·092	2·256	2·362	2·618	2·861	3·095	3·320	3·538	3·749	3·955
(3·7265)	427·56	566·27	730·13	1026·7	1389·0	1669·5	2518·4	3595·6	4922·4	6519·2	8405·7	10601	13123
0·4000	1·724	1·849	1·971	2·146	2·314	2·423	2·686	2·936	3·175	3·406	3·630	3·847	4·058
(3·9227)	438·67	580·98	749·10	1053·4	1425·1	1712·9	2583·8	3689·0	5050·2	6688·6	8624·1	10876	13464
0·4200	1·766	1·895	2·019	2·199	2·371	2·483	2·752	3·008	3·254	3·491	3·720	3·942	4·158
(4·1188)	449·50	595·32	767·60	1079·4	1460·3	1755·2	2647·6	3780·1	5175·0	6853·7	8837·1	11145	13797
0·4400	1·808	1·940	2·067	2·251	2·427	2·542	2·817	3·079	3·330	3·573	3·807	4·034	4·256
(4·3149)	460·08	609·33	785·66	1104·8	1494·6	1796·5	2709·9	3869·0	5296·7	7015·0	9045·0	11407	14121
0·4600	1·849	1·983	2·113	2·301	2·482	2·599	2·880	3·148	3·405	3·653	3·893	4·125	4·351
(4·5111)	470·42	623·03	803·32	1129·6	1528·2	1836·9	2770·8	3956·0	5415·8	7172·7	9248·3	11664	14439
0·4800	1·888	2·026	2·159	2·351	2·535	2·655	2·942	3·216	3·478	3·732	3·976	4·214	4·445
(4·7072)	480·54	636·43	820·60	1153·9	1561·1	1876·4	2830·4	4041·1	5532·3	7326·9	9447·2	11914	14749
0·5000	1·927	2·068	2·203	2·399	2·588	2·709	3·003	3·282	3·550	3·809	4·058	4·301	4·536
(4·9033)	490·45	649·55	837·52	1177·7	1593·3	1915·1	2888·8	4124·4	5646·3	7478·0	9642·0	12160	15053
0·5250	1·975	2·119	2·258	2·458	2·651	2·776	3·077	3·363	3·638	3·903	4·159	4·407	4·649
(5·1485)	502·56	665·59	858·20	1206·8	1632·6	1962·4	2960·1	4226·3	5785·8	7662·7	9880·1	12460	15425
0·5500	2·021	2·169	2·311	2·516	2·714	2·842	3·149	3·442	3·723	3·994	4·256	4·511	4·758
(5·3937)	514·39	681·26	878·40	1235·2	1671·0	2008·6	3029·8	4325·7	5921·9	7843·0	10113	12754	15788
0·5750	2·067	2·217	2·363	2·573	2·775	2·905	3·220	3·520	3·807	4·084	4·352	4·612	4·865
(5·6388)	525·95	696·57	898·14	1263·0	1708·6	2053·7	3097·9	4422·9	6055·0	8019·3	10340	13040	16143
0·6000	2·111	2·265	2·414	2·628	2·834	2·968	3·289	3·595	3·889	4·172	4·446	4·711	4·969
(5·8840)	537·26	711·55	917·46	1290·1	1745·3	2097·9	3164·5	4518·1	6185·3	8191·8	10562	13321	16490
0·6250	2·155	2·312	2·463	2·682	2·893	3·029	3·357	3·669	3·969	4·258	4·537	4·808	5·072
(6·1292)	548·34	726·22	936·37	1316·7	1781·3	2141·1	3229·8	4611·2	6312·8	8360·7	10780	13595	16830
0·6500	2·198	2·357	2·512	2·736	2·950	3·089	3·423	3·742	4·048	4·342	4·627	4·904	5·172
(6·3743)	559·20	740·60	954·92	1342·8	1816·6	2183·5	3293·7	4702·5	6437·8	8526·3	10994	13865	17164
0·6750	2·239	2·402	2·560	2·788	3·006	3·148	3·489	3·813	4·125	4·425	4·715	4·997	5·271
(6·6195)	569·85	754·71	973·11	1368·4	1851·2	2225·1	3356·5	4792·1	6560·5	8688·7	11203	14129	17491
0·7000	2·280	2·446	2·607	2·839	3·062	3·206	3·553	3·883	4·201	4·506	4·802	5·089	5·368
(6·8647)	580·31	768·56	990·97	1393·5	1885·2	2266·0	3418·1	4880·1	6680·8	8848·1	11409	14388	17811
0·7250	2·321	2·490	2·653	2·889	3·116	3·262	3·616	3·952	4·275	4·586	4·887	5·179	5·463
(7·1098)	590·58	782·16	1008·5	1418·2	1918·5	2306·1	3478·6	4966·4	6799·1	9004·7	11611	14643	18127
0·7500	2·361	2·532	2·698	2·938	3·169	3·318	3·677	4·020	4·348	4·664	4·970	5·267	5·556
(7·3550)	600·68	795·54	1025·7	1442·4	1951·3	2345·5	3538·0	5051·3	6915·3	9158·7	11809	14893	18437
0·8000	2·438	2·615	2·787	3·035	3·273	3·427	3·798	4·152	4·491	4·817	5·133	5·440	5·738
(7·8453)	620·37	821·63	1059·4	1489·7	2015·3	2422·4	3654·1	5217·0	7142·1	9459·0	12196	15381	19041
0·8500	2·513	2·696	2·873	3·128	3·374	3·533	3·915	4·279	4·629	4·966	5·291	5·608	5·915
(8·3357)	639·47	846·91	1092·0	1535·6	2077·4	2497·0	3766·5	5377·6	7361·9	9750·2	12572	15855	19627
0·9000	2·586	2·774	2·956	3·219	3·472	3·635	4·028	4·403	4·763	5·110	5·445	5·770	6·086
(8·8260)	658·01	871·47	1123·6	1580·1	2137·6	2569·4	3875·7	5533·5	7575·4	10033	12936	16314	20196
0·9500	2·657	2·850	3·037	3·307	3·567	3·735	4·139	4·524	4·894	5·250	5·594	5·928	6·253
(9·3163)	676·04	895·35	1154·4	1623·4	2196·2	2639·8	3981·9	5685·1	7783·0	10308	13291	16762	20750
S, ($\Delta p/\rho l$)	18	20	22	25	28	30	35	40	45	50	55	60	65

Gradient — (Equivalent) Pipe diameters in mm

S = 0·3000 to 0·9500;

$\Delta p/\rho l$ = (2·9420) to (9·3163) ms^{-2}

m = Manning $n \times 100$
S = 1·0000 to 2·9000

i.e. kin. pr. grad., $\Delta p/\rho l$ =
(9·8067) to (28·439) ms^{-2}

Full bore flow; Tables E give m_C or m_P for Colebrook-White or for laminar solutions resp., to divide mV and/or mQ as follows:

mV to give velocities in ms^{-1}
mQ to give discharges in millilitres/sec

Gradient S, $(\Delta p/\rho l)$	\multicolumn{13}{c}{(Equivalent) Pipe diameters in mm}
	18	20	22	25	28	30	35	40	45	50	55	60	65
1·0000	2·726	2·924	3·116	3·393	3·659	3·832	4·246	4·642	5·021	5·386	5·739	6·082	6·416
(9·8067)	693·60	918·61	1184·4	1665·5	2253·2	2708·4	4085·4	5832·8	7985·1	10576	13636	17197	21289
1·0500	2·793	2·996	3·193	3·477	3·750	3·926	4·351	4·756	5·145	5·519	5·881	6·232	6·574
(10·297)	710·73	941·29	1213·7	1706·7	2308·9	2775·2	4186·3	5976·8	8182·3	10837	13973	17622	21815
1·1000	2·859	3·067	3·268	3·559	3·838	4·019	4·453	4·868	5·266	5·649	6·020	6·379	6·729
(10·787)	727·45	963·44	1242·2	1746·8	2363·2	2840·5	4284·8	6117·5	8374·9	11092	14301	18036	22328
1·1500	2·923	3·136	3·341	3·639	3·924	4·109	4·554	4·978	5·384	5·776	6·155	6·522	6·880
(11·278)	743·80	985·09	1270·2	1786·1	2416·3	2904·4	4381·1	6255·0	8563·1	11341	14623	18442	22830
1·2000	2·986	3·203	3·413	3·717	4·009	4·197	4·652	5·085	5·500	5·900	6·287	6·663	7·028
(11·768)	759·80	1006·3	1297·5	1824·5	2468·3	2966·9	4475·3	6389·5	8747·3	11585	14937	18838	23321
1·2500	3·047	3·269	3·484	3·794	4·091	4·284	4·747	5·189	5·613	6·022	6·417	6·800	7·173
(12·258)	775·47	1027·0	1324·2	1862·1	2519·2	3028·0	4567·6	6521·2	8927·7	11824	15245	19227	23802
1·3000	3·108	3·334	3·553	3·869	4·172	4·369	4·841	5·292	5·725	6·141	6·544	6·935	7·315
(12·749)	790·83	1047·4	1350·5	1899·0	2569·1	3088·0	4658·0	6650·4	9104·5	12058	15547	19608	24273
1·3500	3·167	3·397	3·620	3·942	4·252	4·452	4·934	5·393	5·834	6·258	6·669	7·067	7·454
(13·239)	805·89	1067·3	1376·2	1935·2	2618·0	3146·8	4746·8	6777·1	9277·9	12288	15843	19981	24735
1·4000	3·225	3·460	3·687	4·015	4·330	4·534	5·024	5·492	5·941	6·373	6·791	7·197	7·591
(13·729)	820·68	1086·9	1401·4	1970·7	2666·0	3204·6	4833·9	6901·4	9448·1	12513	16134	20348	25189
1·4500	3·282	3·521	3·752	4·086	4·406	4·614	5·113	5·589	6·046	6·486	6·911	7·324	7·725
(14·220)	835·21	1106·1	1426·2	2005·6	2713·2	3261·3	4919·4	7023·6	9615·4	12735	16420	20708	25635
1·5000	3·338	3·581	3·816	4·156	4·482	4·693	5·201	5·685	6·149	6·597	7·029	7·449	7·857
(14·710)	849·48	1125·1	1450·6	2039·9	2759·6	3317·0	5003·5	7143·7	9779·8	12952	16700	21062	26073
1·6000	3·448	3·699	3·941	4·292	4·629	4·847	5·371	5·871	6·351	6·813	7·260	7·693	8·115
(15·691)	877·34	1162·0	1498·2	2106·8	2850·1	3425·8	5167·6	7378·0	10100	13377	17248	21753	26928
1·7000	3·554	3·812	4·063	4·424	4·771	4·996	5·536	6·052	6·546	7·023	7·483	7·930	8·365
(16·671)	904·34	1197·7	1544·3	2171·6	2937·8	3531·3	5326·7	7605·0	10411	13789	17779	22422	27757
1·8000	3·657	3·923	4·180	4·552	4·909	5·141	5·697	6·227	6·736	7·226	7·700	8·160	8·607
(17·652)	930·56	1232·4	1589·1	2234·6	3023·0	3633·6	5481·1	7825·5	10713	14189	18294	23072	28562
1·9000	3·757	4·030	4·295	4·677	5·044	5·281	5·853	6·398	6·921	7·424	7·911	8·384	8·843
(18·633)	956·06	1266·2	1632·6	2295·8	3105·9	3733·2	5631·3	8039·9	11007	14577	18796	23704	29345
2·0000	3·855	4·135	4·406	4·798	5·175	5·419	6·005	6·564	7·100	7·617	8·117	8·602	9·073
(19·613)	980·90	1299·1	1675·0	2355·4	3186·5	3830·2	5777·6	8248·8	11293	14956	19284	24320	30107
2·1000	3·950	4·237	4·515	4·917	5·303	5·552	6·153	6·726	7·276	7·805	8·317	8·814	9·297
(20·594)	1005·1	1331·2	1716·4	2413·6	3265·2	3924·8	5920·3	8452·5	11572	15325	19760	24921	30850
2·2000	4·043	4·337	4·622	5·033	5·428	5·683	6·298	6·885	7·447	7·989	8·513	9·021	9·516
(21·575)	1028·8	1362·5	1756·8	2470·4	3342·1	4017·1	6059·6	8651·4	11844	15686	20225	25507	31576
2·3000	4·134	4·434	4·725	5·146	5·550	5·811	6·440	7·039	7·614	8·168	8·704	9·224	9·730
(22·555)	1051·9	1393·1	1796·3	2525·9	3417·2	4107·4	6195·8	8845·9	12110	16039	20680	26081	32286
2·4000	4·223	4·530	4·827	5·256	5·669	5·936	6·578	7·191	7·778	8·344	8·891	9·423	9·939
(23·536)	1074·5	1423·1	1834·9	2580·2	3490·7	4195·8	6329·0	9036·1	12371	16384	21125	26641	32980
2·5000	4·310	4·623	4·927	5·365	5·786	6·058	6·714	7·339	7·938	8·516	9·075	9·617	10·14
(24·517)	1096·7	1452·4	1872·7	2633·5	3562·7	4282·3	6459·5	9222·4	12626	16721	21560	27191	33661
2·6000	4·395	4·715	5·024	5·471	5·900	6·178	6·847	7·484	8·096	8·685	9·255	9·807	10·34
(25·497)	1118·4	1481·2	1909·8	2685·6	3633·2	4367·1	6587·4	9405·1	12876	17053	21987	27729	34327
2·7000	4·479	4·805	5·120	5·575	6·013	6·296	6·977	7·627	8·250	8·850	9·431	9·994	10·54
(26·478)	1139·7	1509·4	1946·2	2736·8	3702·4	4450·3	6712·9	9584·2	13121	17377	22406	28258	34981
2·8000	4·561	4·893	5·214	5·678	6·123	6·411	7·105	7·767	8·401	9·013	9·604	10·18	10·74
(27·459)	1160·6	1537·1	1981·9	2787·0	3770·4	4531·9	6836·1	9760·1	13362	17696	22817	28776	35623
2·9000	4·642	4·979	5·306	5·778	6·232	6·525	7·231	7·904	8·550	9·172	9·774	10·36	10·93
(28·439)	1181·2	1564·3	2017·0	2836·3	3837·1	4612·2	6957·1	9932·9	13598	18009	23221	29285	36254
S, $(\Delta p/\rho l)$	18	20	22	25	28	30	35	40	45	50	55	60	65

Gradient (Equivalent) Pipe diameters in mm

S = 1·0000 to 2·9000;

$\Delta p/\rho l$ = (9·8067) to (28·439) ms^{-2}

m = Manning $n \times 100$
S = 3·0000 to 10·000

Full bore flow; Tables E give m_C or m_P for Colebrook-White or for laminar solutions resp., to divide mV and/or mQ as follows:

i.e. kin. pr. grad., $\Delta p/\rho l$ = (29·420) to (98·067) ms^{-2}

mV to give velocities in ms^{-1}
mQ to give discharges in millilitres/sec

Gradient   (Equivalent) Pipe diameters in mm

S, ($\Delta p/\rho l$)	18	20	22	25	28	30	35	40	45	50	55	60	65
3·0000	4·721	5·065	5·397	5·877	6·338	6·636	7·355	8·039	8·696	9·329	9·941	10·53	11·11
(29·420)	1201·4	1591·1	2051·5	2884·8	3902·7	4691·0	7076·1	10103	13831	18317	23618	29786	36873
3·2000	4·876	5·231	5·574	6·070	6·546	6·854	7·596	8·303	8·981	9·635	10·27	10·88	11·48
(31·381)	1240·7	1643·3	2118·8	2979·4	4030·7	4844·9	7308·1	10434	14284	18918	24393	30763	38083
3·4000	5·026	5·392	5·745	6·256	6·747	7·065	7·830	8·559	9·258	9·931	10·58	11·22	11·83
(33·343)	1278·9	1693·8	2184·0	3071·1	4154·7	4994·0	7533·0	10755	14724	19500	25143	31710	39255
3·6000	5·172	5·548	5·912	6·438	6·943	7·270	8·057	8·807	9·526	10·22	10·89	11·54	12·17
(35·304)	1316·0	1742·9	2247·3	3160·1	4275·2	5138·7	7751·4	11067	15151	20066	25872	32629	40393
3·8000	5·313	5·700	6·074	6·614	7·133	7·469	8·277	9·048	9·787	10·50	11·19	11·86	12·51
(37·265)	1352·1	1790·7	2308·9	3246·7	4392·3	5279·6	7963·8	11370	15566	20616	26581	33523	41500
4·0000	5·451	5·848	6·232	6·786	7·319	7·663	8·492	9·283	10·04	10·77	11·48	12·16	12·83
(39·227)	1387·2	1837·2	2368·9	3331·1	4506·4	5416·7	8170·7	11666	15970	21151	27272	34394	42578
4·2000	5·586	5·992	6·386	6·954	7·499	7·852	8·702	9·512	10·29	11·04	11·76	12·46	13·15
(41·188)	1421·5	1882·6	2427·4	3413·3	4617·7	5550·5	8372·5	11954	16365	21673	27945	35243	43629
4·4000	5·717	6·133	6·536	7·117	7·676	8·037	8·907	9·736	10·53	11·30	12·04	12·76	13·46
(43·149)	1454·9	1926·9	2484·5	3493·7	4726·4	5681·1	8569·5	12235	16750	22183	28603	36073	44656
4·6000	5·846	6·271	6·683	7·277	7·848	8·218	9·107	9·955	10·77	11·55	12·31	13·04	13·76
(45·111)	1487·6	1970·2	2540·3	3572·2	4832·6	5808·8	8762·1	12510	17126	22682	29246	36883	45659
4·8000	5·972	6·406	6·826	7·434	8·017	8·394	9·303	10·17	11·00	11·80	12·57	13·33	14·06
(47·072)	1519·6	2012·6	2595·0	3649·0	4936·6	5933·7	8950·6	12779	17495	23170	29875	37677	46641
5·0000	6·095	6·538	6·967	7·587	8·182	8·568	9·495	10·38	11·23	12·04	12·83	13·60	14·35
(49·033)	1550·9	2054·1	2648·5	3724·3	5038·4	6056·1	9135·1	13042	17855	23648	30491	38454	47603
5·2500	6·245	6·700	7·139	7·774	8·385	8·779	9·729	10·64	11·50	12·34	13·15	13·94	14·70
(51·485)	1589·2	2104·8	2713·9	3816·2	5162·8	6205·6	9360·7	13365	18296	24232	31244	39403	48779
5·5000	6·392	6·857	7·307	7·957	8·582	8·986	9·958	10·89	11·77	12·63	13·46	14·26	15·05
(53·937)	1626·6	2154·3	2777·7	3906·0	5284·3	6351·7	9581·0	13679	18727	24802	31979	40331	49927
5·7500	6·536	7·012	7·472	8·136	8·775	9·188	10·18	11·13	12·04	12·92	13·76	14·58	15·38
(56·388)	1663·2	2202·7	2840·2	3993·8	5403·0	6494·4	9796·4	13987	19148	25359	32698	41237	51049
6·0000	6·677	7·162	7·632	8·311	8·963	9·385	10·40	11·37	12·30	13·19	14·06	14·90	15·71
(58·840)	1699·0	2250·1	2901·3	4079·7	5519·2	6634·1	10007	14287	19560	25905	33401	42124	52147
6·2500	6·814	7·310	7·790	8·483	9·148	9·579	10·62	11·60	12·55	13·47	14·35	15·21	16·04
(61·292)	1734·0	2296·5	2961·1	4163·9	5633·1	6770·9	10213	14582	19963	26439	34090	42992	53222
6·5000	6·949	7·455	7·944	8·651	9·329	9·769	10·83	11·83	12·80	13·73	14·63	15·51	16·36
(63·743)	1768·3	2342·0	3019·7	4246·3	5744·6	6905·0	10416	14871	20358	26962	34765	43844	54276
6·7500	7·082	7·597	8·095	8·815	9·507	9·955	11·03	12·06	13·04	13·99	14·91	15·80	16·67
(66·195)	1802·0	2386·6	3077·2	4327·2	5854·0	7036·5	10614	15154	20746	27476	35427	44679	55310
7·0000	7·211	7·736	8·244	8·977	9·682	10·14	11·23	12·28	13·28	14·25	15·19	16·09	16·97
(68·647)	1835·1	2430·4	3133·7	4406·6	5961·5	7165·6	10809	15432	21127	27980	36077	45499	56325
7·2500	7·339	7·873	8·390	9·136	9·853	10·32	11·43	12·50	13·52	14·50	15·45	16·38	17·27
(71·098)	1867·6	2473·4	3189·2	4484·6	6067·0	7292·5	11000	15705	21501	28476	36716	46304	57322
7·5000	7·465	8·008	8·533	9·292	10·02	10·49	11·63	12·71	13·75	14·75	15·72	16·66	17·57
(73·550)	1899·5	2515·7	3243·7	4561·3	6170·7	7417·1	11188	15974	21868	28962	37343	47096	58302
8·0000	7·709	8·270	8·813	9·597	10·35	10·84	12·01	13·13	14·20	15·23	16·23	17·20	18·15
(78·453)	1961·8	2598·2	3350·1	4710·9	6373·1	7660·4	11555	16498	22585	29912	38568	48640	60214
8·5000	7·947	8·525	9·084	9·892	10·67	11·17	12·38	13·53	14·64	15·70	16·73	17·73	18·70
(83·357)	2022·2	2678·2	3453·2	4855·9	6569·2	7896·1	11911	17005	23280	30833	39755	50137	62067
9·0000	8·177	8·772	9·348	10·18	10·98	11·49	12·74	13·92	15·06	16·16	17·22	18·25	19·25
(88·260)	2080·8	2755·8	3553·3	4996·6	6759·7	8125·1	12256	17498	23955	31727	40908	51591	63866
9·5000	8·401	9·012	9·604	10·46	11·28	11·81	13·09	14·31	15·47	16·60	17·69	18·75	19·77
(93·163)	2137·8	2831·3	3650·7	5133·5	6944·9	8347·7	12592	17978	24612	32596	42029	53005	65617
10·000	8·619	9·247	9·853	10·73	11·57	12·12	13·43	14·68	15·88	17·03	18·15	19·23	20·29
(98·067)	2193·4	2904·9	3745·5	5266·9	7125·3	8564·6	12919	18445	25251	33443	43120	54382	67321
S, ($\Delta p/\rho l$)	18	20	22	25	28	30	35	40	45	50	55	60	65

Gradient   (Equivalent) Pipe diameters in mm

S = 3·0000 to 10·000; $\Delta p/\rho l$ = (29·420) to (98·067) ms^{-2}

Tables D1 and D2 before this point show values of mV in black

to give mean velocities V in ms^{-1}

and values of mQ in orange to give discharges Q in millilitres/sec

(i.e. Q in m^{3}s^{-1}/1000000)

Tables D3 to D6 immediately following this point show values of mV in black

to give mean velocities V in ms^{-1}

and values of mQ in green to give discharges Q in litres/sec

(i.e. Q in m^{3}s^{-1}/1000)

m = Manning $n \times 100$
S = 0.000300 to 0.000950

Full bore flow; Tables E give m_C or m_P for Colebrook-White or for laminar solutions resp., to divide mV and/or mQ as follows:

i.e. kin. pr. grad., $\Delta p/\rho l$ = (0.002942) to (0.009316) ms^{-2}

mV to give velocities in ms^{-1}
mQ to give discharges in litres/sec

Gradient — (Equivalent) Pipe diameters in mm

S, ($\Delta p/\rho l$)	65	70	75	80	85	90	95	100	110	125	135	150	160
0.000300	0.111	0.117	0.122	0.128	0.133	0.138	0.143	0.148	0.158	0.172	0.181	0.194	0.203
(0.002942)	0.3687	0.4493	0.5401	0.6415	0.7540	0.8782	1.0144	1.1631	1.4997	2.1088	2.5892	3.4292	4.0731
0.000320	0.115	0.121	0.126	0.132	0.137	0.143	0.148	0.153	0.163	0.177	0.187	0.200	0.209
(0.003138)	0.3808	0.4640	0.5578	0.6625	0.7788	0.9070	1.0477	1.2012	1.5488	2.1780	2.6741	3.5416	4.2067
0.000340	0.118	0.124	0.130	0.136	0.141	0.147	0.152	0.158	0.168	0.183	0.193	0.207	0.216
(0.003334)	0.3925	0.4783	0.5749	0.6829	0.8027	0.9349	1.0799	1.2382	1.5965	2.2450	2.7564	3.6506	4.3362
0.000360	0.122	0.128	0.134	0.140	0.146	0.151	0.157	0.162	0.173	0.188	0.198	0.213	0.222
(0.003530)	0.4039	0.4922	0.5916	0.7027	0.8260	0.9620	1.1112	1.2741	1.6428	2.3101	2.8363	3.7564	4.4619
0.000380	0.125	0.131	0.138	0.144	0.150	0.155	0.161	0.167	0.178	0.193	0.204	0.218	0.228
(0.003727)	0.4150	0.5057	0.6078	0.7220	0.8486	0.9884	1.1417	1.3090	1.6878	2.3734	2.9141	3.8594	4.5842
0.000400	0.128	0.135	0.141	0.147	0.153	0.159	0.165	0.171	0.182	0.198	0.209	0.224	0.234
(0.003923)	0.4258	0.5188	0.6236	0.7407	0.8707	1.0140	1.1713	1.3430	1.7316	2.4350	2.9898	3.9596	4.7033
0.000420	0.131	0.138	0.145	0.151	0.157	0.163	0.169	0.175	0.187	0.203	0.214	0.230	0.240
(0.004119)	0.4363	0.5316	0.6390	0.7590	0.8922	1.0391	1.2002	1.3762	1.7744	2.4952	3.0636	4.0574	4.8194
0.000440	0.135	0.141	0.148	0.155	0.161	0.167	0.173	0.179	0.191	0.208	0.219	0.235	0.245
(0.004315)	0.4466	0.5441	0.6540	0.7769	0.9132	1.0635	1.2285	1.4086	1.8162	2.5539	3.1357	4.1529	4.9328
0.000460	0.138	0.145	0.151	0.158	0.165	0.171	0.177	0.183	0.195	0.213	0.224	0.240	0.251
(0.004511)	0.4566	0.5564	0.6687	0.7943	0.9337	1.0874	1.2561	1.4402	1.8570	2.6113	3.2062	4.2462	5.0437
0.000480	0.141	0.148	0.155	0.161	0.168	0.175	0.181	0.187	0.200	0.217	0.229	0.245	0.256
(0.004707)	0.4664	0.5683	0.6831	0.8114	0.9538	1.1108	1.2831	1.4712	1.8969	2.6675	3.2751	4.3376	5.1522
0.000500	0.143	0.151	0.158	0.165	0.172	0.178	0.185	0.191	0.204	0.222	0.234	0.251	0.262
(0.004903)	0.4760	0.5800	0.6972	0.8281	0.9735	1.1337	1.3096	1.5015	1.9360	2.7225	3.3426	4.4270	5.2584
0.000525	0.147	0.154	0.162	0.169	0.176	0.183	0.189	0.196	0.209	0.227	0.239	0.257	0.268
(0.005148)	0.4878	0.5944	0.7144	0.8486	0.9975	1.1617	1.3419	1.5386	1.9839	2.7897	3.4252	4.5363	5.3883
0.000550	0.150	0.158	0.166	0.173	0.180	0.187	0.194	0.201	0.214	0.233	0.245	0.263	0.274
(0.005394)	0.4993	0.6084	0.7312	0.8686	1.0210	1.1891	1.3735	1.5748	2.0305	2.8553	3.5058	4.6431	5.5151
0.000575	0.154	0.162	0.169	0.177	0.184	0.191	0.198	0.205	0.218	0.238	0.250	0.269	0.280
(0.005639)	0.5105	0.6220	0.7477	0.8881	1.0439	1.2158	1.4044	1.6102	2.0762	2.9195	3.5846	4.7474	5.6390
0.000600	0.157	0.165	0.173	0.180	0.188	0.195	0.202	0.209	0.223	0.243	0.256	0.274	0.286
(0.005884)	0.5215	0.6354	0.7638	0.9072	1.0664	1.2420	1.4346	1.6448	2.1208	2.9823	3.6617	4.8496	5.7603
0.000625	0.160	0.169	0.176	0.184	0.192	0.199	0.207	0.214	0.228	0.248	0.261	0.280	0.292
(0.006129)	0.5322	0.6485	0.7795	0.9259	1.0884	1.2676	1.4641	1.6788	2.1646	3.0438	3.7372	4.9496	5.8791
0.000650	0.164	0.172	0.180	0.188	0.196	0.203	0.211	0.218	0.232	0.253	0.266	0.286	0.298
(0.006374)	0.5428	0.6614	0.7949	0.9442	1.1099	1.2927	1.4931	1.7120	2.2074	3.1041	3.8112	5.0476	5.9955
0.000675	0.167	0.175	0.183	0.191	0.199	0.207	0.215	0.222	0.237	0.258	0.271	0.291	0.304
(0.006619)	0.5531	0.6740	0.8101	0.9622	1.1311	1.3173	1.5216	1.7446	2.2495	3.1632	3.8838	5.1437	6.1097
0.000700	0.170	0.178	0.187	0.195	0.203	0.211	0.219	0.226	0.241	0.262	0.276	0.296	0.309
(0.006865)	0.5632	0.6863	0.8250	0.9799	1.1518	1.3415	1.5495	1.7766	2.2908	3.2213	3.9551	5.2381	6.2218
0.000725	0.173	0.181	0.190	0.198	0.207	0.215	0.222	0.230	0.245	0.267	0.281	0.302	0.315
(0.007110)	0.5732	0.6985	0.8396	0.9972	1.1722	1.3652	1.5769	1.8081	2.3313	3.2783	4.0251	5.3308	6.3320
0.000750	0.176	0.185	0.193	0.202	0.210	0.218	0.226	0.234	0.250	0.272	0.286	0.307	0.320
(0.007355)	0.5830	0.7104	0.8539	1.0143	1.1922	1.3885	1.6039	1.8390	2.3712	3.3343	4.0939	5.4220	6.4402
0.000800	0.181	0.191	0.200	0.208	0.217	0.225	0.234	0.242	0.258	0.281	0.295	0.317	0.331
(0.007845)	0.6021	0.7337	0.8819	1.0475	1.2313	1.4341	1.6565	1.8993	2.4489	3.4437	4.2282	5.5998	6.6514
0.000850	0.187	0.197	0.206	0.215	0.224	0.232	0.241	0.249	0.266	0.289	0.304	0.327	0.341
(0.008336)	0.6207	0.7563	0.9091	1.0798	1.2692	1.4782	1.7075	1.9578	2.5243	3.5496	4.3583	5.7721	6.8561
0.000900	0.192	0.202	0.212	0.221	0.230	0.239	0.248	0.256	0.273	0.298	0.313	0.336	0.351
(0.008826)	0.6387	0.7782	0.9354	1.1111	1.3060	1.5211	1.7570	2.0145	2.5975	3.6526	4.4846	5.9395	7.0549
0.000950	0.198	0.208	0.218	0.227	0.236	0.246	0.255	0.264	0.281	0.306	0.322	0.345	0.360
(0.009316)	0.6562	0.7995	0.9610	1.1415	1.3418	1.5628	1.8051	2.0697	2.6686	3.7526	4.6075	6.1022	7.2482
S, ($\Delta p/\rho l$)	65	70	75	80	85	90	95	100	110	125	135	150	160

Gradient — (Equivalent) Pipe diameters in mm

S = 0.000300 to 0.000950;

$\Delta p/\rho l$ = (0.002942) to (0.009316) ms^{-2}

m = Manning $n \times 100$
S = 0.00100 to 0.00290

i.e. kin. pr. grad., $\Delta p/\rho l$ =
(0.00981) to (0.02844) ms^{-2}

Full bore flow; Tables E give m_C or m_P for Colebrook-White or for laminar solutions resp., to divide mV and/or mQ as follows:

mV to give velocities in ms^{-1}
mQ to give discharges in litres/sec

Gradient (Equivalent) Pipe diameters in mm

S, ($\Delta p/\rho l$)	65	70	75	80	85	90	95	100	110	125	135	150	160
0.00100	0.203	0.213	0.223	0.233	0.243	0.252	0.261	0.270	0.288	0.314	0.330	0.354	0.370
(0.00981)	0.6732	0.8203	0.9860	1.1712	1.3767	1.6034	1.8520	2.1235	2.7380	3.8501	4.7272	6.2607	7.4365
0.00105	0.208	0.218	0.229	0.239	0.249	0.258	0.268	0.277	0.295	0.321	0.338	0.363	0.379
(0.01030)	0.6898	0.8406	1.0104	1.2001	1.4107	1.6429	1.8978	2.1759	2.8056	3.9452	4.8440	6.4154	7.6202
0.00110	0.213	0.224	0.234	0.244	0.254	0.264	0.274	0.284	0.302	0.329	0.346	0.372	0.388
(0.01079)	0.7061	0.8603	1.0341	1.2283	1.4439	1.6816	1.9424	2.2271	2.8716	4.0381	4.9580	6.5663	7.7995
0.00115	0.218	0.229	0.239	0.250	0.260	0.270	0.280	0.290	0.309	0.336	0.354	0.380	0.397
(0.01128)	0.7219	0.8797	1.0574	1.2559	1.4763	1.7194	1.9861	2.2772	2.9362	4.1288	5.0694	6.7139	7.9748
0.00120	0.222	0.233	0.244	0.255	0.266	0.276	0.286	0.296	0.316	0.344	0.362	0.388	0.405
(0.01177)	0.7375	0.8986	1.0801	1.2830	1.5081	1.7564	2.0288	2.3262	2.9993	4.2176	5.1784	6.8583	8.1463
0.00125	0.227	0.238	0.250	0.261	0.271	0.282	0.292	0.302	0.322	0.351	0.369	0.396	0.414
(0.01226)	0.7527	0.9171	1.1024	1.3094	1.5392	1.7926	2.0706	2.3741	3.0612	4.3046	5.2852	6.9997	8.3143
0.00130	0.231	0.243	0.254	0.266	0.277	0.287	0.298	0.308	0.328	0.358	0.377	0.404	0.422
(0.01275)	0.7676	0.9353	1.1242	1.3353	1.5697	1.8281	2.1116	2.4211	3.1218	4.3898	5.3899	7.1383	8.4789
0.00135	0.236	0.248	0.259	0.271	0.282	0.293	0.304	0.314	0.335	0.365	0.384	0.412	0.430
(0.01324)	0.7822	0.9531	1.1456	1.3608	1.5996	1.8629	2.1519	2.4673	3.1812	4.4735	5.4925	7.2743	8.6404
0.00140	0.240	0.252	0.264	0.276	0.287	0.298	0.309	0.320	0.341	0.371	0.391	0.419	0.438
(0.01373)	0.7966	0.9706	1.1667	1.3858	1.6289	1.8971	2.1913	2.5125	3.2396	4.5555	5.5933	7.4078	8.7990
0.00145	0.244	0.257	0.269	0.281	0.292	0.303	0.315	0.326	0.347	0.378	0.398	0.427	0.445
(0.01422)	0.8107	0.9878	1.1873	1.4103	1.6577	1.9307	2.2301	2.5570	3.2970	4.6362	5.6923	7.5389	8.9547
0.00150	0.248	0.261	0.273	0.285	0.297	0.309	0.320	0.331	0.353	0.384	0.404	0.434	0.453
(0.01471)	0.8245	1.0047	1.2076	1.4344	1.6861	1.9637	2.2683	2.6007	3.3533	4.7154	5.7896	7.6678	9.1078
0.00160	0.257	0.270	0.282	0.295	0.307	0.319	0.330	0.342	0.364	0.397	0.418	0.448	0.468
(0.01569)	0.8516	1.0376	1.2472	1.4814	1.7414	2.0281	2.3426	2.6860	3.4633	4.8701	5.9795	7.9193	9.4065
0.00170	0.265	0.278	0.291	0.304	0.316	0.329	0.341	0.353	0.376	0.409	0.431	0.462	0.482
(0.01667)	0.8778	1.0696	1.2856	1.5270	1.7950	2.0905	2.4147	2.7687	3.5699	5.0200	6.1635	8.1630	9.6960
0.00180	0.272	0.286	0.299	0.313	0.325	0.338	0.351	0.363	0.387	0.421	0.443	0.475	0.496
(0.01765)	0.9032	1.1006	1.3229	1.5713	1.8470	2.1511	2.4847	2.8490	3.6734	5.1655	6.3422	8.3997	9.9771
0.00190	0.280	0.294	0.308	0.321	0.334	0.347	0.360	0.373	0.397	0.432	0.455	0.488	0.510
(0.01863)	0.9280	1.1307	1.3591	1.6144	1.8976	2.2101	2.5528	2.9270	3.7740	5.3070	6.5160	8.6298	10.251
0.00200	0.287	0.301	0.316	0.330	0.343	0.356	0.370	0.382	0.407	0.444	0.467	0.501	0.523
(0.01961)	0.9521	1.1601	1.3944	1.6563	1.9469	2.2675	2.6192	3.0031	3.8721	5.4449	6.6853	8.8540	10.517
0.00210	0.294	0.309	0.323	0.338	0.352	0.365	0.379	0.392	0.418	0.455	0.479	0.513	0.536
(0.02059)	0.9756	1.1887	1.4289	1.6972	1.9950	2.3235	2.6838	3.0772	3.9677	5.5794	6.8504	9.0727	10.777
0.00220	0.301	0.316	0.331	0.346	0.360	0.374	0.388	0.401	0.427	0.465	0.490	0.525	0.549
(0.02157)	0.9985	1.2167	1.4625	1.7371	2.0419	2.3782	2.7470	3.1496	4.0611	5.7107	7.0116	9.2862	11.030
0.00230	0.308	0.323	0.338	0.353	0.368	0.382	0.396	0.410	0.437	0.476	0.501	0.537	0.561
(0.02256)	1.0210	1.2441	1.4954	1.7762	2.0878	2.4316	2.8087	3.2204	4.1523	5.8390	7.1692	9.4949	11.278
0.00240	0.314	0.330	0.346	0.361	0.376	0.390	0.405	0.419	0.446	0.486	0.512	0.549	0.573
(0.02354)	1.0429	1.2708	1.5275	1.8144	2.1327	2.4839	2.8691	3.2897	4.2417	5.9646	7.3234	9.6991	11.521
0.00250	0.321	0.337	0.353	0.368	0.384	0.398	0.413	0.427	0.456	0.496	0.522	0.560	0.585
(0.02452)	1.0644	1.2970	1.5590	1.8518	2.1767	2.5351	2.9283	3.3575	4.3291	6.0876	7.4744	9.8991	11.758
0.00260	0.327	0.344	0.360	0.376	0.391	0.406	0.421	0.436	0.465	0.506	0.533	0.571	0.596
(0.02550)	1.0855	1.3227	1.5899	1.8885	2.2198	2.5853	2.9863	3.4240	4.4149	6.2082	7.6224	10.095	11.991
0.00270	0.333	0.350	0.367	0.383	0.399	0.414	0.429	0.444	0.473	0.516	0.543	0.582	0.608
(0.02648)	1.1062	1.3479	1.6202	1.9244	2.2621	2.6346	3.0432	3.4892	4.4990	6.3264	7.7676	10.287	12.219
0.00280	0.339	0.357	0.373	0.390	0.406	0.422	0.437	0.452	0.482	0.525	0.553	0.593	0.619
(0.02746)	1.1265	1.3726	1.6499	1.9598	2.3036	2.6829	3.0990	3.5533	4.5815	6.4425	7.9102	10.476	12.444
0.00290	0.345	0.363	0.380	0.397	0.413	0.429	0.445	0.460	0.491	0.534	0.562	0.603	0.630
(0.02844)	1.1464	1.3969	1.6791	1.9944	2.3444	2.7304	3.1539	3.6162	4.6626	6.5565	8.0502	10.662	12.664
S, ($\Delta p/\rho l$)	65	70	75	80	85	90	95	100	110	125	135	150	160

Gradient (Equivalent) Pipe diameters in mm

S = 0.00100 to 0.00290;

$\Delta p/\rho l$ = (0.00981) to (0.02844) ms^{-2}

m = Manning $n \times 100$
S = 0·00300 to 0·00950

i.e. kin. pr. grad., $\Delta p/\rho l$ = (0·02942) to (0·09316) ms^{-2}

Full bore flow; Tables E give m_C or m_P for Colebrook-White or for laminar solutions resp., to divide mV and/or mQ as follows:

mV to give velocities in ms^{-1}
mQ to give discharges in litres/sec

Gradient (Equivalent) Pipe diameters in mm

S, ($\Delta p/\rho l$)	65	70	75	80	85	90	95	100	110	125	135	150	160
0·00300	0·351	0·369	0·387	0·404	0·420	0·437	0·453	0·468	0·499	0·543	0·572	0·614	0·641
(0·02942)	1·1660	1·4208	1·7078	2·0285	2·3845	2·7771	3·2078	3·6780	4·7423	6·6686	8·1878	10·844	12·880
0·00320	0·363	0·381	0·399	0·417	0·434	0·451	0·467	0·484	0·515	0·561	0·591	0·634	0·662
(0·03138)	1·2043	1·4674	1·7638	2·0951	2·4627	2·8682	3·3130	3·7986	4·8978	6·8873	8·4563	11·200	13·303
0·00340	0·374	0·393	0·412	0·430	0·447	0·465	0·482	0·499	0·531	0·579	0·609	0·653	0·682
(0·03334)	1·2413	1·5126	1·8181	2·1595	2·5385	2·9564	3·4150	3·9155	5·0486	7·0993	8·7166	11·544	13·712
0·00360	0·385	0·404	0·423	0·442	0·460	0·478	0·496	0·513	0·547	0·595	0·627	0·672	0·702
(0·03530)	1·2773	1·5564	1·8708	2·2222	2·6121	3·0421	3·5140	4·0290	5·1949	7·3051	8·9693	11·879	14·110
0·00380	0·395	0·416	0·435	0·454	0·473	0·491	0·509	0·527	0·562	0·612	0·644	0·691	0·721
(0·03727)	1·3123	1·5991	1·9221	2·2830	2·6836	3·1255	3·6102	4·1394	5·3373	7·5053	9·2150	12·204	14·496
0·00400	0·406	0·426	0·446	0·466	0·485	0·504	0·523	0·541	0·576	0·627	0·661	0·709	0·740
(0·03923)	1·3464	1·6406	1·9720	2·3424	2·7534	3·2067	3·7040	4·2470	5·4760	7·7003	9·4544	12·521	14·873
0·00420	0·416	0·437	0·457	0·478	0·497	0·517	0·535	0·554	0·590	0·643	0·677	0·726	0·758
(0·04119)	1·3797	1·6811	2·0207	2·4002	2·8214	3·2859	3·7955	4·3518	5·6112	7·8904	9·6879	12·831	15·240
0·00440	0·426	0·447	0·468	0·489	0·509	0·529	0·548	0·567	0·604	0·658	0·693	0·743	0·776
(0·04315)	1·4121	1·7207	2·0683	2·4567	2·8877	3·3632	3·8848	4·4543	5·7432	8·0761	9·9159	13·133	15·599
0·00460	0·435	0·457	0·479	0·500	0·520	0·541	0·560	0·580	0·618	0·673	0·708	0·760	0·793
(0·04511)	1·4439	1·7594	2·1147	2·5119	2·9526	3·4388	3·9721	4·5544	5·8723	8·2576	10·139	13·428	15·950
0·00480	0·444	0·467	0·489	0·510	0·532	0·552	0·572	0·592	0·631	0·687	0·724	0·776	0·810
(0·04707)	1·4749	1·7972	2·1602	2·5659	3·0162	3·5128	4·0576	4·6523	5·9986	8·4352	10·357	13·717	16·293
0·00500	0·454	0·477	0·499	0·521	0·542	0·564	0·584	0·605	0·644	0·702	0·738	0·792	0·827
(0·04903)	1·5053	1·8343	2·2048	2·6188	3·0783	3·5852	4·1412	4·7483	6·1223	8·6092	10·570	13·999	16·629
0·00525	0·465	0·488	0·511	0·534	0·556	0·577	0·599	0·619	0·660	0·719	0·757	0·812	0·847
(0·05148)	1·5425	1·8796	2·2592	2·6835	3·1544	3·6737	4·2435	4·8655	6·2735	8·8218	10·831	14·345	17·039
0·00550	0·476	0·500	0·523	0·546	0·569	0·591	0·613	0·634	0·676	0·736	0·775	0·831	0·867
(0·05394)	1·5788	1·9238	2·3124	2·7467	3·2286	3·7602	4·3434	4·9800	6·4211	9·0294	11·086	14·683	17·440
0·00575	0·486	0·511	0·535	0·559	0·582	0·604	0·627	0·648	0·691	0·752	0·792	0·850	0·887
(0·05639)	1·6143	1·9670	2·3644	2·8084	3·3012	3·8447	4·4410	5·0919	6·5654	9·2323	11·335	15·013	17·832
0·00600	0·497	0·522	0·547	0·571	0·594	0·617	0·640	0·662	0·706	0·768	0·809	0·868	0·906
(0·05884)	1·6490	2·0093	2·4152	2·8688	3·3722	3·9274	4·5365	5·2015	6·7066	9·4309	11·579	15·336	18·216
0·00625	0·507	0·533	0·558	0·582	0·607	0·630	0·653	0·676	0·720	0·784	0·826	0·886	0·925
(0·06129)	1·6830	2·0508	2·4650	2·9279	3·4417	4·0084	4·6300	5·3087	6·8449	9·6253	11·818	15·652	18·591
0·00650	0·517	0·543	0·569	0·594	0·619	0·643	0·666	0·689	0·735	0·800	0·842	0·903	0·943
(0·06374)	1·7164	2·0914	2·5138	2·9859	3·5099	4·0878	4·7217	5·4138	6·9805	9·8160	12·052	15·962	18·959
0·00675	0·527	0·554	0·580	0·605	0·630	0·655	0·679	0·702	0·749	0·815	0·858	0·920	0·961
(0·06619)	1·7491	2·1312	2·5617	3·0428	3·5767	4·1656	4·8117	5·5170	7·1135	10·003	12·282	16·266	19·321
0·00700	0·537	0·564	0·590	0·616	0·642	0·667	0·691	0·715	0·762	0·830	0·874	0·937	0·979
(0·06865)	1·7811	2·1703	2·6087	3·0986	3·6424	4·2421	4·9000	5·6182	7·2440	10·186	12·507	16·564	19·675
0·00725	0·546	0·574	0·601	0·627	0·653	0·679	0·704	0·728	0·776	0·845	0·889	0·954	0·996
(0·07110)	1·8127	2·2087	2·6549	3·1535	3·7068	4·3172	4·9867	5·7177	7·3722	10·367	12·728	16·858	20·023
0·00750	0·556	0·584	0·611	0·638	0·664	0·690	0·716	0·740	0·789	0·859	0·904	0·970	1·013
(0·07355)	1·8437	2·2465	2·7003	3·2074	3·7702	4·3910	5·0720	5·8154	7·4983	10·544	12·946	17·146	20·366
0·00800	0·574	0·603	0·631	0·659	0·686	0·713	0·739	0·765	0·815	0·887	0·934	1·002	1·046
(0·07845)	1·9041	2·3202	2·7888	3·3126	3·8938	4·5350	5·2383	6·0061	7·7442	10·890	13·371	17·708	21·034
0·00850	0·591	0·621	0·651	0·679	0·707	0·735	0·762	0·788	0·840	0·915	0·963	1·033	1·078
(0·08336)	1·9627	2·3916	2·8747	3·4145	4·0137	4·6745	5·3995	6·1910	7·9825	11·225	13·782	18·253	21·681
0·00900	0·609	0·639	0·670	0·699	0·728	0·756	0·784	0·811	0·864	0·941	0·991	1·063	1·110
(0·08826)	2·0196	2·4609	2·9580	3·5135	4·1300	4·8101	5·5561	6·3705	8·2139	11·550	14·182	18·782	22·310
0·00950	0·625	0·657	0·688	0·718	0·748	0·777	0·805	0·833	0·888	0·967	1·018	1·092	1·140
(0·09316)	2·0750	2·5284	3·0391	3·6098	4·2432	4·9419	5·7083	6·5450	8·4390	11·867	14·570	19·297	22·921
S, ($\Delta p/\rho l$)	65	70	75	80	85	90	95	100	110	125	135	150	160

Gradient (Equivalent) Pipe diameters in mm

S = 0·00300 to 0·00950;

$\Delta p/\rho l$ = (0·02942) to (0·09316) ms^{-2}

m = Manning $n \times 100$
S = 0·0100 to 0·0290

Full bore flow; Tables E give m_C or m_P for Colebrook-White or for laminar solutions resp., to divide mV and/or mQ as follows:

i.e. kin. pr. grad., $\Delta p/\rho l =$ (0·0981) to (0·2844) ms^{-2}

mV to give velocities in ms^{-1}
mQ to give discharges in litres/sec

Gradient S, ($\Delta p/\rho l$)	\multicolumn{13}{c}{(Equivalent) Pipe diameters in mm}

S, ($\Delta p/\rho l$)	65	70	75	80	85	90	95	100	110	125	135	150	160
0·0100	0·642	0·674	0·706	0·737	0·767	0·797	0·826	0·855	0·911	0·992	1·044	1·120	1·170
(0·0981)	2·1289	2·5940	3·1180	3·7036	4·3534	5·0702	5·8566	6·7150	8·6582	12·175	14·949	19·798	23·516
0·0105	0·657	0·691	0·723	0·755	0·786	0·817	0·847	0·876	0·934	1·017	1·070	1·148	1·198
(0·1030)	2·1815	2·6581	3·1950	3·7950	4·4610	5·1955	6·0012	6·8809	8·8721	12·476	15·318	20·287	24·097
0·0110	0·673	0·707	0·740	0·773	0·805	0·836	0·867	0·897	0·956	1·041	1·095	1·175	1·227
(0·1079)	2·2328	2·7207	3·2702	3·8844	4·5659	5·3177	6·1425	7·0428	9·0808	12·769	15·678	20·765	24·664
0·0115	0·688	0·723	0·757	0·790	0·823	0·855	0·886	0·917	0·977	1·064	1·120	1·201	1·254
(0·1128)	2·2830	2·7818	3·3437	3·9717	4·6685	5·4372	6·2805	7·2011	9·2849	13·056	16·031	21·231	25·218
0·0120	0·703	0·738	0·773	0·807	0·840	0·873	0·905	0·937	0·998	1·087	1·144	1·227	1·281
(0·1177)	2·3321	2·8416	3·4156	4·0571	4·7690	5·5542	6·4156	7·3560	9·4846	13·337	16·376	21·688	25·761
0·0125	0·717	0·754	0·789	0·824	0·858	0·891	0·924	0·956	1·019	1·109	1·168	1·253	1·308
(0·1226)	2·3802	2·9002	3·4861	4·1407	4·8673	5·6687	6·5479	7·5077	9·6802	13·612	16·713	22·135	26·292
0·0130	0·731	0·769	0·805	0·840	0·875	0·909	0·942	0·975	1·039	1·131	1·191	1·277	1·334
(0·1275)	2·4273	2·9577	3·5551	4·2227	4·9637	5·7810	6·6775	7·6563	9·8719	13·882	17·044	22·573	26·813
0·0135	0·745	0·783	0·820	0·856	0·891	0·926	0·960	0·993	1·059	1·153	1·213	1·302	1·359
(0·1324)	2·4735	3·0140	3·6228	4·3032	5·0582	5·8911	6·8048	7·8022	10·060	14·146	17·369	23·003	27·323
0·0140	0·759	0·798	0·835	0·872	0·908	0·943	0·978	1·012	1·078	1·174	1·236	1·326	1·384
(0·1373)	2·5189	3·0693	3·6893	4·3821	5·1511	5·9992	6·9296	7·9454	10·245	14·406	17·688	23·426	27·825
0·0145	0·773	0·812	0·850	0·887	0·924	0·960	0·995	1·030	1·097	1·195	1·258	1·349	1·408
(0·1422)	2·5635	3·1236	3·7546	4·4597	5·2422	6·1054	7·0523	8·0860	10·426	14·661	18·001	23·840	28·317
0·0150	0·786	0·826	0·864	0·902	0·940	0·976	1·012	1·047	1·116	1·215	1·279	1·372	1·432
(0·1471)	2·6073	3·1770	3·8188	4·5359	5·3319	6·2098	7·1728	8·2242	10·604	14·912	18·308	24·248	28·801
0·0160	0·812	0·853	0·893	0·932	0·970	1·008	1·045	1·081	1·152	1·255	1·321	1·417	1·479
(0·1569)	2·6928	3·2812	3·9440	4·6847	5·5067	6·4134	7·4081	8·4939	10·952	15·401	18·909	25·043	29·746
0·0170	0·836	0·879	0·920	0·961	1·000	1·039	1·077	1·115	1·188	1·294	1·362	1·461	1·525
(0·1667)	2·7757	3·3822	4·0654	4·8289	5·6762	6·6108	7·6361	8·7554	11·289	15·875	19·491	25·814	30·662
0·0180	0·861	0·904	0·947	0·989	1·029	1·069	1·109	1·147	1·222	1·331	1·401	1·503	1·569
(0·1765)	2·8562	3·4803	4·1833	4·9689	5·8408	6·8025	7·8575	9·0092	11·616	16·335	20·056	26·562	31·550
0·0190	0·884	0·929	0·973	1·016	1·058	1·099	1·139	1·179	1·256	1·368	1·440	1·544	1·612
(0·1863)	2·9345	3·5756	4·2979	5·1050	6·0008	6·9889	8·0728	9·2561	11·935	16·782	20·605	27·290	32·415
0·0200	0·907	0·953	0·998	1·042	1·085	1·127	1·168	1·209	1·288	1·403	1·477	1·584	1·654
(0·1961)	3·0107	3·6685	4·4095	5·2377	6·1567	7·1704	8·2825	9·4965	12·245	17·218	21·141	27·999	33·257
0·0210	0·930	0·977	1·023	1·068	1·112	1·155	1·197	1·239	1·320	1·438	1·513	1·624	1·695
(0·2059)	3·0850	3·7591	4·5184	5·3670	6·3087	7·3475	8·4870	9·7310	12·547	17·644	21·663	28·690	34·078
0·0220	0·952	1·000	1·047	1·093	1·138	1·182	1·226	1·268	1·351	1·472	1·549	1·662	1·735
(0·2157)	3·1576	3·8476	4·6248	5·4933	6·4572	7·5204	8·6867	9·9600	12·842	18·059	22·173	29·365	34·880
0·0230	0·973	1·022	1·070	1·117	1·164	1·209	1·253	1·297	1·382	1·505	1·584	1·699	1·774
(0·2256)	3·2286	3·9341	4·7287	5·6168	6·6023	7·6894	8·8820	10·184	13·131	18·465	22·671	30·025	35·664
0·0240	0·994	1·044	1·093	1·141	1·189	1·235	1·280	1·325	1·411	1·537	1·618	1·736	1·812
(0·2354)	3·2980	4·0187	4·8304	5·7376	6·7443	7·8548	9·0730	10·403	13·413	18·862	23·159	30·671	36·431
0·0250	1·014	1·066	1·116	1·165	1·213	1·260	1·306	1·352	1·441	1·569	1·651	1·771	1·849
(0·2452)	3·3661	4·1015	4·9300	5·8559	6·8834	8·0168	9·2601	10·617	13·690	19·251	23·636	31·304	37·183
0·0260	1·034	1·087	1·138	1·188	1·237	1·285	1·332	1·379	1·469	1·600	1·684	1·807	1·886
(0·2550)	3·4327	4·1828	5·0277	5·9719	7·0197	8·1755	9·4435	10·828	13·961	19·632	24·104	31·924	37·919
0·0270	1·054	1·108	1·160	1·211	1·261	1·310	1·358	1·405	1·497	1·630	1·716	1·841	1·922
(0·2648)	3·4981	4·2624	5·1234	6·0856	7·1534	8·3313	9·6234	11·034	14·227	20·006	24·563	32·532	38·641
0·0280	1·074	1·128	1·181	1·233	1·284	1·334	1·383	1·431	1·525	1·660	1·748	1·875	1·957
(0·2746)	3·5623	4·3407	5·2174	6·1973	7·2847	8·4841	9·8000	11·236	14·488	20·373	25·014	33·129	39·350
0·0290	1·093	1·148	1·202	1·255	1·306	1·357	1·407	1·456	1·552	1·690	1·778	1·908	1·992
(0·2844)	3·6254	4·4175	5·3098	6·3070	7·4136	8·6343	9·9734	11·435	14·744	20·734	25·457	33·715	40·047

S, ($\Delta p/\rho l$)	65	70	75	80	85	90	95	100	110	125	135	150	160

Gradient (Equivalent) Pipe diameters in mm

S = 0·0100 to 0·0290;

$\Delta p/\rho l$ = (0·0981) to (0·2844) ms^{-2}

m = Manning $n \times 100$
S = 0·0300 to 0·0950

i.e. kin. pr. grad., $\Delta p/\rho l$ = (0·2942) to (0·9316) ms^{-2}

Full bore flow; Tables E give m_C or m_P for Colebrook-White or for laminar solutions resp., to divide mV and/or mQ as follows:

mV to give velocities in ms^{-1}
mQ to give discharges in litres/sec

Gradient — (Equivalent) Pipe diameters in mm

S, ($\Delta p/\rho l$)	65	70	75	80	85	90	95	100	110	125	135	150	160
0·0300	1·111	1·167	1·222	1·276	1·329	1·380	1·431	1·481	1·578	1·718	1·809	1·941	2·026
(0·2942)	3·6873	4·4930	5·4006	6·4148	7·5404	8·7819	10·144	11·631	14·997	21·088	25·892	34·292	40·731
0·0320	1·148	1·206	1·263	1·318	1·372	1·426	1·478	1·529	1·630	1·775	1·868	2·004	2·092
(0·3138)	3·8083	4·6404	5·5777	6·6252	7·7877	9·0699	10·477	12·012	15·488	21·780	26·741	35·416	42·067
0·0340	1·183	1·243	1·301	1·359	1·415	1·470	1·524	1·577	1·680	1·829	1·926	2·066	2·157
(0·3334)	3·9255	4·7832	5·7493	6·8291	8·0274	9·3491	10·799	12·382	15·965	22·450	27·564	36·506	43·362
0·0360	1·217	1·279	1·339	1·398	1·456	1·512	1·568	1·622	1·729	1·882	1·982	2·126	2·219
(0·3530)	4·0393	4·9219	5·9160	7·0271	8·2601	9·6201	11·112	12·741	16·428	23·101	28·363	37·564	44·619
0·0380	1·251	1·314	1·376	1·436	1·496	1·554	1·611	1·667	1·776	1·934	2·036	2·184	2·280
(0·3727)	4·1500	5·0567	6·0781	7·2196	8·4864	9·8837	11·417	13·090	16·878	23·734	29·141	38·594	45·842
0·0400	1·283	1·348	1·412	1·474	1·534	1·594	1·652	1·710	1·822	1·984	2·089	2·241	2·339
(0·3923)	4·2578	5·1881	6·2360	7·4072	8·7069	10·140	11·713	13·430	17·316	24·350	29·898	39·596	47·033
0·0420	1·315	1·381	1·446	1·510	1·572	1·633	1·693	1·752	1·867	2·033	2·140	2·296	2·397
(0·4119)	4·3629	5·3162	6·3900	7·5901	8·9219	10·391	12·002	13·762	17·744	24·952	30·636	40·574	48·194
0·0440	1·346	1·414	1·480	1·546	1·609	1·672	1·733	1·793	1·911	2·081	2·191	2·350	2·453
(0·4315)	4·4656	5·4413	6·5404	7·7687	9·1319	10·635	12·285	14·086	18·162	25·539	31·357	41·529	49·328
0·0460	1·376	1·446	1·514	1·580	1·645	1·709	1·772	1·834	1·954	2·128	2·240	2·403	2·509
(0·4511)	4·5659	5·5636	6·6874	7·9433	9·3371	10·874	12·561	14·402	18·570	26·113	32·062	42·462	50·437
0·0480	1·406	1·477	1·546	1·614	1·681	1·746	1·810	1·873	1·996	2·174	2·288	2·455	2·562
(0·4707)	4·6641	5·6833	6·8312	8·1142	9·5379	11·108	12·831	14·712	18·969	26·675	32·751	43·376	51·522
0·0500	1·435	1·507	1·578	1·648	1·715	1·782	1·848	1·912	2·037	2·218	2·335	2·505	2·615
(0·4903)	4·7603	5·8005	6·9721	8·2815	9·7346	11·337	13·096	15·015	19·360	27·225	33·426	44·270	52·584
0·0525	1·470	1·544	1·617	1·688	1·758	1·826	1·893	1·959	2·088	2·273	2·393	2·567	2·680
(0·5148)	4·8779	5·9437	7·1443	8·4860	9·9750	11·617	13·419	15·386	19·839	27·897	34·252	45·363	53·883
0·0550	1·505	1·581	1·655	1·728	1·799	1·869	1·938	2·005	2·137	2·327	2·449	2·627	2·743
(0·5394)	4·9927	6·0836	7·3124	8·6857	10·210	11·891	13·735	15·748	20·305	28·553	35·058	46·431	55·151
0·0575	1·538	1·616	1·692	1·767	1·840	1·911	1·981	2·050	2·185	2·379	2·504	2·687	2·805
(0·5639)	5·1049	6·2203	7·4768	8·8809	10·439	12·158	14·044	16·102	20·762	29·195	35·846	47·474	56·390
0·0600	1·571	1·651	1·729	1·805	1·879	1·952	2·024	2·094	2·232	2·430	2·558	2·744	2·865
(0·5884)	5·2147	6·3541	7·6376	9·0719	10·664	12·420	14·346	16·448	21·208	29·823	36·617	48·496	57·603
0·0625	1·604	1·685	1·764	1·842	1·918	1·992	2·066	2·137	2·278	2·480	2·611	2·801	2·924
(0·6129)	5·3222	6·4851	7·7951	9·2590	10·884	12·676	14·641	16·788	21·646	30·438	37·372	49·496	58·791
0·0650	1·636	1·718	1·799	1·878	1·956	2·032	2·107	2·180	2·323	2·529	2·663	2·856	2·982
(0·6374)	5·4276	6·6135	7·9494	9·4423	11·099	12·927	14·931	17·120	22·074	31·041	38·112	50·476	59·955
0·0675	1·667	1·751	1·834	1·914	1·993	2·071	2·147	2·221	2·367	2·578	2·713	2·911	3·039
(0·6619)	5·5310	6·7395	8·1009	9·6222	11·311	13·173	15·216	17·446	22·495	31·632	38·838	51·437	61·097
0·0700	1·697	1·783	1·867	1·949	2·030	2·109	2·186	2·262	2·410	2·625	2·763	2·964	3·094
(0·6865)	5·6325	6·8632	8·2495	9·7988	11·518	13·415	15·495	17·766	22·908	32·213	39·551	52·381	62·218
0·0725	1·727	1·815	1·900	1·984	2·066	2·146	2·225	2·302	2·453	2·671	2·812	3·017	3·149
(0·7110)	5·7322	6·9847	8·3955	9·9722	11·722	13·652	15·769	18·081	23·313	32·783	40·251	53·308	63·320
0·0750	1·757	1·846	1·933	2·018	2·101	2·183	2·263	2·341	2·495	2·717	2·860	3·068	3·203
(0·7355)	5·8302	7·1041	8·5391	10·143	11·922	13·885	16·039	18·390	23·712	33·343	40·939	54·220	64·402
0·0800	1·815	1·906	1·996	2·084	2·170	2·254	2·337	2·418	2·577	2·806	2·954	3·169	3·308
(0·7845)	6·0214	7·3371	8·8191	10·475	12·313	14·341	16·565	18·993	24·489	34·437	42·282	55·998	66·514
0·0850	1·870	1·965	2·058	2·148	2·237	2·324	2·409	2·493	2·656	2·893	3·045	3·266	3·410
(0·8336)	6·2067	7·5629	9·0905	10·798	12·692	14·782	17·075	19·578	25·243	35·496	43·583	57·721	68·561
0·0900	1·925	2·022	2·117	2·210	2·302	2·391	2·479	2·565	2·733	2·976	3·133	3·361	3·509
(0·8826)	6·3866	7·7821	9·3541	11·111	13·060	15·211	17·570	20·145	25·975	36·526	44·846	59·395	70·549
0·0950	1·977	2·078	2·175	2·271	2·365	2·456	2·547	2·635	2·808	3·058	3·219	3·453	3·605
(0·9316)	6·5617	7·9954	9·6104	11·415	13·418	15·628	18·051	20·697	26·686	37·526	46·075	61·022	72·482

S, ($\Delta p/\rho l$)	65	70	75	80	85	90	95	100	110	125	135	150	160

Gradient — (Equivalent) Pipe diameters in mm

S = 0·0300 to 0·0950;

$\Delta p/\rho l$ = (0·2942) to (0·9316) ms^{-2}

m = Manning $n \times 100$
S = 0·1000 to 0·2900

i.e. kin. pr. grad., $\Delta p/\rho l$ = (0·9807) to (2·8439) ms^{-2}

Full bore flow; Tables E give m_C or m_P for Colebrook-White or for laminar solutions resp., to divide mV and/or mQ as follows:

mV to give velocities in ms^{-1}
mQ to give discharges in litres/sec

Gradient (Equivalent) Pipe diameters in mm

Each cell gives mV (upper) / mQ (lower).

S, ($\Delta p/\rho l$)	65	70	75	80	85	90	95	100	110	125	135	150	160
0·1000 (0·9807)	2·029 / 6·7321	2·132 / 8·2031	2·232 / 9·8601	2·330 / 11·712	2·426 / 13·767	2·520 / 16·034	2·613 / 18·520	2·704 / 21·235	2·881 / 27·380	3·137 / 38·501	3·303 / 47·272	3·543 / 62·607	3·699 / 74·365
0·1050 (1·0297)	2·079 / 6·8984	2·184 / 8·4057	2·287 / 10·104	2·388 / 12·001	2·486 / 14·107	2·583 / 16·429	2·677 / 18·978	2·770 / 21·759	2·952 / 28·056	3·215 / 39·452	3·384 / 48·440	3·630 / 64·154	3·790 / 76·202
0·1100 (1·0787)	2·128 / 7·0607	2·236 / 8·6035	2·341 / 10·341	2·444 / 12·283	2·544 / 14·439	2·643 / 16·816	2·740 / 19·424	2·836 / 22·271	3·022 / 28·716	3·291 / 40·381	3·464 / 49·580	3·716 / 65·663	3·879 / 77·995
0·1150 (1·1278)	2·176 / 7·2194	2·286 / 8·7968	2·393 / 10·574	2·499 / 12·559	2·602 / 14·763	2·703 / 17·194	2·802 / 19·861	2·899 / 22·772	3·090 / 29·362	3·364 / 41·288	3·542 / 50·694	3·799 / 67·139	3·966 / 79·748
0·1200 (1·1768)	2·222 / 7·3747	2·335 / 8·9860	2·445 / 10·801	2·552 / 12·830	2·658 / 15·081	2·761 / 17·564	2·862 / 20·288	2·962 / 23·262	3·156 / 29·993	3·437 / 42·176	3·618 / 51·784	3·881 / 68·583	4·052 / 81·463
0·1250 (1·2258)	2·268 / 7·5267	2·383 / 9·1713	2·495 / 11·024	2·605 / 13·094	2·712 / 15·392	2·818 / 17·926	2·921 / 20·706	3·023 / 23·741	3·221 / 30·612	3·508 / 43·046	3·692 / 52·852	3·961 / 69·997	4·135 / 83·143
0·1300 (1·2749)	2·313 / 7·6758	2·430 / 9·3530	2·545 / 11·242	2·657 / 13·353	2·766 / 15·697	2·874 / 18·281	2·979 / 21·116	3·083 / 24·211	3·285 / 31·218	3·577 / 43·898	3·765 / 53·899	4·039 / 71·383	4·217 / 84·789
0·1350 (1·3239)	2·357 / 7·8220	2·477 / 9·5311	2·593 / 11·456	2·707 / 13·608	2·819 / 15·996	2·928 / 18·629	3·036 / 21·519	3·141 / 24·673	3·348 / 31·812	3·645 / 44·735	3·837 / 54·925	4·116 / 72·743	4·297 / 86·404
0·1400 (1·3729)	2·400 / 7·9655	2·522 / 9·7060	2·641 / 11·667	2·757 / 13·858	2·871 / 16·289	2·982 / 18·971	3·092 / 21·913	3·199 / 25·125	3·409 / 32·396	3·712 / 45·555	3·908 / 55·933	4·192 / 74·078	4·376 / 87·990
0·1450 (1·4220)	2·443 / 8·1065	2·567 / 9·8778	2·688 / 11·873	2·806 / 14·103	2·921 / 16·577	3·035 / 19·307	3·146 / 22·301	3·256 / 25·570	3·469 / 32·970	3·778 / 46·362	3·977 / 56·923	4·266 / 75·389	4·454 / 89·547
0·1500 (1·4710)	2·485 / 8·2451	2·611 / 10·047	2·733 / 12·076	2·854 / 14·344	2·971 / 16·861	3·087 / 19·637	3·200 / 22·683	3·311 / 26·007	3·529 / 33·533	3·842 / 47·154	4·045 / 57·896	4·339 / 76·678	4·530 / 91·078
0·1600 (1·5691)	2·566 / 8·5155	2·696 / 10·376	2·823 / 12·472	2·947 / 14·814	3·069 / 17·414	3·188 / 20·281	3·305 / 23·426	3·420 / 26·860	3·644 / 34·633	3·968 / 48·701	4·177 / 59·795	4·481 / 79·193	4·678 / 94·065
0·1700 (1·6671)	2·645 / 8·7776	2·779 / 10·696	2·910 / 12·856	3·038 / 15·270	3·163 / 17·950	3·286 / 20·905	3·407 / 24·147	3·525 / 27·687	3·756 / 35·699	4·091 / 50·200	4·306 / 61·635	4·619 / 81·630	4·822 / 96·960
0·1800 (1·7652)	2·722 / 9·0321	2·860 / 11·006	2·994 / 13·229	3·126 / 15·713	3·255 / 18·470	3·381 / 21·511	3·505 / 24·847	3·627 / 28·490	3·865 / 36·734	4·209 / 51·655	4·431 / 63·422	4·753 / 83·997	4·962 / 99·771
0·1900 (1·8633)	2·796 / 9·2796	2·938 / 11·307	3·076 / 13·591	3·212 / 16·144	3·344 / 18·976	3·474 / 22·101	3·602 / 25·528	3·727 / 29·270	3·971 / 37·740	4·325 / 53·070	4·552 / 65·160	4·883 / 86·298	5·098 / 102·51
0·2000 (1·9613)	2·869 / 9·5206	3·014 / 11·601	3·156 / 13·944	3·295 / 16·563	3·431 / 19·469	3·564 / 22·675	3·695 / 26·192	3·824 / 30·031	4·074 / 38·721	4·437 / 54·449	4·671 / 66·853	5·010 / 88·540	5·231 / 105·17
0·2100 (2·0594)	2·940 / 9·7558	3·089 / 11·887	3·234 / 14·289	3·376 / 16·972	3·516 / 19·950	3·652 / 23·235	3·786 / 26·838	3·918 / 30·772	4·175 / 39·677	4·546 / 55·794	4·786 / 68·504	5·134 / 90·727	5·360 / 107·77
0·2200 (2·1575)	3·009 / 9·9853	3·162 / 12·167	3·310 / 14·625	3·456 / 17·371	3·598 / 20·419	3·738 / 23·782	3·875 / 27·470	4·010 / 31·496	4·273 / 40·611	4·653 / 57·107	4·898 / 70·116	5·255 / 92·862	5·486 / 110·30
0·2300 (2·2555)	3·077 / 10·210	3·233 / 12·441	3·385 / 14·954	3·534 / 17·762	3·679 / 20·878	3·822 / 24·316	3·963 / 28·087	4·100 / 32·204	4·369 / 41·523	4·758 / 58·390	5·009 / 71·692	5·373 / 94·949	5·609 / 112·78
0·2400 (2·3536)	3·143 / 10·429	3·302 / 12·708	3·458 / 15·275	3·610 / 18·144	3·758 / 21·327	3·904 / 24·839	4·048 / 28·691	4·189 / 32·897	4·463 / 42·417	4·860 / 59·646	5·116 / 73·234	5·489 / 96·991	5·730 / 115·21
0·2500 (2·4517)	3·208 / 10·644	3·370 / 12·970	3·529 / 15·590	3·684 / 18·518	3·836 / 21·767	3·985 / 25·351	4·131 / 29·283	4·275 / 33·575	4·555 / 43·291	4·961 / 60·876	5·222 / 74·744	5·602 / 98·991	5·848 / 117·58
0·2600 (2·5497)	3·271 / 10·855	3·437 / 13·227	3·599 / 15·899	3·757 / 18·885	3·912 / 22·198	4·064 / 25·853	4·213 / 29·863	4·360 / 34·240	4·646 / 44·149	5·059 / 62·082	5·325 / 76·224	5·713 / 100·95	5·964 / 119·91
0·2700 (2·6478)	3·334 / 11·062	3·502 / 13·479	3·667 / 16·202	3·829 / 19·244	3·986 / 22·621	4·141 / 26·346	4·293 / 30·432	4·443 / 34·892	4·734 / 44·990	5·155 / 63·264	5·427 / 77·676	5·822 / 102·87	6·077 / 122·19
0·2800 (2·7459)	3·395 / 11·265	3·567 / 13·726	3·735 / 16·499	3·899 / 19·598	4·060 / 23·036	4·217 / 26·829	4·372 / 30·990	4·524 / 35·533	4·821 / 45·815	5·250 / 64·425	5·526 / 79·102	5·928 / 104·76	6·189 / 124·44
0·2900 (2·8439)	3·455 / 11·464	3·630 / 13·969	3·801 / 16·791	3·968 / 19·944	4·131 / 23·444	4·292 / 27·304	4·449 / 31·539	4·604 / 36·162	4·906 / 46·626	5·343 / 65·565	5·624 / 80·502	6·033 / 106·62	6·299 / 126·64
S, ($\Delta p/\rho l$)	65	70	75	80	85	90	95	100	110	125	135	150	160

Gradient (Equivalent) Pipe diameters in mm

S = 0·1000 to 0·2900; $\Delta p/\rho l$ = (0·9807) to (2·8439) ms^{-2}

m = Manning $n \times 100$
S = 0·3000 to 0·9500

i.e. kin. pr. grad., $\Delta p/\rho l$ = (2·9420) to (9·3163) ms^{-2}

Full bore flow; Tables E give m_C or m_P for Colebrook-White or for laminar solutions resp., to divide mV and/or mQ as follows:

mV to give velocities in ms^{-1}
mQ to give discharges in litres/sec

Gradient (Equivalent) Pipe diameters in mm

S, ($\Delta p/\rho l$)	65	70	75	80	85	90	95	100	110	125	135	150	160
0·3000	3·514	3·692	3·866	4·036	4·202	4·365	4·526	4·683	4·990	5·434	5·720	6·136	6·406
(2·9420)	11·660	14·208	17·078	20·285	23·845	27·771	32·078	36·780	47·423	66·686	81·878	108·44	128·80
0·3200	3·629	3·813	3·992	4·168	4·340	4·508	4·674	4·837	5·154	5·612	5·908	6·338	6·616
(3·1381)	12·043	14·674	17·638	20·951	24·627	28·682	33·130	37·986	48·978	68·873	84·563	112·00	133·03
0·3400	3·741	3·930	4·115	4·296	4·473	4·647	4·818	4·985	5·312	5·785	6·090	6·533	6·820
(3·3343)	12·413	15·126	18·181	21·595	25·385	29·564	34·150	39·155	50·486	70·993	87·166	115·44	137·12
0·3600	3·849	4·044	4·235	4·421	4·603	4·782	4·957	5·130	5·466	5·953	6·266	6·722	7·018
(3·5304)	12·773	15·564	18·708	22·222	26·121	30·421	35·140	40·290	51·949	73·051	89·693	118·79	141·10
0·3800	3·955	4·155	4·351	4·542	4·729	4·913	5·093	5·270	5·616	6·116	6·438	6·906	7·210
(3·7265)	13·123	15·991	19·221	22·830	26·836	31·255	36·102	41·394	53·373	75·053	92·150	122·04	144·96
0·4000	4·058	4·263	4·464	4·660	4·852	5·041	5·226	5·407	5·762	6·275	6·605	7·086	7·397
(3·9227)	13·464	16·406	19·720	23·424	27·534	32·067	37·040	42·470	54·760	77·003	94·544	125·21	148·73
0·4200	4·158	4·368	4·574	4·775	4·972	5·165	5·355	5·541	5·904	6·430	6·768	7·261	7·580
(4·1188)	13·797	16·811	20·207	24·002	28·214	32·859	37·955	43·518	56·112	78·904	96·879	128·31	152·40
0·4400	4·256	4·471	4·682	4·887	5·089	5·287	5·481	5·671	6·043	6·581	6·927	7·432	7·758
(4·3149)	14·121	17·207	20·683	24·567	28·877	33·632	38·848	44·543	57·432	80·761	99·159	131·33	155·99
0·4600	4·351	4·572	4·787	4·997	5·203	5·405	5·604	5·799	6·179	6·729	7·083	7·599	7·933
(4·5111)	14·439	17·594	21·147	25·119	29·526	34·388	39·721	45·544	58·723	82·576	101·39	134·28	159·50
0·4800	4·445	4·670	4·890	5·105	5·315	5·522	5·724	5·924	6·312	6·874	7·236	7·762	8·103
(4·7072)	14·749	17·972	21·602	25·659	30·162	35·128	40·576	46·523	59·986	84·352	103·57	137·17	162·93
0·5000	4·536	4·766	4·991	5·210	5·425	5·636	5·842	6·046	6·442	7·015	7·385	7·922	8·270
(4·9033)	15·053	18·343	22·048	26·188	30·783	35·852	41·412	47·483	61·223	86·092	105·70	139·99	166·29
0·5250	4·649	4·884	5·114	5·339	5·559	5·775	5·987	6·195	6·601	7·189	7·567	8·118	8·475
(5·1485)	15·425	18·796	22·592	26·835	31·544	36·737	42·435	48·655	62·735	88·218	108·31	143·45	170·39
0·5500	4·758	4·999	5·234	5·464	5·690	5·911	6·128	6·341	6·757	7·358	7·745	8·309	8·674
(5·3937)	15·788	19·238	23·124	27·467	32·286	37·602	43·434	49·800	64·211	90·294	110·86	146·83	174·40
0·5750	4·865	5·111	5·352	5·587	5·818	6·044	6·265	6·483	6·909	7·523	7·919	8·495	8·869
(5·6388)	16·143	19·670	23·644	28·084	33·012	38·447	44·410	50·919	65·654	92·323	113·35	150·13	178·32
0·6000	4·969	5·221	5·467	5·707	5·943	6·173	6·400	6·623	7·057	7·685	8·090	8·678	9·060
(5·8840)	16·490	20·093	24·152	28·688	33·722	39·274	45·365	52·015	67·066	94·309	115·79	153·36	182·16
0·6250	5·072	5·329	5·580	5·825	6·065	6·301	6·532	6·759	7·203	7·843	8·256	8·857	9·247
(6·1292)	16·830	20·508	24·650	29·279	34·417	40·084	46·300	53·087	68·449	96·253	118·18	156·52	185·91
0·6500	5·172	5·434	5·690	5·940	6·185	6·426	6·661	6·893	7·345	7·999	8·420	9·033	9·430
(6·3743)	17·164	20·914	25·138	29·859	35·099	40·878	47·217	54·138	69·805	98·160	120·52	159·62	189·59
0·6750	5·271	5·538	5·799	6·053	6·303	6·548	6·788	7·024	7·485	8·151	8·580	9·205	9·609
(6·6195)	17·491	21·312	25·617	30·428	35·767	41·656	48·117	55·170	71·135	100·03	122·82	162·66	193·21
0·7000	5·368	5·640	5·905	6·165	6·419	6·668	6·913	7·153	7·623	8·301	8·738	9·374	9·786
(6·8647)	17·811	21·703	26·087	30·986	36·424	42·421	49·000	56·182	72·440	101·86	125·07	165·64	196·75
0·7250	5·463	5·739	6·009	6·274	6·532	6·786	7·035	7·280	7·758	8·448	8·892	9·539	9·959
(7·1098)	18·127	22·087	26·549	31·535	37·068	43·172	49·867	57·177	73·722	103·67	127·28	168·58	200·23
0·7500	5·556	5·837	6·112	6·381	6·644	6·902	7·155	7·404	7·890	8·592	9·044	9·703	10·13
(7·3550)	18·437	22·465	27·003	32·074	37·702	43·910	50·720	58·154	74·983	105·44	129·46	171·46	203·66
0·8000	5·738	6·029	6·313	6·590	6·862	7·129	7·390	7·647	8·149	8·874	9·341	10·02	10·46
(7·8453)	19·041	23·202	27·888	33·126	38·938	45·350	52·383	60·061	77·442	108·90	133·71	177·08	210·34
0·8500	5·915	6·214	6·507	6·793	7·073	7·348	7·618	7·883	8·400	9·147	9·628	10·33	10·78
(8·3357)	19·627	23·916	28·747	34·145	40·137	46·745	53·995	61·910	79·825	112·25	137·82	182·53	216·81
0·9000	6·086	6·395	6·696	6·990	7·278	7·561	7·838	8·111	8·643	9·412	9·908	10·63	11·10
(8·8260)	20·196	24·609	29·580	35·135	41·300	48·101	55·561	63·705	82·139	115·50	141·82	187·82	223·10
0·9500	6·253	6·570	6·879	7·181	7·478	7·768	8·053	8·333	8·880	9·670	10·18	10·92	11·40
(9·3163)	20·750	25·284	30·391	36·098	42·432	49·419	57·083	65·450	84·390	118·67	145·70	192·97	229·21
S, ($\Delta p/\rho l$)	65	70	75	80	85	90	95	100	110	125	135	150	160

Gradient (Equivalent) Pipe diameters in mm

S = 0·3000 to 0·9500;

$\Delta p/\rho l$ = (2·9420) to (9·3163) ms^{-2}

m = Manning $n \times 100$
S = 1·0000 to 3·0000

i.e. kin. pr. grad., $\Delta p/\rho l$ =
(9·8067) to (29·420) ms^{-2}

Full bore flow; Tables E give m_C or m_P for Colebrook-White or for laminar solutions resp., to divide mV and/or mQ as follows:

mV to give velocities in ms^{-1}
mQ to give discharges in litres/sec

Gradient (Equivalent) Pipe diameters in mm

S, $(\Delta p/\rho l)$	65	70	75	80	85	90	95	100	110	125	135	150	160
1·0000	6·416	6·740	7·058	7·368	7·672	7·970	8·262	8·550	9·111	9·921	10·44	11·20	11·70
(9·8067)	21·289	25·940	31·180	37·036	43·534	50·702	58·566	67·150	86·582	121·75	149·49	197·98	235·16
1·0500	6·574	6·907	7·232	7·550	7·861	8·167	8·466	8·761	9·336	10·17	10·70	11·48	11·98
(10·297)	21·815	26·581	31·950	37·950	44·610	51·955	60·012	68·809	88·721	124·76	153·18	202·87	240·97
1·1000	6·729	7·069	7·402	7·728	8·046	8·359	8·666	8·967	9·555	10·41	10·95	11·75	12·27
(10·787)	22·328	27·207	32·702	38·844	45·659	53·177	61·425	70·428	90·808	127·69	156·78	207·65	246·64
1·1500	6·880	7·228	7·569	7·901	8·227	8·547	8·860	9·169	9·770	10·64	11·20	12·01	12·54
(11·278)	22·830	27·818	33·437	39·717	46·685	54·372	62·805	72·011	92·849	130·56	160·31	212·31	252·18
1·2000	7·028	7·384	7·731	8·071	8·404	8·731	9·051	9·366	9·980	10·87	11·44	12·27	12·81
(11·768)	23·321	28·416	34·156	40·571	47·690	55·542	64·156	73·560	94·846	133·37	163·76	216·88	257·61
1·2500	7·173	7·536	7·891	8·238	8·577	8·911	9·238	9·559	10·19	11·09	11·68	12·53	13·08
(12·258)	23·802	29·002	34·861	41·407	48·673	56·687	65·479	75·077	96·802	136·12	167·13	221·35	262·92
1·3000	7·315	7·685	8·047	8·401	8·747	9·087	9·421	9·748	10·39	11·31	11·91	12·77	13·34
(12·749)	24·273	29·577	35·551	42·227	49·637	57·810	66·775	76·563	98·719	138·82	170·44	225·73	268·13
1·3500	7·454	7·832	8·200	8·561	8·914	9·260	9·600	9·934	10·59	11·53	12·13	13·02	13·59
(13·239)	24·735	30·140	36·228	43·032	50·582	58·911	68·048	78·022	100·60	141·46	173·69	230·03	273·23
1·4000	7·591	7·975	8·351	8·718	9·078	9·430	9·776	10·12	10·78	11·74	12·36	13·26	13·84
(13·729)	25·189	30·693	36·893	43·821	51·511	59·992	69·296	79·454	102·45	144·06	176·88	234·26	278·25
1·4500	7·725	8·117	8·499	8·872	9·238	9·597	9·949	10·30	10·97	11·95	12·58	13·49	14·08
(14·220)	25·635	31·236	37·546	44·597	52·422	61·054	70·523	80·860	104·26	146·61	180·01	238·40	283·17
1·5000	7·857	8·255	8·644	9·024	9·396	9·761	10·12	10·47	11·16	12·15	12·79	13·72	14·32
(14·710)	26·073	31·770	38·188	45·359	53·319	62·098	71·728	82·242	106·04	149·12	183·08	242·48	288·01
1·6000	8·115	8·526	8·927	9·320	9·704	10·08	10·45	10·81	11·52	12·55	13·21	14·17	14·79
(15·691)	26·928	32·812	39·440	46·847	55·067	64·134	74·081	84·939	109·52	154·01	189·09	250·43	297·46
1·7000	8·365	8·789	9·202	9·607	10·00	10·39	10·77	11·15	11·88	12·94	13·62	14·61	15·25
(16·671)	27·757	33·822	40·654	48·289	56·762	66·108	76·361	87·554	112·89	158·75	194·91	258·14	306·62
1·8000	8·607	9·043	9·469	9·885	10·29	10·69	11·09	11·47	12·22	13·31	14·01	15·03	15·69
(17·652)	28·562	34·803	41·833	49·689	58·408	68·025	78·575	90·092	116·16	163·35	200·56	265·62	315·50
1·9000	8·843	9·291	9·728	10·16	10·58	10·99	11·39	11·79	12·56	13·68	14·40	15·44	16·12
(18·633)	29·345	35·756	42·979	51·050	60·008	69·889	80·728	92·561	119·35	167·82	206·05	272·90	324·15
2·0000	9·073	9·532	9·981	10·42	10·85	11·27	11·68	12·09	12·88	14·03	14·77	15·84	16·54
(19·613)	30·107	36·685	44·095	52·377	61·567	71·704	82·825	94·965	122·45	172·18	211·41	279·99	332·57
2·1000	9·297	9·768	10·23	10·68	11·12	11·55	11·97	12·39	13·20	14·38	15·13	16·24	16·95
(20·594)	30·850	37·591	45·184	53·670	63·087	73·475	84·870	97·310	125·47	176·44	216·63	286·90	340·78
2·2000	9·516	9·998	10·47	10·93	11·38	11·82	12·26	12·68	13·51	14·72	15·49	16·62	17·35
(21·575)	31·576	38·476	46·248	54·933	64·572	75·204	86·867	99·600	128·42	180·59	221·73	293·65	348·80
2·3000	9·730	10·22	10·70	11·17	11·64	12·09	12·53	12·97	13·82	15·05	15·84	16·99	17·74
(22·555)	32·286	39·341	47·287	56·168	66·023	76·894	88·820	101·84	131·31	184·65	226·71	300·25	356·64
2·4000	9·939	10·44	10·93	11·41	11·89	12·35	12·80	13·25	14·11	15·37	16·18	17·36	18·12
(23·536)	32·980	40·187	48·304	57·376	67·443	78·548	90·730	104·03	134·13	188·62	231·59	306·71	364·31
2·5000	10·14	10·66	11·16	11·65	12·13	12·60	13·06	13·52	14·41	15·69	16·51	17·71	18·49
(24·517)	33·661	41·015	49·300	58·559	68·834	80·168	92·601	106·17	136·90	192·51	236·36	313·04	371·83
2·6000	10·34	10·87	11·38	11·88	12·37	12·85	13·32	13·79	14·69	16·00	16·84	18·07	18·86
(25·497)	34·327	41·828	50·277	59·719	70·197	81·755	94·435	108·28	139·61	196·32	241·04	319·24	379·19
2·7000	10·54	11·08	11·60	12·11	12·61	13·10	13·58	14·05	14·97	16·30	17·16	18·41	19·22
(26·478)	34·981	42·624	51·234	60·856	71·534	83·313	96·234	110·34	142·27	200·06	245·63	325·32	386·41
2·8000	10·74	11·28	11·81	12·33	12·84	13·34	13·83	14·31	15·25	16·60	17·48	18·75	19·57
(27·459)	35·623	43·407	52·174	61·973	72·847	84·841	98·000	112·36	144·88	203·73	250·14	331·29	393·50
2·9000	10·93	11·48	12·02	12·55	13·06	13·57	14·07	14·56	15·52	16·90	17·78	19·08	19·92
(28·439)	36·254	44·175	53·098	63·070	74·136	86·343	99·734	114·35	147·44	207·34	254·57	337·15	400·47
3·0000	11·11	11·67	12·22	12·76	13·29	13·80	14·31	14·81	15·78	17·18	18·09	19·41	20·26
(29·420)	36·873	44·930	54·006	64·148	75·404	87·819	101·44	116·31	149·97	210·88	258·92	342·92	407·31
S, $(\Delta p/\rho l)$	65	70	75	80	85	90	95	100	110	125	135	150	160

Gradient (Equivalent) Pipe diameters in mm

S = 1·0000 to 3·0000;

$\Delta p/\rho l$ = (9·8067) to (29·420) ms^{-2}

m = Manning $n \times 100$
S = 0·000100 to 0·000290

i.e. kin. pr. grad., $\Delta p/\rho l$ =
(0·000981) to (0·002844) ms^{-2}

Full bore flow; Tables E give m_C or m_P for Colebrook-White or for laminar solutions resp., to divide mV and/or mQ as follows:

mV to give velocities in ms^{-1}
mQ to give discharges in litres/sec

Gradient **(Equivalent) Pipe diameters in mm**

S, $(\Delta p/\rho l)$	160	165	175	200	225	250	275	300	315	330	350	375	400
0·000100	0·117	0·119	0·124	0·136	0·147	0·157	0·168	0·178	0·184	0·190	0·197	0·206	0·215
(0·000981)	2·3516	2·5527	2·9864	4·2638	5·8372	7·7308	9·9679	12·571	14·318	16·209	18·963	22·793	27·073
0·000105	0·120	0·122	0·127	0·139	0·150	0·161	0·172	0·182	0·188	0·194	0·202	0·211	0·221
(0·001030)	2·4097	2·6158	3·0602	4·3691	5·9813	7·9217	10·214	12·882	14·671	16·609	19·431	23·356	27·742
0·000110	0·123	0·125	0·130	0·142	0·154	0·165	0·176	0·187	0·193	0·199	0·207	0·216	0·226
(0·001079)	2·4664	2·6773	3·1322	4·4719	6·1221	8·1081	10·454	13·185	15·017	17·000	19·888	23·905	28·395
0·000115	0·125	0·128	0·133	0·146	0·157	0·169	0·180	0·191	0·197	0·203	0·211	0·221	0·231
(0·001128)	2·5218	2·7375	3·2026	4·5724	6·2597	8·2903	10·689	13·481	15·354	17·382	20·335	24·443	29·033
0·000120	0·128	0·131	0·136	0·149	0·161	0·173	0·184	0·195	0·201	0·208	0·216	0·226	0·236
(0·001177)	2·5761	2·7964	3·2715	4·6707	6·3943	8·4686	10·919	13·771	15·684	17·756	20·772	24·968	29·657
0·000125	0·131	0·133	0·139	0·152	0·164	0·176	0·188	0·199	0·205	0·212	0·220	0·231	0·241
(0·001226)	2·6292	2·8540	3·3389	4·7671	6·5262	8·6433	11·144	14·055	16·008	18·122	21·201	25·483	30·269
0·000130	0·133	0·136	0·142	0·155	0·167	0·180	0·191	0·203	0·209	0·216	0·225	0·235	0·246
(0·001275)	2·6813	2·9106	3·4050	4·8615	6·6554	8·8144	11·365	14·333	16·325	18·481	21·621	25·988	30·868
0·000135	0·136	0·139	0·144	0·158	0·171	0·183	0·195	0·207	0·213	0·220	0·229	0·240	0·250
(0·001324)	2·7323	2·9660	3·4699	4·9541	6·7822	8·9823	11·582	14·606	16·636	18·833	22·033	26·483	31·456
0·000140	0·138	0·141	0·147	0·161	0·174	0·186	0·199	0·210	0·217	0·224	0·233	0·244	0·255
(0·001373)	2·7825	3·0204	3·5336	5·0450	6·9066	9·1472	11·794	14·874	16·941	19·179	22·437	26·969	32·034
0·000145	0·141	0·144	0·150	0·163	0·177	0·190	0·202	0·214	0·221	0·228	0·237	0·249	0·259
(0·001422)	2·8317	3·0739	3·5961	5·1343	7·0289	9·3091	12·003	15·138	17·241	19·518	22·834	27·446	32·601
0·000150	0·143	0·146	0·152	0·166	0·180	0·193	0·206	0·218	0·225	0·232	0·241	0·253	0·264
(0·001471)	2·8801	3·1265	3·6576	5·2221	7·1490	9·4682	12·208	15·396	17·536	19·852	23·224	27·915	33·158
0·000160	0·148	0·151	0·157	0·172	0·186	0·199	0·212	0·225	0·232	0·240	0·249	0·261	0·273
(0·001569)	2·9746	3·2290	3·7776	5·3933	7·3835	9·7787	12·608	15·901	18·111	20·503	23·986	28·831	34·245
0·000170	0·152	0·156	0·162	0·177	0·191	0·205	0·219	0·232	0·240	0·247	0·257	0·269	0·281
(0·001667)	3·0662	3·3284	3·8938	5·5593	7·6107	10·080	12·997	16·391	18·668	21·134	24·724	29·718	35·299
0·000180	0·157	0·160	0·167	0·182	0·197	0·211	0·225	0·239	0·246	0·254	0·264	0·277	0·289
(0·001765)	3·1550	3·4249	4·0067	5·7205	7·8314	10·372	13·373	16·866	19·209	21·746	25·441	30·580	36·323
0·000190	0·161	0·165	0·171	0·187	0·202	0·217	0·231	0·245	0·253	0·261	0·272	0·284	0·297
(0·001863)	3·2415	3·5187	4·1165	5·8772	8·0460	10·656	13·740	17·328	19·736	22·342	26·138	31·418	37·318
0·000200	0·165	0·169	0·176	0·192	0·208	0·223	0·237	0·252	0·260	0·268	0·279	0·292	0·305
(0·001961)	3·3257	3·6101	4·2234	6·0299	8·2550	10·933	14·097	17·778	20·248	22·923	26·817	32·234	38·288
0·000210	0·169	0·173	0·180	0·197	0·213	0·228	0·243	0·258	0·266	0·275	0·286	0·299	0·312
(0·002059)	3·4078	3·6993	4·3277	6·1788	8·4589	11·203	14·445	18·217	20·749	23·489	27·479	33·030	39·233
0·000220	0·173	0·177	0·184	0·201	0·218	0·234	0·249	0·264	0·273	0·281	0·292	0·306	0·320
(0·002157)	3·4880	3·7863	4·4296	6·3242	8·6579	11·467	14·785	18·646	21·237	24·042	28·126	33·807	40·156
0·000230	0·177	0·181	0·188	0·206	0·223	0·239	0·255	0·270	0·279	0·287	0·299	0·313	0·327
(0·002256)	3·5664	3·8714	4·5291	6·4664	8·8525	11·724	15·117	19·065	21·714	24·582	28·758	34·567	41·059
0·000240	0·181	0·185	0·192	0·210	0·227	0·244	0·260	0·276	0·285	0·294	0·305	0·320	0·334
(0·002354)	3·6431	3·9547	4·6265	6·6054	9·0429	11·976	15·442	19·475	22·181	25·111	29·377	35·311	41·942
0·000250	0·185	0·189	0·196	0·215	0·232	0·249	0·265	0·281	0·290	0·300	0·312	0·326	0·341
(0·002452)	3·7183	4·0362	4·7219	6·7416	9·2294	12·223	15·761	19·877	22·638	25·628	29·982	36·039	42·807
0·000260	0·189	0·193	0·200	0·219	0·237	0·254	0·271	0·287	0·296	0·306	0·318	0·333	0·347
(0·002550)	3·7919	4·1162	4·8155	6·8752	9·4122	12·465	16·073	20·270	23·087	26·136	30·576	36·752	43·655
0·000270	0·192	0·196	0·204	0·223	0·241	0·259	0·276	0·292	0·302	0·311	0·324	0·339	0·354
(0·002648)	3·8641	4·1946	4·9072	7·0061	9·5915	12·703	16·379	20·656	23·527	26·634	31·159	37·453	44·486
0·000280	0·196	0·200	0·208	0·227	0·246	0·264	0·281	0·298	0·307	0·317	0·330	0·345	0·361
(0·002746)	3·9350	4·2715	4·9972	7·1347	9·7675	12·936	16·679	21·035	23·958	27·123	31·730	38·140	45·302
0·000290	0·199	0·203	0·211	0·231	0·250	0·268	0·286	0·303	0·313	0·323	0·336	0·351	0·367
(0·002844)	4·0047	4·3472	5·0857	7·2610	9·9403	13·165	16·975	21·408	24·382	27·603	32·292	38·815	46·104

S, $(\Delta p/\rho l)$	160	165	175	200	225	250	275	300	315	330	350	375	400

Gradient **(Equivalent) Pipe diameters in mm**

S = 0·000100 to 0·000290;

$\Delta p/\rho l$ = (0·000981) to (0·002844) ms^{-2}

m = Manning $n \times 100$
S = 0·000300 to 0·000950

i.e. kin. pr. grad., $\Delta p/\rho l$ =
(0·002942) to (0·009316) ms^{-2}

Full bore flow; Tables E give m_C or m_P for Colebrook-White or for laminar solutions resp., to divide mV and/or mQ as follows:

mV to give velocities in ms^{-1}
mQ to give discharges in litres/sec

Gradient (Equivalent) Pipe diameters in mm

S, ($\Delta p/\rho l$)	160	165	175	200	225	250	275	300	315	330	350	375	400
0·000300	0·203	0·207	0·215	0·235	0·254	0·273	0·291	0·308	0·318	0·328	0·341	0·357	0·373
(0·002942)	4·0731	4·4215	5·1726	7·3851	10·110	13·390	17·265	21·774	24·799	28·075	32·844	39·478	46·892
0·000320	0·209	0·214	0·222	0·243	0·263	0·282	0·300	0·318	0·329	0·339	0·353	0·369	0·385
(0·003138)	4·2067	4·5665	5·3423	7·6273	10·442	13·829	17·831	22·488	25·613	28·995	33·921	40·773	48·430
0·000340	0·216	0·220	0·229	0·250	0·271	0·290	0·309	0·328	0·339	0·349	0·363	0·381	0·397
(0·003334)	4·3362	4·7070	5·5067	7·8620	10·763	14·255	18·380	23·180	26·401	29·888	34·965	42·028	49·921
0·000360	0·222	0·227	0·236	0·258	0·279	0·299	0·318	0·337	0·349	0·360	0·374	0·392	0·409
(0·003530)	4·4619	4·8435	5·6663	8·0900	11·075	14·668	18·913	23·852	27·166	30·754	35·979	43·246	51·368
0·000380	0·228	0·233	0·242	0·265	0·286	0·307	0·327	0·347	0·358	0·369	0·384	0·402	0·420
(0·003727)	4·5842	4·9762	5·8216	8·3117	11·379	15·070	19·431	24·506	27·911	31·597	36·965	44·432	52·776
0·000400	0·234	0·239	0·248	0·271	0·294	0·315	0·336	0·356	0·367	0·379	0·394	0·413	0·431
(0·003923)	4·7033	5·1055	5·9728	8·5276	11·674	15·462	19·936	25·142	28·636	32·418	37·925	45·586	54·147
0·000420	0·240	0·245	0·254	0·278	0·301	0·323	0·344	0·364	0·377	0·388	0·404	0·423	0·442
(0·004119)	4·8194	5·2316	6·1203	8·7382	11·963	15·843	20·428	25·763	29·343	33·218	38·862	46·712	55·484
0·000440	0·245	0·250	0·260	0·285	0·308	0·330	0·352	0·373	0·385	0·398	0·413	0·433	0·452
(0·004315)	4·9328	5·3547	6·2644	8·9438	12·244	16·216	20·909	26·369	30·033	34·000	39·776	47·811	56·790
0·000460	0·251	0·256	0·266	0·291	0·315	0·338	0·360	0·381	0·394	0·406	0·423	0·443	0·462
(0·004511)	5·0437	5·4750	6·4052	9·1448	12·519	16·581	21·379	26·962	30·708	34·764	40·670	48·885	58·066
0·000480	0·256	0·262	0·272	0·297	0·322	0·345	0·368	0·390	0·403	0·415	0·432	0·452	0·472
(0·004707)	5·1522	5·5928	6·5429	9·3415	12·789	16·937	21·839	27·542	31·369	35·512	41·545	49·937	59·315
0·000500	0·262	0·267	0·278	0·303	0·328	0·352	0·375	0·398	0·411	0·424	0·441	0·461	0·482
(0·004903)	5·2584	5·7081	6·6778	9·5341	13·052	17·287	22·289	28·110	32·016	36·244	42·402	50·966	60·538
0·000525	0·268	0·274	0·284	0·311	0·336	0·361	0·385	0·407	0·421	0·434	0·452	0·473	0·494
(0·005148)	5·3883	5·8491	6·8427	9·7696	13·375	17·713	22·839	28·804	32·806	37·139	43·449	52·225	62·033
0·000550	0·274	0·280	0·291	0·318	0·344	0·369	0·394	0·417	0·431	0·444	0·462	0·484	0·505
(0·005394)	5·5151	5·9867	7·0038	9·9995	13·689	18·130	23·377	29·482	33·578	38·013	44·471	53·454	63·493
0·000575	0·280	0·286	0·298	0·325	0·352	0·378	0·402	0·426	0·441	0·454	0·473	0·495	0·517
(0·005639)	5·6390	6·1213	7·1612	10·224	13·997	18·538	23·902	30·144	34·333	38·868	45·471	54·655	64·920
0·000600	0·286	0·292	0·304	0·332	0·360	0·386	0·411	0·436	0·450	0·464	0·483	0·506	0·528
(0·005884)	5·7603	6·2529	7·3152	10·444	14·298	18·936	24·416	30·793	35·071	39·703	46·449	55·831	66·316
0·000625	0·292	0·298	0·310	0·339	0·367	0·394	0·420	0·445	0·459	0·474	0·493	0·516	0·539
(0·006129)	5·8791	6·3818	7·4660	10·659	14·593	19·327	24·920	31·428	35·795	40·522	47·406	56·982	67·683
0·000650	0·298	0·304	0·317	0·346	0·374	0·402	0·428	0·453	0·468	0·483	0·502	0·526	0·549
(0·006374)	5·9955	6·5082	7·6139	10·871	14·882	19·710	25·413	32·050	36·503	41·325	48·345	58·111	69·024
0·000675	0·304	0·310	0·323	0·353	0·381	0·409	0·436	0·462	0·477	0·492	0·512	0·536	0·560
(0·006619)	6·1097	6·6322	7·7589	11·078	15·165	20·085	25·897	32·661	37·199	42·112	49·266	59·218	70·339
0·000700	0·309	0·316	0·328	0·359	0·388	0·417	0·444	0·471	0·486	0·501	0·521	0·546	0·570
(0·006865)	6·2218	6·7539	7·9013	11·281	15·444	20·454	26·373	33·260	37·881	42·885	50·170	60·304	71·629
0·000725	0·315	0·321	0·334	0·365	0·395	0·424	0·452	0·479	0·495	0·510	0·531	0·556	0·580
(0·007110)	6·3320	6·8735	8·0412	11·481	15·717	20·816	26·839	33·849	38·552	43·644	51·058	61·372	72·897
0·000750	0·320	0·327	0·340	0·372	0·402	0·431	0·460	0·487	0·503	0·519	0·540	0·565	0·590
(0·007355)	6·4402	6·9910	8·1786	11·677	15·986	21·172	27·298	34·427	39·211	44·390	51·931	62·421	74·144
0·000800	0·331	0·338	0·351	0·384	0·415	0·445	0·475	0·503	0·520	0·536	0·557	0·584	0·609
(0·007845)	6·6514	7·2202	8·4469	12·060	16·510	21·866	28·193	35·556	40·497	45·846	53·634	64·468	76·575
0·000850	0·341	0·348	0·362	0·396	0·428	0·459	0·489	0·519	0·536	0·553	0·575	0·602	0·628
(0·008336)	6·8561	7·4424	8·7068	12·431	17·018	22·539	29·061	36·651	41·743	47·257	55·285	66·452	78·932
0·000900	0·351	0·358	0·372	0·407	0·440	0·472	0·503	0·534	0·551	0·569	0·591	0·619	0·646
(0·008826)	7·0549	7·6582	8·9593	12·791	17·512	23·192	29·904	37·713	42·954	48·627	56·888	68·379	81·220
0·000950	0·360	0·368	0·383	0·418	0·452	0·485	0·517	0·548	0·566	0·584	0·607	0·636	0·664
(0·009316)	7·2482	7·8681	9·2048	13·142	17·991	23·828	30·723	38·747	44·131	49·959	58·447	70·252	83·446
S, ($\Delta p/\rho l$)	160	165	175	200	225	250	275	300	315	330	350	375	400

Gradient (Equivalent) Pipe diameters in mm

S = 0·000300 to 0·000950;

$\Delta p/\rho l$ = (0·002942) to (0·009316) ms^{-2}

$m = \text{Manning } n \times 100$
$S = 0.00100 \text{ to } 0.00290$

i.e. kin. pr. grad., $\Delta p/\rho l =$ (0.00981) to (0.02844) ms^{-2}

Full bore flow; Tables E give m_C or m_P for Colebrook-White or for laminar solutions resp., to divide mV and/or mQ as follows:

mV to give velocities in ms^{-1}
mQ to give discharges in litres/sec

Gradient S, ($\Delta p/\rho l$)	160	165	175	200	225	250	275	300	315	330	350	375	400
0.00100 (0.00981)	0.370	0.378	0.393	0.429	0.464	0.498	0.531	0.562	0.581	0.599	0.623	0.653	0.681
	7.4365	8.0725	9.4439	13.483	18.459	24.447	31.521	39.753	45.277	51.257	59.965	72.077	85.614
0.00105 (0.01030)	0.379	0.387	0.402	0.440	0.476	0.510	0.544	0.576	0.595	0.614	0.639	0.669	0.698
	7.6202	8.2718	9.6771	13.816	18.915	25.051	32.300	40.735	46.395	52.523	61.446	73.857	87.728
0.00110 (0.01079)	0.388	0.396	0.412	0.450	0.487	0.522	0.557	0.590	0.609	0.629	0.654	0.684	0.715
	7.7995	8.4665	9.9048	14.141	19.360	25.640	33.060	41.694	47.487	53.759	62.892	75.595	89.792
0.00115 (0.01128)	0.397	0.405	0.421	0.460	0.498	0.534	0.569	0.603	0.623	0.643	0.668	0.700	0.731
	7.9748	8.6568	10.127	14.459	19.795	26.216	33.803	42.631	48.554	54.967	64.305	77.294	91.810
0.00120 (0.01177)	0.405	0.414	0.430	0.470	0.509	0.546	0.581	0.616	0.636	0.656	0.683	0.715	0.746
	8.1463	8.8429	10.345	14.770	20.221	26.780	34.530	43.547	49.598	56.149	65.688	78.957	93.785
0.00125 (0.01226)	0.414	0.422	0.439	0.480	0.519	0.557	0.593	0.629	0.650	0.670	0.697	0.730	0.762
	8.3143	9.0253	10.559	15.075	20.638	27.332	35.242	44.445	50.621	57.307	67.043	80.585	95.719
0.00130 (0.01275)	0.422	0.430	0.448	0.489	0.529	0.568	0.605	0.641	0.662	0.683	0.711	0.744	0.777
	8.4789	9.2040	10.768	15.373	21.046	27.874	35.940	45.326	51.624	58.442	68.371	82.181	97.614
0.00135 (0.01324)	0.430	0.439	0.456	0.499	0.539	0.579	0.617	0.653	0.675	0.696	0.724	0.758	0.792
	8.6404	9.3794	10.973	15.666	21.447	28.405	36.624	46.189	52.607	59.555	69.673	83.746	99.474
0.00140 (0.01373)	0.438	0.447	0.465	0.508	0.549	0.589	0.628	0.665	0.687	0.709	0.737	0.772	0.806
	8.7990	9.5515	11.174	15.954	21.841	28.926	37.296	47.037	53.572	60.648	70.951	85.283	101.30
0.00145 (0.01422)	0.445	0.455	0.473	0.517	0.559	0.600	0.639	0.677	0.700	0.722	0.751	0.786	0.820
	8.9547	9.7205	11.372	16.236	22.227	29.438	37.957	47.869	54.521	61.722	72.207	86.793	103.09
0.00150 (0.01471)	0.453	0.462	0.481	0.526	0.569	0.610	0.650	0.689	0.712	0.734	0.763	0.799	0.834
	9.1078	9.8867	11.566	16.514	22.607	29.941	38.605	48.688	55.453	62.777	73.442	88.276	104.85
0.00160 (0.01569)	0.468	0.478	0.497	0.543	0.587	0.630	0.671	0.711	0.735	0.758	0.788	0.825	0.862
	9.4065	10.211	11.946	17.055	23.349	30.923	39.872	50.284	57.271	64.836	75.850	91.172	108.29
0.00170 (0.01667)	0.482	0.492	0.512	0.560	0.605	0.649	0.692	0.733	0.758	0.781	0.813	0.851	0.888
	9.6960	10.525	12.313	17.580	24.067	31.875	41.099	51.832	59.034	66.831	78.185	93.977	111.63
0.00180 (0.01765)	0.496	0.507	0.527	0.576	0.623	0.668	0.712	0.755	0.779	0.804	0.836	0.876	0.914
	9.9771	10.830	12.670	18.090	24.765	32.799	42.290	53.335	60.745	68.768	80.451	96.702	114.86
0.00190 (0.01863)	0.510	0.520	0.541	0.592	0.640	0.686	0.732	0.775	0.801	0.826	0.859	0.900	0.939
	10.251	11.127	13.017	18.585	25.444	33.698	43.449	54.796	62.410	70.653	82.656	99.352	118.01
0.00200 (0.01961)	0.523	0.534	0.555	0.607	0.657	0.704	0.751	0.795	0.822	0.848	0.881	0.923	0.963
	10.517	11.416	13.356	19.068	26.105	34.573	44.578	56.220	64.031	72.488	84.803	101.93	121.08
0.00210 (0.02059)	0.536	0.547	0.569	0.622	0.673	0.722	0.769	0.815	0.842	0.868	0.903	0.946	0.987
	10.777	11.698	13.685	19.539	26.749	35.427	45.679	57.608	65.613	74.278	86.897	104.45	124.07
0.00220 (0.02157)	0.549	0.560	0.582	0.637	0.689	0.739	0.787	0.834	0.862	0.889	0.924	0.968	1.011
	11.030	11.973	14.008	19.999	27.379	36.261	46.754	58.964	67.157	76.026	88.942	106.91	126.99
0.00230 (0.02256)	0.561	0.573	0.595	0.651	0.704	0.755	0.805	0.853	0.881	0.909	0.945	0.990	1.033
	11.278	12.243	14.322	20.448	27.994	37.075	47.804	60.289	68.666	77.735	90.941	109.31	129.84
0.00240 (0.02354)	0.573	0.585	0.608	0.665	0.719	0.772	0.822	0.871	0.900	0.928	0.966	1.011	1.055
	11.521	12.506	14.630	20.888	28.596	37.873	48.832	61.585	70.143	79.407	92.897	111.66	132.63
0.00250 (0.02452)	0.585	0.597	0.621	0.679	0.734	0.787	0.839	0.889	0.919	0.948	0.985	1.032	1.077
	11.758	12.764	14.932	21.319	29.186	38.654	49.839	62.855	71.589	81.044	94.813	113.96	135.37
0.00260 (0.02550)	0.596	0.609	0.633	0.692	0.749	0.803	0.856	0.907	0.937	0.966	1.005	1.052	1.099
	11.991	13.016	15.228	21.741	29.764	39.419	50.826	64.100	73.007	82.649	96.691	116.22	138.05
0.00270 (0.02648)	0.608	0.620	0.645	0.705	0.763	0.818	0.872	0.924	0.955	0.985	1.024	1.072	1.119
	12.219	13.264	15.518	22.155	30.331	40.170	51.795	65.321	74.398	84.224	98.532	118.44	140.68
0.00280 (0.02746)	0.619	0.632	0.657	0.718	0.777	0.833	0.888	0.941	0.972	1.003	1.043	1.092	1.140
	12.444	13.508	15.803	22.562	30.887	40.907	52.745	66.520	75.763	85.769	100.34	120.61	143.26
0.00290 (0.02844)	0.630	0.643	0.669	0.731	0.791	0.848	0.904	0.958	0.989	1.021	1.061	1.111	1.160
	12.664	13.747	16.082	22.961	31.434	41.631	53.679	67.697	77.104	87.287	102.12	122.74	145.79
S, ($\Delta p/\rho l$) Gradient	160	165	175	200	225	250	275	300	315	330	350	375	400

(Equivalent) Pipe diameters in mm

$S = 0.00100 \text{ to } 0.00290$;

$\Delta p/\rho l = (0.00981) \text{ to } (0.02844) \text{ ms}^{-2}$

m = Manning $n \times 100$
S = 0·00300 to 0·00950

i.e. kin. pr. grad., $\Delta p/\rho l$ =
(0·02942) to (0·09316) ms^{-2}

Full bore flow; Tables E give m_C or m_P for Colebrook-White or for laminar solutions resp., to divide mV and/or mQ as follows:

mV to give velocities in ms^{-1}
mQ to give discharges in litres/sec

Gradient (Equivalent) Pipe diameters in mm

S, ($\Delta p/\rho l$)	160	165	175	200	225	250	275	300	315	330	350	375	400
0·00300	0·641	0·654	0·680	0·743	0·804	0·863	0·919	0·974	1·006	1·038	1·080	1·130	1·180
(0·02942)	12·880	13·982	16·357	23·354	31·972	42·343	54·596	68·855	78·422	88·780	103·86	124·84	148·29
0·00320	0·662	0·675	0·702	0·768	0·830	0·891	0·949	1·006	1·039	1·072	1·115	1·167	1·219
(0·03138)	13·303	14·440	16·894	24·120	33·020	43·732	56·387	71·113	80·994	91·691	107·27	128·94	153·15
0·00340	0·682	0·696	0·724	0·791	0·856	0·918	0·979	1·037	1·071	1·105	1·149	1·203	1·256
(0·03334)	13·712	14·885	17·414	24·862	34·036	45·078	58·122	73·301	83·487	94·513	110·57	132·90	157·86
0·00360	0·702	0·716	0·745	0·814	0·881	0·945	1·007	1·067	1·102	1·137	1·183	1·238	1·293
(0·03530)	14·110	15·316	17·919	25·583	35·023	46·385	59·807	75·426	85·907	97·253	113·78	136·76	162·44
0·00380	0·721	0·736	0·765	0·837	0·905	0·971	1·035	1·096	1·133	1·168	1·215	1·272	1·328
(0·03727)	14·496	15·736	18·410	26·284	35·983	47·656	61·446	77·493	88·261	99·918	116·89	140·50	166·89
0·00400	0·740	0·755	0·785	0·858	0·928	0·996	1·061	1·125	1·162	1·199	1·247	1·305	1·363
(0·03923)	14·873	16·145	18·888	26·967	36·918	48·894	63·042	79·506	90·554	102·51	119·93	144·15	171·23
0·00420	0·758	0·774	0·805	0·880	0·951	1·021	1·088	1·153	1·191	1·228	1·277	1·337	1·396
(0·04119)	15·240	16·544	19·354	27·633	37·829	50·101	64·599	81·470	92·790	105·05	122·89	147·71	175·46
0·00440	0·776	0·792	0·824	0·900	0·974	1·045	1·113	1·180	1·219	1·257	1·307	1·369	1·429
(0·04315)	15·599	16·933	19·810	28·283	38·719	51·280	66·119	83·387	94·974	107·52	125·78	151·19	179·58
0·00460	0·793	0·810	0·842	0·921	0·996	1·068	1·138	1·206	1·246	1·285	1·337	1·400	1·461
(0·04511)	15·950	17·314	20·255	28·918	39·590	52·433	67·605	85·261	97·108	109·93	128·61	154·59	183·62
0·00480	0·810	0·827	0·860	0·940	1·017	1·091	1·163	1·232	1·273	1·313	1·366	1·430	1·493
(0·04707)	16·293	17·686	20·691	29·540	40·441	53·560	69·060	87·095	99·197	112·30	131·38	157·91	187·57
0·00500	0·827	0·844	0·878	0·960	1·038	1·114	1·187	1·258	1·299	1·340	1·394	1·459	1·523
(0·04903)	16·629	18·051	21·117	30·150	41·275	54·665	70·484	88·891	101·24	114·61	134·09	161·17	191·44
0·00525	0·847	0·865	0·900	0·983	1·064	1·141	1·216	1·289	1·331	1·373	1·428	1·495	1·561
(0·05148)	17·039	18·496	21·639	30·894	42·294	56·015	72·224	91·086	103·74	117·44	137·40	165·15	196·17
0·00550	0·867	0·885	0·921	1·007	1·089	1·168	1·245	1·319	1·363	1·405	1·462	1·530	1·598
(0·05394)	17·440	18·932	22·148	31·621	43·290	57·333	73·924	93·230	106·18	120·21	140·63	169·04	200·78
0·00575	0·887	0·905	0·941	1·029	1·113	1·194	1·273	1·349	1·393	1·437	1·495	1·565	1·634
(0·05639)	17·832	19·357	22·646	32·332	44·263	58·621	75·585	95·325	108·57	122·91	143·79	172·84	205·29
0·00600	0·906	0·925	0·962	1·051	1·137	1·220	1·300	1·378	1·423	1·468	1·527	1·599	1·669
(0·05884)	18·216	19·773	23·133	33·027	45·215	59·882	77·211	97·375	110·91	125·55	146·88	176·55	209·71
0·00625	0·925	0·944	0·982	1·073	1·161	1·245	1·327	1·406	1·452	1·498	1·558	1·632	1·703
(0·06129)	18·591	20·181	23·610	33·708	46·147	61·117	78·803	99·383	113·19	128·14	149·91	180·19	214·03
0·00650	0·943	0·963	1·001	1·094	1·184	1·270	1·353	1·434	1·481	1·528	1·589	1·664	1·737
(0·06374)	18·959	20·581	24·077	34·376	47·061	62·327	80·364	101·35	115·43	130·68	152·88	183·76	218·27
0·00675	0·961	0·981	1·020	1·115	1·206	1·294	1·379	1·461	1·509	1·557	1·619	1·696	1·770
(0·06619)	19·321	20·973	24·536	35·031	47·957	63·515	81·895	103·28	117·63	133·17	155·79	187·26	222·43
0·00700	0·979	0·999	1·039	1·136	1·228	1·318	1·404	1·488	1·537	1·586	1·649	1·727	1·803
(0·06865)	19·675	21·358	24·986	35·673	48·837	64·680	83·397	105·18	119·79	135·61	158·65	190·70	226·51
0·00725	0·996	1·017	1·057	1·156	1·250	1·341	1·429	1·514	1·564	1·614	1·678	1·757	1·834
(0·07110)	20·023	21·736	25·428	36·305	49·702	65·825	84·873	107·04	121·91	138·01	161·46	194·07	230·52
0·00750	1·013	1·034	1·075	1·175	1·271	1·364	1·453	1·540	1·591	1·641	1·707	1·787	1·866
(0·07355)	20·366	22·107	25·863	36·926	50·551	66·950	86·324	108·87	124·00	140·37	164·22	197·39	234·46
0·00800	1·046	1·068	1·111	1·214	1·313	1·409	1·501	1·591	1·643	1·695	1·763	1·846	1·927
(0·07845)	21·034	22·832	26·711	38·137	52·209	69·146	89·155	112·44	128·06	144·98	169·61	203·87	242·15
0·00850	1·078	1·101	1·145	1·251	1·353	1·452	1·547	1·640	1·694	1·747	1·817	1·903	1·986
(0·08336)	21·681	23·535	27·533	39·310	53·816	71·274	91·899	115·90	132·00	149·44	174·83	210·14	249·60
0·00900	1·110	1·133	1·178	1·288	1·393	1·494	1·592	1·687	1·743	1·798	1·870	1·958	2·044
(0·08826)	22·310	24·217	28·332	40·450	55·376	73·340	94·564	119·26	135·83	153·77	179·89	216·23	256·84
0·00950	1·140	1·164	1·210	1·323	1·431	1·535	1·636	1·733	1·791	1·847	1·921	2·011	2·100
(0·09316)	22·921	24·881	29·108	41·558	56·894	75·350	97·155	122·53	139·55	157·98	184·82	222·16	263·88
S, ($\Delta p/\rho l$)	160	165	175	200	225	250	275	300	315	330	350	375	400

Gradient (Equivalent) Pipe diameters in mm

S = 0·00300 to 0·00950;

$\Delta p/\rho l$ = (0·02942) to (0·09316) ms^{-2}

m = Manning $n \times 100$
S = 0·0100 to 0·0290

i.e. kin. pr. grad., $\Delta p/\rho l$ =
(0·0981) to (0·2844) ms^{-2}

Full bore flow; Tables E give m_C or m_P for Colebrook-White or for laminar solutions resp., to divide mV and/or mQ as follows:

mV to give velocities in ms^{-1}
mQ to give discharges in litres/sec

Gradient **(Equivalent) Pipe diameters in mm**

S, ($\Delta p/\rho l$)	160	165	175	200	225	250	275	300	315	330	350	375	400
0·0100	1·170	1·194	1·242	1·357	1·468	1·575	1·678	1·778	1·837	1·895	1·971	2·064	2·154
(0·0981)	23·516	25·527	29·864	42·638	58·372	77·308	99·679	125·71	143·18	162·09	189·63	227·93	270·73
0·0105	1·198	1·223	1·272	1·391	1·504	1·614	1·720	1·822	1·883	1·942	2·020	2·115	2·208
(0·1030)	24·097	26·158	30·602	43·691	59·813	79·217	102·14	128·82	146·71	166·09	194·31	233·56	277·42
0·0110	1·227	1·252	1·302	1·423	1·540	1·652	1·760	1·865	1·927	1·988	2·067	2·164	2·260
(0·1079)	24·664	26·773	31·322	44·719	61·221	81·081	104·54	131·85	150·17	170·00	198·88	239·05	283·95
0·0115	1·254	1·280	1·331	1·455	1·574	1·689	1·800	1·907	1·970	2·032	2·114	2·213	2·310
(0·1128)	25·218	27·375	32·026	45·724	62·597	82·903	106·89	134·81	153·54	173·82	203·35	244·43	290·33
0·0120	1·281	1·308	1·360	1·487	1·608	1·725	1·838	1·948	2·013	2·076	2·159	2·261	2·360
(0·1177)	25·761	27·964	32·715	46·707	63·943	84·686	109·19	137·71	156·84	177·56	207·72	249·68	296·57
0·0125	1·308	1·335	1·388	1·517	1·641	1·761	1·876	1·988	2·054	2·119	2·204	2·307	2·409
(0·1226)	26·292	28·540	33·389	47·671	65·262	86·433	111·44	140·55	160·08	181·22	212·01	254·83	302·69
0·0130	1·334	1·361	1·416	1·547	1·674	1·796	1·913	2·028	2·095	2·161	2·247	2·353	2·456
(0·1275)	26·813	29·106	34·050	48·615	66·554	88·144	113·65	143·33	163·25	184·81	216·21	259·88	308·68
0·0135	1·359	1·387	1·443	1·577	1·706	1·830	1·950	2·066	2·135	2·202	2·290	2·398	2·503
(0·1324)	27·323	29·660	34·699	49·541	67·822	89·823	115·82	146·06	166·36	188·33	220·33	264·83	314·56
0·0140	1·384	1·413	1·469	1·606	1·737	1·863	1·986	2·104	2·174	2·242	2·332	2·442	2·549
(0·1373)	27·825	30·204	35·336	50·450	69·066	91·472	117·94	148·74	169·41	191·79	224·37	269·69	320·34
0·0145	1·408	1·438	1·495	1·634	1·768	1·896	2·021	2·142	2·212	2·282	2·373	2·485	2·594
(0·1422)	28·317	30·739	35·961	51·343	70·289	93·091	120·03	151·38	172·41	195·18	228·34	274·46	326·01
0·0150	1·432	1·462	1·521	1·662	1·798	1·929	2·055	2·178	2·250	2·321	2·414	2·528	2·639
(0·1471)	28·801	31·265	36·576	52·221	71·490	94·682	122·08	153·96	175·36	198·52	232·24	279·15	331·58
0·0160	1·479	1·510	1·571	1·717	1·857	1·992	2·123	2·250	2·324	2·397	2·493	2·610	2·725
(0·1569)	29·746	32·290	37·776	53·933	73·835	97·787	126·08	159·01	181·11	205·03	239·86	288·31	342·45
0·0170	1·525	1·557	1·619	1·770	1·914	2·053	2·188	2·319	2·395	2·471	2·570	2·691	2·809
(0·1667)	30·662	33·284	38·938	55·593	76·107	100·80	129·97	163·91	186·68	211·34	247·24	297·18	352·99
0·0180	1·569	1·602	1·666	1·821	1·970	2·113	2·252	2·386	2·465	2·543	2·644	2·769	2·890
(0·1765)	31·550	34·249	40·067	57·205	78·314	103·72	133·73	168·66	192·09	217·46	254·41	305·80	363·23
0·0190	1·612	1·646	1·711	1·871	2·024	2·171	2·313	2·451	2·532	2·612	2·717	2·845	2·970
(0·1863)	32·415	35·187	41·165	58·772	80·460	106·56	137·40	173·28	197·36	223·42	261·38	314·18	373·18
0·0200	1·654	1·688	1·756	1·919	2·076	2·227	2·373	2·515	2·598	2·680	2·787	2·919	3·047
(0·1961)	33·257	36·101	42·234	60·299	82·550	109·33	140·97	177·78	202·48	229·23	268·17	322·34	382·88
0·0210	1·695	1·730	1·799	1·967	2·127	2·282	2·432	2·577	2·662	2·746	2·856	2·991	3·122
(0·2059)	34·078	36·993	43·277	61·788	84·589	112·03	144·45	182·17	207·49	234·89	274·79	330·30	392·33
0·0220	1·735	1·771	1·842	2·013	2·178	2·336	2·489	2·638	2·725	2·811	2·923	3·061	3·196
(0·2157)	34·880	37·863	44·296	63·242	86·579	114·67	147·85	186·46	212·37	240·42	281·26	338·07	401·56
0·0230	1·774	1·811	1·883	2·058	2·226	2·388	2·545	2·697	2·786	2·874	2·989	3·130	3·267
(0·2256)	35·664	38·714	45·291	64·664	88·525	117·24	151·17	190·65	217·14	245·82	287·58	345·67	410·59
0·0240	1·812	1·850	1·923	2·103	2·274	2·440	2·600	2·755	2·846	2·936	3·053	3·197	3·338
(0·2354)	36·431	39·547	46·265	66·054	90·429	119·76	154·42	194·75	221·81	251·11	293·77	353·11	419·42
0·0250	1·849	1·888	1·963	2·146	2·321	2·490	2·653	2·812	2·905	2·996	3·116	3·263	3·406
(0·2452)	37·183	40·362	47·219	67·416	92·294	122·23	157·61	198·77	226·38	256·28	299·82	360·39	428·07
0·0260	1·886	1·925	2·002	2·188	2·367	2·539	2·706	2·868	2·962	3·056	3·178	3·328	3·474
(0·2550)	37·919	41·162	48·155	68·752	94·122	124·65	160·73	202·70	230·87	261·36	305·76	367·52	436·55
0·0270	1·922	1·962	2·040	2·230	2·412	2·588	2·758	2·922	3·019	3·114	3·239	3·391	3·540
(0·2648)	38·641	41·946	49·072	70·061	95·915	127·03	163·79	206·56	235·27	266·34	311·59	374·53	444·86
0·0280	1·957	1·998	2·078	2·271	2·457	2·635	2·808	2·976	3·074	3·171	3·298	3·453	3·605
(0·2746)	39·350	42·715	49·972	71·347	97·675	129·36	166·79	210·35	239·58	271·23	317·30	381·40	453·02
0·0290	1·992	2·033	2·114	2·311	2·500	2·682	2·858	3·029	3·129	3·227	3·356	3·514	3·669
(0·2844)	40·047	43·472	50·857	72·610	99·403	131·65	169·75	214·08	243·82	276·03	322·92	388·15	461·04
S, ($\Delta p/\rho l$)	160	165	175	200	225	250	275	300	315	330	350	375	400

Gradient **(Equivalent) Pipe diameters in mm**

S = 0·0100 to 0·0290;

$\Delta p/\rho l$ = (0·0981) to (0·2844) ms^{-2}

m = Manning $n \times 100$
S = 0·0300 to 0·0950

i.e. kin. pr. grad., $\Delta p/\rho l$ =
(0·2942) to (0·9316) ms^{-2}

Full bore flow; Tables E give m_C or m_P for Colebrook-White or for laminar solutions resp., to divide mV and/or mQ as follows:

mV to give velocities in ms^{-1}
mQ to give discharges in litres/sec

Gradient $S, (\Delta p/\rho l)$	\(Equivalent\) Pipe diameters in mm												
	160	165	175	200	225	250	275	300	315	330	350	375	400
0·0300 (0·2942)	2·026 / 40·731	2·068 / 44·215	2·151 / 51·726	2·351 / 73·851	2·543 / 101·10	2·728 / 133·90	2·907 / 172·65	3·080 / 217·74	3·182 / 247·99	3·282 / 280·75	3·414 / 328·44	3·574 / 394·78	3·732 / 468·92
0·0320 (0·3138)	2·092 / 42·067	2·136 / 45·665	2·221 / 53·423	2·428 / 76·273	2·626 / 104·42	2·817 / 138·29	3·002 / 178·31	3·181 / 224·88	3·287 / 256·13	3·390 / 289·95	3·526 / 339·21	3·692 / 407·73	3·854 / 484·30
0·0340 (0·3334)	2·157 / 43·362	2·201 / 47·070	2·289 / 55·067	2·503 / 78·620	2·707 / 107·63	2·904 / 142·55	3·094 / 183·80	3·279 / 231·80	3·388 / 264·01	3·494 / 298·88	3·634 / 349·65	3·805 / 420·28	3·973 / 499·21
0·0360 (0·3530)	2·219 / 44·619	2·265 / 48·435	2·356 / 56·663	2·575 / 80·900	2·785 / 110·75	2·988 / 146·68	3·184 / 189·13	3·374 / 238·52	3·486 / 271·66	3·596 / 307·54	3·740 / 359·79	3·916 / 432·46	4·088 / 513·68
0·0380 (0·3727)	2·280 / 45·842	2·327 / 49·762	2·420 / 58·216	2·646 / 83·117	2·862 / 113·79	3·070 / 150·70	3·271 / 194·31	3·467 / 245·06	3·581 / 279·11	3·694 / 315·97	3·842 / 369·65	4·023 / 444·32	4·200 / 527·76
0·0400 (0·3923)	2·339 / 47·033	2·388 / 51·055	2·483 / 59·728	2·714 / 85·276	2·936 / 116·74	3·150 / 154·62	3·356 / 199·36	3·557 / 251·42	3·674 / 286·36	3·790 / 324·18	3·942 / 379·25	4·127 / 455·86	4·309 / 541·47
0·0420 (0·4119)	2·397 / 48·194	2·447 / 52·316	2·545 / 61·203	2·781 / 87·382	3·009 / 119·63	3·228 / 158·43	3·439 / 204·28	3·645 / 257·63	3·765 / 293·43	3·884 / 332·18	4·039 / 388·62	4·229 / 467·12	4·415 / 554·84
0·0440 (0·4315)	2·453 / 49·328	2·504 / 53·547	2·604 / 62·644	2·847 / 89·438	3·079 / 122·44	3·304 / 162·16	3·520 / 209·09	3·730 / 263·69	3·854 / 300·33	3·975 / 340·00	4·134 / 397·76	4·329 / 478·11	4·519 / 567·90
0·0460 (0·4511)	2·509 / 50·437	2·561 / 54·750	2·663 / 64·052	2·911 / 91·448	3·149 / 125·19	3·378 / 165·81	3·599 / 213·79	3·814 / 269·62	3·940 / 307·08	4·065 / 347·64	4·227 / 406·70	4·426 / 488·85	4·621 / 580·66
0·0480 (0·4707)	2·562 / 51·522	2·616 / 55·928	2·720 / 65·429	2·973 / 93·415	3·216 / 127·89	3·450 / 169·37	3·677 / 218·39	3·896 / 275·42	4·025 / 313·69	4·152 / 355·12	4·318 / 415·45	4·521 / 499·37	4·720 / 593·15
0·0500 (0·4903)	2·615 / 52·584	2·670 / 57·081	2·776 / 66·778	3·035 / 95·341	3·283 / 130·52	3·522 / 172·87	3·753 / 222·89	3·977 / 281·10	4·108 / 320·16	4·238 / 362·44	4·407 / 424·02	4·615 / 509·66	4·817 / 605·38
0·0525 (0·5148)	2·680 / 53·883	2·735 / 58·491	2·845 / 68·427	3·110 / 97·696	3·364 / 133·75	3·609 / 177·13	3·845 / 228·39	4·075 / 288·04	4·210 / 328·06	4·342 / 371·39	4·516 / 434·49	4·729 / 522·25	4·936 / 620·33
0·0550 (0·5394)	2·743 / 55·151	2·800 / 59·867	2·912 / 70·038	3·183 / 99·995	3·443 / 136·89	3·693 / 181·30	3·936 / 233·77	4·171 / 294·82	4·309 / 335·78	4·444 / 380·13	4·622 / 444·71	4·840 / 534·54	5·053 / 634·93
0·0575 (0·5639)	2·805 / 56·390	2·863 / 61·213	2·977 / 71·612	3·254 / 102·24	3·520 / 139·97	3·776 / 185·38	4·024 / 239·02	4·265 / 301·44	4·406 / 343·33	4·544 / 388·68	4·726 / 454·71	4·949 / 546·55	5·166 / 649·20
0·0600 (0·5884)	2·865 / 57·603	2·924 / 62·529	3·041 / 73·152	3·324 / 104·44	3·596 / 142·98	3·858 / 189·36	4·111 / 244·16	4·356 / 307·93	4·500 / 350·71	4·642 / 397·03	4·828 / 464·49	5·055 / 558·31	5·277 / 663·16
0·0625 (0·6129)	2·924 / 58·791	2·985 / 63·818	3·104 / 74·660	3·393 / 106·59	3·670 / 145·93	3·937 / 193·27	4·196 / 249·20	4·446 / 314·28	4·593 / 357·95	4·738 / 405·22	4·927 / 474·06	5·159 / 569·82	5·386 / 676·83
0·0650 (0·6374)	2·982 / 59·955	3·044 / 65·082	3·165 / 76·139	3·460 / 108·71	3·743 / 148·82	4·015 / 197·10	4·279 / 254·13	4·534 / 320·50	4·684 / 365·03	4·832 / 413·25	5·025 / 483·45	5·261 / 581·11	5·493 / 690·24
0·0675 (0·6619)	3·039 / 61·097	3·102 / 66·322	3·226 / 77·589	3·526 / 110·78	3·814 / 151·65	4·092 / 200·85	4·360 / 258·97	4·621 / 326·61	4·773 / 371·99	4·924 / 421·12	5·121 / 492·66	5·362 / 592·18	5·597 / 703·39
0·0700 (0·6865)	3·094 / 62·218	3·159 / 67·539	3·285 / 79·013	3·591 / 112·81	3·884 / 154·44	4·167 / 204·54	4·440 / 263·73	4·705 / 332·60	4·861 / 378·81	5·014 / 428·85	5·215 / 501·70	5·460 / 603·04	5·700 / 716·29
0·0725 (0·7110)	3·149 / 63·320	3·215 / 68·735	3·343 / 80·412	3·654 / 114·81	3·953 / 157·17	4·241 / 208·16	4·519 / 268·39	4·789 / 338·49	4·947 / 385·52	5·103 / 436·44	5·307 / 510·58	5·557 / 613·72	5·801 / 728·97
0·0750 (0·7355)	3·203 / 64·402	3·269 / 69·910	3·400 / 81·786	3·717 / 116·77	4·020 / 159·86	4·313 / 211·72	4·596 / 272·98	4·870 / 344·27	5·031 / 392·11	5·190 / 443·90	5·398 / 519·31	5·652 / 624·21	5·900 / 741·44
0·0800 (0·7845)	3·308 / 66·514	3·377 / 72·202	3·512 / 84·469	3·839 / 120·60	4·152 / 165·10	4·454 / 218·66	4·747 / 281·93	5·030 / 355·56	5·197 / 404·97	5·360 / 458·46	5·575 / 536·34	5·837 / 644·68	6·094 / 765·75
0·0850 (0·8336)	3·410 / 68·561	3·481 / 74·424	3·620 / 87·068	3·957 / 124·31	4·280 / 170·18	4·592 / 225·39	4·893 / 290·61	5·185 / 366·51	5·356 / 417·43	5·525 / 472·57	5·746 / 552·85	6·017 / 664·52	6·281 / 789·32
0·0900 (0·8826)	3·509 / 70·549	3·582 / 76·582	3·725 / 89·593	4·072 / 127·91	4·404 / 175·12	4·725 / 231·92	5·035 / 299·04	5·335 / 377·13	5·512 / 429·54	5·685 / 486·27	5·913 / 568·88	6·191 / 683·79	6·463 / 812·20
0·0950 (0·9316)	3·605 / 72·482	3·680 / 78·681	3·827 / 92·048	4·183 / 131·42	4·525 / 179·91	4·854 / 238·28	5·173 / 307·23	5·482 / 387·47	5·663 / 441·31	5·841 / 499·59	6·075 / 584·47	6·361 / 702·52	6·640 / 834·46
$S, (\Delta p/\rho l)$	160	165	175	200	225	250	275	300	315	330	350	375	400
Gradient	\(Equivalent\) Pipe diameters in mm												

S = 0·0300 to 0·0950;

$\Delta p/\rho l$ = (0·2942) to (0·9316) ms^{-2}

m = Manning $n \times 100$
S = 0·1000 to 0·2900

i.e. kin. pr. grad., $\Delta p/\rho l =$
(0·9807) to (2·8439) ms^{-2}

Full bore flow; Tables E give m_C or m_P for Colebrook-White or for laminar solutions resp., to divide mV and/or mQ as follows:

mV to give velocities in ms^{-1}
mQ to give discharges in litres/sec

Gradient — **(Equivalent) Pipe diameters in mm**

$S,(\Delta p/\rho l)$	160	165	175	200	225	250	275	300	315	330	350	375	400
0·1000	3·699	3·775	3·926	4·292	4·642	4·980	5·307	5·624	5·810	5·993	6·233	6·526	6·813
(0·9807)	74·365	80·725	94·439	134·83	184·59	244·47	315·21	397·53	452·77	512·57	599·65	720·77	856·14
0·1050	3·790	3·869	4·023	4·398	4·757	5·103	5·438	5·763	5·953	6·141	6·387	6·687	6·981
(1·0297)	76·202	82·718	96·771	138·16	189·15	250·51	323·00	407·35	463·95	525·23	614·46	738·57	877·28
0·1100	3·879	3·960	4·118	4·501	4·869	5·223	5·566	5·898	6·093	6·285	6·537	6·845	7·145
(1·0787)	77·995	84·665	99·048	141·41	193·60	256·40	330·60	416·94	474·87	537·59	628·92	755·95	897·92
0·1150	3·966	4·049	4·211	4·603	4·978	5·341	5·691	6·031	6·230	6·427	6·684	6·998	7·306
(1·1278)	79·748	86·568	101·27	144·59	197·95	262·16	338·03	426·31	485·54	549·67	643·05	772·94	918·10
0·1200	4·052	4·136	4·301	4·702	5·086	5·456	5·814	6·161	6·364	6·565	6·828	7·149	7·463
(1·1768)	81·463	88·429	103·45	147·70	202·21	267·80	345·30	435·47	495·98	561·49	656·88	789·57	937·85
0·1250	4·135	4·221	4·390	4·798	5·190	5·568	5·933	6·288	6·496	6·700	6·968	7·296	7·617
(1·2258)	83·143	90·253	105·59	150·75	206·38	273·32	352·42	444·45	506·21	573·07	670·43	805·85	957·19
0·1300	4·217	4·304	4·477	4·893	5·293	5·678	6·051	6·412	6·624	6·833	7·106	7·441	7·768
(1·2749)	84·789	92·040	107·68	153·73	210·46	278·74	359·40	453·26	516·24	584·42	683·71	821·81	976·14
0·1350	4·297	4·386	4·562	4·987	5·394	5·787	6·166	6·534	6·750	6·963	7·242	7·583	7·916
(1·3239)	86·404	93·794	109·73	156·66	214·47	284·05	366·24	461·89	526·07	595·55	696·73	837·46	994·74
0·1400	4·376	4·467	4·646	5·078	5·493	5·893	6·279	6·654	6·874	7·091	7·375	7·722	8·061
(1·3729)	87·990	95·515	111·74	159·54	218·41	289·26	372·96	470·37	535·72	606·48	709·51	852·83	1013·0
0·1450	4·454	4·546	4·728	5·168	5·590	5·997	6·390	6·772	6·996	7·216	7·505	7·858	8·204
(1·4220)	89·547	97·205	113·72	162·36	222·27	294·38	379·57	478·69	545·21	617·22	722·07	867·93	1030·9
0·1500	4·530	4·624	4·809	5·256	5·686	6·100	6·500	6·888	7·116	7·340	7·633	7·993	8·344
(1·4710)	91·078	98·867	115·66	165·14	226·07	299·41	386·05	486·88	554·53	627·77	734·42	882·76	1048·5
0·1600	4·678	4·775	4·966	5·429	5·872	6·300	6·713	7·114	7·349	7·580	7·884	8·255	8·618
(1·5691)	94·065	102·11	119·46	170·55	233·49	309·23	398·72	502·84	572·71	648·36	758·50	911·72	1082·9
0·1700	4·822	4·922	5·119	5·596	6·053	6·493	6·919	7·333	7·575	7·814	8·126	8·509	8·883
(1·6671)	96·960	105·25	123·13	175·80	240·67	318·75	410·99	518·32	590·34	668·31	781·85	939·77	1116·3
0·1800	4·962	5·065	5·268	5·758	6·229	6·682	7·120	7·545	7·795	8·040	8·362	8·756	9·140
(1·7652)	99·771	108·30	126·70	180·90	247·65	327·99	422·90	533·35	607·45	687·68	804·51	967·02	1148·6
0·1900	5·098	5·204	5·412	5·916	6·399	6·865	7·315	7·752	8·008	8·261	8·591	8·995	9·391
(1·8633)	102·51	111·27	130·17	185·85	254·44	336·98	434·49	547·96	624·10	706·53	826·56	993·52	1180·1
0·2000	5·231	5·339	5·553	6·070	6·565	7·043	7·505	7·953	8·216	8·475	8·814	9·229	9·635
(1·9613)	105·17	114·16	133·56	190·68	261·05	345·73	445·78	562·20	640·31	724·88	848·03	1019·3	1210·8
0·2100	5·360	5·471	5·690	6·220	6·728	7·217	7·691	8·150	8·419	8·685	9·032	9·457	9·873
(2·0594)	107·77	116·98	136·85	195·39	267·49	354·27	456·79	576·08	656·13	742·78	868·97	1044·5	1240·7
0·2200	5·486	5·600	5·824	6·366	6·886	7·387	7·872	8·342	8·617	8·889	9·244	9·680	10·11
(2·1575)	110·30	119·73	140·08	199·99	273·79	362·61	467·54	589·64	671·57	760·26	889·42	1069·1	1269·9
0·2300	5·609	5·725	5·955	6·509	7·041	7·553	8·048	8·529	8·811	9·089	9·452	9·897	10·33
(2·2555)	112·78	122·43	143·22	204·48	279·94	370·75	478·04	602·89	686·66	777·35	909·41	1093·1	1298·4
0·2400	5·730	5·849	6·083	6·649	7·192	7·715	8·222	8·713	9·001	9·284	9·656	10·11	10·55
(2·3536)	115·21	125·06	146·30	208·88	285·96	378·73	488·32	615·85	701·43	794·07	928·97	1116·6	1326·3
0·2500	5·848	5·969	6·208	6·786	7·340	7·875	8·391	8·892	9·186	9·476	9·855	10·32	10·77
(2·4517)	117·58	127·64	149·32	213·19	291·86	386·54	498·39	628·55	715·89	810·44	948·13	1139·6	1353·7
0·2600	5·964	6·087	6·331	6·920	7·486	8·030	8·557	9·068	9·368	9·663	10·05	10·52	10·99
(2·5497)	119·91	130·16	152·28	217·41	297·64	394·19	508·26	641·00	730·07	826·49	966·91	1162·2	1380·5
0·2700	6·077	6·203	6·452	7·052	7·628	8·183	8·720	9·241	9·547	9·847	10·24	10·72	11·19
(2·6478)	122·19	132·64	155·18	221·55	303·31	401·70	517·95	653·21	743·98	842·24	985·32	1184·4	1406·8
0·2800	6·189	6·317	6·570	7·182	7·768	8·334	8·880	9·411	9·722	10·03	10·43	10·92	11·40
(2·7459)	124·44	135·08	158·03	225·62	308·87	409·07	527·45	665·20	757·63	857·69	1003·4	1206·1	1432·6
0·2900	6·299	6·429	6·686	7·309	7·906	8·481	9·037	9·577	9·894	10·21	10·61	11·11	11·60
(2·8439)	126·64	137·47	160·82	229·61	314·34	416·31	536·79	676·97	771·04	872·87	1021·2	1227·4	1457·9
$S,(\Delta p/\rho l)$	160	165	175	200	225	250	275	300	315	330	350	375	400

Gradient — **(Equivalent) Pipe diameters in mm**

S = 0·1000 to 0·2900;

$\Delta p/\rho l$ = (0·9807) to (2·8439) ms·$^{-2}$

m = Manning $n \times 100$
S = 0·3000 to 1·0000

i.e. kin. pr. grad., $\Delta p/\rho l =$
(2·9420) to (9·8067) ms^{-2}

Full bore flow; Tables E give m_C or m_P for Colebrook-White or for laminar solutions resp., to divide mV and/or mQ as follows:

mV to give velocities in ms^{-1}
mQ to give discharges in litres/sec

Gradient S, ($\Delta p/\rho l$)	160	165	175	200	225	250	275	300	315	330	350	375	400
0·3000	6·406	6·539	6·801	7·434	8·041	8·626	9·192	9·741	10·06	10·38	10·80	11·30	11·80
(2·9420)	128·80	139·82	163·57	233·54	319·72	423·43	545·96	688·55	784·22	887·80	1038·6	1248·4	1482·9
0·3200	6·616	6·753	7·024	7·678	8·305	8·909	9·493	10·06	10·39	10·72	11·15	11·67	12·19
(3·1381)	133·03	144·40	168·94	241·20	330·20	437·32	563·87	711·13	809·94	916·91	1072·7	1289·4	1531·5
0·3400	6·820	6·961	7·240	7·914	8·560	9·183	9·786	10·37	10·71	11·05	11·49	12·03	12·56
(3·3343)	137·12	148·85	174·14	248·62	340·36	450·78	581·22	733·01	834·87	945·13	1105·7	1329·0	1578·6
0·3600	7·018	7·163	7·450	8·143	8·808	9·449	10·07	10·67	11·02	11·37	11·83	12·38	12·93
(3·5304)	141·10	153·16	179·19	255·83	350·23	463·85	598·07	754·26	859·07	972·53	1137·8	1367·6	1624·4
0·3800	7·210	7·359	7·654	8·366	9·050	9·708	10·35	10·96	11·33	11·68	12·15	12·72	13·28
(3·7265)	144·96	157·36	184·10	262·84	359·83	476·56	614·46	774·93	882·61	999·18	1168·9	1405·0	1668·9
0·4000	7·397	7·551	7·853	8·584	9·285	9·961	10·61	11·25	11·62	11·99	12·47	13·05	13·63
(3·9227)	148·73	161·45	188·88	269·67	369·18	488·94	630·42	795·06	905·54	1025·1	1199·3	1441·5	1712·3
0·4200	7·580	7·737	8·047	8·796	9·514	10·21	10·88	11·53	11·91	12·28	12·77	13·37	13·96
(4·1188)	152·40	165·44	193·54	276·33	378·29	501·01	645·99	814·70	927·90	1050·5	1228·9	1477·1	1754·6
0·4400	7·758	7·919	8·236	9·003	9·738	10·45	11·13	11·80	12·19	12·57	13·07	13·69	14·29
(4·3149)	155·99	169·33	198·10	282·83	387·19	512·80	661·19	833·87	949·74	1075·2	1257·8	1511·9	1795·8
0·4600	7·933	8·097	8·421	9·205	9·957	10·68	11·38	12·06	12·46	12·85	13·37	14·00	14·61
(4·5111)	159·50	173·14	202·55	289·18	395·90	524·33	676·05	852·61	971·08	1099·3	1286·1	1545·9	1836·2
0·4800	8·103	8·271	8·602	9·403	10·17	10·91	11·63	12·32	12·73	13·13	13·66	14·30	14·93
(4·7072)	162·93	176·86	206·91	295·40	404·41	535·60	690·60	870·95	991·97	1123·0	1313·8	1579·1	1875·7
0·5000	8·270	8·442	8·780	9·597	10·38	11·14	11·87	12·58	12·99	13·40	13·94	14·59	15·23
(4·9033)	166·29	180·51	211·17	301·50	412·75	546·65	704·84	888·91	1012·4	1146·1	1340·9	1611·7	1914·4
0·5250	8·475	8·650	8·996	9·834	10·64	11·41	12·16	12·89	13·31	13·73	14·28	14·95	15·61
(5·1485)	170·39	184·96	216·39	308·94	422·94	560·15	722·24	910·86	1037·4	1174·4	1374·0	1651·5	1961·7
0·5500	8·674	8·854	9·208	10·07	10·89	11·68	12·45	13·19	13·63	14·05	14·62	15·30	15·98
(5·3937)	174·40	189·32	221·48	316·21	432·90	573·33	739·24	932·30	1061·8	1202·1	1406·3	1690·4	2007·8
0·5750	8·869	9·053	9·415	10·29	11·13	11·94	12·73	13·49	13·93	14·37	14·95	15·65	16·34
(5·6388)	178·32	193·57	226·46	323·32	442·63	586·21	755·85	953·25	1085·7	1229·1	1437·9	1728·4	2052·9
0·6000	9·060	9·248	9·617	10·51	11·37	12·20	13·00	13·78	14·23	14·68	15·27	15·99	16·69
(5·8840)	182·16	197·73	231·33	330·27	452·15	598·82	772·11	973·75	1109·1	1255·5	1468·8	1765·5	2097·1
0·6250	9·247	9·438	9·816	10·73	11·61	12·45	13·27	14·06	14·52	14·98	15·58	16·32	17·03
(6·1292)	185·91	201·81	236·10	337·08	461·47	611·17	788·03	993·83	1131·9	1281·4	1499·1	1801·9	2140·3
0·6500	9·430	9·625	10·01	10·94	11·84	12·70	13·53	14·34	14·81	15·28	15·89	16·64	17·37
(6·3743)	189·59	205·81	240·77	343·76	470·61	623·27	803·64	1013·5	1154·3	1306·8	1528·8	1837·6	2182·7
0·6750	9·609	9·808	10·20	11·15	12·06	12·94	13·79	14·61	15·09	15·57	16·19	16·96	17·70
(6·6195)	193·21	209·73	245·36	350·31	479·57	635·15	818·95	1032·8	1176·3	1331·7	1557·9	1872·6	2224·3
0·7000	9·786	9·988	10·39	11·36	12·28	13·18	14·04	14·88	15·37	15·86	16·49	17·27	18·03
(6·8647)	196·75	213·58	249·86	356·73	488·37	646·80	833·97	1051·8	1197·9	1356·1	1586·5	1907·0	2265·1
0·7250	9·959	10·17	10·57	11·56	12·50	13·41	14·29	15·14	15·64	16·14	16·78	17·57	18·34
(7·1098)	200·23	217·36	254·28	363·05	497·02	658·25	848·73	1070·4	1219·1	1380·1	1614·6	1940·7	2305·2
0·7500	10·13	10·34	10·75	11·75	12·71	13·64	14·53	15·40	15·91	16·41	17·07	17·87	18·66
(7·3550)	203·66	221·07	258·63	369·26	505·51	669·50	863·24	1088·7	1240·0	1403·7	1642·2	1973·9	2344·6
0·8000	10·46	10·68	11·11	12·14	13·13	14·09	15·01	15·91	16·43	16·95	17·63	18·46	19·27
(7·8453)	210·34	228·32	267·11	381·37	522·09	691·46	891·55	1124·4	1280·6	1449·8	1696·1	2038·7	2421·5
0·8500	10·78	11·01	11·45	12·51	13·53	14·52	15·47	16·40	16·94	17·47	18·17	19·03	19·86
(8·3357)	216·81	235·35	275·33	393·10	538·16	712·74	918·99	1159·0	1320·0	1494·4	1748·3	2101·4	2496·0
0·9000	11·10	11·33	11·78	12·88	13·93	14·94	15·92	16·87	17·43	17·98	18·70	19·58	20·44
(8·8260)	223·10	242·17	283·32	404·50	553·76	733·40	945·64	1192·6	1358·3	1537·7	1798·9	2162·3	2568·4
0·9500	11·40	11·64	12·10	13·23	14·31	15·35	16·36	17·33	17·91	18·47	19·21	20·11	21·00
(9·3163)	229·21	248·81	291·08	415·58	568·94	753·50	971·55	1225·3	1395·5	1579·8	1848·2	2221·6	2638·8
1·0000	11·70	11·94	12·42	13·57	14·68	15·75	16·78	17·78	18·37	18·95	19·71	20·64	21·54
(9·8067)	235·16	255·27	298·64	426·38	583·72	773·08	996·79	1257·1	1431·8	1620·9	1896·3	2279·3	2707·3
S, ($\Delta p/\rho l$)	160	165	175	200	225	250	275	300	315	330	350	375	400

Gradient (Equivalent) **Pipe diameters in mm**

S = 0·3000 to 1·0000; $\Delta p/\rho l$ = (2·9420) to (9·8067) ms^{-2}

m = Manning $n \times 100$
S = 0.000100 to 0.000290

i.e. kin. pr. grad., $\Delta p/\rho l$ =
(0.000981) to (0.002844) ms^{-2}

Full bore flow; Tables E give m_C or m_P for Colebrook-White or for laminar solutions resp., to divide mV and/or mQ as follows:

mV to give velocities in ms^{-1}
mQ to give discharges in litres/sec

Gradient — **(Equivalent) Pipe diameters in mm**

S, $(\Delta p/\rho l)$	400	425	450	475	500	525	550	575	600	625	650	675	700
0.000100	0.215	0.224	0.233	0.242	0.250	0.258	0.266	0.274	0.282	0.290	0.298	0.305	0.313
(0.000981)	27.073	31.824	37.064	42.812	49.087	55.908	63.292	71.257	79.821	89.001	98.814	109.28	120.40
0.000105	0.221	0.230	0.239	0.248	0.256	0.265	0.273	0.281	0.289	0.297	0.305	0.313	0.321
(0.001030)	27.742	32.610	37.979	43.869	50.300	57.289	64.855	73.017	81.793	91.199	101.25	111.97	123.38
0.000110	0.226	0.235	0.244	0.253	0.262	0.271	0.279	0.288	0.296	0.304	0.312	0.320	0.328
(0.001079)	28.395	33.377	38.873	44.902	51.483	58.637	66.381	74.735	83.717	93.345	103.64	114.61	126.28
0.000115	0.231	0.241	0.250	0.259	0.268	0.277	0.286	0.294	0.303	0.311	0.319	0.327	0.336
(0.001128)	29.033	34.127	39.746	45.911	52.640	59.955	67.873	76.415	85.599	95.443	105.97	117.19	129.12
0.000120	0.236	0.246	0.255	0.265	0.274	0.283	0.292	0.301	0.309	0.318	0.326	0.335	0.343
(0.001177)	29.657	34.861	40.601	46.898	53.772	61.244	69.333	78.059	87.440	97.496	108.25	119.71	131.90
0.000125	0.241	0.251	0.261	0.270	0.280	0.289	0.298	0.307	0.316	0.324	0.333	0.341	0.350
(0.001226)	30.269	35.580	41.439	47.865	54.881	62.507	70.763	79.668	89.243	99.506	110.48	122.17	134.62
0.000130	0.246	0.256	0.266	0.275	0.285	0.294	0.304	0.313	0.322	0.331	0.340	0.348	0.357
(0.001275)	30.868	36.285	42.259	48.813	55.968	63.745	72.164	81.246	91.010	101.48	112.67	124.59	137.28
0.000135	0.250	0.261	0.271	0.281	0.290	0.300	0.310	0.319	0.328	0.337	0.346	0.355	0.364
(0.001324)	31.456	36.976	43.064	49.743	57.034	64.959	73.539	82.794	92.744	103.41	114.81	126.97	139.90
0.000140	0.255	0.265	0.276	0.286	0.296	0.306	0.315	0.325	0.334	0.343	0.352	0.361	0.370
(0.001373)	32.034	37.655	43.854	50.656	58.081	66.151	74.888	84.313	94.446	105.31	116.92	129.30	142.46
0.000145	0.259	0.270	0.281	0.291	0.301	0.311	0.321	0.330	0.340	0.349	0.359	0.368	0.377
(0.001422)	32.601	38.321	44.631	51.552	59.109	67.322	76.214	85.805	96.118	107.17	118.99	131.59	144.99
0.000150	0.264	0.275	0.285	0.296	0.306	0.316	0.326	0.336	0.346	0.355	0.365	0.374	0.383
(0.001471)	33.158	38.976	45.394	52.434	60.119	68.473	77.517	87.272	97.761	109.00	121.02	133.84	147.47
0.000160	0.273	0.284	0.295	0.306	0.316	0.327	0.337	0.347	0.357	0.367	0.377	0.386	0.396
(0.001569)	34.245	40.254	46.882	54.153	62.091	70.719	80.059	90.134	100.97	112.58	124.99	138.22	152.30
0.000170	0.281	0.292	0.304	0.315	0.326	0.337	0.347	0.358	0.368	0.378	0.388	0.398	0.408
(0.001667)	35.299	41.493	48.325	55.820	64.002	72.895	82.523	92.908	104.07	116.04	128.84	142.48	156.99
0.000180	0.289	0.301	0.313	0.324	0.335	0.346	0.357	0.368	0.379	0.389	0.400	0.410	0.420
(0.001765)	36.323	42.696	49.726	57.438	65.858	75.008	84.915	95.602	107.09	119.41	132.57	146.61	161.54
0.000190	0.297	0.309	0.321	0.333	0.345	0.356	0.367	0.378	0.389	0.400	0.410	0.421	0.431
(0.001863)	37.318	43.866	51.089	59.012	67.662	77.064	87.242	98.222	110.03	122.68	136.21	150.63	165.97
0.000200	0.305	0.317	0.330	0.342	0.354	0.365	0.377	0.388	0.399	0.410	0.421	0.432	0.442
(0.001961)	38.288	45.006	52.416	60.545	69.420	79.066	89.509	100.77	112.88	125.87	139.74	154.54	170.28
0.000210	0.312	0.325	0.338	0.350	0.362	0.374	0.386	0.398	0.409	0.420	0.432	0.443	0.453
(0.002059)	39.233	46.117	53.710	62.040	71.134	81.018	91.719	103.26	115.67	128.98	143.19	158.36	174.48
0.000220	0.320	0.333	0.346	0.358	0.371	0.383	0.395	0.407	0.419	0.430	0.442	0.453	0.464
(0.002157)	40.156	47.202	54.974	63.500	72.808	82.925	93.877	105.69	118.39	132.01	146.56	162.08	178.59
0.000230	0.327	0.340	0.353	0.366	0.379	0.392	0.404	0.416	0.428	0.440	0.452	0.463	0.474
(0.002256)	41.059	48.263	56.210	64.928	74.445	84.789	95.987	108.07	121.06	134.98	149.86	165.73	182.60
0.000240	0.334	0.348	0.361	0.374	0.387	0.400	0.413	0.425	0.437	0.449	0.461	0.473	0.485
(0.002354)	41.942	49.301	57.419	66.324	76.046	86.612	98.052	110.39	123.66	137.88	153.08	169.29	186.53
0.000250	0.341	0.355	0.368	0.382	0.395	0.408	0.421	0.434	0.446	0.459	0.471	0.483	0.495
(0.002452)	42.807	50.318	58.603	67.692	77.614	88.398	100.07	112.67	126.21	140.72	156.24	172.78	190.38
0.000260	0.347	0.362	0.376	0.390	0.403	0.416	0.430	0.442	0.455	0.468	0.480	0.492	0.504
(0.002550)	43.655	51.314	59.763	69.032	79.151	90.149	102.06	114.90	128.71	143.51	159.33	176.20	194.15
0.000270	0.354	0.369	0.383	0.397	0.411	0.424	0.438	0.451	0.464	0.477	0.489	0.502	0.514
(0.002648)	44.486	52.292	60.902	70.347	80.659	91.866	104.00	117.09	131.16	146.24	162.37	179.56	197.85
0.000280	0.361	0.375	0.390	0.404	0.418	0.432	0.446	0.459	0.472	0.485	0.498	0.511	0.524
(0.002746)	45.302	53.252	62.020	71.638	82.139	93.552	105.91	119.24	133.57	148.93	165.35	182.85	201.48
0.000290	0.367	0.382	0.397	0.411	0.426	0.440	0.454	0.467	0.481	0.494	0.507	0.520	0.533
(0.002844)	46.104	54.194	63.117	72.906	83.593	95.208	107.78	121.35	135.93	151.56	168.27	186.09	205.04
S, $(\Delta p/\rho l)$	400	425	450	475	500	525	550	575	600	625	650	675	700

Gradient — **(Equivalent) Pipe diameters in mm**

S = 0.000100 to 0.000290;

$\Delta p/\rho l$ = (0.000981) to (0.002844) ms^{-2}

m = Manning $n \times 100$
S = 0·000300 to 0·000950

Full bore flow; Tables E give m_C or m_P for Colebrook-White or for laminar solutions resp., to divide mV and/or mQ as follows:

i.e. kin. pr. grad., $\Delta p/\rho l$ = (0·002942) to (0·009316) ms^{-2}

mV to give velocities in ms^{-1}
mQ to give discharges in litres/sec

Gradient (Equivalent) Pipe diameters in mm

S, ($\Delta p/\rho l$)	400	425	450	475	500	525	550	575	600	625	650	675	700
0·000300	0·373	0·389	0·404	0·418	0·433	0·447	0·461	0·475	0·489	0·502	0·516	0·529	0·542
(0·002942)	46·892	55·121	64·196	74·153	85·022	96·836	109·63	123·42	138·25	154·15	171·15	189·27	208·55
0·000320	0·385	0·401	0·417	0·432	0·447	0·462	0·477	0·491	0·505	0·519	0·533	0·546	0·560
(0·003138)	48·430	56·928	66·302	76·584	87·810	100·01	113·22	127·47	142·79	159·21	176·76	195·48	215·39
0·000340	0·397	0·414	0·430	0·445	0·461	0·476	0·491	0·506	0·521	0·535	0·549	0·563	0·577
(0·003334)	49·921	58·680	68·342	78·941	90·513	103·09	116·70	131·39	147·18	164·11	182·20	201·50	222·02
0·000360	0·409	0·426	0·442	0·458	0·474	0·490	0·505	0·521	0·536	0·550	0·565	0·579	0·594
(0·003530)	51·368	60·382	70·324	81·230	93·137	106·08	120·09	135·20	151·45	168·87	187·49	207·34	228·45
0·000380	0·420	0·437	0·454	0·471	0·487	0·503	0·519	0·535	0·550	0·566	0·580	0·595	0·610
(0·003727)	52·776	62·036	72·251	83·456	95·689	108·98	123·38	138·91	155·60	173·50	192·62	213·02	234·71
0·000400	0·431	0·449	0·466	0·483	0·500	0·517	0·533	0·549	0·565	0·580	0·596	0·611	0·626
(0·003923)	54·147	63·648	74·127	85·624	98·175	111·82	126·58	142·51	159·64	178·00	197·63	218·55	240·81
0·000420	0·442	0·460	0·478	0·495	0·512	0·529	0·546	0·562	0·579	0·595	0·610	0·626	0·641
(0·004119)	55·484	65·220	75·958	87·738	100·60	114·58	129·71	146·03	163·59	182·40	202·51	223·95	246·76
0·000440	0·452	0·471	0·489	0·507	0·524	0·542	0·559	0·576	0·592	0·609	0·625	0·641	0·656
(0·004315)	56·790	66·754	77·746	89·803	102·97	117·27	132·76	149·47	167·43	186·69	207·27	229·22	252·56
0·000460	0·462	0·481	0·500	0·518	0·536	0·554	0·571	0·589	0·605	0·622	0·639	0·655	0·671
(0·004511)	58·066	68·255	79·493	91·822	105·28	119·91	135·75	152·83	171·20	190·89	211·93	234·37	258·24
0·000480	0·472	0·491	0·511	0·529	0·548	0·566	0·584	0·601	0·619	0·636	0·652	0·669	0·685
(0·004707)	59·315	69·723	81·203	93·796	107·54	122·49	138·67	156·12	174·88	194·99	216·49	239·41	263·79
0·000500	0·482	0·502	0·521	0·540	0·559	0·577	0·596	0·614	0·631	0·649	0·666	0·683	0·700
(0·004903)	60·538	71·160	82·877	95·731	109·76	125·01	141·53	159·34	178·49	199·01	220·95	244·35	269·23
0·000525	0·494	0·514	0·534	0·554	0·573	0·592	0·610	0·629	0·647	0·665	0·682	0·700	0·717
(0·005148)	62·033	72·918	84·924	98·095	112·47	128·10	145·02	163·27	182·89	203·93	226·41	250·38	275·88
0·000550	0·505	0·526	0·547	0·567	0·586	0·606	0·625	0·644	0·662	0·680	0·698	0·716	0·734
(0·005394)	63·493	74·634	86·922	100·40	115·12	131·12	148·43	167·11	187·20	208·73	231·74	256·28	282·37
0·000575	0·517	0·538	0·559	0·579	0·599	0·619	0·639	0·658	0·677	0·696	0·714	0·732	0·750
(0·005639)	64·920	76·311	88·876	102·66	117·71	134·06	151·77	170·87	191·40	213·42	236·95	262·04	288·72
0·000600	0·528	0·549	0·571	0·592	0·612	0·633	0·653	0·672	0·692	0·711	0·729	0·748	0·766
(0·005884)	66·316	77·952	90·787	104·87	120·24	136·95	155·03	174·54	195·52	218·01	242·04	267·67	294·93
0·000625	0·539	0·561	0·583	0·604	0·625	0·646	0·666	0·686	0·706	0·725	0·744	0·763	0·782
(0·006129)	67·683	79·560	92·659	107·03	122·72	139·77	158·23	178·14	199·55	222·50	247·03	273·19	301·01
0·000650	0·549	0·572	0·594	0·616	0·637	0·658	0·679	0·700	0·720	0·740	0·759	0·779	0·798
(0·006374)	69·024	81·135	94·494	109·15	125·15	142·54	161·36	181·67	203·51	226·91	251·93	278·60	306·97
0·000675	0·560	0·583	0·605	0·628	0·650	0·671	0·692	0·713	0·733	0·754	0·774	0·793	0·813
(0·006619)	70·339	82·681	96·294	111·23	127·53	145·25	164·44	185·13	207·38	231·23	256·73	283·91	312·82
0·000700	0·570	0·594	0·617	0·639	0·661	0·683	0·705	0·726	0·747	0·768	0·788	0·808	0·828
(0·006865)	71·629	84·198	98·061	113·27	129·87	147·92	167·46	188·53	211·19	235·48	261·44	289·12	318·56
0·000725	0·580	0·604	0·627	0·651	0·673	0·695	0·717	0·739	0·760	0·781	0·802	0·822	0·842
(0·007110)	72·897	85·688	99·797	115·27	132·17	150·54	170·42	191·87	214·93	239·64	266·06	294·24	324·20
0·000750	0·590	0·614	0·638	0·662	0·685	0·707	0·730	0·752	0·773	0·794	0·816	0·836	0·857
(0·007355)	74·144	87·153	101·50	117·25	134·43	153·11	173·33	195·15	218·60	243·74	270·61	299·27	329·74
0·000800	0·609	0·634	0·659	0·683	0·707	0·730	0·753	0·776	0·798	0·821	0·842	0·864	0·885
(0·007845)	76·575	90·012	104·83	121·09	138·84	158·13	179·02	201·55	225·77	251·73	279·49	309·08	340·56
0·000850	0·628	0·654	0·679	0·704	0·729	0·753	0·777	0·800	0·823	0·846	0·868	0·890	0·912
(0·008336)	78·932	92·782	108·06	124·82	143·11	163·00	184·53	207·75	232·72	259·48	288·09	318·59	351·04
0·000900	0·646	0·673	0·699	0·725	0·750	0·775	0·799	0·823	0·847	0·870	0·893	0·916	0·939
(0·008826)	81·220	95·472	111·19	128·44	147·26	167·72	189·88	213·77	239·46	267·00	296·44	327·83	361·21
0·000950	0·664	0·691	0·718	0·745	0·771	0·796	0·821	0·846	0·870	0·894	0·918	0·941	0·964
(0·009316)	83·446	98·088	114·24	131·96	151·30	172·32	195·08	219·63	246·03	274·32	304·56	336·81	371·11
S, ($\Delta p/\rho l$)	400	425	450	475	500	525	550	575	600	625	650	675	700

Gradient (Equivalent) Pipe diameters in mm

S = 0·000300 to 0·000950;

$\Delta p/\rho l$ = (0·002942) to (0·009316) ms^{-2}

m = Manning $n \times 100$
S = 0·00100 to 0·00290

i.e. kin. pr. grad., $\Delta p/\rho l$ =
(0·00981) to (0·02844) ms^{-2}

Full bore flow; Tables E give m_C or m_P for Colebrook-White or for laminar solutions resp., to divide mV and/or mQ as follows:

mV to give velocities in ms^{-1}
mQ to give discharges in litres/sec

Gradient S, $(\Delta p/\rho l)$	(Equivalent) Pipe diameters in mm												
	400	425	450	475	500	525	550	575	600	625	650	675	700
0·00100	0·681	0·709	0·737	0·764	0·791	0·817	0·842	0·868	0·893	0·917	0·942	0·966	0·989
(0·00981)	85·614	100·64	117·21	135·38	155·23	176·80	200·15	225·34	252·42	281·45	312·48	345·56	380·75
0·00105	0·698	0·727	0·755	0·783	0·810	0·837	0·863	0·889	0·915	0·940	0·965	0·990	1·014
(0·01030)	87·728	103·12	120·10	138·73	159·06	181·16	205·09	230·90	258·65	288·40	320·19	354·10	390·16
0·00110	0·715	0·744	0·773	0·801	0·829	0·857	0·884	0·910	0·936	0·962	0·988	1·013	1·038
(0·01079)	89·792	105·55	122·93	141·99	162·80	185·43	209·92	236·33	264·74	295·18	327·73	362·43	399·34
0·00115	0·731	0·761	0·790	0·819	0·848	0·876	0·903	0·931	0·957	0·984	1·010	1·036	1·061
(0·01128)	91·810	107·92	125·69	145·18	166·46	189·59	214·63	241·65	270·69	301·82	335·09	370·57	408·31
0·00120	0·746	0·777	0·807	0·837	0·866	0·895	0·923	0·951	0·978	1·005	1·032	1·058	1·084
(0·01177)	93·785	110·24	128·39	148·31	170·04	193·67	219·25	246·84	276·51	308·31	342·30	378·54	417·09
0·00125	0·762	0·793	0·824	0·854	0·884	0·913	0·942	0·970	0·998	1·026	1·053	1·080	1·106
(0·01226)	95·719	112·51	131·04	151·36	173·55	197·66	223·77	251·93	282·21	314·67	349·36	386·35	425·70
0·00130	0·777	0·809	0·840	0·871	0·901	0·931	0·961	0·989	1·018	1·046	1·074	1·101	1·128
(0·01275)	97·614	114·74	133·64	154·36	176·99	201·58	228·20	256·92	287·80	320·90	356·28	394·00	434·13
0·00135	0·792	0·824	0·856	0·888	0·919	0·949	0·979	1·008	1·037	1·066	1·094	1·122	1·150
(0·01324)	99·474	116·93	136·18	157·30	180·36	205·42	232·55	261·82	293·28	327·01	363·07	401·51	442·40
0·00140	0·806	0·839	0·872	0·904	0·935	0·966	0·997	1·027	1·056	1·085	1·114	1·143	1·171
(0·01373)	101·30	119·07	138·68	160·19	183·67	209·19	236·82	266·62	298·66	333·01	369·73	408·87	450·51
0·00145	0·820	0·854	0·887	0·920	0·952	0·983	1·014	1·045	1·075	1·105	1·134	1·163	1·191
(0·01422)	103·09	121·18	141·13	163·02	186·92	212·89	241·01	271·34	303·95	338·91	376·27	416·11	458·49
0·00150	0·834	0·869	0·903	0·936	0·968	1·000	1·032	1·063	1·093	1·124	1·153	1·183	1·212
(0·01471)	104·85	123·25	143·55	165·81	190·11	216·53	245·13	275·98	309·15	344·70	382·70	423·23	466·33
0·00160	0·862	0·897	0·932	0·966	1·000	1·033	1·066	1·098	1·129	1·160	1·191	1·221	1·251
(0·01569)	108·29	127·30	148·25	171·25	196·35	223·63	253·17	285·03	319·29	356·00	395·26	437·11	481·62
0·00170	0·888	0·925	0·961	0·996	1·031	1·065	1·098	1·131	1·164	1·196	1·228	1·259	1·290
(0·01667)	111·63	131·21	152·82	176·52	202·39	230·51	260·96	293·80	329·11	366·96	407·42	450·56	496·44
0·00180	0·914	0·952	0·989	1·025	1·061	1·096	1·130	1·164	1·198	1·231	1·263	1·296	1·327
(0·01765)	114·86	135·02	157·25	181·64	208·26	237·20	268·53	302·32	338·65	377·60	419·23	463·62	510·83
0·00190	0·939	0·978	1·016	1·053	1·090	1·126	1·161	1·196	1·231	1·265	1·298	1·331	1·364
(0·01863)	118·01	138·72	161·56	186·61	213·97	243·70	275·88	310·60	347·93	387·95	430·72	476·32	524·83
0·00200	0·963	1·003	1·042	1·080	1·118	1·155	1·191	1·227	1·263	1·297	1·332	1·366	1·399
(0·01961)	121·08	142·32	165·75	191·46	219·53	250·03	283·05	318·67	356·97	398·03	441·91	488·70	538·47
0·00210	0·987	1·028	1·068	1·107	1·146	1·184	1·221	1·258	1·294	1·329	1·365	1·399	1·434
(0·02059)	124·07	145·84	169·85	196·19	224·95	256·20	290·04	326·54	365·79	407·85	452·82	500·77	551·76
0·00220	1·011	1·052	1·093	1·133	1·173	1·211	1·250	1·287	1·324	1·361	1·397	1·432	1·467
(0·02157)	126·99	149·27	173·84	200·81	230·24	262·23	296·87	334·23	374·40	417·45	463·48	512·55	564·75
0·00230	1·033	1·076	1·118	1·159	1·199	1·239	1·278	1·316	1·354	1·391	1·428	1·465	1·500
(0·02256)	129·84	152·62	177·75	205·32	235·41	268·13	303·54	341·74	382·81	426·83	473·89	524·07	577·44
0·00240	1·055	1·099	1·142	1·184	1·225	1·265	1·305	1·344	1·383	1·421	1·459	1·496	1·533
(0·02354)	132·63	155·90	181·57	209·74	240·48	273·89	310·07	349·09	391·04	436·02	484·09	535·34	589·86
0·00250	1·077	1·122	1·165	1·208	1·250	1·291	1·332	1·372	1·412	1·450	1·489	1·527	1·564
(0·02452)	135·37	159·12	185·32	214·06	245·44	279·54	316·46	356·29	399·11	445·01	494·07	546·38	602·02
0·00260	1·099	1·144	1·188	1·232	1·275	1·317	1·358	1·399	1·440	1·479	1·518	1·557	1·595
(0·02550)	138·05	162·27	188·99	218·30	250·30	285·08	322·73	363·34	407·01	453·82	503·85	557·20	613·95
0·00270	1·119	1·166	1·211	1·255	1·299	1·342	1·384	1·426	1·467	1·507	1·547	1·587	1·626
(0·02648)	140·68	165·36	192·59	222·46	255·07	290·51	328·88	370·26	414·76	462·46	513·45	567·82	625·64
0·00280	1·140	1·187	1·233	1·278	1·323	1·367	1·410	1·452	1·494	1·535	1·576	1·616	1·656
(0·02746)	143·26	168·40	196·12	226·54	259·75	295·84	334·91	377·06	422·38	470·95	522·87	578·24	637·12
0·00290	1·160	1·208	1·255	1·301	1·346	1·391	1·435	1·478	1·520	1·562	1·604	1·644	1·685
(0·02844)	145·79	171·38	199·59	230·55	264·34	301·07	340·84	383·73	429·85	479·29	532·13	588·47	648·40
S, $(\Delta p/\rho l)$	400	425	450	475	500	525	550	575	600	625	650	675	700
Gradient	(Equivalent) Pipe diameters in mm												

S = 0·00100 to 0·00290;

$\Delta p/\rho l$ = (0·00981) to (0·02844) ms^{-2}

m = Manning $n \times 100$
S = 0·00300 to 0·00950

i.e. kin. pr. grad., $\Delta p/\rho l$ =
(0·02942) to (0·09316) ms^{-2}

Full bore flow; Tables E give m_C or m_P for Colebrook-White or for laminar solutions resp., to divide mV and/or mQ as follows:

mV to give velocities in ms^{-1}
mQ to give discharges in litres/sec

Gradient (Equivalent) Pipe diameters in mm

S, ($\Delta p/\rho l$)	400	425	450	475	500	525	550	575	600	625	650	675	700
0·00300	1·180	1·229	1·276	1·323	1·369	1·415	1·459	1·503	1·546	1·589	1·631	1·673	1·714
(0·02942)	148·29	174·31	203·01	234·49	268·86	306·22	346·67	390·29	437·20	487·48	541·23	598·53	659·48
0·00320	1·219	1·269	1·318	1·367	1·414	1·461	1·507	1·552	1·597	1·641	1·685	1·727	1·770
(0·03138)	153·15	180·02	209·66	242·18	277·68	316·26	358·03	403·09	451·54	503·47	558·98	618·16	681·11
0·00340	1·256	1·308	1·359	1·409	1·458	1·506	1·553	1·600	1·646	1·692	1·736	1·781	1·824
(0·03334)	157·86	185·56	216·12	249·63	286·23	326·00	369·05	415·50	465·43	518·96	576·18	637·18	702·07
0·00360	1·293	1·346	1·398	1·450	1·500	1·550	1·598	1·646	1·694	1·741	1·787	1·832	1·877
(0·03530)	162·44	190·94	222·38	256·87	294·52	335·45	379·75	427·54	478·93	534·01	592·88	655·66	722·43
0·00380	1·328	1·383	1·437	1·489	1·541	1·592	1·642	1·692	1·740	1·788	1·836	1·882	1·929
(0·03727)	166·89	196·18	228·48	263·91	302·59	344·64	390·16	439·26	492·05	548·64	609·13	673·62	742·23
0·00400	1·363	1·419	1·474	1·528	1·581	1·633	1·685	1·736	1·785	1·835	1·883	1·931	1·979
(0·03923)	171·23	201·27	234·41	270·77	310·46	353·59	400·29	450·67	504·83	562·89	624·95	691·12	761·51
0·00420	1·396	1·454	1·510	1·566	1·620	1·674	1·726	1·778	1·830	1·880	1·930	1·979	2·028
(0·04119)	175·46	206·24	240·20	277·45	318·12	362·33	410·18	461·80	517·30	576·79	640·39	708·19	780·31
0·00440	1·429	1·488	1·546	1·603	1·658	1·713	1·767	1·820	1·873	1·924	1·975	2·026	2·075
(0·04315)	179·58	211·10	245·85	283·98	325·61	370·85	419·83	472·67	529·48	590·37	655·46	724·86	798·68
0·00460	1·461	1·521	1·581	1·639	1·696	1·752	1·807	1·861	1·915	1·968	2·020	2·071	2·122
(0·04511)	183·62	215·84	251·38	290·37	332·93	379·19	429·27	483·29	541·37	603·64	670·19	741·15	816·63
0·00480	1·493	1·554	1·615	1·674	1·732	1·789	1·846	1·901	1·956	2·010	2·063	2·116	2·168
(0·04707)	187·57	220·48	256·79	296·61	340·09	387·34	438·50	493·69	553·02	616·62	684·60	757·09	834·19
0·00500	1·523	1·586	1·648	1·708	1·768	1·826	1·884	1·940	1·996	2·051	2·106	2·159	2·212
(0·04903)	191·44	225·03	262·08	302·73	347·10	395·33	447·54	503·87	564·42	629·33	698·72	772·70	851·39
0·00525	1·561	1·625	1·689	1·751	1·811	1·871	1·930	1·988	2·046	2·102	2·158	2·213	2·267
(0·05148)	196·17	230·59	268·55	310·20	355·67	405·09	458·60	516·31	578·36	644·88	715·97	791·78	872·42
0·00550	1·598	1·664	1·728	1·792	1·854	1·915	1·976	2·035	2·094	2·151	2·208	2·265	2·320
(0·05394)	200·78	236·01	274·87	317·50	364·04	414·62	469·39	528·46	591·97	660·05	732·82	810·41	892·95
0·00575	1·634	1·701	1·767	1·832	1·896	1·958	2·020	2·081	2·141	2·200	2·258	2·316	2·372
(0·05639)	205·29	241·32	281·05	324·64	372·22	423·94	479·94	540·34	605·28	674·89	749·29	828·63	913·02
0·00600	1·669	1·738	1·805	1·871	1·936	2·001	2·064	2·126	2·187	2·247	2·307	2·365	2·423
(0·05884)	209·71	246·51	287·09	331·62	380·23	433·06	490·26	551·96	618·29	689·40	765·41	846·45	932·65
0·00625	1·703	1·773	1·842	1·910	1·976	2·042	2·106	2·169	2·232	2·293	2·354	2·414	2·473
(0·06129)	214·03	251·59	293·01	338·46	388·07	441·99	500·37	563·34	631·04	703·62	781·19	863·90	951·88
0·00650	1·737	1·809	1·879	1·948	2·016	2·082	2·148	2·212	2·276	2·339	2·401	2·462	2·522
(0·06374)	218·27	256·57	298·82	345·16	395·75	450·74	510·28	574·50	643·54	717·55	796·66	881·01	970·73
0·00675	1·770	1·843	1·915	1·985	2·054	2·122	2·189	2·255	2·319	2·383	2·447	2·509	2·570
(0·06619)	222·43	261·46	304·51	351·74	403·29	459·33	520·00	585·44	655·80	731·22	811·84	897·80	989·23
0·00700	1·803	1·877	1·950	2·021	2·092	2·161	2·229	2·296	2·362	2·427	2·491	2·555	2·618
(0·06865)	226·51	266·26	310·10	358·19	410·69	467·76	529·54	596·18	667·83	744·64	826·74	914·27	1007·4
0·00725	1·834	1·910	1·984	2·057	2·129	2·199	2·268	2·337	2·404	2·470	2·536	2·600	2·664
(0·07110)	230·52	270·97	315·59	364·53	417·96	476·04	538·91	606·74	679·65	757·82	841·37	930·45	1025·2
0·00750	1·866	1·943	2·018	2·092	2·165	2·237	2·307	2·376	2·445	2·512	2·579	2·645	2·709
(0·07355)	234·46	275·60	320·98	370·76	425·11	484·18	548·13	617·11	691·27	770·77	855·75	946·36	1042·7
0·00800	1·927	2·006	2·084	2·161	2·236	2·310	2·383	2·454	2·525	2·595	2·663	2·731	2·798
(0·07845)	242·15	284·64	331·51	382·92	439·05	500·06	566·10	637·35	713·94	796·05	883·82	977·40	1076·9
0·00850	1·986	2·068	2·149	2·227	2·305	2·381	2·456	2·530	2·603	2·675	2·745	2·815	2·884
(0·08336)	249·60	293·40	341·71	394·71	452·56	515·45	583·52	656·96	735·92	820·55	911·02	1007·5	1110·1
0·00900	2·044	2·128	2·211	2·292	2·372	2·450	2·527	2·603	2·678	2·752	2·825	2·897	2·968
(0·08826)	256·84	301·91	351·62	406·15	465·68	530·39	600·44	676·01	757·25	844·34	937·43	1036·7	1142·3
0·00950	2·100	2·186	2·271	2·355	2·437	2·517	2·597	2·675	2·752	2·828	2·902	2·976	3·049
(0·09316)	263·88	310·18	361·25	417·28	478·44	544·92	616·90	694·53	778·00	867·48	963·12	1065·1	1173·6
S, ($\Delta p/\rho l$)	400	425	450	475	500	525	550	575	600	625	650	675	700

Gradient (Equivalent) Pipe diameters in mm

S = 0·00300 to 0·00950; $\Delta p/\rho l$ = (0·02942) to (0·09316) ms^{-2}

m = Manning $n \times 100$
S = 0·0100 to 0·0290

i.e. kin. pr. grad., $\Delta p/\rho l$ =
(0·0981) to (0·2844) ms^{-2}

Full bore flow; Tables E give m_C or m_P for Colebrook-White or for laminar solutions resp., to divide mV and/or mQ as follows:

mV to give velocities in ms^{-1}
mQ to give discharges in litres/sec

Gradient (Equivalent) Pipe diameters in mm

S, ($\Delta p/\rho l$)	400	425	450	475	500	525	550	575	600	625	650	675	700
0·0100	2·154	2·243	2·330	2·416	2·500	2·583	2·664	2·744	2·823	2·901	2·978	3·054	3·129
(0·0981)	270·73	318·24	370·64	428·12	490·87	559·08	632·92	712·57	798·21	890·01	988·14	1092·8	1204·0
0·0105	2·208	2·299	2·388	2·476	2·562	2·646	2·730	2·812	2·893	2·973	3·051	3·129	3·206
(0·1030)	277·42	326·10	379·79	438·69	503·00	572·89	648·55	730·17	817·93	911·99	1012·5	1119·7	1233·8
0·0110	2·260	2·353	2·444	2·534	2·622	2·709	2·794	2·878	2·961	3·043	3·123	3·203	3·281
(0·1079)	283·95	333·77	388·73	449·02	514·83	586·37	663·81	747·35	837·17	933·45	1036·4	1146·1	1262·8
0·0115	2·310	2·406	2·499	2·591	2·681	2·770	2·857	2·943	3·027	3·111	3·193	3·275	3·355
(0·1128)	290·33	341·27	397·46	459·11	526·40	599·55	678·73	764·15	855·99	954·43	1059·7	1171·9	1291·2
0·0120	2·360	2·457	2·553	2·647	2·739	2·829	2·918	3·006	3·093	3·178	3·262	3·345	3·427
(0·1177)	296·57	348·61	406·01	468·98	537·72	612·44	693·33	780·59	874·40	974·96	1082·5	1197·1	1319·0
0·0125	2·409	2·508	2·605	2·701	2·795	2·887	2·978	3·068	3·156	3·243	3·329	3·414	3·498
(0·1226)	302·69	355·80	414·39	478·65	548·81	625·07	707·63	796·68	892·43	995·06	1104·8	1221·7	1346·2
0·0130	2·456	2·558	2·657	2·755	2·850	2·945	3·037	3·129	3·219	3·308	3·395	3·482	3·567
(0·1275)	308·68	362·85	422·59	488·13	559·68	637·45	721·64	812·46	910·10	1014·8	1126·7	1245·9	1372·8
0·0135	2·503	2·606	2·708	2·807	2·905	3·001	3·095	3·188	3·280	3·371	3·460	3·548	3·635
(0·1324)	314·56	369·76	430·64	497·43	570·34	649·59	735·39	827·94	927·44	1034·1	1148·1	1269·7	1399·0
0·0140	2·549	2·654	2·757	2·859	2·958	3·056	3·152	3·247	3·340	3·432	3·523	3·613	3·702
(0·1373)	320·34	376·55	438·54	506·56	580·81	661·51	748·88	843·13	944·46	1053·1	1169·2	1293·0	1424·6
0·0145	2·594	2·701	2·806	2·909	3·010	3·110	3·208	3·304	3·399	3·493	3·586	3·677	3·767
(0·1422)	326·01	383·21	446·31	515·52	591·09	673·22	762·14	858·05	961·18	1071·7	1189·9	1315·9	1449·9
0·0150	2·639	2·747	2·854	2·959	3·062	3·163	3·263	3·361	3·458	3·553	3·647	3·740	3·832
(0·1471)	331·58	389·76	453·94	524·34	601·19	684·73	775·17	872·72	977·61	1090·0	1210·2	1338·4	1474·7
0·0160	2·725	2·838	2·948	3·056	3·162	3·267	3·370	3·471	3·571	3·669	3·767	3·863	3·957
(0·1569)	342·45	402·54	468·82	541·53	620·91	707·19	800·59	901·34	1009·7	1125·8	1249·9	1382·2	1523·0
0·0170	2·809	2·925	3·038	3·150	3·260	3·367	3·473	3·578	3·681	3·782	3·883	3·982	4·079
(0·1667)	352·99	414·93	483·25	558·20	640·02	728·95	825·23	929·08	1040·7	1160·4	1288·4	1424·8	1569·9
0·0180	2·890	3·010	3·127	3·241	3·354	3·465	3·574	3·682	3·788	3·892	3·995	4·097	4·198
(0·1765)	363·23	426·96	497·26	574·38	658·58	750·08	849·15	956·02	1070·9	1194·1	1325·7	1466·1	1615·4
0·0190	2·970	3·092	3·212	3·330	3·446	3·560	3·672	3·783	3·891	3·999	4·105	4·209	4·313
(0·1863)	373·18	438·66	510·89	590·12	676·62	770·64	872·42	982·22	1100·3	1226·8	1362·1	1506·3	1659·7
0·0200	3·047	3·172	3·296	3·417	3·536	3·652	3·767	3·881	3·992	4·103	4·211	4·319	4·425
(0·1961)	382·88	450·06	524·16	605·45	694·20	[illegible]	895·09	1007·7	1128·8	1258·7	1397·4	1545·4	1702·8
0·0210	3·122	3·251	3·377	3·501	3·623	3·743	3·861	3·977	4·091	4·204	4·315	4·425	4·534
(0·2059)	392·33	461·17	537·10	620·40	[illegible]	[illegible]	917·19	1032·6	1156·7	1289·8	1431·9	1583·6	1744·8
0·0220	3·196	3·327	3·457	3·583	3·708	3·831	3·951	4·070	4·187	4·303	4·417	4·529	4·641
(0·2157)	401·56	472·02	549·74	635·00	[illegible]	[illegible]	938·77	1056·9	1183·9	1320·1	1465·6	1620·8	1785·9
0·0230	3·267	3·402	3·534	3·664	3·791	3·917	4·040	4·162	4·281	4·400	4·516	4·631	4·745
(0·2256)	410·59	482·63	562·10	649·28	744·45	[illegible]	959·87	1080·7	1210·6	1349·8	1498·6	1657·3	1826·0
0·0240	3·338	3·475	3·610	3·743	3·873	4·001	4·127	4·251	4·374	4·494	4·613	4·731	4·847
(0·2354)	419·42	493·01	574·19	663·24	760·46	866·12	980·52	1103·9	1236·6	1378·8	1530·8	1692·9	1865·3
0·0250	3·406	3·547	3·685	3·820	3·953	4·084	4·212	4·339	4·464	4·587	4·708	4·828	4·947
(0·2452)	428·07	503·18	586·03	676·92	776·14	883·98	1000·7	1126·7	1262·1	1407·2	1562·4	1727·8	1903·8
0·0260	3·474	3·617	3·758	3·896	4·031	4·164	4·296	4·425	4·552	4·678	4·802	4·924	5·045
(0·2550)	436·55	513·14	597·63	690·32	791·51	901·49	1020·6	1149·0	1287·1	1435·1	1593·3	1762·0	1941·5
0·0270	3·540	3·686	3·829	3·970	4·108	4·244	4·377	4·509	4·639	4·767	4·893	5·018	5·141
(0·2648)	444·86	522·92	609·02	703·47	806·59	918·66	1040·0	1170·9	1311·6	1462·4	1623·7	1795·6	1978·5
0·0280	3·605	3·754	3·900	4·043	4·183	4·322	4·458	4·592	4·724	4·854	4·983	5·110	5·235
(0·2746)	453·02	532·52	620·20	716·38	821·39	935·52	1059·1	1192·4	1335·7	1489·3	1653·5	1828·5	2014·8
0·0290	3·669	3·820	3·969	4·114	4·257	4·398	4·537	4·673	4·808	4·940	5·071	5·200	5·328
(0·2844)	461·04	541·94	631·17	729·06	835·93	952·08	1077·8	1213·5	1359·3	1515·6	1682·7	1860·9	2050·4
S, ($\Delta p/\rho l$)	400	425	450	475	500	525	550	575	600	625	650	675	700

Gradient (Equivalent) Pipe diameters in mm

S = 0·0100 to 0·0290; $\Delta p/\rho l$ = (0·0981) to (0·2844) ms^{-2}

m = Manning $n \times 100$
S = 0·0300 to 0·0950

Full bore flow; Tables E give m_C or m_P for Colebrook-White or for laminar solutions resp., to divide mV and/or mQ as follows:

i.e. kin. pr. grad., $\Delta p / \rho l =$
(0·2942) to (0·9316) ms^{-2}

mV to give velocities in ms^{-1}
mQ to give discharges in litres/sec

Gradient   **(Equivalent) Pipe diameters in mm**

S, ($\Delta p/\rho l$)	400	425	450	475	500	525	550	575	600	625	650	675	700
0·0300	3·732	3·885	4·036	4·185	4·330	4·473	4·614	4·753	4·890	5·025	5·158	5·289	5·419
(0·2942)	468·92	551·21	641·96	741·53	850·22	968·36	1096·3	1234·2	1382·5	1541·5	1711·5	1892·7	2085·5
0·0320	3·854	4·013	4·169	4·322	4·472	4·620	4·766	4·909	5·050	5·189	5·327	5·463	5·597
(0·3138)	484·30	569·28	663·02	765·84	878·10	1000·1	1132·2	1274·7	1427·9	1592·1	1767·6	1954·8	2153·9
0·0340	3·973	4·136	4·297	4·455	4·610	4·762	4·912	5·060	5·206	5·349	5·491	5·631	5·769
(0·3334)	499·21	586·80	683·42	789·41	905·13	1030·9	1167·0	1313·9	1471·8	1641·1	1822·0	2015·0	2220·2
0·0360	4·088	4·256	4·422	4·584	4·743	4·900	5·055	5·207	5·356	5·504	5·650	5·794	5·936
(0·3530)	513·68	603·82	703·24	812·30	931·37	1060·8	1200·9	1352·0	1514·5	1688·7	1874·9	2073·4	2284·5
0·0380	4·200	4·373	4·543	4·710	4·873	5·035	5·193	5·349	5·503	5·655	5·805	5·953	6·099
(0·3727)	527·76	620·36	722·51	834·56	956·89	1089·8	1233·8	1389·1	1556·0	1735·0	1926·2	2130·2	2347·1
0·0400	4·309	4·487	4·661	4·832	5·000	5·165	5·328	5·488	5·646	5·802	5·956	6·107	6·257
(0·3923)	541·47	636·48	741·27	856·24	981·75	1118·2	1265·8	1425·1	1596·4	1780·0	1976·3	2185·5	2408·1
0·0420	4·415	4·597	4·776	4·951	5·123	5·293	5·460	5·624	5·786	5·945	6·103	6·258	6·412
(0·4119)	554·84	652·20	759·58	877·38	1006·0	1145·8	1297·1	1460·3	1635·9	1824·0	2025·1	2239·5	2467·6
0·0440	4·519	4·706	4·888	5·068	5·244	5·417	5·588	5·756	5·922	6·085	6·246	6·406	6·563
(0·4315)	567·90	667·54	777·46	898·03	1029·7	1172·7	1327·6	1494·7	1674·3	1866·9	2072·7	2292·2	2525·6
0·0460	4·621	4·811	4·998	5·182	5·362	5·539	5·714	5·886	6·055	6·222	6·387	6·549	6·710
(0·4511)	580·66	682·55	794·93	918·22	1052·8	1199·1	1357·5	1528·3	1712·0	1908·9	2119·3	2343·7	2582·4
0·0480	4·720	4·915	5·106	5·293	5·477	5·658	5·837	6·012	6·185	6·356	6·524	6·690	6·855
(0·4707)	593·15	697·23	812·03	937·96	1075·4	1224·9	1386·7	1561·2	1748·8	1949·9	2164·9	2394·1	2637·9
0·0500	4·817	5·016	5·211	5·402	5·590	5·775	5·957	6·136	6·313	6·487	6·659	6·828	6·996
(0·4903)	605·38	711·60	828·77	957·31	1097·6	1250·1	1415·3	1593·4	1784·9	1990·1	2209·5	2443·5	2692·3
0·0525	4·936	5·140	5·340	5·536	5·728	5·918	6·104	6·288	6·469	6·647	6·823	6·997	7·169
(0·5148)	620·33	729·18	849·24	980·95	1124·7	1281·0	1450·2	1632·7	1828·9	2039·3	2264·1	2503·8	2758·8
0·0550	5·053	5·261	5·465	5·666	5·863	6·057	6·248	6·436	6·621	6·803	6·984	7·162	7·337
(0·5394)	634·93	746·34	869·22	1004·0	1151·2	1311·2	1484·3	1671·1	1872·0	2087·3	2317·4	2562·8	2823·7
0·0575	5·166	5·379	5·588	5·793	5·995	6·193	6·388	6·580	6·770	6·956	7·141	7·323	7·502
(0·5639)	649·20	763·11	888·76	1026·6	1177·1	1340·6	1517·7	1708·7	1914·0	2134·2	2369·5	2620·4	2887·2
0·0600	5·277	5·495	5·708	5·918	6·124	6·326	6·525	6·722	6·915	7·106	7·294	7·480	7·664
(0·5884)	663·16	779·52	907·87	1048·7	1202·4	1369·5	1550·3	1745·4	1955·2	2180·1	2420·4	2676·7	2949·3
0·0625	5·386	5·608	5·826	6·040	6·250	6·457	6·660	6·860	7·058	7·252	7·445	7·634	7·822
(0·6129)	676·83	795·60	926·59	1070·3	1227·2	1397·7	1582·3	1781·4	1995·5	2225·0	2470·3	2731·9	3010·1
0·0650	5·493	5·719	5·941	6·160	6·374	6·584	6·792	6·996	7·198	7·396	7·592	7·785	7·977
(0·6374)	690·24	811·35	944·94	1091·5	1251·5	1425·4	1613·6	1816·7	2035·1	2269·1	2519·3	2786·0	3069·7
0·0675	5·597	5·828	6·055	6·277	6·495	6·710	6·921	7·129	7·335	7·537	7·737	7·934	8·128
(0·6619)	703·39	826·81	962·94	1112·3	1275·3	1452·5	1644·4	1851·3	2073·8	2312·3	2567·3	2839·1	3128·2
0·0700	5·700	5·935	6·166	6·392	6·614	6·833	7·048	7·260	7·469	7·675	7·879	8·079	8·278
(0·6865)	716·29	841·98	980·61	1132·7	1298·7	1479·2	1674·6	1885·3	2111·9	2354·8	2614·4	2891·2	3185·6
0·0725	5·801	6·040	6·275	6·505	6·731	6·954	7·173	7·389	7·601	7·811	8·018	8·222	8·424
(0·7110)	728·97	856·88	997·97	1152·7	1321·7	1505·4	1704·2	1918·7	2149·3	2396·4	2660·6	2942·4	3242·0
0·0750	5·900	6·144	6·382	6·616	6·847	7·073	7·296	7·515	7·731	7·945	8·155	8·363	8·568
(0·7355)	741·44	871·53	1015·0	1172·5	1344·3	1531·1	1733·3	1951·5	2186·0	2437·4	2706·1	2992·7	3297·4
0·0800	6·094	6·345	6·591	6·833	7·071	7·305	7·535	7·762	7·985	8·205	8·423	8·637	8·849
(0·7845)	765·75	900·12	1048·3	1210·9	1388·4	1581·3	1790·2	2015·5	2257·7	2517·3	2794·9	3090·8	3405·6
0·0850	6·281	6·540	6·794	7·044	7·289	7·530	7·767	8·000	8·231	8·458	8·682	8·903	9·122
(0·8336)	789·32	927·82	1080·6	1248·2	1431·1	1630·0	1845·3	2077·5	2327·2	2594·8	2880·9	3185·9	3510·4
0·0900	6·463	6·730	6·991	7·248	7·500	7·748	7·992	8·232	8·469	8·703	8·934	9·161	9·386
(0·8826)	812·20	954·72	1111·9	1284·4	1472·6	1677·2	1898·8	2137·7	2394·6	2670·0	2964·4	3278·3	3612·1
0·0950	6·640	6·914	7·183	7·446	7·706	7·960	8·211	8·458	8·701	8·941	9·178	9·412	9·643
(0·9316)	834·46	980·88	1142·4	1319·6	1513·0	1723·2	1950·8	2196·3	2460·3	2743·2	3045·6	3368·1	3711·1
S, ($\Delta p/\rho l$)	400	425	450	475	500	525	550	575	600	625	650	675	700

Gradient   **(Equivalent) Pipe diameters in mm**

S = 0·0300 to 0·0950;

$\Delta p / \rho l$ = (0·2942) to (0·9316) ms^{-2}

m = Manning $n \times 100$
S = 0·1000 to 0·3000

i.e. kin. pr. grad., $\Delta p/\rho l$ =
(0·9807) to (2·9420) ms^{-2}

Full bore flow; Tables E give m_C or m_P for Colebrook-White or for laminar solutions resp., to divide mV and/or mQ as follows:

mV to give velocities in ms^{-1}
mQ to give discharges in litres/sec

Gradient S, $(\Delta p/\rho l)$	(Equivalent) Pipe diameters in mm												
---	400	425	450	475	500	525	550	575	600	625	650	675	700
0·1000 (0·9807)	6·813 856·14	7·094 1006·4	7·369 1172·1	7·640 1353·8	7·906 1552·3	8·167 1768·0	8·424 2001·5	8·678 2253·4	8·927 2524·2	9·174 2814·5	9·417 3124·8	9·657 3455·6	9·894 3807·5
0·1050 (1·0297)	6·981 877·28	7·269 1031·2	7·551 1201·0	7·829 1387·3	8·101 1590·6	8·369 1811·6	8·632 2050·9	8·892 2309·0	9·148 2586·5	9·400 2884·0	9·649 3201·9	9·895 3541·0	10·14 3901·6
0·1100 (1·0787)	7·145 897·92	7·440 1055·5	7·729 1229·3	8·013 1419·9	8·292 1628·0	8·566 1854·3	8·836 2099·2	9·101 2363·3	9·363 2647·4	9·621 2951·8	9·876 3277·3	10·13 3624·3	10·38 3993·4
0·1150 (1·1278)	7·306 918·10	7·607 1079·2	7·903 1256·9	8·193 1451·8	8·478 1664·6	8·758 1895·9	9·034 2146·3	9·306 2416·5	9·574 2706·9	9·838 3018·2	10·10 3350·9	10·36 3705·7	10·61 4083·1
0·1200 (1·1768)	7·463 937·85	7·771 1102·4	8·073 1283·9	8·369 1483·1	8·660 1700·4	8·947 1936·7	9·228 2192·5	9·506 2468·4	9·780 2765·1	10·05 3083·1	10·32 3423·0	10·58 3785·4	10·84 4170·9
0·1250 (1·2258)	7·617 957·19	7·931 1125·1	8·239 1310·4	8·542 1513·6	8·839 1735·5	9·131 1976·6	9·419 2237·7	9·702 2519·3	9·981 2822·1	10·26 3146·7	10·53 3493·6	10·80 3863·5	11·06 4257·0
0·1300 (1·2749)	7·768 976·14	8·088 1147·4	8·402 1336·4	8·711 1543·6	9·014 1769·9	9·312 2015·8	9·605 2282·0	9·894 2569·2	10·18 2878·0	10·46 3209·0	10·74 3562·8	11·01 3940·0	11·28 4341·3
0·1350 (1·3239)	7·916 994·74	8·242 1169·3	8·563 1361·8	8·877 1573·0	9·186 1803·6	9·489 2054·2	9·788 2325·5	10·08 2618·2	10·37 2932·8	10·66 3270·1	10·94 3630·7	11·22 4015·1	11·50 4424·0
0·1400 (1·3729)	8·061 1013·0	8·394 1190·7	8·720 1386·8	9·040 1601·9	9·354 1836·7	9·663 2091·9	9·968 2368·2	10·27 2666·2	10·56 2986·6	10·85 3330·1	11·14 3697·3	11·43 4088·7	11·71 4505·1
0·1450 (1·4220)	8·204 1030·9	8·542 1211·8	8·874 1411·3	9·200 1630·2	9·520 1869·2	9·834 2128·9	10·14 2410·1	10·45 2713·4	10·75 3039·5	11·05 3389·1	11·34 3762·7	11·63 4161·1	11·91 4584·9
0·1500 (1·4710)	8·344 1048·5	8·688 1232·5	9·026 1435·5	9·357 1658·1	9·682 1901·1	10·00 2165·3	10·32 2451·3	10·63 2759·8	10·93 3091·5	11·24 3447·0	11·53 3827·0	11·83 4232·3	12·12 4663·3
0·1600 (1·5691)	8·618 1082·9	8·973 1273·0	9·322 1482·5	9·664 1712·5	10·00 1963·5	10·33 2236·3	10·66 2531·7	10·98 2850·3	11·29 3192·9	11·60 3560·0	11·91 3952·6	12·21 4371·1	12·51 4816·2
0·1700 (1·6671)	8·883 1116·3	9·249 1312·1	9·609 1528·2	9·961 1765·2	10·31 2023·9	10·65 2305·1	10·98 2609·6	11·31 2938·0	11·64 3291·1	11·96 3669·6	12·28 4074·2	12·59 4505·6	12·90 4964·4
0·1800 (1·7652)	9·140 1148·6	9·517 1350·2	9·887 1572·5	10·25 1816·4	10·61 2082·6	10·96 2372·0	11·30 2685·3	11·64 3023·2	11·98 3386·5	12·31 3776·0	12·63 4192·3	12·96 4636·2	13·27 5108·3
0·1900 (1·8633)	9·391 1180·1	9·778 1387·2	10·16 1615·6	10·53 1866·1	10·90 2139·7	11·26 2437·0	11·61 2758·8	11·96 3106·0	12·31 3479·3	12·65 3879·5	12·98 4307·2	13·31 4763·2	13·64 5248·3
0·2000 (1·9613)	9·635 1210·8	10·03 1423·2	10·42 1657·5	10·80 1914·6	11·18 2195·3	11·55 2500·3	11·91 2830·5	12·27 3186·7	12·63 3569·7	12·97 3980·3	13·32 4419·1	13·66 4887·0	13·99 5384·7
0·2100 (2·0594)	9·873 1240·7	10·28 1458·4	10·68 1698·5	11·07 1961·9	11·46 2249·5	11·84 2562·0	12·21 2900·4	12·58 3265·4	12·94 3657·9	13·29 4078·5	13·65 4528·2	13·99 5007·7	14·34 5517·6
0·2200 (2·1575)	10·11 1269·9	10·52 1492·7	10·93 1738·4	11·33 2008·1	11·73 2302·4	12·11 2622·3	12·50 2968·7	12·87 3342·3	13·24 3744·0	13·61 4174·5	13·97 4634·8	14·32 5125·5	14·67 5647·5
0·2300 (2·2555)	10·33 1298·4	10·76 1526·2	11·18 1777·5	11·59 2053·2	11·99 2354·1	12·39 2681·3	12·78 3035·4	13·16 3417·4	13·54 3828·1	13·91 4268·3	14·28 4738·9	14·65 5240·7	15·00 5774·4
0·2400 (2·3536)	10·55 1326·3	10·99 1559·0	11·42 1815·7	11·84 2097·4	12·25 2404·8	12·65 2738·9	13·05 3100·7	13·44 3490·9	13·83 3910·4	14·21 4360·2	14·59 4840·9	14·96 5353·4	15·33 5898·6
0·2500 (2·4517)	10·77 1353·7	11·22 1591·2	11·65 1853·2	12·08 2140·6	12·50 2454·4	12·91 2795·4	13·32 3164·6	13·72 3562·9	14·12 3991·1	14·50 4450·1	14·89 4940·7	15·27 5463·8	15·64 6020·2
0·2600 (2·5497)	10·99 1380·5	11·44 1622·7	11·88 1889·9	12·32 2183·0	12·75 2503·0	13·17 2850·8	13·58 3227·3	13·99 3633·4	14·40 4070·1	14·79 4538·2	15·18 5038·5	15·57 5572·0	15·95 6139·5
0·2700 (2·6478)	11·19 1406·8	11·66 1653·6	12·11 1925·9	12·55 2224·6	12·99 2550·7	13·42 2905·1	13·84 3288·8	14·26 3702·6	14·67 4147·6	15·07 4624·6	15·47 5134·5	15·87 5678·2	16·26 6256·4
0·2800 (2·7459)	11·40 1432·6	11·87 1684·0	12·33 1961·2	12·78 2265·4	13·23 2597·5	13·67 2958·4	14·10 3349·1	14·52 3770·6	14·94 4223·8	15·35 4709·5	15·76 5228·7	16·16 5782·4	16·56 6371·2
0·2900 (2·8439)	11·60 1457·9	12·08 1713·8	12·55 1995·9	13·01 2305·5	13·46 2643·4	13·91 3010·7	14·35 3408·4	14·78 3837·3	15·20 4298·5	15·62 4792·9	16·04 5321·3	16·44 5884·7	16·85 6484·0
0·3000 (2·9420)	11·80 1482·9	12·29 1743·1	12·76 2030·1	13·23 2344·9	13·69 2688·6	14·15 3062·2	14·59 3466·7	15·03 3902·9	15·46 4372·0	15·89 4874·8	16·31 5412·3	16·73 5985·3	17·14 6594·8

S, $(\Delta p/\rho l)$ Gradient	400	425	450	475	500	525	550	575	600	625	650	675	700
(Equivalent) Pipe diameters in mm													

S = 0·1000 to 0·3000;

$\Delta p/\rho l$ = (0·9807) to (2·9420) ms^{-2}

m = Manning $n \times 100$
S = 0.000100 to 0.000290

i.e. kin. pr. grad., $\Delta p/\rho l$ = (0.000981) to (0.002844) ms^{-2}

Full bore flow; Tables E give m_C or m_P for Colebrook-White or for laminar solutions resp., to divide mV and/or mQ as follows:

mV to give velocities in ms^{-1}
mQ to give discharges in litres/sec

Gradient (Equivalent) Pipe diameters in mm

S, ($\Delta p/\rho l$)	700	725	750	775	800	825	850	875	900	925	950	975	1000
0.000100	0.313	0.320	0.328	0.335	0.342	0.349	0.356	0.363	0.370	0.377	0.384	0.390	0.397
(0.000981)	120.40	132.22	144.73	157.95	171.91	186.61	202.07	218.31	235.34	253.18	271.84	291.34	311.69
0.000105	0.321	0.328	0.336	0.343	0.350	0.358	0.365	0.372	0.379	0.386	0.393	0.400	0.407
(0.001030)	123.38	135.48	148.30	161.85	176.15	191.21	207.06	223.70	241.15	259.43	278.55	298.53	319.38
0.000110	0.328	0.336	0.344	0.351	0.359	0.366	0.373	0.381	0.388	0.395	0.402	0.409	0.416
(0.001079)	126.28	138.67	151.79	165.66	180.30	195.71	211.93	228.96	246.83	265.54	285.11	305.56	326.90
0.000115	0.336	0.343	0.351	0.359	0.367	0.374	0.382	0.389	0.397	0.404	0.411	0.418	0.426
(0.001128)	129.12	141.79	155.20	169.38	184.35	200.11	216.69	234.11	252.37	271.50	291.52	312.42	334.24
0.000120	0.343	0.351	0.359	0.367	0.375	0.382	0.390	0.398	0.405	0.413	0.420	0.427	0.435
(0.001177)	131.90	144.84	158.54	173.03	188.31	204.42	221.36	239.15	257.80	277.34	297.78	319.14	341.43
0.000125	0.350	0.358	0.366	0.374	0.382	0.390	0.398	0.406	0.414	0.421	0.429	0.436	0.444
(0.001226)	134.62	147.82	161.81	176.59	192.20	208.63	225.92	244.08	263.12	283.06	303.93	325.72	348.47
0.000130	0.357	0.365	0.374	0.382	0.390	0.398	0.406	0.414	0.422	0.430	0.437	0.445	0.452
(0.001275)	137.28	150.75	165.01	180.09	196.00	212.76	230.39	248.91	268.33	288.67	309.94	332.17	355.38
0.000135	0.364	0.372	0.381	0.389	0.397	0.406	0.414	0.422	0.430	0.438	0.446	0.453	0.461
(0.001324)	139.90	153.62	168.16	183.52	199.74	216.82	234.78	253.65	273.44	294.17	315.85	338.50	362.15
0.000140	0.370	0.379	0.388	0.396	0.405	0.413	0.421	0.430	0.438	0.446	0.454	0.462	0.470
(0.001373)	142.46	156.44	171.24	186.89	203.40	220.80	239.09	258.31	278.46	299.57	321.64	344.71	368.79
0.000145	0.377	0.386	0.394	0.403	0.412	0.420	0.429	0.437	0.445	0.454	0.462	0.470	0.478
(0.001422)	144.99	159.21	174.27	190.20	207.00	224.70	243.32	262.88	283.39	304.87	327.34	350.82	375.32
0.000150	0.383	0.392	0.401	0.410	0.419	0.428	0.436	0.445	0.453	0.461	0.470	0.478	0.486
(0.001471)	147.47	161.93	177.25	193.45	210.54	228.55	247.48	267.37	288.23	310.08	332.93	356.81	381.73
0.000160	0.396	0.405	0.414	0.424	0.433	0.442	0.450	0.459	0.468	0.477	0.485	0.494	0.502
(0.001569)	152.30	167.24	183.07	199.79	217.44	236.04	255.60	276.14	297.68	320.25	343.85	368.51	394.25
0.000170	0.408	0.418	0.427	0.437	0.446	0.455	0.464	0.473	0.482	0.491	0.500	0.509	0.517
(0.001667)	156.99	172.39	188.70	205.94	224.14	243.30	263.47	284.64	306.85	330.10	354.44	379.86	406.39
0.000180	0.420	0.430	0.440	0.449	0.459	0.468	0.478	0.487	0.496	0.505	0.515	0.524	0.532
(0.001765)	161.54	177.39	194.17	211.91	230.64	250.36	271.10	292.89	315.74	339.67	364.71	390.87	418.17
0.000190	0.431	0.441	0.452	0.462	0.471	0.481	0.491	0.500	0.510	0.519	0.529	0.538	0.547
(0.001863)	165.97	182.25	199.49	217.72	236.96	257.22	278.53	300.92	324.39	348.98	374.70	401.58	429.63
0.000200	0.442	0.453	0.463	0.474	0.484	0.494	0.504	0.513	0.523	0.533	0.542	0.552	0.561
(0.001961)	170.28	186.98	204.67	223.38	243.11	263.90	285.77	308.73	332.82	358.05	384.44	412.01	440.79
0.000210	0.453	0.464	0.475	0.485	0.496	0.506	0.516	0.526	0.536	0.546	0.556	0.565	0.575
(0.002059)	174.48	191.60	209.73	228.89	249.11	270.42	292.83	316.36	341.04	366.89	393.93	422.19	451.67
0.000220	0.464	0.475	0.486	0.497	0.507	0.518	0.528	0.538	0.549	0.559	0.569	0.579	0.589
(0.002157)	178.59	196.11	214.66	234.28	254.98	276.78	299.72	323.80	349.07	375.52	403.20	432.12	462.30
0.000230	0.474	0.486	0.497	0.508	0.519	0.529	0.540	0.551	0.561	0.571	0.582	0.592	0.602
(0.002256)	182.60	200.52	219.49	239.54	260.71	283.00	306.45	331.08	356.91	383.96	412.26	441.83	472.69
0.000240	0.485	0.496	0.508	0.519	0.530	0.541	0.552	0.562	0.573	0.584	0.594	0.605	0.615
(0.002354)	186.53	204.83	224.21	244.70	266.31	289.09	313.04	338.20	364.59	392.22	421.13	451.34	482.86
0.000250	0.495	0.506	0.518	0.529	0.541	0.552	0.563	0.574	0.585	0.596	0.606	0.617	0.627
(0.002452)	190.38	209.05	228.83	249.74	271.81	295.05	319.50	345.18	372.11	400.31	429.82	460.64	492.82
0.000260	0.504	0.516	0.528	0.540	0.551	0.563	0.574	0.585	0.596	0.607	0.618	0.629	0.640
(0.002550)	194.15	213.19	233.36	254.69	277.19	300.89	325.83	352.01	379.47	408.24	438.33	469.77	502.58
0.000270	0.514	0.526	0.538	0.550	0.562	0.574	0.585	0.597	0.608	0.619	0.630	0.641	0.652
(0.002648)	197.85	217.25	237.81	259.54	282.47	306.63	332.03	358.72	386.70	416.02	446.68	478.71	512.15
0.000280	0.524	0.536	0.548	0.560	0.572	0.584	0.596	0.607	0.619	0.630	0.642	0.653	0.664
(0.002746)	201.48	221.24	242.17	264.30	287.65	312.25	338.13	365.30	393.80	423.65	454.87	487.50	521.55
0.000290	0.533	0.545	0.558	0.570	0.582	0.594	0.606	0.618	0.630	0.642	0.653	0.665	0.676
(0.002844)	205.04	225.16	246.46	268.98	292.74	317.78	344.11	371.77	400.77	431.15	462.93	496.13	530.78
S, ($\Delta p/\rho l$)	700	725	750	775	800	825	850	875	900	925	950	975	1000

Gradient (Equivalent) Pipe diameters in mm

S = 0.000100 to 0.000290; $\qquad$ $\Delta p/\rho l$ = (0.000981) to (0.002844) ms^{-2}

m = Manning $n \times 100$
S = 0·000300 to 0·000950

i.e. kin. pr. grad., $\Delta p/\rho l$ = (0·002942) to (0·009316) ms^{-2}

Full bore flow; Tables E give m_C or m_P for Colebrook-White or for laminar solutions resp., to divide mV and/or mQ as follows:

mV to give velocities in ms^{-1}
mQ to give discharges in litres/sec

Gradient — (Equivalent) Pipe diameters in mm

S, ($\Delta p/\rho l$)	700	725	750	775	800	825	850	875	900	925	950	975	1000
0·000300	0·542	0·555	0·567	0·580	0·592	0·605	0·617	0·629	0·641	0·653	0·664	0·676	0·687
(0·002942)	208·55	229·00	250·67	273·58	297·75	323·21	349·99	378·12	407·62	438·52	470·84	504·61	539·85
0·000320	0·560	0·573	0·586	0·599	0·612	0·624	0·637	0·649	0·662	0·674	0·686	0·698	0·710
(0·003138)	215·39	236·52	258·89	282·55	307·51	333·81	361·47	390·52	420·99	452·90	486·28	521·16	557·56
0·000340	0·577	0·591	0·604	0·617	0·631	0·644	0·657	0·669	0·682	0·695	0·707	0·720	0·732
(0·003334)	222·02	243·79	266·86	291·25	316·98	344·09	372·60	402·54	433·95	466·84	501·25	537·20	574·72
0·000360	0·594	0·608	0·622	0·635	0·649	0·662	0·676	0·689	0·702	0·715	0·728	0·740	0·753
(0·003530)	228·45	250·86	274·60	299·69	326·17	354·06	383·40	414·21	446·53	480·37	515·78	552·77	591·38
0·000380	0·610	0·624	0·639	0·653	0·667	0·680	0·694	0·708	0·721	0·734	0·748	0·761	0·774
(0·003727)	234·71	257·74	282·12	307·90	335·11	363·76	393·91	425·56	458·76	493·54	529·91	567·92	607·59
0·000400	0·626	0·641	0·655	0·670	0·684	0·698	0·712	0·726	0·740	0·754	0·767	0·780	0·794
(0·003923)	240·81	264·43	289·45	315·90	343·81	373·21	404·14	436·62	470·68	506·36	543·68	582·67	623·37
0·000420	0·641	0·656	0·671	0·686	0·701	0·715	0·730	0·744	0·758	0·772	0·786	0·800	0·813
(0·004119)	246·76	270·96	296·60	323·70	352·30	382·43	414·12	447·40	482·30	518·86	557·10	597·06	638·76
0·000440	0·656	0·672	0·687	0·702	0·717	0·732	0·747	0·762	0·776	0·790	0·804	0·819	0·832
(0·004315)	252·56	277·34	303·58	331·32	360·59	391·43	423·86	457·93	493·65	531·07	570·21	611·11	653·80
0·000460	0·671	0·687	0·703	0·718	0·733	0·749	0·764	0·779	0·793	0·808	0·823	0·837	0·851
(0·004511)	258·24	283·57	310·40	338·77	368·70	400·23	433·39	468·22	504·75	543·01	583·03	624·85	668·49
0·000480	0·685	0·702	0·718	0·734	0·749	0·765	0·780	0·795	0·810	0·825	0·840	0·855	0·869
(0·004707)	263·79	289·67	317·08	346·05	376·63	408·83	442·71	478·29	515·60	554·69	595·57	638·29	682·87
0·000500	0·700	0·716	0·733	0·749	0·765	0·781	0·796	0·812	0·827	0·842	0·858	0·873	0·887
(0·004903)	269·23	295·64	323·62	353·19	384·39	417·26	451·84	488·15	526·24	566·12	607·85	651·45	696·95
0·000525	0·717	0·734	0·751	0·767	0·784	0·800	0·816	0·832	0·848	0·863	0·879	0·894	0·909
(0·005148)	275·88	302·94	331·61	361·91	393·88	427·57	463·00	500·21	539·23	580·11	622·86	667·54	714·16
0·000550	0·734	0·751	0·768	0·785	0·802	0·819	0·835	0·851	0·868	0·884	0·899	0·915	0·931
(0·005394)	282·37	310·07	339·41	370·43	403·15	437·63	473·89	511·98	551·92	593·76	637·52	683·24	730·97
0·000575	0·750	0·768	0·786	0·803	0·820	0·837	0·854	0·871	0·887	0·903	0·920	0·936	0·952
(0·005639)	288·72	317·04	347·04	378·75	412·21	447·47	484·54	523·49	564·33	607·10	651·85	698·60	747·39
0·000600	0·766	0·785	0·802	0·820	0·838	0·855	0·872	0·889	0·906	0·923	0·939	0·956	0·972
(0·005884)	294·93	323·86	354·50	386·90	421·08	457·09	494·97	534·74	576·46	620·16	665·87	713·63	763·47
0·000625	0·782	0·801	0·819	0·837	0·855	0·873	0·890	0·908	0·925	0·942	0·959	0·976	0·992
(0·006129)	301·01	330·54	361·81	394·88	429·76	466·52	505·17	545·77	588·35	632·95	679·60	728·34	779·21
0·000650	0·798	0·817	0·835	0·854	0·872	0·890	0·908	0·926	0·943	0·961	0·978	0·995	1·012
(0·006374)	306·97	337·09	368·98	402·70	438·27	475·75	515·18	556·58	600·00	645·48	693·06	742·77	794·64
0·000675	0·813	0·832	0·851	0·870	0·889	0·907	0·925	0·943	0·961	0·979	0·996	1·014	1·031
(0·006619)	312·82	343·51	376·01	410·37	446·62	484·82	524·99	567·18	611·43	657·78	706·26	756·91	809·78
0·000700	0·828	0·847	0·867	0·886	0·905	0·924	0·942	0·961	0·979	0·997	1·015	1·032	1·050
(0·006865)	318·56	349·81	382·91	417·90	454·82	493·71	534·62	577·59	622·65	669·85	719·22	770·80	824·64
0·000725	0·842	0·862	0·882	0·902	0·921	0·940	0·959	0·978	0·996	1·014	1·033	1·051	1·069
(0·007110)	324·20	356·00	389·69	425·29	462·87	502·45	544·09	587·81	633·67	681·70	731·95	784·45	839·24
0·000750	0·857	0·877	0·897	0·917	0·937	0·956	0·975	0·994	1·013	1·032	1·050	1·069	1·087
(0·007355)	329·74	362·09	396·35	432·56	470·78	511·04	553·39	597·86	644·51	693·36	744·46	797·86	853·58
0·000800	0·885	0·906	0·927	0·947	0·967	0·987	1·007	1·027	1·046	1·066	1·085	1·104	1·122
(0·007845)	340·56	373·96	409·35	446·75	486·22	527·80	571·54	617·47	665·64	716·10	768·88	824·02	881·58
0·000850	0·912	0·934	0·955	0·976	0·997	1·018	1·038	1·058	1·079	1·098	1·118	1·138	1·157
(0·008336)	351·04	385·47	421·94	460·50	501·19	544·05	589·13	636·47	686·13	738·14	792·54	849·38	908·71
0·000900	0·939	0·961	0·983	1·004	1·026	1·047	1·068	1·089	1·110	1·130	1·151	1·171	1·191
(0·008826)	361·21	396·65	434·18	473·85	515·72	559·82	606·21	654·93	706·02	759·54	815·52	874·01	935·06
0·000950	0·964	0·987	1·010	1·032	1·054	1·076	1·098	1·119	1·140	1·161	1·182	1·203	1·223
(0·009316)	371·11	407·52	446·07	486·84	529·85	575·16	622·82	672·87	725·37	780·35	837·86	897·96	960·68
S, ($\Delta p/\rho l$)	700	725	750	775	800	825	850	875	900	925	950	975	1000

Gradient — (Equivalent) Pipe diameters in mm

S = 0·000300 to 0·000950; $\Delta p/\rho l$ = (0·002942) to (0·009316) ms^{-2}

m = Manning $n \times 100$
S = 0·00100 to 0·00290

i.e. kin. pr. grad., $\Delta p/\rho l$ =
(0·00981) to (0·02844) ms^{-2}

Full bore flow; Tables E give m_C or m_P for Colebrook-White or for laminar solutions resp., to divide mV and/or mQ as follows:

mV to give velocities in ms^{-1}
mQ to give discharges in litres/sec

Gradient   **(Equivalent) Pipe diameters in mm**

S, $(\Delta p/\rho l)$	700	725	750	775	800	825	850	875	900	925	950	975	1000
0·00100	0·989	1·013	1·036	1·059	1·081	1·104	1·126	1·148	1·170	1·191	1·213	1·234	1·255
(0·00981)	380·75	418·10	457·66	499·48	543·61	590·10	639·00	690·35	744·21	800·62	859·63	921·29	985·63
0·00105	1·014	1·038	1·062	1·085	1·108	1·131	1·154	1·176	1·199	1·221	1·243	1·264	1·286
(0·01030)	390·16	428·43	468·97	511·82	557·04	604·67	654·78	707·40	762·59	820·39	880·86	944·04	1010·0
0·00110	1·038	1·062	1·087	1·111	1·134	1·158	1·181	1·204	1·227	1·250	1·272	1·294	1·316
(0·01079)	399·34	438·51	480·00	523·86	570·15	618·90	670·19	724·05	780·53	839·70	901·59	966·25	1033·7
0·00115	1·061	1·086	1·111	1·135	1·160	1·184	1·208	1·231	1·254	1·278	1·301	1·323	1·346
(0·01128)	408·31	448·37	490·79	535·64	582·96	632·81	685·25	740·32	798·08	858·57	921·85	987·97	1057·0
0·00120	1·084	1·109	1·135	1·160	1·185	1·209	1·234	1·258	1·281	1·305	1·329	1·352	1·375
(0·01177)	417·09	458·01	501·34	547·16	595·50	646·42	699·99	756·24	815·24	877·04	941·68	1009·2	1079·7
0·00125	1·106	1·132	1·158	1·184	1·209	1·234	1·259	1·284	1·308	1·332	1·356	1·380	1·403
(0·01226)	425·70	467·45	511·68	558·44	607·78	659·75	714·42	771·84	832·05	895·12	961·10	1030·0	1102·0
0·00130	1·128	1·155	1·181	1·207	1·233	1·259	1·284	1·309	1·334	1·358	1·383	1·407	1·431
(0·01275)	434·13	476·71	521·82	569·50	619·81	672·82	728·57	787·12	848·53	912·85	980·13	1050·4	1123·8
0·00135	1·150	1·177	1·204	1·230	1·257	1·283	1·308	1·334	1·359	1·384	1·409	1·434	1·458
(0·01324)	442·40	485·79	531·76	580·35	631·62	685·64	742·45	802·12	864·69	930·24	998·80	1070·4	1145·2
0·00140	1·171	1·198	1·226	1·253	1·280	1·306	1·332	1·358	1·384	1·410	1·435	1·460	1·485
(0·01373)	450·51	494·71	541·51	591·00	643·21	698·22	756·07	816·84	880·56	947·31	1017·1	1090·1	1166·2
0·00145	1·191	1·220	1·247	1·275	1·302	1·329	1·356	1·382	1·409	1·435	1·460	1·486	1·511
(0·01422)	458·49	503·46	551·10	601·46	654·60	710·58	769·46	831·29	896·15	964·08	1035·1	1109·4	1186·9
0·00150	1·212	1·240	1·269	1·297	1·325	1·352	1·379	1·406	1·433	1·459	1·485	1·511	1·537
(0·01471)	466·33	512·07	560·52	611·74	665·79	722·72	782·61	845·51	911·47	980·56	1052·8	1128·3	1207·2
0·00160	1·251	1·281	1·310	1·339	1·368	1·396	1·424	1·452	1·480	1·507	1·534	1·561	1·587
(0·01569)	481·62	528·86	578·90	631·80	687·62	746·43	808·28	873·23	941·36	1012·7	1087·4	1165·3	1246·7
0·00170	1·290	1·321	1·351	1·381	1·410	1·439	1·468	1·497	1·525	1·553	1·581	1·609	1·636
(0·01667)	496·44	545·14	596·72	651·25	708·78	769·40	833·15	900·11	970·33	1043·9	1120·8	1201·2	1285·1
0·00180	1·327	1·359	1·390	1·421	1·451	1·481	1·511	1·540	1·569	1·598	1·627	1·656	1·684
(0·01765)	510·83	560·94	614·02	670·13	729·33	791·70	857·31	926·20	998·46	1074·1	1153·3	1236·0	1322·4
0·00190	1·364	1·396	1·428	1·460	1·491	1·522	1·552	1·582	1·612	1·642	1·672	1·701	1·730
(0·01863)	524·83	576·32	630·85	688·49	749·32	813·40	880·80	951·58	1025·8	1103·6	1184·9	1269·9	1358·6
0·00200	1·399	1·432	1·465	1·497	1·529	1·561	1·593	1·624	1·654	1·685	1·715	1·745	1·775
(0·01961)	538·47	591·29	647·23	706·38	768·78	834·53	903·68	976·31	1052·5	1132·2	1215·7	1302·9	1393·9
0·00210	1·434	1·468	1·501	1·534	1·567	1·600	1·632	1·664	1·695	1·726	1·757	1·788	1·819
(0·02059)	551·76	605·89	663·22	723·82	787·77	855·14	926·00	1000·4	1078·5	1160·2	1245·7	1335·1	1428·3
0·00220	1·467	1·502	1·537	1·571	1·604	1·637	1·670	1·703	1·735	1·767	1·799	1·830	1·861
(0·02157)	564·75	620·15	678·82	740·85	806·31	875·26	947·79	1024·0	1103·8	1187·5	1275·0	1366·5	1461·9
0·00230	1·500	1·536	1·571	1·606	1·640	1·674	1·708	1·741	1·774	1·807	1·839	1·871	1·903
(0·02256)	577·44	634·09	694·08	757·50	824·43	894·93	969·09	1047·0	1128·7	1214·2	1303·7	1397·2	1494·8
0·00240	1·533	1·569	1·605	1·640	1·675	1·710	1·745	1·779	1·812	1·846	1·879	1·912	1·944
(0·02354)	589·86	647·72	709·01	773·80	842·16	914·18	989·93	1069·5	1152·9	1240·3	1331·7	1427·3	1526·9
0·00250	1·564	1·601	1·638	1·674	1·710	1·745	1·781	1·815	1·850	1·884	1·918	1·951	1·984
(0·02452)	602·02	661·08	723·63	789·75	859·53	933·03	1010·3	1091·5	1176·7	1265·9	1359·2	1456·7	1558·4
0·00260	1·595	1·633	1·670	1·707	1·744	1·780	1·816	1·851	1·886	1·921	1·956	1·990	2·024
(0·02550)	613·95	674·17	737·96	805·39	876·55	951·51	1030·4	1113·2	1200·0	1291·0	1386·1	1485·5	1589·3
0·00270	1·626	1·664	1·702	1·740	1·777	1·814	1·850	1·886	1·922	1·958	1·993	2·028	2·062
(0·02648)	625·64	687·01	752·02	820·73	893·25	969·64	1050·0	1134·4	1222·9	1315·6	1412·5	1513·8	1619·6
0·00280	1·656	1·695	1·733	1·772	1·810	1·847	1·884	1·921	1·957	1·994	2·029	2·065	2·100
(0·02746)	637·12	699·62	765·82	835·79	909·64	987·43	1069·2	1155·2	1245·3	1339·7	1438·4	1541·6	1649·3
0·00290	1·685	1·725	1·764	1·803	1·842	1·880	1·918	1·955	1·992	2·029	2·065	2·101	2·137
(0·02844)	648·40	712·00	779·37	850·59	925·74	1004·9	1088·2	1175·6	1267·3	1363·4	1463·9	1568·9	1678·5
S, $(\Delta p/\rho l)$	700	725	750	775	800	825	850	875	900	925	950	975	1000

Gradient   **(Equivalent) Pipe diameters in mm**

S = 0·00100 to 0·00290; $\Delta p/\rho l$ = (0·00981) to (0·02844) ms^{-2}

m = Manning $n \times 100$
S = 0·00300 to 0·00950

i.e. kin. pr. grad., $\Delta p/\rho l$ =
(0·02942) to (0·09316) ms^{-2}

Full bore flow; Tables E give m_C or m_P for Colebrook-White or for laminar solutions resp., to divide mV and/or mQ as follows:

mV to give velocities in ms^{-1}
mQ to give discharges in litres/sec

Gradient (Equivalent) Pipe diameters in mm

S, ($\Delta p/\rho l$)	700	725	750	775	800	825	850	875	900	925	950	975	1000
0·00300	1·714	1·754	1·794	1·834	1·873	1·912	1·950	1·989	2·026	2·064	2·101	2·137	2·174
(0·02942)	659·48	724·18	792·70	865·13	941·56	1022·1	1106·8	1195·7	1289·0	1386·7	1488·9	1595·7	1707·2
0·00320	1·770	1·812	1·853	1·894	1·935	1·975	2·014	2·054	2·093	2·131	2·169	2·207	2·245
(0·03138)	681·11	747·93	818·69	893·50	972·44	1055·6	1143·1	1234·9	1331·3	1432·2	1537·8	1648·0	1763·2
0·00340	1·824	1·867	1·910	1·952	1·994	2·035	2·076	2·117	2·157	2·197	2·236	2·275	2·314
(0·03334)	702·07	770·94	843·89	921·00	1002·4	1088·1	1178·3	1272·9	1372·3	1476·3	1585·1	1698·8	1817·4
0·00360	1·877	1·922	1·966	2·009	2·052	2·094	2·137	2·178	2·220	2·261	2·301	2·341	2·381
(0·03530)	722·43	793·30	868·35	947·70	1031·4	1119·6	1212·4	1309·9	1412·0	1519·1	1631·0	1748·0	1870·1
0·00380	1·929	1·974	2·019	2·064	2·108	2·152	2·195	2·238	2·280	2·322	2·364	2·405	2·446
(0·03727)	742·23	815·03	892·15	973·67	1059·7	1150·3	1245·6	1345·7	1450·7	1560·7	1675·7	1795·9	1921·4
0·00400	1·979	2·026	2·072	2·118	2·163	2·208	2·252	2·296	2·340	2·383	2·426	2·468	2·510
(0·03923)	761·51	836·21	915·33	998·97	1087·2	1180·2	1278·0	1380·7	1488·4	1601·2	1719·3	1842·6	1971·3
0·00420	2·028	2·076	2·123	2·170	2·216	2·262	2·308	2·353	2·397	2·442	2·485	2·529	2·572
(0·04119)	780·31	856·86	937·93	1023·6	1114·1	1209·3	1309·6	1414·8	1525·2	1640·8	1761·7	1888·1	2019·9
0·00440	2·075	2·124	2·173	2·221	2·269	2·316	2·362	2·408	2·454	2·499	2·544	2·588	2·632
(0·04315)	798·68	877·02	960·00	1047·7	1140·3	1237·8	1340·4	1448·1	1561·1	1679·4	1803·2	1932·5	2067·5
0·00460	2·122	2·172	2·222	2·271	2·320	2·368	2·415	2·462	2·509	2·555	2·601	2·647	2·692
(0·04511)	816·63	896·73	981·58	1071·3	1165·9	1265·6	1370·5	1480·6	1596·2	1717·1	1843·7	1975·9	2114·0
0·00480	2·168	2·219	2·270	2·320	2·369	2·419	2·467	2·515	2·563	2·610	2·657	2·703	2·749
(0·04707)	834·19	916·02	1002·7	1094·3	1191·0	1292·8	1400·0	1512·5	1630·5	1754·1	1883·4	2018·4	2159·4
0·00500	2·212	2·265	2·316	2·368	2·418	2·468	2·518	2·567	2·616	2·664	2·712	2·759	2·806
(0·04903)	851·39	934·91	1023·4	1116·9	1215·6	1319·5	1428·8	1543·7	1664·1	1790·2	1922·2	2060·1	2203·9
0·00525	2·267	2·321	2·374	2·426	2·478	2·529	2·580	2·631	2·680	2·730	2·779	2·827	2·875
(0·05148)	872·42	958·00	1048·6	1144·5	1245·6	1352·1	1464·1	1581·8	1705·2	1834·5	1969·7	2110·9	2258·4
0·00550	2·320	2·375	2·429	2·483	2·536	2·589	2·641	2·692	2·743	2·794	2·844	2·894	2·943
(0·05394)	892·95	980·54	1073·3	1171·4	1274·9	1383·9	1498·6	1619·0	1745·3	1877·6	2016·0	2160·6	2311·5
0·00575	2·372	2·429	2·484	2·539	2·593	2·647	2·700	2·753	2·805	2·857	2·908	2·959	3·009
(0·05639)	913·02	1002·6	1097·4	1197·7	1303·5	1415·0	1532·3	1655·4	1784·6	1919·8	2061·3	2209·2	2363·5
0·00600	2·423	2·481	2·538	2·594	2·649	2·704	2·758	2·812	2·865	2·918	2·971	3·023	3·074
(0·05884)	932·65	1024·1	1121·0	1223·5	1331·6	1445·4	1565·2	1691·0	1822·9	1961·1	2105·7	2256·7	2414·3
0·00625	2·473	2·532	2·590	2·647	2·704	2·760	2·815	2·870	2·925	2·978	3·032	3·085	3·137
(0·06129)	951·88	1045·3	1144·2	1248·7	1359·0	1475·3	1597·5	1725·9	1860·5	2001·6	2149·1	2303·2	2464·1
0·00650	2·522	2·582	2·641	2·700	2·757	2·814	2·871	2·927	2·982	3·037	3·092	3·146	3·200
(0·06374)	970·73	1066·0	1166·8	1273·4	1385·9	1504·5	1629·1	1760·1	1897·4	2041·2	2191·6	2348·8	2512·9
0·00675	2·570	2·631	2·691	2·751	2·810	2·868	2·926	2·983	3·039	3·095	3·151	3·206	3·260
(0·06619)	989·23	1086·3	1189·0	1297·7	1412·3	1533·1	1660·2	1793·6	1933·5	2080·1	2233·4	2393·6	2560·8
0·00700	2·618	2·680	2·741	2·801	2·861	2·921	2·979	3·037	3·095	3·152	3·209	3·265	3·320
(0·06865)	1007·4	1106·2	1210·9	1321·5	1438·3	1561·3	1690·6	1826·5	1969·0	2118·2	2274·4	2437·5	2607·7
0·00725	2·664	2·727	2·789	2·851	2·912	2·972	3·032	3·091	3·150	3·208	3·265	3·323	3·379
(0·07110)	1025·2	1125·8	1232·3	1344·9	1463·7	1588·9	1720·6	1858·8	2003·8	2155·7	2314·6	2480·6	2653·9
0·00750	2·709	2·774	2·837	2·900	2·962	3·023	3·084	3·144	3·204	3·263	3·321	3·379	3·437
(0·07355)	1042·7	1145·0	1253·4	1367·9	1488·7	1616·1	1750·0	1890·6	2038·1	2192·6	2354·2	2523·0	2699·3
0·00800	2·798	2·865	2·930	2·995	3·059	3·122	3·185	3·247	3·309	3·370	3·430	3·490	3·550
(0·07845)	1076·9	1182·6	1294·5	1412·8	1537·6	1669·1	1807·4	1952·6	2104·9	2264·5	2431·4	2605·8	2787·8
0·00850	2·884	2·953	3·020	3·087	3·153	3·218	3·283	3·347	3·411	3·473	3·536	3·598	3·659
(0·08336)	1110·1	1219·0	1334·3	1456·2	1584·9	1720·4	1863·0	2012·7	2169·7	2334·2	2506·2	2686·0	2873·6
0·00900	2·968	3·038	3·108	3·176	3·244	3·312	3·378	3·444	3·509	3·574	3·638	3·702	3·765
(0·08826)	1142·3	1254·3	1373·0	1498·4	1630·8	1770·3	1917·0	2071·1	2232·6	2401·9	2578·9	2763·9	2956·9
0·00950	3·049	3·122	3·193	3·264	3·333	3·402	3·471	3·539	3·606	3·672	3·738	3·803	3·868
(0·09316)	1173·6	1288·7	1410·6	1539·5	1675·5	1818·8	1969·5	2127·8	2293·8	2467·7	2649·6	2839·6	3037·9
S, ($\Delta p/\rho l$)	700	725	750	775	800	825	850	875	900	925	950	975	1000

Gradient (Equivalent) Pipe diameters in mm

S = 0·00300 to 0·00950; $\Delta p/\rho l$ = (0·02942) to (0·09316) ms^{-2}

m = Manning $n \times 100$
S = 0·0100 to 0·0290

i.e. kin. pr. grad., $\Delta p/\rho l$ = (0·0981) to (0·2844) ms^{-2}

Full bore flow; Tables E give m_C or m_P for Colebrook-White or for laminar solutions resp., to divide mV and/or mQ as follows:

mV to give velocities in ms^{-1}
mQ to give discharges in litres/sec

Gradient (Equivalent) Pipe diameters in mm

S, ($\Delta p/\rho l$)	700	725	750	775	800	825	850	875	900	925	950	975	1000
0·0100	3·129	3·203	3·276	3·348	3·420	3·491	3·561	3·630	3·699	3·768	3·835	3·902	3·969
(0·0981)	1204·0	1322·2	1447·3	1579·5	1719·1	1866·1	2020·7	2183·1	2353·4	2531·8	2718·4	2913·4	3116·9
0·0105	3·206	3·282	3·357	3·431	3·504	3·577	3·649	3·720	3·791	3·861	3·930	3·998	4·067
(0·1030)	1233·8	1354·8	1483·0	1618·5	1761·5	1912·1	2070·6	2237·0	2411·5	2594·3	2785·5	2985·3	3193·8
0·0110	3·281	3·359	3·436	3·512	3·587	3·661	3·735	3·808	3·880	3·951	4·022	4·093	4·162
(0·1079)	1262·8	1386·7	1517·9	1656·6	1803·0	1957·1	2119·3	2289·6	2468·3	2655·4	2851·1	3055·6	3269·0
0·0115	3·355	3·435	3·513	3·591	3·667	3·744	3·819	3·893	3·967	4·040	4·113	4·185	4·256
(0·1128)	1291·2	1417·9	1552·0	1693·8	1843·5	2001·1	2166·9	2341·1	2523·7	2715·0	2915·2	3124·2	3342·4
0·0120	3·427	3·508	3·589	3·668	3·746	3·824	3·901	3·977	4·052	4·127	4·201	4·275	4·347
(0·1177)	1319·0	1448·4	1585·4	1730·3	1883·1	2044·2	2213·6	2391·5	2578·0	2773·4	2977·8	3191·4	3414·3
0·0125	3·498	3·581	3·663	3·744	3·824	3·903	3·981	4·059	4·136	4·212	4·288	4·363	4·437
(0·1226)	1346·2	1478·2	1618·1	1765·9	1922·0	2086·3	2259·2	2440·8	2631·2	2830·6	3039·3	3257·2	3484·7
0·0130	3·567	3·652	3·735	3·818	3·899	3·980	4·060	4·139	4·218	4·296	4·373	4·449	4·525
(0·1275)	1372·8	1507·5	1650·1	1800·9	1960·0	2127·6	2303·9	2489·1	2683·3	2886·7	3099·4	3321·7	3553·8
0·0135	3·635	3·721	3·806	3·890	3·974	4·056	4·138	4·218	4·298	4·377	4·456	4·534	4·611
(0·1324)	1399·0	1536·2	1681·6	1835·2	1997·4	2168·2	2347·8	2536·5	2734·4	2941·7	3158·5	3385·0	3621·5
0·0140	3·702	3·790	3·876	3·962	4·047	4·130	4·213	4·296	4·377	4·458	4·538	4·617	4·696
(0·1373)	1424·6	1564·4	1712·4	1868·9	2034·0	2208·0	2390·9	2583·1	2784·6	2995·7	3216·4	3447·1	3687·9
0·0145	3·767	3·857	3·945	4·032	4·118	4·204	4·288	4·372	4·455	4·537	4·618	4·699	4·779
(0·1422)	1449·9	1592·1	1742·7	1902·0	2070·0	2247·0	2433·2	2628·8	2833·9	3048·7	3273·4	3508·2	3753·2
0·0150	3·832	3·923	4·012	4·101	4·189	4·275	4·361	4·446	4·531	4·614	4·697	4·779	4·860
(0·1471)	1474·7	1619·3	1772·5	1934·5	2105·4	2285·5	2474·8	2673·7	2882·3	3100·8	3329·3	3568·1	3817·3
0·0160	3·957	4·051	4·144	4·235	4·326	4·416	4·504	4·592	4·679	4·766	4·851	4·936	5·020
(0·1569)	1523·0	1672·4	1830·7	1997·9	2174·4	2360·4	2556·0	2761·4	2976·8	3202·5	3438·5	3685·1	3942·5
0·0170	4·079	4·176	4·271	4·366	4·459	4·551	4·643	4·734	4·823	4·912	5·000	5·088	5·174
(0·1667)	1569·9	1723·9	1887·0	2059·4	2241·4	2433·0	2634·7	2846·4	3068·5	3301·0	3544·4	3798·6	4063·9
0·0180	4·198	4·297	4·395	4·492	4·588	4·683	4·778	4·871	4·963	5·055	5·145	5·235	5·324
(0·1765)	1615·4	1773·9	1941·7	2119·1	2306·4	2503·6	2711·0	2928·9	3157·4	3396·7	3647·1	3908·7	4181·7
0·0190	4·313	4·415	4·516	4·615	4·714	4·812	4·909	5·004	5·099	5·193	5·286	5·379	5·470
(0·1863)	1659·7	1822·5	1994·9	2177·2	2369·6	2572·2	2785·3	3009·2	3243·9	3489·8	3747·0	4015·8	4296·3
0·0200	4·425	4·529	4·633	4·735	4·837	4·937	5·036	5·134	5·232	5·328	5·424	5·518	5·612
(0·1961)	1702·8	1869·8	2046·7	2233·8	2431·1	2639·0	2857·7	3087·3	3328·2	3580·5	3844·4	4120·1	4407·9
0·0210	4·534	4·641	4·747	4·852	4·956	5·059	5·160	5·261	5·361	5·460	5·558	5·655	5·751
(0·2059)	1744·8	1916·0	2097·3	2288·9	2491·1	2704·2	2928·3	3163·6	3410·4	3668·9	3939·3	4221·9	4516·7
0·0220	4·641	4·750	4·859	4·966	5·073	5·178	5·282	5·385	5·487	5·588	5·688	5·788	5·886
(0·2157)	1785·9	1961·1	2146·6	2342·8	2549·8	2767·8	2997·2	3238·0	3490·7	3755·2	4032·0	4321·2	4623·0
0·0230	4·745	4·857	4·968	5·078	5·187	5·294	5·401	5·506	5·610	5·714	5·816	5·918	6·019
(0·2256)	1826·0	2005·2	2194·9	2395·4	2607·1	2830·0	3064·5	3310·8	3569·1	3839·6	4122·6	4418·3	4726·9
0·0240	4·847	4·962	5·075	5·187	5·298	5·408	5·517	5·624	5·731	5·837	5·941	6·045	6·148
(0·2354)	1865·3	2048·3	2242·1	2447·0	2663·1	2890·9	3130·4	3382·0	3645·9	3922·2	4211·3	4513·4	4828·6
0·0250	4·947	5·064	5·180	5·294	5·407	5·519	5·630	5·740	5·849	5·957	6·064	6·170	6·275
(0·2452)	1903·8	2090·5	2288·3	2497·4	2718·1	2950·5	3195·0	3451·8	3721·1	4003·1	4298·2	4606·4	4928·2
0·0260	5·045	5·164	5·282	5·399	5·515	5·629	5·742	5·854	5·965	6·075	6·184	6·292	6·399
(0·2550)	1941·5	2131·9	2333·6	2546·9	2771·9	3008·9	3258·3	3520·1	3794·7	4082·4	4383·3	4697·7	5025·8
0·0270	5·141	5·263	5·383	5·502	5·620	5·736	5·851	5·966	6·079	6·191	6·302	6·412	6·521
(0·2648)	1978·5	2172·5	2378·1	2595·4	2824·7	3066·3	3320·3	3587·2	3867·0	4160·2	4466·8	4787·1	5121·5
0·0280	5·235	5·359	5·482	5·603	5·723	5·841	5·959	6·075	6·190	6·304	6·417	6·529	6·641
(0·2746)	2014·8	2212·4	2421·7	2643·0	2876·5	3122·5	3381·3	3653·0	3938·0	4236·5	4548·7	4875·0	5215·5
0·0290	5·328	5·454	5·579	5·702	5·824	5·945	6·064	6·182	6·300	6·416	6·531	6·645	6·758
(0·2844)	2050·4	2251·6	2464·6	2689·8	2927·4	3177·8	3441·1	3717·7	4007·7	4311·5	4629·3	4961·3	5307·8
S, ($\Delta p/\rho l$)	700	725	750	775	800	825	850	875	900	925	950	975	1000

Gradient (Equivalent) Pipe diameters in mm

S = 0·0100 to 0·0290; $\qquad$ $\Delta p/\rho l$ = (0·0981) to (0·2844) ms^{-2}

m = Manning $n \times 100$
S = 0·0300 to 0·0950

i.e. kin. pr. grad., $\Delta p/\rho l$ =
(0·2942) to (0·9316) ms^{-2}

Full bore flow; Tables E give m_C or m_P for Colebrook-White or for laminar solutions resp., to divide mV and/or mQ as follows:

mV to give velocities in ms^{-1}
mQ to give discharges in litres/sec

Gradient (Equivalent) Pipe diameters in mm

S, $(\Delta p/\rho l)$	700	725	750	775	800	825	850	875	900	925	950	975	1000
0·0300	5·419	5·547	5·674	5·799	5·924	6·046	6·168	6·288	6·407	6·526	6·643	6·759	6·874
(0·2942)	2085·5	2290·0	2506·7	2735·8	2977·5	3232·1	3499·9	3781·2	4076·2	4385·2	4708·4	5046·1	5398·5
0·0320	5·597	5·729	5·860	5·990	6·118	6·245	6·370	6·494	6·618	6·740	6·860	6·980	7·099
(0·3138)	2153·9	2365·2	2588·9	2825·5	3075·1	3338·1	3614·7	3905·2	4209·9	4529·0	4862·8	5211·6	5575·6
0·0340	5·769	5·906	6·041	6·174	6·306	6·437	6·566	6·694	6·821	6·947	7·072	7·195	7·318
(0·3334)	2220·2	2437·9	2668·6	2912·5	3169·8	3440·9	3726·0	4025·4	4339·5	4668·4	5012·5	5372·0	5747·2
0·0360	5·936	6·077	6·216	6·353	6·489	6·623	6·757	6·888	7·019	7·148	7·277	7·404	7·530
(0·3530)	2284·5	2508·6	2746·0	2996·9	3261·7	3540·6	3834·0	4142·1	4465·3	4803·7	5157·8	5527·7	5913·8
0·0380	6·099	6·243	6·386	6·527	6·667	6·805	6·942	7·077	7·211	7·344	7·476	7·607	7·736
(0·3727)	2347·1	2577·4	2821·2	3079·0	3351·1	3637·6	3939·1	4255·6	4587·6	4935·4	5299·1	5679·2	6075·9
0·0400	6·257	6·405	6·552	6·697	6·840	6·982	7·122	7·261	7·399	7·535	7·670	7·804	7·937
(0·3923)	2408·1	2644·3	2894·5	3159·0	3438·1	3732·1	4041·4	4366·2	4706·8	5063·6	5436·8	5826·7	6233·7
0·0420	6·412	6·564	6·714	6·862	7·009	7·154	7·298	7·440	7·581	7·721	7·860	7·997	8·133
(0·4119)	2467·6	2709·6	2966·0	3237·0	3523·0	3824·3	4141·2	4474·0	4823·0	5188·6	5571·0	5970·6	6387·6
0·0440	6·563	6·718	6·872	7·024	7·174	7·322	7·470	7·615	7·760	7·903	8·045	8·185	8·324
(0·4315)	2525·6	2773·4	3035·8	3313·2	3605·9	3914·3	4238·6	4579·3	4936·5	5310·7	5702·1	6111·1	6538·0
0·0460	6·710	6·869	7·026	7·181	7·335	7·487	7·638	7·787	7·934	8·080	8·225	8·369	8·511
(0·4511)	2582·4	2835·7	3104·0	3387·7	3687·0	4002·3	4333·9	4682·2	5047·5	5430·1	5830·3	6248·5	6684·9
0·0480	6·855	7·017	7·177	7·336	7·493	7·648	7·802	7·954	8·105	8·254	8·402	8·549	8·695
(0·4707)	2637·9	2896·7	3170·8	3460·5	3766·3	4088·3	4427·1	4782·9	5156·0	5546·9	5955·7	6382·9	6828·7
0·0500	6·996	7·161	7·325	7·487	7·647	7·806	7·963	8·118	8·272	8·424	8·576	8·725	8·874
(0·4903)	2692·3	2956·4	3236·2	3531·9	3843·9	4172·6	4518·4	4881·5	5262·4	5661·2	6078·5	6514·5	6969·5
0·0525	7·169	7·338	7·506	7·672	7·836	7·999	8·159	8·318	8·476	8·632	8·787	8·941	9·093
(0·5148)	2758·8	3029·4	3316·1	3619·1	3938·8	4275·7	4630·0	5002·1	5392·3	5801·1	6228·6	6675·4	7141·6
0·0550	7·337	7·511	7·683	7·853	8·020	8·187	8·351	8·514	8·676	8·836	8·994	9·151	9·307
(0·5394)	2823·7	3100·7	3394·1	3704·3	4031·5	4376·3	4738·9	5119·8	5519·2	5937·6	6375·2	6832·4	7309·7
0·0575	7·502	7·680	7·855	8·029	8·201	8·371	8·539	8·706	8·871	9·034	9·196	9·357	9·516
(0·5639)	2887·2	3170·4	3470·4	3787·5	4122·1	4474·7	4845·4	5234·9	5643·3	6071·0	6518·5	6986·0	7473·9
0·0600	7·664	7·845	8·024	8·202	8·377	8·551	8·723	8·893	9·061	9·228	9·394	9·558	9·721
(0·5884)	2949·3	3238·6	3545·0	3869·0	4210·8	4570·9	4949·7	5347·4	5764·6	6201·6	6658·7	7136·3	7634·7
0·0625	7·822	8·007	8·190	8·371	8·550	8·727	8·903	9·076	9·248	9·419	9·588	9·755	9·921
(0·6129)	3010·1	3305·4	3618·1	3948·8	4297·6	4665·2	5051·7	5457·7	5883·5	6329·5	6796·0	7283·4	7792·1
0·0650	7·977	8·165	8·352	8·537	8·719	8·900	9·079	9·256	9·431	9·605	9·778	9·948	10·12
(0·6374)	3069·7	3370·9	3689·8	4027·0	4382·7	4757·5	5151·8	5565·8	6000·0	6454·8	6930·6	7427·7	7946·4
0·0675	8·128	8·321	8·511	8·699	8·885	9·069	9·252	9·432	9·611	9·788	9·964	10·14	10·31
(0·6619)	3128·2	3435·1	3760·1	4103·7	4466·2	4848·2	5249·9	5671·8	6114·3	6577·8	7062·6	7569·1	8097·8
0·0700	8·278	8·474	8·667	8·859	9·048	9·236	9·422	9·605	9·787	9·968	10·15	10·32	10·50
(0·6865)	3185·6	3498·1	3829·1	4179·0	4548·2	4937·1	5346·2	5775·9	6226·5	6698·5	7192·2	7708·0	8246·4
0·0725	8·424	8·624	8·821	9·016	9·208	9·399	9·588	9·775	9·961	10·14	10·33	10·51	10·69
(0·7110)	3242·0	3560·0	3896·9	4252·9	4628·7	5024·5	5440·9	5878·1	6336·7	6817·0	7319·5	7844·5	8392·4
0·0750	8·568	8·771	8·971	9·170	9·366	9·560	9·752	9·943	10·13	10·32	10·50	10·69	10·87
(0·7355)	3297·4	3620·9	3963·5	4325·6	4707·8	5110·4	5533·9	5978·6	6445·1	6933·6	7444·6	7978·6	8535·8
0·0800	8·849	9·059	9·266	9·470	9·673	9·874	10·07	10·27	10·46	10·66	10·85	11·04	11·22
(0·7845)	3405·6	3739·6	4093·5	4467·5	4862·2	5278·0	5715·4	6174·7	6656·4	7161·0	7688·8	8240·2	8815·8
0·0850	9·122	9·337	9·551	9·762	9·971	10·18	10·38	10·58	10·79	10·98	11·18	11·38	11·57
(0·8336)	3510·4	3854·7	4219·4	4605·0	5011·9	5440·5	5891·3	6364·7	6861·3	7381·4	7925·4	8493·8	9087·1
0·0900	9·386	9·608	9·828	10·04	10·26	10·47	10·68	10·89	11·10	11·30	11·51	11·71	11·91
(0·8826)	3612·1	3966·5	4341·8	4738·5	5157·2	5598·2	6062·1	6549·3	7060·2	7595·4	8155·2	8740·1	9350·6
0·0950	9·643	9·871	10·10	10·32	10·54	10·76	10·98	11·19	11·40	11·61	11·82	12·03	12·23
(0·9316)	3711·1	4075·2	4460·7	4868·4	5298·5	5751·6	6228·2	6728·7	7253·7	7803·5	8378·6	8979·6	9606·8
S, $(\Delta p/\rho l)$	700	725	750	775	800	825	850	875	900	925	950	975	1000

Gradient (Equivalent) Pipe diameters in mm

S = 0·0300 to 0·0950; $\Delta p/\rho l$ = (0·2942) to (0·9316) ms^{-2}

m = Manning $n \times 100$
S = 0.1000 to 0.3000

i.e. kin. pr. grad., $\Delta p/\rho l =$
(0.9807) to (2.9420) ms^{-2}

Full bore flow; Tables E give m_C or m_P for Colebrook-White or for laminar solutions resp., to divide mV and/or mQ as follows:

mV to give velocities in ms^{-1}
mQ to give discharges in litres/sec

Gradient   (Equivalent) Pipe diameters in mm

S, ($\Delta p/\rho l$)	700	725	750	775	800	825	850	875	900	925	950	975	1000
0.1000	9.894	10.13	10.36	10.59	10.81	11.04	11.26	11.48	11.70	11.91	12.13	12.34	12.55
(0.9807)	3807.5	4181.0	4576.6	4994.8	5436.1	5901.0	6390.0	6903.5	7442.1	8006.2	8596.3	9212.9	9856.3
0.1050	10.14	10.38	10.62	10.85	11.08	11.31	11.54	11.76	11.99	12.21	12.43	12.64	12.86
(1.0297)	3901.6	4284.3	4689.7	5118.2	5570.4	6046.7	6547.8	7074.0	7625.9	8203.9	8808.6	9440.4	10100
0.1100	10.38	10.62	10.87	11.11	11.34	11.58	11.81	12.04	12.27	12.50	12.72	12.94	13.16
(1.0787)	3993.4	4385.1	4800.0	5238.6	5701.5	6189.0	6701.9	7240.5	7805.3	8397.0	9015.9	9662.5	10337
0.1150	10.61	10.86	11.11	11.35	11.60	11.84	12.08	12.31	12.54	12.78	13.01	13.23	13.46
(1.1278)	4083.1	4483.7	4907.9	5356.4	5829.6	6328.1	6852.5	7403.2	7980.8	8585.7	9218.5	9879.7	10570
0.1200	10.84	11.09	11.35	11.60	11.85	12.09	12.34	12.58	12.81	13.05	13.29	13.52	13.75
(1.1768)	4170.9	4580.1	5013.4	5471.6	5955.0	6464.2	6999.9	7562.4	8152.4	8770.4	9416.8	10092	10797
0.1250	11.06	11.32	11.58	11.84	12.09	12.34	12.59	12.84	13.08	13.32	13.56	13.80	14.03
(1.2258)	4257.0	4674.5	5116.8	5584.4	6077.8	6597.5	7144.2	7718.4	8320.5	8951.2	9611.0	10300	11020
0.1300	11.28	11.55	11.81	12.07	12.33	12.59	12.84	13.09	13.34	13.58	13.83	14.07	14.31
(1.2749)	4341.3	4767.1	5218.2	5695.0	6198.1	6728.2	7285.7	7871.2	8485.3	9128.5	9801.3	10504	11238
0.1350	11.50	11.77	12.04	12.30	12.57	12.83	13.08	13.34	13.59	13.84	14.09	14.34	14.58
(1.3239)	4424.0	4857.9	5317.6	5803.5	6316.2	6856.4	7424.5	8021.2	8646.9	9302.4	9988.0	10704	11452
0.1400	11.71	11.98	12.26	12.53	12.80	13.06	13.32	13.58	13.84	14.10	14.35	14.60	14.85
(1.3729)	4505.1	4947.1	5415.1	5910.0	6432.1	6982.2	7560.7	8168.4	8805.6	9473.1	10171	10901	11662
0.1450	11.91	12.20	12.47	12.75	13.02	13.29	13.56	13.82	14.09	14.35	14.60	14.86	15.11
(1.4220)	4584.9	5034.6	5511.0	6014.6	6546.0	7105.8	7694.6	8312.9	8961.5	9640.8	10351	11094	11869
0.1500	12.12	12.40	12.69	12.97	13.25	13.52	13.79	14.06	14.33	14.59	14.85	15.11	15.37
(1.4710)	4663.3	5120.7	5605.2	6117.4	6657.9	7227.2	7826.1	8455.1	9114.7	9805.6	10528	11283	12072
0.1600	12.51	12.81	13.10	13.39	13.68	13.96	14.24	14.52	14.80	15.07	15.34	15.61	15.87
(1.5691)	4816.2	5288.6	5789.0	6318.0	6876.2	7464.3	8082.8	8732.3	9413.6	10127	10874	11653	12467
0.1700	12.90	13.21	13.51	13.81	14.10	14.39	14.68	14.97	15.25	15.53	15.81	16.09	16.36
(1.6671)	4964.4	5451.4	5967.2	6512.5	7087.8	7694.0	8331.5	9001.1	9703.3	10439	11208	12012	12851
0.1800	13.27	13.59	13.90	14.21	14.51	14.81	15.11	15.40	15.69	15.98	16.27	16.56	16.84
(1.7652)	5108.3	5609.4	6140.2	6701.3	7293.3	7917.0	8573.1	9262.0	9984.6	10741	11533	12360	13224
0.1900	13.64	13.96	14.28	14.60	14.91	15.22	15.52	15.82	16.12	16.42	16.72	17.01	17.30
(1.8633)	5248.3	5763.2	6308.5	6884.9	7493.2	8134.0	8808.0	9515.8	10258	11036	11849	12699	13586
0.2000	13.99	14.32	14.65	14.97	15.29	15.61	15.93	16.24	16.54	16.85	17.15	17.45	17.75
(1.9613)	5384.7	5912.9	6472.3	7063.8	7687.8	8345.3	9036.8	9763.1	10525	11322	12157	13029	13939
0.2100	14.34	14.68	15.01	15.34	15.67	16.00	16.32	16.64	16.95	17.26	17.57	17.88	18.19
(2.0594)	5517.6	6058.9	6632.2	7238.2	7877.7	8551.4	9260.0	10004	10785	11602	12457	13351	14283
0.2200	14.67	15.02	15.37	15.71	16.04	16.37	16.70	17.03	17.35	17.67	17.99	18.30	18.61
(2.1575)	5647.5	6201.5	6788.2	7408.5	8063.1	8752.6	9477.9	10240	11038	11875	12750	13665	14619
0.2300	15.00	15.36	15.71	16.06	16.40	16.74	17.08	17.41	17.74	18.07	18.39	18.71	19.03
(2.2555)	5774.4	6340.9	6940.8	7575.0	8244.3	8949.3	9690.9	10470	11287	12142	13037	13972	14948
0.2400	15.33	15.69	16.05	16.40	16.75	17.10	17.45	17.79	18.12	18.46	18.79	19.12	19.44
(2.3536)	5898.6	6477.2	7090.1	7738.0	8421.6	9141.8	9899.3	10695	11529	12403	13317	14273	15269
0.2500	15.64	16.01	16.38	16.74	17.10	17.45	17.81	18.15	18.50	18.84	19.18	19.51	19.84
(2.4517)	6020.2	6610.8	7236.3	7897.5	8595.3	9330.3	10103	10915	11767	12659	13592	14567	15584
0.2600	15.95	16.33	16.70	17.07	17.44	17.80	18.16	18.51	18.86	19.21	19.56	19.90	20.24
(2.5497)	6139.5	6741.7	7379.6	8053.9	8765.5	9515.1	10304	11132	12000	12910	13861	14855	15893
0.2700	16.26	16.64	17.02	17.40	17.77	18.14	18.50	18.86	19.22	19.58	19.93	20.28	20.62
(2.6478)	6256.4	6870.1	7520.2	8207.3	8932.5	9696.4	10500	11344	12229	13156	14125	15138	16196
0.2800	16.56	16.95	17.33	17.72	18.10	18.47	18.84	19.21	19.57	19.94	20.29	20.65	21.00
(2.7459)	6371.2	6996.2	7658.2	8357.9	9096.4	9874.3	10692	11552	12453	13397	14384	15416	16493
0.2900	16.85	17.25	17.64	18.03	18.42	18.80	19.18	19.55	19.92	20.29	20.65	21.01	21.37
(2.8439)	6484.0	7120.0	7793.7	8505.9	9257.4	10049	10882	11756	12673	13634	14639	15689	16785
0.3000	17.14	17.54	17.94	18.34	18.73	19.12	19.50	19.89	20.26	20.64	21.01	21.37	21.74
(2.9420)	6594.8	7241.8	7927.0	8651.3	9415.6	10221	11068	11957	12890	13867	14889	15957	17072
S, ($\Delta p/\rho l$)	700	725	750	775	800	825	850	875	900	925	950	975	1000

Gradient   (Equivalent) Pipe diameters in mm

S = 0.1000 to 0.3000;

$\Delta p/\rho l$ = (0.9807) to (2.9420) ms^{-2}

Tables D3 to D6 immediately before this point show values of mV in black

to give mean velocities V in ms^{-1}

and values of mQ in green to give discharges Q in litres/sec

(i.e. Q in m^{3}s^{-1}/1000)

Tables D7 to D10 following this point show values of mV in black

to give mean velocities V in ms^{-1}

and values of mQ in blue to give discharges Q in m^{3}s^{-1}

Tables E follow Tables D

Tables E show values of m_C or of m_P in black

to give Colebrook-White or laminar solutions respectively from Tables D

m = Manning $n \times 100$
S = 0·000100 to 0·000290

Full bore flow; Tables E give m_C or m_P for Colebrook-White or for laminar solutions resp., to divide mV and/or mQ as follows:

i.e. kin. pr. grad., $\Delta p/\rho l =$ (0·000981) to (0·002844) ms^{-2}

mV to give velocities in ms^{-1}
mQ to give discharges in m^{3}s^{-1}

Gradient S, ($\Delta p/\rho l$)	1000	1025	1050	1075	1100	1125	1150	1175	1200	1225	1250	1275	1300
(Equivalent) Pipe diameters in mm													
0·000100	0·397	0·403	0·410	0·416	0·423	0·429	0·436	0·442	0·448	0·454	0·461	0·467	0·473
(0·000981)	0·3117	0·3329	0·3550	0·3780	0·4019	0·4267	0·4525	0·4792	0·5068	0·5355	0·5651	0·5958	0·6274
0·000105	0·407	0·413	0·420	0·427	0·433	0·440	0·446	0·453	0·459	0·466	0·472	0·478	0·484
(0·001030)	0·3194	0·3411	0·3638	0·3873	0·4118	0·4372	0·4636	0·4910	0·5194	0·5487	0·5791	0·6105	0·6429
0·000110	0·416	0·423	0·430	0·437	0·444	0·450	0·457	0·463	0·470	0·477	0·483	0·489	0·496
(0·001079)	0·3269	0·3491	0·3723	0·3964	0·4215	0·4475	0·4745	0·5026	0·5316	0·5616	0·5927	0·6248	0·6581
0·000115	0·426	0·433	0·440	0·447	0·453	0·460	0·467	0·474	0·481	0·487	0·494	0·500	0·507
(0·001128)	0·3342	0·3570	0·3807	0·4053	0·4310	0·4576	0·4852	0·5138	0·5435	0·5742	0·6060	0·6389	0·6728
0·000120	0·435	0·442	0·449	0·456	0·463	0·470	0·477	0·484	0·491	0·498	0·504	0·511	0·518
(0·001177)	0·3414	0·3647	0·3889	0·4141	0·4402	0·4674	0·4956	0·5249	0·5552	0·5866	0·6191	0·6526	0·6873
0·000125	0·444	0·451	0·458	0·466	0·473	0·480	0·487	0·494	0·501	0·508	0·515	0·522	0·528
(0·001226)	0·3485	0·3722	0·3969	0·4226	0·4493	0·4771	0·5059	0·5357	0·5667	0·5987	0·6318	0·6661	0·7015
0·000130	0·452	0·460	0·467	0·475	0·482	0·489	0·497	0·504	0·511	0·518	0·525	0·532	0·539
(0·001275)	0·3554	0·3796	0·4048	0·4310	0·4582	0·4865	0·5159	0·5463	0·5779	0·6105	0·6443	0·6793	0·7154
0·000135	0·461	0·469	0·476	0·484	0·491	0·499	0·506	0·513	0·521	0·528	0·535	0·542	0·549
(0·001324)	0·3621	0·3868	0·4125	0·4392	0·4669	0·4958	0·5257	0·5567	0·5889	0·6222	0·6566	0·6922	0·7290
0·000140	0·470	0·477	0·485	0·493	0·500	0·508	0·515	0·523	0·530	0·538	0·545	0·552	0·559
(0·001373)	0·3688	0·3939	0·4200	0·4472	0·4755	0·5049	0·5354	0·5670	0·5997	0·6336	0·6687	0·7049	0·7424
0·000145	0·478	0·486	0·494	0·501	0·509	0·517	0·525	0·532	0·540	0·547	0·555	0·562	0·569
(0·001422)	0·3753	0·4009	0·4275	0·4552	0·4839	0·5138	0·5448	0·5770	0·6103	0·6448	0·6805	0·7174	0·7555
0·000150	0·486	0·494	0·502	0·510	0·518	0·526	0·534	0·541	0·549	0·556	0·564	0·571	0·579
(0·001471)	0·3817	0·4077	0·4348	0·4629	0·4922	0·5226	0·5541	0·5869	0·6207	0·6558	0·6921	0·7297	0·7684
0·000160	0·502	0·510	0·519	0·527	0·535	0·543	0·551	0·559	0·567	0·575	0·582	0·590	0·598
(0·001569)	0·3943	0·4211	0·4490	0·4781	0·5083	0·5397	0·5723	0·6061	0·6411	0·6773	0·7148	0·7536	0·7936
0·000170	0·517	0·526	0·535	0·543	0·551	0·560	0·568	0·576	0·584	0·592	0·600	0·608	0·616
(0·001667)	0·4064	0·4340	0·4629	0·4928	0·5240	0·5563	0·5899	0·6248	0·6608	0·6982	0·7368	0·7768	0·8181
0·000180	0·532	0·541	0·550	0·559	0·567	0·576	0·584	0·593	0·601	0·610	0·618	0·626	0·634
(0·001765)	0·4182	0·4466	0·4763	0·5071	0·5392	0·5725	0·6070	0·6429	0·6800	0·7184	0·7582	0·7993	0·8418
0·000190	0·547	0·556	0·565	0·574	0·583	0·592	0·600	0·609	0·618	0·626	0·635	0·643	0·652
(0·001863)	0·4296	0·4589	0·4893	0·5210	0·5540	0·5882	0·6237	0·6605	0·6986	0·7381	0·7790	0·8212	0·8649
0·000200	0·561	0·571	0·580	0·589	0·598	0·607	0·616	0·625	0·634	0·643	0·651	0·660	0·669
(0·001961)	0·4408	0·4708	0·5020	0·5345	0·5683	0·6034	0·6399	0·6776	0·7168	0·7573	0·7992	0·8425	0·8873
0·000210	0·575	0·585	0·594	0·603	0·613	0·622	0·631	0·640	0·649	0·658	0·667	0·676	0·685
(0·002059)	0·4517	0·4824	0·5144	0·5477	0·5824	0·6183	0·6557	0·6944	0·7345	0·7760	0·8189	0·8633	0·9092
0·000220	0·589	0·598	0·608	0·618	0·627	0·637	0·646	0·655	0·665	0·674	0·683	0·692	0·701
(0·002157)	0·4623	0·4938	0·5265	0·5606	0·5961	0·6329	0·6711	0·7107	0·7518	0·7942	0·8382	0·8837	0·9306
0·000230	0·602	0·612	0·622	0·632	0·641	0·651	0·661	0·670	0·680	0·689	0·698	0·708	0·717
(0·002256)	0·4727	0·5049	0·5384	0·5732	0·6095	0·6471	0·6862	0·7267	0·7687	0·8121	0·8571	0·9035	0·9515
0·000240	0·615	0·625	0·635	0·645	0·655	0·665	0·675	0·685	0·694	0·704	0·713	0·723	0·732
(0·002354)	0·4829	0·5157	0·5500	0·5856	0·6226	0·6610	0·7009	0·7423	0·7852	0·8296	0·8755	0·9230	0·9720
0·000250	0·627	0·638	0·648	0·658	0·669	0·679	0·689	0·699	0·709	0·718	0·728	0·738	0·747
(0·002452)	0·4928	0·5264	0·5613	0·5976	0·6354	0·6747	0·7154	0·7576	0·8014	0·8467	0·8935	0·9420	0·9921
0·000260	0·640	0·651	0·661	0·672	0·682	0·692	0·702	0·713	0·723	0·733	0·743	0·752	0·762
(0·002550)	0·5026	0·5368	0·5724	0·6095	0·6480	0·6880	0·7296	0·7726	0·8172	0·8634	0·9112	0·9606	1·0117
0·000270	0·652	0·663	0·674	0·684	0·695	0·705	0·716	0·726	0·736	0·747	0·757	0·767	0·777
(0·002648)	0·5122	0·5470	0·5833	0·6211	0·6604	0·7011	0·7435	0·7873	0·8328	0·8799	0·9286	0·9789	1·0310
0·000280	0·664	0·675	0·686	0·697	0·708	0·718	0·729	0·739	0·750	0·760	0·771	0·781	0·791
(0·002746)	0·5215	0·5570	0·5940	0·6325	0·6725	0·7140	0·7571	0·8018	0·8481	0·8960	0·9456	0·9969	1·0499
0·000290	0·676	0·687	0·698	0·709	0·720	0·731	0·742	0·753	0·763	0·774	0·784	0·795	0·805
(0·002844)	0·5308	0·5669	0·6045	0·6437	0·6844	0·7266	0·7705	0·8160	0·8631	0·9119	0·9624	1·0146	1·0685
S, ($\Delta p/\rho l$)	1000	1025	1050	1075	1100	1125	1150	1175	1200	1225	1250	1275	1300
Gradient **(Equivalent) Pipe diameters in mm**													

S = 0·000100 to 0·000290; $\Delta p/\rho l$ = (0·000981) to (0·002844) ms^{-2}

m = Manning $n \times 100$
S = 0.000300 to 0.000950

i.e. kin. pr. grad., $\Delta p/\rho l$ = (0.002942) to (0.009316) ms^{-2}

Full bore flow; Tables E give m_C or m_P for Colebrook-White or for laminar solutions resp., to divide mV and/or mQ as follows:

mV to give velocities in ms^{-1}
mQ to give discharges in m^{3}s^{-1}

Gradient   **(Equivalent) Pipe diameters in mm**

S, $(\Delta p/\rho l)$	1000	1025	1050	1075	1100	1125	1150	1175	1200	1225	1250	1275	1300
0.000300	0.687	0.699	0.710	0.721	0.732	0.744	0.754	0.765	0.776	0.787	0.798	0.808	0.819
(0.002942)	0.5399	0.5766	0.6149	0.6547	0.6961	0.7391	0.7837	0.8299	0.8779	0.9275	0.9788	1.0319	1.0867
0.000320	0.710	0.722	0.733	0.745	0.756	0.768	0.779	0.790	0.802	0.813	0.824	0.835	0.846
(0.003138)	0.5576	0.5955	0.6350	0.6762	0.7189	0.7633	0.8094	0.8572	0.9067	0.9579	1.0109	1.0657	1.1224
0.000340	0.732	0.744	0.756	0.768	0.780	0.792	0.803	0.815	0.826	0.838	0.849	0.860	0.872
(0.003334)	0.5747	0.6138	0.6546	0.6970	0.7410	0.7868	0.8343	0.8835	0.9346	0.9874	1.0420	1.0985	1.1569
0.000360	0.753	0.765	0.778	0.790	0.802	0.814	0.827	0.838	0.850	0.862	0.874	0.885	0.897
(0.003530)	0.5914	0.6316	0.6736	0.7172	0.7625	0.8096	0.8585	0.9091	0.9617	1.0160	1.0722	1.1304	1.1905
0.000380	0.774	0.786	0.799	0.812	0.824	0.837	0.849	0.861	0.874	0.886	0.898	0.910	0.921
(0.003727)	0.6076	0.6489	0.6920	0.7368	0.7834	0.8318	0.8820	0.9341	0.9880	1.0438	1.1016	1.1614	1.2231
0.000400	0.794	0.807	0.820	0.833	0.846	0.859	0.871	0.884	0.896	0.909	0.921	0.933	0.945
(0.003923)	0.6234	0.6658	0.7100	0.7560	0.8038	0.8534	0.9049	0.9583	1.0137	1.0710	1.1302	1.1915	1.2549
0.000420	0.813	0.827	0.840	0.853	0.867	0.880	0.893	0.906	0.918	0.931	0.944	0.956	0.969
(0.004119)	0.6388	0.6822	0.7275	0.7746	0.8236	0.8745	0.9273	0.9820	1.0387	1.0974	1.1582	1.2210	1.2858
0.000440	0.832	0.846	0.860	0.874	0.887	0.900	0.914	0.927	0.940	0.953	0.966	0.979	0.992
(0.004315)	0.6538	0.6983	0.7446	0.7929	0.8430	0.8951	0.9491	1.0051	1.0631	1.1232	1.1854	1.2497	1.3161
0.000460	0.851	0.865	0.879	0.893	0.907	0.921	0.934	0.948	0.961	0.974	0.988	1.001	1.014
(0.004511)	0.6685	0.7140	0.7614	0.8107	0.8619	0.9152	0.9704	1.0277	1.0870	1.1485	1.2121	1.2778	1.3457
0.000480	0.869	0.884	0.898	0.912	0.926	0.940	0.954	0.968	0.982	0.995	1.009	1.022	1.036
(0.004707)	0.6829	0.7293	0.7778	0.8281	0.8805	0.9349	0.9913	1.0498	1.1104	1.1732	1.2381	1.3053	1.3746
0.000500	0.887	0.902	0.917	0.931	0.946	0.960	0.974	0.988	1.002	1.016	1.030	1.043	1.057
(0.004903)	0.6969	0.7444	0.7938	0.8452	0.8986	0.9541	1.0117	1.0714	1.1333	1.1974	1.2637	1.3322	1.4030
0.000525	0.909	0.924	0.939	0.954	0.969	0.984	0.998	1.013	1.027	1.041	1.055	1.069	1.083
(0.005148)	0.7142	0.7628	0.8134	0.8661	0.9208	0.9777	1.0367	1.0979	1.1613	1.2269	1.2949	1.3651	1.4376
0.000550	0.931	0.946	0.961	0.977	0.992	1.007	1.022	1.036	1.051	1.066	1.080	1.094	1.109
(0.005394)	0.7310	0.7807	0.8325	0.8864	0.9425	1.0007	1.0611	1.1237	1.1886	1.2558	1.3253	1.3972	1.4715
0.000575	0.952	0.967	0.983	0.999	1.014	1.029	1.045	1.060	1.075	1.089	1.104	1.119	1.134
(0.005639)	0.7474	0.7983	0.8512	0.9064	0.9637	1.0232	1.0850	1.1490	1.2153	1.2840	1.3551	1.4286	1.5045
0.000600	0.972	0.988	1.004	1.020	1.036	1.051	1.067	1.082	1.098	1.113	1.128	1.143	1.158
(0.005884)	0.7635	0.8154	0.8696	0.9259	0.9844	1.0452	1.1083	1.1737	1.2415	1.3117	1.3843	1.4593	1.5369
0.000625	0.992	1.009	1.025	1.041	1.057	1.073	1.089	1.105	1.120	1.136	1.151	1.167	1.182
(0.006129)	0.7792	0.8322	0.8875	0.9450	1.0047	1.0668	1.1311	1.1979	1.2671	1.3387	1.4128	1.4894	1.5686
0.000650	1.012	1.029	1.045	1.062	1.078	1.094	1.111	1.127	1.143	1.158	1.174	1.190	1.205
(0.006374)	0.7946	0.8487	0.9051	0.9637	1.0246	1.0879	1.1535	1.2216	1.2922	1.3652	1.4408	1.5189	1.5996
0.000675	1.031	1.048	1.065	1.082	1.099	1.115	1.132	1.148	1.164	1.180	1.196	1.212	1.228
(0.006619)	0.8098	0.8649	0.9223	0.9820	1.0441	1.1086	1.1755	1.2449	1.3168	1.3912	1.4682	1.5478	1.6301
0.000700	1.050	1.067	1.085	1.102	1.119	1.136	1.152	1.169	1.186	1.202	1.218	1.235	1.251
(0.006865)	0.8246	0.8808	0.9392	1.0000	1.0633	1.1289	1.1971	1.2677	1.3410	1.4168	1.4952	1.5763	1.6600
0.000725	1.069	1.086	1.104	1.121	1.139	1.156	1.173	1.190	1.207	1.223	1.240	1.256	1.273
(0.007110)	0.8392	0.8964	0.9558	1.0177	1.0821	1.1489	1.2183	1.2902	1.3647	1.4418	1.5216	1.6042	1.6894
0.000750	1.087	1.105	1.123	1.141	1.158	1.176	1.193	1.210	1.227	1.244	1.261	1.278	1.295
(0.007355)	0.8536	0.9117	0.9722	1.0351	1.1006	1.1686	1.2391	1.3122	1.3880	1.4665	1.5477	1.6316	1.7183
0.000800	1.122	1.141	1.160	1.178	1.196	1.214	1.232	1.250	1.268	1.285	1.303	1.320	1.337
(0.007845)	0.8816	0.9416	1.0041	1.0691	1.1367	1.2069	1.2797	1.3553	1.4335	1.5146	1.5984	1.6851	1.7746
0.000850	1.157	1.176	1.195	1.214	1.233	1.252	1.270	1.288	1.307	1.325	1.343	1.360	1.378
(0.008336)	0.9087	0.9706	1.0350	1.1020	1.1717	1.2440	1.3191	1.3970	1.4777	1.5612	1.6476	1.7369	1.8293
0.000900	1.191	1.210	1.230	1.249	1.269	1.288	1.307	1.326	1.344	1.363	1.382	1.400	1.418
(0.008826)	0.9351	0.9987	1.0650	1.1339	1.2056	1.2801	1.3574	1.4375	1.5205	1.6064	1.6954	1.7873	1.8823
0.000950	1.223	1.243	1.264	1.284	1.303	1.323	1.343	1.362	1.381	1.400	1.419	1.438	1.457
(0.009316)	0.9607	1.0261	1.0942	1.1650	1.2387	1.3152	1.3946	1.4769	1.5622	1.6505	1.7418	1.8363	1.9339
S, $(\Delta p/\rho l)$	1000	1025	1050	1075	1100	1125	1150	1175	1200	1225	1250	1275	1300

Gradient   **(Equivalent) Pipe diameters in mm**

S = 0.000300 to 0.000950; $\Delta p/\rho l$ = (0.002942) to (0.009316) ms^{-2}

$m = $ Manning $n \times 100$
$S = 0.00100$ to 0.00290

i.e. kin. pr. grad., $\Delta p/\rho l = $
(0.00981) to (0.02844) ms^{-2}

Full bore flow; Tables E give m_C or m_P for Colebrook-White or for laminar solutions resp., to divide mV and/or mQ as follows:

mV to give velocities in ms^{-1}
mQ to give discharges in m^{3}s^{-1}

Gradient   **(Equivalent) Pipe diameters in mm**

S, ($\Delta p/\rho l$)	1000	1025	1050	1075	1100	1125	1150	1175	1200	1225	1250	1275	1300
0.00100	1.255	1.276	1.296	1.317	1.337	1.357	1.378	1.397	1.417	1.437	1.456	1.476	1.495
(0.00981)	0.9856	1.0527	1.1226	1.1953	1.2709	1.3493	1.4308	1.5152	1.6028	1.6933	1.7871	1.8840	1.9841
0.00105	1.286	1.307	1.328	1.349	1.370	1.391	1.412	1.432	1.452	1.472	1.492	1.512	1.532
(0.01030)	1.0100	1.0787	1.1503	1.2248	1.3022	1.3827	1.4661	1.5527	1.6423	1.7352	1.8312	1.9305	2.0331
0.00110	1.316	1.338	1.360	1.381	1.403	1.424	1.445	1.466	1.486	1.507	1.527	1.548	1.568
(0.01079)	1.0337	1.1041	1.1774	1.2536	1.3329	1.4152	1.5006	1.5892	1.6810	1.7760	1.8743	1.9759	2.0809
0.00115	1.346	1.368	1.390	1.412	1.434	1.456	1.477	1.499	1.520	1.541	1.562	1.582	1.603
(0.01128)	1.0570	1.1289	1.2038	1.2818	1.3628	1.4470	1.5344	1.6249	1.7188	1.8159	1.9164	2.0203	2.1277
0.00120	1.375	1.398	1.420	1.443	1.465	1.487	1.509	1.531	1.552	1.574	1.595	1.616	1.637
(0.01177)	1.0797	1.1532	1.2297	1.3094	1.3922	1.4781	1.5674	1.6599	1.7557	1.8550	1.9576	2.0638	2.1735
0.00125	1.403	1.426	1.449	1.472	1.495	1.518	1.540	1.562	1.584	1.606	1.628	1.650	1.671
(0.01226)	1.1020	1.1770	1.2551	1.3364	1.4209	1.5086	1.5997	1.6941	1.7919	1.8932	1.9980	2.1064	2.2183
0.00130	1.431	1.455	1.478	1.502	1.525	1.548	1.571	1.593	1.616	1.638	1.660	1.682	1.704
(0.01275)	1.1238	1.2003	1.2799	1.3628	1.4490	1.5385	1.6314	1.7276	1.8274	1.9307	2.0376	2.1481	2.2622
0.00135	1.458	1.482	1.506	1.530	1.554	1.577	1.601	1.624	1.647	1.669	1.692	1.714	1.737
(0.01324)	1.1452	1.2232	1.3043	1.3888	1.4766	1.5678	1.6624	1.7606	1.8622	1.9675	2.0764	2.1890	2.3053
0.00140	1.485	1.510	1.534	1.558	1.582	1.606	1.630	1.653	1.677	1.700	1.723	1.746	1.769
(0.01373)	1.1662	1.2456	1.3283	1.4143	1.5037	1.5966	1.6929	1.7929	1.8964	2.0036	2.1145	2.2292	2.3476
0.00145	1.511	1.536	1.561	1.586	1.610	1.635	1.659	1.683	1.706	1.730	1.754	1.777	1.800
(0.01422)	1.1869	1.2676	1.3518	1.4393	1.5303	1.6248	1.7229	1.8246	1.9300	2.0391	2.1519	2.2686	2.3892
0.00150	1.537	1.563	1.588	1.613	1.638	1.663	1.687	1.711	1.736	1.760	1.784	1.807	1.831
(0.01471)	1.2072	1.2893	1.3749	1.4639	1.5565	1.6526	1.7524	1.8558	1.9630	2.0739	2.1887	2.3074	2.4300
0.00160	1.587	1.614	1.640	1.666	1.692	1.717	1.742	1.768	1.793	1.817	1.842	1.866	1.891
(0.01569)	1.2467	1.3316	1.4200	1.5119	1.6075	1.7068	1.8098	1.9167	2.0273	2.1419	2.2605	2.3831	2.5097
0.00170	1.636	1.663	1.690	1.717	1.744	1.770	1.796	1.822	1.848	1.873	1.899	1.924	1.949
(0.01667)	1.2851	1.3726	1.4637	1.5585	1.6570	1.7593	1.8655	1.9756	2.0897	2.2079	2.3301	2.4564	2.5870
0.00180	1.684	1.712	1.739	1.767	1.794	1.821	1.848	1.875	1.901	1.928	1.954	1.980	2.006
(0.01765)	1.3224	1.4124	1.5061	1.6036	1.7050	1.8103	1.9196	2.0329	2.1503	2.2719	2.3976	2.5276	2.6620
0.00190	1.730	1.759	1.787	1.815	1.843	1.871	1.899	1.926	1.953	1.980	2.007	2.034	2.060
(0.01863)	1.3586	1.4511	1.5474	1.6476	1.7518	1.8599	1.9722	2.0886	2.2092	2.3341	2.4633	2.5969	2.7349
0.00200	1.775	1.804	1.833	1.862	1.891	1.920	1.948	1.976	2.004	2.032	2.059	2.087	2.114
(0.01961)	1.3939	1.4888	1.5876	1.6904	1.7973	1.9083	2.0234	2.1429	2.2666	2.3948	2.5273	2.6644	2.8059
0.00210	1.819	1.849	1.879	1.908	1.938	1.967	1.996	2.025	2.054	2.082	2.110	2.138	2.166
(0.02059)	1.4283	1.5255	1.6268	1.7321	1.8416	1.9554	2.0734	2.1958	2.3226	2.4539	2.5897	2.7301	2.8752
0.00220	1.861	1.892	1.923	1.953	1.984	2.013	2.043	2.073	2.102	2.131	2.160	2.189	2.217
(0.02157)	1.4619	1.5614	1.6651	1.7729	1.8850	2.0014	2.1222	2.2475	2.3773	2.5116	2.6507	2.7944	2.9429
0.00230	1.903	1.935	1.966	1.997	2.028	2.059	2.089	2.119	2.149	2.179	2.208	2.238	2.267
(0.02256)	1.4948	1.5965	1.7025	1.8127	1.9273	2.0464	2.1699	2.2980	2.4307	2.5681	2.7102	2.8572	3.0090
0.00240	1.944	1.976	2.008	2.040	2.072	2.103	2.134	2.165	2.195	2.226	2.256	2.286	2.316
(0.02354)	1.5269	1.6309	1.7391	1.8517	1.9688	2.0904	2.2166	2.3474	2.4830	2.6233	2.7685	2.9187	3.0738
0.00250	1.984	2.017	2.050	2.082	2.114	2.146	2.178	2.209	2.241	2.272	2.303	2.333	2.364
(0.02452)	1.5584	1.6645	1.7750	1.8899	2.0094	2.1335	2.2623	2.3958	2.5342	2.6774	2.8256	2.9788	3.1371
0.00260	2.024	2.057	2.090	2.123	2.156	2.189	2.221	2.253	2.285	2.317	2.348	2.379	2.410
(0.02550)	1.5893	1.6975	1.8101	1.9273	2.0492	2.1758	2.3071	2.4433	2.5844	2.7304	2.8816	3.0378	3.1993
0.00270	2.062	2.096	2.130	2.164	2.197	2.231	2.263	2.296	2.329	2.361	2.393	2.425	2.456
(0.02648)	1.6196	1.7298	1.8446	1.9641	2.0882	2.2172	2.3510	2.4898	2.6336	2.7825	2.9365	3.0957	3.2602
0.00280	2.100	2.135	2.169	2.204	2.238	2.271	2.305	2.338	2.371	2.404	2.437	2.469	2.501
(0.02746)	1.6493	1.7615	1.8785	2.0001	2.1265	2.2579	2.3942	2.5355	2.6819	2.8335	2.9903	3.1525	3.3200
0.00290	2.137	2.173	2.208	2.243	2.277	2.312	2.346	2.380	2.413	2.447	2.480	2.513	2.546
(0.02844)	1.6785	1.7927	1.9117	2.0355	2.1642	2.2978	2.4366	2.5804	2.7294	2.8837	3.0433	3.2083	3.3788
S, ($\Delta p/\rho l$)	1000	1025	1050	1075	1100	1125	1150	1175	1200	1225	1250	1275	1300

Gradient   **(Equivalent) Pipe diameters in mm**

$S = 0.00100$ to 0.00290; $\Delta p/\rho l = (0.00981)$ to (0.02844) ms^{-2}

m = Manning $n \times 100$
S = 0·00300 to 0·00950

Full bore flow; Tables E give m_C or m_P for Colebrook-White or for laminar solutions resp., to divide mV and/or mQ as follows:

i.e. kin. pr. grad., $\Delta p/\rho l$ = (0·02942) to (0·09316) ms^{-2}

mV to give velocities in ms^{-1}
mQ to give discharges in m^{3}s^{-1}

Gradient   **(Equivalent) Pipe diameters in mm**

S, ($\Delta p/\rho l$)	1000	1025	1050	1075	1100	1125	1150	1175	1200	1225	1250	1275	1300
0·00300	2·174	2·210	2·246	2·281	2·316	2·351	2·386	2·420	2·455	2·489	2·522	2·556	2·589
(0·02942)	1·7072	1·8234	1·9444	2·0703	2·2012	2·3371	2·4782	2·6245	2·7760	2·9330	3·0953	3·2632	3·4366
0·00320	2·245	2·282	2·319	2·356	2·392	2·428	2·464	2·500	2·535	2·570	2·605	2·640	2·674
(0·03138)	1·7632	1·8832	2·0081	2·1382	2·2734	2·4138	2·5595	2·7106	2·8671	3·0291	3·1968	3·3702	3·5493
0·00340	2·314	2·352	2·391	2·428	2·466	2·503	2·540	2·577	2·613	2·649	2·685	2·721	2·756
(0·03334)	1·8174	1·9411	2·0700	2·2040	2·3433	2·4881	2·6383	2·7940	2·9553	3·1224	3·2952	3·4739	3·6585
0·00360	2·381	2·421	2·460	2·499	2·537	2·576	2·614	2·651	2·689	2·726	2·763	2·800	2·836
(0·03530)	1·8701	1·9974	2·1300	2·2679	2·4113	2·5602	2·7147	2·8750	3·0410	3·2129	3·3907	3·5746	3·7646
0·00380	2·446	2·487	2·527	2·567	2·607	2·646	2·685	2·724	2·763	2·801	2·839	2·876	2·914
(0·03727)	1·9214	2·0521	2·1883	2·3300	2·4774	2·6304	2·7891	2·9538	3·1243	3·3009	3·4836	3·6726	3·8677
0·00400	2·510	2·552	2·593	2·634	2·675	2·715	2·755	2·795	2·834	2·874	2·912	2·951	2·990
(0·03923)	1·9713	2·1054	2·2452	2·3906	2·5417	2·6987	2·8616	3·0305	3·2055	3·3867	3·5741	3·7680	3·9682
0·00420	2·572	2·615	2·657	2·699	2·741	2·782	2·823	2·864	2·904	2·944	2·984	3·024	3·063
(0·04119)	2·0199	2·1574	2·3006	2·4496	2·6045	2·7653	2·9323	3·1053	3·2847	3·4703	3·6624	3·8610	4·0662
0·00440	2·632	2·676	2·719	2·762	2·805	2·847	2·889	2·931	2·973	3·014	3·055	3·095	3·136
(0·04315)	2·0675	2·2082	2·3548	2·5073	2·6658	2·8304	3·0013	3·1784	3·3620	3·5520	3·7486	3·9519	4·1619
0·00460	2·692	2·736	2·781	2·825	2·868	2·911	2·954	2·997	3·039	3·082	3·123	3·165	3·206
(0·04511)	2·1140	2·2578	2·4077	2·5636	2·7257	2·8940	3·0687	3·2498	3·4375	3·6318	3·8328	4·0407	4·2554
0·00480	2·749	2·795	2·840	2·885	2·930	2·974	3·018	3·062	3·105	3·148	3·190	3·233	3·275
(0·04707)	2·1594	2·3064	2·4595	2·6187	2·7843	2·9563	3·1347	3·3197	3·5115	3·7099	3·9153	4·1276	4·3470
0·00500	2·806	2·853	2·899	2·945	2·990	3·035	3·080	3·125	3·169	3·213	3·256	3·300	3·343
(0·04903)	2·2039	2·3540	2·5102	2·6727	2·8417	3·0172	3·1994	3·3882	3·5839	3·7864	3·9960	4·2127	4·4366
0·00525	2·875	2·923	2·971	3·017	3·064	3·110	3·156	3·202	3·247	3·292	3·337	3·381	3·425
(0·05148)	2·2584	2·4121	2·5722	2·7387	2·9119	3·0917	3·2784	3·4719	3·6724	3·8799	4·0947	4·3167	4·5462
0·00550	2·943	2·992	3·040	3·088	3·136	3·184	3·231	3·277	3·323	3·369	3·415	3·461	3·506
(0·05394)	2·3115	2·4688	2·6327	2·8032	2·9804	3·1645	3·3555	3·5536	3·7588	3·9712	4·1911	4·4183	4·6531
0·00575	3·009	3·059	3·109	3·158	3·207	3·255	3·303	3·351	3·398	3·445	3·492	3·538	3·584
(0·05639)	2·3635	2·5243	2·6919	2·8662	3·0474	3·2356	3·4309	3·6334	3·8433	4·0605	4·2853	4·5176	4·7577
0·00600	3·074	3·125	3·176	3·226	3·276	3·325	3·374	3·423	3·471	3·519	3·567	3·614	3·662
(0·05884)	2·4143	2·5786	2·7498	2·9278	3·1129	3·3052	3·5047	3·7116	3·9259	4·1478	4·3774	4·6148	4·8600
0·00625	3·137	3·189	3·241	3·292	3·343	3·394	3·444	3·493	3·543	3·592	3·641	3·689	3·737
(0·06129)	2·4641	2·6318	2·8065	2·9882	3·1771	3·3734	3·5770	3·7881	4·0069	4·2334	4·4677	4·7100	4·9603
0·00650	3·200	3·253	3·305	3·358	3·409	3·461	3·512	3·563	3·613	3·663	3·713	3·762	3·811
(0·06374)	2·5129	2·6839	2·8621	3·0474	3·2401	3·4402	3·6478	3·8631	4·0862	4·3172	4·5562	4·8032	5·0585
0·00675	3·260	3·315	3·368	3·422	3·474	3·527	3·579	3·631	3·682	3·733	3·783	3·834	3·884
(0·06619)	2·5608	2·7350	2·9166	3·1054	3·3018	3·5057	3·7173	3·9367	4·1641	4·3994	4·6430	4·8947	5·1549
0·00700	3·320	3·375	3·430	3·484	3·538	3·592	3·645	3·697	3·749	3·801	3·853	3·904	3·955
(0·06865)	2·6077	2·7852	2·9701	3·1624	3·3624	3·5700	3·7855	4·0090	4·2405	4·4802	4·7282	4·9845	5·2495
0·00725	3·379	3·435	3·491	3·546	3·601	3·655	3·709	3·763	3·816	3·869	3·921	3·973	4·025
(0·07110)	2·6539	2·8345	3·0227	3·2184	3·4219	3·6332	3·8525	4·0799	4·3155	4·5595	4·8118	5·0728	5·3424
0·00750	3·437	3·494	3·550	3·607	3·662	3·718	3·772	3·827	3·881	3·935	3·988	4·041	4·094
(0·07355)	2·6993	2·8830	3·0743	3·2734	3·4804	3·6953	3·9184	4·1497	4·3893	4·6374	4·8941	5·1595	5·4337
0·00800	3·550	3·608	3·667	3·725	3·782	3·839	3·896	3·952	4·008	4·064	4·119	4·174	4·228
(0·07845)	2·7878	2·9775	3·1752	3·3808	3·5945	3·8165	4·0469	4·2858	4·5333	4·7895	5·0546	5·3287	5·6119
0·00850	3·659	3·720	3·780	3·840	3·899	3·958	4·016	4·074	4·132	4·189	4·246	4·302	4·358
(0·08336)	2·8736	3·0692	3·2729	3·4848	3·7052	3·9340	4·1714	4·4177	4·6728	4·9369	5·2102	5·4927	5·7846
0·00900	3·765	3·827	3·889	3·951	4·012	4·072	4·133	4·192	4·251	4·310	4·369	4·427	4·484
(0·08826)	2·9569	3·1582	3·3678	3·5859	3·8126	4·0480	4·2924	4·5457	4·8083	5·0800	5·3612	5·6519	5·9523
0·00950	3·868	3·932	3·996	4·059	4·122	4·184	4·246	4·307	4·368	4·428	4·488	4·548	4·607
(0·09316)	3·0379	3·2447	3·4601	3·6841	3·9170	4·1590	4·4100	4·6703	4·9400	5·2192	5·5081	5·8068	6·1154
S, ($\Delta p/\rho l$)	1000	1025	1050	1075	1100	1125	1150	1175	1200	1225	1250	1275	1300

Gradient   **(Equivalent) Pipe diameters in mm**

S = 0·00300 to 0·00950; $\Delta p/\rho l$ = (0·02942) to (0·09316) ms^{-2}

m = Manning $n \times 100$
S = 0·0100 to 0·0290

i.e. kin. pr. grad., $\Delta p/\rho l =$
(0·0981) to (0·2844) ms^{-2}

Full bore flow; Tables E give m_C or m_P for Colebrook-White or for laminar solutions resp., to divide mV and/or mQ as follows:

mV to give velocities in ms^{-1}
mQ to give discharges in m^{3}s^{-1}

Gradient   (Equivalent) Pipe diameters in mm

S, ($\Delta p/\rho l$)	1000	1025	1050	1075	1100	1125	1150	1175	1200	1225	1250	1275	1300
0·0100	3·969	4·034	4·100	4·165	4·229	4·293	4·356	4·419	4·481	4·543	4·605	4·666	4·727
(0·0981)	3·1169	3·3290	3·5499	3·7798	4·0188	4·2670	4·5246	4·7916	5·0683	5·3548	5·6512	5·9577	6·2743
0·0105	4·067	4·134	4·201	4·267	4·333	4·399	4·464	4·528	4·592	4·656	4·719	4·781	4·844
(0·1030)	3·1938	3·4112	3·6376	3·8732	4·1180	4·3724	4·6363	4·9100	5·1935	5·4871	5·7908	6·1048	6·4292
0·0110	4·162	4·231	4·300	4·368	4·435	4·502	4·569	4·635	4·700	4·765	4·830	4·894	4·958
(0·1079)	3·2690	3·4915	3·7232	3·9643	4·2150	4·4753	4·7454	5·0255	5·3157	5·6162	5·9271	6·2485	6·5805
0·0115	4·256	4·326	4·396	4·466	4·535	4·603	4·671	4·739	4·806	4·872	4·938	5·004	5·069
(0·1128)	3·3424	3·5699	3·8069	4·0534	4·3097	4·5758	4·8521	5·1385	5·4352	5·7424	6·0603	6·3889	6·7284
0·0120	4·347	4·419	4·491	4·562	4·632	4·702	4·772	4·841	4·909	4·977	5·045	5·112	5·178
(0·1177)	3·4143	3·6467	3·8888	4·1406	4·4024	4·6743	4·9564	5·2490	5·5521	5·8659	6·1906	6·5263	6·8731
0·0125	4·437	4·511	4·584	4·656	4·728	4·799	4·870	4·941	5·010	5·080	5·149	5·217	5·285
(0·1226)	3·4847	3·7219	3·9690	4·2260	4·4932	4·7707	5·0586	5·3572	5·6666	5·9869	6·3183	6·6609	7·0149
0·0130	4·525	4·600	4·674	4·748	4·822	4·894	4·967	5·038	5·110	5·180	5·251	5·320	5·390
(0·1275)	3·5538	3·7956	4·0476	4·3097	4·5821	4·8651	5·1588	5·4633	5·7788	6·1054	6·4434	6·7928	7·1538
0·0135	4·611	4·688	4·763	4·839	4·913	4·988	5·061	5·134	5·207	5·279	5·351	5·422	5·492
(0·1324)	3·6215	3·8679	4·1247	4·3918	4·6694	4·9578	5·2571	5·5674	5·8889	6·2217	6·5661	6·9222	7·2901
0·0140	4·696	4·774	4·851	4·928	5·004	5·079	5·154	5·229	5·302	5·376	5·449	5·521	5·593
(0·1373)	3·6879	3·9389	4·2003	4·4724	4·7551	5·0488	5·3535	5·6695	5·9969	6·3359	6·6866	7·0492	7·4238
0·0145	4·779	4·858	4·937	5·015	5·092	5·169	5·245	5·321	5·396	5·471	5·545	5·619	5·692
(0·1422)	3·7532	4·0086	4·2747	4·5515	4·8393	5·1381	5·4483	5·7699	6·1031	6·4481	6·8050	7·1740	7·5553
0·0150	4·860	4·941	5·021	5·100	5·179	5·257	5·335	5·412	5·489	5·565	5·640	5·715	5·789
(0·1471)	3·8173	4·0772	4·3478	4·6293	4·9220	5·2260	5·5414	5·8685	6·2074	6·5583	6·9213	7·2966	7·6844
0·0160	5·020	5·103	5·186	5·268	5·349	5·430	5·510	5·590	5·669	5·747	5·825	5·902	5·979
(0·1569)	3·9425	4·2109	4·4904	4·7811	5·0834	5·3974	5·7232	6·0610	6·4110	6·7734	7·1483	7·5359	7·9364
0·0170	5·174	5·260	5·345	5·430	5·514	5·597	5·680	5·762	5·843	5·924	6·004	6·084	6·163
(0·1667)	4·0639	4·3405	4·6286	4·9283	5·2399	5·5635	5·8993	6·2475	6·6083	6·9818	7·3683	7·7679	8·1807
0·0180	5·324	5·413	5·500	5·587	5·674	5·759	5·844	5·929	6·012	6·096	6·178	6·260	6·342
(0·1765)	4·1817	4·4663	4·7627	5·0712	5·3918	5·7248	6·0703	6·4287	6·7999	7·1843	7·5819	7·9931	8·4178
0·0190	5·470	5·561	5·651	5·740	5·829	5·917	6·004	6·091	6·177	6·263	6·348	6·432	6·516
(0·1863)	4·2963	4·5887	4·8933	5·2101	5·5395	5·8817	6·2367	6·6048	6·9862	7·3811	7·7897	8·2121	8·6485
0·0200	5·612	5·705	5·798	5·890	5·980	6·071	6·160	6·249	6·338	6·425	6·513	6·599	6·685
(0·1961)	4·4079	4·7079	5·0204	5·3455	5·6834	6·0344	6·3987	6·7764	7·1677	7·5729	7·9920	8·4254	8·8732
0·0210	5·751	5·846	5·941	6·035	6·128	6·221	6·312	6·404	6·494	6·584	6·673	6·762	6·850
(0·2059)	4·5167	4·8242	5·1443	5·4775	5·8238	6·1835	6·5567	6·9437	7·3447	7·7599	8·1894	8·6335	9·0923
0·0220	5·886	5·984	6·081	6·177	6·272	6·367	6·461	6·554	6·647	6·739	6·830	6·921	7·011
(0·2157)	4·6230	4·9377	5·2654	5·6064	5·9608	6·3290	6·7110	7·1071	7·5176	7·9425	8·3821	8·8367	9·3063
0·0230	6·019	6·118	6·218	6·316	6·413	6·510	6·606	6·702	6·796	6·890	6·984	7·077	7·169
(0·2256)	4·7269	5·0487	5·3837	5·7324	6·0948	6·4712	6·8618	7·2669	7·6865	8·1210	8·5705	9·0353	9·5154
0·0240	6·148	6·250	6·351	6·452	6·551	6·650	6·748	6·846	6·943	7·039	7·134	7·229	7·323
(0·2354)	4·8286	5·1573	5·4995	5·8557	6·2259	6·6104	7·0094	7·4232	7·8518	8·2957	8·7548	9·2296	9·7201
0·0250	6·275	6·379	6·482	6·585	6·686	6·787	6·888	6·987	7·086	7·184	7·281	7·378	7·474
(0·2452)	4·9282	5·2636	5·6129	5·9764	6·3543	6·7467	7·1540	7·5762	8·0138	8·4667	8·9354	9·4199	9·9205
0·0260	6·399	6·505	6·611	6·715	6·819	6·922	7·024	7·125	7·226	7·326	7·425	7·524	7·622
(0·2550)	5·0258	5·3678	5·7241	6·0948	6·4801	6·8803	7·2956	7·7263	8·1725	8·6344	9·1123	9·6065	10·117
0·0270	6·521	6·629	6·737	6·843	6·949	7·054	7·158	7·261	7·364	7·466	7·567	7·667	7·767
(0·2648)	5·1215	5·4701	5·8331	6·2109	6·6036	7·0114	7·4346	7·8735	8·3281	8·7989	9·2859	9·7895	10·310
0·0280	6·641	6·751	6·860	6·969	7·076	7·183	7·289	7·394	7·499	7·603	7·706	7·808	7·910
(0·2746)	5·2155	5·5705	5·9402	6·3249	6·7247	7·1401	7·5710	8·0179	8·4810	8·9603	9·4563	9·9691	10·499
0·0290	6·758	6·870	6·982	7·092	7·201	7·310	7·418	7·525	7·632	7·737	7·842	7·946	8·050
(0·2844)	5·3078	5·6691	6·0453	6·4368	6·8438	7·2664	7·7051	8·1599	8·6311	9·1189	9·6237	10·146	10·685
S, ($\Delta p/\rho l$)	1000	1025	1050	1075	1100	1125	1150	1175	1200	1225	1250	1275	1300

Gradient   (Equivalent) Pipe diameters in mm

S = 0·0100 to 0·0290; $\Delta p/\rho l$ = (0·0981) to (0·2844) ms^{-2}

m = Manning $n \times 100$
S = 0·0300 to 0·1000

Full bore flow; Tables E give m_C or m_P for Colebrook-White or for laminar solutions resp., to divide mV and/or mQ as follows:

i.e. kin. pr. grad., $\Delta p/\rho l$ = (0·2942) to (0·9807) ms^{-2}

mV to give velocities in ms^{-1}
mQ to give discharges in m^{3}s^{-1}

Gradient (Equivalent) **Pipe diameters in mm**

S, ($\Delta p/\rho l$)	1000	1025	1050	1075	1100	1125	1150	1175	1200	1225	1250	1275	1300
0·0300	6·874	6·988	7·101	7·213	7·325	7·435	7·545	7·654	7·762	7·869	7·976	8·082	8·187
(0·2942)	5·3985	5·7660	6·1487	6·5469	6·9608	7·3907	7·8368	8·2994	8·7786	9·2748	9·7882	10·319	10·867
0·0320	7·099	7·217	7·334	7·450	7·565	7·679	7·792	7·905	8·017	8·128	8·238	8·347	8·456
(0·3138)	5·5756	5·9551	6·3503	6·7616	7·1890	7·6330	8·0938	8·5715	9·0665	9·5790	10·109	10·657	11·224
0·0340	7·318	7·439	7·559	7·679	7·798	7·915	8·032	8·148	8·263	8·378	8·491	8·604	8·716
(0·3334)	5·7472	6·1384	6·5458	6·9697	7·4103	7·8680	8·3429	8·8353	9·3456	9·8738	10·420	10·985	11·569
0·0360	7·530	7·655	7·779	7·902	8·024	8·145	8·265	8·384	8·503	8·621	8·737	8·854	8·969
(0·3530)	5·9138	6·3163	6·7355	7·1717	7·6251	8·0961	8·5848	9·0915	9·6165	10·160	10·722	11·304	11·905
0·0380	7·736	7·864	7·992	8·118	8·244	8·368	8·491	8·614	8·736	8·857	8·977	9·096	9·215
(0·3727)	6·0759	6·4894	6·9201	7·3682	7·8341	8·3179	8·8200	9·3406	9·8800	10·438	11·016	11·614	12·231
0·0400	7·937	8·069	8·199	8·329	8·458	8·585	8·712	8·838	8·963	9·087	9·210	9·332	9·454
(0·3923)	6·2337	6·6580	7·0999	7·5597	8·0376	8·5340	9·0491	9·5833	10·137	10·710	11·302	11·915	12·549
0·0420	8·133	8·268	8·402	8·535	8·667	8·797	8·927	9·056	9·184	9·311	9·438	9·563	9·688
(0·4119)	6·3876	6·8224	7·2752	7·7463	8·2361	8·7447	9·2726	9·8199	10·387	10·974	11·582	12·210	12·858
0·0440	8·324	8·463	8·600	8·736	8·870	9·004	9·137	9·269	9·400	9·530	9·660	9·788	9·916
(0·4315)	6·5380	6·9830	7·4464	7·9286	8·4299	8·9505	9·4908	10·051	10·631	11·232	11·854	12·497	13·161
0·0460	8·511	8·653	8·793	8·932	9·070	9·207	9·343	9·478	9·612	9·745	9·877	10·01	10·14
(0·4511)	6·6849	7·1399	7·6138	8·1068	8·6194	9·1517	9·7041	10·277	10·870	11·485	12·121	12·778	13·457
0·0480	8·695	8·839	8·982	9·124	9·265	9·405	9·544	9·681	9·818	9·954	10·09	10·22	10·36
(0·4707)	6·8287	7·2935	7·7775	8·2812	8·8048	9·3485	9·9128	10·498	11·104	11·732	12·381	13·053	13·746
0·0500	8·874	9·021	9·167	9·312	9·456	9·599	9·740	9·881	10·02	10·16	10·30	10·43	10·57
(0·4903)	6·9695	7·4439	7·9379	8·4519	8·9863	9·5413	10·117	10·714	11·333	11·974	12·637	13·322	14·030
0·0525	9·093	9·244	9·394	9·542	9·689	9·836	9·981	10·13	10·27	10·41	10·55	10·69	10·83
(0·5148)	7·1416	7·6277	8·1339	8·6607	9·2082	9·7769	10·367	10·979	11·613	12·269	12·949	13·651	14·376
0·0550	9·307	9·461	9·615	9·767	9·918	10·07	10·22	10·36	10·51	10·66	10·80	10·94	11·09
(0·5394)	7·3097	7·8072	8·3253	8·8645	9·4249	10·007	10·611	11·237	11·886	12·558	13·253	13·972	14·715
0·0575	9·516	9·674	9·831	9·986	10·14	10·29	10·45	10·60	10·75	10·89	11·04	11·19	11·34
(0·5639)	7·4739	7·9826	8·5125	9·0637	9·6367	10·232	10·850	11·490	12·153	12·840	13·551	14·286	15·045
0·0600	9·721	9·882	10·04	10·20	10·36	10·51	10·67	10·82	10·98	11·13	11·28	11·43	11·58
(0·5884)	7·6347	8·1543	8·6955	9·2586	9·8440	10·452	11·083	11·737	12·415	13·117	13·843	14·593	15·369
0·0625	9·921	10·09	10·25	10·41	10·57	10·73	10·89	11·05	11·20	11·36	11·51	11·67	11·82
(0·6129)	7·7921	8·3225	8·8748	9·4496	10·047	10·668	11·311	11·979	12·671	13·387	14·128	14·894	15·686
0·0650	10·12	10·29	10·45	10·62	10·78	10·94	11·11	11·27	11·43	11·58	11·74	11·90	12·05
(0·6374)	7·9464	8·4873	9·0506	9·6367	10·246	10·879	11·535	12·216	12·922	13·652	14·408	15·189	15·996
0·0675	10·31	10·48	10·65	10·82	10·99	11·15	11·32	11·48	11·64	11·80	11·96	12·12	12·28
(0·6619)	8·0978	8·6490	9·2230	9·8203	10·441	11·086	11·755	12·449	13·168	13·912	14·682	15·478	16·301
0·0700	10·50	10·67	10·85	11·02	11·19	11·36	11·52	11·69	11·86	12·02	12·18	12·35	12·51
(0·6865)	8·2464	8·8077	9·3923	10·000	10·633	11·289	11·971	12·677	13·410	14·168	14·952	15·763	16·600
0·0725	10·69	10·86	11·04	11·21	11·39	11·56	11·73	11·90	12·07	12·23	12·40	12·56	12·73
(0·7110)	8·3924	8·9636	9·5585	10·177	10·821	11·489	12·183	12·902	13·647	14·418	15·216	16·042	16·894
0·0750	10·87	11·05	11·23	11·41	11·58	11·76	11·93	12·10	12·27	12·44	12·61	12·78	12·95
(0·7355)	8·5358	9·1168	9·7219	10·351	11·006	11·686	12·391	13·122	13·880	14·665	15·477	16·316	17·183
0·0800	11·22	11·41	11·60	11·78	11·96	12·14	12·32	12·50	12·68	12·85	13·03	13·20	13·37
(0·7845)	8·8158	9·4158	10·041	10·691	11·367	12·069	12·797	13·553	14·335	15·146	15·984	16·851	17·746
0·0850	11·57	11·76	11·95	12·14	12·33	12·52	12·70	12·88	13·07	13·25	13·43	13·60	13·78
(0·8336)	9·0871	9·7056	10·350	11·020	11·717	12·440	13·191	13·970	14·777	15·612	16·476	17·369	18·293
0·0900	11·91	12·10	12·30	12·49	12·69	12·88	13·07	13·26	13·44	13·63	13·82	14·00	14·18
(0·8826)	9·3506	9·9870	10·650	11·339	12·056	12·801	13·574	14·375	15·205	16·064	16·954	17·873	18·823
0·0950	12·23	12·43	12·64	12·84	13·03	13·23	13·43	13·62	13·81	14·00	14·19	14·38	14·57
(0·9316)	9·6068	10·261	10·942	11·650	12·387	13·152	13·946	14·769	15·622	16·505	17·418	18·363	19·339
0·1000	12·55	12·76	12·96	13·17	13·37	13·57	13·78	13·97	14·17	14·37	14·56	14·76	14·95
(0·9807)	9·8563	10·527	11·226	11·953	12·709	13·493	14·308	15·152	16·028	16·933	17·871	18·840	19·841
S, ($\Delta p/\rho l$)	1000	1025	1050	1075	1100	1125	1150	1175	1200	1225	1250	1275	1300

Gradient (Equivalent) **Pipe diameters in mm**

S = 0·0300 to 0·1000; $\Delta p/\rho l$ = (0·2942) to (0·9807) ms^{-2}

m = Manning $n \times 100$
S = 0·000100 to 0·000290

i.e. kin. pr. grad., $\Delta p/\rho l$ =
(0·000981) to (0·002844) ms^{-2}

Full bore flow; Tables E give m_C or m_P for Colebrook-White or for laminar solutions resp., to divide mV and/or mQ as follows:

mV to give velocities in ms^{-1}
mQ to give discharges in m^{3}s^{-1}

Gradient — (Equivalent) Pipe diameters in mm

S, ($\Delta p/\rho l$)	1300	1325	1350	1375	1400	1425	1450	1475	1500	1525	1550	1575	1600
0·000100	0·473	0·479	0·485	0·491	0·497	0·503	0·508	0·514	0·520	0·526	0·532	0·537	0·543
(0·000981)	0·6274	0·6601	0·6939	0·7287	0·7645	0·8015	0·8395	0·8787	0·9190	0·9604	1·0029	1·0466	1·0915
0·000105	0·484	0·491	0·497	0·503	0·509	0·515	0·521	0·527	0·533	0·539	0·545	0·550	0·556
(0·001030)	0·6429	0·6764	0·7110	0·7467	0·7834	0·8213	0·8603	0·9004	0·9416	0·9841	1·0277	1·0725	1·1185
0·000110	0·496	0·502	0·508	0·515	0·521	0·527	0·533	0·539	0·545	0·551	0·557	0·563	0·569
(0·001079)	0·6581	0·6923	0·7277	0·7642	0·8018	0·8406	0·8805	0·9216	0·9638	1·0072	1·0519	1·0977	1·1448
0·000115	0·507	0·513	0·520	0·526	0·533	0·539	0·545	0·551	0·558	0·564	0·570	0·576	0·582
(0·001128)	0·6728	0·7079	0·7441	0·7814	0·8199	0·8595	0·9003	0·9423	0·9855	1·0299	1·0755	1·1224	1·1705
0·000120	0·518	0·524	0·531	0·538	0·544	0·551	0·557	0·563	0·570	0·576	0·582	0·588	0·595
(0·001177)	0·6873	0·7231	0·7601	0·7982	0·8375	0·8780	0·9196	0·9625	1·0067	1·0520	1·0986	1·1465	1·1957
0·000125	0·528	0·535	0·542	0·549	0·555	0·562	0·568	0·575	0·581	0·588	0·594	0·601	0·607
(0·001226)	0·7015	0·7380	0·7758	0·8147	0·8548	0·8961	0·9386	0·9824	1·0274	1·0737	1·1213	1·1702	1·2204
0·000130	0·539	0·546	0·553	0·559	0·566	0·573	0·580	0·586	0·593	0·599	0·606	0·613	0·619
(0·001275)	0·7154	0·7527	0·7911	0·8308	0·8717	0·9138	0·9572	1·0018	1·0478	1·0950	1·1435	1·1934	1·2445
0·000135	0·549	0·556	0·563	0·570	0·577	0·584	0·591	0·597	0·604	0·611	0·618	0·624	0·631
(0·001324)	0·7290	0·7670	0·8062	0·8466	0·8883	0·9312	0·9754	1·0209	1·0677	1·1158	1·1653	1·2161	1·2682
0·000140	0·559	0·566	0·574	0·581	0·588	0·595	0·602	0·608	0·615	0·622	0·629	0·636	0·642
(0·001373)	0·7424	0·7811	0·8210	0·8622	0·9046	0·9483	0·9933	1·0397	1·0873	1·1363	1·1867	1·2384	1·2915
0·000145	0·569	0·576	0·584	0·591	0·598	0·605	0·612	0·619	0·626	0·633	0·640	0·647	0·654
(0·001422)	0·7555	0·7949	0·8355	0·8774	0·9206	0·9651	1·0109	1·0581	1·1066	1·1564	1·2077	1·2603	1·3144
0·000150	0·579	0·586	0·594	0·601	0·608	0·615	0·623	0·630	0·637	0·644	0·651	0·658	0·665
(0·001471)	0·7684	0·8085	0·8498	0·8924	0·9363	0·9816	1·0282	1·0762	1·1255	1·1762	1·2283	1·2819	1·3368
0·000160	0·598	0·606	0·613	0·621	0·628	0·636	0·643	0·650	0·658	0·665	0·672	0·680	0·687
(0·001569)	0·7936	0·8350	0·8777	0·9217	0·9671	1·0138	1·0619	1·1114	1·1624	1·2148	1·2686	1·3239	1·3807
0·000170	0·616	0·624	0·632	0·640	0·648	0·655	0·663	0·670	0·678	0·686	0·693	0·700	0·708
(0·001667)	0·8181	0·8607	0·9047	0·9501	0·9968	1·0450	1·0946	1·1457	1·1982	1·2522	1·3077	1·3647	1·4232
0·000180	0·634	0·642	0·650	0·658	0·666	0·674	0·682	0·690	0·698	0·705	0·713	0·721	0·728
(0·001765)	0·8418	0·8856	0·9309	0·9776	1·0257	1·0753	1·1263	1·1789	1·2329	1·2885	1·3456	1·4042	1·4644
0·000190	0·652	0·660	0·668	0·676	0·685	0·693	0·701	0·709	0·717	0·725	0·733	0·740	0·748
(0·001863)	0·8649	0·9099	0·9564	1·0044	1·0538	1·1048	1·1572	1·2112	1·2667	1·3238	1·3824	1·4427	1·5046
0·000200	0·669	0·677	0·686	0·694	0·702	0·711	0·719	0·727	0·735	0·744	0·752	0·760	0·768
(0·001961)	0·8873	0·9336	0·9813	1·0305	1·0812	1·1335	1·1873	1·2426	1·2996	1·3582	1·4183	1·4802	1·5437
0·000210	0·685	0·694	0·702	0·711	0·720	0·728	0·737	0·745	0·754	0·762	0·770	0·778	0·787
(0·002059)	0·9092	0·9566	1·0055	1·0559	1·1079	1·1614	1·2166	1·2733	1·3317	1·3917	1·4534	1·5167	1·5818
0·000220	0·701	0·710	0·719	0·728	0·737	0·745	0·754	0·763	0·771	0·780	0·788	0·797	0·805
(0·002157)	0·9306	0·9791	1·0292	1·0808	1·1340	1·1888	1·2452	1·3033	1·3630	1·4244	1·4876	1·5524	1·6190
0·000230	0·717	0·726	0·735	0·744	0·753	0·762	0·771	0·780	0·789	0·797	0·806	0·815	0·823
(0·002256)	0·9515	1·0011	1·0523	1·1051	1·1595	1·2155	1·2732	1·3326	1·3937	1·4565	1·5210	1·5873	1·6554
0·000240	0·732	0·742	0·751	0·760	0·769	0·779	0·788	0·797	0·806	0·815	0·823	0·832	0·841
(0·002354)	0·9720	1·0227	1·0749	1·1288	1·1844	1·2416	1·3006	1·3612	1·4236	1·4878	1·5537	1·6214	1·6910
0·000250	0·747	0·757	0·766	0·776	0·785	0·795	0·804	0·813	0·822	0·831	0·840	0·849	0·858
(0·002452)	0·9921	1·0437	1·0971	1·1521	1·2088	1·2672	1·3274	1·3893	1·4530	1·5185	1·5858	1·6549	1·7259
0·000260	0·762	0·772	0·782	0·791	0·801	0·810	0·820	0·829	0·839	0·848	0·857	0·866	0·875
(0·002550)	1·0117	1·0644	1·1188	1·1749	1·2328	1·2923	1·3537	1·4168	1·4818	1·5485	1·6172	1·6877	1·7600
0·000270	0·777	0·787	0·797	0·806	0·816	0·826	0·835	0·845	0·854	0·864	0·873	0·883	0·892
(0·002648)	1·0310	1·0847	1·1401	1·1973	1·2562	1·3170	1·3795	1·4438	1·5100	1·5780	1·6480	1·7198	1·7936
0·000280	0·791	0·801	0·811	0·821	0·831	0·841	0·851	0·860	0·870	0·880	0·889	0·899	0·908
(0·002746)	1·0499	1·1046	1·1611	1·2193	1·2793	1·3411	1·4048	1·4703	1·5377	1·6070	1·6782	1·7514	1·8265
0·000290	0·805	0·815	0·825	0·836	0·846	0·856	0·866	0·876	0·886	0·895	0·905	0·915	0·924
(0·002844)	1·0685	1·1241	1·1816	1·2409	1·3019	1·3649	1·4296	1·4963	1·5649	1·6354	1·7079	1·7824	1·8588
S, ($\Delta p/\rho l$)	1300	1325	1350	1375	1400	1425	1450	1475	1500	1525	1550	1575	1600

Gradient — (Equivalent) Pipe diameters in mm

S = 0·000100 to 0·000290; $\Delta p/\rho l$ = (0·000981) to (0·002844) ms^{-2}

m = Manning $n \times 100$
S = 0·000300 to 0·000950

Full bore flow; Tables E give m_C or m_P for Colebrook-White or for laminar solutions resp., to divide mV and/or mQ as follows:

i.e. kin. pr. grad., $\Delta p/\rho l$ = (0·002942) to (0·009316) ms^{-2}

mV to give velocities in ms^{-1}
mQ to give discharges in m^{3}s^{-1}

Gradient   (Equivalent) Pipe diameters in mm

S, $(\Delta p/\rho l)$	1300	1325	1350	1375	1400	1425	1450	1475	1500	1525	1550	1575	1600
0·000300	0·819	0·829	0·840	0·850	0·860	0·870	0·881	0·891	0·901	0·911	0·921	0·930	0·940
(0·002942)	1·0867	1·1434	1·2018	1·2621	1·3242	1·3882	1·4541	1·5219	1·5917	1·6634	1·7371	1·8128	1·8906
0·000320	0·846	0·856	0·867	0·878	0·888	0·899	0·909	0·920	0·930	0·941	0·951	0·961	0·971
(0·003138)	1·1224	1·1809	1·2412	1·3035	1·3676	1·4337	1·5018	1·5718	1·6439	1·7179	1·7941	1·8723	1·9526
0·000340	0·872	0·883	0·894	0·905	0·916	0·927	0·937	0·948	0·959	0·969	0·980	0·991	1·001
(0·003334)	1·1569	1·2172	1·2794	1·3436	1·4097	1·4778	1·5480	1·6202	1·6945	1·7708	1·8493	1·9299	2·0127
0·000360	0·897	0·908	0·920	0·931	0·942	0·954	0·965	0·976	0·987	0·998	1·008	1·019	1·030
(0·003530)	1·1905	1·2525	1·3165	1·3825	1·4506	1·5207	1·5929	1·6672	1·7436	1·8222	1·9029	1·9859	2·0710
0·000380	0·921	0·933	0·945	0·957	0·968	0·980	0·991	1·002	1·014	1·025	1·036	1·047	1·058
(0·003727)	1·2231	1·2868	1·3526	1·4204	1·4903	1·5624	1·6365	1·7129	1·7914	1·8721	1·9551	2·0403	2·1278
0·000400	0·945	0·957	0·969	0·981	0·993	1·005	1·017	1·028	1·040	1·052	1·063	1·074	1·086
(0·003923)	1·2549	1·3202	1·3877	1·4573	1·5290	1·6029	1·6790	1·7573	1·8379	1·9207	2·0058	2·0933	2·1831
0·000420	0·969	0·981	0·993	1·006	1·018	1·030	1·042	1·054	1·066	1·078	1·089	1·101	1·113
(0·004119)	1·2858	1·3528	1·4220	1·4933	1·5668	1·6425	1·7205	1·8007	1·8833	1·9682	2·0554	2·1450	2·2370
0·000440	0·992	1·004	1·017	1·029	1·042	1·054	1·066	1·079	1·091	1·103	1·115	1·127	1·139
(0·004315)	1·3161	1·3847	1·4555	1·5284	1·6037	1·6812	1·7610	1·8431	1·9276	2·0145	2·1037	2·1955	2·2896
0·000460	1·014	1·027	1·040	1·052	1·065	1·078	1·090	1·103	1·115	1·128	1·140	1·152	1·164
(0·004511)	1·3457	1·4158	1·4882	1·5628	1·6397	1·7190	1·8006	1·8845	1·9709	2·0597	2·1510	2·2448	2·3411
0·000480	1·036	1·049	1·062	1·075	1·088	1·101	1·114	1·127	1·139	1·152	1·164	1·177	1·189
(0·004707)	1·3746	1·4463	1·5202	1·5964	1·6750	1·7559	1·8393	1·9251	2·0133	2·1041	2·1973	2·2931	2·3914
0·000500	1·057	1·071	1·084	1·097	1·111	1·124	1·137	1·150	1·163	1·176	1·189	1·201	1·214
(0·004903)	1·4030	1·4761	1·5515	1·6293	1·7095	1·7921	1·8772	1·9648	2·0548	2·1474	2·2426	2·3404	2·4407
0·000525	1·083	1·097	1·111	1·124	1·138	1·151	1·165	1·178	1·192	1·205	1·218	1·231	1·244
(0·005148)	1·4376	1·5125	1·5898	1·6696	1·7517	1·8364	1·9236	2·0133	2·1056	2·2005	2·2980	2·3982	2·5010
0·000550	1·109	1·123	1·137	1·151	1·165	1·179	1·192	1·206	1·220	1·233	1·247	1·260	1·273
(0·005394)	1·4715	1·5481	1·6272	1·7089	1·7930	1·8796	1·9688	2·0607	2·1551	2·2523	2·3521	2·4546	2·5599
0·000575	1·134	1·148	1·162	1·177	1·191	1·205	1·219	1·233	1·247	1·261	1·275	1·288	1·302
(0·005639)	1·5045	1·5829	1·6638	1·7473	1·8333	1·9219	2·0131	2·1070	2·2036	2·3029	2·4049	2·5098	2·6174
0·000600	1·158	1·173	1·187	1·202	1·217	1·231	1·245	1·260	1·274	1·288	1·302	1·316	1·330
(0·005884)	1·5369	1·6170	1·6996	1·7848	1·8727	1·9632	2·0564	2·1523	2·2510	2·3524	2·4566	2·5637	2·6737
0·000625	1·182	1·197	1·212	1·227	1·242	1·256	1·271	1·286	1·300	1·314	1·329	1·343	1·357
(0·006129)	1·5686	1·6503	1·7347	1·8216	1·9113	2·0037	2·0988	2·1967	2·2974	2·4009	2·5073	2·6166	2·7288
0·000650	1·205	1·221	1·236	1·251	1·266	1·281	1·296	1·311	1·326	1·340	1·355	1·370	1·384
(0·006374)	1·5996	1·6830	1·7690	1·8577	1·9492	2·0434	2·1404	2·2402	2·3429	2·4485	2·5570	2·6684	2·7829
0·000675	1·228	1·244	1·259	1·275	1·290	1·306	1·321	1·336	1·351	1·366	1·381	1·396	1·410
(0·006619)	1·6301	1·7151	1·8027	1·8931	1·9863	2·0823	2·1811	2·2829	2·3875	2·4951	2·6057	2·7193	2·8359
0·000700	1·251	1·267	1·283	1·298	1·314	1·330	1·345	1·361	1·376	1·391	1·406	1·421	1·436
(0·006865)	1·6600	1·7465	1·8358	1·9278	2·0227	2·1205	2·2212	2·3248	2·4313	2·5409	2·6535	2·7691	2·8879
0·000725	1·273	1·289	1·305	1·321	1·337	1·353	1·369	1·385	1·400	1·416	1·431	1·446	1·462
(0·007110)	1·6894	1·7774	1·8683	1·9620	2·0585	2·1580	2·2605	2·3659	2·4744	2·5859	2·7004	2·8182	2·9390
0·000750	1·295	1·311	1·328	1·344	1·360	1·376	1·392	1·408	1·424	1·440	1·456	1·471	1·487
(0·007355)	1·7183	1·8078	1·9002	1·9955	2·0937	2·1949	2·2991	2·4063	2·5167	2·6301	2·7466	2·8663	2·9893
0·000800	1·337	1·354	1·371	1·388	1·405	1·421	1·438	1·454	1·471	1·487	1·503	1·519	1·536
(0·007845)	1·7746	1·8671	1·9625	2·0610	2·1624	2·2669	2·3745	2·4853	2·5992	2·7163	2·8367	2·9603	3·0873
0·000850	1·378	1·396	1·413	1·431	1·448	1·465	1·482	1·499	1·516	1·533	1·550	1·566	1·583
(0·008336)	1·8293	1·9246	2·0229	2·1244	2·2289	2·3367	2·4476	2·5618	2·6792	2·7999	2·9240	3·0515	3·1823
0·000900	1·418	1·436	1·454	1·472	1·490	1·508	1·525	1·543	1·560	1·577	1·595	1·612	1·629
(0·008826)	1·8823	1·9804	2·0816	2·1860	2·2936	2·4044	2·5186	2·6360	2·7569	2·8811	3·0088	3·1399	3·2746
0·000950	1·457	1·476	1·494	1·512	1·531	1·549	1·567	1·585	1·603	1·621	1·638	1·656	1·673
(0·009316)	1·9339	2·0346	2·1386	2·2459	2·3564	2·4703	2·5876	2·7083	2·8324	2·9600	3·0912	3·2260	3·3643
S, $(\Delta p/\rho l)$	1300	1325	1350	1375	1400	1425	1450	1475	1500	1525	1550	1575	1600

Gradient   (Equivalent) Pipe diameters in mm

S = 0·000300 to 0·000950; $\Delta p/\rho l$ = (0·002942) to (0·009316) ms^{-2}

m = Manning $n \times 100$
S = 0·00100 to 0·00290

i.e. kin. pr. grad., $\Delta p/\rho l$ =
(0·00981) to (0·02844) ms^{-2}

Full bore flow; Tables E give m_C or m_P for Colebrook-White or for laminar solutions resp., to divide mV and/or mQ as follows:

mV to give velocities in ms^{-1}
mQ to give discharges in m^{3}s^{-1}

Gradient — **(Equivalent) Pipe diameters in mm**

S, ($\Delta p/\rho l$)	1300	1325	1350	1375	1400	1425	1450	1475	1500	1525	1550	1575	1600
0·00100	1·495	1·514	1·533	1·552	1·571	1·589	1·608	1·626	1·644	1·663	1·681	1·699	1·717
(0·00981)	1·9841	2·0875	2·1942	2·3042	2·4176	2·5345	2·6548	2·7786	2·9060	3·0369	3·1715	3·3098	3·4517
0·00105	1·532	1·551	1·571	1·590	1·609	1·628	1·647	1·666	1·685	1·704	1·722	1·741	1·759
(0·01030)	2·0331	2·1390	2·2484	2·3611	2·4773	2·5971	2·7204	2·8472	2·9777	3·1119	3·2498	3·3915	3·5370
0·00110	1·568	1·588	1·608	1·628	1·647	1·667	1·686	1·705	1·725	1·744	1·763	1·782	1·801
(0·01079)	2·0809	2·1894	2·3013	2·4167	2·5356	2·6582	2·7844	2·9142	3·0478	3·1852	3·3263	3·4713	3·6202
0·00115	1·603	1·624	1·644	1·664	1·684	1·704	1·724	1·744	1·763	1·783	1·802	1·822	1·841
(0·01128)	2·1277	2·2386	2·3530	2·4710	2·5926	2·7179	2·8469	2·9797	3·1163	3·2568	3·4011	3·5493	3·7016
0·00120	1·637	1·658	1·679	1·700	1·720	1·741	1·761	1·781	1·801	1·821	1·841	1·861	1·881
(0·01177)	2·1735	2·2867	2·4036	2·5241	2·6484	2·7764	2·9082	3·0438	3·1833	3·3268	3·4742	3·6257	3·7812
0·00125	1·671	1·693	1·714	1·735	1·756	1·777	1·797	1·818	1·839	1·859	1·879	1·899	1·919
(0·01226)	2·2183	2·3339	2·4532	2·5762	2·7030	2·8336	2·9681	3·1066	3·2490	3·3954	3·5459	3·7004	3·8591
0·00130	1·704	1·726	1·748	1·769	1·791	1·812	1·833	1·854	1·875	1·896	1·916	1·937	1·957
(0·01275)	2·2622	2·3801	2·5018	2·6272	2·7565	2·8898	3·0269	3·1681	3·3133	3·4626	3·6161	3·7737	3·9356
0·00135	1·737	1·759	1·781	1·803	1·825	1·846	1·868	1·889	1·911	1·932	1·953	1·974	1·995
(0·01324)	2·3053	2·4254	2·5494	2·6773	2·8090	2·9448	3·0846	3·2285	3·3764	3·5286	3·6850	3·8456	4·0105
0·00140	1·769	1·791	1·814	1·836	1·858	1·880	1·902	1·924	1·946	1·967	1·989	2·010	2·031
(0·01373)	2·3476	2·4700	2·5962	2·7264	2·8606	2·9988	3·1412	3·2877	3·4384	3·5934	3·7526	3·9162	4·0841
0·00145	1·800	1·823	1·846	1·869	1·891	1·914	1·936	1·958	1·980	2·002	2·024	2·046	2·067
(0·01422)	2·3892	2·5137	2·6421	2·7746	2·9112	3·0519	3·1968	3·3459	3·4993	3·6570	3·8190	3·9855	4·1564
0·00150	1·831	1·854	1·877	1·901	1·923	1·946	1·969	1·992	2·014	2·036	2·059	2·081	2·103
(0·01471)	2·4300	2·5566	2·6873	2·8221	2·9610	3·1041	3·2514	3·4031	3·5591	3·7195	3·8843	4·0536	4·2275
0·00160	1·891	1·915	1·939	1·963	1·987	2·010	2·034	2·057	2·080	2·103	2·126	2·149	2·172
(0·01569)	2·5097	2·6405	2·7754	2·9146	3·0581	3·2059	3·3581	3·5147	3·6758	3·8415	4·0117	4·1866	4·3661
0·00170	1·949	1·974	1·999	2·023	2·048	2·072	2·096	2·120	2·144	2·168	2·191	2·215	2·238
(0·01667)	2·5870	2·7218	2·8609	3·0043	3·1522	3·3046	3·4614	3·6229	3·7889	3·9597	4·1352	4·3154	4·5005
0·00180	2·006	2·031	2·057	2·082	2·107	2·132	2·157	2·182	2·206	2·231	2·255	2·279	2·303
(0·01765)	2·6620	2·8007	2·9438	3·0914	3·2436	3·4004	3·5618	3·7279	3·8988	4·0745	4·2550	4·4405	4·6310
0·00190	2·060	2·087	2·113	2·139	2·165	2·191	2·216	2·241	2·267	2·292	2·317	2·342	2·366
(0·01863)	2·7349	2·8774	3·0245	3·1761	3·3325	3·4935	3·6594	3·8301	4·0056	4·1861	4·3716	4·5622	4·7579
0·00200	2·114	2·141	2·168	2·195	2·221	2·247	2·274	2·300	2·326	2·351	2·377	2·402	2·428
(0·01961)	2·8059	2·9522	3·1030	3·2587	3·4191	3·5843	3·7544	3·9296	4·1097	4·2949	4·4852	4·6807	4·8815
0·00210	2·166	2·194	2·221	2·249	2·276	2·303	2·330	2·356	2·383	2·409	2·436	2·462	2·488
(0·02059)	2·8752	3·0251	3·1797	3·3391	3·5035	3·6728	3·8472	4·0266	4·2112	4·4009	4·5960	4·7963	5·0020
0·00220	2·217	2·246	2·274	2·302	2·329	2·357	2·385	2·412	2·439	2·466	2·493	2·520	2·546
(0·02157)	2·9429	3·0963	3·2545	3·4177	3·5859	3·7592	3·9377	4·1213	4·3103	4·5045	4·7041	4·9092	5·1197
0·00230	2·267	2·296	2·325	2·353	2·382	2·410	2·438	2·466	2·494	2·522	2·549	2·576	2·604
(0·02256)	3·0090	3·1658	3·3276	3·4945	3·6665	3·8437	4·0262	4·2140	4·4071	4·6057	4·8098	5·0195	5·2348
0·00240	2·316	2·345	2·375	2·404	2·433	2·462	2·491	2·519	2·548	2·576	2·604	2·632	2·660
(0·02354)	3·0738	3·2339	3·3992	3·5697	3·7454	3·9264	4·1128	4·3046	4·5019	4·7048	4·9133	5·1275	5·3474
0·00250	2·364	2·394	2·424	2·454	2·483	2·513	2·542	2·571	2·600	2·629	2·658	2·686	2·714
(0·02452)	3·1371	3·3006	3·4693	3·6433	3·8226	4·0074	4·1976	4·3934	4·5948	4·8018	5·0146	5·2332	5·4577
0·00260	2·410	2·441	2·472	2·502	2·532	2·562	2·592	2·622	2·652	2·681	2·710	2·739	2·768
(0·02550)	3·1993	3·3660	3·5380	3·7154	3·8983	4·0867	4·2807	4·4804	4·6858	4·8969	5·1139	5·3368	5·5657
0·00270	2·456	2·488	2·519	2·550	2·581	2·611	2·642	2·672	2·702	2·732	2·762	2·791	2·821
(0·02648)	3·2602	3·4301	3·6054	3·7862	3·9726	4·1646	4·3623	4·5657	4·7750	4·9902	5·2113	5·4385	5·6718
0·00280	2·501	2·533	2·565	2·597	2·628	2·659	2·690	2·721	2·752	2·782	2·813	2·843	2·873
(0·02746)	3·3200	3·4930	3·6716	3·8557	4·0455	4·2410	4·4423	4·6495	4·8626	5·0818	5·3070	5·5383	5·7758
0·00290	2·546	2·578	2·610	2·643	2·675	2·706	2·738	2·769	2·800	2·831	2·862	2·893	2·924
(0·02844)	3·3788	3·5549	3·7366	3·9239	4·1171	4·3161	4·5209	4·7318	4·9487	5·1717	5·4009	5·6363	5·8781
S, ($\Delta p/\rho l$)	1300	1325	1350	1375	1400	1425	1450	1475	1500	1525	1550	1575	1600

Gradient — **(Equivalent) Pipe diameters in mm**

S = 0·00100 to 0·00290;

$\Delta p/\rho l$ = (0·00981) to (0·02844) ms^{-2}

m = Manning $n \times 100$
S = 0.00300 to 0.00950

i.e. kin. pr. grad., $\Delta p/\rho l$ = (0.02942) to (0.09316) ms^{-2}

Full bore flow; Tables E give m_C or m_P for Colebrook-White or for laminar solutions resp., to divide mV and/or mQ as follows:

mV to give velocities in ms^{-1}
mQ to give discharges in m^{3}s^{-1}

Gradient (Equivalent) Pipe diameters in mm

S, $(\Delta p/\rho l)$	1300	1325	1350	1375	1400	1425	1450	1475	1500	1525	1550	1575	1600
0.00300	2.589	2.622	2.655	2.688	2.720	2.753	2.785	2.817	2.848	2.880	2.911	2.942	2.973
(0.02942)	3.4366	3.6156	3.8004	3.9910	4.1875	4.3898	4.5982	4.8127	5.0333	5.2601	5.4932	5.7327	5.9786
0.00320	2.674	2.708	2.742	2.776	2.809	2.843	2.876	2.909	2.942	2.974	3.007	3.039	3.071
(0.03138)	3.5493	3.7342	3.9251	4.1219	4.3248	4.5338	4.7490	4.9705	5.1984	5.4326	5.6734	5.9207	6.1746
0.00340	2.756	2.792	2.827	2.861	2.896	2.930	2.964	2.998	3.032	3.066	3.099	3.132	3.166
(0.03334)	3.6585	3.8491	4.0459	4.2488	4.4579	4.6734	4.8952	5.1235	5.3584	5.5998	5.8480	6.1029	6.3647
0.00360	2.836	2.872	2.908	2.944	2.980	3.015	3.050	3.085	3.120	3.155	3.189	3.223	3.257
(0.03530)	3.7646	3.9607	4.1632	4.3719	4.5871	4.8088	5.0371	5.2720	5.5137	5.7622	6.0175	6.2798	6.5492
0.00380	2.914	2.951	2.988	3.025	3.062	3.098	3.134	3.170	3.206	3.241	3.276	3.312	3.347
(0.03727)	3.8677	4.0693	4.2772	4.4917	4.7128	4.9406	5.1751	5.4165	5.6648	5.9201	6.1824	6.4519	6.7286
0.00400	2.990	3.028	3.066	3.104	3.141	3.178	3.215	3.252	3.289	3.325	3.362	3.398	3.433
(0.03923)	3.9682	4.1750	4.3884	4.6084	4.8353	5.0690	5.3096	5.5572	5.8120	6.0739	6.3430	6.6195	6.9034
0.00420	3.063	3.103	3.142	3.180	3.219	3.257	3.295	3.333	3.370	3.407	3.445	3.482	3.518
(0.04119)	4.0662	4.2781	4.4967	4.7222	4.9547	5.1941	5.4407	5.6945	5.9555	6.2239	6.4997	6.7830	7.0739
0.00440	3.136	3.176	3.215	3.255	3.294	3.333	3.372	3.411	3.449	3.488	3.526	3.563	3.601
(0.04315)	4.1619	4.3788	4.6026	4.8334	5.0713	5.3164	5.5687	5.8285	6.0956	6.3703	6.6526	6.9426	7.2404
0.00460	3.206	3.247	3.288	3.328	3.368	3.408	3.448	3.488	3.527	3.566	3.605	3.644	3.682
(0.04511)	4.2554	4.4772	4.7060	4.9420	5.1852	5.4359	5.6939	5.9595	6.2326	6.5135	6.8021	7.0987	7.4031
0.00480	3.275	3.317	3.358	3.400	3.441	3.482	3.522	3.563	3.603	3.643	3.682	3.722	3.761
(0.04707)	4.3470	4.5735	4.8072	5.0483	5.2968	5.5528	5.8164	6.0876	6.3667	6.6536	6.9484	7.2513	7.5623
0.00500	3.343	3.385	3.428	3.470	3.512	3.553	3.595	3.636	3.677	3.718	3.758	3.799	3.839
(0.04903)	4.4366	4.6678	4.9063	5.1524	5.4060	5.6673	5.9363	6.2132	6.4980	6.7908	7.0917	7.4009	7.7183
0.00525	3.425	3.469	3.512	3.556	3.599	3.641	3.684	3.726	3.768	3.810	3.851	3.892	3.934
(0.05148)	4.5462	4.7830	5.0275	5.2796	5.5395	5.8072	6.0829	6.3666	6.6584	6.9585	7.2669	7.5836	7.9089
0.00550	3.506	3.550	3.595	3.639	3.683	3.727	3.770	3.814	3.857	3.899	3.942	3.984	4.026
(0.05394)	4.6531	4.8956	5.1458	5.4039	5.6699	5.9439	6.2260	6.5164	6.8151	7.1222	7.4379	7.7621	8.0950
0.00575	3.584	3.630	3.676	3.721	3.766	3.811	3.855	3.899	3.943	3.987	4.030	4.074	4.117
(0.05639)	4.7577	5.0056	5.2615	5.5253	5.7973	6.0775	6.3660	6.6629	6.9683	7.2823	7.6050	7.9365	8.2769
0.00600	3.662	3.708	3.755	3.801	3.847	3.893	3.938	3.983	4.028	4.073	4.117	4.161	4.205
(0.05884)	4.8600	5.1133	5.3746	5.6442	5.9220	6.2082	6.5029	6.8062	7.1182	7.4389	7.7686	8.1072	8.4550
0.00625	3.737	3.785	3.832	3.879	3.926	3.973	4.019	4.065	4.111	4.157	4.202	4.247	4.292
(0.06129)	4.9603	5.2187	5.4855	5.7605	6.0441	6.3362	6.6370	6.9465	7.2649	7.5923	7.9288	8.2744	8.6293
0.00650	3.811	3.860	3.908	3.956	4.004	4.052	4.099	4.146	4.193	4.239	4.285	4.331	4.377
(0.06374)	5.0585	5.3221	5.5941	5.8746	6.1638	6.4617	6.7684	7.0841	7.4088	7.7427	8.0858	8.4383	8.8002
0.00675	3.884	3.933	3.983	4.032	4.080	4.129	4.177	4.225	4.272	4.320	4.367	4.414	4.460
(0.06619)	5.1549	5.4235	5.7007	5.9865	6.2812	6.5848	6.8974	7.2190	7.5500	7.8902	8.2398	8.5990	8.9678
0.00700	3.955	4.005	4.056	4.106	4.155	4.205	4.254	4.302	4.351	4.399	4.447	4.495	4.542
(0.06865)	5.2495	5.5230	5.8053	6.0964	6.3965	6.7056	7.0239	7.3515	7.6885	8.0350	8.3910	8.7568	9.1324
0.00725	4.025	4.076	4.127	4.178	4.229	4.279	4.329	4.378	4.428	4.477	4.526	4.574	4.622
(0.07110)	5.3424	5.6207	5.9080	6.2043	6.5097	6.8243	7.1482	7.4816	7.8246	8.1772	8.5396	8.9118	9.2940
0.00750	4.094	4.146	4.198	4.250	4.301	4.352	4.403	4.453	4.504	4.553	4.603	4.652	4.702
(0.07355)	5.4337	5.7168	6.0090	6.3104	6.6210	6.9410	7.2704	7.6095	7.9584	8.3170	8.6856	9.0642	9.4529
0.00800	4.228	4.282	4.336	4.389	4.442	4.495	4.547	4.599	4.651	4.703	4.754	4.805	4.856
(0.07845)	5.6119	5.9043	6.2061	6.5173	6.8381	7.1686	7.5089	7.8591	8.2194	8.5897	8.9704	9.3614	9.7629
0.00850	4.358	4.414	4.469	4.524	4.579	4.633	4.687	4.741	4.794	4.847	4.900	4.953	5.005
(0.08336)	5.7846	6.0860	6.3971	6.7179	7.0486	7.3892	7.7400	8.1010	8.4723	8.8541	9.2465	9.6495	10.063
0.00900	4.484	4.542	4.599	4.655	4.712	4.768	4.823	4.878	4.933	4.988	5.042	5.096	5.150
(0.08826)	5.9523	6.2625	6.5825	6.9126	7.2529	7.6034	7.9644	8.3358	8.7179	9.1108	9.5146	9.9293	10.355
0.00950	4.607	4.666	4.725	4.783	4.841	4.898	4.955	5.012	5.069	5.125	5.181	5.236	5.291
(0.09316)	6.1154	6.4341	6.7629	7.1021	7.4517	7.8118	8.1826	8.5643	8.9568	9.3605	9.7753	10.201	10.639
S, $(\Delta p/\rho l)$	1300	1325	1350	1375	1400	1425	1450	1475	1500	1525	1550	1575	1600

Gradient (Equivalent) Pipe diameters in mm

S = 0.00300 to 0.00950; $\qquad$ $\Delta p/\rho l$ = (0.02942) to (0.09316) ms^{-2}

m = Manning $n \times 100$
S = 0·0100 to 0·0290

i.e. kin. pr. grad., $\Delta p/\rho l$ =
(0·0981) to (0·2844) ms^{-2}

Full bore flow; Tables E give m_C or m_P for Colebrook-White or for laminar solutions resp., to divide mV and/or mQ as follows:

mV to give velocities in ms^{-1}
mQ to give discharges in m^{3}s^{-1}

Gradient — (Equivalent) Pipe diameters in mm

S, ($\Delta p/\rho l$)	1300	1325	1350	1375	1400	1425	1450	1475	1500	1525	1550	1575	1600
0·0100	4·727	4·787	4·847	4·907	4·966	5·025	5·084	5·142	5·200	5·258	5·315	5·372	5·429
(0·0981)	6·2743	6·6012	6·9386	7·2866	7·6452	8·0147	8·3952	8·7867	9·1895	9·6036	10·029	10·466	10·915
0·0105	4·844	4·906	4·967	5·028	5·089	5·149	5·210	5·269	5·329	5·388	5·446	5·505	5·563
(0·1030)	6·4292	6·7642	7·1100	7·4665	7·8340	8·2127	8·6025	9·0037	9·4165	9·8408	10·277	10·725	11·185
0·0110	4·958	5·021	5·084	5·147	5·209	5·271	5·332	5·393	5·454	5·514	5·575	5·634	5·694
(0·1079)	6·5805	6·9234	7·2773	7·6422	8·0184	8·4059	8·8050	9·2156	9·6380	10·072	10·519	10·977	11·448
0·0115	5·069	5·134	5·198	5·262	5·326	5·389	5·452	5·514	5·577	5·638	5·700	5·761	5·822
(0·1128)	6·7284	7·0790	7·4408	7·8140	8·1986	8·5948	9·0028	9·4227	9·8547	10·299	10·755	11·224	11·705
0·0120	5·178	5·244	5·310	5·375	5·440	5·505	5·569	5·633	5·697	5·760	5·822	5·885	5·947
(0·1177)	6·8731	7·2313	7·6009	7·9820	8·3749	8·7797	9·1965	9·6254	10·067	10·520	10·986	11·465	11·957
0·0125	5·285	5·353	5·420	5·486	5·553	5·619	5·684	5·749	5·814	5·878	5·943	6·006	6·070
(0·1226)	7·0149	7·3804	7·7576	8·1466	8·5476	8·9607	9·3861	9·8239	10·274	10·737	11·213	11·702	12·204
0·0130	5·390	5·459	5·527	5·595	5·663	5·730	5·797	5·863	5·929	5·995	6·060	6·125	6·190
(0·1275)	7·1538	7·5266	7·9112	8·3080	8·7169	9·1382	9·5720	10·018	10·478	10·950	11·435	11·934	12·445
0·0135	5·492	5·563	5·632	5·702	5·770	5·839	5·907	5·975	6·042	6·109	6·176	6·242	6·308
(0·1324)	7·2901	7·6699	8·0619	8·4662	8·8830	9·3123	9·7543	10·209	10·677	11·158	11·653	12·161	12·682
0·0140	5·593	5·665	5·736	5·806	5·876	5·946	6·015	6·084	6·153	6·221	6·289	6·356	6·423
(0·1373)	7·4238	7·8107	8·2099	8·6216	9·0460	9·4832	9·9333	10·397	10·873	11·363	11·867	12·384	12·915
0·0145	5·692	5·765	5·837	5·909	5·980	6·051	6·122	6·192	6·262	6·331	6·400	6·469	6·537
(0·1422)	7·5553	7·9489	8·3552	8·7742	9·2061	9·6510	10·109	10·581	11·066	11·564	12·077	12·603	13·144
0·0150	5·789	5·863	5·937	6·010	6·083	6·155	6·227	6·298	6·369	6·439	6·510	6·579	6·649
(0·1471)	7·6844	8·0848	8·4980	8·9242	9·3635	9·8160	10·282	10·762	11·255	11·762	12·283	12·819	13·368
0·0160	5·979	6·056	6·132	6·207	6·282	6·357	6·431	6·505	6·578	6·651	6·723	6·795	6·867
(0·1569)	7·9364	8·3500	8·7767	9·2169	9·6705	10·138	10·619	11·114	11·624	12·148	12·686	13·239	13·807
0·0170	6·163	6·242	6·320	6·398	6·475	6·552	6·629	6·705	6·780	6·855	6·930	7·004	7·078
(0·1667)	8·1807	8·6070	9·0468	9·5005	9·9682	10·450	10·946	11·457	11·982	12·522	13·077	13·647	14·232
0·0180	6·342	6·423	6·504	6·584	6·663	6·742	6·821	6·899	6·977	7·054	7·131	7·207	7·284
(0·1765)	8·4178	8·8565	9·3091	9·7760	10·257	10·753	11·263	11·789	12·329	12·885	13·456	14·042	14·644
0·0190	6·516	6·599	6·682	6·764	6·846	6·927	7·008	7·088	7·168	7·247	7·326	7·405	7·483
(0·1863)	8·6485	9·0992	9·5642	10·044	10·538	11·048	11·572	12·112	12·667	13·238	13·824	14·427	15·046
0·0200	6·685	6·770	6·855	6·940	7·024	7·107	7·190	7·272	7·354	7·436	7·517	7·597	7·678
(0·1961)	8·8732	9·3356	9·8127	10·305	10·812	11·335	11·873	12·426	12·996	13·582	14·183	14·802	15·437
0·0210	6·850	6·938	7·025	7·111	7·197	7·282	7·367	7·452	7·536	7·619	7·702	7·785	7·867
(0·2059)	9·0923	9·5661	10·055	10·559	11·079	11·614	12·166	12·733	13·317	13·917	14·534	15·167	15·818
0·0220	7·011	7·101	7·190	7·278	7·366	7·454	7·541	7·627	7·713	7·799	7·884	7·968	8·052
(0·2157)	9·3063	9·7912	10·292	10·808	11·340	11·888	12·452	13·033	13·630	14·244	14·876	15·524	16·190
0·0230	7·169	7·261	7·352	7·442	7·532	7·621	7·710	7·799	7·887	7·974	8·061	8·147	8·233
(0·2256)	9·5154	10·011	10·523	11·051	11·595	12·155	12·732	13·326	13·937	14·565	15·210	15·873	16·554
0·0240	7·323	7·417	7·510	7·602	7·694	7·785	7·876	7·966	8·056	8·145	8·234	8·322	8·410
(0·2354)	9·7201	10·227	10·749	11·288	11·844	12·416	13·006	13·612	14·236	14·878	15·537	16·214	16·910
0·0250	7·474	7·570	7·665	7·759	7·853	7·946	8·039	8·131	8·222	8·313	8·404	8·494	8·584
(0·2452)	9·9205	10·437	10·971	11·521	12·088	12·672	13·274	13·893	14·530	15·185	15·858	16·549	17·259
0·0260	7·622	7·720	7·816	7·913	8·008	8·103	8·198	8·292	8·385	8·478	8·570	8·662	8·754
(0·2550)	10·117	10·644	11·188	11·749	12·328	12·923	13·537	14·168	14·818	15·485	16·172	16·877	17·600
0·0270	7·767	7·867	7·965	8·063	8·161	8·258	8·354	8·450	8·545	8·639	8·734	8·827	8·920
(0·2648)	10·310	10·847	11·401	11·973	12·562	13·170	13·795	14·438	15·100	15·780	16·480	17·198	17·936
0·0280	7·910	8·011	8·111	8·211	8·310	8·409	8·507	8·605	8·702	8·798	8·894	8·989	9·084
(0·2746)	10·499	11·046	11·611	12·193	12·793	13·411	14·048	14·703	15·377	16·070	16·782	17·514	18·265
0·0290	8·050	8·153	8·255	8·357	8·458	8·558	8·658	8·757	8·856	8·954	9·051	9·148	9·245
(0·2844)	10·685	11·241	11·816	12·409	13·019	13·649	14·296	14·963	15·649	16·354	17·079	17·824	18·588
S, ($\Delta p/\rho l$)	1300	1325	1350	1375	1400	1425	1450	1475	1500	1525	1550	1575	1600

Gradient — (Equivalent) Pipe diameters in mm

S = 0·0100 to 0·0290;

$\Delta p/\rho l$ = (0·0981) to (0·2844) ms^{-2}

m = Manning $n \times 100$
S = 0·0300 to 0·1000

i.e. kin. pr. grad., $\Delta p/\rho l =$
(0·2942) to (0·9807) ms^{-2}

Full bore flow; Tables E give m_C or m_P for Colebrook-White or for laminar solutions resp., to divide mV and/or mQ as follows:

mV to give velocities in ms^{-1}
mQ to give discharges in m^{3}s^{-1}

Gradient — (Equivalent) Pipe diameters in mm

S, ($\Delta p/\rho l$)	1300	1325	1350	1375	1400	1425	1450	1475	1500	1525	1550	1575	1600
0·0300	8·187	8·292	8·396	8·499	8·602	8·704	8·806	8·907	9·007	9·107	9·206	9·305	9·403
(0·2942)	10·867	11·434	12·018	12·621	13·242	13·882	14·541	15·219	15·917	16·634	17·371	18·128	18·906
0·0320	8·456	8·564	8·671	8·778	8·884	8·990	9·095	9·199	9·302	9·405	9·508	9·610	9·711
(0·3138)	11·224	11·809	12·412	13·035	13·676	14·337	15·018	15·718	16·439	17·179	17·941	18·723	19·526
0·0340	8·716	8·828	8·938	9·048	9·158	9·266	9·374	9·482	9·589	9·695	9·801	9·906	10·01
(0·3334)	11·569	12·172	12·794	13·436	14·097	14·778	15·480	16·202	16·945	17·708	18·493	19·299	20·127
0·0360	8·969	9·084	9·197	9·311	9·423	9·535	9·646	9·757	9·867	9·976	10·08	10·19	10·30
(0·3530)	11·905	12·525	13·165	13·825	14·506	15·207	15·929	16·672	17·436	18·222	19·029	19·859	20·710
0·0380	9·215	9·332	9·449	9·566	9·681	9·796	9·911	10·02	10·14	10·25	10·36	10·47	10·58
(0·3727)	12·231	12·868	13·526	14·204	14·903	15·624	16·365	17·129	17·914	18·721	19·551	20·403	21·278
0·0400	9·454	9·575	9·695	9·814	9·933	10·05	10·17	10·28	10·40	10·52	10·63	10·74	10·86
(0·3923)	12·549	13·202	13·877	14·573	15·290	16·029	16·790	17·573	18·379	19·207	20·058	20·933	21·831
0·0420	9·688	9·811	9·934	10·06	10·18	10·30	10·42	10·54	10·66	10·78	10·89	11·01	11·13
(0·4119)	12·858	13·528	14·220	14·933	15·668	16·425	17·205	18·007	18·833	19·682	20·554	21·450	22·370
0·0440	9·916	10·04	10·17	10·29	10·42	10·54	10·66	10·79	10·91	11·03	11·15	11·27	11·39
(0·4315)	13·161	13·847	14·555	15·284	16·037	16·812	17·610	18·431	19·276	20·145	21·037	21·955	22·896
0·0460	10·14	10·27	10·40	10·52	10·65	10·78	10·90	11·03	11·15	11·28	11·40	11·52	11·64
(0·4511)	13·457	14·158	14·882	15·628	16·397	17·190	18·006	18·845	19·709	20·597	21·510	22·448	23·411
0·0480	10·36	10·49	10·62	10·75	10·88	11·01	11·14	11·27	11·39	11·52	11·64	11·77	11·89
(0·4707)	13·746	14·463	15·202	15·964	16·750	17·559	18·393	19·251	20·133	21·041	21·973	22·931	23·914
0·0500	10·57	10·71	10·84	10·97	11·11	11·24	11·37	11·50	11·63	11·76	11·89	12·01	12·14
(0·4903)	14·030	14·761	15·515	16·293	17·095	17·921	18·772	19·648	20·548	21·474	22·426	23·404	24·407
0·0525	10·83	10·97	11·11	11·24	11·38	11·51	11·65	11·78	11·92	12·05	12·18	12·31	12·44
(0·5148)	14·376	15·125	15·898	16·696	17·517	18·364	19·236	20·133	21·056	22·005	22·980	23·982	25·010
0·0550	11·09	11·23	11·37	11·51	11·65	11·79	11·92	12·06	12·20	12·33	12·47	12·60	12·73
(0·5394)	14·715	15·481	16·272	17·089	17·930	18·796	19·688	20·607	21·551	22·523	23·521	24·546	25·599
0·0575	11·34	11·48	11·62	11·77	11·91	12·05	12·19	12·33	12·47	12·61	12·75	12·88	13·02
(0·5639)	15·045	15·829	16·638	17·473	18·333	19·219	20·131	21·070	22·036	23·029	24·049	25·098	26·174
0·0600	11·58	11·73	11·87	12·02	12·17	12·31	12·45	12·60	12·74	12·88	13·02	13·16	13·30
(0·5884)	15·369	16·170	16·996	17·848	18·727	19·632	20·564	21·523	22·510	23·524	24·566	25·637	26·737
0·0625	11·82	11·97	12·12	12·27	12·42	12·56	12·71	12·86	13·00	13·14	13·29	13·43	13·57
(0·6129)	15·686	16·503	17·347	18·216	19·113	20·037	20·988	21·967	22·974	24·009	25·073	26·166	27·288
0·0650	12·05	12·21	12·36	12·51	12·66	12·81	12·96	13·11	13·26	13·40	13·55	13·70	13·84
(0·6374)	15·996	16·830	17·690	18·577	19·492	20·434	21·404	22·402	23·429	24·485	25·570	26·684	27·829
0·0675	12·28	12·44	12·59	12·75	12·90	13·06	13·21	13·36	13·51	13·66	13·81	13·96	14·10
(0·6619)	16·301	17·151	18·027	18·931	19·863	20·823	21·811	22·829	23·875	24·951	26·057	27·193	28·359
0·0700	12·51	12·67	12·83	12·98	13·14	13·30	13·45	13·61	13·76	13·91	14·06	14·21	14·36
(0·6865)	16·600	17·465	18·358	19·278	20·227	21·205	22·212	23·248	24·313	25·409	26·535	27·691	28·879
0·0725	12·73	12·89	13·05	13·21	13·37	13·53	13·69	13·85	14·00	14·16	14·31	14·46	14·62
(0·7110)	16·894	17·774	18·683	19·620	20·585	21·580	22·605	23·659	24·744	25·859	27·004	28·182	29·390
0·0750	12·95	13·11	13·28	13·44	13·60	13·76	13·92	14·08	14·24	14·40	14·56	14·71	14·87
(0·7355)	17·183	18·078	19·002	19·955	20·937	21·949	22·991	24·063	25·167	26·301	27·466	28·663	29·893
0·0800	13·37	13·54	13·71	13·88	14·05	14·21	14·38	14·54	14·71	14·87	15·03	15·19	15·36
(0·7845)	17·746	18·671	19·625	20·610	21·624	22·669	23·745	24·853	25·992	27·163	28·367	29·603	30·873
0·0850	13·78	13·96	14·13	14·31	14·48	14·65	14·82	14·99	15·16	15·33	15·50	15·66	15·83
(0·8336)	18·293	19·246	20·229	21·244	22·289	23·367	24·476	25·618	26·792	27·999	29·240	30·515	31·823
0·0900	14·18	14·36	14·54	14·72	14·90	15·08	15·25	15·43	15·60	15·77	15·95	16·12	16·29
(0·8826)	18·823	19·804	20·816	21·860	22·936	24·044	25·186	26·360	27·569	28·811	30·088	31·399	32·746
0·0950	14·57	14·76	14·94	15·12	15·31	15·49	15·67	15·85	16·03	16·21	16·38	16·56	16·73
(0·9316)	19·339	20·346	21·386	22·459	23·564	24·703	25·876	27·083	28·324	29·600	30·912	32·260	33·643
0·1000	14·95	15·14	15·33	15·52	15·71	15·89	16·08	16·26	16·44	16·63	16·81	16·99	17·17
(0·9807)	19·841	20·875	21·942	23·042	24·176	25·345	26·548	27·786	29·060	30·369	31·715	33·098	34·517
S, ($\Delta p/\rho l$)	1300	1325	1350	1375	1400	1425	1450	1475	1500	1525	1550	1575	1600

Gradient — (Equivalent) Pipe diameters in mm

S = 0·0300 to 0·1000;

$\Delta p/\rho l$ = (0·2942) to (0·9807) ms^{-2}

m = Manning $n \times 100$
S = 0·000100 to 0·000290

i.e. kin. pr. grad., $\Delta p/\rho l$ =
(0·000981) to (0·002844) ms^{-2}

Full bore flow; Tables E give m_C or m_P for Colebrook-White or for laminar solutions resp., to divide mV and/or mQ as follows:

mV to give velocities in ms^{-1}
mQ to give discharges in m^{3}s^{-1}

Gradient **(Equivalent) Pipe diameters in mm**

S, $(\Delta p/\rho l)$	1600	1625	1650	1675	1700	1725	1750	1775	1800	1825	1850	1875	1900
0·000100	0·543	0·549	0·554	0·560	0·565	0·571	0·576	0·582	0·587	0·593	0·598	0·603	0·609
(0·000981)	1·0915	1·1376	1·1849	1·2334	1·2831	1·3340	1·3862	1·4396	1·4943	1·5503	1·6076	1·6662	1·7261
0·000105	0·556	0·562	0·568	0·574	0·579	0·585	0·591	0·596	0·602	0·607	0·613	0·618	0·624
(0·001030)	1·1185	1·1657	1·2141	1·2638	1·3147	1·3669	1·4204	1·4752	1·5312	1·5886	1·6473	1·7073	1·7687
0·000110	0·569	0·575	0·581	0·587	0·593	0·599	0·604	0·610	0·616	0·622	0·627	0·633	0·638
(0·001079)	1·1448	1·1931	1·2427	1·2936	1·3457	1·3991	1·4538	1·5099	1·5673	1·6260	1·6860	1·7475	1·8103
0·000115	0·582	0·588	0·594	0·600	0·606	0·612	0·618	0·624	0·630	0·636	0·641	0·647	0·653
(0·001128)	1·1705	1·2199	1·2706	1·3226	1·3759	1·4305	1·4865	1·5438	1·6025	1·6625	1·7239	1·7868	1·8510
0·000120	0·595	0·601	0·607	0·613	0·619	0·625	0·631	0·637	0·643	0·649	0·655	0·661	0·667
(0·001177)	1·1957	1·2462	1·2980	1·3511	1·4055	1·4613	1·5185	1·5770	1·6369	1·6983	1·7610	1·8252	1·8908
0·000125	0·607	0·613	0·620	0·626	0·632	0·638	0·644	0·650	0·657	0·663	0·669	0·675	0·681
(0·001226)	1·2204	1·2719	1·3247	1·3789	1·4345	1·4914	1·5498	1·6095	1·6707	1·7333	1·7973	1·8628	1·9298
0·000130	0·619	0·625	0·632	0·638	0·645	0·651	0·657	0·663	0·670	0·676	0·682	0·688	0·694
(0·001275)	1·2445	1·2971	1·3510	1·4062	1·4629	1·5210	1·5805	1·6414	1·7038	1·7676	1·8329	1·8997	1·9680
0·000135	0·631	0·637	0·644	0·650	0·657	0·663	0·670	0·676	0·682	0·689	0·695	0·701	0·707
(0·001324)	1·2682	1·3218	1·3767	1·4330	1·4908	1·5500	1·6106	1·6727	1·7362	1·8013	1·8678	1·9359	2·0055
0·000140	0·642	0·649	0·656	0·662	0·669	0·675	0·682	0·688	0·695	0·701	0·708	0·714	0·720
(0·001373)	1·2915	1·3460	1·4020	1·4593	1·5181	1·5784	1·6401	1·7034	1·7681	1·8343	1·9021	1·9714	2·0423
0·000145	0·654	0·661	0·667	0·674	0·681	0·687	0·694	0·701	0·707	0·714	0·720	0·727	0·733
(0·001422)	1·3144	1·3699	1·4268	1·4852	1·5450	1·6063	1·6692	1·7335	1·7994	1·8668	1·9358	2·0063	2·0785
0·000150	0·665	0·672	0·679	0·686	0·692	0·699	0·706	0·713	0·719	0·726	0·732	0·739	0·746
(0·001471)	1·3368	1·3933	1·4512	1·5105	1·5714	1·6338	1·6977	1·7632	1·8302	1·8987	1·9689	2·0406	2·1140
0·000160	0·687	0·694	0·701	0·708	0·715	0·722	0·729	0·736	0·743	0·750	0·756	0·763	0·770
(0·001569)	1·3807	1·4390	1·4988	1·5601	1·6230	1·6874	1·7534	1·8210	1·8902	1·9610	2·0335	2·1076	2·1833
0·000170	0·708	0·715	0·723	0·730	0·737	0·744	0·751	0·759	0·766	0·773	0·780	0·787	0·794
(0·001667)	1·4232	1·4833	1·5449	1·6081	1·6729	1·7393	1·8073	1·8770	1·9484	2·0213	2·0960	2·1724	2·2505
0·000180	0·728	0·736	0·743	0·751	0·758	0·766	0·773	0·781	0·788	0·795	0·802	0·810	0·817
(0·001765)	1·4644	1·5263	1·5897	1·6547	1·7214	1·7897	1·8597	1·9314	2·0048	2·0800	2·1568	2·2354	2·3158
0·000190	0·748	0·756	0·764	0·772	0·779	0·787	0·794	0·802	0·809	0·817	0·824	0·832	0·839
(0·001863)	1·5046	1·5681	1·6332	1·7001	1·7686	1·8388	1·9107	1·9844	2·0598	2·1369	2·2159	2·2967	2·3792
0·000200	0·768	0·776	0·784	0·792	0·799	0·807	0·815	0·823	0·830	0·838	0·846	0·853	0·861
(0·001961)	1·5437	1·6088	1·6757	1·7442	1·8145	1·8866	1·9603	2·0359	2·1133	2·1925	2·2735	2·3563	2·4410
0·000210	0·787	0·795	0·803	0·811	0·819	0·827	0·835	0·843	0·851	0·859	0·867	0·874	0·882
(0·002059)	1·5818	1·6485	1·7170	1·7873	1·8593	1·9331	2·0088	2·0862	2·1655	2·2466	2·3296	2·4145	2·5013
0·000220	0·805	0·814	0·822	0·830	0·838	0·847	0·855	0·863	0·871	0·879	0·887	0·895	0·903
(0·002157)	1·6190	1·6873	1·7575	1·8294	1·9031	1·9786	2·0560	2·1353	2·2164	2·2995	2·3844	2·4713	2·5602
0·000230	0·823	0·832	0·840	0·849	0·857	0·866	0·874	0·882	0·891	0·899	0·907	0·915	0·923
(0·002256)	1·6554	1·7253	1·7970	1·8705	1·9459	2·0231	2·1022	2·1833	2·2662	2·3512	2·4380	2·5269	2·6177
0·000240	0·841	0·850	0·858	0·867	0·876	0·884	0·893	0·901	0·910	0·918	0·927	0·935	0·943
(0·002354)	1·6910	1·7624	1·8356	1·9107	1·9877	2·0666	2·1474	2·2302	2·3150	2·4017	2·4905	2·5812	2·6740
0·000250	0·858	0·867	0·876	0·885	0·894	0·903	0·911	0·920	0·928	0·937	0·946	0·954	0·963
(0·002452)	1·7259	1·7987	1·8735	1·9501	2·0287	2·1092	2·1917	2·2762	2·3627	2·4512	2·5418	2·6344	2·7292
0·000260	0·875	0·884	0·894	0·903	0·911	0·920	0·929	0·938	0·947	0·956	0·964	0·973	0·982
(0·002550)	1·7600	1·8343	1·9106	1·9887	2·0689	2·1510	2·2351	2·3213	2·4095	2·4998	2·5922	2·6866	2·7832
0·000270	0·892	0·901	0·911	0·920	0·929	0·938	0·947	0·956	0·965	0·974	0·983	0·992	1·000
(0·002648)	1·7936	1·8693	1·9470	2·0266	2·1083	2·1920	2·2777	2·3655	2·4554	2·5474	2·6415	2·7378	2·8362
0·000280	0·908	0·918	0·927	0·937	0·946	0·955	0·964	0·974	0·983	0·992	1·001	1·010	1·019
(0·002746)	1·8265	1·9036	1·9827	2·0638	2·1470	2·2322	2·3195	2·4089	2·5005	2·5942	2·6900	2·7880	2·8883
0·000290	0·924	0·934	0·944	0·953	0·963	0·972	0·981	0·991	1·000	1·009	1·018	1·028	1·037
(0·002844)	1·8588	1·9373	2·0178	2·1003	2·1850	2·2717	2·3606	2·4516	2·5447	2·6401	2·7376	2·8374	2·9394
S, $(\Delta p/\rho l)$	1600	1625	1650	1675	1700	1725	1750	1775	1800	1825	1850	1875	1900

Gradient **(Equivalent) Pipe diameters in mm**

S = 0·000100 to 0·000290; $\Delta p/\rho l$ = (0·000981) to (0·002844) ms^{-2}

m = Manning $n \times 100$
S = 0.000300 to 0.000950

Full bore flow; Tables E give m_C or m_P for Colebrook-White or for laminar solutions resp., to divide mV and/or mQ as follows:

i.e. kin. pr. grad., $\Delta p/\rho l$ = (0.002942) to (0.009316) ms^{-2}

mV to give velocities in ms^{-1}
mQ to give discharges in m^{3}s^{-1}

Gradient   **(Equivalent) Pipe diameters in mm**

$S,(\Delta p/\rho l)$	1600	1625	1650	1675	1700	1725	1750	1775	1800	1825	1850	1875	1900
0.000300	0.940	0.950	0.960	0.969	0.979	0.989	0.998	1.008	1.017	1.027	1.036	1.045	1.054
(0.002942)	1.8906	1.9704	2.0523	2.1362	2.2223	2.3105	2.4009	2.4935	2.5882	2.6852	2.7844	2.8859	2.9896
0.000320	0.971	0.981	0.991	1.001	1.011	1.021	1.031	1.041	1.050	1.060	1.070	1.079	1.089
(0.003138)	1.9526	2.0350	2.1196	2.2063	2.2952	2.3863	2.4797	2.5753	2.6731	2.7733	2.8757	2.9805	3.0877
0.000340	1.001	1.011	1.022	1.032	1.042	1.053	1.063	1.073	1.083	1.093	1.103	1.113	1.123
(0.003334)	2.0127	2.0976	2.1848	2.2742	2.3658	2.4598	2.5560	2.6545	2.7554	2.8586	2.9642	3.0723	3.1827
0.000360	1.030	1.041	1.051	1.062	1.073	1.083	1.093	1.104	1.114	1.124	1.135	1.145	1.155
(0.003530)	2.0710	2.1585	2.2481	2.3401	2.4344	2.5311	2.6301	2.7315	2.8353	2.9415	3.0502	3.1613	3.2750
0.000380	1.058	1.069	1.080	1.091	1.102	1.113	1.123	1.134	1.145	1.155	1.166	1.176	1.187
(0.003727)	2.1278	2.2176	2.3097	2.4043	2.5011	2.6004	2.7021	2.8063	2.9130	3.0221	3.1338	3.2480	3.3647
0.000400	1.086	1.097	1.108	1.119	1.131	1.142	1.153	1.164	1.174	1.185	1.196	1.207	1.218
(0.003923)	2.1831	2.2752	2.3698	2.4667	2.5661	2.6680	2.7723	2.8792	2.9886	3.1006	3.2152	3.3323	3.4521
0.000420	1.113	1.124	1.136	1.147	1.158	1.170	1.181	1.192	1.203	1.215	1.226	1.237	1.248
(0.004119)	2.2370	2.3314	2.4283	2.5276	2.6295	2.7339	2.8408	2.9503	3.0624	3.1772	3.2946	3.4146	3.5374
0.000440	1.139	1.151	1.162	1.174	1.186	1.197	1.209	1.220	1.232	1.243	1.254	1.266	1.277
(0.004315)	2.2896	2.3863	2.4854	2.5871	2.6914	2.7982	2.9077	3.0198	3.1345	3.2519	3.3721	3.4950	3.6206
0.000460	1.164	1.176	1.188	1.200	1.212	1.224	1.236	1.248	1.259	1.271	1.283	1.294	1.306
(0.004511)	2.3411	2.4399	2.5413	2.6453	2.7519	2.8611	2.9730	3.0876	3.2050	3.3250	3.4479	3.5735	3.7020
0.000480	1.189	1.202	1.214	1.226	1.238	1.251	1.263	1.275	1.287	1.298	1.310	1.322	1.334
(0.004707)	2.3914	2.4924	2.5959	2.7021	2.8110	2.9226	3.0370	3.1540	3.2739	3.3965	3.5220	3.6504	3.7816
0.000500	1.214	1.227	1.239	1.252	1.264	1.276	1.289	1.301	1.313	1.325	1.337	1.349	1.361
(0.004903)	2.4407	2.5438	2.6495	2.7579	2.8690	2.9829	3.0996	3.2191	3.3414	3.4666	3.5947	3.7257	3.8596
0.000525	1.244	1.257	1.270	1.282	1.295	1.308	1.320	1.333	1.346	1.358	1.370	1.383	1.395
(0.005148)	2.5010	2.6066	2.7149	2.8260	2.9399	3.0566	3.1761	3.2986	3.4239	3.5522	3.6834	3.8177	3.9549
0.000550	1.273	1.286	1.300	1.313	1.326	1.339	1.352	1.364	1.377	1.390	1.403	1.415	1.428
(0.005394)	2.5599	2.6679	2.7788	2.8925	3.0090	3.1285	3.2509	3.3762	3.5045	3.6358	3.7701	3.9075	4.0480
0.000575	1.302	1.315	1.329	1.342	1.355	1.369	1.382	1.395	1.408	1.421	1.434	1.447	1.460
(0.005639)	2.6174	2.7279	2.8412	2.9575	3.0767	3.1988	3.3239	3.4521	3.5832	3.7175	3.8549	3.9953	4.1390
0.000600	1.330	1.344	1.357	1.371	1.385	1.398	1.412	1.425	1.438	1.452	1.465	1.478	1.491
(0.005884)	2.6737	2.7866	2.9023	3.0211	3.1428	3.2676	3.3954	3.5263	3.6603	3.7975	3.9378	4.0813	4.2280
0.000625	1.357	1.371	1.385	1.399	1.413	1.427	1.441	1.454	1.468	1.482	1.495	1.509	1.522
(0.006129)	2.7288	2.8440	2.9622	3.0834	3.2076	3.3350	3.4654	3.5990	3.7358	3.8758	4.0190	4.1654	4.3152
0.000650	1.384	1.398	1.413	1.427	1.441	1.455	1.469	1.483	1.497	1.511	1.525	1.538	1.552
(0.006374)	2.7829	2.9003	3.0209	3.1445	3.2712	3.4010	3.5341	3.6703	3.8098	3.9525	4.0986	4.2479	4.4006
0.000675	1.410	1.425	1.440	1.454	1.469	1.483	1.497	1.512	1.526	1.540	1.554	1.568	1.582
(0.006619)	2.8359	2.9556	3.0784	3.2044	3.3335	3.4658	3.6014	3.7402	3.8823	4.0278	4.1766	4.3288	4.4845
0.000700	1.436	1.451	1.466	1.481	1.496	1.510	1.525	1.539	1.554	1.568	1.582	1.597	1.611
(0.006865)	2.8879	3.0098	3.1349	3.2632	3.3947	3.5294	3.6675	3.8089	3.9536	4.1017	4.2533	4.4083	4.5668
0.000725	1.462	1.477	1.492	1.507	1.522	1.537	1.552	1.566	1.581	1.596	1.610	1.625	1.639
(0.007110)	2.9390	3.0631	3.1904	3.3209	3.4547	3.5919	3.7324	3.8763	4.0236	4.1743	4.3286	4.4863	4.6476
0.000750	1.487	1.502	1.518	1.533	1.548	1.563	1.578	1.593	1.608	1.623	1.638	1.653	1.667
(0.007355)	2.9893	3.1155	3.2449	3.3777	3.5138	3.6533	3.7962	3.9425	4.0924	4.2457	4.4026	4.5630	4.7270
0.000800	1.536	1.551	1.567	1.583	1.599	1.614	1.630	1.646	1.661	1.676	1.692	1.707	1.722
(0.007845)	3.0873	3.2176	3.3513	3.4885	3.6290	3.7731	3.9207	4.0718	4.2266	4.3849	4.5469	4.7126	4.8821
0.000850	1.583	1.599	1.616	1.632	1.648	1.664	1.680	1.696	1.712	1.728	1.744	1.759	1.775
(0.008336)	3.1823	3.3167	3.4545	3.5958	3.7407	3.8892	4.0414	4.1971	4.3566	4.5199	4.6869	4.8577	5.0323
0.000900	1.629	1.646	1.662	1.679	1.696	1.712	1.729	1.745	1.762	1.778	1.794	1.810	1.826
(0.008826)	3.2746	3.4128	3.5546	3.7001	3.8492	4.0020	4.1585	4.3188	4.4829	4.6509	4.8228	4.9985	5.1782
0.000950	1.673	1.691	1.708	1.725	1.742	1.759	1.776	1.793	1.810	1.827	1.843	1.860	1.876
(0.009316)	3.3643	3.5063	3.6520	3.8015	3.9547	4.1116	4.2725	4.4372	4.6058	4.7784	4.9549	5.1355	5.3201
$S,(\Delta p/\rho l)$	1600	1625	1650	1675	1700	1725	1750	1775	1800	1825	1850	1875	1900

Gradient   **(Equivalent) Pipe diameters in mm**

S = 0.000300 to 0.000950; $\Delta p/\rho l$ = (0.002942) to (0.009316) ms^{-2}

m = Manning $n \times 100$
S = 0·00100 to 0·00290

i.e. kin. pr. grad., $\Delta p/\rho l$ = (0·00981) to (0·02844) ms^{-2}

Full bore flow; Tables E give m_C or m_P for Colebrook-White or for laminar solutions resp., to divide mV and/or mQ as follows:

mV to give velocities in ms^{-1}
mQ to give discharges in m^{3}s^{-1}

Gradient (Equivalent) Pipe diameters in mm

S, ($\Delta p/\rho l$)	1600	1625	1650	1675	1700	1725	1750	1775	1800	1825	1850	1875	1900
0·00100	1·717	1·735	1·752	1·770	1·788	1·805	1·822	1·840	1·857	1·874	1·891	1·908	1·925
(0·00981)	3·4517	3·5974	3·7469	3·9002	4·0574	4·2185	4·3835	4·5524	4·7254	4·9025	5·0836	5·2689	5·4583
0·00105	1·759	1·777	1·796	1·814	1·832	1·850	1·867	1·885	1·903	1·920	1·938	1·955	1·973
(0·01030)	3·5370	3·6863	3·8394	3·9965	4·1576	4·3226	4·4917	4·6649	4·8421	5·0236	5·2092	5·3990	5·5931
0·00110	1·801	1·819	1·838	1·856	1·875	1·893	1·911	1·930	1·948	1·966	1·984	2·001	2·019
(0·01079)	3·6202	3·7730	3·9298	4·0906	4·2554	4·4244	4·5974	4·7746	4·9561	5·1418	5·3318	5·5261	5·7247
0·00115	1·841	1·860	1·879	1·898	1·917	1·936	1·954	1·973	1·991	2·010	2·028	2·046	2·064
(0·01128)	3·7016	3·8578	4·0181	4·1825	4·3511	4·5238	4·7007	4·8820	5·0675	5·2573	5·4516	5·6503	5·8534
0·00120	1·881	1·900	1·920	1·939	1·958	1·977	1·996	2·015	2·034	2·053	2·072	2·090	2·109
(0·01177)	3·7812	3·9408	4·1045	4·2725	4·4446	4·6211	4·8018	4·9870	5·1765	5·3704	5·5688	5·7718	5·9793
0·00125	1·919	1·939	1·959	1·979	1·999	2·018	2·038	2·057	2·076	2·095	2·114	2·133	2·152
(0·01226)	3·8591	4·0220	4·1892	4·3606	4·5363	4·7164	4·9009	5·0898	5·2832	5·4812	5·6837	5·8908	6·1026
0·00130	1·957	1·978	1·998	2·018	2·038	2·058	2·078	2·098	2·117	2·137	2·156	2·176	2·195
(0·01275)	3·9356	4·1017	4·2721	4·4469	4·6261	4·8098	4·9979	5·1906	5·3878	5·5897	5·7962	6·0075	6·2234
0·00135	1·995	2·015	2·036	2·057	2·077	2·097	2·117	2·138	2·158	2·178	2·197	2·217	2·237
(0·01324)	4·0105	4·1798	4·3535	4·5316	4·7143	4·9014	5·0931	5·2895	5·4905	5·6962	5·9066	6·1219	6·3420
0·00140	2·031	2·052	2·073	2·094	2·115	2·136	2·156	2·177	2·197	2·218	2·238	2·258	2·278
(0·01373)	4·0841	4·2565	4·4334	4·6148	4·8008	4·9913	5·1866	5·3865	5·5912	5·8007	6·0150	6·2342	6·4584
0·00145	2·067	2·089	2·110	2·131	2·152	2·174	2·195	2·215	2·236	2·257	2·277	2·298	2·318
(0·01422)	4·1564	4·3319	4·5119	4·6965	4·8857	5·0797	5·2784	5·4819	5·6902	5·9034	6·1215	6·3446	6·5727
0·00150	2·103	2·124	2·146	2·168	2·189	2·211	2·232	2·253	2·274	2·295	2·316	2·337	2·358
(0·01471)	4·2275	4·4059	4·5890	4·7768	4·9693	5·1665	5·3686	5·5756	5·7875	6·0043	6·2261	6·4530	6·6850
0·00160	2·172	2·194	2·217	2·239	2·261	2·283	2·305	2·327	2·349	2·371	2·392	2·414	2·435
(0·01569)	4·3661	4·5504	4·7395	4·9334	5·1322	5·3360	5·5447	5·7584	5·9773	6·2012	6·4303	6·6647	6·9043
0·00170	2·238	2·262	2·285	2·308	2·331	2·353	2·376	2·399	2·421	2·444	2·466	2·488	2·510
(0·01667)	4·5005	4·6905	4·8854	5·0853	5·2902	5·5002	5·7153	5·9357	6·1612	6·3921	6·6282	6·8698	7·1168
0·00180	2·303	2·327	2·351	2·375	2·398	2·422	2·445	2·468	2·491	2·514	2·537	2·560	2·583
(0·01765)	4·6310	4·8265	5·0270	5·2327	5·4436	5·6597	5·8810	6·1077	6·3398	6·5774	6·8204	7·0690	7·3231
0·00190	2·366	2·391	2·415	2·440	2·464	2·488	2·512	2·536	2·560	2·583	2·607	2·630	2·654
(0·01863)	4·7579	4·9587	5·1648	5·3761	5·5927	5·8147	6·0422	6·2751	6·5136	6·7576	7·0073	7·2627	7·5238
0·00200	2·428	2·453	2·478	2·503	2·528	2·553	2·577	2·602	2·626	2·650	2·675	2·699	2·723
(0·01961)	4·8815	5·0875	5·2989	5·5157	5·7380	5·9658	6·1992	6·4381	6·6828	6·9332	7·1893	7·4513	7·7192
0·00210	2·488	2·514	2·539	2·565	2·590	2·616	2·641	2·666	2·691	2·716	2·741	2·765	2·790
(0·02059)	5·0020	5·2132	5·4298	5·6520	5·8797	6·1131	6·3522	6·5971	6·8478	7·1044	7·3669	7·6354	7·9099
0·00220	2·546	2·573	2·599	2·625	2·651	2·677	2·703	2·729	2·754	2·780	2·805	2·830	2·855
(0·02157)	5·1197	5·3358	5·5576	5·7850	6·0181	6·2570	6·5017	6·7524	7·0090	7·2716	7·5402	7·8150	8·0960
0·00230	2·604	2·631	2·658	2·684	2·711	2·737	2·764	2·790	2·816	2·842	2·868	2·894	2·920
(0·02256)	5·2348	5·4558	5·6825	5·9150	6·1533	6·3976	6·6479	6·9041	7·1665	7·4350	7·7097	7·9907	8·2779
0·00240	2·660	2·687	2·715	2·742	2·769	2·796	2·823	2·850	2·877	2·903	2·930	2·956	2·982
(0·02354)	5·3474	5·5731	5·8047	6·0422	6·2857	6·5352	6·7908	7·0526	7·3206	7·5949	7·8755	8·1625	8·4560
0·00250	2·714	2·743	2·771	2·799	2·826	2·854	2·882	2·909	2·936	2·963	2·990	3·017	3·044
(0·02452)	5·4577	5·6880	5·9244	6·1668	6·4153	6·6700	6·9309	7·1981	7·4716	7·7515	8·0379	8·3309	8·6304
0·00260	2·768	2·797	2·826	2·854	2·882	2·911	2·939	2·967	2·994	3·022	3·049	3·077	3·104
(0·02550)	5·5657	5·8007	6·0417	6·2889	6·5423	6·8021	7·0681	7·3406	7·6195	7·9050	8·1971	8·4958	8·8013
0·00270	2·821	2·850	2·879	2·908	2·937	2·966	2·995	3·023	3·051	3·080	3·108	3·136	3·163
(0·02648)	5·6718	5·9112	6·1568	6·4087	6·6670	6·9316	7·2028	7·4804	7·7647	8·0556	8·3533	8·6577	8·9689
0·00280	2·873	2·903	2·932	2·962	2·991	3·020	3·050	3·078	3·107	3·136	3·165	3·193	3·221
(0·02746)	5·7758	6·0196	6·2698	6·5263	6·7893	7·0588	7·3349	7·6177	7·9072	8·2034	8·5065	8·8165	9·1335
0·00290	2·924	2·954	2·984	3·014	3·044	3·074	3·103	3·133	3·162	3·192	3·221	3·250	3·278
(0·02844)	5·8781	6·1262	6·3808	6·6418	6·9095	7·1838	7·4648	7·7525	8·0471	8·3486	8·6571	8·9726	9·2952

S, ($\Delta p/\rho l$)	1600	1625	1650	1675	1700	1725	1750	1775	1800	1825	1850	1875	1900

Gradient (Equivalent) Pipe diameters in mm

S = 0·00100 to 0·00290; $\Delta p/\rho l$ = (0·00981) to (0·02844) ms^{-2}

m = Manning $n \times 100$
S = 0·00300 to 0·00950

Full bore flow; Tables E give m_C or m_P for Colebrook-White or for laminar solutions resp., to divide mV and/or mQ as follows:

i.e. kin. pr. grad., $\Delta p/\rho l$ = (0·02942) to (0·09316) ms^{-2}

mV to give velocities in ms^{-1}
mQ to give discharges in m^{3}s^{-1}

Gradient — (Equivalent) Pipe diameters in mm

S, ($\Delta p/\rho l$)	1600	1625	1650	1675	1700	1725	1750	1775	1800	1825	1850	1875	1900
0·00300	2·973	3·004	3·035	3·066	3·096	3·126	3·157	3·187	3·216	3·246	3·276	3·305	3·334
(0·02942)	5·9786	6·2309	6·4898	6·7554	7·0276	7·3066	7·5924	7·8851	8·1847	8·4914	8·8051	9·1260	9·4541
0·00320	3·071	3·103	3·135	3·166	3·198	3·229	3·260	3·291	3·322	3·353	3·383	3·414	3·444
(0·03138)	6·1746	6·4353	6·7027	6·9769	7·2581	7·5462	7·8414	8·1437	8·4531	8·7698	9·0939	9·4253	9·7641
0·00340	3·166	3·198	3·231	3·264	3·296	3·328	3·360	3·392	3·424	3·456	3·487	3·519	3·550
(0·03334)	6·3647	6·6333	6·9090	7·1916	7·4815	7·7784	8·0827	8·3943	8·7133	9·0397	9·3737	9·7154	10·065
0·00360	3·257	3·291	3·325	3·358	3·392	3·425	3·458	3·491	3·523	3·556	3·588	3·621	3·653
(0·03530)	6·5492	6·8256	7·1093	7·4001	7·6983	8·0040	8·3170	8·6377	8·9659	9·3018	9·6455	9·9970	10·356
0·00380	3·347	3·381	3·416	3·450	3·485	3·519	3·553	3·586	3·620	3·653	3·687	3·720	3·753
(0·03727)	6·7286	7·0127	7·3041	7·6029	7·9093	8·2233	8·5449	8·8744	9·2116	9·5567	9·9098	10·271	10·640
0·00400	3·433	3·469	3·505	3·540	3·575	3·610	3·645	3·679	3·714	3·748	3·782	3·816	3·850
(0·03923)	6·9034	7·1948	7·4938	7·8004	8·1148	8·4369	8·7669	9·1049	9·4509	9·8050	10·167	10·538	10·917
0·00420	3·518	3·555	3·591	3·627	3·663	3·699	3·735	3·770	3·806	3·841	3·876	3·911	3·945
(0·04119)	7·0739	7·3725	7·6789	7·9931	8·3152	8·6453	8·9834	9·3297	9·6843	10·047	10·418	10·798	11·186
0·00440	3·601	3·638	3·676	3·713	3·750	3·786	3·823	3·859	3·895	3·931	3·967	4·003	4·038
(0·04315)	7·2404	7·5460	7·8596	8·1812	8·5108	8·8487	9·1948	9·5493	9·9122	10·284	10·664	11·052	11·449
0·00460	3·682	3·720	3·758	3·796	3·834	3·871	3·909	3·946	3·983	4·020	4·056	4·093	4·129
(0·04511)	7·4031	7·7156	8·0362	8·3650	8·7021	9·0476	9·4015	9·7639	10·135	10·515	10·903	11·301	11·707
0·00480	3·761	3·800	3·839	3·878	3·916	3·955	3·993	4·031	4·068	4·106	4·143	4·181	4·218
(0·04707)	7·5623	7·8816	8·2091	8·5449	8·8893	9·2422	9·6037	9·9739	10·353	10·741	11·138	11·544	11·959
0·00500	3·839	3·879	3·918	3·958	3·997	4·036	4·075	4·114	4·152	4·191	4·229	4·267	4·305
(0·04903)	7·7183	8·0441	8·3783	8·7212	9·0726	9·4328	9·8017	10·180	10·566	10·962	11·367	11·782	12·205
0·00525	3·934	3·974	4·015	4·056	4·096	4·136	4·176	4·215	4·255	4·294	4·333	4·372	4·411
(0·05148)	7·9089	8·2427	8·5852	8·9365	9·2966	9·6657	10·044	10·431	10·827	11·233	11·648	12·073	12·507
0·00550	4·026	4·068	4·110	4·151	4·192	4·233	4·274	4·315	4·355	4·395	4·435	4·475	4·515
(0·05394)	8·0950	8·4367	8·7873	9·1468	9·5154	9·8932	10·280	10·676	11·082	11·497	11·922	12·357	12·801
0·00575	4·117	4·159	4·202	4·244	4·286	4·328	4·370	4·412	4·453	4·494	4·535	4·576	4·616
(0·05639)	8·2769	8·6263	8·9848	9·3524	9·7293	10·116	10·511	10·916	11·331	11·756	12·190	12·634	13·089
0·00600	4·205	4·249	4·292	4·336	4·379	4·421	4·464	4·506	4·549	4·591	4·633	4·674	4·716
(0·05884)	8·4550	8·8119	9·1780	9·5535	9·9385	10·333	10·737	11·151	11·575	12·009	12·452	12·906	13·370
0·00625	4·292	4·336	4·381	4·425	4·469	4·513	4·556	4·599	4·642	4·685	4·728	4·771	4·813
(0·06129)	8·6293	8·9936	9·3673	9·7505	10·143	10·546	10·959	11·381	11·814	12·256	12·709	13·172	13·646
0·00650	4·377	4·422	4·468	4·513	4·557	4·602	4·646	4·690	4·734	4·778	4·822	4·865	4·908
(0·06374)	8·8002	9·1717	9·5528	9·9436	10·344	10·755	11·176	11·607	12·048	12·499	12·961	13·433	13·916
0·00675	4·460	4·507	4·553	4·599	4·644	4·690	4·735	4·780	4·825	4·869	4·914	4·958	5·002
(0·06619)	8·9678	9·3464	9·7348	10·133	10·541	10·960	11·389	11·828	12·277	12·737	13·208	13·689	14·181
0·00700	4·542	4·589	4·636	4·683	4·729	4·776	4·822	4·868	4·913	4·959	5·004	5·049	5·093
(0·06865)	9·1324	9·5179	9·9134	10·319	10·735	11·161	11·598	12·045	12·502	12·971	13·450	13·940	14·441
0·00725	4·622	4·671	4·718	4·766	4·813	4·860	4·907	4·954	5·000	5·046	5·092	5·138	5·184
(0·07110)	9·2940	9·6864	10·089	10·502	10·925	11·359	11·803	12·258	12·724	13·200	13·688	14·187	14·697
0·00750	4·702	4·750	4·799	4·847	4·895	4·943	4·991	5·038	5·086	5·133	5·179	5·226	5·272
(0·07355)	9·4529	9·8520	10·261	10·681	11·112	11·553	12·005	12·467	12·941	13·426	13·922	14·429	14·948
0·00800	4·856	4·906	4·956	5·006	5·056	5·105	5·155	5·204	5·252	5·301	5·349	5·397	5·445
(0·07845)	9·7629	10·175	10·598	11·031	11·476	11·932	12·398	12·876	13·366	13·866	14·379	14·903	15·438
0·00850	5·005	5·057	5·109	5·160	5·212	5·263	5·313	5·364	5·414	5·464	5·514	5·563	5·613
(0·08336)	10·063	10·488	10·924	11·371	11·829	12·299	12·780	13·273	13·777	14·293	14·821	15·361	15·914
0·00900	5·150	5·204	5·257	5·310	5·363	5·415	5·467	5·519	5·571	5·622	5·674	5·725	5·775
(0·08826)	10·355	10·792	11·241	11·701	12·172	12·655	13·150	13·657	14·176	14·707	15·251	15·807	16·375
0·00950	5·291	5·346	5·401	5·455	5·510	5·563	5·617	5·670	5·724	5·776	5·829	5·882	5·934
(0·09316)	10·639	11·088	11·549	12·021	12·506	13·002	13·511	14·032	14·565	15·110	15·669	16·240	16·824
S, ($\Delta p/\rho l$)	1600	1625	1650	1675	1700	1725	1750	1775	1800	1825	1850	1875	1900

Gradient — (Equivalent) Pipe diameters in mm

S = 0·00300 to 0·00950;　　　　　　$\Delta p/\rho l$ = (0·02942) to (0·09316) ms^{-2}

m = Manning $n \times 100$
S = 0·0100 to 0·0290

i.e. kin. pr. grad., $\Delta p/\rho l$ =
(0·0981) to (0·2844) ms^{-2}

Full bore flow; Tables E give m_C or m_P for Colebrook-White or for laminar solutions resp., to divide mV and/or mQ as follows:

mV to give velocities in ms^{-1}
mQ to give discharges in m^{3}s^{-1}

Gradient   **(Equivalent) Pipe diameters in mm**

S, $(\Delta p/\rho l)$	1600	1625	1650	1675	1700	1725	1750	1775	1800	1825	1850	1875	1900
0·0100	5·429	5·485	5·541	5·597	5·653	5·708	5·763	5·818	5·872	5·927	5·981	6·034	6·088
(0·0981)	10·915	11·376	11·849	12·334	12·831	13·340	13·862	14·396	14·943	15·503	16·076	16·662	17·261
0·0105	5·563	5·621	5·678	5·735	5·792	5·849	5·905	5·961	6·017	6·073	6·128	6·183	6·238
(0·1030)	11·185	11·657	12·141	12·638	13·147	13·669	14·204	14·752	15·312	15·886	16·473	17·073	17·687
0·0110	5·694	5·753	5·812	5·870	5·929	5·987	6·044	6·102	6·159	6·216	6·272	6·329	6·385
(0·1079)	11·448	11·931	12·427	12·936	13·457	13·991	14·538	15·099	15·673	16·260	16·860	17·475	18·103
0·0115	5·822	5·882	5·942	6·002	6·062	6·121	6·180	6·239	6·297	6·356	6·413	6·471	6·528
(0·1128)	11·705	12·199	12·706	13·226	13·759	14·305	14·865	15·438	16·025	16·625	17·239	17·868	18·510
0·0120	5·947	6·009	6·070	6·131	6·192	6·253	6·313	6·373	6·433	6·492	6·551	6·610	6·669
(0·1177)	11·957	12·462	12·980	13·511	14·055	14·613	15·185	15·770	16·369	16·983	17·610	18·252	18·908
0·0125	6·070	6·133	6·195	6·258	6·320	6·382	6·443	6·504	6·565	6·626	6·686	6·747	6·806
(0·1226)	12·204	12·719	13·247	13·789	14·345	14·914	15·498	16·095	16·707	17·333	17·973	18·628	19·298
0·0130	6·190	6·254	6·318	6·382	6·445	6·508	6·571	6·633	6·695	6·757	6·819	6·880	6·941
(0·1275)	12·445	12·971	13·510	14·062	14·629	15·210	15·805	16·414	17·038	17·676	18·329	18·997	19·680
0·0135	6·308	6·373	6·438	6·503	6·568	6·632	6·696	6·760	6·823	6·886	6·949	7·011	7·073
(0·1324)	12·682	13·218	13·767	14·330	14·908	15·500	16·106	16·727	17·362	18·013	18·678	19·359	20·055
0·0140	6·423	6·490	6·557	6·623	6·688	6·754	6·819	6·884	6·948	7·012	7·076	7·140	7·203
(0·1373)	12·915	13·460	14·020	14·593	15·181	15·784	16·401	17·034	17·681	18·343	19·021	19·714	20·423
0·0145	6·537	6·605	6·673	6·740	6·807	6·873	6·940	7·006	7·071	7·137	7·202	7·266	7·331
(0·1422)	13·144	13·699	14·268	14·852	15·450	16·063	16·692	17·335	17·994	18·668	19·358	20·063	20·785
0·0150	6·649	6·718	6·787	6·855	6·923	6·991	7·058	7·125	7·192	7·259	7·325	7·390	7·456
(0·1471)	13·368	13·933	14·512	15·105	15·714	16·338	16·977	17·632	18·302	18·987	19·689	20·406	21·140
0·0160	6·867	6·938	7·009	7·080	7·150	7·220	7·290	7·359	7·428	7·497	7·565	7·633	7·701
(0·1569)	13·807	14·390	14·988	15·601	16·230	16·874	17·534	18·210	18·902	19·610	20·335	21·076	21·833
0·0170	7·078	7·152	7·225	7·298	7·370	7·442	7·514	7·585	7·657	7·727	7·798	7·868	7·938
(0·1667)	14·232	14·833	15·449	16·081	16·729	17·393	18·073	18·770	19·484	20·213	20·960	21·724	22·505
0·0180	7·284	7·359	7·435	7·509	7·584	7·658	7·732	7·805	7·879	7·951	8·024	8·096	8·168
(0·1765)	14·644	15·263	15·897	16·547	17·214	17·897	18·597	19·314	20·048	20·800	21·568	22·354	23·158
0·0190	7·483	7·561	7·638	7·715	7·792	7·868	7·944	8·019	8·094	8·169	8·244	8·318	8·391
(0·1863)	15·046	15·681	16·332	17·001	17·686	18·388	19·107	19·844	20·598	21·369	22·159	22·967	23·792
0·0200	7·678	7·757	7·837	7·916	7·994	8·072	8·150	8·228	8·305	8·381	8·458	8·534	8·609
(0·1961)	15·437	16·088	16·757	17·442	18·145	18·866	19·603	20·359	21·133	21·925	22·735	23·563	24·410
0·0210	7·867	7·949	8·030	8·111	8·192	8·272	8·351	8·431	8·510	8·588	8·667	8·745	8·822
(0·2059)	15·818	16·485	17·170	17·873	18·593	19·331	20·088	20·862	21·655	22·466	23·296	24·145	25·013
0·0220	8·052	8·136	8·219	8·302	8·384	8·466	8·548	8·629	8·710	8·790	8·871	8·950	9·030
(0·2157)	16·190	16·873	17·575	18·294	19·031	19·786	20·560	21·353	22·164	22·995	23·844	24·713	25·602
0·0230	8·233	8·319	8·404	8·489	8·573	8·657	8·740	8·823	8·906	8·988	9·070	9·151	9·233
(0·2256)	16·554	17·253	17·970	18·705	19·459	20·231	21·022	21·833	22·662	23·512	24·380	25·269	26·177
0·0240	8·410	8·498	8·585	8·671	8·757	8·843	8·928	9·013	9·097	9·181	9·265	9·348	9·431
(0·2354)	16·910	17·624	18·356	19·107	19·877	20·666	21·474	22·302	23·150	24·017	24·905	25·812	26·740
0·0250	8·584	8·673	8·762	8·850	8·938	9·025	9·112	9·199	9·285	9·371	9·456	9·541	9·626
(0·2452)	17·259	17·987	18·735	19·501	20·287	21·092	21·917	22·762	23·627	24·512	25·418	26·344	27·292
0·0260	8·754	8·845	8·935	9·025	9·115	9·204	9·293	9·381	9·469	9·556	9·643	9·730	9·816
(0·2550)	17·600	18·343	19·106	19·887	20·689	21·510	22·351	23·213	24·095	24·998	25·922	26·866	27·832
0·0270	8·920	9·013	9·105	9·197	9·288	9·379	9·470	9·560	9·649	9·738	9·827	9·915	10·00
(0·2648)	17·936	18·693	19·470	20·266	21·083	21·920	22·777	23·655	24·554	25·474	26·415	27·378	28·362
0·0280	9·084	9·179	9·272	9·366	9·459	9·551	9·643	9·735	9·826	9·917	10·01	10·10	10·19
(0·2746)	18·265	19·036	19·827	20·638	21·470	22·322	23·195	24·089	25·005	25·942	26·900	27·880	28·883
0·0290	9·245	9·341	9·437	9·532	9·626	9·720	9·814	9·907	10·00	10·09	10·18	10·28	10·37
(0·2844)	18·588	19·373	20·178	21·003	21·850	22·717	23·606	24·516	25·447	26·401	27·376	28·374	29·394
S, $(\Delta p/\rho l)$	1600	1625	1650	1675	1700	1725	1750	1775	1800	1825	1850	1875	1900

Gradient   **(Equivalent) Pipe diameters in mm**

S = 0·0100 to 0·0290; $\Delta p/\rho l$ = (0·0981) to (0·2844) ms^{-2}

m = Manning $n \times 100$
S = 0·0300 to 0·1000

Full bore flow; Tables E give m_C or m_P for Colebrook-White or for laminar solutions resp., to divide mV and/or mQ as follows:

i.e. kin. pr. grad., $\Delta p/\rho l$ = (0·2942) to (0·9807) ms^{-2}

mV to give velocities in ms^{-1}
mQ to give discharges in m^{3}s^{-1}

Gradient (Equivalent) Pipe diameters in mm

S, ($\Delta p/\rho l$)	1600	1625	1650	1675	1700	1725	1750	1775	1800	1825	1850	1875	1900
0·0300	9·403	9·501	9·598	9·695	9·791	9·887	9·982	10·08	10·17	10·27	10·36	10·45	10·54
(0·2942)	18·906	19·704	20·523	21·362	22·223	23·105	24·009	24·935	25·882	26·852	27·844	28·859	29·896
0·0320	9·711	9·812	9·913	10·01	10·11	10·21	10·31	10·41	10·50	10·60	10·70	10·79	10·89
(0·3138)	19·526	20·350	21·196	22·063	22·952	23·863	24·797	25·753	26·731	27·733	28·757	29·805	30·877
0·0340	10·01	10·11	10·22	10·32	10·42	10·53	10·63	10·73	10·83	10·93	11·03	11·13	11·23
(0·3334)	20·127	20·976	21·848	22·742	23·658	24·598	25·560	26·545	27·554	28·586	29·642	30·723	31·827
0·0360	10·30	10·41	10·51	10·62	10·73	10·83	10·93	11·04	11·14	11·24	11·35	11·45	11·55
(0·3530)	20·710	21·585	22·481	23·401	24·344	25·311	26·301	27·315	28·353	29·415	30·502	31·613	32·750
0·0380	10·58	10·69	10·80	10·91	11·02	11·13	11·23	11·34	11·45	11·55	11·66	11·76	11·87
(0·3727)	21·278	22·176	23·097	24·043	25·011	26·004	27·021	28·063	29·130	30·221	31·338	32·480	33·647
0·0400	10·86	10·97	11·08	11·19	11·31	11·42	11·53	11·64	11·74	11·85	11·96	12·07	12·18
(0·3923)	21·831	22·752	23·698	24·667	25·661	26·680	27·723	28·792	29·886	31·006	32·152	33·323	34·521
0·0420	11·13	11·24	11·36	11·47	11·58	11·70	11·81	11·92	12·03	12·15	12·26	12·37	12·48
(0·4119)	22·370	23·314	24·283	25·276	26·295	27·339	28·408	29·503	30·624	31·772	32·946	34·146	35·374
0·0440	11·39	11·51	11·62	11·74	11·86	11·97	12·09	12·20	12·32	12·43	12·54	12·66	12·77
(0·4315)	22·896	23·863	24·854	25·871	26·914	27·982	29·077	30·198	31·345	32·519	33·721	34·950	36·206
0·0460	11·64	11·76	11·88	12·00	12·12	12·24	12·36	12·48	12·59	12·71	12·83	12·94	13·06
(0·4511)	23·411	24·399	25·413	26·453	27·519	28·611	29·730	30·876	32·050	33·250	34·479	35·735	37·020
0·0480	11·89	12·02	12·14	12·26	12·38	12·51	12·63	12·75	12·87	12·98	13·10	13·22	13·34
(0·4707)	23·914	24·924	25·959	27·021	28·110	29·226	30·370	31·540	32·739	33·965	35·220	36·504	37·816
0·0500	12·14	12·27	12·39	12·52	12·64	12·76	12·89	13·01	13·13	13·25	13·37	13·49	13·61
(0·4903)	24·407	25·438	26·495	27·579	28·690	29·829	30·996	32·191	33·414	34·666	35·947	37·257	38·596
0·0525	12·44	12·57	12·70	12·82	12·95	13·08	13·20	13·33	13·46	13·58	13·70	13·83	13·95
(0·5148)	25·010	26·066	27·149	28·260	29·399	30·566	31·761	32·986	34·239	35·522	36·834	38·177	39·549
0·0550	12·73	12·86	13·00	13·13	13·26	13·39	13·52	13·64	13·77	13·90	14·03	14·15	14·28
(0·5394)	25·599	26·679	27·788	28·925	30·090	31·285	32·509	33·762	35·045	36·358	37·701	39·075	40·480
0·0575	13·02	13·15	13·29	13·42	13·55	13·69	13·82	13·95	14·08	14·21	14·34	14·47	14·60
(0·5639)	26·174	27·279	28·412	29·575	30·767	31·988	33·239	34·521	35·832	37·175	38·549	39·953	41·390
0·0600	13·30	13·44	13·57	13·71	13·85	13·98	14·12	14·25	14·38	14·52	14·65	14·78	14·91
(0·5884)	26·737	27·866	29·023	30·211	31·428	32·676	33·954	35·263	36·603	37·975	39·378	40·813	42·280
0·0625	13·57	13·71	13·85	13·99	14·13	14·27	14·41	14·54	14·68	14·82	14·95	15·09	15·22
(0·6129)	27·288	28·440	29·622	30·834	32·076	33·350	34·654	35·990	37·358	38·758	40·190	41·654	43·152
0·0650	13·84	13·98	14·13	14·27	14·41	14·55	14·69	14·83	14·97	15·11	15·25	15·38	15·52
(0·6374)	27·829	29·003	30·209	31·445	32·712	34·010	35·341	36·703	38·098	39·525	40·986	42·479	44·006
0·0675	14·10	14·25	14·40	14·54	14·69	14·83	14·97	15·12	15·26	15·40	15·54	15·68	15·82
(0·6619)	28·359	29·556	30·784	32·044	33·335	34·658	36·014	37·402	38·823	40·278	41·766	43·288	44·845
0·0700	14·36	14·51	14·66	14·81	14·96	15·10	15·25	15·39	15·54	15·68	15·82	15·97	16·11
(0·6865)	28·879	30·098	31·349	32·632	33·947	35·294	36·675	38·089	39·536	41·017	42·533	44·083	45·668
0·0725	14·62	14·77	14·92	15·07	15·22	15·37	15·52	15·66	15·81	15·96	16·10	16·25	16·39
(0·7110)	29·390	30·631	31·904	33·209	34·547	35·919	37·324	38·763	40·236	41·743	43·286	44·863	46·476
0·0750	14·87	15·02	15·18	15·33	15·48	15·63	15·78	15·93	16·08	16·23	16·38	16·53	16·67
(0·7355)	29·893	31·155	32·449	33·777	35·138	36·533	37·962	39·425	40·924	42·457	44·026	45·630	47·270
0·0800	15·36	15·51	15·67	15·83	15·99	16·14	16·30	16·46	16·61	16·76	16·92	17·07	17·22
(0·7845)	30·873	32·176	33·513	34·885	36·290	37·731	39·207	40·718	42·266	43·849	45·469	47·126	48·821
0·0850	15·83	15·99	16·16	16·32	16·48	16·64	16·80	16·96	17·12	17·28	17·44	17·59	17·75
(0·8336)	31·823	33·167	34·545	35·958	37·407	38·892	40·414	41·971	43·566	45·199	46·869	48·577	50·323
0·0900	16·29	16·46	16·62	16·79	16·96	17·12	17·29	17·45	17·62	17·78	17·94	18·10	18·26
(0·8826)	32·746	34·128	35·546	37·001	38·492	40·020	41·585	43·188	44·829	46·509	48·228	49·985	51·782
0·0950	16·73	16·91	17·08	17·25	17·42	17·59	17·76	17·93	18·10	18·27	18·43	18·60	18·76
(0·9316)	33·643	35·063	36·520	38·015	39·547	41·116	42·725	44·372	46·058	47·784	49·549	51·355	53·201
0·1000	17·17	17·35	17·52	17·70	17·88	18·05	18·22	18·40	18·57	18·74	18·91	19·08	19·25
(0·9807)	34·517	35·974	37·469	39·002	40·574	42·185	43·835	45·524	47·254	49·025	50·836	52·689	54·583
S, ($\Delta p/\rho l$)	1600	1625	1650	1675	1700	1725	1750	1775	1800	1825	1850	1875	1900

Gradient (Equivalent) Pipe diameters in mm

S = 0·0300 to 0·1000;

$\Delta p/\rho l$ = (0·2942) to (0·9807) ms^{-2}

m = Manning $n \times 100$
S = 0.000100 to 0.000290

i.e. kin. pr. grad., $\Delta p/\rho l$ =
(0.000981) to (0.002844) ms^{-2}

Full bore flow; Tables E give m_C or m_P for Colebrook-White or for laminar solutions resp., to divide mV and/or mQ as follows:

mV to give velocities in ms^{-1}
mQ to give discharges in m^{3}s^{-1}

Gradient — (Equivalent) Pipe diameters in mm

S, $(\Delta p/\rho l)$	1900	1950	2000	2050	2100	2150	2200	2250	2300	2350	2400	2450	2500
0.000100	0.609	0.619	0.630	0.640	0.651	0.661	0.671	0.681	0.691	0.701	0.711	0.721	0.731
(0.000981)	1.7261	1.8499	1.9791	2.1138	2.2541	2.4000	2.5518	2.7094	2.8729	3.0425	3.2182	3.4001	3.5883
0.000105	0.624	0.635	0.646	0.656	0.667	0.677	0.688	0.698	0.709	0.719	0.729	0.739	0.749
(0.001030)	1.7687	1.8956	2.0280	2.1660	2.3097	2.4593	2.6148	2.7763	2.9439	3.1176	3.2977	3.4841	3.6769
0.000110	0.638	0.650	0.661	0.672	0.683	0.693	0.704	0.715	0.725	0.736	0.746	0.756	0.767
(0.001079)	1.8103	1.9402	2.0757	2.2170	2.3641	2.5172	2.6763	2.8416	3.0131	3.1910	3.3753	3.5661	3.7634
0.000115	0.653	0.664	0.676	0.687	0.698	0.709	0.720	0.731	0.742	0.752	0.763	0.773	0.784
(0.001128)	1.8510	1.9838	2.1223	2.2668	2.4172	2.5738	2.7365	2.9055	3.0809	3.2627	3.4511	3.6462	3.8480
0.000120	0.667	0.679	0.690	0.702	0.713	0.724	0.735	0.746	0.757	0.768	0.779	0.790	0.801
(0.001177)	1.8908	2.0264	2.1680	2.3155	2.4692	2.6291	2.7953	2.9680	3.1471	3.3329	3.5254	3.7246	3.9308
0.000125	0.681	0.693	0.704	0.716	0.728	0.739	0.751	0.762	0.773	0.784	0.795	0.806	0.817
(0.001226)	1.9298	2.0682	2.2127	2.3633	2.5201	2.6833	2.8530	3.0292	3.2120	3.4016	3.5981	3.8014	4.0118
0.000130	0.694	0.706	0.718	0.730	0.742	0.754	0.765	0.777	0.788	0.800	0.811	0.822	0.833
(0.001275)	1.9680	2.1092	2.2565	2.4101	2.5700	2.7365	2.9095	3.0892	3.2756	3.4690	3.6693	3.8767	4.0913
0.000135	0.707	0.720	0.732	0.744	0.756	0.768	0.780	0.792	0.803	0.815	0.827	0.838	0.849
(0.001324)	2.0055	2.1494	2.2995	2.4560	2.6190	2.7886	2.9649	3.1480	3.3380	3.5351	3.7392	3.9506	4.1692
0.000140	0.720	0.733	0.745	0.758	0.770	0.782	0.794	0.806	0.818	0.830	0.842	0.853	0.865
(0.001373)	2.0423	2.1888	2.3417	2.5011	2.6671	2.8398	3.0193	3.2058	3.3993	3.5999	3.8078	4.0231	4.2457
0.000145	0.733	0.746	0.759	0.771	0.784	0.796	0.808	0.821	0.833	0.845	0.857	0.868	0.880
(0.001422)	2.0785	2.2275	2.3831	2.5453	2.7143	2.8900	3.0727	3.2625	3.4595	3.6637	3.8752	4.0943	4.3209
0.000150	0.746	0.759	0.772	0.784	0.797	0.810	0.822	0.835	0.847	0.859	0.871	0.883	0.895
(0.001471)	2.1140	2.2656	2.4239	2.5888	2.7607	2.9394	3.1253	3.3183	3.5186	3.7263	3.9415	4.1643	4.3948
0.000160	0.770	0.784	0.797	0.810	0.823	0.836	0.849	0.862	0.875	0.887	0.900	0.912	0.925
(0.001569)	2.1833	2.3399	2.5034	2.6737	2.8512	3.0358	3.2278	3.4271	3.6340	3.8485	4.0707	4.3008	4.5389
0.000170	0.794	0.808	0.821	0.835	0.849	0.862	0.875	0.888	0.902	0.915	0.928	0.940	0.953
(0.001667)	2.2505	2.4119	2.5804	2.7560	2.9389	3.1293	3.3271	3.5326	3.7458	3.9669	4.1960	4.4332	4.6786
0.000180	0.817	0.831	0.845	0.859	0.873	0.887	0.901	0.914	0.928	0.941	0.954	0.968	0.981
(0.001765)	2.3158	2.4819	2.6552	2.8359	3.0242	3.2200	3.4236	3.6350	3.8544	4.0819	4.3177	4.5617	4.8142
0.000190	0.839	0.854	0.868	0.883	0.897	0.911	0.925	0.939	0.953	0.967	0.981	0.994	1.008
(0.001863)	2.3792	2.5499	2.7280	2.9136	3.1070	3.3082	3.5174	3.7346	3.9600	4.1938	4.4360	4.6867	4.9461
0.000200	0.861	0.876	0.891	0.906	0.920	0.935	0.949	0.964	0.978	0.992	1.006	1.020	1.034
(0.001961)	2.4410	2.6161	2.7988	2.9893	3.1877	3.3942	3.6088	3.8316	4.0629	4.3027	4.5512	4.8085	5.0746
0.000210	0.882	0.898	0.913	0.928	0.943	0.958	0.973	0.987	1.002	1.017	1.031	1.045	1.059
(0.002059)	2.5013	2.6807	2.8680	3.0632	3.2665	3.4780	3.6979	3.9263	4.1633	4.4090	4.6636	4.9272	5.1999
0.000220	0.903	0.919	0.934	0.950	0.965	0.981	0.996	1.011	1.026	1.040	1.055	1.070	1.084
(0.002157)	2.5602	2.7438	2.9354	3.1352	3.3433	3.5598	3.7849	4.0187	4.2612	4.5128	4.7734	5.0432	5.3223
0.000230	0.923	0.939	0.955	0.971	0.987	1.003	1.018	1.033	1.049	1.064	1.079	1.094	1.109
(0.002256)	2.6177	2.8055	3.0014	3.2057	3.4185	3.6398	3.8700	4.1090	4.3570	4.6142	4.8806	5.1565	5.4419
0.000240	0.943	0.960	0.976	0.992	1.008	1.024	1.040	1.056	1.071	1.087	1.102	1.117	1.132
(0.002354)	2.6740	2.8658	3.0660	3.2747	3.4920	3.7181	3.9532	4.1973	4.4507	4.7134	4.9856	5.2674	5.5590
0.000250	0.963	0.979	0.996	1.013	1.029	1.045	1.061	1.077	1.093	1.109	1.125	1.140	1.156
(0.002452)	2.7292	2.9249	3.1292	3.3422	3.5640	3.7948	4.0347	4.2839	4.5425	4.8106	5.0884	5.3760	5.6736
0.000260	0.982	0.999	1.016	1.033	1.049	1.066	1.082	1.099	1.115	1.131	1.147	1.163	1.179
(0.002550)	2.7832	2.9828	3.1912	3.4084	3.6346	3.8699	4.1146	4.3687	4.6324	4.9059	5.1892	5.4825	5.7860
0.000270	1.000	1.018	1.035	1.052	1.069	1.086	1.103	1.120	1.136	1.153	1.169	1.185	1.201
(0.002648)	2.8362	3.0396	3.2520	3.4733	3.7038	3.9437	4.1930	4.4520	4.7207	4.9993	5.2880	5.5869	5.8962
0.000280	1.019	1.036	1.054	1.072	1.089	1.106	1.123	1.140	1.157	1.174	1.190	1.207	1.223
(0.002746)	2.8883	3.0954	3.3116	3.5370	3.7718	4.0160	4.2699	4.5337	4.8073	5.0911	5.3851	5.6895	6.0044
0.000290	1.037	1.055	1.073	1.091	1.108	1.126	1.143	1.160	1.178	1.195	1.211	1.228	1.245
(0.002844)	2.9394	3.1502	3.3702	3.5996	3.8385	4.0871	4.3455	4.6139	4.8924	5.1812	5.4804	5.7902	6.1107
S, $(\Delta p/\rho l)$	1900	1950	2000	2050	2100	2150	2200	2250	2300	2350	2400	2450	2500

Gradient — (Equivalent) Pipe diameters in mm

S = 0.000100 to 0.000290;

$\Delta p/\rho l$ = (0.000981) to (0.002844) ms^{-2}

m = Manning $n \times 100$
S = 0·000300 to 0·000950

i.e. kin. pr. grad., $\Delta p/\rho l =$
(0·002942) to (0·009316) ms^{-2}

Full bore flow; Tables E give m_C or m_P for Colebrook-White or for laminar solutions resp., to divide mV and/or mQ as follows:

mV to give velocities in ms^{-1}
mQ to give discharges in m^{3}s^{-1}

Gradient (Equivalent) Pipe diameters in mm

S, $(\Delta p/\rho l)$	1900	1950	2000	2050	2100	2150	2200	2250	2300	2350	2400	2450	2500
0·000300	1·054	1·073	1·091	1·109	1·127	1·145	1·163	1·180	1·198	1·215	1·232	1·249	1·266
(0·002942)	2·9896	3·2041	3·4279	3·6612	3·9042	4·1570	4·4198	4·6928	4·9760	5·2698	5·5741	5·8892	6·2151
0·000320	1·089	1·108	1·127	1·146	1·164	1·183	1·201	1·219	1·237	1·255	1·273	1·290	1·308
(0·003138)	3·0877	3·3092	3·5403	3·7812	4·0322	4·2933	4·5648	4·8467	5·1392	5·4426	5·7569	6·0823	6·4190
0·000340	1·123	1·142	1·162	1·181	1·200	1·219	1·238	1·256	1·275	1·293	1·312	1·330	1·348
(0·003334)	3·1827	3·4110	3·6492	3·8976	4·1563	4·4255	4·7052	4·9958	5·2974	5·6101	5·9341	6·2695	6·6165
0·000360	1·155	1·175	1·195	1·215	1·235	1·254	1·274	1·293	1·312	1·331	1·350	1·368	1·387
(0·003530)	3·2750	3·5099	3·7550	4·0106	4·2768	4·5538	4·8417	5·1407	5·4510	5·7727	6·1061	6·4512	6·8083
0·000380	1·187	1·207	1·228	1·248	1·269	1·289	1·309	1·328	1·348	1·367	1·387	1·406	1·425
(0·003727)	3·3647	3·6061	3·8579	4·1205	4·3940	4·6785	4·9743	5·2815	5·6004	5·9309	6·2734	6·6280	6·9949
0·000400	1·218	1·239	1·260	1·281	1·302	1·322	1·343	1·363	1·383	1·403	1·423	1·442	1·462
(0·003923)	3·4521	3·6997	3·9582	4·2276	4·5081	4·8001	5·1036	5·4188	5·7458	6·0850	6·4364	6·8002	7·1766
0·000420	1·248	1·269	1·291	1·312	1·334	1·355	1·376	1·396	1·417	1·438	1·458	1·478	1·498
(0·004119)	3·5374	3·7911	4·0559	4·3320	4·6195	4·9186	5·2296	5·5526	5·8877	6·2353	6·5953	6·9681	7·3538
0·000440	1·277	1·299	1·321	1·343	1·365	1·387	1·408	1·429	1·450	1·471	1·492	1·513	1·533
(0·004315)	3·6206	3·8803	4·1513	4·4339	4·7282	5·0344	5·3527	5·6832	6·0263	6·3820	6·7505	7·1321	7·5269
0·000460	1·306	1·329	1·351	1·374	1·396	1·418	1·440	1·461	1·483	1·504	1·526	1·547	1·568
(0·004511)	3·7020	3·9675	4·2446	4·5336	4·8344	5·1475	5·4730	5·8110	6·1617	6·5254	6·9023	7·2924	7·6961
0·000480	1·334	1·357	1·380	1·403	1·426	1·448	1·471	1·493	1·515	1·537	1·559	1·580	1·602
(0·004707)	3·7816	4·0529	4·3359	4·6311	4·9384	5·2582	5·5907	5·9359	6·2943	6·6658	7·0507	7·4493	7·8616
0·000500	1·361	1·385	1·409	1·432	1·455	1·478	1·501	1·524	1·546	1·569	1·591	1·613	1·635
(0·004903)	3·8596	4·1364	4·4253	4·7266	5·0403	5·3667	5·7060	6·0583	6·4240	6·8032	7·1961	7·6029	8·0237
0·000525	1·395	1·419	1·443	1·467	1·491	1·515	1·538	1·561	1·584	1·607	1·630	1·653	1·675
(0·005148)	3·9549	4·2386	4·5346	4·8433	5·1647	5·4992	5·8469	6·2080	6·5827	6·9712	7·3738	7·7906	8·2218
0·000550	1·428	1·453	1·477	1·502	1·526	1·550	1·574	1·598	1·622	1·645	1·668	1·691	1·714
(0·005394)	4·0480	4·3383	4·6413	4·9573	5·2863	5·6286	5·9845	6·3541	6·7376	7·1353	7·5473	7·9740	8·4153
0·000575	1·460	1·485	1·511	1·536	1·561	1·585	1·610	1·634	1·658	1·682	1·706	1·729	1·753
(0·005639)	4·1390	4·4358	4·7457	5·0687	5·4051	5·7551	6·1190	6·4969	6·8890	7·2957	7·7170	8·1532	8·6045
0·000600	1·491	1·517	1·543	1·569	1·594	1·619	1·644	1·669	1·694	1·718	1·743	1·767	1·791
(0·005884)	4·2280	4·5312	4·8477	5·1777	5·5213	5·8789	6·2506	6·6366	7·0372	7·4526	7·8829	8·3285	8·7895
0·000625	1·522	1·549	1·575	1·601	1·627	1·653	1·678	1·704	1·729	1·754	1·778	1·803	1·828
(0·006129)	4·3152	4·6247	4·9477	5·2844	5·6352	6·0001	6·3794	6·7734	7·1823	7·6062	8·0455	8·5003	8·9708
0·000650	1·552	1·579	1·606	1·633	1·659	1·685	1·711	1·737	1·763	1·788	1·814	1·839	1·864
(0·006374)	4·4006	4·7163	5·0457	5·3891	5·7468	6·1189	6·5058	6·9076	7·3245	7·7569	8·2048	8·6686	9·1484
0·000675	1·582	1·609	1·637	1·664	1·691	1·718	1·744	1·770	1·797	1·822	1·848	1·874	1·899
(0·006619)	4·4845	4·8061	5·1418	5·4918	5·8562	6·2355	6·6297	7·0392	7·4641	7·9046	8·3611	8·8337	9·3227
0·000700	1·611	1·639	1·667	1·694	1·722	1·749	1·776	1·803	1·829	1·856	1·882	1·908	1·934
(0·006865)	4·5668	4·8943	5·2361	5·5925	5·9637	6·3499	6·7514	7·1683	7·6010	8·0497	8·5145	8·9958	9·4938
0·000725	1·639	1·668	1·696	1·724	1·752	1·780	1·807	1·835	1·862	1·889	1·915	1·942	1·968
(0·007110)	4·6476	4·9809	5·3288	5·6915	6·0693	6·4623	6·8709	7·2952	7·7356	8·1922	8·6653	9·1551	9·6618
0·000750	1·667	1·696	1·725	1·754	1·782	1·810	1·838	1·866	1·894	1·921	1·948	1·975	2·002
(0·007355)	4·7270	5·0661	5·4199	5·7888	6·1730	6·5728	6·9883	7·4199	7·8678	8·3322	8·8134	9·3116	9·8270
0·000800	1·722	1·752	1·782	1·811	1·841	1·870	1·899	1·927	1·956	1·984	2·012	2·040	2·068
(0·007845)	4·8821	5·2322	5·5977	5·9787	6·3755	6·7883	7·2175	7·6633	8·1258	8·6055	9·1024	9·6169	10·149
0·000850	1·775	1·806	1·837	1·867	1·897	1·927	1·957	1·987	2·016	2·045	2·074	2·103	2·131
(0·008336)	5·0323	5·3933	5·7700	6·1627	6·5717	6·9973	7·4397	7·8991	8·3759	8·8703	9·3826	9·9129	10·462
0·000900	1·826	1·858	1·890	1·921	1·952	1·983	2·014	2·044	2·074	2·104	2·134	2·164	2·193
(0·008826)	5·1782	5·5496	5·9372	6·3413	6·7622	7·2001	7·6553	8·1281	8·6188	9·1275	9·6546	10·200	10·765
0·000950	1·876	1·909	1·942	1·974	2·006	2·038	2·069	2·100	2·131	2·162	2·193	2·223	2·253
(0·009316)	5·3201	5·7017	6·0999	6·5151	6·9475	7·3974	7·8651	8·3509	8·8549	9·3776	9·9191	10·480	11·060

S, $(\Delta p/\rho l)$	1900	1950	2000	2050	2100	2150	2200	2250	2300	2350	2400	2450	2500

Gradient (Equivalent) Pipe diameters in mm

S = 0·000300 to 0·000950; $\Delta p/\rho l$ = (0·002942) to (0·009316) ms^{-2}

m = Manning $n \times 100$
S = 0·00100 to 0·00290

i.e. kin. pr. grad., $\Delta p/\rho l =$
(0·00981) to (0·02844) ms^{-2}

Full bore flow; Tables E give m_C or m_P for Colebrook-White or for laminar solutions resp., to divide mV and/or mQ as follows:

mV to give velocities in ms^{-1}
mQ to give discharges in m^{3}s^{-1}

Gradient **(Equivalent) Pipe diameters in mm**

S, ($\Delta p/\rho l$)	1900	1950	2000	2050	2100	2150	2200	2250	2300	2350	2400	2450	2500
0·00100	1·925	1·959	1·992	2·025	2·058	2·091	2·123	2·155	2·187	2·218	2·250	2·281	2·312
(0·00981)	5·4583	5·8498	6·2584	6·6844	7·1280	7·5896	8·0694	8·5678	9·0850	9·6212	10·177	10·752	11·347
0·00105	1·973	2·007	2·041	2·075	2·109	2·142	2·175	2·208	2·241	2·273	2·305	2·337	2·369
(0·01030)	5·5931	5·9943	6·4129	6·8494	7·3040	7·7770	8·2687	8·7794	9·3093	9·8588	10·428	11·018	11·627
0·00110	2·019	2·054	2·089	2·124	2·158	2·193	2·226	2·260	2·293	2·326	2·359	2·392	2·424
(0·01079)	5·7247	6·1353	6·5639	7·0106	7·4759	7·9600	8·4633	8·9860	9·5284	10·091	10·674	11·277	11·901
0·00115	2·064	2·101	2·136	2·172	2·207	2·242	2·276	2·311	2·345	2·379	2·412	2·446	2·479
(0·01128)	5·8534	6·2732	6·7114	7·1682	7·6439	8·1389	8·6535	9·1879	9·7425	10·318	10·913	11·530	12·169
0·00120	2·109	2·146	2·182	2·218	2·254	2·290	2·325	2·361	2·395	2·430	2·464	2·498	2·532
(0·01177)	5·9793	6·4081	6·8557	7·3223	7·8083	8·3140	8·8396	9·3856	9·9521	10·540	11·148	11·778	12·430
0·00125	2·152	2·190	2·227	2·264	2·301	2·337	2·373	2·409	2·445	2·480	2·515	2·550	2·584
(0·01226)	6·1026	6·5403	6·9971	7·4733	7·9693	8·4854	9·0219	9·5791	10·157	10·757	11·378	12·021	12·687
0·00130	2·195	2·233	2·271	2·309	2·346	2·384	2·420	2·457	2·493	2·529	2·565	2·600	2·636
(0·01275)	6·2234	6·6698	7·1357	7·6213	8·1272	8·6535	9·2006	9·7688	10·358	10·970	11·603	12·259	12·938
0·00135	2·237	2·276	2·315	2·353	2·391	2·429	2·466	2·504	2·541	2·577	2·614	2·650	2·686
(0·01324)	6·3420	6·7969	7·2716	7·7665	8·2820	8·8183	9·3758	9·9549	10·556	11·179	11·824	12·493	13·184
0·00140	2·278	2·318	2·357	2·396	2·435	2·474	2·512	2·550	2·587	2·625	2·662	2·699	2·735
(0·01373)	6·4584	6·9216	7·4050	7·9090	8·4340	8·9801	9·5479	10·138	10·749	11·384	12·041	12·722	13·426
0·00145	2·318	2·359	2·399	2·439	2·478	2·517	2·556	2·595	2·633	2·671	2·709	2·746	2·784
(0·01422)	6·5727	7·0441	7·5361	8·0490	8·5832	9·1391	9·7169	10·317	10·940	11·585	12·255	12·947	13·664
0·00150	2·358	2·399	2·440	2·480	2·520	2·560	2·600	2·639	2·678	2·717	2·755	2·793	2·831
(0·01471)	6·6850	7·1645	7·6649	8·1866	8·7300	9·2953	9·8830	10·493	11·127	11·784	12·464	13·169	13·897
0·00160	2·435	2·478	2·520	2·562	2·603	2·644	2·685	2·726	2·766	2·806	2·846	2·885	2·924
(0·01569)	6·9043	7·3995	7·9163	8·4551	9·0163	9·6002	10·207	10·838	11·492	12·170	12·873	13·600	14·353
0·00170	2·510	2·554	2·597	2·641	2·683	2·726	2·768	2·810	2·851	2·892	2·933	2·974	3·014
(0·01667)	7·1168	7·6272	8·1599	8·7153	9·2938	9·8956	10·521	11·171	11·845	12·545	13·269	14·019	14·795
0·00180	2·583	2·628	2·673	2·717	2·761	2·805	2·848	2·891	2·934	2·976	3·018	3·060	3·101
(0·01765)	7·3231	7·8483	8·3965	8·9680	9·5632	10·183	10·826	11·495	12·189	12·908	13·654	14·425	15·224
0·00190	2·654	2·700	2·746	2·792	2·837	2·882	2·926	2·970	3·014	3·058	3·101	3·144	3·186
(0·01863)	7·5238	8·0634	8·6266	9·2137	9·8253	10·462	11·123	11·810	12·523	13·262	14·028	14·821	15·641
0·00200	2·723	2·770	2·817	2·864	2·910	2·956	3·002	3·047	3·092	3·137	3·181	3·225	3·269
(0·01961)	7·7192	8·2729	8·8507	9·4531	10·081	10·733	11·412	12·117	12·848	13·606	14·392	15·206	16·047
0·00210	2·790	2·839	2·887	2·935	2·982	3·029	3·076	3·123	3·169	3·215	3·260	3·305	3·350
(0·02059)	7·9099	8·4772	9·0693	9·6866	10·329	10·998	11·694	12·416	13·165	13·942	14·748	15·581	16·444
0·00220	2·855	2·905	2·955	3·004	3·052	3·101	3·149	3·196	3·243	3·290	3·337	3·383	3·429
(0·02157)	8·0960	8·6767	9·2827	9·9145	10·573	11·257	11·969	12·708	13·475	14·271	15·095	15·948	16·831
0·00230	2·920	2·971	3·021	3·071	3·121	3·170	3·219	3·268	3·316	3·364	3·412	3·459	3·506
(0·02256)	8·2779	8·8717	9·4913	10·137	10·810	11·510	12·238	12·994	13·778	14·591	15·434	16·306	17·209
0·00240	2·982	3·035	3·086	3·137	3·188	3·239	3·289	3·338	3·388	3·436	3·485	3·533	3·581
(0·02354)	8·4560	9·0625	9·6955	10·355	11·043	11·758	12·501	13·273	14·074	14·905	15·766	16·657	17·579
0·00250	3·044	3·097	3·150	3·202	3·254	3·305	3·356	3·407	3·457	3·507	3·557	3·606	3·655
(0·02452)	8·6304	9·2494	9·8954	10·569	11·270	12·000	12·759	13·547	14·365	15·212	16·091	17·001	17·942
0·00260	3·104	3·158	3·212	3·265	3·318	3·371	3·423	3·475	3·526	3·577	3·627	3·678	3·727
(0·02550)	8·8013	9·4325	10·091	10·778	11·494	12·238	13·012	13·815	14·649	15·514	16·410	17·337	18·297
0·00270	3·163	3·219	3·273	3·328	3·382	3·435	3·488	3·541	3·593	3·645	3·696	3·748	3·798
(0·02648)	8·9689	9·6122	10·284	10·984	11·712	12·471	13·259	14·078	14·928	15·809	16·722	17·667	18·645
0·00280	3·221	3·278	3·333	3·389	3·444	3·498	3·552	3·606	3·659	3·712	3·764	3·816	3·868
(0·02746)	9·1335	9·7886	10·472	11·185	11·927	12·700	13·503	14·337	15·202	16·099	17·029	17·992	18·988
0·00290	3·278	3·336	3·392	3·449	3·505	3·560	3·615	3·670	3·724	3·777	3·831	3·884	3·937
(0·02844)	9·2952	9·9619	10·658	11·383	12·139	12·925	13·742	14·590	15·471	16·384	17·331	18·310	19·324
S, ($\Delta p/\rho l$)	1900	1950	2000	2050	2100	2150	2200	2250	2300	2350	2400	2450	2500

Gradient **(Equivalent) Pipe diameters in mm**

S = 0·00100 to 0·00290; $\Delta p/\rho l$ = (0·00981) to (0·02844) ms^{-2}

m = Manning $n \times 100$
S = 0.00300 to 0.00950

i.e. kin. pr. grad., $\Delta p/\rho l$ =
(0.02942) to (0.09316) ms^{-2}

Full bore flow; Tables E give m_C or m_P for Colebrook-White or for laminar solutions resp., to divide mV and/or mQ as follows:

mV to give velocities in ms^{-1}
mQ to give discharges in m^{3}s^{-1}

Gradient (Equivalent) Pipe diameters in mm

S, $(\Delta p/\rho l)$	1900	1950	2000	2050	2100	2150	2200	2250	2300	2350	2400	2450	2500
0.00300	3.334	3.393	3.450	3.508	3.565	3.621	3.677	3.732	3.787	3.842	3.896	3.950	4.004
(0.02942)	9.4541	10.132	10.840	11.578	12.346	13.146	13.977	14.840	15.736	16.664	17.627	18.623	19.654
0.00320	3.444	3.504	3.564	3.623	3.681	3.740	3.797	3.855	3.912	3.968	4.024	4.080	4.135
(0.03138)	9.7641	10.464	11.195	11.957	12.751	13.577	14.435	15.327	16.252	17.211	18.205	19.234	20.299
0.00340	3.550	3.612	3.673	3.734	3.795	3.855	3.914	3.973	4.032	4.090	4.148	4.205	4.262
(0.03334)	10.065	10.787	11.540	12.325	13.143	13.995	14.879	15.798	16.752	17.741	18.765	19.826	20.923
0.00360	3.653	3.716	3.780	3.842	3.905	3.966	4.028	4.089	4.149	4.209	4.268	4.327	4.386
(0.03530)	10.356	11.099	11.874	12.683	13.524	14.400	15.311	16.256	17.238	18.255	19.309	20.401	21.530
0.00380	3.753	3.818	3.883	3.948	4.012	4.075	4.138	4.201	4.263	4.324	4.385	4.446	4.506
(0.03727)	10.640	11.403	12.200	13.030	13.895	14.795	15.730	16.702	17.710	18.755	19.838	20.960	22.120
0.00400	3.850	3.918	3.984	4.050	4.116	4.181	4.246	4.310	4.373	4.436	4.499	4.561	4.623
(0.03923)	10.917	11.700	12.517	13.369	14.256	15.179	16.139	17.136	18.170	19.242	20.354	21.504	22.694
0.00420	3.945	4.014	4.083	4.150	4.218	4.284	4.350	4.416	4.481	4.546	4.610	4.674	4.737
(0.04119)	11.186	11.989	12.826	13.699	14.608	15.554	16.537	17.559	18.619	19.718	20.856	22.035	23.255
0.00440	4.038	4.109	4.179	4.248	4.317	4.385	4.453	4.520	4.587	4.653	4.719	4.784	4.849
(0.04315)	11.449	12.271	13.128	14.021	14.952	15.920	16.927	17.972	19.057	20.182	21.347	22.554	23.802
0.00460	4.129	4.201	4.273	4.344	4.414	4.484	4.553	4.622	4.690	4.758	4.825	4.892	4.958
(0.04511)	11.707	12.546	13.423	14.336	15.288	16.278	17.307	18.376	19.485	20.635	21.827	23.061	24.337
0.00480	4.218	4.291	4.364	4.437	4.509	4.580	4.651	4.721	4.791	4.860	4.929	4.997	5.065
(0.04707)	11.959	12.816	13.711	14.645	15.617	16.628	17.679	18.771	19.904	21.079	22.296	23.557	24.861
0.00500	4.305	4.380	4.454	4.528	4.602	4.675	4.747	4.818	4.889	4.960	5.030	5.100	5.169
(0.04903)	12.205	13.081	13.994	14.947	15.939	16.971	18.044	19.158	20.315	21.514	22.756	24.042	25.373
0.00525	4.411	4.488	4.564	4.640	4.715	4.790	4.864	4.937	5.010	5.083	5.154	5.226	5.297
(0.05148)	12.507	13.404	14.340	15.316	16.332	17.390	18.489	19.631	20.816	22.045	23.318	24.636	26.000
0.00550	4.515	4.594	4.672	4.749	4.826	4.903	4.978	5.054	5.128	5.202	5.276	5.349	5.421
(0.05394)	12.801	13.719	14.677	15.676	16.717	17.799	18.925	20.093	21.306	22.564	23.867	25.216	26.612
0.00575	4.616	4.697	4.777	4.856	4.935	5.013	5.090	5.167	5.243	5.319	5.394	5.469	5.543
(0.05639)	13.089	14.027	15.007	16.029	17.092	18.199	19.350	20.545	21.785	23.071	24.403	25.783	27.210
0.00600	4.716	4.798	4.880	4.961	5.041	5.121	5.200	5.278	5.356	5.434	5.510	5.587	5.662
(0.05884)	13.370	14.329	15.330	16.373	17.460	18.591	19.766	20.987	22.254	23.567	24.928	26.337	27.795
0.00625	4.813	4.897	4.980	5.063	5.145	5.226	5.307	5.387	5.467	5.546	5.624	5.702	5.779
(0.06129)	13.646	14.625	15.646	16.711	17.820	18.974	20.174	21.420	22.712	24.053	25.442	26.880	28.368
0.00650	4.908	4.994	5.079	5.163	5.247	5.330	5.412	5.494	5.575	5.655	5.735	5.815	5.894
(0.06374)	13.916	14.914	15.956	17.042	18.173	19.350	20.573	21.844	23.162	24.529	25.946	27.413	28.930
0.00675	5.002	5.089	5.176	5.262	5.347	5.431	5.515	5.598	5.681	5.763	5.845	5.925	6.006
(0.06619)	14.181	15.198	16.260	17.366	18.519	19.718	20.965	22.260	23.603	24.997	26.440	27.935	29.481
0.00700	5.093	5.182	5.271	5.358	5.445	5.531	5.616	5.701	5.785	5.869	5.952	6.034	6.116
(0.06865)	14.441	15.477	16.558	17.685	18.859	20.080	21.350	22.668	24.037	25.455	26.925	28.447	30.022
0.00725	5.184	5.274	5.364	5.453	5.541	5.629	5.716	5.802	5.888	5.973	6.057	6.141	6.224
(0.07110)	14.697	15.751	16.851	17.998	19.193	20.436	21.728	23.070	24.462	25.906	27.402	28.951	30.553
0.00750	5.272	5.364	5.456	5.546	5.636	5.725	5.814	5.901	5.988	6.075	6.161	6.246	6.331
(0.07355)	14.948	16.020	17.139	18.306	19.521	20.785	22.099	23.464	24.880	26.349	27.870	29.446	31.076
0.00800	5.445	5.540	5.635	5.728	5.821	5.913	6.004	6.095	6.185	6.274	6.363	6.451	6.538
(0.07845)	15.438	16.546	17.701	18.906	20.161	21.467	22.824	24.233	25.696	27.213	28.784	30.411	32.095
0.00850	5.613	5.711	5.808	5.904	6.000	6.095	6.189	6.282	6.375	6.467	6.559	6.649	6.740
(0.08336)	15.914	17.055	18.246	19.488	20.781	22.127	23.526	24.979	26.487	28.050	29.670	31.347	33.083
0.00900	5.775	5.876	5.976	6.076	6.174	6.272	6.368	6.465	6.560	6.655	6.749	6.842	6.935
(0.08826)	16.375	17.549	18.775	20.053	21.384	22.769	24.208	25.703	27.255	28.864	30.530	32.256	34.042
0.00950	5.934	6.037	6.140	6.242	6.343	6.443	6.543	6.642	6.740	6.837	6.934	7.030	7.125
(0.09316)	16.824	18.030	19.290	20.603	21.970	23.393	24.872	26.408	28.002	29.655	31.367	33.140	34.974
S, $(\Delta p/\rho l)$	1900	1950	2000	2050	2100	2150	2200	2250	2300	2350	2400	2450	2500

Gradient (Equivalent) Pipe diameters in mm

S = 0.00300 to 0.00950; $\Delta p/\rho l$ = (0.02942) to (0.09316) ms^{-2}

m = Manning $n \times 100$
S = 0·0100 to 0·0290

i.e. kin. pr. grad., $\Delta p/\rho l =$
(0·0981) to (0·2844) ms^{-2}

Full bore flow; Tables E give m_C or m_P for Colebrook-White or for laminar solutions resp., to divide mV and/or mQ as follows:

mV to give velocities in ms^{-1}
mQ to give discharges in m^{3}s^{-1}

Gradient — (Equivalent) Pipe diameters in mm

S, ($\Delta p/\rho l$)	1900	1950	2000	2050	2100	2150	2200	2250	2300	2350	2400	2450	2500
0·0100	6·088	6·194	6·300	6·404	6·508	6·611	6·713	6·814	6·915	7·015	7·114	7·212	7·310
(0·0981)	17·261	18·499	19·791	21·138	22·541	24·000	25·518	27·094	28·729	30·425	32·182	34·001	35·883
0·0105	6·238	6·347	6·455	6·562	6·669	6·774	6·879	6·982	7·086	7·188	7·289	7·390	7·491
(0·1030)	17·687	18·956	20·280	21·660	23·097	24·593	26·148	27·763	29·439	31·176	32·977	34·841	36·769
0·0110	6·385	6·496	6·607	6·717	6·826	6·933	7·041	7·147	7·252	7·357	7·461	7·564	7·667
(0·1079)	18·103	19·402	20·757	22·170	23·641	25·172	26·763	28·416	30·131	31·910	33·753	35·661	37·634
0·0115	6·528	6·643	6·756	6·868	6·979	7·089	7·199	7·307	7·415	7·522	7·629	7·734	7·839
(0·1128)	18·510	19·838	21·223	22·668	24·172	25·738	27·365	29·055	30·809	32·627	34·511	36·462	38·480
0·0120	6·669	6·785	6·901	7·015	7·129	7·242	7·354	7·465	7·575	7·684	7·793	7·901	8·008
(0·1177)	18·908	20·264	21·680	23·155	24·692	26·291	27·953	29·680	31·471	33·329	35·254	37·246	39·308
0·0125	6·806	6·925	7·043	7·160	7·276	7·391	7·505	7·619	7·731	7·843	7·953	8·064	8·173
(0·1226)	19·298	20·682	22·127	23·633	25·201	26·833	28·530	30·292	32·120	34·016	35·981	38·014	40·118
0·0130	6·941	7·062	7·183	7·302	7·420	7·537	7·654	7·769	7·884	7·998	8·111	8·223	8·335
(0·1275)	19·680	21·092	22·565	24·101	25·700	27·365	29·095	30·892	32·756	34·690	36·693	38·767	40·913
0·0135	7·073	7·197	7·319	7·441	7·561	7·681	7·800	7·917	8·034	8·150	8·265	8·380	8·493
(0·1324)	20·055	21·494	22·995	24·560	26·190	27·886	29·649	31·480	33·380	35·351	37·392	39·506	41·692
0·0140	7·203	7·329	7·454	7·578	7·700	7·822	7·943	8·063	8·182	8·300	8·417	8·534	8·649
(0·1373)	20·423	21·888	23·417	25·011	26·671	28·398	30·193	32·058	33·993	35·999	38·078	40·231	42·457
0·0145	7·331	7·459	7·586	7·712	7·837	7·960	8·083	8·205	8·326	8·447	8·566	8·685	8·802
(0·1422)	20·785	22·275	23·831	25·453	27·143	28·900	30·727	32·625	34·595	36·637	38·752	40·943	43·209
0·0150	7·456	7·586	7·715	7·843	7·970	8·097	8·222	8·346	8·469	8·591	8·713	8·833	8·953
(0·1471)	21·140	22·656	24·239	25·888	27·607	29·394	31·253	33·183	35·186	37·263	39·415	41·643	43·948
0·0160	7·701	7·835	7·968	8·101	8·232	8·362	8·491	8·619	8·747	8·873	8·998	9·123	9·247
(0·1569)	21·833	23·399	25·034	26·737	28·512	30·358	32·278	34·271	36·340	38·485	40·707	43·008	45·389
0·0170	7·938	8·076	8·214	8·350	8·485	8·619	8·753	8·885	9·016	9·146	9·275	9·404	9·531
(0·1667)	22·505	24·119	25·804	27·560	29·389	31·293	33·271	35·326	37·458	39·669	41·960	44·332	46·786
0·0180	8·168	8·310	8·452	8·592	8·731	8·869	9·006	9·142	9·277	9·411	9·544	9·676	9·807
(0·1765)	23·158	24·819	26·552	28·359	30·242	32·200	34·236	36·350	38·544	40·819	43·177	45·617	48·142
0·0190	8·391	8·538	8·683	8·828	8·970	9·112	9·253	9·393	9·531	9·669	9·806	9·941	10·08
(0·1863)	23·792	25·499	27·280	29·136	31·070	33·082	35·174	37·346	39·600	41·938	44·360	46·867	49·461
0·0200	8·609	8·760	8·909	9·057	9·204	9·349	9·493	9·637	9·779	9·920	10·06	10·20	10·34
(0·1961)	24·410	26·161	27·988	29·893	31·877	33·942	36·088	38·316	40·629	43·027	45·512	48·085	50·746
0·0210	8·822	8·976	9·129	9·281	9·431	9·580	9·728	9·875	10·02	10·17	10·31	10·45	10·59
(0·2059)	25·013	26·807	28·680	30·632	32·665	34·780	36·979	39·263	41·633	44·090	46·636	49·272	51·999
0·0220	9·030	9·187	9·344	9·499	9·653	9·805	9·957	10·11	10·26	10·40	10·55	10·70	10·84
(0·2157)	25·602	27·438	29·354	31·352	33·433	35·598	37·849	40·187	42·612	45·128	47·734	50·432	53·223
0·0230	9·233	9·394	9·554	9·712	9·870	10·03	10·18	10·33	10·49	10·64	10·79	10·94	11·09
(0·2256)	26·177	28·055	30·014	32·057	34·185	36·398	38·700	41·090	43·570	46·142	48·806	51·565	54·419
0·0240	9·431	9·596	9·759	9·921	10·08	10·24	10·40	10·56	10·71	10·87	11·02	11·17	11·32
(0·2354)	26·740	28·658	30·660	32·747	34·920	37·181	39·532	41·973	44·507	47·134	49·856	52·674	55·590
0·0250	9·626	9·794	9·961	10·13	10·29	10·45	10·61	10·77	10·93	11·09	11·25	11·40	11·56
(0·2452)	27·292	29·249	31·292	33·422	35·640	37·948	40·347	42·839	45·425	48·106	50·884	53·760	56·736
0·0260	9·816	9·988	10·16	10·33	10·49	10·66	10·82	10·99	11·15	11·31	11·47	11·63	11·79
(0·2550)	27·832	29·828	31·912	34·084	36·346	38·699	41·146	43·687	46·324	49·059	51·892	54·825	57·860
0·0270	10·00	10·18	10·35	10·52	10·69	10·86	11·03	11·20	11·36	11·53	11·69	11·85	12·01
(0·2648)	28·362	30·396	32·520	34·733	37·038	39·437	41·930	44·520	47·207	49·993	52·880	55·869	58·962
0·0280	10·19	10·36	10·54	10·72	10·89	11·06	11·23	11·40	11·57	11·74	11·90	12·07	12·23
(0·2746)	28·883	30·954	33·116	35·370	37·718	40·160	42·699	45·337	48·073	50·911	53·851	56·895	60·044
0·0290	10·37	10·55	10·73	10·91	11·08	11·26	11·43	11·60	11·78	11·95	12·11	12·28	12·45
(0·2844)	29·394	31·502	33·702	35·996	38·385	40·871	43·455	46·139	48·924	51·812	54·804	57·902	61·107
S, ($\Delta p/\rho l$)	1900	1950	2000	2050	2100	2150	2200	2250	2300	2350	2400	2450	2500

Gradient — (Equivalent) Pipe diameters in mm

S = 0·0100 to 0·0290; $\Delta p/\rho l$ = (0·0981) to (0·2844) ms^{-2}

m = Manning $n \times 100$
S = 0·0300 to 0·1000

Full bore flow; Tables E give m_C or m_P for Colebrook-White or for laminar solutions resp., to divide mV and/or mQ as follows:

i.e. kin. pr. grad., $\Delta p/\rho l$ =
(0·2942) to (0·9807) ms^{-2}

mV to give velocities in ms^{-1}
mQ to give discharges in m^{3}s^{-1}

Gradient   **(Equivalent) Pipe diameters in mm**

$S, (\Delta p/\rho l)$	1900	1950	2000	2050	2100	2150	2200	2250	2300	2350	2400	2450	2500
0·0300	10·54	10·73	10·91	11·09	11·27	11·45	11·63	11·80	11·98	12·15	12·32	12·49	12·66
(0·2942)	29·896	32·041	34·279	36·612	39·042	41·570	44·198	46·928	49·760	52·698	55·741	58·892	62·151
0·0320	10·89	11·08	11·27	11·46	11·64	11·83	12·01	12·19	12·37	12·55	12·73	12·90	13·08
(0·3138)	30·877	33·092	35·403	37·812	40·322	42·933	45·648	48·467	51·392	54·426	57·569	60·823	64·190
0·0340	11·23	11·42	11·62	11·81	12·00	12·19	12·38	12·56	12·75	12·93	13·12	13·30	13·48
(0·3334)	31·827	34·110	36·492	38·976	41·563	44·255	47·052	49·958	52·974	56·101	59·341	62·695	66·165
0·0360	11·55	11·75	11·95	12·15	12·35	12·54	12·74	12·93	13·12	13·31	13·50	13·68	13·87
(0·3530)	32·750	35·099	37·550	40·106	42·768	45·538	48·417	51·407	54·510	57·727	61·061	64·512	68·083
0·0380	11·87	12·07	12·28	12·48	12·69	12·89	13·09	13·28	13·48	13·67	13·87	14·06	14·25
(0·3727)	33·647	36·061	38·579	41·205	43·940	46·785	49·743	52·815	56·004	59·309	62·734	66·280	69·949
0·0400	12·18	12·39	12·60	12·81	13·02	13·22	13·43	13·63	13·83	14·03	14·23	14·42	14·62
(0·3923)	34·521	36·997	39·582	42·276	45·081	48·001	51·036	54·188	57·458	60·850	64·364	68·002	71·766
0·0420	12·48	12·69	12·91	13·12	13·34	13·55	13·76	13·96	14·17	14·38	14·58	14·78	14·98
(0·4119)	35·374	37·911	40·559	43·320	46·195	49·186	52·296	55·526	58·877	62·353	65·953	69·681	73·538
0·0440	12·77	12·99	13·21	13·43	13·65	13·87	14·08	14·29	14·50	14·71	14·92	15·13	15·33
(0·4315)	36·206	38·803	41·513	44·339	47·282	50·344	53·527	56·832	60·263	63·820	67·505	71·321	75·269
0·0460	13·06	13·29	13·51	13·74	13·96	14·18	14·40	14·61	14·83	15·04	15·26	15·47	15·68
(0·4511)	37·020	39·675	42·446	45·336	48·344	51·475	54·730	58·110	61·617	65·254	69·023	72·924	76·961
0·0480	13·34	13·57	13·80	14·03	14·26	14·48	14·71	14·93	15·15	15·37	15·59	15·80	16·02
(0·4707)	37·816	40·529	43·359	46·311	49·384	52·582	55·907	59·359	62·943	66·658	70·507	74·493	78·616
0·0500	13·61	13·85	14·09	14·32	14·55	14·78	15·01	15·24	15·46	15·69	15·91	16·13	16·35
(0·4903)	38·596	41·364	44·253	47·266	50·403	53·667	57·060	60·583	64·240	68·032	71·961	76·029	80·237
0·0525	13·95	14·19	14·43	14·67	14·91	15·15	15·38	15·61	15·84	16·07	16·30	16·53	16·75
(0·5148)	39·549	42·386	45·346	48·433	51·647	54·992	58·469	62·080	65·827	69·712	73·738	77·906	82·218
0·0550	14·28	14·53	14·77	15·02	15·26	15·50	15·74	15·98	16·22	16·45	16·68	16·91	17·14
(0·5394)	40·480	43·383	46·413	49·573	52·863	56·286	59·845	63·541	67·376	71·353	75·473	79·740	84·153
0·0575	14·60	14·85	15·11	15·36	15·61	15·85	16·10	16·34	16·58	16·82	17·06	17·29	17·53
(0·5639)	41·390	44·358	47·457	50·687	54·051	57·551	61·190	64·969	68·890	72·957	77·170	81·532	86·045
0·0600	14·91	15·17	15·43	15·69	15·94	16·19	16·44	16·69	16·94	17·18	17·43	17·67	17·91
(0·5884)	42·280	45·312	48·477	51·777	55·213	58·789	62·506	66·366	70·372	74·526	78·829	83·285	87·895
0·0625	15·22	15·49	15·75	16·01	16·27	16·53	16·78	17·04	17·29	17·54	17·78	18·03	18·28
(0·6129)	43·152	46·247	49·477	52·844	56·352	60·001	63·794	67·734	71·823	76·062	80·455	85·003	89·708
0·0650	15·52	15·79	16·06	16·33	16·59	16·85	17·11	17·37	17·63	17·88	18·14	18·39	18·64
(0·6374)	44·006	47·163	50·457	53·891	57·468	61·189	65·058	69·076	73·245	77·569	82·048	86·686	91·484
0·0675	15·82	16·09	16·37	16·64	16·91	17·18	17·44	17·70	17·97	18·22	18·48	18·74	18·99
(0·6619)	44·845	48·061	51·418	54·918	58·562	62·355	66·297	70·392	74·641	79·046	83·611	88·337	93·227
0·0700	16·11	16·39	16·67	16·94	17·22	17·49	17·76	18·03	18·29	18·56	18·82	19·08	19·34
(0·6865)	45·668	48·943	52·361	55·925	59·637	63·499	67·514	71·683	76·010	80·497	85·145	89·958	94·938
0·0725	16·39	16·68	16·96	17·24	17·52	17·80	18·07	18·35	18·62	18·89	19·15	19·42	19·68
(0·7110)	46·476	49·809	53·288	56·915	60·693	64·623	68·709	72·952	77·356	81·922	86·653	91·551	96·618
0·0750	16·67	16·96	17·25	17·54	17·82	18·10	18·38	18·66	18·94	19·21	19·48	19·75	20·02
(0·7355)	47·270	50·661	54·199	57·888	61·730	65·728	69·883	74·199	78·678	83·322	88·134	93·116	98·270
0·0800	17·22	17·52	17·82	18·11	18·41	18·70	18·99	19·27	19·56	19·84	20·12	20·40	20·68
(0·7845)	48·821	52·322	55·977	59·787	63·755	67·883	72·175	76·633	81·258	86·055	91·024	96·169	101·49
0·0850	17·75	18·06	18·37	18·67	18·97	19·27	19·57	19·87	20·16	20·45	20·74	21·03	21·31
(0·8336)	50·323	53·933	57·700	61·627	65·717	69·973	74·397	78·991	83·759	88·703	93·826	99·129	104·62
0·0900	18·26	18·58	18·90	19·21	19·52	19·83	20·14	20·44	20·74	21·04	21·34	21·64	21·93
(0·8826)	51·782	55·496	59·372	63·413	67·622	72·001	76·553	81·281	86·188	91·275	96·546	102·00	107·65
0·0950	18·76	19·09	19·42	19·74	20·06	20·38	20·69	21·00	21·31	21·62	21·93	22·23	22·53
(0·9316)	53·201	57·017	60·999	65·151	69·475	73·974	78·651	83·509	88·549	93·776	99·191	104·80	110·60
0·1000	19·25	19·59	19·92	20·25	20·58	20·91	21·23	21·55	21·87	22·18	22·50	22·81	23·12
(0·9807)	54·583	58·498	62·584	66·844	71·280	75·896	80·694	85·678	90·850	96·212	101·77	107·52	113·47
$S, (\Delta p/\rho l)$	1900	1950	2000	2050	2100	2150	2200	2250	2300	2350	2400	2450	2500

Gradient   **(Equivalent) Pipe diameters in mm**

S = 0·0300 to 0·1000;

$\Delta p/\rho l$ = (0·2942) to (0·9807) ms^{-2}

E1

Kin. visc., $\nu = 0.125 \times 10^{-6}$ m^{2}s^{-1};   Roughness size, $k_s = 0.0015$ mm

$S = 0.00010$ to 30.0000

This table shows values of m, as follows

i.e. kin. pr. grad., $\Delta p/\rho l = (0.00098)$ to (294) ms^{-2}

m_C for Colebrook-White solutions; or, where $R \le 2000$, m_P for laminar flow

Grad'nt S	k.p.g. $(\Delta p/\rho l)$	6	8	10	12.5	15	20	25	30	40	50	60	80	100	125
0.00030	(0.00294)	0.857	*	0.872	0.854	0.841	0.824	0.813	0.807	0.798	0.793	0.791	0.788	0.788	0.788
0.00040	(0.00392)	0.742	*	0.850	0.833	0.821	0.806	0.797	0.791	0.784	0.780	0.777	0.776	0.776	0.777
0.00060	(0.00588)	*	0.839	0.821	0.806	0.796	0.783	0.775	0.770	0.764	0.761	0.759	0.759	0.759	0.761
0.00080	(0.00785)	*	0.818	0.802	0.788	0.779	0.767	0.760	0.756	0.751	0.748	0.747	0.747	0.748	0.750
0.00100	(0.00981)	0.826	0.802	0.787	0.775	0.766	0.755	0.749	0.745	0.741	0.739	0.738	0.738	0.740	0.742
0.00150	(0.0147)	0.797	0.775	0.762	0.751	0.744	0.735	0.730	0.727	0.723	0.722	0.722	0.723	0.725	0.728
0.00200	(0.0196)	0.777	0.758	0.746	0.736	0.729	0.721	0.717	0.714	0.712	0.711	0.711	0.713	0.715	0.719
0.00300	(0.0294)	0.750	0.734	0.723	0.715	0.709	0.703	0.699	0.697	0.696	0.696	0.697	0.699	0.702	0.706
0.00400	(0.0392)	0.733	0.718	0.708	0.701	0.696	0.690	0.688	0.686	0.685	0.686	0.687	0.690	0.693	0.697
0.00600	(0.0588)	0.710	0.697	0.689	0.682	0.678	0.674	0.672	0.671	0.671	0.672	0.674	0.677	0.681	0.686
0.00800	(0.0785)	0.694	0.682	0.675	0.670	0.666	0.663	0.661	0.661	0.661	0.663	0.665	0.669	0.673	0.678
0.01000	(0.0981)	0.683	0.672	0.665	0.660	0.657	0.654	0.653	0.653	0.654	0.656	0.658	0.662	0.667	0.672
0.01500	(0.147)	0.663	0.654	0.648	0.644	0.642	0.640	0.639	0.640	0.642	0.644	0.646	0.652	0.656	0.662
0.02000	(0.196)	0.649	0.641	0.637	0.633	0.632	0.630	0.630	0.631	0.633	0.636	0.639	0.644	0.649	0.655
0.03000	(0.294)	0.632	0.625	0.621	0.619	0.618	0.617	0.618	0.619	0.622	0.625	0.628	0.634	0.640	0.646
0.04000	(0.392)	0.620	0.614	0.611	0.609	0.608	0.608	0.609	0.611	0.614	0.618	0.621	0.628	0.634	0.640
0.06000	(0.588)	0.604	0.600	0.597	0.596	0.596	0.597	0.598	0.600	0.604	0.609	0.612	0.619	0.626	0.632
0.08000	(0.785)	0.594	0.590	0.588	0.587	0.588	0.589	0.591	0.593	0.598	0.602	0.606	0.614	0.620	0.627
0.10000	(0.981)	0.586	0.583	0.581	0.581	0.581	0.583	0.586	0.588	0.593	0.598	0.602	0.610	0.616	0.624
0.15000	(1.47)	0.572	0.570	0.570	0.570	0.571	0.573	0.576	0.579	0.585	0.590	0.594	0.602	0.610	0.617
0.20000	(1.96)	0.564	0.562	0.562	0.563	0.564	0.567	0.570	0.573	0.579	0.585	0.589	0.598	0.605	0.613
0.30000	(2.94)	0.552	0.551	0.552	0.553	0.555	0.558	0.562	0.566	0.572	0.578	0.583	0.592	0.600	0.608
0.40000	(3.92)	0.544	0.544	0.545	0.547	0.549	0.553	0.557	0.561	0.567	0.573	0.579	0.588	0.596	0.605
0.60000	(5.88)	0.534	0.535	0.536	0.538	0.541	0.545	0.550	0.554	0.561	0.568	0.574	0.583	0.592	0.601
0.80000	(7.85)	0.527	0.528	0.530	0.533	0.536	0.541	0.546	0.550	0.558	0.564	0.570	0.580	0.589	0.598
1.00000	(9.81)	0.522	0.524	0.526	0.529	0.532	0.537	0.542	0.547	0.555	0.562	0.568	0.578	0.587	0.596
1.50000	(14.7)	0.514	0.516	0.519	0.522	0.526	0.532	0.537	0.542	0.550	0.558	0.564	0.575	0.584	0.593
2.00000	(19.6)	0.508	0.511	0.514	0.518	0.522	0.528	0.534	0.539	0.547	0.555	0.561	0.572	0.581	0.591
3.00000	(29.4)	0.501	0.505	0.509	0.513	0.517	0.523	0.529	0.535	0.544	0.552	0.558	0.570	0.579	0.589
4.00000	(39.2)	0.497	0.501	0.505	0.509	0.513	0.520	0.527	0.532	0.542	0.550	0.556	0.568	0.577	0.587
6.00000	(58.8)	0.491	0.496	0.500	0.505	0.509	0.517	0.523	0.529	0.539	0.547	0.554	0.566	0.575	0.586
8.00000	(78.5)	0.488	0.493	0.497	0.502	0.507	0.515	0.521	0.527	0.537	0.545	0.553	0.564	0.574	0.585
10.0000	(98.1)	0.485	0.490	0.495	0.500	0.505	0.513	0.520	0.526	0.536	0.544	0.552	0.564	0.573	0.584
15.0000	(147)	0.481	0.487	0.492	0.497	0.502	0.510	0.517	0.524	0.534	0.543	0.550	0.562	0.572	0.583
20.0000	(196)	0.478	0.484	0.489	0.495	0.500	0.509	0.516	0.522	0.533	0.541	0.549	0.561	0.571	0.582
30.0000	(294)	0.475	0.481	0.487	0.493	0.498	0.507	0.514	0.521	0.531	0.540	0.548	0.560	0.570	0.581
S	$(\Delta p/\rho l)$	6	8	10	12.5	15	20	25	30	40	50	60	80	100	125

Grad'nt   k.p.g.   (Equivalent) Pipe diameters in mm

Grad'nt S	k.p.g. $(\Delta p/\rho l)$	125	150	200	250	300	400	500	600	800	1000	1250	1500	2000	2500
0.00010	(0.00098)	0.837	0.836	0.838	0.840	0.842	0.848	0.853	0.858	0.866	0.874	0.883	0.890	0.903	0.914
0.00015	(0.00147)	0.818	0.818	0.820	0.823	0.826	0.832	0.838	0.843	0.852	0.860	0.869	0.877	0.890	0.902
0.00020	(0.00196)	0.805	0.806	0.809	0.812	0.815	0.821	0.827	0.833	0.842	0.851	0.860	0.868	0.882	0.893
0.00030	(0.00294)	0.788	0.790	0.793	0.797	0.800	0.807	0.813	0.819	0.829	0.838	0.847	0.856	0.870	0.881
0.00040	(0.00392)	0.777	0.778	0.782	0.786	0.790	0.797	0.804	0.810	0.820	0.829	0.839	0.847	0.862	0.874
0.00060	(0.00588)	0.761	0.763	0.768	0.772	0.776	0.784	0.791	0.797	0.808	0.817	0.827	0.836	0.851	0.863
0.00080	(0.00785)	0.750	0.753	0.758	0.762	0.767	0.775	0.782	0.789	0.800	0.809	0.819	0.828	0.843	0.856
0.00100	(0.00981)	0.742	0.745	0.750	0.755	0.760	0.768	0.776	0.782	0.794	0.803	0.814	0.823	0.838	0.850
0.00150	(0.0147)	0.728	0.731	0.737	0.743	0.748	0.756	0.764	0.771	0.783	0.793	0.804	0.813	0.828	0.841
0.00200	(0.0196)	0.719	0.722	0.728	0.734	0.739	0.748	0.756	0.764	0.776	0.786	0.797	0.806	0.822	0.835
0.00300	(0.0294)	0.706	0.710	0.716	0.722	0.728	0.738	0.746	0.753	0.766	0.776	0.788	0.797	0.814	0.827
0.00400	(0.0392)	0.697	0.701	0.708	0.715	0.720	0.731	0.739	0.747	0.759	0.770	0.782	0.792	0.808	0.822
0.00600	(0.0588)	0.686	0.690	0.698	0.704	0.710	0.721	0.730	0.738	0.751	0.762	0.774	0.784	0.801	0.815
0.00800	(0.0785)	0.678	0.682	0.690	0.697	0.704	0.714	0.724	0.732	0.745	0.757	0.769	0.779	0.796	0.810
0.01000	(0.0981)	0.672	0.677	0.685	0.692	0.699	0.710	0.719	0.727	0.741	0.753	0.765	0.775	0.793	0.807
0.01500	(0.147)	0.662	0.667	0.676	0.683	0.690	0.702	0.711	0.720	0.734	0.746	0.758	0.769	0.787	0.802
0.02000	(0.196)	0.655	0.660	0.669	0.677	0.684	0.696	0.706	0.715	0.729	0.741	0.754	0.765	0.783	0.798
0.03000	(0.294)	0.646	0.652	0.661	0.670	0.677	0.689	0.699	0.708	0.723	0.736	0.749	0.760	0.779	0.794
0.04000	(0.392)	0.640	0.646	0.656	0.665	0.672	0.685	0.695	0.704	0.720	0.732	0.745	0.757	0.776	0.791
0.06000	(0.588)	0.632	0.638	0.649	0.658	0.666	0.679	0.690	0.699	0.715	0.727	0.741	0.753	0.772	0.787
0.08000	(0.785)	0.627	0.634	0.644	0.654	0.661	0.675	0.686	0.696	0.711	0.725	0.738	0.750	0.769	0.785
0.10000	(0.981)	0.624	0.630	0.641	0.650	0.659	0.672	0.683	0.693	0.709	0.722	0.736	0.748	0.768	0.784
0.15000	(1.47)	0.617	0.624	0.636	0.645	0.654	0.668	0.679	0.689	0.706	0.719	0.733	0.745	0.765	0.781
0.20000	(1.96)	0.613	0.620	0.632	0.642	0.650	0.665	0.676	0.687	0.703	0.717	0.731	0.743	0.763	0.780
0.30000	(2.94)	0.608	0.615	0.628	0.638	0.646	0.661	0.673	0.683	0.700	0.714	0.729	0.741	0.761	0.778
0.40000	(3.92)	0.605	0.612	0.625	0.635	0.644	0.659	0.671	0.681	0.699	0.713	0.727	0.740	0.760	0.777
0.60000	(5.88)	0.601	0.608	0.621	0.632	0.641	0.656	0.668	0.679	0.696	0.711	0.725	0.738	0.759	0.775
0.80000	(7.85)	0.598	0.606	0.619	0.630	0.639	0.654	0.667	0.677	0.695	0.709	0.724	0.737	0.758	0.774
1.00000	(9.81)	0.596	0.604	0.617	0.628	0.638	0.653	0.666	0.676	0.694	0.709	0.724	0.736	0.757	0.774
1.50000	(14.7)	0.593	0.601	0.615	0.626	0.635	0.651	0.664	0.675	0.693	0.707	0.722	0.735	0.756	0.773
2.00000	(19.6)	0.591	0.599	0.613	0.624	0.634	0.650	0.663	0.674	0.692	0.706	0.722	0.734	0.755	0.772
3.00000	(29.4)	0.589	0.597	0.611	0.623	0.632	0.648	0.662	0.673	0.691	0.705	0.721	0.734	0.755	0.772
S	$(\Delta p/\rho l)$	125	150	200	250	300	400	500	600	800	1000	1250	1500	2000	2500

Grad'nt   k.p.g.   (Equivalent) Pipe diameters in mm

Kinematic viscosity, $\nu = 0.125 \times 10^{-6}$ m^{2}s^{-1};   Roughness size, $k_s = 0.0015$ mm

Kin. visc., $\nu = 0.125 \times 10^{-6}$ m^{2}s^{-1}; **Roughness size, $k_s = 0.003$ mm** E2
$S = 0.00010$ to 30.0000 This table shows values of m, as follows
i.e. kin. pr. grad., $\Delta p/\rho l =$ m_C for Colebrook-White solutions; or,
(0.00098) to (294) ms^{-2} where $\mathbf{R} \leq 2000$, m_P for laminar flow

Grad'nt S	k.p.g. $(\Delta p/\rho l)$	6	8	10	12·5	15	20	25	30	40	50	60	80	100	125
		(Equivalent) Pipe diameters in mm													
0·00030	(0·00294)	0·857	*	0·874	0·855	0·842	0·825	0·815	0·808	0·800	0·796	0·793	0·791	0·790	0·791
0·00040	(0·00392)	0·742	*	0·852	0·835	0·823	0·808	0·799	0·793	0·786	0·782	0·780	0·778	0·779	0·780
0·00060	(0·00588)	*	0·841	0·823	0·808	0·798	0·785	0·777	0·772	0·767	0·764	0·762	0·762	0·763	0·765
0·00080	(0·00785)	*	0·820	0·804	0·790	0·781	0·770	0·763	0·758	0·754	0·751	0·750	0·751	0·752	0·754
0·00100	(0·00981)	0·829	0·805	0·789	0·777	0·769	0·758	0·752	0·748	0·744	0·742	0·742	0·742	0·744	0·747
0·00150	(0·0147)	0·799	0·778	0·765	0·754	0·747	0·738	0·733	0·730	0·727	0·726	0·726	0·728	0·730	0·733
0·00200	(0·0196)	0·779	0·760	0·748	0·739	0·732	0·725	0·720	0·718	0·716	0·715	0·716	0·718	0·721	0·724
0·00300	(0·0294)	0·753	0·737	0·727	0·718	0·713	0·707	0·703	0·702	0·701	0·701	0·702	0·705	0·708	0·712
0·00400	(0·0392)	0·736	0·721	0·712	0·705	0·700	0·695	0·692	0·691	0·690	0·691	0·693	0·696	0·700	0·704
0·00600	(0·0588)	0·713	0·701	0·693	0·687	0·683	0·679	0·677	0·676	0·677	0·678	0·680	0·685	0·689	0·694
0·00800	(0·0785)	0·698	0·687	0·680	0·675	0·671	0·668	0·667	0·667	0·668	0·670	0·672	0·677	0·681	0·687
0·01000	(0·0981)	0·687	0·676	0·670	0·666	0·663	0·660	0·659	0·659	0·661	0·663	0·666	0·671	0·676	0·682
0·01500	(0·147)	0·668	0·659	0·654	0·650	0·648	0·646	0·646	0·647	0·650	0·652	0·655	0·661	0·667	0·673
0·02000	(0·196)	0·655	0·647	0·643	0·640	0·638	0·637	0·638	0·639	0·642	0·645	0·649	0·655	0·661	0·667
0·03000	(0·294)	0·638	0·632	0·628	0·626	0·625	0·625	0·627	0·628	0·632	0·636	0·639	0·646	0·653	0·660
0·04000	(0·392)	0·627	0·621	0·619	0·617	0·617	0·617	0·619	0·621	0·625	0·629	0·634	0·641	0·647	0·655
0·06000	(0·588)	0·612	0·608	0·606	0·605	0·605	0·607	0·609	0·612	0·617	0·621	0·626	0·634	0·641	0·648
0·08000	(0·785)	0·602	0·599	0·598	0·597	0·598	0·600	0·603	0·606	0·611	0·616	0·621	0·629	0·637	0·644
0·10000	(0·981)	0·595	0·592	0·592	0·592	0·593	0·595	0·598	0·601	0·607	0·612	0·617	0·626	0·633	0·642
0·15000	(1·47)	0·583	0·581	0·581	0·582	0·583	0·587	0·590	0·594	0·600	0·606	0·611	0·620	0·628	0·637
0·20000	(1·96)	0·575	0·574	0·574	0·576	0·577	0·581	0·585	0·589	0·596	0·602	0·607	0·617	0·625	0·634
0·30000	(2·94)	0·564	0·564	0·566	0·567	0·570	0·574	0·579	0·583	0·590	0·597	0·603	0·613	0·621	0·630
0·40000	(3·92)	0·558	0·558	0·560	0·562	0·565	0·570	0·575	0·579	0·587	0·594	0·600	0·610	0·619	0·628
0·60000	(5·88)	0·549	0·550	0·553	0·555	0·558	0·564	0·569	0·574	0·582	0·589	0·596	0·606	0·615	0·625
0·80000	(7·85)	0·543	0·545	0·548	0·551	0·554	0·560	0·566	0·571	0·580	0·587	0·593	0·604	0·614	0·623
1·00000	(9·81)	0·539	0·542	0·545	0·548	0·552	0·558	0·564	0·569	0·578	0·585	0·592	0·603	0·612	0·622
1·50000	(14·7)	0·533	0·536	0·539	0·543	0·547	0·554	0·560	0·565	0·574	0·582	0·589	0·601	0·610	0·620
2·00000	(19·6)	0·528	0·532	0·536	0·540	0·544	0·551	0·557	0·563	0·572	0·580	0·587	0·599	0·609	0·619
3·00000	(29·4)	0·523	0·527	0·531	0·536	0·540	0·548	0·554	0·560	0·570	0·578	0·585	0·597	0·607	0·617
4·00000	(39·2)	0·520	0·524	0·529	0·534	0·538	0·546	0·553	0·559	0·569	0·577	0·584	0·596	0·606	0·617
6·00000	(58·8)	0·515	0·521	0·525	0·531	0·535	0·543	0·550	0·557	0·567	0·575	0·583	0·595	0·605	0·615
8·00000	(78·5)	0·513	0·518	0·523	0·529	0·534	0·542	0·549	0·555	0·566	0·574	0·582	0·594	0·604	0·615
10·0000	(98·1)	0·511	0·517	0·522	0·527	0·532	0·541	0·548	0·554	0·565	0·574	0·581	0·594	0·604	0·614
15·0000	(147)	0·508	0·514	0·519	0·525	0·530	0·539	0·547	0·553	0·564	0·573	0·580	0·593	0·603	0·614
20·0000	(196)	0·506	0·513	0·518	0·524	0·529	0·538	0·546	0·552	0·563	0·572	0·580	0·592	0·603	0·613
30·0000	(294)	0·504	0·511	0·516	0·522	0·528	0·537	0·545	0·551	0·562	0·571	0·579	0·592	0·602	0·613
S	$(\Delta p/\rho l)$	6	8	10	12·5	15	20	25	30	40	50	60	80	100	125

Grad'nt S	k.p.g. $(\Delta p/\rho l)$	125	150	200	250	300	400	500	600	800	1000	1250	1500	2000	2500
		(Equivalent) Pipe diameters in mm													
0·00010	(0·00098)	0·839	0·838	0·840	0·842	0·845	0·850	0·856	0·861	0·870	0·878	0·887	0·895	0·908	0·920
0·00015	(0·00147)	0·820	0·821	0·823	0·826	0·829	0·835	0·841	0·847	0·856	0·865	0·874	0·882	0·896	0·908
0·00020	(0·00196)	0·808	0·809	0·812	0·815	0·818	0·825	0·831	0·837	0·847	0·856	0·865	0·874	0·888	0·900
0·00030	(0·00294)	0·791	0·793	0·796	0·800	0·804	0·811	0·818	0·824	0·835	0·844	0·854	0·862	0·877	0·890
0·00040	(0·00392)	0·780	0·782	0·786	0·790	0·794	0·802	0·809	0·815	0·826	0·836	0·846	0·855	0·870	0·883
0·00060	(0·00588)	0·765	0·767	0·772	0·777	0·781	0·790	0·797	0·804	0·815	0·825	0·835	0·845	0·860	0·873
0·00080	(0·00785)	0·754	0·757	0·762	0·768	0·772	0·781	0·789	0·796	0·808	0·818	0·829	0·838	0·854	0·867
0·00100	(0·00981)	0·747	0·750	0·755	0·761	0·766	0·775	0·783	0·790	0·802	0·812	0·823	0·833	0·849	0·863
0·00150	(0·0147)	0·733	0·737	0·743	0·749	0·754	0·764	0·772	0·780	0·792	0·803	0·815	0·825	0·841	0·855
0·00200	(0·0196)	0·724	0·728	0·735	0·741	0·747	0·757	0·765	0·773	0·786	0·797	0·809	0·819	0·836	0·850
0·00300	(0·0294)	0·712	0·716	0·724	0·731	0·737	0·747	0·756	0·764	0·778	0·789	0·801	0·812	0·829	0·844
0·00400	(0·0392)	0·704	0·709	0·717	0·724	0·730	0·741	0·750	0·758	0·772	0·784	0·796	0·807	0·825	0·840
0·00600	(0·0588)	0·694	0·699	0·707	0·715	0·721	0·733	0·742	0·751	0·765	0·777	0·790	0·801	0·819	0·834
0·00800	(0·0785)	0·687	0·692	0·701	0·709	0·715	0·727	0·737	0·746	0·761	0·773	0·786	0·797	0·816	0·831
0·01000	(0·0981)	0·682	0·687	0·696	0·704	0·711	0·723	0·734	0·742	0·758	0·770	0·783	0·794	0·813	0·829
0·01500	(0·147)	0·673	0·678	0·688	0·697	0·704	0·717	0·727	0·737	0·752	0·765	0·778	0·790	0·809	0·825
0·02000	(0·196)	0·667	0·673	0·683	0·692	0·699	0·713	0·723	0·733	0·749	0·762	0·775	0·787	0·807	0·822
0·03000	(0·294)	0·660	0·666	0·677	0·686	0·694	0·707	0·718	0·728	0·744	0·758	0·772	0·784	0·803	0·819
0·04000	(0·392)	0·655	0·661	0·672	0·682	0·690	0·704	0·715	0·725	0·741	0·755	0·769	0·781	0·801	0·818
0·06000	(0·588)	0·648	0·655	0·667	0·677	0·685	0·699	0·711	0·721	0·738	0·752	0·766	0·779	0·799	0·815
0·08000	(0·785)	0·644	0·651	0·663	0·673	0·682	0·697	0·709	0·719	0·736	0·750	0·764	0·777	0·797	0·814
0·10000	(0·981)	0·642	0·649	0·661	0·671	0·680	0·695	0·707	0·717	0·734	0·748	0·763	0·776	0·796	0·813
0·15000	(1·47)	0·637	0·644	0·657	0·667	0·676	0·692	0·704	0·714	0·732	0·746	0·761	0·774	0·795	0·812
0·20000	(1·96)	0·634	0·642	0·654	0·665	0·674	0·690	0·702	0·713	0·731	0·745	0·760	0·773	0·794	0·811
0·30000	(2·94)	0·630	0·638	0·651	0·662	0·672	0·687	0·700	0·711	0·729	0·743	0·758	0·771	0·792	0·809
0·40000	(3·92)	0·628	0·636	0·649	0·661	0·670	0·686	0·699	0·710	0·728	0·742	0·758	0·770	0·792	0·809
0·60000	(5·88)	0·625	0·633	0·647	0·658	0·668	0·684	0·697	0·708	0·726	0·741	0·756	0·769	0·791	0·808
0·80000	(7·85)	0·623	0·632	0·646	0·657	0·667	0·683	0·696	0·707	0·725	0·740	0·756	0·769	0·790	0·807
1·00000	(9·81)	0·622	0·631	0·645	0·656	0·666	0·682	0·695	0·707	0·725	0·740	0·755	0·768	0·790	0·807
1·50000	(14·7)	0·620	0·629	0·643	0·655	0·665	0·681	0·694	0·706	0·724	0·739	0·755	0·768	0·789	0·807
2·00000	(19·6)	0·619	0·628	0·642	0·654	0·664	0·680	0·694	0·705	0·724	0·739	0·754	0·767	0·789	0·806
3·00000	(29·4)	0·617	0·626	0·641	0·653	0·663	0·679	0·693	0·704	0·723	0·738	0·754	0·767	0·789	0·806
S	$(\Delta p/\rho l)$	125	150	200	250	300	400	500	600	800	1000	1250	1500	2000	2500

Grad'nt k.p.g. (Equivalent) Pipe diameters in mm

Kinematic viscosity, $\nu = 0.125 \times 10^{-6}$ m^{2}s^{-1}; **Roughness size, $k_s = 0.003$ mm**

E3

Kin. visc., $\nu = 0.125 \times 10^{-6}\ \mathrm{m^2 s^{-1}}$;
$S = 0.00010$ to 30.0000

i.e. kin. pr. grad., $\Delta p/\rho l =$
(0.00098) to $(294)\ \mathrm{ms^{-2}}$

Roughness size, $k_s = 0.006$ mm
This table shows values of m, as follows

m_C for Colebrook-White solutions; or,
where $\mathbf{R} \le 2000$, m_P for laminar flow

Grad'nt S	k.p.g. $(\Delta p/\rho l)$	6	8	10	12.5	15	20	25	30	40	50	60	80	100	125
		(Equivalent) Pipe diameters in mm													
0.00030	(0.00294)	0.857	*	0.877	0.858	0.845	0.829	0.819	0.812	0.804	0.800	0.797	0.796	0.796	0.797
0.00040	(0.00392)	0.742	*	0.855	0.838	0.827	0.812	0.803	0.797	0.790	0.787	0.785	0.784	0.784	0.786
0.00060	(0.00588)	*	0.845	0.827	0.812	0.802	0.789	0.782	0.777	0.772	0.769	0.768	0.768	0.769	0.772
0.00080	(0.00785)	*	0.824	0.808	0.795	0.786	0.774	0.768	0.764	0.759	0.757	0.757	0.757	0.759	0.762
0.00100	(0.00981)	*	0.809	0.794	0.782	0.773	0.763	0.757	0.754	0.750	0.749	0.748	0.750	0.752	0.755
0.00150	(0.0147)	0.804	0.783	0.770	0.759	0.752	0.744	0.739	0.736	0.734	0.733	0.734	0.736	0.739	0.743
0.00200	(0.0196)	0.785	0.766	0.754	0.745	0.738	0.731	0.727	0.725	0.723	0.723	0.724	0.727	0.730	0.735
0.00300	(0.0294)	0.759	0.743	0.733	0.725	0.720	0.714	0.711	0.710	0.709	0.710	0.712	0.715	0.719	0.724
0.00400	(0.0392)	0.743	0.728	0.719	0.712	0.708	0.703	0.701	0.700	0.700	0.701	0.703	0.708	0.712	0.717
0.00600	(0.0588)	0.721	0.708	0.701	0.695	0.691	0.688	0.687	0.686	0.688	0.690	0.692	0.697	0.702	0.708
0.00800	(0.0785)	0.706	0.695	0.688	0.684	0.681	0.678	0.677	0.678	0.680	0.682	0.685	0.691	0.696	0.702
0.01000	(0.0981)	0.695	0.685	0.680	0.675	0.673	0.671	0.671	0.671	0.674	0.677	0.680	0.686	0.692	0.698
0.01500	(0.147)	0.677	0.669	0.664	0.661	0.659	0.659	0.659	0.660	0.664	0.667	0.671	0.678	0.684	0.691
0.02000	(0.196)	0.665	0.658	0.654	0.652	0.651	0.651	0.652	0.653	0.657	0.661	0.665	0.673	0.679	0.687
0.03000	(0.294)	0.650	0.644	0.641	0.640	0.639	0.640	0.642	0.644	0.649	0.653	0.658	0.666	0.673	0.681
0.04000	(0.392)	0.640	0.635	0.633	0.632	0.632	0.633	0.636	0.638	0.643	0.648	0.653	0.662	0.669	0.677
0.06000	(0.588)	0.626	0.623	0.622	0.621	0.622	0.625	0.628	0.631	0.637	0.642	0.647	0.656	0.664	0.673
0.08000	(0.785)	0.618	0.615	0.614	0.615	0.616	0.619	0.622	0.626	0.632	0.638	0.643	0.653	0.661	0.670
0.10000	(0.981)	0.611	0.609	0.609	0.610	0.611	0.615	0.619	0.622	0.629	0.635	0.641	0.651	0.659	0.668
0.15000	(1.47)	0.601	0.600	0.601	0.602	0.604	0.608	0.613	0.617	0.624	0.631	0.636	0.647	0.655	0.665
0.20000	(1.96)	0.594	0.594	0.595	0.597	0.599	0.604	0.609	0.613	0.621	0.628	0.634	0.644	0.653	0.663
0.30000	(2.94)	0.586	0.586	0.588	0.591	0.593	0.599	0.604	0.609	0.617	0.624	0.630	0.641	0.650	0.660
0.40000	(3.92)	0.580	0.582	0.584	0.587	0.590	0.596	0.601	0.606	0.614	0.622	0.628	0.639	0.649	0.659
0.60000	(5.88)	0.573	0.576	0.578	0.582	0.585	0.592	0.597	0.602	0.611	0.619	0.626	0.637	0.647	0.657
0.80000	(7.85)	0.569	0.572	0.575	0.579	0.582	0.589	0.595	0.600	0.610	0.617	0.624	0.636	0.646	0.656
1.00000	(9.81)	0.566	0.569	0.572	0.576	0.580	0.587	0.593	0.599	0.608	0.616	0.623	0.635	0.645	0.655
1.50000	(14.7)	0.561	0.565	0.568	0.573	0.577	0.584	0.591	0.597	0.606	0.614	0.622	0.633	0.643	0.654
2.00000	(19.6)	0.558	0.562	0.566	0.571	0.575	0.583	0.589	0.595	0.605	0.613	0.621	0.633	0.643	0.653
3.00000	(29.4)	0.554	0.559	0.563	0.568	0.573	0.581	0.587	0.593	0.604	0.612	0.619	0.632	0.642	0.652
4.00000	(39.2)	0.552	0.557	0.561	0.566	0.571	0.579	0.586	0.592	0.603	0.611	0.619	0.631	0.641	0.652
6.00000	(58.8)	0.549	0.554	0.559	0.564	0.569	0.578	0.585	0.591	0.601	0.610	0.618	0.630	0.640	0.651
8.00000	(78.5)	0.547	0.553	0.558	0.563	0.568	0.577	0.584	0.590	0.601	0.610	0.617	0.630	0.640	0.651
10.0000	(98.1)	0.546	0.552	0.557	0.562	0.567	0.576	0.583	0.590	0.600	0.609	0.617	0.629	0.640	0.650
15.0000	(147)	0.544	0.550	0.555	0.561	0.566	0.575	0.583	0.589	0.600	0.609	0.616	0.629	0.639	0.650
20.0000	(196)	0.543	0.549	0.554	0.560	0.566	0.575	0.582	0.588	0.599	0.608	0.616	0.628	0.639	0.650
30.0000	(294)	0.542	0.548	0.553	0.559	0.565	0.574	0.581	0.588	0.599	0.608	0.615	0.628	0.639	0.649
S	$(\Delta p/\rho l)$	6	8	10	12.5	15	20	25	30	40	50	60	80	100	125

Grad'nt S	k.p.g. $(\Delta p/\rho l)$	125	150	200	250	300	400	500	600	800	1000	1250	1500	2000	2500
		(Equivalent) Pipe diameters in mm													
0.00010	(0.00098)	0.842	0.842	0.844	0.847	0.850	0.856	0.862	0.867	0.877	0.885	0.895	0.903	0.918	0.930
0.00015	(0.00147)	0.825	0.825	0.828	0.831	0.835	0.842	0.848	0.854	0.864	0.873	0.883	0.892	0.907	0.920
0.00020	(0.00196)	0.813	0.814	0.817	0.821	0.825	0.832	0.839	0.845	0.856	0.865	0.875	0.885	0.900	0.913
0.00030	(0.00294)	0.797	0.798	0.803	0.807	0.811	0.819	0.826	0.833	0.845	0.854	0.865	0.875	0.891	0.904
0.00040	(0.00392)	0.786	0.788	0.793	0.798	0.802	0.811	0.818	0.825	0.837	0.847	0.858	0.868	0.884	0.898
0.00060	(0.00588)	0.772	0.774	0.780	0.785	0.790	0.800	0.808	0.815	0.827	0.838	0.850	0.860	0.877	0.891
0.00080	(0.00785)	0.762	0.765	0.771	0.777	0.782	0.792	0.801	0.808	0.821	0.832	0.844	0.854	0.871	0.886
0.00100	(0.00981)	0.755	0.758	0.765	0.771	0.777	0.787	0.795	0.803	0.816	0.828	0.840	0.850	0.868	0.882
0.00150	(0.0147)	0.743	0.747	0.754	0.761	0.767	0.777	0.786	0.795	0.809	0.820	0.833	0.843	0.862	0.877
0.00200	(0.0196)	0.735	0.739	0.747	0.754	0.760	0.771	0.781	0.789	0.803	0.815	0.828	0.839	0.858	0.873
0.00300	(0.0294)	0.724	0.729	0.737	0.745	0.752	0.763	0.773	0.782	0.797	0.809	0.822	0.834	0.853	0.868
0.00400	(0.0392)	0.717	0.722	0.731	0.739	0.746	0.758	0.769	0.778	0.793	0.805	0.819	0.830	0.850	0.865
0.00600	(0.0588)	0.708	0.714	0.723	0.732	0.739	0.752	0.762	0.772	0.788	0.801	0.814	0.826	0.846	0.862
0.00800	(0.0785)	0.702	0.708	0.718	0.727	0.735	0.748	0.759	0.768	0.784	0.798	0.812	0.823	0.843	0.860
0.01000	(0.0981)	0.698	0.704	0.714	0.723	0.731	0.745	0.756	0.766	0.782	0.795	0.809	0.822	0.842	0.858
0.01500	(0.147)	0.691	0.697	0.708	0.718	0.726	0.740	0.751	0.761	0.778	0.792	0.806	0.819	0.839	0.856
0.02000	(0.196)	0.687	0.693	0.704	0.714	0.723	0.737	0.749	0.759	0.776	0.790	0.804	0.817	0.837	0.854
0.03000	(0.294)	0.681	0.688	0.700	0.710	0.718	0.733	0.745	0.755	0.773	0.787	0.802	0.814	0.835	0.852
0.04000	(0.392)	0.677	0.684	0.697	0.707	0.716	0.731	0.743	0.753	0.771	0.785	0.800	0.813	0.834	0.851
0.06000	(0.588)	0.673	0.680	0.693	0.703	0.712	0.728	0.740	0.751	0.769	0.783	0.798	0.811	0.833	0.850
0.08000	(0.785)	0.670	0.677	0.690	0.701	0.710	0.726	0.739	0.749	0.767	0.782	0.797	0.810	0.832	0.849
0.10000	(0.981)	0.668	0.676	0.689	0.700	0.709	0.725	0.737	0.748	0.767	0.781	0.797	0.810	0.831	0.848
0.15000	(1.47)	0.665	0.673	0.686	0.697	0.707	0.723	0.736	0.747	0.765	0.780	0.795	0.809	0.830	0.847
0.20000	(1.96)	0.663	0.671	0.684	0.696	0.705	0.721	0.735	0.746	0.764	0.779	0.795	0.808	0.829	0.847
0.30000	(2.94)	0.660	0.668	0.682	0.694	0.704	0.720	0.733	0.745	0.763	0.778	0.794	0.807	0.829	0.846
0.40000	(3.92)	0.659	0.667	0.681	0.693	0.703	0.719	0.732	0.744	0.762	0.778	0.793	0.807	0.828	0.846
0.60000	(5.88)	0.657	0.665	0.680	0.692	0.702	0.718	0.731	0.743	0.762	0.777	0.793	0.806	0.828	0.845
0.80000	(7.85)	0.656	0.664	0.679	0.691	0.701	0.717	0.731	0.742	0.761	0.776	0.792	0.806	0.828	0.845
1.00000	(9.81)	0.655	0.664	0.678	0.690	0.700	0.717	0.731	0.742	0.761	0.776	0.792	0.805	0.827	0.845
1.50000	(14.7)	0.654	0.663	0.677	0.689	0.699	0.716	0.730	0.741	0.760	0.776	0.792	0.805	0.827	0.845
2.00000	(19.6)	0.653	0.662	0.677	0.689	0.699	0.716	0.730	0.741	0.760	0.775	0.791	0.805	0.827	0.845
3.00000	(29.4)	0.652	0.661	0.676	0.688	0.698	0.715	0.729	0.741	0.760	0.775	0.791	0.805	0.827	0.844
S	$(\Delta p/\rho l)$	125	150	200	250	300	400	500	600	800	1000	1250	1500	2000	2500

Grad'nt k.p.g.   (Equivalent) Pipe diameters in mm

Kinematic viscosity, $\nu = 0.125 \times 10^{-6}\ \mathrm{m^2 s^{-1}}$;   **Roughness size, $k_s = 0.006$ mm**

Kin. visc., $\nu = 0\cdot125\times10^{-6}\ \mathrm{m^2\,s^{-1}}$; Roughness size, $k_s = 0\cdot015$ mm

$S = 0\cdot00010$ to $30\cdot0000$

This table shows values of m, as follows

i.e. kin. pr. grad., $\Delta p/\rho l =$ $(0\cdot00098)$ to $(294)\ \mathrm{ms^{-2}}$

m_C for Colebrook-White solutions; or, where $\mathbf{R} \le 2000$, m_P for laminar flow

Grad'nt S	k.p.g. $(\Delta p/\rho l)$	6	8	10	12·5	15	20	25	30	40	50	60	80	100	125
		(Equivalent) Pipe diameters in mm													
0·00030	(0·00294)	0·857	*	0·886	0·868	0·855	0·839	0·829	0·823	0·815	0·812	0·810	0·809	0·810	0·812
0·00040	(0·00392)	0·742	*	0·865	0·848	0·837	0·823	0·814	0·809	0·803	0·800	0·798	0·798	0·800	0·802
0·00060	(0·00588)	*	0·855	0·838	0·823	0·814	0·802	0·795	0·790	0·786	0·784	0·783	0·785	0·787	0·790
0·00080	(0·00785)	*	0·836	0·820	0·807	0·798	0·788	0·782	0·778	0·775	0·773	0·774	0·775	0·778	0·782
0·00100	(0·00981)	*	0·821	0·807	0·795	0·787	0·777	0·772	0·769	0·766	0·766	0·766	0·769	0·772	0·776
0·00150	(0·0147)	0·817	0·797	0·784	0·774	0·768	0·760	0·756	0·754	0·753	0·753	0·754	0·758	0·762	0·767
0·00200	(0·0196)	0·799	0·781	0·770	0·761	0·755	0·749	0·745	0·744	0·743	0·745	0·746	0·750	0·755	0·760
0·00300	(0·0294)	0·776	0·760	0·751	0·743	0·739	0·734	0·732	0·731	0·732	0·734	0·736	0·741	0·746	0·752
0·00400	(0·0392)	0·761	0·747	0·738	0·732	0·728	0·724	0·723	0·723	0·724	0·727	0·730	0·735	0·741	0·747
0·00600	(0·0588)	0·741	0·729	0·722	0·717	0·714	0·712	0·711	0·712	0·715	0·718	0·721	0·728	0·734	0·741
0·00800	(0·0785)	0·728	0·718	0·712	0·708	0·705	0·704	0·704	0·705	0·708	0·712	0·716	0·723	0·730	0·737
0·01000	(0·0981)	0·719	0·709	0·704	0·701	0·699	0·698	0·699	0·700	0·704	0·708	0·712	0·720	0·726	0·734
0·01500	(0·147)	0·703	0·695	0·691	0·689	0·688	0·688	0·690	0·692	0·697	0·701	0·706	0·714	0·721	0·730
0·02000	(0·196)	0·693	0·686	0·683	0·682	0·681	0·682	0·684	0·687	0·692	0·697	0·702	0·711	0·718	0·727
0·03000	(0·294)	0·680	0·675	0·673	0·672	0·673	0·675	0·677	0·680	0·686	0·692	0·697	0·706	0·714	0·723
0·04000	(0·392)	0·672	0·668	0·667	0·666	0·667	0·670	0·673	0·676	0·683	0·689	0·694	0·704	0·712	0·721
0·06000	(0·588)	0·661	0·659	0·658	0·659	0·660	0·664	0·668	0·671	0·678	0·685	0·690	0·701	0·709	0·718
0·08000	(0·785)	0·655	0·653	0·653	0·654	0·656	0·660	0·664	0·668	0·676	0·682	0·688	0·699	0·707	0·717
0·10000	(0·981)	0·650	0·649	0·650	0·651	0·653	0·657	0·662	0·666	0·674	0·681	0·687	0·697	0·706	0·716
0·15000	(1·47)	0·643	0·643	0·644	0·646	0·648	0·653	0·658	0·663	0·671	0·678	0·684	0·695	0·704	0·714
0·20000	(1·96)	0·638	0·638	0·640	0·642	0·645	0·651	0·656	0·660	0·669	0·676	0·683	0·694	0·703	0·713
0·30000	(2·94)	0·632	0·633	0·635	0·638	0·641	0·647	0·653	0·658	0·667	0·674	0·681	0·692	0·701	0·712
0·40000	(3·92)	0·629	0·630	0·633	0·636	0·639	0·645	0·651	0·656	0·665	0·673	0·680	0·691	0·701	0·711
0·60000	(5·88)	0·624	0·626	0·629	0·633	0·636	0·643	0·649	0·654	0·664	0·671	0·678	0·690	0·700	0·710
0·80000	(7·85)	0·621	0·624	0·627	0·631	0·635	0·642	0·648	0·653	0·663	0·671	0·677	0·689	0·699	0·709
1·00000	(9·81)	0·620	0·623	0·626	0·630	0·634	0·641	0·647	0·652	0·662	0·670	0·677	0·689	0·699	0·709
1·50000	(14·7)	0·617	0·620	0·624	0·628	0·632	0·639	0·645	0·651	0·661	0·669	0·676	0·688	0·698	0·708
2·00000	(19·6)	0·615	0·618	0·622	0·627	0·631	0·638	0·645	0·650	0·660	0·668	0·676	0·688	0·697	0·708
3·00000	(29·4)	0·613	0·617	0·621	0·625	0·629	0·637	0·644	0·649	0·659	0·668	0·675	0·687	0·697	0·708
4·00000	(39·2)	0·611	0·615	0·620	0·624	0·629	0·636	0·643	0·649	0·659	0·667	0·675	0·687	0·697	0·707
6·00000	(58·8)	0·610	0·614	0·618	0·623	0·628	0·636	0·642	0·648	0·658	0·667	0·674	0·686	0·696	0·707
8·00000	(78·5)	0·609	0·613	0·618	0·623	0·627	0·635	0·642	0·648	0·658	0·666	0·674	0·686	0·696	0·707
10·0000	(98·1)	0·608	0·613	0·617	0·622	0·627	0·635	0·642	0·648	0·658	0·666	0·674	0·686	0·696	0·707
15·0000	(147)	0·607	0·612	0·616	0·621	0·626	0·634	0·641	0·647	0·657	0·666	0·673	0·686	0·696	0·706
20·0000	(196)	0·606	0·611	0·616	0·621	0·626	0·634	0·641	0·647	0·657	0·666	0·673	0·685	0·696	0·706
30·0000	(294)	0·606	0·611	0·615	0·621	0·625	0·634	0·641	0·647	0·657	0·666	0·673	0·685	0·696	0·706
S	$(\Delta p/\rho l)$	6	8	10	12·5	15	20	25	30	40	50	60	80	100	125

Grad'nt   k.p.g.   (Equivalent) Pipe diameters in mm

Grad'nt S	k.p.g. $(\Delta p/\rho l)$	125	150	200	250	300	400	500	600	800	1000	1250	1500	2000	2500
0·00010	(0·00098)	0·853	0·853	0·856	0·859	0·863	0·870	0·877	0·883	0·895	0·904	0·915	0·925	0·941	0·955
0·00015	(0·00147)	0·837	0·838	0·842	0·846	0·850	0·858	0·866	0·872	0·884	0·895	0·906	0·916	0·933	0·947
0·00020	(0·00196)	0·826	0·828	0·832	0·837	0·841	0·850	0·858	0·865	0·878	0·888	0·900	0·910	0·928	0·942
0·00030	(0·00294)	0·812	0·814	0·820	0·825	0·830	0·840	0·848	0·856	0·869	0·880	0·892	0·903	0·921	0·936
0·00040	(0·00392)	0·802	0·805	0·811	0·817	0·823	0·833	0·842	0·850	0·863	0·875	0·888	0·899	0·917	0·932
0·00060	(0·00588)	0·790	0·794	0·801	0·807	0·813	0·824	0·834	0·842	0·856	0·868	0·881	0·893	0·912	0·927
0·00080	(0·00785)	0·782	0·786	0·794	0·801	0·807	0·818	0·828	0·837	0·852	0·864	0·878	0·889	0·908	0·924
0·00100	(0·00981)	0·776	0·781	0·789	0·796	0·803	0·814	0·825	0·833	0·849	0·861	0·875	0·887	0·906	0·922
0·00150	(0·0147)	0·767	0·771	0·780	0·788	0·795	0·808	0·818	0·828	0·843	0·856	0·870	0·882	0·902	0·919
0·00200	(0·0196)	0·760	0·766	0·775	0·783	0·791	0·804	0·814	0·824	0·840	0·853	0·868	0·880	0·900	0·917
0·00300	(0·0294)	0·752	0·758	0·768	0·777	0·785	0·798	0·810	0·819	0·836	0·850	0·864	0·877	0·897	0·914
0·00400	(0·0392)	0·747	0·753	0·764	0·773	0·781	0·795	0·807	0·817	0·834	0·847	0·862	0·875	0·895	0·912
0·00600	(0·0588)	0·741	0·747	0·759	0·768	0·777	0·791	0·803	0·813	0·830	0·845	0·860	0·872	0·893	0·910
0·00800	(0·0785)	0·737	0·744	0·755	0·765	0·774	0·788	0·801	0·811	0·829	0·843	0·858	0·871	0·892	0·909
0·01000	(0·0981)	0·734	0·741	0·753	0·763	0·772	0·787	0·799	0·810	0·827	0·842	0·857	0·870	0·891	0·908
0·01500	(0·147)	0·730	0·737	0·749	0·760	0·769	0·784	0·796	0·807	0·825	0·840	0·855	0·868	0·890	0·907
0·02000	(0·196)	0·727	0·734	0·747	0·757	0·767	0·782	0·795	0·806	0·824	0·839	0·854	0·867	0·889	0·906
0·03000	(0·294)	0·723	0·731	0·744	0·755	0·764	0·780	0·793	0·804	0·822	0·837	0·853	0·866	0·888	0·906
0·04000	(0·392)	0·721	0·729	0·742	0·753	0·763	0·779	0·792	0·803	0·821	0·836	0·852	0·865	0·887	0·905
0·06000	(0·588)	0·718	0·726	0·740	0·751	0·761	0·777	0·790	0·802	0·820	0·835	0·851	0·865	0·887	0·904
0·08000	(0·785)	0·717	0·725	0·739	0·750	0·760	0·776	0·790	0·801	0·820	0·835	0·851	0·864	0·886	0·904
0·10000	(0·981)	0·716	0·724	0·738	0·749	0·759	0·776	0·789	0·800	0·819	0·834	0·850	0·864	0·886	0·904
0·15000	(1·47)	0·714	0·722	0·736	0·748	0·758	0·775	0·788	0·799	0·818	0·834	0·850	0·863	0·885	0·903
0·20000	(1·96)	0·713	0·721	0·736	0·747	0·757	0·774	0·787	0·799	0·818	0·833	0·849	0·863	0·885	0·903
0·30000	(2·94)	0·712	0·720	0·735	0·746	0·757	0·773	0·787	0·798	0·817	0·833	0·849	0·862	0·885	0·903
0·40000	(3·92)	0·711	0·719	0·734	0·746	0·756	0·773	0·786	0·798	0·817	0·833	0·849	0·862	0·884	0·902
0·60000	(5·88)	0·710	0·719	0·733	0·745	0·755	0·772	0·786	0·798	0·817	0·832	0·848	0·862	0·884	0·902
0·80000	(7·85)	0·709	0·718	0·733	0·745	0·755	0·772	0·786	0·797	0·817	0·832	0·848	0·862	0·884	0·902
1·00000	(9·81)	0·709	0·718	0·732	0·745	0·755	0·772	0·786	0·797	0·816	0·832	0·848	0·862	0·884	0·902
1·50000	(14·7)	0·708	0·717	0·732	0·744	0·754	0·771	0·785	0·797	0·816	0·832	0·848	0·862	0·884	0·902
2·00000	(19·6)	0·708	0·717	0·732	0·744	0·754	0·771	0·785	0·797	0·816	0·832	0·848	0·861	0·884	0·902
3·00000	(29·4)	0·708	0·716	0·731	0·744	0·754	0·771	0·785	0·797	0·816	0·831	0·848	0·861	0·884	0·902
S	$(\Delta p/\rho l)$	125	150	200	250	300	400	500	600	800	1000	1250	1500	2000	2500

Grad'nt   k.p.g.   (Equivalent) Pipe diameters in mm

Kinematic viscosity, $\nu = 0\cdot125\times10^{-6}\ \mathrm{m^2\,s^{-1}}$; Roughness size, $k_s = 0\cdot015$ mm

E5

Kin. visc., $\nu = 0{\cdot}125 \times 10^{-6}\ \mathrm{m^2\,s^{-1}}$; **Roughness size, $k_s = 0{\cdot}030$ mm**

$S = 0{\cdot}00010$ to $3{\cdot}00000$

This table shows values of m, as follows

i.e. kin. pr. grad., $\Delta p/\rho l = (0{\cdot}00098)$ to $(29{\cdot}4)\ \mathrm{ms^{-2}}$

m_C for Colebrook-White solutions

Grad'nt S	k.p.g. $(\Delta p/\rho l)$	125	150	200	250	300	400	500	600	800	1000	1250	1500	2000	2500
							(Equivalent) Pipe diameters in mm								
0·00010	(0·00098)	0·869	0·870	0·874	0·878	0·882	0·891	0·899	0·906	0·919	0·930	0·942	0·952	0·970	0·985
0·00015	(0·00147)	0·854	0·856	0·861	0·866	0·871	0·881	0·889	0·897	0·911	0·922	0·935	0·946	0·964	0·980
0·00020	(0·00196)	0·845	0·848	0·853	0·859	0·864	0·875	0·884	0·892	0·906	0·918	0·930	0·942	0·961	0·976
0·00030	(0·00294)	0·833	0·836	0·843	0·849	0·855	0·866	0·876	0·884	0·899	0·912	0·925	0·937	0·956	0·972
0·00040	(0·00392)	0·825	0·829	0·836	0·843	0·850	0·861	0·871	0·880	0·895	0·908	0·922	0·933	0·953	0·969
0·00060	(0·00588)	0·815	0·820	0·828	0·836	0·843	0·855	0·865	0·874	0·890	0·903	0·917	0·930	0·950	0·966
0·00080	(0·00785)	0·809	0·814	0·823	0·831	0·838	0·851	0·861	0·871	0·887	0·901	0·915	0·927	0·948	0·964
0·00100	(0·00981)	0·805	0·810	0·819	0·827	0·835	0·848	0·859	0·869	0·885	0·899	0·913	0·925	0·946	0·963
0·00150	(0·0147)	0·797	0·803	0·813	0·822	0·830	0·843	0·855	0·865	0·881	0·895	0·910	0·923	0·944	0·961
0·00200	(0·0196)	0·792	0·798	0·809	0·818	0·826	0·840	0·852	0·862	0·879	0·893	0·908	0·921	0·942	0·960
0·00300	(0·0294)	0·787	0·793	0·804	0·814	0·822	0·837	0·849	0·859	0·877	0·891	0·906	0·919	0·941	0·958
0·00400	(0·0392)	0·783	0·790	0·801	0·811	0·820	0·835	0·847	0·857	0·875	0·890	0·905	0·918	0·940	0·957
0·00600	(0·0588)	0·779	0·786	0·798	0·808	0·817	0·832	0·844	0·855	0·873	0·888	0·903	0·917	0·938	0·956
0·00800	(0·0785)	0·776	0·783	0·795	0·806	0·815	0·830	0·843	0·854	0·872	0·887	0·902	0·916	0·938	0·955
0·01000	(0·0981)	0·774	0·781	0·794	0·805	0·814	0·829	0·842	0·853	0·871	0·886	0·902	0·915	0·937	0·955
0·01500	(0·147)	0·771	0·779	0·792	0·802	0·812	0·827	0·840	0·852	0·870	0·885	0·901	0·914	0·936	0·954
0·02000	(0·196)	0·769	0·777	0·790	0·801	0·811	0·826	0·840	0·851	0·869	0·884	0·900	0·914	0·936	0·954
0·03000	(0·294)	0·767	0·775	0·788	0·799	0·809	0·825	0·838	0·850	0·868	0·884	0·899	0·913	0·935	0·953
0·04000	(0·392)	0·766	0·774	0·787	0·798	0·808	0·824	0·838	0·849	0·868	0·883	0·899	0·913	0·935	0·953
0·06000	(0·588)	0·764	0·772	0·786	0·797	0·807	0·823	0·837	0·848	0·867	0·883	0·899	0·912	0·934	0·952
0·08000	(0·785)	0·763	0·771	0·785	0·797	0·806	0·823	0·836	0·848	0·867	0·882	0·898	0·912	0·934	0·952
0·10000	(0·981)	0·762	0·771	0·785	0·796	0·806	0·823	0·836	0·848	0·866	0·882	0·898	0·912	0·934	0·952
0·15000	(1·47)	0·761	0·770	0·784	0·795	0·805	0·822	0·835	0·847	0·866	0·882	0·898	0·911	0·934	0·952
0·20000	(1·96)	0·761	0·769	0·783	0·795	0·805	0·822	0·835	0·847	0·866	0·881	0·897	0·911	0·934	0·952
0·30000	(2·94)	0·760	0·768	0·783	0·794	0·804	0·821	0·835	0·846	0·866	0·881	0·897	0·911	0·933	0·951
0·40000	(3·92)	0·759	0·768	0·782	0·794	0·804	0·821	0·835	0·846	0·865	0·881	0·897	0·911	0·933	0·951
0·60000	(5·88)	0·759	0·767	0·782	0·794	0·804	0·821	0·834	0·846	0·865	0·881	0·897	0·911	0·933	0·951
0·80000	(7·85)	0·758	0·767	0·782	0·793	0·804	0·820	0·834	0·846	0·865	0·881	0·897	0·911	0·933	0·951
1·00000	(9·81)	0·758	0·767	0·781	0·793	0·803	0·820	0·834	0·846	0·865	0·881	0·897	0·910	0·933	0·951
1·50000	(14·7)	0·758	0·767	0·781	0·793	0·803	0·820	0·834	0·846	0·865	0·880	0·897	0·910	0·933	0·951
2·00000	(19·6)	0·758	0·766	0·781	0·793	0·803	0·820	0·834	0·845	0·865	0·880	0·897	0·910	0·933	0·951
3·00000	(29·4)	0·757	0·766	0·781	0·793	0·803	0·820	0·834	0·845	0·865	0·880	0·896	0·910	0·933	0·951
S	$(\Delta p/\rho l)$	125	150	200	250	300	400	500	600	800	1000	1250	1500	2000	2500
Grad'nt	k.p.g.				(Equivalent) Pipe diameters in mm										

Roughness size, $k_s = 0{\cdot}030$ mm

E6

Kin. visc., $\nu = 0{\cdot}125 \times 10^{-6}\ \mathrm{m^2\,s^{-1}}$; **Roughness size, $k_s = 0{\cdot}060$ mm**

$S = 0{\cdot}00010$ to $3{\cdot}00000$

This table shows values of m, as follows

i.e. kin. pr. grad., $\Delta p/\rho l = (0{\cdot}00098)$ to $(29{\cdot}4)\ \mathrm{ms^{-2}}$

m_C for Colebrook-White solutions

Grad'nt S	k.p.g. $(\Delta p/\rho l)$	125	150	200	250	300	400	500	600	800	1000	1250	1500	2000	2500
							(Equivalent) Pipe diameters in mm								
0·00010	(0·00098)	0·895	0·898	0·903	0·908	0·913	0·923	0·932	0·941	0·955	0·967	0·980	0·992	1·011	1·027
0·00015	(0·00147)	0·884	0·887	0·893	0·899	0·905	0·916	0·926	0·934	0·949	0·962	0·975	0·987	1·007	1·023
0·00020	(0·00196)	0·876	0·880	0·887	0·894	0·900	0·911	0·921	0·930	0·946	0·959	0·972	0·984	1·005	1·021
0·00030	(0·00294)	0·867	0·871	0·879	0·887	0·893	0·906	0·916	0·925	0·941	0·955	0·969	0·981	1·002	1·019
0·00040	(0·00392)	0·861	0·866	0·874	0·882	0·889	0·902	0·913	0·922	0·939	0·952	0·967	0·979	1·000	1·017
0·00060	(0·00588)	0·854	0·859	0·868	0·877	0·884	0·897	0·909	0·919	0·935	0·949	0·964	0·977	0·998	1·015
0·00080	(0·00785)	0·849	0·855	0·865	0·873	0·881	0·895	0·906	0·916	0·933	0·948	0·962	0·975	0·996	1·014
0·00100	(0·00981)	0·846	0·852	0·862	0·871	0·879	0·893	0·905	0·915	0·932	0·946	0·961	0·974	0·996	1·013
0·00150	(0·0147)	0·841	0·847	0·858	0·867	0·876	0·890	0·902	0·912	0·930	0·944	0·960	0·973	0·994	1·012
0·00200	(0·0196)	0·838	0·844	0·855	0·865	0·873	0·888	0·900	0·911	0·929	0·943	0·958	0·972	0·993	1·011
0·00300	(0·0294)	0·834	0·840	0·852	0·862	0·871	0·886	0·898	0·909	0·927	0·942	0·957	0·970	0·992	1·010
0·00400	(0·0392)	0·831	0·838	0·850	0·860	0·869	0·884	0·897	0·908	0·926	0·941	0·956	0·970	0·992	1·010
0·00600	(0·0588)	0·829	0·836	0·848	0·858	0·868	0·883	0·896	0·907	0·925	0·940	0·956	0·969	0·991	1·009
0·00800	(0·0785)	0·827	0·834	0·847	0·857	0·866	0·882	0·895	0·906	0·924	0·939	0·955	0·968	0·990	1·008
0·01000	(0·0981)	0·826	0·833	0·846	0·856	0·866	0·881	0·894	0·905	0·924	0·939	0·955	0·968	0·990	1·008
0·01500	(0·147)	0·824	0·831	0·844	0·855	0·864	0·880	0·893	0·904	0·923	0·938	0·954	0·967	0·990	1·008
0·02000	(0·196)	0·823	0·830	0·843	0·854	0·864	0·880	0·893	0·904	0·923	0·938	0·954	0·967	0·989	1·008
0·03000	(0·294)	0·821	0·829	0·842	0·853	0·863	0·879	0·892	0·903	0·922	0·937	0·953	0·967	0·989	1·007
0·04000	(0·392)	0·820	0·828	0·842	0·853	0·862	0·878	0·892	0·903	0·922	0·937	0·953	0·967	0·989	1·007
0·06000	(0·588)	0·819	0·827	0·841	0·852	0·862	0·878	0·891	0·902	0·921	0·937	0·953	0·966	0·989	1·007
0·08000	(0·785)	0·819	0·827	0·840	0·852	0·861	0·878	0·891	0·902	0·921	0·936	0·952	0·966	0·989	1·007
0·10000	(0·981)	0·818	0·826	0·840	0·851	0·861	0·877	0·891	0·902	0·921	0·936	0·952	0·966	0·988	1·007
0·15000	(1·47)	0·818	0·826	0·839	0·851	0·861	0·877	0·890	0·902	0·921	0·936	0·952	0·966	0·988	1·006
0·20000	(1·96)	0·817	0·825	0·839	0·851	0·860	0·877	0·890	0·902	0·921	0·936	0·952	0·966	0·988	1·006
0·30000	(2·94)	0·817	0·825	0·839	0·850	0·860	0·876	0·890	0·901	0·920	0·936	0·952	0·966	0·988	1·006
0·40000	(3·92)	0·817	0·825	0·839	0·850	0·860	0·876	0·890	0·901	0·920	0·936	0·952	0·966	0·988	1·006
0·60000	(5·88)	0·816	0·825	0·838	0·850	0·860	0·876	0·890	0·901	0·920	0·936	0·952	0·965	0·988	1·006
0·80000	(7·85)	0·816	0·824	0·838	0·850	0·860	0·876	0·890	0·901	0·920	0·936	0·952	0·965	0·988	1·006
1·00000	(9·81)	0·816	0·824	0·838	0·850	0·860	0·876	0·889	0·901	0·920	0·936	0·952	0·965	0·988	1·006
1·50000	(14·7)	0·816	0·824	0·838	0·850	0·859	0·876	0·889	0·901	0·920	0·935	0·952	0·965	0·988	1·006
2·00000	(19·6)	0·816	0·824	0·838	0·849	0·859	0·876	0·889	0·901	0·920	0·935	0·952	0·965	0·988	1·006
3·00000	(29·4)	0·815	0·824	0·838	0·849	0·859	0·876	0·889	0·901	0·920	0·935	0·952	0·965	0·988	1·006
S	$(\Delta p/\rho l)$	125	150	200	250	300	400	500	600	800	1000	1250	1500	2000	2500
Grad'nt	k.p.g.				(Equivalent) Pipe diameters in mm										

Roughness size, $k_s = 0{\cdot}060$ mm

Kin. visc., $\nu = 0\cdot125\times10^{-6}\ \mathrm{m^2\,s^{-1}}$;   Roughness size, $k_s = 0\cdot150$ mm

$S = 0\cdot00010$ to $3\cdot00000$

This table shows values of m, as follows

i.e. kin. pr. grad., $\Delta p/\rho l = (0\cdot00098)$ to $(29\cdot4)\ \mathrm{ms^{-2}}$

m_C for Colebrook-White solutions

Grad'nt S	k.p.g. $(\Delta p/\rho l)$	125	150	200	250	300	400	500	600	800	1000	1250	1500	2000	2500
							(Equivalent) Pipe diameters in mm								
0·00010	(0·00098)	0·956	0·959	0·965	0·971	0·978	0·989	0·999	1·008	1·023	1·036	1·050	1·063	1·083	1·100
0·00015	(0·00147)	0·948	0·951	0·959	0·966	0·972	0·984	0·995	1·004	1·020	1·034	1·048	1·060	1·081	1·098
0·00020	(0·00196)	0·943	0·947	0·955	0·962	0·969	0·982	0·992	1·002	1·018	1·032	1·046	1·059	1·080	1·097
0·00030	(0·00294)	0·937	0·942	0·950	0·958	0·966	0·978	0·989	0·999	1·016	1·030	1·044	1·057	1·078	1·096
0·00040	(0·00392)	0·933	0·938	0·947	0·956	0·963	0·976	0·988	0·998	1·014	1·028	1·043	1·056	1·077	1·095
0·00060	(0·00588)	0·929	0·934	0·944	0·953	0·960	0·974	0·985	0·996	1·013	1·027	1·042	1·055	1·076	1·094
0·00080	(0·00785)	0·926	0·932	0·942	0·951	0·959	0·972	0·984	0·994	1·012	1·026	1·041	1·054	1·076	1·093
0·00100	(0·00981)	0·924	0·930	0·940	0·949	0·958	0·971	0·983	0·994	1·011	1·025	1·041	1·054	1·075	1·093
0·00150	(0·0147)	0·922	0·927	0·938	0·947	0·956	0·970	0·982	0·992	1·010	1·024	1·040	1·053	1·074	1·092
0·00200	(0·0196)	0·920	0·926	0·937	0·946	0·955	0·969	0·981	0·992	1·009	1·024	1·039	1·052	1·074	1·092
0·00300	(0·0294)	0·918	0·924	0·935	0·945	0·953	0·968	0·980	0·991	1·008	1·023	1·038	1·052	1·074	1·091
0·00400	(0·0392)	0·916	0·923	0·934	0·944	0·952	0·967	0·979	0·990	1·008	1·023	1·038	1·051	1·073	1·091
0·00600	(0·0588)	0·915	0·921	0·933	0·943	0·951	0·966	0·979	0·989	1·007	1·022	1·038	1·051	1·073	1·091
0·00800	(0·0785)	0·914	0·921	0·932	0·942	0·951	0·966	0·978	0·989	1·007	1·022	1·037	1·051	1·073	1·091
0·01000	(0·0981)	0·913	0·920	0·932	0·942	0·951	0·965	0·978	0·989	1·007	1·022	1·037	1·051	1·073	1·091
0·01500	(0·147)	0·912	0·919	0·931	0·941	0·950	0·965	0·977	0·988	1·006	1·021	1·037	1·050	1·072	1·090
0·02000	(0·196)	0·912	0·918	0·930	0·941	0·950	0·965	0·977	0·988	1·006	1·021	1·037	1·050	1·072	1·090
0·03000	(0·294)	0·911	0·918	0·930	0·940	0·949	0·964	0·977	0·988	1·006	1·021	1·037	1·050	1·072	1·090
0·04000	(0·392)	0·911	0·917	0·930	0·940	0·949	0·964	0·977	0·988	1·006	1·021	1·036	1·050	1·072	1·090
0·06000	(0·588)	0·910	0·917	0·929	0·940	0·949	0·964	0·976	0·987	1·006	1·021	1·036	1·050	1·072	1·090
0·08000	(0·785)	0·910	0·917	0·929	0·939	0·948	0·964	0·976	0·987	1·005	1·020	1·036	1·050	1·072	1·090
0·10000	(0·981)	0·909	0·917	0·929	0·939	0·948	0·964	0·976	0·987	1·005	1·020	1·036	1·050	1·072	1·090
0·15000	(1·47)	0·909	0·916	0·929	0·939	0·948	0·963	0·976	0·987	1·005	1·020	1·036	1·049	1·072	1·090
0·20000	(1·96)	0·909	0·916	0·928	0·939	0·948	0·963	0·976	0·987	1·005	1·020	1·036	1·049	1·072	1·090
0·30000	(2·94)	0·909	0·916	0·928	0·939	0·948	0·963	0·976	0·987	1·005	1·020	1·036	1·049	1·072	1·090
0·40000	(3·92)	0·909	0·916	0·928	0·939	0·948	0·963	0·976	0·987	1·005	1·020	1·036	1·049	1·072	1·090
0·60000	(5·88)	0·908	0·916	0·928	0·938	0·948	0·963	0·976	0·987	1·005	1·020	1·036	1·049	1·071	1·090
0·80000	(7·85)	0·908	0·916	0·928	0·938	0·948	0·963	0·976	0·987	1·005	1·020	1·036	1·049	1·071	1·090
1·00000	(9·81)	0·908	0·915	0·928	0·938	0·947	0·963	0·976	0·987	1·005	1·020	1·036	1·049	1·071	1·090
1·50000	(14·7)	0·908	0·915	0·928	0·938	0·947	0·963	0·976	0·987	1·005	1·020	1·036	1·049	1·071	1·090
2·00000	(19·6)	0·908	0·915	0·928	0·938	0·947	0·963	0·976	0·987	1·005	1·020	1·036	1·049	1·071	1·090
3·00000	(29·4)	0·908	0·915	0·928	0·938	0·947	0·963	0·976	0·987	1·005	1·020	1·036	1·049	1·071	1·089
S	$(\Delta p/\rho l)$	125	150	200	250	300	400	500	600	800	1000	1250	1500	2000	2500

Grad'nt   k.p.g.   (Equivalent) Pipe diameters in mm   Roughness size, $k_s = 0\cdot150$ mm

Kin. visc., $\nu = 0\cdot125\times10^{-6}\ \mathrm{m^2\,s^{-1}}$;   Roughness size, $k_s = 0\cdot30$ mm

$S = 0\cdot00010$ to $3\cdot00000$

This table shows values of m, as follows

i.e. kin. pr. grad., $\Delta p/\rho l = (0\cdot00098)$ to $(29\cdot4)\ \mathrm{ms^{-2}}$

m_C for Colebrook-White solutions

Grad'nt S	k.p.g. $(\Delta p/\rho l)$	125	150	200	250	300	400	500	600	800	1000	1250	1500	2000	2500
							(Equivalent) Pipe diameters in mm								
0·00010	(0·00098)	1·024	1·027	1·033	1·040	1·046	1·057	1·067	1·076	1·092	1·105	1·119	1·131	1·152	1·169
0·00015	(0·00147)	1·019	1·022	1·029	1·036	1·043	1·054	1·065	1·074	1·090	1·103	1·117	1·130	1·150	1·167
0·00020	(0·00196)	1·016	1·019	1·027	1·034	1·041	1·053	1·063	1·072	1·088	1·102	1·116	1·129	1·150	1·167
0·00030	(0·00294)	1·012	1·016	1·024	1·031	1·038	1·051	1·061	1·071	1·087	1·101	1·115	1·128	1·149	1·166
0·00040	(0·00392)	1·009	1·014	1·022	1·030	1·037	1·049	1·060	1·070	1·086	1·100	1·114	1·127	1·148	1·166
0·00060	(0·00588)	1·007	1·011	1·020	1·028	1·035	1·048	1·059	1·069	1·085	1·099	1·114	1·126	1·148	1·165
0·00080	(0·00785)	1·005	1·010	1·018	1·027	1·034	1·047	1·058	1·068	1·084	1·098	1·113	1·126	1·147	1·165
0·00100	(0·00981)	1·004	1·009	1·018	1·026	1·033	1·046	1·058	1·067	1·084	1·098	1·113	1·126	1·147	1·164
0·00150	(0·0147)	1·002	1·007	1·016	1·025	1·032	1·045	1·057	1·067	1·083	1·097	1·112	1·125	1·147	1·164
0·00200	(0·0196)	1·001	1·006	1·015	1·024	1·031	1·045	1·056	1·066	1·083	1·097	1·112	1·125	1·146	1·164
0·00300	(0·0294)	0·999	1·005	1·014	1·023	1·031	1·044	1·056	1·066	1·083	1·097	1·112	1·125	1·146	1·164
0·00400	(0·0392)	0·999	1·004	1·014	1·022	1·030	1·044	1·055	1·065	1·082	1·096	1·111	1·124	1·146	1·163
0·00600	(0·0588)	0·998	1·003	1·013	1·022	1·030	1·043	1·055	1·065	1·082	1·096	1·111	1·124	1·146	1·163
0·00800	(0·0785)	0·997	1·003	1·013	1·021	1·029	1·043	1·055	1·065	1·082	1·096	1·111	1·124	1·146	1·163
0·01000	(0·0981)	0·997	1·002	1·012	1·021	1·029	1·043	1·054	1·064	1·082	1·096	1·111	1·124	1·145	1·163
0·01500	(0·147)	0·996	1·002	1·012	1·021	1·029	1·042	1·054	1·064	1·081	1·096	1·111	1·124	1·145	1·163
0·02000	(0·196)	0·996	1·001	1·012	1·020	1·028	1·042	1·054	1·064	1·081	1·096	1·111	1·124	1·145	1·163
0·03000	(0·294)	0·995	1·001	1·011	1·020	1·028	1·042	1·054	1·064	1·081	1·095	1·111	1·124	1·145	1·163
0·04000	(0·392)	0·995	1·001	1·011	1·020	1·028	1·042	1·054	1·064	1·081	1·095	1·111	1·123	1·145	1·163
0·06000	(0·588)	0·995	1·000	1·011	1·020	1·028	1·042	1·053	1·064	1·081	1·095	1·110	1·123	1·145	1·163
0·08000	(0·785)	0·995	1·000	1·011	1·020	1·028	1·042	1·053	1·064	1·081	1·095	1·110	1·123	1·145	1·163
0·10000	(0·981)	0·994	1·000	1·011	1·020	1·028	1·042	1·053	1·064	1·081	1·095	1·110	1·123	1·145	1·163
0·15000	(1·47)	0·994	1·000	1·010	1·019	1·028	1·041	1·053	1·063	1·081	1·095	1·110	1·123	1·145	1·163
0·20000	(1·96)	0·994	1·000	1·010	1·019	1·027	1·041	1·053	1·063	1·081	1·095	1·110	1·123	1·145	1·163
0·30000	(2·94)	0·994	1·000	1·010	1·019	1·027	1·041	1·053	1·063	1·081	1·095	1·110	1·123	1·145	1·163
0·40000	(3·92)	0·994	1·000	1·010	1·019	1·027	1·041	1·053	1·063	1·081	1·095	1·110	1·123	1·145	1·163
0·60000	(5·88)	0·994	1·000	1·010	1·019	1·027	1·041	1·053	1·063	1·081	1·095	1·110	1·123	1·145	1·163
0·80000	(7·85)	0·994	1·000	1·010	1·019	1·027	1·041	1·053	1·063	1·081	1·095	1·110	1·123	1·145	1·163
1·00000	(9·81)	0·994	1·000	1·010	1·019	1·027	1·041	1·053	1·063	1·081	1·095	1·110	1·123	1·145	1·163
1·50000	(14·7)	0·994	0·999	1·010	1·019	1·027	1·041	1·053	1·063	1·081	1·095	1·110	1·123	1·145	1·163
2·00000	(19·6)	0·994	0·999	1·010	1·019	1·027	1·041	1·053	1·063	1·081	1·095	1·110	1·123	1·145	1·163
3·00000	(29·4)	0·994	0·999	1·010	1·019	1·027	1·041	1·053	1·063	1·081	1·095	1·110	1·123	1·145	1·163
S	$(\Delta p/\rho l)$	125	150	200	250	300	400	500	600	800	1000	1250	1500	2000	2500

Grad'nt   k.p.g.   (Equivalent) Pipe diameters in mm   Roughness size, $k_s = 0\cdot30$ mm

E9

Kin. visc., $\nu = 0.15 \times 10^{-6}$ m^{2}s^{-1};
$S = 0.00010$ to 30.0000
i.e. kin. pr. grad., $\Delta p/\rho l =$
(0.00098) to (294) ms^{-2}

Roughness size, $k_s = 0.0015$ mm
This table shows values of m, as follows
m_C for Colebrook-White solutions; or,
where $\mathbf{R} \le 2000$, m_P for laminar flow

Grad'nt S	k.p.g. $(\Delta p/\rho l)$	6	8	10	12·5	15	20	25	30	40	50	60	80	100	125
						(Equivalent) Pipe diameters in mm									
0·00030	(0·00294)	1·029	*	*	0·881	0·866	0·847	0·835	0·827	0·818	0·812	0·808	0·805	0·804	0·804
0·00040	(0·00392)	0·891	*	0·878	0·859	0·846	0·829	0·818	0·811	0·802	0·797	0·794	0·792	0·791	0·792
0·00060	(0·00588)	0·727	*	0·847	0·830	0·819	0·804	0·795	0·789	0·782	0·778	0·776	0·774	0·774	0·775
0·00080	(0·00785)	*	0·845	0·826	0·811	0·801	0·787	0·779	0·774	0·768	0·764	0·763	0·762	0·762	0·764
0·00100	(0·00981)	*	0·828	0·811	0·797	0·787	0·775	0·767	0·763	0·757	0·754	0·753	0·753	0·754	0·756
0·00150	(0·0147)	0·823	0·800	0·785	0·772	0·764	0·753	0·747	0·743	0·739	0·737	0·736	0·737	0·738	0·741
0·00200	(0·0196)	0·802	0·780	0·767	0·756	0·748	0·739	0·733	0·730	0·727	0·725	0·725	0·726	0·728	0·731
0·00300	(0·0294)	0·774	0·755	0·743	0·734	0·727	0·719	0·715	0·712	0·710	0·709	0·710	0·712	0·714	0·717
0·00400	(0·0392)	0·755	0·738	0·727	0·719	0·713	0·706	0·703	0·701	0·699	0·699	0·699	0·702	0·705	0·708
0·00600	(0·0588)	0·730	0·716	0·706	0·699	0·694	0·689	0·686	0·685	0·684	0·684	0·686	0·689	0·692	0·696
0·00800	(0·0785)	0·714	0·700	0·692	0·686	0·682	0·677	0·675	0·674	0·674	0·675	0·676	0·680	0·683	0·688
0·01000	(0·0981)	0·702	0·689	0·682	0·676	0·672	0·668	0·666	0·666	0·666	0·667	0·669	0·673	0·677	0·682
0·01500	(0·147)	0·681	0·670	0·664	0·659	0·656	0·653	0·652	0·652	0·653	0·655	0·657	0·661	0·666	0·671
0·02000	(0·196)	0·666	0·657	0·651	0·647	0·645	0·643	0·642	0·642	0·644	0·646	0·649	0·654	0·658	0·664
0·03000	(0·294)	0·648	0·640	0·635	0·632	0·630	0·629	0·629	0·630	0·632	0·635	0·638	0·643	0·648	0·654
0·04000	(0·392)	0·635	0·628	0·624	0·621	0·620	0·619	0·620	0·621	0·624	0·627	0·630	0·636	0·642	0·648
0·06000	(0·588)	0·618	0·613	0·610	0·608	0·607	0·607	0·608	0·610	0·613	0·617	0·621	0·627	0·633	0·639
0·08000	(0·785)	0·607	0·602	0·600	0·598	0·598	0·599	0·600	0·602	0·606	0·610	0·614	0·621	0·627	0·634
0·10000	(0·981)	0·599	0·595	0·593	0·592	0·592	0·593	0·595	0·597	0·601	0·605	0·609	0·616	0·623	0·630
0·15000	(1·47)	0·584	0·581	0·580	0·580	0·580	0·582	0·585	0·587	0·592	0·597	0·601	0·609	0·616	0·623
0·20000	(1·96)	0·575	0·572	0·572	0·572	0·573	0·575	0·578	0·581	0·586	0·591	0·596	0·604	0·611	0·618
0·30000	(2·94)	0·562	0·561	0·561	0·562	0·563	0·566	0·569	0·572	0·578	0·584	0·589	0·597	0·605	0·613
0·40000	(3·92)	0·554	0·553	0·553	0·555	0·556	0·560	0·564	0·567	0·573	0·579	0·584	0·593	0·601	0·609
0·60000	(5·88)	0·543	0·543	0·544	0·546	0·548	0·552	0·556	0·560	0·567	0·573	0·578	0·588	0·596	0·604
0·80000	(7·85)	0·536	0·536	0·538	0·540	0·542	0·547	0·551	0·555	0·563	0·569	0·575	0·584	0·593	0·601
1·00000	(9·81)	0·530	0·531	0·533	0·536	0·538	0·543	0·548	0·552	0·559	0·566	0·572	0·582	0·590	0·599
1·50000	(14·7)	0·521	0·523	0·525	0·528	0·531	0·537	0·542	0·546	0·554	0·561	0·567	0·578	0·586	0·596
2·00000	(19·6)	0·515	0·518	0·520	0·524	0·527	0·533	0·538	0·543	0·551	0·558	0·565	0·575	0·584	0·594
3·00000	(29·4)	0·508	0·511	0·514	0·517	0·521	0·528	0·533	0·538	0·547	0·555	0·561	0·572	0·581	0·591
4·00000	(39·2)	0·503	0·506	0·510	0·514	0·517	0·524	0·530	0·535	0·545	0·552	0·559	0·570	0·579	0·589
6·00000	(58·8)	0·496	0·500	0·504	0·509	0·513	0·520	0·526	0·532	0·541	0·549	0·556	0·568	0·577	0·587
8·00000	(78·5)	0·492	0·497	0·501	0·506	0·510	0·517	0·524	0·530	0·539	0·547	0·554	0·566	0·576	0·586
10·0000	(98·1)	0·489	0·494	0·498	0·503	0·508	0·516	0·522	0·528	0·538	0·546	0·553	0·565	0·575	0·585
15·0000	(147)	0·485	0·490	0·495	0·500	0·505	0·513	0·520	0·526	0·536	0·544	0·551	0·563	0·573	0·584
20·0000	(196)	0·482	0·487	0·492	0·498	0·502	0·511	0·518	0·524	0·534	0·543	0·550	0·562	0·572	0·583
30·0000	(294)	0·478	0·484	0·489	0·495	0·500	0·509	0·516	0·522	0·533	0·541	0·549	0·561	0·571	0·582
S	$(\Delta p/\rho l)$	6	8	10	12·5	15	20	25	30	40	50	60	80	100	125

Grad'nt   k.p.g.   (Equivalent) Pipe diameters in mm

S	$(\Delta p/\rho l)$	125	150	200	250	300	400	500	600	800	1000	1250	1500	2000	2500
0·00010	(0·00098)	0·854	0·853	0·854	0·855	0·858	0·862	0·867	0·871	0·880	0·887	0·895	0·903	0·915	0·926
0·00015	(0·00147)	0·835	0·835	0·836	0·838	0·841	0·846	0·851	0·856	0·865	0·873	0·881	0·889	0·902	0·913
0·00020	(0·00196)	0·822	0·822	0·824	0·826	0·829	0·835	0·840	0·846	0·855	0·863	0·872	0·879	0·893	0·904
0·00030	(0·00294)	0·804	0·804	0·807	0·810	0·814	0·820	0·826	0·831	0·841	0·849	0·859	0·867	0·880	0·892
0·00040	(0·00392)	0·792	0·793	0·796	0·799	0·803	0·810	0·816	0·822	0·832	0·840	0·850	0·858	0·872	0·884
0·00060	(0·00588)	0·775	0·777	0·781	0·785	0·789	0·796	0·803	0·809	0·819	0·828	0·838	0·846	0·860	0·872
0·00080	(0·00785)	0·764	0·766	0·770	0·775	0·779	0·787	0·793	0·800	0·810	0·820	0·829	0·838	0·853	0·865
0·00100	(0·00981)	0·756	0·758	0·762	0·767	0·771	0·779	0·787	0·793	0·804	0·813	0·823	0·832	0·847	0·859
0·00150	(0·0147)	0·741	0·743	0·749	0·754	0·759	0·767	0·774	0·781	0·793	0·802	0·813	0·822	0·837	0·849
0·00200	(0·0196)	0·731	0·734	0·740	0·745	0·750	0·759	0·766	0·773	0·785	0·795	0·805	0·815	0·830	0·843
0·00300	(0·0294)	0·717	0·721	0·727	0·733	0·738	0·747	0·755	0·762	0·775	0·785	0·796	0·805	0·821	0·834
0·00400	(0·0392)	0·708	0·712	0·719	0·725	0·730	0·740	0·748	0·755	0·768	0·778	0·789	0·799	0·815	0·829
0·00600	(0·0588)	0·696	0·700	0·707	0·714	0·719	0·730	0·738	0·746	0·759	0·769	0·781	0·791	0·807	0·821
0·00800	(0·0785)	0·688	0·692	0·700	0·706	0·712	0·723	0·732	0·739	0·753	0·764	0·775	0·785	0·802	0·816
0·01000	(0·0981)	0·682	0·686	0·694	0·701	0·707	0·718	0·727	0·735	0·748	0·759	0·771	0·781	0·798	0·813
0·01500	(0·147)	0·671	0·676	0·684	0·691	0·698	0·709	0·718	0·726	0·740	0·752	0·764	0·775	0·792	0·807
0·02000	(0·196)	0·664	0·669	0·677	0·685	0·692	0·703	0·713	0·721	0·735	0·747	0·759	0·770	0·788	0·803
0·03000	(0·294)	0·654	0·659	0·669	0·677	0·684	0·695	0·705	0·714	0·729	0·741	0·754	0·765	0·783	0·798
0·04000	(0·392)	0·648	0·653	0·663	0·671	0·678	0·690	0·701	0·710	0·724	0·737	0·750	0·761	0·779	0·794
0·06000	(0·588)	0·639	0·645	0·655	0·664	0·671	0·684	0·695	0·704	0·719	0·732	0·745	0·756	0·775	0·791
0·08000	(0·785)	0·634	0·640	0·650	0·659	0·667	0·680	0·691	0·700	0·715	0·728	0·742	0·753	0·772	0·788
0·10000	(0·981)	0·630	0·636	0·647	0·656	0·663	0·677	0·688	0·697	0·713	0·726	0·740	0·751	0·771	0·786
0·15000	(1·47)	0·623	0·629	0·641	0·650	0·658	0·672	0·683	0·693	0·709	0·722	0·736	0·748	0·767	0·783
0·20000	(1·96)	0·618	0·625	0·637	0·646	0·654	0·668	0·680	0·690	0·706	0·720	0·734	0·746	0·766	0·782
0·30000	(2·94)	0·613	0·620	0·632	0·641	0·650	0·664	0·676	0·686	0·703	0·717	0·731	0·743	0·763	0·779
0·40000	(3·92)	0·609	0·616	0·628	0·638	0·647	0·662	0·674	0·684	0·701	0·715	0·729	0·741	0·762	0·778
0·60000	(5·88)	0·604	0·612	0·624	0·635	0·644	0·659	0·671	0·681	0·698	0·712	0·727	0·739	0·760	0·776
0·80000	(7·85)	0·601	0·609	0·622	0·632	0·641	0·657	0·669	0·679	0·697	0·711	0·726	0·738	0·759	0·775
1·00000	(9·81)	0·599	0·607	0·620	0·631	0·640	0·655	0·668	0·678	0·696	0·710	0·725	0·737	0·758	0·775
1·50000	(14·7)	0·596	0·604	0·617	0·628	0·637	0·653	0·666	0·676	0·694	0·708	0·723	0·736	0·757	0·774
2·00000	(19·6)	0·594	0·602	0·615	0·626	0·636	0·651	0·664	0·675	0·693	0·707	0·723	0·735	0·756	0·773
3·00000	(29·4)	0·591	0·599	0·613	0·624	0·634	0·650	0·663	0·674	0·692	0·706	0·721	0·734	0·755	0·772
S	$(\Delta p/\rho l)$	125	150	200	250	300	400	500	600	800	1000	1250	1500	2000	2500

Grad'nt   k.p.g.   (Equivalent) Pipe diameters in mm

Kinematic viscosity, $\nu = 0.15 \times 10^{-6}$ m^{2}s^{-1};

Roughness size, $k_s = 0.0015$ mm

Kin. visc., $\nu = 0.15\times10^{-6}\ \mathrm{m^2 s^{-1}}$;
$S = 0.00010$ to 30.0000

i.e. kin. pr. grad., $\Delta p/\rho l =$
(0.00098) to $(294)\ \mathrm{ms^{-2}}$

Roughness size, $k_s = 0.003$ mm
This table shows values of m, as follows

m_C for Colebrook-White solutions; or,
where $\mathbf{R} \le 2000$, m_P for laminar flow

Grad'nt S	k.p.g. $(\Delta p/\rho l)$	6	8	10	12.5	15	20	25	30	40	50	60	80	100	125
		(Equivalent) Pipe diameters in mm													
0.00030	(0.00294)	1.029	*	*	0.882	0.868	0.849	0.837	0.829	0.819	0.814	0.810	0.807	0.806	0.806
0.00040	(0.00392)	0.891	*	0.880	0.861	0.847	0.830	0.820	0.813	0.804	0.799	0.797	0.794	0.794	0.794
0.00060	(0.00588)	0.727	*	0.849	0.832	0.821	0.806	0.797	0.791	0.784	0.780	0.778	0.777	0.777	0.778
0.00080	(0.00785)	*	0.847	0.828	0.813	0.803	0.789	0.781	0.776	0.770	0.767	0.766	0.765	0.766	0.767
0.00100	(0.00981)	*	0.830	0.813	0.799	0.789	0.777	0.770	0.765	0.760	0.757	0.756	0.756	0.757	0.759
0.00150	(0.0147)	0.826	0.802	0.787	0.775	0.766	0.756	0.750	0.746	0.742	0.740	0.740	0.741	0.743	0.745
0.00200	(0.0196)	0.804	0.783	0.769	0.758	0.751	0.742	0.736	0.733	0.730	0.729	0.729	0.730	0.733	0.736
0.00300	(0.0294)	0.777	0.758	0.746	0.737	0.730	0.723	0.719	0.716	0.714	0.714	0.714	0.717	0.719	0.723
0.00400	(0.0392)	0.758	0.741	0.731	0.722	0.717	0.710	0.707	0.705	0.703	0.704	0.705	0.707	0.711	0.715
0.00600	(0.0588)	0.734	0.719	0.710	0.703	0.698	0.693	0.691	0.689	0.689	0.690	0.691	0.695	0.699	0.703
0.00800	(0.0785)	0.718	0.704	0.696	0.690	0.686	0.682	0.680	0.679	0.679	0.681	0.683	0.687	0.691	0.696
0.01000	(0.0981)	0.706	0.693	0.686	0.680	0.677	0.673	0.672	0.671	0.672	0.674	0.676	0.681	0.685	0.690
0.01500	(0.147)	0.685	0.675	0.669	0.664	0.661	0.659	0.658	0.658	0.660	0.662	0.665	0.670	0.675	0.681
0.02000	(0.196)	0.671	0.662	0.657	0.653	0.651	0.649	0.649	0.649	0.652	0.654	0.657	0.663	0.668	0.674
0.03000	(0.294)	0.653	0.646	0.641	0.638	0.637	0.636	0.637	0.638	0.641	0.644	0.648	0.654	0.660	0.666
0.04000	(0.392)	0.641	0.635	0.631	0.629	0.628	0.628	0.629	0.630	0.634	0.637	0.641	0.648	0.654	0.661
0.06000	(0.588)	0.625	0.620	0.617	0.616	0.616	0.616	0.618	0.620	0.624	0.629	0.633	0.640	0.647	0.654
0.08000	(0.785)	0.615	0.610	0.608	0.607	0.608	0.609	0.611	0.613	0.618	0.623	0.627	0.635	0.642	0.650
0.10000	(0.981)	0.607	0.603	0.602	0.601	0.602	0.603	0.606	0.609	0.614	0.619	0.623	0.631	0.639	0.646
0.15000	(1.47)	0.594	0.591	0.590	0.591	0.592	0.594	0.597	0.600	0.606	0.612	0.617	0.625	0.633	0.641
0.20000	(1.96)	0.585	0.583	0.583	0.584	0.585	0.588	0.592	0.595	0.601	0.607	0.612	0.621	0.629	0.638
0.30000	(2.94)	0.574	0.573	0.573	0.575	0.577	0.581	0.585	0.588	0.595	0.601	0.607	0.617	0.625	0.634
0.40000	(3.92)	0.566	0.566	0.567	0.569	0.571	0.576	0.580	0.584	0.591	0.598	0.603	0.613	0.622	0.631
0.60000	(5.88)	0.557	0.557	0.559	0.562	0.564	0.569	0.574	0.578	0.586	0.593	0.599	0.610	0.618	0.628
0.80000	(7.85)	0.550	0.552	0.554	0.557	0.560	0.565	0.570	0.575	0.583	0.590	0.596	0.607	0.616	0.626
1.00000	(9.81)	0.546	0.548	0.550	0.553	0.556	0.562	0.568	0.572	0.581	0.588	0.595	0.605	0.614	0.624
1.50000	(14.7)	0.539	0.541	0.544	0.548	0.551	0.557	0.563	0.568	0.577	0.585	0.591	0.603	0.612	0.622
2.00000	(19.6)	0.534	0.537	0.540	0.544	0.548	0.554	0.560	0.566	0.575	0.583	0.590	0.601	0.610	0.620
3.00000	(29.4)	0.528	0.531	0.535	0.539	0.544	0.551	0.557	0.563	0.572	0.580	0.587	0.599	0.609	0.619
4.00000	(39.2)	0.524	0.528	0.532	0.537	0.541	0.548	0.555	0.561	0.570	0.579	0.586	0.598	0.607	0.618
6.00000	(58.8)	0.519	0.524	0.528	0.533	0.538	0.546	0.552	0.558	0.568	0.577	0.584	0.596	0.606	0.616
8.00000	(78.5)	0.516	0.521	0.526	0.531	0.536	0.544	0.551	0.557	0.567	0.576	0.583	0.595	0.605	0.616
10.0000	(98.1)	0.514	0.519	0.524	0.530	0.534	0.543	0.550	0.556	0.566	0.575	0.582	0.595	0.605	0.615
15.0000	(147)	0.511	0.516	0.521	0.527	0.532	0.541	0.548	0.554	0.565	0.574	0.581	0.593	0.604	0.614
20.0000	(196)	0.509	0.515	0.520	0.526	0.531	0.540	0.547	0.553	0.564	0.573	0.580	0.593	0.603	0.614
30.0000	(294)	0.506	0.512	0.518	0.524	0.529	0.538	0.546	0.552	0.563	0.572	0.579	0.592	0.602	0.613
S	$(\Delta p/\rho l)$	6	8	10	12.5	15	20	25	30	40	50	60	80	100	125

Grad'nt S	k.p.g. $(\Delta p/\rho l)$	125	150	200	250	300	400	500	600	800	1000	1250	1500	2000	2500
		(Equivalent) Pipe diameters in mm													
0.00010	(0.00098)	0.856	0.855	0.856	0.857	0.860	0.865	0.870	0.874	0.883	0.891	0.899	0.907	0.920	0.931
0.00015	(0.00147)	0.837	0.837	0.838	0.841	0.843	0.849	0.854	0.859	0.869	0.877	0.886	0.894	0.907	0.919
0.00020	(0.00196)	0.824	0.824	0.826	0.829	0.832	0.838	0.844	0.849	0.859	0.867	0.876	0.885	0.899	0.910
0.00030	(0.00294)	0.806	0.807	0.810	0.813	0.817	0.824	0.830	0.836	0.846	0.855	0.864	0.873	0.887	0.899
0.00040	(0.00392)	0.794	0.796	0.799	0.803	0.807	0.814	0.820	0.826	0.837	0.846	0.856	0.865	0.879	0.892
0.00060	(0.00588)	0.778	0.780	0.784	0.789	0.793	0.801	0.808	0.814	0.825	0.835	0.845	0.854	0.869	0.882
0.00080	(0.00785)	0.767	0.770	0.774	0.779	0.784	0.792	0.799	0.806	0.817	0.827	0.837	0.847	0.862	0.875
0.00100	(0.00981)	0.759	0.762	0.767	0.772	0.777	0.785	0.793	0.799	0.811	0.821	0.832	0.841	0.857	0.870
0.00150	(0.0147)	0.745	0.748	0.754	0.760	0.765	0.774	0.782	0.789	0.801	0.811	0.822	0.832	0.848	0.862
0.00200	(0.0196)	0.736	0.739	0.745	0.751	0.757	0.766	0.774	0.782	0.794	0.805	0.816	0.826	0.843	0.856
0.00300	(0.0294)	0.723	0.727	0.734	0.740	0.746	0.756	0.764	0.772	0.785	0.796	0.808	0.818	0.835	0.849
0.00400	(0.0392)	0.715	0.719	0.726	0.733	0.739	0.749	0.758	0.766	0.779	0.791	0.803	0.813	0.830	0.845
0.00600	(0.0588)	0.703	0.708	0.716	0.723	0.729	0.740	0.749	0.758	0.772	0.783	0.796	0.806	0.824	0.839
0.00800	(0.0785)	0.696	0.700	0.709	0.716	0.723	0.734	0.744	0.752	0.767	0.779	0.791	0.802	0.820	0.835
0.01000	(0.0981)	0.690	0.695	0.704	0.711	0.718	0.730	0.740	0.748	0.763	0.775	0.788	0.799	0.817	0.833
0.01500	(0.147)	0.681	0.686	0.695	0.703	0.710	0.723	0.733	0.742	0.757	0.769	0.783	0.794	0.813	0.828
0.02000	(0.196)	0.674	0.680	0.690	0.698	0.705	0.718	0.729	0.738	0.753	0.766	0.779	0.791	0.810	0.826
0.03000	(0.294)	0.666	0.672	0.682	0.691	0.699	0.712	0.723	0.732	0.748	0.761	0.775	0.787	0.806	0.822
0.04000	(0.392)	0.661	0.667	0.678	0.687	0.695	0.708	0.719	0.729	0.745	0.758	0.772	0.784	0.804	0.820
0.06000	(0.588)	0.654	0.661	0.672	0.681	0.689	0.703	0.715	0.725	0.741	0.755	0.769	0.781	0.801	0.817
0.08000	(0.785)	0.650	0.656	0.668	0.678	0.686	0.700	0.712	0.722	0.739	0.752	0.767	0.779	0.799	0.816
0.10000	(0.981)	0.646	0.653	0.665	0.675	0.683	0.698	0.710	0.720	0.737	0.751	0.765	0.778	0.798	0.815
0.15000	(1.47)	0.641	0.648	0.661	0.671	0.680	0.694	0.706	0.717	0.734	0.748	0.763	0.776	0.796	0.813
0.20000	(1.96)	0.638	0.645	0.658	0.668	0.677	0.692	0.704	0.715	0.732	0.747	0.762	0.774	0.795	0.812
0.30000	(2.94)	0.634	0.641	0.654	0.665	0.674	0.689	0.702	0.713	0.730	0.745	0.760	0.773	0.793	0.810
0.40000	(3.92)	0.631	0.639	0.652	0.663	0.672	0.688	0.700	0.711	0.729	0.744	0.759	0.772	0.793	0.810
0.60000	(5.88)	0.628	0.636	0.649	0.660	0.670	0.686	0.698	0.709	0.727	0.742	0.757	0.770	0.792	0.809
0.80000	(7.85)	0.626	0.634	0.647	0.659	0.668	0.684	0.697	0.708	0.727	0.741	0.757	0.770	0.791	0.808
1.00000	(9.81)	0.624	0.632	0.646	0.658	0.667	0.683	0.696	0.708	0.726	0.741	0.756	0.769	0.790	0.808
1.50000	(14.7)	0.622	0.630	0.644	0.656	0.666	0.682	0.695	0.706	0.725	0.740	0.755	0.768	0.790	0.807
2.00000	(19.6)	0.620	0.629	0.643	0.655	0.665	0.681	0.694	0.706	0.724	0.739	0.755	0.768	0.789	0.807
3.00000	(29.4)	0.619	0.627	0.642	0.654	0.664	0.680	0.694	0.705	0.723	0.739	0.754	0.767	0.789	0.806
S	$(\Delta p/\rho l)$	125	150	200	250	300	400	500	600	800	1000	1250	1500	2000	2500

Grad'nt   k.p.g.   (Equivalent) Pipe diameters in mm

Kinematic viscosity, $\nu = 0.15\times10^{-6}\ \mathrm{m^2 s^{-1}}$;   **Roughness size, $k_s = 0.003$ mm**

E11

Kin. visc., $\nu = 0.15\times10^{-6}$ m^{2}s^{-1};   **Roughness size, k_s = 0.006 mm**

S = 0.00010 to 30.0000   This table shows values of m, as follows

i.e. kin. pr. grad., $\Delta p/\rho l$ = (0.00098) to (294) ms^{-2}

m_C for Colebrook-White solutions; or, where $\mathbf{R} \leq 2000$, m_P for laminar flow

Grad'nt S	k.p.g. $(\Delta p/\rho l)$	6	8	10	12.5	15	20	25	30	40	50	60	80	100	125
0.00030	(0.00294)	1.029	*	*	0.885	0.871	0.852	0.840	0.832	0.823	0.817	0.814	0.811	0.811	0.811
0.00040	(0.00392)	0.891	*	0.883	0.864	0.851	0.834	0.823	0.816	0.808	0.803	0.801	0.799	0.799	0.800
0.00060	(0.00588)	0.727	*	0.852	0.836	0.824	0.810	0.801	0.795	0.788	0.785	0.783	0.782	0.783	0.784
0.00080	(0.00785)	*	0.850	0.832	0.817	0.807	0.794	0.786	0.781	0.775	0.772	0.771	0.771	0.772	0.774
0.00100	(0.00981)	*	0.834	0.817	0.803	0.794	0.782	0.775	0.770	0.765	0.763	0.762	0.763	0.764	0.767
0.00150	(0.0147)	0.830	0.806	0.791	0.779	0.771	0.761	0.755	0.752	0.748	0.747	0.747	0.748	0.750	0.754
0.00200	(0.0196)	0.809	0.788	0.774	0.764	0.756	0.747	0.742	0.740	0.737	0.736	0.737	0.739	0.741	0.745
0.00300	(0.0294)	0.782	0.763	0.752	0.743	0.737	0.729	0.725	0.723	0.722	0.722	0.723	0.726	0.729	0.734
0.00400	(0.0392)	0.764	0.747	0.737	0.729	0.723	0.717	0.714	0.713	0.712	0.713	0.714	0.718	0.721	0.726
0.00600	(0.0588)	0.740	0.726	0.717	0.710	0.706	0.701	0.699	0.698	0.699	0.700	0.702	0.706	0.711	0.716
0.00800	(0.0785)	0.725	0.712	0.704	0.698	0.694	0.691	0.689	0.689	0.690	0.692	0.694	0.699	0.704	0.710
0.01000	(0.0981)	0.713	0.701	0.694	0.689	0.686	0.683	0.682	0.682	0.683	0.686	0.689	0.694	0.699	0.705
0.01500	(0.147)	0.694	0.684	0.678	0.674	0.671	0.670	0.669	0.670	0.673	0.676	0.679	0.685	0.691	0.697
0.02000	(0.196)	0.681	0.672	0.667	0.664	0.662	0.661	0.661	0.662	0.666	0.669	0.673	0.679	0.685	0.692
0.03000	(0.294)	0.664	0.657	0.653	0.650	0.649	0.649	0.651	0.652	0.656	0.660	0.664	0.672	0.678	0.686
0.04000	(0.392)	0.653	0.647	0.644	0.642	0.641	0.642	0.644	0.646	0.650	0.655	0.659	0.667	0.674	0.682
0.06000	(0.588)	0.638	0.634	0.631	0.631	0.631	0.632	0.635	0.637	0.643	0.648	0.653	0.661	0.668	0.677
0.08000	(0.785)	0.629	0.625	0.624	0.623	0.624	0.626	0.629	0.632	0.638	0.643	0.648	0.657	0.665	0.673
0.10000	(0.981)	0.622	0.619	0.618	0.618	0.619	0.622	0.625	0.628	0.634	0.640	0.645	0.655	0.662	0.671
0.15000	(1.47)	0.610	0.608	0.608	0.609	0.611	0.614	0.618	0.622	0.629	0.635	0.640	0.650	0.658	0.667
0.20000	(1.96)	0.603	0.602	0.602	0.604	0.605	0.610	0.614	0.618	0.625	0.631	0.637	0.647	0.656	0.665
0.30000	(2.94)	0.593	0.593	0.594	0.596	0.599	0.604	0.608	0.613	0.620	0.627	0.633	0.644	0.653	0.662
0.40000	(3.92)	0.587	0.588	0.589	0.592	0.595	0.600	0.605	0.609	0.618	0.625	0.631	0.642	0.651	0.661
0.60000	(5.88)	0.579	0.581	0.583	0.586	0.589	0.595	0.601	0.605	0.614	0.622	0.628	0.639	0.649	0.658
0.80000	(7.85)	0.575	0.577	0.579	0.583	0.586	0.592	0.598	0.603	0.612	0.620	0.626	0.638	0.647	0.657
1.00000	(9.81)	0.571	0.574	0.576	0.580	0.584	0.590	0.596	0.601	0.610	0.618	0.625	0.637	0.646	0.656
1.50000	(14.7)	0.565	0.569	0.572	0.576	0.580	0.587	0.593	0.599	0.608	0.616	0.623	0.635	0.645	0.655
2.00000	(19.6)	0.562	0.565	0.569	0.574	0.578	0.585	0.591	0.597	0.607	0.615	0.622	0.634	0.644	0.654
3.00000	(29.4)	0.558	0.562	0.566	0.570	0.575	0.582	0.589	0.595	0.605	0.613	0.620	0.632	0.642	0.653
4.00000	(39.2)	0.555	0.559	0.564	0.569	0.573	0.581	0.588	0.594	0.604	0.612	0.620	0.632	0.642	0.652
6.00000	(58.8)	0.551	0.556	0.561	0.566	0.571	0.579	0.586	0.592	0.602	0.611	0.618	0.631	0.641	0.652
8.00000	(78.5)	0.549	0.555	0.559	0.565	0.570	0.578	0.585	0.591	0.602	0.610	0.618	0.630	0.640	0.651
10.0000	(98.1)	0.548	0.553	0.558	0.564	0.569	0.577	0.584	0.591	0.601	0.610	0.617	0.630	0.640	0.651
15.0000	(147)	0.546	0.552	0.557	0.562	0.567	0.576	0.583	0.590	0.600	0.609	0.617	0.629	0.640	0.650
20.0000	(196)	0.544	0.550	0.556	0.561	0.567	0.575	0.583	0.589	0.600	0.609	0.616	0.629	0.639	0.650
30.0000	(294)	0.543	0.549	0.554	0.560	0.566	0.574	0.582	0.588	0.599	0.608	0.616	0.628	0.639	0.650
S	$(\Delta p/\rho l)$	6	8	10	12.5	15	20	25	30	40	50	60	80	100	125

Grad'nt   k.p.g.   (Equivalent) Pipe diameters in mm

Grad'nt S	k.p.g. $(\Delta p/\rho l)$	125	150	200	250	300	400	500	600	800	1000	1250	1500	2000	2500
0.00010	(0.00098)	0.859	0.859	0.859	0.861	0.864	0.869	0.875	0.880	0.889	0.897	0.906	0.914	0.928	0.940
0.00015	(0.00147)	0.840	0.841	0.842	0.845	0.848	0.854	0.860	0.866	0.875	0.884	0.894	0.902	0.917	0.929
0.00020	(0.00196)	0.828	0.828	0.831	0.834	0.838	0.844	0.850	0.856	0.866	0.876	0.885	0.894	0.909	0.922
0.00030	(0.00294)	0.811	0.812	0.816	0.819	0.823	0.831	0.837	0.844	0.855	0.864	0.874	0.883	0.899	0.912
0.00040	(0.00392)	0.800	0.801	0.805	0.810	0.814	0.822	0.829	0.835	0.847	0.856	0.867	0.876	0.892	0.906
0.00060	(0.00588)	0.784	0.787	0.791	0.796	0.801	0.810	0.817	0.824	0.836	0.846	0.858	0.867	0.884	0.897
0.00080	(0.00785)	0.774	0.777	0.782	0.788	0.793	0.802	0.810	0.817	0.829	0.840	0.851	0.861	0.878	0.892
0.00100	(0.00981)	0.767	0.770	0.775	0.781	0.786	0.796	0.804	0.811	0.824	0.835	0.847	0.857	0.874	0.888
0.00150	(0.0147)	0.754	0.757	0.764	0.770	0.776	0.786	0.794	0.802	0.816	0.827	0.839	0.849	0.867	0.882
0.00200	(0.0196)	0.745	0.749	0.756	0.762	0.768	0.779	0.788	0.796	0.810	0.822	0.834	0.845	0.863	0.878
0.00300	(0.0294)	0.734	0.738	0.746	0.753	0.759	0.770	0.780	0.788	0.803	0.815	0.828	0.839	0.857	0.872
0.00400	(0.0392)	0.726	0.731	0.739	0.746	0.753	0.765	0.775	0.783	0.798	0.810	0.824	0.835	0.854	0.869
0.00600	(0.0588)	0.716	0.721	0.730	0.738	0.745	0.758	0.768	0.777	0.792	0.805	0.818	0.830	0.849	0.865
0.00800	(0.0785)	0.710	0.715	0.725	0.733	0.740	0.753	0.764	0.773	0.788	0.801	0.815	0.827	0.846	0.862
0.01000	(0.0981)	0.705	0.711	0.721	0.729	0.737	0.750	0.760	0.770	0.786	0.799	0.813	0.825	0.845	0.861
0.01500	(0.147)	0.697	0.703	0.714	0.723	0.731	0.744	0.755	0.765	0.781	0.795	0.809	0.821	0.841	0.858
0.02000	(0.196)	0.692	0.699	0.709	0.719	0.727	0.741	0.752	0.762	0.779	0.792	0.807	0.819	0.839	0.856
0.03000	(0.294)	0.686	0.693	0.704	0.714	0.722	0.736	0.748	0.758	0.775	0.789	0.804	0.816	0.837	0.854
0.04000	(0.392)	0.682	0.689	0.700	0.710	0.719	0.734	0.746	0.756	0.773	0.787	0.802	0.815	0.836	0.853
0.06000	(0.588)	0.677	0.684	0.696	0.706	0.715	0.730	0.743	0.753	0.771	0.785	0.800	0.813	0.834	0.851
0.08000	(0.785)	0.673	0.681	0.693	0.704	0.713	0.728	0.741	0.751	0.769	0.784	0.799	0.812	0.833	0.850
0.10000	(0.981)	0.671	0.679	0.691	0.702	0.711	0.727	0.739	0.750	0.768	0.783	0.798	0.811	0.832	0.849
0.15000	(1.47)	0.667	0.675	0.688	0.699	0.709	0.724	0.737	0.748	0.766	0.781	0.796	0.810	0.831	0.848
0.20000	(1.96)	0.665	0.673	0.687	0.698	0.707	0.723	0.736	0.747	0.765	0.780	0.796	0.809	0.830	0.848
0.30000	(2.94)	0.662	0.670	0.684	0.696	0.705	0.721	0.734	0.746	0.764	0.779	0.795	0.808	0.829	0.847
0.40000	(3.92)	0.661	0.669	0.683	0.694	0.704	0.720	0.733	0.745	0.763	0.778	0.794	0.807	0.829	0.846
0.60000	(5.88)	0.658	0.667	0.681	0.693	0.703	0.719	0.732	0.744	0.762	0.777	0.793	0.806	0.828	0.846
0.80000	(7.85)	0.657	0.666	0.680	0.692	0.702	0.718	0.732	0.743	0.762	0.777	0.793	0.806	0.828	0.846
1.00000	(9.81)	0.656	0.665	0.679	0.691	0.701	0.718	0.731	0.743	0.761	0.777	0.792	0.806	0.828	0.845
1.50000	(14.7)	0.655	0.664	0.678	0.690	0.700	0.717	0.730	0.742	0.761	0.776	0.792	0.805	0.827	0.845
2.00000	(19.6)	0.654	0.663	0.677	0.689	0.700	0.716	0.730	0.742	0.760	0.776	0.792	0.805	0.827	0.845
3.00000	(29.4)	0.653	0.662	0.677	0.689	0.699	0.716	0.729	0.741	0.760	0.775	0.791	0.805	0.827	0.844
S	$(\Delta p/\rho l)$	125	150	200	250	300	400	500	600	800	1000	1250	1500	2000	2500

Grad'nt   k.p.g.   (Equivalent) Pipe diameters in mm

Kinematic viscosity, $\nu = 0.15\times10^{-6}$ m^{2}s^{-1} ;   **Roughness size, k_s = 0.006 mm**

Kin. visc., $\nu = 0.15 \times 10^{-6}$ m^{2}s^{-1};
$S = 0.00010$ to 30.0000

i.e. kin. pr. grad., $\Delta p/\rho l =$
(0.00098) to (294) ms^{-2}

Roughness size, $k_s = 0.015$ mm
This table shows values of m, as follows

m_C for Colebrook-White solutions; or,
where $\mathbf{R} \leq 2000$, m_P for laminar flow

Grad'nt S	k.p.g. $(\Delta p/\rho l)$	6	8	10	12·5	15	20	25	30	40	50	60	80	100	125
0·00030	(0·00294)	1·029	*	*	0·893	0·879	0·860	0·849	0·842	0·833	0·828	0·825	0·823	0·823	0·824
0·00040	(0·00392)	0·891	*	0·892	0·873	0·860	0·843	0·833	0·827	0·819	0·815	0·813	0·812	0·812	0·814
0·00060	(0·00588)	0·727	*	0·862	0·846	0·835	0·821	0·812	0·807	0·801	0·798	0·797	0·797	0·798	0·801
0·00080	(0·00785)	*	0·861	0·843	0·828	0·818	0·806	0·798	0·794	0·789	0·787	0·786	0·787	0·789	0·792
0·00100	(0·00981)	*	0·845	0·829	0·815	0·806	0·794	0·788	0·784	0·780	0·779	0·778	0·780	0·782	0·786
0·00150	(0·0147)	0·842	0·819	0·804	0·793	0·785	0·776	0·770	0·768	0·765	0·764	0·765	0·768	0·771	0·775
0·00200	(0·0196)	0·823	0·801	0·789	0·778	0·771	0·763	0·759	0·757	0·755	0·755	0·756	0·760	0·764	0·768
0·00300	(0·0294)	0·797	0·779	0·768	0·759	0·753	0·747	0·744	0·743	0·742	0·743	0·745	0·750	0·754	0·760
0·00400	(0·0392)	0·780	0·764	0·754	0·747	0·742	0·737	0·734	0·733	0·734	0·736	0·738	0·743	0·748	0·754
0·00600	(0·0588)	0·759	0·745	0·737	0·730	0·727	0·723	0·722	0·722	0·723	0·726	0·729	0·734	0·740	0·747
0·00800	(0·0785)	0·744	0·732	0·725	0·720	0·717	0·714	0·714	0·714	0·716	0·719	0·723	0·729	0·735	0·742
0·01000	(0·0981)	0·734	0·723	0·717	0·712	0·710	0·708	0·708	0·708	0·711	0·715	0·718	0·725	0·732	0·739
0·01500	(0·147)	0·717	0·708	0·703	0·699	0·698	0·697	0·698	0·699	0·703	0·707	0·711	0·719	0·726	0·734
0·02000	(0·196)	0·706	0·698	0·694	0·691	0·690	0·690	0·692	0·694	0·698	0·703	0·707	0·715	0·722	0·730
0·03000	(0·294)	0·692	0·685	0·682	0·681	0·680	0·682	0·684	0·686	0·691	0·697	0·701	0·710	0·718	0·726
0·04000	(0·392)	0·682	0·677	0·675	0·674	0·674	0·676	0·679	0·682	0·687	0·693	0·698	0·707	0·715	0·724
0·06000	(0·588)	0·671	0·667	0·666	0·666	0·666	0·669	0·672	0·676	0·682	0·688	0·694	0·703	0·712	0·721
0·08000	(0·785)	0·663	0·660	0·660	0·660	0·662	0·665	0·669	0·672	0·679	0·685	0·691	0·701	0·710	0·719
0·10000	(0·981)	0·658	0·656	0·656	0·657	0·658	0·662	0·666	0·670	0·677	0·683	0·689	0·699	0·708	0·718
0·15000	(1·47)	0·649	0·648	0·649	0·650	0·652	0·657	0·661	0·666	0·673	0·680	0·686	0·697	0·706	0·715
0·20000	(1·96)	0·644	0·644	0·645	0·647	0·649	0·654	0·659	0·663	0·671	0·678	0·685	0·695	0·704	0·714
0·30000	(2·94)	0·637	0·638	0·639	0·642	0·645	0·650	0·655	0·660	0·669	0·676	0·682	0·693	0·703	0·713
0·40000	(3·92)	0·633	0·634	0·636	0·639	0·642	0·648	0·653	0·658	0·667	0·674	0·681	0·692	0·702	0·712
0·60000	(5·88)	0·628	0·630	0·632	0·636	0·639	0·645	0·651	0·656	0·665	0·673	0·679	0·691	0·701	0·711
0·80000	(7·85)	0·625	0·627	0·630	0·633	0·637	0·644	0·649	0·655	0·664	0·672	0·679	0·690	0·700	0·710
1·00000	(9·81)	0·623	0·625	0·628	0·632	0·636	0·642	0·648	0·654	0·663	0·671	0·678	0·690	0·699	0·710
1·50000	(14·7)	0·619	0·622	0·626	0·630	0·633	0·640	0·647	0·652	0·662	0·670	0·677	0·689	0·698	0·709
2·00000	(19·6)	0·617	0·620	0·624	0·628	0·632	0·639	0·646	0·651	0·661	0·669	0·676	0·688	0·698	0·708
3·00000	(29·4)	0·614	0·618	0·622	0·626	0·631	0·638	0·645	0·650	0·660	0·668	0·675	0·687	0·697	0·708
4·00000	(39·2)	0·613	0·617	0·621	0·625	0·630	0·637	0·644	0·650	0·659	0·668	0·675	0·687	0·697	0·708
6·00000	(58·8)	0·611	0·615	0·619	0·624	0·629	0·636	0·643	0·649	0·659	0·667	0·674	0·687	0·697	0·707
8·00000	(78·5)	0·610	0·614	0·619	0·623	0·628	0·636	0·642	0·648	0·658	0·667	0·674	0·686	0·696	0·707
10·0000	(98·1)	0·609	0·614	0·618	0·623	0·627	0·635	0·642	0·648	0·658	0·667	0·674	0·686	0·696	0·707
15·0000	(147)	0·608	0·613	0·617	0·622	0·627	0·635	0·642	0·648	0·658	0·666	0·674	0·686	0·696	0·707
20·0000	(196)	0·607	0·612	0·616	0·622	0·626	0·634	0·641	0·647	0·657	0·666	0·673	0·686	0·696	0·706
30·0000	(294)	0·606	0·611	0·616	0·621	0·626	0·634	0·641	0·647	0·657	0·666	0·673	0·685	0·696	0·706
S	$(\Delta p/\rho l)$	6	8	10	12·5	15	20	25	30	40	50	60	80	100	125

Grad'nt S	k.p.g. $(\Delta p/\rho l)$	125	150	200	250	300	400	500	600	800	1000	1250	1500	2000	2500
0·00010	(0·00098)	0·868	0·868	0·870	0·873	0·876	0·882	0·888	0·894	0·905	0·914	0·924	0·933	0·949	0·962
0·00015	(0·00147)	0·851	0·852	0·855	0·858	0·862	0·869	0·876	0·882	0·894	0·903	0·914	0·924	0·940	0·954
0·00020	(0·00196)	0·840	0·841	0·844	0·848	0·852	0·860	0·868	0·874	0·886	0·897	0·908	0·918	0·934	0·948
0·00030	(0·00294)	0·824	0·826	0·831	0·836	0·840	0·849	0·857	0·864	0·877	0·888	0·899	0·910	0·927	0·942
0·00040	(0·00392)	0·814	0·817	0·822	0·827	0·832	0·842	0·850	0·857	0·871	0·882	0·894	0·904	0·922	0·937
0·00060	(0·00588)	0·801	0·804	0·810	0·816	0·822	0·832	0·841	0·849	0·863	0·874	0·887	0·898	0·916	0·932
0·00080	(0·00785)	0·792	0·796	0·803	0·809	0·815	0·826	0·835	0·843	0·858	0·870	0·883	0·894	0·913	0·928
0·00100	(0·00981)	0·786	0·790	0·797	0·804	0·810	0·821	0·831	0·839	0·854	0·866	0·879	0·891	0·910	0·926
0·00150	(0·0147)	0·775	0·780	0·788	0·795	0·802	0·814	0·824	0·833	0·848	0·861	0·874	0·886	0·906	0·922
0·00200	(0·0196)	0·768	0·773	0·782	0·790	0·797	0·809	0·819	0·829	0·844	0·857	0·871	0·883	0·903	0·919
0·00300	(0·0294)	0·760	0·765	0·774	0·783	0·790	0·803	0·814	0·824	0·840	0·853	0·867	0·879	0·900	0·916
0·00400	(0·0392)	0·754	0·759	0·769	0·778	0·786	0·799	0·811	0·820	0·837	0·850	0·865	0·877	0·898	0·914
0·00600	(0·0588)	0·747	0·753	0·763	0·773	0·781	0·795	0·806	0·816	0·833	0·847	0·862	0·874	0·895	0·912
0·00800	(0·0785)	0·742	0·748	0·760	0·769	0·777	0·792	0·804	0·814	0·831	0·845	0·860	0·873	0·894	0·911
0·01000	(0·0981)	0·739	0·745	0·757	0·767	0·775	0·790	0·802	0·812	0·829	0·844	0·859	0·871	0·893	0·910
0·01500	(0·147)	0·734	0·741	0·753	0·763	0·771	0·786	0·799	0·809	0·827	0·841	0·857	0·870	0·891	0·908
0·02000	(0·196)	0·730	0·738	0·750	0·760	0·769	0·784	0·797	0·808	0·825	0·840	0·855	0·868	0·890	0·907
0·03000	(0·294)	0·726	0·734	0·746	0·757	0·766	0·782	0·795	0·806	0·824	0·838	0·854	0·867	0·889	0·906
0·04000	(0·392)	0·724	0·731	0·744	0·755	0·765	0·780	0·793	0·804	0·823	0·837	0·853	0·866	0·888	0·906
0·06000	(0·588)	0·721	0·729	0·742	0·753	0·763	0·779	0·792	0·803	0·821	0·836	0·852	0·865	0·887	0·905
0·08000	(0·785)	0·719	0·727	0·740	0·752	0·761	0·777	0·791	0·802	0·820	0·836	0·851	0·865	0·887	0·904
0·10000	(0·981)	0·718	0·726	0·739	0·751	0·760	0·777	0·790	0·801	0·820	0·835	0·851	0·864	0·886	0·904
0·15000	(1·47)	0·715	0·724	0·738	0·749	0·759	0·775	0·789	0·800	0·819	0·834	0·850	0·864	0·886	0·904
0·20000	(1·96)	0·714	0·723	0·737	0·748	0·758	0·775	0·788	0·800	0·819	0·834	0·850	0·863	0·885	0·903
0·30000	(2·94)	0·713	0·721	0·735	0·747	0·757	0·774	0·787	0·799	0·818	0·833	0·849	0·863	0·885	0·903
0·40000	(3·92)	0·712	0·720	0·735	0·747	0·757	0·773	0·787	0·798	0·818	0·833	0·849	0·863	0·885	0·903
0·60000	(5·88)	0·711	0·719	0·734	0·746	0·756	0·773	0·786	0·798	0·817	0·833	0·849	0·862	0·884	0·902
0·80000	(7·85)	0·710	0·719	0·733	0·745	0·756	0·772	0·786	0·798	0·817	0·832	0·848	0·862	0·884	0·902
1·00000	(9·81)	0·710	0·718	0·733	0·745	0·755	0·772	0·786	0·797	m_P0·817	0·832	0·848	0·862	0·884	0·902
1·50000	(14·7)	0·709	0·718	0·732	0·744	0·755	0·772	0·785	0·797	0·816	0·832	0·848	0·862	0·884	0·902
2·00000	(19·6)	0·708	0·717	0·732	0·744	0·754	0·771	0·785	0·797	0·816	0·832	0·848	0·862	0·884	0·902
3·00000	(29·4)	0·708	0·717	0·732	0·744	0·754	0·771	0·785	0·797	0·816	0·832	0·848	0·861	0·884	0·902
S	$(\Delta p/\rho l)$	125	150	200	250	300	400	500	600	800	1000	1250	1500	2000	2500

Grad'nt k.p.g. (Equivalent) Pipe diameters in mm

Kinematic viscosity, $\nu = 0.15 \times 10^{-6}$ m^{2}s^{-1} ;   **Roughness size, $k_s = 0.015$ mm**

E13

Kin. visc., $\nu = 0.15 \times 10^{-6}$ m²s⁻¹;
$S = 0.00010$ to 3.00000

i.e. kin. pr. grad., $\Delta p/\rho l =$ (0.00098) to (29.4) ms⁻²

Roughness size, $k_s = 0.030$ mm
This table shows values of m, as follows

m_C for Colebrook-White solutions

Grad'nt S	k.p.g. $(\Delta p/\rho l)$	125	150	200	250	300	400	500	600	800	1000	1250	1500	2000	2500
0.00010	(0.00098)	0.883	0.883	0.886	0.889	0.893	0.901	0.908	0.915	0.927	0.937	0.949	0.959	0.976	0.991
0.00015	(0.00147)	0.867	0.868	0.872	0.877	0.881	0.890	0.898	0.905	0.918	0.929	0.941	0.952	0.970	0.985
0.00020	(0.00196)	0.857	0.859	0.863	0.869	0.873	0.883	0.891	0.899	0.912	0.924	0.936	0.947	0.965	0.981
0.00030	(0.00294)	0.844	0.846	0.852	0.858	0.863	0.874	0.883	0.891	0.905	0.917	0.930	0.941	0.960	0.976
0.00040	(0.00392)	0.835	0.838	0.845	0.851	0.857	0.868	0.877	0.886	0.900	0.913	0.926	0.938	0.957	0.973
0.00060	(0.00588)	0.824	0.828	0.835	0.842	0.849	0.861	0.871	0.879	0.895	0.907	0.921	0.933	0.953	0.969
0.00080	(0.00785)	0.817	0.821	0.830	0.837	0.844	0.856	0.866	0.875	0.891	0.904	0.918	0.930	0.950	0.967
0.00100	(0.00981)	0.812	0.817	0.825	0.833	0.840	0.853	0.863	0.873	0.889	0.902	0.916	0.928	0.949	0.965
0.00150	(0.0147)	0.804	0.809	0.818	0.827	0.834	0.847	0.858	0.868	0.885	0.898	0.913	0.925	0.946	0.963
0.00200	(0.0196)	0.798	0.804	0.814	0.823	0.830	0.844	0.855	0.865	0.882	0.896	0.911	0.923	0.944	0.961
0.00300	(0.0294)	0.792	0.798	0.808	0.818	0.826	0.840	0.852	0.862	0.879	0.893	0.908	0.921	0.942	0.959
0.00400	(0.0392)	0.788	0.794	0.805	0.815	0.823	0.837	0.849	0.860	0.877	0.892	0.907	0.920	0.941	0.958
0.00600	(0.0588)	0.783	0.789	0.801	0.811	0.820	0.834	0.847	0.857	0.875	0.890	0.905	0.918	0.939	0.957
0.00800	(0.0785)	0.779	0.786	0.798	0.808	0.817	0.832	0.845	0.856	0.874	0.888	0.904	0.917	0.939	0.956
0.01000	(0.0981)	0.777	0.784	0.797	0.807	0.816	0.831	0.844	0.855	0.873	0.887	0.903	0.916	0.938	0.956
0.01500	(0.147)	0.774	0.781	0.794	0.804	0.814	0.829	0.842	0.853	0.871	0.886	0.902	0.915	0.937	0.955
0.02000	(0.196)	0.772	0.779	0.792	0.803	0.812	0.828	0.841	0.852	0.870	0.885	0.901	0.914	0.936	0.954
0.03000	(0.294)	0.769	0.777	0.790	0.801	0.810	0.826	0.839	0.851	0.869	0.884	0.900	0.914	0.936	0.954
0.04000	(0.392)	0.767	0.775	0.789	0.800	0.809	0.825	0.839	0.850	0.868	0.884	0.900	0.913	0.935	0.953
0.06000	(0.588)	0.765	0.773	0.787	0.798	0.808	0.824	0.838	0.849	0.868	0.883	0.899	0.912	0.935	0.953
0.08000	(0.785)	0.764	0.772	0.786	0.798	0.807	0.824	0.837	0.848	0.867	0.883	0.899	0.912	0.934	0.952
0.10000	(0.981)	0.763	0.772	0.785	0.797	0.807	0.823	0.837	0.848	0.867	0.882	0.898	0.912	0.934	0.952
0.15000	(1.47)	0.762	0.770	0.784	0.796	0.806	0.822	0.836	0.847	0.866	0.882	0.898	0.912	0.934	0.952
0.20000	(1.96)	0.761	0.770	0.784	0.796	0.805	0.822	0.836	0.847	0.866	0.882	0.898	0.911	0.934	0.952
0.30000	(2.94)	0.760	0.769	0.783	0.795	0.805	0.822	0.835	0.847	0.866	0.881	0.897	0.911	0.933	0.952
0.40000	(3.92)	0.760	0.768	0.783	0.794	0.805	0.821	0.835	0.846	0.866	0.881	0.897	0.911	0.933	0.952
0.60000	(5.88)	0.759	0.768	0.782	0.794	0.804	0.821	0.835	0.846	0.865	0.881	0.897	0.911	0.933	0.952
0.80000	(7.85)	0.759	0.767	0.782	0.794	0.804	0.821	0.834	0.846	0.865	0.881	0.897	0.911	0.933	0.951
1.00000	(9.81)	0.759	0.767	0.782	0.794	0.804	0.821	0.834	0.846	0.865	0.881	0.897	0.911	0.933	0.951
1.50000	(14.7)	0.758	0.767	0.781	0.793	0.803	0.820	0.834	0.846	0.865	0.880	0.897	0.910	0.933	0.951
2.00000	(19.6)	0.758	0.767	0.781	0.793	0.803	0.820	0.834	0.846	0.865	0.880	0.897	0.910	0.933	0.951
3.00000	(29.4)	0.758	0.766	0.781	0.793	0.803	0.820	0.834	0.845	0.865	0.880	0.897	0.910	0.933	0.951
S	$(\Delta p/\rho l)$	125	150	200	250	300	400	500	600	800	1000	1250	1500	2000	2500

Grad'nt k.p.g. (Equivalent) Pipe diameters in mm — Roughness size, $k_s = 0.030$ mm

E14

Kin. visc., $\nu = 0.15 \times 10^{-6}$ m²s⁻¹;
$S = 0.00010$ to 3.00000

i.e. kin. pr. grad., $\Delta p/\rho l =$ (0.00098) to (29.4) ms⁻²

Roughness size, $k_s = 0.060$ mm
This table shows values of m, as follows

m_C for Colebrook-White solutions

Grad'nt S	k.p.g. $(\Delta p/\rho l)$	125	150	200	250	300	400	500	600	800	1000	1250	1500	2000	2500
0.00010	(0.00098)	0.907	0.908	0.913	0.917	0.922	0.931	0.939	0.947	0.961	0.972	0.985	0.996	1.015	1.031
0.00015	(0.00147)	0.894	0.896	0.902	0.907	0.913	0.923	0.932	0.940	0.954	0.966	0.980	0.991	1.010	1.027
0.00020	(0.00196)	0.886	0.889	0.895	0.901	0.907	0.917	0.927	0.935	0.950	0.963	0.976	0.988	1.008	1.024
0.00030	(0.00294)	0.875	0.879	0.886	0.893	0.899	0.911	0.921	0.930	0.945	0.958	0.972	0.984	1.004	1.021
0.00040	(0.00392)	0.869	0.873	0.880	0.888	0.895	0.907	0.917	0.926	0.942	0.955	0.970	0.982	1.002	1.019
0.00060	(0.00588)	0.860	0.865	0.874	0.882	0.889	0.901	0.912	0.922	0.938	0.952	0.966	0.979	1.000	1.017
0.00080	(0.00785)	0.855	0.860	0.869	0.878	0.885	0.898	0.909	0.919	0.936	0.950	0.965	0.977	0.998	1.015
0.00100	(0.00981)	0.851	0.857	0.866	0.875	0.883	0.896	0.907	0.917	0.934	0.948	0.963	0.976	0.997	1.014
0.00150	(0.0147)	0.846	0.851	0.861	0.871	0.879	0.893	0.904	0.915	0.932	0.946	0.961	0.974	0.995	1.013
0.00200	(0.0196)	0.842	0.848	0.859	0.868	0.876	0.890	0.902	0.913	0.930	0.945	0.960	0.973	0.994	1.012
0.00300	(0.0294)	0.837	0.844	0.855	0.865	0.873	0.888	0.900	0.911	0.928	0.943	0.958	0.971	0.993	1.011
0.00400	(0.0392)	0.835	0.841	0.853	0.863	0.871	0.886	0.899	0.909	0.927	0.942	0.957	0.971	0.992	1.010
0.00600	(0.0588)	0.831	0.838	0.850	0.860	0.869	0.884	0.897	0.908	0.926	0.941	0.956	0.970	0.992	1.009
0.00800	(0.0785)	0.829	0.836	0.848	0.859	0.868	0.883	0.896	0.907	0.925	0.940	0.956	0.969	0.991	1.009
0.01000	(0.0981)	0.828	0.835	0.847	0.858	0.867	0.882	0.895	0.906	0.924	0.939	0.955	0.969	0.991	1.009
0.01500	(0.147)	0.825	0.833	0.845	0.856	0.865	0.881	0.894	0.905	0.924	0.939	0.955	0.968	0.990	1.008
0.02000	(0.196)	0.824	0.832	0.844	0.855	0.865	0.880	0.893	0.905	0.923	0.938	0.954	0.968	0.990	1.008
0.03000	(0.294)	0.822	0.830	0.843	0.854	0.864	0.879	0.893	0.904	0.922	0.938	0.954	0.967	0.989	1.007
0.04000	(0.392)	0.821	0.829	0.842	0.853	0.863	0.879	0.892	0.903	0.922	0.937	0.953	0.967	0.989	1.007
0.06000	(0.588)	0.820	0.828	0.841	0.853	0.862	0.878	0.892	0.903	0.922	0.937	0.953	0.967	0.989	1.007
0.08000	(0.785)	0.820	0.827	0.841	0.852	0.862	0.878	0.891	0.903	0.921	0.937	0.953	0.966	0.989	1.007
0.10000	(0.981)	0.819	0.827	0.841	0.852	0.861	0.878	0.891	0.902	0.921	0.937	0.953	0.966	0.989	1.007
0.15000	(1.47)	0.818	0.826	0.840	0.851	0.861	0.877	0.891	0.902	0.921	0.936	0.952	0.966	0.988	1.007
0.20000	(1.96)	0.818	0.826	0.840	0.851	0.861	0.877	0.890	0.902	0.921	0.936	0.952	0.966	0.988	1.007
0.30000	(2.94)	0.817	0.825	0.839	0.851	0.860	0.877	0.890	0.902	0.921	0.936	0.952	0.966	0.988	1.006
0.40000	(3.92)	0.817	0.825	0.839	0.850	0.860	0.877	0.890	0.901	0.920	0.936	0.952	0.966	0.988	1.006
0.60000	(5.88)	0.817	0.825	0.839	0.850	0.860	0.876	0.890	0.901	0.920	0.936	0.952	0.966	0.988	1.006
0.80000	(7.85)	0.816	0.825	0.838	0.850	0.860	0.876	0.890	0.901	0.920	0.936	0.952	0.965	0.988	1.006
1.00000	(9.81)	0.816	0.824	0.838	0.850	0.860	0.876	0.890	0.901	0.920	0.936	0.952	0.965	0.988	1.006
1.50000	(14.7)	0.816	0.824	0.838	0.850	0.860	0.876	0.889	0.901	0.920	0.935	0.952	0.965	0.988	1.006
2.00000	(19.6)	0.816	0.824	0.838	0.850	0.859	0.876	0.889	0.901	0.920	0.935	0.952	0.965	0.988	1.006
3.00000	(29.4)	0.816	0.824	0.838	0.849	0.859	0.876	0.889	0.901	0.920	0.935	0.952	0.965	0.988	1.006
S	$(\Delta p/\rho l)$	125	150	200	250	300	400	500	600	800	1000	1250	1500	2000	2500

Grad'nt k.p.g. (Equivalent) Pipe diameters in mm — Roughness size, $k_s = 0.060$ mm

Kin. visc., $\nu = 0.15 \times 10^{-6}$ m^{2}s^{-1};
$S = 0.00010$ to 3.00000
i.e. kin. pr. grad., $\Delta p/\rho l =$ (0.00098) to (29.4) ms^{-2}

Roughness size, $k_s = 0.150$ mm
This table shows values of m, as follows
m_C for Colebrook-White solutions

Grad'nt S	k.p.g. $(\Delta p/\rho l)$	125	150	200	250	300	400	500	600	800	1000	1250	1500	2000	2500
0.00010	(0.00098)	0.964	0.966	0.971	0.977	0.983	0.994	1.003	1.012	1.027	1.039	1.053	1.065	1.085	1.102
0.00015	(0.00147)	0.955	0.958	0.964	0.971	0.977	0.988	0.998	1.007	1.023	1.036	1.050	1.062	1.083	1.100
0.00020	(0.00196)	0.949	0.953	0.960	0.967	0.973	0.985	0.996	1.005	1.021	1.034	1.048	1.061	1.081	1.098
0.00030	(0.00294)	0.942	0.946	0.954	0.962	0.969	0.981	0.992	1.002	1.018	1.032	1.046	1.059	1.080	1.097
0.00040	(0.00392)	0.938	0.943	0.951	0.959	0.966	0.979	0.990	1.000	1.016	1.030	1.045	1.057	1.078	1.096
0.00060	(0.00588)	0.933	0.938	0.947	0.955	0.963	0.976	0.987	0.997	1.014	1.028	1.043	1.056	1.077	1.095
0.00080	(0.00785)	0.930	0.935	0.945	0.953	0.961	0.974	0.986	0.996	1.013	1.027	1.042	1.055	1.076	1.094
0.00100	(0.00981)	0.928	0.933	0.943	0.952	0.959	0.973	0.985	0.995	1.012	1.026	1.041	1.054	1.076	1.094
0.00150	(0.0147)	0.924	0.930	0.940	0.949	0.957	0.971	0.983	0.993	1.011	1.025	1.040	1.053	1.075	1.093
0.00200	(0.0196)	0.922	0.928	0.938	0.948	0.956	0.970	0.982	0.993	1.010	1.024	1.040	1.053	1.075	1.092
0.00300	(0.0294)	0.920	0.926	0.937	0.946	0.954	0.969	0.981	0.991	1.009	1.024	1.039	1.052	1.074	1.092
0.00400	(0.0392)	0.918	0.924	0.935	0.945	0.953	0.968	0.980	0.991	1.008	1.023	1.039	1.052	1.074	1.091
0.00600	(0.0588)	0.916	0.923	0.934	0.944	0.952	0.967	0.979	0.990	1.008	1.023	1.038	1.051	1.073	1.091
0.00800	(0.0785)	0.915	0.922	0.933	0.943	0.952	0.966	0.979	0.989	1.007	1.022	1.038	1.051	1.073	1.091
0.01000	(0.0981)	0.914	0.921	0.932	0.942	0.951	0.966	0.978	0.989	1.007	1.022	1.038	1.051	1.073	1.091
0.01500	(0.147)	0.913	0.920	0.932	0.942	0.950	0.965	0.978	0.989	1.007	1.022	1.037	1.050	1.073	1.090
0.02000	(0.196)	0.912	0.919	0.931	0.941	0.950	0.965	0.978	0.988	1.006	1.021	1.037	1.050	1.072	1.090
0.03000	(0.294)	0.912	0.918	0.930	0.941	0.949	0.965	0.977	0.988	1.006	1.021	1.037	1.050	1.072	1.090
0.04000	(0.392)	0.911	0.918	0.930	0.940	0.949	0.964	0.977	0.988	1.006	1.021	1.037	1.050	1.072	1.090
0.06000	(0.588)	0.910	0.917	0.930	0.940	0.949	0.964	0.977	0.988	1.006	1.021	1.036	1.050	1.072	1.090
0.08000	(0.785)	0.910	0.917	0.929	0.940	0.949	0.964	0.976	0.987	1.006	1.021	1.036	1.050	1.072	1.090
0.10000	(0.981)	0.910	0.917	0.929	0.939	0.948	0.964	0.976	0.987	1.006	1.021	1.036	1.050	1.072	1.090
0.15000	(1.47)	0.909	0.917	0.929	0.939	0.948	0.963	0.976	0.987	1.005	1.020	1.036	1.050	1.072	1.090
0.20000	(1.96)	0.909	0.916	0.929	0.939	0.948	0.963	0.976	0.987	1.005	1.020	1.036	1.049	1.072	1.090
0.30000	(2.94)	0.909	0.916	0.928	0.939	0.948	0.963	0.976	0.987	1.005	1.020	1.036	1.049	1.072	1.090
0.40000	(3.92)	0.909	0.916	0.928	0.939	0.948	0.963	0.976	0.987	1.005	1.020	1.036	1.049	1.072	1.090
0.60000	(5.88)	0.909	0.916	0.928	0.939	0.948	0.963	0.976	0.987	1.005	1.020	1.036	1.049	1.072	1.090
0.80000	(7.85)	0.908	0.916	0.928	0.938	0.948	0.963	0.976	0.987	1.005	1.020	1.036	1.049	1.071	1.090
1.00000	(9.81)	0.908	0.916	0.928	0.938	0.948	0.963	0.976	0.987	1.005	1.020	1.036	1.049	1.071	1.090
1.50000	(14.7)	0.908	0.915	0.928	0.938	0.947	0.963	0.976	0.987	1.005	1.020	1.036	1.049	1.071	1.090
2.00000	(19.6)	0.908	0.915	0.928	0.938	0.947	0.963	0.976	0.987	1.005	1.020	1.036	1.049	1.071	1.090
3.00000	(29.4)	0.908	0.915	0.928	0.938	0.947	0.963	0.976	0.987	1.005	1.020	1.036	1.049	1.071	1.090
S	$(\Delta p/\rho l)$	125	150	200	250	300	400	500	600	800	1000	1250	1500	2000	2500

Grad'nt k.p.g. (Equivalent) Pipe diameters in mm Roughness size, $k_s = 0.150$ mm

Kin. visc., $\nu = 0.15 \times 10^{-6}$ m^{2}s^{-1};
$S = 0.00010$ to 3.00000
i.e. kin. pr. grad., $\Delta p/\rho l =$ (0.00098) to (29.4) ms^{-2}

Roughness size, $k_s = 0.30$ mm
This table shows values of m, as follows
m_C for Colebrook-White solutions

Grad'nt S	k.p.g. $(\Delta p/\rho l)$	125	150	200	250	300	400	500	600	800	1000	1250	1500	2000	2500
0.00010	(0.00098)	1.030	1.032	1.038	1.044	1.049	1.060	1.070	1.079	1.094	1.107	1.120	1.133	1.153	1.170
0.00015	(0.00147)	1.024	1.027	1.033	1.039	1.046	1.057	1.067	1.076	1.091	1.105	1.119	1.131	1.151	1.168
0.00020	(0.00196)	1.020	1.023	1.030	1.037	1.043	1.055	1.065	1.074	1.090	1.103	1.118	1.130	1.151	1.168
0.00030	(0.00294)	1.015	1.019	1.026	1.034	1.040	1.052	1.063	1.072	1.088	1.102	1.116	1.129	1.150	1.167
0.00040	(0.00392)	1.013	1.016	1.024	1.032	1.039	1.051	1.062	1.071	1.087	1.101	1.115	1.128	1.149	1.166
0.00060	(0.00588)	1.009	1.013	1.022	1.029	1.037	1.049	1.060	1.070	1.086	1.100	1.114	1.127	1.148	1.165
0.00080	(0.00785)	1.007	1.012	1.020	1.028	1.035	1.048	1.059	1.069	1.085	1.099	1.114	1.127	1.148	1.165
0.00100	(0.00981)	1.006	1.010	1.019	1.027	1.034	1.047	1.058	1.068	1.085	1.099	1.113	1.126	1.147	1.165
0.00150	(0.0147)	1.004	1.008	1.017	1.026	1.033	1.046	1.057	1.067	1.084	1.098	1.113	1.126	1.147	1.164
0.00200	(0.0196)	1.002	1.007	1.016	1.025	1.032	1.046	1.057	1.067	1.084	1.098	1.112	1.125	1.147	1.164
0.00300	(0.0294)	1.001	1.006	1.015	1.024	1.031	1.045	1.056	1.066	1.083	1.097	1.112	1.125	1.146	1.164
0.00400	(0.0392)	1.000	1.005	1.014	1.023	1.031	1.044	1.056	1.066	1.083	1.097	1.112	1.125	1.146	1.164
0.00600	(0.0588)	0.999	1.004	1.014	1.022	1.030	1.044	1.055	1.065	1.082	1.096	1.111	1.124	1.146	1.163
0.00800	(0.0785)	0.998	1.003	1.013	1.022	1.030	1.043	1.055	1.065	1.082	1.096	1.111	1.124	1.146	1.163
0.01000	(0.0981)	0.997	1.003	1.013	1.022	1.029	1.043	1.055	1.065	1.082	1.096	1.111	1.124	1.146	1.163
0.01500	(0.147)	0.997	1.002	1.012	1.021	1.029	1.043	1.054	1.064	1.082	1.096	1.111	1.124	1.145	1.163
0.02000	(0.196)	0.996	1.002	1.012	1.021	1.029	1.042	1.054	1.064	1.081	1.096	1.111	1.124	1.145	1.163
0.03000	(0.294)	0.996	1.001	1.011	1.020	1.028	1.042	1.054	1.064	1.081	1.096	1.111	1.124	1.145	1.163
0.04000	(0.392)	0.995	1.001	1.011	1.020	1.028	1.042	1.054	1.064	1.081	1.095	1.111	1.124	1.145	1.163
0.06000	(0.588)	0.995	1.001	1.011	1.020	1.028	1.042	1.054	1.064	1.081	1.095	1.110	1.123	1.145	1.163
0.08000	(0.785)	0.995	1.001	1.011	1.020	1.028	1.042	1.053	1.064	1.081	1.095	1.110	1.123	1.145	1.163
0.10000	(0.981)	0.995	1.000	1.011	1.020	1.028	1.042	1.053	1.064	1.081	1.095	1.110	1.123	1.145	1.163
0.15000	(1.47)	0.994	1.000	1.011	1.020	1.028	1.042	1.053	1.064	1.081	1.095	1.110	1.123	1.145	1.163
0.20000	(1.96)	0.994	1.000	1.010	1.019	1.028	1.041	1.053	1.064	1.081	1.095	1.110	1.123	1.145	1.163
0.30000	(2.94)	0.994	1.000	1.010	1.019	1.027	1.041	1.053	1.063	1.081	1.095	1.110	1.123	1.145	1.163
0.40000	(3.92)	0.994	1.000	1.010	1.019	1.027	1.041	1.053	1.063	1.081	1.095	1.110	1.123	1.145	1.163
0.60000	(5.88)	0.994	1.000	1.010	1.019	1.027	1.041	1.053	1.063	1.081	1.095	1.110	1.123	1.145	1.163
0.80000	(7.85)	0.994	1.000	1.010	1.019	1.027	1.041	1.053	1.063	1.081	1.095	1.110	1.123	1.145	1.163
1.00000	(9.81)	0.994	1.000	1.010	1.019	1.027	1.041	1.053	1.063	1.081	1.095	1.110	1.123	1.145	1.163
1.50000	(14.7)	0.994	1.000	1.010	1.019	1.027	1.041	1.053	1.063	1.081	1.095	1.110	1.123	1.145	1.163
2.00000	(19.6)	0.994	1.000	1.010	1.019	1.027	1.041	1.053	1.063	1.081	1.095	1.110	1.123	1.145	1.163
3.00000	(29.4)	0.994	0.999	1.010	1.019	1.027	1.041	1.053	1.063	1.081	1.095	1.110	1.123	1.145	1.163
S	$(\Delta p/\rho l)$	125	150	200	250	300	400	500	600	800	1000	1250	1500	2000	2500

Grad'nt k.p.g. (Equivalent) Pipe diameters in mm Roughness size, $k_s = 0.30$ mm

E17

Kin. visc., $\nu = 0.20 \times 10^{-6}\ \text{m}^2\text{s}^{-1}$;
$S = 0.00010$ to 30.0000

i.e. kin. pr. grad., $\Delta p/\rho l =$
(0.00098) to $(294)\ \text{ms}^{-2}$

Roughness size, $k_s = 0.0015$ mm
This table shows values of m, as follows

m_C for Colebrook-White solutions; or,
where $R \leq 2000$, m_P for laminar flow

Grad'nt S	k.p.g. $(\Delta p/\rho l)$	(Equivalent) Pipe diameters in mm													
		6	8	10	12.5	15	20	25	30	40	50	60	80	100	125
0.00030	(0.00294)	1.371	0.935	*	0.928	0.910	0.887	0.872	0.863	0.850	0.843	0.838	0.833	0.830	0.829
0.00040	(0.00392)	1.188	0.809	*	0.904	0.888	0.867	0.853	0.845	0.834	0.827	0.823	0.819	0.817	0.816
0.00060	(0.00588)	0.970	*	0.892	0.872	0.858	0.839	0.828	0.821	0.811	0.806	0.803	0.799	0.798	0.799
0.00080	(0.00785)	0.840	*	0.869	0.851	0.838	0.821	0.811	0.804	0.796	0.791	0.789	0.786	0.786	0.787
0.00100	(0.00981)	0.751	*	0.852	0.835	0.823	0.808	0.798	0.792	0.785	0.781	0.779	0.777	0.777	0.778
0.00150	(0.0147)	*	0.841	0.823	0.808	0.797	0.784	0.776	0.771	0.765	0.762	0.760	0.760	0.760	0.762
0.00200	(0.0196)	*	0.820	0.803	0.790	0.780	0.768	0.761	0.757	0.752	0.749	0.748	0.748	0.749	0.751
0.00300	(0.0294)	0.815	0.792	0.777	0.765	0.757	0.747	0.741	0.738	0.734	0.732	0.732	0.732	0.734	0.737
0.00400	(0.0392)	0.794	0.773	0.760	0.749	0.742	0.733	0.728	0.725	0.722	0.721	0.721	0.722	0.724	0.727
0.00600	(0.0588)	0.766	0.748	0.737	0.728	0.721	0.714	0.710	0.708	0.705	0.705	0.706	0.708	0.710	0.714
0.00800	(0.0785)	0.748	0.731	0.721	0.713	0.708	0.701	0.698	0.696	0.694	0.695	0.695	0.698	0.701	0.705
0.01000	(0.0981)	0.734	0.719	0.710	0.702	0.697	0.691	0.689	0.687	0.686	0.687	0.688	0.691	0.694	0.698
0.01500	(0.147)	0.711	0.698	0.690	0.683	0.679	0.675	0.673	0.672	0.672	0.673	0.674	0.678	0.682	0.686
0.02000	(0.196)	0.695	0.683	0.676	0.671	0.667	0.664	0.662	0.662	0.662	0.664	0.665	0.670	0.674	0.678
0.03000	(0.294)	0.675	0.664	0.659	0.654	0.651	0.649	0.648	0.648	0.649	0.651	0.653	0.658	0.663	0.668
0.04000	(0.392)	0.661	0.652	0.647	0.643	0.640	0.638	0.638	0.638	0.640	0.643	0.645	0.650	0.655	0.661
0.06000	(0.588)	0.642	0.635	0.631	0.628	0.626	0.625	0.625	0.626	0.629	0.632	0.635	0.640	0.646	0.651
0.08000	(0.785)	0.630	0.623	0.620	0.617	0.616	0.616	0.617	0.618	0.621	0.624	0.627	0.633	0.639	0.645
0.10000	(0.981)	0.621	0.615	0.612	0.610	0.609	0.609	0.610	0.612	0.615	0.619	0.622	0.628	0.634	0.641
0.15000	(1.47)	0.605	0.600	0.598	0.597	0.597	0.597	0.599	0.601	0.605	0.609	0.613	0.620	0.626	0.633
0.20000	(1.96)	0.594	0.591	0.589	0.588	0.588	0.590	0.592	0.594	0.598	0.603	0.607	0.614	0.621	0.628
0.30000	(2.94)	0.580	0.578	0.577	0.577	0.577	0.579	0.582	0.584	0.590	0.594	0.599	0.607	0.614	0.621
0.40000	(3.92)	0.571	0.569	0.568	0.569	0.570	0.572	0.575	0.578	0.584	0.589	0.594	0.602	0.609	0.617
0.60000	(5.88)	0.559	0.558	0.558	0.559	0.560	0.563	0.567	0.570	0.576	0.582	0.587	0.596	0.603	0.611
0.80000	(7.85)	0.551	0.550	0.551	0.552	0.554	0.558	0.561	0.565	0.571	0.577	0.582	0.591	0.599	0.608
1.00000	(9.81)	0.545	0.544	0.545	0.547	0.549	0.553	0.557	0.561	0.568	0.574	0.579	0.589	0.596	0.605
1.50000	(14.7)	0.534	0.535	0.537	0.539	0.541	0.546	0.550	0.554	0.562	0.568	0.574	0.584	0.592	0.601
2.00000	(19.6)	0.528	0.529	0.531	0.533	0.536	0.541	0.546	0.550	0.558	0.565	0.570	0.581	0.589	0.598
3.00000	(29.4)	0.519	0.521	0.523	0.526	0.529	0.535	0.540	0.545	0.553	0.560	0.566	0.577	0.585	0.595
4.00000	(39.2)	0.513	0.516	0.518	0.522	0.525	0.531	0.537	0.541	0.550	0.557	0.564	0.574	0.583	0.593
6.00000	(58.8)	0.506	0.509	0.512	0.516	0.520	0.526	0.532	0.537	0.546	0.554	0.560	0.571	0.580	0.590
8.00000	(78.5)	0.501	0.504	0.508	0.512	0.516	0.523	0.529	0.534	0.544	0.551	0.558	0.569	0.579	0.589
10.0000	(98.1)	0.497	0.501	0.505	0.509	0.514	0.521	0.527	0.532	0.542	0.550	0.557	0.568	0.577	0.587
15.0000	(147)	0.492	0.496	0.500	0.505	0.509	0.517	0.524	0.529	0.539	0.547	0.554	0.566	0.576	0.586
20.0000	(196)	0.488	0.493	0.497	0.502	0.507	0.515	0.521	0.527	0.537	0.546	0.553	0.565	0.574	0.585
30.0000	(294)	0.483	0.489	0.494	0.499	0.504	0.512	0.519	0.525	0.535	0.544	0.551	0.563	0.573	0.583
S	$(\Delta p/\rho l)$	6	8	10	12.5	15	20	25	30	40	50	60	80	100	125

Grad'nt   k.p.g.   (Equivalent) Pipe diameters in mm

S	$(\Delta p/\rho l)$	125	150	200	250	300	400	500	600	800	1000	1250	1500	2000	2500
0.00010	(0.00098)	0.884	0.882	0.881	0.881	0.883	0.886	0.890	0.894	0.902	0.909	0.916	0.923	0.936	0.946
0.00015	(0.00147)	0.863	0.862	0.862	0.863	0.865	0.869	0.874	0.878	0.886	0.893	0.902	0.909	0.921	0.932
0.00020	(0.00196)	0.849	0.848	0.849	0.850	0.853	0.857	0.862	0.867	0.875	0.883	0.891	0.899	0.911	0.922
0.00030	(0.00294)	0.829	0.829	0.831	0.833	0.836	0.841	0.847	0.852	0.861	0.869	0.877	0.885	0.898	0.909
0.00040	(0.00392)	0.816	0.817	0.819	0.822	0.825	0.831	0.836	0.841	0.851	0.859	0.868	0.876	0.889	0.900
0.00060	(0.00588)	0.799	0.800	0.802	0.806	0.809	0.816	0.822	0.827	0.837	0.846	0.855	0.863	0.877	0.888
0.00080	(0.00785)	0.787	0.788	0.791	0.795	0.799	0.806	0.812	0.818	0.828	0.837	0.846	0.854	0.868	0.880
0.00100	(0.00981)	0.778	0.779	0.783	0.787	0.791	0.798	0.805	0.811	0.821	0.830	0.839	0.848	0.862	0.874
0.00150	(0.0147)	0.762	0.764	0.768	0.773	0.777	0.785	0.792	0.798	0.809	0.818	0.828	0.837	0.851	0.864
0.00200	(0.0196)	0.751	0.754	0.758	0.763	0.768	0.776	0.783	0.789	0.800	0.810	0.820	0.829	0.844	0.856
0.00300	(0.0294)	0.737	0.739	0.745	0.750	0.755	0.764	0.771	0.778	0.789	0.799	0.810	0.819	0.834	0.847
0.00400	(0.0392)	0.727	0.730	0.736	0.741	0.746	0.755	0.763	0.770	0.782	0.792	0.803	0.812	0.827	0.840
0.00600	(0.0588)	0.714	0.717	0.724	0.729	0.735	0.744	0.752	0.760	0.772	0.782	0.793	0.803	0.819	0.832
0.00800	(0.0785)	0.705	0.708	0.715	0.721	0.727	0.737	0.745	0.752	0.765	0.776	0.787	0.797	0.813	0.826
0.01000	(0.0981)	0.698	0.702	0.709	0.715	0.721	0.731	0.740	0.747	0.760	0.771	0.782	0.792	0.809	0.822
0.01500	(0.147)	0.686	0.690	0.698	0.705	0.711	0.721	0.730	0.738	0.751	0.762	0.774	0.784	0.801	0.815
0.02000	(0.196)	0.678	0.683	0.691	0.698	0.704	0.715	0.724	0.732	0.746	0.757	0.769	0.779	0.797	0.811
0.03000	(0.294)	0.668	0.673	0.681	0.689	0.695	0.706	0.716	0.724	0.738	0.750	0.762	0.773	0.790	0.805
0.04000	(0.392)	0.661	0.666	0.675	0.682	0.689	0.701	0.711	0.719	0.733	0.745	0.758	0.768	0.786	0.801
0.06000	(0.588)	0.651	0.657	0.666	0.674	0.681	0.693	0.703	0.712	0.727	0.739	0.752	0.763	0.781	0.796
0.08000	(0.785)	0.645	0.651	0.661	0.669	0.676	0.689	0.699	0.708	0.723	0.735	0.748	0.759	0.778	0.793
0.10000	(0.981)	0.641	0.646	0.656	0.665	0.672	0.685	0.695	0.705	0.720	0.732	0.746	0.757	0.776	0.791
0.15000	(1.47)	0.633	0.639	0.649	0.658	0.666	0.679	0.690	0.699	0.715	0.728	0.741	0.753	0.772	0.788
0.20000	(1.96)	0.628	0.634	0.645	0.654	0.662	0.675	0.686	0.696	0.712	0.725	0.738	0.750	0.770	0.785
0.30000	(2.94)	0.621	0.628	0.639	0.648	0.657	0.670	0.682	0.691	0.708	0.721	0.735	0.747	0.767	0.783
0.40000	(3.92)	0.617	0.623	0.635	0.645	0.653	0.667	0.679	0.689	0.705	0.719	0.733	0.745	0.765	0.781
0.60000	(5.88)	0.611	0.618	0.630	0.640	0.649	0.663	0.675	0.685	0.702	0.716	0.730	0.742	0.763	0.779
0.80000	(7.85)	0.608	0.615	0.627	0.637	0.646	0.661	0.673	0.683	0.700	0.714	0.729	0.741	0.761	0.778
1.00000	(9.81)	0.605	0.612	0.625	0.635	0.644	0.659	0.671	0.682	0.699	0.713	0.727	0.740	0.760	0.777
1.50000	(14.7)	0.601	0.608	0.621	0.632	0.641	0.656	0.669	0.679	0.697	0.711	0.726	0.738	0.759	0.775
2.00000	(19.6)	0.598	0.606	0.619	0.630	0.639	0.654	0.667	0.678	0.695	0.710	0.724	0.737	0.758	0.774
3.00000	(29.4)	0.595	0.603	0.616	0.627	0.637	0.652	0.665	0.676	0.694	0.708	0.723	0.736	0.757	0.773
S	$(\Delta p/\rho l)$	125	150	200	250	300	400	500	600	800	1000	1250	1500	2000	2500

Grad'nt   k.p.g.   (Equivalent) Pipe diameters in mm

Kinematic viscosity, $\nu = 0.20 \times 10^{-6}\ \text{m}^2\text{s}^{-1}$;

Roughness size, $k_s = 0.0015$ mm

Kin. visc., $\nu = 0.20\times10^{-6}\ \mathrm{m^2 s^{-1}}$; Roughness size, $k_s = 0.003$ mm E18

$S = 0.00010$ to 30.0000

i.e. kin. pr. grad., $\Delta p/\rho l =$ (0.00098) to (294) ms^{-2}

This table shows values of m, as follows m_C for Colebrook-White solutions; or, where $\mathbf{R} \le 2000$, m_P for laminar flow

Grad'nt S **k.p.g.** $(\Delta p/\rho l)$ **(Equivalent) Pipe diameters in mm**

S	$(\Delta p/\rho l)$	6	8	10	12.5	15	20	25	30	40	50	60	80	100	125
0.00030	(0.00294)	1.371	0.935	*	0.929	0.911	0.888	0.874	0.864	0.852	0.844	0.840	0.835	0.832	0.831
0.00040	(0.00392)	1.188	0.809	*	0.905	0.889	0.868	0.855	0.846	0.835	0.829	0.825	0.821	0.819	0.819
0.00060	(0.00588)	0.970	*	0.894	0.873	0.859	0.841	0.830	0.822	0.813	0.808	0.805	0.802	0.801	0.801
0.00080	(0.00785)	0.840	*	0.871	0.852	0.839	0.823	0.813	0.806	0.798	0.794	0.791	0.789	0.789	0.790
0.00100	(0.00981)	0.751	*	0.854	0.837	0.825	0.810	0.800	0.794	0.787	0.783	0.781	0.779	0.780	0.781
0.00150	(0.0147)	*	0.843	0.825	0.810	0.799	0.786	0.779	0.773	0.768	0.765	0.763	0.763	0.764	0.766
0.00200	(0.0196)	*	0.822	0.805	0.792	0.782	0.771	0.764	0.759	0.755	0.752	0.751	0.752	0.753	0.755
0.00300	(0.0294)	0.817	0.794	0.780	0.768	0.760	0.750	0.744	0.741	0.737	0.736	0.735	0.737	0.738	0.741
0.00400	(0.0392)	0.796	0.775	0.762	0.752	0.745	0.736	0.731	0.728	0.725	0.725	0.725	0.726	0.729	0.732
0.00600	(0.0588)	0.769	0.751	0.740	0.731	0.725	0.717	0.714	0.711	0.710	0.710	0.710	0.713	0.716	0.720
0.00800	(0.0785)	0.751	0.735	0.725	0.716	0.711	0.705	0.702	0.700	0.699	0.700	0.701	0.704	0.707	0.711
0.01000	(0.0981)	0.738	0.722	0.713	0.706	0.701	0.696	0.693	0.692	0.691	0.692	0.693	0.697	0.701	0.705
0.01500	(0.147)	0.715	0.702	0.694	0.688	0.684	0.680	0.678	0.677	0.678	0.679	0.681	0.685	0.689	0.694
0.02000	(0.196)	0.699	0.688	0.681	0.676	0.672	0.669	0.668	0.667	0.669	0.670	0.673	0.677	0.682	0.687
0.03000	(0.294)	0.679	0.669	0.664	0.659	0.657	0.655	0.654	0.655	0.656	0.659	0.662	0.667	0.672	0.678
0.04000	(0.392)	0.666	0.657	0.652	0.649	0.647	0.645	0.645	0.646	0.648	0.651	0.654	0.660	0.666	0.672
0.06000	(0.588)	0.648	0.641	0.637	0.634	0.633	0.633	0.633	0.635	0.638	0.641	0.645	0.651	0.657	0.664
0.08000	(0.785)	0.636	0.630	0.627	0.625	0.624	0.624	0.625	0.627	0.631	0.635	0.639	0.646	0.652	0.659
0.10000	(0.981)	0.628	0.622	0.619	0.618	0.617	0.618	0.620	0.622	0.626	0.630	0.634	0.641	0.648	0.655
0.15000	(1.47)	0.613	0.609	0.607	0.606	0.606	0.608	0.610	0.612	0.617	0.622	0.626	0.634	0.641	0.649
0.20000	(1.96)	0.603	0.600	0.598	0.598	0.599	0.601	0.603	0.606	0.611	0.617	0.621	0.630	0.637	0.645
0.30000	(2.94)	0.590	0.588	0.587	0.587	0.589	0.592	0.595	0.598	0.604	0.610	0.615	0.624	0.631	0.640
0.40000	(3.92)	0.582	0.580	0.580	0.581	0.583	0.586	0.590	0.593	0.600	0.605	0.611	0.620	0.628	0.636
0.60000	(5.88)	0.571	0.570	0.571	0.572	0.574	0.578	0.583	0.586	0.594	0.600	0.605	0.615	0.624	0.632
0.80000	(7.85)	0.563	0.564	0.565	0.567	0.569	0.574	0.578	0.582	0.590	0.596	0.602	0.612	0.621	0.630
1.00000	(9.81)	0.558	0.559	0.560	0.563	0.565	0.570	0.575	0.579	0.587	0.594	0.600	0.610	0.619	0.628
1.50000	(14.7)	0.549	0.551	0.553	0.556	0.559	0.564	0.570	0.574	0.583	0.590	0.596	0.607	0.616	0.625
2.00000	(19.6)	0.544	0.546	0.548	0.552	0.555	0.561	0.566	0.571	0.580	0.587	0.594	0.605	0.614	0.623
3.00000	(29.4)	0.537	0.539	0.542	0.546	0.550	0.556	0.562	0.567	0.576	0.584	0.591	0.602	0.611	0.621
4.00000	(39.2)	0.532	0.535	0.539	0.543	0.546	0.553	0.559	0.565	0.574	0.582	0.589	0.600	0.610	0.620
6.00000	(58.8)	0.526	0.530	0.534	0.538	0.542	0.550	0.556	0.562	0.571	0.580	0.587	0.598	0.608	0.618
8.00000	(78.5)	0.522	0.527	0.531	0.536	0.540	0.548	0.554	0.560	0.570	0.578	0.585	0.597	0.607	0.617
10.0000	(98.1)	0.520	0.524	0.529	0.534	0.538	0.546	0.553	0.559	0.569	0.577	0.584	0.596	0.606	0.617
15.0000	(147)	0.516	0.521	0.525	0.531	0.535	0.544	0.551	0.557	0.567	0.575	0.583	0.595	0.605	0.616
20.0000	(196)	0.513	0.518	0.523	0.529	0.534	0.542	0.549	0.555	0.566	0.574	0.582	0.594	0.604	0.615
30.0000	(294)	0.510	0.516	0.521	0.526	0.532	0.540	0.548	0.554	0.564	0.573	0.581	0.593	0.603	0.614
S	$(\Delta p/\rho l)$	6	8	10	12.5	15	20	25	30	40	50	60	80	100	125

Grad'nt S **k.p.g.** $(\Delta p/\rho l)$ **(Equivalent) Pipe diameters in mm**

S	$(\Delta p/\rho l)$	125	150	200	250	300	400	500	600	800	1000	1250	1500	2000	2500
0.00010	(0.00098)	0.885	0.883	0.882	0.883	0.884	0.888	0.892	0.897	0.904	0.911	0.919	0.927	0.939	0.950
0.00015	(0.00147)	0.864	0.863	0.863	0.865	0.867	0.871	0.876	0.881	0.889	0.897	0.905	0.912	0.925	0.936
0.00020	(0.00196)	0.850	0.850	0.851	0.852	0.855	0.860	0.865	0.870	0.879	0.886	0.895	0.903	0.916	0.927
0.00030	(0.00294)	0.831	0.831	0.833	0.836	0.839	0.844	0.850	0.855	0.865	0.873	0.882	0.890	0.904	0.915
0.00040	(0.00392)	0.819	0.819	0.821	0.824	0.828	0.834	0.840	0.845	0.855	0.864	0.873	0.881	0.895	0.907
0.00060	(0.00588)	0.801	0.802	0.806	0.809	0.813	0.820	0.826	0.832	0.842	0.851	0.861	0.869	0.884	0.896
0.00080	(0.00785)	0.790	0.791	0.795	0.799	0.803	0.810	0.817	0.823	0.833	0.843	0.853	0.861	0.876	0.889
0.00100	(0.00981)	0.781	0.783	0.787	0.791	0.795	0.803	0.810	0.816	0.827	0.836	0.847	0.855	0.871	0.883
0.00150	(0.0147)	0.766	0.768	0.773	0.777	0.782	0.790	0.798	0.804	0.816	0.825	0.836	0.845	0.861	0.874
0.00200	(0.0196)	0.755	0.758	0.763	0.768	0.773	0.782	0.789	0.796	0.808	0.818	0.829	0.839	0.854	0.868
0.00300	(0.0294)	0.741	0.744	0.750	0.756	0.761	0.771	0.779	0.786	0.798	0.809	0.820	0.830	0.846	0.860
0.00400	(0.0392)	0.732	0.735	0.742	0.748	0.753	0.763	0.771	0.779	0.792	0.802	0.814	0.824	0.840	0.854
0.00600	(0.0588)	0.720	0.723	0.731	0.737	0.743	0.753	0.762	0.770	0.783	0.794	0.806	0.816	0.833	0.848
0.00800	(0.0785)	0.711	0.715	0.723	0.730	0.736	0.746	0.755	0.763	0.777	0.788	0.801	0.811	0.829	0.843
0.01000	(0.0981)	0.705	0.709	0.717	0.724	0.731	0.741	0.751	0.759	0.773	0.784	0.797	0.807	0.825	0.840
0.01500	(0.147)	0.694	0.699	0.708	0.715	0.722	0.733	0.743	0.751	0.766	0.778	0.790	0.801	0.820	0.835
0.02000	(0.196)	0.687	0.692	0.701	0.709	0.716	0.728	0.738	0.746	0.761	0.773	0.786	0.798	0.816	0.831
0.03000	(0.294)	0.678	0.683	0.693	0.701	0.708	0.721	0.731	0.740	0.755	0.768	0.781	0.793	0.812	0.827
0.04000	(0.392)	0.672	0.678	0.687	0.696	0.703	0.716	0.727	0.736	0.752	0.764	0.778	0.789	0.809	0.824
0.06000	(0.588)	0.664	0.670	0.680	0.689	0.697	0.710	0.721	0.731	0.747	0.760	0.774	0.786	0.805	0.821
0.08000	(0.785)	0.659	0.665	0.676	0.685	0.693	0.707	0.718	0.728	0.744	0.757	0.771	0.783	0.803	0.819
0.10000	(0.981)	0.655	0.662	0.673	0.682	0.690	0.704	0.715	0.725	0.742	0.755	0.769	0.781	0.801	0.818
0.15000	(1.47)	0.649	0.656	0.667	0.677	0.685	0.700	0.711	0.721	0.738	0.752	0.766	0.779	0.799	0.815
0.20000	(1.96)	0.645	0.652	0.664	0.674	0.682	0.697	0.709	0.719	0.736	0.750	0.765	0.777	0.797	0.814
0.30000	(2.94)	0.640	0.647	0.659	0.670	0.679	0.693	0.706	0.716	0.733	0.748	0.762	0.775	0.796	0.812
0.40000	(3.92)	0.636	0.644	0.657	0.667	0.676	0.691	0.704	0.714	0.732	0.746	0.761	0.774	0.794	0.811
0.60000	(5.88)	0.632	0.640	0.653	0.664	0.673	0.689	0.701	0.712	0.730	0.744	0.759	0.772	0.793	0.810
0.80000	(7.85)	0.630	0.638	0.651	0.662	0.671	0.687	0.700	0.711	0.729	0.743	0.758	0.771	0.792	0.809
1.00000	(9.81)	0.628	0.636	0.650	0.661	0.670	0.686	0.699	0.710	0.728	0.742	0.758	0.771	0.792	0.809
1.50000	(14.7)	0.625	0.633	0.647	0.658	0.668	0.684	0.697	0.708	0.726	0.741	0.757	0.770	0.791	0.808
2.00000	(19.6)	0.623	0.632	0.646	0.657	0.667	0.683	0.696	0.707	0.726	0.740	0.756	0.769	0.790	0.808
3.00000	(29.4)	0.621	0.630	0.644	0.656	0.665	0.682	0.695	0.706	0.725	0.740	0.755	0.768	0.790	0.807
S	$(\Delta p/\rho l)$	125	150	200	250	300	400	500	600	800	1000	1250	1500	2000	2500

Grad'nt   **k.p.g.**   **(Equivalent) Pipe diameters in mm**

Kinematic viscosity, $\nu = 0.20\times10^{-6}\ \mathrm{m^2 s^{-1}}$; **Roughness size,** $k_s = 0.003$ mm

E19

Kin. visc., $\nu = 0.20 \times 10^{-6}$ m^{2}s^{-1};
$S = 0.00010$ to 30.0000

i.e. kin. pr. grad., $\Delta p/\rho l =$
(0.00098) to (294) ms^{-2}

Roughness size, $k_s = 0.006$ mm
This table shows values of m, as follows

m_C for Colebrook-White solutions; or,
where $\mathbf{R} \leq 2000$, m_P for laminar flow

Grad'nt **k.p.g.** **(Equivalent) Pipe diameters in mm**

S	$(\Delta p/\rho l)$	6	8	10	12.5	15	20	25	30	40	50	60	80	100	125
0.00030	(0.00294)	1.371	0.935	*	0.931	0.914	0.891	0.876	0.867	0.855	0.847	0.843	0.838	0.836	0.835
0.00040	(0.00392)	1.188	0.809	*	0.908	0.892	0.871	0.858	0.849	0.838	0.832	0.828	0.824	0.823	0.823
0.00060	(0.00588)	0.970	*	0.897	0.876	0.862	0.844	0.833	0.826	0.817	0.812	0.809	0.806	0.806	0.806
0.00080	(0.00785)	0.840	*	0.874	0.855	0.843	0.826	0.817	0.810	0.802	0.798	0.796	0.794	0.794	0.795
0.00100	(0.00981)	0.751	*	0.857	0.840	0.828	0.813	0.804	0.798	0.791	0.788	0.786	0.785	0.785	0.787
0.00150	(0.0147)	*	0.846	0.828	0.814	0.803	0.791	0.783	0.778	0.773	0.770	0.769	0.769	0.770	0.772
0.00200	(0.0196)	*	0.826	0.809	0.796	0.787	0.775	0.769	0.765	0.760	0.758	0.758	0.758	0.760	0.763
0.00300	(0.0294)	0.821	0.798	0.784	0.773	0.765	0.755	0.750	0.747	0.743	0.742	0.743	0.744	0.747	0.750
0.00400	(0.0392)	0.801	0.780	0.768	0.757	0.750	0.742	0.737	0.735	0.732	0.732	0.733	0.735	0.738	0.742
0.00600	(0.0588)	0.774	0.757	0.746	0.737	0.731	0.724	0.721	0.719	0.718	0.718	0.719	0.722	0.726	0.731
0.00800	(0.0785)	0.757	0.741	0.731	0.723	0.718	0.712	0.710	0.708	0.708	0.709	0.710	0.714	0.718	0.723
0.01000	(0.0981)	0.744	0.729	0.720	0.713	0.709	0.704	0.701	0.700	0.701	0.702	0.704	0.708	0.713	0.718
0.01500	(0.147)	0.722	0.709	0.702	0.696	0.692	0.689	0.687	0.687	0.688	0.690	0.693	0.698	0.703	0.709
0.02000	(0.196)	0.707	0.696	0.689	0.685	0.682	0.679	0.678	0.678	0.680	0.683	0.686	0.691	0.697	0.703
0.03000	(0.294)	0.688	0.679	0.673	0.670	0.667	0.666	0.666	0.667	0.670	0.673	0.676	0.683	0.689	0.695
0.04000	(0.392)	0.676	0.667	0.663	0.660	0.658	0.657	0.658	0.659	0.663	0.666	0.670	0.677	0.683	0.690
0.06000	(0.588)	0.659	0.652	0.649	0.647	0.646	0.646	0.648	0.650	0.654	0.658	0.662	0.670	0.677	0.684
0.08000	(0.785)	0.648	0.643	0.640	0.638	0.638	0.639	0.641	0.643	0.648	0.653	0.657	0.665	0.672	0.680
0.10000	(0.981)	0.640	0.635	0.633	0.632	0.632	0.634	0.636	0.639	0.644	0.649	0.654	0.662	0.669	0.677
0.15000	(1.47)	0.627	0.623	0.622	0.622	0.623	0.625	0.628	0.631	0.637	0.642	0.648	0.657	0.664	0.673
0.20000	(1.96)	0.618	0.616	0.615	0.615	0.616	0.619	0.623	0.626	0.633	0.638	0.644	0.653	0.661	0.670
0.30000	(2.94)	0.607	0.606	0.606	0.607	0.608	0.612	0.616	0.620	0.627	0.633	0.639	0.649	0.657	0.666
0.40000	(3.92)	0.600	0.599	0.600	0.601	0.603	0.608	0.612	0.616	0.624	0.630	0.636	0.646	0.655	0.664
0.60000	(5.88)	0.591	0.591	0.592	0.595	0.597	0.602	0.607	0.611	0.619	0.626	0.632	0.643	0.652	0.662
0.80000	(7.85)	0.585	0.586	0.587	0.590	0.593	0.598	0.604	0.608	0.617	0.624	0.630	0.641	0.650	0.660
1.00000	(9.81)	0.581	0.582	0.584	0.587	0.590	0.596	0.601	0.606	0.615	0.622	0.629	0.640	0.649	0.659
1.50000	(14.7)	0.574	0.576	0.579	0.582	0.585	0.592	0.597	0.603	0.612	0.619	0.626	0.637	0.647	0.657
2.00000	(19.6)	0.569	0.572	0.575	0.579	0.583	0.589	0.595	0.601	0.610	0.618	0.624	0.636	0.646	0.656
3.00000	(29.4)	0.564	0.567	0.571	0.575	0.579	0.586	0.592	0.598	0.607	0.616	0.623	0.634	0.644	0.654
4.00000	(39.2)	0.561	0.564	0.568	0.573	0.577	0.584	0.591	0.596	0.606	0.614	0.621	0.633	0.643	0.654
6.00000	(58.8)	0.556	0.561	0.565	0.570	0.574	0.582	0.588	0.594	0.604	0.613	0.620	0.632	0.642	0.653
8.00000	(78.5)	0.554	0.558	0.563	0.568	0.572	0.580	0.587	0.593	0.603	0.612	0.619	0.631	0.641	0.652
10.0000	(98.1)	0.552	0.557	0.561	0.567	0.571	0.579	0.586	0.592	0.603	0.611	0.619	0.631	0.641	0.652
15.0000	(147)	0.549	0.554	0.559	0.565	0.569	0.578	0.585	0.591	0.602	0.610	0.618	0.630	0.640	0.651
20.0000	(196)	0.547	0.553	0.558	0.563	0.568	0.577	0.584	0.590	0.601	0.610	0.617	0.630	0.640	0.651
30.0000	(294)	0.545	0.551	0.556	0.562	0.567	0.576	0.583	0.589	0.600	0.609	0.616	0.629	0.639	0.650
S	$(\Delta p/\rho l)$	6	8	10	12.5	15	20	25	30	40	50	60	80	100	125

Grad'nt **k.p.g.** **(Equivalent) Pipe diameters in mm**

S	$(\Delta p/\rho l)$	125	150	200	250	300	400	500	600	800	1000	1250	1500	2000	2500
0.00010	(0.00098)	0.888	0.886	0.885	0.886	0.888	0.892	0.897	0.901	0.909	0.917	0.925	0.933	0.946	0.957
0.00015	(0.00147)	0.867	0.866	0.867	0.869	0.871	0.876	0.881	0.886	0.895	0.903	0.912	0.919	0.933	0.945
0.00020	(0.00196)	0.854	0.853	0.854	0.857	0.859	0.865	0.870	0.875	0.885	0.893	0.902	0.911	0.925	0.937
0.00030	(0.00294)	0.835	0.836	0.838	0.841	0.844	0.850	0.856	0.862	0.872	0.881	0.890	0.899	0.913	0.926
0.00040	(0.00392)	0.823	0.824	0.826	0.830	0.833	0.840	0.847	0.852	0.863	0.872	0.882	0.891	0.906	0.919
0.00060	(0.00588)	0.806	0.808	0.811	0.815	0.819	0.827	0.834	0.840	0.851	0.861	0.871	0.881	0.896	0.909
0.00080	(0.00785)	0.795	0.797	0.801	0.806	0.810	0.818	0.825	0.832	0.843	0.853	0.864	0.874	0.890	0.903
0.00100	(0.00981)	0.787	0.789	0.794	0.798	0.803	0.811	0.819	0.826	0.838	0.848	0.859	0.869	0.885	0.899
0.00150	(0.0147)	0.772	0.775	0.781	0.786	0.791	0.800	0.808	0.815	0.828	0.839	0.850	0.860	0.877	0.891
0.00200	(0.0196)	0.763	0.766	0.772	0.778	0.783	0.793	0.801	0.809	0.822	0.833	0.844	0.855	0.872	0.886
0.00300	(0.0294)	0.750	0.754	0.760	0.767	0.773	0.783	0.792	0.800	0.813	0.825	0.837	0.847	0.865	0.880
0.00400	(0.0392)	0.742	0.746	0.753	0.760	0.766	0.776	0.786	0.794	0.808	0.819	0.832	0.843	0.861	0.876
0.00600	(0.0588)	0.731	0.735	0.743	0.750	0.757	0.768	0.778	0.786	0.801	0.813	0.826	0.837	0.856	0.871
0.00800	(0.0785)	0.723	0.728	0.736	0.744	0.751	0.763	0.773	0.781	0.796	0.809	0.822	0.833	0.852	0.868
0.01000	(0.0981)	0.718	0.723	0.732	0.740	0.747	0.759	0.769	0.778	0.793	0.806	0.819	0.831	0.850	0.866
0.01500	(0.147)	0.709	0.714	0.724	0.732	0.739	0.752	0.763	0.772	0.788	0.801	0.815	0.826	0.846	0.862
0.02000	(0.196)	0.703	0.709	0.719	0.727	0.735	0.748	0.759	0.768	0.784	0.798	0.812	0.824	0.844	0.860
0.03000	(0.294)	0.695	0.701	0.712	0.721	0.729	0.743	0.754	0.764	0.780	0.794	0.808	0.820	0.841	0.857
0.04000	(0.392)	0.690	0.697	0.708	0.717	0.725	0.739	0.751	0.761	0.778	0.791	0.806	0.818	0.839	0.855
0.06000	(0.588)	0.684	0.691	0.702	0.712	0.721	0.735	0.747	0.757	0.774	0.789	0.803	0.816	0.836	0.853
0.08000	(0.785)	0.680	0.687	0.699	0.709	0.718	0.733	0.745	0.755	0.773	0.787	0.802	0.814	0.835	0.852
0.10000	(0.981)	0.677	0.685	0.697	0.707	0.716	0.731	0.743	0.754	0.771	0.785	0.800	0.813	0.834	0.851
0.15000	(1.47)	0.673	0.680	0.693	0.703	0.713	0.728	0.740	0.751	0.769	0.783	0.799	0.811	0.833	0.850
0.20000	(1.96)	0.670	0.678	0.691	0.701	0.711	0.726	0.739	0.750	0.768	0.782	0.797	0.810	0.832	0.849
0.30000	(2.94)	0.666	0.674	0.688	0.699	0.708	0.724	0.737	0.748	0.766	0.781	0.796	0.809	0.831	0.848
0.40000	(3.92)	0.664	0.672	0.686	0.697	0.707	0.722	0.736	0.747	0.765	0.780	0.795	0.808	0.830	0.847
0.60000	(5.88)	0.662	0.670	0.684	0.695	0.705	0.721	0.734	0.745	0.764	0.779	0.794	0.808	0.829	0.847
0.80000	(7.85)	0.660	0.668	0.682	0.694	0.704	0.720	0.733	0.744	0.763	0.778	0.794	0.807	0.829	0.846
1.00000	(9.81)	0.659	0.667	0.681	0.693	0.703	0.719	0.733	0.744	0.762	0.778	0.793	0.807	0.828	0.846
1.50000	(14.7)	0.657	0.665	0.680	0.692	0.702	0.718	0.732	0.743	0.762	0.777	0.793	0.806	0.828	0.845
2.00000	(19.6)	0.656	0.664	0.679	0.691	0.701	0.717	0.731	0.742	0.761	0.776	0.792	0.806	0.828	0.845
3.00000	(29.4)	0.654	0.663	0.678	0.690	0.700	0.717	0.730	0.742	0.761	0.776	0.792	0.805	0.827	0.845
S	$(\Delta p/\rho l)$	125	150	200	250	300	400	500	600	800	1000	1250	1500	2000	2500

Grad'nt **k.p.g.** **(Equivalent) Pipe diameters in mm**

Kinematic viscosity, $\nu = 0.20 \times 10^{-6}$ m^{2}s^{-1} ; Roughness size, $k_s = 0.006$ mm

Kin. visc., $\nu = 0.20\times10^{-6}\ \mathrm{m^2 s^{-1}}$;
$S = 0.00010$ to 30.0000
i.e. kin. pr. grad., $\Delta p/\rho l =$
(0.00098) to $(294)\ \mathrm{ms^{-2}}$

Roughness size, $k_\mathrm{s} = 0.015$ mm
This table shows values of m, as follows
m_C for Colebrook-White solutions; or,
where $\mathbf{R} \leq 2000$, m_P for laminar flow

Grad'nt S	k.p.g. $(\Delta p/\rho l)$	(Equivalent) Pipe diameters in mm													
		6	8	10	12.5	15	20	25	30	40	50	60	80	100	125
0.00030	(0.00294)	1.371	0.935	*	0.938	0.921	0.898	0.884	0.875	0.863	0.856	0.852	0.848	0.846	0.846
0.00040	(0.00392)	1.188	0.809	*	0.915	0.899	0.879	0.866	0.858	0.847	0.842	0.838	0.835	0.834	0.835
0.00060	(0.00588)	0.970	*	0.905	0.885	0.871	0.853	0.843	0.836	0.827	0.823	0.820	0.818	0.819	0.820
0.00080	(0.00785)	0.840	*	0.883	0.865	0.852	0.836	0.827	0.821	0.814	0.810	0.808	0.807	0.808	0.810
0.00100	(0.00981)	0.751	*	0.867	0.850	0.839	0.824	0.815	0.810	0.804	0.801	0.799	0.799	0.801	0.803
0.00150	(0.0147)	*	0.857	0.839	0.825	0.815	0.803	0.796	0.791	0.787	0.785	0.784	0.785	0.787	0.791
0.00200	(0.0196)	*	0.837	0.821	0.808	0.800	0.789	0.783	0.779	0.775	0.774	0.774	0.776	0.779	0.783
0.00300	(0.0294)	0.834	0.812	0.798	0.787	0.779	0.770	0.766	0.763	0.761	0.761	0.761	0.764	0.768	0.772
0.00400	(0.0392)	0.815	0.795	0.782	0.772	0.766	0.758	0.755	0.752	0.751	0.752	0.753	0.757	0.761	0.766
0.00600	(0.0588)	0.790	0.773	0.762	0.754	0.749	0.743	0.740	0.739	0.739	0.740	0.742	0.747	0.751	0.757
0.00800	(0.0785)	0.774	0.758	0.749	0.742	0.737	0.732	0.730	0.730	0.731	0.733	0.735	0.740	0.746	0.752
0.01000	(0.0981)	0.762	0.748	0.739	0.733	0.729	0.725	0.724	0.723	0.725	0.727	0.730	0.736	0.741	0.748
0.01500	(0.147)	0.742	0.730	0.723	0.718	0.715	0.712	0.712	0.713	0.715	0.718	0.722	0.728	0.734	0.741
0.02000	(0.196)	0.729	0.718	0.713	0.708	0.706	0.704	0.705	0.706	0.709	0.713	0.716	0.723	0.730	0.737
0.03000	(0.294)	0.712	0.704	0.699	0.696	0.695	0.694	0.695	0.697	0.701	0.705	0.710	0.717	0.724	0.732
0.04000	(0.392)	0.701	0.694	0.690	0.688	0.687	0.688	0.689	0.691	0.696	0.701	0.705	0.714	0.721	0.729
0.06000	(0.588)	0.688	0.682	0.679	0.678	0.678	0.679	0.682	0.684	0.690	0.695	0.700	0.709	0.717	0.725
0.08000	(0.785)	0.679	0.674	0.672	0.671	0.672	0.674	0.677	0.680	0.686	0.691	0.697	0.706	0.714	0.723
0.10000	(0.981)	0.672	0.669	0.667	0.667	0.668	0.670	0.673	0.677	0.683	0.689	0.694	0.704	0.712	0.721
0.15000	(1.47)	0.662	0.659	0.659	0.659	0.661	0.664	0.668	0.672	0.679	0.685	0.691	0.701	0.709	0.719
0.20000	(1.96)	0.655	0.654	0.654	0.655	0.656	0.660	0.664	0.668	0.676	0.682	0.688	0.699	0.707	0.717
0.30000	(2.94)	0.647	0.646	0.647	0.649	0.651	0.656	0.660	0.665	0.672	0.679	0.686	0.696	0.705	0.715
0.40000	(3.92)	0.642	0.642	0.643	0.645	0.648	0.653	0.658	0.662	0.670	0.678	0.684	0.695	0.704	0.714
0.60000	(5.88)	0.636	0.636	0.638	0.641	0.644	0.649	0.655	0.659	0.668	0.675	0.682	0.693	0.702	0.712
0.80000	(7.85)	0.632	0.633	0.635	0.638	0.641	0.647	0.653	0.658	0.666	0.674	0.681	0.692	0.701	0.711
1.00000	(9.81)	0.629	0.631	0.633	0.636	0.639	0.646	0.651	0.656	0.665	0.673	0.680	0.691	0.701	0.711
1.50000	(14.7)	0.624	0.627	0.629	0.633	0.637	0.643	0.649	0.654	0.664	0.671	0.678	0.690	0.700	0.710
2.00000	(19.6)	0.622	0.624	0.627	0.631	0.635	0.642	0.648	0.653	0.663	0.671	0.678	0.689	0.699	0.709
3.00000	(29.4)	0.618	0.621	0.625	0.629	0.633	0.640	0.646	0.652	0.661	0.670	0.677	0.688	0.698	0.709
4.00000	(39.2)	0.616	0.620	0.623	0.628	0.632	0.639	0.645	0.651	0.661	0.669	0.676	0.688	0.698	0.708
6.00000	(58.8)	0.614	0.618	0.621	0.626	0.630	0.638	0.644	0.650	0.660	0.668	0.675	0.687	0.697	0.708
8.00000	(78.5)	0.612	0.616	0.620	0.625	0.629	0.637	0.644	0.649	0.659	0.668	0.675	0.687	0.697	0.707
10.0000	(98.1)	0.611	0.616	0.620	0.624	0.629	0.636	0.643	0.649	0.659	0.667	0.675	0.687	0.697	0.707
15.0000	(147)	0.610	0.614	0.618	0.623	0.628	0.636	0.642	0.648	0.658	0.667	0.674	0.686	0.696	0.707
20.0000	(196)	0.609	0.613	0.618	0.623	0.627	0.635	0.642	0.648	0.658	0.666	0.674	0.686	0.696	0.707
30.0000	(294)	0.608	0.612	0.617	0.622	0.626	0.635	0.641	0.647	0.658	0.666	0.673	0.686	0.696	0.707
S	$(\Delta p/\rho l)$	6	8	10	12.5	15	20	25	30	40	50	60	80	100	125

Grad'nt k.p.g. (Equivalent) Pipe diameters in mm

Grad'nt S	k.p.g. $(\Delta p/\rho l)$	(Equivalent) Pipe diameters in mm													
		125	150	200	250	300	400	500	600	800	1000	1250	1500	2000	2500
0.00010	(0.00098)	0.895	0.894	0.894	0.895	0.898	0.903	0.908	0.913	0.922	0.931	0.940	0.949	0.963	0.976
0.00015	(0.00147)	0.876	0.876	0.877	0.879	0.882	0.888	0.894	0.900	0.910	0.919	0.929	0.938	0.953	0.966
0.00020	(0.00196)	0.863	0.863	0.865	0.868	0.872	0.878	0.885	0.891	0.901	0.911	0.921	0.931	0.946	0.960
0.00030	(0.00294)	0.846	0.847	0.850	0.854	0.858	0.865	0.872	0.879	0.890	0.901	0.912	0.921	0.938	0.952
0.00040	(0.00392)	0.835	0.836	0.840	0.844	0.849	0.857	0.864	0.871	0.883	0.894	0.905	0.915	0.932	0.946
0.00060	(0.00588)	0.820	0.822	0.827	0.832	0.837	0.846	0.854	0.861	0.874	0.885	0.897	0.907	0.925	0.940
0.00080	(0.00785)	0.810	0.813	0.818	0.824	0.829	0.839	0.847	0.855	0.868	0.880	0.892	0.902	0.920	0.935
0.00100	(0.00981)	0.803	0.806	0.812	0.818	0.823	0.833	0.842	0.850	0.864	0.875	0.888	0.899	0.917	0.932
0.00150	(0.0147)	0.791	0.794	0.801	0.808	0.814	0.825	0.834	0.842	0.857	0.869	0.882	0.893	0.912	0.928
0.00200	(0.0196)	0.783	0.787	0.794	0.801	0.808	0.819	0.829	0.837	0.852	0.865	0.878	0.889	0.909	0.924
0.00300	(0.0294)	0.772	0.777	0.785	0.793	0.800	0.812	0.822	0.831	0.846	0.859	0.873	0.885	0.904	0.921
0.00400	(0.0392)	0.766	0.771	0.780	0.788	0.795	0.807	0.818	0.827	0.843	0.856	0.870	0.882	0.902	0.918
0.00600	(0.0588)	0.757	0.763	0.772	0.781	0.788	0.801	0.813	0.822	0.838	0.852	0.866	0.878	0.899	0.915
0.00800	(0.0785)	0.752	0.757	0.768	0.776	0.784	0.798	0.809	0.819	0.836	0.849	0.864	0.876	0.897	0.914
0.01000	(0.0981)	0.748	0.754	0.764	0.773	0.781	0.795	0.807	0.817	0.834	0.848	0.862	0.875	0.896	0.912
0.01500	(0.147)	0.741	0.748	0.759	0.768	0.777	0.791	0.803	0.813	0.831	0.845	0.860	0.872	0.893	0.911
0.02000	(0.196)	0.737	0.744	0.756	0.765	0.774	0.789	0.801	0.811	0.829	0.843	0.858	0.871	0.892	0.909
0.03000	(0.294)	0.732	0.739	0.751	0.762	0.770	0.785	0.798	0.809	0.826	0.841	0.856	0.869	0.891	0.908
0.04000	(0.392)	0.729	0.736	0.749	0.759	0.768	0.784	0.796	0.807	0.825	0.840	0.855	0.868	0.890	0.907
0.06000	(0.588)	0.725	0.733	0.746	0.756	0.766	0.781	0.794	0.805	0.823	0.838	0.854	0.867	0.888	0.906
0.08000	(0.785)	0.723	0.731	0.744	0.755	0.764	0.780	0.793	0.804	0.822	0.837	0.853	0.866	0.888	0.905
0.10000	(0.981)	0.721	0.729	0.742	0.753	0.763	0.779	0.792	0.803	0.821	0.836	0.852	0.865	0.887	0.905
0.15000	(1.47)	0.719	0.727	0.740	0.751	0.761	0.777	0.790	0.802	0.820	0.835	0.851	0.865	0.887	0.904
0.20000	(1.96)	0.717	0.725	0.739	0.750	0.760	0.776	0.790	0.801	0.820	0.835	0.851	0.864	0.886	0.904
0.30000	(2.94)	0.715	0.723	0.737	0.749	0.759	0.775	0.789	0.800	0.819	0.834	0.850	0.863	0.886	0.903
0.40000	(3.92)	0.714	0.722	0.736	0.748	0.758	0.774	0.788	0.799	0.818	0.834	0.850	0.863	0.885	0.903
0.60000	(5.88)	0.712	0.721	0.735	0.747	0.757	0.774	0.787	0.799	0.818	0.833	0.849	0.863	0.885	0.903
0.80000	(7.85)	0.711	0.720	0.734	0.746	0.756	0.773	0.787	0.798	0.817	0.833	0.849	0.862	0.885	0.903
1.00000	(9.81)	0.711	0.720	0.734	0.746	0.756	0.773	0.786	0.798	0.817	0.833	0.849	0.862	0.885	0.902
1.50000	(14.7)	0.710	0.719	0.733	0.745	0.755	0.772	0.786	0.798	0.817	0.832	0.848	0.862	0.884	0.902
2.00000	(19.6)	0.709	0.718	0.733	0.745	0.755	0.772	0.786	0.797	0.817	0.832	0.848	0.862	0.884	0.902
3.00000	(29.4)	0.709	0.718	0.732	0.744	0.755	0.772	0.785	0.797	0.816	0.832	0.848	0.862	0.884	0.902
S	$(\Delta p/\rho l)$	125	150	200	250	300	400	500	600	800	1000	1250	1500	2000	2500

Grad'nt k.p.g. (Equivalent) Pipe diameters in mm

Kinematic viscosity, $\nu = 0.20\times10^{-6}\ \mathrm{m^2 s^{-1}}$;   **Roughness size, $k_\mathrm{s} = 0.015$ mm**

E21

Kin. visc., $\nu = 0.20 \times 10^{-6}$ m²s⁻¹;
$S = 0.00010$ to 3.00000
i.e. kin. pr. grad., $\Delta p/\rho l =$ (0.00098) to (29.4) ms⁻²

Roughness size, $k_s = 0.030$ mm
This table shows values of m, as follows
m_C for Colebrook-White solutions

Grad'nt S	k.p.g. $(\Delta p/\rho l)$	125	150	200	250	300	400	500	600	800	1000	1250	1500	2000	2500
0.00010	(0.00098)	0.907	0.906	0.907	0.909	0.912	0.919	0.925	0.931	0.941	0.951	0.962	0.971	0.987	1.001
0.00015	(0.00147)	0.889	0.890	0.892	0.895	0.899	0.906	0.913	0.919	0.931	0.941	0.952	0.962	0.979	0.994
0.00020	(0.00196)	0.878	0.879	0.882	0.886	0.890	0.898	0.905	0.912	0.924	0.935	0.946	0.957	0.974	0.989
0.00030	(0.00294)	0.863	0.864	0.869	0.873	0.878	0.887	0.895	0.902	0.915	0.927	0.939	0.950	0.968	0.983
0.00040	(0.00392)	0.853	0.855	0.860	0.865	0.870	0.880	0.889	0.896	0.910	0.922	0.934	0.945	0.964	0.979
0.00060	(0.00588)	0.840	0.843	0.849	0.855	0.861	0.871	0.880	0.889	0.903	0.915	0.928	0.940	0.959	0.975
0.00080	(0.00785)	0.832	0.835	0.842	0.849	0.855	0.866	0.875	0.884	0.899	0.911	0.925	0.936	0.956	0.972
0.00100	(0.00981)	0.826	0.830	0.837	0.844	0.850	0.862	0.872	0.880	0.896	0.908	0.922	0.934	0.953	0.970
0.00150	(0.0147)	0.816	0.820	0.828	0.836	0.843	0.855	0.866	0.875	0.890	0.904	0.918	0.930	0.950	0.966
0.00200	(0.0196)	0.810	0.814	0.823	0.831	0.838	0.851	0.862	0.871	0.887	0.901	0.915	0.927	0.948	0.965
0.00300	(0.0294)	0.802	0.807	0.816	0.825	0.833	0.846	0.857	0.867	0.884	0.897	0.912	0.924	0.945	0.962
0.00400	(0.0392)	0.796	0.802	0.812	0.821	0.829	0.843	0.854	0.864	0.881	0.895	0.910	0.923	0.944	0.961
0.00600	(0.0588)	0.790	0.796	0.807	0.816	0.825	0.839	0.851	0.861	0.878	0.893	0.908	0.920	0.942	0.959
0.00800	(0.0785)	0.786	0.793	0.804	0.813	0.822	0.836	0.849	0.859	0.876	0.891	0.906	0.919	0.940	0.958
0.01000	(0.0981)	0.783	0.790	0.801	0.811	0.820	0.835	0.847	0.858	0.875	0.890	0.905	0.918	0.940	0.957
0.01500	(0.147)	0.779	0.786	0.798	0.808	0.817	0.832	0.845	0.855	0.873	0.888	0.904	0.917	0.938	0.956
0.02000	(0.196)	0.776	0.783	0.796	0.806	0.815	0.830	0.843	0.854	0.872	0.887	0.903	0.916	0.938	0.955
0.03000	(0.294)	0.773	0.780	0.793	0.804	0.813	0.828	0.841	0.852	0.871	0.886	0.901	0.915	0.937	0.954
0.04000	(0.392)	0.771	0.778	0.791	0.802	0.812	0.827	0.840	0.851	0.870	0.885	0.901	0.914	0.936	0.954
0.06000	(0.588)	0.768	0.776	0.789	0.800	0.810	0.826	0.839	0.850	0.869	0.884	0.900	0.913	0.935	0.953
0.08000	(0.785)	0.767	0.775	0.788	0.799	0.809	0.825	0.838	0.850	0.868	0.884	0.899	0.913	0.935	0.953
0.10000	(0.981)	0.766	0.774	0.787	0.799	0.808	0.824	0.838	0.849	0.868	0.883	0.899	0.913	0.935	0.953
0.15000	(1.47)	0.764	0.772	0.786	0.797	0.807	0.824	0.837	0.848	0.867	0.883	0.899	0.912	0.934	0.952
0.20000	(1.96)	0.763	0.771	0.785	0.797	0.807	0.823	0.836	0.848	0.867	0.882	0.898	0.912	0.934	0.952
0.30000	(2.94)	0.762	0.770	0.784	0.796	0.806	0.822	0.836	0.847	0.866	0.882	0.898	0.911	0.934	0.952
0.40000	(3.92)	0.761	0.770	0.784	0.795	0.805	0.822	0.835	0.847	0.866	0.882	0.898	0.911	0.934	0.952
0.60000	(5.88)	0.760	0.769	0.783	0.795	0.805	0.821	0.835	0.847	0.866	0.881	0.897	0.911	0.933	0.952
0.80000	(7.85)	0.760	0.768	0.783	0.794	0.804	0.821	0.835	0.846	0.865	0.881	0.897	0.911	0.933	0.951
1.00000	(9.81)	0.759	0.768	0.782	0.794	0.804	0.821	0.835	0.846	0.865	0.881	0.897	0.911	0.933	0.951
1.50000	(14.7)	0.759	0.767	0.782	0.794	0.804	0.821	0.834	0.846	0.865	0.881	0.897	0.911	0.933	0.951
2.00000	(19.6)	0.758	0.767	0.782	0.793	0.804	0.820	0.834	0.846	0.865	0.881	0.897	0.911	0.933	0.951
3.00000	(29.4)	0.758	0.767	0.781	0.793	0.803	0.820	0.834	0.846	0.865	0.880	0.897	0.910	0.933	0.951

S — $(\Delta p/\rho l)$ — Grad'nt — k.p.g. — (Equivalent) Pipe diameters in mm — Roughness size, $k_s = 0.030$ mm

E22

Kin. visc., $\nu = 0.20 \times 10^{-6}$ m²s⁻¹;
$S = 0.00010$ to 3.00000
i.e. kin. pr. grad., $\Delta p/\rho l =$ (0.00098) to (29.4) ms⁻²

Roughness size, $k_s = 0.060$ mm
This table shows values of m, as follows
m_C for Colebrook-White solutions

Grad'nt S	k.p.g. $(\Delta p/\rho l)$	125	150	200	250	300	400	500	600	800	1000	1250	1500	2000	2500
0.00010	(0.00098)	0.928	0.928	0.930	0.934	0.937	0.945	0.952	0.959	0.972	0.982	0.994	1.005	1.023	1.038
0.00015	(0.00147)	0.913	0.914	0.918	0.922	0.926	0.935	0.943	0.950	0.964	0.975	0.988	0.999	1.017	1.033
0.00020	(0.00196)	0.903	0.905	0.909	0.914	0.919	0.929	0.937	0.945	0.959	0.971	0.983	0.995	1.014	1.029
0.00030	(0.00294)	0.891	0.893	0.899	0.904	0.910	0.920	0.930	0.938	0.952	0.965	0.978	0.990	1.009	1.025
0.00040	(0.00392)	0.883	0.886	0.892	0.898	0.904	0.915	0.925	0.934	0.949	0.961	0.975	0.987	1.007	1.023
0.00060	(0.00588)	0.873	0.876	0.884	0.891	0.897	0.909	0.919	0.928	0.944	0.957	0.971	0.983	1.003	1.020
0.00080	(0.00785)	0.866	0.870	0.878	0.886	0.893	0.905	0.916	0.925	0.941	0.954	0.969	0.981	1.001	1.018
0.00100	(0.00981)	0.862	0.866	0.875	0.882	0.890	0.902	0.913	0.923	0.939	0.952	0.967	0.979	1.000	1.017
0.00150	(0.0147)	0.854	0.859	0.869	0.877	0.885	0.898	0.909	0.919	0.936	0.949	0.964	0.977	0.998	1.015
0.00200	(0.0196)	0.850	0.855	0.865	0.874	0.881	0.895	0.907	0.917	0.934	0.948	0.963	0.975	0.997	1.014
0.00300	(0.0294)	0.844	0.850	0.860	0.869	0.878	0.892	0.904	0.914	0.931	0.945	0.961	0.974	0.995	1.012
0.00400	(0.0392)	0.840	0.847	0.857	0.867	0.875	0.890	0.902	0.912	0.930	0.944	0.959	0.972	0.994	1.012
0.00600	(0.0588)	0.836	0.843	0.854	0.864	0.872	0.887	0.899	0.910	0.928	0.943	0.958	0.971	0.993	1.011
0.00800	(0.0785)	0.833	0.840	0.852	0.862	0.871	0.886	0.898	0.909	0.927	0.942	0.957	0.970	0.992	1.010
0.01000	(0.0981)	0.832	0.838	0.850	0.861	0.869	0.885	0.897	0.908	0.926	0.941	0.956	0.970	0.992	1.010
0.01500	(0.147)	0.829	0.836	0.848	0.859	0.868	0.883	0.896	0.907	0.925	0.940	0.956	0.969	0.991	1.009
0.02000	(0.196)	0.827	0.834	0.847	0.857	0.866	0.882	0.895	0.906	0.924	0.939	0.955	0.968	0.991	1.008
0.03000	(0.294)	0.825	0.832	0.845	0.856	0.865	0.881	0.894	0.905	0.923	0.939	0.954	0.968	0.990	1.008
0.04000	(0.392)	0.824	0.831	0.844	0.855	0.864	0.880	0.893	0.904	0.923	0.938	0.954	0.967	0.990	1.008
0.06000	(0.588)	0.822	0.830	0.843	0.854	0.863	0.879	0.892	0.904	0.922	0.938	0.953	0.967	0.989	1.007
0.08000	(0.785)	0.821	0.829	0.842	0.853	0.863	0.879	0.892	0.903	0.922	0.937	0.953	0.967	0.989	1.007
0.10000	(0.981)	0.820	0.828	0.842	0.853	0.862	0.878	0.892	0.903	0.922	0.937	0.953	0.967	0.989	1.007
0.15000	(1.47)	0.819	0.827	0.841	0.852	0.862	0.878	0.891	0.902	0.921	0.937	0.953	0.966	0.989	1.007
0.20000	(1.96)	0.819	0.827	0.840	0.852	0.861	0.878	0.891	0.902	0.921	0.936	0.953	0.966	0.989	1.007
0.30000	(2.94)	0.818	0.826	0.840	0.851	0.861	0.877	0.891	0.902	0.921	0.936	0.952	0.966	0.988	1.007
0.40000	(3.92)	0.818	0.826	0.839	0.851	0.861	0.877	0.890	0.902	0.921	0.936	0.952	0.966	0.988	1.006
0.60000	(5.88)	0.817	0.825	0.839	0.850	0.860	0.877	0.890	0.902	0.920	0.936	0.952	0.966	0.988	1.006
0.80000	(7.85)	0.817	0.825	0.839	0.850	0.860	0.876	0.890	0.901	0.920	0.936	0.952	0.966	0.988	1.006
1.00000	(9.81)	0.817	0.825	0.839	0.850	0.860	0.876	0.890	0.901	0.920	0.936	0.952	0.966	0.988	1.006
1.50000	(14.7)	0.816	0.825	0.838	0.850	0.860	0.876	0.890	0.901	0.920	0.936	0.952	0.965	0.988	1.006
2.00000	(19.6)	0.816	0.824	0.838	0.850	0.860	0.876	0.890	0.901	0.920	0.936	0.952	0.965	0.988	1.006
3.00000	(29.4)	0.816	0.824	0.838	0.850	0.859	0.876	0.889	0.901	0.920	0.935	0.952	0.965	0.988	1.006

S — $(\Delta p/\rho l)$ — Grad'nt — k.p.g. — (Equivalent) Pipe diameters in mm — Roughness size, $k_s = 0.060$ mm

Kin. visc., $\nu = 0.20 \times 10^{-6}$ m^{2}s^{-1};
$S = 0.00010$ to 3.00000

i.e. kin. pr. grad., $\Delta p/\rho l =$
(0.00098) to (29.4) ms^{-2}

Roughness size, $k_s = 0.150$ mm
This table shows values of m, as follows

m_C for Colebrook-White solutions

Grad'nt S	k.p.g. $(\Delta p/\rho l)$	125	150	200	250	300	400	500	600	800	1000	1250	1500	2000	2500
0·00010	(0·00098)	0·979	0·980	0·984	0·988	0·993	1·003	1·011	1·019	1·033	1·045	1·059	1·070	1·090	1·106
0·00015	(0·00147)	0·968	0·970	0·975	0·980	0·986	0·996	1·005	1·014	1·028	1·041	1·055	1·066	1·086	1·103
0·00020	(0·00196)	0·961	0·963	0·969	0·975	0·981	0·992	1·002	1·010	1·026	1·038	1·052	1·064	1·084	1·101
0·00030	(0·00294)	0·952	0·956	0·962	0·969	0·975	0·987	0·997	1·006	1·022	1·035	1·049	1·062	1·082	1·099
0·00040	(0·00392)	0·947	0·951	0·958	0·965	0·972	0·984	0·994	1·004	1·020	1·033	1·048	1·060	1·081	1·098
0·00060	(0·00588)	0·941	0·945	0·953	0·961	0·968	0·980	0·991	1·001	1·017	1·031	1·045	1·058	1·079	1·096
0·00080	(0·00785)	0·937	0·941	0·950	0·958	0·965	0·978	0·989	0·999	1·016	1·029	1·044	1·057	1·078	1·095
0·00100	(0·00981)	0·934	0·939	0·948	0·956	0·963	0·977	0·988	0·998	1·014	1·028	1·043	1·056	1·077	1·095
0·00150	(0·0147)	0·929	0·934	0·944	0·953	0·961	0·974	0·986	0·996	1·013	1·027	1·042	1·055	1·076	1·094
0·00200	(0·0196)	0·927	0·932	0·942	0·951	0·959	0·973	0·984	0·994	1·012	1·026	1·041	1·054	1·076	1·093
0·00300	(0·0294)	0·923	0·929	0·939	0·949	0·957	0·971	0·983	0·993	1·010	1·025	1·040	1·053	1·075	1·093
0·00400	(0·0392)	0·921	0·927	0·938	0·947	0·955	0·970	0·982	0·992	1·010	1·024	1·040	1·053	1·074	1·092
0·00600	(0·0588)	0·919	0·925	0·936	0·946	0·954	0·968	0·981	0·991	1·009	1·023	1·039	1·052	1·074	1·092
0·00800	(0·0785)	0·917	0·924	0·935	0·945	0·953	0·968	0·980	0·991	1·008	1·023	1·038	1·052	1·074	1·091
0·01000	(0·0981)	0·916	0·923	0·934	0·944	0·952	0·967	0·979	0·990	1·008	1·023	1·038	1·051	1·073	1·091
0·01500	(0·147)	0·915	0·921	0·933	0·943	0·952	0·966	0·979	0·989	1·007	1·022	1·038	1·051	1·073	1·091
0·02000	(0·196)	0·914	0·921	0·932	0·942	0·951	0·966	0·978	0·989	1·007	1·022	1·037	1·051	1·073	1·091
0·03000	(0·294)	0·913	0·920	0·931	0·941	0·950	0·965	0·978	0·989	1·007	1·021	1·037	1·050	1·072	1·090
0·04000	(0·392)	0·912	0·919	0·931	0·941	0·950	0·965	0·977	0·988	1·006	1·021	1·037	1·050	1·072	1·090
0·06000	(0·588)	0·911	0·918	0·930	0·940	0·949	0·964	0·977	0·988	1·006	1·021	1·037	1·050	1·072	1·090
0·08000	(0·785)	0·911	0·918	0·930	0·940	0·949	0·964	0·977	0·988	1·006	1·021	1·037	1·050	1·072	1·090
0·10000	(0·981)	0·911	0·918	0·930	0·940	0·949	0·964	0·977	0·988	1·006	1·021	1·036	1·050	1·072	1·090
0·15000	(1·47)	0·910	0·917	0·929	0·940	0·949	0·964	0·976	0·987	1·006	1·021	1·036	1·050	1·072	1·090
0·20000	(1·96)	0·910	0·917	0·929	0·939	0·948	0·964	0·976	0·987	1·005	1·020	1·036	1·050	1·072	1·090
0·30000	(2·94)	0·909	0·916	0·929	0·939	0·948	0·963	0·976	0·987	1·005	1·020	1·036	1·050	1·072	1·090
0·40000	(3·92)	0·909	0·916	0·929	0·939	0·948	0·963	0·976	0·987	1·005	1·020	1·036	1·049	1·072	1·090
0·60000	(5·88)	0·909	0·916	0·928	0·939	0·948	0·963	0·976	0·987	1·005	1·020	1·036	1·049	1·072	1·090
0·80000	(7·85)	0·909	0·916	0·928	0·939	0·948	0·963	0·976	0·987	1·005	1·020	1·036	1·049	1·072	1·090
1·00000	(9·81)	0·909	0·916	0·928	0·939	0·948	0·963	0·976	0·987	1·005	1·020	1·036	1·049	1·072	1·090
1·50000	(14·7)	0·908	0·916	0·928	0·938	0·948	0·963	0·976	0·987	1·005	1·020	1·036	1·049	1·071	1·090
2·00000	(19·6)	0·908	0·916	0·928	0·938	0·948	0·963	0·976	0·987	1·005	1·020	1·036	1·049	1·071	1·090
3·00000	(29·4)	0·908	0·915	0·928	0·938	0·947	0·963	0·976	0·987	1·005	1·020	1·036	1·049	1·071	1·090
S ($\Delta p/\rho l$)		125	150	200	250	300	400	500	600	800	1000	1250	1500	2000	2500

Grad'nt   k.p.g.   (Equivalent) Pipe diameters in mm   Roughness size, $k_s = 0.150$ mm

Kin. visc., $\nu = 0.20 \times 10^{-6}$ m^{2}s^{-1};
$S = 0.00010$ to 3.00000

i.e. kin. pr. grad., $\Delta p/\rho l =$
(0.00098) to (29.4) ms^{-2}

Roughness size, $k_s = 0.30$ mm
This table shows values of m, as follows

m_C for Colebrook-White solutions

Grad'nt S	k.p.g. $(\Delta p/\rho l)$	125	150	200	250	300	400	500	600	800	1000	1250	1500	2000	2500
0·00010	(0·00098)	1·041	1·042	1·046	1·051	1·056	1·066	1·075	1·083	1·098	1·110	1·124	1·136	1·156	1·172
0·00015	(0·00147)	1·033	1·035	1·040	1·046	1·051	1·062	1·071	1·080	1·095	1·108	1·121	1·133	1·154	1·170
0·00020	(0·00196)	1·028	1·030	1·036	1·042	1·048	1·059	1·069	1·078	1·093	1·106	1·120	1·132	1·152	1·169
0·00030	(0·00294)	1·022	1·025	1·032	1·038	1·045	1·056	1·066	1·075	1·091	1·104	1·118	1·130	1·151	1·168
0·00040	(0·00392)	1·018	1·022	1·029	1·036	1·042	1·054	1·064	1·074	1·089	1·103	1·117	1·129	1·150	1·167
0·00060	(0·00588)	1·014	1·018	1·026	1·033	1·040	1·052	1·062	1·072	1·088	1·101	1·116	1·128	1·149	1·166
0·00080	(0·00785)	1·011	1·016	1·023	1·031	1·038	1·050	1·061	1·071	1·087	1·101	1·115	1·128	1·149	1·166
0·00100	(0·00981)	1·010	1·014	1·022	1·030	1·037	1·049	1·060	1·070	1·086	1·100	1·115	1·127	1·148	1·166
0·00150	(0·0147)	1·007	1·011	1·020	1·028	1·035	1·048	1·059	1·069	1·085	1·099	1·114	1·126	1·148	1·165
0·00200	(0·0196)	1·005	1·010	1·019	1·027	1·034	1·047	1·058	1·068	1·085	1·098	1·113	1·126	1·147	1·165
0·00300	(0·0294)	1·003	1·008	1·017	1·025	1·033	1·046	1·057	1·067	1·084	1·098	1·113	1·125	1·147	1·164
0·00400	(0·0392)	1·002	1·007	1·016	1·024	1·032	1·045	1·057	1·067	1·083	1·097	1·112	1·125	1·146	1·164
0·00600	(0·0588)	1·000	1·005	1·015	1·023	1·031	1·044	1·056	1·066	1·083	1·097	1·112	1·125	1·146	1·164
0·00800	(0·0785)	0·999	1·005	1·014	1·023	1·031	1·044	1·055	1·065	1·082	1·096	1·111	1·124	1·146	1·163
0·01000	(0·0981)	0·999	1·004	1·014	1·022	1·030	1·044	1·055	1·065	1·082	1·096	1·111	1·124	1·146	1·163
0·01500	(0·147)	0·998	1·003	1·013	1·022	1·030	1·043	1·055	1·065	1·082	1·096	1·111	1·124	1·146	1·163
0·02000	(0·196)	0·997	1·003	1·013	1·021	1·029	1·043	1·055	1·065	1·082	1·096	1·111	1·124	1·146	1·163
0·03000	(0·294)	0·996	1·002	1·012	1·021	1·029	1·043	1·054	1·064	1·082	1·096	1·111	1·124	1·145	1·163
0·04000	(0·392)	0·996	1·002	1·012	1·021	1·029	1·042	1·054	1·064	1·081	1·096	1·111	1·124	1·145	1·163
0·06000	(0·588)	0·996	1·001	1·011	1·020	1·028	1·042	1·054	1·064	1·081	1·096	1·111	1·124	1·145	1·163
0·08000	(0·785)	0·995	1·001	1·011	1·020	1·028	1·042	1·054	1·064	1·081	1·095	1·111	1·123	1·145	1·163
0·10000	(0·981)	0·995	1·001	1·011	1·020	1·028	1·042	1·054	1·064	1·081	1·095	1·110	1·123	1·145	1·163
0·15000	(1·47)	0·995	1·000	1·011	1·020	1·028	1·042	1·053	1·064	1·081	1·095	1·110	1·123	1·145	1·163
0·20000	(1·96)	0·995	1·000	1·011	1·020	1·028	1·042	1·053	1·064	1·081	1·095	1·110	1·123	1·145	1·163
0·30000	(2·94)	0·994	1·000	1·010	1·020	1·028	1·042	1·053	1·064	1·081	1·095	1·110	1·123	1·145	1·163
0·40000	(3·92)	0·994	1·000	1·010	1·019	1·028	1·041	1·053	1·063	1·081	1·095	1·110	1·123	1·145	1·163
0·60000	(5·88)	0·994	1·000	1·010	1·019	1·027	1·041	1·053	1·063	1·081	1·095	1·110	1·123	1·145	1·163
0·80000	(7·85)	0·994	1·000	1·010	1·019	1·027	1·041	1·053	1·063	1·081	1·095	1·110	1·123	1·145	1·163
1·00000	(9·81)	0·994	1·000	1·010	1·019	1·027	1·041	1·053	1·063	1·081	1·095	1·110	1·123	1·145	1·163
1·50000	(14·7)	0·994	1·000	1·010	1·019	1·027	1·041	1·053	1·063	1·081	1·095	1·110	1·123	1·145	1·163
2·00000	(19·6)	0·994	1·000	1·010	1·019	1·027	1·041	1·053	1·063	1·081	1·095	1·110	1·123	1·145	1·163
3·00000	(29·4)	0·994	1·000	1·010	1·019	1·027	1·041	1·053	1·063	1·081	1·095	1·110	1·123	1·145	1·163
S ($\Delta p/\rho l$)		125	150	200	250	300	400	500	600	800	1000	1250	1500	2000	2500

Grad'nt   k.p.g.   (Equivalent) Pipe diameters in mm   Roughness size, $k_s = 0.30$ mm

E25

Kin. visc., $\nu = 0.25 \times 10^{-6}$ m^{2}s^{-1};
$S = 0.00010$ to 30.0000

i.e. kin. pr. grad., $\Delta p/\rho l =$
(0.00098) to (294) ms^{-2}

Roughness size, $k_s = 0.0015$ mm
This table shows values of m, as follows

m_C for Colebrook-White solutions; or,
where $R \leq 2000$, m_P for laminar flow

Grad'nt S	k.p.g. $(\Delta p/\rho l)$	6	8	10	12.5	15	20	25	30	40	50	60	80	100	125
0.00030	(0.00294)	1.714	1.168	0.868	*	0.947	0.921	0.904	0.892	0.877	0.869	0.863	0.856	0.853	0.850
0.00040	(0.00392)	1.485	1.012	0.751	*	0.923	0.899	0.883	0.873	0.860	0.852	0.847	0.841	0.838	0.837
0.00060	(0.00588)	1.212	0.826	*	0.907	0.891	0.869	0.856	0.847	0.836	0.829	0.825	0.821	0.819	0.818
0.00080	(0.00785)	1.050	*	*	0.884	0.869	0.850	0.838	0.830	0.820	0.814	0.810	0.807	0.806	0.805
0.00100	(0.00981)	0.939	*	0.887	0.867	0.853	0.835	0.824	0.817	0.808	0.803	0.800	0.797	0.796	0.796
0.00150	(0.0147)	0.767	*	0.855	0.838	0.826	0.810	0.801	0.794	0.787	0.783	0.780	0.778	0.778	0.779
0.00200	(0.0196)	*	0.853	0.834	0.818	0.807	0.793	0.785	0.779	0.773	0.769	0.767	0.766	0.767	0.768
0.00300	(0.0294)	*	0.823	0.806	0.792	0.783	0.771	0.763	0.759	0.754	0.751	0.750	0.750	0.751	0.753
0.00400	(0.0392)	0.826	0.802	0.787	0.775	0.766	0.755	0.749	0.745	0.741	0.739	0.738	0.738	0.740	0.742
0.00600	(0.0588)	0.797	0.775	0.762	0.751	0.744	0.735	0.730	0.727	0.723	0.722	0.722	0.723	0.725	0.728
0.00800	(0.0785)	0.777	0.758	0.746	0.736	0.729	0.721	0.717	0.714	0.712	0.711	0.711	0.713	0.715	0.719
0.01000	(0.0981)	0.762	0.744	0.733	0.724	0.718	0.711	0.707	0.705	0.703	0.703	0.703	0.705	0.708	0.712
0.01500	(0.147)	0.737	0.721	0.712	0.704	0.699	0.693	0.690	0.689	0.688	0.688	0.689	0.692	0.695	0.699
0.02000	(0.196)	0.720	0.706	0.697	0.691	0.686	0.681	0.679	0.678	0.677	0.678	0.680	0.683	0.686	0.691
0.03000	(0.294)	0.698	0.685	0.678	0.673	0.669	0.665	0.664	0.663	0.663	0.665	0.667	0.671	0.675	0.679
0.04000	(0.392)	0.683	0.672	0.665	0.660	0.657	0.654	0.653	0.653	0.654	0.656	0.658	0.662	0.667	0.672
0.06000	(0.588)	0.663	0.654	0.648	0.644	0.642	0.640	0.639	0.640	0.642	0.644	0.646	0.652	0.656	0.662
0.08000	(0.785)	0.649	0.641	0.637	0.633	0.632	0.630	0.630	0.631	0.633	0.636	0.639	0.644	0.649	0.655
0.10000	(0.981)	0.640	0.632	0.628	0.625	0.624	0.623	0.623	0.624	0.627	0.630	0.633	0.639	0.644	0.650
0.15000	(1.47)	0.622	0.616	0.613	0.611	0.610	0.610	0.611	0.613	0.616	0.620	0.623	0.629	0.635	0.641
0.20000	(1.96)	0.611	0.606	0.603	0.602	0.601	0.602	0.603	0.605	0.609	0.613	0.616	0.623	0.629	0.636
0.30000	(2.94)	0.596	0.592	0.590	0.589	0.589	0.591	0.593	0.595	0.599	0.604	0.608	0.615	0.621	0.628
0.40000	(3.92)	0.586	0.583	0.581	0.581	0.581	0.583	0.586	0.588	0.593	0.598	0.602	0.610	0.616	0.624
0.60000	(5.88)	0.572	0.570	0.570	0.570	0.571	0.573	0.576	0.579	0.585	0.590	0.594	0.602	0.610	0.617
0.80000	(7.85)	0.564	0.562	0.562	0.563	0.564	0.567	0.570	0.573	0.579	0.585	0.589	0.598	0.605	0.613
1.00000	(9.81)	0.557	0.556	0.556	0.557	0.559	0.562	0.566	0.569	0.575	0.581	0.586	0.595	0.602	0.610
1.50000	(14.7)	0.546	0.545	0.546	0.548	0.550	0.554	0.558	0.562	0.568	0.574	0.580	0.589	0.597	0.605
2.00000	(19.6)	0.538	0.539	0.540	0.542	0.544	0.549	0.553	0.557	0.564	0.570	0.576	0.586	0.594	0.602
3.00000	(29.4)	0.529	0.530	0.531	0.534	0.537	0.542	0.546	0.551	0.558	0.565	0.571	0.581	0.589	0.598
4.00000	(39.2)	0.522	0.524	0.526	0.529	0.532	0.537	0.542	0.547	0.555	0.562	0.568	0.578	0.587	0.596
6.00000	(58.8)	0.514	0.516	0.519	0.522	0.526	0.532	0.537	0.542	0.550	0.558	0.564	0.575	0.584	0.593
8.00000	(78.5)	0.508	0.511	0.514	0.518	0.522	0.528	0.534	0.539	0.547	0.555	0.561	0.572	0.581	0.591
10.0000	(98.1)	0.505	0.508	0.511	0.515	0.519	0.525	0.531	0.536	0.545	0.553	0.560	0.571	0.580	0.590
15.0000	(147)	0.498	0.502	0.506	0.510	0.514	0.521	0.527	0.533	0.542	0.550	0.557	0.568	0.578	0.588
20.0000	(196)	0.494	0.498	0.502	0.507	0.511	0.518	0.525	0.530	0.540	0.548	0.555	0.567	0.576	0.586
30.0000	(294)	0.488	0.493	0.498	0.503	0.507	0.515	0.522	0.528	0.537	0.546	0.553	0.565	0.575	0.585
S	$(\Delta p/\rho l)$	6	8	10	12.5	15	20	25	30	40	50	60	80	100	125

Grad'nt k.p.g. (Equivalent) Pipe diameters in mm

Grad'nt S	k.p.g. $(\Delta p/\rho l)$	125	150	200	250	300	400	500	600	800	1000	1250	1500	2000	2500
0.00010	(0.00098)	0.908	0.905	0.903	0.903	0.903	0.906	0.910	0.913	0.920	0.926	0.934	0.940	0.952	0.962
0.00015	(0.00147)	0.886	0.884	0.883	0.883	0.885	0.888	0.892	0.896	0.903	0.910	0.918	0.925	0.937	0.947
0.00020	(0.00196)	0.871	0.869	0.869	0.870	0.872	0.876	0.880	0.884	0.892	0.899	0.907	0.914	0.927	0.937
0.00030	(0.00294)	0.850	0.850	0.850	0.852	0.854	0.859	0.864	0.868	0.877	0.884	0.893	0.900	0.913	0.924
0.00040	(0.00392)	0.837	0.836	0.838	0.840	0.842	0.848	0.853	0.858	0.866	0.874	0.883	0.890	0.903	0.914
0.00060	(0.00588)	0.818	0.818	0.820	0.823	0.826	0.832	0.838	0.843	0.852	0.860	0.869	0.877	0.890	0.902
0.00080	(0.00785)	0.805	0.806	0.809	0.812	0.815	0.821	0.827	0.833	0.842	0.851	0.860	0.868	0.882	0.893
0.00100	(0.00981)	0.796	0.797	0.800	0.803	0.807	0.813	0.820	0.825	0.835	0.844	0.853	0.861	0.875	0.887
0.00150	(0.0147)	0.779	0.781	0.785	0.788	0.792	0.799	0.806	0.812	0.822	0.831	0.841	0.849	0.863	0.875
0.00200	(0.0196)	0.768	0.770	0.774	0.778	0.782	0.790	0.797	0.803	0.813	0.823	0.832	0.841	0.855	0.868
0.00300	(0.0294)	0.753	0.755	0.760	0.765	0.769	0.777	0.784	0.791	0.802	0.811	0.821	0.830	0.845	0.857
0.00400	(0.0392)	0.742	0.745	0.750	0.755	0.760	0.768	0.776	0.782	0.794	0.803	0.814	0.823	0.838	0.850
0.00600	(0.0588)	0.728	0.731	0.737	0.743	0.748	0.756	0.764	0.771	0.783	0.793	0.804	0.813	0.828	0.841
0.00800	(0.0785)	0.719	0.722	0.728	0.734	0.739	0.748	0.756	0.764	0.776	0.786	0.797	0.806	0.822	0.835
0.01000	(0.0981)	0.712	0.715	0.722	0.728	0.733	0.742	0.751	0.758	0.770	0.781	0.792	0.801	0.817	0.831
0.01500	(0.147)	0.699	0.703	0.710	0.716	0.722	0.732	0.741	0.748	0.761	0.772	0.783	0.793	0.809	0.823
0.02000	(0.196)	0.691	0.695	0.702	0.709	0.715	0.725	0.734	0.742	0.755	0.766	0.777	0.787	0.804	0.818
0.03000	(0.294)	0.679	0.684	0.692	0.699	0.705	0.716	0.725	0.733	0.746	0.758	0.770	0.780	0.797	0.811
0.04000	(0.392)	0.672	0.677	0.685	0.692	0.699	0.710	0.719	0.727	0.741	0.753	0.765	0.775	0.793	0.807
0.06000	(0.588)	0.662	0.667	0.676	0.683	0.690	0.702	0.711	0.720	0.734	0.746	0.758	0.769	0.787	0.802
0.08000	(0.785)	0.655	0.660	0.669	0.677	0.684	0.696	0.706	0.715	0.729	0.741	0.754	0.765	0.783	0.798
0.10000	(0.981)	0.650	0.655	0.665	0.673	0.680	0.692	0.702	0.711	0.726	0.738	0.751	0.762	0.781	0.796
0.15000	(1.47)	0.641	0.647	0.657	0.666	0.673	0.686	0.696	0.705	0.720	0.733	0.746	0.757	0.776	0.791
0.20000	(1.96)	0.636	0.642	0.652	0.661	0.668	0.681	0.692	0.701	0.717	0.729	0.743	0.754	0.773	0.789
0.30000	(2.94)	0.628	0.635	0.645	0.655	0.662	0.676	0.687	0.696	0.712	0.725	0.739	0.751	0.770	0.786
0.40000	(3.92)	0.624	0.630	0.641	0.650	0.659	0.672	0.683	0.693	0.709	0.722	0.736	0.748	0.768	0.784
0.60000	(5.88)	0.617	0.624	0.636	0.645	0.654	0.668	0.679	0.689	0.706	0.719	0.733	0.745	0.765	0.781
0.80000	(7.85)	0.613	0.620	0.632	0.642	0.650	0.665	0.676	0.687	0.703	0.717	0.731	0.743	0.763	0.780
1.00000	(9.81)	0.610	0.617	0.629	0.640	0.648	0.663	0.675	0.685	0.702	0.715	0.730	0.742	0.762	0.779
1.50000	(14.7)	0.605	0.613	0.625	0.636	0.645	0.659	0.671	0.682	0.699	0.713	0.728	0.740	0.760	0.777
2.00000	(19.6)	0.602	0.610	0.623	0.633	0.642	0.657	0.670	0.680	0.697	0.712	0.726	0.739	0.759	0.776
3.00000	(29.4)	0.598	0.606	0.619	0.630	0.639	0.655	0.667	0.678	0.695	0.710	0.725	0.737	0.758	0.775
S	$(\Delta p/\rho l)$	125	150	200	250	300	400	500	600	800	1000	1250	1500	2000	2500

Grad'nt k.p.g. (Equivalent) Pipe diameters in mm

Kinematic viscosity, $\nu = 0.25 \times 10^{-6}$ m^{2}s^{-1} ;

Roughness size, $k_s = 0.0015$ mm

Kin. visc., $\nu = 0{\cdot}25\times10^{-6}$ m^{2}s^{-1};
$S = 0{\cdot}00010$ to $30{\cdot}0000$

i.e. kin. pr. grad., $\Delta p/\rho l =$
$(0{\cdot}00098)$ to (294) ms^{-2}

Roughness size, $k_s = 0{\cdot}003$ mm
This table shows values of m, as follows
m_C for Colebrook-White solutions; or,
where $\mathbf{R} \le 2000$, m_P for laminar flow

Grad'nt	k.p.g.	(Equivalent) Pipe diameters in mm													
S	$(\Delta p/\rho l)$	6	8	10	12·5	15	20	25	30	40	50	60	80	100	125
0·00030	(0·00294)	1·714	1·168	0·868	*	0·948	0·922	0·905	0·893	0·879	0·870	0·864	0·857	0·854	0·852
0·00040	(0·00392)	1·485	1·012	0·751	*	0·924	0·900	0·885	0·874	0·861	0·853	0·848	0·843	0·840	0·839
0·00060	(0·00588)	1·212	0·826	*	0·908	0·892	0·871	0·858	0·849	0·837	0·831	0·827	0·823	0·821	0·820
0·00080	(0·00785)	1·050	*	*	0·885	0·871	0·851	0·839	0·831	0·822	0·816	0·812	0·809	0·808	0·808
0·00100	(0·00981)	0·939	*	0·888	0·868	0·855	0·837	0·826	0·819	0·810	0·805	0·802	0·799	0·798	0·799
0·00150	(0·0147)	0·767	*	0·857	0·839	0·827	0·812	0·803	0·796	0·789	0·785	0·783	0·781	0·781	0·782
0·00200	(0·0196)	*	0·855	0·836	0·820	0·809	0·795	0·787	0·781	0·775	0·772	0·770	0·769	0·770	0·771
0·00300	(0·0294)	*	0·825	0·808	0·794	0·785	0·773	0·766	0·761	0·756	0·754	0·753	0·753	0·754	0·757
0·00400	(0·0392)	0·829	0·805	0·789	0·777	0·769	0·758	0·752	0·748	0·744	0·742	0·742	0·742	0·744	0·747
0·00600	(0·0588)	0·799	0·778	0·765	0·754	0·747	0·738	0·733	0·730	0·727	0·726	0·726	0·728	0·730	0·733
0·00800	(0·0785)	0·779	0·760	0·748	0·739	0·732	0·725	0·720	0·718	0·716	0·715	0·716	0·718	0·721	0·724
0·01000	(0·0981)	0·765	0·747	0·736	0·727	0·722	0·715	0·711	0·709	0·707	0·707	0·708	0·711	0·714	0·718
0·01500	(0·147)	0·740	0·725	0·715	0·708	0·703	0·697	0·695	0·693	0·693	0·693	0·695	0·698	0·702	0·706
0·02000	(0·196)	0·723	0·710	0·701	0·695	0·690	0·686	0·684	0·683	0·683	0·684	0·686	0·690	0·694	0·699
0·03000	(0·294)	0·702	0·690	0·683	0·677	0·674	0·670	0·669	0·669	0·670	0·672	0·674	0·678	0·683	0·688
0·04000	(0·392)	0·687	0·676	0·670	0·666	0·663	0·660	0·659	0·659	0·661	0·663	0·666	0·671	0·676	0·682
0·06000	(0·588)	0·668	0·659	0·654	0·650	0·648	0·646	0·646	0·647	0·650	0·652	0·655	0·661	0·667	0·673
0·08000	(0·785)	0·655	0·647	0·643	0·640	0·638	0·637	0·638	0·639	0·642	0·645	0·649	0·655	0·661	0·667
0·10000	(0·981)	0·645	0·638	0·635	0·632	0·631	0·631	0·631	0·633	0·636	0·640	0·643	0·650	0·656	0·663
0·15000	(1·47)	0·629	0·624	0·621	0·619	0·619	0·619	0·621	0·623	0·627	0·631	0·635	0·642	0·649	0·656
0·20000	(1·96)	0·618	0·614	0·612	0·611	0·610	0·612	0·614	0·616	0·620	0·625	0·629	0·637	0·644	0·651
0·30000	(2·94)	0·604	0·601	0·599	0·599	0·600	0·602	0·604	0·607	0·612	0·617	0·622	0·630	0·637	0·645
0·40000	(3·92)	0·595	0·592	0·592	0·592	0·593	0·595	0·598	0·601	0·607	0·612	0·617	0·626	0·633	0·642
0·60000	(5·88)	0·583	0·581	0·581	0·582	0·583	0·587	0·590	0·594	0·600	0·606	0·611	0·620	0·628	0·637
0·80000	(7·85)	0·575	0·574	0·574	0·576	0·577	0·581	0·585	0·589	0·596	0·602	0·607	0·617	0·625	0·634
1·00000	(9·81)	0·569	0·569	0·569	0·571	0·573	0·577	0·582	0·585	0·593	0·599	0·605	0·615	0·623	0·632
1·50000	(14·7)	0·559	0·560	0·561	0·563	0·566	0·571	0·575	0·580	0·587	0·594	0·600	0·610	0·619	0·628
2·00000	(19·6)	0·553	0·554	0·556	0·558	0·561	0·567	0·572	0·576	0·584	0·591	0·597	0·608	0·617	0·626
3·00000	(29·4)	0·545	0·546	0·549	0·552	0·555	0·561	0·567	0·572	0·580	0·587	0·594	0·605	0·614	0·624
4·00000	(39·2)	0·539	0·542	0·545	0·548	0·552	0·558	0·564	0·569	0·578	0·585	0·592	0·603	0·612	0·622
6·00000	(58·8)	0·533	0·536	0·539	0·543	0·547	0·554	0·560	0·565	0·574	0·582	0·589	0·601	0·610	0·620
8·00000	(78·5)	0·528	0·532	0·536	0·540	0·544	0·551	0·557	0·563	0·572	0·580	0·587	0·599	0·609	0·619
10·0000	(98·1)	0·525	0·529	0·533	0·538	0·542	0·549	0·556	0·561	0·571	0·579	0·586	0·598	0·608	0·618
15·0000	(147)	0·520	0·525	0·529	0·534	0·539	0·546	0·553	0·559	0·569	0·577	0·584	0·596	0·606	0·617
20·0000	(196)	0·517	0·522	0·527	0·532	0·536	0·545	0·551	0·557	0·568	0·576	0·583	0·595	0·605	0·616
30·0000	(294)	0·513	0·519	0·524	0·529	0·534	0·542	0·549	0·556	0·566	0·575	0·582	0·594	0·604	0·615
S	$(\Delta p/\rho l)$	6	8	10	12·5	15	20	25	30	40	50	60	80	100	125

Grad'nt   k.p.g.   (Equivalent) Pipe diameters in mm

S	$(\Delta p/\rho l)$	125	150	200	250	300	400	500	600	800	1000	1250	1500	2000	2500
0·00010	(0·00098)	0·909	0·906	0·904	0·904	0·905	0·908	0·911	0·915	0·922	0·929	0·936	0·943	0·955	0·965
0·00015	(0·00147)	0·887	0·885	0·884	0·885	0·886	0·890	0·894	0·898	0·906	0·913	0·921	0·928	0·940	0·951
0·00020	(0·00196)	0·872	0·871	0·871	0·872	0·874	0·878	0·882	0·887	0·895	0·902	0·911	0·918	0·931	0·942
0·00030	(0·00294)	0·852	0·852	0·852	0·854	0·857	0·862	0·867	0·871	0·880	0·888	0·897	0·904	0·917	0·929
0·00040	(0·00392)	0·839	0·838	0·840	0·842	0·845	0·850	0·856	0·861	0·870	0·878	0·887	0·895	0·908	0·920
0·00060	(0·00588)	0·820	0·821	0·823	0·826	0·829	0·835	0·841	0·847	0·856	0·865	0·874	0·882	0·896	0·908
0·00080	(0·00785)	0·808	0·809	0·812	0·815	0·818	0·825	0·831	0·837	0·847	0·856	0·865	0·874	0·888	0·900
0·00100	(0·00981)	0·799	0·800	0·803	0·807	0·810	0·817	0·824	0·830	0·840	0·849	0·859	0·867	0·882	0·894
0·00150	(0·0147)	0·782	0·784	0·788	0·792	0·796	0·804	0·811	0·817	0·828	0·837	0·848	0·857	0·872	0·884
0·00200	(0·0196)	0·771	0·773	0·778	0·783	0·787	0·795	0·802	0·809	0·820	0·830	0·840	0·849	0·865	0·877
0·00300	(0·0294)	0·757	0·759	0·764	0·770	0·774	0·783	0·791	0·797	0·809	0·819	0·830	0·839	0·855	0·869
0·00400	(0·0392)	0·747	0·750	0·755	0·761	0·766	0·775	0·783	0·790	0·802	0·812	0·823	0·833	0·849	0·863
0·00600	(0·0588)	0·733	0·737	0·743	0·749	0·754	0·764	0·772	0·780	0·792	0·803	0·815	0·825	0·841	0·855
0·00800	(0·0785)	0·724	0·728	0·735	0·741	0·747	0·757	0·765	0·773	0·786	0·797	0·809	0·819	0·836	0·850
0·01000	(0·0981)	0·718	0·722	0·729	0·735	0·741	0·751	0·760	0·768	0·781	0·793	0·805	0·815	0·832	0·846
0·01500	(0·147)	0·706	0·710	0·718	0·725	0·731	0·742	0·752	0·760	0·774	0·785	0·797	0·808	0·826	0·840
0·02000	(0·196)	0·699	0·703	0·711	0·719	0·725	0·736	0·746	0·754	0·768	0·780	0·793	0·804	0·822	0·837
0·03000	(0·294)	0·688	0·693	0·702	0·710	0·717	0·728	0·738	0·747	0·762	0·774	0·787	0·798	0·817	0·832
0·04000	(0·392)	0·682	0·687	0·696	0·704	0·711	0·723	0·734	0·742	0·758	0·770	0·783	0·794	0·813	0·829
0·06000	(0·588)	0·673	0·678	0·688	0·697	0·704	0·717	0·727	0·737	0·752	0·765	0·778	0·790	0·809	0·825
0·08000	(0·785)	0·667	0·673	0·683	0·692	0·699	0·713	0·723	0·733	0·749	0·762	0·775	0·787	0·807	0·822
0·10000	(0·981)	0·663	0·669	0·679	0·688	0·696	0·709	0·721	0·730	0·746	0·759	0·773	0·785	0·805	0·821
0·15000	(1·47)	0·656	0·662	0·673	0·683	0·691	0·704	0·716	0·726	0·742	0·756	0·770	0·782	0·802	0·818
0·20000	(1·96)	0·651	0·658	0·669	0·679	0·687	0·701	0·713	0·723	0·739	0·753	0·768	0·780	0·800	0·816
0·30000	(2·94)	0·645	0·652	0·664	0·674	0·683	0·697	0·709	0·719	0·736	0·750	0·765	0·777	0·798	0·814
0·40000	(3·92)	0·642	0·649	0·661	0·671	0·680	0·695	0·707	0·717	0·734	0·748	0·763	0·776	0·796	0·813
0·60000	(5·88)	0·637	0·644	0·657	0·667	0·676	0·692	0·704	0·714	0·732	0·746	0·761	0·774	0·795	0·812
0·80000	(7·85)	0·634	0·642	0·654	0·665	0·674	0·690	0·702	0·713	0·731	0·745	0·760	0·773	0·794	0·811
1·00000	(9·81)	0·632	0·640	0·653	0·663	0·673	0·688	0·701	0·712	0·729	0·744	0·759	0·772	0·793	0·810
1·50000	(14·7)	0·628	0·636	0·650	0·661	0·670	0·686	0·699	0·710	0·728	0·743	0·758	0·771	0·792	0·809
2·00000	(19·6)	0·626	0·634	0·648	0·659	0·669	0·685	0·698	0·709	0·727	0·742	0·757	0·770	0·791	0·808
3·00000	(29·4)	0·624	0·632	0·646	0·657	0·667	0·683	0·696	0·707	0·726	0·741	0·756	0·769	0·790	0·808
S	$(\Delta p/\rho l)$	125	150	200	250	300	400	500	600	800	1000	1250	1500	2000	2500

Grad'nt   k.p.g.   (Equivalent) Pipe diameters in mm

Kinematic viscosity, $\nu = 0{\cdot}25\times10^{-6}$ m^{2}s^{-1} ;   **Roughness size, $k_s = 0{\cdot}003$ mm**

E27

Kin. visc., $\nu = 0.25 \times 10^{-6}$ m^{2}s^{-1};
$S = 0.00010$ to 30.0000

i.e. kin. pr. grad., $\Delta p/\rho l =$
(0.00098) to (294) ms^{-2}

Roughness size, $k_s = 0.006$ mm
This table shows values of m, as follows

m_C for Colebrook-White solutions; or,
where $R \leq 2000$, m_P for laminar flow

Grad'nt S	k.p.g. $(\Delta p/\rho l)$	6	8	10	12.5	15	20	25	30	40	50	60	80	100	125
		(Equivalent) Pipe diameters in mm													
0.00030	(0.00294)	1.714	1.168	0.868	*	0.950	0.924	0.907	0.896	0.881	0.872	0.867	0.860	0.857	0.856
0.00040	(0.00392)	1.485	1.012	0.751	*	0.926	0.902	0.887	0.877	0.864	0.856	0.851	0.846	0.843	0.842
0.00060	(0.00588)	1.212	0.826	*	0.911	0.895	0.873	0.860	0.852	0.841	0.834	0.830	0.826	0.825	0.825
0.00080	(0.00785)	1.050	*	*	0.888	0.873	0.854	0.843	0.835	0.825	0.820	0.816	0.813	0.812	0.813
0.00100	(0.00981)	0.939	*	0.891	0.871	0.858	0.840	0.829	0.822	0.813	0.809	0.806	0.803	0.803	0.804
0.00150	(0.0147)	0.767	*	0.860	0.843	0.831	0.816	0.806	0.800	0.793	0.790	0.788	0.786	0.787	0.788
0.00200	(0.0196)	*	0.858	0.839	0.824	0.813	0.799	0.791	0.786	0.780	0.777	0.775	0.775	0.776	0.778
0.00300	(0.0294)	*	0.829	0.812	0.798	0.789	0.778	0.771	0.767	0.762	0.760	0.759	0.760	0.761	0.764
0.00400	(0.0392)	*	0.809	0.794	0.782	0.773	0.763	0.757	0.754	0.750	0.749	0.748	0.750	0.752	0.755
0.00600	(0.0588)	0.804	0.783	0.770	0.759	0.752	0.744	0.739	0.736	0.734	0.733	0.734	0.736	0.739	0.743
0.00800	(0.0785)	0.785	0.766	0.754	0.745	0.738	0.731	0.727	0.725	0.723	0.723	0.724	0.727	0.730	0.735
0.01000	(0.0981)	0.770	0.753	0.742	0.734	0.728	0.722	0.718	0.716	0.715	0.716	0.717	0.721	0.724	0.729
0.01500	(0.147)	0.746	0.731	0.722	0.715	0.710	0.705	0.703	0.702	0.702	0.703	0.705	0.709	0.714	0.719
0.02000	(0.196)	0.730	0.717	0.709	0.703	0.699	0.694	0.693	0.692	0.693	0.695	0.697	0.702	0.707	0.712
0.03000	(0.294)	0.709	0.698	0.691	0.686	0.683	0.680	0.679	0.680	0.681	0.684	0.687	0.692	0.698	0.704
0.04000	(0.392)	0.695	0.685	0.680	0.675	0.673	0.671	0.671	0.671	0.674	0.677	0.680	0.686	0.692	0.698
0.06000	(0.588)	0.677	0.669	0.664	0.661	0.659	0.659	0.659	0.660	0.664	0.667	0.671	0.678	0.684	0.691
0.08000	(0.785)	0.665	0.658	0.654	0.652	0.651	0.651	0.652	0.653	0.657	0.661	0.665	0.673	0.679	0.687
0.10000	(0.981)	0.657	0.650	0.647	0.645	0.644	0.645	0.646	0.648	0.652	0.657	0.661	0.669	0.676	0.683
0.15000	(1.47)	0.642	0.637	0.634	0.633	0.633	0.635	0.637	0.639	0.645	0.650	0.654	0.663	0.670	0.678
0.20000	(1.96)	0.632	0.628	0.626	0.626	0.626	0.628	0.631	0.634	0.640	0.645	0.650	0.659	0.666	0.675
0.30000	(2.94)	0.620	0.617	0.616	0.616	0.617	0.620	0.624	0.627	0.633	0.639	0.644	0.654	0.662	0.670
0.40000	(3.92)	0.611	0.609	0.609	0.610	0.611	0.615	0.619	0.622	0.629	0.635	0.641	0.651	0.659	0.668
0.60000	(5.88)	0.601	0.600	0.601	0.602	0.604	0.608	0.613	0.617	0.624	0.631	0.636	0.647	0.655	0.665
0.80000	(7.85)	0.594	0.594	0.595	0.597	0.599	0.604	0.609	0.613	0.621	0.628	0.634	0.644	0.653	0.663
1.00000	(9.81)	0.589	0.590	0.591	0.593	0.596	0.601	0.606	0.611	0.619	0.626	0.632	0.643	0.652	0.661
1.50000	(14.7)	0.581	0.583	0.585	0.588	0.591	0.596	0.602	0.606	0.615	0.622	0.629	0.640	0.649	0.659
2.00000	(19.6)	0.576	0.578	0.581	0.584	0.587	0.593	0.599	0.604	0.613	0.620	0.627	0.638	0.648	0.658
3.00000	(29.4)	0.570	0.573	0.576	0.579	0.583	0.590	0.595	0.601	0.610	0.618	0.625	0.636	0.646	0.656
4.00000	(39.2)	0.566	0.569	0.572	0.576	0.580	0.587	0.593	0.599	0.608	0.616	0.623	0.635	0.645	0.655
6.00000	(58.8)	0.561	0.565	0.568	0.573	0.577	0.584	0.591	0.597	0.606	0.614	0.622	0.633	0.643	0.654
8.00000	(78.5)	0.558	0.562	0.566	0.571	0.575	0.583	0.589	0.595	0.605	0.613	0.621	0.633	0.643	0.653
10.0000	(98.1)	0.556	0.560	0.564	0.569	0.574	0.581	0.588	0.594	0.604	0.613	0.620	0.632	0.642	0.653
15.0000	(147)	0.552	0.557	0.562	0.567	0.571	0.580	0.586	0.593	0.603	0.611	0.619	0.631	0.641	0.652
20.0000	(196)	0.550	0.555	0.560	0.565	0.570	0.578	0.585	0.592	0.602	0.611	0.618	0.630	0.641	0.651
30.0000	(294)	0.548	0.553	0.558	0.564	0.568	0.577	0.584	0.590	0.601	0.610	0.617	0.630	0.640	0.651
S	$(\Delta p/\rho l)$	6	8	10	12.5	15	20	25	30	40	50	60	80	100	125

Grad'nt   k.p.g.   (Equivalent) Pipe diameters in mm

S	$(\Delta p/\rho l)$	125	150	200	250	300	400	500	600	800	1000	1250	1500	2000	2500
0.00010	(0.00098)	0.911	0.909	0.907	0.907	0.908	0.911	0.915	0.919	0.926	0.933	0.941	0.948	0.961	0.971
0.00015	(0.00147)	0.890	0.888	0.887	0.888	0.890	0.894	0.898	0.903	0.911	0.918	0.927	0.934	0.947	0.958
0.00020	(0.00196)	0.875	0.874	0.874	0.875	0.877	0.882	0.887	0.892	0.900	0.908	0.917	0.925	0.938	0.949
0.00030	(0.00294)	0.856	0.855	0.856	0.858	0.861	0.866	0.872	0.877	0.886	0.895	0.904	0.912	0.926	0.938
0.00040	(0.00392)	0.842	0.842	0.844	0.847	0.850	0.856	0.862	0.867	0.877	0.885	0.895	0.903	0.918	0.930
0.00060	(0.00588)	0.825	0.825	0.828	0.831	0.835	0.842	0.848	0.854	0.864	0.873	0.883	0.892	0.907	0.920
0.00080	(0.00785)	0.813	0.814	0.817	0.821	0.825	0.832	0.839	0.845	0.856	0.865	0.875	0.885	0.900	0.913
0.00100	(0.00981)	0.804	0.805	0.809	0.813	0.817	0.825	0.832	0.838	0.849	0.859	0.870	0.879	0.895	0.908
0.00150	(0.0147)	0.788	0.790	0.795	0.800	0.804	0.813	0.820	0.827	0.839	0.849	0.860	0.870	0.886	0.899
0.00200	(0.0196)	0.778	0.780	0.786	0.791	0.796	0.804	0.812	0.819	0.832	0.842	0.853	0.863	0.880	0.894
0.00300	(0.0294)	0.764	0.767	0.773	0.779	0.784	0.794	0.802	0.810	0.822	0.833	0.845	0.855	0.873	0.887
0.00400	(0.0392)	0.755	0.758	0.765	0.771	0.777	0.787	0.795	0.803	0.816	0.828	0.840	0.850	0.868	0.882
0.00600	(0.0588)	0.743	0.747	0.754	0.761	0.767	0.777	0.786	0.795	0.809	0.820	0.833	0.843	0.862	0.877
0.00800	(0.0785)	0.735	0.739	0.747	0.754	0.760	0.771	0.781	0.789	0.803	0.815	0.828	0.839	0.858	0.873
0.01000	(0.0981)	0.729	0.733	0.741	0.749	0.755	0.767	0.777	0.785	0.800	0.812	0.825	0.836	0.855	0.870
0.01500	(0.147)	0.719	0.724	0.733	0.740	0.747	0.759	0.770	0.779	0.794	0.806	0.820	0.831	0.850	0.866
0.02000	(0.196)	0.712	0.717	0.727	0.735	0.742	0.755	0.765	0.774	0.790	0.803	0.816	0.828	0.847	0.863
0.03000	(0.294)	0.704	0.709	0.719	0.728	0.735	0.749	0.759	0.769	0.785	0.798	0.812	0.824	0.844	0.860
0.04000	(0.392)	0.698	0.704	0.714	0.723	0.731	0.745	0.756	0.766	0.782	0.795	0.809	0.822	0.842	0.858
0.06000	(0.588)	0.691	0.697	0.708	0.718	0.726	0.740	0.751	0.761	0.778	0.792	0.806	0.819	0.839	0.856
0.08000	(0.785)	0.687	0.693	0.704	0.714	0.723	0.737	0.749	0.759	0.776	0.790	0.804	0.817	0.837	0.854
0.10000	(0.981)	0.683	0.690	0.702	0.712	0.720	0.735	0.747	0.757	0.774	0.788	0.803	0.815	0.836	0.853
0.15000	(1.47)	0.678	0.685	0.697	0.707	0.716	0.731	0.743	0.754	0.771	0.786	0.801	0.813	0.834	0.851
0.20000	(1.96)	0.675	0.682	0.694	0.705	0.714	0.729	0.741	0.752	0.770	0.784	0.799	0.812	0.833	0.850
0.30000	(2.94)	0.670	0.678	0.691	0.702	0.711	0.726	0.739	0.750	0.768	0.782	0.798	0.811	0.832	0.849
0.40000	(3.92)	0.668	0.676	0.689	0.700	0.709	0.725	0.737	0.748	0.767	0.781	0.797	0.810	0.831	0.848
0.60000	(5.88)	0.665	0.673	0.686	0.697	0.707	0.723	0.736	0.747	0.765	0.780	0.795	0.809	0.830	0.847
0.80000	(7.85)	0.663	0.671	0.684	0.696	0.705	0.721	0.735	0.746	0.764	0.779	0.795	0.808	0.829	0.847
1.00000	(9.81)	0.661	0.669	0.683	0.695	0.704	0.721	0.734	0.745	0.764	0.779	0.794	0.807	0.829	0.847
1.50000	(14.7)	0.659	0.667	0.681	0.693	0.703	0.719	0.733	0.744	0.763	0.778	0.793	0.807	0.828	0.846
2.00000	(19.6)	0.658	0.666	0.680	0.692	0.702	0.718	0.732	0.743	0.762	0.777	0.793	0.806	0.828	0.846
3.00000	(29.4)	0.656	0.665	0.679	0.691	0.701	0.718	0.731	0.742	0.761	0.777	0.792	0.806	0.828	0.845
S	$(\Delta p/\rho l)$	125	150	200	250	300	400	500	600	800	1000	1250	1500	2000	2500

Grad'nt   k.p.g.   (Equivalent) Pipe diameters in mm

Kinematic viscosity, $\nu = 0.25 \times 10^{-6}$ m^{2}s^{-1} ;

Roughness size, $k_s = 0.006$ mm

Kin. visc., $\nu = 0.25 \times 10^{-6}$ m^{2}s^{-1};
$S = 0.00010$ to 30.0000
i.e. kin. pr. grad., $\Delta p/\rho l = (0.00098)$ to (294) ms^{-2}

Roughness size, $k_s = 0.015$ mm
This table shows values of m, as follows
m_C for Colebrook-White solutions; or,
where $\mathbf{R} \leq 2000$, m_P for laminar flow

Grad'nt S	k.p.g. $(\Delta p/\rho l)$	(Equivalent) Pipe diameters in mm													
		6	8	10	12.5	15	20	25	30	40	50	60	80	100	125
0.00030	(0.00294)	1.714	1.168	0.868	*	0.957	0.930	0.914	0.903	0.888	0.880	0.875	0.869	0.866	0.865
0.00040	(0.00392)	1.485	1.012	0.751	*	0.933	0.909	0.894	0.884	0.872	0.864	0.860	0.855	0.853	0.853
0.00060	(0.00588)	1.212	0.826	*	0.918	0.902	0.881	0.869	0.860	0.850	0.844	0.840	0.837	0.836	0.837
0.00080	(0.00785)	1.050	*	*	0.896	0.882	0.863	0.852	0.844	0.835	0.830	0.827	0.825	0.825	0.826
0.00100	(0.00981)	0.939	*	0.900	0.880	0.867	0.849	0.839	0.832	0.824	0.820	0.818	0.816	0.816	0.818
0.00150	(0.0147)	0.767	*	0.870	0.853	0.841	0.826	0.817	0.812	0.805	0.802	0.801	0.801	0.802	0.804
0.00200	(0.0196)	*	0.869	0.850	0.834	0.824	0.811	0.803	0.798	0.793	0.791	0.790	0.791	0.792	0.795
0.00300	(0.0294)	*	0.840	0.824	0.811	0.802	0.791	0.785	0.781	0.777	0.776	0.776	0.777	0.780	0.784
0.00400	(0.0392)	*	0.821	0.807	0.795	0.787	0.777	0.772	0.769	0.766	0.766	0.766	0.769	0.772	0.776
0.00600	(0.0588)	0.817	0.797	0.784	0.774	0.768	0.760	0.756	0.754	0.753	0.753	0.754	0.758	0.762	0.767
0.00800	(0.0785)	0.799	0.781	0.770	0.761	0.755	0.749	0.745	0.744	0.743	0.745	0.746	0.750	0.755	0.760
0.01000	(0.0981)	0.786	0.769	0.759	0.751	0.746	0.740	0.738	0.737	0.737	0.738	0.741	0.745	0.750	0.756
0.01500	(0.147)	0.764	0.750	0.741	0.734	0.730	0.726	0.725	0.725	0.726	0.728	0.731	0.737	0.742	0.748
0.02000	(0.196)	0.749	0.737	0.729	0.724	0.720	0.717	0.716	0.717	0.719	0.722	0.725	0.731	0.737	0.744
0.03000	(0.294)	0.731	0.720	0.714	0.710	0.707	0.706	0.706	0.707	0.710	0.713	0.717	0.724	0.731	0.738
0.04000	(0.392)	0.719	0.709	0.704	0.701	0.699	0.698	0.699	0.700	0.704	0.708	0.712	0.720	0.726	0.734
0.06000	(0.588)	0.703	0.695	0.691	0.689	0.688	0.688	0.690	0.692	0.697	0.701	0.706	0.714	0.721	0.730
0.08000	(0.785)	0.693	0.686	0.683	0.682	0.681	0.682	0.684	0.687	0.692	0.697	0.702	0.711	0.718	0.727
0.10000	(0.981)	0.686	0.680	0.678	0.676	0.676	0.678	0.680	0.683	0.689	0.694	0.699	0.708	0.716	0.725
0.15000	(1.47)	0.674	0.670	0.668	0.668	0.668	0.671	0.674	0.677	0.684	0.689	0.695	0.704	0.713	0.721
0.20000	(1.96)	0.666	0.663	0.662	0.662	0.663	0.666	0.670	0.673	0.680	0.686	0.692	0.702	0.710	0.720
0.30000	(2.94)	0.656	0.654	0.654	0.655	0.657	0.661	0.665	0.669	0.676	0.683	0.689	0.699	0.708	0.717
0.40000	(3.92)	0.650	0.649	0.650	0.651	0.653	0.657	0.662	0.666	0.674	0.681	0.687	0.697	0.706	0.716
0.60000	(5.88)	0.643	0.643	0.644	0.646	0.648	0.653	0.658	0.663	0.671	0.678	0.684	0.695	0.704	0.714
0.80000	(7.85)	0.638	0.638	0.640	0.642	0.645	0.651	0.656	0.660	0.669	0.676	0.683	0.694	0.703	0.713
1.00000	(9.81)	0.635	0.636	0.637	0.640	0.643	0.649	0.654	0.659	0.668	0.675	0.682	0.693	0.702	0.712
1.50000	(14.7)	0.629	0.631	0.633	0.636	0.640	0.646	0.651	0.657	0.665	0.673	0.680	0.691	0.701	0.711
2.00000	(19.6)	0.626	0.628	0.631	0.634	0.638	0.644	0.650	0.655	0.664	0.672	0.679	0.690	0.700	0.710
3.00000	(29.4)	0.622	0.625	0.628	0.632	0.635	0.642	0.648	0.653	0.663	0.671	0.678	0.689	0.699	0.709
4.00000	(39.2)	0.620	0.623	0.626	0.630	0.634	0.641	0.647	0.652	0.662	0.670	0.677	0.689	0.699	0.709
6.00000	(58.8)	0.617	0.620	0.624	0.628	0.632	0.639	0.645	0.651	0.661	0.669	0.676	0.688	0.698	0.708
8.00000	(78.5)	0.615	0.618	0.622	0.627	0.631	0.638	0.645	0.650	0.660	0.668	0.676	0.688	0.697	0.708
10.0000	(98.1)	0.613	0.617	0.621	0.626	0.630	0.638	0.644	0.650	0.660	0.668	0.675	0.687	0.697	0.708
15.0000	(147)	0.611	0.616	0.620	0.624	0.629	0.637	0.643	0.649	0.659	0.667	0.675	0.687	0.697	0.707
20.0000	(196)	0.610	0.615	0.619	0.624	0.628	0.636	0.643	0.649	0.659	0.667	0.674	0.686	0.696	0.707
30.0000	(294)	0.609	0.613	0.618	0.623	0.627	0.635	0.642	0.648	0.658	0.667	0.674	0.686	0.696	0.707
S	$(\Delta p/\rho l)$	6	8	10	12.5	15	20	25	30	40	50	60	80	100	125

Grad'nt S	k.p.g. $(\Delta p/\rho l)$	(Equivalent) Pipe diameters in mm													
		125	150	200	250	300	400	500	600	800	1000	1250	1500	2000	2500
0.00010	(0.00098)	0.918	0.916	0.914	0.915	0.916	0.920	0.925	0.929	0.938	0.945	0.954	0.962	0.976	0.988
0.00015	(0.00147)	0.897	0.896	0.896	0.897	0.899	0.904	0.909	0.914	0.924	0.932	0.942	0.950	0.965	0.977
0.00020	(0.00196)	0.883	0.883	0.883	0.885	0.888	0.894	0.899	0.905	0.915	0.923	0.933	0.942	0.957	0.970
0.00030	(0.00294)	0.865	0.865	0.865	0.870	0.873	0.880	0.886	0.892	0.903	0.912	0.922	0.932	0.947	0.961
0.00040	(0.00392)	0.853	0.853	0.856	0.859	0.863	0.870	0.877	0.883	0.895	0.904	0.915	0.925	0.941	0.955
0.00060	(0.00588)	0.837	0.838	0.842	0.846	0.850	0.858	0.866	0.872	0.884	0.895	0.906	0.916	0.933	0.947
0.00080	(0.00785)	0.826	0.828	0.832	0.837	0.841	0.850	0.858	0.865	0.878	0.888	0.900	0.910	0.928	0.942
0.00100	(0.00981)	0.818	0.820	0.825	0.830	0.835	0.844	0.852	0.860	0.873	0.884	0.896	0.906	0.924	0.939
0.00150	(0.0147)	0.804	0.807	0.813	0.819	0.824	0.834	0.843	0.851	0.865	0.876	0.889	0.900	0.918	0.933
0.00200	(0.0196)	0.795	0.799	0.805	0.812	0.817	0.828	0.837	0.845	0.859	0.871	0.884	0.895	0.914	0.929
0.00300	(0.0294)	0.784	0.788	0.795	0.802	0.808	0.820	0.829	0.838	0.853	0.865	0.878	0.890	0.909	0.925
0.00400	(0.0392)	0.776	0.781	0.789	0.796	0.803	0.814	0.825	0.833	0.849	0.861	0.875	0.887	0.906	0.922
0.00600	(0.0588)	0.767	0.771	0.780	0.788	0.795	0.808	0.818	0.828	0.843	0.856	0.870	0.882	0.902	0.919
0.00800	(0.0785)	0.760	0.766	0.775	0.783	0.791	0.804	0.814	0.824	0.840	0.853	0.868	0.880	0.900	0.917
0.01000	(0.0981)	0.756	0.761	0.771	0.780	0.787	0.801	0.812	0.821	0.838	0.851	0.866	0.878	0.898	0.915
0.01500	(0.147)	0.748	0.754	0.765	0.774	0.782	0.796	0.807	0.817	0.834	0.848	0.863	0.875	0.896	0.913
0.02000	(0.196)	0.744	0.750	0.761	0.770	0.779	0.793	0.804	0.815	0.832	0.846	0.861	0.873	0.894	0.911
0.03000	(0.294)	0.738	0.744	0.756	0.766	0.774	0.789	0.801	0.811	0.829	0.843	0.858	0.871	0.892	0.910
0.04000	(0.392)	0.734	0.741	0.753	0.763	0.772	0.787	0.799	0.810	0.827	0.842	0.857	0.870	0.891	0.908
0.06000	(0.588)	0.730	0.737	0.749	0.760	0.769	0.784	0.796	0.807	0.825	0.840	0.855	0.868	0.890	0.907
0.08000	(0.785)	0.727	0.734	0.747	0.757	0.767	0.782	0.795	0.806	0.824	0.839	0.854	0.867	0.889	0.906
0.10000	(0.981)	0.725	0.732	0.745	0.756	0.765	0.781	0.794	0.805	0.823	0.838	0.853	0.867	0.888	0.906
0.15000	(1.47)	0.721	0.729	0.743	0.754	0.763	0.779	0.792	0.803	0.822	0.837	0.852	0.866	0.887	0.905
0.20000	(1.96)	0.720	0.727	0.741	0.752	0.762	0.778	0.791	0.802	0.821	0.836	0.852	0.865	0.887	0.905
0.30000	(2.94)	0.717	0.725	0.739	0.750	0.760	0.776	0.790	0.801	0.820	0.835	0.851	0.864	0.886	0.904
0.40000	(3.92)	0.716	0.724	0.738	0.749	0.759	0.776	0.789	0.800	0.819	0.834	0.850	0.864	0.886	0.904
0.60000	(5.88)	0.714	0.722	0.736	0.748	0.758	0.775	0.788	0.799	0.818	0.834	0.850	0.863	0.885	0.903
0.80000	(7.85)	0.713	0.721	0.736	0.747	0.757	0.774	0.787	0.799	0.818	0.833	0.849	0.863	0.885	0.903
1.00000	(9.81)	0.712	0.721	0.735	0.747	0.757	0.774	0.787	0.799	0.818	0.833	0.849	0.862	0.885	0.903
1.50000	(14.7)	0.711	0.720	0.734	0.746	0.756	0.773	0.787	0.798	0.817	0.833	0.849	0.862	0.885	0.903
2.00000	(19.6)	0.710	0.719	0.734	0.745	0.756	0.772	0.786	0.798	0.817	0.832	0.848	0.862	0.884	0.902
3.00000	(29.4)	0.709	0.718	0.733	0.745	0.755	0.772	0.786	0.797	0.817	0.832	0.848	0.862	0.884	0.902
S	$(\Delta p/\rho l)$	125	150	200	250	300	400	500	600	800	1000	1250	1500	2000	2500

Grad'nt   k.p.g.   (Equivalent) Pipe diameters in mm

Kinematic viscosity, $\nu = 0.25 \times 10^{-6}$ m^{2}s^{-1} ;   **Roughness size, $k_s = 0.015$ mm**

E29

Kin. visc., $\nu = 0.25\times10^{-6}$ m^{2}s^{-1};
$S = 0.00010$ to 3.00000

i.e. kin. pr. grad., $\Delta p/\rho l =$
(0.00098) to (29.4) ms^{-2}

Roughness size, $k_s = 0.030$ mm
This table shows values of m, as follows

m_C for Colebrook-White solutions

Grad'nt S	k.p.g. $(\Delta p/\rho l)$	125	150	200	250	300	400	500	600	800	1000	1250	1500	2000	2500
		(Equivalent) Pipe diameters in mm													
0.00010	(0.00098)	0.928	0.926	0.926	0.927	0.929	0.934	0.939	0.944	0.954	0.963	0.973	0.982	0.997	1.010
0.00015	(0.00147)	0.909	0.908	0.909	0.911	0.914	0.920	0.926	0.932	0.942	0.952	0.963	0.972	0.988	1.002
0.00020	(0.00196)	0.896	0.896	0.898	0.901	0.904	0.911	0.917	0.924	0.935	0.945	0.956	0.966	0.982	0.996
0.00030	(0.00294)	0.879	0.880	0.883	0.887	0.891	0.899	0.906	0.913	0.925	0.936	0.947	0.957	0.975	0.989
0.00040	(0.00392)	0.869	0.870	0.874	0.878	0.882	0.891	0.899	0.906	0.919	0.930	0.942	0.952	0.970	0.985
0.00060	(0.00588)	0.854	0.856	0.861	0.866	0.871	0.881	0.889	0.897	0.911	0.922	0.935	0.946	0.964	0.980
0.00080	(0.00785)	0.845	0.848	0.853	0.859	0.864	0.875	0.884	0.892	0.906	0.918	0.930	0.942	0.961	0.976
0.00100	(0.00981)	0.838	0.841	0.847	0.854	0.859	0.870	0.879	0.888	0.902	0.914	0.927	0.939	0.958	0.974
0.00150	(0.0147)	0.827	0.831	0.838	0.845	0.851	0.862	0.872	0.881	0.896	0.909	0.922	0.934	0.954	0.970
0.00200	(0.0196)	0.820	0.824	0.832	0.839	0.846	0.858	0.868	0.877	0.892	0.905	0.919	0.931	0.951	0.968
0.00300	(0.0294)	0.810	0.815	0.824	0.832	0.839	0.852	0.862	0.872	0.888	0.901	0.915	0.928	0.948	0.965
0.00400	(0.0392)	0.805	0.810	0.819	0.827	0.835	0.848	0.859	0.869	0.885	0.899	0.913	0.925	0.946	0.963
0.00600	(0.0588)	0.797	0.803	0.813	0.822	0.830	0.843	0.855	0.865	0.881	0.895	0.910	0.923	0.944	0.961
0.00800	(0.0785)	0.792	0.798	0.809	0.818	0.826	0.840	0.852	0.862	0.879	0.893	0.908	0.921	0.942	0.960
0.01000	(0.0981)	0.789	0.795	0.806	0.816	0.824	0.838	0.850	0.861	0.878	0.892	0.907	0.920	0.941	0.959
0.01500	(0.147)	0.784	0.790	0.802	0.812	0.820	0.835	0.847	0.858	0.875	0.890	0.905	0.918	0.940	0.957
0.02000	(0.196)	0.781	0.787	0.799	0.809	0.818	0.833	0.845	0.856	0.874	0.889	0.904	0.917	0.939	0.956
0.03000	(0.294)	0.777	0.784	0.796	0.806	0.815	0.831	0.843	0.854	0.872	0.887	0.903	0.916	0.938	0.955
0.04000	(0.392)	0.774	0.781	0.794	0.805	0.814	0.829	0.842	0.853	0.871	0.886	0.902	0.915	0.937	0.955
0.06000	(0.588)	0.771	0.779	0.792	0.802	0.812	0.827	0.840	0.852	0.870	0.885	0.901	0.915	0.937	0.955
0.08000	(0.785)	0.769	0.777	0.790	0.801	0.811	0.826	0.840	0.851	0.869	0.885	0.901	0.914	0.936	0.954
0.10000	(0.981)	0.768	0.776	0.789	0.800	0.810	0.826	0.839	0.850	0.869	0.884	0.900	0.914	0.936	0.954
0.15000	(1.47)	0.766	0.774	0.787	0.799	0.808	0.825	0.838	0.849	0.868	0.883	0.899	0.913	0.935	0.953
0.20000	(1.96)	0.765	0.773	0.786	0.798	0.808	0.824	0.837	0.849	0.867	0.883	0.899	0.912	0.935	0.953
0.30000	(2.94)	0.763	0.771	0.785	0.797	0.807	0.823	0.836	0.848	0.867	0.882	0.898	0.912	0.935	0.953
0.40000	(3.92)	0.762	0.771	0.785	0.796	0.806	0.823	0.836	0.848	0.866	0.882	0.898	0.912	0.934	0.952
0.60000	(5.88)	0.761	0.770	0.784	0.795	0.805	0.822	0.835	0.847	0.866	0.882	0.898	0.911	0.934	0.952
0.80000	(7.85)	0.761	0.769	0.783	0.795	0.805	0.822	0.835	0.847	0.866	0.881	0.897	0.911	0.934	0.952
1.00000	(9.81)	0.760	0.769	0.783	0.795	0.805	0.822	0.835	0.847	0.866	0.881	0.897	0.911	0.934	0.952
1.50000	(14.7)	0.759	0.768	0.782	0.794	0.804	0.821	0.835	0.847	0.866	0.881	0.897	0.911	0.933	0.952
2.00000	(19.6)	0.759	0.768	0.782	0.794	0.804	0.821	0.834	0.846	0.865	0.881	0.897	0.911	0.933	0.951
3.00000	(29.4)	0.758	0.767	0.782	0.794	0.804	0.820	0.834	0.846	0.865	0.881	0.897	0.911	0.933	0.951
S ($\Delta p/\rho l$)		125	150	200	250	300	400	500	600	800	1000	1250	1500	2000	2500

Grad'nt   k.p.g.   (Equivalent) Pipe diameters in mm   **Roughness size, $k_s = 0.030$ mm**

E30

Kin. visc., $\nu = 0.25\times10^{-6}$ m^{2}s^{-1};
$S = 0.00010$ to 3.00000

i.e. kin. pr. grad., $\Delta p/\rho l =$
(0.00098) to (29.4) ms^{-2}

Roughness size, $k_s = 0.060$ mm
This table shows values of m, as follows

m_C for Colebrook-White solutions

Grad'nt S	k.p.g. $(\Delta p/\rho l)$	125	150	200	250	300	400	500	600	800	1000	1250	1500	2000	2500
		(Equivalent) Pipe diameters in mm													
0.00010	(0.00098)	0.947	0.946	0.946	0.948	0.951	0.958	0.964	0.970	0.982	0.992	1.003	1.013	1.030	1.044
0.00015	(0.00147)	0.930	0.930	0.932	0.935	0.939	0.946	0.953	0.960	0.972	0.983	0.995	1.005	1.023	1.038
0.00020	(0.00196)	0.919	0.919	0.922	0.926	0.930	0.939	0.947	0.954	0.967	0.978	0.990	1.001	1.019	1.035
0.00030	(0.00294)	0.905	0.906	0.910	0.915	0.920	0.929	0.938	0.946	0.959	0.971	0.984	0.995	1.014	1.030
0.00040	(0.00392)	0.895	0.898	0.903	0.908	0.913	0.923	0.932	0.941	0.955	0.967	0.980	0.992	1.011	1.027
0.00060	(0.00588)	0.884	0.887	0.893	0.899	0.905	0.916	0.926	0.934	0.949	0.962	0.975	0.987	1.007	1.023
0.00080	(0.00785)	0.876	0.880	0.887	0.894	0.900	0.911	0.921	0.930	0.946	0.959	0.972	0.984	1.005	1.021
0.00100	(0.00981)	0.871	0.875	0.882	0.890	0.896	0.908	0.918	0.927	0.943	0.956	0.970	0.983	1.003	1.020
0.00150	(0.0147)	0.862	0.867	0.875	0.883	0.890	0.903	0.913	0.923	0.939	0.953	0.967	0.980	1.000	1.017
0.00200	(0.0196)	0.857	0.862	0.871	0.879	0.886	0.899	0.911	0.920	0.937	0.951	0.965	0.978	0.999	1.016
0.00300	(0.0294)	0.850	0.856	0.865	0.874	0.882	0.895	0.907	0.917	0.934	0.948	0.963	0.976	0.997	1.014
0.00400	(0.0392)	0.846	0.852	0.862	0.871	0.879	0.893	0.905	0.915	0.932	0.946	0.961	0.974	0.996	1.013
0.00600	(0.0588)	0.841	0.847	0.858	0.867	0.876	0.890	0.902	0.912	0.930	0.944	0.960	0.973	0.994	1.012
0.00800	(0.0785)	0.838	0.844	0.855	0.865	0.873	0.888	0.900	0.911	0.929	0.943	0.958	0.972	0.993	1.011
0.01000	(0.0981)	0.835	0.842	0.853	0.863	0.872	0.887	0.899	0.910	0.928	0.942	0.958	0.971	0.993	1.010
0.01500	(0.147)	0.832	0.839	0.851	0.861	0.870	0.885	0.897	0.908	0.926	0.941	0.957	0.970	0.992	1.010
0.02000	(0.196)	0.830	0.837	0.849	0.859	0.868	0.884	0.896	0.907	0.925	0.940	0.956	0.969	0.991	1.009
0.03000	(0.294)	0.827	0.834	0.847	0.857	0.867	0.882	0.895	0.906	0.924	0.939	0.955	0.968	0.991	1.009
0.04000	(0.392)	0.826	0.833	0.846	0.856	0.866	0.881	0.894	0.905	0.924	0.939	0.955	0.968	0.990	1.008
0.06000	(0.588)	0.824	0.831	0.844	0.855	0.864	0.880	0.893	0.904	0.923	0.938	0.954	0.967	0.990	1.008
0.08000	(0.785)	0.823	0.830	0.843	0.854	0.864	0.880	0.893	0.904	0.923	0.938	0.954	0.967	0.989	1.008
0.10000	(0.981)	0.822	0.829	0.843	0.854	0.863	0.879	0.892	0.904	0.922	0.937	0.953	0.967	0.989	1.007
0.15000	(1.47)	0.821	0.828	0.842	0.853	0.862	0.878	0.892	0.903	0.922	0.937	0.953	0.967	0.989	1.007
0.20000	(1.96)	0.820	0.828	0.841	0.852	0.862	0.878	0.891	0.903	0.921	0.937	0.953	0.966	0.989	1.007
0.30000	(2.94)	0.819	0.827	0.840	0.852	0.861	0.878	0.891	0.902	0.921	0.936	0.953	0.966	0.989	1.007
0.40000	(3.92)	0.818	0.826	0.840	0.851	0.861	0.877	0.891	0.902	0.921	0.936	0.952	0.966	0.988	1.007
0.60000	(5.88)	0.818	0.826	0.839	0.851	0.861	0.877	0.890	0.902	0.921	0.936	0.952	0.966	0.988	1.006
0.80000	(7.85)	0.817	0.825	0.839	0.851	0.860	0.877	0.890	0.902	0.921	0.936	0.952	0.966	0.988	1.006
1.00000	(9.81)	0.817	0.825	0.839	0.850	0.860	0.877	0.890	0.901	0.920	0.936	0.952	0.966	0.988	1.006
1.50000	(14.7)	0.817	0.825	0.839	0.850	0.860	0.876	0.890	0.901	0.920	0.936	0.952	0.966	0.988	1.006
2.00000	(19.6)	0.816	0.825	0.838	0.850	0.860	0.876	0.890	0.901	0.920	0.936	0.952	0.966	0.988	1.006
3.00000	(29.4)	0.816	0.824	0.838	0.850	0.860	0.876	0.890	0.901	0.920	0.936	0.952	0.965	0.988	1.006
S ($\Delta p/\rho l$)		125	150	200	250	300	400	500	600	800	1000	1250	1500	2000	2500

Grad'nt   k.p.g.   (Equivalent) Pipe diameters in mm   **Roughness size, $k_s = 0.060$ mm**

Kin. visc., $\nu = 0.25 \times 10^{-6}$ m^{2}s^{-1};
$S = 0.00010$ to 3.00000
i.e. kin. pr. grad., $\Delta p/\rho l =$ (0.00098) to (29.4) ms^{-2}

Roughness size, $k_s = 0.150$ mm
This table shows values of m, as follows
m_C for Colebrook-White solutions

Grad'nt S	k.p.g. $(\Delta p/\rho l)$	\multicolumn{14}{c}{(Equivalent) Pipe diameters in mm}
		125	150	200	250	300	400	500	600	800	1000	1250	1500	2000	2500
0·00010	(0·00098)	0·993	0·993	0·995	0·999	1·003	1·011	1·019	1·026	1·040	1·051	1·064	1·075	1·094	1·110
0·00015	(0·00147)	0·980	0·981	0·985	0·989	0·994	1·003	1·012	1·020	1·034	1·046	1·059	1·070	1·090	1·106
0·00020	(0·00196)	0·972	0·974	0·978	0·983	0·989	0·998	1·008	1·016	1·030	1·043	1·056	1·068	1·088	1·104
0·00030	(0·00294)	0·962	0·964	0·970	0·976	0·982	0·993	1·002	1·011	1·026	1·039	1·053	1·065	1·085	1·101
0·00040	(0·00392)	0·956	0·959	0·965	0·971	0·978	0·989	0·999	1·008	1·023	1·036	1·050	1·063	1·083	1·100
0·00060	(0·00588)	0·948	0·951	0·959	0·966	0·972	0·984	0·995	1·004	1·020	1·034	1·048	1·060	1·081	1·098
0·00080	(0·00785)	0·943	0·947	0·955	0·962	0·969	0·982	0·992	1·002	1·018	1·032	1·046	1·059	1·080	1·097
0·00100	(0·00981)	0·940	0·944	0·952	0·960	0·967	0·980	0·991	1·000	1·017	1·031	1·045	1·058	1·079	1·096
0·00150	(0·0147)	0·934	0·939	0·948	0·956	0·964	0·977	0·988	0·998	1·015	1·029	1·043	1·056	1·078	1·095
0·00200	(0·0196)	0·931	0·936	0·945	0·954	0·962	0·975	0·986	0·996	1·013	1·028	1·042	1·055	1·077	1·094
0·00300	(0·0294)	0·927	0·932	0·942	0·951	0·959	0·973	0·984	0·995	1·012	1·026	1·041	1·054	1·076	1·093
0·00400	(0·0392)	0·924	0·930	0·940	0·949	0·958	0·971	0·983	0·994	1·011	1·025	1·041	1·054	1·075	1·093
0·00600	(0·0588)	0·922	0·927	0·938	0·947	0·956	0·970	0·982	0·992	1·010	1·024	1·040	1·053	1·074	1·092
0·00800	(0·0785)	0·920	0·926	0·937	0·946	0·955	0·969	0·981	0·992	1·009	1·024	1·039	1·052	1·074	1·092
0·01000	(0·0981)	0·919	0·925	0·936	0·945	0·954	0·968	0·980	0·991	1·009	1·023	1·039	1·052	1·074	1·092
0·01500	(0·147)	0·917	0·923	0·934	0·944	0·953	0·967	0·980	0·990	1·008	1·023	1·038	1·051	1·073	1·091
0·02000	(0·196)	0·915	0·922	0·933	0·943	0·952	0·967	0·979	0·990	1·008	1·022	1·038	1·051	1·073	1·091
0·03000	(0·294)	0·914	0·921	0·932	0·942	0·951	0·966	0·978	0·989	1·007	1·022	1·037	1·051	1·073	1·091
0·04000	(0·392)	0·913	0·920	0·932	0·942	0·951	0·965	0·978	0·989	1·007	1·022	1·037	1·051	1·073	1·091
0·06000	(0·588)	0·912	0·919	0·931	0·941	0·950	0·965	0·977	0·988	1·006	1·021	1·037	1·050	1·072	1·090
0·08000	(0·785)	0·912	0·918	0·930	0·941	0·950	0·965	0·977	0·988	1·006	1·021	1·037	1·050	1·072	1·090
0·10000	(0·981)	0·911	0·918	0·930	0·940	0·949	0·964	0·977	0·988	1·006	1·021	1·037	1·050	1·072	1·090
0·15000	(1·47)	0·911	0·918	0·930	0·940	0·949	0·964	0·977	0·988	1·006	1·021	1·036	1·050	1·072	1·090
0·20000	(1·96)	0·910	0·917	0·929	0·940	0·949	0·964	0·977	0·987	1·006	1·021	1·036	1·050	1·072	1·090
0·30000	(2·94)	0·910	0·917	0·929	0·939	0·948	0·964	0·976	0·987	1·005	1·020	1·036	1·050	1·072	1·090
0·40000	(3·92)	0·909	0·917	0·929	0·939	0·948	0·964	0·976	0·987	1·005	1·020	1·036	1·050	1·072	1·090
0·60000	(5·88)	0·909	0·916	0·929	0·939	0·948	0·963	0·976	0·987	1·005	1·020	1·036	1·049	1·072	1·090
0·80000	(7·85)	0·909	0·916	0·928	0·939	0·948	0·963	0·976	0·987	1·005	1·020	1·036	1·049	1·072	1·090
1·00000	(9·81)	0·909	0·916	0·928	0·939	0·948	0·963	0·976	0·987	1·005	1·020	1·036	1·049	1·072	1·090
1·50000	(14·7)	0·909	0·916	0·928	0·939	0·948	0·963	0·976	0·987	1·005	1·020	1·036	1·049	1·072	1·090
2·00000	(19·6)	0·909	0·916	0·928	0·939	0·948	0·963	0·976	0·987	1·005	1·020	1·036	1·049	1·071	1·090
3·00000	(29·4)	0·908	0·916	0·928	0·938	0·948	0·963	0·976	0·987	1·005	1·020	1·036	1·049	1·071	1·090
S	$(\Delta p/\rho l)$	125	150	200	250	300	400	500	600	800	1000	1250	1500	2000	2500

Grad'nt k.p.g. (Equivalent) Pipe diameters in mm **Roughness size, $k_s = 0.150$ mm**

Kin. visc., $\nu = 0.25 \times 10^{-6}$ m^{2}s^{-1};
$S = 0.00010$ to 3.00000
i.e. kin. pr. grad., $\Delta p/\rho l =$ (0.00098) to (29.4) ms^{-2}

Roughness size, $k_s = 0.30$ mm
This table shows values of m, as follows
m_C for Colebrook-White solutions

Grad'nt S	k.p.g. $(\Delta p/\rho l)$	\multicolumn{14}{c}{(Equivalent) Pipe diameters in mm}
		125	150	200	250	300	400	500	600	800	1000	1250	1500	2000	2500
0·00010	(0·00098)	1·051	1·051	1·054	1·058	1·063	1·072	1·080	1·088	1·102	1·114	1·127	1·139	1·158	1·174
0·00015	(0·00147)	1·042	1·043	1·047	1·052	1·057	1·067	1·076	1·084	1·098	1·111	1·124	1·136	1·156	1·172
0·00020	(0·00196)	1·036	1·038	1·042	1·048	1·053	1·063	1·073	1·081	1·096	1·109	1·122	1·134	1·154	1·171
0·00030	(0·00294)	1·029	1·031	1·037	1·043	1·049	1·060	1·069	1·078	1·093	1·106	1·120	1·132	1·153	1·169
0·00040	(0·00392)	1·024	1·027	1·033	1·040	1·046	1·057	1·067	1·076	1·092	1·105	1·119	1·131	1·152	1·169
0·00060	(0·00588)	1·019	1·022	1·029	1·036	1·043	1·054	1·065	1·074	1·090	1·103	1·117	1·130	1·150	1·167
0·00080	(0·00785)	1·016	1·019	1·027	1·034	1·041	1·053	1·063	1·072	1·088	1·102	1·116	1·129	1·150	1·167
0·00100	(0·00981)	1·014	1·017	1·025	1·032	1·039	1·051	1·062	1·071	1·088	1·101	1·116	1·128	1·149	1·166
0·00150	(0·0147)	1·010	1·014	1·022	1·030	1·037	1·050	1·060	1·070	1·086	1·100	1·115	1·127	1·148	1·166
0·00200	(0·0196)	1·008	1·012	1·021	1·029	1·036	1·048	1·059	1·069	1·086	1·099	1·114	1·127	1·148	1·165
0·00300	(0·0294)	1·005	1·010	1·019	1·027	1·034	1·047	1·058	1·068	1·085	1·099	1·113	1·126	1·147	1·165
0·00400	(0·0392)	1·004	1·009	1·018	1·026	1·033	1·046	1·058	1·067	1·084	1·098	1·113	1·126	1·147	1·164
0·00600	(0·0588)	1·002	1·007	1·016	1·025	1·032	1·045	1·057	1·067	1·083	1·097	1·112	1·125	1·147	1·164
0·00800	(0·0785)	1·001	1·006	1·015	1·024	1·031	1·045	1·056	1·066	1·083	1·097	1·112	1·125	1·146	1·164
0·01000	(0·0981)	1·000	1·005	1·015	1·023	1·031	1·044	1·056	1·066	1·083	1·097	1·112	1·125	1·146	1·164
0·01500	(0·147)	0·999	1·004	1·014	1·022	1·030	1·044	1·055	1·065	1·082	1·097	1·112	1·124	1·146	1·163
0·02000	(0·196)	0·998	1·003	1·013	1·022	1·030	1·043	1·055	1·065	1·082	1·096	1·111	1·124	1·146	1·163
0·03000	(0·294)	0·997	1·003	1·013	1·021	1·029	1·043	1·055	1·065	1·082	1·096	1·111	1·124	1·146	1·163
0·04000	(0·392)	0·997	1·002	1·012	1·021	1·029	1·043	1·054	1·064	1·082	1·096	1·111	1·124	1·145	1·163
0·06000	(0·588)	0·996	1·002	1·012	1·021	1·029	1·042	1·054	1·064	1·081	1·096	1·111	1·124	1·145	1·163
0·08000	(0·785)	0·996	1·001	1·012	1·020	1·028	1·042	1·054	1·064	1·081	1·096	1·111	1·124	1·145	1·163
0·10000	(0·981)	0·996	1·001	1·011	1·020	1·028	1·042	1·054	1·064	1·081	1·096	1·111	1·124	1·145	1·163
0·15000	(1·47)	0·995	1·001	1·011	1·020	1·028	1·042	1·054	1·064	1·081	1·095	1·111	1·124	1·145	1·163
0·20000	(1·96)	0·995	1·001	1·011	1·020	1·028	1·042	1·054	1·064	1·081	1·095	1·110	1·123	1·145	1·163
0·30000	(2·94)	0·995	1·000	1·011	1·020	1·028	1·042	1·053	1·064	1·081	1·095	1·110	1·123	1·145	1·163
0·40000	(3·92)	0·994	1·000	1·011	1·020	1·028	1·042	1·053	1·064	1·081	1·095	1·110	1·123	1·145	1·163
0·60000	(5·88)	0·994	1·000	1·010	1·019	1·028	1·041	1·053	1·063	1·081	1·095	1·110	1·123	1·145	1·163
0·80000	(7·85)	0·994	1·000	1·010	1·019	1·027	1·041	1·053	1·063	1·081	1·095	1·110	1·123	1·145	1·163
1·00000	(9·81)	0·994	1·000	1·010	1·019	1·027	1·041	1·053	1·063	1·081	1·095	1·110	1·123	1·145	1·163
1·50000	(14·7)	0·994	1·000	1·010	1·019	1·027	1·041	1·053	1·063	1·081	1·095	1·110	1·123	1·145	1·163
2·00000	(19·6)	0·994	1·000	1·010	1·019	1·027	1·041	1·053	1·063	1·081	1·095	1·110	1·123	1·145	1·163
3·00000	(29·4)	0·994	1·000	1·010	1·019	1·027	1·041	1·053	1·063	1·081	1·095	1·110	1·123	1·145	1·163
S	$(\Delta p/\rho l)$	125	150	200	250	300	400	500	600	800	1000	1250	1500	2000	2500

Grad'nt k.p.g. (Equivalent) Pipe diameters in mm **Roughness size, $k_s = 0.30$ mm**

E33

Kin. visc., $\nu = 0.30 \times 10^{-6}$ m^{2}s^{-1};
$S = 0.00010$ to 30.0000

i.e. kin. pr. grad., $\Delta p/\rho l =$
(0.00098) to (294) ms^{-2}

Roughness size, $k_s = 0.0015$ mm
This table shows values of m, as follows

m_C for Colebrook-White solutions; or,
where $R \leq 2000$, m_P for laminar flow

Grad'nt S	k.p.g. $(\Delta p/\rho l)$	6	8	10	12.5	15	20	25	30	40	50	60	80	100	125
0.00030	(0.00294)	2.057	1.402	1.041	0.773	*	0.950	0.931	0.918	0.901	0.891	0.884	0.876	0.872	0.869
0.00040	(0.00392)	1.782	1.214	0.902	*	0.954	0.927	0.909	0.898	0.882	0.873	0.867	0.860	0.856	0.854
0.00060	(0.00588)	1.455	0.991	*	*	0.920	0.896	0.881	0.870	0.857	0.849	0.844	0.839	0.836	0.835
0.00080	(0.00785)	1.260	0.858	*	0.913	0.897	0.875	0.861	0.852	0.840	0.833	0.829	0.824	0.822	0.822
0.00100	(0.00981)	1.127	0.768	*	0.895	0.880	0.859	0.847	0.838	0.828	0.821	0.818	0.814	0.812	0.812
0.00150	(0.0147)	0.920	*	0.884	0.864	0.850	0.833	0.822	0.814	0.806	0.801	0.798	0.795	0.794	0.794
0.00200	(0.0196)	0.797	*	0.861	0.843	0.831	0.815	0.805	0.799	0.791	0.786	0.784	0.782	0.782	0.782
0.00300	(0.0294)	*	0.850	0.831	0.815	0.805	0.791	0.783	0.777	0.771	0.767	0.766	0.765	0.765	0.766
0.00400	(0.0392)	*	0.828	0.811	0.797	0.787	0.775	0.767	0.763	0.757	0.754	0.753	0.753	0.754	0.756
0.00600	(0.0588)	0.823	0.800	0.785	0.772	0.764	0.753	0.747	0.743	0.739	0.737	0.736	0.737	0.738	0.741
0.00800	(0.0785)	0.802	0.780	0.767	0.756	0.748	0.739	0.733	0.730	0.727	0.725	0.725	0.726	0.728	0.731
0.01000	(0.0981)	0.786	0.766	0.754	0.743	0.736	0.728	0.723	0.720	0.717	0.717	0.717	0.718	0.720	0.723
0.01500	(0.147)	0.759	0.742	0.731	0.722	0.716	0.709	0.705	0.703	0.701	0.701	0.702	0.704	0.707	0.710
0.02000	(0.196)	0.741	0.726	0.716	0.708	0.703	0.697	0.693	0.692	0.690	0.691	0.692	0.694	0.698	0.701
0.03000	(0.294)	0.718	0.704	0.695	0.689	0.684	0.680	0.677	0.676	0.676	0.677	0.678	0.682	0.685	0.690
0.04000	(0.392)	0.702	0.689	0.682	0.676	0.672	0.668	0.666	0.666	0.666	0.667	0.669	0.673	0.677	0.682
0.06000	(0.588)	0.681	0.670	0.664	0.659	0.656	0.653	0.652	0.652	0.653	0.655	0.657	0.661	0.666	0.671
0.08000	(0.785)	0.666	0.657	0.651	0.647	0.645	0.643	0.642	0.642	0.644	0.646	0.649	0.654	0.658	0.664
0.10000	(0.981)	0.656	0.647	0.642	0.639	0.637	0.635	0.635	0.635	0.637	0.640	0.642	0.648	0.653	0.658
0.15000	(1.47)	0.638	0.631	0.627	0.624	0.622	0.622	0.622	0.623	0.626	0.629	0.632	0.638	0.643	0.649
0.20000	(1.96)	0.626	0.619	0.616	0.614	0.613	0.613	0.613	0.615	0.618	0.621	0.625	0.631	0.637	0.643
0.30000	(2.94)	0.610	0.604	0.602	0.601	0.600	0.601	0.602	0.604	0.608	0.612	0.615	0.622	0.628	0.635
0.40000	(3.92)	0.599	0.595	0.593	0.592	0.592	0.593	0.595	0.597	0.601	0.605	0.609	0.616	0.623	0.630
0.60000	(5.88)	0.584	0.581	0.580	0.580	0.580	0.582	0.585	0.587	0.592	0.597	0.601	0.609	0.616	0.623
0.80000	(7.85)	0.575	0.572	0.572	0.572	0.573	0.575	0.578	0.581	0.586	0.591	0.596	0.604	0.611	0.618
1.00000	(9.81)	0.568	0.566	0.566	0.566	0.567	0.570	0.573	0.576	0.582	0.587	0.592	0.600	0.607	0.615
1.50000	(14.7)	0.556	0.555	0.555	0.556	0.558	0.561	0.565	0.568	0.574	0.580	0.585	0.594	0.602	0.610
2.00000	(19.6)	0.548	0.547	0.548	0.550	0.552	0.555	0.559	0.563	0.570	0.576	0.581	0.590	0.598	0.606
3.00000	(29.4)	0.537	0.538	0.539	0.541	0.543	0.548	0.552	0.556	0.563	0.570	0.575	0.585	0.593	0.602
4.00000	(39.2)	0.530	0.531	0.533	0.536	0.538	0.543	0.548	0.552	0.559	0.566	0.572	0.582	0.590	0.599
6.00000	(58.8)	0.521	0.523	0.525	0.528	0.531	0.537	0.542	0.546	0.554	0.561	0.567	0.578	0.586	0.596
8.00000	(78.5)	0.515	0.518	0.520	0.524	0.527	0.533	0.538	0.543	0.551	0.558	0.565	0.575	0.584	0.594
10.0000	(98.1)	0.511	0.514	0.517	0.520	0.524	0.530	0.535	0.540	0.549	0.556	0.563	0.573	0.582	0.592
15.0000	(147)	0.504	0.507	0.510	0.514	0.518	0.525	0.531	0.536	0.545	0.553	0.559	0.570	0.580	0.590
20.0000	(196)	0.499	0.503	0.507	0.511	0.515	0.522	0.528	0.533	0.543	0.551	0.557	0.569	0.578	0.588
30.0000	(294)	0.493	0.498	0.502	0.506	0.511	0.518	0.524	0.530	0.540	0.548	0.555	0.566	0.576	0.586
S	$(\Delta p/\rho l)$	6	8	10	12.5	15	20	25	30	40	50	60	80	100	125

Grad'nt   k.p.g.   (Equivalent) Pipe diameters in mm

Grad'nt S	k.p.g. $(\Delta p/\rho l)$	125	150	200	250	300	400	500	600	800	1000	1250	1500	2000	2500
0.00010	(0.00098)	0.929	0.925	0.922	0.921	0.921	0.923	0.926	0.929	0.935	0.941	0.948	0.955	0.966	0.976
0.00015	(0.00147)	0.906	0.903	0.901	0.901	0.902	0.904	0.908	0.911	0.918	0.925	0.932	0.939	0.950	0.960
0.00020	(0.00196)	0.890	0.888	0.886	0.887	0.888	0.892	0.895	0.899	0.907	0.913	0.921	0.928	0.940	0.950
0.00030	(0.00294)	0.869	0.867	0.867	0.868	0.870	0.874	0.878	0.883	0.891	0.898	0.906	0.913	0.925	0.936
0.00040	(0.00392)	0.854	0.853	0.854	0.855	0.858	0.862	0.867	0.871	0.880	0.887	0.895	0.903	0.915	0.926
0.00060	(0.00588)	0.835	0.835	0.836	0.838	0.841	0.846	0.851	0.856	0.865	0.873	0.881	0.889	0.902	0.913
0.00080	(0.00785)	0.822	0.822	0.824	0.826	0.829	0.835	0.840	0.846	0.855	0.863	0.872	0.879	0.893	0.904
0.00100	(0.00981)	0.812	0.812	0.814	0.817	0.821	0.827	0.832	0.838	0.847	0.855	0.864	0.872	0.886	0.897
0.00150	(0.0147)	0.794	0.795	0.798	0.802	0.805	0.812	0.818	0.824	0.834	0.842	0.852	0.860	0.874	0.885
0.00200	(0.0196)	0.782	0.784	0.787	0.791	0.795	0.802	0.809	0.814	0.825	0.833	0.843	0.851	0.865	0.877
0.00300	(0.0294)	0.766	0.768	0.773	0.777	0.781	0.789	0.795	0.802	0.812	0.821	0.831	0.840	0.854	0.867
0.00400	(0.0392)	0.756	0.758	0.762	0.767	0.771	0.779	0.787	0.793	0.804	0.813	0.823	0.832	0.847	0.859
0.00600	(0.0588)	0.741	0.743	0.749	0.754	0.759	0.767	0.774	0.781	0.793	0.802	0.813	0.822	0.837	0.849
0.00800	(0.0785)	0.731	0.734	0.740	0.745	0.750	0.759	0.766	0.773	0.785	0.795	0.805	0.815	0.830	0.843
0.01000	(0.0981)	0.723	0.727	0.733	0.738	0.743	0.752	0.760	0.767	0.779	0.789	0.800	0.809	0.825	0.838
0.01500	(0.147)	0.710	0.714	0.720	0.726	0.732	0.741	0.750	0.757	0.769	0.780	0.791	0.800	0.817	0.830
0.02000	(0.196)	0.701	0.705	0.712	0.718	0.724	0.734	0.743	0.750	0.763	0.773	0.785	0.794	0.811	0.824
0.03000	(0.294)	0.690	0.694	0.701	0.708	0.714	0.724	0.733	0.741	0.754	0.765	0.776	0.787	0.803	0.817
0.04000	(0.392)	0.682	0.686	0.694	0.701	0.707	0.718	0.727	0.735	0.748	0.759	0.771	0.781	0.798	0.813
0.06000	(0.588)	0.671	0.676	0.684	0.691	0.698	0.709	0.718	0.726	0.740	0.752	0.764	0.775	0.792	0.807
0.08000	(0.785)	0.664	0.669	0.677	0.685	0.692	0.703	0.713	0.721	0.735	0.747	0.759	0.770	0.788	0.803
0.10000	(0.981)	0.658	0.663	0.672	0.680	0.687	0.699	0.709	0.717	0.732	0.744	0.756	0.767	0.785	0.800
0.15000	(1.47)	0.649	0.655	0.664	0.672	0.679	0.692	0.702	0.711	0.725	0.738	0.751	0.762	0.780	0.795
0.20000	(1.96)	0.643	0.649	0.659	0.667	0.674	0.687	0.697	0.706	0.721	0.734	0.747	0.758	0.777	0.792
0.30000	(2.94)	0.635	0.641	0.651	0.660	0.668	0.681	0.691	0.701	0.716	0.729	0.742	0.754	0.773	0.789
0.40000	(3.92)	0.630	0.636	0.647	0.656	0.663	0.677	0.688	0.697	0.713	0.726	0.740	0.751	0.771	0.786
0.60000	(5.88)	0.623	0.629	0.641	0.650	0.658	0.672	0.683	0.693	0.709	0.722	0.736	0.748	0.767	0.783
0.80000	(7.85)	0.618	0.625	0.637	0.646	0.654	0.668	0.680	0.690	0.706	0.720	0.734	0.746	0.766	0.782
1.00000	(9.81)	0.615	0.622	0.634	0.644	0.652	0.666	0.678	0.688	0.704	0.718	0.732	0.744	0.764	0.780
1.50000	(14.7)	0.610	0.617	0.629	0.639	0.648	0.662	0.674	0.684	0.701	0.715	0.730	0.742	0.762	0.778
2.00000	(19.6)	0.606	0.614	0.626	0.636	0.645	0.660	0.672	0.682	0.699	0.713	0.728	0.740	0.761	0.777
3.00000	(29.4)	0.602	0.610	0.622	0.633	0.642	0.657	0.669	0.680	0.697	0.711	0.726	0.739	0.759	0.776
S	$(\Delta p/\rho l)$	125	150	200	250	300	400	500	600	800	1000	1250	1500	2000	2500

Grad'nt   k.p.g.   (Equivalent) Pipe diameters in mm

Kinematic viscosity, $\nu = 0.30 \times 10^{-6}$ m^{2}s^{-1} ;

Roughness size, $k_s = 0.0015$ mm

Kin. visc., $\nu = 0.30 \times 10^{-6} \text{ m}^2\text{s}^{-1}$;
$S = 0.00010$ to 30.0000

i.e. kin. pr. grad., $\Delta p/\rho l =$
(0.00098) to (294) ms^{-2}

Roughness size, $k_s = 0.003$ mm
This table shows values of m, as follows

m_C for Colebrook-White solutions; or,
where $R \leq 2000$, m_P for laminar flow

Grad'nt S	k.p.g. $(\Delta p/\rho l)$	6	8	10	12·5	15	20	25	30	40	50	60	80	100	125
0.00030	(0.00294)	2·057	1·402	1·041	0·773	*	0·951	0·932	0·919	0·902	0·892	0·885	0·877	0·873	0·870
0.00040	(0.00392)	1·782	1·214	0·902	*	0·955	0·928	0·910	0·899	0·884	0·874	0·869	0·862	0·858	0·856
0.00060	(0.00588)	1·455	0·991	*	*	0·921	0·897	0·882	0·872	0·859	0·851	0·846	0·841	0·838	0·837
0.00080	(0.00785)	1·260	0·858	*	0·915	0·898	0·876	0·863	0·853	0·842	0·835	0·831	0·826	0·824	0·824
0.00100	(0.00981)	1·127	0·768	*	0·897	0·881	0·861	0·848	0·840	0·829	0·823	0·819	0·816	0·814	0·814
0.00150	(0.0147)	0·920	*	0·885	0·865	0·852	0·834	0·824	0·816	0·807	0·803	0·800	0·797	0·796	0·797
0.00200	(0.0196)	0·797	*	0·863	0·845	0·832	0·817	0·807	0·800	0·793	0·789	0·786	0·784	0·784	0·785
0.00300	(0.0294)	*	0·852	0·833	0·817	0·807	0·793	0·785	0·779	0·773	0·770	0·768	0·768	0·768	0·770
0.00400	(0.0392)	*	0·830	0·813	0·799	0·789	0·777	0·770	0·765	0·760	0·757	0·756	0·756	0·757	0·759
0.00600	(0.0588)	0·826	0·802	0·787	0·775	0·766	0·756	0·750	0·746	0·742	0·740	0·740	0·741	0·743	0·745
0.00800	(0.0785)	0·804	0·783	0·769	0·758	0·751	0·742	0·736	0·733	0·730	0·729	0·729	0·730	0·733	0·736
0.01000	(0.0981)	0·789	0·769	0·756	0·746	0·739	0·731	0·726	0·724	0·721	0·721	0·721	0·723	0·725	0·729
0.01500	(0.147)	0·762	0·745	0·734	0·725	0·720	0·713	0·709	0·707	0·706	0·706	0·707	0·709	0·713	0·717
0.02000	(0.196)	0·745	0·729	0·719	0·711	0·706	0·701	0·698	0·696	0·695	0·696	0·697	0·700	0·704	0·708
0.03000	(0.294)	0·721	0·708	0·699	0·693	0·689	0·684	0·682	0·681	0·681	0·683	0·685	0·688	0·693	0·697
0.04000	(0.392)	0·706	0·693	0·686	0·680	0·677	0·673	0·672	0·671	0·672	0·674	0·676	0·681	0·685	0·690
0.06000	(0.588)	0·685	0·675	0·669	0·664	0·661	0·659	0·658	0·658	0·660	0·662	0·665	0·670	0·675	0·681
0.08000	(0.785)	0·671	0·662	0·657	0·653	0·651	0·649	0·649	0·649	0·652	0·654	0·657	0·663	0·668	0·674
0.10000	(0.981)	0·661	0·653	0·648	0·645	0·643	0·642	0·642	0·643	0·646	0·649	0·652	0·658	0·664	0·670
0.15000	(1.47)	0·644	0·637	0·633	0·631	0·630	0·629	0·630	0·632	0·635	0·639	0·643	0·649	0·655	0·662
0.20000	(1.96)	0·632	0·626	0·623	0·622	0·621	0·621	0·623	0·624	0·628	0·633	0·636	0·644	0·650	0·657
0.30000	(2.94)	0·617	0·612	0·610	0·609	0·609	0·611	0·613	0·615	0·620	0·624	0·628	0·636	0·643	0·651
0.40000	(3.92)	0·607	0·603	0·602	0·601	0·602	0·603	0·606	0·609	0·614	0·619	0·623	0·631	0·639	0·646
0.60000	(5.88)	0·594	0·591	0·591	0·590	0·591	0·592	0·594	0·597	0·600	0·606	0·612	0·617	0·625	0·633
0.80000	(7.85)	0·585	0·583	0·583	0·584	0·585	0·588	0·592	0·595	0·601	0·607	0·612	0·621	0·629	0·638
1.00000	(9.81)	0·579	0·577	0·578	0·579	0·580	0·584	0·588	0·591	0·598	0·604	0·609	0·619	0·627	0·635
1.50000	(14.7)	0·568	0·568	0·568	0·570	0·572	0·577	0·581	0·585	0·592	0·599	0·604	0·614	0·622	0·631
2.00000	(19.6)	0·561	0·561	0·563	0·565	0·567	0·572	0·577	0·581	0·588	0·595	0·601	0·611	0·620	0·629
3.00000	(29.4)	0·552	0·553	0·555	0·558	0·561	0·566	0·571	0·576	0·584	0·591	0·597	0·608	0·617	0·626
4.00000	(39.2)	0·546	0·548	0·550	0·553	0·556	0·562	0·568	0·572	0·581	0·588	0·595	0·605	0·614	0·624
6.00000	(58.8)	0·539	0·541	0·544	0·548	0·551	0·557	0·563	0·568	0·577	0·585	0·591	0·603	0·612	0·622
8.00000	(78.5)	0·534	0·537	0·540	0·544	0·548	0·554	0·560	0·566	0·575	0·583	0·590	0·601	0·610	0·620
10.0000	(98.1)	0·530	0·534	0·537	0·541	0·545	0·552	0·559	0·564	0·573	0·581	0·588	0·600	0·609	0·619
15.0000	(147)	0·525	0·529	0·533	0·537	0·542	0·549	0·555	0·561	0·571	0·579	0·586	0·598	0·608	0·618
20.0000	(196)	0·521	0·526	0·530	0·535	0·539	0·547	0·554	0·559	0·569	0·578	0·585	0·597	0·607	0·617
30.0000	(294)	0·517	0·522	0·526	0·532	0·536	0·544	0·551	0·557	0·567	0·576	0·583	0·595	0·605	0·616
S	$(\Delta p/\rho l)$	6	8	10	12·5	15	20	25	30	40	50	60	80	100	125

Grad'nt   k.p.g.   (Equivalent) Pipe diameters in mm

S	$(\Delta p/\rho l)$	125	150	200	250	300	400	500	600	800	1000	1250	1500	2000	2500
0.00010	(0.00098)	0·930	0·926	0·923	0·922	0·922	0·925	0·928	0·931	0·937	0·943	0·951	0·957	0·968	0·978
0.00015	(0.00147)	0·907	0·904	0·902	0·902	0·903	0·906	0·910	0·913	0·920	0·927	0·935	0·941	0·953	0·964
0.00020	(0.00196)	0·891	0·889	0·888	0·888	0·890	0·893	0·897	0·901	0·909	0·916	0·924	0·931	0·943	0·954
0.00030	(0.00294)	0·870	0·869	0·869	0·870	0·872	0·876	0·881	0·885	0·893	0·901	0·909	0·916	0·929	0·940
0.00040	(0.00392)	0·856	0·855	0·856	0·857	0·860	0·865	0·870	0·874	0·883	0·891	0·899	0·907	0·920	0·931
0.00060	(0.00588)	0·837	0·837	0·838	0·841	0·843	0·849	0·854	0·859	0·869	0·877	0·886	0·894	0·907	0·919
0.00080	(0.00785)	0·824	0·824	0·826	0·829	0·832	0·838	0·844	0·849	0·859	0·867	0·876	0·885	0·899	0·910
0.00100	(0.00981)	0·814	0·815	0·817	0·820	0·824	0·830	0·836	0·842	0·852	0·860	0·870	0·878	0·892	0·904
0.00150	(0.0147)	0·797	0·798	0·802	0·805	0·809	0·816	0·823	0·828	0·839	0·848	0·858	0·866	0·881	0·893
0.00200	(0.0196)	0·785	0·787	0·791	0·795	0·799	0·807	0·813	0·820	0·830	0·840	0·850	0·859	0·874	0·886
0.00300	(0.0294)	0·770	0·772	0·777	0·781	0·786	0·794	0·801	0·808	0·819	0·829	0·839	0·848	0·864	0·876
0.00400	(0.0392)	0·759	0·762	0·767	0·772	0·777	0·785	0·793	0·799	0·811	0·821	0·832	0·841	0·857	0·870
0.00600	(0.0588)	0·745	0·748	0·754	0·760	0·765	0·774	0·782	0·789	0·801	0·811	0·822	0·832	0·848	0·862
0.00800	(0.0785)	0·736	0·739	0·745	0·751	0·757	0·766	0·774	0·782	0·794	0·805	0·816	0·826	0·843	0·856
0.01000	(0.0981)	0·729	0·732	0·739	0·745	0·750	0·760	0·769	0·776	0·789	0·800	0·812	0·822	0·839	0·852
0.01500	(0.147)	0·717	0·720	0·728	0·734	0·740	0·751	0·759	0·767	0·781	0·792	0·804	0·814	0·832	0·846
0.02000	(0.196)	0·708	0·713	0·720	0·727	0·733	0·744	0·753	0·761	0·775	0·787	0·799	0·809	0·827	0·842
0.03000	(0.294)	0·697	0·702	0·710	0·718	0·724	0·735	0·745	0·753	0·768	0·780	0·792	0·803	0·821	0·836
0.04000	(0.392)	0·690	0·695	0·704	0·711	0·718	0·730	0·740	0·748	0·763	0·775	0·788	0·799	0·817	0·833
0.06000	(0.588)	0·681	0·686	0·695	0·703	0·710	0·723	0·733	0·742	0·757	0·769	0·783	0·794	0·813	0·828
0.08000	(0.785)	0·674	0·680	0·690	0·698	0·705	0·718	0·729	0·738	0·753	0·766	0·779	0·791	0·810	0·826
0.10000	(0.981)	0·670	0·676	0·686	0·694	0·702	0·715	0·725	0·735	0·750	0·763	0·777	0·788	0·808	0·824
0.15000	(1.47)	0·662	0·668	0·679	0·688	0·696	0·709	0·720	0·730	0·746	0·759	0·773	0·785	0·804	0·820
0.20000	(1.96)	0·657	0·663	0·674	0·684	0·692	0·705	0·717	0·726	0·743	0·756	0·770	0·782	0·802	0·819
0.30000	(2.94)	0·651	0·657	0·669	0·678	0·687	0·701	0·712	0·722	0·739	0·753	0·767	0·779	0·800	0·816
0.40000	(3.92)	0·646	0·653	0·665	0·675	0·683	0·698	0·710	0·720	0·737	0·751	0·765	0·778	0·798	0·815
0.60000	(5.88)	0·641	0·648	0·661	0·671	0·680	0·694	0·706	0·717	0·734	0·748	0·763	0·776	0·796	0·813
0.80000	(7.85)	0·638	0·645	0·658	0·668	0·677	0·692	0·704	0·715	0·732	0·747	0·762	0·774	0·795	0·812
1.00000	(9.81)	0·635	0·643	0·656	0·666	0·675	0·691	0·703	0·714	0·731	0·746	0·761	0·773	0·794	0·811
1.50000	(14.7)	0·631	0·639	0·652	0·663	0·673	0·688	0·701	0·711	0·729	0·744	0·759	0·772	0·793	0·810
2.00000	(19.6)	0·629	0·637	0·650	0·661	0·671	0·686	0·699	0·710	0·728	0·743	0·758	0·771	0·792	0·809
3.00000	(29.4)	0·626	0·634	0·648	0·659	0·669	0·685	0·698	0·709	0·727	0·741	0·757	0·770	0·791	0·808
S	$(\Delta p/\rho l)$	125	150	200	250	300	400	500	600	800	1000	1250	1500	2000	2500

Grad'nt   k.p.g.   (Equivalent) Pipe diameters in mm

Kinematic viscosity, $\nu = 0.30 \times 10^{-6} \text{ m}^2\text{s}^{-1}$;   **Roughness size, $k_s = 0.003$ mm**

E35

Kin. visc., $\nu = 0.30 \times 10^{-6}$ m^{2}s^{-1}; $S = 0.00010$ to 30.0000

i.e. kin. pr. grad., $\Delta p/\rho l =$ (0.00098) to (294) ms^{-2}

Roughness size, $k_s = 0.006$ mm

This table shows values of m, as follows

m_C for Colebrook-White solutions; or, where $R \le 2000$, m_P for laminar flow

Grad'nt S	k.p.g. $(\Delta p/\rho l)$	(Equivalent) Pipe diameters in mm													
		6	8	10	12·5	15	20	25	30	40	50	60	80	100	125
0·00030	(0·00294)	2·057	1·402	1·041	0·773	*	0·953	0·934	0·921	0·904	0·894	0·888	0·880	0·876	0·873
0·00040	(0·00392)	1·782	1·214	0·902	*	0·957	0·930	0·913	0·901	0·886	0·877	0·871	0·864	0·861	0·859
0·00060	(0·00588)	1·455	0·991	*	*	0·923	0·899	0·884	0·874	0·861	0·854	0·849	0·844	0·841	0·840
0·00080	(0·00785)	1·260	0·858	*	0·917	0·900	0·879	0·865	0·856	0·845	0·838	0·834	0·830	0·828	0·828
0·00100	(0·00981)	1·127	0·768	*	0·899	0·884	0·864	0·851	0·843	0·833	0·827	0·823	0·820	0·818	0·818
0·00150	(0·0147)	0·920	*	0·888	0·868	0·855	0·838	0·827	0·820	0·811	0·807	0·804	0·802	0·801	0·802
0·00200	(0·0196)	0·797	*	0·866	0·848	0·836	0·820	0·811	0·804	0·797	0·793	0·791	0·789	0·790	0·791
0·00300	(0·0294)	*	0·855	0·837	0·821	0·810	0·797	0·789	0·784	0·778	0·775	0·774	0·773	0·774	0·777
0·00400	(0·0392)	*	0·834	0·817	0·803	0·794	0·782	0·775	0·770	0·765	0·763	0·762	0·763	0·764	0·767
0·00600	(0·0588)	0·830	0·806	0·791	0·779	0·771	0·761	0·755	0·752	0·748	0·747	0·747	0·748	0·750	0·754
0·00800	(0·0785)	0·809	0·788	0·774	0·764	0·756	0·747	0·742	0·740	0·737	0·736	0·737	0·739	0·741	0·745
0·01000	(0·0981)	0·794	0·774	0·762	0·752	0·745	0·737	0·733	0·730	0·728	0·728	0·729	0·731	0·735	0·739
0·01500	(0·147)	0·768	0·751	0·740	0·732	0·726	0·720	0·717	0·715	0·714	0·715	0·716	0·719	0·723	0·728
0·02000	(0·196)	0·751	0·735	0·726	0·718	0·714	0·708	0·706	0·705	0·704	0·706	0·707	0·711	0·716	0·721
0·03000	(0·294)	0·728	0·715	0·707	0·701	0·697	0·693	0·691	0·691	0·692	0·694	0·696	0·701	0·706	0·711
0·04000	(0·392)	0·713	0·701	0·694	0·689	0·686	0·683	0·682	0·682	0·683	0·686	0·689	0·694	0·699	0·705
0·06000	(0·588)	0·694	0·684	0·678	0·674	0·671	0·670	0·669	0·670	0·673	0·676	0·679	0·685	0·691	0·697
0·08000	(0·785)	0·681	0·672	0·667	0·664	0·662	0·661	0·661	0·662	0·666	0·669	0·673	0·679	0·685	0·692
0·10000	(0·981)	0·671	0·663	0·659	0·656	0·655	0·654	0·655	0·657	0·660	0·664	0·668	0·675	0·681	0·689
0·15000	(1·47)	0·655	0·649	0·646	0·644	0·643	0·644	0·645	0·647	0·652	0·656	0·660	0·668	0·675	0·683
0·20000	(1·96)	0·644	0·639	0·637	0·636	0·635	0·637	0·639	0·641	0·646	0·651	0·655	0·664	0·671	0·679
0·30000	(2·94)	0·631	0·627	0·625	0·625	0·625	0·628	0·630	0·633	0·639	0·644	0·649	0·658	0·666	0·674
0·40000	(3·92)	0·622	0·619	0·618	0·618	0·619	0·622	0·625	0·628	0·634	0·640	0·645	0·655	0·662	0·671
0·60000	(5·88)	0·610	0·608	0·608	0·609	0·611	0·614	0·618	0·622	0·629	0·635	0·640	0·650	0·658	0·667
0·80000	(7·85)	0·603	0·602	0·602	0·604	0·605	0·610	0·614	0·618	0·625	0·631	0·637	0·647	0·656	0·665
1·00000	(9·81)	0·597	0·597	0·598	0·600	0·602	0·606	0·611	0·615	0·622	0·629	0·635	0·645	0·654	0·663
1·50000	(14·7)	0·588	0·589	0·590	0·593	0·595	0·601	0·606	0·610	0·618	0·625	0·632	0·642	0·651	0·661
2·00000	(19·6)	0·583	0·584	0·586	0·589	0·592	0·597	0·602	0·607	0·616	0·623	0·629	0·640	0·650	0·659
3·00000	(29·4)	0·576	0·578	0·580	0·583	0·587	0·593	0·599	0·604	0·612	0·620	0·627	0·638	0·647	0·657
4·00000	(39·2)	0·571	0·574	0·576	0·580	0·584	0·590	0·596	0·601	0·610	0·618	0·625	0·637	0·646	0·656
6·00000	(58·8)	0·565	0·569	0·572	0·576	0·580	0·587	0·593	0·599	0·608	0·616	0·623	0·635	0·645	0·655
8·00000	(78·5)	0·562	0·565	0·569	0·574	0·578	0·585	0·591	0·597	0·607	0·615	0·622	0·634	0·644	0·654
10·0000	(98·1)	0·559	0·563	0·567	0·572	0·576	0·583	0·590	0·596	0·606	0·614	0·621	0·633	0·643	0·653
15·0000	(147)	0·555	0·560	0·564	0·569	0·573	0·581	0·588	0·594	0·604	0·612	0·620	0·632	0·642	0·652
20·0000	(196)	0·553	0·558	0·562	0·567	0·572	0·580	0·587	0·593	0·603	0·612	0·619	0·631	0·641	0·652
30·0000	(294)	0·550	0·555	0·560	0·565	0·570	0·578	0·585	0·591	0·602	0·611	0·618	0·630	0·641	0·651
S	$(\Delta p/\rho l)$	6	8	10	12·5	15	20	25	30	40	50	60	80	100	125

Grad'nt   k.p.g.   (Equivalent) Pipe diameters in mm

Grad'nt S	k.p.g. $(\Delta p/\rho l)$	(Equivalent) Pipe diameters in mm													
		125	150	200	250	300	400	500	600	800	1000	1250	1500	2000	2500
0·00010	(0·00098)	0·932	0·928	0·925	0·925	0·925	0·927	0·931	0·934	0·941	0·947	0·955	0·961	0·973	0·984
0·00015	(0·00147)	0·909	0·907	0·905	0·905	0·906	0·909	0·913	0·917	0·925	0·932	0·939	0·947	0·959	0·970
0·00020	(0·00196)	0·894	0·892	0·891	0·892	0·893	0·897	0·901	0·906	0·914	0·921	0·929	0·937	0·950	0·961
0·00030	(0·00294)	0·873	0·872	0·872	0·874	0·876	0·880	0·885	0·890	0·899	0·907	0·915	0·923	0·937	0·948
0·00040	(0·00392)	0·859	0·859	0·859	0·861	0·864	0·869	0·875	0·880	0·889	0·897	0·906	0·914	0·928	0·940
0·00060	(0·00588)	0·840	0·841	0·842	0·845	0·848	0·854	0·860	0·866	0·875	0·884	0·894	0·902	0·917	0·929
0·00080	(0·00785)	0·828	0·828	0·831	0·834	0·838	0·844	0·850	0·856	0·866	0·876	0·885	0·894	0·909	0·922
0·00100	(0·00981)	0·818	0·819	0·822	0·826	0·830	0·837	0·843	0·849	0·860	0·869	0·879	0·888	0·903	0·916
0·00150	(0·0147)	0·802	0·804	0·808	0·812	0·816	0·824	0·831	0·837	0·848	0·858	0·869	0·878	0·894	0·907
0·00200	(0·0196)	0·791	0·793	0·798	0·802	0·807	0·815	0·822	0·829	0·841	0·851	0·862	0·871	0·887	0·901
0·00300	(0·0294)	0·777	0·779	0·784	0·789	0·794	0·803	0·811	0·818	0·831	0·841	0·853	0·863	0·879	0·893
0·00400	(0·0392)	0·767	0·770	0·775	0·781	0·786	0·796	0·804	0·811	0·824	0·835	0·847	0·857	0·874	0·888
0·00600	(0·0588)	0·754	0·757	0·764	0·770	0·776	0·786	0·794	0·802	0·816	0·827	0·839	0·849	0·867	0·882
0·00800	(0·0785)	0·745	0·749	0·756	0·762	0·768	0·779	0·788	0·796	0·810	0·822	0·834	0·845	0·863	0·878
0·01000	(0·0981)	0·739	0·743	0·750	0·757	0·763	0·774	0·784	0·792	0·806	0·818	0·830	0·841	0·860	0·875
0·01500	(0·147)	0·728	0·732	0·741	0·748	0·754	0·766	0·776	0·784	0·799	0·811	0·824	0·836	0·854	0·870
0·02000	(0·196)	0·721	0·725	0·734	0·742	0·749	0·761	0·771	0·780	0·795	0·807	0·821	0·832	0·851	0·867
0·03000	(0·294)	0·711	0·717	0·726	0·734	0·741	0·754	0·765	0·774	0·789	0·802	0·816	0·828	0·847	0·863
0·04000	(0·392)	0·705	0·711	0·721	0·729	0·737	0·750	0·760	0·770	0·786	0·799	0·813	0·825	0·845	0·861
0·06000	(0·588)	0·697	0·703	0·714	0·723	0·731	0·744	0·755	0·765	0·781	0·795	0·809	0·821	0·841	0·858
0·08000	(0·785)	0·692	0·699	0·709	0·719	0·727	0·741	0·752	0·762	0·779	0·792	0·807	0·819	0·839	0·856
0·10000	(0·981)	0·689	0·695	0·706	0·716	0·724	0·738	0·750	0·760	0·777	0·791	0·805	0·818	0·838	0·855
0·15000	(1·47)	0·683	0·689	0·701	0·711	0·720	0·734	0·746	0·757	0·774	0·788	0·803	0·815	0·836	0·853
0·20000	(1·96)	0·679	0·686	0·698	0·708	0·717	0·732	0·744	0·754	0·772	0·786	0·801	0·814	0·835	0·852
0·30000	(2·94)	0·674	0·681	0·694	0·704	0·713	0·729	0·741	0·752	0·770	0·784	0·799	0·812	0·833	0·850
0·40000	(3·92)	0·671	0·679	0·691	0·702	0·711	0·727	0·739	0·750	0·768	0·783	0·798	0·811	0·832	0·849
0·60000	(5·88)	0·667	0·675	0·688	0·699	0·709	0·724	0·737	0·748	0·766	0·781	0·796	0·810	0·831	0·848
0·80000	(7·85)	0·665	0·673	0·687	0·698	0·707	0·723	0·736	0·747	0·765	0·780	0·796	0·809	0·830	0·848
1·00000	(9·81)	0·663	0·672	0·685	0·696	0·706	0·722	0·735	0·746	0·765	0·780	0·795	0·808	0·830	0·847
1·50000	(14·7)	0·661	0·669	0·683	0·695	0·704	0·720	0·734	0·745	0·763	0·778	0·794	0·807	0·829	0·846
2·00000	(19·6)	0·659	0·668	0·682	0·693	0·703	0·720	0·733	0·744	0·763	0·778	0·794	0·807	0·829	0·846
3·00000	(29·4)	0·657	0·666	0·680	0·692	0·702	0·718	0·732	0·743	0·762	0·777	0·793	0·806	0·828	0·846
S	$(\Delta p/\rho l)$	125	150	200	250	300	400	500	600	800	1000	1250	1500	2000	2500

Grad'nt   k.p.g.   (Equivalent) Pipe diameters in mm

Kinematic viscosity, $\nu = 0.30 \times 10^{-6}$ m^{2}s^{-1};   **Roughness size, $k_s = 0.006$ mm**

Kin. visc., $\nu = 0.30 \times 10^{-6}$ m^{2}s^{-1};
$S = 0.00010$ to 30.0000
i.e. kin. pr. grad., $\Delta p/\rho l =$
(0.00098) to (294) ms^{-2}

Roughness size, $k_s = 0.015$ mm
This table shows values of m, as follows
m_C for Colebrook-White solutions; or,
where $R \leq 2000$, m_P for laminar flow

Grad'nt S	k.p.g. $(\Delta p/\rho l)$	6	8	10	12·5	15	20	25	30	40	50	60	80	100	125
0.00030	(0·00294)	2·057	1·402	1·041	0·773	*	0·959	0·940	0·927	0·911	0·901	0·895	0·887	0·884	0·881
0.00040	(0·00392)	1·782	1·214	0·902	*	0·963	0·936	0·919	0·908	0·893	0·884	0·879	0·873	0·870	0·868
0.00060	(0·00588)	1·455	0·991	*	*	0·930	0·906	0·892	0·882	0·869	0·862	0·858	0·853	0·851	0·851
0.00080	(0·00785)	1·260	0·858	*	0·924	0·908	0·887	0·873	0·865	0·854	0·848	0·844	0·840	0·839	0·840
0.00100	(0·00981)	1·127	0·768	*	0·907	0·892	0·872	0·860	0·852	0·842	0·837	0·834	0·831	0·830	0·831
0.00150	(0·0147)	0·920	*	0·897	0·877	0·864	0·847	0·837	0·830	0·822	0·818	0·816	0·814	0·815	0·816
0.00200	(0·0196)	0·797	*	0·875	0·858	0·846	0·831	0·821	0·816	0·809	0·806	0·804	0·803	0·805	0·807
0.00300	(0·0294)	*	0·866	0·847	0·832	0·822	0·809	0·801	0·797	0·791	0·789	0·789	0·789	0·791	0·794
0.00400	(0·0392)	*	0·845	0·829	0·815	0·806	0·794	0·788	0·784	0·780	0·779	0·778	0·780	0·782	0·786
0.00600	(0·0588)	0·842	0·819	0·804	0·793	0·785	0·776	0·770	0·768	0·765	0·764	0·765	0·768	0·771	0·775
0.00800	(0·0785)	0·823	0·801	0·789	0·778	0·771	0·763	0·759	0·757	0·755	0·755	0·756	0·760	0·764	0·768
0.01000	(0·0981)	0·808	0·789	0·777	0·767	0·761	0·754	0·751	0·749	0·748	0·749	0·750	0·754	0·758	0·763
0.01500	(0·147)	0·784	0·767	0·757	0·749	0·744	0·739	0·736	0·735	0·736	0·737	0·739	0·744	0·749	0·755
0.02000	(0·196)	0·768	0·753	0·744	0·737	0·733	0·729	0·727	0·727	0·728	0·730	0·733	0·738	0·744	0·750
0.03000	(0·294)	0·748	0·735	0·728	0·722	0·719	0·716	0·715	0·716	0·718	0·721	0·724	0·730	0·736	0·743
0.04000	(0·392)	0·734	0·723	0·717	0·712	0·710	0·708	0·708	0·708	0·711	0·715	0·718	0·725	0·732	0·739
0.06000	(0·588)	0·717	0·708	0·703	0·699	0·698	0·697	0·698	0·699	0·703	0·707	0·711	0·719	0·726	0·734
0.08000	(0·785)	0·706	0·698	0·694	0·691	0·690	0·690	0·692	0·694	0·698	0·703	0·707	0·715	0·722	0·730
0.10000	(0·981)	0·698	0·691	0·687	0·685	0·685	0·685	0·687	0·689	0·694	0·699	0·704	0·712	0·720	0·728
0.15000	(1·47)	0·684	0·679	0·677	0·675	0·676	0·677	0·680	0·683	0·688	0·694	0·699	0·708	0·716	0·724
0.20000	(1·96)	0·676	0·671	0·670	0·669	0·670	0·672	0·675	0·678	0·684	0·690	0·696	0·705	0·713	0·722
0.30000	(2·94)	0·665	0·662	0·661	0·661	0·663	0·666	0·669	0·673	0·680	0·686	0·692	0·702	0·710	0·719
0.40000	(3·92)	0·658	0·656	0·656	0·657	0·658	0·662	0·666	0·670	0·677	0·683	0·689	0·699	0·708	0·718
0.60000	(5·88)	0·649	0·648	0·649	0·650	0·652	0·657	0·661	0·666	0·673	0·680	0·686	0·697	0·706	0·715
0.80000	(7·85)	0·644	0·644	0·645	0·647	0·649	0·654	0·659	0·663	0·671	0·678	0·685	0·695	0·704	0·714
1.00000	(9·81)	0·640	0·640	0·642	0·644	0·647	0·652	0·657	0·661	0·670	0·677	0·683	0·694	0·703	0·713
1.50000	(14·7)	0·634	0·635	0·637	0·640	0·643	0·648	0·654	0·659	0·667	0·675	0·681	0·693	0·702	0·712
2.00000	(19·6)	0·630	0·632	0·634	0·637	0·640	0·646	0·652	0·657	0·666	0·673	0·680	0·692	0·701	0·711
3.00000	(29·4)	0·626	0·628	0·630	0·634	0·637	0·644	0·650	0·655	0·664	0·672	0·679	0·690	0·700	0·710
4.00000	(39·2)	0·623	0·625	0·628	0·632	0·636	0·642	0·648	0·654	0·663	0·671	0·678	0·690	0·699	0·710
6.00000	(58·8)	0·619	0·622	0·626	0·630	0·633	0·640	0·647	0·652	0·662	0·670	0·677	0·689	0·698	0·709
8.00000	(78·5)	0·617	0·620	0·624	0·628	0·632	0·639	0·646	0·651	0·661	0·669	0·676	0·688	0·698	0·708
10.0000	(98·1)	0·616	0·619	0·623	0·627	0·631	0·639	0·645	0·651	0·660	0·669	0·676	0·688	0·698	0·708
15.0000	(147)	0·613	0·617	0·621	0·626	0·630	0·637	0·644	0·650	0·660	0·668	0·675	0·687	0·697	0·708
20.0000	(196)	0·612	0·616	0·620	0·625	0·629	0·637	0·643	0·649	0·659	0·667	0·675	0·687	0·697	0·707
30.0000	(294)	0·610	0·615	0·619	0·624	0·628	0·636	0·643	0·648	0·659	0·667	0·674	0·686	0·696	0·707
S	$(\Delta p/\rho l)$	6	8	10	12·5	15	20	25	30	40	50	60	80	100	125

Grad'nt k.p.g. (Equivalent) Pipe diameters in mm

S	$(\Delta p/\rho l)$	125	150	200	250	300	400	500	600	800	1000	1250	1500	2000	2500
0.00010	(0·00098)	0·937	0·934	0·932	0·932	0·932	0·935	0·939	0·943	0·951	0·958	0·966	0·974	0·987	0·998
0.00015	(0·00147)	0·916	0·913	0·912	0·913	0·914	0·918	0·923	0·928	0·936	0·944	0·953	0·961	0·975	0·987
0.00020	(0·00196)	0·901	0·899	0·899	0·900	0·902	0·907	0·912	0·917	0·926	0·935	0·944	0·952	0·967	0·979
0.00030	(0·00294)	0·881	0·881	0·882	0·884	0·886	0·892	0·898	0·903	0·913	0·922	0·932	0·941	0·956	0·969
0.00040	(0·00392)	0·868	0·868	0·870	0·873	0·876	0·882	0·888	0·894	0·905	0·914	0·924	0·933	0·949	0·962
0.00060	(0·00588)	0·851	0·852	0·855	0·858	0·862	0·869	0·876	0·882	0·894	0·903	0·914	0·924	0·940	0·954
0.00080	(0·00785)	0·840	0·841	0·844	0·848	0·852	0·860	0·868	0·874	0·886	0·897	0·908	0·918	0·934	0·948
0.00100	(0·00981)	0·831	0·833	0·837	0·841	0·846	0·854	0·862	0·869	0·881	0·891	0·903	0·913	0·930	0·945
0.00150	(0·0147)	0·816	0·819	0·824	0·829	0·834	0·843	0·851	0·859	0·872	0·883	0·895	0·906	0·923	0·938
0.00200	(0·0196)	0·807	0·810	0·815	0·821	0·826	0·836	0·845	0·853	0·866	0·878	0·890	0·901	0·919	0·934
0.00300	(0·0294)	0·794	0·798	0·804	0·811	0·816	0·827	0·836	0·845	0·859	0·871	0·884	0·895	0·913	0·929
0.00400	(0·0392)	0·786	0·790	0·797	0·804	0·810	0·821	0·831	0·839	0·854	0·866	0·879	0·891	0·910	0·926
0.00600	(0·0588)	0·775	0·780	0·788	0·795	0·802	0·814	0·824	0·833	0·848	0·861	0·874	0·886	0·906	0·922
0.00800	(0·0785)	0·768	0·773	0·782	0·790	0·797	0·809	0·819	0·829	0·844	0·857	0·871	0·883	0·903	0·919
0.01000	(0·0981)	0·763	0·768	0·778	0·786	0·793	0·806	0·816	0·826	0·842	0·855	0·869	0·881	0·901	0·918
0.01500	(0·147)	0·755	0·761	0·770	0·779	0·787	0·800	0·811	0·821	0·837	0·851	0·865	0·878	0·898	0·915
0.02000	(0·196)	0·750	0·756	0·766	0·775	0·783	0·797	0·808	0·818	0·835	0·849	0·863	0·876	0·896	0·913
0.03000	(0·294)	0·743	0·749	0·760	0·770	0·778	0·792	0·804	0·814	0·831	0·846	0·860	0·873	0·894	0·911
0.04000	(0·392)	0·739	0·745	0·757	0·767	0·775	0·790	0·802	0·812	0·829	0·844	0·859	0·871	0·893	0·910
0.06000	(0·588)	0·734	0·741	0·753	0·763	0·771	0·786	0·799	0·809	0·827	0·841	0·857	0·870	0·891	0·908
0.08000	(0·785)	0·730	0·738	0·750	0·760	0·769	0·784	0·797	0·808	0·825	0·840	0·855	0·868	0·890	0·907
0.10000	(0·981)	0·728	0·735	0·748	0·758	0·768	0·783	0·796	0·806	0·824	0·839	0·855	0·868	0·889	0·907
0.15000	(1·47)	0·724	0·732	0·745	0·756	0·765	0·781	0·794	0·805	0·823	0·838	0·853	0·866	0·888	0·906
0.20000	(1·96)	0·722	0·730	0·743	0·754	0·763	0·779	0·792	0·803	0·822	0·837	0·852	0·866	0·888	0·905
0.30000	(2·94)	0·719	0·727	0·741	0·752	0·762	0·778	0·791	0·802	0·821	0·836	0·851	0·865	0·887	0·905
0.40000	(3·92)	0·718	0·726	0·739	0·751	0·760	0·777	0·790	0·801	0·820	0·835	0·851	0·864	0·886	0·904
0.60000	(5·88)	0·715	0·724	0·738	0·749	0·759	0·775	0·789	0·800	0·819	0·834	0·850	0·864	0·886	0·904
0.80000	(7·85)	0·714	0·723	0·737	0·748	0·758	0·775	0·788	0·800	0·819	0·834	0·850	0·863	0·885	0·903
1.00000	(9·81)	0·713	0·722	0·736	0·748	0·758	0·774	0·788	0·799	0·818	0·834	0·849	0·863	0·885	0·903
1.50000	(14·7)	0·712	0·721	0·735	0·747	0·757	0·773	0·787	0·799	0·818	0·833	0·849	0·863	0·885	0·903
2.00000	(19·6)	0·711	0·720	0·734	0·746	0·756	0·773	0·787	0·798	0·817	0·833	0·849	0·862	0·885	0·903
3.00000	(29·4)	0·710	0·719	0·733	0·745	0·756	0·772	0·786	0·798	0·817	0·832	0·848	0·862	0·884	0·902
S	$(\Delta p/\rho l)$	125	150	200	250	300	400	500	600	800	1000	1250	1500	2000	2500

Grad'nt k.p.g. (Equivalent) Pipe diameters in mm

Kinematic viscosity, $\nu = 0.30 \times 10^{-6}$ m^{2}s^{-1} ;   **Roughness size, $k_s = 0.015$ mm**

Kin. visc., $\nu = 0.30 \times 10^{-6}$ m^{2}s^{-1};   Roughness size, $k_s = 0.030$ mm
$S = 0.00010$ to 30.0000 This table shows values of m, as follows

i.e. kin. pr. grad., $\Delta p/\rho l =$ m_C for Colebrook-White solutions; or,
(0.00098) to (294) ms^{-2} where $R \le 2000$, m_P for laminar flow

Grad'nt S	k.p.g. ($\Delta p/\rho l$)	(Equivalent) Pipe diameters in mm 6	8	10	12·5	15	20	25	30	40	50	60	80	100	125
0.00030	(0.00294)	2.057	1.402	1.041	0.773	*	0.968	0.949	0.937	0.921	0.912	0.906	0.899	0.896	0.894
0.00040	(0.00392)	1.782	1.214	0.902	*	0.973	0.946	0.929	0.918	0.904	0.896	0.891	0.885	0.883	0.883
0.00060	(0.00588)	1.455	0.991	*	*	0.941	0.918	0.903	0.894	0.882	0.875	0.871	0.868	0.867	0.867
0.00080	(0.00785)	1.260	0.858	*	0.936	0.920	0.899	0.886	0.878	0.867	0.862	0.859	0.856	0.856	0.857
0.00100	(0.00981)	1.127	0.768	*	0.920	0.904	0.885	0.873	0.866	0.857	0.852	0.849	0.848	0.848	0.849
0.00150	(0.0147)	0.920	*	0.910	0.891	0.878	0.862	0.852	0.846	0.839	0.835	0.834	0.833	0.834	0.837
0.00200	(0.0196)	0.797	*	0.890	0.873	0.861	0.847	0.838	0.833	0.827	0.824	0.823	0.824	0.826	0.829
0.00300	(0.0294)	*	0.882	0.864	0.849	0.839	0.827	0.820	0.816	0.811	0.810	0.810	0.812	0.815	0.819
0.00400	(0.0392)	*	0.863	0.847	0.833	0.825	0.814	0.808	0.805	0.802	0.801	0.801	0.804	0.807	0.812
0.00600	(0.0588)	*	0.839	0.824	0.813	0.806	0.797	0.793	0.790	0.789	0.789	0.790	0.794	0.798	0.804
0.00800	(0.0785)	0.843	0.823	0.810	0.800	0.794	0.786	0.783	0.781	0.781	0.782	0.783	0.788	0.793	0.798
0.01000	(0.0981)	0.830	0.811	0.800	0.791	0.785	0.779	0.776	0.775	0.775	0.776	0.778	0.783	0.789	0.795
0.01500	(0.147)	0.808	0.792	0.782	0.775	0.770	0.766	0.764	0.764	0.765	0.767	0.770	0.776	0.782	0.789
0.02000	(0.196)	0.794	0.780	0.771	0.765	0.761	0.758	0.757	0.757	0.759	0.762	0.765	0.772	0.778	0.785
0.03000	(0.294)	0.776	0.764	0.757	0.752	0.749	0.747	0.747	0.748	0.751	0.755	0.759	0.766	0.773	0.780
0.04000	(0.392)	0.765	0.754	0.748	0.744	0.742	0.741	0.741	0.743	0.746	0.750	0.755	0.762	0.769	0.777
0.06000	(0.588)	0.750	0.741	0.737	0.734	0.732	0.732	0.734	0.736	0.740	0.745	0.750	0.758	0.765	0.774
0.08000	(0.785)	0.741	0.733	0.729	0.727	0.727	0.727	0.729	0.731	0.737	0.742	0.747	0.755	0.763	0.772
0.10000	(0.981)	0.734	0.728	0.724	0.723	0.722	0.724	0.726	0.729	0.734	0.739	0.744	0.753	0.761	0.770
0.15000	(1.47)	0.724	0.718	0.716	0.715	0.716	0.718	0.721	0.724	0.730	0.736	0.741	0.750	0.759	0.768
0.20000	(1.96)	0.717	0.713	0.711	0.711	0.712	0.714	0.717	0.721	0.727	0.733	0.739	0.749	0.757	0.766
0.30000	(2.94)	0.709	0.705	0.705	0.705	0.706	0.710	0.713	0.717	0.724	0.731	0.736	0.746	0.755	0.764
0.40000	(3.92)	0.703	0.701	0.701	0.702	0.703	0.707	0.711	0.715	0.722	0.729	0.735	0.745	0.754	0.763
0.60000	(5.88)	0.697	0.696	0.696	0.697	0.699	0.704	0.708	0.712	0.720	0.727	0.733	0.744	0.752	0.762
0.80000	(7.85)	0.693	0.692	0.693	0.695	0.697	0.702	0.706	0.711	0.719	0.726	0.732	0.743	0.752	0.761
1.00000	(9.81)	0.690	0.690	0.691	0.693	0.695	0.700	0.705	0.710	0.718	0.725	0.731	0.742	0.751	0.761
1.50000	(14.7)	0.686	0.686	0.688	0.690	0.693	0.698	0.703	0.708	0.716	0.723	0.730	0.741	0.750	0.760
2.00000	(19.6)	0.684	0.684	0.686	0.689	0.691	0.697	0.702	0.707	0.715	0.723	0.729	0.740	0.750	0.760
3.00000	(29.4)	0.680	0.682	0.684	0.686	0.689	0.695	0.701	0.706	0.714	0.722	0.728	0.739	0.749	0.759
4.00000	(39.2)	0.678	0.680	0.682	0.685	0.688	0.694	0.700	0.705	0.714	0.721	0.728	0.739	0.749	0.759
6.00000	(58.8)	0.676	0.678	0.680	0.684	0.687	0.693	0.699	0.704	0.713	0.720	0.727	0.739	0.748	0.758
8.00000	(78.5)	0.675	0.677	0.679	0.683	0.686	0.692	0.698	0.703	0.712	0.720	0.727	0.738	0.748	0.758
10.0000	(98.1)	0.674	0.676	0.679	0.682	0.686	0.692	0.698	0.703	0.712	0.720	0.726	0.738	0.748	0.758
15.0000	(147)	0.672	0.675	0.678	0.681	0.685	0.691	0.697	0.702	0.711	0.719	0.726	0.738	0.747	0.757
20.0000	(196)	0.671	0.674	0.677	0.681	0.684	0.691	0.697	0.702	0.711	0.719	0.726	0.737	0.747	0.757
30.0000	(294)	0.670	0.673	0.676	0.680	0.684	0.690	0.696	0.702	0.711	0.719	0.726	0.737	0.747	0.757
S	($\Delta p/\rho l$)	6	8	10	12·5	15	20	25	30	40	50	60	80	100	125

Grad'nt k.p.g. (Equivalent) Pipe diameters in mm

S	($\Delta p/\rho l$)	125	150	200	250	300	400	500	600	800	1000	1250	1500	2000	2500
0.00010	(0.00098)	0.946	0.944	0.942	0.942	0.944	0.948	0.952	0.957	0.966	0.974	0.983	0.992	1.006	1.019
0.00015	(0.00147)	0.926	0.924	0.924	0.925	0.927	0.932	0.938	0.943	0.953	0.962	0.972	0.981	0.996	1.009
0.00020	(0.00196)	0.912	0.911	0.912	0.914	0.917	0.922	0.928	0.934	0.945	0.954	0.964	0.974	0.990	1.003
0.00030	(0.00294)	0.894	0.894	0.896	0.899	0.902	0.909	0.916	0.922	0.934	0.944	0.955	0.965	0.981	0.996
0.00040	(0.00392)	0.883	0.883	0.886	0.889	0.893	0.901	0.908	0.915	0.927	0.937	0.949	0.959	0.976	0.991
0.00060	(0.00588)	0.867	0.868	0.872	0.877	0.881	0.890	0.898	0.905	0.918	0.929	0.941	0.952	0.970	0.985
0.00080	(0.00785)	0.857	0.859	0.863	0.869	0.873	0.883	0.891	0.899	0.912	0.924	0.936	0.947	0.965	0.981
0.00100	(0.00981)	0.849	0.852	0.857	0.863	0.868	0.878	0.886	0.894	0.908	0.920	0.933	0.944	0.962	0.978
0.00150	(0.0147)	0.837	0.840	0.846	0.853	0.858	0.869	0.878	0.887	0.901	0.914	0.927	0.938	0.958	0.973
0.00200	(0.0196)	0.829	0.832	0.840	0.846	0.852	0.864	0.873	0.882	0.897	0.910	0.923	0.935	0.955	0.971
0.00300	(0.0294)	0.819	0.823	0.831	0.838	0.845	0.857	0.867	0.876	0.892	0.905	0.919	0.931	0.951	0.967
0.00400	(0.0392)	0.812	0.817	0.825	0.833	0.840	0.853	0.863	0.873	0.889	0.902	0.916	0.928	0.949	0.965
0.00600	(0.0588)	0.804	0.809	0.818	0.827	0.834	0.847	0.858	0.868	0.885	0.898	0.913	0.925	0.946	0.963
0.00800	(0.0785)	0.798	0.804	0.814	0.823	0.830	0.844	0.855	0.865	0.882	0.896	0.911	0.923	0.944	0.961
0.01000	(0.0981)	0.795	0.800	0.811	0.820	0.828	0.842	0.853	0.863	0.880	0.894	0.909	0.922	0.943	0.960
0.01500	(0.147)	0.789	0.795	0.806	0.815	0.824	0.838	0.850	0.860	0.878	0.892	0.907	0.920	0.941	0.959
0.02000	(0.196)	0.785	0.791	0.803	0.812	0.821	0.836	0.848	0.858	0.876	0.890	0.906	0.919	0.940	0.958
0.03000	(0.294)	0.780	0.787	0.799	0.809	0.818	0.833	0.845	0.856	0.874	0.889	0.904	0.917	0.939	0.956
0.04000	(0.392)	0.777	0.784	0.797	0.807	0.816	0.831	0.844	0.855	0.873	0.887	0.903	0.916	0.938	0.956
0.06000	(0.588)	0.774	0.781	0.794	0.804	0.814	0.829	0.842	0.853	0.871	0.886	0.902	0.915	0.937	0.955
0.08000	(0.785)	0.772	0.779	0.792	0.803	0.812	0.828	0.841	0.852	0.870	0.885	0.901	0.914	0.936	0.954
0.10000	(0.981)	0.770	0.778	0.791	0.802	0.811	0.827	0.840	0.851	0.870	0.885	0.900	0.914	0.936	0.954
0.15000	(1.47)	0.768	0.775	0.789	0.800	0.810	0.826	0.839	0.850	0.869	0.884	0.900	0.913	0.935	0.953
0.20000	(1.96)	0.766	0.774	0.788	0.799	0.809	0.825	0.838	0.849	0.868	0.883	0.899	0.913	0.935	0.953
0.30000	(2.94)	0.764	0.773	0.786	0.798	0.807	0.824	0.837	0.849	0.867	0.883	0.899	0.912	0.934	0.953
0.40000	(3.92)	0.763	0.772	0.785	0.797	0.807	0.823	0.837	0.848	0.867	0.882	0.898	0.912	0.934	0.952
0.60000	(5.88)	0.762	0.770	0.784	0.796	0.806	0.822	0.836	0.847	0.866	0.882	0.898	0.912	0.934	0.952
0.80000	(7.85)	0.761	0.770	0.784	0.796	0.805	0.822	0.836	0.847	0.866	0.882	0.898	0.911	0.934	0.952
1.00000	(9.81)	0.761	0.769	0.783	0.795	0.805	0.822	0.835	0.847	0.866	0.881	0.898	0.911	0.934	0.952
1.50000	(14.7)	0.760	0.769	0.783	0.795	0.805	0.821	0.835	0.847	0.866	0.881	0.897	0.911	0.933	0.952
2.00000	(19.6)	0.760	0.768	0.782	0.794	0.804	0.821	0.835	0.846	0.865	0.881	0.897	0.911	0.933	0.951
3.00000	(29.4)	0.759	0.768	0.782	0.794	0.804	0.821	0.834	0.846	0.865	0.881	0.897	0.911	0.933	0.951
S	($\Delta p/\rho l$)	125	150	200	250	300	400	500	600	800	1000	1250	1500	2000	2500

Grad'nt k.p.g. (Equivalent) Pipe diameters in mm

Kinematic viscosity, $\nu = 0.30 \times 10^{-6}$ m^{2}s^{-1} ;   **Roughness size, $k_s = 0.030$ mm**

Kin. visc., $\nu = 0.30 \times 10^{-6}$ m^{2}s^{-1};
$S = 0.00010$ to 30.0000
i.e. kin. pr. grad., $\Delta p/\rho l =$
(0.00098) to (294) ms^{-2}

Roughness size, $k_s = 0.060$ mm
This table shows values of m, as follows
m_C for Colebrook-White solutions; or,
where $R \le 2000$, m_P for laminar flow

Grad'nt S	k.p.g. $(\Delta p/\rho l)$	(Equivalent) Pipe diameters in mm													
		6	8	10	12.5	15	20	25	30	40	50	60	80	100	125
0.00030	(0.00294)	2.057	1.402	1.041	0.773	*	0.986	0.967	0.955	0.940	0.931	0.926	0.920	0.918	0.917
0.00040	(0.00392)	1.782	1.214	0.902	*	0.992	0.965	0.949	0.938	0.925	0.917	0.913	0.908	0.907	0.907
0.00060	(0.00588)	1.455	0.991	*	*	0.962	0.939	0.925	0.916	0.905	0.899	0.896	0.893	0.893	0.894
0.00080	(0.00785)	1.260	0.858	*	0.958	0.942	0.922	0.909	0.901	0.892	0.887	0.885	0.883	0.884	0.886
0.00100	(0.00981)	1.127	0.768	*	0.943	0.928	0.909	0.898	0.891	0.883	0.879	0.877	0.876	0.877	0.880
0.00150	(0.0147)	0.920	*	0.936	0.917	0.905	0.889	0.879	0.874	0.867	0.865	0.864	0.864	0.867	0.870
0.00200	(0.0196)	0.797	*	0.918	0.901	0.889	0.875	0.867	0.862	0.857	0.856	0.855	0.857	0.860	0.864
0.00300	(0.0294)	*	0.912	0.894	0.880	0.870	0.858	0.852	0.848	0.845	0.844	0.845	0.848	0.851	0.856
0.00400	(0.0392)	*	0.895	0.879	0.866	0.857	0.847	0.842	0.839	0.837	0.837	0.838	0.842	0.846	0.851
0.00600	(0.0588)	*	0.874	0.860	0.849	0.842	0.834	0.830	0.828	0.827	0.828	0.830	0.835	0.839	0.846
0.00800	(0.0785)	0.881	0.860	0.848	0.838	0.832	0.825	0.822	0.821	0.821	0.822	0.825	0.830	0.835	0.842
0.01000	(0.0981)	0.869	0.850	0.839	0.830	0.825	0.819	0.816	0.816	0.816	0.818	0.821	0.827	0.833	0.839
0.01500	(0.147)	0.851	0.834	0.825	0.817	0.813	0.809	0.807	0.807	0.809	0.812	0.815	0.822	0.828	0.835
0.02000	(0.196)	0.839	0.824	0.816	0.809	0.806	0.803	0.802	0.802	0.805	0.808	0.812	0.819	0.825	0.833
0.03000	(0.294)	0.824	0.811	0.804	0.799	0.797	0.795	0.795	0.796	0.799	0.803	0.807	0.815	0.822	0.830
0.04000	(0.392)	0.815	0.804	0.797	0.793	0.791	0.790	0.791	0.792	0.796	0.800	0.804	0.812	0.820	0.828
0.06000	(0.588)	0.803	0.794	0.789	0.786	0.784	0.784	0.785	0.787	0.792	0.797	0.801	0.810	0.817	0.825
0.08000	(0.785)	0.796	0.788	0.784	0.781	0.780	0.780	0.782	0.784	0.789	0.794	0.799	0.808	0.816	0.824
0.10000	(0.981)	0.791	0.784	0.780	0.778	0.777	0.778	0.780	0.782	0.788	0.793	0.798	0.807	0.814	0.823
0.15000	(1.47)	0.783	0.777	0.774	0.773	0.772	0.774	0.776	0.779	0.785	0.790	0.796	0.805	0.813	0.822
0.20000	(1.96)	0.778	0.773	0.770	0.769	0.770	0.772	0.774	0.777	0.783	0.789	0.794	0.804	0.812	0.821
0.30000	(2.94)	0.772	0.768	0.766	0.766	0.766	0.769	0.772	0.775	0.781	0.787	0.793	0.802	0.811	0.820
0.40000	(3.92)	0.769	0.765	0.763	0.763	0.764	0.767	0.770	0.774	0.780	0.786	0.792	0.801	0.810	0.819
0.60000	(5.88)	0.764	0.761	0.760	0.760	0.762	0.765	0.768	0.772	0.779	0.785	0.791	0.801	0.809	0.818
0.80000	(7.85)	0.762	0.759	0.758	0.759	0.760	0.763	0.767	0.771	0.778	0.784	0.790	0.800	0.809	0.818
1.00000	(9.81)	0.760	0.757	0.757	0.758	0.759	0.763	0.766	0.770	0.777	0.784	0.789	0.800	0.808	0.818
1.50000	(14.7)	0.757	0.755	0.755	0.756	0.757	0.761	0.765	0.769	0.776	0.783	0.789	0.799	0.808	0.817
2.00000	(19.6)	0.755	0.753	0.753	0.755	0.756	0.760	0.764	0.768	0.776	0.782	0.788	0.799	0.807	0.817
3.00000	(29.4)	0.753	0.752	0.752	0.753	0.755	0.759	0.764	0.768	0.775	0.782	0.788	0.798	0.807	0.816
4.00000	(39.2)	0.752	0.751	0.751	0.753	0.754	0.759	0.763	0.767	0.775	0.781	0.787	0.798	0.807	0.816
6.00000	(58.8)	0.750	0.749	0.750	0.752	0.754	0.758	0.762	0.767	0.774	0.781	0.787	0.798	0.806	0.816
8.00000	(78.5)	0.749	0.749	0.749	0.751	0.753	0.758	0.762	0.766	0.774	0.781	0.787	0.797	0.806	0.816
10.0000	(98.1)	0.749	0.748	0.749	0.751	0.753	0.757	0.762	0.766	0.774	0.781	0.787	0.797	0.806	0.816
15.0000	(147)	0.748	0.747	0.748	0.750	0.752	0.757	0.761	0.766	0.774	0.780	0.786	0.797	0.806	0.816
20.0000	(196)	0.747	0.747	0.748	0.750	0.752	0.757	0.761	0.766	0.773	0.780	0.786	0.797	0.806	0.815
30.0000	(294)	0.746	0.746	0.747	0.749	0.752	0.756	0.761	0.765	0.773	0.780	0.786	0.797	0.806	0.815
S	$(\Delta p/\rho l)$	6	8	10	12.5	15	20	25	30	40	50	60	80	100	125
Grad'nt	k.p.g.	(Equivalent) Pipe diameters in mm													

Grad'nt S	k.p.g. $(\Delta p/\rho l)$	(Equivalent) Pipe diameters in mm													
		125	150	200	250	300	400	500	600	800	1000	1250	1500	2000	2500
0.00010	(0.00098)	0.963	0.961	0.961	0.962	0.964	0.969	0.975	0.980	0.991	1.000	1.011	1.020	1.037	1.051
0.00015	(0.00147)	0.945	0.944	0.945	0.947	0.950	0.956	0.963	0.969	0.981	0.991	1.002	1.012	1.029	1.044
0.00020	(0.00196)	0.933	0.933	0.934	0.937	0.941	0.948	0.955	0.962	0.974	0.985	0.996	1.007	1.025	1.039
0.00030	(0.00294)	0.917	0.918	0.921	0.925	0.929	0.938	0.946	0.953	0.966	0.977	0.989	1.000	1.019	1.034
0.00040	(0.00392)	0.907	0.908	0.913	0.917	0.922	0.931	0.939	0.947	0.961	0.972	0.985	0.996	1.015	1.031
0.00060	(0.00588)	0.894	0.896	0.902	0.907	0.913	0.923	0.932	0.940	0.954	0.966	0.980	0.991	1.010	1.027
0.00080	(0.00785)	0.886	0.889	0.895	0.901	0.907	0.917	0.927	0.935	0.950	0.963	0.976	0.988	1.008	1.024
0.00100	(0.00981)	0.880	0.883	0.890	0.896	0.902	0.914	0.923	0.932	0.947	0.960	0.974	0.986	1.006	1.022
0.00150	(0.0147)	0.870	0.874	0.882	0.889	0.896	0.907	0.918	0.927	0.943	0.956	0.970	0.982	1.003	1.019
0.00200	(0.0196)	0.864	0.868	0.877	0.884	0.891	0.904	0.914	0.924	0.940	0.953	0.968	0.980	1.001	1.018
0.00300	(0.0294)	0.856	0.861	0.870	0.878	0.886	0.899	0.910	0.920	0.936	0.950	0.965	0.977	0.998	1.016
0.00400	(0.0392)	0.851	0.857	0.866	0.875	0.883	0.896	0.907	0.917	0.934	0.948	0.963	0.976	0.997	1.014
0.00600	(0.0588)	0.846	0.851	0.861	0.871	0.879	0.893	0.904	0.915	0.932	0.946	0.961	0.974	0.995	1.013
0.00800	(0.0785)	0.842	0.848	0.859	0.868	0.876	0.890	0.902	0.913	0.930	0.945	0.960	0.973	0.994	1.012
0.01000	(0.0981)	0.839	0.845	0.856	0.866	0.874	0.889	0.901	0.912	0.929	0.944	0.959	0.972	0.994	1.011
0.01500	(0.147)	0.835	0.842	0.853	0.863	0.872	0.887	0.899	0.910	0.927	0.942	0.958	0.971	0.993	1.010
0.02000	(0.196)	0.833	0.839	0.851	0.861	0.870	0.885	0.898	0.908	0.926	0.941	0.957	0.970	0.992	1.010
0.03000	(0.294)	0.830	0.837	0.849	0.859	0.868	0.883	0.896	0.907	0.925	0.940	0.956	0.969	0.991	1.009
0.04000	(0.392)	0.828	0.835	0.847	0.858	0.867	0.882	0.895	0.906	0.924	0.939	0.955	0.969	0.991	1.009
0.06000	(0.588)	0.825	0.833	0.845	0.856	0.865	0.881	0.894	0.905	0.924	0.939	0.955	0.968	0.990	1.008
0.08000	(0.785)	0.824	0.832	0.844	0.855	0.865	0.880	0.893	0.905	0.923	0.938	0.954	0.968	0.990	1.008
0.10000	(0.981)	0.823	0.831	0.844	0.855	0.864	0.880	0.893	0.904	0.923	0.938	0.954	0.967	0.990	1.008
0.15000	(1.47)	0.822	0.829	0.843	0.854	0.863	0.879	0.892	0.903	0.922	0.937	0.953	0.967	0.989	1.007
0.20000	(1.96)	0.821	0.829	0.842	0.853	0.863	0.879	0.892	0.903	0.922	0.937	0.953	0.967	0.989	1.007
0.30000	(2.94)	0.820	0.828	0.841	0.852	0.862	0.878	0.891	0.903	0.921	0.937	0.953	0.966	0.989	1.007
0.40000	(3.92)	0.819	0.827	0.841	0.852	0.861	0.878	0.891	0.902	0.921	0.937	0.953	0.966	0.989	1.007
0.60000	(5.88)	0.818	0.826	0.840	0.851	0.861	0.877	0.891	0.902	0.921	0.936	0.952	0.966	0.988	1.007
0.80000	(7.85)	0.818	0.826	0.840	0.851	0.861	0.877	0.890	0.902	0.921	0.936	0.952	0.966	0.988	1.007
1.00000	(9.81)	0.818	0.826	0.839	0.851	0.861	0.877	0.890	0.902	0.921	0.936	0.952	0.966	0.988	1.006
1.50000	(14.7)	0.817	0.825	0.839	0.850	0.860	0.877	0.890	0.901	0.920	0.936	0.952	0.966	0.988	1.006
2.00000	(19.6)	0.817	0.825	0.839	0.850	0.860	0.876	0.890	0.901	0.920	0.936	0.952	0.965	0.988	1.006
3.00000	(29.4)	0.816	0.825	0.838	0.850	0.860	0.876	0.890	0.901	0.920	0.936	0.952	0.966	0.988	1.006
S	$(\Delta p/\rho l)$	125	150	200	250	300	400	500	600	800	1000	1250	1500	2000	2500
Grad'nt	k.p.g.	(Equivalent) Pipe diameters in mm													

Kinematic viscosity, $\nu = 0.30 \times 10^{-6}$ m^{2}s^{-1} ;
Roughness size, $k_s = 0.060$ mm

E39

Kin. visc., $\nu = 0.30 \times 10^{-6}$ m^{2}s^{-1};
$S = 0.00010$ to 3.00000

i.e. kin. pr. grad., $\Delta p/\rho l =$
(0.00098) to (29.4) ms^{-2}

Roughness size, $k_s = 0.150$ mm
This table shows values of m, as follows

m_C for Colebrook-White solutions

Grad'nt S	k.p.g. $(\Delta p/\rho l)$	125	150	200	250	300	400	500	600	800	1000	1250	1500	2000	2500
							(Equivalent) Pipe diameters in mm								
0.00010	(0.00098)	1.006	1.005	1.006	1.009	1.012	1.019	1.026	1.033	1.046	1.057	1.069	1.079	1.098	1.113
0.00015	(0.00147)	0.991	0.992	0.994	0.998	1.002	1.010	1.018	1.026	1.039	1.050	1.063	1.074	1.093	1.109
0.00020	(0.00196)	0.982	0.983	0.987	0.991	0.996	1.005	1.013	1.021	1.035	1.047	1.060	1.071	1.091	1.107
0.00030	(0.00294)	0.971	0.973	0.977	0.983	0.988	0.998	1.007	1.015	1.030	1.042	1.056	1.067	1.087	1.104
0.00040	(0.00392)	0.964	0.966	0.971	0.977	0.983	0.994	1.003	1.012	1.027	1.039	1.053	1.065	1.085	1.102
0.00060	(0.00588)	0.955	0.958	0.964	0.971	0.977	0.988	0.998	1.007	1.023	1.036	1.050	1.062	1.083	1.100
0.00080	(0.00785)	0.949	0.953	0.960	0.967	0.973	0.985	0.996	1.005	1.021	1.034	1.048	1.061	1.081	1.098
0.00100	(0.00981)	0.945	0.949	0.957	0.964	0.971	0.983	0.994	1.003	1.019	1.033	1.047	1.059	1.080	1.097
0.00150	(0.0147)	0.939	0.943	0.952	0.960	0.967	0.979	0.990	1.000	1.017	1.030	1.045	1.058	1.079	1.096
0.00200	(0.0196)	0.935	0.940	0.949	0.957	0.964	0.977	0.989	0.998	1.015	1.029	1.044	1.057	1.078	1.095
0.00300	(0.0294)	0.930	0.936	0.945	0.954	0.961	0.975	0.986	0.996	1.013	1.027	1.042	1.055	1.077	1.094
0.00400	(0.0392)	0.928	0.933	0.943	0.952	0.959	0.973	0.985	0.995	1.012	1.026	1.041	1.054	1.076	1.094
0.00600	(0.0588)	0.924	0.930	0.940	0.949	0.957	0.971	0.983	0.993	1.011	1.025	1.040	1.053	1.075	1.093
0.00800	(0.0785)	0.922	0.928	0.938	0.948	0.956	0.970	0.982	0.993	1.010	1.024	1.040	1.053	1.075	1.092
0.01000	(0.0981)	0.921	0.927	0.937	0.947	0.955	0.969	0.981	0.992	1.009	1.024	1.039	1.052	1.074	1.092
0.01500	(0.147)	0.918	0.925	0.936	0.945	0.954	0.968	0.980	0.991	1.009	1.023	1.039	1.052	1.074	1.092
0.02000	(0.196)	0.917	0.923	0.935	0.944	0.953	0.967	0.980	0.990	1.008	1.023	1.038	1.051	1.073	1.091
0.03000	(0.294)	0.915	0.922	0.933	0.943	0.952	0.967	0.979	0.990	1.008	1.022	1.038	1.051	1.073	1.091
0.04000	(0.392)	0.914	0.921	0.932	0.942	0.951	0.966	0.978	0.989	1.007	1.022	1.038	1.051	1.073	1.091
0.06000	(0.588)	0.913	0.920	0.932	0.942	0.950	0.965	0.978	0.989	1.007	1.022	1.037	1.050	1.073	1.090
0.08000	(0.785)	0.912	0.919	0.931	0.941	0.950	0.965	0.978	0.988	1.006	1.021	1.037	1.050	1.072	1.090
0.10000	(0.981)	0.912	0.919	0.931	0.941	0.950	0.965	0.977	0.988	1.006	1.021	1.037	1.050	1.072	1.090
0.15000	(1.47)	0.911	0.918	0.930	0.940	0.949	0.964	0.977	0.988	1.006	1.021	1.037	1.050	1.072	1.090
0.20000	(1.96)	0.911	0.918	0.930	0.940	0.949	0.964	0.977	0.988	1.006	1.021	1.036	1.050	1.072	1.090
0.30000	(2.94)	0.910	0.917	0.929	0.940	0.949	0.964	0.977	0.987	1.006	1.021	1.036	1.050	1.072	1.090
0.40000	(3.92)	0.910	0.917	0.929	0.939	0.948	0.964	0.976	0.987	1.006	1.021	1.036	1.050	1.072	1.090
0.60000	(5.88)	0.909	0.917	0.929	0.939	0.948	0.963	0.976	0.987	1.005	1.020	1.036	1.050	1.072	1.090
0.80000	(7.85)	0.909	0.916	0.929	0.939	0.948	0.963	0.976	0.987	1.005	1.020	1.036	1.049	1.072	1.090
1.00000	(9.81)	0.909	0.916	0.928	0.939	0.948	0.963	0.976	0.987	1.005	1.020	1.036	1.049	1.072	1.090
1.50000	(14.7)	0.909	0.916	0.928	0.939	0.948	0.963	0.976	0.987	1.005	1.020	1.036	1.049	1.072	1.090
2.00000	(19.6)	0.909	0.916	0.928	0.939	0.948	0.963	0.976	0.987	1.005	1.020	1.036	1.049	1.072	1.090
3.00000	(29.4)	0.909	0.916	0.928	0.939	0.948	0.963	0.976	0.987	1.005	1.020	1.036	1.049	1.071	1.090
S Grad'nt	$(\Delta p/\rho l)$ k.p.g.	125	150	200	250	300	400	500	600	800	1000	1250	1500	2000	2500

Roughness size, $k_s = 0.150$ mm

E40

Kin. visc., $\nu = 0.30 \times 10^{-6}$ m^{2}s^{-1};
$S = 0.00010$ to 3.00000

i.e. kin. pr. grad., $\Delta p/\rho l =$
(0.00098) to (29.4) ms^{-2}

Roughness size, $k_s = 0.30$ mm
This table shows values of m, as follows

m_C for Colebrook-White solutions

Grad'nt S	k.p.g. $(\Delta p/\rho l)$	125	150	200	250	300	400	500	600	800	1000	1250	1500	2000	2500
							(Equivalent) Pipe diameters in mm								
0.00010	(0.00098)	1.061	1.060	1.062	1.065	1.069	1.077	1.085	1.093	1.106	1.117	1.130	1.141	1.161	1.177
0.00015	(0.00147)	1.050	1.050	1.054	1.058	1.062	1.071	1.080	1.088	1.102	1.114	1.127	1.138	1.158	1.174
0.00020	(0.00196)	1.043	1.044	1.048	1.053	1.058	1.068	1.076	1.085	1.099	1.111	1.125	1.136	1.156	1.173
0.00030	(0.00294)	1.035	1.037	1.042	1.047	1.053	1.063	1.072	1.081	1.096	1.108	1.122	1.134	1.154	1.171
0.00040	(0.00392)	1.030	1.032	1.038	1.044	1.049	1.060	1.070	1.079	1.094	1.107	1.120	1.133	1.153	1.170
0.00060	(0.00588)	1.024	1.027	1.033	1.039	1.046	1.057	1.067	1.076	1.091	1.105	1.119	1.131	1.151	1.168
0.00080	(0.00785)	1.020	1.023	1.030	1.037	1.043	1.055	1.065	1.074	1.090	1.103	1.118	1.130	1.151	1.168
0.00100	(0.00981)	1.017	1.021	1.028	1.035	1.042	1.053	1.064	1.073	1.089	1.102	1.117	1.129	1.150	1.167
0.00150	(0.0147)	1.013	1.017	1.025	1.032	1.039	1.051	1.062	1.071	1.087	1.101	1.116	1.128	1.149	1.166
0.00200	(0.0196)	1.011	1.015	1.023	1.030	1.037	1.050	1.061	1.070	1.087	1.100	1.115	1.127	1.148	1.166
0.00300	(0.0294)	1.008	1.012	1.020	1.028	1.036	1.048	1.059	1.069	1.085	1.099	1.114	1.127	1.148	1.165
0.00400	(0.0392)	1.006	1.010	1.019	1.027	1.034	1.047	1.058	1.068	1.085	1.099	1.113	1.126	1.147	1.165
0.00600	(0.0588)	1.004	1.008	1.017	1.026	1.033	1.046	1.057	1.067	1.084	1.098	1.113	1.126	1.147	1.164
0.00800	(0.0785)	1.002	1.007	1.016	1.025	1.032	1.046	1.057	1.067	1.084	1.098	1.112	1.125	1.147	1.164
0.01000	(0.0981)	1.001	1.006	1.016	1.024	1.032	1.045	1.056	1.066	1.083	1.097	1.112	1.125	1.146	1.164
0.01500	(0.147)	1.000	1.005	1.015	1.023	1.031	1.044	1.056	1.066	1.083	1.097	1.112	1.125	1.146	1.164
0.02000	(0.196)	0.999	1.004	1.014	1.023	1.030	1.044	1.055	1.065	1.082	1.097	1.112	1.124	1.146	1.164
0.03000	(0.294)	0.998	1.003	1.013	1.022	1.030	1.043	1.055	1.065	1.082	1.096	1.111	1.124	1.146	1.163
0.04000	(0.392)	0.997	1.003	1.013	1.022	1.029	1.043	1.055	1.065	1.082	1.096	1.111	1.124	1.146	1.163
0.06000	(0.588)	0.997	1.002	1.012	1.021	1.029	1.043	1.054	1.064	1.082	1.096	1.111	1.124	1.145	1.163
0.08000	(0.785)	0.996	1.002	1.012	1.021	1.029	1.042	1.054	1.064	1.081	1.096	1.111	1.124	1.145	1.163
0.10000	(0.981)	0.996	1.002	1.012	1.021	1.029	1.042	1.054	1.064	1.081	1.096	1.111	1.124	1.145	1.163
0.15000	(1.47)	0.995	1.001	1.011	1.020	1.028	1.042	1.054	1.064	1.081	1.095	1.111	1.124	1.145	1.163
0.20000	(1.96)	0.995	1.001	1.011	1.020	1.028	1.042	1.054	1.064	1.081	1.095	1.111	1.124	1.145	1.163
0.30000	(2.94)	0.995	1.001	1.011	1.020	1.028	1.042	1.054	1.064	1.081	1.095	1.110	1.123	1.145	1.163
0.40000	(3.92)	0.995	1.000	1.011	1.020	1.028	1.042	1.053	1.064	1.081	1.095	1.110	1.123	1.145	1.163
0.60000	(5.88)	0.994	1.000	1.011	1.020	1.028	1.042	1.053	1.064	1.081	1.095	1.110	1.123	1.145	1.163
0.80000	(7.85)	0.994	1.000	1.010	1.019	1.028	1.041	1.053	1.064	1.081	1.095	1.110	1.123	1.145	1.163
1.00000	(9.81)	0.994	1.000	1.010	1.019	1.028	1.041	1.053	1.063	1.081	1.095	1.110	1.123	1.145	1.163
1.50000	(14.7)	0.994	1.000	1.010	1.019	1.027	1.041	1.053	1.063	1.081	1.095	1.110	1.123	1.145	1.163
2.00000	(19.6)	0.994	1.000	1.010	1.019	1.027	1.041	1.053	1.063	1.081	1.095	1.110	1.123	1.145	1.163
3.00000	(29.4)	0.994	1.000	1.010	1.019	1.027	1.041	1.053	1.063	1.081	1.095	1.110	1.123	1.145	1.163
S Grad'nt	$(\Delta p/\rho l)$ k.p.g.	125	150	200	250	300	400	500	600	800	1000	1250	1500	2000	2500

Roughness size, $k_s = 0.30$ mm

Kin. visc., $\nu = 0.30 \times 10^{-6}$ m^{2}s^{-1};
S = 0.00010 to 3.00000

i.e. kin. pr. grad., $\Delta p/\rho l =$
(0.00098) to (29.4) ms^{-2}

Roughness size, $k_s = 0.60$ mm
This table shows values of m, as follows

m_C for Colebrook-White solutions

E41

Grad'nt S	k.p.g. $(\Delta p/\rho l)$	125	150	200	250	300	400	500	600	800	1000	1250	1500	2000	2500
0.00010	(0.00098)	1.142	1.141	1.142	1.145	1.148	1.156	1.164	1.171	1.184	1.195	1.208	1.219	1.238	1.254
0.00015	(0.00147)	1.134	1.134	1.136	1.140	1.144	1.152	1.160	1.168	1.181	1.193	1.206	1.217	1.237	1.253
0.00020	(0.00196)	1.129	1.130	1.132	1.137	1.141	1.150	1.158	1.166	1.180	1.192	1.205	1.216	1.236	1.252
0.00030	(0.00294)	1.124	1.125	1.128	1.133	1.138	1.147	1.156	1.163	1.178	1.190	1.203	1.215	1.234	1.251
0.00040	(0.00392)	1.120	1.122	1.126	1.130	1.135	1.145	1.154	1.162	1.176	1.189	1.202	1.214	1.234	1.250
0.00060	(0.00588)	1.116	1.118	1.122	1.128	1.133	1.143	1.152	1.160	1.175	1.188	1.201	1.213	1.233	1.249
0.00080	(0.00785)	1.114	1.116	1.121	1.126	1.131	1.142	1.151	1.159	1.174	1.187	1.200	1.212	1.232	1.249
0.00100	(0.00981)	1.112	1.114	1.119	1.125	1.130	1.141	1.150	1.159	1.174	1.186	1.200	1.212	1.232	1.249
0.00150	(0.0147)	1.109	1.112	1.117	1.123	1.129	1.140	1.149	1.158	1.173	1.185	1.199	1.211	1.231	1.248
0.00200	(0.0196)	1.108	1.110	1.116	1.122	1.128	1.139	1.148	1.157	1.172	1.185	1.199	1.211	1.231	1.248
0.00300	(0.0294)	1.106	1.108	1.115	1.121	1.127	1.138	1.147	1.156	1.171	1.184	1.198	1.210	1.231	1.248
0.00400	(0.0392)	1.105	1.107	1.114	1.120	1.126	1.137	1.147	1.156	1.171	1.184	1.198	1.210	1.230	1.247
0.00600	(0.0588)	1.103	1.106	1.113	1.119	1.125	1.136	1.146	1.155	1.171	1.184	1.198	1.210	1.230	1.247
0.00800	(0.0785)	1.102	1.105	1.112	1.119	1.125	1.136	1.146	1.155	1.170	1.183	1.197	1.210	1.230	1.247
0.01000	(0.0981)	1.102	1.105	1.112	1.118	1.124	1.136	1.146	1.155	1.170	1.183	1.197	1.209	1.230	1.247
0.01500	(0.147)	1.101	1.104	1.111	1.118	1.124	1.135	1.145	1.154	1.170	1.183	1.197	1.209	1.230	1.247
0.02000	(0.196)	1.100	1.104	1.111	1.117	1.124	1.135	1.145	1.154	1.170	1.183	1.197	1.209	1.230	1.247
0.03000	(0.294)	1.100	1.103	1.110	1.117	1.123	1.135	1.145	1.154	1.169	1.183	1.197	1.209	1.229	1.246
0.04000	(0.392)	1.099	1.103	1.110	1.117	1.123	1.135	1.145	1.154	1.169	1.183	1.197	1.209	1.229	1.246
0.06000	(0.588)	1.099	1.102	1.110	1.116	1.123	1.134	1.145	1.154	1.169	1.182	1.197	1.209	1.229	1.246
0.08000	(0.785)	1.099	1.102	1.109	1.116	1.123	1.134	1.144	1.153	1.169	1.182	1.196	1.209	1.229	1.246
0.10000	(0.981)	1.099	1.102	1.109	1.116	1.123	1.134	1.144	1.153	1.169	1.182	1.196	1.209	1.229	1.246
0.15000	(1.47)	1.098	1.102	1.109	1.116	1.122	1.134	1.144	1.153	1.169	1.182	1.196	1.209	1.229	1.246
0.20000	(1.96)	1.098	1.102	1.109	1.116	1.122	1.134	1.144	1.153	1.169	1.182	1.196	1.209	1.229	1.246
0.30000	(2.94)	1.098	1.101	1.109	1.116	1.122	1.134	1.144	1.153	1.169	1.182	1.196	1.209	1.229	1.246
0.40000	(3.92)	1.098	1.101	1.109	1.116	1.122	1.134	1.144	1.153	1.169	1.182	1.196	1.208	1.229	1.246
0.60000	(5.88)	1.098	1.101	1.109	1.115	1.122	1.134	1.144	1.153	1.169	1.182	1.196	1.208	1.229	1.246
0.80000	(7.85)	1.098	1.101	1.108	1.115	1.122	1.134	1.144	1.153	1.169	1.182	1.196	1.208	1.229	1.246
1.00000	(9.81)	1.097	1.101	1.108	1.115	1.122	1.134	1.144	1.153	1.169	1.182	1.196	1.208	1.229	1.246
1.50000	(14.7)	1.097	1.101	1.108	1.115	1.122	1.134	1.144	1.153	1.169	1.182	1.196	1.208	1.229	1.246
2.00000	(19.6)	1.097	1.101	1.108	1.115	1.122	1.134	1.144	1.153	1.169	1.182	1.196	1.208	1.229	1.246
3.00000	(29.4)	1.097	1.101	1.108	1.115	1.122	1.134	1.144	1.153	1.169	1.182	1.196	1.208	1.229	1.246
S	$(\Delta p/\rho l)$	125	150	200	250	300	400	500	600	800	1000	1250	1500	2000	2500

Grad'nt k.p.g.   (Equivalent) Pipe diameters in mm   **Roughness size, $k_s = 0.60$ mm**

Kin. visc., $\nu = 0.30 \times 10^{-6}$ m^{2}s^{-1};
S = 0.00010 to 3.00000

i.e. kin. pr. grad., $\Delta p/\rho l =$
(0.00098) to (29.4) ms^{-2}

Roughness size, $k_s = 1.50$ mm
This table shows values of m, as follows

m_C for Colebrook-White solutions

E42

Grad'nt S	k.p.g. $(\Delta p/\rho l)$	125	150	200	250	300	400	500	600	800	1000	1250	1500	2000	2500
0.00010	(0.00098)	1.298	1.294	1.290	1.290	1.292	1.296	1.301	1.307	1.318	1.327	1.339	1.349	1.366	1.381
0.00015	(0.00147)	1.293	1.290	1.287	1.287	1.289	1.294	1.300	1.305	1.316	1.326	1.337	1.348	1.365	1.380
0.00020	(0.00196)	1.290	1.287	1.285	1.286	1.287	1.293	1.298	1.304	1.315	1.325	1.337	1.347	1.365	1.380
0.00030	(0.00294)	1.287	1.284	1.283	1.284	1.286	1.291	1.297	1.303	1.314	1.324	1.336	1.346	1.364	1.379
0.00040	(0.00392)	1.285	1.283	1.281	1.282	1.284	1.290	1.296	1.302	1.314	1.324	1.335	1.346	1.364	1.379
0.00060	(0.00588)	1.283	1.280	1.279	1.281	1.283	1.289	1.295	1.301	1.313	1.323	1.335	1.345	1.363	1.378
0.00080	(0.00785)	1.281	1.279	1.278	1.280	1.282	1.288	1.295	1.301	1.312	1.323	1.335	1.345	1.363	1.378
0.00100	(0.00981)	1.280	1.278	1.278	1.279	1.282	1.288	1.294	1.300	1.312	1.323	1.334	1.345	1.363	1.378
0.00150	(0.0147)	1.279	1.277	1.277	1.278	1.281	1.287	1.294	1.300	1.312	1.322	1.334	1.344	1.363	1.378
0.00200	(0.0196)	1.278	1.276	1.276	1.278	1.280	1.287	1.293	1.300	1.311	1.322	1.334	1.344	1.362	1.378
0.00300	(0.0294)	1.277	1.275	1.275	1.277	1.280	1.286	1.293	1.299	1.311	1.322	1.334	1.344	1.362	1.378
0.00400	(0.0392)	1.276	1.275	1.275	1.277	1.279	1.286	1.293	1.299	1.311	1.322	1.333	1.344	1.362	1.377
0.00600	(0.0588)	1.276	1.274	1.274	1.276	1.279	1.286	1.292	1.299	1.311	1.321	1.333	1.344	1.362	1.377
0.00800	(0.0785)	1.275	1.274	1.274	1.276	1.279	1.285	1.292	1.299	1.310	1.321	1.333	1.344	1.362	1.377
0.01000	(0.0981)	1.275	1.273	1.274	1.276	1.279	1.285	1.292	1.298	1.310	1.321	1.333	1.344	1.362	1.377
0.01500	(0.147)	1.274	1.273	1.273	1.275	1.278	1.285	1.292	1.298	1.310	1.321	1.333	1.343	1.362	1.377
0.02000	(0.196)	1.274	1.273	1.273	1.275	1.278	1.285	1.292	1.298	1.310	1.321	1.333	1.343	1.362	1.377
0.03000	(0.294)	1.274	1.272	1.273	1.275	1.278	1.285	1.291	1.298	1.310	1.321	1.333	1.343	1.362	1.377
0.04000	(0.392)	1.274	1.272	1.273	1.275	1.278	1.285	1.291	1.298	1.310	1.321	1.333	1.343	1.362	1.377
0.06000	(0.588)	1.273	1.272	1.272	1.275	1.278	1.284	1.291	1.298	1.310	1.321	1.333	1.343	1.362	1.377
0.08000	(0.785)	1.273	1.272	1.272	1.275	1.278	1.284	1.291	1.298	1.310	1.321	1.333	1.343	1.361	1.377
0.10000	(0.981)	1.273	1.272	1.272	1.275	1.278	1.284	1.291	1.298	1.310	1.321	1.333	1.343	1.361	1.377
0.15000	(1.47)	1.273	1.272	1.272	1.274	1.278	1.284	1.291	1.298	1.310	1.321	1.333	1.343	1.361	1.377
0.20000	(1.96)	1.273	1.272	1.272	1.274	1.277	1.284	1.291	1.298	1.310	1.321	1.333	1.343	1.361	1.377
0.30000	(2.94)	1.273	1.272	1.272	1.274	1.277	1.284	1.291	1.298	1.310	1.321	1.332	1.343	1.361	1.377
0.40000	(3.92)	1.273	1.271	1.272	1.274	1.277	1.284	1.291	1.298	1.310	1.320	1.332	1.343	1.361	1.377
0.60000	(5.88)	1.273	1.271	1.272	1.274	1.277	1.284	1.291	1.298	1.310	1.320	1.332	1.343	1.361	1.377
0.80000	(7.85)	1.273	1.271	1.272	1.274	1.277	1.284	1.291	1.298	1.310	1.320	1.332	1.343	1.361	1.377
1.00000	(9.81)	1.273	1.271	1.272	1.274	1.277	1.284	1.291	1.298	1.310	1.320	1.332	1.343	1.361	1.377
1.50000	(14.7)	1.272	1.271	1.272	1.274	1.277	1.284	1.291	1.298	1.310	1.320	1.332	1.343	1.361	1.377
2.00000	(19.6)	1.272	1.271	1.272	1.274	1.277	1.284	1.291	1.298	1.310	1.320	1.332	1.343	1.361	1.377
3.00000	(29.4)	1.272	1.271	1.272	1.274	1.277	1.284	1.291	1.298	1.310	1.320	1.332	1.343	1.361	1.377
S	$(\Delta p/\rho l)$	125	150	200	250	300	400	500	600	800	1000	1250	1500	2000	2500

Grad'nt k.p.g.   (Equivalent) Pipe diameters in mm   **Roughness size, $k_s = 1.50$ mm**

Kin. visc., $\nu = 0.40 \times 10^{-6}$ m^{2}s^{-1};
$S = 0.00010$ to 30.0000

i.e. kin. pr. grad., $\Delta p/\rho l =$
(0.00098) to (294) ms^{-2}

Roughness size, $k_s = 0.0015$ mm
This table shows values of m, as follows

m_C for Colebrook-White solutions; or,
where $\mathbf{R} \le 2000$, m_P for laminar flow

Grad'nt S	k.p.g. $(\Delta p/\rho l)$	6	8	10	12.5	15	20	25	30	40	50	60	80	100	125
0.00030	(0.00294)	2.743	1.869	1.388	1.031	0.808	1.001	0.978	0.962	0.941	0.929	0.920	0.909	0.903	0.899
0.00040	(0.00392)	2.375	1.619	1.202	0.893	*	0.975	0.954	0.939	0.921	0.909	0.902	0.892	0.887	0.884
0.00060	(0.00588)	1.940	1.322	0.982	*	*	0.940	0.922	0.910	0.893	0.884	0.877	0.869	0.865	0.863
0.00080	(0.00785)	1.680	1.145	0.850	*	0.944	0.917	0.901	0.889	0.875	0.866	0.860	0.854	0.850	0.849
0.00100	(0.00981)	1.502	1.024	0.760	*	0.925	0.900	0.885	0.874	0.861	0.853	0.848	0.842	0.839	0.838
0.00150	(0.0147)	1.227	0.836	*	0.909	0.892	0.871	0.858	0.849	0.837	0.831	0.826	0.822	0.820	0.819
0.00200	(0.0196)	1.062	0.724	*	0.886	0.871	0.851	0.839	0.831	0.821	0.815	0.812	0.808	0.807	0.806
0.00300	(0.0294)	0.867	*	0.874	0.855	0.842	0.825	0.815	0.808	0.799	0.795	0.792	0.789	0.789	0.789
0.00400	(0.0392)	0.751	*	0.852	0.835	0.823	0.808	0.798	0.792	0.785	0.781	0.779	0.777	0.777	0.778
0.00600	(0.0588)	*	0.841	0.823	0.808	0.797	0.784	0.776	0.771	0.765	0.762	0.760	0.760	0.760	0.762
0.00800	(0.0785)	*	0.820	0.803	0.790	0.780	0.768	0.761	0.757	0.752	0.749	0.748	0.748	0.749	0.751
0.01000	(0.0981)	*	0.804	0.789	0.776	0.767	0.757	0.750	0.746	0.742	0.740	0.739	0.739	0.741	0.743
0.01500	(0.147)	0.798	0.777	0.764	0.753	0.745	0.736	0.731	0.728	0.724	0.723	0.723	0.724	0.726	0.729
0.02000	(0.196)	0.778	0.759	0.747	0.737	0.730	0.722	0.718	0.715	0.713	0.712	0.712	0.714	0.716	0.719
0.03000	(0.294)	0.752	0.735	0.725	0.716	0.711	0.704	0.700	0.698	0.697	0.697	0.698	0.700	0.703	0.707
0.04000	(0.392)	0.734	0.719	0.710	0.702	0.697	0.691	0.689	0.687	0.686	0.687	0.688	0.691	0.694	0.698
0.06000	(0.588)	0.711	0.698	0.690	0.683	0.679	0.675	0.673	0.672	0.672	0.673	0.674	0.678	0.682	0.686
0.08000	(0.785)	0.695	0.683	0.676	0.671	0.667	0.664	0.662	0.662	0.662	0.664	0.665	0.670	0.674	0.678
0.10000	(0.981)	0.684	0.673	0.666	0.661	0.658	0.655	0.654	0.654	0.655	0.657	0.659	0.663	0.667	0.673
0.15000	(1.47)	0.664	0.655	0.649	0.645	0.643	0.641	0.640	0.641	0.642	0.645	0.647	0.652	0.657	0.662
0.20000	(1.96)	0.650	0.642	0.638	0.634	0.632	0.631	0.631	0.632	0.634	0.637	0.639	0.645	0.650	0.656
0.30000	(2.94)	0.633	0.626	0.622	0.620	0.618	0.618	0.618	0.620	0.623	0.626	0.629	0.635	0.640	0.647
0.40000	(3.92)	0.621	0.615	0.612	0.610	0.609	0.609	0.610	0.612	0.615	0.619	0.622	0.628	0.634	0.641
0.60000	(5.88)	0.605	0.600	0.598	0.597	0.597	0.597	0.599	0.601	0.605	0.609	0.613	0.620	0.626	0.633
0.80000	(7.85)	0.594	0.591	0.589	0.588	0.588	0.590	0.592	0.594	0.598	0.603	0.607	0.614	0.621	0.628
1.00000	(9.81)	0.587	0.583	0.582	0.582	0.582	0.584	0.586	0.589	0.593	0.598	0.602	0.610	0.617	0.624
1.50000	(14.7)	0.573	0.571	0.570	0.571	0.571	0.574	0.577	0.580	0.585	0.590	0.595	0.603	0.610	0.618
2.00000	(19.6)	0.564	0.563	0.562	0.563	0.564	0.567	0.571	0.574	0.580	0.585	0.590	0.598	0.606	0.614
3.00000	(29.4)	0.552	0.552	0.552	0.553	0.555	0.559	0.562	0.566	0.572	0.578	0.583	0.592	0.600	0.608
4.00000	(39.2)	0.545	0.544	0.545	0.547	0.549	0.553	0.557	0.561	0.568	0.574	0.579	0.589	0.596	0.605
6.00000	(58.8)	0.534	0.535	0.537	0.539	0.541	0.546	0.550	0.554	0.562	0.568	0.574	0.584	0.592	0.601
8.00000	(78.5)	0.528	0.529	0.531	0.533	0.536	0.541	0.546	0.550	0.558	0.565	0.570	0.581	0.589	0.598
10.0000	(98.1)	0.523	0.524	0.526	0.529	0.532	0.538	0.543	0.547	0.555	0.562	0.568	0.578	0.587	0.596
15.0000	(147)	0.514	0.517	0.519	0.523	0.526	0.532	0.537	0.542	0.551	0.558	0.564	0.575	0.584	0.593
20.0000	(196)	0.509	0.512	0.515	0.518	0.522	0.528	0.534	0.539	0.548	0.555	0.562	0.573	0.582	0.591
30.0000	(294)	0.502	0.505	0.509	0.513	0.517	0.524	0.530	0.535	0.544	0.552	0.558	0.570	0.579	0.589
S	$(\Delta p/\rho l)$	6	8	10	12.5	15	20	25	30	40	50	60	80	100	125

Grad'nt k.p.g. (Equivalent) Pipe diameters in mm

S	$(\Delta p/\rho l)$	125	150	200	250	300	400	500	600	800	1000	1250	1500	2000	2500
0.00010	(0.00098)	0.964	0.959	0.954	0.951	0.951	0.951	0.953	0.956	0.961	0.966	0.973	0.978	0.989	0.998
0.00015	(0.00147)	0.939	0.935	0.931	0.930	0.930	0.931	0.934	0.937	0.943	0.949	0.955	0.962	0.973	0.982
0.00020	(0.00196)	0.922	0.919	0.916	0.915	0.915	0.918	0.921	0.924	0.930	0.937	0.944	0.950	0.961	0.971
0.00030	(0.00294)	0.899	0.897	0.895	0.895	0.896	0.899	0.903	0.906	0.913	0.920	0.927	0.934	0.946	0.956
0.00040	(0.00392)	0.884	0.882	0.881	0.881	0.883	0.886	0.890	0.894	0.902	0.909	0.916	0.923	0.936	0.946
0.00060	(0.00588)	0.863	0.862	0.862	0.863	0.865	0.869	0.874	0.878	0.886	0.893	0.902	0.909	0.921	0.932
0.00080	(0.00785)	0.849	0.848	0.849	0.850	0.853	0.857	0.862	0.867	0.875	0.883	0.891	0.899	0.911	0.922
0.00100	(0.00981)	0.838	0.838	0.839	0.841	0.843	0.849	0.854	0.858	0.867	0.875	0.884	0.891	0.904	0.915
0.00150	(0.0147)	0.819	0.819	0.821	0.824	0.827	0.833	0.839	0.844	0.853	0.861	0.870	0.878	0.891	0.902
0.00200	(0.0196)	0.806	0.807	0.810	0.813	0.816	0.822	0.828	0.834	0.843	0.852	0.861	0.869	0.882	0.894
0.00300	(0.0294)	0.789	0.791	0.794	0.797	0.801	0.808	0.814	0.820	0.830	0.839	0.848	0.856	0.870	0.882
0.00400	(0.0392)	0.778	0.779	0.783	0.787	0.791	0.798	0.805	0.811	0.821	0.830	0.839	0.848	0.862	0.874
0.00600	(0.0588)	0.762	0.764	0.768	0.773	0.777	0.785	0.792	0.798	0.809	0.818	0.828	0.837	0.851	0.864
0.00800	(0.0785)	0.751	0.754	0.758	0.763	0.768	0.776	0.783	0.789	0.800	0.810	0.820	0.829	0.844	0.856
0.01000	(0.0981)	0.743	0.746	0.751	0.756	0.761	0.769	0.776	0.783	0.794	0.804	0.814	0.823	0.838	0.851
0.01500	(0.147)	0.729	0.732	0.738	0.743	0.748	0.757	0.765	0.772	0.784	0.793	0.804	0.813	0.829	0.842
0.02000	(0.196)	0.719	0.723	0.729	0.735	0.740	0.749	0.757	0.764	0.776	0.786	0.797	0.807	0.823	0.836
0.03000	(0.294)	0.707	0.710	0.717	0.723	0.729	0.738	0.747	0.754	0.767	0.777	0.788	0.798	0.814	0.828
0.04000	(0.392)	0.698	0.702	0.709	0.715	0.721	0.731	0.740	0.747	0.760	0.771	0.782	0.792	0.809	0.822
0.06000	(0.588)	0.686	0.690	0.698	0.705	0.711	0.721	0.730	0.738	0.751	0.762	0.774	0.784	0.801	0.815
0.08000	(0.785)	0.678	0.683	0.691	0.698	0.704	0.715	0.724	0.732	0.746	0.757	0.769	0.779	0.797	0.811
0.10000	(0.981)	0.673	0.677	0.685	0.693	0.699	0.710	0.720	0.728	0.742	0.753	0.765	0.776	0.793	0.807
0.15000	(1.47)	0.662	0.667	0.676	0.684	0.691	0.702	0.712	0.720	0.734	0.746	0.759	0.769	0.787	0.802
0.20000	(1.96)	0.656	0.661	0.670	0.678	0.685	0.697	0.707	0.715	0.730	0.742	0.754	0.765	0.784	0.798
0.30000	(2.94)	0.647	0.652	0.662	0.670	0.677	0.690	0.700	0.709	0.724	0.736	0.749	0.760	0.779	0.794
0.40000	(3.92)	0.641	0.646	0.656	0.665	0.672	0.685	0.695	0.705	0.720	0.732	0.746	0.757	0.776	0.791
0.60000	(5.88)	0.633	0.639	0.649	0.658	0.666	0.679	0.690	0.699	0.715	0.728	0.741	0.753	0.772	0.788
0.80000	(7.85)	0.628	0.634	0.645	0.654	0.662	0.675	0.686	0.696	0.712	0.725	0.738	0.750	0.770	0.785
1.00000	(9.81)	0.624	0.630	0.641	0.651	0.659	0.672	0.684	0.693	0.709	0.723	0.737	0.748	0.768	0.784
1.50000	(14.7)	0.618	0.624	0.636	0.646	0.654	0.668	0.679	0.689	0.706	0.719	0.733	0.745	0.765	0.781
2.00000	(19.6)	0.614	0.621	0.632	0.642	0.651	0.665	0.677	0.687	0.703	0.717	0.731	0.743	0.763	0.780
3.00000	(29.4)	0.608	0.616	0.628	0.638	0.647	0.661	0.673	0.684	0.701	0.714	0.729	0.741	0.761	0.778
S	$(\Delta p/\rho l)$	125	150	200	250	300	400	500	600	800	1000	1250	1500	2000	2500

Grad'nt k.p.g. (Equivalent) Pipe diameters in mm

Kinematic viscosity, $\nu = 0.40 \times 10^{-6}$ m^{2}s^{-1} ;   **Roughness size, $k_s = 0.0015$ mm**

Kin. visc., $\nu = 0{\cdot}40 \times 10^{-6}$ m^{2}s^{-1};
$S = 0{\cdot}00010$ to $30{\cdot}0000$

i.e. kin. pr. grad., $\Delta p/\rho l = (0{\cdot}00098)$ to (294) ms^{-2}

Roughness size, $k_s = 0{\cdot}003$ mm
This table shows values of m, as follows

m_C for Colebrook-White solutions; or, where $\mathbf{R} \le 2000$, m_P for laminar flow

Grad'nt **k.p.g.** (Equivalent) Pipe diameters in mm

S	$(\Delta p/\rho l)$	6	8	10	12·5	15	20	25	30	40	50	60	80	100	125
0·00030	(0·00294)	2·743	1·869	1·388	1·031	0·808	1·001	0·979	0·963	0·942	0·930	0·921	0·910	0·905	0·900
0·00040	(0·00392)	2·375	1·619	1·202	0·893	*	0·976	0·955	0·940	0·922	0·910	0·903	0·894	0·888	0·885
0·00060	(0·00588)	1·940	1·322	0·982	*	*	0·941	0·923	0·911	0·895	0·885	0·878	0·871	0·867	0·864
0·00080	(0·00785)	1·680	1·145	0·850	*	0·945	0·918	0·902	0·891	0·876	0·867	0·862	0·855	0·852	0·850
0·00100	(0·00981)	1·502	1·024	0·760	*	0·926	0·902	0·886	0·876	0·862	0·855	0·850	0·844	0·841	0·840
0·00150	(0·0147)	1·227	0·836	*	0·910	0·894	0·872	0·859	0·850	0·839	0·832	0·828	0·824	0·822	0·821
0·00200	(0·0196)	1·062	0·724	*	0·887	0·872	0·853	0·841	0·833	0·823	0·817	0·814	0·810	0·809	0·809
0·00300	(0·0294)	0·867	*	0·876	0·857	0·844	0·827	0·817	0·810	0·801	0·797	0·794	0·792	0·791	0·792
0·00400	(0·0392)	0·751	*	0·854	0·837	0·825	0·810	0·800	0·794	0·787	0·783	0·781	0·779	0·780	0·781
0·00600	(0·0588)	*	0·843	0·825	0·810	0·799	0·786	0·779	0·773	0·768	0·765	0·763	0·763	0·764	0·766
0·00800	(0·0785)	*	0·822	0·805	0·792	0·782	0·771	0·764	0·759	0·755	0·752	0·751	0·752	0·753	0·755
0·01000	(0·0981)	*	0·806	0·791	0·778	0·770	0·759	0·753	0·749	0·745	0·743	0·743	0·743	0·745	0·748
0·01500	(0·147)	0·801	0·779	0·766	0·755	0·748	0·739	0·734	0·731	0·728	0·727	0·727	0·729	0·731	0·734
0·02000	(0·196)	0·781	0·762	0·750	0·740	0·734	0·726	0·721	0·719	0·717	0·716	0·717	0·719	0·721	0·725
0·03000	(0·294)	0·755	0·738	0·728	0·720	0·714	0·708	0·704	0·703	0·701	0·702	0·703	0·706	0·709	0·713
0·04000	(0·392)	0·738	0·722	0·713	0·706	0·701	0·696	0·693	0·692	0·691	0·692	0·693	0·697	0·701	0·705
0·06000	(0·588)	0·715	0·702	0·694	0·688	0·684	0·680	0·678	0·677	0·678	0·679	0·681	0·685	0·689	0·694
0·08000	(0·785)	0·699	0·688	0·681	0·676	0·672	0·669	0·668	0·667	0·669	0·670	0·673	0·677	0·682	0·687
0·10000	(0·981)	0·688	0·678	0·671	0·667	0·664	0·661	0·660	0·660	0·662	0·664	0·667	0·672	0·677	0·682
0·15000	(1·47)	0·669	0·660	0·655	0·651	0·649	0·647	0·647	0·648	0·650	0·653	0·656	0·662	0·667	0·673
0·20000	(1·96)	0·656	0·648	0·644	0·641	0·639	0·638	0·639	0·640	0·643	0·646	0·649	0·655	0·661	0·668
0·30000	(2·94)	0·639	0·633	0·629	0·627	0·626	0·626	0·627	0·629	0·632	0·636	0·640	0·647	0·653	0·660
0·40000	(3·92)	0·628	0·622	0·619	0·618	0·617	0·618	0·620	0·622	0·626	0·630	0·634	0·641	0·648	0·655
0·60000	(5·88)	0·613	0·609	0·607	0·606	0·606	0·608	0·610	0·612	0·617	0·622	0·626	0·634	0·641	0·649
0·80000	(7·85)	0·603	0·600	0·598	0·598	0·599	0·601	0·603	0·606	0·611	0·617	0·621	0·630	0·637	0·645
1·00000	(9·81)	0·596	0·593	0·592	0·592	0·593	0·596	0·599	0·602	0·607	0·613	0·618	0·626	0·634	0·642
1·50000	(14·7)	0·583	0·582	0·582	0·583	0·584	0·587	0·591	0·594	0·601	0·606	0·612	0·621	0·629	0·637
2·00000	(19·6)	0·575	0·574	0·575	0·576	0·578	0·582	0·586	0·589	0·596	0·602	0·608	0·617	0·625	0·634
3·00000	(29·4)	0·565	0·565	0·566	0·568	0·570	0·575	0·579	0·583	0·591	0·597	0·603	0·613	0·621	0·630
4·00000	(39·2)	0·558	0·559	0·560	0·563	0·565	0·570	0·575	0·579	0·587	0·594	0·600	0·610	0·619	0·628
6·00000	(58·8)	0·549	0·551	0·553	0·556	0·559	0·564	0·570	0·574	0·583	0·590	0·596	0·607	0·616	0·625
8·00000	(78·5)	0·544	0·546	0·548	0·552	0·555	0·561	0·566	0·571	0·580	0·587	0·594	0·605	0·614	0·623
10·0000	(98·1)	0·540	0·542	0·545	0·548	0·552	0·558	0·564	0·569	0·578	0·585	0·592	0·603	0·612	0·622
15·0000	(147)	0·533	0·536	0·539	0·543	0·547	0·554	0·560	0·565	0·575	0·582	0·589	0·601	0·610	0·620
20·0000	(196)	0·529	0·532	0·536	0·540	0·544	0·551	0·558	0·563	0·573	0·581	0·588	0·599	0·609	0·619
30·0000	(294)	0·523	0·527	0·532	0·536	0·540	0·548	0·555	0·560	0·570	0·578	0·585	0·597	0·607	0·618
S	$(\Delta p/\rho l)$	6	8	10	12·5	15	20	25	30	40	50	60	80	100	125

Grad'nt **k.p.g.** (Equivalent) Pipe diameters in mm

S	$(\Delta p/\rho l)$	125	150	200	250	300	400	500	600	800	1000	1250	1500	2000	2500
0·00010	(0·00098)	0·965	0·960	0·955	0·952	0·952	0·952	0·954	0·957	0·962	0·968	0·974	0·980	0·991	1·000
0·00015	(0·00147)	0·940	0·936	0·932	0·931	0·931	0·933	0·935	0·938	0·944	0·950	0·957	0·964	0·975	0·985
0·00020	(0·00196)	0·923	0·920	0·917	0·916	0·917	0·919	0·922	0·926	0·932	0·939	0·946	0·952	0·964	0·974
0·00030	(0·00294)	0·900	0·898	0·896	0·896	0·898	0·901	0·905	0·908	0·916	0·922	0·930	0·937	0·949	0·960
0·00040	(0·00392)	0·885	0·883	0·882	0·883	0·884	0·888	0·892	0·897	0·904	0·911	0·919	0·927	0·939	0·950
0·00060	(0·00588)	0·864	0·863	0·863	0·865	0·867	0·871	0·876	0·881	0·889	0·897	0·905	0·912	0·925	0·936
0·00080	(0·00785)	0·850	0·850	0·851	0·852	0·855	0·860	0·865	0·870	0·879	0·886	0·895	0·903	0·916	0·927
0·00100	(0·00981)	0·840	0·840	0·841	0·843	0·846	0·851	0·857	0·862	0·871	0·879	0·888	0·896	0·909	0·921
0·00150	(0·0147)	0·821	0·822	0·824	0·827	0·830	0·836	0·842	0·847	0·857	0·866	0·875	0·883	0·897	0·909
0·00200	(0·0196)	0·809	0·810	0·813	0·816	0·819	0·826	0·832	0·838	0·848	0·857	0·866	0·874	0·889	0·901
0·00300	(0·0294)	0·792	0·794	0·797	0·801	0·805	0·812	0·819	0·825	0·835	0·844	0·854	0·863	0·878	0·890
0·00400	(0·0392)	0·781	0·783	0·787	0·791	0·795	0·803	0·810	0·816	0·827	0·836	0·847	0·855	0·871	0·883
0·00600	(0·0588)	0·766	0·768	0·773	0·777	0·782	0·790	0·798	0·804	0·816	0·825	0·836	0·845	0·861	0·874
0·00800	(0·0785)	0·755	0·758	0·763	0·768	0·773	0·782	0·789	0·796	0·808	0·818	0·829	0·839	0·854	0·868
0·01000	(0·0981)	0·748	0·750	0·756	0·761	0·766	0·775	0·783	0·790	0·803	0·813	0·824	0·834	0·850	0·863
0·01500	(0·147)	0·734	0·737	0·744	0·750	0·755	0·765	0·773	0·780	0·793	0·804	0·815	0·825	0·842	0·855
0·02000	(0·196)	0·725	0·729	0·736	0·742	0·747	0·757	0·766	0·774	0·787	0·798	0·809	0·819	0·836	0·850
0·03000	(0·294)	0·713	0·717	0·725	0·731	0·737	0·748	0·757	0·765	0·778	0·790	0·802	0·812	0·830	0·844
0·04000	(0·392)	0·705	0·709	0·717	0·724	0·731	0·741	0·751	0·759	0·773	0·784	0·797	0·807	0·825	0·840
0·06000	(0·588)	0·694	0·699	0·708	0·715	0·722	0·733	0·743	0·751	0·766	0·778	0·790	0·801	0·820	0·835
0·08000	(0·785)	0·687	0·692	0·701	0·709	0·716	0·728	0·738	0·746	0·761	0·773	0·786	0·798	0·816	0·831
0·10000	(0·981)	0·682	0·687	0·697	0·705	0·712	0·724	0·734	0·743	0·758	0·770	0·783	0·795	0·814	0·829
0·15000	(1·47)	0·673	0·679	0·689	0·697	0·704	0·717	0·728	0·737	0·752	0·765	0·779	0·790	0·809	0·825
0·20000	(1·96)	0·668	0·673	0·684	0·692	0·700	0·713	0·724	0·733	0·749	0·762	0·776	0·787	0·807	0·823
0·30000	(2·94)	0·660	0·666	0·677	0·686	0·694	0·707	0·719	0·728	0·744	0·758	0·772	0·784	0·803	0·820
0·40000	(3·92)	0·655	0·662	0·673	0·682	0·690	0·704	0·715	0·725	0·742	0·755	0·769	0·781	0·801	0·818
0·60000	(5·88)	0·649	0·656	0·667	0·677	0·685	0·700	0·711	0·721	0·738	0·752	0·766	0·779	0·799	0·815
0·80000	(7·85)	0·645	0·652	0·664	0·674	0·682	0·697	0·709	0·719	0·736	0·750	0·765	0·777	0·797	0·814
1·00000	(9·81)	0·642	0·649	0·661	0·671	0·680	0·695	0·707	0·717	0·735	0·749	0·763	0·776	0·796	0·813
1·50000	(14·7)	0·637	0·645	0·657	0·668	0·677	0·692	0·704	0·715	0·732	0·746	0·761	0·774	0·795	0·812
2·00000	(19·6)	0·634	0·642	0·655	0·665	0·674	0·690	0·702	0·713	0·731	0·745	0·760	0·773	0·794	0·811
3·00000	(29·4)	0·630	0·638	0·651	0·662	0·672	0·687	0·700	0·711	0·729	0·743	0·759	0·771	0·792	0·810
S	$(\Delta p/\rho l)$	125	150	200	250	300	400	500	600	800	1000	1250	1500	2000	2500

Grad'nt **k.p.g.** (Equivalent) Pipe diameters in mm

Kinematic viscosity, $\nu = 0{\cdot}40 \times 10^{-6}$ m^{2}s^{-1};

Roughness size, $k_s = 0{\cdot}003$ mm

E45

Kin. visc., $\nu = 0.40 \times 10^{-6}\ m^2 s^{-1}$;
$S = 0.00010$ to 30.0000

i.e. kin. pr. grad., $\Delta p/\rho l =$
(0.00098) to (294) ms^{-2}

Roughness size, $k_s = 0.006$ mm
This table shows values of m, as follows

m_C for Colebrook-White solutions; or,
where $\mathbf{R} \le 2000$, m_P for laminar flow

Grad'nt S	k.p.g. $(\Delta p/\rho l)$	6	8	10	12.5	15	20	25	30	40	50	60	80	100	125
0.00030	(0.00294)	2.743	1.869	1.388	1.031	0.808	1.003	0.980	0.964	0.944	0.931	0.923	0.913	0.907	0.903
0.00040	(0.00392)	2.375	1.619	1.202	0.893	*	0.977	0.957	0.942	0.924	0.912	0.905	0.896	0.891	0.888
0.00060	(0.00588)	1.940	1.322	0.982	*	*	0.943	0.925	0.913	0.897	0.887	0.881	0.873	0.870	0.867
0.00080	(0.00785)	1.680	1.145	0.850	*	0.947	0.921	0.904	0.893	0.879	0.870	0.865	0.858	0.855	0.854
0.00100	(0.00981)	1.502	1.024	0.760	*	0.928	0.904	0.889	0.878	0.865	0.857	0.852	0.847	0.844	0.843
0.00150	(0.0147)	1.227	0.836	*	0.913	0.896	0.875	0.862	0.853	0.842	0.835	0.832	0.827	0.826	0.826
0.00200	(0.0196)	1.062	0.724	*	0.890	0.875	0.856	0.844	0.836	0.826	0.821	0.817	0.814	0.813	0.814
0.00300	(0.0294)	0.867	*	0.879	0.860	0.847	0.830	0.820	0.813	0.805	0.801	0.799	0.797	0.796	0.798
0.00400	(0.0392)	0.751	*	0.857	0.840	0.828	0.813	0.804	0.798	0.791	0.788	0.786	0.785	0.785	0.787
0.00600	(0.0588)	*	0.846	0.828	0.814	0.803	0.791	0.783	0.778	0.773	0.770	0.769	0.769	0.770	0.772
0.00800	(0.0785)	*	0.826	0.809	0.796	0.787	0.775	0.769	0.765	0.760	0.758	0.758	0.758	0.760	0.763
0.01000	(0.0981)	*	0.810	0.795	0.783	0.775	0.764	0.758	0.755	0.751	0.749	0.749	0.750	0.753	0.756
0.01500	(0.147)	0.805	0.784	0.771	0.761	0.754	0.745	0.740	0.737	0.735	0.734	0.735	0.737	0.740	0.744
0.02000	(0.196)	0.786	0.767	0.755	0.746	0.740	0.732	0.728	0.726	0.724	0.724	0.725	0.728	0.731	0.735
0.03000	(0.294)	0.761	0.744	0.734	0.726	0.721	0.715	0.712	0.711	0.710	0.711	0.712	0.716	0.720	0.725
0.04000	(0.392)	0.744	0.729	0.720	0.713	0.709	0.704	0.701	0.700	0.701	0.702	0.704	0.708	0.713	0.718
0.06000	(0.588)	0.722	0.709	0.702	0.696	0.692	0.689	0.687	0.687	0.688	0.690	0.693	0.698	0.703	0.709
0.08000	(0.785)	0.707	0.696	0.689	0.685	0.682	0.679	0.678	0.678	0.680	0.683	0.686	0.691	0.697	0.703
0.10000	(0.981)	0.697	0.686	0.680	0.676	0.674	0.672	0.671	0.672	0.674	0.677	0.680	0.686	0.692	0.699
0.15000	(1.47)	0.678	0.670	0.665	0.662	0.660	0.659	0.660	0.661	0.664	0.668	0.671	0.678	0.684	0.691
0.20000	(1.96)	0.666	0.659	0.655	0.653	0.651	0.651	0.652	0.654	0.658	0.662	0.666	0.673	0.680	0.687
0.30000	(2.94)	0.651	0.645	0.642	0.640	0.640	0.641	0.642	0.645	0.649	0.654	0.658	0.666	0.673	0.681
0.40000	(3.92)	0.640	0.635	0.633	0.632	0.632	0.634	0.636	0.639	0.644	0.649	0.654	0.662	0.669	0.677
0.60000	(5.88)	0.627	0.623	0.622	0.622	0.623	0.625	0.628	0.631	0.637	0.642	0.648	0.657	0.664	0.673
0.80000	(7.85)	0.618	0.616	0.615	0.615	0.616	0.619	0.623	0.626	0.633	0.638	0.644	0.653	0.661	0.670
1.00000	(9.81)	0.612	0.610	0.610	0.611	0.612	0.615	0.619	0.623	0.629	0.636	0.641	0.651	0.659	0.668
1.50000	(14.7)	0.601	0.601	0.601	0.603	0.605	0.609	0.613	0.617	0.624	0.631	0.637	0.647	0.655	0.665
2.00000	(19.6)	0.595	0.595	0.596	0.597	0.600	0.604	0.609	0.613	0.621	0.628	0.634	0.644	0.653	0.663
3.00000	(29.4)	0.586	0.587	0.589	0.591	0.594	0.599	0.604	0.609	0.617	0.624	0.631	0.641	0.651	0.660
4.00000	(39.2)	0.581	0.582	0.584	0.587	0.590	0.596	0.601	0.606	0.615	0.622	0.629	0.640	0.649	0.659
6.00000	(58.8)	0.574	0.576	0.579	0.582	0.585	0.592	0.597	0.603	0.612	0.619	0.626	0.637	0.647	0.657
8.00000	(78.5)	0.569	0.572	0.575	0.579	0.583	0.589	0.595	0.601	0.610	0.618	0.624	0.636	0.646	0.656
10.0000	(98.1)	0.566	0.569	0.573	0.577	0.580	0.587	0.594	0.599	0.608	0.616	0.623	0.635	0.645	0.655
15.0000	(147)	0.561	0.565	0.569	0.573	0.577	0.585	0.591	0.597	0.606	0.615	0.622	0.634	0.643	0.654
20.0000	(196)	0.558	0.562	0.566	0.571	0.575	0.583	0.589	0.595	0.605	0.613	0.621	0.633	0.643	0.653
30.0000	(294)	0.554	0.559	0.563	0.568	0.573	0.581	0.587	0.593	0.604	0.612	0.619	0.632	0.642	0.652
S	$(\Delta p/\rho l)$	6	8	10	12.5	15	20	25	30	40	50	60	80	100	125

Grad'nt   k.p.g.   (Equivalent) Pipe diameters in mm

Grad'nt S	k.p.g. $(\Delta p/\rho l)$	125	150	200	250	300	400	500	600	800	1000	1250	1500	2000	2500
0.00010	(0.00098)	0.966	0.961	0.956	0.954	0.954	0.955	0.957	0.960	0.965	0.971	0.978	0.984	0.995	1.005
0.00015	(0.00147)	0.942	0.938	0.934	0.933	0.933	0.935	0.938	0.941	0.948	0.954	0.961	0.968	0.980	0.990
0.00020	(0.00196)	0.925	0.922	0.919	0.919	0.919	0.922	0.925	0.929	0.936	0.943	0.950	0.957	0.969	0.980
0.00030	(0.00294)	0.903	0.900	0.899	0.899	0.901	0.904	0.908	0.912	0.920	0.927	0.935	0.943	0.955	0.966
0.00040	(0.00392)	0.888	0.886	0.885	0.886	0.888	0.892	0.897	0.901	0.909	0.917	0.925	0.933	0.946	0.957
0.00060	(0.00588)	0.867	0.866	0.867	0.869	0.871	0.876	0.881	0.886	0.895	0.903	0.912	0.919	0.933	0.945
0.00080	(0.00785)	0.854	0.853	0.854	0.857	0.859	0.865	0.870	0.875	0.885	0.893	0.902	0.911	0.925	0.937
0.00100	(0.00981)	0.843	0.843	0.845	0.848	0.851	0.857	0.862	0.868	0.878	0.886	0.896	0.904	0.918	0.931
0.00150	(0.0147)	0.826	0.826	0.829	0.832	0.836	0.842	0.849	0.854	0.865	0.874	0.884	0.893	0.908	0.920
0.00200	(0.0196)	0.814	0.815	0.818	0.822	0.825	0.833	0.839	0.846	0.856	0.866	0.876	0.885	0.900	0.913
0.00300	(0.0294)	0.798	0.799	0.803	0.808	0.812	0.820	0.827	0.834	0.845	0.855	0.866	0.875	0.891	0.904
0.00400	(0.0392)	0.787	0.789	0.794	0.798	0.803	0.811	0.819	0.826	0.838	0.848	0.859	0.869	0.885	0.899
0.00600	(0.0588)	0.772	0.775	0.781	0.786	0.791	0.800	0.808	0.815	0.828	0.839	0.850	0.860	0.877	0.891
0.00800	(0.0785)	0.763	0.766	0.772	0.778	0.783	0.793	0.801	0.809	0.822	0.833	0.844	0.855	0.872	0.886
0.01000	(0.0981)	0.756	0.759	0.766	0.772	0.777	0.787	0.796	0.804	0.817	0.828	0.840	0.851	0.868	0.883
0.01500	(0.147)	0.744	0.747	0.755	0.761	0.767	0.778	0.787	0.795	0.809	0.821	0.833	0.844	0.862	0.877
0.02000	(0.196)	0.735	0.740	0.747	0.754	0.761	0.772	0.781	0.790	0.804	0.816	0.829	0.840	0.858	0.873
0.03000	(0.294)	0.725	0.729	0.738	0.745	0.752	0.764	0.774	0.782	0.797	0.810	0.823	0.834	0.853	0.869
0.04000	(0.392)	0.718	0.723	0.732	0.740	0.747	0.759	0.769	0.778	0.793	0.806	0.819	0.831	0.850	0.866
0.06000	(0.588)	0.709	0.714	0.724	0.732	0.739	0.752	0.763	0.772	0.788	0.801	0.815	0.826	0.846	0.862
0.08000	(0.785)	0.703	0.709	0.719	0.727	0.735	0.748	0.759	0.768	0.784	0.798	0.812	0.824	0.844	0.860
0.10000	(0.981)	0.699	0.704	0.715	0.724	0.732	0.745	0.756	0.766	0.782	0.796	0.810	0.822	0.842	0.858
0.15000	(1.47)	0.691	0.698	0.709	0.718	0.726	0.740	0.752	0.762	0.778	0.792	0.806	0.819	0.839	0.856
0.20000	(1.96)	0.687	0.693	0.705	0.714	0.723	0.737	0.749	0.759	0.776	0.790	0.804	0.817	0.837	0.854
0.30000	(2.94)	0.681	0.688	0.700	0.710	0.719	0.733	0.745	0.756	0.773	0.787	0.802	0.815	0.835	0.852
0.40000	(3.92)	0.677	0.685	0.697	0.707	0.716	0.731	0.743	0.754	0.771	0.785	0.800	0.813	0.834	0.851
0.60000	(5.88)	0.673	0.680	0.693	0.703	0.713	0.728	0.740	0.751	0.769	0.783	0.799	0.811	0.833	0.850
0.80000	(7.85)	0.670	0.678	0.691	0.701	0.711	0.726	0.739	0.750	0.768	0.782	0.797	0.810	0.832	0.849
1.00000	(9.81)	0.668	0.676	0.689	0.700	0.709	0.725	0.738	0.749	0.767	0.781	0.797	0.810	0.831	0.848
1.50000	(14.7)	0.665	0.673	0.686	0.697	0.707	0.723	0.736	0.747	0.765	0.780	0.795	0.809	0.830	0.848
2.00000	(19.6)	0.663	0.671	0.685	0.696	0.705	0.722	0.735	0.746	0.764	0.779	0.795	0.808	0.830	0.847
3.00000	(29.4)	0.660	0.669	0.683	0.694	0.704	0.720	0.733	0.745	0.763	0.778	0.794	0.807	0.829	0.846
S	$(\Delta p/\rho l)$	125	150	200	250	300	400	500	600	800	1000	1250	1500	2000	2500

Grad'nt   k.p.g.   (Equivalent) Pipe diameters in mm

Kinematic viscosity, $\nu = 0.40 \times 10^{-6}\ m^2 s^{-1}$;   **Roughness size, $k_s = 0.006$ mm**

Kin. visc., $\nu = 0.40 \times 10^{-6}$ m^{2}s^{-1};
$S = 0.00010$ to 30.0000
i.e. kin. pr. grad., $\Delta p/\rho l =$
(0.00098) to (294) ms^{-2}

Roughness size, $k_s = 0.015$ mm
This table shows values of m, as follows
m_C for Colebrook-White solutions; or,
where $\mathbf{R} \leq 2000$, m_P for laminar flow

Grad'nt S	k.p.g. $(\Delta p/\rho l)$	6	8	10	12·5	15	20	25	30	40	50	60	80	100	125
0·00030	(0·00294)	2·743	1·869	1·388	1·031	0·808	1·008	0·985	0·969	0·949	0·937	0·929	0·919	0·913	0·909
0·00040	(0·00392)	2·375	1·619	1·202	0·893	*	0·983	0·962	0·948	0·930	0·919	0·911	0·903	0·898	0·895
0·00060	(0·00588)	1·940	1·322	0·982	*	*	0·949	0·931	0·919	0·903	0·894	0·888	0·881	0·878	0·876
0·00080	(0·00785)	1·680	1·145	0·850	*	0·953	0·927	0·911	0·900	0·886	0·878	0·873	0·867	0·864	0·863
0·00100	(0·00981)	1·502	1·024	0·760	*	0·935	0·911	0·896	0·886	0·873	0·866	0·861	0·856	0·854	0·854
0·00150	(0·0147)	1·227	0·836	*	0·920	0·904	0·883	0·870	0·861	0·851	0·845	0·841	0·838	0·837	0·837
0·00200	(0·0196)	1·062	0·724	*	0·898	0·883	0·865	0·853	0·845	0·836	0·831	0·828	0·826	0·826	0·827
0·00300	(0·0294)	0·867	*	0·888	0·869	0·856	0·840	0·830	0·824	0·817	0·813	0·811	0·810	0·810	0·812
0·00400	(0·0392)	0·751	*	0·867	0·850	0·839	0·824	0·815	0·810	0·804	0·801	0·799	0·799	0·801	0·803
0·00600	(0·0588)	*	0·857	0·839	0·825	0·815	0·803	0·796	0·791	0·787	0·785	0·784	0·785	0·787	0·791
0·00800	(0·0785)	*	0·837	0·821	0·808	0·800	0·789	0·783	0·779	0·775	0·774	0·774	0·776	0·779	0·783
0·01000	(0·0981)	*	0·823	0·808	0·796	0·788	0·779	0·773	0·770	0·767	0·767	0·767	0·770	0·773	0·777
0·01500	(0·147)	0·819	0·798	0·786	0·776	0·769	0·761	0·757	0·755	0·753	0·754	0·755	0·758	0·762	0·767
0·02000	(0·196)	0·801	0·782	0·771	0·762	0·756	0·750	0·746	0·745	0·744	0·745	0·747	0·751	0·755	0·761
0·03000	(0·294)	0·777	0·761	0·752	0·744	0·740	0·735	0·733	0·732	0·732	0·734	0·737	0·742	0·747	0·753
0·04000	(0·392)	0·762	0·748	0·739	0·733	0·729	0·725	0·724	0·723	0·725	0·727	0·730	0·736	0·741	0·748
0·06000	(0·588)	0·742	0·730	0·723	0·718	0·715	0·712	0·712	0·713	0·715	0·718	0·722	0·728	0·734	0·741
0·08000	(0·785)	0·729	0·718	0·713	0·708	0·706	0·704	0·705	0·706	0·709	0·713	0·716	0·723	0·730	0·737
0·10000	(0·981)	0·720	0·710	0·705	0·701	0·700	0·699	0·699	0·701	0·704	0·708	0·712	0·720	0·727	0·734
0·15000	(1·47)	0·704	0·696	0·692	0·690	0·689	0·689	0·690	0·692	0·697	0·702	0·706	0·714	0·722	0·730
0·20000	(1·96)	0·694	0·687	0·684	0·682	0·682	0·683	0·685	0·687	0·692	0·697	0·702	0·711	0·719	0·727
0·30000	(2·94)	0·681	0·676	0·674	0·673	0·673	0·675	0·678	0·681	0·687	0·692	0·697	0·707	0·715	0·723
0·40000	(3·92)	0·672	0·669	0·667	0·667	0·668	0·670	0·673	0·677	0·683	0·689	0·694	0·704	0·712	0·721
0·60000	(5·88)	0·662	0·659	0·659	0·659	0·661	0·664	0·668	0·672	0·679	0·685	0·691	0·701	0·709	0·719
0·80000	(7·85)	0·655	0·654	0·654	0·655	0·656	0·660	0·664	0·668	0·676	0·682	0·688	0·699	0·707	0·717
1·00000	(9·81)	0·651	0·650	0·650	0·651	0·653	0·658	0·662	0·666	0·674	0·681	0·687	0·697	0·706	0·716
1·50000	(14·7)	0·643	0·643	0·644	0·646	0·648	0·653	0·658	0·663	0·671	0·678	0·684	0·695	0·704	0·714
2·00000	(19·6)	0·638	0·639	0·640	0·643	0·645	0·651	0·656	0·661	0·669	0·676	0·683	0·694	0·703	0·713
3·00000	(29·4)	0·632	0·634	0·636	0·639	0·642	0·648	0·653	0·658	0·667	0·674	0·681	0·692	0·702	0·712
4·00000	(39·2)	0·629	0·631	0·633	0·636	0·639	0·646	0·651	0·656	0·665	0·673	0·680	0·691	0·701	0·711
6·00000	(58·8)	0·624	0·627	0·629	0·633	0·637	0·643	0·649	0·654	0·664	0·671	0·678	0·690	0·700	0·710
8·00000	(78·5)	0·622	0·624	0·627	0·631	0·635	0·642	0·648	0·653	0·663	0·671	0·678	0·689	0·699	0·709
10·0000	(98·1)	0·620	0·623	0·626	0·630	0·634	0·641	0·647	0·652	0·662	0·670	0·677	0·689	0·699	0·709
15·0000	(147)	0·617	0·620	0·624	0·628	0·632	0·639	0·646	0·651	0·661	0·669	0·676	0·688	0·698	0·708
20·0000	(196)	0·615	0·619	0·622	0·627	0·631	0·638	0·645	0·650	0·660	0·668	0·676	0·688	0·697	0·708
30·0000	(294)	0·613	0·617	0·621	0·625	0·630	0·637	0·644	0·649	0·659	0·668	0·675	0·687	0·697	0·708
S	$(\Delta p/\rho l)$	6	8	10	12·5	15	20	25	30	40	50	60	80	100	125

Grad'nt k.p.g. (Equivalent) Pipe diameters in mm

Grad'nt S	k.p.g. $(\Delta p/\rho l)$	125	150	200	250	300	400	500	600	800	1000	1250	1500	2000	2500
0·00010	(0·00098)	0·971	0·966	0·962	0·960	0·960	0·961	0·964	0·967	0·973	0·980	0·987	0·994	1·006	1·017
0·00015	(0·00147)	0·947	0·944	0·940	0·940	0·940	0·943	0·946	0·950	0·957	0·964	0·972	0·980	0·992	1·004
0·00020	(0·00196)	0·931	0·928	0·926	0·926	0·927	0·930	0·934	0·938	0·946	0·954	0·962	0·970	0·983	0·995
0·00030	(0·00294)	0·909	0·908	0·907	0·908	0·909	0·914	0·919	0·923	0·932	0·940	0·949	0·957	0·971	0·983
0·00040	(0·00392)	0·895	0·894	0·894	0·895	0·898	0·903	0·908	0·913	0·922	0·931	0·940	0·949	0·963	0·976
0·00060	(0·00588)	0·876	0·876	0·877	0·879	0·882	0·888	0·894	0·900	0·910	0·919	0·929	0·938	0·953	0·966
0·00080	(0·00785)	0·863	0·863	0·865	0·868	0·872	0·878	0·885	0·891	0·901	0·911	0·921	0·931	0·946	0·960
0·00100	(0·00981)	0·854	0·854	0·857	0·860	0·864	0·871	0·878	0·884	0·895	0·905	0·916	0·925	0·942	0·955
0·00150	(0·0147)	0·837	0·839	0·842	0·847	0·851	0·859	0·866	0·873	0·885	0·895	0·907	0·916	0·933	0·947
0·00200	(0·0196)	0·827	0·828	0·833	0·837	0·842	0·851	0·859	0·866	0·878	0·889	0·901	0·911	0·928	0·943
0·00300	(0·0294)	0·812	0·815	0·820	0·826	0·831	0·840	0·849	0·856	0·869	0·881	0·893	0·904	0·921	0·936
0·00400	(0·0392)	0·803	0·806	0·812	0·818	0·823	0·833	0·842	0·850	0·864	0·875	0·888	0·899	0·917	0·932
0·00600	(0·0588)	0·791	0·794	0·801	0·808	0·814	0·825	0·834	0·842	0·857	0·869	0·882	0·893	0·912	0·928
0·00800	(0·0785)	0·783	0·787	0·794	0·801	0·808	0·819	0·829	0·837	0·852	0·865	0·878	0·889	0·909	0·924
0·01000	(0·0981)	0·777	0·781	0·789	0·797	0·803	0·815	0·825	0·834	0·849	0·862	0·875	0·887	0·906	0·922
0·01500	(0·147)	0·767	0·772	0·781	0·789	0·796	0·808	0·819	0·828	0·844	0·857	0·871	0·883	0·902	0·919
0·02000	(0·196)	0·761	0·766	0·775	0·784	0·791	0·804	0·815	0·824	0·840	0·854	0·868	0·880	0·900	0·917
0·03000	(0·294)	0·753	0·759	0·769	0·777	0·785	0·799	0·810	0·820	0·836	0·850	0·864	0·877	0·897	0·914
0·04000	(0·392)	0·748	0·754	0·764	0·773	0·781	0·795	0·807	0·817	0·834	0·848	0·862	0·875	0·896	0·912
0·06000	(0·588)	0·741	0·748	0·759	0·768	0·777	0·791	0·803	0·813	0·831	0·845	0·860	0·872	0·893	0·911
0·08000	(0·785)	0·737	0·744	0·756	0·765	0·774	0·789	0·801	0·811	0·829	0·843	0·858	0·871	0·892	0·909
0·10000	(0·981)	0·734	0·741	0·753	0·763	0·772	0·787	0·799	0·810	0·827	0·842	0·857	0·870	0·891	0·909
0·15000	(1·47)	0·730	0·737	0·749	0·760	0·769	0·784	0·797	0·807	0·825	0·840	0·855	0·868	0·890	0·907
0·20000	(1·96)	0·727	0·734	0·747	0·758	0·767	0·782	0·795	0·806	0·824	0·839	0·854	0·867	0·889	0·906
0·30000	(2·94)	0·723	0·731	0·744	0·755	0·764	0·780	0·793	0·804	0·822	0·837	0·853	0·866	0·888	0·906
0·40000	(3·92)	0·721	0·729	0·742	0·753	0·763	0·779	0·792	0·803	0·821	0·836	0·852	0·865	0·887	0·905
0·60000	(5·88)	0·719	0·727	0·740	0·751	0·761	0·777	0·790	0·802	0·820	0·835	0·851	0·865	0·887	0·904
0·80000	(7·85)	0·717	0·725	0·739	0·750	0·760	0·776	0·790	0·801	0·820	0·835	0·851	0·864	0·886	0·904
1·00000	(9·81)	0·716	0·724	0·738	0·749	0·759	0·776	0·789	0·800	0·819	0·834	0·850	0·864	0·886	0·904
1·50000	(14·7)	0·714	0·722	0·737	0·748	0·758	0·775	0·788	0·800	0·818	0·834	0·850	0·863	0·885	0·903
2·00000	(19·6)	0·713	0·721	0·736	0·747	0·757	0·774	0·788	0·799	0·818	0·833	0·849	0·863	0·885	0·903
3·00000	(29·4)	0·712	0·720	0·735	0·746	0·757	0·773	0·787	0·798	0·817	0·833	0·849	0·862	0·885	0·903
S	$(\Delta p/\rho l)$	125	150	200	250	300	400	500	600	800	1000	1250	1500	2000	2500

Grad'nt k.p.g. (Equivalent) Pipe diameters in mm

Kinematic viscosity, $\nu = 0.40 \times 10^{-6}$ m^{2}s^{-1};　　　　**Roughness size, $k_s = 0.015$ mm**

E47

Kin. visc., $\nu = 0.40 \times 10^{-6}$ m^2 s^{-1};
$S = 0.00010$ to 30.0000
i.e. kin. pr. grad., $\Delta p/\rho l =$ (0.00098) to (294) ms^{-2}

Roughness size, $k_s = 0.030$ mm
This table shows values of m, as follows
m_C for Colebrook-White solutions; or,
where $R \leq 2000$, m_P for laminar flow

Grad'nt S	k.p.g. $(\Delta p/\rho l)$	6	8	10	12.5	15	20	25	30	40	50	60	80	100	125
		(Equivalent) Pipe diameters in mm													
0.00030	(0.00294)	2.743	1.869	1.388	1.031	0.808	1.016	0.993	0.978	0.958	0.946	0.938	0.928	0.923	0.920
0.00040	(0.00392)	2.375	1.619	1.202	0.893	*	0.991	0.971	0.957	0.939	0.928	0.921	0.913	0.909	0.907
0.00060	(0.00588)	1.940	1.322	0.982	*	*	0.959	0.941	0.929	0.914	0.905	0.899	0.893	0.891	0.889
0.00080	(0.00785)	1.680	1.145	0.850	*	0.963	0.938	0.922	0.911	0.898	0.890	0.885	0.880	0.878	0.878
0.00100	(0.00981)	1.502	1.024	0.760	*	0.946	0.922	0.907	0.898	0.885	0.879	0.874	0.870	0.869	0.869
0.00150	(0.0147)	1.227	0.836	*	0.932	0.916	0.896	0.883	0.875	0.865	0.859	0.856	0.854	0.854	0.855
0.00200	(0.0196)	1.062	0.724	*	0.911	0.897	0.878	0.867	0.860	0.851	0.847	0.845	0.843	0.844	0.846
0.00300	(0.0294)	0.867	*	0.902	0.884	0.871	0.855	0.846	0.840	0.834	0.831	0.829	0.829	0.831	0.834
0.00400	(0.0392)	0.751	*	0.882	0.866	0.855	0.841	0.833	0.828	0.822	0.820	0.819	0.820	0.822	0.826
0.00600	(0.0588)	*	0.874	0.857	0.842	0.833	0.821	0.815	0.811	0.807	0.806	0.806	0.809	0.812	0.816
0.00800	(0.0785)	*	0.856	0.840	0.827	0.819	0.809	0.803	0.800	0.798	0.797	0.798	0.801	0.805	0.810
0.01000	(0.0981)	*	0.842	0.828	0.816	0.809	0.800	0.795	0.793	0.791	0.791	0.792	0.796	0.800	0.805
0.01500	(0.147)	0.840	0.820	0.808	0.798	0.792	0.785	0.781	0.780	0.779	0.780	0.782	0.787	0.792	0.798
0.02000	(0.196)	0.824	0.805	0.794	0.786	0.781	0.775	0.772	0.771	0.772	0.774	0.776	0.781	0.787	0.793
0.03000	(0.294)	0.802	0.787	0.778	0.771	0.767	0.762	0.761	0.761	0.762	0.765	0.768	0.774	0.780	0.787
0.04000	(0.392)	0.789	0.775	0.767	0.761	0.758	0.754	0.754	0.754	0.756	0.760	0.763	0.770	0.776	0.783
0.06000	(0.588)	0.771	0.760	0.753	0.749	0.746	0.744	0.745	0.746	0.749	0.753	0.757	0.764	0.771	0.779
0.08000	(0.785)	0.760	0.750	0.745	0.741	0.739	0.738	0.739	0.740	0.744	0.749	0.753	0.761	0.768	0.776
0.10000	(0.981)	0.752	0.743	0.738	0.735	0.734	0.734	0.735	0.737	0.741	0.746	0.750	0.759	0.766	0.774
0.15000	(1.47)	0.739	0.732	0.728	0.726	0.726	0.726	0.728	0.731	0.736	0.741	0.746	0.755	0.763	0.771
0.20000	(1.96)	0.731	0.725	0.722	0.720	0.720	0.722	0.724	0.727	0.733	0.738	0.743	0.752	0.761	0.769
0.30000	(2.94)	0.721	0.716	0.714	0.713	0.714	0.716	0.719	0.722	0.729	0.735	0.740	0.750	0.758	0.767
0.40000	(3.92)	0.714	0.710	0.709	0.709	0.710	0.713	0.716	0.720	0.726	0.732	0.738	0.748	0.756	0.766
0.60000	(5.88)	0.706	0.704	0.703	0.704	0.705	0.709	0.712	0.716	0.723	0.730	0.736	0.746	0.755	0.764
0.80000	(7.85)	0.701	0.699	0.699	0.700	0.702	0.706	0.710	0.714	0.722	0.728	0.734	0.745	0.753	0.763
1.00000	(9.81)	0.698	0.696	0.697	0.698	0.700	0.704	0.709	0.713	0.720	0.727	0.733	0.744	0.753	0.762
1.50000	(14.7)	0.693	0.692	0.693	0.694	0.697	0.701	0.706	0.710	0.718	0.725	0.732	0.742	0.751	0.761
2.00000	(19.6)	0.689	0.689	0.690	0.692	0.695	0.700	0.705	0.709	0.717	0.724	0.731	0.742	0.751	0.761
3.00000	(29.4)	0.685	0.685	0.687	0.689	0.692	0.698	0.703	0.707	0.716	0.723	0.730	0.741	0.750	0.760
4.00000	(39.2)	0.683	0.683	0.685	0.688	0.691	0.696	0.702	0.706	0.715	0.722	0.729	0.740	0.749	0.759
6.00000	(58.8)	0.680	0.681	0.683	0.686	0.689	0.695	0.700	0.705	0.714	0.721	0.728	0.739	0.749	0.759
8.00000	(78.5)	0.678	0.679	0.682	0.685	0.688	0.694	0.699	0.704	0.713	0.721	0.728	0.739	0.748	0.758
10.0000	(98.1)	0.676	0.678	0.681	0.684	0.687	0.693	0.699	0.704	0.713	0.721	0.727	0.739	0.748	0.758
15.0000	(147)	0.674	0.677	0.679	0.683	0.686	0.692	0.698	0.703	0.712	0.720	0.727	0.738	0.748	0.758
20.0000	(196)	0.673	0.676	0.678	0.682	0.685	0.692	0.698	0.703	0.712	0.720	0.726	0.738	0.747	0.758
30.0000	(294)	0.672	0.674	0.677	0.681	0.685	0.691	0.697	0.702	0.711	0.719	0.726	0.738	0.747	0.757
S	$(\Delta p/\rho l)$	6	8	10	12.5	15	20	25	30	40	50	60	80	100	125

Grad'nt k.p.g. (Equivalent) Pipe diameters in mm

S	$(\Delta p/\rho l)$	125	150	200	250	300	400	500	600	800	1000	1250	1500	2000	2500
0.00010	(0.00098)	0.978	0.974	0.970	0.969	0.969	0.971	0.975	0.978	0.986	0.993	1.001	1.009	1.022	1.034
0.00015	(0.00147)	0.956	0.952	0.950	0.950	0.951	0.954	0.959	0.963	0.971	0.979	0.988	0.996	1.011	1.023
0.00020	(0.00196)	0.940	0.938	0.937	0.937	0.939	0.943	0.948	0.953	0.962	0.970	0.980	0.988	1.003	1.016
0.00030	(0.00294)	0.920	0.919	0.919	0.921	0.923	0.928	0.934	0.939	0.949	0.959	0.969	0.978	0.994	1.007
0.00040	(0.00392)	0.907	0.906	0.907	0.909	0.912	0.919	0.925	0.931	0.941	0.951	0.962	0.971	0.987	1.001
0.00060	(0.00588)	0.889	0.890	0.892	0.895	0.899	0.906	0.913	0.919	0.931	0.941	0.952	0.962	0.979	0.994
0.00080	(0.00785)	0.878	0.879	0.882	0.886	0.890	0.898	0.905	0.912	0.924	0.935	0.946	0.957	0.974	0.989
0.00100	(0.00981)	0.869	0.871	0.874	0.879	0.883	0.892	0.899	0.907	0.919	0.930	0.942	0.953	0.971	0.985
0.00150	(0.0147)	0.855	0.857	0.862	0.867	0.872	0.881	0.890	0.898	0.911	0.923	0.935	0.946	0.965	0.980
0.00200	(0.0196)	0.846	0.848	0.854	0.860	0.865	0.875	0.884	0.892	0.906	0.918	0.931	0.942	0.961	0.977
0.00300	(0.0294)	0.834	0.837	0.844	0.850	0.856	0.867	0.876	0.885	0.900	0.912	0.925	0.937	0.956	0.972
0.00400	(0.0392)	0.826	0.830	0.837	0.844	0.850	0.862	0.872	0.880	0.896	0.908	0.922	0.934	0.953	0.970
0.00600	(0.0588)	0.816	0.820	0.828	0.836	0.843	0.855	0.866	0.875	0.890	0.904	0.918	0.930	0.950	0.966
0.00800	(0.0785)	0.810	0.814	0.823	0.831	0.838	0.851	0.862	0.871	0.887	0.901	0.915	0.927	0.948	0.965
0.01000	(0.0981)	0.805	0.810	0.819	0.828	0.835	0.848	0.859	0.869	0.885	0.899	0.913	0.926	0.946	0.963
0.01500	(0.147)	0.798	0.803	0.813	0.822	0.830	0.843	0.855	0.865	0.882	0.896	0.910	0.923	0.944	0.961
0.02000	(0.196)	0.793	0.799	0.809	0.818	0.827	0.840	0.852	0.862	0.879	0.894	0.909	0.921	0.942	0.960
0.03000	(0.294)	0.787	0.793	0.804	0.814	0.823	0.837	0.849	0.859	0.877	0.891	0.906	0.919	0.941	0.958
0.04000	(0.392)	0.783	0.790	0.801	0.811	0.820	0.835	0.847	0.858	0.875	0.890	0.905	0.918	0.940	0.957
0.06000	(0.588)	0.779	0.786	0.798	0.808	0.817	0.832	0.845	0.855	0.873	0.888	0.904	0.917	0.938	0.956
0.08000	(0.785)	0.776	0.783	0.796	0.806	0.815	0.830	0.843	0.854	0.872	0.887	0.903	0.916	0.938	0.955
0.10000	(0.981)	0.774	0.782	0.794	0.805	0.814	0.829	0.842	0.853	0.871	0.886	0.902	0.915	0.937	0.955
0.15000	(1.47)	0.771	0.779	0.792	0.802	0.812	0.828	0.841	0.852	0.870	0.885	0.901	0.914	0.936	0.954
0.20000	(1.96)	0.769	0.777	0.790	0.801	0.811	0.826	0.840	0.851	0.869	0.884	0.900	0.914	0.936	0.954
0.30000	(2.94)	0.767	0.775	0.788	0.800	0.809	0.825	0.838	0.850	0.868	0.884	0.900	0.913	0.935	0.953
0.40000	(3.92)	0.766	0.774	0.787	0.799	0.808	0.824	0.838	0.849	0.868	0.883	0.899	0.913	0.935	0.953
0.60000	(5.88)	0.764	0.772	0.786	0.797	0.807	0.824	0.837	0.848	0.867	0.883	0.899	0.912	0.934	0.952
0.80000	(7.85)	0.763	0.771	0.785	0.797	0.807	0.823	0.836	0.848	0.867	0.882	0.898	0.912	0.934	0.952
1.00000	(9.81)	0.762	0.771	0.785	0.796	0.806	0.823	0.836	0.848	0.867	0.882	0.898	0.912	0.934	0.952
1.50000	(14.7)	0.761	0.770	0.784	0.795	0.805	0.822	0.836	0.847	0.866	0.882	0.898	0.911	0.934	0.952
2.00000	(19.6)	0.761	0.769	0.783	0.795	0.805	0.822	0.835	0.847	0.866	0.881	0.897	0.911	0.934	0.952
3.00000	(29.4)	0.760	0.768	0.783	0.794	0.804	0.821	0.835	0.846	0.866	0.881	0.897	0.911	0.933	0.951
S	$(\Delta p/\rho l)$	125	150	200	250	300	400	500	600	800	1000	1250	1500	2000	2500

Grad'nt k.p.g. (Equivalent) Pipe diameters in mm

Kinematic viscosity, $\nu = 0.40 \times 10^{-6}$ m^2 s^{-1};   **Roughness size, $k_s = 0.030$ mm**

Kin. visc., $\nu = 0.40\times10^{-6}$ m^{2}s^{-1};
$S = 0.00010$ to 30.0000

i.e. kin. pr. grad., $\Delta p/\rho l =$
(0.00098) to (294) ms^{-2}

Roughness size, $k_s = 0.060$ mm
This table shows values of m, as follows
m_C for Colebrook-White solutions; or,
where $R \leq 2000$, m_P for laminar flow

E48

Grad'nt S	k.p.g. $(\Delta p/\rho l)$	6	8	10	12·5	15	20	25	30	40	50	60	80	100	125
0·00030	(0·00294)	2·743	1·869	1·388	1·031	0·808	1·031	1·008	0·993	0·974	0·962	0·955	0·946	0·942	0·940
0·00040	(0·00392)	2·375	1·619	1·202	0·893	*	1·007	0·987	0·974	0·956	0·946	0·940	0·933	0·930	0·928
0·00060	(0·00588)	1·940	1·322	0·982	*	*	0·977	0·960	0·948	0·934	0·925	0·920	0·915	0·913	0·913
0·00080	(0·00785)	1·680	1·145	0·850	*	0·983	0·957	0·942	0·931	0·919	0·912	0·908	0·904	0·903	0·903
0·00100	(0·00981)	1·502	1·024	0·760	*	0·966	0·943	0·929	0·919	0·908	0·902	0·898	0·895	0·895	0·896
0·00150	(0·0147)	1·227	0·836	*	0·955	0·939	0·919	0·907	0·899	0·890	0·885	0·883	0·881	0·882	0·884
0·00200	(0·0196)	1·062	0·724	*	0·935	0·921	0·903	0·892	0·886	0·878	0·875	0·873	0·873	0·874	0·877
0·00300	(0·0294)	0·867	*	0·928	0·910	0·898	0·883	0·874	0·869	0·863	0·861	0·860	0·861	0·864	0·868
0·00400	(0·0392)	0·751	*	0·911	0·894	0·883	0·870	0·863	0·858	0·854	0·852	0·852	0·854	0·857	0·862
0·00600	(0·0588)	*	0·905	0·888	0·874	0·865	0·854	0·848	0·844	0·842	0·841	0·842	0·845	0·849	0·854
0·00800	(0·0785)	*	0·889	0·873	0·861	0·853	0·843	0·838	0·836	0·834	0·834	0·836	0·840	0·844	0·850
0·01000	(0·0981)	*	0·877	0·863	0·851	0·844	0·836	0·832	0·830	0·828	0·829	0·831	0·836	0·840	0·846
0·01500	(0·147)	0·878	0·858	0·845	0·836	0·830	0·823	0·821	0·819	0·820	0·821	0·824	0·829	0·835	0·841
0·02000	(0·196)	0·864	0·845	0·834	0·826	0·821	0·816	0·814	0·813	0·814	0·816	0·819	0·825	0·831	0·838
0·03000	(0·294)	0·846	0·830	0·821	0·814	0·810	0·806	0·805	0·805	0·807	0·810	0·814	0·820	0·827	0·834
0·04000	(0·392)	0·834	0·820	0·812	0·806	0·803	0·800	0·800	0·800	0·803	0·807	0·810	0·817	0·824	0·832
0·06000	(0·588)	0·820	0·808	0·801	0·797	0·794	0·793	0·793	0·794	0·798	0·802	0·806	0·814	0·821	0·829
0·08000	(0·785)	0·811	0·801	0·795	0·791	0·789	0·788	0·789	0·791	0·795	0·799	0·803	0·812	0·819	0·827
0·10000	(0·981)	0·805	0·795	0·790	0·787	0·785	0·785	0·786	0·788	0·792	0·797	0·802	0·810	0·817	0·826
0·15000	(1·47)	0·795	0·787	0·783	0·780	0·779	0·780	0·782	0·784	0·789	0·794	0·799	0·808	0·815	0·824
0·20000	(1·96)	0·789	0·782	0·778	0·776	0·776	0·777	0·779	0·781	0·787	0·792	0·797	0·806	0·814	0·823
0·30000	(2·94)	0·781	0·775	0·772	0·771	0·771	0·773	0·776	0·778	0·784	0·790	0·795	0·804	0·812	0·821
0·40000	(3·92)	0·776	0·771	0·769	0·768	0·769	0·771	0·774	0·777	0·783	0·788	0·794	0·803	0·811	0·820
0·60000	(5·88)	0·771	0·766	0·765	0·765	0·765	0·768	0·771	0·774	0·781	0·787	0·792	0·802	0·810	0·819
0·80000	(7·85)	0·767	0·764	0·762	0·762	0·763	0·766	0·770	0·773	0·780	0·786	0·791	0·801	0·810	0·819
1·00000	(9·81)	0·765	0·762	0·761	0·761	0·762	0·765	0·769	0·772	0·779	0·785	0·791	0·801	0·809	0·818
1·50000	(14·7)	0·761	0·758	0·758	0·758	0·760	0·763	0·767	0·771	0·778	0·784	0·790	0·800	0·808	0·818
2·00000	(19·6)	0·759	0·756	0·756	0·757	0·758	0·762	0·766	0·770	0·777	0·783	0·789	0·799	0·808	0·817
3·00000	(29·4)	0·756	0·754	0·754	0·755	0·757	0·761	0·765	0·769	0·776	0·783	0·789	0·799	0·807	0·817
4·00000	(39·2)	0·754	0·753	0·753	0·754	0·756	0·760	0·764	0·768	0·776	0·782	0·788	0·798	0·807	0·817
6·00000	(58·8)	0·752	0·751	0·752	0·753	0·755	0·759	0·763	0·767	0·775	0·782	0·788	0·798	0·807	0·816
8·00000	(78·5)	0·751	0·750	0·751	0·752	0·754	0·759	0·763	0·767	0·775	0·781	0·787	0·798	0·807	0·816
10·0000	(98·1)	0·750	0·750	0·750	0·752	0·754	0·758	0·763	0·767	0·774	0·781	0·787	0·798	0·806	0·816
15·0000	(147)	0·749	0·748	0·749	0·751	0·753	0·758	0·762	0·766	0·774	0·781	0·787	0·797	0·806	0·816
20·0000	(196)	0·748	0·748	0·749	0·750	0·753	0·757	0·762	0·766	0·774	0·781	0·787	0·797	0·806	0·816
30·0000	(294)	0·747	0·747	0·748	0·750	0·752	0·757	0·761	0·766	0·773	0·780	0·786	0·797	0·806	0·815
S	$(\Delta p/\rho l)$	6	8	10	12·5	15	20	25	30	40	50	60	80	100	125

Grad'nt   k.p.g.   (Equivalent) Pipe diameters in mm

Grad'nt S	k.p.g. $(\Delta p/\rho l)$	125	150	200	250	300	400	500	600	800	1000	1250	1500	2000	2500
0·00010	(0·00098)	0·992	0·989	0·986	0·985	0·986	0·990	0·994	0·999	1·007	1·016	1·025	1·034	1·049	1·062
0·00015	(0·00147)	0·972	0·969	0·968	0·968	0·970	0·975	0·980	0·985	0·995	1·005	1·015	1·024	1·040	1·054
0·00020	(0·00196)	0·958	0·956	0·956	0·957	0·960	0·965	0·971	0·977	0·988	0·997	1·008	1·018	1·034	1·049
0·00030	(0·00294)	0·940	0·939	0·940	0·943	0·946	0·953	0·960	0·966	0·978	0·988	1·000	1·010	1·027	1·042
0·00040	(0·00392)	0·928	0·928	0·930	0·934	0·937	0·945	0·952	0·959	0·972	0·982	0·994	1·005	1·023	1·038
0·00060	(0·00588)	0·913	0·914	0·918	0·922	0·926	0·935	0·943	0·950	0·964	0·975	0·988	0·999	1·017	1·033
0·00080	(0·00785)	0·903	0·905	0·909	0·914	0·919	0·929	0·937	0·945	0·959	0·971	0·983	0·995	1·014	1·029
0·00100	(0·00981)	0·896	0·898	0·903	0·909	0·914	0·924	0·933	0·941	0·955	0·967	0·980	0·992	1·011	1·027
0·00150	(0·0147)	0·884	0·887	0·893	0·900	0·906	0·916	0·926	0·935	0·949	0·962	0·976	0·987	1·007	1·024
0·00200	(0·0196)	0·877	0·880	0·887	0·894	0·900	0·912	0·922	0·931	0·946	0·959	0·973	0·985	1·005	1·021
0·00300	(0·0294)	0·868	0·872	0·880	0·887	0·894	0·906	0·916	0·926	0·942	0·955	0·969	0·981	1·002	1·019
0·00400	(0·0392)	0·862	0·866	0·875	0·882	0·890	0·902	0·913	0·923	0·939	0·952	0·967	0·979	1·000	1·017
0·00600	(0·0588)	0·854	0·859	0·869	0·877	0·885	0·898	0·909	0·919	0·936	0·949	0·964	0·977	0·998	1·015
0·00800	(0·0785)	0·850	0·855	0·865	0·874	0·881	0·895	0·907	0·917	0·934	0·948	0·963	0·975	0·997	1·014
0·01000	(0·0981)	0·846	0·852	0·862	0·871	0·879	0·893	0·905	0·915	0·932	0·946	0·961	0·974	0·996	1·013
0·01500	(0·147)	0·841	0·847	0·858	0·867	0·876	0·890	0·902	0·912	0·930	0·944	0·960	0·973	0·994	1·012
0·02000	(0·196)	0·838	0·844	0·855	0·865	0·874	0·888	0·900	0·911	0·929	0·943	0·959	0·972	0·993	1·011
0·03000	(0·294)	0·834	0·841	0·852	0·862	0·871	0·886	0·898	0·909	0·927	0·942	0·957	0·970	0·992	1·010
0·04000	(0·392)	0·832	0·838	0·850	0·861	0·869	0·885	0·897	0·908	0·926	0·941	0·956	0·970	0·992	1·010
0·06000	(0·588)	0·829	0·836	0·848	0·859	0·868	0·883	0·896	0·907	0·925	0·940	0·956	0·969	0·991	1·009
0·08000	(0·785)	0·827	0·834	0·847	0·857	0·866	0·882	0·895	0·906	0·924	0·939	0·955	0·968	0·991	1·008
0·10000	(0·981)	0·826	0·833	0·846	0·856	0·866	0·881	0·894	0·905	0·924	0·939	0·955	0·968	0·990	1·008
0·15000	(1·47)	0·824	0·831	0·844	0·855	0·864	0·880	0·893	0·904	0·923	0·938	0·954	0·968	0·990	1·008
0·20000	(1·96)	0·823	0·830	0·843	0·854	0·864	0·880	0·893	0·904	0·923	0·938	0·954	0·967	0·989	1·008
0·30000	(2·94)	0·821	0·829	0·842	0·853	0·863	0·879	0·892	0·903	0·922	0·937	0·953	0·967	0·989	1·007
0·40000	(3·92)	0·820	0·828	0·842	0·853	0·862	0·878	0·892	0·903	0·922	0·937	0·953	0·967	0·989	1·007
0·60000	(5·88)	0·819	0·827	0·841	0·852	0·862	0·878	0·891	0·902	0·921	0·937	0·953	0·966	0·989	1·007
0·80000	(7·85)	0·819	0·827	0·840	0·852	0·861	0·878	0·891	0·902	0·921	0·936	0·953	0·966	0·989	1·007
1·00000	(9·81)	0·818	0·826	0·840	0·851	0·861	0·877	0·891	0·902	0·921	0·936	0·952	0·966	0·988	1·007
1·50000	(14·7)	0·818	0·826	0·840	0·851	0·861	0·877	0·890	0·902	0·921	0·936	0·952	0·966	0·988	1·006
2·00000	(19·6)	0·817	0·825	0·839	0·851	0·860	0·877	0·890	0·902	0·921	0·936	0·952	0·966	0·988	1·006
3·00000	(29·4)	0·817	0·825	0·839	0·850	0·860	0·877	0·890	0·901	0·920	0·936	0·952	0·966	0·988	1·006
S	$(\Delta p/\rho l)$	125	150	200	250	300	400	500	600	800	1000	1250	1500	2000	2500

Grad'nt   k.p.g.   (Equivalent) Pipe diameters in mm

Kinematic viscosity, $\nu = 0.40\times10^{-6}$ m^{2}s^{-1} ;   **Roughness size, $k_s = 0.060$ mm**

Kin. visc., $\nu = 0.40 \times 10^{-6}\ \mathrm{m^2 s^{-1}}$;
$S = 0.00010$ to 30.0000
i.e. kin. pr. grad., $\Delta p/\rho l =$
(0.00098) to (294) $\mathrm{ms^{-2}}$

Roughness size, $k_s = 0.150$ mm
This table shows values of m, as follows
m_C for Colebrook-White solutions; or,
where $R \le 2000$, m_P for laminar flow

Grad'nt S	k.p.g. $(\Delta p/\rho l)$	(Equivalent) Pipe diameters in mm													
		6	8	10	12.5	15	20	25	30	40	50	60	80	100	125
0.00030	(0.00294)	2.743	1.869	1.388	1.031	0.808	*	1.050	1.035	1.017	1.006	0.999	0.992	0.989	0.988
0.00040	(0.00392)	2.375	1.619	1.202	0.893	*	1.052	1.032	1.018	1.002	0.993	0.987	0.981	0.979	0.979
0.00060	(0.00588)	1.940	1.322	0.982	*	*	1.026	1.008	0.997	0.984	0.976	0.972	0.968	0.967	0.968
0.00080	(0.00785)	1.680	1.145	0.850	*	1.034	1.009	0.994	0.984	0.972	0.966	0.962	0.959	0.959	0.961
0.00100	(0.00981)	1.502	1.024	0.760	*	1.020	0.997	0.983	0.974	0.963	0.958	0.955	0.953	0.954	0.956
0.00150	(0.0147)	1.227	0.836	*	1.014	0.998	0.977	0.966	0.958	0.950	0.946	0.944	0.944	0.945	0.948
0.00200	(0.0196)	1.062	0.724	*	0.998	0.983	0.965	0.955	0.948	0.941	0.938	0.937	0.938	0.940	0.943
0.00300	(0.0294)	0.867	*	0.996	0.978	0.965	0.950	0.941	0.936	0.931	0.929	0.928	0.930	0.933	0.937
0.00400	(0.0392)	0.751	*	0.982	0.965	0.954	0.940	0.933	0.928	0.924	0.923	0.923	0.925	0.929	0.934
0.00600	(0.0588)	*	0.983	0.964	0.949	0.940	0.928	0.922	0.919	0.916	0.916	0.916	0.920	0.924	0.929
0.00800	(0.0785)	*	0.970	0.953	0.940	0.931	0.921	0.916	0.913	0.911	0.911	0.912	0.916	0.921	0.927
0.01000	(0.0981)	*	0.961	0.945	0.933	0.925	0.916	0.911	0.909	0.907	0.908	0.910	0.914	0.919	0.925
0.01500	(0.147)	0.970	0.947	0.933	0.922	0.915	0.907	0.904	0.902	0.902	0.903	0.905	0.910	0.915	0.922
0.02000	(0.196)	0.959	0.938	0.925	0.915	0.909	0.902	0.899	0.898	0.898	0.900	0.902	0.908	0.913	0.920
0.03000	(0.294)	0.946	0.927	0.916	0.907	0.902	0.896	0.894	0.893	0.894	0.896	0.899	0.905	0.911	0.918
0.04000	(0.392)	0.937	0.920	0.910	0.902	0.897	0.892	0.891	0.890	0.892	0.894	0.897	0.903	0.909	0.916
0.06000	(0.588)	0.928	0.912	0.903	0.896	0.892	0.888	0.887	0.887	0.889	0.892	0.895	0.901	0.908	0.915
0.08000	(0.785)	0.922	0.907	0.898	0.892	0.889	0.885	0.884	0.885	0.887	0.890	0.893	0.900	0.907	0.914
0.10000	(0.981)	0.917	0.903	0.895	0.890	0.886	0.883	0.883	0.883	0.886	0.889	0.892	0.899	0.906	0.913
0.15000	(1.47)	0.911	0.898	0.891	0.886	0.883	0.880	0.880	0.881	0.884	0.887	0.891	0.898	0.905	0.912
0.20000	(1.96)	0.907	0.895	0.888	0.883	0.880	0.878	0.878	0.879	0.883	0.886	0.890	0.897	0.904	0.912
0.30000	(2.94)	0.902	0.891	0.885	0.880	0.878	0.876	0.877	0.878	0.881	0.885	0.889	0.896	0.903	0.911
0.40000	(3.92)	0.899	0.888	0.883	0.878	0.876	0.875	0.875	0.877	0.880	0.884	0.888	0.896	0.903	0.911
0.60000	(5.88)	0.896	0.886	0.880	0.876	0.874	0.873	0.874	0.876	0.879	0.883	0.887	0.895	0.902	0.910
0.80000	(7.85)	0.893	0.884	0.879	0.875	0.873	0.872	0.873	0.875	0.879	0.883	0.887	0.895	0.902	0.910
1.00000	(9.81)	0.892	0.883	0.878	0.874	0.873	0.872	0.873	0.874	0.878	0.883	0.887	0.895	0.902	0.910
1.50000	(14.7)	0.890	0.881	0.876	0.873	0.871	0.871	0.872	0.874	0.878	0.882	0.886	0.894	0.901	0.909
2.00000	(19.6)	0.888	0.880	0.875	0.872	0.871	0.870	0.871	0.873	0.877	0.882	0.886	0.894	0.901	0.909
3.00000	(29.4)	0.887	0.878	0.874	0.871	0.870	0.870	0.871	0.873	0.877	0.881	0.886	0.894	0.901	0.909
4.00000	(39.2)	0.886	0.878	0.873	0.871	0.869	0.869	0.870	0.872	0.877	0.881	0.885	0.893	0.901	0.909
6.00000	(58.8)	0.885	0.877	0.873	0.870	0.869	0.869	0.870	0.872	0.876	0.881	0.885	0.893	0.900	0.908
8.00000	(78.5)	0.884	0.876	0.872	0.869	0.868	0.868	0.870	0.872	0.876	0.881	0.885	0.893	0.900	0.908
10.0000	(98.1)	0.884	0.876	0.872	0.869	0.868	0.868	0.870	0.871	0.876	0.880	0.885	0.893	0.900	0.908
15.0000	(147)	0.883	0.875	0.871	0.869	0.868	0.868	0.869	0.871	0.876	0.880	0.885	0.893	0.900	0.908
20.0000	(196)	0.882	0.875	0.871	0.868	0.867	0.868	0.869	0.871	0.876	0.880	0.885	0.893	0.900	0.908
30.0000	(294)	0.882	0.874	0.871	0.868	0.867	0.867	0.869	0.871	0.875	0.880	0.884	0.893	0.900	0.908
S	$(\Delta p/\rho l)$	6	8	10	12.5	15	20	25	30	40	50	60	80	100	125

Grad'nt   k.p.g.   (Equivalent) Pipe diameters in mm

S	$(\Delta p/\rho l)$	125	150	200	250	300	400	500	600	800	1000	1250	1500	2000	2500
0.00010	(0.00098)	1.029	1.027	1.025	1.026	1.029	1.034	1.040	1.046	1.057	1.067	1.078	1.088	1.105	1.120
0.00015	(0.00147)	1.012	1.011	1.011	1.013	1.016	1.023	1.030	1.037	1.049	1.059	1.071	1.082	1.100	1.115
0.00020	(0.00196)	1.001	1.001	1.002	1.005	1.009	1.016	1.024	1.031	1.043	1.055	1.067	1.078	1.096	1.112
0.00030	(0.00294)	0.988	0.988	0.991	0.995	0.999	1.008	1.016	1.024	1.037	1.049	1.062	1.073	1.092	1.108
0.00040	(0.00392)	0.979	0.980	0.984	0.988	0.993	1.003	1.011	1.019	1.033	1.045	1.059	1.070	1.090	1.106
0.00060	(0.00588)	0.968	0.970	0.975	0.980	0.986	0.996	1.005	1.014	1.028	1.041	1.055	1.066	1.086	1.103
0.00080	(0.00785)	0.961	0.963	0.969	0.975	0.981	0.992	1.002	1.010	1.026	1.038	1.052	1.064	1.084	1.101
0.00100	(0.00981)	0.956	0.959	0.965	0.972	0.978	0.989	0.999	1.008	1.023	1.037	1.051	1.063	1.083	1.100
0.00150	(0.0147)	0.948	0.952	0.959	0.966	0.973	0.985	0.995	1.004	1.020	1.034	1.048	1.060	1.081	1.098
0.00200	(0.0196)	0.943	0.947	0.955	0.963	0.970	0.982	0.993	1.002	1.018	1.032	1.046	1.059	1.080	1.097
0.00300	(0.0294)	0.937	0.942	0.950	0.958	0.966	0.979	0.990	0.999	1.016	1.030	1.044	1.057	1.078	1.096
0.00400	(0.0392)	0.934	0.939	0.948	0.956	0.963	0.977	0.988	0.998	1.014	1.028	1.043	1.056	1.077	1.095
0.00600	(0.0588)	0.929	0.934	0.944	0.953	0.961	0.974	0.986	0.996	1.013	1.027	1.042	1.055	1.076	1.094
0.00800	(0.0785)	0.927	0.932	0.942	0.951	0.959	0.973	0.984	0.994	1.012	1.026	1.041	1.054	1.076	1.093
0.01000	(0.0981)	0.925	0.930	0.941	0.950	0.958	0.972	0.983	0.994	1.011	1.025	1.041	1.054	1.075	1.093
0.01500	(0.147)	0.922	0.928	0.938	0.948	0.956	0.970	0.982	0.992	1.010	1.024	1.040	1.053	1.074	1.092
0.02000	(0.196)	0.920	0.926	0.937	0.946	0.955	0.969	0.981	0.992	1.009	1.024	1.039	1.052	1.074	1.092
0.03000	(0.294)	0.918	0.924	0.935	0.945	0.953	0.968	0.980	0.991	1.008	1.023	1.039	1.052	1.074	1.091
0.04000	(0.392)	0.916	0.923	0.934	0.944	0.952	0.967	0.979	0.990	1.008	1.023	1.038	1.051	1.073	1.091
0.06000	(0.588)	0.915	0.921	0.933	0.943	0.952	0.966	0.979	0.989	1.007	1.022	1.038	1.051	1.073	1.091
0.08000	(0.785)	0.914	0.921	0.932	0.942	0.951	0.966	0.978	0.989	1.007	1.022	1.037	1.051	1.073	1.091
0.10000	(0.981)	0.913	0.920	0.932	0.942	0.951	0.965	0.978	0.989	1.007	1.022	1.037	1.051	1.073	1.091
0.15000	(1.47)	0.912	0.919	0.931	0.941	0.950	0.965	0.978	0.988	1.006	1.021	1.037	1.050	1.072	1.090
0.20000	(1.96)	0.912	0.919	0.930	0.941	0.950	0.965	0.977	0.988	1.006	1.021	1.037	1.050	1.072	1.090
0.30000	(2.94)	0.911	0.918	0.930	0.940	0.949	0.964	0.977	0.988	1.006	1.021	1.037	1.050	1.072	1.090
0.40000	(3.92)	0.911	0.918	0.930	0.940	0.949	0.964	0.977	0.988	1.006	1.021	1.036	1.050	1.072	1.090
0.60000	(5.88)	0.910	0.917	0.929	0.940	0.949	0.964	0.976	0.987	1.006	1.021	1.036	1.050	1.072	1.090
0.80000	(7.85)	0.910	0.917	0.929	0.939	0.948	0.964	0.976	0.987	1.005	1.020	1.036	1.050	1.072	1.090
1.00000	(9.81)	0.910	0.917	0.929	0.939	0.948	0.964	0.976	0.987	1.005	1.020	1.036	1.050	1.072	1.090
1.50000	(14.7)	0.909	0.916	0.929	0.939	0.948	0.963	0.976	0.987	1.005	1.020	1.036	1.049	1.072	1.090
2.00000	(19.6)	0.909	0.916	0.928	0.939	0.948	0.963	0.976	0.987	1.005	1.020	1.036	1.049	1.072	1.090
3.00000	(29.4)	0.909	0.916	0.928	0.939	0.948	0.963	0.976	0.987	1.005	1.020	1.036	1.049	1.072	1.090
S	$(\Delta p/\rho l)$	125	150	200	250	300	400	500	600	800	1000	1250	1500	2000	2500

Grad'nt   k.p.g.   (Equivalent) Pipe diameters in mm

Kinematic viscosity, $\nu = 0.40 \times 10^{-6}\ \mathrm{m^2 s^{-1}}$;   **Roughness size, $k_s = 0.150$ mm**

Kin. visc., $\nu = 0.40\times10^{-6}\ \mathrm{m^2s^{-1}}$;
$S = 0.00010$ to 30.0000
i.e. kin. pr. grad., $\Delta p/\rho l =$
(0.00098) to $(294)\ \mathrm{ms^{-2}}$

Roughness size, $k_s = 0.30$ mm
This table shows values of m, as follows
m_C for Colebrook-White solutions; or,
where $\mathbf{R} \leq 2000$, m_P for laminar flow

Grad'nt S	k.p.g. $(\Delta p/\rho l)$	(Equivalent) Pipe diameters in mm													
		6	8	10	12.5	15	20	25	30	40	50	60	80	100	125
0.00030	(0.00294)	2.743	1.869	1.388	1.031	0.808	*	1.109	1.094	1.075	1.064	1.058	1.051	1.048	1.047
0.00040	(0.00392)	2.375	1.619	1.202	0.893	*	1.115	1.094	1.080	1.064	1.054	1.048	1.043	1.041	1.041
0.00060	(0.00588)	1.940	1.322	0.982	*	*	1.093	1.075	1.063	1.049	1.041	1.037	1.033	1.032	1.033
0.00080	(0.00785)	1.680	1.145	0.850	*	*	1.079	1.063	1.053	1.040	1.033	1.030	1.027	1.026	1.028
0.00100	(0.00981)	1.502	1.024	0.760	*	1.095	1.070	1.055	1.045	1.034	1.028	1.025	1.022	1.023	1.025
0.00150	(0.0147)	1.227	0.836	*	*	1.077	1.054	1.041	1.033	1.023	1.019	1.016	1.015	1.017	1.019
0.00200	(0.0196)	1.062	0.724	*	1.081	1.065	1.045	1.033	1.026	1.017	1.013	1.012	1.011	1.013	1.016
0.00300	(0.0294)	0.867	*	*	1.066	1.051	1.033	1.023	1.017	1.010	1.007	1.006	1.006	1.008	1.012
0.00400	(0.0392)	0.751	*	1.076	1.056	1.043	1.026	1.017	1.011	1.005	1.003	1.002	1.003	1.006	1.010
0.00600	(0.0588)	*	*	1.062	1.044	1.032	1.017	1.009	1.004	0.999	0.998	0.997	0.999	1.002	1.007
0.00800	(0.0785)	*	1.074	1.054	1.037	1.026	1.012	1.005	1.000	0.996	0.995	0.995	0.997	1.000	1.005
0.01000	(0.0981)	*	1.068	1.048	1.032	1.021	1.008	1.001	0.997	0.993	0.993	0.993	0.996	0.999	1.004
0.01500	(0.147)	*	1.057	1.038	1.024	1.014	1.003	0.996	0.993	0.990	0.989	0.990	0.993	0.997	1.002
0.02000	(0.196)	1.078	1.050	1.033	1.019	1.010	0.999	0.993	0.990	0.988	0.987	0.988	0.992	0.996	1.001
0.03000	(0.294)	1.068	1.042	1.026	1.013	1.005	0.995	0.990	0.987	0.985	0.985	0.986	0.990	0.994	1.000
0.04000	(0.392)	1.062	1.037	1.022	1.010	1.002	0.992	0.988	0.985	0.983	0.984	0.985	0.989	0.993	0.999
0.06000	(0.588)	1.055	1.031	1.017	1.005	0.998	0.989	0.985	0.983	0.981	0.982	0.983	0.988	0.992	0.998
0.08000	(0.785)	1.050	1.028	1.014	1.003	0.996	0.988	0.983	0.981	0.980	0.981	0.983	0.987	0.991	0.997
0.10000	(0.981)	1.047	1.025	1.012	1.001	0.994	0.986	0.982	0.980	0.979	0.980	0.982	0.986	0.991	0.997
0.15000	(1.47)	1.043	1.021	1.009	0.998	0.992	0.984	0.981	0.979	0.978	0.979	0.981	0.985	0.990	0.996
0.20000	(1.96)	1.040	1.019	1.007	0.997	0.990	0.983	0.980	0.978	0.977	0.979	0.980	0.985	0.990	0.996
0.30000	(2.94)	1.037	1.017	1.004	0.995	0.989	0.982	0.978	0.977	0.976	0.978	0.980	0.984	0.989	0.995
0.40000	(3.92)	1.034	1.015	1.003	0.994	0.988	0.981	0.978	0.976	0.976	0.977	0.979	0.984	0.989	0.995
0.60000	(5.88)	1.032	1.013	1.001	0.992	0.986	0.980	0.977	0.975	0.975	0.977	0.979	0.984	0.989	0.995
0.80000	(7.85)	1.031	1.012	1.000	0.991	0.986	0.979	0.976	0.975	0.975	0.976	0.979	0.983	0.989	0.995
1.00000	(9.81)	1.030	1.011	1.000	0.991	0.985	0.979	0.976	0.975	0.975	0.976	0.978	0.983	0.988	0.994
1.50000	(14.7)	1.028	1.010	0.999	0.990	0.984	0.978	0.975	0.974	0.974	0.976	0.978	0.983	0.988	0.994
2.00000	(19.6)	1.027	1.009	0.998	0.989	0.984	0.978	0.975	0.974	0.974	0.976	0.978	0.983	0.988	0.994
3.00000	(29.4)	1.026	1.008	0.997	0.989	0.983	0.977	0.975	0.974	0.974	0.975	0.978	0.983	0.988	0.994
4.00000	(39.2)	1.025	1.008	0.997	0.988	0.983	0.977	0.974	0.973	0.974	0.975	0.977	0.983	0.988	0.994
6.00000	(58.8)	1.025	1.007	0.996	0.988	0.983	0.977	0.974	0.973	0.973	0.975	0.977	0.982	0.988	0.994
8.00000	(78.5)	1.024	1.007	0.996	0.988	0.982	0.977	0.974	0.973	0.973	0.975	0.977	0.982	0.988	0.994
10.0000	(98.1)	1.024	1.006	0.996	0.988	0.982	0.976	0.974	0.973	0.973	0.975	0.977	0.982	0.988	0.994
15.0000	(147)	1.023	1.006	0.995	0.987	0.982	0.976	0.974	0.973	0.973	0.975	0.977	0.982	0.987	0.994
20.0000	(196)	1.023	1.006	0.995	0.987	0.982	0.976	0.974	0.973	0.973	0.975	0.977	0.982	0.987	0.994
30.0000	(294)	1.023	1.005	0.995	0.987	0.982	0.976	0.973	0.973	0.973	0.975	0.977	0.982	0.987	0.994
S	$(\Delta p/\rho l)$	6	8	10	12.5	15	20	25	30	40	50	60	80	100	125

Grad'nt S	k.p.g. $(\Delta p/\rho l)$	(Equivalent) Pipe diameters in mm													
		125	150	200	250	300	400	500	600	800	1000	1250	1500	2000	2500
0.00010	(0.00098)	1.079	1.077	1.077	1.079	1.081	1.088	1.095	1.101	1.113	1.124	1.136	1.147	1.166	1.181
0.00015	(0.00147)	1.066	1.065	1.066	1.069	1.073	1.080	1.088	1.095	1.108	1.119	1.132	1.143	1.162	1.178
0.00020	(0.00196)	1.058	1.057	1.059	1.063	1.067	1.076	1.084	1.091	1.105	1.116	1.129	1.140	1.160	1.176
0.00030	(0.00294)	1.047	1.048	1.051	1.056	1.060	1.070	1.078	1.086	1.100	1.113	1.126	1.137	1.157	1.174
0.00040	(0.00392)	1.041	1.042	1.046	1.051	1.056	1.066	1.075	1.083	1.098	1.110	1.124	1.136	1.156	1.172
0.00060	(0.00588)	1.033	1.035	1.040	1.046	1.051	1.062	1.071	1.080	1.095	1.108	1.121	1.133	1.154	1.170
0.00080	(0.00785)	1.028	1.030	1.036	1.042	1.048	1.059	1.069	1.078	1.093	1.106	1.120	1.132	1.152	1.169
0.00100	(0.00981)	1.025	1.027	1.034	1.040	1.046	1.057	1.067	1.076	1.092	1.105	1.119	1.131	1.152	1.169
0.00150	(0.0147)	1.019	1.023	1.030	1.036	1.043	1.054	1.065	1.074	1.090	1.103	1.117	1.130	1.150	1.168
0.00200	(0.0196)	1.016	1.020	1.027	1.034	1.041	1.053	1.063	1.073	1.088	1.102	1.116	1.129	1.150	1.167
0.00300	(0.0294)	1.012	1.016	1.024	1.031	1.038	1.051	1.061	1.071	1.087	1.101	1.115	1.128	1.149	1.166
0.00400	(0.0392)	1.010	1.014	1.022	1.030	1.037	1.049	1.060	1.070	1.086	1.100	1.115	1.127	1.148	1.166
0.00600	(0.0588)	1.007	1.011	1.020	1.028	1.035	1.048	1.059	1.069	1.085	1.099	1.114	1.126	1.148	1.165
0.00800	(0.0785)	1.005	1.010	1.019	1.027	1.034	1.047	1.058	1.068	1.085	1.098	1.113	1.126	1.147	1.165
0.01000	(0.0981)	1.004	1.009	1.018	1.026	1.033	1.046	1.058	1.067	1.084	1.098	1.113	1.126	1.147	1.164
0.01500	(0.147)	1.002	1.007	1.016	1.025	1.032	1.045	1.057	1.067	1.083	1.097	1.112	1.125	1.147	1.164
0.02000	(0.196)	1.001	1.006	1.015	1.024	1.031	1.045	1.056	1.066	1.083	1.097	1.112	1.125	1.146	1.164
0.03000	(0.294)	1.000	1.005	1.014	1.023	1.031	1.044	1.056	1.066	1.083	1.097	1.112	1.125	1.146	1.164
0.04000	(0.392)	0.999	1.004	1.014	1.022	1.030	1.044	1.055	1.065	1.082	1.096	1.111	1.124	1.146	1.163
0.06000	(0.588)	0.998	1.003	1.013	1.022	1.030	1.043	1.055	1.065	1.082	1.096	1.111	1.124	1.146	1.163
0.08000	(0.785)	0.997	1.003	1.013	1.021	1.029	1.043	1.055	1.065	1.082	1.096	1.111	1.124	1.146	1.163
0.10000	(0.981)	0.997	1.002	1.012	1.021	1.029	1.043	1.054	1.064	1.082	1.096	1.111	1.124	1.145	1.163
0.15000	(1.47)	0.996	1.002	1.012	1.021	1.029	1.042	1.054	1.064	1.081	1.096	1.111	1.124	1.145	1.163
0.20000	(1.96)	0.996	1.001	1.012	1.020	1.028	1.042	1.054	1.064	1.081	1.096	1.111	1.124	1.145	1.163
0.30000	(2.94)	0.995	1.001	1.011	1.020	1.028	1.042	1.054	1.064	1.081	1.095	1.111	1.124	1.145	1.163
0.40000	(3.92)	0.995	1.001	1.011	1.020	1.028	1.042	1.054	1.064	1.081	1.095	1.111	1.123	1.145	1.163
0.60000	(5.88)	0.995	1.000	1.011	1.020	1.028	1.042	1.053	1.064	1.081	1.095	1.110	1.123	1.145	1.163
0.80000	(7.85)	0.995	1.000	1.011	1.020	1.028	1.042	1.053	1.064	1.081	1.095	1.110	1.123	1.145	1.163
1.00000	(9.81)	0.994	1.000	1.011	1.020	1.028	1.042	1.053	1.064	1.081	1.095	1.110	1.123	1.145	1.163
1.50000	(14.7)	0.994	1.000	1.010	1.019	1.028	1.041	1.053	1.063	1.081	1.095	1.110	1.123	1.145	1.163
2.00000	(19.6)	0.994	1.000	1.010	1.019	1.027	1.041	1.053	1.063	1.081	1.095	1.110	1.123	1.145	1.163
3.00000	(29.4)	0.994	1.000	1.010	1.019	1.027	1.041	1.053	1.063	1.081	1.095	1.110	1.123	1.145	1.163
S	$(\Delta p/\rho l)$	125	150	200	250	300	400	500	600	800	1000	1250	1500	2000	2500

Grad'nt   k.p.g.   (Equivalent) Pipe diameters in mm

Kinematic viscosity, $\nu = 0.40\times10^{-6}\ \mathrm{m^2s^{-1}}$;   **Roughness size, $k_s = 0.30$ mm**

E51

Kin. visc., $\nu = 0.40 \times 10^{-6}$ m²s⁻¹; $S = 0.00010$ to 3.00000

i.e. kin. pr. grad., $\Delta p/\rho l =$ (0.00098) to (29.4) ms⁻²

Roughness size, $k_s = 0.60$ mm

This table shows values of m, as follows

m_C for Colebrook-White solutions

Grad'nt S	k.p.g. $(\Delta p/\rho l)$	125	150	200	250	300	400	500	600	800	1000	1250	1500	2000	2500
							(Equivalent) Pipe diameters in mm								
0.00010	(0.00098)	1.155	1.153	1.152	1.154	1.157	1.163	1.170	1.177	1.189	1.200	1.212	1.223	1.241	1.257
0.00015	(0.00147)	1.145	1.144	1.145	1.147	1.151	1.158	1.165	1.172	1.185	1.197	1.209	1.220	1.239	1.255
0.00020	(0.00196)	1.139	1.139	1.140	1.143	1.147	1.155	1.163	1.170	1.183	1.195	1.207	1.219	1.238	1.254
0.00030	(0.00294)	1.132	1.132	1.135	1.138	1.143	1.151	1.159	1.167	1.181	1.192	1.205	1.217	1.236	1.252
0.00040	(0.00392)	1.128	1.128	1.131	1.135	1.140	1.149	1.157	1.165	1.179	1.191	1.204	1.216	1.235	1.252
0.00060	(0.00588)	1.122	1.123	1.127	1.132	1.137	1.146	1.155	1.163	1.177	1.189	1.203	1.214	1.234	1.251
0.00080	(0.00785)	1.119	1.120	1.125	1.130	1.135	1.144	1.153	1.162	1.176	1.188	1.202	1.214	1.233	1.250
0.00100	(0.00981)	1.117	1.118	1.123	1.128	1.133	1.143	1.152	1.161	1.175	1.188	1.201	1.213	1.233	1.250
0.00150	(0.0147)	1.113	1.115	1.120	1.126	1.131	1.142	1.151	1.159	1.174	1.187	1.200	1.212	1.232	1.249
0.00200	(0.0196)	1.111	1.113	1.119	1.124	1.130	1.140	1.150	1.158	1.173	1.186	1.200	1.212	1.232	1.249
0.00300	(0.0294)	1.109	1.111	1.117	1.123	1.128	1.139	1.149	1.157	1.172	1.185	1.199	1.211	1.231	1.248
0.00400	(0.0392)	1.107	1.110	1.116	1.122	1.128	1.138	1.148	1.157	1.172	1.185	1.199	1.211	1.231	1.248
0.00600	(0.0588)	1.105	1.108	1.114	1.120	1.126	1.137	1.147	1.156	1.171	1.184	1.198	1.210	1.231	1.247
0.00800	(0.0785)	1.104	1.107	1.113	1.120	1.126	1.137	1.147	1.156	1.171	1.184	1.198	1.210	1.230	1.247
0.01000	(0.0981)	1.103	1.106	1.113	1.119	1.125	1.137	1.146	1.155	1.171	1.184	1.198	1.210	1.230	1.247
0.01500	(0.147)	1.102	1.105	1.112	1.119	1.125	1.136	1.146	1.155	1.170	1.183	1.197	1.210	1.230	1.247
0.02000	(0.196)	1.102	1.105	1.111	1.118	1.124	1.136	1.146	1.155	1.170	1.183	1.197	1.209	1.230	1.247
0.03000	(0.294)	1.101	1.104	1.111	1.118	1.124	1.135	1.145	1.154	1.170	1.183	1.197	1.209	1.230	1.247
0.04000	(0.392)	1.100	1.104	1.110	1.117	1.124	1.135	1.145	1.154	1.170	1.183	1.197	1.209	1.230	1.247
0.06000	(0.588)	1.100	1.103	1.110	1.117	1.123	1.135	1.145	1.154	1.169	1.183	1.197	1.209	1.229	1.246
0.08000	(0.785)	1.099	1.103	1.110	1.117	1.123	1.135	1.145	1.154	1.169	1.182	1.197	1.209	1.229	1.246
0.10000	(0.981)	1.099	1.102	1.110	1.116	1.123	1.134	1.145	1.154	1.169	1.182	1.197	1.209	1.229	1.246
0.15000	(1.47)	1.099	1.102	1.109	1.116	1.123	1.134	1.144	1.153	1.169	1.182	1.196	1.209	1.229	1.246
0.20000	(1.96)	1.098	1.102	1.109	1.116	1.122	1.134	1.144	1.153	1.169	1.182	1.196	1.209	1.229	1.246
0.30000	(2.94)	1.098	1.102	1.109	1.116	1.122	1.134	1.144	1.153	1.169	1.182	1.196	1.209	1.229	1.246
0.40000	(3.92)	1.098	1.102	1.109	1.116	1.122	1.134	1.144	1.153	1.169	1.182	1.196	1.209	1.229	1.246
0.60000	(5.88)	1.098	1.101	1.109	1.116	1.122	1.134	1.144	1.153	1.169	1.182	1.196	1.208	1.229	1.246
0.80000	(7.85)	1.098	1.101	1.109	1.116	1.122	1.134	1.144	1.153	1.169	1.182	1.196	1.208	1.229	1.246
1.00000	(9.81)	1.098	1.101	1.109	1.115	1.122	1.134	1.144	1.153	1.169	1.182	1.196	1.208	1.229	1.246
1.50000	(14.7)	1.097	1.101	1.108	1.115	1.122	1.134	1.144	1.153	1.169	1.182	1.196	1.208	1.229	1.246
2.00000	(19.6)	1.097	1.101	1.108	1.115	1.122	1.134	1.144	1.153	1.169	1.182	1.196	1.208	1.229	1.246
3.00000	(29.4)	1.097	1.101	1.108	1.115	1.122	1.134	1.144	1.153	1.169	1.182	1.196	1.208	1.229	1.246
S	$(\Delta p/\rho l)$	125	150	200	250	300	400	500	600	800	1000	1250	1500	2000	2500

Grad'nt k.p.g.   (Equivalent) Pipe diameters in mm   Roughness size, $k_s = 0.60$ mm

E52

Kin. visc., $\nu = 0.40 \times 10^{-6}$ m²s⁻¹; $S = 0.00010$ to 3.00000

i.e. kin. pr. grad., $\Delta p/\rho l =$ (0.00098) to (29.4) ms⁻²

Roughness size, $k_s = 1.50$ mm

This table shows values of m, as follows

m_C for Colebrook-White solutions

Grad'nt S	k.p.g. $(\Delta p/\rho l)$	125	150	200	250	300	400	500	600	800	1000	1250	1500	2000	2500
							(Equivalent) Pipe diameters in mm								
0.00010	(0.00098)	1.306	1.301	1.296	1.295	1.296	1.300	1.305	1.310	1.320	1.330	1.341	1.350	1.368	1.382
0.00015	(0.00147)	1.300	1.296	1.292	1.292	1.293	1.297	1.302	1.308	1.318	1.328	1.339	1.349	1.366	1.381
0.00020	(0.00196)	1.296	1.292	1.289	1.289	1.291	1.295	1.301	1.306	1.317	1.327	1.338	1.348	1.366	1.381
0.00030	(0.00294)	1.292	1.289	1.286	1.287	1.288	1.293	1.299	1.305	1.316	1.326	1.337	1.347	1.365	1.380
0.00040	(0.00392)	1.289	1.286	1.284	1.285	1.287	1.292	1.298	1.304	1.315	1.325	1.336	1.347	1.364	1.380
0.00060	(0.00588)	1.286	1.284	1.282	1.283	1.285	1.291	1.297	1.303	1.314	1.324	1.336	1.346	1.364	1.379
0.00080	(0.00785)	1.284	1.282	1.281	1.282	1.284	1.290	1.296	1.302	1.313	1.324	1.335	1.346	1.364	1.379
0.00100	(0.00981)	1.283	1.281	1.280	1.281	1.283	1.289	1.295	1.301	1.313	1.323	1.335	1.345	1.363	1.379
0.00150	(0.0147)	1.281	1.279	1.278	1.280	1.282	1.288	1.295	1.301	1.312	1.323	1.335	1.345	1.363	1.378
0.00200	(0.0196)	1.280	1.278	1.277	1.279	1.282	1.288	1.294	1.300	1.312	1.322	1.334	1.345	1.363	1.378
0.00300	(0.0294)	1.279	1.277	1.276	1.278	1.281	1.287	1.293	1.300	1.312	1.322	1.334	1.344	1.362	1.378
0.00400	(0.0392)	1.278	1.276	1.276	1.278	1.280	1.287	1.293	1.299	1.311	1.322	1.334	1.344	1.362	1.378
0.00600	(0.0588)	1.277	1.275	1.275	1.277	1.280	1.286	1.293	1.299	1.311	1.322	1.333	1.344	1.362	1.378
0.00800	(0.0785)	1.276	1.275	1.275	1.277	1.279	1.286	1.292	1.299	1.311	1.321	1.333	1.344	1.362	1.377
0.01000	(0.0981)	1.276	1.274	1.274	1.276	1.279	1.286	1.292	1.299	1.311	1.321	1.333	1.344	1.362	1.377
0.01500	(0.147)	1.275	1.274	1.274	1.276	1.279	1.285	1.292	1.298	1.310	1.321	1.333	1.344	1.362	1.377
0.02000	(0.196)	1.275	1.273	1.273	1.276	1.279	1.285	1.292	1.298	1.310	1.321	1.333	1.344	1.362	1.377
0.03000	(0.294)	1.274	1.273	1.273	1.275	1.278	1.285	1.292	1.298	1.310	1.321	1.333	1.343	1.362	1.377
0.04000	(0.392)	1.274	1.273	1.273	1.275	1.278	1.285	1.292	1.298	1.310	1.321	1.333	1.343	1.362	1.377
0.06000	(0.588)	1.274	1.272	1.273	1.275	1.278	1.285	1.291	1.298	1.310	1.321	1.333	1.343	1.362	1.377
0.08000	(0.785)	1.273	1.272	1.273	1.275	1.278	1.285	1.291	1.298	1.310	1.321	1.333	1.343	1.362	1.377
0.10000	(0.981)	1.273	1.272	1.272	1.275	1.278	1.284	1.291	1.298	1.310	1.321	1.333	1.343	1.362	1.377
0.15000	(1.47)	1.273	1.272	1.272	1.275	1.278	1.284	1.291	1.298	1.310	1.321	1.333	1.343	1.361	1.377
0.20000	(1.96)	1.273	1.272	1.272	1.275	1.278	1.284	1.291	1.298	1.310	1.321	1.333	1.343	1.361	1.377
0.30000	(2.94)	1.273	1.272	1.272	1.274	1.277	1.284	1.291	1.298	1.310	1.321	1.333	1.343	1.361	1.377
0.40000	(3.92)	1.273	1.272	1.272	1.274	1.277	1.284	1.291	1.298	1.310	1.321	1.332	1.343	1.361	1.377
0.60000	(5.88)	1.273	1.271	1.272	1.274	1.277	1.284	1.291	1.298	1.310	1.321	1.332	1.343	1.361	1.377
0.80000	(7.85)	1.273	1.271	1.272	1.274	1.277	1.284	1.291	1.298	1.310	1.320	1.332	1.343	1.361	1.377
1.00000	(9.81)	1.273	1.271	1.272	1.274	1.277	1.284	1.291	1.298	1.310	1.320	1.332	1.343	1.361	1.377
1.50000	(14.7)	1.273	1.271	1.272	1.274	1.277	1.284	1.291	1.298	1.310	1.320	1.332	1.343	1.361	1.377
2.00000	(19.6)	1.273	1.271	1.272	1.274	1.277	1.284	1.291	1.298	1.310	1.320	1.332	1.343	1.361	1.377
3.00000	(29.4)	1.272	1.271	1.272	1.274	1.277	1.284	1.291	1.298	1.310	1.320	1.332	1.343	1.361	1.377
S	$(\Delta p/\rho l)$	125	150	200	250	300	400	500	600	800	1000	1250	1500	2000	2500

Grad'nt k.p.g.   (Equivalent) Pipe diameters in mm   Roughness size, $k_s = 1.50$ mm

Kin. visc., $v = 0.50 \times 10^{-6}$ m^{2}s^{-1}; **Roughness size, $k_s = 0.0015$ mm** # E53

$S = 0.00010$ to 30.0000

i.e. kin. pr. grad., $\Delta p / \rho l = (0.00098)$ to (294) ms^{-2}

This table shows values of m, as follows
m_C for Colebrook-White solutions; or,
where $R \leq 2000$, m_P for laminar flow

Grad'nt S	k.p.g. $(\Delta p/\rho l)$	6	8	10	12·5	15	20	25	30	40	50	60	80	100	125
0·00030	(0·00294)	3·429	2·336	1·735	1·289	1·011	*	1·017	0·999	0·975	0·960	0·950	0·937	0·930	0·924
0·00040	(0·00392)	2·969	2·023	1·503	1·116	0·875	*	0·991	0·975	0·953	0·940	0·930	0·919	0·913	0·908
0·00060	(0·00588)	2·424	1·652	1·227	0·911	*	0·978	0·957	0·943	0·924	0·912	0·904	0·895	0·889	0·886
0·00080	(0·00785)	2·100	1·431	1·063	0·789	*	0·953	0·934	0·921	0·904	0·893	0·887	0·878	0·874	0·871
0·00100	(0·00981)	1·878	1·280	0·950	*	*	0·935	0·917	0·905	0·889	0·880	0·873	0·866	0·862	0·859
0·00150	(0·0147)	1·533	1·045	0·776	*	0·928	0·903	0·888	0·877	0·864	0·855	0·850	0·844	0·841	0·840
0·00200	(0·0196)	1·328	0·905	*	0·922	0·905	0·882	0·868	0·859	0·846	0·839	0·835	0·830	0·827	0·826
0·00300	(0·0294)	1·084	0·739	*	0·889	0·874	0·854	0·842	0·834	0·823	0·817	0·814	0·810	0·808	0·808
0·00400	(0·0392)	0·939	*	0·887	0·867	0·853	0·835	0·824	0·817	0·808	0·803	0·800	0·797	0·796	0·796
0·00600	(0·0588)	0·767	*	0·855	0·838	0·826	0·810	0·801	0·794	0·787	0·783	0·780	0·778	0·778	0·779
0·00800	(0·0785)	*	0·853	0·834	0·818	0·807	0·793	0·785	0·779	0·773	0·769	0·767	0·766	0·767	0·768
0·01000	(0·0981)	*	0·836	0·818	0·804	0·793	0·781	0·773	0·768	0·762	0·759	0·758	0·757	0·758	0·759
0·01500	(0·147)	*	0·807	0·791	0·778	0·770	0·759	0·752	0·748	0·744	0·742	0·741	0·741	0·742	0·745
0·02000	(0·196)	0·810	0·787	0·773	0·762	0·754	0·744	0·738	0·735	0·731	0·730	0·729	0·730	0·732	0·734
0·03000	(0·294)	0·781	0·761	0·749	0·739	0·733	0·724	0·720	0·717	0·714	0·714	0·714	0·715	0·718	0·721
0·04000	(0·392)	0·762	0·744	0·733	0·724	0·718	0·711	0·707	0·705	0·703	0·703	0·703	0·705	0·708	0·712
0·06000	(0·588)	0·737	0·721	0·712	0·704	0·699	0·693	0·690	0·689	0·688	0·688	0·689	0·692	0·695	0·699
0·08000	(0·785)	0·720	0·706	0·697	0·691	0·686	0·681	0·679	0·678	0·677	0·678	0·680	0·683	0·686	0·691
0·10000	(0·981)	0·707	0·695	0·687	0·681	0·677	0·672	0·670	0·669	0·670	0·671	0·672	0·676	0·680	0·684
0·15000	(1·47)	0·686	0·675	0·668	0·663	0·660	0·657	0·655	0·655	0·656	0·658	0·660	0·664	0·669	0·674
0·20000	(1·96)	0·671	0·662	0·656	0·651	0·649	0·646	0·646	0·646	0·647	0·649	0·652	0·656	0·661	0·666
0·30000	(2·94)	0·652	0·644	0·639	0·636	0·634	0·632	0·632	0·633	0·635	0·638	0·640	0·646	0·651	0·657
0·40000	(3·92)	0·640	0·632	0·628	0·625	0·624	0·623	0·623	0·624	0·627	0·630	0·633	0·639	0·644	0·650
0·60000	(5·88)	0·622	0·616	0·613	0·611	0·610	0·610	0·611	0·613	0·616	0·620	0·623	0·629	0·635	0·641
0·80000	(7·85)	0·611	0·606	0·603	0·602	0·601	0·602	0·603	0·605	0·609	0·613	0·616	0·623	0·629	0·636
1·00000	(9·81)	0·603	0·598	0·596	0·595	0·595	0·596	0·597	0·599	0·603	0·608	0·611	0·619	0·625	0·632
1·50000	(14·7)	0·588	0·585	0·583	0·583	0·583	0·585	0·587	0·590	0·594	0·599	0·603	0·611	0·617	0·625
2·00000	(19·6)	0·578	0·576	0·575	0·575	0·575	0·578	0·580	0·583	0·588	0·593	0·598	0·606	0·612	0·620
3·00000	(29·4)	0·566	0·564	0·564	0·564	0·565	0·568	0·571	0·575	0·580	0·586	0·590	0·599	0·606	0·614
4·00000	(39·2)	0·557	0·556	0·556	0·557	0·559	0·562	0·566	0·569	0·575	0·581	0·586	0·595	0·602	0·610
6·00000	(58·8)	0·546	0·545	0·546	0·548	0·550	0·554	0·558	0·562	0·568	0·574	0·580	0·589	0·597	0·605
8·00000	(78·5)	0·538	0·539	0·540	0·542	0·544	0·549	0·553	0·557	0·564	0·570	0·576	0·586	0·594	0·602
10·0000	(98·1)	0·533	0·534	0·535	0·538	0·540	0·545	0·549	0·553	0·561	0·567	0·573	0·583	0·591	0·600
15·0000	(147)	0·524	0·525	0·527	0·530	0·533	0·538	0·543	0·548	0·556	0·562	0·568	0·579	0·587	0·597
20·0000	(196)	0·517	0·520	0·522	0·525	0·528	0·534	0·539	0·544	0·552	0·559	0·566	0·576	0·585	0·594
30·0000	(294)	0·510	0·512	0·515	0·519	0·522	0·529	0·534	0·539	0·548	0·555	0·562	0·573	0·582	0·592
S	$(\Delta p/\rho l)$	6	8	10	12·5	15	20	25	30	40	50	60	80	100	125

Grad'nt k.p.g.   (Equivalent) Pipe diameters in mm

Grad'nt S	k.p.g. $(\Delta p/\rho l)$	125	150	200	250	300	400	500	600	800	1000	1250	1500	2000	2500
0·00010	(0·00098)	0·993	0·987	0·980	0·977	0·975	0·974	0·976	0·977	0·982	0·987	0·992	0·998	1·008	1·017
0·00015	(0·00147)	0·966	0·961	0·956	0·954	0·953	0·953	0·955	0·958	0·963	0·968	0·974	0·980	0·991	1·000
0·00020	(0·00196)	0·948	0·944	0·940	0·938	0·938	0·939	0·941	0·944	0·950	0·955	0·962	0·968	0·979	0·988
0·00030	(0·00294)	0·924	0·921	0·918	0·917	0·917	0·920	0·922	0·926	0·932	0·938	0·945	0·952	0·963	0·973
0·00040	(0·00392)	0·908	0·905	0·903	0·903	0·903	0·906	0·910	0·913	0·920	0·926	0·934	0·940	0·952	0·962
0·00060	(0·00588)	0·886	0·884	0·883	0·883	0·885	0·888	0·892	0·896	0·903	0·910	0·918	0·925	0·937	0·947
0·00080	(0·00785)	0·871	0·869	0·869	0·870	0·872	0·876	0·880	0·884	0·892	0·899	0·907	0·914	0·927	0·937
0·00100	(0·00981)	0·859	0·858	0·859	0·860	0·862	0·866	0·871	0·875	0·884	0·891	0·899	0·906	0·919	0·930
0·00150	(0·0147)	0·840	0·839	0·840	0·843	0·845	0·850	0·855	0·860	0·869	0·876	0·885	0·892	0·905	0·916
0·00200	(0·0196)	0·826	0·826	0·828	0·831	0·833	0·839	0·844	0·849	0·858	0·866	0·875	0·883	0·896	0·907
0·00300	(0·0294)	0·808	0·809	0·811	0·814	0·818	0·824	0·830	0·835	0·845	0·853	0·862	0·870	0·884	0·895
0·00400	(0·0392)	0·796	0·797	0·800	0·803	0·807	0·813	0·820	0·825	0·835	0·844	0·853	0·861	0·875	0·887
0·00600	(0·0588)	0·779	0·781	0·785	0·788	0·792	0·799	0·806	0·812	0·822	0·831	0·841	0·849	0·863	0·875
0·00800	(0·0785)	0·768	0·770	0·774	0·778	0·782	0·790	0·797	0·803	0·813	0·823	0·832	0·841	0·855	0·868
0·01000	(0·0981)	0·759	0·762	0·766	0·771	0·775	0·783	0·790	0·796	0·807	0·816	0·826	0·835	0·850	0·862
0·01500	(0·147)	0·745	0·747	0·752	0·757	0·762	0·770	0·778	0·784	0·795	0·805	0·815	0·824	0·839	0·852
0·02000	(0·196)	0·734	0·737	0·743	0·748	0·753	0·762	0·769	0·776	0·788	0·797	0·808	0·817	0·833	0·845
0·03000	(0·294)	0·721	0·724	0·730	0·736	0·741	0·750	0·758	0·765	0·777	0·787	0·798	0·808	0·823	0·837
0·04000	(0·392)	0·712	0·715	0·722	0·728	0·733	0·742	0·751	0·758	0·770	0·781	0·792	0·801	0·817	0·831
0·06000	(0·588)	0·699	0·703	0·710	0·716	0·722	0·732	0·741	0·748	0·761	0·772	0·783	0·793	0·809	0·823
0·08000	(0·785)	0·691	0·695	0·702	0·709	0·715	0·725	0·734	0·742	0·755	0·766	0·777	0·787	0·804	0·818
0·10000	(0·981)	0·684	0·689	0·696	0·703	0·709	0·720	0·729	0·737	0·750	0·761	0·773	0·783	0·800	0·814
0·15000	(1·47)	0·674	0·678	0·686	0·694	0·700	0·711	0·720	0·729	0·742	0·754	0·766	0·776	0·794	0·808
0·20000	(1·96)	0·666	0·671	0·680	0·687	0·694	0·705	0·715	0·723	0·737	0·749	0·761	0·772	0·789	0·804
0·30000	(2·94)	0·657	0·662	0·671	0·679	0·686	0·697	0·707	0·716	0·730	0·742	0·755	0·766	0·784	0·799
0·40000	(3·92)	0·650	0·655	0·665	0·673	0·680	0·692	0·702	0·711	0·726	0·738	0·751	0·762	0·781	0·796
0·60000	(5·88)	0·641	0·647	0·657	0·666	0·673	0·686	0·696	0·705	0·720	0·733	0·746	0·757	0·776	0·791
0·80000	(7·85)	0·636	0·642	0·652	0·661	0·668	0·681	0·692	0·701	0·717	0·729	0·743	0·754	0·773	0·789
1·00000	(9·81)	0·632	0·638	0·648	0·657	0·665	0·678	0·689	0·698	0·714	0·727	0·741	0·752	0·771	0·787
1·50000	(14·7)	0·625	0·631	0·642	0·651	0·659	0·673	0·684	0·694	0·710	0·723	0·737	0·749	0·768	0·784
2·00000	(19·6)	0·620	0·627	0·638	0·647	0·656	0·670	0·681	0·691	0·707	0·721	0·735	0·746	0·766	0·782
3·00000	(29·4)	0·614	0·621	0·633	0·643	0·651	0·665	0·677	0·687	0·704	0·717	0·732	0·744	0·764	0·780
S	$(\Delta p/\rho l)$	125	150	200	250	300	400	500	600	800	1000	1250	1500	2000	2500

Grad'nt k.p.g.   (Equivalent) Pipe diameters in mm

Kinematic viscosity, $v = 0.50 \times 10^{-6}$ m^{2}s^{-1}; **Roughness size, $k_s = 0.0015$ mm**

E54

Kin. visc., $\nu = 0{\cdot}50\times10^{-6}$ m^{2}s^{-1};
$S = 0{\cdot}00010$ to $30{\cdot}0000$

i.e. kin. pr. grad., $\Delta p/\rho l =$
$(0{\cdot}00098)$ to (294) ms^{-2}

Roughness size, $k_s = 0{\cdot}003$ mm
This table shows values of m, as follows

m_C for Colebrook-White solutions; or,
where $\mathbf{R} \le 2000$, m_P for laminar flow

Grad'nt S	k.p.g. $(\Delta p/\rho l)$	6	8	10	12·5	15	20	25	30	40	50	60	80	100	125
0·00030	(0·00294)	3·429	2·336	1·735	1·289	1·011	*	1·018	1·000	0·976	0·961	0·951	0·938	0·931	0·925
0·00040	(0·00392)	2·969	2·023	1·503	1·116	0·875	*	0·992	0·976	0·954	0·940	0·931	0·920	0·914	0·909
0·00060	(0·00588)	2·424	1·652	1·227	0·911	*	0·979	0·958	0·943	0·925	0·913	0·905	0·896	0·891	0·887
0·00080	(0·00785)	2·100	1·431	1·063	0·789	*	0·954	0·935	0·922	0·905	0·895	0·888	0·879	0·875	0·872
0·00100	(0·00981)	1·878	1·280	0·950	*	*	0·936	0·918	0·906	0·890	0·881	0·875	0·867	0·863	0·861
0·00150	(0·0147)	1·533	1·045	0·776	*	0·929	0·905	0·889	0·878	0·865	0·857	0·852	0·846	0·843	0·842
0·00200	(0·0196)	1·328	0·905	*	0·923	0·906	0·884	0·870	0·860	0·848	0·841	0·836	0·831	0·829	0·828
0·00300	(0·0294)	1·084	0·739	*	0·890	0·875	0·856	0·843	0·835	0·825	0·819	0·816	0·812	0·811	0·811
0·00400	(0·0392)	0·939	*	0·888	0·868	0·855	0·837	0·826	0·819	0·810	0·805	0·802	0·799	0·798	0·799
0·00600	(0·0588)	0·767	*	0·857	0·839	0·827	0·812	0·803	0·796	0·789	0·785	0·783	0·781	0·781	0·782
0·00800	(0·0785)	*	0·855	0·836	0·820	0·809	0·795	0·787	0·781	0·775	0·772	0·770	0·769	0·770	0·771
0·01000	(0·0981)	*	0·838	0·820	0·806	0·796	0·783	0·775	0·770	0·765	0·762	0·761	0·760	0·761	0·763
0·01500	(0·147)	*	0·809	0·794	0·781	0·772	0·761	0·755	0·751	0·747	0·745	0·744	0·745	0·746	0·749
0·02000	(0·196)	0·812	0·790	0·776	0·764	0·756	0·747	0·741	0·738	0·734	0·733	0·733	0·734	0·736	0·739
0·03000	(0·294)	0·784	0·764	0·752	0·742	0·736	0·728	0·723	0·721	0·718	0·718	0·718	0·720	0·723	0·726
0·04000	(0·392)	0·765	0·747	0·736	0·727	0·722	0·715	0·711	0·709	0·707	0·707	0·708	0·711	0·714	0·718
0·06000	(0·588)	0·740	0·725	0·715	0·708	0·703	0·697	0·695	0·693	0·693	0·693	0·695	0·698	0·702	0·706
0·08000	(0·785)	0·723	0·710	0·701	0·695	0·690	0·686	0·684	0·683	0·683	0·684	0·686	0·690	0·694	0·699
0·10000	(0·981)	0·711	0·699	0·691	0·685	0·681	0·677	0·676	0·675	0·676	0·677	0·679	0·683	0·688	0·693
0·15000	(1·47)	0·690	0·679	0·673	0·668	0·665	0·662	0·661	0·662	0·663	0·665	0·668	0·673	0·678	0·683
0·20000	(1·96)	0·676	0·667	0·661	0·657	0·655	0·652	0·652	0·653	0·655	0·657	0·660	0·666	0·671	0·677
0·30000	(2·94)	0·658	0·650	0·645	0·642	0·640	0·639	0·640	0·641	0·644	0·647	0·650	0·656	0·662	0·668
0·40000	(3·92)	0·645	0·638	0·635	0·632	0·631	0·631	0·631	0·633	0·636	0·640	0·643	0·650	0·656	0·663
0·60000	(5·88)	0·629	0·624	0·621	0·619	0·619	0·619	0·621	0·623	0·627	0·631	0·635	0·642	0·649	0·656
0·80000	(7·85)	0·618	0·614	0·612	0·611	0·610	0·612	0·614	0·616	0·620	0·625	0·629	0·637	0·644	0·651
1·00000	(9·81)	0·611	0·607	0·605	0·604	0·604	0·606	0·608	0·611	0·616	0·621	0·625	0·633	0·640	0·648
1·50000	(14·7)	0·597	0·594	0·593	0·593	0·594	0·597	0·599	0·602	0·608	0·613	0·618	0·627	0·634	0·642
2·00000	(19·6)	0·588	0·586	0·586	0·586	0·587	0·590	0·594	0·597	0·603	0·609	0·614	0·623	0·631	0·639
3·00000	(29·4)	0·577	0·575	0·576	0·577	0·579	0·582	0·586	0·590	0·597	0·603	0·608	0·618	0·626	0·635
4·00000	(39·2)	0·569	0·569	0·569	0·571	0·573	0·577	0·582	0·585	0·593	0·599	0·605	0·615	0·623	0·632
6·00000	(58·8)	0·559	0·560	0·561	0·563	0·566	0·571	0·575	0·580	0·587	0·594	0·600	0·610	0·619	0·628
8·00000	(78·5)	0·553	0·554	0·556	0·558	0·561	0·567	0·572	0·576	0·584	0·591	0·597	0·608	0·617	0·626
10·0000	(98·1)	0·548	0·550	0·552	0·555	0·558	0·564	0·569	0·574	0·582	0·589	0·595	0·606	0·615	0·625
15·0000	(147)	0·540	0·543	0·545	0·549	0·552	0·559	0·564	0·569	0·578	0·586	0·592	0·603	0·613	0·622
20·0000	(196)	0·535	0·538	0·541	0·545	0·549	0·556	0·561	0·567	0·576	0·583	0·590	0·602	0·611	0·621
30·0000	(294)	0·529	0·533	0·536	0·541	0·545	0·552	0·558	0·563	0·573	0·581	0·588	0·599	0·609	0·619
S	$(\Delta p/\rho l)$	6	8	10	12·5	15	20	25	30	40	50	60	80	100	125

Grad'nt   k.p.g.   (Equivalent) Pipe diameters in mm

Grad'nt S	k.p.g. $(\Delta p/\rho l)$	125	150	200	250	300	400	500	600	800	1000	1250	1500	2000	2500
0·00010	(0·00098)	0·994	0·988	0·981	0·977	0·976	0·975	0·977	0·978	0·983	0·988	0·994	0·999	1·009	1·018
0·00015	(0·00147)	0·967	0·962	0·957	0·955	0·954	0·954	0·956	0·959	0·964	0·970	0·976	0·982	0·993	1·002
0·00020	(0·00196)	0·949	0·945	0·941	0·939	0·939	0·940	0·943	0·945	0·951	0·957	0·964	0·970	0·981	0·991
0·00030	(0·00294)	0·925	0·922	0·919	0·918	0·919	0·921	0·924	0·927	0·934	0·940	0·947	0·954	0·966	0·976
0·00040	(0·00392)	0·909	0·906	0·904	0·904	0·905	0·908	0·911	0·915	0·922	0·929	0·936	0·943	0·955	0·965
0·00060	(0·00588)	0·887	0·885	0·884	0·885	0·886	0·890	0·894	0·898	0·906	0·913	0·921	0·928	0·940	0·951
0·00080	(0·00785)	0·872	0·871	0·871	0·872	0·874	0·878	0·882	0·887	0·895	0·902	0·911	0·918	0·931	0·942
0·00100	(0·00981)	0·861	0·860	0·860	0·862	0·864	0·869	0·874	0·878	0·887	0·894	0·903	0·910	0·923	0·934
0·00150	(0·0147)	0·842	0·841	0·843	0·845	0·847	0·853	0·858	0·863	0·872	0·880	0·889	0·897	0·910	0·922
0·00200	(0·0196)	0·828	0·829	0·831	0·833	0·836	0·842	0·848	0·853	0·862	0·871	0·880	0·888	0·902	0·913
0·00300	(0·0294)	0·811	0·811	0·814	0·817	0·821	0·827	0·833	0·839	0·849	0·858	0·867	0·876	0·890	0·902
0·00400	(0·0392)	0·799	0·800	0·803	0·807	0·810	0·817	0·824	0·830	0·840	0·849	0·859	0·867	0·882	0·894
0·00600	(0·0588)	0·782	0·784	0·788	0·792	0·796	0·804	0·811	0·817	0·828	0·837	0·848	0·857	0·872	0·884
0·00800	(0·0785)	0·771	0·773	0·778	0·783	0·787	0·795	0·802	0·809	0·820	0·830	0·840	0·849	0·865	0·877
0·01000	(0·0981)	0·763	0·766	0·770	0·775	0·780	0·788	0·796	0·802	0·814	0·824	0·834	0·844	0·859	0·872
0·01500	(0·147)	0·749	0·752	0·757	0·763	0·768	0·777	0·784	0·791	0·804	0·814	0·825	0·834	0·851	0·864
0·02000	(0·196)	0·739	0·742	0·748	0·754	0·759	0·769	0·777	0·784	0·797	0·807	0·818	0·828	0·845	0·858
0·03000	(0·294)	0·726	0·730	0·737	0·743	0·748	0·758	0·767	0·775	0·787	0·798	0·810	0·820	0·837	0·851
0·04000	(0·392)	0·718	0·722	0·729	0·735	0·741	0·751	0·760	0·768	0·781	0·793	0·805	0·815	0·832	0·846
0·06000	(0·588)	0·706	0·710	0·718	0·725	0·731	0·742	0·752	0·760	0·774	0·785	0·797	0·808	0·826	0·840
0·08000	(0·785)	0·699	0·703	0·711	0·719	0·725	0·736	0·746	0·754	0·768	0·780	0·793	0·804	0·822	0·837
0·10000	(0·981)	0·693	0·698	0·706	0·714	0·720	0·732	0·742	0·750	0·765	0·777	0·790	0·800	0·819	0·834
0·15000	(1·47)	0·683	0·688	0·697	0·705	0·712	0·724	0·735	0·744	0·758	0·771	0·784	0·795	0·814	0·829
0·20000	(1·96)	0·677	0·682	0·692	0·700	0·707	0·720	0·730	0·739	0·754	0·767	0·780	0·792	0·811	0·827
0·30000	(2·94)	0·668	0·674	0·684	0·693	0·700	0·713	0·724	0·734	0·749	0·762	0·776	0·788	0·807	0·823
0·40000	(3·92)	0·663	0·669	0·679	0·688	0·696	0·709	0·721	0·730	0·746	0·759	0·773	0·785	0·805	0·821
0·60000	(5·88)	0·656	0·662	0·673	0·683	0·691	0·704	0·716	0·726	0·742	0·756	0·770	0·782	0·802	0·818
0·80000	(7·85)	0·651	0·658	0·669	0·679	0·687	0·701	0·713	0·723	0·739	0·753	0·768	0·780	0·800	0·816
1·00000	(9·81)	0·648	0·655	0·666	0·676	0·685	0·699	0·711	0·721	0·738	0·752	0·766	0·778	0·799	0·815
1·50000	(14·7)	0·642	0·650	0·662	0·672	0·681	0·695	0·707	0·718	0·735	0·749	0·764	0·776	0·797	0·813
2·00000	(19·6)	0·639	0·646	0·659	0·669	0·678	0·693	0·705	0·716	0·733	0·747	0·762	0·775	0·795	0·812
3·00000	(29·4)	0·635	0·642	0·655	0·666	0·675	0·690	0·703	0·713	0·731	0·745	0·760	0·773	0·794	0·811
S	$(\Delta p/\rho l)$	125	150	200	250	300	400	500	600	800	1000	1250	1500	2000	2500

Grad'nt   k.p.g.   (Equivalent) Pipe diameters in mm

Kinematic viscosity, $\nu = 0{\cdot}50\times10^{-6}$ m^{2}s^{-1} ;   **Roughness size, $k_s = 0{\cdot}003$ mm**

Kin. visc., $\nu = 0.50 \times 10^{-6}\ \mathrm{m^2 s^{-1}}$; **Roughness size, $k_s = 0.006$ mm** # E55

$S = 0.00010$ to 30.0000
i.e. kin. pr. grad., $\Delta p/\rho l =$
(0.00098) to $(294)\ \mathrm{ms^{-2}}$

This table shows values of m, as follows
m_C for Colebrook-White solutions; or,
where $\mathbf{R} \le 2000$, m_P for laminar flow

Grad'nt S	k.p.g. $(\Delta p/\rho l)$	6	8	10	12.5	15	20	25	30	40	50	60	80	100	125
0.00030	(0.00294)	3.429	2.336	1.735	1.289	1.011	*	1.020	1.001	0.977	0.963	0.953	0.940	0.933	0.927
0.00040	(0.00392)	2.969	2.023	1.503	1.116	0.875	*	0.994	0.977	0.956	0.942	0.933	0.922	0.916	0.911
0.00060	(0.00588)	2.424	1.652	1.227	0.911	*	0.981	0.960	0.945	0.927	0.915	0.907	0.898	0.893	0.890
0.00080	(0.00785)	2.100	1.431	1.063	0.789	*	0.956	0.937	0.924	0.907	0.897	0.890	0.882	0.878	0.875
0.00100	(0.00981)	1.878	1.280	0.950	*	*	0.938	0.920	0.908	0.893	0.883	0.877	0.870	0.866	0.864
0.00150	(0.0147)	1.533	1.045	0.776	*	0.932	0.907	0.891	0.881	0.868	0.860	0.855	0.849	0.846	0.845
0.00200	(0.0196)	1.328	0.905	*	0.926	0.909	0.886	0.872	0.863	0.851	0.844	0.840	0.835	0.833	0.832
0.00300	(0.0294)	1.084	0.739	*	0.893	0.878	0.859	0.846	0.838	0.828	0.823	0.819	0.816	0.815	0.815
0.00400	(0.0392)	0.939	*	0.891	0.871	0.858	0.840	0.829	0.822	0.813	0.809	0.806	0.803	0.803	0.804
0.00600	(0.0588)	0.767	*	0.860	0.843	0.831	0.816	0.806	0.800	0.793	0.790	0.788	0.786	0.787	0.788
0.00800	(0.0785)	*	0.858	0.839	0.824	0.813	0.799	0.791	0.786	0.780	0.777	0.775	0.775	0.776	0.778
0.01000	(0.0981)	*	0.842	0.824	0.810	0.800	0.787	0.780	0.775	0.770	0.767	0.766	0.766	0.768	0.770
0.01500	(0.147)	*	0.813	0.798	0.785	0.777	0.766	0.760	0.756	0.753	0.751	0.751	0.752	0.754	0.757
0.02000	(0.196)	0.817	0.794	0.781	0.769	0.762	0.752	0.747	0.744	0.741	0.740	0.740	0.742	0.745	0.748
0.03000	(0.294)	0.789	0.769	0.758	0.748	0.742	0.734	0.730	0.727	0.726	0.726	0.726	0.729	0.732	0.737
0.04000	(0.392)	0.770	0.753	0.742	0.734	0.728	0.722	0.718	0.716	0.715	0.716	0.717	0.721	0.724	0.729
0.06000	(0.588)	0.746	0.731	0.722	0.715	0.710	0.705	0.703	0.702	0.702	0.703	0.705	0.709	0.714	0.719
0.08000	(0.785)	0.730	0.717	0.709	0.703	0.699	0.694	0.693	0.692	0.693	0.695	0.697	0.702	0.707	0.712
0.10000	(0.981)	0.719	0.706	0.699	0.693	0.690	0.686	0.685	0.685	0.686	0.689	0.691	0.696	0.701	0.707
0.15000	(1.47)	0.699	0.688	0.682	0.678	0.675	0.673	0.673	0.673	0.675	0.678	0.681	0.687	0.693	0.699
0.20000	(1.96)	0.685	0.676	0.671	0.667	0.665	0.664	0.664	0.665	0.668	0.671	0.675	0.681	0.687	0.694
0.30000	(2.94)	0.668	0.660	0.656	0.654	0.653	0.652	0.653	0.655	0.659	0.663	0.666	0.674	0.680	0.687
0.40000	(3.92)	0.657	0.650	0.647	0.645	0.644	0.645	0.646	0.648	0.652	0.657	0.661	0.669	0.676	0.683
0.60000	(5.88)	0.642	0.637	0.634	0.633	0.633	0.635	0.637	0.639	0.645	0.650	0.654	0.663	0.670	0.678
0.80000	(7.85)	0.632	0.628	0.626	0.626	0.626	0.628	0.631	0.634	0.640	0.645	0.650	0.659	0.666	0.675
1.00000	(9.81)	0.625	0.622	0.621	0.620	0.621	0.624	0.627	0.630	0.636	0.642	0.647	0.656	0.664	0.672
1.50000	(14.7)	0.613	0.611	0.611	0.611	0.613	0.616	0.620	0.623	0.630	0.636	0.642	0.651	0.659	0.668
2.00000	(19.6)	0.605	0.604	0.604	0.606	0.607	0.611	0.615	0.619	0.626	0.633	0.638	0.648	0.657	0.666
3.00000	(29.4)	0.596	0.595	0.596	0.598	0.600	0.605	0.610	0.614	0.622	0.628	0.634	0.645	0.654	0.663
4.00000	(39.2)	0.589	0.590	0.591	0.593	0.596	0.601	0.606	0.611	0.619	0.626	0.632	0.643	0.652	0.661
6.00000	(58.8)	0.581	0.583	0.585	0.588	0.591	0.596	0.602	0.606	0.615	0.622	0.629	0.640	0.649	0.659
8.00000	(78.5)	0.576	0.578	0.581	0.584	0.587	0.593	0.599	0.604	0.613	0.620	0.627	0.638	0.648	0.658
10.0000	(98.1)	0.573	0.575	0.578	0.581	0.585	0.591	0.597	0.602	0.611	0.619	0.626	0.637	0.647	0.657
15.0000	(147)	0.567	0.570	0.573	0.577	0.581	0.588	0.594	0.599	0.609	0.617	0.624	0.635	0.645	0.655
20.0000	(196)	0.563	0.567	0.570	0.574	0.578	0.586	0.592	0.598	0.607	0.615	0.622	0.634	0.644	0.654
30.0000	(294)	0.559	0.563	0.567	0.571	0.575	0.583	0.590	0.595	0.605	0.614	0.621	0.633	0.643	0.653
S	$(\Delta p/\rho l)$	6	8	10	12.5	15	20	25	30	40	50	60	80	100	125

Grad'nt   k.p.g.   (Equivalent) Pipe diameters in mm

Grad'nt S	k.p.g. $(\Delta p/\rho l)$	125	150	200	250	300	400	500	600	800	1000	1250	1500	2000	2500
0.00010	(0.00098)	0.995	0.989	0.982	0.979	0.978	0.977	0.979	0.981	0.985	0.990	0.997	1.002	1.013	1.022
0.00015	(0.00147)	0.969	0.964	0.959	0.957	0.956	0.957	0.959	0.961	0.967	0.973	0.979	0.986	0.997	1.006
0.00020	(0.00196)	0.951	0.947	0.943	0.941	0.941	0.943	0.945	0.948	0.955	0.961	0.968	0.974	0.986	0.996
0.00030	(0.00294)	0.927	0.924	0.921	0.921	0.921	0.924	0.927	0.931	0.938	0.944	0.952	0.959	0.971	0.981
0.00040	(0.00392)	0.911	0.909	0.907	0.907	0.908	0.911	0.915	0.919	0.926	0.933	0.941	0.948	0.961	0.971
0.00060	(0.00588)	0.890	0.888	0.887	0.888	0.890	0.894	0.898	0.903	0.911	0.918	0.927	0.934	0.947	0.958
0.00080	(0.00785)	0.875	0.874	0.874	0.875	0.877	0.882	0.887	0.892	0.900	0.908	0.917	0.925	0.938	0.949
0.00100	(0.00981)	0.864	0.863	0.864	0.866	0.868	0.873	0.878	0.883	0.892	0.900	0.909	0.917	0.931	0.943
0.00150	(0.0147)	0.845	0.845	0.847	0.849	0.852	0.858	0.864	0.869	0.879	0.887	0.897	0.905	0.920	0.932
0.00200	(0.0196)	0.832	0.833	0.835	0.838	0.841	0.848	0.854	0.860	0.870	0.879	0.888	0.897	0.912	0.924
0.00300	(0.0294)	0.815	0.816	0.820	0.823	0.827	0.834	0.841	0.847	0.858	0.867	0.877	0.886	0.901	0.914
0.00400	(0.0392)	0.804	0.805	0.809	0.813	0.817	0.825	0.832	0.838	0.849	0.859	0.870	0.879	0.895	0.908
0.00600	(0.0588)	0.788	0.790	0.795	0.800	0.804	0.813	0.820	0.827	0.839	0.849	0.860	0.870	0.886	0.899
0.00800	(0.0785)	0.778	0.780	0.786	0.791	0.796	0.804	0.812	0.819	0.832	0.842	0.853	0.863	0.880	0.894
0.01000	(0.0981)	0.770	0.773	0.779	0.784	0.789	0.798	0.807	0.814	0.826	0.837	0.849	0.859	0.876	0.890
0.01500	(0.147)	0.757	0.760	0.767	0.773	0.778	0.788	0.797	0.804	0.818	0.829	0.841	0.851	0.869	0.883
0.02000	(0.196)	0.748	0.752	0.759	0.765	0.771	0.781	0.790	0.798	0.812	0.823	0.836	0.846	0.864	0.879
0.03000	(0.294)	0.737	0.741	0.748	0.755	0.762	0.773	0.782	0.790	0.805	0.816	0.829	0.840	0.859	0.874
0.04000	(0.392)	0.729	0.733	0.741	0.749	0.755	0.767	0.777	0.785	0.800	0.812	0.825	0.836	0.855	0.870
0.06000	(0.588)	0.719	0.724	0.733	0.740	0.747	0.759	0.770	0.779	0.794	0.806	0.820	0.831	0.850	0.866
0.08000	(0.785)	0.712	0.717	0.727	0.735	0.742	0.755	0.765	0.774	0.790	0.803	0.816	0.828	0.847	0.863
0.10000	(0.981)	0.707	0.713	0.723	0.731	0.738	0.751	0.762	0.771	0.787	0.800	0.814	0.826	0.845	0.862
0.15000	(1.47)	0.699	0.705	0.715	0.724	0.732	0.745	0.757	0.766	0.783	0.796	0.810	0.822	0.842	0.859
0.20000	(1.96)	0.694	0.700	0.711	0.720	0.728	0.742	0.753	0.763	0.780	0.793	0.808	0.820	0.840	0.857
0.30000	(2.94)	0.687	0.694	0.705	0.715	0.723	0.737	0.749	0.759	0.776	0.790	0.805	0.817	0.838	0.854
0.40000	(3.92)	0.683	0.690	0.702	0.712	0.720	0.735	0.747	0.757	0.774	0.788	0.803	0.815	0.836	0.853
0.60000	(5.88)	0.678	0.685	0.697	0.707	0.716	0.731	0.743	0.754	0.771	0.786	0.801	0.813	0.834	0.851
0.80000	(7.85)	0.675	0.682	0.694	0.705	0.714	0.729	0.741	0.752	0.770	0.784	0.799	0.812	0.833	0.850
1.00000	(9.81)	0.672	0.680	0.692	0.703	0.712	0.727	0.740	0.751	0.769	0.783	0.798	0.811	0.832	0.850
1.50000	(14.7)	0.668	0.676	0.689	0.700	0.709	0.725	0.738	0.749	0.767	0.781	0.797	0.810	0.831	0.849
2.00000	(19.6)	0.666	0.674	0.687	0.698	0.708	0.724	0.736	0.747	0.766	0.780	0.796	0.809	0.830	0.848
3.00000	(29.4)	0.663	0.671	0.685	0.696	0.706	0.722	0.735	0.746	0.764	0.779	0.795	0.808	0.830	0.847
S	$(\Delta p/\rho l)$	125	150	200	250	300	400	500	600	800	1000	1250	1500	2000	2500

Grad'nt   k.p.g.   (Equivalent) Pipe diameters in mm

Kinematic viscosity, $\nu = 0.50 \times 10^{-6}\ \mathrm{m^2 s^{-1}}$; **Roughness size, $k_s = 0.006$ mm**

E56

Kin. visc., $\nu = 0.50 \times 10^{-6}$ m^{2}s^{-1};
$S = 0.00010$ to 30.0000

i.e. kin. pr. grad., $\Delta p/\rho l =$
(0.00098) to (294) ms^{-2}

Roughness size, $k_s = 0.015$ mm
This table shows values of m, as follows

m_C for Colebrook-White solutions; or,
where $R \leq 2000$, m_P for laminar flow

Grad'nt S	k.p.g. $(\Delta p/\rho l)$	\multicolumn{14}{c}{(Equivalent) Pipe diameters in mm}
		6	8	10	12.5	15	20	25	30	40	50	60	80	100	125
0.00030	(0.00294)	3.429	2.336	1.735	1.289	1.011	*	1.024	1.006	0.982	0.967	0.957	0.945	0.938	0.933
0.00040	(0.00392)	2.969	2.023	1.503	1.116	0.875	*	0.998	0.982	0.961	0.947	0.939	0.928	0.922	0.918
0.00060	(0.00588)	2.424	1.652	1.227	0.911	*	0.986	0.965	0.951	0.932	0.921	0.914	0.905	0.900	0.897
0.00080	(0.00785)	2.100	1.431	1.063	0.789	*	0.962	0.943	0.930	0.913	0.903	0.897	0.889	0.886	0.883
0.00100	(0.00981)	1.878	1.280	0.950	*	*	0.944	0.927	0.915	0.899	0.890	0.885	0.878	0.875	0.873
0.00150	(0.0147)	1.533	1.045	0.776	*	0.938	0.914	0.899	0.888	0.875	0.868	0.863	0.858	0.856	0.855
0.00200	(0.0196)	1.328	0.905	*	0.933	0.916	0.894	0.880	0.871	0.859	0.853	0.849	0.845	0.844	0.844
0.00300	(0.0294)	1.084	0.739	*	0.901	0.886	0.867	0.855	0.848	0.838	0.833	0.830	0.828	0.827	0.828
0.00400	(0.0392)	0.939	*	0.900	0.880	0.867	0.849	0.839	0.832	0.824	0.820	0.818	0.816	0.816	0.818
0.00600	(0.0588)	0.767	*	0.870	0.853	0.841	0.826	0.817	0.812	0.805	0.802	0.801	0.801	0.802	0.804
0.00800	(0.0785)	*	0.869	0.850	0.834	0.824	0.811	0.803	0.798	0.793	0.791	0.790	0.791	0.792	0.795
0.01000	(0.0981)	*	0.853	0.835	0.821	0.812	0.800	0.793	0.789	0.784	0.782	0.782	0.783	0.786	0.789
0.01500	(0.147)	*	0.826	0.811	0.798	0.790	0.780	0.775	0.772	0.769	0.768	0.768	0.771	0.774	0.778
0.02000	(0.196)	0.830	0.808	0.794	0.783	0.776	0.768	0.763	0.761	0.759	0.759	0.760	0.763	0.766	0.771
0.03000	(0.294)	0.803	0.785	0.773	0.764	0.758	0.751	0.748	0.746	0.745	0.746	0.748	0.752	0.756	0.762
0.04000	(0.392)	0.786	0.769	0.759	0.751	0.746	0.740	0.738	0.737	0.737	0.738	0.741	0.745	0.750	0.756
0.06000	(0.588)	0.764	0.750	0.741	0.734	0.730	0.726	0.725	0.725	0.726	0.728	0.731	0.737	0.742	0.748
0.08000	(0.785)	0.749	0.737	0.729	0.724	0.720	0.717	0.716	0.717	0.719	0.722	0.725	0.731	0.737	0.744
0.10000	(0.981)	0.739	0.727	0.721	0.716	0.713	0.711	0.710	0.711	0.714	0.717	0.720	0.727	0.733	0.740
0.15000	(1.47)	0.721	0.712	0.706	0.703	0.701	0.700	0.700	0.702	0.705	0.709	0.713	0.721	0.727	0.735
0.20000	(1.96)	0.710	0.701	0.697	0.694	0.693	0.693	0.694	0.696	0.700	0.704	0.709	0.716	0.724	0.732
0.30000	(2.94)	0.695	0.688	0.685	0.683	0.683	0.684	0.686	0.688	0.693	0.698	0.703	0.711	0.719	0.727
0.40000	(3.92)	0.686	0.680	0.678	0.676	0.676	0.678	0.680	0.683	0.689	0.694	0.699	0.708	0.716	0.725
0.60000	(5.88)	0.674	0.670	0.668	0.668	0.668	0.671	0.674	0.677	0.684	0.689	0.695	0.704	0.713	0.721
0.80000	(7.85)	0.666	0.663	0.662	0.662	0.663	0.666	0.670	0.673	0.680	0.686	0.692	0.702	0.710	0.720
1.00000	(9.81)	0.661	0.658	0.658	0.658	0.660	0.663	0.667	0.671	0.678	0.684	0.690	0.700	0.709	0.718
1.50000	(14.7)	0.652	0.650	0.651	0.652	0.654	0.658	0.662	0.667	0.674	0.681	0.687	0.698	0.706	0.716
2.00000	(19.6)	0.646	0.645	0.646	0.648	0.650	0.655	0.660	0.664	0.672	0.679	0.685	0.696	0.705	0.715
3.00000	(29.4)	0.639	0.639	0.641	0.643	0.646	0.651	0.656	0.661	0.669	0.676	0.683	0.694	0.703	0.713
4.00000	(39.2)	0.635	0.636	0.637	0.640	0.643	0.649	0.654	0.659	0.668	0.675	0.682	0.693	0.702	0.712
6.00000	(58.8)	0.629	0.631	0.633	0.636	0.640	0.646	0.651	0.657	0.665	0.673	0.680	0.691	0.701	0.711
8.00000	(78.5)	0.626	0.628	0.631	0.634	0.638	0.644	0.650	0.655	0.664	0.672	0.679	0.690	0.700	0.710
10.0000	(98.1)	0.624	0.626	0.629	0.633	0.636	0.643	0.649	0.654	0.663	0.671	0.678	0.690	0.699	0.710
15.0000	(147)	0.620	0.623	0.626	0.630	0.634	0.641	0.647	0.653	0.662	0.670	0.677	0.689	0.699	0.709
20.0000	(196)	0.618	0.621	0.625	0.629	0.633	0.640	0.646	0.652	0.661	0.669	0.676	0.688	0.698	0.709
30.0000	(294)	0.615	0.619	0.622	0.627	0.631	0.638	0.645	0.650	0.660	0.668	0.676	0.688	0.698	0.708
S	$(\Delta p/\rho l)$	6	8	10	12.5	15	20	25	30	40	50	60	80	100	125

Grad'nt   k.p.g.   (Equivalent) Pipe diameters in mm

Grad'nt S	k.p.g. $(\Delta p/\rho l)$	\multicolumn{14}{c}{(Equivalent) Pipe diameters in mm}
		125	150	200	250	300	400	500	600	800	1000	1250	1500	2000	2500
0.00010	(0.00098)	0.999	0.993	0.987	0.984	0.983	0.983	0.985	0.987	0.992	0.998	1.005	1.011	1.022	1.032
0.00015	(0.00147)	0.973	0.969	0.964	0.962	0.962	0.963	0.966	0.969	0.975	0.981	0.989	0.995	1.008	1.018
0.00020	(0.00196)	0.956	0.952	0.949	0.947	0.948	0.950	0.953	0.956	0.963	0.970	0.978	0.985	0.998	1.009
0.00030	(0.00294)	0.933	0.930	0.928	0.928	0.929	0.932	0.936	0.940	0.948	0.955	0.964	0.971	0.985	0.996
0.00040	(0.00392)	0.918	0.916	0.914	0.915	0.916	0.920	0.925	0.929	0.938	0.945	0.954	0.962	0.976	0.988
0.00060	(0.00588)	0.897	0.896	0.896	0.897	0.899	0.904	0.909	0.914	0.924	0.932	0.942	0.950	0.965	0.977
0.00080	(0.00785)	0.883	0.883	0.883	0.885	0.888	0.894	0.899	0.905	0.915	0.923	0.933	0.942	0.957	0.970
0.00100	(0.00981)	0.873	0.873	0.874	0.877	0.880	0.886	0.892	0.898	0.908	0.917	0.927	0.936	0.952	0.965
0.00150	(0.0147)	0.855	0.856	0.858	0.862	0.865	0.872	0.879	0.885	0.896	0.906	0.917	0.926	0.942	0.956
0.00200	(0.0196)	0.844	0.845	0.848	0.852	0.856	0.863	0.871	0.877	0.889	0.899	0.910	0.920	0.936	0.950
0.00300	(0.0294)	0.828	0.830	0.834	0.839	0.843	0.852	0.860	0.867	0.879	0.890	0.901	0.912	0.929	0.943
0.00400	(0.0392)	0.818	0.820	0.825	0.830	0.835	0.844	0.852	0.860	0.873	0.884	0.896	0.906	0.924	0.939
0.00600	(0.0588)	0.804	0.807	0.813	0.819	0.824	0.834	0.843	0.851	0.865	0.876	0.889	0.900	0.918	0.933
0.00800	(0.0785)	0.795	0.799	0.805	0.812	0.817	0.828	0.837	0.845	0.859	0.871	0.884	0.895	0.914	0.929
0.01000	(0.0981)	0.789	0.793	0.800	0.806	0.812	0.823	0.833	0.841	0.856	0.868	0.881	0.892	0.911	0.927
0.01500	(0.147)	0.778	0.782	0.790	0.797	0.804	0.816	0.826	0.834	0.850	0.862	0.876	0.887	0.907	0.923
0.02000	(0.196)	0.771	0.775	0.784	0.792	0.799	0.811	0.821	0.830	0.846	0.859	0.872	0.884	0.904	0.920
0.03000	(0.294)	0.762	0.767	0.776	0.784	0.792	0.804	0.815	0.825	0.841	0.854	0.868	0.880	0.900	0.917
0.04000	(0.392)	0.756	0.761	0.771	0.780	0.787	0.801	0.812	0.821	0.838	0.851	0.866	0.878	0.898	0.915
0.06000	(0.588)	0.748	0.754	0.765	0.774	0.782	0.796	0.807	0.817	0.834	0.848	0.863	0.875	0.896	0.913
0.08000	(0.785)	0.744	0.750	0.761	0.770	0.779	0.793	0.804	0.815	0.832	0.846	0.861	0.873	0.894	0.911
0.10000	(0.981)	0.740	0.747	0.758	0.768	0.776	0.791	0.803	0.813	0.830	0.844	0.859	0.872	0.893	0.910
0.15000	(1.47)	0.735	0.742	0.754	0.764	0.772	0.787	0.799	0.810	0.828	0.842	0.857	0.870	0.891	0.909
0.20000	(1.96)	0.732	0.739	0.751	0.761	0.770	0.785	0.797	0.808	0.826	0.841	0.856	0.869	0.890	0.908
0.30000	(2.94)	0.727	0.735	0.747	0.758	0.767	0.782	0.795	0.806	0.824	0.839	0.854	0.867	0.889	0.907
0.40000	(3.92)	0.725	0.732	0.745	0.756	0.765	0.781	0.794	0.805	0.823	0.838	0.853	0.867	0.888	0.906
0.60000	(5.88)	0.721	0.729	0.743	0.754	0.763	0.779	0.792	0.803	0.822	0.837	0.852	0.866	0.887	0.905
0.80000	(7.85)	0.720	0.727	0.741	0.752	0.762	0.778	0.791	0.802	0.821	0.836	0.852	0.865	0.887	0.905
1.00000	(9.81)	0.718	0.726	0.740	0.751	0.761	0.777	0.790	0.802	0.820	0.835	0.851	0.864	0.886	0.904
1.50000	(14.7)	0.716	0.724	0.738	0.750	0.759	0.776	0.789	0.800	0.819	0.834	0.850	0.864	0.886	0.904
2.00000	(19.6)	0.715	0.723	0.737	0.749	0.759	0.775	0.788	0.800	0.819	0.834	0.850	0.863	0.886	0.903
3.00000	(29.4)	0.713	0.722	0.736	0.747	0.757	0.774	0.788	0.799	0.818	0.833	0.849	0.863	0.885	0.903
S	$(\Delta p/\rho l)$	125	150	200	250	300	400	500	600	800	1000	1250	1500	2000	2500

Grad'nt   k.p.g.   (Equivalent) Pipe diameters in mm

Kinematic viscosity, $\nu = 0.50 \times 10^{-6}$ m^{2}s^{-1} ;   **Roughness size, $k_s = 0.015$ mm**

Kin. visc., $\nu = 0.50\times10^{-6}\ m^2 s^{-1}$;   Roughness size, $k_s = 0.030\ mm$

$S = 0.00010$ to 30.0000

This table shows values of m, as follows

i.e. kin. pr. grad., $\Delta p/\rho l = (0.00098)$ to $(294)\ ms^{-2}$

m_C for Colebrook-White solutions; or, where $R \leq 2000$, m_P for laminar flow

Grad'nt S	k.p.g. $(\Delta p/\rho l)$	6	8	10	12.5	15	20	25	30	40	50	60	80	100	125
0.00030	(0.00294)	3.429	2.336	1.735	1.289	1.011	*	1.031	1.013	0.989	0.975	0.965	0.954	0.947	0.942
0.00040	(0.00392)	2.969	2.023	1.503	1.116	0.875	*	1.006	0.990	0.969	0.956	0.947	0.937	0.932	0.928
0.00060	(0.00588)	2.424	1.652	1.227	0.911	*	0.994	0.974	0.960	0.942	0.931	0.924	0.915	0.911	0.909
0.00080	(0.00785)	2.100	1.431	1.063	0.789	*	0.971	0.952	0.940	0.923	0.914	0.908	0.901	0.898	0.896
0.00100	(0.00981)	1.878	1.280	0.950	*	*	0.954	0.937	0.925	0.910	0.902	0.896	0.890	0.888	0.887
0.00150	(0.0147)	1.533	1.045	0.776	*	0.949	0.925	0.910	0.900	0.888	0.881	0.876	0.872	0.871	0.871
0.00200	(0.0196)	1.328	0.905	*	0.944	0.927	0.906	0.892	0.883	0.873	0.867	0.863	0.860	0.860	0.860
0.00300	(0.0294)	1.084	0.739	*	0.914	0.899	0.881	0.869	0.862	0.853	0.849	0.846	0.845	0.845	0.847
0.00400	(0.0392)	0.939	*	0.913	0.894	0.881	0.864	0.854	0.848	0.840	0.837	0.835	0.835	0.836	0.838
0.00600	(0.0588)	0.767	*	0.885	0.868	0.857	0.843	0.834	0.829	0.824	0.821	0.821	0.821	0.824	0.827
0.00800	(0.0785)	*	0.885	0.866	0.851	0.841	0.829	0.822	0.817	0.813	0.811	0.811	0.813	0.816	0.820
0.01000	(0.0981)	*	0.870	0.853	0.839	0.830	0.818	0.812	0.809	0.805	0.804	0.804	0.807	0.810	0.814
0.01500	(0.147)	*	0.845	0.830	0.818	0.811	0.801	0.797	0.794	0.792	0.792	0.793	0.797	0.801	0.806
0.02000	(0.196)	0.850	0.828	0.815	0.805	0.798	0.790	0.786	0.784	0.783	0.784	0.786	0.790	0.795	0.800
0.03000	(0.294)	0.826	0.807	0.796	0.788	0.782	0.776	0.773	0.772	0.773	0.774	0.777	0.782	0.787	0.793
0.04000	(0.392)	0.810	0.794	0.784	0.777	0.772	0.767	0.765	0.765	0.766	0.768	0.771	0.777	0.783	0.789
0.06000	(0.588)	0.791	0.777	0.768	0.762	0.759	0.756	0.755	0.755	0.757	0.760	0.764	0.770	0.777	0.784
0.08000	(0.785)	0.778	0.765	0.758	0.753	0.750	0.748	0.748	0.749	0.752	0.755	0.759	0.766	0.773	0.781
0.10000	(0.981)	0.769	0.757	0.751	0.747	0.745	0.743	0.743	0.745	0.748	0.752	0.756	0.764	0.770	0.778
0.15000	(1.47)	0.754	0.744	0.739	0.736	0.735	0.734	0.736	0.737	0.742	0.746	0.751	0.759	0.766	0.775
0.20000	(1.96)	0.744	0.736	0.732	0.730	0.729	0.729	0.731	0.733	0.738	0.743	0.748	0.756	0.764	0.772
0.30000	(2.94)	0.732	0.726	0.723	0.721	0.721	0.722	0.725	0.728	0.733	0.739	0.744	0.753	0.761	0.770
0.40000	(3.92)	0.725	0.719	0.717	0.716	0.716	0.718	0.721	0.724	0.730	0.736	0.741	0.751	0.759	0.768
0.60000	(5.88)	0.715	0.711	0.710	0.710	0.710	0.713	0.717	0.720	0.727	0.733	0.738	0.748	0.757	0.766
0.80000	(7.85)	0.709	0.706	0.705	0.706	0.707	0.710	0.714	0.718	0.724	0.731	0.737	0.747	0.755	0.765
1.00000	(9.81)	0.705	0.703	0.702	0.703	0.704	0.708	0.712	0.716	0.723	0.729	0.735	0.746	0.754	0.764
1.50000	(14.7)	0.699	0.697	0.697	0.698	0.700	0.705	0.709	0.713	0.721	0.727	0.733	0.744	0.753	0.762
2.00000	(19.6)	0.695	0.694	0.694	0.696	0.698	0.702	0.707	0.711	0.719	0.726	0.732	0.743	0.752	0.762
3.00000	(29.4)	0.690	0.689	0.690	0.692	0.695	0.700	0.705	0.709	0.717	0.724	0.731	0.742	0.751	0.761
4.00000	(39.2)	0.687	0.687	0.688	0.690	0.693	0.698	0.703	0.708	0.716	0.724	0.730	0.741	0.750	0.760
6.00000	(58.8)	0.683	0.684	0.685	0.688	0.691	0.697	0.702	0.707	0.715	0.722	0.729	0.740	0.749	0.759
8.00000	(78.5)	0.681	0.682	0.684	0.687	0.690	0.695	0.701	0.706	0.714	0.722	0.728	0.740	0.749	0.759
10.0000	(98.1)	0.679	0.680	0.683	0.686	0.689	0.695	0.700	0.705	0.714	0.721	0.728	0.739	0.749	0.759
15.0000	(147)	0.677	0.678	0.681	0.684	0.687	0.693	0.699	0.704	0.713	0.721	0.727	0.739	0.748	0.758
20.0000	(196)	0.675	0.677	0.680	0.683	0.686	0.693	0.698	0.703	0.712	0.720	0.727	0.738	0.748	0.758
30.0000	(294)	0.673	0.676	0.678	0.682	0.685	0.692	0.698	0.703	0.712	0.720	0.726	0.738	0.748	0.758
S	$(\Delta p/\rho l)$	6	8	10	12.5	15	20	25	30	40	50	60	80	100	125

Grad'nt   k.p.g.   (Equivalent) Pipe diameters in mm

Grad'nt S	k.p.g. $(\Delta p/\rho l)$	125	150	200	250	300	400	500	600	800	1000	1250	1500	2000	2500
0.00010	(0.00098)	1.005	1.000	0.994	0.992	0.991	0.992	0.994	0.997	1.003	1.009	1.017	1.024	1.037	1.048
0.00015	(0.00147)	0.981	0.976	0.972	0.971	0.971	0.973	0.976	0.980	0.987	0.994	1.003	1.010	1.024	1.035
0.00020	(0.00196)	0.964	0.961	0.958	0.957	0.958	0.961	0.965	0.969	0.977	0.985	0.993	1.001	1.015	1.027
0.00030	(0.00294)	0.942	0.940	0.938	0.939	0.940	0.945	0.949	0.954	0.963	0.972	0.981	0.989	1.004	1.017
0.00040	(0.00392)	0.928	0.926	0.926	0.927	0.929	0.934	0.939	0.944	0.954	0.963	0.973	0.982	0.997	1.010
0.00060	(0.00588)	0.909	0.908	0.909	0.911	0.914	0.920	0.926	0.932	0.942	0.952	0.963	0.972	0.988	1.002
0.00080	(0.00785)	0.896	0.896	0.898	0.901	0.904	0.911	0.917	0.924	0.935	0.945	0.956	0.966	0.982	0.996
0.00100	(0.00981)	0.887	0.887	0.889	0.893	0.897	0.904	0.911	0.918	0.929	0.940	0.951	0.961	0.978	0.992
0.00150	(0.0147)	0.871	0.872	0.876	0.880	0.884	0.893	0.900	0.907	0.920	0.931	0.943	0.953	0.971	0.986
0.00200	(0.0196)	0.860	0.862	0.867	0.871	0.876	0.885	0.894	0.901	0.914	0.926	0.938	0.949	0.967	0.982
0.00300	(0.0294)	0.847	0.849	0.855	0.861	0.866	0.876	0.885	0.893	0.907	0.919	0.931	0.943	0.961	0.977
0.00400	(0.0392)	0.838	0.841	0.847	0.854	0.859	0.870	0.879	0.888	0.902	0.914	0.927	0.939	0.958	0.974
0.00600	(0.0588)	0.827	0.831	0.838	0.845	0.851	0.862	0.872	0.881	0.896	0.909	0.922	0.934	0.954	0.970
0.00800	(0.0785)	0.820	0.824	0.832	0.839	0.846	0.858	0.868	0.877	0.892	0.905	0.919	0.931	0.951	0.968
0.01000	(0.0981)	0.814	0.819	0.827	0.835	0.842	0.854	0.865	0.874	0.890	0.903	0.917	0.929	0.949	0.966
0.01500	(0.147)	0.806	0.811	0.820	0.828	0.836	0.849	0.860	0.869	0.886	0.899	0.914	0.926	0.947	0.963
0.02000	(0.196)	0.800	0.806	0.815	0.824	0.832	0.845	0.856	0.866	0.883	0.897	0.911	0.924	0.945	0.962
0.03000	(0.294)	0.793	0.799	0.810	0.819	0.827	0.841	0.853	0.863	0.880	0.894	0.909	0.922	0.943	0.960
0.04000	(0.392)	0.789	0.795	0.806	0.816	0.824	0.838	0.850	0.861	0.878	0.892	0.907	0.920	0.941	0.959
0.06000	(0.588)	0.784	0.790	0.802	0.812	0.820	0.835	0.847	0.858	0.875	0.890	0.905	0.918	0.940	0.957
0.08000	(0.785)	0.781	0.787	0.799	0.809	0.818	0.833	0.845	0.856	0.874	0.889	0.904	0.917	0.939	0.956
0.10000	(0.981)	0.778	0.785	0.797	0.808	0.817	0.832	0.844	0.855	0.873	0.888	0.903	0.916	0.938	0.956
0.15000	(1.47)	0.775	0.782	0.794	0.805	0.814	0.830	0.842	0.853	0.871	0.886	0.902	0.915	0.937	0.955
0.20000	(1.96)	0.772	0.780	0.793	0.803	0.813	0.828	0.841	0.852	0.871	0.886	0.901	0.915	0.937	0.954
0.30000	(2.94)	0.770	0.777	0.790	0.801	0.811	0.827	0.840	0.851	0.869	0.885	0.900	0.914	0.936	0.954
0.40000	(3.92)	0.768	0.776	0.789	0.800	0.810	0.826	0.839	0.850	0.869	0.884	0.900	0.913	0.935	0.953
0.60000	(5.88)	0.766	0.774	0.787	0.799	0.808	0.825	0.838	0.849	0.868	0.883	0.899	0.913	0.935	0.953
0.80000	(7.85)	0.765	0.773	0.786	0.798	0.808	0.824	0.837	0.849	0.867	0.883	0.899	0.912	0.935	0.953
1.00000	(9.81)	0.764	0.772	0.786	0.797	0.807	0.823	0.837	0.848	0.867	0.882	0.898	0.912	0.934	0.952
1.50000	(14.7)	0.762	0.771	0.785	0.796	0.806	0.823	0.836	0.848	0.867	0.882	0.898	0.912	0.934	0.952
2.00000	(19.6)	0.762	0.770	0.784	0.796	0.806	0.822	0.836	0.847	0.866	0.882	0.898	0.911	0.934	0.952
3.00000	(29.4)	0.761	0.769	0.783	0.795	0.805	0.822	0.835	0.847	0.866	0.881	0.897	0.911	0.934	0.952
S	$(\Delta p/\rho l)$	125	150	200	250	300	400	500	600	800	1000	1250	1500	2000	2500

Grad'nt   k.p.g.   (Equivalent) Pipe diameters in mm

Kinematic viscosity, $\nu = 0.50\times10^{-6}\ m^2 s^{-1}$;   Roughness size, $k_s = 0.030\ mm$

E58

Kin. visc., $\nu = 0.50 \times 10^{-6}$ m^{2}s^{-1};
$S = 0.00010$ to 30.0000
i.e. kin. pr. grad., $\Delta p/\rho l =$
(0.00098) to (294) ms^{-2}

Roughness size, $k_s = 0.060$ mm
This table shows values of m, as follows
m_C for Colebrook-White solutions; or,
where $\mathbf{R} \leq 2000$, m_P for laminar flow

Grad'nt S	k.p.g. $(\Delta p/\rho l)$	6	8	10	12.5	15	20	25	30	40	50	60	80	100	125
															(Equivalent) Pipe diameters in mm
0.00030	(0.00294)	3.429	2.336	1.735	1.289	1.011	*	1.044	1.026	1.004	0.990	0.980	0.969	0.964	0.960
0.00040	(0.00392)	2.969	2.023	1.503	1.116	0.875	*	1.021	1.005	0.984	0.972	0.964	0.954	0.950	0.947
0.00060	(0.00588)	2.424	1.652	1.227	0.911	*	1.011	0.990	0.976	0.959	0.949	0.942	0.935	0.931	0.930
0.00080	(0.00785)	2.100	1.431	1.063	0.789	*	0.989	0.970	0.958	0.942	0.933	0.928	0.922	0.919	0.919
0.00100	(0.00981)	1.878	1.280	0.950	*	*	0.972	0.956	0.944	0.930	0.922	0.917	0.912	0.911	0.911
0.00150	(0.0147)	1.533	1.045	0.776	*	0.969	0.945	0.931	0.921	0.910	0.904	0.900	0.897	0.896	0.897
0.00200	(0.0196)	1.328	0.905	*	0.966	0.949	0.928	0.915	0.907	0.897	0.891	0.889	0.887	0.887	0.889
0.00300	(0.0294)	1.084	0.739	*	0.938	0.923	0.905	0.894	0.888	0.880	0.876	0.874	0.874	0.875	0.878
0.00400	(0.0392)	0.939	*	0.939	0.920	0.907	0.891	0.881	0.875	0.869	0.866	0.865	0.866	0.868	0.871
0.00600	(0.0588)	0.767	*	0.913	0.896	0.885	0.872	0.864	0.860	0.855	0.853	0.853	0.855	0.858	0.862
0.00800	(0.0785)	*	0.915	0.896	0.882	0.872	0.860	0.853	0.850	0.846	0.845	0.846	0.849	0.852	0.857
0.01000	(0.0981)	*	0.901	0.884	0.871	0.862	0.851	0.846	0.842	0.840	0.840	0.841	0.844	0.848	0.853
0.01500	(0.147)	*	0.879	0.865	0.853	0.846	0.837	0.833	0.831	0.829	0.830	0.832	0.836	0.841	0.847
0.02000	(0.196)	0.886	0.865	0.852	0.842	0.835	0.828	0.825	0.823	0.823	0.824	0.827	0.832	0.837	0.843
0.03000	(0.294)	0.866	0.847	0.836	0.828	0.822	0.817	0.815	0.814	0.815	0.817	0.820	0.826	0.832	0.838
0.04000	(0.392)	0.852	0.836	0.826	0.819	0.814	0.810	0.808	0.808	0.810	0.813	0.816	0.822	0.828	0.835
0.06000	(0.588)	0.836	0.822	0.813	0.807	0.804	0.801	0.800	0.801	0.804	0.807	0.811	0.818	0.824	0.832
0.08000	(0.785)	0.825	0.813	0.805	0.800	0.798	0.795	0.796	0.797	0.800	0.804	0.808	0.815	0.822	0.830
0.10000	(0.981)	0.818	0.806	0.800	0.795	0.793	0.792	0.792	0.793	0.797	0.801	0.805	0.813	0.820	0.828
0.15000	(1.47)	0.806	0.796	0.791	0.788	0.786	0.786	0.787	0.788	0.793	0.797	0.802	0.810	0.818	0.826
0.20000	(1.96)	0.799	0.790	0.785	0.783	0.782	0.782	0.783	0.785	0.790	0.795	0.800	0.808	0.816	0.825
0.30000	(2.94)	0.789	0.782	0.779	0.777	0.776	0.777	0.779	0.782	0.787	0.792	0.797	0.806	0.814	0.823
0.40000	(3.92)	0.784	0.778	0.775	0.773	0.773	0.774	0.777	0.780	0.785	0.791	0.796	0.805	0.813	0.822
0.60000	(5.88)	0.777	0.772	0.769	0.769	0.769	0.771	0.774	0.777	0.783	0.789	0.794	0.803	0.812	0.821
0.80000	(7.85)	0.773	0.768	0.766	0.766	0.766	0.769	0.772	0.775	0.782	0.787	0.793	0.802	0.811	0.820
1.00000	(9.81)	0.770	0.766	0.764	0.764	0.765	0.767	0.771	0.774	0.781	0.787	0.792	0.802	0.810	0.819
1.50000	(14.7)	0.765	0.762	0.761	0.761	0.762	0.765	0.769	0.772	0.779	0.785	0.791	0.801	0.809	0.818
2.00000	(19.6)	0.762	0.760	0.759	0.759	0.761	0.764	0.768	0.771	0.778	0.784	0.790	0.800	0.809	0.818
3.00000	(29.4)	0.759	0.757	0.756	0.757	0.759	0.762	0.766	0.770	0.777	0.783	0.789	0.799	0.808	0.817
4.00000	(39.2)	0.757	0.755	0.755	0.756	0.758	0.761	0.765	0.769	0.776	0.783	0.789	0.799	0.808	0.817
6.00000	(58.8)	0.755	0.753	0.753	0.754	0.756	0.760	0.764	0.768	0.776	0.782	0.788	0.798	0.807	0.817
8.00000	(78.5)	0.753	0.752	0.752	0.753	0.755	0.759	0.764	0.768	0.775	0.782	0.788	0.798	0.807	0.816
10.0000	(98.1)	0.752	0.751	0.751	0.753	0.755	0.759	0.763	0.767	0.775	0.782	0.788	0.798	0.807	0.816
15.0000	(147)	0.751	0.750	0.750	0.752	0.754	0.758	0.763	0.767	0.774	0.781	0.787	0.798	0.806	0.816
20.0000	(196)	0.750	0.749	0.750	0.751	0.753	0.758	0.762	0.766	0.774	0.781	0.787	0.797	0.806	0.816
30.0000	(294)	0.748	0.748	0.749	0.751	0.753	0.757	0.762	0.766	0.774	0.781	0.787	0.797	0.806	0.816
S	$(\Delta p/\rho l)$	6	8	10	12.5	15	20	25	30	40	50	60	80	100	125

Grad'nt k.p.g. (Equivalent) Pipe diameters in mm

Grad'nt S	k.p.g. $(\Delta p/\rho l)$	125	150	200	250	300	400	500	600	800	1000	1250	1500	2000	2500
0.00010	(0.00098)	1.018	1.013	1.008	1.006	1.006	1.008	1.011	1.015	1.022	1.030	1.038	1.046	1.061	1.073
0.00015	(0.00147)	0.995	0.991	0.988	0.987	0.988	0.991	0.995	1.000	1.009	1.017	1.026	1.035	1.050	1.063
0.00020	(0.00196)	0.979	0.977	0.975	0.975	0.976	0.981	0.985	0.990	1.000	1.009	1.019	1.028	1.044	1.057
0.00030	(0.00294)	0.960	0.958	0.957	0.959	0.961	0.967	0.973	0.978	0.989	0.998	1.009	1.019	1.035	1.049
0.00040	(0.00392)	0.947	0.946	0.946	0.948	0.951	0.958	0.964	0.970	0.982	0.992	1.003	1.013	1.030	1.044
0.00060	(0.00588)	0.930	0.930	0.932	0.935	0.939	0.946	0.953	0.960	0.972	0.983	0.995	1.005	1.023	1.038
0.00080	(0.00785)	0.919	0.919	0.922	0.926	0.930	0.939	0.947	0.954	0.967	0.978	0.990	1.001	1.019	1.035
0.00100	(0.00981)	0.911	0.912	0.916	0.920	0.925	0.933	0.942	0.949	0.963	0.974	0.987	0.998	1.016	1.032
0.00150	(0.0147)	0.897	0.899	0.904	0.910	0.915	0.925	0.934	0.942	0.956	0.968	0.981	0.992	1.012	1.028
0.00200	(0.0196)	0.889	0.891	0.897	0.903	0.909	0.919	0.929	0.937	0.952	0.964	0.977	0.989	1.009	1.025
0.00300	(0.0294)	0.878	0.881	0.888	0.895	0.901	0.912	0.922	0.931	0.946	0.959	0.973	0.985	1.005	1.022
0.00400	(0.0392)	0.871	0.875	0.882	0.890	0.896	0.908	0.918	0.927	0.943	0.956	0.970	0.983	1.003	1.020
0.00600	(0.0588)	0.862	0.867	0.875	0.883	0.890	0.903	0.913	0.923	0.939	0.953	0.967	0.980	1.000	1.017
0.00800	(0.0785)	0.857	0.862	0.871	0.879	0.886	0.899	0.911	0.920	0.937	0.951	0.965	0.978	0.999	1.016
0.01000	(0.0981)	0.853	0.858	0.868	0.876	0.884	0.897	0.908	0.918	0.935	0.949	0.964	0.976	0.998	1.015
0.01500	(0.147)	0.847	0.853	0.863	0.872	0.880	0.893	0.905	0.915	0.932	0.947	0.962	0.974	0.996	1.013
0.02000	(0.196)	0.843	0.849	0.860	0.869	0.877	0.891	0.903	0.913	0.931	0.945	0.960	0.973	0.995	1.012
0.03000	(0.294)	0.838	0.845	0.856	0.865	0.874	0.888	0.901	0.911	0.929	0.943	0.959	0.972	0.993	1.011
0.04000	(0.392)	0.835	0.842	0.853	0.863	0.872	0.887	0.899	0.910	0.928	0.942	0.958	0.971	0.993	1.010
0.06000	(0.588)	0.832	0.839	0.851	0.861	0.870	0.885	0.897	0.908	0.926	0.941	0.957	0.970	0.992	1.010
0.08000	(0.785)	0.830	0.837	0.849	0.859	0.868	0.884	0.896	0.907	0.925	0.940	0.956	0.969	0.991	1.009
0.10000	(0.981)	0.828	0.835	0.848	0.858	0.867	0.883	0.895	0.906	0.925	0.940	0.955	0.969	0.991	1.009
0.15000	(1.47)	0.826	0.833	0.846	0.857	0.866	0.881	0.894	0.905	0.924	0.939	0.955	0.968	0.990	1.008
0.20000	(1.96)	0.825	0.832	0.845	0.856	0.865	0.881	0.894	0.905	0.923	0.938	0.954	0.968	0.990	1.008
0.30000	(2.94)	0.823	0.830	0.843	0.854	0.864	0.880	0.893	0.904	0.923	0.938	0.954	0.967	0.989	1.008
0.40000	(3.92)	0.822	0.829	0.843	0.854	0.863	0.879	0.892	0.904	0.922	0.937	0.953	0.967	0.989	1.007
0.60000	(5.88)	0.821	0.828	0.842	0.853	0.862	0.878	0.892	0.903	0.922	0.937	0.953	0.967	0.989	1.007
0.80000	(7.85)	0.820	0.828	0.841	0.852	0.862	0.878	0.891	0.903	0.921	0.937	0.953	0.966	0.989	1.007
1.00000	(9.81)	0.819	0.827	0.841	0.852	0.862	0.878	0.891	0.902	0.921	0.937	0.953	0.966	0.989	1.007
1.50000	(14.7)	0.818	0.826	0.840	0.851	0.861	0.877	0.891	0.902	0.921	0.936	0.952	0.966	0.988	1.007
2.00000	(19.6)	0.818	0.826	0.840	0.851	0.861	0.877	0.890	0.902	0.921	0.936	0.952	0.966	0.988	1.007
3.00000	(29.4)	0.817	0.826	0.839	0.851	0.860	0.877	0.890	0.902	0.921	0.936	0.952	0.966	0.988	1.006
S	$(\Delta p/\rho l)$	125	150	200	250	300	400	500	600	800	1000	1250	1500	2000	2500

Grad'nt k.p.g. (Equivalent) Pipe diameters in mm

Kinematic viscosity, $\nu = 0.50 \times 10^{-6}$ m^{2}s^{-1} ;   **Roughness size, $k_s = 0.060$ mm**

Kin. visc., $\nu = 0.50 \times 10^{-6}$ m^{2}s^{-1};
$S = 0.00010$ to 30.0000
i.e. kin. pr. grad., $\Delta p/\rho l =$
(0.00098) to (294) ms^{-2}

Roughness size, $k_s = 0.150$ mm
This table shows values of m, as follows
m_C for Colebrook-White solutions; or,
where $\mathbf{R} \leq 2000$, m_P for laminar flow

Grad'nt S	k.p.g. $(\Delta p/\rho l)$	(Equivalent) Pipe diameters in mm													
		6	8	10	12.5	15	20	25	30	40	50	60	80	100	125
0.00030	(0.00294)	3.429	2.336	1.735	1.289	1.011	*	1.082	1.064	1.042	1.029	1.020	1.010	1.006	1.003
0.00040	(0.00392)	2.969	2.023	1.503	1.116	0.875	*	1.061	1.045	1.025	1.014	1.006	0.998	0.995	0.993
0.00060	(0.00588)	2.424	1.652	1.227	0.911	*	1.054	1.034	1.021	1.004	0.995	0.989	0.983	0.980	0.980
0.00080	(0.00785)	2.100	1.431	1.063	0.789	*	1.035	1.017	1.005	0.991	0.982	0.978	0.973	0.972	0.972
0.00100	(0.00981)	1.878	1.280	0.950	*	*	1.022	1.005	0.994	0.981	0.974	0.969	0.966	0.965	0.966
0.00150	(0.0147)	1.533	1.045	0.776	*	1.023	0.999	0.985	0.976	0.965	0.959	0.956	0.954	0.955	0.957
0.00200	(0.0196)	1.328	0.905	*	*	1.006	0.985	0.972	0.964	0.955	0.950	0.948	0.947	0.948	0.951
0.00300	(0.0294)	1.084	0.739	*	1.000	0.985	0.967	0.956	0.950	0.942	0.939	0.938	0.938	0.940	0.944
0.00400	(0.0392)	0.939	*	*	0.985	0.972	0.956	0.946	0.940	0.934	0.932	0.932	0.933	0.936	0.940
0.00600	(0.0588)	0.767	*	0.984	0.967	0.955	0.941	0.934	0.929	0.925	0.924	0.924	0.926	0.930	0.934
0.00800	(0.0785)	*	*	0.971	0.955	0.945	0.933	0.926	0.922	0.919	0.918	0.919	0.922	0.926	0.931
0.01000	(0.0981)	*	0.980	0.962	0.947	0.938	0.927	0.921	0.917	0.915	0.914	0.915	0.919	0.923	0.929
0.01500	(0.147)	*	0.963	0.947	0.934	0.926	0.917	0.912	0.909	0.908	0.909	0.910	0.914	0.919	0.925
0.02000	(0.196)	*	0.952	0.937	0.926	0.919	0.910	0.906	0.905	0.904	0.905	0.907	0.912	0.917	0.923
0.03000	(0.294)	0.960	0.939	0.926	0.916	0.910	0.903	0.900	0.899	0.899	0.900	0.903	0.908	0.914	0.920
0.04000	(0.392)	0.951	0.931	0.919	0.910	0.904	0.898	0.896	0.895	0.896	0.898	0.900	0.906	0.912	0.919
0.06000	(0.588)	0.939	0.921	0.911	0.903	0.898	0.893	0.891	0.891	0.892	0.895	0.898	0.904	0.910	0.917
0.08000	(0.785)	0.931	0.915	0.905	0.898	0.894	0.890	0.888	0.888	0.890	0.893	0.896	0.902	0.908	0.915
0.10000	(0.981)	0.926	0.911	0.902	0.895	0.891	0.887	0.886	0.886	0.888	0.891	0.895	0.901	0.907	0.915
0.15000	(1.47)	0.918	0.904	0.896	0.890	0.887	0.884	0.883	0.883	0.886	0.889	0.893	0.900	0.906	0.913
0.20000	(1.96)	0.913	0.900	0.893	0.887	0.884	0.881	0.881	0.882	0.884	0.888	0.892	0.899	0.905	0.913
0.30000	(2.94)	0.907	0.895	0.888	0.883	0.881	0.879	0.879	0.880	0.883	0.886	0.890	0.897	0.904	0.912
0.40000	(3.92)	0.904	0.892	0.886	0.881	0.879	0.877	0.877	0.878	0.882	0.885	0.889	0.897	0.904	0.911
0.60000	(5.88)	0.899	0.889	0.883	0.879	0.877	0.875	0.876	0.877	0.880	0.884	0.888	0.896	0.903	0.911
0.80000	(7.85)	0.897	0.887	0.881	0.877	0.875	0.874	0.875	0.876	0.880	0.884	0.888	0.895	0.902	0.910
1.00000	(9.81)	0.895	0.885	0.880	0.876	0.874	0.873	0.874	0.875	0.879	0.883	0.887	0.895	0.902	0.910
1.50000	(14.7)	0.892	0.883	0.878	0.874	0.873	0.872	0.873	0.874	0.878	0.883	0.887	0.895	0.902	0.910
2.00000	(19.6)	0.891	0.882	0.877	0.873	0.872	0.871	0.872	0.874	0.878	0.882	0.886	0.894	0.901	0.909
3.00000	(29.4)	0.889	0.880	0.875	0.872	0.871	0.870	0.871	0.873	0.877	0.882	0.886	0.894	0.901	0.909
4.00000	(39.2)	0.887	0.879	0.874	0.871	0.870	0.870	0.871	0.873	0.877	0.881	0.886	0.894	0.901	0.909
6.00000	(58.8)	0.886	0.878	0.873	0.871	0.869	0.869	0.870	0.872	0.877	0.881	0.885	0.893	0.901	0.909
8.00000	(78.5)	0.885	0.877	0.873	0.870	0.869	0.869	0.870	0.872	0.876	0.881	0.885	0.893	0.900	0.909
10.0000	(98.1)	0.885	0.877	0.872	0.870	0.869	0.869	0.870	0.872	0.876	0.881	0.885	0.893	0.900	0.908
15.0000	(147)	0.884	0.876	0.872	0.869	0.868	0.868	0.870	0.871	0.876	0.880	0.885	0.893	0.900	0.908
20.0000	(196)	0.883	0.875	0.871	0.869	0.868	0.868	0.869	0.871	0.876	0.880	0.885	0.893	0.900	0.908
30.0000	(294)	0.882	0.875	0.871	0.868	0.868	0.868	0.869	0.871	0.876	0.880	0.885	0.893	0.900	0.908
S	$(\Delta p/\rho l)$	6	8	10	12.5	15	20	25	30	40	50	60	80	100	125

Grad'nt S	k.p.g. $(\Delta p/\rho l)$	(Equivalent) Pipe diameters in mm													
		125	150	200	250	300	400	500	600	800	1000	1250	1500	2000	2500
0.00010	(0.00098)	1.051	1.047	1.043	1.043	1.044	1.048	1.052	1.057	1.067	1.076	1.087	1.096	1.113	1.127
0.00015	(0.00147)	1.031	1.029	1.027	1.028	1.030	1.035	1.041	1.047	1.058	1.068	1.079	1.089	1.106	1.121
0.00020	(0.00196)	1.019	1.017	1.017	1.018	1.021	1.027	1.034	1.040	1.052	1.062	1.074	1.084	1.102	1.117
0.00030	(0.00294)	1.003	1.002	1.003	1.006	1.010	1.017	1.025	1.032	1.044	1.055	1.068	1.078	1.097	1.112
0.00040	(0.00392)	0.993	0.993	0.995	0.999	1.003	1.011	1.019	1.026	1.040	1.051	1.064	1.075	1.094	1.110
0.00060	(0.00588)	0.980	0.981	0.985	0.989	0.994	1.003	1.012	1.020	1.034	1.046	1.059	1.070	1.090	1.106
0.00080	(0.00785)	0.972	0.974	0.978	0.983	0.989	0.998	1.008	1.016	1.030	1.043	1.056	1.068	1.088	1.104
0.00100	(0.00981)	0.966	0.968	0.974	0.979	0.985	0.995	1.004	1.013	1.028	1.040	1.054	1.066	1.086	1.103
0.00150	(0.0147)	0.957	0.960	0.966	0.972	0.978	0.990	1.000	1.008	1.024	1.037	1.051	1.063	1.083	1.100
0.00200	(0.0196)	0.951	0.954	0.961	0.968	0.975	0.986	0.997	1.006	1.021	1.035	1.049	1.061	1.082	1.099
0.00300	(0.0294)	0.944	0.948	0.956	0.963	0.970	0.982	0.993	1.002	1.019	1.032	1.047	1.059	1.080	1.097
0.00400	(0.0392)	0.940	0.944	0.952	0.960	0.967	0.980	0.991	1.000	1.017	1.031	1.045	1.058	1.079	1.096
0.00600	(0.0588)	0.934	0.939	0.948	0.956	0.964	0.977	0.988	0.998	1.015	1.029	1.043	1.056	1.078	1.095
0.00800	(0.0785)	0.931	0.936	0.945	0.954	0.962	0.975	0.986	0.996	1.013	1.028	1.042	1.055	1.077	1.094
0.01000	(0.0981)	0.929	0.934	0.944	0.952	0.960	0.974	0.985	0.995	1.012	1.027	1.042	1.055	1.076	1.094
0.01500	(0.147)	0.925	0.931	0.941	0.950	0.958	0.972	0.984	0.994	1.011	1.025	1.041	1.054	1.075	1.093
0.02000	(0.196)	0.923	0.929	0.939	0.948	0.956	0.971	0.982	0.993	1.010	1.025	1.040	1.053	1.075	1.092
0.03000	(0.294)	0.920	0.926	0.937	0.946	0.955	0.969	0.981	0.992	1.009	1.024	1.039	1.052	1.074	1.092
0.04000	(0.392)	0.919	0.925	0.936	0.945	0.954	0.968	0.980	0.991	1.009	1.023	1.039	1.052	1.074	1.092
0.06000	(0.588)	0.917	0.923	0.934	0.944	0.953	0.967	0.980	0.990	1.008	1.023	1.038	1.051	1.073	1.091
0.08000	(0.785)	0.915	0.922	0.933	0.943	0.952	0.967	0.979	0.990	1.008	1.022	1.038	1.051	1.073	1.091
0.10000	(0.981)	0.915	0.921	0.933	0.943	0.951	0.966	0.979	0.989	1.007	1.022	1.038	1.051	1.073	1.091
0.15000	(1.47)	0.913	0.920	0.932	0.942	0.951	0.966	0.978	0.989	1.007	1.022	1.037	1.051	1.073	1.091
0.20000	(1.96)	0.913	0.919	0.931	0.941	0.950	0.965	0.978	0.988	1.007	1.021	1.037	1.050	1.072	1.090
0.30000	(2.94)	0.912	0.919	0.931	0.941	0.950	0.965	0.977	0.988	1.006	1.021	1.037	1.050	1.072	1.090
0.40000	(3.92)	0.911	0.918	0.930	0.940	0.949	0.964	0.977	0.988	1.006	1.021	1.037	1.050	1.072	1.090
0.60000	(5.88)	0.911	0.918	0.930	0.940	0.949	0.964	0.977	0.988	1.006	1.021	1.036	1.050	1.072	1.090
0.80000	(7.85)	0.910	0.917	0.929	0.940	0.949	0.964	0.977	0.987	1.006	1.021	1.036	1.050	1.072	1.090
1.00000	(9.81)	0.910	0.917	0.929	0.939	0.949	0.964	0.976	0.987	1.006	1.021	1.036	1.050	1.072	1.090
1.50000	(14.7)	0.910	0.917	0.929	0.939	0.948	0.964	0.976	0.987	1.005	1.020	1.036	1.050	1.072	1.090
2.00000	(19.6)	0.909	0.916	0.929	0.939	0.948	0.963	0.976	0.987	1.005	1.020	1.036	1.049	1.072	1.090
3.00000	(29.4)	0.909	0.916	0.928	0.939	0.948	0.963	0.976	0.987	1.005	1.020	1.036	1.049	1.072	1.090
S	$(\Delta p/\rho l)$	125	150	200	250	300	400	500	600	800	1000	1250	1500	2000	2500

Grad'nt   k.p.g.   (Equivalent) Pipe diameters in mm

Kinematic viscosity, $\nu = 0.50 \times 10^{-6}$ m^{2}s^{-1} ;   **Roughness size, $k_s = 0.150$ mm**

E60

Kin. visc., $\nu = 0.50 \times 10^{-6}$ m^{2}s^{-1}; $S = 0.00010$ to 30.0000

i.e. kin. pr. grad., $\Delta p/\rho l =$ (0.00098) to (294) ms^{-2}

Roughness size, $k_s = 0.30$ mm

This table shows values of m, as follows

m_C for Colebrook-White solutions; or, where **R** ≤ 2000, m_P for laminar flow

Grad'nt	k.p.g.	(Equivalent) Pipe diameters in mm													
S	$(\Delta p/\rho l)$	6	8	10	12.5	15	20	25	30	40	50	60	80	100	125
0.00030	(0.00294)	3.429	2.336	1.735	1.289	1.011	*	1.136	1.118	1.096	1.083	1.075	1.065	1.061	1.059
0.00040	(0.00392)	2.969	2.023	1.503	1.116	0.875	*	1.118	1.102	1.082	1.071	1.063	1.056	1.052	1.051
0.00060	(0.00588)	2.424	1.652	1.227	0.911	*	1.117	1.096	1.082	1.065	1.056	1.050	1.044	1.042	1.042
0.00080	(0.00785)	2.100	1.431	1.063	0.789	*	1.101	1.082	1.070	1.054	1.046	1.041	1.036	1.035	1.036
0.00100	(0.00981)	1.878	1.280	0.950	*	*	1.090	1.072	1.061	1.047	1.039	1.035	1.031	1.031	1.032
0.00150	(0.0147)	1.533	1.045	0.776	*	1.097	1.072	1.056	1.046	1.035	1.029	1.025	1.023	1.023	1.025
0.00200	(0.0196)	1.328	0.905	*	*	1.084	1.060	1.046	1.037	1.027	1.022	1.019	1.018	1.019	1.021
0.00300	(0.0294)	1.084	0.739	*	1.083	1.067	1.046	1.034	1.027	1.018	1.014	1.012	1.012	1.013	1.016
0.00400	(0.0392)	0.939	*	*	1.072	1.056	1.038	1.027	1.020	1.012	1.009	1.008	1.008	1.010	1.014
0.00600	(0.0588)	0.767	*	1.077	1.057	1.044	1.027	1.018	1.012	1.006	1.003	1.002	1.003	1.006	1.010
0.00800	(0.0785)	*	*	1.067	1.048	1.036	1.021	1.012	1.007	1.001	0.999	0.999	1.001	1.004	1.008
0.01000	(0.0981)	*	*	1.060	1.042	1.031	1.016	1.008	1.003	0.998	0.997	0.997	0.999	1.002	1.006
0.01500	(0.147)	*	1.069	1.049	1.033	1.022	1.009	1.002	0.998	0.994	0.993	0.993	0.996	0.999	1.004
0.02000	(0.196)	*	1.061	1.042	1.027	1.017	1.005	0.998	0.995	0.991	0.990	0.991	0.994	0.998	1.003
0.03000	(0.294)	1.079	1.051	1.034	1.020	1.010	1.000	0.994	0.991	0.988	0.988	0.989	0.992	0.996	1.001
0.04000	(0.392)	1.072	1.045	1.028	1.015	1.007	0.996	0.991	0.988	0.986	0.986	0.987	0.991	0.995	1.000
0.06000	(0.588)	1.063	1.038	1.022	1.010	1.002	0.993	0.988	0.985	0.983	0.984	0.985	0.989	0.993	0.999
0.08000	(0.785)	1.058	1.033	1.019	1.007	0.999	0.990	0.986	0.984	0.982	0.983	0.984	0.988	0.992	0.998
0.10000	(0.981)	1.054	1.030	1.016	1.005	0.997	0.989	0.985	0.982	0.981	0.982	0.983	0.987	0.992	0.998
0.15000	(1.47)	1.048	1.026	1.012	1.001	0.994	0.986	0.982	0.980	0.979	0.980	0.982	0.986	0.991	0.997
0.20000	(1.96)	1.044	1.023	1.010	0.999	0.993	0.985	0.981	0.979	0.979	0.980	0.981	0.986	0.991	0.996
0.30000	(2.94)	1.040	1.020	1.007	0.997	0.991	0.983	0.980	0.978	0.977	0.979	0.980	0.985	0.990	0.996
0.40000	(3.92)	1.038	1.018	1.005	0.996	0.989	0.982	0.979	0.977	0.977	0.978	0.980	0.985	0.990	0.996
0.60000	(5.88)	1.035	1.015	1.003	0.994	0.988	0.981	0.978	0.976	0.976	0.977	0.979	0.984	0.989	0.995
0.80000	(7.85)	1.033	1.014	1.002	0.993	0.987	0.980	0.977	0.976	0.976	0.977	0.979	0.984	0.989	0.995
1.00000	(9.81)	1.032	1.013	1.001	0.992	0.986	0.980	0.977	0.975	0.975	0.977	0.979	0.984	0.989	0.995
1.50000	(14.7)	1.030	1.011	1.000	0.991	0.985	0.979	0.976	0.975	0.975	0.976	0.978	0.983	0.988	0.995
2.00000	(19.6)	1.029	1.010	0.999	0.990	0.985	0.978	0.976	0.974	0.974	0.976	0.978	0.983	0.988	0.994
3.00000	(29.4)	1.027	1.009	0.998	0.989	0.984	0.978	0.975	0.974	0.974	0.976	0.978	0.983	0.988	0.994
4.00000	(39.2)	1.026	1.008	0.998	0.989	0.984	0.978	0.975	0.974	0.974	0.975	0.978	0.983	0.988	0.994
6.00000	(58.8)	1.025	1.008	0.997	0.988	0.983	0.977	0.974	0.973	0.974	0.975	0.978	0.983	0.988	0.994
8.00000	(78.5)	1.025	1.007	0.997	0.988	0.983	0.977	0.974	0.973	0.973	0.975	0.977	0.982	0.988	0.994
10.0000	(98.1)	1.025	1.007	0.996	0.988	0.983	0.977	0.974	0.973	0.973	0.975	0.977	0.982	0.988	0.994
15.0000	(147)	1.024	1.006	0.996	0.988	0.982	0.976	0.974	0.973	0.973	0.975	0.977	0.982	0.988	0.994
20.0000	(196)	1.024	1.006	0.996	0.987	0.982	0.976	0.974	0.973	0.973	0.975	0.977	0.982	0.987	0.994
30.0000	(294)	1.023	1.006	0.995	0.987	0.982	0.976	0.974	0.973	0.973	0.975	0.977	0.982	0.987	0.994
S	$(\Delta p/\rho l)$	6	8	10	12.5	15	20	25	30	40	50	60	80	100	125
Grad'nt	k.p.g.	(Equivalent) Pipe diameters in mm													

Grad'nt	k.p.g.	(Equivalent) Pipe diameters in mm													
S	$(\Delta p/\rho l)$	125	150	200	250	300	400	500	600	800	1000	1250	1500	2000	2500
0.00010	(0.00098)	1.096	1.093	1.090	1.091	1.093	1.098	1.104	1.110	1.121	1.131	1.142	1.153	1.170	1.185
0.00015	(0.00147)	1.081	1.079	1.078	1.080	1.082	1.089	1.096	1.102	1.114	1.125	1.137	1.148	1.166	1.181
0.00020	(0.00196)	1.071	1.070	1.070	1.073	1.076	1.083	1.090	1.097	1.110	1.121	1.134	1.145	1.163	1.179
0.00030	(0.00294)	1.059	1.058	1.060	1.064	1.068	1.076	1.084	1.092	1.105	1.117	1.130	1.141	1.160	1.176
0.00040	(0.00392)	1.051	1.051	1.054	1.058	1.063	1.072	1.080	1.088	1.102	1.114	1.127	1.139	1.158	1.174
0.00060	(0.00588)	1.042	1.043	1.047	1.052	1.057	1.067	1.076	1.084	1.098	1.111	1.124	1.136	1.156	1.172
0.00080	(0.00785)	1.036	1.038	1.042	1.048	1.053	1.063	1.073	1.081	1.096	1.109	1.122	1.134	1.154	1.171
0.00100	(0.00981)	1.032	1.034	1.039	1.045	1.051	1.061	1.071	1.079	1.094	1.107	1.121	1.133	1.153	1.170
0.00150	(0.0147)	1.025	1.028	1.034	1.040	1.046	1.058	1.068	1.076	1.092	1.105	1.119	1.131	1.152	1.169
0.00200	(0.0196)	1.021	1.024	1.031	1.038	1.044	1.056	1.066	1.075	1.090	1.104	1.118	1.130	1.151	1.168
0.00300	(0.0294)	1.016	1.020	1.027	1.034	1.041	1.053	1.063	1.073	1.089	1.102	1.116	1.129	1.150	1.167
0.00400	(0.0392)	1.014	1.017	1.025	1.032	1.039	1.051	1.062	1.071	1.088	1.101	1.116	1.128	1.149	1.166
0.00600	(0.0588)	1.010	1.014	1.022	1.030	1.037	1.050	1.060	1.070	1.086	1.100	1.115	1.127	1.148	1.166
0.00800	(0.0785)	1.008	1.012	1.021	1.029	1.036	1.048	1.059	1.069	1.086	1.099	1.114	1.127	1.148	1.165
0.01000	(0.0981)	1.006	1.011	1.020	1.028	1.035	1.048	1.059	1.068	1.085	1.099	1.114	1.126	1.148	1.165
0.01500	(0.147)	1.004	1.009	1.018	1.026	1.033	1.046	1.058	1.067	1.084	1.098	1.113	1.126	1.147	1.164
0.02000	(0.196)	1.003	1.008	1.017	1.025	1.033	1.046	1.057	1.067	1.084	1.098	1.113	1.125	1.147	1.164
0.03000	(0.294)	1.001	1.006	1.015	1.024	1.032	1.045	1.056	1.066	1.083	1.097	1.112	1.125	1.146	1.164
0.04000	(0.392)	1.000	1.005	1.015	1.023	1.031	1.044	1.056	1.066	1.083	1.097	1.112	1.125	1.146	1.164
0.06000	(0.588)	0.999	1.004	1.014	1.022	1.030	1.044	1.055	1.065	1.082	1.097	1.112	1.124	1.146	1.163
0.08000	(0.785)	0.998	1.003	1.013	1.022	1.030	1.043	1.055	1.065	1.082	1.096	1.111	1.124	1.146	1.163
0.10000	(0.981)	0.998	1.003	1.013	1.022	1.030	1.043	1.055	1.065	1.082	1.096	1.111	1.124	1.146	1.163
0.15000	(1.47)	0.997	1.002	1.012	1.021	1.029	1.043	1.054	1.065	1.082	1.096	1.111	1.124	1.145	1.163
0.20000	(1.96)	0.996	1.002	1.012	1.021	1.029	1.043	1.054	1.064	1.082	1.096	1.111	1.124	1.145	1.163
0.30000	(2.94)	0.996	1.001	1.012	1.021	1.028	1.042	1.054	1.064	1.081	1.096	1.111	1.124	1.145	1.163
0.40000	(3.92)	0.996	1.001	1.011	1.020	1.028	1.042	1.054	1.064	1.081	1.096	1.111	1.124	1.145	1.163
0.60000	(5.88)	0.995	1.001	1.011	1.020	1.028	1.042	1.054	1.064	1.081	1.095	1.111	1.124	1.145	1.163
0.80000	(7.85)	0.995	1.001	1.011	1.020	1.028	1.042	1.054	1.064	1.081	1.095	1.110	1.123	1.145	1.163
1.00000	(9.81)	0.995	1.000	1.011	1.020	1.028	1.042	1.053	1.064	1.081	1.095	1.110	1.123	1.145	1.163
1.50000	(14.7)	0.995	1.000	1.011	1.020	1.028	1.042	1.053	1.064	1.081	1.095	1.110	1.123	1.145	1.163
2.00000	(19.6)	0.994	1.000	1.010	1.020	1.028	1.042	1.053	1.064	1.081	1.095	1.110	1.123	1.145	1.163
3.00000	(29.4)	0.994	1.000	1.010	1.019	1.027	1.041	1.053	1.063	1.081	1.095	1.110	1.123	1.145	1.163
S	$(\Delta p/\rho l)$	125	150	200	250	300	400	500	600	800	1000	1250	1500	2000	2500
Grad'nt	k.p.g.	(Equivalent) Pipe diameters in mm													

Kinematic viscosity, $\nu = 0.50 \times 10^{-6}$ m^{2}s^{-1} ;   Roughness size, $k_s = 0.30$ mm

Kin. visc., $\nu = 0.50 \times 10^{-6}\ \mathrm{m^2 s^{-1}}$;
S = 0.00010 to 3.00000

i.e. kin. pr. grad., $\Delta p/\rho l =$
(0.00098) to (29.4) $\mathrm{ms^{-2}}$

Roughness size, k_s = 0.60 mm
This table shows values of m, as follows

m_C for Colebrook-White solutions

Grad'nt S	k.p.g. $(\Delta p/\rho l)$	125	150	200	250	300	400	500	600	800	1000	1250	1500	2000	2500
		(Equivalent) Pipe diameters in mm													
0.00010	(0.00098)	1.168	1.164	1.162	1.163	1.165	1.170	1.176	1.182	1.194	1.204	1.216	1.226	1.244	1.260
0.00015	(0.00147)	1.156	1.154	1.153	1.155	1.157	1.164	1.171	1.177	1.189	1.200	1.212	1.223	1.242	1.257
0.00020	(0.00196)	1.149	1.147	1.147	1.150	1.153	1.160	1.167	1.174	1.187	1.198	1.210	1.221	1.240	1.256
0.00030	(0.00294)	1.140	1.139	1.141	1.144	1.147	1.155	1.163	1.170	1.183	1.195	1.208	1.219	1.238	1.254
0.00040	(0.00392)	1.135	1.134	1.137	1.140	1.144	1.153	1.161	1.168	1.181	1.193	1.206	1.218	1.237	1.253
0.00060	(0.00588)	1.128	1.129	1.132	1.136	1.140	1.149	1.158	1.165	1.179	1.191	1.204	1.216	1.235	1.252
0.00080	(0.00785)	1.124	1.125	1.129	1.133	1.138	1.147	1.156	1.164	1.178	1.190	1.203	1.215	1.235	1.251
0.00100	(0.00981)	1.122	1.123	1.127	1.131	1.136	1.146	1.155	1.163	1.177	1.189	1.202	1.214	1.234	1.250
0.00150	(0.0147)	1.117	1.119	1.123	1.128	1.134	1.143	1.153	1.161	1.175	1.188	1.201	1.213	1.233	1.250
0.00200	(0.0196)	1.115	1.116	1.121	1.127	1.132	1.142	1.151	1.160	1.174	1.187	1.201	1.212	1.233	1.249
0.00300	(0.0294)	1.111	1.114	1.119	1.125	1.130	1.141	1.150	1.159	1.173	1.186	1.200	1.212	1.232	1.249
0.00400	(0.0392)	1.110	1.112	1.117	1.123	1.129	1.140	1.149	1.158	1.173	1.186	1.199	1.211	1.231	1.248
0.00600	(0.0588)	1.107	1.110	1.116	1.122	1.128	1.139	1.148	1.157	1.172	1.185	1.199	1.211	1.231	1.248
0.00800	(0.0785)	1.106	1.109	1.115	1.121	1.127	1.138	1.148	1.156	1.172	1.184	1.198	1.210	1.231	1.248
0.01000	(0.0981)	1.105	1.108	1.114	1.120	1.126	1.137	1.147	1.156	1.171	1.184	1.198	1.210	1.231	1.247
0.01500	(0.147)	1.104	1.107	1.113	1.119	1.125	1.137	1.147	1.155	1.171	1.184	1.198	1.210	1.230	1.247
0.02000	(0.196)	1.103	1.106	1.112	1.119	1.125	1.136	1.146	1.155	1.170	1.183	1.197	1.210	1.230	1.247
0.03000	(0.294)	1.102	1.105	1.112	1.118	1.124	1.136	1.146	1.155	1.170	1.183	1.197	1.209	1.230	1.247
0.04000	(0.392)	1.101	1.104	1.111	1.118	1.124	1.135	1.145	1.154	1.170	1.183	1.197	1.209	1.230	1.247
0.06000	(0.588)	1.100	1.104	1.111	1.117	1.124	1.135	1.145	1.154	1.170	1.183	1.197	1.209	1.230	1.247
0.08000	(0.785)	1.100	1.103	1.110	1.117	1.123	1.135	1.145	1.154	1.169	1.183	1.197	1.209	1.230	1.247
0.10000	(0.981)	1.100	1.103	1.110	1.117	1.123	1.135	1.145	1.154	1.169	1.183	1.197	1.209	1.229	1.246
0.15000	(1.47)	1.099	1.102	1.110	1.116	1.123	1.134	1.145	1.154	1.169	1.182	1.197	1.209	1.229	1.246
0.20000	(1.96)	1.099	1.102	1.109	1.116	1.123	1.134	1.144	1.154	1.169	1.182	1.196	1.209	1.229	1.246
0.30000	(2.94)	1.098	1.102	1.109	1.116	1.122	1.134	1.144	1.153	1.169	1.182	1.196	1.209	1.229	1.246
0.40000	(3.92)	1.098	1.102	1.109	1.116	1.122	1.134	1.144	1.153	1.169	1.182	1.196	1.209	1.229	1.246
0.60000	(5.88)	1.098	1.102	1.109	1.116	1.122	1.134	1.144	1.153	1.169	1.182	1.196	1.209	1.229	1.246
0.80000	(7.85)	1.098	1.101	1.109	1.116	1.122	1.134	1.144	1.153	1.169	1.182	1.196	1.209	1.229	1.246
1.00000	(9.81)	1.098	1.101	1.109	1.116	1.122	1.134	1.144	1.153	1.169	1.182	1.196	1.208	1.229	1.246
1.50000	(14.7)	1.098	1.101	1.109	1.116	1.122	1.134	1.144	1.153	1.169	1.182	1.196	1.208	1.229	1.246
2.00000	(19.6)	1.098	1.101	1.108	1.115	1.122	1.134	1.144	1.153	1.169	1.182	1.196	1.208	1.229	1.246
3.00000	(29.4)	1.097	1.101	1.108	1.115	1.122	1.134	1.144	1.153	1.169	1.182	1.196	1.208	1.229	1.246
S $(\Delta p/\rho l)$		125	150	200	250	300	400	500	600	800	1000	1250	1500	2000	2500

Grad'nt   k.p.g.   (Equivalent) Pipe diameters in mm   Roughness size, k_s = 0.60 mm

Kin. visc., $\nu = 0.50 \times 10^{-6}\ \mathrm{m^2 s^{-1}}$;
S = 0.00010 to 3.00000

i.e. kin. pr. grad., $\Delta p/\rho l =$
(0.00098) to (29.4) $\mathrm{ms^{-2}}$

Roughness size, k_s = 1.50 mm
This table shows values of m, as follows

m_C for Colebrook-White solutions

Grad'nt S	k.p.g. $(\Delta p/\rho l)$	125	150	200	250	300	400	500	600	800	1000	1250	1500	2000	2500
		(Equivalent) Pipe diameters in mm													
0.00010	(0.00098)	1.314	1.308	1.302	1.301	1.301	1.304	1.308	1.313	1.323	1.332	1.343	1.352	1.369	1.384
0.00015	(0.00147)	1.306	1.301	1.297	1.296	1.297	1.300	1.305	1.310	1.320	1.330	1.341	1.351	1.368	1.382
0.00020	(0.00196)	1.302	1.297	1.294	1.293	1.294	1.298	1.303	1.309	1.319	1.329	1.340	1.350	1.367	1.382
0.00030	(0.00294)	1.297	1.293	1.290	1.290	1.291	1.296	1.301	1.307	1.317	1.327	1.338	1.348	1.366	1.381
0.00040	(0.00392)	1.294	1.290	1.287	1.288	1.289	1.294	1.300	1.305	1.316	1.326	1.338	1.348	1.365	1.380
0.00060	(0.00588)	1.290	1.287	1.285	1.285	1.287	1.292	1.298	1.304	1.315	1.325	1.337	1.347	1.365	1.380
0.00080	(0.00785)	1.287	1.285	1.283	1.284	1.286	1.291	1.297	1.303	1.314	1.325	1.336	1.346	1.364	1.379
0.00100	(0.00981)	1.286	1.283	1.282	1.283	1.285	1.290	1.296	1.302	1.314	1.324	1.336	1.346	1.364	1.379
0.00150	(0.0147)	1.283	1.281	1.280	1.281	1.283	1.289	1.295	1.302	1.313	1.323	1.335	1.345	1.363	1.379
0.00200	(0.0196)	1.282	1.280	1.279	1.280	1.283	1.289	1.295	1.301	1.313	1.323	1.335	1.345	1.363	1.378
0.00300	(0.0294)	1.280	1.278	1.277	1.279	1.282	1.288	1.294	1.300	1.312	1.323	1.334	1.345	1.363	1.378
0.00400	(0.0392)	1.279	1.277	1.277	1.278	1.281	1.287	1.294	1.300	1.312	1.322	1.334	1.344	1.363	1.378
0.00600	(0.0588)	1.278	1.276	1.276	1.278	1.280	1.287	1.293	1.300	1.311	1.322	1.334	1.344	1.362	1.378
0.00800	(0.0785)	1.277	1.275	1.275	1.277	1.280	1.286	1.293	1.299	1.311	1.322	1.334	1.344	1.362	1.378
0.01000	(0.0981)	1.277	1.275	1.275	1.277	1.280	1.286	1.293	1.299	1.311	1.322	1.333	1.344	1.362	1.377
0.01500	(0.147)	1.276	1.274	1.274	1.276	1.279	1.286	1.292	1.299	1.311	1.321	1.333	1.344	1.362	1.377
0.02000	(0.196)	1.275	1.274	1.274	1.276	1.279	1.285	1.292	1.299	1.311	1.321	1.333	1.344	1.362	1.377
0.03000	(0.294)	1.275	1.273	1.274	1.276	1.279	1.285	1.292	1.298	1.310	1.321	1.333	1.344	1.362	1.377
0.04000	(0.392)	1.274	1.273	1.273	1.275	1.278	1.285	1.292	1.298	1.310	1.321	1.333	1.343	1.362	1.377
0.06000	(0.588)	1.274	1.273	1.273	1.275	1.278	1.285	1.292	1.298	1.310	1.321	1.333	1.343	1.362	1.377
0.08000	(0.785)	1.274	1.272	1.273	1.275	1.278	1.285	1.291	1.298	1.310	1.321	1.333	1.343	1.362	1.377
0.10000	(0.981)	1.274	1.272	1.273	1.275	1.278	1.285	1.291	1.298	1.310	1.321	1.333	1.343	1.362	1.377
0.15000	(1.47)	1.273	1.272	1.273	1.275	1.278	1.284	1.291	1.298	1.310	1.321	1.333	1.343	1.362	1.377
0.20000	(1.96)	1.273	1.272	1.272	1.275	1.278	1.284	1.291	1.298	1.310	1.321	1.333	1.343	1.361	1.377
0.30000	(2.94)	1.273	1.272	1.272	1.275	1.278	1.284	1.291	1.298	1.310	1.321	1.333	1.343	1.361	1.377
0.40000	(3.92)	1.273	1.272	1.272	1.274	1.278	1.284	1.291	1.298	1.310	1.321	1.333	1.343	1.361	1.377
0.60000	(5.88)	1.273	1.272	1.272	1.274	1.277	1.284	1.291	1.298	1.310	1.321	1.333	1.343	1.361	1.377
0.80000	(7.85)	1.273	1.272	1.272	1.274	1.277	1.284	1.291	1.298	1.310	1.321	1.332	1.343	1.361	1.377
1.00000	(9.81)	1.273	1.271	1.272	1.274	1.277	1.284	1.291	1.298	1.310	1.321	1.332	1.343	1.361	1.377
1.50000	(14.7)	1.273	1.271	1.272	1.274	1.277	1.284	1.291	1.298	1.310	1.320	1.332	1.343	1.361	1.377
2.00000	(19.6)	1.273	1.271	1.272	1.274	1.277	1.284	1.291	1.298	1.310	1.320	1.332	1.343	1.361	1.377
3.00000	(29.4)	1.273	1.271	1.272	1.274	1.277	1.284	1.291	1.298	1.310	1.320	1.332	1.343	1.361	1.377
S $(\Delta p/\rho l)$		125	150	200	250	300	400	500	600	800	1000	1250	1500	2000	2500

Grad'nt   k.p.g.   (Equivalent) Pipe diameters in mm   Roughness size, k_s = 1.50 mm

E63

Kin. visc., $\nu = 0.60 \times 10^{-6}$ m^{2}s^{-1};
$S = 0.00010$ to 30.0000

i.e. kin. pr. grad., $\Delta p/\rho l =$
(0.00098) to (294) ms^{-2}

Roughness size, $k_s = 0.0015$ mm
This table shows values of m, as follows

m_C for Colebrook-White solutions; or,
where $R \le 2000$, m_P for laminar flow

Grad'nt S	k.p.g. $(\Delta p/\rho l)$	6	8	10	12.5	15	20	25	30	40	50	60	80	100	125
0.00030	(0.00294)	4.114	2.804	2.082	1.546	1.213	*	*	1.032	1.005	0.988	0.976	0.961	0.953	0.946
0.00040	(0.00392)	3.563	2.428	1.803	1.339	1.050	*	1.025	1.006	0.981	0.966	0.955	0.942	0.935	0.929
0.00060	(0.00588)	2.909	1.982	1.472	1.093	0.857	*	0.988	0.971	0.950	0.937	0.928	0.917	0.910	0.906
0.00080	(0.00785)	2.520	1.717	1.275	0.947	*	0.985	0.963	0.948	0.929	0.917	0.909	0.899	0.894	0.890
0.00100	(0.00981)	2.254	1.536	1.140	0.847	*	0.966	0.945	0.931	0.914	0.902	0.895	0.886	0.881	0.878
0.00150	(0.0147)	1.840	1.254	0.931	*	0.960	0.932	0.914	0.902	0.887	0.877	0.871	0.864	0.860	0.857
0.00200	(0.0196)	1.594	1.086	0.806	*	0.935	0.909	0.893	0.882	0.868	0.860	0.855	0.848	0.845	0.843
0.00300	(0.0294)	1.301	0.887	*	0.919	0.902	0.879	0.865	0.856	0.844	0.837	0.833	0.828	0.825	0.825
0.00400	(0.0392)	1.127	0.768	*	0.895	0.880	0.859	0.847	0.838	0.828	0.821	0.818	0.814	0.812	0.812
0.00600	(0.0588)	0.920	*	0.884	0.864	0.850	0.833	0.822	0.814	0.806	0.801	0.798	0.795	0.794	0.794
0.00800	(0.0785)	0.797	*	0.861	0.843	0.831	0.815	0.805	0.799	0.791	0.786	0.784	0.782	0.782	0.782
0.01000	(0.0981)	0.713	0.864	0.844	0.828	0.816	0.801	0.792	0.787	0.780	0.776	0.774	0.772	0.772	0.773
0.01500	(0.147)	*	0.833	0.815	0.801	0.791	0.778	0.771	0.766	0.760	0.757	0.756	0.755	0.756	0.758
0.02000	(0.196)	*	0.812	0.796	0.783	0.774	0.763	0.756	0.752	0.747	0.745	0.744	0.744	0.745	0.747
0.03000	(0.294)	0.807	0.785	0.771	0.759	0.752	0.742	0.736	0.733	0.729	0.728	0.728	0.729	0.730	0.733
0.04000	(0.392)	0.786	0.766	0.754	0.743	0.736	0.728	0.723	0.720	0.717	0.717	0.717	0.718	0.720	0.723
0.06000	(0.588)	0.759	0.742	0.731	0.722	0.716	0.709	0.705	0.703	0.701	0.701	0.702	0.704	0.707	0.710
0.08000	(0.785)	0.741	0.726	0.716	0.708	0.703	0.697	0.693	0.692	0.690	0.691	0.692	0.694	0.698	0.701
0.10000	(0.981)	0.728	0.713	0.704	0.697	0.693	0.687	0.684	0.683	0.682	0.683	0.684	0.687	0.691	0.695
0.15000	(1.47)	0.705	0.692	0.685	0.679	0.675	0.671	0.669	0.668	0.668	0.669	0.671	0.675	0.679	0.683
0.20000	(1.96)	0.690	0.678	0.672	0.666	0.663	0.660	0.658	0.658	0.659	0.660	0.662	0.666	0.671	0.676
0.30000	(2.94)	0.669	0.660	0.654	0.650	0.647	0.645	0.644	0.644	0.646	0.648	0.650	0.655	0.660	0.665
0.40000	(3.92)	0.656	0.647	0.642	0.639	0.637	0.635	0.635	0.635	0.637	0.640	0.642	0.648	0.653	0.658
0.60000	(5.88)	0.638	0.631	0.627	0.624	0.622	0.622	0.622	0.623	0.626	0.629	0.632	0.638	0.643	0.649
0.80000	(7.85)	0.626	0.619	0.616	0.614	0.613	0.613	0.613	0.615	0.618	0.621	0.625	0.631	0.637	0.643
1.00000	(9.81)	0.617	0.611	0.608	0.606	0.606	0.606	0.607	0.609	0.612	0.616	0.620	0.626	0.632	0.639
1.50000	(14.7)	0.601	0.597	0.595	0.594	0.593	0.594	0.596	0.598	0.603	0.607	0.611	0.618	0.624	0.631
2.00000	(19.6)	0.591	0.587	0.586	0.585	0.585	0.587	0.589	0.591	0.596	0.600	0.605	0.612	0.619	0.626
3.00000	(29.4)	0.577	0.574	0.574	0.574	0.574	0.577	0.579	0.582	0.587	0.592	0.597	0.605	0.612	0.619
4.00000	(39.2)	0.568	0.566	0.566	0.566	0.567	0.570	0.573	0.576	0.582	0.587	0.592	0.600	0.607	0.615
6.00000	(58.8)	0.556	0.555	0.555	0.556	0.558	0.561	0.565	0.568	0.574	0.580	0.585	0.594	0.602	0.610
8.00000	(78.5)	0.548	0.547	0.548	0.550	0.552	0.555	0.559	0.563	0.570	0.576	0.581	0.590	0.598	0.606
10.0000	(98.1)	0.542	0.542	0.543	0.545	0.547	0.551	0.555	0.559	0.566	0.572	0.578	0.587	0.595	0.604
15.0000	(147)	0.532	0.533	0.534	0.537	0.539	0.544	0.549	0.553	0.560	0.567	0.573	0.583	0.591	0.600
20.0000	(196)	0.525	0.527	0.529	0.531	0.534	0.539	0.544	0.549	0.557	0.563	0.569	0.580	0.588	0.597
30.0000	(294)	0.517	0.519	0.521	0.525	0.528	0.534	0.539	0.544	0.552	0.559	0.565	0.576	0.585	0.594
S	$(\Delta p/\rho l)$	6	8	10	12.5	15	20	25	30	40	50	60	80	100	125

Grad'nt k.p.g. (Equivalent) Pipe diameters in mm

Grad'nt S	k.p.g. $(\Delta p/\rho l)$	125	150	200	250	300	400	500	600	800	1000	1250	1500	2000	2500
0.00010	(0.00098)	1.018	1.011	1.003	0.998	0.996	0.994	0.995	0.996	1.000	1.004	1.009	1.014	1.024	1.032
0.00015	(0.00147)	0.990	0.984	0.977	0.974	0.973	0.972	0.973	0.975	0.980	0.985	0.990	0.996	1.006	1.015
0.00020	(0.00196)	0.971	0.966	0.960	0.958	0.957	0.957	0.959	0.961	0.966	0.972	0.978	0.983	0.994	1.003
0.00030	(0.00294)	0.946	0.942	0.938	0.936	0.936	0.937	0.939	0.942	0.948	0.954	0.960	0.966	0.977	0.987
0.00040	(0.00392)	0.929	0.925	0.922	0.921	0.921	0.923	0.926	0.929	0.935	0.941	0.948	0.955	0.966	0.976
0.00060	(0.00588)	0.906	0.903	0.901	0.901	0.902	0.904	0.908	0.911	0.918	0.925	0.932	0.939	0.950	0.960
0.00080	(0.00785)	0.890	0.888	0.886	0.887	0.888	0.892	0.895	0.899	0.907	0.913	0.921	0.928	0.940	0.950
0.00100	(0.00981)	0.878	0.876	0.876	0.876	0.878	0.882	0.886	0.890	0.898	0.905	0.913	0.920	0.932	0.942
0.00150	(0.0147)	0.857	0.857	0.857	0.858	0.860	0.865	0.869	0.874	0.882	0.890	0.898	0.905	0.918	0.928
0.00200	(0.0196)	0.843	0.843	0.844	0.846	0.848	0.853	0.858	0.863	0.871	0.879	0.888	0.895	0.908	0.919
0.00300	(0.0294)	0.825	0.825	0.826	0.829	0.832	0.837	0.843	0.848	0.857	0.865	0.874	0.882	0.895	0.906
0.00400	(0.0392)	0.812	0.812	0.814	0.817	0.821	0.827	0.832	0.838	0.847	0.855	0.864	0.872	0.886	0.897
0.00600	(0.0588)	0.794	0.795	0.798	0.802	0.805	0.812	0.818	0.824	0.834	0.842	0.852	0.860	0.874	0.885
0.00800	(0.0785)	0.782	0.784	0.787	0.791	0.795	0.802	0.809	0.814	0.825	0.833	0.843	0.851	0.865	0.877
0.01000	(0.0981)	0.773	0.775	0.779	0.783	0.787	0.795	0.801	0.807	0.818	0.827	0.836	0.845	0.859	0.871
0.01500	(0.147)	0.758	0.760	0.765	0.769	0.774	0.781	0.788	0.795	0.806	0.815	0.825	0.834	0.849	0.861
0.02000	(0.196)	0.747	0.750	0.755	0.760	0.764	0.773	0.780	0.786	0.798	0.807	0.817	0.826	0.841	0.854
0.03000	(0.294)	0.733	0.736	0.742	0.747	0.752	0.761	0.768	0.775	0.787	0.796	0.807	0.816	0.832	0.844
0.04000	(0.392)	0.723	0.727	0.733	0.738	0.743	0.752	0.760	0.767	0.779	0.789	0.800	0.809	0.825	0.838
0.06000	(0.588)	0.710	0.714	0.720	0.726	0.732	0.741	0.750	0.757	0.769	0.780	0.791	0.800	0.817	0.830
0.08000	(0.785)	0.701	0.705	0.712	0.718	0.724	0.734	0.743	0.750	0.763	0.773	0.785	0.794	0.811	0.824
0.10000	(0.981)	0.695	0.699	0.706	0.713	0.718	0.729	0.737	0.745	0.758	0.769	0.780	0.790	0.807	0.820
0.15000	(1.47)	0.683	0.688	0.695	0.702	0.708	0.719	0.728	0.736	0.749	0.760	0.772	0.782	0.800	0.814
0.20000	(1.96)	0.676	0.680	0.688	0.695	0.702	0.713	0.722	0.730	0.744	0.755	0.767	0.778	0.795	0.809
0.30000	(2.94)	0.665	0.670	0.679	0.686	0.693	0.704	0.714	0.722	0.736	0.748	0.760	0.771	0.789	0.803
0.40000	(3.92)	0.658	0.663	0.672	0.680	0.687	0.699	0.709	0.717	0.732	0.744	0.756	0.767	0.785	0.800
0.60000	(5.88)	0.649	0.655	0.664	0.672	0.679	0.692	0.702	0.711	0.725	0.738	0.751	0.762	0.780	0.795
0.80000	(7.85)	0.643	0.649	0.659	0.667	0.674	0.687	0.697	0.706	0.721	0.734	0.747	0.758	0.777	0.792
1.00000	(9.81)	0.639	0.644	0.655	0.663	0.671	0.683	0.694	0.703	0.718	0.731	0.744	0.756	0.775	0.790
1.50000	(14.7)	0.631	0.637	0.648	0.657	0.664	0.678	0.689	0.698	0.714	0.727	0.740	0.752	0.771	0.787
2.00000	(19.6)	0.626	0.632	0.643	0.652	0.660	0.674	0.685	0.695	0.711	0.724	0.738	0.749	0.769	0.785
3.00000	(29.4)	0.619	0.626	0.637	0.647	0.655	0.669	0.681	0.690	0.707	0.720	0.734	0.746	0.766	0.782
S	$(\Delta p/\rho l)$	125	150	200	250	300	400	500	600	800	1000	1250	1500	2000	2500

Grad'nt k.p.g. (Equivalent) Pipe diameters in mm

Kinematic viscosity, $\nu = 0.60 \times 10^{-6}$ m^{2}s^{-1} ; **Roughness size, $k_s = 0.0015$ mm**

Kin. visc., $\nu = 0.60 \times 10^{-6}\ \mathrm{m^2 s^{-1}}$;
$S = 0.00010$ to 30.0000

i.e. kin. pr. grad., $\Delta p/\rho l =$
(0.00098) to $(294)\ \mathrm{ms^{-2}}$

Roughness size, $k_s = 0.003$ mm
This table shows values of m, as follows

m_C for Colebrook-White solutions; or,
where $\mathbf{R} \le 2000$, m_P for laminar flow

| Grad'nt | k.p.g. | (Equivalent) Pipe diameters in mm | | | | | | | | | | | | | |
S	$(\Delta p/\rho l)$	6	8	10	12·5	15	20	25	30	40	50	60	80	100	125
0·00030	(0·00294)	4·114	2·804	2·082	1·546	1·213	*	*	1·032	1·005	0·988	0·977	0·962	0·953	0·947
0·00040	(0·00392)	3·563	2·428	1·803	1·339	1·050	*	1·025	1·006	0·982	0·967	0·956	0·943	0·935	0·930
0·00060	(0·00588)	2·909	1·982	1·472	1·093	0·857	*	0·989	0·972	0·951	0·938	0·929	0·918	0·911	0·907
0·00080	(0·00785)	2·520	1·717	1·275	0·947	*	0·986	0·964	0·949	0·930	0·918	0·910	0·900	0·895	0·891
0·00100	(0·00981)	2·254	1·536	1·140	0·847	*	0·966	0·946	0·932	0·915	0·904	0·896	0·887	0·883	0·879
0·00150	(0·0147)	1·840	1·254	0·931	*	0·961	0·933	0·915	0·903	0·888	0·878	0·872	0·865	0·861	0·859
0·00200	(0·0196)	1·594	1·086	0·806	*	0·936	0·910	0·894	0·884	0·870	0·861	0·856	0·850	0·847	0·845
0·00300	(0·0294)	1·301	0·887	*	0·920	0·903	0·881	0·867	0·857	0·846	0·839	0·834	0·829	0·827	0·827
0·00400	(0·0392)	1·127	0·768	*	0·897	0·881	0·861	0·848	0·840	0·829	0·823	0·819	0·816	0·814	0·814
0·00600	(0·0588)	0·920	*	0·885	0·865	0·852	0·834	0·824	0·816	0·807	0·803	0·800	0·797	0·796	0·797
0·00800	(0·0785)	0·797	*	0·863	0·845	0·832	0·817	0·807	0·800	0·793	0·789	0·786	0·784	0·784	0·785
0·01000	(0·0981)	0·713	0·866	0·846	0·829	0·818	0·803	0·795	0·789	0·782	0·778	0·776	0·775	0·775	0·777
0·01500	(0·147)	*	0·835	0·817	0·803	0·793	0·781	0·773	0·768	0·763	0·760	0·759	0·759	0·760	0·762
0·02000	(0·196)	*	0·814	0·798	0·785	0·776	0·765	0·759	0·755	0·750	0·748	0·747	0·748	0·749	0·752
0·03000	(0·294)	0·809	0·787	0·773	0·762	0·754	0·745	0·739	0·736	0·733	0·732	0·731	0·733	0·735	0·738
0·04000	(0·392)	0·789	0·769	0·756	0·746	0·739	0·731	0·726	0·724	0·721	0·721	0·721	0·723	0·725	0·729
0·06000	(0·588)	0·762	0·745	0·734	0·725	0·720	0·713	0·709	0·707	0·706	0·706	0·707	0·709	0·713	0·717
0·08000	(0·785)	0·745	0·729	0·719	0·711	0·706	0·701	0·698	0·696	0·695	0·696	0·697	0·700	0·704	0·708
0·10000	(0·981)	0·731	0·717	0·708	0·701	0·697	0·691	0·689	0·688	0·688	0·689	0·690	0·694	0·698	0·702
0·15000	(1·47)	0·709	0·697	0·689	0·683	0·680	0·676	0·674	0·674	0·674	0·676	0·678	0·682	0·687	0·692
0·20000	(1·96)	0·694	0·683	0·676	0·671	0·668	0·665	0·664	0·664	0·665	0·667	0·670	0·675	0·679	0·685
0·30000	(2·94)	0·674	0·665	0·659	0·655	0·653	0·651	0·651	0·651	0·653	0·656	0·659	0·665	0·670	0·676
0·40000	(3·92)	0·661	0·653	0·648	0·645	0·643	0·642	0·642	0·643	0·646	0·649	0·652	0·658	0·664	0·670
0·60000	(5·88)	0·644	0·637	0·633	0·631	0·630	0·629	0·630	0·632	0·635	0·639	0·643	0·649	0·655	0·662
0·80000	(7·85)	0·632	0·626	0·623	0·622	0·621	0·621	0·623	0·624	0·628	0·633	0·636	0·644	0·650	0·657
1·00000	(9·81)	0·624	0·619	0·616	0·615	0·614	0·615	0·617	0·619	0·623	0·628	0·632	0·639	0·646	0·653
1·50000	(14·7)	0·609	0·605	0·604	0·603	0·603	0·605	0·607	0·610	0·615	0·620	0·624	0·632	0·640	0·647
2·00000	(19·6)	0·600	0·596	0·595	0·595	0·596	0·598	0·601	0·604	0·610	0·615	0·619	0·628	0·635	0·643
3·00000	(29·4)	0·587	0·585	0·585	0·585	0·586	0·590	0·593	0·596	0·602	0·608	0·613	0·622	0·630	0·638
4·00000	(39·2)	0·579	0·577	0·578	0·579	0·580	0·584	0·588	0·591	0·598	0·604	0·609	0·619	0·627	0·635
6·00000	(58·8)	0·568	0·568	0·568	0·570	0·572	0·577	0·581	0·585	0·592	0·599	0·604	0·614	0·622	0·631
8·00000	(78·5)	0·561	0·561	0·563	0·565	0·567	0·572	0·577	0·581	0·588	0·595	0·601	0·611	0·620	0·629
10·0000	(98·1)	0·556	0·557	0·558	0·561	0·563	0·569	0·573	0·578	0·586	0·593	0·599	0·609	0·618	0·627
15·0000	(147)	0·547	0·549	0·551	0·554	0·557	0·563	0·568	0·573	0·581	0·589	0·595	0·606	0·615	0·625
20·0000	(196)	0·542	0·544	0·547	0·550	0·553	0·560	0·565	0·570	0·579	0·586	0·593	0·604	0·613	0·623
30·0000	(294)	0·535	0·538	0·541	0·545	0·548	0·555	0·561	0·566	0·575	0·583	0·590	0·601	0·611	0·621
S	$(\Delta p/\rho l)$	6	8	10	12·5	15	20	25	30	40	50	60	80	100	125

Grad'nt   k.p.g.   (Equivalent) Pipe diameters in mm

S	$(\Delta p/\rho l)$	125	150	200	250	300	400	500	600	800	1000	1250	1500	2000	2500
0·00010	(0·00098)	1·019	1·012	1·003	0·999	0·997	0·995	0·996	0·997	1·001	1·005	1·010	1·016	1·025	1·034
0·00015	(0·00147)	0·991	0·985	0·978	0·975	0·974	0·973	0·974	0·976	0·981	0·986	0·992	0·998	1·008	1·017
0·00020	(0·00196)	0·972	0·967	0·961	0·959	0·958	0·958	0·960	0·963	0·968	0·973	0·979	0·985	0·996	1·005
0·00030	(0·00294)	0·947	0·943	0·939	0·937	0·937	0·938	0·941	0·944	0·950	0·955	0·962	0·968	0·980	0·989
0·00040	(0·00392)	0·930	0·926	0·923	0·922	0·922	0·925	0·928	0·931	0·937	0·943	0·951	0·957	0·968	0·978
0·00060	(0·00588)	0·907	0·904	0·902	0·902	0·903	0·906	0·910	0·913	0·920	0·927	0·935	0·941	0·953	0·964
0·00080	(0·00785)	0·891	0·889	0·888	0·888	0·890	0·893	0·897	0·901	0·909	0·916	0·924	0·931	0·943	0·954
0·00100	(0·00981)	0·879	0·878	0·877	0·878	0·880	0·884	0·888	0·892	0·900	0·908	0·916	0·923	0·935	0·946
0·00150	(0·0147)	0·859	0·858	0·859	0·860	0·862	0·867	0·872	0·877	0·885	0·893	0·901	0·909	0·922	0·933
0·00200	(0·0196)	0·845	0·845	0·846	0·848	0·851	0·856	0·861	0·866	0·875	0·883	0·892	0·899	0·913	0·924
0·00300	(0·0294)	0·827	0·827	0·829	0·832	0·835	0·841	0·846	0·851	0·861	0·869	0·879	0·887	0·900	0·912
0·00400	(0·0392)	0·814	0·815	0·817	0·820	0·824	0·830	0·836	0·842	0·852	0·860	0·870	0·878	0·892	0·904
0·00600	(0·0588)	0·797	0·798	0·802	0·805	0·809	0·816	0·823	0·828	0·839	0·848	0·858	0·866	0·881	0·893
0·00800	(0·0785)	0·785	0·787	0·791	0·795	0·799	0·807	0·813	0·820	0·830	0·840	0·850	0·859	0·874	0·886
0·01000	(0·0981)	0·777	0·779	0·783	0·787	0·792	0·799	0·806	0·813	0·824	0·833	0·844	0·853	0·868	0·881
0·01500	(0·147)	0·762	0·764	0·769	0·774	0·779	0·787	0·795	0·801	0·813	0·823	0·833	0·843	0·858	0·872
0·02000	(0·196)	0·752	0·754	0·760	0·765	0·770	0·779	0·787	0·793	0·805	0·816	0·827	0·836	0·852	0·866
0·03000	(0·294)	0·738	0·741	0·747	0·753	0·758	0·768	0·776	0·783	0·796	0·806	0·818	0·827	0·844	0·858
0·04000	(0·392)	0·729	0·732	0·739	0·745	0·750	0·760	0·769	0·776	0·789	0·800	0·812	0·822	0·839	0·852
0·06000	(0·588)	0·717	0·720	0·728	0·734	0·740	0·751	0·759	0·767	0·781	0·792	0·804	0·814	0·832	0·846
0·08000	(0·785)	0·708	0·713	0·720	0·727	0·733	0·744	0·753	0·761	0·775	0·787	0·799	0·809	0·827	0·842
0·10000	(0·981)	0·702	0·707	0·715	0·722	0·728	0·739	0·749	0·757	0·771	0·783	0·795	0·806	0·824	0·838
0·15000	(1·47)	0·692	0·697	0·705	0·713	0·720	0·731	0·741	0·750	0·764	0·776	0·789	0·800	0·818	0·833
0·20000	(1·96)	0·685	0·690	0·699	0·707	0·714	0·726	0·736	0·745	0·760	0·772	0·785	0·796	0·815	0·830
0·30000	(2·94)	0·676	0·681	0·691	0·699	0·706	0·719	0·729	0·739	0·754	0·767	0·780	0·791	0·811	0·826
0·40000	(3·92)	0·670	0·676	0·686	0·694	0·702	0·715	0·725	0·735	0·750	0·763	0·777	0·788	0·808	0·824
0·60000	(5·88)	0·662	0·668	0·679	0·688	0·696	0·709	0·720	0·730	0·746	0·759	0·773	0·785	0·804	0·820
0·80000	(7·85)	0·657	0·663	0·674	0·684	0·692	0·705	0·717	0·726	0·743	0·756	0·770	0·782	0·802	0·819
1·00000	(9·81)	0·653	0·660	0·671	0·681	0·689	0·703	0·714	0·724	0·741	0·754	0·769	0·781	0·801	0·817
1·50000	(14·7)	0·647	0·654	0·666	0·676	0·684	0·698	0·710	0·720	0·737	0·751	0·766	0·778	0·798	0·815
2·00000	(19·6)	0·643	0·650	0·662	0·673	0·681	0·696	0·708	0·718	0·735	0·749	0·764	0·776	0·797	0·814
3·00000	(29·4)	0·638	0·646	0·658	0·669	0·678	0·693	0·705	0·715	0·733	0·747	0·762	0·774	0·795	0·812
S	$(\Delta p/\rho l)$	125	150	200	250	300	400	500	600	800	1000	1250	1500	2000	2500

Grad'nt   k.p.g.   (Equivalent) Pipe diameters in mm

Kinematic viscosity, $\nu = 0.60 \times 10^{-6}\ \mathrm{m^2 s^{-1}}$;   **Roughness size, $k_s = 0.003$ mm**

Kin. visc., $\nu = 0.60\times10^{-6}$ m^{2}s^{-1};
$S = 0.00010$ to 30.0000

i.e. kin. pr. grad., $\Delta p/\rho l =$
(0.00098) to (294) ms^{-2}

Roughness size, $k_s = 0.006$ mm
This table shows values of m, as follows

m_C for Colebrook-White solutions; or,
where $R \leq 2000$, m_P for laminar flow

Grad'nt S	k.p.g. $(\Delta p/\rho l)$	6	8	10	12.5	15	20	25	30	40	50	60	80	100	125
0.00030	(0.00294)	4.114	2.804	2.082	1.546	1.213	*	*	1.034	1.007	0.990	0.978	0.964	0.955	0.949
0.00040	(0.00392)	3.563	2.428	1.803	1.339	1.050	*	1.027	1.008	0.983	0.968	0.958	0.945	0.937	0.932
0.00060	(0.00588)	2.909	1.982	1.472	1.093	0.857	*	0.990	0.974	0.953	0.939	0.931	0.920	0.913	0.909
0.00080	(0.00785)	2.520	1.717	1.275	0.947	*	0.988	0.966	0.951	0.932	0.920	0.912	0.903	0.897	0.894
0.00100	(0.00981)	2.254	1.536	1.140	0.847	*	0.968	0.948	0.934	0.917	0.906	0.899	0.890	0.885	0.882
0.00150	(0.0147)	1.840	1.254	0.931	*	0.963	0.935	0.917	0.905	0.890	0.881	0.875	0.868	0.864	0.862
0.00200	(0.0196)	1.594	1.086	0.806	*	0.938	0.913	0.897	0.886	0.872	0.864	0.859	0.853	0.850	0.849
0.00300	(0.0294)	1.301	0.887	*	0.922	0.905	0.883	0.869	0.860	0.848	0.842	0.837	0.833	0.831	0.831
0.00400	(0.0392)	1.127	0.768	*	0.899	0.884	0.864	0.851	0.843	0.833	0.827	0.823	0.820	0.818	0.818
0.00600	(0.0588)	0.920	*	0.888	0.868	0.855	0.838	0.827	0.820	0.811	0.807	0.804	0.802	0.801	0.802
0.00800	(0.0785)	0.797	*	0.866	0.848	0.836	0.820	0.811	0.804	0.797	0.793	0.791	0.789	0.790	0.791
0.01000	(0.0981)	0.713	0.869	0.849	0.833	0.822	0.807	0.799	0.793	0.786	0.783	0.781	0.780	0.781	0.783
0.01500	(0.147)	*	0.839	0.821	0.807	0.797	0.785	0.778	0.773	0.768	0.766	0.765	0.765	0.766	0.769
0.02000	(0.196)	*	0.818	0.803	0.790	0.781	0.770	0.764	0.760	0.756	0.754	0.754	0.755	0.756	0.759
0.03000	(0.294)	0.814	0.792	0.778	0.767	0.760	0.750	0.745	0.742	0.739	0.739	0.739	0.741	0.743	0.747
0.04000	(0.392)	0.794	0.774	0.762	0.752	0.745	0.737	0.733	0.730	0.728	0.728	0.729	0.731	0.735	0.739
0.06000	(0.588)	0.768	0.751	0.740	0.732	0.726	0.720	0.717	0.715	0.714	0.715	0.716	0.719	0.723	0.728
0.08000	(0.785)	0.751	0.735	0.726	0.718	0.714	0.708	0.706	0.705	0.704	0.706	0.707	0.711	0.716	0.721
0.10000	(0.981)	0.738	0.724	0.715	0.709	0.704	0.700	0.698	0.697	0.697	0.699	0.701	0.705	0.710	0.715
0.15000	(1.47)	0.716	0.704	0.697	0.692	0.688	0.685	0.684	0.684	0.685	0.688	0.690	0.695	0.701	0.707
0.20000	(1.96)	0.702	0.691	0.685	0.681	0.678	0.675	0.675	0.675	0.677	0.680	0.683	0.689	0.694	0.701
0.30000	(2.94)	0.683	0.674	0.669	0.666	0.664	0.663	0.663	0.664	0.667	0.670	0.674	0.680	0.687	0.693
0.40000	(3.92)	0.671	0.663	0.659	0.656	0.655	0.654	0.655	0.657	0.660	0.664	0.668	0.675	0.681	0.689
0.60000	(5.88)	0.655	0.649	0.646	0.644	0.643	0.644	0.645	0.647	0.652	0.656	0.660	0.668	0.675	0.683
0.80000	(7.85)	0.644	0.639	0.637	0.636	0.635	0.637	0.639	0.641	0.646	0.651	0.655	0.664	0.671	0.679
1.00000	(9.81)	0.637	0.632	0.630	0.630	0.630	0.631	0.634	0.637	0.642	0.647	0.652	0.660	0.668	0.676
1.50000	(14.7)	0.624	0.621	0.619	0.620	0.620	0.623	0.626	0.629	0.635	0.641	0.646	0.655	0.663	0.672
2.00000	(19.6)	0.615	0.613	0.612	0.613	0.614	0.618	0.621	0.625	0.631	0.637	0.642	0.652	0.660	0.669
3.00000	(29.4)	0.604	0.603	0.603	0.605	0.607	0.611	0.615	0.619	0.626	0.632	0.638	0.648	0.656	0.666
4.00000	(39.2)	0.597	0.597	0.598	0.600	0.602	0.606	0.611	0.615	0.622	0.629	0.635	0.645	0.654	0.663
6.00000	(58.8)	0.588	0.589	0.590	0.593	0.595	0.601	0.606	0.610	0.618	0.625	0.632	0.642	0.651	0.661
8.00000	(78.5)	0.583	0.584	0.586	0.589	0.592	0.597	0.602	0.607	0.616	0.623	0.629	0.640	0.650	0.659
10.0000	(98.1)	0.579	0.580	0.583	0.586	0.589	0.595	0.600	0.605	0.614	0.621	0.628	0.639	0.648	0.658
15.0000	(147)	0.572	0.574	0.577	0.581	0.584	0.591	0.597	0.602	0.611	0.619	0.625	0.637	0.646	0.656
20.0000	(196)	0.568	0.571	0.574	0.578	0.582	0.588	0.594	0.600	0.609	0.617	0.624	0.636	0.645	0.655
30.0000	(294)	0.563	0.566	0.570	0.574	0.578	0.585	0.592	0.597	0.607	0.615	0.622	0.634	0.644	0.654
S	$(\Delta p/\rho l)$	6	8	10	12.5	15	20	25	30	40	50	60	80	100	125

Grad'nt   k.p.g.   (Equivalent) Pipe diameters in mm

Grad'nt S	k.p.g. $(\Delta p/\rho l)$	125	150	200	250	300	400	500	600	800	1000	1250	1500	2000	2500
0.00010	(0.00098)	1.020	1.013	1.005	1.000	0.998	0.997	0.997	0.999	1.003	1.007	1.013	1.018	1.028	1.037
0.00015	(0.00147)	0.992	0.986	0.980	0.977	0.975	0.975	0.977	0.979	0.984	0.989	0.995	1.001	1.011	1.020
0.00020	(0.00196)	0.974	0.969	0.963	0.961	0.960	0.961	0.962	0.965	0.970	0.976	0.983	0.989	1.000	1.009
0.00030	(0.00294)	0.949	0.945	0.941	0.939	0.939	0.941	0.943	0.946	0.953	0.959	0.966	0.972	0.984	0.994
0.00040	(0.00392)	0.932	0.928	0.925	0.925	0.925	0.927	0.931	0.934	0.941	0.947	0.955	0.961	0.973	0.984
0.00060	(0.00588)	0.909	0.907	0.905	0.905	0.906	0.909	0.913	0.917	0.925	0.932	0.939	0.947	0.959	0.970
0.00080	(0.00785)	0.894	0.892	0.891	0.892	0.893	0.897	0.901	0.906	0.914	0.921	0.929	0.937	0.950	0.961
0.00100	(0.00981)	0.882	0.881	0.880	0.882	0.883	0.888	0.892	0.897	0.905	0.913	0.922	0.929	0.942	0.954
0.00150	(0.0147)	0.862	0.862	0.862	0.864	0.867	0.872	0.877	0.882	0.891	0.899	0.908	0.916	0.930	0.942
0.00200	(0.0196)	0.849	0.849	0.850	0.852	0.855	0.861	0.867	0.872	0.881	0.890	0.899	0.907	0.922	0.934
0.00300	(0.0294)	0.831	0.831	0.834	0.837	0.840	0.846	0.853	0.858	0.868	0.877	0.887	0.896	0.911	0.923
0.00400	(0.0392)	0.818	0.819	0.822	0.826	0.830	0.837	0.843	0.849	0.860	0.869	0.879	0.888	0.903	0.916
0.00600	(0.0588)	0.802	0.804	0.808	0.812	0.816	0.824	0.831	0.837	0.848	0.858	0.869	0.878	0.894	0.907
0.00800	(0.0785)	0.791	0.793	0.798	0.802	0.807	0.815	0.822	0.829	0.841	0.851	0.862	0.871	0.887	0.901
0.01000	(0.0981)	0.783	0.785	0.790	0.795	0.800	0.808	0.816	0.823	0.835	0.845	0.857	0.866	0.883	0.897
0.01500	(0.147)	0.769	0.772	0.777	0.783	0.788	0.797	0.806	0.813	0.826	0.836	0.848	0.858	0.875	0.889
0.02000	(0.196)	0.759	0.763	0.769	0.775	0.780	0.790	0.799	0.806	0.819	0.830	0.842	0.853	0.870	0.885
0.03000	(0.294)	0.747	0.751	0.758	0.764	0.770	0.780	0.789	0.797	0.811	0.823	0.835	0.846	0.864	0.878
0.04000	(0.392)	0.739	0.743	0.750	0.757	0.763	0.774	0.784	0.792	0.806	0.818	0.830	0.841	0.860	0.875
0.06000	(0.588)	0.728	0.732	0.741	0.748	0.754	0.766	0.776	0.784	0.799	0.811	0.824	0.836	0.854	0.870
0.08000	(0.785)	0.721	0.725	0.734	0.742	0.749	0.761	0.771	0.780	0.795	0.807	0.821	0.832	0.851	0.867
0.10000	(0.981)	0.715	0.720	0.730	0.738	0.745	0.757	0.767	0.776	0.792	0.804	0.818	0.830	0.849	0.865
0.15000	(1.47)	0.707	0.712	0.722	0.730	0.738	0.751	0.761	0.771	0.787	0.800	0.814	0.825	0.845	0.861
0.20000	(1.96)	0.701	0.707	0.717	0.726	0.733	0.747	0.758	0.767	0.783	0.797	0.811	0.823	0.843	0.859
0.30000	(2.94)	0.693	0.700	0.710	0.720	0.728	0.741	0.753	0.763	0.779	0.793	0.807	0.820	0.840	0.856
0.40000	(3.92)	0.689	0.695	0.706	0.716	0.724	0.738	0.750	0.760	0.777	0.791	0.805	0.818	0.838	0.855
0.60000	(5.88)	0.683	0.689	0.701	0.711	0.720	0.734	0.746	0.757	0.774	0.788	0.803	0.815	0.836	0.853
0.80000	(7.85)	0.679	0.686	0.698	0.708	0.717	0.732	0.744	0.754	0.772	0.786	0.801	0.814	0.835	0.852
1.00000	(9.81)	0.676	0.683	0.696	0.706	0.715	0.730	0.742	0.753	0.771	0.785	0.800	0.813	0.834	0.851
1.50000	(14.7)	0.672	0.679	0.692	0.703	0.712	0.727	0.740	0.751	0.768	0.783	0.798	0.811	0.832	0.850
2.00000	(19.6)	0.669	0.677	0.690	0.701	0.710	0.725	0.738	0.749	0.767	0.782	0.797	0.810	0.831	0.849
3.00000	(29.4)	0.666	0.674	0.687	0.698	0.708	0.723	0.736	0.747	0.766	0.780	0.796	0.809	0.830	0.848
S	$(\Delta p/\rho l)$	125	150	200	250	300	400	500	600	800	1000	1250	1500	2000	2500

Grad'nt   k.p.g.   (Equivalent) Pipe diameters in mm

Kinematic viscosity, $\nu = 0.60\times10^{-6}$ m^{2}s^{-1} ;

Roughness size, $k_s = 0.006$ mm

Kin. visc., $\nu = 0.60 \times 10^{-6} \ \mathrm{m^2 s^{-1}}$;
$S = 0.00010$ to 30.0000
i.e. kin. pr. grad., $\Delta p/\rho l =$
(0.00098) to $(294) \ \mathrm{ms^{-2}}$

Roughness size, $k_s = 0.015$ mm
This table shows values of m, as follows
m_C for Colebrook-White solutions; or,
where $R \leq 2000$, m_P for laminar flow

Grad'nt   k.p.g.   (Equivalent) Pipe diameters in mm

S	$(\Delta p/\rho l)$	6	8	10	12.5	15	20	25	30	40	50	60	80	100	125
0.00030	(0.00294)	4.114	2.804	2.082	1.546	1.213	*	*	1.037	1.011	0.994	0.983	0.968	0.960	0.954
0.00040	(0.00392)	3.563	2.428	1.803	1.339	1.050	*	1.031	1.012	0.988	0.973	0.963	0.950	0.943	0.937
0.00060	(0.00588)	2.909	1.982	1.472	1.093	0.857	*	0.995	0.979	0.958	0.945	0.936	0.926	0.920	0.916
0.00080	(0.00785)	2.520	1.717	1.275	0.947	*	0.993	0.971	0.957	0.938	0.926	0.918	0.909	0.904	0.901
0.00100	(0.00981)	2.254	1.536	1.140	0.847	*	0.974	0.954	0.940	0.923	0.912	0.905	0.897	0.893	0.890
0.00150	(0.0147)	1.840	1.254	0.931	*	0.969	0.941	0.924	0.912	0.897	0.888	0.882	0.876	0.873	0.871
0.00200	(0.0196)	1.594	1.086	0.806	*	0.945	0.919	0.904	0.893	0.880	0.872	0.867	0.862	0.860	0.859
0.00300	(0.0294)	1.301	0.887	*	0.930	0.913	0.891	0.877	0.868	0.857	0.851	0.847	0.843	0.842	0.842
0.00400	(0.0392)	1.127	0.768	*	0.907	0.892	0.872	0.860	0.852	0.842	0.837	0.834	0.831	0.830	0.831
0.00600	(0.0588)	0.920	*	0.897	0.877	0.864	0.847	0.837	0.830	0.822	0.818	0.816	0.814	0.815	0.816
0.00800	(0.0785)	0.797	*	0.875	0.858	0.846	0.831	0.821	0.816	0.809	0.806	0.804	0.803	0.805	0.807
0.01000	(0.0981)	0.713	0.879	0.859	0.843	0.832	0.818	0.810	0.805	0.799	0.796	0.795	0.795	0.797	0.800
0.01500	(0.147)	*	0.850	0.833	0.819	0.809	0.798	0.791	0.787	0.782	0.781	0.781	0.782	0.784	0.788
0.02000	(0.196)	*	0.830	0.815	0.803	0.794	0.784	0.778	0.775	0.772	0.771	0.771	0.773	0.776	0.780
0.03000	(0.294)	0.827	0.805	0.792	0.781	0.774	0.766	0.762	0.759	0.757	0.757	0.758	0.761	0.765	0.770
0.04000	(0.392)	0.808	0.789	0.777	0.767	0.761	0.754	0.751	0.749	0.748	0.749	0.750	0.754	0.758	0.763
0.06000	(0.588)	0.784	0.767	0.757	0.749	0.744	0.739	0.736	0.735	0.736	0.737	0.739	0.744	0.749	0.755
0.08000	(0.785)	0.768	0.753	0.744	0.737	0.733	0.729	0.727	0.727	0.728	0.730	0.733	0.738	0.744	0.750
0.10000	(0.981)	0.756	0.743	0.735	0.729	0.725	0.722	0.720	0.721	0.722	0.725	0.728	0.734	0.739	0.746
0.15000	(1.47)	0.737	0.726	0.719	0.714	0.712	0.709	0.709	0.710	0.713	0.716	0.720	0.726	0.733	0.740
0.20000	(1.96)	0.724	0.714	0.709	0.705	0.703	0.702	0.702	0.703	0.707	0.711	0.714	0.722	0.728	0.736
0.30000	(2.94)	0.708	0.700	0.696	0.693	0.692	0.692	0.693	0.695	0.699	0.704	0.708	0.716	0.723	0.731
0.40000	(3.92)	0.698	0.691	0.687	0.685	0.685	0.685	0.687	0.689	0.694	0.699	0.704	0.712	0.720	0.728
0.60000	(5.88)	0.684	0.679	0.677	0.675	0.676	0.677	0.680	0.683	0.688	0.694	0.699	0.708	0.716	0.724
0.80000	(7.85)	0.676	0.671	0.670	0.669	0.670	0.672	0.675	0.678	0.684	0.690	0.696	0.705	0.713	0.722
1.00000	(9.81)	0.670	0.666	0.665	0.665	0.666	0.669	0.672	0.675	0.682	0.688	0.693	0.703	0.711	0.720
1.50000	(14.7)	0.660	0.657	0.657	0.658	0.659	0.663	0.667	0.670	0.678	0.684	0.690	0.700	0.709	0.718
2.00000	(19.6)	0.653	0.652	0.652	0.653	0.655	0.659	0.663	0.667	0.675	0.682	0.688	0.698	0.707	0.716
3.00000	(29.4)	0.645	0.645	0.646	0.647	0.650	0.655	0.659	0.664	0.672	0.679	0.685	0.696	0.705	0.714
4.00000	(39.2)	0.640	0.640	0.642	0.644	0.647	0.652	0.657	0.661	0.670	0.677	0.683	0.694	0.703	0.713
6.00000	(58.8)	0.634	0.635	0.637	0.640	0.643	0.648	0.654	0.659	0.667	0.675	0.681	0.693	0.702	0.712
8.00000	(78.5)	0.630	0.632	0.634	0.637	0.640	0.646	0.652	0.657	0.666	0.673	0.680	0.692	0.701	0.711
10.0000	(98.1)	0.628	0.629	0.632	0.635	0.639	0.645	0.651	0.656	0.665	0.673	0.679	0.691	0.700	0.711
15.0000	(147)	0.623	0.626	0.629	0.632	0.636	0.643	0.649	0.654	0.663	0.671	0.678	0.690	0.699	0.710
20.0000	(196)	0.621	0.624	0.627	0.631	0.634	0.641	0.647	0.653	0.662	0.670	0.677	0.689	0.699	0.709
30.0000	(294)	0.618	0.621	0.624	0.628	0.632	0.640	0.646	0.651	0.661	0.669	0.676	0.688	0.698	0.708
S	$(\Delta p/\rho l)$	6	8	10	12.5	15	20	25	30	40	50	60	80	100	125

Grad'nt   k.p.g.   (Equivalent) Pipe diameters in mm

S	$(\Delta p/\rho l)$	125	150	200	250	300	400	500	600	800	1000	1250	1500	2000	2500
0.00010	(0.00098)	1.023	1.016	1.008	1.004	1.002	1.002	1.002	1.004	1.009	1.014	1.020	1.026	1.036	1.046
0.00015	(0.00147)	0.996	0.991	0.984	0.982	0.980	0.981	0.983	0.985	0.991	0.996	1.003	1.009	1.021	1.031
0.00020	(0.00196)	0.978	0.973	0.968	0.966	0.966	0.967	0.969	0.972	0.978	0.984	0.992	0.998	1.010	1.021
0.00030	(0.00294)	0.954	0.950	0.946	0.945	0.946	0.948	0.951	0.955	0.962	0.969	0.976	0.984	0.996	1.007
0.00040	(0.00392)	0.937	0.934	0.932	0.932	0.932	0.935	0.939	0.943	0.951	0.958	0.966	0.974	0.987	0.998
0.00060	(0.00588)	0.916	0.913	0.912	0.913	0.914	0.918	0.923	0.928	0.936	0.944	0.953	0.961	0.975	0.987
0.00080	(0.00785)	0.901	0.899	0.899	0.900	0.902	0.907	0.912	0.917	0.926	0.935	0.944	0.952	0.967	0.979
0.00100	(0.00981)	0.890	0.889	0.889	0.891	0.893	0.899	0.904	0.909	0.919	0.928	0.937	0.946	0.961	0.973
0.00150	(0.0147)	0.871	0.871	0.873	0.875	0.878	0.884	0.890	0.896	0.907	0.916	0.926	0.935	0.951	0.964
0.00200	(0.0196)	0.859	0.859	0.861	0.865	0.868	0.875	0.881	0.887	0.898	0.908	0.919	0.928	0.944	0.958
0.00300	(0.0294)	0.842	0.843	0.847	0.850	0.854	0.862	0.869	0.876	0.888	0.898	0.909	0.919	0.936	0.950
0.00400	(0.0392)	0.831	0.833	0.837	0.841	0.846	0.854	0.862	0.869	0.881	0.891	0.903	0.913	0.930	0.945
0.00600	(0.0588)	0.816	0.819	0.824	0.829	0.834	0.843	0.851	0.859	0.872	0.883	0.895	0.906	0.923	0.938
0.00800	(0.0785)	0.807	0.810	0.815	0.821	0.826	0.836	0.845	0.853	0.866	0.878	0.890	0.901	0.919	0.934
0.01000	(0.0981)	0.800	0.803	0.809	0.815	0.821	0.831	0.840	0.848	0.862	0.874	0.886	0.897	0.916	0.931
0.01500	(0.147)	0.788	0.791	0.799	0.805	0.811	0.822	0.832	0.840	0.855	0.867	0.880	0.892	0.911	0.926
0.02000	(0.196)	0.780	0.784	0.792	0.799	0.805	0.817	0.827	0.836	0.851	0.863	0.877	0.888	0.907	0.923
0.03000	(0.294)	0.770	0.775	0.783	0.791	0.798	0.810	0.820	0.830	0.845	0.858	0.872	0.884	0.904	0.920
0.04000	(0.392)	0.763	0.768	0.778	0.786	0.793	0.806	0.816	0.826	0.842	0.855	0.869	0.881	0.901	0.918
0.06000	(0.588)	0.755	0.761	0.770	0.779	0.787	0.800	0.811	0.821	0.837	0.851	0.865	0.878	0.898	0.915
0.08000	(0.785)	0.750	0.756	0.766	0.775	0.783	0.797	0.808	0.818	0.835	0.849	0.863	0.876	0.896	0.913
0.10000	(0.981)	0.746	0.752	0.763	0.772	0.780	0.794	0.806	0.816	0.833	0.847	0.862	0.874	0.895	0.912
0.15000	(1.47)	0.740	0.746	0.758	0.767	0.776	0.790	0.802	0.813	0.830	0.844	0.859	0.872	0.893	0.910
0.20000	(1.96)	0.736	0.743	0.754	0.764	0.773	0.788	0.800	0.810	0.828	0.842	0.858	0.870	0.892	0.909
0.30000	(2.94)	0.731	0.738	0.750	0.761	0.770	0.785	0.797	0.808	0.826	0.840	0.856	0.869	0.890	0.908
0.40000	(3.92)	0.728	0.735	0.748	0.758	0.768	0.783	0.796	0.806	0.824	0.839	0.855	0.868	0.889	0.907
0.60000	(5.88)	0.724	0.732	0.745	0.756	0.765	0.781	0.794	0.805	0.823	0.838	0.853	0.866	0.888	0.906
0.80000	(7.85)	0.722	0.730	0.743	0.754	0.763	0.779	0.792	0.803	0.822	0.837	0.852	0.866	0.888	0.905
1.00000	(9.81)	0.720	0.728	0.742	0.753	0.762	0.778	0.791	0.803	0.821	0.836	0.852	0.865	0.887	0.905
1.50000	(14.7)	0.718	0.726	0.740	0.751	0.761	0.777	0.790	0.801	0.820	0.835	0.851	0.864	0.886	0.904
2.00000	(19.6)	0.716	0.725	0.738	0.750	0.760	0.776	0.789	0.801	0.819	0.835	0.850	0.864	0.886	0.904
3.00000	(29.4)	0.714	0.723	0.737	0.749	0.758	0.775	0.788	0.800	0.819	0.834	0.850	0.863	0.885	0.903
S	$(\Delta p/\rho l)$	125	150	200	250	300	400	500	600	800	1000	1250	1500	2000	2500

Grad'nt   k.p.g.   (Equivalent) Pipe diameters in mm

Kinematic viscosity, $\nu = 0.60 \times 10^{-6} \ \mathrm{m^2 s^{-1}}$;   **Roughness size, $k_s = 0.015$ mm**

E67

Kin. visc., $\nu = 0.60 \times 10^{-6}$ m²s⁻¹; $S = 0.00010$ to 30.0000

i.e. kin. pr. grad., $\Delta p/\rho l =$ (0.00098) to (294) ms⁻²

Roughness size, $k_s = 0.030$ mm

This table shows values of m, as follows

m_C for Colebrook-White solutions; or, where $\mathbf{R} \leq 2000$, m_P for laminar flow

Grad'nt S	k.p.g. $(\Delta p/\rho l)$	6	8	10	12.5	15	20	25	30	40	50	60	80	100	125
0.00030	(0.00294)	4.114	2.804	2.082	1.546	1.213	*	*	1.044	1.017	1.001	0.990	0.976	0.968	0.962
0.00040	(0.00392)	3.563	2.428	1.803	1.339	1.050	*	1.038	1.019	0.995	0.980	0.970	0.958	0.951	0.946
0.00060	(0.00588)	2.909	1.982	1.472	1.093	0.857	*	1.003	0.987	0.966	0.953	0.945	0.935	0.929	0.926
0.00080	(0.00785)	2.520	1.717	1.275	0.947	*	1.001	0.980	0.965	0.947	0.935	0.928	0.919	0.915	0.912
0.00100	(0.00981)	2.254	1.536	1.140	0.847	*	0.982	0.963	0.949	0.932	0.922	0.915	0.908	0.904	0.902
0.00150	(0.0147)	1.840	1.254	0.931	*	*	0.951	0.934	0.922	0.908	0.899	0.894	0.888	0.886	0.885
0.00200	(0.0196)	1.594	1.086	0.806	*	0.955	0.930	0.915	0.904	0.892	0.884	0.880	0.875	0.874	0.874
0.00300	(0.0294)	1.301	0.887	*	0.941	0.924	0.903	0.890	0.881	0.871	0.865	0.862	0.859	0.858	0.859
0.00400	(0.0392)	1.127	0.768	*	0.920	0.904	0.885	0.873	0.866	0.857	0.852	0.849	0.848	0.848	0.849
0.00600	(0.0588)	0.920	*	0.910	0.891	0.878	0.862	0.852	0.846	0.839	0.835	0.834	0.833	0.834	0.837
0.00800	(0.0785)	0.797	*	0.890	0.873	0.861	0.847	0.838	0.833	0.827	0.824	0.823	0.824	0.826	0.829
0.01000	(0.0981)	0.713	*	0.875	0.859	0.849	0.835	0.828	0.823	0.818	0.816	0.816	0.817	0.819	0.823
0.01500	(0.147)	*	0.867	0.850	0.837	0.828	0.817	0.811	0.807	0.804	0.803	0.803	0.806	0.809	0.814
0.02000	(0.196)	*	0.849	0.834	0.822	0.814	0.805	0.799	0.797	0.794	0.794	0.795	0.798	0.802	0.807
0.03000	(0.294)	0.847	0.826	0.813	0.803	0.796	0.789	0.785	0.783	0.782	0.783	0.785	0.789	0.794	0.800
0.04000	(0.392)	0.830	0.811	0.800	0.791	0.785	0.779	0.776	0.775	0.775	0.776	0.778	0.783	0.789	0.795
0.06000	(0.588)	0.808	0.792	0.782	0.775	0.770	0.766	0.764	0.764	0.765	0.767	0.770	0.776	0.782	0.789
0.08000	(0.785)	0.794	0.780	0.771	0.765	0.761	0.758	0.757	0.757	0.759	0.762	0.765	0.772	0.778	0.785
0.10000	(0.981)	0.784	0.771	0.763	0.758	0.754	0.752	0.751	0.752	0.754	0.758	0.761	0.768	0.775	0.782
0.15000	(1.47)	0.767	0.756	0.750	0.746	0.743	0.742	0.742	0.744	0.747	0.751	0.755	0.763	0.770	0.778
0.20000	(1.96)	0.756	0.747	0.742	0.738	0.737	0.736	0.737	0.739	0.743	0.747	0.752	0.760	0.767	0.775
0.30000	(2.94)	0.743	0.735	0.731	0.729	0.728	0.728	0.730	0.732	0.737	0.742	0.747	0.756	0.764	0.772
0.40000	(3.92)	0.734	0.728	0.724	0.723	0.722	0.724	0.726	0.729	0.734	0.739	0.744	0.753	0.761	0.770
0.60000	(5.88)	0.724	0.718	0.716	0.715	0.716	0.718	0.721	0.724	0.730	0.736	0.741	0.750	0.759	0.768
0.80000	(7.85)	0.717	0.713	0.711	0.711	0.712	0.714	0.717	0.721	0.727	0.733	0.739	0.749	0.757	0.766
1.00000	(9.81)	0.712	0.709	0.707	0.708	0.709	0.712	0.715	0.719	0.726	0.732	0.737	0.747	0.756	0.765
1.50000	(14.7)	0.704	0.702	0.702	0.702	0.704	0.708	0.712	0.715	0.723	0.729	0.735	0.745	0.754	0.764
2.00000	(19.6)	0.700	0.698	0.698	0.699	0.701	0.705	0.709	0.713	0.721	0.728	0.734	0.744	0.753	0.763
3.00000	(29.4)	0.694	0.693	0.694	0.695	0.698	0.702	0.707	0.711	0.719	0.726	0.732	0.743	0.752	0.762
4.00000	(39.2)	0.690	0.690	0.691	0.693	0.695	0.700	0.705	0.710	0.718	0.725	0.731	0.742	0.751	0.761
6.00000	(58.8)	0.686	0.686	0.688	0.690	0.693	0.698	0.703	0.708	0.716	0.723	0.730	0.741	0.750	0.760
8.00000	(78.5)	0.684	0.684	0.686	0.689	0.691	0.697	0.702	0.707	0.715	0.723	0.729	0.740	0.750	0.760
10.0000	(98.1)	0.682	0.683	0.685	0.687	0.690	0.696	0.701	0.706	0.715	0.722	0.729	0.740	0.749	0.759
15.0000	(147)	0.679	0.680	0.682	0.685	0.689	0.694	0.700	0.705	0.714	0.721	0.728	0.739	0.749	0.759
20.0000	(196)	0.677	0.679	0.681	0.684	0.688	0.694	0.699	0.704	0.713	0.721	0.727	0.739	0.748	0.758
30.0000	(294)	0.675	0.677	0.680	0.683	0.686	0.693	0.698	0.703	0.712	0.720	0.727	0.738	0.748	0.758
S	$(\Delta p/\rho l)$	6	8	10	12.5	15	20	25	30	40	50	60	80	100	125

Grad'nt — k.p.g. — (Equivalent) Pipe diameters in mm

Grad'nt S	k.p.g. $(\Delta p/\rho l)$	125	150	200	250	300	400	500	600	800	1000	1250	1500	2000	2500
0.00010	(0.00098)	1.029	1.022	1.015	1.011	1.010	1.009	1.011	1.013	1.018	1.024	1.031	1.037	1.049	1.059
0.00015	(0.00147)	1.003	0.997	0.992	0.989	0.989	0.990	0.992	0.995	1.001	1.008	1.015	1.023	1.035	1.046
0.00020	(0.00196)	0.985	0.981	0.976	0.975	0.975	0.977	0.980	0.983	0.990	0.997	1.005	1.013	1.026	1.038
0.00030	(0.00294)	0.962	0.959	0.956	0.955	0.956	0.959	0.963	0.967	0.975	0.983	0.992	1.000	1.014	1.026
0.00040	(0.00392)	0.946	0.944	0.942	0.942	0.944	0.948	0.952	0.957	0.966	0.974	0.983	0.992	1.006	1.019
0.00060	(0.00588)	0.926	0.924	0.924	0.925	0.927	0.932	0.938	0.943	0.953	0.962	0.972	0.981	0.996	1.009
0.00080	(0.00785)	0.912	0.911	0.912	0.914	0.917	0.922	0.928	0.934	0.945	0.954	0.964	0.974	0.990	1.003
0.00100	(0.00981)	0.902	0.902	0.903	0.906	0.909	0.915	0.921	0.928	0.939	0.948	0.959	0.969	0.985	0.999
0.00150	(0.0147)	0.885	0.886	0.888	0.892	0.895	0.903	0.910	0.916	0.928	0.939	0.950	0.960	0.977	0.992
0.00200	(0.0196)	0.874	0.875	0.878	0.882	0.886	0.895	0.902	0.909	0.922	0.933	0.944	0.955	0.972	0.987
0.00300	(0.0294)	0.859	0.861	0.865	0.870	0.875	0.884	0.893	0.900	0.913	0.925	0.937	0.948	0.966	0.981
0.00400	(0.0392)	0.849	0.852	0.857	0.863	0.868	0.878	0.886	0.894	0.908	0.920	0.933	0.944	0.962	0.978
0.00600	(0.0588)	0.837	0.840	0.846	0.853	0.858	0.869	0.878	0.887	0.901	0.914	0.927	0.938	0.958	0.973
0.00800	(0.0785)	0.829	0.832	0.840	0.846	0.852	0.864	0.873	0.882	0.897	0.910	0.923	0.935	0.955	0.971
0.01000	(0.0981)	0.823	0.827	0.835	0.842	0.848	0.860	0.870	0.879	0.894	0.907	0.921	0.933	0.952	0.969
0.01500	(0.147)	0.814	0.818	0.826	0.834	0.841	0.854	0.864	0.873	0.889	0.903	0.917	0.929	0.949	0.966
0.02000	(0.196)	0.807	0.812	0.821	0.829	0.837	0.850	0.860	0.870	0.886	0.900	0.914	0.927	0.947	0.964
0.03000	(0.294)	0.800	0.805	0.815	0.824	0.831	0.845	0.856	0.866	0.883	0.896	0.911	0.924	0.945	0.962
0.04000	(0.392)	0.795	0.800	0.811	0.820	0.828	0.842	0.853	0.863	0.880	0.894	0.909	0.922	0.943	0.960
0.06000	(0.588)	0.789	0.795	0.806	0.815	0.824	0.838	0.850	0.860	0.878	0.892	0.907	0.920	0.941	0.959
0.08000	(0.785)	0.785	0.791	0.803	0.812	0.821	0.836	0.848	0.858	0.876	0.890	0.906	0.919	0.940	0.958
0.10000	(0.981)	0.782	0.789	0.800	0.810	0.819	0.834	0.846	0.857	0.875	0.889	0.905	0.918	0.939	0.957
0.15000	(1.47)	0.778	0.785	0.797	0.807	0.816	0.831	0.844	0.855	0.873	0.888	0.903	0.916	0.938	0.956
0.20000	(1.96)	0.775	0.782	0.795	0.805	0.815	0.830	0.843	0.854	0.872	0.887	0.902	0.915	0.937	0.955
0.30000	(2.94)	0.772	0.779	0.792	0.803	0.812	0.828	0.841	0.852	0.870	0.885	0.901	0.914	0.936	0.954
0.40000	(3.92)	0.770	0.778	0.791	0.802	0.811	0.827	0.840	0.851	0.870	0.885	0.900	0.914	0.936	0.954
0.60000	(5.88)	0.768	0.775	0.789	0.800	0.810	0.826	0.839	0.850	0.869	0.884	0.900	0.913	0.935	0.953
0.80000	(7.85)	0.766	0.774	0.788	0.799	0.809	0.825	0.838	0.849	0.868	0.883	0.899	0.913	0.935	0.953
1.00000	(9.81)	0.765	0.773	0.787	0.798	0.808	0.824	0.837	0.849	0.868	0.883	0.899	0.912	0.935	0.953
1.50000	(14.7)	0.764	0.772	0.786	0.797	0.807	0.823	0.837	0.848	0.867	0.882	0.898	0.912	0.934	0.952
2.00000	(19.6)	0.763	0.771	0.785	0.796	0.806	0.823	0.836	0.848	0.867	0.882	0.898	0.912	0.934	0.952
3.00000	(29.4)	0.762	0.770	0.784	0.796	0.806	0.822	0.836	0.847	0.866	0.882	0.898	0.911	0.934	0.952
S	$(\Delta p/\rho l)$	125	150	200	250	300	400	500	600	800	1000	1250	1500	2000	2500

Grad'nt — k.p.g. — (Equivalent) Pipe diameters in mm

Kinematic viscosity, $\nu = 0.60 \times 10^{-6}$ m²s⁻¹;   **Roughness size, $k_s = 0.030$ mm**

Kin. visc., $\nu = 0.60 \times 10^{-6}\ \text{m}^2\text{s}^{-1}$;
$S = 0.00010$ to 30.0000
i.e. kin. pr. grad., $\Delta p/\rho l =$
(0.00098) to $(294)\ \text{ms}^{-2}$

Roughness size, $k_s = 0.060$ mm
This table shows values of m, as follows
m_C for Colebrook-White solutions; or,
where $\mathbf{R} \leq 2000$, m_P for laminar flow

Grad'nt S	k.p.g. $(\Delta p/\rho l)$	6	8	10	12·5	15	20	25	30	40	50	60	80	100	125
		(Equivalent) Pipe diameters in mm													
0·00030	(0·00294)	4·114	2·804	2·082	1·546	1·213	*	*	1·056	1·030	1·014	1·003	0·990	0·983	0·977
0·00040	(0·00392)	3·563	2·428	1·803	1·339	1·050	*	1·051	1·032	1·009	0·995	0·985	0·974	0·967	0·963
0·00060	(0·00588)	2·909	1·982	1·472	1·093	0·857	*	1·018	1·002	0·981	0·969	0·961	0·952	0·948	0·945
0·00080	(0·00785)	2·520	1·717	1·275	0·947	*	1·017	0·996	0·981	0·963	0·953	0·946	0·938	0·935	0·933
0·00100	(0·00981)	2·254	1·536	1·140	0·847	*	0·999	0·980	0·967	0·950	0·941	0·935	0·928	0·925	0·924
0·00150	(0·0147)	1·840	1·254	0·931	*	*	0·970	0·953	0·942	0·928	0·920	0·916	0·911	0·909	0·909
0·00200	(0·0196)	1·594	1·086	0·806	*	0·975	0·950	0·935	0·926	0·914	0·907	0·903	0·900	0·899	0·900
0·00300	(0·0294)	1·301	0·887	*	0·963	0·946	0·925	0·913	0·905	0·895	0·890	0·887	0·885	0·886	0·888
0·00400	(0·0392)	1·127	0·768	*	0·943	0·928	0·909	0·898	0·891	0·883	0·879	0·877	0·876	0·877	0·880
0·00600	(0·0588)	0·920	*	0·936	0·917	0·905	0·889	0·879	0·874	0·867	0·865	0·864	0·864	0·867	0·870
0·00800	(0·0785)	0·797	*	0·918	0·901	0·889	0·875	0·867	0·862	0·857	0·856	0·855	0·857	0·860	0·864
0·01000	(0·0981)	0·713	*	0·904	0·889	0·878	0·866	0·859	0·854	0·850	0·849	0·849	0·852	0·855	0·860
0·01500	(0·147)	*	0·899	0·882	0·869	0·860	0·850	0·844	0·841	0·839	0·839	0·840	0·843	0·847	0·853
0·02000	(0·196)	*	0·883	0·868	0·856	0·849	0·840	0·835	0·833	0·831	0·832	0·834	0·838	0·842	0·848
0·03000	(0·294)	0·884	0·863	0·850	0·840	0·834	0·827	0·824	0·822	0·822	0·824	0·826	0·831	0·836	0·843
0·04000	(0·392)	0·869	0·850	0·839	0·830	0·825	0·819	0·816	0·816	0·816	0·818	0·821	0·827	0·833	0·839
0·06000	(0·588)	0·851	0·834	0·825	0·817	0·813	0·809	0·807	0·807	0·809	0·812	0·815	0·822	0·828	0·835
0·08000	(0·785)	0·839	0·824	0·816	0·809	0·806	0·803	0·802	0·802	0·805	0·808	0·812	0·819	0·825	0·833
0·10000	(0·981)	0·830	0·817	0·809	0·804	0·801	0·798	0·798	0·799	0·802	0·805	0·809	0·816	0·823	0·831
0·15000	(1·47)	0·817	0·805	0·799	0·795	0·792	0·791	0·792	0·793	0·797	0·801	0·805	0·813	0·820	0·828
0·20000	(1·96)	0·808	0·798	0·793	0·789	0·787	0·787	0·788	0·789	0·794	0·798	0·803	0·811	0·818	0·826
0·30000	(2·94)	0·798	0·789	0·785	0·782	0·781	0·781	0·783	0·785	0·790	0·795	0·800	0·808	0·816	0·824
0·40000	(3·92)	0·791	0·784	0·780	0·778	0·777	0·778	0·780	0·782	0·788	0·793	0·798	0·807	0·814	0·823
0·60000	(5·88)	0·783	0·777	0·774	0·773	0·772	0·774	0·776	0·779	0·785	0·790	0·796	0·805	0·813	0·822
0·80000	(7·85)	0·778	0·773	0·770	0·769	0·770	0·772	0·774	0·777	0·783	0·789	0·794	0·804	0·812	0·821
1·00000	(9·81)	0·775	0·770	0·768	0·767	0·768	0·770	0·773	0·776	0·782	0·788	0·793	0·803	0·811	0·820
1·50000	(14·7)	0·769	0·765	0·764	0·764	0·765	0·767	0·771	0·774	0·780	0·786	0·792	0·802	0·810	0·819
2·00000	(19·6)	0·766	0·763	0·761	0·762	0·763	0·766	0·769	0·773	0·779	0·785	0·791	0·801	0·809	0·819
3·00000	(29·4)	0·762	0·759	0·759	0·759	0·760	0·764	0·767	0·771	0·778	0·784	0·790	0·800	0·809	0·818
4·00000	(39·2)	0·760	0·757	0·757	0·758	0·759	0·763	0·766	0·770	0·777	0·784	0·789	0·800	0·808	0·818
6·00000	(58·8)	0·757	0·755	0·755	0·756	0·757	0·761	0·765	0·769	0·776	0·783	0·789	0·799	0·808	0·817
8·00000	(78·5)	0·755	0·753	0·753	0·755	0·756	0·760	0·764	0·768	0·776	0·782	0·788	0·799	0·807	0·817
10·0000	(98·1)	0·754	0·752	0·753	0·754	0·756	0·760	0·764	0·768	0·775	0·782	0·788	0·798	0·807	0·817
15·0000	(147)	0·752	0·751	0·751	0·753	0·755	0·759	0·763	0·767	0·775	0·782	0·787	0·798	0·807	0·816
20·0000	(196)	0·751	0·750	0·750	0·752	0·754	0·758	0·763	0·767	0·774	0·781	0·787	0·798	0·807	0·816
30·0000	(294)	0·749	0·749	0·749	0·751	0·753	0·758	0·762	0·766	0·774	0·781	0·787	0·797	0·806	0·816
S	$(\Delta p/\rho l)$	6	8	10	12·5	15	20	25	30	40	50	60	80	100	125

Grad'nt k.p.g. (Equivalent) Pipe diameters in mm

Grad'nt S	k.p.g. $(\Delta p/\rho l)$	125	150	200	250	300	400	500	600	800	1000	1250	1500	2000	2500
0·00010	(0·00098)	1·040	1·034	1·027	1·024	1·023	1·024	1·026	1·029	1·035	1·042	1·050	1·057	1·071	1·083
0·00015	(0·00147)	1·015	1·010	1·006	1·004	1·004	1·006	1·009	1·013	1·021	1·028	1·037	1·045	1·059	1·072
0·00020	(0·00196)	0·999	0·995	0·991	0·991	0·991	0·994	0·998	1·003	1·011	1·019	1·029	1·037	1·052	1·065
0·00030	(0·00294)	0·977	0·975	0·973	0·973	0·975	0·979	0·984	0·989	0·999	1·008	1·018	1·027	1·043	1·056
0·00040	(0·00392)	0·963	0·961	0·961	0·962	0·964	0·969	0·975	0·980	0·991	1·000	1·011	1·020	1·037	1·051
0·00060	(0·00588)	0·945	0·944	0·945	0·947	0·950	0·956	0·963	0·969	0·981	0·991	1·002	1·012	1·029	1·044
0·00080	(0·00785)	0·933	0·933	0·934	0·937	0·941	0·948	0·955	0·962	0·974	0·985	0·996	1·007	1·025	1·039
0·00100	(0·00981)	0·924	0·924	0·927	0·930	0·934	0·942	0·950	0·957	0·969	0·980	0·992	1·003	1·021	1·036
0·00150	(0·0147)	0·909	0·911	0·914	0·919	0·924	0·933	0·941	0·948	0·962	0·973	0·986	0·997	1·016	1·031
0·00200	(0·0196)	0·900	0·902	0·906	0·912	0·917	0·926	0·935	0·943	0·957	0·969	0·982	0·993	1·012	1·028
0·00300	(0·0294)	0·888	0·890	0·896	0·902	0·908	0·918	0·928	0·936	0·951	0·964	0·977	0·989	1·008	1·025
0·00400	(0·0392)	0·880	0·883	0·890	0·896	0·902	0·914	0·923	0·932	0·947	0·960	0·974	0·986	1·006	1·022
0·00600	(0·0588)	0·870	0·874	0·882	0·889	0·896	0·907	0·918	0·927	0·943	0·956	0·970	0·982	1·003	1·019
0·00800	(0·0785)	0·864	0·868	0·877	0·884	0·891	0·904	0·914	0·924	0·940	0·953	0·968	0·980	1·001	1·018
0·01000	(0·0981)	0·860	0·864	0·873	0·881	0·888	0·901	0·912	0·922	0·938	0·952	0·966	0·979	0·999	1·017
0·01500	(0·147)	0·853	0·858	0·867	0·876	0·883	0·897	0·908	0·918	0·935	0·949	0·964	0·976	0·997	1·015
0·02000	(0·196)	0·848	0·854	0·864	0·872	0·880	0·894	0·906	0·916	0·933	0·947	0·962	0·975	0·996	1·014
0·03000	(0·294)	0·843	0·849	0·859	0·868	0·877	0·891	0·903	0·913	0·931	0·945	0·960	0·973	0·995	1·012
0·04000	(0·392)	0·839	0·845	0·856	0·866	0·874	0·889	0·901	0·912	0·929	0·944	0·959	0·972	0·994	1·011
0·06000	(0·588)	0·835	0·842	0·853	0·863	0·872	0·887	0·899	0·910	0·927	0·942	0·958	0·971	0·993	1·010
0·08000	(0·785)	0·833	0·839	0·851	0·861	0·870	0·885	0·898	0·908	0·926	0·941	0·957	0·970	0·992	1·010
0·10000	(0·981)	0·831	0·838	0·850	0·860	0·869	0·884	0·897	0·908	0·926	0·941	0·956	0·970	0·991	1·009
0·15000	(1·47)	0·828	0·835	0·848	0·858	0·867	0·883	0·895	0·906	0·925	0·940	0·955	0·969	0·991	1·009
0·20000	(1·96)	0·826	0·834	0·846	0·857	0·866	0·882	0·895	0·906	0·924	0·939	0·955	0·968	0·990	1·008
0·30000	(2·94)	0·824	0·832	0·845	0·855	0·865	0·881	0·894	0·905	0·923	0·938	0·954	0·968	0·990	1·008
0·40000	(3·92)	0·823	0·831	0·844	0·855	0·864	0·880	0·893	0·904	0·923	0·938	0·954	0·967	0·990	1·008
0·60000	(5·88)	0·822	0·829	0·843	0·854	0·863	0·879	0·892	0·903	0·922	0·937	0·953	0·967	0·989	1·007
0·80000	(7·85)	0·821	0·829	0·842	0·853	0·863	0·879	0·892	0·903	0·922	0·937	0·953	0·967	0·989	1·007
1·00000	(9·81)	0·820	0·828	0·841	0·853	0·862	0·878	0·891	0·903	0·922	0·937	0·953	0·966	0·989	1·007
1·50000	(14·7)	0·819	0·827	0·841	0·852	0·862	0·878	0·891	0·902	0·921	0·937	0·953	0·966	0·989	1·007
2·00000	(19·6)	0·819	0·827	0·840	0·851	0·861	0·877	0·891	0·902	0·921	0·936	0·952	0·966	0·988	1·007
3·00000	(29·4)	0·818	0·826	0·840	0·851	0·861	0·877	0·890	0·902	0·921	0·936	0·952	0·966	0·988	1·007
S	$(\Delta p/\rho l)$	125	150	200	250	300	400	500	600	800	1000	1250	1500	2000	2500

Grad'nt k.p.g. (Equivalent) Pipe diameters in mm

Kinematic viscosity, $\nu = 0.60 \times 10^{-6}\ \text{m}^2\text{s}^{-1}$;

Roughness size, $k_s = 0.060$ mm

E69

Kin. visc., $\nu = 0.60 \times 10^{-6}$ m^{2}s^{-1};
$S = 0.00010$ to 30.0000

i.e. kin. pr. grad., $\Delta p/\rho l =$
(0.00098) to (294) ms^{-2}

Roughness size, $k_s = 0.150$ mm
This table shows values of m, as follows

m_C for Colebrook-White solutions; or,
where $\mathbf{R} \le 2000$, m_P for laminar flow

Grad'nt S	k.p.g. $(\Delta p/\rho l)$	6	8	10	12.5	15	20	25	30	40	50	60	80	100	125
		(Equivalent) Pipe diameters in mm													
0.00030	(0.00294)	4.114	2.804	2.082	1.546	1.213	*	*	1.091	1.065	1.050	1.039	1.028	1.021	1.017
0.00040	(0.00392)	3.563	2.428	1.803	1.339	1.050	*	1.088	1.069	1.047	1.033	1.024	1.014	1.009	1.006
0.00060	(0.00588)	2.909	1.982	1.472	1.093	0.857	*	1.058	1.042	1.023	1.012	1.004	0.997	0.993	0.991
0.00080	(0.00785)	2.520	1.717	1.275	0.947	*	1.060	1.039	1.025	1.008	0.998	0.992	0.985	0.983	0.982
0.00100	(0.00981)	2.254	1.536	1.140	0.847	*	1.045	1.025	1.013	0.997	0.988	0.983	0.978	0.976	0.976
0.00150	(0.0147)	1.840	1.254	0.931	*	*	1.019	1.003	0.992	0.979	0.972	0.968	0.965	0.964	0.965
0.00200	(0.0196)	1.594	1.086	0.806	*	1.028	1.003	0.989	0.979	0.968	0.962	0.959	0.956	0.957	0.959
0.00300	(0.0294)	1.301	0.887	*	*	1.004	0.983	0.970	0.963	0.954	0.949	0.947	0.946	0.948	0.950
0.00400	(0.0392)	1.127	0.768	*	1.004	0.989	0.970	0.959	0.952	0.945	0.941	0.940	0.940	0.942	0.945
0.00600	(0.0588)	0.920	*	*	0.983	0.970	0.954	0.945	0.939	0.933	0.931	0.931	0.932	0.935	0.939
0.00800	(0.0785)	0.797	*	0.988	0.970	0.958	0.944	0.936	0.931	0.927	0.925	0.925	0.927	0.931	0.935
0.01000	(0.0981)	0.713	*	0.977	0.961	0.950	0.937	0.930	0.926	0.922	0.921	0.921	0.924	0.928	0.932
0.01500	(0.147)	*	0.978	0.960	0.946	0.936	0.925	0.920	0.916	0.914	0.914	0.915	0.919	0.923	0.928
0.02000	(0.196)	*	0.966	0.949	0.936	0.928	0.918	0.913	0.911	0.909	0.910	0.911	0.915	0.920	0.926
0.03000	(0.294)	*	0.951	0.936	0.925	0.918	0.910	0.906	0.904	0.903	0.904	0.906	0.911	0.916	0.923
0.04000	(0.392)	0.963	0.941	0.928	0.918	0.911	0.904	0.901	0.900	0.900	0.901	0.904	0.909	0.914	0.921
0.06000	(0.588)	0.949	0.930	0.918	0.909	0.904	0.898	0.895	0.895	0.895	0.897	0.900	0.906	0.912	0.918
0.08000	(0.785)	0.941	0.923	0.912	0.904	0.899	0.894	0.892	0.891	0.893	0.895	0.898	0.904	0.910	0.917
0.10000	(0.981)	0.935	0.918	0.908	0.900	0.896	0.891	0.890	0.889	0.891	0.894	0.897	0.903	0.909	0.916
0.15000	(1.47)	0.925	0.910	0.901	0.894	0.891	0.887	0.886	0.886	0.888	0.891	0.894	0.901	0.907	0.915
0.20000	(1.96)	0.920	0.905	0.897	0.891	0.887	0.884	0.883	0.884	0.886	0.890	0.893	0.900	0.906	0.914
0.30000	(2.94)	0.913	0.900	0.892	0.887	0.884	0.881	0.881	0.881	0.884	0.888	0.891	0.898	0.905	0.913
0.40000	(3.92)	0.908	0.896	0.889	0.884	0.881	0.879	0.879	0.880	0.883	0.887	0.890	0.898	0.904	0.912
0.60000	(5.88)	0.903	0.892	0.886	0.881	0.879	0.877	0.877	0.878	0.882	0.885	0.889	0.897	0.903	0.911
0.80000	(7.85)	0.900	0.889	0.883	0.879	0.877	0.875	0.876	0.877	0.881	0.885	0.889	0.896	0.903	0.911
1.00000	(9.81)	0.898	0.888	0.882	0.878	0.876	0.875	0.875	0.876	0.880	0.884	0.888	0.896	0.903	0.910
1.50000	(14.7)	0.895	0.885	0.880	0.876	0.874	0.873	0.874	0.875	0.879	0.883	0.887	0.895	0.902	0.910
2.00000	(19.6)	0.893	0.883	0.878	0.875	0.873	0.872	0.873	0.875	0.878	0.883	0.887	0.895	0.902	0.910
3.00000	(29.4)	0.890	0.881	0.876	0.873	0.872	0.871	0.872	0.874	0.878	0.882	0.886	0.894	0.901	0.909
4.00000	(39.2)	0.889	0.880	0.875	0.872	0.871	0.870	0.872	0.873	0.877	0.882	0.886	0.894	0.901	0.909
6.00000	(58.8)	0.887	0.879	0.874	0.871	0.870	0.870	0.871	0.873	0.877	0.881	0.886	0.894	0.901	0.909
8.00000	(78.5)	0.886	0.878	0.874	0.871	0.870	0.869	0.871	0.872	0.877	0.881	0.885	0.893	0.901	0.909
10.0000	(98.1)	0.886	0.877	0.873	0.870	0.869	0.869	0.870	0.872	0.876	0.881	0.885	0.893	0.901	0.909
15.0000	(147)	0.884	0.877	0.872	0.870	0.869	0.869	0.870	0.872	0.876	0.881	0.885	0.893	0.900	0.908
20.0000	(196)	0.884	0.876	0.872	0.869	0.868	0.868	0.870	0.872	0.876	0.880	0.885	0.893	0.900	0.908
30.0000	(294)	0.883	0.875	0.871	0.869	0.868	0.868	0.869	0.871	0.876	0.880	0.885	0.893	0.900	0.908
S	$(\Delta p/\rho l)$	6	8	10	12.5	15	20	25	30	40	50	60	80	100	125

Grad'nt S	k.p.g. $(\Delta p/\rho l)$	125	150	200	250	300	400	500	600	800	1000	1250	1500	2000	2500
		(Equivalent) Pipe diameters in mm													
0.00010	(0.00098)	1.070	1.064	1.059	1.058	1.058	1.060	1.064	1.068	1.077	1.085	1.095	1.104	1.120	1.133
0.00015	(0.00147)	1.049	1.045	1.041	1.041	1.042	1.046	1.051	1.056	1.066	1.075	1.086	1.095	1.112	1.126
0.00020	(0.00196)	1.035	1.032	1.030	1.030	1.032	1.037	1.043	1.049	1.059	1.069	1.080	1.090	1.107	1.122
0.00030	(0.00294)	1.017	1.015	1.015	1.017	1.020	1.026	1.033	1.039	1.051	1.061	1.073	1.083	1.101	1.117
0.00040	(0.00392)	1.006	1.005	1.006	1.009	1.012	1.019	1.026	1.033	1.046	1.057	1.069	1.079	1.098	1.113
0.00060	(0.00588)	0.991	0.992	0.994	0.998	1.002	1.010	1.018	1.026	1.039	1.050	1.063	1.074	1.093	1.109
0.00080	(0.00785)	0.982	0.983	0.987	0.991	0.996	1.005	1.013	1.021	1.035	1.047	1.060	1.071	1.091	1.107
0.00100	(0.00981)	0.976	0.977	0.981	0.986	0.991	1.001	1.010	1.018	1.032	1.044	1.057	1.069	1.089	1.105
0.00150	(0.0147)	0.965	0.967	0.973	0.978	0.984	0.995	1.004	1.012	1.027	1.040	1.054	1.066	1.086	1.102
0.00200	(0.0196)	0.959	0.961	0.967	0.974	0.980	0.991	1.000	1.009	1.025	1.038	1.051	1.064	1.084	1.101
0.00300	(0.0294)	0.950	0.954	0.961	0.968	0.974	0.986	0.996	1.005	1.021	1.034	1.049	1.061	1.082	1.099
0.00400	(0.0392)	0.945	0.949	0.957	0.964	0.971	0.983	0.994	1.003	1.019	1.033	1.047	1.059	1.080	1.097
0.00600	(0.0588)	0.939	0.943	0.952	0.960	0.967	0.979	0.990	1.000	1.017	1.030	1.045	1.058	1.079	1.096
0.00800	(0.0785)	0.935	0.940	0.949	0.957	0.964	0.977	0.989	0.998	1.015	1.029	1.044	1.057	1.078	1.095
0.01000	(0.0981)	0.932	0.937	0.947	0.955	0.963	0.976	0.987	0.997	1.014	1.028	1.043	1.056	1.077	1.095
0.01500	(0.147)	0.928	0.934	0.943	0.952	0.960	0.974	0.985	0.995	1.012	1.027	1.042	1.055	1.076	1.094
0.02000	(0.196)	0.926	0.931	0.941	0.950	0.958	0.972	0.984	0.994	1.011	1.026	1.041	1.054	1.075	1.093
0.03000	(0.294)	0.923	0.928	0.939	0.948	0.956	0.970	0.982	0.993	1.010	1.025	1.040	1.053	1.075	1.092
0.04000	(0.392)	0.921	0.927	0.937	0.947	0.955	0.969	0.981	0.992	1.009	1.024	1.039	1.052	1.074	1.092
0.06000	(0.588)	0.918	0.925	0.936	0.945	0.954	0.968	0.980	0.991	1.009	1.023	1.039	1.052	1.074	1.092
0.08000	(0.785)	0.917	0.923	0.935	0.944	0.953	0.967	0.980	0.990	1.008	1.023	1.038	1.051	1.073	1.091
0.10000	(0.981)	0.916	0.922	0.934	0.944	0.952	0.967	0.979	0.990	1.008	1.022	1.038	1.051	1.073	1.091
0.15000	(1.47)	0.915	0.921	0.933	0.943	0.951	0.966	0.979	0.989	1.007	1.022	1.038	1.051	1.073	1.091
0.20000	(1.96)	0.914	0.920	0.932	0.942	0.951	0.966	0.978	0.989	1.007	1.022	1.037	1.051	1.073	1.091
0.30000	(2.94)	0.913	0.919	0.931	0.941	0.950	0.965	0.978	0.988	1.007	1.021	1.037	1.050	1.072	1.090
0.40000	(3.92)	0.912	0.919	0.931	0.941	0.950	0.965	0.977	0.988	1.006	1.021	1.037	1.050	1.072	1.090
0.60000	(5.88)	0.911	0.918	0.930	0.940	0.949	0.964	0.977	0.988	1.006	1.021	1.037	1.050	1.072	1.090
0.80000	(7.85)	0.911	0.918	0.930	0.940	0.949	0.964	0.977	0.988	1.006	1.021	1.036	1.050	1.072	1.090
1.00000	(9.81)	0.910	0.917	0.929	0.940	0.949	0.964	0.977	0.988	1.006	1.021	1.036	1.050	1.072	1.090
1.50000	(14.7)	0.910	0.917	0.929	0.939	0.948	0.964	0.976	0.987	1.006	1.021	1.036	1.050	1.072	1.090
2.00000	(19.6)	0.910	0.917	0.929	0.939	0.948	0.964	0.976	0.987	1.005	1.020	1.036	1.050	1.072	1.090
3.00000	(29.4)	0.909	0.916	0.929	0.939	0.948	0.963	0.976	0.987	1.005	1.020	1.036	1.049	1.072	1.090
S	$(\Delta p/\rho l)$	125	150	200	250	300	400	500	600	800	1000	1250	1500	2000	2500

Grad'nt k.p.g. (Equivalent) Pipe diameters in mm

Kinematic viscosity, $\nu = 0.60 \times 10^{-6}$ m^{2}s^{-1};   **Roughness size, $k_s = 0.150$ mm**

Kin. visc., $\nu = 0.60 \times 10^{-6}\ m^2 s^{-1}$;
$S = 0.00010$ to 30.0000

i.e. kin. pr. grad., $\Delta p/\rho l =$
(0.00098) to (294) ms^{-2}

Roughness size, $k_s = 0.30$ mm
This table shows values of m, as follows

m_C for Colebrook-White solutions; or,
where $\mathbf{R} \leq 2000$, m_P for laminar flow

Grad'nt S	k.p.g. $(\Delta p/\rho l)$	6	8	10	12·5	15	20	25	30	40	50	60	80	100	125
		(Equivalent) Pipe diameters in mm													
0·00030	(0·00294)	4·114	2·804	2·082	1·546	1·213	*	*	1·141	1·115	1·100	1·090	1·079	1·073	1·070
0·00040	(0·00392)	3·563	2·428	1·803	1·339	1·050	*	1·141	1·123	1·100	1·086	1·078	1·068	1·063	1·061
0·00060	(0·00588)	2·909	1·982	1·472	1·093	0·857	*	1·116	1·100	1·080	1·069	1·062	1·054	1·051	1·050
0·00080	(0·00785)	2·520	1·717	1·275	0·947	*	1·122	1·100	1·086	1·068	1·058	1·052	1·046	1·044	1·043
0·00100	(0·00981)	2·254	1·536	1·140	0·847	*	1·109	1·089	1·076	1·060	1·051	1·045	1·040	1·038	1·039
0·00150	(0·0147)	1·840	1·254	0·931	*	*	1·088	1·071	1·059	1·046	1·038	1·034	1·030	1·030	1·031
0·00200	(0·0196)	1·594	1·086	0·806	*	1·101	1·075	1·059	1·049	1·037	1·031	1·027	1·025	1·025	1·026
0·00300	(0·0294)	1·301	0·887	*	*	1·082	1·059	1·045	1·036	1·026	1·021	1·019	1·017	1·018	1·021
0·00400	(0·0392)	1·127	0·768	*	1·087	1·070	1·049	1·036	1·029	1·020	1·015	1·013	1·013	1·014	1·017
0·00600	(0·0588)	0·920	*	*	1·070	1·055	1·036	1·026	1·019	1·012	1·008	1·007	1·008	1·010	1·013
0·00800	(0·0785)	0·797	*	1·080	1·060	1·046	1·029	1·019	1·013	1·007	1·004	1·003	1·004	1·007	1·011
0·01000	(0·0981)	0·713	*	1·072	1·053	1·040	1·024	1·015	1·009	1·003	1·001	1·001	1·002	1·005	1·009
0·01500	(0·147)	*	*	1·059	1·041	1·030	1·015	1·007	1·003	0·998	0·996	0·996	0·998	1·002	1·006
0·02000	(0·196)	*	1·071	1·051	1·034	1·024	1·010	1·003	0·999	0·995	0·994	0·994	0·996	1·000	1·004
0·03000	(0·294)	*	1·060	1·041	1·026	1·016	1·004	0·998	0·994	0·991	0·990	0·991	0·994	0·998	1·002
0·04000	(0·392)	1·081	1·053	1·035	1·021	1·012	1·001	0·995	0·991	0·988	0·988	0·989	0·992	0·996	1·001
0·06000	(0·588)	1·071	1·044	1·028	1·015	1·006	0·996	0·991	0·988	0·986	0·986	0·987	0·990	0·995	1·000
0·08000	(0·785)	1·065	1·039	1·023	1·011	1·003	0·993	0·988	0·986	0·984	0·984	0·985	0·989	0·994	0·999
0·10000	(0·981)	1·060	1·035	1·020	1·008	1·001	0·992	0·987	0·984	0·983	0·983	0·985	0·988	0·993	0·998
0·15000	(1·47)	1·053	1·030	1·016	1·004	0·997	0·989	0·984	0·982	0·981	0·982	0·983	0·987	0·992	0·998
0·20000	(1·96)	1·049	1·027	1·013	1·002	0·995	0·987	0·983	0·981	0·980	0·981	0·982	0·987	0·991	0·997
0·30000	(2·94)	1·044	1·023	1·009	0·999	0·992	0·985	0·981	0·979	0·978	0·979	0·981	0·986	0·990	0·996
0·40000	(3·92)	1·041	1·020	1·007	0·997	0·991	0·984	0·980	0·978	0·978	0·979	0·981	0·985	0·990	0·996
0·60000	(5·88)	1·037	1·017	1·005	0·995	0·989	0·982	0·979	0·977	0·977	0·978	0·980	0·985	0·990	0·995
0·80000	(7·85)	1·035	1·016	1·004	0·994	0·988	0·981	0·978	0·976	0·976	0·977	0·979	0·984	0·989	0·995
1·00000	(9·81)	1·034	1·014	1·003	0·993	0·987	0·981	0·977	0·976	0·976	0·977	0·979	0·984	0·989	0·995
1·50000	(14·7)	1·032	1·012	1·001	0·992	0·986	0·980	0·977	0·975	0·975	0·977	0·979	0·984	0·989	0·995
2·00000	(19·6)	1·030	1·011	1·000	0·991	0·985	0·979	0·976	0·975	0·975	0·976	0·978	0·983	0·988	0·995
3·00000	(29·4)	1·028	1·010	0·999	0·990	0·985	0·978	0·976	0·974	0·974	0·976	0·978	0·983	0·988	0·994
4·00000	(39·2)	1·028	1·009	0·998	0·990	0·984	0·978	0·975	0·974	0·974	0·976	0·978	0·983	0·988	0·994
6·00000	(58·8)	1·026	1·008	0·997	0·989	0·984	0·978	0·975	0·974	0·974	0·975	0·978	0·983	0·988	0·994
8·00000	(78·5)	1·026	1·008	0·997	0·989	0·983	0·977	0·975	0·973	0·974	0·975	0·978	0·983	0·988	0·994
10·0000	(98·1)	1·025	1·007	0·997	0·988	0·983	0·977	0·974	0·973	0·974	0·975	0·977	0·983	0·988	0·994
15·0000	(147)	1·024	1·007	0·996	0·988	0·983	0·977	0·974	0·973	0·973	0·975	0·977	0·982	0·988	0·994
20·0000	(196)	1·024	1·006	0·996	0·988	0·982	0·977	0·974	0·973	0·973	0·975	0·977	0·982	0·988	0·994
30·0000	(294)	1·023	1·006	0·996	0·987	0·982	0·976	0·974	0·973	0·973	0·975	0·977	0·982	0·987	0·994
S	$(\Delta p/\rho l)$	6	8	10	12·5	15	20	25	30	40	50	60	80	100	125

Grad'nt **k.p.g.** (Equivalent) Pipe diameters in mm

Grad'nt S	k.p.g. $(\Delta p/\rho l)$	125	150	200	250	300	400	500	600	800	1000	1250	1500	2000	2500
0·00010	(0·00098)	1·112	1·107	1·103	1·103	1·104	1·107	1·112	1·118	1·128	1·137	1·148	1·158	1·175	1·190
0·00015	(0·00147)	1·095	1·091	1·089	1·090	1·092	1·097	1·103	1·109	1·120	1·130	1·142	1·152	1·170	1·185
0·00020	(0·00196)	1·083	1·081	1·080	1·082	1·084	1·090	1·097	1·103	1·115	1·126	1·138	1·148	1·167	1·182
0·00030	(0·00294)	1·070	1·068	1·069	1·072	1·075	1·082	1·090	1·097	1·109	1·121	1·133	1·144	1·163	1·179
0·00040	(0·00392)	1·061	1·060	1·062	1·065	1·069	1·077	1·085	1·093	1·106	1·117	1·130	1·141	1·161	1·177
0·00060	(0·00588)	1·050	1·050	1·054	1·058	1·062	1·071	1·080	1·088	1·102	1·114	1·127	1·138	1·158	1·174
0·00080	(0·00785)	1·043	1·044	1·048	1·053	1·058	1·068	1·076	1·085	1·099	1·111	1·125	1·136	1·156	1·173
0·00100	(0·00981)	1·039	1·040	1·045	1·050	1·055	1·065	1·074	1·082	1·097	1·110	1·123	1·135	1·155	1·172
0·00150	(0·0147)	1·031	1·033	1·039	1·044	1·050	1·061	1·070	1·079	1·094	1·107	1·121	1·133	1·153	1·170
0·00200	(0·0196)	1·026	1·029	1·035	1·041	1·047	1·058	1·068	1·077	1·092	1·105	1·119	1·132	1·152	1·169
0·00300	(0·0294)	1·021	1·024	1·031	1·037	1·044	1·055	1·065	1·075	1·090	1·104	1·118	1·130	1·151	1·168
0·00400	(0·0392)	1·017	1·021	1·028	1·035	1·042	1·053	1·064	1·073	1·089	1·102	1·117	1·129	1·150	1·167
0·00600	(0·0588)	1·013	1·017	1·025	1·032	1·039	1·051	1·062	1·071	1·087	1·101	1·116	1·128	1·149	1·166
0·00800	(0·0785)	1·011	1·015	1·023	1·030	1·037	1·050	1·061	1·070	1·087	1·100	1·115	1·127	1·148	1·166
0·01000	(0·0981)	1·009	1·013	1·021	1·029	1·036	1·049	1·060	1·069	1·086	1·100	1·114	1·127	1·148	1·165
0·01500	(0·147)	1·006	1·011	1·019	1·027	1·035	1·048	1·059	1·068	1·085	1·099	1·114	1·126	1·147	1·165
0·02000	(0·196)	1·004	1·009	1·018	1·026	1·034	1·047	1·058	1·068	1·084	1·098	1·113	1·126	1·147	1·165
0·03000	(0·294)	1·002	1·007	1·017	1·025	1·032	1·046	1·057	1·067	1·084	1·098	1·113	1·125	1·147	1·164
0·04000	(0·392)	1·001	1·006	1·016	1·024	1·032	1·045	1·056	1·066	1·083	1·097	1·112	1·125	1·146	1·164
0·06000	(0·588)	1·000	1·005	1·015	1·023	1·031	1·044	1·056	1·066	1·083	1·097	1·112	1·125	1·146	1·164
0·08000	(0·785)	0·999	1·004	1·014	1·023	1·030	1·044	1·055	1·065	1·082	1·097	1·112	1·124	1·146	1·164
0·10000	(0·981)	0·998	1·004	1·014	1·022	1·030	1·044	1·055	1·065	1·082	1·096	1·111	1·124	1·146	1·163
0·15000	(1·47)	0·998	1·003	1·013	1·022	1·029	1·043	1·055	1·065	1·082	1·096	1·111	1·124	1·146	1·163
0·20000	(1·96)	0·997	1·002	1·012	1·021	1·029	1·043	1·054	1·065	1·082	1·096	1·111	1·124	1·145	1·163
0·30000	(2·94)	0·996	1·002	1·012	1·021	1·029	1·043	1·054	1·064	1·081	1·096	1·111	1·124	1·145	1·163
0·40000	(3·92)	0·996	1·002	1·012	1·021	1·029	1·042	1·054	1·064	1·081	1·096	1·111	1·124	1·145	1·163
0·60000	(5·88)	0·995	1·001	1·011	1·020	1·028	1·042	1·054	1·064	1·081	1·095	1·111	1·124	1·145	1·163
0·80000	(7·85)	0·995	1·001	1·011	1·020	1·028	1·042	1·054	1·064	1·081	1·095	1·111	1·124	1·145	1·163
1·00000	(9·81)	0·995	1·001	1·011	1·020	1·028	1·042	1·054	1·064	1·081	1·095	1·110	1·123	1·145	1·163
1·50000	(14·7)	0·995	1·000	1·011	1·020	1·028	1·042	1·053	1·064	1·081	1·095	1·110	1·123	1·145	1·163
2·00000	(19·6)	0·995	1·000	1·011	1·020	1·028	1·042	1·053	1·064	1·081	1·095	1·110	1·123	1·145	1·163
3·00000	(29·4)	0·994	1·000	1·010	1·020	1·028	1·042	1·053	1·064	1·081	1·095	1·110	1·123	1·145	1·163
S	$(\Delta p/\rho l)$	125	150	200	250	300	400	500	600	800	1000	1250	1500	2000	2500

Grad'nt **k.p.g.** (Equivalent) Pipe diameters in mm

Kinematic viscosity, $\nu = 0.60 \times 10^{-6}\ m^2 s^{-1}$;

Roughness size, $k_s = 0.30$ mm

E71

Kin. visc., $\nu = 0.60 \times 10^{-6}$ m^{2}s^{-1}; $S = 0.00010$ to 3.00000

i.e. kin. pr. grad., $\Delta p/\rho l =$ (0.00098) to (29.4) ms^{-2}

Roughness size, $k_s = 0.60$ mm

This table shows values of m, as follows

m_C for Colebrook-White solutions

Grad'nt S	k.p.g. $(\Delta p/\rho l)$	\multicolumn{14}{c}{(Equivalent) Pipe diameters in mm}

Grad'nt S	k.p.g. $(\Delta p/\rho l)$	125	150	200	250	300	400	500	600	800	1000	1250	1500	2000	2500
0.00010	(0.00098)	1.180	1.175	1.171	1.171	1.172	1.177	1.182	1.188	1.198	1.208	1.219	1.230	1.247	1.262
0.00015	(0.00147)	1.166	1.163	1.161	1.162	1.164	1.169	1.175	1.182	1.193	1.204	1.215	1.226	1.244	1.259
0.00020	(0.00196)	1.158	1.156	1.155	1.156	1.159	1.165	1.171	1.178	1.190	1.201	1.213	1.224	1.242	1.258
0.00030	(0.00294)	1.148	1.146	1.147	1.149	1.152	1.159	1.167	1.174	1.186	1.198	1.210	1.221	1.240	1.256
0.00040	(0.00392)	1.142	1.141	1.142	1.145	1.148	1.156	1.164	1.171	1.184	1.195	1.208	1.219	1.238	1.254
0.00060	(0.00588)	1.134	1.134	1.136	1.140	1.144	1.152	1.160	1.168	1.181	1.193	1.206	1.217	1.237	1.253
0.00080	(0.00785)	1.129	1.130	1.132	1.137	1.141	1.150	1.158	1.166	1.180	1.192	1.205	1.216	1.236	1.252
0.00100	(0.00981)	1.126	1.127	1.130	1.134	1.139	1.148	1.157	1.164	1.178	1.191	1.204	1.215	1.235	1.251
0.00150	(0.0147)	1.121	1.122	1.126	1.131	1.136	1.145	1.154	1.162	1.177	1.189	1.202	1.214	1.234	1.250
0.00200	(0.0196)	1.118	1.119	1.124	1.129	1.134	1.144	1.153	1.161	1.176	1.188	1.202	1.213	1.233	1.250
0.00300	(0.0294)	1.114	1.116	1.121	1.126	1.132	1.142	1.151	1.160	1.174	1.187	1.201	1.212	1.232	1.249
0.00400	(0.0392)	1.112	1.114	1.119	1.125	1.130	1.141	1.150	1.159	1.174	1.186	1.200	1.212	1.232	1.249
0.00600	(0.0588)	1.109	1.112	1.117	1.123	1.129	1.140	1.149	1.158	1.173	1.185	1.199	1.211	1.231	1.248
0.00800	(0.0785)	1.108	1.110	1.116	1.122	1.128	1.139	1.148	1.157	1.172	1.185	1.199	1.211	1.231	1.248
0.01000	(0.0981)	1.107	1.109	1.115	1.121	1.127	1.138	1.148	1.157	1.172	1.185	1.198	1.211	1.231	1.248
0.01500	(0.147)	1.105	1.108	1.114	1.120	1.126	1.137	1.147	1.156	1.171	1.184	1.198	1.210	1.231	1.247
0.02000	(0.196)	1.104	1.107	1.113	1.120	1.126	1.137	1.147	1.155	1.171	1.184	1.198	1.210	1.230	1.247
0.03000	(0.294)	1.103	1.106	1.112	1.119	1.125	1.136	1.146	1.155	1.170	1.183	1.197	1.210	1.230	1.247
0.04000	(0.392)	1.102	1.105	1.112	1.118	1.124	1.136	1.146	1.155	1.170	1.183	1.197	1.209	1.230	1.247
0.06000	(0.588)	1.101	1.104	1.111	1.118	1.124	1.135	1.145	1.154	1.170	1.183	1.197	1.209	1.230	1.247
0.08000	(0.785)	1.100	1.104	1.111	1.117	1.124	1.135	1.145	1.154	1.170	1.183	1.197	1.209	1.230	1.247
0.10000	(0.981)	1.100	1.103	1.110	1.117	1.123	1.135	1.145	1.154	1.170	1.183	1.197	1.209	1.230	1.247
0.15000	(1.47)	1.099	1.103	1.110	1.117	1.123	1.135	1.145	1.154	1.169	1.183	1.197	1.209	1.229	1.246
0.20000	(1.96)	1.099	1.103	1.110	1.116	1.123	1.134	1.145	1.154	1.169	1.182	1.197	1.209	1.229	1.246
0.30000	(2.94)	1.099	1.102	1.109	1.116	1.123	1.134	1.144	1.153	1.169	1.182	1.196	1.209	1.229	1.246
0.40000	(3.92)	1.099	1.102	1.109	1.116	1.123	1.134	1.144	1.153	1.169	1.182	1.196	1.209	1.229	1.246
0.60000	(5.88)	1.098	1.102	1.109	1.116	1.122	1.134	1.144	1.153	1.169	1.182	1.196	1.209	1.229	1.246
0.80000	(7.85)	1.098	1.102	1.109	1.116	1.122	1.134	1.144	1.153	1.169	1.182	1.196	1.209	1.229	1.246
1.00000	(9.81)	1.098	1.101	1.109	1.116	1.122	1.134	1.144	1.153	1.169	1.182	1.196	1.209	1.229	1.246
1.50000	(14.7)	1.098	1.101	1.109	1.116	1.122	1.134	1.144	1.153	1.169	1.182	1.196	1.208	1.229	1.246
2.00000	(19.6)	1.098	1.101	1.109	1.116	1.122	1.134	1.144	1.153	1.169	1.182	1.196	1.208	1.229	1.246
3.00000	(29.4)	1.098	1.101	1.108	1.115	1.122	1.134	1.144	1.153	1.169	1.182	1.196	1.208	1.229	1.246
S	$(\Delta p/\rho l)$	125	150	200	250	300	400	500	600	800	1000	1250	1500	2000	2500

Grad'nt k.p.g. (Equivalent) Pipe diameters in mm

Roughness size, $k_s = 0.60$ mm

E72

Kin. visc., $\nu = 0.60 \times 10^{-6}$ m^{2}s^{-1}; $S = 0.00010$ to 3.00000

i.e. kin. pr. grad., $\Delta p/\rho l =$ (0.00098) to (29.4) ms^{-2}

Roughness size, $k_s = 1.50$ mm

This table shows values of m, as follows

m_C for Colebrook-White solutions

Grad'nt S	k.p.g. $(\Delta p/\rho l)$	125	150	200	250	300	400	500	600	800	1000	1250	1500	2000	2500
0.00010	(0.00098)	1.321	1.315	1.308	1.306	1.305	1.308	1.312	1.316	1.325	1.334	1.345	1.354	1.371	1.385
0.00015	(0.00147)	1.313	1.307	1.302	1.300	1.300	1.303	1.308	1.313	1.322	1.332	1.342	1.352	1.369	1.384
0.00020	(0.00196)	1.308	1.303	1.298	1.297	1.297	1.301	1.306	1.311	1.321	1.330	1.341	1.351	1.368	1.383
0.00030	(0.00294)	1.301	1.297	1.293	1.293	1.294	1.298	1.303	1.308	1.319	1.328	1.339	1.349	1.367	1.382
0.00040	(0.00392)	1.298	1.294	1.290	1.290	1.292	1.296	1.301	1.307	1.318	1.327	1.339	1.349	1.366	1.381
0.00060	(0.00588)	1.293	1.290	1.287	1.287	1.289	1.294	1.300	1.305	1.316	1.326	1.337	1.348	1.365	1.380
0.00080	(0.00785)	1.290	1.287	1.285	1.286	1.287	1.293	1.298	1.304	1.315	1.325	1.337	1.347	1.365	1.380
0.00100	(0.00981)	1.289	1.286	1.284	1.284	1.286	1.292	1.298	1.303	1.315	1.325	1.336	1.347	1.364	1.379
0.00150	(0.0147)	1.286	1.283	1.281	1.283	1.285	1.290	1.296	1.302	1.314	1.324	1.336	1.346	1.364	1.379
0.00200	(0.0196)	1.284	1.281	1.280	1.281	1.284	1.289	1.296	1.302	1.313	1.324	1.335	1.346	1.363	1.379
0.00300	(0.0294)	1.282	1.279	1.279	1.280	1.282	1.288	1.295	1.301	1.313	1.323	1.335	1.345	1.363	1.378
0.00400	(0.0392)	1.280	1.278	1.278	1.279	1.282	1.288	1.294	1.300	1.312	1.323	1.334	1.345	1.363	1.378
0.00600	(0.0588)	1.279	1.277	1.277	1.278	1.281	1.287	1.294	1.300	1.312	1.322	1.334	1.344	1.363	1.378
0.00800	(0.0785)	1.278	1.276	1.276	1.278	1.280	1.287	1.293	1.300	1.311	1.322	1.334	1.344	1.362	1.378
0.01000	(0.0981)	1.277	1.276	1.276	1.277	1.280	1.286	1.293	1.299	1.311	1.322	1.334	1.344	1.362	1.378
0.01500	(0.147)	1.277	1.275	1.275	1.277	1.280	1.286	1.293	1.299	1.311	1.322	1.333	1.344	1.362	1.377
0.02000	(0.196)	1.276	1.274	1.274	1.276	1.279	1.286	1.292	1.299	1.311	1.321	1.333	1.344	1.362	1.377
0.03000	(0.294)	1.275	1.274	1.274	1.276	1.279	1.285	1.292	1.299	1.311	1.321	1.333	1.344	1.362	1.377
0.04000	(0.392)	1.275	1.273	1.274	1.276	1.279	1.285	1.292	1.298	1.310	1.321	1.333	1.344	1.362	1.377
0.06000	(0.588)	1.274	1.273	1.273	1.275	1.278	1.285	1.292	1.298	1.310	1.321	1.333	1.343	1.362	1.377
0.08000	(0.785)	1.274	1.273	1.273	1.275	1.278	1.285	1.292	1.298	1.310	1.321	1.333	1.343	1.362	1.377
0.10000	(0.981)	1.274	1.273	1.273	1.275	1.278	1.285	1.292	1.298	1.310	1.321	1.333	1.343	1.362	1.377
0.15000	(1.47)	1.274	1.272	1.273	1.275	1.278	1.285	1.291	1.298	1.310	1.321	1.333	1.343	1.362	1.377
0.20000	(1.96)	1.273	1.272	1.273	1.275	1.278	1.285	1.291	1.298	1.310	1.321	1.333	1.343	1.362	1.377
0.30000	(2.94)	1.273	1.272	1.272	1.275	1.278	1.284	1.291	1.298	1.310	1.321	1.333	1.343	1.361	1.377
0.40000	(3.92)	1.273	1.272	1.272	1.275	1.278	1.284	1.291	1.298	1.310	1.321	1.333	1.343	1.361	1.377
0.60000	(5.88)	1.273	1.272	1.272	1.274	1.278	1.284	1.291	1.298	1.310	1.321	1.333	1.343	1.361	1.377
0.80000	(7.85)	1.273	1.272	1.272	1.274	1.277	1.284	1.291	1.298	1.310	1.321	1.333	1.343	1.361	1.377
1.00000	(9.81)	1.273	1.272	1.272	1.274	1.277	1.284	1.291	1.298	1.310	1.321	1.332	1.343	1.361	1.377
1.50000	(14.7)	1.273	1.271	1.272	1.274	1.277	1.284	1.291	1.298	1.310	1.321	1.332	1.343	1.361	1.377
2.00000	(19.6)	1.273	1.271	1.272	1.274	1.277	1.284	1.291	1.298	1.310	1.320	1.332	1.343	1.361	1.377
3.00000	(29.4)	1.273	1.271	1.272	1.274	1.277	1.284	1.291	1.298	1.310	1.320	1.332	1.343	1.361	1.377
S	$(\Delta p/\rho l)$	125	150	200	250	300	400	500	600	800	1000	1250	1500	2000	2500

Grad'nt k.p.g. (Equivalent) Pipe diameters in mm

Roughness size, $k_s = 1.50$ mm

Kin. visc., $\nu = 0{\cdot}60\times10^{-6}\ \mathrm{m^2 s^{-1}}$;
$S = 0{\cdot}00010$ to $3{\cdot}00000$

i.e. kin. pr. grad., $\Delta p/\rho l =$
$(0{\cdot}00098)$ to $(29{\cdot}4)\ \mathrm{ms^{-2}}$

Roughness size, $k_s = 3{\cdot}0$ mm
This table shows values of m, as follows

m_C for Colebrook-White solutions

Grad'nt S	k.p.g. $(\Delta p/\rho l)$	125	150	200	250	300	400	500	600	800	1000	1250	1500	2000	2500
0·00010	(0·00098)	1·480	1·469	1·455	1·448	1·445	1·442	1·443	1·445	1·451	1·457	1·465	1·473	1·488	1·500
0·00015	(0·00147)	1·474	1·464	1·451	1·445	1·442	1·440	1·441	1·443	1·449	1·456	1·464	1·472	1·487	1·500
0·00020	(0·00196)	1·471	1·460	1·449	1·443	1·440	1·438	1·439	1·442	1·448	1·455	1·463	1·471	1·486	1·499
0·00030	(0·00294)	1·467	1·457	1·446	1·440	1·437	1·436	1·438	1·440	1·447	1·454	1·462	1·471	1·485	1·498
0·00040	(0·00392)	1·464	1·454	1·444	1·438	1·436	1·435	1·437	1·439	1·446	1·453	1·462	1·470	1·485	1·498
0·00060	(0·00588)	1·461	1·452	1·441	1·437	1·434	1·434	1·435	1·438	1·445	1·452	1·461	1·469	1·484	1·498
0·00080	(0·00785)	1·459	1·450	1·440	1·435	1·433	1·433	1·435	1·438	1·445	1·452	1·461	1·469	1·484	1·497
0·00100	(0·00981)	1·458	1·449	1·439	1·435	1·433	1·432	1·434	1·437	1·444	1·452	1·460	1·469	1·484	1·497
0·00150	(0·0147)	1·456	1·447	1·438	1·433	1·431	1·431	1·433	1·437	1·444	1·451	1·460	1·468	1·484	1·497
0·00200	(0·0196)	1·455	1·446	1·437	1·433	1·431	1·431	1·433	1·436	1·443	1·451	1·460	1·468	1·483	1·497
0·00300	(0·0294)	1·453	1·445	1·436	1·432	1·430	1·430	1·432	1·436	1·443	1·450	1·459	1·468	1·483	1·496
0·00400	(0·0392)	1·453	1·444	1·435	1·431	1·430	1·430	1·432	1·435	1·443	1·450	1·459	1·468	1·483	1·496
0·00600	(0·0588)	1·452	1·444	1·435	1·431	1·429	1·429	1·432	1·435	1·442	1·450	1·459	1·467	1·483	1·496
0·00800	(0·0785)	1·451	1·443	1·434	1·430	1·429	1·429	1·432	1·435	1·442	1·450	1·459	1·467	1·483	1·496
0·01000	(0·0981)	1·451	1·443	1·434	1·430	1·429	1·429	1·431	1·435	1·442	1·450	1·459	1·467	1·483	1·496
0·01500	(0·147)	1·450	1·442	1·434	1·430	1·428	1·429	1·431	1·434	1·442	1·450	1·459	1·467	1·483	1·496
0·02000	(0·196)	1·450	1·442	1·433	1·430	1·428	1·428	1·431	1·434	1·442	1·450	1·459	1·467	1·482	1·496
0·03000	(0·294)	1·449	1·441	1·433	1·429	1·428	1·428	1·431	1·434	1·442	1·449	1·459	1·467	1·482	1·496
0·04000	(0·392)	1·449	1·441	1·433	1·429	1·428	1·428	1·431	1·434	1·442	1·449	1·459	1·467	1·482	1·496
0·06000	(0·588)	1·449	1·441	1·433	1·429	1·428	1·428	1·431	1·434	1·442	1·449	1·458	1·467	1·482	1·496
0·08000	(0·785)	1·448	1·441	1·432	1·429	1·427	1·428	1·431	1·434	1·442	1·449	1·458	1·467	1·482	1·496
0·10000	(0·981)	1·448	1·441	1·432	1·429	1·427	1·428	1·430	1·434	1·442	1·449	1·458	1·467	1·482	1·496
0·15000	(1·47)	1·448	1·441	1·432	1·429	1·427	1·428	1·430	1·434	1·441	1·449	1·458	1·467	1·482	1·496
0·20000	(1·96)	1·448	1·440	1·432	1·429	1·427	1·428	1·430	1·434	1·441	1·449	1·458	1·467	1·482	1·496
0·30000	(2·94)	1·448	1·440	1·432	1·428	1·427	1·428	1·430	1·434	1·441	1·449	1·458	1·467	1·482	1·496
0·40000	(3·92)	1·448	1·440	1·432	1·428	1·427	1·428	1·430	1·434	1·441	1·449	1·458	1·467	1·482	1·496
0·60000	(5·88)	1·448	1·440	1·432	1·428	1·427	1·428	1·430	1·434	1·441	1·449	1·458	1·467	1·482	1·496
0·80000	(7·85)	1·448	1·440	1·432	1·428	1·427	1·428	1·430	1·434	1·441	1·449	1·458	1·467	1·482	1·496
1·00000	(9·81)	1·448	1·440	1·432	1·428	1·427	1·428	1·430	1·434	1·441	1·449	1·458	1·467	1·482	1·496
1·50000	(14·7)	1·447	1·440	1·432	1·428	1·427	1·428	1·430	1·434	1·441	1·449	1·458	1·467	1·482	1·496
2·00000	(19·6)	1·447	1·440	1·432	1·428	1·427	1·428	1·430	1·434	1·441	1·449	1·458	1·467	1·482	1·496
3·00000	(29·4)	1·447	1·440	1·432	1·428	1·427	1·428	1·430	1·434	1·441	1·449	1·458	1·467	1·482	1·496
S	$(\Delta p/\rho l)$	125	150	200	250	300	400	500	600	800	1000	1250	1500	2000	2500

Grad'nt k.p.g.   (Equivalent) Pipe diameters in mm   Roughness size, $k_s = 3{\cdot}0$ mm

Kin. visc., $\nu = 0{\cdot}60\times10^{-6}\ \mathrm{m^2 s^{-1}}$;
$S = 0{\cdot}00010$ to $3{\cdot}00000$

i.e. kin. pr. grad., $\Delta p/\rho l =$
$(0{\cdot}00098)$ to $(29{\cdot}4)\ \mathrm{ms^{-2}}$

Roughness size, $k_s = 6{\cdot}0$ mm
This table shows values of m, as follows

m_C for Colebrook-White solutions

Grad'nt S	k.p.g. $(\Delta p/\rho l)$	125	150	200	250	300	400	500	600	800	1000	1250	1500	2000	2500
0·00010	(0·00098)	1·701	1·680	1·653	1·638	1·628	1·617	1·611	1·609	1·608	1·610	1·615	1·620	1·630	1·640
0·00015	(0·00147)	1·696	1·676	1·651	1·635	1·626	1·615	1·610	1·607	1·607	1·609	1·614	1·619	1·629	1·639
0·00020	(0·00196)	1·694	1·674	1·649	1·634	1·624	1·614	1·609	1·607	1·606	1·609	1·613	1·618	1·629	1·639
0·00030	(0·00294)	1·691	1·671	1·647	1·632	1·623	1·613	1·608	1·606	1·606	1·608	1·613	1·618	1·628	1·638
0·00040	(0·00392)	1·689	1·670	1·646	1·631	1·622	1·612	1·607	1·605	1·605	1·608	1·612	1·618	1·628	1·638
0·00060	(0·00588)	1·687	1·668	1·644	1·630	1·621	1·611	1·606	1·604	1·605	1·607	1·612	1·617	1·628	1·638
0·00080	(0·00785)	1·686	1·667	1·643	1·629	1·620	1·610	1·606	1·604	1·604	1·607	1·612	1·617	1·628	1·638
0·00100	(0·00981)	1·685	1·666	1·643	1·629	1·620	1·610	1·606	1·604	1·604	1·607	1·612	1·617	1·627	1·638
0·00150	(0·0147)	1·684	1·665	1·642	1·628	1·619	1·609	1·605	1·603	1·604	1·607	1·611	1·617	1·627	1·637
0·00200	(0·0196)	1·683	1·665	1·641	1·627	1·619	1·609	1·605	1·603	1·604	1·606	1·611	1·616	1·627	1·637
0·00300	(0·0294)	1·682	1·664	1·641	1·627	1·618	1·609	1·604	1·603	1·603	1·606	1·611	1·616	1·627	1·637
0·00400	(0·0392)	1·682	1·663	1·640	1·627	1·618	1·608	1·604	1·603	1·603	1·606	1·611	1·616	1·627	1·637
0·00600	(0·0588)	1·681	1·663	1·640	1·626	1·618	1·608	1·604	1·602	1·603	1·606	1·611	1·616	1·627	1·637
0·00800	(0·0785)	1·680	1·662	1·639	1·626	1·617	1·608	1·604	1·602	1·603	1·606	1·611	1·616	1·627	1·637
0·01000	(0·0981)	1·680	1·662	1·639	1·626	1·617	1·608	1·604	1·602	1·603	1·606	1·611	1·616	1·627	1·637
0·01500	(0·147)	1·680	1·662	1·639	1·626	1·617	1·608	1·604	1·602	1·603	1·606	1·610	1·616	1·627	1·637
0·02000	(0·196)	1·680	1·661	1·639	1·625	1·617	1·608	1·604	1·602	1·603	1·606	1·610	1·616	1·627	1·637
0·03000	(0·294)	1·679	1·661	1·639	1·625	1·617	1·608	1·603	1·602	1·603	1·605	1·610	1·616	1·627	1·637
0·04000	(0·392)	1·679	1·661	1·638	1·625	1·617	1·607	1·603	1·602	1·603	1·605	1·610	1·616	1·627	1·637
0·06000	(0·588)	1·679	1·661	1·638	1·625	1·617	1·607	1·603	1·602	1·602	1·605	1·610	1·616	1·627	1·637
0·08000	(0·785)	1·679	1·661	1·638	1·625	1·617	1·607	1·603	1·602	1·602	1·605	1·610	1·616	1·626	1·637
0·10000	(0·981)	1·679	1·661	1·638	1·625	1·616	1·607	1·603	1·602	1·602	1·605	1·610	1·616	1·626	1·637
0·15000	(1·47)	1·679	1·661	1·638	1·625	1·616	1·607	1·603	1·602	1·602	1·605	1·610	1·616	1·626	1·637
0·20000	(1·96)	1·678	1·661	1·638	1·625	1·616	1·607	1·603	1·602	1·602	1·605	1·610	1·616	1·626	1·637
0·30000	(2·94)	1·678	1·660	1·638	1·625	1·616	1·607	1·603	1·602	1·602	1·605	1·610	1·616	1·626	1·637
0·40000	(3·92)	1·678	1·660	1·638	1·625	1·616	1·607	1·603	1·602	1·602	1·605	1·610	1·616	1·626	1·637
0·60000	(5·88)	1·678	1·660	1·638	1·625	1·616	1·607	1·603	1·602	1·602	1·605	1·610	1·616	1·626	1·637
0·80000	(7·85)	1·678	1·660	1·638	1·625	1·616	1·607	1·603	1·602	1·602	1·605	1·610	1·616	1·626	1·637
1·00000	(9·81)	1·678	1·660	1·638	1·625	1·616	1·607	1·603	1·602	1·602	1·605	1·610	1·616	1·626	1·637
1·50000	(14·7)	1·678	1·660	1·638	1·625	1·616	1·607	1·603	1·602	1·602	1·605	1·610	1·616	1·626	1·637
2·00000	(19·6)	1·678	1·660	1·638	1·625	1·616	1·607	1·603	1·602	1·602	1·605	1·610	1·616	1·626	1·637
3·00000	(29·4)	1·678	1·660	1·638	1·625	1·616	1·607	1·603	1·602	1·602	1·605	1·610	1·616	1·626	1·637
S	$(\Delta p/\rho l)$	125	150	200	250	300	400	500	600	800	1000	1250	1500	2000	2500

Grad'nt k.p.g.   (Equivalent) Pipe diameters in mm   Roughness size, $k_s = 6{\cdot}0$ mm

E75

Kin. visc., $\nu = 0.80 \times 10^{-6}$ m^{2}s^{-1};
$S = 0.00010$ to 30.0000

i.e. kin. pr. grad., $\Delta p / \rho l =$
(0.00098) to (294) ms^{-2}

Roughness size, $k_s = 0.0015$ mm
This table shows values of m, as follows

m_C for Colebrook-White solutions; or,
where $\mathbf{R} \leq 2000$, m_P for laminar flow

Grad'nt — **k.p.g.** — (Equivalent) Pipe diameters in mm

S	$(\Delta p/\rho l)$	6	8	10	12.5	15	20	25	30	40	50	60	80	100	125
0.00030	(0.00294)	5.486	3.738	2.776	2.062	1.617	1.102	*	*	1.055	1.034	1.020	1.002	0.991	0.982
0.00040	(0.00392)	4.751	3.237	2.404	1.786	1.400	0.954	*	1.059	1.029	1.010	0.998	0.981	0.971	0.964
0.00060	(0.00588)	3.879	2.643	1.963	1.458	1.143	*	1.041	1.021	0.995	0.979	0.968	0.953	0.945	0.939
0.00080	(0.00785)	3.359	2.289	1.700	1.263	0.990	*	1.014	0.995	0.972	0.957	0.947	0.935	0.927	0.922
0.00100	(0.00981)	3.005	2.048	1.521	1.129	0.886	*	0.994	0.977	0.955	0.941	0.932	0.921	0.914	0.909
0.00150	(0.0147)	2.453	1.672	1.242	0.922	*	0.980	0.959	0.944	0.925	0.914	0.906	0.896	0.891	0.887
0.00200	(0.0196)	2.125	1.448	1.075	0.799	*	0.955	0.936	0.923	0.905	0.895	0.888	0.880	0.875	0.872
0.00300	(0.0294)	1.735	1.182	0.878	*	0.949	0.922	0.905	0.894	0.879	0.870	0.864	0.857	0.854	0.852
0.00400	(0.0392)	1.502	1.024	0.760	*	0.925	0.900	0.885	0.874	0.861	0.853	0.848	0.842	0.839	0.838
0.00600	(0.0588)	1.227	0.836	*	0.909	0.892	0.871	0.858	0.849	0.837	0.831	0.826	0.822	0.820	0.819
0.00800	(0.0785)	1.062	0.724	*	0.886	0.871	0.851	0.839	0.831	0.821	0.815	0.812	0.808	0.807	0.806
0.01000	(0.0981)	0.950	*	0.889	0.869	0.855	0.837	0.826	0.818	0.809	0.804	0.801	0.798	0.797	0.797
0.01500	(0.147)	0.776	*	0.857	0.839	0.827	0.812	0.802	0.796	0.788	0.784	0.781	0.780	0.779	0.780
0.02000	(0.196)	*	0.855	0.836	0.820	0.809	0.795	0.786	0.780	0.774	0.770	0.768	0.767	0.768	0.769
0.03000	(0.294)	*	0.824	0.807	0.794	0.784	0.772	0.765	0.760	0.755	0.752	0.751	0.751	0.752	0.754
0.04000	(0.392)	*	0.804	0.789	0.776	0.767	0.757	0.750	0.746	0.742	0.740	0.739	0.739	0.741	0.743
0.06000	(0.588)	0.798	0.777	0.764	0.753	0.745	0.736	0.731	0.728	0.724	0.723	0.723	0.724	0.726	0.729
0.08000	(0.785)	0.778	0.759	0.747	0.737	0.730	0.722	0.718	0.715	0.713	0.712	0.712	0.714	0.716	0.719
0.10000	(0.981)	0.764	0.746	0.734	0.725	0.719	0.712	0.708	0.706	0.704	0.704	0.704	0.706	0.709	0.712
0.15000	(1.47)	0.738	0.723	0.713	0.705	0.700	0.694	0.691	0.690	0.688	0.689	0.690	0.693	0.696	0.700
0.20000	(1.96)	0.721	0.707	0.699	0.692	0.687	0.682	0.680	0.679	0.678	0.679	0.680	0.684	0.687	0.691
0.30000	(2.94)	0.699	0.687	0.679	0.674	0.670	0.666	0.664	0.664	0.664	0.666	0.667	0.671	0.675	0.680
0.40000	(3.92)	0.684	0.673	0.666	0.661	0.658	0.655	0.654	0.654	0.655	0.657	0.659	0.663	0.667	0.673
0.60000	(5.88)	0.664	0.655	0.649	0.645	0.643	0.641	0.640	0.641	0.642	0.645	0.647	0.652	0.657	0.662
0.80000	(7.85)	0.650	0.642	0.638	0.634	0.632	0.631	0.631	0.632	0.634	0.637	0.639	0.645	0.650	0.656
1.00000	(9.81)	0.641	0.633	0.629	0.626	0.625	0.624	0.624	0.625	0.628	0.631	0.634	0.639	0.645	0.651
1.50000	(14.7)	0.623	0.617	0.614	0.612	0.611	0.611	0.612	0.613	0.617	0.620	0.624	0.630	0.636	0.642
2.00000	(19.6)	0.612	0.607	0.604	0.603	0.602	0.603	0.604	0.606	0.609	0.613	0.617	0.624	0.630	0.636
3.00000	(29.4)	0.597	0.593	0.591	0.590	0.590	0.591	0.593	0.595	0.600	0.604	0.608	0.615	0.622	0.629
4.00000	(39.2)	0.587	0.583	0.582	0.582	0.582	0.584	0.586	0.589	0.593	0.598	0.602	0.610	0.617	0.624
6.00000	(58.8)	0.573	0.571	0.570	0.571	0.571	0.574	0.577	0.580	0.585	0.590	0.595	0.603	0.610	0.618
8.00000	(78.5)	0.564	0.563	0.562	0.563	0.564	0.567	0.571	0.574	0.580	0.585	0.590	0.598	0.606	0.614
10.0000	(98.1)	0.558	0.557	0.557	0.558	0.559	0.563	0.566	0.569	0.576	0.581	0.586	0.595	0.602	0.611
15.0000	(147)	0.546	0.546	0.547	0.549	0.550	0.554	0.558	0.562	0.569	0.575	0.580	0.589	0.597	0.606
20.0000	(196)	0.539	0.539	0.540	0.542	0.545	0.549	0.553	0.557	0.564	0.571	0.576	0.586	0.594	0.603
30.0000	(294)	0.529	0.530	0.532	0.535	0.537	0.542	0.547	0.551	0.559	0.565	0.571	0.581	0.590	0.599
S	$(\Delta p/\rho l)$	6	8	10	12.5	15	20	25	30	40	50	60	80	100	125

Grad'nt — **k.p.g.** — (Equivalent) Pipe diameters in mm

S	$(\Delta p/\rho l)$	125	150	200	250	300	400	500	600	800	1000	1250	1500	2000	2500
0.00010	(0.00098)	1.060	1.051	1.040	1.034	1.031	1.027	1.026	1.027	1.029	1.032	1.037	1.041	1.050	1.058
0.00015	(0.00147)	1.030	1.023	1.013	1.009	1.006	1.004	1.004	1.005	1.008	1.012	1.017	1.022	1.031	1.040
0.00020	(0.00196)	1.010	1.003	0.995	0.991	0.989	0.988	0.988	0.990	0.994	0.998	1.004	1.009	1.018	1.027
0.00030	(0.00294)	0.982	0.977	0.970	0.967	0.966	0.966	0.967	0.970	0.974	0.979	0.985	0.991	1.001	1.010
0.00040	(0.00392)	0.964	0.959	0.954	0.951	0.951	0.951	0.953	0.956	0.961	0.966	0.973	0.978	0.989	0.998
0.00060	(0.00588)	0.939	0.935	0.931	0.930	0.930	0.931	0.934	0.937	0.943	0.949	0.955	0.962	0.973	0.982
0.00080	(0.00785)	0.922	0.919	0.916	0.915	0.915	0.918	0.921	0.924	0.930	0.937	0.944	0.950	0.961	0.971
0.00100	(0.00981)	0.909	0.906	0.904	0.904	0.905	0.907	0.911	0.914	0.921	0.927	0.935	0.941	0.953	0.963
0.00150	(0.0147)	0.887	0.885	0.884	0.884	0.886	0.889	0.893	0.897	0.904	0.911	0.919	0.926	0.938	0.948
0.00200	(0.0196)	0.872	0.871	0.870	0.871	0.873	0.877	0.881	0.885	0.893	0.900	0.908	0.915	0.928	0.938
0.00300	(0.0294)	0.852	0.851	0.851	0.853	0.855	0.860	0.865	0.869	0.878	0.885	0.894	0.901	0.914	0.924
0.00400	(0.0392)	0.838	0.838	0.839	0.841	0.843	0.849	0.854	0.858	0.867	0.875	0.884	0.891	0.904	0.915
0.00600	(0.0588)	0.819	0.819	0.821	0.824	0.827	0.833	0.839	0.844	0.853	0.861	0.870	0.878	0.891	0.902
0.00800	(0.0785)	0.806	0.807	0.810	0.813	0.816	0.822	0.828	0.834	0.843	0.852	0.861	0.869	0.882	0.894
0.01000	(0.0981)	0.797	0.798	0.801	0.804	0.808	0.814	0.820	0.826	0.836	0.844	0.854	0.862	0.876	0.887
0.01500	(0.147)	0.780	0.782	0.785	0.789	0.793	0.800	0.807	0.813	0.823	0.832	0.841	0.850	0.864	0.876
0.02000	(0.196)	0.769	0.771	0.775	0.779	0.783	0.791	0.797	0.804	0.814	0.823	0.833	0.842	0.856	0.868
0.03000	(0.294)	0.754	0.756	0.761	0.765	0.770	0.778	0.785	0.791	0.802	0.812	0.822	0.831	0.846	0.858
0.04000	(0.392)	0.743	0.746	0.751	0.756	0.761	0.769	0.776	0.783	0.794	0.804	0.814	0.823	0.838	0.851
0.06000	(0.588)	0.729	0.732	0.738	0.743	0.748	0.757	0.765	0.772	0.784	0.793	0.804	0.813	0.829	0.842
0.08000	(0.785)	0.719	0.723	0.729	0.735	0.740	0.749	0.757	0.764	0.776	0.786	0.797	0.807	0.823	0.836
0.10000	(0.981)	0.712	0.716	0.722	0.728	0.734	0.743	0.751	0.758	0.771	0.781	0.792	0.802	0.818	0.831
0.15000	(1.47)	0.700	0.704	0.711	0.717	0.723	0.733	0.741	0.749	0.761	0.772	0.784	0.793	0.810	0.823
0.20000	(1.96)	0.691	0.695	0.703	0.710	0.715	0.726	0.734	0.742	0.755	0.766	0.778	0.788	0.804	0.818
0.30000	(2.94)	0.680	0.685	0.692	0.699	0.706	0.716	0.726	0.733	0.747	0.758	0.770	0.780	0.798	0.812
0.40000	(3.92)	0.673	0.677	0.685	0.693	0.699	0.710	0.720	0.728	0.742	0.753	0.765	0.776	0.793	0.807
0.60000	(5.88)	0.662	0.667	0.676	0.684	0.691	0.702	0.712	0.720	0.734	0.746	0.759	0.769	0.787	0.802
0.80000	(7.85)	0.656	0.661	0.670	0.678	0.685	0.697	0.707	0.715	0.730	0.742	0.754	0.765	0.784	0.798
1.00000	(9.81)	0.651	0.656	0.665	0.673	0.681	0.693	0.703	0.712	0.726	0.739	0.751	0.762	0.781	0.796
1.50000	(14.7)	0.642	0.648	0.658	0.666	0.673	0.686	0.696	0.705	0.721	0.733	0.746	0.758	0.776	0.792
2.00000	(19.6)	0.636	0.642	0.652	0.661	0.669	0.682	0.692	0.702	0.717	0.730	0.743	0.755	0.774	0.789
3.00000	(29.4)	0.629	0.635	0.646	0.655	0.663	0.676	0.687	0.697	0.712	0.725	0.739	0.751	0.770	0.786
S	$(\Delta p/\rho l)$	125	150	200	250	300	400	500	600	800	1000	1250	1500	2000	2500

Grad'nt — **k.p.g.** — (Equivalent) Pipe diameters in mm

Kinematic viscosity, $\nu = 0.80 \times 10^{-6}$ m^{2}s^{-1};

Roughness size, $k_s = 0.0015$ mm

Kin. visc., $\nu = 0.80 \times 10^{-6}$ m²s⁻¹;   Roughness size, $k_s = 0.003$ mm
$S = 0.00010$ to 30.0000 This table shows values of m, as follows
i.e. kin. pr. grad., $\Delta p/\rho l =$ m_C for Colebrook-White solutions; or,
(0.00098) to (294) ms⁻²   where $\mathbf{R} \le 2000$, m_P for laminar flow

Grad'nt S	k.p.g. $(\Delta p/\rho l)$	6	8	10	12·5	15	20	25	30	40	50	60	80	100	125
0·00030	(0·00294)	5·486	3·738	2·776	2·062	1·617	1·102	*	*	1·055	1·035	1·021	1·003	0·992	0·983
0·00040	(0·00392)	4·751	3·237	2·404	1·786	1·400	0·954	*	1·059	1·030	1·011	0·998	0·982	0·972	0·965
0·00060	(0·00588)	3·879	2·643	1·963	1·458	1·143	*	1·041	1·021	0·996	0·979	0·968	0·954	0·946	0·940
0·00080	(0·00785)	3·359	2·289	1·700	1·263	0·990	*	1·014	0·996	0·973	0·958	0·948	0·936	0·928	0·923
0·00100	(0·00981)	3·005	2·048	1·521	1·129	0·886	*	0·994	0·978	0·956	0·942	0·933	0·922	0·915	0·910
0·00150	(0·0147)	2·453	1·672	1·242	0·922	*	0·981	0·960	0·945	0·926	0·915	0·907	0·897	0·892	0·888
0·00200	(0·0196)	2·125	1·448	1·075	0·799	*	0·956	0·937	0·924	0·907	0·896	0·889	0·881	0·876	0·873
0·00300	(0·0294)	1·735	1·182	0·878	*	0·950	0·924	0·907	0·895	0·880	0·871	0·865	0·859	0·855	0·853
0·00400	(0·0392)	1·502	1·024	0·760	*	0·926	0·902	0·886	0·876	0·862	0·855	0·850	0·844	0·841	0·840
0·00600	(0·0588)	1·227	0·836	*	0·910	0·894	0·872	0·859	0·850	0·839	0·832	0·828	0·824	0·822	0·821
0·00800	(0·0785)	1·062	0·724	*	0·887	0·872	0·853	0·841	0·833	0·823	0·817	0·814	0·810	0·809	0·809
0·01000	(0·0981)	0·950	*	0·890	0·870	0·856	0·838	0·827	0·820	0·811	0·806	0·803	0·800	0·799	0·800
0·01500	(0·147)	0·776	*	0·859	0·841	0·829	0·813	0·804	0·798	0·790	0·786	0·784	0·782	0·782	0·783
0·02000	(0·196)	*	0·857	0·837	0·822	0·811	0·797	0·788	0·783	0·776	0·773	0·771	0·770	0·771	0·772
0·03000	(0·294)	*	0·826	0·810	0·796	0·786	0·774	0·767	0·763	0·757	0·755	0·754	0·754	0·755	0·757
0·04000	(0·392)	*	0·806	0·791	0·778	0·770	0·759	0·753	0·749	0·745	0·743	0·743	0·743	0·745	0·748
0·06000	(0·588)	0·801	0·779	0·766	0·755	0·748	0·739	0·734	0·731	0·728	0·727	0·727	0·729	0·731	0·734
0·08000	(0·785)	0·781	0·762	0·750	0·740	0·734	0·726	0·721	0·719	0·717	0·716	0·717	0·719	0·721	0·725
0·10000	(0·981)	0·766	0·749	0·738	0·729	0·723	0·716	0·712	0·710	0·708	0·708	0·709	0·711	0·714	0·718
0·15000	(1·47)	0·741	0·726	0·716	0·709	0·704	0·698	0·696	0·694	0·693	0·694	0·695	0·699	0·702	0·707
0·20000	(1·96)	0·725	0·711	0·702	0·696	0·692	0·687	0·685	0·684	0·684	0·685	0·686	0·690	0·694	0·699
0·30000	(2·94)	0·703	0·691	0·684	0·678	0·675	0·671	0·670	0·670	0·671	0·672	0·675	0·679	0·684	0·689
0·40000	(3·92)	0·688	0·678	0·671	0·667	0·664	0·661	0·660	0·660	0·662	0·664	0·667	0·672	0·677	0·682
0·60000	(5·88)	0·669	0·660	0·655	0·651	0·649	0·647	0·647	0·648	0·650	0·653	0·656	0·662	0·667	0·673
0·80000	(7·85)	0·656	0·648	0·644	0·641	0·639	0·638	0·639	0·640	0·643	0·646	0·649	0·655	0·661	0·668
1·00000	(9·81)	0·646	0·639	0·636	0·633	0·632	0·631	0·632	0·634	0·637	0·640	0·644	0·651	0·657	0·663
1·50000	(14·7)	0·630	0·624	0·622	0·620	0·619	0·620	0·621	0·623	0·627	0·631	0·635	0·643	0·649	0·656
2·00000	(19·6)	0·619	0·615	0·612	0·611	0·611	0·612	0·614	0·616	0·621	0·625	0·630	0·637	0·644	0·652
3·00000	(29·4)	0·605	0·602	0·600	0·600	0·600	0·602	0·605	0·607	0·613	0·618	0·622	0·631	0·638	0·646
4·00000	(39·2)	0·596	0·593	0·592	0·592	0·593	0·596	0·599	0·602	0·607	0·613	0·618	0·626	0·634	0·642
6·00000	(58·8)	0·583	0·582	0·582	0·583	0·584	0·587	0·591	0·594	0·601	0·606	0·612	0·621	0·629	0·637
8·00000	(78·5)	0·575	0·574	0·575	0·576	0·578	0·582	0·586	0·589	0·596	0·602	0·608	0·617	0·625	0·634
10·0000	(98·1)	0·570	0·569	0·570	0·572	0·574	0·578	0·582	0·586	0·593	0·599	0·605	0·615	0·623	0·632
15·0000	(147)	0·560	0·560	0·562	0·564	0·566	0·571	0·576	0·580	0·588	0·594	0·600	0·611	0·619	0·629
20·0000	(196)	0·553	0·554	0·556	0·559	0·562	0·567	0·572	0·576	0·584	0·591	0·598	0·608	0·617	0·626
30·0000	(294)	0·545	0·547	0·549	0·552	0·556	0·562	0·567	0·572	0·580	0·588	0·594	0·605	0·614	0·624
S	$(\Delta p/\rho l)$	6	8	10	12·5	15	20	25	30	40	50	60	80	100	125

Grad'nt   k.p.g.   (Equivalent) Pipe diameters in mm

S	$(\Delta p/\rho l)$	125	150	200	250	300	400	500	600	800	1000	1250	1500	2000	2500
0·00010	(0·00098)	1·061	1·052	1·041	1·035	1·031	1·028	1·027	1·027	1·030	1·033	1·038	1·042	1·051	1·059
0·00015	(0·00147)	1·031	1·023	1·014	1·009	1·007	1·004	1·005	1·006	1·009	1·013	1·018	1·023	1·033	1·041
0·00020	(0·00196)	1·010	1·004	0·996	0·992	0·990	0·989	0·989	0·991	0·995	0·999	1·005	1·010	1·020	1·029
0·00030	(0·00294)	0·983	0·978	0·971	0·968	0·967	0·967	0·969	0·971	0·976	0·981	0·987	0·992	1·003	1·012
0·00040	(0·00392)	0·965	0·960	0·955	0·952	0·952	0·952	0·954	0·957	0·962	0·968	0·974	0·980	0·991	1·000
0·00060	(0·00588)	0·940	0·936	0·932	0·931	0·931	0·933	0·935	0·938	0·944	0·950	0·957	0·964	0·975	0·985
0·00080	(0·00785)	0·923	0·920	0·917	0·916	0·917	0·919	0·922	0·926	0·932	0·939	0·946	0·952	0·964	0·974
0·00100	(0·00981)	0·910	0·908	0·905	0·905	0·906	0·909	0·912	0·916	0·923	0·930	0·937	0·944	0·956	0·966
0·00150	(0·0147)	0·888	0·886	0·885	0·886	0·887	0·891	0·895	0·899	0·907	0·914	0·922	0·929	0·941	0·952
0·00200	(0·0196)	0·873	0·872	0·872	0·873	0·875	0·879	0·883	0·888	0·896	0·903	0·911	0·919	0·931	0·942
0·00300	(0·0294)	0·853	0·853	0·853	0·855	0·857	0·862	0·867	0·872	0·881	0·889	0·897	0·905	0·918	0·929
0·00400	(0·0392)	0·840	0·840	0·841	0·843	0·846	0·851	0·857	0·862	0·871	0·879	0·888	0·896	0·909	0·921
0·00600	(0·0588)	0·821	0·822	0·824	0·827	0·830	0·836	0·842	0·847	0·857	0·866	0·875	0·883	0·897	0·909
0·00800	(0·0785)	0·809	0·810	0·813	0·816	0·819	0·826	0·832	0·838	0·848	0·857	0·866	0·874	0·889	0·901
0·01000	(0·0981)	0·800	0·801	0·804	0·808	0·811	0·818	0·825	0·831	0·841	0·850	0·860	0·868	0·883	0·895
0·01500	(0·147)	0·783	0·785	0·789	0·793	0·797	0·805	0·812	0·818	0·829	0·838	0·848	0·857	0·872	0·885
0·02000	(0·196)	0·772	0·774	0·779	0·783	0·788	0·796	0·803	0·809	0·821	0·830	0·841	0·850	0·865	0·878
0·03000	(0·294)	0·757	0·760	0·765	0·770	0·775	0·784	0·791	0·798	0·810	0·820	0·831	0·840	0·856	0·869
0·04000	(0·392)	0·748	0·750	0·756	0·761	0·766	0·775	0·783	0·790	0·803	0·813	0·824	0·834	0·850	0·863
0·06000	(0·588)	0·734	0·737	0·744	0·750	0·755	0·765	0·773	0·780	0·793	0·804	0·815	0·825	0·842	0·855
0·08000	(0·785)	0·725	0·729	0·736	0·742	0·747	0·757	0·766	0·774	0·787	0·798	0·809	0·819	0·836	0·850
0·10000	(0·981)	0·718	0·722	0·729	0·736	0·742	0·752	0·761	0·769	0·782	0·793	0·805	0·815	0·833	0·847
0·15000	(1·47)	0·707	0·711	0·719	0·726	0·732	0·743	0·752	0·760	0·774	0·786	0·798	0·808	0·826	0·841
0·20000	(1·96)	0·699	0·704	0·712	0·719	0·726	0·737	0·746	0·755	0·769	0·781	0·793	0·804	0·822	0·837
0·30000	(2·94)	0·689	0·694	0·703	0·710	0·717	0·729	0·739	0·748	0·762	0·774	0·787	0·798	0·817	0·832
0·40000	(3·92)	0·682	0·687	0·697	0·705	0·712	0·724	0·734	0·743	0·758	0·770	0·783	0·795	0·814	0·829
0·60000	(5·88)	0·673	0·679	0·689	0·697	0·704	0·717	0·728	0·737	0·752	0·765	0·779	0·790	0·809	0·825
0·80000	(7·85)	0·668	0·673	0·684	0·692	0·700	0·713	0·724	0·733	0·749	0·762	0·776	0·787	0·807	0·823
1·00000	(9·81)	0·663	0·669	0·680	0·689	0·697	0·710	0·721	0·730	0·746	0·760	0·773	0·785	0·805	0·821
1·50000	(14·7)	0·656	0·663	0·674	0·683	0·691	0·705	0·716	0·726	0·742	0·756	0·770	0·782	0·802	0·818
2·00000	(19·6)	0·652	0·658	0·670	0·679	0·687	0·701	0·713	0·723	0·740	0·753	0·768	0·780	0·800	0·816
3·00000	(29·4)	0·646	0·653	0·664	0·674	0·683	0·697	0·709	0·720	0·737	0·750	0·765	0·777	0·798	0·814
S	$(\Delta p/\rho l)$	125	150	200	250	300	400	500	600	800	1000	1250	1500	2000	2500

Grad'nt   k.p.g.   (Equivalent) Pipe diameters in mm

Kinematic viscosity, $\nu = 0.80 \times 10^{-6}$ m²s⁻¹ ;   **Roughness size, $k_s = 0.003$ mm**

E77

Kin. visc., $\nu = 0.80 \times 10^{-6}$ m^{2}s^{-1}; $\quad$ Roughness size, $k_s = 0.006$ mm
S = 0.00010 to 30.0000

This table shows values of m, as follows

i.e. kin. pr. grad., $\Delta p/\rho l =$ (0.00098) to (294) ms^{-2}

m_C for Colebrook-White solutions; or, where $\mathbf{R} \leq 2000$, m_P for laminar flow

Grad'nt S	k.p.g. $(\Delta p/\rho l)$	6	8	10	12.5	15	20	25	30	40	50	60	80	100	125
									(Equivalent) Pipe diameters in mm					
0.00030	(0.00294)	5.486	3.738	2.776	2.062	1.617	1.102	*	*	1.057	1.036	1.022	1.004	0.993	0.984
0.00040	(0.00392)	4.751	3.237	2.404	1.786	1.400	0.954	*	1.061	1.031	1.012	1.000	0.983	0.974	0.966
0.00060	(0.00588)	3.879	2.643	1.963	1.458	1.143	*	1.043	1.023	0.997	0.981	0.970	0.956	0.948	0.942
0.00080	(0.00785)	3.359	2.289	1.700	1.263	0.990	*	1.016	0.998	0.974	0.960	0.950	0.937	0.930	0.925
0.00100	(0.00981)	3.005	2.048	1.521	1.129	0.886	*	0.996	0.979	0.957	0.944	0.935	0.924	0.917	0.913
0.00150	(0.0147)	2.453	1.672	1.242	0.922	*	0.983	0.962	0.947	0.928	0.917	0.909	0.900	0.894	0.891
0.00200	(0.0196)	2.125	1.448	1.075	0.799	*	0.958	0.939	0.926	0.909	0.898	0.891	0.883	0.879	0.876
0.00300	(0.0294)	1.735	1.182	0.878	*	0.953	0.926	0.909	0.897	0.883	0.874	0.868	0.862	0.858	0.857
0.00400	(0.0392)	1.502	1.024	0.760	*	0.928	0.904	0.889	0.878	0.865	0.857	0.852	0.847	0.844	0.843
0.00600	(0.0588)	1.227	0.836	*	0.913	0.896	0.875	0.862	0.853	0.842	0.835	0.832	0.827	0.826	0.826
0.00800	(0.0785)	1.062	0.724	*	0.890	0.875	0.856	0.844	0.836	0.826	0.821	0.817	0.814	0.813	0.814
0.01000	(0.0981)	0.950	*	0.893	0.873	0.859	0.842	0.831	0.823	0.815	0.810	0.807	0.804	0.804	0.805
0.01500	(0.147)	0.776	*	0.862	0.844	0.832	0.817	0.808	0.802	0.794	0.791	0.789	0.787	0.788	0.789
0.02000	(0.196)	*	0.860	0.841	0.825	0.814	0.801	0.792	0.787	0.781	0.778	0.776	0.776	0.777	0.779
0.03000	(0.294)	*	0.830	0.814	0.800	0.791	0.779	0.772	0.768	0.763	0.761	0.760	0.761	0.762	0.765
0.04000	(0.392)	*	0.810	0.795	0.783	0.775	0.764	0.758	0.755	0.751	0.749	0.749	0.750	0.753	0.756
0.06000	(0.588)	0.805	0.784	0.771	0.761	0.754	0.745	0.740	0.737	0.735	0.734	0.735	0.737	0.740	0.744
0.08000	(0.785)	0.786	0.767	0.755	0.746	0.740	0.732	0.728	0.726	0.724	0.724	0.725	0.728	0.731	0.735
0.10000	(0.981)	0.772	0.754	0.743	0.735	0.729	0.723	0.719	0.717	0.716	0.717	0.718	0.721	0.725	0.729
0.15000	(1.47)	0.748	0.732	0.723	0.716	0.711	0.706	0.704	0.703	0.703	0.704	0.706	0.710	0.714	0.719
0.20000	(1.96)	0.732	0.718	0.710	0.704	0.700	0.695	0.694	0.693	0.694	0.696	0.698	0.702	0.707	0.713
0.30000	(2.94)	0.711	0.699	0.692	0.687	0.684	0.681	0.680	0.680	0.682	0.684	0.687	0.693	0.698	0.704
0.40000	(3.92)	0.697	0.686	0.680	0.676	0.674	0.672	0.671	0.672	0.674	0.677	0.680	0.686	0.692	0.699
0.60000	(5.88)	0.678	0.670	0.665	0.662	0.660	0.659	0.660	0.661	0.664	0.668	0.671	0.678	0.684	0.691
0.80000	(7.85)	0.666	0.659	0.655	0.653	0.651	0.651	0.652	0.654	0.658	0.662	0.666	0.673	0.680	0.687
1.00000	(9.81)	0.658	0.651	0.648	0.646	0.645	0.645	0.647	0.649	0.653	0.657	0.661	0.669	0.676	0.684
1.50000	(14.7)	0.643	0.638	0.635	0.634	0.634	0.635	0.638	0.640	0.645	0.650	0.655	0.663	0.670	0.678
2.00000	(19.6)	0.633	0.629	0.627	0.627	0.627	0.629	0.632	0.634	0.640	0.645	0.650	0.659	0.667	0.675
3.00000	(29.4)	0.620	0.617	0.617	0.617	0.618	0.621	0.624	0.627	0.634	0.639	0.645	0.654	0.662	0.671
4.00000	(39.2)	0.612	0.610	0.610	0.611	0.612	0.615	0.619	0.623	0.629	0.636	0.641	0.651	0.659	0.668
6.00000	(58.8)	0.601	0.601	0.601	0.603	0.605	0.609	0.613	0.617	0.624	0.631	0.637	0.647	0.655	0.665
8.00000	(78.5)	0.595	0.595	0.596	0.597	0.600	0.604	0.609	0.613	0.621	0.628	0.634	0.644	0.653	0.663
10.0000	(98.1)	0.590	0.590	0.592	0.594	0.596	0.601	0.606	0.611	0.619	0.626	0.632	0.643	0.652	0.661
15.0000	(147)	0.582	0.583	0.585	0.588	0.591	0.597	0.602	0.607	0.615	0.623	0.629	0.640	0.649	0.659
20.0000	(196)	0.577	0.579	0.581	0.584	0.587	0.594	0.599	0.604	0.613	0.620	0.627	0.638	0.648	0.658
30.0000	(294)	0.570	0.573	0.576	0.580	0.583	0.590	0.596	0.601	0.610	0.618	0.625	0.636	0.646	0.656
S	$(\Delta p/\rho l)$	6	8	10	12.5	15	20	25	30	40	50	60	80	100	125

Grad'nt $\quad$ k.p.g. $\quad$ (Equivalent) Pipe diameters in mm

Grad'nt S	k.p.g. $(\Delta p/\rho l)$	125	150	200	250	300	400	500	600	800	1000	1250	1500	2000	2500
								(Equivalent) Pipe diameters in mm							
0.00010	(0.00098)	1.062	1.053	1.042	1.036	1.032	1.029	1.028	1.029	1.032	1.035	1.040	1.045	1.054	1.062
0.00015	(0.00147)	1.032	1.024	1.015	1.011	1.008	1.006	1.006	1.007	1.011	1.015	1.021	1.026	1.035	1.044
0.00020	(0.00196)	1.012	1.005	0.997	0.993	0.991	0.990	0.991	0.993	0.997	1.002	1.008	1.013	1.023	1.032
0.00030	(0.00294)	0.984	0.979	0.973	0.970	0.969	0.969	0.971	0.973	0.978	0.983	0.990	0.996	1.006	1.016
0.00040	(0.00392)	0.966	0.961	0.956	0.954	0.954	0.955	0.957	0.960	0.965	0.971	0.978	0.984	0.995	1.005
0.00060	(0.00588)	0.942	0.938	0.934	0.933	0.933	0.935	0.938	0.941	0.948	0.954	0.961	0.968	0.980	0.990
0.00080	(0.00785)	0.925	0.922	0.919	0.919	0.919	0.922	0.925	0.929	0.936	0.943	0.950	0.957	0.969	0.980
0.00100	(0.00981)	0.913	0.910	0.908	0.908	0.909	0.912	0.916	0.920	0.927	0.934	0.942	0.949	0.961	0.972
0.00150	(0.0147)	0.891	0.889	0.888	0.889	0.891	0.895	0.899	0.903	0.912	0.919	0.927	0.935	0.948	0.959
0.00200	(0.0196)	0.876	0.875	0.875	0.876	0.878	0.883	0.888	0.892	0.901	0.909	0.918	0.925	0.939	0.950
0.00300	(0.0294)	0.857	0.856	0.857	0.859	0.862	0.867	0.873	0.878	0.887	0.895	0.904	0.913	0.927	0.938
0.00400	(0.0392)	0.843	0.843	0.845	0.848	0.851	0.857	0.862	0.868	0.878	0.886	0.896	0.904	0.918	0.931
0.00600	(0.0588)	0.826	0.826	0.829	0.832	0.836	0.842	0.849	0.854	0.865	0.874	0.884	0.893	0.908	0.920
0.00800	(0.0785)	0.814	0.815	0.818	0.822	0.825	0.833	0.839	0.846	0.856	0.866	0.876	0.885	0.900	0.913
0.01000	(0.0981)	0.805	0.806	0.810	0.814	0.818	0.826	0.833	0.839	0.850	0.860	0.870	0.880	0.895	0.908
0.01500	(0.147)	0.789	0.791	0.796	0.800	0.805	0.813	0.821	0.828	0.839	0.850	0.860	0.870	0.886	0.900
0.02000	(0.196)	0.779	0.781	0.786	0.791	0.796	0.805	0.813	0.820	0.832	0.843	0.854	0.864	0.880	0.894
0.03000	(0.294)	0.765	0.768	0.774	0.779	0.785	0.794	0.803	0.810	0.823	0.834	0.846	0.856	0.873	0.887
0.04000	(0.392)	0.756	0.759	0.766	0.772	0.777	0.787	0.796	0.804	0.817	0.828	0.840	0.851	0.868	0.883
0.06000	(0.588)	0.744	0.747	0.755	0.761	0.767	0.778	0.787	0.795	0.809	0.821	0.833	0.844	0.862	0.877
0.08000	(0.785)	0.735	0.740	0.747	0.754	0.761	0.772	0.781	0.790	0.804	0.816	0.829	0.840	0.858	0.873
0.10000	(0.981)	0.729	0.734	0.742	0.749	0.756	0.767	0.777	0.786	0.800	0.812	0.825	0.836	0.855	0.871
0.15000	(1.47)	0.719	0.724	0.733	0.741	0.748	0.760	0.770	0.779	0.794	0.807	0.820	0.831	0.851	0.866
0.20000	(1.96)	0.713	0.718	0.727	0.735	0.743	0.755	0.765	0.775	0.790	0.803	0.817	0.828	0.848	0.864
0.30000	(2.94)	0.704	0.710	0.720	0.728	0.736	0.749	0.760	0.769	0.785	0.798	0.812	0.824	0.844	0.860
0.40000	(3.92)	0.699	0.704	0.715	0.724	0.732	0.745	0.756	0.766	0.782	0.796	0.810	0.822	0.842	0.858
0.60000	(5.88)	0.691	0.698	0.709	0.718	0.726	0.740	0.752	0.762	0.778	0.792	0.806	0.819	0.839	0.856
0.80000	(7.85)	0.687	0.693	0.705	0.714	0.723	0.737	0.749	0.759	0.776	0.790	0.804	0.817	0.837	0.854
1.00000	(9.81)	0.684	0.690	0.702	0.712	0.720	0.735	0.747	0.757	0.774	0.788	0.803	0.816	0.836	0.853
1.50000	(14.7)	0.678	0.685	0.697	0.708	0.716	0.731	0.744	0.754	0.772	0.786	0.801	0.813	0.834	0.851
2.00000	(19.6)	0.675	0.682	0.695	0.705	0.714	0.729	0.742	0.752	0.770	0.784	0.799	0.812	0.833	0.850
3.00000	(29.4)	0.671	0.678	0.691	0.702	0.711	0.726	0.739	0.750	0.768	0.782	0.798	0.811	0.832	0.849
S	$(\Delta p/\rho l)$	125	150	200	250	300	400	500	600	800	1000	1250	1500	2000	2500

Grad'nt $\quad$ k.p.g. $\quad$ (Equivalent) Pipe diameters in mm

Kinematic viscosity, $\nu = 0.80 \times 10^{-6}$ m^{2}s^{-1}; $\qquad$ **Roughness size, $k_s = 0.006$ mm**

Kin. visc., $\nu = 0.80 \times 10^{-6}$ m^{2}s^{-1};
$S = 0.00010$ to 30.0000

i.e. kin. pr. grad., $\Delta p/\rho l =$
(0.00098) to (294) ms^{-2}

Roughness size, $k_s = 0.015$ mm
This table shows values of m, as follows

m_C for Colebrook-White solutions; or,
where $\mathbf{R} \leq 2000$, m_P for laminar flow

Grad'nt S	k.p.g. $(\Delta p/\rho l)$	6	8	10	12·5	15	20	25	30	40	50	60	80	100	125
		\(Equivalent) Pipe diameters in mm													
0·00030	(0·00294)	5·486	3·738	2·776	2·062	1·617	1·102	*	*	1·060	1·040	1·026	1·008	0·997	0·989
0·00040	(0·00392)	4·751	3·237	2·404	1·786	1·400	0·954	*	1·064	1·035	1·016	1·004	0·988	0·978	0·971
0·00060	(0·00588)	3·879	2·643	1·963	1·458	1·143	*	1·047	1·027	1·001	0·985	0·974	0·961	0·953	0·947
0·00080	(0·00785)	3·359	2·289	1·700	1·263	0·990	*	1·020	1·002	0·979	0·964	0·955	0·943	0·936	0·931
0·00100	(0·00981)	3·005	2·048	1·521	1·129	0·886	*	1·001	0·984	0·962	0·949	0·940	0·929	0·923	0·919
0·00150	(0·0147)	2·453	1·672	1·242	0·922	*	0·988	0·967	0·953	0·934	0·923	0·915	0·906	0·901	0·898
0·00200	(0·0196)	2·125	1·448	1·075	0·799	*	0·964	0·945	0·932	0·915	0·905	0·898	0·891	0·887	0·884
0·00300	(0·0294)	1·735	1·182	0·878	*	0·959	0·932	0·915	0·904	0·890	0·881	0·876	0·870	0·867	0·866
0·00400	(0·0392)	1·502	1·024	0·760	*	0·935	0·911	0·896	0·886	0·873	0·866	0·861	0·856	0·854	0·854
0·00600	(0·0588)	1·227	0·836	*	0·920	0·904	0·883	0·870	0·861	0·851	0·845	0·841	0·838	0·837	0·837
0·00800	(0·0785)	1·062	0·724	*	0·898	0·883	0·865	0·853	0·845	0·836	0·831	0·828	0·826	0·826	0·827
0·01000	(0·0981)	0·950	*	0·902	0·882	0·868	0·851	0·840	0·833	0·825	0·821	0·819	0·817	0·817	0·819
0·01500	(0·147)	0·776	*	0·871	0·854	0·842	0·828	0·819	0·813	0·806	0·803	0·802	0·802	0·803	0·805
0·02000	(0·196)	*	0·870	0·851	0·836	0·825	0·812	0·804	0·799	0·794	0·792	0·791	0·791	0·793	0·796
0·03000	(0·294)	*	0·842	0·825	0·812	0·803	0·792	0·786	0·782	0·778	0·777	0·777	0·778	0·781	0·785
0·04000	(0·392)	*	0·823	0·808	0·796	0·788	0·779	0·773	0·770	0·767	0·767	0·767	0·770	0·773	0·777
0·06000	(0·588)	0·819	0·798	0·786	0·776	0·769	0·761	0·757	0·755	0·753	0·754	0·755	0·758	0·762	0·767
0·08000	(0·785)	0·801	0·782	0·771	0·762	0·756	0·750	0·746	0·745	0·744	0·745	0·747	0·751	0·755	0·761
0·10000	(0·981)	0·788	0·771	0·760	0·752	0·747	0·741	0·739	0·737	0·738	0·739	0·741	0·746	0·751	0·756
0·15000	(1·47)	0·765	0·751	0·742	0·735	0·731	0·727	0·726	0·725	0·727	0·729	0·731	0·737	0·743	0·749
0·20000	(1·96)	0·751	0·738	0·730	0·724	0·721	0·718	0·717	0·717	0·719	0·722	0·725	0·731	0·737	0·744
0·30000	(2·94)	0·732	0·721	0·715	0·710	0·708	0·706	0·706	0·707	0·710	0·714	0·717	0·724	0·731	0·738
0·40000	(3·92)	0·720	0·710	0·705	0·701	0·700	0·699	0·699	0·701	0·704	0·708	0·712	0·720	0·727	0·734
0·60000	(5·88)	0·704	0·696	0·692	0·690	0·689	0·689	0·690	0·692	0·697	0·702	0·706	0·714	0·722	0·730
0·80000	(7·85)	0·694	0·687	0·684	0·682	0·682	0·683	0·685	0·687	0·692	0·697	0·702	0·711	0·719	0·727
1·00000	(9·81)	0·686	0·681	0·678	0·677	0·677	0·678	0·681	0·684	0·689	0·694	0·699	0·708	0·716	0·725
1·50000	(14·7)	0·674	0·670	0·668	0·668	0·669	0·671	0·674	0·678	0·684	0·690	0·695	0·705	0·713	0·722
2·00000	(19·6)	0·667	0·663	0·662	0·663	0·664	0·667	0·670	0·674	0·681	0·687	0·692	0·702	0·711	0·720
3·00000	(29·4)	0·657	0·655	0·655	0·656	0·657	0·661	0·665	0·669	0·676	0·683	0·689	0·699	0·708	0·717
4·00000	(39·2)	0·651	0·650	0·650	0·651	0·653	0·658	0·662	0·666	0·674	0·681	0·687	0·697	0·706	0·716
6·00000	(58·8)	0·643	0·643	0·644	0·646	0·648	0·653	0·658	0·663	0·671	0·678	0·684	0·695	0·704	0·714
8·00000	(78·5)	0·638	0·639	0·640	0·643	0·645	0·651	0·656	0·661	0·669	0·676	0·683	0·694	0·703	0·713
10·0000	(98·1)	0·635	0·636	0·638	0·640	0·643	0·649	0·654	0·659	0·668	0·675	0·682	0·693	0·702	0·712
15·0000	(147)	0·630	0·631	0·633	0·637	0·640	0·646	0·652	0·657	0·666	0·673	0·680	0·691	0·701	0·711
20·0000	(196)	0·626	0·628	0·631	0·634	0·638	0·644	0·650	0·655	0·664	0·672	0·679	0·690	0·700	0·710
30·0000	(294)	0·622	0·625	0·628	0·632	0·635	0·642	0·648	0·653	0·663	0·671	0·678	0·689	0·699	0·709
S	$(\Delta p/\rho l)$	6	8	10	12·5	15	20	25	30	40	50	60	80	100	125

Grad'nt   k.p.g.   (Equivalent) Pipe diameters in mm

Grad'nt S	k.p.g. $(\Delta p/\rho l)$	125	150	200	250	300	400	500	600	800	1000	1250	1500	2000	2500
0·00010	(0·00098)	1·065	1·056	1·045	1·039	1·036	1·033	1·033	1·033	1·037	1·040	1·046	1·051	1·060	1·069
0·00015	(0·00147)	1·035	1·028	1·019	1·014	1·012	1·011	1·011	1·013	1·017	1·021	1·027	1·033	1·043	1·053
0·00020	(0·00196)	1·015	1·009	1·001	0·998	0·996	0·995	0·997	0·999	1·003	1·009	1·015	1·021	1·032	1·041
0·00030	(0·00294)	0·989	0·983	0·978	0·975	0·974	0·975	0·977	0·980	0·985	0·991	0·998	1·005	1·016	1·027
0·00040	(0·00392)	0·971	0·966	0·962	0·960	0·960	0·961	0·964	0·967	0·973	0·980	0·987	0·994	1·006	1·017
0·00060	(0·00588)	0·947	0·944	0·940	0·940	0·940	0·943	0·946	0·950	0·957	0·964	0·972	0·980	0·992	1·004
0·00080	(0·00785)	0·931	0·928	0·926	0·926	0·927	0·930	0·934	0·938	0·946	0·954	0·962	0·970	0·983	0·995
0·00100	(0·00981)	0·919	0·917	0·915	0·916	0·917	0·921	0·925	0·930	0·938	0·946	0·955	0·963	0·977	0·989
0·00150	(0·0147)	0·898	0·897	0·897	0·898	0·900	0·905	0·910	0·915	0·925	0·933	0·942	0·951	0·965	0·978
0·00200	(0·0196)	0·884	0·884	0·884	0·886	0·889	0·895	0·900	0·905	0·915	0·924	0·934	0·943	0·958	0·970
0·00300	(0·0294)	0·866	0·866	0·868	0·871	0·874	0·880	0·887	0·893	0·903	0·913	0·923	0·932	0·948	0·961
0·00400	(0·0392)	0·854	0·854	0·857	0·860	0·864	0·871	0·878	0·884	0·895	0·905	0·916	0·925	0·942	0·955
0·00600	(0·0588)	0·837	0·839	0·842	0·847	0·851	0·859	0·866	0·873	0·885	0·895	0·907	0·916	0·933	0·947
0·00800	(0·0785)	0·827	0·828	0·833	0·837	0·842	0·851	0·859	0·866	0·878	0·889	0·901	0·911	0·928	0·943
0·01000	(0·0981)	0·819	0·821	0·826	0·831	0·836	0·845	0·853	0·860	0·873	0·884	0·896	0·907	0·924	0·939
0·01500	(0·147)	0·805	0·808	0·814	0·820	0·825	0·835	0·844	0·851	0·865	0·877	0·889	0·900	0·918	0·933
0·02000	(0·196)	0·796	0·799	0·806	0·812	0·818	0·828	0·838	0·846	0·860	0·872	0·885	0·896	0·914	0·930
0·03000	(0·294)	0·785	0·788	0·796	0·803	0·809	0·820	0·830	0·838	0·853	0·866	0·879	0·890	0·909	0·925
0·04000	(0·392)	0·777	0·781	0·789	0·797	0·803	0·815	0·825	0·834	0·849	0·862	0·875	0·887	0·906	0·922
0·06000	(0·588)	0·767	0·772	0·781	0·789	0·796	0·808	0·819	0·828	0·844	0·857	0·871	0·883	0·902	0·919
0·08000	(0·785)	0·761	0·766	0·775	0·784	0·791	0·804	0·815	0·824	0·840	0·854	0·868	0·880	0·900	0·917
0·10000	(0·981)	0·756	0·762	0·772	0·780	0·788	0·801	0·812	0·822	0·838	0·852	0·866	0·878	0·899	0·915
0·15000	(1·47)	0·749	0·755	0·765	0·774	0·782	0·796	0·807	0·817	0·834	0·848	0·863	0·875	0·896	0·913
0·20000	(1·96)	0·744	0·750	0·761	0·771	0·779	0·793	0·805	0·815	0·832	0·846	0·861	0·873	0·894	0·911
0·30000	(2·94)	0·738	0·745	0·756	0·766	0·775	0·789	0·801	0·812	0·829	0·843	0·858	0·871	0·892	0·910
0·40000	(3·92)	0·734	0·741	0·753	0·763	0·772	0·787	0·799	0·810	0·827	0·842	0·857	0·870	0·891	0·909
0·60000	(5·88)	0·730	0·737	0·749	0·760	0·769	0·784	0·797	0·807	0·825	0·840	0·855	0·868	0·890	0·907
0·80000	(7·85)	0·727	0·734	0·747	0·758	0·767	0·782	0·795	0·806	0·824	0·839	0·854	0·867	0·889	0·906
1·00000	(9·81)	0·725	0·732	0·745	0·756	0·765	0·781	0·794	0·805	0·823	0·838	0·853	0·867	0·888	0·906
1·50000	(14·7)	0·722	0·729	0·742	0·754	0·763	0·779	0·792	0·803	0·822	0·837	0·852	0·866	0·887	0·905
2·00000	(19·6)	0·720	0·728	0·741	0·752	0·762	0·778	0·791	0·802	0·821	0·836	0·852	0·865	0·887	0·905
3·00000	(29·4)	0·717	0·725	0·739	0·751	0·760	0·776	0·790	0·801	0·820	0·835	0·851	0·864	0·886	0·904
S	$(\Delta p/\rho l)$	125	150	200	250	300	400	500	600	800	1000	1250	1500	2000	2500

Grad'nt   k.p.g.   (Equivalent) Pipe diameters in mm

Kinematic viscosity, $\nu = 0.80 \times 10^{-6}$ m^{2}s^{-1} ;   **Roughness size, $k_s = 0.015$ mm**

E79

Kin. visc., $\nu = 0.80 \times 10^{-6}$ m²s⁻¹;
$S = 0.00010$ to 30.0000

i.e. kin. pr. grad., $\Delta p/\rho l =$
(0.00098) to (294) ms⁻²

Roughness size, $k_s = 0.030$ mm
This table shows values of m, as follows

m_C for Colebrook-White solutions; or,
where $\mathbf{R} \leq 2000$, m_P for laminar flow

Grad'nt S	k.p.g. $(\Delta p/\rho l)$	6	8	10	12.5	15	20	25	30	40	50	60	80	100	125
		(Equivalent) Pipe diameters in mm													
0.00030	(0.00294)	5.486	3.738	2.776	2.062	1.617	1.102	*	*	1.065	1.045	1.032	1.014	1.004	0.996
0.00040	(0.00392)	4.751	3.237	2.404	1.786	1.400	0.954	*	1.070	1.041	1.022	1.010	0.994	0.985	0.978
0.00060	(0.00588)	3.879	2.643	1.963	1.458	1.143	*	1.053	1.033	1.008	0.992	0.982	0.969	0.961	0.956
0.00080	(0.00785)	3.359	2.289	1.700	1.263	0.990	*	1.027	1.009	0.986	0.972	0.963	0.951	0.945	0.940
0.00100	(0.00981)	3.005	2.048	1.521	1.129	0.886	*	1.008	0.992	0.970	0.957	0.949	0.938	0.933	0.929
0.00150	(0.0147)	2.453	1.672	1.242	0.922	*	0.996	0.976	0.961	0.943	0.932	0.925	0.917	0.912	0.910
0.00200	(0.0196)	2.125	1.448	1.075	0.799	*	0.973	0.954	0.941	0.925	0.915	0.909	0.902	0.899	0.897
0.00300	(0.0294)	1.735	1.182	0.878	*	0.969	0.942	0.926	0.915	0.901	0.893	0.888	0.883	0.881	0.880
0.00400	(0.0392)	1.502	1.024	0.760	*	0.946	0.922	0.907	0.898	0.885	0.879	0.874	0.870	0.869	0.869
0.00600	(0.0588)	1.227	0.836	*	0.932	0.916	0.896	0.883	0.875	0.865	0.859	0.856	0.854	0.854	0.855
0.00800	(0.0785)	1.062	0.724	*	0.911	0.897	0.878	0.867	0.860	0.851	0.847	0.845	0.843	0.844	0.846
0.01000	(0.0981)	0.950	*	0.915	0.896	0.882	0.865	0.855	0.849	0.841	0.838	0.836	0.835	0.836	0.839
0.01500	(0.147)	0.776	*	0.887	0.870	0.858	0.844	0.836	0.830	0.825	0.822	0.821	0.822	0.824	0.828
0.02000	(0.196)	*	0.886	0.868	0.853	0.842	0.830	0.823	0.818	0.814	0.812	0.812	0.814	0.816	0.820
0.03000	(0.294)	*	0.860	0.844	0.831	0.822	0.812	0.806	0.803	0.800	0.799	0.800	0.803	0.806	0.811
0.04000	(0.392)	*	0.842	0.828	0.816	0.809	0.800	0.795	0.793	0.791	0.791	0.792	0.796	0.800	0.805
0.06000	(0.588)	0.840	0.820	0.808	0.798	0.792	0.785	0.781	0.780	0.779	0.780	0.782	0.787	0.792	0.798
0.08000	(0.785)	0.824	0.805	0.794	0.786	0.781	0.775	0.772	0.771	0.772	0.774	0.776	0.781	0.787	0.793
0.10000	(0.981)	0.812	0.795	0.785	0.777	0.773	0.768	0.766	0.765	0.766	0.769	0.771	0.777	0.783	0.790
0.15000	(1.47)	0.792	0.777	0.769	0.763	0.760	0.756	0.755	0.756	0.758	0.761	0.764	0.771	0.777	0.784
0.20000	(1.96)	0.779	0.766	0.759	0.754	0.751	0.749	0.749	0.749	0.752	0.756	0.760	0.767	0.773	0.781
0.30000	(2.94)	0.763	0.752	0.746	0.743	0.741	0.739	0.740	0.742	0.745	0.750	0.754	0.762	0.769	0.777
0.40000	(3.92)	0.752	0.743	0.738	0.735	0.734	0.734	0.735	0.737	0.741	0.746	0.750	0.759	0.766	0.774
0.60000	(5.88)	0.739	0.732	0.728	0.726	0.726	0.726	0.728	0.731	0.736	0.741	0.746	0.755	0.763	0.771
0.80000	(7.85)	0.731	0.725	0.722	0.720	0.720	0.722	0.724	0.727	0.733	0.738	0.743	0.752	0.761	0.769
1.00000	(9.81)	0.725	0.720	0.717	0.716	0.717	0.719	0.721	0.724	0.730	0.736	0.741	0.751	0.759	0.768
1.50000	(14.7)	0.716	0.712	0.710	0.710	0.711	0.714	0.717	0.720	0.727	0.733	0.739	0.748	0.757	0.766
2.00000	(19.6)	0.710	0.707	0.706	0.706	0.707	0.710	0.714	0.718	0.725	0.731	0.737	0.747	0.755	0.765
3.00000	(29.4)	0.702	0.700	0.700	0.701	0.703	0.707	0.711	0.715	0.722	0.729	0.735	0.745	0.754	0.763
4.00000	(39.2)	0.698	0.696	0.697	0.698	0.700	0.704	0.709	0.713	0.720	0.727	0.733	0.744	0.753	0.762
6.00000	(58.8)	0.693	0.692	0.693	0.694	0.697	0.701	0.706	0.710	0.718	0.725	0.732	0.742	0.751	0.761
8.00000	(78.5)	0.689	0.689	0.690	0.692	0.695	0.700	0.705	0.709	0.717	0.724	0.731	0.742	0.751	0.761
10.0000	(98.1)	0.687	0.687	0.688	0.691	0.693	0.698	0.703	0.708	0.716	0.724	0.730	0.741	0.750	0.760
15.0000	(147)	0.683	0.684	0.686	0.688	0.691	0.697	0.702	0.707	0.715	0.722	0.729	0.740	0.749	0.759
20.0000	(196)	0.681	0.682	0.684	0.687	0.690	0.695	0.701	0.706	0.714	0.722	0.728	0.740	0.749	0.759
30.0000	(294)	0.678	0.680	0.682	0.685	0.688	0.694	0.700	0.705	0.713	0.721	0.728	0.739	0.748	0.759
S	$(\Delta p/\rho l)$	6	8	10	12.5	15	20	25	30	40	50	60	80	100	125

Grad'nt **k.p.g.** (Equivalent) Pipe diameters in mm

Grad'nt S	k.p.g. $(\Delta p/\rho l)$	125	150	200	250	300	400	500	600	800	1000	1250	1500	2000	2500
0.00010	(0.00098)	1.069	1.061	1.050	1.045	1.042	1.039	1.039	1.041	1.044	1.049	1.054	1.060	1.071	1.080
0.00015	(0.00147)	1.041	1.033	1.025	1.021	1.019	1.018	1.019	1.021	1.026	1.031	1.038	1.044	1.055	1.065
0.00020	(0.00196)	1.021	1.015	1.008	1.005	1.003	1.003	1.005	1.008	1.013	1.019	1.026	1.033	1.045	1.056
0.00030	(0.00294)	0.996	0.991	0.985	0.983	0.983	0.984	0.987	0.990	0.997	1.003	1.011	1.018	1.031	1.043
0.00040	(0.00392)	0.978	0.974	0.970	0.969	0.969	0.971	0.975	0.978	0.986	0.993	1.001	1.009	1.022	1.034
0.00060	(0.00588)	0.956	0.952	0.950	0.950	0.951	0.954	0.959	0.963	0.971	0.979	0.988	0.996	1.011	1.023
0.00080	(0.00785)	0.940	0.938	0.937	0.937	0.939	0.943	0.948	0.953	0.962	0.970	0.980	0.988	1.003	1.016
0.00100	(0.00981)	0.929	0.927	0.927	0.928	0.930	0.935	0.940	0.945	0.955	0.964	0.974	0.982	0.998	1.011
0.00150	(0.0147)	0.910	0.909	0.910	0.912	0.915	0.921	0.927	0.932	0.943	0.953	0.963	0.972	0.989	1.002
0.00200	(0.0196)	0.897	0.897	0.899	0.901	0.905	0.911	0.918	0.924	0.935	0.945	0.956	0.966	0.983	0.997
0.00300	(0.0294)	0.880	0.881	0.884	0.888	0.892	0.899	0.907	0.913	0.926	0.936	0.948	0.958	0.975	0.990
0.00400	(0.0392)	0.869	0.871	0.874	0.879	0.883	0.892	0.899	0.907	0.919	0.930	0.942	0.953	0.971	0.985
0.00600	(0.0588)	0.855	0.857	0.862	0.867	0.872	0.881	0.890	0.898	0.911	0.923	0.935	0.946	0.965	0.980
0.00800	(0.0785)	0.846	0.848	0.854	0.860	0.865	0.875	0.884	0.892	0.906	0.918	0.931	0.942	0.961	0.977
0.01000	(0.0981)	0.839	0.842	0.848	0.854	0.860	0.870	0.880	0.888	0.902	0.915	0.928	0.939	0.958	0.974
0.01500	(0.147)	0.828	0.831	0.838	0.845	0.851	0.863	0.873	0.881	0.896	0.909	0.923	0.934	0.954	0.970
0.02000	(0.196)	0.820	0.824	0.832	0.839	0.846	0.858	0.868	0.877	0.893	0.906	0.919	0.931	0.951	0.968
0.03000	(0.294)	0.811	0.816	0.824	0.832	0.839	0.852	0.863	0.872	0.888	0.901	0.916	0.928	0.948	0.965
0.04000	(0.392)	0.805	0.810	0.819	0.828	0.835	0.848	0.859	0.869	0.885	0.899	0.913	0.926	0.946	0.963
0.06000	(0.588)	0.798	0.803	0.813	0.822	0.830	0.843	0.855	0.865	0.882	0.896	0.910	0.923	0.944	0.961
0.08000	(0.785)	0.793	0.799	0.809	0.818	0.827	0.840	0.852	0.862	0.879	0.894	0.909	0.921	0.942	0.960
0.10000	(0.981)	0.790	0.796	0.807	0.816	0.824	0.838	0.850	0.861	0.878	0.892	0.907	0.920	0.941	0.959
0.15000	(1.47)	0.784	0.791	0.802	0.812	0.821	0.835	0.847	0.858	0.876	0.890	0.905	0.918	0.940	0.957
0.20000	(1.96)	0.781	0.788	0.799	0.809	0.818	0.833	0.846	0.856	0.874	0.889	0.904	0.917	0.939	0.957
0.30000	(2.94)	0.777	0.784	0.796	0.807	0.816	0.831	0.843	0.854	0.872	0.887	0.903	0.916	0.938	0.955
0.40000	(3.92)	0.774	0.782	0.794	0.805	0.814	0.829	0.842	0.853	0.871	0.886	0.902	0.915	0.937	0.955
0.60000	(5.88)	0.771	0.779	0.792	0.802	0.812	0.828	0.841	0.852	0.870	0.885	0.901	0.914	0.936	0.954
0.80000	(7.85)	0.769	0.777	0.790	0.801	0.811	0.826	0.840	0.851	0.869	0.884	0.900	0.914	0.936	0.954
1.00000	(9.81)	0.768	0.776	0.789	0.800	0.810	0.826	0.839	0.850	0.869	0.884	0.900	0.913	0.935	0.953
1.50000	(14.7)	0.766	0.774	0.787	0.799	0.808	0.825	0.838	0.849	0.868	0.883	0.899	0.913	0.935	0.953
2.00000	(19.6)	0.765	0.773	0.787	0.798	0.808	0.824	0.837	0.849	0.867	0.883	0.899	0.912	0.935	0.953
3.00000	(29.4)	0.763	0.771	0.785	0.797	0.807	0.823	0.837	0.848	0.867	0.882	0.898	0.912	0.934	0.952
S	$(\Delta p/\rho l)$	125	150	200	250	300	400	500	600	800	1000	1250	1500	2000	2500

Grad'nt **k.p.g.** (Equivalent) Pipe diameters in mm

Kinematic viscosity, $\nu = 0.80 \times 10^{-6}$ m²s⁻¹ ;

Roughness size, $k_s = 0.030$ mm

Kin. visc., $\nu = 0.80 \times 10^{-6}$ m^{2}s^{-1};
$S = 0.00010$ to 30.0000
i.e. kin. pr. grad., $\Delta p/\rho l =$
(0.00098) to (294) ms^{-2}

Roughness size, $k_s = 0.060$ mm
This table shows values of m, as follows
m_C for Colebrook-White solutions; or,
where $\mathbf{R} \leq 2000$, m_P for laminar flow

Grad'nt S	k.p.g. $(\Delta p/\rho l)$	6	8	10	12·5	15	20	25	30	40	50	60	80	100	125
0·00030	(0·00294)	5·486	3·738	2·776	2·062	1·617	1·102	*	*	1·076	1·056	1·043	1·026	1·016	1·008
0·00040	(0·00392)	4·751	3·237	2·404	1·786	1·400	0·954	*	1·081	1·052	1·035	1·022	1·007	0·999	0·992
0·00060	(0·00588)	3·879	2·643	1·963	1·458	1·143	*	1·066	1·046	1·021	1·006	0·996	0·983	0·976	0·972
0·00080	(0·00785)	3·359	2·289	1·700	1·263	0·990	*	1·041	1·023	1·001	0·987	0·978	0·967	0·961	0·958
0·00100	(0·00981)	3·005	2·048	1·521	1·129	0·886	*	1·023	1·006	0·986	0·973	0·965	0·956	0·951	0·948
0·00150	(0·0147)	2·453	1·672	1·242	0·922	*	1·012	0·992	0·978	0·960	0·950	0·943	0·936	0·932	0·931
0·00200	(0·0196)	2·125	1·448	1·075	0·799	*	0·990	0·972	0·959	0·944	0·935	0·929	0·923	0·920	0·920
0·00300	(0·0294)	1·735	1·182	0·878	*	0·988	0·962	0·946	0·935	0·922	0·915	0·910	0·906	0·905	0·905
0·00400	(0·0392)	1·502	1·024	0·760	*	0·966	0·943	0·929	0·919	0·908	0·902	0·898	0·895	0·895	0·896
0·00600	(0·0588)	1·227	0·836	*	0·955	0·939	0·919	0·907	0·899	0·890	0·885	0·883	0·881	0·882	0·884
0·00800	(0·0785)	1·062	0·724	*	0·935	0·921	0·903	0·892	0·886	0·878	0·875	0·873	0·873	0·874	0·877
0·01000	(0·0981)	0·950	*	0·941	0·921	0·908	0·892	0·882	0·876	0·870	0·867	0·866	0·866	0·868	0·872
0·01500	(0·147)	0·776	*	0·914	0·898	0·887	0·873	0·865	0·860	0·856	0·854	0·854	0·856	0·859	0·863
0·02000	(0·196)	*	0·916	0·898	0·883	0·873	0·861	0·854	0·850	0·847	0·846	0·846	0·849	0·853	0·857
0·03000	(0·294)	*	0·892	0·876	0·864	0·855	0·845	0·840	0·838	0·836	0·836	0·837	0·841	0·845	0·851
0·04000	(0·392)	*	0·877	0·863	0·851	0·844	0·836	0·832	0·830	0·828	0·829	0·831	0·836	0·840	0·846
0·06000	(0·588)	0·878	0·858	0·845	0·836	0·830	0·823	0·821	0·819	0·820	0·821	0·824	0·829	0·835	0·841
0·08000	(0·785)	0·864	0·845	0·834	0·826	0·821	0·816	0·814	0·813	0·814	0·816	0·819	0·825	0·831	0·838
0·10000	(0·981)	0·853	0·837	0·827	0·819	0·815	0·810	0·809	0·809	0·810	0·813	0·816	0·822	0·829	0·836
0·15000	(1·47)	0·837	0·822	0·814	0·808	0·805	0·801	0·801	0·801	0·804	0·807	0·811	0·818	0·825	0·832
0·20000	(1·96)	0·826	0·813	0·806	0·801	0·798	0·796	0·796	0·797	0·800	0·804	0·808	0·815	0·822	0·830
0·30000	(2·94)	0·813	0·802	0·796	0·792	0·790	0·789	0·790	0·791	0·795	0·800	0·804	0·812	0·819	0·827
0·40000	(3·92)	0·805	0·795	0·790	0·787	0·785	0·785	0·786	0·788	0·792	0·797	0·802	0·810	0·817	0·826
0·60000	(5·88)	0·795	0·787	0·783	0·780	0·779	0·780	0·782	0·784	0·789	0·794	0·799	0·808	0·815	0·824
0·80000	(7·85)	0·789	0·782	0·778	0·776	0·776	0·777	0·779	0·781	0·787	0·792	0·797	0·806	0·814	0·823
1·00000	(9·81)	0·784	0·778	0·775	0·773	0·773	0·775	0·777	0·780	0·785	0·791	0·796	0·805	0·813	0·822
1·50000	(14·7)	0·777	0·772	0·770	0·769	0·769	0·771	0·774	0·777	0·783	0·789	0·794	0·803	0·812	0·821
2·00000	(19·6)	0·773	0·768	0·767	0·766	0·767	0·769	0·772	0·775	0·782	0·787	0·793	0·802	0·811	0·820
3·00000	(29·4)	0·768	0·764	0·763	0·763	0·764	0·767	0·770	0·773	0·780	0·786	0·792	0·801	0·810	0·819
4·00000	(39·2)	0·765	0·762	0·761	0·761	0·762	0·765	0·769	0·772	0·779	0·785	0·791	0·801	0·809	0·818
6·00000	(58·8)	0·761	0·758	0·758	0·758	0·760	0·763	0·767	0·771	0·778	0·784	0·790	0·800	0·808	0·818
8·00000	(78·5)	0·759	0·756	0·756	0·757	0·758	0·762	0·766	0·770	0·777	0·783	0·789	0·799	0·808	0·817
10·0000	(98·1)	0·757	0·755	0·755	0·756	0·758	0·761	0·765	0·769	0·776	0·783	0·789	0·799	0·808	0·817
15·0000	(147)	0·755	0·753	0·753	0·754	0·756	0·760	0·764	0·768	0·776	0·782	0·788	0·798	0·807	0·817
20·0000	(196)	0·753	0·752	0·752	0·754	0·755	0·760	0·764	0·768	0·775	0·782	0·788	0·798	0·807	0·816
30·0000	(294)	0·751	0·750	0·751	0·752	0·754	0·759	0·763	0·767	0·775	0·781	0·787	0·798	0·807	0·816
S	$(\Delta p/\rho l)$	6	8	10	12·5	15	20	25	30	40	50	60	80	100	125

Grad'nt k.p.g. (Equivalent) Pipe diameters in mm

S	$(\Delta p/\rho l)$	125	150	200	250	300	400	500	600	800	1000	1250	1500	2000	2500
0·00010	(0·00098)	1·079	1·070	1·061	1·055	1·053	1·051	1·052	1·054	1·059	1·064	1·071	1·077	1·089	1·100
0·00015	(0·00147)	1·051	1·044	1·037	1·033	1·031	1·031	1·033	1·036	1·042	1·048	1·056	1·063	1·076	1·087
0·00020	(0·00196)	1·033	1·027	1·021	1·018	1·017	1·018	1·021	1·024	1·031	1·038	1·046	1·054	1·067	1·079
0·00030	(0·00294)	1·008	1·004	1·000	0·998	0·999	1·001	1·005	1·009	1·017	1·025	1·034	1·042	1·056	1·069
0·00040	(0·00392)	0·992	0·989	0·986	0·985	0·986	0·990	0·994	0·999	1·007	1·016	1·025	1·034	1·049	1·062
0·00060	(0·00588)	0·972	0·969	0·968	0·968	0·970	0·975	0·980	0·985	0·995	1·005	1·015	1·024	1·040	1·054
0·00080	(0·00785)	0·958	0·956	0·956	0·957	0·960	0·965	0·971	0·977	0·988	0·997	1·008	1·018	1·034	1·049
0·00100	(0·00981)	0·948	0·947	0·947	0·949	0·952	0·958	0·965	0·971	0·982	0·992	1·003	1·013	1·030	1·045
0·00150	(0·0147)	0·931	0·931	0·933	0·936	0·939	0·947	0·954	0·961	0·973	0·984	0·995	1·006	1·024	1·039
0·00200	(0·0196)	0·920	0·920	0·923	0·927	0·931	0·939	0·947	0·954	0·967	0·978	0·991	1·001	1·020	1·035
0·00300	(0·0294)	0·905	0·907	0·911	0·916	0·921	0·930	0·938	0·946	0·960	0·972	0·984	0·995	1·014	1·030
0·00400	(0·0392)	0·896	0·898	0·903	0·909	0·914	0·924	0·933	0·941	0·955	0·967	0·980	0·992	1·011	1·027
0·00600	(0·0588)	0·884	0·887	0·893	0·900	0·906	0·916	0·926	0·935	0·949	0·962	0·976	0·987	1·007	1·024
0·00800	(0·0785)	0·877	0·880	0·887	0·894	0·900	0·912	0·922	0·931	0·946	0·959	0·973	0·985	1·005	1·021
0·01000	(0·0981)	0·872	0·875	0·883	0·890	0·897	0·908	0·919	0·928	0·943	0·957	0·971	0·983	1·003	1·020
0·01500	(0·147)	0·863	0·867	0·876	0·883	0·890	0·903	0·914	0·923	0·939	0·953	0·967	0·980	1·000	1·017
0·02000	(0·196)	0·857	0·862	0·871	0·879	0·887	0·900	0·911	0·920	0·937	0·951	0·965	0·978	0·999	1·016
0·03000	(0·294)	0·851	0·856	0·866	0·874	0·882	0·896	0·907	0·917	0·934	0·948	0·963	0·976	0·997	1·014
0·04000	(0·392)	0·846	0·852	0·862	0·871	0·879	0·893	0·905	0·915	0·932	0·946	0·961	0·974	0·996	1·013
0·06000	(0·588)	0·841	0·847	0·858	0·867	0·876	0·890	0·902	0·912	0·930	0·944	0·960	0·973	0·994	1·012
0·08000	(0·785)	0·838	0·844	0·855	0·865	0·874	0·888	0·900	0·911	0·929	0·943	0·959	0·972	0·993	1·011
0·10000	(0·981)	0·836	0·842	0·854	0·864	0·872	0·887	0·899	0·910	0·928	0·942	0·958	0·971	0·993	1·010
0·15000	(1·47)	0·832	0·839	0·851	0·861	0·870	0·885	0·897	0·908	0·926	0·941	0·957	0·970	0·992	1·010
0·20000	(1·96)	0·830	0·837	0·849	0·859	0·868	0·884	0·896	0·907	0·925	0·940	0·956	0·969	0·991	1·009
0·30000	(2·94)	0·827	0·835	0·847	0·858	0·867	0·882	0·895	0·906	0·924	0·939	0·955	0·969	0·991	1·009
0·40000	(3·92)	0·826	0·833	0·846	0·856	0·866	0·881	0·894	0·905	0·924	0·939	0·955	0·968	0·990	1·008
0·60000	(5·88)	0·824	0·831	0·844	0·855	0·864	0·880	0·893	0·904	0·923	0·938	0·954	0·967	0·990	1·008
0·80000	(7·85)	0·823	0·830	0·843	0·854	0·864	0·880	0·893	0·904	0·923	0·938	0·954	0·967	0·989	1·008
1·00000	(9·81)	0·822	0·830	0·843	0·854	0·863	0·879	0·892	0·904	0·922	0·937	0·953	0·967	0·989	1·007
1·50000	(14·7)	0·821	0·828	0·842	0·853	0·862	0·878	0·892	0·903	0·922	0·937	0·953	0·966	0·989	1·007
2·00000	(19·6)	0·820	0·828	0·841	0·852	0·861	0·878	0·891	0·903	0·921	0·936	0·953	0·966	0·989	1·007
3·00000	(29·4)	0·819	0·827	0·840	0·852	0·861	0·878	0·891	0·902	0·921	0·936	0·953	0·966	0·989	1·007
S	$(\Delta p/\rho l)$	125	150	200	250	300	400	500	600	800	1000	1250	1500	2000	2500

Grad'nt k.p.g. (Equivalent) Pipe diameters in mm

Kinematic viscosity, $\nu = 0.80 \times 10^{-6}$ m^{2}s^{-1} ;   **Roughness size, $k_s = 0.060$ mm**

E81

Kin. visc., $\nu = 0.80 \times 10^{-6}$ m^{2}s^{-1}; $S = 0.00010$ to 30.0000

i.e. kin. pr. grad., $\Delta p/\rho l = (0.00098)$ to (294) ms^{-2}

Roughness size, $k_s = 0.150$ mm

This table shows values of m, as follows

m_C for Colebrook-White solutions; or, where $\mathbf{R} \le 2000$, m_P for laminar flow

Grad'nt S	k.p.g. $(\Delta p/\rho l)$	6	8	10	12.5	15	20	25	30	40	50	60	80	100	125
0.00030	(0.00294)	5.486	3.738	2.776	2.062	1.617	1.102	*	*	1.106	1.087	1.074	1.058	1.049	1.043
0.00040	(0.00392)	4.751	3.237	2.404	1.786	1.400	0.954	*	1.113	1.085	1.068	1.056	1.042	1.035	1.029
0.00060	(0.00588)	3.879	2.643	1.963	1.458	1.143	*	*	1.082	1.057	1.043	1.033	1.022	1.016	1.012
0.00080	(0.00785)	3.359	2.289	1.700	1.263	0.990	*	1.079	1.061	1.039	1.026	1.018	1.009	1.004	1.001
0.00100	(0.00981)	3.005	2.048	1.521	1.129	0.886	*	1.062	1.047	1.027	1.015	1.007	0.999	0.995	0.994
0.00150	(0.0147)	2.453	1.672	1.242	0.922	*	1.056	1.036	1.022	1.005	0.996	0.990	0.984	0.981	0.981
0.00200	(0.0196)	2.125	1.448	1.075	0.799	*	1.037	1.019	1.006	0.992	0.983	0.978	0.974	0.972	0.973
0.00300	(0.0294)	1.735	1.182	0.878	*	*	1.012	0.997	0.987	0.974	0.968	0.964	0.961	0.961	0.962
0.00400	(0.0392)	1.502	1.024	0.760	*	1.020	0.997	0.983	0.974	0.963	0.958	0.955	0.953	0.954	0.956
0.00600	(0.0588)	1.227	0.836	*	1.014	0.998	0.977	0.966	0.958	0.950	0.946	0.944	0.944	0.945	0.948
0.00800	(0.0785)	1.062	0.724	*	0.998	0.983	0.965	0.955	0.948	0.941	0.938	0.937	0.938	0.940	0.943
0.01000	(0.0981)	0.950	*	*	0.986	0.973	0.956	0.947	0.941	0.935	0.933	0.932	0.933	0.936	0.940
0.01500	(0.147)	0.776	*	0.985	0.968	0.956	0.942	0.934	0.930	0.925	0.924	0.924	0.926	0.930	0.935
0.02000	(0.196)	*	*	0.972	0.956	0.946	0.933	0.927	0.923	0.919	0.919	0.919	0.922	0.926	0.931
0.03000	(0.294)	*	0.973	0.956	0.942	0.933	0.922	0.917	0.914	0.912	0.912	0.913	0.917	0.922	0.927
0.04000	(0.392)	*	0.961	0.945	0.933	0.925	0.916	0.911	0.909	0.907	0.908	0.910	0.914	0.919	0.925
0.06000	(0.588)	0.970	0.947	0.933	0.922	0.915	0.907	0.904	0.902	0.902	0.903	0.905	0.910	0.915	0.922
0.08000	(0.785)	0.959	0.938	0.925	0.915	0.909	0.902	0.899	0.898	0.898	0.900	0.902	0.908	0.913	0.920
0.10000	(0.981)	0.951	0.931	0.920	0.911	0.905	0.899	0.896	0.895	0.896	0.898	0.901	0.906	0.912	0.919
0.15000	(1.47)	0.939	0.921	0.911	0.903	0.898	0.893	0.891	0.891	0.892	0.895	0.898	0.904	0.910	0.917
0.20000	(1.96)	0.932	0.915	0.906	0.898	0.894	0.890	0.888	0.888	0.890	0.893	0.896	0.902	0.908	0.916
0.30000	(2.94)	0.923	0.908	0.899	0.893	0.889	0.886	0.885	0.885	0.887	0.890	0.894	0.901	0.907	0.914
0.40000	(3.92)	0.917	0.903	0.895	0.890	0.886	0.883	0.883	0.883	0.886	0.889	0.892	0.899	0.906	0.913
0.60000	(5.88)	0.911	0.898	0.891	0.886	0.883	0.880	0.880	0.881	0.884	0.887	0.891	0.898	0.905	0.912
0.80000	(7.85)	0.907	0.895	0.888	0.883	0.880	0.878	0.878	0.879	0.883	0.886	0.890	0.897	0.904	0.912
1.00000	(9.81)	0.904	0.892	0.886	0.881	0.879	0.877	0.877	0.878	0.882	0.886	0.889	0.897	0.904	0.911
1.50000	(14.7)	0.900	0.889	0.883	0.879	0.877	0.875	0.876	0.877	0.880	0.884	0.888	0.896	0.903	0.911
2.00000	(19.6)	0.897	0.887	0.881	0.877	0.875	0.874	0.875	0.876	0.880	0.884	0.888	0.895	0.902	0.910
3.00000	(29.4)	0.894	0.884	0.879	0.875	0.874	0.873	0.873	0.875	0.879	0.883	0.887	0.895	0.902	0.910
4.00000	(39.2)	0.892	0.883	0.878	0.874	0.873	0.872	0.873	0.874	0.878	0.883	0.887	0.895	0.902	0.910
6.00000	(58.8)	0.890	0.881	0.876	0.873	0.871	0.871	0.872	0.874	0.878	0.882	0.886	0.894	0.901	0.909
8.00000	(78.5)	0.888	0.880	0.875	0.872	0.871	0.870	0.871	0.873	0.877	0.882	0.886	0.894	0.901	0.909
10.0000	(98.1)	0.888	0.879	0.874	0.872	0.870	0.870	0.871	0.873	0.877	0.881	0.886	0.894	0.901	0.909
15.0000	(147)	0.886	0.878	0.873	0.871	0.869	0.869	0.870	0.872	0.877	0.881	0.885	0.893	0.901	0.909
20.0000	(196)	0.885	0.877	0.873	0.870	0.869	0.869	0.870	0.872	0.876	0.881	0.885	0.893	0.901	0.909
30.0000	(294)	0.884	0.876	0.872	0.870	0.868	0.868	0.870	0.872	0.876	0.881	0.885	0.893	0.900	0.909
S	$(\Delta p/\rho l)$	6	8	10	12.5	15	20	25	30	40	50	60	80	100	125

Grad'nt — k.p.g. — (Equivalent) Pipe diameters in mm

Grad'nt S	k.p.g. $(\Delta p/\rho l)$	125	150	200	250	300	400	500	600	800	1000	1250	1500	2000	2500
0.00010	(0.00098)	1.104	1.096	1.088	1.084	1.072	1.083	1.085	1.088	1.095	1.102	1.110	1.118	1.132	1.145
0.00015	(0.00147)	1.079	1.073	1.067	1.065	1.065	1.066	1.070	1.074	1.082	1.090	1.099	1.108	1.123	1.136
0.00020	(0.00196)	1.063	1.058	1.054	1.053	1.053	1.056	1.060	1.065	1.074	1.082	1.092	1.101	1.117	1.131
0.00030	(0.00294)	1.043	1.039	1.037	1.037	1.038	1.042	1.048	1.053	1.063	1.073	1.084	1.093	1.110	1.124
0.00040	(0.00392)	1.029	1.027	1.025	1.026	1.029	1.034	1.040	1.046	1.057	1.067	1.078	1.088	1.105	1.120
0.00060	(0.00588)	1.012	1.011	1.011	1.013	1.016	1.023	1.030	1.037	1.049	1.059	1.071	1.082	1.100	1.115
0.00080	(0.00785)	1.001	1.001	1.002	1.005	1.009	1.016	1.024	1.031	1.043	1.055	1.067	1.078	1.096	1.112
0.00100	(0.00981)	0.994	0.994	0.996	0.999	1.003	1.012	1.019	1.027	1.040	1.051	1.064	1.075	1.094	1.110
0.00150	(0.0147)	0.981	0.982	0.985	0.990	0.995	1.004	1.012	1.020	1.034	1.046	1.059	1.071	1.090	1.106
0.00200	(0.0196)	0.973	0.974	0.979	0.984	0.989	0.999	1.008	1.016	1.031	1.043	1.056	1.068	1.088	1.104
0.00300	(0.0294)	0.962	0.965	0.970	0.976	0.982	0.993	1.002	1.011	1.026	1.039	1.053	1.065	1.085	1.102
0.00400	(0.0392)	0.956	0.959	0.965	0.972	0.978	0.989	0.999	1.008	1.023	1.037	1.051	1.063	1.083	1.100
0.00600	(0.0588)	0.948	0.952	0.959	0.966	0.973	0.985	0.995	1.004	1.020	1.034	1.048	1.060	1.081	1.098
0.00800	(0.0785)	0.943	0.947	0.955	0.963	0.970	0.982	0.993	1.002	1.018	1.032	1.046	1.059	1.080	1.097
0.01000	(0.0981)	0.940	0.944	0.953	0.960	0.967	0.980	0.991	1.001	1.017	1.031	1.045	1.058	1.079	1.096
0.01500	(0.147)	0.935	0.939	0.948	0.956	0.964	0.977	0.988	0.998	1.015	1.029	1.044	1.056	1.078	1.095
0.02000	(0.196)	0.931	0.936	0.946	0.954	0.962	0.975	0.987	0.997	1.013	1.028	1.043	1.055	1.077	1.094
0.03000	(0.294)	0.927	0.933	0.942	0.951	0.959	0.973	0.985	0.995	1.012	1.026	1.041	1.054	1.076	1.093
0.04000	(0.392)	0.925	0.930	0.941	0.950	0.958	0.972	0.983	0.994	1.011	1.025	1.041	1.054	1.075	1.093
0.06000	(0.588)	0.922	0.928	0.938	0.948	0.956	0.970	0.982	0.992	1.010	1.024	1.040	1.053	1.074	1.092
0.08000	(0.785)	0.920	0.926	0.937	0.946	0.955	0.969	0.981	0.992	1.009	1.024	1.039	1.052	1.074	1.092
0.10000	(0.981)	0.919	0.925	0.936	0.945	0.954	0.968	0.981	0.991	1.009	1.023	1.039	1.052	1.074	1.092
0.15000	(1.47)	0.917	0.923	0.934	0.944	0.953	0.967	0.980	0.990	1.008	1.023	1.038	1.051	1.073	1.091
0.20000	(1.96)	0.916	0.922	0.933	0.943	0.952	0.967	0.979	0.990	1.008	1.022	1.038	1.051	1.073	1.091
0.30000	(2.94)	0.914	0.921	0.932	0.942	0.951	0.966	0.978	0.989	1.007	1.022	1.037	1.051	1.073	1.091
0.40000	(3.92)	0.913	0.920	0.932	0.942	0.951	0.965	0.978	0.989	1.007	1.022	1.037	1.051	1.073	1.091
0.60000	(5.88)	0.912	0.919	0.931	0.941	0.950	0.965	0.978	0.988	1.006	1.021	1.037	1.050	1.072	1.090
0.80000	(7.85)	0.912	0.919	0.930	0.941	0.950	0.965	0.977	0.988	1.006	1.021	1.037	1.050	1.072	1.090
1.00000	(9.81)	0.911	0.918	0.930	0.940	0.949	0.964	0.977	0.988	1.006	1.021	1.037	1.050	1.072	1.090
1.50000	(14.7)	0.911	0.918	0.930	0.940	0.949	0.964	0.977	0.988	1.006	1.021	1.036	1.050	1.072	1.090
2.00000	(19.6)	0.910	0.917	0.929	0.940	0.949	0.964	0.977	0.987	1.006	1.021	1.036	1.050	1.072	1.090
3.00000	(29.4)	0.910	0.917	0.929	0.939	0.948	0.964	0.976	0.987	1.006	1.020	1.036	1.050	1.072	1.090
S	$(\Delta p/\rho l)$	125	150	200	250	300	400	500	600	800	1000	1250	1500	2000	2500

Grad'nt — k.p.g. — (Equivalent) Pipe diameters in mm

Kinematic viscosity, $\nu = 0.80 \times 10^{-6}$ m^{2}s^{-1} ;

Roughness size, $k_s = 0.150$ mm

Kin. visc., $v = 0.80 \times 10^{-6} \ m^2 s^{-1}$; $S = 0.00010$ to 30.0000

i.e. kin. pr. grad., $\Delta p/\rho l =$ (0.00098) to (294) ms^{-2}

Roughness size, $k_s = 0.30$ mm

This table shows values of m, as follows

m_C for Colebrook-White solutions; or, where $R \leq 2000$, m_P for laminar flow

Grad'nt S	k.p.g. $(\Delta p/\rho l)$	6	8	10	12.5	15	20	25	30	40	50	60	80	100	125
0.00030	(0.00294)	5.486	3.738	2.776	2.062	1.617	1.102	*	*	1.151	1.132	1.119	1.104	1.096	1.090
0.00040	(0.00392)	4.751	3.237	2.404	1.786	1.400	0.954	*	1.161	1.132	1.115	1.104	1.091	1.084	1.079
0.00060	(0.00588)	3.879	2.643	1.963	1.458	1.143	*	*	1.133	1.109	1.094	1.085	1.074	1.069	1.066
0.00080	(0.00785)	3.359	2.289	1.700	1.263	0.990	*	1.134	1.116	1.094	1.081	1.073	1.064	1.060	1.058
0.00100	(0.00981)	3.005	2.048	1.521	1.129	0.886	*	1.120	1.104	1.083	1.072	1.064	1.056	1.053	1.052
0.00150	(0.0147)	2.453	1.672	1.242	0.922	*	1.118	1.097	1.083	1.066	1.056	1.050	1.044	1.042	1.042
0.00200	(0.0196)	2.125	1.448	1.075	0.799	*	1.102	1.083	1.071	1.055	1.047	1.042	1.037	1.036	1.036
0.00300	(0.0294)	1.735	1.182	0.878	*	*	1.082	1.066	1.055	1.042	1.035	1.031	1.028	1.028	1.029
0.00400	(0.0392)	1.502	1.024	0.760	*	1.095	1.070	1.055	1.045	1.034	1.028	1.025	1.022	1.023	1.025
0.00600	(0.0588)	1.227	0.836	*	*	1.077	1.054	1.041	1.033	1.023	1.019	1.016	1.015	1.017	1.019
0.00800	(0.0785)	1.062	0.724	*	1.081	1.065	1.045	1.033	1.026	1.017	1.013	1.012	1.011	1.013	1.016
0.01000	(0.0981)	0.950	*	*	1.072	1.057	1.038	1.027	1.020	1.013	1.009	1.008	1.008	1.010	1.014
0.01500	(0.147)	0.776	*	1.078	1.058	1.044	1.028	1.018	1.012	1.006	1.003	1.003	1.004	1.006	1.010
0.02000	(0.196)	*	*	1.068	1.049	1.037	1.021	1.012	1.007	1.002	1.000	0.999	1.001	1.004	1.008
0.03000	(0.294)	*	1.077	1.055	1.038	1.027	1.013	1.006	1.001	0.997	0.995	0.995	0.998	1.001	1.005
0.04000	(0.392)	*	1.068	1.048	1.032	1.021	1.008	1.001	0.997	0.993	0.993	0.993	0.996	0.999	1.004
0.06000	(0.588)	*	1.057	1.038	1.024	1.014	1.003	0.996	0.993	0.990	0.989	0.990	0.993	0.997	1.002
0.08000	(0.785)	1.078	1.050	1.033	1.019	1.010	0.999	0.993	0.990	0.988	0.987	0.988	0.992	0.996	1.001
0.10000	(0.981)	1.072	1.045	1.029	1.016	1.007	0.997	0.991	0.988	0.986	0.986	0.987	0.991	0.995	1.000
0.15000	(1.47)	1.063	1.038	1.023	1.010	1.002	0.993	0.988	0.985	0.984	0.984	0.985	0.989	0.993	0.999
0.20000	(1.96)	1.058	1.034	1.019	1.007	1.000	0.991	0.986	0.984	0.982	0.983	0.984	0.988	0.993	0.998
0.30000	(2.94)	1.051	1.028	1.014	1.003	0.996	0.988	0.984	0.982	0.980	0.981	0.983	0.987	0.992	0.997
0.40000	(3.92)	1.047	1.025	1.012	1.001	0.994	0.986	0.982	0.980	0.979	0.980	0.982	0.986	0.991	0.997
0.60000	(5.88)	1.043	1.021	1.009	0.998	0.992	0.984	0.981	0.979	0.978	0.979	0.981	0.985	0.990	0.996
0.80000	(7.85)	1.040	1.019	1.007	0.997	0.990	0.983	0.980	0.978	0.977	0.979	0.980	0.985	0.990	0.996
1.00000	(9.81)	1.038	1.018	1.005	0.996	0.989	0.982	0.979	0.977	0.977	0.978	0.980	0.985	0.990	0.996
1.50000	(14.7)	1.035	1.015	1.003	0.994	0.988	0.981	0.978	0.976	0.976	0.977	0.979	0.984	0.989	0.995
2.00000	(19.6)	1.033	1.014	1.002	0.993	0.987	0.980	0.977	0.976	0.976	0.977	0.979	0.984	0.989	0.995
3.00000	(29.4)	1.031	1.012	1.001	0.992	0.986	0.979	0.976	0.975	0.975	0.976	0.979	0.983	0.989	0.995
4.00000	(39.2)	1.030	1.011	1.000	0.991	0.985	0.979	0.976	0.975	0.975	0.976	0.978	0.983	0.988	0.994
6.00000	(58.8)	1.028	1.010	0.999	0.990	0.984	0.978	0.975	0.974	0.974	0.976	0.978	0.983	0.988	0.994
8.00000	(78.5)	1.027	1.009	0.998	0.989	0.984	0.978	0.975	0.974	0.974	0.976	0.978	0.983	0.988	0.994
10.0000	(98.1)	1.027	1.009	0.998	0.989	0.984	0.978	0.975	0.974	0.974	0.975	0.978	0.983	0.988	0.994
15.0000	(147)	1.026	1.008	0.997	0.988	0.983	0.977	0.974	0.973	0.974	0.975	0.978	0.983	0.988	0.994
20.0000	(196)	1.025	1.007	0.997	0.988	0.983	0.977	0.974	0.973	0.973	0.975	0.977	0.983	0.988	0.994
30.0000	(294)	1.024	1.007	0.996	0.988	0.982	0.977	0.974	0.973	0.973	0.975	0.977	0.982	0.988	0.994
S	$(\Delta p/\rho l)$	6	8	10	12.5	15	20	25	30	40	50	60	80	100	125

Grad'nt k.p.g. (Equivalent) Pipe diameters in mm

Grad'nt S	k.p.g. $(\Delta p/\rho l)$	125	150	200	250	300	400	500	600	800	1000	1250	1500	2000	2500
0.00010	(0.00098)	1.141	1.134	1.127	1.124	1.123	1.125	1.128	1.132	1.141	1.149	1.159	1.168	1.184	1.198
0.00015	(0.00147)	1.120	1.115	1.110	1.109	1.109	1.112	1.117	1.122	1.131	1.140	1.151	1.161	1.177	1.192
0.00020	(0.00196)	1.107	1.103	1.099	1.099	1.100	1.104	1.109	1.115	1.125	1.135	1.146	1.156	1.173	1.188
0.00030	(0.00294)	1.090	1.087	1.085	1.086	1.089	1.094	1.100	1.106	1.118	1.128	1.140	1.151	1.169	1.184
0.00040	(0.00392)	1.079	1.077	1.077	1.079	1.081	1.088	1.095	1.101	1.113	1.124	1.136	1.147	1.166	1.181
0.00060	(0.00588)	1.066	1.065	1.066	1.069	1.073	1.080	1.088	1.095	1.108	1.119	1.132	1.143	1.162	1.178
0.00080	(0.00785)	1.058	1.057	1.059	1.063	1.067	1.076	1.084	1.091	1.105	1.116	1.129	1.140	1.160	1.176
0.00100	(0.00981)	1.052	1.052	1.055	1.059	1.063	1.072	1.081	1.088	1.102	1.114	1.127	1.139	1.158	1.175
0.00150	(0.0147)	1.042	1.043	1.047	1.052	1.057	1.067	1.076	1.084	1.098	1.111	1.124	1.136	1.156	1.172
0.00200	(0.0196)	1.036	1.038	1.043	1.048	1.053	1.064	1.073	1.081	1.096	1.109	1.122	1.134	1.154	1.171
0.00300	(0.0294)	1.029	1.031	1.037	1.043	1.049	1.060	1.069	1.078	1.093	1.106	1.120	1.132	1.153	1.170
0.00400	(0.0392)	1.025	1.027	1.034	1.040	1.046	1.057	1.067	1.076	1.092	1.105	1.119	1.131	1.152	1.169
0.00600	(0.0588)	1.019	1.023	1.030	1.036	1.043	1.054	1.065	1.074	1.090	1.103	1.117	1.130	1.150	1.168
0.00800	(0.0785)	1.016	1.020	1.027	1.034	1.041	1.053	1.063	1.073	1.088	1.102	1.116	1.129	1.150	1.167
0.01000	(0.0981)	1.014	1.018	1.025	1.033	1.039	1.052	1.062	1.072	1.088	1.101	1.116	1.128	1.149	1.166
0.01500	(0.147)	1.010	1.014	1.022	1.030	1.037	1.050	1.060	1.070	1.086	1.100	1.115	1.127	1.148	1.166
0.02000	(0.196)	1.008	1.012	1.021	1.029	1.036	1.049	1.059	1.069	1.086	1.099	1.114	1.127	1.148	1.165
0.03000	(0.294)	1.005	1.010	1.019	1.027	1.034	1.047	1.058	1.068	1.085	1.099	1.113	1.126	1.147	1.165
0.04000	(0.392)	1.004	1.009	1.018	1.026	1.033	1.046	1.058	1.067	1.084	1.098	1.113	1.126	1.147	1.164
0.06000	(0.588)	1.002	1.007	1.016	1.025	1.032	1.045	1.057	1.067	1.083	1.097	1.112	1.125	1.147	1.164
0.08000	(0.785)	1.001	1.006	1.015	1.024	1.031	1.045	1.056	1.066	1.083	1.097	1.112	1.125	1.146	1.164
0.10000	(0.981)	1.000	1.005	1.015	1.023	1.031	1.044	1.056	1.066	1.083	1.097	1.112	1.125	1.146	1.164
0.15000	(1.47)	0.999	1.004	1.014	1.023	1.030	1.044	1.055	1.065	1.082	1.097	1.112	1.124	1.146	1.163
0.20000	(1.96)	0.998	1.004	1.013	1.022	1.030	1.043	1.055	1.065	1.082	1.096	1.111	1.124	1.146	1.163
0.30000	(2.94)	0.997	1.003	1.013	1.021	1.029	1.043	1.055	1.065	1.082	1.096	1.111	1.124	1.146	1.163
0.40000	(3.92)	0.997	1.002	1.012	1.021	1.029	1.043	1.054	1.064	1.082	1.096	1.111	1.124	1.145	1.163
0.60000	(5.88)	0.996	1.002	1.012	1.021	1.029	1.042	1.054	1.064	1.081	1.096	1.111	1.124	1.145	1.163
0.80000	(7.85)	0.996	1.001	1.012	1.020	1.028	1.042	1.054	1.064	1.081	1.096	1.111	1.124	1.145	1.163
1.00000	(9.81)	0.996	1.001	1.011	1.020	1.028	1.042	1.054	1.064	1.081	1.096	1.111	1.124	1.145	1.163
1.50000	(14.7)	0.995	1.001	1.011	1.020	1.028	1.042	1.054	1.064	1.081	1.095	1.111	1.124	1.145	1.163
2.00000	(19.6)	0.995	1.001	1.011	1.020	1.028	1.042	1.054	1.064	1.081	1.095	1.110	1.123	1.145	1.163
3.00000	(29.4)	0.995	1.000	1.011	1.020	1.028	1.042	1.053	1.064	1.081	1.095	1.110	1.123	1.145	1.163
S	$(\Delta p/\rho l)$	125	150	200	250	300	400	500	600	800	1000	1250	1500	2000	2500

Grad'nt k.p.g. (Equivalent) Pipe diameters in mm

Kinematic viscosity, $v = 0.80 \times 10^{-6} \ m^2 s^{-1}$;

Roughness size, $k_s = 0.30$ mm

E83

Kin. visc., $\nu = 0{\cdot}80 \times 10^{-6}$ m²s⁻¹;   Roughness size, $k_s = 0{\cdot}60$ mm

$S = 0{\cdot}00010$ to $3{\cdot}00000$

i.e. kin. pr. grad., $\Delta p/\rho l = (0{\cdot}00098)$ to $(29{\cdot}4)$ ms⁻²

This table shows values of m, as follows

m_C for Colebrook-White solutions

Grad'nt S	k.p.g. $(\Delta p/\rho l)$	\multicolumn{14}{c}{(Equivalent) Pipe diameters in mm}													
		125	150	200	250	300	400	500	600	800	1000	1250	1500	2000	2500
0·00010	(0·00098)	1·203	1·196	1·189	1·187	1·187	1·189	1·193	1·198	1·207	1·216	1·227	1·236	1·253	1·267
0·00015	(0·00147)	1·186	1·181	1·176	1·175	1·176	1·180	1·185	1·190	1·201	1·210	1·221	1·231	1·249	1·264
0·00020	(0·00196)	1·176	1·171	1·168	1·168	1·170	1·174	1·180	1·186	1·197	1·207	1·218	1·228	1·246	1·261
0·00030	(0·00294)	1·163	1·160	1·158	1·159	1·162	1·167	1·174	1·180	1·192	1·202	1·214	1·225	1·243	1·259
0·00040	(0·00392)	1·155	1·153	1·152	1·154	1·157	1·163	1·170	1·177	1·189	1·200	1·212	1·223	1·241	1·257
0·00060	(0·00588)	1·145	1·144	1·145	1·147	1·151	1·158	1·165	1·172	1·185	1·197	1·209	1·220	1·239	1·255
0·00080	(0·00785)	1·139	1·139	1·140	1·143	1·147	1·155	1·163	1·170	1·183	1·195	1·207	1·219	1·238	1·254
0·00100	(0·00981)	1·135	1·135	1·137	1·140	1·144	1·153	1·161	1·168	1·182	1·193	1·206	1·218	1·237	1·253
0·00150	(0·0147)	1·129	1·129	1·132	1·136	1·140	1·149	1·158	1·165	1·179	1·191	1·204	1·216	1·235	1·252
0·00200	(0·0196)	1·125	1·125	1·129	1·133	1·138	1·147	1·156	1·164	1·178	1·190	1·203	1·215	1·235	1·251
0·00300	(0·0294)	1·120	1·121	1·125	1·130	1·135	1·145	1·154	1·162	1·176	1·189	1·202	1·214	1·234	1·250
0·00400	(0·0392)	1·117	1·118	1·123	1·128	1·133	1·143	1·152	1·161	1·175	1·188	1·201	1·213	1·233	1·250
0·00600	(0·0588)	1·113	1·115	1·120	1·126	1·131	1·142	1·151	1·159	1·174	1·187	1·200	1·212	1·232	1·249
0·00800	(0·0785)	1·111	1·113	1·119	1·124	1·130	1·140	1·150	1·158	1·173	1·186	1·200	1·212	1·232	1·249
0·01000	(0·0981)	1·110	1·112	1·118	1·123	1·129	1·140	1·149	1·158	1·173	1·186	1·199	1·211	1·232	1·248
0·01500	(0·147)	1·107	1·110	1·116	1·122	1·128	1·139	1·148	1·157	1·172	1·185	1·199	1·211	1·231	1·248
0·02000	(0·196)	1·106	1·109	1·115	1·121	1·127	1·138	1·148	1·156	1·172	1·184	1·198	1·210	1·231	1·248
0·03000	(0·294)	1·104	1·107	1·114	1·120	1·126	1·137	1·147	1·156	1·171	1·184	1·198	1·210	1·230	1·247
0·04000	(0·392)	1·103	1·106	1·113	1·119	1·125	1·137	1·146	1·155	1·171	1·184	1·198	1·210	1·230	1·247
0·06000	(0·588)	1·102	1·105	1·112	1·119	1·125	1·136	1·146	1·155	1·170	1·183	1·197	1·210	1·230	1·247
0·08000	(0·785)	1·102	1·105	1·111	1·118	1·124	1·136	1·146	1·155	1·170	1·183	1·197	1·209	1·230	1·247
0·10000	(0·981)	1·101	1·104	1·111	1·118	1·124	1·135	1·145	1·154	1·170	1·183	1·197	1·209	1·230	1·247
0·15000	(1·47)	1·100	1·104	1·111	1·117	1·124	1·135	1·145	1·154	1·170	1·183	1·197	1·209	1·230	1·247
0·20000	(1·96)	1·100	1·103	1·110	1·117	1·123	1·135	1·145	1·154	1·169	1·183	1·197	1·209	1·230	1·247
0·30000	(2·94)	1·099	1·103	1·110	1·117	1·123	1·135	1·145	1·154	1·169	1·182	1·197	1·209	1·229	1·246
0·40000	(3·92)	1·099	1·102	1·110	1·116	1·123	1·134	1·145	1·154	1·169	1·182	1·197	1·209	1·229	1·246
0·60000	(5·88)	1·099	1·102	1·109	1·116	1·123	1·134	1·144	1·153	1·169	1·182	1·196	1·209	1·229	1·246
0·80000	(7·85)	1·098	1·102	1·109	1·116	1·122	1·134	1·144	1·153	1·169	1·182	1·196	1·209	1·229	1·246
1·00000	(9·81)	1·098	1·102	1·109	1·116	1·122	1·134	1·144	1·153	1·169	1·182	1·196	1·209	1·229	1·246
1·50000	(14·7)	1·098	1·102	1·109	1·116	1·122	1·134	1·144	1·153	1·169	1·182	1·196	1·209	1·229	1·246
2·00000	(19·6)	1·098	1·101	1·109	1·116	1·122	1·134	1·144	1·153	1·169	1·182	1·196	1·209	1·229	1·246
3·00000	(29·4)	1·098	1·101	1·109	1·116	1·122	1·134	1·144	1·153	1·169	1·182	1·196	1·208	1·229	1·246
S	$(\Delta p/\rho l)$	125	150	200	250	300	400	500	600	800	1000	1250	1500	2000	2500

Grad'nt   k.p.g.   (Equivalent) Pipe diameters in mm   Roughness size, $k_s = 0{\cdot}60$ mm

E84

Kin. visc., $\nu = 0{\cdot}80 \times 10^{-6}$ m²s⁻¹;   Roughness size, $k_s = 1{\cdot}50$ mm

$S = 0{\cdot}00010$ to $3{\cdot}00000$

i.e. kin. pr. grad., $\Delta p/\rho l = (0{\cdot}00098)$ to $(29{\cdot}4)$ ms⁻²

This table shows values of m, as follows

m_C for Colebrook-White solutions

Grad'nt S	k.p.g. $(\Delta p/\rho l)$	125	150	200	250	300	400	500	600	800	1000	1250	1500	2000	2500
0·00010	(0·00098)	1·336	1·328	1·319	1·315	1·314	1·315	1·318	1·322	1·330	1·339	1·348	1·357	1·374	1·388
0·00015	(0·00147)	1·325	1·318	1·311	1·308	1·308	1·310	1·313	1·318	1·327	1·335	1·346	1·355	1·371	1·386
0·00020	(0·00196)	1·319	1·312	1·306	1·304	1·304	1·306	1·310	1·315	1·324	1·333	1·344	1·353	1·370	1·385
0·00030	(0·00294)	1·311	1·305	1·300	1·299	1·299	1·302	1·307	1·312	1·322	1·331	1·342	1·351	1·369	1·383
0·00040	(0·00392)	1·306	1·301	1·296	1·295	1·296	1·300	1·305	1·310	1·320	1·330	1·341	1·350	1·368	1·382
0·00060	(0·00588)	1·300	1·296	1·292	1·292	1·293	1·297	1·302	1·308	1·318	1·328	1·339	1·349	1·366	1·381
0·00080	(0·00785)	1·296	1·292	1·289	1·289	1·291	1·295	1·301	1·306	1·317	1·327	1·338	1·348	1·366	1·381
0·00100	(0·00981)	1·294	1·290	1·288	1·288	1·289	1·294	1·300	1·305	1·316	1·326	1·338	1·348	1·365	1·380
0·00150	(0·0147)	1·290	1·287	1·285	1·285	1·287	1·292	1·298	1·304	1·315	1·325	1·337	1·347	1·365	1·380
0·00200	(0·0196)	1·288	1·285	1·283	1·284	1·286	1·291	1·297	1·303	1·314	1·325	1·336	1·346	1·364	1·379
0·00300	(0·0294)	1·285	1·282	1·281	1·282	1·284	1·290	1·296	1·302	1·314	1·324	1·335	1·346	1·364	1·379
0·00400	(0·0392)	1·283	1·281	1·280	1·281	1·283	1·289	1·295	1·301	1·313	1·323	1·335	1·345	1·363	1·379
0·00600	(0·0588)	1·281	1·279	1·278	1·280	1·282	1·288	1·295	1·301	1·312	1·323	1·335	1·345	1·363	1·378
0·00800	(0·0785)	1·280	1·278	1·277	1·279	1·282	1·288	1·294	1·300	1·312	1·322	1·334	1·345	1·363	1·378
0·01000	(0·0981)	1·279	1·277	1·277	1·278	1·281	1·287	1·294	1·300	1·312	1·322	1·334	1·344	1·363	1·378
0·01500	(0·147)	1·278	1·276	1·276	1·278	1·280	1·287	1·293	1·300	1·311	1·322	1·334	1·344	1·362	1·378
0·02000	(0·196)	1·277	1·275	1·275	1·277	1·280	1·286	1·293	1·299	1·311	1·322	1·334	1·344	1·362	1·378
0·03000	(0·294)	1·276	1·275	1·275	1·277	1·279	1·286	1·293	1·299	1·311	1·321	1·333	1·344	1·362	1·377
0·04000	(0·392)	1·276	1·274	1·274	1·276	1·279	1·286	1·292	1·299	1·311	1·321	1·333	1·344	1·362	1·377
0·06000	(0·588)	1·275	1·274	1·274	1·276	1·279	1·285	1·292	1·298	1·310	1·321	1·333	1·344	1·362	1·377
0·08000	(0·785)	1·275	1·273	1·273	1·276	1·279	1·285	1·292	1·298	1·310	1·321	1·333	1·344	1·362	1·377
0·10000	(0·981)	1·274	1·273	1·273	1·275	1·278	1·285	1·292	1·298	1·310	1·321	1·333	1·343	1·362	1·377
0·15000	(1·47)	1·274	1·273	1·273	1·275	1·278	1·285	1·292	1·298	1·310	1·321	1·333	1·343	1·362	1·377
0·20000	(1·96)	1·274	1·272	1·273	1·275	1·278	1·285	1·291	1·298	1·310	1·321	1·333	1·343	1·362	1·377
0·30000	(2·94)	1·274	1·272	1·273	1·275	1·278	1·285	1·291	1·298	1·310	1·321	1·333	1·343	1·362	1·377
0·40000	(3·92)	1·273	1·272	1·272	1·275	1·278	1·284	1·291	1·298	1·310	1·321	1·333	1·343	1·362	1·377
0·60000	(5·88)	1·273	1·272	1·272	1·275	1·278	1·284	1·291	1·298	1·310	1·321	1·333	1·343	1·361	1·377
0·80000	(7·85)	1·273	1·272	1·272	1·275	1·278	1·284	1·291	1·298	1·310	1·321	1·333	1·343	1·361	1·377
1·00000	(9·81)	1·273	1·272	1·272	1·274	1·278	1·284	1·291	1·298	1·310	1·321	1·333	1·343	1·361	1·377
1·50000	(14·7)	1·273	1·272	1·272	1·274	1·277	1·284	1·291	1·298	1·310	1·321	1·333	1·343	1·361	1·377
2·00000	(19·6)	1·273	1·272	1·272	1·274	1·277	1·284	1·291	1·298	1·310	1·321	1·332	1·343	1·361	1·377
3·00000	(29·4)	1·273	1·271	1·272	1·274	1·277	1·284	1·291	1·298	1·310	1·320	1·332	1·343	1·361	1·377
S	$(\Delta p/\rho l)$	125	150	200	250	300	400	500	600	800	1000	1250	1500	2000	2500

Grad'nt   k.p.g.   (Equivalent) Pipe diameters in mm   Roughness size, $k_s = 1{\cdot}50$ mm

Kin. visc., $\nu = 0.80 \times 10^{-6}$ m^{2}s^{-1};
$S = 0.00010$ to 3.00000

i.e. kin. pr. grad., $\Delta p/\rho l =$
(0.00098) to (29.4) ms^{-2}

Roughness size, $k_s = 3.0$ mm
This table shows values of m, as follows

m_C for Colebrook-White solutions

E85

| Grad'nt | k.p.g. | (Equivalent) Pipe diameters in mm | | | | | | | | | | | | | |
S	$(\Delta p/\rho l)$	125	150	200	250	300	400	500	600	800	1000	1250	1500	2000	2500
0·00010	(0·00098)	1·491	1·478	1·463	1·455	1·451	1·447	1·447	1·449	1·454	1·460	1·468	1·475	1·489	1·502
0·00015	(0·00147)	1·483	1·471	1·457	1·450	1·446	1·444	1·444	1·446	1·452	1·458	1·466	1·474	1·488	1·501
0·00020	(0·00196)	1·478	1·467	1·454	1·447	1·444	1·442	1·442	1·444	1·450	1·457	1·465	1·473	1·487	1·500
0·00030	(0·00294)	1·473	1·462	1·450	1·444	1·441	1·439	1·440	1·442	1·449	1·455	1·464	1·472	1·486	1·499
0·00040	(0·00392)	1·469	1·459	1·448	1·442	1·439	1·437	1·439	1·441	1·448	1·455	1·463	1·471	1·486	1·499
0·00060	(0·00588)	1·465	1·456	1·445	1·439	1·437	1·436	1·437	1·440	1·446	1·454	1·462	1·470	1·485	1·498
0·00080	(0·00785)	1·463	1·454	1·443	1·438	1·435	1·435	1·436	1·439	1·446	1·453	1·462	1·470	1·485	1·498
0·00100	(0·00981)	1·461	1·452	1·442	1·437	1·434	1·434	1·436	1·438	1·445	1·453	1·461	1·470	1·484	1·498
0·00150	(0·0147)	1·459	1·450	1·440	1·435	1·433	1·433	1·435	1·438	1·445	1·452	1·461	1·469	1·484	1·497
0·00200	(0·0196)	1·457	1·449	1·439	1·434	1·432	1·432	1·434	1·437	1·444	1·451	1·460	1·469	1·484	1·497
0·00300	(0·0294)	1·455	1·447	1·438	1·433	1·431	1·431	1·433	1·436	1·444	1·451	1·460	1·468	1·483	1·497
0·00400	(0·0392)	1·454	1·446	1·437	1·432	1·431	1·431	1·433	1·436	1·443	1·451	1·460	1·468	1·483	1·497
0·00600	(0·0588)	1·453	1·445	1·436	1·432	1·430	1·430	1·432	1·436	1·443	1·450	1·459	1·468	1·483	1·496
0·00800	(0·0785)	1·452	1·444	1·435	1·431	1·429	1·430	1·432	1·435	1·443	1·450	1·459	1·468	1·483	1·496
0·01000	(0·0981)	1·452	1·444	1·435	1·431	1·429	1·429	1·432	1·435	1·443	1·450	1·459	1·468	1·483	1·496
0·01500	(0·147)	1·451	1·443	1·434	1·430	1·429	1·429	1·432	1·435	1·442	1·450	1·459	1·467	1·483	1·496
0·02000	(0·196)	1·450	1·443	1·434	1·430	1·428	1·429	1·431	1·435	1·442	1·450	1·459	1·467	1·483	1·496
0·03000	(0·294)	1·450	1·442	1·434	1·430	1·428	1·429	1·431	1·434	1·442	1·450	1·459	1·467	1·483	1·496
0·04000	(0·392)	1·449	1·442	1·433	1·429	1·428	1·428	1·431	1·434	1·442	1·449	1·459	1·467	1·482	1·496
0·06000	(0·588)	1·449	1·441	1·433	1·429	1·428	1·428	1·431	1·434	1·442	1·449	1·459	1·467	1·482	1·496
0·08000	(0·785)	1·449	1·441	1·433	1·429	1·428	1·428	1·431	1·434	1·442	1·449	1·458	1·467	1·482	1·496
0·10000	(0·981)	1·449	1·441	1·433	1·429	1·428	1·428	1·431	1·434	1·442	1·449	1·458	1·467	1·482	1·496
0·15000	(1·47)	1·448	1·441	1·432	1·429	1·427	1·428	1·431	1·434	1·442	1·449	1·458	1·467	1·482	1·496
0·20000	(1·96)	1·448	1·441	1·432	1·429	1·427	1·428	1·430	1·434	1·441	1·449	1·458	1·467	1·482	1·496
0·30000	(2·94)	1·448	1·441	1·432	1·429	1·427	1·428	1·430	1·434	1·441	1·449	1·458	1·467	1·482	1·496
0·40000	(3·92)	1·448	1·440	1·432	1·428	1·427	1·428	1·430	1·434	1·441	1·449	1·458	1·467	1·482	1·496
0·60000	(5·88)	1·448	1·440	1·432	1·428	1·427	1·428	1·430	1·434	1·441	1·449	1·458	1·467	1·482	1·496
0·80000	(7·85)	1·448	1·440	1·432	1·428	1·427	1·428	1·430	1·434	1·441	1·449	1·458	1·467	1·482	1·496
1·00000	(9·81)	1·448	1·440	1·432	1·428	1·427	1·428	1·430	1·434	1·441	1·449	1·458	1·467	1·482	1·496
1·50000	(14·7)	1·448	1·440	1·432	1·428	1·427	1·428	1·430	1·434	1·441	1·449	1·458	1·467	1·482	1·496
2·00000	(19·6)	1·448	1·440	1·432	1·428	1·427	1·428	1·430	1·434	1·441	1·449	1·458	1·467	1·482	1·496
3·00000	(29·4)	1·447	1·440	1·432	1·428	1·427	1·428	1·430	1·434	1·441	1·449	1·458	1·467	1·482	1·496
S	$(\Delta p/\rho l)$	125	150	200	250	300	400	500	600	800	1000	1250	1500	2000	2500
Grad'nt	k.p.g.	(Equivalent) Pipe diameters in mm													

Roughness size, $k_s = 3.0$ mm

Kin. visc., $\nu = 0.80 \times 10^{-6}$ m^{2}s^{-1};
$S = 0.00010$ to 3.00000

i.e. kin. pr. grad., $\Delta p/\rho l =$
(0.00098) to (29.4) ms^{-2}

Roughness size, $k_s = 6.0$ mm
This table shows values of m, as follows

m_C for Colebrook-White solutions

E86

| Grad'nt | k.p.g. | (Equivalent) Pipe diameters in mm | | | | | | | | | | | | | |
S	$(\Delta p/\rho l)$	125	150	200	250	300	400	500	600	800	1000	1250	1500	2000	2500
0·00010	(0·00098)	1·708	1·686	1·659	1·642	1·632	1·620	1·614	1·611	1·610	1·612	1·616	1·621	1·631	1·641
0·00015	(0·00147)	1·702	1·681	1·655	1·639	1·629	1·617	1·612	1·609	1·609	1·611	1·615	1·620	1·630	1·640
0·00020	(0·00196)	1·699	1·679	1·652	1·637	1·627	1·616	1·611	1·608	1·608	1·610	1·614	1·619	1·630	1·639
0·00030	(0·00294)	1·695	1·675	1·650	1·635	1·625	1·614	1·609	1·607	1·607	1·609	1·614	1·619	1·629	1·639
0·00040	(0·00392)	1·693	1·673	1·648	1·633	1·624	1·613	1·608	1·606	1·606	1·609	1·613	1·618	1·629	1·639
0·00060	(0·00588)	1·690	1·671	1·646	1·632	1·623	1·612	1·607	1·605	1·606	1·608	1·613	1·618	1·628	1·638
0·00080	(0·00785)	1·689	1·669	1·645	1·631	1·622	1·612	1·607	1·605	1·605	1·608	1·612	1·617	1·628	1·638
0·00100	(0·00981)	1·688	1·668	1·644	1·630	1·621	1·611	1·606	1·605	1·605	1·607	1·612	1·617	1·628	1·638
0·00150	(0·0147)	1·686	1·667	1·643	1·629	1·620	1·610	1·606	1·604	1·604	1·607	1·612	1·617	1·628	1·638
0·00200	(0·0196)	1·685	1·666	1·642	1·628	1·620	1·610	1·605	1·604	1·604	1·607	1·611	1·617	1·627	1·638
0·00300	(0·0294)	1·683	1·665	1·641	1·628	1·619	1·609	1·605	1·603	1·604	1·606	1·611	1·616	1·627	1·637
0·00400	(0·0392)	1·683	1·664	1·641	1·627	1·619	1·609	1·605	1·603	1·603	1·606	1·611	1·616	1·627	1·637
0·00600	(0·0588)	1·682	1·663	1·640	1·627	1·618	1·609	1·604	1·603	1·603	1·606	1·611	1·616	1·627	1·637
0·00800	(0·0785)	1·681	1·663	1·640	1·626	1·618	1·608	1·604	1·603	1·603	1·606	1·611	1·616	1·627	1·637
0·01000	(0·0981)	1·681	1·663	1·640	1·626	1·618	1·608	1·604	1·602	1·603	1·606	1·611	1·616	1·627	1·637
0·01500	(0·147)	1·680	1·662	1·639	1·626	1·617	1·608	1·604	1·602	1·603	1·606	1·611	1·616	1·627	1·637
0·02000	(0·196)	1·680	1·662	1·639	1·626	1·617	1·608	1·604	1·602	1·603	1·606	1·611	1·616	1·627	1·637
0·03000	(0·294)	1·680	1·662	1·639	1·625	1·617	1·608	1·604	1·602	1·603	1·606	1·610	1·616	1·627	1·637
0·04000	(0·392)	1·679	1·661	1·639	1·625	1·617	1·608	1·603	1·602	1·603	1·606	1·610	1·616	1·627	1·637
0·06000	(0·588)	1·679	1·661	1·638	1·625	1·617	1·607	1·603	1·602	1·603	1·605	1·610	1·616	1·627	1·637
0·08000	(0·785)	1·679	1·661	1·638	1·625	1·617	1·607	1·603	1·602	1·603	1·605	1·610	1·616	1·627	1·637
0·10000	(0·981)	1·679	1·661	1·638	1·625	1·617	1·607	1·603	1·602	1·602	1·605	1·610	1·616	1·627	1·637
0·15000	(1·47)	1·679	1·661	1·638	1·625	1·617	1·607	1·603	1·602	1·602	1·605	1·610	1·616	1·626	1·637
0·20000	(1·96)	1·679	1·661	1·638	1·625	1·616	1·607	1·603	1·602	1·602	1·605	1·610	1·616	1·626	1·637
0·30000	(2·94)	1·678	1·661	1·638	1·625	1·616	1·607	1·603	1·602	1·602	1·605	1·610	1·616	1·626	1·637
0·40000	(3·92)	1·678	1·661	1·638	1·625	1·616	1·607	1·603	1·602	1·602	1·605	1·610	1·616	1·626	1·637
0·60000	(5·88)	1·678	1·660	1·638	1·625	1·616	1·607	1·603	1·602	1·602	1·605	1·610	1·616	1·626	1·637
0·80000	(7·85)	1·678	1·660	1·638	1·625	1·616	1·607	1·603	1·602	1·602	1·605	1·610	1·616	1·626	1·637
1·00000	(9·81)	1·678	1·660	1·638	1·625	1·616	1·607	1·603	1·602	1·602	1·605	1·610	1·616	1·626	1·637
1·50000	(14·7)	1·678	1·660	1·638	1·625	1·616	1·607	1·603	1·602	1·602	1·605	1·610	1·616	1·626	1·637
2·00000	(19·6)	1·678	1·660	1·638	1·625	1·616	1·607	1·603	1·602	1·602	1·605	1·610	1·616	1·626	1·637
3·00000	(29·4)	1·678	1·660	1·638	1·625	1·616	1·607	1·603	1·602	1·602	1·605	1·610	1·616	1·626	1·637
S	$(\Delta p/\rho l)$	125	150	200	250	300	400	500	600	800	1000	1250	1500	2000	2500
Grad'nt	k.p.g.	(Equivalent) Pipe diameters in mm													

Roughness size, $k_s = 6.0$ mm

Kin. visc., $\nu = 1.0 \times 10^{-6}$ m^{2}s^{-1}; $\quad$ **Roughness size, k_s = 0.0015 mm**

S = 0.00010 to 30.0000 $\quad$ This table shows values of m, as follows

i.e. kin. pr. grad., $\Delta p/\rho l$ = $\quad$ m_C for Colebrook-White solutions; or,

(0.00098) to (294) ms^{-2} $\quad$ where $\mathbf{R} \le 2000$, m_P for laminar flow

Grad'nt $\quad$ **k.p.g.** $\quad$ **(Equivalent) Pipe diameters in mm**

S	$(\Delta p/\rho l)$	6	8	10	12.5	15	20	25	30	40	50	60	80	100	125
0.00030	(0.00294)	6.857	4.673	3.470	2.577	2.021	1.377	1.023	*	1.098	1.074	1.057	1.036	1.023	1.013
0.00040	(0.00392)	5.939	4.047	3.005	2.232	1.750	1.193	0.886	*	1.070	1.048	1.033	1.014	1.002	0.993
0.00060	(0.00588)	4.849	3.304	2.454	1.822	1.429	0.974	*	1.063	1.033	1.014	1.001	0.984	0.974	0.966
0.00080	(0.00785)	4.199	2.861	2.125	1.578	1.238	0.843	*	1.035	1.008	0.991	0.979	0.964	0.955	0.948
0.00100	(0.00981)	3.756	2.559	1.901	1.412	1.107	*	1.035	1.015	0.990	0.974	0.963	0.949	0.941	0.935
0.00150	(0.0147)	3.067	2.090	1.552	1.153	0.904	*	0.997	0.980	0.958	0.944	0.935	0.923	0.916	0.911
0.00200	(0.0196)	2.656	1.810	1.344	0.998	*	0.995	0.972	0.957	0.937	0.924	0.916	0.906	0.900	0.896
0.00300	(0.0294)	2.169	1.478	1.097	0.815	*	0.959	0.939	0.926	0.908	0.898	0.890	0.882	0.877	0.874
0.00400	(0.0392)	1.878	1.280	0.950	*	*	0.935	0.917	0.905	0.889	0.880	0.873	0.866	0.862	0.859
0.00600	(0.0588)	1.533	1.045	0.776	*	0.928	0.903	0.888	0.877	0.864	0.855	0.850	0.844	0.841	0.840
0.00800	(0.0785)	1.328	0.905	*	0.922	0.905	0.882	0.868	0.859	0.846	0.839	0.835	0.830	0.827	0.826
0.01000	(0.0981)	1.188	0.809	*	0.904	0.888	0.867	0.853	0.845	0.834	0.827	0.823	0.819	0.817	0.816
0.01500	(0.147)	0.970	*	0.892	0.872	0.858	0.839	0.828	0.821	0.811	0.806	0.803	0.799	0.798	0.799
0.02000	(0.196)	0.840	*	0.869	0.851	0.838	0.821	0.811	0.804	0.796	0.791	0.789	0.786	0.786	0.787
0.03000	(0.294)	0.686	0.858	0.839	0.822	0.811	0.797	0.788	0.783	0.776	0.772	0.770	0.769	0.769	0.770
0.04000	(0.392)	*	0.836	0.818	0.804	0.793	0.781	0.773	0.768	0.762	0.759	0.758	0.757	0.758	0.759
0.06000	(0.588)	*	0.807	0.791	0.778	0.770	0.759	0.752	0.748	0.744	0.742	0.741	0.741	0.742	0.745
0.08000	(0.785)	0.810	0.787	0.773	0.762	0.754	0.744	0.738	0.735	0.731	0.730	0.729	0.730	0.732	0.734
0.10000	(0.981)	0.794	0.773	0.760	0.749	0.742	0.733	0.728	0.725	0.722	0.721	0.721	0.722	0.724	0.727
0.15000	(1.47)	0.766	0.748	0.737	0.728	0.721	0.714	0.710	0.708	0.705	0.705	0.706	0.708	0.710	0.714
0.20000	(1.96)	0.748	0.731	0.721	0.713	0.708	0.701	0.698	0.696	0.694	0.695	0.695	0.698	0.701	0.705
0.30000	(2.94)	0.724	0.709	0.701	0.694	0.689	0.684	0.681	0.680	0.680	0.680	0.682	0.685	0.688	0.693
0.40000	(3.92)	0.707	0.695	0.687	0.681	0.677	0.672	0.670	0.669	0.670	0.671	0.672	0.676	0.680	0.684
0.60000	(5.88)	0.686	0.675	0.668	0.663	0.660	0.657	0.655	0.655	0.656	0.658	0.660	0.664	0.669	0.674
0.80000	(7.85)	0.671	0.662	0.656	0.651	0.649	0.646	0.646	0.646	0.647	0.649	0.652	0.656	0.661	0.666
1.00000	(9.81)	0.661	0.652	0.647	0.643	0.640	0.638	0.638	0.638	0.640	0.643	0.645	0.650	0.655	0.661
1.50000	(14.7)	0.642	0.635	0.631	0.628	0.626	0.625	0.625	0.626	0.629	0.632	0.635	0.640	0.646	0.651
2.00000	(19.6)	0.630	0.623	0.620	0.617	0.616	0.616	0.617	0.618	0.621	0.624	0.627	0.633	0.639	0.645
3.00000	(29.4)	0.614	0.608	0.606	0.604	0.603	0.604	0.605	0.607	0.610	0.614	0.618	0.624	0.630	0.637
4.00000	(39.2)	0.603	0.598	0.596	0.595	0.595	0.596	0.597	0.599	0.603	0.608	0.611	0.619	0.625	0.632
6.00000	(58.8)	0.588	0.585	0.583	0.583	0.583	0.585	0.587	0.590	0.594	0.599	0.603	0.611	0.617	0.625
8.00000	(78.5)	0.578	0.576	0.575	0.575	0.575	0.578	0.580	0.583	0.588	0.593	0.598	0.606	0.612	0.620
10.0000	(98.1)	0.571	0.569	0.568	0.569	0.570	0.572	0.575	0.578	0.584	0.589	0.594	0.602	0.609	0.617
15.0000	(147)	0.559	0.558	0.558	0.559	0.560	0.563	0.567	0.570	0.576	0.582	0.587	0.596	0.603	0.611
20.0000	(196)	0.551	0.550	0.551	0.552	0.554	0.558	0.561	0.565	0.571	0.577	0.582	0.591	0.599	0.608
30.0000	(294)	0.540	0.540	0.541	0.543	0.545	0.550	0.554	0.558	0.565	0.571	0.577	0.586	0.594	0.603
S	$(\Delta p/\rho l)$	6	8	10	12.5	15	20	25	30	40	50	60	80	100	125

Grad'nt $\quad$ **k.p.g.** $\quad$ **(Equivalent) Pipe diameters in mm**

S	$(\Delta p/\rho l)$	125	150	200	250	300	400	500	600	800	1000	1250	1500	2000	2500
0.00010	(0.00098)	1.096	1.085	1.072	1.064	1.059	1.054	1.052	1.052	1.053	1.056	1.060	1.064	1.071	1.079
0.00015	(0.00147)	1.064	1.054	1.043	1.037	1.033	1.030	1.029	1.029	1.031	1.035	1.039	1.043	1.052	1.060
0.00020	(0.00196)	1.042	1.034	1.024	1.018	1.015	1.013	1.012	1.013	1.016	1.020	1.025	1.030	1.039	1.047
0.00030	(0.00294)	1.013	1.006	0.998	0.994	0.991	0.990	0.991	0.992	0.996	1.000	1.006	1.011	1.020	1.029
0.00040	(0.00392)	0.993	0.987	0.980	0.977	0.975	0.974	0.976	0.977	0.982	0.987	0.992	0.998	1.008	1.017
0.00060	(0.00588)	0.966	0.961	0.956	0.954	0.953	0.953	0.955	0.958	0.963	0.968	0.974	0.980	0.991	1.000
0.00080	(0.00785)	0.948	0.944	0.940	0.938	0.938	0.939	0.941	0.944	0.950	0.955	0.962	0.968	0.979	0.988
0.00100	(0.00981)	0.935	0.931	0.928	0.926	0.926	0.928	0.931	0.934	0.940	0.946	0.953	0.959	0.970	0.980
0.00150	(0.0147)	0.911	0.909	0.906	0.906	0.907	0.909	0.912	0.916	0.923	0.929	0.936	0.943	0.954	0.964
0.00200	(0.0196)	0.896	0.893	0.892	0.892	0.893	0.896	0.900	0.904	0.911	0.917	0.925	0.932	0.944	0.954
0.00300	(0.0294)	0.874	0.873	0.872	0.873	0.875	0.879	0.883	0.887	0.895	0.902	0.910	0.917	0.929	0.939
0.00400	(0.0392)	0.859	0.858	0.859	0.860	0.862	0.866	0.871	0.875	0.884	0.891	0.899	0.906	0.919	0.930
0.00600	(0.0588)	0.840	0.839	0.840	0.843	0.845	0.850	0.855	0.860	0.869	0.876	0.885	0.892	0.905	0.916
0.00800	(0.0785)	0.826	0.826	0.828	0.831	0.833	0.839	0.844	0.849	0.858	0.866	0.875	0.883	0.896	0.907
0.01000	(0.0981)	0.816	0.817	0.819	0.822	0.825	0.831	0.836	0.841	0.851	0.859	0.868	0.876	0.889	0.900
0.01500	(0.147)	0.799	0.800	0.802	0.806	0.809	0.816	0.822	0.827	0.837	0.846	0.855	0.863	0.877	0.888
0.02000	(0.196)	0.787	0.788	0.791	0.795	0.799	0.806	0.812	0.818	0.828	0.837	0.846	0.854	0.868	0.880
0.03000	(0.294)	0.770	0.772	0.776	0.781	0.785	0.792	0.799	0.805	0.815	0.824	0.834	0.843	0.857	0.869
0.04000	(0.392)	0.759	0.762	0.766	0.771	0.775	0.783	0.790	0.796	0.807	0.816	0.826	0.835	0.850	0.862
0.06000	(0.588)	0.745	0.747	0.752	0.757	0.762	0.770	0.778	0.784	0.795	0.805	0.815	0.824	0.839	0.852
0.08000	(0.785)	0.734	0.737	0.743	0.748	0.753	0.762	0.769	0.776	0.788	0.797	0.808	0.817	0.833	0.845
0.10000	(0.981)	0.727	0.730	0.736	0.741	0.746	0.755	0.763	0.770	0.782	0.792	0.803	0.812	0.827	0.840
0.15000	(1.47)	0.714	0.717	0.724	0.729	0.735	0.744	0.752	0.760	0.772	0.782	0.793	0.803	0.819	0.832
0.20000	(1.96)	0.705	0.708	0.715	0.721	0.727	0.737	0.745	0.752	0.765	0.776	0.787	0.797	0.813	0.826
0.30000	(2.94)	0.693	0.697	0.704	0.711	0.716	0.727	0.735	0.743	0.756	0.767	0.779	0.789	0.805	0.819
0.40000	(3.92)	0.684	0.689	0.696	0.703	0.709	0.720	0.729	0.737	0.750	0.761	0.773	0.783	0.800	0.814
0.60000	(5.88)	0.674	0.678	0.686	0.694	0.700	0.711	0.720	0.729	0.742	0.754	0.766	0.776	0.794	0.808
0.80000	(7.85)	0.666	0.671	0.680	0.687	0.694	0.705	0.715	0.723	0.737	0.749	0.761	0.772	0.789	0.804
1.00000	(9.81)	0.661	0.666	0.675	0.682	0.689	0.701	0.711	0.719	0.733	0.745	0.758	0.768	0.786	0.801
1.50000	(14.7)	0.651	0.657	0.666	0.674	0.681	0.693	0.703	0.712	0.727	0.739	0.752	0.763	0.781	0.796
2.00000	(19.6)	0.645	0.651	0.661	0.669	0.676	0.689	0.699	0.708	0.723	0.735	0.748	0.759	0.778	0.793
3.00000	(29.4)	0.637	0.643	0.653	0.662	0.669	0.682	0.693	0.702	0.718	0.730	0.744	0.755	0.774	0.789
S	$(\Delta p/\rho l)$	125	150	200	250	300	400	500	600	800	1000	1250	1500	2000	2500

Grad'nt $\quad$ **k.p.g.** $\quad$ **(Equivalent) Pipe diameters in mm**

Kinematic viscosity, $\nu = 1.0 \times 10^{-6}$ m^{2}s^{-1} ; $\qquad$ **Roughness size, k_s = 0.0015 mm**

Kin. visc., $\nu = 1.00 \times 10^{-6}$ m^{2}s^{-1};
$S = 0.00010$ to 30.0000

i.e. kin. pr. grad., $\Delta p/\rho l =$
(0.00098) to (294) ms^{-2}

Roughness size, $k_s = 0.003$ mm
This table shows values of m, as follows

m_C for Colebrook-White solutions; or,
where $R \leq 2000$, m_P for laminar flow

Grad'nt S	k.p.g. $(\Delta p/\rho l)$	6	8	10	12.5	15	20	25	30	40	50	60	80	100	125
0.00030	(0.00294)	6.857	4.673	3.470	2.577	2.021	1.377	1.023	*	1.098	1.074	1.058	1.037	1.024	1.013
0.00040	(0.00392)	5.939	4.047	3.005	2.232	1.750	1.193	0.886	*	1.070	1.049	1.034	1.015	1.003	0.994
0.00060	(0.00588)	4.849	3.304	2.454	1.822	1.429	0.974	*	1.063	1.033	1.014	1.001	0.985	0.975	0.967
0.00080	(0.00785)	4.199	2.861	2.125	1.578	1.238	0.843	*	1.036	1.009	0.992	0.980	0.965	0.956	0.949
0.00100	(0.00981)	3.756	2.559	1.901	1.412	1.107	*	1.035	1.016	0.990	0.974	0.964	0.950	0.942	0.936
0.00150	(0.0147)	3.067	2.090	1.552	1.153	0.904	*	0.998	0.981	0.959	0.945	0.936	0.924	0.917	0.913
0.00200	(0.0196)	2.656	1.810	1.344	0.998	*	0.996	0.973	0.958	0.938	0.925	0.917	0.907	0.901	0.897
0.00300	(0.0294)	2.169	1.478	1.097	0.815	*	0.960	0.940	0.927	0.909	0.899	0.892	0.883	0.879	0.875
0.00400	(0.0392)	1.878	1.280	0.950	*	*	0.936	0.918	0.906	0.890	0.881	0.875	0.867	0.863	0.861
0.00600	(0.0588)	1.533	1.045	0.776	*	0.929	0.905	0.889	0.878	0.865	0.857	0.852	0.846	0.843	0.842
0.00800	(0.0785)	1.328	0.905	*	0.923	0.906	0.884	0.870	0.860	0.848	0.841	0.836	0.831	0.829	0.828
0.01000	(0.0981)	1.188	0.809	*	0.905	0.889	0.868	0.855	0.846	0.835	0.829	0.825	0.821	0.819	0.819
0.01500	(0.147)	0.970	*	0.894	0.873	0.859	0.841	0.830	0.822	0.813	0.808	0.805	0.802	0.801	0.801
0.02000	(0.196)	0.840	*	0.871	0.852	0.839	0.823	0.813	0.806	0.798	0.794	0.791	0.789	0.789	0.790
0.03000	(0.294)	0.686	0.860	0.840	0.824	0.813	0.799	0.790	0.785	0.778	0.775	0.773	0.772	0.772	0.774
0.04000	(0.392)	*	0.838	0.820	0.806	0.796	0.783	0.775	0.770	0.765	0.762	0.761	0.760	0.761	0.763
0.06000	(0.588)	*	0.809	0.794	0.781	0.772	0.761	0.755	0.751	0.747	0.745	0.744	0.745	0.746	0.749
0.08000	(0.785)	0.812	0.790	0.776	0.764	0.756	0.747	0.741	0.738	0.734	0.733	0.733	0.734	0.736	0.739
0.10000	(0.981)	0.796	0.775	0.762	0.752	0.745	0.736	0.731	0.728	0.725	0.725	0.725	0.726	0.729	0.732
0.15000	(1.47)	0.769	0.751	0.740	0.731	0.725	0.717	0.714	0.711	0.710	0.710	0.710	0.713	0.716	0.720
0.20000	(1.96)	0.751	0.735	0.725	0.716	0.711	0.705	0.702	0.700	0.699	0.700	0.701	0.704	0.707	0.711
0.30000	(2.94)	0.727	0.713	0.704	0.698	0.693	0.688	0.686	0.685	0.685	0.686	0.688	0.692	0.695	0.700
0.40000	(3.92)	0.711	0.699	0.691	0.685	0.681	0.677	0.676	0.675	0.676	0.677	0.679	0.683	0.688	0.693
0.60000	(5.88)	0.690	0.679	0.673	0.668	0.665	0.662	0.661	0.662	0.663	0.665	0.668	0.673	0.678	0.683
0.80000	(7.85)	0.676	0.667	0.661	0.657	0.655	0.652	0.652	0.653	0.655	0.657	0.660	0.666	0.671	0.677
1.00000	(9.81)	0.666	0.657	0.652	0.649	0.647	0.645	0.645	0.646	0.648	0.651	0.654	0.660	0.666	0.672
1.50000	(14.7)	0.648	0.641	0.637	0.634	0.633	0.633	0.633	0.635	0.638	0.641	0.645	0.651	0.657	0.664
2.00000	(19.6)	0.636	0.630	0.627	0.625	0.624	0.624	0.625	0.627	0.631	0.635	0.639	0.646	0.652	0.659
3.00000	(29.4)	0.621	0.616	0.614	0.612	0.612	0.613	0.615	0.617	0.622	0.626	0.630	0.638	0.645	0.652
4.00000	(39.2)	0.611	0.607	0.605	0.604	0.604	0.606	0.608	0.611	0.616	0.621	0.625	0.633	0.640	0.648
6.00000	(58.8)	0.597	0.594	0.593	0.593	0.594	0.594	0.597	0.599	0.602	0.608	0.613	0.618	0.627	0.634
8.00000	(78.5)	0.588	0.586	0.586	0.586	0.587	0.590	0.594	0.597	0.603	0.609	0.614	0.623	0.631	0.639
10.0000	(98.1)	0.582	0.580	0.580	0.581	0.583	0.586	0.590	0.593	0.600	0.605	0.611	0.620	0.628	0.636
15.0000	(147)	0.571	0.570	0.571	0.572	0.574	0.578	0.583	0.586	0.594	0.600	0.605	0.615	0.624	0.632
20.0000	(196)	0.563	0.564	0.565	0.567	0.569	0.574	0.578	0.582	0.590	0.596	0.602	0.612	0.621	0.630
30.0000	(294)	0.554	0.555	0.557	0.559	0.562	0.567	0.572	0.577	0.585	0.592	0.598	0.609	0.617	0.627
S	$(\Delta p/\rho l)$	6	8	10	12.5	15	20	25	30	40	50	60	80	100	125

Grad'nt k.p.g. (Equivalent) Pipe diameters in mm

Grad'nt S	k.p.g. $(\Delta p/\rho l)$	125	150	200	250	300	400	500	600	800	1000	1250	1500	2000	2500
0.00010	(0.00098)	1.096	1.086	1.072	1.065	1.060	1.055	1.053	1.053	1.054	1.057	1.060	1.065	1.072	1.080
0.00015	(0.00147)	1.064	1.055	1.044	1.037	1.034	1.030	1.029	1.030	1.032	1.035	1.040	1.044	1.053	1.061
0.00020	(0.00196)	1.042	1.034	1.024	1.019	1.016	1.013	1.013	1.014	1.017	1.021	1.026	1.031	1.040	1.048
0.00030	(0.00294)	1.013	1.006	0.998	0.994	0.992	0.991	0.991	0.993	0.997	1.001	1.007	1.012	1.022	1.030
0.00040	(0.00392)	0.994	0.988	0.981	0.977	0.976	0.975	0.977	0.978	0.983	0.988	0.994	0.999	1.009	1.018
0.00060	(0.00588)	0.967	0.962	0.957	0.955	0.954	0.954	0.956	0.959	0.964	0.970	0.976	0.982	0.993	1.002
0.00080	(0.00785)	0.949	0.945	0.941	0.939	0.939	0.940	0.943	0.945	0.951	0.957	0.964	0.970	0.981	0.991
0.00100	(0.00981)	0.936	0.932	0.929	0.928	0.928	0.930	0.932	0.935	0.942	0.948	0.955	0.961	0.973	0.982
0.00150	(0.0147)	0.913	0.910	0.907	0.907	0.908	0.911	0.914	0.918	0.925	0.931	0.939	0.945	0.957	0.967
0.00200	(0.0196)	0.897	0.895	0.893	0.893	0.895	0.898	0.902	0.906	0.913	0.920	0.928	0.935	0.947	0.957
0.00300	(0.0294)	0.875	0.874	0.874	0.875	0.876	0.881	0.885	0.889	0.897	0.905	0.913	0.920	0.933	0.944
0.00400	(0.0392)	0.861	0.860	0.860	0.862	0.864	0.869	0.874	0.878	0.887	0.894	0.903	0.910	0.923	0.934
0.00600	(0.0588)	0.842	0.841	0.843	0.845	0.847	0.853	0.858	0.863	0.872	0.880	0.889	0.897	0.910	0.922
0.00800	(0.0785)	0.828	0.829	0.831	0.833	0.836	0.842	0.848	0.853	0.862	0.871	0.880	0.888	0.902	0.913
0.01000	(0.0981)	0.819	0.819	0.821	0.824	0.828	0.834	0.840	0.845	0.855	0.864	0.873	0.881	0.895	0.907
0.01500	(0.147)	0.801	0.802	0.806	0.809	0.813	0.820	0.826	0.832	0.842	0.851	0.861	0.869	0.884	0.896
0.02000	(0.196)	0.790	0.791	0.795	0.799	0.803	0.810	0.817	0.823	0.833	0.843	0.853	0.861	0.876	0.889
0.03000	(0.294)	0.774	0.776	0.780	0.785	0.789	0.797	0.804	0.811	0.822	0.831	0.842	0.851	0.866	0.879
0.04000	(0.392)	0.763	0.766	0.770	0.775	0.780	0.788	0.796	0.802	0.814	0.824	0.834	0.844	0.859	0.872
0.06000	(0.588)	0.749	0.752	0.757	0.763	0.768	0.777	0.784	0.791	0.804	0.814	0.825	0.834	0.851	0.864
0.08000	(0.785)	0.739	0.742	0.748	0.754	0.759	0.769	0.777	0.784	0.797	0.807	0.818	0.828	0.845	0.858
0.10000	(0.981)	0.732	0.735	0.742	0.748	0.753	0.763	0.771	0.779	0.792	0.802	0.814	0.824	0.840	0.854
0.15000	(1.47)	0.720	0.723	0.731	0.737	0.743	0.753	0.762	0.770	0.783	0.794	0.806	0.816	0.833	0.848
0.20000	(1.96)	0.711	0.715	0.723	0.730	0.736	0.746	0.755	0.763	0.777	0.788	0.801	0.811	0.829	0.843
0.30000	(2.94)	0.700	0.705	0.713	0.720	0.726	0.738	0.747	0.755	0.770	0.781	0.794	0.805	0.823	0.837
0.40000	(3.92)	0.693	0.698	0.706	0.714	0.720	0.732	0.742	0.750	0.765	0.777	0.790	0.800	0.819	0.834
0.60000	(5.88)	0.683	0.688	0.697	0.705	0.712	0.724	0.735	0.744	0.758	0.771	0.784	0.795	0.814	0.829
0.80000	(7.85)	0.677	0.682	0.692	0.700	0.707	0.720	0.730	0.739	0.754	0.767	0.780	0.792	0.811	0.827
1.00000	(9.81)	0.672	0.678	0.687	0.696	0.703	0.716	0.727	0.736	0.752	0.764	0.778	0.789	0.809	0.824
1.50000	(14.7)	0.664	0.670	0.680	0.689	0.697	0.710	0.721	0.731	0.747	0.760	0.774	0.786	0.805	0.821
2.00000	(19.6)	0.659	0.665	0.676	0.685	0.693	0.707	0.718	0.728	0.744	0.757	0.771	0.783	0.803	0.819
3.00000	(29.4)	0.652	0.659	0.670	0.680	0.688	0.702	0.713	0.723	0.740	0.754	0.768	0.780	0.800	0.817
S	$(\Delta p/\rho l)$	125	150	200	250	300	400	500	600	800	1000	1250	1500	2000	2500

Grad'nt k.p.g. (Equivalent) Pipe diameters in mm

Kinematic viscosity, $\nu = 1.00 \times 10^{-6}$ m^{2}s^{-1};

Roughness size, $k_s = 0.003$ mm

E89

Kin. visc., $\nu = 1.00 \times 10^{-6}$ m^{2}s^{-1};
$S = 0.00010$ to 30.0000

i.e. kin. pr. grad., $\Delta p/\rho l =$
(0.00098) to (294) ms^{-2}

Roughness size, $k_s = 0.006$ mm
This table shows values of m, as follows

m_C for Colebrook-White solutions; or,
where $\mathbf{R} \leq 2000$, m_P for laminar flow

Grad'nt S	k.p.g. $(\Delta p/\rho l)$	6	8	10	12.5	15	20	25	30	40	50	60	80	100	125
0.00030	(0.00294)	6.857	4.673	3.470	2.577	2.021	1.377	1.023	*	1.099	1.075	1.059	1.038	1.025	1.014
0.00040	(0.00392)	5.939	4.047	3.005	2.232	1.750	1.193	0.886	*	1.071	1.050	1.035	1.016	1.004	0.995
0.00060	(0.00588)	4.849	3.304	2.454	1.822	1.429	0.974	*	1.065	1.035	1.016	1.003	0.986	0.976	0.969
0.00080	(0.00785)	4.199	2.861	2.125	1.578	1.238	0.843	*	1.037	1.010	0.993	0.981	0.967	0.958	0.951
0.00100	(0.00981)	3.756	2.559	1.901	1.412	1.107	*	1.037	1.017	0.992	0.976	0.965	0.952	0.944	0.938
0.00150	(0.0147)	3.067	2.090	1.552	1.153	0.904	*	0.999	0.982	0.960	0.947	0.937	0.926	0.920	0.915
0.00200	(0.0196)	2.656	1.810	1.344	0.998	*	0.997	0.975	0.959	0.939	0.927	0.919	0.909	0.903	0.899
0.00300	(0.0294)	2.169	1.478	1.097	0.815	*	0.962	0.942	0.929	0.911	0.901	0.894	0.886	0.881	0.878
0.00400	(0.0392)	1.878	1.280	0.950	*	*	0.938	0.920	0.908	0.893	0.883	0.877	0.870	0.866	0.864
0.00600	(0.0588)	1.533	1.045	0.776	*	0.932	0.907	0.891	0.881	0.868	0.860	0.855	0.849	0.846	0.845
0.00800	(0.0785)	1.328	0.905	*	0.926	0.909	0.886	0.872	0.863	0.851	0.844	0.840	0.835	0.833	0.832
0.01000	(0.0981)	1.188	0.809	*	0.908	0.892	0.871	0.858	0.849	0.838	0.832	0.828	0.824	0.823	0.823
0.01500	(0.147)	0.970	*	0.897	0.876	0.862	0.844	0.833	0.826	0.817	0.812	0.809	0.806	0.806	0.806
0.02000	(0.196)	0.840	*	0.874	0.855	0.843	0.826	0.817	0.810	0.802	0.798	0.796	0.794	0.794	0.795
0.03000	(0.294)	0.686	0.863	0.844	0.828	0.817	0.803	0.795	0.789	0.783	0.780	0.778	0.777	0.778	0.780
0.04000	(0.392)	*	0.842	0.824	0.810	0.800	0.787	0.780	0.775	0.770	0.767	0.766	0.766	0.768	0.770
0.06000	(0.588)	*	0.813	0.798	0.785	0.777	0.766	0.760	0.756	0.753	0.751	0.751	0.752	0.754	0.757
0.08000	(0.785)	0.817	0.794	0.781	0.769	0.762	0.752	0.747	0.744	0.741	0.740	0.740	0.742	0.745	0.748
0.10000	(0.981)	0.801	0.780	0.768	0.757	0.750	0.742	0.737	0.735	0.732	0.732	0.733	0.735	0.738	0.742
0.15000	(1.47)	0.774	0.757	0.746	0.737	0.731	0.724	0.721	0.719	0.718	0.718	0.719	0.722	0.726	0.731
0.20000	(1.96)	0.757	0.741	0.731	0.723	0.718	0.712	0.710	0.708	0.708	0.709	0.710	0.714	0.718	0.723
0.30000	(2.94)	0.734	0.720	0.712	0.705	0.701	0.697	0.695	0.694	0.695	0.697	0.699	0.703	0.708	0.714
0.40000	(3.92)	0.719	0.706	0.699	0.693	0.690	0.686	0.685	0.685	0.686	0.689	0.691	0.696	0.701	0.707
0.60000	(5.88)	0.699	0.688	0.682	0.678	0.675	0.673	0.673	0.673	0.675	0.678	0.681	0.687	0.693	0.699
0.80000	(7.85)	0.685	0.676	0.671	0.667	0.665	0.664	0.664	0.665	0.668	0.671	0.675	0.681	0.687	0.694
1.00000	(9.81)	0.676	0.667	0.663	0.660	0.658	0.657	0.658	0.659	0.663	0.666	0.670	0.677	0.683	0.690
1.50000	(14.7)	0.659	0.652	0.649	0.647	0.646	0.646	0.648	0.650	0.654	0.658	0.662	0.670	0.677	0.684
2.00000	(19.6)	0.648	0.643	0.640	0.638	0.638	0.639	0.641	0.643	0.648	0.653	0.657	0.665	0.672	0.680
3.00000	(29.4)	0.634	0.630	0.628	0.628	0.628	0.630	0.632	0.635	0.641	0.646	0.651	0.659	0.667	0.675
4.00000	(39.2)	0.625	0.622	0.621	0.620	0.621	0.624	0.627	0.630	0.636	0.642	0.647	0.656	0.664	0.672
6.00000	(58.8)	0.613	0.611	0.611	0.611	0.613	0.616	0.620	0.623	0.630	0.636	0.642	0.651	0.659	0.668
8.00000	(78.5)	0.605	0.604	0.604	0.606	0.607	0.611	0.615	0.619	0.626	0.633	0.638	0.648	0.657	0.666
10.0000	(98.1)	0.600	0.599	0.600	0.601	0.603	0.608	0.612	0.616	0.624	0.630	0.636	0.646	0.655	0.664
15.0000	(147)	0.591	0.591	0.592	0.595	0.597	0.602	0.607	0.611	0.619	0.626	0.632	0.643	0.652	0.662
20.0000	(196)	0.585	0.586	0.587	0.590	0.593	0.598	0.604	0.608	0.617	0.624	0.630	0.641	0.650	0.660
30.0000	(294)	0.577	0.579	0.581	0.585	0.588	0.594	0.599	0.604	0.613	0.621	0.627	0.639	0.648	0.658
S	$(\Delta p/\rho l)$	6	8	10	12.5	15	20	25	30	40	50	60	80	100	125

Grad'nt k.p.g. (Equivalent) Pipe diameters in mm

S	$(\Delta p/\rho l)$	125	150	200	250	300	400	500	600	800	1000	1250	1500	2000	2500
0.00010	(0.00098)	1.097	1.086	1.073	1.066	1.061	1.056	1.054	1.054	1.055	1.058	1.062	1.066	1.074	1.082
0.00015	(0.00147)	1.065	1.056	1.045	1.039	1.035	1.032	1.031	1.031	1.034	1.037	1.042	1.047	1.055	1.063
0.00020	(0.00196)	1.043	1.035	1.026	1.020	1.017	1.015	1.015	1.016	1.019	1.023	1.028	1.033	1.042	1.051
0.00030	(0.00294)	1.014	1.008	1.000	0.996	0.994	0.993	0.993	0.995	0.999	1.004	1.009	1.015	1.025	1.034
0.00040	(0.00392)	0.995	0.989	0.982	0.979	0.978	0.977	0.979	0.981	0.985	0.990	0.997	1.002	1.013	1.022
0.00060	(0.00588)	0.969	0.964	0.959	0.957	0.956	0.957	0.959	0.961	0.967	0.973	0.979	0.986	0.997	1.006
0.00080	(0.00785)	0.951	0.947	0.943	0.941	0.941	0.943	0.945	0.948	0.955	0.961	0.968	0.974	0.986	0.996
0.00100	(0.00981)	0.938	0.934	0.931	0.930	0.930	0.932	0.935	0.938	0.945	0.951	0.959	0.965	0.977	0.988
0.00150	(0.0147)	0.915	0.912	0.910	0.910	0.911	0.914	0.918	0.921	0.929	0.936	0.943	0.950	0.963	0.974
0.00200	(0.0196)	0.899	0.897	0.896	0.896	0.898	0.901	0.906	0.910	0.918	0.925	0.933	0.940	0.953	0.964
0.00300	(0.0294)	0.878	0.877	0.877	0.878	0.880	0.885	0.889	0.894	0.903	0.910	0.919	0.927	0.940	0.951
0.00400	(0.0392)	0.864	0.863	0.864	0.866	0.868	0.873	0.878	0.883	0.892	0.900	0.909	0.917	0.931	0.943
0.00600	(0.0588)	0.845	0.845	0.847	0.849	0.852	0.858	0.864	0.869	0.879	0.887	0.897	0.905	0.920	0.932
0.00800	(0.0785)	0.832	0.833	0.835	0.838	0.841	0.848	0.854	0.860	0.870	0.879	0.888	0.897	0.912	0.924
0.01000	(0.0981)	0.823	0.824	0.826	0.830	0.833	0.840	0.847	0.852	0.863	0.872	0.882	0.891	0.906	0.919
0.01500	(0.147)	0.806	0.808	0.811	0.815	0.819	0.827	0.834	0.840	0.851	0.861	0.871	0.881	0.896	0.909
0.02000	(0.196)	0.795	0.797	0.801	0.806	0.810	0.818	0.825	0.832	0.843	0.853	0.864	0.874	0.890	0.903
0.03000	(0.294)	0.780	0.783	0.788	0.793	0.797	0.806	0.814	0.821	0.833	0.844	0.855	0.865	0.881	0.895
0.04000	(0.392)	0.770	0.773	0.779	0.784	0.789	0.798	0.807	0.814	0.826	0.837	0.849	0.859	0.876	0.890
0.06000	(0.588)	0.757	0.760	0.767	0.773	0.778	0.788	0.797	0.804	0.818	0.829	0.841	0.851	0.869	0.883
0.08000	(0.785)	0.748	0.752	0.759	0.765	0.771	0.781	0.790	0.798	0.812	0.823	0.836	0.846	0.864	0.879
0.10000	(0.981)	0.742	0.746	0.753	0.760	0.766	0.776	0.786	0.794	0.808	0.819	0.832	0.843	0.861	0.876
0.15000	(1.47)	0.731	0.735	0.743	0.750	0.757	0.768	0.778	0.786	0.801	0.813	0.826	0.837	0.856	0.871
0.20000	(1.96)	0.723	0.728	0.736	0.744	0.751	0.763	0.773	0.781	0.796	0.809	0.822	0.833	0.852	0.868
0.30000	(2.94)	0.714	0.719	0.728	0.736	0.743	0.756	0.766	0.775	0.791	0.803	0.817	0.829	0.848	0.864
0.40000	(3.92)	0.707	0.713	0.723	0.731	0.738	0.751	0.762	0.771	0.787	0.800	0.814	0.826	0.845	0.862
0.60000	(5.88)	0.699	0.705	0.715	0.724	0.732	0.745	0.757	0.766	0.783	0.796	0.810	0.822	0.842	0.859
0.80000	(7.85)	0.694	0.700	0.711	0.720	0.728	0.742	0.753	0.763	0.780	0.793	0.808	0.820	0.840	0.857
1.00000	(9.81)	0.690	0.697	0.708	0.717	0.725	0.739	0.751	0.761	0.778	0.791	0.806	0.818	0.839	0.855
1.50000	(14.7)	0.684	0.691	0.702	0.712	0.721	0.735	0.747	0.757	0.774	0.789	0.803	0.816	0.836	0.853
2.00000	(19.6)	0.680	0.687	0.699	0.709	0.718	0.733	0.745	0.755	0.773	0.787	0.802	0.814	0.835	0.852
3.00000	(29.4)	0.675	0.683	0.695	0.705	0.714	0.729	0.742	0.752	0.770	0.785	0.800	0.812	0.833	0.851
S	$(\Delta p/\rho l)$	125	150	200	250	300	400	500	600	800	1000	1250	1500	2000	2500

Grad'nt k.p.g. (Equivalent) Pipe diameters in mm

Kinematic viscosity, $\nu = 1.00 \times 10^{-6}$ m^{2}s^{-1} ;

Roughness size, $k_s = 0.006$ mm

Kin. visc., $\nu = 1.00\times10^{-6}\ \mathrm{m^2\,s^{-1}}$; **Roughness size, $k_s = 0.015$ mm**
$S = 0.00010$ to 30.0000 This table shows values of m, as follows

i.e. kin. pr. grad., $\Delta p/\rho l =$ m_C for Colebrook-White solutions; or,
(0.00098) to $(294)\ \mathrm{ms^{-2}}$ where $\mathbf{R} \leq 2000$, m_P for laminar flow

Grad'nt S	k.p.g. $(\Delta p/\rho l)$	\(6\)	\(8\)	\(10\)	\(12.5\)	\(15\)	\(20\)	\(25\)	\(30\)	\(40\)	\(50\)	\(60\)	\(80\)	\(100\)	\(125\)
		(Equivalent) Pipe diameters in mm													
0.00030	(0.00294)	6.857	4.673	3.470	2.577	2.021	1.377	1.023	*	1.102	1.078	1.062	1.041	1.028	1.018
0.00040	(0.00392)	5.939	4.047	3.005	2.232	1.750	1.193	0.886	*	1.075	1.053	1.038	1.019	1.008	0.999
0.00060	(0.00588)	4.849	3.304	2.454	1.822	1.429	0.974	*	1.068	1.038	1.019	1.007	0.990	0.981	0.973
0.00080	(0.00785)	4.199	2.861	2.125	1.578	1.238	0.843	*	1.041	1.014	0.997	0.986	0.971	0.963	0.956
0.00100	(0.00981)	3.756	2.559	1.901	1.412	1.107	*	1.041	1.021	0.996	0.980	0.970	0.957	0.949	0.943
0.00150	(0.0147)	3.067	2.090	1.552	1.153	0.904	*	1.004	0.987	0.965	0.952	0.943	0.932	0.925	0.921
0.00200	(0.0196)	2.656	1.810	1.344	0.998	*	1.002	0.980	0.965	0.945	0.933	0.925	0.915	0.910	0.906
0.00300	(0.0294)	2.169	1.478	1.097	0.815	*	0.967	0.948	0.935	0.918	0.907	0.901	0.893	0.889	0.886
0.00400	(0.0392)	1.878	1.280	0.950	*	*	0.944	0.927	0.915	0.899	0.890	0.885	0.878	0.875	0.873
0.00600	(0.0588)	1.533	1.045	0.776	*	0.938	0.914	0.899	0.888	0.875	0.868	0.863	0.858	0.856	0.855
0.00800	(0.0785)	1.328	0.905	*	0.933	0.916	0.894	0.880	0.871	0.859	0.853	0.849	0.845	0.844	0.844
0.01000	(0.0981)	1.188	0.809	*	0.915	0.899	0.879	0.866	0.858	0.847	0.842	0.838	0.835	0.834	0.835
0.01500	(0.147)	0.970	*	0.905	0.885	0.871	0.853	0.843	0.836	0.827	0.823	0.820	0.818	0.819	0.820
0.02000	(0.196)	0.840	*	0.883	0.865	0.852	0.836	0.827	0.821	0.814	0.810	0.808	0.807	0.808	0.810
0.03000	(0.294)	0.686	0.873	0.854	0.838	0.828	0.814	0.806	0.801	0.796	0.793	0.792	0.793	0.795	0.797
0.04000	(0.392)	*	0.853	0.835	0.821	0.812	0.800	0.793	0.789	0.784	0.782	0.782	0.783	0.786	0.789
0.06000	(0.588)	*	0.826	0.811	0.798	0.790	0.780	0.775	0.772	0.769	0.768	0.768	0.771	0.774	0.778
0.08000	(0.785)	0.830	0.808	0.794	0.783	0.776	0.768	0.763	0.761	0.759	0.759	0.760	0.763	0.766	0.771
0.10000	(0.981)	0.815	0.795	0.782	0.772	0.766	0.758	0.755	0.752	0.751	0.752	0.753	0.757	0.761	0.766
0.15000	(1.47)	0.790	0.773	0.762	0.754	0.749	0.743	0.740	0.739	0.739	0.740	0.742	0.747	0.751	0.757
0.20000	(1.96)	0.774	0.758	0.749	0.742	0.737	0.732	0.730	0.730	0.731	0.733	0.735	0.740	0.746	0.752
0.30000	(2.94)	0.753	0.739	0.732	0.726	0.723	0.719	0.718	0.718	0.720	0.723	0.726	0.732	0.738	0.745
0.40000	(3.92)	0.739	0.727	0.721	0.716	0.713	0.711	0.710	0.711	0.714	0.717	0.720	0.727	0.733	0.740
0.60000	(5.88)	0.721	0.712	0.706	0.703	0.701	0.700	0.700	0.702	0.705	0.709	0.713	0.721	0.727	0.735
0.80000	(7.85)	0.710	0.701	0.697	0.694	0.693	0.693	0.694	0.696	0.700	0.704	0.709	0.716	0.724	0.732
1.00000	(9.81)	0.701	0.694	0.690	0.688	0.687	0.688	0.689	0.691	0.696	0.701	0.705	0.714	0.721	0.729
1.50000	(14.7)	0.688	0.682	0.679	0.678	0.678	0.679	0.682	0.684	0.690	0.695	0.700	0.709	0.717	0.725
2.00000	(19.6)	0.679	0.674	0.672	0.671	0.672	0.674	0.677	0.680	0.686	0.691	0.697	0.706	0.714	0.723
3.00000	(29.4)	0.668	0.664	0.663	0.663	0.664	0.667	0.671	0.674	0.681	0.687	0.693	0.702	0.711	0.720
4.00000	(39.2)	0.661	0.658	0.658	0.658	0.660	0.663	0.667	0.671	0.678	0.684	0.690	0.700	0.709	0.718
6.00000	(58.8)	0.652	0.650	0.651	0.652	0.654	0.658	0.662	0.667	0.674	0.681	0.687	0.698	0.706	0.716
8.00000	(78.5)	0.646	0.645	0.646	0.648	0.650	0.655	0.660	0.664	0.672	0.679	0.685	0.696	0.705	0.715
10.0000	(98.1)	0.642	0.642	0.643	0.645	0.648	0.653	0.658	0.662	0.670	0.678	0.684	0.695	0.704	0.714
15.0000	(147)	0.636	0.636	0.638	0.641	0.644	0.649	0.655	0.659	0.668	0.675	0.682	0.693	0.702	0.712
20.0000	(196)	0.632	0.633	0.635	0.638	0.641	0.647	0.653	0.658	0.666	0.674	0.681	0.692	0.701	0.711
30.0000	(294)	0.627	0.629	0.631	0.635	0.638	0.644	0.650	0.655	0.665	0.672	0.679	0.691	0.700	0.710
S	$(\Delta p/\rho l)$	6	8	10	12.5	15	20	25	30	40	50	60	80	100	125

Grad'nt k.p.g. (Equivalent) Pipe diameters in mm

S	$(\Delta p/\rho l)$	\(125\)	\(150\)	\(200\)	\(250\)	\(300\)	\(400\)	\(500\)	\(600\)	\(800\)	\(1000\)	\(1250\)	\(1500\)	\(2000\)	\(2500\)
0.00010	(0.00098)	1.100	1.089	1.076	1.068	1.064	1.059	1.058	1.058	1.060	1.063	1.067	1.071	1.080	1.088
0.00015	(0.00147)	1.068	1.059	1.048	1.042	1.038	1.035	1.035	1.036	1.039	1.042	1.047	1.052	1.062	1.071
0.00020	(0.00196)	1.047	1.039	1.029	1.024	1.021	1.019	1.019	1.021	1.024	1.029	1.034	1.040	1.050	1.059
0.00030	(0.00294)	1.018	1.011	1.004	1.000	0.998	0.997	0.999	1.001	1.005	1.010	1.017	1.023	1.033	1.043
0.00040	(0.00392)	0.999	0.993	0.987	0.984	0.983	0.983	0.985	0.987	0.992	0.998	1.005	1.011	1.022	1.032
0.00060	(0.00588)	0.973	0.969	0.964	0.962	0.962	0.963	0.966	0.969	0.975	0.981	0.989	0.995	1.008	1.018
0.00080	(0.00785)	0.956	0.952	0.949	0.947	0.948	0.950	0.953	0.956	0.963	0.970	0.978	0.985	0.998	1.009
0.00100	(0.00981)	0.943	0.940	0.937	0.937	0.937	0.940	0.943	0.947	0.955	0.962	0.970	0.977	0.990	1.002
0.00150	(0.0147)	0.921	0.919	0.917	0.918	0.919	0.923	0.927	0.931	0.940	0.948	0.956	0.964	0.978	0.990
0.00200	(0.0196)	0.906	0.905	0.904	0.905	0.907	0.911	0.916	0.921	0.930	0.938	0.947	0.955	0.970	0.982
0.00300	(0.0294)	0.886	0.886	0.886	0.888	0.891	0.896	0.902	0.907	0.917	0.925	0.935	0.944	0.959	0.971
0.00400	(0.0392)	0.873	0.873	0.874	0.877	0.880	0.886	0.892	0.898	0.908	0.917	0.927	0.936	0.952	0.965
0.00600	(0.0588)	0.855	0.856	0.858	0.862	0.865	0.872	0.879	0.885	0.896	0.906	0.917	0.926	0.942	0.956
0.00800	(0.0785)	0.844	0.845	0.848	0.852	0.856	0.863	0.871	0.877	0.889	0.899	0.910	0.920	0.936	0.950
0.01000	(0.0981)	0.835	0.836	0.840	0.844	0.849	0.857	0.864	0.871	0.883	0.894	0.905	0.915	0.932	0.946
0.01500	(0.147)	0.820	0.822	0.827	0.832	0.837	0.846	0.854	0.861	0.874	0.885	0.897	0.907	0.925	0.940
0.02000	(0.196)	0.810	0.813	0.818	0.824	0.829	0.839	0.847	0.855	0.868	0.880	0.892	0.902	0.920	0.935
0.03000	(0.294)	0.797	0.801	0.807	0.813	0.819	0.829	0.838	0.847	0.861	0.872	0.885	0.896	0.915	0.930
0.04000	(0.392)	0.789	0.793	0.800	0.806	0.812	0.823	0.833	0.841	0.856	0.868	0.881	0.892	0.911	0.927
0.06000	(0.588)	0.778	0.782	0.790	0.797	0.804	0.816	0.826	0.834	0.850	0.862	0.876	0.887	0.907	0.923
0.08000	(0.785)	0.771	0.775	0.784	0.792	0.799	0.811	0.821	0.830	0.846	0.859	0.872	0.884	0.904	0.920
0.10000	(0.981)	0.766	0.771	0.780	0.788	0.795	0.807	0.818	0.827	0.843	0.856	0.870	0.882	0.902	0.918
0.15000	(1.47)	0.757	0.763	0.772	0.781	0.788	0.801	0.813	0.822	0.838	0.852	0.866	0.878	0.899	0.915
0.20000	(1.96)	0.752	0.757	0.768	0.776	0.784	0.798	0.809	0.819	0.836	0.849	0.864	0.876	0.897	0.914
0.30000	(2.94)	0.745	0.751	0.762	0.771	0.779	0.793	0.805	0.815	0.832	0.846	0.861	0.874	0.895	0.912
0.40000	(3.92)	0.740	0.747	0.758	0.768	0.776	0.791	0.803	0.813	0.830	0.844	0.859	0.872	0.893	0.910
0.60000	(5.88)	0.735	0.742	0.754	0.764	0.772	0.787	0.799	0.810	0.828	0.842	0.857	0.870	0.891	0.909
0.80000	(7.85)	0.732	0.739	0.751	0.761	0.770	0.785	0.797	0.808	0.826	0.841	0.856	0.869	0.890	0.908
1.00000	(9.81)	0.729	0.736	0.749	0.759	0.768	0.784	0.796	0.807	0.825	0.840	0.855	0.868	0.890	0.907
1.50000	(14.7)	0.725	0.733	0.746	0.756	0.766	0.781	0.794	0.805	0.823	0.838	0.854	0.867	0.888	0.906
2.00000	(19.6)	0.723	0.731	0.744	0.755	0.764	0.780	0.793	0.804	0.822	0.837	0.853	0.866	0.888	0.905
3.00000	(29.4)	0.720	0.728	0.741	0.752	0.762	0.778	0.791	0.802	0.821	0.836	0.852	0.865	0.887	0.905
S	$(\Delta p/\rho l)$	125	150	200	250	300	400	500	600	800	1000	1250	1500	2000	2500

Grad'nt k.p.g. (Equivalent) Pipe diameters in mm

Kinematic viscosity, $\nu = 1.00\times10^{-6}\ \mathrm{m^2\,s^{-1}}$; **Roughness size, $k_s = 0.015$ mm**

E91

Kin. visc., $\nu = 1.00 \times 10^{-6}\ \mathrm{m^2 s^{-1}}$;
$S = 0.00010$ to 30.0000

i.e. kin. pr. grad., $\Delta p/\rho l =$
(0.00098) to $(294)\ \mathrm{ms^{-2}}$

Roughness size, $k_s = 0.030$ mm
This table shows values of m, as follows

m_C for Colebrook-White solutions; or,
where $\mathbf{R} \leq 2000$, m_P for laminar flow

Grad'nt S	k.p.g. $(\Delta p/\rho l)$	6	8	10	12.5	15	20	25	30	40	50	60	80	100	125
		(Equivalent) Pipe diameters in mm													
0.00030	(0.00294)	6.857	4.673	3.470	2.577	2.021	1.377	1.023	*	1.107	1.083	1.067	1.046	1.034	1.024
0.00040	(0.00392)	5.939	4.047	3.005	2.232	1.750	1.193	0.886	*	1.080	1.058	1.044	1.025	1.014	1.005
0.00060	(0.00588)	4.849	3.304	2.454	1.822	1.429	0.974	*	1.074	1.044	1.026	1.013	0.997	0.988	0.981
0.00080	(0.00785)	4.199	2.861	2.125	1.578	1.238	0.843	*	1.047	1.021	1.004	0.993	0.978	0.970	0.964
0.00100	(0.00981)	3.756	2.559	1.901	1.412	1.107	*	1.047	1.028	1.003	0.988	0.977	0.965	0.957	0.952
0.00150	(0.0147)	3.067	2.090	1.552	1.153	0.904	*	1.012	0.995	0.973	0.960	0.951	0.941	0.935	0.931
0.00200	(0.0196)	2.656	1.810	1.344	0.998	*	1.010	0.988	0.973	0.953	0.942	0.934	0.925	0.920	0.917
0.00300	(0.0294)	2.169	1.478	1.097	0.815	*	0.976	0.957	0.944	0.927	0.918	0.911	0.904	0.901	0.899
0.00400	(0.0392)	1.878	1.280	0.950	*	*	0.954	0.937	0.925	0.910	0.902	0.896	0.890	0.888	0.887
0.00600	(0.0588)	1.533	1.045	0.776	*	0.949	0.925	0.910	0.900	0.888	0.881	0.876	0.872	0.871	0.871
0.00800	(0.0785)	1.328	0.905	*	0.944	0.927	0.906	0.892	0.883	0.873	0.867	0.863	0.860	0.860	0.860
0.01000	(0.0981)	1.188	0.809	*	0.927	0.912	0.892	0.879	0.871	0.862	0.857	0.854	0.851	0.851	0.853
0.01500	(0.147)	0.970	*	0.918	0.898	0.885	0.868	0.857	0.851	0.843	0.839	0.838	0.837	0.838	0.840
0.02000	(0.196)	0.840	*	0.897	0.880	0.867	0.852	0.843	0.837	0.831	0.828	0.827	0.827	0.829	0.832
0.03000	(0.294)	0.686	0.889	0.870	0.855	0.845	0.832	0.824	0.820	0.815	0.814	0.813	0.815	0.817	0.821
0.04000	(0.392)	*	0.870	0.853	0.839	0.830	0.818	0.812	0.809	0.805	0.804	0.804	0.807	0.810	0.814
0.06000	(0.588)	*	0.845	0.830	0.818	0.811	0.801	0.797	0.794	0.792	0.792	0.793	0.797	0.801	0.806
0.08000	(0.785)	0.850	0.828	0.815	0.805	0.798	0.790	0.786	0.784	0.783	0.784	0.786	0.790	0.795	0.800
0.10000	(0.981)	0.836	0.817	0.805	0.795	0.789	0.782	0.779	0.778	0.777	0.779	0.781	0.786	0.790	0.796
0.15000	(1.47)	0.814	0.797	0.787	0.779	0.774	0.769	0.767	0.766	0.767	0.770	0.772	0.778	0.784	0.790
0.20000	(1.96)	0.799	0.784	0.775	0.769	0.764	0.761	0.759	0.759	0.761	0.764	0.767	0.773	0.779	0.786
0.30000	(2.94)	0.781	0.768	0.761	0.755	0.752	0.750	0.749	0.750	0.753	0.757	0.760	0.767	0.774	0.781
0.40000	(3.92)	0.769	0.757	0.751	0.747	0.745	0.743	0.743	0.745	0.748	0.752	0.756	0.764	0.770	0.778
0.60000	(5.88)	0.754	0.744	0.739	0.736	0.735	0.734	0.736	0.737	0.742	0.746	0.751	0.759	0.766	0.775
0.80000	(7.85)	0.744	0.736	0.732	0.730	0.729	0.729	0.731	0.733	0.738	0.743	0.748	0.756	0.764	0.772
1.00000	(9.81)	0.737	0.730	0.727	0.725	0.724	0.725	0.727	0.730	0.735	0.740	0.745	0.754	0.762	0.771
1.50000	(14.7)	0.726	0.721	0.718	0.717	0.717	0.719	0.722	0.725	0.731	0.737	0.742	0.751	0.759	0.768
2.00000	(19.6)	0.719	0.715	0.713	0.712	0.713	0.715	0.719	0.722	0.728	0.734	0.740	0.749	0.758	0.767
3.00000	(29.4)	0.711	0.707	0.706	0.707	0.708	0.711	0.714	0.718	0.725	0.731	0.737	0.747	0.756	0.765
4.00000	(39.2)	0.705	0.703	0.702	0.703	0.704	0.708	0.712	0.716	0.723	0.729	0.735	0.746	0.754	0.765
6.00000	(58.8)	0.699	0.697	0.697	0.698	0.700	0.705	0.709	0.713	0.721	0.727	0.733	0.744	0.753	0.762
8.00000	(78.5)	0.695	0.694	0.694	0.696	0.698	0.702	0.707	0.711	0.719	0.726	0.732	0.743	0.752	0.762
10.0000	(98.1)	0.692	0.691	0.692	0.694	0.696	0.701	0.706	0.710	0.718	0.725	0.731	0.742	0.751	0.761
15.0000	(147)	0.687	0.687	0.689	0.691	0.693	0.699	0.704	0.708	0.717	0.724	0.730	0.741	0.750	0.760
20.0000	(196)	0.684	0.685	0.687	0.689	0.692	0.697	0.702	0.707	0.716	0.723	0.729	0.740	0.750	0.760
30.0000	(294)	0.681	0.682	0.684	0.687	0.690	0.696	0.701	0.706	0.714	0.722	0.728	0.740	0.749	0.759
S	$(\Delta p/\rho l)$	6	8	10	12.5	15	20	25	30	40	50	60	80	100	125
Grad'nt	k.p.g.	(Equivalent) Pipe diameters in mm													

Grad'nt S	k.p.g. $(\Delta p/\rho l)$	125	150	200	250	300	400	500	600	800	1000	1250	1500	2000	2500
0.00010	(0.00098)	1.104	1.093	1.080	1.073	1.069	1.065	1.063	1.064	1.066	1.070	1.075	1.079	1.089	1.098
0.00015	(0.00147)	1.073	1.064	1.053	1.047	1.044	1.042	1.042	1.043	1.046	1.051	1.056	1.062	1.072	1.082
0.00020	(0.00196)	1.052	1.044	1.035	1.030	1.028	1.026	1.027	1.028	1.033	1.038	1.044	1.050	1.061	1.071
0.00030	(0.00294)	1.024	1.018	1.010	1.007	1.006	1.005	1.007	1.009	1.015	1.021	1.028	1.034	1.046	1.057
0.00040	(0.00392)	1.005	1.000	0.994	0.992	0.991	0.992	0.994	0.997	1.003	1.009	1.017	1.024	1.037	1.048
0.00060	(0.00588)	0.981	0.976	0.972	0.971	0.971	0.973	0.976	0.980	0.987	0.994	1.003	1.010	1.024	1.035
0.00080	(0.00785)	0.964	0.961	0.958	0.957	0.958	0.961	0.965	0.969	0.977	0.985	0.993	1.001	1.015	1.027
0.00100	(0.00981)	0.952	0.949	0.947	0.947	0.948	0.952	0.956	0.961	0.969	0.977	0.986	0.995	1.009	1.022
0.00150	(0.0147)	0.931	0.929	0.928	0.930	0.931	0.936	0.941	0.947	0.956	0.965	0.975	0.983	0.999	1.012
0.00200	(0.0196)	0.917	0.916	0.916	0.918	0.920	0.926	0.932	0.937	0.948	0.957	0.967	0.976	0.992	1.006
0.00300	(0.0294)	0.899	0.899	0.900	0.903	0.906	0.913	0.919	0.925	0.936	0.946	0.957	0.967	0.984	0.998
0.00400	(0.0392)	0.887	0.887	0.889	0.893	0.897	0.904	0.911	0.918	0.929	0.940	0.951	0.961	0.978	0.992
0.00600	(0.0588)	0.871	0.872	0.876	0.880	0.884	0.893	0.900	0.907	0.920	0.931	0.943	0.953	0.971	0.992
0.00800	(0.0785)	0.860	0.862	0.867	0.871	0.876	0.885	0.894	0.901	0.914	0.926	0.938	0.949	0.967	0.986
0.01000	(0.0981)	0.853	0.855	0.860	0.865	0.870	0.880	0.889	0.896	0.910	0.922	0.934	0.945	0.964	0.979
0.01500	(0.147)	0.840	0.843	0.849	0.855	0.861	0.871	0.880	0.889	0.903	0.915	0.928	0.940	0.959	0.975
0.02000	(0.196)	0.832	0.835	0.842	0.849	0.855	0.866	0.875	0.884	0.899	0.911	0.925	0.936	0.956	0.972
0.03000	(0.294)	0.821	0.825	0.833	0.840	0.847	0.859	0.869	0.878	0.893	0.906	0.920	0.932	0.952	0.968
0.04000	(0.392)	0.814	0.819	0.827	0.835	0.842	0.854	0.865	0.874	0.890	0.903	0.917	0.929	0.949	0.966
0.06000	(0.588)	0.806	0.811	0.820	0.828	0.836	0.849	0.860	0.869	0.886	0.899	0.914	0.926	0.947	0.963
0.08000	(0.785)	0.800	0.806	0.815	0.824	0.832	0.845	0.856	0.866	0.883	0.897	0.911	0.924	0.945	0.962
0.10000	(0.981)	0.796	0.802	0.812	0.821	0.829	0.843	0.854	0.864	0.881	0.895	0.910	0.923	0.944	0.961
0.15000	(1.47)	0.790	0.796	0.807	0.816	0.825	0.839	0.851	0.861	0.878	0.893	0.908	0.920	0.942	0.959
0.20000	(1.96)	0.786	0.793	0.804	0.813	0.822	0.836	0.849	0.859	0.876	0.891	0.906	0.919	0.940	0.958
0.30000	(2.94)	0.781	0.788	0.800	0.810	0.819	0.833	0.846	0.857	0.874	0.889	0.904	0.917	0.939	0.957
0.40000	(3.92)	0.778	0.785	0.797	0.808	0.817	0.832	0.844	0.855	0.873	0.888	0.903	0.916	0.938	0.956
0.60000	(5.88)	0.775	0.782	0.794	0.805	0.814	0.830	0.842	0.853	0.871	0.886	0.902	0.915	0.937	0.955
0.80000	(7.85)	0.772	0.780	0.793	0.803	0.813	0.828	0.841	0.852	0.871	0.886	0.901	0.915	0.937	0.954
1.00000	(9.81)	0.771	0.778	0.791	0.802	0.812	0.827	0.840	0.851	0.870	0.885	0.901	0.914	0.936	0.954
1.50000	(14.7)	0.768	0.776	0.789	0.800	0.810	0.826	0.839	0.850	0.869	0.884	0.900	0.913	0.935	0.953
2.00000	(19.6)	0.767	0.775	0.788	0.799	0.809	0.825	0.838	0.850	0.868	0.884	0.899	0.913	0.935	0.953
3.00000	(29.4)	0.765	0.773	0.787	0.798	0.808	0.824	0.837	0.849	0.868	0.883	0.899	0.912	0.935	0.953
S	$(\Delta p/\rho l)$	125	150	200	250	300	400	500	600	800	1000	1250	1500	2000	2500
Grad'nt	k.p.g.	(Equivalent) Pipe diameters in mm													

Kinematic viscosity, $\nu = 1.00 \times 10^{-6}\ \mathrm{m^2 s^{-1}}$;

Roughness size, $k_s = 0.030$ mm

Kin. visc., $\nu = 1.00\times10^{-6}$ m^{2}s^{-1};
$S = 0.00010$ to 30.0000

i.e. kin. pr. grad., $\Delta p/\rho l =$
(0.00098) to (294) ms^{-2}

Roughness size, $k_s = 0.060$ mm
This table shows values of m, as follows

m_C for Colebrook-White solutions; or,
where $\mathbf{R} \le 2000$, m_P for laminar flow

| Grad'nt | k.p.g. | (Equivalent) Pipe diameters in mm | | | | | | | | | | | | | |
S	$(\Delta p/\rho l)$	6	8	10	12.5	15	20	25	30	40	50	60	80	100	125
0.00030	(0.00294)	6.857	4.673	3.470	2.577	2.021	1.377	1.023	*	1.116	1.093	1.077	1.057	1.045	1.035
0.00040	(0.00392)	5.939	4.047	3.005	2.232	1.750	1.193	0.886	*	1.090	1.069	1.055	1.037	1.026	1.018
0.00060	(0.00588)	4.849	3.304	2.454	1.822	1.429	0.974	*	1.085	1.056	1.038	1.025	1.010	1.001	0.995
0.00080	(0.00785)	4.199	2.861	2.125	1.578	1.238	0.843	*	1.060	1.033	1.017	1.006	0.992	0.985	0.979
0.00100	(0.00981)	3.756	2.559	1.901	1.412	1.107	*	1.060	1.041	1.017	1.002	0.992	0.980	0.973	0.968
0.00150	(0.0147)	3.067	2.090	1.552	1.153	0.904	*	1.026	1.009	0.988	0.976	0.967	0.958	0.953	0.949
0.00200	(0.0196)	2.656	1.810	1.344	0.998	*	1.025	1.004	0.989	0.970	0.959	0.952	0.943	0.939	0.937
0.00300	(0.0294)	2.169	1.478	1.097	0.815	*	0.993	0.974	0.962	0.946	0.937	0.931	0.925	0.922	0.921
0.00400	(0.0392)	1.878	1.280	0.950	*	*	0.972	0.956	0.944	0.930	0.922	0.917	0.912	0.911	0.911
0.00600	(0.0588)	1.533	1.045	0.776	*	0.969	0.945	0.931	0.921	0.910	0.904	0.900	0.897	0.896	0.897
0.00800	(0.0785)	1.328	0.905	*	0.966	0.949	0.928	0.915	0.907	0.897	0.891	0.889	0.887	0.887	0.889
0.01000	(0.0981)	1.188	0.809	*	0.950	0.935	0.915	0.903	0.896	0.887	0.883	0.881	0.879	0.880	0.883
0.01500	(0.147)	0.970	*	*	0.924	0.910	0.894	0.884	0.878	0.871	0.868	0.867	0.867	0.869	0.873
0.02000	(0.196)	0.840	*	0.924	0.907	0.895	0.880	0.872	0.866	0.861	0.859	0.858	0.860	0.862	0.866
0.03000	(0.294)	0.686	*	0.900	0.885	0.875	0.862	0.856	0.852	0.848	0.847	0.847	0.850	0.854	0.858
0.04000	(0.392)	*	0.901	0.884	0.871	0.862	0.851	0.846	0.842	0.840	0.840	0.841	0.844	0.848	0.853
0.06000	(0.588)	*	0.879	0.865	0.853	0.846	0.837	0.833	0.831	0.829	0.830	0.832	0.836	0.841	0.847
0.08000	(0.785)	0.886	0.865	0.852	0.842	0.835	0.828	0.825	0.823	0.823	0.824	0.827	0.832	0.837	0.843
0.10000	(0.981)	0.875	0.855	0.843	0.834	0.828	0.822	0.819	0.818	0.818	0.820	0.823	0.828	0.834	0.840
0.15000	(1.47)	0.855	0.838	0.828	0.821	0.816	0.811	0.810	0.809	0.811	0.814	0.817	0.823	0.829	0.836
0.20000	(1.96)	0.843	0.828	0.819	0.812	0.808	0.805	0.804	0.804	0.806	0.809	0.813	0.820	0.826	0.833
0.30000	(2.94)	0.828	0.815	0.807	0.802	0.799	0.797	0.797	0.797	0.801	0.804	0.808	0.816	0.823	0.830
0.40000	(3.92)	0.818	0.806	0.800	0.795	0.793	0.792	0.792	0.793	0.797	0.801	0.805	0.813	0.820	0.828
0.60000	(5.88)	0.806	0.796	0.791	0.788	0.786	0.786	0.787	0.788	0.793	0.797	0.802	0.810	0.818	0.826
0.80000	(7.85)	0.799	0.790	0.785	0.783	0.782	0.782	0.783	0.785	0.790	0.795	0.800	0.808	0.816	0.825
1.00000	(9.81)	0.793	0.786	0.782	0.779	0.779	0.779	0.781	0.783	0.788	0.794	0.798	0.807	0.815	0.824
1.50000	(14.7)	0.785	0.779	0.775	0.774	0.774	0.775	0.777	0.780	0.786	0.791	0.796	0.805	0.813	0.822
2.00000	(19.6)	0.780	0.774	0.772	0.771	0.771	0.772	0.775	0.778	0.784	0.789	0.795	0.804	0.812	0.821
3.00000	(29.4)	0.774	0.769	0.767	0.767	0.767	0.769	0.772	0.776	0.782	0.788	0.793	0.803	0.811	0.820
4.00000	(39.2)	0.770	0.766	0.764	0.764	0.765	0.767	0.771	0.774	0.781	0.787	0.792	0.802	0.810	0.819
6.00000	(58.8)	0.765	0.762	0.761	0.761	0.762	0.765	0.769	0.772	0.779	0.785	0.791	0.801	0.809	0.818
8.00000	(78.5)	0.762	0.760	0.759	0.759	0.761	0.764	0.768	0.771	0.778	0.784	0.790	0.800	0.809	0.818
10.0000	(98.1)	0.761	0.758	0.757	0.758	0.759	0.763	0.767	0.771	0.778	0.784	0.790	0.800	0.808	0.818
15.0000	(147)	0.757	0.755	0.755	0.756	0.758	0.762	0.765	0.769	0.777	0.783	0.789	0.799	0.808	0.817
20.0000	(196)	0.756	0.754	0.754	0.755	0.757	0.761	0.765	0.769	0.776	0.783	0.788	0.799	0.807	0.817
30.0000	(294)	0.753	0.752	0.752	0.754	0.755	0.760	0.764	0.768	0.775	0.782	0.788	0.798	0.807	0.816
S	$(\Delta p/\rho l)$	6	8	10	12.5	15	20	25	30	40	50	60	80	100	125

Grad'nt   k.p.g.   (Equivalent) Pipe diameters in mm

| Grad'nt | k.p.g. | (Equivalent) Pipe diameters in mm | | | | | | | | | | | | | |
S	$(\Delta p/\rho l)$	125	150	200	250	300	400	500	600	800	1000	1250	1500	2000	2500
0.00010	(0.00098)	1.112	1.101	1.089	1.082	1.078	1.075	1.074	1.075	1.079	1.083	1.089	1.094	1.105	1.115
0.00015	(0.00147)	1.081	1.073	1.063	1.058	1.055	1.053	1.054	1.056	1.060	1.066	1.072	1.079	1.090	1.101
0.00020	(0.00196)	1.062	1.054	1.046	1.042	1.040	1.039	1.040	1.043	1.048	1.054	1.062	1.068	1.081	1.092
0.00030	(0.00294)	1.035	1.029	1.023	1.020	1.019	1.020	1.023	1.026	1.033	1.039	1.047	1.055	1.069	1.080
0.00040	(0.00392)	1.018	1.013	1.008	1.006	1.006	1.008	1.011	1.015	1.022	1.030	1.038	1.046	1.061	1.073
0.00060	(0.00588)	0.995	0.991	0.988	0.987	0.988	0.991	0.995	1.000	1.009	1.017	1.026	1.035	1.050	1.063
0.00080	(0.00785)	0.979	0.977	0.975	0.975	0.976	0.981	0.985	0.990	1.000	1.009	1.019	1.028	1.044	1.057
0.00100	(0.00981)	0.968	0.966	0.965	0.966	0.968	0.973	0.978	0.984	0.994	1.003	1.013	1.023	1.039	1.053
0.00150	(0.0147)	0.949	0.948	0.949	0.951	0.953	0.960	0.966	0.972	0.983	0.993	1.004	1.014	1.031	1.045
0.00200	(0.0196)	0.937	0.937	0.938	0.941	0.944	0.951	0.958	0.965	0.976	0.987	0.998	1.009	1.026	1.041
0.00300	(0.0294)	0.921	0.922	0.924	0.928	0.932	0.940	0.948	0.955	0.968	0.979	0.991	1.002	1.020	1.035
0.00400	(0.0392)	0.911	0.912	0.916	0.920	0.925	0.933	0.942	0.949	0.963	0.974	0.987	0.998	1.016	1.032
0.00600	(0.0588)	0.897	0.899	0.904	0.910	0.915	0.925	0.934	0.942	0.956	0.968	0.981	0.992	1.012	1.028
0.00800	(0.0785)	0.889	0.891	0.897	0.903	0.909	0.919	0.929	0.937	0.952	0.964	0.977	0.989	1.009	1.025
0.01000	(0.0981)	0.883	0.886	0.892	0.898	0.904	0.915	0.925	0.934	0.949	0.961	0.975	0.987	1.007	1.023
0.01500	(0.147)	0.873	0.876	0.884	0.891	0.897	0.909	0.919	0.928	0.944	0.957	0.971	0.983	1.003	1.020
0.02000	(0.196)	0.866	0.870	0.878	0.886	0.893	0.905	0.916	0.925	0.941	0.954	0.969	0.981	1.001	1.018
0.03000	(0.294)	0.858	0.863	0.872	0.880	0.887	0.900	0.911	0.921	0.937	0.951	0.966	0.978	0.999	1.016
0.04000	(0.392)	0.853	0.858	0.868	0.876	0.884	0.897	0.908	0.918	0.935	0.949	0.964	0.976	0.998	1.015
0.06000	(0.588)	0.847	0.853	0.863	0.872	0.880	0.893	0.905	0.915	0.932	0.947	0.962	0.974	0.996	1.013
0.08000	(0.785)	0.843	0.849	0.860	0.869	0.877	0.891	0.903	0.913	0.931	0.945	0.960	0.973	0.995	1.012
0.10000	(0.981)	0.840	0.847	0.857	0.867	0.875	0.890	0.902	0.912	0.930	0.944	0.959	0.972	0.994	1.012
0.15000	(1.47)	0.836	0.843	0.854	0.864	0.872	0.887	0.899	0.910	0.928	0.943	0.958	0.971	0.993	1.011
0.20000	(1.96)	0.833	0.840	0.852	0.862	0.871	0.886	0.898	0.909	0.927	0.942	0.957	0.970	0.992	1.010
0.30000	(2.94)	0.830	0.837	0.849	0.860	0.869	0.884	0.896	0.907	0.926	0.940	0.956	0.969	0.991	1.009
0.40000	(3.92)	0.828	0.835	0.848	0.858	0.867	0.883	0.895	0.906	0.925	0.940	0.955	0.969	0.991	1.009
0.60000	(5.88)	0.826	0.833	0.846	0.857	0.866	0.881	0.894	0.905	0.924	0.939	0.955	0.968	0.990	1.008
0.80000	(7.85)	0.825	0.832	0.845	0.856	0.865	0.881	0.894	0.905	0.923	0.938	0.954	0.968	0.990	1.008
1.00000	(9.81)	0.824	0.831	0.844	0.855	0.864	0.880	0.893	0.904	0.923	0.938	0.954	0.967	0.990	1.008
1.50000	(14.7)	0.822	0.830	0.843	0.854	0.863	0.879	0.892	0.904	0.922	0.938	0.953	0.967	0.989	1.007
2.00000	(19.6)	0.821	0.829	0.842	0.853	0.863	0.879	0.892	0.903	0.922	0.937	0.953	0.967	0.989	1.007
3.00000	(29.4)	0.820	0.828	0.841	0.852	0.862	0.878	0.891	0.903	0.922	0.937	0.953	0.966	0.989	1.007
S	$(\Delta p/\rho l)$	125	150	200	250	300	400	500	600	800	1000	1250	1500	2000	2500

Grad'nt   k.p.g.   (Equivalent) Pipe diameters in mm

Kinematic viscosity, $\nu = 1.00\times10^{-6}$ m^{2}s^{-1} ;   **Roughness size, $k_s = 0.060$ mm**

E93

Kin. visc., $\nu = 1.00 \times 10^{-6}$ m^{2}s^{-1}; $S = 0.00010$ to 30.0000

i.e. kin. pr. grad., $\Delta p/\rho l =$ (0.00098) to (294) ms^{-2}

Roughness size, $k_s = 0.150$ mm

This table shows values of m, as follows

m_C for Colebrook-White solutions; or, where $\mathbf{R} \leq 2000$, m_P for laminar flow

Grad'nt	k.p.g.	(Equivalent) Pipe diameters in mm													
S	$(\Delta p/\rho l)$	6	8	10	12.5	15	20	25	30	40	50	60	80	100	125
0.00030	(0.00294)	6.857	4.673	3.470	2.577	2.021	1.377	1.023	*	1.143	1.120	1.105	1.085	1.074	1.066
0.00040	(0.00392)	5.939	4.047	3.005	2.232	1.750	1.193	0.886	*	1.119	1.098	1.085	1.068	1.058	1.051
0.00060	(0.00588)	4.849	3.304	2.454	1.822	1.429	0.974	*	1.117	1.088	1.070	1.059	1.044	1.037	1.031
0.00080	(0.00785)	4.199	2.861	2.125	1.578	1.238	0.843	*	1.094	1.068	1.052	1.042	1.030	1.023	1.019
0.00100	(0.00981)	3.756	2.559	1.901	1.412	1.107	*	1.096	1.077	1.053	1.039	1.030	1.019	1.013	1.010
0.00150	(0.0147)	3.067	2.090	1.552	1.153	0.904	*	1.065	1.049	1.029	1.017	1.009	1.001	0.997	0.995
0.00200	(0.0196)	2.656	1.810	1.344	0.998	*	1.067	1.046	1.031	1.013	1.003	0.996	0.989	0.987	0.986
0.00300	(0.0294)	2.169	1.478	1.097	0.815	*	1.039	1.021	1.009	0.993	0.985	0.980	0.975	0.973	0.974
0.00400	(0.0392)	1.878	1.280	0.950	*	*	1.022	1.005	0.994	0.981	0.974	0.969	0.966	0.965	0.966
0.00600	(0.0588)	1.533	1.045	0.776	*	1.023	0.999	0.985	0.976	0.965	0.959	0.956	0.954	0.955	0.957
0.00800	(0.0785)	1.328	0.905	*	*	1.006	0.985	0.972	0.964	0.955	0.950	0.948	0.947	0.948	0.951
0.01000	(0.0981)	1.188	0.809	*	1.010	0.994	0.975	0.963	0.956	0.948	0.944	0.942	0.942	0.944	0.947
0.01500	(0.147)	0.970	*	*	0.988	0.975	0.958	0.948	0.942	0.936	0.934	0.933	0.934	0.937	0.941
0.02000	(0.196)	0.840	*	0.993	0.975	0.962	0.948	0.939	0.934	0.929	0.927	0.927	0.929	0.932	0.937
0.03000	(0.294)	0.686	*	0.974	0.958	0.947	0.935	0.928	0.924	0.920	0.919	0.920	0.923	0.927	0.932
0.04000	(0.392)	*	0.980	0.962	0.947	0.938	0.927	0.921	0.917	0.915	0.914	0.915	0.919	0.923	0.929
0.06000	(0.588)	*	0.963	0.947	0.934	0.926	0.917	0.912	0.909	0.908	0.909	0.910	0.914	0.919	0.925
0.08000	(0.785)	*	0.952	0.937	0.926	0.919	0.910	0.906	0.905	0.904	0.905	0.907	0.912	0.917	0.923
0.10000	(0.981)	0.967	0.944	0.931	0.920	0.914	0.906	0.903	0.901	0.901	0.902	0.905	0.910	0.915	0.921
0.15000	(1.47)	0.953	0.932	0.921	0.911	0.906	0.899	0.897	0.896	0.896	0.898	0.901	0.907	0.912	0.919
0.20000	(1.96)	0.944	0.925	0.914	0.906	0.901	0.895	0.893	0.893	0.894	0.896	0.899	0.905	0.911	0.917
0.30000	(2.94)	0.933	0.916	0.906	0.899	0.895	0.890	0.889	0.889	0.890	0.893	0.896	0.903	0.909	0.916
0.40000	(3.92)	0.926	0.911	0.902	0.895	0.891	0.887	0.886	0.886	0.888	0.891	0.895	0.901	0.907	0.915
0.60000	(5.88)	0.918	0.904	0.896	0.890	0.887	0.884	0.883	0.883	0.886	0.889	0.893	0.900	0.906	0.913
0.80000	(7.85)	0.913	0.900	0.893	0.887	0.884	0.881	0.881	0.882	0.884	0.888	0.892	0.899	0.905	0.913
1.00000	(9.81)	0.910	0.897	0.890	0.885	0.882	0.880	0.880	0.880	0.883	0.887	0.891	0.898	0.905	0.912
1.50000	(14.7)	0.904	0.893	0.886	0.882	0.879	0.877	0.878	0.879	0.882	0.886	0.890	0.897	0.904	0.911
2.00000	(19.6)	0.901	0.890	0.884	0.880	0.878	0.876	0.876	0.877	0.881	0.885	0.889	0.896	0.903	0.911
3.00000	(29.4)	0.897	0.887	0.881	0.877	0.875	0.874	0.875	0.876	0.880	0.884	0.888	0.896	0.902	0.910
4.00000	(39.2)	0.895	0.885	0.880	0.876	0.874	0.873	0.874	0.875	0.879	0.883	0.887	0.895	0.902	0.910
6.00000	(58.8)	0.892	0.883	0.878	0.874	0.873	0.872	0.873	0.874	0.878	0.883	0.887	0.895	0.902	0.910
8.00000	(78.5)	0.891	0.882	0.877	0.873	0.872	0.871	0.872	0.874	0.878	0.882	0.886	0.894	0.901	0.910
10.0000	(98.1)	0.889	0.881	0.876	0.873	0.871	0.871	0.872	0.873	0.878	0.882	0.886	0.894	0.901	0.909
15.0000	(147)	0.888	0.879	0.875	0.872	0.870	0.870	0.871	0.873	0.877	0.881	0.886	0.894	0.901	0.909
20.0000	(196)	0.887	0.878	0.874	0.871	0.870	0.869	0.871	0.872	0.877	0.881	0.885	0.894	0.901	0.909
30.0000	(294)	0.885	0.877	0.873	0.870	0.869	0.869	0.870	0.872	0.876	0.881	0.885	0.893	0.900	0.909
S	$(\Delta p/\rho l)$	6	8	10	12.5	15	20	25	30	40	50	60	80	100	125

Grad'nt	k.p.g.	(Equivalent) Pipe diameters in mm													
S	$(\Delta p/\rho l)$	125	150	200	250	300	400	500	600	800	1000	1250	1500	2000	2500
0.00010	(0.00098)	1.134	1.124	1.113	1.107	1.104	1.103	1.103	1.105	1.111	1.117	1.124	1.131	1.144	1.156
0.00015	(0.00147)	1.107	1.099	1.090	1.086	1.084	1.084	1.086	1.089	1.096	1.103	1.111	1.119	1.134	1.146
0.00020	(0.00196)	1.089	1.082	1.075	1.072	1.071	1.073	1.075	1.079	1.087	1.094	1.103	1.112	1.127	1.140
0.00030	(0.00294)	1.066	1.061	1.056	1.054	1.055	1.057	1.061	1.066	1.075	1.083	1.093	1.102	1.118	1.132
0.00040	(0.00392)	1.051	1.047	1.043	1.043	1.044	1.048	1.052	1.057	1.067	1.076	1.087	1.096	1.113	1.127
0.00060	(0.00588)	1.031	1.029	1.027	1.028	1.030	1.035	1.041	1.047	1.058	1.068	1.079	1.089	1.106	1.121
0.00080	(0.00785)	1.019	1.017	1.017	1.018	1.021	1.027	1.034	1.040	1.052	1.062	1.074	1.084	1.102	1.117
0.00100	(0.00981)	1.010	1.009	1.009	1.012	1.015	1.022	1.029	1.035	1.047	1.058	1.070	1.081	1.099	1.114
0.00150	(0.0147)	0.995	0.995	0.997	1.000	1.004	1.012	1.020	1.027	1.041	1.052	1.065	1.076	1.094	1.110
0.00200	(0.0196)	0.986	0.986	0.989	0.993	0.998	1.007	1.015	1.023	1.036	1.048	1.061	1.072	1.092	1.108
0.00300	(0.0294)	0.974	0.975	0.980	0.985	0.990	1.000	1.008	1.017	1.031	1.043	1.057	1.068	1.088	1.104
0.00400	(0.0392)	0.966	0.968	0.974	0.979	0.985	0.995	1.004	1.013	1.028	1.040	1.054	1.066	1.086	1.103
0.00600	(0.0588)	0.957	0.960	0.966	0.972	0.978	0.990	1.000	1.008	1.024	1.037	1.051	1.063	1.083	1.100
0.00800	(0.0785)	0.951	0.954	0.961	0.968	0.975	0.986	0.997	1.006	1.021	1.035	1.049	1.061	1.082	1.099
0.01000	(0.0981)	0.947	0.951	0.958	0.965	0.972	0.984	0.994	1.004	1.020	1.033	1.048	1.060	1.081	1.098
0.01500	(0.147)	0.941	0.945	0.953	0.961	0.968	0.980	0.991	1.001	1.017	1.031	1.045	1.058	1.079	1.096
0.02000	(0.196)	0.937	0.941	0.950	0.958	0.965	0.978	0.989	0.999	1.016	1.029	1.044	1.057	1.078	1.095
0.03000	(0.294)	0.932	0.937	0.946	0.954	0.962	0.975	0.987	0.997	1.014	1.028	1.043	1.056	1.077	1.094
0.04000	(0.392)	0.929	0.934	0.944	0.952	0.960	0.974	0.985	0.995	1.012	1.027	1.042	1.055	1.076	1.094
0.06000	(0.588)	0.925	0.931	0.941	0.950	0.958	0.972	0.984	0.994	1.011	1.025	1.041	1.054	1.075	1.093
0.08000	(0.785)	0.923	0.929	0.939	0.948	0.956	0.971	0.982	0.993	1.010	1.025	1.040	1.053	1.075	1.092
0.10000	(0.981)	0.921	0.927	0.938	0.947	0.955	0.970	0.982	0.992	1.010	1.024	1.040	1.053	1.074	1.092
0.15000	(1.47)	0.919	0.925	0.936	0.946	0.954	0.968	0.981	0.991	1.009	1.023	1.039	1.052	1.074	1.092
0.20000	(1.96)	0.917	0.924	0.935	0.945	0.953	0.968	0.980	0.991	1.008	1.023	1.038	1.052	1.074	1.091
0.30000	(2.94)	0.916	0.922	0.934	0.943	0.952	0.967	0.979	0.990	1.008	1.022	1.038	1.051	1.073	1.091
0.40000	(3.92)	0.915	0.921	0.933	0.943	0.951	0.966	0.979	0.989	1.007	1.022	1.038	1.051	1.073	1.091
0.60000	(5.88)	0.913	0.920	0.932	0.942	0.951	0.966	0.978	0.989	1.007	1.022	1.037	1.051	1.073	1.091
0.80000	(7.85)	0.913	0.919	0.931	0.941	0.950	0.965	0.978	0.988	1.007	1.021	1.037	1.050	1.072	1.090
1.00000	(9.81)	0.912	0.919	0.931	0.941	0.950	0.965	0.977	0.988	1.006	1.021	1.037	1.050	1.072	1.090
1.50000	(14.7)	0.911	0.918	0.930	0.940	0.949	0.964	0.977	0.988	1.006	1.021	1.037	1.050	1.072	1.090
2.00000	(19.6)	0.911	0.918	0.930	0.940	0.949	0.964	0.977	0.988	1.006	1.021	1.037	1.050	1.072	1.090
3.00000	(29.4)	0.910	0.917	0.929	0.940	0.949	0.964	0.977	0.987	1.006	1.021	1.036	1.050	1.072	1.090
S	$(\Delta p/\rho l)$	125	150	200	250	300	400	500	600	800	1000	1250	1500	2000	2500

Grad'nt   **k.p.g.**   (Equivalent) Pipe diameters in mm

Kinematic viscosity, $\nu = 1.00 \times 10^{-6}$ m^{2}s^{-1} ;   **Roughness size, $k_s = 0.150$ mm**

Kin. visc., $\nu = 1.00 \times 10^{-6}$ m^{2}s^{-1};
$S = 0.00010$ to 30.0000

i.e. kin. pr. grad., $\Delta p/\rho l =$
(0.00098) to (294) ms^{-2}

Roughness size, $k_s = 0.30$ mm
This table shows values of m, as follows

m_C for Colebrook-White solutions; or,
where $R \leq 2000$, m_P for laminar flow

Grad'nt S	k.p.g. $(\Delta p/\rho l)$	(Equivalent) Pipe diameters in mm													
		6	8	10	12·5	15	20	25	30	40	50	60	80	100	125
0.00030	(0.00294)	6.857	4.673	3.470	2.577	2.021	1.377	1.023	*	1.184	1.161	1.146	1.127	1.116	1.109
0.00040	(0.00392)	5.939	4.047	3.005	2.232	1.750	1.193	0.886	*	1.162	1.142	1.128	1.112	1.103	1.096
0.00060	(0.00588)	4.849	3.304	2.454	1.822	1.429	0.974	*	1.164	1.135	1.118	1.106	1.093	1.085	1.081
0.00080	(0.00785)	4.199	2.861	2.125	1.578	1.238	0.843	*	1.144	1.118	1.102	1.092	1.080	1.074	1.071
0.00100	(0.00981)	3.756	2.559	1.901	1.412	1.107	*	*	1.129	1.105	1.091	1.082	1.072	1.067	1.064
0.00150	(0.0147)	3.067	2.090	1.552	1.153	0.904	*	1.122	1.106	1.085	1.073	1.066	1.058	1.054	1.053
0.00200	(0.0196)	2.656	1.810	1.344	0.998	*	*	1.106	1.091	1.073	1.062	1.056	1.049	1.046	1.046
0.00300	(0.0294)	2.169	1.478	1.097	0.815	*	1.105	1.085	1.072	1.057	1.048	1.043	1.038	1.037	1.037
0.00400	(0.0392)	1.878	1.280	0.950	*	*	1.090	1.072	1.061	1.047	1.039	1.035	1.031	1.031	1.032
0.00600	(0.0588)	1.533	1.045	0.776	*	1.097	1.072	1.056	1.046	1.035	1.029	1.025	1.023	1.023	1.025
0.00800	(0.0785)	1.328	0.905	*	*	1.084	1.060	1.046	1.037	1.027	1.022	1.019	1.018	1.019	1.021
0.01000	(0.0981)	1.188	0.809	*	1.091	1.074	1.052	1.039	1.031	1.022	1.017	1.015	1.014	1.016	1.018
0.01500	(0.147)	0.970	*	*	1.074	1.059	1.039	1.028	1.021	1.014	1.010	1.009	1.009	1.011	1.014
0.02000	(0.196)	0.840	*	1.085	1.063	1.049	1.032	1.021	1.015	1.008	1.006	1.005	1.005	1.008	1.011
0.03000	(0.294)	0.686	*	1.069	1.050	1.038	1.022	1.013	1.008	1.002	1.000	1.000	1.001	1.004	1.008
0.04000	(0.392)	*	*	1.060	1.042	1.031	1.016	1.008	1.003	0.998	0.997	0.997	0.999	1.002	1.006
0.06000	(0.588)	*	1.069	1.049	1.033	1.022	1.009	1.002	0.998	0.994	0.993	0.993	0.996	0.999	1.004
0.08000	(0.785)	*	1.061	1.042	1.027	1.017	1.005	0.998	0.995	0.991	0.990	0.991	0.994	0.998	1.003
0.10000	(0.981)	*	1.055	1.037	1.023	1.013	1.002	0.996	0.992	0.989	0.989	0.990	0.993	0.997	1.002
0.15000	(1.47)	1.073	1.046	1.030	1.016	1.007	0.997	0.992	0.989	0.986	0.986	0.987	0.991	0.995	1.000
0.20000	(1.96)	1.067	1.041	1.025	1.012	1.004	0.994	0.989	0.986	0.984	0.985	0.986	0.990	0.994	0.999
0.30000	(2.94)	1.059	1.034	1.019	1.008	1.000	0.991	0.986	0.984	0.982	0.983	0.984	0.988	0.993	0.998
0.40000	(3.92)	1.054	1.030	1.016	1.005	0.997	0.989	0.985	0.982	0.981	0.982	0.983	0.987	0.992	0.998
0.60000	(5.88)	1.048	1.026	1.012	1.001	0.994	0.986	0.982	0.980	0.979	0.980	0.982	0.986	0.991	0.997
0.80000	(7.85)	1.044	1.023	1.010	0.999	0.993	0.985	0.981	0.979	0.979	0.980	0.981	0.986	0.991	0.996
1.00000	(9.81)	1.042	1.021	1.008	0.998	0.991	0.984	0.980	0.979	0.978	0.979	0.981	0.985	0.990	0.996
1.50000	(14.7)	1.038	1.018	1.006	0.996	0.990	0.982	0.979	0.977	0.977	0.978	0.980	0.985	0.990	0.996
2.00000	(19.6)	1.036	1.016	1.004	0.995	0.988	0.981	0.978	0.977	0.976	0.978	0.980	0.984	0.989	0.995
3.00000	(29.4)	1.033	1.014	1.002	0.993	0.987	0.980	0.977	0.976	0.976	0.977	0.979	0.984	0.989	0.995
4.00000	(39.2)	1.032	1.013	1.001	0.992	0.986	0.980	0.977	0.975	0.975	0.977	0.979	0.984	0.989	0.995
6.00000	(58.8)	1.030	1.011	1.000	0.991	0.985	0.979	0.976	0.975	0.975	0.976	0.978	0.983	0.988	0.995
8.00000	(78.5)	1.029	1.010	0.999	0.990	0.985	0.978	0.976	0.974	0.974	0.976	0.978	0.983	0.988	0.994
10.0000	(98.1)	1.028	1.010	0.999	0.990	0.984	0.978	0.975	0.974	0.974	0.976	0.978	0.983	0.988	0.994
15.0000	(147)	1.027	1.009	0.998	0.989	0.984	0.978	0.975	0.974	0.974	0.976	0.978	0.983	0.988	0.994
20.0000	(196)	1.026	1.008	0.997	0.989	0.983	0.977	0.975	0.974	0.974	0.975	0.978	0.983	0.988	0.994
30.0000	(294)	1.025	1.007	0.997	0.988	0.983	0.977	0.974	0.973	0.973	0.975	0.977	0.983	0.988	0.994
S	$(\Delta p/\rho l)$	6	8	10	12·5	15	20	25	30	40	50	60	80	100	125

Grad'nt k.p.g. (Equivalent) Pipe diameters in mm

S	$(\Delta p/\rho l)$	125	150	200	250	300	400	500	600	800	1000	1250	1500	2000	2500
0.00010	(0.00098)	1.167	1.158	1.148	1.144	1.141	1.141	1.143	1.146	1.153	1.160	1.169	1.177	1.192	1.205
0.00015	(0.00147)	1.143	1.136	1.129	1.126	1.125	1.126	1.130	1.134	1.142	1.150	1.160	1.169	1.185	1.198
0.00020	(0.00196)	1.128	1.122	1.116	1.114	1.114	1.117	1.121	1.126	1.135	1.144	1.154	1.163	1.180	1.194
0.00030	(0.00294)	1.109	1.104	1.100	1.100	1.101	1.105	1.110	1.116	1.126	1.136	1.147	1.157	1.174	1.189
0.00040	(0.00392)	1.096	1.093	1.090	1.091	1.093	1.098	1.104	1.110	1.121	1.131	1.142	1.153	1.170	1.185
0.00060	(0.00588)	1.081	1.079	1.078	1.080	1.082	1.089	1.096	1.102	1.114	1.125	1.137	1.148	1.166	1.181
0.00080	(0.00785)	1.071	1.070	1.070	1.073	1.076	1.083	1.090	1.097	1.110	1.121	1.134	1.145	1.163	1.179
0.00100	(0.00981)	1.064	1.063	1.065	1.068	1.071	1.079	1.087	1.094	1.107	1.119	1.131	1.142	1.161	1.177
0.00150	(0.0147)	1.053	1.053	1.056	1.060	1.064	1.073	1.081	1.089	1.103	1.115	1.128	1.139	1.159	1.175
0.00200	(0.0196)	1.046	1.047	1.050	1.055	1.059	1.069	1.078	1.086	1.100	1.112	1.125	1.137	1.157	1.173
0.00300	(0.0294)	1.037	1.039	1.043	1.049	1.054	1.064	1.073	1.082	1.096	1.109	1.123	1.135	1.155	1.171
0.00400	(0.0392)	1.032	1.034	1.039	1.045	1.051	1.061	1.071	1.079	1.094	1.107	1.121	1.133	1.153	1.170
0.00600	(0.0588)	1.025	1.028	1.034	1.040	1.046	1.058	1.068	1.076	1.092	1.105	1.119	1.131	1.152	1.169
0.00800	(0.0785)	1.021	1.024	1.031	1.038	1.044	1.056	1.066	1.075	1.090	1.104	1.118	1.130	1.151	1.168
0.01000	(0.0981)	1.018	1.022	1.029	1.036	1.042	1.054	1.064	1.074	1.089	1.103	1.117	1.129	1.150	1.167
0.01500	(0.147)	1.014	1.018	1.026	1.033	1.040	1.052	1.062	1.072	1.088	1.101	1.116	1.128	1.149	1.166
0.02000	(0.196)	1.011	1.016	1.023	1.031	1.038	1.050	1.061	1.071	1.087	1.101	1.115	1.128	1.149	1.166
0.03000	(0.294)	1.008	1.013	1.021	1.029	1.036	1.049	1.060	1.069	1.086	1.099	1.114	1.127	1.148	1.165
0.04000	(0.392)	1.006	1.011	1.020	1.028	1.035	1.048	1.059	1.068	1.085	1.099	1.114	1.126	1.148	1.165
0.06000	(0.588)	1.004	1.009	1.018	1.026	1.033	1.046	1.058	1.067	1.084	1.098	1.113	1.126	1.147	1.164
0.08000	(0.785)	1.003	1.008	1.017	1.025	1.033	1.046	1.057	1.067	1.084	1.098	1.113	1.125	1.147	1.164
0.10000	(0.981)	1.002	1.007	1.016	1.024	1.032	1.045	1.057	1.067	1.083	1.097	1.112	1.125	1.146	1.164
0.15000	(1.47)	1.000	1.005	1.015	1.023	1.031	1.044	1.056	1.066	1.083	1.097	1.112	1.125	1.146	1.164
0.20000	(1.96)	0.999	1.005	1.014	1.023	1.031	1.044	1.056	1.066	1.083	1.097	1.112	1.125	1.146	1.164
0.30000	(2.94)	0.998	1.004	1.013	1.022	1.030	1.043	1.055	1.065	1.082	1.096	1.111	1.124	1.146	1.163
0.40000	(3.92)	0.998	1.003	1.013	1.022	1.030	1.043	1.055	1.065	1.082	1.096	1.111	1.124	1.146	1.163
0.60000	(5.88)	0.997	1.002	1.012	1.021	1.029	1.043	1.054	1.065	1.082	1.096	1.111	1.124	1.145	1.163
0.80000	(7.85)	0.996	1.002	1.012	1.021	1.029	1.043	1.054	1.064	1.082	1.096	1.111	1.124	1.145	1.163
1.00000	(9.81)	0.996	1.002	1.012	1.021	1.029	1.042	1.054	1.064	1.081	1.096	1.111	1.124	1.145	1.163
1.50000	(14.7)	0.996	1.001	1.011	1.020	1.028	1.042	1.054	1.064	1.081	1.096	1.111	1.124	1.145	1.163
2.00000	(19.6)	0.995	1.001	1.011	1.020	1.028	1.042	1.054	1.064	1.081	1.095	1.111	1.124	1.145	1.163
3.00000	(29.4)	0.995	1.001	1.011	1.020	1.028	1.042	1.054	1.064	1.081	1.095	1.110	1.123	1.145	1.163
S	$(\Delta p/\rho l)$	125	150	200	250	300	400	500	600	800	1000	1250	1500	2000	2500

Grad'nt k.p.g. (Equivalent) Pipe diameters in mm

Kinematic viscosity, $\nu = 1.00 \times 10^{-6}$ m^{2}s^{-1} ; **Roughness size, $k_s = 0.30$ mm**

E95

$\text{Kin. visc., } \nu = 1.00 \times 10^{-6} \text{ m}^2\text{s}^{-1};$
$S = 0.00010 \text{ to } 30.0000$

i.e. kin. pr. grad., $\Delta p/\rho l = $ (0.00098) to (294) ms^{-2}

Roughness size, $k_s = 0.60$ mm
This table shows values of m, as follows

m_C for Colebrook-White solutions; or,
where $\mathbf{R} \leq 2000$, m_P for laminar flow

Grad'nt S	k.p.g. $(\Delta p/\rho l)$	6	8	10	12·5	15	20	25	30	40	50	60	80	100	125
															(Equivalent) Pipe diameters in mm
0·00030	(0·00294)	6·857	4·673	3·470	2·577	2·021	1·377	1·023	*	1·254	1·230	1·215	1·196	1·185	1·177
0·00040	(0·00392)	5·939	4·047	3·005	2·232	1·750	1·193	0·886	*	1·236	1·215	1·201	1·184	1·174	1·168
0·00060	(0·00588)	4·849	3·304	2·454	1·822	1·429	0·974	*	*	1·214	1·195	1·183	1·169	1·161	1·156
0·00080	(0·00785)	4·199	2·861	2·125	1·578	1·238	0·843	*	1·228	1·200	1·183	1·172	1·159	1·153	1·149
0·00100	(0·00981)	3·756	2·559	1·901	1·412	1·107	*	*	1·216	1·190	1·175	1·165	1·153	1·147	1·144
0·00150	(0·0147)	3·067	2·090	1·552	1·153	0·904	*	1·216	1·197	1·174	1·161	1·152	1·143	1·138	1·136
0·00200	(0·0196)	2·656	1·810	1·344	0·998	*	*	1·203	1·186	1·165	1·152	1·145	1·136	1·133	1·131
0·00300	(0·0294)	2·169	1·478	1·097	0·815	*	1·209	1·187	1·172	1·153	1·142	1·136	1·129	1·126	1·125
0·00400	(0·0392)	1·878	1·280	0·950	*	*	1·198	1·177	1·163	1·146	1·136	1·130	1·124	1·122	1·122
0·00600	(0·0588)	1·533	1·045	0·776	*	*	1·184	1·165	1·152	1·137	1·128	1·123	1·118	1·117	1·117
0·00800	(0·0785)	1·328	0·905	*	*	1·205	1·175	1·158	1·146	1·132	1·124	1·119	1·115	1·114	1·115
0·01000	(0·0981)	1·188	0·809	*	*	1·198	1·169	1·152	1·141	1·128	1·121	1·116	1·113	1·112	1·113
0·01500	(0·147)	0·970	*	*	1·206	1·186	1·160	1·144	1·134	1·122	1·115	1·112	1·109	1·109	1·110
0·02000	(0·196)	0·840	*	*	1·198	1·179	1·154	1·140	1·130	1·119	1·112	1·109	1·107	1·107	1·108
0·03000	(0·294)	0·686	*	1·215	1·189	1·171	1·148	1·134	1·125	1·114	1·109	1·106	1·104	1·104	1·106
0·04000	(0·392)	*	*	1·208	1·183	1·165	1·143	1·130	1·122	1·112	1·107	1·104	1·102	1·103	1·105
0·06000	(0·588)	*	1·229	1·200	1·176	1·159	1·138	1·126	1·118	1·109	1·104	1·102	1·100	1·101	1·104
0·08000	(0·785)	*	1·223	1·195	1·171	1·156	1·135	1·124	1·116	1·107	1·102	1·100	1·099	1·100	1·103
0·10000	(0·981)	*	1·219	1·191	1·169	1·153	1·133	1·122	1·114	1·106	1·101	1·099	1·098	1·099	1·102
0·15000	(1·47)	1·255	1·213	1·186	1·164	1·149	1·130	1·119	1·112	1·104	1·100	1·098	1·097	1·098	1·101
0·20000	(1·96)	1·250	1·209	1·183	1·161	1·147	1·128	1·117	1·110	1·102	1·099	1·097	1·096	1·098	1·101
0·30000	(2·94)	1·244	1·204	1·179	1·158	1·144	1·126	1·115	1·109	1·101	1·097	1·096	1·096	1·097	1·100
0·40000	(3·92)	1·241	1·201	1·176	1·156	1·142	1·125	1·114	1·108	1·100	1·097	1·095	1·095	1·097	1·100
0·60000	(5·88)	1·236	1·198	1·174	1·154	1·140	1·123	1·113	1·106	1·099	1·096	1·094	1·094	1·096	1·099
0·80000	(7·85)	1·234	1·196	1·172	1·152	1·139	1·122	1·112	1·106	1·099	1·095	1·094	1·094	1·096	1·099
1·00000	(9·81)	1·232	1·194	1·171	1·151	1·138	1·121	1·112	1·105	1·098	1·095	1·094	1·094	1·095	1·099
1·50000	(14·7)	1·229	1·192	1·169	1·150	1·137	1·120	1·111	1·104	1·098	1·094	1·093	1·093	1·095	1·098
2·00000	(19·6)	1·228	1·191	1·168	1·149	1·136	1·120	1·110	1·104	1·097	1·094	1·093	1·093	1·095	1·098
3·00000	(29·4)	1·226	1·189	1·167	1·148	1·135	1·119	1·110	1·103	1·097	1·094	1·093	1·093	1·095	1·098
4·00000	(39·2)	1·225	1·189	1·166	1·147	1·135	1·119	1·109	1·103	1·097	1·094	1·092	1·093	1·095	1·098
6·00000	(58·8)	1·223	1·187	1·165	1·147	1·134	1·118	1·109	1·103	1·096	1·093	1·092	1·092	1·094	1·098
8·00000	(78·5)	1·222	1·187	1·164	1·146	1·134	1·118	1·108	1·103	1·096	1·093	1·092	1·092	1·094	1·098
10·0000	(98·1)	1·222	1·186	1·164	1·146	1·133	1·118	1·108	1·102	1·096	1·093	1·092	1·092	1·094	1·097
15·0000	(147)	1·221	1·186	1·164	1·145	1·133	1·117	1·108	1·102	1·096	1·093	1·092	1·092	1·094	1·097
20·0000	(196)	1·220	1·185	1·163	1·145	1·133	1·117	1·108	1·102	1·096	1·093	1·092	1·092	1·094	1·097
30·0000	(294)	1·220	1·185	1·163	1·145	1·132	1·117	1·108	1·102	1·095	1·093	1·092	1·092	1·094	1·097
S	$(\Delta p/\rho l)$	6	8	10	12·5	15	20	25	30	40	50	60	80	100	125

Grad'nt S	k.p.g. $(\Delta p/\rho l)$	125	150	200	250	300	400	500	600	800	1000	1250	1500	2000	2500
															(Equivalent) Pipe diameters in mm
0·00010	(0·00098)	1·224	1·215	1·206	1·202	1·201	1·201	1·204	1·208	1·216	1·224	1·234	1·243	1·259	1·272
0·00015	(0·00147)	1·204	1·198	1·191	1·188	1·188	1·190	1·194	1·199	1·208	1·217	1·227	1·237	1·253	1·268
0·00020	(0·00196)	1·192	1·187	1·181	1·180	1·180	1·183	1·188	1·193	1·203	1·213	1·223	1·233	1·250	1·265
0·00030	(0·00294)	1·177	1·173	1·169	1·169	1·171	1·175	1·181	1·186	1·197	1·207	1·219	1·229	1·247	1·262
0·00040	(0·00392)	1·168	1·164	1·162	1·163	1·165	1·170	1·176	1·182	1·194	1·204	1·216	1·226	1·244	1·260
0·00060	(0·00588)	1·156	1·154	1·153	1·155	1·157	1·164	1·171	1·177	1·189	1·200	1·212	1·223	1·242	1·257
0·00080	(0·00785)	1·149	1·147	1·147	1·150	1·153	1·160	1·167	1·174	1·187	1·198	1·210	1·221	1·240	1·256
0·00100	(0·00981)	1·144	1·143	1·144	1·146	1·150	1·157	1·165	1·172	1·185	1·196	1·209	1·220	1·239	1·255
0·00150	(0·0147)	1·136	1·136	1·137	1·141	1·145	1·153	1·161	1·169	1·182	1·194	1·206	1·218	1·237	1·253
0·00200	(0·0196)	1·131	1·131	1·134	1·138	1·142	1·151	1·159	1·166	1·180	1·192	1·205	1·217	1·236	1·252
0·00300	(0·0294)	1·125	1·126	1·129	1·134	1·138	1·148	1·156	1·164	1·178	1·190	1·203	1·215	1·235	1·251
0·00400	(0·0392)	1·122	1·123	1·127	1·131	1·136	1·146	1·155	1·163	1·177	1·189	1·202	1·214	1·234	1·250
0·00600	(0·0588)	1·117	1·119	1·123	1·128	1·134	1·143	1·153	1·161	1·175	1·188	1·201	1·213	1·233	1·250
0·00800	(0·0785)	1·115	1·116	1·121	1·127	1·132	1·142	1·151	1·160	1·174	1·187	1·201	1·212	1·233	1·249
0·01000	(0·0981)	1·113	1·115	1·120	1·125	1·131	1·141	1·151	1·159	1·174	1·186	1·200	1·212	1·232	1·249
0·01500	(0·147)	1·110	1·112	1·118	1·124	1·129	1·140	1·149	1·158	1·173	1·186	1·199	1·211	1·232	1·248
0·02000	(0·196)	1·108	1·111	1·116	1·122	1·128	1·139	1·149	1·157	1·172	1·185	1·199	1·211	1·231	1·248
0·03000	(0·294)	1·106	1·109	1·115	1·121	1·127	1·138	1·148	1·156	1·172	1·185	1·198	1·210	1·231	1·248
0·04000	(0·392)	1·105	1·108	1·114	1·120	1·126	1·137	1·147	1·156	1·171	1·184	1·198	1·210	1·231	1·247
0·06000	(0·588)	1·104	1·107	1·113	1·119	1·125	1·137	1·147	1·155	1·171	1·184	1·198	1·210	1·230	1·247
0·08000	(0·785)	1·103	1·106	1·112	1·119	1·125	1·136	1·146	1·155	1·170	1·183	1·197	1·210	1·230	1·247
0·10000	(0·981)	1·102	1·105	1·112	1·118	1·125	1·136	1·146	1·155	1·170	1·183	1·197	1·209	1·230	1·247
0·15000	(1·47)	1·101	1·104	1·111	1·118	1·124	1·135	1·145	1·154	1·170	1·183	1·197	1·209	1·230	1·247
0·20000	(1·96)	1·101	1·104	1·111	1·117	1·124	1·135	1·145	1·154	1·170	1·183	1·197	1·209	1·230	1·247
0·30000	(2·94)	1·100	1·103	1·110	1·117	1·123	1·135	1·145	1·154	1·169	1·183	1·197	1·209	1·230	1·247
0·40000	(3·92)	1·100	1·103	1·110	1·117	1·123	1·135	1·145	1·154	1·169	1·183	1·197	1·209	1·229	1·246
0·60000	(5·88)	1·099	1·102	1·110	1·116	1·123	1·134	1·145	1·154	1·169	1·182	1·197	1·209	1·229	1·246
0·80000	(7·85)	1·099	1·102	1·109	1·116	1·123	1·134	1·144	1·154	1·169	1·182	1·196	1·209	1·229	1·246
1·00000	(9·81)	1·099	1·102	1·109	1·116	1·123	1·134	1·144	1·153	1·169	1·182	1·196	1·209	1·229	1·246
1·50000	(14·7)	1·098	1·102	1·109	1·116	1·122	1·134	1·144	1·153	1·169	1·182	1·196	1·209	1·229	1·246
2·00000	(19·6)	1·098	1·102	1·109	1·116	1·122	1·134	1·144	1·153	1·169	1·182	1·196	1·209	1·229	1·246
3·00000	(29·4)	1·098	1·101	1·109	1·116	1·122	1·134	1·144	1·153	1·169	1·182	1·196	1·209	1·229	1·246
S	$(\Delta p/\rho l)$	125	150	200	250	300	400	500	600	800	1000	1250	1500	2000	2500

Grad'nt k.p.g. (Equivalent) Pipe diameters in mm

Kinematic viscosity, $\nu = 1.00 \times 10^{-6}$ m^{2}s^{-1} ; **Roughness size, $k_s = 0.60$ mm**

Kin. visc., $\nu = 1{\cdot}00 \times 10^{-6}\ \mathrm{m^2\,s^{-1}}$;
$S = 0{\cdot}00010$ to $30{\cdot}0000$

i.e. kin. pr. grad., $\Delta p/\rho l =$
$(0{\cdot}00098)$ to $(294)\ \mathrm{ms^{-2}}$

Roughness size, $k_s = 1{\cdot}50$ mm
This table shows values of m, as follows

m_C for Colebrook-White solutions; or,
where $\mathbf{R} \le 2000$, m_P for laminar flow

Grad'nt S	k.p.g. $(\Delta p/\rho l)$	(Equivalent) Pipe diameters in mm													
		6	8	10	12·5	15	20	25	30	40	50	60	80	100	125
0·00030	(0·00294)		4·673	3·470	2·577	2·021	1·377	1·023	*	*	1·389	1·369	1·345	1·330	1·320
0·00040	(0·00392)		4·047	3·005	2·232	1·750	1·193	0·886	*	1·405	1·378	1·359	1·337	1·324	1·314
0·00060	(0·00588)		3·304	2·454	1·822	1·429	0·974	*	*	1·389	1·364	1·348	1·327	1·315	1·306
0·00080	(0·00785)		2·861	2·125	1·578	1·238	0·843	*	*	1·380	1·356	1·340	1·321	1·310	1·302
0·00100	(0·00981)		2·559	1·901	1·412	1·107	*	*	1·410	1·373	1·350	1·335	1·317	1·307	1·299
0·00150	(0·0147)		2·090	1·552	1·153	0·904	*	*	1·397	1·362	1·342	1·328	1·311	1·301	1·294
0·00200	(0·0196)		1·810	1·344	0·998	*	*	1·415	1·389	1·356	1·336	1·323	1·307	1·298	1·291
0·00300	(0·0294)		1·478	1·097	0·815	*	*	1·404	1·379	1·348	1·330	1·317	1·302	1·294	1·288
0·00400	(0·0392)		1·280	0·950	*	*	1·431	1·397	1·374	1·344	1·326	1·314	1·299	1·291	1·286
0·00600	(0·0588)		1·045	0·776	*	*	1·422	1·389	1·367	1·338	1·321	1·310	1·296	1·288	1·283
0·00800	(0·0785)		0·905	*	*	*	1·416	1·384	1·363	1·335	1·318	1·307	1·294	1·287	1·282
0·01000	(0·0981)		0·809	*	*	1·462	1·412	1·381	1·360	1·333	1·316	1·306	1·293	1·286	1·281
0·01500	(0·147)		*	*	*	1·454	1·406	1·376	1·355	1·329	1·313	1·303	1·290	1·284	1·279
0·02000	(0·196)		*	*	1·485	1·449	1·402	1·373	1·353	1·327	1·312	1·301	1·289	1·283	1·278
0·03000	(0·294)		*	*	1·479	1·443	1·398	1·369	1·350	1·325	1·309	1·299	1·288	1·281	1·277
0·04000	(0·392)		*	1·525	1·475	1·440	1·395	1·367	1·348	1·323	1·308	1·298	1·287	1·281	1·277
0·06000	(0·588)		*	1·519	1·470	1·436	1·392	1·364	1·345	1·321	1·307	1·297	1·286	1·280	1·276
0·08000	(0·785)		*	1·515	1·467	1·434	1·390	1·363	1·344	1·320	1·306	1·296	1·285	1·279	1·275
0·10000	(0·981)		1·571	1·513	1·465	1·432	1·389	1·362	1·343	1·319	1·305	1·296	1·285	1·279	1·275
0·15000	(1·47)		1·566	1·509	1·462	1·429	1·387	1·360	1·342	1·318	1·304	1·295	1·284	1·278	1·275
0·20000	(1·96)		1·564	1·507	1·460	1·428	1·386	1·359	1·341	1·318	1·304	1·294	1·283	1·278	1·274
0·30000	(2·94)		1·560	1·504	1·458	1·426	1·384	1·358	1·340	1·317	1·303	1·294	1·283	1·277	1·274
0·40000	(3·92)		1·558	1·503	1·457	1·425	1·383	1·357	1·339	1·316	1·302	1·293	1·283	1·277	1·274
0·60000	(5·88)		1·556	1·501	1·455	1·424	1·382	1·356	1·339	1·316	1·302	1·293	1·282	1·277	1·273
0·80000	(7·85)		1·555	1·500	1·454	1·423	1·382	1·356	1·338	1·315	1·302	1·293	1·282	1·277	1·273
1·00000	(9·81)		1·554	1·499	1·454	1·422	1·381	1·355	1·338	1·315	1·301	1·292	1·282	1·276	1·273
1·50000	(14·7)		1·552	1·498	1·453	1·421	1·381	1·355	1·337	1·315	1·301	1·292	1·282	1·276	1·273
2·00000	(19·6)		1·551	1·497	1·452	1·421	1·380	1·355	1·337	1·315	1·301	1·292	1·282	1·276	1·273
3·00000	(29·4)		1·550	1·496	1·451	1·420	1·380	1·354	1·337	1·314	1·301	1·292	1·281	1·276	1·273
4·00000	(39·2)		1·550	1·496	1·451	1·420	1·379	1·354	1·337	1·314	1·301	1·292	1·281	1·276	1·273
6·00000	(58·8)		1·549	1·495	1·451	1·420	1·379	1·354	1·336	1·314	1·300	1·292	1·281	1·276	1·273
8·00000	(78·5)		1·548	1·495	1·450	1·419	1·379	1·354	1·336	1·314	1·300	1·292	1·281	1·276	1·273
10·0000	(98·1)		1·548	1·495	1·450	1·419	1·379	1·353	1·336	1·314	1·300	1·291	1·281	1·276	1·273
15·0000	(147)		1·548	1·494	1·450	1·419	1·379	1·353	1·336	1·314	1·300	1·291	1·281	1·276	1·272
20·0000	(196)		1·547	1·494	1·450	1·419	1·379	1·353	1·336	1·314	1·300	1·291	1·281	1·276	1·272
30·0000	(294)		1·547	1·494	1·449	1·419	1·378	1·353	1·336	1·314	1·300	1·291	1·281	1·276	1·272
S	$(\Delta p/\rho l)$	6	8	10	12·5	15	20	25	30	40	50	60	80	100	125

Grad'nt k.p.g.   (Equivalent) Pipe diameters in mm

Grad'nt S	k.p.g. $(\Delta p/\rho l)$	125	150	200	250	300	400	500	600	800	1000	1250	1500	2000	2500
0·00010	(0·00098)	1·351	1·341	1·330	1·325	1·323	1·322	1·324	1·328	1·335	1·343	1·352	1·361	1·377	1·390
0·00015	(0·00147)	1·338	1·329	1·320	1·316	1·315	1·316	1·319	1·322	1·331	1·339	1·349	1·358	1·374	1·388
0·00020	(0·00196)	1·330	1·322	1·314	1·311	1·310	1·312	1·315	1·319	1·328	1·337	1·347	1·356	1·372	1·386
0·00030	(0·00294)	1·320	1·313	1·307	1·304	1·304	1·307	1·311	1·315	1·325	1·334	1·344	1·354	1·370	1·385
0·00040	(0·00392)	1·314	1·308	1·302	1·301	1·301	1·304	1·308	1·313	1·323	1·332	1·343	1·352	1·369	1·384
0·00060	(0·00588)	1·306	1·301	1·297	1·296	1·297	1·300	1·305	1·310	1·320	1·330	1·341	1·351	1·368	1·382
0·00080	(0·00785)	1·302	1·297	1·294	1·293	1·294	1·298	1·303	1·309	1·319	1·329	1·340	1·350	1·367	1·382
0·00100	(0·00981)	1·299	1·295	1·291	1·291	1·292	1·297	1·302	1·307	1·318	1·328	1·339	1·349	1·366	1·381
0·00150	(0·0147)	1·294	1·291	1·288	1·288	1·290	1·294	1·300	1·306	1·316	1·326	1·338	1·348	1·365	1·380
0·00200	(0·0196)	1·291	1·288	1·286	1·286	1·288	1·293	1·299	1·305	1·316	1·326	1·337	1·347	1·365	1·380
0·00300	(0·0294)	1·288	1·285	1·283	1·284	1·286	1·291	1·297	1·303	1·314	1·325	1·336	1·346	1·364	1·379
0·00400	(0·0392)	1·286	1·283	1·282	1·283	1·285	1·290	1·296	1·302	1·314	1·324	1·336	1·346	1·364	1·379
0·00600	(0·0588)	1·283	1·281	1·280	1·281	1·283	1·289	1·295	1·302	1·313	1·323	1·335	1·345	1·363	1·379
0·00800	(0·0785)	1·282	1·280	1·279	1·280	1·283	1·289	1·295	1·301	1·313	1·323	1·335	1·345	1·363	1·378
0·01000	(0·0981)	1·281	1·279	1·278	1·280	1·282	1·288	1·294	1·301	1·312	1·323	1·334	1·345	1·363	1·378
0·01500	(0·147)	1·279	1·277	1·277	1·279	1·281	1·287	1·294	1·300	1·312	1·322	1·334	1·345	1·363	1·378
0·02000	(0·196)	1·278	1·277	1·276	1·278	1·281	1·287	1·293	1·300	1·311	1·322	1·334	1·344	1·362	1·378
0·03000	(0·294)	1·277	1·276	1·275	1·277	1·280	1·286	1·293	1·299	1·311	1·322	1·333	1·344	1·362	1·377
0·04000	(0·392)	1·277	1·275	1·275	1·277	1·280	1·286	1·293	1·299	1·311	1·321	1·333	1·344	1·362	1·377
0·06000	(0·588)	1·276	1·274	1·274	1·276	1·279	1·286	1·292	1·299	1·311	1·321	1·333	1·344	1·362	1·377
0·08000	(0·785)	1·275	1·274	1·274	1·276	1·279	1·285	1·292	1·299	1·311	1·321	1·333	1·344	1·362	1·377
0·10000	(0·981)	1·275	1·274	1·274	1·276	1·279	1·285	1·292	1·298	1·310	1·321	1·333	1·344	1·362	1·377
0·15000	(1·47)	1·275	1·273	1·273	1·275	1·278	1·285	1·292	1·298	1·310	1·321	1·333	1·343	1·362	1·377
0·20000	(1·96)	1·274	1·273	1·273	1·275	1·278	1·285	1·292	1·298	1·310	1·321	1·333	1·343	1·362	1·377
0·30000	(2·94)	1·274	1·273	1·273	1·275	1·278	1·285	1·292	1·298	1·310	1·321	1·333	1·343	1·362	1·377
0·40000	(3·92)	1·274	1·272	1·273	1·275	1·278	1·285	1·291	1·298	1·310	1·321	1·333	1·343	1·362	1·377
0·60000	(5·88)	1·273	1·272	1·273	1·275	1·278	1·284	1·291	1·298	1·310	1·321	1·333	1·343	1·362	1·377
0·80000	(7·85)	1·273	1·272	1·272	1·275	1·278	1·284	1·291	1·298	1·310	1·321	1·333	1·343	1·361	1·377
1·00000	(9·81)	1·273	1·272	1·272	1·275	1·278	1·284	1·291	1·298	1·310	1·321	1·333	1·343	1·361	1·377
1·50000	(14·7)	1·273	1·272	1·272	1·274	1·278	1·284	1·291	1·298	1·310	1·321	1·333	1·343	1·361	1·377
2·00000	(19·6)	1·273	1·272	1·272	1·274	1·277	1·284	1·291	1·298	1·310	1·321	1·332	1·343	1·361	1·377
3·00000	(29·4)	1·273	1·272	1·272	1·274	1·277	1·284	1·291	1·298	1·310	1·321	1·332	1·343	1·361	1·377
S	$(\Delta p/\rho l)$	125	150	200	250	300	400	500	600	800	1000	1250	1500	2000	2500

Grad'nt k.p.g.   (Equivalent) Pipe diameters in mm

Kinematic viscosity, $\nu = 1{\cdot}00 \times 10^{-6}\ \mathrm{m^2\,s^{-1}}$;   **Roughness size, $k_s = 1{\cdot}50$ mm**

E97

Kin. visc., $\nu = 1 \cdot 00 \times 10^{-6}$ m^{2}s^{-1};
$S = 0 \cdot 00010$ to $3 \cdot 00000$
i.e. kin. pr. grad., $\Delta p/\rho l =$ $(0 \cdot 00098)$ to $(29 \cdot 4)$ ms^{-2}

Roughness size, $k_s = 3 \cdot 0$ mm
This table shows values of m, as follows
m_C for Colebrook-White solutions

Grad'nt S	k.p.g. $(\Delta p/\rho l)$	125	150	200	250	300	400	500	600	800	1000	1250	1500	2000	2500
0·00010	(0·00098)	1·501	1·487	1·470	1·461	1·456	1·452	1·451	1·453	1·457	1·463	1·470	1·478	1·491	1·504
0·00015	(0·00147)	1·492	1·479	1·464	1·456	1·451	1·448	1·448	1·449	1·454	1·460	1·468	1·476	1·490	1·502
0·00020	(0·00196)	1·486	1·474	1·459	1·452	1·448	1·445	1·445	1·447	1·453	1·459	1·467	1·474	1·489	1·501
0·00030	(0·00294)	1·479	1·468	1·454	1·448	1·444	1·442	1·443	1·445	1·451	1·457	1·465	1·473	1·487	1·500
0·00040	(0·00392)	1·475	1·464	1·452	1·445	1·442	1·440	1·441	1·443	1·449	1·456	1·464	1·472	1·487	1·500
0·00060	(0·00588)	1·470	1·460	1·448	1·442	1·439	1·438	1·439	1·441	1·448	1·455	1·463	1·471	1·486	1·499
0·00080	(0·00785)	1·467	1·457	1·446	1·440	1·438	1·436	1·438	1·440	1·447	1·454	1·462	1·471	1·485	1·498
0·00100	(0·00981)	1·465	1·455	1·444	1·439	1·436	1·435	1·437	1·440	1·446	1·453	1·462	1·470	1·485	1·498
0·00150	(0·0147)	1·462	1·452	1·442	1·437	1·435	1·434	1·436	1·439	1·445	1·453	1·461	1·470	1·485	1·498
0·00200	(0·0196)	1·460	1·451	1·441	1·436	1·434	1·433	1·435	1·438	1·445	1·452	1·461	1·469	1·484	1·497
0·00300	(0·0294)	1·457	1·449	1·439	1·434	1·432	1·432	1·434	1·437	1·444	1·452	1·460	1·469	1·484	1·497
0·00400	(0·0392)	1·456	1·448	1·438	1·434	1·432	1·431	1·434	1·437	1·444	1·451	1·460	1·468	1·484	1·497
0·00600	(0·0588)	1·454	1·446	1·437	1·433	1·431	1·431	1·433	1·436	1·443	1·451	1·460	1·468	1·484	1·497
0·00800	(0·0785)	1·454	1·445	1·436	1·432	1·430	1·430	1·433	1·436	1·443	1·451	1·460	1·468	1·483	1·496
0·01000	(0·0981)	1·453	1·445	1·436	1·432	1·430	1·430	1·432	1·435	1·443	1·450	1·459	1·468	1·483	1·496
0·01500	(0·147)	1·452	1·444	1·435	1·431	1·429	1·429	1·432	1·435	1·443	1·450	1·459	1·468	1·483	1·496
0·02000	(0·196)	1·451	1·443	1·435	1·431	1·429	1·429	1·432	1·435	1·442	1·450	1·459	1·467	1·483	1·496
0·03000	(0·294)	1·450	1·443	1·434	1·430	1·429	1·429	1·431	1·435	1·442	1·450	1·459	1·467	1·483	1·496
0·04000	(0·392)	1·450	1·442	1·434	1·430	1·428	1·429	1·431	1·434	1·442	1·450	1·459	1·467	1·483	1·496
0·06000	(0·588)	1·450	1·442	1·433	1·429	1·428	1·428	1·431	1·434	1·442	1·450	1·459	1·467	1·483	1·496
0·08000	(0·785)	1·449	1·442	1·433	1·429	1·428	1·428	1·431	1·434	1·442	1·449	1·459	1·467	1·482	1·496
0·10000	(0·981)	1·449	1·441	1·433	1·429	1·428	1·428	1·431	1·434	1·442	1·449	1·459	1·467	1·482	1·496
0·15000	(1·47)	1·449	1·441	1·433	1·429	1·428	1·428	1·431	1·434	1·442	1·449	1·458	1·467	1·482	1·496
0·20000	(1·96)	1·448	1·441	1·433	1·429	1·427	1·428	1·431	1·434	1·442	1·449	1·458	1·467	1·482	1·496
0·30000	(2·94)	1·448	1·441	1·432	1·429	1·427	1·428	1·430	1·434	1·441	1·449	1·458	1·467	1·482	1·496
0·40000	(3·92)	1·448	1·441	1·432	1·429	1·427	1·428	1·430	1·434	1·441	1·449	1·458	1·467	1·482	1·496
0·60000	(5·88)	1·448	1·440	1·432	1·429	1·427	1·428	1·430	1·434	1·441	1·449	1·458	1·467	1·482	1·496
0·80000	(7·85)	1·448	1·440	1·432	1·428	1·427	1·428	1·430	1·434	1·441	1·449	1·458	1·467	1·482	1·496
1·00000	(9·81)	1·448	1·440	1·432	1·428	1·427	1·428	1·430	1·434	1·441	1·449	1·458	1·467	1·482	1·496
1·50000	(14·7)	1·448	1·440	1·432	1·428	1·427	1·428	1·430	1·434	1·441	1·449	1·458	1·467	1·482	1·496
2·00000	(19·6)	1·448	1·440	1·432	1·428	1·427	1·428	1·430	1·434	1·441	1·449	1·458	1·467	1·482	1·496
3·00000	(29·4)	1·448	1·440	1·432	1·428	1·427	1·428	1·430	1·434	1·441	1·449	1·458	1·467	1·482	1·496
S	$(\Delta p/\rho l)$	125	150	200	250	300	400	500	600	800	1000	1250	1500	2000	2500

Grad'nt k.p.g. (Equivalent) Pipe diameters in mm   Roughness size, $k_s = 3 \cdot 0$ mm

E98

Kin. visc., $\nu = 1 \cdot 00 \times 10^{-6}$ m^{2}s^{-1};
$S = 0 \cdot 00010$ to $3 \cdot 00000$
i.e. kin. pr. grad., $\Delta p/\rho l =$ $(0 \cdot 00098)$ to $(29 \cdot 4)$ ms^{-2}

Roughness size, $k_s = 6 \cdot 0$ mm
This table shows values of m, as follows
m_C for Colebrook-White solutions

Grad'nt S	k.p.g. $(\Delta p/\rho l)$	125	150	200	250	300	400	500	600	800	1000	1250	1500	2000	2500
0·00010	(0·00098)	1·715	1·692	1·664	1·647	1·636	1·623	1·617	1·614	1·612	1·614	1·618	1·622	1·632	1·642
0·00015	(0·00147)	1·709	1·687	1·659	1·643	1·632	1·620	1·614	1·611	1·610	1·612	1·616	1·621	1·631	1·641
0·00020	(0·00196)	1·704	1·683	1·656	1·640	1·630	1·618	1·613	1·610	1·609	1·611	1·615	1·620	1·630	1·640
0·00030	(0·00294)	1·700	1·679	1·653	1·637	1·627	1·616	1·611	1·608	1·608	1·610	1·614	1·619	1·630	1·640
0·00040	(0·00392)	1·697	1·676	1·651	1·636	1·626	1·615	1·610	1·608	1·607	1·610	1·614	1·619	1·629	1·639
0·00060	(0·00588)	1·693	1·673	1·648	1·634	1·624	1·614	1·609	1·606	1·606	1·609	1·613	1·618	1·629	1·639
0·00080	(0·00785)	1·691	1·672	1·647	1·632	1·623	1·613	1·608	1·606	1·606	1·608	1·613	1·618	1·628	1·638
0·00100	(0·00981)	1·690	1·670	1·646	1·632	1·622	1·612	1·607	1·605	1·605	1·608	1·613	1·618	1·628	1·638
0·00150	(0·0147)	1·688	1·669	1·644	1·630	1·621	1·611	1·607	1·605	1·605	1·607	1·612	1·617	1·628	1·638
0·00200	(0·0196)	1·686	1·667	1·644	1·629	1·620	1·611	1·606	1·604	1·604	1·607	1·612	1·617	1·628	1·638
0·00300	(0·0294)	1·685	1·666	1·642	1·629	1·620	1·610	1·605	1·604	1·604	1·607	1·611	1·617	1·628	1·638
0·00400	(0·0392)	1·684	1·665	1·642	1·628	1·619	1·610	1·605	1·603	1·604	1·607	1·611	1·617	1·627	1·638
0·00600	(0·0588)	1·683	1·664	1·641	1·627	1·619	1·609	1·605	1·603	1·604	1·606	1·611	1·616	1·627	1·637
0·00800	(0·0785)	1·682	1·664	1·641	1·627	1·618	1·609	1·604	1·603	1·603	1·606	1·611	1·616	1·627	1·637
0·01000	(0·0981)	1·682	1·663	1·640	1·627	1·618	1·609	1·604	1·603	1·603	1·606	1·611	1·616	1·627	1·637
0·01500	(0·147)	1·681	1·663	1·640	1·626	1·618	1·608	1·604	1·602	1·603	1·606	1·611	1·616	1·627	1·637
0·02000	(0·196)	1·681	1·662	1·639	1·626	1·617	1·608	1·604	1·602	1·603	1·606	1·611	1·616	1·627	1·637
0·03000	(0·294)	1·680	1·662	1·639	1·626	1·617	1·608	1·604	1·602	1·603	1·606	1·611	1·616	1·627	1·637
0·04000	(0·392)	1·680	1·662	1·639	1·626	1·617	1·608	1·604	1·602	1·603	1·606	1·610	1·616	1·627	1·637
0·06000	(0·588)	1·679	1·661	1·639	1·625	1·617	1·608	1·604	1·602	1·603	1·606	1·610	1·616	1·627	1·637
0·08000	(0·785)	1·679	1·661	1·639	1·625	1·617	1·608	1·603	1·602	1·603	1·606	1·610	1·616	1·627	1·637
0·10000	(0·981)	1·679	1·661	1·638	1·625	1·617	1·607	1·603	1·602	1·603	1·605	1·610	1·616	1·627	1·637
0·15000	(1·47)	1·679	1·661	1·638	1·625	1·617	1·607	1·603	1·602	1·602	1·605	1·610	1·616	1·627	1·637
0·20000	(1·96)	1·679	1·661	1·638	1·625	1·617	1·607	1·603	1·602	1·602	1·605	1·610	1·616	1·626	1·637
0·30000	(2·94)	1·679	1·661	1·638	1·625	1·616	1·607	1·603	1·602	1·602	1·605	1·610	1·616	1·626	1·637
0·40000	(3·92)	1·679	1·661	1·638	1·625	1·616	1·607	1·603	1·602	1·602	1·605	1·610	1·616	1·626	1·637
0·60000	(5·88)	1·678	1·661	1·638	1·625	1·616	1·607	1·603	1·602	1·602	1·605	1·610	1·616	1·626	1·637
0·80000	(7·85)	1·678	1·660	1·638	1·625	1·616	1·607	1·603	1·602	1·602	1·605	1·610	1·616	1·626	1·637
1·00000	(9·81)	1·678	1·660	1·638	1·625	1·616	1·607	1·603	1·602	1·602	1·605	1·610	1·616	1·626	1·637
1·50000	(14·7)	1·678	1·660	1·638	1·625	1·616	1·607	1·603	1·602	1·602	1·605	1·610	1·616	1·626	1·637
2·00000	(19·6)	1·678	1·660	1·638	1·625	1·616	1·607	1·603	1·602	1·602	1·605	1·610	1·616	1·626	1·637
3·00000	(29·4)	1·678	1·660	1·638	1·625	1·616	1·607	1·603	1·602	1·602	1·605	1·610	1·616	1·626	1·637
S	$(\Delta p/\rho l)$	125	150	200	250	300	400	500	600	800	1000	1250	1500	2000	2500

Grad'nt k.p.g. (Equivalent) Pipe diameters in mm   Roughness size, $k_s = 6 \cdot 0$ mm

Kin. visc., $\nu = 1.00 \times 10^{-6}$ m^{2}s^{-1};
$S = 0.00010$ to 3.00000

i.e. kin. pr. grad., $\Delta p/\rho l =$
(0.00098) to (29.4) ms^{-2}

Roughness size, $k_s = 15.0$ mm
This table shows values of m, as follows

m_C for Colebrook-White solutions

Grad'nt S	k.p.g. $(\Delta p/\rho l)$	125	150	200	250	300	400	500	600	800	1000	1250	1500	2000	2500
0.00010	(0.00098)	2.150	2.102	2.038	1.999	1.971	1.937	1.916	1.902	1.885	1.877	1.871	1.869	1.870	1.873
0.00015	(0.00147)	2.146	2.098	2.035	1.996	1.969	1.935	1.914	1.901	1.884	1.876	1.871	1.869	1.869	1.872
0.00020	(0.00196)	2.143	2.096	2.034	1.995	1.968	1.934	1.913	1.900	1.884	1.875	1.870	1.868	1.869	1.872
0.00030	(0.00294)	2.140	2.093	2.032	1.993	1.967	1.933	1.912	1.899	1.883	1.875	1.870	1.868	1.868	1.872
0.00040	(0.00392)	2.138	2.091	2.030	1.992	1.966	1.932	1.912	1.898	1.883	1.874	1.869	1.868	1.868	1.871
0.00060	(0.00588)	2.136	2.090	2.029	1.991	1.965	1.931	1.911	1.898	1.882	1.874	1.869	1.867	1.868	1.871
0.00080	(0.00785)	2.135	2.088	2.028	1.990	1.964	1.931	1.911	1.897	1.882	1.874	1.869	1.867	1.868	1.871
0.00100	(0.00981)	2.134	2.088	2.027	1.990	1.964	1.930	1.910	1.897	1.882	1.874	1.869	1.867	1.868	1.871
0.00150	(0.0147)	2.132	2.086	2.026	1.989	1.963	1.930	1.910	1.897	1.881	1.873	1.868	1.867	1.867	1.871
0.00200	(0.0196)	2.131	2.086	2.026	1.988	1.963	1.930	1.910	1.896	1.881	1.873	1.868	1.867	1.867	1.871
0.00300	(0.0294)	2.130	2.085	2.025	1.988	1.962	1.929	1.909	1.896	1.881	1.873	1.868	1.866	1.867	1.871
0.00400	(0.0392)	2.130	2.084	2.025	1.987	1.962	1.929	1.909	1.896	1.881	1.873	1.868	1.866	1.867	1.870
0.00600	(0.0588)	2.129	2.084	2.024	1.987	1.961	1.929	1.909	1.896	1.880	1.873	1.868	1.866	1.867	1.870
0.00800	(0.0785)	2.129	2.083	2.024	1.987	1.961	1.928	1.909	1.896	1.880	1.873	1.868	1.866	1.867	1.870
0.01000	(0.0981)	2.128	2.083	2.024	1.987	1.961	1.928	1.909	1.896	1.880	1.873	1.868	1.866	1.867	1.870
0.01500	(0.147)	2.128	2.083	2.024	1.986	1.961	1.928	1.908	1.896	1.880	1.872	1.868	1.866	1.867	1.870
0.02000	(0.196)	2.128	2.082	2.023	1.986	1.961	1.928	1.908	1.895	1.880	1.872	1.868	1.866	1.867	1.870
0.03000	(0.294)	2.127	2.082	2.023	1.986	1.961	1.928	1.908	1.895	1.880	1.872	1.868	1.866	1.867	1.870
0.04000	(0.392)	2.127	2.082	2.023	1.986	1.961	1.928	1.908	1.895	1.880	1.872	1.868	1.866	1.867	1.870
0.06000	(0.588)	2.127	2.082	2.023	1.986	1.960	1.928	1.908	1.895	1.880	1.872	1.868	1.866	1.867	1.870
0.08000	(0.785)	2.127	2.082	2.023	1.986	1.960	1.928	1.908	1.895	1.880	1.872	1.868	1.866	1.867	1.870
0.10000	(0.981)	2.127	2.082	2.023	1.986	1.960	1.928	1.908	1.895	1.880	1.872	1.868	1.866	1.867	1.870
0.15000	(1.47)	2.127	2.082	2.023	1.986	1.960	1.928	1.908	1.895	1.880	1.872	1.868	1.866	1.867	1.870
0.20000	(1.96)	2.127	2.081	2.023	1.986	1.960	1.928	1.908	1.895	1.880	1.872	1.868	1.866	1.867	1.870
0.30000	(2.94)	2.126	2.081	2.023	1.986	1.960	1.928	1.908	1.895	1.880	1.872	1.867	1.866	1.867	1.870
0.40000	(3.92)	2.126	2.081	2.022	1.986	1.960	1.928	1.908	1.895	1.880	1.872	1.867	1.866	1.867	1.870
0.60000	(5.88)	2.126	2.081	2.022	1.985	1.960	1.928	1.908	1.895	1.880	1.872	1.867	1.866	1.867	1.870
0.80000	(7.85)	2.126	2.081	2.022	1.985	1.960	1.928	1.908	1.895	1.880	1.872	1.867	1.866	1.867	1.870
1.00000	(9.81)	2.126	2.081	2.022	1.985	1.960	1.928	1.908	1.895	1.880	1.872	1.867	1.866	1.867	1.870
1.50000	(14.7)	2.126	2.081	2.022	1.985	1.960	1.928	1.908	1.895	1.880	1.872	1.867	1.866	1.867	1.870
2.00000	(19.6)	2.126	2.081	2.022	1.985	1.960	1.928	1.908	1.895	1.880	1.872	1.867	1.866	1.867	1.870
3.00000	(29.4)	2.126	2.081	2.022	1.985	1.960	1.927	1.908	1.895	1.880	1.872	1.867	1.866	1.867	1.870
S	$(\Delta p/\rho l)$	125	150	200	250	300	400	500	600	800	1000	1250	1500	2000	2500

Grad'nt   k.p.g.   (Equivalent) Pipe diameters in mm   Roughness size, $k_s = 15.0$ mm

Kin. visc., $\nu = 1.00 \times 10^{-6}$ m^{2}s^{-1};
$S = 0.00010$ to 3.00000

i.e. kin. pr. grad., $\Delta p/\rho l =$
(0.00098) to (29.4) ms^{-2}

Roughness size, $k_s = 30.0$ mm
This table shows values of m, as follows

m_C for Colebrook-White solutions

Grad'nt S	k.p.g. $(\Delta p/\rho l)$	125	150	200	250	300	400	500	600	800	1000	1250	1500	2000	2500
0.00010	(0.00098)		2.591	2.471	2.396	2.344	2.276	2.234	2.205	2.167	2.145	2.127	2.116	2.103	2.098
0.00015	(0.00147)		2.588	2.469	2.394	2.343	2.275	2.233	2.204	2.166	2.144	2.126	2.115	2.103	2.097
0.00020	(0.00196)		2.586	2.468	2.393	2.342	2.274	2.232	2.203	2.166	2.144	2.126	2.115	2.103	2.097
0.00030	(0.00294)		2.584	2.466	2.392	2.341	2.274	2.232	2.203	2.166	2.143	2.126	2.115	2.102	2.097
0.00040	(0.00392)		2.583	2.465	2.391	2.340	2.273	2.231	2.202	2.165	2.143	2.126	2.114	2.102	2.097
0.00060	(0.00588)		2.581	2.464	2.390	2.339	2.273	2.231	2.202	2.165	2.143	2.125	2.114	2.102	2.097
0.00080	(0.00785)		2.581	2.463	2.390	2.339	2.272	2.230	2.202	2.165	2.143	2.125	2.114	2.102	2.097
0.00100	(0.00981)		2.580	2.463	2.389	2.338	2.272	2.230	2.201	2.165	2.142	2.125	2.114	2.102	2.097
0.00150	(0.0147)		2.579	2.462	2.389	2.338	2.272	2.230	2.201	2.164	2.142	2.125	2.114	2.102	2.097
0.00200	(0.0196)		2.578	2.462	2.389	2.338	2.271	2.230	2.201	2.164	2.142	2.125	2.114	2.102	2.096
0.00300	(0.0294)		2.578	2.461	2.388	2.337	2.271	2.229	2.201	2.164	2.142	2.125	2.114	2.102	2.096
0.00400	(0.0392)		2.577	2.461	2.388	2.337	2.271	2.229	2.201	2.164	2.142	2.125	2.114	2.102	2.096
0.00600	(0.0588)		2.577	2.461	2.388	2.337	2.271	2.229	2.201	2.164	2.142	2.125	2.114	2.102	2.096
0.00800	(0.0785)		2.577	2.461	2.387	2.337	2.271	2.229	2.200	2.164	2.142	2.124	2.114	2.102	2.096
0.01000	(0.0981)		2.577	2.460	2.387	2.337	2.271	2.229	2.200	2.164	2.142	2.124	2.114	2.102	2.096
0.01500	(0.147)		2.576	2.460	2.387	2.337	2.270	2.229	2.200	2.164	2.142	2.124	2.113	2.102	2.096
0.02000	(0.196)		2.576	2.460	2.387	2.336	2.270	2.229	2.200	2.164	2.142	2.124	2.113	2.101	2.096
0.03000	(0.294)		2.576	2.460	2.387	2.336	2.270	2.229	2.200	2.164	2.142	2.124	2.113	2.101	2.096
0.04000	(0.392)		2.576	2.460	2.387	2.336	2.270	2.229	2.200	2.164	2.142	2.124	2.113	2.101	2.096
0.06000	(0.588)		2.576	2.460	2.387	2.336	2.270	2.229	2.200	2.164	2.142	2.124	2.113	2.101	2.096
0.08000	(0.785)		2.575	2.460	2.387	2.336	2.270	2.229	2.200	2.164	2.142	2.124	2.113	2.101	2.096
0.10000	(0.981)		2.575	2.460	2.387	2.336	2.270	2.229	2.200	2.164	2.142	2.124	2.113	2.101	2.096
0.15000	(1.47)		2.575	2.459	2.387	2.336	2.270	2.229	2.200	2.164	2.142	2.124	2.113	2.101	2.096
0.20000	(1.96)		2.575	2.459	2.387	2.336	2.270	2.229	2.200	2.164	2.142	2.124	2.113	2.101	2.096
0.30000	(2.94)		2.575	2.459	2.387	2.336	2.270	2.229	2.200	2.164	2.142	2.124	2.113	2.101	2.096
0.40000	(3.92)		2.575	2.459	2.387	2.336	2.270	2.229	2.200	2.164	2.141	2.124	2.113	2.101	2.096
0.60000	(5.88)		2.575	2.459	2.386	2.336	2.270	2.229	2.200	2.164	2.141	2.124	2.113	2.101	2.096
0.80000	(7.85)		2.575	2.459	2.386	2.336	2.270	2.228	2.200	2.164	2.141	2.124	2.113	2.101	2.096
1.00000	(9.81)		2.575	2.459	2.386	2.336	2.270	2.228	2.200	2.164	2.141	2.124	2.113	2.101	2.096
1.50000	(14.7)		2.575	2.459	2.386	2.336	2.270	2.228	2.200	2.164	2.141	2.124	2.113	2.101	2.096
2.00000	(19.6)		2.575	2.459	2.386	2.336	2.270	2.228	2.200	2.164	2.141	2.124	2.113	2.101	2.096
3.00000	(29.4)		2.575	2.459	2.386	2.336	2.270	2.228	2.200	2.163	2.141	2.124	2.113	2.101	2.096
S	$(\Delta p/\rho l)$	125	150	200	250	300	400	500	600	800	1000	1250	1500	2000	2500

Grad'nt   k.p.g.   (Equivalent) Pipe diameters in mm   Roughness size, $k_s = 30.0$ mm

E101

$\nu = 1{\cdot}25 \times 10^{-6}\ \mathrm{m^2\,s^{-1}}$; $S = 0{\cdot}00010$ to $30{\cdot}0000$ i.e. kin. pr. grad., $\Delta p/\rho l = (0{\cdot}00098)$ to $(294)\ \mathrm{ms^{-2}}$

Kin. visc., $\nu = 1{\cdot}25 \times 10^{-6}\ \mathrm{m^2\,s^{-1}}$; $S = 0{\cdot}00010$ to $30{\cdot}0000$

Roughness size, $k_s = 0{\cdot}0015$ mm

This table shows values of m, as follows m_C for Colebrook-White solutions; or, where $R \leq 2000$, m_P for laminar flow

Grad'nt — k.p.g. — (Equivalent) Pipe diameters in mm

S	$(\Delta p/\rho l)$	6	8	10	12·5	15	20	25	30	40	50	60	80	100	125
0·00030	(0·00294)	8·572	5·841	4·338	3·222	2·526	1·721	1·278	1·003	*	1·116	1·097	1·073	1·057	1·045
0·00040	(0·00392)	7·423	5·058	3·757	2·790	2·188	1·491	1·107	*	1·114	1·089	1·071	1·049	1·035	1·024
0·00060	(0·00588)	6·061	4·130	3·067	2·278	1·786	1·217	0·904	*	1·074	1·052	1·037	1·017	1·005	0·996
0·00080	(0·00785)	5·249	3·577	2·656	1·973	1·547	1·054	*	1·079	1·047	1·027	1·013	0·996	0·985	0·977
0·00100	(0·00981)	4·695	3·199	2·376	1·764	1·384	0·943	*	1·056	1·027	1·009	0·996	0·980	0·970	0·962
0·00150	(0·0147)	3·833	2·612	1·940	1·441	1·130	*	1·038	1·019	0·993	0·977	0·966	0·952	0·944	0·937
0·00200	(0·0196)	3·320	2·262	1·680	1·248	0·978	*	1·011	0·993	0·970	0·955	0·946	0·933	0·926	0·921
0·00300	(0·0294)	2·711	1·847	1·372	1·019	0·799	0·998	0·976	0·960	0·940	0·927	0·918	0·908	0·902	0·898
0·00400	(0·0392)	2·347	1·600	1·188	0·882	*	0·973	0·952	0·938	0·919	0·908	0·900	0·891	0·886	0·882
0·00600	(0·0588)	1·917	1·306	0·970	*	*	0·938	0·920	0·908	0·892	0·882	0·876	0·868	0·864	0·862
0·00800	(0·0785)	1·660	1·131	0·840	*	0·942	0·916	0·899	0·888	0·873	0·865	0·859	0·853	0·855	0·853
0·01000	(0·0981)	1·485	1·012	0·751	*	0·923	0·899	0·883	0·873	0·860	0·852	0·847	0·847	0·849	0·847
0·01500	(0·147)	1·212	0·826	*	0·907	0·891	0·869	0·856	0·847	0·836	0·829	0·825	0·821	0·819	0·818
0·02000	(0·196)	1·050	*	*	0·884	0·869	0·850	0·838	0·830	0·820	0·814	0·810	0·807	0·806	0·805
0·03000	(0·294)	0·857	*	0·872	0·854	0·841	0·824	0·813	0·807	0·798	0·793	0·791	0·788	0·788	0·788
0·04000	(0·392)	0·742	*	0·850	0·833	0·821	0·806	0·797	0·791	0·784	0·780	0·777	0·776	0·776	0·777
0·06000	(0·588)	*	0·839	0·821	0·806	0·796	0·783	0·775	0·770	0·764	0·761	0·759	0·759	0·759	0·761
0·08000	(0·785)	*	0·818	0·802	0·788	0·779	0·767	0·760	0·756	0·751	0·748	0·747	0·747	0·748	0·750
0·10000	(0·981)	0·826	0·802	0·787	0·775	0·766	0·755	0·749	0·745	0·741	0·739	0·738	0·738	0·740	0·742
0·15000	(1·47)	0·797	0·775	0·762	0·751	0·744	0·735	0·730	0·727	0·723	0·722	0·722	0·723	0·725	0·728
0·20000	(1·96)	0·777	0·758	0·746	0·736	0·729	0·721	0·717	0·714	0·712	0·711	0·711	0·713	0·715	0·719
0·30000	(2·94)	0·750	0·734	0·723	0·715	0·709	0·703	0·699	0·697	0·696	0·696	0·697	0·699	0·702	0·706
0·40000	(3·92)	0·733	0·718	0·708	0·701	0·696	0·690	0·688	0·686	0·685	0·686	0·687	0·690	0·693	0·697
0·60000	(5·88)	0·710	0·697	0·689	0·682	0·678	0·674	0·672	0·671	0·671	0·672	0·674	0·677	0·681	0·686
0·80000	(7·85)	0·694	0·682	0·675	0·670	0·666	0·663	0·661	0·661	0·661	0·663	0·665	0·669	0·673	0·678
1·00000	(9·81)	0·683	0·672	0·665	0·660	0·657	0·654	0·653	0·653	0·654	0·656	0·658	0·662	0·667	0·672
1·50000	(14·7)	0·663	0·654	0·648	0·644	0·642	0·640	0·639	0·640	0·642	0·644	0·646	0·652	0·656	0·662
2·00000	(19·6)	0·649	0·641	0·637	0·633	0·632	0·630	0·630	0·631	0·633	0·636	0·639	0·644	0·649	0·655
3·00000	(29·4)	0·632	0·625	0·621	0·619	0·618	0·617	0·618	0·619	0·622	0·625	0·628	0·634	0·640	0·646
4·00000	(39·2)	0·620	0·614	0·611	0·609	0·608	0·608	0·609	0·611	0·614	0·618	0·621	0·628	0·634	0·640
6·00000	(58·8)	0·604	0·600	0·597	0·596	0·596	0·597	0·598	0·600	0·604	0·609	0·612	0·619	0·626	0·632
8·00000	(78·5)	0·594	0·590	0·588	0·587	0·588	0·589	0·591	0·593	0·598	0·602	0·606	0·614	0·620	0·627
10·0000	(98·1)	0·586	0·583	0·581	0·581	0·581	0·583	0·586	0·588	0·593	0·598	0·602	0·610	0·616	0·624
15·0000	(147)	0·572	0·570	0·570	0·570	0·571	0·573	0·576	0·579	0·585	0·590	0·594	0·602	0·610	0·617
20·0000	(196)	0·564	0·562	0·562	0·563	0·564	0·567	0·570	0·573	0·579	0·585	0·589	0·598	0·605	0·613
30·0000	(294)	0·552	0·551	0·552	0·553	0·555	0·558	0·562	0·566	0·572	0·578	0·583	0·592	0·600	0·608

| S | $(\Delta p/\rho l)$ | 6 | 8 | 10 | 12·5 | 15 | 20 | 25 | 30 | 40 | 50 | 60 | 80 | 100 | 125 |

Grad'nt — k.p.g. — (Equivalent) Pipe diameters in mm

S	$(\Delta p/\rho l)$	125	150	200	250	300	400	500	600	800	1000	1250	1500	2000	2500
0·00010	(0·00098)	1·134	1·121	1·105	1·096	1·090	1·083	1·080	1·079	1·079	1·080	1·083	1·087	1·094	1·101
0·00015	(0·00147)	1·099	1·088	1·075	1·067	1·062	1·057	1·055	1·054	1·056	1·058	1·062	1·066	1·073	1·081
0·00020	(0·00196)	1·076	1·066	1·054	1·047	1·043	1·039	1·038	1·038	1·040	1·043	1·047	1·051	1·059	1·067
0·00030	(0·00294)	1·045	1·037	1·026	1·021	1·018	1·015	1·015	1·015	1·018	1·022	1·027	1·032	1·040	1·048
0·00040	(0·00392)	1·024	1·016	1·008	1·003	1·001	0·999	0·999	1·000	1·004	1·008	1·013	1·018	1·027	1·036
0·00060	(0·00588)	0·996	0·989	0·982	0·979	0·977	0·977	0·978	0·979	0·984	0·988	0·994	1·000	1·009	1·018
0·00080	(0·00785)	0·977	0·971	0·965	0·963	0·961	0·961	0·963	0·965	0·970	0·975	0·981	0·987	0·997	1·006
0·00100	(0·00981)	0·962	0·958	0·952	0·950	0·949	0·950	0·952	0·955	0·960	0·965	0·972	0·977	0·988	0·997
0·00150	(0·0147)	0·937	0·934	0·930	0·929	0·928	0·930	0·933	0·936	0·942	0·948	0·954	0·961	0·972	0·981
0·00200	(0·0196)	0·921	0·917	0·914	0·914	0·914	0·916	0·920	0·923	0·929	0·936	0·943	0·949	0·960	0·970
0·00300	(0·0294)	0·898	0·895	0·894	0·894	0·895	0·898	0·902	0·905	0·912	0·919	0·927	0·933	0·945	0·955
0·00400	(0·0392)	0·882	0·881	0·880	0·880	0·882	0·885	0·889	0·893	0·901	0·908	0·916	0·923	0·935	0·945
0·00600	(0·0588)	0·862	0·860	0·860	0·862	0·864	0·868	0·873	0·877	0·885	0·893	0·901	0·908	0·920	0·931
0·00800	(0·0785)	0·847	0·847	0·847	0·849	0·852	0·856	0·861	0·866	0·874	0·882	0·890	0·898	0·911	0·921
0·01000	(0·0981)	0·837	0·836	0·838	0·840	0·842	0·848	0·853	0·858	0·866	0·874	0·883	0·890	0·903	0·914
0·01500	(0·147)	0·818	0·818	0·820	0·823	0·826	0·832	0·838	0·843	0·852	0·860	0·869	0·877	0·890	0·902
0·02000	(0·196)	0·805	0·806	0·809	0·812	0·815	0·821	0·827	0·833	0·842	0·851	0·860	0·868	0·882	0·893
0·03000	(0·294)	0·788	0·790	0·793	0·797	0·800	0·807	0·813	0·819	0·829	0·838	0·847	0·856	0·870	0·881
0·04000	(0·392)	0·777	0·778	0·782	0·786	0·790	0·797	0·804	0·810	0·820	0·829	0·839	0·847	0·862	0·874
0·06000	(0·588)	0·761	0·763	0·768	0·772	0·776	0·784	0·791	0·797	0·808	0·817	0·827	0·836	0·851	0·863
0·08000	(0·785)	0·750	0·753	0·758	0·762	0·767	0·775	0·782	0·789	0·800	0·809	0·819	0·828	0·843	0·856
0·10000	(0·981)	0·742	0·745	0·750	0·755	0·760	0·768	0·776	0·782	0·794	0·803	0·814	0·823	0·838	0·850
0·15000	(1·47)	0·728	0·731	0·737	0·743	0·748	0·756	0·764	0·771	0·783	0·793	0·804	0·813	0·828	0·841
0·20000	(1·96)	0·719	0·722	0·728	0·734	0·739	0·748	0·756	0·764	0·776	0·786	0·797	0·806	0·822	0·835
0·30000	(2·94)	0·706	0·710	0·716	0·722	0·728	0·738	0·746	0·753	0·766	0·776	0·788	0·797	0·814	0·827
0·40000	(3·92)	0·697	0·701	0·708	0·715	0·720	0·731	0·739	0·747	0·759	0·770	0·782	0·792	0·808	0·822
0·60000	(5·88)	0·686	0·690	0·698	0·704	0·710	0·721	0·730	0·738	0·751	0·762	0·774	0·784	0·801	0·815
0·80000	(7·85)	0·678	0·682	0·690	0·697	0·704	0·714	0·724	0·732	0·745	0·757	0·769	0·779	0·796	0·810
1·00000	(9·81)	0·672	0·677	0·685	0·692	0·699	0·710	0·719	0·727	0·741	0·753	0·765	0·775	0·793	0·807
1·50000	(14·7)	0·662	0·667	0·676	0·683	0·690	0·702	0·711	0·720	0·734	0·746	0·758	0·769	0·787	0·802
2·00000	(19·6)	0·655	0·660	0·669	0·677	0·684	0·696	0·706	0·715	0·729	0·741	0·754	0·765	0·783	0·798
3·00000	(29·4)	0·646	0·652	0·661	0·670	0·677	0·689	0·699	0·708	0·723	0·736	0·749	0·760	0·779	0·794

| S | $(\Delta p/\rho l)$ | 125 | 150 | 200 | 250 | 300 | 400 | 500 | 600 | 800 | 1000 | 1250 | 1500 | 2000 | 2500 |

Grad'nt — k.p.g. — (Equivalent) Pipe diameters in mm

Kinematic viscosity, $\nu = 1{\cdot}25 \times 10^{-6}\ \mathrm{m^2\,s^{-1}}$;

Roughness size, $k_s = 0{\cdot}0015$ mm

Kin. visc., $\nu = 1{\cdot}25\times10^{-6}$ m^{2}s^{-1}; $S = 0{\cdot}00010$ to $30{\cdot}0000$
i.e. kin. pr. grad., $\Delta p/\rho l = (0{\cdot}00098)$ to (294) ms^{-2}

Roughness size, $k_s = 0{\cdot}003$ mm
This table shows values of m, as follows
m_C for Colebrook-White solutions; or, where $\mathbf{R} \leq 2000$, m_P for laminar flow

Grad'nt S	k.p.g. $(\Delta p/\rho l)$	_(Equivalent) Pipe diameters in mm_ 6	8	10	12·5	15	20	25	30	40	50	60	80	100	125
0·00030	(0·00294)	8·572	5·841	4·338	3·222	2·526	1·721	1·278	1·003	*	1·117	1·098	1·073	1·058	1·045
0·00040	(0·00392)	7·423	5·058	3·757	2·790	2·188	1·491	1·107	*	1·114	1·089	1·072	1·049	1·036	1·024
0·00060	(0·00588)	6·061	4·130	3·067	2·278	1·786	1·217	0·904	*	1·074	1·052	1·037	1·018	1·006	0·996
0·00080	(0·00785)	5·249	3·577	2·656	1·973	1·547	1·054	*	1·079	1·047	1·028	1·014	0·996	0·986	0·977
0·00100	(0·00981)	4·695	3·199	2·376	1·764	1·384	0·943	*	1·057	1·028	1·009	0·997	0·980	0·971	0·963
0·00150	(0·0147)	3·833	2·612	1·940	1·441	1·130	*	1·039	1·019	0·994	0·978	0·967	0·953	0·945	0·938
0·00200	(0·0196)	3·320	2·262	1·680	1·248	0·978	*	1·012	0·994	0·971	0·956	0·946	0·934	0·927	0·922
0·00300	(0·0294)	2·711	1·847	1·372	1·019	0·799	0·999	0·977	0·961	0·940	0·928	0·919	0·909	0·903	0·899
0·00400	(0·0392)	2·347	1·600	1·188	0·882	*	0·974	0·953	0·939	0·920	0·909	0·901	0·892	0·887	0·884
0·00600	(0·0588)	1·917	1·306	0·970	*	*	0·939	0·921	0·909	0·893	0·883	0·877	0·869	0·866	0·863
0·00800	(0·0785)	1·660	1·131	0·840	*	0·943	0·917	0·900	0·889	0·875	0·866	0·861	0·854	0·851	0·849
0·01000	(0·0981)	1·485	1·012	0·751	*	0·924	0·900	0·885	0·874	0·861	0·853	0·848	0·843	0·840	0·839
0·01500	(0·147)	1·212	0·826	*	0·908	0·892	0·871	0·858	0·849	0·837	0·831	0·827	0·823	0·821	0·820
0·02000	(0·196)	1·050	*	*	0·885	0·871	0·851	0·839	0·831	0·822	0·816	0·812	0·809	0·808	0·808
0·03000	(0·294)	0·857	*	0·874	0·855	0·842	0·825	0·815	0·808	0·800	0·796	0·793	0·791	0·790	0·791
0·04000	(0·392)	0·742	*	0·852	0·835	0·823	0·808	0·799	0·793	0·786	0·782	0·780	0·778	0·779	0·780
0·06000	(0·588)	*	0·841	0·823	0·808	0·798	0·785	0·777	0·772	0·767	0·764	0·762	0·762	0·763	0·765
0·08000	(0·785)	*	0·820	0·804	0·790	0·781	0·770	0·763	0·758	0·754	0·751	0·750	0·751	0·752	0·754
0·10000	(0·981)	0·829	0·805	0·789	0·777	0·769	0·758	0·752	0·748	0·744	0·742	0·742	0·742	0·744	0·747
0·15000	(1·47)	0·799	0·778	0·765	0·754	0·747	0·738	0·733	0·730	0·727	0·726	0·726	0·728	0·730	0·733
0·20000	(1·96)	0·779	0·760	0·748	0·739	0·732	0·725	0·720	0·718	0·716	0·715	0·716	0·718	0·721	0·724
0·30000	(2·94)	0·753	0·737	0·727	0·718	0·713	0·707	0·703	0·702	0·701	0·701	0·702	0·705	0·708	0·712
0·40000	(3·92)	0·736	0·721	0·712	0·705	0·700	0·695	0·692	0·691	0·690	0·691	0·693	0·696	0·700	0·704
0·60000	(5·88)	0·713	0·701	0·693	0·687	0·683	0·679	0·677	0·676	0·677	0·678	0·680	0·685	0·689	0·694
0·80000	(7·85)	0·698	0·687	0·680	0·675	0·671	0·668	0·667	0·667	0·668	0·670	0·672	0·677	0·681	0·687
1·00000	(9·81)	0·687	0·676	0·670	0·666	0·663	0·660	0·659	0·659	0·661	0·663	0·666	0·671	0·676	0·682
1·50000	(14·7)	0·668	0·659	0·654	0·650	0·648	0·646	0·646	0·647	0·650	0·652	0·655	0·661	0·667	0·673
2·00000	(19·6)	0·655	0·647	0·643	0·640	0·638	0·637	0·638	0·639	0·642	0·645	0·649	0·655	0·661	0·667
3·00000	(29·4)	0·638	0·632	0·628	0·626	0·625	0·625	0·627	0·628	0·632	0·636	0·639	0·646	0·653	0·660
4·00000	(39·2)	0·627	0·621	0·619	0·617	0·617	0·617	0·619	0·621	0·625	0·629	0·634	0·641	0·647	0·655
6·00000	(58·8)	0·612	0·608	0·606	0·605	0·605	0·607	0·609	0·612	0·617	0·621	0·626	0·634	0·641	0·648
8·00000	(78·5)	0·602	0·599	0·598	0·597	0·598	0·600	0·603	0·606	0·611	0·616	0·621	0·629	0·637	0·644
10·0000	(98·1)	0·595	0·592	0·592	0·592	0·593	0·595	0·598	0·601	0·607	0·612	0·617	0·626	0·633	0·642
15·0000	(147)	0·583	0·581	0·581	0·582	0·583	0·587	0·590	0·594	0·600	0·606	0·611	0·620	0·628	0·637
20·0000	(196)	0·575	0·574	0·574	0·576	0·577	0·581	0·585	0·589	0·596	0·602	0·607	0·617	0·625	0·634
30·0000	(294)	0·564	0·564	0·566	0·567	0·570	0·574	0·579	0·583	0·590	0·597	0·603	0·613	0·621	0·630
S	$(\Delta p/\rho l)$	6	8	10	12·5	15	20	25	30	40	50	60	80	100	125

Grad'nt   k.p.g.   (Equivalent) Pipe diameters in mm

S	$(\Delta p/\rho l)$	125	150	200	250	300	400	500	600	800	1000	1250	1500	2000	2500
0·00010	(0·00098)	1·134	1·122	1·106	1·096	1·090	1·083	1·080	1·079	1·079	1·081	1·084	1·088	1·095	1·101
0·00015	(0·00147)	1·100	1·089	1·075	1·067	1·063	1·057	1·055	1·055	1·056	1·059	1·063	1·067	1·074	1·082
0·00020	(0·00196)	1·076	1·067	1·055	1·048	1·044	1·040	1·038	1·039	1·041	1·044	1·048	1·052	1·061	1·068
0·00030	(0·00294)	1·045	1·037	1·027	1·022	1·019	1·016	1·016	1·016	1·019	1·023	1·028	1·033	1·042	1·050
0·00040	(0·00392)	1·024	1·017	1·008	1·004	1·001	1·000	1·000	1·001	1·005	1·009	1·014	1·019	1·029	1·037
0·00060	(0·00588)	0·996	0·990	0·983	0·980	0·978	0·978	0·978	0·979	0·980	0·985	0·990	0·996	1·001	1·011	1·020
0·00080	(0·00785)	0·977	0·972	0·966	0·963	0·962	0·963	0·964	0·966	0·972	0·977	0·983	0·989	0·999	1·008
0·00100	(0·00981)	0·963	0·958	0·953	0·951	0·950	0·951	0·953	0·956	0·961	0·967	0·973	0·979	0·990	0·999
0·00150	(0·0147)	0·938	0·935	0·931	0·930	0·930	0·931	0·934	0·937	0·943	0·949	0·956	0·963	0·974	0·984
0·00200	(0·0196)	0·922	0·918	0·916	0·915	0·916	0·918	0·921	0·925	0·931	0·938	0·945	0·952	0·963	0·973
0·00300	(0·0294)	0·899	0·897	0·895	0·895	0·896	0·900	0·904	0·907	0·915	0·922	0·929	0·936	0·948	0·959
0·00400	(0·0392)	0·884	0·882	0·881	0·882	0·883	0·887	0·891	0·896	0·903	0·911	0·919	0·926	0·938	0·949
0·00600	(0·0588)	0·863	0·862	0·862	0·864	0·866	0·870	0·875	0·880	0·888	0·896	0·904	0·912	0·925	0·936
0·00800	(0·0785)	0·849	0·849	0·849	0·851	0·854	0·859	0·864	0·869	0·878	0·886	0·894	0·902	0·915	0·927
0·01000	(0·0981)	0·839	0·838	0·840	0·842	0·845	0·850	0·856	0·861	0·870	0·878	0·887	0·895	0·908	0·920
0·01500	(0·147)	0·820	0·821	0·823	0·826	0·829	0·835	0·841	0·847	0·856	0·865	0·874	0·882	0·896	0·908
0·02000	(0·196)	0·808	0·809	0·812	0·815	0·818	0·825	0·831	0·837	0·847	0·856	0·865	0·874	0·888	0·900
0·03000	(0·294)	0·791	0·793	0·796	0·800	0·804	0·811	0·818	0·824	0·835	0·844	0·854	0·862	0·877	0·890
0·04000	(0·392)	0·780	0·782	0·786	0·790	0·794	0·802	0·809	0·815	0·826	0·836	0·846	0·855	0·870	0·883
0·06000	(0·588)	0·765	0·767	0·772	0·777	0·781	0·790	0·797	0·804	0·815	0·825	0·835	0·845	0·860	0·873
0·08000	(0·785)	0·754	0·757	0·762	0·768	0·772	0·781	0·789	0·796	0·808	0·818	0·829	0·838	0·854	0·867
0·10000	(0·981)	0·747	0·750	0·755	0·761	0·766	0·775	0·783	0·790	0·802	0·812	0·823	0·833	0·849	0·863
0·15000	(1·47)	0·733	0·737	0·743	0·749	0·754	0·764	0·772	0·780	0·792	0·803	0·815	0·825	0·841	0·855
0·20000	(1·96)	0·724	0·728	0·735	0·741	0·747	0·757	0·765	0·773	0·786	0·797	0·809	0·819	0·836	0·850
0·30000	(2·94)	0·712	0·716	0·724	0·731	0·737	0·747	0·756	0·764	0·778	0·789	0·801	0·812	0·829	0·844
0·40000	(3·92)	0·704	0·709	0·717	0·724	0·730	0·741	0·750	0·758	0·772	0·784	0·796	0·807	0·825	0·840
0·60000	(5·88)	0·694	0·699	0·707	0·715	0·721	0·733	0·742	0·751	0·765	0·777	0·790	0·801	0·819	0·834
0·80000	(7·85)	0·687	0·692	0·701	0·709	0·715	0·727	0·737	0·746	0·761	0·773	0·786	0·797	0·816	0·831
1·00000	(9·81)	0·682	0·687	0·696	0·704	0·711	0·723	0·734	0·742	0·758	0·770	0·783	0·794	0·813	0·829
1·50000	(14·7)	0·673	0·678	0·688	0·697	0·704	0·717	0·727	0·737	0·752	0·765	0·778	0·790	0·809	0·825
2·00000	(19·6)	0·667	0·673	0·683	0·692	0·699	0·713	0·723	0·733	0·749	0·762	0·775	0·787	0·807	0·822
3·00000	(29·4)	0·660	0·666	0·677	0·686	0·694	0·707	0·718	0·728	0·744	0·758	0·772	0·784	0·803	0·819
S	$(\Delta p/\rho l)$	125	150	200	250	300	400	500	600	800	1000	1250	1500	2000	2500

Grad'nt   k.p.g.   (Equivalent) Pipe diameters in mm

Kinematic viscosity, $\nu = 1{\cdot}25\times10^{-6}$ m^{2}s^{-1} ;   **Roughness size, $k_s = 0{\cdot}003$ mm**

E103

Kin. visc., $\nu = 1.25 \times 10^{-6}$ m^{2}s^{-1}; $\quad$ **Roughness size, $k_s = 0.006$ mm**

$S = 0.00010$ to 30.0000

This table shows values of m, as follows

i.e. kin. pr. grad., $\Delta p/\rho l = (0.00098)$ to (294) ms^{-2}

m_C for Colebrook-White solutions; or, where $\mathbf{R} \le 2000$, m_P for laminar flow

Grad'nt S	k.p.g. $(\Delta p/\rho l)$	6	8	10	12.5	15	20	25	30	40	50	60	80	100	125
								(Equivalent) Pipe diameters in mm							
0.00030	(0.00294)	8.572	5.841	4.338	3.222	2.526	1.721	1.278	1.003	*	1.118	1.099	1.074	1.059	1.046
0.00040	(0.00392)	7.423	5.058	3.757	2.790	2.188	1.491	1.107	*	1.115	1.090	1.073	1.050	1.037	1.026
0.00060	(0.00588)	6.061	4.130	3.067	2.278	1.786	1.217	0.904	*	1.075	1.053	1.038	1.019	1.007	0.998
0.00080	(0.00785)	5.249	3.577	2.656	1.973	1.547	1.054	*	1.080	1.049	1.029	1.015	0.998	0.987	0.979
0.00100	(0.00981)	4.695	3.199	2.376	1.764	1.384	0.943	*	1.058	1.029	1.011	0.998	0.982	0.972	0.965
0.00150	(0.0147)	3.833	2.612	1.940	1.441	1.130	*	1.040	1.021	0.995	0.979	0.968	0.954	0.946	0.940
0.00200	(0.0196)	3.320	2.262	1.680	1.248	0.978	*	1.014	0.996	0.972	0.958	0.948	0.936	0.929	0.924
0.00300	(0.0294)	2.711	1.847	1.372	1.019	0.799	1.001	0.978	0.963	0.942	0.930	0.921	0.911	0.905	0.901
0.00400	(0.0392)	2.347	1.600	1.188	0.882	*	0.975	0.955	0.940	0.922	0.911	0.903	0.894	0.890	0.886
0.00600	(0.0588)	1.917	1.306	0.970	*	*	0.941	0.923	0.911	0.895	0.886	0.879	0.872	0.868	0.866
0.00800	(0.0785)	1.660	1.131	0.840	*	0.945	0.919	0.903	0.891	0.877	0.869	0.863	0.857	0.854	0.852
0.01000	(0.0981)	1.485	1.012	0.751	*	0.926	0.902	0.887	0.877	0.864	0.856	0.851	0.846	0.843	0.842
0.01500	(0.147)	1.212	0.826	*	0.911	0.895	0.873	0.860	0.852	0.841	0.834	0.830	0.826	0.825	0.825
0.02000	(0.196)	1.050	*	*	0.888	0.873	0.854	0.843	0.835	0.825	0.820	0.816	0.813	0.812	0.813
0.03000	(0.294)	0.857	*	0.877	0.858	0.845	0.829	0.819	0.812	0.804	0.800	0.797	0.796	0.796	0.797
0.04000	(0.392)	0.742	*	0.855	0.838	0.827	0.812	0.803	0.797	0.790	0.787	0.785	0.784	0.784	0.786
0.06000	(0.588)	*	0.845	0.827	0.812	0.802	0.789	0.782	0.777	0.772	0.769	0.768	0.768	0.769	0.772
0.08000	(0.785)	*	0.824	0.808	0.795	0.786	0.774	0.768	0.764	0.759	0.757	0.757	0.757	0.759	0.762
0.10000	(0.981)	*	0.809	0.794	0.782	0.773	0.763	0.757	0.754	0.750	0.749	0.748	0.750	0.752	0.755
0.15000	(1.47)	0.804	0.783	0.770	0.759	0.752	0.744	0.739	0.736	0.734	0.733	0.734	0.736	0.739	0.743
0.20000	(1.96)	0.785	0.766	0.754	0.745	0.738	0.731	0.727	0.725	0.723	0.723	0.724	0.727	0.730	0.735
0.30000	(2.94)	0.759	0.743	0.733	0.725	0.720	0.714	0.711	0.710	0.709	0.710	0.712	0.715	0.719	0.724
0.40000	(3.92)	0.743	0.728	0.719	0.712	0.708	0.703	0.701	0.700	0.700	0.701	0.703	0.708	0.712	0.717
0.60000	(5.88)	0.721	0.708	0.701	0.695	0.691	0.688	0.687	0.686	0.688	0.690	0.692	0.697	0.702	0.708
0.80000	(7.85)	0.706	0.695	0.688	0.684	0.681	0.678	0.677	0.678	0.680	0.682	0.685	0.691	0.696	0.702
1.00000	(9.81)	0.695	0.685	0.680	0.675	0.673	0.671	0.671	0.671	0.674	0.677	0.680	0.686	0.692	0.698
1.50000	(14.7)	0.677	0.669	0.664	0.661	0.659	0.659	0.659	0.660	0.664	0.667	0.671	0.678	0.684	0.691
2.00000	(19.6)	0.665	0.658	0.654	0.652	0.651	0.651	0.652	0.653	0.657	0.661	0.665	0.673	0.679	0.687
3.00000	(29.4)	0.650	0.644	0.641	0.640	0.639	0.640	0.642	0.644	0.649	0.653	0.658	0.666	0.673	0.681
4.00000	(39.2)	0.640	0.635	0.633	0.632	0.632	0.633	0.636	0.638	0.643	0.648	0.653	0.662	0.669	0.677
6.00000	(58.8)	0.626	0.623	0.622	0.621	0.622	0.625	0.628	0.631	0.637	0.642	0.647	0.656	0.664	0.673
8.00000	(78.5)	0.618	0.615	0.614	0.615	0.616	0.619	0.622	0.626	0.632	0.638	0.643	0.653	0.661	0.670
10.0000	(98.1)	0.611	0.609	0.609	0.610	0.611	0.615	0.619	0.622	0.629	0.635	0.641	0.651	0.659	0.668
15.0000	(147)	0.601	0.600	0.601	0.602	0.604	0.608	0.613	0.617	0.624	0.631	0.636	0.647	0.655	0.665
20.0000	(196)	0.594	0.594	0.595	0.597	0.599	0.604	0.609	0.613	0.621	0.628	0.634	0.644	0.653	0.663
30.0000	(294)	0.586	0.586	0.588	0.591	0.593	0.599	0.604	0.609	0.617	0.624	0.630	0.641	0.650	0.660
S	$(\Delta p/\rho l)$	6	8	10	12.5	15	20	25	30	40	50	60	80	100	125
Grad'nt	k.p.g.						(Equivalent) Pipe diameters in mm								

Grad'nt S	k.p.g. $(\Delta p/\rho l)$	125	150	200	250	300	400	500	600	800	1000	1250	1500	2000	2500
								(Equivalent) Pipe diameters in mm							
0.00010	(0.00098)	1.135	1.122	1.106	1.097	1.091	1.084	1.081	1.080	1.080	1.082	1.085	1.089	1.096	1.103
0.00015	(0.00147)	1.101	1.090	1.076	1.068	1.064	1.059	1.057	1.056	1.058	1.060	1.064	1.068	1.076	1.084
0.00020	(0.00196)	1.077	1.068	1.056	1.049	1.045	1.041	1.040	1.040	1.042	1.045	1.050	1.054	1.063	1.071
0.00030	(0.00294)	1.046	1.038	1.028	1.023	1.020	1.017	1.017	1.018	1.021	1.025	1.030	1.035	1.044	1.053
0.00040	(0.00392)	1.026	1.018	1.010	1.005	1.003	1.001	1.002	1.003	1.007	1.011	1.017	1.022	1.032	1.040
0.00060	(0.00588)	0.998	0.992	0.985	0.981	0.980	0.979	0.981	0.983	0.987	0.992	0.998	1.004	1.014	1.024
0.00080	(0.00785)	0.979	0.974	0.968	0.965	0.964	0.965	0.966	0.969	0.974	0.980	0.986	0.992	1.003	1.012
0.00100	(0.00981)	0.965	0.960	0.955	0.953	0.953	0.954	0.956	0.958	0.964	0.970	0.977	0.983	0.994	1.004
0.00150	(0.0147)	0.940	0.937	0.933	0.932	0.932	0.934	0.937	0.940	0.947	0.953	0.960	0.967	0.979	0.989
0.00200	(0.0196)	0.924	0.921	0.918	0.918	0.918	0.921	0.924	0.928	0.935	0.942	0.949	0.956	0.968	0.979
0.00300	(0.0294)	0.901	0.899	0.898	0.898	0.900	0.903	0.907	0.911	0.919	0.926	0.934	0.942	0.954	0.965
0.00400	(0.0392)	0.886	0.885	0.884	0.885	0.887	0.891	0.896	0.900	0.908	0.916	0.924	0.932	0.945	0.956
0.00600	(0.0588)	0.866	0.865	0.866	0.868	0.870	0.875	0.880	0.885	0.894	0.902	0.911	0.919	0.932	0.944
0.00800	(0.0785)	0.852	0.852	0.853	0.856	0.858	0.864	0.869	0.875	0.884	0.892	0.902	0.910	0.924	0.936
0.01000	(0.0981)	0.842	0.842	0.844	0.847	0.850	0.856	0.862	0.867	0.877	0.885	0.895	0.903	0.918	0.930
0.01500	(0.147)	0.825	0.825	0.828	0.831	0.835	0.842	0.848	0.854	0.864	0.873	0.883	0.892	0.907	0.920
0.02000	(0.196)	0.813	0.814	0.817	0.821	0.825	0.832	0.839	0.845	0.856	0.865	0.875	0.885	0.900	0.913
0.03000	(0.294)	0.797	0.798	0.803	0.807	0.811	0.819	0.826	0.833	0.845	0.854	0.865	0.875	0.891	0.904
0.04000	(0.392)	0.786	0.788	0.793	0.798	0.802	0.811	0.818	0.825	0.837	0.847	0.858	0.868	0.884	0.898
0.06000	(0.588)	0.772	0.774	0.780	0.785	0.790	0.800	0.808	0.815	0.827	0.838	0.850	0.860	0.877	0.891
0.08000	(0.785)	0.762	0.765	0.771	0.777	0.782	0.792	0.801	0.808	0.821	0.832	0.844	0.854	0.871	0.886
0.10000	(0.981)	0.755	0.758	0.765	0.771	0.777	0.787	0.795	0.803	0.816	0.828	0.840	0.850	0.868	0.882
0.15000	(1.47)	0.743	0.747	0.754	0.761	0.767	0.777	0.786	0.795	0.809	0.820	0.833	0.843	0.862	0.877
0.20000	(1.96)	0.735	0.739	0.747	0.754	0.760	0.771	0.781	0.789	0.803	0.815	0.828	0.839	0.858	0.873
0.30000	(2.94)	0.724	0.729	0.737	0.745	0.752	0.763	0.773	0.782	0.797	0.809	0.822	0.834	0.853	0.868
0.40000	(3.92)	0.717	0.722	0.731	0.739	0.746	0.758	0.769	0.778	0.793	0.805	0.819	0.830	0.850	0.865
0.60000	(5.88)	0.708	0.714	0.723	0.732	0.739	0.752	0.762	0.772	0.788	0.801	0.814	0.826	0.846	0.862
0.80000	(7.85)	0.702	0.708	0.718	0.727	0.735	0.748	0.759	0.768	0.784	0.798	0.812	0.823	0.843	0.860
1.00000	(9.81)	0.698	0.704	0.714	0.723	0.731	0.745	0.756	0.766	0.782	0.795	0.809	0.822	0.842	0.858
1.50000	(14.7)	0.691	0.697	0.708	0.718	0.726	0.740	0.751	0.761	0.778	0.792	0.806	0.819	0.839	0.856
2.00000	(19.6)	0.687	0.693	0.704	0.714	0.723	0.737	0.749	0.759	0.776	0.790	0.804	0.817	0.837	0.854
3.00000	(29.4)	0.681	0.688	0.700	0.710	0.718	0.733	0.745	0.755	0.773	0.787	0.802	0.814	0.835	0.852
S	$(\Delta p/\rho l)$	125	150	200	250	300	400	500	600	800	1000	1250	1500	2000	2500
Grad'nt	k.p.g.						(Equivalent) Pipe diameters in mm								

Kinematic viscosity, $\nu = 1.25 \times 10^{-6}$ m^{2}s^{-1}; $\qquad$ **Roughness size, $k_s = 0.006$ mm**

Kin. visc., $\nu = 1{\cdot}25\times10^{-6}\ \mathrm{m^2\,s^{-1}}$;
$S = 0{\cdot}00010$ to $30{\cdot}0000$

i.e. kin. pr. grad., $\Delta p/\rho l =$
$(0{\cdot}00098)$ to $(294)\ \mathrm{ms^{-2}}$

Roughness size, $k_s = 0{\cdot}015$ mm
This table shows values of m, as follows

m_C for Colebrook-White solutions; or,
where $\mathbf{R} \le 2000$, m_P for laminar flow

Grad'nt S	k.p.g. $(\Delta p/\rho l)$	(Equivalent) Pipe diameters in mm													
		6	8	10	12·5	15	20	25	30	40	50	60	80	100	125
0·00030	(0·00294)	8·572	5·841	4·338	3·222	2·526	1·721	1·278	1·003	*	1·120	1·101	1·077	1·062	1·050
0·00040	(0·00392)	7·423	5·058	3·757	2·790	2·188	1·491	1·107	*	1·118	1·093	1·076	1·054	1·040	1·029
0·00060	(0·00588)	6·061	4·130	3·067	2·278	1·786	1·217	0·904	*	1·078	1·057	1·042	1·022	1·011	1·002
0·00080	(0·00785)	5·249	3·577	2·656	1·973	1·547	1·054	*	1·084	1·052	1·032	1·019	1·002	0·991	0·983
0·00100	(0·00981)	4·695	3·199	2·376	1·764	1·384	0·943	*	1·062	1·033	1·014	1·002	0·986	0·977	0·969
0·00150	(0·0147)	3·833	2·612	1·940	1·441	1·130	*	1·044	1·025	0·999	0·983	0·973	0·959	0·951	0·946
0·00200	(0·0196)	3·320	2·262	1·680	1·248	0·978	*	1·018	1·000	0·977	0·963	0·953	0·941	0·934	0·930
0·00300	(0·0294)	2·711	1·847	1·372	1·019	0·799	1·006	0·983	0·968	0·948	0·935	0·927	0·917	0·912	0·908
0·00400	(0·0392)	2·347	1·600	1·188	0·882	*	0·981	0·960	0·946	0·928	0·917	0·910	0·901	0·897	0·894
0·00600	(0·0588)	1·917	1·306	0·970	*	*	0·947	0·930	0·917	0·902	0·893	0·887	0·880	0·877	0·875
0·00800	(0·0785)	1·660	1·131	0·840	*	0·951	0·925	0·909	0·898	0·885	0·876	0·871	0·866	0·863	0·862
0·01000	(0·0981)	1·485	1·012	0·751	*	0·933	0·909	0·894	0·884	0·872	0·864	0·860	0·855	0·853	0·853
0·01500	(0·147)	1·212	0·826	*	0·918	0·902	0·881	0·869	0·860	0·850	0·844	0·840	0·837	0·836	0·837
0·02000	(0·196)	1·050	*	*	0·896	0·882	0·863	0·852	0·844	0·835	0·830	0·827	0·825	0·825	0·826
0·03000	(0·294)	0·857	*	0·886	0·868	0·855	0·839	0·829	0·823	0·815	0·812	0·810	0·809	0·810	0·812
0·04000	(0·392)	0·742	*	0·865	0·848	0·837	0·823	0·814	0·809	0·803	0·800	0·798	0·798	0·800	0·802
0·06000	(0·588)	*	0·855	0·838	0·823	0·814	0·802	0·795	0·790	0·786	0·784	0·783	0·785	0·787	0·790
0·08000	(0·785)	*	0·836	0·820	0·807	0·798	0·788	0·782	0·778	0·775	0·773	0·774	0·775	0·778	0·782
0·10000	(0·981)	*	0·821	0·807	0·795	0·787	0·777	0·772	0·769	0·766	0·766	0·766	0·769	0·772	0·776
0·15000	(1·47)	0·817	0·797	0·784	0·774	0·768	0·760	0·756	0·754	0·753	0·753	0·754	0·758	0·762	0·767
0·20000	(1·96)	0·799	0·781	0·770	0·761	0·755	0·749	0·745	0·744	0·743	0·745	0·746	0·750	0·755	0·760
0·30000	(2·94)	0·776	0·760	0·751	0·743	0·739	0·734	0·732	0·731	0·732	0·734	0·736	0·741	0·746	0·752
0·40000	(3·92)	0·761	0·747	0·738	0·732	0·728	0·724	0·723	0·723	0·724	0·727	0·730	0·735	0·741	0·747
0·60000	(5·88)	0·741	0·729	0·722	0·717	0·714	0·712	0·711	0·712	0·715	0·718	0·721	0·728	0·734	0·741
0·80000	(7·85)	0·728	0·718	0·712	0·708	0·705	0·704	0·704	0·705	0·708	0·712	0·716	0·723	0·730	0·737
1·00000	(9·81)	0·719	0·709	0·704	0·701	0·699	0·698	0·699	0·700	0·704	0·708	0·712	0·720	0·726	0·734
1·50000	(14·7)	0·703	0·695	0·691	0·689	0·688	0·688	0·690	0·692	0·697	0·701	0·706	0·714	0·721	0·730
2·00000	(19·6)	0·693	0·686	0·683	0·682	0·681	0·682	0·684	0·687	0·692	0·697	0·702	0·711	0·718	0·727
3·00000	(29·4)	0·680	0·675	0·673	0·672	0·673	0·675	0·677	0·680	0·686	0·692	0·697	0·706	0·714	0·723
4·00000	(39·2)	0·672	0·668	0·667	0·666	0·667	0·670	0·673	0·676	0·683	0·689	0·694	0·704	0·712	0·721
6·00000	(58·8)	0·661	0·659	0·658	0·659	0·660	0·664	0·668	0·671	0·678	0·685	0·690	0·701	0·709	0·718
8·00000	(78·5)	0·655	0·653	0·653	0·654	0·656	0·660	0·664	0·668	0·676	0·682	0·688	0·699	0·707	0·717
10·0000	(98·1)	0·650	0·649	0·650	0·651	0·653	0·657	0·662	0·666	0·674	0·681	0·687	0·697	0·706	0·716
15·0000	(147)	0·643	0·643	0·644	0·646	0·648	0·653	0·658	0·663	0·671	0·678	0·684	0·695	0·704	0·714
20·0000	(196)	0·638	0·638	0·640	0·642	0·645	0·651	0·656	0·660	0·669	0·676	0·683	0·694	0·703	0·713
30·0000	(294)	0·632	0·633	0·635	0·638	0·641	0·647	0·653	0·658	0·667	0·674	0·681	0·692	0·701	0·712
S	$(\Delta p/\rho l)$	6	8	10	12·5	15	20	25	30	40	50	60	80	100	125

Grad'nt S	k.p.g. $(\Delta p/\rho l)$	(Equivalent) Pipe diameters in mm													
		125	150	200	250	300	400	500	600	800	1000	1250	1500	2000	2500
0·00010	(0·00098)	1·137	1·124	1·109	1·099	1·094	1·087	1·084	1·083	1·084	1·086	1·090	1·093	1·101	1·108
0·00015	(0·00147)	1·103	1·092	1·079	1·071	1·067	1·062	1·060	1·060	1·062	1·065	1·069	1·073	1·082	1·090
0·00020	(0·00196)	1·080	1·070	1·059	1·052	1·048	1·045	1·044	1·044	1·047	1·050	1·055	1·060	1·069	1·077
0·00030	(0·00294)	1·050	1·041	1·032	1·027	1·024	1·021	1·022	1·023	1·026	1·031	1·036	1·041	1·051	1·060
0·00040	(0·00392)	1·029	1·022	1·014	1·009	1·007	1·006	1·007	1·008	1·013	1·017	1·023	1·029	1·040	1·049
0·00060	(0·00588)	1·002	0·996	0·989	0·986	0·985	0·985	0·986	0·989	0·994	1·000	1·006	1·013	1·024	1·034
0·00080	(0·00785)	0·983	0·978	0·973	0·970	0·970	0·971	0·973	0·976	0·982	0·988	0·995	1·001	1·013	1·024
0·00100	(0·00981)	0·969	0·965	0·960	0·959	0·959	0·960	0·963	0·966	0·972	0·979	0·986	0·993	1·005	1·016
0·00150	(0·0147)	0·946	0·942	0·939	0·939	0·939	0·942	0·945	0·949	0·956	0·963	0·971	0·979	0·992	1·003
0·00200	(0·0196)	0·930	0·927	0·925	0·925	0·926	0·929	0·933	0·938	0·946	0·953	0·961	0·969	0·983	0·994
0·00300	(0·0294)	0·908	0·907	0·906	0·907	0·908	0·913	0·918	0·922	0·931	0·939	0·948	0·957	0·971	0·983
0·00400	(0·0392)	0·894	0·893	0·893	0·894	0·897	0·902	0·907	0·912	0·922	0·930	0·940	0·948	0·963	0·975
0·00600	(0·0588)	0·875	0·875	0·876	0·878	0·881	0·887	0·893	0·899	0·909	0·918	0·928	0·937	0·953	0·966
0·00800	(0·0785)	0·862	0·862	0·864	0·867	0·871	0·877	0·884	0·890	0·901	0·910	0·921	0·930	0·946	0·959
0·01000	(0·0981)	0·853	0·853	0·856	0·859	0·863	0·870	0·877	0·883	0·895	0·904	0·915	0·925	0·941	0·955
0·01500	(0·147)	0·837	0·838	0·842	0·846	0·850	0·858	0·866	0·872	0·884	0·895	0·906	0·916	0·933	0·947
0·02000	(0·196)	0·826	0·828	0·832	0·837	0·841	0·850	0·858	0·865	0·878	0·888	0·900	0·910	0·928	0·942
0·03000	(0·294)	0·812	0·814	0·820	0·825	0·830	0·840	0·848	0·856	0·869	0·880	0·892	0·903	0·921	0·936
0·04000	(0·392)	0·802	0·805	0·811	0·817	0·823	0·833	0·842	0·850	0·863	0·875	0·888	0·899	0·917	0·932
0·06000	(0·588)	0·790	0·794	0·801	0·807	0·813	0·824	0·834	0·842	0·856	0·868	0·881	0·893	0·912	0·927
0·08000	(0·785)	0·782	0·786	0·794	0·801	0·807	0·818	0·828	0·837	0·852	0·864	0·878	0·889	0·908	0·924
0·10000	(0·981)	0·776	0·781	0·789	0·796	0·803	0·814	0·825	0·833	0·849	0·861	0·875	0·887	0·906	0·922
0·15000	(1·47)	0·767	0·771	0·780	0·788	0·795	0·808	0·818	0·828	0·843	0·856	0·870	0·882	0·902	0·919
0·20000	(1·96)	0·760	0·766	0·775	0·783	0·791	0·804	0·814	0·824	0·840	0·853	0·868	0·880	0·900	0·917
0·30000	(2·94)	0·752	0·758	0·768	0·777	0·785	0·798	0·810	0·819	0·836	0·850	0·864	0·877	0·897	0·914
0·40000	(3·92)	0·747	0·753	0·764	0·773	0·781	0·795	0·807	0·817	0·834	0·847	0·862	0·875	0·895	0·912
0·60000	(5·88)	0·741	0·747	0·759	0·768	0·777	0·791	0·803	0·813	0·830	0·845	0·860	0·872	0·893	0·910
0·80000	(7·85)	0·737	0·744	0·755	0·765	0·774	0·788	0·801	0·811	0·829	0·843	0·858	0·871	0·892	0·909
1·00000	(9·81)	0·734	0·741	0·753	0·763	0·772	0·787	0·799	0·810	0·827	0·842	0·857	0·870	0·891	0·908
1·50000	(14·7)	0·730	0·737	0·749	0·760	0·769	0·784	0·796	0·807	0·825	0·840	0·855	0·868	0·890	0·907
2·00000	(19·6)	0·727	0·734	0·747	0·757	0·767	0·782	0·795	0·806	0·824	0·839	0·854	0·867	0·889	0·906
3·00000	(29·4)	0·723	0·731	0·744	0·755	0·764	0·780	0·793	0·804	0·822	0·837	0·853	0·866	0·888	0·906
S	$(\Delta p/\rho l)$	125	150	200	250	300	400	500	600	800	1000	1250	1500	2000	2500

Grad'nt   k.p.g.   (Equivalent) Pipe diameters in mm

Kinematic viscosity, $\nu = 1{\cdot}25\times10^{-6}\ \mathrm{m^2\,s^{-1}}$;   **Roughness size,** $k_s = 0{\cdot}015$ mm

E105

Kin. visc., $\nu = 1.25\times10^{-6}$ m^{2}s^{-1}; $S = 0.00010$ to 30.0000

i.e. kin. pr. grad., $\Delta p/\rho l =$ (0.00098) to (294) ms^{-2}

Roughness size, $k_s = 0.030$ mm

This table shows values of m, as follows

m_C for Colebrook-White solutions; or, where $\mathbf{R} \leq 2000$, m_P for laminar flow

Grad'nt S	k.p.g. $(\Delta p/\rho l)$	6	8	10	12·5	15	20	25	30	40	50	60	80	100	125
0·00030	(0·00294)	8·572	5·841	4·338	3·222	2·526	1·721	1·278	1·003	*	1·125	1·106	1·082	1·067	1·055
0·00040	(0·00392)	7·423	5·058	3·757	2·790	2·188	1·491	1·107	*	1·122	1·098	1·081	1·059	1·045	1·035
0·00060	(0·00588)	6·061	4·130	3·067	2·278	1·786	1·217	0·904	*	1·084	1·062	1·047	1·028	1·017	1·008
0·00080	(0·00785)	5·249	3·577	2·656	1·973	1·547	1·054	*	*	1·058	1·038	1·025	1·008	0·998	0·990
0·00100	(0·00981)	4·695	3·199	2·376	1·764	1·384	0·943	*	1·068	1·039	1·021	1·008	0·993	0·984	0·977
0·00150	(0·0147)	3·833	2·612	1·940	1·441	1·130	*	1·051	1·031	1·006	0·991	0·980	0·967	0·960	0·954
0·00200	(0·0196)	3·320	2·262	1·680	1·248	0·978	*	1·025	1·007	0·985	0·971	0·961	0·950	0·944	0·939
0·00300	(0·0294)	2·711	1·847	1·372	1·019	0·799	1·014	0·991	0·976	0·956	0·944	0·936	0·927	0·922	0·919
0·00400	(0·0392)	2·347	1·600	1·188	0·882	*	0·989	0·969	0·955	0·937	0·927	0·920	0·912	0·908	0·906
0·00600	(0·0588)	1·917	1·306	0·970	*	*	0·957	0·939	0·928	0·913	0·904	0·898	0·892	0·889	0·888
0·00800	(0·0785)	1·660	1·131	0·840	*	0·961	0·936	0·920	0·909	0·896	0·889	0·884	0·879	0·877	0·877
0·01000	(0·0981)	1·485	1·012	0·751	*	0·944	0·920	0·906	0·896	0·884	0·877	0·873	0·869	0·868	0·869
0·01500	(0·147)	1·212	0·826	*	0·930	0·914	0·894	0·882	0·873	0·864	0·858	0·855	0·853	0·853	0·854
0·02000	(0·196)	1·050	*	*	0·909	0·895	0·877	0·866	0·859	0·850	0·846	0·844	0·842	0·843	0·845
0·03000	(0·294)	0·857	*	0·900	0·882	0·870	0·854	0·845	0·839	0·833	0·830	0·828	0·828	0·830	0·833
0·04000	(0·392)	0·742	*	0·881	0·864	0·853	0·839	0·831	0·827	0·821	0·819	0·818	0·819	0·822	0·825
0·06000	(0·588)	*	0·873	0·855	0·841	0·832	0·820	0·814	0·810	0·806	0·805	0·806	0·808	0·811	0·815
0·08000	(0·785)	*	0·854	0·839	0·826	0·818	0·808	0·802	0·799	0·797	0·797	0·797	0·800	0·804	0·809
0·10000	(0·981)	*	0·841	0·827	0·815	0·808	0·799	0·794	0·792	0·790	0·790	0·792	0·795	0·799	0·805
0·15000	(1·47)	0·839	0·819	0·806	0·797	0·791	0·784	0·780	0·779	0·778	0·780	0·782	0·786	0·791	0·797
0·20000	(1·96)	0·822	0·804	0·793	0·785	0·780	0·774	0·772	0·771	0·771	0·773	0·775	0·781	0·786	0·792
0·30000	(2·94)	0·801	0·786	0·777	0·770	0·766	0·762	0·760	0·760	0·762	0·765	0·768	0·774	0·780	0·787
0·40000	(3·92)	0·788	0·774	0·766	0·760	0·757	0·754	0·753	0·754	0·756	0·759	0·763	0·769	0·776	0·783
0·60000	(5·88)	0·770	0·759	0·752	0·748	0·746	0·744	0·744	0·745	0·749	0·753	0·757	0·764	0·771	0·779
0·80000	(7·85)	0·759	0·749	0·744	0·740	0·738	0·738	0·738	0·740	0·744	0·748	0·753	0·761	0·768	0·776
1·00000	(9·81)	0·751	0·742	0·738	0·735	0·733	0·733	0·734	0·736	0·741	0·746	0·750	0·758	0·766	0·774
1·50000	(14·7)	0·738	0·731	0·728	0·726	0·725	0·726	0·728	0·730	0·736	0·741	0·746	0·755	0·762	0·771
2·00000	(19·6)	0·730	0·724	0·721	0·720	0·720	0·721	0·724	0·727	0·732	0·738	0·743	0·752	0·760	0·769
3·00000	(29·4)	0·720	0·715	0·714	0·713	0·714	0·716	0·719	0·722	0·729	0·734	0·740	0·750	0·758	0·767
4·00000	(39·2)	0·714	0·710	0·709	0·709	0·710	0·713	0·716	0·719	0·726	0·732	0·738	0·748	0·756	0·766
6·00000	(58·8)	0·706	0·703	0·703	0·703	0·705	0·708	0·712	0·716	0·723	0·730	0·736	0·746	0·754	0·764
8·00000	(78·5)	0·701	0·699	0·699	0·700	0·702	0·706	0·710	0·714	0·721	0·728	0·734	0·745	0·753	0·763
10·0000	(98·1)	0·698	0·696	0·696	0·698	0·700	0·704	0·708	0·713	0·720	0·727	0·733	0·744	0·753	0·762
15·0000	(147)	0·692	0·692	0·692	0·694	0·696	0·701	0·706	0·710	0·718	0·725	0·732	0·742	0·751	0·761
20·0000	(196)	0·689	0·689	0·690	0·692	0·694	0·700	0·704	0·709	0·717	0·724	0·731	0·742	0·751	0·761
30·0000	(294)	0·685	0·685	0·687	0·689	0·692	0·697	0·703	0·707	0·716	0·723	0·729	0·741	0·750	0·760
S Grad'nt	$(\Delta p/\rho l)$ k.p.g.	6	8	10	12·5	15	20	25	30	40	50	60	80	100	125

(Equivalent) Pipe diameters in mm

Grad'nt S	k.p.g. $(\Delta p/\rho l)$	125	150	200	250	300	400	500	600	800	1000	1250	1500	2000	2500
0·00010	(0·00098)	1·140	1·128	1·113	1·104	1·098	1·092	1·089	1·089	1·090	1·092	1·096	1·100	1·109	1·117
0·00015	(0·00147)	1·107	1·096	1·083	1·076	1·071	1·067	1·066	1·066	1·068	1·072	1·076	1·081	1·091	1·099
0·00020	(0·00196)	1·084	1·075	1·064	1·057	1·054	1·050	1·050	1·051	1·054	1·058	1·063	1·069	1·079	1·088
0·00030	(0·00294)	1·055	1·047	1·037	1·032	1·030	1·028	1·029	1·030	1·035	1·040	1·046	1·052	1·063	1·072
0·00040	(0·00392)	1·035	1·028	1·020	1·016	1·014	1·013	1·015	1·017	1·022	1·027	1·034	1·040	1·052	1·062
0·00060	(0·00588)	1·008	1·002	0·996	0·994	0·993	0·993	0·996	0·999	1·005	1·011	1·018	1·025	1·038	1·049
0·00080	(0·00785)	0·990	0·985	0·981	0·979	0·979	0·980	0·983	0·986	0·993	1·000	1·008	1·015	1·029	1·040
0·00100	(0·00981)	0·977	0·973	0·969	0·968	0·968	0·970	0·974	0·977	0·985	0·992	1·001	1·008	1·022	1·033
0·00150	(0·0147)	0·954	0·951	0·949	0·949	0·950	0·953	0·958	0·962	0·971	0·979	0·988	0·996	1·010	1·023
0·00200	(0·0196)	0·939	0·937	0·936	0·936	0·938	0·942	0·947	0·952	0·961	0·970	0·979	0·988	1·003	1·015
0·00300	(0·0294)	0·919	0·918	0·918	0·920	0·922	0·927	0·933	0·939	0·949	0·958	0·968	0·977	0·993	1·006
0·00400	(0·0392)	0·906	0·905	0·906	0·909	0·912	0·918	0·924	0·930	0·941	0·950	0·961	0·970	0·987	1·001
0·00600	(0·0588)	0·888	0·889	0·891	0·894	0·898	0·905	0·912	0·919	0·930	0·941	0·952	0·962	0·979	0·993
0·00800	(0·0785)	0·877	0·878	0·881	0·885	0·889	0·897	0·904	0·911	0·924	0·934	0·946	0·956	0·974	0·988
0·01000	(0·0981)	0·869	0·870	0·874	0·878	0·882	0·891	0·899	0·906	0·919	0·930	0·942	0·952	0·970	0·985
0·01500	(0·147)	0·854	0·856	0·861	0·866	0·871	0·881	0·889	0·897	0·911	0·922	0·935	0·946	0·964	0·980
0·02000	(0·196)	0·845	0·848	0·853	0·859	0·864	0·875	0·884	0·892	0·906	0·918	0·930	0·942	0·961	0·976
0·03000	(0·294)	0·833	0·836	0·843	0·849	0·855	0·866	0·876	0·884	0·899	0·912	0·925	0·937	0·956	0·972
0·04000	(0·392)	0·825	0·829	0·836	0·843	0·850	0·861	0·871	0·880	0·895	0·908	0·922	0·933	0·953	0·969
0·06000	(0·588)	0·815	0·820	0·828	0·836	0·843	0·855	0·865	0·874	0·890	0·903	0·917	0·930	0·950	0·966
0·08000	(0·785)	0·809	0·814	0·823	0·831	0·838	0·851	0·861	0·871	0·887	0·901	0·915	0·927	0·948	0·964
0·10000	(0·981)	0·805	0·810	0·819	0·827	0·835	0·848	0·859	0·869	0·885	0·899	0·913	0·925	0·946	0·963
0·15000	(1·47)	0·797	0·803	0·813	0·822	0·830	0·843	0·855	0·865	0·881	0·895	0·910	0·923	0·944	0·961
0·20000	(1·96)	0·792	0·798	0·809	0·818	0·826	0·840	0·852	0·862	0·879	0·893	0·908	0·921	0·942	0·960
0·30000	(2·94)	0·787	0·793	0·804	0·814	0·822	0·837	0·849	0·859	0·877	0·891	0·906	0·919	0·941	0·958
0·40000	(3·92)	0·783	0·790	0·801	0·811	0·820	0·835	0·847	0·857	0·875	0·890	0·905	0·918	0·940	0·957
0·60000	(5·88)	0·779	0·786	0·798	0·808	0·817	0·832	0·844	0·855	0·873	0·888	0·903	0·917	0·938	0·956
0·80000	(7·85)	0·776	0·783	0·795	0·806	0·815	0·830	0·843	0·854	0·872	0·887	0·902	0·916	0·938	0·956
1·00000	(9·81)	0·774	0·781	0·794	0·805	0·814	0·829	0·842	0·853	0·871	0·886	0·902	0·915	0·937	0·955
1·50000	(14·7)	0·771	0·779	0·792	0·802	0·812	0·827	0·840	0·852	0·870	0·885	0·901	0·914	0·937	0·955
2·00000	(19·6)	0·769	0·777	0·790	0·801	0·811	0·826	0·840	0·851	0·869	0·884	0·900	0·914	0·936	0·954
3·00000	(29·4)	0·767	0·775	0·788	0·799	0·809	0·825	0·838	0·850	0·868	0·884	0·899	0·913	0·935	0·953
S Grad'nt	$(\Delta p/\rho l)$ k.p.g.	125	150	200	250	300	400	500	600	800	1000	1250	1500	2000	2500

(Equivalent) Pipe diameters in mm

Kinematic viscosity, $\nu = 1.25\times10^{-6}$ m^{2}s^{-1};

Roughness size, $k_s = 0.030$ mm

Kin. visc., $v = 1.25 \times 10^{-6}$ m^{2}s^{-1}; Roughness size, $k_s = 0.060$ mm # E106

$S = 0.00010$ to 30.0000

This table shows values of m, as follows

i.e. kin. pr. grad., $\Delta p/\rho l =$ (0.00098) to (294) ms^{-2}

m_C for Colebrook-White solutions; or, where $\mathbf{R} \leq 2000$, m_P for laminar flow

| Grad'nt | k.p.g. | (Equivalent) Pipe diameters in mm | | | | | | | | | | | | | |
S	($\Delta p/\rho l$)	6	8	10	12·5	15	20	25	30	40	50	60	80	100	125
0.00030	(0.00294)	8·572	5·841	4·338	3·222	2·526	1·721	1·278	1·003	*	1·133	1·115	1·091	1·076	1·064
0.00040	(0.00392)	7·423	5·058	3·757	2·790	2·188	1·491	1·107	*	1·131	1·107	1·090	1·068	1·055	1·045
0.00060	(0.00588)	6·061	4·130	3·067	2·278	1·786	1·217	0·904	*	1·094	1·072	1·058	1·039	1·028	1·020
0.00080	(0.00785)	5·249	3·577	2·656	1·973	1·547	1·054	*	*	1·069	1·050	1·036	1·020	1·011	1·003
0.00100	(0.00981)	4·695	3·199	2·376	1·764	1·384	0·943	*	1·079	1·051	1·033	1·021	1·006	0·997	0·991
0.00150	(0.0147)	3·833	2·612	1·940	1·441	1·130	*	1·064	1·044	1·019	1·004	0·994	0·982	0·975	0·970
0.00200	(0.0196)	3·320	2·262	1·680	1·248	0·978	*	1·039	1·021	0·999	0·986	0·977	0·966	0·960	0·957
0.00300	(0.0294)	2·711	1·847	1·372	1·019	0·799	1·029	1·007	0·992	0·972	0·961	0·954	0·945	0·941	0·939
0.00400	(0.0392)	2·347	1·600	1·188	0·882	*	1·005	0·986	0·972	0·955	0·945	0·939	0·932	0·929	0·927
0.00600	(0.0588)	1·917	1·306	0·970	*	*	0·975	0·958	0·947	0·932	0·924	0·919	0·914	0·912	0·912
0.00800	(0.0785)	1·660	1·131	0·840	*	0·981	0·956	0·940	0·930	0·918	0·911	0·907	0·903	0·902	0·902
0.01000	(0.0981)	1·485	1·012	0·751	*	0·965	0·941	0·927	0·918	0·907	0·901	0·897	0·894	0·894	0·895
0.01500	(0.147)	1·212	0·826	*	0·953	0·937	0·917	0·905	0·898	0·889	0·884	0·882	0·881	0·882	0·884
0.02000	(0.196)	1·050	*	*	0·934	0·920	0·902	0·891	0·885	0·877	0·874	0·872	0·872	0·873	0·876
0.03000	(0.294)	0·857	*	0·927	0·909	0·897	0·882	0·873	0·868	0·862	0·860	0·860	0·861	0·863	0·867
0.04000	(0.392)	0·742	*	0·909	0·893	0·882	0·869	0·862	0·857	0·853	0·852	0·852	0·854	0·857	0·861
0.06000	(0.588)	*	0·904	0·887	0·873	0·864	0·853	0·847	0·844	0·841	0·841	0·842	0·845	0·849	0·854
0.08000	(0.785)	*	0·887	0·872	0·860	0·852	0·842	0·838	0·835	0·833	0·834	0·835	0·839	0·844	0·849
0.10000	(0.981)	*	0·876	0·862	0·850	0·843	0·835	0·831	0·829	0·828	0·829	0·831	0·835	0·840	0·846
0.15000	(1.47)	0·877	0·857	0·844	0·835	0·829	0·823	0·820	0·819	0·819	0·821	0·823	0·829	0·834	0·841
0.20000	(1.96)	0·863	0·844	0·834	0·825	0·820	0·815	0·813	0·813	0·814	0·816	0·819	0·825	0·831	0·838
0.30000	(2.94)	0·845	0·829	0·820	0·813	0·809	0·806	0·805	0·805	0·807	0·810	0·813	0·820	0·826	0·834
0.40000	(3.92)	0·833	0·819	0·812	0·806	0·802	0·800	0·799	0·800	0·803	0·806	0·810	0·817	0·824	0·831
0.60000	(5.88)	0·819	0·807	0·801	0·796	0·794	0·792	0·793	0·794	0·798	0·802	0·806	0·814	0·821	0·829
0.80000	(7.85)	0·810	0·800	0·794	0·791	0·789	0·788	0·789	0·790	0·794	0·799	0·803	0·811	0·819	0·827
1.00000	(9.81)	0·804	0·795	0·790	0·786	0·785	0·785	0·786	0·788	0·792	0·797	0·801	0·810	0·817	0·826
1.50000	(14.7)	0·794	0·786	0·782	0·780	0·779	0·780	0·781	0·784	0·789	0·794	0·799	0·807	0·815	0·824
2.00000	(19.6)	0·788	0·781	0·778	0·776	0·775	0·776	0·779	0·781	0·787	0·792	0·797	0·806	0·814	0·823
3.00000	(29.4)	0·781	0·775	0·772	0·771	0·771	0·773	0·775	0·778	0·784	0·790	0·795	0·804	0·812	0·821
4.00000	(39.2)	0·776	0·771	0·769	0·768	0·768	0·770	0·773	0·776	0·783	0·788	0·794	0·803	0·811	0·820
6.00000	(58.8)	0·770	0·766	0·765	0·764	0·765	0·768	0·771	0·774	0·781	0·787	0·792	0·802	0·810	0·819
8.00000	(78.5)	0·767	0·763	0·762	0·762	0·763	0·766	0·769	0·773	0·780	0·786	0·791	0·801	0·810	0·819
10.0000	(98.1)	0·765	0·761	0·760	0·761	0·762	0·765	0·768	0·772	0·779	0·785	0·791	0·801	0·809	0·818
15.0000	(147)	0·761	0·758	0·758	0·758	0·760	0·763	0·767	0·771	0·778	0·784	0·790	0·800	0·808	0·818
20.0000	(196)	0·759	0·756	0·756	0·757	0·758	0·762	0·766	0·770	0·777	0·783	0·789	0·799	0·808	0·817
30.0000	(294)	0·756	0·754	0·754	0·755	0·757	0·761	0·765	0·769	0·776	0·783	0·788	0·799	0·807	0·817
S	($\Delta p/\rho l$)	6	8	10	12·5	15	20	25	30	40	50	60	80	100	125

| Grad'nt | k.p.g. | (Equivalent) Pipe diameters in mm | | | | | | | | | | | | | |
S	($\Delta p/\rho l$)	125	150	200	250	300	400	500	600	800	1000	1250	1500	2000	2500
0.00010	(0.00098)	1·147	1·135	1·120	1·111	1·106	1·101	1·099	1·098	1·100	1·103	1·108	1·113	1·123	1·132
0.00015	(0.00147)	1·115	1·104	1·092	1·085	1·081	1·077	1·077	1·077	1·080	1·085	1·090	1·096	1·107	1·116
0.00020	(0.00196)	1·093	1·084	1·073	1·067	1·064	1·062	1·062	1·063	1·067	1·072	1·079	1·085	1·096	1·106
0.00030	(0.00294)	1·064	1·057	1·048	1·044	1·042	1·041	1·042	1·045	1·050	1·056	1·063	1·070	1·082	1·093
0.00040	(0.00392)	1·045	1·039	1·032	1·028	1·027	1·027	1·029	1·032	1·039	1·045	1·053	1·060	1·073	1·085
0.00060	(0.00588)	1·020	1·015	1·010	1·008	1·008	1·009	1·012	1·016	1·024	1·031	1·040	1·047	1·062	1·074
0.00080	(0.00785)	1·003	0·999	0·995	0·994	0·995	0·997	1·001	1·005	1·014	1·022	1·031	1·039	1·054	1·067
0.00100	(0.00981)	0·991	0·988	0·985	0·984	0·985	0·989	0·993	0·998	1·007	1·015	1·025	1·033	1·049	1·062
0.00150	(0.0147)	0·970	0·968	0·967	0·967	0·969	0·974	0·979	0·985	0·995	1·004	1·014	1·024	1·040	1·053
0.00200	(0.0196)	0·957	0·955	0·955	0·956	0·959	0·965	0·971	0·976	0·987	0·997	1·008	1·017	1·034	1·048
0.00300	(0.0294)	0·939	0·938	0·940	0·942	0·945	0·952	0·959	0·966	0·977	0·988	0·999	1·009	1·027	1·042
0.00400	(0.0392)	0·927	0·927	0·930	0·933	0·937	0·944	0·952	0·959	0·971	0·982	0·994	1·004	1·022	1·037
0.00600	(0.0588)	0·912	0·913	0·917	0·921	0·926	0·934	0·943	0·950	0·963	0·975	0·987	0·998	1·017	1·032
0.00800	(0.0785)	0·902	0·904	0·909	0·914	0·919	0·928	0·937	0·944	0·958	0·970	0·983	0·994	1·013	1·029
0.01000	(0.0981)	0·895	0·898	0·903	0·908	0·913	0·923	0·932	0·941	0·955	0·967	0·980	0·992	1·011	1·027
0.01500	(0.147)	0·884	0·887	0·893	0·899	0·905	0·916	0·926	0·934	0·949	0·962	0·975	0·987	1·007	1·023
0.02000	(0.196)	0·876	0·880	0·887	0·894	0·900	0·911	0·921	0·930	0·946	0·959	0·972	0·984	1·005	1·021
0.03000	(0.294)	0·867	0·871	0·879	0·887	0·893	0·906	0·916	0·925	0·941	0·955	0·969	0·981	1·002	1·019
0.04000	(0.392)	0·861	0·866	0·874	0·882	0·889	0·902	0·913	0·922	0·939	0·952	0·967	0·979	1·000	1·017
0.06000	(0.588)	0·854	0·859	0·868	0·877	0·884	0·897	0·909	0·919	0·935	0·949	0·964	0·977	0·998	1·015
0.08000	(0.785)	0·849	0·855	0·865	0·873	0·881	0·895	0·906	0·916	0·933	0·948	0·962	0·975	0·996	1·014
0.10000	(0.981)	0·846	0·852	0·862	0·871	0·879	0·893	0·905	0·915	0·932	0·946	0·961	0·974	0·996	1·013
0.15000	(1.47)	0·841	0·847	0·858	0·867	0·876	0·890	0·902	0·912	0·930	0·944	0·960	0·973	0·994	1·012
0.20000	(1.96)	0·838	0·844	0·855	0·865	0·873	0·888	0·900	0·911	0·929	0·943	0·958	0·972	0·993	1·011
0.30000	(2.94)	0·834	0·840	0·852	0·862	0·871	0·886	0·898	0·909	0·927	0·942	0·957	0·970	0·992	1·010
0.40000	(3.92)	0·831	0·838	0·850	0·860	0·869	0·884	0·897	0·908	0·926	0·941	0·956	0·970	0·992	1·010
0.60000	(5.88)	0·829	0·836	0·848	0·858	0·868	0·883	0·896	0·907	0·925	0·940	0·956	0·969	0·991	1·009
0.80000	(7.85)	0·827	0·834	0·847	0·857	0·866	0·882	0·895	0·906	0·924	0·939	0·955	0·968	0·990	1·008
1.00000	(9.81)	0·826	0·833	0·846	0·856	0·866	0·881	0·894	0·905	0·924	0·939	0·955	0·968	0·990	1·008
1.50000	(14.7)	0·824	0·831	0·844	0·855	0·864	0·880	0·893	0·904	0·923	0·938	0·954	0·967	0·990	1·008
2.00000	(19.6)	0·823	0·830	0·843	0·854	0·864	0·880	0·893	0·904	0·923	0·938	0·954	0·967	0·989	1·008
3.00000	(29.4)	0·821	0·829	0·842	0·853	0·863	0·879	0·892	0·903	0·922	0·937	0·953	0·967	0·989	1·007
S	($\Delta p/\rho l$)	125	150	200	250	300	400	500	600	800	1000	1250	1500	2000	2500

Grad'nt k.p.g. (Equivalent) Pipe diameters in mm

Kinematic viscosity, $v = 1.25 \times 10^{-6}$ m^{2}s^{-1} ; **Roughness size, $k_s = 0.060$ mm**

E107

Kin. visc., $\nu = 1.25 \times 10^{-6}$ m^{2}s^{-1};
$S = 0.00010$ to 30.0000

i.e. kin. pr. grad., $\Delta p/\rho l =$
(0.00098) to (294) ms^{-2}

Roughness size, $k_s = 0.150$ mm
This table shows values of m, as follows

m_C for Colebrook-White solutions; or,
where $\mathbf{R} \leq 2000$, m_P for laminar flow

Grad'nt S	k.p.g. $(\Delta p/\rho l)$	6	8	10	12.5	15	20	25	30	40	50	60	80	100	125
0.00030	(0.00294)	8.572	5.841	4.338	3.222	2.526	1.721	1.278	1.003	*	1.157	1.139	1.116	1.102	1.091
0.00040	(0.00392)	7.423	5.058	3.757	2.790	2.188	1.491	1.107	*	1.157	1.133	1.116	1.096	1.084	1.074
0.00060	(0.00588)	6.061	4.130	3.067	2.278	1.786	1.217	0.904	*	1.122	1.101	1.087	1.070	1.060	1.053
0.00080	(0.00785)	5.249	3.577	2.656	1.973	1.547	1.054	*	*	1.100	1.081	1.068	1.053	1.045	1.039
0.00100	(0.00981)	4.695	3.199	2.376	1.764	1.384	0.943	*	1.111	1.083	1.066	1.055	1.041	1.033	1.028
0.00150	(0.0147)	3.833	2.612	1.940	1.441	1.130	*	*	1.080	1.056	1.041	1.032	1.021	1.015	1.011
0.00200	(0.0196)	3.320	2.262	1.680	1.248	0.978	*	1.077	1.060	1.038	1.025	1.017	1.008	1.003	1.001
0.00300	(0.0294)	2.711	1.847	1.372	1.019	0.799	*	1.048	1.034	1.015	1.005	0.998	0.991	0.988	0.987
0.00400	(0.0392)	2.347	1.600	1.188	0.882	*	1.050	1.030	1.017	1.001	0.992	0.986	0.980	0.978	0.978
0.00600	(0.0588)	1.917	1.306	0.970	*	*	1.024	1.007	0.996	0.983	0.975	0.971	0.967	0.966	0.967
0.00800	(0.0785)	1.660	1.131	0.840	*	1.033	1.008	0.992	0.983	0.971	0.965	0.961	0.959	0.959	0.960
0.01000	(0.0981)	1.485	1.012	0.751	*	1.019	0.996	0.982	0.973	0.963	0.957	0.954	0.953	0.953	0.956
0.01500	(0.147)	1.212	0.826	*	1.012	0.996	0.976	0.965	0.957	0.949	0.945	0.943	0.943	0.945	0.948
0.02000	(0.196)	1.050	*	*	0.996	0.982	0.964	0.954	0.947	0.940	0.937	0.936	0.937	0.939	0.943
0.03000	(0.294)	0.857	*	0.995	0.977	0.964	0.949	0.940	0.935	0.930	0.928	0.928	0.930	0.933	0.937
0.04000	(0.392)	0.742	*	0.981	0.964	0.953	0.939	0.932	0.928	0.923	0.922	0.923	0.925	0.929	0.933
0.06000	(0.588)	*	0.982	0.963	0.949	0.939	0.928	0.922	0.918	0.915	0.915	0.916	0.920	0.924	0.929
0.08000	(0.785)	*	0.969	0.952	0.939	0.930	0.920	0.915	0.912	0.910	0.911	0.912	0.916	0.921	0.926
0.10000	(0.981)	*	0.960	0.945	0.932	0.924	0.915	0.911	0.908	0.907	0.908	0.909	0.914	0.919	0.924
0.15000	(1.47)	0.969	0.946	0.932	0.921	0.915	0.907	0.903	0.902	0.901	0.903	0.905	0.910	0.915	0.922
0.20000	(1.96)	0.958	0.937	0.924	0.915	0.909	0.902	0.899	0.898	0.898	0.900	0.902	0.908	0.913	0.920
0.30000	(2.94)	0.945	0.926	0.915	0.907	0.901	0.896	0.894	0.893	0.894	0.896	0.899	0.905	0.911	0.918
0.40000	(3.92)	0.937	0.919	0.909	0.902	0.897	0.892	0.890	0.890	0.892	0.894	0.897	0.903	0.909	0.916
0.60000	(5.88)	0.927	0.911	0.902	0.896	0.892	0.888	0.886	0.887	0.889	0.892	0.895	0.901	0.908	0.915
0.80000	(7.85)	0.921	0.907	0.898	0.892	0.888	0.885	0.884	0.884	0.887	0.890	0.893	0.900	0.907	0.914
1.00000	(9.81)	0.917	0.903	0.895	0.889	0.886	0.883	0.882	0.883	0.886	0.889	0.892	0.899	0.906	0.913
1.50000	(14.7)	0.910	0.898	0.891	0.885	0.882	0.880	0.880	0.881	0.884	0.887	0.891	0.898	0.905	0.912
2.00000	(19.6)	0.906	0.895	0.888	0.883	0.880	0.878	0.878	0.879	0.882	0.886	0.890	0.897	0.904	0.912
3.00000	(29.4)	0.902	0.891	0.884	0.880	0.878	0.876	0.876	0.878	0.881	0.885	0.889	0.896	0.903	0.911
4.00000	(39.2)	0.899	0.888	0.882	0.878	0.876	0.875	0.875	0.877	0.880	0.884	0.888	0.896	0.903	0.911
6.00000	(58.8)	0.895	0.885	0.880	0.876	0.874	0.873	0.874	0.875	0.879	0.883	0.887	0.895	0.902	0.910
8.00000	(78.5)	0.893	0.884	0.879	0.875	0.873	0.872	0.873	0.875	0.879	0.883	0.887	0.895	0.902	0.910
10.0000	(98.1)	0.892	0.883	0.878	0.874	0.873	0.872	0.873	0.874	0.878	0.882	0.887	0.894	0.902	0.909
15.0000	(147)	0.890	0.881	0.876	0.873	0.871	0.871	0.872	0.873	0.878	0.882	0.886	0.894	0.901	0.909
20.0000	(196)	0.888	0.880	0.875	0.872	0.871	0.870	0.871	0.873	0.877	0.882	0.886	0.894	0.901	0.909
30.0000	(294)	0.887	0.878	0.874	0.871	0.870	0.870	0.871	0.873	0.877	0.881	0.886	0.894	0.901	0.909
S	$(\Delta p/\rho l)$	6	8	10	12.5	15	20	25	30	40	50	60	80	100	125

Grad'nt   k.p.g.   (Equivalent) Pipe diameters in mm

Grad'nt S	k.p.g. $(\Delta p/\rho l)$	125	150	200	250	300	400	500	600	800	1000	1250	1500	2000	2500
0.00010	(0.00098)	1.167	1.155	1.141	1.133	1.129	1.125	1.124	1.125	1.129	1.133	1.140	1.146	1.158	1.168
0.00015	(0.00147)	1.137	1.127	1.116	1.110	1.107	1.105	1.105	1.107	1.112	1.118	1.125	1.132	1.145	1.157
0.00020	(0.00196)	1.117	1.109	1.099	1.094	1.092	1.091	1.093	1.095	1.102	1.108	1.116	1.124	1.138	1.150
0.00030	(0.00294)	1.091	1.084	1.077	1.074	1.073	1.074	1.077	1.080	1.088	1.096	1.104	1.113	1.128	1.141
0.00040	(0.00392)	1.074	1.069	1.063	1.061	1.061	1.063	1.067	1.071	1.079	1.087	1.097	1.106	1.121	1.135
0.00060	(0.00588)	1.053	1.048	1.045	1.044	1.045	1.049	1.054	1.059	1.068	1.077	1.088	1.097	1.113	1.128
0.00080	(0.00785)	1.039	1.035	1.033	1.033	1.035	1.040	1.045	1.051	1.061	1.071	1.082	1.092	1.109	1.123
0.00100	(0.00981)	1.028	1.026	1.025	1.026	1.028	1.033	1.039	1.045	1.056	1.066	1.078	1.088	1.105	1.120
0.00150	(0.0147)	1.011	1.010	1.010	1.013	1.016	1.023	1.029	1.036	1.048	1.059	1.071	1.081	1.100	1.115
0.00200	(0.0196)	1.001	1.000	1.002	1.005	1.008	1.016	1.023	1.030	1.043	1.054	1.067	1.078	1.096	1.112
0.00300	(0.0294)	0.987	0.987	0.990	0.994	0.999	1.007	1.016	1.023	1.037	1.049	1.061	1.073	1.092	1.108
0.00400	(0.0392)	0.978	0.979	0.983	0.988	0.993	1.002	1.011	1.019	1.033	1.045	1.058	1.070	1.089	1.106
0.00600	(0.0588)	0.967	0.969	0.974	0.980	0.985	0.996	1.005	1.013	1.028	1.041	1.054	1.066	1.086	1.103
0.00800	(0.0785)	0.960	0.963	0.969	0.975	0.981	0.992	1.001	1.010	1.025	1.038	1.052	1.064	1.084	1.101
0.01000	(0.0981)	0.956	0.959	0.965	0.971	0.978	0.989	0.999	1.008	1.023	1.036	1.050	1.063	1.083	1.100
0.01500	(0.147)	0.948	0.951	0.959	0.966	0.972	0.984	0.995	1.004	1.020	1.034	1.048	1.060	1.081	1.098
0.02000	(0.196)	0.943	0.947	0.955	0.962	0.969	0.982	0.992	1.002	1.018	1.032	1.046	1.059	1.080	1.097
0.03000	(0.294)	0.937	0.942	0.950	0.958	0.966	0.978	0.989	0.999	1.016	1.030	1.044	1.057	1.078	1.096
0.04000	(0.392)	0.933	0.938	0.947	0.956	0.963	0.976	0.988	0.998	1.014	1.028	1.043	1.056	1.077	1.095
0.06000	(0.588)	0.929	0.934	0.944	0.953	0.960	0.974	0.985	0.996	1.013	1.027	1.042	1.055	1.076	1.094
0.08000	(0.785)	0.926	0.932	0.942	0.951	0.959	0.972	0.984	0.994	1.012	1.026	1.041	1.054	1.076	1.093
0.10000	(0.981)	0.924	0.930	0.940	0.949	0.958	0.971	0.983	0.994	1.011	1.025	1.041	1.054	1.075	1.093
0.15000	(1.47)	0.922	0.927	0.938	0.947	0.956	0.970	0.982	0.992	1.010	1.024	1.040	1.053	1.074	1.092
0.20000	(1.96)	0.920	0.926	0.937	0.946	0.955	0.969	0.981	0.992	1.009	1.024	1.039	1.052	1.074	1.092
0.30000	(2.94)	0.918	0.924	0.935	0.945	0.953	0.968	0.980	0.991	1.008	1.023	1.038	1.052	1.074	1.091
0.40000	(3.92)	0.916	0.923	0.934	0.944	0.952	0.967	0.979	0.990	1.008	1.023	1.038	1.051	1.073	1.091
0.60000	(5.88)	0.915	0.921	0.933	0.943	0.951	0.966	0.979	0.989	1.007	1.022	1.038	1.051	1.073	1.091
0.80000	(7.85)	0.914	0.921	0.932	0.942	0.951	0.966	0.978	0.989	1.007	1.022	1.037	1.051	1.073	1.091
1.00000	(9.81)	0.913	0.920	0.932	0.942	0.951	0.965	0.978	0.989	1.007	1.022	1.037	1.051	1.073	1.091
1.50000	(14.7)	0.912	0.919	0.931	0.941	0.950	0.965	0.977	0.988	1.006	1.021	1.037	1.050	1.072	1.090
2.00000	(19.6)	0.912	0.918	0.930	0.941	0.950	0.965	0.977	0.988	1.006	1.021	1.037	1.050	1.072	1.090
3.00000	(29.4)	0.911	0.918	0.930	0.940	0.949	0.964	0.977	0.988	1.006	1.021	1.037	1.050	1.072	1.090
S	$(\Delta p/\rho l)$	125	150	200	250	300	400	500	600	800	1000	1250	1500	2000	2500

Grad'nt   k.p.g.   (Equivalent) Pipe diameters in mm

Kinematic viscosity, $\nu = 1.25 \times 10^{-6}$ m^{2}s^{-1} ;   **Roughness size, $k_s = 0.150$ mm**

Grad'nt S	k.p.g. $(\Delta p/\rho l)$	(Equivalent) Pipe diameters in mm													
		6	8	10	12·5	15	20	25	30	40	50	60	80	100	125
0·00030	(0·00294)	8·572	5·841	4·338	3·222	2·526	1·721	1·278	1·003	*	1·194	1·176	1·153	1·140	1·130
0·00040	(0·00392)	7·423	5·058	3·757	2·790	2·188	1·491	1·107	*	1·196	1·172	1·156	1·136	1·124	1·116
0·00060	(0·00588)	6·061	4·130	3·067	2·278	1·786	1·217	0·904	*	1·165	1·144	1·131	1·114	1·105	1·098
0·00080	(0·00785)	5·249	3·577	2·656	1·973	1·547	1·054	*	*	1·145	1·127	1·114	1·100	1·092	1·087
0·00100	(0·00981)	4·695	3·199	2·376	1·764	1·384	0·943	*	1·159	1·131	1·114	1·103	1·090	1·083	1·078
0·00150	(0·0147)	3·833	2·612	1·940	1·441	1·130	*	*	1·132	1·108	1·093	1·084	1·073	1·068	1·065
0·00200	(0·0196)	3·320	2·262	1·680	1·248	0·978	*	1·132	1·115	1·093	1·080	1·072	1·063	1·059	1·057
0·00300	(0·0294)	2·711	1·847	1·372	1·019	0·799	*	1·108	1·093	1·074	1·064	1·057	1·050	1·047	1·047
0·00400	(0·0392)	2·347	1·600	1·188	0·882	*	1·113	1·093	1·079	1·063	1·053	1·048	1·042	1·040	1·040
0·00600	(0·0588)	1·917	1·306	0·970	*	*	1·092	1·074	1·062	1·048	1·041	1·036	1·032	1·031	1·032
0·00800	(0·0785)	1·660	1·131	0·840	*	*	1·078	1·062	1·052	1·039	1·033	1·029	1·026	1·026	1·028
0·01000	(0·0981)	1·485	1·012	0·751	*	1·094	1·069	1·054	1·044	1·033	1·027	1·024	1·022	1·022	1·024
0·01500	(0·147)	1·212	0·826	*	*	1·076	1·054	1·041	1·032	1·023	1·018	1·016	1·015	1·016	1·019
0·02000	(0·196)	1·050	*	*	1·080	1·064	1·044	1·032	1·025	1·017	1·013	1·011	1·011	1·013	1·016
0·03000	(0·294)	0·857	*	*	1·065	1·050	1·033	1·022	1·016	1·009	1·006	1·005	1·006	1·008	1·012
0·04000	(0·392)	0·742	*	1·075	1·055	1·042	1·026	1·016	1·011	1·005	1·002	1·002	1·003	1·005	1·009
0·06000	(0·588)	*	*	1·061	1·043	1·032	1·017	1·009	1·004	0·999	0·997	0·997	0·999	1·002	1·007
0·08000	(0·785)	*	1·074	1·053	1·036	1·025	1·012	1·004	1·000	0·996	0·994	0·995	0·997	1·000	1·005
0·10000	(0·981)	*	1·067	1·047	1·031	1·021	1·008	1·001	0·997	0·993	0·992	0·993	0·995	0·999	1·004
0·15000	(1·47)	*	1·056	1·038	1·023	1·014	1·002	0·996	0·993	0·990	0·989	0·990	0·993	0·997	1·002
0·20000	(1·96)	1·078	1·049	1·032	1·019	1·010	0·999	0·993	0·990	0·987	0·987	0·988	0·992	0·996	1·001
0·30000	(2·94)	1·068	1·041	1·026	1·013	1·005	0·995	0·990	0·987	0·985	0·985	0·986	0·990	0·994	0·999
0·40000	(3·92)	1·062	1·037	1·021	1·009	1·001	0·992	0·987	0·985	0·983	0·983	0·985	0·989	0·993	0·999
0·60000	(5·88)	1·054	1·031	1·017	1·005	0·998	0·989	0·985	0·983	0·981	0·982	0·983	0·987	0·992	0·998
0·80000	(7·85)	1·050	1·027	1·014	1·003	0·996	0·987	0·983	0·981	0·980	0·981	0·982	0·987	0·991	0·997
1·00000	(9·81)	1·047	1·025	1·012	1·001	0·994	0·986	0·982	0·980	0·979	0·980	0·982	0·986	0·991	0·997
1·50000	(14·7)	1·043	1·021	1·008	0·998	0·992	0·984	0·980	0·979	0·978	0·979	0·981	0·985	0·990	0·996
2·00000	(19·6)	1·040	1·019	1·007	0·997	0·990	0·983	0·979	0·978	0·977	0·978	0·980	0·985	0·990	0·996
3·00000	(29·4)	1·036	1·016	1·004	0·995	0·989	0·982	0·978	0·977	0·976	0·978	0·980	0·984	0·989	0·995
4·00000	(39·2)	1·034	1·015	1·003	0·994	0·988	0·981	0·978	0·976	0·976	0·977	0·979	0·984	0·989	0·995
6·00000	(58·8)	1·032	1·013	1·001	0·992	0·986	0·980	0·977	0·975	0·975	0·977	0·979	0·984	0·989	0·995
8·00000	(78·5)	1·031	1·012	1·000	0·991	0·986	0·979	0·976	0·975	0·975	0·976	0·979	0·983	0·988	0·995
10·0000	(98·1)	1·030	1·011	1·000	0·991	0·985	0·979	0·976	0·975	0·975	0·976	0·978	0·983	0·988	0·994
15·0000	(147)	1·028	1·010	0·999	0·990	0·984	0·978	0·975	0·974	0·974	0·976	0·978	0·983	0·988	0·994
20·0000	(196)	1·027	1·009	0·998	0·989	0·984	0·978	0·975	0·974	0·974	0·976	0·978	0·983	0·988	0·994
30·0000	(294)	1·026	1·008	0·997	0·989	0·983	0·977	0·975	0·974	0·974	0·975	0·978	0·983	0·988	0·994
S	$(\Delta p/\rho l)$	6	8	10	12·5	15	20	25	30	40	50	60	80	100	125

Grad'nt k.p.g. (Equivalent) Pipe diameters in mm

S	$(\Delta p/\rho l)$	125	150	200	250	300	400	500	600	800	1000	1250	1500	2000	2500
0·00010	(0·00098)	1·197	1·186	1·173	1·166	1·162	1·160	1·160	1·162	1·167	1·173	1·181	1·188	1·202	1·214
0·00015	(0·00147)	1·170	1·161	1·150	1·145	1·143	1·143	1·144	1·147	1·154	1·161	1·170	1·178	1·193	1·206
0·00020	(0·00196)	1·152	1·145	1·136	1·132	1·131	1·132	1·135	1·138	1·146	1·154	1·163	1·172	1·187	1·201
0·00030	(0·00294)	1·130	1·124	1·118	1·116	1·116	1·118	1·122	1·127	1·136	1·145	1·155	1·164	1·180	1·195
0·00040	(0·00392)	1·116	1·111	1·106	1·105	1·106	1·110	1·114	1·119	1·129	1·139	1·149	1·159	1·176	1·191
0·00060	(0·00588)	1·098	1·094	1·092	1·092	1·094	1·099	1·105	1·110	1·121	1·132	1·143	1·153	1·171	1·186
0·00080	(0·00785)	1·087	1·084	1·083	1·084	1·086	1·092	1·099	1·105	1·117	1·127	1·139	1·149	1·168	1·183
0·00100	(0·00981)	1·078	1·076	1·076	1·078	1·081	1·087	1·094	1·101	1·113	1·124	1·136	1·147	1·165	1·181
0·00150	(0·0147)	1·065	1·064	1·066	1·069	1·072	1·080	1·087	1·095	1·108	1·119	1·132	1·143	1·162	1·178
0·00200	(0·0196)	1·057	1·057	1·059	1·063	1·067	1·075	1·083	1·091	1·104	1·116	1·129	1·140	1·160	1·176
0·00300	(0·0294)	1·047	1·047	1·051	1·055	1·060	1·069	1·078	1·086	1·100	1·112	1·126	1·137	1·157	1·173
0·00400	(0·0392)	1·040	1·042	1·046	1·051	1·056	1·066	1·075	1·083	1·098	1·110	1·124	1·135	1·155	1·172
0·00600	(0·0588)	1·032	1·034	1·040	1·045	1·051	1·062	1·071	1·080	1·095	1·107	1·121	1·133	1·154	1·170
0·00800	(0·0785)	1·028	1·030	1·036	1·042	1·048	1·059	1·069	1·078	1·093	1·106	1·120	1·132	1·152	1·169
0·01000	(0·0981)	1·024	1·027	1·033	1·040	1·046	1·057	1·067	1·076	1·092	1·105	1·119	1·131	1·152	1·169
0·01500	(0·147)	1·019	1·022	1·029	1·036	1·043	1·054	1·065	1·074	1·090	1·103	1·117	1·130	1·150	1·167
0·02000	(0·196)	1·016	1·019	1·027	1·034	1·041	1·053	1·063	1·072	1·088	1·102	1·116	1·129	1·150	1·167
0·03000	(0·294)	1·012	1·016	1·024	1·031	1·038	1·051	1·061	1·071	1·087	1·101	1·115	1·128	1·149	1·166
0·04000	(0·392)	1·009	1·014	1·022	1·030	1·037	1·049	1·060	1·070	1·086	1·100	1·114	1·127	1·148	1·166
0·06000	(0·588)	1·007	1·011	1·020	1·028	1·035	1·048	1·059	1·069	1·085	1·099	1·114	1·126	1·148	1·165
0·08000	(0·785)	1·005	1·010	1·018	1·027	1·034	1·047	1·058	1·068	1·084	1·098	1·113	1·126	1·147	1·165
0·10000	(0·981)	1·004	1·009	1·018	1·026	1·033	1·046	1·058	1·067	1·084	1·098	1·113	1·126	1·147	1·164
0·15000	(1·47)	1·002	1·007	1·016	1·025	1·032	1·045	1·057	1·067	1·083	1·097	1·112	1·125	1·147	1·164
0·20000	(1·96)	1·001	1·006	1·015	1·024	1·031	1·045	1·056	1·066	1·083	1·097	1·112	1·125	1·146	1·164
0·30000	(2·94)	0·999	1·005	1·014	1·023	1·031	1·044	1·056	1·066	1·083	1·097	1·112	1·125	1·146	1·164
0·40000	(3·92)	0·999	1·004	1·014	1·022	1·030	1·044	1·055	1·065	1·082	1·096	1·111	1·124	1·146	1·163
0·60000	(5·88)	0·998	1·003	1·013	1·022	1·030	1·043	1·055	1·065	1·082	1·096	1·111	1·124	1·146	1·163
0·80000	(7·85)	0·997	1·003	1·013	1·021	1·029	1·043	1·055	1·065	1·082	1·096	1·111	1·124	1·146	1·163
1·00000	(9·81)	0·997	1·002	1·012	1·021	1·029	1·043	1·054	1·064	1·082	1·096	1·111	1·124	1·145	1·163
1·50000	(14·7)	0·996	1·002	1·012	1·021	1·029	1·042	1·054	1·064	1·081	1·096	1·111	1·124	1·145	1·163
2·00000	(19·6)	0·996	1·001	1·012	1·020	1·028	1·042	1·054	1·064	1·081	1·096	1·111	1·124	1·145	1·163
3·00000	(29·4)	0·995	1·001	1·011	1·020	1·028	1·042	1·054	1·064	1·081	1·095	1·111	1·124	1·145	1·163
S	$(\Delta p/\rho l)$	125	150	200	250	300	400	500	600	800	1000	1250	1500	2000	2500

Grad'nt k.p.g. (Equivalent) Pipe diameters in mm

Kinematic viscosity, $\nu = 1 \cdot 25 \times 10^{-6}\ \mathrm{m^2 s^{-1}}$;   **Roughness size, k_s = 0·30 mm**

E109

Kin. visc., $\nu = 1.25 \times 10^{-6}$ m^{2}s^{-1};
$S = 0.00010$ to 30.0000

i.e. kin. pr. grad., $\Delta p/\rho l =$
(0.00098) to (294) ms^{-2}

Roughness size, $k_s = 0.60$ mm
This table shows values of m, as follows

m_C for Colebrook-White solutions; or,
where $\mathbf{R} \leq 2000$, m_P for laminar flow

Grad'nt S	k.p.g. $(\Delta p/\rho l)$	6	8	10	12·5	15	20	25	30	40	50	60	80	100	125
											(Equivalent) Pipe diameters in mm				
0·00030	(0·00294)	8·572	5·841	4·338	3·222	2·526	1·721	1·278	1·003	*	1·258	1·240	1·217	1·204	1·194
0·00040	(0·00392)	7·423	5·058	3·757	2·790	2·188	1·491	1·107	*	*	1·240	1·223	1·203	1·191	1·183
0·00060	(0·00588)	6·061	4·130	3·067	2·278	1·786	1·217	0·904	*	1·238	1·217	1·202	1·185	1·176	1·169
0·00080	(0·00785)	5·249	3·577	2·656	1·973	1·547	1·054	*	*	1·222	1·202	1·190	1·174	1·166	1·160
0·00100	(0·00981)	4·695	3·199	2·376	1·764	1·384	0·943	*	*	1·210	1·192	1·181	1·167	1·159	1·154
0·00150	(0·0147)	3·833	2·612	1·940	1·441	1·130	*	*	1·218	1·192	1·176	1·166	1·154	1·148	1·145
0·00200	(0·0196)	3·320	2·262	1·680	1·248	0·978	*	1·224	1·204	1·180	1·166	1·157	1·147	1·142	1·139
0·00300	(0·0294)	2·711	1·847	1·372	1·019	0·799	*	1·205	1·187	1·166	1·154	1·146	1·137	1·133	1·132
0·00400	(0·0392)	2·347	1·600	1·188	0·882	*	1·216	1·193	1·177	1·157	1·146	1·139	1·132	1·128	1·127
0·00600	(0·0588)	1·917	1·306	0·970	*	*	1·199	1·178	1·164	1·147	1·137	1·131	1·125	1·122	1·122
0·00800	(0·0785)	1·660	1·131	0·840	*	*	1·189	1·169	1·156	1·140	1·131	1·126	1·120	1·119	1·119
0·01000	(0·0981)	1·485	1·012	0·751	*	*	1·182	1·163	1·151	1·136	1·127	1·122	1·118	1·116	1·117
0·01500	(0·147)	1·212	0·826	*	*	1·199	1·170	1·153	1·142	1·128	1·121	1·117	1·113	1·112	1·113
0·02000	(0·196)	1·050	*	*	1·211	1·190	1·164	1·147	1·137	1·124	1·117	1·113	1·110	1·110	1·111
0·03000	(0·294)	0·857	*	*	1·199	1·180	1·155	1·140	1·131	1·119	1·113	1·109	1·107	1·107	1·108
0·04000	(0·392)	0·742	*	1·220	1·192	1·174	1·150	1·136	1·127	1·116	1·110	1·107	1·105	1·105	1·107
0·06000	(0·588)	*	*	1·209	1·184	1·166	1·144	1·131	1·122	1·112	1·107	1·104	1·102	1·103	1·105
0·08000	(0·785)	*	*	1·203	1·178	1·162	1·140	1·128	1·119	1·110	1·105	1·102	1·101	1·102	1·104
0·10000	(0·981)	*	1·228	1·199	1·175	1·158	1·138	1·125	1·117	1·108	1·104	1·101	1·100	1·101	1·103
0·15000	(1·47)	*	1·220	1·192	1·169	1·153	1·134	1·122	1·115	1·106	1·102	1·099	1·099	1·100	1·102
0·20000	(1·96)	*	1·215	1·188	1·166	1·151	1·131	1·120	1·113	1·104	1·100	1·098	1·098	1·099	1·101
0·30000	(2·94)	1·251	1·209	1·183	1·162	1·147	1·129	1·118	1·111	1·103	1·099	1·097	1·097	1·098	1·101
0·40000	(3·92)	1·247	1·206	1·180	1·159	1·145	1·127	1·116	1·109	1·102	1·098	1·096	1·096	1·097	1·100
0·60000	(5·88)	1·241	1·201	1·177	1·156	1·142	1·125	1·114	1·108	1·100	1·097	1·095	1·095	1·097	1·100
0·80000	(7·85)	1·238	1·199	1·175	1·155	1·141	1·124	1·113	1·107	1·100	1·096	1·095	1·095	1·096	1·099
1·00000	(9·81)	1·236	1·197	1·173	1·153	1·140	1·123	1·113	1·106	1·099	1·096	1·094	1·094	1·096	1·099
1·50000	(14·7)	1·232	1·195	1·171	1·152	1·138	1·122	1·112	1·105	1·098	1·095	1·094	1·094	1·096	1·099
2·00000	(19·6)	1·230	1·193	1·170	1·150	1·137	1·121	1·111	1·105	1·098	1·095	1·093	1·093	1·095	1·098
3·00000	(29·4)	1·228	1·191	1·168	1·149	1·136	1·120	1·110	1·104	1·097	1·094	1·093	1·093	1·095	1·098
4·00000	(39·2)	1·226	1·190	1·167	1·148	1·135	1·119	1·110	1·104	1·097	1·094	1·093	1·093	1·095	1·098
6·00000	(58·8)	1·225	1·189	1·166	1·147	1·135	1·119	1·109	1·103	1·097	1·094	1·092	1·093	1·095	1·098
8·00000	(78·5)	1·224	1·188	1·165	1·147	1·134	1·118	1·109	1·103	1·096	1·093	1·092	1·093	1·094	1·098
10·0000	(98·1)	1·223	1·187	1·165	1·146	1·134	1·118	1·109	1·103	1·096	1·093	1·092	1·092	1·094	1·098
15·0000	(147)	1·222	1·186	1·164	1·146	1·133	1·118	1·108	1·102	1·096	1·093	1·092	1·092	1·094	1·097
20·0000	(196)	1·221	1·186	1·164	1·145	1·133	1·117	1·108	1·102	1·096	1·093	1·092	1·092	1·094	1·097
30·0000	(294)	1·220	1·185	1·163	1·145	1·133	1·117	1·108	1·102	1·096	1·093	1·092	1·092	1·094	1·097
S	$(\Delta p/\rho l)$	6	8	10	12·5	15	20	25	30	40	50	60	80	100	125

Grad'nt   k.p.g.   (Equivalent) Pipe diameters in mm

S	$(\Delta p/\rho l)$	125	150	200	250	300	400	500	600	800	1000	1250	1500	2000	2500
0·00010	(0·00098)	1·248	1·238	1·226	1·220	1·217	1·215	1·217	1·219	1·226	1·233	1·242	1·250	1·265	1·278
0·00015	(0·00147)	1·226	1·217	1·208	1·204	1·202	1·202	1·205	1·209	1·217	1·225	1·234	1·243	1·259	1·273
0·00020	(0·00196)	1·212	1·204	1·196	1·193	1·193	1·194	1·198	1·202	1·211	1·219	1·230	1·239	1·255	1·270
0·00030	(0·00294)	1·194	1·188	1·182	1·181	1·181	1·184	1·189	1·194	1·204	1·213	1·224	1·234	1·251	1·265
0·00040	(0·00392)	1·183	1·178	1·174	1·173	1·174	1·178	1·183	1·189	1·199	1·209	1·220	1·230	1·248	1·263
0·00060	(0·00588)	1·169	1·165	1·163	1·164	1·165	1·171	1·177	1·183	1·194	1·204	1·216	1·227	1·245	1·260
0·00080	(0·00785)	1·160	1·158	1·156	1·158	1·160	1·166	1·173	1·179	1·191	1·202	1·214	1·224	1·243	1·258
0·00100	(0·00981)	1·154	1·152	1·152	1·153	1·156	1·163	1·170	1·176	1·189	1·200	1·212	1·223	1·241	1·257
0·00150	(0·0147)	1·145	1·144	1·144	1·147	1·150	1·158	1·165	1·172	1·185	1·196	1·209	1·220	1·239	1·255
0·00200	(0·0196)	1·139	1·138	1·140	1·143	1·147	1·155	1·162	1·170	1·183	1·195	1·207	1·219	1·238	1·254
0·00300	(0·0294)	1·132	1·132	1·134	1·138	1·142	1·151	1·159	1·167	1·180	1·192	1·205	1·217	1·236	1·252
0·00400	(0·0392)	1·127	1·128	1·131	1·135	1·140	1·149	1·157	1·165	1·179	1·191	1·204	1·216	1·235	1·252
0·00600	(0·0588)	1·122	1·123	1·127	1·132	1·136	1·146	1·155	1·163	1·177	1·189	1·203	1·214	1·234	1·251
0·00800	(0·0785)	1·119	1·120	1·124	1·129	1·135	1·144	1·153	1·161	1·176	1·188	1·202	1·214	1·233	1·250
0·01000	(0·0981)	1·117	1·118	1·123	1·128	1·133	1·143	1·152	1·161	1·175	1·188	1·201	1·213	1·233	1·250
0·01500	(0·147)	1·113	1·115	1·120	1·126	1·131	1·141	1·151	1·159	1·174	1·187	1·200	1·212	1·232	1·249
0·02000	(0·196)	1·111	1·113	1·119	1·124	1·130	1·140	1·150	1·158	1·173	1·186	1·200	1·212	1·232	1·249
0·03000	(0·294)	1·108	1·111	1·117	1·123	1·128	1·139	1·149	1·157	1·172	1·185	1·199	1·211	1·231	1·248
0·04000	(0·392)	1·107	1·110	1·116	1·122	1·127	1·138	1·148	1·157	1·172	1·185	1·199	1·211	1·231	1·248
0·06000	(0·588)	1·105	1·108	1·114	1·120	1·126	1·137	1·147	1·156	1·171	1·184	1·198	1·210	1·231	1·247
0·08000	(0·785)	1·104	1·107	1·113	1·120	1·126	1·137	1·147	1·156	1·171	1·184	1·198	1·210	1·230	1·247
0·10000	(0·981)	1·103	1·106	1·113	1·119	1·125	1·137	1·146	1·155	1·171	1·184	1·198	1·210	1·230	1·247
0·15000	(1·47)	1·102	1·105	1·112	1·118	1·125	1·136	1·146	1·155	1·170	1·183	1·197	1·210	1·230	1·247
0·20000	(1·96)	1·101	1·105	1·111	1·118	1·124	1·136	1·146	1·155	1·170	1·183	1·197	1·209	1·230	1·247
0·30000	(2·94)	1·101	1·104	1·111	1·117	1·124	1·135	1·145	1·154	1·170	1·183	1·197	1·209	1·230	1·247
0·40000	(3·92)	1·100	1·103	1·110	1·117	1·123	1·135	1·145	1·154	1·170	1·183	1·197	1·209	1·230	1·247
0·60000	(5·88)	1·100	1·103	1·110	1·117	1·123	1·135	1·145	1·154	1·169	1·183	1·197	1·209	1·229	1·246
0·80000	(7·85)	1·099	1·103	1·110	1·117	1·123	1·135	1·145	1·154	1·169	1·182	1·197	1·209	1·229	1·246
1·00000	(9·81)	1·099	1·102	1·110	1·116	1·123	1·134	1·145	1·154	1·169	1·182	1·197	1·209	1·229	1·246
1·50000	(14·7)	1·099	1·102	1·109	1·116	1·123	1·134	1·144	1·153	1·169	1·182	1·196	1·209	1·229	1·246
2·00000	(19·6)	1·098	1·102	1·109	1·116	1·122	1·134	1·144	1·153	1·169	1·182	1·196	1·209	1·229	1·246
3·00000	(29·4)	1·098	1·102	1·109	1·116	1·122	1·134	1·144	1·153	1·169	1·182	1·196	1·209	1·229	1·246
S	$(\Delta p/\rho l)$	125	150	200	250	300	400	500	600	800	1000	1250	1500	2000	2500

Grad'nt   k.p.g.   (Equivalent) Pipe diameters in mm

Kinematic viscosity, $\nu = 1.25 \times 10^{-6}$ m^{2}s^{-1} ;

Roughness size, $k_s = 0.60$ mm

Kin. visc., $\nu = 1.25 \times 10^{-6}$ m^{2}s^{-1};
S = 0.00010 to 30.0000

i.e. kin. pr. grad., $\Delta p/\rho l$ =
(0.00098) to (294) ms^{-2}

Roughness size, k_s = 1.50 mm
This table shows values of m, as follows

m_C for Colebrook-White solutions; or,
where $R \leq 2000$, m_P for laminar flow

Grad'nt **k.p.g.** **(Equivalent) Pipe diameters in mm**

S	$(\Delta p/\rho l)$	6	8	10	12.5	15	20	25	30	40	50	60	80	100	125
0.00030	(0.00294)		5.841	4.338	3.222	2.526	1.721	1.278	1.003	*	1.409	1.387	1.359	1.343	1.331
0.00040	(0.00392)		5.058	3.757	2.790	2.188	1.491	1.107	*	*	1.396	1.375	1.350	1.335	1.323
0.00060	(0.00588)		4.130	3.067	2.278	1.786	1.217	0.904	*	1.407	1.379	1.361	1.338	1.324	1.314
0.00080	(0.00785)		3.577	2.656	1.973	1.547	1.054	*	*	1.395	1.369	1.352	1.331	1.318	1.309
0.00100	(0.00981)		3.199	2.376	1.764	1.384	0.943	*	*	1.387	1.362	1.346	1.326	1.314	1.305
0.00150	(0.0147)		2.612	1.940	1.441	1.130	*	*	1.411	1.374	1.351	1.336	1.318	1.307	1.300
0.00200	(0.0196)		2.262	1.680	1.248	0.978	*	*	1.402	1.366	1.345	1.331	1.313	1.303	1.296
0.00300	(0.0294)		1.847	1.372	1.019	0.799	*	1.416	1.390	1.357	1.337	1.324	1.307	1.298	1.292
0.00400	(0.0392)		1.600	1.188	0.882	*	*	1.408	1.383	1.351	1.332	1.319	1.304	1.295	1.289
0.00600	(0.0588)		1.306	0.970	*	*	1.432	1.398	1.374	1.345	1.326	1.314	1.300	1.292	1.286
0.00800	(0.0785)		1.131	0.840	*	*	1.425	1.392	1.369	1.340	1.323	1.311	1.297	1.290	1.284
0.01000	(0.0981)		1.012	0.751	*	*	1.420	1.388	1.366	1.338	1.320	1.309	1.295	1.288	1.283
0.01500	(0.147)		0.826	*	*	1.462	1.413	1.382	1.360	1.333	1.317	1.306	1.293	1.286	1.281
0.02000	(0.196)		*	*	*	1.457	1.408	1.378	1.357	1.331	1.314	1.304	1.291	1.284	1.280
0.03000	(0.294)		*	*	1.486	1.450	1.403	1.373	1.353	1.327	1.312	1.302	1.289	1.283	1.279
0.04000	(0.392)		*	*	1.481	1.445	1.399	1.370	1.351	1.326	1.310	1.300	1.288	1.282	1.278
0.06000	(0.588)		*	*	1.475	1.440	1.395	1.367	1.348	1.323	1.308	1.298	1.287	1.281	1.277
0.08000	(0.785)		*	1.521	1.472	1.437	1.393	1.365	1.346	1.322	1.307	1.298	1.286	1.280	1.276
0.10000	(0.981)		*	1.518	1.469	1.435	1.391	1.364	1.345	1.321	1.306	1.297	1.285	1.279	1.276
0.15000	(1.47)		1.572	1.513	1.465	1.432	1.389	1.362	1.343	1.320	1.305	1.296	1.285	1.279	1.275
0.20000	(1.96)		1.568	1.511	1.463	1.430	1.387	1.361	1.342	1.319	1.304	1.295	1.284	1.278	1.275
0.30000	(2.94)		1.564	1.507	1.460	1.428	1.386	1.359	1.341	1.318	1.304	1.294	1.283	1.278	1.274
0.40000	(3.92)		1.562	1.505	1.459	1.427	1.385	1.358	1.340	1.317	1.303	1.294	1.283	1.277	1.274
0.60000	(5.88)		1.559	1.503	1.457	1.425	1.383	1.357	1.339	1.316	1.302	1.293	1.283	1.277	1.274
0.80000	(7.85)		1.557	1.502	1.456	1.424	1.383	1.357	1.339	1.316	1.302	1.293	1.282	1.277	1.273
1.00000	(9.81)		1.556	1.501	1.455	1.423	1.382	1.356	1.338	1.316	1.302	1.293	1.282	1.277	1.273
1.50000	(14.7)		1.554	1.499	1.454	1.422	1.381	1.356	1.338	1.315	1.301	1.293	1.282	1.276	1.273
2.00000	(19.6)		1.553	1.498	1.453	1.422	1.381	1.355	1.337	1.315	1.301	1.292	1.282	1.276	1.273
3.00000	(29.4)		1.551	1.497	1.452	1.421	1.380	1.355	1.337	1.315	1.301	1.292	1.282	1.276	1.273
4.00000	(39.2)		1.551	1.497	1.452	1.421	1.380	1.354	1.337	1.314	1.301	1.292	1.281	1.276	1.273
6.00000	(58.8)		1.550	1.496	1.451	1.420	1.380	1.354	1.337	1.314	1.301	1.292	1.281	1.276	1.273
8.00000	(78.5)		1.549	1.495	1.451	1.420	1.379	1.354	1.336	1.314	1.301	1.292	1.281	1.276	1.273
10.0000	(98.1)		1.549	1.495	1.451	1.420	1.379	1.354	1.336	1.314	1.300	1.292	1.281	1.276	1.273
15.0000	(147)		1.548	1.495	1.450	1.419	1.379	1.354	1.336	1.314	1.300	1.291	1.281	1.276	1.273
20.0000	(196)		1.548	1.494	1.450	1.419	1.379	1.353	1.336	1.314	1.300	1.291	1.281	1.276	1.272
30.0000	(294)		1.547	1.494	1.450	1.419	1.379	1.353	1.336	1.314	1.300	1.291	1.281	1.276	1.272
S	$(\Delta p/\rho l)$	6	8	10	12.5	15	20	25	30	40	50	60	80	100	125

Grad'nt **k.p.g.** **(Equivalent) Pipe diameters in mm**

S	$(\Delta p/\rho l)$	125	150	200	250	300	400	500	600	800	1000	1250	1500	2000	2500
0.00010	(0.00098)	1.368	1.357	1.343	1.337	1.333	1.331	1.332	1.335	1.341	1.348	1.357	1.365	1.380	1.394
0.00015	(0.00147)	1.352	1.342	1.331	1.326	1.324	1.323	1.325	1.328	1.336	1.343	1.353	1.361	1.377	1.391
0.00020	(0.00196)	1.343	1.334	1.324	1.319	1.318	1.318	1.321	1.324	1.332	1.340	1.350	1.359	1.375	1.389
0.00030	(0.00294)	1.331	1.323	1.315	1.312	1.311	1.312	1.316	1.320	1.328	1.337	1.347	1.356	1.372	1.387
0.00040	(0.00392)	1.323	1.316	1.309	1.307	1.306	1.309	1.312	1.317	1.326	1.335	1.345	1.354	1.371	1.385
0.00060	(0.00588)	1.314	1.309	1.303	1.301	1.301	1.304	1.309	1.313	1.323	1.332	1.343	1.352	1.369	1.384
0.00080	(0.00785)	1.309	1.304	1.299	1.298	1.298	1.302	1.306	1.311	1.321	1.331	1.341	1.351	1.368	1.383
0.00100	(0.00981)	1.305	1.300	1.296	1.295	1.296	1.300	1.305	1.310	1.320	1.330	1.340	1.350	1.367	1.382
0.00150	(0.0147)	1.300	1.295	1.292	1.291	1.293	1.297	1.302	1.308	1.318	1.328	1.339	1.349	1.366	1.381
0.00200	(0.0196)	1.296	1.292	1.289	1.289	1.291	1.295	1.301	1.306	1.317	1.327	1.338	1.348	1.366	1.381
0.00300	(0.0294)	1.292	1.288	1.286	1.286	1.288	1.293	1.299	1.305	1.316	1.326	1.337	1.347	1.365	1.380
0.00400	(0.0392)	1.289	1.286	1.284	1.285	1.287	1.292	1.298	1.304	1.315	1.325	1.336	1.347	1.364	1.380
0.00600	(0.0588)	1.286	1.283	1.282	1.283	1.285	1.291	1.297	1.303	1.314	1.324	1.336	1.346	1.364	1.379
0.00800	(0.0785)	1.284	1.282	1.281	1.282	1.284	1.290	1.296	1.302	1.313	1.324	1.335	1.346	1.364	1.379
0.01000	(0.0981)	1.283	1.281	1.280	1.281	1.283	1.289	1.295	1.301	1.313	1.323	1.335	1.345	1.363	1.379
0.01500	(0.147)	1.281	1.279	1.278	1.280	1.282	1.288	1.294	1.301	1.312	1.323	1.334	1.345	1.363	1.378
0.02000	(0.196)	1.280	1.278	1.277	1.279	1.281	1.288	1.294	1.300	1.312	1.322	1.334	1.345	1.363	1.378
0.03000	(0.294)	1.279	1.277	1.276	1.278	1.281	1.287	1.293	1.300	1.312	1.322	1.334	1.344	1.362	1.378
0.04000	(0.392)	1.278	1.276	1.276	1.277	1.280	1.287	1.293	1.299	1.311	1.322	1.334	1.344	1.362	1.378
0.06000	(0.588)	1.277	1.275	1.275	1.277	1.280	1.286	1.293	1.299	1.311	1.322	1.333	1.344	1.362	1.377
0.08000	(0.785)	1.276	1.275	1.275	1.276	1.279	1.286	1.292	1.299	1.311	1.321	1.333	1.344	1.362	1.377
0.10000	(0.981)	1.276	1.274	1.274	1.276	1.279	1.286	1.292	1.299	1.311	1.321	1.333	1.344	1.362	1.377
0.15000	(1.47)	1.275	1.274	1.274	1.276	1.279	1.285	1.292	1.298	1.310	1.321	1.333	1.344	1.362	1.377
0.20000	(1.96)	1.275	1.273	1.273	1.276	1.278	1.285	1.292	1.298	1.310	1.321	1.333	1.344	1.362	1.377
0.30000	(2.94)	1.274	1.273	1.273	1.275	1.278	1.285	1.292	1.298	1.310	1.321	1.333	1.343	1.362	1.377
0.40000	(3.92)	1.274	1.273	1.273	1.275	1.278	1.285	1.292	1.298	1.310	1.321	1.333	1.343	1.362	1.377
0.60000	(5.88)	1.274	1.272	1.273	1.275	1.278	1.285	1.291	1.298	1.310	1.321	1.333	1.343	1.362	1.377
0.80000	(7.85)	1.273	1.272	1.273	1.275	1.278	1.285	1.291	1.298	1.310	1.321	1.333	1.343	1.362	1.377
1.00000	(9.81)	1.273	1.272	1.272	1.275	1.278	1.284	1.291	1.298	1.310	1.321	1.333	1.343	1.362	1.377
1.50000	(14.7)	1.273	1.272	1.272	1.275	1.278	1.284	1.291	1.298	1.310	1.321	1.333	1.343	1.361	1.377
2.00000	(19.6)	1.273	1.272	1.272	1.275	1.278	1.284	1.291	1.298	1.310	1.321	1.333	1.343	1.361	1.377
3.00000	(29.4)	1.273	1.272	1.272	1.274	1.277	1.284	1.291	1.298	1.310	1.321	1.333	1.343	1.361	1.377
S	$(\Delta p/\rho l)$	125	150	200	250	300	400	500	600	800	1000	1250	1500	2000	2500

Grad'nt **k.p.g.** **(Equivalent) Pipe diameters in mm**

Kinematic viscosity, $\nu = 1.25 \times 10^{-6}$ m^{2}s^{-1} ;   **Roughness size,** k_s = 1.50 mm

E111

Kin. visc., $\nu = 1.25 \times 10^{-6}$ m^{2}s^{-1};
$S = 0.00010$ to 3.00000

i.e. kin. pr. grad., $\Delta p/\rho l =$ (0.00098) to (29.4) ms^{-2}

Roughness size, $k_s = 3.0$ mm
This table shows values of m, as follows

m_C for Colebrook-White solutions

Grad'nt S	k.p.g. $(\Delta p/\rho l)$	125	150	200	250	300	400	500	600	800	1000	1250	1500	2000	2500
		(Equivalent) Pipe diameters in mm													
0.00010	(0.00098)	1.514	1.498	1.480	1.469	1.463	1.458	1.457	1.457	1.461	1.466	1.473	1.480	1.494	1.506
0.00015	(0.00147)	1.502	1.488	1.471	1.462	1.457	1.453	1.452	1.453	1.457	1.463	1.471	1.478	1.492	1.504
0.00020	(0.00196)	1.495	1.482	1.466	1.458	1.453	1.449	1.449	1.450	1.455	1.461	1.469	1.476	1.490	1.503
0.00030	(0.00294)	1.487	1.474	1.460	1.452	1.448	1.445	1.446	1.447	1.453	1.459	1.467	1.475	1.489	1.501
0.00040	(0.00392)	1.482	1.470	1.456	1.449	1.446	1.443	1.444	1.446	1.451	1.458	1.466	1.474	1.488	1.501
0.00060	(0.00588)	1.475	1.465	1.452	1.445	1.442	1.440	1.441	1.443	1.449	1.456	1.464	1.472	1.487	1.500
0.00080	(0.00785)	1.472	1.461	1.449	1.443	1.440	1.439	1.440	1.442	1.448	1.455	1.464	1.472	1.486	1.499
0.00100	(0.00981)	1.469	1.459	1.447	1.442	1.439	1.437	1.439	1.441	1.448	1.455	1.463	1.471	1.486	1.499
0.00150	(0.0147)	1.465	1.456	1.445	1.439	1.437	1.436	1.437	1.440	1.446	1.453	1.462	1.470	1.485	1.498
0.00200	(0.0196)	1.463	1.454	1.443	1.438	1.435	1.434	1.436	1.439	1.446	1.453	1.462	1.470	1.485	1.498
0.00300	(0.0294)	1.460	1.451	1.441	1.436	1.434	1.433	1.435	1.438	1.445	1.452	1.461	1.469	1.484	1.497
0.00400	(0.0392)	1.458	1.450	1.440	1.435	1.433	1.432	1.434	1.437	1.444	1.452	1.461	1.469	1.484	1.497
0.00600	(0.0588)	1.456	1.448	1.438	1.434	1.432	1.432	1.434	1.437	1.444	1.451	1.460	1.468	1.484	1.497
0.00800	(0.0785)	1.455	1.447	1.437	1.433	1.431	1.431	1.433	1.436	1.443	1.451	1.460	1.468	1.483	1.497
0.01000	(0.0981)	1.454	1.446	1.437	1.432	1.431	1.431	1.433	1.436	1.443	1.451	1.460	1.468	1.483	1.497
0.01500	(0.147)	1.453	1.445	1.436	1.432	1.430	1.430	1.432	1.436	1.443	1.450	1.459	1.468	1.483	1.496
0.02000	(0.196)	1.452	1.444	1.435	1.431	1.429	1.430	1.432	1.435	1.443	1.450	1.459	1.468	1.483	1.496
0.03000	(0.294)	1.451	1.443	1.435	1.431	1.429	1.429	1.432	1.435	1.442	1.450	1.459	1.468	1.483	1.496
0.04000	(0.392)	1.451	1.443	1.434	1.430	1.429	1.429	1.431	1.435	1.442	1.450	1.459	1.467	1.483	1.496
0.06000	(0.588)	1.450	1.442	1.434	1.430	1.428	1.429	1.431	1.435	1.442	1.450	1.459	1.467	1.483	1.496
0.08000	(0.785)	1.450	1.442	1.433	1.430	1.428	1.429	1.431	1.434	1.442	1.450	1.459	1.467	1.482	1.496
0.10000	(0.981)	1.449	1.442	1.433	1.429	1.428	1.428	1.431	1.434	1.442	1.449	1.459	1.467	1.482	1.496
0.15000	(1.47)	1.449	1.441	1.433	1.429	1.428	1.428	1.431	1.434	1.442	1.449	1.459	1.467	1.482	1.496
0.20000	(1.96)	1.449	1.441	1.433	1.429	1.428	1.428	1.431	1.434	1.442	1.449	1.458	1.467	1.482	1.496
0.30000	(2.94)	1.448	1.441	1.433	1.429	1.427	1.428	1.431	1.434	1.442	1.449	1.458	1.467	1.482	1.496
0.40000	(3.92)	1.448	1.441	1.432	1.429	1.427	1.428	1.430	1.434	1.442	1.449	1.458	1.467	1.482	1.496
0.60000	(5.88)	1.448	1.441	1.432	1.429	1.427	1.428	1.430	1.434	1.441	1.449	1.458	1.467	1.482	1.496
0.80000	(7.85)	1.448	1.440	1.432	1.429	1.427	1.428	1.430	1.434	1.441	1.449	1.458	1.467	1.482	1.496
1.00000	(9.81)	1.448	1.440	1.432	1.428	1.427	1.428	1.430	1.434	1.441	1.449	1.458	1.467	1.482	1.496
1.50000	(14.7)	1.448	1.440	1.432	1.428	1.427	1.428	1.430	1.434	1.441	1.449	1.458	1.467	1.482	1.496
2.00000	(19.6)	1.448	1.440	1.432	1.428	1.427	1.428	1.430	1.434	1.441	1.449	1.458	1.467	1.482	1.496
3.00000	(29.4)	1.448	1.440	1.432	1.428	1.427	1.428	1.430	1.434	1.441	1.449	1.458	1.467	1.482	1.496
S	$(\Delta p/\rho l)$	125	150	200	250	300	400	500	600	800	1000	1250	1500	2000	2500

Grad'nt k.p.g.   (Equivalent) Pipe diameters in mm   Roughness size, $k_s = 3.0$ mm

E112

Kin. visc., $\nu = 1.25 \times 10^{-6}$ m^{2}s^{-1};
$S = 0.00010$ to 3.00000

i.e. kin. pr. grad., $\Delta p/\rho l =$ (0.00098) to (29.4) ms^{-2}

Roughness size, $k_s = 6.0$ mm
This table shows values of m, as follows

m_C for Colebrook-White solutions

Grad'nt S	k.p.g. $(\Delta p/\rho l)$	125	150	200	250	300	400	500	600	800	1000	1250	1500	2000	2500
		(Equivalent) Pipe diameters in mm													
0.00010	(0.00098)	1.724	1.700	1.670	1.652	1.640	1.627	1.620	1.617	1.615	1.616	1.619	1.624	1.633	1.643
0.00015	(0.00147)	1.716	1.693	1.664	1.647	1.636	1.623	1.617	1.614	1.612	1.614	1.618	1.622	1.632	1.642
0.00020	(0.00196)	1.711	1.689	1.661	1.644	1.633	1.621	1.615	1.612	1.611	1.613	1.617	1.621	1.631	1.641
0.00030	(0.00294)	1.705	1.684	1.657	1.640	1.630	1.618	1.613	1.610	1.609	1.611	1.616	1.620	1.630	1.640
0.00040	(0.00392)	1.701	1.680	1.654	1.638	1.628	1.617	1.612	1.609	1.608	1.611	1.615	1.620	1.630	1.640
0.00060	(0.00588)	1.697	1.677	1.651	1.636	1.626	1.615	1.610	1.608	1.607	1.610	1.614	1.619	1.629	1.639
0.00080	(0.00785)	1.695	1.675	1.649	1.634	1.625	1.614	1.609	1.607	1.607	1.609	1.613	1.618	1.629	1.639
0.00100	(0.00981)	1.693	1.673	1.648	1.633	1.624	1.613	1.608	1.606	1.606	1.609	1.613	1.618	1.629	1.639
0.00150	(0.0147)	1.690	1.671	1.646	1.632	1.622	1.612	1.607	1.605	1.605	1.608	1.613	1.618	1.628	1.638
0.00200	(0.0196)	1.689	1.669	1.645	1.631	1.622	1.611	1.607	1.605	1.605	1.608	1.612	1.617	1.628	1.638
0.00300	(0.0294)	1.687	1.668	1.644	1.630	1.621	1.611	1.606	1.604	1.605	1.607	1.612	1.617	1.628	1.638
0.00400	(0.0392)	1.685	1.667	1.643	1.629	1.620	1.610	1.606	1.604	1.604	1.607	1.612	1.617	1.627	1.638
0.00600	(0.0588)	1.684	1.665	1.642	1.628	1.619	1.610	1.605	1.603	1.604	1.607	1.611	1.617	1.627	1.637
0.00800	(0.0785)	1.683	1.665	1.641	1.628	1.619	1.609	1.605	1.603	1.604	1.606	1.611	1.616	1.627	1.637
0.01000	(0.0981)	1.683	1.664	1.641	1.627	1.619	1.609	1.605	1.603	1.603	1.606	1.611	1.616	1.627	1.637
0.01500	(0.147)	1.682	1.663	1.640	1.627	1.618	1.609	1.604	1.603	1.603	1.606	1.611	1.616	1.627	1.637
0.02000	(0.196)	1.681	1.663	1.640	1.626	1.618	1.608	1.604	1.603	1.603	1.606	1.611	1.616	1.627	1.637
0.03000	(0.294)	1.681	1.662	1.640	1.626	1.618	1.608	1.604	1.602	1.603	1.606	1.611	1.616	1.627	1.637
0.04000	(0.392)	1.680	1.662	1.639	1.626	1.617	1.608	1.604	1.602	1.603	1.606	1.611	1.616	1.627	1.637
0.06000	(0.588)	1.680	1.662	1.639	1.626	1.617	1.608	1.604	1.602	1.603	1.606	1.610	1.616	1.627	1.637
0.08000	(0.785)	1.680	1.662	1.639	1.625	1.617	1.608	1.604	1.602	1.603	1.606	1.610	1.616	1.627	1.637
0.10000	(0.981)	1.679	1.661	1.639	1.625	1.617	1.608	1.603	1.602	1.603	1.606	1.610	1.616	1.627	1.637
0.15000	(1.47)	1.679	1.661	1.638	1.625	1.617	1.607	1.603	1.602	1.603	1.605	1.610	1.616	1.627	1.637
0.20000	(1.96)	1.679	1.661	1.638	1.625	1.617	1.607	1.603	1.602	1.603	1.605	1.610	1.616	1.627	1.637
0.30000	(2.94)	1.679	1.661	1.638	1.625	1.617	1.607	1.603	1.602	1.602	1.605	1.610	1.616	1.626	1.637
0.40000	(3.92)	1.679	1.661	1.638	1.625	1.616	1.607	1.603	1.602	1.602	1.605	1.610	1.616	1.626	1.637
0.60000	(5.88)	1.679	1.661	1.638	1.625	1.616	1.607	1.603	1.602	1.602	1.605	1.610	1.616	1.626	1.637
0.80000	(7.85)	1.678	1.661	1.638	1.625	1.616	1.607	1.603	1.602	1.602	1.605	1.610	1.616	1.626	1.637
1.00000	(9.81)	1.678	1.660	1.638	1.625	1.616	1.607	1.603	1.602	1.602	1.605	1.610	1.616	1.626	1.637
1.50000	(14.7)	1.678	1.660	1.638	1.625	1.616	1.607	1.603	1.602	1.602	1.605	1.610	1.616	1.626	1.637
2.00000	(19.6)	1.678	1.660	1.638	1.625	1.616	1.607	1.603	1.602	1.602	1.605	1.610	1.616	1.626	1.637
3.00000	(29.4)	1.678	1.660	1.638	1.625	1.616	1.607	1.603	1.602	1.602	1.605	1.610	1.616	1.626	1.637
S	$(\Delta p/\rho l)$	125	150	200	250	300	400	500	600	800	1000	1250	1500	2000	2500

Grad'nt k.p.g.   (Equivalent) Pipe diameters in mm   Roughness size, $k_s = 6.0$ mm

Kin. visc., $\nu = 1\cdot25\times10^{-6}$ m^{2}s^{-1}; $S = 0\cdot00010$ to $3\cdot00000$
i.e. kin. pr. grad., $\Delta p/\rho l = (0\cdot00098)$ to $(29\cdot4)$ ms^{-2}

Roughness size, $k_s = 15\cdot0$ mm
This table shows values of m, as follows
m_C for Colebrook-White solutions

Grad'nt S	k.p.g. $(\Delta p/\rho l)$	125	150	200	250	300	400	500	600	800	1000	1250	1500	2000	2500
0·00010	(0·00098)	2·156	2·107	2·042	2·002	1·974	1·939	1·918	1·903	1·887	1·878	1·872	1·870	1·870	1·873
0·00015	(0·00147)	2·151	2·102	2·039	1·999	1·972	1·937	1·916	1·902	1·886	1·877	1·872	1·869	1·870	1·873
0·00020	(0·00196)	2·148	2·099	2·037	1·997	1·970	1·936	1·915	1·901	1·885	1·876	1·871	1·869	1·869	1·872
0·00030	(0·00294)	2·144	2·096	2·034	1·995	1·968	1·934	1·913	1·900	1·884	1·876	1·870	1·868	1·869	1·872
0·00040	(0·00392)	2·141	2·094	2·032	1·994	1·967	1·933	1·913	1·899	1·883	1·875	1·870	1·868	1·868	1·872
0·00060	(0·00588)	2·138	2·092	2·030	1·992	1·966	1·932	1·912	1·898	1·883	1·875	1·869	1·868	1·868	1·871
0·00080	(0·00785)	2·137	2·090	2·029	1·991	1·965	1·932	1·911	1·898	1·882	1·874	1·869	1·867	1·868	1·871
0·00100	(0·00981)	2·136	2·089	2·029	1·991	1·965	1·931	1·911	1·898	1·882	1·874	1·869	1·867	1·868	1·871
0·00150	(0·0147)	2·134	2·088	2·027	1·990	1·964	1·930	1·910	1·897	1·882	1·874	1·869	1·867	1·868	1·871
0·00200	(0·0196)	2·133	2·087	2·027	1·989	1·963	1·930	1·910	1·897	1·881	1·873	1·869	1·867	1·867	1·871
0·00300	(0·0294)	2·132	2·086	2·026	1·988	1·963	1·930	1·910	1·897	1·881	1·873	1·868	1·867	1·867	1·871
0·00400	(0·0392)	2·131	2·085	2·025	1·988	1·962	1·929	1·909	1·896	1·881	1·873	1·868	1·866	1·867	1·871
0·00600	(0·0588)	2·130	2·084	2·025	1·987	1·962	1·929	1·909	1·896	1·881	1·873	1·868	1·866	1·867	1·870
0·00800	(0·0785)	2·129	2·084	2·025	1·987	1·962	1·929	1·909	1·896	1·881	1·873	1·868	1·866	1·867	1·870
0·01000	(0·0981)	2·129	2·084	2·024	1·987	1·961	1·929	1·909	1·896	1·881	1·873	1·868	1·866	1·867	1·870
0·01500	(0·147)	2·129	2·083	2·024	1·987	1·961	1·928	1·909	1·896	1·880	1·873	1·868	1·866	1·867	1·870
0·02000	(0·196)	2·128	2·083	2·024	1·987	1·961	1·928	1·909	1·896	1·880	1·873	1·868	1·866	1·867	1·870
0·03000	(0·294)	2·128	2·083	2·023	1·986	1·961	1·928	1·908	1·895	1·880	1·872	1·868	1·866	1·867	1·870
0·04000	(0·392)	2·128	2·082	2·023	1·986	1·961	1·928	1·908	1·895	1·880	1·872	1·868	1·866	1·867	1·870
0·06000	(0·588)	2·127	2·082	2·023	1·986	1·961	1·928	1·908	1·895	1·880	1·872	1·868	1·866	1·867	1·870
0·08000	(0·785)	2·127	2·082	2·023	1·986	1·960	1·928	1·908	1·895	1·880	1·872	1·868	1·866	1·867	1·870
0·10000	(0·981)	2·127	2·082	2·023	1·986	1·960	1·928	1·908	1·895	1·880	1·872	1·868	1·866	1·867	1·870
0·15000	(1·47)	2·127	2·082	2·023	1·986	1·960	1·928	1·908	1·895	1·880	1·872	1·868	1·866	1·867	1·870
0·20000	(1·96)	2·127	2·082	2·023	1·986	1·960	1·928	1·908	1·895	1·880	1·872	1·868	1·866	1·867	1·870
0·30000	(2·94)	2·127	2·081	2·023	1·986	1·960	1·928	1·908	1·895	1·880	1·872	1·868	1·866	1·867	1·870
0·40000	(3·92)	2·126	2·081	2·023	1·986	1·960	1·928	1·908	1·895	1·880	1·872	1·867	1·866	1·867	1·870
0·60000	(5·88)	2·126	2·081	2·023	1·986	1·960	1·928	1·908	1·895	1·880	1·872	1·867	1·866	1·867	1·870
0·80000	(7·85)	2·126	2·081	2·022	1·985	1·960	1·928	1·908	1·895	1·880	1·872	1·867	1·866	1·867	1·870
1·00000	(9·81)	2·126	2·081	2·022	1·985	1·960	1·928	1·908	1·895	1·880	1·872	1·867	1·866	1·867	1·870
1·50000	(14·7)	2·126	2·081	2·022	1·985	1·960	1·928	1·908	1·895	1·880	1·872	1·867	1·866	1·867	1·870
2·00000	(19·6)	2·126	2·081	2·022	1·985	1·960	1·928	1·908	1·895	1·880	1·872	1·867	1·866	1·867	1·870
3·00000	(29·4)	2·126	2·081	2·022	1·985	1·960	1·928	1·908	1·895	1·880	1·872	1·867	1·866	1·867	1·870
S	$(\Delta p/\rho l)$	125	150	200	250	300	400	500	600	800	1000	1250	1500	2000	2500

Grad'nt   k.p.g.   (Equivalent) Pipe diameters in mm   Roughness size, $k_s = 15\cdot0$ mm

Kin. visc., $\nu = 1\cdot25\times10^{-6}$ m^{2}s^{-1}; $S = 0\cdot00010$ to $3\cdot00000$
i.e. kin. pr. grad., $\Delta p/\rho l = (0\cdot00098)$ to $(29\cdot4)$ ms^{-2}

Roughness size, $k_s = 30\cdot0$ mm
This table shows values of m, as follows
m_C for Colebrook-White solutions

Grad'nt S	k.p.g. $(\Delta p/\rho l)$	125	150	200	250	300	400	500	600	800	1000	1250	1500	2000	2500
0·00010	(0·00098)		2·595	2·474	2·398	2·346	2·278	2·235	2·206	2·168	2·145	2·127	2·116	2·104	2·098
0·00015	(0·00147)		2·591	2·471	2·396	2·344	2·276	2·234	2·205	2·167	2·145	2·127	2·116	2·103	2·098
0·00020	(0·00196)		2·589	2·470	2·395	2·343	2·276	2·233	2·204	2·167	2·144	2·127	2·115	2·103	2·098
0·00030	(0·00294)		2·586	2·468	2·393	2·342	2·275	2·232	2·203	2·166	2·144	2·126	2·115	2·103	2·097
0·00040	(0·00392)		2·585	2·467	2·392	2·341	2·274	2·232	2·203	2·166	2·143	2·126	2·115	2·103	2·097
0·00060	(0·00588)		2·583	2·465	2·391	2·340	2·273	2·231	2·202	2·165	2·143	2·126	2·114	2·102	2·097
0·00080	(0·00785)		2·582	2·464	2·391	2·340	2·273	2·231	2·202	2·165	2·143	2·125	2·114	2·102	2·097
0·00100	(0·00981)		2·581	2·464	2·390	2·339	2·272	2·231	2·202	2·165	2·143	2·125	2·114	2·102	2·097
0·00150	(0·0147)		2·580	2·463	2·390	2·339	2·272	2·230	2·201	2·165	2·142	2·125	2·114	2·102	2·097
0·00200	(0·0196)		2·579	2·463	2·389	2·338	2·272	2·230	2·201	2·165	2·142	2·125	2·114	2·102	2·097
0·00300	(0·0294)		2·579	2·462	2·389	2·338	2·271	2·230	2·201	2·164	2·142	2·125	2·114	2·102	2·096
0·00400	(0·0392)		2·578	2·462	2·388	2·338	2·271	2·229	2·201	2·164	2·142	2·125	2·114	2·102	2·096
0·00600	(0·0588)		2·578	2·461	2·388	2·337	2·271	2·229	2·201	2·164	2·142	2·125	2·114	2·102	2·096
0·00800	(0·0785)		2·577	2·461	2·388	2·337	2·271	2·229	2·201	2·164	2·142	2·125	2·114	2·102	2·096
0·01000	(0·0981)		2·577	2·461	2·388	2·337	2·271	2·229	2·201	2·164	2·142	2·125	2·114	2·102	2·096
0·01500	(0·147)		2·577	2·460	2·387	2·337	2·271	2·229	2·200	2·164	2·142	2·124	2·114	2·102	2·096
0·02000	(0·196)		2·576	2·460	2·387	2·337	2·270	2·229	2·200	2·164	2·142	2·124	2·114	2·102	2·096
0·03000	(0·294)		2·576	2·460	2·387	2·336	2·270	2·229	2·200	2·164	2·142	2·124	2·113	2·101	2·096
0·04000	(0·392)		2·576	2·460	2·387	2·336	2·270	2·229	2·200	2·164	2·142	2·124	2·113	2·101	2·096
0·06000	(0·588)		2·576	2·460	2·387	2·336	2·270	2·229	2·200	2·164	2·142	2·124	2·113	2·101	2·096
0·08000	(0·785)		2·576	2·460	2·387	2·336	2·270	2·229	2·200	2·164	2·142	2·124	2·113	2·101	2·096
0·10000	(0·981)		2·576	2·460	2·387	2·336	2·270	2·229	2·200	2·164	2·142	2·124	2·113	2·101	2·096
0·15000	(1·47)		2·575	2·460	2·387	2·336	2·270	2·229	2·200	2·164	2·142	2·124	2·113	2·101	2·096
0·20000	(1·96)		2·575	2·459	2·387	2·336	2·270	2·229	2·200	2·164	2·142	2·124	2·113	2·101	2·096
0·30000	(2·94)		2·575	2·459	2·387	2·336	2·270	2·229	2·200	2·164	2·142	2·124	2·113	2·101	2·096
0·40000	(3·92)		2·575	2·459	2·387	2·336	2·270	2·229	2·200	2·164	2·142	2·124	2·113	2·101	2·096
0·60000	(5·88)		2·575	2·459	2·387	2·336	2·270	2·229	2·200	2·164	2·141	2·124	2·113	2·101	2·096
0·80000	(7·85)		2·575	2·459	2·386	2·336	2·270	2·229	2·200	2·164	2·141	2·124	2·113	2·101	2·096
1·00000	(9·81)		2·575	2·459	2·386	2·336	2·270	2·229	2·200	2·164	2·141	2·124	2·113	2·101	2·096
1·50000	(14·7)		2·575	2·459	2·386	2·336	2·270	2·228	2·200	2·164	2·141	2·124	2·113	2·101	2·096
2·00000	(19·6)		2·575	2·459	2·386	2·336	2·270	2·228	2·200	2·164	2·141	2·124	2·113	2·101	2·096
3·00000	(29·4)		2·575	2·459	2·386	2·336	2·270	2·228	2·200	2·164	2·141	2·124	2·113	2·101	2·096
S	$(\Delta p/\rho l)$	125	150	200	250	300	400	500	600	800	1000	1250	1500	2000	2500

Grad'nt   k.p.g.   (Equivalent) Pipe diameters in mm   Roughness size, $k_s = 30\cdot0$ mm

E115

Kin. visc., $\nu = 1.5\times10^{-6}\ \mathrm{m^2\,s^{-1}}$;
$S = 0.00010$ to 30.0000

i.e. kin. pr. grad., $\Delta p/\rho l =$
(0.00098) to $(294)\ \mathrm{ms^{-2}}$

Roughness size, $k_s = 0.0015$ mm
This table shows values of m, as follows

m_C for Colebrook-White solutions; or,
where $\mathbf{R} \le 2000$, m_P for laminar flow

Grad'nt   k.p.g.   (Equivalent) Pipe diameters in mm

S	$(\Delta p/\rho l)$	6	8	10	12·5	15	20	25	30	40	50	60	80	100	125
0·00030	(0·00294)	10·29	7·009	5·205	3·866	3·032	2·066	1·534	1·203	*	1·154	1·132	1·105	1·087	1·073
0·00040	(0·00392)	8·908	6·070	4·508	3·348	2·625	1·789	1·329	1·042	*	1·124	1·105	1·079	1·063	1·051
0·00060	(0·00588)	7·273	4·956	3·681	2·734	2·144	1·461	1·085	*	1·109	1·085	1·068	1·046	1·032	1·021
0·00080	(0·00785)	6·299	4·292	3·188	2·367	1·856	1·265	0·939	*	1·081	1·058	1·043	1·023	1·011	1·001
0·00100	(0·00981)	5·634	3·839	2·851	2·117	1·660	1·131	*	*	1·060	1·039	1·025	1·006	0·995	0·986
0·00150	(0·0147)	4·600	3·135	2·328	1·729	1·356	0·924	*	1·053	1·023	1·005	0·993	0·977	0·967	0·960
0·00200	(0·0196)	3·984	2·715	2·016	1·497	1·174	*	1·046	1·026	0·999	0·983	0·971	0·957	0·949	0·942
0·00300	(0·0294)	3·253	2·216	1·646	1·222	0·959	*	1·008	0·990	0·967	0·953	0·943	0·931	0·923	0·918
0·00400	(0·0392)	2·817	1·920	1·426	1·059	0·830	1·006	0·982	0·966	0·945	0·932	0·924	0·913	0·906	0·902
0·00600	(0·0588)	2·300	1·567	1·164	0·864	*	0·969	0·949	0·935	0·916	0·905	0·898	0·889	0·884	0·880
0·00800	(0·0785)	1·992	1·357	1·008	*	*	0·945	0·926	0·913	0·897	0·887	0·880	0·872	0·868	0·865
0·01000	(0·0981)	1·782	1·214	0·902	*	0·954	0·927	0·909	0·898	0·882	0·873	0·867	0·860	0·856	0·854
0·01500	(0·147)	1·455	0·991	*	*	0·920	0·896	0·881	0·870	0·857	0·849	0·844	0·839	0·836	0·835
0·02000	(0·196)	1·260	0·858	*	0·913	0·897	0·875	0·861	0·852	0·840	0·833	0·829	0·824	0·822	0·822
0·03000	(0·294)	1·029	*	*	0·881	0·866	0·847	0·835	0·827	0·818	0·812	0·808	0·805	0·804	0·804
0·04000	(0·392)	0·891	*	0·878	0·859	0·846	0·829	0·818	0·811	0·802	0·797	0·794	0·792	0·791	0·792
0·06000	(0·588)	0·727	*	0·847	0·830	0·819	0·804	0·795	0·789	0·782	0·778	0·776	0·774	0·774	0·775
0·08000	(0·785)	*	0·845	0·826	0·811	0·801	0·787	0·779	0·774	0·768	0·764	0·763	0·762	0·762	0·764
0·10000	(0·981)	*	0·828	0·811	0·797	0·787	0·775	0·767	0·763	0·757	0·754	0·753	0·753	0·754	0·756
0·15000	(1·47)	0·823	0·800	0·785	0·772	0·764	0·753	0·747	0·743	0·739	0·737	0·736	0·737	0·738	0·741
0·20000	(1·96)	0·802	0·780	0·767	0·756	0·748	0·739	0·733	0·730	0·727	0·725	0·725	0·726	0·728	0·731
0·30000	(2·94)	0·774	0·755	0·743	0·734	0·727	0·719	0·715	0·712	0·710	0·709	0·710	0·712	0·714	0·717
0·40000	(3·92)	0·755	0·738	0·727	0·719	0·713	0·706	0·703	0·701	0·699	0·699	0·699	0·702	0·705	0·708
0·60000	(5·88)	0·730	0·716	0·706	0·699	0·694	0·689	0·686	0·685	0·684	0·684	0·686	0·689	0·692	0·696
0·80000	(7·85)	0·714	0·700	0·692	0·686	0·682	0·677	0·675	0·674	0·674	0·675	0·676	0·680	0·683	0·688
1·00000	(9·81)	0·702	0·689	0·682	0·676	0·672	0·668	0·666	0·666	0·666	0·667	0·669	0·673	0·677	0·682
1·50000	(14·7)	0·681	0·670	0·664	0·659	0·656	0·653	0·652	0·652	0·653	0·655	0·657	0·661	0·666	0·671
2·00000	(19·6)	0·666	0·657	0·651	0·647	0·645	0·643	0·642	0·642	0·644	0·646	0·649	0·654	0·658	0·664
3·00000	(29·4)	0·648	0·640	0·635	0·632	0·630	0·629	0·629	0·630	0·632	0·635	0·638	0·643	0·648	0·654
4·00000	(39·2)	0·635	0·628	0·624	0·621	0·620	0·619	0·620	0·621	0·624	0·627	0·630	0·636	0·642	0·648
6·00000	(58·8)	0·618	0·613	0·610	0·608	0·607	0·607	0·608	0·610	0·613	0·617	0·621	0·627	0·633	0·639
8·00000	(78·5)	0·607	0·602	0·600	0·598	0·598	0·599	0·600	0·602	0·606	0·610	0·614	0·621	0·627	0·634
10·0000	(98·1)	0·599	0·595	0·593	0·592	0·592	0·593	0·595	0·597	0·601	0·605	0·609	0·616	0·623	0·630
15·0000	(147)	0·584	0·581	0·580	0·580	0·580	0·582	0·585	0·587	0·592	0·597	0·601	0·609	0·616	0·623
20·0000	(196)	0·575	0·572	0·572	0·572	0·573	0·575	0·578	0·581	0·586	0·591	0·596	0·604	0·611	0·618
30·0000	(294)	0·562	0·561	0·561	0·562	0·563	0·566	0·569	0·572	0·578	0·584	0·589	0·597	0·605	0·613
S	$(\Delta p/\rho l)$	6	8	10	12·5	15	20	25	30	40	50	60	80	100	125

Grad'nt   k.p.g.   (Equivalent) Pipe diameters in mm

S	$(\Delta p/\rho l)$	125	150	200	250	300	400	500	600	800	1000	1250	1500	2000	2500
0·00010	(0·00098)	1·167	1·152	1·134	1·123	1·116	1·108	1·103	1·101	1·100	1·101	1·104	1·107	1·113	1·119
0·00015	(0·00147)	1·130	1·118	1·102	1·093	1·087	1·080	1·077	1·076	1·076	1·078	1·081	1·085	1·092	1·099
0·00020	(0·00196)	1·106	1·094	1·080	1·072	1·067	1·062	1·060	1·059	1·060	1·062	1·066	1·070	1·077	1·084
0·00030	(0·00294)	1·073	1·063	1·051	1·045	1·041	1·037	1·035	1·036	1·038	1·041	1·045	1·049	1·058	1·065
0·00040	(0·00392)	1·051	1·042	1·032	1·026	1·023	1·020	1·019	1·020	1·022	1·026	1·031	1·035	1·044	1·052
0·00060	(0·00588)	1·021	1·014	1·005	1·001	0·998	0·996	0·997	0·998	1·002	1·006	1·011	1·016	1·026	1·034
0·00080	(0·00785)	1·001	0·995	0·987	0·983	0·982	0·981	0·982	0·983	0·988	0·992	0·998	1·003	1·013	1·022
0·00100	(0·00981)	0·986	0·980	0·974	0·971	0·969	0·969	0·970	0·972	0·977	0·982	0·988	0·993	1·003	1·012
0·00150	(0·0147)	0·960	0·955	0·950	0·948	0·947	0·948	0·950	0·953	0·958	0·963	0·970	0·976	0·986	0·996
0·00200	(0·0196)	0·942	0·938	0·934	0·933	0·932	0·934	0·936	0·939	0·945	0·951	0·958	0·964	0·975	0·984
0·00300	(0·0294)	0·918	0·915	0·912	0·912	0·912	0·915	0·918	0·921	0·928	0·934	0·941	0·947	0·959	0·969
0·00400	(0·0392)	0·902	0·900	0·898	0·898	0·898	0·901	0·905	0·909	0·916	0·922	0·930	0·936	0·948	0·958
0·00600	(0·0588)	0·880	0·879	0·878	0·878	0·880	0·884	0·888	0·892	0·899	0·906	0·914	0·921	0·933	0·944
0·00800	(0·0785)	0·865	0·864	0·864	0·865	0·867	0·871	0·876	0·880	0·888	0·895	0·903	0·911	0·923	0·934
0·01000	(0·0981)	0·854	0·853	0·854	0·855	0·858	0·862	0·867	0·871	0·880	0·887	0·895	0·903	0·915	0·926
0·01500	(0·147)	0·835	0·835	0·836	0·838	0·841	0·846	0·851	0·856	0·865	0·873	0·881	0·889	0·902	0·913
0·02000	(0·196)	0·822	0·822	0·824	0·826	0·829	0·835	0·840	0·846	0·855	0·863	0·872	0·879	0·893	0·904
0·03000	(0·294)	0·804	0·804	0·807	0·810	0·814	0·820	0·826	0·831	0·841	0·849	0·859	0·867	0·880	0·892
0·04000	(0·392)	0·792	0·793	0·796	0·799	0·803	0·810	0·816	0·822	0·832	0·840	0·850	0·858	0·872	0·884
0·06000	(0·588)	0·775	0·777	0·781	0·785	0·789	0·796	0·803	0·809	0·819	0·828	0·838	0·846	0·860	0·872
0·08000	(0·785)	0·764	0·766	0·770	0·775	0·779	0·787	0·793	0·800	0·810	0·820	0·829	0·838	0·853	0·865
0·10000	(0·981)	0·756	0·758	0·762	0·767	0·771	0·779	0·787	0·793	0·804	0·813	0·823	0·832	0·847	0·859
0·15000	(1·47)	0·741	0·743	0·749	0·754	0·759	0·767	0·774	0·781	0·793	0·802	0·813	0·822	0·837	0·849
0·20000	(1·96)	0·731	0·734	0·740	0·745	0·750	0·759	0·766	0·773	0·785	0·795	0·805	0·815	0·830	0·843
0·30000	(2·94)	0·717	0·721	0·727	0·733	0·738	0·747	0·755	0·762	0·775	0·785	0·796	0·805	0·821	0·834
0·40000	(3·92)	0·708	0·712	0·719	0·725	0·730	0·740	0·748	0·755	0·768	0·778	0·789	0·799	0·815	0·829
0·60000	(5·88)	0·696	0·700	0·707	0·714	0·719	0·730	0·738	0·746	0·759	0·769	0·781	0·791	0·807	0·821
0·80000	(7·85)	0·688	0·692	0·700	0·706	0·712	0·723	0·732	0·739	0·753	0·764	0·775	0·785	0·802	0·816
1·00000	(9·81)	0·682	0·686	0·694	0·701	0·707	0·718	0·727	0·735	0·748	0·759	0·771	0·781	0·798	0·813
1·50000	(14·7)	0·671	0·676	0·684	0·691	0·698	0·709	0·718	0·726	0·740	0·752	0·764	0·775	0·792	0·807
2·00000	(19·6)	0·664	0·669	0·677	0·685	0·692	0·703	0·713	0·721	0·735	0·747	0·759	0·770	0·788	0·803
3·00000	(29·4)	0·654	0·659	0·669	0·677	0·684	0·695	0·705	0·714	0·729	0·741	0·754	0·765	0·783	0·798
S	$(\Delta p/\rho l)$	125	150	200	250	300	400	500	600	800	1000	1250	1500	2000	2500

Grad'nt   k.p.g.   (Equivalent) Pipe diameters in mm

Kinematic viscosity, $\nu = 1.5\times10^{-6}\ \mathrm{m^2\,s^{-1}}$;   **Roughness size, $k_s = 0.0015$ mm**

Kin. visc., $\nu = 1.5 \times 10^{-6}$ m^{2}s^{-1}; $S = 0.00010$ to 30.0000

i.e. kin. pr. grad., $\Delta p/\rho l =$ (0.00098) to (294) ms^{-2}

Roughness size, $k_s = 0.003$ mm

This table shows values of m, as follows m_C for Colebrook-White solutions; or, where $\mathbf{R} \leq 2000$, m_P for laminar flow

Grad'nt S	k.p.g. $(\Delta p/\rho l)$	6	8	10	12·5	15	20	25	30	40	50	60	80	100	125
0.00030	(0.00294)	10.29	7.009	5.205	3.866	3.032	2.066	1.534	1.203	*	1.154	1.133	1.105	1.088	1.073
0.00040	(0.00392)	8.908	6.070	4.508	3.348	2.625	1.789	1.329	1.042	*	1.125	1.105	1.080	1.064	1.051
0.00060	(0.00588)	7.273	4.956	3.681	2.734	2.144	1.461	1.085	*	1.110	1.085	1.068	1.046	1.033	1.022
0.00080	(0.00785)	6.299	4.292	3.188	2.367	1.856	1.265	0.939	*	1.081	1.059	1.044	1.024	1.011	1.002
0.00100	(0.00981)	5.634	3.839	2.851	2.117	1.660	1.131	*	*	1.060	1.040	1.025	1.007	0.995	0.987
0.00150	(0.0147)	4.600	3.135	2.328	1.729	1.356	0.924	*	1.053	1.024	1.006	0.993	0.978	0.968	0.961
0.00200	(0.0196)	3.984	2.715	2.016	1.497	1.174	*	1.047	1.026	1.000	0.983	0.972	0.958	0.949	0.943
0.00300	(0.0294)	3.253	2.216	1.646	1.222	0.959	*	1.008	0.991	0.968	0.953	0.944	0.931	0.924	0.919
0.00400	(0.0392)	2.817	1.920	1.426	1.059	0.830	1.006	0.983	0.967	0.946	0.933	0.924	0.914	0.908	0.903
0.00600	(0.0588)	2.300	1.567	1.164	0.864	*	0.970	0.950	0.935	0.917	0.906	0.899	0.890	0.885	0.882
0.00800	(0.0785)	1.992	1.357	1.008	*	*	0.946	0.927	0.914	0.898	0.888	0.881	0.874	0.870	0.867
0.01000	(0.0981)	1.782	1.214	0.902	*	0.955	0.928	0.910	0.899	0.884	0.874	0.869	0.862	0.858	0.856
0.01500	(0.147)	1.455	0.991	*	*	0.921	0.897	0.882	0.872	0.859	0.851	0.846	0.841	0.838	0.837
0.02000	(0.196)	1.260	0.858	*	0.915	0.898	0.876	0.863	0.853	0.842	0.835	0.831	0.826	0.824	0.824
0.03000	(0.294)	1.029	*	*	0.882	0.868	0.849	0.837	0.829	0.819	0.814	0.810	0.807	0.806	0.806
0.04000	(0.392)	0.891	*	0.880	0.861	0.847	0.830	0.820	0.813	0.804	0.799	0.797	0.794	0.794	0.794
0.06000	(0.588)	0.727	*	0.849	0.832	0.821	0.806	0.797	0.791	0.784	0.780	0.778	0.777	0.777	0.778
0.08000	(0.785)	*	0.847	0.828	0.813	0.803	0.789	0.781	0.776	0.770	0.767	0.766	0.765	0.766	0.767
0.10000	(0.981)	*	0.830	0.813	0.799	0.789	0.777	0.770	0.765	0.760	0.757	0.756	0.756	0.757	0.759
0.15000	(1.47)	0.826	0.802	0.787	0.775	0.766	0.756	0.750	0.746	0.742	0.740	0.740	0.741	0.743	0.745
0.20000	(1.96)	0.804	0.783	0.769	0.758	0.751	0.742	0.736	0.733	0.730	0.729	0.729	0.730	0.733	0.736
0.30000	(2.94)	0.777	0.758	0.746	0.737	0.730	0.723	0.719	0.716	0.714	0.714	0.714	0.717	0.719	0.723
0.40000	(3.92)	0.758	0.741	0.731	0.722	0.717	0.710	0.707	0.705	0.703	0.704	0.705	0.707	0.711	0.715
0.60000	(5.88)	0.734	0.719	0.710	0.703	0.698	0.693	0.691	0.689	0.689	0.690	0.691	0.695	0.699	0.703
0.80000	(7.85)	0.718	0.704	0.696	0.690	0.686	0.682	0.680	0.679	0.679	0.681	0.683	0.687	0.691	0.696
1.00000	(9.81)	0.706	0.693	0.686	0.680	0.677	0.673	0.672	0.671	0.672	0.674	0.676	0.681	0.685	0.690
1.50000	(14.7)	0.685	0.675	0.669	0.664	0.661	0.659	0.658	0.658	0.660	0.662	0.665	0.670	0.675	0.681
2.00000	(19.6)	0.671	0.662	0.657	0.653	0.651	0.649	0.649	0.649	0.652	0.654	0.657	0.663	0.668	0.674
3.00000	(29.4)	0.653	0.646	0.641	0.638	0.637	0.636	0.637	0.638	0.641	0.644	0.648	0.654	0.660	0.666
4.00000	(39.2)	0.641	0.635	0.631	0.629	0.628	0.628	0.629	0.630	0.634	0.637	0.641	0.648	0.654	0.661
6.00000	(58.8)	0.625	0.620	0.617	0.616	0.616	0.616	0.618	0.620	0.624	0.629	0.633	0.640	0.647	0.654
8.00000	(78.5)	0.615	0.610	0.608	0.607	0.608	0.609	0.611	0.613	0.618	0.623	0.627	0.635	0.642	0.650
10.0000	(98.1)	0.607	0.603	0.602	0.601	0.602	0.603	0.606	0.609	0.614	0.619	0.623	0.631	0.639	0.646
15.0000	(147)	0.594	0.591	0.590	0.591	0.592	0.594	0.597	0.600	0.606	0.612	0.617	0.625	0.633	0.641
20.0000	(196)	0.585	0.583	0.583	0.584	0.585	0.588	0.592	0.595	0.601	0.607	0.612	0.621	0.629	0.638
30.0000	(294)	0.574	0.573	0.573	0.575	0.577	0.581	0.585	0.588	0.595	0.601	0.607	0.617	0.625	0.634
S $(\Delta p/\rho l)$		6	8	10	12·5	15	20	25	30	40	50	60	80	100	125

Grad'nt   k.p.g.   (Equivalent) Pipe diameters in mm

S	$(\Delta p/\rho l)$	125	150	200	250	300	400	500	600	800	1000	1250	1500	2000	2500
0.00010	(0.00098)	1.167	1.153	1.134	1.123	1.116	1.108	1.104	1.102	1.101	1.102	1.104	1.107	1.114	1.120
0.00015	(0.00147)	1.131	1.118	1.102	1.093	1.087	1.081	1.078	1.077	1.077	1.079	1.082	1.085	1.093	1.099
0.00020	(0.00196)	1.106	1.095	1.081	1.073	1.068	1.062	1.060	1.059	1.061	1.063	1.067	1.070	1.078	1.085
0.00030	(0.00294)	1.073	1.064	1.052	1.045	1.041	1.037	1.036	1.036	1.038	1.041	1.046	1.050	1.059	1.066
0.00040	(0.00392)	1.051	1.043	1.032	1.026	1.023	1.020	1.020	1.020	1.023	1.027	1.032	1.036	1.045	1.053
0.00060	(0.00588)	1.022	1.014	1.006	1.001	0.999	0.997	0.998	0.999	1.003	1.007	1.012	1.017	1.027	1.035
0.00080	(0.00785)	1.002	0.995	0.988	0.984	0.982	0.982	0.983	0.984	0.989	0.993	0.999	1.005	1.014	1.023
0.00100	(0.00981)	0.987	0.981	0.974	0.971	0.970	0.970	0.971	0.973	0.978	0.983	0.989	0.995	1.005	1.014
0.00150	(0.0147)	0.961	0.956	0.951	0.949	0.948	0.949	0.951	0.954	0.959	0.965	0.972	0.978	0.988	0.998
0.00200	(0.0196)	0.943	0.939	0.935	0.934	0.934	0.935	0.938	0.941	0.947	0.953	0.960	0.966	0.977	0.987
0.00300	(0.0294)	0.919	0.916	0.914	0.913	0.914	0.916	0.919	0.923	0.930	0.936	0.943	0.950	0.962	0.972
0.00400	(0.0392)	0.903	0.901	0.899	0.899	0.900	0.903	0.907	0.911	0.918	0.925	0.932	0.939	0.951	0.961
0.00600	(0.0588)	0.882	0.880	0.879	0.880	0.882	0.886	0.890	0.894	0.902	0.909	0.917	0.924	0.937	0.948
0.00800	(0.0785)	0.867	0.866	0.866	0.867	0.869	0.874	0.878	0.883	0.891	0.898	0.907	0.914	0.927	0.938
0.01000	(0.0981)	0.856	0.855	0.856	0.857	0.860	0.865	0.870	0.874	0.883	0.891	0.899	0.907	0.920	0.931
0.01500	(0.147)	0.837	0.837	0.838	0.841	0.843	0.849	0.854	0.859	0.869	0.877	0.886	0.894	0.907	0.919
0.02000	(0.196)	0.824	0.824	0.826	0.829	0.832	0.838	0.844	0.849	0.859	0.867	0.876	0.885	0.899	0.910
0.03000	(0.294)	0.806	0.807	0.810	0.813	0.817	0.824	0.830	0.836	0.846	0.855	0.864	0.873	0.887	0.899
0.04000	(0.392)	0.794	0.796	0.799	0.803	0.807	0.814	0.820	0.826	0.837	0.846	0.856	0.865	0.879	0.892
0.06000	(0.588)	0.778	0.780	0.784	0.789	0.793	0.801	0.808	0.814	0.825	0.835	0.845	0.854	0.869	0.882
0.08000	(0.785)	0.767	0.770	0.774	0.779	0.784	0.792	0.799	0.806	0.817	0.827	0.837	0.847	0.862	0.875
0.10000	(0.981)	0.759	0.762	0.767	0.772	0.777	0.785	0.793	0.799	0.811	0.821	0.832	0.841	0.857	0.870
0.15000	(1.47)	0.745	0.748	0.754	0.760	0.765	0.774	0.782	0.789	0.801	0.811	0.822	0.832	0.848	0.862
0.20000	(1.96)	0.736	0.739	0.745	0.751	0.757	0.766	0.774	0.782	0.794	0.805	0.816	0.826	0.843	0.856
0.30000	(2.94)	0.723	0.727	0.734	0.740	0.746	0.756	0.764	0.772	0.785	0.796	0.808	0.818	0.835	0.849
0.40000	(3.92)	0.715	0.719	0.726	0.733	0.739	0.749	0.758	0.766	0.779	0.791	0.803	0.813	0.830	0.845
0.60000	(5.88)	0.703	0.708	0.716	0.723	0.729	0.740	0.749	0.758	0.772	0.783	0.796	0.806	0.824	0.839
0.80000	(7.85)	0.696	0.700	0.709	0.716	0.723	0.734	0.744	0.752	0.767	0.779	0.791	0.802	0.820	0.835
1.00000	(9.81)	0.690	0.695	0.704	0.711	0.718	0.730	0.740	0.748	0.763	0.775	0.788	0.799	0.817	0.833
1.50000	(14.7)	0.681	0.686	0.695	0.703	0.710	0.723	0.733	0.742	0.757	0.769	0.783	0.794	0.813	0.828
2.00000	(19.6)	0.674	0.680	0.690	0.698	0.705	0.718	0.729	0.738	0.753	0.766	0.779	0.791	0.810	0.826
3.00000	(29.4)	0.666	0.672	0.682	0.691	0.699	0.712	0.723	0.732	0.748	0.761	0.775	0.787	0.806	0.822
S $(\Delta p/\rho l)$		125	150	200	250	300	400	500	600	800	1000	1250	1500	2000	2500

Grad'nt   k.p.g.   (Equivalent) Pipe diameters in mm

Kinematic viscosity, $\nu = 1.5 \times 10^{-6}$ m^{2}s^{-1} ;   **Roughness size, $k_s = 0.003$ mm**

E117

Kin. visc., $\nu = 1.5 \times 10^{-6}$ m^{2}s^{-1};
S = 0.00010 to 30.0000

i.e. kin. pr. grad., $\Delta p/\rho l$ =
(0.00098) to (294) ms^{-2}

Roughness size, k_s = 0.006 mm
This table shows values of m, as follows

m_C for Colebrook-White solutions; or,
where $\mathbf{R} \leq 2000$, m_P for laminar flow

Grad'nt S	k.p.g. $(\Delta p/\rho l)$	6	8	10	12.5	15	20	25	30	40	50	60	80	100	125
		(Equivalent) Pipe diameters in mm													
0.00030	(0.00294)	10.29	7.009	5.205	3.866	3.032	2.066	1.534	1.203	*	1.155	1.134	1.106	1.088	1.074
0.00040	(0.00392)	8.908	6.070	4.508	3.348	2.625	1.789	1.329	1.042	*	1.125	1.106	1.081	1.065	1.052
0.00060	(0.00588)	7.273	4.956	3.681	2.734	2.144	1.461	1.085	*	1.111	1.086	1.069	1.047	1.034	1.023
0.00080	(0.00785)	6.299	4.292	3.188	2.367	1.856	1.265	0.939	*	1.082	1.060	1.045	1.025	1.013	1.003
0.00100	(0.00981)	5.634	3.839	2.851	2.117	1.660	1.131	*	*	1.061	1.041	1.026	1.008	0.997	0.988
0.00150	(0.0147)	4.600	3.135	2.328	1.729	1.356	0.924	*	1.054	1.025	1.007	0.995	0.979	0.969	0.962
0.00200	(0.0196)	3.984	2.715	2.016	1.497	1.174	*	1.048	1.028	1.001	0.985	0.974	0.959	0.951	0.945
0.00300	(0.0294)	3.253	2.216	1.646	1.222	0.959	*	1.010	0.992	0.969	0.955	0.945	0.933	0.926	0.921
0.00400	(0.0392)	2.817	1.920	1.426	1.059	0.830	1.008	0.985	0.969	0.948	0.935	0.926	0.916	0.910	0.905
0.00600	(0.0588)	2.300	1.567	1.164	0.864	*	0.972	0.951	0.937	0.919	0.908	0.901	0.892	0.887	0.884
0.00800	(0.0785)	1.992	1.357	1.008	*	*	0.948	0.929	0.916	0.900	0.890	0.884	0.876	0.872	0.870
0.01000	(0.0981)	1.782	1.214	0.902	*	0.957	0.930	0.913	0.901	0.886	0.877	0.871	0.864	0.861	0.859
0.01500	(0.147)	1.455	0.991	*	*	0.923	0.899	0.884	0.874	0.861	0.854	0.849	0.844	0.841	0.840
0.02000	(0.196)	1.260	0.858	*	0.917	0.900	0.879	0.865	0.856	0.845	0.838	0.834	0.830	0.828	0.828
0.03000	(0.294)	1.029	*	*	0.885	0.871	0.852	0.840	0.832	0.823	0.817	0.814	0.811	0.811	0.811
0.04000	(0.392)	0.891	*	0.883	0.864	0.851	0.834	0.823	0.816	0.808	0.803	0.801	0.799	0.799	0.800
0.06000	(0.588)	0.727	*	0.852	0.836	0.824	0.810	0.801	0.795	0.788	0.785	0.783	0.782	0.783	0.784
0.08000	(0.785)	*	0.850	0.832	0.817	0.807	0.794	0.786	0.781	0.775	0.772	0.771	0.771	0.772	0.774
0.10000	(0.981)	*	0.834	0.817	0.803	0.794	0.782	0.775	0.770	0.765	0.763	0.762	0.763	0.764	0.767
0.15000	(1.47)	0.830	0.806	0.791	0.779	0.771	0.761	0.755	0.752	0.748	0.747	0.747	0.748	0.750	0.754
0.20000	(1.96)	0.809	0.788	0.774	0.764	0.756	0.747	0.742	0.740	0.737	0.736	0.737	0.739	0.741	0.745
0.30000	(2.94)	0.782	0.763	0.752	0.743	0.737	0.729	0.725	0.723	0.722	0.722	0.723	0.726	0.729	0.734
0.40000	(3.92)	0.764	0.747	0.737	0.729	0.723	0.717	0.714	0.713	0.712	0.713	0.714	0.718	0.721	0.726
0.60000	(5.88)	0.740	0.726	0.717	0.710	0.706	0.701	0.699	0.698	0.699	0.700	0.702	0.706	0.711	0.716
0.80000	(7.85)	0.725	0.712	0.704	0.698	0.694	0.691	0.689	0.689	0.690	0.692	0.694	0.699	0.704	0.710
1.00000	(9.81)	0.713	0.701	0.694	0.689	0.686	0.683	0.682	0.682	0.683	0.686	0.689	0.694	0.699	0.705
1.50000	(14.7)	0.694	0.684	0.678	0.674	0.671	0.670	0.669	0.670	0.673	0.676	0.679	0.685	0.691	0.697
2.00000	(19.6)	0.681	0.672	0.667	0.664	0.662	0.661	0.661	0.662	0.666	0.669	0.673	0.679	0.685	0.692
3.00000	(29.4)	0.664	0.657	0.653	0.650	0.649	0.649	0.651	0.652	0.656	0.660	0.664	0.672	0.678	0.686
4.00000	(39.2)	0.653	0.647	0.644	0.642	0.641	0.642	0.644	0.646	0.650	0.655	0.659	0.667	0.674	0.682
6.00000	(58.8)	0.638	0.634	0.631	0.631	0.631	0.632	0.635	0.637	0.643	0.648	0.653	0.661	0.668	0.677
8.00000	(78.5)	0.629	0.625	0.624	0.623	0.624	0.626	0.629	0.632	0.638	0.643	0.648	0.657	0.665	0.673
10.0000	(98.1)	0.622	0.619	0.618	0.618	0.619	0.622	0.625	0.628	0.634	0.640	0.645	0.655	0.662	0.671
15.0000	(147)	0.610	0.608	0.608	0.609	0.611	0.614	0.618	0.622	0.629	0.635	0.640	0.650	0.658	0.667
20.0000	(196)	0.603	0.602	0.602	0.604	0.605	0.610	0.614	0.618	0.625	0.631	0.637	0.647	0.656	0.665
30.0000	(294)	0.593	0.593	0.594	0.596	0.599	0.604	0.608	0.613	0.620	0.627	0.633	0.644	0.653	0.662
S	$(\Delta p/\rho l)$	6	8	10	12.5	15	20	25	30	40	50	60	80	100	125

Grad'nt S	k.p.g. $(\Delta p/\rho l)$	125	150	200	250	300	400	500	600	800	1000	1250	1500	2000	2500
		(Equivalent) Pipe diameters in mm													
0.00010	(0.00098)	1.168	1.153	1.135	1.124	1.117	1.109	1.105	1.103	1.102	1.103	1.105	1.108	1.115	1.121
0.00015	(0.00147)	1.131	1.119	1.103	1.094	1.088	1.082	1.079	1.078	1.078	1.080	1.083	1.087	1.094	1.101
0.00020	(0.00196)	1.107	1.096	1.082	1.074	1.069	1.063	1.061	1.061	1.062	1.064	1.068	1.072	1.080	1.087
0.00030	(0.00294)	1.074	1.065	1.053	1.046	1.042	1.039	1.037	1.038	1.040	1.043	1.048	1.052	1.061	1.069
0.00040	(0.00392)	1.052	1.044	1.033	1.028	1.024	1.022	1.021	1.022	1.025	1.029	1.034	1.039	1.048	1.056
0.00060	(0.00588)	1.023	1.016	1.007	1.003	1.000	0.999	0.999	1.001	1.005	1.009	1.015	1.020	1.030	1.039
0.00080	(0.00785)	1.003	0.997	0.989	0.986	0.984	0.983	0.985	0.986	0.991	0.996	1.002	1.007	1.018	1.027
0.00100	(0.00981)	0.988	0.982	0.976	0.973	0.972	0.972	0.973	0.976	0.981	0.986	0.992	0.998	1.009	1.018
0.00150	(0.0147)	0.962	0.958	0.953	0.951	0.950	0.952	0.954	0.957	0.962	0.968	0.975	0.981	0.993	1.002
0.00200	(0.0196)	0.945	0.941	0.937	0.936	0.936	0.938	0.941	0.944	0.950	0.956	0.963	0.970	0.982	0.992
0.00300	(0.0294)	0.921	0.918	0.916	0.916	0.916	0.919	0.923	0.926	0.933	0.940	0.948	0.955	0.967	0.978
0.00400	(0.0392)	0.905	0.903	0.902	0.902	0.903	0.906	0.910	0.914	0.922	0.929	0.937	0.944	0.957	0.968
0.00600	(0.0588)	0.884	0.883	0.882	0.883	0.885	0.889	0.894	0.898	0.907	0.914	0.923	0.931	0.944	0.955
0.00800	(0.0785)	0.870	0.869	0.869	0.871	0.873	0.878	0.883	0.888	0.897	0.904	0.913	0.921	0.935	0.946
0.01000	(0.0981)	0.859	0.859	0.859	0.861	0.864	0.869	0.875	0.880	0.889	0.897	0.906	0.914	0.928	0.940
0.01500	(0.147)	0.840	0.841	0.842	0.845	0.848	0.854	0.860	0.866	0.875	0.884	0.894	0.902	0.917	0.929
0.02000	(0.196)	0.828	0.828	0.831	0.834	0.838	0.844	0.850	0.856	0.866	0.876	0.885	0.894	0.909	0.922
0.03000	(0.294)	0.811	0.812	0.816	0.819	0.823	0.831	0.837	0.844	0.855	0.864	0.874	0.883	0.899	0.912
0.04000	(0.392)	0.800	0.801	0.805	0.810	0.814	0.822	0.829	0.835	0.847	0.856	0.867	0.876	0.892	0.906
0.06000	(0.588)	0.784	0.787	0.791	0.796	0.801	0.810	0.817	0.824	0.836	0.846	0.858	0.867	0.884	0.897
0.08000	(0.785)	0.774	0.777	0.782	0.788	0.793	0.802	0.810	0.817	0.829	0.840	0.851	0.861	0.878	0.892
0.10000	(0.981)	0.767	0.770	0.775	0.781	0.786	0.796	0.804	0.811	0.824	0.835	0.847	0.857	0.874	0.888
0.15000	(1.47)	0.754	0.757	0.764	0.770	0.776	0.786	0.794	0.802	0.816	0.827	0.839	0.849	0.867	0.882
0.20000	(1.96)	0.745	0.749	0.756	0.762	0.768	0.779	0.788	0.796	0.810	0.822	0.834	0.845	0.863	0.878
0.30000	(2.94)	0.734	0.738	0.746	0.753	0.759	0.770	0.780	0.788	0.803	0.815	0.828	0.839	0.857	0.872
0.40000	(3.92)	0.726	0.731	0.739	0.746	0.753	0.765	0.775	0.783	0.798	0.810	0.824	0.835	0.854	0.869
0.60000	(5.88)	0.716	0.721	0.730	0.738	0.745	0.758	0.768	0.777	0.792	0.805	0.818	0.830	0.849	0.865
0.80000	(7.85)	0.710	0.715	0.725	0.733	0.740	0.753	0.764	0.773	0.788	0.801	0.815	0.827	0.846	0.862
1.00000	(9.81)	0.705	0.711	0.721	0.729	0.737	0.750	0.760	0.770	0.786	0.799	0.813	0.825	0.845	0.861
1.50000	(14.7)	0.697	0.703	0.714	0.723	0.731	0.744	0.755	0.765	0.781	0.795	0.809	0.821	0.841	0.858
2.00000	(19.6)	0.692	0.699	0.709	0.719	0.727	0.741	0.752	0.762	0.779	0.792	0.807	0.819	0.839	0.856
3.00000	(29.4)	0.686	0.693	0.704	0.714	0.722	0.736	0.748	0.758	0.775	0.789	0.804	0.816	0.837	0.854
S	$(\Delta p/\rho l)$	125	150	200	250	300	400	500	600	800	1000	1250	1500	2000	2500

Grad'nt **k.p.g.** (Equivalent) Pipe diameters in mm

Kinematic viscosity, $\nu = 1.5 \times 10^{-6}$ m^{2}s^{-1}; **Roughness size, k_s = 0.006 mm**

Kin. visc., $\nu = 1.5 \times 10^{-6}$ m^{2}s^{-1}; S = 0.00010 to 30.0000

i.e. kin. pr. grad., $\Delta p/\rho l$ = (0.00098) to (294) ms^{-2}

Roughness size, k_s = 0.015 mm

This table shows values of m, as follows m_C for Colebrook-White solutions; or, where $R \le 2000$, m_P for laminar flow

Grad'nt S	k.p.g. $(\Delta p/\rho l)$	6	8	10	12.5	15	20	25	30	40	50	60	80	100	125
0.00030	(0.00294)	10.29	7.009	5.205	3.866	3.032	2.066	1.534	1.203	*	1.157	1.136	1.108	1.091	1.077
0.00040	(0.00392)	8.908	6.070	4.508	3.348	2.625	1.789	1.329	1.042	*	1.128	1.109	1.083	1.068	1.055
0.00060	(0.00588)	7.273	4.956	3.681	2.734	2.144	1.461	1.085	*	1.114	1.089	1.072	1.050	1.037	1.026
0.00080	(0.00785)	6.299	4.292	3.188	2.367	1.856	1.265	0.939	*	1.086	1.063	1.048	1.028	1.016	1.007
0.00100	(0.00981)	5.634	3.839	2.851	2.117	1.660	1.131	*	*	1.065	1.044	1.030	1.012	1.001	0.992
0.00150	(0.0147)	4.600	3.135	2.328	1.729	1.356	0.924	*	1.058	1.029	1.011	0.999	0.983	0.974	0.967
0.00200	(0.0196)	3.984	2.715	2.016	1.497	1.174	*	1.052	1.032	1.005	0.989	0.978	0.964	0.956	0.950
0.00300	(0.0294)	3.253	2.216	1.646	1.222	0.959	*	1.014	0.997	0.974	0.960	0.950	0.939	0.932	0.927
0.00400	(0.0392)	2.817	1.920	1.426	1.059	0.830	1.013	0.990	0.974	0.953	0.940	0.932	0.922	0.916	0.912
0.00600	(0.0588)	2.300	1.567	1.164	0.864	*	0.977	0.957	0.943	0.925	0.915	0.907	0.899	0.895	0.892
0.00800	(0.0785)	1.992	1.357	1.008	*	*	0.953	0.935	0.923	0.907	0.897	0.891	0.884	0.880	0.878
0.01000	(0.0981)	1.782	1.214	0.902	*	0.963	0.936	0.919	0.908	0.893	0.884	0.879	0.873	0.870	0.868
0.01500	(0.147)	1.455	0.991	*	*	0.930	0.906	0.892	0.882	0.869	0.862	0.858	0.853	0.851	0.851
0.02000	(0.196)	1.260	0.858	*	0.924	0.908	0.887	0.873	0.865	0.854	0.848	0.844	0.840	0.839	0.840
0.03000	(0.294)	1.029	*	*	0.893	0.879	0.860	0.849	0.842	0.833	0.828	0.825	0.823	0.823	0.824
0.04000	(0.392)	0.891	*	0.892	0.873	0.860	0.843	0.833	0.827	0.819	0.815	0.813	0.812	0.812	0.814
0.06000	(0.588)	0.727	*	0.862	0.846	0.835	0.821	0.812	0.807	0.801	0.798	0.797	0.797	0.798	0.801
0.08000	(0.785)	*	0.861	0.843	0.828	0.818	0.806	0.798	0.794	0.789	0.787	0.786	0.787	0.789	0.792
0.10000	(0.981)	*	0.845	0.829	0.815	0.806	0.794	0.788	0.784	0.780	0.779	0.778	0.780	0.782	0.786
0.15000	(1.47)	0.842	0.819	0.804	0.793	0.785	0.776	0.770	0.768	0.765	0.764	0.765	0.768	0.771	0.775
0.20000	(1.96)	0.823	0.801	0.789	0.778	0.771	0.763	0.759	0.757	0.755	0.755	0.756	0.760	0.764	0.768
0.30000	(2.94)	0.797	0.779	0.768	0.759	0.753	0.747	0.744	0.743	0.742	0.743	0.745	0.750	0.754	0.760
0.40000	(3.92)	0.780	0.764	0.754	0.747	0.742	0.737	0.734	0.733	0.734	0.736	0.738	0.743	0.748	0.754
0.60000	(5.88)	0.759	0.745	0.737	0.730	0.727	0.723	0.722	0.722	0.723	0.726	0.729	0.734	0.740	0.747
0.80000	(7.85)	0.744	0.732	0.725	0.720	0.717	0.714	0.714	0.714	0.716	0.719	0.723	0.729	0.735	0.742
1.00000	(9.81)	0.734	0.723	0.717	0.712	0.710	0.708	0.708	0.708	0.711	0.715	0.718	0.725	0.732	0.739
1.50000	(14.7)	0.717	0.708	0.703	0.699	0.698	0.697	0.698	0.699	0.703	0.707	0.711	0.719	0.726	0.734
2.00000	(19.6)	0.706	0.698	0.694	0.691	0.690	0.690	0.692	0.694	0.698	0.703	0.707	0.715	0.722	0.730
3.00000	(29.4)	0.692	0.685	0.682	0.681	0.680	0.682	0.684	0.686	0.691	0.697	0.701	0.710	0.718	0.726
4.00000	(39.2)	0.682	0.677	0.675	0.674	0.674	0.676	0.679	0.682	0.687	0.693	0.698	0.707	0.715	0.724
6.00000	(58.8)	0.671	0.667	0.666	0.666	0.666	0.669	0.672	0.676	0.682	0.688	0.694	0.703	0.712	0.721
8.00000	(78.5)	0.663	0.660	0.660	0.660	0.662	0.665	0.669	0.672	0.679	0.685	0.691	0.701	0.710	0.719
10.0000	(98.1)	0.658	0.656	0.656	0.657	0.658	0.662	0.666	0.670	0.677	0.683	0.689	0.699	0.708	0.718
15.0000	(147)	0.649	0.648	0.649	0.650	0.652	0.657	0.661	0.666	0.673	0.680	0.686	0.697	0.706	0.715
20.0000	(196)	0.644	0.644	0.645	0.647	0.649	0.654	0.659	0.663	0.671	0.678	0.685	0.695	0.704	0.714
30.0000	(294)	0.637	0.638	0.639	0.642	0.645	0.650	0.655	0.660	0.669	0.676	0.682	0.693	0.703	0.713
S	$(\Delta p/\rho l)$	6	8	10	12.5	15	20	25	30	40	50	60	80	100	125

Grad'nt k.p.g. (Equivalent) Pipe diameters in mm

S	$(\Delta p/\rho l)$	125	150	200	250	300	400	500	600	800	1000	1250	1500	2000	2500
0.00010	(0.00098)	1.170	1.155	1.137	1.126	1.119	1.111	1.107	1.106	1.105	1.106	1.109	1.112	1.119	1.126
0.00015	(0.00147)	1.133	1.121	1.106	1.097	1.091	1.085	1.082	1.081	1.082	1.084	1.087	1.091	1.099	1.107
0.00020	(0.00196)	1.109	1.098	1.084	1.076	1.072	1.067	1.065	1.064	1.066	1.069	1.073	1.077	1.085	1.093
0.00030	(0.00294)	1.077	1.067	1.056	1.049	1.046	1.042	1.041	1.042	1.045	1.048	1.053	1.058	1.067	1.076
0.00040	(0.00392)	1.055	1.047	1.037	1.031	1.028	1.026	1.026	1.027	1.030	1.034	1.040	1.045	1.055	1.064
0.00060	(0.00588)	1.026	1.019	1.011	1.007	1.005	1.004	1.005	1.006	1.011	1.016	1.022	1.027	1.038	1.047
0.00080	(0.00785)	1.007	1.001	0.994	0.990	0.989	0.989	0.990	0.992	0.998	1.003	1.009	1.016	1.027	1.037
0.00100	(0.00981)	0.992	0.987	0.981	0.978	0.977	0.978	0.980	0.982	0.988	0.994	1.000	1.007	1.018	1.029
0.00150	(0.0147)	0.967	0.963	0.958	0.957	0.957	0.958	0.961	0.964	0.971	0.977	0.985	0.992	1.004	1.015
0.00200	(0.0196)	0.950	0.946	0.943	0.942	0.943	0.945	0.948	0.952	0.959	0.966	0.974	0.981	0.994	1.005
0.00300	(0.0294)	0.927	0.925	0.923	0.923	0.924	0.928	0.932	0.936	0.944	0.952	0.960	0.968	0.981	0.993
0.00400	(0.0392)	0.912	0.910	0.909	0.910	0.912	0.916	0.921	0.925	0.934	0.942	0.951	0.959	0.973	0.985
0.00600	(0.0588)	0.892	0.891	0.891	0.893	0.895	0.900	0.906	0.911	0.920	0.929	0.938	0.947	0.962	0.974
0.00800	(0.0785)	0.878	0.878	0.879	0.881	0.884	0.890	0.896	0.901	0.911	0.920	0.930	0.939	0.954	0.967
0.01000	(0.0981)	0.868	0.868	0.870	0.873	0.876	0.882	0.888	0.894	0.905	0.914	0.924	0.933	0.949	0.962
0.01500	(0.147)	0.851	0.852	0.855	0.858	0.862	0.869	0.876	0.882	0.894	0.903	0.914	0.924	0.940	0.954
0.02000	(0.196)	0.840	0.841	0.844	0.848	0.852	0.860	0.868	0.874	0.886	0.897	0.908	0.918	0.934	0.948
0.03000	(0.294)	0.824	0.826	0.831	0.836	0.840	0.849	0.857	0.864	0.877	0.888	0.899	0.910	0.927	0.942
0.04000	(0.392)	0.814	0.817	0.822	0.827	0.832	0.842	0.850	0.857	0.871	0.882	0.894	0.904	0.922	0.937
0.06000	(0.588)	0.801	0.804	0.810	0.816	0.822	0.832	0.841	0.849	0.863	0.874	0.887	0.898	0.916	0.932
0.08000	(0.785)	0.792	0.796	0.803	0.809	0.815	0.826	0.835	0.843	0.858	0.870	0.883	0.894	0.913	0.928
0.10000	(0.981)	0.786	0.790	0.797	0.804	0.810	0.821	0.831	0.839	0.854	0.866	0.879	0.891	0.910	0.926
0.15000	(1.47)	0.775	0.780	0.788	0.795	0.802	0.814	0.824	0.833	0.848	0.861	0.874	0.886	0.906	0.922
0.20000	(1.96)	0.768	0.773	0.782	0.790	0.797	0.809	0.819	0.829	0.844	0.857	0.871	0.883	0.903	0.919
0.30000	(2.94)	0.760	0.765	0.774	0.783	0.790	0.803	0.814	0.824	0.840	0.853	0.867	0.879	0.900	0.916
0.40000	(3.92)	0.754	0.759	0.769	0.778	0.786	0.799	0.811	0.820	0.837	0.850	0.865	0.877	0.898	0.914
0.60000	(5.88)	0.747	0.753	0.763	0.773	0.781	0.795	0.806	0.816	0.833	0.847	0.862	0.874	0.895	0.912
0.80000	(7.85)	0.742	0.748	0.760	0.769	0.777	0.792	0.804	0.814	0.831	0.845	0.860	0.873	0.894	0.911
1.00000	(9.81)	0.739	0.745	0.757	0.767	0.775	0.790	0.802	0.812	0.829	0.844	0.859	0.871	0.893	0.910
1.50000	(14.7)	0.734	0.741	0.753	0.763	0.771	0.786	0.799	0.809	0.827	0.841	0.857	0.870	0.891	0.908
2.00000	(19.6)	0.730	0.738	0.750	0.760	0.769	0.784	0.797	0.808	0.825	0.840	0.855	0.868	0.890	0.907
3.00000	(29.4)	0.726	0.734	0.746	0.757	0.766	0.782	0.795	0.806	0.824	0.838	0.854	0.867	0.889	0.906
S	$(\Delta p/\rho l)$	125	150	200	250	300	400	500	600	800	1000	1250	1500	2000	2500

Grad'nt k.p.g. (Equivalent) Pipe diameters in mm

Kinematic viscosity, $\nu = 1.5 \times 10^{-6}$ m^{2}s^{-1} ;   Roughness size, k_s = 0.015 mm

E119

Kin. visc., $\nu = 1.5\times10^{-6}$ m^{2}s^{-1};
$S = 0.00010$ to 30.0000

i.e. kin. pr. grad., $\Delta p/\rho l =$
(0.00098) to (294) ms^{-2}

Roughness size, $k_s = 0.030$ mm
This table shows values of m, as follows

m_C for Colebrook-White solutions; or,
where $\mathbf{R} \leq 2000$, m_P for laminar flow

Grad'nt **k.p.g.** **(Equivalent) Pipe diameters in mm**

S	$(\Delta p/\rho l)$	6	8	10	12·5	15	20	25	30	40	50	60	80	100	125
0·00030	(0·00294)	10·29	7·009	5·205	3·866	3·032	2·066	1·534	1·203	*	1·161	1·140	1·112	1·095	1·081
0·00040	(0·00392)	8·908	6·070	4·508	3·348	2·625	1·789	1·329	1·042	*	1·132	1·113	1·088	1·072	1·060
0·00060	(0·00588)	7·273	4·956	3·681	2·734	2·144	1·461	1·085	*	1·118	1·094	1·077	1·055	1·042	1·032
0·00080	(0·00785)	6·299	4·292	3·188	2·367	1·856	1·265	0·939	*	1·091	1·068	1·053	1·034	1·022	1·013
0·00100	(0·00981)	5·634	3·839	2·851	2·117	1·660	1·131	*	*	1·070	1·050	1·036	1·018	1·007	0·999
0·00150	(0·0147)	4·600	3·135	2·328	1·729	1·356	0·924	*	1·064	1·035	1·018	1·005	0·990	0·981	0·975
0·00200	(0·0196)	3·984	2·715	2·016	1·497	1·174	*	*	1·038	1·012	0·996	0·985	0·972	0·964	0·958
0·00300	(0·0294)	3·253	2·216	1·646	1·222	0·959	*	1·022	1·004	0·982	0·968	0·959	0·948	0·941	0·937
0·00400	(0·0392)	2·817	1·920	1·426	1·059	0·830	1·020	0·997	0·982	0·961	0·949	0·941	0·931	0·926	0·923
0·00600	(0·0588)	2·300	1·567	1·164	0·864	*	0·986	0·966	0·952	0·935	0·924	0·918	0·910	0·906	0·904
0·00800	(0·0785)	1·992	1·357	1·008	*	*	0·963	0·945	0·933	0·917	0·908	0·902	0·896	0·893	0·892
0·01000	(0·0981)	1·782	1·214	0·902	*	0·973	0·946	0·929	0·918	0·904	0·896	0·891	0·885	0·883	0·883
0·01500	(0·147)	1·455	0·991	*	*	0·941	0·918	0·903	0·894	0·882	0·875	0·871	0·868	0·867	0·867
0·02000	(0·196)	1·260	0·858	*	0·936	0·920	0·899	0·886	0·878	0·867	0·862	0·859	0·856	0·856	0·857
0·03000	(0·294)	1·029	*	*	0·907	0·892	0·874	0·864	0·857	0·848	0·844	0·842	0·841	0·842	0·844
0·04000	(0·392)	0·891	*	0·906	0·887	0·874	0·858	0·849	0·843	0·836	0·833	0·831	0·831	0·832	0·835
0·06000	(0·588)	0·727	*	0·878	0·862	0·851	0·837	0·830	0·825	0·820	0·818	0·817	0·818	0·821	0·824
0·08000	(0·785)	*	0·878	0·860	0·845	0·836	0·824	0·817	0·813	0·809	0·808	0·808	0·810	0·813	0·817
0·10000	(0·981)	*	0·863	0·847	0·833	0·825	0·814	0·808	0·805	0·802	0·801	0·801	0·804	0·807	0·812
0·15000	(1·47)	*	0·839	0·824	0·813	0·806	0·797	0·793	0·790	0·789	0·789	0·790	0·794	0·798	0·804
0·20000	(1·96)	0·843	0·823	0·810	0·800	0·794	0·786	0·783	0·781	0·781	0·782	0·783	0·788	0·793	0·798
0·30000	(2·94)	0·820	0·802	0·792	0·783	0·778	0·773	0·770	0·770	0·770	0·772	0·775	0·780	0·785	0·792
0·40000	(3·92)	0·805	0·789	0·780	0·773	0·768	0·764	0·762	0·762	0·764	0·766	0·769	0·775	0·781	0·788
0·60000	(5·88)	0·786	0·772	0·765	0·759	0·756	0·753	0·752	0·753	0·755	0·759	0·762	0·769	0·775	0·783
0·80000	(7·85)	0·773	0·762	0·755	0·750	0·748	0·746	0·746	0·747	0·750	0·754	0·758	0·765	0·772	0·779
1·00000	(9·81)	0·765	0·754	0·748	0·744	0·742	0·741	0·741	0·743	0·746	0·750	0·755	0·762	0·769	0·777
1·50000	(14·7)	0·750	0·741	0·737	0·734	0·732	0·732	0·734	0·736	0·740	0·745	0·750	0·758	0·765	0·774
2·00000	(19·6)	0·741	0·733	0·729	0·727	0·727	0·727	0·729	0·731	0·737	0·742	0·747	0·755	0·763	0·772
3·00000	(29·4)	0·729	0·723	0·721	0·719	0·719	0·721	0·723	0·726	0·732	0·738	0·743	0·752	0·760	0·769
4·00000	(39·2)	0·722	0·717	0·715	0·714	0·715	0·717	0·720	0·723	0·729	0·735	0·740	0·750	0·758	0·767
6·00000	(58·8)	0·713	0·709	0·708	0·708	0·709	0·712	0·716	0·719	0·726	0·732	0·738	0·748	0·756	0·765
8·00000	(78·5)	0·707	0·704	0·704	0·704	0·706	0·709	0·713	0·717	0·724	0·730	0·736	0·746	0·755	0·764
10·0000	(98·1)	0·703	0·701	0·701	0·702	0·703	0·707	0·711	0·715	0·722	0·729	0·735	0·745	0·754	0·763
15·0000	(147)	0·697	0·696	0·696	0·697	0·699	0·704	0·708	0·712	0·720	0·727	0·733	0·744	0·752	0·762
20·0000	(196)	0·693	0·692	0·693	0·695	0·697	0·702	0·706	0·711	0·719	0·726	0·732	0·743	0·752	0·761
30·0000	(294)	0·688	0·688	0·690	0·692	0·694	0·699	0·704	0·709	0·717	0·724	0·730	0·741	0·751	0·760
S	$(\Delta p/\rho l)$	6	8	10	12·5	15	20	25	30	40	50	60	80	100	125

Grad'nt **k.p.g.** **(Equivalent) Pipe diameters in mm**

S	$(\Delta p/\rho l)$	125	150	200	250	300	400	500	600	800	1000	1250	1500	2000	2500
0·00010	(0·00098)	1·173	1·159	1·141	1·130	1·123	1·115	1·112	1·110	1·110	1·111	1·115	1·118	1·126	1·133
0·00015	(0·00147)	1·137	1·125	1·110	1·101	1·095	1·089	1·087	1·086	1·087	1·090	1·094	1·098	1·107	1·115
0·00020	(0·00196)	1·113	1·102	1·089	1·081	1·076	1·072	1·070	1·070	1·072	1·075	1·080	1·085	1·094	1·103
0·00030	(0·00294)	1·081	1·072	1·061	1·055	1·051	1·048	1·048	1·049	1·052	1·056	1·061	1·067	1·077	1·086
0·00040	(0·00392)	1·060	1·052	1·042	1·037	1·034	1·032	1·033	1·034	1·038	1·043	1·049	1·055	1·066	1·075
0·00060	(0·00588)	1·032	1·025	1·017	1·014	1·012	1·011	1·013	1·015	1·020	1·026	1·032	1·039	1·050	1·061
0·00080	(0·00785)	1·013	1·007	1·001	0·998	0·997	0·997	0·999	1·002	1·008	1·014	1·021	1·028	1·040	1·051
0·00100	(0·00981)	0·999	0·994	0·988	0·986	0·985	0·987	0·989	0·992	0·999	1·005	1·013	1·020	1·033	1·044
0·00150	(0·0147)	0·975	0·971	0·967	0·966	0·966	0·969	0·972	0·976	0·983	0·991	0·999	1·007	1·021	1·032
0·00200	(0·0196)	0·958	0·955	0·953	0·952	0·953	0·957	0·961	0·965	0·973	0·981	0·990	0·998	1·012	1·025
0·00300	(0·0294)	0·937	0·935	0·934	0·934	0·936	0·941	0·946	0·950	0·960	0·968	0·978	0·987	1·002	1·015
0·00400	(0·0392)	0·923	0·921	0·921	0·923	0·925	0·930	0·936	0·941	0·951	0·960	0·970	0·979	0·995	1·008
0·00600	(0·0588)	0·904	0·904	0·905	0·907	0·910	0·916	0·923	0·929	0·940	0·949	0·960	0·970	0·986	1·000
0·00800	(0·0785)	0·892	0·892	0·894	0·897	0·900	0·907	0·914	0·921	0·932	0·942	0·953	0·963	0·980	0·994
0·01000	(0·0981)	0·883	0·883	0·886	0·889	0·893	0·901	0·908	0·915	0·927	0·937	0·949	0·959	0·976	0·991
0·01500	(0·147)	0·867	0·868	0·872	0·877	0·881	0·890	0·898	0·905	0·918	0·929	0·941	0·952	0·970	0·985
0·02000	(0·196)	0·857	0·859	0·863	0·869	0·873	0·883	0·891	0·899	0·912	0·924	0·936	0·947	0·965	0·981
0·03000	(0·294)	0·844	0·846	0·852	0·858	0·863	0·874	0·883	0·891	0·905	0·917	0·930	0·941	0·960	0·976
0·04000	(0·392)	0·835	0·838	0·845	0·851	0·857	0·868	0·877	0·886	0·900	0·913	0·926	0·938	0·957	0·973
0·06000	(0·588)	0·824	0·828	0·835	0·842	0·849	0·861	0·871	0·879	0·895	0·907	0·921	0·933	0·953	0·969
0·08000	(0·785)	0·817	0·821	0·830	0·837	0·844	0·856	0·866	0·875	0·891	0·904	0·918	0·930	0·950	0·967
0·10000	(0·981)	0·812	0·817	0·825	0·833	0·840	0·853	0·863	0·873	0·889	0·902	0·916	0·928	0·949	0·965
0·15000	(1·47)	0·804	0·809	0·818	0·827	0·834	0·847	0·858	0·868	0·885	0·898	0·913	0·925	0·946	0·963
0·20000	(1·96)	0·798	0·804	0·814	0·823	0·830	0·844	0·855	0·865	0·882	0·896	0·911	0·923	0·944	0·961
0·30000	(2·94)	0·792	0·798	0·808	0·818	0·826	0·840	0·852	0·862	0·879	0·893	0·908	0·921	0·942	0·959
0·40000	(3·92)	0·788	0·794	0·805	0·815	0·823	0·837	0·849	0·860	0·877	0·892	0·907	0·920	0·941	0·958
0·60000	(5·88)	0·783	0·789	0·801	0·811	0·820	0·834	0·847	0·857	0·875	0·890	0·905	0·918	0·939	0·957
0·80000	(7·85)	0·779	0·786	0·798	0·808	0·817	0·832	0·845	0·856	0·874	0·888	0·904	0·917	0·939	0·956
1·00000	(9·81)	0·777	0·784	0·797	0·807	0·816	0·831	0·844	0·855	0·873	0·887	0·903	0·916	0·938	0·956
1·50000	(14·7)	0·774	0·781	0·794	0·804	0·814	0·829	0·842	0·853	0·871	0·886	0·902	0·915	0·937	0·955
2·00000	(19·6)	0·772	0·779	0·792	0·803	0·812	0·828	0·841	0·852	0·870	0·885	0·901	0·914	0·936	0·954
3·00000	(29·4)	0·769	0·777	0·790	0·801	0·810	0·826	0·839	0·851	0·869	0·884	0·900	0·914	0·936	0·954
S	$(\Delta p/\rho l)$	125	150	200	250	300	400	500	600	800	1000	1250	1500	2000	2500

Grad'nt **k.p.g.** **(Equivalent) Pipe diameters in mm**

Kinematic viscosity, $\nu = 1.5\times10^{-6}$ m^{2}s^{-1} ;

Roughness size, $k_s = 0.030$ mm

Kin. visc., $\nu = 1.5 \times 10^{-6}\ \mathrm{m^2\,s^{-1}}$;
$S = 0.00010$ to 30.0000
i.e. kin. pr. grad., $\Delta p/\rho l =$
(0.00098) to $(294)\ \mathrm{ms^{-2}}$

Roughness size, $k_s = 0.060$ mm
This table shows values of m, as follows
m_C for Colebrook-White solutions; or,
where $\mathbf{R} \le 2000$, m_P for laminar flow

Grad'nt S	k.p.g. $(\Delta p/\rho l)$	6	8	10	12·5	15	20	25	30	40	50	60	80	100	125
		(Equivalent) Pipe diameters in mm													
0·00030	(0·00294)	10·29	7·009	5·205	3·866	3·032	2·066	1·534	1·203	*	1·169	1·148	1·120	1·104	1·090
0·00040	(0·00392)	8·908	6·070	4·508	3·348	2·625	1·789	1·329	1·042	*	1·140	1·121	1·097	1·082	1·070
0·00060	(0·00588)	7·273	4·956	3·681	2·734	2·144	1·461	1·085	*	1·127	1·103	1·087	1·065	1·053	1·043
0·00080	(0·00785)	6·299	4·292	3·188	2·367	1·856	1·265	0·939	*	1·101	1·079	1·064	1·045	1·033	1·025
0·00100	(0·00981)	5·634	3·839	2·851	2·117	1·660	1·131	*	*	1·081	1·061	1·047	1·029	1·019	1·012
0·00150	(0·0147)	4·600	3·135	2·328	1·729	1·356	0·924	*	1·076	1·047	1·030	1·018	1·003	0·995	0·989
0·00200	(0·0196)	3·984	2·715	2·016	1·497	1·174	*	*	1·051	1·025	1·010	0·999	0·986	0·979	0·974
0·00300	(0·0294)	3·253	2·216	1·646	1·222	0·959	*	1·035	1·018	0·996	0·983	0·974	0·964	0·958	0·955
0·00400	(0·0392)	2·817	1·920	1·426	1·059	0·830	*	1·013	0·997	0·977	0·966	0·958	0·949	0·945	0·942
0·00600	(0·0588)	2·300	1·567	1·164	0·864	*	1·002	0·983	0·969	0·953	0·943	0·937	0·930	0·927	0·926
0·00800	(0·0785)	1·992	1·357	1·008	*	*	0·981	0·963	0·951	0·936	0·928	0·923	0·917	0·915	0·915
0·01000	(0·0981)	1·782	1·214	0·902	*	0·992	0·965	0·949	0·938	0·925	0·917	0·913	0·908	0·907	0·907
0·01500	(0·147)	1·455	0·991	*	*	0·962	0·939	0·925	0·916	0·905	0·899	0·896	0·893	0·893	0·894
0·02000	(0·196)	1·260	0·858	*	0·958	0·942	0·922	0·909	0·901	0·892	0·887	0·885	0·883	0·884	0·886
0·03000	(0·294)	1·029	*	*	0·931	0·917	0·900	0·889	0·883	0·876	0·872	0·871	0·871	0·872	0·875
0·04000	(0·392)	0·891	*	0·932	0·913	0·901	0·886	0·877	0·871	0·865	0·863	0·862	0·863	0·865	0·869
0·06000	(0·588)	0·727	*	0·907	0·891	0·880	0·867	0·860	0·856	0·852	0·850	0·851	0·853	0·856	0·860
0·08000	(0·785)	*	0·908	0·891	0·876	0·867	0·856	0·850	0·846	0·843	0·843	0·843	0·846	0·850	0·855
0·10000	(0·981)	*	0·895	0·879	0·866	0·857	0·847	0·842	0·839	0·837	0·837	0·838	0·842	0·846	0·851
0·15000	(1·47)	*	0·874	0·860	0·849	0·842	0·834	0·830	0·828	0·827	0·828	0·830	0·835	0·839	0·846
0·20000	(1·96)	0·881	0·860	0·848	0·838	0·832	0·825	0·822	0·821	0·821	0·822	0·825	0·830	0·835	0·842
0·30000	(2·94)	0·861	0·843	0·832	0·824	0·819	0·814	0·812	0·812	0·813	0·815	0·818	0·824	0·830	0·837
0·40000	(3·92)	0·848	0·832	0·822	0·816	0·811	0·807	0·806	0·806	0·808	0·811	0·814	0·821	0·827	0·835
0·60000	(5·88)	0·832	0·818	0·810	0·805	0·802	0·799	0·799	0·799	0·802	0·806	0·809	0·817	0·823	0·831
0·80000	(7·85)	0·822	0·810	0·803	0·798	0·795	0·794	0·794	0·795	0·798	0·802	0·806	0·814	0·821	0·829
1·00000	(9·81)	0·815	0·804	0·797	0·793	0·791	0·790	0·791	0·792	0·796	0·800	0·804	0·812	0·820	0·828
1·50000	(14·7)	0·803	0·794	0·789	0·786	0·784	0·784	0·785	0·787	0·792	0·797	0·801	0·810	0·817	0·825
2·00000	(19·6)	0·796	0·788	0·784	0·781	0·780	0·780	0·782	0·784	0·789	0·794	0·799	0·808	0·816	0·824
3·00000	(29·4)	0·787	0·780	0·777	0·775	0·775	0·776	0·778	0·781	0·786	0·792	0·797	0·806	0·814	0·822
4·00000	(39·2)	0·782	0·776	0·773	0·772	0·772	0·773	0·776	0·779	0·785	0·790	0·795	0·805	0·813	0·821
6·00000	(58·8)	0·775	0·770	0·768	0·768	0·768	0·770	0·773	0·776	0·782	0·788	0·794	0·803	0·811	0·820
8·00000	(78·5)	0·771	0·767	0·765	0·765	0·766	0·768	0·771	0·775	0·781	0·787	0·792	0·802	0·810	0·820
10·0000	(98·1)	0·769	0·765	0·763	0·763	0·764	0·767	0·770	0·774	0·780	0·786	0·792	0·801	0·810	0·819
15·0000	(147)	0·764	0·761	0·760	0·760	0·762	0·765	0·768	0·772	0·779	0·785	0·791	0·801	0·809	0·818
20·0000	(196)	0·762	0·759	0·758	0·759	0·760	0·763	0·767	0·771	0·778	0·784	0·790	0·800	0·809	0·818
30·0000	(294)	0·758	0·756	0·756	0·757	0·758	0·762	0·766	0·770	0·777	0·783	0·789	0·799	0·808	0·817
S	$(\Delta p/\rho l)$	6	8	10	12·5	15	20	25	30	40	50	60	80	100	125

Grad'nt **k.p.g.** (Equivalent) Pipe diameters in mm

S	$(\Delta p/\rho l)$	125	150	200	250	300	400	500	600	800	1000	1250	1500	2000	2500
0·00010	(0·00098)	1·179	1·165	1·147	1·137	1·130	1·123	1·120	1·119	1·119	1·122	1·125	1·130	1·138	1·146
0·00015	(0·00147)	1·144	1·132	1·117	1·109	1·103	1·098	1·096	1·096	1·098	1·101	1·106	1·111	1·121	1·130
0·00020	(0·00196)	1·121	1·110	1·097	1·090	1·086	1·082	1·081	1·081	1·084	1·088	1·094	1·099	1·110	1·119
0·00030	(0·00294)	1·090	1·081	1·070	1·065	1·062	1·059	1·060	1·061	1·065	1·070	1·077	1·083	1·095	1·105
0·00040	(0·00392)	1·070	1·062	1·053	1·048	1·046	1·045	1·046	1·048	1·053	1·059	1·066	1·073	1·085	1·096
0·00060	(0·00588)	1·043	1·036	1·029	1·026	1·025	1·025	1·028	1·031	1·037	1·044	1·051	1·059	1·072	1·084
0·00080	(0·00785)	1·025	1·019	1·014	1·012	1·011	1·013	1·016	1·019	1·026	1·034	1·042	1·050	1·064	1·076
0·00100	(0·00981)	1·012	1·007	1·002	1·001	1·001	1·003	1·007	1·011	1·019	1·026	1·035	1·043	1·058	1·070
0·00150	(0·0147)	0·989	0·986	0·983	0·983	0·984	0·987	0·992	0·996	1·005	1·014	1·024	1·032	1·048	1·061
0·00200	(0·0196)	0·974	0·972	0·970	0·971	0·972	0·977	0·982	0·987	0·997	1·006	1·016	1·025	1·041	1·055
0·00300	(0·0294)	0·955	0·953	0·953	0·955	0·957	0·963	0·969	0·975	0·986	0·996	1·007	1·016	1·033	1·047
0·00400	(0·0392)	0·942	0·941	0·942	0·945	0·948	0·955	0·961	0·968	0·979	0·989	1·001	1·011	1·028	1·043
0·00600	(0·0588)	0·926	0·926	0·928	0·932	0·936	0·943	0·951	0·958	0·970	0·981	0·993	1·004	1·022	1·037
0·00800	(0·0785)	0·915	0·916	0·919	0·923	0·928	0·936	0·944	0·952	0·965	0·976	0·988	0·999	1·018	1·033
0·01000	(0·0981)	0·907	0·908	0·913	0·917	0·922	0·931	0·939	0·947	0·961	0·972	0·985	0·996	1·015	1·031
0·01500	(0·147)	0·894	0·896	0·902	0·907	0·913	0·923	0·932	0·940	0·954	0·966	0·980	0·991	1·010	1·027
0·02000	(0·196)	0·886	0·889	0·895	0·901	0·907	0·917	0·927	0·935	0·950	0·963	0·976	0·988	1·008	1·024
0·03000	(0·294)	0·875	0·879	0·886	0·893	0·899	0·911	0·921	0·930	0·945	0·958	0·972	0·984	1·004	1·021
0·04000	(0·392)	0·869	0·873	0·880	0·888	0·895	0·907	0·917	0·926	0·942	0·955	0·970	0·982	1·002	1·019
0·06000	(0·588)	0·860	0·865	0·874	0·882	0·889	0·901	0·912	0·922	0·938	0·952	0·966	0·979	1·000	1·017
0·08000	(0·785)	0·855	0·860	0·869	0·878	0·885	0·898	0·909	0·919	0·936	0·950	0·965	0·977	0·998	1·015
0·10000	(0·981)	0·851	0·857	0·866	0·875	0·883	0·896	0·907	0·917	0·934	0·948	0·963	0·976	0·997	1·014
0·15000	(1·47)	0·846	0·851	0·861	0·871	0·879	0·893	0·904	0·915	0·932	0·946	0·961	0·974	0·995	1·013
0·20000	(1·96)	0·842	0·848	0·859	0·868	0·876	0·890	0·902	0·913	0·930	0·945	0·960	0·973	0·994	1·012
0·30000	(2·94)	0·837	0·844	0·855	0·865	0·873	0·888	0·900	0·911	0·928	0·943	0·958	0·971	0·993	1·011
0·40000	(3·92)	0·835	0·841	0·853	0·863	0·871	0·886	0·899	0·909	0·927	0·942	0·957	0·971	0·992	1·010
0·60000	(5·88)	0·831	0·838	0·850	0·860	0·869	0·884	0·897	0·908	0·926	0·941	0·956	0·970	0·992	1·009
0·80000	(7·85)	0·829	0·836	0·848	0·859	0·868	0·883	0·896	0·907	0·925	0·940	0·956	0·969	0·991	1·009
1·00000	(9·81)	0·828	0·835	0·847	0·858	0·867	0·882	0·895	0·906	0·924	0·939	0·955	0·969	0·991	1·009
1·50000	(14·7)	0·825	0·833	0·845	0·856	0·865	0·881	0·894	0·905	0·924	0·939	0·955	0·968	0·990	1·008
2·00000	(19·6)	0·824	0·832	0·844	0·855	0·865	0·880	0·893	0·905	0·923	0·938	0·954	0·968	0·990	1·008
3·00000	(29·4)	0·822	0·830	0·843	0·854	0·864	0·879	0·893	0·904	0·922	0·938	0·954	0·967	0·989	1·007
S	$(\Delta p/\rho l)$	125	150	200	250	300	400	500	600	800	1000	1250	1500	2000	2500

Grad'nt **k.p.g.** (Equivalent) Pipe diameters in mm

Kinematic viscosity, $\nu = 1.5 \times 10^{-6}\ \mathrm{m^2\,s^{-1}}$;   **Roughness size, $k_s = 0.060$ mm**

E121

Kin. visc., $\nu = 1.5 \times 10^{-6}\ \mathrm{m^2 s^{-1}}$;
S = 0.00010 to 30.0000

i.e. kin. pr. grad., $\Delta p/\rho l$ = (0.00098) to (294) ms^{-2}

Roughness size, k_s = 0.150 mm
This table shows values of m, as follows
m_C for Colebrook-White solutions; or,
where $\mathbf{R} \leq 2000$, m_P for laminar flow

Grad'nt S	k.p.g. $(\Delta p/\rho l)$	\(\)				(Equivalent) Pipe diameters in mm									
		6	8	10	12.5	15	20	25	30	40	50	60	80	100	125
0.00030	(0.00294)	10.29	7.009	5.205	3.866	3.032	2.066	1.534	1.203	*	1.191	1.170	1.143	1.127	1.114
0.00040	(0.00392)	8.908	6.070	4.508	3.348	2.625	1.789	1.329	1.042	*	1.164	1.145	1.121	1.107	1.096
0.00060	(0.00588)	7.273	4.956	3.681	2.734	2.144	1.461	1.085	*	1.153	1.130	1.113	1.093	1.081	1.072
0.00080	(0.00785)	6.299	4.292	3.188	2.367	1.856	1.265	0.939	*	1.129	1.107	1.093	1.075	1.064	1.057
0.00100	(0.00981)	5.634	3.839	2.851	2.117	1.660	1.131	*	*	1.111	1.091	1.078	1.061	1.052	1.045
0.00150	(0.0147)	4.600	3.135	2.328	1.729	1.356	0.924	*	1.108	1.080	1.063	1.052	1.039	1.032	1.027
0.00200	(0.0196)	3.984	2.715	2.016	1.497	1.174	*	*	1.086	1.061	1.046	1.036	1.024	1.018	1.015
0.00300	(0.0294)	3.253	2.216	1.646	1.222	0.959	*	1.074	1.057	1.036	1.023	1.015	1.006	1.001	0.999
0.00400	(0.0392)	2.817	1.920	1.426	1.059	0.830	*	1.054	1.038	1.019	1.008	1.002	0.994	0.991	0.989
0.00600	(0.0588)	2.300	1.567	1.164	0.864	*	1.047	1.028	1.015	0.999	0.990	0.984	0.979	0.977	0.977
0.00800	(0.0785)	1.992	1.357	1.008	*	*	1.029	1.011	1.000	0.986	0.978	0.974	0.969	0.968	0.969
0.01000	(0.0981)	1.782	1.214	0.902	*	*	1.016	1.000	0.989	0.976	0.970	0.966	0.963	0.962	0.964
0.01500	(0.147)	1.455	0.991	*	*	1.017	0.994	0.980	0.971	0.961	0.956	0.953	0.952	0.952	0.955
0.02000	(0.196)	1.260	0.858	*	1.017	1.000	0.980	0.968	0.960	0.951	0.947	0.945	0.945	0.946	0.949
0.03000	(0.294)	1.029	*	*	0.994	0.980	0.963	0.952	0.946	0.939	0.936	0.936	0.936	0.939	0.942
0.04000	(0.392)	0.891	*	0.999	0.980	0.967	0.952	0.943	0.937	0.932	0.930	0.929	0.931	0.934	0.938
0.06000	(0.588)	0.727	*	0.979	0.962	0.951	0.938	0.931	0.927	0.923	0.921	0.922	0.924	0.928	0.933
0.08000	(0.785)	*	0.985	0.966	0.951	0.941	0.930	0.923	0.920	0.917	0.916	0.917	0.921	0.925	0.930
0.10000	(0.981)	*	0.975	0.958	0.944	0.934	0.924	0.918	0.915	0.913	0.913	0.914	0.918	0.922	0.928
0.15000	(1.47)	*	0.959	0.943	0.931	0.923	0.914	0.910	0.908	0.906	0.907	0.909	0.913	0.918	0.924
0.20000	(1.96)	0.972	0.948	0.934	0.923	0.916	0.908	0.905	0.903	0.902	0.904	0.906	0.911	0.916	0.922
0.30000	(2.94)	0.957	0.936	0.923	0.914	0.908	0.901	0.898	0.897	0.898	0.899	0.902	0.907	0.913	0.920
0.40000	(3.92)	0.947	0.928	0.917	0.908	0.903	0.897	0.895	0.894	0.895	0.897	0.900	0.905	0.911	0.918
0.60000	(5.88)	0.936	0.919	0.909	0.901	0.896	0.892	0.890	0.890	0.891	0.894	0.897	0.903	0.909	0.916
0.80000	(7.85)	0.929	0.913	0.904	0.897	0.892	0.888	0.887	0.887	0.889	0.892	0.895	0.902	0.908	0.915
1.00000	(9.81)	0.924	0.909	0.900	0.894	0.890	0.886	0.885	0.885	0.888	0.891	0.894	0.901	0.907	0.914
1.50000	(14.7)	0.916	0.903	0.895	0.889	0.886	0.883	0.882	0.883	0.885	0.889	0.892	0.899	0.906	0.913
2.00000	(19.6)	0.912	0.899	0.891	0.886	0.883	0.881	0.880	0.881	0.884	0.887	0.891	0.898	0.905	0.912
3.00000	(29.4)	0.906	0.894	0.887	0.883	0.880	0.878	0.878	0.879	0.882	0.886	0.890	0.897	0.904	0.912
4.00000	(39.2)	0.903	0.891	0.885	0.881	0.878	0.877	0.877	0.878	0.881	0.885	0.889	0.897	0.903	0.911
6.00000	(58.8)	0.898	0.888	0.882	0.878	0.876	0.875	0.875	0.877	0.880	0.884	0.888	0.896	0.903	0.910
8.00000	(78.5)	0.896	0.886	0.880	0.877	0.875	0.874	0.874	0.876	0.879	0.884	0.888	0.895	0.902	0.910
10.0000	(98.1)	0.894	0.885	0.879	0.876	0.874	0.873	0.874	0.875	0.879	0.883	0.887	0.895	0.902	0.910
15.0000	(147)	0.892	0.882	0.877	0.874	0.872	0.872	0.873	0.874	0.878	0.882	0.887	0.894	0.902	0.909
20.0000	(196)	0.890	0.881	0.876	0.873	0.872	0.871	0.872	0.874	0.878	0.882	0.886	0.894	0.901	0.909
30.0000	(294)	0.888	0.880	0.875	0.872	0.871	0.870	0.871	0.873	0.877	0.882	0.886	0.894	0.901	0.909
S	$(\Delta p/\rho l)$	6	8	10	12.5	15	20	25	30	40	50	60	80	100	125

Grad'nt k.p.g. (Equivalent) Pipe diameters in mm

Grad'nt S	k.p.g. $(\Delta p/\rho l)$	(Equivalent) Pipe diameters in mm													
		125	150	200	250	300	400	500	600	800	1000	1250	1500	2000	2500
0.00010	(0.00098)	1.196	1.183	1.166	1.157	1.151	1.145	1.143	1.143	1.145	1.148	1.154	1.159	1.170	1.180
0.00015	(0.00147)	1.164	1.152	1.138	1.131	1.127	1.123	1.122	1.123	1.127	1.132	1.138	1.144	1.156	1.167
0.00020	(0.00196)	1.142	1.132	1.120	1.114	1.111	1.108	1.109	1.110	1.115	1.121	1.128	1.135	1.148	1.159
0.00030	(0.00294)	1.114	1.106	1.097	1.092	1.090	1.089	1.091	1.094	1.100	1.107	1.115	1.123	1.136	1.149
0.00040	(0.00392)	1.096	1.089	1.081	1.078	1.077	1.077	1.080	1.083	1.090	1.098	1.107	1.115	1.129	1.142
0.00060	(0.00588)	1.072	1.067	1.061	1.059	1.059	1.062	1.065	1.069	1.078	1.086	1.096	1.105	1.120	1.134
0.00080	(0.00785)	1.057	1.052	1.048	1.047	1.048	1.051	1.056	1.061	1.070	1.079	1.089	1.099	1.115	1.129
0.00100	(0.00981)	1.045	1.042	1.039	1.039	1.040	1.044	1.049	1.054	1.065	1.074	1.085	1.094	1.111	1.125
0.00150	(0.0147)	1.027	1.024	1.023	1.024	1.026	1.032	1.038	1.044	1.055	1.066	1.077	1.087	1.105	1.119
0.00200	(0.0196)	1.015	1.013	1.013	1.015	1.018	1.024	1.031	1.038	1.050	1.060	1.072	1.083	1.101	1.116
0.00300	(0.0294)	0.999	0.999	1.000	1.003	1.007	1.015	1.022	1.030	1.042	1.054	1.066	1.077	1.096	1.111
0.00400	(0.0392)	0.989	0.990	0.992	0.996	1.000	1.009	1.017	1.025	1.038	1.050	1.062	1.074	1.093	1.109
0.00600	(0.0588)	0.977	0.978	0.982	0.987	0.992	1.001	1.010	1.018	1.032	1.045	1.058	1.069	1.089	1.105
0.00800	(0.0785)	0.969	0.971	0.976	0.981	0.987	0.997	1.006	1.014	1.029	1.042	1.055	1.067	1.087	1.103
0.01000	(0.0981)	0.964	0.966	0.971	0.977	0.983	0.994	1.003	1.012	1.027	1.039	1.053	1.065	1.085	1.102
0.01500	(0.147)	0.955	0.958	0.964	0.971	0.977	0.988	0.998	1.007	1.023	1.036	1.050	1.062	1.083	1.100
0.02000	(0.196)	0.949	0.953	0.960	0.967	0.973	0.985	0.996	1.005	1.021	1.034	1.048	1.061	1.081	1.098
0.03000	(0.294)	0.942	0.946	0.954	0.962	0.969	0.981	0.992	1.002	1.018	1.032	1.046	1.059	1.080	1.097
0.04000	(0.392)	0.938	0.943	0.951	0.959	0.966	0.979	0.990	1.000	1.016	1.030	1.045	1.057	1.078	1.096
0.06000	(0.588)	0.933	0.938	0.947	0.955	0.963	0.976	0.987	0.997	1.014	1.028	1.043	1.056	1.077	1.095
0.08000	(0.785)	0.930	0.935	0.945	0.953	0.961	0.974	0.986	0.996	1.013	1.027	1.042	1.055	1.076	1.094
0.10000	(0.981)	0.928	0.933	0.943	0.952	0.959	0.973	0.985	0.995	1.012	1.026	1.041	1.054	1.076	1.094
0.15000	(1.47)	0.924	0.930	0.940	0.949	0.957	0.971	0.983	0.993	1.011	1.025	1.040	1.053	1.075	1.093
0.20000	(1.96)	0.922	0.928	0.938	0.948	0.956	0.970	0.982	0.993	1.010	1.024	1.040	1.053	1.075	1.092
0.30000	(2.94)	0.920	0.926	0.937	0.946	0.954	0.969	0.981	0.991	1.009	1.024	1.039	1.052	1.074	1.092
0.40000	(3.92)	0.918	0.924	0.935	0.945	0.953	0.968	0.980	0.991	1.008	1.023	1.039	1.052	1.074	1.091
0.60000	(5.88)	0.916	0.923	0.934	0.944	0.952	0.967	0.979	0.990	1.008	1.023	1.038	1.051	1.073	1.091
0.80000	(7.85)	0.915	0.922	0.933	0.943	0.952	0.966	0.979	0.989	1.007	1.022	1.038	1.051	1.073	1.091
1.00000	(9.81)	0.914	0.921	0.932	0.942	0.951	0.966	0.978	0.989	1.007	1.022	1.038	1.051	1.073	1.091
1.50000	(14.7)	0.913	0.920	0.932	0.942	0.950	0.965	0.978	0.989	1.007	1.022	1.037	1.050	1.073	1.090
2.00000	(19.6)	0.912	0.919	0.931	0.941	0.950	0.965	0.978	0.988	1.006	1.021	1.037	1.050	1.072	1.090
3.00000	(29.4)	0.912	0.918	0.930	0.941	0.949	0.965	0.977	0.988	1.006	1.021	1.037	1.050	1.072	1.090
S	$(\Delta p/\rho l)$	125	150	200	250	300	400	500	600	800	1000	1250	1500	2000	2500

Grad'nt k.p.g. (Equivalent) Pipe diameters in mm

Kinematic viscosity, $\nu = 1.5 \times 10^{-6}\ \mathrm{m^2 s^{-1}}$;   **Roughness size, k_s = 0.150 mm**

Kin. visc., $\nu = 1.50 \times 10^{-6}$ m^2 s^{-1};
$S = 0.00010$ to 30.0000

i.e. kin. pr. grad., $\Delta p/\rho l =$
(0.00098) to (294) ms^{-2}

Roughness size, $k_s = 0.30$ mm
This table shows values of m, as follows

m_C for Colebrook-White solutions; or,
where $R \leq 2000$, m_P for laminar flow

Grad'nt	k.p.g.	(Equivalent) Pipe diameters in mm													
S	$(\Delta p/\rho l)$	6	8	10	12·5	15	20	25	30	40	50	60	80	100	125
0·00030	(0·00294)	10·29	7·009	5·205	3·866	3·032	2·066	1·534	1·203	*	1·224	1·204	1·178	1·162	1·150
0·00040	(0·00392)	8·908	6·070	4·508	3·348	2·625	1·789	1·329	1·042	*	1·200	1·182	1·158	1·145	1·134
0·00060	(0·00588)	7·273	4·956	3·681	2·734	2·144	1·461	1·085	*	1·193	1·169	1·153	1·134	1·122	1·114
0·00080	(0·00785)	6·299	4·292	3·188	2·367	1·856	1·265	0·939	*	1·171	1·149	1·135	1·118	1·108	1·101
0·00100	(0·00981)	5·634	3·839	2·851	2·117	1·660	1·131	*	*	1·155	1·135	1·122	1·107	1·098	1·092
0·00150	(0·0147)	4·600	3·135	2·328	1·729	1·356	0·924	*	1·156	1·128	1·112	1·101	1·088	1·081	1·077
0·00200	(0·0196)	3·984	2·715	2·016	1·497	1·174	*	*	1·137	1·112	1·097	1·087	1·076	1·071	1·068
0·00300	(0·0294)	3·253	2·216	1·646	1·222	0·959	*	1·129	1·112	1·091	1·078	1·070	1·062	1·058	1·056
0·00400	(0·0392)	2·817	1·920	1·426	1·059	0·830	*	1·112	1·097	1·078	1·067	1·060	1·052	1·049	1·048
0·00600	(0·0588)	2·300	1·567	1·164	0·864	*	1·111	1·091	1·077	1·061	1·052	1·046	1·041	1·039	1·039
0·00800	(0·0785)	1·992	1·357	1·008	*	*	1·096	1·077	1·065	1·051	1·043	1·038	1·034	1·033	1·034
0·01000	(0·0981)	1·782	1·214	0·902	*	*	1·085	1·068	1·057	1·044	1·036	1·032	1·029	1·029	1·030
0·01500	(0·147)	1·455	0·991	*	*	1·092	1·067	1·053	1·043	1·032	1·026	1·023	1·021	1·022	1·024
0·02000	(0·196)	1·260	0·858	*	*	1·079	1·056	1·043	1·034	1·025	1·020	1·017	1·016	1·017	1·020
0·03000	(0·294)	1·029	*	*	1·079	1·063	1·043	1·031	1·024	1·016	1·012	1·010	1·010	1·012	1·015
0·04000	(0·392)	0·891	*	*	1·068	1·053	1·035	1·024	1·018	1·010	1·007	1·006	1·007	1·009	1·013
0·06000	(0·588)	0·727	*	1·074	1·054	1·041	1·025	1·015	1·010	1·004	1·002	1·001	1·002	1·005	1·009
0·08000	(0·785)	*	*	1·064	1·045	1·033	1·018	1·010	1·005	1·000	0·998	0·998	1·000	1·003	1·007
0·10000	(0·981)	*	1·078	1·057	1·040	1·028	1·014	1·006	1·002	0·997	0·996	0·996	0·998	1·001	1·006
0·15000	(1·47)	*	1·066	1·046	1·030	1·020	1·007	1·001	0·997	0·993	0·992	0·992	0·995	0·999	1·004
0·20000	(1·96)	*	1·058	1·040	1·025	1·015	1·003	0·997	0·993	0·990	0·990	0·990	0·993	0·997	1·002
0·30000	(2·94)	1·076	1·049	1·032	1·018	1·009	0·998	0·993	0·990	0·987	0·987	0·988	0·991	0·995	1·001
0·40000	(3·92)	1·069	1·043	1·027	1·014	1·005	0·995	0·990	0·987	0·985	0·985	0·986	0·990	0·994	1·000
0·60000	(5·88)	1·061	1·036	1·021	1·009	1·001	0·992	0·987	0·985	0·983	0·983	0·985	0·989	0·993	0·999
0·80000	(7·85)	1·056	1·032	1·017	1·006	0·998	0·990	0·985	0·983	0·982	0·982	0·984	0·988	0·992	0·998
1·00000	(9·81)	1·052	1·029	1·015	1·004	0·997	0·988	0·984	0·982	0·981	0·981	0·983	0·987	0·992	0·997
1·50000	(14·7)	1·047	1·025	1·011	1·001	0·994	0·986	0·982	0·980	0·979	0·980	0·982	0·986	0·991	0·997
2·00000	(19·6)	1·043	1·022	1·009	0·999	0·992	0·985	0·981	0·979	0·978	0·979	0·981	0·986	0·990	0·996
3·00000	(29·4)	1·039	1·019	1·006	0·996	0·990	0·983	0·979	0·978	0·977	0·978	0·980	0·985	0·990	0·996
4·00000	(39·2)	1·037	1·017	1·005	0·995	0·989	0·982	0·979	0·977	0·977	0·978	0·980	0·985	0·989	0·995
6·00000	(58·8)	1·034	1·015	1·003	0·993	0·987	0·981	0·977	0·976	0·976	0·977	0·979	0·984	0·989	0·995
8·00000	(78·5)	1·032	1·013	1·002	0·992	0·987	0·980	0·977	0·976	0·975	0·977	0·979	0·984	0·989	0·995
10·0000	(98·1)	1·031	1·012	1·001	0·992	0·986	0·979	0·976	0·975	0·975	0·977	0·979	0·984	0·989	0·995
15·0000	(147)	1·029	1·011	1·000	0·991	0·985	0·979	0·976	0·975	0·975	0·976	0·978	0·983	0·988	0·994
20·0000	(196)	1·028	1·010	0·999	0·990	0·984	0·978	0·975	0·974	0·974	0·976	0·978	0·983	0·988	0·994
30·0000	(294)	1·027	1·009	0·998	0·989	0·984	0·978	0·975	0·974	0·974	0·976	0·978	0·983	0·988	0·994
S	$(\Delta p/\rho l)$	6	8	10	12·5	15	20	25	30	40	50	60	80	100	125
Grad'nt	k.p.g.	(Equivalent) Pipe diameters in mm													

S	$(\Delta p/\rho l)$	125	150	200	250	300	400	500	600	800	1000	1250	1500	2000	2500
0·00010	(0·00098)	1·223	1·210	1·195	1·186	1·181	1·176	1·176	1·176	1·180	1·185	1·192	1·199	1·211	1·223
0·00015	(0·00147)	1·194	1·183	1·170	1·164	1·160	1·158	1·158	1·160	1·166	1·172	1·180	1·187	1·201	1·213
0·00020	(0·00196)	1·175	1·165	1·154	1·149	1·147	1·146	1·147	1·150	1·157	1·163	1·172	1·180	1·195	1·207
0·00030	(0·00294)	1·150	1·142	1·134	1·131	1·130	1·130	1·133	1·137	1·145	1·153	1·162	1·171	1·187	1·200
0·00040	(0·00392)	1·134	1·128	1·121	1·119	1·119	1·121	1·124	1·129	1·138	1·146	1·156	1·165	1·182	1·196
0·00060	(0·00588)	1·114	1·109	1·105	1·104	1·105	1·109	1·113	1·118	1·129	1·138	1·149	1·158	1·176	1·190
0·00080	(0·00785)	1·101	1·097	1·094	1·095	1·096	1·101	1·106	1·112	1·123	1·133	1·144	1·154	1·172	1·187
0·00100	(0·00981)	1·092	1·089	1·087	1·088	1·090	1·095	1·101	1·108	1·119	1·129	1·141	1·151	1·169	1·184
0·00150	(0·0147)	1·077	1·075	1·075	1·077	1·080	1·087	1·094	1·100	1·113	1·123	1·136	1·146	1·165	1·181
0·00200	(0·0196)	1·068	1·066	1·067	1·070	1·074	1·081	1·089	1·096	1·109	1·120	1·132	1·144	1·162	1·178
0·00300	(0·0294)	1·056	1·056	1·058	1·062	1·066	1·075	1·083	1·090	1·104	1·116	1·129	1·140	1·159	1·175
0·00400	(0·0392)	1·048	1·049	1·052	1·057	1·061	1·070	1·079	1·087	1·101	1·113	1·126	1·138	1·157	1·174
0·00600	(0·0588)	1·039	1·041	1·045	1·050	1·055	1·065	1·074	1·083	1·097	1·110	1·123	1·135	1·155	1·172
0·00800	(0·0785)	1·034	1·036	1·041	1·046	1·052	1·062	1·072	1·080	1·095	1·108	1·122	1·134	1·154	1·171
0·01000	(0·0981)	1·030	1·032	1·038	1·044	1·049	1·060	1·070	1·079	1·094	1·107	1·120	1·133	1·153	1·170
0·01500	(0·147)	1·024	1·027	1·033	1·039	1·046	1·057	1·067	1·076	1·091	1·105	1·119	1·131	1·151	1·168
0·02000	(0·196)	1·020	1·023	1·030	1·037	1·043	1·055	1·065	1·074	1·090	1·103	1·118	1·130	1·151	1·168
0·03000	(0·294)	1·015	1·019	1·026	1·034	1·040	1·052	1·063	1·072	1·088	1·102	1·116	1·129	1·150	1·167
0·04000	(0·392)	1·013	1·016	1·024	1·032	1·039	1·051	1·062	1·071	1·087	1·101	1·115	1·128	1·149	1·166
0·06000	(0·588)	1·009	1·013	1·022	1·029	1·037	1·049	1·060	1·070	1·086	1·100	1·114	1·127	1·148	1·165
0·08000	(0·785)	1·007	1·012	1·020	1·028	1·035	1·048	1·059	1·069	1·085	1·099	1·114	1·127	1·148	1·165
0·10000	(0·981)	1·006	1·010	1·019	1·027	1·034	1·047	1·058	1·068	1·085	1·099	1·113	1·126	1·147	1·165
0·15000	(1·47)	1·004	1·008	1·017	1·026	1·033	1·046	1·057	1·067	1·084	1·098	1·113	1·126	1·147	1·164
0·20000	(1·96)	1·002	1·007	1·016	1·025	1·032	1·046	1·057	1·067	1·084	1·098	1·112	1·125	1·147	1·164
0·30000	(2·94)	1·001	1·006	1·015	1·024	1·031	1·045	1·056	1·066	1·083	1·097	1·112	1·125	1·146	1·164
0·40000	(3·92)	1·000	1·005	1·014	1·023	1·031	1·044	1·056	1·066	1·083	1·097	1·112	1·125	1·146	1·164
0·60000	(5·88)	0·999	1·004	1·014	1·022	1·030	1·044	1·055	1·065	1·082	1·096	1·111	1·124	1·146	1·163
0·80000	(7·85)	0·998	1·003	1·013	1·022	1·030	1·043	1·055	1·065	1·082	1·096	1·111	1·124	1·146	1·163
1·00000	(9·81)	0·997	1·003	1·013	1·022	1·029	1·043	1·055	1·065	1·082	1·096	1·111	1·124	1·146	1·163
1·50000	(14·7)	0·997	1·002	1·012	1·021	1·029	1·043	1·054	1·064	1·082	1·096	1·111	1·124	1·145	1·163
2·00000	(19·6)	0·996	1·002	1·012	1·021	1·029	1·042	1·054	1·064	1·081	1·096	1·111	1·124	1·145	1·163
3·00000	(29·4)	0·996	1·001	1·011	1·020	1·028	1·042	1·054	1·064	1·081	1·096	1·111	1·124	1·145	1·163
S	$(\Delta p/\rho l)$	125	150	200	250	300	400	500	600	800	1000	1250	1500	2000	2500
Grad'nt	k.p.g.	(Equivalent) Pipe diameters in mm													

Kinematic viscosity, $\nu = 1.50 \times 10^{-6}$ m^2 s^{-1} ;   **Roughness size, $k_s = 0.30$ mm**

E123

Kin. visc., $\nu = 1.5 \times 10^{-6}$ m^{2}s^{-1};
$S = 0.00010$ to 3.00000

i.e. kin. pr. grad., $\Delta p/\rho l =$
(0.00098) to (29.4) ms^{-2}

Roughness size, $k_s = 0.60$ mm
This table shows values of m, as follows

m_C for Colebrook-White solutions

Grad'nt S	k.p.g. $(\Delta p/\rho l)$	125	150	200	250	300	400	500	600	800	1000	1250	1500	2000	2500
0.00010	(0.00098)	1.271	1.259	1.244	1.236	1.232	1.229	1.229	1.230	1.236	1.242	1.250	1.257	1.272	1.284
0.00015	(0.00147)	1.246	1.235	1.224	1.218	1.215	1.214	1.215	1.218	1.225	1.232	1.241	1.249	1.265	1.278
0.00020	(0.00196)	1.230	1.221	1.211	1.206	1.205	1.205	1.207	1.211	1.218	1.226	1.236	1.244	1.260	1.274
0.00030	(0.00294)	1.210	1.202	1.195	1.192	1.191	1.193	1.197	1.201	1.210	1.219	1.229	1.238	1.255	1.269
0.00040	(0.00392)	1.197	1.191	1.185	1.183	1.183	1.186	1.190	1.195	1.205	1.214	1.225	1.235	1.252	1.266
0.00060	(0.00588)	1.181	1.177	1.173	1.172	1.173	1.177	1.183	1.188	1.199	1.209	1.220	1.230	1.248	1.263
0.00080	(0.00785)	1.171	1.168	1.165	1.165	1.167	1.172	1.178	1.184	1.195	1.205	1.217	1.227	1.245	1.261
0.00100	(0.00981)	1.164	1.161	1.159	1.160	1.163	1.168	1.175	1.181	1.192	1.203	1.215	1.225	1.244	1.259
0.00150	(0.0147)	1.153	1.151	1.151	1.153	1.156	1.162	1.169	1.176	1.188	1.199	1.212	1.222	1.241	1.257
0.00200	(0.0196)	1.146	1.145	1.146	1.148	1.151	1.159	1.166	1.173	1.186	1.197	1.210	1.221	1.239	1.255
0.00300	(0.0294)	1.138	1.137	1.139	1.142	1.146	1.154	1.162	1.169	1.183	1.194	1.207	1.218	1.238	1.254
0.00400	(0.0392)	1.133	1.133	1.135	1.139	1.143	1.152	1.160	1.167	1.181	1.193	1.206	1.217	1.236	1.253
0.00600	(0.0588)	1.127	1.127	1.130	1.135	1.139	1.148	1.157	1.165	1.179	1.191	1.204	1.215	1.235	1.251
0.00800	(0.0785)	1.123	1.124	1.128	1.132	1.137	1.146	1.155	1.163	1.177	1.190	1.203	1.215	1.234	1.251
0.01000	(0.0981)	1.120	1.122	1.126	1.130	1.135	1.145	1.154	1.162	1.176	1.189	1.202	1.214	1.234	1.250
0.01500	(0.147)	1.116	1.118	1.122	1.128	1.133	1.143	1.152	1.160	1.175	1.188	1.201	1.213	1.233	1.249
0.02000	(0.196)	1.114	1.116	1.121	1.126	1.131	1.142	1.151	1.159	1.174	1.187	1.200	1.212	1.232	1.249
0.03000	(0.294)	1.111	1.113	1.118	1.124	1.130	1.140	1.150	1.158	1.173	1.186	1.200	1.212	1.232	1.248
0.04000	(0.392)	1.109	1.111	1.117	1.123	1.129	1.139	1.149	1.158	1.173	1.185	1.199	1.211	1.231	1.248
0.06000	(0.588)	1.107	1.109	1.115	1.121	1.127	1.138	1.148	1.157	1.172	1.185	1.199	1.211	1.231	1.248
0.08000	(0.785)	1.105	1.108	1.114	1.121	1.127	1.138	1.147	1.156	1.171	1.184	1.198	1.210	1.231	1.248
0.10000	(0.981)	1.105	1.107	1.114	1.120	1.126	1.137	1.147	1.156	1.171	1.184	1.198	1.210	1.230	1.247
0.15000	(1.47)	1.103	1.106	1.113	1.119	1.125	1.136	1.146	1.155	1.171	1.184	1.198	1.210	1.230	1.247
0.20000	(1.96)	1.102	1.105	1.112	1.119	1.125	1.136	1.146	1.155	1.170	1.183	1.197	1.210	1.230	1.247
0.30000	(2.94)	1.101	1.105	1.111	1.118	1.124	1.136	1.146	1.155	1.170	1.183	1.197	1.209	1.230	1.247
0.40000	(3.92)	1.101	1.104	1.111	1.118	1.124	1.135	1.145	1.154	1.170	1.183	1.197	1.209	1.230	1.247
0.60000	(5.88)	1.100	1.103	1.110	1.117	1.123	1.135	1.145	1.154	1.170	1.183	1.197	1.209	1.230	1.247
0.80000	(7.85)	1.100	1.103	1.110	1.117	1.123	1.135	1.145	1.154	1.169	1.183	1.197	1.209	1.229	1.246
1.00000	(9.81)	1.099	1.103	1.110	1.117	1.123	1.135	1.145	1.154	1.169	1.183	1.197	1.209	1.229	1.246
1.50000	(14.7)	1.099	1.102	1.110	1.116	1.123	1.134	1.145	1.154	1.169	1.182	1.197	1.209	1.229	1.246
2.00000	(19.6)	1.099	1.102	1.109	1.116	1.123	1.134	1.144	1.153	1.169	1.182	1.196	1.209	1.229	1.246
3.00000	(29.4)	1.098	1.102	1.109	1.116	1.122	1.134	1.144	1.153	1.169	1.182	1.196	1.209	1.229	1.246
S	$(\Delta p/\rho l)$	125	150	200	250	300	400	500	600	800	1000	1250	1500	2000	2500

Grad'nt k.p.g. (Equivalent) Pipe diameters in mm **Roughness size, $k_s = 0.60$ mm**

E124

Kin. visc., $\nu = 1.5 \times 10^{-6}$ m^{2}s^{-1};
$S = 0.00010$ to 3.00000

i.e. kin. pr. grad., $\Delta p/\rho l =$
(0.00098) to (29.4) ms^{-2}

Roughness size, $k_s = 1.50$ mm
This table shows values of m, as follows

m_C for Colebrook-White solutions

Grad'nt S	k.p.g. $(\Delta p/\rho l)$	125	150	200	250	300	400	500	600	800	1000	1250	1500	2000	2500
0.00010	(0.00098)	1.385	1.372	1.356	1.348	1.343	1.340	1.340	1.342	1.347	1.353	1.362	1.369	1.384	1.397
0.00015	(0.00147)	1.367	1.355	1.342	1.335	1.332	1.330	1.331	1.334	1.341	1.348	1.356	1.365	1.380	1.393
0.00020	(0.00196)	1.355	1.345	1.333	1.328	1.325	1.325	1.326	1.329	1.337	1.344	1.353	1.362	1.378	1.391
0.00030	(0.00294)	1.341	1.332	1.323	1.319	1.317	1.318	1.320	1.324	1.332	1.340	1.350	1.359	1.375	1.389
0.00040	(0.00392)	1.333	1.325	1.316	1.313	1.312	1.313	1.316	1.320	1.329	1.337	1.347	1.357	1.373	1.387
0.00060	(0.00588)	1.322	1.316	1.309	1.306	1.306	1.308	1.312	1.316	1.326	1.334	1.345	1.354	1.371	1.385
0.00080	(0.00785)	1.316	1.310	1.304	1.302	1.302	1.305	1.309	1.314	1.324	1.333	1.343	1.353	1.370	1.384
0.00100	(0.00981)	1.312	1.306	1.301	1.299	1.300	1.303	1.307	1.312	1.322	1.331	1.342	1.352	1.369	1.383
0.00150	(0.0147)	1.305	1.300	1.296	1.295	1.296	1.299	1.304	1.310	1.320	1.329	1.340	1.350	1.367	1.382
0.00200	(0.0196)	1.301	1.296	1.293	1.292	1.293	1.297	1.303	1.308	1.318	1.328	1.339	1.349	1.367	1.381
0.00300	(0.0294)	1.296	1.292	1.289	1.289	1.290	1.295	1.301	1.306	1.317	1.327	1.338	1.348	1.366	1.381
0.00400	(0.0392)	1.293	1.289	1.287	1.287	1.289	1.294	1.299	1.305	1.316	1.326	1.337	1.347	1.365	1.380
0.00600	(0.0588)	1.289	1.286	1.284	1.285	1.287	1.292	1.298	1.304	1.315	1.325	1.336	1.347	1.364	1.379
0.00800	(0.0785)	1.287	1.284	1.282	1.283	1.285	1.291	1.297	1.303	1.314	1.324	1.336	1.346	1.364	1.379
0.01000	(0.0981)	1.285	1.283	1.281	1.282	1.284	1.290	1.296	1.302	1.314	1.324	1.335	1.346	1.364	1.379
0.01500	(0.147)	1.283	1.280	1.279	1.281	1.283	1.289	1.295	1.301	1.313	1.323	1.335	1.345	1.363	1.378
0.02000	(0.196)	1.281	1.279	1.278	1.280	1.282	1.288	1.295	1.301	1.312	1.323	1.335	1.345	1.363	1.378
0.03000	(0.294)	1.280	1.278	1.277	1.279	1.281	1.288	1.294	1.300	1.312	1.322	1.334	1.345	1.363	1.378
0.04000	(0.392)	1.279	1.277	1.276	1.278	1.281	1.287	1.294	1.300	1.312	1.322	1.334	1.344	1.363	1.378
0.06000	(0.588)	1.278	1.276	1.276	1.277	1.280	1.286	1.293	1.299	1.311	1.322	1.334	1.344	1.362	1.378
0.08000	(0.785)	1.277	1.275	1.275	1.277	1.280	1.286	1.293	1.299	1.311	1.322	1.333	1.344	1.362	1.378
0.10000	(0.981)	1.276	1.275	1.275	1.277	1.279	1.286	1.293	1.299	1.311	1.322	1.333	1.344	1.362	1.377
0.15000	(1.47)	1.276	1.274	1.274	1.276	1.279	1.286	1.292	1.299	1.311	1.321	1.333	1.344	1.362	1.377
0.20000	(1.96)	1.275	1.274	1.274	1.276	1.279	1.285	1.292	1.299	1.310	1.321	1.333	1.344	1.362	1.377
0.30000	(2.94)	1.275	1.273	1.273	1.276	1.278	1.285	1.292	1.298	1.310	1.321	1.333	1.344	1.362	1.377
0.40000	(3.92)	1.274	1.273	1.273	1.275	1.278	1.285	1.292	1.298	1.310	1.321	1.333	1.343	1.362	1.377
0.60000	(5.88)	1.274	1.273	1.273	1.275	1.278	1.285	1.292	1.298	1.310	1.321	1.333	1.343	1.362	1.377
0.80000	(7.85)	1.274	1.272	1.273	1.275	1.278	1.285	1.291	1.298	1.310	1.321	1.333	1.343	1.362	1.377
1.00000	(9.81)	1.274	1.272	1.273	1.275	1.278	1.285	1.291	1.298	1.310	1.321	1.333	1.343	1.362	1.377
1.50000	(14.7)	1.273	1.272	1.272	1.275	1.278	1.284	1.291	1.298	1.310	1.321	1.333	1.343	1.362	1.377
2.00000	(19.6)	1.273	1.272	1.272	1.275	1.278	1.284	1.291	1.298	1.310	1.321	1.333	1.343	1.361	1.377
3.00000	(29.4)	1.273	1.272	1.272	1.275	1.278	1.284	1.291	1.298	1.310	1.321	1.333	1.343	1.361	1.377
S	$(\Delta p/\rho l)$	125	150	200	250	300	400	500	600	800	1000	1250	1500	2000	2500

Grad'nt k.p.g. (Equivalent) Pipe diameters in mm **Roughness size, $k_s = 1.50$ mm**

Kin. visc., $\nu = 1.5 \times 10^{-6}$ m^{2}s^{-1};
$S = 0.00010$ to 3.00000

i.e. kin. pr. grad., $\Delta p/\rho l =$
(0.00098) to (29.4) ms^{-2}

Roughness size, $k_s = 3.0$ mm
This table shows values of m, as follows
m_C for Colebrook-White solutions

Grad'nt S	k.p.g. $(\Delta p/\rho l)$	125	150	200	250	300	400	500	600	800	1000	1250	1500	2000	2500
							(Equivalent) Pipe diameters in mm								
0.00010	(0.00098)	1.526	1.509	1.489	1.477	1.470	1.464	1.462	1.462	1.465	1.470	1.476	1.483	1.496	1.508
0.00015	(0.00147)	1.512	1.497	1.479	1.469	1.463	1.457	1.456	1.457	1.461	1.466	1.473	1.480	1.493	1.505
0.00020	(0.00196)	1.504	1.490	1.473	1.463	1.458	1.453	1.453	1.454	1.458	1.464	1.471	1.478	1.492	1.504
0.00030	(0.00294)	1.494	1.481	1.465	1.457	1.453	1.449	1.449	1.450	1.455	1.461	1.469	1.476	1.490	1.503
0.00040	(0.00392)	1.488	1.476	1.461	1.453	1.449	1.446	1.446	1.448	1.453	1.459	1.467	1.475	1.489	1.502
0.00060	(0.00588)	1.481	1.469	1.456	1.449	1.445	1.443	1.443	1.445	1.451	1.458	1.466	1.473	1.488	1.501
0.00080	(0.00785)	1.476	1.465	1.453	1.446	1.443	1.441	1.442	1.444	1.450	1.456	1.465	1.473	1.487	1.500
0.00100	(0.00981)	1.473	1.463	1.451	1.444	1.441	1.439	1.440	1.443	1.449	1.456	1.464	1.472	1.487	1.499
0.00150	(0.0147)	1.469	1.459	1.447	1.441	1.439	1.437	1.438	1.441	1.447	1.454	1.463	1.471	1.486	1.499
0.00200	(0.0196)	1.466	1.456	1.445	1.440	1.437	1.436	1.437	1.440	1.447	1.454	1.462	1.470	1.485	1.498
0.00300	(0.0294)	1.463	1.453	1.443	1.437	1.435	1.434	1.436	1.439	1.446	1.453	1.462	1.470	1.485	1.498
0.00400	(0.0392)	1.461	1.451	1.441	1.436	1.434	1.433	1.435	1.438	1.445	1.452	1.461	1.469	1.484	1.497
0.00600	(0.0588)	1.458	1.449	1.439	1.435	1.433	1.432	1.434	1.437	1.444	1.452	1.461	1.469	1.484	1.497
0.00800	(0.0785)	1.457	1.448	1.438	1.434	1.432	1.432	1.434	1.437	1.444	1.451	1.460	1.469	1.484	1.497
0.01000	(0.0981)	1.456	1.447	1.438	1.433	1.431	1.431	1.433	1.436	1.444	1.451	1.460	1.468	1.483	1.497
0.01500	(0.147)	1.454	1.446	1.437	1.432	1.431	1.431	1.433	1.436	1.443	1.451	1.460	1.468	1.483	1.497
0.02000	(0.196)	1.453	1.445	1.436	1.432	1.430	1.430	1.432	1.436	1.443	1.450	1.459	1.468	1.483	1.496
0.03000	(0.294)	1.452	1.444	1.435	1.431	1.429	1.430	1.432	1.435	1.443	1.450	1.459	1.468	1.483	1.496
0.04000	(0.392)	1.451	1.443	1.435	1.431	1.429	1.429	1.432	1.435	1.442	1.450	1.459	1.468	1.483	1.496
0.06000	(0.588)	1.451	1.443	1.434	1.430	1.429	1.429	1.431	1.435	1.442	1.450	1.459	1.467	1.483	1.496
0.08000	(0.785)	1.450	1.442	1.434	1.430	1.428	1.429	1.431	1.435	1.442	1.450	1.459	1.467	1.483	1.496
0.10000	(0.981)	1.450	1.442	1.434	1.430	1.428	1.429	1.431	1.434	1.442	1.450	1.459	1.467	1.483	1.496
0.15000	(1.47)	1.449	1.442	1.433	1.429	1.428	1.428	1.431	1.434	1.442	1.449	1.459	1.467	1.482	1.496
0.20000	(1.96)	1.449	1.441	1.433	1.429	1.428	1.428	1.431	1.434	1.442	1.449	1.459	1.467	1.482	1.496
0.30000	(2.94)	1.449	1.441	1.433	1.429	1.428	1.428	1.431	1.434	1.442	1.449	1.458	1.467	1.482	1.496
0.40000	(3.92)	1.449	1.441	1.433	1.429	1.427	1.428	1.431	1.434	1.442	1.449	1.458	1.467	1.482	1.496
0.60000	(5.88)	1.448	1.441	1.432	1.429	1.427	1.428	1.430	1.434	1.442	1.449	1.458	1.467	1.482	1.496
0.80000	(7.85)	1.448	1.441	1.432	1.429	1.427	1.428	1.430	1.434	1.441	1.449	1.458	1.467	1.482	1.496
1.00000	(9.81)	1.448	1.441	1.432	1.429	1.427	1.428	1.430	1.434	1.441	1.449	1.458	1.467	1.482	1.496
1.50000	(14.7)	1.448	1.440	1.432	1.428	1.427	1.428	1.430	1.434	1.441	1.449	1.458	1.467	1.482	1.496
2.00000	(19.6)	1.448	1.440	1.432	1.428	1.427	1.428	1.430	1.434	1.441	1.449	1.458	1.467	1.482	1.496
3.00000	(29.4)	1.448	1.440	1.432	1.428	1.427	1.428	1.430	1.434	1.441	1.449	1.458	1.467	1.482	1.496
S	$(\Delta p/\rho l)$	125	150	200	250	300	400	500	600	800	1000	1250	1500	2000	2500

Grad'nt   k.p.g.   (Equivalent) Pipe diameters in mm   **Roughness size, $k_s = 3.0$ mm**

Kin. visc., $\nu = 1.5 \times 10^{-6}$ m^{2}s^{-1};
$S = 0.00010$ to 3.00000

i.e. kin. pr. grad., $\Delta p/\rho l =$
(0.00098) to (29.4) ms^{-2}

Roughness size, $k_s = 6.0$ mm
This table shows values of m, as follows
m_C for Colebrook-White solutions

Grad'nt S	k.p.g. $(\Delta p/\rho l)$	125	150	200	250	300	400	500	600	800	1000	1250	1500	2000	2500
							(Equivalent) Pipe diameters in mm								
0.00010	(0.00098)	1.733	1.708	1.676	1.657	1.645	1.631	1.623	1.619	1.617	1.618	1.621	1.625	1.635	1.644
0.00015	(0.00147)	1.723	1.700	1.669	1.651	1.640	1.626	1.620	1.616	1.614	1.616	1.619	1.624	1.633	1.643
0.00020	(0.00196)	1.717	1.694	1.665	1.648	1.637	1.624	1.617	1.614	1.613	1.614	1.618	1.623	1.632	1.642
0.00030	(0.00294)	1.710	1.688	1.660	1.644	1.633	1.621	1.615	1.612	1.611	1.613	1.617	1.621	1.631	1.641
0.00040	(0.00392)	1.706	1.684	1.657	1.641	1.631	1.619	1.613	1.611	1.610	1.612	1.616	1.621	1.631	1.640
0.00060	(0.00588)	1.701	1.680	1.654	1.638	1.628	1.617	1.611	1.609	1.608	1.611	1.615	1.620	1.630	1.640
0.00080	(0.00785)	1.698	1.677	1.652	1.636	1.626	1.615	1.610	1.608	1.608	1.610	1.614	1.619	1.629	1.639
0.00100	(0.00981)	1.696	1.676	1.650	1.635	1.625	1.615	1.609	1.607	1.607	1.609	1.614	1.619	1.629	1.639
0.00150	(0.0147)	1.693	1.673	1.648	1.633	1.624	1.613	1.608	1.606	1.606	1.609	1.613	1.618	1.629	1.639
0.00200	(0.0196)	1.691	1.671	1.646	1.632	1.623	1.612	1.608	1.606	1.606	1.608	1.613	1.618	1.628	1.638
0.00300	(0.0294)	1.688	1.669	1.645	1.631	1.621	1.611	1.607	1.605	1.605	1.608	1.612	1.617	1.628	1.638
0.00400	(0.0392)	1.687	1.668	1.644	1.630	1.621	1.611	1.606	1.604	1.605	1.607	1.612	1.617	1.628	1.638
0.00600	(0.0588)	1.685	1.666	1.643	1.629	1.620	1.610	1.606	1.604	1.604	1.607	1.612	1.617	1.627	1.638
0.00800	(0.0785)	1.684	1.666	1.642	1.628	1.619	1.610	1.605	1.604	1.604	1.607	1.611	1.617	1.627	1.638
0.01000	(0.0981)	1.684	1.665	1.642	1.628	1.619	1.609	1.605	1.603	1.604	1.607	1.611	1.617	1.627	1.637
0.01500	(0.147)	1.683	1.664	1.641	1.627	1.619	1.609	1.605	1.603	1.603	1.606	1.611	1.616	1.627	1.637
0.02000	(0.196)	1.682	1.664	1.640	1.627	1.618	1.609	1.604	1.603	1.603	1.606	1.611	1.616	1.627	1.637
0.03000	(0.294)	1.681	1.663	1.640	1.626	1.618	1.608	1.604	1.603	1.603	1.606	1.611	1.616	1.627	1.637
0.04000	(0.392)	1.681	1.663	1.640	1.626	1.618	1.608	1.604	1.602	1.603	1.606	1.611	1.616	1.627	1.637
0.06000	(0.588)	1.680	1.662	1.639	1.626	1.617	1.608	1.604	1.602	1.603	1.606	1.611	1.616	1.627	1.637
0.08000	(0.785)	1.680	1.662	1.639	1.626	1.617	1.608	1.604	1.602	1.603	1.606	1.611	1.616	1.627	1.637
0.10000	(0.981)	1.680	1.662	1.639	1.625	1.617	1.608	1.604	1.602	1.603	1.606	1.610	1.616	1.627	1.637
0.15000	(1.47)	1.679	1.661	1.639	1.625	1.617	1.608	1.603	1.602	1.603	1.606	1.610	1.616	1.627	1.637
0.20000	(1.96)	1.679	1.661	1.639	1.625	1.617	1.607	1.603	1.602	1.603	1.605	1.610	1.616	1.627	1.637
0.30000	(2.94)	1.679	1.661	1.638	1.625	1.617	1.607	1.603	1.602	1.602	1.605	1.610	1.616	1.627	1.637
0.40000	(3.92)	1.679	1.661	1.638	1.625	1.617	1.607	1.603	1.602	1.602	1.605	1.610	1.616	1.626	1.637
0.60000	(5.88)	1.679	1.661	1.638	1.625	1.616	1.607	1.603	1.602	1.602	1.605	1.610	1.616	1.626	1.637
0.80000	(7.85)	1.679	1.661	1.638	1.625	1.616	1.607	1.603	1.602	1.602	1.605	1.610	1.616	1.626	1.637
1.00000	(9.81)	1.679	1.661	1.638	1.625	1.616	1.607	1.603	1.602	1.602	1.605	1.610	1.616	1.626	1.637
1.50000	(14.7)	1.678	1.660	1.638	1.625	1.616	1.607	1.603	1.602	1.602	1.605	1.610	1.616	1.626	1.637
2.00000	(19.6)	1.678	1.660	1.638	1.625	1.616	1.607	1.603	1.602	1.602	1.605	1.610	1.616	1.626	1.637
3.00000	(29.4)	1.678	1.660	1.638	1.625	1.616	1.607	1.603	1.602	1.602	1.605	1.610	1.616	1.626	1.637
S	$(\Delta p/\rho l)$	125	150	200	250	300	400	500	600	800	1000	1250	1500	2000	2500

Grad'nt   k.p.g.   (Equivalent) Pipe diameters in mm   **Roughness size, $k_s = 6.0$ mm**

E127

Kin. visc., $\nu = 2{\cdot}0 \times 10^{-6}\ \mathrm{m^2 s^{-1}}$;
$S = 0{\cdot}00010$ to $30{\cdot}0000$

i.e. kin. pr. grad., $\Delta p/\rho l =$
$(0{\cdot}00098)$ to $(294)\ \mathrm{ms^{-2}}$

Roughness size, $k_s = 0{\cdot}0015$ mm
This table shows values of m, as follows

m_C for Colebrook-White solutions; or,
where $R \le 2000$, m_P for laminar flow

Grad'nt S	k.p.g. $(\Delta p/\rho l)$	6	8	10	12·5	15	20	25	30	40	50	60	80	100	125
		(Equivalent) Pipe diameters in mm													
0·00030	(0·00294)	13·71	9·346	6·941	5·154	4·042	2·754	2·046	1·604	1·093	*	1·193	1·159	1·138	1·120
0·00040	(0·00392)	11·88	8·093	6·011	4·464	3·501	2·385	1·771	1·389	0·947	*	1·162	1·131	1·112	1·096
0·00060	(0·00588)	9·698	6·608	4·908	3·645	2·858	1·948	1·446	1·134	*	1·142	1·121	1·094	1·077	1·064
0·00080	(0·00785)	8·399	5·723	4·250	3·156	2·475	1·687	1·253	0·982	*	1·112	1·094	1·069	1·054	1·042
0·00100	(0·00981)	7·512	5·119	3·801	2·823	2·214	1·509	1·120	*	1·116	1·091	1·073	1·051	1·037	1·026
0·00150	(0·0147)	6·133	4·179	3·104	2·305	1·808	1·232	0·915	*	1·076	1·054	1·039	1·019	1·007	0·997
0·00200	(0·0196)	5·312	3·620	2·688	1·996	1·565	1·067	*	*	1·049	1·029	1·015	0·997	0·987	0·978
0·00300	(0·0294)	4·337	2·955	2·195	1·630	1·278	0·871	*	1·041	1·013	0·996	0·984	0·969	0·959	0·952
0·00400	(0·0392)	3·756	2·559	1·901	1·412	1·107	*	1·035	1·015	0·990	0·974	0·963	0·949	0·941	0·935
0·00600	(0·0588)	3·067	2·090	1·552	1·153	0·904	*	0·997	0·980	0·958	0·944	0·935	0·923	0·916	0·911
0·00800	(0·0785)	2·656	1·810	1·344	0·998	*	0·995	0·972	0·957	0·937	0·924	0·916	0·906	0·900	0·896
0·01000	(0·0981)	2·375	1·619	1·202	0·893	*	0·975	0·954	0·939	0·921	0·909	0·902	0·892	0·887	0·884
0·01500	(0·147)	1·940	1·322	0·982	*	*	0·940	0·922	0·910	0·893	0·884	0·877	0·869	0·865	0·863
0·02000	(0·196)	1·680	1·145	0·850	*	0·944	0·917	0·901	0·889	0·875	0·866	0·860	0·854	0·850	0·849
0·03000	(0·294)	1·371	0·935	*	0·928	0·910	0·887	0·872	0·863	0·850	0·843	0·838	0·833	0·830	0·829
0·04000	(0·392)	1·188	0·809	*	0·904	0·888	0·867	0·853	0·845	0·834	0·827	0·823	0·819	0·817	0·816
0·06000	(0·588)	0·970	*	0·892	0·872	0·858	0·839	0·828	0·821	0·811	0·806	0·803	0·799	0·798	0·799
0·08000	(0·785)	0·840	*	0·869	0·851	0·838	0·821	0·811	0·804	0·796	0·791	0·789	0·786	0·786	0·787
0·10000	(0·981)	0·751	*	0·852	0·835	0·823	0·808	0·798	0·792	0·785	0·781	0·779	0·777	0·777	0·778
0·15000	(1·47)	*	0·841	0·823	0·808	0·797	0·784	0·776	0·771	0·765	0·762	0·760	0·760	0·760	0·762
0·20000	(1·96)	*	0·820	0·803	0·790	0·780	0·768	0·761	0·757	0·752	0·749	0·748	0·748	0·749	0·751
0·30000	(2·94)	0·815	0·792	0·777	0·765	0·757	0·747	0·741	0·738	0·734	0·732	0·732	0·732	0·734	0·737
0·40000	(3·92)	0·794	0·773	0·760	0·749	0·742	0·733	0·728	0·725	0·722	0·721	0·721	0·722	0·724	0·727
0·60000	(5·88)	0·766	0·748	0·737	0·728	0·721	0·714	0·710	0·708	0·705	0·705	0·706	0·708	0·710	0·714
0·80000	(7·85)	0·748	0·731	0·721	0·713	0·708	0·701	0·698	0·696	0·694	0·695	0·695	0·698	0·701	0·705
1·00000	(9·81)	0·734	0·719	0·710	0·702	0·697	0·691	0·689	0·687	0·686	0·687	0·688	0·691	0·694	0·698
1·50000	(14·7)	0·711	0·698	0·690	0·683	0·679	0·675	0·673	0·672	0·672	0·673	0·674	0·678	0·682	0·686
2·00000	(19·6)	0·695	0·683	0·676	0·671	0·667	0·664	0·662	0·662	0·662	0·664	0·665	0·670	0·674	0·678
3·00000	(29·4)	0·675	0·664	0·659	0·654	0·651	0·649	0·648	0·648	0·649	0·651	0·653	0·658	0·663	0·668
4·00000	(39·2)	0·661	0·652	0·647	0·643	0·640	0·638	0·638	0·638	0·640	0·643	0·645	0·650	0·655	0·661
6·00000	(58·8)	0·642	0·635	0·631	0·628	0·626	0·625	0·625	0·626	0·629	0·632	0·635	0·640	0·646	0·651
8·00000	(78·5)	0·630	0·623	0·620	0·617	0·616	0·616	0·617	0·618	0·621	0·624	0·627	0·633	0·639	0·645
10·0000	(98·1)	0·621	0·615	0·612	0·610	0·609	0·609	0·610	0·612	0·615	0·619	0·622	0·628	0·634	0·641
15·0000	(147)	0·605	0·600	0·598	0·597	0·597	0·597	0·599	0·601	0·605	0·609	0·613	0·620	0·626	0·633
20·0000	(196)	0·594	0·591	0·589	0·588	0·588	0·590	0·592	0·594	0·598	0·603	0·607	0·614	0·621	0·628
30·0000	(294)	0·580	0·578	0·577	0·577	0·577	0·579	0·582	0·584	0·590	0·594	0·599	0·607	0·614	0·621
S	$(\Delta p/\rho l)$	6	8	10	12·5	15	20	25	30	40	50	60	80	100	125

Grad'nt S	k.p.g. $(\Delta p/\rho l)$	125	150	200	250	300	400	500	600	800	1000	1250	1500	2000	2500
		(Equivalent) Pipe diameters in mm													
0·00010	(0·00098)	1·223	1·206	1·183	1·169	1·160	1·149	1·143	1·139	1·136	1·136	1·137	1·139	1·144	1·150
0·00015	(0·00147)	1·183	1·168	1·148	1·136	1·129	1·119	1·115	1·112	1·111	1·111	1·113	1·116	1·122	1·128
0·00020	(0·00196)	1·156	1·142	1·125	1·114	1·107	1·100	1·096	1·094	1·093	1·094	1·097	1·100	1·107	1·113
0·00030	(0·00294)	1·120	1·108	1·093	1·084	1·079	1·073	1·070	1·069	1·069	1·072	1·075	1·078	1·086	1·093
0·00040	(0·00392)	1·096	1·085	1·072	1·064	1·059	1·054	1·052	1·052	1·053	1·056	1·060	1·064	1·071	1·079
0·00060	(0·00588)	1·064	1·054	1·043	1·037	1·033	1·030	1·029	1·029	1·031	1·035	1·039	1·043	1·052	1·060
0·00080	(0·00785)	1·042	1·034	1·024	1·018	1·015	1·013	1·012	1·013	1·016	1·020	1·025	1·030	1·039	1·047
0·00100	(0·00981)	1·026	1·018	1·009	1·005	1·002	1·000	1·000	1·001	1·005	1·009	1·014	1·019	1·028	1·037
0·00150	(0·0147)	0·997	0·991	0·984	0·980	0·979	0·978	0·979	0·981	0·985	0·990	0·995	1·001	1·011	1·019
0·00200	(0·0196)	0·978	0·973	0·967	0·964	0·963	0·963	0·964	0·966	0·971	0·976	0·982	0·988	0·998	1·007
0·00300	(0·0294)	0·952	0·948	0·943	0·942	0·941	0·942	0·944	0·947	0·953	0·958	0·965	0·971	0·982	0·991
0·00400	(0·0392)	0·935	0·931	0·928	0·926	0·926	0·928	0·931	0·934	0·940	0·946	0·953	0·959	0·970	0·980
0·00600	(0·0588)	0·911	0·909	0·906	0·906	0·907	0·909	0·912	0·916	0·923	0·929	0·936	0·943	0·954	0·964
0·00800	(0·0785)	0·896	0·893	0·892	0·892	0·893	0·896	0·900	0·904	0·911	0·917	0·925	0·932	0·944	0·954
0·01000	(0·0981)	0·884	0·882	0·881	0·881	0·883	0·886	0·890	0·894	0·902	0·909	0·916	0·923	0·936	0·946
0·01500	(0·147)	0·863	0·862	0·862	0·863	0·865	0·869	0·874	0·878	0·886	0·893	0·902	0·909	0·921	0·932
0·02000	(0·196)	0·849	0·848	0·849	0·850	0·853	0·857	0·862	0·867	0·875	0·883	0·891	0·899	0·911	0·922
0·03000	(0·294)	0·829	0·829	0·831	0·833	0·836	0·841	0·847	0·852	0·861	0·869	0·877	0·885	0·898	0·909
0·04000	(0·392)	0·816	0·817	0·819	0·822	0·825	0·831	0·836	0·841	0·851	0·859	0·868	0·876	0·889	0·900
0·06000	(0·588)	0·799	0·800	0·802	0·806	0·809	0·816	0·822	0·827	0·837	0·846	0·855	0·863	0·877	0·888
0·08000	(0·785)	0·787	0·788	0·791	0·795	0·799	0·806	0·812	0·818	0·828	0·837	0·846	0·854	0·868	0·880
0·10000	(0·981)	0·778	0·779	0·783	0·787	0·791	0·798	0·805	0·811	0·821	0·830	0·839	0·848	0·862	0·874
0·15000	(1·47)	0·762	0·764	0·768	0·773	0·777	0·785	0·792	0·798	0·809	0·818	0·828	0·837	0·851	0·864
0·20000	(1·96)	0·751	0·754	0·758	0·763	0·768	0·776	0·783	0·789	0·800	0·810	0·820	0·829	0·844	0·856
0·30000	(2·94)	0·737	0·739	0·745	0·750	0·755	0·764	0·771	0·778	0·789	0·799	0·810	0·819	0·834	0·847
0·40000	(3·92)	0·727	0·730	0·736	0·741	0·746	0·755	0·763	0·770	0·782	0·792	0·803	0·812	0·827	0·840
0·60000	(5·88)	0·714	0·717	0·724	0·729	0·735	0·744	0·752	0·760	0·772	0·782	0·793	0·803	0·819	0·832
0·80000	(7·85)	0·705	0·708	0·715	0·721	0·727	0·737	0·745	0·752	0·765	0·776	0·787	0·797	0·813	0·826
1·00000	(9·81)	0·698	0·702	0·709	0·715	0·721	0·731	0·740	0·747	0·760	0·771	0·782	0·792	0·809	0·822
1·50000	(14·7)	0·686	0·690	0·698	0·705	0·711	0·721	0·730	0·738	0·751	0·762	0·774	0·784	0·801	0·815
2·00000	(19·6)	0·678	0·683	0·691	0·698	0·704	0·715	0·724	0·732	0·746	0·757	0·769	0·779	0·797	0·811
3·00000	(29·4)	0·668	0·673	0·681	0·689	0·695	0·706	0·716	0·724	0·738	0·750	0·762	0·773	0·790	0·805
S	$(\Delta p/\rho l)$	125	150	200	250	300	400	500	600	800	1000	1250	1500	2000	2500

Grad'nt k.p.g. (Equivalent) Pipe diameters in mm

Kinematic viscosity, $\nu = 2{\cdot}0 \times 10^{-6}\ \mathrm{m^2 s^{-1}}$;   Roughness size, $k_s = 0{\cdot}0015$ mm

Kin. visc., $\nu = 2{\cdot}0\times10^{-6}\ \mathrm{m^2\,s^{-1}}$; $S = 0{\cdot}00010$ to $30{\cdot}0000$

i.e. kin. pr. grad., $\Delta p/\rho l = (0{\cdot}00098)$ to $(294)\ \mathrm{ms^{-2}}$

Roughness size, $k_s = 0{\cdot}003$ mm

This table shows values of m, as follows

m_C for Colebrook-White solutions; or, where $R \le 2000$, m_P for laminar flow

Grad'nt S	k.p.g. $(\Delta p/\rho l)$	6	8	10	12·5	15	20	25	30	40	50	60	80	100	125
						(Equivalent) Pipe diameters in mm									
0·00030	(0·00294)	13·71	9·346	6·941	5·154	4·042	2·754	2·046	1·604	1·093	*	1·193	1·159	1·138	1·120
0·00040	(0·00392)	11·88	8·093	6·011	4·464	3·501	2·385	1·771	1·389	0·947	*	1·162	1·131	1·112	1·096
0·00060	(0·00588)	9·698	6·608	4·908	3·645	2·858	1·948	1·446	1·134	*	1·142	1·121	1·094	1·078	1·064
0·00080	(0·00785)	8·399	5·723	4·250	3·156	2·475	1·687	1·253	0·982	*	1·113	1·094	1·070	1·055	1·042
0·00100	(0·00981)	7·512	5·119	3·801	2·823	2·214	1·509	1·120	*	1·117	1·091	1·074	1·051	1·037	1·026
0·00150	(0·0147)	6·133	4·179	3·104	2·305	1·808	1·232	0·915	*	1·076	1·054	1·039	1·019	1·007	0·998
0·00200	(0·0196)	5·312	3·620	2·688	1·996	1·565	1·067	*	*	1·050	1·030	1·016	0·998	0·987	0·979
0·00300	(0·0294)	4·337	2·955	2·195	1·630	1·278	0·871	*	1·042	1·014	0·997	0·985	0·969	0·960	0·953
0·00400	(0·0392)	3·756	2·559	1·901	1·412	1·107	*	1·035	1·016	0·990	0·974	0·964	0·950	0·942	0·936
0·00600	(0·0588)	3·067	2·090	1·552	1·153	0·904	*	0·998	0·981	0·959	0·945	0·936	0·924	0·917	0·913
0·00800	(0·0785)	2·656	1·810	1·344	0·998	*	0·996	0·973	0·958	0·938	0·925	0·917	0·907	0·901	0·897
0·01000	(0·0981)	2·375	1·619	1·202	0·893	*	0·976	0·955	0·940	0·922	0·910	0·903	0·894	0·888	0·885
0·01500	(0·147)	1·940	1·322	0·982	*	*	0·941	0·923	0·911	0·895	0·885	0·878	0·871	0·867	0·864
0·02000	(0·196)	1·680	1·145	0·850	*	0·945	0·918	0·902	0·891	0·876	0·867	0·862	0·855	0·852	0·850
0·03000	(0·294)	1·371	0·935	*	0·929	0·911	0·888	0·874	0·864	0·852	0·844	0·840	0·835	0·832	0·831
0·04000	(0·392)	1·188	0·809	*	0·905	0·889	0·868	0·855	0·846	0·835	0·829	0·825	0·821	0·819	0·819
0·06000	(0·588)	0·970	*	0·894	0·873	0·859	0·841	0·830	0·822	0·813	0·808	0·805	0·802	0·801	0·801
0·08000	(0·785)	0·840	*	0·871	0·852	0·839	0·823	0·813	0·806	0·798	0·794	0·791	0·789	0·789	0·790
0·10000	(0·981)	0·751	*	0·854	0·837	0·825	0·810	0·800	0·794	0·787	0·783	0·781	0·779	0·780	0·781
0·15000	(1·47)	*	0·843	0·825	0·810	0·799	0·786	0·779	0·773	0·768	0·765	0·763	0·763	0·764	0·766
0·20000	(1·96)	*	0·822	0·805	0·792	0·782	0·771	0·764	0·759	0·755	0·752	0·751	0·752	0·753	0·755
0·30000	(2·94)	0·817	0·794	0·780	0·768	0·760	0·750	0·744	0·741	0·737	0·736	0·735	0·737	0·738	0·741
0·40000	(3·92)	0·796	0·775	0·762	0·752	0·745	0·736	0·731	0·728	0·725	0·725	0·725	0·726	0·729	0·732
0·60000	(5·88)	0·769	0·751	0·740	0·731	0·725	0·717	0·714	0·711	0·710	0·710	0·710	0·713	0·716	0·720
0·80000	(7·85)	0·751	0·735	0·725	0·716	0·711	0·705	0·702	0·700	0·699	0·700	0·701	0·704	0·707	0·711
1·00000	(9·81)	0·738	0·722	0·713	0·706	0·701	0·696	0·693	0·692	0·691	0·692	0·693	0·697	0·701	0·705
1·50000	(14·7)	0·715	0·702	0·694	0·688	0·684	0·680	0·678	0·677	0·678	0·679	0·681	0·685	0·689	0·694
2·00000	(19·6)	0·699	0·688	0·681	0·676	0·672	0·669	0·668	0·667	0·669	0·670	0·673	0·677	0·682	0·687
3·00000	(29·4)	0·679	0·669	0·664	0·659	0·657	0·655	0·654	0·655	0·656	0·659	0·662	0·667	0·672	0·678
4·00000	(39·2)	0·666	0·657	0·652	0·649	0·647	0·645	0·645	0·646	0·648	0·651	0·654	0·660	0·666	0·672
6·00000	(58·8)	0·648	0·641	0·637	0·634	0·633	0·633	0·633	0·635	0·638	0·641	0·645	0·651	0·657	0·664
8·00000	(78·5)	0·636	0·630	0·627	0·625	0·624	0·624	0·625	0·627	0·631	0·635	0·639	0·646	0·652	0·659
10·0000	(98·1)	0·628	0·622	0·619	0·618	0·617	0·618	0·620	0·622	0·626	0·630	0·634	0·641	0·648	0·655
15·0000	(147)	0·613	0·609	0·607	0·606	0·606	0·608	0·610	0·612	0·617	0·622	0·626	0·634	0·641	0·649
20·0000	(196)	0·603	0·600	0·598	0·598	0·599	0·601	0·603	0·606	0·611	0·617	0·621	0·630	0·637	0·645
30·0000	(294)	0·590	0·588	0·587	0·588	0·589	0·592	0·595	0·598	0·604	0·610	0·615	0·624	0·631	0·640
S	$(\Delta p/\rho l)$	6	8	10	12·5	15	20	25	30	40	50	60	80	100	125

Grad'nt   k.p.g.   (Equivalent) Pipe diameters in mm

Grad'nt S	k.p.g. $(\Delta p/\rho l)$	125	150	200	250	300	400	500	600	800	1000	1250	1500	2000	2500
						(Equivalent) Pipe diameters in mm									
0·00010	(0·00098)	1·223	1·206	1·183	1·169	1·160	1·149	1·143	1·140	1·137	1·136	1·138	1·140	1·145	1·150
0·00015	(0·00147)	1·183	1·168	1·148	1·137	1·129	1·120	1·115	1·113	1·111	1·112	1·114	1·117	1·123	1·129
0·00020	(0·00196)	1·156	1·143	1·125	1·114	1·108	1·100	1·096	1·094	1·094	1·095	1·098	1·101	1·107	1·114
0·00030	(0·00294)	1·120	1·109	1·094	1·085	1·079	1·073	1·070	1·070	1·070	1·072	1·076	1·079	1·087	1·094
0·00040	(0·00392)	1·096	1·086	1·072	1·065	1·060	1·055	1·053	1·053	1·054	1·057	1·060	1·065	1·072	1·080
0·00060	(0·00588)	1·064	1·055	1·044	1·037	1·034	1·030	1·029	1·030	1·032	1·035	1·040	1·044	1·053	1·061
0·00080	(0·00785)	1·042	1·034	1·024	1·019	1·016	1·013	1·013	1·014	1·017	1·021	1·026	1·031	1·040	1·048
0·00100	(0·00981)	1·026	1·019	1·010	1·005	1·003	1·001	1·001	1·002	1·006	1·010	1·015	1·020	1·030	1·038
0·00150	(0·0147)	0·998	0·992	0·985	0·981	0·979	0·979	0·980	0·982	0·986	0·991	0·997	1·002	1·012	1·021
0·00200	(0·0196)	0·979	0·973	0·967	0·965	0·964	0·964	0·965	0·968	0·973	0·978	0·984	0·990	1·000	1·009
0·00300	(0·0294)	0·953	0·949	0·944	0·943	0·942	0·943	0·946	0·948	0·954	0·960	0·967	0·973	0·984	0·993
0·00400	(0·0392)	0·936	0·932	0·929	0·928	0·928	0·930	0·932	0·935	0·942	0·948	0·955	0·961	0·973	0·982
0·00600	(0·0588)	0·913	0·910	0·907	0·907	0·908	0·911	0·914	0·918	0·925	0·931	0·939	0·945	0·957	0·967
0·00800	(0·0785)	0·897	0·895	0·893	0·893	0·895	0·898	0·902	0·906	0·913	0·920	0·928	0·935	0·947	0·957
0·01000	(0·0981)	0·885	0·883	0·882	0·883	0·884	0·888	0·892	0·897	0·904	0·911	0·919	0·927	0·939	0·950
0·01500	(0·147)	0·864	0·863	0·863	0·865	0·867	0·871	0·876	0·881	0·889	0·897	0·905	0·912	0·925	0·936
0·02000	(0·196)	0·850	0·850	0·851	0·852	0·855	0·860	0·865	0·870	0·879	0·886	0·895	0·903	0·916	0·927
0·03000	(0·294)	0·831	0·831	0·833	0·836	0·839	0·844	0·850	0·855	0·865	0·873	0·882	0·890	0·904	0·915
0·04000	(0·392)	0·819	0·819	0·821	0·824	0·828	0·834	0·840	0·845	0·855	0·864	0·873	0·881	0·895	0·907
0·06000	(0·588)	0·801	0·802	0·806	0·809	0·813	0·820	0·826	0·832	0·842	0·851	0·861	0·869	0·884	0·896
0·08000	(0·785)	0·790	0·791	0·795	0·799	0·803	0·810	0·817	0·823	0·833	0·843	0·853	0·861	0·876	0·889
0·10000	(0·981)	0·781	0·783	0·787	0·791	0·795	0·803	0·810	0·816	0·827	0·836	0·847	0·855	0·871	0·883
0·15000	(1·47)	0·766	0·768	0·773	0·777	0·782	0·790	0·798	0·804	0·816	0·825	0·836	0·845	0·861	0·874
0·20000	(1·96)	0·755	0·758	0·763	0·768	0·773	0·782	0·789	0·796	0·808	0·818	0·829	0·839	0·854	0·868
0·30000	(2·94)	0·741	0·744	0·750	0·756	0·761	0·771	0·779	0·786	0·798	0·809	0·820	0·830	0·846	0·860
0·40000	(3·92)	0·732	0·735	0·742	0·748	0·753	0·763	0·771	0·779	0·792	0·802	0·814	0·824	0·840	0·854
0·60000	(5·88)	0·720	0·723	0·731	0·737	0·743	0·753	0·762	0·770	0·783	0·794	0·806	0·816	0·833	0·848
0·80000	(7·85)	0·711	0·715	0·723	0·730	0·736	0·746	0·755	0·763	0·777	0·788	0·801	0·811	0·829	0·843
1·00000	(9·81)	0·705	0·709	0·717	0·724	0·731	0·741	0·751	0·759	0·773	0·784	0·797	0·807	0·825	0·840
1·50000	(14·7)	0·694	0·699	0·708	0·715	0·722	0·733	0·743	0·751	0·766	0·778	0·790	0·801	0·820	0·835
2·00000	(19·6)	0·687	0·692	0·701	0·709	0·716	0·728	0·738	0·746	0·761	0·773	0·786	0·798	0·816	0·831
3·00000	(29·4)	0·678	0·683	0·693	0·701	0·708	0·721	0·731	0·740	0·755	0·768	0·781	0·793	0·812	0·827
S	$(\Delta p/\rho l)$	125	150	200	250	300	400	500	600	800	1000	1250	1500	2000	2500

Grad'nt   k.p.g.   (Equivalent) Pipe diameters in mm

Kinematic viscosity, $\nu = 2{\cdot}0\times10^{-6}\ \mathrm{m^2\,s^{-1}}$;

Roughness size, $k_s = 0{\cdot}003$ mm

E129

Kin. visc., $\nu = 2.0 \times 10^{-6}$ m^{2}s^{-1};
$S = 0.00010$ to 30.0000

i.e. kin. pr. grad., $\Delta p/\rho l =$
(0.00098) to (294) ms^{-2}

Roughness size, $k_s = 0.006$ mm
This table shows values of m, as follows

m_C for Colebrook-White solutions; or,
where $\mathbf{R} \le 2000$, m_P for laminar flow

Grad'nt S	k.p.g. $(\Delta p/\rho l)$	6	8	10	12·5	15	20	25	30	40	50	60	80	100	125
		(Equivalent) Pipe diameters in mm														
0·00030	(0·00294)	13·71	9·346	6·941	5·154	4·042	2·754	2·046	1·604	1·093	*	1·194	1·160	1·139	1·121
0·00040	(0·00392)	11·88	8·093	6·011	4·464	3·501	2·385	1·771	1·389	0·947	*	1·163	1·132	1·113	1·097
0·00060	(0·00588)	9·698	6·608	4·908	3·645	2·858	1·948	1·446	1·134	*	1·143	1·122	1·095	1·079	1·065
0·00080	(0·00785)	8·399	5·723	4·250	3·156	2·475	1·687	1·253	0·982	*	1·114	1·095	1·071	1·056	1·043
0·00100	(0·00981)	7·512	5·119	3·801	2·823	2·214	1·509	1·120	*	1·117	1·092	1·075	1·052	1·038	1·027
0·00150	(0·0147)	6·133	4·179	3·104	2·305	1·808	1·232	0·915	*	1·077	1·055	1·040	1·021	1·009	0·999
0·00200	(0·0196)	5·312	3·620	2·688	1·996	1·565	1·067	*	*	1·051	1·031	1·017	0·999	0·989	0·980
0·00300	(0·0294)	4·337	2·955	2·195	1·630	1·278	0·871	*	1·043	1·015	0·998	0·986	0·971	0·962	0·955
0·00400	(0·0392)	3·756	2·559	1·901	1·412	1·107	*	1·037	1·017	0·992	0·976	0·965	0·952	0·944	0·938
0·00600	(0·0588)	3·067	2·090	1·552	1·153	0·904	*	0·999	0·982	0·960	0·947	0·937	0·926	0·920	0·915
0·00800	(0·0785)	2·656	1·810	1·344	0·998	*	0·997	0·975	0·959	0·939	0·927	0·919	0·909	0·903	0·899
0·01000	(0·0981)	2·375	1·619	1·202	0·893	*	0·977	0·957	0·942	0·924	0·912	0·905	0·896	0·891	0·888
0·01500	(0·147)	1·940	1·322	0·982	*	*	0·943	0·925	0·913	0·897	0·887	0·881	0·873	0·870	0·867
0·02000	(0·196)	1·680	1·145	0·850	*	0·947	0·921	0·904	0·893	0·879	0·870	0·865	0·858	0·855	0·854
0·03000	(0·294)	1·371	0·935	*	0·931	0·914	0·891	0·876	0·867	0·855	0·847	0·843	0·838	0·836	0·835
0·04000	(0·392)	1·188	0·809	*	0·908	0·892	0·871	0·858	0·849	0·838	0·832	0·828	0·824	0·823	0·823
0·06000	(0·588)	0·970	*	0·897	0·876	0·862	0·844	0·833	0·826	0·817	0·812	0·809	0·806	0·806	0·806
0·08000	(0·785)	0·840	*	0·874	0·855	0·843	0·826	0·817	0·810	0·802	0·798	0·796	0·794	0·794	0·795
0·10000	(0·981)	0·751	*	0·857	0·840	0·828	0·813	0·804	0·798	0·791	0·788	0·786	0·785	0·785	0·787
0·15000	(1·47)	*	0·846	0·828	0·814	0·803	0·791	0·783	0·778	0·773	0·770	0·769	0·769	0·770	0·772
0·20000	(1·96)	*	0·826	0·809	0·796	0·787	0·775	0·769	0·765	0·760	0·758	0·758	0·758	0·760	0·763
0·30000	(2·94)	0·821	0·798	0·784	0·773	0·765	0·755	0·750	0·747	0·743	0·742	0·743	0·744	0·747	0·750
0·40000	(3·92)	0·801	0·780	0·768	0·757	0·750	0·742	0·737	0·735	0·732	0·732	0·733	0·735	0·738	0·742
0·60000	(5·88)	0·774	0·757	0·746	0·737	0·731	0·724	0·721	0·719	0·718	0·718	0·719	0·722	0·726	0·731
0·80000	(7·85)	0·757	0·741	0·731	0·723	0·718	0·712	0·710	0·708	0·708	0·709	0·710	0·714	0·718	0·723
1·00000	(9·81)	0·744	0·729	0·720	0·713	0·709	0·704	0·701	0·700	0·701	0·702	0·704	0·708	0·713	0·718
1·50000	(14·7)	0·722	0·709	0·702	0·696	0·692	0·689	0·687	0·687	0·688	0·690	0·693	0·698	0·703	0·709
2·00000	(19·6)	0·707	0·696	0·689	0·685	0·682	0·679	0·678	0·678	0·680	0·683	0·686	0·691	0·697	0·703
3·00000	(29·4)	0·688	0·679	0·673	0·670	0·667	0·666	0·666	0·667	0·670	0·673	0·676	0·683	0·689	0·695
4·00000	(39·2)	0·676	0·667	0·663	0·660	0·658	0·657	0·658	0·659	0·663	0·666	0·670	0·677	0·683	0·690
6·00000	(58·8)	0·659	0·652	0·649	0·647	0·646	0·646	0·648	0·650	0·654	0·658	0·662	0·670	0·677	0·684
8·00000	(78·5)	0·648	0·643	0·640	0·638	0·638	0·639	0·641	0·643	0·648	0·653	0·657	0·665	0·672	0·680
10·0000	(98·1)	0·640	0·635	0·633	0·632	0·632	0·634	0·636	0·639	0·644	0·649	0·654	0·662	0·669	0·677
15·0000	(147)	0·627	0·623	0·622	0·622	0·623	0·625	0·628	0·631	0·637	0·642	0·648	0·657	0·664	0·673
20·0000	(196)	0·618	0·616	0·615	0·615	0·616	0·619	0·623	0·626	0·633	0·638	0·644	0·653	0·661	0·670
30·0000	(294)	0·607	0·606	0·606	0·607	0·608	0·612	0·616	0·620	0·627	0·633	0·639	0·649	0·657	0·666
S	$(\Delta p/\rho l)$	6	8	10	12·5	15	20	25	30	40	50	60	80	100	125

Grad'nt S	k.p.g. $(\Delta p/\rho l)$	125	150	200	250	300	400	500	600	800	1000	1250	1500	2000	2500
		(Equivalent) Pipe diameters in mm														
0·00010	(0·00098)	1·224	1·206	1·184	1·170	1·161	1·150	1·144	1·140	1·138	1·137	1·139	1·141	1·146	1·152
0·00015	(0·00147)	1·184	1·169	1·149	1·137	1·130	1·121	1·116	1·114	1·112	1·113	1·115	1·118	1·124	1·130
0·00020	(0·00196)	1·157	1·143	1·126	1·115	1·109	1·101	1·097	1·095	1·095	1·096	1·099	1·102	1·109	1·115
0·00030	(0·00294)	1·121	1·109	1·094	1·086	1·080	1·074	1·072	1·071	1·071	1·074	1·077	1·081	1·088	1·096
0·00040	(0·00392)	1·097	1·086	1·073	1·066	1·061	1·056	1·054	1·054	1·055	1·058	1·062	1·066	1·074	1·082
0·00060	(0·00588)	1·065	1·056	1·045	1·039	1·035	1·032	1·031	1·031	1·034	1·037	1·042	1·047	1·055	1·063
0·00080	(0·00785)	1·043	1·035	1·026	1·020	1·017	1·015	1·015	1·016	1·019	1·023	1·028	1·033	1·042	1·051
0·00100	(0·00981)	1·027	1·020	1·011	1·007	1·004	1·002	1·003	1·004	1·008	1·012	1·018	1·023	1·033	1·041
0·00150	(0·0147)	0·999	0·993	0·986	0·983	0·981	0·981	0·982	0·984	0·988	0·993	0·999	1·005	1·015	1·025
0·00200	(0·0196)	0·980	0·975	0·969	0·967	0·965	0·966	0·968	0·970	0·975	0·981	0·987	0·993	1·004	1·013
0·00300	(0·0294)	0·955	0·951	0·946	0·945	0·944	0·946	0·948	0·951	0·957	0·963	0·970	0·977	0·988	0·998
0·00400	(0·0392)	0·938	0·934	0·931	0·930	0·930	0·932	0·935	0·938	0·945	0·951	0·959	0·965	0·977	0·988
0·00600	(0·0588)	0·915	0·912	0·910	0·910	0·911	0·914	0·918	0·921	0·929	0·936	0·943	0·950	0·963	0·974
0·00800	(0·0785)	0·899	0·897	0·896	0·896	0·898	0·901	0·906	0·910	0·918	0·925	0·933	0·940	0·953	0·964
0·01000	(0·0981)	0·888	0·886	0·885	0·886	0·888	0·892	0·897	0·901	0·909	0·917	0·925	0·933	0·946	0·957
0·01500	(0·147)	0·867	0·866	0·867	0·869	0·871	0·876	0·881	0·886	0·895	0·903	0·912	0·919	0·933	0·945
0·02000	(0·196)	0·854	0·853	0·854	0·857	0·859	0·865	0·870	0·875	0·885	0·893	0·902	0·911	0·925	0·937
0·03000	(0·294)	0·835	0·836	0·838	0·841	0·844	0·850	0·856	0·862	0·872	0·881	0·890	0·899	0·913	0·926
0·04000	(0·392)	0·823	0·824	0·826	0·830	0·833	0·840	0·847	0·852	0·863	0·872	0·882	0·891	0·906	0·919
0·06000	(0·588)	0·806	0·808	0·811	0·815	0·819	0·827	0·834	0·840	0·851	0·861	0·871	0·881	0·896	0·909
0·08000	(0·785)	0·795	0·797	0·801	0·806	0·810	0·818	0·825	0·832	0·843	0·853	0·864	0·874	0·890	0·903
0·10000	(0·981)	0·787	0·789	0·794	0·798	0·803	0·811	0·819	0·826	0·838	0·848	0·859	0·869	0·885	0·899
0·15000	(1·47)	0·772	0·775	0·781	0·786	0·791	0·800	0·808	0·815	0·828	0·839	0·850	0·860	0·877	0·891
0·20000	(1·96)	0·763	0·766	0·772	0·778	0·783	0·793	0·801	0·809	0·822	0·833	0·844	0·855	0·872	0·886
0·30000	(2·94)	0·750	0·754	0·760	0·767	0·773	0·783	0·792	0·800	0·813	0·825	0·837	0·847	0·865	0·880
0·40000	(3·92)	0·742	0·746	0·753	0·760	0·766	0·776	0·786	0·794	0·808	0·819	0·832	0·843	0·861	0·876
0·60000	(5·88)	0·731	0·735	0·743	0·750	0·757	0·768	0·778	0·786	0·801	0·813	0·826	0·837	0·856	0·871
0·80000	(7·85)	0·723	0·728	0·736	0·744	0·751	0·763	0·773	0·781	0·796	0·809	0·822	0·833	0·852	0·868
1·00000	(9·81)	0·718	0·723	0·732	0·740	0·747	0·759	0·769	0·778	0·793	0·806	0·819	0·831	0·850	0·866
1·50000	(14·7)	0·709	0·714	0·724	0·732	0·739	0·752	0·763	0·772	0·788	0·801	0·815	0·826	0·846	0·862
2·00000	(19·6)	0·703	0·709	0·719	0·727	0·735	0·748	0·759	0·768	0·784	0·798	0·812	0·824	0·844	0·860
3·00000	(29·4)	0·695	0·701	0·712	0·721	0·729	0·743	0·754	0·764	0·780	0·794	0·808	0·820	0·841	0·857
S	$(\Delta p/\rho l)$	125	150	200	250	300	400	500	600	800	1000	1250	1500	2000	2500

Grad'nt k.p.g. (Equivalent) Pipe diameters in mm

Kinematic viscosity, $\nu = 2.0 \times 10^{-6}$ m^{2}s^{-1} ;

Roughness size, $k_s = 0.006$ mm

Kin. visc., $\nu = 2 \cdot 0 \times 10^{-6}$ m^{2}s^{-1}; Roughness size, $k_s = 0 \cdot 015$ mm E130

$S = 0 \cdot 00010$ to $30 \cdot 0000$

i.e. kin. pr. grad., $\Delta p / \rho l =$ ($0 \cdot 00098$) to (294) ms^{-2}

This table shows values of m, as follows

m_C for Colebrook-White solutions; or, where $\mathbf{R} \leq 2000$, m_P for laminar flow

Grad'nt S	k.p.g. ($\Delta p/\rho l$)	6	8	10	12·5	15	20	25	30	40	50	60	80	100	125
		(Equivalent) Pipe diameters in mm													
0·00030	(0·00294)	13·71	9·346	6·941	5·154	4·042	2·754	2·046	1·604	1·093	*	1·196	1·162	1·141	1·123
0·00040	(0·00392)	11·88	8·093	6·011	4·464	3·501	2·385	1·771	1·389	0·947	*	1·165	1·134	1·115	1·100
0·00060	(0·00588)	9·698	6·608	4·908	3·645	2·858	1·948	1·446	1·134	*	1·145	1·125	1·098	1·081	1·068
0·00080	(0·00785)	8·399	5·723	4·250	3·156	2·475	1·687	1·253	0·982	*	1·116	1·098	1·074	1·059	1·047
0·00100	(0·00981)	7·512	5·119	3·801	2·823	2·214	1·509	1·120	*	1·120	1·095	1·078	1·055	1·042	1·031
0·00150	(0·0147)	6·133	4·179	3·104	2·305	1·808	1·232	0·915	*	1·081	1·059	1·044	1·024	1·012	1·003
0·00200	(0·0196)	5·312	3·620	2·688	1·996	1·565	1·067	*	*	1·054	1·034	1·021	1·003	0·993	0·985
0·00300	(0·0294)	4·337	2·955	2·195	1·630	1·278	0·871	*	1·047	1·019	1·002	0·990	0·975	0·967	0·960
0·00400	(0·0392)	3·756	2·559	1·901	1·412	1·107	*	1·041	1·021	0·996	0·980	0·970	0·957	0·949	0·943
0·00600	(0·0588)	3·067	2·090	1·552	1·153	0·904	*	1·004	0·987	0·965	0·952	0·943	0·932	0·925	0·921
0·00800	(0·0785)	2·656	1·810	1·344	0·998	*	1·002	0·980	0·965	0·945	0·933	0·925	0·915	0·910	0·906
0·01000	(0·0981)	2·375	1·619	1·202	0·893	*	0·983	0·962	0·948	0·930	0·919	0·911	0·903	0·898	0·895
0·01500	(0·147)	1·940	1·322	0·982	*	*	0·949	0·931	0·919	0·903	0·894	0·888	0·881	0·878	0·876
0·02000	(0·196)	1·680	1·145	0·850	*	0·953	0·927	0·911	0·900	0·886	0·878	0·873	0·867	0·864	0·863
0·03000	(0·294)	1·371	0·935	*	0·938	0·921	0·898	0·884	0·875	0·863	0·856	0·852	0·848	0·846	0·846
0·04000	(0·392)	1·188	0·809	*	0·915	0·899	0·879	0·866	0·858	0·847	0·842	0·838	0·835	0·834	0·835
0·06000	(0·588)	0·970	*	0·905	0·885	0·871	0·853	0·843	0·836	0·827	0·823	0·820	0·818	0·819	0·820
0·08000	(0·785)	0·840	*	0·883	0·865	0·852	0·836	0·827	0·821	0·814	0·810	0·808	0·807	0·808	0·810
0·10000	(0·981)	0·751	*	0·867	0·850	0·839	0·824	0·815	0·810	0·804	0·801	0·799	0·799	0·801	0·803
0·15000	(1·47)	*	0·857	0·839	0·825	0·815	0·803	0·796	0·791	0·787	0·785	0·784	0·785	0·787	0·791
0·20000	(1·96)	*	0·837	0·821	0·808	0·800	0·789	0·783	0·779	0·775	0·774	0·774	0·776	0·779	0·783
0·30000	(2·94)	0·834	0·812	0·798	0·787	0·779	0·770	0·766	0·763	0·761	0·761	0·761	0·764	0·768	0·772
0·40000	(3·92)	0·815	0·795	0·782	0·772	0·766	0·758	0·755	0·752	0·751	0·752	0·753	0·757	0·761	0·766
0·60000	(5·88)	0·790	0·773	0·762	0·754	0·749	0·743	0·740	0·739	0·739	0·740	0·742	0·747	0·751	0·757
0·80000	(7·85)	0·774	0·758	0·749	0·742	0·737	0·732	0·730	0·730	0·731	0·733	0·735	0·740	0·746	0·752
1·00000	(9·81)	0·762	0·748	0·739	0·733	0·729	0·725	0·724	0·723	0·725	0·727	0·730	0·736	0·741	0·748
1·50000	(14·7)	0·742	0·730	0·723	0·718	0·715	0·712	0·712	0·713	0·715	0·718	0·722	0·728	0·734	0·741
2·00000	(19·6)	0·729	0·718	0·713	0·708	0·706	0·704	0·705	0·706	0·709	0·713	0·716	0·723	0·730	0·737
3·00000	(29·4)	0·712	0·704	0·699	0·696	0·695	0·694	0·695	0·697	0·701	0·705	0·710	0·717	0·724	0·732
4·00000	(39·2)	0·701	0·694	0·690	0·688	0·687	0·688	0·689	0·691	0·696	0·701	0·705	0·714	0·721	0·729
6·00000	(58·8)	0·688	0·682	0·679	0·678	0·678	0·679	0·682	0·684	0·690	0·695	0·700	0·709	0·717	0·725
8·00000	(78·5)	0·679	0·674	0·672	0·671	0·672	0·674	0·677	0·680	0·686	0·691	0·697	0·706	0·714	0·723
10·0000	(98·1)	0·672	0·669	0·667	0·667	0·668	0·670	0·673	0·677	0·683	0·689	0·694	0·704	0·712	0·721
15·0000	(147)	0·662	0·659	0·659	0·659	0·661	0·664	0·668	0·672	0·679	0·685	0·691	0·701	0·709	0·719
20·0000	(196)	0·655	0·654	0·654	0·655	0·656	0·660	0·664	0·668	0·676	0·682	0·688	0·699	0·707	0·717
30·0000	(294)	0·647	0·646	0·647	0·649	0·651	0·656	0·660	0·665	0·672	0·679	0·686	0·696	0·705	0·715
S	($\Delta p/\rho l$)	6	8	10	12·5	15	20	25	30	40	50	60	80	100	125

Grad'nt k.p.g. (Equivalent) Pipe diameters in mm

Grad'nt S	k.p.g. ($\Delta p/\rho l$)	125	150	200	250	300	400	500	600	800	1000	1250	1500	2000	2500
0·00010	(0·00098)	1·225	1·208	1·185	1·172	1·163	1·152	1·146	1·143	1·140	1·140	1·141	1·144	1·149	1·155
0·00015	(0·00147)	1·186	1·170	1·151	1·139	1·132	1·123	1·118	1·116	1·115	1·116	1·118	1·121	1·128	1·134
0·00020	(0·00196)	1·159	1·145	1·128	1·118	1·111	1·103	1·100	1·098	1·098	1·100	1·103	1·106	1·113	1·120
0·00030	(0·00294)	1·123	1·112	1·097	1·088	1·083	1·077	1·075	1·074	1·075	1·078	1·081	1·085	1·093	1·101
0·00040	(0·00392)	1·100	1·089	1·076	1·068	1·064	1·059	1·058	1·058	1·060	1·063	1·067	1·071	1·080	1·088
0·00060	(0·00588)	1·068	1·059	1·048	1·042	1·038	1·035	1·035	1·036	1·039	1·042	1·047	1·052	1·062	1·071
0·00080	(0·00785)	1·047	1·039	1·029	1·024	1·021	1·019	1·019	1·021	1·024	1·029	1·034	1·040	1·050	1·059
0·00100	(0·00981)	1·031	1·023	1·015	1·011	1·008	1·007	1·008	1·009	1·014	1·018	1·024	1·030	1·041	1·050
0·00150	(0·0147)	1·003	0·997	0·991	0·987	0·986	0·986	0·988	0·990	0·995	1·001	1·007	1·013	1·025	1·035
0·00200	(0·0196)	0·985	0·980	0·974	0·972	0·971	0·972	0·974	0·977	0·983	0·989	0·996	1·002	1·014	1·024
0·00300	(0·0294)	0·960	0·956	0·952	0·951	0·951	0·953	0·956	0·959	0·966	0·973	0·980	0·987	1·000	1·011
0·00400	(0·0392)	0·943	0·940	0·937	0·937	0·937	0·940	0·943	0·947	0·955	0·962	0·970	0·977	0·990	1·002
0·00600	(0·0588)	0·921	0·919	0·917	0·918	0·919	0·923	0·927	0·931	0·940	0·948	0·956	0·964	0·978	0·990
0·00800	(0·0785)	0·906	0·905	0·904	0·905	0·907	0·911	0·916	0·921	0·930	0·938	0·947	0·955	0·970	0·982
0·01000	(0·0981)	0·895	0·894	0·894	0·895	0·898	0·903	0·908	0·913	0·922	0·931	0·940	0·949	0·963	0·976
0·01500	(0·147)	0·876	0·876	0·877	0·879	0·882	0·888	0·894	0·900	0·910	0·919	0·929	0·938	0·953	0·966
0·02000	(0·196)	0·863	0·863	0·865	0·868	0·872	0·878	0·885	0·891	0·901	0·911	0·921	0·931	0·946	0·960
0·03000	(0·294)	0·846	0·847	0·850	0·854	0·858	0·865	0·872	0·879	0·890	0·901	0·912	0·921	0·938	0·952
0·04000	(0·392)	0·835	0·836	0·840	0·844	0·849	0·857	0·864	0·871	0·883	0·894	0·905	0·915	0·932	0·946
0·06000	(0·588)	0·820	0·822	0·827	0·832	0·837	0·846	0·854	0·861	0·874	0·885	0·897	0·907	0·925	0·940
0·08000	(0·785)	0·810	0·813	0·818	0·824	0·829	0·839	0·847	0·855	0·868	0·880	0·892	0·902	0·920	0·935
0·10000	(0·981)	0·803	0·806	0·812	0·818	0·823	0·833	0·842	0·850	0·864	0·875	0·888	0·899	0·917	0·932
0·15000	(1·47)	0·791	0·794	0·801	0·808	0·814	0·825	0·834	0·842	0·857	0·869	0·882	0·893	0·912	0·928
0·20000	(1·96)	0·783	0·787	0·794	0·801	0·808	0·819	0·829	0·837	0·852	0·865	0·878	0·889	0·909	0·924
0·30000	(2·94)	0·772	0·777	0·785	0·793	0·800	0·812	0·822	0·831	0·846	0·859	0·873	0·885	0·904	0·921
0·40000	(3·92)	0·766	0·771	0·780	0·788	0·795	0·807	0·818	0·827	0·843	0·856	0·870	0·882	0·902	0·918
0·60000	(5·88)	0·757	0·763	0·772	0·781	0·788	0·801	0·813	0·822	0·838	0·852	0·866	0·878	0·899	0·915
0·80000	(7·85)	0·752	0·757	0·768	0·776	0·784	0·798	0·809	0·819	0·836	0·849	0·864	0·876	0·897	0·914
1·00000	(9·81)	0·748	0·754	0·764	0·773	0·781	0·795	0·807	0·817	0·834	0·848	0·862	0·875	0·896	0·912
1·50000	(14·7)	0·741	0·748	0·759	0·768	0·777	0·791	0·803	0·813	0·831	0·845	0·860	0·872	0·893	0·911
2·00000	(19·6)	0·737	0·744	0·756	0·765	0·774	0·789	0·801	0·811	0·829	0·843	0·858	0·871	0·892	0·909
3·00000	(29·4)	0·732	0·739	0·751	0·762	0·770	0·785	0·798	0·809	0·826	0·841	0·856	0·869	0·891	0·908
S	($\Delta p/\rho l$)	125	150	200	250	300	400	500	600	800	1000	1250	1500	2000	2500

Grad'nt k.p.g. (Equivalent) Pipe diameters in mm

Kinematic viscosity, $\nu = 2 \cdot 0 \times 10^{-6}$ m^{2}s^{-1}; Roughness size, $k_s = 0 \cdot 015$ mm

E131

Kin. visc., $\nu = 2.0 \times 10^{-6}\ \mathrm{m^2 s^{-1}}$;
$S = 0.00010$ to 30.0000

i.e. kin. pr. grad., $\Delta p/\rho l =$
(0.00098) to $(294)\ \mathrm{ms^{-2}}$

Roughness size, $k_s = 0.030$ mm
This table shows values of m, as follows

m_C for Colebrook-White solutions; or,
where $\mathbf{R} \le 2000$, m_P for laminar flow

Grad'nt	k.p.g.	(Equivalent) Pipe diameters in mm													
S	$(\Delta p/\rho l)$	6	8	10	12·5	15	20	25	30	40	50	60	80	100	125
0·00030	(0·00294)	13·71	9·346	6·941	5·154	4·042	2·754	2·046	1·604	1·093	*	1·199	1·165	1·144	1·127
0·00040	(0·00392)	11·88	8·093	6·011	4·464	3·501	2·385	1·771	1·389	0·947	*	1·169	1·138	1·119	1·104
0·00060	(0·00588)	9·698	6·608	4·908	3·645	2·858	1·948	1·446	1·134	*	1·149	1·129	1·102	1·086	1·073
0·00080	(0·00785)	8·399	5·723	4·250	3·156	2·475	1·687	1·253	0·982	*	1·121	1·102	1·078	1·063	1·052
0·00100	(0·00981)	7·512	5·119	3·801	2·823	2·214	1·509	1·120	*	1·125	1·100	1·083	1·060	1·047	1·036
0·00150	(0·0147)	6·133	4·179	3·104	2·305	1·808	1·232	0·915	*	1·086	1·064	1·049	1·030	1·018	1·009
0·00200	(0·0196)	5·312	3·620	2·688	1·996	1·565	1·067	*	*	1·060	1·040	1·027	1·010	0·999	0·992
0·00300	(0·0294)	4·337	2·955	2·195	1·630	1·278	0·871	*	1·053	1·026	1·009	0·997	0·983	0·974	0·968
0·00400	(0·0392)	3·756	2·559	1·901	1·412	1·107	*	1·047	1·028	1·003	0·988	0·977	0·965	0·957	0·952
0·00600	(0·0588)	3·067	2·090	1·552	1·153	0·904	*	1·012	0·995	0·973	0·960	0·951	0·941	0·935	0·931
0·00800	(0·0785)	2·656	1·810	1·344	0·998	*	1·010	0·988	0·973	0·953	0·942	0·934	0·925	0·920	0·917
0·01000	(0·0981)	2·375	1·619	1·202	0·893	*	0·991	0·971	0·957	0·939	0·928	0·921	0·913	0·909	0·907
0·01500	(0·147)	1·940	1·322	0·982	*	*	0·959	0·941	0·929	0·914	0·905	0·899	0·893	0·891	0·889
0·02000	(0·196)	1·680	1·145	0·850	*	0·963	0·938	0·922	0·911	0·898	0·890	0·885	0·880	0·878	0·878
0·03000	(0·294)	1·371	0·935	*	0·949	0·932	0·910	0·896	0·887	0·876	0·870	0·866	0·863	0·862	0·863
0·04000	(0·392)	1·188	0·809	*	0·927	0·912	0·892	0·879	0·871	0·862	0·857	0·854	0·851	0·851	0·853
0·06000	(0·588)	0·970	*	0·918	0·898	0·885	0·868	0·857	0·851	0·843	0·839	0·838	0·837	0·838	0·840
0·08000	(0·785)	0·840	*	0·897	0·880	0·867	0·852	0·843	0·837	0·831	0·828	0·827	0·827	0·829	0·832
0·10000	(0·981)	0·751	*	0·882	0·866	0·855	0·841	0·833	0·828	0·822	0·820	0·819	0·820	0·822	0·826
0·15000	(1·47)	*	0·874	0·857	0·842	0·833	0·821	0·815	0·811	0·807	0·806	0·806	0·809	0·812	0·816
0·20000	(1·96)	*	0·856	0·840	0·827	0·819	0·809	0·803	0·800	0·798	0·797	0·798	0·801	0·805	0·810
0·30000	(2·94)	0·854	0·832	0·818	0·808	0·801	0·793	0·789	0·787	0·785	0·786	0·788	0·792	0·796	0·802
0·40000	(3·92)	0·836	0·817	0·805	0·795	0·789	0·782	0·779	0·778	0·777	0·779	0·781	0·786	0·790	0·796
0·60000	(5·88)	0·814	0·797	0·787	0·779	0·774	0·769	0·767	0·766	0·767	0·770	0·772	0·778	0·784	0·790
0·80000	(7·85)	0·799	0·784	0·775	0·769	0·764	0·761	0·759	0·759	0·761	0·764	0·767	0·773	0·779	0·786
1·00000	(9·81)	0·789	0·775	0·767	0·761	0·758	0·754	0·754	0·754	0·756	0·760	0·763	0·770	0·776	0·783
1·50000	(14·7)	0·771	0·760	0·753	0·749	0·746	0·744	0·745	0·746	0·749	0·753	0·757	0·764	0·771	0·779
2·00000	(19·6)	0·760	0·750	0·745	0·741	0·739	0·738	0·739	0·740	0·744	0·749	0·753	0·761	0·768	0·776
3·00000	(29·4)	0·746	0·738	0·734	0·731	0·730	0·730	0·732	0·734	0·739	0·744	0·748	0·757	0·764	0·773
4·00000	(39·2)	0·737	0·730	0·727	0·725	0·724	0·725	0·727	0·730	0·735	0·740	0·745	0·754	0·762	0·771
6·00000	(58·8)	0·726	0·721	0·718	0·717	0·717	0·719	0·722	0·725	0·731	0·737	0·742	0·751	0·759	0·768
8·00000	(78·5)	0·719	0·715	0·713	0·712	0·713	0·715	0·719	0·722	0·728	0·734	0·740	0·749	0·758	0·767
10·0000	(98·1)	0·714	0·710	0·709	0·709	0·710	0·713	0·716	0·720	0·726	0·732	0·738	0·748	0·756	0·766
15·0000	(147)	0·706	0·704	0·703	0·704	0·705	0·709	0·712	0·716	0·723	0·730	0·736	0·746	0·755	0·764
20·0000	(196)	0·701	0·699	0·699	0·700	0·702	0·706	0·710	0·714	0·722	0·728	0·734	0·745	0·753	0·763
30·0000	(294)	0·695	0·694	0·695	0·696	0·698	0·703	0·707	0·712	0·719	0·726	0·732	0·743	0·752	0·762
S	$(\Delta p/\rho l)$	6	8	10	12·5	15	20	25	30	40	50	60	80	100	125

Grad'nt k.p.g. (Equivalent) Pipe diameters in mm

S	$(\Delta p/\rho l)$	125	150	200	250	300	400	500	600	800	1000	1250	1500	2000	2500
0·00010	(0·00098)	1·228	1·211	1·188	1·175	1·166	1·155	1·149	1·146	1·144	1·144	1·146	1·149	1·155	1·161
0·00015	(0·00147)	1·188	1·173	1·154	1·143	1·135	1·127	1·122	1·120	1·120	1·121	1·124	1·127	1·134	1·141
0·00020	(0·00196)	1·162	1·149	1·131	1·121	1·115	1·108	1·104	1·103	1·103	1·105	1·109	1·112	1·120	1·128
0·00030	(0·00294)	1·127	1·115	1·101	1·093	1·087	1·082	1·080	1·080	1·081	1·084	1·088	1·093	1·101	1·110
0·00040	(0·00392)	1·104	1·093	1·080	1·073	1·069	1·065	1·063	1·064	1·066	1·070	1·075	1·079	1·089	1·098
0·00060	(0·00588)	1·073	1·064	1·053	1·047	1·044	1·042	1·042	1·043	1·046	1·051	1·056	1·062	1·072	1·082
0·00080	(0·00785)	1·052	1·044	1·035	1·030	1·028	1·026	1·027	1·028	1·033	1·038	1·044	1·050	1·061	1·071
0·00100	(0·00981)	1·036	1·029	1·021	1·017	1·015	1·015	1·016	1·018	1·023	1·028	1·035	1·041	1·053	1·063
0·00150	(0·0147)	1·009	1·004	0·998	0·995	0·994	0·995	0·997	1·000	1·006	1·012	1·019	1·026	1·039	1·050
0·00200	(0·0196)	0·992	0·987	0·982	0·980	0·980	0·981	0·984	0·987	0·994	1·001	1·009	1·016	1·029	1·041
0·00300	(0·0294)	0·968	0·964	0·961	0·960	0·961	0·964	0·967	0·971	0·979	0·987	0·995	1·003	1·017	1·029
0·00400	(0·0392)	0·952	0·949	0·947	0·947	0·948	0·952	0·956	0·961	0·969	0·977	0·986	0·995	1·009	1·022
0·00600	(0·0588)	0·931	0·929	0·928	0·930	0·931	0·936	0·941	0·947	0·956	0·965	0·975	0·983	0·999	1·012
0·00800	(0·0785)	0·917	0·916	0·916	0·918	0·920	0·926	0·932	0·937	0·948	0·957	0·967	0·976	0·992	1·006
0·01000	(0·0981)	0·907	0·906	0·907	0·909	0·912	0·919	0·925	0·931	0·941	0·951	0·962	0·971	0·987	1·001
0·01500	(0·147)	0·889	0·890	0·892	0·895	0·899	0·906	0·913	0·919	0·931	0·941	0·952	0·962	0·979	0·994
0·02000	(0·196)	0·878	0·879	0·882	0·886	0·890	0·898	0·905	0·912	0·924	0·935	0·946	0·957	0·974	0·989
0·03000	(0·294)	0·863	0·864	0·869	0·873	0·878	0·887	0·895	0·902	0·915	0·927	0·939	0·950	0·968	0·983
0·04000	(0·392)	0·853	0·855	0·860	0·865	0·870	0·880	0·889	0·896	0·910	0·922	0·934	0·945	0·964	0·979
0·06000	(0·588)	0·840	0·843	0·849	0·855	0·861	0·871	0·880	0·889	0·903	0·915	0·928	0·940	0·959	0·975
0·08000	(0·785)	0·832	0·835	0·842	0·849	0·855	0·866	0·875	0·884	0·899	0·911	0·925	0·936	0·956	0·972
0·10000	(0·981)	0·826	0·830	0·837	0·844	0·850	0·862	0·872	0·880	0·896	0·908	0·922	0·934	0·953	0·970
0·15000	(1·47)	0·816	0·820	0·828	0·836	0·843	0·855	0·866	0·875	0·890	0·904	0·918	0·930	0·950	0·966
0·20000	(1·96)	0·810	0·814	0·823	0·831	0·838	0·851	0·862	0·871	0·887	0·901	0·915	0·927	0·948	0·965
0·30000	(2·94)	0·802	0·807	0·816	0·825	0·833	0·846	0·857	0·867	0·884	0·897	0·912	0·924	0·945	0·962
0·40000	(3·92)	0·796	0·802	0·812	0·821	0·829	0·843	0·854	0·864	0·881	0·895	0·910	0·923	0·944	0·961
0·60000	(5·88)	0·790	0·796	0·807	0·816	0·825	0·839	0·851	0·861	0·878	0·893	0·908	0·920	0·942	0·959
0·80000	(7·85)	0·786	0·793	0·804	0·813	0·822	0·836	0·849	0·859	0·876	0·891	0·906	0·919	0·940	0·958
1·00000	(9·81)	0·783	0·790	0·801	0·811	0·820	0·835	0·847	0·858	0·875	0·890	0·905	0·918	0·940	0·957
1·50000	(14·7)	0·779	0·786	0·798	0·808	0·817	0·832	0·845	0·855	0·873	0·888	0·904	0·917	0·938	0·956
2·00000	(19·6)	0·776	0·783	0·796	0·806	0·815	0·830	0·843	0·854	0·872	0·887	0·903	0·916	0·938	0·955
3·00000	(29·4)	0·773	0·780	0·793	0·804	0·813	0·828	0·841	0·852	0·871	0·886	0·901	0·915	0·937	0·954
S	$(\Delta p/\rho l)$	125	150	200	250	300	400	500	600	800	1000	1250	1500	2000	2500

Grad'nt k.p.g. (Equivalent) Pipe diameters in mm

Kinematic viscosity, $\nu = 2.0 \times 10^{-6}\ \mathrm{m^2 s^{-1}}$;   **Roughness size, $k_s = 0.030$ mm**

Kin. visc., $\nu = 2{\cdot}0\times10^{-6}\ \mathrm{m^2\,s^{-1}}$;
$S = 0{\cdot}00010$ to $30{\cdot}0000$

i.e. kin. pr. grad., $\Delta p/\rho l =$
$(0{\cdot}00098)$ to $(294)\ \mathrm{ms^{-2}}$

Roughness size, $k_s = 0{\cdot}060$ mm
This table shows values of m, as follows

m_C for Colebrook-White solutions; or,
where $\mathbf{R} \le 2000$, m_P for laminar flow

E132

Grad'nt S	k.p.g. $(\Delta p/\rho l)$	6	8	10	12·5	15	20	25	30	40	50	60	80	100	125
0·00030	(0·00294)	13·71	9·346	6·941	5·154	4·042	2·754	2·046	1·604	1·093	*	1·205	1·172	1·151	1·134
0·00040	(0·00392)	11·88	8·093	6·011	4·464	3·501	2·385	1·771	1·389	0·947	*	1·176	1·145	1·127	1·112
0·00060	(0·00588)	9·698	6·608	4·908	3·645	2·858	1·948	1·446	1·134	*	1·157	1·137	1·111	1·094	1·081
0·00080	(0·00785)	8·399	5·723	4·250	3·156	2·475	1·687	1·253	0·982	*	1·129	1·111	1·087	1·073	1·062
0·00100	(0·00981)	7·512	5·119	3·801	2·823	2·214	1·509	1·120	*	1·134	1·109	1·092	1·070	1·057	1·047
0·00150	(0·0147)	6·133	4·179	3·104	2·305	1·808	1·232	0·915	*	1·096	1·074	1·060	1·041	1·030	1·021
0·00200	(0·0196)	5·312	3·620	2·688	1·996	1·565	1·067	*	*	1·071	1·051	1·038	1·022	1·012	1·005
0·00300	(0·0294)	4·337	2·955	2·195	1·630	1·278	0·871	*	1·065	1·038	1·021	1·010	0·996	0·988	0·983
0·00400	(0·0392)	3·756	2·559	1·901	1·412	1·107	*	1·060	1·041	1·017	1·002	0·992	0·980	0·973	0·968
0·00600	(0·0588)	3·067	2·090	1·552	1·153	0·904	*	1·026	1·009	0·988	0·976	0·967	0·958	0·953	0·949
0·00800	(0·0785)	2·656	1·810	1·344	0·998	*	1·025	1·004	0·989	0·970	0·959	0·952	0·943	0·939	0·937
0·01000	(0·0981)	2·375	1·619	1·202	0·893	*	1·007	0·987	0·974	0·956	0·946	0·940	0·933	0·930	0·928
0·01500	(0·147)	1·940	1·322	0·982	*	*	0·977	0·960	0·948	0·934	0·925	0·920	0·915	0·913	0·913
0·02000	(0·196)	1·680	1·145	0·850	*	0·983	0·957	0·942	0·931	0·919	0·912	0·908	0·904	0·903	0·903
0·03000	(0·294)	1·371	0·935	*	*	0·954	0·932	0·918	0·910	0·900	0·894	0·891	0·889	0·889	0·891
0·04000	(0·392)	1·188	0·809	*	0·950	0·935	0·915	0·903	0·896	0·887	0·883	0·881	0·879	0·880	0·883
0·06000	(0·588)	0·970	*	*	0·924	0·910	0·894	0·884	0·878	0·871	0·868	0·867	0·867	0·869	0·873
0·08000	(0·785)	0·840	*	0·924	0·907	0·895	0·880	0·872	0·866	0·861	0·859	0·858	0·860	0·862	0·866
0·10000	(0·981)	0·751	*	0·911	0·894	0·883	0·870	0·863	0·858	0·854	0·852	0·852	0·854	0·857	0·862
0·15000	(1·47)	*	0·905	0·888	0·874	0·865	0·854	0·848	0·844	0·842	0·841	0·842	0·845	0·849	0·854
0·20000	(1·96)	*	0·889	0·873	0·861	0·853	0·843	0·838	0·836	0·834	0·834	0·836	0·840	0·844	0·850
0·30000	(2·94)	0·890	0·868	0·855	0·844	0·838	0·830	0·826	0·825	0·824	0·826	0·828	0·833	0·838	0·844
0·40000	(3·92)	0·875	0·855	0·843	0·834	0·828	0·822	0·819	0·818	0·818	0·820	0·823	0·828	0·834	0·840
0·60000	(5·88)	0·855	0·838	0·828	0·821	0·816	0·811	0·810	0·809	0·811	0·814	0·817	0·823	0·829	0·836
0·80000	(7·85)	0·843	0·828	0·819	0·812	0·808	0·805	0·804	0·804	0·806	0·809	0·813	0·820	0·826	0·833
1·00000	(9·81)	0·834	0·820	0·812	0·806	0·803	0·800	0·800	0·800	0·803	0·807	0·810	0·817	0·824	0·832
1·50000	(14·7)	0·820	0·808	0·801	0·797	0·794	0·793	0·793	0·794	0·798	0·802	0·806	0·814	0·821	0·829
2·00000	(19·6)	0·811	0·801	0·795	0·791	0·789	0·788	0·789	0·791	0·795	0·799	0·803	0·812	0·819	0·827
3·00000	(29·4)	0·800	0·791	0·787	0·784	0·783	0·783	0·784	0·786	0·791	0·796	0·800	0·809	0·816	0·825
4·00000	(39·2)	0·793	0·786	0·782	0·779	0·779	0·779	0·781	0·783	0·788	0·794	0·798	0·807	0·815	0·824
6·00000	(58·8)	0·785	0·779	0·775	0·774	0·774	0·775	0·777	0·780	0·786	0·791	0·796	0·805	0·813	0·822
8·00000	(78·5)	0·780	0·774	0·772	0·771	0·771	0·772	0·775	0·778	0·784	0·789	0·795	0·804	0·812	0·821
10·0000	(98·1)	0·776	0·771	0·769	0·768	0·769	0·771	0·774	0·777	0·783	0·788	0·794	0·803	0·811	0·820
15·0000	(147)	0·771	0·766	0·765	0·765	0·765	0·768	0·771	0·774	0·781	0·787	0·792	0·802	0·810	0·819
20·0000	(196)	0·767	0·764	0·762	0·762	0·763	0·766	0·770	0·773	0·780	0·786	0·791	0·801	0·810	0·819
30·0000	(294)	0·763	0·760	0·759	0·760	0·761	0·764	0·768	0·771	0·778	0·785	0·790	0·800	0·809	0·818
S	$(\Delta p/\rho l)$	6	8	10	12·5	15	20	25	30	40	50	60	80	100	125

Grad'nt   k.p.g.   (Equivalent) Pipe diameters in mm

Grad'nt S	k.p.g. $(\Delta p/\rho l)$	125	150	200	250	300	400	500	600	800	1000	1250	1500	2000	2500
0·00010	(0·00098)	1·233	1·216	1·194	1·180	1·172	1·162	1·156	1·153	1·152	1·152	1·155	1·158	1·165	1·172
0·00015	(0·00147)	1·194	1·179	1·160	1·149	1·142	1·134	1·130	1·129	1·129	1·130	1·134	1·138	1·146	1·154
0·00020	(0·00196)	1·168	1·155	1·138	1·128	1·122	1·116	1·113	1·112	1·113	1·116	1·120	1·124	1·133	1·141
0·00030	(0·00294)	1·134	1·123	1·109	1·101	1·096	1·091	1·090	1·090	1·092	1·096	1·101	1·106	1·116	1·125
0·00040	(0·00392)	1·112	1·101	1·089	1·082	1·078	1·075	1·074	1·075	1·079	1·083	1·089	1·094	1·105	1·115
0·00060	(0·00588)	1·081	1·073	1·063	1·058	1·055	1·053	1·054	1·056	1·060	1·066	1·072	1·079	1·090	1·101
0·00080	(0·00785)	1·062	1·054	1·046	1·042	1·040	1·039	1·040	1·043	1·048	1·054	1·062	1·068	1·081	1·092
0·00100	(0·00981)	1·047	1·040	1·033	1·030	1·028	1·028	1·030	1·033	1·039	1·046	1·054	1·061	1·074	1·085
0·00150	(0·0147)	1·021	1·016	1·011	1·009	1·009	1·010	1·013	1·017	1·024	1·032	1·040	1·048	1·062	1·074
0·00200	(0·0196)	1·005	1·000	0·996	0·995	0·996	0·998	1·002	1·006	1·015	1·023	1·032	1·040	1·055	1·067
0·00300	(0·0294)	0·983	0·980	0·977	0·978	0·979	0·983	0·988	0·993	1·002	1·011	1·021	1·029	1·045	1·058
0·00400	(0·0392)	0·968	0·966	0·965	0·966	0·968	0·973	0·978	0·984	0·994	1·003	1·013	1·023	1·039	1·053
0·00600	(0·0588)	0·949	0·948	0·949	0·951	0·953	0·960	0·966	0·972	0·983	0·993	1·004	1·014	1·031	1·045
0·00800	(0·0785)	0·937	0·937	0·938	0·941	0·944	0·951	0·958	0·965	0·976	0·987	0·998	1·009	1·026	1·041
0·01000	(0·0981)	0·928	0·928	0·930	0·934	0·937	0·945	0·952	0·959	0·972	0·982	0·994	1·005	1·023	1·038
0·01500	(0·147)	0·913	0·914	0·918	0·922	0·926	0·935	0·943	0·950	0·964	0·975	0·988	0·999	1·017	1·033
0·02000	(0·196)	0·903	0·905	0·909	0·914	0·919	0·929	0·937	0·945	0·959	0·971	0·983	0·995	1·014	1·029
0·03000	(0·294)	0·891	0·893	0·899	0·904	0·910	0·920	0·930	0·938	0·952	0·965	0·978	0·990	1·009	1·025
0·04000	(0·392)	0·883	0·886	0·892	0·898	0·904	0·915	0·925	0·934	0·949	0·961	0·975	0·987	1·007	1·023
0·06000	(0·588)	0·873	0·876	0·884	0·891	0·897	0·909	0·919	0·928	0·944	0·957	0·971	0·983	1·003	1·020
0·08000	(0·785)	0·866	0·870	0·878	0·886	0·893	0·905	0·916	0·925	0·941	0·954	0·969	0·981	1·001	1·018
0·10000	(0·981)	0·862	0·866	0·875	0·882	0·890	0·902	0·913	0·923	0·939	0·952	0·967	0·979	1·000	1·017
0·15000	(1·47)	0·854	0·859	0·869	0·877	0·885	0·898	0·909	0·919	0·936	0·949	0·964	0·977	0·998	1·015
0·20000	(1·96)	0·850	0·855	0·865	0·874	0·881	0·895	0·907	0·917	0·934	0·948	0·963	0·975	0·997	1·014
0·30000	(2·94)	0·844	0·850	0·860	0·869	0·878	0·892	0·904	0·914	0·931	0·945	0·961	0·974	0·995	1·012
0·40000	(3·92)	0·840	0·847	0·857	0·867	0·875	0·890	0·902	0·912	0·930	0·944	0·959	0·972	0·994	1·012
0·60000	(5·88)	0·836	0·843	0·854	0·864	0·872	0·887	0·899	0·910	0·928	0·943	0·958	0·971	0·993	1·011
0·80000	(7·85)	0·833	0·840	0·852	0·862	0·871	0·886	0·898	0·909	0·927	0·942	0·957	0·970	0·992	1·010
1·00000	(9·81)	0·832	0·838	0·850	0·861	0·869	0·885	0·897	0·908	0·926	0·941	0·956	0·970	0·992	1·010
1·50000	(14·7)	0·829	0·836	0·848	0·859	0·868	0·883	0·896	0·907	0·925	0·940	0·956	0·969	0·991	1·009
2·00000	(19·6)	0·827	0·834	0·847	0·857	0·866	0·882	0·895	0·906	0·924	0·939	0·955	0·968	0·991	1·008
3·00000	(29·4)	0·825	0·832	0·845	0·856	0·865	0·881	0·894	0·905	0·923	0·939	0·954	0·968	0·990	1·008
S	$(\Delta p/\rho l)$	125	150	200	250	300	400	500	600	800	1000	1250	1500	2000	2500

Grad'nt   k.p.g.   (Equivalent) Pipe diameters in mm

Kinematic viscosity, $\nu = 2{\cdot}0\times10^{-6}\ \mathrm{m^2\,s^{-1}}$;   Roughness size, $k_s = 0{\cdot}060$ mm

E133

Kin. visc., $\nu = 2.0 \times 10^{-6}$ m^{2}s^{-1};
S = 0.00010 to 30.0000

i.e. kin. pr. grad., $\Delta p/\rho l =$
(0.00098) to (294) ms^{-2}

Roughness size, k_s = 0.150 mm
This table shows values of m, as follows

m_C for Colebrook-White solutions; or,
where $\mathbf{R} \leq 2000$, m_P for laminar flow

Grad'nt   **k.p.g.**   **(Equivalent) Pipe diameters in mm**

S	$(\Delta p/\rho l)$	6	8	10	12.5	15	20	25	30	40	50	60	80	100	125
0.00030	(0.00294)	13.71	9.346	6.941	5.154	4.042	2.754	2.046	1.604	1.093	*	1.224	1.191	1.171	1.155
0.00040	(0.00392)	11.88	8.093	6.011	4.464	3.501	2.385	1.771	1.389	0.947	*	1.196	1.166	1.148	1.134
0.00060	(0.00588)	9.698	6.608	4.908	3.645	2.858	1.948	1.446	1.134	*	1.180	1.160	1.134	1.119	1.107
0.00080	(0.00785)	8.399	5.723	4.250	3.156	2.475	1.687	1.253	0.982	*	1.154	1.136	1.113	1.099	1.089
0.00100	(0.00981)	7.512	5.119	3.801	2.823	2.214	1.509	1.120	*	1.159	1.135	1.118	1.097	1.085	1.076
0.00150	(0.0147)	6.133	4.179	3.104	2.305	1.808	1.232	0.915	*	1.124	1.103	1.089	1.071	1.061	1.054
0.00200	(0.0196)	5.312	3.620	2.688	1.996	1.565	1.067	*	*	1.101	1.083	1.070	1.055	1.046	1.040
0.00300	(0.0294)	4.337	2.955	2.195	1.630	1.278	0.871	*	1.099	1.072	1.056	1.045	1.033	1.026	1.022
0.00400	(0.0392)	3.756	2.559	1.901	1.412	1.107	*	1.096	1.077	1.053	1.039	1.030	1.019	1.013	1.010
0.00600	(0.0588)	3.067	2.090	1.552	1.153	0.904	*	1.065	1.049	1.029	1.017	1.009	1.001	0.997	0.995
0.00800	(0.0785)	2.656	1.810	1.344	0.998	*	1.067	1.046	1.031	1.013	1.003	0.996	0.989	0.987	0.986
0.01000	(0.0981)	2.375	1.619	1.202	0.893	*	1.052	1.032	1.018	1.002	0.993	0.987	0.981	0.979	0.979
0.01500	(0.147)	1.940	1.322	0.982	*	*	1.026	1.008	0.997	0.984	0.976	0.972	0.968	0.967	0.968
0.02000	(0.196)	1.680	1.145	0.850	*	1.034	1.009	0.994	0.984	0.972	0.966	0.962	0.959	0.959	0.961
0.03000	(0.294)	1.371	0.935	*	*	1.010	0.988	0.975	0.967	0.957	0.952	0.950	0.949	0.950	0.952
0.04000	(0.392)	1.188	0.809	*	1.010	0.994	0.975	0.963	0.956	0.948	0.944	0.942	0.942	0.944	0.947
0.06000	(0.588)	0.970	*	*	0.988	0.975	0.958	0.948	0.942	0.936	0.934	0.933	0.934	0.937	0.941
0.08000	(0.785)	0.840	*	0.993	0.975	0.962	0.948	0.939	0.934	0.929	0.927	0.927	0.929	0.932	0.937
0.10000	(0.981)	0.751	*	0.982	0.965	0.954	0.940	0.933	0.928	0.924	0.923	0.923	0.925	0.929	0.934
0.15000	(1.47)	*	0.983	0.964	0.949	0.940	0.928	0.922	0.919	0.916	0.916	0.916	0.920	0.924	0.929
0.20000	(1.96)	*	0.970	0.953	0.940	0.931	0.921	0.916	0.913	0.911	0.911	0.912	0.916	0.921	0.927
0.30000	(2.94)	*	0.954	0.939	0.928	0.920	0.912	0.908	0.906	0.905	0.906	0.908	0.912	0.917	0.923
0.40000	(3.92)	0.967	0.944	0.931	0.920	0.914	0.906	0.903	0.901	0.901	0.902	0.905	0.910	0.915	0.921
0.60000	(5.88)	0.953	0.932	0.921	0.911	0.906	0.899	0.897	0.896	0.896	0.898	0.901	0.907	0.912	0.919
0.80000	(7.85)	0.944	0.925	0.914	0.906	0.901	0.895	0.893	0.893	0.894	0.896	0.899	0.905	0.911	0.917
1.00000	(9.81)	0.937	0.920	0.910	0.902	0.897	0.892	0.891	0.890	0.892	0.894	0.897	0.903	0.909	0.916
1.50000	(14.7)	0.928	0.912	0.903	0.896	0.892	0.888	0.887	0.887	0.889	0.892	0.895	0.901	0.908	0.915
2.00000	(19.6)	0.922	0.907	0.898	0.892	0.889	0.885	0.884	0.885	0.887	0.890	0.893	0.900	0.907	0.914
3.00000	(29.4)	0.914	0.901	0.893	0.888	0.885	0.882	0.881	0.882	0.885	0.888	0.892	0.899	0.905	0.913
4.00000	(39.2)	0.910	0.897	0.890	0.885	0.882	0.880	0.880	0.880	0.883	0.887	0.891	0.898	0.905	0.912
6.00000	(58.8)	0.904	0.893	0.886	0.882	0.879	0.877	0.878	0.879	0.882	0.886	0.890	0.897	0.904	0.911
8.00000	(78.5)	0.901	0.890	0.884	0.880	0.878	0.876	0.876	0.877	0.881	0.885	0.889	0.896	0.903	0.911
10.0000	(98.1)	0.899	0.888	0.883	0.878	0.876	0.875	0.875	0.877	0.880	0.884	0.888	0.896	0.903	0.911
15.0000	(147)	0.896	0.886	0.880	0.876	0.874	0.873	0.874	0.876	0.879	0.883	0.887	0.895	0.902	0.910
20.0000	(196)	0.893	0.884	0.879	0.875	0.873	0.872	0.873	0.875	0.879	0.883	0.887	0.895	0.902	0.910
30.0000	(294)	0.891	0.882	0.877	0.874	0.872	0.871	0.872	0.874	0.878	0.882	0.886	0.894	0.901	0.909
S	$(\Delta p/\rho l)$	6	8	10	12.5	15	20	25	30	40	50	60	80	100	125

Grad'nt   **k.p.g.**   **(Equivalent) Pipe diameters in mm**

S	$(\Delta p/\rho l)$	125	150	200	250	300	400	500	600	800	1000	1250	1500	2000	2500
0.00010	(0.00098)	1.248	1.231	1.209	1.197	1.189	1.180	1.175	1.173	1.173	1.175	1.179	1.183	1.192	1.200
0.00015	(0.00147)	1.211	1.196	1.178	1.168	1.162	1.155	1.152	1.151	1.153	1.156	1.161	1.166	1.176	1.186
0.00020	(0.00196)	1.186	1.174	1.158	1.149	1.143	1.138	1.137	1.137	1.139	1.143	1.149	1.155	1.166	1.176
0.00030	(0.00294)	1.155	1.144	1.131	1.124	1.120	1.117	1.116	1.118	1.122	1.127	1.134	1.140	1.153	1.164
0.00040	(0.00392)	1.134	1.124	1.113	1.107	1.104	1.103	1.103	1.105	1.111	1.117	1.124	1.131	1.144	1.156
0.00060	(0.00588)	1.107	1.099	1.090	1.086	1.084	1.084	1.086	1.089	1.096	1.103	1.111	1.119	1.134	1.146
0.00080	(0.00785)	1.089	1.082	1.075	1.072	1.071	1.073	1.075	1.079	1.087	1.094	1.103	1.112	1.127	1.140
0.00100	(0.00981)	1.076	1.070	1.064	1.062	1.062	1.064	1.067	1.072	1.080	1.088	1.098	1.106	1.122	1.135
0.00150	(0.0147)	1.054	1.050	1.046	1.045	1.046	1.050	1.054	1.059	1.069	1.078	1.088	1.098	1.114	1.128
0.00200	(0.0196)	1.040	1.036	1.034	1.034	1.036	1.040	1.046	1.051	1.062	1.071	1.082	1.092	1.109	1.123
0.00300	(0.0294)	1.022	1.019	1.019	1.020	1.023	1.029	1.035	1.041	1.053	1.063	1.075	1.085	1.103	1.118
0.00400	(0.0392)	1.010	1.009	1.009	1.012	1.015	1.022	1.029	1.035	1.047	1.058	1.070	1.081	1.099	1.114
0.00600	(0.0588)	0.995	0.995	0.997	1.000	1.004	1.012	1.020	1.027	1.041	1.052	1.065	1.076	1.094	1.110
0.00800	(0.0785)	0.986	0.986	0.989	0.993	0.998	1.007	1.015	1.023	1.036	1.048	1.061	1.072	1.092	1.108
0.01000	(0.0981)	0.979	0.980	0.984	0.988	0.993	1.003	1.011	1.019	1.033	1.045	1.059	1.070	1.090	1.106
0.01500	(0.147)	0.968	0.970	0.975	0.980	0.986	0.996	1.005	1.014	1.028	1.041	1.055	1.066	1.086	1.103
0.02000	(0.196)	0.961	0.963	0.969	0.975	0.981	0.992	1.002	1.010	1.026	1.038	1.052	1.064	1.084	1.101
0.03000	(0.294)	0.952	0.956	0.962	0.969	0.975	0.987	0.997	1.006	1.022	1.035	1.049	1.062	1.082	1.099
0.04000	(0.392)	0.947	0.951	0.958	0.965	0.972	0.984	0.994	1.004	1.020	1.033	1.048	1.060	1.081	1.098
0.06000	(0.588)	0.941	0.945	0.953	0.961	0.968	0.980	0.991	1.001	1.017	1.031	1.045	1.058	1.079	1.096
0.08000	(0.785)	0.937	0.941	0.950	0.958	0.965	0.978	0.989	0.999	1.016	1.029	1.044	1.057	1.078	1.095
0.10000	(0.981)	0.934	0.939	0.948	0.956	0.963	0.977	0.988	0.998	1.014	1.028	1.043	1.056	1.077	1.095
0.15000	(1.47)	0.929	0.934	0.944	0.953	0.961	0.974	0.986	0.996	1.013	1.027	1.042	1.055	1.076	1.094
0.20000	(1.96)	0.927	0.932	0.942	0.951	0.959	0.973	0.984	0.994	1.012	1.026	1.041	1.054	1.076	1.093
0.30000	(2.94)	0.923	0.929	0.939	0.949	0.957	0.971	0.983	0.993	1.010	1.025	1.040	1.053	1.075	1.093
0.40000	(3.92)	0.921	0.927	0.938	0.947	0.955	0.970	0.982	0.992	1.010	1.024	1.040	1.053	1.074	1.092
0.60000	(5.88)	0.919	0.925	0.936	0.946	0.954	0.968	0.981	0.991	1.009	1.023	1.039	1.052	1.074	1.092
0.80000	(7.85)	0.917	0.924	0.935	0.945	0.953	0.968	0.980	0.991	1.008	1.023	1.038	1.052	1.074	1.091
1.00000	(9.81)	0.916	0.923	0.934	0.944	0.952	0.967	0.979	0.990	1.008	1.023	1.038	1.051	1.073	1.091
1.50000	(14.7)	0.915	0.921	0.933	0.943	0.952	0.966	0.979	0.989	1.007	1.022	1.038	1.051	1.073	1.091
2.00000	(19.6)	0.914	0.921	0.932	0.942	0.951	0.966	0.978	0.989	1.007	1.022	1.037	1.051	1.073	1.091
3.00000	(29.4)	0.913	0.920	0.931	0.941	0.950	0.965	0.978	0.989	1.007	1.021	1.037	1.050	1.072	1.090
S	$(\Delta p/\rho l)$	125	150	200	250	300	400	500	600	800	1000	1250	1500	2000	2500

Grad'nt   **k.p.g.**   **(Equivalent) Pipe diameters in mm**

Kinematic viscosity, $\nu = 2.0 \times 10^{-6}$ m^{2}s^{-1} ;
     Roughness size, k_s = 0.150 mm

Kin. visc., $\nu = 2.0 \times 10^{-6}$ m^{2}s^{-1}; S = 0·00010 to 30·0000
i.e. kin. pr. grad., $\Delta p/\rho l$ = (0·00098) to (294) ms^{-2}

Roughness size, k_s = 0·30 mm
This table shows values of m, as follows
m_C for Colebrook-White solutions; or, where **R** $\leq$ 2000, m_P for laminar flow

Grad'nt S	k.p.g. $(\Delta p/\rho l)$	6	8	10	12·5	15	20	25	30	40	50	60	80	100	125
									(Equivalent) Pipe diameters in mm						
0·00030	(0·00294)	13·71	9·346	6·941	5·154	4·042	2·754	2·046	1·604	1·093	*	1·253	1·221	1·201	1·186
0·00040	(0·00392)	11·88	8·093	6·011	4·464	3·501	2·385	1·771	1·389	0·947	*	1·228	1·198	1·181	1·167
0·00060	(0·00588)	9·698	6·608	4·908	3·645	2·858	1·948	1·446	1·134	*	1·214	1·194	1·169	1·155	1·143
0·00080	(0·00785)	8·399	5·723	4·250	3·156	2·475	1·687	1·253	0·982	*	1·191	1·173	1·151	1·138	1·128
0·00100	(0·00981)	7·512	5·119	3·801	2·823	2·214	1·509	1·120	*	1·198	1·174	1·158	1·137	1·126	1·117
0·00150	(0·0147)	6·133	4·179	3·104	2·305	1·808	1·232	0·915	*	1·167	1·146	1·132	1·115	1·106	1·099
0·00200	(0·0196)	5·312	3·620	2·688	1·996	1·565	1·067	*	*	1·147	1·128	1·116	1·101	1·093	1·087
0·00300	(0·0294)	4·337	2·955	2·195	1·630	1·278	0·871	*	1·148	1·121	1·106	1·095	1·083	1·077	1·073
0·00400	(0·0392)	3·756	2·559	1·901	1·412	1·107	*	*	1·129	1·105	1·091	1·082	1·072	1·067	1·064
0·00600	(0·0588)	3·067	2·090	1·552	1·153	0·904	*	1·122	1·106	1·085	1·073	1·066	1·058	1·054	1·053
0·00800	(0·0785)	2·656	1·810	1·344	0·998	*	*	1·106	1·091	1·073	1·062	1·056	1·049	1·046	1·046
0·01000	(0·0981)	2·375	1·619	1·202	0·893	*	1·115	1·094	1·080	1·064	1·054	1·048	1·043	1·041	1·041
0·01500	(0·147)	1·940	1·322	0·982	*	*	1·093	1·075	1·063	1·049	1·041	1·037	1·033	1·032	1·033
0·02000	(0·196)	1·680	1·145	0·850	*	*	1·079	1·063	1·053	1·040	1·033	1·030	1·027	1·026	1·028
0·03000	(0·294)	1·371	0·935	*	*	1·086	1·063	1·049	1·039	1·029	1·023	1·021	1·019	1·020	1·022
0·04000	(0·392)	1·188	0·809	*	1·091	1·074	1·052	1·039	1·031	1·022	1·017	1·015	1·014	1·016	1·018
0·06000	(0·588)	0·970	*	*	1·074	1·059	1·039	1·028	1·021	1·014	1·010	1·009	1·009	1·011	1·014
0·08000	(0·785)	0·840	*	1·085	1·063	1·049	1·032	1·021	1·015	1·008	1·006	1·005	1·005	1·008	1·011
0·10000	(0·981)	0·751	*	1·076	1·056	1·043	1·026	1·017	1·011	1·005	1·003	1·002	1·003	1·006	1·010
0·15000	(1·47)	*	*	1·062	1·044	1·032	1·017	1·009	1·004	0·999	0·998	0·997	0·999	1·002	1·007
0·20000	(1·96)	*	1·074	1·054	1·037	1·026	1·012	1·005	1·000	0·996	0·995	0·995	0·997	1·000	1·005
0·30000	(2·94)	*	1·062	1·043	1·028	1·018	1·006	0·999	0·995	0·992	0·991	0·992	0·994	0·998	1·003
0·40000	(3·92)	*	1·055	1·037	1·023	1·013	1·002	0·996	0·992	0·989	0·989	0·990	0·993	0·997	1·002
0·60000	(5·88)	1·073	1·046	1·030	1·016	1·007	0·997	0·992	0·989	0·986	0·986	0·987	0·991	0·995	1·000
0·80000	(7·85)	1·067	1·041	1·025	1·012	1·004	0·994	0·989	0·986	0·984	0·985	0·986	0·990	0·994	0·999
1·00000	(9·81)	1·062	1·037	1·022	1·010	1·002	0·992	0·988	0·985	0·983	0·984	0·985	0·989	0·993	0·999
1·50000	(14·7)	1·055	1·031	1·017	1·005	0·998	0·989	0·985	0·983	0·981	0·982	0·983	0·988	0·992	0·998
2·00000	(19·6)	1·050	1·028	1·014	1·003	0·996	0·988	0·983	0·981	0·980	0·981	0·983	0·987	0·991	0·997
3·00000	(29·4)	1·045	1·023	1·010	1·000	0·993	0·985	0·981	0·980	0·979	0·980	0·981	0·986	0·991	0·996
4·00000	(39·2)	1·042	1·021	1·008	0·998	0·991	0·984	0·980	0·979	0·978	0·979	0·981	0·985	0·990	0·996
6·00000	(58·8)	1·038	1·018	1·006	0·996	0·990	0·982	0·979	0·977	0·977	0·978	0·980	0·985	0·990	0·996
8·00000	(78·5)	1·036	1·016	1·004	0·995	0·988	0·981	0·978	0·977	0·976	0·978	0·980	0·984	0·989	0·995
10·0000	(98·1)	1·034	1·015	1·003	0·994	0·988	0·981	0·978	0·976	0·976	0·977	0·979	0·984	0·989	0·995
15·0000	(147)	1·032	1·013	1·001	0·992	0·986	0·980	0·977	0·975	0·975	0·977	0·979	0·984	0·989	0·995
20·0000	(196)	1·031	1·012	1·000	0·991	0·986	0·979	0·976	0·975	0·975	0·976	0·979	0·983	0·989	0·995
30·0000	(294)	1·029	1·010	0·999	0·990	0·985	0·979	0·976	0·974	0·974	0·976	0·978	0·983	0·988	0·994
S	$(\Delta p/\rho l)$	6	8	10	12·5	15	20	25	30	40	50	60	80	100	125

Grad'nt k.p.g. (Equivalent) Pipe diameters in mm

Grad'nt S	k.p.g. $(\Delta p/\rho l)$	125	150	200	250	300	400	500	600	800	1000	1250	1500	2000	2500
0·00010	(0·00098)	1·271	1·254	1·234	1·222	1·215	1·207	1·204	1·203	1·204	1·207	1·212	1·218	1·228	1·239
0·00015	(0·00147)	1·237	1·223	1·206	1·196	1·190	1·185	1·183	1·184	1·187	1·191	1·198	1·204	1·216	1·227
0·00020	(0·00196)	1·214	1·202	1·187	1·179	1·175	1·171	1·170	1·171	1·176	1·181	1·188	1·195	1·208	1·220
0·00030	(0·00294)	1·186	1·175	1·164	1·158	1·154	1·153	1·154	1·156	1·162	1·168	1·176	1·184	1·198	1·211
0·00040	(0·00392)	1·167	1·158	1·148	1·144	1·141	1·141	1·143	1·146	1·153	1·160	1·169	1·177	1·192	1·205
0·00060	(0·00588)	1·143	1·136	1·129	1·126	1·125	1·126	1·130	1·134	1·142	1·150	1·160	1·169	1·185	1·198
0·00080	(0·00785)	1·128	1·122	1·116	1·114	1·114	1·117	1·121	1·126	1·135	1·144	1·154	1·163	1·180	1·194
0·00100	(0·00981)	1·117	1·112	1·107	1·106	1·107	1·110	1·115	1·120	1·130	1·139	1·150	1·160	1·176	1·191
0·00150	(0·0147)	1·099	1·095	1·093	1·093	1·095	1·100	1·105	1·111	1·122	1·132	1·143	1·153	1·171	1·186
0·00200	(0·0196)	1·087	1·085	1·085	1·085	1·087	1·093	1·099	1·105	1·117	1·127	1·139	1·150	1·168	1·183
0·00300	(0·0294)	1·073	1·071	1·072	1·074	1·077	1·084	1·091	1·098	1·111	1·122	1·134	1·145	1·164	1·180
0·00400	(0·0392)	1·064	1·063	1·065	1·068	1·071	1·079	1·087	1·094	1·107	1·119	1·131	1·142	1·161	1·177
0·00600	(0·0588)	1·053	1·053	1·056	1·060	1·064	1·073	1·081	1·089	1·103	1·115	1·128	1·139	1·159	1·175
0·00800	(0·0785)	1·046	1·047	1·050	1·055	1·059	1·069	1·078	1·086	1·100	1·112	1·125	1·137	1·157	1·173
0·01000	(0·0981)	1·041	1·042	1·046	1·051	1·056	1·066	1·075	1·083	1·098	1·110	1·124	1·136	1·156	1·172
0·01500	(0·147)	1·033	1·035	1·040	1·046	1·051	1·062	1·071	1·080	1·095	1·108	1·121	1·133	1·154	1·170
0·02000	(0·196)	1·028	1·030	1·036	1·042	1·048	1·059	1·069	1·078	1·093	1·106	1·120	1·132	1·152	1·169
0·03000	(0·294)	1·022	1·025	1·032	1·038	1·045	1·056	1·066	1·075	1·091	1·104	1·118	1·130	1·151	1·168
0·04000	(0·392)	1·018	1·022	1·029	1·036	1·042	1·054	1·064	1·074	1·089	1·103	1·117	1·129	1·150	1·167
0·06000	(0·588)	1·014	1·018	1·026	1·033	1·040	1·052	1·062	1·072	1·088	1·101	1·116	1·128	1·149	1·166
0·08000	(0·785)	1·011	1·016	1·023	1·031	1·038	1·050	1·061	1·071	1·087	1·101	1·115	1·128	1·149	1·166
0·10000	(0·981)	1·010	1·014	1·022	1·030	1·037	1·049	1·060	1·070	1·086	1·100	1·115	1·127	1·148	1·166
0·15000	(1·47)	1·007	1·011	1·020	1·028	1·035	1·048	1·059	1·069	1·085	1·099	1·114	1·126	1·148	1·165
0·20000	(1·96)	1·005	1·010	1·019	1·027	1·034	1·047	1·058	1·068	1·085	1·098	1·113	1·126	1·147	1·165
0·30000	(2·94)	1·003	1·008	1·017	1·025	1·033	1·046	1·057	1·067	1·084	1·098	1·113	1·125	1·147	1·164
0·40000	(3·92)	1·002	1·007	1·016	1·024	1·032	1·045	1·057	1·067	1·083	1·097	1·112	1·125	1·146	1·164
0·60000	(5·88)	1·000	1·005	1·015	1·023	1·031	1·044	1·056	1·066	1·083	1·097	1·112	1·125	1·146	1·164
0·80000	(7·85)	0·999	1·005	1·014	1·023	1·031	1·044	1·056	1·066	1·083	1·097	1·112	1·125	1·146	1·164
1·00000	(9·81)	0·999	1·004	1·014	1·022	1·030	1·044	1·055	1·065	1·082	1·096	1·111	1·124	1·146	1·163
1·50000	(14·7)	0·998	1·003	1·013	1·022	1·030	1·043	1·055	1·065	1·082	1·096	1·111	1·124	1·146	1·163
2·00000	(19·6)	0·997	1·003	1·013	1·021	1·029	1·043	1·055	1·065	1·082	1·096	1·111	1·124	1·146	1·163
3·00000	(29·4)	0·996	1·002	1·012	1·021	1·029	1·043	1·054	1·064	1·082	1·096	1·111	1·124	1·145	1·163
S	$(\Delta p/\rho l)$	125	150	200	250	300	400	500	600	800	1000	1250	1500	2000	2500

Grad'nt k.p.g. (Equivalent) Pipe diameters in mm

Kinematic viscosity, $\nu = 2.0 \times 10^{-6}$ m^{2}s^{-1} ; **Roughness size, k_s = 0·30 mm**

E135

Kin. visc., $\nu = 2{\cdot}0 \times 10^{-6}$ $m^2 s^{-1}$; $S = 0{\cdot}00010$ to $3{\cdot}00000$

i.e. kin. pr. grad., $\Delta p/\rho l =$ ($0{\cdot}00098$) to ($29{\cdot}4$) ms^{-2}

Roughness size, $k_s = 0{\cdot}60$ mm

This table shows values of m, as follows

m_C for Colebrook-White solutions

Grad'nt S	k.p.g. $(\Delta p/\rho l)$	125	150	200	250	300	400	500	600	800	1000	1250	1500	2000	2500
0·00010	(0·00098)	1·313	1·297	1·277	1·266	1·260	1·253	1·251	1·251	1·254	1·258	1·265	1·271	1·284	1·295
0·00015	(0·00147)	1·283	1·269	1·253	1·244	1·239	1·235	1·235	1·236	1·241	1·246	1·254	1·261	1·275	1·287
0·00020	(0·00196)	1·263	1·252	1·238	1·230	1·227	1·224	1·225	1·227	1·232	1·239	1·247	1·255	1·269	1·282
0·00030	(0·00294)	1·239	1·229	1·218	1·213	1·211	1·210	1·212	1·215	1·222	1·230	1·239	1·247	1·263	1·276
0·00040	(0·00392)	1·224	1·215	1·206	1·202	1·201	1·201	1·204	1·208	1·216	1·224	1·234	1·243	1·259	1·272
0·00060	(0·00588)	1·204	1·198	1·191	1·188	1·188	1·190	1·194	1·199	1·208	1·217	1·227	1·237	1·253	1·268
0·00080	(0·00785)	1·192	1·187	1·181	1·180	1·180	1·183	1·188	1·193	1·203	1·213	1·223	1·233	1·250	1·265
0·00100	(0·00981)	1·184	1·179	1·174	1·174	1·175	1·179	1·184	1·189	1·200	1·210	1·221	1·231	1·248	1·263
0·00150	(0·0147)	1·170	1·166	1·164	1·164	1·166	1·171	1·177	1·183	1·194	1·205	1·216	1·227	1·245	1·260
0·00200	(0·0196)	1·161	1·158	1·157	1·158	1·160	1·166	1·173	1·179	1·191	1·202	1·214	1·224	1·243	1·258
0·00300	(0·0294)	1·150	1·149	1·149	1·151	1·154	1·161	1·168	1·175	1·187	1·198	1·211	1·222	1·240	1·256
0·00400	(0·0392)	1·144	1·143	1·144	1·146	1·150	1·157	1·165	1·172	1·185	1·196	1·209	1·220	1·239	1·255
0·00600	(0·0588)	1·136	1·136	1·137	1·141	1·145	1·153	1·161	1·169	1·182	1·194	1·206	1·218	1·237	1·253
0·00800	(0·0785)	1·131	1·131	1·134	1·138	1·142	1·151	1·159	1·166	1·180	1·192	1·205	1·217	1·236	1·252
0·01000	(0·0981)	1·128	1·128	1·131	1·135	1·140	1·149	1·157	1·165	1·179	1·191	1·204	1·216	1·235	1·252
0·01500	(0·147)	1·122	1·123	1·127	1·132	1·137	1·146	1·155	1·163	1·177	1·189	1·203	1·214	1·234	1·251
0·02000	(0·196)	1·119	1·120	1·125	1·130	1·135	1·144	1·153	1·162	1·176	1·188	1·202	1·214	1·233	1·250
0·03000	(0·294)	1·115	1·117	1·122	1·127	1·132	1·142	1·152	1·160	1·175	1·187	1·201	1·213	1·233	1·249
0·04000	(0·392)	1·113	1·115	1·120	1·125	1·131	1·141	1·151	1·159	1·174	1·186	1·200	1·212	1·232	1·249
0·06000	(0·588)	1·110	1·112	1·118	1·124	1·129	1·140	1·149	1·158	1·173	1·186	1·199	1·211	1·232	1·248
0·08000	(0·785)	1·108	1·111	1·116	1·122	1·128	1·139	1·149	1·157	1·172	1·185	1·199	1·211	1·231	1·248
0·10000	(0·981)	1·107	1·110	1·116	1·122	1·128	1·138	1·148	1·157	1·172	1·185	1·199	1·211	1·231	1·248
0·15000	(1·47)	1·105	1·108	1·114	1·120	1·126	1·137	1·147	1·156	1·171	1·184	1·198	1·210	1·231	1·247
0·20000	(1·96)	1·104	1·107	1·113	1·120	1·126	1·137	1·147	1·156	1·171	1·184	1·198	1·210	1·231	1·247
0·30000	(2·94)	1·103	1·106	1·112	1·119	1·125	1·136	1·146	1·155	1·170	1·184	1·198	1·210	1·230	1·247
0·40000	(3·92)	1·102	1·105	1·112	1·118	1·125	1·136	1·146	1·155	1·170	1·183	1·197	1·209	1·230	1·247
0·60000	(5·88)	1·101	1·104	1·111	1·118	1·124	1·135	1·145	1·154	1·170	1·183	1·197	1·209	1·230	1·247
0·80000	(7·85)	1·101	1·104	1·111	1·117	1·124	1·135	1·145	1·154	1·170	1·183	1·197	1·209	1·230	1·247
1·00000	(9·81)	1·100	1·104	1·110	1·117	1·124	1·135	1·145	1·154	1·170	1·183	1·197	1·209	1·230	1·247
1·50000	(14·7)	1·100	1·103	1·110	1·117	1·123	1·135	1·145	1·154	1·169	1·183	1·197	1·209	1·229	1·246
2·00000	(19·6)	1·099	1·103	1·110	1·117	1·123	1·135	1·145	1·154	1·169	1·182	1·197	1·209	1·229	1·246
3·00000	(29·4)	1·099	1·102	1·109	1·116	1·123	1·134	1·144	1·154	1·169	1·182	1·196	1·209	1·229	1·246
S	$(\Delta p/\rho l)$	125	150	200	250	300	400	500	600	800	1000	1250	1500	2000	2500

Grad'nt   k.p.g.   (Equivalent) Pipe diameters in mm   **Roughness size,** $k_s = 0{\cdot}60$ mm

E136

Kin. visc., $\nu = 2{\cdot}0 \times 10^{-6}$ $m^2 s^{-1}$; $S = 0{\cdot}00010$ to $3{\cdot}00000$

i.e. kin. pr. grad., $\Delta p/\rho l =$ ($0{\cdot}00098$) to ($29{\cdot}4$) ms^{-2}

Roughness size, $k_s = 1{\cdot}50$ mm

This table shows values of m, as follows

m_C for Colebrook-White solutions

Grad'nt S	k.p.g. $(\Delta p/\rho l)$	125	150	200	250	300	400	500	600	800	1000	1250	1500	2000	2500
0·00010	(0·00098)	1·416	1·400	1·380	1·369	1·362	1·356	1·354	1·355	1·358	1·364	1·371	1·378	1·391	1·403
0·00015	(0·00147)	1·394	1·379	1·362	1·353	1·348	1·344	1·344	1·345	1·350	1·356	1·364	1·372	1·386	1·398
0·00020	(0·00196)	1·379	1·367	1·352	1·344	1·340	1·337	1·337	1·339	1·345	1·352	1·360	1·368	1·383	1·396
0·00030	(0·00294)	1·362	1·351	1·338	1·332	1·329	1·328	1·329	1·332	1·339	1·346	1·355	1·364	1·379	1·392
0·00040	(0·00392)	1·351	1·341	1·330	1·325	1·323	1·322	1·324	1·328	1·335	1·343	1·352	1·361	1·377	1·390
0·00060	(0·00588)	1·338	1·329	1·320	1·316	1·315	1·316	1·319	1·322	1·331	1·339	1·349	1·358	1·374	1·388
0·00080	(0·00785)	1·330	1·322	1·314	1·311	1·310	1·312	1·315	1·319	1·328	1·337	1·347	1·356	1·372	1·386
0·00100	(0·00981)	1·324	1·317	1·310	1·307	1·307	1·309	1·313	1·317	1·326	1·335	1·345	1·355	1·371	1·385
0·00150	(0·0147)	1·315	1·309	1·303	1·301	1·302	1·304	1·309	1·314	1·323	1·332	1·343	1·352	1·369	1·384
0·00200	(0·0196)	1·310	1·304	1·299	1·298	1·298	1·302	1·306	1·311	1·321	1·331	1·341	1·351	1·368	1·383
0·00300	(0·0294)	1·303	1·298	1·294	1·294	1·295	1·299	1·304	1·309	1·319	1·329	1·340	1·350	1·367	1·382
0·00400	(0·0392)	1·299	1·295	1·291	1·291	1·292	1·297	1·302	1·307	1·318	1·328	1·339	1·349	1·366	1·381
0·00600	(0·0588)	1·294	1·291	1·288	1·288	1·290	1·294	1·300	1·306	1·316	1·326	1·338	1·348	1·365	1·380
0·00800	(0·0785)	1·291	1·288	1·286	1·286	1·288	1·293	1·299	1·305	1·316	1·326	1·337	1·347	1·365	1·380
0·01000	(0·0981)	1·289	1·286	1·284	1·285	1·287	1·292	1·298	1·304	1·315	1·325	1·336	1·347	1·364	1·380
0·01500	(0·147)	1·286	1·284	1·282	1·283	1·285	1·291	1·297	1·303	1·314	1·324	1·336	1·346	1·364	1·379
0·02000	(0·196)	1·284	1·282	1·281	1·282	1·284	1·290	1·296	1·302	1·313	1·324	1·335	1·346	1·364	1·379
0·03000	(0·294)	1·282	1·280	1·279	1·280	1·283	1·289	1·295	1·301	1·313	1·323	1·335	1·345	1·363	1·378
0·04000	(0·392)	1·281	1·279	1·278	1·280	1·282	1·288	1·294	1·301	1·312	1·323	1·334	1·345	1·363	1·378
0·06000	(0·588)	1·279	1·277	1·277	1·279	1·281	1·287	1·294	1·300	1·312	1·322	1·334	1·345	1·363	1·378
0·08000	(0·785)	1·278	1·277	1·276	1·278	1·281	1·287	1·293	1·300	1·311	1·322	1·334	1·344	1·362	1·378
0·10000	(0·981)	1·278	1·276	1·276	1·278	1·280	1·287	1·293	1·299	1·311	1·322	1·334	1·344	1·362	1·378
0·15000	(1·47)	1·277	1·275	1·275	1·277	1·280	1·286	1·293	1·299	1·311	1·322	1·333	1·344	1·362	1·378
0·20000	(1·96)	1·276	1·275	1·275	1·277	1·279	1·286	1·292	1·299	1·311	1·321	1·333	1·344	1·362	1·377
0·30000	(2·94)	1·275	1·274	1·274	1·276	1·279	1·285	1·292	1·299	1·311	1·321	1·333	1·344	1·362	1·377
0·40000	(3·92)	1·275	1·274	1·274	1·276	1·279	1·285	1·292	1·298	1·310	1·321	1·333	1·344	1·362	1·377
0·60000	(5·88)	1·275	1·273	1·273	1·275	1·278	1·285	1·292	1·298	1·310	1·321	1·333	1·343	1·362	1·377
0·80000	(7·85)	1·274	1·273	1·273	1·275	1·278	1·285	1·292	1·298	1·310	1·321	1·333	1·343	1·362	1·377
1·00000	(9·81)	1·274	1·273	1·273	1·275	1·278	1·285	1·292	1·298	1·310	1·321	1·333	1·343	1·362	1·377
1·50000	(14·7)	1·274	1·272	1·273	1·275	1·278	1·285	1·291	1·298	1·310	1·321	1·333	1·343	1·362	1·377
2·00000	(19·6)	1·273	1·272	1·273	1·275	1·278	1·285	1·291	1·298	1·310	1·321	1·333	1·343	1·362	1·377
3·00000	(29·4)	1·273	1·272	1·272	1·275	1·278	1·284	1·291	1·298	1·310	1·321	1·333	1·343	1·362	1·377
S	$(\Delta p/\rho l)$	125	150	200	250	300	400	500	600	800	1000	1250	1500	2000	2500

Grad'nt   k.p.g.   (Equivalent) Pipe diameters in mm   **Roughness size,** $k_s = 1{\cdot}50$ mm

Kin. visc., $\nu = 2.5 \times 10^{-6}\ m^2 s^{-1}$; $S = 0.00010$ to 30.0000 i.e. kin. pr. grad., $\Delta p/\rho l = (0.00098)$ to $(294)\ ms^{-2}$

Roughness size, $k_s = 0.0015$ mm
This table shows values of m, as follows
m_C for Colebrook-White solutions; or,
where $\mathbf{R} \le 2000$, m_P for laminar flow

Grad'nt S	k.p.g. $(\Delta p/\rho l)$	\(Equivalent\) Pipe diameters in mm													
		6	8	10	12.5	15	20	25	30	40	50	60	80	100	125
0.00030	(0.00294)	17.14	11.68	8.676	6.443	5.053	3.443	2.557	2.005	1.366	1.015	*	1.205	1.180	1.160
0.00040	(0.00392)	14.85	10.12	7.513	5.580	4.376	2.982	2.214	1.736	1.183	*	1.210	1.175	1.152	1.134
0.00060	(0.00588)	12.12	8.260	6.135	4.556	3.573	2.435	1.808	1.418	0.966	*	1.166	1.135	1.115	1.099
0.00080	(0.00785)	10.50	7.154	5.313	3.946	3.094	2.108	1.566	1.228	*	1.158	1.137	1.108	1.091	1.076
0.00100	(0.00981)	9.390	6.398	4.752	3.529	2.767	1.886	1.400	1.098	*	1.135	1.115	1.088	1.072	1.059
0.00150	(0.0147)	7.667	5.224	3.880	2.881	2.260	1.540	1.143	0.897	1.120	1.095	1.077	1.054	1.040	1.028
0.00200	(0.0196)	6.640	4.524	3.360	2.495	1.957	1.333	0.990	*	1.091	1.068	1.052	1.031	1.018	1.008
0.00300	(0.0294)	5.421	3.694	2.743	2.037	1.598	1.089	*	*	1.053	1.032	1.018	1.000	0.989	0.981
0.00400	(0.0392)	4.695	3.199	2.376	1.764	1.384	0.943	*	1.056	1.027	1.009	0.996	0.980	0.970	0.962
0.00600	(0.0588)	3.833	2.612	1.940	1.441	1.130	*	1.038	1.019	0.993	0.977	0.966	0.952	0.944	0.937
0.00800	(0.0785)	3.320	2.262	1.680	1.248	0.978	*	1.011	0.993	0.970	0.955	0.946	0.933	0.926	0.921
0.01000	(0.0981)	2.969	2.023	1.503	1.116	0.875	*	0.991	0.975	0.953	0.940	0.930	0.919	0.913	0.908
0.01500	(0.147)	2.424	1.652	1.227	0.911	*	0.978	0.957	0.943	0.924	0.912	0.904	0.895	0.889	0.886
0.02000	(0.196)	2.100	1.431	1.063	0.789	*	0.953	0.934	0.921	0.904	0.893	0.887	0.878	0.874	0.871
0.03000	(0.294)	1.714	1.168	0.868	*	0.947	0.921	0.904	0.892	0.877	0.869	0.863	0.856	0.853	0.850
0.04000	(0.392)	1.485	1.012	0.751	*	0.923	0.899	0.883	0.873	0.860	0.852	0.847	0.841	0.838	0.837
0.06000	(0.588)	1.212	0.826	*	0.907	0.891	0.869	0.856	0.847	0.836	0.829	0.825	0.821	0.819	0.818
0.08000	(0.785)	1.050	*	*	0.884	0.869	0.850	0.838	0.830	0.820	0.814	0.810	0.807	0.806	0.805
0.10000	(0.981)	0.939	*	0.887	0.867	0.853	0.835	0.824	0.817	0.808	0.803	0.800	0.797	0.796	0.796
0.15000	(1.47)	0.767	*	0.855	0.838	0.826	0.810	0.801	0.794	0.787	0.783	0.780	0.778	0.778	0.779
0.20000	(1.96)	*	0.853	0.834	0.818	0.807	0.793	0.785	0.779	0.773	0.769	0.767	0.766	0.767	0.768
0.30000	(2.94)	*	0.823	0.806	0.792	0.783	0.771	0.763	0.759	0.754	0.751	0.750	0.750	0.751	0.753
0.40000	(3.92)	0.826	0.802	0.787	0.775	0.766	0.755	0.749	0.745	0.741	0.739	0.738	0.738	0.740	0.742
0.60000	(5.88)	0.797	0.775	0.762	0.751	0.744	0.735	0.730	0.727	0.723	0.722	0.722	0.723	0.725	0.728
0.80000	(7.85)	0.777	0.758	0.746	0.736	0.729	0.721	0.717	0.714	0.712	0.711	0.711	0.713	0.715	0.719
1.00000	(9.81)	0.762	0.744	0.733	0.724	0.718	0.711	0.707	0.705	0.703	0.703	0.703	0.705	0.708	0.712
1.50000	(14.7)	0.737	0.721	0.712	0.704	0.699	0.693	0.690	0.689	0.688	0.688	0.689	0.692	0.695	0.699
2.00000	(19.6)	0.720	0.706	0.697	0.691	0.686	0.681	0.679	0.678	0.677	0.678	0.680	0.683	0.686	0.691
3.00000	(29.4)	0.698	0.685	0.678	0.673	0.669	0.665	0.664	0.663	0.663	0.665	0.667	0.671	0.675	0.679
4.00000	(39.2)	0.683	0.672	0.665	0.660	0.657	0.654	0.653	0.653	0.654	0.656	0.658	0.662	0.667	0.672
6.00000	(58.8)	0.663	0.654	0.648	0.644	0.642	0.640	0.639	0.640	0.642	0.644	0.646	0.652	0.656	0.662
8.00000	(78.5)	0.649	0.641	0.637	0.633	0.632	0.630	0.630	0.631	0.633	0.636	0.639	0.644	0.649	0.655
10.0000	(98.1)	0.640	0.632	0.628	0.625	0.624	0.623	0.623	0.624	0.627	0.630	0.633	0.639	0.644	0.650
15.0000	(147)	0.622	0.616	0.613	0.611	0.610	0.610	0.611	0.613	0.616	0.620	0.623	0.629	0.635	0.641
20.0000	(196)	0.611	0.606	0.603	0.602	0.601	0.602	0.603	0.605	0.609	0.613	0.616	0.623	0.629	0.636
30.0000	(294)	0.596	0.592	0.590	0.589	0.589	0.591	0.593	0.595	0.599	0.604	0.608	0.615	0.621	0.628
S	$(\Delta p/\rho l)$	6	8	10	12.5	15	20	25	30	40	50	60	80	100	125

Grad'nt k.p.g. \(Equivalent\) Pipe diameters in mm

S	$(\Delta p/\rho l)$	125	150	200	250	300	400	500	600	800	1000	1250	1500	2000	2500
0.00010	(0.00098)	1.271	1.250	1.224	1.208	1.197	1.183	1.175	1.171	1.166	1.165	1.165	1.166	1.170	1.175
0.00015	(0.00147)	1.227	1.209	1.187	1.173	1.163	1.152	1.146	1.142	1.139	1.139	1.140	1.142	1.147	1.152
0.00020	(0.00196)	1.198	1.182	1.161	1.149	1.141	1.131	1.126	1.123	1.121	1.121	1.123	1.125	1.131	1.136
0.00030	(0.00294)	1.160	1.146	1.128	1.117	1.110	1.102	1.098	1.096	1.096	1.097	1.099	1.102	1.109	1.115
0.00040	(0.00392)	1.134	1.121	1.105	1.096	1.090	1.083	1.080	1.079	1.079	1.080	1.083	1.087	1.094	1.101
0.00060	(0.00588)	1.099	1.088	1.075	1.067	1.062	1.057	1.055	1.054	1.056	1.058	1.062	1.066	1.073	1.081
0.00080	(0.00785)	1.076	1.066	1.054	1.047	1.043	1.039	1.038	1.038	1.040	1.043	1.047	1.051	1.059	1.067
0.00100	(0.00981)	1.059	1.050	1.039	1.033	1.029	1.026	1.025	1.025	1.028	1.031	1.036	1.040	1.049	1.057
0.00150	(0.0147)	1.028	1.021	1.012	1.007	1.004	1.002	1.002	1.004	1.007	1.011	1.016	1.021	1.030	1.039
0.00200	(0.0196)	1.008	1.001	0.994	0.990	0.988	0.986	0.987	0.989	0.993	0.997	1.003	1.008	1.017	1.026
0.00300	(0.0294)	0.981	0.975	0.969	0.966	0.965	0.965	0.966	0.968	0.973	0.978	0.984	0.990	1.000	1.009
0.00400	(0.0392)	0.962	0.958	0.952	0.950	0.949	0.950	0.952	0.955	0.960	0.965	0.972	0.977	0.988	0.997
0.00600	(0.0588)	0.937	0.934	0.930	0.929	0.928	0.930	0.933	0.936	0.942	0.948	0.954	0.961	0.972	0.981
0.00800	(0.0785)	0.921	0.917	0.914	0.914	0.914	0.916	0.920	0.923	0.929	0.936	0.943	0.949	0.960	0.970
0.01000	(0.0981)	0.908	0.905	0.903	0.903	0.903	0.906	0.910	0.913	0.920	0.926	0.934	0.940	0.952	0.962
0.01500	(0.147)	0.886	0.884	0.883	0.883	0.885	0.888	0.892	0.896	0.903	0.910	0.918	0.925	0.937	0.947
0.02000	(0.196)	0.871	0.869	0.869	0.870	0.872	0.876	0.880	0.884	0.892	0.899	0.907	0.914	0.927	0.937
0.03000	(0.294)	0.850	0.850	0.850	0.852	0.854	0.859	0.864	0.868	0.877	0.884	0.893	0.900	0.913	0.924
0.04000	(0.392)	0.837	0.836	0.838	0.840	0.842	0.848	0.853	0.858	0.866	0.874	0.883	0.890	0.903	0.914
0.06000	(0.588)	0.818	0.818	0.820	0.823	0.826	0.832	0.838	0.843	0.852	0.860	0.869	0.877	0.890	0.902
0.08000	(0.785)	0.805	0.806	0.809	0.812	0.815	0.821	0.827	0.833	0.842	0.851	0.860	0.868	0.882	0.893
0.10000	(0.981)	0.796	0.797	0.800	0.803	0.807	0.813	0.820	0.825	0.835	0.844	0.853	0.861	0.875	0.887
0.15000	(1.47)	0.779	0.781	0.785	0.788	0.792	0.799	0.806	0.812	0.822	0.831	0.841	0.849	0.863	0.875
0.20000	(1.96)	0.768	0.770	0.774	0.778	0.782	0.790	0.797	0.803	0.813	0.823	0.832	0.841	0.855	0.868
0.30000	(2.94)	0.753	0.755	0.760	0.765	0.769	0.777	0.784	0.791	0.802	0.811	0.821	0.830	0.845	0.857
0.40000	(3.92)	0.742	0.745	0.750	0.755	0.760	0.768	0.776	0.782	0.794	0.803	0.814	0.823	0.838	0.850
0.60000	(5.88)	0.728	0.731	0.737	0.743	0.748	0.756	0.764	0.771	0.783	0.793	0.804	0.813	0.828	0.841
0.80000	(7.85)	0.719	0.722	0.728	0.734	0.739	0.748	0.756	0.764	0.776	0.786	0.797	0.806	0.822	0.835
1.00000	(9.81)	0.712	0.715	0.722	0.728	0.733	0.742	0.751	0.758	0.770	0.781	0.792	0.801	0.817	0.831
1.50000	(14.7)	0.699	0.703	0.710	0.716	0.722	0.732	0.741	0.748	0.761	0.772	0.783	0.793	0.809	0.823
2.00000	(19.6)	0.691	0.695	0.702	0.709	0.715	0.725	0.734	0.742	0.755	0.766	0.777	0.787	0.804	0.818
3.00000	(29.4)	0.679	0.684	0.692	0.699	0.705	0.716	0.725	0.733	0.746	0.758	0.770	0.780	0.797	0.811
S	$(\Delta p/\rho l)$	125	150	200	250	300	400	500	600	800	1000	1250	1500	2000	2500

Grad'nt k.p.g. \(Equivalent\) Pipe diameters in mm

Kinematic viscosity, $\nu = 2.5 \times 10^{-6}\ m^2 s^{-1}$;   **Roughness size, $k_s = 0.0015$ mm**

E138

Kin. visc., $\nu = 2.5 \times 10^{-6}$ m^{2}s^{-1};
$S = 0.00010$ to 30.0000
i.e. kin. pr. grad., $\Delta p/\rho l =$ (0.00098) to (294) ms^{-2}

Roughness size, $k_s = 0.003$ mm
This table shows values of m, as follows
m_C for Colebrook-White solutions; or,
where $R \le 2000$, m_P for laminar flow

Grad'nt S	k.p.g. $(\Delta p/\rho l)$	6	8	10	12.5	15	20	25	30	40	50	60	80	100	125
0.00030	(0.00294)	17.14	11.68	8.676	6.443	5.053	3.443	2.557	2.005	1.366	1.015	*	1.205	1.180	1.160
0.00040	(0.00392)	14.85	10.12	7.513	5.580	4.376	2.982	2.214	1.736	1.183	*	1.211	1.175	1.153	1.134
0.00060	(0.00588)	12.12	8.260	6.135	4.556	3.573	2.435	1.808	1.418	0.966	*	1.166	1.135	1.116	1.100
0.00080	(0.00785)	10.50	7.154	5.313	3.946	3.094	2.108	1.566	1.228	*	1.159	1.137	1.109	1.091	1.076
0.00100	(0.00981)	9.390	6.398	4.752	3.529	2.767	1.886	1.400	1.098	*	1.135	1.115	1.089	1.072	1.059
0.00150	(0.0147)	7.667	5.224	3.880	2.881	2.260	1.540	1.143	0.897	1.121	1.095	1.078	1.055	1.040	1.029
0.00200	(0.0196)	6.640	4.524	3.360	2.495	1.957	1.333	0.990	*	1.092	1.068	1.052	1.032	1.019	1.009
0.00300	(0.0294)	5.421	3.694	2.743	2.037	1.598	1.089	*	*	1.053	1.033	1.019	1.001	0.990	0.982
0.00400	(0.0392)	4.695	3.199	2.376	1.764	1.384	0.943	*	1.057	1.028	1.009	0.997	0.980	0.971	0.963
0.00600	(0.0588)	3.833	2.612	1.940	1.441	1.130	*	1.039	1.019	0.994	0.978	0.967	0.953	0.945	0.938
0.00800	(0.0785)	3.320	2.262	1.680	1.248	0.978	*	1.012	0.994	0.971	0.956	0.946	0.934	0.927	0.922
0.01000	(0.0981)	2.969	2.023	1.503	1.116	0.875	*	0.992	0.976	0.954	0.940	0.931	0.920	0.914	0.909
0.01500	(0.147)	2.424	1.652	1.227	0.911	*	0.979	0.958	0.943	0.925	0.913	0.905	0.896	0.891	0.887
0.02000	(0.196)	2.100	1.431	1.063	0.789	*	0.954	0.935	0.922	0.905	0.895	0.888	0.879	0.875	0.872
0.03000	(0.294)	1.714	1.168	0.868	*	0.948	0.922	0.905	0.893	0.879	0.870	0.864	0.857	0.854	0.852
0.04000	(0.392)	1.485	1.012	0.751	*	0.924	0.900	0.885	0.874	0.861	0.853	0.848	0.843	0.840	0.839
0.06000	(0.588)	1.212	0.826	*	0.908	0.892	0.871	0.858	0.849	0.837	0.831	0.827	0.823	0.821	0.820
0.08000	(0.785)	1.050	*	*	0.885	0.871	0.851	0.839	0.831	0.822	0.816	0.812	0.809	0.808	0.808
0.10000	(0.981)	0.939	*	0.888	0.868	0.855	0.837	0.826	0.819	0.810	0.805	0.802	0.799	0.798	0.799
0.15000	(1.47)	0.767	*	0.857	0.839	0.827	0.812	0.803	0.796	0.789	0.785	0.783	0.781	0.781	0.782
0.20000	(1.96)	*	0.855	0.836	0.820	0.809	0.795	0.787	0.781	0.775	0.772	0.770	0.769	0.770	0.771
0.30000	(2.94)	*	0.825	0.808	0.794	0.785	0.773	0.766	0.761	0.756	0.754	0.753	0.753	0.754	0.757
0.40000	(3.92)	0.829	0.805	0.789	0.777	0.769	0.758	0.752	0.748	0.744	0.742	0.742	0.742	0.744	0.747
0.60000	(5.88)	0.799	0.778	0.765	0.754	0.747	0.738	0.733	0.730	0.727	0.726	0.726	0.728	0.730	0.733
0.80000	(7.85)	0.779	0.760	0.748	0.739	0.732	0.725	0.720	0.718	0.716	0.715	0.716	0.718	0.721	0.724
1.00000	(9.81)	0.765	0.747	0.736	0.727	0.722	0.715	0.711	0.709	0.707	0.707	0.708	0.711	0.714	0.718
1.50000	(14.7)	0.740	0.725	0.715	0.708	0.703	0.697	0.695	0.693	0.693	0.693	0.695	0.698	0.702	0.706
2.00000	(19.6)	0.723	0.710	0.701	0.695	0.690	0.686	0.684	0.683	0.683	0.684	0.686	0.690	0.694	0.699
3.00000	(29.4)	0.702	0.690	0.683	0.677	0.674	0.670	0.669	0.669	0.670	0.672	0.674	0.678	0.683	0.688
4.00000	(39.2)	0.687	0.676	0.670	0.666	0.663	0.660	0.659	0.659	0.661	0.663	0.666	0.671	0.676	0.682
6.00000	(58.8)	0.668	0.659	0.654	0.650	0.648	0.646	0.646	0.647	0.650	0.652	0.655	0.661	0.667	0.673
8.00000	(78.5)	0.655	0.647	0.643	0.640	0.638	0.637	0.638	0.639	0.642	0.645	0.649	0.655	0.661	0.667
10.0000	(98.1)	0.645	0.638	0.635	0.632	0.631	0.631	0.631	0.633	0.636	0.640	0.643	0.650	0.656	0.663
15.0000	(147)	0.629	0.624	0.621	0.619	0.619	0.619	0.621	0.623	0.627	0.631	0.635	0.642	0.649	0.656
20.0000	(196)	0.618	0.614	0.612	0.611	0.610	0.612	0.614	0.616	0.620	0.625	0.629	0.635	0.642	0.649
30.0000	(294)	0.604	0.601	0.599	0.599	0.600	0.602	0.604	0.607	0.612	0.617	0.622	0.630	0.637	0.645
S	$(\Delta p/\rho l)$	6	8	10	12.5	15	20	25	30	40	50	60	80	100	125

Grad'nt k.p.g. (Equivalent) Pipe diameters in mm

Grad'nt S	k.p.g. $(\Delta p/\rho l)$	125	150	200	250	300	400	500	600	800	1000	1250	1500	2000	2500
0.00010	(0.00098)	1.271	1.250	1.224	1.208	1.197	1.183	1.176	1.171	1.166	1.165	1.165	1.166	1.170	1.175
0.00015	(0.00147)	1.227	1.210	1.187	1.173	1.163	1.152	1.146	1.142	1.139	1.139	1.140	1.142	1.147	1.153
0.00020	(0.00196)	1.199	1.182	1.162	1.149	1.141	1.131	1.126	1.123	1.121	1.121	1.123	1.126	1.131	1.137
0.00030	(0.00294)	1.160	1.146	1.128	1.118	1.111	1.103	1.099	1.097	1.096	1.097	1.100	1.103	1.110	1.116
0.00040	(0.00392)	1.134	1.122	1.106	1.096	1.090	1.083	1.080	1.079	1.079	1.081	1.084	1.088	1.095	1.101
0.00060	(0.00588)	1.100	1.089	1.075	1.067	1.063	1.057	1.055	1.055	1.056	1.059	1.063	1.067	1.074	1.082
0.00080	(0.00785)	1.076	1.067	1.055	1.048	1.044	1.040	1.038	1.039	1.041	1.044	1.048	1.052	1.061	1.068
0.00100	(0.00981)	1.059	1.050	1.039	1.033	1.030	1.026	1.026	1.026	1.029	1.032	1.037	1.041	1.050	1.058
0.00150	(0.0147)	1.029	1.022	1.013	1.008	1.005	1.003	1.003	1.004	1.008	1.012	1.017	1.022	1.032	1.040
0.00200	(0.0196)	1.009	1.002	0.994	0.990	0.988	0.987	0.988	0.990	0.994	0.998	1.004	1.009	1.019	1.028
0.00300	(0.0294)	0.982	0.976	0.970	0.967	0.966	0.966	0.967	0.970	0.974	0.980	0.986	0.991	1.002	1.011
0.00400	(0.0392)	0.963	0.958	0.953	0.951	0.950	0.951	0.953	0.956	0.961	0.967	0.973	0.979	0.990	0.999
0.00600	(0.0588)	0.938	0.935	0.931	0.930	0.930	0.931	0.934	0.937	0.943	0.949	0.956	0.963	0.974	0.984
0.00800	(0.0785)	0.922	0.918	0.916	0.915	0.916	0.918	0.921	0.925	0.931	0.938	0.945	0.952	0.963	0.973
0.01000	(0.0981)	0.909	0.906	0.904	0.904	0.905	0.908	0.911	0.915	0.922	0.929	0.936	0.943	0.955	0.965
0.01500	(0.147)	0.887	0.885	0.884	0.885	0.886	0.890	0.894	0.898	0.906	0.913	0.921	0.928	0.940	0.951
0.02000	(0.196)	0.872	0.871	0.871	0.872	0.874	0.878	0.882	0.887	0.895	0.902	0.911	0.918	0.931	0.942
0.03000	(0.294)	0.852	0.852	0.852	0.854	0.857	0.862	0.867	0.871	0.880	0.888	0.897	0.904	0.917	0.929
0.04000	(0.392)	0.839	0.838	0.840	0.842	0.845	0.850	0.856	0.861	0.870	0.878	0.887	0.895	0.908	0.920
0.06000	(0.588)	0.820	0.821	0.823	0.826	0.829	0.835	0.841	0.847	0.856	0.865	0.874	0.882	0.896	0.908
0.08000	(0.785)	0.808	0.809	0.812	0.815	0.818	0.825	0.831	0.837	0.847	0.856	0.865	0.874	0.888	0.900
0.10000	(0.981)	0.799	0.800	0.803	0.807	0.810	0.817	0.824	0.830	0.840	0.849	0.859	0.867	0.882	0.894
0.15000	(1.47)	0.782	0.784	0.788	0.792	0.796	0.804	0.811	0.817	0.828	0.837	0.848	0.857	0.872	0.884
0.20000	(1.96)	0.771	0.773	0.778	0.783	0.787	0.795	0.802	0.809	0.820	0.830	0.840	0.849	0.865	0.877
0.30000	(2.94)	0.757	0.759	0.764	0.770	0.774	0.783	0.791	0.797	0.809	0.819	0.830	0.839	0.855	0.869
0.40000	(3.92)	0.747	0.750	0.755	0.761	0.766	0.775	0.783	0.790	0.802	0.812	0.823	0.833	0.849	0.863
0.60000	(5.88)	0.733	0.737	0.743	0.749	0.754	0.764	0.772	0.780	0.792	0.803	0.815	0.825	0.841	0.855
0.80000	(7.85)	0.724	0.728	0.735	0.741	0.747	0.757	0.765	0.773	0.786	0.797	0.809	0.819	0.836	0.850
1.00000	(9.81)	0.718	0.722	0.729	0.735	0.741	0.751	0.760	0.768	0.781	0.793	0.805	0.815	0.832	0.846
1.50000	(14.7)	0.706	0.710	0.718	0.725	0.731	0.742	0.752	0.760	0.774	0.785	0.797	0.808	0.826	0.840
2.00000	(19.6)	0.699	0.703	0.711	0.719	0.725	0.736	0.746	0.754	0.768	0.780	0.793	0.804	0.822	0.837
3.00000	(29.4)	0.688	0.693	0.702	0.710	0.717	0.728	0.738	0.747	0.762	0.774	0.787	0.798	0.817	0.832
S	$(\Delta p/\rho l)$	125	150	200	250	300	400	500	600	800	1000	1250	1500	2000	2500

Grad'nt k.p.g. (Equivalent) Pipe diameters in mm

Kinematic viscosity, $\nu = 2.5 \times 10^{-6}$ m^{2}s^{-1};

Roughness size, $k_s = 0.003$ mm

Kin. visc., $\nu = 2.5 \times 10^{-6}$ m^2 s^{-1};
$S = 0.00010$ to 30.0000

i.e. kin. pr. grad., $\Delta p/\rho l =$
(0.00098) to (294) ms^{-2}

Roughness size, $k_s = 0.006$ mm
This table shows values of m, as follows
m_C for Colebrook-White solutions; or,
where $\mathbf{R} \leq 2000$, m_P for laminar flow

E139

Grad'nt S	k.p.g. $(\Delta p/\rho l)$	6	8	10	12·5	15	20	25	30	40	50	60	80	100	125
		(Equivalent) Pipe diameters in mm													
0·00030	(0·00294)	17·14	11·68	8·676	6·443	5·053	3·443	2·557	2·005	1·366	1·015	*	1·206	1·181	1·161
0·00040	(0·00392)	14·85	10·12	7·513	5·580	4·376	2·982	2·214	1·736	1·183	*	1·211	1·176	1·153	1·135
0·00060	(0·00588)	12·12	8·260	6·135	4·556	3·573	2·435	1·808	1·418	0·966	*	1·167	1·136	1·117	1·101
0·00080	(0·00785)	10·50	7·154	5·313	3·946	3·094	2·108	1·566	1·228	*	1·159	1·138	1·109	1·092	1·077
0·00100	(0·00981)	9·390	6·398	4·752	3·529	2·767	1·886	1·400	1·098	*	1·136	1·116	1·090	1·073	1·060
0·00150	(0·0147)	7·667	5·224	3·880	2·881	2·260	1·540	1·143	0·897	1·122	1·096	1·079	1·056	1·042	1·030
0·00200	(0·0196)	6·640	4·524	3·360	2·495	1·957	1·333	0·990	*	1·093	1·070	1·053	1·033	1·020	1·010
0·00300	(0·0294)	5·421	3·694	2·743	2·037	1·598	1·089	*	*	1·054	1·034	1·020	1·002	0·991	0·983
0·00400	(0·0392)	4·695	3·199	2·376	1·764	1·384	0·943	*	1·058	1·029	1·011	0·998	0·982	0·972	0·965
0·00600	(0·0588)	3·833	2·612	1·940	1·441	1·130	*	1·040	1·021	0·995	0·979	0·968	0·954	0·946	0·940
0·00800	(0·0785)	3·320	2·262	1·680	1·248	0·978	*	1·014	0·996	0·972	0·958	0·948	0·936	0·929	0·924
0·01000	(0·0981)	2·969	2·023	1·503	1·116	0·875	*	0·994	0·977	0·956	0·942	0·933	0·922	0·916	0·911
0·01500	(0·147)	2·424	1·652	1·227	0·911	*	0·981	0·960	0·945	0·927	0·915	0·907	0·898	0·893	0·890
0·02000	(0·196)	2·100	1·431	1·063	0·789	*	0·956	0·937	0·924	0·907	0·897	0·890	0·882	0·878	0·875
0·03000	(0·294)	1·714	1·168	0·868	*	0·950	0·924	0·907	0·896	0·881	0·872	0·867	0·860	0·857	0·856
0·04000	(0·392)	1·485	1·012	0·751	*	0·926	0·902	0·887	0·877	0·864	0·856	0·851	0·846	0·843	0·842
0·06000	(0·588)	1·212	0·826	*	0·911	0·895	0·873	0·860	0·852	0·841	0·834	0·830	0·826	0·825	0·825
0·08000	(0·785)	1·050	*	*	0·888	0·873	0·854	0·843	0·835	0·825	0·820	0·816	0·813	0·812	0·813
0·10000	(0·981)	0·939	*	0·891	0·871	0·858	0·840	0·829	0·822	0·813	0·809	0·806	0·803	0·803	0·804
0·15000	(1·47)	0·767	*	0·860	0·843	0·831	0·816	0·806	0·800	0·793	0·790	0·788	0·786	0·787	0·788
0·20000	(1·96)	*	0·858	0·839	0·824	0·813	0·799	0·791	0·786	0·780	0·777	0·775	0·775	0·776	0·778
0·30000	(2·94)	*	0·829	0·812	0·798	0·789	0·778	0·771	0·767	0·762	0·760	0·759	0·760	0·761	0·764
0·40000	(3·92)	*	0·809	0·794	0·782	0·773	0·763	0·757	0·754	0·750	0·749	0·748	0·750	0·752	0·755
0·60000	(5·88)	0·804	0·783	0·770	0·759	0·752	0·744	0·739	0·736	0·734	0·733	0·734	0·736	0·739	0·743
0·80000	(7·85)	0·785	0·766	0·754	0·745	0·738	0·731	0·727	0·725	0·723	0·723	0·724	0·727	0·730	0·735
1·00000	(9·81)	0·770	0·753	0·742	0·734	0·728	0·722	0·718	0·716	0·715	0·716	0·717	0·721	0·724	0·729
1·50000	(14·7)	0·746	0·731	0·722	0·715	0·710	0·705	0·703	0·702	0·702	0·703	0·705	0·709	0·714	0·719
2·00000	(19·6)	0·730	0·717	0·709	0·703	0·699	0·694	0·693	0·692	0·693	0·695	0·697	0·702	0·707	0·712
3·00000	(29·4)	0·709	0·698	0·691	0·686	0·683	0·680	0·679	0·680	0·681	0·684	0·687	0·692	0·698	0·704
4·00000	(39·2)	0·695	0·685	0·680	0·675	0·673	0·671	0·671	0·671	0·674	0·677	0·680	0·686	0·692	0·698
6·00000	(58·8)	0·677	0·669	0·664	0·661	0·659	0·659	0·659	0·660	0·664	0·667	0·671	0·678	0·684	0·691
8·00000	(78·5)	0·665	0·658	0·654	0·652	0·651	0·651	0·652	0·653	0·657	0·661	0·665	0·673	0·679	0·687
10·0000	(98·1)	0·657	0·650	0·647	0·645	0·644	0·645	0·646	0·648	0·652	0·657	0·661	0·669	0·676	0·683
15·0000	(147)	0·642	0·637	0·634	0·633	0·633	0·635	0·637	0·639	0·645	0·650	0·654	0·663	0·670	0·678
20·0000	(196)	0·632	0·628	0·626	0·626	0·626	0·628	0·631	0·634	0·640	0·645	0·650	0·659	0·666	0·675
30·0000	(294)	0·620	0·617	0·616	0·616	0·617	0·620	0·624	0·627	0·633	0·639	0·644	0·654	0·662	0·670
S	$(\Delta p/\rho l)$	6	8	10	12·5	15	20	25	30	40	50	60	80	100	125
Grad'nt	k.p.g.	(Equivalent) Pipe diameters in mm													

Grad'nt S	k.p.g. $(\Delta p/\rho l)$	125	150	200	250	300	400	500	600	800	1000	1250	1500	2000	2500
0·00010	(0·00098)	1·271	1·251	1·225	1·208	1·197	1·184	1·176	1·172	1·167	1·166	1·166	1·167	1·171	1·176
0·00015	(0·00147)	1·228	1·210	1·187	1·173	1·164	1·153	1·147	1·143	1·140	1·140	1·141	1·143	1·148	1·154
0·00020	(0·00196)	1·199	1·183	1·162	1·150	1·142	1·132	1·127	1·124	1·122	1·122	1·124	1·127	1·132	1·138
0·00030	(0·00294)	1·161	1·147	1·129	1·118	1·111	1·104	1·100	1·098	1·097	1·098	1·101	1·104	1·111	1·118
0·00040	(0·00392)	1·135	1·122	1·106	1·097	1·091	1·084	1·081	1·080	1·080	1·082	1·085	1·089	1·096	1·103
0·00060	(0·00588)	1·101	1·090	1·076	1·068	1·064	1·059	1·057	1·056	1·058	1·060	1·064	1·068	1·076	1·084
0·00080	(0·00785)	1·077	1·068	1·056	1·049	1·045	1·041	1·040	1·040	1·042	1·045	1·050	1·054	1·063	1·071
0·00100	(0·00981)	1·060	1·051	1·040	1·034	1·031	1·028	1·027	1·028	1·030	1·034	1·039	1·043	1·052	1·061
0·00150	(0·0147)	1·030	1·023	1·014	1·009	1·007	1·005	1·005	1·006	1·010	1·014	1·020	1·025	1·034	1·043
0·00200	(0·0196)	1·010	1·003	0·996	0·992	0·990	0·989	0·990	0·992	0·996	1·001	1·006	1·012	1·022	1·031
0·00300	(0·0294)	0·983	0·978	0·972	0·969	0·968	0·968	0·970	0·972	0·977	0·982	0·989	0·995	1·005	1·015
0·00400	(0·0392)	0·965	0·960	0·955	0·953	0·953	0·954	0·956	0·958	0·964	0·970	0·977	0·983	0·994	1·004
0·00600	(0·0588)	0·940	0·937	0·933	0·932	0·932	0·934	0·937	0·940	0·947	0·953	0·960	0·967	0·979	0·989
0·00800	(0·0785)	0·924	0·921	0·918	0·918	0·918	0·921	0·924	0·928	0·935	0·942	0·949	0·956	0·968	0·979
0·01000	(0·0981)	0·911	0·909	0·907	0·907	0·908	0·911	0·915	0·919	0·926	0·933	0·941	0·948	0·961	0·971
0·01500	(0·147)	0·890	0·888	0·887	0·888	0·890	0·894	0·898	0·903	0·911	0·918	0·927	0·934	0·947	0·958
0·02000	(0·196)	0·875	0·874	0·874	0·875	0·877	0·882	0·887	0·892	0·900	0·908	0·917	0·925	0·938	0·949
0·03000	(0·294)	0·856	0·855	0·856	0·858	0·861	0·866	0·872	0·877	0·886	0·895	0·904	0·912	0·926	0·938
0·04000	(0·392)	0·842	0·842	0·844	0·847	0·850	0·856	0·862	0·867	0·877	0·885	0·895	0·903	0·918	0·930
0·06000	(0·588)	0·825	0·825	0·828	0·831	0·835	0·842	0·848	0·854	0·864	0·873	0·883	0·892	0·907	0·920
0·08000	(0·785)	0·813	0·814	0·817	0·821	0·825	0·832	0·839	0·845	0·856	0·865	0·875	0·885	0·900	0·913
0·10000	(0·981)	0·804	0·805	0·809	0·813	0·817	0·825	0·832	0·838	0·849	0·859	0·870	0·879	0·895	0·908
0·15000	(1·47)	0·788	0·790	0·795	0·800	0·804	0·813	0·820	0·827	0·839	0·849	0·860	0·870	0·886	0·899
0·20000	(1·96)	0·778	0·780	0·786	0·791	0·796	0·804	0·812	0·819	0·832	0·842	0·853	0·863	0·880	0·894
0·30000	(2·94)	0·764	0·767	0·773	0·779	0·784	0·794	0·802	0·810	0·822	0·833	0·845	0·855	0·873	0·887
0·40000	(3·92)	0·755	0·758	0·765	0·771	0·777	0·787	0·795	0·803	0·816	0·828	0·840	0·850	0·868	0·882
0·60000	(5·88)	0·743	0·747	0·754	0·761	0·767	0·777	0·786	0·795	0·809	0·820	0·833	0·843	0·862	0·877
0·80000	(7·85)	0·735	0·739	0·747	0·754	0·760	0·771	0·781	0·789	0·803	0·815	0·828	0·839	0·858	0·873
1·00000	(9·81)	0·729	0·733	0·741	0·749	0·755	0·767	0·777	0·785	0·800	0·812	0·825	0·836	0·855	0·870
1·50000	(14·7)	0·719	0·724	0·733	0·740	0·747	0·759	0·770	0·779	0·794	0·806	0·820	0·831	0·850	0·866
2·00000	(19·6)	0·712	0·717	0·727	0·735	0·742	0·755	0·765	0·774	0·790	0·803	0·816	0·828	0·847	0·863
3·00000	(29·4)	0·704	0·709	0·719	0·728	0·735	0·749	0·759	0·769	0·785	0·798	0·812	0·824	0·844	0·860
S	$(\Delta p/\rho l)$	125	150	200	250	300	400	500	600	800	1000	1250	1500	2000	2500
Grad'nt	k.p.g.	(Equivalent) Pipe diameters in mm													

Kinematic viscosity, $\nu = 2.5 \times 10^{-6}$ m^2 s^{-1} ;

Roughness size, $k_s = 0.006$ mm

Kin. visc., $\nu = 2{\cdot}5\times10^{-6}$ m^{2}s^{-1};
$S = 0{\cdot}00010$ to $30{\cdot}0000$

i.e. kin. pr. grad., $\Delta p/\rho l =$
$(0{\cdot}00098)$ to (294) ms^{-2}

Roughness size, $k_s = 0{\cdot}015$ mm
This table shows values of m, as follows

m_C for Colebrook-White solutions; or,
where $\mathbf{R} \le 2000$, m_P for laminar flow

Grad'nt S	k.p.g. $(\Delta p/\rho l)$	(Equivalent) Pipe diameters in mm													
		6	8	10	12·5	15	20	25	30	40	50	60	80	100	125
0·00030	(0·00294)	17·14	11·68	8·676	6·443	5·053	3·443	2·557	2·005	1·366	1·015	*	1·207	1·183	1·163
0·00040	(0·00392)	14·85	10·12	7·513	5·580	4·376	2·982	2·214	1·736	1·183	*	1·213	1·178	1·155	1·137
0·00060	(0·00588)	12·12	8·260	6·135	4·556	3·573	2·435	1·808	1·418	0·966	*	1·169	1·138	1·119	1·103
0·00080	(0·00785)	10·50	7·154	5·313	3·946	3·094	2·108	1·566	1·228	*	1·162	1·140	1·112	1·094	1·080
0·00100	(0·00981)	9·390	6·398	4·752	3·529	2·767	1·886	1·400	1·098	*	1·139	1·118	1·092	1·076	1·063
0·00150	(0·0147)	7·667	5·224	3·880	2·881	2·260	1·540	1·143	0·897	1·124	1·099	1·081	1·059	1·045	1·034
0·00200	(0·0196)	6·640	4·524	3·360	2·495	1·957	1·333	0·990	*	1·096	1·073	1·057	1·036	1·024	1·014
0·00300	(0·0294)	5·421	3·694	2·743	2·037	1·598	1·089	*	*	1·058	1·038	1·024	1·006	0·996	0·987
0·00400	(0·0392)	4·695	3·199	2·376	1·764	1·384	0·943	*	1·062	1·033	1·014	1·002	0·986	0·977	0·969
0·00600	(0·0588)	3·833	2·612	1·940	1·441	1·130	*	1·044	1·025	0·999	0·983	0·973	0·959	0·951	0·946
0·00800	(0·0785)	3·320	2·262	1·680	1·248	0·978	*	1·018	1·000	0·977	0·963	0·953	0·941	0·934	0·930
0·01000	(0·0981)	2·969	2·023	1·503	1·116	0·875	*	0·998	0·982	0·961	0·947	0·939	0·928	0·922	0·918
0·01500	(0·147)	2·424	1·652	1·227	0·911	*	0·986	0·965	0·951	0·932	0·921	0·914	0·905	0·900	0·897
0·02000	(0·196)	2·100	1·431	1·063	0·789	*	0·962	0·943	0·930	0·913	0·903	0·897	0·889	0·886	0·883
0·03000	(0·294)	1·714	1·168	0·868	*	0·957	0·930	0·914	0·903	0·888	0·880	0·875	0·869	0·866	0·865
0·04000	(0·392)	1·485	1·012	0·751	*	0·933	0·909	0·894	0·884	0·872	0·864	0·860	0·855	0·853	0·853
0·06000	(0·588)	1·212	0·826	*	0·918	0·902	0·881	0·869	0·860	0·850	0·844	0·840	0·837	0·836	0·837
0·08000	(0·785)	1·050	*	*	0·896	0·882	0·863	0·852	0·844	0·835	0·830	0·827	0·825	0·825	0·826
0·10000	(0·981)	0·939	*	0·900	0·880	0·867	0·849	0·839	0·832	0·824	0·820	0·818	0·816	0·816	0·818
0·15000	(1·47)	0·767	*	0·870	0·853	0·841	0·826	0·817	0·812	0·805	0·802	0·801	0·801	0·802	0·804
0·20000	(1·96)	*	0·869	0·850	0·834	0·824	0·811	0·803	0·798	0·793	0·791	0·790	0·791	0·792	0·795
0·30000	(2·94)	*	0·840	0·824	0·811	0·802	0·791	0·785	0·781	0·777	0·776	0·776	0·777	0·780	0·784
0·40000	(3·92)	*	0·821	0·807	0·795	0·787	0·777	0·772	0·769	0·766	0·766	0·766	0·769	0·772	0·776
0·60000	(5·88)	0·817	0·797	0·784	0·774	0·768	0·760	0·756	0·754	0·753	0·753	0·754	0·758	0·762	0·767
0·80000	(7·85)	0·799	0·781	0·770	0·761	0·755	0·749	0·745	0·744	0·743	0·745	0·746	0·750	0·755	0·760
1·00000	(9·81)	0·786	0·769	0·759	0·751	0·746	0·740	0·738	0·737	0·737	0·738	0·741	0·745	0·750	0·756
1·50000	(14·7)	0·764	0·750	0·741	0·734	0·730	0·726	0·725	0·725	0·726	0·728	0·731	0·737	0·742	0·748
2·00000	(19·6)	0·749	0·737	0·729	0·724	0·720	0·717	0·716	0·717	0·719	0·722	0·725	0·731	0·737	0·744
3·00000	(29·4)	0·731	0·720	0·714	0·710	0·707	0·706	0·706	0·707	0·710	0·713	0·717	0·724	0·731	0·738
4·00000	(39·2)	0·719	0·709	0·704	0·701	0·699	0·698	0·699	0·700	0·704	0·708	0·712	0·720	0·726	0·734
6·00000	(58·8)	0·703	0·695	0·691	0·689	0·688	0·688	0·690	0·692	0·697	0·701	0·706	0·714	0·721	0·730
8·00000	(78·5)	0·693	0·686	0·683	0·682	0·681	0·682	0·684	0·687	0·692	0·697	0·702	0·711	0·718	0·727
10·0000	(98·1)	0·686	0·680	0·678	0·676	0·676	0·678	0·680	0·683	0·689	0·694	0·699	0·708	0·716	0·725
15·0000	(147)	0·674	0·670	0·668	0·668	0·668	0·671	0·674	0·677	0·684	0·689	0·695	0·704	0·713	0·721
20·0000	(196)	0·666	0·663	0·662	0·662	0·663	0·666	0·670	0·673	0·680	0·686	0·692	0·702	0·710	0·720
30·0000	(294)	0·656	0·654	0·654	0·655	0·657	0·661	0·665	0·669	0·676	0·683	0·689	0·699	0·708	0·717
S	$(\Delta p/\rho l)$	6	8	10	12·5	15	20	25	30	40	50	60	80	100	125

Grad'nt   k.p.g.   (Equivalent) Pipe diameters in mm

Grad'nt S	k.p.g. $(\Delta p/\rho l)$	(Equivalent) Pipe diameters in mm													
		125	150	200	250	300	400	500	600	800	1000	1250	1500	2000	2500
0·00010	(0·00098)	1·273	1·252	1·226	1·210	1·199	1·186	1·178	1·174	1·169	1·168	1·168	1·170	1·174	1·179
0·00015	(0·00147)	1·230	1·212	1·189	1·175	1·166	1·155	1·149	1·145	1·143	1·142	1·144	1·146	1·152	1·157
0·00020	(0·00196)	1·201	1·185	1·164	1·152	1·144	1·134	1·129	1·126	1·125	1·125	1·127	1·130	1·136	1·142
0·00030	(0·00294)	1·163	1·149	1·131	1·121	1·114	1·106	1·102	1·101	1·100	1·102	1·105	1·108	1·115	1·122
0·00040	(0·00392)	1·137	1·124	1·109	1·099	1·094	1·087	1·084	1·083	1·084	1·086	1·090	1·093	1·101	1·108
0·00060	(0·00588)	1·103	1·092	1·079	1·071	1·067	1·062	1·060	1·060	1·062	1·065	1·069	1·073	1·082	1·090
0·00080	(0·00785)	1·080	1·070	1·059	1·052	1·048	1·045	1·044	1·044	1·047	1·050	1·055	1·060	1·069	1·077
0·00100	(0·00981)	1·063	1·054	1·044	1·038	1·035	1·032	1·031	1·032	1·035	1·039	1·044	1·050	1·059	1·068
0·00150	(0·0147)	1·034	1·026	1·018	1·013	1·011	1·009	1·010	1·011	1·016	1·020	1·026	1·032	1·042	1·052
0·00200	(0·0196)	1·014	1·007	1·000	0·996	0·995	0·994	0·995	0·997	1·002	1·008	1·014	1·020	1·031	1·041
0·00300	(0·0294)	0·987	0·982	0·976	0·974	0·973	0·974	0·976	0·979	0·984	0·990	0·997	1·004	1·016	1·026
0·00400	(0·0392)	0·969	0·965	0·960	0·959	0·959	0·960	0·963	0·966	0·972	0·979	0·986	0·993	1·005	1·016
0·00600	(0·0588)	0·946	0·942	0·939	0·939	0·939	0·942	0·945	0·949	0·956	0·963	0·971	0·979	0·992	1·003
0·00800	(0·0785)	0·930	0·927	0·925	0·925	0·926	0·929	0·933	0·938	0·946	0·953	0·961	0·969	0·983	0·994
0·01000	(0·0981)	0·918	0·916	0·914	0·915	0·916	0·920	0·925	0·929	0·938	0·945	0·954	0·962	0·976	0·988
0·01500	(0·147)	0·897	0·896	0·896	0·897	0·899	0·904	0·909	0·914	0·924	0·932	0·942	0·950	0·965	0·977
0·02000	(0·196)	0·883	0·883	0·883	0·885	0·888	0·894	0·899	0·905	0·915	0·923	0·933	0·942	0·957	0·970
0·03000	(0·294)	0·865	0·865	0·867	0·870	0·873	0·880	0·886	0·892	0·903	0·912	0·922	0·932	0·947	0·961
0·04000	(0·392)	0·853	0·853	0·856	0·859	0·863	0·870	0·877	0·883	0·895	0·904	0·915	0·925	0·941	0·955
0·06000	(0·588)	0·837	0·838	0·842	0·846	0·850	0·858	0·866	0·872	0·884	0·895	0·906	0·916	0·933	0·947
0·08000	(0·785)	0·826	0·828	0·832	0·837	0·841	0·850	0·858	0·865	0·878	0·888	0·900	0·910	0·928	0·942
0·10000	(0·981)	0·818	0·820	0·825	0·830	0·835	0·844	0·852	0·860	0·873	0·884	0·896	0·906	0·924	0·939
0·15000	(1·47)	0·804	0·807	0·813	0·819	0·824	0·834	0·843	0·851	0·865	0·876	0·889	0·900	0·918	0·933
0·20000	(1·96)	0·795	0·799	0·805	0·812	0·817	0·828	0·837	0·845	0·859	0·871	0·884	0·895	0·914	0·929
0·30000	(2·94)	0·784	0·788	0·795	0·802	0·808	0·820	0·829	0·838	0·853	0·865	0·878	0·890	0·909	0·925
0·40000	(3·92)	0·776	0·781	0·789	0·796	0·803	0·814	0·825	0·833	0·849	0·861	0·875	0·887	0·906	0·922
0·60000	(5·88)	0·767	0·771	0·780	0·788	0·795	0·808	0·818	0·828	0·843	0·856	0·870	0·882	0·902	0·919
0·80000	(7·85)	0·760	0·766	0·775	0·783	0·791	0·804	0·814	0·824	0·840	0·853	0·868	0·880	0·900	0·917
1·00000	(9·81)	0·756	0·761	0·771	0·780	0·787	0·801	0·812	0·821	0·838	0·851	0·866	0·878	0·898	0·915
1·50000	(14·7)	0·748	0·754	0·765	0·774	0·782	0·796	0·807	0·817	0·834	0·848	0·863	0·875	0·896	0·913
2·00000	(19·6)	0·744	0·750	0·761	0·770	0·779	0·793	0·804	0·815	0·832	0·846	0·861	0·873	0·894	0·911
3·00000	(29·4)	0·738	0·744	0·756	0·766	0·774	0·789	0·801	0·811	0·829	0·843	0·858	0·871	0·892	0·910
S	$(\Delta p/\rho l)$	125	150	200	250	300	400	500	600	800	1000	1250	1500	2000	2500

Grad'nt   k.p.g.   (Equivalent) Pipe diameters in mm

Kinematic viscosity, $\nu = 2{\cdot}5\times10^{-6}$ m^{2}s^{-1} ;   **Roughness size, $k_s = 0{\cdot}015$ mm**

Kin. visc., $\nu = 2.5\times10^{-6}\ \mathrm{m^2\,s^{-1}}$;
$S = 0.00010$ to 30.0000

i.e. kin. pr. grad., $\Delta p/\rho l =$
(0.00098) to $(294)\ \mathrm{ms^{-2}}$

Roughness size, $k_s = 0.030$ mm
This table shows values of m, as follows

m_C for Colebrook-White solutions; or,
where $\mathbf{R} \leq 2000$, m_P for laminar flow

Grad'nt S	k.p.g. $(\Delta p/\rho l)$	(Equivalent) Pipe diameters in mm													
		6	8	10	12·5	15	20	25	30	40	50	60	80	100	125
0·00030	(0·00294)	17·14	11·68	8·676	6·443	5·053	3·443	2·557	2·005	1·366	1·015	*	1·210	1·186	1·166
0·00040	(0·00392)	14·85	10·12	7·513	5·580	4·376	2·982	2·214	1·736	1·183	*	1·216	1·181	1·159	1·140
0·00060	(0·00588)	12·12	8·260	6·135	4·556	3·573	2·435	1·808	1·418	0·966	*	1·173	1·142	1·123	1·107
0·00080	(0·00785)	10·50	7·154	5·313	3·946	3·094	2·108	1·566	1·228	*	1·166	1·144	1·116	1·099	1·084
0·00100	(0·00981)	9·390	6·398	4·752	3·529	2·767	1·886	1·400	1·098	*	1·143	1·123	1·097	1·081	1·068
0·00150	(0·0147)	7·667	5·224	3·880	2·881	2·260	1·540	1·143	0·897	1·129	1·104	1·086	1·064	1·050	1·039
0·00200	(0·0196)	6·640	4·524	3·360	2·495	1·957	1·333	0·990	*	1·101	1·078	1·062	1·042	1·029	1·020
0·00300	(0·0294)	5·421	3·694	2·743	2·037	1·598	1·089	*	*	1·063	1·043	1·030	1·012	1·002	0·994
0·00400	(0·0392)	4·695	3·199	2·376	1·764	1·384	0·943	*	1·068	1·039	1·021	1·008	0·993	0·984	0·977
0·00600	(0·0588)	3·833	2·612	1·940	1·441	1·130	*	1·051	1·031	1·006	0·991	0·980	0·967	0·960	0·954
0·00800	(0·0785)	3·320	2·262	1·680	1·248	0·978	*	1·025	1·007	0·985	0·971	0·961	0·950	0·944	0·939
0·01000	(0·0981)	2·969	2·023	1·503	1·116	0·875	*	1·006	0·990	0·969	0·956	0·947	0·937	0·932	0·928
0·01500	(0·147)	2·424	1·652	1·227	0·911	*	0·994	0·974	0·960	0·942	0·931	0·924	0·915	0·911	0·909
0·02000	(0·196)	2·100	1·431	1·063	0·789	*	0·971	0·952	0·940	0·923	0·914	0·908	0·901	0·898	0·896
0·03000	(0·294)	1·714	1·168	0·868	*	0·967	0·941	0·924	0·913	0·900	0·892	0·887	0·882	0·880	0·879
0·04000	(0·392)	1·485	1·012	0·751	*	0·944	0·920	0·906	0·896	0·884	0·877	0·873	0·869	0·868	0·869
0·06000	(0·588)	1·212	0·826	*	0·930	0·914	0·894	0·882	0·873	0·864	0·858	0·855	0·853	0·853	0·854
0·08000	(0·785)	1·050	*	*	0·909	0·895	0·877	0·866	0·859	0·850	0·846	0·844	0·842	0·843	0·845
0·10000	(0·981)	0·939	*	0·913	0·894	0·881	0·864	0·854	0·848	0·840	0·837	0·835	0·835	0·836	0·838
0·15000	(1·47)	0·767	*	0·885	0·868	0·857	0·843	0·834	0·829	0·824	0·821	0·821	0·821	0·824	0·827
0·20000	(1·96)	*	0·885	0·866	0·851	0·841	0·829	0·822	0·817	0·813	0·811	0·811	0·813	0·816	0·820
0·30000	(2·94)	*	0·858	0·842	0·829	0·821	0·811	0·805	0·802	0·799	0·799	0·799	0·802	0·806	0·810
0·40000	(3·92)	*	0·841	0·827	0·815	0·808	0·799	0·794	0·792	0·790	0·790	0·792	0·795	0·799	0·805
0·60000	(5·88)	0·839	0·819	0·806	0·797	0·791	0·784	0·780	0·779	0·778	0·780	0·782	0·786	0·791	0·797
0·80000	(7·85)	0·822	0·804	0·793	0·785	0·780	0·774	0·772	0·771	0·771	0·773	0·775	0·781	0·786	0·792
1·00000	(9·81)	0·810	0·794	0·784	0·777	0·772	0·767	0·765	0·765	0·766	0·768	0·771	0·777	0·783	0·789
1·50000	(14·7)	0·791	0·777	0·768	0·762	0·759	0·756	0·755	0·755	0·757	0·760	0·764	0·770	0·777	0·784
2·00000	(19·6)	0·778	0·765	0·758	0·753	0·750	0·748	0·748	0·749	0·752	0·755	0·759	0·766	0·773	0·781
3·00000	(29·4)	0·762	0·751	0·746	0·742	0·740	0·739	0·740	0·741	0·745	0·749	0·754	0·761	0·769	0·777
4·00000	(39·2)	0·751	0·742	0·738	0·735	0·733	0·733	0·734	0·736	0·741	0·746	0·750	0·758	0·766	0·774
6·00000	(58·8)	0·738	0·731	0·728	0·726	0·725	0·726	0·728	0·730	0·736	0·741	0·746	0·755	0·762	0·771
8·00000	(78·5)	0·730	0·724	0·721	0·720	0·720	0·721	0·724	0·727	0·732	0·738	0·743	0·752	0·760	0·769
10·0000	(98·1)	0·725	0·719	0·717	0·716	0·716	0·718	0·721	0·724	0·730	0·736	0·741	0·751	0·759	0·768
15·0000	(147)	0·715	0·711	0·710	0·710	0·710	0·713	0·717	0·720	0·727	0·733	0·738	0·748	0·757	0·766
20·0000	(196)	0·709	0·706	0·705	0·706	0·707	0·710	0·714	0·718	0·724	0·731	0·737	0·747	0·755	0·765
30·0000	(294)	0·702	0·700	0·700	0·701	0·702	0·706	0·710	0·714	0·722	0·728	0·734	0·745	0·754	0·763
S	$(\Delta p/\rho l)$	6	8	10	12·5	15	20	25	30	40	50	60	80	100	125

Grad'nt S	k.p.g. $(\Delta p/\rho l)$	(Equivalent) Pipe diameters in mm													
		125	150	200	250	300	400	500	600	800	1000	1250	1500	2000	2500
0·00010	(0·00098)	1·275	1·255	1·228	1·212	1·202	1·188	1·181	1·177	1·173	1·171	1·172	1·174	1·179	1·184
0·00015	(0·00147)	1·232	1·214	1·192	1·178	1·169	1·158	1·152	1·149	1·147	1·147	1·148	1·151	1·157	1·163
0·00020	(0·00196)	1·204	1·188	1·167	1·155	1·147	1·138	1·133	1·130	1·129	1·130	1·132	1·135	1·142	1·149
0·00030	(0·00294)	1·166	1·152	1·135	1·124	1·118	1·110	1·107	1·105	1·105	1·107	1·111	1·114	1·122	1·130
0·00040	(0·00392)	1·140	1·128	1·113	1·104	1·098	1·092	1·089	1·089	1·090	1·092	1·096	1·100	1·109	1·117
0·00060	(0·00588)	1·107	1·096	1·083	1·076	1·071	1·067	1·066	1·066	1·068	1·072	1·076	1·081	1·091	1·099
0·00080	(0·00785)	1·084	1·075	1·064	1·057	1·054	1·050	1·050	1·051	1·054	1·058	1·063	1·069	1·079	1·088
0·00100	(0·00981)	1·068	1·059	1·049	1·043	1·040	1·038	1·038	1·039	1·043	1·048	1·053	1·059	1·070	1·079
0·00150	(0·0147)	1·039	1·032	1·024	1·020	1·018	1·017	1·018	1·020	1·025	1·030	1·037	1·043	1·054	1·064
0·00200	(0·0196)	1·020	1·014	1·007	1·003	1·002	1·002	1·004	1·007	1·012	1·018	1·025	1·032	1·044	1·055
0·00300	(0·0294)	0·994	0·989	0·984	0·982	0·982	0·983	0·986	0·989	0·996	1·003	1·010	1·018	1·031	1·042
0·00400	(0·0392)	0·977	0·973	0·969	0·968	0·968	0·970	0·974	0·977	0·985	0·992	1·001	1·008	1·022	1·033
0·00600	(0·0588)	0·954	0·951	0·949	0·949	0·950	0·953	0·958	0·962	0·971	0·979	0·988	0·996	1·003	1·015
0·00800	(0·0785)	0·939	0·937	0·936	0·936	0·938	0·942	0·947	0·952	0·961	0·970	0·979	0·988	1·003	1·015
0·01000	(0·0981)	0·928	0·926	0·926	0·927	0·929	0·934	0·939	0·944	0·954	0·963	0·973	0·982	0·997	1·010
0·01500	(0·147)	0·909	0·908	0·909	0·911	0·914	0·920	0·926	0·932	0·942	0·952	0·963	0·972	0·988	1·002
0·02000	(0·196)	0·896	0·896	0·898	0·901	0·904	0·911	0·917	0·924	0·935	0·945	0·956	0·966	0·982	0·996
0·03000	(0·294)	0·879	0·880	0·883	0·887	0·891	0·899	0·906	0·913	0·925	0·936	0·947	0·957	0·975	0·989
0·04000	(0·392)	0·869	0·870	0·874	0·878	0·882	0·891	0·899	0·906	0·919	0·930	0·942	0·952	0·970	0·985
0·06000	(0·588)	0·854	0·856	0·861	0·866	0·871	0·881	0·889	0·897	0·911	0·922	0·935	0·946	0·964	0·980
0·08000	(0·785)	0·845	0·848	0·853	0·859	0·864	0·875	0·884	0·892	0·906	0·918	0·930	0·942	0·961	0·976
0·10000	(0·981)	0·838	0·841	0·847	0·854	0·859	0·870	0·879	0·888	0·902	0·914	0·927	0·939	0·958	0·974
0·15000	(1·47)	0·827	0·831	0·838	0·845	0·851	0·862	0·872	0·881	0·896	0·909	0·922	0·934	0·954	0·970
0·20000	(1·96)	0·820	0·824	0·832	0·839	0·846	0·858	0·868	0·877	0·892	0·905	0·919	0·931	0·951	0·968
0·30000	(2·94)	0·810	0·815	0·824	0·832	0·839	0·852	0·862	0·872	0·888	0·901	0·915	0·928	0·948	0·965
0·40000	(3·92)	0·805	0·810	0·819	0·827	0·835	0·848	0·859	0·869	0·885	0·899	0·913	0·925	0·946	0·963
0·60000	(5·88)	0·797	0·803	0·813	0·822	0·830	0·843	0·855	0·865	0·881	0·895	0·910	0·923	0·944	0·961
0·80000	(7·85)	0·792	0·798	0·809	0·818	0·826	0·840	0·852	0·862	0·879	0·893	0·908	0·921	0·942	0·960
1·00000	(9·81)	0·789	0·795	0·806	0·816	0·824	0·838	0·850	0·861	0·878	0·892	0·907	0·920	0·941	0·959
1·50000	(14·7)	0·784	0·790	0·802	0·812	0·820	0·835	0·847	0·858	0·875	0·890	0·905	0·918	0·940	0·957
2·00000	(19·6)	0·781	0·787	0·799	0·809	0·818	0·833	0·845	0·856	0·874	0·889	0·904	0·917	0·939	0·956
3·00000	(29·4)	0·777	0·784	0·796	0·806	0·815	0·831	0·843	0·854	0·872	0·887	0·903	0·916	0·938	0·955
S	$(\Delta p/\rho l)$	125	150	200	250	300	400	500	600	800	1000	1250	1500	2000	2500

Grad'nt   k.p.g.   (Equivalent) Pipe diameters in mm

Kinematic viscosity, $\nu = 2.5\times10^{-6}\ \mathrm{m^2\,s^{-1}}$;

Roughness size, $k_s = 0.030$ mm

E142

Kin. visc., $\nu = 2.5 \times 10^{-6}$ m^{2}s^{-1}; $S = 0.00010$ to 30.0000

i.e. kin. pr. grad., $\Delta p / \rho l =$ (0.00098) to (294) ms^{-2}

Roughness size, $k_s = 0.060$ mm

This table shows values of m, as follows

m_C for Colebrook-White solutions; or, where $\mathbf{R} \leq 2000$, m_P for laminar flow

Grad'nt S	k.p.g. $(\Delta p/\rho l)$	6	8	10	12·5	15	20	25	30	40	50	60	80	100	125
0·00030	(0·00294)	17·14	11·68	8·676	6·443	5·053	3·443	2·557	2·005	1·366	1·015	*	1·216	1·192	1·172
0·00040	(0·00392)	14·85	10·12	7·513	5·580	4·376	2·982	2·214	1·736	1·183	*	*	1·187	1·165	1·147
0·00060	(0·00588)	12·12	8·260	6·135	4·556	3·573	2·435	1·808	1·418	0·966	*	1·180	1·149	1·130	1·115
0·00080	(0·00785)	10·50	7·154	5·313	3·946	3·094	2·108	1·566	1·228	*	1·173	1·152	1·124	1·107	1·093
0·00100	(0·00981)	9·390	6·398	4·752	3·529	2·767	1·886	1·400	1·098	*	1·151	1·131	1·105	1·090	1·077
0·00150	(0·0147)	7·667	5·224	3·880	2·881	2·260	1·540	1·143	0·897	1·138	1·113	1·095	1·073	1·060	1·049
0·00200	(0·0196)	6·640	4·524	3·360	2·495	1·957	1·333	0·990	*	1·110	1·087	1·072	1·052	1·040	1·031
0·00300	(0·0294)	5·421	3·694	2·743	2·037	1·598	1·089	*	*	1·074	1·055	1·041	1·024	1·014	1·007
0·00400	(0·0392)	4·695	3·199	2·376	1·764	1·384	0·943	*	1·079	1·051	1·033	1·021	1·006	0·997	0·991
0·00600	(0·0588)	3·833	2·612	1·940	1·441	1·130	*	1·064	1·044	1·019	1·004	0·994	0·982	0·975	0·970
0·00800	(0·0785)	3·320	2·262	1·680	1·248	0·978	*	1·039	1·021	0·999	0·986	0·977	0·966	0·960	0·957
0·01000	(0·0981)	2·969	2·023	1·503	1·116	0·875	*	1·021	1·005	0·984	0·972	0·964	0·954	0·950	0·947
0·01500	(0·147)	2·424	1·652	1·227	0·911	*	1·011	0·990	0·976	0·959	0·949	0·942	0·935	0·931	0·930
0·02000	(0·196)	2·100	1·431	1·063	0·789	*	0·989	0·970	0·958	0·942	0·933	0·928	0·922	0·919	0·919
0·03000	(0·294)	1·714	1·168	0·868	*	0·986	0·960	0·944	0·934	0·921	0·914	0·909	0·905	0·904	0·905
0·04000	(0·392)	1·485	1·012	0·751	*	0·965	0·941	0·927	0·918	0·907	0·901	0·897	0·894	0·894	0·895
0·06000	(0·588)	1·212	0·826	*	0·953	0·937	0·917	0·905	0·898	0·889	0·884	0·882	0·881	0·882	0·884
0·08000	(0·785)	1·050	*	*	0·934	0·920	0·902	0·891	0·885	0·877	0·874	0·872	0·872	0·873	0·876
0·10000	(0·981)	0·939	*	0·939	0·920	0·907	0·891	0·881	0·875	0·869	0·866	0·865	0·866	0·868	0·871
0·15000	(1·47)	0·767	*	0·913	0·896	0·885	0·872	0·864	0·860	0·855	0·853	0·853	0·855	0·858	0·862
0·20000	(1·96)	*	0·915	0·896	0·882	0·872	0·860	0·853	0·850	0·846	0·845	0·846	0·849	0·852	0·857
0·30000	(2·94)	*	0·891	0·875	0·863	0·854	0·845	0·840	0·837	0·835	0·835	0·837	0·840	0·845	0·850
0·40000	(3·92)	*	0·876	0·862	0·850	0·843	0·835	0·831	0·829	0·828	0·829	0·831	0·835	0·840	0·846
0·60000	(5·88)	0·877	0·857	0·844	0·835	0·829	0·823	0·820	0·819	0·819	0·821	0·823	0·829	0·834	0·841
0·80000	(7·85)	0·863	0·844	0·834	0·825	0·820	0·815	0·813	0·813	0·814	0·816	0·819	0·825	0·831	0·838
1·00000	(9·81)	0·852	0·836	0·826	0·819	0·814	0·810	0·808	0·808	0·810	0·813	0·816	0·822	0·828	0·835
1·50000	(14·7)	0·836	0·822	0·813	0·807	0·804	0·801	0·800	0·801	0·804	0·807	0·811	0·818	0·824	0·832
2·00000	(19·6)	0·825	0·813	0·805	0·800	0·798	0·795	0·796	0·797	0·800	0·804	0·808	0·815	0·822	0·830
3·00000	(29·4)	0·812	0·802	0·796	0·792	0·790	0·789	0·790	0·791	0·795	0·799	0·804	0·812	0·819	0·827
4·00000	(39·2)	0·804	0·795	0·790	0·786	0·785	0·785	0·786	0·788	0·792	0·797	0·801	0·810	0·817	0·826
6·00000	(58·8)	0·794	0·786	0·782	0·780	0·779	0·780	0·781	0·784	0·789	0·794	0·799	0·807	0·815	0·824
8·00000	(78·5)	0·788	0·781	0·778	0·776	0·775	0·776	0·779	0·781	0·787	0·792	0·797	0·806	0·814	0·823
10·0000	(98·1)	0·784	0·778	0·775	0·773	0·773	0·774	0·777	0·780	0·785	0·791	0·796	0·805	0·813	0·822
15·0000	(147)	0·777	0·772	0·769	0·769	0·769	0·771	0·774	0·777	0·783	0·789	0·794	0·803	0·812	0·821
20·0000	(196)	0·773	0·768	0·766	0·766	0·766	0·769	0·772	0·775	0·782	0·787	0·793	0·802	0·811	0·820
30·0000	(294)	0·768	0·764	0·763	0·763	0·764	0·766	0·770	0·773	0·780	0·786	0·791	0·801	0·810	0·819
S	$(\Delta p/\rho l)$	6	8	10	12·5	15	20	25	30	40	50	60	80	100	125

Grad'nt k.p.g. (Equivalent) Pipe diameters in mm

Grad'nt S	k.p.g. $(\Delta p/\rho l)$	125	150	200	250	300	400	500	600	800	1000	1250	1500	2000	2500
0·00010	(0·00098)	1·279	1·259	1·233	1·217	1·207	1·194	1·187	1·183	1·179	1·179	1·180	1·182	1·187	1·193
0·00015	(0·00147)	1·237	1·220	1·197	1·184	1·175	1·164	1·159	1·156	1·154	1·155	1·157	1·160	1·167	1·174
0·00020	(0·00196)	1·209	1·193	1·173	1·161	1·153	1·145	1·140	1·138	1·137	1·139	1·142	1·145	1·153	1·161
0·00030	(0·00294)	1·172	1·158	1·141	1·131	1·125	1·118	1·115	1·114	1·115	1·118	1·122	1·126	1·135	1·143
0·00040	(0·00392)	1·147	1·135	1·120	1·111	1·106	1·101	1·099	1·098	1·100	1·103	1·108	1·113	1·123	1·132
0·00060	(0·00588)	1·115	1·104	1·092	1·085	1·081	1·077	1·077	1·077	1·080	1·085	1·090	1·096	1·107	1·116
0·00080	(0·00785)	1·093	1·084	1·073	1·067	1·064	1·062	1·062	1·063	1·067	1·072	1·079	1·085	1·096	1·106
0·00100	(0·00981)	1·077	1·069	1·059	1·054	1·052	1·050	1·051	1·053	1·058	1·063	1·070	1·076	1·088	1·099
0·00150	(0·0147)	1·049	1·043	1·035	1·032	1·030	1·030	1·032	1·035	1·041	1·047	1·055	1·062	1·075	1·087
0·00200	(0·0196)	1·031	1·025	1·019	1·017	1·016	1·017	1·020	1·023	1·030	1·037	1·045	1·053	1·067	1·079
0·00300	(0·0294)	1·007	1·003	0·998	0·997	0·998	1·000	1·004	1·008	1·016	1·024	1·033	1·041	1·056	1·068
0·00400	(0·0392)	0·991	0·988	0·985	0·984	0·985	0·989	0·993	0·998	1·007	1·015	1·025	1·033	1·049	1·062
0·00600	(0·0588)	0·970	0·968	0·967	0·967	0·969	0·974	0·979	0·985	0·995	1·004	1·014	1·024	1·040	1·053
0·00800	(0·0785)	0·957	0·955	0·955	0·956	0·959	0·965	0·971	0·976	0·987	0·997	1·008	1·017	1·034	1·048
0·01000	(0·0981)	0·947	0·946	0·946	0·948	0·951	0·958	0·964	0·970	0·982	0·992	1·003	1·013	1·030	1·044
0·01500	(0·147)	0·930	0·930	0·932	0·935	0·939	0·946	0·953	0·960	0·972	0·983	0·995	1·005	1·023	1·038
0·02000	(0·196)	0·919	0·919	0·922	0·926	0·930	0·939	0·947	0·954	0·967	0·978	0·990	1·001	1·019	1·035
0·03000	(0·294)	0·905	0·906	0·910	0·915	0·920	0·929	0·938	0·946	0·959	0·971	0·984	0·995	1·014	1·030
0·04000	(0·392)	0·895	0·898	0·903	0·908	0·913	0·923	0·932	0·941	0·955	0·967	0·980	0·992	1·011	1·027
0·06000	(0·588)	0·884	0·887	0·893	0·899	0·905	0·916	0·926	0·934	0·949	0·962	0·975	0·987	1·007	1·023
0·08000	(0·785)	0·876	0·880	0·887	0·894	0·900	0·911	0·921	0·930	0·946	0·959	0·972	0·984	1·005	1·021
0·10000	(0·981)	0·871	0·875	0·882	0·890	0·896	0·908	0·918	0·927	0·943	0·956	0·970	0·983	1·003	1·020
0·15000	(1·47)	0·862	0·867	0·875	0·883	0·890	0·903	0·913	0·923	0·939	0·953	0·967	0·980	1·000	1·017
0·20000	(1·96)	0·857	0·862	0·871	0·879	0·886	0·899	0·911	0·920	0·937	0·951	0·965	0·978	0·999	1·016
0·30000	(2·94)	0·850	0·856	0·865	0·874	0·882	0·895	0·907	0·917	0·934	0·948	0·963	0·976	0·997	1·014
0·40000	(3·92)	0·846	0·852	0·862	0·871	0·879	0·893	0·905	0·915	0·932	0·946	0·961	0·974	0·996	1·013
0·60000	(5·88)	0·841	0·847	0·858	0·867	0·876	0·890	0·902	0·912	0·930	0·944	0·960	0·973	0·994	1·012
0·80000	(7·85)	0·838	0·844	0·855	0·865	0·873	0·888	0·900	0·911	0·929	0·943	0·958	0·972	0·993	1·011
1·00000	(9·81)	0·835	0·842	0·853	0·863	0·872	0·887	0·899	0·910	0·928	0·942	0·958	0·971	0·993	1·010
1·50000	(14·7)	0·832	0·839	0·851	0·861	0·870	0·885	0·897	0·908	0·926	0·941	0·957	0·970	0·992	1·010
2·00000	(19·6)	0·830	0·837	0·849	0·859	0·868	0·884	0·896	0·907	0·925	0·940	0·956	0·969	0·991	1·009
3·00000	(29·4)	0·827	0·834	0·847	0·857	0·867	0·882	0·895	0·906	0·924	0·939	0·955	0·968	0·991	1·009
S	$(\Delta p/\rho l)$	125	150	200	250	300	400	500	600	800	1000	1250	1500	2000	2500

Grad'nt k.p.g. (Equivalent) Pipe diameters in mm

Kinematic viscosity, $\nu = 2.5 \times 10^{-6}$ m^{2}s^{-1};

Roughness size, $k_s = 0.060$ mm

Kin. visc., $\nu = 2{\cdot}5\times10^{-6}$ m²s⁻¹;
$S = 0{\cdot}00010$ to $30{\cdot}0000$
i.e. kin. pr. grad., $\Delta p/\rho l =$
$(0{\cdot}00098)$ to (294) ms⁻²

Roughness size, $k_s = 0{\cdot}150$ mm
This table shows values of m, as follows
m_C for Colebrook-White solutions; or,
where $\mathbf{R} \le 2000$, m_P for laminar flow

Grad'nt **k.p.g.** (Equivalent) Pipe diameters in mm

S	$(\Delta p/\rho l)$	6	8	10	12·5	15	20	25	30	40	50	60	80	100	125
0·00030	(0·00294)	17·14	11·68	8·676	6·443	5·053	3·443	2·557	2·005	1·366	1·015	*	1·233	1·209	1·190
0·00040	(0·00392)	14·85	10·12	7·513	5·580	4·376	2·982	2·214	1·736	1·183	*	*	1·206	1·184	1·167
0·00060	(0·00588)	12·12	8·260	6·135	4·556	3·573	2·435	1·808	1·418	0·966	*	1·200	1·170	1·151	1·137
0·00080	(0·00785)	10·50	7·154	5·313	3·946	3·094	2·108	1·566	1·228	*	1·195	1·173	1·146	1·130	1·117
0·00100	(0·00981)	9·390	6·398	4·752	3·529	2·767	1·886	1·400	1·098	*	1·174	1·154	1·129	1·114	1·102
0·00150	(0·0147)	7·667	5·224	3·880	2·881	2·260	1·540	1·143	0·897	1·163	1·138	1·121	1·100	1·088	1·078
0·00200	(0·0196)	6·640	4·524	3·360	2·495	1·957	1·333	0·990	*	1·137	1·115	1·100	1·081	1·070	1·062
0·00300	(0·0294)	5·421	3·694	2·743	2·037	1·598	1·089	*	*	1·105	1·085	1·073	1·057	1·048	1·042
0·00400	(0·0392)	4·695	3·199	2·376	1·764	1·384	0·943	*	1·111	1·083	1·066	1·055	1·041	1·033	1·028
0·00600	(0·0588)	3·833	2·612	1·940	1·441	1·130	*	*	1·080	1·056	1·041	1·032	1·021	1·015	1·011
0·00800	(0·0785)	3·320	2·262	1·680	1·248	0·978	*	1·077	1·060	1·038	1·025	1·017	1·008	1·003	1·001
0·01000	(0·0981)	2·969	2·023	1·503	1·116	0·875	*	1·061	1·045	1·025	1·014	1·006	0·998	0·995	0·993
0·01500	(0·147)	2·424	1·652	1·227	0·911	*	1·054	1·034	1·021	1·004	0·995	0·989	0·983	0·980	0·980
0·02000	(0·196)	2·100	1·431	1·063	0·789	*	1·035	1·017	1·005	0·991	0·982	0·978	0·973	0·972	0·972
0·03000	(0·294)	1·714	1·168	0·868	*	1·037	1·011	0·996	0·985	0·973	0·967	0·963	0·960	0·960	0·962
0·04000	(0·392)	1·485	1·012	0·751	*	1·019	0·996	0·982	0·973	0·963	0·957	0·954	0·953	0·953	0·956
0·06000	(0·588)	1·212	0·826	*	1·012	0·996	0·976	0·965	0·957	0·949	0·945	0·943	0·943	0·945	0·948
0·08000	(0·785)	1·050	*	*	0·996	0·982	0·964	0·954	0·947	0·940	0·937	0·936	0·937	0·939	0·943
0·10000	(0·981)	0·939	*	*	0·985	0·972	0·956	0·946	0·940	0·934	0·932	0·932	0·933	0·936	0·940
0·15000	(1·47)	0·767	*	0·984	0·967	0·955	0·941	0·934	0·929	0·925	0·924	0·924	0·926	0·930	0·934
0·20000	(1·96)	*	*	0·971	0·955	0·945	0·933	0·926	0·922	0·919	0·918	0·919	0·922	0·926	0·931
0·30000	(2·94)	*	0·972	0·955	0·941	0·932	0·922	0·916	0·914	0·911	0·912	0·913	0·917	0·921	0·927
0·40000	(3·92)	*	0·960	0·945	0·932	0·924	0·915	0·911	0·908	0·907	0·908	0·909	0·914	0·919	0·924
0·60000	(5·88)	0·969	0·946	0·932	0·921	0·915	0·907	0·903	0·902	0·901	0·903	0·905	0·910	0·915	0·922
0·80000	(7·85)	0·958	0·937	0·924	0·915	0·909	0·902	0·899	0·898	0·898	0·900	0·902	0·908	0·913	0·920
1·00000	(9·81)	0·951	0·931	0·919	0·910	0·904	0·898	0·896	0·895	0·896	0·898	0·900	0·906	0·912	0·919
1·50000	(14·7)	0·939	0·921	0·911	0·903	0·898	0·893	0·891	0·891	0·892	0·895	0·898	0·904	0·910	0·917
2·00000	(19·6)	0·931	0·915	0·905	0·898	0·894	0·890	0·888	0·888	0·890	0·893	0·896	0·902	0·908	0·915
3·00000	(29·4)	0·922	0·908	0·899	0·893	0·889	0·885	0·885	0·885	0·887	0·890	0·894	0·900	0·907	0·914
4·00000	(39·2)	0·917	0·903	0·895	0·889	0·886	0·883	0·882	0·883	0·886	0·889	0·892	0·899	0·906	0·913
6·00000	(58·8)	0·910	0·898	0·891	0·885	0·882	0·880	0·880	0·881	0·884	0·887	0·891	0·898	0·905	0·912
8·00000	(78·5)	0·906	0·895	0·888	0·883	0·880	0·878	0·878	0·879	0·882	0·886	0·890	0·897	0·904	0·912
10·0000	(98·1)	0·904	0·892	0·886	0·881	0·879	0·877	0·877	0·878	0·882	0·885	0·889	0·897	0·904	0·911
15·0000	(147)	0·899	0·889	0·883	0·879	0·877	0·875	0·876	0·877	0·880	0·884	0·888	0·896	0·903	0·911
20·0000	(196)	0·897	0·887	0·881	0·877	0·875	0·874	0·875	0·876	0·880	0·884	0·888	0·895	0·902	0·910
30·0000	(294)	0·894	0·884	0·879	0·875	0·874	0·873	0·873	0·875	0·879	0·883	0·887	0·895	0·902	0·910
S	$(\Delta p/\rho l)$	6	8	10	12·5	15	20	25	30	40	50	60	80	100	125

Grad'nt **k.p.g.** (Equivalent) Pipe diameters in mm

S	$(\Delta p/\rho l)$	125	150	200	250	300	400	500	600	800	1000	1250	1500	2000	2500
0·00010	(0·00098)	1·292	1·272	1·247	1·231	1·221	1·210	1·203	1·200	1·198	1·198	1·200	1·204	1·211	1·218
0·00015	(0·00147)	1·252	1·234	1·213	1·200	1·192	1·182	1·178	1·176	1·175	1·177	1·181	1·185	1·193	1·202
0·00020	(0·00196)	1·225	1·209	1·190	1·179	1·172	1·164	1·161	1·160	1·160	1·163	1·168	1·172	1·182	1·191
0·00030	(0·00294)	1·190	1·177	1·161	1·152	1·146	1·140	1·139	1·139	1·141	1·145	1·151	1·156	1·167	1·177
0·00040	(0·00392)	1·167	1·155	1·141	1·133	1·129	1·125	1·124	1·125	1·129	1·133	1·140	1·146	1·158	1·168
0·00060	(0·00588)	1·137	1·127	1·116	1·110	1·107	1·105	1·105	1·107	1·112	1·118	1·125	1·132	1·145	1·157
0·00080	(0·00785)	1·117	1·109	1·099	1·094	1·092	1·091	1·093	1·095	1·102	1·108	1·116	1·124	1·138	1·150
0·00100	(0·00981)	1·102	1·095	1·087	1·083	1·081	1·082	1·084	1·087	1·094	1·101	1·110	1·118	1·132	1·145
0·00150	(0·0147)	1·078	1·072	1·066	1·064	1·064	1·065	1·069	1·073	1·081	1·089	1·099	1·107	1·123	1·136
0·00200	(0·0196)	1·062	1·057	1·053	1·052	1·052	1·055	1·059	1·064	1·073	1·082	1·092	1·101	1·117	1·131
0·00300	(0·0294)	1·042	1·038	1·036	1·036	1·037	1·042	1·047	1·052	1·063	1·072	1·083	1·093	1·110	1·124
0·00400	(0·0392)	1·028	1·026	1·025	1·026	1·028	1·033	1·039	1·045	1·056	1·066	1·078	1·088	1·105	1·120
0·00600	(0·0588)	1·011	1·010	1·010	1·013	1·016	1·023	1·029	1·036	1·048	1·059	1·071	1·081	1·100	1·115
0·00800	(0·0785)	1·001	1·000	1·002	1·005	1·008	1·016	1·023	1·030	1·043	1·054	1·067	1·078	1·096	1·112
0·01000	(0·0981)	0·993	0·993	0·995	0·999	1·003	1·011	1·019	1·026	1·040	1·051	1·064	1·075	1·094	1·110
0·01500	(0·147)	0·980	0·981	0·985	0·989	0·994	1·003	1·012	1·020	1·034	1·046	1·059	1·070	1·090	1·106
0·02000	(0·196)	0·972	0·974	0·978	0·983	0·989	0·998	1·008	1·016	1·030	1·043	1·056	1·068	1·088	1·104
0·03000	(0·294)	0·962	0·964	0·970	0·976	0·982	0·993	1·002	1·011	1·026	1·039	1·053	1·065	1·085	1·101
0·04000	(0·392)	0·956	0·959	0·965	0·971	0·978	0·989	0·999	1·008	1·023	1·036	1·050	1·063	1·083	1·100
0·06000	(0·588)	0·948	0·951	0·959	0·966	0·972	0·984	0·995	1·004	1·020	1·034	1·048	1·060	1·081	1·098
0·08000	(0·785)	0·943	0·947	0·955	0·962	0·969	0·982	0·992	1·002	1·018	1·032	1·046	1·059	1·080	1·097
0·10000	(0·981)	0·940	0·944	0·952	0·960	0·967	0·980	0·991	1·000	1·017	1·031	1·045	1·058	1·079	1·096
0·15000	(1·47)	0·934	0·939	0·948	0·956	0·964	0·977	0·988	0·998	1·015	1·029	1·043	1·056	1·078	1·095
0·20000	(1·96)	0·931	0·936	0·945	0·954	0·962	0·975	0·986	0·996	1·013	1·028	1·042	1·055	1·077	1·094
0·30000	(2·94)	0·927	0·932	0·942	0·951	0·959	0·973	0·984	0·995	1·012	1·026	1·041	1·054	1·076	1·093
0·40000	(3·92)	0·924	0·930	0·940	0·949	0·958	0·971	0·983	0·994	1·011	1·025	1·041	1·054	1·075	1·093
0·60000	(5·88)	0·922	0·927	0·938	0·947	0·956	0·970	0·982	0·992	1·010	1·024	1·040	1·053	1·074	1·092
0·80000	(7·85)	0·920	0·926	0·937	0·946	0·955	0·969	0·981	0·992	1·009	1·024	1·039	1·052	1·074	1·092
1·00000	(9·81)	0·919	0·925	0·936	0·945	0·954	0·968	0·980	0·991	1·009	1·023	1·039	1·052	1·074	1·092
1·50000	(14·7)	0·917	0·923	0·934	0·944	0·953	0·967	0·980	0·990	1·008	1·023	1·038	1·051	1·073	1·091
2·00000	(19·6)	0·915	0·922	0·933	0·943	0·952	0·967	0·979	0·990	1·008	1·022	1·038	1·051	1·073	1·091
3·00000	(29·4)	0·914	0·921	0·932	0·942	0·951	0·966	0·978	0·989	1·007	1·022	1·037	1·051	1·073	1·091
S	$(\Delta p/\rho l)$	125	150	200	250	300	400	500	600	800	1000	1250	1500	2000	2500

Grad'nt **k.p.g.** (Equivalent) Pipe diameters in mm

Kinematic viscosity, $\nu = 2{\cdot}5\times10^{-6}$ m²s⁻¹ ; **Roughness size, $k_s = 0{\cdot}150$ mm**

E144

Kin. visc., $\nu = 2.5 \times 10^{-6}\ \mathrm{m^2 s^{-1}}$;
$S = 0.00010$ to 30.0000

i.e. kin. pr. grad., $\Delta p/\rho l =$
(0.00098) to $(294)\ \mathrm{ms^{-2}}$

Roughness size, $k_s = 0.30$ mm

This table shows values of m, as follows

m_C for Colebrook-White solutions; or,
where $\mathbf{R} \leq 2000$, m_P for laminar flow

Grad'nt S	k.p.g. $(\Delta p/\rho l)$	6	8	10	12.5	15	20	25	30	40	50	60	80	100	125
		(Equivalent) Pipe diameters in mm													
0.00030	(0.00294)	17.14	11.68	8.676	6.443	5.053	3.443	2.557	2.005	1.366	1.015	*	1.260	1.236	1.217
0.00040	(0.00392)	14.85	10.12	7.513	5.580	4.376	2.982	2.214	1.736	1.183	*	*	1.234	1.213	1.197
0.00060	(0.00588)	12.12	8.260	6.135	4.556	3.573	2.435	1.808	1.418	0.966	*	1.231	1.202	1.184	1.170
0.00080	(0.00785)	10.50	7.154	5.313	3.946	3.094	2.108	1.566	1.228	*	1.228	1.207	1.180	1.165	1.152
0.00100	(0.00981)	9.390	6.398	4.752	3.529	2.767	1.886	1.400	1.098	*	1.209	1.189	1.165	1.151	1.140
0.00150	(0.0147)	7.667	5.224	3.880	2.881	2.260	1.540	1.143	0.897	*	1.177	1.160	1.140	1.128	1.119
0.00200	(0.0196)	6.640	4.524	3.360	2.495	1.957	1.333	0.990	*	1.179	1.156	1.142	1.123	1.113	1.106
0.00300	(0.0294)	5.421	3.694	2.743	2.037	1.598	1.089	*	*	1.149	1.131	1.118	1.103	1.095	1.089
0.00400	(0.0392)	4.695	3.199	2.376	1.764	1.384	0.943	*	1.159	1.131	1.114	1.103	1.090	1.083	1.078
0.00600	(0.0588)	3.833	2.612	1.940	1.441	1.130	*	*	1.132	1.108	1.093	1.084	1.073	1.068	1.065
0.00800	(0.0785)	3.320	2.262	1.680	1.248	0.978	*	1.132	1.115	1.093	1.080	1.072	1.063	1.059	1.057
0.01000	(0.0981)	2.969	2.023	1.503	1.116	0.875	*	1.118	1.102	1.082	1.071	1.063	1.056	1.052	1.051
0.01500	(0.147)	2.424	1.652	1.227	0.911	*	1.117	1.096	1.082	1.065	1.056	1.050	1.044	1.042	1.042
0.02000	(0.196)	2.100	1.431	1.063	0.789	*	1.101	1.082	1.070	1.054	1.046	1.041	1.036	1.035	1.036
0.03000	(0.294)	1.714	1.168	0.868	*	*	1.081	1.065	1.054	1.041	1.034	1.031	1.027	1.027	1.029
0.04000	(0.392)	1.485	1.012	0.751	*	1.094	1.069	1.054	1.044	1.033	1.027	1.024	1.022	1.022	1.024
0.06000	(0.588)	1.212	0.826	*	*	1.076	1.054	1.041	1.032	1.023	1.018	1.016	1.015	1.016	1.019
0.08000	(0.785)	1.050	*	*	1.080	1.064	1.044	1.032	1.025	1.017	1.013	1.011	1.011	1.013	1.016
0.10000	(0.981)	0.939	*	*	1.072	1.056	1.038	1.027	1.020	1.012	1.009	1.008	1.008	1.010	1.014
0.15000	(1.47)	0.767	*	1.077	1.057	1.044	1.027	1.018	1.012	1.006	1.003	1.002	1.003	1.006	1.010
0.20000	(1.96)	*	*	1.067	1.048	1.036	1.021	1.012	1.007	1.001	0.999	0.999	1.001	1.004	1.008
0.30000	(2.94)	*	1.076	1.055	1.038	1.027	1.013	1.005	1.001	0.996	0.995	0.995	0.997	1.001	1.005
0.40000	(3.92)	*	1.067	1.047	1.031	1.021	1.008	1.001	0.997	0.993	0.992	0.993	0.995	0.999	1.004
0.60000	(5.88)	*	1.056	1.038	1.023	1.014	1.002	0.996	0.993	0.990	0.989	0.990	0.993	0.997	1.002
0.80000	(7.85)	1.078	1.049	1.032	1.019	1.010	0.999	0.993	0.990	0.987	0.987	0.988	0.992	0.996	1.001
1.00000	(9.81)	1.072	1.045	1.028	1.015	1.007	0.996	0.991	0.988	0.986	0.986	0.987	0.991	0.995	1.000
1.50000	(14.7)	1.063	1.038	1.022	1.010	1.002	0.993	0.988	0.985	0.983	0.984	0.985	0.989	0.993	0.999
2.00000	(19.6)	1.058	1.033	1.019	1.007	0.999	0.990	0.986	0.984	0.982	0.983	0.984	0.988	0.992	0.998
3.00000	(29.4)	1.051	1.028	1.014	1.003	0.996	0.988	0.984	0.981	0.980	0.981	0.983	0.987	0.992	0.997
4.00000	(39.2)	1.047	1.025	1.012	1.001	0.994	0.986	0.982	0.980	0.979	0.980	0.982	0.986	0.991	0.997
6.00000	(58.8)	1.043	1.021	1.008	0.998	0.992	0.984	0.980	0.979	0.978	0.979	0.981	0.985	0.990	0.996
8.00000	(78.5)	1.040	1.019	1.007	0.997	0.990	0.983	0.979	0.978	0.977	0.978	0.980	0.985	0.990	0.996
10.0000	(98.1)	1.038	1.018	1.005	0.996	0.989	0.982	0.979	0.977	0.977	0.978	0.980	0.985	0.990	0.996
15.0000	(147)	1.035	1.015	1.003	0.994	0.988	0.981	0.978	0.976	0.976	0.977	0.979	0.984	0.989	0.995
20.0000	(196)	1.033	1.014	1.002	0.993	0.987	0.980	0.977	0.976	0.976	0.977	0.979	0.984	0.989	0.995
30.0000	(294)	1.031	1.012	1.001	0.992	0.986	0.979	0.976	0.975	0.975	0.976	0.979	0.983	0.989	0.995
S	$(\Delta p/\rho l)$	6	8	10	12.5	15	20	25	30	40	50	60	80	100	125

Grad'nt S	k.p.g. $(\Delta p/\rho l)$	125	150	200	250	300	400	500	600	800	1000	1250	1500	2000	2500
		(Equivalent) Pipe diameters in mm													
0.00010	(0.00098)	1.312	1.293	1.268	1.254	1.244	1.233	1.228	1.226	1.225	1.227	1.230	1.235	1.244	1.253
0.00015	(0.00147)	1.274	1.258	1.237	1.225	1.217	1.209	1.206	1.205	1.206	1.209	1.214	1.219	1.230	1.240
0.00020	(0.00196)	1.250	1.235	1.216	1.206	1.200	1.193	1.191	1.191	1.193	1.197	1.203	1.209	1.221	1.231
0.00030	(0.00294)	1.217	1.205	1.190	1.182	1.177	1.173	1.172	1.173	1.177	1.183	1.190	1.196	1.209	1.221
0.00040	(0.00392)	1.197	1.186	1.173	1.166	1.162	1.160	1.160	1.162	1.167	1.173	1.181	1.188	1.202	1.214
0.00060	(0.00588)	1.170	1.161	1.150	1.145	1.143	1.143	1.144	1.147	1.154	1.161	1.170	1.178	1.193	1.206
0.00080	(0.00785)	1.152	1.145	1.136	1.132	1.131	1.132	1.135	1.138	1.146	1.154	1.163	1.172	1.187	1.201
0.00100	(0.00981)	1.140	1.133	1.126	1.123	1.122	1.124	1.128	1.132	1.140	1.149	1.158	1.167	1.183	1.197
0.00150	(0.0147)	1.119	1.114	1.109	1.108	1.108	1.112	1.116	1.121	1.131	1.140	1.151	1.160	1.177	1.191
0.00200	(0.0196)	1.106	1.102	1.098	1.098	1.099	1.104	1.109	1.114	1.125	1.135	1.146	1.156	1.173	1.188
0.00300	(0.0294)	1.089	1.086	1.085	1.086	1.088	1.094	1.100	1.106	1.118	1.128	1.140	1.150	1.168	1.184
0.00400	(0.0392)	1.078	1.076	1.076	1.078	1.081	1.087	1.094	1.101	1.113	1.124	1.136	1.147	1.165	1.181
0.00600	(0.0588)	1.065	1.064	1.066	1.069	1.072	1.080	1.087	1.095	1.108	1.119	1.132	1.143	1.162	1.178
0.00800	(0.0785)	1.057	1.057	1.059	1.063	1.067	1.075	1.083	1.091	1.104	1.116	1.129	1.140	1.160	1.176
0.01000	(0.0981)	1.051	1.051	1.054	1.058	1.063	1.072	1.080	1.088	1.102	1.114	1.127	1.139	1.158	1.174
0.01500	(0.147)	1.042	1.043	1.047	1.052	1.057	1.067	1.076	1.084	1.098	1.111	1.124	1.136	1.156	1.172
0.02000	(0.196)	1.036	1.038	1.042	1.048	1.053	1.063	1.073	1.081	1.096	1.109	1.122	1.134	1.154	1.171
0.03000	(0.294)	1.029	1.031	1.037	1.043	1.049	1.060	1.069	1.078	1.093	1.106	1.120	1.132	1.153	1.169
0.04000	(0.392)	1.024	1.027	1.033	1.040	1.046	1.057	1.067	1.076	1.092	1.105	1.119	1.131	1.152	1.169
0.06000	(0.588)	1.019	1.022	1.029	1.036	1.043	1.054	1.065	1.074	1.090	1.103	1.117	1.130	1.150	1.167
0.08000	(0.785)	1.016	1.019	1.027	1.034	1.041	1.053	1.063	1.072	1.088	1.102	1.116	1.129	1.150	1.167
0.10000	(0.981)	1.014	1.017	1.025	1.032	1.039	1.051	1.062	1.071	1.088	1.101	1.116	1.128	1.149	1.166
0.15000	(1.47)	1.010	1.014	1.022	1.030	1.037	1.050	1.060	1.070	1.086	1.100	1.115	1.127	1.148	1.166
0.20000	(1.96)	1.008	1.012	1.021	1.029	1.036	1.048	1.059	1.069	1.086	1.099	1.114	1.127	1.148	1.165
0.30000	(2.94)	1.005	1.010	1.019	1.027	1.034	1.047	1.058	1.068	1.085	1.099	1.113	1.126	1.147	1.165
0.40000	(3.92)	1.004	1.009	1.018	1.026	1.033	1.046	1.058	1.067	1.084	1.098	1.113	1.126	1.147	1.164
0.60000	(5.88)	1.002	1.007	1.016	1.025	1.032	1.045	1.057	1.067	1.083	1.097	1.112	1.125	1.147	1.164
0.80000	(7.85)	1.001	1.006	1.015	1.024	1.031	1.045	1.056	1.066	1.083	1.097	1.112	1.125	1.146	1.164
1.00000	(9.81)	1.000	1.005	1.015	1.023	1.031	1.044	1.056	1.066	1.083	1.097	1.112	1.125	1.146	1.164
1.50000	(14.7)	0.999	1.004	1.014	1.022	1.030	1.044	1.055	1.065	1.082	1.097	1.112	1.124	1.146	1.163
2.00000	(19.6)	0.998	1.003	1.013	1.022	1.030	1.043	1.055	1.065	1.082	1.096	1.111	1.124	1.146	1.163
3.00000	(29.4)	0.997	1.003	1.013	1.021	1.029	1.043	1.055	1.065	1.082	1.096	1.111	1.124	1.146	1.163
S	$(\Delta p/\rho l)$	125	150	200	250	300	400	500	600	800	1000	1250	1500	2000	2500

Grad'nt   k.p.g.   (Equivalent) Pipe diameters in mm

Kinematic viscosity, $\nu = 2.5 \times 10^{-6}\ \mathrm{m^2 s^{-1}}$;   **Roughness size, $k_s = 0.30$ mm**

Kin. visc., $\nu = 2.5 \times 10^{-6}$ m^{2}s^{-1};   Roughness size, $k_s = 0.60$ mm

$S = 0.00010$ to 3.00000

i.e. kin. pr. grad., $\Delta p / \rho l =$ (0.00098) to (29.4) ms^{-2}

This table shows values of m, as follows

m_C for Colebrook-White solutions

E145

Grad'nt S	k.p.g. $(\Delta p / \rho l)$	125	150	200	250	300	400	500	600	800	1000	1250	1500	2000	2500
							(Equivalent) Pipe diameters in mm								
0.00010	(0.00098)	1.350	1.331	1.307	1.293	1.285	1.275	1.271	1.270	1.271	1.274	1.279	1.284	1.295	1.306
0.00015	(0.00147)	1.316	1.300	1.280	1.268	1.262	1.255	1.253	1.252	1.255	1.260	1.266	1.272	1.285	1.296
0.00020	(0.00196)	1.294	1.280	1.262	1.253	1.247	1.242	1.241	1.242	1.246	1.251	1.258	1.265	1.278	1.290
0.00030	(0.00294)	1.266	1.254	1.240	1.232	1.228	1.226	1.226	1.228	1.234	1.240	1.248	1.256	1.270	1.283
0.00040	(0.00392)	1.248	1.238	1.226	1.220	1.217	1.215	1.217	1.219	1.226	1.233	1.242	1.250	1.265	1.278
0.00060	(0.00588)	1.226	1.217	1.208	1.204	1.202	1.202	1.205	1.209	1.217	1.225	1.234	1.243	1.259	1.273
0.00080	(0.00785)	1.212	1.204	1.196	1.193	1.193	1.194	1.198	1.202	1.211	1.219	1.230	1.239	1.255	1.270
0.00100	(0.00981)	1.202	1.195	1.189	1.186	1.186	1.189	1.193	1.197	1.207	1.216	1.226	1.236	1.253	1.267
0.00150	(0.0147)	1.185	1.180	1.176	1.175	1.176	1.180	1.185	1.190	1.200	1.210	1.221	1.231	1.249	1.264
0.00200	(0.0196)	1.175	1.171	1.168	1.168	1.169	1.174	1.180	1.185	1.196	1.207	1.218	1.228	1.246	1.261
0.00300	(0.0294)	1.162	1.159	1.158	1.159	1.161	1.167	1.173	1.180	1.192	1.202	1.214	1.225	1.243	1.259
0.00400	(0.0392)	1.154	1.152	1.152	1.153	1.156	1.163	1.170	1.176	1.189	1.200	1.212	1.223	1.241	1.257
0.00600	(0.0588)	1.145	1.144	1.144	1.147	1.150	1.158	1.165	1.172	1.185	1.196	1.209	1.220	1.239	1.255
0.00800	(0.0785)	1.139	1.138	1.140	1.143	1.147	1.155	1.162	1.170	1.183	1.195	1.207	1.219	1.238	1.254
0.01000	(0.0981)	1.135	1.134	1.137	1.140	1.144	1.153	1.161	1.168	1.181	1.193	1.206	1.218	1.237	1.253
0.01500	(0.147)	1.128	1.129	1.132	1.136	1.140	1.149	1.158	1.165	1.179	1.191	1.204	1.216	1.235	1.252
0.02000	(0.196)	1.124	1.125	1.129	1.133	1.138	1.147	1.156	1.164	1.178	1.190	1.203	1.215	1.235	1.251
0.03000	(0.294)	1.119	1.121	1.125	1.130	1.135	1.145	1.154	1.162	1.176	1.189	1.202	1.214	1.234	1.250
0.04000	(0.392)	1.117	1.118	1.123	1.128	1.133	1.143	1.152	1.161	1.175	1.188	1.201	1.213	1.233	1.250
0.06000	(0.588)	1.113	1.115	1.120	1.126	1.131	1.141	1.151	1.159	1.174	1.187	1.200	1.212	1.232	1.249
0.08000	(0.785)	1.111	1.113	1.119	1.124	1.130	1.140	1.150	1.158	1.173	1.186	1.200	1.212	1.232	1.249
0.10000	(0.981)	1.110	1.112	1.117	1.123	1.129	1.140	1.149	1.158	1.173	1.186	1.199	1.211	1.231	1.248
0.15000	(1.47)	1.107	1.110	1.116	1.122	1.128	1.139	1.148	1.157	1.172	1.185	1.199	1.211	1.231	1.248
0.20000	(1.96)	1.106	1.109	1.115	1.121	1.127	1.138	1.148	1.156	1.172	1.184	1.198	1.210	1.231	1.248
0.30000	(2.94)	1.104	1.107	1.114	1.120	1.126	1.137	1.147	1.156	1.171	1.184	1.198	1.210	1.230	1.247
0.40000	(3.92)	1.103	1.106	1.113	1.119	1.125	1.137	1.146	1.155	1.171	1.184	1.198	1.210	1.230	1.247
0.60000	(5.88)	1.102	1.105	1.112	1.118	1.125	1.136	1.146	1.155	1.170	1.183	1.197	1.210	1.230	1.247
0.80000	(7.85)	1.101	1.105	1.111	1.118	1.124	1.136	1.146	1.155	1.170	1.183	1.197	1.209	1.230	1.247
1.00000	(9.81)	1.101	1.104	1.111	1.118	1.124	1.135	1.145	1.154	1.170	1.183	1.197	1.209	1.230	1.247
1.50000	(14.7)	1.100	1.104	1.111	1.117	1.124	1.135	1.145	1.154	1.170	1.183	1.197	1.209	1.230	1.247
2.00000	(19.6)	1.100	1.103	1.110	1.117	1.123	1.135	1.145	1.154	1.169	1.183	1.197	1.209	1.230	1.247
3.00000	(29.4)	1.099	1.103	1.110	1.117	1.123	1.135	1.145	1.154	1.169	1.182	1.197	1.209	1.229	1.246
S	$(\Delta p / \rho l)$	125	150	200	250	300	400	500	600	800	1000	1250	1500	2000	2500

Grad'nt   k.p.g.   (Equivalent) Pipe diameters in mm   Roughness size, $k_s = 0.60$ mm

Kin. visc., $\nu = 2.5 \times 10^{-6}$ m^{2}s^{-1};   Roughness size, $k_s = 1.50$ mm

$S = 0.00010$ to 3.00000

i.e. kin. pr. grad., $\Delta p / \rho l =$ (0.00098) to (29.4) ms^{-2}

This table shows values of m, as follows

m_C for Colebrook-White solutions

E146

Grad'nt S	k.p.g. $(\Delta p / \rho l)$	125	150	200	250	300	400	500	600	800	1000	1250	1500	2000	2500
							(Equivalent) Pipe diameters in mm								
0.00010	(0.00098)	1.446	1.426	1.402	1.389	1.380	1.372	1.368	1.367	1.369	1.373	1.379	1.385	1.398	1.409
0.00015	(0.00147)	1.419	1.402	1.382	1.371	1.364	1.357	1.356	1.356	1.359	1.364	1.371	1.378	1.391	1.404
0.00020	(0.00196)	1.402	1.387	1.369	1.359	1.354	1.349	1.348	1.349	1.353	1.359	1.366	1.374	1.388	1.400
0.00030	(0.00294)	1.381	1.368	1.353	1.345	1.341	1.338	1.338	1.340	1.346	1.352	1.361	1.368	1.383	1.396
0.00040	(0.00392)	1.368	1.357	1.343	1.337	1.333	1.331	1.332	1.335	1.341	1.348	1.357	1.365	1.380	1.394
0.00060	(0.00588)	1.352	1.342	1.331	1.326	1.324	1.323	1.325	1.328	1.336	1.343	1.353	1.361	1.377	1.391
0.00080	(0.00785)	1.343	1.334	1.324	1.319	1.318	1.318	1.321	1.324	1.332	1.340	1.350	1.359	1.375	1.389
0.00100	(0.00981)	1.336	1.328	1.319	1.315	1.314	1.315	1.318	1.322	1.330	1.338	1.348	1.357	1.373	1.388
0.00150	(0.0147)	1.325	1.318	1.311	1.308	1.307	1.309	1.313	1.317	1.326	1.335	1.345	1.355	1.371	1.386
0.00200	(0.0196)	1.318	1.312	1.306	1.304	1.303	1.306	1.310	1.315	1.324	1.333	1.344	1.353	1.370	1.384
0.00300	(0.0294)	1.310	1.305	1.300	1.298	1.299	1.302	1.307	1.312	1.322	1.331	1.342	1.351	1.368	1.383
0.00400	(0.0392)	1.305	1.300	1.296	1.295	1.296	1.300	1.305	1.310	1.320	1.330	1.340	1.350	1.367	1.382
0.00600	(0.0588)	1.300	1.295	1.292	1.291	1.293	1.297	1.302	1.308	1.318	1.328	1.339	1.349	1.366	1.381
0.00800	(0.0785)	1.296	1.292	1.289	1.289	1.291	1.295	1.301	1.306	1.317	1.327	1.338	1.348	1.366	1.381
0.01000	(0.0981)	1.294	1.290	1.287	1.288	1.289	1.294	1.300	1.305	1.316	1.326	1.338	1.348	1.365	1.380
0.01500	(0.147)	1.290	1.287	1.285	1.285	1.287	1.292	1.298	1.304	1.315	1.325	1.337	1.347	1.365	1.380
0.02000	(0.196)	1.287	1.285	1.283	1.284	1.286	1.291	1.297	1.303	1.314	1.325	1.336	1.346	1.364	1.379
0.03000	(0.294)	1.285	1.282	1.281	1.282	1.284	1.290	1.296	1.302	1.313	1.324	1.335	1.346	1.364	1.379
0.04000	(0.392)	1.283	1.281	1.280	1.281	1.283	1.289	1.295	1.301	1.313	1.323	1.335	1.345	1.363	1.379
0.06000	(0.588)	1.281	1.279	1.278	1.280	1.282	1.288	1.294	1.301	1.312	1.323	1.334	1.345	1.363	1.378
0.08000	(0.785)	1.280	1.278	1.277	1.279	1.281	1.288	1.294	1.300	1.312	1.322	1.334	1.345	1.363	1.378
0.10000	(0.981)	1.279	1.277	1.277	1.278	1.281	1.287	1.294	1.300	1.312	1.322	1.334	1.344	1.363	1.378
0.15000	(1.47)	1.278	1.276	1.276	1.278	1.280	1.287	1.293	1.300	1.311	1.322	1.334	1.344	1.362	1.378
0.20000	(1.96)	1.277	1.275	1.275	1.277	1.280	1.286	1.293	1.299	1.311	1.322	1.334	1.344	1.362	1.378
0.30000	(2.94)	1.276	1.275	1.275	1.277	1.279	1.286	1.292	1.299	1.311	1.321	1.333	1.344	1.362	1.377
0.40000	(3.92)	1.276	1.274	1.274	1.276	1.279	1.286	1.292	1.299	1.311	1.321	1.333	1.344	1.362	1.377
0.60000	(5.88)	1.275	1.274	1.274	1.276	1.279	1.285	1.292	1.298	1.310	1.321	1.333	1.344	1.362	1.377
0.80000	(7.85)	1.275	1.273	1.273	1.276	1.278	1.285	1.292	1.298	1.310	1.321	1.333	1.344	1.362	1.377
1.00000	(9.81)	1.274	1.273	1.273	1.275	1.278	1.285	1.292	1.298	1.310	1.321	1.333	1.343	1.362	1.377
1.50000	(14.7)	1.274	1.273	1.273	1.275	1.278	1.285	1.292	1.298	1.310	1.321	1.333	1.343	1.362	1.377
2.00000	(19.6)	1.274	1.272	1.273	1.275	1.278	1.285	1.291	1.298	1.310	1.321	1.333	1.343	1.362	1.377
3.00000	(29.4)	1.274	1.272	1.273	1.275	1.278	1.285	1.291	1.298	1.310	1.321	1.333	1.343	1.362	1.377
S	$(\Delta p / \rho l)$	125	150	200	250	300	400	500	600	800	1000	1250	1500	2000	2500

Grad'nt   k.p.g.   (Equivalent) Pipe diameters in mm   Roughness size, $k_s = 1.50$ mm

Kin. visc., $\nu = 3{\cdot}0 \times 10^{-6}$ m^{2}s^{-1};
$S = 0{\cdot}00010$ to $30{\cdot}0000$

i.e. kin. pr. grad., $\Delta p/\rho l =$
$(0{\cdot}00098)$ to (294) ms^{-2}

Roughness size, $k_s = 0{\cdot}0015$ mm
This table shows values of m, as follows

m_C for Colebrook-White solutions; or,
where $\mathbf{R} \leq 2000$, m_P for laminar flow

Grad'nt S	k.p.g. $(\Delta p/\rho l)$	(Equivalent) Pipe diameters in mm													
		6	8	10	12·5	15	20	25	30	40	50	60	80	100	125
0·00030	(0·00294)	20·57	14·02	10·41	7·732	6·063	4·132	3·068	2·406	1·640	1·218	*	1·245	1·217	1·194
0·00040	(0·00392)	17·82	12·14	9·016	6·696	5·251	3·578	2·657	2·084	1·420	1·055	*	1·213	1·188	1·167
0·00060	(0·00588)	14·55	9·912	7·362	5·467	4·287	2·921	2·170	1·701	1·159	*	1·206	1·171	1·149	1·130
0·00080	(0·00785)	12·60	8·584	6·375	4·735	3·713	2·530	1·879	1·473	1·004	*	1·174	1·142	1·122	1·106
0·00100	(0·00981)	11·27	7·678	5·702	4·235	3·321	2·263	1·681	1·318	*	1·174	1·151	1·121	1·103	1·087
0·00150	(0·0147)	9·200	6·269	4·656	3·458	2·712	1·848	1·372	1·076	*	1·131	1·111	1·085	1·069	1·056
0·00200	(0·0196)	7·968	5·429	4·032	2·994	2·348	1·600	1·188	0·932	1·128	1·102	1·084	1·060	1·046	1·034
0·00300	(0·0294)	6·506	4·433	3·292	2·445	1·917	1·307	0·970	*	1·087	1·064	1·048	1·028	1·015	1·005
0·00400	(0·0392)	5·634	3·839	2·851	2·117	1·660	1·131	*	*	1·060	1·039	1·025	1·006	0·995	0·986
0·00600	(0·0588)	4·600	3·135	2·328	1·729	1·356	0·924	*	1·053	1·023	1·005	0·993	0·977	0·967	0·960
0·00800	(0·0785)	3·984	2·715	2·016	1·497	1·174	*	1·046	1·026	0·999	0·983	0·971	0·957	0·949	0·942
0·01000	(0·0981)	3·563	2·428	1·803	1·339	1·050	*	1·025	1·006	0·981	0·966	0·955	0·942	0·935	0·929
0·01500	(0·147)	2·909	1·982	1·472	1·093	0·857	*	0·988	0·971	0·950	0·937	0·928	0·917	0·910	0·906
0·02000	(0·196)	2·520	1·717	1·275	0·947	*	0·985	0·963	0·948	0·929	0·917	0·909	0·899	0·894	0·890
0·03000	(0·294)	2·057	1·402	1·041	0·773	*	0·950	0·931	0·918	0·901	0·891	0·884	0·876	0·872	0·869
0·04000	(0·392)	1·782	1·214	0·902	*	0·954	0·927	0·909	0·898	0·882	0·873	0·867	0·860	0·856	0·854
0·06000	(0·588)	1·455	0·991	*	*	0·920	0·896	0·881	0·870	0·857	0·849	0·844	0·839	0·836	0·835
0·08000	(0·785)	1·260	0·858	*	0·913	0·897	0·875	0·861	0·852	0·840	0·833	0·829	0·824	0·822	0·822
0·10000	(0·981)	1·127	0·768	*	0·895	0·880	0·859	0·847	0·838	0·828	0·821	0·818	0·814	0·812	0·812
0·15000	(1·47)	0·920	*	0·884	0·864	0·850	0·833	0·822	0·814	0·806	0·801	0·798	0·795	0·794	0·794
0·20000	(1·96)	0·797	*	0·861	0·843	0·831	0·815	0·805	0·799	0·791	0·786	0·784	0·782	0·782	0·782
0·30000	(2·94)	*	0·850	0·831	0·815	0·805	0·791	0·783	0·777	0·771	0·767	0·766	0·765	0·765	0·766
0·40000	(3·92)	*	0·828	0·811	0·797	0·787	0·775	0·767	0·763	0·757	0·754	0·753	0·753	0·754	0·756
0·60000	(5·88)	0·823	0·800	0·785	0·772	0·764	0·753	0·747	0·743	0·739	0·737	0·736	0·737	0·738	0·741
0·80000	(7·85)	0·802	0·780	0·767	0·756	0·748	0·739	0·733	0·730	0·727	0·725	0·725	0·726	0·728	0·731
1·00000	(9·81)	0·786	0·766	0·754	0·743	0·736	0·728	0·723	0·720	0·717	0·717	0·717	0·718	0·720	0·723
1·50000	(14·7)	0·759	0·742	0·731	0·722	0·716	0·709	0·705	0·703	0·701	0·701	0·702	0·704	0·707	0·710
2·00000	(19·6)	0·741	0·726	0·716	0·708	0·703	0·697	0·693	0·692	0·690	0·691	0·692	0·694	0·698	0·701
3·00000	(29·4)	0·718	0·704	0·695	0·689	0·684	0·680	0·677	0·676	0·676	0·677	0·678	0·682	0·685	0·690
4·00000	(39·2)	0·702	0·689	0·682	0·676	0·672	0·668	0·666	0·666	0·666	0·667	0·669	0·673	0·677	0·682
6·00000	(58·8)	0·681	0·670	0·664	0·659	0·656	0·653	0·652	0·652	0·653	0·655	0·657	0·661	0·666	0·671
8·00000	(78·5)	0·666	0·657	0·651	0·647	0·645	0·643	0·642	0·642	0·644	0·646	0·649	0·654	0·658	0·664
10·0000	(98·1)	0·656	0·647	0·642	0·639	0·637	0·635	0·635	0·635	0·637	0·640	0·642	0·648	0·653	0·658
15·0000	(147)	0·638	0·631	0·627	0·624	0·622	0·622	0·622	0·623	0·626	0·629	0·632	0·638	0·643	0·649
20·0000	(196)	0·626	0·619	0·616	0·614	0·613	0·613	0·613	0·615	0·618	0·621	0·625	0·631	0·637	0·643
30·0000	(294)	0·610	0·604	0·602	0·601	0·600	0·601	0·602	0·604	0·608	0·612	0·615	0·622	0·628	0·635
S	$(\Delta p/\rho l)$	6	8	10	12·5	15	20	25	30	40	50	60	80	100	125

Grad'nt k.p.g. **(Equivalent) Pipe diameters in mm**

Grad'nt S	k.p.g. $(\Delta p/\rho l)$	(Equivalent) Pipe diameters in mm													
		125	150	200	250	300	400	500	600	800	1000	1250	1500	2000	2500
0·00010	(0·00098)	1·312	1·289	1·259	1·241	1·228	1·212	1·203	1·198	1·191	1·189	1·188	1·189	1·192	1·196
0·00015	(0·00147)	1·266	1·246	1·220	1·204	1·193	1·180	1·172	1·168	1·163	1·162	1·162	1·163	1·168	1·172
0·00020	(0·00196)	1·235	1·217	1·193	1·179	1·169	1·158	1·151	1·147	1·144	1·143	1·144	1·146	1·151	1·156
0·00030	(0·00294)	1·194	1·178	1·158	1·146	1·137	1·128	1·123	1·120	1·118	1·118	1·120	1·123	1·128	1·134
0·00040	(0·00392)	1·167	1·152	1·134	1·123	1·116	1·108	1·103	1·101	1·100	1·101	1·104	1·107	1·113	1·119
0·00060	(0·00588)	1·130	1·118	1·102	1·093	1·087	1·080	1·077	1·076	1·076	1·078	1·081	1·085	1·092	1·099
0·00080	(0·00785)	1·106	1·094	1·080	1·072	1·067	1·062	1·060	1·059	1·060	1·062	1·066	1·070	1·077	1·084
0·00100	(0·00981)	1·087	1·077	1·064	1·057	1·052	1·048	1·046	1·046	1·047	1·050	1·054	1·058	1·066	1·074
0·00150	(0·0147)	1·056	1·047	1·036	1·030	1·027	1·023	1·023	1·023	1·026	1·029	1·034	1·038	1·047	1·055
0·00200	(0·0196)	1·034	1·026	1·017	1·012	1·009	1·007	1·007	1·008	1·011	1·015	1·020	1·025	1·034	1·042
0·00300	(0·0294)	1·005	0·999	0·991	0·987	0·985	0·984	0·985	0·987	0·991	0·995	1·001	1·006	1·016	1·024
0·00400	(0·0392)	0·986	0·980	0·974	0·971	0·969	0·969	0·970	0·972	0·977	0·982	0·988	0·993	1·003	1·012
0·00600	(0·0588)	0·960	0·955	0·950	0·948	0·947	0·948	0·950	0·953	0·958	0·963	0·970	0·976	0·986	0·996
0·00800	(0·0785)	0·942	0·938	0·934	0·933	0·932	0·934	0·936	0·939	0·945	0·951	0·958	0·964	0·975	0·984
0·01000	(0·0981)	0·929	0·925	0·922	0·921	0·921	0·923	0·926	0·929	0·935	0·941	0·948	0·955	0·966	0·976
0·01500	(0·147)	0·906	0·903	0·901	0·901	0·902	0·904	0·908	0·911	0·918	0·925	0·932	0·939	0·950	0·960
0·02000	(0·196)	0·890	0·888	0·886	0·887	0·888	0·892	0·895	0·899	0·907	0·913	0·921	0·928	0·940	0·950
0·03000	(0·294)	0·869	0·867	0·867	0·868	0·870	0·874	0·878	0·883	0·891	0·898	0·906	0·913	0·925	0·936
0·04000	(0·392)	0·854	0·853	0·854	0·855	0·858	0·862	0·867	0·871	0·880	0·887	0·895	0·903	0·915	0·926
0·06000	(0·588)	0·835	0·835	0·836	0·838	0·841	0·846	0·851	0·856	0·865	0·873	0·881	0·889	0·902	0·913
0·08000	(0·785)	0·822	0·822	0·824	0·826	0·829	0·835	0·840	0·846	0·855	0·863	0·872	0·879	0·893	0·904
0·10000	(0·981)	0·812	0·812	0·814	0·817	0·821	0·827	0·832	0·838	0·847	0·855	0·864	0·872	0·886	0·897
0·15000	(1·47)	0·794	0·795	0·798	0·802	0·805	0·812	0·818	0·824	0·834	0·842	0·852	0·860	0·874	0·885
0·20000	(1·96)	0·782	0·784	0·787	0·791	0·795	0·802	0·809	0·814	0·825	0·833	0·843	0·851	0·865	0·877
0·30000	(2·94)	0·766	0·768	0·773	0·777	0·781	0·789	0·795	0·802	0·812	0·821	0·831	0·840	0·854	0·867
0·40000	(3·92)	0·756	0·758	0·762	0·767	0·771	0·779	0·787	0·793	0·804	0·813	0·823	0·832	0·847	0·859
0·60000	(5·88)	0·741	0·743	0·749	0·754	0·759	0·767	0·774	0·781	0·793	0·802	0·813	0·822	0·837	0·849
0·80000	(7·85)	0·731	0·734	0·740	0·745	0·750	0·759	0·766	0·773	0·785	0·795	0·805	0·815	0·830	0·843
1·00000	(9·81)	0·723	0·727	0·733	0·738	0·743	0·752	0·760	0·767	0·779	0·789	0·800	0·809	0·825	0·838
1·50000	(14·7)	0·710	0·714	0·720	0·726	0·732	0·741	0·750	0·757	0·769	0·780	0·791	0·800	0·817	0·830
2·00000	(19·6)	0·701	0·705	0·712	0·718	0·724	0·734	0·743	0·750	0·763	0·773	0·785	0·794	0·811	0·824
3·00000	(29·4)	0·690	0·694	0·701	0·708	0·714	0·724	0·733	0·741	0·754	0·765	0·776	0·787	0·803	0·817
S	$(\Delta p/\rho l)$	125	150	200	250	300	400	500	600	800	1000	1250	1500	2000	2500

Grad'nt k.p.g. **(Equivalent) Pipe diameters in mm**

Kinematic viscosity, $\nu = 3{\cdot}0 \times 10^{-6}$ m^{2}s^{-1} ;   **Roughness size, $k_s = 0{\cdot}0015$ mm**

Kin. visc., $\nu = 3 \cdot 0 \times 10^{-6}$ m^{2}s^{-1}; $S = 0 \cdot 00010$ to $30 \cdot 0000$

i.e. kin. pr. grad., $\Delta p / \rho l = (0 \cdot 00098)$ to (294) ms^{-2}

Roughness size, $k_s = 0 \cdot 003$ mm

This table shows values of m, as follows

m_C for Colebrook-White solutions; or, where $\mathbf{R} \leq 2000$, m_P for laminar flow

Grad'nt S	k.p.g. $(\Delta p/\rho l)$	6	8	10	12·5	15	20	25	30	40	50	60	80	100	125
						(Equivalent) Pipe diameters in mm									
0·00030	(0·00294)	20·57	14·02	10·41	7·732	6·063	4·132	3·068	2·406	1·640	1·218	*	1·245	1·218	1·195
0·00040	(0·00392)	17·82	12·14	9·016	6·696	5·251	3·578	2·657	2·084	1·420	1·055	*	1·213	1·188	1·167
0·00060	(0·00588)	14·55	9·912	7·362	5·467	4·287	2·921	2·170	1·701	1·159	*	1·206	1·171	1·149	1·131
0·00080	(0·00785)	12·60	8·584	6·375	4·735	3·713	2·530	1·879	1·473	1·004	*	1·175	1·143	1·123	1·106
0·00100	(0·00981)	11·27	7·678	5·702	4·235	3·321	2·263	1·681	1·318	*	1·174	1·151	1·122	1·103	1·088
0·00150	(0·0147)	9·200	6·269	4·656	3·458	2·712	1·848	1·372	1·076	*	1·131	1·111	1·085	1·069	1·056
0·00200	(0·0196)	7·968	5·429	4·032	2·994	2·348	1·600	1·188	0·932	1·129	1·103	1·084	1·061	1·046	1·035
0·00300	(0·0294)	6·506	4·433	3·292	2·445	1·917	1·307	0·970	*	1·088	1·065	1·049	1·029	1·016	1·006
0·00400	(0·0392)	5·634	3·839	2·851	2·117	1·660	1·131	*	*	1·060	1·040	1·025	1·007	0·995	0·987
0·00600	(0·0588)	4·600	3·135	2·328	1·729	1·356	0·924	*	1·053	1·024	1·006	0·993	0·978	0·968	0·961
0·00800	(0·0785)	3·984	2·715	2·016	1·497	1·174	*	1·047	1·026	1·000	0·983	0·972	0·958	0·949	0·943
0·01000	(0·0981)	3·563	2·428	1·803	1·339	1·050	*	1·025	1·006	0·982	0·967	0·956	0·943	0·935	0·930
0·01500	(0·147)	2·909	1·982	1·472	1·093	0·857	*	0·989	0·972	0·951	0·938	0·929	0·918	0·911	0·907
0·02000	(0·196)	2·520	1·717	1·275	0·947	*	0·986	0·964	0·949	0·930	0·918	0·910	0·900	0·895	0·891
0·03000	(0·294)	2·057	1·402	1·041	0·773	*	0·951	0·932	0·919	0·902	0·892	0·885	0·877	0·873	0·870
0·04000	(0·392)	1·782	1·214	0·902	*	0·955	0·928	0·910	0·899	0·884	0·874	0·869	0·862	0·858	0·856
0·06000	(0·588)	1·455	0·991	*	*	0·921	0·897	0·882	0·872	0·859	0·851	0·846	0·841	0·838	0·837
0·08000	(0·785)	1·260	0·858	*	0·915	0·898	0·876	0·863	0·853	0·842	0·835	0·831	0·826	0·824	0·824
0·10000	(0·981)	1·127	0·768	*	0·897	0·881	0·861	0·848	0·840	0·829	0·823	0·819	0·816	0·814	0·814
0·15000	(1·47)	0·920	*	0·885	0·865	0·852	0·834	0·824	0·816	0·807	0·803	0·800	0·797	0·796	0·797
0·20000	(1·96)	0·797	*	0·863	0·845	0·832	0·817	0·807	0·800	0·793	0·789	0·786	0·784	0·784	0·785
0·30000	(2·94)	*	0·852	0·833	0·817	0·807	0·793	0·785	0·779	0·773	0·770	0·768	0·768	0·768	0·770
0·40000	(3·92)	*	0·830	0·813	0·799	0·789	0·777	0·770	0·765	0·760	0·757	0·756	0·756	0·757	0·759
0·60000	(5·88)	0·826	0·802	0·787	0·775	0·766	0·756	0·750	0·746	0·742	0·740	0·740	0·741	0·743	0·745
0·80000	(7·85)	0·804	0·783	0·769	0·758	0·751	0·742	0·736	0·733	0·730	0·729	0·729	0·730	0·733	0·736
1·00000	(9·81)	0·789	0·769	0·756	0·746	0·739	0·731	0·726	0·724	0·721	0·721	0·721	0·723	0·725	0·729
1·50000	(14·7)	0·762	0·745	0·734	0·725	0·720	0·713	0·709	0·707	0·706	0·706	0·707	0·709	0·713	0·717
2·00000	(19·6)	0·745	0·729	0·719	0·711	0·706	0·701	0·698	0·696	0·695	0·696	0·697	0·700	0·704	0·708
3·00000	(29·4)	0·721	0·708	0·699	0·693	0·689	0·684	0·682	0·681	0·681	0·683	0·685	0·688	0·693	0·697
4·00000	(39·2)	0·706	0·693	0·686	0·680	0·677	0·673	0·672	0·671	0·672	0·674	0·676	0·681	0·685	0·690
6·00000	(58·8)	0·685	0·675	0·669	0·664	0·661	0·659	0·658	0·658	0·660	0·662	0·665	0·670	0·675	0·681
8·00000	(78·5)	0·671	0·662	0·657	0·653	0·651	0·649	0·649	0·649	0·652	0·654	0·657	0·663	0·668	0·674
10·0000	(98·1)	0·661	0·653	0·648	0·645	0·643	0·642	0·642	0·643	0·646	0·649	0·652	0·658	0·664	0·670
15·0000	(147)	0·644	0·637	0·633	0·631	0·630	0·629	0·630	0·632	0·635	0·639	0·643	0·649	0·655	0·662
20·0000	(196)	0·632	0·626	0·623	0·622	0·621	0·621	0·623	0·624	0·628	0·633	0·636	0·644	0·650	0·657
30·0000	(294)	0·617	0·612	0·610	0·609	0·609	0·611	0·613	0·615	0·620	0·624	0·628	0·636	0·643	0·651
S	$(\Delta p/\rho l)$	6	8	10	12·5	15	20	25	30	40	50	60	80	100	125

Grad'nt S	k.p.g. $(\Delta p/\rho l)$	125	150	200	250	300	400	500	600	800	1000	1250	1500	2000	2500
						(Equivalent) Pipe diameters in mm									
0·00010	(0·00098)	1·312	1·289	1·260	1·241	1·228	1·213	1·204	1·198	1·192	1·189	1·189	1·189	1·192	1·196
0·00015	(0·00147)	1·266	1·246	1·220	1·204	1·193	1·180	1·173	1·168	1·164	1·162	1·163	1·164	1·168	1·173
0·00020	(0·00196)	1·235	1·217	1·194	1·179	1·170	1·158	1·152	1·148	1·144	1·144	1·145	1·147	1·152	1·157
0·00030	(0·00294)	1·195	1·179	1·158	1·146	1·138	1·128	1·123	1·120	1·119	1·119	1·121	1·123	1·129	1·135
0·00040	(0·00392)	1·167	1·153	1·134	1·123	1·116	1·108	1·104	1·102	1·101	1·102	1·104	1·107	1·114	1·120
0·00060	(0·00588)	1·131	1·118	1·102	1·093	1·087	1·081	1·078	1·077	1·077	1·079	1·082	1·085	1·093	1·099
0·00080	(0·00785)	1·106	1·095	1·081	1·073	1·068	1·062	1·060	1·059	1·061	1·063	1·067	1·070	1·078	1·085
0·00100	(0·00981)	1·088	1·077	1·065	1·057	1·053	1·048	1·047	1·047	1·048	1·051	1·055	1·059	1·067	1·075
0·00150	(0·0147)	1·056	1·047	1·037	1·031	1·027	1·024	1·023	1·024	1·027	1·030	1·035	1·039	1·048	1·056
0·00200	(0·0196)	1·035	1·027	1·017	1·012	1·010	1·007	1·007	1·008	1·012	1·016	1·021	1·026	1·035	1·043
0·00300	(0·0294)	1·006	0·999	0·992	0·988	0·986	0·985	0·986	0·988	0·992	0·996	1·002	1·007	1·017	1·026
0·00400	(0·0392)	0·987	0·981	0·974	0·971	0·970	0·970	0·971	0·973	0·978	0·983	0·989	0·995	1·005	1·014
0·00600	(0·0588)	0·961	0·956	0·951	0·949	0·948	0·949	0·951	0·954	0·959	0·965	0·972	0·978	0·988	0·998
0·00800	(0·0785)	0·943	0·939	0·935	0·934	0·934	0·935	0·938	0·941	0·947	0·953	0·960	0·966	0·977	0·987
0·01000	(0·0981)	0·930	0·926	0·923	0·922	0·922	0·925	0·928	0·931	0·937	0·943	0·951	0·957	0·968	0·978
0·01500	(0·147)	0·907	0·904	0·902	0·902	0·903	0·906	0·910	0·913	0·920	0·927	0·935	0·941	0·953	0·964
0·02000	(0·196)	0·891	0·889	0·888	0·888	0·890	0·893	0·897	0·901	0·909	0·916	0·924	0·931	0·943	0·954
0·03000	(0·294)	0·870	0·869	0·869	0·870	0·872	0·876	0·881	0·885	0·893	0·901	0·909	0·916	0·929	0·940
0·04000	(0·392)	0·856	0·855	0·856	0·857	0·860	0·865	0·870	0·874	0·883	0·891	0·899	0·907	0·920	0·931
0·06000	(0·588)	0·837	0·837	0·838	0·841	0·843	0·849	0·854	0·859	0·869	0·877	0·886	0·894	0·907	0·919
0·08000	(0·785)	0·824	0·824	0·826	0·829	0·832	0·838	0·844	0·849	0·859	0·867	0·876	0·885	0·899	0·910
0·10000	(0·981)	0·814	0·815	0·817	0·820	0·824	0·830	0·836	0·842	0·852	0·860	0·870	0·878	0·892	0·904
0·15000	(1·47)	0·797	0·798	0·802	0·805	0·809	0·816	0·823	0·828	0·839	0·848	0·858	0·866	0·881	0·893
0·20000	(1·96)	0·785	0·787	0·791	0·795	0·799	0·807	0·813	0·820	0·830	0·840	0·850	0·859	0·874	0·886
0·30000	(2·94)	0·770	0·772	0·777	0·781	0·786	0·794	0·801	0·808	0·819	0·829	0·839	0·848	0·864	0·876
0·40000	(3·92)	0·759	0·762	0·767	0·772	0·777	0·785	0·793	0·799	0·811	0·821	0·832	0·841	0·857	0·870
0·60000	(5·88)	0·745	0·748	0·754	0·760	0·765	0·774	0·782	0·789	0·801	0·811	0·822	0·832	0·848	0·862
0·80000	(7·85)	0·736	0·739	0·745	0·751	0·757	0·766	0·774	0·782	0·794	0·805	0·816	0·826	0·843	0·856
1·00000	(9·81)	0·729	0·732	0·739	0·745	0·750	0·760	0·769	0·776	0·789	0·800	0·812	0·822	0·839	0·852
1·50000	(14·7)	0·717	0·720	0·728	0·734	0·740	0·751	0·759	0·767	0·781	0·792	0·804	0·814	0·832	0·846
2·00000	(19·6)	0·708	0·713	0·720	0·727	0·733	0·744	0·753	0·761	0·775	0·787	0·799	0·809	0·827	0·842
3·00000	(29·4)	0·697	0·702	0·710	0·718	0·724	0·735	0·745	0·753	0·768	0·780	0·792	0·803	0·821	0·836
S	$(\Delta p/\rho l)$	125	150	200	250	300	400	500	600	800	1000	1250	1500	2000	2500

Grad'nt k.p.g. (Equivalent) Pipe diameters in mm

Kinematic viscosity, $\nu = 3 \cdot 0 \times 10^{-6}$ m^{2}s^{-1} ; Roughness size, $k_s = 0 \cdot 003$ mm

E149

Kin. visc., $\nu = 3.0 \times 10^{-6}$ m^{2}s^{-1};
$S = 0.00010$ to 30.0000

i.e. kin. pr. grad., $\Delta p/\rho l =$
(0.00098) to (294) ms^{-2}

Roughness size, $k_s = 0.006$ mm
This table shows values of m, as follows

m_C for Colebrook-White solutions; or,
where $\mathbf{R} \le 2000$, m_P for laminar flow

Grad'nt	k.p.g.	(Equivalent) Pipe diameters in mm													
S	$(\Delta p/\rho l)$	6	8	10	12.5	15	20	25	30	40	50	60	80	100	125
0.00030	(0.00294)	20.57	14.02	10.41	7.732	6.063	4.132	3.068	2.406	1.640	1.218	*	1.246	1.218	1.195
0.00040	(0.00392)	17.82	12.14	9.016	6.696	5.251	3.578	2.657	2.084	1.420	1.055	*	1.214	1.189	1.168
0.00060	(0.00588)	14.55	9.912	7.362	5.467	4.287	2.921	2.170	1.701	1.159	*	1.207	1.172	1.150	1.131
0.00080	(0.00785)	12.60	8.584	6.375	4.735	3.713	2.530	1.879	1.473	1.004	*	1.175	1.143	1.123	1.107
0.00100	(0.00981)	11.27	7.678	5.702	4.235	3.321	2.263	1.681	1.318	*	1.175	1.152	1.122	1.104	1.089
0.00150	(0.0147)	9.200	6.269	4.656	3.458	2.712	1.848	1.372	1.076	*	1.132	1.112	1.086	1.070	1.057
0.00200	(0.0196)	7.968	5.429	4.032	2.994	2.348	1.600	1.188	0.932	1.130	1.104	1.085	1.062	1.047	1.036
0.00300	(0.0294)	6.506	4.433	3.292	2.445	1.917	1.307	0.970	*	1.089	1.066	1.050	1.030	1.017	1.007
0.00400	(0.0392)	5.634	3.839	2.851	2.117	1.660	1.131	*	*	1.061	1.041	1.026	1.008	0.997	0.988
0.00600	(0.0588)	4.600	3.135	2.328	1.729	1.356	0.924	*	1.054	1.025	1.007	0.995	0.979	0.969	0.962
0.00800	(0.0785)	3.984	2.715	2.016	1.497	1.174	*	1.048	1.028	1.001	0.985	0.974	0.959	0.951	0.945
0.01000	(0.0981)	3.563	2.428	1.803	1.339	1.050	*	1.027	1.008	0.983	0.968	0.958	0.945	0.937	0.932
0.01500	(0.147)	2.909	1.982	1.472	1.093	0.857	*	0.990	0.974	0.953	0.939	0.931	0.920	0.913	0.909
0.02000	(0.196)	2.520	1.717	1.275	0.947	*	0.988	0.966	0.951	0.932	0.920	0.912	0.903	0.897	0.894
0.03000	(0.294)	2.057	1.402	1.041	0.773	*	0.953	0.934	0.921	0.904	0.894	0.888	0.880	0.876	0.873
0.04000	(0.392)	1.782	1.214	0.902	*	0.957	0.930	0.913	0.901	0.886	0.877	0.871	0.864	0.861	0.859
0.06000	(0.588)	1.455	0.991	*	*	0.923	0.899	0.884	0.874	0.861	0.854	0.849	0.844	0.841	0.840
0.08000	(0.785)	1.260	0.858	*	0.917	0.900	0.879	0.865	0.856	0.845	0.838	0.834	0.830	0.828	0.828
0.10000	(0.981)	1.127	0.768	*	0.899	0.884	0.864	0.851	0.843	0.833	0.827	0.823	0.820	0.818	0.818
0.15000	(1.47)	0.920	*	0.888	0.868	0.855	0.838	0.827	0.820	0.811	0.807	0.804	0.802	0.801	0.802
0.20000	(1.96)	0.797	*	0.866	0.848	0.836	0.820	0.811	0.804	0.797	0.793	0.791	0.789	0.790	0.791
0.30000	(2.94)	*	0.855	0.837	0.821	0.810	0.797	0.789	0.784	0.778	0.775	0.774	0.773	0.774	0.777
0.40000	(3.92)	*	0.834	0.817	0.803	0.794	0.782	0.775	0.770	0.765	0.763	0.762	0.763	0.764	0.767
0.60000	(5.88)	0.830	0.806	0.791	0.779	0.771	0.761	0.755	0.752	0.748	0.747	0.747	0.748	0.750	0.754
0.80000	(7.85)	0.809	0.788	0.774	0.764	0.756	0.747	0.742	0.740	0.737	0.736	0.737	0.739	0.741	0.745
1.00000	(9.81)	0.794	0.774	0.762	0.752	0.745	0.737	0.733	0.730	0.728	0.728	0.729	0.731	0.735	0.739
1.50000	(14.7)	0.768	0.751	0.740	0.732	0.726	0.720	0.717	0.715	0.714	0.715	0.716	0.719	0.723	0.728
2.00000	(19.6)	0.751	0.735	0.726	0.718	0.714	0.708	0.706	0.705	0.704	0.706	0.707	0.711	0.716	0.721
3.00000	(29.4)	0.728	0.715	0.707	0.701	0.697	0.693	0.691	0.691	0.692	0.694	0.696	0.701	0.706	0.711
4.00000	(39.2)	0.713	0.701	0.694	0.689	0.686	0.683	0.682	0.682	0.683	0.686	0.689	0.694	0.699	0.705
6.00000	(58.8)	0.694	0.684	0.678	0.674	0.671	0.670	0.669	0.670	0.673	0.676	0.679	0.685	0.691	0.697
8.00000	(78.5)	0.681	0.672	0.667	0.664	0.662	0.661	0.661	0.662	0.666	0.669	0.673	0.679	0.685	0.692
10.0000	(98.1)	0.671	0.663	0.659	0.656	0.655	0.654	0.655	0.657	0.660	0.664	0.668	0.675	0.681	0.689
15.0000	(147)	0.655	0.649	0.646	0.644	0.643	0.644	0.645	0.647	0.652	0.656	0.660	0.668	0.675	0.683
20.0000	(196)	0.644	0.639	0.637	0.636	0.635	0.637	0.639	0.641	0.646	0.651	0.655	0.664	0.671	0.679
30.0000	(294)	0.631	0.627	0.625	0.625	0.625	0.628	0.630	0.633	0.639	0.644	0.649	0.658	0.666	0.674
S	$(\Delta p/\rho l)$	6	8	10	12.5	15	20	25	30	40	50	60	80	100	125

Grad'nt	k.p.g.	(Equivalent) Pipe diameters in mm													
S	$(\Delta p/\rho l)$	125	150	200	250	300	400	500	600	800	1000	1250	1500	2000	2500
0.00010	(0.00098)	1.313	1.290	1.260	1.242	1.229	1.213	1.204	1.198	1.192	1.190	1.189	1.190	1.193	1.197
0.00015	(0.00147)	1.267	1.247	1.221	1.205	1.194	1.181	1.173	1.169	1.164	1.163	1.163	1.165	1.169	1.174
0.00020	(0.00196)	1.236	1.218	1.194	1.180	1.170	1.159	1.152	1.149	1.145	1.145	1.146	1.148	1.153	1.158
0.00030	(0.00294)	1.195	1.179	1.159	1.147	1.139	1.129	1.124	1.121	1.119	1.120	1.122	1.124	1.130	1.136
0.00040	(0.00392)	1.168	1.153	1.135	1.124	1.117	1.109	1.105	1.103	1.102	1.103	1.105	1.108	1.115	1.121
0.00060	(0.00588)	1.131	1.119	1.103	1.094	1.088	1.082	1.079	1.078	1.078	1.080	1.083	1.087	1.094	1.101
0.00080	(0.00785)	1.107	1.096	1.082	1.074	1.069	1.063	1.061	1.061	1.062	1.064	1.068	1.072	1.080	1.087
0.00100	(0.00981)	1.089	1.078	1.066	1.058	1.054	1.050	1.048	1.048	1.050	1.053	1.057	1.061	1.069	1.077
0.00150	(0.0147)	1.057	1.048	1.038	1.032	1.028	1.025	1.025	1.025	1.028	1.032	1.037	1.042	1.051	1.059
0.00200	(0.0196)	1.036	1.028	1.019	1.014	1.011	1.009	1.009	1.010	1.014	1.018	1.023	1.028	1.038	1.046
0.00300	(0.0294)	1.007	1.001	0.993	0.990	0.988	0.987	0.988	0.990	0.994	0.999	1.005	1.010	1.020	1.029
0.00400	(0.0392)	0.988	0.982	0.976	0.973	0.972	0.972	0.973	0.976	0.981	0.986	0.992	0.998	1.009	1.018
0.00600	(0.0588)	0.962	0.958	0.953	0.951	0.950	0.952	0.954	0.957	0.962	0.968	0.975	0.981	0.993	1.002
0.00800	(0.0785)	0.945	0.941	0.937	0.936	0.936	0.938	0.941	0.944	0.950	0.956	0.963	0.970	0.982	0.992
0.01000	(0.0981)	0.932	0.928	0.925	0.925	0.925	0.927	0.931	0.934	0.941	0.947	0.955	0.961	0.973	0.984
0.01500	(0.147)	0.909	0.907	0.905	0.905	0.906	0.909	0.913	0.917	0.925	0.932	0.939	0.947	0.959	0.970
0.02000	(0.196)	0.894	0.892	0.891	0.892	0.893	0.897	0.901	0.906	0.914	0.921	0.929	0.937	0.950	0.961
0.03000	(0.294)	0.873	0.872	0.872	0.874	0.876	0.880	0.885	0.890	0.899	0.907	0.915	0.923	0.937	0.948
0.04000	(0.392)	0.859	0.859	0.859	0.861	0.864	0.869	0.875	0.880	0.889	0.897	0.906	0.914	0.928	0.940
0.06000	(0.588)	0.840	0.841	0.842	0.845	0.848	0.854	0.860	0.866	0.875	0.884	0.894	0.902	0.917	0.929
0.08000	(0.785)	0.828	0.828	0.831	0.834	0.838	0.844	0.850	0.856	0.866	0.876	0.885	0.894	0.909	0.922
0.10000	(0.981)	0.818	0.819	0.822	0.826	0.830	0.837	0.843	0.849	0.860	0.869	0.879	0.888	0.903	0.916
0.15000	(1.47)	0.802	0.804	0.808	0.812	0.816	0.824	0.831	0.837	0.848	0.858	0.869	0.878	0.894	0.907
0.20000	(1.96)	0.791	0.793	0.798	0.802	0.807	0.815	0.822	0.829	0.841	0.851	0.862	0.871	0.887	0.901
0.30000	(2.94)	0.777	0.779	0.784	0.789	0.794	0.803	0.811	0.818	0.831	0.841	0.853	0.863	0.879	0.893
0.40000	(3.92)	0.767	0.770	0.775	0.781	0.786	0.796	0.804	0.811	0.824	0.835	0.847	0.857	0.874	0.888
0.60000	(5.88)	0.754	0.757	0.764	0.770	0.776	0.786	0.794	0.802	0.816	0.827	0.839	0.849	0.867	0.882
0.80000	(7.85)	0.745	0.749	0.756	0.762	0.768	0.779	0.788	0.796	0.810	0.822	0.834	0.845	0.863	0.878
1.00000	(9.81)	0.739	0.743	0.750	0.757	0.763	0.774	0.784	0.792	0.806	0.818	0.830	0.841	0.860	0.875
1.50000	(14.7)	0.728	0.732	0.741	0.748	0.754	0.766	0.776	0.784	0.799	0.811	0.824	0.836	0.854	0.870
2.00000	(19.6)	0.721	0.725	0.734	0.742	0.749	0.761	0.771	0.780	0.795	0.807	0.821	0.832	0.851	0.867
3.00000	(29.4)	0.711	0.717	0.726	0.734	0.741	0.754	0.765	0.774	0.789	0.802	0.816	0.828	0.847	0.863
S	$(\Delta p/\rho l)$	125	150	200	250	300	400	500	600	800	1000	1250	1500	2000	2500

Grad'nt   k.p.g.   (Equivalent) Pipe diameters in mm

Kinematic viscosity, $\nu = 3.0 \times 10^{-6}$ m^{2}s^{-1} ;

Roughness size, $k_s = 0.006$ mm

Kin. visc., $\nu = 3.0 \times 10^{-6}\ \mathrm{m^2 s^{-1}}$;
$S = 0.00010$ to 30.0000

i.e. kin. pr. grad., $\Delta p/\rho l = (0.00098)$ to $(294)\ \mathrm{ms^{-2}}$

Roughness size, $k_s = 0.015$ mm
This table shows values of m, as follows
m_C for Colebrook-White solutions; or,
where $\mathbf{R} \le 2000$, m_P for laminar flow

Grad'nt S	k.p.g. $(\Delta p/\rho l)$	6	8	10	12.5	15	20	25	30	40	50	60	80	100	125
		\(Equivalent\) Pipe diameters in mm													
0.00030	(0.00294)	20.57	14.02	10.41	7.732	6.063	4.132	3.068	2.406	1.640	1.218	*	1.247	1.220	1.197
0.00040	(0.00392)	17.82	12.14	9.016	6.696	5.251	3.578	2.657	2.084	1.420	1.055	*	1.216	1.191	1.170
0.00060	(0.00588)	14.55	9.912	7.362	5.467	4.287	2.921	2.170	1.701	1.159	*	1.209	1.174	1.152	1.133
0.00080	(0.00785)	12.60	8.584	6.375	4.735	3.713	2.530	1.879	1.473	1.004	*	1.177	1.146	1.126	1.109
0.00100	(0.00981)	11.27	7.678	5.702	4.235	3.321	2.263	1.681	1.318	*	1.177	1.154	1.125	1.106	1.091
0.00150	(0.0147)	9.200	6.269	4.656	3.458	2.712	1.848	1.372	1.076	*	1.134	1.115	1.089	1.073	1.060
0.00200	(0.0196)	7.968	5.429	4.032	2.994	2.348	1.600	1.188	0.932	1.132	1.106	1.088	1.065	1.051	1.039
0.00300	(0.0294)	6.506	4.433	3.292	2.445	1.917	1.307	0.970	*	1.092	1.069	1.053	1.033	1.021	1.011
0.00400	(0.0392)	5.634	3.839	2.851	2.117	1.660	1.131	*	*	1.065	1.044	1.030	1.012	1.001	0.992
0.00600	(0.0588)	4.600	3.135	2.328	1.729	1.356	0.924	*	1.058	1.029	1.011	0.999	0.983	0.974	0.967
0.00800	(0.0785)	3.984	2.715	2.016	1.497	1.174	*	1.052	1.032	1.005	0.989	0.978	0.964	0.956	0.950
0.01000	(0.0981)	3.563	2.428	1.803	1.339	1.050	*	1.031	1.012	0.988	0.973	0.963	0.950	0.943	0.937
0.01500	(0.147)	2.909	1.982	1.472	1.093	0.857	*	0.995	0.979	0.958	0.945	0.936	0.926	0.920	0.916
0.02000	(0.196)	2.520	1.717	1.275	0.947	*	0.993	0.971	0.957	0.938	0.926	0.918	0.909	0.904	0.901
0.03000	(0.294)	2.057	1.402	1.041	0.773	*	0.959	0.940	0.927	0.911	0.901	0.895	0.887	0.884	0.881
0.04000	(0.392)	1.782	1.214	0.902	*	0.963	0.936	0.919	0.908	0.893	0.884	0.879	0.873	0.870	0.868
0.06000	(0.588)	1.455	0.991	*	*	0.930	0.906	0.892	0.882	0.869	0.862	0.858	0.853	0.851	0.851
0.08000	(0.785)	1.260	0.858	*	0.924	0.908	0.887	0.873	0.865	0.854	0.848	0.844	0.840	0.839	0.840
0.10000	(0.981)	1.127	0.768	*	0.907	0.892	0.872	0.860	0.852	0.842	0.837	0.834	0.831	0.830	0.831
0.15000	(1.47)	0.920	*	0.897	0.877	0.864	0.847	0.837	0.830	0.822	0.818	0.816	0.814	0.815	0.816
0.20000	(1.96)	0.797	*	0.875	0.858	0.846	0.831	0.821	0.816	0.809	0.806	0.804	0.803	0.805	0.807
0.30000	(2.94)	*	0.866	0.847	0.832	0.822	0.809	0.801	0.797	0.791	0.789	0.789	0.789	0.791	0.794
0.40000	(3.92)	*	0.845	0.829	0.815	0.806	0.794	0.788	0.784	0.780	0.779	0.778	0.780	0.782	0.786
0.60000	(5.88)	0.842	0.819	0.804	0.793	0.785	0.776	0.770	0.768	0.765	0.764	0.765	0.768	0.771	0.775
0.80000	(7.85)	0.823	0.801	0.789	0.778	0.771	0.763	0.759	0.757	0.755	0.755	0.756	0.760	0.764	0.768
1.00000	(9.81)	0.808	0.789	0.777	0.767	0.761	0.754	0.751	0.749	0.748	0.749	0.750	0.754	0.758	0.763
1.50000	(14.7)	0.784	0.767	0.757	0.749	0.744	0.739	0.736	0.735	0.736	0.737	0.739	0.744	0.749	0.755
2.00000	(19.6)	0.768	0.753	0.744	0.737	0.733	0.729	0.727	0.727	0.728	0.730	0.733	0.738	0.744	0.750
3.00000	(29.4)	0.748	0.735	0.728	0.722	0.719	0.716	0.715	0.716	0.718	0.721	0.724	0.730	0.736	0.743
4.00000	(39.2)	0.734	0.723	0.717	0.712	0.710	0.708	0.708	0.708	0.711	0.715	0.718	0.725	0.732	0.739
6.00000	(58.8)	0.717	0.708	0.703	0.699	0.698	0.697	0.698	0.699	0.703	0.707	0.711	0.719	0.726	0.734
8.00000	(78.5)	0.706	0.698	0.694	0.691	0.690	0.690	0.692	0.694	0.698	0.703	0.707	0.715	0.722	0.730
10.0000	(98.1)	0.698	0.691	0.687	0.685	0.685	0.685	0.687	0.689	0.694	0.699	0.704	0.712	0.720	0.728
15.0000	(147)	0.684	0.679	0.677	0.675	0.676	0.677	0.680	0.683	0.688	0.694	0.699	0.708	0.716	0.724
20.0000	(196)	0.676	0.671	0.670	0.669	0.670	0.672	0.675	0.678	0.684	0.690	0.696	0.705	0.713	0.722
30.0000	(294)	0.665	0.662	0.661	0.661	0.663	0.666	0.669	0.673	0.680	0.686	0.692	0.702	0.710	0.719
S	$(\Delta p/\rho l)$	6	8	10	12.5	15	20	25	30	40	50	60	80	100	125

Grad'nt k.p.g. (Equivalent) Pipe diameters in mm

Grad'nt S	k.p.g. $(\Delta p/\rho l)$	125	150	200	250	300	400	500	600	800	1000	1250	1500	2000	2500
0.00010	(0.00098)	1.314	1.291	1.261	1.243	1.230	1.215	1.206	1.200	1.194	1.192	1.191	1.192	1.196	1.200
0.00015	(0.00147)	1.268	1.248	1.222	1.206	1.196	1.182	1.175	1.171	1.166	1.165	1.166	1.167	1.172	1.177
0.00020	(0.00196)	1.237	1.219	1.196	1.182	1.172	1.161	1.154	1.151	1.148	1.147	1.148	1.151	1.156	1.161
0.00030	(0.00294)	1.197	1.181	1.161	1.149	1.141	1.131	1.126	1.124	1.122	1.123	1.125	1.128	1.134	1.140
0.00040	(0.00392)	1.170	1.155	1.137	1.126	1.119	1.111	1.107	1.106	1.105	1.106	1.109	1.112	1.119	1.126
0.00060	(0.00588)	1.133	1.121	1.106	1.097	1.091	1.085	1.082	1.081	1.082	1.084	1.087	1.091	1.099	1.107
0.00080	(0.00785)	1.109	1.098	1.084	1.076	1.072	1.067	1.065	1.064	1.066	1.069	1.073	1.077	1.085	1.093
0.00100	(0.00981)	1.091	1.081	1.068	1.061	1.057	1.053	1.052	1.052	1.054	1.057	1.062	1.066	1.075	1.083
0.00150	(0.0147)	1.060	1.051	1.041	1.035	1.032	1.029	1.029	1.030	1.033	1.037	1.043	1.048	1.057	1.066
0.00200	(0.0196)	1.039	1.031	1.022	1.018	1.015	1.013	1.014	1.015	1.019	1.024	1.030	1.035	1.045	1.055
0.00300	(0.0294)	1.011	1.005	0.997	0.994	0.992	0.992	0.993	0.995	1.000	1.006	1.012	1.018	1.029	1.039
0.00400	(0.0392)	0.992	0.987	0.981	0.978	0.977	0.978	0.980	0.982	0.988	0.994	1.000	1.007	1.018	1.029
0.00600	(0.0588)	0.967	0.963	0.958	0.957	0.957	0.958	0.961	0.964	0.971	0.977	0.985	0.992	1.004	1.015
0.00800	(0.0785)	0.950	0.946	0.943	0.942	0.943	0.945	0.948	0.952	0.959	0.966	0.974	0.981	0.994	1.005
0.01000	(0.0981)	0.937	0.934	0.932	0.932	0.932	0.935	0.939	0.943	0.951	0.958	0.966	0.974	0.987	0.998
0.01500	(0.147)	0.916	0.913	0.912	0.913	0.914	0.918	0.923	0.928	0.936	0.944	0.953	0.961	0.975	0.987
0.02000	(0.196)	0.901	0.899	0.899	0.900	0.902	0.907	0.912	0.917	0.926	0.935	0.944	0.952	0.967	0.979
0.03000	(0.294)	0.881	0.881	0.882	0.884	0.886	0.892	0.898	0.903	0.913	0.922	0.932	0.941	0.956	0.969
0.04000	(0.392)	0.868	0.868	0.870	0.873	0.876	0.882	0.888	0.894	0.905	0.914	0.924	0.933	0.949	0.962
0.06000	(0.588)	0.851	0.852	0.855	0.858	0.862	0.869	0.876	0.882	0.894	0.903	0.914	0.924	0.940	0.954
0.08000	(0.785)	0.840	0.841	0.844	0.848	0.852	0.860	0.868	0.874	0.886	0.897	0.908	0.918	0.934	0.948
0.10000	(0.981)	0.831	0.833	0.837	0.841	0.846	0.854	0.862	0.869	0.881	0.891	0.903	0.913	0.930	0.945
0.15000	(1.47)	0.816	0.819	0.824	0.829	0.834	0.843	0.851	0.859	0.872	0.883	0.895	0.906	0.923	0.938
0.20000	(1.96)	0.807	0.810	0.815	0.821	0.826	0.836	0.845	0.853	0.866	0.878	0.890	0.901	0.919	0.934
0.30000	(2.94)	0.794	0.798	0.804	0.811	0.816	0.827	0.836	0.845	0.859	0.871	0.884	0.895	0.913	0.929
0.40000	(3.92)	0.786	0.790	0.797	0.804	0.810	0.821	0.831	0.839	0.854	0.866	0.879	0.891	0.910	0.926
0.60000	(5.88)	0.775	0.780	0.788	0.795	0.802	0.814	0.824	0.833	0.848	0.861	0.874	0.886	0.906	0.922
0.80000	(7.85)	0.768	0.773	0.782	0.790	0.797	0.809	0.819	0.829	0.844	0.857	0.871	0.883	0.903	0.919
1.00000	(9.81)	0.763	0.768	0.778	0.786	0.793	0.806	0.816	0.826	0.842	0.855	0.869	0.881	0.901	0.918
1.50000	(14.7)	0.755	0.761	0.770	0.779	0.787	0.800	0.811	0.821	0.837	0.851	0.865	0.878	0.898	0.915
2.00000	(19.6)	0.750	0.756	0.766	0.775	0.783	0.797	0.808	0.818	0.835	0.849	0.863	0.876	0.896	0.913
3.00000	(29.4)	0.743	0.749	0.760	0.770	0.778	0.792	0.804	0.814	0.831	0.846	0.860	0.873	0.894	0.911
S	$(\Delta p/\rho l)$	125	150	200	250	300	400	500	600	800	1000	1250	1500	2000	2500

Grad'nt k.p.g. (Equivalent) Pipe diameters in mm

Kinematic viscosity, $\nu = 3.0 \times 10^{-6}\ \mathrm{m^2 s^{-1}}$;
Roughness size, $k_s = 0.015$ mm

E151

Kin. visc., $\nu = 3.0 \times 10^{-6}$ m^{2}s^{-1};
$S = 0.00010$ to 3.00000

i.e. kin. pr. grad., $\Delta p/\rho l =$
(0.00098) to (29.4) ms^{-2}

Roughness size, $k_s = 0.030$ mm
This table shows values of m, as follows

m_C for Colebrook-White solutions

Grad'nt S	k.p.g. $(\Delta p/\rho l)$	125	150	200	250	300	400	500	600	800	1000	1250	1500	2000	2500
0.00010	(0.00098)	1.316	1.293	1.263	1.245	1.233	1.217	1.208	1.203	1.197	1.195	1.195	1.196	1.199	1.204
0.00015	(0.00147)	1.270	1.250	1.225	1.209	1.198	1.185	1.178	1.174	1.170	1.169	1.170	1.172	1.176	1.182
0.00020	(0.00196)	1.240	1.222	1.199	1.184	1.175	1.164	1.158	1.154	1.151	1.151	1.153	1.155	1.161	1.167
0.00030	(0.00294)	1.200	1.184	1.164	1.152	1.144	1.135	1.130	1.128	1.127	1.128	1.130	1.133	1.140	1.147
0.00040	(0.00392)	1.173	1.159	1.141	1.130	1.123	1.115	1.112	1.110	1.110	1.111	1.115	1.118	1.126	1.133
0.00060	(0.00588)	1.137	1.125	1.110	1.101	1.095	1.089	1.087	1.086	1.087	1.090	1.094	1.098	1.107	1.115
0.00080	(0.00785)	1.113	1.102	1.089	1.081	1.076	1.072	1.070	1.070	1.072	1.075	1.080	1.085	1.094	1.103
0.00100	(0.00981)	1.095	1.085	1.073	1.066	1.062	1.059	1.058	1.058	1.061	1.065	1.070	1.075	1.085	1.093
0.00150	(0.0147)	1.065	1.056	1.046	1.041	1.038	1.036	1.036	1.037	1.041	1.046	1.052	1.057	1.068	1.078
0.00200	(0.0196)	1.044	1.037	1.028	1.024	1.022	1.021	1.022	1.023	1.028	1.033	1.040	1.046	1.057	1.067
0.00300	(0.0294)	1.017	1.011	1.004	1.001	1.000	1.000	1.002	1.005	1.011	1.017	1.024	1.030	1.043	1.053
0.00400	(0.0392)	0.999	0.994	0.988	0.986	0.985	0.987	0.989	0.992	0.999	1.005	1.013	1.020	1.033	1.044
0.00600	(0.0588)	0.975	0.971	0.967	0.966	0.966	0.969	0.972	0.976	0.983	0.991	0.999	1.007	1.021	1.032
0.00800	(0.0785)	0.958	0.955	0.953	0.952	0.953	0.957	0.961	0.965	0.973	0.981	0.990	0.998	1.012	1.025
0.01000	(0.0981)	0.946	0.944	0.942	0.942	0.944	0.948	0.952	0.957	0.966	0.974	0.983	0.992	1.006	1.019
0.01500	(0.147)	0.926	0.924	0.924	0.925	0.927	0.932	0.938	0.943	0.953	0.962	0.972	0.981	0.996	1.009
0.02000	(0.196)	0.912	0.911	0.912	0.914	0.917	0.922	0.928	0.934	0.945	0.954	0.964	0.974	0.990	1.003
0.03000	(0.294)	0.894	0.894	0.896	0.899	0.902	0.909	0.916	0.922	0.934	0.944	0.955	0.965	0.981	0.996
0.04000	(0.392)	0.883	0.883	0.886	0.889	0.893	0.901	0.908	0.915	0.927	0.937	0.949	0.959	0.976	0.991
0.06000	(0.588)	0.867	0.868	0.872	0.877	0.881	0.890	0.898	0.905	0.918	0.929	0.941	0.952	0.970	0.985
0.08000	(0.785)	0.857	0.859	0.863	0.869	0.873	0.883	0.891	0.899	0.912	0.924	0.936	0.947	0.965	0.981
0.10000	(0.981)	0.849	0.852	0.857	0.863	0.868	0.878	0.886	0.894	0.908	0.920	0.933	0.944	0.962	0.978
0.15000	(1.47)	0.837	0.840	0.846	0.853	0.858	0.869	0.878	0.887	0.901	0.914	0.927	0.938	0.958	0.973
0.20000	(1.96)	0.829	0.832	0.840	0.846	0.852	0.864	0.873	0.882	0.897	0.910	0.923	0.935	0.955	0.971
0.30000	(2.94)	0.819	0.823	0.831	0.838	0.845	0.857	0.867	0.876	0.892	0.905	0.919	0.931	0.951	0.967
0.40000	(3.92)	0.812	0.817	0.825	0.833	0.840	0.853	0.863	0.873	0.889	0.902	0.916	0.928	0.949	0.965
0.60000	(5.88)	0.804	0.809	0.818	0.827	0.834	0.847	0.858	0.868	0.885	0.898	0.913	0.925	0.946	0.963
0.80000	(7.85)	0.798	0.804	0.814	0.823	0.830	0.844	0.855	0.865	0.882	0.896	0.911	0.923	0.944	0.961
1.00000	(9.81)	0.795	0.800	0.811	0.820	0.828	0.842	0.853	0.863	0.880	0.894	0.909	0.922	0.943	0.960
1.50000	(14.7)	0.789	0.795	0.806	0.815	0.824	0.838	0.850	0.860	0.878	0.892	0.907	0.920	0.941	0.959
2.00000	(19.6)	0.785	0.791	0.803	0.812	0.821	0.836	0.848	0.858	0.876	0.890	0.906	0.919	0.940	0.958
3.00000	(29.4)	0.780	0.787	0.799	0.809	0.818	0.833	0.845	0.856	0.874	0.889	0.904	0.917	0.939	0.956
S	$(\Delta p/\rho l)$	125	150	200	250	300	400	500	600	800	1000	1250	1500	2000	2500
Grad'nt	k.p.g.	(Equivalent) Pipe diameters in mm													

Roughness size, $k_s = 0.030$ mm

E152

Kin. visc., $\nu = 3.0 \times 10^{-6}$ m^{2}s^{-1};
$S = 0.00010$ to 3.00000

i.e. kin. pr. grad., $\Delta p/\rho l =$
(0.00098) to (29.4) ms^{-2}

Roughness size, $k_s = 0.060$ mm
This table shows values of m, as follows

m_C for Colebrook-White solutions

Grad'nt S	k.p.g. $(\Delta p/\rho l)$	125	150	200	250	300	400	500	600	800	1000	1250	1500	2000	2500
0.00010	(0.00098)	1.320	1.297	1.268	1.249	1.237	1.222	1.213	1.208	1.203	1.201	1.201	1.203	1.207	1.212
0.00015	(0.00147)	1.275	1.255	1.229	1.214	1.203	1.191	1.184	1.180	1.177	1.176	1.177	1.180	1.185	1.191
0.00020	(0.00196)	1.245	1.227	1.204	1.190	1.181	1.170	1.164	1.161	1.159	1.159	1.161	1.164	1.171	1.177
0.00030	(0.00294)	1.205	1.190	1.170	1.158	1.151	1.142	1.138	1.136	1.135	1.137	1.140	1.143	1.151	1.159
0.00040	(0.00392)	1.179	1.165	1.147	1.137	1.130	1.123	1.120	1.119	1.119	1.122	1.125	1.130	1.138	1.146
0.00060	(0.00588)	1.144	1.132	1.117	1.109	1.103	1.098	1.096	1.096	1.098	1.101	1.106	1.111	1.121	1.130
0.00080	(0.00785)	1.121	1.110	1.097	1.090	1.086	1.082	1.081	1.081	1.084	1.088	1.094	1.099	1.110	1.119
0.00100	(0.00981)	1.104	1.094	1.082	1.076	1.072	1.069	1.069	1.070	1.074	1.078	1.084	1.090	1.101	1.111
0.00150	(0.0147)	1.074	1.066	1.057	1.052	1.049	1.048	1.049	1.051	1.056	1.061	1.068	1.075	1.087	1.098
0.00200	(0.0196)	1.054	1.047	1.040	1.036	1.034	1.034	1.036	1.038	1.044	1.050	1.058	1.065	1.078	1.089
0.00300	(0.0294)	1.029	1.023	1.017	1.015	1.014	1.015	1.018	1.022	1.029	1.036	1.044	1.052	1.066	1.078
0.00400	(0.0392)	1.012	1.007	1.002	1.001	1.001	1.003	1.007	1.011	1.019	1.026	1.035	1.043	1.058	1.070
0.00600	(0.0588)	0.989	0.986	0.983	0.983	0.984	0.987	0.992	0.996	1.005	1.014	1.024	1.032	1.048	1.061
0.00800	(0.0785)	0.974	0.972	0.970	0.971	0.972	0.977	0.982	0.987	0.997	1.006	1.016	1.025	1.041	1.055
0.01000	(0.0981)	0.963	0.961	0.961	0.962	0.964	0.969	0.975	0.980	0.991	1.000	1.011	1.020	1.037	1.051
0.01500	(0.147)	0.945	0.944	0.945	0.947	0.950	0.956	0.963	0.969	0.981	0.991	1.002	1.012	1.029	1.044
0.02000	(0.196)	0.933	0.933	0.934	0.937	0.941	0.948	0.955	0.962	0.974	0.985	0.996	1.007	1.025	1.039
0.03000	(0.294)	0.917	0.918	0.921	0.925	0.929	0.938	0.946	0.953	0.966	0.977	0.989	1.000	1.019	1.034
0.04000	(0.392)	0.907	0.908	0.913	0.917	0.922	0.931	0.939	0.947	0.961	0.972	0.985	0.996	1.015	1.031
0.06000	(0.588)	0.894	0.896	0.902	0.907	0.913	0.923	0.932	0.940	0.954	0.966	0.980	0.991	1.010	1.027
0.08000	(0.785)	0.886	0.889	0.895	0.901	0.907	0.917	0.927	0.935	0.950	0.963	0.976	0.988	1.008	1.024
0.10000	(0.981)	0.880	0.883	0.890	0.896	0.902	0.914	0.923	0.932	0.947	0.960	0.974	0.986	1.006	1.022
0.15000	(1.47)	0.870	0.874	0.882	0.889	0.896	0.907	0.918	0.927	0.943	0.956	0.970	0.982	1.003	1.019
0.20000	(1.96)	0.864	0.868	0.877	0.884	0.891	0.904	0.914	0.924	0.940	0.953	0.968	0.980	1.001	1.018
0.30000	(2.94)	0.856	0.861	0.870	0.878	0.886	0.899	0.910	0.920	0.936	0.950	0.965	0.977	0.998	1.016
0.40000	(3.92)	0.851	0.857	0.866	0.875	0.883	0.896	0.907	0.917	0.934	0.948	0.963	0.976	0.997	1.014
0.60000	(5.88)	0.846	0.851	0.861	0.871	0.879	0.893	0.904	0.915	0.932	0.946	0.961	0.974	0.995	1.013
0.80000	(7.85)	0.842	0.848	0.859	0.868	0.876	0.890	0.902	0.913	0.930	0.945	0.960	0.973	0.994	1.012
1.00000	(9.81)	0.839	0.845	0.856	0.866	0.874	0.889	0.901	0.912	0.929	0.944	0.959	0.972	0.994	1.011
1.50000	(14.7)	0.835	0.842	0.853	0.863	0.872	0.887	0.899	0.910	0.927	0.942	0.958	0.971	0.993	1.010
2.00000	(19.6)	0.833	0.839	0.851	0.861	0.870	0.885	0.898	0.908	0.926	0.941	0.957	0.970	0.992	1.010
3.00000	(29.4)	0.830	0.837	0.849	0.859	0.868	0.883	0.896	0.907	0.925	0.940	0.956	0.969	0.991	1.009
S	$(\Delta p/\rho l)$	125	150	200	250	300	400	500	600	800	1000	1250	1500	2000	2500
Grad'nt	k.p.g.	(Equivalent) Pipe diameters in mm													

Roughness size, $k_s = 0.060$ mm

Kin. visc., $\nu = 3.0 \times 10^{-6}$ m^2 s^{-1};
$S = 0.00010$ to 3.00000

i.e. kin. pr. grad., $\Delta p/\rho l =$
(0.00098) to (29.4) ms^{-2}

Roughness size, $k_s = 0.150$ mm
This table shows values of m, as follows

m_C for Colebrook-White solutions

Grad'nt S	k.p.g. $(\Delta p/\rho l)$	125	150	200	250	300	400	500	600	800	1000	1250	1500	2000	2500
0.00010	(0.00098)	1.331	1.309	1.280	1.262	1.250	1.236	1.228	1.223	1.219	1.219	1.220	1.222	1.228	1.235
0.00015	(0.00147)	1.288	1.268	1.243	1.228	1.218	1.207	1.201	1.197	1.195	1.196	1.198	1.202	1.209	1.217
0.00020	(0.00196)	1.259	1.241	1.219	1.206	1.197	1.187	1.182	1.180	1.179	1.181	1.184	1.188	1.197	1.205
0.00030	(0.00294)	1.221	1.206	1.187	1.176	1.169	1.162	1.159	1.157	1.158	1.161	1.166	1.171	1.180	1.190
0.00040	(0.00392)	1.196	1.183	1.166	1.157	1.151	1.145	1.143	1.143	1.145	1.148	1.154	1.159	1.170	1.180
0.00060	(0.00588)	1.164	1.152	1.138	1.131	1.127	1.123	1.122	1.123	1.127	1.132	1.138	1.144	1.156	1.167
0.00080	(0.00785)	1.142	1.132	1.120	1.114	1.111	1.108	1.109	1.110	1.115	1.121	1.128	1.135	1.148	1.159
0.00100	(0.00981)	1.126	1.117	1.107	1.102	1.099	1.098	1.099	1.101	1.107	1.113	1.121	1.128	1.141	1.153
0.00150	(0.0147)	1.100	1.093	1.085	1.081	1.080	1.080	1.082	1.085	1.093	1.100	1.108	1.116	1.131	1.144
0.00200	(0.0196)	1.082	1.076	1.070	1.067	1.067	1.068	1.072	1.075	1.083	1.091	1.101	1.109	1.124	1.138
0.00300	(0.0294)	1.060	1.055	1.051	1.050	1.050	1.054	1.058	1.063	1.072	1.081	1.091	1.100	1.116	1.130
0.00400	(0.0392)	1.045	1.042	1.039	1.039	1.040	1.044	1.049	1.054	1.065	1.074	1.085	1.094	1.111	1.125
0.00600	(0.0588)	1.027	1.024	1.023	1.024	1.026	1.032	1.038	1.044	1.055	1.066	1.077	1.087	1.105	1.119
0.00800	(0.0785)	1.015	1.013	1.013	1.015	1.018	1.024	1.031	1.038	1.050	1.060	1.072	1.083	1.101	1.116
0.01000	(0.0981)	1.006	1.005	1.006	1.009	1.012	1.019	1.026	1.033	1.046	1.057	1.069	1.079	1.098	1.113
0.01500	(0.147)	0.991	0.992	0.994	0.998	1.002	1.010	1.018	1.026	1.039	1.050	1.063	1.074	1.093	1.109
0.02000	(0.196)	0.982	0.983	0.987	0.991	0.996	1.005	1.013	1.021	1.035	1.047	1.060	1.071	1.091	1.107
0.03000	(0.294)	0.971	0.973	0.977	0.983	0.988	0.998	1.007	1.015	1.030	1.042	1.056	1.067	1.087	1.104
0.04000	(0.392)	0.964	0.966	0.971	0.977	0.983	0.994	1.003	1.012	1.027	1.039	1.053	1.065	1.085	1.102
0.06000	(0.588)	0.955	0.958	0.964	0.971	0.977	0.988	0.998	1.007	1.023	1.036	1.050	1.062	1.083	1.100
0.08000	(0.785)	0.949	0.953	0.960	0.967	0.973	0.985	0.996	1.005	1.021	1.034	1.048	1.061	1.081	1.098
0.10000	(0.981)	0.945	0.949	0.957	0.964	0.971	0.983	0.994	1.003	1.019	1.033	1.047	1.059	1.080	1.097
0.15000	(1.47)	0.939	0.943	0.952	0.960	0.967	0.979	0.990	1.000	1.017	1.030	1.045	1.058	1.079	1.096
0.20000	(1.96)	0.935	0.940	0.949	0.957	0.964	0.977	0.989	0.998	1.015	1.029	1.044	1.057	1.078	1.095
0.30000	(2.94)	0.930	0.936	0.945	0.954	0.961	0.975	0.986	0.996	1.013	1.027	1.042	1.055	1.077	1.094
0.40000	(3.92)	0.928	0.933	0.943	0.952	0.959	0.973	0.985	0.995	1.012	1.026	1.041	1.054	1.076	1.094
0.60000	(5.88)	0.924	0.930	0.940	0.949	0.957	0.971	0.983	0.993	1.011	1.025	1.040	1.053	1.075	1.093
0.80000	(7.85)	0.922	0.928	0.938	0.948	0.956	0.970	0.982	0.993	1.010	1.024	1.040	1.053	1.075	1.092
1.00000	(9.81)	0.921	0.927	0.937	0.947	0.955	0.969	0.981	0.992	1.009	1.024	1.039	1.052	1.074	1.092
1.50000	(14.7)	0.918	0.925	0.936	0.945	0.954	0.968	0.980	0.991	1.009	1.023	1.039	1.052	1.074	1.092
2.00000	(19.6)	0.917	0.923	0.935	0.944	0.953	0.967	0.980	0.990	1.008	1.023	1.038	1.051	1.073	1.091
3.00000	(29.4)	0.915	0.922	0.933	0.943	0.952	0.967	0.979	0.990	1.008	1.022	1.038	1.051	1.073	1.091
S	$(\Delta p/\rho l)$	125	150	200	250	300	400	500	600	800	1000	1250	1500	2000	2500

Grad'nt k.p.g. (Equivalent) Pipe diameters in mm Roughness size, $k_s = 0.150$ mm

Kin. visc., $\nu = 3.0 \times 10^{-6}$ m^2 s^{-1};
$S = 0.00010$ to 3.00000

i.e. kin. pr. grad., $\Delta p/\rho l =$
(0.00098) to (29.4) ms^{-2}

Roughness size, $k_s = 0.30$ mm
This table shows values of m, as follows

m_C for Colebrook-White solutions

Grad'nt S	k.p.g. $(\Delta p/\rho l)$	125	150	200	250	300	400	500	600	800	1000	1250	1500	2000	2500
0.00010	(0.00098)	1.350	1.327	1.299	1.282	1.271	1.258	1.250	1.247	1.244	1.244	1.247	1.250	1.258	1.266
0.00015	(0.00147)	1.308	1.289	1.265	1.251	1.241	1.231	1.226	1.224	1.223	1.225	1.229	1.233	1.242	1.251
0.00020	(0.00196)	1.281	1.264	1.243	1.230	1.222	1.214	1.210	1.208	1.209	1.212	1.217	1.222	1.232	1.242
0.00030	(0.00294)	1.246	1.232	1.214	1.203	1.197	1.191	1.189	1.189	1.192	1.196	1.202	1.208	1.219	1.230
0.00040	(0.00392)	1.223	1.210	1.195	1.186	1.181	1.176	1.176	1.176	1.180	1.185	1.192	1.199	1.211	1.223
0.00060	(0.00588)	1.194	1.183	1.170	1.164	1.160	1.158	1.158	1.160	1.166	1.172	1.180	1.187	1.201	1.213
0.00080	(0.00785)	1.175	1.165	1.154	1.149	1.147	1.146	1.147	1.150	1.157	1.163	1.172	1.180	1.195	1.207
0.00100	(0.00981)	1.161	1.152	1.143	1.139	1.137	1.137	1.139	1.143	1.150	1.157	1.166	1.175	1.190	1.203
0.00150	(0.0147)	1.138	1.131	1.124	1.121	1.121	1.123	1.126	1.131	1.139	1.148	1.158	1.167	1.183	1.197
0.00200	(0.0196)	1.123	1.117	1.112	1.110	1.111	1.114	1.118	1.123	1.132	1.142	1.152	1.161	1.178	1.192
0.00300	(0.0294)	1.104	1.100	1.097	1.097	1.098	1.103	1.108	1.113	1.124	1.134	1.145	1.155	1.173	1.187
0.00400	(0.0392)	1.092	1.089	1.087	1.088	1.090	1.095	1.101	1.108	1.119	1.129	1.141	1.151	1.169	1.184
0.00600	(0.0588)	1.077	1.075	1.075	1.077	1.080	1.087	1.094	1.100	1.113	1.123	1.136	1.146	1.165	1.181
0.00800	(0.0785)	1.068	1.066	1.067	1.070	1.074	1.081	1.089	1.096	1.109	1.120	1.132	1.144	1.162	1.178
0.01000	(0.0981)	1.061	1.060	1.062	1.065	1.069	1.077	1.085	1.093	1.106	1.117	1.130	1.141	1.161	1.177
0.01500	(0.147)	1.050	1.050	1.054	1.058	1.062	1.071	1.080	1.088	1.102	1.114	1.127	1.138	1.158	1.174
0.02000	(0.196)	1.043	1.044	1.048	1.053	1.058	1.068	1.076	1.085	1.099	1.111	1.125	1.136	1.156	1.173
0.03000	(0.294)	1.035	1.037	1.042	1.047	1.053	1.063	1.072	1.081	1.096	1.108	1.122	1.134	1.154	1.171
0.04000	(0.392)	1.030	1.032	1.038	1.044	1.049	1.060	1.070	1.079	1.094	1.107	1.120	1.133	1.153	1.170
0.06000	(0.588)	1.024	1.027	1.033	1.039	1.046	1.057	1.067	1.076	1.091	1.105	1.119	1.131	1.151	1.168
0.08000	(0.785)	1.020	1.023	1.030	1.037	1.043	1.055	1.065	1.074	1.090	1.103	1.118	1.130	1.151	1.168
0.10000	(0.981)	1.017	1.021	1.028	1.035	1.042	1.053	1.064	1.073	1.089	1.102	1.117	1.129	1.150	1.167
0.15000	(1.47)	1.013	1.017	1.025	1.032	1.039	1.051	1.062	1.071	1.087	1.101	1.116	1.128	1.149	1.166
0.20000	(1.96)	1.011	1.015	1.023	1.030	1.037	1.050	1.061	1.070	1.087	1.100	1.115	1.127	1.148	1.166
0.30000	(2.94)	1.008	1.012	1.020	1.028	1.036	1.048	1.059	1.069	1.085	1.099	1.114	1.127	1.148	1.165
0.40000	(3.92)	1.006	1.010	1.019	1.027	1.034	1.047	1.058	1.068	1.085	1.099	1.113	1.126	1.147	1.165
0.60000	(5.88)	1.004	1.008	1.017	1.026	1.033	1.046	1.057	1.067	1.084	1.098	1.113	1.126	1.147	1.164
0.80000	(7.85)	1.002	1.007	1.016	1.025	1.032	1.046	1.057	1.067	1.084	1.098	1.112	1.125	1.147	1.164
1.00000	(9.81)	1.001	1.006	1.016	1.024	1.032	1.045	1.056	1.066	1.083	1.097	1.112	1.125	1.146	1.164
1.50000	(14.7)	1.000	1.005	1.015	1.023	1.031	1.044	1.056	1.066	1.083	1.097	1.112	1.125	1.146	1.164
2.00000	(19.6)	0.999	1.004	1.014	1.023	1.030	1.044	1.055	1.065	1.082	1.097	1.112	1.124	1.146	1.164
3.00000	(29.4)	0.998	1.003	1.013	1.022	1.030	1.043	1.055	1.065	1.082	1.096	1.111	1.124	1.146	1.163
S	$(\Delta p/\rho l)$	125	150	200	250	300	400	500	600	800	1000	1250	1500	2000	2500

Grad'nt k.p.g. (Equivalent) Pipe diameters in mm Roughness size, $k_s = 0.30$ mm

E155

Kin. visc., $\nu = 3{\cdot}0 \times 10^{-6}$ m^{2}s^{-1};
$S = 0{\cdot}00010$ to $3{\cdot}00000$
i.e. kin. pr. grad., $\Delta p/\rho l =$
$(0{\cdot}00098)$ to $(29{\cdot}4)$ ms^{-2}

Roughness size, $k_s = 0{\cdot}60$ mm
This table shows values of m, as follows
m_C for Colebrook-White solutions

Grad'nt S	k.p.g. $(\Delta p/\rho l)$	125	150	200	250	300	400	500	600	800	1000	1250	1500	2000	2500
0·00010	(0·00098)	1·384	1·362	1·335	1·318	1·308	1·296	1·290	1·287	1·286	1·288	1·292	1·296	1·306	1·316
0·00015	(0·00147)	1·346	1·328	1·304	1·291	1·282	1·273	1·269	1·268	1·269	1·272	1·277	1·283	1·294	1·305
0·00020	(0·00196)	1·322	1·305	1·285	1·273	1·266	1·259	1·256	1·256	1·258	1·262	1·268	1·274	1·287	1·298
0·00030	(0·00294)	1·291	1·277	1·260	1·250	1·245	1·240	1·239	1·240	1·244	1·250	1·257	1·264	1·277	1·289
0·00040	(0·00392)	1·271	1·259	1·244	1·236	1·232	1·229	1·229	1·230	1·236	1·242	1·250	1·257	1·272	1·284
0·00060	(0·00588)	1·246	1·235	1·224	1·218	1·215	1·214	1·215	1·218	1·225	1·232	1·241	1·249	1·265	1·278
0·00080	(0·00785)	1·230	1·221	1·211	1·206	1·205	1·205	1·207	1·211	1·218	1·226	1·236	1·244	1·260	1·274
0·00100	(0·00981)	1·218	1·210	1·202	1·198	1·197	1·198	1·201	1·205	1·214	1·222	1·232	1·241	1·257	1·271
0·00150	(0·0147)	1·200	1·193	1·187	1·185	1·185	1·188	1·192	1·197	1·206	1·215	1·226	1·235	1·252	1·267
0·00200	(0·0196)	1·188	1·183	1·178	1·177	1·178	1·181	1·186	1·191	1·201	1·211	1·222	1·232	1·249	1·264
0·00300	(0·0294)	1·173	1·170	1·167	1·167	1·168	1·173	1·179	1·185	1·196	1·206	1·218	1·228	1·246	1·261
0·00400	(0·0392)	1·164	1·161	1·159	1·160	1·163	1·168	1·175	1·181	1·192	1·203	1·215	1·225	1·244	1·259
0·00600	(0·0588)	1·153	1·151	1·151	1·153	1·156	1·162	1·169	1·176	1·188	1·199	1·212	1·222	1·241	1·257
0·00800	(0·0785)	1·146	1·145	1·146	1·148	1·151	1·159	1·166	1·173	1·186	1·197	1·210	1·221	1·239	1·255
0·01000	(0·0981)	1·142	1·141	1·142	1·145	1·148	1·156	1·164	1·171	1·184	1·195	1·208	1·219	1·238	1·254
0·01500	(0·147)	1·134	1·134	1·136	1·140	1·144	1·152	1·160	1·168	1·181	1·193	1·206	1·217	1·237	1·253
0·02000	(0·196)	1·129	1·130	1·132	1·137	1·141	1·150	1·158	1·166	1·180	1·192	1·205	1·216	1·236	1·252
0·03000	(0·294)	1·124	1·125	1·128	1·133	1·138	1·147	1·156	1·163	1·178	1·190	1·203	1·215	1·234	1·251
0·04000	(0·392)	1·120	1·122	1·126	1·130	1·135	1·145	1·154	1·162	1·176	1·189	1·202	1·214	1·234	1·250
0·06000	(0·588)	1·116	1·118	1·122	1·128	1·133	1·143	1·152	1·160	1·175	1·188	1·201	1·213	1·233	1·249
0·08000	(0·785)	1·114	1·116	1·121	1·126	1·131	1·142	1·151	1·159	1·174	1·187	1·200	1·212	1·232	1·249
0·10000	(0·981)	1·112	1·114	1·119	1·125	1·130	1·141	1·150	1·159	1·174	1·186	1·200	1·212	1·232	1·249
0·15000	(1·47)	1·109	1·112	1·117	1·123	1·129	1·140	1·149	1·158	1·173	1·185	1·199	1·211	1·231	1·248
0·20000	(1·96)	1·108	1·110	1·116	1·122	1·128	1·139	1·148	1·157	1·172	1·185	1·199	1·211	1·231	1·248
0·30000	(2·94)	1·106	1·108	1·115	1·121	1·127	1·138	1·147	1·156	1·171	1·184	1·198	1·210	1·231	1·248
0·40000	(3·92)	1·105	1·107	1·114	1·120	1·126	1·137	1·147	1·156	1·171	1·184	1·198	1·210	1·230	1·247
0·60000	(5·88)	1·103	1·106	1·113	1·119	1·125	1·136	1·146	1·155	1·171	1·184	1·198	1·210	1·230	1·247
0·80000	(7·85)	1·102	1·105	1·112	1·119	1·125	1·136	1·146	1·155	1·170	1·183	1·197	1·210	1·230	1·247
1·00000	(9·81)	1·102	1·105	1·112	1·118	1·124	1·136	1·146	1·155	1·170	1·183	1·197	1·209	1·230	1·247
1·50000	(14·7)	1·101	1·104	1·111	1·118	1·124	1·135	1·145	1·154	1·170	1·183	1·197	1·209	1·230	1·247
2·00000	(19·6)	1·100	1·104	1·111	1·117	1·124	1·135	1·145	1·154	1·170	1·183	1·197	1·209	1·230	1·247
3·00000	(29·4)	1·100	1·103	1·110	1·117	1·123	1·135	1·145	1·154	1·169	1·183	1·197	1·209	1·229	1·246
S	$(\Delta p/\rho l)$	125	150	200	250	300	400	500	600	800	1000	1250	1500	2000	2500

Grad'nt k.p.g. (Equivalent) Pipe diameters in mm Roughness size, $k_s = 0{\cdot}60$ mm

E156

Kin. visc., $\nu = 3{\cdot}0 \times 10^{-6}$ m^{2}s^{-1};
$S = 0{\cdot}00010$ to $3{\cdot}00000$
i.e. kin. pr. grad., $\Delta p/\rho l =$
$(0{\cdot}00098)$ to $(29{\cdot}4)$ ms^{-2}

Roughness size, $k_s = 1{\cdot}50$ mm
This table shows values of m, as follows
m_C for Colebrook-White solutions

Grad'nt S	k.p.g. $(\Delta p/\rho l)$	125	150	200	250	300	400	500	600	800	1000	1250	1500	2000	2500
0·00010	(0·00098)	1·474	1·451	1·424	1·408	1·398	1·386	1·381	1·379	1·380	1·382	1·387	1·393	1·404	1·415
0·00015	(0·00147)	1·443	1·424	1·400	1·387	1·379	1·370	1·367	1·366	1·368	1·372	1·378	1·385	1·397	1·408
0·00020	(0·00196)	1·424	1·407	1·386	1·374	1·367	1·360	1·358	1·358	1·361	1·366	1·373	1·379	1·393	1·405
0·00030	(0·00294)	1·400	1·385	1·367	1·358	1·352	1·347	1·347	1·348	1·352	1·358	1·366	1·373	1·387	1·400
0·00040	(0·00392)	1·385	1·372	1·356	1·348	1·343	1·340	1·340	1·342	1·347	1·353	1·362	1·369	1·384	1·397
0·00060	(0·00588)	1·367	1·355	1·342	1·335	1·332	1·330	1·331	1·334	1·341	1·348	1·356	1·365	1·380	1·393
0·00080	(0·00785)	1·355	1·345	1·333	1·328	1·325	1·325	1·326	1·329	1·337	1·344	1·353	1·362	1·378	1·391
0·00100	(0·00981)	1·347	1·338	1·327	1·323	1·321	1·320	1·323	1·326	1·334	1·342	1·351	1·360	1·376	1·390
0·00150	(0·0147)	1·335	1·327	1·318	1·314	1·313	1·314	1·317	1·321	1·330	1·338	1·348	1·357	1·373	1·387
0·00200	(0·0196)	1·327	1·320	1·312	1·309	1·308	1·310	1·314	1·318	1·327	1·336	1·346	1·355	1·372	1·386
0·00300	(0·0294)	1·317	1·311	1·305	1·303	1·303	1·306	1·310	1·314	1·324	1·333	1·343	1·353	1·370	1·384
0·00400	(0·0392)	1·312	1·306	1·301	1·299	1·300	1·303	1·307	1·312	1·322	1·331	1·342	1·352	1·369	1·383
0·00600	(0·0588)	1·305	1·300	1·296	1·295	1·296	1·299	1·304	1·310	1·320	1·329	1·340	1·350	1·367	1·382
0·00800	(0·0785)	1·301	1·296	1·293	1·292	1·293	1·297	1·303	1·308	1·318	1·328	1·339	1·349	1·367	1·381
0·01000	(0·0981)	1·298	1·294	1·290	1·290	1·292	1·296	1·301	1·307	1·318	1·327	1·339	1·349	1·366	1·381
0·01500	(0·147)	1·293	1·290	1·287	1·287	1·289	1·294	1·300	1·305	1·316	1·326	1·337	1·348	1·365	1·380
0·02000	(0·196)	1·290	1·287	1·285	1·286	1·287	1·293	1·298	1·304	1·315	1·325	1·337	1·347	1·365	1·380
0·03000	(0·294)	1·287	1·284	1·283	1·284	1·286	1·291	1·297	1·303	1·314	1·324	1·336	1·346	1·364	1·379
0·04000	(0·392)	1·285	1·283	1·281	1·282	1·284	1·290	1·296	1·302	1·314	1·324	1·335	1·346	1·364	1·379
0·06000	(0·588)	1·283	1·280	1·279	1·281	1·283	1·289	1·295	1·301	1·313	1·323	1·335	1·345	1·363	1·378
0·08000	(0·785)	1·281	1·279	1·278	1·280	1·282	1·288	1·295	1·301	1·312	1·323	1·335	1·345	1·363	1·378
0·10000	(0·981)	1·280	1·278	1·278	1·279	1·282	1·288	1·294	1·300	1·312	1·323	1·334	1·345	1·363	1·378
0·15000	(1·47)	1·279	1·277	1·277	1·278	1·281	1·287	1·294	1·300	1·312	1·322	1·334	1·344	1·363	1·378
0·20000	(1·96)	1·278	1·276	1·276	1·278	1·280	1·287	1·293	1·300	1·311	1·322	1·334	1·344	1·362	1·378
0·30000	(2·94)	1·277	1·275	1·275	1·277	1·280	1·286	1·293	1·299	1·311	1·322	1·334	1·344	1·362	1·378
0·40000	(3·92)	1·276	1·275	1·275	1·277	1·279	1·286	1·293	1·299	1·311	1·322	1·333	1·344	1·362	1·377
0·60000	(5·88)	1·276	1·274	1·274	1·276	1·279	1·286	1·292	1·299	1·311	1·321	1·333	1·344	1·362	1·377
0·80000	(7·85)	1·275	1·274	1·274	1·276	1·279	1·285	1·292	1·299	1·310	1·321	1·333	1·344	1·362	1·377
1·00000	(9·81)	1·275	1·273	1·274	1·276	1·279	1·285	1·292	1·298	1·310	1·321	1·333	1·344	1·362	1·377
1·50000	(14·7)	1·274	1·273	1·273	1·275	1·278	1·285	1·292	1·298	1·310	1·321	1·333	1·343	1·362	1·377
2·00000	(19·6)	1·274	1·273	1·273	1·275	1·278	1·285	1·292	1·298	1·310	1·321	1·333	1·343	1·362	1·377
3·00000	(29·4)	1·274	1·272	1·273	1·275	1·278	1·285	1·291	1·298	1·310	1·321	1·333	1·343	1·362	1·377
S	$(\Delta p/\rho l)$	125	150	200	250	300	400	500	600	800	1000	1250	1500	2000	2500

Grad'nt k.p.g. (Equivalent) Pipe diameters in mm Roughness size, $k_s = 1{\cdot}50$ mm

Kin. visc., $\nu = 4{\cdot}0\times10^{-6}\ \mathrm{m^2 s^{-1}}$; $S = 0{\cdot}00010$ to $30{\cdot}0000$

i.e. kin. pr. grad., $\Delta p/\rho l =$ $(0{\cdot}00098)$ to $(294)\ \mathrm{ms^{-2}}$

Roughness size, $k_s = 0{\cdot}003$ mm. This table shows values of m, as follows m_C for Colebrook-White solutions; or, where $R \le 2000$, m_P for laminar flow

Grad'nt S	k.p.g. $(\Delta p/\rho l)$	\(Equivalent\) Pipe diameters in mm													
		6	8	10	12·5	15	20	25	30	40	50	60	80	100	125
0·00030	(0·00294)	27·43	18·69	13·88	10·31	8·084	5·509	4·091	3·208	2·186	1·624	1·273	*	1·281	1·253
0·00040	(0·00392)	23·75	16·19	12·02	8·928	7·001	4·771	3·543	2·778	1·893	1·406	1·103	*	1·249	1·223
0·00060	(0·00588)	19·40	13·22	9·815	7·289	5·716	3·895	2·893	2·269	1·546	1·148	*	1·232	1·205	1·183
0·00080	(0·00785)	16·80	11·45	8·500	6·313	4·951	3·373	2·505	1·965	1·339	0·994	*	1·201	1·176	1·156
0·00100	(0·00981)	15·02	10·24	7·603	5·646	4·428	3·017	2·241	1·757	1·197	*	*	1·177	1·155	1·136
0·00150	(0·0147)	12·27	8·359	6·208	4·610	3·615	2·464	1·830	1·435	0·978	*	1·169	1·138	1·118	1·102
0·00200	(0·0196)	10·62	7·239	5·376	3·993	3·131	2·134	1·584	1·243	*	1·161	1·139	1·111	1·093	1·078
0·00300	(0·0294)	8·674	5·911	4·390	3·260	2·556	1·742	1·294	1·015	*	1·119	1·100	1·075	1·060	1·047
0·00400	(0·0392)	7·512	5·119	3·801	2·823	2·214	1·509	1·120	*	1·117	1·091	1·074	1·051	1·037	1·026
0·00600	(0·0588)	6·133	4·179	3·104	2·305	1·808	1·232	0·915	*	1·076	1·054	1·039	1·019	1·007	0·998
0·00800	(0·0785)	5·312	3·620	2·688	1·996	1·565	1·067	*	*	1·050	1·030	1·016	0·998	0·987	0·979
0·01000	(0·0981)	4·751	3·237	2·404	1·786	1·400	0·954	*	1·059	1·030	1·011	0·998	0·982	0·972	0·965
0·01500	(0·147)	3·879	2·643	1·963	1·458	1·143	*	1·041	1·021	0·996	0·979	0·968	0·954	0·946	0·940
0·02000	(0·196)	3·359	2·289	1·700	1·263	0·990	*	1·014	0·996	0·973	0·958	0·948	0·936	0·928	0·923
0·03000	(0·294)	2·743	1·869	1·388	1·031	0·808	1·001	0·979	0·963	0·942	0·930	0·921	0·910	0·905	0·900
0·04000	(0·392)	2·375	1·619	1·202	0·893	*	0·976	0·955	0·940	0·922	0·910	0·903	0·894	0·888	0·885
0·06000	(0·588)	1·940	1·322	0·982	*	*	0·941	0·923	0·911	0·895	0·885	0·878	0·871	0·867	0·864
0·08000	(0·785)	1·680	1·145	0·850	*	0·945	0·918	0·902	0·891	0·876	0·867	0·862	0·855	0·852	0·850
0·10000	(0·981)	1·502	1·024	0·760	*	0·926	0·902	0·886	0·876	0·862	0·855	0·850	0·844	0·841	0·840
0·15000	(1·47)	1·227	0·836	*	0·910	0·894	0·872	0·859	0·850	0·839	0·832	0·828	0·824	0·822	0·821
0·20000	(1·96)	1·062	0·724	*	0·887	0·872	0·853	0·841	0·833	0·823	0·817	0·814	0·810	0·809	0·809
0·30000	(2·94)	0·867	*	0·876	0·857	0·844	0·827	0·817	0·810	0·801	0·797	0·794	0·792	0·791	0·792
0·40000	(3·92)	0·751	*	0·854	0·837	0·825	0·810	0·800	0·794	0·787	0·783	0·781	0·779	0·780	0·781
0·60000	(5·88)	*	0·843	0·825	0·810	0·799	0·786	0·779	0·773	0·768	0·765	0·763	0·763	0·764	0·766
0·80000	(7·85)	*	0·822	0·805	0·792	0·782	0·771	0·764	0·759	0·755	0·752	0·751	0·752	0·753	0·755
1·00000	(9·81)	*	0·806	0·791	0·778	0·770	0·759	0·753	0·749	0·745	0·743	0·743	0·743	0·745	0·748
1·50000	(14·7)	0·801	0·779	0·766	0·755	0·748	0·739	0·734	0·731	0·728	0·727	0·727	0·729	0·731	0·734
2·00000	(19·6)	0·781	0·762	0·750	0·740	0·734	0·726	0·721	0·719	0·717	0·716	0·717	0·719	0·721	0·725
3·00000	(29·4)	0·755	0·738	0·728	0·720	0·714	0·708	0·704	0·703	0·701	0·702	0·703	0·706	0·709	0·713
4·00000	(39·2)	0·738	0·722	0·713	0·706	0·701	0·696	0·693	0·692	0·691	0·692	0·693	0·697	0·701	0·705
6·00000	(58·8)	0·715	0·702	0·694	0·688	0·684	0·680	0·678	0·677	0·678	0·679	0·681	0·685	0·689	0·694
8·00000	(78·5)	0·699	0·688	0·681	0·676	0·672	0·669	0·668	0·667	0·669	0·670	0·673	0·677	0·682	0·687
10·0000	(98·1)	0·688	0·678	0·671	0·667	0·664	0·661	0·660	0·660	0·662	0·664	0·667	0·672	0·677	0·682
15·0000	(147)	0·669	0·660	0·655	0·651	0·649	0·647	0·647	0·648	0·650	0·653	0·656	0·662	0·667	0·673
20·0000	(196)	0·656	0·648	0·644	0·641	0·639	0·638	0·639	0·640	0·643	0·646	0·649	0·655	0·661	0·668
30·0000	(294)	0·639	0·633	0·629	0·627	0·626	0·626	0·627	0·629	0·632	0·636	0·640	0·647	0·653	0·660
S $(\Delta p/\rho l)$		6	8	10	12·5	15	20	25	30	40	50	60	80	100	125

Grad'nt k.p.g.   \(Equivalent\) Pipe diameters in mm

Grad'nt S	k.p.g. $(\Delta p/\rho l)$	\(Equivalent\) Pipe diameters in mm													
		125	150	200	250	300	400	500	600	800	1000	1250	1500	2000	2500
0·00010	(0·00098)	*	1·356	1·320	1·298	1·282	1·262	1·251	1·243	1·234	1·230	1·228	1·227	1·228	1·231
0·00015	(0·00147)	1·333	1·308	1·277	1·257	1·244	1·227	1·217	1·211	1·204	1·201	1·200	1·200	1·203	1·206
0·00020	(0·00196)	1·299	1·277	1·248	1·230	1·218	1·203	1·194	1·189	1·183	1·181	1·181	1·182	1·185	1·189
0·00030	(0·00294)	1·253	1·234	1·209	1·194	1·184	1·171	1·164	1·160	1·156	1·155	1·155	1·157	1·161	1·166
0·00040	(0·00392)	1·223	1·206	1·183	1·169	1·160	1·149	1·143	1·140	1·137	1·136	1·138	1·140	1·145	1·150
0·00060	(0·00588)	1·183	1·168	1·148	1·137	1·129	1·120	1·115	1·113	1·111	1·112	1·114	1·117	1·123	1·129
0·00080	(0·00785)	1·156	1·143	1·125	1·114	1·108	1·100	1·096	1·094	1·094	1·095	1·098	1·101	1·107	1·114
0·00100	(0·00981)	1·136	1·124	1·107	1·098	1·092	1·085	1·082	1·081	1·081	1·082	1·085	1·089	1·096	1·103
0·00150	(0·0147)	1·102	1·091	1·077	1·069	1·064	1·059	1·057	1·056	1·058	1·060	1·064	1·068	1·076	1·083
0·00200	(0·0196)	1·078	1·069	1·056	1·049	1·045	1·041	1·040	1·040	1·042	1·045	1·049	1·053	1·062	1·069
0·00300	(0·0294)	1·047	1·039	1·029	1·023	1·020	1·017	1·017	1·018	1·020	1·024	1·029	1·034	1·043	1·051
0·00400	(0·0392)	1·026	1·019	1·010	1·005	1·003	1·001	1·001	1·002	1·006	1·010	1·015	1·020	1·030	1·038
0·00600	(0·0588)	0·998	0·992	0·985	0·981	0·979	0·979	0·980	0·982	0·986	0·991	0·997	1·002	1·012	1·021
0·00800	(0·0785)	0·979	0·973	0·967	0·965	0·964	0·964	0·965	0·968	0·973	0·978	0·984	0·990	1·000	1·009
0·01000	(0·0981)	0·965	0·960	0·955	0·952	0·952	0·952	0·954	0·957	0·962	0·968	0·974	0·980	0·991	1·000
0·01500	(0·147)	0·940	0·936	0·932	0·931	0·931	0·933	0·935	0·938	0·944	0·950	0·957	0·964	0·975	0·985
0·02000	(0·196)	0·923	0·920	0·917	0·916	0·917	0·919	0·922	0·926	0·932	0·939	0·946	0·952	0·964	0·974
0·03000	(0·294)	0·900	0·898	0·896	0·896	0·898	0·901	0·905	0·908	0·916	0·922	0·930	0·937	0·949	0·960
0·04000	(0·392)	0·885	0·883	0·882	0·883	0·884	0·888	0·892	0·897	0·904	0·911	0·919	0·927	0·939	0·950
0·06000	(0·588)	0·864	0·863	0·863	0·865	0·867	0·871	0·876	0·881	0·889	0·897	0·905	0·912	0·925	0·936
0·08000	(0·785)	0·850	0·850	0·851	0·852	0·855	0·860	0·865	0·870	0·879	0·886	0·895	0·903	0·916	0·927
0·10000	(0·981)	0·840	0·840	0·841	0·843	0·846	0·851	0·857	0·862	0·871	0·879	0·888	0·896	0·909	0·921
0·15000	(1·47)	0·821	0·822	0·824	0·827	0·830	0·836	0·842	0·847	0·857	0·866	0·875	0·883	0·897	0·909
0·20000	(1·96)	0·809	0·810	0·813	0·816	0·819	0·826	0·832	0·838	0·848	0·857	0·866	0·874	0·889	0·901
0·30000	(2·94)	0·792	0·794	0·797	0·801	0·805	0·812	0·819	0·825	0·835	0·844	0·854	0·863	0·878	0·890
0·40000	(3·92)	0·781	0·783	0·787	0·791	0·795	0·803	0·810	0·816	0·827	0·836	0·847	0·855	0·871	0·883
0·60000	(5·88)	0·766	0·768	0·773	0·777	0·782	0·790	0·798	0·804	0·816	0·825	0·836	0·845	0·861	0·874
0·80000	(7·85)	0·755	0·758	0·763	0·768	0·773	0·782	0·789	0·796	0·808	0·818	0·829	0·839	0·854	0·868
1·00000	(9·81)	0·748	0·750	0·756	0·761	0·766	0·775	0·783	0·790	0·803	0·813	0·824	0·834	0·850	0·863
1·50000	(14·7)	0·734	0·737	0·744	0·750	0·755	0·765	0·773	0·780	0·793	0·804	0·815	0·825	0·842	0·855
2·00000	(19·6)	0·725	0·729	0·736	0·742	0·747	0·757	0·766	0·774	0·787	0·798	0·809	0·819	0·836	0·850
3·00000	(29·4)	0·713	0·717	0·725	0·731	0·737	0·748	0·757	0·765	0·778	0·790	0·802	0·812	0·830	0·844
S $(\Delta p/\rho l)$		125	150	200	250	300	400	500	600	800	1000	1250	1500	2000	2500

Grad'nt k.p.g.   \(Equivalent\) Pipe diameters in mm

Kinematic viscosity, $\nu = 4{\cdot}0\times10^{-6}\ \mathrm{m^2 s^{-1}}$;

Roughness size, $k_s = 0{\cdot}003$ mm

Kin. visc., $\nu = 4.0 \times 10^{-6}$ m^{2}s^{-1};
$S = 0.00010$ to 30.0000

i.e. kin. pr. grad., $\Delta p/\rho l =$
(0.00098) to (294) ms^{-2}

Roughness size, $k_s = 0.006$ mm
This table shows values of m, as follows

m_C for Colebrook-White solutions; or,
where $\mathbf{R} \le 2000$, m_P for laminar flow

Grad'nt S	k.p.g. $(\Delta p/\rho l)$	6	8	10	12.5	15	20	25	30	40	50	60	80	100	125
		\multicolumn — (Equivalent) Pipe diameters in mm													
0.00030	(0.00294)	27.43	18.69	13.88	10.31	8.084	5.509	4.091	3.208	2.186	1.624	1.273	*	1.282	1.254
0.00040	(0.00392)	23.75	16.19	12.02	8.928	7.001	4.771	3.543	2.778	1.893	1.406	1.103	*	1.249	1.224
0.00060	(0.00588)	19.40	13.22	9.815	7.289	5.716	3.895	2.893	2.269	1.546	1.148	*	1.233	1.206	1.184
0.00080	(0.00785)	16.80	11.45	8.500	6.313	4.951	3.373	2.505	1.965	1.339	0.994	*	1.201	1.177	1.157
0.00100	(0.00981)	15.02	10.24	7.603	5.646	4.428	3.017	2.241	1.757	1.197	*	*	1.178	1.156	1.137
0.00150	(0.0147)	12.27	8.359	6.208	4.610	3.615	2.464	1.830	1.435	0.978	*	1.170	1.138	1.119	1.102
0.00200	(0.0196)	10.62	7.239	5.376	3.993	3.131	2.134	1.584	1.243	*	1.162	1.140	1.112	1.094	1.079
0.00300	(0.0294)	8.674	5.911	4.390	3.260	2.556	1.742	1.294	1.015	*	1.120	1.101	1.076	1.061	1.048
0.00400	(0.0392)	7.512	5.119	3.801	2.823	2.214	1.509	1.120	*	1.117	1.092	1.075	1.052	1.038	1.027
0.00600	(0.0588)	6.133	4.179	3.104	2.305	1.808	1.232	0.915	*	1.077	1.055	1.040	1.021	1.009	0.999
0.00800	(0.0785)	5.312	3.620	2.688	1.996	1.565	1.067	*	*	1.051	1.031	1.017	0.999	0.989	0.980
0.01000	(0.0981)	4.751	3.237	2.404	1.786	1.400	0.954	*	1.061	1.031	1.012	1.000	0.983	0.974	0.966
0.01500	(0.147)	3.879	2.643	1.963	1.458	1.143	*	1.043	1.023	0.997	0.981	0.970	0.956	0.948	0.942
0.02000	(0.196)	3.359	2.289	1.700	1.263	0.990	*	1.016	0.998	0.974	0.960	0.950	0.937	0.930	0.925
0.03000	(0.294)	2.743	1.869	1.388	1.031	0.808	1.003	0.980	0.964	0.944	0.931	0.923	0.913	0.907	0.903
0.04000	(0.392)	2.375	1.619	1.202	0.893	*	0.977	0.957	0.942	0.924	0.912	0.905	0.896	0.891	0.888
0.06000	(0.588)	1.940	1.322	0.982	*	*	0.943	0.925	0.913	0.897	0.887	0.881	0.873	0.870	0.867
0.08000	(0.785)	1.680	1.145	0.850	*	0.947	0.921	0.904	0.893	0.879	0.870	0.865	0.858	0.855	0.854
0.10000	(0.981)	1.502	1.024	0.760	*	0.928	0.904	0.889	0.878	0.865	0.857	0.852	0.847	0.844	0.843
0.15000	(1.47)	1.227	0.836	*	0.913	0.896	0.875	0.862	0.853	0.842	0.835	0.832	0.827	0.826	0.826
0.20000	(1.96)	1.062	0.724	*	0.890	0.875	0.856	0.844	0.836	0.826	0.821	0.817	0.814	0.813	0.814
0.30000	(2.94)	0.867	*	0.879	0.860	0.847	0.830	0.820	0.813	0.805	0.801	0.799	0.797	0.796	0.798
0.40000	(3.92)	0.751	*	0.857	0.840	0.828	0.813	0.804	0.798	0.791	0.788	0.786	0.785	0.785	0.787
0.60000	(5.88)	*	0.846	0.828	0.814	0.803	0.791	0.783	0.778	0.773	0.770	0.769	0.769	0.770	0.772
0.80000	(7.85)	*	0.826	0.809	0.796	0.787	0.775	0.769	0.765	0.760	0.758	0.758	0.758	0.760	0.763
1.00000	(9.81)	*	0.810	0.795	0.783	0.775	0.764	0.758	0.755	0.751	0.749	0.749	0.750	0.753	0.756
1.50000	(14.7)	0.805	0.784	0.771	0.761	0.754	0.745	0.740	0.737	0.735	0.734	0.735	0.737	0.740	0.744
2.00000	(19.6)	0.786	0.767	0.755	0.746	0.740	0.732	0.728	0.726	0.724	0.724	0.725	0.728	0.731	0.735
3.00000	(29.4)	0.761	0.744	0.734	0.726	0.721	0.715	0.712	0.711	0.710	0.711	0.712	0.716	0.720	0.725
4.00000	(39.2)	0.744	0.729	0.720	0.713	0.709	0.704	0.701	0.700	0.701	0.702	0.704	0.708	0.713	0.718
6.00000	(58.8)	0.722	0.709	0.702	0.696	0.692	0.689	0.687	0.687	0.688	0.690	0.693	0.698	0.703	0.709
8.00000	(78.5)	0.707	0.696	0.689	0.685	0.682	0.679	0.678	0.678	0.680	0.683	0.686	0.691	0.697	0.703
10.0000	(98.1)	0.697	0.686	0.680	0.676	0.674	0.672	0.671	0.672	0.674	0.677	0.680	0.686	0.692	0.699
15.0000	(147)	0.678	0.670	0.665	0.662	0.660	0.659	0.660	0.661	0.664	0.668	0.671	0.678	0.684	0.691
20.0000	(196)	0.666	0.659	0.655	0.653	0.651	0.651	0.652	0.654	0.658	0.662	0.666	0.673	0.680	0.687
30.0000	(294)	0.651	0.645	0.642	0.640	0.640	0.641	0.642	0.645	0.649	0.654	0.658	0.666	0.673	0.681
S	$(\Delta p/\rho l)$	6	8	10	12.5	15	20	25	30	40	50	60	80	100	125

Grad'nt k.p.g. (Equivalent) Pipe diameters in mm

S	$(\Delta p/\rho l)$	125	150	200	250	300	400	500	600	800	1000	1250	1500	2000	2500
0.00010	(0.00098)	*	1.357	1.321	1.298	1.282	1.263	1.251	1.243	1.235	1.230	1.228	1.228	1.229	1.232
0.00015	(0.00147)	1.333	1.309	1.277	1.258	1.244	1.227	1.218	1.211	1.205	1.202	1.200	1.201	1.203	1.207
0.00020	(0.00196)	1.299	1.277	1.248	1.231	1.219	1.204	1.195	1.190	1.184	1.182	1.182	1.182	1.186	1.190
0.00030	(0.00294)	1.254	1.235	1.210	1.194	1.184	1.172	1.164	1.160	1.156	1.155	1.156	1.158	1.162	1.167
0.00040	(0.00392)	1.224	1.206	1.184	1.170	1.161	1.150	1.144	1.140	1.138	1.137	1.139	1.141	1.146	1.152
0.00060	(0.00588)	1.184	1.169	1.149	1.137	1.130	1.121	1.116	1.114	1.112	1.113	1.115	1.118	1.124	1.130
0.00080	(0.00785)	1.157	1.143	1.126	1.115	1.109	1.101	1.097	1.095	1.095	1.096	1.099	1.102	1.109	1.115
0.00100	(0.00981)	1.137	1.124	1.108	1.099	1.093	1.086	1.083	1.082	1.082	1.084	1.087	1.090	1.097	1.104
0.00150	(0.0147)	1.102	1.092	1.078	1.070	1.065	1.060	1.058	1.058	1.059	1.062	1.065	1.069	1.077	1.085
0.00200	(0.0196)	1.079	1.069	1.057	1.051	1.046	1.042	1.041	1.041	1.043	1.046	1.051	1.055	1.064	1.072
0.00300	(0.0294)	1.048	1.040	1.030	1.024	1.021	1.019	1.018	1.019	1.022	1.026	1.031	1.036	1.045	1.054
0.00400	(0.0392)	1.027	1.020	1.011	1.007	1.004	1.002	1.003	1.004	1.008	1.012	1.018	1.023	1.033	1.041
0.00600	(0.0588)	0.999	0.993	0.986	0.983	0.981	0.981	0.982	0.984	0.988	0.993	0.999	1.005	1.015	1.025
0.00800	(0.0785)	0.980	0.975	0.969	0.967	0.965	0.966	0.968	0.970	0.975	0.981	0.987	0.993	1.004	1.013
0.01000	(0.0981)	0.966	0.961	0.956	0.954	0.954	0.955	0.957	0.960	0.965	0.971	0.978	0.984	0.995	1.005
0.01500	(0.147)	0.942	0.938	0.934	0.933	0.933	0.935	0.938	0.941	0.948	0.954	0.961	0.968	0.980	0.990
0.02000	(0.196)	0.925	0.922	0.919	0.919	0.919	0.922	0.925	0.929	0.936	0.943	0.950	0.957	0.969	0.980
0.03000	(0.294)	0.903	0.900	0.899	0.899	0.901	0.904	0.908	0.912	0.920	0.927	0.935	0.943	0.955	0.966
0.04000	(0.392)	0.888	0.886	0.885	0.886	0.888	0.892	0.897	0.901	0.909	0.917	0.925	0.933	0.946	0.957
0.06000	(0.588)	0.867	0.866	0.867	0.869	0.871	0.876	0.881	0.886	0.895	0.903	0.912	0.919	0.933	0.945
0.08000	(0.785)	0.854	0.853	0.854	0.857	0.859	0.865	0.870	0.875	0.885	0.893	0.902	0.911	0.925	0.937
0.10000	(0.981)	0.843	0.843	0.845	0.848	0.851	0.857	0.862	0.868	0.878	0.886	0.896	0.904	0.918	0.931
0.15000	(1.47)	0.826	0.826	0.829	0.832	0.836	0.842	0.849	0.854	0.865	0.874	0.884	0.893	0.908	0.920
0.20000	(1.96)	0.814	0.815	0.818	0.822	0.825	0.833	0.839	0.846	0.856	0.866	0.876	0.885	0.900	0.913
0.30000	(2.94)	0.798	0.799	0.803	0.808	0.812	0.820	0.827	0.834	0.845	0.855	0.866	0.875	0.891	0.904
0.40000	(3.92)	0.787	0.789	0.794	0.798	0.803	0.811	0.819	0.826	0.838	0.848	0.859	0.869	0.885	0.899
0.60000	(5.88)	0.772	0.775	0.781	0.786	0.791	0.800	0.808	0.815	0.828	0.839	0.850	0.860	0.877	0.891
0.80000	(7.85)	0.763	0.766	0.772	0.778	0.783	0.793	0.801	0.809	0.822	0.833	0.844	0.855	0.872	0.886
1.00000	(9.81)	0.756	0.759	0.766	0.772	0.777	0.787	0.796	0.804	0.817	0.828	0.840	0.851	0.868	0.883
1.50000	(14.7)	0.744	0.747	0.755	0.761	0.767	0.778	0.787	0.795	0.809	0.821	0.833	0.844	0.862	0.877
2.00000	(19.6)	0.735	0.740	0.747	0.754	0.761	0.772	0.781	0.790	0.804	0.816	0.829	0.840	0.858	0.873
3.00000	(29.4)	0.725	0.729	0.738	0.745	0.752	0.764	0.774	0.782	0.797	0.810	0.823	0.834	0.853	0.869
S	$(\Delta p/\rho l)$	125	150	200	250	300	400	500	600	800	1000	1250	1500	2000	2500

Grad'nt k.p.g. (Equivalent) Pipe diameters in mm

Kinematic viscosity, $\nu = 4.0 \times 10^{-6}$ m^{2}s^{-1};

Roughness size, $k_s = 0.006$ mm

Kin. visc., $\nu = 4{\cdot}0\times10^{-6}\ \mathrm{m^2\,s^{-1}}$;
$S = 0{\cdot}00010$ to $30{\cdot}0000$

i.e. kin. pr. grad., $\Delta p/\rho l =$
$(0{\cdot}00098)$ to $(294)\ \mathrm{ms^{-2}}$

Roughness size, $k_s = 0{\cdot}015$ mm
This table shows values of m, as follows

m_C for Colebrook-White solutions; or,
where $\mathbf{R} \le 2000$, m_P for laminar flow

E159

Grad'nt S	k.p.g. $(\Delta p/\rho l)$	\(Equivalent\) Pipe diameters in mm													
		6	8	10	12·5	15	20	25	30	40	50	60	80	100	125
0·00030	(0·00294)	27·43	18·69	13·88	10·31	8·084	5·509	4·091	3·208	2·186	1·624	1·273	*	1·283	1·255
0·00040	(0·00392)	23·75	16·19	12·02	8·928	7·001	4·771	3·543	2·778	1·893	1·406	1·103	*	1·251	1·225
0·00060	(0·00588)	19·40	13·22	9·815	7·289	5·716	3·895	2·893	2·269	1·546	1·148	*	1·234	1·208	1·186
0·00080	(0·00785)	16·80	11·45	8·500	6·313	4·951	3·373	2·505	1·965	1·339	0·994	*	1·203	1·179	1·159
0·00100	(0·00981)	15·02	10·24	7·603	5·646	4·428	3·017	2·241	1·757	1·197	*	*	1·180	1·158	1·139
0·00150	(0·0147)	12·27	8·359	6·208	4·610	3·615	2·464	1·830	1·435	0·978	*	1·172	1·140	1·121	1·105
0·00200	(0·0196)	10·62	7·239	5·376	3·993	3·131	2·134	1·584	1·243	*	1·164	1·142	1·114	1·096	1·082
0·00300	(0·0294)	8·674	5·911	4·390	3·260	2·556	1·742	1·294	1·015	*	1·123	1·104	1·079	1·064	1·051
0·00400	(0·0392)	7·512	5·119	3·801	2·823	2·214	1·509	1·120	*	1·120	1·095	1·078	1·055	1·042	1·031
0·00600	(0·0588)	6·133	4·179	3·104	2·305	1·808	1·232	0·915	*	1·081	1·059	1·044	1·024	1·012	1·003
0·00800	(0·0785)	5·312	3·620	2·688	1·996	1·565	1·067	*	*	1·054	1·034	1·021	1·003	0·993	0·985
0·01000	(0·0981)	4·751	3·237	2·404	1·786	1·400	0·954	*	1·064	1·035	1·016	1·004	0·988	0·978	0·971
0·01500	(0·147)	3·879	2·643	1·963	1·458	1·143	*	1·047	1·027	1·001	0·985	0·974	0·961	0·953	0·947
0·02000	(0·196)	3·359	2·289	1·700	1·263	0·990	*	1·020	1·002	0·979	0·964	0·955	0·943	0·936	0·931
0·03000	(0·294)	2·743	1·869	1·388	1·031	0·808	1·008	0·985	0·969	0·949	0·937	0·929	0·919	0·913	0·909
0·04000	(0·392)	2·375	1·619	1·202	0·893	*	0·983	0·962	0·948	0·930	0·919	0·911	0·903	0·898	0·895
0·06000	(0·588)	1·940	1·322	0·982	*	*	0·949	0·931	0·919	0·903	0·894	0·888	0·881	0·878	0·876
0·08000	(0·785)	1·680	1·145	0·850	*	0·953	0·927	0·911	0·900	0·886	0·878	0·873	0·867	0·864	0·863
0·10000	(0·981)	1·502	1·024	0·760	*	0·935	0·911	0·896	0·886	0·873	0·866	0·861	0·856	0·854	0·854
0·15000	(1·47)	1·227	0·836	*	0·920	0·904	0·883	0·870	0·861	0·851	0·845	0·841	0·838	0·837	0·837
0·20000	(1·96)	1·062	0·724	*	0·898	0·883	0·865	0·853	0·845	0·836	0·831	0·828	0·826	0·826	0·827
0·30000	(2·94)	0·867	*	0·888	0·869	0·856	0·840	0·830	0·824	0·817	0·813	0·811	0·810	0·810	0·812
0·40000	(3·92)	0·751	*	0·867	0·850	0·839	0·824	0·815	0·810	0·804	0·801	0·799	0·799	0·801	0·803
0·60000	(5·88)	*	0·857	0·839	0·825	0·815	0·803	0·796	0·791	0·787	0·785	0·784	0·785	0·787	0·791
0·80000	(7·85)	*	0·837	0·821	0·808	0·800	0·789	0·783	0·779	0·775	0·774	0·774	0·776	0·779	0·783
1·00000	(9·81)	*	0·823	0·808	0·796	0·788	0·779	0·773	0·770	0·767	0·767	0·767	0·770	0·773	0·777
1·50000	(14·7)	0·819	0·798	0·786	0·776	0·769	0·761	0·757	0·755	0·753	0·754	0·755	0·758	0·762	0·767
2·00000	(19·6)	0·801	0·782	0·771	0·762	0·756	0·750	0·746	0·745	0·744	0·745	0·747	0·751	0·755	0·761
3·00000	(29·4)	0·777	0·761	0·752	0·744	0·740	0·735	0·733	0·732	0·732	0·734	0·737	0·742	0·747	0·753
4·00000	(39·2)	0·762	0·748	0·739	0·733	0·729	0·725	0·724	0·723	0·725	0·727	0·730	0·736	0·741	0·748
6·00000	(58·8)	0·742	0·730	0·723	0·718	0·715	0·712	0·712	0·713	0·715	0·718	0·722	0·728	0·734	0·741
8·00000	(78·5)	0·729	0·718	0·713	0·708	0·706	0·704	0·705	0·706	0·709	0·713	0·716	0·723	0·730	0·737
10·0000	(98·1)	0·720	0·710	0·705	0·701	0·700	0·699	0·699	0·701	0·704	0·708	0·712	0·720	0·727	0·734
15·0000	(147)	0·704	0·696	0·692	0·690	0·689	0·689	0·690	0·692	0·697	0·702	0·706	0·714	0·722	0·730
20·0000	(196)	0·694	0·687	0·684	0·682	0·682	0·683	0·685	0·687	0·692	0·697	0·702	0·711	0·719	0·727
30·0000	(294)	0·681	0·676	0·674	0·673	0·673	0·675	0·678	0·681	0·687	0·692	0·697	0·707	0·715	0·723
S	$(\Delta p/\rho l)$	6	8	10	12·5	15	20	25	30	40	50	60	80	100	125

Grad'nt   k.p.g.   (Equivalent) Pipe diameters in mm

Grad'nt S	k.p.g. $(\Delta p/\rho l)$	\(Equivalent\) Pipe diameters in mm													
		125	150	200	250	300	400	500	600	800	1000	1250	1500	2000	2500
0·00010	(0·00098)	*	1·358	1·322	1·299	1·284	1·264	1·252	1·245	1·236	1·232	1·230	1·229	1·231	1·234
0·00015	(0·00147)	1·334	1·310	1·279	1·259	1·246	1·229	1·219	1·213	1·206	1·203	1·202	1·203	1·206	1·210
0·00020	(0·00196)	1·300	1·278	1·250	1·232	1·220	1·205	1·197	1·191	1·186	1·184	1·184	1·185	1·189	1·193
0·00030	(0·00294)	1·255	1·236	1·211	1·196	1·186	1·173	1·166	1·162	1·159	1·158	1·159	1·160	1·165	1·171
0·00040	(0·00392)	1·225	1·208	1·185	1·172	1·163	1·152	1·146	1·143	1·140	1·140	1·141	1·144	1·149	1·155
0·00060	(0·00588)	1·186	1·170	1·151	1·139	1·132	1·123	1·118	1·116	1·115	1·116	1·118	1·121	1·128	1·134
0·00080	(0·00785)	1·159	1·145	1·128	1·118	1·111	1·103	1·100	1·098	1·098	1·100	1·103	1·106	1·113	1·120
0·00100	(0·00981)	1·139	1·126	1·111	1·101	1·095	1·089	1·086	1·085	1·085	1·087	1·091	1·095	1·102	1·110
0·00150	(0·0147)	1·105	1·094	1·081	1·073	1·068	1·063	1·062	1·061	1·063	1·066	1·070	1·074	1·083	1·091
0·00200	(0·0196)	1·082	1·072	1·060	1·054	1·050	1·046	1·045	1·045	1·048	1·051	1·056	1·061	1·070	1·078
0·00300	(0·0294)	1·051	1·043	1·033	1·028	1·025	1·023	1·023	1·024	1·027	1·032	1·037	1·043	1·052	1·061
0·00400	(0·0392)	1·031	1·023	1·015	1·011	1·008	1·007	1·008	1·009	1·014	1·018	1·024	1·030	1·041	1·050
0·00600	(0·0588)	1·003	0·997	0·991	0·987	0·986	0·986	0·988	0·990	0·995	1·001	1·007	1·013	1·025	1·035
0·00800	(0·0785)	0·985	0·980	0·974	0·972	0·971	0·972	0·974	0·977	0·983	0·989	0·996	1·002	1·014	1·024
0·01000	(0·0981)	0·971	0·966	0·962	0·960	0·960	0·961	0·964	0·967	0·973	0·980	0·987	0·994	1·006	1·017
0·01500	(0·147)	0·947	0·944	0·940	0·940	0·940	0·943	0·946	0·950	0·957	0·964	0·972	0·980	0·992	1·004
0·02000	(0·196)	0·931	0·928	0·926	0·926	0·927	0·930	0·934	0·938	0·946	0·954	0·962	0·970	0·983	0·995
0·03000	(0·294)	0·909	0·908	0·907	0·908	0·909	0·914	0·919	0·923	0·932	0·940	0·949	0·957	0·971	0·983
0·04000	(0·392)	0·895	0·894	0·894	0·895	0·898	0·903	0·908	0·913	0·922	0·931	0·940	0·949	0·963	0·976
0·06000	(0·588)	0·876	0·876	0·877	0·879	0·882	0·888	0·894	0·900	0·910	0·919	0·929	0·938	0·953	0·966
0·08000	(0·785)	0·863	0·863	0·865	0·868	0·872	0·878	0·885	0·891	0·901	0·911	0·921	0·931	0·946	0·960
0·10000	(0·981)	0·854	0·854	0·857	0·860	0·864	0·871	0·878	0·884	0·895	0·905	0·916	0·925	0·942	0·955
0·15000	(1·47)	0·837	0·839	0·842	0·847	0·851	0·859	0·866	0·873	0·885	0·895	0·907	0·916	0·933	0·947
0·20000	(1·96)	0·827	0·828	0·833	0·837	0·842	0·851	0·859	0·866	0·878	0·889	0·901	0·911	0·928	0·943
0·30000	(2·94)	0·812	0·815	0·820	0·826	0·831	0·840	0·849	0·856	0·869	0·881	0·893	0·904	0·921	0·936
0·40000	(3·92)	0·803	0·806	0·812	0·818	0·823	0·833	0·842	0·850	0·864	0·875	0·888	0·899	0·917	0·932
0·60000	(5·88)	0·791	0·794	0·801	0·808	0·814	0·825	0·834	0·842	0·857	0·869	0·882	0·893	0·912	0·928
0·80000	(7·85)	0·783	0·787	0·794	0·801	0·808	0·819	0·829	0·837	0·852	0·865	0·878	0·889	0·909	0·924
1·00000	(9·81)	0·777	0·781	0·789	0·797	0·803	0·815	0·825	0·834	0·849	0·862	0·875	0·887	0·906	0·922
1·50000	(14·7)	0·767	0·772	0·781	0·789	0·796	0·808	0·819	0·828	0·844	0·857	0·871	0·883	0·902	0·919
2·00000	(19·6)	0·761	0·766	0·775	0·784	0·791	0·804	0·815	0·824	0·840	0·854	0·868	0·880	0·900	0·917
3·00000	(29·4)	0·753	0·759	0·769	0·777	0·785	0·799	0·810	0·820	0·836	0·850	0·864	0·877	0·897	0·914
S	$(\Delta p/\rho l)$	125	150	200	250	300	400	500	600	800	1000	1250	1500	2000	2500

Grad'nt   k.p.g.   (Equivalent) Pipe diameters in mm

Kinematic viscosity, $\nu = 4{\cdot}0\times10^{-6}\ \mathrm{m^2\,s^{-1}}$;

Roughness size, $k_s = 0{\cdot}015$ mm

E160

Kin. visc., $\nu = 4.0 \times 10^{-6}\ \text{m}^2\text{s}^{-1}$;
$S = 0.00010$ to 3.00000

i.e. kin. pr. grad., $\Delta p/\rho l = (0.00098)$ to $(29.4)\ \text{ms}^{-2}$

Roughness size, $k_s = 0.030$ mm
This table shows values of m, as follows

m_C for Colebrook-White solutions

Grad'nt S	k.p.g. $(\Delta p/\rho l)$	125	150	200	250	300	400	500	600	800	1000	1250	1500	2000	2500
		(Equivalent) Pipe diameters in mm													
0.00010	(0.00098)	*	1.359	1.323	1.301	1.285	1.266	1.254	1.247	1.239	1.235	1.233	1.232	1.234	1.238
0.00015	(0.00147)	1.336	1.312	1.281	1.261	1.248	1.231	1.221	1.215	1.209	1.206	1.206	1.206	1.209	1.214
0.00020	(0.00196)	1.302	1.280	1.252	1.234	1.222	1.208	1.199	1.194	1.189	1.187	1.187	1.189	1.193	1.198
0.00030	(0.00294)	1.258	1.239	1.214	1.199	1.189	1.176	1.170	1.166	1.162	1.162	1.163	1.165	1.170	1.176
0.00040	(0.00392)	1.228	1.211	1.188	1.175	1.166	1.155	1.149	1.146	1.144	1.144	1.146	1.149	1.155	1.161
0.00060	(0.00588)	1.188	1.173	1.154	1.143	1.135	1.127	1.122	1.120	1.120	1.121	1.124	1.127	1.134	1.141
0.00080	(0.00785)	1.162	1.149	1.131	1.121	1.115	1.108	1.104	1.103	1.103	1.105	1.109	1.112	1.120	1.128
0.00100	(0.00981)	1.143	1.130	1.114	1.105	1.099	1.093	1.091	1.090	1.091	1.093	1.097	1.101	1.110	1.118
0.00150	(0.0147)	1.109	1.098	1.085	1.077	1.073	1.069	1.067	1.067	1.069	1.073	1.078	1.082	1.092	1.100
0.00200	(0.0196)	1.086	1.077	1.065	1.059	1.055	1.052	1.051	1.052	1.055	1.059	1.064	1.070	1.080	1.089
0.00300	(0.0294)	1.056	1.048	1.039	1.034	1.031	1.030	1.030	1.031	1.036	1.041	1.047	1.053	1.064	1.073
0.00400	(0.0392)	1.036	1.029	1.021	1.017	1.015	1.015	1.016	1.018	1.023	1.028	1.035	1.041	1.053	1.063
0.00600	(0.0588)	1.009	1.004	0.998	0.995	0.994	0.995	0.997	1.000	1.006	1.012	1.019	1.026	1.039	1.050
0.00800	(0.0785)	0.992	0.987	0.982	0.980	0.980	0.981	0.984	0.987	0.994	1.001	1.009	1.016	1.029	1.041
0.01000	(0.0981)	0.978	0.974	0.970	0.969	0.969	0.971	0.975	0.978	0.986	0.993	1.001	1.009	1.022	1.034
0.01500	(0.147)	0.956	0.952	0.950	0.950	0.951	0.954	0.959	0.963	0.971	0.979	0.988	0.996	1.011	1.023
0.02000	(0.196)	0.940	0.938	0.937	0.937	0.939	0.943	0.948	0.953	0.962	0.970	0.980	0.988	1.003	1.016
0.03000	(0.294)	0.920	0.919	0.919	0.921	0.923	0.928	0.934	0.939	0.949	0.959	0.969	0.978	0.994	1.007
0.04000	(0.392)	0.907	0.906	0.907	0.909	0.912	0.919	0.925	0.931	0.941	0.951	0.962	0.971	0.987	1.001
0.06000	(0.588)	0.889	0.890	0.892	0.895	0.899	0.906	0.913	0.919	0.931	0.941	0.952	0.962	0.979	0.994
0.08000	(0.785)	0.878	0.879	0.882	0.886	0.890	0.898	0.905	0.912	0.924	0.935	0.946	0.957	0.974	0.989
0.10000	(0.981)	0.869	0.871	0.874	0.879	0.883	0.892	0.899	0.907	0.919	0.930	0.942	0.953	0.971	0.985
0.15000	(1.47)	0.855	0.857	0.862	0.867	0.872	0.881	0.890	0.898	0.911	0.923	0.935	0.946	0.965	0.980
0.20000	(1.96)	0.846	0.848	0.854	0.860	0.865	0.875	0.884	0.892	0.906	0.918	0.931	0.942	0.961	0.977
0.30000	(2.94)	0.834	0.837	0.844	0.850	0.856	0.867	0.876	0.885	0.900	0.912	0.925	0.937	0.956	0.972
0.40000	(3.92)	0.826	0.830	0.837	0.844	0.850	0.862	0.872	0.880	0.896	0.908	0.922	0.934	0.953	0.970
0.60000	(5.88)	0.816	0.820	0.828	0.836	0.843	0.855	0.866	0.875	0.890	0.904	0.918	0.930	0.950	0.966
0.80000	(7.85)	0.810	0.814	0.823	0.831	0.838	0.851	0.862	0.871	0.887	0.901	0.915	0.927	0.948	0.965
1.00000	(9.81)	0.805	0.810	0.819	0.828	0.835	0.848	0.859	0.869	0.885	0.899	0.913	0.926	0.946	0.963
1.50000	(14.7)	0.798	0.803	0.813	0.822	0.830	0.843	0.855	0.865	0.882	0.896	0.910	0.923	0.944	0.961
2.00000	(19.6)	0.793	0.799	0.809	0.818	0.827	0.840	0.852	0.862	0.879	0.894	0.909	0.921	0.942	0.960
3.00000	(29.4)	0.787	0.793	0.804	0.814	0.823	0.837	0.849	0.859	0.877	0.891	0.906	0.919	0.941	0.958
S	$(\Delta p/\rho l)$	125	150	200	250	300	400	500	600	800	1000	1250	1500	2000	2500
Grad'nt	k.p.g.	(Equivalent) Pipe diameters in mm													

Roughness size, $k_s = 0.030$ mm

E161

Kin. visc., $\nu = 4.0 \times 10^{-6}\ \text{m}^2\text{s}^{-1}$;
$S = 0.00010$ to 3.00000

i.e. kin. pr. grad., $\Delta p/\rho l = (0.00098)$ to $(29.4)\ \text{ms}^{-2}$

Roughness size, $k_s = 0.060$ mm
This table shows values of m, as follows

m_C for Colebrook-White solutions

Grad'nt S	k.p.g. $(\Delta p/\rho l)$	125	150	200	250	300	400	500	600	800	1000	1250	1500	2000	2500
		(Equivalent) Pipe diameters in mm													
0.00010	(0.00098)	*	1.363	1.327	1.304	1.289	1.270	1.259	1.251	1.243	1.240	1.238	1.238	1.240	1.244
0.00015	(0.00147)	1.340	1.316	1.284	1.265	1.252	1.236	1.226	1.221	1.214	1.212	1.212	1.213	1.217	1.221
0.00020	(0.00196)	1.306	1.285	1.256	1.239	1.227	1.213	1.205	1.200	1.195	1.194	1.194	1.196	1.201	1.206
0.00030	(0.00294)	1.262	1.243	1.219	1.204	1.194	1.182	1.176	1.172	1.169	1.169	1.171	1.173	1.179	1.186
0.00040	(0.00392)	1.233	1.216	1.194	1.180	1.172	1.162	1.156	1.153	1.152	1.152	1.155	1.158	1.165	1.172
0.00060	(0.00588)	1.194	1.179	1.160	1.149	1.142	1.134	1.130	1.129	1.129	1.130	1.134	1.138	1.146	1.154
0.00080	(0.00785)	1.168	1.155	1.138	1.128	1.122	1.116	1.113	1.112	1.113	1.116	1.120	1.124	1.133	1.141
0.00100	(0.00981)	1.149	1.137	1.122	1.113	1.108	1.102	1.100	1.100	1.101	1.105	1.109	1.114	1.124	1.133
0.00150	(0.0147)	1.117	1.106	1.093	1.086	1.082	1.079	1.078	1.078	1.082	1.086	1.091	1.097	1.108	1.117
0.00200	(0.0196)	1.095	1.086	1.075	1.069	1.065	1.063	1.063	1.064	1.068	1.073	1.079	1.086	1.097	1.107
0.00300	(0.0294)	1.066	1.058	1.049	1.045	1.043	1.042	1.043	1.046	1.051	1.057	1.064	1.071	1.083	1.094
0.00400	(0.0392)	1.047	1.040	1.033	1.030	1.028	1.028	1.030	1.033	1.039	1.046	1.054	1.061	1.074	1.085
0.00600	(0.0588)	1.021	1.016	1.011	1.009	1.009	1.010	1.013	1.017	1.024	1.032	1.040	1.048	1.062	1.074
0.00800	(0.0785)	1.005	1.000	0.996	0.995	0.996	0.998	1.002	1.006	1.015	1.023	1.032	1.040	1.055	1.067
0.01000	(0.0981)	0.992	0.989	0.986	0.985	0.986	0.990	0.994	0.999	1.007	1.016	1.025	1.034	1.049	1.062
0.01500	(0.147)	0.972	0.969	0.968	0.968	0.970	0.975	0.980	0.985	0.995	1.005	1.015	1.024	1.040	1.054
0.02000	(0.196)	0.958	0.956	0.956	0.957	0.960	0.965	0.971	0.977	0.988	0.997	1.008	1.018	1.034	1.049
0.03000	(0.294)	0.940	0.939	0.940	0.943	0.946	0.953	0.960	0.966	0.978	0.988	1.000	1.010	1.027	1.042
0.04000	(0.392)	0.928	0.928	0.930	0.934	0.937	0.945	0.952	0.959	0.972	0.982	0.994	1.005	1.023	1.038
0.06000	(0.588)	0.913	0.914	0.918	0.922	0.926	0.935	0.943	0.950	0.964	0.975	0.988	0.999	1.017	1.033
0.08000	(0.785)	0.903	0.905	0.909	0.914	0.919	0.929	0.937	0.945	0.959	0.971	0.983	0.995	1.014	1.029
0.10000	(0.981)	0.896	0.898	0.903	0.909	0.914	0.924	0.933	0.941	0.955	0.967	0.980	0.992	1.011	1.027
0.15000	(1.47)	0.884	0.887	0.893	0.900	0.906	0.916	0.926	0.935	0.949	0.962	0.976	0.987	1.007	1.024
0.20000	(1.96)	0.877	0.880	0.887	0.894	0.900	0.912	0.922	0.931	0.946	0.959	0.973	0.985	1.005	1.021
0.30000	(2.94)	0.868	0.872	0.880	0.887	0.894	0.906	0.916	0.926	0.942	0.955	0.969	0.981	1.002	1.019
0.40000	(3.92)	0.862	0.866	0.875	0.882	0.890	0.902	0.913	0.923	0.939	0.952	0.967	0.979	1.000	1.017
0.60000	(5.88)	0.854	0.859	0.869	0.877	0.885	0.898	0.909	0.919	0.936	0.949	0.964	0.977	0.998	1.015
0.80000	(7.85)	0.850	0.855	0.865	0.874	0.881	0.895	0.907	0.917	0.934	0.948	0.963	0.975	0.997	1.014
1.00000	(9.81)	0.846	0.852	0.862	0.871	0.879	0.893	0.905	0.915	0.932	0.946	0.961	0.974	0.996	1.013
1.50000	(14.7)	0.841	0.847	0.858	0.867	0.876	0.890	0.902	0.912	0.930	0.944	0.960	0.973	0.994	1.012
2.00000	(19.6)	0.838	0.844	0.855	0.865	0.874	0.888	0.900	0.911	0.929	0.943	0.959	0.972	0.993	1.011
3.00000	(29.4)	0.834	0.841	0.852	0.862	0.871	0.886	0.898	0.909	0.927	0.942	0.957	0.970	0.992	1.010
S	$(\Delta p/\rho l)$	125	150	200	250	300	400	500	600	800	1000	1250	1500	2000	2500
Grad'nt	k.p.g.	(Equivalent) Pipe diameters in mm													

Roughness size, $k_s = 0.060$ mm

Kin. visc., $\nu = 4{\cdot}0 \times 10^{-6}$ m²s⁻¹ ;
$S = 0{\cdot}00010$ to $3{\cdot}00000$

i.e. kin. pr. grad., $\Delta p/\rho l =$
$(0{\cdot}00098)$ to $(29{\cdot}4)$ ms⁻²

Roughness size, $k_s = 0{\cdot}150$ mm
This table shows values of m, as follows

m_C for Colebrook-White solutions

Grad'nt S	k.p.g. $(\Delta p/\rho l)$	125	150	200	250	300	400	500	600	800	1000	1250	1500	2000	2500
0·00010	(0·00098)	*	1·372	1·337	1·315	1·300	1·281	1·271	1·264	1·257	1·254	1·253	1·254	1·258	1·263
0·00015	(0·00147)	1·351	1·327	1·296	1·277	1·264	1·249	1·240	1·235	1·230	1·229	1·229	1·231	1·237	1·243
0·00020	(0·00196)	1·318	1·297	1·269	1·252	1·241	1·227	1·220	1·216	1·212	1·212	1·213	1·216	1·222	1·229
0·00030	(0·00294)	1·276	1·257	1·233	1·219	1·209	1·199	1·193	1·190	1·189	1·190	1·192	1·196	1·204	1·212
0·00040	(0·00392)	1·248	1·231	1·209	1·197	1·189	1·180	1·175	1·173	1·173	1·175	1·179	1·183	1·192	1·200
0·00060	(0·00588)	1·211	1·196	1·178	1·168	1·162	1·155	1·152	1·151	1·153	1·156	1·161	1·166	1·176	1·186
0·00080	(0·00785)	1·186	1·174	1·158	1·149	1·143	1·138	1·137	1·137	1·139	1·143	1·149	1·155	1·166	1·176
0·00100	(0·00981)	1·169	1·157	1·143	1·135	1·130	1·126	1·125	1·126	1·130	1·134	1·140	1·147	1·158	1·169
0·00150	(0·0147)	1·138	1·129	1·117	1·111	1·108	1·106	1·106	1·108	1·113	1·119	1·126	1·133	1·146	1·158
0·00200	(0·0196)	1·118	1·110	1·100	1·095	1·093	1·092	1·094	1·096	1·102	1·109	1·117	1·124	1·138	1·150
0·00300	(0·0294)	1·093	1·086	1·078	1·075	1·074	1·075	1·078	1·081	1·089	1·096	1·105	1·113	1·128	1·141
0·00400	(0·0392)	1·076	1·070	1·064	1·062	1·062	1·064	1·067	1·072	1·080	1·088	1·098	1·106	1·122	1·135
0·00600	(0·0588)	1·054	1·050	1·046	1·045	1·046	1·050	1·054	1·059	1·069	1·078	1·088	1·098	1·114	1·128
0·00800	(0·0785)	1·040	1·036	1·034	1·034	1·036	1·040	1·046	1·051	1·062	1·071	1·082	1·092	1·109	1·123
0·01000	(0·0981)	1·029	1·027	1·025	1·026	1·029	1·034	1·040	1·046	1·057	1·067	1·078	1·088	1·105	1·120
0·01500	(0·147)	1·012	1·011	1·011	1·013	1·016	1·023	1·030	1·037	1·049	1·059	1·071	1·082	1·100	1·115
0·02000	(0·196)	1·001	1·001	1·002	1·005	1·009	1·016	1·024	1·031	1·043	1·055	1·067	1·078	1·096	1·112
0·03000	(0·294)	0·988	0·988	0·991	0·995	0·999	1·008	1·016	1·024	1·037	1·049	1·062	1·073	1·092	1·108
0·04000	(0·392)	0·979	0·980	0·984	0·988	0·993	1·003	1·011	1·019	1·033	1·045	1·059	1·070	1·090	1·106
0·06000	(0·588)	0·968	0·970	0·975	0·980	0·986	0·996	1·005	1·014	1·028	1·041	1·055	1·066	1·086	1·103
0·08000	(0·785)	0·961	0·963	0·969	0·975	0·981	0·992	1·002	1·010	1·026	1·038	1·052	1·064	1·084	1·101
0·10000	(0·981)	0·956	0·959	0·965	0·972	0·978	0·989	0·999	1·008	1·023	1·037	1·051	1·063	1·083	1·100
0·15000	(1·47)	0·948	0·952	0·959	0·966	0·973	0·985	0·995	1·004	1·020	1·034	1·048	1·060	1·081	1·098
0·20000	(1·96)	0·943	0·947	0·955	0·963	0·970	0·982	0·993	1·002	1·018	1·032	1·046	1·059	1·080	1·097
0·30000	(2·94)	0·937	0·942	0·950	0·958	0·966	0·979	0·990	0·999	1·016	1·030	1·044	1·057	1·078	1·096
0·40000	(3·92)	0·934	0·939	0·948	0·956	0·963	0·977	0·988	0·998	1·014	1·028	1·043	1·056	1·077	1·095
0·60000	(5·88)	0·929	0·934	0·944	0·953	0·961	0·974	0·986	0·996	1·013	1·027	1·042	1·055	1·076	1·094
0·80000	(7·85)	0·927	0·932	0·942	0·951	0·959	0·973	0·984	0·994	1·012	1·026	1·041	1·054	1·076	1·093
1·00000	(9·81)	0·925	0·930	0·941	0·950	0·958	0·972	0·983	0·994	1·011	1·025	1·041	1·054	1·075	1·093
1·50000	(14·7)	0·922	0·928	0·938	0·948	0·956	0·970	0·982	0·992	1·010	1·024	1·040	1·053	1·074	1·092
2·00000	(19·6)	0·920	0·926	0·937	0·946	0·955	0·969	0·981	0·992	1·009	1·024	1·039	1·052	1·074	1·092
3·00000	(29·4)	0·918	0·924	0·935	0·945	0·953	0·968	0·980	0·991	1·008	1·023	1·039	1·052	1·074	1·091
S	$(\Delta p/\rho l)$	125	150	200	250	300	400	500	600	800	1000	1250	1500	2000	2500

Grad'nt k.p.g. (Equivalent) Pipe diameters in mm   Roughness size, $k_s = 0{\cdot}150$ mm

---

Kin. visc., $\nu = 4{\cdot}0 \times 10^{-6}$ m²s⁻¹ ;
$S = 0{\cdot}00010$ to $3{\cdot}00000$

i.e. kin. pr. grad., $\Delta p/\rho l =$
$(0{\cdot}00098)$ to $(29{\cdot}4)$ ms⁻²

Roughness size, $k_s = 0{\cdot}30$ mm
This table shows values of m, as follows

m_C for Colebrook-White solutions

Grad'nt S	k.p.g. $(\Delta p/\rho l)$	125	150	200	250	300	400	500	600	800	1000	1250	1500	2000	2500
0·00010	(0·00098)	*	1·388	1·353	1·332	1·317	1·300	1·290	1·284	1·278	1·276	1·276	1·278	1·283	1·290
0·00015	(0·00147)	1·368	1·345	1·314	1·296	1·284	1·269	1·262	1·257	1·253	1·253	1·255	1·258	1·265	1·273
0·00020	(0·00196)	1·337	1·316	1·289	1·273	1·262	1·250	1·243	1·240	1·238	1·238	1·241	1·245	1·253	1·262
0·00030	(0·00294)	1·297	1·279	1·256	1·242	1·233	1·224	1·219	1·217	1·217	1·219	1·224	1·228	1·238	1·248
0·00040	(0·00392)	1·271	1·254	1·234	1·222	1·215	1·207	1·204	1·203	1·204	1·207	1·212	1·218	1·228	1·239
0·00060	(0·00588)	1·237	1·223	1·206	1·196	1·190	1·185	1·183	1·184	1·187	1·191	1·198	1·204	1·216	1·227
0·00080	(0·00785)	1·214	1·202	1·187	1·179	1·175	1·171	1·170	1·171	1·176	1·181	1·188	1·195	1·208	1·220
0·00100	(0·00981)	1·198	1·187	1·174	1·167	1·163	1·161	1·161	1·163	1·168	1·174	1·182	1·189	1·203	1·215
0·00150	(0·0147)	1·171	1·162	1·152	1·147	1·144	1·144	1·145	1·148	1·155	1·162	1·171	1·179	1·193	1·206
0·00200	(0·0196)	1·154	1·146	1·137	1·134	1·132	1·133	1·135	1·139	1·147	1·154	1·164	1·172	1·188	1·201
0·00300	(0·0294)	1·131	1·125	1·119	1·117	1·117	1·119	1·123	1·127	1·136	1·145	1·155	1·164	1·181	1·195
0·00400	(0·0392)	1·117	1·112	1·107	1·106	1·107	1·110	1·115	1·120	1·130	1·139	1·150	1·160	1·176	1·191
0·00600	(0·0588)	1·099	1·095	1·093	1·093	1·095	1·100	1·105	1·111	1·122	1·132	1·143	1·153	1·171	1·186
0·00800	(0·0785)	1·087	1·085	1·083	1·085	1·087	1·093	1·099	1·105	1·117	1·127	1·139	1·150	1·168	1·183
0·01000	(0·0981)	1·079	1·077	1·077	1·079	1·081	1·088	1·095	1·101	1·113	1·124	1·136	1·147	1·166	1·181
0·01500	(0·147)	1·066	1·065	1·066	1·069	1·073	1·080	1·088	1·095	1·108	1·119	1·132	1·143	1·162	1·178
0·02000	(0·196)	1·058	1·057	1·059	1·063	1·067	1·076	1·084	1·091	1·105	1·116	1·129	1·140	1·160	1·176
0·03000	(0·294)	1·047	1·048	1·051	1·056	1·060	1·070	1·078	1·086	1·100	1·113	1·126	1·137	1·157	1·174
0·04000	(0·392)	1·041	1·042	1·046	1·051	1·056	1·066	1·075	1·083	1·098	1·110	1·124	1·136	1·156	1·172
0·06000	(0·588)	1·033	1·035	1·040	1·046	1·051	1·062	1·071	1·080	1·095	1·108	1·121	1·133	1·154	1·170
0·08000	(0·785)	1·028	1·030	1·036	1·042	1·048	1·059	1·069	1·078	1·093	1·106	1·120	1·132	1·152	1·169
0·10000	(0·981)	1·025	1·027	1·034	1·040	1·046	1·057	1·067	1·076	1·092	1·105	1·119	1·131	1·152	1·169
0·15000	(1·47)	1·019	1·023	1·030	1·036	1·043	1·054	1·065	1·074	1·090	1·103	1·117	1·130	1·150	1·168
0·20000	(1·96)	1·016	1·020	1·027	1·034	1·041	1·053	1·063	1·073	1·088	1·102	1·116	1·129	1·150	1·167
0·30000	(2·94)	1·012	1·016	1·024	1·031	1·038	1·051	1·061	1·071	1·087	1·101	1·115	1·128	1·149	1·166
0·40000	(3·92)	1·010	1·014	1·022	1·030	1·037	1·049	1·060	1·070	1·086	1·100	1·115	1·127	1·148	1·166
0·60000	(5·88)	1·007	1·011	1·020	1·028	1·035	1·048	1·059	1·069	1·085	1·099	1·114	1·126	1·148	1·165
0·80000	(7·85)	1·005	1·010	1·019	1·027	1·034	1·047	1·058	1·068	1·085	1·098	1·113	1·126	1·147	1·165
1·00000	(9·81)	1·004	1·009	1·018	1·026	1·033	1·046	1·058	1·067	1·084	1·098	1·113	1·126	1·147	1·164
1·50000	(14·7)	1·002	1·007	1·016	1·025	1·032	1·045	1·057	1·067	1·083	1·097	1·112	1·125	1·147	1·164
2·00000	(19·6)	1·001	1·006	1·015	1·024	1·031	1·045	1·056	1·066	1·083	1·097	1·112	1·125	1·146	1·164
3·00000	(29·4)	1·000	1·005	1·014	1·023	1·031	1·044	1·056	1·066	1·083	1·097	1·112	1·125	1·146	1·164
S	$(\Delta p/\rho l)$	125	150	200	250	300	400	500	600	800	1000	1250	1500	2000	2500

Grad'nt k.p.g. (Equivalent) Pipe diameters in mm   Roughness size, $k_s = 0{\cdot}30$ mm

E164

Kin. visc., $\nu = 4.0 \times 10^{-6}$ m^{2}s^{-1};
$S = 0.00010$ to 3.00000

i.e. kin. pr. grad., $\Delta p/\rho l =$ (0.00098) to (29.4) ms^{-2}

Roughness size, $k_s = 0.60$ mm
This table shows values of m, as follows
m_C for Colebrook-White solutions

Grad'nt S	k.p.g. $(\Delta p/\rho l)$	125	150	200	250	300	400	500	600	800	1000	1250	1500	2000	2500
0.00010	(0.00098)	*	1.418	1.384	1.363	1.349	1.333	1.324	1.319	1.314	1.314	1.315	1.318	1.326	1.334
0.00015	(0.00147)	1.401	1.378	1.348	1.331	1.319	1.306	1.299	1.296	1.294	1.295	1.298	1.302	1.311	1.321
0.00020	(0.00196)	1.373	1.352	1.326	1.310	1.300	1.289	1.284	1.281	1.281	1.283	1.287	1.292	1.302	1.312
0.00030	(0.00294)	1.336	1.318	1.296	1.283	1.275	1.267	1.264	1.263	1.264	1.268	1.273	1.279	1.291	1.302
0.00040	(0.00392)	1.313	1.297	1.277	1.266	1.260	1.253	1.251	1.251	1.254	1.258	1.265	1.271	1.284	1.295
0.00060	(0.00588)	1.283	1.269	1.253	1.244	1.239	1.235	1.235	1.236	1.241	1.246	1.254	1.261	1.275	1.287
0.00080	(0.00785)	1.263	1.252	1.238	1.230	1.227	1.224	1.225	1.227	1.232	1.239	1.247	1.255	1.269	1.282
0.00100	(0.00981)	1.250	1.239	1.227	1.221	1.218	1.216	1.217	1.220	1.227	1.234	1.242	1.251	1.266	1.279
0.00150	(0.0147)	1.227	1.218	1.209	1.204	1.203	1.203	1.206	1.209	1.217	1.225	1.235	1.244	1.259	1.273
0.00200	(0.0196)	1.213	1.205	1.197	1.194	1.193	1.195	1.198	1.203	1.211	1.220	1.230	1.239	1.256	1.270
0.00300	(0.0294)	1.195	1.189	1.183	1.182	1.182	1.185	1.189	1.194	1.204	1.213	1.224	1.234	1.251	1.266
0.00400	(0.0392)	1.184	1.179	1.174	1.174	1.175	1.179	1.184	1.189	1.200	1.210	1.221	1.231	1.248	1.263
0.00600	(0.0588)	1.170	1.166	1.164	1.164	1.166	1.171	1.177	1.183	1.194	1.205	1.216	1.227	1.245	1.260
0.00800	(0.0785)	1.161	1.158	1.157	1.158	1.160	1.166	1.173	1.179	1.191	1.202	1.214	1.224	1.243	1.258
0.01000	(0.0981)	1.155	1.153	1.152	1.154	1.157	1.163	1.170	1.177	1.189	1.200	1.212	1.223	1.241	1.257
0.01500	(0.147)	1.145	1.144	1.145	1.147	1.151	1.158	1.165	1.172	1.185	1.197	1.209	1.220	1.239	1.255
0.02000	(0.196)	1.139	1.139	1.140	1.143	1.147	1.155	1.163	1.170	1.183	1.195	1.207	1.219	1.238	1.254
0.03000	(0.294)	1.132	1.132	1.135	1.138	1.143	1.151	1.159	1.167	1.181	1.192	1.205	1.217	1.236	1.252
0.04000	(0.392)	1.128	1.128	1.131	1.135	1.140	1.149	1.157	1.165	1.179	1.191	1.204	1.216	1.235	1.252
0.06000	(0.588)	1.122	1.123	1.127	1.132	1.137	1.146	1.155	1.163	1.177	1.189	1.203	1.214	1.234	1.251
0.08000	(0.785)	1.119	1.120	1.125	1.130	1.135	1.144	1.153	1.162	1.176	1.188	1.202	1.214	1.233	1.250
0.10000	(0.981)	1.117	1.118	1.123	1.128	1.133	1.143	1.152	1.161	1.175	1.188	1.201	1.213	1.233	1.250
0.15000	(1.47)	1.113	1.115	1.120	1.126	1.131	1.142	1.151	1.159	1.174	1.187	1.200	1.212	1.232	1.249
0.20000	(1.96)	1.111	1.113	1.119	1.124	1.130	1.140	1.150	1.158	1.173	1.186	1.200	1.212	1.232	1.249
0.30000	(2.94)	1.109	1.111	1.117	1.123	1.128	1.139	1.149	1.157	1.172	1.185	1.199	1.211	1.231	1.248
0.40000	(3.92)	1.107	1.110	1.116	1.122	1.128	1.138	1.148	1.157	1.172	1.185	1.199	1.211	1.231	1.248
0.60000	(5.88)	1.105	1.108	1.114	1.120	1.126	1.137	1.147	1.156	1.171	1.184	1.198	1.210	1.231	1.247
0.80000	(7.85)	1.104	1.107	1.113	1.120	1.126	1.137	1.147	1.156	1.171	1.184	1.198	1.210	1.230	1.247
1.00000	(9.81)	1.103	1.106	1.113	1.119	1.125	1.137	1.146	1.155	1.171	1.184	1.198	1.210	1.230	1.247
1.50000	(14.7)	1.102	1.105	1.112	1.119	1.125	1.136	1.146	1.155	1.170	1.183	1.197	1.210	1.230	1.247
2.00000	(19.6)	1.102	1.105	1.111	1.118	1.124	1.136	1.146	1.155	1.170	1.183	1.197	1.210	1.230	1.247
3.00000	(29.4)	1.101	1.104	1.111	1.118	1.124	1.135	1.145	1.154	1.170	1.183	1.197	1.209	1.230	1.247

S	$(\Delta p/\rho l)$	125	150	200	250	300	400	500	600	800	1000	1250	1500	2000	2500

Grad'nt k.p.g. (Equivalent) Pipe diameters in mm **Roughness size, $k_s = 0.60$ mm**

E165

Kin. visc., $\nu = 4.0 \times 10^{-6}$ m^{2}s^{-1};
$S = 0.00010$ to 3.00000

i.e. kin. pr. grad., $\Delta p/\rho l =$ (0.00098) to (29.4) ms^{-2}

Roughness size, $k_s = 1.50$ mm
This table shows values of m, as follows
m_C for Colebrook-White solutions

Grad'nt S	k.p.g. $(\Delta p/\rho l)$	125	150	200	250	300	400	500	600	800	1000	1250	1500	2000	2500
0.00010	(0.00098)	*	1.498	1.463	1.443	1.430	1.414	1.406	1.402	1.399	1.400	1.403	1.408	1.417	1.426
0.00015	(0.00147)	1.488	1.464	1.435	1.417	1.406	1.394	1.388	1.386	1.385	1.387	1.392	1.397	1.408	1.418
0.00020	(0.00196)	1.464	1.443	1.417	1.401	1.392	1.381	1.377	1.375	1.376	1.379	1.385	1.390	1.402	1.413
0.00030	(0.00294)	1.435	1.417	1.394	1.381	1.374	1.366	1.363	1.363	1.365	1.370	1.376	1.382	1.395	1.407
0.00040	(0.00392)	1.416	1.400	1.380	1.369	1.362	1.356	1.354	1.355	1.358	1.364	1.371	1.378	1.391	1.403
0.00060	(0.00588)	1.394	1.379	1.362	1.353	1.348	1.344	1.344	1.345	1.350	1.356	1.364	1.372	1.386	1.398
0.00080	(0.00785)	1.379	1.367	1.352	1.344	1.340	1.337	1.337	1.339	1.345	1.352	1.360	1.368	1.383	1.396
0.00100	(0.00981)	1.369	1.358	1.344	1.337	1.334	1.332	1.333	1.335	1.342	1.349	1.357	1.365	1.380	1.394
0.00150	(0.0147)	1.353	1.343	1.332	1.327	1.324	1.324	1.326	1.329	1.336	1.344	1.353	1.362	1.377	1.391
0.00200	(0.0196)	1.343	1.334	1.324	1.320	1.318	1.319	1.321	1.325	1.333	1.341	1.350	1.359	1.375	1.389
0.00300	(0.0294)	1.331	1.324	1.315	1.312	1.311	1.313	1.316	1.320	1.329	1.337	1.347	1.356	1.373	1.387
0.00400	(0.0392)	1.324	1.317	1.310	1.307	1.307	1.309	1.313	1.317	1.326	1.335	1.345	1.355	1.371	1.385
0.00600	(0.0588)	1.315	1.309	1.303	1.301	1.302	1.304	1.309	1.314	1.323	1.332	1.343	1.352	1.369	1.384
0.00800	(0.0785)	1.310	1.304	1.299	1.298	1.298	1.302	1.306	1.311	1.321	1.331	1.341	1.351	1.368	1.383
0.01000	(0.0981)	1.306	1.301	1.296	1.295	1.296	1.300	1.305	1.310	1.320	1.330	1.341	1.350	1.368	1.382
0.01500	(0.147)	1.300	1.296	1.292	1.292	1.293	1.297	1.302	1.308	1.318	1.328	1.339	1.349	1.366	1.381
0.02000	(0.196)	1.296	1.292	1.289	1.289	1.291	1.295	1.301	1.306	1.317	1.327	1.338	1.348	1.366	1.381
0.03000	(0.294)	1.292	1.289	1.286	1.287	1.288	1.293	1.299	1.305	1.316	1.326	1.337	1.347	1.365	1.380
0.04000	(0.392)	1.289	1.286	1.284	1.285	1.287	1.292	1.298	1.304	1.315	1.325	1.336	1.347	1.364	1.380
0.06000	(0.588)	1.286	1.284	1.282	1.283	1.285	1.291	1.297	1.303	1.314	1.324	1.336	1.346	1.364	1.379
0.08000	(0.785)	1.284	1.282	1.281	1.282	1.284	1.290	1.296	1.302	1.313	1.324	1.335	1.346	1.364	1.379
0.10000	(0.981)	1.283	1.281	1.280	1.281	1.283	1.289	1.295	1.301	1.313	1.323	1.335	1.345	1.363	1.379
0.15000	(1.47)	1.281	1.279	1.278	1.280	1.282	1.288	1.295	1.301	1.312	1.323	1.335	1.345	1.363	1.378
0.20000	(1.96)	1.280	1.278	1.277	1.279	1.282	1.288	1.294	1.300	1.312	1.322	1.334	1.345	1.363	1.378
0.30000	(2.94)	1.279	1.277	1.276	1.278	1.281	1.287	1.293	1.300	1.312	1.322	1.334	1.344	1.362	1.378
0.40000	(3.92)	1.278	1.276	1.276	1.278	1.280	1.287	1.293	1.299	1.311	1.322	1.334	1.344	1.362	1.378
0.60000	(5.88)	1.277	1.275	1.275	1.277	1.280	1.286	1.293	1.299	1.311	1.322	1.333	1.344	1.362	1.378
0.80000	(7.85)	1.276	1.275	1.275	1.277	1.279	1.286	1.292	1.299	1.311	1.321	1.333	1.344	1.362	1.377
1.00000	(9.81)	1.276	1.274	1.274	1.276	1.279	1.286	1.292	1.299	1.311	1.321	1.333	1.344	1.362	1.377
1.50000	(14.7)	1.275	1.274	1.274	1.276	1.279	1.285	1.292	1.298	1.310	1.321	1.333	1.344	1.362	1.377
2.00000	(19.6)	1.275	1.273	1.273	1.276	1.279	1.285	1.292	1.298	1.310	1.321	1.333	1.344	1.362	1.377
3.00000	(29.4)	1.274	1.273	1.273	1.275	1.278	1.285	1.292	1.298	1.310	1.321	1.333	1.343	1.362	1.377

S	$(\Delta p/\rho l)$	125	150	200	250	300	400	500	600	800	1000	1250	1500	2000	2500

Grad'nt k.p.g. (Equivalent) Pipe diameters in mm **Roughness size, $k_s = 1.50$ mm**

Kin. visc., $\nu = 5.0 \times 10^{-6}$ m^{2}s^{-1};
$S = 0.00010$ to 30.0000
i.e. kin. pr. grad., $\Delta p/\rho l =$
(0.00098) to (294) ms^{-2}

Roughness size, $k_s = 0.003$ mm
This table shows values of m, as follows
m_C for Colebrook-White solutions; or,
where $R \leq 2000$, m_P for laminar flow

Grad'nt	k.p.g.	(Equivalent) Pipe diameters in mm													
S	$(\Delta p/\rho l)$	6	8	10	12.5	15	20	25	30	40	50	60	80	100	125
0.00030	(0.00294)	34.29	23.36	17.35	12.89	10.11	6.886	5.114	4.010	2.733	2.029	1.591	1.084	*	1.303
0.00040	(0.00392)	29.69	20.23	15.03	11.16	8.751	5.963	4.429	3.473	2.367	1.758	1.378	*	1.300	1.271
0.00060	(0.00588)	24.24	16.52	12.27	9.112	7.145	4.869	3.616	2.836	1.932	1.435	1.125	*	1.253	1.227
0.00080	(0.00785)	21.00	14.31	10.63	7.891	6.188	4.217	3.132	2.456	1.673	1.243	*	1.250	1.222	1.199
0.00100	(0.00981)	18.78	12.80	9.504	7.058	5.535	3.772	2.801	2.196	1.497	1.112	*	1.225	1.199	1.177
0.00150	(0.0147)	15.33	10.45	7.760	5.763	4.519	3.079	2.287	1.793	1.222	*	*	1.182	1.159	1.140
0.00200	(0.0196)	13.28	9.049	6.720	4.991	3.914	2.667	1.981	1.553	1.058	*	1.186	1.153	1.132	1.115
0.00300	(0.0294)	10.84	7.388	5.487	4.075	3.196	2.178	1.617	1.268	*	1.166	1.143	1.114	1.096	1.082
0.00400	(0.0392)	9.390	6.398	4.752	3.529	2.767	1.886	1.400	1.098	*	1.135	1.115	1.089	1.072	1.059
0.00600	(0.0588)	7.667	5.224	3.880	2.881	2.260	1.540	1.143	0.897	1.121	1.095	1.078	1.055	1.040	1.029
0.00800	(0.0785)	6.640	4.524	3.360	2.495	1.957	1.333	0.990	*	1.092	1.068	1.052	1.032	1.019	1.009
0.01000	(0.0981)	5.939	4.047	3.005	2.232	1.750	1.193	0.886	*	1.070	1.049	1.034	1.015	1.003	0.994
0.01500	(0.147)	4.849	3.304	2.454	1.822	1.429	0.974	*	1.063	1.033	1.014	1.001	0.985	0.975	0.967
0.02000	(0.196)	4.199	2.861	2.125	1.578	1.238	0.843	*	1.036	1.009	0.992	0.980	0.965	0.956	0.949
0.03000	(0.294)	3.429	2.336	1.735	1.289	1.011	*	1.018	1.000	0.976	0.961	0.951	0.938	0.931	0.925
0.04000	(0.392)	2.969	2.023	1.503	1.116	0.875	*	0.992	0.976	0.954	0.940	0.931	0.920	0.914	0.909
0.06000	(0.588)	2.424	1.652	1.227	0.911	*	0.979	0.958	0.943	0.925	0.913	0.905	0.896	0.891	0.887
0.08000	(0.785)	2.100	1.431	1.063	0.789	*	0.954	0.935	0.922	0.905	0.895	0.888	0.879	0.875	0.872
0.10000	(0.981)	1.878	1.280	0.950	*	*	0.936	0.918	0.906	0.890	0.881	0.875	0.867	0.863	0.861
0.15000	(1.47)	1.533	1.045	0.776	*	0.929	0.905	0.889	0.878	0.865	0.857	0.852	0.846	0.843	0.842
0.20000	(1.96)	1.328	0.905	*	0.923	0.906	0.884	0.870	0.860	0.848	0.841	0.836	0.831	0.829	0.828
0.30000	(2.94)	1.084	0.739	*	0.890	0.875	0.856	0.843	0.835	0.825	0.819	0.816	0.812	0.811	0.811
0.40000	(3.92)	0.939	*	0.888	0.868	0.855	0.837	0.826	0.819	0.810	0.805	0.802	0.799	0.798	0.799
0.60000	(5.88)	0.767	*	0.857	0.839	0.827	0.812	0.803	0.796	0.789	0.785	0.783	0.781	0.781	0.782
0.80000	(7.85)	*	0.855	0.836	0.820	0.809	0.795	0.787	0.781	0.775	0.772	0.770	0.769	0.770	0.771
1.00000	(9.81)	*	0.838	0.820	0.806	0.796	0.783	0.775	0.770	0.765	0.762	0.761	0.760	0.761	0.763
1.50000	(14.7)	*	0.809	0.794	0.781	0.772	0.761	0.755	0.751	0.747	0.745	0.744	0.745	0.746	0.749
2.00000	(19.6)	0.812	0.790	0.776	0.764	0.756	0.747	0.741	0.738	0.734	0.733	0.733	0.734	0.736	0.739
3.00000	(29.4)	0.784	0.764	0.752	0.742	0.736	0.728	0.723	0.721	0.718	0.718	0.718	0.720	0.723	0.726
4.00000	(39.2)	0.765	0.747	0.736	0.727	0.722	0.715	0.711	0.709	0.707	0.707	0.708	0.711	0.714	0.718
6.00000	(58.8)	0.740	0.725	0.715	0.708	0.703	0.697	0.695	0.693	0.693	0.693	0.695	0.698	0.702	0.706
8.00000	(78.5)	0.723	0.710	0.701	0.695	0.690	0.686	0.684	0.683	0.683	0.684	0.686	0.690	0.694	0.699
10.0000	(98.1)	0.711	0.699	0.691	0.685	0.681	0.677	0.676	0.675	0.676	0.677	0.679	0.683	0.688	0.693
15.0000	(147)	0.690	0.679	0.673	0.668	0.665	0.662	0.661	0.662	0.663	0.665	0.668	0.673	0.678	0.683
20.0000	(196)	0.676	0.667	0.661	0.657	0.655	0.652	0.652	0.653	0.655	0.657	0.660	0.666	0.671	0.677
30.0000	(294)	0.658	0.650	0.645	0.642	0.640	0.639	0.640	0.641	0.644	0.647	0.650	0.656	0.662	0.668
S	$(\Delta p/\rho l)$	6	8	10	12.5	15	20	25	30	40	50	60	80	100	125

Grad'nt   k.p.g.   (Equivalent) Pipe diameters in mm

Grad'nt	k.p.g.	(Equivalent) Pipe diameters in mm													
S	$(\Delta p/\rho l)$	125	150	200	250	300	400	500	600	800	1000	1250	1500	2000	2500
0.00010	(0.00098)	*	*	1.371	1.345	1.327	1.304	1.290	1.280	1.269	1.263	1.260	1.258	1.258	1.260
0.00015	(0.00147)	*	1.361	1.325	1.302	1.286	1.266	1.254	1.246	1.237	1.233	1.230	1.230	1.231	1.234
0.00020	(0.00196)	1.352	1.327	1.293	1.273	1.259	1.241	1.230	1.223	1.216	1.212	1.211	1.210	1.213	1.216
0.00030	(0.00294)	1.303	1.281	1.252	1.234	1.222	1.206	1.198	1.192	1.186	1.184	1.184	1.184	1.188	1.192
0.00040	(0.00392)	1.271	1.250	1.224	1.208	1.197	1.183	1.176	1.171	1.166	1.165	1.165	1.166	1.170	1.175
0.00060	(0.00588)	1.227	1.210	1.187	1.173	1.163	1.152	1.146	1.142	1.139	1.139	1.140	1.142	1.147	1.153
0.00080	(0.00785)	1.199	1.182	1.162	1.149	1.141	1.131	1.126	1.123	1.121	1.121	1.123	1.126	1.131	1.137
0.00100	(0.00981)	1.177	1.162	1.143	1.132	1.124	1.115	1.111	1.109	1.107	1.108	1.110	1.113	1.119	1.125
0.00150	(0.0147)	1.140	1.127	1.111	1.101	1.095	1.088	1.084	1.083	1.083	1.085	1.088	1.091	1.098	1.105
0.00200	(0.0196)	1.115	1.103	1.089	1.080	1.075	1.069	1.067	1.066	1.066	1.069	1.072	1.076	1.083	1.091
0.00300	(0.0294)	1.082	1.072	1.059	1.052	1.048	1.044	1.042	1.042	1.044	1.047	1.051	1.055	1.064	1.071
0.00400	(0.0392)	1.059	1.050	1.039	1.033	1.030	1.026	1.026	1.026	1.029	1.032	1.037	1.041	1.050	1.058
0.00600	(0.0588)	1.029	1.022	1.013	1.008	1.005	1.003	1.003	1.004	1.008	1.012	1.017	1.022	1.032	1.040
0.00800	(0.0785)	1.009	1.002	0.994	0.990	0.988	0.987	0.988	0.990	0.994	0.998	1.004	1.009	1.019	1.028
0.01000	(0.0981)	0.994	0.988	0.981	0.977	0.976	0.975	0.977	0.978	0.983	0.988	0.994	0.999	1.009	1.018
0.01500	(0.147)	0.967	0.962	0.957	0.955	0.954	0.954	0.956	0.959	0.964	0.970	0.976	0.982	0.993	1.002
0.02000	(0.196)	0.949	0.945	0.941	0.939	0.939	0.940	0.943	0.945	0.951	0.957	0.964	0.970	0.981	0.991
0.03000	(0.294)	0.925	0.922	0.919	0.918	0.919	0.921	0.924	0.927	0.934	0.940	0.947	0.954	0.966	0.976
0.04000	(0.392)	0.909	0.906	0.904	0.904	0.905	0.908	0.911	0.915	0.922	0.929	0.936	0.943	0.955	0.965
0.06000	(0.588)	0.887	0.885	0.884	0.885	0.886	0.890	0.894	0.898	0.906	0.913	0.921	0.928	0.940	0.951
0.08000	(0.785)	0.872	0.871	0.871	0.872	0.874	0.878	0.882	0.887	0.895	0.902	0.911	0.918	0.931	0.942
0.10000	(0.981)	0.861	0.860	0.860	0.862	0.864	0.869	0.874	0.878	0.887	0.894	0.903	0.910	0.923	0.934
0.15000	(1.47)	0.842	0.841	0.843	0.845	0.847	0.853	0.858	0.863	0.872	0.880	0.889	0.897	0.910	0.922
0.20000	(1.96)	0.828	0.829	0.831	0.833	0.836	0.842	0.848	0.853	0.862	0.871	0.880	0.888	0.902	0.913
0.30000	(2.94)	0.811	0.811	0.814	0.817	0.821	0.827	0.833	0.839	0.849	0.858	0.867	0.876	0.890	0.902
0.40000	(3.92)	0.799	0.800	0.803	0.807	0.810	0.817	0.824	0.830	0.840	0.849	0.859	0.867	0.882	0.894
0.60000	(5.88)	0.782	0.784	0.788	0.792	0.796	0.804	0.811	0.817	0.828	0.837	0.848	0.857	0.872	0.884
0.80000	(7.85)	0.771	0.773	0.778	0.783	0.787	0.795	0.802	0.809	0.820	0.830	0.840	0.849	0.865	0.877
1.00000	(9.81)	0.763	0.766	0.770	0.775	0.780	0.788	0.796	0.802	0.814	0.824	0.834	0.844	0.859	0.872
1.50000	(14.7)	0.749	0.752	0.757	0.763	0.768	0.777	0.784	0.791	0.804	0.814	0.825	0.834	0.851	0.864
2.00000	(19.6)	0.739	0.742	0.748	0.754	0.759	0.769	0.777	0.784	0.797	0.807	0.818	0.828	0.845	0.858
3.00000	(29.4)	0.726	0.730	0.737	0.743	0.748	0.758	0.767	0.775	0.787	0.798	0.810	0.820	0.837	0.851
S	$(\Delta p/\rho l)$	125	150	200	250	300	400	500	600	800	1000	1250	1500	2000	2500

Grad'nt   k.p.g.   (Equivalent) Pipe diameters in mm

Kinematic viscosity, $\nu = 5.0 \times 10^{-6}$ m^{2}s^{-1} ;   **Roughness size, $k_s = 0.003$ mm**

E167

Kin. visc., $\nu = 5.0 \times 10^{-6}$ m^{2}s^{-1};
S = 0·00010 to 30·0000
i.e. kin. pr. grad., $\Delta p/\rho l =$
(0·00098) to (294) ms^{-2}

Roughness size, k_s = 0·006 mm
This table shows values of m, as follows
m_C for Colebrook-White solutions; or,
where $R \leq 2000$, m_P for laminar flow

Grad'nt S	k.p.g. $(\Delta p/\rho l)$	6	8	10	12·5	15	20	25	30	40	50	60	80	100	125
0·00030	(0·00294)	34·29	23·36	17·35	12·89	10·11	6·886	5·114	4·010	2·733	2·029	1·591	1·084	*	1·304
0·00040	(0·00392)	29·69	20·23	15·03	11·16	8·751	5·963	4·429	3·473	2·367	1·758	1·378	*	1·301	1·271
0·00060	(0·00588)	24·24	16·52	12·27	9·112	7·145	4·869	3·616	2·836	1·932	1·435	1·125	*	1·254	1·228
0·00080	(0·00785)	21·00	14·31	10·63	7·891	6·188	4·217	3·132	2·456	1·673	1·243	*	1·251	1·222	1·199
0·00100	(0·00981)	18·78	12·80	9·504	7·058	5·535	3·772	2·801	2·196	1·497	1·112	*	1·225	1·199	1·178
0·00150	(0·0147)	15·33	10·45	7·760	5·763	4·519	3·079	2·287	1·793	1·222	*	*	1·182	1·159	1·141
0·00200	(0·0196)	13·28	9·049	6·720	4·991	3·914	2·667	1·981	1·553	1·058	*	1·187	1·154	1·133	1·116
0·00300	(0·0294)	10·84	7·388	5·487	4·075	3·196	2·178	1·617	1·268	*	1·166	1·144	1·115	1·097	1·082
0·00400	(0·0392)	9·390	6·398	4·752	3·529	2·767	1·886	1·400	1·098	*	1·136	1·116	1·090	1·073	1·060
0·00600	(0·0588)	7·667	5·224	3·880	2·881	2·260	1·540	1·143	0·897	1·122	1·096	1·079	1·056	1·042	1·030
0·00800	(0·0785)	6·640	4·524	3·360	2·495	1·957	1·333	0·990	*	1·093	1·070	1·053	1·033	1·020	1·010
0·01000	(0·0981)	5·939	4·047	3·005	2·232	1·750	1·193	0·886	*	1·071	1·050	1·035	1·016	1·004	0·995
0·01500	(0·147)	4·849	3·304	2·454	1·822	1·429	0·974	*	1·065	1·035	1·016	1·003	0·986	0·976	0·969
0·02000	(0·196)	4·199	2·861	2·125	1·578	1·238	0·843	*	1·037	1·010	0·993	0·981	0·967	0·958	0·951
0·03000	(0·294)	3·429	2·336	1·735	1·289	1·011	*	1·020	1·001	0·977	0·963	0·953	0·940	0·933	0·927
0·04000	(0·392)	2·969	2·023	1·503	1·116	0·875	*	0·994	0·977	0·956	0·942	0·933	0·922	0·916	0·911
0·06000	(0·588)	2·424	1·652	1·227	0·911	*	0·981	0·960	0·945	0·927	0·915	0·907	0·898	0·893	0·890
0·08000	(0·785)	2·100	1·431	1·063	0·789	*	0·956	0·937	0·924	0·907	0·897	0·890	0·882	0·878	0·875
0·10000	(0·981)	1·878	1·280	0·950	*	*	0·938	0·920	0·908	0·893	0·883	0·877	0·870	0·866	0·864
0·15000	(1·47)	1·533	1·045	0·776	*	0·932	0·907	0·891	0·881	0·868	0·860	0·855	0·849	0·846	0·845
0·20000	(1·96)	1·328	0·905	*	0·926	0·909	0·886	0·872	0·863	0·851	0·844	0·840	0·835	0·833	0·832
0·30000	(2·94)	1·084	0·739	*	0·893	0·878	0·859	0·846	0·838	0·828	0·823	0·819	0·816	0·815	0·815
0·40000	(3·92)	0·939	*	0·891	0·871	0·858	0·840	0·829	0·822	0·813	0·809	0·806	0·803	0·803	0·804
0·60000	(5·88)	0·767	*	0·860	0·843	0·831	0·816	0·806	0·800	0·793	0·790	0·788	0·786	0·787	0·788
0·80000	(7·85)	*	0·858	0·839	0·824	0·813	0·799	0·791	0·786	0·780	0·777	0·775	0·775	0·776	0·778
1·00000	(9·81)	*	0·842	0·824	0·810	0·800	0·787	0·780	0·775	0·770	0·767	0·766	0·766	0·768	0·770
1·50000	(14·7)	*	0·813	0·798	0·785	0·777	0·766	0·760	0·756	0·753	0·751	0·751	0·752	0·754	0·757
2·00000	(19·6)	0·817	0·794	0·781	0·769	0·762	0·752	0·747	0·744	0·741	0·740	0·740	0·742	0·745	0·748
3·00000	(29·4)	0·789	0·769	0·758	0·748	0·742	0·734	0·730	0·727	0·726	0·726	0·726	0·729	0·732	0·737
4·00000	(39·2)	0·770	0·753	0·742	0·734	0·728	0·722	0·718	0·716	0·715	0·716	0·717	0·721	0·724	0·729
6·00000	(58·8)	0·746	0·731	0·722	0·715	0·710	0·705	0·703	0·702	0·702	0·703	0·705	0·709	0·714	0·719
8·00000	(78·5)	0·730	0·717	0·709	0·703	0·699	0·694	0·693	0·692	0·693	0·695	0·697	0·702	0·707	0·712
10·0000	(98·1)	0·719	0·706	0·699	0·693	0·690	0·686	0·685	0·685	0·686	0·689	0·691	0·696	0·701	0·707
15·0000	(147)	0·699	0·688	0·682	0·678	0·675	0·673	0·673	0·673	0·675	0·678	0·681	0·687	0·693	0·699
20·0000	(196)	0·685	0·676	0·671	0·667	0·665	0·664	0·664	0·665	0·668	0·671	0·675	0·681	0·687	0·694
30·0000	(294)	0·668	0·660	0·656	0·654	0·653	0·652	0·653	0·655	0·659	0·663	0·666	0·674	0·680	0·687
S	$(\Delta p/\rho l)$	6	8	10	12·5	15	20	25	30	40	50	60	80	100	125

Grad'nt k.p.g. (Equivalent) Pipe diameters in mm

S	$(\Delta p/\rho l)$	125	150	200	250	300	400	500	600	800	1000	1250	1500	2000	2500
0·00010	(0·00098)	*	*	1·372	1·345	1·327	1·304	1·290	1·281	1·270	1·264	1·260	1·259	1·259	1·261
0·00015	(0·00147)	*	1·362	1·325	1·302	1·286	1·266	1·254	1·247	1·238	1·233	1·231	1·230	1·232	1·235
0·00020	(0·00196)	1·353	1·327	1·294	1·273	1·259	1·241	1·230	1·224	1·216	1·213	1·211	1·211	1·213	1·217
0·00030	(0·00294)	1·304	1·281	1·252	1·234	1·222	1·207	1·198	1·193	1·187	1·185	1·184	1·185	1·188	1·193
0·00040	(0·00392)	1·271	1·251	1·225	1·208	1·197	1·184	1·176	1·172	1·167	1·166	1·166	1·167	1·171	1·176
0·00060	(0·00588)	1·228	1·210	1·187	1·173	1·164	1·153	1·147	1·143	1·140	1·140	1·141	1·143	1·148	1·154
0·00080	(0·00785)	1·199	1·183	1·162	1·150	1·142	1·132	1·127	1·124	1·122	1·122	1·124	1·127	1·132	1·138
0·00100	(0·00981)	1·178	1·163	1·144	1·132	1·125	1·116	1·112	1·109	1·108	1·109	1·111	1·114	1·121	1·127
0·00150	(0·0147)	1·141	1·128	1·111	1·102	1·096	1·089	1·085	1·084	1·084	1·086	1·089	1·092	1·100	1·106
0·00200	(0·0196)	1·116	1·104	1·090	1·081	1·076	1·070	1·068	1·067	1·068	1·070	1·074	1·077	1·085	1·092
0·00300	(0·0294)	1·082	1·073	1·060	1·053	1·049	1·045	1·044	1·044	1·046	1·049	1·053	1·057	1·066	1·074
0·00400	(0·0392)	1·060	1·051	1·040	1·034	1·031	1·028	1·027	1·028	1·030	1·034	1·039	1·043	1·052	1·061
0·00600	(0·0588)	1·030	1·023	1·014	1·009	1·007	1·005	1·005	1·006	1·010	1·014	1·020	1·025	1·034	1·043
0·00800	(0·0785)	1·010	1·003	0·996	0·992	0·990	0·989	0·990	0·992	0·996	1·001	1·006	1·012	1·022	1·031
0·01000	(0·0981)	0·995	0·989	0·982	0·979	0·978	0·977	0·979	0·981	0·985	0·990	0·997	1·002	1·013	1·022
0·01500	(0·147)	0·969	0·964	0·959	0·957	0·956	0·957	0·959	0·961	0·967	0·973	0·979	0·986	0·997	1·006
0·02000	(0·196)	0·951	0·947	0·943	0·941	0·941	0·943	0·945	0·948	0·955	0·961	0·968	0·974	0·986	0·996
0·03000	(0·294)	0·927	0·924	0·921	0·921	0·921	0·924	0·927	0·931	0·938	0·944	0·952	0·959	0·971	0·981
0·04000	(0·392)	0·911	0·909	0·907	0·907	0·908	0·911	0·915	0·919	0·926	0·933	0·941	0·948	0·961	0·971
0·06000	(0·588)	0·890	0·888	0·887	0·888	0·890	0·894	0·898	0·903	0·911	0·918	0·927	0·934	0·947	0·958
0·08000	(0·785)	0·875	0·874	0·874	0·875	0·877	0·882	0·887	0·892	0·900	0·908	0·917	0·925	0·938	0·949
0·10000	(0·981)	0·864	0·863	0·864	0·866	0·868	0·873	0·878	0·883	0·892	0·900	0·909	0·917	0·931	0·943
0·15000	(1·47)	0·845	0·845	0·847	0·849	0·852	0·858	0·864	0·869	0·879	0·887	0·897	0·905	0·920	0·932
0·20000	(1·96)	0·832	0·833	0·835	0·838	0·841	0·848	0·854	0·860	0·870	0·879	0·888	0·897	0·912	0·924
0·30000	(2·94)	0·815	0·816	0·820	0·823	0·827	0·834	0·841	0·847	0·858	0·867	0·877	0·886	0·901	0·914
0·40000	(3·92)	0·804	0·805	0·809	0·813	0·817	0·825	0·832	0·838	0·849	0·859	0·870	0·879	0·895	0·908
0·60000	(5·88)	0·788	0·790	0·795	0·800	0·804	0·813	0·820	0·827	0·839	0·849	0·860	0·870	0·886	0·899
0·80000	(7·85)	0·778	0·780	0·786	0·791	0·796	0·804	0·812	0·819	0·832	0·842	0·853	0·863	0·880	0·894
1·00000	(9·81)	0·770	0·773	0·779	0·784	0·789	0·798	0·807	0·814	0·826	0·837	0·849	0·859	0·876	0·890
1·50000	(14·7)	0·757	0·760	0·767	0·773	0·778	0·788	0·797	0·804	0·818	0·829	0·841	0·851	0·869	0·883
2·00000	(19·6)	0·748	0·752	0·759	0·765	0·771	0·781	0·790	0·798	0·812	0·823	0·836	0·846	0·864	0·879
3·00000	(29·4)	0·737	0·741	0·748	0·755	0·762	0·773	0·782	0·790	0·805	0·816	0·829	0·840	0·859	0·874
S	$(\Delta p/\rho l)$	125	150	200	250	300	400	500	600	800	1000	1250	1500	2000	2500

Grad'nt k.p.g. (Equivalent) Pipe diameters in mm

Kinematic viscosity, $\nu = 5.0 \times 10^{-6}$ m^{2}s^{-1} ; **Roughness size, k_s = 0·006 mm**

Kin. visc., $\nu = 5.0 \times 10^{-6}$ m^{2}s^{-1};
$S = 0.00010$ to 30.0000
i.e. kin. pr. grad., $\Delta p/\rho l =$
(0.00098) to (294) ms^{-2}

Roughness size, $k_s = 0.015$ mm
This table shows values of m, as follows
m_C for Colebrook-White solutions; or,
where $R \le 2000$, m_P for laminar flow

Grad'nt S	k.p.g. $(\Delta p/\rho l)$	6	8	10	12·5	15	20	25	30	40	50	60	80	100	125
0·00030	(0·00294)	34·29	23·36	17·35	12·89	10·11	6·886	5·114	4·010	2·733	2·029	1·591	1·084	*	1·305
0·00040	(0·00392)	29·69	20·23	15·03	11·16	8·751	5·963	4·429	3·473	2·367	1·758	1·378	*	1·302	1·273
0·00060	(0·00588)	24·24	16·52	12·27	9·112	7·145	4·869	3·616	2·836	1·932	1·435	1·125	*	1·255	1·230
0·00080	(0·00785)	21·00	14·31	10·63	7·891	6·188	4·217	3·132	2·456	1·673	1·243	*	1·252	1·224	1·201
0·00100	(0·00981)	18·78	12·80	9·504	7·058	5·535	3·772	2·801	2·196	1·497	1·112	*	1·227	1·201	1·179
0·00150	(0·0147)	15·33	10·45	7·760	5·763	4·519	3·079	2·287	1·793	1·222	*	*	1·184	1·161	1·143
0·00200	(0·0196)	13·28	9·049	6·720	4·991	3·914	2·667	1·981	1·553	1·058	*	1·189	1·156	1·135	1·118
0·00300	(0·0294)	10·84	7·388	5·487	4·075	3·196	2·178	1·617	1·268	*	1·169	1·146	1·118	1·100	1·085
0·00400	(0·0392)	9·390	6·398	4·752	3·529	2·767	1·886	1·400	1·098	*	1·139	1·118	1·092	1·076	1·063
0·00600	(0·0588)	7·667	5·224	3·880	2·881	2·260	1·540	1·143	0·897	1·124	1·099	1·081	1·059	1·045	1·034
0·00800	(0·0785)	6·640	4·524	3·360	2·495	1·957	1·333	0·990	*	1·096	1·073	1·057	1·036	1·024	1·014
0·01000	(0·0981)	5·939	4·047	3·005	2·232	1·750	1·193	0·886	*	1·075	1·053	1·038	1·019	1·008	0·999
0·01500	(0·147)	4·849	3·304	2·454	1·822	1·429	0·974	*	1·068	1·038	1·019	1·007	0·990	0·981	0·973
0·02000	(0·196)	4·199	2·861	2·125	1·578	1·238	0·843	*	1·041	1·014	0·997	0·986	0·971	0·963	0·956
0·03000	(0·294)	3·429	2·336	1·735	1·289	1·011	*	1·024	1·006	0·982	0·967	0·957	0·945	0·938	0·933
0·04000	(0·392)	2·969	2·023	1·503	1·116	0·875	*	0·998	0·982	0·961	0·947	0·939	0·928	0·922	0·918
0·06000	(0·588)	2·424	1·652	1·227	0·911	*	0·986	0·965	0·951	0·932	0·921	0·914	0·905	0·900	0·897
0·08000	(0·785)	2·100	1·431	1·063	0·789	*	0·962	0·943	0·930	0·913	0·903	0·897	0·889	0·886	0·883
0·10000	(0·981)	1·878	1·280	0·950	*	*	0·944	0·927	0·915	0·899	0·890	0·885	0·878	0·875	0·873
0·15000	(1·47)	1·533	1·045	0·776	*	0·938	0·914	0·899	0·888	0·875	0·868	0·863	0·858	0·856	0·855
0·20000	(1·96)	1·328	0·905	*	0·933	0·916	0·894	0·880	0·871	0·859	0·853	0·849	0·845	0·844	0·844
0·30000	(2·94)	1·084	0·739	*	0·901	0·886	0·867	0·855	0·848	0·838	0·833	0·830	0·828	0·827	0·828
0·40000	(3·92)	0·939	*	0·900	0·880	0·867	0·849	0·839	0·832	0·824	0·820	0·818	0·816	0·816	0·818
0·60000	(5·88)	0·767	*	0·870	0·853	0·841	0·826	0·817	0·812	0·805	0·802	0·801	0·801	0·802	0·804
0·80000	(7·85)	*	0·869	0·850	0·834	0·824	0·811	0·803	0·798	0·793	0·791	0·790	0·791	0·792	0·795
1·00000	(9·81)	*	0·853	0·835	0·821	0·812	0·800	0·793	0·789	0·784	0·782	0·782	0·783	0·786	0·789
1·50000	(14·7)	*	0·826	0·811	0·798	0·790	0·780	0·775	0·772	0·769	0·768	0·768	0·771	0·774	0·778
2·00000	(19·6)	0·830	0·808	0·794	0·783	0·776	0·768	0·763	0·761	0·759	0·759	0·760	0·763	0·766	0·771
3·00000	(29·4)	0·803	0·785	0·773	0·764	0·758	0·751	0·748	0·746	0·745	0·746	0·748	0·752	0·756	0·762
4·00000	(39·2)	0·786	0·769	0·759	0·751	0·746	0·740	0·738	0·737	0·737	0·738	0·741	0·745	0·750	0·756
6·00000	(58·8)	0·764	0·750	0·741	0·734	0·730	0·726	0·725	0·725	0·726	0·728	0·731	0·737	0·742	0·748
8·00000	(78·5)	0·749	0·737	0·729	0·724	0·720	0·717	0·716	0·717	0·719	0·722	0·725	0·731	0·737	0·744
10·0000	(98·1)	0·739	0·727	0·721	0·716	0·713	0·711	0·710	0·711	0·714	0·717	0·720	0·727	0·733	0·740
15·0000	(147)	0·721	0·712	0·706	0·703	0·701	0·700	0·700	0·702	0·705	0·709	0·713	0·721	0·727	0·735
20·0000	(196)	0·710	0·701	0·697	0·694	0·693	0·693	0·694	0·696	0·700	0·704	0·709	0·716	0·724	0·732
30·0000	(294)	0·695	0·688	0·685	0·683	0·683	0·684	0·686	0·688	0·693	0·698	0·703	0·711	0·719	0·727
S	$(\Delta p/\rho l)$	6	8	10	12·5	15	20	25	30	40	50	60	80	100	125

Grad'nt k.p.g. (Equivalent) Pipe diameters in mm

S	$(\Delta p/\rho l)$	125	150	200	250	300	400	500	600	800	1000	1250	1500	2000	2500
0·00010	(0·00098)	*	*	1·372	1·346	1·328	1·305	1·291	1·282	1·271	1·265	1·262	1·260	1·260	1·262
0·00015	(0·00147)	*	1·363	1·326	1·303	1·288	1·268	1·256	1·248	1·239	1·235	1·233	1·232	1·234	1·237
0·00020	(0·00196)	1·354	1·328	1·295	1·274	1·260	1·242	1·232	1·225	1·218	1·214	1·213	1·213	1·216	1·219
0·00030	(0·00294)	1·305	1·283	1·254	1·236	1·224	1·208	1·200	1·194	1·189	1·187	1·186	1·187	1·191	1·195
0·00040	(0·00392)	1·273	1·252	1·226	1·210	1·199	1·186	1·178	1·174	1·169	1·168	1·168	1·170	1·174	1·179
0·00060	(0·00588)	1·230	1·212	1·189	1·175	1·166	1·155	1·149	1·145	1·143	1·142	1·144	1·146	1·152	1·157
0·00080	(0·00785)	1·201	1·185	1·164	1·152	1·144	1·134	1·129	1·126	1·125	1·125	1·127	1·130	1·136	1·142
0·00100	(0·00981)	1·179	1·165	1·146	1·134	1·127	1·118	1·114	1·112	1·111	1·112	1·115	1·118	1·125	1·131
0·00150	(0·0147)	1·143	1·130	1·114	1·104	1·098	1·091	1·088	1·087	1·088	1·090	1·093	1·097	1·104	1·111
0·00200	(0·0196)	1·118	1·106	1·092	1·084	1·079	1·073	1·071	1·070	1·072	1·074	1·078	1·082	1·090	1·098
0·00300	(0·0294)	1·085	1·075	1·063	1·056	1·052	1·048	1·047	1·048	1·050	1·053	1·058	1·063	1·072	1·080
0·00400	(0·0392)	1·063	1·054	1·044	1·038	1·035	1·032	1·031	1·032	1·035	1·039	1·044	1·050	1·059	1·068
0·00600	(0·0588)	1·034	1·026	1·018	1·013	1·011	1·009	1·010	1·011	1·016	1·020	1·026	1·032	1·042	1·052
0·00800	(0·0785)	1·014	1·007	1·000	0·996	0·995	0·994	0·995	0·997	1·002	1·008	1·014	1·020	1·031	1·041
0·01000	(0·0981)	0·999	0·993	0·987	0·984	0·983	0·983	0·985	0·987	0·992	0·998	1·005	1·011	1·022	1·032
0·01500	(0·147)	0·973	0·969	0·964	0·962	0·962	0·963	0·966	0·969	0·975	0·981	0·989	0·995	1·008	1·018
0·02000	(0·196)	0·956	0·952	0·949	0·947	0·948	0·950	0·953	0·956	0·963	0·970	0·978	0·985	0·998	1·009
0·03000	(0·294)	0·933	0·930	0·928	0·928	0·929	0·932	0·936	0·940	0·948	0·955	0·964	0·971	0·985	0·996
0·04000	(0·392)	0·918	0·916	0·914	0·915	0·916	0·920	0·925	0·929	0·938	0·945	0·954	0·962	0·976	0·988
0·06000	(0·588)	0·897	0·896	0·896	0·897	0·899	0·904	0·909	0·914	0·924	0·932	0·942	0·950	0·965	0·977
0·08000	(0·785)	0·883	0·883	0·883	0·885	0·888	0·894	0·899	0·905	0·915	0·923	0·933	0·942	0·957	0·970
0·10000	(0·981)	0·873	0·873	0·874	0·877	0·880	0·886	0·892	0·898	0·908	0·917	0·927	0·936	0·952	0·965
0·15000	(1·47)	0·855	0·856	0·858	0·862	0·865	0·872	0·879	0·885	0·896	0·906	0·917	0·926	0·942	0·956
0·20000	(1·96)	0·844	0·845	0·848	0·852	0·856	0·863	0·871	0·877	0·889	0·899	0·910	0·920	0·936	0·950
0·30000	(2·94)	0·828	0·830	0·834	0·839	0·843	0·852	0·860	0·867	0·879	0·890	0·901	0·912	0·929	0·943
0·40000	(3·92)	0·818	0·820	0·825	0·830	0·835	0·844	0·852	0·860	0·873	0·884	0·896	0·906	0·924	0·939
0·60000	(5·88)	0·804	0·807	0·813	0·819	0·824	0·834	0·843	0·851	0·865	0·876	0·889	0·900	0·918	0·933
0·80000	(7·85)	0·795	0·799	0·805	0·812	0·817	0·828	0·837	0·845	0·859	0·871	0·884	0·895	0·914	0·929
1·00000	(9·81)	0·789	0·793	0·800	0·806	0·812	0·823	0·833	0·841	0·856	0·868	0·881	0·892	0·911	0·927
1·50000	(14·7)	0·778	0·782	0·790	0·797	0·804	0·816	0·826	0·834	0·850	0·862	0·876	0·887	0·907	0·923
2·00000	(19·6)	0·771	0·775	0·784	0·792	0·799	0·811	0·821	0·830	0·846	0·859	0·872	0·884	0·904	0·920
3·00000	(29·4)	0·762	0·767	0·776	0·784	0·792	0·804	0·815	0·825	0·841	0·854	0·868	0·880	0·900	0·917
S	$(\Delta p/\rho l)$	125	150	200	250	300	400	500	600	800	1000	1250	1500	2000	2500

Grad'nt k.p.g. (Equivalent) Pipe diameters in mm

Kinematic viscosity, $\nu = 5.0 \times 10^{-6}$ m^{2}s^{-1} ; Roughness size, $k_s = 0.015$ mm

E169

Kin. visc., $\nu = 5.0 \times 10^{-6}$ m^{2}s^{-1};
$S = 0.00010$ to 3.00000

i.e. kin. pr. grad., $\Delta p/\rho l =$
(0.00098) to (29.4) ms^{-2}

Roughness size, $k_s = 0.030$ mm
This table shows values of m, as follows

m_C for Colebrook-White solutions

Grad'nt S	k.p.g. $(\Delta p/\rho l)$														
		125	150	200	250	300	400	500	600	800	1000	1250	1500	2000	2500
0.00010	(0.00098)	*	*	1.374	1.348	1.330	1.307	1.293	1.284	1.273	1.267	1.264	1.263	1.263	1.265
0.00015	(0.00147)	*	1.364	1.328	1.305	1.289	1.270	1.258	1.250	1.242	1.237	1.235	1.235	1.237	1.240
0.00020	(0.00196)	1.355	1.330	1.297	1.276	1.262	1.245	1.234	1.228	1.220	1.217	1.216	1.216	1.219	1.223
0.00030	(0.00294)	1.307	1.285	1.256	1.238	1.226	1.211	1.202	1.197	1.192	1.190	1.190	1.191	1.195	1.200
0.00040	(0.00392)	1.275	1.255	1.228	1.212	1.202	1.188	1.181	1.177	1.173	1.171	1.172	1.174	1.179	1.184
0.00060	(0.00588)	1.232	1.214	1.192	1.178	1.169	1.158	1.152	1.149	1.147	1.147	1.148	1.151	1.157	1.163
0.00080	(0.00785)	1.204	1.188	1.167	1.155	1.147	1.138	1.133	1.130	1.129	1.130	1.132	1.135	1.142	1.149
0.00100	(0.00981)	1.182	1.168	1.149	1.138	1.131	1.122	1.118	1.116	1.116	1.117	1.120	1.124	1.131	1.138
0.00150	(0.0147)	1.146	1.133	1.117	1.108	1.102	1.096	1.093	1.092	1.093	1.095	1.099	1.103	1.112	1.119
0.00200	(0.0196)	1.122	1.110	1.096	1.088	1.083	1.078	1.076	1.076	1.078	1.081	1.085	1.090	1.099	1.107
0.00300	(0.0294)	1.089	1.080	1.068	1.061	1.058	1.054	1.053	1.054	1.057	1.061	1.066	1.071	1.081	1.090
0.00400	(0.0392)	1.068	1.059	1.049	1.043	1.040	1.038	1.038	1.039	1.043	1.048	1.053	1.059	1.070	1.079
0.00600	(0.0588)	1.039	1.032	1.024	1.020	1.018	1.017	1.018	1.020	1.025	1.030	1.037	1.043	1.054	1.064
0.00800	(0.0785)	1.020	1.014	1.007	1.003	1.002	1.002	1.004	1.007	1.012	1.018	1.025	1.032	1.044	1.055
0.01000	(0.0981)	1.005	1.000	0.994	0.992	0.991	0.992	0.994	0.997	1.003	1.009	1.017	1.024	1.037	1.048
0.01500	(0.147)	0.981	0.976	0.972	0.971	0.971	0.973	0.976	0.980	0.987	0.994	1.003	1.010	1.024	1.035
0.02000	(0.196)	0.964	0.961	0.958	0.957	0.958	0.961	0.965	0.969	0.977	0.985	0.993	1.001	1.015	1.027
0.03000	(0.294)	0.942	0.940	0.938	0.939	0.940	0.945	0.949	0.954	0.963	0.972	0.981	0.989	1.004	1.017
0.04000	(0.392)	0.928	0.926	0.926	0.927	0.929	0.934	0.939	0.944	0.954	0.963	0.973	0.982	0.997	1.010
0.06000	(0.588)	0.909	0.908	0.909	0.911	0.914	0.920	0.926	0.932	0.942	0.952	0.963	0.972	0.988	1.002
0.08000	(0.785)	0.896	0.896	0.898	0.901	0.904	0.911	0.917	0.924	0.935	0.945	0.956	0.966	0.982	0.996
0.10000	(0.981)	0.887	0.887	0.889	0.893	0.897	0.904	0.911	0.918	0.929	0.940	0.951	0.961	0.978	0.992
0.15000	(1.47)	0.871	0.872	0.876	0.880	0.884	0.893	0.900	0.907	0.920	0.931	0.943	0.953	0.971	0.986
0.20000	(1.96)	0.860	0.862	0.867	0.871	0.876	0.885	0.894	0.901	0.914	0.926	0.938	0.949	0.967	0.982
0.30000	(2.94)	0.847	0.849	0.855	0.861	0.866	0.876	0.885	0.893	0.907	0.919	0.931	0.943	0.961	0.977
0.40000	(3.92)	0.838	0.841	0.847	0.854	0.859	0.870	0.879	0.888	0.902	0.914	0.927	0.939	0.958	0.974
0.60000	(5.88)	0.827	0.831	0.838	0.845	0.851	0.862	0.872	0.881	0.896	0.909	0.922	0.934	0.954	0.970
0.80000	(7.85)	0.820	0.824	0.832	0.839	0.846	0.858	0.868	0.877	0.892	0.905	0.919	0.931	0.951	0.968
1.00000	(9.81)	0.814	0.819	0.827	0.835	0.842	0.854	0.865	0.874	0.890	0.903	0.917	0.929	0.949	0.966
1.50000	(14.7)	0.806	0.811	0.820	0.828	0.836	0.849	0.860	0.869	0.886	0.899	0.914	0.926	0.947	0.963
2.00000	(19.6)	0.800	0.806	0.815	0.824	0.832	0.845	0.856	0.866	0.883	0.897	0.911	0.924	0.945	0.962
3.00000	(29.4)	0.793	0.799	0.810	0.819	0.827	0.841	0.853	0.863	0.880	0.894	0.909	0.922	0.943	0.960
S	$(\Delta p/\rho l)$	125	150	200	250	300	400	500	600	800	1000	1250	1500	2000	2500

Grad'nt   k.p.g.   (Equivalent) Pipe diameters in mm

Roughness size, $k_s = 0.030$ mm

E170

Kin. visc., $\nu = 5.0 \times 10^{-6}$ m^{2}s^{-1};
$S = 0.00010$ to 3.00000

i.e. kin. pr. grad., $\Delta p/\rho l =$
(0.00098) to (29.4) ms^{-2}

Roughness size, $k_s = 0.060$ mm
This table shows values of m, as follows

m_C for Colebrook-White solutions

Grad'nt S	k.p.g. $(\Delta p/\rho l)$	125	150	200	250	300	400	500	600	800	1000	1250	1500	2000	2500
0.00010	(0.00098)	*	*	1.377	1.351	1.333	1.310	1.296	1.287	1.277	1.272	1.268	1.267	1.268	1.271
0.00015	(0.00147)	*	1.368	1.331	1.309	1.293	1.273	1.262	1.255	1.246	1.242	1.241	1.241	1.243	1.247
0.00020	(0.00196)	1.359	1.334	1.301	1.280	1.266	1.249	1.239	1.232	1.226	1.223	1.222	1.222	1.226	1.230
0.00030	(0.00294)	1.311	1.289	1.260	1.242	1.231	1.216	1.208	1.203	1.198	1.196	1.197	1.198	1.203	1.208
0.00040	(0.00392)	1.279	1.259	1.233	1.217	1.207	1.194	1.187	1.183	1.179	1.179	1.180	1.182	1.187	1.193
0.00060	(0.00588)	1.237	1.220	1.197	1.184	1.175	1.164	1.159	1.156	1.154	1.155	1.157	1.160	1.167	1.174
0.00080	(0.00785)	1.209	1.193	1.173	1.161	1.153	1.145	1.140	1.138	1.137	1.139	1.142	1.145	1.153	1.161
0.00100	(0.00981)	1.188	1.174	1.155	1.145	1.138	1.130	1.126	1.125	1.125	1.127	1.131	1.135	1.143	1.151
0.00150	(0.0147)	1.153	1.140	1.125	1.116	1.110	1.105	1.102	1.102	1.104	1.107	1.111	1.116	1.125	1.134
0.00200	(0.0196)	1.129	1.118	1.104	1.097	1.092	1.088	1.086	1.087	1.089	1.093	1.098	1.103	1.114	1.123
0.00300	(0.0294)	1.098	1.088	1.077	1.071	1.068	1.065	1.065	1.066	1.070	1.075	1.081	1.087	1.098	1.109
0.00400	(0.0392)	1.077	1.069	1.059	1.054	1.052	1.050	1.051	1.053	1.058	1.063	1.070	1.076	1.088	1.099
0.00600	(0.0588)	1.049	1.043	1.035	1.032	1.030	1.030	1.032	1.035	1.041	1.047	1.055	1.062	1.075	1.087
0.00800	(0.0785)	1.031	1.025	1.019	1.017	1.016	1.017	1.020	1.023	1.030	1.037	1.045	1.053	1.067	1.079
0.01000	(0.0981)	1.018	1.013	1.008	1.006	1.006	1.008	1.011	1.015	1.022	1.030	1.038	1.046	1.061	1.073
0.01500	(0.147)	0.995	0.991	0.988	0.987	0.988	0.991	0.995	1.000	1.009	1.017	1.026	1.035	1.050	1.063
0.02000	(0.196)	0.979	0.977	0.975	0.975	0.976	0.981	0.985	0.990	1.000	1.009	1.019	1.028	1.044	1.057
0.03000	(0.294)	0.960	0.958	0.957	0.959	0.961	0.967	0.973	0.978	0.989	0.998	1.009	1.019	1.035	1.049
0.04000	(0.392)	0.947	0.946	0.946	0.948	0.951	0.958	0.964	0.970	0.982	0.992	1.003	1.013	1.030	1.044
0.06000	(0.588)	0.930	0.930	0.932	0.935	0.939	0.946	0.953	0.960	0.972	0.983	0.995	1.005	1.023	1.038
0.08000	(0.785)	0.919	0.919	0.922	0.926	0.930	0.939	0.947	0.954	0.967	0.978	0.990	1.001	1.019	1.035
0.10000	(0.981)	0.911	0.912	0.916	0.920	0.925	0.933	0.942	0.949	0.963	0.974	0.987	0.998	1.016	1.032
0.15000	(1.47)	0.897	0.899	0.904	0.910	0.915	0.925	0.934	0.942	0.956	0.968	0.981	0.992	1.012	1.028
0.20000	(1.96)	0.889	0.891	0.897	0.903	0.909	0.919	0.929	0.937	0.952	0.964	0.977	0.989	1.009	1.025
0.30000	(2.94)	0.878	0.881	0.888	0.895	0.901	0.912	0.922	0.931	0.946	0.959	0.973	0.985	1.005	1.022
0.40000	(3.92)	0.871	0.875	0.882	0.890	0.896	0.908	0.918	0.927	0.943	0.956	0.970	0.983	1.003	1.020
0.60000	(5.88)	0.862	0.867	0.875	0.883	0.890	0.903	0.913	0.923	0.939	0.953	0.967	0.980	1.000	1.017
0.80000	(7.85)	0.857	0.862	0.871	0.879	0.886	0.899	0.911	0.920	0.937	0.951	0.965	0.978	0.999	1.016
1.00000	(9.81)	0.853	0.858	0.868	0.876	0.884	0.897	0.908	0.918	0.935	0.949	0.964	0.976	0.998	1.015
1.50000	(14.7)	0.847	0.853	0.863	0.872	0.880	0.893	0.905	0.915	0.932	0.947	0.962	0.974	0.996	1.013
2.00000	(19.6)	0.843	0.849	0.860	0.869	0.877	0.891	0.903	0.913	0.931	0.945	0.960	0.973	0.995	1.012
3.00000	(29.4)	0.838	0.845	0.856	0.865	0.874	0.888	0.901	0.911	0.929	0.943	0.959	0.972	0.993	1.011
S	$(\Delta p/\rho l)$	125	150	200	250	300	400	500	600	800	1000	1250	1500	2000	2500

Grad'nt   k.p.g.   (Equivalent) Pipe diameters in mm

Roughness size, $k_s = 0.060$ mm

Kin. visc., $\nu = 5{\cdot}0 \times 10^{-6}$ m^{2}s^{-1};
$S = 0{\cdot}00010$ to $3{\cdot}00000$
i.e. kin. pr. grad., $\Delta p/\rho l =$
$(0{\cdot}00098)$ to $(29{\cdot}4)$ ms^{-2}

Roughness size, $k_s = 0{\cdot}150$ mm
This table shows values of m, as follows
m_C for Colebrook-White solutions

Grad'nt S	k.p.g. $(\Delta p/\rho l)$	125	150	200	250	300	400	500	600	800	1000	1250	1500	2000	2500
0·00010	(0·00098)	*	*	1·386	1·360	1·342	1·320	1·307	1·298	1·289	1·284	1·282	1·281	1·283	1·287
0·00015	(0·00147)	*	1·377	1·341	1·319	1·304	1·285	1·274	1·267	1·260	1·257	1·256	1·256	1·260	1·265
0·00020	(0·00196)	1·369	1·344	1·312	1·292	1·278	1·261	1·252	1·246	1·240	1·238	1·238	1·240	1·245	1·250
0·00030	(0·00294)	1·323	1·301	1·273	1·255	1·244	1·230	1·223	1·218	1·215	1·214	1·216	1·218	1·224	1·231
0·00040	(0·00392)	1·292	1·272	1·247	1·231	1·221	1·210	1·203	1·200	1·198	1·198	1·200	1·204	1·211	1·218
0·00060	(0·00588)	1·252	1·234	1·213	1·200	1·192	1·182	1·178	1·176	1·175	1·177	1·181	1·185	1·193	1·202
0·00080	(0·00785)	1·225	1·209	1·190	1·179	1·172	1·164	1·161	1·160	1·160	1·163	1·168	1·172	1·182	1·191
0·00100	(0·00981)	1·205	1·191	1·174	1·164	1·157	1·151	1·148	1·148	1·150	1·153	1·158	1·163	1·174	1·183
0·00150	(0·0147)	1·172	1·160	1·145	1·137	1·133	1·128	1·127	1·128	1·131	1·136	1·142	1·148	1·160	1·170
0·00200	(0·0196)	1·150	1·139	1·127	1·120	1·116	1·113	1·113	1·115	1·119	1·125	1·132	1·138	1·151	1·162
0·00300	(0·0294)	1·121	1·113	1·103	1·098	1·095	1·094	1·096	1·098	1·104	1·110	1·118	1·126	1·139	1·151
0·00400	(0·0392)	1·102	1·095	1·087	1·083	1·081	1·082	1·084	1·087	1·094	1·101	1·110	1·118	1·132	1·145
0·00600	(0·0588)	1·078	1·072	1·066	1·064	1·064	1·065	1·069	1·073	1·081	1·089	1·099	1·107	1·123	1·136
0·00800	(0·0785)	1·062	1·057	1·053	1·052	1·052	1·055	1·059	1·064	1·073	1·082	1·092	1·101	1·117	1·131
0·01000	(0·0981)	1·051	1·047	1·043	1·043	1·044	1·048	1·052	1·057	1·067	1·076	1·087	1·096	1·113	1·127
0·01500	(0·147)	1·031	1·029	1·027	1·028	1·030	1·035	1·041	1·047	1·058	1·068	1·079	1·089	1·106	1·121
0·02000	(0·196)	1·019	1·017	1·017	1·018	1·021	1·027	1·034	1·040	1·052	1·062	1·074	1·084	1·102	1·117
0·03000	(0·294)	1·003	1·002	1·003	1·006	1·010	1·017	1·025	1·032	1·044	1·055	1·068	1·078	1·097	1·112
0·04000	(0·392)	0·993	0·993	0·995	0·999	1·003	1·011	1·019	1·026	1·040	1·051	1·064	1·075	1·094	1·110
0·06000	(0·588)	0·980	0·981	0·985	0·989	0·994	1·003	1·012	1·020	1·034	1·046	1·059	1·070	1·090	1·106
0·08000	(0·785)	0·972	0·974	0·978	0·983	0·989	0·998	1·008	1·016	1·030	1·043	1·056	1·068	1·088	1·104
0·10000	(0·981)	0·966	0·968	0·974	0·979	0·985	0·995	1·004	1·013	1·028	1·040	1·054	1·066	1·086	1·103
0·15000	(1·47)	0·957	0·960	0·966	0·972	0·978	0·990	1·000	1·008	1·024	1·037	1·051	1·063	1·083	1·100
0·20000	(1·96)	0·951	0·954	0·961	0·968	0·975	0·986	0·997	1·006	1·021	1·035	1·049	1·061	1·082	1·099
0·30000	(2·94)	0·944	0·948	0·956	0·963	0·970	0·982	0·993	1·002	1·019	1·032	1·047	1·059	1·080	1·097
0·40000	(3·92)	0·940	0·944	0·952	0·960	0·967	0·980	0·991	1·000	1·017	1·031	1·045	1·058	1·079	1·096
0·60000	(5·88)	0·934	0·939	0·948	0·956	0·964	0·977	0·988	0·998	1·015	1·029	1·043	1·056	1·078	1·095
0·80000	(7·85)	0·931	0·936	0·945	0·954	0·962	0·975	0·986	0·996	1·013	1·028	1·042	1·055	1·077	1·094
1·00000	(9·81)	0·929	0·934	0·944	0·952	0·960	0·974	0·985	0·995	1·012	1·027	1·042	1·055	1·076	1·094
1·50000	(14·7)	0·925	0·931	0·941	0·950	0·958	0·972	0·984	0·994	1·011	1·025	1·041	1·054	1·075	1·093
2·00000	(19·6)	0·923	0·929	0·939	0·948	0·956	0·971	0·982	0·993	1·010	1·025	1·040	1·053	1·075	1·092
3·00000	(29·4)	0·920	0·926	0·937	0·946	0·955	0·969	0·981	0·992	1·009	1·024	1·039	1·052	1·074	1·092
S	$(\Delta p/\rho l)$	125	150	200	250	300	400	500	600	800	1000	1250	1500	2000	2500

Grad'nt k.p.g. (Equivalent) Pipe diameters in mm Roughness size, $k_s = 0{\cdot}150$ mm

Kin. visc., $\nu = 5{\cdot}0 \times 10^{-6}$ m^{2}s^{-1};
$S = 0{\cdot}00010$ to $3{\cdot}00000$
i.e. kin. pr. grad., $\Delta p/\rho l =$
$(0{\cdot}00098)$ to $(29{\cdot}4)$ ms^{-2}

Roughness size, $k_s = 0{\cdot}30$ mm
This table shows values of m, as follows
m_C for Colebrook-White solutions

Grad'nt S	k.p.g. $(\Delta p/\rho l)$	125	150	200	250	300	400	500	600	800	1000	1250	1500	2000	2500
0·00010	(0·00098)	*	*	1·400	1·375	1·358	1·336	1·323	1·315	1·307	1·303	1·302	1·302	1·306	1·311
0·00015	(0·00147)	*	1·393	1·357	1·336	1·321	1·303	1·293	1·286	1·280	1·278	1·278	1·280	1·285	1·292
0·00020	(0·00196)	1·386	1·361	1·329	1·310	1·297	1·281	1·272	1·267	1·263	1·262	1·263	1·266	1·272	1·279
0·00030	(0·00294)	1·342	1·320	1·292	1·276	1·265	1·252	1·246	1·242	1·240	1·241	1·243	1·247	1·255	1·263
0·00040	(0·00392)	1·312	1·293	1·268	1·254	1·244	1·233	1·228	1·226	1·225	1·227	1·230	1·235	1·244	1·253
0·00060	(0·00588)	1·274	1·258	1·237	1·225	1·217	1·209	1·206	1·205	1·206	1·209	1·214	1·219	1·230	1·240
0·00080	(0·00785)	1·250	1·235	1·216	1·206	1·200	1·193	1·191	1·191	1·193	1·197	1·203	1·209	1·221	1·231
0·00100	(0·00981)	1·231	1·218	1·201	1·192	1·187	1·182	1·180	1·181	1·184	1·189	1·195	1·202	1·214	1·225
0·00150	(0·0147)	1·201	1·190	1·176	1·169	1·165	1·162	1·163	1·164	1·169	1·175	1·183	1·190	1·204	1·216
0·00200	(0·0196)	1·181	1·171	1·160	1·154	1·151	1·150	1·151	1·154	1·160	1·166	1·175	1·183	1·197	1·210
0·00300	(0·0294)	1·156	1·148	1·139	1·135	1·134	1·134	1·137	1·140	1·148	1·156	1·165	1·173	1·189	1·202
0·00400	(0·0392)	1·140	1·133	1·126	1·123	1·122	1·124	1·128	1·132	1·140	1·149	1·158	1·167	1·183	1·197
0·00600	(0·0588)	1·119	1·114	1·109	1·108	1·108	1·112	1·116	1·121	1·131	1·140	1·151	1·160	1·177	1·191
0·00800	(0·0785)	1·106	1·102	1·098	1·098	1·099	1·104	1·109	1·114	1·125	1·135	1·146	1·156	1·173	1·188
0·01000	(0·0981)	1·096	1·093	1·090	1·091	1·093	1·098	1·104	1·110	1·121	1·131	1·142	1·153	1·170	1·185
0·01500	(0·147)	1·081	1·079	1·078	1·080	1·082	1·089	1·096	1·102	1·114	1·125	1·137	1·148	1·166	1·181
0·02000	(0·196)	1·071	1·070	1·070	1·073	1·076	1·083	1·090	1·097	1·110	1·121	1·134	1·145	1·163	1·179
0·03000	(0·294)	1·059	1·058	1·060	1·064	1·068	1·076	1·084	1·092	1·105	1·117	1·130	1·141	1·160	1·176
0·04000	(0·392)	1·051	1·051	1·054	1·058	1·063	1·072	1·080	1·088	1·102	1·114	1·127	1·139	1·158	1·174
0·06000	(0·588)	1·042	1·043	1·047	1·052	1·057	1·067	1·076	1·084	1·098	1·111	1·124	1·136	1·156	1·172
0·08000	(0·785)	1·036	1·038	1·042	1·048	1·053	1·063	1·073	1·081	1·096	1·109	1·122	1·134	1·154	1·171
0·10000	(0·981)	1·032	1·034	1·039	1·045	1·051	1·061	1·071	1·079	1·094	1·107	1·121	1·133	1·153	1·170
0·15000	(1·47)	1·025	1·028	1·034	1·040	1·046	1·058	1·068	1·076	1·092	1·105	1·119	1·131	1·152	1·169
0·20000	(1·96)	1·021	1·024	1·031	1·038	1·044	1·056	1·066	1·075	1·090	1·104	1·118	1·130	1·151	1·168
0·30000	(2·94)	1·016	1·020	1·027	1·034	1·041	1·053	1·063	1·073	1·089	1·102	1·116	1·129	1·150	1·167
0·40000	(3·92)	1·014	1·017	1·025	1·032	1·039	1·051	1·062	1·071	1·088	1·101	1·116	1·128	1·149	1·166
0·60000	(5·88)	1·010	1·014	1·022	1·030	1·037	1·050	1·060	1·070	1·086	1·100	1·115	1·127	1·148	1·166
0·80000	(7·85)	1·008	1·012	1·021	1·029	1·036	1·048	1·059	1·069	1·086	1·099	1·114	1·127	1·148	1·165
1·00000	(9·81)	1·006	1·011	1·020	1·028	1·035	1·048	1·059	1·068	1·085	1·099	1·114	1·126	1·148	1·165
1·50000	(14·7)	1·004	1·009	1·018	1·026	1·033	1·046	1·058	1·067	1·084	1·098	1·113	1·126	1·147	1·164
2·00000	(19·6)	1·003	1·008	1·017	1·025	1·033	1·046	1·057	1·067	1·084	1·098	1·113	1·125	1·147	1·164
3·00000	(29·4)	1·001	1·006	1·015	1·024	1·032	1·045	1·056	1·066	1·083	1·097	1·112	1·125	1·146	1·164
S	$(\Delta p/\rho l)$	125	150	200	250	300	400	500	600	800	1000	1250	1500	2000	2500

Grad'nt k.p.g. (Equivalent) Pipe diameters in mm Roughness size, $k_s = 0{\cdot}30$ mm

E173

Kin. visc., $\nu = 5.0 \times 10^{-6}$ m^{2}s^{-1};
$S = 0.00010$ to 3.00000
i.e. kin. pr. grad., $\Delta p/\rho l =$ (0.00098) to (29.4) ms^{-2}

Roughness size, $k_s = 0.60$ mm
This table shows values of m, as follows
m_C for Colebrook-White solutions

Grad'nt S	k.p.g. $(\Delta p/\rho l)$	125	150	200	250	300	400	500	600	800	1000	1250	1500	2000	2500
		\multicolumn — (Equivalent) Pipe diameters in mm													
0.00010	(0.00098)	*	*	1.427	1.403	1.386	1.365	1.354	1.347	1.339	1.337	1.337	1.338	1.344	1.350
0.00015	(0.00147)	*	1.423	1.388	1.366	1.352	1.335	1.326	1.321	1.316	1.316	1.317	1.320	1.327	1.335
0.00020	(0.00196)	1.418	1.393	1.362	1.343	1.331	1.316	1.309	1.304	1.302	1.302	1.305	1.308	1.317	1.326
0.00030	(0.00294)	1.377	1.355	1.329	1.313	1.303	1.291	1.286	1.283	1.283	1.285	1.289	1.294	1.304	1.313
0.00040	(0.00392)	1.350	1.331	1.307	1.293	1.285	1.275	1.271	1.270	1.271	1.274	1.279	1.284	1.295	1.306
0.00060	(0.00588)	1.316	1.300	1.280	1.268	1.262	1.255	1.253	1.252	1.255	1.260	1.266	1.272	1.285	1.296
0.00080	(0.00785)	1.294	1.280	1.262	1.253	1.247	1.242	1.241	1.242	1.246	1.251	1.258	1.265	1.278	1.290
0.00100	(0.00981)	1.278	1.265	1.249	1.241	1.236	1.233	1.232	1.234	1.239	1.245	1.252	1.260	1.274	1.286
0.00150	(0.0147)	1.252	1.241	1.229	1.222	1.219	1.217	1.219	1.221	1.228	1.235	1.243	1.251	1.266	1.279
0.00200	(0.0196)	1.236	1.226	1.215	1.210	1.208	1.208	1.210	1.213	1.221	1.228	1.238	1.246	1.262	1.275
0.00300	(0.0294)	1.215	1.207	1.199	1.196	1.195	1.196	1.199	1.203	1.212	1.221	1.231	1.240	1.256	1.270
0.00400	(0.0392)	1.202	1.195	1.189	1.186	1.186	1.189	1.193	1.197	1.207	1.216	1.226	1.236	1.253	1.267
0.00600	(0.0588)	1.185	1.180	1.176	1.175	1.176	1.180	1.185	1.190	1.200	1.210	1.221	1.231	1.249	1.264
0.00800	(0.0785)	1.175	1.171	1.168	1.168	1.169	1.174	1.180	1.185	1.196	1.207	1.218	1.228	1.246	1.261
0.01000	(0.0981)	1.168	1.164	1.162	1.163	1.165	1.170	1.176	1.182	1.194	1.204	1.216	1.226	1.244	1.260
0.01500	(0.147)	1.156	1.154	1.153	1.155	1.157	1.164	1.171	1.177	1.189	1.200	1.212	1.223	1.242	1.257
0.02000	(0.196)	1.149	1.147	1.147	1.150	1.153	1.160	1.167	1.174	1.187	1.198	1.210	1.221	1.240	1.256
0.03000	(0.294)	1.140	1.139	1.141	1.144	1.147	1.155	1.163	1.170	1.183	1.195	1.208	1.219	1.238	1.254
0.04000	(0.392)	1.135	1.134	1.137	1.140	1.144	1.153	1.161	1.168	1.181	1.193	1.206	1.218	1.237	1.253
0.06000	(0.588)	1.128	1.129	1.132	1.136	1.140	1.149	1.158	1.165	1.179	1.191	1.204	1.216	1.235	1.252
0.08000	(0.785)	1.124	1.125	1.129	1.133	1.138	1.147	1.156	1.164	1.178	1.190	1.203	1.215	1.235	1.251
0.10000	(0.981)	1.122	1.123	1.127	1.131	1.136	1.146	1.155	1.163	1.177	1.189	1.202	1.214	1.234	1.250
0.15000	(1.47)	1.117	1.119	1.123	1.128	1.134	1.143	1.153	1.161	1.175	1.188	1.201	1.213	1.233	1.250
0.20000	(1.96)	1.115	1.116	1.121	1.127	1.132	1.142	1.151	1.160	1.174	1.187	1.201	1.212	1.233	1.249
0.30000	(2.94)	1.111	1.114	1.119	1.125	1.130	1.141	1.150	1.159	1.173	1.186	1.200	1.212	1.232	1.249
0.40000	(3.92)	1.110	1.112	1.117	1.123	1.129	1.140	1.149	1.158	1.173	1.186	1.199	1.211	1.231	1.248
0.60000	(5.88)	1.107	1.110	1.116	1.122	1.128	1.139	1.148	1.157	1.172	1.185	1.199	1.211	1.231	1.248
0.80000	(7.85)	1.106	1.109	1.115	1.121	1.127	1.138	1.148	1.156	1.172	1.184	1.198	1.210	1.231	1.248
1.00000	(9.81)	1.105	1.108	1.114	1.120	1.126	1.137	1.147	1.156	1.171	1.184	1.198	1.210	1.231	1.247
1.50000	(14.7)	1.104	1.107	1.113	1.119	1.125	1.137	1.147	1.155	1.171	1.184	1.198	1.210	1.230	1.247
2.00000	(19.6)	1.103	1.106	1.112	1.119	1.125	1.136	1.146	1.155	1.170	1.183	1.197	1.210	1.230	1.247
3.00000	(29.4)	1.102	1.105	1.112	1.118	1.124	1.136	1.146	1.155	1.170	1.183	1.197	1.209	1.230	1.247
S	$(\Delta p/\rho l)$	125	150	200	250	300	400	500	600	800	1000	1250	1500	2000	2500

Grad'nt k.p.g. (Equivalent) Pipe diameters in mm
Roughness size, $k_s = 0.60$ mm

E174

Kin. visc., $\nu = 5.0 \times 10^{-6}$ m^{2}s^{-1};
$S = 0.00010$ to 3.00000
i.e. kin. pr. grad., $\Delta p/\rho l =$ (0.00098) to (29.4) ms^{-2}

Roughness size, $k_s = 1.50$ mm
This table shows values of m, as follows
m_C for Colebrook-White solutions

Grad'nt S	k.p.g. $(\Delta p/\rho l)$	125	150	200	250	300	400	500	600	800	1000	1250	1500	2000	2500
0.00010	(0.00098)	*	*	1.500	1.475	1.459	1.440	1.429	1.423	1.418	1.417	1.418	1.421	1.429	1.437
0.00015	(0.00147)	*	1.501	1.466	1.446	1.432	1.416	1.408	1.404	1.401	1.402	1.405	1.409	1.418	1.427
0.00020	(0.00196)	*	1.477	1.445	1.427	1.415	1.402	1.395	1.392	1.390	1.392	1.396	1.401	1.411	1.421
0.00030	(0.00294)	1.468	1.446	1.419	1.404	1.394	1.383	1.378	1.377	1.377	1.380	1.386	1.391	1.403	1.414
0.00040	(0.00392)	1.446	1.426	1.402	1.389	1.380	1.372	1.368	1.367	1.369	1.373	1.379	1.385	1.398	1.409
0.00060	(0.00588)	1.419	1.402	1.382	1.371	1.364	1.357	1.356	1.356	1.359	1.364	1.371	1.378	1.391	1.404
0.00080	(0.00785)	1.402	1.387	1.369	1.359	1.354	1.349	1.348	1.349	1.353	1.359	1.366	1.374	1.388	1.400
0.00100	(0.00981)	1.390	1.376	1.360	1.351	1.346	1.342	1.342	1.344	1.349	1.355	1.363	1.371	1.385	1.398
0.00150	(0.0147)	1.371	1.359	1.345	1.338	1.335	1.333	1.334	1.336	1.342	1.349	1.358	1.366	1.381	1.394
0.00200	(0.0196)	1.359	1.349	1.336	1.330	1.328	1.327	1.328	1.331	1.338	1.345	1.354	1.363	1.378	1.392
0.00300	(0.0294)	1.345	1.336	1.325	1.321	1.319	1.319	1.322	1.325	1.333	1.341	1.351	1.359	1.375	1.389
0.00400	(0.0392)	1.336	1.328	1.319	1.315	1.314	1.315	1.318	1.322	1.330	1.338	1.348	1.357	1.373	1.388
0.00600	(0.0588)	1.325	1.318	1.311	1.308	1.307	1.309	1.313	1.317	1.326	1.335	1.345	1.355	1.371	1.386
0.00800	(0.0785)	1.318	1.312	1.306	1.304	1.303	1.306	1.310	1.315	1.324	1.333	1.344	1.353	1.370	1.384
0.01000	(0.0981)	1.314	1.308	1.302	1.301	1.301	1.304	1.308	1.313	1.323	1.332	1.343	1.352	1.369	1.384
0.01500	(0.147)	1.306	1.301	1.297	1.296	1.297	1.300	1.305	1.310	1.320	1.330	1.341	1.351	1.368	1.382
0.02000	(0.196)	1.302	1.297	1.294	1.293	1.294	1.298	1.303	1.309	1.319	1.329	1.340	1.350	1.367	1.382
0.03000	(0.294)	1.297	1.293	1.290	1.290	1.291	1.296	1.301	1.307	1.317	1.327	1.338	1.348	1.366	1.381
0.04000	(0.392)	1.294	1.290	1.287	1.288	1.289	1.294	1.300	1.305	1.316	1.326	1.338	1.348	1.365	1.380
0.06000	(0.588)	1.290	1.287	1.285	1.285	1.287	1.292	1.298	1.304	1.315	1.325	1.337	1.347	1.365	1.380
0.08000	(0.785)	1.287	1.285	1.283	1.284	1.286	1.291	1.297	1.303	1.314	1.325	1.336	1.346	1.364	1.379
0.10000	(0.981)	1.286	1.283	1.282	1.283	1.285	1.290	1.296	1.302	1.314	1.324	1.336	1.346	1.364	1.379
0.15000	(1.47)	1.283	1.281	1.280	1.281	1.283	1.289	1.295	1.302	1.313	1.323	1.335	1.345	1.363	1.379
0.20000	(1.96)	1.282	1.280	1.279	1.280	1.283	1.289	1.295	1.301	1.313	1.323	1.335	1.345	1.363	1.378
0.30000	(2.94)	1.280	1.278	1.277	1.279	1.282	1.288	1.294	1.300	1.312	1.323	1.334	1.345	1.363	1.378
0.40000	(3.92)	1.279	1.277	1.277	1.278	1.281	1.287	1.294	1.300	1.312	1.322	1.334	1.344	1.363	1.378
0.60000	(5.88)	1.278	1.276	1.276	1.278	1.280	1.287	1.293	1.300	1.311	1.322	1.334	1.344	1.362	1.378
0.80000	(7.85)	1.277	1.275	1.275	1.277	1.280	1.286	1.293	1.299	1.311	1.322	1.334	1.344	1.362	1.378
1.00000	(9.81)	1.277	1.275	1.275	1.277	1.280	1.286	1.293	1.299	1.311	1.322	1.333	1.344	1.362	1.377
1.50000	(14.7)	1.276	1.274	1.274	1.276	1.279	1.286	1.292	1.299	1.311	1.321	1.333	1.344	1.362	1.377
2.00000	(19.6)	1.275	1.274	1.274	1.276	1.279	1.285	1.292	1.299	1.311	1.321	1.333	1.344	1.362	1.377
3.00000	(29.4)	1.275	1.273	1.274	1.276	1.279	1.285	1.292	1.298	1.310	1.321	1.333	1.344	1.362	1.377
S	$(\Delta p/\rho l)$	125	150	200	250	300	400	500	600	800	1000	1250	1500	2000	2500

Grad'nt k.p.g. (Equivalent) Pipe diameters in mm
Roughness size, $k_s = 1.50$ mm

Kin. visc., $\nu = 6.0\times10^{-6}$ m^{2}s^{-1}; $S = 0.00010$ to 30.0000
i.e. kin. pr. grad., $\Delta p/\rho l =$ (0.00098) to (294) ms^{-2}

Roughness size, $k_s = 0.003$ mm
This table shows values of m, as follows
m_C for Colebrook-White solutions; or, where $\mathbf{R} \leq 2000$, m_P for laminar flow

E175

Grad'nt S	k.p.g. $(\Delta p/\rho l)$	6	8	10	12·5	15	20	25	30	40	50	60	80	100	125
		(Equivalent) Pipe diameters in mm													
0·00030	(0·00294)	41·14	28·04	20·82	15·46	12·13	8·263	6·137	4·812	3·279	2·435	1·910	1·301	*	1·347
0·00040	(0·00392)	35·63	24·28	18·03	13·39	10·50	7·156	5·314	4·168	2·840	2·109	1·654	1·127	*	1·312
0·00060	(0·00588)	29·09	19·82	14·72	10·93	8·575	5·843	4·339	3·403	2·319	1·722	1·350	*	1·295	1·266
0·00080	(0·00785)	25·20	17·17	12·75	9·469	7·426	5·060	3·758	2·947	2·008	1·491	1·169	*	1·262	1·235
0·00100	(0·00981)	22·54	15·36	11·40	8·470	6·642	4·526	3·361	2·636	1·796	1·334	1·046	1·267	1·237	1·213
0·00150	(0·0147)	18·40	12·54	9·312	6·915	5·423	3·695	2·744	2·152	1·466	1·089	*	1·220	1·195	1·173
0·00200	(0·0196)	15·94	10·86	8·064	5·989	4·696	3·200	2·377	1·864	1·270	*	*	1·190	1·166	1·147
0·00300	(0·0294)	13·01	8·866	6·584	4·890	3·835	2·613	1·941	1·522	1·037	*	1·181	1·149	1·128	1·111
0·00400	(0·0392)	11·27	7·678	5·702	4·235	3·321	2·263	1·681	1·318	*	1·174	1·151	1·122	1·103	1·088
0·00600	(0·0588)	9·200	6·269	4·656	3·458	2·712	1·848	1·372	1·076	*	1·131	1·111	1·085	1·069	1·056
0·00800	(0·0785)	7·968	5·429	4·032	2·994	2·348	1·600	1·188	0·932	1·129	1·103	1·084	1·061	1·046	1·035
0·01000	(0·0981)	7·126	4·856	3·606	2·678	2·100	1·431	1·063	*	1·106	1·081	1·065	1·043	1·029	1·019
0·01500	(0·147)	5·819	3·965	2·945	2·187	1·715	1·169	0·868	*	1·066	1·045	1·030	1·012	1·000	0·991
0·02000	(0·196)	5·039	3·434	2·550	1·894	1·485	1·012	*	1·071	1·040	1·021	1·007	0·990	0·980	0·972
0·03000	(0·294)	4·114	2·804	2·082	1·546	1·213	*	*	1·032	1·005	0·988	0·977	0·962	0·953	0·947
0·04000	(0·392)	3·563	2·428	1·803	1·339	1·050	*	1·025	1·006	0·982	0·967	0·956	0·943	0·935	0·930
0·06000	(0·588)	2·909	1·982	1·472	1·093	0·857	*	0·989	0·972	0·951	0·938	0·929	0·918	0·911	0·907
0·08000	(0·785)	2·520	1·717	1·275	0·947	*	0·986	0·964	0·949	0·930	0·918	0·910	0·900	0·895	0·891
0·10000	(0·981)	2·254	1·536	1·140	0·847	*	0·966	0·946	0·932	0·915	0·904	0·896	0·887	0·883	0·879
0·15000	(1·47)	1·840	1·254	0·931	*	0·961	0·933	0·915	0·903	0·888	0·878	0·872	0·865	0·861	0·859
0·20000	(1·96)	1·594	1·086	0·806	*	0·936	0·910	0·894	0·884	0·870	0·861	0·856	0·850	0·847	0·845
0·30000	(2·94)	1·301	0·887	*	0·920	0·903	0·881	0·867	0·857	0·846	0·839	0·834	0·829	0·827	0·827
0·40000	(3·92)	1·127	0·768	*	0·897	0·881	0·861	0·848	0·840	0·829	0·823	0·819	0·816	0·814	0·814
0·60000	(5·88)	0·920	*	0·885	0·865	0·852	0·834	0·824	0·816	0·807	0·803	0·800	0·797	0·796	0·797
0·80000	(7·85)	0·797	*	0·863	0·845	0·832	0·817	0·807	0·800	0·793	0·789	0·786	0·784	0·784	0·785
1·00000	(9·81)	0·713	0·866	0·846	0·829	0·818	0·803	0·795	0·789	0·782	0·778	0·776	0·775	0·775	0·777
1·50000	(14·7)	*	0·835	0·817	0·803	0·793	0·781	0·773	0·768	0·763	0·760	0·759	0·759	0·760	0·762
2·00000	(19·6)	*	0·814	0·798	0·785	0·776	0·765	0·759	0·755	0·750	0·748	0·747	0·748	0·749	0·752
3·00000	(29·4)	0·809	0·787	0·773	0·762	0·754	0·745	0·739	0·736	0·733	0·732	0·731	0·733	0·735	0·738
4·00000	(39·2)	0·789	0·769	0·756	0·746	0·739	0·731	0·726	0·724	0·721	0·721	0·721	0·723	0·725	0·729
6·00000	(58·8)	0·762	0·745	0·734	0·725	0·720	0·713	0·709	0·707	0·706	0·706	0·707	0·709	0·713	0·717
8·00000	(78·5)	0·745	0·729	0·719	0·711	0·706	0·701	0·698	0·696	0·695	0·696	0·697	0·700	0·704	0·708
10·0000	(98·1)	0·731	0·717	0·708	0·701	0·697	0·691	0·689	0·688	0·688	0·689	0·690	0·694	0·698	0·702
15·0000	(147)	0·709	0·697	0·689	0·683	0·680	0·676	0·674	0·674	0·674	0·676	0·678	0·682	0·687	0·692
20·0000	(196)	0·694	0·683	0·676	0·671	0·668	0·665	0·664	0·664	0·665	0·667	0·670	0·675	0·679	0·685
30·0000	(294)	0·674	0·665	0·659	0·655	0·653	0·651	0·651	0·651	0·653	0·656	0·659	0·665	0·670	0·676
S	$(\Delta p/\rho l)$	6	8	10	12·5	15	20	25	30	40	50	60	80	100	125

Grad'nt S	k.p.g. $(\Delta p/\rho l)$	125	150	200	250	300	400	500	600	800	1000	1250	1500	2000	2500
		(Equivalent) Pipe diameters in mm													
0·00010	(0·00098)	1·243	*	1·416	1·386	1·366	1·340	1·323	1·313	1·299	1·292	1·287	1·285	1·283	1·284
0·00015	(0·00147)	*	1·408	1·366	1·341	1·323	1·300	1·286	1·277	1·266	1·260	1·257	1·255	1·255	1·257
0·00020	(0·00196)	*	1·371	1·333	1·310	1·294	1·273	1·261	1·253	1·243	1·239	1·236	1·235	1·236	1·239
0·00030	(0·00294)	1·347	1·322	1·289	1·269	1·255	1·237	1·227	1·220	1·213	1·209	1·208	1·208	1·210	1·214
0·00040	(0·00392)	1·312	1·289	1·260	1·241	1·228	1·213	1·204	1·198	1·192	1·189	1·189	1·189	1·192	1·196
0·00060	(0·00588)	1·266	1·246	1·220	1·204	1·193	1·180	1·173	1·168	1·164	1·162	1·163	1·164	1·168	1·173
0·00080	(0·00785)	1·235	1·217	1·194	1·179	1·170	1·158	1·152	1·148	1·144	1·144	1·145	1·147	1·152	1·157
0·00100	(0·00981)	1·213	1·196	1·174	1·161	1·152	1·141	1·136	1·133	1·130	1·130	1·131	1·134	1·139	1·145
0·00150	(0·0147)	1·173	1·158	1·140	1·128	1·121	1·112	1·108	1·106	1·105	1·106	1·108	1·111	1·117	1·123
0·00200	(0·0196)	1·147	1·133	1·117	1·107	1·100	1·093	1·089	1·088	1·088	1·089	1·092	1·095	1·102	1·109
0·00300	(0·0294)	1·111	1·100	1·086	1·077	1·072	1·066	1·064	1·063	1·064	1·066	1·070	1·074	1·081	1·089
0·00400	(0·0392)	1·088	1·077	1·065	1·057	1·053	1·048	1·047	1·047	1·048	1·051	1·055	1·059	1·067	1·075
0·00600	(0·0588)	1·056	1·047	1·037	1·031	1·027	1·024	1·023	1·024	1·027	1·030	1·035	1·039	1·048	1·056
0·00800	(0·0785)	1·035	1·027	1·017	1·012	1·010	1·007	1·007	1·008	1·012	1·016	1·021	1·026	1·035	1·043
0·01000	(0·0981)	1·019	1·012	1·003	0·999	0·997	0·995	0·996	0·997	1·001	1·005	1·010	1·016	1·025	1·034
0·01500	(0·147)	0·991	0·985	0·978	0·975	0·974	0·973	0·974	0·976	0·981	0·986	0·992	0·998	1·008	1·017
0·02000	(0·196)	0·972	0·967	0·961	0·959	0·958	0·958	0·960	0·963	0·968	0·973	0·979	0·985	0·996	1·005
0·03000	(0·294)	0·947	0·943	0·939	0·937	0·937	0·938	0·941	0·944	0·950	0·955	0·962	0·968	0·980	0·989
0·04000	(0·392)	0·930	0·926	0·923	0·922	0·922	0·925	0·928	0·931	0·937	0·943	0·951	0·957	0·968	0·978
0·06000	(0·588)	0·907	0·904	0·902	0·902	0·903	0·906	0·910	0·913	0·920	0·927	0·935	0·941	0·953	0·964
0·08000	(0·785)	0·891	0·889	0·888	0·888	0·890	0·893	0·897	0·901	0·909	0·916	0·924	0·931	0·943	0·954
0·10000	(0·981)	0·879	0·878	0·877	0·878	0·880	0·884	0·888	0·892	0·900	0·908	0·916	0·923	0·935	0·946
0·15000	(1·47)	0·859	0·858	0·859	0·860	0·862	0·867	0·872	0·877	0·885	0·893	0·901	0·909	0·922	0·933
0·20000	(1·96)	0·845	0·845	0·846	0·848	0·851	0·856	0·861	0·866	0·875	0·883	0·892	0·899	0·913	0·924
0·30000	(2·94)	0·827	0·827	0·829	0·832	0·835	0·841	0·846	0·851	0·861	0·869	0·879	0·887	0·900	0·912
0·40000	(3·92)	0·814	0·815	0·817	0·820	0·824	0·830	0·836	0·842	0·852	0·860	0·870	0·878	0·892	0·904
0·60000	(5·88)	0·797	0·798	0·802	0·805	0·809	0·816	0·823	0·828	0·839	0·848	0·858	0·866	0·881	0·893
0·80000	(7·85)	0·785	0·787	0·791	0·795	0·799	0·807	0·813	0·820	0·830	0·840	0·850	0·859	0·874	0·886
1·00000	(9·81)	0·777	0·779	0·783	0·787	0·792	0·799	0·806	0·813	0·824	0·833	0·844	0·853	0·868	0·881
1·50000	(14·7)	0·762	0·764	0·769	0·774	0·779	0·787	0·795	0·801	0·813	0·823	0·833	0·843	0·858	0·872
2·00000	(19·6)	0·752	0·754	0·760	0·765	0·770	0·779	0·787	0·793	0·805	0·816	0·827	0·836	0·852	0·866
3·00000	(29·4)	0·738	0·741	0·747	0·753	0·758	0·768	0·776	0·783	0·796	0·806	0·818	0·827	0·844	0·858
S	$(\Delta p/\rho l)$	125	150	200	250	300	400	500	600	800	1000	1250	1500	2000	2500

Grad'nt   k.p.g.   (Equivalent) Pipe diameters in mm

Kinematic viscosity, $\nu = 6.0\times10^{-6}$ m^{2}s^{-1} ;   **Roughness size, $k_s = 0.003$ mm**

E176

Kin. visc., $\nu = 6.0\times10^{-6}\ \mathrm{m^2 s^{-1}}$;
$S = 0.00010$ to 30.0000

i.e. kin. pr. grad., $\Delta p/\rho l =$
(0.00098) to $(294)\ \mathrm{ms^{-2}}$

Roughness size, $k_s = 0.006$ mm
This table shows values of m, as follows

m_C for Colebrook-White solutions; or,
where $\mathbf{R} \leq 2000$, m_P for laminar flow

Grad'nt S	k.p.g. $(\Delta p/\rho l)$	6	8	10	12.5	15	20	25	30	40	50	60	80	100	125
								(Equivalent) Pipe diameters in mm							
0.00030	(0.00294)	41.14	28.04	20.82	15.46	12.13	8.263	6.137	4.812	3.279	2.435	1.910	1.301	*	1.348
0.00040	(0.00392)	35.63	24.28	18.03	13.39	10.50	7.156	5.314	4.168	2.840	2.109	1.654	1.127	*	1.313
0.00060	(0.00588)	29.09	19.82	14.72	10.93	8.575	5.843	4.339	3.403	2.319	1.722	1.350	*	1.296	1.267
0.00080	(0.00785)	25.20	17.17	12.75	9.469	7.426	5.060	3.758	2.947	2.008	1.491	1.169	*	1.262	1.236
0.00100	(0.00981)	22.54	15.36	11.40	8.470	6.642	4.526	3.361	2.636	1.796	1.334	1.046	1.267	1.238	1.213
0.00150	(0.0147)	18.40	12.54	9.312	6.915	5.423	3.695	2.744	2.152	1.466	1.089	*	1.221	1.195	1.174
0.00200	(0.0196)	15.94	10.86	8.064	5.989	4.696	3.200	2.377	1.864	1.270	*	*	1.190	1.167	1.147
0.00300	(0.0294)	13.01	8.866	6.584	4.890	3.835	2.613	1.941	1.522	1.037	*	1.182	1.150	1.129	1.112
0.00400	(0.0392)	11.27	7.678	5.702	4.235	3.321	2.263	1.681	1.318	*	1.175	1.152	1.122	1.104	1.089
0.00600	(0.0588)	9.200	6.269	4.656	3.458	2.712	1.848	1.372	1.076	*	1.132	1.112	1.086	1.070	1.057
0.00800	(0.0785)	7.968	5.429	4.032	2.994	2.348	1.600	1.188	0.932	1.130	1.104	1.085	1.062	1.047	1.036
0.01000	(0.0981)	7.126	4.856	3.606	2.678	2.100	1.431	1.063	*	1.107	1.082	1.066	1.044	1.031	1.020
0.01500	(0.147)	5.819	3.965	2.945	2.187	1.715	1.169	0.868	*	1.067	1.046	1.032	1.013	1.001	0.992
0.02000	(0.196)	5.039	3.434	2.550	1.894	1.485	1.012	*	1.072	1.041	1.022	1.009	0.992	0.982	0.974
0.03000	(0.294)	4.114	2.804	2.082	1.546	1.213	*	*	1.034	1.007	0.990	0.978	0.964	0.955	0.949
0.04000	(0.392)	3.563	2.428	1.803	1.339	1.050	*	1.027	1.008	0.983	0.968	0.958	0.945	0.937	0.932
0.06000	(0.588)	2.909	1.982	1.472	1.093	0.857	*	0.990	0.974	0.953	0.939	0.931	0.920	0.913	0.909
0.08000	(0.785)	2.520	1.717	1.275	0.947	*	0.988	0.966	0.951	0.932	0.920	0.912	0.903	0.897	0.894
0.10000	(0.981)	2.254	1.536	1.140	0.847	*	0.968	0.948	0.934	0.917	0.906	0.899	0.890	0.885	0.882
0.15000	(1.47)	1.840	1.254	0.931	*	0.963	0.935	0.917	0.905	0.890	0.881	0.875	0.868	0.864	0.862
0.20000	(1.96)	1.594	1.086	0.806	*	0.938	0.913	0.897	0.886	0.872	0.864	0.859	0.853	0.850	0.849
0.30000	(2.94)	1.301	0.887	*	0.922	0.905	0.883	0.869	0.860	0.848	0.842	0.837	0.833	0.831	0.831
0.40000	(3.92)	1.127	0.768	*	0.899	0.884	0.864	0.851	0.843	0.833	0.827	0.823	0.820	0.818	0.818
0.60000	(5.88)	0.920	*	0.888	0.868	0.855	0.838	0.827	0.820	0.811	0.807	0.804	0.802	0.801	0.802
0.80000	(7.85)	0.797	*	0.866	0.848	0.836	0.820	0.811	0.804	0.797	0.793	0.791	0.789	0.790	0.791
1.00000	(9.81)	0.713	0.869	0.849	0.833	0.822	0.807	0.799	0.793	0.786	0.783	0.781	0.780	0.781	0.783
1.50000	(14.7)	*	0.839	0.821	0.807	0.797	0.785	0.778	0.773	0.768	0.766	0.765	0.765	0.766	0.769
2.00000	(19.6)	*	0.818	0.803	0.790	0.781	0.770	0.764	0.760	0.756	0.754	0.754	0.755	0.756	0.759
3.00000	(29.4)	0.814	0.792	0.778	0.767	0.760	0.750	0.745	0.742	0.739	0.739	0.739	0.741	0.743	0.747
4.00000	(39.2)	0.794	0.774	0.762	0.752	0.745	0.737	0.733	0.730	0.728	0.728	0.729	0.731	0.735	0.739
6.00000	(58.8)	0.768	0.751	0.740	0.732	0.726	0.720	0.717	0.715	0.714	0.715	0.716	0.719	0.723	0.728
8.00000	(78.5)	0.751	0.735	0.726	0.718	0.714	0.708	0.706	0.705	0.704	0.706	0.707	0.711	0.716	0.721
10.0000	(98.1)	0.738	0.724	0.715	0.709	0.704	0.700	0.698	0.697	0.697	0.699	0.701	0.705	0.710	0.715
15.0000	(147)	0.716	0.704	0.697	0.692	0.688	0.685	0.684	0.684	0.685	0.688	0.690	0.695	0.701	0.707
20.0000	(196)	0.702	0.691	0.685	0.681	0.678	0.675	0.675	0.675	0.677	0.680	0.683	0.689	0.694	0.701
30.0000	(294)	0.683	0.674	0.669	0.666	0.664	0.663	0.663	0.664	0.667	0.670	0.674	0.680	0.687	0.693
S	$(\Delta p/\rho l)$	6	8	10	12.5	15	20	25	30	40	50	60	80	100	125

Grad'nt k.p.g. (Equivalent) Pipe diameters in mm

Grad'nt S	k.p.g. $(\Delta p/\rho l)$	125	150	200	250	300	400	500	600	800	1000	1250	1500	2000	2500
								(Equivalent) Pipe diameters in mm							
0.00010	(0.00098)	1.243	*	1.416	1.387	1.366	1.340	1.324	1.313	1.300	1.293	1.288	1.285	1.284	1.285
0.00015	(0.00147)	*	1.408	1.367	1.341	1.323	1.300	1.286	1.277	1.266	1.261	1.257	1.256	1.256	1.258
0.00020	(0.00196)	*	1.371	1.334	1.310	1.294	1.273	1.261	1.253	1.244	1.239	1.236	1.236	1.237	1.239
0.00030	(0.00294)	1.348	1.322	1.290	1.269	1.255	1.237	1.227	1.220	1.213	1.210	1.208	1.208	1.211	1.214
0.00040	(0.00392)	1.313	1.290	1.260	1.242	1.229	1.213	1.204	1.198	1.192	1.190	1.189	1.190	1.193	1.197
0.00060	(0.00588)	1.267	1.247	1.221	1.205	1.194	1.181	1.173	1.169	1.164	1.163	1.163	1.165	1.169	1.174
0.00080	(0.00785)	1.236	1.218	1.194	1.180	1.170	1.159	1.152	1.149	1.145	1.145	1.146	1.148	1.153	1.158
0.00100	(0.00981)	1.213	1.196	1.175	1.161	1.153	1.142	1.136	1.133	1.131	1.131	1.132	1.135	1.140	1.146
0.00150	(0.0147)	1.174	1.159	1.140	1.129	1.122	1.113	1.109	1.107	1.106	1.107	1.109	1.112	1.118	1.125
0.00200	(0.0196)	1.147	1.134	1.117	1.107	1.101	1.094	1.090	1.089	1.089	1.090	1.093	1.096	1.103	1.110
0.00300	(0.0294)	1.112	1.101	1.087	1.078	1.073	1.067	1.065	1.064	1.065	1.068	1.072	1.075	1.083	1.091
0.00400	(0.0392)	1.089	1.078	1.066	1.058	1.054	1.050	1.048	1.048	1.050	1.053	1.057	1.061	1.069	1.077
0.00600	(0.0588)	1.057	1.048	1.038	1.032	1.028	1.025	1.025	1.025	1.028	1.032	1.037	1.042	1.051	1.059
0.00800	(0.0785)	1.036	1.028	1.019	1.014	1.011	1.009	1.009	1.010	1.014	1.018	1.023	1.028	1.038	1.046
0.01000	(0.0981)	1.020	1.013	1.005	1.000	0.998	0.997	0.997	0.999	1.003	1.007	1.013	1.018	1.028	1.037
0.01500	(0.147)	0.992	0.986	0.980	0.977	0.975	0.975	0.977	0.979	0.984	0.989	0.995	1.001	1.011	1.020
0.02000	(0.196)	0.974	0.969	0.963	0.961	0.960	0.961	0.962	0.965	0.970	0.976	0.983	0.989	1.000	1.009
0.03000	(0.294)	0.949	0.945	0.941	0.939	0.939	0.941	0.943	0.946	0.953	0.959	0.966	0.972	0.984	0.994
0.04000	(0.392)	0.932	0.928	0.925	0.925	0.925	0.927	0.931	0.934	0.941	0.947	0.955	0.961	0.973	0.984
0.06000	(0.588)	0.909	0.907	0.905	0.905	0.906	0.909	0.913	0.917	0.925	0.932	0.939	0.947	0.959	0.970
0.08000	(0.785)	0.894	0.892	0.891	0.892	0.893	0.897	0.901	0.906	0.914	0.921	0.929	0.937	0.950	0.961
0.10000	(0.981)	0.882	0.881	0.880	0.882	0.883	0.888	0.892	0.897	0.905	0.913	0.922	0.929	0.942	0.954
0.15000	(1.47)	0.862	0.862	0.862	0.864	0.867	0.872	0.877	0.882	0.891	0.899	0.908	0.916	0.930	0.942
0.20000	(1.96)	0.849	0.849	0.850	0.852	0.855	0.861	0.867	0.872	0.881	0.890	0.899	0.907	0.922	0.934
0.30000	(2.94)	0.831	0.831	0.834	0.837	0.840	0.846	0.853	0.858	0.868	0.877	0.887	0.896	0.911	0.923
0.40000	(3.92)	0.818	0.819	0.822	0.826	0.830	0.837	0.843	0.849	0.860	0.869	0.879	0.888	0.903	0.916
0.60000	(5.88)	0.802	0.804	0.808	0.812	0.816	0.824	0.831	0.837	0.848	0.858	0.869	0.878	0.894	0.907
0.80000	(7.85)	0.791	0.793	0.798	0.802	0.807	0.815	0.822	0.829	0.841	0.851	0.862	0.871	0.887	0.901
1.00000	(9.81)	0.783	0.785	0.790	0.795	0.800	0.808	0.816	0.823	0.835	0.845	0.857	0.866	0.883	0.897
1.50000	(14.7)	0.769	0.772	0.777	0.783	0.788	0.797	0.806	0.813	0.826	0.836	0.848	0.858	0.875	0.889
2.00000	(19.6)	0.759	0.763	0.769	0.775	0.780	0.790	0.799	0.806	0.819	0.830	0.842	0.853	0.870	0.885
3.00000	(29.4)	0.747	0.751	0.758	0.764	0.770	0.780	0.789	0.797	0.811	0.823	0.835	0.846	0.864	0.878
S	$(\Delta p/\rho l)$	125	150	200	250	300	400	500	600	800	1000	1250	1500	2000	2500

Grad'nt k.p.g. (Equivalent) Pipe diameters in mm

Kinematic viscosity, $\nu = 6.0\times10^{-6}\ \mathrm{m^2 s^{-1}}$;   **Roughness size, $k_s = 0.006$ mm**

Kin. visc., $\nu = 6{\cdot}0 \times 10^{-6}\ \text{m}^2\text{s}^{-1}$; $S = 0{\cdot}00010$ to $30{\cdot}0000$ i.e. kin. pr. grad., $\Delta p/\rho l = (0{\cdot}00098)$ to $(294)\ \text{ms}^{-2}$

Roughness size, $k_s = 0{\cdot}015$ mm. This table shows values of m, as follows m_C for Colebrook-White solutions; or, where $\mathbf{R} \leq 2000$, m_P for laminar flow

Grad'nt S	k.p.g. $(\Delta p/\rho l)$	6	8	10	12·5	15	20	25	30	40	50	60	80	100	125
0·00030	(0·00294)	41·14	28·04	20·82	15·46	12·13	8·263	6·137	4·812	3·279	2·435	1·910	1·301	*	1·349
0·00040	(0·00392)	35·63	24·28	18·03	13·39	10·50	7·156	5·314	4·168	2·840	2·109	1·654	1·127	*	1·314
0·00060	(0·00588)	29·09	19·82	14·72	10·93	8·575	5·843	4·339	3·403	2·319	1·722	1·350	*	1·297	1·268
0·00080	(0·00785)	25·20	17·17	12·75	9·469	7·426	5·060	3·758	2·947	2·008	1·491	1·169	*	1·264	1·237
0·00100	(0·00981)	22·54	15·36	11·40	8·470	6·642	4·526	3·361	2·636	1·796	1·334	1·046	1·269	1·239	1·215
0·00150	(0·0147)	18·40	12·54	9·312	6·915	5·423	3·695	2·744	2·152	1·466	1·089	*	1·223	1·197	1·176
0·00200	(0·0196)	15·94	10·86	8·064	5·989	4·696	3·200	2·377	1·864	1·270	*	*	1·192	1·169	1·149
0·00300	(0·0294)	13·01	8·866	6·584	4·890	3·835	2·613	1·941	1·522	1·037	*	1·184	1·152	1·131	1·115
0·00400	(0·0392)	11·27	7·678	5·702	4·235	3·321	2·263	1·681	1·318	*	1·177	1·154	1·125	1·106	1·091
0·00600	(0·0588)	9·200	6·269	4·656	3·458	2·712	1·848	1·372	1·076	*	1·134	1·115	1·089	1·073	1·060
0·00800	(0·0785)	7·968	5·429	4·032	2·994	2·348	1·600	1·188	0·932	1·132	1·106	1·088	1·065	1·051	1·039
0·01000	(0·0981)	7·126	4·856	3·606	2·678	2·100	1·431	1·063	*	1·110	1·085	1·069	1·047	1·034	1·023
0·01500	(0·147)	5·819	3·965	2·945	2·187	1·715	1·169	0·868	*	1·071	1·050	1·035	1·016	1·005	0·996
0·02000	(0·196)	5·039	3·434	2·550	1·894	1·485	1·012	*	1·076	1·045	1·026	1·013	0·996	0·986	0·978
0·03000	(0·294)	4·114	2·804	2·082	1·546	1·213	*	*	1·037	1·011	0·994	0·983	0·968	0·960	0·954
0·04000	(0·392)	3·563	2·428	1·803	1·339	1·050	*	1·031	1·012	0·988	0·973	0·963	0·950	0·943	0·937
0·06000	(0·588)	2·909	1·982	1·472	1·093	0·857	*	0·995	0·979	0·958	0·945	0·936	0·926	0·920	0·916
0·08000	(0·785)	2·520	1·717	1·275	0·947	*	0·993	0·971	0·957	0·938	0·926	0·918	0·909	0·904	0·901
0·10000	(0·981)	2·254	1·536	1·140	0·847	*	0·974	0·954	0·940	0·923	0·912	0·905	0·897	0·893	0·890
0·15000	(1·47)	1·840	1·254	0·931	*	0·969	0·941	0·924	0·912	0·897	0·888	0·882	0·876	0·873	0·871
0·20000	(1·96)	1·594	1·086	0·806	*	0·945	0·919	0·904	0·893	0·880	0·872	0·867	0·862	0·860	0·859
0·30000	(2·94)	1·301	0·887	*	0·930	0·913	0·891	0·877	0·868	0·857	0·851	0·847	0·843	0·842	0·842
0·40000	(3·92)	1·127	0·768	*	0·907	0·892	0·872	0·860	0·852	0·842	0·837	0·834	0·831	0·830	0·831
0·60000	(5·88)	0·920	*	0·897	0·877	0·864	0·847	0·837	0·830	0·822	0·818	0·816	0·814	0·815	0·816
0·80000	(7·85)	0·797	*	0·875	0·858	0·846	0·831	0·821	0·821	0·816	0·809	0·806	0·804	0·803	0·805
1·00000	(9·81)	0·713	0·879	0·859	0·843	0·832	0·818	0·810	0·805	0·799	0·796	0·795	0·795	0·797	0·800
1·50000	(14·7)	*	0·850	0·833	0·819	0·809	0·798	0·791	0·787	0·782	0·781	0·781	0·782	0·784	0·788
2·00000	(19·6)	*	0·830	0·815	0·803	0·794	0·784	0·778	0·775	0·772	0·771	0·771	0·773	0·776	0·780
3·00000	(29·4)	0·827	0·805	0·792	0·781	0·774	0·766	0·762	0·759	0·757	0·757	0·758	0·761	0·765	0·770
4·00000	(39·2)	0·808	0·789	0·777	0·767	0·761	0·754	0·751	0·749	0·748	0·749	0·750	0·754	0·758	0·763
6·00000	(58·8)	0·784	0·767	0·757	0·749	0·744	0·739	0·736	0·735	0·736	0·737	0·739	0·744	0·749	0·755
8·00000	(78·5)	0·768	0·753	0·744	0·737	0·733	0·729	0·727	0·727	0·728	0·730	0·733	0·738	0·744	0·750
10·0000	(98·1)	0·756	0·743	0·735	0·729	0·725	0·722	0·720	0·721	0·722	0·725	0·728	0·734	0·739	0·746
15·0000	(147)	0·737	0·726	0·719	0·714	0·712	0·709	0·709	0·710	0·713	0·716	0·720	0·726	0·733	0·740
20·0000	(196)	0·724	0·714	0·709	0·705	0·703	0·702	0·702	0·703	0·707	0·711	0·714	0·722	0·728	0·736
30·0000	(294)	0·708	0·700	0·696	0·693	0·692	0·692	0·693	0·695	0·699	0·704	0·708	0·716	0·723	0·731
S	$(\Delta p/\rho l)$	6	8	10	12·5	15	20	25	30	40	50	60	80	100	125

Grad'nt k.p.g. (Equivalent) Pipe diameters in mm

Grad'nt S	k.p.g. $(\Delta p/\rho l)$	125	150	200	250	300	400	500	600	800	1000	1250	1500	2000	2500
0·00010	(0·00098)	1·243	*	1·417	1·388	1·367	1·341	1·325	1·314	1·301	1·294	1·289	1·286	1·285	1·287
0·00015	(0·00147)	*	1·409	1·368	1·342	1·324	1·301	1·287	1·278	1·268	1·262	1·259	1·257	1·258	1·260
0·00020	(0·00196)	*	1·372	1·335	1·311	1·295	1·275	1·262	1·254	1·245	1·241	1·238	1·237	1·239	1·241
0·00030	(0·00294)	1·349	1·323	1·291	1·270	1·256	1·239	1·228	1·222	1·215	1·212	1·210	1·210	1·213	1·217
0·00040	(0·00392)	1·314	1·291	1·261	1·243	1·230	1·215	1·206	1·200	1·194	1·192	1·191	1·192	1·196	1·200
0·00060	(0·00588)	1·268	1·248	1·222	1·206	1·196	1·182	1·175	1·171	1·166	1·165	1·166	1·167	1·172	1·177
0·00080	(0·00785)	1·237	1·219	1·196	1·182	1·172	1·161	1·154	1·151	1·148	1·147	1·148	1·151	1·156	1·161
0·00100	(0·00981)	1·215	1·198	1·176	1·163	1·154	1·144	1·139	1·136	1·133	1·134	1·135	1·138	1·144	1·150
0·00150	(0·0147)	1·176	1·161	1·142	1·131	1·124	1·116	1·112	1·110	1·109	1·110	1·112	1·116	1·122	1·129
0·00200	(0·0196)	1·149	1·136	1·120	1·110	1·103	1·096	1·093	1·092	1·092	1·094	1·097	1·101	1·108	1·115
0·00300	(0·0294)	1·115	1·103	1·089	1·081	1·076	1·071	1·068	1·068	1·069	1·072	1·076	1·080	1·088	1·096
0·00400	(0·0392)	1·091	1·081	1·068	1·061	1·057	1·053	1·052	1·052	1·054	1·057	1·062	1·066	1·075	1·083
0·00600	(0·0588)	1·060	1·051	1·041	1·035	1·032	1·029	1·029	1·030	1·033	1·037	1·043	1·048	1·057	1·066
0·00800	(0·0785)	1·039	1·031	1·022	1·018	1·015	1·013	1·014	1·015	1·019	1·024	1·030	1·035	1·045	1·055
0·01000	(0·0981)	1·023	1·016	1·008	1·004	1·002	1·002	1·002	1·004	1·009	1·014	1·020	1·026	1·036	1·046
0·01500	(0·147)	0·996	0·991	0·984	0·982	0·980	0·981	0·983	0·985	0·991	0·996	1·003	1·009	1·021	1·031
0·02000	(0·196)	0·978	0·973	0·968	0·966	0·966	0·967	0·969	0·972	0·978	0·984	0·992	0·998	1·010	1·021
0·03000	(0·294)	0·954	0·950	0·946	0·945	0·946	0·948	0·951	0·955	0·962	0·969	0·976	0·984	0·996	1·007
0·04000	(0·392)	0·937	0·934	0·932	0·932	0·932	0·935	0·939	0·943	0·951	0·958	0·966	0·974	0·987	0·998
0·06000	(0·588)	0·916	0·913	0·912	0·913	0·914	0·918	0·923	0·928	0·936	0·944	0·953	0·961	0·975	0·987
0·08000	(0·785)	0·901	0·899	0·899	0·900	0·902	0·907	0·912	0·917	0·926	0·935	0·944	0·952	0·967	0·979
0·10000	(0·981)	0·890	0·889	0·889	0·891	0·893	0·899	0·904	0·909	0·919	0·928	0·937	0·946	0·961	0·973
0·15000	(1·47)	0·871	0·871	0·873	0·875	0·878	0·884	0·890	0·896	0·907	0·916	0·926	0·935	0·951	0·964
0·20000	(1·96)	0·859	0·859	0·861	0·865	0·868	0·875	0·881	0·887	0·898	0·908	0·919	0·928	0·944	0·958
0·30000	(2·94)	0·842	0·843	0·847	0·850	0·854	0·862	0·869	0·876	0·888	0·898	0·909	0·919	0·936	0·950
0·40000	(3·92)	0·831	0·833	0·837	0·841	0·846	0·854	0·862	0·869	0·881	0·891	0·903	0·913	0·930	0·945
0·60000	(5·88)	0·816	0·819	0·824	0·829	0·834	0·843	0·851	0·859	0·872	0·883	0·895	0·906	0·923	0·938
0·80000	(7·85)	0·807	0·810	0·815	0·821	0·826	0·836	0·845	0·853	0·866	0·878	0·890	0·901	0·919	0·934
1·00000	(9·81)	0·800	0·803	0·809	0·815	0·821	0·831	0·840	0·848	0·862	0·874	0·886	0·897	0·916	0·931
1·50000	(14·7)	0·788	0·791	0·799	0·805	0·811	0·822	0·832	0·840	0·855	0·867	0·880	0·892	0·911	0·926
2·00000	(19·6)	0·780	0·784	0·792	0·799	0·805	0·817	0·827	0·836	0·851	0·863	0·877	0·888	0·907	0·923
3·00000	(29·4)	0·770	0·775	0·783	0·791	0·798	0·810	0·820	0·830	0·845	0·858	0·872	0·884	0·904	0·920
S	$(\Delta p/\rho l)$	125	150	200	250	300	400	500	600	800	1000	1250	1500	2000	2500

Grad'nt k.p.g. (Equivalent) Pipe diameters in mm

Kinematic viscosity, $\nu = 6{\cdot}0 \times 10^{-6}\ \text{m}^2\text{s}^{-1}$;   Roughness size, $k_s = 0{\cdot}015$ mm

E178

Kin. visc., $\nu = 6.0 \times 10^{-6}$ m^{2}s^{-1}; $S = 0.00010$ to 3.00000

i.e. kin. pr. grad., $\Delta p/\rho l =$ (0.00098) to (29.4) ms^{-2}

Roughness size, $k_s = 0.030$ mm
This table shows values of m, as follows

m_C for Colebrook-White solutions; or, where $R \leq 2000$, m_P for laminar flow

Grad'nt S	k.p.g. $(\Delta p/\rho l)$	125	150	200	250	300	400	500	600	800	1000	1250	1500	2000	2500
0.00010	(0.00098)	1.243	*	1.419	1.389	1.369	1.342	1.326	1.316	1.303	1.296	1.291	1.289	1.288	1.289
0.00015	(0.00147)	*	1.410	1.369	1.343	1.326	1.303	1.289	1.280	1.270	1.264	1.261	1.260	1.260	1.263
0.00020	(0.00196)	*	1.374	1.336	1.313	1.297	1.276	1.264	1.256	1.247	1.243	1.241	1.240	1.242	1.245
0.00030	(0.00294)	1.350	1.325	1.293	1.272	1.258	1.241	1.231	1.224	1.217	1.214	1.213	1.214	1.217	1.221
0.00040	(0.00392)	1.316	1.293	1.263	1.245	1.233	1.217	1.208	1.203	1.197	1.195	1.195	1.196	1.199	1.204
0.00060	(0.00588)	1.270	1.250	1.225	1.209	1.198	1.185	1.178	1.174	1.170	1.169	1.170	1.172	1.176	1.182
0.00080	(0.00785)	1.240	1.222	1.199	1.184	1.175	1.164	1.158	1.154	1.151	1.151	1.153	1.155	1.161	1.167
0.00100	(0.00981)	1.217	1.201	1.179	1.166	1.158	1.148	1.142	1.139	1.138	1.138	1.140	1.143	1.149	1.156
0.00150	(0.0147)	1.179	1.164	1.146	1.135	1.128	1.120	1.116	1.114	1.114	1.115	1.118	1.122	1.129	1.136
0.00200	(0.0196)	1.153	1.140	1.123	1.114	1.107	1.101	1.098	1.097	1.097	1.099	1.103	1.107	1.115	1.123
0.00300	(0.0294)	1.118	1.107	1.093	1.085	1.080	1.076	1.074	1.074	1.076	1.079	1.083	1.088	1.097	1.105
0.00400	(0.0392)	1.095	1.085	1.073	1.066	1.062	1.059	1.058	1.058	1.061	1.065	1.070	1.075	1.085	1.093
0.00600	(0.0588)	1.065	1.056	1.046	1.041	1.038	1.036	1.036	1.037	1.041	1.046	1.052	1.057	1.068	1.078
0.00800	(0.0785)	1.044	1.037	1.028	1.024	1.022	1.021	1.022	1.023	1.028	1.033	1.040	1.046	1.057	1.067
0.01000	(0.0981)	1.029	1.022	1.015	1.011	1.010	1.009	1.011	1.013	1.018	1.024	1.031	1.037	1.049	1.059
0.01500	(0.147)	1.003	0.997	0.992	0.989	0.989	0.990	0.992	0.995	1.001	1.008	1.015	1.023	1.035	1.046
0.02000	(0.196)	0.985	0.981	0.976	0.975	0.975	0.977	0.980	0.983	0.990	0.997	1.005	1.013	1.026	1.038
0.03000	(0.294)	0.962	0.959	0.956	0.955	0.956	0.959	0.963	0.967	0.975	0.983	0.992	1.000	1.014	1.026
0.04000	(0.392)	0.946	0.944	0.942	0.942	0.944	0.948	0.952	0.957	0.966	0.974	0.983	0.992	1.006	1.019
0.06000	(0.588)	0.926	0.924	0.924	0.925	0.927	0.932	0.938	0.943	0.953	0.962	0.972	0.981	0.996	1.009
0.08000	(0.785)	0.912	0.911	0.912	0.914	0.917	0.922	0.928	0.934	0.945	0.954	0.964	0.974	0.990	1.003
0.10000	(0.981)	0.902	0.902	0.903	0.906	0.909	0.915	0.921	0.928	0.939	0.948	0.959	0.969	0.985	0.999
0.15000	(1.47)	0.885	0.886	0.888	0.892	0.895	0.903	0.910	0.916	0.928	0.939	0.950	0.960	0.977	0.992
0.20000	(1.96)	0.874	0.875	0.878	0.882	0.886	0.895	0.902	0.909	0.922	0.933	0.944	0.955	0.972	0.987
0.30000	(2.94)	0.859	0.861	0.865	0.870	0.875	0.884	0.893	0.900	0.913	0.925	0.937	0.948	0.966	0.981
0.40000	(3.92)	0.849	0.852	0.857	0.863	0.868	0.878	0.886	0.894	0.908	0.920	0.933	0.944	0.962	0.978
0.60000	(5.88)	0.837	0.840	0.846	0.853	0.858	0.869	0.878	0.887	0.901	0.914	0.927	0.938	0.958	0.973
0.80000	(7.85)	0.829	0.832	0.840	0.846	0.852	0.864	0.873	0.882	0.897	0.910	0.923	0.935	0.955	0.971
1.00000	(9.81)	0.823	0.827	0.835	0.842	0.848	0.860	0.870	0.879	0.894	0.907	0.921	0.933	0.952	0.969
1.50000	(14.7)	0.814	0.818	0.826	0.834	0.841	0.854	0.864	0.873	0.889	0.903	0.917	0.929	0.949	0.966
2.00000	(19.6)	0.807	0.812	0.821	0.829	0.837	0.850	0.860	0.870	0.886	0.900	0.914	0.927	0.947	0.964
3.00000	(29.4)	0.800	0.805	0.815	0.824	0.831	0.845	0.856	0.866	0.883	0.896	0.911	0.924	0.945	0.962
S	$(\Delta p/\rho l)$	125	150	200	250	300	400	500	600	800	1000	1250	1500	2000	2500

Grad'nt k.p.g. (Equivalent) Pipe diameters in mm — **Roughness size, $k_s = 0.030$ mm**

E179

Kin. visc., $\nu = 6.0 \times 10^{-6}$ m^{2}s^{-1}; $S = 0.00010$ to 3.00000

i.e. kin. pr. grad., $\Delta p/\rho l =$ (0.00098) to (29.4) ms^{-2}

Roughness size, $k_s = 0.060$ mm
This table shows values of m, as follows

m_C for Colebrook-White solutions; or, where $R \leq 2000$, m_P for laminar flow

Grad'nt S	k.p.g. $(\Delta p/\rho l)$	125	150	200	250	300	400	500	600	800	1000	1250	1500	2000	2500
0.00010	(0.00098)	1.243	*	1.421	1.392	1.371	1.345	1.329	1.319	1.306	1.299	1.295	1.293	1.292	1.294
0.00015	(0.00147)	*	1.413	1.372	1.347	1.329	1.306	1.293	1.284	1.274	1.269	1.266	1.265	1.266	1.268
0.00020	(0.00196)	*	1.377	1.340	1.316	1.300	1.280	1.268	1.261	1.252	1.248	1.246	1.246	1.248	1.251
0.00030	(0.00294)	1.354	1.329	1.296	1.276	1.263	1.245	1.235	1.229	1.223	1.220	1.219	1.220	1.223	1.228
0.00040	(0.00392)	1.320	1.297	1.268	1.249	1.237	1.222	1.213	1.208	1.203	1.201	1.201	1.203	1.207	1.212
0.00060	(0.00588)	1.275	1.255	1.229	1.214	1.203	1.191	1.184	1.180	1.177	1.176	1.177	1.180	1.185	1.191
0.00080	(0.00785)	1.245	1.227	1.204	1.190	1.181	1.170	1.164	1.161	1.159	1.159	1.161	1.164	1.171	1.177
0.00100	(0.00981)	1.223	1.206	1.185	1.172	1.164	1.154	1.149	1.147	1.146	1.147	1.149	1.153	1.160	1.167
0.00150	(0.0147)	1.185	1.170	1.152	1.142	1.135	1.127	1.124	1.122	1.123	1.125	1.129	1.133	1.141	1.149
0.00200	(0.0196)	1.159	1.146	1.130	1.121	1.115	1.109	1.107	1.106	1.107	1.110	1.115	1.119	1.129	1.137
0.00300	(0.0294)	1.126	1.115	1.101	1.094	1.089	1.085	1.084	1.084	1.087	1.091	1.096	1.102	1.112	1.122
0.00400	(0.0392)	1.104	1.094	1.082	1.076	1.072	1.069	1.069	1.070	1.074	1.078	1.084	1.090	1.101	1.111
0.00600	(0.0588)	1.074	1.066	1.057	1.052	1.049	1.048	1.049	1.051	1.056	1.061	1.068	1.075	1.087	1.098
0.00800	(0.0785)	1.054	1.047	1.040	1.036	1.034	1.034	1.036	1.038	1.044	1.050	1.058	1.065	1.078	1.089
0.01000	(0.0981)	1.040	1.034	1.027	1.024	1.023	1.024	1.026	1.029	1.035	1.042	1.050	1.057	1.071	1.083
0.01500	(0.147)	1.015	1.010	1.006	1.004	1.004	1.006	1.009	1.013	1.021	1.028	1.037	1.045	1.059	1.072
0.02000	(0.196)	0.999	0.995	0.991	0.991	0.991	0.994	0.998	1.003	1.011	1.019	1.029	1.037	1.052	1.065
0.03000	(0.294)	0.977	0.975	0.973	0.973	0.975	0.979	0.984	0.989	0.999	1.008	1.018	1.027	1.043	1.056
0.04000	(0.392)	0.963	0.961	0.961	0.962	0.964	0.969	0.975	0.980	0.991	1.000	1.011	1.020	1.037	1.051
0.06000	(0.588)	0.945	0.944	0.945	0.947	0.950	0.956	0.963	0.969	0.981	0.991	1.002	1.012	1.029	1.044
0.08000	(0.785)	0.933	0.933	0.934	0.937	0.941	0.948	0.955	0.962	0.974	0.985	0.996	1.007	1.025	1.039
0.10000	(0.981)	0.924	0.924	0.927	0.930	0.934	0.942	0.950	0.957	0.969	0.980	0.992	1.003	1.021	1.036
0.15000	(1.47)	0.909	0.911	0.914	0.919	0.924	0.933	0.941	0.948	0.962	0.973	0.986	0.997	1.016	1.031
0.20000	(1.96)	0.900	0.902	0.906	0.912	0.917	0.926	0.935	0.943	0.957	0.969	0.982	0.993	1.012	1.028
0.30000	(2.94)	0.888	0.890	0.896	0.902	0.908	0.918	0.928	0.936	0.951	0.964	0.977	0.989	1.008	1.025
0.40000	(3.92)	0.880	0.883	0.890	0.896	0.902	0.914	0.923	0.932	0.947	0.960	0.974	0.986	1.006	1.022
0.60000	(5.88)	0.870	0.874	0.882	0.889	0.896	0.907	0.918	0.927	0.943	0.956	0.970	0.982	1.003	1.019
0.80000	(7.85)	0.864	0.868	0.877	0.884	0.891	0.904	0.914	0.924	0.940	0.953	0.968	0.980	1.001	1.018
1.00000	(9.81)	0.860	0.864	0.873	0.881	0.888	0.901	0.912	0.922	0.938	0.952	0.966	0.979	0.999	1.017
1.50000	(14.7)	0.853	0.858	0.867	0.876	0.883	0.897	0.908	0.918	0.935	0.949	0.964	0.976	0.997	1.015
2.00000	(19.6)	0.848	0.854	0.864	0.872	0.880	0.894	0.906	0.916	0.933	0.947	0.962	0.975	0.996	1.014
3.00000	(29.4)	0.843	0.849	0.859	0.868	0.877	0.891	0.903	0.913	0.931	0.945	0.960	0.973	0.995	1.012
S	$(\Delta p/\rho l)$	125	150	200	250	300	400	500	600	800	1000	1250	1500	2000	2500

Grad'nt k.p.g. (Equivalent) Pipe diameters in mm — **Roughness size, $k_s = 0.060$ mm**

Kin. visc., $\nu = 6.0 \times 10^{-6}$ m²s⁻¹; $S = 0.00010$ to 3.00000

i.e. kin. pr. grad., $\Delta p/\rho l =$ (0.00098) to (29.4) ms⁻²

Roughness size, $k_s = 0.150$ mm
This table shows values of m, as follows

m_C for Colebrook-White solutions; or, where $R \leq 2000$, m_P for laminar flow

Grad'nt S	k.p.g. $(\Delta p/\rho l)$	125	150	200	250	300	400	500	600	800	1000	1250	1500	2000	2500
		(Equivalent) Pipe diameters in mm													
0.00010	(0.00098)	1.243	*	1.429	1.400	1.380	1.354	1.339	1.328	1.316	1.310	1.306	1.305	1.305	1.308
0.00015	(0.00147)	*	1.422	1.381	1.356	1.338	1.316	1.303	1.295	1.286	1.281	1.279	1.279	1.281	1.285
0.00020	(0.00196)	*	1.386	1.349	1.327	1.311	1.291	1.280	1.273	1.265	1.262	1.260	1.261	1.264	1.269
0.00030	(0.00294)	1.364	1.340	1.308	1.288	1.275	1.258	1.249	1.243	1.238	1.236	1.236	1.238	1.243	1.248
0.00040	(0.00392)	1.331	1.309	1.280	1.262	1.250	1.236	1.228	1.223	1.219	1.219	1.220	1.222	1.228	1.235
0.00060	(0.00588)	1.288	1.268	1.243	1.228	1.218	1.207	1.201	1.197	1.195	1.196	1.198	1.202	1.209	1.217
0.00080	(0.00785)	1.259	1.241	1.219	1.206	1.197	1.187	1.182	1.180	1.179	1.181	1.184	1.188	1.197	1.205
0.00100	(0.00981)	1.238	1.222	1.201	1.189	1.181	1.173	1.169	1.167	1.168	1.170	1.174	1.178	1.188	1.196
0.00150	(0.0147)	1.202	1.188	1.171	1.161	1.155	1.149	1.146	1.146	1.148	1.151	1.156	1.162	1.172	1.182
0.00200	(0.0196)	1.178	1.166	1.151	1.142	1.137	1.132	1.131	1.132	1.135	1.139	1.145	1.151	1.162	1.173
0.00300	(0.0294)	1.147	1.137	1.124	1.118	1.114	1.111	1.112	1.113	1.118	1.123	1.130	1.137	1.150	1.161
0.00400	(0.0392)	1.126	1.117	1.107	1.102	1.099	1.098	1.099	1.101	1.107	1.113	1.121	1.128	1.141	1.153
0.00600	(0.0588)	1.100	1.093	1.085	1.081	1.080	1.080	1.082	1.085	1.093	1.100	1.108	1.116	1.131	1.144
0.00800	(0.0785)	1.082	1.076	1.070	1.067	1.067	1.068	1.072	1.075	1.083	1.091	1.101	1.109	1.124	1.138
0.01000	(0.0981)	1.070	1.064	1.059	1.058	1.058	1.060	1.064	1.068	1.077	1.085	1.095	1.104	1.120	1.133
0.01500	(0.147)	1.049	1.045	1.041	1.041	1.042	1.046	1.051	1.056	1.066	1.075	1.086	1.095	1.112	1.126
0.02000	(0.196)	1.035	1.032	1.030	1.030	1.032	1.037	1.043	1.049	1.059	1.069	1.080	1.090	1.107	1.122
0.03000	(0.294)	1.017	1.015	1.015	1.017	1.020	1.026	1.033	1.039	1.051	1.061	1.073	1.083	1.101	1.117
0.04000	(0.392)	1.006	1.005	1.006	1.009	1.012	1.019	1.026	1.033	1.046	1.057	1.069	1.079	1.098	1.113
0.06000	(0.588)	0.991	0.992	0.994	0.998	1.002	1.010	1.018	1.026	1.039	1.050	1.063	1.074	1.093	1.109
0.08000	(0.785)	0.982	0.983	0.987	0.991	0.996	1.005	1.013	1.021	1.035	1.047	1.060	1.071	1.091	1.107
0.10000	(0.981)	0.976	0.977	0.981	0.986	0.991	1.001	1.010	1.018	1.032	1.044	1.057	1.069	1.089	1.105
0.15000	(1.47)	0.965	0.967	0.973	0.978	0.984	0.995	1.004	1.012	1.027	1.040	1.054	1.066	1.086	1.102
0.20000	(1.96)	0.959	0.961	0.967	0.974	0.980	0.991	1.000	1.009	1.025	1.038	1.051	1.064	1.084	1.101
0.30000	(2.94)	0.950	0.954	0.961	0.968	0.974	0.986	0.996	1.005	1.021	1.034	1.049	1.061	1.082	1.099
0.40000	(3.92)	0.945	0.949	0.957	0.964	0.971	0.983	0.994	1.003	1.019	1.033	1.047	1.059	1.080	1.097
0.60000	(5.88)	0.939	0.943	0.952	0.960	0.967	0.979	0.990	1.000	1.017	1.030	1.045	1.058	1.079	1.096
0.80000	(7.85)	0.935	0.940	0.949	0.957	0.964	0.977	0.989	0.998	1.015	1.029	1.044	1.057	1.078	1.095
1.00000	(9.81)	0.932	0.937	0.947	0.955	0.963	0.976	0.987	0.997	1.014	1.028	1.043	1.056	1.077	1.095
1.50000	(14.7)	0.928	0.934	0.943	0.952	0.960	0.974	0.985	0.995	1.012	1.027	1.042	1.055	1.076	1.094
2.00000	(19.6)	0.926	0.931	0.941	0.950	0.958	0.972	0.984	0.994	1.011	1.026	1.041	1.054	1.075	1.093
3.00000	(29.4)	0.923	0.928	0.939	0.948	0.956	0.970	0.982	0.993	1.010	1.025	1.040	1.053	1.075	1.092
S	$(\Delta p/\rho l)$	125	150	200	250	300	400	500	600	800	1000	1250	1500	2000	2500
Grad'nt	k.p.g.	(Equivalent) Pipe diameters in mm												Roughness size, $k_s = 0.150$ mm	

Kin. visc., $\nu = 6.0 \times 10^{-6}$ m²s⁻¹; $S = 0.00010$ to 3.00000

i.e. kin. pr. grad., $\Delta p/\rho l =$ (0.00098) to (29.4) ms⁻²

Roughness size, $k_s = 0.30$ mm
This table shows values of m, as follows

m_C for Colebrook-White solutions; or, where $R \leq 2000$, m_P for laminar flow

Grad'nt S	k.p.g. $(\Delta p/\rho l)$	125	150	200	250	300	400	500	600	800	1000	1250	1500	2000	2500
		(Equivalent) Pipe diameters in mm													
0.00010	(0.00098)	1.243	*	1.442	1.413	1.393	1.368	1.353	1.344	1.333	1.327	1.324	1.324	1.326	1.329
0.00015	(0.00147)	*	*	1.396	1.371	1.354	1.333	1.320	1.312	1.304	1.300	1.299	1.300	1.303	1.309
0.00020	(0.00196)	*	1.402	1.365	1.343	1.328	1.309	1.298	1.292	1.285	1.283	1.283	1.284	1.289	1.295
0.00030	(0.00294)	1.381	1.357	1.325	1.306	1.293	1.278	1.269	1.265	1.260	1.260	1.261	1.264	1.270	1.277
0.00040	(0.00392)	1.350	1.327	1.299	1.282	1.271	1.258	1.250	1.247	1.244	1.244	1.247	1.250	1.258	1.266
0.00060	(0.00588)	1.308	1.289	1.265	1.251	1.241	1.231	1.226	1.224	1.223	1.225	1.229	1.233	1.242	1.251
0.00080	(0.00785)	1.281	1.264	1.243	1.230	1.222	1.214	1.210	1.208	1.209	1.212	1.217	1.222	1.232	1.242
0.00100	(0.00981)	1.261	1.246	1.226	1.215	1.208	1.201	1.198	1.197	1.199	1.203	1.208	1.214	1.225	1.235
0.00150	(0.0147)	1.228	1.215	1.199	1.190	1.185	1.180	1.178	1.179	1.183	1.188	1.194	1.201	1.213	1.224
0.00200	(0.0196)	1.207	1.195	1.181	1.173	1.169	1.166	1.166	1.167	1.172	1.178	1.185	1.192	1.205	1.217
0.00300	(0.0294)	1.179	1.169	1.158	1.152	1.150	1.148	1.150	1.152	1.159	1.165	1.174	1.182	1.196	1.209
0.00400	(0.0392)	1.161	1.152	1.143	1.139	1.137	1.137	1.139	1.143	1.150	1.157	1.166	1.175	1.190	1.203
0.00600	(0.0588)	1.138	1.131	1.124	1.121	1.121	1.123	1.126	1.131	1.139	1.148	1.158	1.167	1.183	1.197
0.00800	(0.0785)	1.123	1.117	1.112	1.110	1.111	1.114	1.118	1.123	1.132	1.142	1.152	1.161	1.178	1.192
0.01000	(0.0981)	1.112	1.107	1.103	1.103	1.104	1.107	1.112	1.118	1.128	1.137	1.148	1.158	1.175	1.190
0.01500	(0.147)	1.095	1.091	1.089	1.090	1.092	1.097	1.103	1.109	1.120	1.130	1.142	1.152	1.170	1.185
0.02000	(0.196)	1.083	1.081	1.080	1.082	1.084	1.090	1.097	1.103	1.115	1.126	1.138	1.148	1.167	1.182
0.03000	(0.294)	1.070	1.068	1.069	1.072	1.075	1.082	1.090	1.097	1.109	1.121	1.133	1.144	1.163	1.179
0.04000	(0.392)	1.061	1.060	1.062	1.065	1.069	1.077	1.085	1.093	1.106	1.117	1.130	1.141	1.161	1.177
0.06000	(0.588)	1.050	1.050	1.054	1.058	1.062	1.071	1.080	1.088	1.102	1.114	1.127	1.138	1.158	1.174
0.08000	(0.785)	1.043	1.044	1.048	1.053	1.058	1.068	1.076	1.085	1.099	1.111	1.125	1.136	1.156	1.173
0.10000	(0.981)	1.039	1.040	1.045	1.050	1.055	1.065	1.074	1.082	1.097	1.110	1.123	1.135	1.155	1.172
0.15000	(1.47)	1.031	1.033	1.039	1.044	1.050	1.061	1.070	1.079	1.094	1.107	1.121	1.133	1.153	1.170
0.20000	(1.96)	1.026	1.029	1.035	1.041	1.047	1.058	1.068	1.077	1.092	1.105	1.119	1.132	1.152	1.169
0.30000	(2.94)	1.021	1.024	1.031	1.037	1.044	1.055	1.065	1.075	1.090	1.104	1.118	1.130	1.151	1.168
0.40000	(3.92)	1.017	1.021	1.028	1.035	1.042	1.053	1.064	1.073	1.089	1.102	1.117	1.129	1.150	1.167
0.60000	(5.88)	1.013	1.017	1.025	1.032	1.039	1.051	1.062	1.071	1.087	1.101	1.116	1.128	1.149	1.166
0.80000	(7.85)	1.011	1.015	1.023	1.030	1.037	1.050	1.061	1.070	1.087	1.100	1.115	1.127	1.148	1.166
1.00000	(9.81)	1.009	1.013	1.021	1.029	1.036	1.049	1.060	1.069	1.086	1.100	1.114	1.127	1.148	1.165
1.50000	(14.7)	1.006	1.011	1.019	1.027	1.035	1.048	1.059	1.068	1.085	1.099	1.114	1.126	1.147	1.165
2.00000	(19.6)	1.004	1.009	1.018	1.026	1.034	1.047	1.058	1.068	1.084	1.098	1.113	1.126	1.147	1.165
3.00000	(29.4)	1.002	1.007	1.017	1.025	1.032	1.046	1.057	1.067	1.084	1.098	1.113	1.125	1.147	1.164
S	$(\Delta p/\rho l)$	125	150	200	250	300	400	500	600	800	1000	1250	1500	2000	2500
Grad'nt	k.p.g.	(Equivalent) Pipe diameters in mm												Roughness size, $k_s = 0.30$ mm	

E182

Kin. visc., $v = 6\cdot0\times10^{-6}$ m^{2}s^{-1};
$S = 0\cdot00010$ to $3\cdot00000$

i.e. kin. pr. grad., $\Delta p/\rho l =$
($0\cdot00098$) to ($29\cdot4$) ms^{-2}

Roughness size, $k_s = 0\cdot60$ mm
This table shows values of m, as follows

m_C for Colebrook-White solutions; or,
where $R \leq 2000$, m_P for laminar flow

Grad'nt S	k.p.g. $(\Delta p/\rho l)$	125	150	200	250	300	400	500	600	800	1000	1250	1500	2000	2500
0·00010	(0·00098)	1·243	*	1·467	1·438	1·419	1·395	1·381	1·372	1·362	1·358	1·356	1·357	1·360	1·366
0·00015	(0·00147)	*	*	1·423	1·399	1·382	1·362	1·351	1·344	1·337	1·335	1·335	1·336	1·342	1·349
0·00020	(0·00196)	*	1·431	1·395	1·373	1·359	1·341	1·331	1·326	1·321	1·319	1·321	1·323	1·330	1·338
0·00030	(0·00294)	1·413	1·389	1·358	1·340	1·328	1·314	1·306	1·302	1·300	1·300	1·303	1·307	1·315	1·324
0·00040	(0·00392)	1·384	1·362	1·335	1·318	1·308	1·296	1·290	1·287	1·286	1·288	1·292	1·296	1·306	1·316
0·00060	(0·00588)	1·346	1·328	1·304	1·291	1·282	1·273	1·269	1·268	1·269	1·272	1·277	1·283	1·294	1·305
0·00080	(0·00785)	1·322	1·305	1·285	1·273	1·266	1·259	1·256	1·256	1·258	1·262	1·268	1·274	1·287	1·298
0·00100	(0·00981)	1·304	1·289	1·271	1·260	1·254	1·248	1·246	1·247	1·250	1·255	1·262	1·269	1·281	1·293
0·00150	(0·0147)	1·275	1·263	1·247	1·239	1·235	1·231	1·231	1·232	1·238	1·244	1·251	1·259	1·273	1·285
0·00200	(0·0196)	1·257	1·246	1·232	1·226	1·222	1·220	1·221	1·223	1·230	1·236	1·245	1·253	1·268	1·281
0·00300	(0·0294)	1·233	1·224	1·214	1·209	1·207	1·207	1·209	1·212	1·220	1·227	1·237	1·245	1·261	1·275
0·00400	(0·0392)	1·218	1·210	1·202	1·198	1·197	1·198	1·201	1·205	1·214	1·222	1·232	1·241	1·257	1·271
0·00600	(0·0588)	1·200	1·193	1·187	1·185	1·185	1·188	1·192	1·197	1·206	1·215	1·226	1·235	1·252	1·267
0·00800	(0·0785)	1·188	1·183	1·178	1·177	1·178	1·181	1·186	1·191	1·201	1·211	1·222	1·232	1·249	1·264
0·01000	(0·0981)	1·180	1·175	1·171	1·171	1·172	1·177	1·182	1·188	1·198	1·208	1·219	1·230	1·247	1·262
0·01500	(0·147)	1·166	1·163	1·161	1·162	1·164	1·169	1·175	1·182	1·193	1·204	1·215	1·226	1·244	1·259
0·02000	(0·196)	1·158	1·156	1·155	1·156	1·159	1·165	1·171	1·178	1·190	1·201	1·213	1·224	1·242	1·258
0·03000	(0·294)	1·148	1·146	1·147	1·149	1·152	1·159	1·167	1·174	1·186	1·198	1·210	1·221	1·240	1·256
0·04000	(0·392)	1·142	1·141	1·142	1·145	1·148	1·156	1·164	1·171	1·184	1·195	1·208	1·219	1·238	1·254
0·06000	(0·588)	1·134	1·134	1·136	1·140	1·144	1·152	1·160	1·168	1·181	1·193	1·206	1·217	1·237	1·253
0·08000	(0·785)	1·129	1·130	1·132	1·137	1·141	1·150	1·158	1·166	1·180	1·192	1·205	1·216	1·236	1·252
0·10000	(0·981)	1·126	1·127	1·130	1·134	1·139	1·148	1·157	1·164	1·178	1·191	1·204	1·215	1·235	1·251
0·15000	(1·47)	1·121	1·122	1·126	1·131	1·136	1·145	1·154	1·162	1·177	1·189	1·202	1·214	1·234	1·250
0·20000	(1·96)	1·118	1·119	1·124	1·129	1·134	1·144	1·153	1·161	1·176	1·188	1·202	1·213	1·233	1·250
0·30000	(2·94)	1·114	1·116	1·121	1·126	1·132	1·142	1·151	1·160	1·174	1·187	1·201	1·212	1·233	1·249
0·40000	(3·92)	1·112	1·114	1·119	1·125	1·130	1·141	1·150	1·159	1·174	1·186	1·200	1·212	1·232	1·249
0·60000	(5·88)	1·109	1·112	1·117	1·123	1·129	1·140	1·149	1·158	1·173	1·185	1·199	1·211	1·231	1·248
0·80000	(7·85)	1·108	1·110	1·116	1·122	1·128	1·139	1·148	1·157	1·172	1·185	1·199	1·211	1·231	1·248
1·00000	(9·81)	1·107	1·109	1·115	1·121	1·127	1·138	1·148	1·157	1·172	1·185	1·198	1·211	1·231	1·248
1·50000	(14·7)	1·105	1·108	1·114	1·120	1·126	1·137	1·147	1·156	1·171	1·184	1·198	1·210	1·231	1·247
2·00000	(19·6)	1·104	1·107	1·113	1·120	1·126	1·137	1·147	1·155	1·171	1·184	1·198	1·210	1·230	1·247
3·00000	(29·4)	1·103	1·106	1·112	1·119	1·125	1·136	1·146	1·155	1·170	1·183	1·197	1·210	1·230	1·247
S $(\Delta p/\rho l)$		125	150	200	250	300	400	500	600	800	1000	1250	1500	2000	2500

Grad'nt k.p.g. (Equivalent) Pipe diameters in mm **Roughness size, $k_s = 0\cdot60$ mm**

E183

Kin. visc., $v = 6\cdot0\times10^{-6}$ m^{2}s^{-1};
$S = 0\cdot00010$ to $3\cdot00000$

i.e. kin. pr. grad., $\Delta p/\rho l =$
($0\cdot00098$) to ($29\cdot4$) ms^{-2}

Roughness size, $k_s = 1\cdot50$ mm
This table shows values of m, as follows

m_C for Colebrook-White solutions; or,
where $R \leq 2000$, m_P for laminar flow

Grad'nt S	k.p.g. $(\Delta p/\rho l)$	125	150	200	250	300	400	500	600	800	1000	1250	1500	2000	2500
0·00010	(0·00098)	1·243	*	1·533	1·505	1·486	1·463	1·450	1·443	1·435	1·432	1·432	1·434	1·440	1·448
0·00015	(0·00147)	*	*	1·496	1·472	1·456	1·437	1·427	1·421	1·416	1·415	1·417	1·420	1·428	1·436
0·00020	(0·00196)	*	1·508	1·472	1·451	1·437	1·420	1·412	1·407	1·404	1·404	1·407	1·411	1·420	1·429
0·00030	(0·00294)	*	1·473	1·443	1·424	1·413	1·400	1·393	1·390	1·389	1·391	1·395	1·400	1·410	1·420
0·00040	(0·00392)	1·474	1·451	1·424	1·408	1·398	1·386	1·381	1·379	1·380	1·382	1·387	1·393	1·404	1·415
0·00060	(0·00588)	1·443	1·424	1·400	1·387	1·379	1·370	1·367	1·366	1·368	1·372	1·378	1·385	1·397	1·408
0·00080	(0·00785)	1·424	1·407	1·386	1·374	1·367	1·360	1·358	1·358	1·361	1·366	1·373	1·379	1·393	1·405
0·00100	(0·00981)	1·410	1·394	1·375	1·365	1·358	1·353	1·351	1·352	1·356	1·361	1·369	1·376	1·389	1·402
0·00150	(0·0147)	1·388	1·375	1·358	1·350	1·345	1·341	1·341	1·343	1·348	1·354	1·362	1·370	1·385	1·397
0·00200	(0·0196)	1·374	1·362	1·348	1·341	1·337	1·334	1·335	1·337	1·343	1·350	1·359	1·367	1·382	1·395
0·00300	(0·0294)	1·358	1·347	1·335	1·329	1·327	1·326	1·327	1·330	1·338	1·345	1·354	1·363	1·378	1·392
0·00400	(0·0392)	1·347	1·338	1·327	1·323	1·321	1·320	1·323	1·326	1·334	1·342	1·351	1·360	1·376	1·390
0·00600	(0·0588)	1·335	1·327	1·318	1·314	1·313	1·314	1·317	1·321	1·330	1·338	1·348	1·357	1·373	1·387
0·00800	(0·0785)	1·327	1·320	1·312	1·309	1·308	1·310	1·314	1·318	1·327	1·336	1·346	1·355	1·372	1·386
0·01000	(0·0981)	1·321	1·315	1·308	1·306	1·305	1·308	1·312	1·316	1·325	1·334	1·345	1·354	1·371	1·385
0·01500	(0·147)	1·313	1·307	1·302	1·300	1·300	1·303	1·308	1·313	1·322	1·332	1·342	1·352	1·369	1·384
0·02000	(0·196)	1·308	1·303	1·298	1·297	1·297	1·301	1·306	1·311	1·321	1·330	1·341	1·351	1·368	1·383
0·03000	(0·294)	1·301	1·297	1·293	1·293	1·294	1·298	1·303	1·308	1·319	1·328	1·339	1·349	1·367	1·382
0·04000	(0·392)	1·298	1·294	1·290	1·290	1·292	1·296	1·301	1·307	1·318	1·327	1·339	1·349	1·366	1·381
0·06000	(0·588)	1·293	1·290	1·287	1·287	1·289	1·294	1·300	1·305	1·316	1·326	1·337	1·348	1·365	1·380
0·08000	(0·785)	1·290	1·287	1·285	1·286	1·287	1·293	1·298	1·304	1·315	1·325	1·337	1·347	1·365	1·380
0·10000	(0·981)	1·289	1·286	1·284	1·284	1·286	1·292	1·298	1·303	1·315	1·325	1·336	1·347	1·364	1·379
0·15000	(1·47)	1·286	1·283	1·281	1·283	1·285	1·290	1·296	1·302	1·314	1·324	1·336	1·346	1·364	1·379
0·20000	(1·96)	1·284	1·281	1·280	1·281	1·284	1·289	1·296	1·302	1·313	1·324	1·335	1·346	1·363	1·379
0·30000	(2·94)	1·282	1·279	1·279	1·280	1·282	1·288	1·295	1·301	1·313	1·323	1·335	1·345	1·363	1·378
0·40000	(3·92)	1·280	1·278	1·278	1·279	1·282	1·288	1·294	1·300	1·312	1·323	1·334	1·345	1·363	1·378
0·60000	(5·88)	1·279	1·277	1·277	1·278	1·281	1·287	1·294	1·300	1·312	1·322	1·334	1·344	1·363	1·378
0·80000	(7·85)	1·278	1·276	1·276	1·278	1·280	1·287	1·293	1·300	1·311	1·322	1·334	1·344	1·362	1·378
1·00000	(9·81)	1·277	1·276	1·276	1·277	1·280	1·286	1·293	1·299	1·311	1·322	1·334	1·344	1·362	1·378
1·50000	(14·7)	1·277	1·275	1·275	1·277	1·280	1·286	1·293	1·299	1·311	1·322	1·333	1·344	1·362	1·377
2·00000	(19·6)	1·276	1·274	1·274	1·276	1·279	1·286	1·292	1·299	1·311	1·321	1·333	1·344	1·362	1·377
3·00000	(29·4)	1·275	1·274	1·274	1·276	1·279	1·285	1·292	1·299	1·311	1·321	1·333	1·344	1·362	1·377
S $(\Delta p/\rho l)$		125	150	200	250	300	400	500	600	800	1000	1250	1500	2000	2500

Grad'nt k.p.g. (Equivalent) Pipe diameters in mm **Roughness size, $k_s = 1\cdot50$ mm**

Kin. visc., $\nu = 8{\cdot}0 \times 10^{-6}\ \mathrm{m^2\,s^{-1}}$;
$S = 0{\cdot}00010$ to $30{\cdot}0000$
i.e. kin. pr. grad., $\Delta p/\rho l = (0{\cdot}00098)$ to $(294)\ \mathrm{ms^{-2}}$

Roughness size, $k_s = 0{\cdot}003$ mm
This table shows values of m, as follows
m_C for Colebrook-White solutions; or,
where $\mathbf{R} \leq 2000$, m_P for laminar flow

Grad'nt S	k.p.g. $(\Delta p/\rho l)$	(Equivalent) Pipe diameters in mm													
		6	8	10	12·5	15	20	25	30	40	50	60	80	100	125
0·00030	(0·00294)	54·86	37·38	27·76	20·62	16·17	11·02	8·182	6·416	4·372	3·247	2·546	1·735	1·289	*
0·00040	(0·00392)	47·51	32·37	24·04	17·86	14·00	9·541	7·086	5·557	3·786	2·812	2·205	1·503	1·116	*
0·00060	(0·00588)	38·79	26·43	19·63	14·58	11·43	7·790	5·786	4·537	3·092	2·296	1·801	1·227	*	1·333
0·00080	(0·00785)	33·59	22·89	17·00	12·63	9·901	6·747	5·010	3·929	2·677	1·988	1·559	1·063	*	1·299
0·00100	(0·00981)	30·05	20·48	15·21	11·29	8·856	6·034	4·482	3·514	2·395	1·778	1·395	*	1·303	1·273
0·00150	(0·0147)	24·53	16·72	12·42	9·220	7·231	4·927	3·659	2·869	1·955	1·452	1·139	*	1·256	1·230
0·00200	(0·0196)	21·25	14·48	10·75	7·985	6·262	4·267	3·169	2·485	1·693	1·258	*	1·253	1·224	1·201
0·00300	(0·0294)	17·35	11·82	8·779	6·520	5·113	3·484	2·587	2·029	1·383	1·027	*	1·208	1·183	1·162
0·00400	(0·0392)	15·02	10·24	7·603	5·646	4·428	3·017	2·241	1·757	1·197	*	*	1·177	1·155	1·136
0·00600	(0·0588)	12·27	8·359	6·208	4·610	3·615	2·464	1·830	1·435	0·978	*	1·169	1·138	1·118	1·102
0·00800	(0·0785)	10·62	7·239	5·376	3·993	3·131	2·134	1·584	1·243	*	1·161	1·139	1·111	1·093	1·078
0·01000	(0·0981)	9·502	6·475	4·809	3·571	2·800	1·908	1·417	1·111	*	1·138	1·117	1·091	1·074	1·061
0·01500	(0·147)	7·758	5·287	3·926	2·916	2·287	1·558	1·157	0·907	1·123	1·097	1·080	1·057	1·042	1·031
0·02000	(0·196)	6·719	4·578	3·400	2·525	1·980	1·349	1·002	*	1·094	1·071	1·054	1·034	1·021	1·010
0·03000	(0·294)	5·486	3·738	2·776	2·062	1·617	1·102	*	*	1·055	1·035	1·021	1·003	0·992	0·983
0·04000	(0·392)	4·751	3·237	2·404	1·786	1·400	0·954	*	1·059	1·030	1·011	0·998	0·982	0·972	0·965
0·06000	(0·588)	3·879	2·643	1·963	1·458	1·143	*	1·041	1·021	0·996	0·979	0·968	0·954	0·946	0·940
0·08000	(0·785)	3·359	2·289	1·700	1·263	0·990	*	1·014	0·996	0·973	0·958	0·948	0·936	0·928	0·923
0·10000	(0·981)	3·005	2·048	1·521	1·129	0·886	*	0·994	0·978	0·956	0·942	0·933	0·922	0·915	0·910
0·15000	(1·47)	2·453	1·672	1·242	0·922	*	0·981	0·960	0·945	0·926	0·915	0·907	0·897	0·892	0·888
0·20000	(1·96)	2·125	1·448	1·075	0·799	*	0·956	0·937	0·924	0·907	0·896	0·889	0·881	0·876	0·873
0·30000	(2·94)	1·735	1·182	0·878	*	0·950	0·924	0·907	0·895	0·880	0·871	0·865	0·859	0·855	0·853
0·40000	(3·92)	1·502	1·024	0·760	*	0·926	0·902	0·886	0·876	0·862	0·855	0·850	0·844	0·841	0·840
0·60000	(5·88)	1·227	0·836	*	0·910	0·894	0·872	0·859	0·850	0·839	0·832	0·828	0·824	0·822	0·821
0·80000	(7·85)	1·062	0·724	*	0·887	0·872	0·853	0·841	0·833	0·823	0·817	0·814	0·810	0·809	0·809
1·00000	(9·81)	0·950	*	0·890	0·870	0·856	0·838	0·827	0·820	0·811	0·806	0·803	0·800	0·799	0·800
1·50000	(14·7)	0·776	*	0·859	0·841	0·829	0·813	0·804	0·798	0·790	0·786	0·784	0·782	0·782	0·783
2·00000	(19·6)	*	0·857	0·837	0·822	0·811	0·797	0·788	0·783	0·776	0·773	0·771	0·770	0·771	0·772
3·00000	(29·4)	*	0·826	0·810	0·796	0·786	0·774	0·767	0·763	0·757	0·755	0·754	0·754	0·755	0·757
4·00000	(39·2)	*	0·806	0·791	0·778	0·770	0·759	0·753	0·749	0·745	0·743	0·743	0·743	0·745	0·748
6·00000	(58·8)	0·801	0·779	0·766	0·755	0·748	0·739	0·734	0·731	0·728	0·727	0·727	0·729	0·731	0·734
8·00000	(78·5)	0·781	0·762	0·750	0·740	0·734	0·726	0·721	0·719	0·717	0·716	0·717	0·719	0·721	0·725
10·0000	(98·1)	0·766	0·749	0·738	0·729	0·723	0·716	0·712	0·710	0·708	0·708	0·709	0·711	0·714	0·718
15·0000	(147)	0·741	0·726	0·716	0·709	0·704	0·698	0·696	0·694	0·693	0·694	0·695	0·699	0·702	0·707
20·0000	(196)	0·725	0·711	0·702	0·696	0·692	0·687	0·685	0·684	0·684	0·685	0·686	0·690	0·694	0·699
30·0000	(294)	0·703	0·691	0·684	0·678	0·675	0·671	0·670	0·670	0·671	0·672	0·675	0·679	0·684	0·689
S	$(\Delta p/\rho l)$	6	8	10	12·5	15	20	25	30	40	50	60	80	100	125

Grad'nt   k.p.g.   (Equivalent) Pipe diameters in mm

Grad'nt S	k.p.g. $(\Delta p/\rho l)$	(Equivalent) Pipe diameters in mm													
		125	150	200	250	300	400	500	600	800	1000	1250	1500	2000	2500
0·00010	(0·00098)	1·658	1·300	*	1·457	1·433	1·400	1·381	1·367	1·350	1·340	1·333	1·329	1·326	1·325
0·00015	(0·00147)	1·353	*	1·438	1·407	1·385	1·357	1·340	1·328	1·314	1·306	1·300	1·298	1·296	1·296
0·00020	(0·00196)	1·172	*	1·401	1·373	1·353	1·328	1·312	1·302	1·290	1·283	1·278	1·276	1·275	1·276
0·00030	(0·00294)	*	1·392	1·353	1·328	1·311	1·289	1·275	1·267	1·257	1·251	1·248	1·247	1·247	1·250
0·00040	(0·00392)	*	1·356	1·320	1·298	1·282	1·262	1·251	1·243	1·234	1·230	1·228	1·227	1·228	1·231
0·00060	(0·00588)	1·333	1·308	1·277	1·257	1·244	1·227	1·217	1·211	1·204	1·201	1·200	1·200	1·203	1·206
0·00080	(0·00785)	1·299	1·277	1·248	1·230	1·218	1·203	1·194	1·189	1·183	1·181	1·181	1·182	1·185	1·189
0·00100	(0·00981)	1·273	1·253	1·226	1·210	1·199	1·185	1·177	1·173	1·168	1·166	1·167	1·168	1·172	1·177
0·00150	(0·0147)	1·230	1·212	1·189	1·175	1·165	1·154	1·148	1·144	1·141	1·140	1·142	1·144	1·148	1·154
0·00200	(0·0196)	1·201	1·185	1·164	1·151	1·143	1·133	1·128	1·125	1·123	1·123	1·124	1·127	1·133	1·138
0·00300	(0·0294)	1·162	1·148	1·130	1·119	1·112	1·104	1·100	1·098	1·098	1·099	1·101	1·104	1·111	1·117
0·00400	(0·0392)	1·136	1·124	1·107	1·098	1·092	1·085	1·082	1·081	1·081	1·082	1·085	1·089	1·096	1·103
0·00600	(0·0588)	1·102	1·091	1·077	1·069	1·064	1·059	1·057	1·056	1·058	1·060	1·064	1·068	1·076	1·083
0·00800	(0·0785)	1·078	1·069	1·056	1·049	1·045	1·041	1·040	1·040	1·042	1·045	1·049	1·053	1·062	1·069
0·01000	(0·0981)	1·061	1·052	1·041	1·035	1·031	1·028	1·027	1·027	1·030	1·033	1·038	1·042	1·051	1·059
0·01500	(0·147)	1·031	1·023	1·014	1·009	1·007	1·004	1·005	1·006	1·009	1·013	1·018	1·023	1·033	1·041
0·02000	(0·196)	1·010	1·004	0·996	0·992	0·990	0·989	0·989	0·991	0·995	0·999	1·005	1·010	1·020	1·029
0·03000	(0·294)	0·983	0·978	0·971	0·968	0·967	0·967	0·969	0·971	0·976	0·981	0·987	0·992	1·003	1·012
0·04000	(0·392)	0·965	0·960	0·955	0·952	0·952	0·952	0·954	0·957	0·962	0·968	0·974	0·980	0·991	1·000
0·06000	(0·588)	0·940	0·936	0·932	0·931	0·931	0·933	0·935	0·938	0·944	0·950	0·957	0·964	0·975	0·985
0·08000	(0·785)	0·923	0·920	0·917	0·916	0·917	0·919	0·922	0·926	0·932	0·939	0·946	0·952	0·964	0·974
0·10000	(0·981)	0·910	0·908	0·905	0·905	0·906	0·909	0·912	0·916	0·923	0·930	0·937	0·944	0·956	0·966
0·15000	(1·47)	0·888	0·886	0·885	0·886	0·887	0·891	0·895	0·899	0·907	0·914	0·922	0·929	0·941	0·952
0·20000	(1·96)	0·873	0·872	0·872	0·873	0·875	0·879	0·883	0·888	0·896	0·903	0·911	0·919	0·931	0·942
0·30000	(2·94)	0·853	0·853	0·853	0·855	0·857	0·862	0·867	0·872	0·881	0·889	0·897	0·905	0·918	0·929
0·40000	(3·92)	0·840	0·840	0·841	0·843	0·846	0·851	0·857	0·862	0·871	0·879	0·888	0·896	0·909	0·921
0·60000	(5·88)	0·821	0·822	0·824	0·827	0·830	0·836	0·842	0·847	0·857	0·866	0·875	0·883	0·897	0·909
0·80000	(7·85)	0·809	0·810	0·813	0·816	0·819	0·826	0·832	0·838	0·848	0·857	0·866	0·874	0·889	0·901
1·00000	(9·81)	0·800	0·801	0·804	0·808	0·811	0·818	0·825	0·831	0·841	0·850	0·860	0·868	0·883	0·895
1·50000	(14·7)	0·783	0·785	0·789	0·793	0·797	0·805	0·812	0·818	0·829	0·838	0·848	0·857	0·872	0·885
2·00000	(19·6)	0·772	0·774	0·779	0·783	0·788	0·796	0·803	0·809	0·821	0·830	0·841	0·850	0·865	0·878
3·00000	(29·4)	0·757	0·760	0·765	0·770	0·775	0·784	0·791	0·798	0·810	0·820	0·831	0·840	0·856	0·869
S	$(\Delta p/\rho l)$	125	150	200	250	300	400	500	600	800	1000	1250	1500	2000	2500

Grad'nt   k.p.g.   (Equivalent) Pipe diameters in mm

Kinematic viscosity, $\nu = 8{\cdot}0 \times 10^{-6}\ \mathrm{m^2\,s^{-1}}$;   Roughness size, $k_s = 0{\cdot}003$ mm

E185

Kin. visc., $\nu = 8\cdot0\times10^{-6}\ \mathrm{m^2 s^{-1}}$;   Roughness size, $k_s = 0\cdot006$ mm

$S = 0\cdot00010$ to $30\cdot0000$

This table shows values of m, as follows

i.e. kin. pr. grad., $\Delta p/\rho l =$ ($0\cdot00098$) to (294) $\mathrm{ms^{-2}}$

m_C for Colebrook-White solutions; or, where $\mathbf{R} \leq 2000$, m_P for laminar flow

Grad'nt S **k.p.g.** ($\Delta p/\rho l$)   **(Equivalent) Pipe diameters in mm**

S	($\Delta p/\rho l$)	6	8	10	12.5	15	20	25	30	40	50	60	80	100	125
0.00030	(0.00294)	54.86	37.38	27.76	20.62	16.17	11.02	8.182	6.416	4.372	3.247	2.546	1.735	1.289	*
0.00040	(0.00392)	47.51	32.37	24.04	17.86	14.00	9.541	7.086	5.557	3.786	2.812	2.205	1.503	1.116	*
0.00060	(0.00588)	38.79	26.43	19.63	14.58	11.43	7.790	5.786	4.537	3.092	2.296	1.801	1.227	*	1.333
0.00080	(0.00785)	33.59	22.89	17.00	12.63	9.901	6.747	5.010	3.929	2.677	1.988	1.559	1.063	*	1.299
0.00100	(0.00981)	30.05	20.48	15.21	11.29	8.856	6.034	4.482	3.514	2.395	1.778	1.395	*	1.303	1.274
0.00150	(0.0147)	24.53	16.72	12.42	9.220	7.231	4.927	3.659	2.869	1.955	1.452	1.139	*	1.256	1.230
0.00200	(0.0196)	21.25	14.48	10.75	7.985	6.262	4.267	3.169	2.485	1.693	1.258	*	1.253	1.225	1.201
0.00300	(0.0294)	17.35	11.82	8.779	6.520	5.113	3.484	2.587	2.029	1.383	1.027	*	1.208	1.183	1.163
0.00400	(0.0392)	15.02	10.24	7.603	5.646	4.428	3.017	2.241	1.757	1.197	*	*	1.178	1.156	1.137
0.00600	(0.0588)	12.27	8.359	6.208	4.610	3.615	2.464	1.830	1.435	0.978	*	1.170	1.138	1.119	1.102
0.00800	(0.0785)	10.62	7.239	5.376	3.993	3.131	2.134	1.584	1.243	*	1.162	1.140	1.112	1.094	1.079
0.01000	(0.0981)	9.502	6.475	4.809	3.571	2.800	1.908	1.417	1.111	*	1.139	1.118	1.092	1.075	1.062
0.01500	(0.147)	7.758	5.287	3.926	2.916	2.287	1.558	1.157	0.907	1.124	1.098	1.081	1.058	1.043	1.032
0.02000	(0.196)	6.719	4.578	3.400	2.525	1.980	1.349	1.002	*	1.095	1.072	1.056	1.035	1.022	1.012
0.03000	(0.294)	5.486	3.738	2.776	2.062	1.617	1.102	*	*	1.057	1.036	1.022	1.004	0.993	0.984
0.04000	(0.392)	4.751	3.237	2.404	1.786	1.400	0.954	*	1.061	1.031	1.012	1.000	0.983	0.974	0.966
0.06000	(0.588)	3.879	2.643	1.963	1.458	1.143	*	1.043	1.023	0.997	0.981	0.970	0.956	0.948	0.942
0.08000	(0.785)	3.359	2.289	1.700	1.263	0.990	*	1.016	0.998	0.974	0.960	0.950	0.937	0.930	0.925
0.10000	(0.981)	3.005	2.048	1.521	1.129	0.886	*	0.996	0.979	0.957	0.944	0.935	0.924	0.917	0.913
0.15000	(1.47)	2.453	1.672	1.242	0.922	*	0.983	0.962	0.947	0.928	0.917	0.909	0.900	0.894	0.891
0.20000	(1.96)	2.125	1.448	1.075	0.799	*	0.958	0.939	0.926	0.909	0.898	0.891	0.883	0.879	0.876
0.30000	(2.94)	1.735	1.182	0.878	*	0.953	0.926	0.909	0.897	0.883	0.874	0.868	0.862	0.858	0.857
0.40000	(3.92)	1.502	1.024	0.760	*	0.928	0.904	0.889	0.878	0.865	0.857	0.852	0.847	0.844	0.843
0.60000	(5.88)	1.227	0.836	*	0.913	0.896	0.875	0.862	0.853	0.842	0.835	0.832	0.827	0.826	0.826
0.80000	(7.85)	1.062	0.724	*	0.890	0.875	0.856	0.844	0.836	0.826	0.821	0.817	0.814	0.813	0.814
1.00000	(9.81)	0.950	*	0.893	0.873	0.859	0.842	0.831	0.823	0.815	0.810	0.807	0.804	0.804	0.805
1.50000	(14.7)	0.776	*	0.862	0.844	0.832	0.817	0.808	0.802	0.794	0.791	0.789	0.787	0.788	0.789
2.00000	(19.6)	*	0.860	0.841	0.825	0.814	0.801	0.792	0.787	0.781	0.778	0.776	0.776	0.777	0.779
3.00000	(29.4)	*	0.830	0.814	0.800	0.791	0.779	0.772	0.768	0.763	0.761	0.760	0.761	0.762	0.765
4.00000	(39.2)	*	0.810	0.795	0.783	0.775	0.764	0.758	0.755	0.751	0.749	0.749	0.750	0.753	0.756
6.00000	(58.8)	0.805	0.784	0.771	0.761	0.754	0.745	0.740	0.737	0.735	0.734	0.735	0.737	0.740	0.744
8.00000	(78.5)	0.786	0.767	0.755	0.746	0.740	0.732	0.728	0.726	0.724	0.724	0.725	0.728	0.731	0.735
10.0000	(98.1)	0.772	0.754	0.743	0.735	0.729	0.723	0.719	0.717	0.716	0.717	0.718	0.721	0.725	0.729
15.0000	(147)	0.748	0.732	0.723	0.716	0.711	0.706	0.704	0.703	0.703	0.704	0.706	0.710	0.714	0.719
20.0000	(196)	0.732	0.718	0.710	0.704	0.700	0.695	0.694	0.693	0.694	0.696	0.698	0.702	0.707	0.713
30.0000	(294)	0.711	0.699	0.692	0.687	0.684	0.681	0.680	0.680	0.682	0.684	0.687	0.693	0.698	0.704
S	($\Delta p/\rho l$)	6	8	10	12.5	15	20	25	30	40	50	60	80	100	125

Grad'nt S **k.p.g.** ($\Delta p/\rho l$)   **(Equivalent) Pipe diameters in mm**

S	($\Delta p/\rho l$)	125	150	200	250	300	400	500	600	800	1000	1250	1500	2000	2500
0.00010	(0.00098)	1.658	1.300	*	1.458	1.433	1.401	1.381	1.367	1.350	1.341	1.334	1.329	1.326	1.325
0.00015	(0.00147)	1.353	*	1.438	1.407	1.385	1.357	1.340	1.328	1.314	1.306	1.301	1.298	1.296	1.297
0.00020	(0.00196)	1.172	*	1.402	1.373	1.353	1.328	1.313	1.302	1.290	1.283	1.279	1.276	1.276	1.277
0.00030	(0.00294)	*	1.393	1.353	1.328	1.311	1.289	1.276	1.267	1.257	1.252	1.249	1.247	1.248	1.250
0.00040	(0.00392)	*	1.357	1.321	1.298	1.282	1.263	1.251	1.243	1.235	1.230	1.228	1.228	1.229	1.232
0.00060	(0.00588)	1.333	1.309	1.277	1.258	1.244	1.227	1.218	1.211	1.205	1.202	1.200	1.201	1.203	1.207
0.00080	(0.00785)	1.299	1.277	1.248	1.231	1.219	1.204	1.195	1.190	1.184	1.182	1.182	1.182	1.186	1.190
0.00100	(0.00981)	1.274	1.253	1.227	1.210	1.199	1.186	1.178	1.173	1.169	1.167	1.167	1.169	1.173	1.178
0.00150	(0.0147)	1.230	1.213	1.190	1.175	1.166	1.155	1.148	1.145	1.142	1.141	1.142	1.145	1.150	1.155
0.00200	(0.0196)	1.201	1.185	1.164	1.152	1.143	1.134	1.128	1.126	1.123	1.124	1.125	1.128	1.134	1.140
0.00300	(0.0294)	1.163	1.149	1.131	1.120	1.113	1.105	1.101	1.099	1.099	1.100	1.102	1.106	1.112	1.119
0.00400	(0.0392)	1.137	1.124	1.108	1.099	1.093	1.086	1.083	1.082	1.082	1.084	1.087	1.090	1.097	1.104
0.00600	(0.0588)	1.102	1.092	1.078	1.070	1.065	1.060	1.058	1.058	1.059	1.062	1.065	1.069	1.077	1.085
0.00800	(0.0785)	1.079	1.069	1.057	1.051	1.046	1.042	1.041	1.041	1.043	1.046	1.051	1.055	1.064	1.072
0.01000	(0.0981)	1.062	1.053	1.042	1.036	1.032	1.029	1.028	1.029	1.032	1.035	1.040	1.045	1.054	1.062
0.01500	(0.147)	1.032	1.024	1.015	1.011	1.008	1.006	1.006	1.007	1.011	1.015	1.021	1.026	1.035	1.044
0.02000	(0.196)	1.012	1.005	0.997	0.993	0.991	0.990	0.991	0.993	0.997	1.002	1.008	1.013	1.023	1.032
0.03000	(0.294)	0.984	0.979	0.973	0.970	0.969	0.969	0.971	0.973	0.978	0.983	0.990	0.996	1.006	1.016
0.04000	(0.392)	0.966	0.961	0.956	0.954	0.954	0.955	0.957	0.960	0.965	0.971	0.978	0.984	0.995	1.005
0.06000	(0.588)	0.942	0.938	0.934	0.933	0.933	0.935	0.938	0.941	0.948	0.954	0.961	0.968	0.980	0.990
0.08000	(0.785)	0.925	0.922	0.919	0.919	0.919	0.922	0.925	0.929	0.936	0.943	0.950	0.957	0.969	0.980
0.10000	(0.981)	0.913	0.910	0.908	0.908	0.909	0.912	0.916	0.920	0.927	0.934	0.942	0.949	0.961	0.972
0.15000	(1.47)	0.891	0.889	0.888	0.889	0.891	0.895	0.899	0.903	0.912	0.919	0.927	0.935	0.948	0.959
0.20000	(1.96)	0.876	0.875	0.875	0.876	0.878	0.883	0.888	0.892	0.901	0.909	0.918	0.925	0.939	0.950
0.30000	(2.94)	0.857	0.856	0.857	0.859	0.862	0.867	0.873	0.878	0.887	0.895	0.904	0.913	0.927	0.938
0.40000	(3.92)	0.843	0.843	0.845	0.848	0.851	0.857	0.862	0.868	0.878	0.886	0.896	0.904	0.918	0.931
0.60000	(5.88)	0.826	0.826	0.829	0.832	0.836	0.842	0.849	0.854	0.865	0.874	0.884	0.893	0.908	0.920
0.80000	(7.85)	0.814	0.815	0.818	0.822	0.825	0.833	0.839	0.846	0.856	0.866	0.876	0.885	0.900	0.913
1.00000	(9.81)	0.805	0.806	0.810	0.814	0.818	0.826	0.833	0.839	0.850	0.860	0.870	0.880	0.895	0.908
1.50000	(14.7)	0.789	0.791	0.796	0.800	0.805	0.813	0.821	0.828	0.839	0.850	0.860	0.870	0.886	0.900
2.00000	(19.6)	0.779	0.781	0.786	0.791	0.796	0.805	0.813	0.820	0.832	0.843	0.854	0.864	0.880	0.894
3.00000	(29.4)	0.765	0.768	0.774	0.779	0.785	0.794	0.803	0.810	0.823	0.834	0.846	0.856	0.873	0.887
S	($\Delta p/\rho l$)	125	150	200	250	300	400	500	600	800	1000	1250	1500	2000	2500

Grad'nt   **k.p.g.**   **(Equivalent) Pipe diameters in mm**

Kinematic viscosity, $\nu = 8\cdot0\times10^{-6}\ \mathrm{m^2 s^{-1}}$;   **Roughness size,** $k_s = 0\cdot006$ mm

Kin. visc., $\nu = 8{\cdot}0\times10^{-6}$ m^2 s^{-1};
$S = 0{\cdot}00010$ to $30{\cdot}0000$

i.e. kin. pr. grad., $\Delta p/\rho l =$
$(0{\cdot}00098)$ to (294) ms^{-2}

Roughness size, $k_s = 0{\cdot}015$ mm
This table shows values of m, as follows

m_C for Colebrook-White solutions; or,
where $R \le 2000$, m_P for laminar flow

Grad'nt S	k.p.g. $(\Delta p/\rho l)$	6	8	10	12·5	15	20	25	30	40	50	60	80	100	125
				(Equivalent) Pipe diameters in mm											
0·00030	(0·00294)	54·86	37·38	27·76	20·62	16·17	11·02	8·182	6·416	4·372	3·247	2·546	1·735	1·289	*
0·00040	(0·00392)	47·51	32·37	24·04	17·86	14·00	9·541	7·086	5·557	3·786	2·812	2·205	1·503	1·116	*
0·00060	(0·00588)	38·79	26·43	19·63	14·58	11·43	7·790	5·786	4·537	3·092	2·296	1·801	1·227	*	1·334
0·00080	(0·00785)	33·59	22·89	17·00	12·63	9·901	6·747	5·010	3·929	2·677	1·988	1·559	1·063	*	1·300
0·00100	(0·00981)	30·05	20·48	15·21	11·29	8·856	6·034	4·482	3·514	2·395	1·778	1·395	*	1·305	1·275
0·00150	(0·0147)	24·53	16·72	12·42	9·220	7·231	4·927	3·659	2·869	1·955	1·452	1·139	*	1·258	1·232
0·00200	(0·0196)	21·25	14·48	10·75	7·985	6·262	4·267	3·169	2·485	1·693	1·258	*	1·255	1·227	1·203
0·00300	(0·0294)	17·35	11·82	8·779	6·520	5·113	3·484	2·587	2·029	1·383	1·027	*	1·210	1·185	1·165
0·00400	(0·0392)	15·02	10·24	7·603	5·646	4·428	3·017	2·241	1·757	1·197	*	*	1·180	1·158	1·139
0·00600	(0·0588)	12·27	8·359	6·208	4·610	3·615	2·464	1·830	1·435	0·978	*	1·172	1·140	1·121	1·105
0·00800	(0·0785)	10·62	7·239	5·376	3·993	3·131	2·134	1·584	1·243	*	1·164	1·142	1·114	1·096	1·082
0·01000	(0·0981)	9·502	6·475	4·809	3·571	2·800	1·908	1·417	1·111	*	1·141	1·121	1·094	1·078	1·065
0·01500	(0·147)	7·758	5·287	3·926	2·916	2·287	1·558	1·157	0·907	1·127	1·101	1·084	1·061	1·047	1·035
0·02000	(0·196)	6·719	4·578	3·400	2·525	1·980	1·349	1·002	*	1·098	1·075	1·059	1·038	1·025	1·015
0·03000	(0·294)	5·486	3·738	2·776	2·062	1·617	1·102	*	*	1·060	1·040	1·026	1·008	0·997	0·989
0·04000	(0·392)	4·751	3·237	2·404	1·786	1·400	0·954	*	1·064	1·035	1·016	1·004	0·988	0·978	0·971
0·06000	(0·588)	3·879	2·643	1·963	1·458	1·143	*	1·047	1·027	1·001	0·985	0·974	0·961	0·953	0·947
0·08000	(0·785)	3·359	2·289	1·700	1·263	0·990	*	1·020	1·002	0·979	0·964	0·955	0·943	0·936	0·931
0·10000	(0·981)	3·005	2·048	1·521	1·129	0·886	*	1·001	0·984	0·962	0·949	0·940	0·929	0·923	0·919
0·15000	(1·47)	2·453	1·672	1·242	0·922	*	0·988	0·967	0·953	0·934	0·923	0·915	0·906	0·901	0·898
0·20000	(1·96)	2·125	1·448	1·075	0·799	*	0·964	0·945	0·932	0·915	0·905	0·898	0·891	0·887	0·884
0·30000	(2·94)	1·735	1·182	0·878	*	0·959	0·932	0·915	0·904	0·890	0·881	0·876	0·870	0·867	0·866
0·40000	(3·92)	1·502	1·024	0·760	*	0·935	0·911	0·896	0·886	0·873	0·866	0·861	0·856	0·854	0·854
0·60000	(5·88)	1·227	0·836	*	0·920	0·904	0·883	0·870	0·861	0·851	0·845	0·841	0·838	0·837	0·837
0·80000	(7·85)	1·062	0·724	*	0·898	0·883	0·865	0·853	0·845	0·836	0·831	0·828	0·826	0·826	0·827
1·00000	(9·81)	0·950	*	0·902	0·882	0·868	0·851	0·840	0·833	0·825	0·821	0·819	0·817	0·817	0·819
1·50000	(14·7)	0·776	*	0·871	0·854	0·842	0·828	0·819	0·813	0·806	0·803	0·802	0·802	0·803	0·805
2·00000	(19·6)	*	0·870	0·851	0·836	0·825	0·812	0·804	0·799	0·794	0·792	0·791	0·791	0·793	0·796
3·00000	(29·4)	*	0·842	0·825	0·812	0·803	0·792	0·786	0·782	0·778	0·777	0·777	0·778	0·781	0·785
4·00000	(39·2)	*	0·823	0·808	0·796	0·788	0·779	0·773	0·770	0·767	0·767	0·767	0·770	0·773	0·777
6·00000	(58·8)	0·819	0·798	0·786	0·776	0·769	0·761	0·757	0·755	0·753	0·754	0·755	0·758	0·762	0·767
8·00000	(78·5)	0·801	0·782	0·771	0·762	0·756	0·750	0·746	0·745	0·744	0·745	0·747	0·751	0·755	0·761
10·0000	(98·1)	0·788	0·771	0·760	0·752	0·747	0·741	0·739	0·737	0·738	0·739	0·741	0·746	0·751	0·756
15·0000	(147)	0·765	0·751	0·742	0·735	0·731	0·727	0·726	0·725	0·727	0·729	0·731	0·737	0·743	0·749
20·0000	(196)	0·751	0·738	0·730	0·724	0·721	0·718	0·717	0·717	0·719	0·722	0·725	0·731	0·737	0·744
30·0000	(294)	0·732	0·721	0·715	0·710	0·708	0·706	0·706	0·707	0·710	0·714	0·717	0·724	0·731	0·738
S	$(\Delta p/\rho l)$	6	8	10	12·5	15	20	25	30	40	50	60	80	100	125

Grad'nt S	k.p.g. $(\Delta p/\rho l)$	125	150	200	250	300	400	500	600	800	1000	1250	1500	2000	2500
				(Equivalent) Pipe diameters in mm											
0·00010	(0·00098)	1·658	1·300	*	1·458	1·434	1·401	1·382	1·368	1·351	1·342	1·335	1·330	1·327	1·327
0·00015	(0·00147)	1·353	*	1·439	1·408	1·386	1·358	1·341	1·329	1·315	1·307	1·302	1·299	1·297	1·298
0·00020	(0·00196)	1·172	*	1·402	1·374	1·354	1·329	1·314	1·303	1·291	1·284	1·280	1·278	1·277	1·279
0·00030	(0·00294)	*	1·394	1·354	1·329	1·312	1·290	1·277	1·268	1·258	1·253	1·250	1·249	1·250	1·252
0·00040	(0·00392)	*	1·358	1·322	1·299	1·284	1·264	1·252	1·245	1·236	1·232	1·230	1·229	1·231	1·234
0·00060	(0·00588)	1·334	1·310	1·279	1·259	1·246	1·229	1·219	1·213	1·206	1·203	1·202	1·203	1·206	1·210
0·00080	(0·00785)	1·300	1·278	1·250	1·232	1·220	1·205	1·197	1·191	1·186	1·184	1·184	1·185	1·189	1·193
0·00100	(0·00981)	1·275	1·255	1·228	1·212	1·201	1·187	1·180	1·175	1·171	1·169	1·170	1·171	1·176	1·181
0·00150	(0·0147)	1·232	1·214	1·191	1·177	1·168	1·157	1·150	1·147	1·144	1·144	1·145	1·147	1·153	1·159
0·00200	(0·0196)	1·203	1·187	1·166	1·154	1·145	1·136	1·131	1·128	1·126	1·127	1·129	1·131	1·137	1·144
0·00300	(0·0294)	1·165	1·151	1·133	1·122	1·115	1·108	1·104	1·102	1·102	1·103	1·106	1·109	1·116	1·123
0·00400	(0·0392)	1·139	1·126	1·111	1·101	1·095	1·089	1·086	1·085	1·085	1·087	1·091	1·095	1·102	1·110
0·00600	(0·0588)	1·105	1·094	1·081	1·073	1·068	1·063	1·062	1·061	1·063	1·066	1·070	1·074	1·083	1·091
0·00800	(0·0785)	1·082	1·072	1·060	1·054	1·050	1·046	1·045	1·045	1·048	1·051	1·056	1·061	1·070	1·078
0·01000	(0·0981)	1·065	1·056	1·045	1·039	1·036	1·033	1·033	1·033	1·037	1·040	1·046	1·051	1·060	1·069
0·01500	(0·147)	1·035	1·028	1·019	1·014	1·012	1·011	1·011	1·013	1·017	1·021	1·027	1·033	1·043	1·053
0·02000	(0·196)	1·015	1·009	1·001	0·998	0·996	0·995	0·997	0·999	1·003	1·009	1·015	1·021	1·032	1·041
0·03000	(0·294)	0·989	0·983	0·978	0·975	0·974	0·975	0·977	0·980	0·985	0·991	0·998	1·005	1·016	1·027
0·04000	(0·392)	0·971	0·966	0·962	0·960	0·960	0·961	0·964	0·967	0·973	0·980	0·987	0·994	1·006	1·017
0·06000	(0·588)	0·947	0·944	0·940	0·940	0·940	0·943	0·946	0·950	0·957	0·964	0·972	0·980	0·992	1·004
0·08000	(0·785)	0·931	0·928	0·926	0·926	0·927	0·930	0·934	0·938	0·946	0·954	0·962	0·970	0·983	0·995
0·10000	(0·981)	0·919	0·917	0·915	0·916	0·917	0·921	0·925	0·930	0·938	0·946	0·955	0·963	0·977	0·989
0·15000	(1·47)	0·898	0·897	0·897	0·898	0·900	0·905	0·910	0·915	0·925	0·933	0·942	0·951	0·965	0·978
0·20000	(1·96)	0·884	0·884	0·884	0·886	0·889	0·895	0·900	0·905	0·915	0·924	0·934	0·943	0·958	0·970
0·30000	(2·94)	0·866	0·866	0·868	0·871	0·874	0·880	0·887	0·893	0·903	0·913	0·923	0·932	0·948	0·961
0·40000	(3·92)	0·854	0·854	0·857	0·860	0·864	0·871	0·878	0·884	0·895	0·905	0·916	0·925	0·942	0·955
0·60000	(5·88)	0·837	0·839	0·842	0·847	0·851	0·859	0·866	0·873	0·885	0·895	0·907	0·916	0·933	0·947
0·80000	(7·85)	0·827	0·828	0·833	0·837	0·842	0·851	0·859	0·866	0·878	0·889	0·901	0·911	0·928	0·943
1·00000	(9·81)	0·819	0·821	0·826	0·831	0·836	0·845	0·853	0·860	0·873	0·884	0·896	0·907	0·924	0·939
1·50000	(14·7)	0·805	0·808	0·814	0·820	0·825	0·835	0·844	0·851	0·865	0·877	0·889	0·900	0·918	0·933
2·00000	(19·6)	0·796	0·799	0·806	0·812	0·818	0·828	0·838	0·846	0·860	0·872	0·885	0·896	0·914	0·930
3·00000	(29·4)	0·785	0·788	0·796	0·803	0·809	0·820	0·830	0·838	0·853	0·866	0·879	0·890	0·909	0·925
S	$(\Delta p/\rho l)$	125	150	200	250	300	400	500	600	800	1000	1250	1500	2000	2500

Grad'nt   k.p.g.   (Equivalent) Pipe diameters in mm

Kinematic viscosity, $\nu = 8{\cdot}0\times10^{-6}$ m^2 s^{-1} ;

Roughness size, $k_s = 0{\cdot}015$ mm

E187

Kin. visc., $\nu = 8{\cdot}0 \times 10^{-6}$ m²s⁻¹;
S = 0·00010 to 3·00000

i.e. kin. pr. grad., $\Delta p/\rho l$ =
(0·00098) to (29·4) ms⁻²

Roughness size, k_s = 0·030 mm
This table shows values of m, as follows

m_C for Colebrook-White solutions; or,
where $\mathbf{R} \le 2000$, m_P for laminar flow

Grad'nt S	k.p.g. $(\Delta p/\rho l)$	\(Equivalent) Pipe diameters in mm												
		125	150	200	250	300	400	500	600	800	1000	1250	1500	2000	2500
0·00010	(0·00098)	1·658	1·300	*	1·459	1·435	1·403	1·383	1·369	1·353	1·343	1·336	1·332	1·329	1·329
0·00015	(0·00147)	1·353	*	1·440	1·409	1·387	1·360	1·342	1·331	1·317	1·309	1·304	1·301	1·300	1·300
0·00020	(0·00196)	1·172	*	1·404	1·375	1·356	1·331	1·315	1·305	1·293	1·286	1·282	1·280	1·280	1·281
0·00030	(0·00294)	*	1·395	1·356	1·331	1·314	1·292	1·279	1·270	1·260	1·255	1·253	1·252	1·253	1·255
0·00040	(0·00392)	*	1·359	1·323	1·301	1·285	1·266	1·254	1·247	1·239	1·235	1·233	1·232	1·234	1·238
0·00060	(0·00588)	1·336	1·312	1·281	1·261	1·248	1·231	1·221	1·215	1·209	1·206	1·206	1·206	1·209	1·214
0·00080	(0·00785)	1·302	1·280	1·252	1·234	1·222	1·208	1·199	1·194	1·189	1·187	1·187	1·189	1·193	1·198
0·00100	(0·00981)	1·277	1·257	1·231	1·214	1·204	1·190	1·183	1·178	1·174	1·173	1·174	1·175	1·180	1·185
0·00150	(0·0147)	1·234	1·217	1·194	1·180	1·171	1·160	1·154	1·151	1·148	1·148	1·150	1·152	1·158	1·164
0·00200	(0·0196)	1·206	1·190	1·169	1·157	1·149	1·139	1·134	1·132	1·130	1·131	1·134	1·137	1·143	1·150
0·00300	(0·0294)	1·168	1·154	1·136	1·126	1·119	1·112	1·108	1·107	1·107	1·109	1·112	1·116	1·123	1·131
0·00400	(0·0392)	1·143	1·130	1·114	1·105	1·099	1·093	1·091	1·090	1·091	1·093	1·097	1·101	1·110	1·118
0·00600	(0·0588)	1·109	1·098	1·085	1·077	1·073	1·069	1·067	1·067	1·069	1·073	1·078	1·082	1·092	1·100
0·00800	(0·0785)	1·086	1·077	1·065	1·059	1·055	1·052	1·051	1·052	1·055	1·059	1·064	1·070	1·080	1·089
0·01000	(0·0981)	1·069	1·061	1·050	1·045	1·042	1·039	1·039	1·041	1·044	1·049	1·054	1·060	1·071	1·080
0·01500	(0·147)	1·041	1·033	1·025	1·021	1·019	1·018	1·019	1·021	1·026	1·031	1·038	1·044	1·055	1·065
0·02000	(0·196)	1·021	1·015	1·008	1·005	1·003	1·003	1·005	1·008	1·013	1·019	1·026	1·033	1·045	1·056
0·03000	(0·294)	0·996	0·991	0·985	0·983	0·983	0·984	0·987	0·990	0·997	1·003	1·011	1·018	1·031	1·043
0·04000	(0·392)	0·978	0·974	0·970	0·969	0·969	0·971	0·975	0·978	0·986	0·993	1·001	1·009	1·022	1·034
0·06000	(0·588)	0·956	0·952	0·950	0·950	0·951	0·954	0·959	0·963	0·971	0·979	0·988	0·996	1·011	1·023
0·08000	(0·785)	0·940	0·938	0·937	0·937	0·939	0·943	0·948	0·953	0·962	0·970	0·980	0·988	1·003	1·016
0·10000	(0·981)	0·929	0·927	0·927	0·928	0·930	0·935	0·940	0·945	0·955	0·964	0·974	0·982	0·998	1·011
0·15000	(1·47)	0·910	0·909	0·910	0·912	0·915	0·921	0·927	0·932	0·943	0·953	0·963	0·972	0·989	1·002
0·20000	(1·96)	0·897	0·897	0·899	0·901	0·905	0·911	0·918	0·924	0·935	0·945	0·956	0·966	0·983	0·997
0·30000	(2·94)	0·880	0·881	0·884	0·888	0·892	0·899	0·907	0·913	0·926	0·936	0·948	0·958	0·975	0·990
0·40000	(3·92)	0·869	0·871	0·874	0·879	0·883	0·892	0·899	0·907	0·919	0·930	0·942	0·953	0·971	0·985
0·60000	(5·88)	0·855	0·857	0·862	0·867	0·872	0·881	0·890	0·898	0·911	0·923	0·935	0·946	0·965	0·980
0·80000	(7·85)	0·846	0·848	0·854	0·860	0·865	0·875	0·884	0·892	0·906	0·918	0·931	0·942	0·961	0·977
1·00000	(9·81)	0·839	0·842	0·848	0·854	0·860	0·870	0·880	0·888	0·902	0·915	0·928	0·939	0·958	0·974
1·50000	(14·7)	0·828	0·831	0·838	0·845	0·851	0·863	0·873	0·881	0·896	0·909	0·923	0·934	0·954	0·970
2·00000	(19·6)	0·820	0·824	0·832	0·839	0·846	0·858	0·868	0·877	0·893	0·906	0·919	0·931	0·951	0·968
3·00000	(29·4)	0·811	0·816	0·824	0·832	0·839	0·852	0·863	0·872	0·888	0·901	0·916	0·928	0·948	0·965
S	$(\Delta p/\rho l)$	125	150	200	250	300	400	500	600	800	1000	1250	1500	2000	2500

Grad'nt k.p.g. (Equivalent) Pipe diameters in mm **Roughness size, k_s = 0·030 mm**

E188

Kin. visc., $\nu = 8{\cdot}0 \times 10^{-6}$ m²s⁻¹;
S = 0·00010 to 3·00000

i.e. kin. pr. grad., $\Delta p/\rho l$ =
(0·00098) to (29·4) ms⁻²

Roughness size, k_s = 0·060 mm
This table shows values of m, as follows

m_C for Colebrook-White solutions; or,
where $\mathbf{R} \le 2000$, m_P for laminar flow

Grad'nt S	k.p.g. $(\Delta p/\rho l)$	\(Equivalent) Pipe diameters in mm												
		125	150	200	250	300	400	500	600	800	1000	1250	1500	2000	2500
0·00010	(0·00098)	1·658	1·300	*	1·462	1·437	1·405	1·385	1·372	1·356	1·346	1·339	1·336	1·333	1·333
0·00015	(0·00147)	1·353	*	1·443	1·412	1·390	1·362	1·345	1·334	1·320	1·313	1·308	1·305	1·304	1·305
0·00020	(0·00196)	1·172	*	1·407	1·378	1·359	1·334	1·319	1·309	1·297	1·290	1·286	1·284	1·284	1·286
0·00030	(0·00294)	*	1·398	1·359	1·334	1·317	1·296	1·283	1·274	1·265	1·260	1·257	1·257	1·258	1·261
0·00040	(0·00392)	*	1·363	1·327	1·304	1·289	1·270	1·259	1·251	1·243	1·240	1·238	1·238	1·240	1·244
0·00060	(0·00588)	1·340	1·316	1·284	1·265	1·252	1·236	1·226	1·221	1·214	1·212	1·212	1·213	1·217	1·221
0·00080	(0·00785)	1·306	1·285	1·256	1·239	1·227	1·213	1·205	1·200	1·195	1·194	1·194	1·196	1·201	1·206
0·00100	(0·00981)	1·282	1·261	1·235	1·219	1·209	1·196	1·188	1·184	1·181	1·180	1·181	1·183	1·189	1·195
0·00150	(0·0147)	1·239	1·222	1·199	1·186	1·177	1·166	1·161	1·158	1·156	1·156	1·158	1·161	1·168	1·175
0·00200	(0·0196)	1·211	1·195	1·175	1·163	1·155	1·146	1·142	1·140	1·139	1·140	1·143	1·147	1·154	1·162
0·00300	(0·0294)	1·174	1·160	1·143	1·133	1·127	1·120	1·117	1·116	1·116	1·119	1·123	1·127	1·136	1·144
0·00400	(0·0392)	1·149	1·137	1·122	1·113	1·108	1·102	1·100	1·100	1·101	1·105	1·109	1·114	1·124	1·133
0·00600	(0·0588)	1·117	1·106	1·093	1·086	1·082	1·079	1·078	1·078	1·082	1·086	1·091	1·097	1·108	1·117
0·00800	(0·0785)	1·095	1·086	1·075	1·069	1·065	1·063	1·063	1·064	1·068	1·073	1·079	1·086	1·097	1·107
0·01000	(0·0981)	1·079	1·070	1·061	1·055	1·053	1·051	1·052	1·054	1·059	1·064	1·071	1·077	1·089	1·100
0·01500	(0·147)	1·051	1·044	1·037	1·033	1·031	1·031	1·033	1·036	1·042	1·048	1·056	1·063	1·076	1·087
0·02000	(0·196)	1·033	1·027	1·021	1·018	1·017	1·018	1·021	1·024	1·031	1·038	1·046	1·054	1·067	1·079
0·03000	(0·294)	1·008	1·004	1·000	0·998	0·999	1·001	1·005	1·009	1·017	1·025	1·034	1·042	1·056	1·069
0·04000	(0·392)	0·992	0·989	0·986	0·985	0·986	0·990	0·994	0·999	1·007	1·016	1·025	1·034	1·049	1·062
0·06000	(0·588)	0·972	0·969	0·968	0·968	0·970	0·975	0·980	0·985	0·995	1·005	1·015	1·024	1·040	1·054
0·08000	(0·785)	0·958	0·956	0·956	0·957	0·960	0·965	0·971	0·977	0·988	0·997	1·008	1·018	1·034	1·049
0·10000	(0·981)	0·948	0·947	0·947	0·949	0·952	0·958	0·965	0·971	0·982	0·992	1·003	1·013	1·030	1·045
0·15000	(1·47)	0·931	0·931	0·933	0·936	0·939	0·947	0·954	0·961	0·973	0·984	0·995	1·006	1·024	1·039
0·20000	(1·96)	0·920	0·920	0·923	0·927	0·931	0·939	0·947	0·954	0·967	0·978	0·991	1·001	1·020	1·035
0·30000	(2·94)	0·905	0·907	0·911	0·916	0·921	0·930	0·938	0·946	0·960	0·972	0·984	0·995	1·014	1·030
0·40000	(3·92)	0·896	0·898	0·903	0·909	0·914	0·924	0·933	0·941	0·955	0·967	0·980	0·992	1·011	1·027
0·60000	(5·88)	0·884	0·887	0·893	0·900	0·906	0·916	0·926	0·935	0·949	0·962	0·976	0·987	1·007	1·024
0·80000	(7·85)	0·877	0·880	0·887	0·894	0·900	0·912	0·922	0·931	0·946	0·959	0·973	0·985	1·005	1·021
1·00000	(9·81)	0·872	0·875	0·883	0·890	0·897	0·908	0·919	0·928	0·943	0·957	0·971	0·983	1·003	1·020
1·50000	(14·7)	0·863	0·867	0·876	0·883	0·890	0·903	0·914	0·923	0·939	0·953	0·967	0·980	1·000	1·017
2·00000	(19·6)	0·857	0·862	0·871	0·879	0·887	0·900	0·911	0·920	0·937	0·951	0·965	0·978	0·999	1·016
3·00000	(29·4)	0·851	0·856	0·866	0·874	0·882	0·896	0·907	0·917	0·934	0·948	0·963	0·976	0·997	1·014
S	$(\Delta p/\rho l)$	125	150	200	250	300	400	500	600	800	1000	1250	1500	2000	2500

Grad'nt k.p.g. (Equivalent) Pipe diameters in mm **Roughness size, k_s = 0·060 mm**

Kin. visc., $\nu = 8{\cdot}0\times10^{-6}$ m²s⁻¹;
$S = 0{\cdot}00010$ to $3{\cdot}00000$

i.e. kin. pr. grad., $\Delta p/\rho l =$
$(0{\cdot}00098)$ to $(29{\cdot}4)$ ms⁻²

Roughness size, $k_s = 0{\cdot}150$ mm
This table shows values of m, as follows

m_C for Colebrook-White solutions; or,
where $\mathbf{R} \le 2000$, m_P for laminar flow

Grad'nt S	k.p.g. $(\Delta p/\rho l)$	125	150	200	250	300	400	500	600	800	1000	1250	1500	2000	2500
						(Equivalent) Pipe diameters in mm									
0·00010	(0·00098)	1·658	1·300	*	1·469	1·444	1·412	1·393	1·380	1·364	1·355	1·349	1·346	1·343	1·344
0·00015	(0·00147)	1·353	*	1·450	1·419	1·398	1·371	1·354	1·343	1·330	1·323	1·319	1·317	1·316	1·318
0·00020	(0·00196)	1·172	*	1·415	1·387	1·367	1·343	1·328	1·319	1·307	1·302	1·298	1·297	1·298	1·301
0·00030	(0·00294)	*	1·407	1·368	1·344	1·327	1·306	1·294	1·286	1·277	1·273	1·271	1·271	1·274	1·278
0·00040	(0·00392)	*	1·372	1·337	1·315	1·300	1·281	1·271	1·264	1·257	1·254	1·253	1·254	1·258	1·263
0·00060	(0·00588)	1·351	1·327	1·296	1·277	1·264	1·249	1·240	1·235	1·230	1·229	1·229	1·231	1·237	1·243
0·00080	(0·00785)	1·318	1·297	1·269	1·252	1·241	1·227	1·220	1·216	1·212	1·212	1·213	1·216	1·222	1·229
0·00100	(0·00981)	1·294	1·274	1·249	1·233	1·223	1·211	1·205	1·201	1·199	1·199	1·202	1·205	1·212	1·219
0·00150	(0·0147)	1·254	1·237	1·215	1·202	1·193	1·184	1·179	1·177	1·176	1·178	1·182	1·186	1·194	1·203
0·00200	(0·0196)	1·227	1·211	1·192	1·181	1·173	1·166	1·162	1·161	1·162	1·164	1·169	1·173	1·183	1·192
0·00300	(0·0294)	1·192	1·179	1·162	1·153	1·147	1·142	1·140	1·140	1·142	1·146	1·152	1·157	1·168	1·178
0·00400	(0·0392)	1·169	1·157	1·143	1·135	1·130	1·126	1·125	1·126	1·130	1·134	1·140	1·147	1·158	1·169
0·00600	(0·0588)	1·138	1·129	1·117	1·111	1·108	1·106	1·106	1·108	1·113	1·119	1·126	1·133	1·146	1·158
0·00800	(0·0785)	1·118	1·110	1·100	1·095	1·093	1·092	1·094	1·096	1·102	1·109	1·117	1·124	1·138	1·150
0·01000	(0·0981)	1·104	1·096	1·088	1·084	1·082	1·083	1·085	1·088	1·095	1·102	1·110	1·118	1·132	1·145
0·01500	(0·147)	1·079	1·073	1·067	1·065	1·065	1·066	1·070	1·074	1·082	1·090	1·099	1·108	1·123	1·136
0·02000	(0·196)	1·063	1·058	1·054	1·053	1·053	1·056	1·060	1·065	1·074	1·082	1·092	1·101	1·117	1·131
0·03000	(0·294)	1·043	1·039	1·037	1·037	1·038	1·042	1·048	1·053	1·063	1·073	1·084	1·093	1·110	1·124
0·04000	(0·392)	1·029	1·027	1·025	1·026	1·029	1·034	1·040	1·046	1·057	1·067	1·078	1·088	1·105	1·120
0·06000	(0·588)	1·012	1·011	1·011	1·013	1·016	1·023	1·030	1·037	1·049	1·059	1·071	1·082	1·100	1·158
0·08000	(0·785)	1·001	1·001	1·002	1·005	1·009	1·016	1·024	1·031	1·043	1·055	1·067	1·078	1·096	1·112
0·10000	(0·981)	0·994	0·994	0·996	0·999	1·003	1·012	1·019	1·027	1·040	1·051	1·064	1·075	1·094	1·110
0·15000	(1·47)	0·981	0·982	0·985	0·990	0·995	1·004	1·012	1·020	1·034	1·046	1·059	1·071	1·090	1·106
0·20000	(1·96)	0·973	0·974	0·979	0·984	0·989	0·999	1·008	1·016	1·031	1·043	1·056	1·068	1·088	1·104
0·30000	(2·94)	0·962	0·965	0·970	0·976	0·982	0·993	1·002	1·011	1·026	1·039	1·053	1·065	1·085	1·102
0·40000	(3·92)	0·956	0·959	0·965	0·972	0·978	0·989	0·999	1·008	1·023	1·037	1·051	1·063	1·083	1·100
0·60000	(5·88)	0·948	0·952	0·959	0·966	0·973	0·985	0·995	1·004	1·020	1·034	1·048	1·060	1·081	1·098
0·80000	(7·85)	0·943	0·947	0·955	0·963	0·970	0·982	0·993	1·002	1·018	1·032	1·046	1·059	1·080	1·097
1·00000	(9·81)	0·940	0·944	0·953	0·960	0·967	0·980	0·991	1·001	1·017	1·031	1·045	1·058	1·079	1·096
1·50000	(14·7)	0·935	0·939	0·948	0·956	0·964	0·977	0·988	0·998	1·015	1·029	1·044	1·056	1·078	1·095
2·00000	(19·6)	0·931	0·936	0·946	0·954	0·962	0·975	0·987	0·997	1·013	1·028	1·043	1·055	1·077	1·094
3·00000	(29·4)	0·927	0·933	0·942	0·951	0·959	0·973	0·985	0·995	1·012	1·026	1·041	1·054	1·076	1·093
S	$(\Delta p/\rho l)$	125	150	200	250	300	400	500	600	800	1000	1250	1500	2000	2500

Grad'nt   k.p.g.   (Equivalent) Pipe diameters in mm   **Roughness size, $k_s = 0{\cdot}150$ mm**

Kin. visc., $\nu = 8{\cdot}0\times10^{-6}$ m²s⁻¹;
$S = 0{\cdot}00010$ to $3{\cdot}00000$

i.e. kin. pr. grad., $\Delta p/\rho l =$
$(0{\cdot}00098)$ to $(29{\cdot}4)$ ms⁻²

Roughness size, $k_s = 0{\cdot}30$ mm
This table shows values of m, as follows

m_C for Colebrook-White solutions; or,
where $\mathbf{R} \le 2000$, m_P for laminar flow

Grad'nt S	k.p.g. $(\Delta p/\rho l)$	125	150	200	250	300	400	500	600	800	1000	1250	1500	2000	2500
						(Equivalent) Pipe diameters in mm									
0·00010	(0·00098)	1·658	1·300	*	1·480	1·455	1·424	1·405	1·393	1·377	1·369	1·364	1·361	1·360	1·362
0·00015	(0·00147)	1·353	*	1·463	1·432	1·411	1·384	1·368	1·358	1·345	1·339	1·335	1·334	1·335	1·339
0·00020	(0·00196)	1·172	*	1·428	1·400	1·382	1·358	1·343	1·334	1·324	1·319	1·317	1·316	1·319	1·323
0·00030	(0·00294)	*	1·422	1·383	1·359	1·343	1·323	1·311	1·304	1·296	1·293	1·292	1·293	1·297	1·303
0·00040	(0·00392)	*	1·388	1·353	1·332	1·317	1·300	1·290	1·284	1·278	1·276	1·276	1·278	1·283	1·290
0·00060	(0·00588)	1·368	1·345	1·314	1·296	1·284	1·269	1·262	1·257	1·253	1·253	1·255	1·258	1·265	1·273
0·00080	(0·00785)	1·337	1·316	1·289	1·273	1·262	1·250	1·243	1·240	1·238	1·238	1·241	1·245	1·253	1·262
0·00100	(0·00981)	1·315	1·295	1·270	1·255	1·246	1·235	1·230	1·227	1·226	1·228	1·231	1·236	1·245	1·254
0·00150	(0·0147)	1·276	1·260	1·239	1·226	1·219	1·210	1·207	1·206	1·207	1·210	1·215	1·220	1·230	1·240
0·00200	(0·0196)	1·251	1·236	1·218	1·207	1·201	1·194	1·192	1·192	1·194	1·198	1·204	1·210	1·221	1·232
0·00300	(0·0294)	1·219	1·207	1·191	1·183	1·178	1·174	1·173	1·174	1·178	1·183	1·190	1·197	1·210	1·221
0·00400	(0·0392)	1·198	1·187	1·174	1·167	1·163	1·161	1·161	1·163	1·168	1·174	1·182	1·189	1·203	1·215
0·00600	(0·0588)	1·171	1·162	1·152	1·147	1·144	1·144	1·145	1·148	1·155	1·162	1·171	1·179	1·193	1·206
0·00800	(0·0785)	1·154	1·146	1·137	1·134	1·132	1·133	1·135	1·139	1·147	1·154	1·164	1·172	1·188	1·201
0·01000	(0·0981)	1·141	1·134	1·127	1·124	1·123	1·125	1·128	1·132	1·141	1·149	1·159	1·168	1·184	1·198
0·01500	(0·147)	1·120	1·115	1·110	1·109	1·109	1·112	1·117	1·122	1·131	1·140	1·151	1·161	1·177	1·192
0·02000	(0·196)	1·107	1·103	1·099	1·099	1·100	1·104	1·109	1·115	1·125	1·135	1·146	1·156	1·173	1·188
0·03000	(0·294)	1·090	1·087	1·085	1·086	1·089	1·094	1·100	1·106	1·118	1·128	1·140	1·151	1·169	1·184
0·04000	(0·392)	1·079	1·077	1·077	1·079	1·081	1·088	1·095	1·101	1·113	1·124	1·136	1·147	1·166	1·181
0·06000	(0·588)	1·066	1·065	1·066	1·069	1·073	1·080	1·088	1·095	1·108	1·119	1·132	1·143	1·162	1·178
0·08000	(0·785)	1·058	1·057	1·059	1·063	1·067	1·076	1·084	1·091	1·105	1·116	1·129	1·140	1·160	1·176
0·10000	(0·981)	1·052	1·052	1·055	1·059	1·063	1·072	1·081	1·088	1·102	1·114	1·127	1·139	1·158	1·175
0·15000	(1·47)	1·042	1·043	1·047	1·052	1·057	1·067	1·076	1·084	1·098	1·111	1·124	1·136	1·156	1·172
0·20000	(1·96)	1·036	1·038	1·043	1·048	1·053	1·064	1·073	1·081	1·096	1·109	1·122	1·134	1·154	1·171
0·30000	(2·94)	1·029	1·031	1·037	1·043	1·049	1·060	1·069	1·078	1·093	1·106	1·120	1·132	1·153	1·170
0·40000	(3·92)	1·025	1·027	1·034	1·040	1·046	1·057	1·067	1·076	1·092	1·105	1·119	1·131	1·152	1·169
0·60000	(5·88)	1·019	1·023	1·030	1·036	1·043	1·054	1·065	1·074	1·090	1·103	1·117	1·130	1·150	1·168
0·80000	(7·85)	1·016	1·020	1·027	1·034	1·041	1·053	1·063	1·073	1·088	1·102	1·116	1·129	1·150	1·167
1·00000	(9·81)	1·014	1·018	1·025	1·033	1·039	1·052	1·062	1·072	1·088	1·101	1·116	1·128	1·149	1·166
1·50000	(14·7)	1·010	1·014	1·022	1·030	1·037	1·050	1·060	1·070	1·086	1·100	1·115	1·127	1·148	1·166
2·00000	(19·6)	1·008	1·012	1·021	1·029	1·036	1·049	1·059	1·069	1·086	1·099	1·114	1·127	1·148	1·165
3·00000	(29·4)	1·005	1·010	1·019	1·027	1·034	1·047	1·058	1·068	1·085	1·099	1·113	1·126	1·147	1·165
S	$(\Delta p/\rho l)$	125	150	200	250	300	400	500	600	800	1000	1250	1500	2000	2500

Grad'nt   k.p.g.   (Equivalent) Pipe diameters in mm   **Roughness size, $k_s = 0{\cdot}30$ mm**

E191

Kin. visc., $\nu = 8{\cdot}0\times10^{-6}\ \mathrm{m^2 s^{-1}}$;
S = 0·00010 to 3·00000

i.e. kin. pr. grad., $\Delta p/\rho l$ = (0·00098) to (29·4) ms^{-2}

Roughness size, k_s = 0·60 mm
This table shows values of m, as follows

m_C for Colebrook-White solutions; or, where $\mathbf{R} \leq 2000$, m_P for laminar flow

Grad'nt S	k.p.g. ($\Delta p/\rho l$)	(Equivalent) Pipe diameters in mm													
		125	150	200	250	300	400	500	600	800	1000	1250	1500	2000	2500
0·00010	(0·00098)	1·658	1·300	*	1·501	1·477	1·447	1·428	1·417	1·403	1·395	1·391	1·389	1·390	1·393
0·00015	(0·00147)	1·353	*	1·486	1·456	1·435	1·409	1·394	1·384	1·374	1·368	1·366	1·366	1·369	1·373
0·00020	(0·00196)	1·172	*	1·454	1·426	1·408	1·385	1·372	1·363	1·355	1·351	1·350	1·351	1·355	1·361
0·00030	(0·00294)	*	1·450	1·411	1·388	1·372	1·353	1·343	1·336	1·330	1·328	1·329	1·331	1·337	1·344
0·00040	(0·00392)	*	1·418	1·384	1·363	1·349	1·333	1·324	1·319	1·314	1·314	1·315	1·318	1·326	1·334
0·00060	(0·00588)	1·401	1·378	1·348	1·331	1·319	1·306	1·299	1·296	1·294	1·295	1·298	1·302	1·311	1·321
0·00080	(0·00785)	1·373	1·352	1·326	1·310	1·300	1·289	1·284	1·281	1·281	1·283	1·287	1·292	1·302	1·312
0·00100	(0·00981)	1·352	1·333	1·309	1·295	1·286	1·277	1·272	1·271	1·271	1·274	1·279	1·285	1·296	1·306
0·00150	(0·0147)	1·318	1·301	1·281	1·270	1·263	1·256	1·254	1·253	1·256	1·260	1·267	1·273	1·285	1·297
0·00200	(0·0196)	1·296	1·281	1·264	1·254	1·248	1·243	1·242	1·242	1·246	1·252	1·259	1·266	1·279	1·291
0·00300	(0·0294)	1·268	1·255	1·241	1·233	1·229	1·226	1·227	1·229	1·234	1·241	1·249	1·256	1·271	1·283
0·00400	(0·0392)	1·250	1·239	1·227	1·221	1·218	1·216	1·217	1·220	1·227	1·234	1·242	1·251	1·266	1·279
0·00600	(0·0588)	1·227	1·218	1·209	1·204	1·203	1·203	1·206	1·209	1·217	1·225	1·235	1·244	1·259	1·273
0·00800	(0·0785)	1·213	1·205	1·197	1·194	1·193	1·195	1·198	1·203	1·211	1·220	1·230	1·239	1·256	1·270
0·01000	(0·0981)	1·203	1·196	1·189	1·187	1·187	1·189	1·193	1·198	1·207	1·216	1·227	1·236	1·253	1·267
0·01500	(0·147)	1·186	1·181	1·176	1·175	1·176	1·180	1·185	1·190	1·201	1·210	1·221	1·231	1·249	1·264
0·02000	(0·196)	1·176	1·171	1·168	1·168	1·170	1·174	1·180	1·186	1·197	1·207	1·218	1·228	1·246	1·261
0·03000	(0·294)	1·163	1·160	1·158	1·159	1·162	1·167	1·174	1·180	1·192	1·202	1·214	1·225	1·243	1·259
0·04000	(0·392)	1·155	1·153	1·152	1·154	1·157	1·163	1·170	1·177	1·189	1·200	1·212	1·223	1·241	1·257
0·06000	(0·588)	1·145	1·144	1·145	1·147	1·151	1·158	1·165	1·172	1·185	1·197	1·209	1·220	1·239	1·255
0·08000	(0·785)	1·139	1·139	1·140	1·143	1·147	1·155	1·163	1·170	1·183	1·195	1·207	1·219	1·238	1·254
0·10000	(0·981)	1·135	1·135	1·137	1·140	1·144	1·153	1·161	1·168	1·182	1·193	1·206	1·218	1·237	1·253
0·15000	(1·47)	1·129	1·129	1·132	1·136	1·140	1·149	1·158	1·165	1·179	1·191	1·204	1·216	1·235	1·252
0·20000	(1·96)	1·125	1·125	1·129	1·133	1·138	1·147	1·156	1·164	1·178	1·190	1·203	1·215	1·235	1·251
0·30000	(2·94)	1·120	1·121	1·125	1·130	1·135	1·145	1·154	1·162	1·176	1·189	1·202	1·214	1·234	1·250
0·40000	(3·92)	1·117	1·118	1·123	1·128	1·133	1·143	1·152	1·161	1·175	1·188	1·201	1·213	1·233	1·250
0·60000	(5·88)	1·113	1·115	1·120	1·126	1·131	1·142	1·151	1·159	1·174	1·187	1·200	1·212	1·232	1·249
0·80000	(7·85)	1·111	1·113	1·119	1·124	1·130	1·140	1·150	1·158	1·173	1·186	1·200	1·212	1·232	1·249
1·00000	(9·81)	1·110	1·112	1·118	1·123	1·129	1·140	1·149	1·158	1·173	1·186	1·199	1·211	1·232	1·248
1·50000	(14·7)	1·107	1·110	1·116	1·122	1·128	1·139	1·148	1·157	1·172	1·185	1·199	1·211	1·231	1·248
2·00000	(19·6)	1·106	1·109	1·115	1·121	1·127	1·138	1·148	1·156	1·172	1·184	1·198	1·210	1·231	1·248
3·00000	(29·4)	1·104	1·107	1·114	1·120	1·126	1·137	1·147	1·156	1·171	1·184	1·198	1·210	1·230	1·247
S ($\Delta p/\rho l$)		125	150	200	250	300	400	500	600	800	1000	1250	1500	2000	2500

Grad'nt k.p.g. (Equivalent) Pipe diameters in mm — Roughness size, k_s = 0·60 mm

E192

Kin. visc., $\nu = 8{\cdot}0\times10^{-6}\ \mathrm{m^2 s^{-1}}$;
S = 0·00010 to 3·00000

i.e. kin. pr. grad., $\Delta p/\rho l$ = (0·00098) to (29·4) ms^{-2}

Roughness size, k_s = 1·50 mm
This table shows values of m, as follows

m_C for Colebrook-White solutions; or, where $\mathbf{R} \leq 2000$, m_P for laminar flow

Grad'nt S	k.p.g. ($\Delta p/\rho l$)	(Equivalent) Pipe diameters in mm													
		125	150	200	250	300	400	500	600	800	1000	1250	1500	2000	2500
0·00010	(0·00098)	1·658	1·300	*	1·560	1·536	1·507	1·490	1·479	1·467	1·461	1·458	1·458	1·462	1·467
0·00015	(0·00147)	1·353	*	1·550	1·520	1·500	1·475	1·461	1·453	1·444	1·440	1·440	1·441	1·446	1·453
0·00020	(0·00196)	1·172	*	1·522	1·495	1·477	1·455	1·443	1·436	1·429	1·427	1·428	1·430	1·436	1·444
0·00030	(0·00294)	*	*	1·486	1·463	1·448	1·430	1·421	1·415	1·411	1·411	1·413	1·416	1·424	1·433
0·00040	(0·00392)	*	1·498	1·463	1·443	1·430	1·414	1·406	1·402	1·399	1·400	1·403	1·408	1·417	1·426
0·00060	(0·00588)	1·488	1·464	1·435	1·417	1·406	1·394	1·388	1·386	1·385	1·387	1·392	1·397	1·408	1·418
0·00080	(0·00785)	1·464	1·443	1·417	1·401	1·392	1·381	1·377	1·375	1·376	1·379	1·385	1·390	1·402	1·413
0·00100	(0·00981)	1·448	1·428	1·404	1·390	1·381	1·373	1·369	1·368	1·370	1·374	1·380	1·386	1·398	1·409
0·00150	(0·0147)	1·420	1·404	1·383	1·372	1·365	1·358	1·356	1·356	1·360	1·365	1·372	1·379	1·392	1·404
0·00200	(0·0196)	1·403	1·388	1·370	1·360	1·354	1·349	1·348	1·349	1·354	1·359	1·367	1·374	1·388	1·400
0·00300	(0·0294)	1·382	1·369	1·354	1·346	1·342	1·338	1·339	1·341	1·346	1·353	1·361	1·369	1·383	1·396
0·00400	(0·0392)	1·369	1·358	1·344	1·337	1·334	1·332	1·333	1·335	1·342	1·349	1·357	1·365	1·380	1·394
0·00600	(0·0588)	1·353	1·343	1·332	1·327	1·324	1·324	1·326	1·329	1·336	1·344	1·353	1·362	1·377	1·391
0·00800	(0·0785)	1·343	1·334	1·324	1·320	1·318	1·319	1·321	1·325	1·333	1·341	1·350	1·359	1·375	1·389
0·01000	(0·0981)	1·336	1·328	1·319	1·315	1·314	1·315	1·318	1·322	1·330	1·339	1·348	1·357	1·374	1·388
0·01500	(0·147)	1·325	1·318	1·311	1·308	1·308	1·310	1·313	1·318	1·327	1·335	1·346	1·355	1·371	1·386
0·02000	(0·196)	1·319	1·312	1·306	1·304	1·304	1·306	1·310	1·315	1·324	1·333	1·344	1·353	1·370	1·385
0·03000	(0·294)	1·311	1·305	1·300	1·299	1·299	1·302	1·307	1·312	1·322	1·331	1·342	1·351	1·369	1·383
0·04000	(0·392)	1·306	1·301	1·296	1·295	1·296	1·300	1·305	1·310	1·320	1·330	1·341	1·350	1·368	1·382
0·06000	(0·588)	1·300	1·296	1·292	1·292	1·293	1·297	1·302	1·308	1·318	1·328	1·339	1·349	1·366	1·381
0·08000	(0·785)	1·296	1·292	1·289	1·289	1·291	1·295	1·301	1·306	1·317	1·327	1·338	1·348	1·366	1·381
0·10000	(0·981)	1·294	1·290	1·288	1·288	1·289	1·294	1·300	1·305	1·316	1·326	1·338	1·348	1·365	1·380
0·15000	(1·47)	1·290	1·287	1·285	1·285	1·287	1·292	1·298	1·304	1·315	1·325	1·337	1·347	1·365	1·380
0·20000	(1·96)	1·288	1·285	1·283	1·284	1·286	1·291	1·297	1·303	1·314	1·325	1·336	1·346	1·364	1·379
0·30000	(2·94)	1·285	1·282	1·281	1·282	1·284	1·290	1·296	1·302	1·314	1·324	1·335	1·346	1·364	1·379
0·40000	(3·92)	1·283	1·281	1·280	1·281	1·283	1·289	1·295	1·301	1·313	1·323	1·335	1·345	1·363	1·379
0·60000	(5·88)	1·281	1·279	1·278	1·280	1·282	1·288	1·295	1·301	1·312	1·323	1·335	1·345	1·363	1·378
0·80000	(7·85)	1·280	1·278	1·277	1·279	1·282	1·288	1·294	1·300	1·312	1·322	1·334	1·345	1·363	1·378
1·00000	(9·81)	1·279	1·277	1·277	1·278	1·281	1·287	1·294	1·300	1·312	1·322	1·334	1·344	1·363	1·378
1·50000	(14·7)	1·278	1·276	1·276	1·278	1·280	1·287	1·293	1·300	1·311	1·322	1·334	1·344	1·362	1·378
2·00000	(19·6)	1·277	1·275	1·275	1·277	1·280	1·286	1·293	1·299	1·311	1·322	1·334	1·344	1·362	1·378
3·00000	(29·4)	1·276	1·275	1·275	1·277	1·279	1·286	1·293	1·299	1·311	1·321	1·333	1·344	1·362	1·377
S ($\Delta p/\rho l$)		125	150	200	250	300	400	500	600	800	1000	1250	1500	2000	2500

Grad'nt k.p.g. (Equivalent) Pipe diameters in mm — Roughness size, k_s = 1·50 mm

Kin. visc., $\nu = 10{\cdot}0 \times 10^{-6}\ \mathrm{m^2 s^{-1}}$;
$S = 0{\cdot}00010$ to $30{\cdot}0000$

i.e. kin. pr. grad., $\Delta p/\rho l =$
$(0{\cdot}00098)$ to $(294)\ \mathrm{ms^{-2}}$

Roughness size, $k_\mathrm{s} = 0{\cdot}003$ mm
This table shows values of m, as follows

m_C for Colebrook-White solutions; or,
where $\mathbf{R} \le 2000$, m_P for laminar flow

Grad'nt S	k.p.g. $(\Delta p/\rho l)$	6	8	10	12·5	15	20	25	30	40	50	60	80	100	125
		(Equivalent) Pipe diameters in mm													
0·00030	(0·00294)	68·57	46·73	34·70	25·77	20·21	13·77	10·23	8·020	5·465	4·059	3·183	2·169	1·611	1·196
0·00040	(0·00392)	59·39	40·47	30·05	22·32	17·50	11·93	8·857	6·946	4·733	3·515	2·756	1·878	1·395	*
0·00060	(0·00588)	48·49	33·04	24·54	18·22	14·29	9·738	7·232	5·671	3·865	2·870	2·251	1·534	1·139	*
0·00080	(0·00785)	41·99	28·61	21·25	15·78	12·38	8·433	6·263	4·912	3·347	2·486	1·949	1·328	*	1·352
0·00100	(0·00981)	37·56	25·59	19·01	14·12	11·07	7·543	5·602	4·393	2·993	2·223	1·743	1·188	*	1·325
0·00150	(0·0147)	30·67	20·90	15·52	11·53	9·038	6·159	4·574	3·587	2·444	1·815	1·423	*	1·308	1·278
0·00200	(0·0196)	26·56	18·10	13·44	9·981	7·827	5·334	3·961	3·106	2·117	1·572	1·233	*	1·274	1·247
0·00300	(0·0294)	21·69	14·78	10·97	8·150	6·391	4·355	3·234	2·536	1·728	1·284	1·007	1·257	1·229	1·205
0·00400	(0·0392)	18·78	12·80	9·504	7·058	5·535	3·772	2·801	2·196	1·497	1·112	*	1·225	1·199	1·177
0·00600	(0·0588)	15·33	10·45	7·760	5·763	4·519	3·079	2·287	1·793	1·222	*	*	1·182	1·159	1·140
0·00800	(0·0785)	13·28	9·049	6·720	4·991	3·914	2·667	1·981	1·553	1·058	*	1·186	1·153	1·132	1·115
0·01000	(0·0981)	11·88	8·093	6·011	4·464	3·501	2·385	1·771	1·389	0·947	*	1·162	1·131	1·112	1·096
0·01500	(0·147)	9·698	6·608	4·908	3·645	2·858	1·948	1·446	1·134	*	1·142	1·121	1·094	1·078	1·064
0·02000	(0·196)	8·399	5·723	4·250	3·156	2·475	1·687	1·253	0·982	*	1·113	1·094	1·070	1·055	1·042
0·03000	(0·294)	6·857	4·673	3·470	2·577	2·021	1·377	1·023	*	1·098	1·074	1·058	1·037	1·024	1·013
0·04000	(0·392)	5·939	4·047	3·005	2·232	1·750	1·193	0·886	*	1·070	1·049	1·034	1·015	1·003	0·994
0·06000	(0·588)	4·849	3·304	2·454	1·822	1·429	0·974	*	1·063	1·033	1·014	1·001	0·985	0·975	0·967
0·08000	(0·785)	4·199	2·861	2·125	1·578	1·238	0·843	*	1·036	1·009	0·992	0·980	0·965	0·956	0·949
0·10000	(0·981)	3·756	2·559	1·901	1·412	1·107	*	1·035	1·016	0·990	0·974	0·964	0·950	0·942	0·936
0·15000	(1·47)	3·067	2·090	1·552	1·153	0·904	*	0·998	0·981	0·959	0·945	0·936	0·924	0·917	0·913
0·20000	(1·96)	2·656	1·810	1·344	0·998	*	0·996	0·973	0·958	0·938	0·925	0·917	0·907	0·901	0·897
0·30000	(2·94)	2·169	1·478	1·097	0·815	*	0·960	0·940	0·927	0·909	0·899	0·892	0·883	0·879	0·875
0·40000	(3·92)	1·878	1·280	0·950	*	*	0·936	0·918	0·906	0·890	0·881	0·875	0·867	0·863	0·861
0·60000	(5·88)	1·533	1·045	0·776	*	0·929	0·905	0·889	0·878	0·865	0·857	0·852	0·846	0·843	0·842
0·80000	(7·85)	1·328	0·905	*	0·923	0·906	0·884	0·870	0·860	0·848	0·841	0·836	0·831	0·829	0·828
1·00000	(9·81)	1·188	0·809	*	0·905	0·889	0·868	0·855	0·846	0·835	0·829	0·825	0·821	0·819	0·819
1·50000	(14·7)	0·970	*	0·894	0·873	0·859	0·841	0·830	0·822	0·813	0·808	0·805	0·802	0·801	0·801
2·00000	(19·6)	0·840	*	0·871	0·852	0·839	0·823	0·813	0·806	0·798	0·794	0·791	0·789	0·789	0·790
3·00000	(29·4)	0·686	0·860	0·840	0·824	0·813	0·799	0·790	0·785	0·778	0·775	0·773	0·772	0·772	0·774
4·00000	(39·2)	*	0·838	0·820	0·806	0·796	0·783	0·775	0·770	0·765	0·762	0·761	0·760	0·761	0·763
6·00000	(58·8)	*	0·809	0·794	0·781	0·772	0·761	0·755	0·751	0·747	0·745	0·744	0·745	0·746	0·749
8·00000	(78·5)	0·812	0·790	0·776	0·764	0·756	0·747	0·741	0·738	0·734	0·733	0·733	0·734	0·736	0·739
10·0000	(98·1)	0·796	0·775	0·762	0·752	0·745	0·736	0·731	0·728	0·725	0·725	0·725	0·726	0·729	0·732
15·0000	(147)	0·769	0·751	0·740	0·731	0·725	0·717	0·714	0·711	0·710	0·710	0·710	0·713	0·716	0·720
20·0000	(196)	0·751	0·735	0·725	0·716	0·711	0·705	0·702	0·700	0·699	0·700	0·701	0·704	0·707	0·711
30·0000	(294)	0·727	0·713	0·704	0·698	0·693	0·688	0·686	0·685	0·685	0·686	0·688	0·692	0·695	0·700
S	$(\Delta p/\rho l)$	6	8	10	12·5	15	20	25	30	40	50	60	80	100	125
Grad'nt	k.p.g.	(Equivalent) Pipe diameters in mm													

Grad'nt S	k.p.g. $(\Delta p/\rho l)$	125	150	200	250	300	400	500	600	800	1000	1250	1500	2000	2500
0·00010	(0·00098)	2·072	1·625	*	1·518	1·489	1·452	1·428	1·412	1·392	1·380	1·371	1·366	1·360	1·358
0·00015	(0·00147)	1·692	1·327	*	1·463	1·438	1·405	1·385	1·371	1·354	1·344	1·337	1·332	1·329	1·328
0·00020	(0·00196)	1·465	*	1·459	1·426	1·403	1·374	1·355	1·343	1·328	1·319	1·313	1·310	1·307	1·307
0·00030	(0·00294)	1·196	*	1·406	1·378	1·358	1·332	1·316	1·306	1·293	1·286	1·281	1·279	1·278	1·279
0·00040	(0·00392)	*	*	1·371	1·345	1·327	1·304	1·290	1·280	1·269	1·263	1·260	1·258	1·258	1·260
0·00060	(0·00588)	*	1·361	1·325	1·302	1·286	1·266	1·254	1·246	1·237	1·233	1·230	1·230	1·231	1·234
0·00080	(0·00785)	1·352	1·327	1·293	1·273	1·259	1·241	1·230	1·223	1·216	1·212	1·211	1·210	1·213	1·216
0·00100	(0·00981)	1·325	1·301	1·270	1·251	1·238	1·222	1·212	1·206	1·199	1·196	1·195	1·196	1·199	1·203
0·00150	(0·0147)	1·278	1·257	1·230	1·214	1·202	1·188	1·180	1·176	1·171	1·169	1·169	1·170	1·174	1·179
0·00200	(0·0196)	1·247	1·228	1·203	1·188	1·178	1·166	1·159	1·155	1·151	1·150	1·151	1·153	1·158	1·163
0·00300	(0·0294)	1·205	1·188	1·167	1·154	1·146	1·136	1·130	1·127	1·125	1·125	1·127	1·129	1·135	1·140
0·00400	(0·0392)	1·177	1·162	1·143	1·132	1·124	1·115	1·111	1·109	1·107	1·108	1·110	1·113	1·119	1·125
0·00600	(0·0588)	1·140	1·127	1·111	1·101	1·095	1·088	1·084	1·083	1·083	1·085	1·088	1·091	1·098	1·105
0·00800	(0·0785)	1·115	1·103	1·089	1·080	1·075	1·069	1·067	1·066	1·066	1·069	1·072	1·076	1·083	1·091
0·01000	(0·0981)	1·096	1·086	1·072	1·065	1·060	1·055	1·053	1·053	1·054	1·057	1·060	1·065	1·072	1·080
0·01500	(0·147)	1·064	1·055	1·044	1·037	1·034	1·030	1·029	1·030	1·032	1·035	1·040	1·044	1·053	1·061
0·02000	(0·196)	1·042	1·034	1·024	1·019	1·016	1·013	1·013	1·014	1·017	1·021	1·026	1·031	1·040	1·048
0·03000	(0·294)	1·013	1·006	0·998	0·994	0·992	0·991	0·991	0·993	0·997	1·001	1·007	1·012	1·022	1·030
0·04000	(0·392)	0·994	0·988	0·981	0·977	0·976	0·975	0·977	0·978	0·983	0·988	0·994	0·999	1·009	1·018
0·06000	(0·588)	0·967	0·962	0·957	0·955	0·954	0·954	0·956	0·959	0·964	0·970	0·976	0·982	0·993	1·002
0·08000	(0·785)	0·949	0·945	0·941	0·939	0·939	0·940	0·943	0·945	0·951	0·957	0·964	0·970	0·981	0·991
0·10000	(0·981)	0·936	0·932	0·929	0·928	0·928	0·930	0·932	0·935	0·942	0·948	0·955	0·961	0·973	0·982
0·15000	(1·47)	0·913	0·910	0·907	0·907	0·908	0·911	0·914	0·918	0·925	0·931	0·939	0·945	0·957	0·967
0·20000	(1·96)	0·897	0·895	0·893	0·893	0·895	0·898	0·902	0·906	0·913	0·920	0·928	0·935	0·947	0·957
0·30000	(2·94)	0·875	0·874	0·874	0·875	0·876	0·881	0·885	0·889	0·897	0·905	0·913	0·920	0·933	0·944
0·40000	(3·92)	0·861	0·860	0·860	0·862	0·864	0·869	0·874	0·878	0·887	0·894	0·903	0·910	0·923	0·934
0·60000	(5·88)	0·842	0·841	0·843	0·845	0·847	0·853	0·858	0·863	0·872	0·880	0·889	0·897	0·910	0·922
0·80000	(7·85)	0·828	0·829	0·831	0·833	0·836	0·842	0·848	0·853	0·862	0·871	0·880	0·888	0·902	0·913
1·00000	(9·81)	0·819	0·819	0·821	0·824	0·828	0·834	0·840	0·845	0·855	0·864	0·873	0·881	0·895	0·907
1·50000	(14·7)	0·801	0·802	0·806	0·809	0·813	0·820	0·826	0·832	0·842	0·851	0·861	0·869	0·884	0·896
2·00000	(19·6)	0·790	0·791	0·795	0·799	0·803	0·810	0·817	0·823	0·833	0·843	0·853	0·861	0·876	0·889
3·00000	(29·4)	0·774	0·776	0·780	0·785	0·789	0·797	0·804	0·811	0·822	0·831	0·842	0·851	0·866	0·879
S	$(\Delta p/\rho l)$	125	150	200	250	300	400	500	600	800	1000	1250	1500	2000	2500
Grad'nt	k.p.g.	(Equivalent) Pipe diameters in mm													

Kinematic viscosity, $\nu = 10 \times 10^{-6}\ \mathrm{m^2 s^{-1}}$;   **Roughness size, $k_\mathrm{s} = 0{\cdot}003$ mm**

E194

Kin. visc., $\nu = 10 \cdot 0 \times 10^{-6}$ m^{2}s^{-1};
$S = 0 \cdot 00010$ to $30 \cdot 0000$
i.e. kin. pr. grad., $\Delta p/\rho l =$
$(0 \cdot 00098)$ to (294) ms^{-2}

Roughness size, $k_s = 0 \cdot 006$ mm
This table shows values of m, as follows

m_C for Colebrook-White solutions; or,
where $R \le 2000$, m_P for laminar flow

Grad'nt S	k.p.g. $(\Delta p/\rho l)$	6	8	10	12·5	15	20	25	30	40	50	60	80	100	125
		\(Equivalent\) Pipe diameters in mm													
0·00030	(0·00294)	68·57	46·73	34·70	25·77	20·21	13·77	10·23	8·020	5·465	4·059	3·183	2·169	1·611	1·196
0·00040	(0·00392)	59·39	40·47	30·05	22·32	17·50	11·93	8·857	6·946	4·733	3·515	2·756	1·878	1·395	*
0·00060	(0·00588)	48·49	33·04	24·54	18·22	14·29	9·738	7·232	5·671	3·865	2·870	2·251	1·534	1·139	*
0·00080	(0·00785)	41·99	28·61	21·25	15·78	12·38	8·433	6·263	4·912	3·347	2·486	1·949	1·328	*	1·353
0·00100	(0·00981)	37·56	25·59	19·01	14·12	11·07	7·543	5·602	4·393	2·993	2·223	1·743	1·188	*	1·325
0·00150	(0·0147)	30·67	20·90	15·52	11·53	9·038	6·159	4·574	3·587	2·444	1·815	1·423	*	1·308	1·278
0·00200	(0·0196)	26·56	18·10	13·44	9·981	7·827	5·334	3·961	3·106	2·117	1·572	1·233	*	1·274	1·247
0·00300	(0·0294)	21·69	14·78	10·97	8·150	6·391	4·355	3·234	2·536	1·728	1·284	1·007	1·258	1·229	1·205
0·00400	(0·0392)	18·78	12·80	9·504	7·058	5·535	3·772	2·801	2·196	1·497	1·112	*	1·225	1·199	1·178
0·00600	(0·0588)	15·33	10·45	7·760	5·763	4·519	3·079	2·287	1·793	1·222	*	*	1·182	1·159	1·141
0·00800	(0·0785)	13·28	9·049	6·720	4·991	3·914	2·667	1·981	1·553	1·058	*	1·187	1·154	1·133	1·116
0·01000	(0·0981)	11·88	8·093	6·011	4·464	3·501	2·385	1·771	1·389	0·947	*	1·163	1·132	1·113	1·097
0·01500	(0·147)	9·698	6·608	4·908	3·645	2·858	1·948	1·446	1·134	*	1·143	1·122	1·095	1·079	1·065
0·02000	(0·196)	8·399	5·723	4·250	3·156	2·475	1·687	1·253	0·982	*	1·114	1·095	1·071	1·056	1·043
0·03000	(0·294)	6·857	4·673	3·470	2·577	2·021	1·377	1·023	*	1·099	1·075	1·059	1·038	1·025	1·014
0·04000	(0·392)	5·939	4·047	3·005	2·232	1·750	1·193	0·886	*	1·071	1·050	1·035	1·016	1·004	0·995
0·06000	(0·588)	4·849	3·304	2·454	1·822	1·429	0·974	*	1·065	1·035	1·016	1·003	0·986	0·976	0·969
0·08000	(0·785)	4·199	2·861	2·125	1·578	1·238	0·843	*	1·037	1·010	0·993	0·981	0·967	0·958	0·951
0·10000	(0·981)	3·756	2·559	1·901	1·412	1·107	*	1·037	1·017	0·992	0·976	0·965	0·952	0·944	0·938
0·15000	(1·47)	3·067	2·090	1·552	1·153	0·904	*	0·999	0·982	0·960	0·947	0·937	0·926	0·920	0·915
0·20000	(1·96)	2·656	1·810	1·344	0·998	*	0·997	0·975	0·959	0·939	0·927	0·919	0·909	0·903	0·899
0·30000	(2·94)	2·169	1·478	1·097	0·815	*	0·962	0·942	0·929	0·911	0·901	0·894	0·886	0·881	0·878
0·40000	(3·92)	1·878	1·280	0·950	*	*	0·938	0·920	0·908	0·893	0·883	0·877	0·870	0·866	0·864
0·60000	(5·88)	1·533	1·045	0·776	*	0·932	0·907	0·891	0·881	0·868	0·860	0·855	0·849	0·846	0·845
0·80000	(7·85)	1·328	0·905	*	0·926	0·909	0·886	0·872	0·863	0·851	0·844	0·840	0·835	0·833	0·832
1·00000	(9·81)	1·188	0·809	*	0·908	0·892	0·871	0·858	0·849	0·838	0·832	0·828	0·824	0·823	0·823
1·50000	(14·7)	0·970	*	0·897	0·876	0·862	0·844	0·833	0·826	0·817	0·812	0·809	0·806	0·806	0·806
2·00000	(19·6)	0·840	*	0·874	0·855	0·843	0·826	0·817	0·810	0·802	0·798	0·796	0·794	0·794	0·795
3·00000	(29·4)	0·686	0·863	0·844	0·828	0·817	0·803	0·795	0·789	0·783	0·780	0·778	0·777	0·778	0·780
4·00000	(39·2)	*	0·842	0·824	0·810	0·800	0·787	0·780	0·775	0·770	0·767	0·766	0·766	0·768	0·770
6·00000	(58·8)	*	0·813	0·798	0·785	0·777	0·766	0·760	0·756	0·753	0·751	0·751	0·752	0·754	0·757
8·00000	(78·5)	0·817	0·794	0·781	0·769	0·762	0·752	0·747	0·744	0·741	0·740	0·740	0·742	0·745	0·748
10·0000	(98·1)	0·801	0·780	0·768	0·757	0·750	0·742	0·737	0·735	0·732	0·732	0·733	0·735	0·738	0·742
15·0000	(147)	0·774	0·757	0·746	0·737	0·731	0·724	0·721	0·719	0·718	0·718	0·719	0·722	0·726	0·731
20·0000	(196)	0·757	0·741	0·731	0·723	0·718	0·712	0·710	0·708	0·708	0·709	0·710	0·714	0·718	0·723
30·0000	(294)	0·734	0·720	0·712	0·705	0·701	0·697	0·695	0·694	0·695	0·697	0·699	0·703	0·708	0·714
S	$(\Delta p/\rho l)$	6	8	10	12·5	15	20	25	30	40	50	60	80	100	125

Grad'nt k.p.g. (Equivalent) Pipe diameters in mm

S	$(\Delta p/\rho l)$	125	150	200	250	300	400	500	600	800	1000	1250	1500	2000	2500
0·00010	(0·00098)	2·072	1·625	*	1·518	1·489	1·452	1·429	1·413	1·392	1·381	1·372	1·366	1·361	1·359
0·00015	(0·00147)	1·692	1·327	*	1·463	1·438	1·405	1·385	1·371	1·354	1·344	1·337	1·333	1·329	1·328
0·00020	(0·00196)	1·465	*	1·459	1·426	1·404	1·374	1·356	1·343	1·328	1·320	1·313	1·310	1·308	1·308
0·00030	(0·00294)	1·196	*	1·407	1·378	1·358	1·332	1·316	1·306	1·293	1·286	1·282	1·279	1·278	1·280
0·00040	(0·00392)	*	*	1·372	1·345	1·327	1·304	1·290	1·281	1·270	1·264	1·260	1·259	1·259	1·261
0·00060	(0·00588)	*	1·362	1·325	1·302	1·286	1·266	1·254	1·247	1·238	1·233	1·231	1·230	1·232	1·235
0·00080	(0·00785)	1·353	1·327	1·294	1·273	1·259	1·241	1·230	1·224	1·216	1·213	1·211	1·211	1·213	1·217
0·00100	(0·00981)	1·325	1·302	1·271	1·251	1·238	1·222	1·212	1·206	1·200	1·197	1·196	1·197	1·200	1·203
0·00150	(0·0147)	1·278	1·258	1·231	1·214	1·203	1·189	1·181	1·176	1·171	1·170	1·170	1·171	1·175	1·180
0·00200	(0·0196)	1·247	1·228	1·204	1·189	1·179	1·167	1·160	1·156	1·152	1·151	1·152	1·154	1·159	1·164
0·00300	(0·0294)	1·205	1·189	1·168	1·155	1·147	1·137	1·131	1·128	1·126	1·126	1·128	1·130	1·136	1·142
0·00400	(0·0392)	1·178	1·163	1·144	1·132	1·125	1·116	1·112	1·109	1·108	1·109	1·111	1·114	1·121	1·127
0·00600	(0·0588)	1·141	1·128	1·111	1·102	1·096	1·089	1·085	1·084	1·084	1·086	1·089	1·092	1·100	1·106
0·00800	(0·0785)	1·116	1·104	1·090	1·081	1·076	1·070	1·068	1·067	1·068	1·070	1·074	1·077	1·085	1·092
0·01000	(0·0981)	1·097	1·086	1·073	1·066	1·061	1·056	1·054	1·054	1·055	1·058	1·062	1·066	1·074	1·082
0·01500	(0·147)	1·065	1·056	1·045	1·039	1·035	1·032	1·031	1·031	1·034	1·037	1·042	1·047	1·055	1·063
0·02000	(0·196)	1·043	1·035	1·026	1·020	1·017	1·015	1·015	1·016	1·019	1·023	1·028	1·033	1·042	1·051
0·03000	(0·294)	1·014	1·008	1·000	0·996	0·994	0·993	0·993	0·995	0·999	1·004	1·009	1·015	1·025	1·034
0·04000	(0·392)	0·995	0·989	0·982	0·979	0·978	0·977	0·979	0·981	0·985	0·990	0·997	1·002	1·013	1·022
0·06000	(0·588)	0·969	0·964	0·959	0·957	0·956	0·957	0·959	0·961	0·967	0·973	0·979	0·986	0·997	1·006
0·08000	(0·785)	0·951	0·947	0·943	0·941	0·941	0·943	0·945	0·948	0·955	0·961	0·968	0·974	0·986	0·996
0·10000	(0·981)	0·938	0·934	0·931	0·930	0·930	0·932	0·935	0·938	0·945	0·951	0·959	0·965	0·977	0·988
0·15000	(1·47)	0·915	0·912	0·910	0·910	0·911	0·914	0·918	0·921	0·929	0·936	0·943	0·950	0·963	0·974
0·20000	(1·96)	0·899	0·897	0·896	0·896	0·898	0·901	0·906	0·910	0·918	0·925	0·933	0·940	0·953	0·964
0·30000	(2·94)	0·878	0·877	0·877	0·878	0·880	0·885	0·889	0·894	0·903	0·910	0·919	0·927	0·940	0·951
0·40000	(3·92)	0·864	0·863	0·864	0·866	0·868	0·873	0·878	0·883	0·892	0·900	0·909	0·917	0·931	0·943
0·60000	(5·88)	0·845	0·845	0·847	0·849	0·852	0·858	0·864	0·869	0·879	0·887	0·897	0·905	0·920	0·932
0·80000	(7·85)	0·832	0·833	0·835	0·838	0·841	0·848	0·854	0·860	0·870	0·879	0·888	0·897	0·912	0·924
1·00000	(9·81)	0·823	0·824	0·826	0·830	0·833	0·840	0·847	0·852	0·863	0·872	0·882	0·891	0·906	0·919
1·50000	(14·7)	0·806	0·808	0·811	0·815	0·819	0·827	0·834	0·840	0·851	0·861	0·871	0·881	0·896	0·909
2·00000	(19·6)	0·795	0·797	0·801	0·806	0·810	0·818	0·825	0·832	0·843	0·853	0·864	0·874	0·890	0·903
3·00000	(29·4)	0·780	0·783	0·788	0·793	0·797	0·806	0·814	0·821	0·833	0·844	0·855	0·865	0·881	0·895
S	$(\Delta p/\rho l)$	125	150	200	250	300	400	500	600	800	1000	1250	1500	2000	2500

Grad'nt k.p.g. (Equivalent) Pipe diameters in mm

Kinematic viscosity, $\nu = 10 \cdot 0 \times 10^{-6}$ m^{2}s^{-1} ; Roughness size, $k_s = 0 \cdot 006$ mm

Kin. visc., $\nu = 10.0 \times 10^{-6}$ m^{2}s^{-1};
$S = 0.00010$ to 30.0000
i.e. kin. pr. grad., $\Delta p/\rho l =$
(0.00098) to (294) ms^{-2}

Roughness size, $k_s = 0.015$ mm
This table shows values of m, as follows
m_C for Colebrook-White solutions; or,
where $\mathbf{R} \leq 2000$, m_P for laminar flow

Grad'nt S	k.p.g. $(\Delta p/\rho l)$	6	8	10	12·5	15	20	25	30	40	50	60	80	100	125
0·00030	(0·00294)	68·57	46·73	34·70	25·77	20·21	13·77	10·23	8·020	5·465	4·059	3·183	2·169	1·611	1·196
0·00040	(0·00392)	59·39	40·47	30·05	22·32	17·50	11·93	8·857	6·946	4·733	3·515	2·756	1·878	1·395	*
0·00060	(0·00588)	48·49	33·04	24·54	18·22	14·29	9·738	7·232	5·671	3·865	2·870	2·251	1·534	1·139	*
0·00080	(0·00785)	41·99	28·61	21·25	15·78	12·38	8·433	6·263	4·912	3·347	2·486	1·949	1·328	*	1·354
0·00100	(0·00981)	37·56	25·59	19·01	14·12	11·07	7·543	5·602	4·393	2·993	2·223	1·743	1·188	*	1·326
0·00150	(0·0147)	30·67	20·90	15·52	11·53	9·038	6·159	4·574	3·587	2·444	1·815	1·423	*	1·310	1·280
0·00200	(0·0196)	26·56	18·10	13·44	9·981	7·827	5·334	3·961	3·106	2·117	1·572	1·233	*	1·276	1·248
0·00300	(0·0294)	21·69	14·78	10·97	8·150	6·391	4·355	3·234	2·536	1·728	1·284	1·007	1·260	1·231	1·207
0·00400	(0·0392)	18·78	12·80	9·504	7·058	5·535	3·772	2·801	2·196	1·497	1·112	*	1·227	1·201	1·179
0·00600	(0·0588)	15·33	10·45	7·760	5·763	4·519	3·079	2·287	1·793	1·222	*	*	1·184	1·161	1·143
0·00800	(0·0785)	13·28	9·049	6·720	4·991	3·914	2·667	1·981	1·553	1·058	*	1·189	1·156	1·135	1·118
0·01000	(0·0981)	11·88	8·093	6·011	4·464	3·501	2·385	1·771	1·389	0·947	*	1·165	1·134	1·115	1·100
0·01500	(0·147)	9·698	6·608	4·908	3·645	2·858	1·948	1·446	1·134	*	1·145	1·125	1·098	1·081	1·068
0·02000	(0·196)	8·399	5·723	4·250	3·156	2·475	1·687	1·253	0·982	*	1·116	1·098	1·074	1·059	1·047
0·03000	(0·294)	6·857	4·673	3·470	2·577	2·021	1·377	1·023	*	1·102	1·078	1·062	1·041	1·028	1·018
0·04000	(0·392)	5·939	4·047	3·005	2·232	1·750	1·193	0·886	*	1·075	1·053	1·038	1·019	1·008	0·999
0·06000	(0·588)	4·849	3·304	2·454	1·822	1·429	0·974	*	1·068	1·038	1·019	1·007	0·990	0·981	0·973
0·08000	(0·785)	4·199	2·861	2·125	1·578	1·238	0·843	*	1·041	1·014	0·997	0·986	0·971	0·963	0·956
0·10000	(0·981)	3·756	2·559	1·901	1·412	1·107	*	1·041	1·021	0·996	0·980	0·970	0·957	0·949	0·943
0·15000	(1·47)	3·067	2·090	1·552	1·153	0·904	*	1·004	0·987	0·965	0·952	0·943	0·932	0·925	0·921
0·20000	(1·96)	2·656	1·810	1·344	0·998	*	1·002	0·980	0·965	0·945	0·933	0·925	0·915	0·910	0·906
0·30000	(2·94)	2·169	1·478	1·097	0·815	*	0·967	0·948	0·935	0·918	0·907	0·901	0·893	0·889	0·886
0·40000	(3·92)	1·878	1·280	0·950	*	*	0·944	0·927	0·915	0·899	0·890	0·885	0·878	0·875	0·873
0·60000	(5·88)	1·533	1·045	0·776	*	0·938	0·914	0·899	0·888	0·875	0·868	0·863	0·858	0·856	0·855
0·80000	(7·85)	1·328	0·905	*	0·933	0·916	0·894	0·880	0·871	0·859	0·853	0·849	0·845	0·844	0·844
1·00000	(9·81)	1·188	0·809	*	0·915	0·899	0·879	0·866	0·858	0·847	0·842	0·838	0·835	0·834	0·835
1·50000	(14·7)	0·970	*	0·905	0·885	0·871	0·853	0·843	0·836	0·827	0·823	0·820	0·818	0·819	0·820
2·00000	(19·6)	0·840	*	0·883	0·865	0·852	0·836	0·827	0·821	0·814	0·810	0·808	0·807	0·808	0·810
3·00000	(29·4)	0·686	0·873	0·854	0·838	0·828	0·814	0·806	0·801	0·796	0·793	0·792	0·793	0·795	0·797
4·00000	(39·2)	*	0·853	0·835	0·821	0·812	0·800	0·793	0·789	0·784	0·782	0·782	0·783	0·786	0·789
6·00000	(58·8)	*	0·826	0·811	0·798	0·790	0·780	0·775	0·772	0·769	0·768	0·768	0·771	0·774	0·778
8·00000	(78·5)	0·830	0·808	0·794	0·783	0·776	0·768	0·763	0·761	0·759	0·759	0·760	0·763	0·766	0·771
10·0000	(98·1)	0·815	0·795	0·782	0·772	0·766	0·758	0·755	0·752	0·751	0·752	0·753	0·757	0·761	0·766
15·0000	(147)	0·790	0·773	0·762	0·754	0·749	0·743	0·740	0·739	0·739	0·740	0·742	0·747	0·751	0·757
20·0000	(196)	0·774	0·758	0·749	0·742	0·737	0·732	0·730	0·730	0·731	0·733	0·735	0·740	0·746	0·752
30·0000	(294)	0·753	0·739	0·732	0·726	0·723	0·719	0·718	0·718	0·720	0·723	0·726	0·732	0·738	0·745
S	$(\Delta p/\rho l)$	6	8	10	12·5	15	20	25	30	40	50	60	80	100	125

Grad'nt k.p.g. (Equivalent) Pipe diameters in mm

S	$(\Delta p/\rho l)$	125	150	200	250	300	400	500	600	800	1000	1250	1500	2000	2500
0·00010	(0·00098)	2·072	1·625	*	1·518	1·490	1·453	1·429	1·413	1·393	1·381	1·372	1·367	1·362	1·360
0·00015	(0·00147)	1·692	1·327	*	1·464	1·439	1·406	1·386	1·372	1·355	1·345	1·338	1·334	1·330	1·330
0·00020	(0·00196)	1·465	*	1·460	1·427	1·404	1·375	1·357	1·344	1·329	1·321	1·315	1·311	1·309	1·309
0·00030	(0·00294)	1·196	*	1·408	1·379	1·359	1·333	1·317	1·307	1·294	1·288	1·283	1·281	1·280	1·281
0·00040	(0·00392)	*	*	1·372	1·346	1·328	1·305	1·291	1·282	1·271	1·265	1·262	1·260	1·260	1·262
0·00060	(0·00588)	*	1·363	1·326	1·303	1·288	1·268	1·256	1·248	1·239	1·235	1·233	1·232	1·234	1·237
0·00080	(0·00785)	1·354	1·328	1·295	1·274	1·260	1·242	1·232	1·225	1·218	1·214	1·213	1·213	1·216	1·219
0·00100	(0·00981)	1·326	1·303	1·272	1·253	1·240	1·223	1·214	1·208	1·202	1·199	1·198	1·199	1·202	1·206
0·00150	(0·0147)	1·280	1·259	1·232	1·216	1·204	1·191	1·183	1·178	1·173	1·172	1·172	1·174	1·178	1·183
0·00200	(0·0196)	1·248	1·230	1·205	1·190	1·181	1·168	1·162	1·158	1·154	1·154	1·155	1·157	1·162	1·167
0·00300	(0·0294)	1·207	1·191	1·170	1·157	1·149	1·139	1·133	1·131	1·129	1·129	1·131	1·134	1·140	1·146
0·00400	(0·0392)	1·179	1·165	1·146	1·134	1·127	1·118	1·114	1·112	1·111	1·112	1·115	1·118	1·125	1·131
0·00600	(0·0588)	1·143	1·130	1·114	1·104	1·098	1·091	1·088	1·087	1·088	1·090	1·093	1·097	1·104	1·111
0·00800	(0·0785)	1·118	1·106	1·092	1·084	1·079	1·073	1·071	1·070	1·072	1·074	1·078	1·082	1·090	1·098
0·01000	(0·0981)	1·100	1·089	1·076	1·068	1·064	1·059	1·058	1·058	1·060	1·063	1·067	1·071	1·080	1·088
0·01500	(0·147)	1·068	1·059	1·048	1·042	1·038	1·035	1·035	1·036	1·039	1·042	1·047	1·052	1·062	1·071
0·02000	(0·196)	1·047	1·039	1·029	1·024	1·021	1·019	1·019	1·021	1·024	1·029	1·034	1·040	1·050	1·059
0·03000	(0·294)	1·018	1·011	1·004	1·000	0·998	0·997	0·999	1·001	1·005	1·010	1·017	1·023	1·033	1·043
0·04000	(0·392)	0·999	0·993	0·987	0·984	0·983	0·983	0·985	0·987	0·992	0·998	1·005	1·011	1·022	1·032
0·06000	(0·588)	0·973	0·969	0·964	0·962	0·962	0·963	0·966	0·969	0·975	0·981	0·989	0·995	1·008	1·018
0·08000	(0·785)	0·956	0·952	0·949	0·947	0·948	0·950	0·953	0·956	0·963	0·970	0·978	0·985	0·998	1·009
0·10000	(0·981)	0·943	0·940	0·937	0·937	0·937	0·940	0·943	0·947	0·955	0·962	0·970	0·977	0·990	1·002
0·15000	(1·47)	0·921	0·919	0·917	0·918	0·919	0·923	0·927	0·931	0·940	0·948	0·956	0·964	0·978	0·990
0·20000	(1·96)	0·906	0·905	0·904	0·905	0·907	0·911	0·916	0·921	0·930	0·938	0·947	0·955	0·970	0·982
0·30000	(2·94)	0·886	0·886	0·886	0·888	0·891	0·896	0·902	0·907	0·917	0·925	0·935	0·944	0·959	0·971
0·40000	(3·92)	0·873	0·873	0·874	0·877	0·880	0·886	0·892	0·898	0·908	0·917	0·927	0·936	0·952	0·965
0·60000	(5·88)	0·855	0·856	0·858	0·862	0·865	0·872	0·879	0·885	0·896	0·906	0·917	0·926	0·942	0·956
0·80000	(7·85)	0·844	0·845	0·848	0·852	0·856	0·863	0·871	0·877	0·889	0·899	0·910	0·920	0·936	0·950
1·00000	(9·81)	0·835	0·836	0·840	0·844	0·849	0·857	0·864	0·871	0·883	0·894	0·905	0·915	0·932	0·946
1·50000	(14·7)	0·820	0·822	0·827	0·832	0·837	0·846	0·854	0·861	0·874	0·885	0·897	0·907	0·925	0·940
2·00000	(19·6)	0·810	0·813	0·818	0·824	0·829	0·839	0·847	0·855	0·868	0·880	0·892	0·902	0·920	0·935
3·00000	(29·4)	0·797	0·801	0·807	0·813	0·819	0·829	0·838	0·847	0·861	0·872	0·885	0·896	0·915	0·930
S	$(\Delta p/\rho l)$	125	150	200	250	300	400	500	600	800	1000	1250	1500	2000	2500

Grad'nt k.p.g. (Equivalent) Pipe diameters in mm

Kinematic viscosity, $\nu = 10.0 \times 10^{-6}$ m^{2}s^{-1} ;

Roughness size, $k_s = 0.015$ mm

E196

Kin. visc., $\nu = 10 \cdot 0 \times 10^{-6}$ m^{2}s^{-1}; $S = 0 \cdot 00010$ to $3 \cdot 00000$

i.e. kin. pr. grad., $\Delta p / \rho l =$ ($0 \cdot 00098$) to ($29 \cdot 4$) ms^{-2}

Roughness size, $k_s = 0 \cdot 030$ mm

This table shows values of m, as follows

m_C for Colebrook-White solutions; or, where $\mathbf{R} \leq 2000$, m_P for laminar flow

Grad'nt S	k.p.g. $(\Delta p / \rho l)$	125	150	200	250	300	400	500	600	800	1000	1250	1500	2000	2500
0·00010	(0·00098)	2·072	1·625	*	1·519	1·491	1·454	1·430	1·414	1·394	1·383	1·374	1·368	1·363	1·361
0·00015	(0·00147)	1·692	1·327	*	1·465	1·440	1·407	1·387	1·373	1·356	1·347	1·339	1·335	1·332	1·332
0·00020	(0·00196)	1·465	*	1·461	1·428	1·406	1·376	1·358	1·346	1·331	1·322	1·316	1·313	1·311	1·311
0·00030	(0·00294)	1·196	*	1·409	1·380	1·360	1·335	1·319	1·309	1·296	1·290	1·285	1·283	1·282	1·284
0·00040	(0·00392)	*	*	1·374	1·348	1·330	1·307	1·293	1·284	1·273	1·267	1·264	1·263	1·263	1·265
0·00060	(0·00588)	*	1·364	1·328	1·305	1·289	1·270	1·258	1·250	1·242	1·237	1·235	1·235	1·237	1·240
0·00080	(0·00785)	1·355	1·330	1·297	1·276	1·262	1·245	1·234	1·228	1·220	1·217	1·216	1·216	1·219	1·223
0·00100	(0·00981)	1·328	1·305	1·274	1·255	1·242	1·226	1·216	1·211	1·204	1·202	1·201	1·202	1·206	1·210
0·00150	(0·0147)	1·282	1·261	1·234	1·218	1·207	1·193	1·186	1·181	1·177	1·176	1·176	1·178	1·182	1·188
0·00200	(0·0196)	1·251	1·232	1·208	1·193	1·183	1·171	1·165	1·161	1·158	1·158	1·159	1·161	1·167	1·172
0·00300	(0·0294)	1·210	1·194	1·173	1·160	1·152	1·142	1·137	1·135	1·133	1·134	1·136	1·139	1·145	1·152
0·00400	(0·0392)	1·182	1·168	1·149	1·138	1·131	1·122	1·118	1·116	1·116	1·117	1·120	1·124	1·131	1·138
0·00600	(0·0588)	1·146	1·133	1·117	1·108	1·102	1·096	1·093	1·092	1·093	1·095	1·099	1·103	1·112	1·119
0·00800	(0·0785)	1·122	1·110	1·096	1·088	1·083	1·078	1·076	1·076	1·078	1·081	1·085	1·090	1·099	1·107
0·01000	(0·0981)	1·104	1·093	1·080	1·073	1·069	1·065	1·063	1·064	1·066	1·070	1·075	1·079	1·089	1·098
0·01500	(0·147)	1·073	1·064	1·053	1·047	1·044	1·042	1·042	1·043	1·046	1·051	1·056	1·062	1·072	1·082
0·02000	(0·196)	1·052	1·044	1·035	1·030	1·028	1·026	1·027	1·028	1·033	1·038	1·044	1·050	1·061	1·071
0·03000	(0·294)	1·024	1·018	1·010	1·007	1·006	1·005	1·007	1·009	1·015	1·021	1·028	1·034	1·046	1·057
0·04000	(0·392)	1·005	1·000	0·994	0·992	0·991	0·992	0·994	0·997	1·003	1·009	1·017	1·024	1·037	1·048
0·06000	(0·588)	0·981	0·976	0·972	0·971	0·971	0·973	0·976	0·980	0·987	0·994	1·003	1·010	1·024	1·035
0·08000	(0·785)	0·964	0·961	0·958	0·957	0·958	0·961	0·965	0·969	0·977	0·985	0·993	1·001	1·015	1·027
0·10000	(0·981)	0·952	0·949	0·947	0·947	0·948	0·952	0·956	0·961	0·969	0·977	0·986	0·995	1·009	1·022
0·15000	(1·47)	0·931	0·929	0·928	0·930	0·931	0·936	0·941	0·947	0·956	0·965	0·975	0·983	0·999	1·012
0·20000	(1·96)	0·917	0·916	0·916	0·918	0·920	0·926	0·932	0·937	0·948	0·957	0·967	0·976	0·992	1·006
0·30000	(2·94)	0·899	0·899	0·900	0·903	0·906	0·913	0·919	0·925	0·936	0·946	0·957	0·967	0·984	0·998
0·40000	(3·92)	0·887	0·887	0·889	0·893	0·897	0·904	0·911	0·918	0·929	0·940	0·951	0·961	0·978	0·992
0·60000	(5·88)	0·871	0·872	0·876	0·880	0·884	0·893	0·900	0·907	0·920	0·931	0·943	0·953	0·971	0·986
0·80000	(7·85)	0·860	0·862	0·867	0·871	0·876	0·885	0·894	0·901	0·914	0·926	0·938	0·949	0·967	0·982
1·00000	(9·81)	0·853	0·855	0·860	0·865	0·870	0·880	0·889	0·896	0·910	0·922	0·934	0·945	0·964	0·979
1·50000	(14·7)	0·840	0·843	0·849	0·855	0·861	0·871	0·880	0·889	0·903	0·915	0·928	0·940	0·959	0·975
2·00000	(19·6)	0·832	0·835	0·842	0·849	0·855	0·866	0·875	0·884	0·899	0·911	0·925	0·936	0·956	0·972
3·00000	(29·4)	0·821	0·825	0·833	0·840	0·847	0·859	0·869	0·878	0·893	0·906	0·920	0·932	0·952	0·968
S	$(\Delta p / \rho l)$	125	150	200	250	300	400	500	600	800	1000	1250	1500	2000	2500

Grad'nt   k.p.g.   (Equivalent) Pipe diameters in mm   Roughness size, $k_s = 0 \cdot 030$ mm

E197

Kin. visc., $\nu = 10 \cdot 0 \times 10^{-6}$ m^{2}s^{-1}; $S = 0 \cdot 00010$ to $3 \cdot 00000$

i.e. kin. pr. grad., $\Delta p / \rho l =$ ($0 \cdot 00098$) to ($29 \cdot 4$) ms^{-2}

Roughness size, $k_s = 0 \cdot 060$ mm

This table shows values of m, as follows

m_C for Colebrook-White solutions; or, where $\mathbf{R} \leq 2000$, m_P for laminar flow

Grad'nt S	k.p.g. $(\Delta p / \rho l)$	125	150	200	250	300	400	500	600	800	1000	1250	1500	2000	2500
0·00010	(0·00098)	2·072	1·625	*	1·521	1·493	1·456	1·433	1·417	1·397	1·385	1·376	1·371	1·366	1·365
0·00015	(0·00147)	1·692	1·327	*	1·467	1·442	1·410	1·390	1·376	1·359	1·350	1·343	1·339	1·336	1·336
0·00020	(0·00196)	1·465	*	1·464	1·431	1·408	1·379	1·361	1·349	1·334	1·326	1·320	1·317	1·315	1·316
0·00030	(0·00294)	1·196	*	1·412	1·383	1·363	1·338	1·322	1·312	1·300	1·293	1·289	1·287	1·287	1·289
0·00040	(0·00392)	*	*	1·377	1·351	1·333	1·310	1·296	1·287	1·277	1·272	1·268	1·267	1·268	1·271
0·00060	(0·00588)	*	1·368	1·331	1·309	1·293	1·273	1·262	1·255	1·246	1·242	1·241	1·241	1·243	1·247
0·00080	(0·00785)	1·359	1·334	1·301	1·280	1·266	1·249	1·239	1·232	1·226	1·223	1·222	1·222	1·226	1·230
0·00100	(0·00981)	1·332	1·309	1·278	1·259	1·246	1·230	1·221	1·216	1·210	1·208	1·208	1·209	1·213	1·218
0·00150	(0·0147)	1·286	1·266	1·239	1·223	1·212	1·199	1·191	1·187	1·183	1·182	1·183	1·185	1·191	1·197
0·00200	(0·0196)	1·256	1·237	1·213	1·198	1·189	1·177	1·171	1·168	1·165	1·165	1·167	1·170	1·176	1·182
0·00300	(0·0294)	1·215	1·199	1·178	1·166	1·158	1·149	1·144	1·142	1·141	1·142	1·145	1·149	1·156	1·164
0·00400	(0·0392)	1·188	1·174	1·155	1·145	1·138	1·130	1·126	1·125	1·125	1·127	1·131	1·135	1·143	1·151
0·00600	(0·0588)	1·153	1·140	1·125	1·116	1·110	1·105	1·102	1·102	1·104	1·107	1·111	1·116	1·125	1·134
0·00800	(0·0785)	1·129	1·118	1·104	1·097	1·092	1·088	1·086	1·087	1·089	1·093	1·098	1·103	1·114	1·123
0·01000	(0·0981)	1·112	1·101	1·089	1·082	1·078	1·075	1·074	1·075	1·079	1·083	1·089	1·094	1·105	1·115
0·01500	(0·147)	1·081	1·073	1·063	1·058	1·055	1·053	1·054	1·056	1·060	1·066	1·072	1·079	1·090	1·101
0·02000	(0·196)	1·062	1·054	1·046	1·042	1·040	1·039	1·040	1·043	1·048	1·054	1·062	1·068	1·081	1·092
0·03000	(0·294)	1·035	1·029	1·023	1·020	1·019	1·020	1·023	1·026	1·033	1·039	1·047	1·055	1·069	1·080
0·04000	(0·392)	1·018	1·013	1·008	1·006	1·006	1·008	1·011	1·015	1·022	1·030	1·038	1·046	1·061	1·073
0·06000	(0·588)	0·995	0·991	0·988	0·987	0·988	0·991	0·995	1·000	1·009	1·017	1·026	1·035	1·050	1·063
0·08000	(0·785)	0·979	0·977	0·975	0·975	0·976	0·981	0·985	0·990	1·000	1·009	1·019	1·028	1·044	1·057
0·10000	(0·981)	0·968	0·966	0·965	0·966	0·968	0·973	0·978	0·984	0·994	1·003	1·013	1·023	1·039	1·053
0·15000	(1·47)	0·949	0·948	0·949	0·951	0·953	0·960	0·966	0·972	0·983	0·993	1·004	1·014	1·031	1·045
0·20000	(1·96)	0·937	0·937	0·938	0·941	0·944	0·951	0·958	0·965	0·976	0·987	0·998	1·009	1·026	1·041
0·30000	(2·94)	0·921	0·922	0·924	0·928	0·932	0·940	0·948	0·955	0·968	0·979	0·991	1·002	1·020	1·035
0·40000	(3·92)	0·911	0·912	0·916	0·920	0·925	0·933	0·942	0·949	0·963	0·974	0·987	0·998	1·016	1·032
0·60000	(5·88)	0·897	0·899	0·904	0·910	0·915	0·925	0·934	0·942	0·956	0·968	0·981	0·992	1·012	1·028
0·80000	(7·85)	0·889	0·891	0·897	0·903	0·909	0·919	0·929	0·937	0·952	0·964	0·977	0·989	1·009	1·025
1·00000	(9·81)	0·883	0·886	0·892	0·898	0·904	0·915	0·925	0·934	0·949	0·961	0·975	0·987	1·007	1·023
1·50000	(14·7)	0·873	0·876	0·884	0·891	0·897	0·909	0·919	0·928	0·944	0·957	0·971	0·983	1·003	1·020
2·00000	(19·6)	0·866	0·870	0·878	0·886	0·893	0·905	0·916	0·925	0·941	0·954	0·969	0·981	1·001	1·018
3·00000	(29·4)	0·858	0·863	0·872	0·880	0·887	0·900	0·911	0·921	0·937	0·951	0·966	0·978	0·999	1·016
S	$(\Delta p / \rho l)$	125	150	200	250	300	400	500	600	800	1000	1250	1500	2000	2500

Grad'nt   k.p.g.   (Equivalent) Pipe diameters in mm   Roughness size, $k_s = 0 \cdot 060$ mm

Kin. visc., $\nu = 10 \cdot 0 \times 10^{-6}$ m^{2}s^{-1};
$S = 0 \cdot 00010$ to $3 \cdot 00000$

i.e. kin. pr. grad., $\Delta p/\rho l =$
$(0 \cdot 00098)$ to $(29 \cdot 4)$ ms^{-2}

Roughness size, $k_s = 0 \cdot 150$ mm
This table shows values of m, as follows

m_C for Colebrook-White solutions; or,
where $R \le 2000$, m_P for laminar flow

E198

Grad'nt S	k.p.g. $(\Delta p/\rho l)$	125	150	200	250	300	400	500	600	800	1000	1250	1500	2000	2500
0·00010	(0·00098)	2·072	1·625	*	1·527	1·499	1·462	1·439	1·424	1·404	1·393	1·385	1·380	1·375	1·375
0·00015	(0·00147)	1·692	1·327	*	1·474	1·449	1·417	1·397	1·384	1·368	1·358	1·352	1·349	1·346	1·347
0·00020	(0·00196)	1·465	*	1·471	1·438	1·416	1·387	1·369	1·357	1·343	1·335	1·330	1·328	1·327	1·328
0·00030	(0·00294)	1·196	*	1·420	1·391	1·372	1·347	1·332	1·322	1·310	1·305	1·301	1·300	1·301	1·303
0·00040	(0·00392)	*	*	1·386	1·360	1·342	1·320	1·307	1·298	1·289	1·284	1·282	1·281	1·283	1·287
0·00060	(0·00588)	*	1·377	1·341	1·319	1·304	1·285	1·274	1·267	1·260	1·257	1·256	1·256	1·260	1·265
0·00080	(0·00785)	1·369	1·344	1·312	1·292	1·278	1·261	1·252	1·246	1·240	1·238	1·238	1·240	1·245	1·250
0·00100	(0·00981)	1·343	1·320	1·290	1·271	1·259	1·244	1·235	1·231	1·226	1·225	1·226	1·228	1·233	1·240
0·00150	(0·0147)	1·299	1·278	1·252	1·237	1·226	1·214	1·208	1·204	1·201	1·202	1·204	1·207	1·214	1·221
0·00200	(0·0196)	1·269	1·251	1·228	1·214	1·205	1·194	1·189	1·186	1·185	1·186	1·189	1·193	1·201	1·209
0·00300	(0·0294)	1·231	1·215	1·195	1·184	1·176	1·168	1·165	1·163	1·164	1·166	1·170	1·175	1·185	1·194
0·00400	(0·0392)	1·205	1·191	1·174	1·164	1·157	1·151	1·148	1·148	1·150	1·153	1·158	1·163	1·174	1·183
0·00600	(0·0588)	1·172	1·160	1·145	1·137	1·133	1·128	1·127	1·128	1·131	1·136	1·142	1·148	1·160	1·170
0·00800	(0·0785)	1·150	1·139	1·127	1·120	1·116	1·113	1·113	1·115	1·119	1·125	1·132	1·138	1·151	1·162
0·01000	(0·0981)	1·134	1·124	1·113	1·107	1·104	1·103	1·103	1·105	1·111	1·117	1·124	1·131	1·144	1·156
0·01500	(0·147)	1·107	1·099	1·090	1·086	1·084	1·084	1·086	1·089	1·096	1·103	1·111	1·119	1·134	1·146
0·02000	(0·196)	1·089	1·082	1·075	1·072	1·071	1·073	1·075	1·079	1·087	1·094	1·103	1·112	1·127	1·140
0·03000	(0·294)	1·066	1·061	1·056	1·054	1·055	1·057	1·061	1·066	1·075	1·083	1·093	1·102	1·118	1·132
0·04000	(0·392)	1·051	1·047	1·043	1·043	1·044	1·048	1·052	1·057	1·067	1·076	1·087	1·096	1·113	1·127
0·06000	(0·588)	1·031	1·029	1·027	1·028	1·030	1·035	1·041	1·047	1·058	1·068	1·079	1·089	1·106	1·121
0·08000	(0·785)	1·019	1·017	1·017	1·018	1·021	1·027	1·034	1·040	1·052	1·062	1·074	1·084	1·102	1·117
0·10000	(0·981)	1·010	1·009	1·009	1·012	1·015	1·022	1·029	1·035	1·047	1·058	1·070	1·081	1·099	1·114
0·15000	(1·47)	0·995	0·995	0·997	1·000	1·004	1·012	1·020	1·027	1·041	1·052	1·065	1·076	1·094	1·110
0·20000	(1·96)	0·986	0·986	0·989	0·993	0·998	1·007	1·015	1·023	1·036	1·048	1·061	1·072	1·092	1·108
0·30000	(2·94)	0·974	0·975	0·980	0·985	0·990	1·000	1·008	1·017	1·031	1·043	1·057	1·068	1·088	1·104
0·40000	(3·92)	0·966	0·968	0·974	0·979	0·985	0·995	1·004	1·013	1·028	1·040	1·054	1·066	1·086	1·103
0·60000	(5·88)	0·957	0·960	0·966	0·972	0·978	0·990	1·000	1·008	1·024	1·037	1·051	1·063	1·083	1·100
0·80000	(7·85)	0·951	0·954	0·961	0·968	0·975	0·986	0·997	1·006	1·021	1·035	1·049	1·061	1·082	1·099
1·00000	(9·81)	0·947	0·951	0·958	0·965	0·972	0·984	0·994	1·004	1·020	1·033	1·048	1·060	1·081	1·098
1·50000	(14·7)	0·941	0·945	0·953	0·961	0·968	0·980	0·991	1·001	1·017	1·031	1·045	1·058	1·079	1·096
2·00000	(19·6)	0·937	0·941	0·950	0·958	0·965	0·978	0·989	0·999	1·016	1·029	1·044	1·057	1·078	1·095
3·00000	(29·4)	0·932	0·937	0·946	0·954	0·962	0·975	0·987	0·997	1·014	1·028	1·043	1·056	1·077	1·094
S	$(\Delta p/\rho l)$	125	150	200	250	300	400	500	600	800	1000	1250	1500	2000	2500

Grad'nt k.p.g. (Equivalent) Pipe diameters in mm Roughness size, $k_s = 0 \cdot 150$ mm

---

Kin. visc., $\nu = 10 \cdot 0 \times 10^{-6}$ m^{2}s^{-1};
$S = 0 \cdot 00010$ to $3 \cdot 00000$

i.e. kin. pr. grad., $\Delta p/\rho l =$
$(0 \cdot 00098)$ to $(29 \cdot 4)$ ms^{-2}

Roughness size, $k_s = 0 \cdot 30$ mm
This table shows values of m, as follows

m_C for Colebrook-White solutions; or,
where $R \le 2000$, m_P for laminar flow

E199

Grad'nt S	k.p.g. $(\Delta p/\rho l)$	125	150	200	250	300	400	500	600	800	1000	1250	1500	2000	2500
0·00010	(0·00098)	2·072	1·625	*	1·537	1·509	1·472	1·450	1·435	1·416	1·405	1·397	1·393	1·390	1·390
0·00015	(0·00147)	1·692	1·327	*	1·485	1·460	1·429	1·409	1·396	1·381	1·372	1·367	1·364	1·363	1·364
0·00020	(0·00196)	1·465	*	1·483	1·450	1·428	1·400	1·382	1·371	1·358	1·351	1·346	1·345	1·345	1·347
0·00030	(0·00294)	1·196	*	1·433	1·405	1·386	1·361	1·347	1·338	1·327	1·322	1·319	1·319	1·321	1·325
0·00040	(0·00392)	*	*	1·400	1·375	1·358	1·336	1·323	1·315	1·307	1·303	1·302	1·302	1·306	1·311
0·00060	(0·00588)	*	1·393	1·357	1·336	1·321	1·303	1·293	1·286	1·280	1·278	1·278	1·280	1·285	1·292
0·00080	(0·00785)	1·386	1·361	1·329	1·310	1·297	1·281	1·272	1·267	1·263	1·262	1·263	1·266	1·272	1·279
0·00100	(0·00981)	1·361	1·338	1·309	1·291	1·279	1·265	1·257	1·253	1·250	1·250	1·252	1·255	1·262	1·270
0·00150	(0·0147)	1·319	1·299	1·274	1·259	1·249	1·238	1·232	1·229	1·228	1·230	1·233	1·237	1·246	1·255
0·00200	(0·0196)	1·291	1·273	1·251	1·237	1·229	1·220	1·216	1·214	1·214	1·217	1·221	1·226	1·236	1·245
0·00300	(0·0294)	1·255	1·240	1·221	1·210	1·203	1·197	1·194	1·194	1·196	1·200	1·205	1·211	1·223	1·233
0·00400	(0·0392)	1·231	1·218	1·201	1·192	1·187	1·182	1·180	1·181	1·184	1·189	1·195	1·202	1·214	1·225
0·00600	(0·0588)	1·201	1·190	1·176	1·169	1·165	1·162	1·163	1·164	1·169	1·175	1·183	1·190	1·204	1·216
0·00800	(0·0785)	1·181	1·171	1·160	1·154	1·151	1·150	1·151	1·154	1·160	1·166	1·175	1·183	1·197	1·210
0·01000	(0·0981)	1·167	1·158	1·148	1·144	1·141	1·141	1·143	1·146	1·153	1·160	1·169	1·177	1·192	1·205
0·01500	(0·147)	1·143	1·136	1·129	1·126	1·125	1·126	1·130	1·134	1·142	1·150	1·160	1·169	1·185	1·198
0·02000	(0·196)	1·128	1·122	1·116	1·114	1·114	1·117	1·121	1·126	1·135	1·144	1·154	1·163	1·180	1·194
0·03000	(0·294)	1·109	1·104	1·100	1·100	1·101	1·105	1·110	1·116	1·126	1·136	1·147	1·157	1·174	1·189
0·04000	(0·392)	1·096	1·093	1·090	1·091	1·093	1·098	1·104	1·110	1·121	1·131	1·142	1·153	1·170	1·185
0·06000	(0·588)	1·081	1·079	1·078	1·080	1·082	1·089	1·096	1·102	1·114	1·125	1·137	1·148	1·166	1·181
0·08000	(0·785)	1·071	1·070	1·070	1·073	1·076	1·083	1·090	1·097	1·110	1·121	1·134	1·145	1·163	1·179
0·10000	(0·981)	1·064	1·063	1·065	1·068	1·071	1·079	1·087	1·094	1·107	1·119	1·131	1·142	1·161	1·177
0·15000	(1·47)	1·053	1·053	1·056	1·060	1·064	1·073	1·081	1·089	1·103	1·115	1·128	1·139	1·159	1·175
0·20000	(1·96)	1·046	1·047	1·050	1·055	1·059	1·069	1·078	1·086	1·100	1·112	1·125	1·137	1·157	1·173
0·30000	(2·94)	1·037	1·039	1·043	1·049	1·054	1·064	1·073	1·082	1·096	1·109	1·123	1·135	1·155	1·171
0·40000	(3·92)	1·032	1·034	1·039	1·045	1·051	1·061	1·071	1·079	1·094	1·107	1·121	1·133	1·153	1·170
0·60000	(5·88)	1·025	1·028	1·034	1·040	1·046	1·058	1·068	1·076	1·092	1·105	1·119	1·131	1·152	1·169
0·80000	(7·85)	1·021	1·024	1·031	1·038	1·044	1·056	1·066	1·075	1·090	1·104	1·118	1·130	1·151	1·168
1·00000	(9·81)	1·018	1·022	1·029	1·036	1·042	1·054	1·064	1·074	1·089	1·103	1·117	1·129	1·150	1·167
1·50000	(14·7)	1·014	1·018	1·026	1·033	1·040	1·052	1·062	1·072	1·088	1·101	1·116	1·128	1·149	1·166
2·00000	(19·6)	1·011	1·016	1·023	1·031	1·038	1·050	1·061	1·071	1·087	1·101	1·115	1·128	1·149	1·166
3·00000	(29·4)	1·008	1·013	1·021	1·029	1·036	1·049	1·060	1·069	1·086	1·099	1·114	1·127	1·148	1·165
S	$(\Delta p/\rho l)$	125	150	200	250	300	400	500	600	800	1000	1250	1500	2000	2500

Grad'nt k.p.g. (Equivalent) Pipe diameters in mm Roughness size, $k_s = 0 \cdot 30$ mm

E200

Kin. visc., $\nu = 10.0 \times 10^{-6}$ m^{2}s^{-1}; Roughness size, $k_s = 0.60$ mm

$S = 0.00010$ to 3.00000

i.e. kin. pr. grad., $\Delta p/\rho l =$ (0·00098) to (29·4) ms^{-2}

This table shows values of m, as follows m_C for Colebrook-White solutions; or, where $\mathbf{R} \leq 2000$, m_P for laminar flow

Grad'nt S	k.p.g. $(\Delta p/\rho l)$	125	150	200	250	300	400	500	600	800	1000	1250	1500	2000	2500
						(Equivalent) Pipe diameters in mm									
0·00010	(0·00098)	2·072	1·625	*	1·556	1·528	1·492	1·470	1·455	1·438	1·428	1·421	1·418	1·416	1·418
0·00015	(0·00147)	1·692	1·327	*	1·506	1·482	1·451	1·432	1·420	1·406	1·398	1·394	1·392	1·392	1·395
0·00020	(0·00196)	1·465	*	1·505	1·473	1·451	1·424	1·407	1·397	1·385	1·379	1·375	1·375	1·377	1·381
0·00030	(0·00294)	1·196	*	1·458	1·430	1·412	1·388	1·375	1·366	1·357	1·353	1·352	1·353	1·357	1·362
0·00040	(0·00392)	*	*	1·427	1·403	1·386	1·365	1·354	1·347	1·339	1·337	1·337	1·338	1·344	1·350
0·00060	(0·00588)	*	1·423	1·388	1·366	1·352	1·335	1·326	1·321	1·316	1·316	1·317	1·320	1·327	1·335
0·00080	(0·00785)	1·418	1·393	1·362	1·343	1·331	1·316	1·309	1·304	1·302	1·302	1·305	1·308	1·317	1·326
0·00100	(0·00981)	1·395	1·372	1·343	1·326	1·315	1·302	1·296	1·292	1·291	1·292	1·296	1·300	1·309	1·319
0·00150	(0·0147)	1·356	1·336	1·312	1·298	1·289	1·279	1·274	1·273	1·273	1·276	1·281	1·286	1·297	1·307
0·00200	(0·0196)	1·331	1·313	1·292	1·279	1·272	1·264	1·261	1·260	1·262	1·266	1·271	1·277	1·289	1·300
0·00300	(0·0294)	1·299	1·284	1·266	1·256	1·250	1·245	1·243	1·244	1·248	1·253	1·260	1·267	1·280	1·291
0·00400	(0·0392)	1·278	1·265	1·249	1·241	1·236	1·233	1·232	1·234	1·239	1·245	1·252	1·260	1·274	1·286
0·00600	(0·0588)	1·252	1·241	1·229	1·222	1·219	1·217	1·219	1·221	1·228	1·235	1·243	1·251	1·266	1·279
0·00800	(0·0785)	1·236	1·226	1·215	1·210	1·208	1·208	1·210	1·213	1·221	1·228	1·238	1·246	1·262	1·275
0·01000	(0·0981)	1·224	1·215	1·206	1·202	1·201	1·201	1·204	1·208	1·216	1·224	1·234	1·243	1·259	1·272
0·01500	(0·147)	1·204	1·198	1·191	1·188	1·188	1·190	1·194	1·199	1·208	1·217	1·227	1·237	1·253	1·268
0·02000	(0·196)	1·192	1·187	1·181	1·180	1·180	1·183	1·188	1·193	1·203	1·213	1·223	1·233	1·250	1·265
0·03000	(0·294)	1·177	1·173	1·169	1·169	1·171	1·175	1·181	1·186	1·197	1·207	1·219	1·229	1·247	1·262
0·04000	(0·392)	1·168	1·164	1·162	1·163	1·165	1·170	1·176	1·182	1·194	1·204	1·216	1·226	1·244	1·260
0·06000	(0·588)	1·156	1·154	1·153	1·155	1·157	1·164	1·171	1·177	1·189	1·200	1·212	1·223	1·242	1·257
0·08000	(0·785)	1·149	1·147	1·147	1·150	1·153	1·160	1·167	1·174	1·187	1·198	1·210	1·221	1·240	1·256
0·10000	(0·981)	1·144	1·143	1·144	1·146	1·150	1·157	1·165	1·172	1·185	1·196	1·209	1·220	1·239	1·255
0·15000	(1·47)	1·136	1·136	1·137	1·141	1·145	1·153	1·161	1·169	1·182	1·194	1·206	1·218	1·237	1·253
0·20000	(1·96)	1·131	1·131	1·134	1·138	1·142	1·151	1·159	1·166	1·180	1·192	1·205	1·217	1·236	1·252
0·30000	(2·94)	1·125	1·126	1·129	1·134	1·138	1·148	1·156	1·164	1·178	1·190	1·203	1·215	1·235	1·251
0·40000	(3·92)	1·122	1·123	1·127	1·131	1·136	1·146	1·155	1·163	1·177	1·189	1·202	1·214	1·234	1·250
0·60000	(5·88)	1·117	1·119	1·123	1·128	1·134	1·143	1·153	1·161	1·175	1·188	1·201	1·213	1·233	1·250
0·80000	(7·85)	1·115	1·116	1·121	1·127	1·132	1·142	1·151	1·160	1·174	1·187	1·201	1·212	1·233	1·249
1·00000	(9·81)	1·113	1·115	1·120	1·125	1·131	1·141	1·151	1·159	1·174	1·186	1·200	1·212	1·232	1·249
1·50000	(14·7)	1·110	1·112	1·118	1·124	1·129	1·140	1·149	1·158	1·173	1·186	1·199	1·211	1·232	1·248
2·00000	(19·6)	1·108	1·111	1·116	1·122	1·128	1·139	1·149	1·157	1·172	1·185	1·199	1·211	1·231	1·248
3·00000	(29·4)	1·106	1·109	1·115	1·121	1·127	1·138	1·148	1·156	1·172	1·185	1·198	1·210	1·231	1·248
S	$(\Delta p/\rho l)$	125	150	200	250	300	400	500	600	800	1000	1250	1500	2000	2500

Grad'nt k.p.g. (Equivalent) Pipe diameters in mm Roughness size, $k_s = 0.60$ mm

E201

Kin. visc., $\nu = 10.0 \times 10^{-6}$ m^{2}s^{-1}; Roughness size, $k_s = 1.50$ mm

$S = 0.00010$ to 3.00000

i.e. kin. pr. grad., $\Delta p/\rho l =$ (0·00098) to (29·4) ms^{-2}

This table shows values of m, as follows m_C for Colebrook-White solutions; or, where $\mathbf{R} \leq 2000$, m_P for laminar flow

Grad'nt S	k.p.g. $(\Delta p/\rho l)$	125	150	200	250	300	400	500	600	800	1000	1250	1500	2000	2500
						(Equivalent) Pipe diameters in mm									
0·00010	(0·00098)	2·072	1·625	*	1·608	1·581	1·546	1·525	1·511	1·495	1·487	1·482	1·480	1·481	1·485
0·00015	(0·00147)	1·692	1·327	*	1·564	1·540	1·510	1·493	1·482	1·469	1·463	1·460	1·460	1·463	1·468
0·00020	(0·00196)	1·465	*	1·567	1·535	1·514	1·487	1·472	1·462	1·452	1·448	1·447	1·447	1·452	1·458
0·00030	(0·00294)	1·196	*	1·526	1·498	1·480	1·458	1·446	1·438	1·431	1·429	1·429	1·431	1·438	1·445
0·00040	(0·00392)	*	*	1·500	1·475	1·459	1·440	1·429	1·423	1·418	1·417	1·418	1·421	1·429	1·437
0·00060	(0·00588)	*	1·501	1·466	1·446	1·432	1·416	1·408	1·404	1·401	1·402	1·405	1·409	1·418	1·427
0·00080	(0·00785)	*	1·477	1·445	1·427	1·415	1·402	1·395	1·392	1·390	1·392	1·396	1·401	1·411	1·421
0·00100	(0·00981)	1·482	1·459	1·430	1·414	1·403	1·391	1·386	1·383	1·383	1·385	1·390	1·395	1·406	1·417
0·00150	(0·0147)	1·451	1·431	1·406	1·392	1·383	1·374	1·370	1·369	1·371	1·375	1·381	1·387	1·399	1·410
0·00200	(0·0196)	1·431	1·413	1·391	1·378	1·371	1·364	1·361	1·361	1·364	1·368	1·375	1·381	1·394	1·406
0·00300	(0·0294)	1·406	1·390	1·372	1·362	1·356	1·350	1·349	1·350	1·354	1·360	1·367	1·375	1·388	1·401
0·00400	(0·0392)	1·390	1·376	1·360	1·351	1·346	1·342	1·342	1·344	1·349	1·355	1·363	1·371	1·385	1·398
0·00600	(0·0588)	1·371	1·359	1·345	1·338	1·335	1·333	1·334	1·336	1·342	1·349	1·358	1·366	1·381	1·394
0·00800	(0·0785)	1·359	1·349	1·336	1·330	1·328	1·327	1·328	1·331	1·338	1·345	1·354	1·363	1·378	1·392
0·01000	(0·0981)	1·351	1·341	1·330	1·325	1·323	1·322	1·324	1·328	1·335	1·343	1·352	1·361	1·377	1·390
0·01500	(0·147)	1·338	1·329	1·320	1·316	1·315	1·316	1·319	1·322	1·331	1·339	1·349	1·358	1·374	1·388
0·02000	(0·196)	1·330	1·322	1·314	1·311	1·310	1·312	1·315	1·319	1·328	1·337	1·347	1·356	1·372	1·386
0·03000	(0·294)	1·320	1·313	1·307	1·304	1·304	1·307	1·311	1·315	1·325	1·334	1·344	1·354	1·370	1·385
0·04000	(0·392)	1·314	1·308	1·302	1·301	1·301	1·304	1·308	1·313	1·323	1·332	1·343	1·352	1·369	1·384
0·06000	(0·588)	1·306	1·301	1·297	1·296	1·297	1·300	1·305	1·310	1·320	1·330	1·341	1·351	1·368	1·382
0·08000	(0·785)	1·302	1·297	1·294	1·293	1·294	1·298	1·303	1·309	1·319	1·329	1·340	1·350	1·367	1·382
0·10000	(0·981)	1·299	1·295	1·291	1·291	1·292	1·297	1·302	1·307	1·318	1·328	1·339	1·349	1·366	1·381
0·15000	(1·47)	1·294	1·291	1·288	1·288	1·290	1·294	1·300	1·306	1·316	1·326	1·338	1·348	1·365	1·380
0·20000	(1·96)	1·291	1·288	1·286	1·286	1·288	1·293	1·299	1·305	1·316	1·326	1·337	1·347	1·365	1·380
0·30000	(2·94)	1·288	1·285	1·283	1·284	1·286	1·291	1·297	1·303	1·314	1·325	1·336	1·346	1·364	1·379
0·40000	(3·92)	1·286	1·283	1·282	1·283	1·285	1·290	1·296	1·302	1·314	1·324	1·336	1·346	1·364	1·379
0·60000	(5·88)	1·283	1·281	1·280	1·281	1·283	1·289	1·295	1·302	1·313	1·323	1·335	1·345	1·363	1·379
0·80000	(7·85)	1·282	1·280	1·279	1·280	1·283	1·289	1·295	1·301	1·313	1·323	1·335	1·345	1·363	1·378
1·00000	(9·81)	1·281	1·279	1·278	1·280	1·282	1·288	1·294	1·301	1·312	1·323	1·334	1·345	1·363	1·378
1·50000	(14·7)	1·279	1·277	1·277	1·279	1·281	1·287	1·294	1·300	1·312	1·322	1·334	1·345	1·363	1·378
2·00000	(19·6)	1·278	1·277	1·276	1·278	1·281	1·287	1·293	1·300	1·311	1·322	1·334	1·344	1·362	1·378
3·00000	(29·4)	1·277	1·276	1·275	1·277	1·280	1·286	1·293	1·299	1·311	1·322	1·334	1·344	1·362	1·378
S	$(\Delta p/\rho l)$	125	150	200	250	300	400	500	600	800	1000	1250	1500	2000	2500

Grad'nt k.p.g. (Equivalent) Pipe diameters in mm Roughness size, $k_s = 1.50$ mm

Kin. visc., $\nu = 12.5 \times 10^{-6}$ m^{2}s^{-1};
$S = 0.00010$ to 30.0000
i.e. kin. pr. grad., $\Delta p/\rho l =$
(0.00098) to (294) ms^{-2}

Roughness size, $k_s = 0.003$ mm
This table shows values of m, as follows
m_C for Colebrook-White solutions; or,
where $R \leq 2000$, m_P for laminar flow

Grad'nt	k.p.g.	(Equivalent) Pipe diameters in mm													
S	$(\Delta p/\rho l)$	6	8	10	12.5	15	20	25	30	40	50	60	80	100	125
0.00030	(0.00294)	85.72	58.41	43.38	32.22	25.26	17.21	12.78	10.03	6.832	5.074	3.979	2.711	2.013	1.495
0.00040	(0.00392)	74.23	50.58	37.57	27.90	21.88	14.91	11.07	8.682	5.916	4.394	3.446	2.348	1.744	1.295
0.00060	(0.00588)	60.61	41.30	30.67	22.78	17.86	12.17	9.040	7.089	4.831	3.588	2.813	1.917	1.424	*
0.00080	(0.00785)	52.49	35.77	26.56	19.73	15.47	10.54	7.829	6.139	4.184	3.107	2.436	1.660	1.233	*
0.00100	(0.00981)	46.95	31.99	23.76	17.64	13.84	9.429	7.002	5.491	3.742	2.779	2.179	1.485	1.103	*
0.00150	(0.0147)	38.33	26.12	19.40	14.41	11.30	7.699	5.717	4.484	3.055	2.269	1.779	1.212	*	1.330
0.00200	(0.0196)	33.20	22.62	16.80	12.48	9.784	6.667	4.951	3.883	2.646	1.965	1.541	*	*	1.296
0.00300	(0.0294)	27.11	18.47	13.72	10.19	7.989	5.444	4.043	3.170	2.160	1.604	1.258	*	1.279	1.251
0.00400	(0.0392)	23.47	16.00	11.88	8.822	6.919	4.714	3.501	2.746	1.871	1.389	1.090	*	1.246	1.221
0.00600	(0.0588)	19.17	13.06	9.700	7.204	5.649	3.849	2.859	2.242	1.528	1.134	*	1.229	1.203	1.181
0.00800	(0.0785)	16.60	11.31	8.400	6.238	4.892	3.334	2.476	1.941	1.323	0.982	*	1.198	1.174	1.154
0.01000	(0.0981)	14.85	10.12	7.513	5.580	4.376	2.982	2.214	1.736	1.183	*	1.211	1.175	1.153	1.134
0.01500	(0.147)	12.12	8.260	6.135	4.556	3.573	2.435	1.808	1.418	0.966	*	1.166	1.135	1.116	1.100
0.02000	(0.196)	10.50	7.154	5.313	3.946	3.094	2.108	1.566	1.228	*	1.159	1.137	1.109	1.091	1.076
0.03000	(0.294)	8.572	5.841	4.338	3.222	2.526	1.721	1.278	1.003	*	1.117	1.098	1.073	1.058	1.045
0.04000	(0.392)	7.423	5.058	3.757	2.790	2.188	1.491	1.107	*	1.114	1.089	1.072	1.049	1.036	1.024
0.06000	(0.588)	6.061	4.130	3.067	2.278	1.786	1.217	0.904	*	1.074	1.052	1.037	1.018	1.006	0.996
0.08000	(0.785)	5.249	3.577	2.656	1.973	1.547	1.054	*	1.079	1.047	1.028	1.014	0.996	0.986	0.977
0.10000	(0.981)	4.695	3.199	2.376	1.764	1.384	0.943	*	1.057	1.028	1.009	0.997	0.980	0.971	0.963
0.15000	(1.47)	3.833	2.612	1.940	1.441	1.130	*	1.039	1.019	0.994	0.978	0.967	0.953	0.945	0.938
0.20000	(1.96)	3.320	2.262	1.680	1.248	0.978	*	1.012	0.994	0.971	0.956	0.946	0.934	0.927	0.922
0.30000	(2.94)	2.711	1.847	1.372	1.019	0.799	0.999	0.977	0.961	0.940	0.928	0.919	0.909	0.903	0.899
0.40000	(3.92)	2.347	1.600	1.188	0.882	*	0.974	0.953	0.939	0.920	0.909	0.901	0.892	0.887	0.884
0.60000	(5.88)	1.917	1.306	0.970	*	*	0.939	0.921	0.909	0.893	0.883	0.877	0.869	0.866	0.863
0.80000	(7.85)	1.660	1.131	0.840	*	0.943	0.917	0.900	0.889	0.875	0.866	0.861	0.854	0.851	0.849
1.00000	(9.81)	1.485	1.012	0.751	*	0.924	0.900	0.885	0.874	0.861	0.853	0.848	0.843	0.840	0.839
1.50000	(14.7)	1.212	0.826	*	0.908	0.892	0.871	0.858	0.849	0.837	0.831	0.827	0.823	0.821	0.820
2.00000	(19.6)	1.050	*	*	0.885	0.871	0.851	0.839	0.831	0.822	0.816	0.812	0.809	0.808	0.808
3.00000	(29.4)	0.857	*	0.874	0.855	0.842	0.825	0.815	0.808	0.800	0.796	0.793	0.791	0.790	0.791
4.00000	(39.2)	0.742	*	0.852	0.835	0.823	0.808	0.799	0.793	0.786	0.782	0.780	0.778	0.779	0.780
6.00000	(58.8)	*	0.841	0.823	0.808	0.798	0.785	0.777	0.772	0.767	0.764	0.762	0.762	0.763	0.765
8.00000	(78.5)	*	0.820	0.804	0.790	0.781	0.770	0.763	0.758	0.754	0.751	0.750	0.751	0.752	0.754
10.0000	(98.1)	0.829	0.805	0.789	0.777	0.769	0.758	0.752	0.748	0.744	0.742	0.742	0.742	0.744	0.747
15.0000	(147)	0.799	0.778	0.765	0.754	0.747	0.738	0.733	0.730	0.727	0.726	0.726	0.728	0.730	0.733
20.0000	(196)	0.779	0.760	0.748	0.739	0.732	0.725	0.720	0.718	0.716	0.715	0.716	0.718	0.721	0.724
30.0000	(294)	0.753	0.737	0.727	0.718	0.713	0.707	0.703	0.702	0.701	0.701	0.702	0.705	0.708	0.712
S	$(\Delta p/\rho l)$	6	8	10	12.5	15	20	25	30	40	50	60	80	100	125

Grad'nt   k.p.g.   (Equivalent) Pipe diameters in mm

Grad'nt	k.p.g.	(Equivalent) Pipe diameters in mm													
S	$(\Delta p/\rho l)$	125	150	200	250	300	400	500	600	800	1000	1250	1500	2000	2500
0.00010	(0.00098)	2.590	2.031	1.384	*	1.550	1.507	1.480	1.461	1.437	1.423	1.412	1.404	1.397	1.393
0.00015	(0.00147)	2.115	1.658	*	1.523	1.494	1.457	1.433	1.417	1.396	1.384	1.375	1.369	1.363	1.362
0.00020	(0.00196)	1.831	1.436	*	1.484	1.457	1.423	1.401	1.387	1.369	1.358	1.350	1.345	1.341	1.340
0.00030	(0.00294)	1.495	1.173	1.465	1.431	1.408	1.378	1.359	1.347	1.331	1.323	1.316	1.313	1.310	1.310
0.00040	(0.00392)	1.295	*	1.427	1.396	1.375	1.348	1.331	1.320	1.306	1.299	1.294	1.291	1.289	1.290
0.00060	(0.00588)	*	*	1.376	1.350	1.331	1.308	1.293	1.284	1.273	1.267	1.263	1.261	1.261	1.263
0.00080	(0.00785)	*	1.381	1.343	1.318	1.302	1.280	1.268	1.259	1.250	1.245	1.242	1.241	1.242	1.244
0.00100	(0.00981)	*	1.353	1.318	1.295	1.280	1.260	1.249	1.241	1.232	1.228	1.226	1.225	1.227	1.230
0.00150	(0.0147)	1.330	1.306	1.274	1.255	1.242	1.225	1.215	1.209	1.202	1.199	1.198	1.199	1.201	1.205
0.00200	(0.0196)	1.296	1.274	1.246	1.228	1.216	1.201	1.193	1.187	1.182	1.180	1.179	1.180	1.184	1.188
0.00300	(0.0294)	1.251	1.232	1.207	1.192	1.182	1.169	1.162	1.158	1.154	1.153	1.154	1.155	1.160	1.165
0.00400	(0.0392)	1.221	1.204	1.181	1.167	1.158	1.147	1.141	1.138	1.135	1.135	1.136	1.138	1.144	1.149
0.00600	(0.0588)	1.181	1.166	1.146	1.135	1.127	1.118	1.114	1.111	1.110	1.110	1.113	1.115	1.121	1.127
0.00800	(0.0785)	1.154	1.140	1.123	1.113	1.106	1.098	1.095	1.093	1.092	1.094	1.096	1.099	1.106	1.113
0.01000	(0.0981)	1.134	1.122	1.106	1.096	1.090	1.083	1.080	1.079	1.079	1.081	1.084	1.088	1.095	1.101
0.01500	(0.147)	1.100	1.089	1.075	1.067	1.063	1.057	1.055	1.055	1.056	1.059	1.063	1.067	1.074	1.082
0.02000	(0.196)	1.076	1.067	1.055	1.048	1.044	1.040	1.038	1.039	1.041	1.044	1.048	1.052	1.061	1.068
0.03000	(0.294)	1.045	1.037	1.027	1.022	1.019	1.016	1.016	1.016	1.019	1.023	1.028	1.033	1.042	1.050
0.04000	(0.392)	1.024	1.017	1.008	1.004	1.001	1.000	1.000	1.001	1.005	1.009	1.014	1.019	1.029	1.037
0.06000	(0.588)	0.996	0.990	0.983	0.980	0.978	0.978	0.979	0.980	0.985	0.990	0.996	1.001	1.011	1.020
0.08000	(0.785)	0.977	0.972	0.966	0.963	0.962	0.963	0.964	0.966	0.972	0.977	0.983	0.989	0.999	1.008
0.10000	(0.981)	0.963	0.958	0.953	0.951	0.950	0.951	0.953	0.956	0.961	0.967	0.973	0.979	0.990	0.999
0.15000	(1.47)	0.938	0.935	0.931	0.930	0.930	0.931	0.934	0.937	0.943	0.949	0.956	0.963	0.974	0.984
0.20000	(1.96)	0.922	0.918	0.916	0.915	0.916	0.918	0.921	0.925	0.931	0.938	0.945	0.952	0.963	0.973
0.30000	(2.94)	0.899	0.897	0.895	0.895	0.896	0.900	0.904	0.907	0.915	0.922	0.929	0.936	0.948	0.959
0.40000	(3.92)	0.884	0.882	0.881	0.882	0.883	0.887	0.891	0.896	0.903	0.911	0.919	0.926	0.938	0.949
0.60000	(5.88)	0.863	0.862	0.862	0.864	0.866	0.870	0.875	0.880	0.888	0.896	0.904	0.912	0.925	0.936
0.80000	(7.85)	0.849	0.849	0.849	0.851	0.854	0.859	0.864	0.869	0.878	0.886	0.894	0.902	0.915	0.927
1.00000	(9.81)	0.839	0.838	0.840	0.842	0.845	0.850	0.856	0.861	0.870	0.878	0.887	0.895	0.908	0.920
1.50000	(14.7)	0.820	0.821	0.823	0.826	0.829	0.835	0.841	0.847	0.856	0.865	0.874	0.882	0.896	0.908
2.00000	(19.6)	0.808	0.809	0.812	0.815	0.818	0.825	0.831	0.837	0.847	0.856	0.865	0.874	0.888	0.900
3.00000	(29.4)	0.791	0.793	0.796	0.800	0.804	0.811	0.818	0.824	0.835	0.844	0.854	0.862	0.877	0.890
S	$(\Delta p/\rho l)$	125	150	200	250	300	400	500	600	800	1000	1250	1500	2000	2500

Grad'nt   k.p.g.   (Equivalent) Pipe diameters in mm

Kinematic viscosity, $\nu = 12.5 \times 10^{-6}$ m^{2}s^{-1};   **Roughness size, $k_s = 0.003$ mm**

E203

Kin. visc., $\nu = 12.5 \times 10^{-6} \ m^2 s^{-1}$;
$S = 0.00010$ to 30.0000

i.e. kin. pr. grad., $\Delta p/\rho l =$
(0.00098) to $(294) \ ms^{-2}$

Roughness size, $k_s = 0.006$ mm
This table shows values of m, as follows

m_C for Colebrook-White solutions; or,
where $\mathbf{R} \leq 2000$, m_P for laminar flow

Grad'nt S	k.p.g. $(\Delta p/\rho l)$	6	8	10	12.5	15	20	25	30	40	50	60	80	100	125
		(Equivalent) Pipe diameters in mm													
0.00030	(0.00294)	85.72	58.41	43.38	32.22	25.26	17.21	12.78	10.03	6.832	5.074	3.979	2.711	2.013	1.495
0.00040	(0.00392)	74.23	50.58	37.57	27.90	21.88	14.91	11.07	8.682	5.916	4.394	3.446	2.348	1.744	1.295
0.00060	(0.00588)	60.61	41.30	30.67	22.78	17.86	12.17	9.040	7.089	4.831	3.588	2.813	1.917	1.424	*
0.00080	(0.00785)	52.49	35.77	26.56	19.73	15.47	10.54	7.829	6.139	4.184	3.107	2.436	1.660	1.233	*
0.00100	(0.00981)	46.95	31.99	23.76	17.64	13.84	9.429	7.002	5.491	3.742	2.779	2.179	1.485	1.103	*
0.00150	(0.0147)	38.33	26.12	19.40	14.41	11.30	7.699	5.717	4.484	3.055	2.269	1.779	1.212	*	1.330
0.00200	(0.0196)	33.20	22.62	16.80	12.48	9.784	6.667	4.951	3.883	2.646	1.965	1.541	*	*	1.296
0.00300	(0.0294)	27.11	18.47	13.72	10.19	7.989	5.444	4.043	3.170	2.160	1.604	1.258	*	1.279	1.251
0.00400	(0.0392)	23.47	16.00	11.88	8.822	6.919	4.714	3.501	2.746	1.871	1.389	1.090	*	1.247	1.221
0.00600	(0.0588)	19.17	13.06	9.700	7.204	5.649	3.849	2.859	2.242	1.528	1.134	*	1.230	1.204	1.182
0.00800	(0.0785)	16.60	11.31	8.400	6.238	4.892	3.334	2.476	1.941	1.323	0.982	*	1.199	1.175	1.155
0.01000	(0.0981)	14.85	10.12	7.513	5.580	4.376	2.982	2.214	1.736	1.183	*	1.211	1.176	1.153	1.135
0.01500	(0.147)	12.12	8.260	6.135	4.556	3.573	2.435	1.808	1.418	0.966	*	1.167	1.136	1.117	1.101
0.02000	(0.196)	10.50	7.154	5.313	3.946	3.094	2.108	1.566	1.228	*	1.159	1.138	1.109	1.092	1.077
0.03000	(0.294)	8.572	5.841	4.338	3.222	2.526	1.721	1.278	1.003	*	1.118	1.099	1.074	1.059	1.046
0.04000	(0.392)	7.423	5.058	3.757	2.790	2.188	1.491	1.107	*	1.115	1.090	1.073	1.050	1.037	1.026
0.06000	(0.588)	6.061	4.130	3.067	2.278	1.786	1.217	0.904	*	1.075	1.053	1.038	1.019	1.007	0.998
0.08000	(0.785)	5.249	3.577	2.656	1.973	1.547	1.054	*	1.080	1.049	1.029	1.015	0.998	0.987	0.979
0.10000	(0.981)	4.695	3.199	2.376	1.764	1.384	0.943	*	1.058	1.029	1.011	0.998	0.982	0.972	0.965
0.15000	(1.47)	3.833	2.612	1.940	1.441	1.130	*	1.040	1.021	0.995	0.979	0.968	0.954	0.946	0.940
0.20000	(1.96)	3.320	2.262	1.680	1.248	0.978	*	1.014	0.996	0.972	0.958	0.948	0.936	0.929	0.924
0.30000	(2.94)	2.711	1.847	1.372	1.019	0.799	1.001	0.978	0.963	0.942	0.930	0.921	0.911	0.905	0.901
0.40000	(3.92)	2.347	1.600	1.188	0.882	*	0.975	0.955	0.940	0.922	0.911	0.903	0.894	0.890	0.886
0.60000	(5.88)	1.917	1.306	0.970	*	*	0.941	0.923	0.911	0.895	0.886	0.879	0.872	0.868	0.866
0.80000	(7.85)	1.660	1.131	0.840	*	0.945	0.919	0.903	0.891	0.877	0.869	0.863	0.857	0.854	0.852
1.00000	(9.81)	1.485	1.012	0.751	*	0.926	0.902	0.887	0.877	0.864	0.856	0.851	0.846	0.843	0.842
1.50000	(14.7)	1.212	0.826	*	0.911	0.895	0.873	0.860	0.852	0.841	0.834	0.830	0.826	0.825	0.825
2.00000	(19.6)	1.050	*	*	0.888	0.873	0.854	0.843	0.835	0.825	0.820	0.816	0.813	0.812	0.813
3.00000	(29.4)	0.857	*	0.877	0.858	0.845	0.829	0.819	0.812	0.804	0.800	0.797	0.796	0.796	0.797
4.00000	(39.2)	0.742	*	0.855	0.838	0.827	0.812	0.803	0.797	0.790	0.787	0.785	0.784	0.784	0.786
6.00000	(58.8)	*	0.845	0.827	0.812	0.802	0.789	0.782	0.777	0.772	0.769	0.768	0.768	0.769	0.772
8.00000	(78.5)	*	0.824	0.808	0.795	0.786	0.774	0.768	0.764	0.759	0.757	0.757	0.757	0.759	0.762
10.0000	(98.1)	*	0.809	0.794	0.782	0.773	0.763	0.757	0.754	0.750	0.749	0.748	0.750	0.752	0.755
15.0000	(147)	0.804	0.783	0.770	0.759	0.752	0.744	0.739	0.736	0.734	0.733	0.734	0.736	0.739	0.743
20.0000	(196)	0.785	0.766	0.754	0.745	0.738	0.731	0.727	0.725	0.723	0.723	0.724	0.727	0.730	0.735
30.0000	(294)	0.759	0.743	0.733	0.725	0.720	0.714	0.711	0.710	0.709	0.710	0.712	0.715	0.719	0.724
S	$(\Delta p/\rho l)$	6	8	10	12.5	15	20	25	30	40	50	60	80	100	125

Grad'nt S	k.p.g. $(\Delta p/\rho l)$	125	150	200	250	300	400	500	600	800	1000	1250	1500	2000	2500
		(Equivalent) Pipe diameters in mm													
0.00010	(0.00098)	2.590	2.031	1.384	*	1.550	1.507	1.480	1.461	1.437	1.423	1.412	1.405	1.397	1.394
0.00015	(0.00147)	2.115	1.658	*	1.524	1.495	1.457	1.433	1.417	1.396	1.384	1.375	1.369	1.364	1.362
0.00020	(0.00196)	1.831	1.436	*	1.484	1.458	1.423	1.402	1.387	1.369	1.358	1.350	1.346	1.341	1.340
0.00030	(0.00294)	1.495	1.173	1.465	1.431	1.408	1.378	1.360	1.347	1.332	1.323	1.317	1.313	1.311	1.311
0.00040	(0.00392)	1.295	*	1.427	1.396	1.375	1.348	1.332	1.320	1.307	1.299	1.294	1.291	1.290	1.291
0.00060	(0.00588)	*	*	1.376	1.350	1.332	1.308	1.294	1.284	1.273	1.267	1.263	1.262	1.261	1.263
0.00080	(0.00785)	*	1.381	1.343	1.319	1.302	1.281	1.268	1.260	1.250	1.245	1.242	1.241	1.242	1.245
0.00100	(0.00981)	*	1.354	1.318	1.295	1.280	1.261	1.249	1.241	1.233	1.229	1.226	1.226	1.228	1.231
0.00150	(0.0147)	1.330	1.306	1.275	1.255	1.242	1.225	1.216	1.210	1.203	1.200	1.199	1.199	1.202	1.206
0.00200	(0.0196)	1.296	1.274	1.246	1.228	1.216	1.202	1.193	1.188	1.182	1.180	1.180	1.181	1.185	1.189
0.00300	(0.0294)	1.251	1.232	1.208	1.192	1.182	1.170	1.163	1.159	1.155	1.154	1.155	1.156	1.161	1.166
0.00400	(0.0392)	1.221	1.204	1.182	1.168	1.159	1.148	1.142	1.139	1.136	1.136	1.137	1.139	1.145	1.150
0.00600	(0.0588)	1.182	1.166	1.147	1.135	1.128	1.119	1.114	1.112	1.111	1.111	1.114	1.116	1.123	1.129
0.00800	(0.0785)	1.155	1.141	1.124	1.113	1.107	1.099	1.096	1.094	1.093	1.095	1.098	1.101	1.108	1.114
0.01000	(0.0981)	1.135	1.122	1.106	1.097	1.091	1.084	1.081	1.080	1.080	1.082	1.085	1.089	1.096	1.103
0.01500	(0.147)	1.101	1.090	1.076	1.068	1.064	1.059	1.057	1.056	1.058	1.060	1.064	1.068	1.076	1.084
0.02000	(0.196)	1.077	1.068	1.056	1.049	1.045	1.041	1.040	1.040	1.042	1.045	1.050	1.054	1.063	1.071
0.03000	(0.294)	1.046	1.038	1.028	1.023	1.020	1.017	1.017	1.018	1.021	1.025	1.030	1.035	1.044	1.053
0.04000	(0.392)	1.026	1.018	1.010	1.005	1.003	1.001	1.002	1.003	1.007	1.011	1.017	1.022	1.032	1.040
0.06000	(0.588)	0.998	0.992	0.985	0.981	0.980	0.979	0.981	0.983	0.987	0.992	0.998	1.004	1.014	1.024
0.08000	(0.785)	0.979	0.974	0.968	0.965	0.964	0.965	0.966	0.969	0.974	0.980	0.986	0.992	1.003	1.012
0.10000	(0.981)	0.965	0.960	0.955	0.953	0.953	0.954	0.956	0.958	0.964	0.970	0.977	0.983	0.994	1.004
0.15000	(1.47)	0.940	0.937	0.933	0.932	0.932	0.934	0.937	0.940	0.947	0.953	0.960	0.967	0.979	0.989
0.20000	(1.96)	0.924	0.921	0.918	0.918	0.918	0.921	0.924	0.928	0.935	0.942	0.949	0.956	0.968	0.979
0.30000	(2.94)	0.901	0.899	0.898	0.898	0.900	0.903	0.907	0.911	0.919	0.926	0.934	0.942	0.954	0.965
0.40000	(3.92)	0.886	0.885	0.884	0.885	0.887	0.891	0.896	0.900	0.908	0.916	0.924	0.932	0.945	0.956
0.60000	(5.88)	0.866	0.865	0.866	0.868	0.870	0.875	0.880	0.885	0.894	0.902	0.911	0.919	0.932	0.944
0.80000	(7.85)	0.852	0.852	0.853	0.856	0.858	0.864	0.869	0.875	0.884	0.892	0.902	0.910	0.924	0.936
1.00000	(9.81)	0.842	0.842	0.844	0.847	0.850	0.856	0.862	0.867	0.877	0.885	0.895	0.903	0.918	0.930
1.50000	(14.7)	0.825	0.825	0.828	0.831	0.835	0.842	0.848	0.854	0.864	0.873	0.883	0.892	0.907	0.920
2.00000	(19.6)	0.813	0.814	0.817	0.821	0.825	0.832	0.839	0.845	0.856	0.865	0.875	0.885	0.900	0.913
3.00000	(29.4)	0.797	0.798	0.803	0.807	0.811	0.819	0.826	0.833	0.845	0.854	0.865	0.875	0.891	0.904
S	$(\Delta p/\rho l)$	125	150	200	250	300	400	500	600	800	1000	1250	1500	2000	2500

Grad'nt   k.p.g.   (Equivalent) Pipe diameters in mm

Kinematic viscosity, $\nu = 12.5 \times 10^{-6} \ m^2 s^{-1}$;

Roughness size, $k_s = 0.006$ mm

Kin. visc., $\nu = 12{\cdot}5\times10^{-6}$ m^{2}s^{-1};
$S = 0{\cdot}00010$ to $30{\cdot}0000$
i.e. kin. pr. grad., $\Delta p/\rho l =$
$(0{\cdot}00098)$ to (294) ms^{-2}

Roughness size, $k_s = 0{\cdot}015$ mm
This table shows values of m, as follows
m_C for Colebrook-White solutions; or,
where $\mathbf{R} \le 2000$, m_P for laminar flow

| Grad'nt | k.p.g. | (Equivalent) Pipe diameters in mm | | | | | | | | | | | | | |
S	$(\Delta p/\rho l)$	6	8	10	12·5	15	20	25	30	40	50	60	80	100	125
0·00030	(0·00294)	85·72	58·41	43·38	32·22	25·26	17·21	12·78	10·03	6·832	5·074	3·979	2·711	2·013	1·495
0·00040	(0·00392)	74·23	50·58	37·57	27·90	21·88	14·91	11·07	8·682	5·916	4·394	3·446	2·348	1·744	1·295
0·00060	(0·00588)	60·61	41·30	30·67	22·78	17·86	12·17	9·040	7·089	4·831	3·588	2·813	1·917	1·424	*
0·00080	(0·00785)	52·49	35·77	26·56	19·73	15·47	10·54	7·829	6·139	4·184	3·107	2·436	1·660	1·233	*
0·00100	(0·00981)	46·95	31·99	23·76	17·64	13·84	9·429	7·002	5·491	3·742	2·779	2·179	1·485	1·103	*
0·00150	(0·0147)	38·33	26·12	19·40	14·41	11·30	7·699	5·717	4·484	3·055	2·269	1·779	1·212	*	1·331
0·00200	(0·0196)	33·20	22·62	16·80	12·48	9·784	6·667	4·951	3·883	2·646	1·965	1·541	*	*	1·298
0·00300	(0·0294)	27·11	18·47	13·72	10·19	7·989	5·444	4·043	3·170	2·160	1·604	1·258	*	1·280	1·253
0·00400	(0·0392)	23·47	16·00	11·88	8·822	6·919	4·714	3·501	2·746	1·871	1·389	1·090	*	1·248	1·223
0·00600	(0·0588)	19·17	13·06	9·700	7·204	5·649	3·849	2·859	2·242	1·528	1·134	*	1·232	1·205	1·183
0·00800	(0·0785)	16·60	11·31	8·400	6·238	4·892	3·334	2·476	1·941	1·323	0·982	*	1·201	1·177	1·157
0·01000	(0·0981)	14·85	10·12	7·513	5·580	4·376	2·982	2·214	1·736	1·183	*	1·213	1·178	1·155	1·137
0·01500	(0·147)	12·12	8·260	6·135	4·556	3·573	2·435	1·808	1·418	0·966	*	1·169	1·138	1·119	1·103
0·02000	(0·196)	10·50	7·154	5·313	3·946	3·094	2·108	1·566	1·228	*	1·162	1·140	1·112	1·094	1·080
0·03000	(0·294)	8·572	5·841	4·338	3·222	2·526	1·721	1·278	1·003	*	1·120	1·101	1·077	1·062	1·050
0·04000	(0·392)	7·423	5·058	3·757	2·790	2·188	1·491	1·107	*	1·118	1·093	1·076	1·054	1·040	1·029
0·06000	(0·588)	6·061	4·130	3·067	2·278	1·786	1·217	0·904	*	1·078	1·057	1·042	1·022	1·011	1·002
0·08000	(0·785)	5·249	3·577	2·656	1·973	1·547	1·054	*	1·084	1·052	1·032	1·019	1·002	0·991	0·983
0·10000	(0·981)	4·695	3·199	2·376	1·764	1·384	0·943	*	1·062	1·033	1·014	1·002	0·986	0·977	0·969
0·15000	(1·47)	3·833	2·612	1·940	1·441	1·130	*	1·044	1·025	0·999	0·983	0·973	0·959	0·951	0·946
0·20000	(1·96)	3·320	2·262	1·680	1·248	0·978	*	1·018	1·000	0·977	0·963	0·953	0·941	0·934	0·930
0·30000	(2·94)	2·711	1·847	1·372	1·019	0·799	1·006	0·983	0·968	0·948	0·935	0·927	0·917	0·912	0·908
0·40000	(3·92)	2·347	1·600	1·188	0·882	*	0·981	0·960	0·946	0·928	0·917	0·910	0·901	0·897	0·894
0·60000	(5·88)	1·917	1·306	0·970	*	*	0·947	0·930	0·917	0·902	0·893	0·887	0·880	0·877	0·875
0·80000	(7·85)	1·660	1·131	0·840	*	0·951	0·925	0·909	0·898	0·885	0·876	0·871	0·866	0·863	0·862
1·00000	(9·81)	1·485	1·012	0·751	*	0·933	0·909	0·894	0·884	0·872	0·864	0·860	0·855	0·853	0·853
1·50000	(14·7)	1·212	0·826	*	0·918	0·902	0·881	0·869	0·860	0·850	0·844	0·840	0·837	0·836	0·837
2·00000	(19·6)	1·050	*	*	0·896	0·882	0·863	0·852	0·844	0·835	0·830	0·827	0·825	0·825	0·826
3·00000	(29·4)	0·857	*	0·886	0·868	0·855	0·839	0·829	0·823	0·815	0·812	0·810	0·809	0·810	0·812
4·00000	(39·2)	0·742	*	0·865	0·848	0·837	0·823	0·814	0·809	0·803	0·800	0·798	0·798	0·800	0·802
6·00000	(58·8)	*	0·855	0·838	0·823	0·814	0·802	0·795	0·790	0·786	0·784	0·783	0·785	0·787	0·790
8·00000	(78·5)	*	0·836	0·820	0·807	0·798	0·788	0·782	0·778	0·775	0·773	0·774	0·775	0·778	0·782
10·0000	(98·1)	*	0·821	0·807	0·795	0·787	0·777	0·772	0·769	0·766	0·766	0·766	0·769	0·772	0·776
15·0000	(147)	0·817	0·797	0·784	0·774	0·768	0·760	0·756	0·754	0·753	0·753	0·754	0·758	0·762	0·767
20·0000	(196)	0·799	0·781	0·770	0·761	0·755	0·749	0·745	0·744	0·743	0·745	0·746	0·750	0·755	0·760
30·0000	(294)	0·776	0·760	0·751	0·743	0·739	0·734	0·732	0·731	0·732	0·734	0·736	0·741	0·746	0·752
S	$(\Delta p/\rho l)$	6	8	10	12·5	15	20	25	30	40	50	60	80	100	125

| Grad'nt | k.p.g. | (Equivalent) Pipe diameters in mm | | | | | | | | | | | | | |
S	$(\Delta p/\rho l)$	125	150	200	250	300	400	500	600	800	1000	1250	1500	2000	2500
0·00010	(0·00098)	2·590	2·031	1·384	*	1·551	1·507	1·480	1·462	1·438	1·424	1·412	1·405	1·398	1·395
0·00015	(0·00147)	2·115	1·658	*	1·524	1·495	1·457	1·434	1·418	1·397	1·385	1·376	1·370	1·365	1·363
0·00020	(0·00196)	1·831	1·436	*	1·485	1·458	1·424	1·402	1·388	1·370	1·359	1·351	1·347	1·342	1·341
0·00030	(0·00294)	1·495	1·173	1·466	1·432	1·409	1·379	1·361	1·348	1·333	1·324	1·318	1·314	1·312	1·312
0·00040	(0·00392)	1·295	*	1·428	1·397	1·376	1·349	1·332	1·321	1·308	1·300	1·295	1·293	1·291	1·292
0·00060	(0·00588)	*	*	1·377	1·351	1·332	1·309	1·295	1·285	1·274	1·268	1·265	1·263	1·263	1·265
0·00080	(0·00785)	*	1·382	1·344	1·320	1·303	1·282	1·269	1·261	1·251	1·247	1·244	1·243	1·244	1·247
0·00100	(0·00981)	*	1·355	1·319	1·297	1·281	1·262	1·250	1·243	1·234	1·230	1·228	1·228	1·230	1·233
0·00150	(0·0147)	1·331	1·307	1·276	1·257	1·243	1·227	1·217	1·211	1·205	1·202	1·201	1·201	1·204	1·208
0·00200	(0·0196)	1·298	1·276	1·247	1·230	1·218	1·203	1·195	1·190	1·184	1·182	1·182	1·183	1·187	1·192
0·00300	(0·0294)	1·253	1·234	1·209	1·194	1·184	1·172	1·165	1·161	1·157	1·156	1·157	1·159	1·164	1·169
0·00400	(0·0392)	1·223	1·206	1·183	1·170	1·161	1·150	1·144	1·141	1·139	1·139	1·140	1·142	1·148	1·154
0·00600	(0·0588)	1·183	1·168	1·149	1·138	1·130	1·121	1·117	1·115	1·114	1·115	1·117	1·120	1·127	1·133
0·00800	(0·0785)	1·157	1·143	1·126	1·116	1·109	1·102	1·098	1·097	1·097	1·098	1·101	1·105	1·112	1·119
0·01000	(0·0981)	1·137	1·124	1·109	1·099	1·094	1·087	1·084	1·083	1·084	1·086	1·090	1·093	1·101	1·108
0·01500	(0·147)	1·103	1·092	1·079	1·071	1·067	1·062	1·060	1·060	1·062	1·065	1·069	1·073	1·082	1·090
0·02000	(0·196)	1·080	1·070	1·059	1·052	1·048	1·045	1·044	1·044	1·047	1·050	1·055	1·060	1·069	1·077
0·03000	(0·294)	1·050	1·041	1·032	1·027	1·024	1·021	1·022	1·023	1·026	1·031	1·036	1·041	1·051	1·060
0·04000	(0·392)	1·029	1·022	1·014	1·009	1·007	1·006	1·007	1·008	1·013	1·017	1·023	1·029	1·040	1·049
0·06000	(0·588)	1·002	0·996	0·989	0·986	0·985	0·985	0·986	0·989	0·994	1·000	1·006	1·013	1·024	1·034
0·08000	(0·785)	0·983	0·978	0·973	0·970	0·970	0·971	0·973	0·976	0·982	0·988	0·995	1·001	1·013	1·024
0·10000	(0·981)	0·969	0·965	0·960	0·959	0·959	0·960	0·963	0·966	0·972	0·979	0·986	0·993	1·005	1·016
0·15000	(1·47)	0·946	0·942	0·939	0·939	0·939	0·942	0·945	0·949	0·956	0·963	0·971	0·979	0·992	1·003
0·20000	(1·96)	0·930	0·927	0·925	0·925	0·926	0·929	0·933	0·938	0·946	0·953	0·961	0·969	0·983	0·994
0·30000	(2·94)	0·908	0·907	0·906	0·907	0·908	0·913	0·918	0·922	0·931	0·939	0·948	0·957	0·971	0·983
0·40000	(3·92)	0·894	0·893	0·893	0·894	0·897	0·902	0·907	0·912	0·922	0·930	0·940	0·948	0·963	0·975
0·60000	(5·88)	0·875	0·875	0·876	0·878	0·881	0·887	0·893	0·899	0·909	0·918	0·928	0·937	0·953	0·966
0·80000	(7·85)	0·862	0·862	0·864	0·867	0·871	0·877	0·884	0·890	0·901	0·910	0·921	0·930	0·946	0·959
1·00000	(9·81)	0·853	0·853	0·856	0·859	0·863	0·870	0·877	0·883	0·895	0·904	0·915	0·925	0·941	0·955
1·50000	(14·7)	0·837	0·838	0·842	0·846	0·850	0·858	0·866	0·872	0·884	0·895	0·906	0·916	0·933	0·947
2·00000	(19·6)	0·826	0·828	0·832	0·837	0·841	0·850	0·858	0·865	0·878	0·888	0·900	0·910	0·928	0·942
3·00000	(29·4)	0·812	0·814	0·820	0·825	0·830	0·840	0·848	0·856	0·869	0·880	0·892	0·903	0·921	0·936
S	$(\Delta p/\rho l)$	125	150	200	250	300	400	500	600	800	1000	1250	1500	2000	2500

Grad'nt   k.p.g.   (Equivalent) Pipe diameters in mm

Kinematic viscosity, $\nu = 12{\cdot}5\times10^{-6}$ m^{2}s^{-1} ;   **Roughness size, $k_s = 0{\cdot}015$ mm**

E205

Kin. visc., $\nu = 12{\cdot}5\times10^{-6}$ $m^2 s^{-1}$;
$S = 0{\cdot}00010$ to $3{\cdot}00000$

i.e. kin. pr. grad., $\Delta p/\rho l =$
$(0{\cdot}00098)$ to $(29{\cdot}4)$ ms^{-2}

Roughness size, $k_s = 0{\cdot}030$ mm
This table shows values of m, as follows

m_C for Colebrook-White solutions; or,
where $\mathbf{R} \leq 2000$, m_P for laminar flow

Grad'nt S	k.p.g. $(\Delta p/\rho l)$	125	150	200	250	300	400	500	600	800	1000	1250	1500	2000	2500
0·00010	(0·00098)	2·590	2·031	1·384	*	1·552	1·508	1·481	1·463	1·439	1·425	1·414	1·407	1·399	1·396
0·00015	(0·00147)	2·115	1·658	*	1·525	1·496	1·458	1·435	1·419	1·398	1·386	1·377	1·372	1·366	1·365
0·00020	(0·00196)	1·831	1·436	*	1·486	1·459	1·425	1·404	1·389	1·371	1·360	1·353	1·348	1·344	1·343
0·00030	(0·00294)	1·495	1·173	1·467	1·433	1·410	1·380	1·362	1·350	1·334	1·326	1·320	1·316	1·314	1·314
0·00040	(0·00392)	1·295	*	1·429	1·399	1·378	1·351	1·334	1·323	1·310	1·302	1·297	1·295	1·293	1·294
0·00060	(0·00588)	*	*	1·379	1·352	1·334	1·311	1·297	1·287	1·276	1·270	1·267	1·265	1·266	1·268
0·00080	(0·00785)	*	1·384	1·345	1·321	1·305	1·284	1·271	1·263	1·254	1·249	1·246	1·246	1·247	1·250
0·00100	(0·00981)	*	1·356	1·321	1·298	1·283	1·264	1·252	1·245	1·237	1·233	1·231	1·231	1·233	1·236
0·00150	(0·0147)	1·333	1·309	1·278	1·259	1·246	1·229	1·220	1·214	1·207	1·205	1·204	1·205	1·208	1·212
0·00200	(0·0196)	1·300	1·278	1·250	1·232	1·220	1·206	1·198	1·192	1·187	1·186	1·186	1·187	1·191	1·196
0·00300	(0·0294)	1·255	1·236	1·212	1·197	1·187	1·174	1·168	1·164	1·161	1·160	1·161	1·163	1·169	1·174
0·00400	(0·0392)	1·226	1·208	1·186	1·173	1·164	1·153	1·148	1·145	1·143	1·143	1·145	1·147	1·153	1·160
0·00600	(0·0588)	1·186	1·171	1·152	1·141	1·134	1·125	1·121	1·119	1·118	1·120	1·122	1·126	1·133	1·140
0·00800	(0·0785)	1·160	1·147	1·130	1·120	1·113	1·106	1·103	1·102	1·102	1·104	1·107	1·111	1·119	1·127
0·01000	(0·0981)	1·140	1·128	1·113	1·104	1·098	1·092	1·089	1·089	1·090	1·092	1·096	1·100	1·109	1·117
0·01500	(0·147)	1·107	1·096	1·083	1·076	1·071	1·067	1·066	1·066	1·068	1·072	1·076	1·081	1·091	1·099
0·02000	(0·196)	1·084	1·075	1·064	1·057	1·054	1·050	1·050	1·051	1·054	1·058	1·063	1·069	1·079	1·088
0·03000	(0·294)	1·055	1·047	1·037	1·032	1·030	1·028	1·029	1·030	1·035	1·040	1·046	1·052	1·063	1·072
0·04000	(0·392)	1·035	1·028	1·020	1·016	1·014	1·013	1·015	1·017	1·022	1·027	1·034	1·040	1·052	1·062
0·06000	(0·588)	1·008	1·002	0·996	0·994	0·993	0·993	0·996	0·999	1·005	1·011	1·018	1·025	1·038	1·049
0·08000	(0·785)	0·990	0·985	0·981	0·979	0·979	0·980	0·983	0·986	0·993	1·000	1·008	1·015	1·029	1·040
0·10000	(0·981)	0·977	0·973	0·969	0·968	0·968	0·970	0·974	0·977	0·985	0·992	1·001	1·008	1·022	1·033
0·15000	(1·47)	0·954	0·951	0·949	0·949	0·950	0·953	0·958	0·962	0·971	0·979	0·988	0·996	1·010	1·023
0·20000	(1·96)	0·939	0·937	0·936	0·936	0·938	0·942	0·947	0·952	0·961	0·970	0·979	0·988	1·003	1·015
0·30000	(2·94)	0·919	0·918	0·918	0·920	0·922	0·927	0·933	0·939	0·949	0·958	0·968	0·977	0·993	1·006
0·40000	(3·92)	0·906	0·905	0·906	0·909	0·912	0·918	0·924	0·930	0·941	0·950	0·961	0·970	0·987	1·001
0·60000	(5·88)	0·888	0·889	0·891	0·894	0·898	0·905	0·912	0·919	0·930	0·941	0·952	0·962	0·979	0·993
0·80000	(7·85)	0·877	0·878	0·881	0·885	0·889	0·897	0·904	0·911	0·924	0·934	0·946	0·956	0·974	0·988
1·00000	(9·81)	0·869	0·870	0·874	0·878	0·882	0·891	0·899	0·906	0·919	0·930	0·942	0·952	0·970	0·985
1·50000	(14·7)	0·854	0·856	0·861	0·866	0·871	0·881	0·889	0·897	0·911	0·922	0·935	0·946	0·964	0·980
2·00000	(19·6)	0·845	0·848	0·853	0·859	0·864	0·875	0·884	0·892	0·906	0·918	0·930	0·942	0·961	0·976
3·00000	(29·4)	0·833	0·836	0·843	0·849	0·855	0·866	0·876	0·884	0·899	0·912	0·925	0·937	0·956	0·972
S	$(\Delta p/\rho l)$	125	150	200	250	300	400	500	600	800	1000	1250	1500	2000	2500

Grad'nt   k.p.g.   (Equivalent) Pipe diameters in mm   **Roughness size, $k_s = 0{\cdot}030$ mm**

E206

Kin. visc., $\nu = 12{\cdot}5\times10^{-6}$ $m^2 s^{-1}$;
$S = 0{\cdot}00010$ to $3{\cdot}00000$

i.e. kin. pr. grad., $\Delta p/\rho l =$
$(0{\cdot}00098)$ to $(29{\cdot}4)$ ms^{-2}

Roughness size, $k_s = 0{\cdot}060$ mm
This table shows values of m, as follows

m_C for Colebrook-White solutions; or,
where $\mathbf{R} \leq 2000$, m_P for laminar flow

Grad'nt S	k.p.g. $(\Delta p/\rho l)$	125	150	200	250	300	400	500	600	800	1000	1250	1500	2000	2500
0·00010	(0·00098)	2·590	2·031	1·384	*	1·553	1·510	1·483	1·465	1·441	1·427	1·416	1·409	1·402	1·399
0·00015	(0·00147)	2·115	1·658	*	1·527	1·498	1·461	1·437	1·421	1·401	1·389	1·380	1·375	1·369	1·368
0·00020	(0·00196)	1·831	1·436	*	1·488	1·462	1·427	1·406	1·392	1·374	1·363	1·356	1·351	1·347	1·347
0·00030	(0·00294)	1·495	1·173	1·469	1·436	1·413	1·383	1·365	1·352	1·337	1·329	1·323	1·320	1·318	1·318
0·00040	(0·00392)	1·295	*	1·432	1·401	1·380	1·353	1·337	1·326	1·313	1·306	1·301	1·299	1·298	1·299
0·00060	(0·00588)	*	*	1·382	1·355	1·337	1·314	1·300	1·291	1·280	1·275	1·271	1·270	1·271	1·273
0·00080	(0·00785)	*	1·387	1·349	1·325	1·308	1·288	1·275	1·267	1·258	1·254	1·251	1·251	1·253	1·256
0·00100	(0·00981)	*	1·360	1·324	1·302	1·287	1·268	1·257	1·249	1·241	1·238	1·236	1·236	1·239	1·243
0·00150	(0·0147)	1·337	1·313	1·282	1·263	1·250	1·234	1·224	1·219	1·213	1·211	1·210	1·211	1·215	1·220
0·00200	(0·0196)	1·304	1·282	1·254	1·237	1·225	1·211	1·203	1·198	1·194	1·192	1·193	1·194	1·199	1·205
0·00300	(0·0294)	1·260	1·241	1·217	1·202	1·192	1·180	1·174	1·171	1·168	1·168	1·169	1·172	1·178	1·184
0·00400	(0·0392)	1·231	1·214	1·192	1·179	1·170	1·160	1·155	1·152	1·150	1·151	1·154	1·157	1·164	1·171
0·00600	(0·0588)	1·192	1·177	1·159	1·148	1·140	1·133	1·129	1·127	1·127	1·129	1·133	1·137	1·145	1·153
0·00800	(0·0785)	1·166	1·153	1·136	1·127	1·121	1·114	1·112	1·111	1·112	1·114	1·119	1·123	1·132	1·141
0·01000	(0·0981)	1·147	1·135	1·120	1·111	1·106	1·101	1·099	1·098	1·100	1·103	1·108	1·113	1·123	1·132
0·01500	(0·147)	1·115	1·104	1·092	1·085	1·081	1·077	1·077	1·077	1·080	1·085	1·090	1·096	1·107	1·116
0·02000	(0·196)	1·093	1·084	1·073	1·067	1·064	1·062	1·062	1·063	1·067	1·072	1·079	1·085	1·096	1·106
0·03000	(0·294)	1·064	1·057	1·048	1·044	1·042	1·041	1·042	1·045	1·050	1·056	1·063	1·070	1·082	1·093
0·04000	(0·392)	1·045	1·039	1·032	1·028	1·027	1·027	1·029	1·032	1·039	1·045	1·053	1·060	1·073	1·085
0·06000	(0·588)	1·020	1·015	1·010	1·008	1·008	1·009	1·012	1·016	1·024	1·031	1·040	1·047	1·062	1·074
0·08000	(0·785)	1·003	0·999	0·995	0·994	0·995	0·997	1·001	1·005	1·014	1·022	1·031	1·039	1·054	1·067
0·10000	(0·981)	0·991	0·988	0·985	0·984	0·985	0·989	0·993	0·998	1·007	1·015	1·025	1·033	1·049	1·062
0·15000	(1·47)	0·970	0·968	0·967	0·967	0·969	0·974	0·979	0·985	0·995	1·004	1·014	1·024	1·040	1·053
0·20000	(1·96)	0·957	0·955	0·955	0·956	0·959	0·965	0·971	0·976	0·987	0·997	1·008	1·017	1·034	1·048
0·30000	(2·94)	0·939	0·938	0·940	0·942	0·945	0·952	0·959	0·966	0·977	0·988	0·999	1·009	1·027	1·042
0·40000	(3·92)	0·927	0·927	0·930	0·933	0·937	0·944	0·952	0·959	0·971	0·982	0·994	1·004	1·022	1·037
0·60000	(5·88)	0·912	0·913	0·917	0·921	0·926	0·934	0·943	0·950	0·963	0·975	0·987	0·998	1·017	1·032
0·80000	(7·85)	0·902	0·904	0·909	0·914	0·919	0·928	0·937	0·944	0·958	0·970	0·983	0·994	1·013	1·029
1·00000	(9·81)	0·895	0·898	0·903	0·908	0·913	0·923	0·932	0·941	0·955	0·967	0·980	0·992	1·011	1·027
1·50000	(14·7)	0·884	0·887	0·893	0·899	0·905	0·916	0·926	0·934	0·949	0·962	0·975	0·987	1·007	1·023
2·00000	(19·6)	0·876	0·880	0·887	0·894	0·900	0·911	0·921	0·930	0·946	0·959	0·972	0·984	1·005	1·021
3·00000	(29·4)	0·867	0·871	0·879	0·887	0·893	0·906	0·916	0·925	0·941	0·955	0·969	0·981	1·002	1·019
S	$(\Delta p/\rho l)$	125	150	200	250	300	400	500	600	800	1000	1250	1500	2000	2500

Grad'nt   k.p.g.   (Equivalent) Pipe diameters in mm   **Roughness size, $k_s = 0{\cdot}060$ mm**

Kin. visc., $\nu = 12.5 \times 10^{-6}$ m^{2}s^{-1};
$S = 0.00010$ to 3.00000
i.e. kin. pr. grad., $\Delta p/\rho l =$
(0.00098) to (29.4) ms^{-2}

Roughness size, $k_s = 0.150$ mm
This table shows values of m, as follows
m_C for Colebrook-White solutions; or,
where $\mathbf{R} \leq 2000$, m_P for laminar flow

Grad'nt S	k.p.g. ($\Delta p/\rho l$)	\multicolumn{14}{c}{(Equivalent) Pipe diameters in mm}
		125	150	200	250	300	400	500	600	800	1000	1250	1500	2000	2500
0.00010	(0.00098)	2.590	2.031	1.384	*	1.559	1.516	1.489	1.471	1.447	1.433	1.423	1.416	1.410	1.407
0.00015	(0.00147)	2.115	1.658	*	1.533	1.504	1.467	1.443	1.428	1.408	1.396	1.388	1.383	1.378	1.378
0.00020	(0.00196)	1.831	1.436	*	1.494	1.468	1.434	1.413	1.399	1.382	1.372	1.364	1.361	1.357	1.358
0.00030	(0.00294)	1.495	1.173	1.476	1.443	1.420	1.391	1.373	1.361	1.346	1.339	1.333	1.331	1.329	1.331
0.00040	(0.00392)	1.295	*	1.439	1.409	1.389	1.362	1.346	1.335	1.323	1.316	1.312	1.310	1.311	1.313
0.00060	(0.00588)	*	*	1.391	1.364	1.347	1.324	1.310	1.302	1.292	1.287	1.284	1.284	1.286	1.289
0.00080	(0.00785)	*	1.396	1.358	1.335	1.319	1.298	1.286	1.279	1.271	1.267	1.266	1.266	1.269	1.273
0.00100	(0.00981)	*	1.370	1.334	1.313	1.298	1.279	1.269	1.262	1.255	1.252	1.252	1.253	1.256	1.262
0.00150	(0.0147)	1.348	1.324	1.294	1.275	1.262	1.247	1.238	1.233	1.228	1.227	1.228	1.230	1.235	1.241
0.00200	(0.0196)	1.316	1.294	1.267	1.250	1.239	1.225	1.218	1.214	1.211	1.210	1.212	1.215	1.221	1.228
0.00300	(0.0294)	1.273	1.255	1.231	1.217	1.208	1.197	1.192	1.189	1.187	1.188	1.191	1.195	1.203	1.211
0.00400	(0.0392)	1.245	1.229	1.208	1.195	1.187	1.178	1.174	1.172	1.172	1.174	1.178	1.182	1.191	1.200
0.00600	(0.0588)	1.209	1.194	1.177	1.166	1.160	1.153	1.151	1.150	1.152	1.155	1.160	1.165	1.175	1.185
0.00800	(0.0785)	1.185	1.172	1.156	1.147	1.142	1.137	1.135	1.135	1.138	1.142	1.148	1.154	1.165	1.175
0.01000	(0.0981)	1.167	1.155	1.141	1.133	1.129	1.125	1.124	1.125	1.129	1.133	1.140	1.146	1.158	1.168
0.01500	(0.147)	1.137	1.127	1.116	1.110	1.107	1.105	1.105	1.107	1.112	1.118	1.125	1.132	1.145	1.157
0.02000	(0.196)	1.117	1.109	1.099	1.094	1.092	1.091	1.093	1.095	1.102	1.108	1.116	1.124	1.138	1.150
0.03000	(0.294)	1.091	1.084	1.077	1.074	1.073	1.074	1.077	1.080	1.088	1.096	1.104	1.113	1.128	1.141
0.04000	(0.392)	1.074	1.069	1.063	1.061	1.061	1.063	1.067	1.071	1.079	1.087	1.097	1.106	1.121	1.135
0.06000	(0.588)	1.053	1.048	1.045	1.044	1.045	1.049	1.054	1.059	1.068	1.077	1.088	1.097	1.113	1.128
0.08000	(0.785)	1.039	1.035	1.033	1.033	1.035	1.040	1.045	1.051	1.061	1.071	1.082	1.092	1.109	1.123
0.10000	(0.981)	1.028	1.026	1.025	1.026	1.028	1.033	1.039	1.045	1.056	1.066	1.078	1.088	1.105	1.120
0.15000	(1.47)	1.011	1.010	1.010	1.013	1.016	1.023	1.029	1.036	1.048	1.059	1.071	1.081	1.100	1.115
0.20000	(1.96)	1.001	1.000	1.002	1.005	1.008	1.016	1.023	1.030	1.043	1.054	1.067	1.078	1.096	1.112
0.30000	(2.94)	0.987	0.987	0.990	0.994	0.999	1.007	1.016	1.023	1.037	1.049	1.061	1.073	1.092	1.108
0.40000	(3.92)	0.978	0.979	0.983	0.988	0.993	1.002	1.011	1.019	1.033	1.045	1.058	1.070	1.089	1.106
0.60000	(5.88)	0.967	0.969	0.974	0.980	0.985	0.996	1.005	1.013	1.028	1.041	1.054	1.066	1.086	1.103
0.80000	(7.85)	0.960	0.963	0.969	0.975	0.981	0.992	1.001	1.010	1.025	1.038	1.052	1.064	1.084	1.101
1.00000	(9.81)	0.956	0.959	0.965	0.971	0.978	0.989	0.999	1.008	1.023	1.036	1.050	1.063	1.083	1.100
1.50000	(14.7)	0.948	0.951	0.959	0.966	0.972	0.984	0.995	1.004	1.020	1.034	1.048	1.060	1.081	1.098
2.00000	(19.6)	0.943	0.947	0.955	0.962	0.969	0.982	0.992	1.002	1.018	1.032	1.046	1.059	1.080	1.097
3.00000	(29.4)	0.937	0.942	0.950	0.958	0.966	0.978	0.989	0.999	1.016	1.030	1.044	1.057	1.078	1.096
S	($\Delta p/\rho l$)	125	150	200	250	300	400	500	600	800	1000	1250	1500	2000	2500

Grad'nt k.p.g. (Equivalent) Pipe diameters in mm   Roughness size, $k_s = 0.150$ mm

---

Kin. visc., $\nu = 12.5 \times 10^{-6}$ m^{2}s^{-1};
$S = 0.00010$ to 3.00000
i.e. kin. pr. grad., $\Delta p/\rho l =$
(0.00098) to (29.4) ms^{-2}

Roughness size, $k_s = 0.30$ mm
This table shows values of m, as follows
m_C for Colebrook-White solutions; or,
where $\mathbf{R} \leq 2000$, m_P for laminar flow

Grad'nt S	k.p.g. ($\Delta p/\rho l$)	\multicolumn{14}{c}{(Equivalent) Pipe diameters in mm}
		125	150	200	250	300	400	500	600	800	1000	1250	1500	2000	2500
0.00010	(0.00098)	2.590	2.031	1.384	*	1.567	1.525	1.498	1.480	1.457	1.444	1.434	1.428	1.422	1.420
0.00015	(0.00147)	2.115	1.658	*	1.543	1.514	1.477	1.454	1.439	1.419	1.409	1.401	1.396	1.393	1.393
0.00020	(0.00196)	1.831	1.436	*	1.505	1.479	1.445	1.425	1.411	1.394	1.385	1.378	1.375	1.373	1.374
0.00030	(0.00294)	1.495	1.173	1.488	1.455	1.432	1.404	1.386	1.375	1.361	1.354	1.349	1.347	1.347	1.350
0.00040	(0.00392)	1.295	*	1.452	1.422	1.402	1.376	1.360	1.350	1.339	1.333	1.330	1.329	1.330	1.334
0.00060	(0.00588)	*	*	1.405	1.379	1.361	1.340	1.327	1.318	1.310	1.306	1.304	1.304	1.308	1.313
0.00080	(0.00785)	*	1.411	1.374	1.351	1.335	1.315	1.304	1.297	1.290	1.288	1.287	1.288	1.293	1.299
0.00100	(0.00981)	*	1.386	1.351	1.330	1.315	1.298	1.288	1.282	1.276	1.274	1.275	1.277	1.282	1.289
0.00150	(0.0147)	1.366	1.342	1.312	1.294	1.282	1.268	1.260	1.256	1.252	1.252	1.254	1.257	1.264	1.272
0.00200	(0.0196)	1.335	1.314	1.287	1.271	1.260	1.248	1.242	1.238	1.236	1.237	1.240	1.244	1.252	1.261
0.00300	(0.0294)	1.295	1.277	1.254	1.240	1.232	1.222	1.218	1.216	1.216	1.218	1.223	1.227	1.237	1.247
0.00400	(0.0392)	1.269	1.252	1.232	1.220	1.213	1.205	1.202	1.201	1.203	1.206	1.211	1.217	1.228	1.238
0.00600	(0.0588)	1.235	1.221	1.204	1.195	1.189	1.184	1.182	1.183	1.186	1.190	1.197	1.203	1.215	1.226
0.00800	(0.0785)	1.213	1.200	1.186	1.178	1.173	1.170	1.169	1.170	1.175	1.180	1.188	1.195	1.208	1.219
0.01000	(0.0981)	1.197	1.186	1.173	1.166	1.162	1.160	1.160	1.162	1.167	1.173	1.181	1.188	1.202	1.214
0.01500	(0.147)	1.170	1.161	1.150	1.145	1.143	1.143	1.144	1.147	1.154	1.161	1.170	1.178	1.193	1.206
0.02000	(0.196)	1.152	1.145	1.136	1.132	1.131	1.132	1.135	1.138	1.146	1.154	1.163	1.172	1.187	1.201
0.03000	(0.294)	1.130	1.124	1.118	1.116	1.116	1.118	1.122	1.127	1.136	1.145	1.155	1.164	1.180	1.195
0.04000	(0.392)	1.116	1.111	1.106	1.105	1.106	1.110	1.114	1.119	1.129	1.139	1.149	1.159	1.176	1.191
0.06000	(0.588)	1.098	1.094	1.092	1.092	1.094	1.099	1.105	1.110	1.121	1.132	1.143	1.153	1.171	1.186
0.08000	(0.785)	1.087	1.084	1.083	1.084	1.086	1.092	1.099	1.105	1.117	1.127	1.139	1.149	1.168	1.183
0.10000	(0.981)	1.078	1.076	1.076	1.078	1.081	1.087	1.094	1.101	1.113	1.124	1.136	1.147	1.165	1.181
0.15000	(1.47)	1.065	1.064	1.066	1.069	1.072	1.080	1.087	1.095	1.108	1.119	1.132	1.143	1.162	1.178
0.20000	(1.96)	1.057	1.057	1.059	1.063	1.067	1.075	1.083	1.091	1.104	1.116	1.129	1.140	1.160	1.176
0.30000	(2.94)	1.047	1.047	1.051	1.055	1.060	1.069	1.078	1.086	1.100	1.112	1.126	1.137	1.157	1.173
0.40000	(3.92)	1.040	1.042	1.046	1.051	1.056	1.066	1.075	1.083	1.098	1.110	1.124	1.135	1.155	1.172
0.60000	(5.88)	1.032	1.034	1.040	1.045	1.051	1.062	1.071	1.080	1.095	1.107	1.121	1.133	1.154	1.170
0.80000	(7.85)	1.028	1.030	1.036	1.042	1.048	1.059	1.069	1.078	1.093	1.106	1.120	1.132	1.152	1.169
1.00000	(9.81)	1.024	1.027	1.033	1.040	1.046	1.057	1.067	1.076	1.092	1.105	1.119	1.131	1.152	1.169
1.50000	(14.7)	1.019	1.022	1.029	1.036	1.043	1.054	1.065	1.074	1.090	1.103	1.117	1.130	1.150	1.167
2.00000	(19.6)	1.016	1.019	1.027	1.034	1.041	1.053	1.063	1.072	1.088	1.102	1.116	1.129	1.150	1.167
3.00000	(29.4)	1.012	1.016	1.024	1.031	1.038	1.051	1.061	1.071	1.087	1.101	1.115	1.128	1.149	1.166
S	($\Delta p/\rho l$)	125	150	200	250	300	400	500	600	800	1000	1250	1500	2000	2500

Grad'nt k.p.g. (Equivalent) Pipe diameters in mm   Roughness size, $k_s = 0.30$ mm

Kin. visc., $\nu = 15 \times 10^{-6}\ m^2 s^{-1}$;
$S = 0.00010$ to 30.0000

i.e. kin. pr. grad., $\Delta p/\rho l =$
(0.00098) to (294) ms^{-2}

Roughness size, $k_s = 0.003$ mm
This table shows values of m, as follows

m_C for Colebrook-White solutions; or,
where $R \le 2000$, m_P for laminar flow

Grad'nt S	k.p.g. $(\Delta p/\rho l)$	6	8	10	12.5	15	20	25	30	40	50	60	80	100	125
0.00030	(0.00294)	102.9	70.09	52.05	38.66	30.32	20.66	15.34	12.03	8.198	6.088	4.774	3.253	2.416	1.794
0.00040	(0.00392)	89.08	60.70	45.08	33.48	26.25	17.89	13.29	10.42	7.100	5.273	4.135	2.817	2.092	1.554
0.00060	(0.00588)	72.73	49.56	36.81	27.34	21.44	14.61	10.85	8.507	5.797	4.305	3.376	2.300	1.708	1.269
0.00080	(0.00785)	62.99	42.92	31.88	23.67	18.56	12.65	9.395	7.367	5.020	3.728	2.924	1.992	1.480	*
0.00100	(0.00981)	56.34	38.39	28.51	21.17	16.60	11.31	8.403	6.589	4.490	3.335	2.615	1.782	1.323	*
0.00150	(0.0147)	46.00	31.35	23.28	17.29	13.56	9.238	6.861	5.380	3.666	2.723	2.135	1.455	*	*
0.00200	(0.0196)	39.84	27.15	20.16	14.97	11.74	8.001	5.942	4.659	3.175	2.358	1.849	1.260	*	1.339
0.00300	(0.0294)	32.53	22.16	16.46	12.22	9.587	6.533	4.851	3.804	2.592	1.925	1.510	*	*	1.291
0.00400	(0.0392)	28.17	19.20	14.26	10.59	8.302	5.657	4.201	3.295	2.245	1.667	1.308	*	1.288	1.259
0.00600	(0.0588)	23.00	15.67	11.64	8.644	6.779	4.619	3.430	2.690	1.833	1.361	1.068	*	1.242	1.217
0.00800	(0.0785)	19.92	13.57	10.08	7.486	5.871	4.000	2.971	2.330	1.588	1.179	*	1.238	1.211	1.188
0.01000	(0.0981)	17.82	12.14	9.016	6.696	5.251	3.578	2.657	2.084	1.420	1.055	*	1.213	1.188	1.167
0.01500	(0.147)	14.55	9.912	7.362	5.467	4.287	2.921	2.170	1.701	1.159	*	1.206	1.171	1.149	1.131
0.02000	(0.196)	12.60	8.584	6.375	4.735	3.713	2.530	1.879	1.473	1.004	*	1.175	1.143	1.123	1.106
0.03000	(0.294)	10.29	7.009	5.205	3.866	3.032	2.066	1.534	1.203	*	1.154	1.133	1.105	1.088	1.073
0.04000	(0.392)	8.908	6.070	4.508	3.348	2.625	1.789	1.329	1.042	*	1.125	1.105	1.080	1.064	1.051
0.06000	(0.588)	7.273	4.956	3.681	2.734	2.144	1.461	1.085	*	1.110	1.085	1.068	1.046	1.033	1.022
0.08000	(0.785)	6.299	4.292	3.188	2.367	1.856	1.265	0.939	*	1.081	1.059	1.044	1.024	1.011	1.002
0.10000	(0.981)	5.634	3.839	2.851	2.117	1.660	1.131	*	*	1.060	1.040	1.025	1.007	0.995	0.987
0.15000	(1.47)	4.600	3.135	2.328	1.729	1.356	0.924	*	1.053	1.024	1.006	0.993	0.978	0.968	0.961
0.20000	(1.96)	3.984	2.715	2.016	1.497	1.174	*	1.047	1.026	1.000	0.983	0.972	0.958	0.949	0.943
0.30000	(2.94)	3.253	2.216	1.646	1.222	0.959	*	1.008	0.991	0.968	0.953	0.944	0.931	0.924	0.919
0.40000	(3.92)	2.817	1.920	1.426	1.059	0.830	1.006	0.983	0.967	0.946	0.933	0.924	0.914	0.908	0.903
0.60000	(5.88)	2.300	1.567	1.164	0.864	*	0.970	0.950	0.935	0.917	0.906	0.899	0.890	0.885	0.882
0.80000	(7.85)	1.992	1.357	1.008	*	*	0.946	0.927	0.914	0.898	0.888	0.881	0.874	0.870	0.867
1.00000	(9.81)	1.782	1.214	0.902	*	0.955	0.928	0.910	0.899	0.884	0.874	0.869	0.862	0.858	0.856
1.50000	(14.7)	1.455	0.991	*	*	0.921	0.897	0.882	0.872	0.859	0.851	0.846	0.841	0.838	0.837
2.00000	(19.6)	1.260	0.858	*	0.915	0.898	0.876	0.863	0.853	0.842	0.835	0.831	0.826	0.824	0.824
3.00000	(29.4)	1.029	*	*	0.882	0.868	0.849	0.837	0.829	0.819	0.814	0.810	0.807	0.806	0.806
4.00000	(39.2)	0.891	*	0.880	0.861	0.847	0.830	0.820	0.813	0.804	0.799	0.797	0.794	0.794	0.794
6.00000	(58.8)	0.727	*	0.849	0.832	0.821	0.806	0.797	0.791	0.784	0.780	0.778	0.777	0.777	0.778
8.00000	(78.5)	*	0.847	0.828	0.813	0.803	0.789	0.781	0.776	0.770	0.767	0.766	0.765	0.766	0.767
10.0000	(98.1)	*	0.830	0.813	0.799	0.789	0.777	0.770	0.765	0.760	0.757	0.756	0.756	0.757	0.759
15.0000	(147)	0.826	0.802	0.787	0.775	0.766	0.756	0.750	0.746	0.742	0.740	0.740	0.741	0.743	0.745
20.0000	(196)	0.804	0.783	0.769	0.758	0.751	0.742	0.736	0.733	0.730	0.729	0.729	0.730	0.733	0.736
30.0000	(294)	0.777	0.758	0.746	0.737	0.730	0.723	0.719	0.716	0.714	0.714	0.714	0.717	0.719	0.723
S	$(\Delta p/\rho l)$	6	8	10	12.5	15	20	25	30	40	50	60	80	100	125

Grad'nt k.p.g. (Equivalent) Pipe diameters in mm

Grad'nt S	k.p.g. $(\Delta p/\rho l)$	125	150	200	250	300	400	500	600	800	1000	1250	1500	2000	2500
0.00010	(0.00098)	3.108	2.437	1.661	*	*	1.555	1.524	1.503	1.476	1.459	1.446	1.438	1.428	1.424
0.00015	(0.00147)	2.538	1.990	1.356	*	1.544	1.502	1.475	1.456	1.433	1.419	1.408	1.401	1.393	1.390
0.00020	(0.00196)	2.198	1.723	*	1.534	1.505	1.466	1.442	1.425	1.404	1.391	1.382	1.376	1.370	1.367
0.00030	(0.00294)	1.794	1.407	*	1.478	1.452	1.418	1.397	1.383	1.365	1.354	1.346	1.342	1.338	1.337
0.00040	(0.00392)	1.554	1.219	1.475	1.441	1.417	1.386	1.367	1.354	1.338	1.329	1.323	1.319	1.316	1.316
0.00060	(0.00588)	1.269	*	1.421	1.391	1.371	1.344	1.327	1.316	1.303	1.295	1.290	1.288	1.286	1.287
0.00080	(0.00785)	*	*	1.385	1.358	1.339	1.315	1.300	1.291	1.279	1.273	1.268	1.267	1.266	1.268
0.00100	(0.00981)	*	1.399	1.359	1.333	1.316	1.294	1.280	1.271	1.261	1.255	1.252	1.251	1.251	1.253
0.00150	(0.0147)	*	1.348	1.313	1.291	1.276	1.257	1.245	1.238	1.229	1.225	1.223	1.223	1.224	1.227
0.00200	(0.0196)	1.339	1.315	1.282	1.262	1.249	1.231	1.221	1.215	1.208	1.205	1.203	1.204	1.206	1.210
0.00300	(0.0294)	1.291	1.270	1.242	1.224	1.212	1.198	1.189	1.184	1.179	1.177	1.177	1.178	1.181	1.186
0.00400	(0.0392)	1.259	1.240	1.214	1.199	1.188	1.175	1.168	1.163	1.159	1.158	1.159	1.160	1.164	1.169
0.00600	(0.0588)	1.217	1.200	1.178	1.164	1.155	1.144	1.139	1.135	1.133	1.132	1.134	1.136	1.141	1.147
0.00800	(0.0785)	1.188	1.173	1.153	1.141	1.133	1.124	1.119	1.116	1.114	1.115	1.117	1.120	1.125	1.131
0.01000	(0.0981)	1.167	1.153	1.134	1.123	1.116	1.108	1.104	1.102	1.101	1.102	1.104	1.107	1.114	1.120
0.01500	(0.147)	1.131	1.118	1.102	1.093	1.087	1.081	1.078	1.077	1.077	1.079	1.082	1.085	1.093	1.099
0.02000	(0.196)	1.106	1.095	1.081	1.073	1.068	1.062	1.060	1.059	1.061	1.063	1.067	1.070	1.078	1.085
0.03000	(0.294)	1.073	1.064	1.052	1.045	1.041	1.037	1.036	1.036	1.038	1.041	1.046	1.050	1.059	1.066
0.04000	(0.392)	1.051	1.043	1.032	1.026	1.023	1.020	1.020	1.020	1.023	1.027	1.032	1.036	1.045	1.053
0.06000	(0.588)	1.022	1.014	1.006	1.001	0.999	0.997	0.998	0.999	1.003	1.007	1.012	1.017	1.027	1.035
0.08000	(0.785)	1.002	0.995	0.988	0.984	0.982	0.982	0.983	0.984	0.989	0.993	0.999	1.005	1.014	1.023
0.10000	(0.981)	0.987	0.981	0.974	0.971	0.970	0.970	0.971	0.973	0.978	0.983	0.989	0.995	1.005	1.014
0.15000	(1.47)	0.961	0.956	0.951	0.949	0.948	0.949	0.951	0.954	0.959	0.965	0.972	0.978	0.988	0.998
0.20000	(1.96)	0.943	0.939	0.935	0.934	0.934	0.935	0.938	0.941	0.947	0.953	0.960	0.966	0.977	0.987
0.30000	(2.94)	0.919	0.916	0.914	0.913	0.914	0.916	0.919	0.923	0.930	0.936	0.943	0.950	0.962	0.972
0.40000	(3.92)	0.903	0.901	0.899	0.899	0.900	0.903	0.907	0.911	0.918	0.925	0.932	0.939	0.951	0.961
0.60000	(5.88)	0.882	0.880	0.879	0.880	0.882	0.886	0.890	0.894	0.902	0.909	0.917	0.924	0.937	0.948
0.80000	(7.85)	0.867	0.866	0.866	0.867	0.869	0.874	0.878	0.883	0.891	0.898	0.907	0.914	0.927	0.938
1.00000	(9.81)	0.856	0.855	0.856	0.857	0.860	0.865	0.870	0.874	0.883	0.891	0.899	0.907	0.920	0.931
1.50000	(14.7)	0.837	0.837	0.838	0.841	0.843	0.849	0.854	0.859	0.869	0.877	0.886	0.894	0.907	0.919
2.00000	(19.6)	0.824	0.824	0.826	0.829	0.832	0.838	0.844	0.849	0.859	0.867	0.876	0.885	0.899	0.910
3.00000	(29.4)	0.806	0.807	0.810	0.813	0.817	0.824	0.830	0.836	0.846	0.855	0.864	0.873	0.887	0.899
S	$(\Delta p/\rho l)$	125	150	200	250	300	400	500	600	800	1000	1250	1500	2000	2500

Grad'nt k.p.g. (Equivalent) Pipe diameters in mm

Kinematic viscosity, $\nu = 15 \times 10^{-6}\ m^2 s^{-1}$;   **Roughness size, $k_s = 0.003$ mm**

Kin. visc., $\nu = 15\times10^{-6}\ \text{m}^2\,\text{s}^{-1}$;
$S = 0\cdot00010$ to $30\cdot0000$
i.e. kin. pr. grad., $\Delta p/\rho l =$
$(0\cdot00098)$ to $(294)\ \text{ms}^{-2}$

Roughness size, $k_s = 0\cdot006$ mm
This table shows values of m, as follows
m_C for Colebrook-White solutions; or,
where $R \le 2000$, m_P for laminar flow

Grad'nt	k.p.g.	(Equivalent) Pipe diameters in mm													
S	$(\Delta p/\rho l)$	6	8	10	12·5	15	20	25	30	40	50	60	80	100	125
0·00030	(0·00294)	102·9	70·09	52·05	38·66	30·32	20·66	15·34	12·03	8·198	6·088	4·774	3·253	2·416	1·794
0·00040	(0·00392)	89·08	60·70	45·08	33·48	26·25	17·89	13·29	10·42	7·100	5·273	4·135	2·817	2·092	1·554
0·00060	(0·00588)	72·73	49·56	36·81	27·34	21·44	14·61	10·85	8·507	5·797	4·305	3·376	2·300	1·708	1·269
0·00080	(0·00785)	62·99	42·92	31·88	23·67	18·56	12·65	9·395	7·367	5·020	3·728	2·924	1·992	1·480	*
0·00100	(0·00981)	56·34	38·39	28·51	21·17	16·60	11·31	8·403	6·589	4·490	3·335	2·615	1·782	1·323	*
0·00150	(0·0147)	46·00	31·35	23·28	17·29	13·56	9·238	6·861	5·380	3·666	2·723	2·135	1·455	*	*
0·00200	(0·0196)	39·84	27·15	20·16	14·97	11·74	8·001	5·942	4·659	3·175	2·358	1·849	1·260	*	1·340
0·00300	(0·0294)	32·53	22·16	16·46	12·22	9·587	6·533	4·851	3·804	2·592	1·925	1·510	*	*	1·292
0·00400	(0·0392)	28·17	19·20	14·26	10·59	8·302	5·657	4·201	3·295	2·245	1·667	1·308	*	1·288	1·260
0·00600	(0·0588)	23·00	15·67	11·64	8·644	6·779	4·619	3·430	2·690	1·833	1·361	1·068	*	1·242	1·217
0·00800	(0·0785)	19·92	13·57	10·08	7·486	5·871	4·000	2·971	2·330	1·588	1·179	*	1·239	1·211	1·189
0·01000	(0·0981)	17·82	12·14	9·016	6·696	5·251	3·578	2·657	2·084	1·420	1·055	*	1·214	1·189	1·168
0·01500	(0·147)	14·55	9·912	7·362	5·467	4·287	2·921	2·170	1·701	1·159	*	1·207	1·172	1·150	1·131
0·02000	(0·196)	12·60	8·584	6·375	4·735	3·713	2·530	1·879	1·473	1·004	*	1·175	1·143	1·123	1·107
0·03000	(0·294)	10·29	7·009	5·205	3·866	3·032	2·066	1·534	1·203	*	1·155	1·134	1·106	1·088	1·074
0·04000	(0·392)	8·908	6·070	4·508	3·348	2·625	1·789	1·329	1·042	*	1·125	1·106	1·081	1·065	1·052
0·06000	(0·588)	7·273	4·956	3·681	2·734	2·144	1·461	1·085	*	1·111	1·086	1·069	1·047	1·034	1·023
0·08000	(0·785)	6·299	4·292	3·188	2·367	1·856	1·265	0·939	*	1·082	1·060	1·045	1·025	1·013	1·003
0·10000	(0·981)	5·634	3·839	2·851	2·117	1·660	1·131	*	*	1·061	1·041	1·026	1·008	0·997	0·988
0·15000	(1·47)	4·600	3·135	2·328	1·729	1·356	0·924	*	1·054	1·025	1·007	0·995	0·979	0·969	0·962
0·20000	(1·96)	3·984	2·715	2·016	1·497	1·174	*	1·048	1·028	1·001	0·985	0·974	0·959	0·951	0·945
0·30000	(2·94)	3·253	2·216	1·646	1·222	0·959	*	1·010	0·992	0·969	0·955	0·945	0·933	0·926	0·921
0·40000	(3·92)	2·817	1·920	1·426	1·059	0·830	1·008	0·985	0·969	0·948	0·935	0·926	0·916	0·910	0·905
0·60000	(5·88)	2·300	1·567	1·164	0·864	*	0·972	0·951	0·937	0·919	0·908	0·901	0·892	0·887	0·884
0·80000	(7·85)	1·992	1·357	1·008	*	*	0·948	0·929	0·916	0·900	0·890	0·884	0·876	0·872	0·870
1·00000	(9·81)	1·782	1·214	0·902	*	0·957	0·930	0·913	0·901	0·886	0·877	0·871	0·864	0·861	0·859
1·50000	(14·7)	1·455	0·991	*	*	0·923	0·899	0·884	0·874	0·861	0·854	0·849	0·844	0·841	0·840
2·00000	(19·6)	1·260	0·858	*	0·917	0·900	0·879	0·865	0·856	0·845	0·838	0·834	0·830	0·828	0·828
3·00000	(29·4)	1·029	*	*	0·885	0·871	0·852	0·840	0·832	0·823	0·817	0·814	0·811	0·811	0·811
4·00000	(39·2)	0·891	*	0·883	0·864	0·851	0·834	0·823	0·816	0·808	0·803	0·801	0·799	0·799	0·800
6·00000	(58·8)	0·727	*	0·852	0·836	0·824	0·810	0·801	0·795	0·788	0·785	0·783	0·782	0·783	0·784
8·00000	(78·5)	*	0·850	0·832	0·817	0·807	0·794	0·786	0·781	0·775	0·772	0·771	0·771	0·772	0·774
10·0000	(98·1)	*	0·834	0·817	0·803	0·794	0·782	0·775	0·770	0·765	0·763	0·762	0·763	0·764	0·767
15·0000	(147)	0·830	0·806	0·791	0·779	0·771	0·761	0·755	0·752	0·748	0·747	0·747	0·748	0·750	0·754
20·0000	(196)	0·809	0·788	0·774	0·764	0·756	0·747	0·742	0·740	0·737	0·736	0·737	0·739	0·741	0·745
30·0000	(294)	0·782	0·763	0·752	0·743	0·737	0·729	0·725	0·723	0·722	0·722	0·723	0·726	0·729	0·734
S	$(\Delta p/\rho l)$	6	8	10	12·5	15	20	25	30	40	50	60	80	100	125

Grad'nt	k.p.g.	(Equivalent) Pipe diameters in mm													
S	$(\Delta p/\rho l)$	125	150	200	250	300	400	500	600	800	1000	1250	1500	2000	2500
0·00010	(0·00098)	3·108	2·437	1·661	*	*	1·555	1·524	1·503	1·476	1·460	1·446	1·438	1·428	1·424
0·00015	(0·00147)	2·538	1·990	1·356	*	1·544	1·502	1·475	1·457	1·433	1·419	1·408	1·401	1·394	1·390
0·00020	(0·00196)	2·198	1·723	*	1·535	1·505	1·466	1·442	1·425	1·404	1·392	1·382	1·376	1·370	1·368
0·00030	(0·00294)	1·794	1·407	*	1·478	1·452	1·418	1·397	1·383	1·365	1·355	1·347	1·342	1·338	1·337
0·00040	(0·00392)	1·554	1·219	1·475	1·441	1·417	1·387	1·368	1·355	1·339	1·330	1·323	1·319	1·316	1·316
0·00060	(0·00588)	1·269	*	1·422	1·392	1·371	1·344	1·328	1·317	1·303	1·296	1·291	1·288	1·287	1·288
0·00080	(0·00785)	*	*	1·386	1·358	1·340	1·315	1·301	1·291	1·279	1·273	1·269	1·267	1·267	1·268
0·00100	(0·00981)	*	1·399	1·359	1·334	1·316	1·294	1·281	1·272	1·261	1·256	1·252	1·251	1·252	1·254
0·00150	(0·0147)	*	1·349	1·313	1·291	1·276	1·257	1·246	1·238	1·230	1·226	1·224	1·223	1·225	1·228
0·00200	(0·0196)	1·340	1·315	1·283	1·263	1·249	1·232	1·222	1·215	1·208	1·205	1·204	1·204	1·207	1·210
0·00300	(0·0294)	1·292	1·270	1·242	1·225	1·213	1·198	1·190	1·185	1·180	1·178	1·177	1·178	1·182	1·187
0·00400	(0·0392)	1·260	1·240	1·215	1·199	1·189	1·176	1·168	1·164	1·160	1·159	1·159	1·161	1·165	1·170
0·00600	(0·0588)	1·217	1·200	1·178	1·165	1·156	1·145	1·139	1·136	1·133	1·133	1·135	1·137	1·142	1·148
0·00800	(0·0785)	1·189	1·173	1·154	1·142	1·134	1·124	1·120	1·117	1·115	1·116	1·118	1·121	1·127	1·133
0·01000	(0·0981)	1·168	1·153	1·135	1·124	1·117	1·109	1·105	1·103	1·102	1·103	1·105	1·108	1·115	1·121
0·01500	(0·147)	1·131	1·119	1·103	1·094	1·088	1·082	1·079	1·078	1·078	1·080	1·083	1·087	1·094	1·101
0·02000	(0·196)	1·107	1·096	1·082	1·074	1·069	1·063	1·061	1·061	1·062	1·064	1·068	1·072	1·080	1·087
0·03000	(0·294)	1·074	1·065	1·053	1·046	1·042	1·039	1·037	1·038	1·040	1·043	1·048	1·052	1·061	1·069
0·04000	(0·392)	1·052	1·044	1·033	1·028	1·024	1·022	1·021	1·022	1·025	1·029	1·034	1·039	1·048	1·056
0·06000	(0·588)	1·023	1·016	1·007	1·003	1·000	0·999	0·999	1·001	1·005	1·009	1·015	1·020	1·030	1·039
0·08000	(0·785)	1·003	0·997	0·989	0·986	0·984	0·983	0·985	0·986	0·991	0·996	1·002	1·007	1·018	1·027
0·10000	(0·981)	0·988	0·982	0·976	0·973	0·972	0·972	0·973	0·976	0·981	0·986	0·992	0·998	1·009	1·018
0·15000	(1·47)	0·962	0·958	0·953	0·951	0·950	0·952	0·954	0·957	0·962	0·968	0·975	0·981	0·993	1·002
0·20000	(1·96)	0·945	0·941	0·937	0·936	0·936	0·938	0·941	0·944	0·950	0·956	0·963	0·970	0·982	0·992
0·30000	(2·94)	0·921	0·918	0·916	0·916	0·916	0·919	0·923	0·926	0·933	0·940	0·948	0·955	0·967	0·978
0·40000	(3·92)	0·905	0·903	0·902	0·902	0·903	0·906	0·910	0·914	0·922	0·929	0·937	0·944	0·957	0·968
0·60000	(5·88)	0·884	0·883	0·882	0·883	0·885	0·889	0·894	0·898	0·907	0·914	0·923	0·931	0·944	0·955
0·80000	(7·85)	0·870	0·869	0·869	0·871	0·873	0·878	0·883	0·888	0·897	0·904	0·913	0·921	0·935	0·946
1·00000	(9·81)	0·859	0·859	0·859	0·861	0·864	0·869	0·875	0·880	0·889	0·897	0·906	0·914	0·928	0·940
1·50000	(14·7)	0·840	0·841	0·842	0·845	0·848	0·854	0·860	0·866	0·875	0·884	0·894	0·902	0·917	0·929
2·00000	(19·6)	0·828	0·828	0·831	0·834	0·838	0·844	0·850	0·856	0·866	0·876	0·885	0·894	0·909	0·922
3·00000	(29·4)	0·811	0·812	0·816	0·819	0·823	0·831	0·837	0·844	0·855	0·864	0·874	0·883	0·899	0·912
S	$(\Delta p/\rho l)$	125	150	200	250	300	400	500	600	800	1000	1250	1500	2000	2500

Grad'nt **k.p.g.** **(Equivalent) Pipe diameters in mm**

Kinematic viscosity, $\nu = 15\times10^{-6}\ \text{m}^2\,\text{s}^{-1}$;

Roughness size, $k_s = 0\cdot006$ mm

E211

Kin. visc., $\nu = 15 \times 10^{-6}$ m^{2}s^{-1};
S = 0·00010 to 30·0000

i.e. kin. pr. grad., $\Delta p / \rho l =$
(0·00098) to (294) ms^{-2}

Roughness size, k_s = 0·015 mm
This table shows values of m, as follows

m_C for Colebrook-White solutions; or,
where $R \leq 2000$, m_P for laminar flow

Grad'nt S	k.p.g. $(\Delta p / \rho l)$	6	8	10	12·5	15	20	25	30	40	50	60	80	100	125
		(Equivalent) Pipe diameters in mm													
0·00030	(0·00294)	102·9	70·09	52·05	38·66	30·32	20·66	15·34	12·03	8·198	6·088	4·774	3·253	2·416	1·794
0·00040	(0·00392)	89·08	60·70	45·08	33·48	26·25	17·89	13·29	10·42	7·100	5·273	4·135	2·817	2·092	1·554
0·00060	(0·00588)	72·73	49·56	36·81	27·34	21·44	14·61	10·85	8·507	5·797	4·305	3·376	2·300	1·708	1·269
0·00080	(0·00785)	62·99	42·92	31·88	23·67	18·56	12·65	9·395	7·367	5·020	3·728	2·924	1·992	1·480	*
0·00100	(0·00981)	56·34	38·39	28·51	21·17	16·60	11·31	8·403	6·589	4·490	3·335	2·615	1·782	1·323	*
0·00150	(0·0147)	46·00	31·35	23·28	17·29	13·56	9·238	6·861	5·380	3·666	2·723	2·135	1·455	*	*
0·00200	(0·0196)	39·84	27·15	20·16	14·97	11·74	8·001	5·942	4·659	3·175	2·358	1·849	1·260	*	1·341
0·00300	(0·0294)	32·53	22·16	16·46	12·22	9·587	6·533	4·851	3·804	2·592	1·925	1·510	*	*	1·293
0·00400	(0·0392)	28·17	19·20	14·26	10·59	8·302	5·657	4·201	3·295	2·245	1·667	1·308	*	1·289	1·261
0·00600	(0·0588)	23·00	15·67	11·64	8·644	6·779	4·619	3·430	2·690	1·833	1·361	1·068	*	1·244	1·219
0·00800	(0·0785)	19·92	13·57	10·08	7·486	5·871	4·000	2·971	2·330	1·588	1·179	*	1·240	1·213	1·191
0·01000	(0·0981)	17·82	12·14	9·016	6·696	5·251	3·578	2·657	2·084	1·420	1·055	*	1·216	1·191	1·170
0·01500	(0·147)	14·55	9·912	7·362	5·467	4·287	2·921	2·170	1·701	1·159	*	1·209	1·174	1·152	1·133
0·02000	(0·196)	12·60	8·584	6·375	4·735	3·713	2·530	1·879	1·473	1·004	*	1·177	1·146	1·126	1·109
0·03000	(0·294)	10·29	7·009	5·205	3·866	3·032	2·066	1·534	1·203	*	1·157	1·136	1·108	1·091	1·077
0·04000	(0·392)	8·908	6·070	4·508	3·348	2·625	1·789	1·329	1·042	*	1·128	1·109	1·083	1·068	1·055
0·06000	(0·588)	7·273	4·956	3·681	2·734	2·144	1·461	1·085	*	1·114	1·089	1·072	1·050	1·037	1·026
0·08000	(0·785)	6·299	4·292	3·188	2·367	1·856	1·265	0·939	*	1·086	1·063	1·048	1·028	1·016	1·007
0·10000	(0·981)	5·634	3·839	2·851	2·117	1·660	1·131	*	*	1·065	1·044	1·030	1·012	1·001	0·992
0·15000	(1·47)	4·600	3·135	2·328	1·729	1·356	0·924	*	1·058	1·029	1·011	0·999	0·983	0·974	0·967
0·20000	(1·96)	3·984	2·715	2·016	1·497	1·174	*	1·052	1·032	1·005	0·989	0·978	0·964	0·956	0·950
0·30000	(2·94)	3·253	2·216	1·646	1·222	0·959	*	1·014	0·997	0·974	0·960	0·950	0·939	0·932	0·927
0·40000	(3·92)	2·817	1·920	1·426	1·059	0·830	1·013	0·990	0·974	0·953	0·940	0·932	0·922	0·916	0·912
0·60000	(5·88)	2·300	1·567	1·164	0·864	*	0·977	0·957	0·943	0·925	0·915	0·907	0·899	0·895	0·892
0·80000	(7·85)	1·992	1·357	1·008	*	*	0·953	0·935	0·923	0·907	0·897	0·891	0·884	0·880	0·878
1·00000	(9·81)	1·782	1·214	0·902	*	0·963	0·936	0·919	0·908	0·893	0·884	0·879	0·873	0·870	0·868
1·50000	(14·7)	1·455	0·991	*	*	0·930	0·906	0·892	0·882	0·869	0·862	0·858	0·853	0·851	0·851
2·00000	(19·6)	1·260	0·858	*	0·924	0·908	0·887	0·873	0·865	0·854	0·848	0·844	0·840	0·839	0·840
3·00000	(29·4)	1·029	*	*	0·893	0·879	0·860	0·849	0·842	0·833	0·828	0·825	0·823	0·823	0·824
4·00000	(39·2)	0·891	*	0·892	0·873	0·860	0·843	0·833	0·827	0·819	0·815	0·813	0·812	0·812	0·814
6·00000	(58·8)	0·727	*	0·862	0·846	0·835	0·821	0·812	0·807	0·801	0·798	0·797	0·797	0·798	0·801
8·00000	(78·5)	*	0·861	0·843	0·828	0·818	0·806	0·798	0·794	0·789	0·787	0·786	0·787	0·789	0·792
10·0000	(98·1)	*	0·845	0·829	0·815	0·806	0·794	0·788	0·784	0·780	0·779	0·778	0·780	0·782	0·786
15·0000	(147)	0·842	0·819	0·804	0·793	0·785	0·776	0·770	0·768	0·765	0·764	0·765	0·768	0·771	0·775
20·0000	(196)	0·823	0·801	0·789	0·778	0·771	0·763	0·759	0·757	0·755	0·755	0·756	0·760	0·764	0·768
30·0000	(294)	0·797	0·779	0·768	0·759	0·753	0·747	0·744	0·743	0·742	0·743	0·745	0·750	0·754	0·760
S	$(\Delta p / \rho l)$	6	8	10	12·5	15	20	25	30	40	50	60	80	100	125

Grad'nt k.p.g. (Equivalent) Pipe diameters in mm

S	$(\Delta p / \rho l)$	125	150	200	250	300	400	500	600	800	1000	1250	1500	2000	2500
0·00010	(0·00098)	3·108	2·437	1·661	*	*	1·556	1·525	1·504	1·477	1·460	1·447	1·439	1·429	1·425
0·00015	(0·00147)	2·538	1·990	1·356	*	1·545	1·502	1·476	1·457	1·434	1·420	1·409	1·402	1·394	1·391
0·00020	(0·00196)	2·198	1·723	*	1·535	1·505	1·467	1·442	1·426	1·405	1·392	1·383	1·377	1·371	1·369
0·00030	(0·00294)	1·794	1·407	*	1·479	1·453	1·419	1·398	1·384	1·366	1·355	1·348	1·343	1·339	1·338
0·00040	(0·00392)	1·554	1·219	1·476	1·442	1·418	1·387	1·368	1·356	1·340	1·331	1·324	1·320	1·317	1·317
0·00060	(0·00588)	1·269	*	1·422	1·392	1·372	1·345	1·329	1·318	1·304	1·297	1·292	1·289	1·288	1·289
0·00080	(0·00785)	*	*	1·387	1·359	1·341	1·316	1·302	1·292	1·280	1·274	1·270	1·268	1·268	1·270
0·00100	(0·00981)	*	1·400	1·360	1·335	1·317	1·295	1·282	1·273	1·262	1·257	1·254	1·253	1·253	1·256
0·00150	(0·0147)	*	1·350	1·315	1·292	1·277	1·258	1·247	1·240	1·231	1·227	1·225	1·225	1·227	1·230
0·00200	(0·0196)	1·341	1·316	1·284	1·264	1·250	1·233	1·223	1·217	1·210	1·207	1·206	1·206	1·209	1·213
0·00300	(0·0294)	1·293	1·271	1·243	1·226	1·214	1·200	1·192	1·187	1·182	1·180	1·180	1·181	1·185	1·189
0·00400	(0·0392)	1·261	1·241	1·216	1·201	1·190	1·177	1·170	1·166	1·162	1·161	1·162	1·164	1·168	1·173
0·00600	(0·0588)	1·219	1·202	1·180	1·166	1·158	1·147	1·141	1·138	1·136	1·136	1·138	1·140	1·146	1·152
0·00800	(0·0785)	1·191	1·175	1·155	1·144	1·136	1·127	1·122	1·120	1·118	1·119	1·121	1·124	1·131	1·137
0·01000	(0·0981)	1·170	1·155	1·137	1·126	1·119	1·111	1·107	1·106	1·105	1·106	1·109	1·112	1·119	1·126
0·01500	(0·147)	1·133	1·121	1·106	1·097	1·091	1·085	1·082	1·081	1·082	1·084	1·087	1·091	1·099	1·107
0·02000	(0·196)	1·109	1·098	1·084	1·076	1·072	1·067	1·065	1·064	1·066	1·069	1·073	1·077	1·085	1·093
0·03000	(0·294)	1·077	1·067	1·056	1·049	1·046	1·042	1·041	1·042	1·045	1·048	1·053	1·058	1·067	1·076
0·04000	(0·392)	1·055	1·047	1·037	1·031	1·028	1·026	1·026	1·027	1·030	1·034	1·040	1·045	1·055	1·064
0·06000	(0·588)	1·026	1·019	1·011	1·007	1·005	1·004	1·005	1·006	1·011	1·016	1·022	1·027	1·038	1·047
0·08000	(0·785)	1·007	1·001	0·994	0·990	0·989	0·989	0·990	0·992	0·998	1·003	1·009	1·016	1·027	1·037
0·10000	(0·981)	0·992	0·987	0·981	0·978	0·977	0·978	0·980	0·982	0·988	0·994	1·000	1·007	1·018	1·029
0·15000	(1·47)	0·967	0·963	0·958	0·957	0·957	0·958	0·961	0·964	0·971	0·977	0·985	0·992	1·004	1·015
0·20000	(1·96)	0·950	0·946	0·943	0·942	0·943	0·945	0·948	0·952	0·959	0·966	0·974	0·981	0·994	1·005
0·30000	(2·94)	0·927	0·925	0·923	0·923	0·924	0·928	0·932	0·936	0·944	0·952	0·960	0·968	0·981	0·993
0·40000	(3·92)	0·912	0·910	0·909	0·910	0·912	0·916	0·921	0·925	0·934	0·942	0·951	0·959	0·973	0·985
0·60000	(5·88)	0·892	0·891	0·891	0·893	0·895	0·900	0·906	0·911	0·920	0·929	0·938	0·947	0·962	0·974
0·80000	(7·85)	0·878	0·878	0·879	0·881	0·884	0·890	0·896	0·901	0·911	0·920	0·930	0·939	0·954	0·967
1·00000	(9·81)	0·868	0·868	0·870	0·873	0·876	0·882	0·888	0·894	0·905	0·914	0·924	0·933	0·949	0·962
1·50000	(14·7)	0·851	0·852	0·855	0·858	0·862	0·869	0·876	0·882	0·894	0·903	0·914	0·924	0·940	0·954
2·00000	(19·6)	0·840	0·841	0·844	0·848	0·852	0·860	0·868	0·874	0·886	0·897	0·908	0·918	0·934	0·948
3·00000	(29·4)	0·824	0·826	0·831	0·836	0·840	0·849	0·857	0·864	0·877	0·888	0·899	0·910	0·927	0·942
S	$(\Delta p / \rho l)$	125	150	200	250	300	400	500	600	800	1000	1250	1500	2000	2500

Grad'nt k.p.g. (Equivalent) Pipe diameters in mm

Kinematic viscosity, $\nu = 15 \times 10^{-6}$ m^2 s^{-1} ;
Roughness size, k_s = 0·015 mm

Kin. visc., $\nu = 15 \times 10^{-6}$ m^{2}s^{-1};
$S = 0.00010$ to 3.00000

i.e. kin. pr. grad., $\Delta p/\rho l =$
(0.00098) to (29.4) ms^{-2}

Roughness size, $k_s = 0.030$ mm
This table shows values of m, as follows

m_C for Colebrook-White solutions; or,
where $\mathbf{R} \le 2000$, m_P for laminar flow

Grad'nt S	k.p.g. $(\Delta p/\rho l)$	125	150	200	250	300	400	500	600	800	1000	1250	1500	2000	2500
0.00010	(0.00098)	3.108	2.437	1.661	*	*	1.556	1.526	1.505	1.478	1.461	1.448	1.440	1.430	1.426
0.00015	(0.00147)	2.538	1.990	1.356	*	1.546	1.503	1.476	1.458	1.435	1.421	1.410	1.403	1.396	1.393
0.00020	(0.00196)	2.198	1.723	*	1.536	1.506	1.468	1.443	1.427	1.406	1.394	1.384	1.378	1.372	1.370
0.00030	(0.00294)	1.794	1.407	*	1.480	1.454	1.420	1.399	1.385	1.367	1.357	1.349	1.345	1.341	1.340
0.00040	(0.00392)	1.554	1.219	1.477	1.443	1.419	1.389	1.370	1.357	1.341	1.332	1.326	1.322	1.319	1.320
0.00060	(0.00588)	1.269	*	1.424	1.394	1.373	1.346	1.330	1.319	1.306	1.299	1.294	1.292	1.290	1.292
0.00080	(0.00785)	*	*	1.388	1.361	1.342	1.318	1.303	1.294	1.282	1.276	1.272	1.271	1.271	1.273
0.00100	(0.00981)	*	1.402	1.362	1.336	1.319	1.297	1.284	1.275	1.265	1.259	1.256	1.255	1.256	1.259
0.00150	(0.0147)	*	1.351	1.316	1.294	1.279	1.260	1.249	1.242	1.234	1.230	1.228	1.228	1.230	1.234
0.00200	(0.0196)	1.342	1.318	1.286	1.266	1.253	1.236	1.226	1.219	1.213	1.210	1.209	1.210	1.213	1.217
0.00300	(0.0294)	1.295	1.273	1.246	1.228	1.217	1.203	1.194	1.190	1.185	1.183	1.183	1.185	1.189	1.194
0.00400	(0.0392)	1.263	1.244	1.219	1.203	1.193	1.180	1.173	1.169	1.166	1.165	1.166	1.168	1.173	1.179
0.00600	(0.0588)	1.221	1.204	1.183	1.169	1.161	1.150	1.145	1.142	1.140	1.140	1.142	1.145	1.151	1.158
0.00800	(0.0785)	1.193	1.178	1.159	1.147	1.139	1.130	1.126	1.124	1.123	1.124	1.127	1.130	1.137	1.144
0.01000	(0.0981)	1.173	1.159	1.141	1.130	1.123	1.115	1.112	1.110	1.110	1.111	1.115	1.118	1.126	1.133
0.01500	(0.147)	1.137	1.125	1.110	1.101	1.095	1.089	1.087	1.086	1.087	1.090	1.094	1.098	1.107	1.115
0.02000	(0.196)	1.113	1.102	1.089	1.081	1.076	1.072	1.070	1.070	1.072	1.075	1.080	1.085	1.094	1.103
0.03000	(0.294)	1.081	1.072	1.061	1.055	1.051	1.048	1.048	1.049	1.052	1.056	1.061	1.067	1.077	1.086
0.04000	(0.392)	1.060	1.052	1.042	1.037	1.034	1.032	1.033	1.034	1.038	1.043	1.049	1.055	1.066	1.075
0.06000	(0.588)	1.032	1.025	1.017	1.014	1.012	1.011	1.013	1.015	1.020	1.026	1.032	1.039	1.050	1.061
0.08000	(0.785)	1.013	1.007	1.001	0.998	0.997	0.997	0.999	1.002	1.008	1.014	1.021	1.028	1.040	1.051
0.10000	(0.981)	0.999	0.994	0.988	0.986	0.985	0.987	0.989	0.992	0.999	1.005	1.013	1.020	1.033	1.044
0.15000	(1.47)	0.975	0.971	0.967	0.966	0.966	0.969	0.972	0.976	0.983	0.991	0.999	1.007	1.021	1.032
0.20000	(1.96)	0.958	0.955	0.953	0.952	0.953	0.957	0.961	0.965	0.973	0.981	0.990	0.998	1.012	1.025
0.30000	(2.94)	0.937	0.935	0.934	0.934	0.936	0.941	0.946	0.950	0.960	0.968	0.978	0.987	1.002	1.015
0.40000	(3.92)	0.923	0.921	0.921	0.923	0.925	0.930	0.936	0.941	0.951	0.960	0.970	0.979	0.995	1.008
0.60000	(5.88)	0.904	0.904	0.905	0.907	0.910	0.916	0.923	0.929	0.940	0.949	0.960	0.970	0.986	1.000
0.80000	(7.85)	0.892	0.892	0.894	0.897	0.900	0.907	0.914	0.921	0.932	0.942	0.953	0.963	0.980	0.994
1.00000	(9.81)	0.883	0.883	0.886	0.889	0.893	0.901	0.908	0.915	0.927	0.937	0.949	0.959	0.976	0.991
1.50000	(14.7)	0.867	0.868	0.872	0.877	0.881	0.890	0.898	0.905	0.918	0.929	0.941	0.952	0.970	0.985
2.00000	(19.6)	0.857	0.859	0.863	0.869	0.873	0.883	0.891	0.899	0.912	0.924	0.936	0.947	0.965	0.981
3.00000	(29.4)	0.844	0.846	0.852	0.858	0.863	0.874	0.883	0.891	0.905	0.917	0.930	0.941	0.960	0.976
S	$(\Delta p/\rho l)$	125	150	200	250	300	400	500	600	800	1000	1250	1500	2000	2500

Grad'nt k.p.g.   (Equivalent) Pipe diameters in mm   **Roughness size, $k_s = 0.030$ mm**

Kin. visc., $\nu = 15 \times 10^{-6}$ m^{2}s^{-1};
$S = 0.00010$ to 3.00000

i.e. kin. pr. grad., $\Delta p/\rho l =$
(0.00098) to (29.4) ms^{-2}

Roughness size, $k_s = 0.060$ mm
This table shows values of m, as follows

m_C for Colebrook-White solutions; or,
where $\mathbf{R} \le 2000$, m_P for laminar flow

Grad'nt S	k.p.g. $(\Delta p/\rho l)$	125	150	200	250	300	400	500	600	800	1000	1250	1500	2000	2500
0.00010	(0.00098)	3.108	2.437	1.661	*	*	1.558	1.528	1.506	1.479	1.463	1.450	1.442	1.433	1.428
0.00015	(0.00147)	2.538	1.990	1.356	*	1.548	1.505	1.478	1.460	1.437	1.423	1.412	1.406	1.399	1.396
0.00020	(0.00196)	2.198	1.723	*	1.538	1.508	1.470	1.446	1.429	1.408	1.396	1.387	1.381	1.375	1.374
0.00030	(0.00294)	1.794	1.407	*	1.482	1.456	1.423	1.402	1.388	1.370	1.360	1.352	1.348	1.344	1.344
0.00040	(0.00392)	1.554	1.219	1.480	1.445	1.422	1.391	1.372	1.360	1.344	1.335	1.329	1.326	1.323	1.324
0.00060	(0.00588)	1.269	*	1.426	1.396	1.376	1.349	1.333	1.322	1.310	1.303	1.298	1.296	1.295	1.297
0.00080	(0.00785)	*	*	1.391	1.364	1.345	1.321	1.307	1.297	1.286	1.280	1.277	1.275	1.276	1.278
0.00100	(0.00981)	*	1.405	1.365	1.340	1.322	1.300	1.287	1.279	1.269	1.264	1.261	1.260	1.262	1.264
0.00150	(0.0147)	*	1.355	1.320	1.298	1.283	1.264	1.253	1.246	1.239	1.235	1.234	1.234	1.237	1.241
0.00200	(0.0196)	1.346	1.322	1.290	1.270	1.257	1.240	1.230	1.224	1.218	1.216	1.215	1.216	1.220	1.224
0.00300	(0.0294)	1.299	1.278	1.250	1.233	1.222	1.208	1.200	1.195	1.191	1.190	1.190	1.192	1.197	1.203
0.00400	(0.0392)	1.268	1.248	1.224	1.208	1.198	1.186	1.179	1.176	1.172	1.172	1.174	1.176	1.182	1.188
0.00600	(0.0588)	1.227	1.210	1.188	1.175	1.167	1.157	1.152	1.149	1.148	1.149	1.151	1.155	1.162	1.169
0.00800	(0.0785)	1.199	1.184	1.165	1.153	1.146	1.138	1.134	1.132	1.132	1.133	1.136	1.140	1.148	1.156
0.01000	(0.0981)	1.179	1.165	1.147	1.137	1.130	1.123	1.120	1.119	1.119	1.122	1.125	1.130	1.138	1.146
0.01500	(0.147)	1.144	1.132	1.117	1.109	1.103	1.098	1.096	1.096	1.098	1.101	1.106	1.111	1.121	1.130
0.02000	(0.196)	1.121	1.110	1.097	1.090	1.086	1.082	1.081	1.081	1.084	1.088	1.094	1.099	1.110	1.119
0.03000	(0.294)	1.090	1.081	1.070	1.065	1.062	1.059	1.060	1.061	1.065	1.070	1.077	1.083	1.095	1.105
0.04000	(0.392)	1.070	1.062	1.053	1.048	1.046	1.045	1.046	1.048	1.053	1.059	1.066	1.073	1.085	1.096
0.06000	(0.588)	1.043	1.036	1.029	1.026	1.025	1.025	1.028	1.031	1.037	1.044	1.051	1.059	1.072	1.084
0.08000	(0.785)	1.025	1.019	1.014	1.012	1.011	1.013	1.016	1.019	1.026	1.034	1.042	1.050	1.064	1.076
0.10000	(0.981)	1.012	1.007	1.002	1.001	1.001	1.003	1.007	1.011	1.019	1.026	1.035	1.043	1.058	1.070
0.15000	(1.47)	0.989	0.986	0.983	0.983	0.984	0.987	0.992	0.996	1.005	1.014	1.024	1.032	1.048	1.061
0.20000	(1.96)	0.974	0.972	0.970	0.971	0.972	0.977	0.982	0.987	0.997	1.006	1.016	1.025	1.041	1.055
0.30000	(2.94)	0.955	0.953	0.953	0.955	0.957	0.963	0.969	0.975	0.986	0.996	1.007	1.016	1.033	1.047
0.40000	(3.92)	0.942	0.941	0.942	0.945	0.948	0.955	0.961	0.968	0.979	0.989	1.001	1.011	1.028	1.043
0.60000	(5.88)	0.926	0.926	0.928	0.932	0.936	0.943	0.951	0.958	0.970	0.981	0.993	1.004	1.022	1.037
0.80000	(7.85)	0.915	0.916	0.919	0.923	0.928	0.936	0.944	0.952	0.965	0.976	0.988	0.999	1.018	1.033
1.00000	(9.81)	0.907	0.908	0.913	0.917	0.922	0.931	0.939	0.947	0.961	0.972	0.985	0.996	1.015	1.031
1.50000	(14.7)	0.894	0.896	0.902	0.907	0.913	0.923	0.932	0.940	0.954	0.966	0.980	0.991	1.010	1.027
2.00000	(19.6)	0.886	0.889	0.895	0.901	0.907	0.917	0.927	0.935	0.950	0.963	0.976	0.988	1.008	1.024
3.00000	(29.4)	0.875	0.879	0.886	0.893	0.899	0.911	0.921	0.930	0.945	0.958	0.972	0.984	1.004	1.021
S	$(\Delta p/\rho l)$	125	150	200	250	300	400	500	600	800	1000	1250	1500	2000	2500

Grad'nt k.p.g.   (Equivalent) Pipe diameters in mm   **Roughness size, $k_s = 0.060$ mm**

E214

Kin. visc., $\nu = 15\times10^{-6}$ m^{2}s^{-1};
$S = 0{\cdot}00010$ to $3{\cdot}00000$

i.e. kin. pr. grad., $\Delta p/\rho l =$
$(0{\cdot}00098)$ to $(29{\cdot}4)$ ms^{-2}

Roughness size, $k_s = 0{\cdot}150$ mm
This table shows values of m, as follows

m_C for Colebrook-White solutions; or,
where $\mathbf{R} \leq 2000$, m_P for laminar flow

Grad'nt S	k.p.g. $(\Delta p/\rho l)$	125	150	200	250	300	400	500	600	800	1000	1250	1500	2000	2500
0·00010	(0·00098)	3·108	2·437	1·661	*	*	1·563	1·533	1·512	1·485	1·469	1·456	1·448	1·439	1·436
0·00015	(0·00147)	2·538	1·990	1·356	*	1·553	1·511	1·484	1·466	1·443	1·430	1·419	1·413	1·406	1·404
0·00020	(0·00196)	2·198	1·723	*	1·544	1·514	1·476	1·452	1·436	1·415	1·403	1·394	1·389	1·384	1·383
0·00030	(0·00294)	1·794	1·407	*	1·489	1·463	1·430	1·409	1·395	1·378	1·368	1·361	1·357	1·355	1·355
0·00040	(0·00392)	1·554	1·219	1·486	1·453	1·429	1·399	1·380	1·368	1·353	1·345	1·339	1·336	1·335	1·336
0·00060	(0·00588)	1·269	*	1·434	1·405	1·384	1·358	1·342	1·332	1·320	1·313	1·309	1·308	1·308	1·310
0·00080	(0·00785)	*	*	1·400	1·373	1·354	1·331	1·317	1·308	1·297	1·292	1·289	1·289	1·290	1·294
0·00100	(0·00981)	*	1·414	1·374	1·349	1·332	1·311	1·298	1·290	1·281	1·277	1·275	1·275	1·277	1·281
0·00150	(0·0147)	*	1·365	1·330	1·309	1·294	1·276	1·266	1·259	1·252	1·250	1·249	1·250	1·254	1·259
0·00200	(0·0196)	1·357	1·333	1·301	1·282	1·269	1·253	1·244	1·239	1·233	1·232	1·232	1·234	1·239	1·245
0·00300	(0·0294)	1·311	1·290	1·263	1·247	1·236	1·223	1·215	1·211	1·208	1·208	1·210	1·213	1·219	1·226
0·00400	(0·0392)	1·281	1·262	1·238	1·223	1·213	1·202	1·196	1·193	1·192	1·192	1·195	1·199	1·206	1·214
0·00600	(0·0588)	1·242	1·225	1·204	1·192	1·184	1·175	1·171	1·170	1·170	1·172	1·176	1·180	1·189	1·198
0·00800	(0·0785)	1·215	1·201	1·182	1·172	1·165	1·158	1·155	1·154	1·155	1·158	1·163	1·168	1·178	1·187
0·01000	(0·0981)	1·196	1·183	1·166	1·157	1·151	1·145	1·143	1·143	1·145	1·148	1·154	1·159	1·170	1·180
0·01500	(0·147)	1·164	1·152	1·138	1·131	1·127	1·123	1·122	1·123	1·127	1·132	1·138	1·144	1·156	1·167
0·02000	(0·196)	1·142	1·132	1·120	1·114	1·111	1·108	1·109	1·110	1·115	1·121	1·128	1·135	1·148	1·159
0·03000	(0·294)	1·114	1·106	1·097	1·092	1·090	1·089	1·091	1·094	1·100	1·107	1·115	1·123	1·136	1·149
0·04000	(0·392)	1·096	1·089	1·081	1·078	1·077	1·077	1·080	1·083	1·090	1·098	1·107	1·115	1·129	1·142
0·06000	(0·588)	1·072	1·067	1·061	1·059	1·059	1·062	1·065	1·069	1·078	1·086	1·096	1·105	1·120	1·134
0·08000	(0·785)	1·057	1·052	1·048	1·047	1·048	1·051	1·056	1·061	1·070	1·079	1·089	1·099	1·115	1·129
0·10000	(0·981)	1·045	1·042	1·039	1·039	1·040	1·044	1·049	1·054	1·065	1·074	1·085	1·094	1·111	1·125
0·15000	(1·47)	1·027	1·024	1·023	1·024	1·026	1·032	1·038	1·044	1·055	1·066	1·077	1·087	1·105	1·119
0·20000	(1·96)	1·015	1·013	1·013	1·015	1·018	1·024	1·031	1·038	1·050	1·060	1·072	1·083	1·101	1·116
0·30000	(2·94)	0·999	0·999	1·000	1·003	1·007	1·015	1·022	1·030	1·042	1·054	1·066	1·077	1·096	1·111
0·40000	(3·92)	0·989	0·990	0·992	0·996	1·000	1·009	1·017	1·025	1·038	1·050	1·062	1·074	1·093	1·109
0·60000	(5·88)	0·977	0·978	0·982	0·987	0·992	1·001	1·010	1·018	1·032	1·045	1·058	1·069	1·089	1·105
0·80000	(7·85)	0·969	0·971	0·976	0·981	0·987	0·997	1·006	1·014	1·029	1·042	1·055	1·067	1·087	1·103
1·00000	(9·81)	0·964	0·966	0·971	0·977	0·983	0·994	1·003	1·012	1·027	1·039	1·053	1·065	1·085	1·102
1·50000	(14·7)	0·955	0·958	0·964	0·971	0·977	0·988	0·998	1·007	1·023	1·036	1·050	1·062	1·083	1·100
2·00000	(19·6)	0·949	0·953	0·960	0·967	0·973	0·985	0·996	1·005	1·021	1·034	1·048	1·061	1·081	1·098
3·00000	(29·4)	0·942	0·946	0·954	0·962	0·969	0·981	0·992	1·002	1·018	1·032	1·046	1·059	1·080	1·097
S	$(\Delta p/\rho l)$	125	150	200	250	300	400	500	600	800	1000	1250	1500	2000	2500

Grad'nt   k.p.g.   (Equivalent) Pipe diameters in mm   **Roughness size, $k_s = 0{\cdot}150$ mm**

E215

Kin. visc., $\nu = 15\times10^{-6}$ m^{2}s^{-1};
$S = 0{\cdot}00010$ to $3{\cdot}00000$

i.e. kin. pr. grad., $\Delta p/\rho l =$
$(0{\cdot}00098)$ to $(29{\cdot}4)$ ms^{-2}

Roughness size, $k_s = 0{\cdot}30$ mm
This table shows values of m, as follows

m_C for Colebrook-White solutions; or,
where $\mathbf{R} \leq 2000$, m_P for laminar flow

Grad'nt S	k.p.g. $(\Delta p/\rho l)$	125	150	200	250	300	400	500	600	800	1000	1250	1500	2000	2500
0·00010	(0·00098)	3·108	2·437	1·661	*	*	1·571	1·541	1·520	1·494	1·478	1·466	1·458	1·450	1·447
0·00015	(0·00147)	2·538	1·990	1·356	*	1·562	1·520	1·494	1·476	1·453	1·440	1·431	1·425	1·419	1·418
0·00020	(0·00196)	2·198	1·723	*	*	1·524	1·486	1·462	1·446	1·426	1·415	1·407	1·402	1·398	1·398
0·00030	(0·00294)	1·794	1·407	*	1·499	1·474	1·441	1·421	1·407	1·391	1·382	1·375	1·372	1·370	1·372
0·00040	(0·00392)	1·554	1·219	1·498	1·464	1·441	1·411	1·393	1·381	1·367	1·359	1·355	1·352	1·352	1·354
0·00060	(0·00588)	1·269	*	1·447	1·418	1·398	1·372	1·357	1·347	1·336	1·330	1·327	1·326	1·328	1·331
0·00080	(0·00785)	*	*	1·413	1·387	1·369	1·346	1·333	1·324	1·315	1·311	1·309	1·309	1·312	1·316
0·00100	(0·00981)	*	1·428	1·389	1·364	1·348	1·327	1·315	1·308	1·300	1·296	1·295	1·296	1·300	1·305
0·00150	(0·0147)	*	1·381	1·347	1·326	1·312	1·295	1·285	1·279	1·274	1·272	1·273	1·275	1·280	1·287
0·00200	(0·0196)	1·374	1·350	1·319	1·301	1·288	1·273	1·265	1·260	1·257	1·256	1·258	1·260	1·267	1·275
0·00300	(0·0294)	1·331	1·310	1·283	1·268	1·257	1·245	1·239	1·236	1·234	1·235	1·238	1·242	1·251	1·259
0·00400	(0·0392)	1·302	1·283	1·260	1·246	1·237	1·227	1·222	1·220	1·220	1·222	1·226	1·230	1·240	1·249
0·00600	(0·0588)	1·265	1·249	1·229	1·218	1·211	1·203	1·200	1·199	1·201	1·204	1·210	1·215	1·226	1·237
0·00800	(0·0785)	1·241	1·227	1·209	1·199	1·193	1·188	1·186	1·186	1·189	1·193	1·199	1·206	1·218	1·229
0·01000	(0·0981)	1·223	1·210	1·195	1·186	1·181	1·176	1·176	1·176	1·180	1·185	1·192	1·199	1·211	1·223
0·01500	(0·147)	1·194	1·183	1·170	1·164	1·160	1·158	1·158	1·160	1·166	1·172	1·180	1·187	1·201	1·213
0·02000	(0·196)	1·175	1·165	1·154	1·149	1·147	1·146	1·147	1·150	1·157	1·163	1·172	1·180	1·195	1·207
0·03000	(0·294)	1·150	1·142	1·134	1·131	1·130	1·130	1·133	1·137	1·145	1·153	1·162	1·171	1·187	1·200
0·04000	(0·392)	1·134	1·128	1·121	1·119	1·119	1·121	1·124	1·129	1·138	1·146	1·156	1·165	1·182	1·196
0·06000	(0·588)	1·114	1·109	1·105	1·104	1·105	1·109	1·113	1·118	1·129	1·138	1·149	1·158	1·176	1·190
0·08000	(0·785)	1·101	1·097	1·094	1·095	1·096	1·101	1·106	1·112	1·123	1·133	1·144	1·154	1·172	1·187
0·10000	(0·981)	1·092	1·089	1·087	1·088	1·090	1·095	1·101	1·108	1·119	1·129	1·141	1·151	1·169	1·184
0·15000	(1·47)	1·077	1·075	1·075	1·077	1·080	1·087	1·094	1·100	1·113	1·123	1·136	1·146	1·165	1·181
0·20000	(1·96)	1·068	1·066	1·067	1·070	1·074	1·081	1·089	1·096	1·109	1·120	1·132	1·144	1·162	1·178
0·30000	(2·94)	1·056	1·056	1·058	1·062	1·066	1·075	1·083	1·090	1·104	1·116	1·129	1·140	1·159	1·175
0·40000	(3·92)	1·048	1·049	1·052	1·057	1·061	1·070	1·079	1·087	1·101	1·113	1·126	1·138	1·157	1·174
0·60000	(5·88)	1·039	1·041	1·045	1·050	1·055	1·065	1·074	1·083	1·097	1·110	1·123	1·135	1·155	1·172
0·80000	(7·85)	1·034	1·036	1·041	1·046	1·052	1·062	1·072	1·080	1·095	1·108	1·122	1·134	1·154	1·171
1·00000	(9·81)	1·030	1·032	1·038	1·044	1·049	1·060	1·070	1·079	1·094	1·107	1·120	1·133	1·153	1·170
1·50000	(14·7)	1·024	1·027	1·033	1·039	1·046	1·057	1·067	1·076	1·091	1·105	1·119	1·131	1·151	1·168
2·00000	(19·6)	1·020	1·023	1·030	1·037	1·043	1·055	1·065	1·074	1·090	1·103	1·118	1·130	1·151	1·168
3·00000	(29·4)	1·015	1·019	1·026	1·034	1·040	1·052	1·063	1·072	1·088	1·102	1·116	1·129	1·150	1·167
S	$(\Delta p/\rho l)$	125	150	200	250	300	400	500	600	800	1000	1250	1500	2000	2500

Grad'nt   k.p.g.   (Equivalent) Pipe diameters in mm   **Roughness size, $k_s = 0{\cdot}30$ mm**

Kin. visc., $\nu = 20 \times 10^{-6}$ m^2 s^{-1};
S = 0·00010 to 30·0000

i.e. kin. pr. grad., $\Delta p/\rho l$ =
(0·00098) to (294) ms^{-2}

Roughness size, k_s = 0·003 mm
This table shows values of m, as follows

m_C for Colebrook-White solutions; or,
where $\mathbf{R} \le 2000$, m_P for laminar flow

Grad'nt S	k.p.g. $(\Delta p/\rho l)$	6	8	10	12·5	15	20	25	30	40	50	60	80	100	125
0·00030	(0·00294)	137·1	93·46	69·41	51·54	40·42	27·54	20·46	16·04	10·93	8·118	6·366	4·338	3·222	2·392
0·00040	(0·00392)	118·8	80·93	60·11	44·64	35·01	23·85	17·71	13·89	9·466	7·030	5·513	3·757	2·790	2·072
0·00060	(0·00588)	96·98	66·08	49·08	36·45	28·58	19·48	14·46	11·34	7·729	5·740	4·501	3·067	2·278	1·692
0·00080	(0·00785)	83·99	57·23	42·50	31·56	24·75	16·87	12·53	9·823	6·694	4·971	3·898	2·656	1·973	1·465
0·00100	(0·00981)	75·12	51·19	38·01	28·23	22·14	15·09	11·20	8·786	5·987	4·446	3·487	2·376	1·764	1·310
0·00150	(0·0147)	61·33	41·79	31·04	23·05	18·08	12·32	9·148	7·174	4·888	3·630	2·847	1·940	1·441	*
0·00200	(0·0196)	53·12	36·20	26·88	19·96	15·65	10·67	7·922	6·213	4·233	3·144	2·465	1·680	1·248	*
0·00300	(0·0294)	43·37	29·55	21·95	16·30	12·78	8·710	6·469	5·073	3·457	2·567	2·013	1·372	*	1·360
0·00400	(0·0392)	37·56	25·59	19·01	14·12	11·07	7·543	5·602	4·393	2·993	2·223	1·743	1·188	*	1·325
0·00600	(0·0588)	30·67	20·90	15·52	11·53	9·038	6·159	4·574	3·587	2·444	1·815	1·423	*	1·308	1·278
0·00800	(0·0785)	26·56	18·10	13·44	9·981	7·827	5·334	3·961	3·106	2·117	1·572	1·233	*	1·274	1·247
0·01000	(0·0981)	23·75	16·19	12·02	8·928	7·001	4·771	3·543	2·778	1·893	1·406	1·103	*	1·249	1·223
0·01500	(0·147)	19·40	13·22	9·815	7·289	5·716	3·895	2·893	2·269	1·546	1·148	*	1·232	1·205	1·183
0·02000	(0·196)	16·80	11·45	8·500	6·313	4·951	3·373	2·505	1·965	1·339	0·994	*	1·201	1·176	1·156
0·03000	(0·294)	13·71	9·346	6·941	5·154	4·042	2·754	2·046	1·604	1·093	*	1·193	1·159	1·138	1·120
0·04000	(0·392)	11·88	8·093	6·011	4·464	3·501	2·385	1·771	1·389	0·947	*	1·162	1·131	1·112	1·096
0·06000	(0·588)	9·698	6·608	4·908	3·645	2·858	1·948	1·446	1·134	*	1·142	1·121	1·094	1·078	1·064
0·08000	(0·785)	8·399	5·723	4·250	3·156	2·475	1·687	1·253	0·982	*	1·113	1·094	1·070	1·055	1·042
0·10000	(0·981)	7·512	5·119	3·801	2·823	2·214	1·509	1·120	*	1·117	1·091	1·074	1·051	1·037	1·026
0·15000	(1·47)	6·133	4·179	3·104	2·305	1·808	1·232	0·915	*	1·076	1·054	1·039	1·019	1·007	0·998
0·20000	(1·96)	5·312	3·620	2·688	1·996	1·565	1·067	*	*	1·050	1·030	1·016	0·998	0·987	0·979
0·30000	(2·94)	4·337	2·955	2·195	1·630	1·278	0·871	*	1·042	1·014	0·997	0·985	0·969	0·960	0·953
0·40000	(3·92)	3·756	2·559	1·901	1·412	1·107	*	1·035	1·016	0·990	0·974	0·964	0·950	0·942	0·936
0·60000	(5·88)	3·067	2·090	1·552	1·153	0·904	*	0·998	0·981	0·959	0·945	0·936	0·924	0·917	0·913
0·80000	(7·85)	2·656	1·810	1·344	0·998	*	0·996	0·973	0·958	0·938	0·925	0·917	0·907	0·901	0·897
1·00000	(9·81)	2·375	1·619	1·202	0·893	*	0·976	0·955	0·940	0·922	0·910	0·903	0·894	0·888	0·885
1·50000	(14·7)	1·940	1·322	0·982	*	*	0·941	0·923	0·911	0·895	0·885	0·878	0·871	0·867	0·864
2·00000	(19·6)	1·680	1·145	0·850	*	0·945	0·918	0·902	0·891	0·876	0·867	0·862	0·855	0·852	0·850
·3·00000	(29·4)	1·371	0·935	*	0·929	0·911	0·888	0·874	0·864	0·852	0·844	0·840	0·835	0·832	0·831
4·00000	(39·2)	1·188	0·809	*	0·905	0·889	0·868	0·855	0·846	0·835	0·829	0·825	0·821	0·819	0·819
6·00000	(58·8)	0·970	*	0·894	0·873	0·859	0·841	0·830	0·822	0·813	0·808	0·805	0·802	0·801	0·801
8·00000	(78·5)	0·840	*	0·871	0·852	0·839	0·823	0·813	0·806	0·798	0·794	0·791	0·789	0·789	0·790
10·0000	(98·1)	0·751	*	0·854	0·837	0·825	0·810	0·800	0·794	0·787	0·783	0·781	0·779	0·780	0·781
15·0000	(147)	*	0·843	0·825	0·810	0·799	0·786	0·779	0·773	0·768	0·765	0·763	0·763	0·764	0·766
20·0000	(196)	*	0·822	0·805	0·792	0·782	0·771	0·764	0·759	0·755	0·752	0·751	0·752	0·753	0·755
30·0000	(294)	0·817	0·794	0·780	0·768	0·760	0·750	0·744	0·741	0·737	0·736	0·735	0·737	0·738	0·741
S	$(\Delta p/\rho l)$	6	8	10	12·5	15	20	25	30	40	50	60	80	100	125

Grad'nt   k.p.g.   (Equivalent) Pipe diameters in mm

Grad'nt S	k.p.g. $(\Delta p/\rho l)$	125	150	200	250	300	400	500	600	800	1000	1250	1500	2000	2500
0·00010	(0·00098)	4·144	3·250	2·214	1·644	*	1·638	1·601	1·575	1·542	1·521	1·505	1·494	1·481	1·474
0·00015	(0·00147)	3·383	2·653	1·808	1·343	*	1·578	1·546	1·524	1·495	1·477	1·463	1·454	1·443	1·438
0·00020	(0·00196)	2·930	2·298	1·566	*	*	1·539	1·510	1·489	1·463	1·447	1·435	1·427	1·418	1·414
0·00030	(0·00294)	2·392	1·876	1·278	*	1·528	1·487	1·461	1·443	1·421	1·407	1·397	1·390	1·384	1·381
0·00040	(0·00392)	2·072	1·625	*	1·518	1·489	1·452	1·428	1·412	1·392	1·380	1·371	1·366	1·360	1·358
0·00060	(0·00588)	1·692	1·327	*	1·463	1·438	1·405	1·385	1·371	1·354	1·344	1·337	1·332	1·329	1·328
0·00080	(0·00785)	1·465	*	1·459	1·426	1·403	1·374	1·355	1·343	1·328	1·319	1·313	1·310	1·307	1·307
0·00100	(0·00981)	1·310	*	1·430	1·399	1·378	1·350	1·334	1·322	1·308	1·301	1·295	1·293	1·291	1·292
0·00150	(0·0147)	*	*	1·379	1·352	1·334	1·310	1·295	1·286	1·274	1·268	1·264	1·263	1·263	1·264
0·00200	(0·0196)	*	1·384	1·345	1·321	1·304	1·283	1·270	1·261	1·251	1·246	1·243	1·242	1·243	1·246
0·00300	(0·0294)	1·360	1·334	1·300	1·279	1·265	1·246	1·235	1·228	1·220	1·217	1·215	1·215	1·217	1·220
0·00400	(0·0392)	1·325	1·301	1·270	1·251	1·238	1·222	1·212	1·206	1·199	1·196	1·195	1·196	1·199	1·203
0·00600	(0·0588)	1·278	1·257	1·230	1·214	1·202	1·188	1·180	1·176	1·171	1·169	1·169	1·170	1·174	1·179
0·00800	(0·0785)	1·247	1·228	1·203	1·188	1·178	1·166	1·159	1·155	1·151	1·150	1·151	1·153	1·158	1·163
0·01000	(0·0981)	1·223	1·206	1·183	1·169	1·160	1·149	1·143	1·140	1·137	1·136	1·138	1·140	1·145	1·150
0·01500	(0·147)	1·183	1·168	1·148	1·137	1·129	1·120	1·115	1·113	1·111	1·112	1·114	1·117	1·123	1·129
0·02000	(0·196)	1·156	1·143	1·125	1·114	1·108	1·100	1·096	1·094	1·094	1·095	1·098	1·101	1·107	1·114
0·03000	(0·294)	1·120	1·109	1·094	1·085	1·079	1·073	1·070	1·070	1·070	1·072	1·076	1·079	1·087	1·094
0·04000	(0·392)	1·096	1·086	1·072	1·065	1·060	1·055	1·053	1·053	1·054	1·057	1·060	1·065	1·072	1·080
0·06000	(0·588)	1·064	1·055	1·044	1·037	1·034	1·030	1·029	1·030	1·032	1·035	1·040	1·044	1·053	1·061
0·08000	(0·785)	1·042	1·034	1·024	1·019	1·016	1·013	1·013	1·014	1·017	1·021	1·026	1·031	1·040	1·048
0·10000	(0·981)	1·026	1·019	1·010	1·005	1·003	1·001	1·001	1·002	1·006	1·010	1·015	1·020	1·030	1·038
0·15000	(1·47)	0·998	0·992	0·985	0·981	0·979	0·979	0·980	0·982	0·986	0·991	0·997	1·002	1·012	1·021
0·20000	(1·96)	0·979	0·973	0·967	0·965	0·964	0·964	0·965	0·968	0·973	0·978	0·984	0·990	1·000	1·009
0·30000	(2·94)	0·953	0·949	0·944	0·943	0·942	0·943	0·946	0·948	0·954	0·960	0·967	0·973	0·984	0·993
0·40000	(3·92)	0·936	0·932	0·929	0·928	0·928	0·930	0·932	0·935	0·942	0·948	0·955	0·961	0·973	0·982
0·60000	(5·88)	0·913	0·910	0·907	0·907	0·908	0·911	0·914	0·918	0·925	0·931	0·939	0·945	0·957	0·967
0·80000	(7·85)	0·897	0·895	0·893	0·893	0·895	0·898	0·902	0·906	0·913	0·920	0·928	0·935	0·947	0·957
1·00000	(9·81)	0·885	0·883	0·882	0·883	0·884	0·888	0·892	0·897	0·904	0·911	0·919	0·927	0·939	0·950
1·50000	(14·7)	0·864	0·863	0·863	0·865	0·867	0·871	0·876	0·881	0·889	0·897	0·905	0·912	0·925	0·936
2·00000	(19·6)	0·850	0·850	0·851	0·852	0·855	0·860	0·865	0·870	0·879	0·886	0·895	0·903	0·916	0·927
3·00000	(29·4)	0·831	0·831	0·833	0·836	0·839	0·844	0·850	0·855	0·865	0·873	0·882	0·890	0·904	0·915
S	$(\Delta p/\rho l)$	125	150	200	250	300	400	500	600	800	1000	1250	1500	2000	2500

Grad'nt   k.p.g.   (Equivalent) Pipe diameters in mm

Kinematic viscosity, $\nu = 20 \times 10^{-6}$ m^{2}s^{-1} ;   **Roughness size, k_s = 0·003 mm**

E217

Kin. visc., $\nu = 20 \times 10^{-6}$ m^{2}s^{-1};
S = 0·00010 to 30·0000
i.e. kin. pr. grad., $\Delta p/\rho l$ = (0·00098) to (294) ms^{-2}

Roughness size, k_s = 0·006 mm
This table shows values of m, as follows
m_C for Colebrook-White solutions; or, where $\mathbf{R} \leq 2000$, m_P for laminar flow

Grad'nt **k.p.g.** (Equivalent) Pipe diameters in mm

S	$(\Delta p/\rho l)$	6	8	10	12·5	15	20	25	30	40	50	60	80	100	125
0·00030	(0·00294)	137·1	93·46	69·41	51·54	40·42	27·54	20·46	16·04	10·93	8·118	6·366	4·338	3·222	2·392
0·00040	(0·00392)	118·8	80·93	60·11	44·64	35·01	23·85	17·71	13·89	9·466	7·030	5·513	3·757	2·790	2·072
0·00060	(0·00588)	96·98	66·08	49·08	36·45	28·58	19·48	14·46	11·34	7·729	5·740	4·501	3·067	2·278	1·692
0·00080	(0·00785)	83·99	57·23	42·50	31·56	24·75	16·87	12·53	9·823	6·694	4·971	3·898	2·656	1·973	1·465
0·00100	(0·00981)	75·12	51·19	38·01	28·23	22·14	15·09	11·20	8·786	5·987	4·446	3·487	2·376	1·764	1·310
0·00150	(0·0147)	61·33	41·79	31·04	23·05	18·08	12·32	9·148	7·174	4·888	3·630	2·847	1·940	1·441	*
0·00200	(0·0196)	53·12	36·20	26·88	19·96	15·65	10·67	7·922	6·213	4·233	3·144	2·465	1·680	1·248	*
0·00300	(0·0294)	43·37	29·55	21·95	16·30	12·78	8·710	6·469	5·073	3·457	2·567	2·013	1·372	*	1·361
0·00400	(0·0392)	37·56	25·59	19·01	14·12	11·07	7·543	5·602	4·393	2·993	2·223	1·743	1·188	*	1·325
0·00600	(0·0588)	30·67	20·90	15·52	11·53	9·038	6·159	4·574	3·587	2·444	1·815	1·423	*	1·308	1·278
0·00800	(0·0785)	26·56	18·10	13·44	9·981	7·827	5·334	3·961	3·106	2·117	1·572	1·233	*	1·274	1·247
0·01000	(0·0981)	23·75	16·19	12·02	8·928	7·001	4·771	3·543	2·778	1·893	1·406	1·103	*	1·249	1·224
0·01500	(0·147)	19·40	13·22	9·815	7·289	5·716	3·895	2·893	2·269	1·546	1·148	*	1·233	1·206	1·184
0·02000	(0·196)	16·80	11·45	8·500	6·313	4·951	3·373	2·505	1·965	1·339	0·994	*	1·201	1·177	1·157
0·03000	(0·294)	13·71	9·346	6·941	5·154	4·042	2·754	2·046	1·604	1·093	*	1·194	1·160	1·139	1·121
0·04000	(0·392)	11·88	8·093	6·011	4·464	3·501	2·385	1·771	1·389	0·947	*	1·163	1·132	1·113	1·097
0·06000	(0·588)	9·698	6·608	4·908	3·645	2·858	1·948	1·446	1·134	*	1·143	1·122	1·095	1·079	1·065
0·08000	(0·785)	8·399	5·723	4·250	3·156	2·475	1·687	1·253	0·982	*	1·114	1·095	1·071	1·056	1·043
0·10000	(0·981)	7·512	5·119	3·801	2·823	2·214	1·509	1·120	*	1·117	1·092	1·075	1·052	1·038	1·027
0·15000	(1·47)	6·133	4·179	3·104	2·305	1·808	1·232	0·915	*	1·077	1·055	1·040	1·021	1·009	0·999
0·20000	(1·96)	5·312	3·620	2·688	1·996	1·565	1·067	*	*	1·051	1·031	1·017	0·999	0·989	0·980
0·30000	(2·94)	4·337	2·955	2·195	1·630	1·278	0·871	*	1·043	1·015	0·998	0·986	0·971	0·962	0·955
0·40000	(3·92)	3·756	2·559	1·901	1·412	1·107	*	1·037	1·017	0·992	0·976	0·965	0·952	0·944	0·938
0·60000	(5·88)	3·067	2·090	1·552	1·153	0·904	*	0·999	0·982	0·960	0·947	0·937	0·926	0·920	0·915
0·80000	(7·85)	2·656	1·810	1·344	0·998	*	0·997	0·975	0·959	0·939	0·927	0·919	0·909	0·903	0·899
1·00000	(9·81)	2·375	1·619	1·202	0·893	*	0·977	0·957	0·942	0·924	0·912	0·905	0·896	0·891	0·888
1·50000	(14·7)	1·940	1·322	0·982	*	*	0·943	0·925	0·913	0·897	0·887	0·881	0·873	0·870	0·867
2·00000	(19·6)	1·680	1·145	0·850	*	0·947	0·921	0·904	0·893	0·879	0·870	0·865	0·858	0·855	0·854
3·00000	(29·4)	1·371	0·935	*	0·931	0·914	0·891	0·876	0·867	0·855	0·847	0·843	0·838	0·836	0·835
4·00000	(39·2)	1·188	0·809	*	0·908	0·892	0·871	0·858	0·849	0·838	0·832	0·828	0·824	0·823	0·823
6·00000	(58·8)	0·970	*	0·897	0·876	0·862	0·844	0·833	0·826	0·817	0·812	0·809	0·806	0·806	0·806
8·00000	(78·5)	0·840	*	0·874	0·855	0·843	0·826	0·817	0·810	0·802	0·798	0·796	0·794	0·794	0·795
10·0000	(98·1)	0·751	*	0·857	0·840	0·828	0·813	0·804	0·798	0·791	0·788	0·786	0·785	0·785	0·787
15·0000	(147)	*	0·846	0·828	0·814	0·803	0·791	0·783	0·778	0·773	0·770	0·769	0·769	0·770	0·772
20·0000	(196)	*	0·826	0·809	0·796	0·787	0·775	0·769	0·765	0·760	0·758	0·758	0·758	0·760	0·763
30·0000	(294)	0·821	0·798	0·784	0·773	0·765	0·755	0·750	0·747	0·743	0·742	0·743	0·744	0·747	0·750
S	$(\Delta p/\rho l)$	6	8	10	12·5	15	20	25	30	40	50	60	80	100	125

Grad'nt **k.p.g.** (Equivalent) Pipe diameters in mm

S	$(\Delta p/\rho l)$	125	150	200	250	300	400	500	600	800	1000	1250	1500	2000	2500
0·00010	(0·00098)	4·144	3·250	2·214	1·644	*	1·638	1·601	1·575	1·542	1·521	1·505	1·494	1·481	1·474
0·00015	(0·00147)	3·383	2·653	1·808	1·343	*	1·579	1·546	1·524	1·495	1·477	1·463	1·454	1·443	1·438
0·00020	(0·00196)	2·930	2·298	1·566	*	*	1·539	1·510	1·489	1·463	1·448	1·435	1·427	1·418	1·414
0·00030	(0·00294)	2·392	1·876	1·278	*	1·528	1·487	1·461	1·444	1·421	1·408	1·397	1·391	1·384	1·381
0·00040	(0·00392)	2·072	1·625	*	1·518	1·489	1·452	1·429	1·413	1·392	1·381	1·372	1·366	1·361	1·359
0·00060	(0·00588)	1·692	1·327	*	1·463	1·438	1·405	1·385	1·371	1·354	1·344	1·337	1·333	1·329	1·328
0·00080	(0·00785)	1·465	*	1·459	1·426	1·404	1·374	1·356	1·343	1·328	1·320	1·313	1·310	1·308	1·308
0·00100	(0·00981)	1·310	*	1·430	1·399	1·378	1·351	1·334	1·323	1·309	1·301	1·296	1·293	1·291	1·292
0·00150	(0·0147)	*	*	1·379	1·352	1·334	1·310	1·296	1·286	1·275	1·269	1·265	1·263	1·263	1·265
0·00200	(0·0196)	*	1·384	1·346	1·321	1·304	1·283	1·270	1·262	1·252	1·247	1·244	1·243	1·244	1·246
0·00300	(0·0294)	1·361	1·335	1·301	1·279	1·265	1·247	1·236	1·229	1·221	1·217	1·216	1·215	1·217	1·221
0·00400	(0·0392)	1·325	1·302	1·271	1·251	1·238	1·222	1·212	1·206	1·200	1·197	1·196	1·197	1·200	1·203
0·00600	(0·0588)	1·278	1·258	1·231	1·214	1·203	1·189	1·181	1·176	1·171	1·170	1·170	1·171	1·175	1·180
0·00800	(0·0785)	1·247	1·228	1·204	1·189	1·179	1·167	1·160	1·156	1·152	1·151	1·152	1·154	1·159	1·164
0·01000	(0·0981)	1·224	1·206	1·184	1·170	1·161	1·150	1·144	1·140	1·138	1·137	1·139	1·141	1·146	1·152
0·01500	(0·147)	1·184	1·169	1·149	1·137	1·130	1·121	1·116	1·114	1·112	1·113	1·115	1·118	1·124	1·130
0·02000	(0·196)	1·157	1·143	1·126	1·115	1·109	1·101	1·097	1·095	1·095	1·096	1·099	1·102	1·109	1·115
0·03000	(0·294)	1·121	1·109	1·094	1·086	1·080	1·074	1·072	1·071	1·071	1·074	1·077	1·081	1·088	1·096
0·04000	(0·392)	1·097	1·086	1·073	1·066	1·061	1·056	1·054	1·054	1·055	1·058	1·062	1·066	1·074	1·082
0·06000	(0·588)	1·065	1·056	1·045	1·039	1·035	1·032	1·031	1·031	1·034	1·037	1·042	1·047	1·055	1·063
0·08000	(0·785)	1·043	1·035	1·026	1·020	1·017	1·015	1·015	1·016	1·019	1·023	1·028	1·033	1·042	1·051
0·10000	(0·981)	1·027	1·020	1·011	1·007	1·004	1·002	1·003	1·004	1·008	1·012	1·018	1·023	1·033	1·041
0·15000	(1·47)	0·999	0·993	0·986	0·983	0·981	0·981	0·982	0·984	0·988	0·993	0·999	1·005	1·015	1·025
0·20000	(1·96)	0·980	0·975	0·969	0·967	0·965	0·966	0·968	0·970	0·975	0·981	0·987	0·993	1·004	1·013
0·30000	(2·94)	0·955	0·951	0·946	0·945	0·944	0·946	0·948	0·951	0·957	0·963	0·970	0·977	0·988	0·998
0·40000	(3·92)	0·938	0·934	0·931	0·930	0·930	0·932	0·935	0·938	0·945	0·951	0·959	0·965	0·977	0·988
0·60000	(5·88)	0·915	0·912	0·910	0·910	0·911	0·914	0·918	0·921	0·929	0·936	0·943	0·950	0·963	0·974
0·80000	(7·85)	0·899	0·897	0·896	0·896	0·898	0·901	0·906	0·910	0·918	0·925	0·933	0·940	0·953	0·964
1·00000	(9·81)	0·888	0·886	0·885	0·886	0·888	0·892	0·897	0·901	0·909	0·917	0·925	0·933	0·946	0·957
1·50000	(14·7)	0·867	0·866	0·867	0·869	0·871	0·876	0·881	0·886	0·895	0·903	0·912	0·919	0·933	0·945
2·00000	(19·6)	0·854	0·853	0·854	0·857	0·859	0·865	0·870	0·875	0·885	0·893	0·902	0·911	0·925	0·937
3·00000	(29·4)	0·835	0·836	0·838	0·841	0·844	0·850	0·856	0·862	0·872	0·881	0·890	0·899	0·913	0·926
S	$(\Delta p/\rho l)$	125	150	200	250	300	400	500	600	800	1000	1250	1500	2000	2500

Grad'nt **k.p.g.** (Equivalent) Pipe diameters in mm

Kinematic viscosity, $\nu = 20 \times 10^{-6}$ m^{2}s^{-1} ;

Roughness size, k_s = 0·006 mm

Kin. visc., $\nu = 20\times10^{-6}$ m^2 s^{-1}; $S = 0.00010$ to 30.0000
i.e. kin. pr. grad., $\Delta p/\rho l =$ (0.00098) to (294) ms^{-2}

Roughness size, $k_s = 0.015$ mm
This table shows values of m, as follows
m_C for Colebrook-White solutions; or, where $\mathbf{R} \leq 2000$, m_P for laminar flow

E218

Grad'nt S	k.p.g. $(\Delta p/\rho l)$	6	8	10	12·5	15	20	25	30	40	50	60	80	100	125
0·00030	(0·00294)	137·1	93·46	69·41	51·54	40·42	27·54	20·46	16·04	10·93	8·118	6·366	4·338	3·222	2·392
0·00040	(0·00392)	118·8	80·93	60·11	44·64	35·01	23·85	17·71	13·89	9·466	7·030	5·513	3·757	2·790	2·072
0·00060	(0·00588)	96·98	66·08	49·08	36·45	28·58	19·48	14·46	11·34	7·729	5·740	4·501	3·067	2·278	1·692
0·00080	(0·00785)	83·99	57·23	42·50	31·56	24·75	16·87	12·53	9·823	6·694	4·971	3·898	2·656	1·973	1·465
0·00100	(0·00981)	75·12	51·19	38·01	28·23	22·14	15·09	11·20	8·786	5·987	4·446	3·487	2·376	1·764	1·310
0·00150	(0·0147)	61·33	41·79	31·04	23·05	18·08	12·32	9·148	7·174	4·888	3·630	2·847	1·940	1·441	*
0·00200	(0·0196)	53·12	36·20	26·88	19·96	15·65	10·67	7·922	6·213	4·233	3·144	2·465	1·680	1·248	*
0·00300	(0·0294)	43·37	29·55	21·95	16·30	12·78	8·710	6·469	5·073	3·457	2·567	2·013	1·372	*	1·362
0·00400	(0·0392)	37·56	25·59	19·01	14·12	11·07	7·543	5·602	4·393	2·993	2·223	1·743	1·188	*	1·326
0·00600	(0·0588)	30·67	20·90	15·52	11·53	9·038	6·159	4·574	3·587	2·444	1·815	1·423	*	1·310	1·280
0·00800	(0·0785)	26·56	18·10	13·44	9·981	7·827	5·334	3·961	3·106	2·117	1·572	1·233	*	1·276	1·248
0·01000	(0·0981)	23·75	16·19	12·02	8·928	7·001	4·771	3·543	2·778	1·893	1·406	1·103	*	1·251	1·225
0·01500	(0·147)	19·40	13·22	9·815	7·289	5·716	3·895	2·893	2·269	1·546	1·148	*	1·234	1·208	1·186
0·02000	(0·196)	16·80	11·45	8·500	6·313	4·951	3·373	2·505	1·965	1·339	0·994	*	1·203	1·179	1·159
0·03000	(0·294)	13·71	9·346	6·941	5·154	4·042	2·754	2·046	1·604	1·093	*	1·196	1·162	1·141	1·123
0·04000	(0·392)	11·88	8·093	6·011	4·464	3·501	2·385	1·771	1·389	0·947	*	1·165	1·134	1·115	1·100
0·06000	(0·588)	9·698	6·608	4·908	3·645	2·858	1·948	1·446	1·134	*	1·145	1·125	1·098	1·081	1·068
0·08000	(0·785)	8·399	5·723	4·250	3·156	2·475	1·687	1·253	0·982	*	1·116	1·098	1·074	1·059	1·047
0·10000	(0·981)	7·512	5·119	3·801	2·823	2·214	1·509	1·120	*	1·120	1·095	1·078	1·055	1·042	1·031
0·15000	(1·47)	6·133	4·179	3·104	2·305	1·808	1·232	0·915	*	1·081	1·059	1·044	1·024	1·012	1·003
0·20000	(1·96)	5·312	3·620	2·688	1·996	1·565	1·067	*	*	1·054	1·034	1·021	1·003	0·993	0·985
0·30000	(2·94)	4·337	2·955	2·195	1·630	1·278	0·871	*	1·047	1·019	1·002	0·990	0·975	0·967	0·960
0·40000	(3·92)	3·756	2·559	1·901	1·412	1·107	*	1·041	1·021	0·996	0·980	0·970	0·957	0·949	0·943
0·60000	(5·88)	3·067	2·090	1·552	1·153	0·904	*	1·004	0·987	0·965	0·952	0·943	0·932	0·925	0·921
0·80000	(7·85)	2·656	1·810	1·344	0·998	*	1·002	0·980	0·965	0·945	0·933	0·925	0·915	0·910	0·906
1·00000	(9·81)	2·375	1·619	1·202	0·893	*	0·983	0·962	0·948	0·930	0·919	0·911	0·903	0·898	0·895
1·50000	(14·7)	1·940	1·322	0·982	*	*	0·949	0·931	0·919	0·903	0·894	0·888	0·881	0·878	0·876
2·00000	(19·6)	1·680	1·145	0·850	*	0·953	0·927	0·911	0·900	0·886	0·878	0·873	0·867	0·864	0·863
3·00000	(29·4)	1·371	0·935	*	0·938	0·921	0·898	0·884	0·875	0·863	0·856	0·852	0·848	0·846	0·846
4·00000	(39·2)	1·188	0·809	*	0·915	0·899	0·879	0·866	0·858	0·847	0·842	0·838	0·835	0·834	0·835
6·00000	(58·8)	0·970	*	0·905	0·885	0·871	0·853	0·843	0·836	0·827	0·823	0·820	0·818	0·819	0·820
8·00000	(78·5)	0·840	*	0·883	0·865	0·852	0·836	0·827	0·821	0·814	0·810	0·808	0·807	0·808	0·810
10·0000	(98·1)	0·751	*	0·867	0·850	0·839	0·824	0·815	0·810	0·804	0·801	0·799	0·799	0·801	0·803
15·0000	(147)	*	0·857	0·839	0·825	0·815	0·803	0·796	0·791	0·787	0·785	0·784	0·785	0·787	0·791
20·0000	(196)	*	0·837	0·821	0·808	0·800	0·789	0·783	0·779	0·775	0·774	0·774	0·776	0·779	0·783
30·0000	(294)	0·834	0·812	0·798	0·787	0·779	0·770	0·766	0·763	0·761	0·761	0·761	0·764	0·768	0·772
S	$(\Delta p/\rho l)$	6	8	10	12·5	15	20	25	30	40	50	60	80	100	125

Grad'nt k.p.g. (Equivalent) Pipe diameters in mm

Grad'nt S	k.p.g. $(\Delta p/\rho l)$	125	150	200	250	300	400	500	600	800	1000	1250	1500	2000	2500
0·00010	(0·00098)	4·144	3·250	2·214	1·644	*	1·638	1·601	1·576	1·542	1·522	1·505	1·494	1·481	1·474
0·00015	(0·00147)	3·383	2·653	1·808	1·343	*	1·579	1·547	1·524	1·496	1·478	1·464	1·455	1·444	1·439
0·00020	(0·00196)	2·930	2·298	1·566	*	*	1·540	1·510	1·490	1·464	1·448	1·436	1·428	1·419	1·415
0·00030	(0·00294)	2·392	1·876	1·278	*	1·528	1·487	1·462	1·444	1·422	1·408	1·398	1·391	1·385	1·382
0·00040	(0·00392)	2·072	1·625	*	1·518	1·490	1·453	1·429	1·413	1·393	1·381	1·372	1·367	1·362	1·360
0·00060	(0·00588)	1·692	1·327	*	1·464	1·439	1·406	1·386	1·372	1·355	1·345	1·338	1·334	1·330	1·330
0·00080	(0·00785)	1·465	*	1·460	1·427	1·404	1·375	1·357	1·344	1·329	1·321	1·315	1·311	1·309	1·309
0·00100	(0·00981)	1·310	*	1·431	1·400	1·379	1·352	1·335	1·323	1·310	1·302	1·297	1·294	1·293	1·294
0·00150	(0·0147)	*	*	1·380	1·353	1·335	1·311	1·297	1·287	1·276	1·270	1·266	1·265	1·265	1·267
0·00200	(0·0196)	*	1·385	1·347	1·322	1·306	1·284	1·271	1·263	1·253	1·248	1·245	1·245	1·245	1·248
0·00300	(0·0294)	1·362	1·336	1·302	1·281	1·266	1·248	1·237	1·230	1·222	1·219	1·217	1·217	1·220	1·223
0·00400	(0·0392)	1·326	1·303	1·272	1·253	1·240	1·223	1·214	1·208	1·202	1·199	1·198	1·199	1·202	1·206
0·00600	(0·0588)	1·280	1·259	1·232	1·216	1·204	1·191	1·183	1·178	1·173	1·172	1·172	1·174	1·178	1·183
0·00800	(0·0785)	1·248	1·230	1·205	1·190	1·181	1·168	1·162	1·158	1·154	1·154	1·155	1·157	1·162	1·167
0·01000	(0·0981)	1·225	1·208	1·185	1·172	1·163	1·152	1·146	1·143	1·140	1·140	1·141	1·144	1·149	1·155
0·01500	(0·147)	1·186	1·170	1·151	1·139	1·132	1·123	1·118	1·116	1·115	1·116	1·118	1·121	1·128	1·134
0·02000	(0·196)	1·159	1·145	1·128	1·118	1·111	1·103	1·100	1·098	1·098	1·100	1·103	1·106	1·113	1·120
0·03000	(0·294)	1·123	1·112	1·097	1·088	1·083	1·077	1·075	1·074	1·075	1·078	1·081	1·085	1·093	1·101
0·04000	(0·392)	1·100	1·089	1·076	1·068	1·064	1·059	1·058	1·058	1·060	1·063	1·067	1·071	1·080	1·088
0·06000	(0·588)	1·068	1·059	1·048	1·042	1·038	1·035	1·035	1·036	1·039	1·042	1·047	1·052	1·062	1·071
0·08000	(0·785)	1·047	1·039	1·029	1·024	1·021	1·019	1·019	1·021	1·024	1·029	1·034	1·040	1·050	1·059
0·10000	(0·981)	1·031	1·023	1·015	1·011	1·008	1·007	1·008	1·009	1·014	1·018	1·024	1·030	1·041	1·050
0·15000	(1·47)	1·003	0·997	0·991	0·987	0·986	0·986	0·988	0·990	0·995	1·001	1·007	1·013	1·025	1·035
0·20000	(1·96)	0·985	0·980	0·974	0·972	0·971	0·972	0·974	0·977	0·983	0·989	0·996	1·002	1·014	1·024
0·30000	(2·94)	0·960	0·956	0·952	0·951	0·951	0·953	0·956	0·959	0·966	0·973	0·980	0·987	1·000	1·011
0·40000	(3·92)	0·943	0·940	0·937	0·937	0·937	0·940	0·943	0·947	0·955	0·962	0·970	0·977	0·990	1·002
0·60000	(5·88)	0·921	0·919	0·917	0·918	0·919	0·923	0·927	0·931	0·940	0·948	0·956	0·964	0·978	0·990
0·80000	(7·85)	0·906	0·905	0·904	0·905	0·907	0·911	0·916	0·921	0·930	0·938	0·947	0·955	0·970	0·982
1·00000	(9·81)	0·895	0·894	0·894	0·895	0·898	0·903	0·908	0·913	0·922	0·931	0·940	0·949	0·963	0·976
1·50000	(14·7)	0·876	0·876	0·877	0·879	0·882	0·888	0·894	0·900	0·910	0·919	0·929	0·938	0·953	0·966
2·00000	(19·6)	0·863	0·863	0·865	0·868	0·872	0·878	0·885	0·891	0·901	0·911	0·921	0·931	0·946	0·960
3·00000	(29·4)	0·846	0·847	0·850	0·854	0·858	0·865	0·872	0·879	0·890	0·901	0·912	0·921	0·938	0·952
S	$(\Delta p/\rho l)$	125	150	200	250	300	400	500	600	800	1000	1250	1500	2000	2500

Grad'nt k.p.g. (Equivalent) Pipe diameters in mm

Kinematic viscosity, $\nu = 20\times10^{-6}$ m^2 s^{-1} ;   **Roughness size, $k_s = 0.015$ mm**

E219

Kin. visc., $\nu = 20 \times 10^{-6}$ m^{2}s^{-1};
$S = 0.00010$ to 3.00000

i.e. kin. pr. grad., $\Delta p/\rho l =$
(0.00098) to (29.4) ms^{-2}

Roughness size, $k_s = 0.030$ mm
This table shows values of m, as follows

m_C for Colebrook-White solutions; or,
where $\mathbf{R} \leq 2000$, m_P for laminar flow

Grad'nt S	k.p.g. $(\Delta p/\rho l)$	125	150	200	250	300	400	500	600	800	1000	1250	1500	2000	2500
0.00010	(0.00098)	4.144	3.250	2.214	1.644	*	1.639	1.602	1.576	1.543	1.523	1.506	1.495	1.482	1.475
0.00015	(0.00147)	3.383	2.653	1.808	1.343	*	1.580	1.548	1.525	1.496	1.479	1.465	1.456	1.445	1.440
0.00020	(0.00196)	2.930	2.298	1.566	*	*	1.541	1.511	1.491	1.465	1.449	1.437	1.429	1.420	1.416
0.00030	(0.00294)	2.392	1.876	1.278	*	1.529	1.488	1.463	1.445	1.423	1.409	1.399	1.393	1.386	1.384
0.00040	(0.00392)	2.072	1.625	*	1.519	1.491	1.454	1.430	1.414	1.394	1.383	1.374	1.368	1.363	1.361
0.00060	(0.00588)	1.692	1.327	*	1.465	1.440	1.407	1.387	1.373	1.356	1.347	1.339	1.335	1.332	1.332
0.00080	(0.00785)	1.465	*	1.461	1.428	1.406	1.376	1.358	1.346	1.331	1.322	1.316	1.313	1.311	1.311
0.00100	(0.00981)	1.310	*	1.432	1.401	1.380	1.353	1.336	1.325	1.312	1.304	1.299	1.296	1.295	1.296
0.00150	(0.0147)	*	*	1.382	1.355	1.337	1.313	1.299	1.289	1.278	1.272	1.269	1.267	1.267	1.269
0.00200	(0.0196)	*	1.387	1.348	1.324	1.307	1.286	1.273	1.265	1.255	1.251	1.248	1.247	1.248	1.251
0.00300	(0.0294)	1.364	1.338	1.304	1.283	1.268	1.250	1.239	1.233	1.225	1.222	1.220	1.220	1.223	1.227
0.00400	(0.0392)	1.328	1.305	1.274	1.255	1.242	1.226	1.216	1.211	1.204	1.202	1.201	1.202	1.206	1.210
0.00600	(0.0588)	1.282	1.261	1.234	1.218	1.207	1.193	1.186	1.181	1.177	1.176	1.176	1.178	1.182	1.188
0.00800	(0.0785)	1.251	1.232	1.208	1.193	1.183	1.171	1.165	1.161	1.158	1.158	1.159	1.161	1.167	1.172
0.01000	(0.0981)	1.228	1.211	1.188	1.175	1.166	1.155	1.149	1.146	1.144	1.144	1.146	1.149	1.155	1.161
0.01500	(0.147)	1.188	1.173	1.154	1.143	1.135	1.127	1.122	1.120	1.120	1.121	1.124	1.127	1.134	1.141
0.02000	(0.196)	1.162	1.149	1.131	1.121	1.115	1.108	1.104	1.103	1.103	1.105	1.109	1.112	1.120	1.128
0.03000	(0.294)	1.127	1.115	1.101	1.093	1.087	1.082	1.080	1.080	1.081	1.084	1.088	1.093	1.101	1.110
0.04000	(0.392)	1.104	1.093	1.080	1.073	1.069	1.065	1.063	1.064	1.066	1.070	1.075	1.079	1.089	1.098
0.06000	(0.588)	1.073	1.064	1.053	1.047	1.044	1.042	1.042	1.043	1.046	1.051	1.056	1.062	1.072	1.082
0.08000	(0.785)	1.052	1.044	1.035	1.030	1.028	1.026	1.027	1.028	1.033	1.038	1.044	1.050	1.061	1.071
0.10000	(0.981)	1.036	1.029	1.021	1.017	1.015	1.015	1.016	1.018	1.023	1.028	1.035	1.041	1.053	1.063
0.15000	(1.47)	1.009	1.004	0.998	0.995	0.994	0.995	0.997	1.000	1.006	1.012	1.019	1.026	1.039	1.050
0.20000	(1.96)	0.992	0.987	0.982	0.980	0.980	0.981	0.984	0.987	0.994	1.001	1.009	1.016	1.029	1.041
0.30000	(2.94)	0.968	0.964	0.961	0.960	0.961	0.964	0.967	0.971	0.979	0.987	0.995	1.003	1.017	1.029
0.40000	(3.92)	0.952	0.949	0.947	0.947	0.948	0.952	0.956	0.961	0.969	0.977	0.986	0.995	1.009	1.022
0.60000	(5.88)	0.931	0.929	0.928	0.930	0.931	0.936	0.941	0.947	0.956	0.965	0.975	0.983	0.999	1.012
0.80000	(7.85)	0.917	0.916	0.916	0.918	0.920	0.926	0.932	0.937	0.948	0.957	0.967	0.976	0.992	1.006
1.00000	(9.81)	0.907	0.906	0.907	0.909	0.912	0.919	0.925	0.931	0.941	0.951	0.962	0.971	0.987	1.001
1.50000	(14.7)	0.889	0.890	0.892	0.895	0.899	0.906	0.913	0.919	0.931	0.941	0.952	0.962	0.979	0.994
2.00000	(19.6)	0.878	0.879	0.882	0.886	0.890	0.898	0.905	0.912	0.924	0.935	0.946	0.957	0.974	0.989
3.00000	(29.4)	0.863	0.864	0.869	0.873	0.878	0.887	0.895	0.902	0.915	0.927	0.939	0.950	0.968	0.983
S	$(\Delta p/\rho l)$	125	150	200	250	300	400	500	600	800	1000	1250	1500	2000	2500

Grad'nt k.p.g. (Equivalent) Pipe diameters in mm — **Roughness size, $k_s = 0.030$ mm**

E220

Kin. visc., $\nu = 20 \times 10^{-6}$ m^{2}s^{-1};
$S = 0.00010$ to 3.00000

i.e. kin. pr. grad., $\Delta p/\rho l =$
(0.00098) to (29.4) ms^{-2}

Roughness size, $k_s = 0.060$ mm
This table shows values of m, as follows

m_C for Colebrook-White solutions; or,
where $\mathbf{R} \leq 2000$, m_P for laminar flow

Grad'nt S	k.p.g. $(\Delta p/\rho l)$	125	150	200	250	300	400	500	600	800	1000	1250	1500	2000	2500
0.00010	(0.00098)	4.144	3.250	2.214	1.644	*	1.640	1.603	1.578	1.545	1.524	1.508	1.497	1.484	1.477
0.00015	(0.00147)	3.383	2.653	1.808	1.343	*	1.581	1.549	1.527	1.498	1.481	1.467	1.458	1.447	1.442
0.00020	(0.00196)	2.930	2.298	1.566	*	*	1.542	1.513	1.493	1.467	1.451	1.439	1.431	1.423	1.419
0.00030	(0.00294)	2.392	1.876	1.278	*	1.531	1.490	1.465	1.447	1.425	1.412	1.402	1.395	1.389	1.387
0.00040	(0.00392)	2.072	1.625	*	1.521	1.493	1.456	1.433	1.417	1.397	1.385	1.376	1.371	1.366	1.365
0.00060	(0.00588)	1.692	1.327	*	1.467	1.442	1.410	1.390	1.376	1.359	1.350	1.343	1.339	1.336	1.336
0.00080	(0.00785)	1.465	*	1.464	1.431	1.408	1.379	1.361	1.349	1.334	1.326	1.320	1.317	1.315	1.316
0.00100	(0.00981)	1.310	*	1.435	1.404	1.383	1.356	1.339	1.328	1.315	1.308	1.303	1.300	1.299	1.301
0.00150	(0.0147)	*	*	1.385	1.358	1.340	1.316	1.302	1.293	1.282	1.276	1.273	1.272	1.272	1.275
0.00200	(0.0196)	*	1.390	1.351	1.327	1.311	1.290	1.277	1.269	1.260	1.255	1.253	1.252	1.254	1.257
0.00300	(0.0294)	1.367	1.341	1.307	1.286	1.272	1.254	1.244	1.237	1.230	1.227	1.226	1.226	1.230	1.234
0.00400	(0.0392)	1.332	1.309	1.278	1.259	1.246	1.230	1.221	1.216	1.210	1.208	1.208	1.209	1.213	1.218
0.00600	(0.0588)	1.286	1.266	1.239	1.223	1.212	1.199	1.191	1.187	1.183	1.182	1.183	1.185	1.191	1.197
0.00800	(0.0785)	1.256	1.237	1.213	1.198	1.189	1.177	1.171	1.168	1.165	1.165	1.167	1.170	1.176	1.182
0.01000	(0.0981)	1.233	1.216	1.194	1.180	1.172	1.162	1.156	1.153	1.152	1.152	1.155	1.158	1.165	1.172
0.01500	(0.147)	1.194	1.179	1.160	1.149	1.142	1.134	1.130	1.129	1.129	1.130	1.134	1.138	1.146	1.154
0.02000	(0.196)	1.168	1.155	1.138	1.128	1.122	1.116	1.113	1.112	1.113	1.116	1.120	1.124	1.133	1.141
0.03000	(0.294)	1.134	1.123	1.109	1.101	1.096	1.091	1.090	1.090	1.092	1.096	1.101	1.106	1.116	1.125
0.04000	(0.392)	1.112	1.101	1.089	1.082	1.078	1.075	1.074	1.075	1.079	1.083	1.089	1.094	1.105	1.115
0.06000	(0.588)	1.081	1.073	1.063	1.058	1.055	1.053	1.054	1.056	1.060	1.066	1.072	1.079	1.090	1.101
0.08000	(0.785)	1.062	1.054	1.046	1.042	1.040	1.039	1.040	1.043	1.048	1.054	1.062	1.068	1.081	1.092
0.10000	(0.981)	1.047	1.040	1.033	1.030	1.028	1.028	1.030	1.033	1.039	1.046	1.054	1.061	1.074	1.085
0.15000	(1.47)	1.021	1.016	1.011	1.009	1.009	1.010	1.013	1.017	1.024	1.032	1.040	1.048	1.062	1.074
0.20000	(1.96)	1.005	1.000	0.996	0.995	0.996	0.998	1.002	1.006	1.015	1.023	1.032	1.040	1.055	1.067
0.30000	(2.94)	0.983	0.980	0.977	0.978	0.979	0.983	0.988	0.993	1.002	1.011	1.021	1.029	1.045	1.058
0.40000	(3.92)	0.968	0.966	0.965	0.966	0.968	0.973	0.978	0.984	0.994	1.003	1.013	1.023	1.039	1.053
0.60000	(5.88)	0.949	0.948	0.949	0.951	0.953	0.960	0.966	0.972	0.983	0.993	1.004	1.014	1.031	1.045
0.80000	(7.85)	0.937	0.937	0.938	0.941	0.944	0.951	0.958	0.965	0.976	0.987	0.998	1.009	1.026	1.041
1.00000	(9.81)	0.928	0.928	0.930	0.934	0.937	0.945	0.952	0.959	0.972	0.982	0.994	1.005	1.023	1.038
1.50000	(14.7)	0.913	0.914	0.918	0.922	0.926	0.935	0.943	0.950	0.964	0.975	0.988	0.999	1.017	1.033
2.00000	(19.6)	0.903	0.905	0.909	0.914	0.919	0.929	0.937	0.945	0.959	0.971	0.983	0.995	1.014	1.029
3.00000	(29.4)	0.891	0.893	0.899	0.904	0.910	0.920	0.930	0.938	0.952	0.965	0.978	0.990	1.009	1.025
S	$(\Delta p/\rho l)$	125	150	200	250	300	400	500	600	800	1000	1250	1500	2000	2500

Grad'nt k.p.g. (Equivalent) Pipe diameters in mm — **Roughness size, $k_s = 0.060$ mm**

Kin. visc., $\nu = 20 \times 10^{-6}$ m^{2}s^{-1};
S = 0.00010 to 3.00000

i.e. kin. pr. grad., $\Delta p/\rho l$ = (0.00098) to (29.4) ms^{-2}

Roughness size, k_s = 0.150 mm
This table shows values of m, as follows
m_C for Colebrook-White solutions; or, where **R** $\leq$ 2000, m_P for laminar flow

Grad'nt S	k.p.g. $(\Delta p/\rho l)$	125	150	200	250	300	400	500	600	800	1000	1250	1500	2000	2500
0.00010	(0.00098)	4.144	3.250	2.214	1.644	*	1.644	1.607	1.582	1.549	1.529	1.513	1.502	1.490	1.483
0.00015	(0.00147)	3.383	2.653	1.808	1.343	*	1.586	1.554	1.532	1.503	1.486	1.472	1.464	1.454	1.449
0.00020	(0.00196)	2.930	2.298	1.566	*	*	1.547	1.518	1.498	1.473	1.457	1.445	1.438	1.430	1.426
0.00030	(0.00294)	2.392	1.876	1.278	*	1.537	1.496	1.471	1.453	1.432	1.419	1.409	1.403	1.397	1.395
0.00040	(0.00392)	2.072	1.625	*	1.527	1.499	1.462	1.439	1.424	1.404	1.393	1.385	1.380	1.375	1.375
0.00060	(0.00588)	1.692	1.327	*	1.474	1.449	1.417	1.397	1.384	1.368	1.358	1.352	1.349	1.346	1.347
0.00080	(0.00785)	1.465	*	1.471	1.438	1.416	1.387	1.369	1.357	1.343	1.335	1.330	1.328	1.327	1.328
0.00100	(0.00981)	1.310	*	1.442	1.412	1.391	1.364	1.348	1.338	1.325	1.318	1.314	1.312	1.312	1.314
0.00150	(0.0147)	*	*	1.393	1.367	1.349	1.326	1.312	1.304	1.293	1.288	1.286	1.285	1.287	1.290
0.00200	(0.0196)	*	1.399	1.361	1.337	1.321	1.300	1.288	1.281	1.272	1.269	1.267	1.267	1.270	1.275
0.00300	(0.0294)	1.377	1.351	1.318	1.298	1.284	1.267	1.257	1.251	1.244	1.242	1.242	1.243	1.248	1.253
0.00400	(0.0392)	1.343	1.320	1.290	1.271	1.259	1.244	1.235	1.231	1.226	1.225	1.226	1.228	1.233	1.240
0.00600	(0.0588)	1.299	1.278	1.252	1.237	1.226	1.214	1.208	1.204	1.201	1.202	1.204	1.207	1.214	1.221
0.00800	(0.0785)	1.269	1.251	1.228	1.214	1.205	1.194	1.189	1.186	1.185	1.186	1.189	1.193	1.201	1.209
0.01000	(0.0981)	1.248	1.231	1.209	1.197	1.189	1.180	1.175	1.173	1.173	1.175	1.179	1.183	1.192	1.200
0.01500	(0.147)	1.211	1.196	1.178	1.168	1.162	1.155	1.152	1.151	1.153	1.156	1.161	1.166	1.176	1.186
0.02000	(0.196)	1.186	1.174	1.158	1.149	1.143	1.138	1.137	1.137	1.139	1.143	1.149	1.155	1.166	1.176
0.03000	(0.294)	1.155	1.144	1.131	1.124	1.120	1.117	1.116	1.118	1.122	1.127	1.134	1.140	1.153	1.164
0.04000	(0.392)	1.134	1.124	1.113	1.107	1.104	1.103	1.103	1.105	1.111	1.117	1.124	1.131	1.144	1.156
0.06000	(0.588)	1.107	1.099	1.090	1.086	1.084	1.084	1.086	1.089	1.096	1.103	1.111	1.119	1.134	1.146
0.08000	(0.785)	1.089	1.082	1.075	1.072	1.071	1.073	1.075	1.079	1.087	1.094	1.103	1.112	1.127	1.140
0.10000	(0.981)	1.076	1.070	1.064	1.062	1.062	1.064	1.067	1.072	1.080	1.088	1.098	1.106	1.122	1.135
0.15000	(1.47)	1.054	1.050	1.046	1.045	1.046	1.050	1.054	1.059	1.069	1.078	1.088	1.098	1.114	1.128
0.20000	(1.96)	1.040	1.036	1.034	1.034	1.036	1.040	1.046	1.051	1.062	1.071	1.082	1.092	1.109	1.123
0.30000	(2.94)	1.022	1.019	1.019	1.020	1.023	1.029	1.035	1.041	1.053	1.063	1.075	1.085	1.103	1.118
0.40000	(3.92)	1.010	1.009	1.009	1.012	1.015	1.022	1.029	1.035	1.047	1.058	1.070	1.081	1.099	1.114
0.60000	(5.88)	0.995	0.995	0.997	1.000	1.004	1.012	1.020	1.027	1.041	1.052	1.065	1.076	1.094	1.110
0.80000	(7.85)	0.986	0.986	0.989	0.993	0.998	1.007	1.015	1.023	1.036	1.048	1.061	1.072	1.092	1.108
1.00000	(9.81)	0.979	0.980	0.984	0.988	0.993	1.003	1.011	1.019	1.033	1.045	1.059	1.070	1.090	1.106
1.50000	(14.7)	0.968	0.970	0.975	0.980	0.986	0.996	1.005	1.014	1.028	1.041	1.055	1.066	1.086	1.103
2.00000	(19.6)	0.961	0.963	0.969	0.975	0.981	0.992	1.002	1.010	1.026	1.038	1.052	1.064	1.084	1.101
3.00000	(29.4)	0.952	0.956	0.962	0.969	0.975	0.987	0.997	1.006	1.022	1.035	1.049	1.062	1.082	1.099
S	$(\Delta p/\rho l)$	125	150	200	250	300	400	500	600	800	1000	1250	1500	2000	2500

Grad'nt k.p.g. (Equivalent) Pipe diameters in mm — Roughness size, k_s = 0.150 mm

---

Kin. visc., $\nu = 20 \times 10^{-6}$ m^{2}s^{-1};
S = 0.00010 to 3.00000

i.e. kin. pr. grad., $\Delta p/\rho l$ = (0.00098) to (29.4) ms^{-2}

Roughness size, k_s = 0.30 mm
This table shows values of m, as follows
m_C for Colebrook-White solutions; or, where **R** $\leq$ 2000, m_P for laminar flow

Grad'nt S	k.p.g. $(\Delta p/\rho l)$	125	150	200	250	300	400	500	600	800	1000	1250	1500	2000	2500
0.00010	(0.00098)	4.144	3.250	2.214	1.644	*	1.651	1.614	1.589	1.557	1.537	1.521	1.511	1.499	1.493
0.00015	(0.00147)	3.383	2.653	1.808	1.343	*	1.594	1.562	1.540	1.512	1.495	1.482	1.473	1.464	1.460
0.00020	(0.00196)	2.930	2.298	1.566	*	*	1.556	1.527	1.507	1.482	1.467	1.455	1.448	1.441	1.438
0.00030	(0.00294)	2.392	1.876	1.278	*	1.546	1.506	1.480	1.463	1.442	1.430	1.421	1.415	1.410	1.409
0.00040	(0.00392)	2.072	1.625	*	1.537	1.509	1.472	1.450	1.435	1.416	1.405	1.397	1.393	1.390	1.390
0.00060	(0.00588)	1.692	1.327	*	1.485	1.460	1.429	1.409	1.396	1.381	1.372	1.367	1.364	1.363	1.364
0.00080	(0.00785)	1.465	*	1.483	1.450	1.428	1.400	1.382	1.371	1.358	1.351	1.346	1.345	1.345	1.347
0.00100	(0.00981)	1.310	*	1.455	1.425	1.404	1.378	1.362	1.352	1.340	1.335	1.331	1.330	1.332	1.335
0.00150	(0.0147)	*	*	1.407	1.381	1.364	1.342	1.329	1.320	1.311	1.307	1.305	1.306	1.309	1.314
0.00200	(0.0196)	*	1.414	1.376	1.353	1.337	1.317	1.306	1.299	1.292	1.289	1.289	1.290	1.294	1.300
0.00300	(0.0294)	1.394	1.368	1.335	1.315	1.302	1.286	1.277	1.271	1.266	1.265	1.266	1.269	1.275	1.282
0.00400	(0.0392)	1.361	1.338	1.309	1.291	1.279	1.265	1.257	1.253	1.250	1.250	1.252	1.255	1.262	1.270
0.00600	(0.0588)	1.319	1.299	1.274	1.259	1.249	1.238	1.232	1.229	1.228	1.230	1.233	1.237	1.246	1.255
0.00800	(0.0785)	1.291	1.273	1.251	1.237	1.229	1.220	1.216	1.214	1.214	1.217	1.221	1.226	1.236	1.245
0.01000	(0.0981)	1.271	1.254	1.234	1.222	1.215	1.207	1.204	1.203	1.204	1.207	1.212	1.218	1.228	1.239
0.01500	(0.147)	1.237	1.223	1.206	1.196	1.190	1.185	1.183	1.184	1.187	1.191	1.198	1.204	1.216	1.227
0.02000	(0.196)	1.214	1.202	1.187	1.179	1.175	1.171	1.170	1.171	1.176	1.181	1.188	1.195	1.208	1.220
0.03000	(0.294)	1.186	1.175	1.164	1.158	1.154	1.153	1.154	1.156	1.162	1.168	1.176	1.184	1.198	1.211
0.04000	(0.392)	1.167	1.158	1.148	1.144	1.141	1.141	1.143	1.146	1.153	1.160	1.169	1.177	1.192	1.205
0.06000	(0.588)	1.143	1.136	1.129	1.126	1.125	1.126	1.130	1.134	1.142	1.150	1.160	1.169	1.185	1.198
0.08000	(0.785)	1.128	1.122	1.116	1.114	1.114	1.117	1.121	1.126	1.135	1.144	1.154	1.163	1.180	1.194
0.10000	(0.981)	1.117	1.112	1.107	1.106	1.107	1.110	1.115	1.120	1.130	1.139	1.150	1.160	1.176	1.191
0.15000	(1.47)	1.099	1.095	1.093	1.093	1.095	1.100	1.105	1.111	1.122	1.132	1.143	1.153	1.171	1.186
0.20000	(1.96)	1.087	1.085	1.083	1.085	1.087	1.093	1.099	1.105	1.117	1.127	1.139	1.150	1.168	1.183
0.30000	(2.94)	1.073	1.071	1.072	1.074	1.077	1.084	1.091	1.098	1.111	1.122	1.134	1.145	1.164	1.180
0.40000	(3.92)	1.064	1.063	1.065	1.068	1.071	1.079	1.087	1.094	1.107	1.119	1.131	1.142	1.161	1.177
0.60000	(5.88)	1.053	1.053	1.056	1.060	1.064	1.073	1.081	1.089	1.103	1.115	1.128	1.139	1.159	1.175
0.80000	(7.85)	1.046	1.047	1.050	1.055	1.059	1.069	1.078	1.086	1.100	1.112	1.125	1.137	1.157	1.173
1.00000	(9.81)	1.041	1.042	1.046	1.051	1.056	1.066	1.075	1.083	1.098	1.110	1.124	1.136	1.156	1.172
1.50000	(14.7)	1.033	1.035	1.040	1.046	1.051	1.062	1.071	1.080	1.095	1.108	1.121	1.133	1.154	1.170
2.00000	(19.6)	1.028	1.030	1.036	1.042	1.048	1.059	1.069	1.078	1.093	1.106	1.120	1.132	1.152	1.169
3.00000	(29.4)	1.022	1.025	1.032	1.038	1.045	1.056	1.066	1.075	1.091	1.104	1.118	1.130	1.151	1.168
S	$(\Delta p/\rho l)$	125	150	200	250	300	400	500	600	800	1000	1250	1500	2000	2500

Grad'nt k.p.g. (Equivalent) Pipe diameters in mm — Roughness size, k_s = 0.30 mm

E223

Kin. visc., $\nu = 25\times10^{-6}$ m^{2}s^{-1};
S = 0·00010 to 30·0000

i.e. kin. pr. grad., $\Delta p/\rho l =$
(0·00098) to (294) ms^{-2}

Roughness size, k_s = 0·006 mm
This table shows values of m, as follows

m_C for Colebrook-White solutions; or,
where $R \leq 2000$, m_P for laminar flow

Grad'nt S	k.p.g. $(\Delta p/\rho l)$	6	8	10	12·5	15	20	25	30	40	50	60	80	100	125
0·00030	(0·00294)	171·4	116·8	86·76	64·43	50·53	34·43	25·57	20·05	13·66	10·15	7·957	5·422	4·027	2·991
0·00040	(0·00392)	148·5	101·2	75·13	55·80	43·76	29·82	22·14	17·36	11·83	8·788	6·891	4·696	3·487	2·590
0·00060	(0·00588)	121·2	82·60	61·35	45·56	35·73	24·35	18·08	14·18	9·661	7·175	5·627	3·834	2·847	2·115
0·00080	(0·00785)	105·0	71·54	53·13	39·46	30·94	21·08	15·66	12·28	8·367	6·214	4·873	3·320	2·466	1·831
0·00100	(0·00981)	93·90	63·98	47·52	35·29	27·67	18·86	14·00	10·98	7·484	5·558	4·358	2·970	2·206	1·638
0·00150	(0·0147)	76·67	52·24	38·80	28·81	22·60	15·40	11·43	8·967	6·110	4·538	3·559	2·425	1·801	1·337
0·00200	(0·0196)	66·40	45·24	33·60	24·95	19·57	13·33	9·903	7·766	5·292	3·930	3·082	2·100	1·560	1·158
0·00300	(0·0294)	54·21	36·94	27·43	20·37	15·98	10·89	8·086	6·341	4·321	3·209	2·516	1·715	1·273	*
0·00400	(0·0392)	46·95	31·99	23·76	17·64	13·84	9·429	7·002	5·491	3·742	2·779	2·179	1·485	1·103	*
0·00600	(0·0588)	38·33	26·12	19·40	14·41	11·30	7·699	5·717	4·484	3·055	2·269	1·779	1·212	*	1·330
0·00800	(0·0785)	33·20	22·62	16·80	12·48	9·784	6·667	4·951	3·883	2·646	1·965	1·541	*	*	1·296
0·01000	(0·0981)	29·69	20·23	15·03	11·16	8·751	5·963	4·429	3·473	2·367	1·758	1·378	*	1·301	1·271
0·01500	(0·147)	24·24	16·52	12·27	9·112	7·145	4·869	3·616	2·836	1·932	1·435	1·125	*	1·254	1·228
0·02000	(0·196)	21·00	14·31	10·63	7·891	6·188	4·217	3·132	2·456	1·673	1·243	*	1·251	1·222	1·199
0·03000	(0·294)	17·14	11·68	8·676	6·443	5·053	3·443	2·557	2·005	1·366	1·015	*	1·206	1·181	1·161
0·04000	(0·392)	14·85	10·12	7·513	5·580	4·376	2·982	2·214	1·736	1·183	*	1·211	1·176	1·153	1·135
0·06000	(0·588)	12·12	8·260	6·135	4·556	3·573	2·435	1·808	1·418	0·966	*	1·167	1·136	1·117	1·101
0·08000	(0·785)	10·50	7·154	5·313	3·946	3·094	2·108	1·566	1·228	*	1·159	1·138	1·109	1·092	1·077
0·10000	(0·981)	9·390	6·398	4·752	3·529	2·767	1·886	1·400	1·098	*	1·136	1·116	1·090	1·073	1·060
0·15000	(1·47)	7·667	5·224	3·880	2·881	2·260	1·540	1·143	0·897	1·122	1·096	1·079	1·056	1·042	1·030
0·20000	(1·96)	6·640	4·524	3·360	2·495	1·957	1·333	0·990	*	1·093	1·070	1·053	1·033	1·020	1·010
0·30000	(2·94)	5·421	3·694	2·743	2·037	1·598	1·089	*	*	1·054	1·034	1·020	1·002	0·991	0·983
0·40000	(3·92)	4·695	3·199	2·376	1·764	1·384	0·943	*	1·058	1·029	1·011	0·998	0·982	0·972	0·965
0·60000	(5·88)	3·833	2·612	1·940	1·441	1·130	*	1·040	1·021	0·995	0·979	0·968	0·954	0·946	0·940
0·80000	(7·85)	3·320	2·262	1·680	1·248	0·978	*	1·014	0·996	0·972	0·958	0·948	0·936	0·929	0·924
1·00000	(9·81)	2·969	2·023	1·503	1·116	0·875	*	0·994	0·977	0·956	0·942	0·933	0·922	0·916	0·911
1·50000	(14·7)	2·424	1·652	1·227	0·911	*	0·981	0·960	0·945	0·927	0·915	0·907	0·898	0·893	0·890
2·00000	(19·6)	2·100	1·431	1·063	0·789	*	0·956	0·937	0·924	0·907	0·897	0·890	0·882	0·878	0·875
3·00000	(29·4)	1·714	1·168	0·868	*	0·950	0·924	0·907	0·896	0·881	0·872	0·867	0·860	0·857	0·856
4·00000	(39·2)	1·485	1·012	0·751	*	0·926	0·902	0·887	0·877	0·864	0·856	0·851	0·846	0·843	0·842
6·00000	(58·8)	1·212	0·826	*	0·911	0·895	0·873	0·860	0·852	0·841	0·834	0·830	0·826	0·825	0·825
8·00000	(78·5)	1·050	*	*	0·888	0·873	0·854	0·843	0·835	0·825	0·820	0·816	0·813	0·812	0·813
10·0000	(98·1)	0·939	*	0·891	0·871	0·858	0·840	0·829	0·822	0·813	0·809	0·806	0·803	0·803	0·804
15·0000	(147)	0·767	*	0·860	0·843	0·831	0·816	0·806	0·800	0·793	0·790	0·788	0·786	0·787	0·788
20·0000	(196)	*	0·858	0·839	0·824	0·813	0·799	0·791	0·786	0·780	0·777	0·775	0·775	0·776	0·778
30·0000	(294)	*	0·829	0·812	0·798	0·789	0·778	0·771	0·767	0·762	0·760	0·759	0·760	0·761	0·764
S	**$(\Delta p/\rho l)$**	**6**	**8**	**10**	**12·5**	**15**	**20**	**25**	**30**	**40**	**50**	**60**	**80**	**100**	**125**

Grad'nt k.p.g. (Equivalent) Pipe diameters in mm

Grad'nt S	k.p.g. $(\Delta p/\rho l)$	125	150	200	250	300	400	500	600	800	1000	1250	1500	2000	2500
0·00010	(0·00098)	5·180	4·062	2·768	2·056	1·612	*	1·665	1·636	1·597	1·573	1·553	1·540	1·524	1·515
0·00015	(0·00147)	4·229	3·317	2·260	1·678	1·316	1·644	1·606	1·581	1·547	1·526	1·509	1·498	1·485	1·478
0·00020	(0·00196)	3·663	2·872	1·957	1·454	*	1·601	1·567	1·544	1·513	1·494	1·479	1·469	1·458	1·452
0·00030	(0·00294)	2·991	2·345	1·598	*	*	1·545	1·515	1·494	1·468	1·452	1·439	1·431	1·422	1·417
0·00040	(0·00392)	2·590	2·031	1·384	*	1·550	1·507	1·480	1·461	1·437	1·423	1·412	1·405	1·397	1·394
0·00060	(0·00588)	2·115	1·658	*	1·524	1·495	1·457	1·433	1·417	1·396	1·384	1·375	1·369	1·364	1·362
0·00080	(0·00785)	1·831	1·436	*	1·484	1·458	1·423	1·402	1·387	1·369	1·358	1·350	1·346	1·341	1·340
0·00100	(0·00981)	1·638	1·285	*	1·455	1·430	1·398	1·378	1·365	1·348	1·339	1·332	1·328	1·324	1·324
0·00150	(0·0147)	1·337	*	1·435	1·404	1·383	1·355	1·338	1·326	1·312	1·304	1·299	1·296	1·294	1·295
0·00200	(0·0196)	1·158	*	1·399	1·370	1·351	1·326	1·310	1·300	1·288	1·281	1·277	1·275	1·274	1·275
0·00300	(0·0294)	*	1·390	1·350	1·326	1·309	1·287	1·274	1·265	1·255	1·250	1·247	1·246	1·246	1·249
0·00400	(0·0392)	*	1·354	1·318	1·295	1·280	1·261	1·249	1·241	1·233	1·229	1·226	1·226	1·228	1·231
0·00600	(0·0588)	1·330	1·306	1·275	1·255	1·242	1·225	1·216	1·210	1·203	1·200	1·199	1·199	1·202	1·206
0·00800	(0·0785)	1·296	1·274	1·246	1·228	1·216	1·202	1·193	1·188	1·182	1·180	1·180	1·181	1·185	1·189
0·01000	(0·0981)	1·271	1·251	1·225	1·208	1·197	1·184	1·176	1·172	1·167	1·166	1·166	1·167	1·171	1·176
0·01500	(0·147)	1·228	1·210	1·187	1·173	1·164	1·153	1·147	1·143	1·140	1·140	1·141	1·143	1·148	1·154
0·02000	(0·196)	1·199	1·183	1·162	1·150	1·142	1·132	1·127	1·124	1·122	1·122	1·124	1·127	1·132	1·138
0·03000	(0·294)	1·161	1·147	1·129	1·118	1·111	1·104	1·100	1·098	1·097	1·098	1·101	1·104	1·111	1·118
0·04000	(0·392)	1·135	1·122	1·106	1·097	1·091	1·084	1·081	1·080	1·080	1·082	1·085	1·089	1·096	1·103
0·06000	(0·588)	1·101	1·090	1·076	1·068	1·064	1·059	1·057	1·056	1·058	1·060	1·064	1·068	1·076	1·084
0·08000	(0·785)	1·077	1·068	1·056	1·049	1·045	1·041	1·040	1·040	1·042	1·045	1·050	1·054	1·063	1·071
0·10000	(0·981)	1·060	1·051	1·040	1·034	1·031	1·028	1·027	1·028	1·030	1·034	1·039	1·043	1·052	1·061
0·15000	(1·47)	1·030	1·023	1·014	1·009	1·007	1·005	1·005	1·006	1·010	1·014	1·020	1·025	1·034	1·043
0·20000	(1·96)	1·010	1·003	0·996	0·992	0·990	0·989	0·990	0·992	0·996	1·001	1·006	1·012	1·022	1·031
0·30000	(2·94)	0·983	0·978	0·972	0·969	0·968	0·968	0·970	0·972	0·977	0·982	0·989	0·995	1·005	1·015
0·40000	(3·92)	0·965	0·960	0·955	0·953	0·953	0·954	0·956	0·958	0·964	0·970	0·977	0·983	0·994	1·004
0·60000	(5·88)	0·940	0·937	0·933	0·932	0·932	0·934	0·937	0·940	0·947	0·953	0·960	0·967	0·979	0·989
0·80000	(7·85)	0·924	0·921	0·918	0·918	0·918	0·921	0·924	0·928	0·935	0·942	0·949	0·956	0·968	0·979
1·00000	(9·81)	0·911	0·909	0·907	0·907	0·908	0·911	0·915	0·919	0·926	0·933	0·941	0·948	0·961	0·971
1·50000	(14·7)	0·890	0·888	0·887	0·888	0·890	0·894	0·898	0·903	0·911	0·918	0·927	0·934	0·947	0·958
2·00000	(19·6)	0·875	0·874	0·874	0·875	0·877	0·882	0·887	0·892	0·900	0·908	0·917	0·925	0·938	0·949
3·00000	(29·4)	0·856	0·855	0·856	0·858	0·861	0·866	0·872	0·877	0·886	0·895	0·904	0·912	0·926	0·938
S	**$(\Delta p/\rho l)$**	**125**	**150**	**200**	**250**	**300**	**400**	**500**	**600**	**800**	**1000**	**1250**	**1500**	**2000**	**2500**

Grad'nt k.p.g. (Equivalent) Pipe diameters in mm

Kinematic viscosity, $\nu = 25\times10^{-6}$ m^{2}s^{-1};

Roughness size, k_s = 0·006 mm

Kin. visc., $\nu = 25 \times 10^{-6}$ m^2 s^{-1};
$S = 0.00010$ to 30.0000
i.e. kin. pr. grad., $\Delta p/\rho l =$
(0.00098) to (294) ms^{-2}

Roughness size, $k_s = 0.015$ mm
This table shows values of m, as follows
m_C for Colebrook-White solutions; or,
where $\mathbf{R} \leq 2000$, m_P for laminar flow

E224

Grad'nt S	k.p.g. $(\Delta p/\rho l)$	(Equivalent) Pipe diameters in mm													
		6	8	10	12·5	15	20	25	30	40	50	60	80	100	125
0·00030	(0·00294)	171·4	116·8	86·76	64·43	50·53	34·43	25·57	20·05	13·66	10·15	7·957	5·422	4·027	2·991
0·00040	(0·00392)	148·5	101·2	75·13	55·80	43·76	29·82	22·14	17·36	11·83	8·788	6·891	4·696	3·487	2·590
0·00060	(0·00588)	121·2	82·60	61·35	45·56	35·73	24·35	18·08	14·18	9·661	7·175	5·627	3·834	2·847	2·115
0·00080	(0·00785)	105·0	71·54	53·13	39·46	30·94	21·08	15·66	12·28	8·367	6·214	4·873	3·320	2·466	1·831
0·00100	(0·00981)	93·90	63·98	47·52	35·29	27·67	18·86	14·00	10·98	7·484	5·558	4·358	2·970	2·206	1·638
0·00150	(0·0147)	76·67	52·24	38·80	28·81	22·60	15·40	11·43	8·967	6·110	4·538	3·559	2·425	1·801	1·337
0·00200	(0·0196)	66·40	45·24	33·60	24·95	19·57	13·33	9·903	7·766	5·292	3·930	3·082	2·100	1·560	1·158
0·00300	(0·0294)	54·21	36·94	27·43	20·37	15·98	10·89	8·086	6·341	4·321	3·209	2·516	1·715	1·273	*
0·00400	(0·0392)	46·95	31·99	23·76	17·64	13·84	9·429	7·002	5·491	3·742	2·779	2·179	1·485	1·103	*
0·00600	(0·0588)	38·33	26·12	19·40	14·41	11·30	7·699	5·717	4·484	3·055	2·269	1·779	1·212	*	1·331
0·00800	(0·0785)	33·20	22·62	16·80	12·48	9·784	6·667	4·951	3·883	2·646	1·965	1·541	*	*	1·298
0·01000	(0·0981)	29·69	20·23	15·03	11·16	8·751	5·963	4·429	3·473	2·367	1·758	1·378	*	1·302	1·273
0·01500	(0·147)	24·24	16·52	12·27	9·112	7·145	4·869	3·616	2·836	1·932	1·435	1·125	*	1·255	1·230
0·02000	(0·196)	21·00	14·31	10·63	7·891	6·188	4·217	3·132	2·456	1·673	1·243	*	1·252	1·224	1·201
0·03000	(0·294)	17·14	11·68	8·676	6·443	5·053	3·443	2·557	2·005	1·366	1·015	*	1·207	1·183	1·163
0·04000	(0·392)	14·85	10·12	7·513	5·580	4·376	2·982	2·214	1·736	1·183	*	1·213	1·178	1·155	1·137
0·06000	(0·588)	12·12	8·260	6·135	4·556	3·573	2·435	1·808	1·418	0·966	*	1·169	1·138	1·119	1·103
0·08000	(0·785)	10·50	7·154	5·313	3·946	3·094	2·108	1·566	1·228	*	1·162	1·140	1·112	1·094	1·080
0·10000	(0·981)	9·390	6·398	4·752	3·529	2·767	1·886	1·400	1·098	*	1·139	1·118	1·092	1·076	1·063
0·15000	(1·47)	7·667	5·224	3·880	2·881	2·260	1·540	1·143	0·897	1·124	1·099	1·081	1·059	1·045	1·034
0·20000	(1·96)	6·640	4·524	3·360	2·495	1·957	1·333	0·990	*	1·096	1·073	1·057	1·036	1·024	1·014
0·30000	(2·94)	5·421	3·694	2·743	2·037	1·598	1·089	*	*	1·058	1·038	1·024	1·006	0·996	0·987
0·40000	(3·92)	4·695	3·199	2·376	1·764	1·384	0·943	*	1·062	1·033	1·014	1·002	0·986	0·977	0·969
0·60000	(5·88)	3·833	2·612	1·940	1·441	1·130	*	1·044	1·025	0·999	0·983	0·973	0·959	0·951	0·946
0·80000	(7·85)	3·320	2·262	1·680	1·248	0·978	*	1·018	1·000	0·977	0·963	0·953	0·941	0·934	0·930
1·00000	(9·81)	2·969	2·023	1·503	1·116	0·875	*	0·998	0·982	0·961	0·947	0·939	0·928	0·922	0·918
1·50000	(14·7)	2·424	1·652	1·227	0·911	*	0·986	0·965	0·951	0·932	0·921	0·914	0·905	0·900	0·897
2·00000	(19·6)	2·100	1·431	1·063	0·789	*	0·962	0·943	0·930	0·913	0·903	0·897	0·889	0·886	0·883
3·00000	(29·4)	1·714	1·168	0·868	*	0·957	0·930	0·914	0·903	0·888	0·880	0·875	0·869	0·866	0·865
4·00000	(39·2)	1·485	1·012	0·751	*	0·933	0·909	0·894	0·884	0·872	0·864	0·860	0·855	0·853	0·853
6·00000	(58·8)	1·212	0·826	*	0·918	0·902	0·881	0·869	0·860	0·850	0·844	0·840	0·837	0·836	0·837
8·00000	(78·5)	1·050	*	*	0·896	0·882	0·863	0·852	0·844	0·835	0·830	0·827	0·825	0·825	0·826
10·0000	(98·1)	0·939	*	0·900	0·880	0·867	0·849	0·839	0·832	0·824	0·820	0·818	0·816	0·816	0·818
15·0000	(147)	0·767	*	0·870	0·853	0·841	0·826	0·817	0·812	0·805	0·802	0·801	0·801	0·802	0·804
20·0000	(196)	*	0·869	0·850	0·834	0·824	0·811	0·803	0·798	0·793	0·791	0·790	0·791	0·792	0·795
30·0000	(294)	*	0·840	0·824	0·811	0·802	0·791	0·785	0·781	0·777	0·776	0·776	0·777	0·780	0·784
S	$(\Delta p/\rho l)$	6	8	10	12·5	15	20	25	30	40	50	60	80	100	125

Grad'nt k.p.g. (Equivalent) Pipe diameters in mm

S	$(\Delta p/\rho l)$	125	150	200	250	300	400	500	600	800	1000	1250	1500	2000	2500
0·00010	(0·00098)	5·180	4·062	2·768	2·056	1·612	*	1·666	1·636	1·598	1·574	1·554	1·541	1·525	1·516
0·00015	(0·00147)	4·229	3·317	2·260	1·678	1·316	1·644	1·607	1·581	1·547	1·526	1·510	1·499	1·485	1·478
0·00020	(0·00196)	3·663	2·872	1·957	1·454	*	1·602	1·568	1·544	1·514	1·495	1·480	1·470	1·458	1·453
0·00030	(0·00294)	2·991	2·345	1·598	*	*	1·545	1·515	1·495	1·468	1·452	1·440	1·431	1·422	1·418
0·00040	(0·00392)	2·590	2·031	1·384	*	1·551	1·507	1·480	1·462	1·438	1·424	1·412	1·405	1·398	1·395
0·00060	(0·00588)	2·115	1·658	*	1·524	1·495	1·457	1·434	1·418	1·397	1·385	1·376	1·370	1·365	1·363
0·00080	(0·00785)	1·831	1·436	*	1·485	1·458	1·424	1·402	1·388	1·370	1·359	1·351	1·347	1·342	1·341
0·00100	(0·00981)	1·638	1·285	*	1·455	1·431	1·399	1·379	1·366	1·349	1·340	1·333	1·329	1·325	1·325
0·00150	(0·0147)	1·337	*	1·436	1·405	1·383	1·356	1·339	1·327	1·313	1·306	1·300	1·297	1·296	1·296
0·00200	(0·0196)	1·158	*	1·399	1·371	1·352	1·327	1·311	1·301	1·289	1·282	1·278	1·276	1·275	1·277
0·00300	(0·0294)	*	1·390	1·351	1·327	1·310	1·288	1·275	1·266	1·256	1·251	1·248	1·247	1·248	1·251
0·00400	(0·0392)	*	1·355	1·319	1·297	1·281	1·262	1·250	1·243	1·234	1·230	1·228	1·228	1·230	1·233
0·00600	(0·0588)	1·331	1·307	1·276	1·257	1·243	1·227	1·217	1·211	1·205	1·202	1·201	1·201	1·204	1·208
0·00800	(0·0785)	1·298	1·276	1·247	1·230	1·218	1·203	1·195	1·190	1·184	1·182	1·182	1·183	1·187	1·192
0·01000	(0·0981)	1·273	1·252	1·226	1·210	1·199	1·186	1·178	1·174	1·169	1·168	1·168	1·170	1·174	1·179
0·01500	(0·147)	1·230	1·212	1·189	1·175	1·166	1·155	1·149	1·145	1·143	1·142	1·144	1·146	1·152	1·157
0·02000	(0·196)	1·201	1·185	1·164	1·152	1·144	1·134	1·129	1·126	1·125	1·125	1·127	1·130	1·136	1·142
0·03000	(0·294)	1·163	1·149	1·131	1·121	1·114	1·106	1·102	1·101	1·100	1·102	1·105	1·108	1·115	1·122
0·04000	(0·392)	1·137	1·124	1·109	1·099	1·094	1·087	1·084	1·083	1·084	1·086	1·090	1·093	1·101	1·108
0·06000	(0·588)	1·103	1·092	1·079	1·071	1·067	1·062	1·060	1·060	1·062	1·065	1·069	1·073	1·082	1·090
0·08000	(0·785)	1·080	1·070	1·059	1·052	1·048	1·045	1·044	1·044	1·047	1·050	1·055	1·060	1·069	1·077
0·10000	(0·981)	1·063	1·054	1·044	1·038	1·035	1·032	1·031	1·032	1·035	1·039	1·044	1·050	1·059	1·068
0·15000	(1·47)	1·034	1·026	1·018	1·013	1·011	1·009	1·010	1·011	1·016	1·020	1·026	1·032	1·042	1·052
0·20000	(1·96)	1·014	1·007	1·000	0·996	0·995	0·994	0·995	0·997	1·002	1·008	1·014	1·020	1·031	1·041
0·30000	(2·94)	0·987	0·982	0·976	0·974	0·973	0·974	0·976	0·979	0·984	0·990	0·997	1·004	1·016	1·026
0·40000	(3·92)	0·969	0·965	0·960	0·959	0·959	0·960	0·963	0·966	0·972	0·979	0·986	0·993	1·005	1·016
0·60000	(5·88)	0·946	0·942	0·939	0·939	0·939	0·942	0·945	0·949	0·956	0·963	0·971	0·979	0·992	1·003
0·80000	(7·85)	0·930	0·927	0·925	0·925	0·926	0·929	0·933	0·938	0·946	0·953	0·961	0·969	0·983	0·994
1·00000	(9·81)	0·918	0·916	0·914	0·915	0·916	0·920	0·925	0·929	0·938	0·945	0·954	0·962	0·976	0·988
1·50000	(14·7)	0·897	0·896	0·896	0·897	0·899	0·904	0·909	0·914	0·924	0·932	0·942	0·950	0·965	0·977
2·00000	(19·6)	0·883	0·883	0·883	0·885	0·888	0·894	0·899	0·905	0·915	0·923	0·933	0·942	0·957	0·970
3·00000	(29·4)	0·865	0·865	0·867	0·870	0·873	0·880	0·886	0·892	0·903	0·912	0·922	0·932	0·947	0·961
S	$(\Delta p/\rho l)$	125	150	200	250	300	400	500	600	800	1000	1250	1500	2000	2500

Grad'nt k.p.g. (Equivalent) Pipe diameters in mm

Kinematic viscosity, $\nu = 25 \times 10^{-6}$ m^2 s^{-1};   Roughness size, $k_s = 0.015$ mm

E225

Kin. visc., $\nu = 25\times10^{-6}\ \mathrm{m^2 s^{-1}}$;
$S = 0\cdot00010$ to $3\cdot00000$

i.e. kin. pr. grad., $\Delta p/\rho l =$
$(0\cdot00098)$ to $(29\cdot4)\ \mathrm{ms^{-2}}$

Roughness size, $k_s = 0\cdot030$ mm
This table shows values of m, as follows

m_C for Colebrook-White solutions; or,
where $\mathbf{R} \leq 2000$, m_P for laminar flow

Grad'nt S	k.p.g. $(\Delta p/\rho l)$	125	150	200	250	300	400	500	600	800	1000	1250	1500	2000	2500
								(Equivalent) Pipe diameters in mm							
0·00010	(0·00098)	5·180	4·062	2·768	2·056	1·612	*	1·666	1·637	1·598	1·574	1·555	1·541	1·525	1·517
0·00015	(0·00147)	4·229	3·317	2·260	1·678	1·316	1·645	1·608	1·582	1·548	1·527	1·510	1·499	1·486	1·479
0·00020	(0·00196)	3·663	2·872	1·957	1·454	*	1·602	1·568	1·545	1·514	1·496	1·481	1·471	1·459	1·454
0·00030	(0·00294)	2·991	2·345	1·598	*	*	1·546	1·516	1·496	1·469	1·453	1·441	1·433	1·424	1·419
0·00040	(0·00392)	2·590	2·031	1·384	*	1·552	1·508	1·481	1·463	1·439	1·425	1·414	1·407	1·399	1·396
0·00060	(0·00588)	2·115	1·658	*	1·525	1·496	1·458	1·435	1·419	1·398	1·386	1·377	1·372	1·366	1·365
0·00080	(0·00785)	1·831	1·436	*	1·486	1·459	1·425	1·404	1·389	1·371	1·360	1·353	1·348	1·344	1·343
0·00100	(0·00981)	1·638	1·285	*	1·456	1·432	1·400	1·380	1·367	1·351	1·341	1·334	1·330	1·327	1·327
0·00150	(0·0147)	1·337	*	1·437	1·406	1·385	1·357	1·340	1·329	1·315	1·307	1·302	1·299	1·298	1·299
0·00200	(0·0196)	1·158	*	1·401	1·373	1·353	1·328	1·313	1·303	1·291	1·284	1·280	1·278	1·278	1·280
0·00300	(0·0294)	*	1·392	1·353	1·328	1·311	1·290	1·277	1·268	1·259	1·254	1·251	1·250	1·251	1·254
0·00400	(0·0392)	*	1·356	1·321	1·298	1·283	1·264	1·252	1·245	1·237	1·233	1·231	1·231	1·233	1·236
0·00600	(0·0588)	1·333	1·309	1·278	1·259	1·246	1·229	1·220	1·214	1·207	1·205	1·204	1·205	1·208	1·212
0·00800	(0·0785)	1·300	1·278	1·250	1·232	1·220	1·206	1·198	1·192	1·187	1·186	1·186	1·187	1·191	1·196
0·01000	(0·0981)	1·275	1·255	1·228	1·212	1·202	1·188	1·181	1·177	1·173	1·171	1·172	1·174	1·179	1·184
0·01500	(0·147)	1·232	1·214	1·192	1·178	1·169	1·158	1·152	1·149	1·147	1·147	1·148	1·151	1·157	1·163
0·02000	(0·196)	1·204	1·188	1·167	1·155	1·147	1·138	1·133	1·130	1·129	1·130	1·132	1·135	1·142	1·149
0·03000	(0·294)	1·166	1·152	1·135	1·124	1·118	1·110	1·107	1·105	1·105	1·107	1·111	1·114	1·122	1·130
0·04000	(0·392)	1·140	1·128	1·113	1·104	1·098	1·092	1·089	1·089	1·090	1·092	1·096	1·100	1·109	1·117
0·06000	(0·588)	1·107	1·096	1·083	1·076	1·071	1·067	1·066	1·066	1·068	1·072	1·076	1·081	1·091	1·099
0·08000	(0·785)	1·084	1·075	1·064	1·057	1·054	1·050	1·050	1·051	1·054	1·058	1·063	1·069	1·079	1·088
0·10000	(0·981)	1·068	1·059	1·049	1·043	1·040	1·038	1·038	1·039	1·043	1·048	1·053	1·059	1·070	1·079
0·15000	(1·47)	1·039	1·032	1·024	1·020	1·018	1·017	1·018	1·020	1·025	1·030	1·037	1·043	1·054	1·064
0·20000	(1·96)	1·020	1·014	1·007	1·003	1·002	1·002	1·004	1·007	1·012	1·018	1·025	1·032	1·044	1·055
0·30000	(2·94)	0·994	0·989	0·984	0·982	0·982	0·983	0·986	0·989	0·996	1·003	1·010	1·018	1·031	1·042
0·40000	(3·92)	0·977	0·973	0·969	0·968	0·968	0·970	0·974	0·977	0·985	0·992	1·001	1·008	1·022	1·033
0·60000	(5·88)	0·954	0·951	0·949	0·949	0·950	0·953	0·958	0·962	0·971	0·979	0·988	0·996	1·010	1·023
0·80000	(7·85)	0·939	0·937	0·936	0·936	0·938	0·942	0·947	0·952	0·961	0·970	0·979	0·988	1·003	1·015
1·00000	(9·81)	0·928	0·926	0·926	0·927	0·929	0·934	0·939	0·944	0·954	0·963	0·973	0·982	0·997	1·010
1·50000	(14·7)	0·909	0·908	0·909	0·911	0·914	0·920	0·926	0·932	0·942	0·952	0·963	0·972	0·988	1·002
2·00000	(19·6)	0·896	0·896	0·898	0·901	0·904	0·911	0·917	0·924	0·935	0·945	0·956	0·966	0·982	0·996
3·00000	(29·4)	0·879	0·880	0·883	0·887	0·891	0·899	0·906	0·913	0·925	0·936	0·947	0·957	0·975	0·989
S	$(\Delta p/\rho l)$	125	150	200	250	300	400	500	600	800	1000	1250	1500	2000	2500

Grad'nt   k.p.g.   (Equivalent) Pipe diameters in mm   **Roughness size, $k_s = 0\cdot030$ mm**

E226

Kin. visc., $\nu = 25\times10^{-6}\ \mathrm{m^2 s^{-1}}$;
$S = 0\cdot00010$ to $3\cdot00000$

i.e. kin. pr. grad., $\Delta p/\rho l =$
$(0\cdot00098)$ to $(29\cdot4)\ \mathrm{ms^{-2}}$

Roughness size, $k_s = 0\cdot060$ mm
This table shows values of m, as follows

m_C for Colebrook-White solutions; or,
where $\mathbf{R} \leq 2000$, m_P for laminar flow

Grad'nt S	k.p.g. $(\Delta p/\rho l)$	125	150	200	250	300	400	500	600	800	1000	1250	1500	2000	2500
								(Equivalent) Pipe diameters in mm							
0·00010	(0·00098)	5·180	4·062	2·768	2·056	1·612	*	1·668	1·638	1·600	1·576	1·556	1·543	1·527	1·518
0·00015	(0·00147)	4·229	3·317	2·260	1·678	1·316	1·646	1·609	1·583	1·550	1·529	1·512	1·501	1·488	1·481
0·00020	(0·00196)	3·663	2·872	1·957	1·454	*	1·604	1·570	1·546	1·516	1·497	1·482	1·473	1·462	1·456
0·00030	(0·00294)	2·991	2·345	1·598	*	*	1·548	1·518	1·497	1·471	1·455	1·443	1·435	1·426	1·422
0·00040	(0·00392)	2·590	2·031	1·384	*	1·553	1·510	1·483	1·465	1·441	1·427	1·416	1·409	1·402	1·399
0·00060	(0·00588)	2·115	1·658	*	1·527	1·498	1·461	1·437	1·421	1·401	1·389	1·380	1·375	1·369	1·368
0·00080	(0·00785)	1·831	1·436	*	1·488	1·462	1·427	1·406	1·392	1·374	1·363	1·356	1·351	1·347	1·347
0·00100	(0·00981)	1·638	1·285	*	1·459	1·434	1·403	1·383	1·370	1·353	1·344	1·337	1·334	1·331	1·331
0·00150	(0·0147)	1·337	*	1·440	1·409	1·388	1·360	1·343	1·332	1·318	1·311	1·306	1·303	1·302	1·304
0·00200	(0·0196)	1·158	*	1·404	1·376	1·356	1·331	1·316	1·306	1·295	1·288	1·284	1·283	1·283	1·285
0·00300	(0·0294)	*	1·395	1·356	1·332	1·315	1·293	1·281	1·272	1·263	1·258	1·256	1·255	1·257	1·260
0·00400	(0·0392)	*	1·360	1·324	1·302	1·287	1·268	1·257	1·249	1·241	1·238	1·236	1·236	1·239	1·243
0·00600	(0·0588)	1·337	1·313	1·282	1·263	1·250	1·234	1·224	1·219	1·213	1·211	1·210	1·211	1·215	1·220
0·00800	(0·0785)	1·304	1·282	1·254	1·237	1·225	1·211	1·203	1·198	1·194	1·192	1·193	1·194	1·199	1·205
0·01000	(0·0981)	1·279	1·259	1·233	1·217	1·207	1·194	1·187	1·183	1·179	1·179	1·180	1·182	1·187	1·193
0·01500	(0·147)	1·237	1·220	1·197	1·184	1·175	1·164	1·159	1·156	1·154	1·155	1·157	1·160	1·167	1·174
0·02000	(0·196)	1·209	1·193	1·173	1·161	1·153	1·145	1·140	1·138	1·137	1·139	1·142	1·145	1·153	1·161
0·03000	(0·294)	1·172	1·158	1·141	1·131	1·125	1·118	1·115	1·114	1·115	1·118	1·122	1·126	1·135	1·143
0·04000	(0·392)	1·147	1·135	1·120	1·111	1·106	1·101	1·099	1·098	1·100	1·103	1·108	1·113	1·123	1·132
0·06000	(0·588)	1·115	1·104	1·092	1·085	1·081	1·077	1·077	1·077	1·080	1·085	1·090	1·096	1·107	1·116
0·08000	(0·785)	1·093	1·084	1·073	1·067	1·064	1·062	1·062	1·063	1·067	1·072	1·079	1·085	1·096	1·106
0·10000	(0·981)	1·077	1·069	1·059	1·054	1·052	1·050	1·051	1·053	1·058	1·063	1·070	1·076	1·088	1·099
0·15000	(1·47)	1·049	1·043	1·035	1·032	1·030	1·030	1·032	1·035	1·041	1·047	1·055	1·062	1·075	1·087
0·20000	(1·96)	1·031	1·025	1·019	1·017	1·016	1·017	1·020	1·023	1·030	1·037	1·045	1·053	1·067	1·079
0·30000	(2·94)	1·007	1·003	0·998	0·997	0·998	1·000	1·004	1·008	1·016	1·024	1·033	1·041	1·056	1·068
0·40000	(3·92)	0·991	0·988	0·985	0·984	0·985	0·989	0·993	0·998	1·007	1·015	1·025	1·033	1·049	1·062
0·60000	(5·88)	0·970	0·968	0·967	0·967	0·969	0·974	0·979	0·985	0·995	1·004	1·014	1·024	1·040	1·053
0·80000	(7·85)	0·957	0·955	0·955	0·956	0·959	0·965	0·971	0·976	0·987	0·997	1·008	1·017	1·034	1·048
1·00000	(9·81)	0·947	0·946	0·946	0·948	0·951	0·958	0·964	0·970	0·982	0·992	1·003	1·013	1·030	1·044
1·50000	(14·7)	0·930	0·930	0·932	0·935	0·939	0·946	0·953	0·960	0·972	0·983	0·995	1·005	1·023	1·038
2·00000	(19·6)	0·919	0·919	0·922	0·926	0·930	0·939	0·947	0·954	0·967	0·978	0·990	1·001	1·019	1·035
3·00000	(29·4)	0·905	0·906	0·910	0·915	0·920	0·929	0·938	0·946	0·959	0·971	0·984	0·995	1·014	1·030
S	$(\Delta p/\rho l)$	125	150	200	250	300	400	500	600	800	1000	1250	1500	2000	2500

Grad'nt   k.p.g.   (Equivalent) Pipe diameters in mm   **Roughness size, $k_s = 0\cdot060$ mm**

Kin. visc., $\nu = 25 \times 10^{-6}\ m^2 s^{-1}$; S = 0·00010 to 3·00000

i.e. kin. pr. grad., $\Delta p/\rho l$ = (0·00098) to (29·4) ms^{-2}

Roughness size, k_s = 0·150 mm
This table shows values of m, as follows

m_C for Colebrook-White solutions; or, where $\mathbf{R} \leq 2000$, m_P for laminar flow

Grad'nt S	k.p.g. $(\Delta p/\rho l)$	125	150	200	250	300	400	500	600	800	1000	1250	1500	2000	2500
							(Equivalent) Pipe diameters in mm								
0·00010	(0·00098)	5·180	4·062	2·768	2·056	1·612	*	1·671	1·642	1·603	1·580	1·560	1·547	1·532	1·523
0·00015	(0·00147)	4·229	3·317	2·260	1·678	1·316	1·650	1·613	1·587	1·554	1·533	1·517	1·506	1·494	1·487
0·00020	(0·00196)	3·663	2·872	1·957	1·454	*	1·608	1·574	1·551	1·521	1·502	1·488	1·478	1·468	1·462
0·00030	(0·00294)	2·991	2·345	1·598	*	*	1·553	1·523	1·503	1·477	1·461	1·449	1·441	1·433	1·429
0·00040	(0·00392)	2·590	2·031	1·384	*	1·559	1·516	1·489	1·471	1·447	1·433	1·423	1·416	1·410	1·407
0·00060	(0·00588)	2·115	1·658	*	1·533	1·504	1·467	1·443	1·428	1·408	1·396	1·388	1·383	1·378	1·378
0·00080	(0·00785)	1·831	1·436	*	1·494	1·468	1·434	1·413	1·399	1·382	1·372	1·364	1·361	1·357	1·358
0·00100	(0·00981)	1·638	1·285	*	1·466	1·441	1·410	1·391	1·378	1·362	1·353	1·347	1·344	1·342	1·343
0·00150	(0·0147)	1·337	*	1·447	1·417	1·396	1·368	1·352	1·341	1·328	1·321	1·317	1·315	1·315	1·317
0·00200	(0·0196)	1·158	*	1·412	1·384	1·365	1·341	1·326	1·317	1·305	1·300	1·297	1·296	1·297	1·300
0·00300	(0·0294)	*	1·404	1·365	1·341	1·325	1·304	1·292	1·284	1·275	1·271	1·270	1·270	1·273	1·277
0·00400	(0·0392)	*	1·370	1·334	1·313	1·298	1·279	1·269	1·262	1·255	1·252	1·252	1·253	1·256	1·262
0·00600	(0·0588)	1·348	1·324	1·294	1·275	1·262	1·247	1·238	1·233	1·228	1·227	1·228	1·230	1·235	1·241
0·00800	(0·0785)	1·316	1·294	1·267	1·250	1·239	1·225	1·218	1·214	1·211	1·210	1·212	1·215	1·221	1·228
0·01000	(0·0981)	1·292	1·272	1·247	1·231	1·221	1·210	1·203	1·200	1·198	1·198	1·200	1·204	1·211	1·218
0·01500	(0·147)	1·252	1·234	1·213	1·200	1·192	1·182	1·178	1·176	1·175	1·177	1·181	1·185	1·193	1·202
0·02000	(0·196)	1·225	1·209	1·190	1·179	1·172	1·164	1·161	1·160	1·160	1·163	1·168	1·172	1·182	1·191
0·03000	(0·294)	1·190	1·177	1·161	1·152	1·146	1·140	1·139	1·139	1·141	1·145	1·151	1·156	1·167	1·177
0·04000	(0·392)	1·167	1·155	1·141	1·133	1·129	1·125	1·124	1·125	1·129	1·133	1·140	1·146	1·158	1·168
0·06000	(0·588)	1·137	1·127	1·116	1·110	1·107	1·105	1·105	1·107	1·112	1·118	1·125	1·132	1·145	1·157
0·08000	(0·785)	1·117	1·109	1·099	1·094	1·092	1·091	1·093	1·095	1·102	1·108	1·116	1·124	1·138	1·150
0·10000	(0·981)	1·102	1·095	1·087	1·083	1·081	1·082	1·084	1·087	1·094	1·101	1·110	1·118	1·132	1·145
0·15000	(1·47)	1·078	1·072	1·066	1·064	1·064	1·065	1·069	1·073	1·081	1·089	1·099	1·107	1·123	1·136
0·20000	(1·96)	1·062	1·057	1·053	1·052	1·052	1·055	1·059	1·064	1·073	1·082	1·092	1·101	1·117	1·131
0·30000	(2·94)	1·042	1·038	1·036	1·036	1·037	1·042	1·047	1·052	1·063	1·072	1·083	1·093	1·110	1·124
0·40000	(3·92)	1·028	1·026	1·025	1·026	1·028	1·033	1·039	1·045	1·056	1·066	1·078	1·088	1·105	1·120
0·60000	(5·88)	1·011	1·010	1·010	1·013	1·016	1·023	1·029	1·036	1·048	1·059	1·071	1·081	1·100	1·115
0·80000	(7·85)	1·001	1·000	1·002	1·005	1·008	1·016	1·023	1·030	1·043	1·054	1·067	1·078	1·096	1·112
1·00000	(9·81)	0·993	0·993	0·995	0·999	1·003	1·011	1·019	1·026	1·040	1·051	1·064	1·075	1·094	1·110
1·50000	(14·7)	0·980	0·981	0·985	0·989	0·994	1·003	1·012	1·020	1·034	1·046	1·059	1·070	1·090	1·106
2·00000	(19·6)	0·972	0·974	0·978	0·983	0·989	0·998	1·008	1·016	1·030	1·043	1·056	1·068	1·088	1·104
3·00000	(29·4)	0·962	0·964	0·970	0·976	0·982	0·993	1·002	1·011	1·026	1·039	1·053	1·065	1·085	1·101
S	$(\Delta p/\rho l)$	125	150	200	250	300	400	500	600	800	1000	1250	1500	2000	2500

Grad'nt k.p.g. (Equivalent) Pipe diameters in mm   Roughness size, k_s = 0·150 mm

---

Kin. visc., $\nu = 25 \times 10^{-6}\ m^2 s^{-1}$; S = 0·00010 to 3·00000

i.e. kin. pr. grad., $\Delta p/\rho l$ = (0·00098) to (29·4) ms^{-2}

Roughness size, k_s = 0·30 mm
This table shows values of m, as follows

m_C for Colebrook-White solutions; or, where $\mathbf{R} \leq 2000$, m_P for laminar flow

Grad'nt S	k.p.g. $(\Delta p/\rho l)$	125	150	200	250	300	400	500	600	800	1000	1250	1500	2000	2500
							(Equivalent) Pipe diameters in mm								
0·00010	(0·00098)	5·180	4·062	2·768	2·056	1·612	*	1·677	1·648	1·610	1·586	1·567	1·555	1·540	1·532
0·00015	(0·00147)	4·229	3·317	2·260	1·678	1·316	1·657	1·620	1·594	1·561	1·541	1·525	1·515	1·502	1·496
0·00020	(0·00196)	3·663	2·872	1·957	1·454	*	1·615	1·582	1·559	1·529	1·511	1·497	1·488	1·478	1·473
0·00030	(0·00294)	2·991	2·345	1·598	*	*	1·561	1·532	1·511	1·486	1·471	1·459	1·452	1·444	1·441
0·00040	(0·00392)	2·590	2·031	1·384	*	1·567	1·525	1·498	1·480	1·457	1·444	1·434	1·428	1·422	1·420
0·00060	(0·00588)	2·115	1·658	*	1·543	1·514	1·477	1·454	1·439	1·419	1·409	1·401	1·396	1·393	1·393
0·00080	(0·00785)	1·831	1·436	*	1·505	1·479	1·445	1·425	1·411	1·394	1·385	1·378	1·375	1·373	1·374
0·00100	(0·00981)	1·638	1·285	*	1·477	1·453	1·422	1·403	1·391	1·376	1·367	1·362	1·359	1·359	1·361
0·00150	(0·0147)	1·337	*	1·460	1·429	1·408	1·382	1·366	1·356	1·344	1·337	1·334	1·333	1·334	1·337
0·00200	(0·0196)	1·158	*	1·425	1·398	1·379	1·355	1·341	1·332	1·322	1·318	1·315	1·315	1·318	1·322
0·00300	(0·0294)	*	1·419	1·380	1·357	1·341	1·321	1·309	1·302	1·294	1·292	1·291	1·292	1·296	1·302
0·00400	(0·0392)	*	1·386	1·351	1·330	1·315	1·298	1·288	1·282	1·276	1·274	1·275	1·277	1·282	1·289
0·00600	(0·0588)	1·366	1·342	1·312	1·294	1·282	1·268	1·260	1·256	1·252	1·252	1·254	1·257	1·264	1·272
0·00800	(0·0785)	1·335	1·314	1·287	1·271	1·260	1·248	1·242	1·238	1·236	1·237	1·240	1·244	1·252	1·261
0·01000	(0·0981)	1·312	1·293	1·268	1·254	1·244	1·233	1·228	1·226	1·225	1·227	1·230	1·235	1·244	1·253
0·01500	(0·147)	1·274	1·258	1·237	1·225	1·217	1·209	1·206	1·205	1·206	1·209	1·214	1·219	1·230	1·240
0·02000	(0·196)	1·250	1·235	1·216	1·206	1·200	1·193	1·191	1·191	1·193	1·197	1·203	1·209	1·221	1·231
0·03000	(0·294)	1·217	1·205	1·190	1·182	1·177	1·173	1·172	1·173	1·177	1·183	1·190	1·196	1·209	1·221
0·04000	(0·392)	1·197	1·186	1·173	1·166	1·162	1·160	1·160	1·162	1·167	1·173	1·181	1·188	1·202	1·214
0·06000	(0·588)	1·170	1·161	1·150	1·145	1·143	1·143	1·144	1·147	1·154	1·161	1·170	1·178	1·193	1·206
0·08000	(0·785)	1·152	1·145	1·136	1·132	1·131	1·132	1·135	1·138	1·146	1·154	1·163	1·172	1·187	1·201
0·10000	(0·981)	1·140	1·133	1·126	1·123	1·122	1·124	1·128	1·132	1·140	1·149	1·158	1·167	1·183	1·197
0·15000	(1·47)	1·119	1·114	1·109	1·108	1·108	1·112	1·116	1·121	1·131	1·140	1·151	1·160	1·177	1·191
0·20000	(1·96)	1·106	1·102	1·098	1·098	1·099	1·104	1·109	1·114	1·125	1·135	1·146	1·156	1·173	1·188
0·30000	(2·94)	1·089	1·086	1·085	1·086	1·088	1·094	1·100	1·106	1·118	1·128	1·140	1·150	1·168	1·184
0·40000	(3·92)	1·078	1·076	1·076	1·078	1·081	1·087	1·094	1·101	1·113	1·124	1·136	1·147	1·165	1·181
0·60000	(5·88)	1·065	1·064	1·066	1·069	1·072	1·080	1·087	1·095	1·108	1·119	1·132	1·143	1·162	1·178
0·80000	(7·85)	1·057	1·057	1·059	1·063	1·067	1·075	1·083	1·091	1·104	1·116	1·129	1·140	1·160	1·176
1·00000	(9·81)	1·051	1·051	1·054	1·058	1·063	1·072	1·080	1·088	1·102	1·114	1·127	1·139	1·158	1·174
1·50000	(14·7)	1·042	1·043	1·047	1·052	1·057	1·067	1·076	1·084	1·098	1·111	1·124	1·136	1·156	1·172
2·00000	(19·6)	1·036	1·038	1·042	1·048	1·053	1·063	1·073	1·081	1·096	1·109	1·122	1·134	1·154	1·171
3·00000	(29·4)	1·029	1·031	1·037	1·043	1·049	1·060	1·069	1·078	1·093	1·106	1·120	1·132	1·153	1·169
S	$(\Delta p/\rho l)$	125	150	200	250	300	400	500	600	800	1000	1250	1500	2000	2500

Grad'nt k.p.g. (Equivalent) Pipe diameters in mm   Roughness size, k_s = 0·30 mm

E229

Kin. visc., $\nu = 30\times10^{-6}$ m^2 s^{-1};
$S = 0.00010$ to 30.0000

i.e. kin. pr. grad., $\Delta p/\rho l =$
(0.00098) to (294) ms^{-2}

Roughness size, $k_s = 0.006$ mm
This table shows values of m, as follows

m_C for Colebrook-White solutions; or,
where $\mathbf{R} \leq 2000$, m_P for laminar flow

Grad'nt S	k.p.g. $(\Delta p/\rho l)$	6	8	10	12·5	15	20	25	30	40	50	60	80	100	125
						(Equivalent) Pipe diameters in mm									
0·00030	(0·00294)	205·7	140·2	104·1	77·32	60·63	41·32	30·68	24·06	16·40	12·18	9·549	6·507	4·832	3·589
0·00040	(0·00392)	178·2	121·4	90·16	66·96	52·51	35·78	26·57	20·84	14·20	10·55	8·269	5·635	4·185	3·108
0·00060	(0·00588)	145·5	99·12	73·62	54·67	42·87	29·21	21·70	17·01	11·59	8·610	6·752	4·601	3·417	2·538
0·00080	(0·00785)	126·0	85·84	63·75	47·35	37·13	25·30	18·79	14·73	10·04	7·457	5·847	3·985	2·959	2·198
0·00100	(0·00981)	112·7	76·78	57·02	42·35	33·21	22·63	16·81	13·18	8·980	6·669	5·230	3·564	2·647	1·966
0·00150	(0·0147)	92·00	62·69	46·56	34·58	27·12	18·48	13·72	10·76	7·332	5·446	4·270	2·910	2·161	1·605
0·00200	(0·0196)	79·68	54·29	40·32	29·94	23·48	16·00	11·88	9·319	6·350	4·716	3·698	2·520	1·872	1·390
0·00300	(0·0294)	65·06	44·33	32·92	24·45	19·17	13·07	9·703	7·609	5·185	3·851	3·020	2·058	1·528	1·135
0·00400	(0·0392)	56·34	38·39	28·51	21·17	16·60	11·31	8·403	6·589	4·490	3·335	2·615	1·782	1·323	*
0·00600	(0·0588)	46·00	31·35	23·28	17·29	13·56	9·238	6·861	5·380	3·666	2·723	2·135	1·455	*	*
0·00800	(0·0785)	39·84	27·15	20·16	14·97	11·74	8·001	5·942	4·659	3·175	2·358	1·849	1·260	*	1·340
0·01000	(0·0981)	35·63	24·28	18·03	13·39	10·50	7·156	5·314	4·168	2·840	2·109	1·654	1·127	*	1·313
0·01500	(0·147)	29·09	19·82	14·72	10·93	8·575	5·843	4·339	3·403	2·319	1·722	1·350	*	1·296	1·267
0·02000	(0·196)	25·20	17·17	12·75	9·469	7·426	5·060	3·758	2·947	2·008	1·491	1·169	*	1·262	1·236
0·03000	(0·294)	20·57	14·02	10·41	7·732	6·063	4·132	3·068	2·406	1·640	1·218	*	1·246	1·218	1·195
0·04000	(0·392)	17·82	12·14	9·016	6·696	5·251	3·578	2·657	2·084	1·420	1·055	*	1·214	1·189	1·168
0·06000	(0·588)	14·55	9·912	7·362	5·467	4·287	2·921	2·170	1·701	1·159	*	1·207	1·172	1·150	1·131
0·08000	(0·785)	12·60	8·584	6·375	4·735	3·713	2·530	1·879	1·473	1·004	*	1·175	1·143	1·123	1·107
0·10000	(0·981)	11·27	7·678	5·702	4·235	3·321	2·263	1·681	1·318	*	1·175	1·152	1·122	1·104	1·089
0·15000	(1·47)	9·200	6·269	4·656	3·458	2·712	1·848	1·372	1·076	*	1·132	1·112	1·086	1·070	1·057
0·20000	(1·96)	7·968	5·429	4·032	2·994	2·348	1·600	1·188	0·932	1·130	1·104	1·085	1·062	1·047	1·036
0·30000	(2·94)	6·506	4·433	3·292	2·445	1·917	1·307	0·970	*	1·089	1·066	1·050	1·030	1·017	1·007
0·40000	(3·92)	5·634	3·839	2·851	2·117	1·660	1·131	*	*	1·061	1·041	1·026	1·008	0·997	0·988
0·60000	(5·88)	4·600	3·135	2·328	1·729	1·356	0·924	*	1·054	1·025	1·007	0·995	0·979	0·969	0·962
0·80000	(7·85)	3·984	2·715	2·016	1·497	1·174	*	1·048	1·028	1·001	0·985	0·974	0·959	0·951	0·945
1·00000	(9·81)	3·563	2·428	1·803	1·339	1·050	*	1·027	1·008	0·983	0·968	0·958	0·945	0·937	0·932
1·50000	(14·7)	2·909	1·982	1·472	1·093	0·857	*	0·990	0·974	0·953	0·939	0·931	0·920	0·913	0·909
2·00000	(19·6)	2·520	1·717	1·275	0·947	*	0·988	0·966	0·951	0·932	0·920	0·912	0·903	0·897	0·894
3·00000	(29·4)	2·057	1·402	1·041	0·773	*	0·953	0·934	0·921	0·904	0·894	0·888	0·880	0·876	0·873
4·00000	(39·2)	1·782	1·214	0·902	*	0·957	0·930	0·913	0·901	0·886	0·877	0·871	0·864	0·861	0·859
6·00000	(58·8)	1·455	0·991	*	*	0·923	0·899	0·884	0·874	0·861	0·854	0·849	0·844	0·841	0·840
8·00000	(78·5)	1·260	0·858	*	0·917	0·900	0·879	0·865	0·856	0·845	0·838	0·834	0·830	0·828	0·828
10·0000	(98·1)	1·127	0·768	*	0·899	0·884	0·864	0·851	0·843	0·833	0·827	0·823	0·820	0·818	0·818
15·0000	(147)	0·920	*	0·888	0·868	0·855	0·838	0·827	0·820	0·811	0·807	0·804	0·802	0·801	0·802
20·0000	(196)	0·797	*	0·866	0·848	0·836	0·820	0·811	0·804	0·797	0·793	0·791	0·789	0·790	0·791
30·0000	(294)	*	0·855	0·837	0·821	0·810	0·797	0·789	0·784	0·778	0·775	0·774	0·773	0·774	0·777
S	$(\Delta p/\rho l)$	6	8	10	12·5	15	20	25	30	40	50	60	80	100	125

Grad'nt — k.p.g. — (Equivalent) Pipe diameters in mm

Grad'nt S	k.p.g. $(\Delta p/\rho l)$	125	150	200	250	300	400	500	600	800	1000	1250	1500	2000	2500
0·00010	(0·00098)	6·216	4·874	3·322	2·467	1·934	*	1·722	1·689	1·645	1·618	1·596	1·580	1·562	1·551
0·00015	(0·00147)	5·075	3·980	2·712	2·014	1·579	*	1·659	1·630	1·592	1·568	1·549	1·536	1·520	1·511
0·00020	(0·00196)	4·395	3·447	2·349	1·744	1·368	1·656	1·617	1·591	1·556	1·535	1·517	1·506	1·492	1·485
0·00030	(0·00294)	3·589	2·814	1·918	1·424	*	1·595	1·562	1·538	1·508	1·490	1·475	1·465	1·454	1·448
0·00040	(0·00392)	3·108	2·437	1·661	*	*	1·555	1·524	1·503	1·476	1·460	1·446	1·438	1·428	1·424
0·00060	(0·00588)	2·538	1·990	1·356	*	1·544	1·502	1·475	1·457	1·433	1·419	1·408	1·401	1·394	1·390
0·00080	(0·00785)	2·198	1·723	*	1·535	1·505	1·466	1·442	1·425	1·404	1·392	1·382	1·376	1·370	1·368
0·00100	(0·00981)	1·966	1·541	*	1·503	1·476	1·439	1·417	1·402	1·382	1·371	1·362	1·357	1·352	1·351
0·00150	(0·0147)	1·605	1·259	*	1·449	1·425	1·394	1·374	1·361	1·345	1·335	1·328	1·324	1·321	1·321
0·00200	(0·0196)	1·390	*	1·445	1·413	1·391	1·363	1·345	1·333	1·319	1·311	1·305	1·302	1·300	1·300
0·00300	(0·0294)	1·135	*	1·394	1·366	1·347	1·322	1·307	1·297	1·285	1·278	1·274	1·272	1·271	1·273
0·00400	(0·0392)	*	1·399	1·359	1·334	1·316	1·294	1·281	1·272	1·261	1·256	1·252	1·251	1·252	1·254
0·00600	(0·0588)	*	1·349	1·313	1·291	1·276	1·257	1·246	1·238	1·230	1·226	1·224	1·223	1·225	1·228
0·00800	(0·0785)	1·340	1·315	1·283	1·263	1·249	1·232	1·222	1·215	1·208	1·205	1·204	1·204	1·207	1·210
0·01000	(0·0981)	1·313	1·290	1·260	1·242	1·229	1·213	1·204	1·198	1·192	1·190	1·189	1·190	1·193	1·197
0·01500	(0·147)	1·267	1·247	1·221	1·205	1·194	1·181	1·173	1·169	1·164	1·163	1·163	1·165	1·169	1·174
0·02000	(0·196)	1·236	1·218	1·194	1·180	1·170	1·159	1·152	1·149	1·145	1·145	1·146	1·148	1·153	1·158
0·03000	(0·294)	1·195	1·179	1·159	1·147	1·139	1·129	1·124	1·121	1·119	1·120	1·122	1·124	1·130	1·136
0·04000	(0·392)	1·168	1·153	1·135	1·124	1·117	1·109	1·105	1·103	1·102	1·103	1·105	1·108	1·115	1·121
0·06000	(0·588)	1·131	1·119	1·103	1·094	1·088	1·082	1·079	1·078	1·078	1·080	1·083	1·087	1·094	1·101
0·08000	(0·785)	1·107	1·096	1·082	1·074	1·069	1·063	1·061	1·061	1·062	1·064	1·068	1·072	1·080	1·087
0·10000	(0·981)	1·089	1·078	1·066	1·058	1·054	1·050	1·048	1·048	1·050	1·053	1·057	1·061	1·069	1·077
0·15000	(1·47)	1·057	1·048	1·038	1·032	1·028	1·025	1·025	1·025	1·028	1·032	1·037	1·042	1·051	1·059
0·20000	(1·96)	1·036	1·028	1·019	1·014	1·011	1·009	1·009	1·010	1·014	1·018	1·023	1·028	1·038	1·046
0·30000	(2·94)	1·007	1·001	0·993	0·990	0·988	0·987	0·988	0·990	0·994	0·999	1·005	1·010	1·020	1·029
0·40000	(3·92)	0·988	0·982	0·976	0·973	0·972	0·972	0·973	0·976	0·981	0·986	0·992	0·998	1·009	1·018
0·60000	(5·88)	0·962	0·958	0·953	0·951	0·950	0·952	0·954	0·957	0·962	0·968	0·975	0·981	0·993	1·002
0·80000	(7·85)	0·945	0·941	0·937	0·936	0·936	0·938	0·941	0·944	0·950	0·956	0·963	0·970	0·982	0·992
1·00000	(9·81)	0·932	0·928	0·925	0·925	0·925	0·927	0·931	0·934	0·941	0·947	0·955	0·961	0·973	0·984
1·50000	(14·7)	0·909	0·907	0·905	0·905	0·906	0·909	0·913	0·917	0·925	0·932	0·939	0·947	0·959	0·970
2·00000	(19·6)	0·894	0·892	0·891	0·892	0·893	0·897	0·901	0·906	0·914	0·921	0·929	0·937	0·950	0·961
3·00000	(29·4)	0·873	0·872	0·872	0·874	0·876	0·880	0·885	0·890	0·899	0·907	0·915	0·923	0·937	0·948
S	$(\Delta p/\rho l)$	125	150	200	250	300	400	500	600	800	1000	1250	1500	2000	2500

Grad'nt — k.p.g. — (Equivalent) Pipe diameters in mm

Kinematic viscosity, $\nu = 30\times10^{-6}$ m^2 s^{-1};   **Roughness size, $k_s = 0.006$ mm**

Kin. visc., $\nu = 30 \times 10^{-6}$ m^2 s^{-1};
$S = 0.00010$ to 30.0000

i.e. kin. pr. grad., $\Delta p/\rho l =$
(0.00098) to (294) ms^{-2}

Roughness size, $k_s = 0.015$ mm
This table shows values of m, as follows

m_C for Colebrook-White solutions; or,
where $\mathbf{R} \leq 2000$, m_P for laminar flow

Grad'nt S	k.p.g. $(\Delta p/\rho l)$	6	8	10	12·5	15	20	25	30	40	50	60	80	100	125
0·00030	(0·00294)	205·7	140·2	104·1	77·32	60·63	41·32	30·68	24·06	16·40	12·18	9·549	6·507	4·832	3·589
0·00040	(0·00392)	178·2	121·4	90·16	66·96	52·51	35·78	26·57	20·84	14·20	10·55	8·269	5·635	4·185	3·108
0·00060	(0·00588)	145·5	99·12	73·62	54·67	42·87	29·21	21·70	17·01	11·59	8·610	6·752	4·601	3·417	2·538
0·00080	(0·00785)	126·0	85·84	63·75	47·35	37·13	25·30	18·79	14·73	10·04	7·457	5·847	3·985	2·959	2·198
0·00100	(0·00981)	112·7	76·78	57·02	42·35	33·21	22·63	16·81	13·18	8·980	6·669	5·230	3·564	2·647	1·966
0·00150	(0·0147)	92·00	62·69	46·56	34·58	27·12	18·48	13·72	10·76	7·332	5·446	4·270	2·910	2·161	1·605
0·00200	(0·0196)	79·68	54·29	40·32	29·94	23·48	16·00	11·88	9·319	6·350	4·716	3·698	2·520	1·872	1·390
0·00300	(0·0294)	65·06	44·33	32·92	24·45	19·17	13·07	9·703	7·609	5·185	3·851	3·020	2·058	1·528	1·135
0·00400	(0·0392)	56·34	38·39	28·51	21·17	16·60	11·31	8·403	6·589	4·490	3·335	2·615	1·782	1·323	*
0·00600	(0·0588)	46·00	31·35	23·28	17·29	13·56	9·238	6·861	5·380	3·666	2·723	2·135	1·455	*	*
0·00800	(0·0785)	39·84	27·15	20·16	14·97	11·74	8·001	5·942	4·659	3·175	2·358	1·849	1·260	*	1·341
0·01000	(0·0981)	35·63	24·28	18·03	13·39	10·50	7·156	5·314	4·168	2·840	2·109	1·654	1·127	*	1·314
0·01500	(0·147)	29·09	19·82	14·72	10·93	8·575	5·843	4·339	3·403	2·319	1·722	1·350	*	1·297	1·268
0·02000	(0·196)	25·20	17·17	12·75	9·469	7·426	5·060	3·758	2·947	2·008	1·491	1·169	*	1·264	1·237
0·03000	(0·294)	20·57	14·02	10·41	7·732	6·063	4·132	3·068	2·406	1·640	1·218	*	1·247	1·220	1·197
0·04000	(0·392)	17·82	12·14	9·016	6·696	5·251	3·578	2·657	2·084	1·420	1·055	*	1·216	1·191	1·170
0·06000	(0·588)	14·55	9·912	7·362	5·467	4·287	2·921	2·170	1·701	1·159	*	1·209	1·174	1·152	1·133
0·08000	(0·785)	12·60	8·584	6·375	4·735	3·713	2·530	1·879	1·473	1·004	*	1·177	1·146	1·126	1·109
0·10000	(0·981)	11·27	7·678	5·702	4·235	3·321	2·263	1·681	1·318	*	1·177	1·154	1·125	1·106	1·091
0·15000	(1·47)	9·200	6·269	4·656	3·458	2·712	1·848	1·372	1·076	*	1·134	1·115	1·089	1·073	1·060
0·20000	(1·96)	7·968	5·429	4·032	2·994	2·348	1·600	1·188	0·932	1·132	1·106	1·088	1·065	1·051	1·039
0·30000	(2·94)	6·506	4·433	3·292	2·445	1·917	1·307	0·970	*	1·092	1·069	1·053	1·033	1·021	1·011
0·40000	(3·92)	5·634	3·839	2·851	2·117	1·660	1·131	*	*	1·065	1·044	1·030	1·012	1·001	0·992
0·60000	(5·88)	4·600	3·135	2·328	1·729	1·356	0·924	*	1·058	1·029	1·011	0·999	0·983	0·974	0·967
0·80000	(7·85)	3·984	2·715	2·016	1·497	1·174	*	1·052	1·032	1·005	0·989	0·978	0·964	0·956	0·950
1·00000	(9·81)	3·563	2·428	1·803	1·339	1·050	*	1·031	1·012	0·988	0·973	0·963	0·950	0·943	0·937
1·50000	(14·7)	2·909	1·982	1·472	1·093	0·857	*	0·995	0·979	0·958	0·945	0·936	0·926	0·920	0·916
2·00000	(19·6)	2·520	1·717	1·275	0·947	*	0·993	0·971	0·957	0·938	0·926	0·918	0·909	0·904	0·901
3·00000	(29·4)	2·057	1·402	1·041	0·773	*	0·959	0·940	0·927	0·911	0·901	0·895	0·887	0·884	0·881
4·00000	(39·2)	1·782	1·214	0·902	*	0·963	0·936	0·919	0·908	0·893	0·884	0·879	0·873	0·870	0·868
6·00000	(58·8)	1·455	0·991	*	*	0·930	0·906	0·892	0·882	0·869	0·862	0·858	0·853	0·851	0·851
8·00000	(78·5)	1·260	0·858	*	0·924	0·908	0·887	0·873	0·865	0·854	0·848	0·844	0·840	0·839	0·840
10·0000	(98·1)	1·127	0·768	*	0·907	0·892	0·872	0·860	0·852	0·842	0·837	0·834	0·831	0·830	0·831
15·0000	(147)	0·920	*	0·897	0·877	0·864	0·847	0·837	0·830	0·822	0·818	0·816	0·814	0·815	0·816
20·0000	(196)	0·797	*	0·875	0·858	0·846	0·831	0·821	0·816	0·809	0·806	0·804	0·803	0·805	0·807
30·0000	(294)	*	0·866	0·847	0·832	0·822	0·809	0·801	0·797	0·791	0·789	0·789	0·789	0·791	0·794
S	$(\Delta p/\rho l)$	6	8	10	12·5	15	20	25	30	40	50	60	80	100	125

Grad'nt k.p.g. (Equivalent) Pipe diameters in mm

Grad'nt S	k.p.g. $(\Delta p/\rho l)$	125	150	200	250	300	400	500	600	800	1000	1250	1500	2000	2500
0·00010	(0·00098)	6·216	4·874	3·322	2·467	1·934	*	1·723	1·689	1·646	1·618	1·596	1·581	1·562	1·551
0·00015	(0·00147)	5·075	3·980	2·712	2·014	1·579	*	1·660	1·630	1·592	1·569	1·549	1·536	1·521	1·512
0·00020	(0·00196)	4·395	3·447	2·349	1·744	1·368	1·656	1·618	1·591	1·557	1·535	1·518	1·506	1·493	1·485
0·00030	(0·00294)	3·589	2·814	1·918	1·424	*	1·596	1·562	1·539	1·509	1·490	1·476	1·466	1·455	1·449
0·00040	(0·00392)	3·108	2·437	1·661	*	*	1·556	1·525	1·504	1·477	1·460	1·447	1·439	1·429	1·425
0·00060	(0·00588)	2·538	1·990	1·356	*	1·545	1·502	1·476	1·457	1·434	1·420	1·409	1·402	1·394	1·391
0·00080	(0·00785)	2·198	1·723	*	1·535	1·505	1·467	1·442	1·426	1·405	1·392	1·383	1·377	1·371	1·369
0·00100	(0·00981)	1·966	1·541	*	1·504	1·476	1·440	1·418	1·402	1·383	1·372	1·363	1·358	1·353	1·352
0·00150	(0·0147)	1·605	1·259	*	1·450	1·426	1·394	1·375	1·362	1·345	1·336	1·329	1·325	1·322	1·322
0·00200	(0·0196)	1·390	*	1·446	1·414	1·392	1·364	1·346	1·334	1·320	1·312	1·306	1·303	1·301	1·302
0·00300	(0·0294)	1·135	*	1·394	1·367	1·347	1·323	1·308	1·298	1·286	1·279	1·275	1·273	1·273	1·274
0·00400	(0·0392)	*	1·400	1·360	1·335	1·317	1·295	1·282	1·273	1·262	1·257	1·254	1·253	1·253	1·256
0·00600	(0·0588)	*	1·350	1·315	1·292	1·277	1·258	1·247	1·240	1·231	1·227	1·225	1·225	1·227	1·230
0·00800	(0·0785)	1·341	1·316	1·284	1·264	1·250	1·233	1·223	1·217	1·210	1·207	1·206	1·206	1·209	1·213
0·01000	(0·0981)	1·314	1·291	1·261	1·243	1·230	1·215	1·206	1·200	1·194	1·192	1·191	1·192	1·196	1·200
0·01500	(0·147)	1·268	1·248	1·222	1·206	1·196	1·182	1·175	1·171	1·166	1·165	1·166	1·167	1·172	1·177
0·02000	(0·196)	1·237	1·219	1·196	1·182	1·172	1·161	1·154	1·151	1·148	1·147	1·148	1·151	1·156	1·161
0·03000	(0·294)	1·197	1·181	1·161	1·149	1·141	1·131	1·126	1·124	1·122	1·123	1·125	1·128	1·134	1·140
0·04000	(0·392)	1·170	1·155	1·137	1·126	1·119	1·111	1·107	1·106	1·105	1·106	1·109	1·112	1·119	1·126
0·06000	(0·588)	1·133	1·121	1·106	1·097	1·091	1·085	1·082	1·081	1·082	1·084	1·087	1·091	1·099	1·107
0·08000	(0·785)	1·109	1·098	1·084	1·076	1·072	1·067	1·065	1·064	1·066	1·069	1·073	1·077	1·085	1·093
0·10000	(0·981)	1·091	1·081	1·068	1·061	1·057	1·053	1·052	1·052	1·054	1·057	1·062	1·066	1·075	1·083
0·15000	(1·47)	1·060	1·051	1·041	1·035	1·032	1·029	1·029	1·030	1·033	1·037	1·043	1·048	1·057	1·066
0·20000	(1·96)	1·039	1·031	1·022	1·018	1·015	1·013	1·014	1·015	1·019	1·024	1·030	1·035	1·045	1·055
0·30000	(2·94)	1·011	1·005	0·997	0·994	0·992	0·992	0·993	0·995	1·000	1·006	1·012	1·018	1·029	1·039
0·40000	(3·92)	0·992	0·987	0·981	0·978	0·977	0·978	0·980	0·982	0·988	0·994	1·000	1·007	1·018	1·029
0·60000	(5·88)	0·967	0·963	0·958	0·957	0·957	0·958	0·961	0·964	0·971	0·977	0·985	0·992	1·004	1·015
0·80000	(7·85)	0·950	0·946	0·943	0·942	0·943	0·945	0·948	0·952	0·959	0·966	0·974	0·981	0·994	1·005
1·00000	(9·81)	0·937	0·934	0·932	0·932	0·932	0·935	0·939	0·943	0·951	0·958	0·966	0·974	0·987	0·998
1·50000	(14·7)	0·916	0·913	0·912	0·913	0·914	0·918	0·923	0·928	0·936	0·944	0·953	0·961	0·975	0·987
2·00000	(19·6)	0·901	0·899	0·899	0·900	0·902	0·907	0·912	0·917	0·926	0·935	0·944	0·952	0·967	0·979
3·00000	(29·4)	0·881	0·881	0·882	0·884	0·886	0·892	0·898	0·903	0·913	0·922	0·932	0·941	0·956	0·969
S	$(\Delta p/\rho l)$	125	150	200	250	300	400	500	600	800	1000	1250	1500	2000	2500

Grad'nt k.p.g. (Equivalent) Pipe diameters in mm

Kinematic viscosity, $\nu = 30 \times 10^{-6}$ m^2 s^{-1};

Roughness size, $k_s = 0.015$ mm

E231

Kin. visc., $\nu = 30\times10^{-6}$ m^2 s^{-1};
$S = 0.00010$ to 3.00000

i.e. kin. pr. grad., $\Delta p/\rho l =$
(0.00098) to (29.4) ms^{-2}

Roughness size, $k_s = 0.030$ mm
This table shows values of m, as follows

m_C for Colebrook-White solutions; or,
where $\mathbf{R} \le 2000$, m_P for laminar flow

Grad'nt S	k.p.g. $(\Delta p/\rho l)$	125	150	200	250	300	400	500	600	800	1000	1250	1500	2000	2500
0.00010	(0.00098)	6.216	4.874	3.322	2.467	1.934	*	1.723	1.690	1.646	1.619	1.597	1.582	1.563	1.552
0.00015	(0.00147)	5.075	3.980	2.712	2.014	1.579	*	1.660	1.631	1.593	1.569	1.550	1.537	1.521	1.513
0.00020	(0.00196)	4.395	3.447	2.349	1.744	1.368	1.657	1.618	1.592	1.557	1.536	1.519	1.507	1.493	1.486
0.00030	(0.00294)	3.589	2.814	1.918	1.424	*	1.597	1.563	1.540	1.510	1.491	1.477	1.467	1.456	1.450
0.00040	(0.00392)	3.108	2.437	1.661	*	*	1.556	1.526	1.505	1.478	1.461	1.448	1.440	1.430	1.426
0.00060	(0.00588)	2.538	1.990	1.356	*	1.546	1.503	1.476	1.458	1.435	1.421	1.410	1.403	1.396	1.393
0.00080	(0.00785)	2.198	1.723	*	1.536	1.506	1.468	1.443	1.427	1.406	1.394	1.384	1.378	1.372	1.370
0.00100	(0.00981)	1.966	1.541	*	1.505	1.477	1.441	1.419	1.404	1.384	1.373	1.365	1.360	1.355	1.354
0.00150	(0.0147)	1.605	1.259	*	1.451	1.427	1.396	1.376	1.363	1.347	1.338	1.331	1.327	1.324	1.324
0.00200	(0.0196)	1.390	*	1.447	1.415	1.394	1.365	1.348	1.336	1.322	1.314	1.308	1.305	1.303	1.304
0.00300	(0.0294)	1.135	*	1.396	1.368	1.349	1.324	1.309	1.299	1.288	1.281	1.277	1.275	1.275	1.277
0.00400	(0.0392)	*	1.402	1.362	1.336	1.319	1.297	1.284	1.275	1.265	1.259	1.256	1.255	1.256	1.259
0.00600	(0.0588)	*	1.351	1.316	1.294	1.279	1.260	1.249	1.242	1.234	1.230	1.228	1.228	1.230	1.234
0.00800	(0.0785)	1.342	1.318	1.286	1.266	1.253	1.236	1.226	1.219	1.213	1.210	1.209	1.210	1.213	1.217
0.01000	(0.0981)	1.316	1.293	1.263	1.245	1.233	1.217	1.208	1.203	1.197	1.195	1.195	1.196	1.199	1.204
0.01500	(0.147)	1.270	1.250	1.225	1.209	1.198	1.185	1.178	1.174	1.170	1.169	1.170	1.172	1.176	1.182
0.02000	(0.196)	1.240	1.222	1.199	1.184	1.175	1.164	1.158	1.154	1.151	1.151	1.153	1.155	1.161	1.167
0.03000	(0.294)	1.200	1.184	1.164	1.152	1.144	1.135	1.130	1.128	1.127	1.128	1.130	1.133	1.140	1.147
0.04000	(0.392)	1.173	1.159	1.141	1.130	1.123	1.115	1.112	1.110	1.110	1.111	1.115	1.118	1.126	1.133
0.06000	(0.588)	1.137	1.125	1.110	1.101	1.095	1.089	1.087	1.086	1.087	1.090	1.094	1.098	1.107	1.115
0.08000	(0.785)	1.113	1.102	1.089	1.081	1.076	1.072	1.070	1.070	1.072	1.075	1.080	1.085	1.094	1.103
0.10000	(0.981)	1.095	1.085	1.073	1.066	1.062	1.059	1.058	1.058	1.061	1.065	1.070	1.075	1.085	1.093
0.15000	(1.47)	1.065	1.056	1.046	1.041	1.038	1.036	1.036	1.037	1.041	1.046	1.052	1.057	1.068	1.078
0.20000	(1.96)	1.044	1.037	1.028	1.024	1.022	1.021	1.022	1.023	1.028	1.033	1.040	1.046	1.057	1.067
0.30000	(2.94)	1.017	1.011	1.004	1.001	1.000	1.000	1.002	1.005	1.011	1.017	1.024	1.030	1.043	1.053
0.40000	(3.92)	0.999	0.994	0.988	0.986	0.985	0.987	0.989	0.992	0.999	1.005	1.013	1.020	1.033	1.044
0.60000	(5.88)	0.975	0.971	0.967	0.966	0.966	0.969	0.972	0.976	0.983	0.991	0.999	1.007	1.021	1.032
0.80000	(7.85)	0.958	0.955	0.953	0.952	0.953	0.957	0.961	0.965	0.973	0.981	0.990	0.998	1.012	1.025
1.00000	(9.81)	0.946	0.944	0.942	0.942	0.944	0.948	0.952	0.957	0.966	0.974	0.983	0.992	1.006	1.019
1.50000	(14.7)	0.926	0.924	0.924	0.925	0.927	0.932	0.938	0.943	0.953	0.962	0.972	0.981	0.996	1.009
2.00000	(19.6)	0.912	0.911	0.912	0.914	0.917	0.922	0.928	0.934	0.945	0.954	0.964	0.974	0.990	1.003
3.00000	(29.4)	0.894	0.894	0.896	0.899	0.902	0.909	0.916	0.922	0.934	0.944	0.955	0.965	0.981	0.996
S	$(\Delta p/\rho l)$	125	150	200	250	300	400	500	600	800	1000	1250	1500	2000	2500

Grad'nt   k.p.g.   (Equivalent) Pipe diameters in mm   **Roughness size, $k_s = 0.030$ mm**

E232

Kin. visc., $\nu = 30\times10^{-6}$ m^2 s^{-1};
$S = 0.00010$ to 3.00000

i.e. kin. pr. grad., $\Delta p/\rho l =$
(0.00098) to (29.4) ms^{-2}

Roughness size, $k_s = 0.060$ mm
This table shows values of m, as follows

m_C for Colebrook-White solutions; or,
where $\mathbf{R} \le 2000$, m_P for laminar flow

Grad'nt S	k.p.g. $(\Delta p/\rho l)$	125	150	200	250	300	400	500	600	800	1000	1250	1500	2000	2500
0.00010	(0.00098)	6.216	4.874	3.322	2.467	1.934	*	1.724	1.691	1.647	1.620	1.598	1.583	1.564	1.554
0.00015	(0.00147)	5.075	3.980	2.712	2.014	1.579	*	1.662	1.632	1.594	1.571	1.551	1.539	1.523	1.514
0.00020	(0.00196)	4.395	3.447	2.349	1.744	1.368	1.658	1.620	1.593	1.559	1.537	1.520	1.509	1.495	1.488
0.00030	(0.00294)	3.589	2.814	1.918	1.424	*	1.598	1.564	1.541	1.511	1.493	1.478	1.469	1.458	1.452
0.00040	(0.00392)	3.108	2.437	1.661	*	*	1.558	1.528	1.506	1.479	1.463	1.450	1.442	1.433	1.428
0.00060	(0.00588)	2.538	1.990	1.356	*	1.548	1.505	1.478	1.460	1.437	1.423	1.412	1.406	1.399	1.396
0.00080	(0.00785)	2.198	1.723	*	1.538	1.508	1.470	1.446	1.429	1.408	1.396	1.387	1.381	1.375	1.374
0.00100	(0.00981)	1.966	1.541	*	1.507	1.479	1.443	1.421	1.406	1.387	1.376	1.368	1.363	1.358	1.357
0.00150	(0.0147)	1.605	1.259	*	1.454	1.429	1.398	1.379	1.366	1.350	1.341	1.334	1.331	1.328	1.328
0.00200	(0.0196)	1.390	*	1.450	1.418	1.396	1.368	1.351	1.339	1.325	1.317	1.312	1.309	1.308	1.309
0.00300	(0.0294)	1.135	*	1.399	1.371	1.352	1.328	1.313	1.303	1.291	1.285	1.281	1.280	1.280	1.282
0.00400	(0.0392)	*	1.405	1.365	1.340	1.322	1.300	1.287	1.279	1.269	1.264	1.261	1.260	1.262	1.264
0.00600	(0.0588)	*	1.355	1.320	1.298	1.283	1.264	1.253	1.246	1.239	1.235	1.234	1.234	1.237	1.241
0.00800	(0.0785)	1.346	1.322	1.290	1.270	1.257	1.240	1.230	1.224	1.218	1.216	1.215	1.216	1.220	1.224
0.01000	(0.0981)	1.320	1.297	1.268	1.249	1.237	1.222	1.213	1.208	1.203	1.201	1.201	1.203	1.207	1.212
0.01500	(0.147)	1.275	1.255	1.229	1.214	1.203	1.191	1.184	1.180	1.177	1.176	1.177	1.180	1.185	1.191
0.02000	(0.196)	1.245	1.227	1.204	1.190	1.181	1.170	1.164	1.161	1.159	1.159	1.161	1.164	1.171	1.177
0.03000	(0.294)	1.205	1.190	1.170	1.158	1.151	1.142	1.138	1.136	1.135	1.137	1.140	1.143	1.151	1.159
0.04000	(0.392)	1.179	1.165	1.147	1.137	1.130	1.123	1.120	1.119	1.119	1.122	1.125	1.130	1.138	1.146
0.06000	(0.588)	1.144	1.132	1.117	1.109	1.103	1.098	1.096	1.096	1.098	1.101	1.106	1.111	1.121	1.130
0.08000	(0.785)	1.121	1.110	1.097	1.090	1.086	1.082	1.081	1.081	1.084	1.088	1.094	1.099	1.110	1.119
0.10000	(0.981)	1.104	1.094	1.082	1.076	1.072	1.069	1.069	1.070	1.074	1.078	1.084	1.090	1.101	1.111
0.15000	(1.47)	1.074	1.066	1.057	1.052	1.049	1.048	1.049	1.051	1.056	1.061	1.068	1.075	1.087	1.098
0.20000	(1.96)	1.054	1.047	1.040	1.036	1.034	1.034	1.036	1.038	1.044	1.050	1.058	1.065	1.078	1.089
0.30000	(2.94)	1.029	1.023	1.017	1.015	1.014	1.015	1.018	1.022	1.029	1.036	1.044	1.052	1.066	1.078
0.40000	(3.92)	1.012	1.007	1.002	1.001	1.001	1.003	1.007	1.011	1.019	1.026	1.035	1.043	1.058	1.070
0.60000	(5.88)	0.989	0.986	0.983	0.983	0.984	0.987	0.992	0.996	1.005	1.014	1.024	1.032	1.048	1.061
0.80000	(7.85)	0.974	0.972	0.970	0.971	0.972	0.977	0.982	0.987	0.997	1.006	1.016	1.025	1.041	1.055
1.00000	(9.81)	0.963	0.961	0.961	0.962	0.964	0.969	0.975	0.980	0.991	1.000	1.011	1.020	1.037	1.051
1.50000	(14.7)	0.945	0.944	0.945	0.947	0.950	0.956	0.963	0.969	0.981	0.991	1.002	1.012	1.029	1.044
2.00000	(19.6)	0.933	0.933	0.934	0.937	0.941	0.948	0.955	0.962	0.974	0.985	0.996	1.007	1.025	1.039
3.00000	(29.4)	0.917	0.918	0.921	0.925	0.929	0.938	0.946	0.953	0.966	0.977	0.989	1.000	1.019	1.034
S	$(\Delta p/\rho l)$	125	150	200	250	300	400	500	600	800	1000	1250	1500	2000	2500

Grad'nt   k.p.g.   (Equivalent) Pipe diameters in mm   **Roughness size, $k_s = 0.060$ mm**

Kin. visc., $\nu = 30 \times 10^{-6}$ m^{2}s^{-1};
$S = 0.00010$ to 3.00000

i.e. kin. pr. grad., $\Delta p/\rho l =$
(0.00098) to (29.4) ms^{-2}

Roughness size, $k_s = 0.150$ mm
This table shows values of m, as follows

m_C for Colebrook-White solutions; or,
where $\mathbf{R} \le 2000$, m_P for laminar flow

Grad'nt S	k.p.g. $(\Delta p/\rho l)$	125	150	200	250	300	400	500	600	800	1000	1250	1500	2000	2500
0.00010	(0.00098)	6.216	4.874	3.322	2.467	1.934	*	1.727	1.694	1.651	1.624	1.602	1.587	1.568	1.558
0.00015	(0.00147)	5.075	3.980	2.712	2.014	1.579	*	1.665	1.636	1.598	1.575	1.556	1.543	1.528	1.520
0.00020	(0.00196)	4.395	3.447	2.349	1.744	1.368	1.662	1.624	1.597	1.563	1.542	1.525	1.514	1.501	1.494
0.00030	(0.00294)	3.589	2.814	1.918	1.424	*	1.602	1.569	1.546	1.516	1.498	1.484	1.475	1.464	1.459
0.00040	(0.00392)	3.108	2.437	1.661	*	*	1.563	1.533	1.512	1.485	1.469	1.456	1.448	1.439	1.436
0.00060	(0.00588)	2.538	1.990	1.356	*	1.553	1.511	1.484	1.466	1.443	1.430	1.419	1.413	1.406	1.404
0.00080	(0.00785)	2.198	1.723	*	1.544	1.514	1.476	1.452	1.436	1.415	1.403	1.394	1.389	1.384	1.383
0.00100	(0.00981)	1.966	1.541	*	1.513	1.486	1.450	1.428	1.413	1.394	1.384	1.376	1.371	1.368	1.367
0.00150	(0.0147)	1.605	1.259	*	1.460	1.437	1.406	1.387	1.374	1.358	1.350	1.344	1.341	1.339	1.340
0.00200	(0.0196)	1.390	*	1.457	1.426	1.404	1.376	1.359	1.348	1.334	1.327	1.322	1.320	1.320	1.322
0.00300	(0.0294)	1.135	*	1.407	1.380	1.361	1.337	1.323	1.313	1.302	1.297	1.294	1.293	1.294	1.297
0.00400	(0.0392)	*	1.414	1.374	1.349	1.332	1.311	1.298	1.290	1.281	1.277	1.275	1.275	1.277	1.281
0.00600	(0.0588)	*	1.365	1.330	1.309	1.294	1.276	1.266	1.259	1.252	1.250	1.249	1.250	1.254	1.259
0.00800	(0.0785)	1.357	1.333	1.301	1.282	1.269	1.253	1.244	1.239	1.233	1.232	1.232	1.234	1.239	1.245
0.01000	(0.0981)	1.331	1.309	1.280	1.262	1.250	1.236	1.228	1.223	1.219	1.219	1.220	1.222	1.228	1.235
0.01500	(0.147)	1.288	1.268	1.243	1.228	1.218	1.207	1.201	1.197	1.195	1.196	1.198	1.202	1.209	1.217
0.02000	(0.196)	1.259	1.241	1.219	1.206	1.197	1.187	1.182	1.180	1.179	1.181	1.184	1.188	1.197	1.205
0.03000	(0.294)	1.221	1.206	1.187	1.176	1.169	1.162	1.159	1.157	1.158	1.161	1.166	1.171	1.180	1.190
0.04000	(0.392)	1.196	1.183	1.166	1.157	1.151	1.145	1.143	1.143	1.145	1.148	1.154	1.159	1.170	1.180
0.06000	(0.588)	1.164	1.152	1.138	1.131	1.127	1.123	1.122	1.123	1.127	1.132	1.138	1.144	1.156	1.167
0.08000	(0.785)	1.142	1.132	1.120	1.114	1.111	1.108	1.109	1.110	1.115	1.121	1.128	1.135	1.148	1.159
0.10000	(0.981)	1.126	1.117	1.107	1.102	1.099	1.098	1.099	1.101	1.107	1.113	1.121	1.128	1.141	1.153
0.15000	(1.47)	1.100	1.093	1.085	1.081	1.080	1.080	1.082	1.085	1.093	1.100	1.108	1.116	1.131	1.144
0.20000	(1.96)	1.082	1.076	1.070	1.067	1.067	1.068	1.072	1.075	1.083	1.091	1.101	1.109	1.124	1.138
0.30000	(2.94)	1.060	1.055	1.051	1.050	1.050	1.054	1.058	1.063	1.072	1.081	1.091	1.100	1.116	1.130
0.40000	(3.92)	1.045	1.042	1.039	1.039	1.040	1.044	1.049	1.054	1.065	1.074	1.085	1.094	1.111	1.125
0.60000	(5.88)	1.027	1.024	1.023	1.024	1.026	1.032	1.038	1.044	1.055	1.066	1.077	1.087	1.105	1.119
0.80000	(7.85)	1.015	1.013	1.013	1.015	1.018	1.024	1.031	1.038	1.050	1.060	1.072	1.083	1.101	1.116
1.00000	(9.81)	1.006	1.005	1.006	1.009	1.012	1.019	1.026	1.033	1.046	1.057	1.069	1.079	1.098	1.113
1.50000	(14.7)	0.991	0.992	0.994	0.998	1.002	1.010	1.018	1.026	1.039	1.050	1.063	1.074	1.093	1.109
2.00000	(19.6)	0.982	0.983	0.987	0.991	0.996	1.005	1.013	1.021	1.035	1.047	1.060	1.071	1.091	1.107
3.00000	(29.4)	0.971	0.973	0.977	0.983	0.988	0.998	1.007	1.015	1.030	1.042	1.056	1.067	1.087	1.104
S	$(\Delta p/\rho l)$	125	150	200	250	300	400	500	600	800	1000	1250	1500	2000	2500

Grad'nt   k.p.g.   (Equivalent) Pipe diameters in mm   **Roughness size, $k_s = 0.150$ mm**

Kin. visc., $\nu = 30 \times 10^{-6}$ m^{2}s^{-1};
$S = 0.00010$ to 3.00000

i.e. kin. pr. grad., $\Delta p/\rho l =$
(0.00098) to (29.4) ms^{-2}

Roughness size, $k_s = 0.30$ mm
This table shows values of m, as follows

m_C for Colebrook-White solutions; or,
where $\mathbf{R} \le 2000$, m_P for laminar flow

Grad'nt S	k.p.g. $(\Delta p/\rho l)$	125	150	200	250	300	400	500	600	800	1000	1250	1500	2000	2500
0.00010	(0.00098)	6.216	4.874	3.322	2.467	1.934	*	1.733	1.700	1.656	1.630	1.608	1.593	1.575	1.565
0.00015	(0.00147)	5.075	3.980	2.712	2.014	1.579	*	1.671	1.642	1.605	1.582	1.563	1.550	1.536	1.528
0.00020	(0.00196)	4.395	3.447	2.349	1.744	1.368	1.669	1.630	1.604	1.570	1.549	1.533	1.522	1.509	1.503
0.00030	(0.00294)	3.589	2.814	1.918	1.424	*	1.610	1.577	1.554	1.524	1.507	1.493	1.484	1.474	1.470
0.00040	(0.00392)	3.108	2.437	1.661	*	*	1.571	1.541	1.520	1.494	1.478	1.466	1.458	1.450	1.447
0.00060	(0.00588)	2.538	1.990	1.356	*	1.562	1.520	1.494	1.476	1.453	1.440	1.431	1.425	1.419	1.418
0.00080	(0.00785)	2.198	1.723	*	*	1.524	1.486	1.462	1.446	1.426	1.415	1.407	1.402	1.398	1.398
0.00100	(0.00981)	1.966	1.541	*	1.523	1.496	1.461	1.439	1.424	1.406	1.396	1.389	1.385	1.383	1.383
0.00150	(0.0147)	1.605	1.259	*	1.472	1.448	1.418	1.399	1.387	1.372	1.364	1.359	1.357	1.356	1.358
0.00200	(0.0196)	1.390	*	1.469	1.438	1.417	1.389	1.373	1.362	1.349	1.343	1.339	1.338	1.338	1.341
0.00300	(0.0294)	1.135	*	1.421	1.394	1.375	1.352	1.338	1.329	1.319	1.315	1.313	1.313	1.315	1.320
0.00400	(0.0392)	*	1.428	1.389	1.364	1.348	1.327	1.315	1.308	1.300	1.296	1.295	1.296	1.300	1.305
0.00600	(0.0588)	*	1.381	1.347	1.326	1.312	1.295	1.285	1.279	1.274	1.272	1.273	1.275	1.280	1.287
0.00800	(0.0785)	1.374	1.350	1.319	1.301	1.288	1.273	1.265	1.260	1.257	1.256	1.258	1.260	1.267	1.275
0.01000	(0.0981)	1.350	1.327	1.299	1.282	1.271	1.258	1.250	1.247	1.244	1.244	1.247	1.250	1.258	1.266
0.01500	(0.147)	1.308	1.289	1.265	1.251	1.241	1.231	1.226	1.224	1.223	1.225	1.229	1.233	1.242	1.251
0.02000	(0.196)	1.281	1.264	1.243	1.230	1.222	1.214	1.210	1.208	1.209	1.212	1.217	1.222	1.232	1.242
0.03000	(0.294)	1.246	1.232	1.214	1.203	1.197	1.191	1.189	1.189	1.192	1.196	1.202	1.208	1.219	1.230
0.04000	(0.392)	1.223	1.210	1.195	1.186	1.181	1.176	1.176	1.176	1.180	1.185	1.192	1.199	1.211	1.223
0.06000	(0.588)	1.194	1.183	1.170	1.164	1.160	1.158	1.158	1.160	1.166	1.172	1.180	1.187	1.201	1.213
0.08000	(0.785)	1.175	1.165	1.154	1.149	1.147	1.146	1.147	1.150	1.157	1.163	1.172	1.180	1.195	1.207
0.10000	(0.981)	1.161	1.152	1.143	1.139	1.137	1.137	1.139	1.143	1.150	1.157	1.166	1.175	1.190	1.203
0.15000	(1.47)	1.138	1.131	1.124	1.121	1.121	1.123	1.126	1.131	1.139	1.148	1.158	1.167	1.183	1.197
0.20000	(1.96)	1.123	1.117	1.112	1.110	1.111	1.114	1.118	1.123	1.132	1.142	1.152	1.161	1.178	1.192
0.30000	(2.94)	1.104	1.100	1.097	1.097	1.098	1.103	1.108	1.113	1.124	1.134	1.145	1.155	1.173	1.187
0.40000	(3.92)	1.092	1.089	1.087	1.088	1.090	1.095	1.101	1.108	1.119	1.129	1.141	1.151	1.169	1.184
0.60000	(5.88)	1.077	1.075	1.075	1.077	1.080	1.087	1.094	1.100	1.113	1.123	1.136	1.146	1.165	1.181
0.80000	(7.85)	1.068	1.066	1.067	1.070	1.074	1.081	1.089	1.096	1.109	1.120	1.132	1.144	1.162	1.178
1.00000	(9.81)	1.061	1.060	1.062	1.065	1.069	1.077	1.085	1.093	1.106	1.117	1.130	1.141	1.161	1.177
1.50000	(14.7)	1.050	1.050	1.054	1.058	1.062	1.071	1.080	1.088	1.102	1.114	1.127	1.138	1.158	1.174
2.00000	(19.6)	1.043	1.044	1.048	1.053	1.058	1.068	1.076	1.085	1.099	1.111	1.125	1.136	1.156	1.173
3.00000	(29.4)	1.035	1.037	1.042	1.047	1.053	1.063	1.072	1.081	1.096	1.108	1.122	1.134	1.154	1.171
S	$(\Delta p/\rho l)$	125	150	200	250	300	400	500	600	800	1000	1250	1500	2000	2500

Grad'nt   k.p.g.   (Equivalent) Pipe diameters in mm   **Roughness size, $k_s = 0.30$ mm**

E235

Kin. visc., $\nu = 40\times10^{-6}\ \mathrm{m^2 s^{-1}}$;
$S = 0{\cdot}00010$ to $30{\cdot}0000$

i.e. kin. pr. grad., $\Delta p/\rho l =$
$(0{\cdot}00098)$ to $(294)\ \mathrm{ms^{-2}}$

Roughness size, $k_s = 0{\cdot}006$ mm
This table shows values of m, as follows

m_C for Colebrook-White solutions; or,
where $\mathbf{R} \leq 2000$, m_P for laminar flow

Grad'nt S	k.p.g. $(\Delta p/\rho l)$	6	8	10	12.5	15	20	25	30	40	50	60	80	100	125
															(Equivalent) Pipe diameters in mm
0.00030	(0.00294)	274.3	186.9	138.8	103.1	80.84	55.09	40.91	32.08	21.86	16.24	12.73	8.676	6.443	4.785
0.00040	(0.00392)	237.5	161.9	120.2	89.28	70.01	47.71	35.43	27.78	18.93	14.06	11.03	7.513	5.580	4.144
0.00060	(0.00588)	194.0	132.2	98.15	72.89	57.16	38.95	28.93	22.69	15.46	11.48	9.003	6.135	4.556	3.383
0.00080	(0.00785)	168.0	114.5	85.00	63.13	49.51	33.73	25.05	19.65	13.39	9.942	7.797	5.313	3.946	2.930
0.00100	(0.00981)	150.2	102.4	76.03	56.46	44.28	30.17	22.41	17.57	11.97	8.892	6.973	4.752	3.529	2.621
0.00150	(0.0147)	122.7	83.59	62.08	46.10	36.15	24.64	18.30	14.35	9.777	7.261	5.694	3.880	2.881	2.140
0.00200	(0.0196)	106.2	72.39	53.76	39.93	31.31	21.34	15.84	12.43	8.467	6.288	4.931	3.360	2.495	1.853
0.00300	(0.0294)	86.74	59.11	43.90	32.60	25.56	17.42	12.94	10.15	6.913	5.134	4.026	2.743	2.037	1.513
0.00400	(0.0392)	75.12	51.19	38.01	28.23	22.14	15.09	11.20	8.786	5.987	4.446	3.487	2.376	1.764	1.310
0.00600	(0.0588)	61.33	41.79	31.04	23.05	18.08	12.32	9.148	7.174	4.888	3.630	2.847	1.940	1.441	*
0.00800	(0.0785)	53.12	36.20	26.88	19.96	15.65	10.67	7.922	6.213	4.233	3.144	2.465	1.680	1.248	*
0.01000	(0.0981)	47.51	32.37	24.04	17.86	14.00	9.541	7.086	5.557	3.786	2.812	2.205	1.503	1.116	*
0.01500	(0.147)	38.79	26.43	19.63	14.58	11.43	7.790	5.786	4.537	3.092	2.296	1.801	1.227	*	1.333
0.02000	(0.196)	33.59	22.89	17.00	12.63	9.901	6.747	5.010	3.929	2.677	1.988	1.559	1.063	*	1.299
0.03000	(0.294)	27.43	18.69	13.88	10.31	8.084	5.509	4.091	3.208	2.186	1.624	1.273	*	1.282	1.254
0.04000	(0.392)	23.75	16.19	12.02	8.928	7.001	4.771	3.543	2.778	1.893	1.406	1.103	*	1.249	1.224
0.06000	(0.588)	19.40	13.22	9.815	7.289	5.716	3.895	2.893	2.269	1.546	1.148	*	1.233	1.206	1.184
0.08000	(0.785)	16.80	11.45	8.500	6.313	4.951	3.373	2.505	1.965	1.339	0.994	*	1.201	1.177	1.157
0.10000	(0.981)	15.02	10.24	7.603	5.646	4.428	3.017	2.241	1.757	1.197	*	*	1.178	1.156	1.137
0.15000	(1.47)	12.27	8.359	6.208	4.610	3.615	2.464	1.830	1.435	0.978	*	1.170	1.138	1.119	1.102
0.20000	(1.96)	10.62	7.239	5.376	3.993	3.131	2.134	1.584	1.243	*	1.162	1.140	1.112	1.094	1.079
0.30000	(2.94)	8.674	5.911	4.390	3.260	2.556	1.742	1.294	1.015	*	1.120	1.101	1.076	1.061	1.048
0.40000	(3.92)	7.512	5.119	3.801	2.823	2.214	1.509	1.120	*	1.117	1.092	1.075	1.052	1.038	1.027
0.60000	(5.88)	6.133	4.179	3.104	2.305	1.808	1.232	0.915	*	1.077	1.055	1.040	1.021	1.009	0.999
0.80000	(7.85)	5.312	3.620	2.688	1.996	1.565	1.067	*	*	1.051	1.031	1.017	0.999	0.989	0.980
1.00000	(9.81)	4.751	3.237	2.404	1.786	1.400	0.954	*	1.061	1.031	1.012	1.000	0.983	0.974	0.966
1.50000	(14.7)	3.879	2.643	1.963	1.458	1.143	*	1.043	1.023	0.997	0.981	0.970	0.956	0.948	0.942
2.00000	(19.6)	3.359	2.289	1.700	1.263	0.990	*	1.016	0.998	0.974	0.960	0.950	0.937	0.930	0.925
3.00000	(29.4)	2.743	1.869	1.388	1.031	0.808	1.003	0.980	0.964	0.944	0.931	0.923	0.913	0.907	0.903
4.00000	(39.2)	2.375	1.619	1.202	0.893	*	0.977	0.957	0.942	0.924	0.912	0.905	0.896	0.891	0.888
6.00000	(58.8)	1.940	1.322	0.982	*	*	0.943	0.925	0.913	0.897	0.887	0.881	0.873	0.870	0.867
8.00000	(78.5)	1.680	1.145	0.850	*	0.947	0.921	0.904	0.893	0.879	0.870	0.865	0.858	0.855	0.854
10.0000	(98.1)	1.502	1.024	0.760	*	0.928	0.904	0.889	0.878	0.865	0.857	0.852	0.847	0.844	0.843
15.0000	(147)	1.227	0.836	*	0.913	0.896	0.875	0.862	0.853	0.842	0.835	0.832	0.827	0.826	0.826
20.0000	(196)	1.062	0.724	*	0.890	0.875	0.856	0.844	0.836	0.826	0.821	0.817	0.814	0.813	0.814
30.0000	(294)	0.867	*	0.879	0.860	0.847	0.830	0.820	0.813	0.805	0.801	0.799	0.797	0.796	0.798
S	$(\Delta p/\rho l)$	6	8	10	12.5	15	20	25	30	40	50	60	80	100	125

Grad'nt k.p.g. (Equivalent) Pipe diameters in mm

Grad'nt S	k.p.g. $(\Delta p/\rho l)$	125	150	200	250	300	400	500	600	800	1000	1250	1500	2000	2500
0.00010	(0.00098)	8.288	6.499	4.429	3.289	2.579	1.758	*	*	1.728	1.694	1.667	1.648	1.625	1.611
0.00015	(0.00147)	6.767	5.307	3.616	2.685	2.106	1.435	*	1.715	1.669	1.640	1.616	1.600	1.580	1.568
0.00020	(0.00196)	5.860	4.596	3.132	2.326	1.824	*	1.703	1.671	1.629	1.603	1.582	1.567	1.549	1.539
0.00030	(0.00294)	4.785	3.752	2.557	1.899	1.489	*	1.642	1.614	1.577	1.554	1.536	1.523	1.508	1.500
0.00040	(0.00392)	4.144	3.250	2.214	1.644	*	1.638	1.601	1.575	1.542	1.521	1.505	1.494	1.481	1.474
0.00060	(0.00588)	3.383	2.653	1.808	1.343	*	1.579	1.546	1.524	1.495	1.477	1.463	1.454	1.443	1.438
0.00080	(0.00785)	2.930	2.298	1.566	*	*	1.539	1.510	1.489	1.463	1.448	1.435	1.427	1.418	1.414
0.00100	(0.00981)	2.621	2.055	1.400	*	1.554	1.510	1.483	1.464	1.440	1.425	1.414	1.407	1.399	1.396
0.00150	(0.0147)	2.140	1.678	*	1.527	1.498	1.460	1.436	1.419	1.399	1.387	1.377	1.371	1.366	1.364
0.00200	(0.0196)	1.853	1.453	*	1.487	1.461	1.426	1.404	1.390	1.371	1.360	1.352	1.347	1.343	1.342
0.00300	(0.0294)	1.513	1.187	1.468	1.434	1.411	1.381	1.362	1.350	1.334	1.325	1.319	1.315	1.312	1.312
0.00400	(0.0392)	1.310	*	1.430	1.399	1.378	1.351	1.334	1.323	1.309	1.301	1.296	1.293	1.291	1.292
0.00600	(0.0588)	*	*	1.379	1.352	1.334	1.310	1.296	1.286	1.275	1.269	1.265	1.263	1.263	1.265
0.00800	(0.0785)	*	1.384	1.346	1.321	1.304	1.283	1.270	1.262	1.252	1.247	1.244	1.243	1.244	1.246
0.01000	(0.0981)	*	1.357	1.321	1.298	1.282	1.263	1.251	1.243	1.235	1.230	1.228	1.228	1.229	1.232
0.01500	(0.147)	1.333	1.309	1.277	1.258	1.244	1.227	1.218	1.211	1.205	1.202	1.200	1.201	1.203	1.207
0.02000	(0.196)	1.299	1.277	1.248	1.231	1.219	1.204	1.195	1.190	1.184	1.182	1.182	1.182	1.186	1.190
0.03000	(0.294)	1.254	1.235	1.210	1.194	1.184	1.172	1.164	1.160	1.156	1.155	1.156	1.158	1.162	1.167
0.04000	(0.392)	1.224	1.206	1.184	1.170	1.161	1.150	1.144	1.140	1.138	1.137	1.139	1.141	1.146	1.152
0.06000	(0.588)	1.184	1.169	1.149	1.137	1.130	1.121	1.116	1.114	1.112	1.113	1.115	1.118	1.124	1.130
0.08000	(0.785)	1.157	1.143	1.126	1.115	1.109	1.101	1.097	1.095	1.095	1.096	1.099	1.102	1.109	1.115
0.10000	(0.981)	1.137	1.124	1.108	1.099	1.093	1.086	1.083	1.082	1.082	1.084	1.087	1.090	1.097	1.104
0.15000	(1.47)	1.102	1.092	1.078	1.070	1.065	1.060	1.058	1.058	1.059	1.062	1.065	1.069	1.077	1.085
0.20000	(1.96)	1.079	1.069	1.057	1.051	1.046	1.042	1.041	1.041	1.043	1.046	1.051	1.055	1.064	1.072
0.30000	(2.94)	1.048	1.040	1.030	1.024	1.021	1.019	1.018	1.019	1.022	1.026	1.031	1.036	1.045	1.054
0.40000	(3.92)	1.027	1.020	1.011	1.007	1.004	1.002	1.003	1.004	1.008	1.012	1.018	1.023	1.033	1.041
0.60000	(5.88)	0.999	0.993	0.986	0.983	0.981	0.981	0.982	0.984	0.988	0.993	0.999	1.005	1.015	1.025
0.80000	(7.85)	0.980	0.975	0.969	0.967	0.965	0.966	0.968	0.970	0.975	0.981	0.987	0.993	1.004	1.013
1.00000	(9.81)	0.966	0.961	0.956	0.954	0.954	0.955	0.957	0.960	0.965	0.971	0.978	0.984	0.995	1.005
1.50000	(14.7)	0.942	0.938	0.934	0.933	0.933	0.935	0.938	0.941	0.948	0.954	0.961	0.968	0.980	0.990
2.00000	(19.6)	0.925	0.922	0.919	0.919	0.919	0.922	0.925	0.929	0.936	0.943	0.950	0.957	0.969	0.980
3.00000	(29.4)	0.903	0.900	0.899	0.899	0.901	0.904	0.908	0.912	0.920	0.927	0.935	0.943	0.955	0.966
S	$(\Delta p/\rho l)$	125	150	200	250	300	400	500	600	800	1000	1250	1500	2000	2500

Grad'nt k.p.g. (Equivalent) Pipe diameters in mm

Kinematic viscosity, $\nu = 40\times10^{-6}\ \mathrm{m^2 s^{-1}}$;

Roughness size, $k_s = 0{\cdot}006$ mm

Kin. visc., $\nu = 40\times10^{-6}\ \text{m}^2\text{s}^{-1}$;
S = 0·00010 to 30·0000
i.e. kin. pr. grad., $\Delta p/\rho l =$
(0·00098) to (294) ms^{-2}

Roughness size, k_s = 0·015 mm
This table shows values of m, as follows
m_C for Colebrook-White solutions; or,
where $\mathbf{R} \le 2000$, m_P for laminar flow

Grad'nt S	k.p.g. $(\Delta p/\rho l)$	6	8	10	12·5	15	20	25	30	40	50	60	80	100	125
							(Equivalent) Pipe diameters in mm								
0·00030	(0·00294)	274·3	186·9	138·8	103·1	80·84	55·09	40·91	32·08	21·86	16·24	12·73	8·676	6·443	4·785
0·00040	(0·00392)	237·5	161·9	120·2	89·28	70·01	47·71	35·43	27·78	18·93	14·06	11·03	7·513	5·580	4·144
0·00060	(0·00588)	194·0	132·2	98·15	72·89	57·16	38·95	28·93	22·69	15·46	11·48	9·003	6·135	4·556	3·383
0·00080	(0·00785)	168·0	114·5	85·00	63·13	49·51	33·73	25·05	19·65	13·39	9·942	7·797	5·313	3·946	2·930
0·00100	(0·00981)	150·2	102·4	76·03	56·46	44·28	30·17	22·41	17·57	11·97	8·892	6·973	4·752	3·529	2·621
0·00150	(0·0147)	122·7	83·59	62·08	46·10	36·15	24·64	18·30	14·35	9·777	7·261	5·694	3·880	2·881	2·140
0·00200	(0·0196)	106·2	72·39	53·76	39·93	31·31	21·34	15·84	12·43	8·467	6·288	4·931	3·360	2·495	1·853
0·00300	(0·0294)	86·74	59·11	43·90	32·60	25·56	17·42	12·94	10·15	6·913	5·134	4·026	2·743	2·037	1·513
0·00400	(0·0392)	75·12	51·19	38·01	28·23	22·14	15·09	11·20	8·786	5·987	4·446	3·487	2·376	1·764	1·310
0·00600	(0·0588)	61·33	41·79	31·04	23·05	18·08	12·32	9·148	7·174	4·888	3·630	2·847	1·940	1·441	*
0·00800	(0·0785)	53·12	36·20	26·88	19·96	15·65	10·67	7·922	6·213	4·233	3·144	2·465	1·680	1·248	*
0·01000	(0·0981)	47·51	32·37	24·04	17·86	14·00	9·541	7·086	5·557	3·786	2·812	2·205	1·503	1·116	*
0·01500	(0·147)	38·79	26·43	19·63	14·58	11·43	7·790	5·786	4·537	3·092	2·296	1·801	1·227	*	1·334
0·02000	(0·196)	33·59	22·89	17·00	12·63	9·901	6·747	5·010	3·929	2·677	1·988	1·559	1·063	*	1·300
0·03000	(0·294)	27·43	18·69	13·88	10·31	8·084	5·509	4·091	3·208	2·186	1·624	1·273	*	1·283	1·255
0·04000	(0·392)	23·75	16·19	12·02	8·928	7·001	4·771	3·543	2·778	1·893	1·406	1·103	*	1·251	1·225
0·06000	(0·588)	19·40	13·22	9·815	7·289	5·716	3·895	2·893	2·269	1·546	1·148	*	1·234	1·208	1·186
0·08000	(0·785)	16·80	11·45	8·500	6·313	4·951	3·373	2·505	1·965	1·339	0·994	*	1·203	1·179	1·159
0·10000	(0·981)	15·02	10·24	7·603	5·646	4·428	3·017	2·241	1·757	1·197	*	*	1·180	1·158	1·139
0·15000	(1·47)	12·27	8·359	6·208	4·610	3·615	2·464	1·830	1·435	0·978	*	1·172	1·140	1·121	1·105
0·20000	(1·96)	10·62	7·239	5·376	3·993	3·131	2·134	1·584	1·243	*	1·164	1·142	1·114	1·096	1·082
0·30000	(2·94)	8·674	5·911	4·390	3·260	2·556	1·742	1·294	1·015	*	1·123	1·104	1·079	1·064	1·051
0·40000	(3·92)	7·512	5·119	3·801	2·823	2·214	1·509	1·120	*	1·120	1·095	1·078	1·055	1·042	1·031
0·60000	(5·88)	6·133	4·179	3·104	2·305	1·808	1·232	0·915	*	1·081	1·059	1·044	1·024	1·012	1·003
0·80000	(7·85)	5·312	3·620	2·688	1·996	1·565	1·067	*	*	1·054	1·034	1·021	1·003	0·993	0·985
1·00000	(9·81)	4·751	3·237	2·404	1·786	1·400	0·954	*	1·064	1·035	1·016	1·004	0·988	0·978	0·971
1·50000	(14·7)	3·879	2·643	1·963	1·458	1·143	*	1·047	1·027	1·001	0·985	0·974	0·961	0·953	0·947
2·00000	(19·6)	3·359	2·289	1·700	1·263	0·990	*	1·020	1·002	0·979	0·964	0·955	0·943	0·936	0·931
3·00000	(29·4)	2·743	1·869	1·388	1·031	0·808	1·008	0·985	0·969	0·949	0·937	0·929	0·919	0·913	0·909
4·00000	(39·2)	2·375	1·619	1·202	0·893	*	0·983	0·962	0·948	0·930	0·919	0·911	0·903	0·898	0·895
6·00000	(58·8)	1·940	1·322	0·982	*	*	0·949	0·931	0·919	0·903	0·894	0·888	0·881	0·878	0·876
8·00000	(78·5)	1·680	1·145	0·850	*	0·953	0·927	0·911	0·900	0·886	0·878	0·873	0·867	0·864	0·863
10·0000	(98·1)	1·502	1·024	0·760	*	0·935	0·911	0·896	0·886	0·873	0·866	0·861	0·856	0·854	0·854
15·0000	(147)	1·227	0·836	*	0·920	0·904	0·883	0·870	0·861	0·851	0·845	0·841	0·838	0·837	0·837
20·0000	(196)	1·062	0·724	*	0·898	0·883	0·865	0·853	0·845	0·836	0·831	0·828	0·826	0·826	0·827
30·0000	(294)	0·867	*	0·888	0·869	0·856	0·840	0·830	0·824	0·817	0·813	0·811	0·810	0·810	0·812
S	$(\Delta p/\rho l)$	6	8	10	12·5	15	20	25	30	40	50	60	80	100	125

Grad'nt k.p.g. (Equivalent) Pipe diameters in mm

Grad'nt S	k.p.g. $(\Delta p/\rho l)$	125	150	200	250	300	400	500	600	800	1000	1250	1500	2000	2500
0·00010	(0·00098)	8·288	6·499	4·429	3·289	2·579	1·758	*	*	1·728	1·695	1·667	1·649	1·625	1·611
0·00015	(0·00147)	6·767	5·307	3·616	2·685	2·106	1·435	*	1·715	1·669	1·640	1·616	1·600	1·580	1·568
0·00020	(0·00196)	5·860	4·596	3·132	2·326	1·824	*	1·704	1·672	1·630	1·604	1·582	1·568	1·550	1·540
0·00030	(0·00294)	4·785	3·752	2·557	1·899	1·489	*	1·642	1·614	1·577	1·555	1·536	1·524	1·509	1·501
0·00040	(0·00392)	4·144	3·250	2·214	1·644	*	1·638	1·601	1·576	1·542	1·522	1·505	1·494	1·481	1·474
0·00060	(0·00588)	3·383	2·653	1·808	1·343	*	1·579	1·547	1·524	1·496	1·478	1·464	1·455	1·444	1·439
0·00080	(0·00785)	2·930	2·298	1·566	*	*	1·540	1·510	1·490	1·464	1·448	1·436	1·428	1·419	1·415
0·00100	(0·00981)	2·621	2·055	1·400	*	1·554	1·511	1·483	1·464	1·440	1·426	1·415	1·408	1·400	1·397
0·00150	(0·0147)	2·140	1·678	*	1·528	1·498	1·460	1·436	1·420	1·400	1·387	1·378	1·372	1·367	1·365
0·00200	(0·0196)	1·853	1·453	*	1·488	1·461	1·427	1·405	1·390	1·372	1·361	1·353	1·348	1·344	1·343
0·00300	(0·0294)	1·513	1·187	1·469	1·435	1·412	1·382	1·363	1·350	1·335	1·326	1·320	1·316	1·314	1·314
0·00400	(0·0392)	1·310	*	1·431	1·400	1·379	1·352	1·335	1·323	1·310	1·302	1·297	1·294	1·293	1·294
0·00600	(0·0588)	*	*	1·380	1·353	1·335	1·311	1·297	1·287	1·276	1·270	1·266	1·265	1·265	1·267
0·00800	(0·0785)	*	1·385	1·347	1·322	1·306	1·284	1·271	1·263	1·253	1·248	1·245	1·245	1·245	1·248
0·01000	(0·0981)	*	1·358	1·322	1·299	1·284	1·264	1·252	1·245	1·236	1·232	1·230	1·229	1·231	1·234
0·01500	(0·147)	1·334	1·310	1·279	1·259	1·246	1·229	1·219	1·213	1·206	1·203	1·202	1·203	1·206	1·210
0·02000	(0·196)	1·300	1·278	1·250	1·232	1·220	1·205	1·197	1·191	1·186	1·184	1·184	1·185	1·189	1·193
0·03000	(0·294)	1·255	1·236	1·211	1·196	1·186	1·173	1·166	1·162	1·159	1·158	1·159	1·160	1·165	1·171
0·04000	(0·392)	1·225	1·208	1·185	1·172	1·163	1·152	1·146	1·143	1·140	1·140	1·141	1·144	1·149	1·155
0·06000	(0·588)	1·186	1·170	1·151	1·139	1·132	1·123	1·118	1·116	1·115	1·116	1·118	1·121	1·128	1·134
0·08000	(0·785)	1·159	1·145	1·128	1·118	1·111	1·103	1·100	1·098	1·098	1·100	1·103	1·106	1·113	1·120
0·10000	(0·981)	1·139	1·126	1·111	1·101	1·095	1·089	1·086	1·085	1·085	1·087	1·091	1·095	1·102	1·110
0·15000	(1·47)	1·105	1·094	1·081	1·073	1·068	1·063	1·062	1·061	1·063	1·066	1·070	1·074	1·083	1·091
0·20000	(1·96)	1·082	1·072	1·060	1·054	1·050	1·046	1·045	1·045	1·048	1·051	1·056	1·061	1·070	1·078
0·30000	(2·94)	1·051	1·043	1·033	1·028	1·025	1·023	1·023	1·024	1·027	1·032	1·037	1·043	1·052	1·061
0·40000	(3·92)	1·031	1·023	1·015	1·011	1·008	1·007	1·008	1·009	1·014	1·018	1·024	1·030	1·041	1·050
0·60000	(5·88)	1·003	0·997	0·991	0·987	0·986	0·986	0·988	0·990	0·995	1·001	1·007	1·013	1·025	1·035
0·80000	(7·85)	0·985	0·980	0·974	0·972	0·971	0·972	0·974	0·977	0·983	0·989	0·996	1·002	1·014	1·024
1·00000	(9·81)	0·971	0·966	0·962	0·960	0·960	0·961	0·964	0·967	0·973	0·980	0·987	0·994	1·006	1·017
1·50000	(14·7)	0·947	0·944	0·940	0·940	0·940	0·943	0·946	0·950	0·957	0·964	0·972	0·980	0·992	1·004
2·00000	(19·6)	0·931	0·928	0·926	0·926	0·927	0·930	0·934	0·938	0·946	0·954	0·962	0·970	0·983	0·995
3·00000	(29·4)	0·909	0·908	0·907	0·908	0·909	0·914	0·919	0·923	0·932	0·940	0·949	0·957	0·971	0·983
S	$(\Delta p/\rho l)$	125	150	200	250	300	400	500	600	800	1000	1250	1500	2000	2500

Grad'nt k.p.g. (Equivalent) Pipe diameters in mm

Kinematic viscosity, $\nu = 40\times10^{-6}\ \text{m}^2\text{s}^{-1}$;
Roughness size, k_s = 0·015 mm

E237

Kin. visc., $\nu = 40 \times 10^{-6}$ m^{2}s^{-1}; $S = 0.00010$ to 3.00000

i.e. kin. pr. grad., $\Delta p/\rho l =$ (0·00098) to (29·4) ms^{-2}

Roughness size, $k_s = 0.030$ mm
This table shows values of m, as follows

m_C for Colebrook-White solutions; or, where $\mathbf{R} \le 2000$, m_P for laminar flow

Grad'nt S	k.p.g. $(\Delta p/\rho l)$	125	150	200	250	300	400	500	600	800	1000	1250	1500	2000	2500
0·00010	(0·00098)	8·288	6·499	4·429	3·289	2·579	1·758	*	*	1·728	1·695	1·668	1·649	1·625	1·612
0·00015	(0·00147)	6·767	5·307	3·616	2·685	2·106	1·435	*	1·716	1·670	1·641	1·617	1·601	1·581	1·569
0·00020	(0·00196)	5·860	4·596	3·132	2·326	1·824	*	1·704	1·672	1·630	1·604	1·583	1·568	1·551	1·540
0·00030	(0·00294)	4·785	3·752	2·557	1·899	1·489	*	1·643	1·615	1·578	1·555	1·537	1·525	1·510	1·502
0·00040	(0·00392)	4·144	3·250	2·214	1·644	*	1·639	1·602	1·576	1·543	1·523	1·506	1·495	1·482	1·475
0·00060	(0·00588)	3·383	2·653	1·808	1·343	*	1·580	1·548	1·525	1·496	1·479	1·465	1·456	1·445	1·440
0·00080	(0·00785)	2·930	2·298	1·566	*	*	1·541	1·511	1·491	1·465	1·449	1·437	1·429	1·420	1·416
0·00100	(0·00981)	2·621	2·055	1·400	*	1·555	1·511	1·484	1·465	1·441	1·427	1·416	1·409	1·401	1·398
0·00150	(0·0147)	2·140	1·678	*	1·529	1·499	1·461	1·437	1·421	1·401	1·389	1·379	1·374	1·368	1·366
0·00200	(0·0196)	1·853	1·453	*	1·489	1·462	1·428	1·406	1·392	1·373	1·363	1·355	1·350	1·346	1·345
0·00300	(0·0294)	1·513	1·187	1·470	1·436	1·413	1·383	1·364	1·352	1·336	1·328	1·321	1·318	1·316	1·316
0·00400	(0·0392)	1·310	*	1·432	1·401	1·380	1·353	1·336	1·325	1·312	1·304	1·299	1·296	1·295	1·296
0·00600	(0·0588)	*	*	1·382	1·355	1·337	1·313	1·299	1·289	1·278	1·272	1·269	1·267	1·267	1·269
0·00800	(0·0785)	*	1·387	1·348	1·324	1·307	1·286	1·273	1·265	1·255	1·251	1·248	1·247	1·247	1·251
0·01000	(0·0981)	*	1·359	1·323	1·301	1·285	1·266	1·254	1·247	1·239	1·235	1·233	1·232	1·234	1·238
0·01500	(0·147)	1·336	1·312	1·281	1·261	1·248	1·231	1·221	1·215	1·209	1·206	1·206	1·206	1·209	1·214
0·02000	(0·196)	1·302	1·280	1·252	1·234	1·222	1·208	1·199	1·194	1·189	1·187	1·187	1·189	1·193	1·198
0·03000	(0·294)	1·258	1·239	1·214	1·199	1·189	1·176	1·170	1·166	1·162	1·162	1·163	1·165	1·170	1·176
0·04000	(0·392)	1·228	1·211	1·188	1·175	1·166	1·155	1·149	1·146	1·144	1·144	1·146	1·149	1·155	1·161
0·06000	(0·588)	1·188	1·173	1·154	1·143	1·135	1·127	1·122	1·120	1·120	1·121	1·124	1·127	1·134	1·141
0·08000	(0·785)	1·162	1·149	1·131	1·121	1·115	1·108	1·104	1·103	1·103	1·105	1·109	1·112	1·120	1·128
0·10000	(0·981)	1·143	1·130	1·114	1·105	1·099	1·093	1·091	1·090	1·091	1·093	1·097	1·101	1·110	1·118
0·15000	(1·47)	1·109	1·098	1·085	1·077	1·073	1·069	1·067	1·067	1·069	1·073	1·078	1·082	1·092	1·100
0·20000	(1·96)	1·086	1·077	1·065	1·059	1·055	1·052	1·051	1·052	1·055	1·059	1·064	1·070	1·080	1·089
0·30000	(2·94)	1·056	1·048	1·039	1·034	1·031	1·030	1·030	1·031	1·036	1·041	1·047	1·053	1·064	1·073
0·40000	(3·92)	1·036	1·029	1·021	1·017	1·015	1·015	1·016	1·018	1·023	1·028	1·035	1·041	1·053	1·063
0·60000	(5·88)	1·009	1·004	0·998	0·995	0·994	0·995	0·997	1·000	1·006	1·012	1·019	1·026	1·039	1·050
0·80000	(7·85)	0·992	0·987	0·982	0·980	0·980	0·981	0·984	0·987	0·994	1·001	1·009	1·016	1·029	1·041
1·00000	(9·81)	0·978	0·974	0·970	0·969	0·969	0·971	0·975	0·978	0·986	0·993	1·001	1·009	1·022	1·034
1·50000	(14·7)	0·956	0·952	0·950	0·950	0·951	0·954	0·959	0·963	0·971	0·979	0·988	0·996	1·011	1·023
2·00000	(19·6)	0·940	0·938	0·937	0·937	0·939	0·943	0·948	0·953	0·962	0·970	0·980	0·988	1·003	1·016
3·00000	(29·4)	0·920	0·919	0·919	0·921	0·923	0·928	0·934	0·939	0·949	0·959	0·969	0·978	0·994	1·007
S Grad'nt	$(\Delta p/\rho l)$ k.p.g.	125	150	200	250	300	400	500	600	800	1000	1250	1500	2000	2500

(Equivalent) Pipe diameters in mm — Roughness size, $k_s = 0.030$ mm

E238

Kin. visc., $\nu = 40 \times 10^{-6}$ m^{2}s^{-1}; $S = 0.00010$ to 3.00000

i.e. kin. pr. grad., $\Delta p/\rho l =$ (0·00098) to (29·4) ms^{-2}

Roughness size, $k_s = 0.060$ mm
This table shows values of m, as follows

m_C for Colebrook-White solutions; or, where $\mathbf{R} \le 2000$, m_P for laminar flow

Grad'nt S	k.p.g. $(\Delta p/\rho l)$	125	150	200	250	300	400	500	600	800	1000	1250	1500	2000	2500
0·00010	(0·00098)	8·288	6·499	4·429	3·289	2·579	1·758	*	*	1·729	1·696	1·669	1·650	1·627	1·613
0·00015	(0·00147)	6·767	5·307	3·616	2·685	2·106	1·435	*	1·717	1·671	1·642	1·618	1·602	1·582	1·571
0·00020	(0·00196)	5·860	4·596	3·132	2·326	1·824	*	1·706	1·673	1·632	1·605	1·584	1·570	1·552	1·542
0·00030	(0·00294)	4·785	3·752	2·557	1·899	1·489	*	1·644	1·616	1·580	1·557	1·539	1·526	1·512	1·504
0·00040	(0·00392)	4·144	3·250	2·214	1·644	*	1·640	1·603	1·578	1·545	1·524	1·508	1·497	1·484	1·477
0·00060	(0·00588)	3·383	2·653	1·808	1·343	*	1·581	1·549	1·527	1·498	1·481	1·467	1·458	1·447	1·442
0·00080	(0·00785)	2·930	2·298	1·566	*	*	1·542	1·513	1·493	1·467	1·451	1·439	1·431	1·423	1·419
0·00100	(0·00981)	2·621	2·055	1·400	*	1·557	1·513	1·486	1·467	1·444	1·429	1·418	1·411	1·404	1·401
0·00150	(0·0147)	2·140	1·678	*	1·530	1·501	1·463	1·440	1·423	1·403	1·391	1·382	1·377	1·371	1·370
0·00200	(0·0196)	1·853	1·453	*	1·491	1·464	1·430	1·409	1·394	1·376	1·365	1·358	1·353	1·349	1·349
0·00300	(0·0294)	1·513	1·187	1·472	1·439	1·416	1·386	1·367	1·355	1·339	1·331	1·325	1·322	1·320	1·320
0·00400	(0·0392)	1·310	*	1·435	1·404	1·383	1·356	1·339	1·328	1·315	1·308	1·303	1·300	1·299	1·301
0·00600	(0·0588)	*	*	1·385	1·358	1·340	1·316	1·302	1·293	1·282	1·276	1·273	1·272	1·272	1·275
0·00800	(0·0785)	*	1·390	1·351	1·327	1·311	1·290	1·277	1·269	1·260	1·255	1·253	1·252	1·254	1·257
0·01000	(0·0981)	*	1·363	1·327	1·304	1·289	1·270	1·259	1·251	1·243	1·240	1·238	1·238	1·240	1·244
0·01500	(0·147)	1·340	1·316	1·284	1·265	1·252	1·236	1·226	1·221	1·214	1·212	1·212	1·213	1·217	1·221
0·02000	(0·196)	1·306	1·285	1·256	1·239	1·227	1·213	1·205	1·200	1·195	1·194	1·194	1·196	1·201	1·206
0·03000	(0·294)	1·262	1·243	1·219	1·204	1·194	1·182	1·176	1·172	1·169	1·169	1·171	1·173	1·179	1·186
0·04000	(0·392)	1·233	1·216	1·194	1·180	1·172	1·162	1·156	1·153	1·152	1·152	1·155	1·158	1·165	1·172
0·06000	(0·588)	1·194	1·179	1·160	1·149	1·142	1·134	1·130	1·129	1·129	1·130	1·134	1·138	1·146	1·154
0·08000	(0·785)	1·168	1·155	1·138	1·128	1·122	1·116	1·113	1·112	1·113	1·116	1·120	1·124	1·133	1·141
0·10000	(0·981)	1·149	1·137	1·122	1·113	1·108	1·102	1·100	1·100	1·101	1·105	1·109	1·114	1·124	1·133
0·15000	(1·47)	1·117	1·106	1·093	1·086	1·082	1·079	1·078	1·078	1·082	1·086	1·091	1·097	1·108	1·117
0·20000	(1·96)	1·095	1·086	1·075	1·069	1·065	1·063	1·063	1·064	1·068	1·073	1·079	1·086	1·097	1·107
0·30000	(2·94)	1·066	1·058	1·049	1·045	1·043	1·042	1·043	1·046	1·051	1·057	1·064	1·071	1·083	1·094
0·40000	(3·92)	1·047	1·040	1·033	1·030	1·028	1·028	1·030	1·033	1·039	1·046	1·054	1·061	1·074	1·085
0·60000	(5·88)	1·021	1·016	1·011	1·009	1·009	1·010	1·013	1·017	1·024	1·032	1·040	1·048	1·062	1·074
0·80000	(7·85)	1·005	1·000	0·996	0·995	0·996	0·998	1·002	1·006	1·015	1·023	1·032	1·040	1·055	1·067
1·00000	(9·81)	0·992	0·989	0·986	0·985	0·986	0·990	0·994	0·999	1·007	1·016	1·025	1·034	1·049	1·062
1·50000	(14·7)	0·972	0·969	0·968	0·968	0·970	0·975	0·980	0·985	0·995	1·005	1·015	1·024	1·040	1·054
2·00000	(19·6)	0·958	0·956	0·956	0·957	0·960	0·965	0·971	0·977	0·988	0·997	1·008	1·018	1·034	1·049
3·00000	(29·4)	0·940	0·939	0·940	0·943	0·946	0·953	0·960	0·966	0·978	0·988	1·000	1·010	1·027	1·042
S Grad'nt	$(\Delta p/\rho l)$ k.p.g.	125	150	200	250	300	400	500	600	800	1000	1250	1500	2000	2500

(Equivalent) Pipe diameters in mm — Roughness size, $k_s = 0.060$ mm

Kin. visc., $\nu = 40 \times 10^{-6}$ m^2 s^{-1};
$S = 0.00010$ to 3.00000

i.e. kin. pr. grad., $\Delta p/\rho l =$ (0.00098) to (29.4) ms^{-2}

Roughness size, $k_s = 0.150$ mm
This table shows values of m, as follows

m_C for Colebrook-White solutions; or, where $\mathbf{R} \leq 2000$, m_P for laminar flow

E239

Grad'nt S	k.p.g. $(\Delta p/\rho l)$	125	150	200	250	300	400	500	600	800	1000	1250	1500	2000	2500
0.00010	(0.00098)	8.288	6.499	4.429	3.289	2.579	1.758	*	*	1.732	1.699	1.672	1.653	1.630	1.616
0.00015	(0.00147)	6.767	5.307	3.616	2.685	2.106	1.435	*	1.720	1.674	1.645	1.622	1.606	1.586	1.575
0.00020	(0.00196)	5.860	4.596	3.132	2.326	1.824	*	1.709	1.677	1.635	1.609	1.588	1.574	1.556	1.547
0.00030	(0.00294)	4.785	3.752	2.557	1.899	1.489	*	1.648	1.620	1.584	1.561	1.543	1.531	1.517	1.509
0.00040	(0.00392)	4.144	3.250	2.214	1.644	*	1.644	1.607	1.582	1.549	1.529	1.513	1.502	1.490	1.483
0.00060	(0.00588)	3.383	2.653	1.808	1.343	*	1.586	1.554	1.532	1.503	1.486	1.472	1.464	1.454	1.449
0.00080	(0.00785)	2.930	2.298	1.566	*	*	1.547	1.518	1.498	1.473	1.457	1.445	1.438	1.430	1.426
0.00100	(0.00981)	2.621	2.055	1.400	*	1.562	1.519	1.492	1.473	1.450	1.436	1.425	1.418	1.412	1.409
0.00150	(0.0147)	2.140	1.678	*	1.536	1.507	1.470	1.446	1.430	1.410	1.399	1.390	1.385	1.380	1.379
0.00200	(0.0196)	1.853	1.453	*	1.497	1.471	1.437	1.416	1.401	1.384	1.374	1.366	1.362	1.359	1.359
0.00300	(0.0294)	1.513	1.187	1.479	1.446	1.423	1.393	1.375	1.363	1.348	1.340	1.335	1.332	1.331	1.332
0.00400	(0.0392)	1.310	*	1.442	1.412	1.391	1.364	1.348	1.338	1.325	1.318	1.314	1.312	1.312	1.314
0.00600	(0.0588)	*	*	1.393	1.367	1.349	1.326	1.312	1.304	1.293	1.288	1.286	1.285	1.287	1.290
0.00800	(0.0785)	*	1.399	1.361	1.337	1.321	1.300	1.288	1.281	1.272	1.269	1.267	1.267	1.270	1.275
0.01000	(0.0981)	*	1.372	1.337	1.315	1.300	1.281	1.271	1.264	1.257	1.254	1.253	1.254	1.258	1.263
0.01500	(0.147)	1.351	1.327	1.296	1.277	1.264	1.249	1.240	1.235	1.230	1.229	1.229	1.231	1.237	1.243
0.02000	(0.196)	1.318	1.297	1.269	1.252	1.241	1.227	1.220	1.216	1.212	1.212	1.213	1.216	1.222	1.229
0.03000	(0.294)	1.276	1.257	1.233	1.219	1.209	1.199	1.193	1.190	1.189	1.190	1.192	1.196	1.204	1.212
0.04000	(0.392)	1.248	1.231	1.209	1.197	1.189	1.180	1.175	1.173	1.173	1.175	1.179	1.183	1.192	1.200
0.06000	(0.588)	1.211	1.196	1.178	1.168	1.162	1.155	1.152	1.151	1.153	1.156	1.161	1.166	1.176	1.186
0.08000	(0.785)	1.186	1.174	1.158	1.149	1.143	1.138	1.137	1.137	1.139	1.143	1.149	1.155	1.166	1.176
0.10000	(0.981)	1.169	1.157	1.143	1.135	1.130	1.126	1.125	1.126	1.130	1.134	1.140	1.147	1.158	1.169
0.15000	(1.47)	1.138	1.129	1.117	1.111	1.108	1.106	1.106	1.108	1.113	1.119	1.126	1.133	1.146	1.158
0.20000	(1.96)	1.118	1.110	1.100	1.095	1.093	1.092	1.094	1.096	1.102	1.109	1.117	1.124	1.138	1.150
0.30000	(2.94)	1.093	1.086	1.078	1.075	1.074	1.075	1.078	1.081	1.089	1.096	1.105	1.113	1.128	1.141
0.40000	(3.92)	1.076	1.070	1.064	1.062	1.062	1.064	1.067	1.072	1.080	1.088	1.098	1.106	1.122	1.135
0.60000	(5.88)	1.054	1.050	1.046	1.045	1.046	1.050	1.054	1.059	1.069	1.078	1.088	1.098	1.114	1.128
0.80000	(7.85)	1.040	1.036	1.034	1.034	1.036	1.040	1.046	1.051	1.062	1.071	1.082	1.092	1.109	1.123
1.00000	(9.81)	1.029	1.027	1.025	1.026	1.029	1.034	1.040	1.046	1.057	1.067	1.078	1.088	1.105	1.120
1.50000	(14.7)	1.012	1.011	1.011	1.013	1.016	1.023	1.030	1.037	1.049	1.059	1.071	1.082	1.100	1.115
2.00000	(19.6)	1.001	1.001	1.002	1.005	1.009	1.016	1.024	1.031	1.043	1.055	1.067	1.078	1.096	1.112
3.00000	(29.4)	0.988	0.988	0.991	0.995	0.999	1.008	1.016	1.024	1.037	1.049	1.062	1.073	1.092	1.108
S Grad'nt	$(\Delta p/\rho l)$ k.p.g.	125	150	200	250	300	400	500	600	800	1000	1250	1500	2000	2500

(Equivalent) Pipe diameters in mm — Roughness size, $k_s = 0.150$ mm

---

Kin. visc., $\nu = 40 \times 10^{-6}$ m^2 s^{-1};
$S = 0.00010$ to 3.00000

i.e. kin. pr. grad., $\Delta p/\rho l =$ (0.00098) to (29.4) ms^{-2}

Roughness size, $k_s = 0.30$ mm
This table shows values of m, as follows

m_C for Colebrook-White solutions; or, where $\mathbf{R} \leq 2000$, m_P for laminar flow

E240

Grad'nt S	k.p.g. $(\Delta p/\rho l)$	125	150	200	250	300	400	500	600	800	1000	1250	1500	2000	2500
0.00010	(0.00098)	8.288	6.499	4.429	3.289	2.579	1.758	*	*	1.737	1.704	1.677	1.659	1.636	1.622
0.00015	(0.00147)	6.767	5.307	3.616	2.685	2.106	1.435	*	1.725	1.679	1.651	1.628	1.612	1.592	1.581
0.00020	(0.00196)	5.860	4.596	3.132	2.326	1.824	*	1.714	1.682	1.641	1.615	1.594	1.580	1.563	1.554
0.00030	(0.00294)	4.785	3.752	2.557	1.899	1.489	*	1.654	1.626	1.590	1.568	1.550	1.539	1.525	1.518
0.00040	(0.00392)	4.144	3.250	2.214	1.644	*	1.651	1.614	1.589	1.557	1.537	1.521	1.511	1.499	1.493
0.00060	(0.00588)	3.383	2.653	1.808	1.343	*	1.594	1.562	1.540	1.512	1.495	1.482	1.473	1.464	1.460
0.00080	(0.00785)	2.930	2.298	1.566	*	*	1.556	1.527	1.507	1.482	1.467	1.455	1.448	1.441	1.438
0.00100	(0.00981)	2.621	2.055	1.400	*	1.571	1.528	1.501	1.483	1.460	1.446	1.436	1.430	1.424	1.422
0.00150	(0.0147)	2.140	1.678	*	1.546	1.517	1.480	1.456	1.441	1.422	1.411	1.403	1.398	1.394	1.394
0.00200	(0.0196)	1.853	1.453	*	1.508	1.482	1.448	1.427	1.413	1.396	1.387	1.380	1.377	1.375	1.376
0.00300	(0.0294)	1.513	1.187	1.491	1.458	1.435	1.406	1.388	1.377	1.363	1.355	1.351	1.349	1.349	1.351
0.00400	(0.0392)	1.310	*	1.455	1.425	1.404	1.378	1.362	1.352	1.340	1.335	1.331	1.330	1.332	1.335
0.00600	(0.0588)	*	*	1.407	1.381	1.364	1.342	1.329	1.320	1.311	1.307	1.305	1.306	1.309	1.314
0.00800	(0.0785)	*	1.414	1.376	1.353	1.337	1.317	1.306	1.299	1.292	1.289	1.289	1.290	1.294	1.300
0.01000	(0.0981)	*	1.388	1.353	1.332	1.317	1.300	1.290	1.284	1.278	1.276	1.276	1.278	1.283	1.290
0.01500	(0.147)	1.368	1.345	1.314	1.296	1.284	1.269	1.262	1.257	1.253	1.253	1.255	1.258	1.265	1.273
0.02000	(0.196)	1.337	1.316	1.289	1.273	1.262	1.250	1.243	1.240	1.238	1.238	1.241	1.245	1.253	1.262
0.03000	(0.294)	1.297	1.279	1.256	1.242	1.233	1.224	1.219	1.217	1.217	1.219	1.224	1.228	1.238	1.248
0.04000	(0.392)	1.271	1.254	1.234	1.222	1.215	1.207	1.204	1.203	1.204	1.207	1.212	1.218	1.228	1.239
0.06000	(0.588)	1.237	1.223	1.206	1.196	1.190	1.185	1.183	1.184	1.187	1.191	1.198	1.204	1.216	1.227
0.08000	(0.785)	1.214	1.202	1.187	1.179	1.175	1.171	1.170	1.171	1.176	1.181	1.188	1.195	1.208	1.220
0.10000	(0.981)	1.198	1.187	1.174	1.167	1.163	1.161	1.161	1.163	1.168	1.174	1.182	1.189	1.203	1.215
0.15000	(1.47)	1.171	1.162	1.152	1.147	1.144	1.144	1.145	1.148	1.155	1.162	1.171	1.179	1.193	1.206
0.20000	(1.96)	1.154	1.146	1.137	1.134	1.132	1.133	1.135	1.139	1.147	1.154	1.164	1.172	1.188	1.201
0.30000	(2.94)	1.131	1.125	1.119	1.117	1.117	1.119	1.123	1.127	1.136	1.145	1.155	1.164	1.181	1.195
0.40000	(3.92)	1.117	1.112	1.107	1.106	1.107	1.110	1.115	1.120	1.130	1.139	1.150	1.160	1.176	1.191
0.60000	(5.88)	1.099	1.095	1.093	1.093	1.095	1.100	1.105	1.111	1.122	1.132	1.143	1.153	1.171	1.186
0.80000	(7.85)	1.087	1.085	1.083	1.085	1.087	1.093	1.099	1.105	1.117	1.127	1.139	1.150	1.168	1.183
1.00000	(9.81)	1.079	1.077	1.077	1.079	1.081	1.088	1.095	1.101	1.113	1.124	1.136	1.147	1.166	1.181
1.50000	(14.7)	1.066	1.065	1.066	1.069	1.073	1.080	1.088	1.095	1.108	1.119	1.132	1.143	1.162	1.178
2.00000	(19.6)	1.058	1.057	1.059	1.063	1.067	1.076	1.084	1.091	1.105	1.116	1.129	1.140	1.160	1.176
3.00000	(29.4)	1.047	1.048	1.051	1.056	1.060	1.070	1.078	1.086	1.100	1.113	1.126	1.137	1.157	1.174
S Grad'nt	$(\Delta p/\rho l)$ k.p.g.	125	150	200	250	300	400	500	600	800	1000	1250	1500	2000	2500

(Equivalent) Pipe diameters in mm — Roughness size, $k_s = 0.30$ mm

E241

Kin. visc., $\nu = 50 \times 10^{-6}$ m^{2}s^{-1}; S = 0·00010 to 30·0000

i.e. kin. pr. grad., $\Delta p/\rho l$ = (0·00098) to (294) ms^{-2}

Roughness size, k_s = 0·006 mm

This table shows values of m, as follows

m_C for Colebrook-White solutions; or, where **R** $\leq$ 2000, m_P for laminar flow

Grad'nt S	k.p.g. $(\Delta p/\rho l)$	6	8	10	12·5	15	20	25	30	40	50	60	80	100	125
0·00030	(0·00294)	342·9	233·6	173·5	128·9	101·1	68·86	51·14	40·10	27·33	20·29	15·91	10·84	8·054	5·981
0·00040	(0·00392)	296·9	202·3	150·3	111·6	87·51	59·63	44·29	34·73	23·67	17·58	13·78	9·392	6·975	5·180
0·00060	(0·00588)	242·4	165·2	122·7	91·12	71·45	48·69	36·16	28·36	19·32	14·35	11·25	7·668	5·695	4·229
0·00080	(0·00785)	210·0	143·1	106·3	78·91	61·88	42·17	31·32	24·56	16·73	12·43	9·746	6·641	4·932	3·663
0·00100	(0·00981)	187·8	128·0	95·04	70·58	55·35	37·72	28·01	21·96	14·97	11·12	8·717	5·940	4·411	3·276
0·00150	(0·0147)	153·3	104·5	77·60	57·63	45·19	30·79	22·87	17·93	12·22	9·076	7·117	4·850	3·602	2·675
0·00200	(0·0196)	132·8	90·49	67·20	49·91	39·14	26·67	19·81	15·53	10·58	7·860	6·164	4·200	3·119	2·316
0·00300	(0·0294)	108·4	73·88	54·87	40·75	31·96	21·78	16·17	12·68	8·641	6·418	5·033	3·429	2·547	1·891
0·00400	(0·0392)	93·90	63·98	47·52	35·29	27·67	18·86	14·00	10·98	7·484	5·558	4·358	2·970	2·206	1·638
0·00600	(0·0588)	76·67	52·24	38·80	28·81	22·60	15·40	11·43	8·967	6·110	4·538	3·559	2·425	1·801	1·337
0·00800	(0·0785)	66·40	45·24	33·60	24·95	19·57	13·33	9·903	7·766	5·292	3·930	3·082	2·100	1·560	1·158
0·01000	(0·0981)	59·39	40·47	30·05	22·32	17·50	11·93	8·857	6·946	4·733	3·515	2·756	1·878	1·395	*
0·01500	(0·147)	48·49	33·04	24·54	18·22	14·29	9·738	7·232	5·671	3·865	2·870	2·251	1·534	1·139	*
0·02000	(0·196)	41·99	28·61	21·25	15·78	12·38	8·433	6·263	4·912	3·347	2·486	1·949	1·328	*	1·353
0·03000	(0·294)	34·29	23·36	17·35	12·89	10·11	6·886	5·114	4·010	2·733	2·029	1·591	1·084	*	1·304
0·04000	(0·392)	29·69	20·23	15·03	11·16	8·751	5·963	4·429	3·473	2·367	1·758	1·378	*	1·301	1·271
0·06000	(0·588)	24·24	16·52	12·27	9·112	7·145	4·869	3·616	2·836	1·932	1·435	1·125	*	1·254	1·228
0·08000	(0·785)	21·00	14·31	10·63	7·891	6·188	4·217	3·132	2·456	1·673	1·243	*	1·251	1·222	1·199
0·10000	(0·981)	18·78	12·80	9·504	7·058	5·535	3·772	2·801	2·196	1·497	1·112	*	1·225	1·199	1·178
0·15000	(1·47)	15·33	10·45	7·760	5·763	4·519	3·079	2·287	1·793	1·222	*	*	1·182	1·159	1·141
0·20000	(1·96)	13·28	9·049	6·720	4·991	3·914	2·667	1·981	1·553	1·058	*	1·187	1·154	1·133	1·116
0·30000	(2·94)	10·84	7·388	5·487	4·075	3·196	2·178	1·617	1·268	*	1·166	1·144	1·115	1·097	1·082
0·40000	(3·92)	9·390	6·398	4·752	3·529	2·767	1·886	1·400	1·098	*	1·136	1·116	1·090	1·073	1·060
0·60000	(5·88)	7·667	5·224	3·880	2·881	2·260	1·540	1·143	0·897	1·122	1·096	1·079	1·056	1·042	1·030
0·80000	(7·85)	6·640	4·524	3·360	2·495	1·957	1·333	0·990	*	1·093	1·070	1·053	1·033	1·020	1·010
1·00000	(9·81)	5·939	4·047	3·005	2·232	1·750	1·193	0·886	*	1·071	1·050	1·035	1·016	1·004	0·995
1·50000	(14·7)	4·849	3·304	2·454	1·822	1·429	0·974	*	1·065	1·035	1·016	1·003	0·986	0·976	0·969
2·00000	(19·6)	4·199	2·861	2·125	1·578	1·238	0·843	*	1·037	1·010	0·993	0·981	0·967	0·958	0·951
3·00000	(29·4)	3·429	2·336	1·735	1·289	1·011	*	1·020	1·001	0·977	0·963	0·953	0·940	0·933	0·927
4·00000	(39·2)	2·969	2·023	1·503	1·116	0·875	*	0·994	0·977	0·956	0·942	0·933	0·922	0·916	0·911
6·00000	(58·8)	2·424	1·652	1·227	0·911	*	0·981	0·960	0·945	0·927	0·915	0·907	0·898	0·893	0·890
8·00000	(78·5)	2·100	1·431	1·063	0·789	*	0·956	0·937	0·924	0·907	0·897	0·890	0·882	0·878	0·875
10·0000	(98·1)	1·878	1·280	0·950	*	*	0·938	0·920	0·908	0·893	0·883	0·877	0·870	0·866	0·864
15·0000	(147)	1·533	1·045	0·776	*	0·932	0·907	0·891	0·881	0·868	0·860	0·855	0·849	0·846	0·845
20·0000	(196)	1·328	0·905	*	0·926	0·909	0·886	0·872	0·863	0·851	0·844	0·840	0·835	0·833	0·832
30·0000	(294)	1·084	0·739	*	0·893	0·878	0·859	0·846	0·838	0·828	0·823	0·819	0·816	0·815	0·815
S	$(\Delta p/\rho l)$	6	8	10	12·5	15	20	25	30	40	50	60	80	100	125

Grad'nt k.p.g. **(Equivalent) Pipe diameters in mm**

S	$(\Delta p/\rho l)$	125	150	200	250	300	400	500	600	800	1000	1250	1500	2000	2500
0·00010	(0·00098)	10·36	8·124	5·536	4·111	3·224	2·197	1·632	*	1·797	1·759	1·727	1·705	1·677	1·660
0·00015	(0·00147)	8·459	6·633	4·520	3·357	2·632	1·794	*	*	1·734	1·700	1·672	1·653	1·629	1·615
0·00020	(0·00196)	7·325	5·745	3·914	2·907	2·280	1·553	*	1·740	1·691	1·661	1·636	1·618	1·597	1·584
0·00030	(0·00294)	5·981	4·690	3·196	2·374	1·861	*	1·710	1·677	1·635	1·608	1·587	1·572	1·554	1·543
0·00040	(0·00392)	5·180	4·062	2·768	2·056	1·612	*	1·665	1·636	1·597	1·573	1·553	1·540	1·524	1·515
0·00060	(0·00588)	4·229	3·317	2·260	1·678	1·316	1·644	1·606	1·581	1·547	1·526	1·509	1·498	1·485	1·478
0·00080	(0·00785)	3·663	2·872	1·957	1·454	*	1·601	1·567	1·544	1·513	1·494	1·479	1·469	1·458	1·452
0·00100	(0·00981)	3·276	2·569	1·751	1·300	*	1·570	1·538	1·516	1·488	1·471	1·457	1·448	1·438	1·433
0·00150	(0·0147)	2·675	2·098	1·429	*	1·559	1·515	1·487	1·468	1·444	1·429	1·418	1·410	1·402	1·399
0·00200	(0·0196)	2·316	1·817	1·238	*	1·519	1·479	1·454	1·436	1·415	1·401	1·391	1·385	1·379	1·376
0·00300	(0·0294)	1·891	1·483	*	1·493	1·466	1·431	1·409	1·394	1·375	1·364	1·356	1·351	1·346	1·345
0·00400	(0·0392)	1·638	1·285	*	1·455	1·430	1·398	1·378	1·365	1·348	1·339	1·332	1·328	1·324	1·324
0·00600	(0·0588)	1·337	*	1·435	1·404	1·383	1·355	1·338	1·326	1·312	1·304	1·299	1·296	1·294	1·295
0·00800	(0·0785)	1·158	*	1·399	1·370	1·351	1·326	1·310	1·300	1·288	1·281	1·277	1·275	1·274	1·275
0·01000	(0·0981)	*	*	1·372	1·345	1·327	1·304	1·290	1·281	1·270	1·264	1·260	1·259	1·259	1·261
0·01500	(0·147)	*	1·362	1·325	1·302	1·286	1·266	1·254	1·247	1·238	1·233	1·231	1·230	1·232	1·235
0·02000	(0·196)	1·353	1·327	1·294	1·273	1·259	1·241	1·230	1·224	1·216	1·213	1·211	1·211	1·213	1·217
0·03000	(0·294)	1·304	1·281	1·252	1·234	1·222	1·207	1·198	1·193	1·187	1·185	1·184	1·185	1·188	1·193
0·04000	(0·392)	1·271	1·251	1·225	1·208	1·197	1·184	1·176	1·172	1·167	1·166	1·166	1·167	1·171	1·176
0·06000	(0·588)	1·228	1·210	1·187	1·173	1·164	1·153	1·147	1·143	1·140	1·140	1·141	1·143	1·148	1·154
0·08000	(0·785)	1·199	1·183	1·162	1·150	1·142	1·132	1·127	1·124	1·122	1·122	1·124	1·127	1·132	1·138
0·10000	(0·981)	1·178	1·163	1·144	1·132	1·125	1·116	1·112	1·109	1·108	1·109	1·111	1·114	1·121	1·127
0·15000	(1·47)	1·141	1·128	1·111	1·102	1·096	1·089	1·085	1·084	1·084	1·086	1·089	1·092	1·100	1·106
0·20000	(1·96)	1·116	1·104	1·090	1·081	1·076	1·070	1·068	1·067	1·068	1·070	1·074	1·077	1·085	1·092
0·30000	(2·94)	1·082	1·073	1·060	1·053	1·049	1·045	1·044	1·044	1·046	1·049	1·053	1·057	1·066	1·074
0·40000	(3·92)	1·060	1·051	1·040	1·034	1·031	1·028	1·027	1·028	1·030	1·034	1·039	1·043	1·052	1·061
0·60000	(5·88)	1·030	1·023	1·014	1·009	1·007	1·005	1·005	1·006	1·010	1·014	1·020	1·025	1·034	1·043
0·80000	(7·85)	1·010	1·003	0·996	0·992	0·990	0·989	0·990	0·992	0·996	1·001	1·006	1·012	1·022	1·031
1·00000	(9·81)	0·995	0·989	0·982	0·979	0·978	0·977	0·979	0·981	0·985	0·990	0·997	1·002	1·013	1·022
1·50000	(14·7)	0·969	0·964	0·959	0·957	0·956	0·957	0·959	0·961	0·967	0·973	0·979	0·986	0·997	1·006
2·00000	(19·6)	0·951	0·947	0·943	0·941	0·941	0·943	0·945	0·948	0·955	0·961	0·968	0·974	0·986	0·996
3·00000	(29·4)	0·927	0·924	0·921	0·921	0·921	0·924	0·927	0·931	0·938	0·944	0·952	0·959	0·971	0·981
S	$(\Delta p/\rho l)$	125	150	200	250	300	400	500	600	800	1000	1250	1500	2000	2500

Grad'nt k.p.g. **(Equivalent) Pipe diameters in mm**

Kinematic viscosity, $\nu = 50 \times 10^{-6}$ m^{2}s^{-1} ;

Roughness size, k_s = 0·006 mm

Kin. visc., $\nu = 50 \times 10^{-6}\ \text{m}^2\text{s}^{-1}$;
$S = 0{\cdot}00010$ to $30{\cdot}0000$
i.e. kin. pr. grad., $\Delta p/\rho l =$
$(0{\cdot}00098)$ to (294) ms^{-2}

Roughness size, $k_s = 0{\cdot}015$ mm
This table shows values of m, as follows
m_C for Colebrook-White solutions; or,
where $R \leq 2000$, m_P for laminar flow

Grad'nt	k.p.g.	(Equivalent) Pipe diameters in mm													
S	$(\Delta p/\rho l)$	6	8	10	12·5	15	20	25	30	40	50	60	80	100	125
0·00030	(0·00294)	342·9	233·6	173·5	128·9	101·1	68·86	51·14	40·10	27·33	20·29	15·91	10·84	8·054	5·981
0·00040	(0·00392)	296·9	202·3	150·3	111·6	87·51	59·63	44·29	34·73	23·67	17·58	13·78	9·392	6·975	5·180
0·00060	(0·00588)	242·4	165·2	122·7	91·12	71·45	48·69	36·16	28·36	19·32	14·35	11·25	7·668	5·695	4·229
0·00080	(0·00785)	210·0	143·1	106·3	78·91	61·88	42·17	31·32	24·56	16·73	12·43	9·746	6·641	4·932	3·663
0·00100	(0·00981)	187·8	128·0	95·04	70·58	55·35	37·72	28·01	21·96	14·97	11·12	8·717	5·940	4·411	3·276
0·00150	(0·0147)	153·3	104·5	77·60	57·63	45·19	30·79	22·87	17·93	12·22	9·076	7·117	4·850	3·602	2·675
0·00200	(0·0196)	132·8	90·49	67·20	49·91	39·14	26·67	19·81	15·53	10·58	7·860	6·164	4·200	3·119	2·316
0·00300	(0·0294)	108·4	73·88	54·87	40·75	31·96	21·78	16·17	12·68	8·641	6·418	5·033	3·429	2·547	1·891
0·00400	(0·0392)	93·90	63·98	47·52	35·29	27·67	18·86	14·00	10·98	7·484	5·558	4·358	2·970	2·206	1·638
0·00600	(0·0588)	76·67	52·24	38·80	28·81	22·60	15·40	11·43	8·967	6·110	4·538	3·559	2·425	1·801	1·337
0·00800	(0·0785)	66·40	45·24	33·60	24·95	19·57	13·33	9·903	7·766	5·292	3·930	3·082	2·100	1·560	1·158
0·01000	(0·0981)	59·39	40·47	30·05	22·32	17·50	11·93	8·857	6·946	4·733	3·515	2·756	1·878	1·395	*
0·01500	(0·147)	48·49	33·04	24·54	18·22	14·29	9·738	7·232	5·671	3·865	2·870	2·251	1·534	1·139	*
0·02000	(0·196)	41·99	28·61	21·25	15·78	12·38	8·433	6·263	4·912	3·347	2·486	1·949	1·328	*	1·354
0·03000	(0·294)	34·29	23·36	17·35	12·89	10·11	6·886	5·114	4·010	2·733	2·029	1·591	1·084	*	1·305
0·04000	(0·392)	29·69	20·23	15·03	11·16	8·751	5·963	4·429	3·473	2·367	1·758	1·378	*	1·302	1·273
0·06000	(0·588)	24·24	16·52	12·27	9·112	7·145	4·869	3·616	2·836	1·932	1·435	1·125	*	1·255	1·230
0·08000	(0·785)	21·00	14·31	10·63	7·891	6·188	4·217	3·132	2·456	1·673	1·243	*	1·252	1·224	1·201
0·10000	(0·981)	18·78	12·80	9·504	7·058	5·535	3·772	2·801	2·196	1·497	1·112	*	1·227	1·201	1·179
0·15000	(1·47)	15·33	10·45	7·760	5·763	4·519	3·079	2·287	1·793	1·222	*	*	1·184	1·161	1·143
0·20000	(1·96)	13·28	9·049	6·720	4·991	3·914	2·667	1·981	1·553	1·058	*	1·189	1·156	1·135	1·118
0·30000	(2·94)	10·84	7·388	5·487	4·075	3·196	2·178	1·617	1·268	*	1·169	1·146	1·118·	1·100	1·085
0·40000	(3·92)	9·390	6·398	4·752	3·529	2·767	1·886	1·400	1·098	*	1·139	1·118	1·092	1·076	1·063
0·60000	(5·88)	7·667	5·224	3·880	2·881	2·260	1·540	1·143	0·897	1·124	1·099	1·081	1·059	1·045	1·034
0·80000	(7·85)	6·640	4·524	3·360	2·495	1·957	1·333	0·990	*	1·096	1·073	1·057	1·036	1·024	1·014
1·00000	(9·81)	5·939	4·047	3·005	2·232	1·750	1·193	0·886	*	1·075	1·053	1·038	1·019	1·008	0·999
1·50000	(14·7)	4·849	3·304	2·454	1·822	1·429	0·974	*	1·068	1·038	1·019	1·007	0·990	0·981	0·973
2·00000	(19·6)	4·199	2·861	2·125	1·578	1·238	0·843	*	1·041	1·014	0·997	0·986	0·971	0·963	0·956
3·00000	(29·4)	3·429	2·336	1·735	1·289	1·011	*	1·024	1·006	0·982	0·967	0·957	0·945	0·938	0·933
4·00000	(39·2)	2·969	2·023	1·503	1·116	0·875	*	0·998	0·982	0·961	0·947	0·939	0·928	0·922	0·918
6·00000	(58·8)	2·424	1·652	1·227	0·911	*	0·986	0·965	0·951	0·932	0·921	0·914	0·905	0·900	0·897
8·00000	(78·5)	2·100	1·431	1·063	0·789	*	0·962	0·943	0·930	0·913	0·903	0·897	0·889	0·886	0·883
10·0000	(98·1)	1·878	1·280	0·950	*	*	0·944	0·927	0·915	0·899	0·890	0·885	0·878	0·875	0·873
15·0000	(147)	1·533	1·045	0·776	*	0·938	0·914	0·899	0·888	0·875	0·868	0·863	0·858	0·856	0·855
20·0000	(196)	1·328	0·905	*	0·933	0·916	0·894	0·880	0·871	0·859	0·853	0·849	0·845	0·844	0·844
30·0000	(294)	1·084	0·739	*	0·901	0·886	0·867	0·855	0·848	0·838	0·833	0·830	0·828	0·827	0·828
S	$(\Delta p/\rho l)$	6	8	10	12·5	15	20	25	30	40	50	60	80	100	125

Grad'nt   k.p.g.   (Equivalent) Pipe diameters in mm

S	$(\Delta p/\rho l)$	125	150	200	250	300	400	500	600	800	1000	1250	1500	2000	2500
0·00010	(0·00098)	10·36	8·124	5·536	4·111	3·224	2·197	1·632	*	1·797	1·759	1·727	1·705	1·677	1·660
0·00015	(0·00147)	8·459	6·633	4·520	3·357	2·632	1·794	*	*	1·734	1·700	1·673	1·654	1·630	1·615
0·00020	(0·00196)	7·325	5·745	3·914	2·907	2·280	1·553	*	1·740	1·692	1·661	1·636	1·619	1·597	1·585
0·00030	(0·00294)	5·981	4·690	3·196	2·374	1·861	*	1·710	1·678	1·635	1·609	1·587	1·572	1·554	1·544
0·00040	(0·00392)	5·180	4·062	2·768	2·056	1·612	*	1·666	1·636	1·598	1·574	1·554	1·541	1·525	1·516
0·00060	(0·00588)	4·229	3·317	2·260	1·678	1·316	1·644	1·607	1·581	1·547	1·526	1·510	1·499	1·485	1·478
0·00080	(0·00785)	3·663	2·872	1·957	1·454	*	1·602	1·568	1·544	1·514	1·495	1·480	1·470	1·458	1·453
0·00100	(0·00981)	3·276	2·569	1·751	1·300	*	1·570	1·538	1·516	1·488	1·471	1·457	1·449	1·438	1·433
0·00150	(0·0147)	2·675	2·098	1·429	*	1·560	1·516	1·488	1·469	1·445	1·430	1·418	1·411	1·403	1·400
0·00200	(0·0196)	2·316	1·817	1·238	*	1·520	1·479	1·454	1·437	1·415	1·402	1·392	1·386	1·379	1·377
0·00300	(0·0294)	1·891	1·483	*	1·493	1·466	1·431	1·409	1·394	1·376	1·365	1·357	1·352	1·347	1·346
0·00400	(0·0392)	1·638	1·285	*	1·455	1·431	1·399	1·379	1·366	1·349	1·340	1·333	1·329	1·325	1·325
0·00600	(0·0588)	1·337	*	1·436	1·405	1·383	1·356	1·339	1·327	1·313	1·306	1·300	1·297	1·296	1·296
0·00800	(0·0785)	1·158	*	1·399	1·371	1·352	1·327	1·311	1·301	1·289	1·282	1·278	1·276	1·275	1·277
0·01000	(0·0981)	*	*	1·372	1·346	1·328	1·305	1·291	1·282	1·271	1·265	1·262	1·260	1·260	1·262
0·01500	(0·147)	*	1·363	1·326	1·303	1·288	1·268	1·256	1·248	1·239	1·235	1·233	1·232	1·234	1·237
0·02000	(0·196)	1·354	1·328	1·295	1·274	1·260	1·242	1·232	1·225	1·218	1·214	1·213	1·213	1·216	1·219
0·03000	(0·294)	1·305	1·283	1·254	1·236	1·224	1·208	1·200	1·194	1·189	1·187	1·186	1·187	1·191	1·195
0·04000	(0·392)	1·273	1·252	1·226	1·210	1·199	1·186	1·178	1·174	1·169	1·168	1·168	1·170	1·174	1·179
0·06000	(0·588)	1·230	1·212	1·189	1·175	1·166	1·155	1·149	1·145	1·143	1·142	1·144	1·146	1·152	1·157
0·08000	(0·785)	1·201	1·185	1·164	1·152	1·144	1·134	1·129	1·126	1·125	1·125	1·127	1·130	1·136	1·142
0·10000	(0·981)	1·179	1·165	1·146	1·134	1·127	1·118	1·114	1·112	1·111	1·112	1·115	1·118	1·125	1·131
0·15000	(1·47)	1·143	1·130	1·114	1·104	1·098	1·091	1·088	1·087	1·088	1·090	1·093	1·097	1·104	1·111
0·20000	(1·96)	1·118	1·106	1·092	1·084	1·079	1·073	1·071	1·070	1·072	1·074	1·078	1·082	1·090	1·098
0·30000	(2·94)	1·085	1·075	1·063	1·056	1·052	1·048	1·047	1·048	1·050	1·053	1·058	1·063	1·072	1·080
0·40000	(3·92)	1·063	1·054	1·044	1·038	1·035	1·032	1·031	1·032	1·035	1·039	1·044	1·050	1·059	1·068
0·60000	(5·88)	1·034	1·026	1·018	1·013	1·011	1·009	1·010	1·011	1·016	1·020	1·026	1·032	1·042	1·052
0·80000	(7·85)	1·014	1·007	1·000	0·996	0·995	0·994	0·995	0·997	1·002	1·008	1·014	1·020	1·031	1·041
1·00000	(9·81)	0·999	0·993	0·987	0·984	0·983	0·983	0·985	0·987	0·992	0·998	1·005	1·011	1·022	1·032
1·50000	(14·7)	0·973	0·969	0·964	0·962	0·962	0·963	0·966	0·969	0·975	0·981	0·989	0·995	1·008	1·018
2·00000	(19·6)	0·956	0·952	0·949	0·947	0·948	0·950	0·953	0·956	0·963	0·970	0·978	0·985	0·998	1·009
3·00000	(29·4)	0·933	0·930	0·928	0·928	0·929	0·932	0·936	0·940	0·948	0·955	0·964	0·971	0·985	0·996
S	$(\Delta p/\rho l)$	125	150	200	250	300	400	500	600	800	1000	1250	1500	2000	2500

Grad'nt   k.p.g.   (Equivalent) Pipe diameters in mm

Kinematic viscosity, $\nu = 50 \times 10^{-6}\ \text{m}^2\text{s}^{-1}$;   **Roughness size, $k_s = 0{\cdot}015$ mm**

E243

Kin. visc., $\nu = 50\times10^{-6}$ m^2 s^{-1};
$S = 0\cdot00010$ to $3\cdot00000$

i.e. kin. pr. grad., $\Delta p/\rho l =$
($0\cdot00098$) to ($29\cdot4$) ms^{-2}

Roughness size, $k_s = 0\cdot030$ mm
This table shows values of m, as follows

m_C for Colebrook-White solutions; or,
where $R \le 2000$, m_P for laminar flow

Grad'nt S	k.p.g. ($\Delta p/\rho l$)	(Equivalent) Pipe diameters in mm													
		125	150	200	250	300	400	500	600	800	1000	1250	1500	2000	2500
0·00010	(0·00098)	10·36	8·124	5·536	4·111	3·224	2·197	1·632	*	1·798	1·760	1·728	1·706	1·678	1·661
0·00015	(0·00147)	8·459	6·633	4·520	3·357	2·632	1·794	*	*	1·734	1·701	1·673	1·654	1·630	1·616
0·00020	(0·00196)	7·325	5·745	3·914	2·907	2·280	1·553	*	1·741	1·692	1·662	1·637	1·619	1·598	1·585
0·00030	(0·00294)	5·981	4·690	3·196	2·374	1·861	*	1·711	1·678	1·636	1·609	1·588	1·573	1·555	1·544
0·00040	(0·00392)	5·180	4·062	2·768	2·056	1·612	*	1·666	1·637	1·598	1·574	1·555	1·541	1·525	1·517
0·00060	(0·00588)	4·229	3·317	2·260	1·678	1·316	1·645	1·608	1·582	1·548	1·527	1·510	1·499	1·486	1·479
0·00080	(0·00785)	3·663	2·872	1·957	1·454	*	1·602	1·568	1·545	1·514	1·496	1·481	1·471	1·459	1·454
0·00100	(0·00981)	3·276	2·569	1·751	1·300	*	1·571	1·539	1·517	1·489	1·472	1·458	1·450	1·439	1·435
0·00150	(0·0147)	2·675	2·098	1·429	*	1·561	1·517	1·489	1·470	1·446	1·431	1·420	1·412	1·405	1·401
0·00200	(0·0196)	2·316	1·817	1·238	*	1·521	1·480	1·455	1·438	1·416	1·403	1·393	1·387	1·381	1·379
0·00300	(0·0294)	1·891	1·483	*	1·494	1·467	1·432	1·411	1·396	1·377	1·366	1·358	1·353	1·349	1·348
0·00400	(0·0392)	1·638	1·285	*	1·456	1·432	1·400	1·380	1·367	1·351	1·341	1·334	1·330	1·327	1·327
0·00600	(0·0588)	1·337	*	1·437	1·406	1·385	1·357	1·340	1·329	1·315	1·307	1·302	1·299	1·298	1·299
0·00800	(0·0785)	1·158	*	1·401	1·373	1·353	1·328	1·313	1·303	1·291	1·284	1·280	1·278	1·278	1·280
0·01000	(0·0981)	*	*	1·374	1·348	1·330	1·307	1·293	1·284	1·273	1·267	1·264	1·263	1·263	1·265
0·01500	(0·147)	*	1·364	1·328	1·305	1·289	1·270	1·258	1·250	1·242	1·237	1·235	1·235	1·237	1·240
0·02000	(0·196)	1·355	1·330	1·297	1·276	1·262	1·245	1·234	1·228	1·220	1·217	1·216	1·216	1·219	1·223
0·03000	(0·294)	1·307	1·285	1·256	1·238	1·226	1·211	1·202	1·197	1·192	1·190	1·190	1·191	1·195	1·200
0·04000	(0·392)	1·275	1·255	1·228	1·212	1·202	1·188	1·181	1·177	1·173	1·171	1·172	1·174	1·179	1·184
0·06000	(0·588)	1·232	1·214	1·192	1·178	1·169	1·158	1·152	1·149	1·147	1·147	1·148	1·151	1·157	1·163
0·08000	(0·785)	1·204	1·188	1·167	1·155	1·147	1·138	1·133	1·130	1·129	1·130	1·132	1·135	1·142	1·149
0·10000	(0·981)	1·182	1·168	1·149	1·138	1·131	1·122	1·118	1·116	1·116	1·117	1·120	1·124	1·131	1·138
0·15000	(1·47)	1·146	1·133	1·117	1·108	1·102	1·096	1·093	1·092	1·093	1·095	1·099	1·103	1·112	1·119
0·20000	(1·96)	1·122	1·110	1·096	1·088	1·083	1·078	1·076	1·076	1·078	1·081	1·085	1·090	1·099	1·107
0·30000	(2·94)	1·089	1·080	1·068	1·061	1·058	1·054	1·053	1·054	1·057	1·061	1·066	1·071	1·081	1·090
0·40000	(3·92)	1·068	1·059	1·049	1·043	1·040	1·038	1·038	1·039	1·043	1·048	1·053	1·059	1·070	1·079
0·60000	(5·88)	1·039	1·032	1·024	1·020	1·018	1·017	1·018	1·020	1·025	1·030	1·037	1·043	1·054	1·064
0·80000	(7·85)	1·020	1·014	1·007	1·003	1·002	1·002	1·004	1·007	1·012	1·018	1·025	1·032	1·044	1·055
1·00000	(9·81)	1·005	1·000	0·994	0·992	0·991	0·992	0·994	0·997	1·003	1·009	1·017	1·024	1·037	1·048
1·50000	(14·7)	0·981	0·976	0·972	0·971	0·971	0·973	0·976	0·980	0·987	0·994	1·003	1·010	1·024	1·035
2·00000	(19·6)	0·964	0·961	0·958	0·957	0·958	0·961	0·965	0·969	0·977	0·985	0·993	1·001	1·015	1·027
3·00000	(29·4)	0·942	0·940	0·938	0·939	0·940	0·945	0·949	0·954	0·963	0·972	0·981	0·989	1·004	1·017
S	($\Delta p/\rho l$)	125	150	200	250	300	400	500	600	800	1000	1250	1500	2000	2500

Grad'nt k.p.g. (Equivalent) Pipe diameters in mm **Roughness size, $k_s = 0\cdot030$ mm**

E244

Kin. visc., $\nu = 50\times10^{-6}$ m^2 s^{-1};
$S = 0\cdot00010$ to $3\cdot00000$

i.e. kin. pr. grad., $\Delta p/\rho l =$
($0\cdot00098$) to ($29\cdot4$) ms^{-2}

Roughness size, $k_s = 0\cdot060$ mm
This table shows values of m, as follows

m_C for Colebrook-White solutions; or,
where $R \le 2000$, m_P for laminar flow

Grad'nt S	k.p.g. ($\Delta p/\rho l$)	(Equivalent) Pipe diameters in mm													
		125	150	200	250	300	400	500	600	800	1000	1250	1500	2000	2500
0·00010	(0·00098)	10·36	8·124	5·536	4·111	3·224	2·197	1·632	*	1·799	1·760	1·729	1·707	1·679	1·662
0·00015	(0·00147)	8·459	6·633	4·520	3·357	2·632	1·794	*	*	1·735	1·702	1·674	1·655	1·631	1·617
0·00020	(0·00196)	7·325	5·745	3·914	2·907	2·280	1·553	*	1·742	1·693	1·663	1·638	1·621	1·599	1·587
0·00030	(0·00294)	5·981	4·690	3·196	2·374	1·861	*	1·712	1·679	1·637	1·611	1·589	1·574	1·556	1·546
0·00040	(0·00392)	5·180	4·062	2·768	2·056	1·612	*	1·668	1·638	1·600	1·576	1·556	1·543	1·527	1·518
0·00060	(0·00588)	4·229	3·317	2·260	1·678	1·316	1·646	1·609	1·583	1·550	1·529	1·512	1·501	1·488	1·481
0·00080	(0·00785)	3·663	2·872	1·957	1·454	*	1·604	1·570	1·546	1·516	1·497	1·482	1·473	1·462	1·456
0·00100	(0·00981)	3·276	2·569	1·751	1·300	*	1·572	1·541	1·519	1·491	1·474	1·460	1·452	1·442	1·437
0·00150	(0·0147)	2·675	2·098	1·429	*	1·563	1·518	1·491	1·472	1·448	1·433	1·422	1·415	1·407	1·404
0·00200	(0·0196)	2·316	1·817	1·238	*	1·522	1·482	1·457	1·440	1·419	1·406	1·396	1·390	1·384	1·382
0·00300	(0·0294)	1·891	1·483	*	1·496	1·470	1·435	1·413	1·398	1·380	1·369	1·361	1·356	1·352	1·351
0·00400	(0·0392)	1·638	1·285	*	1·459	1·434	1·403	1·383	1·370	1·353	1·344	1·337	1·334	1·331	1·331
0·00600	(0·0588)	1·337	*	1·440	1·409	1·388	1·360	1·343	1·332	1·318	1·311	1·306	1·303	1·302	1·304
0·00800	(0·0785)	1·158	*	1·404	1·376	1·356	1·331	1·316	1·306	1·295	1·288	1·284	1·283	1·283	1·285
0·01000	(0·0981)	*	*	1·377	1·351	1·333	1·310	1·296	1·287	1·277	1·272	1·268	1·267	1·268	1·271
0·01500	(0·147)	*	1·368	1·331	1·309	1·293	1·273	1·262	1·255	1·246	1·242	1·241	1·241	1·243	1·247
0·02000	(0·196)	1·359	1·334	1·301	1·280	1·266	1·249	1·239	1·232	1·226	1·223	1·222	1·222	1·226	1·230
0·03000	(0·294)	1·311	1·289	1·260	1·242	1·231	1·216	1·208	1·203	1·198	1·196	1·197	1·198	1·203	1·208
0·04000	(0·392)	1·279	1·259	1·233	1·217	1·207	1·194	1·187	1·183	1·179	1·179	1·180	1·182	1·187	1·193
0·06000	(0·588)	1·237	1·220	1·197	1·184	1·175	1·164	1·159	1·156	1·154	1·155	1·157	1·160	1·167	1·174
0·08000	(0·785)	1·209	1·193	1·173	1·161	1·153	1·145	1·140	1·138	1·137	1·139	1·142	1·145	1·153	1·161
0·10000	(0·981)	1·188	1·174	1·155	1·145	1·138	1·130	1·126	1·125	1·125	1·127	1·131	1·135	1·143	1·151
0·15000	(1·47)	1·153	1·140	1·125	1·116	1·110	1·105	1·102	1·102	1·104	1·107	1·111	1·116	1·125	1·134
0·20000	(1·96)	1·129	1·118	1·104	1·097	1·092	1·088	1·086	1·087	1·089	1·093	1·098	1·103	1·114	1·123
0·30000	(2·94)	1·098	1·088	1·077	1·071	1·068	1·065	1·065	1·066	1·070	1·075	1·081	1·087	1·098	1·109
0·40000	(3·92)	1·077	1·069	1·059	1·054	1·052	1·050	1·051	1·053	1·058	1·063	1·070	1·076	1·088	1·099
0·60000	(5·88)	1·049	1·043	1·035	1·032	1·030	1·030	1·032	1·035	1·041	1·047	1·055	1·062	1·075	1·087
0·80000	(7·85)	1·031	1·025	1·019	1·017	1·016	1·017	1·020	1·023	1·030	1·037	1·045	1·053	1·067	1·079
1·00000	(9·81)	1·018	1·013	1·008	1·006	1·006	1·008	1·011	1·015	1·022	1·030	1·038	1·046	1·061	1·073
1·50000	(14·7)	0·995	0·991	0·988	0·987	0·988	0·991	0·995	1·000	1·009	1·017	1·026	1·035	1·050	1·063
2·00000	(19·6)	0·979	0·977	0·975	0·975	0·976	0·981	0·985	0·990	1·000	1·009	1·019	1·028	1·044	1·057
3·00000	(29·4)	0·960	0·958	0·957	0·959	0·961	0·967	0·973	0·978	0·989	0·998	1·009	1·019	1·035	1·049
S	($\Delta p/\rho l$)	125	150	200	250	300	400	500	600	800	1000	1250	1500	2000	2500

Grad'nt k.p.g. (Equivalent) Pipe diameters in mm **Roughness size, $k_s = 0\cdot060$ mm**

Kin. visc., $\nu = 50 \times 10^{-6}$ m^{2}s^{-1};
$S = 0.00010$ to 3.00000

i.e. kin. pr. grad., $\Delta p/\rho l =$ (0.00098) to (29.4) ms^{-2}

Roughness size, $k_s = 0.150$ mm
This table shows values of m, as follows
m_C for Colebrook-White solutions; or, where $\mathbf{R} \le 2000$, m_P for laminar flow

E245

Grad'nt S	k.p.g. $(\Delta p/\rho l)$	125	150	200	250	300	400	500	600	800	1000	1250	1500	2000	2500
0.00010	(0.00098)	10.36	8.124	5.536	4.111	3.224	2.197	1.632	*	1.801	1.763	1.731	1.709	1.682	1.665
0.00015	(0.00147)	8.459	6.633	4.520	3.357	2.632	1.794	*	*	1.738	1.705	1.677	1.658	1.635	1.621
0.00020	(0.00196)	7.325	5.745	3.914	2.907	2.280	1.553	*	1.745	1.696	1.666	1.641	1.624	1.603	1.591
0.00030	(0.00294)	5.981	4.690	3.196	2.374	1.861	*	1.715	1.683	1.641	1.614	1.593	1.578	1.560	1.550
0.00040	(0.00392)	5.180	4.062	2.768	2.056	1.612	*	1.671	1.642	1.603	1.580	1.560	1.547	1.532	1.523
0.00060	(0.00588)	4.229	3.317	2.260	1.678	1.316	1.650	1.613	1.587	1.554	1.533	1.517	1.506	1.494	1.487
0.00080	(0.00785)	3.663	2.872	1.957	1.454	*	1.608	1.574	1.551	1.521	1.502	1.488	1.478	1.468	1.462
0.00100	(0.00981)	3.276	2.569	1.751	1.300	*	1.577	1.546	1.524	1.496	1.479	1.466	1.458	1.448	1.444
0.00150	(0.0147)	2.675	2.098	1.429	*	1.568	1.524	1.496	1.478	1.454	1.440	1.429	1.422	1.415	1.412
0.00200	(0.0196)	2.316	1.817	1.238	*	1.528	1.488	1.464	1.447	1.425	1.413	1.403	1.398	1.392	1.391
0.00300	(0.0294)	1.891	1.483	*	1.503	1.476	1.441	1.420	1.405	1.387	1.377	1.370	1.365	1.362	1.362
0.00400	(0.0392)	1.638	1.285	*	1.466	1.441	1.410	1.391	1.378	1.362	1.353	1.347	1.344	1.342	1.343
0.00600	(0.0588)	1.337	*	1.447	1.417	1.396	1.368	1.352	1.341	1.328	1.321	1.317	1.315	1.315	1.317
0.00800	(0.0785)	1.158	*	1.412	1.384	1.365	1.341	1.326	1.317	1.305	1.300	1.297	1.296	1.297	1.300
0.01000	(0.0981)	*	*	1.386	1.360	1.342	1.320	1.307	1.298	1.289	1.284	1.282	1.281	1.283	1.287
0.01500	(0.147)	*	1.377	1.341	1.319	1.304	1.285	1.274	1.267	1.260	1.257	1.256	1.256	1.260	1.265
0.02000	(0.196)	1.369	1.344	1.312	1.292	1.278	1.261	1.252	1.246	1.240	1.238	1.238	1.240	1.245	1.250
0.03000	(0.294)	1.323	1.301	1.273	1.255	1.244	1.230	1.223	1.218	1.215	1.214	1.216	1.218	1.224	1.231
0.04000	(0.392)	1.292	1.272	1.247	1.231	1.221	1.210	1.203	1.200	1.198	1.198	1.200	1.204	1.211	1.218
0.06000	(0.588)	1.252	1.234	1.213	1.200	1.192	1.182	1.178	1.176	1.175	1.177	1.181	1.185	1.193	1.202
0.08000	(0.785)	1.225	1.209	1.190	1.179	1.172	1.164	1.161	1.160	1.160	1.163	1.168	1.172	1.182	1.191
0.10000	(0.981)	1.205	1.191	1.174	1.164	1.157	1.151	1.148	1.148	1.150	1.153	1.158	1.163	1.174	1.183
0.15000	(1.47)	1.172	1.160	1.145	1.137	1.133	1.128	1.127	1.128	1.131	1.136	1.142	1.148	1.160	1.170
0.20000	(1.96)	1.150	1.139	1.127	1.120	1.116	1.113	1.113	1.115	1.119	1.125	1.132	1.138	1.151	1.162
0.30000	(2.94)	1.121	1.113	1.103	1.098	1.095	1.094	1.096	1.098	1.104	1.110	1.118	1.126	1.139	1.151
0.40000	(3.92)	1.102	1.095	1.087	1.083	1.081	1.082	1.084	1.087	1.094	1.101	1.110	1.118	1.132	1.145
0.60000	(5.88)	1.078	1.072	1.066	1.064	1.064	1.065	1.069	1.073	1.081	1.089	1.099	1.107	1.123	1.136
0.80000	(7.85)	1.062	1.057	1.053	1.052	1.052	1.055	1.059	1.064	1.073	1.082	1.092	1.101	1.117	1.131
1.00000	(9.81)	1.051	1.047	1.043	1.043	1.044	1.048	1.052	1.057	1.067	1.076	1.087	1.096	1.113	1.127
1.50000	(14.7)	1.031	1.029	1.027	1.028	1.030	1.035	1.041	1.047	1.058	1.068	1.079	1.089	1.106	1.121
2.00000	(19.6)	1.019	1.017	1.017	1.018	1.021	1.027	1.034	1.040	1.052	1.062	1.074	1.084	1.102	1.117
3.00000	(29.4)	1.003	1.002	1.003	1.006	1.010	1.017	1.025	1.032	1.044	1.055	1.068	1.078	1.097	1.112
S	$(\Delta p/\rho l)$	125	150	200	250	300	400	500	600	800	1000	1250	1500	2000	2500

Grad'nt   k.p.g.   (Equivalent) Pipe diameters in mm   Roughness size, $k_s = 0.150$ mm

---

Kin. visc., $\nu = 50 \times 10^{-6}$ m^{2}s^{-1};
$S = 0.00010$ to 3.00000

i.e. kin. pr. grad., $\Delta p/\rho l =$ (0.00098) to (29.4) ms^{-2}

Roughness size, $k_s = 0.30$ mm
This table shows values of m, as follows
m_C for Colebrook-White solutions; or, where $\mathbf{R} \le 2000$, m_P for laminar flow

E246

Grad'nt S	k.p.g. $(\Delta p/\rho l)$	125	150	200	250	300	400	500	600	800	1000	1250	1500	2000	2500
0.00010	(0.00098)	10.36	8.124	5.536	4.111	3.224	2.197	1.632	*	1.805	1.767	1.736	1.714	1.686	1.670
0.00015	(0.00147)	8.459	6.633	4.520	3.357	2.632	1.794	*	*	1.743	1.710	1.682	1.664	1.640	1.626
0.00020	(0.00196)	7.325	5.745	3.914	2.907	2.280	1.553	*	1.749	1.701	1.671	1.646	1.630	1.609	1.597
0.00030	(0.00294)	5.981	4.690	3.196	2.374	1.861	*	1.721	1.688	1.646	1.620	1.599	1.585	1.567	1.558
0.00040	(0.00392)	5.180	4.062	2.768	2.056	1.612	*	1.677	1.648	1.610	1.586	1.567	1.555	1.540	1.532
0.00060	(0.00588)	4.229	3.317	2.260	1.678	1.316	1.657	1.620	1.594	1.561	1.541	1.525	1.515	1.502	1.496
0.00080	(0.00785)	3.663	2.872	1.957	1.454	*	1.615	1.582	1.559	1.529	1.511	1.497	1.488	1.478	1.473
0.00100	(0.00981)	3.276	2.569	1.751	1.300	*	1.585	1.554	1.532	1.505	1.488	1.476	1.468	1.459	1.455
0.00150	(0.0147)	2.675	2.098	1.429	*	1.576	1.533	1.505	1.487	1.464	1.450	1.439	1.433	1.427	1.425
0.00200	(0.0196)	2.316	1.817	1.238	*	1.537	1.498	1.473	1.457	1.436	1.424	1.415	1.410	1.406	1.405
0.00300	(0.0294)	1.891	1.483	*	1.513	1.486	1.452	1.431	1.417	1.400	1.390	1.383	1.380	1.377	1.378
0.00400	(0.0392)	1.638	1.285	*	1.477	1.453	1.422	1.403	1.391	1.376	1.367	1.362	1.359	1.359	1.361
0.00600	(0.0588)	1.337	*	1.460	1.429	1.408	1.382	1.366	1.356	1.344	1.337	1.334	1.333	1.334	1.337
0.00800	(0.0785)	1.158	*	1.425	1.398	1.379	1.355	1.341	1.332	1.322	1.318	1.315	1.315	1.318	1.322
0.01000	(0.0981)	*	*	1.400	1.375	1.358	1.336	1.323	1.315	1.307	1.303	1.302	1.302	1.306	1.311
0.01500	(0.147)	*	1.393	1.357	1.336	1.321	1.303	1.293	1.286	1.280	1.278	1.278	1.280	1.285	1.292
0.02000	(0.196)	1.386	1.361	1.329	1.310	1.297	1.281	1.272	1.267	1.263	1.262	1.263	1.266	1.272	1.279
0.03000	(0.294)	1.342	1.320	1.292	1.276	1.265	1.252	1.246	1.242	1.240	1.241	1.243	1.247	1.255	1.263
0.04000	(0.392)	1.312	1.293	1.268	1.254	1.244	1.233	1.228	1.226	1.225	1.227	1.230	1.235	1.244	1.253
0.06000	(0.588)	1.274	1.258	1.237	1.225	1.217	1.209	1.206	1.205	1.206	1.209	1.214	1.219	1.230	1.240
0.08000	(0.785)	1.250	1.235	1.216	1.206	1.200	1.193	1.191	1.191	1.193	1.197	1.203	1.209	1.221	1.231
0.10000	(0.981)	1.231	1.218	1.201	1.192	1.187	1.182	1.180	1.181	1.184	1.189	1.195	1.202	1.214	1.225
0.15000	(1.47)	1.201	1.190	1.176	1.169	1.165	1.162	1.163	1.164	1.169	1.175	1.183	1.190	1.204	1.216
0.20000	(1.96)	1.181	1.171	1.160	1.154	1.151	1.150	1.151	1.154	1.160	1.166	1.175	1.183	1.197	1.210
0.30000	(2.94)	1.156	1.148	1.139	1.135	1.134	1.134	1.137	1.140	1.148	1.156	1.165	1.173	1.189	1.202
0.40000	(3.92)	1.140	1.133	1.126	1.123	1.122	1.124	1.128	1.132	1.140	1.149	1.158	1.167	1.183	1.197
0.60000	(5.88)	1.119	1.114	1.109	1.108	1.108	1.112	1.116	1.121	1.131	1.140	1.151	1.160	1.177	1.191
0.80000	(7.85)	1.106	1.102	1.098	1.098	1.099	1.104	1.109	1.114	1.125	1.135	1.146	1.156	1.173	1.188
1.00000	(9.81)	1.096	1.093	1.090	1.091	1.093	1.098	1.104	1.110	1.121	1.131	1.142	1.153	1.170	1.185
1.50000	(14.7)	1.081	1.079	1.078	1.080	1.082	1.089	1.096	1.102	1.114	1.125	1.137	1.148	1.166	1.181
2.00000	(19.6)	1.071	1.070	1.070	1.073	1.076	1.083	1.090	1.097	1.110	1.121	1.134	1.145	1.163	1.179
3.00000	(29.4)	1.059	1.058	1.060	1.064	1.068	1.076	1.084	1.092	1.105	1.117	1.130	1.141	1.160	1.176
S	$(\Delta p/\rho l)$	125	150	200	250	300	400	500	600	800	1000	1250	1500	2000	2500

Grad'nt   k.p.g.   (Equivalent) Pipe diameters in mm   Roughness size, $k_s = 0.30$ mm

E247

Kin. visc., $\nu = 60\times10^{-6}\ \text{m}^2\text{s}^{-1}$; $S = 0.00010$ to 30.0000

i.e. kin. pr. grad., $\Delta p/\rho l = (0.00098)$ to $(294)\ \text{ms}^{-2}$

Roughness size, $k_s = 0.006$ mm

This table shows values of m, as follows

m_C for Colebrook-White solutions; or,
where $\mathbf{R} \le 2000$, m_P for laminar flow

Grad'nt S	k.p.g. $(\Delta p/\rho l)$	(Equivalent) Pipe diameters in mm													
		6	8	10	12.5	15	20	25	30	40	50	60	80	100	125
0.00030	(0.00294)	411.4	280.4	208.2	154.6	121.3	82.63	61.37	48.12	32.79	24.35	19.10	13.01	9.665	7.177
0.00040	(0.00392)	356.3	242.8	180.3	133.9	105.0	71.56	53.14	41.68	28.40	21.09	16.54	11.27	8.370	6.216
0.00060	(0.00588)	290.9	198.2	147.2	109.3	85.75	58.43	43.39	34.03	23.19	17.22	13.50	9.202	6.834	5.075
0.00080	(0.00785)	252.0	171.7	127.5	94.69	74.26	50.60	37.58	29.47	20.08	14.91	11.69	7.969	5.918	4.395
0.00100	(0.00981)	225.4	153.6	114.0	84.70	66.42	45.26	33.61	26.36	17.96	13.34	10.46	7.128	5.293	3.931
0.00150	(0.0147)	184.0	125.4	93.12	69.15	54.23	36.95	27.44	21.52	14.66	10.89	8.541	5.820	4.322	3.210
0.00200	(0.0196)	159.4	108.6	80.64	59.89	46.96	32.00	23.77	18.64	12.70	9.432	7.396	5.040	3.743	2.780
0.00300	(0.0294)	130.1	88.66	65.84	48.90	38.35	26.13	19.41	15.22	10.37	7.701	6.039	4.115	3.056	2.270
0.00400	(0.0392)	112.7	76.78	57.02	42.35	33.21	22.63	16.81	13.18	8.980	6.669	5.230	3.564	2.647	1.966
0.00600	(0.0588)	92.00	62.69	46.56	34.58	27.12	18.48	13.72	10.76	7.332	5.446	4.270	2.910	2.161	1.605
0.00800	(0.0785)	79.68	54.29	40.32	29.94	23.48	16.00	11.88	9.319	6.350	4.716	3.698	2.520	1.872	1.390
0.01000	(0.0981)	71.26	48.56	36.06	26.78	21.00	14.31	10.63	8.335	5.680	4.218	3.308	2.254	1.674	1.243
0.01500	(0.147)	58.19	39.65	29.45	21.87	17.15	11.69	8.678	6.806	4.637	3.444	2.701	1.840	1.367	*
0.02000	(0.196)	50.39	34.34	25.50	18.94	14.85	10.12	7.516	5.894	4.016	2.983	2.339	1.594	1.184	*
0.03000	(0.294)	41.14	28.04	20.82	15.46	12.13	8.263	6.137	4.812	3.279	2.435	1.910	1.301	*	1.348
0.04000	(0.392)	35.63	24.28	18.03	13.39	10.50	7.156	5.314	4.168	2.840	2.109	1.654	1.127	*	1.313
0.06000	(0.588)	29.09	19.82	14.72	10.93	8.575	5.843	4.339	3.403	2.319	1.722	1.350	*	1.296	1.267
0.08000	(0.785)	25.20	17.17	12.75	9.469	7.426	5.060	3.758	2.947	2.008	1.491	1.169	*	1.262	1.236
0.10000	(0.981)	22.54	15.36	11.40	8.470	6.642	4.526	3.361	2.636	1.796	1.334	1.046	1.267	1.238	1.213
0.15000	(1.47)	18.40	12.54	9.312	6.915	5.423	3.695	2.744	2.152	1.466	1.089	*	1.221	1.195	1.174
0.20000	(1.96)	15.94	10.86	8.064	5.989	4.696	3.200	2.377	1.864	1.270	*	*	1.190	1.167	1.147
0.30000	(2.94)	13.01	8.866	6.584	4.890	3.835	2.613	1.941	1.522	1.037	*	1.182	1.150	1.129	1.112
0.40000	(3.92)	11.27	7.678	5.702	4.235	3.321	2.263	1.681	1.318	*	1.175	1.152	1.122	1.104	1.089
0.60000	(5.88)	9.200	6.269	4.656	3.458	2.712	1.848	1.372	1.076	*	1.132	1.112	1.086	1.070	1.057
0.80000	(7.85)	7.968	5.429	4.032	2.994	2.348	1.600	1.188	0.932	1.130	1.104	1.085	1.062	1.047	1.036
1.00000	(9.81)	7.126	4.856	3.606	2.678	2.100	1.431	1.063	*	1.107	1.082	1.066	1.044	1.031	1.020
1.50000	(14.7)	5.819	3.965	2.945	2.187	1.715	1.169	0.868	*	1.067	1.046	1.032	1.013	1.001	0.992
2.00000	(19.6)	5.039	3.434	2.550	1.894	1.485	1.012	*	1.072	1.041	1.022	1.009	0.992	0.982	0.974
3.00000	(29.4)	4.114	2.804	2.082	1.546	1.213	*	*	1.034	1.007	0.990	0.978	0.964	0.955	0.949
4.00000	(39.2)	3.563	2.428	1.803	1.339	1.050	*	1.027	1.008	0.983	0.968	0.958	0.945	0.937	0.932
6.00000	(58.8)	2.909	1.982	1.472	1.093	0.857	*	0.990	0.974	0.953	0.939	0.931	0.920	0.913	0.909
8.00000	(78.5)	2.520	1.717	1.275	0.947	*	0.988	0.966	0.951	0.932	0.920	0.912	0.903	0.897	0.894
10.0000	(98.1)	2.254	1.536	1.140	0.847	*	0.968	0.948	0.934	0.917	0.906	0.899	0.890	0.885	0.882
15.0000	(147)	1.840	1.254	0.931	*	0.963	0.935	0.917	0.905	0.890	0.881	0.875	0.868	0.864	0.862
20.0000	(196)	1.594	1.086	0.806	*	0.938	0.913	0.897	0.886	0.872	0.864	0.859	0.853	0.850	0.849
30.0000	(294)	1.301	0.887	*	0.922	0.905	0.883	0.869	0.860	0.848	0.842	0.837	0.833	0.831	0.831
S	$(\Delta p/\rho l)$	6	8	10	12.5	15	20	25	30	40	50	60	80	100	125

Grad'nt   k.p.g.   (Equivalent) Pipe diameters in mm

Grad'nt S	k.p.g. $(\Delta p/\rho l)$	(Equivalent) Pipe diameters in mm													
		125	150	200	250	300	400	500	600	800	1000	1250	1500	2000	2500
0.00010	(0.00098)	12.43	9.749	6.643	4.933	3.869	2.636	1.958	1.535	1.858	1.815	1.779	1.754	1.722	1.703
0.00015	(0.00147)	10.15	7.960	5.424	4.028	3.159	2.153	1.599	*	1.791	1.753	1.721	1.700	1.672	1.655
0.00020	(0.00196)	8.790	6.893	4.697	3.489	2.736	1.864	*	*	1.745	1.711	1.682	1.663	1.638	1.623
0.00030	(0.00294)	7.177	5.628	3.835	2.848	2.234	1.522	*	1.733	1.685	1.655	1.631	1.614	1.592	1.580
0.00040	(0.00392)	6.216	4.874	3.322	2.467	1.934	*	1.722	1.689	1.645	1.618	1.596	1.580	1.562	1.551
0.00060	(0.00588)	5.075	3.980	2.712	2.014	1.579	*	1.659	1.630	1.592	1.568	1.549	1.536	1.520	1.511
0.00080	(0.00785)	4.395	3.447	2.349	1.744	1.368	1.656	1.617	1.591	1.556	1.535	1.517	1.506	1.492	1.485
0.00100	(0.00981)	3.931	3.083	2.101	1.560	*	1.622	1.586	1.561	1.529	1.510	1.494	1.483	1.471	1.464
0.00150	(0.0147)	3.210	2.517	1.715	1.274	*	1.564	1.533	1.511	1.483	1.466	1.453	1.444	1.434	1.429
0.00200	(0.0196)	2.780	2.180	1.485	*	1.571	1.525	1.497	1.477	1.452	1.437	1.425	1.417	1.409	1.405
0.00300	(0.0294)	2.270	1.780	*	*	1.514	1.474	1.449	1.432	1.410	1.398	1.388	1.382	1.375	1.373
0.00400	(0.0392)	1.966	1.541	*	1.503	1.476	1.439	1.417	1.402	1.382	1.371	1.362	1.357	1.352	1.351
0.00600	(0.0588)	1.605	1.259	*	1.449	1.425	1.394	1.374	1.361	1.345	1.335	1.328	1.324	1.321	1.321
0.00800	(0.0785)	1.390	*	1.445	1.413	1.391	1.363	1.345	1.333	1.319	1.311	1.305	1.302	1.300	1.300
0.01000	(0.0981)	1.243	*	1.416	1.387	1.366	1.340	1.324	1.313	1.300	1.293	1.288	1.285	1.284	1.285
0.01500	(0.147)	*	1.408	1.367	1.341	1.323	1.300	1.286	1.277	1.266	1.261	1.257	1.256	1.256	1.258
0.02000	(0.196)	*	1.371	1.334	1.310	1.294	1.273	1.261	1.253	1.244	1.239	1.236	1.236	1.237	1.239
0.03000	(0.294)	1.348	1.322	1.290	1.269	1.255	1.237	1.227	1.220	1.213	1.210	1.208	1.208	1.211	1.214
0.04000	(0.392)	1.313	1.290	1.260	1.242	1.229	1.213	1.204	1.198	1.192	1.190	1.189	1.190	1.193	1.197
0.06000	(0.588)	1.267	1.247	1.221	1.205	1.194	1.181	1.173	1.169	1.164	1.163	1.163	1.165	1.169	1.174
0.08000	(0.785)	1.236	1.218	1.194	1.180	1.170	1.159	1.152	1.149	1.145	1.145	1.146	1.148	1.153	1.158
0.10000	(0.981)	1.213	1.196	1.175	1.161	1.153	1.142	1.136	1.133	1.131	1.131	1.132	1.135	1.140	1.146
0.15000	(1.47)	1.174	1.159	1.140	1.129	1.122	1.113	1.109	1.107	1.106	1.107	1.109	1.112	1.118	1.125
0.20000	(1.96)	1.147	1.134	1.117	1.107	1.101	1.094	1.090	1.089	1.089	1.090	1.093	1.096	1.103	1.110
0.30000	(2.94)	1.112	1.101	1.087	1.078	1.073	1.067	1.065	1.064	1.065	1.068	1.072	1.075	1.083	1.091
0.40000	(3.92)	1.089	1.078	1.066	1.058	1.054	1.050	1.048	1.048	1.050	1.053	1.057	1.061	1.069	1.077
0.60000	(5.88)	1.057	1.048	1.038	1.032	1.028	1.025	1.025	1.025	1.028	1.032	1.037	1.042	1.051	1.059
0.80000	(7.85)	1.036	1.028	1.019	1.014	1.011	1.009	1.009	1.010	1.014	1.018	1.023	1.028	1.038	1.046
1.00000	(9.81)	1.020	1.013	1.005	1.000	0.998	0.997	0.997	0.999	1.003	1.007	1.013	1.018	1.028	1.037
1.50000	(14.7)	0.992	0.986	0.980	0.977	0.975	0.975	0.977	0.979	0.984	0.989	0.995	1.001	1.011	1.020
2.00000	(19.6)	0.974	0.969	0.963	0.961	0.960	0.961	0.962	0.965	0.970	0.976	0.983	0.989	1.000	1.009
3.00000	(29.4)	0.949	0.945	0.941	0.939	0.939	0.941	0.943	0.946	0.953	0.959	0.966	0.972	0.984	0.994
S	$(\Delta p/\rho l)$	125	150	200	250	300	400	500	600	800	1000	1250	1500	2000	2500

Grad'nt   k.p.g.   (Equivalent) Pipe diameters in mm

Kinematic viscosity, $\nu = 60\times10^{-6}\ \text{m}^2\text{s}^{-1}$;   **Roughness size, $k_s = 0.006$ mm**

Kin. visc., $\nu = 60\times10^{-6}$ m^2 s^{-1};
S = 0·00010 to 30·0000

i.e. kin. pr. grad., $\Delta p/\rho l$ =
(0·00098) to (294) ms^{-2}

Roughness size, k_s = 0·015 mm
This table shows values of m, as follows

m_C for Colebrook-White solutions; or,
where $R \leq 2000$, m_P for laminar flow

E248

Grad'nt S	k.p.g. $(\Delta p/\rho l)$	6	8	10	12·5	15	20	25	30	40	50	60	80	100	125
0·00030	(0·00294)	411·4	280·4	208·2	154·6	121·3	82·63	61·37	48·12	32·79	24·35	19·10	13·01	9·665	7·177
0·00040	(0·00392)	356·3	242·8	180·3	133·9	105·0	71·56	53·14	41·68	28·40	21·09	16·54	11·27	8·370	6·216
0·00060	(0·00588)	290·9	198·2	147·2	109·3	85·75	58·43	43·39	34·03	23·19	17·22	13·50	9·202	6·834	5·075
0·00080	(0·00785)	252·0	171·7	127·5	94·69	74·26	50·60	37·58	29·47	20·08	14·91	11·69	7·969	5·918	4·395
0·00100	(0·00981)	225·4	153·6	114·0	84·70	66·42	45·26	33·61	26·36	17·96	13·34	10·46	7·128	5·293	3·931
0·00150	(0·0147)	184·0	125·4	93·12	69·15	54·23	36·95	27·44	21·52	14·66	10·89	8·541	5·820	4·322	3·210
0·00200	(0·0196)	159·4	108·6	80·64	59·89	46·96	32·00	23·77	18·64	12·70	9·432	7·396	5·040	3·743	2·780
0·00300	(0·0294)	130·1	88·66	65·84	48·90	38·35	26·13	19·41	15·22	10·37	7·701	6·039	4·115	3·056	2·270
0·00400	(0·0392)	112·7	76·78	57·02	42·35	33·21	22·63	16·81	13·18	8·980	6·669	5·230	3·564	2·647	1·966
0·00600	(0·0588)	92·00	62·69	46·56	34·58	27·12	18·48	13·72	10·76	7·332	5·446	4·270	2·910	2·161	1·605
0·00800	(0·0785)	79·68	54·29	40·32	29·94	23·48	16·00	11·88	9·319	6·350	4·716	3·698	2·520	1·872	1·390
0·01000	(0·0981)	71·26	48·56	36·06	26·78	21·00	14·31	10·63	8·335	5·680	4·218	3·308	2·254	1·674	1·243
0·01500	(0·147)	58·19	39·65	29·45	21·87	17·15	11·69	8·678	6·806	4·637	3·444	2·701	1·840	1·367	*
0·02000	(0·196)	50·39	34·34	25·50	18·94	14·85	10·12	7·516	5·894	4·016	2·983	2·339	1·594	1·184	*
0·03000	(0·294)	41·14	28·04	20·82	15·46	12·13	8·263	6·137	4·812	3·279	2·435	1·910	1·301	*	1·349
0·04000	(0·392)	35·63	24·28	18·03	13·39	10·50	7·156	5·314	4·168	2·840	2·109	1·654	1·127	*	1·314
0·06000	(0·588)	29·09	19·82	14·72	10·93	8·575	5·843	4·339	3·403	2·319	1·722	1·350	*	1·297	1·268
0·08000	(0·785)	25·20	17·17	12·75	9·469	7·426	5·060	3·758	2·947	2·008	1·491	1·169	*	1·264	1·237
0·10000	(0·981)	22·54	15·36	11·40	8·470	6·642	4·526	3·361	2·636	1·796	1·334	1·046	1·269	1·239	1·215
0·15000	(1·47)	18·40	12·54	9·312	6·915	5·423	3·695	2·744	2·152	1·466	1·089	*	1·223	1·197	1·176
0·20000	(1·96)	15·94	10·86	8·064	5·989	4·696	3·200	2·377	1·864	1·270	*	*	1·192	1·169	1·149
0·30000	(2·94)	13·01	8·866	6·584	4·890	3·835	2·613	1·941	1·522	1·037	*	1·184	1·152	1·131	1·115
0·40000	(3·92)	11·27	7·678	5·702	4·235	3·321	2·263	1·681	1·318	*	1·177	1·154	1·125	1·106	1·091
0·60000	(5·88)	9·200	6·269	4·656	3·458	2·712	1·848	1·372	1·076	*	1·134	1·115	1·089	1·073	1·060
0·80000	(7·85)	7·968	5·429	4·032	2·994	2·348	1·600	1·188	0·932	1·132	1·106	1·088	1·065	1·051	1·039
1·00000	(9·81)	7·126	4·856	3·606	2·678	2·100	1·431	1·063	*	1·110	1·085	1·069	1·047	1·034	1·023
1·50000	(14·7)	5·819	3·965	2·945	2·187	1·715	1·169	0·868	*	1·071	1·050	1·035	1·016	1·005	0·996
2·00000	(19·6)	5·039	3·434	2·550	1·894	1·485	1·012	*	1·076	1·045	1·026	1·013	0·996	0·986	0·978
3·00000	(29·4)	4·114	2·804	2·082	1·546	1·213	*	*	1·037	1·011	0·994	0·983	0·968	0·960	0·954
4·00000	(39·2)	3·563	2·428	1·803	1·339	1·050	*	1·031	1·012	0·988	0·973	0·963	0·950	0·943	0·937
6·00000	(58·8)	2·909	1·982	1·472	1·093	0·857	*	0·995	0·979	0·958	0·945	0·936	0·926	0·920	0·916
8·00000	(78·5)	2·520	1·717	1·275	0·947	*	0·993	0·971	0·957	0·938	0·926	0·918	0·909	0·904	0·901
10·0000	(98·1)	2·254	1·536	1·140	0·847	*	0·974	0·954	0·940	0·923	0·912	0·905	0·897	0·893	0·890
15·0000	(147)	1·840	1·254	0·931	*	0·969	0·941	0·924	0·912	0·897	0·888	0·882	0·876	0·873	0·871
20·0000	(196)	1·594	1·086	0·806	*	0·945	0·919	0·904	0·893	0·880	0·872	0·867	0·862	0·860	0·859
30·0000	(294)	1·301	0·887	*	0·930	0·913	0·891	0·877	0·868	0·857	0·851	0·847	0·843	0·842	0·842
S	$(\Delta p/\rho l)$	6	8	10	12·5	15	20	25	30	40	50	60	80	100	125

Grad'nt k.p.g. (Equivalent) Pipe diameters in mm

Grad'nt S	k.p.g. $(\Delta p/\rho l)$	125	150	200	250	300	400	500	600	800	1000	1250	1500	2000	2500
0·00010	(0·00098)	12·43	9·749	6·643	4·933	3·869	2·636	1·958	1·535	1·859	1·815	1·779	1·755	1·723	1·703
0·00015	(0·00147)	10·15	7·960	5·424	4·028	3·159	2·153	1·599	*	1·791	1·753	1·722	1·700	1·672	1·656
0·00020	(0·00196)	8·790	6·893	4·697	3·489	2·736	1·864	*	*	1·746	1·711	1·683	1·663	1·638	1·624
0·00030	(0·00294)	7·177	5·628	3·835	2·848	2·234	1·522	*	1·734	1·686	1·656	1·631	1·614	1·593	1·581
0·00040	(0·00392)	6·216	4·874	3·322	2·467	1·934	*	1·723	1·689	1·646	1·618	1·596	1·581	1·562	1·551
0·00060	(0·00588)	5·075	3·980	2·712	2·014	1·579	*	1·660	1·630	1·592	1·569	1·549	1·536	1·521	1·512
0·00080	(0·00785)	4·395	3·447	2·349	1·744	1·368	1·656	1·618	1·591	1·557	1·535	1·518	1·506	1·493	1·485
0·00100	(0·00981)	3·931	3·083	2·101	1·560	*	1·622	1·587	1·562	1·530	1·510	1·494	1·484	1·471	1·465
0·00150	(0·0147)	3·210	2·517	1·715	1·274	*	1·564	1·533	1·512	1·484	1·467	1·453	1·445	1·435	1·430
0·00200	(0·0196)	2·780	2·180	1·485	*	1·571	1·526	1·497	1·478	1·453	1·438	1·426	1·418	1·410	1·406
0·00300	(0·0294)	2·270	1·780	*	*	1·514	1·474	1·450	1·433	1·411	1·398	1·388	1·382	1·376	1·374
0·00400	(0·0392)	1·966	1·541	*	1·504	1·476	1·440	1·418	1·402	1·383	1·372	1·363	1·358	1·353	1·352
0·00600	(0·0588)	1·605	1·259	*	1·450	1·426	1·394	1·375	1·362	1·345	1·336	1·329	1·325	1·322	1·322
0·00800	(0·0785)	1·390	*	1·446	1·414	1·392	1·364	1·346	1·334	1·320	1·312	1·306	1·303	1·301	1·302
0·01000	(0·0981)	1·243	*	1·417	1·388	1·367	1·341	1·325	1·314	1·301	1·294	1·289	1·286	1·285	1·287
0·01500	(0·147)	*	1·409	1·368	1·342	1·324	1·301	1·287	1·278	1·268	1·262	1·259	1·257	1·258	1·260
0·02000	(0·196)	*	1·372	1·335	1·311	1·295	1·275	1·262	1·254	1·245	1·241	1·238	1·237	1·239	1·241
0·03000	(0·294)	1·349	1·323	1·291	1·270	1·256	1·239	1·228	1·222	1·215	1·212	1·210	1·210	1·213	1·217
0·04000	(0·392)	1·314	1·291	1·261	1·243	1·230	1·215	1·206	1·200	1·194	1·192	1·191	1·192	1·196	1·200
0·06000	(0·588)	1·268	1·248	1·222	1·206	1·196	1·182	1·175	1·171	1·166	1·165	1·166	1·167	1·172	1·177
0·08000	(0·785)	1·237	1·219	1·196	1·182	1·172	1·161	1·154	1·151	1·148	1·147	1·148	1·151	1·156	1·161
0·10000	(0·981)	1·215	1·198	1·176	1·163	1·154	1·144	1·139	1·136	1·133	1·134	1·135	1·138	1·144	1·150
0·15000	(1·47)	1·176	1·161	1·142	1·131	1·124	1·116	1·112	1·110	1·109	1·110	1·112	1·116	1·122	1·129
0·20000	(1·96)	1·149	1·136	1·120	1·110	1·103	1·096	1·093	1·092	1·092	1·094	1·097	1·101	1·108	1·115
0·30000	(2·94)	1·115	1·103	1·089	1·081	1·076	1·071	1·068	1·068	1·069	1·072	1·076	1·080	1·088	1·096
0·40000	(3·92)	1·091	1·081	1·068	1·061	1·057	1·053	1·052	1·052	1·054	1·057	1·062	1·066	1·075	1·083
0·60000	(5·88)	1·060	1·051	1·041	1·035	1·032	1·029	1·029	1·030	1·033	1·037	1·043	1·048	1·057	1·066
0·80000	(7·85)	1·039	1·031	1·022	1·018	1·015	1·013	1·014	1·015	1·019	1·024	1·030	1·035	1·045	1·055
1·00000	(9·81)	1·023	1·016	1·008	1·004	1·002	1·002	1·002	1·004	1·009	1·014	1·020	1·026	1·036	1·046
1·50000	(14·7)	0·996	0·991	0·984	0·982	0·980	0·981	0·983	0·985	0·991	0·996	1·003	1·009	1·021	1·031
2·00000	(19·6)	0·978	0·973	0·968	0·966	0·966	0·967	0·969	0·972	0·978	0·984	0·992	0·998	1·010	1·021
3·00000	(29·4)	0·954	0·950	0·946	0·945	0·946	0·948	0·951	0·955	0·962	0·969	0·976	0·984	0·996	1·007
S	$(\Delta p/\rho l)$	125	150	200	250	300	400	500	600	800	1000	1250	1500	2000	2500

Grad'nt k.p.g. (Equivalent) Pipe diameters in mm

Kinematic viscosity, $\nu = 60\times10^{-6}$ m^2 s^{-1}; **Roughness size, k_s = 0·015 mm**

E249

Kin. visc., $\nu = 60 \times 10^{-6}\ \mathrm{m^2 s^{-1}}$;
$S = 0.00010$ to 3.00000

i.e. kin. pr. grad., $\Delta p/\rho l =$
(0.00098) to $(29.4)\ \mathrm{ms^{-2}}$

Roughness size, $k_s = 0.030$ mm
This table shows values of m, as follows

m_C for Colebrook-White solutions; or,
where $\mathbf{R} \leq 2000$, m_P for laminar flow

Grad'nt S	k.p.g. $(\Delta p/\rho l)$	125	150	200	250	300	400	500	600	800	1000	1250	1500	2000	2500
0.00010	(0.00098)	12.43	9.749	6.643	4.933	3.869	2.636	1.958	1.535	1.859	1.816	1.780	1.755	1.723	1.704
0.00015	(0.00147)	10.15	7.960	5.424	4.028	3.159	2.153	1.599	*	1.791	1.753	1.722	1.700	1.673	1.656
0.00020	(0.00196)	8.790	6.893	4.697	3.489	2.736	1.864	*	*	1.746	1.712	1.683	1.664	1.639	1.624
0.00030	(0.00294)	7.177	5.628	3.835	2.848	2.234	1.522	*	1.734	1.686	1.656	1.631	1.615	1.593	1.581
0.00040	(0.00392)	6.216	4.874	3.322	2.467	1.934	*	1.723	1.690	1.646	1.619	1.597	1.582	1.563	1.552
0.00060	(0.00588)	5.075	3.980	2.712	2.014	1.579	*	1.660	1.631	1.593	1.569	1.550	1.537	1.521	1.513
0.00080	(0.00785)	4.395	3.447	2.349	1.744	1.368	1.657	1.618	1.592	1.557	1.536	1.519	1.507	1.493	1.486
0.00100	(0.00981)	3.931	3.083	2.101	1.560	*	1.623	1.587	1.563	1.531	1.511	1.495	1.485	1.472	1.466
0.00150	(0.0147)	3.210	2.517	1.715	1.274	*	1.565	1.534	1.512	1.485	1.468	1.454	1.446	1.436	1.431
0.00200	(0.0196)	2.780	2.180	1.485	*	1.572	1.527	1.498	1.479	1.454	1.439	1.427	1.419	1.411	1.407
0.00300	(0.0294)	2.270	1.780	*	*	1.515	1.475	1.451	1.434	1.412	1.400	1.390	1.384	1.378	1.375
0.00400	(0.0392)	1.966	1.541	*	1.505	1.477	1.441	1.419	1.404	1.384	1.373	1.365	1.360	1.355	1.354
0.00600	(0.0588)	1.605	1.259	*	1.451	1.427	1.396	1.376	1.363	1.347	1.338	1.331	1.327	1.324	1.324
0.00800	(0.0785)	1.390	*	1.447	1.415	1.394	1.365	1.348	1.336	1.322	1.314	1.308	1.305	1.303	1.304
0.01000	(0.0981)	1.243	*	1.419	1.389	1.369	1.342	1.326	1.316	1.303	1.296	1.291	1.289	1.288	1.289
0.01500	(0.147)	*	1.410	1.369	1.343	1.326	1.303	1.289	1.280	1.270	1.264	1.261	1.260	1.260	1.263
0.02000	(0.196)	*	1.374	1.336	1.313	1.297	1.276	1.264	1.256	1.247	1.243	1.241	1.240	1.242	1.245
0.03000	(0.294)	1.350	1.325	1.293	1.272	1.258	1.241	1.231	1.224	1.217	1.214	1.213	1.214	1.217	1.221
0.04000	(0.392)	1.316	1.293	1.263	1.245	1.233	1.217	1.208	1.203	1.197	1.195	1.195	1.196	1.199	1.204
0.06000	(0.588)	1.270	1.250	1.225	1.209	1.198	1.185	1.178	1.174	1.170	1.169	1.170	1.172	1.176	1.182
0.08000	(0.785)	1.240	1.222	1.199	1.184	1.175	1.164	1.158	1.154	1.151	1.151	1.153	1.155	1.161	1.167
0.10000	(0.981)	1.217	1.201	1.179	1.166	1.158	1.148	1.142	1.139	1.138	1.138	1.140	1.143	1.149	1.156
0.15000	(1.47)	1.179	1.164	1.146	1.135	1.128	1.120	1.116	1.114	1.114	1.115	1.118	1.122	1.129	1.136
0.20000	(1.96)	1.153	1.140	1.123	1.114	1.107	1.101	1.098	1.097	1.097	1.099	1.103	1.107	1.115	1.123
0.30000	(2.94)	1.118	1.107	1.093	1.085	1.080	1.076	1.074	1.074	1.076	1.079	1.083	1.088	1.097	1.105
0.40000	(3.92)	1.095	1.085	1.073	1.066	1.062	1.059	1.058	1.058	1.061	1.065	1.070	1.075	1.085	1.093
0.60000	(5.88)	1.065	1.056	1.046	1.041	1.038	1.036	1.036	1.037	1.041	1.046	1.052	1.057	1.068	1.078
0.80000	(7.85)	1.044	1.037	1.028	1.024	1.022	1.021	1.022	1.023	1.028	1.033	1.040	1.046	1.057	1.067
1.00000	(9.81)	1.029	1.022	1.015	1.011	1.010	1.009	1.011	1.013	1.018	1.024	1.031	1.037	1.049	1.059
1.50000	(14.7)	1.003	0.997	0.992	0.989	0.989	0.990	0.992	0.995	1.001	1.008	1.015	1.023	1.035	1.046
2.00000	(19.6)	0.985	0.981	0.976	0.975	0.975	0.977	0.980	0.983	0.990	0.997	1.005	1.013	1.026	1.038
3.00000	(29.4)	0.962	0.959	0.956	0.955	0.956	0.959	0.963	0.967	0.975	0.983	0.992	1.000	1.014	1.026
S $(\Delta p/\rho l)$		125	150	200	250	300	400	500	600	800	1000	1250	1500	2000	2500

Grad'nt k.p.g.   (Equivalent) Pipe diameters in mm

Roughness size, $k_s = 0.030$ mm

E250

Kin. visc., $\nu = 60 \times 10^{-6}\ \mathrm{m^2 s^{-1}}$;
$S = 0.00010$ to 3.00000

i.e. kin. pr. grad., $\Delta p/\rho l =$
(0.00098) to $(29.4)\ \mathrm{ms^{-2}}$

Roughness size, $k_s = 0.060$ mm
This table shows values of m, as follows

m_C for Colebrook-White solutions; or,
where $\mathbf{R} \leq 2000$, m_P for laminar flow

Grad'nt S	k.p.g. $(\Delta p/\rho l)$	125	150	200	250	300	400	500	600	800	1000	1250	1500	2000	2500
0.00010	(0.00098)	12.43	9.749	6.643	4.933	3.869	2.636	1.958	1.535	1.860	1.817	1.781	1.756	1.724	1.705
0.00015	(0.00147)	10.15	7.960	5.424	4.028	3.159	2.153	1.599	*	1.792	1.754	1.723	1.701	1.674	1.657
0.00020	(0.00196)	8.790	6.893	4.697	3.489	2.736	1.864	*	*	1.747	1.713	1.684	1.665	1.640	1.625
0.00030	(0.00294)	7.177	5.628	3.835	2.848	2.234	1.522	*	1.735	1.687	1.657	1.633	1.616	1.595	1.583
0.00040	(0.00392)	6.216	4.874	3.322	2.467	.934	*	1.724	1.691	1.647	1.620	1.598	1.583	1.564	1.554
0.00060	(0.00588)	5.075	3.980	2.712	2.014	1.579	*	1.662	1.632	1.594	1.571	1.551	1.539	1.523	1.514
0.00080	(0.00785)	4.395	3.447	2.349	1.744	1.368	1.658	1.620	1.593	1.559	1.537	1.520	1.509	1.495	1.488
0.00100	(0.00981)	3.931	3.083	2.101	1.560	*	1.625	1.589	1.564	1.532	1.513	1.497	1.487	1.474	1.468
0.00150	(0.0147)	3.210	2.517	1.715	1.274	*	1.567	1.536	1.514	1.487	1.470	1.456	1.448	1.438	1.434
0.00200	(0.0196)	2.780	2.180	1.485	*	1.574	1.528	1.500	1.481	1.456	1.441	1.429	1.422	1.414	1.410
0.00300	(0.0294)	2.270	1.780	*	*	1.517	1.477	1.453	1.436	1.415	1.402	1.392	1.386	1.381	1.379
0.00400	(0.0392)	1.966	1.541	*	1.507	1.479	1.443	1.421	1.406	1.387	1.376	1.368	1.363	1.358	1.357
0.00600	(0.0588)	1.605	1.259	*	1.454	1.429	1.398	1.379	1.366	1.350	1.341	1.334	1.331	1.328	1.328
0.00800	(0.0785)	1.390	*	1.450	1.418	1.396	1.368	1.351	1.339	1.325	1.317	1.312	1.309	1.308	1.309
0.01000	(0.0981)	1.243	*	1.421	1.392	1.371	1.345	1.329	1.319	1.306	1.299	1.295	1.293	1.292	1.294
0.01500	(0.147)	*	1.413	1.372	1.347	1.329	1.306	1.293	1.284	1.274	1.269	1.266	1.265	1.266	1.268
0.02000	(0.196)	*	1.377	1.340	1.316	1.300	1.280	1.268	1.261	1.252	1.248	1.246	1.246	1.248	1.251
0.03000	(0.294)	1.354	1.329	1.296	1.276	1.263	1.245	1.235	1.229	1.223	1.220	1.219	1.220	1.223	1.228
0.04000	(0.392)	1.320	1.297	1.268	1.249	1.237	1.222	1.213	1.208	1.203	1.201	1.201	1.203	1.207	1.212
0.06000	(0.588)	1.275	1.255	1.229	1.214	1.203	1.191	1.184	1.180	1.177	1.176	1.177	1.180	1.185	1.191
0.08000	(0.785)	1.245	1.227	1.204	1.190	1.181	1.170	1.164	1.161	1.159	1.159	1.161	1.164	1.171	1.177
0.10000	(0.981)	1.223	1.206	1.185	1.172	1.164	1.154	1.149	1.147	1.146	1.147	1.149	1.153	1.160	1.167
0.15000	(1.47)	1.185	1.170	1.152	1.142	1.135	1.127	1.124	1.122	1.123	1.125	1.129	1.133	1.141	1.149
0.20000	(1.96)	1.159	1.146	1.130	1.121	1.115	1.109	1.107	1.106	1.107	1.110	1.115	1.119	1.129	1.137
0.30000	(2.94)	1.126	1.115	1.101	1.094	1.089	1.085	1.084	1.084	1.087	1.091	1.096	1.102	1.112	1.122
0.40000	(3.92)	1.104	1.094	1.082	1.076	1.072	1.069	1.069	1.070	1.074	1.078	1.084	1.090	1.101	1.111
0.60000	(5.88)	1.074	1.066	1.057	1.052	1.049	1.048	1.049	1.051	1.056	1.061	1.068	1.075	1.087	1.098
0.80000	(7.85)	1.054	1.047	1.040	1.036	1.034	1.034	1.036	1.038	1.044	1.050	1.058	1.065	1.078	1.089
1.00000	(9.81)	1.040	1.034	1.027	1.024	1.023	1.024	1.026	1.029	1.035	1.042	1.050	1.057	1.071	1.083
1.50000	(14.7)	1.015	1.010	1.006	1.004	1.004	1.006	1.009	1.013	1.021	1.028	1.037	1.045	1.059	1.072
2.00000	(19.6)	0.999	0.995	0.991	0.991	0.991	0.994	0.998	1.003	1.011	1.019	1.029	1.037	1.052	1.065
3.00000	(29.4)	0.977	0.975	0.973	0.973	0.975	0.979	0.984	0.989	0.999	1.008	1.018	1.027	1.043	1.056
S $(\Delta p/\rho l)$		125	150	200	250	300	400	500	600	800	1000	1250	1500	2000	2500

Grad'nt k.p.g.   (Equivalent) Pipe diameters in mm

Roughness size, $k_s = 0.060$ mm

Kin. visc., $\nu = 60 \times 10^{-6}$ m^{2}s^{-1};
$S = 0.00010$ to 3.00000

i.e. kin. pr. grad., $\Delta p/\rho l =$
(0.00098) to (29.4) ms^{-2}

Roughness size, $k_s = 0.150$ mm
This table shows values of m, as follows

m_C for Colebrook-White solutions; or,
where $\mathbf{R} \leq 2000$, m_P for laminar flow

Grad'nt S	k.p.g. $(\Delta p/\rho l)$	125	150	200	250	300	400	500	600	800	1000	1250	1500	2000	2500
0.00010	(0.00098)	12.43	9.749	6.643	4.933	3.869	2.636	1.958	1.535	1.862	1.819	1.783	1.758	1.727	1.707
0.00015	(0.00147)	10.15	7.960	5.424	4.028	3.159	2.153	1.599	*	1.795	1.757	1.726	1.704	1.677	1.660
0.00020	(0.00196)	8.790	6.893	4.697	3.489	2.736	1.864	*	*	1.750	1.716	1.687	1.668	1.643	1.629
0.00030	(0.00294)	7.177	5.628	3.835	2.848	2.234	1.522	*	1.738	1.690	1.661	1.636	1.619	1.599	1.587
0.00040	(0.00392)	6.216	4.874	3.322	2.467	1.934	*	1.727	1.694	1.651	1.624	1.602	1.587	1.568	1.558
0.00060	(0.00588)	5.075	3.980	2.712	2.014	1.579	*	1.665	1.636	1.598	1.575	1.556	1.543	1.528	1.520
0.00080	(0.00785)	4.395	3.447	2.349	1.744	1.368	1.662	1.624	1.597	1.563	1.542	1.525	1.514	1.501	1.494
0.00100	(0.00981)	3.931	3.083	2.101	1.560	*	1.629	1.593	1.569	1.537	1.518	1.502	1.492	1.480	1.474
0.00150	(0.0147)	3.210	2.517	1.715	1.274	*	1.572	1.541	1.519	1.492	1.475	1.462	1.454	1.445	1.441
0.00200	(0.0196)	2.780	2.180	1.485	*	1.579	1.534	1.505	1.486	1.462	1.447	1.436	1.429	1.421	1.418
0.00300	(0.0294)	2.270	1.780	*	*	1.523	1.483	1.459	1.442	1.421	1.409	1.400	1.394	1.389	1.388
0.00400	(0.0392)	1.966	1.541	*	1.513	1.486	1.450	1.428	1.413	1.394	1.384	1.376	1.371	1.368	1.367
0.00600	(0.0588)	1.605	1.259	*	1.460	1.437	1.406	1.387	1.374	1.358	1.350	1.344	1.341	1.339	1.340
0.00800	(0.0785)	1.390	*	1.457	1.426	1.404	1.376	1.359	1.348	1.334	1.327	1.322	1.320	1.320	1.322
0.01000	(0.0981)	1.243	*	1.429	1.400	1.380	1.354	1.339	1.328	1.316	1.310	1.306	1.305	1.305	1.308
0.01500	(0.147)	*	1.422	1.381	1.356	1.338	1.316	1.303	1.295	1.286	1.281	1.279	1.279	1.281	1.285
0.02000	(0.196)	*	1.386	1.349	1.327	1.311	1.291	1.280	1.273	1.265	1.262	1.260	1.261	1.264	1.269
0.03000	(0.294)	1.364	1.340	1.308	1.288	1.275	1.258	1.249	1.243	1.238	1.236	1.236	1.238	1.243	1.248
0.04000	(0.392)	1.331	1.309	1.280	1.262	1.250	1.236	1.228	1.223	1.219	1.219	1.220	1.222	1.228	1.235
0.06000	(0.588)	1.288	1.268	1.243	1.228	1.218	1.207	1.201	1.197	1.195	1.196	1.198	1.202	1.209	1.217
0.08000	(0.785)	1.259	1.241	1.219	1.206	1.197	1.187	1.182	1.180	1.179	1.181	1.184	1.188	1.197	1.205
0.10000	(0.981)	1.238	1.222	1.201	1.189	1.181	1.173	1.169	1.167	1.168	1.170	1.174	1.178	1.188	1.196
0.15000	(1.47)	1.202	1.188	1.171	1.161	1.155	1.149	1.146	1.146	1.148	1.151	1.156	1.162	1.172	1.182
0.20000	(1.96)	1.178	1.166	1.151	1.142	1.137	1.132	1.131	1.132	1.135	1.139	1.145	1.151	1.162	1.173
0.30000	(2.94)	1.147	1.137	1.124	1.118	1.114	1.111	1.112	1.113	1.118	1.123	1.130	1.137	1.150	1.161
0.40000	(3.92)	1.126	1.117	1.107	1.102	1.099	1.098	1.099	1.101	1.107	1.113	1.121	1.128	1.141	1.153
0.60000	(5.88)	1.100	1.093	1.085	1.081	1.080	1.080	1.082	1.085	1.093	1.100	1.108	1.116	1.131	1.144
0.80000	(7.85)	1.082	1.076	1.070	1.067	1.067	1.068	1.072	1.075	1.083	1.091	1.101	1.109	1.124	1.138
1.00000	(9.81)	1.070	1.064	1.059	1.058	1.058	1.060	1.064	1.068	1.077	1.085	1.095	1.104	1.120	1.133
1.50000	(14.7)	1.049	1.045	1.041	1.041	1.042	1.046	1.051	1.056	1.066	1.075	1.086	1.095	1.112	1.126
2.00000	(19.6)	1.035	1.032	1.030	1.030	1.032	1.037	1.043	1.049	1.059	1.069	1.080	1.090	1.107	1.122
3.00000	(29.4)	1.017	1.015	1.015	1.017	1.020	1.026	1.033	1.039	1.051	1.061	1.073	1.083	1.101	1.117
S	$(\Delta p/\rho l)$	125	150	200	250	300	400	500	600	800	1000	1250	1500	2000	2500

Grad'nt k.p.g. (Equivalent) Pipe diameters in mm — Roughness size, $k_s = 0.150$ mm

Kin. visc., $\nu = 60 \times 10^{-6}$ m^{2}s^{-1};
$S = 0.00010$ to 3.00000

i.e. kin. pr. grad., $\Delta p/\rho l =$
(0.00098) to (29.4) ms^{-2}

Roughness size, $k_s = 0.30$ mm
This table shows values of m, as follows

m_C for Colebrook-White solutions; or,
where $\mathbf{R} \leq 2000$, m_P for laminar flow

Grad'nt S	k.p.g. $(\Delta p/\rho l)$	125	150	200	250	300	400	500	600	800	1000	1250	1500	2000	2500
0.00010	(0.00098)	12.43	9.749	6.643	4.933	3.869	2.636	1.958	1.535	1.866	1.823	1.787	1.762	1.731	1.712
0.00015	(0.00147)	10.15	7.960	5.424	4.028	3.159	2.153	1.599	*	1.799	1.761	1.730	1.709	1.682	1.666
0.00020	(0.00196)	8.790	6.893	4.697	3.489	2.736	1.864	*	*	1.754	1.720	1.692	1.673	1.649	1.635
0.00030	(0.00294)	7.177	5.628	3.835	2.848	2.234	1.522	*	1.743	1.696	1.666	1.642	1.625	1.605	1.593
0.00040	(0.00392)	6.216	4.874	3.322	2.467	1.934	*	1.733	1.700	1.656	1.630	1.608	1.593	1.575	1.565
0.00060	(0.00588)	5.075	3.980	2.712	2.014	1.579	*	1.671	1.642	1.605	1.582	1.563	1.550	1.536	1.528
0.00080	(0.00785)	4.395	3.447	2.349	1.744	1.368	1.669	1.630	1.604	1.570	1.549	1.533	1.522	1.509	1.503
0.00100	(0.00981)	3.931	3.083	2.101	1.560	*	1.636	1.600	1.576	1.545	1.526	1.510	1.501	1.490	1.484
0.00150	(0.0147)	3.210	2.517	1.715	1.274	*	1.579	1.549	1.528	1.501	1.484	1.472	1.464	1.456	1.452
0.00200	(0.0196)	2.780	2.180	1.485	*	1.587	1.542	1.514	1.495	1.471	1.457	1.446	1.440	1.433	1.431
0.00300	(0.0294)	2.270	1.780	*	*	1.532	1.493	1.469	1.453	1.432	1.421	1.412	1.407	1.403	1.402
0.00400	(0.0392)	1.966	1.541	*	1.523	1.496	1.461	1.439	1.424	1.406	1.396	1.389	1.385	1.383	1.383
0.00600	(0.0588)	1.605	1.259	*	1.472	1.448	1.418	1.399	1.387	1.372	1.364	1.359	1.357	1.356	1.358
0.00800	(0.0785)	1.390	*	1.469	1.438	1.417	1.389	1.373	1.362	1.349	1.343	1.339	1.338	1.338	1.341
0.01000	(0.0981)	1.243	*	1.442	1.413	1.393	1.368	1.353	1.344	1.333	1.327	1.324	1.324	1.326	1.329
0.01500	(0.147)	*	*	1.396	1.371	1.354	1.333	1.320	1.312	1.304	1.300	1.299	1.300	1.303	1.309
0.02000	(0.196)	*	1.402	1.365	1.343	1.328	1.309	1.298	1.292	1.285	1.283	1.283	1.284	1.289	1.295
0.03000	(0.294)	1.381	1.357	1.325	1.306	1.293	1.278	1.269	1.265	1.260	1.260	1.261	1.264	1.270	1.277
0.04000	(0.392)	1.350	1.327	1.299	1.282	1.271	1.258	1.250	1.247	1.244	1.244	1.247	1.250	1.258	1.266
0.06000	(0.588)	1.308	1.289	1.265	1.251	1.241	1.231	1.226	1.224	1.223	1.225	1.229	1.233	1.242	1.251
0.08000	(0.785)	1.281	1.264	1.243	1.230	1.222	1.214	1.210	1.208	1.209	1.212	1.217	1.222	1.232	1.242
0.10000	(0.981)	1.261	1.246	1.226	1.215	1.208	1.201	1.198	1.197	1.199	1.203	1.208	1.214	1.225	1.235
0.15000	(1.47)	1.228	1.215	1.199	1.190	1.185	1.180	1.178	1.179	1.183	1.188	1.194	1.201	1.213	1.224
0.20000	(1.96)	1.207	1.195	1.181	1.173	1.169	1.166	1.166	1.167	1.172	1.178	1.185	1.192	1.205	1.217
0.30000	(2.94)	1.179	1.169	1.158	1.152	1.150	1.148	1.150	1.152	1.159	1.165	1.174	1.182	1.196	1.209
0.40000	(3.92)	1.161	1.152	1.143	1.139	1.137	1.137	1.139	1.143	1.150	1.157	1.166	1.175	1.190	1.203
0.60000	(5.88)	1.138	1.131	1.124	1.121	1.121	1.123	1.126	1.131	1.139	1.148	1.158	1.167	1.183	1.197
0.80000	(7.85)	1.123	1.117	1.112	1.110	1.111	1.114	1.118	1.123	1.132	1.142	1.152	1.161	1.178	1.192
1.00000	(9.81)	1.112	1.107	1.103	1.103	1.104	1.107	1.112	1.118	1.128	1.137	1.148	1.158	1.175	1.190
1.50000	(14.7)	1.095	1.091	1.089	1.090	1.092	1.097	1.103	1.109	1.120	1.130	1.142	1.152	1.170	1.185
2.00000	(19.6)	1.083	1.081	1.080	1.082	1.084	1.090	1.097	1.103	1.115	1.126	1.138	1.148	1.167	1.182
3.00000	(29.4)	1.070	1.068	1.069	1.072	1.075	1.082	1.090	1.097	1.109	1.121	1.133	1.144	1.163	1.179
S	$(\Delta p/\rho l)$	125	150	200	250	300	400	500	600	800	1000	1250	1500	2000	2500

Grad'nt k.p.g. (Equivalent) Pipe diameters in mm — Roughness size, $k_s = 0.30$ mm

E253

Kin. visc., $\nu = 80 \times 10^{-6}$ m^{2}s^{-1};
$S = 0.00010$ to 30.0000

i.e. kin. pr. grad., $\Delta p/\rho l =$
(0.00098) to (294) ms^{-2}

Roughness size, $k_s = 0.006$ mm
This table shows values of m, as follows

m_C for Colebrook-White solutions; or,
where $\mathbf{R} \leq 2000$, m_P for laminar flow

Grad'nt S	k.p.g. $(\Delta p/\rho l)$	6	8	10	12·5	15	20	25	30	40	50	60	80	100	125
0·00030	(0·00294)	548·6	373·8	277·6	206·2	161·7	110·2	81·82	64·16	43·72	32·47	25·46	17·35	12·89	9·570
0·00040	(0·00392)	475·1	323·7	240·4	178·6	140·0	95·41	70·86	55·57	37·86	28·12	22·05	15·03	11·16	8·288
0·00060	(0·00588)	387·9	264·3	196·3	145·8	114·3	77·90	57·86	45·37	30·92	22·96	18·01	12·27	9·112	6·767
0·00080	(0·00785)	335·9	228·9	170·0	126·3	99·01	67·47	50·10	39·29	26·77	19·88	15·59	10·63	7·891	5·860
0·00100	(0·00981)	300·5	204·8	152·1	112·9	88·56	60·34	44·82	35·14	23·95	17·78	13·95	9·504	7·058	5·242
0·00150	(0·0147)	245·3	167·2	124·2	92·20	72·31	49·27	36·59	28·69	19·55	14·52	11·39	7·760	5·763	4·280
0·00200	(0·0196)	212·5	144·8	107·5	79·85	62·62	42·67	31·69	24·85	16·93	12·58	9·862	6·720	4·991	3·706
0·00300	(0·0294)	173·5	118·2	87·79	65·20	51·13	34·84	25·87	20·29	13·83	10·27	8·052	5·487	4·075	3·026
0·00400	(0·0392)	150·2	102·4	76·03	56·46	44·28	30·17	22·41	17·57	11·97	8·892	6·973	4·752	3·529	2·621
0·00600	(0·0588)	122·7	83·59	62·08	46·10	36·15	24·64	18·30	14·35	9·777	7·261	5·694	3·880	2·881	2·140
0·00800	(0·0785)	106·2	72·39	53·76	39·93	31·31	21·34	15·84	12·43	8·467	6·288	4·931	3·360	2·495	1·853
0·01000	(0·0981)	95·02	64·75	48·09	35·71	28·00	19·08	14·17	11·11	7·573	5·624	4·410	3·005	2·232	1·658
0·01500	(0·147)	77·58	52·87	39·26	29·16	22·87	15·58	11·57	9·074	6·183	4·592	3·601	2·454	1·822	1·353
0·02000	(0·196)	67·19	45·78	34·00	25·25	19·80	13·49	10·02	7·858	5·355	3·977	3·119	2·125	1·578	1·172
0·03000	(0·294)	54·86	37·38	27·76	20·62	16·17	11·02	8·182	6·416	4·372	3·247	2·546	1·735	1·289	*
0·04000	(0·392)	47·51	32·37	24·04	17·86	14·00	9·541	7·086	5·557	3·786	2·812	2·205	1·503	1·116	*
0·06000	(0·588)	38·79	26·43	19·63	14·58	11·43	7·790	5·786	4·537	3·092	2·296	1·801	1·227	*	1·333
0·08000	(0·785)	33·59	22·89	17·00	12·63	9·901	6·747	5·010	3·929	2·677	1·988	1·559	1·063	*	1·299
0·10000	(0·981)	30·05	20·48	15·21	11·29	8·856	6·034	4·482	3·514	2·395	1·778	1·395	*	1·303	1·274
0·15000	(1·47)	24·53	16·72	12·42	9·220	7·231	4·927	3·659	2·869	1·955	1·452	1·139	*	1·256	1·230
0·20000	(1·96)	21·25	14·48	10·75	7·985	6·262	4·267	3·169	2·485	1·693	1·258	*	1·253	1·225	1·201
0·30000	(2·94)	17·35	11·82	8·779	6·520	5·113	3·484	2·587	2·029	1·383	1·027	*	1·208	1·183	1·163
0·40000	(3·92)	15·02	10·24	7·603	5·646	4·428	3·017	2·241	1·757	1·197	*	*	1·178	1·156	1·137
0·60000	(5·88)	12·27	8·359	6·208	4·610	3·615	2·464	1·830	1·435	0·978	*	1·170	1·138	1·119	1·137
0·80000	(7·85)	10·62	7·239	5·376	3·993	3·131	2·134	1·584	1·243	*	1·162	1·140	1·112	1·094	1·079
1·00000	(9·81)	9·502	6·475	4·809	3·571	2·800	1·908	1·417	1·111	*	1·139	1·118	1·092	1·075	1·062
1·50000	(14·7)	7·758	5·287	3·926	2·916	2·287	1·558	1·157	0·907	1·124	1·098	1·081	1·058	1·043	1·032
2·00000	(19·6)	6·719	4·578	3·400	2·525	1·980	1·349	1·002	*	1·095	1·072	1·056	1·035	1·022	1·012
3·00000	(29·4)	5·486	3·738	2·776	2·062	1·617	1·102	*	*	1·057	1·036	1·022	1·004	0·993	0·984
4·00000	(39·2)	4·751	3·237	2·404	1·786	1·400	0·954	*	1·061	1·031	1·012	1·000	0·983	0·974	0·966
6·00000	(58·8)	3·879	2·643	1·963	1·458	1·143	*	1·043	1·023	0·997	0·981	0·970	0·956	0·948	0·942
8·00000	(78·5)	3·359	2·289	1·700	1·263	0·990	*	1·016	0·998	0·974	0·960	0·950	0·937	0·930	0·925
10·0000	(98·1)	3·005	2·048	1·521	1·129	0·886	*	0·996	0·979	0·957	0·944	0·935	0·924	0·917	0·913
15·0000	(147)	2·453	1·672	1·242	0·922	*	0·983	0·962	0·947	0·928	0·917	0·909	0·900	0·894	0·891
20·0000	(196)	2·125	1·448	1·075	0·799	*	0·958	0·939	0·926	0·909	0·898	0·891	0·883	0·879	0·876
30·0000	(294)	1·735	1·182	0·878	*	0·953	0·926	0·909	0·897	0·883	0·874	0·868	0·862	0·858	0·857
S	$(\Delta p/\rho l)$	6	8	10	12·5	15	20	25	30	40	50	60	80	100	125

Grad'nt   k.p.g.   (Equivalent) Pipe diameters in mm

Grad'nt S	k.p.g. $(\Delta p/\rho l)$	125	150	200	250	300	400	500	600	800	1000	1250	1500	2000	2500
0·00010	(0·00098)	16·58	13·00	8·857	6·578	5·158	3·515	2·610	2·047	*	1·912	1·869	1·838	1·799	1·775
0·00015	(0·00147)	13·53	10·61	7·232	5·371	4·212	2·870	2·131	1·671	*	1·843	1·805	1·778	1·745	1·724
0·00020	(0·00196)	11·72	9·191	6·263	4·651	3·648	2·486	1·846	*	1·838	1·797	1·762	1·738	1·708	1·689
0·00030	(0·00294)	9·570	7·505	5·114	3·798	2·978	2·029	1·507	*	1·772	1·735	1·705	1·684	1·658	1·642
0·00040	(0·00392)	8·288	6·499	4·429	3·289	2·579	1·758	*	*	1·728	1·694	1·667	1·648	1·625	1·611
0·00060	(0·00588)	6·767	5·307	3·616	2·685	2·106	1·435	*	1·715	1·669	1·640	1·616	1·600	1·580	1·568
0·00080	(0·00785)	5·860	4·596	3·132	2·326	1·824	*	1·703	1·671	1·629	1·603	1·582	1·567	1·549	1·539
0·00100	(0·00981)	5·242	4·110	2·801	2·080	1·631	*	1·669	1·639	1·600	1·576	1·556	1·543	1·527	1·518
0·00150	(0·0147)	4·280	3·356	2·287	1·698	1·332	1·648	1·610	1·584	1·550	1·529	1·512	1·500	1·487	1·480
0·00200	(0·0196)	3·706	2·907	1·981	1·471	*	1·605	1·570	1·546	1·516	1·497	1·482	1·472	1·460	1·454
0·00300	(0·0294)	3·026	2·373	1·617	*	*	1·548	1·518	1·497	1·470	1·454	1·441	1·433	1·424	1·419
0·00400	(0·0392)	2·621	2·055	1·400	*	1·554	1·510	1·483	1·464	1·440	1·425	1·414	1·407	1·399	1·396
0·00600	(0·0588)	2·140	1·678	*	1·527	1·498	1·460	1·436	1·419	1·399	1·387	1·377	1·371	1·366	1·364
0·00800	(0·0785)	1·853	1·453	*	1·487	1·461	1·426	1·404	1·390	1·371	1·360	1·352	1·347	1·343	1·342
0·01000	(0·0981)	1·658	1·300	*	1·458	1·433	1·401	1·381	1·367	1·350	1·341	1·334	1·329	1·326	1·325
0·01500	(0·147)	1·353	*	1·438	1·407	1·385	1·357	1·340	1·328	1·314	1·306	1·301	1·298	1·296	1·297
0·02000	(0·196)	1·172	*	1·402	1·373	1·353	1·328	1·313	1·302	1·290	1·283	1·279	1·276	1·276	1·277
0·03000	(0·294)	*	1·393	1·353	1·328	1·311	1·289	1·276	1·267	1·257	1·252	1·249	1·247	1·248	1·250
0·04000	(0·392)	*	1·357	1·321	1·298	1·282	1·263	1·251	1·243	1·235	1·230	1·228	1·228	1·229	1·232
0·06000	(0·588)	1·333	1·309	1·277	1·258	1·244	1·227	1·218	1·211	1·205	1·202	1·200	1·201	1·203	1·207
0·08000	(0·785)	1·299	1·277	1·248	1·231	1·219	1·204	1·195	1·190	1·184	1·182	1·182	1·182	1·186	1·190
0·10000	(0·981)	1·274	1·253	1·227	1·210	1·199	1·186	1·178	1·173	1·169	1·167	1·167	1·169	1·173	1·178
0·15000	(1·47)	1·230	1·213	1·190	1·175	1·166	1·155	1·148	1·145	1·142	1·141	1·142	1·145	1·150	1·155
0·20000	(1·96)	1·201	1·185	1·164	1·152	1·143	1·134	1·128	1·126	1·123	1·124	1·125	1·128	1·134	1·140
0·30000	(2·94)	1·163	1·149	1·131	1·120	1·113	1·105	1·101	1·099	1·099	1·100	1·102	1·106	1·112	1·119
0·40000	(3·92)	1·137	1·124	1·108	1·099	1·093	1·086	1·083	1·082	1·082	1·084	1·087	1·090	1·097	1·104
0·60000	(5·88)	1·102	1·092	1·078	1·070	1·065	1·060	1·058	1·058	1·059	1·062	1·065	1·069	1·077	1·085
0·80000	(7·85)	1·079	1·069	1·057	1·051	1·046	1·042	1·041	1·041	1·043	1·046	1·051	1·055	1·064	1·072
1·00000	(9·81)	1·062	1·053	1·042	1·036	1·032	1·029	1·028	1·029	1·032	1·035	1·040	1·045	1·054	1·062
1·50000	(14·7)	1·032	1·024	1·015	1·011	1·008	1·006	1·006	1·007	1·011	1·015	1·021	1·026	1·035	1·044
2·00000	(19·6)	1·012	1·005	0·997	0·993	0·991	0·990	0·991	0·993	0·997	1·002	1·008	1·013	1·023	1·032
3·00000	(29·4)	0·984	0·979	0·973	0·970	0·969	0·969	0·971	0·973	0·978	0·983	0·990	0·996	1·006	1·016
S	$(\Delta p/\rho l)$	125	150	200	250	300	400	500	600	800	1000	1250	1500	2000	2500

Grad'nt   k.p.g.   (Equivalent) Pipe diameters in mm

Kinematic viscosity, $\nu = 80 \times 10^{-6}$ m^{2}s^{-1} ;   **Roughness size, $k_s = 0.006$ mm**

Kin. visc., $\nu = 80 \times 10^{-6}$ m^{2}s^{-1};
$S = 0.00010$ to 30.0000

i.e. kin. pr. grad., $\Delta p/\rho l =$
(0·00098) to (294) ms^{-2}

Roughness size, $k_s = 0.015$ mm
This table shows values of m, as follows

m_C for Colebrook-White solutions; or,
where $\mathbf{R} \leq 2000$, m_P for laminar flow

E254

Grad'nt S	k.p.g. $(\Delta p/\rho l)$	6	8	10	12·5	15	20	25	30	40	50	60	80	100	125
0·00030	(0·00294)	548·6	373·8	277·6	206·2	161·7	110·2	81·82	64·16	43·72	32·47	25·46	17·35	12·89	9·570
0·00040	(0·00392)	475·1	323·7	240·4	178·6	140·0	95·41	70·86	55·57	37·86	28·12	22·05	15·03	11·16	8·288
0·00060	(0·00588)	387·9	264·3	196·3	145·8	114·3	77·90	57·86	45·37	30·92	22·96	18·01	12·27	9·112	6·767
0·00080	(0·00785)	335·9	228·9	170·0	126·3	99·01	67·47	50·10	39·29	26·77	19·88	15·59	10·63	7·891	5·860
0·00100	(0·00981)	300·5	204·8	152·1	112·9	88·56	60·34	44·82	35·14	23·95	17·78	13·95	9·504	7·058	5·242
0·00150	(0·0147)	245·3	167·2	124·2	92·20	72·31	49·27	36·59	28·69	19·55	14·52	11·39	7·760	5·763	4·280
0·00200	(0·0196)	212·5	144·8	107·5	79·85	62·62	42·67	31·69	24·85	16·93	12·58	9·862	6·720	4·991	3·706
0·00300	(0·0294)	173·5	118·2	87·79	65·20	51·13	34·84	25·87	20·29	13·83	10·27	8·052	5·487	4·075	3·026
0·00400	(0·0392)	150·2	102·4	76·03	56·46	44·28	30·17	22·41	17·57	11·97	8·892	6·973	4·752	3·529	2·621
0·00600	(0·0588)	122·7	83·59	62·08	46·10	36·15	24·64	18·30	14·35	9·777	7·261	5·694	3·880	2·881	2·140
0·00800	(0·0785)	106·2	72·39	53·76	39·93	31·31	21·34	15·84	12·43	8·467	6·288	4·931	3·360	2·495	1·853
0·01000	(0·0981)	95·02	64·75	48·09	35·71	28·00	19·08	14·17	11·11	7·573	5·624	4·410	3·005	2·232	1·658
0·01500	(0·147)	77·58	52·87	39·26	29·16	22·87	15·58	11·57	9·074	6·183	4·592	3·601	2·454	1·822	1·353
0·02000	(0·196)	67·19	45·78	34·00	25·25	19·80	13·49	10·02	7·858	5·355	3·977	3·119	2·125	1·578	1·172
0·03000	(0·294)	54·86	37·38	27·76	20·62	16·17	11·02	8·182	6·416	4·372	3·247	2·546	1·735	1·289	*
0·04000	(0·392)	47·51	32·37	24·04	17·86	14·00	9·541	7·086	5·557	3·786	2·812	2·205	1·503	1·116	*
0·06000	(0·588)	38·79	26·43	19·63	14·58	11·43	7·790	5·786	4·537	3·092	2·296	1·801	1·227	*	1·334
0·08000	(0·785)	33·59	22·89	17·00	12·63	9·901	6·747	5·010	3·929	2·677	1·988	1·559	1·063	*	1·300
0·10000	(0·981)	30·05	20·48	15·21	11·29	8·856	6·034	4·482	3·514	2·395	1·778	1·395	*	1·305	1·275
0·15000	(1·47)	24·53	16·72	12·42	9·220	7·231	4·927	3·659	2·869	1·955	1·452	1·139	*	1·258	1·232
0·20000	(1·96)	21·25	14·48	10·75	7·985	6·262	4·267	3·169	2·485	1·693	1·258	*	1·255	1·227	1·203
0·30000	(2·94)	17·35	11·82	8·779	6·520	5·113	3·484	2·587	2·029	1·383	1·027	*	1·210	1·185	1·165
0·40000	(3·92)	15·02	10·24	7·603	5·646	4·428	3·017	2·241	1·757	1·197	*	*	1·180	1·158	1·139
0·60000	(5·88)	12·27	8·359	6·208	4·610	3·615	2·464	1·830	1·435	0·978	*	1·172	1·140	1·121	1·105
0·80000	(7·85)	10·62	7·239	5·376	3·993	3·131	2·134	1·584	1·243	*	1·164	1·142	1·114	1·096	1·082
1·00000	(9·81)	9·502	6·475	4·809	3·571	2·800	1·908	1·417	1·111	*	1·141	1·121	1·094	1·078	1·065
1·50000	(14·7)	7·758	5·287	3·926	2·916	2·287	1·558	1·157	0·907	1·127	1·101	1·084	1·061	1·047	1·035
2·00000	(19·6)	6·719	4·578	3·400	2·525	1·980	1·349	1·002	*	1·098	1·075	1·059	1·038	1·025	1·015
3·00000	(29·4)	5·486	3·738	2·776	2·062	1·617	1·102	*	*	1·060	1·040	1·026	1·008	0·997	0·989
4·00000	(39·2)	4·751	3·237	2·404	1·786	1·400	0·954	*	1·064	1·035	1·016	1·004	0·988	0·978	0·971
6·00000	(58·8)	3·879	2·643	1·963	1·458	1·143	*	1·047	1·027	1·001	0·985	0·974	0·961	0·953	0·947
8·00000	(78·5)	3·359	2·289	1·700	1·263	0·990	*	1·020	1·002	0·979	0·964	0·955	0·943	0·936	0·931
10·0000	(98·1)	3·005	2·048	1·521	1·129	0·886	*	1·001	0·984	0·962	0·949	0·940	0·929	0·923	0·919
15·0000	(147)	2·453	1·672	1·242	0·922	*	0·988	0·967	0·953	0·934	0·923	0·915	0·906	0·901	0·898
20·0000	(196)	2·125	1·448	1·075	0·799	*	0·964	0·945	0·932	0·915	0·905	0·898	0·891	0·887	0·884
30·0000	(294)	1·735	1·182	0·878	*	0·959	0·932	0·915	0·904	0·890	0·881	0·876	0·870	0·867	0·866
S	$(\Delta p/\rho l)$	6	8	10	12·5	15	20	25	30	40	50	60	80	100	125

Grad'nt   k.p.g.   (Equivalent) Pipe diameters in mm

Grad'nt S	k.p.g. $(\Delta p/\rho l)$	125	150	200	250	300	400	500	600	800	1000	1250	1500	2000	2500
0·00010	(0·00098)	16·58	13·00	8·857	6·578	5·158	3·515	2·610	2·047	*	1·912	1·869	1·839	1·800	1·776
0·00015	(0·00147)	13·53	10·61	7·232	5·371	4·212	2·870	2·131	1·671	*	1·843	1·805	1·779	1·745	1·724
0·00020	(0·00196)	11·72	9·191	6·263	4·651	3·648	2·486	1·846	*	1·838	1·797	1·762	1·738	1·708	1·689
0·00030	(0·00294)	9·570	7·505	5·114	3·798	2·978	2·029	1·507	*	1·772	1·736	1·705	1·685	1·658	1·642
0·00040	(0·00392)	8·288	6·499	4·429	3·289	2·579	1·758	*	*	1·728	1·695	1·667	1·649	1·625	1·611
0·00060	(0·00588)	6·767	5·307	3·616	2·685	2·106	1·435	*	1·715	1·669	1·640	1·616	1·600	1·580	1·568
0·00080	(0·00785)	5·860	4·596	3·132	2·326	1·824	*	1·704	1·672	1·630	1·604	1·582	1·568	1·550	1·540
0·00100	(0·00981)	5·242	4·110	2·801	2·080	1·631	*	1·669	1·639	1·601	1·576	1·557	1·543	1·527	1·518
0·00150	(0·0147)	4·280	3·356	2·287	1·698	1·332	1·648	1·610	1·584	1·550	1·529	1·512	1·501	1·487	1·480
0·00200	(0·0196)	3·706	2·907	1·981	1·471	*	1·605	1·571	1·547	1·516	1·497	1·482	1·472	1·461	1·455
0·00300	(0·0294)	3·026	2·373	1·617	*	*	1·548	1·518	1·498	1·471	1·455	1·442	1·434	1·424	1·420
0·00400	(0·0392)	2·621	2·055	1·400	*	1·554	1·511	1·483	1·464	1·440	1·426	1·415	1·408	1·400	1·397
0·00600	(0·0588)	2·140	1·678	*	1·528	1·498	1·460	1·436	1·420	1·400	1·387	1·378	1·372	1·367	1·365
0·00800	(0·0785)	1·853	1·453	*	1·488	1·461	1·427	1·405	1·390	1·372	1·361	1·353	1·348	1·344	1·343
0·01000	(0·0981)	1·658	1·300	*	1·458	1·434	1·401	1·382	1·368	1·351	1·342	1·335	1·330	1·327	1·327
0·01500	(0·147)	1·353	*	1·439	1·408	1·386	1·358	1·341	1·329	1·315	1·307	1·302	1·299	1·297	1·298
0·02000	(0·196)	1·172	*	1·402	1·374	1·354	1·329	1·314	1·303	1·291	1·284	1·280	1·278	1·277	1·279
0·03000	(0·294)	*	1·394	1·354	1·329	1·312	1·290	1·277	1·268	1·258	1·253	1·250	1·249	1·250	1·252
0·04000	(0·392)	*	1·358	1·322	1·299	1·284	1·264	1·252	1·245	1·236	1·232	1·230	1·229	1·231	1·234
0·06000	(0·588)	1·334	1·310	1·279	1·259	1·246	1·229	1·219	1·213	1·206	1·203	1·202	1·203	1·206	1·210
0·08000	(0·785)	1·300	1·278	1·250	1·232	1·220	1·205	1·197	1·191	1·186	1·184	1·184	1·185	1·189	1·193
0·10000	(0·981)	1·275	1·255	1·228	1·212	1·201	1·187	1·180	1·175	1·171	1·169	1·170	1·171	1·176	1·181
0·15000	(1·47)	1·232	1·214	1·191	1·177	1·168	1·157	1·150	1·147	1·144	1·144	1·145	1·147	1·153	1·159
0·20000	(1·96)	1·203	1·187	1·166	1·154	1·145	1·136	1·131	1·128	1·126	1·127	1·129	1·131	1·137	1·144
0·30000	(2·94)	1·165	1·151	1·133	1·122	1·115	1·108	1·104	1·102	1·102	1·103	1·106	1·109	1·116	1·123
0·40000	(3·92)	1·139	1·126	1·111	1·101	1·095	1·089	1·086	1·085	1·085	1·087	1·091	1·095	1·102	1·110
0·60000	(5·88)	1·105	1·094	1·081	1·073	1·068	1·063	1·062	1·061	1·063	1·066	1·070	1·074	1·083	1·091
0·80000	(7·85)	1·082	1·072	1·060	1·054	1·050	1·046	1·045	1·045	1·048	1·051	1·056	1·061	1·070	1·078
1·00000	(9·81)	1·065	1·056	1·045	1·039	1·036	1·033	1·033	1·033	1·037	1·040	1·046	1·051	1·060	1·069
1·50000	(14·7)	1·035	1·028	1·019	1·014	1·012	1·011	1·011	1·013	1·017	1·021	1·027	1·033	1·043	1·053
2·00000	(19·6)	1·015	1·009	1·001	0·998	0·996	0·995	0·997	0·999	1·003	1·009	1·015	1·021	1·032	1·041
3·00000	(29·4)	0·989	0·983	0·978	0·975	0·974	0·975	0·977	0·980	0·985	0·991	0·998	1·005	1·016	1·027
S	$(\Delta p/\rho l)$	125	150	200	250	300	400	500	600	800	1000	1250	1500	2000	2500

Grad'nt   k.p.g.   (Equivalent) Pipe diameters in mm

Kinematic viscosity, $\nu = 80 \times 10^{-6}$ m^{2}s^{-1} ;   Roughness size, $k_s = 0.015$ mm

E255

Kin. visc., $\nu = 80 \times 10^{-6}\ \mathrm{m^2\,s^{-1}}$;
$S = 0{\cdot}00010$ to $3{\cdot}00000$

i.e. kin. pr. grad., $\Delta p/\rho l =$ ($0{\cdot}00098$) to ($29{\cdot}4$) $\mathrm{ms^{-2}}$

Roughness size, $k_s = 0{\cdot}030$ mm
This table shows values of m, as follows

m_C for Colebrook-White solutions; or, where $\mathbf{R} \le 2000$, m_P for laminar flow

Grad'nt S	k.p.g. $(\Delta p/\rho l)$	\(Equivalent) Pipe diameters in mm 125	150	200	250	300	400	500	600	800	1000	1250	1500	2000	2500
0·00010	(0·00098)	16·58	13·00	8·857	6·578	5·158	3·515	2·610	2·047	*	1·912	1·869	1·839	1·800	1·776
0·00015	(0·00147)	13·53	10·61	7·232	5·371	4·212	2·870	2·131	1·671	*	1·843	1·805	1·779	1·745	1·724
0·00020	(0·00196)	11·72	9·191	6·263	4·651	3·648	2·486	1·846	*	1·839	1·797	1·763	1·739	1·708	1·690
0·00030	(0·00294)	9·570	7·505	5·114	3·798	2·978	2·029	1·507	*	1·772	1·736	1·706	1·685	1·659	1·643
0·00040	(0·00392)	8·288	6·499	4·429	3·289	2·579	1·758	*	*	1·728	1·695	1·668	1·649	1·625	1·612
0·00060	(0·00588)	6·767	5·307	3·616	2·685	2·106	1·435	*	1·716	1·670	1·641	1·617	1·601	1·581	1·569
0·00080	(0·00785)	5·860	4·596	3·132	2·326	1·824	*	1·704	1·672	1·630	1·604	1·583	1·568	1·551	1·540
0·00100	(0·00981)	5·242	4·110	2·801	2·080	1·631	*	1·670	1·640	1·601	1·577	1·557	1·544	1·528	1·519
0·00150	(0·0147)	4·280	3·356	2·287	1·698	1·332	1·649	1·611	1·585	1·551	1·530	1·513	1·502	1·488	1·481
0·00200	(0·0196)	3·706	2·907	1·981	1·471	*	1·606	1·571	1·548	1·517	1·498	1·483	1·473	1·462	1·456
0·00300	(0·0294)	3·026	2·373	1·617	*	*	1·549	1·519	1·498	1·472	1·456	1·443	1·435	1·426	1·421
0·00400	(0·0392)	2·621	2·055	1·400	*	1·555	1·511	1·484	1·465	1·441	1·427	1·416	1·409	1·401	1·398
0·00600	(0·0588)	2·140	1·678	*	1·529	1·499	1·461	1·437	1·421	1·401	1·389	1·379	1·374	1·368	1·366
0·00800	(0·0785)	1·853	1·453	*	1·489	1·462	1·428	1·406	1·392	1·373	1·363	1·355	1·350	1·346	1·345
0·01000	(0·0981)	1·658	1·300	*	1·459	1·435	1·403	1·383	1·369	1·353	1·343	1·336	1·332	1·329	1·329
0·01500	(0·147)	1·353	*	1·440	1·409	1·387	1·360	1·342	1·331	1·317	1·309	1·304	1·301	1·300	1·300
0·02000	(0·196)	1·172	*	1·404	1·375	1·356	1·331	1·315	1·305	1·293	1·286	1·282	1·280	1·280	1·281
0·03000	(0·294)	*	1·395	1·356	1·331	1·314	1·292	1·279	1·270	1·260	1·255	1·253	1·252	1·253	1·255
0·04000	(0·392)	*	1·359	1·323	1·301	1·285	1·266	1·254	1·247	1·239	1·235	1·233	1·232	1·234	1·238
0·06000	(0·588)	1·336	1·312	1·281	1·261	1·248	1·231	1·221	1·215	1·209	1·206	1·206	1·206	1·209	1·214
0·08000	(0·785)	1·302	1·280	1·252	1·234	1·222	1·208	1·199	1·194	1·189	1·187	1·187	1·189	1·193	1·198
0·10000	(0·981)	1·277	1·257	1·231	1·214	1·204	1·190	1·183	1·178	1·174	1·173	1·174	1·175	1·180	1·185
0·15000	(1·47)	1·234	1·217	1·194	1·180	1·171	1·160	1·154	1·151	1·148	1·148	1·150	1·152	1·158	1·164
0·20000	(1·96)	1·206	1·190	1·169	1·157	1·149	1·139	1·134	1·132	1·130	1·131	1·134	1·137	1·143	1·150
0·30000	(2·94)	1·168	1·154	1·136	1·126	1·119	1·112	1·108	1·107	1·107	1·109	1·112	1·116	1·123	1·131
0·40000	(3·92)	1·143	1·130	1·114	1·105	1·099	1·093	1·091	1·090	1·091	1·093	1·097	1·101	1·110	1·118
0·60000	(5·88)	1·109	1·098	1·085	1·077	1·073	1·069	1·067	1·067	1·069	1·073	1·078	1·082	1·092	1·100
0·80000	(7·85)	1·086	1·077	1·065	1·059	1·055	1·052	1·051	1·052	1·055	1·059	1·064	1·070	1·080	1·089
1·00000	(9·81)	1·069	1·061	1·050	1·045	1·042	1·039	1·039	1·041	1·044	1·049	1·054	1·060	1·071	1·080
1·50000	(14·7)	1·041	1·033	1·025	1·021	1·019	1·018	1·019	1·021	1·026	1·031	1·038	1·044	1·055	1·065
2·00000	(19·6)	1·021	1·015	1·008	1·005	1·003	1·003	1·005	1·008	1·013	1·019	1·026	1·033	1·045	1·056
3·00000	(29·4)	0·996	0·991	0·985	0·983	0·983	0·984	0·987	0·990	0·997	1·003	1·011	1·018	1·031	1·043
S $(\Delta p/\rho l)$		125	150	200	250	300	400	500	600	800	1000	1250	1500	2000	2500

Grad'nt   k.p.g.   (Equivalent) Pipe diameters in mm   Roughness size, $k_s = 0{\cdot}030$ mm

E256

Kin. visc., $\nu = 80 \times 10^{-6}\ \mathrm{m^2\,s^{-1}}$;
$S = 0{\cdot}00010$ to $3{\cdot}00000$

i.e. kin. pr. grad., $\Delta p/\rho l =$ ($0{\cdot}00098$) to ($29{\cdot}4$) $\mathrm{ms^{-2}}$

Roughness size, $k_s = 0{\cdot}060$ mm
This table shows values of m, as follows

m_C for Colebrook-White solutions; or, where $\mathbf{R} \le 2000$, m_P for laminar flow

Grad'nt S	k.p.g. $(\Delta p/\rho l)$	\(Equivalent) Pipe diameters in mm 125	150	200	250	300	400	500	600	800	1000	1250	1500	2000	2500
0·00010	(0·00098)	16·58	13·00	8·857	6·578	5·158	3·515	2·610	2·047	*	1·913	1·870	1·840	1·801	1·777
0·00015	(0·00147)	13·53	10·61	7·232	5·371	4·212	2·870	2·131	1·671	*	1·844	1·806	1·780	1·746	1·725
0·00020	(0·00196)	11·72	9·191	6·263	4·651	3·648	2·486	1·846	*	1·840	1·798	1·763	1·740	1·709	1·691
0·00030	(0·00294)	9·570	7·505	5·114	3·798	2·978	2·029	1·507	*	1·773	1·737	1·707	1·686	1·660	1·644
0·00040	(0·00392)	8·288	6·499	4·429	3·289	2·579	1·758	*	*	1·729	1·696	1·669	1·650	1·627	1·613
0·00060	(0·00588)	6·767	5·307	3·616	2·685	2·106	1·435	*	1·717	1·671	1·642	1·618	1·602	1·582	1·571
0·00080	(0·00785)	5·860	4·596	3·132	2·326	1·824	*	1·706	1·673	1·632	1·605	1·584	1·570	1·552	1·542
0·00100	(0·00981)	5·242	4·110	2·801	2·080	1·631	*	1·671	1·641	1·603	1·578	1·559	1·546	1·529	1·521
0·00150	(0·0147)	4·280	3·356	2·287	1·698	1·332	1·650	1·612	1·586	1·552	1·531	1·515	1·503	1·490	1·483
0·00200	(0·0196)	3·706	2·907	1·981	1·471	*	1·607	1·573	1·549	1·519	1·500	1·485	1·475	1·464	1·458
0·00300	(0·0294)	3·026	2·373	1·617	*	*	1·551	1·521	1·500	1·474	1·458	1·445	1·437	1·428	1·424
0·00400	(0·0392)	2·621	2·055	1·400	*	1·557	1·513	1·486	1·467	1·444	1·429	1·418	1·411	1·404	1·401
0·00600	(0·0588)	2·140	1·678	*	1·530	1·501	1·463	1·440	1·423	1·403	1·391	1·382	1·377	1·371	1·370
0·00800	(0·0785)	1·853	1·453	*	1·491	1·464	1·430	1·409	1·394	1·376	1·365	1·358	1·353	1·349	1·349
0·01000	(0·0981)	1·658	1·300	*	1·462	1·437	1·405	1·385	1·372	1·356	1·346	1·339	1·336	1·333	1·333
0·01500	(0·147)	1·353	*	1·443	1·412	1·390	1·362	1·345	1·334	1·320	1·313	1·308	1·305	1·304	1·305
0·02000	(0·196)	1·172	*	1·407	1·378	1·359	1·334	1·319	1·309	1·297	1·290	1·286	1·284	1·284	1·286
0·03000	(0·294)	*	1·398	1·359	1·334	1·317	1·296	1·283	1·274	1·265	1·260	1·257	1·257	1·258	1·261
0·04000	(0·392)	*	1·363	1·327	1·304	1·289	1·270	1·259	1·251	1·243	1·240	1·238	1·238	1·240	1·244
0·06000	(0·588)	1·340	1·316	1·284	1·265	1·252	1·236	1·226	1·221	1·214	1·212	1·212	1·213	1·217	1·221
0·08000	(0·785)	1·306	1·285	1·256	1·239	1·227	1·213	1·205	1·200	1·195	1·194	1·194	1·196	1·201	1·206
0·10000	(0·981)	1·282	1·261	1·235	1·219	1·209	1·196	1·188	1·184	1·181	1·180	1·181	1·183	1·189	1·195
0·15000	(1·47)	1·239	1·222	1·199	1·186	1·177	1·166	1·161	1·158	1·156	1·156	1·158	1·161	1·168	1·175
0·20000	(1·96)	1·211	1·195	1·175	1·163	1·155	1·146	1·142	1·140	1·139	1·140	1·143	1·147	1·154	1·162
0·30000	(2·94)	1·174	1·160	1·143	1·133	1·127	1·120	1·117	1·116	1·116	1·119	1·123	1·127	1·136	1·144
0·40000	(3·92)	1·149	1·137	1·122	1·113	1·108	1·102	1·100	1·100	1·101	1·105	1·109	1·114	1·124	1·133
0·60000	(5·88)	1·117	1·106	1·093	1·086	1·082	1·079	1·078	1·078	1·082	1·086	1·091	1·097	1·108	1·117
0·80000	(7·85)	1·095	1·086	1·075	1·069	1·065	1·063	1·063	1·064	1·068	1·073	1·079	1·086	1·097	1·107
1·00000	(9·81)	1·079	1·070	1·061	1·055	1·053	1·051	1·052	1·054	1·059	1·064	1·071	1·077	1·089	1·100
1·50000	(14·7)	1·051	1·044	1·037	1·033	1·031	1·031	1·033	1·036	1·042	1·048	1·056	1·063	1·076	1·087
2·00000	(19·6)	1·033	1·027	1·021	1·018	1·017	1·018	1·021	1·024	1·031	1·038	1·046	1·054	1·067	1·079
3·00000	(29·4)	1·008	1·004	1·000	0·998	0·999	1·001	1·005	1·009	1·017	1·025	1·034	1·042	1·056	1·069
S $(\Delta p/\rho l)$		125	150	200	250	300	400	500	600	800	1000	1250	1500	2000	2500

Grad'nt   k.p.g.   (Equivalent) Pipe diameters in mm   Roughness size, $k_s = 0{\cdot}060$ mm

Kin. visc., $\nu = 80 \times 10^{-6}$ m^{2}s^{-1};
$S = 0.00010$ to 3.00000

i.e. kin. pr. grad., $\Delta p/\rho l =$
(0.00098) to (29.4) ms^{-2}

Roughness size, $k_s = 0.150$ mm
This table shows values of m, as follows

m_C for Colebrook-White solutions; or,
where $R \leq 2000$, m_P for laminar flow

E257

Grad'nt S	k.p.g. $(\Delta p/\rho l)$	125	150	200	250	300	400	500	600	800	1000	1250	1500	2000	2500
0.00010	(0.00098)	16.58	13.00	8.857	6.578	5.158	3.515	2.610	2.047	*	1.915	1.872	1.842	1.803	1.779
0.00015	(0.00147)	13.53	10.61	7.232	5.371	4.212	2.870	2.131	1.671	*	1.846	1.808	1.782	1.748	1.728
0.00020	(0.00196)	11.72	9.191	6.263	4.651	3.648	2.486	1.846	*	1.842	1.800	1.766	1.742	1.712	1.693
0.00030	(0.00294)	9.570	7.505	5.114	3.798	2.978	2.029	1.507	*	1.776	1.740	1.710	1.689	1.663	1.647
0.00040	(0.00392)	8.288	6.499	4.429	3.289	2.579	1.758	*	*	1.732	1.699	1.672	1.653	1.630	1.616
0.00060	(0.00588)	6.767	5.307	3.616	2.685	2.106	1.435	*	1.720	1.674	1.645	1.622	1.606	1.586	1.575
0.00080	(0.00785)	5.860	4.596	3.132	2.326	1.824	*	1.709	1.677	1.635	1.609	1.588	1.574	1.556	1.547
0.00100	(0.00981)	5.242	4.110	2.801	2.080	1.631	*	1.675	1.645	1.606	1.582	1.563	1.550	1.534	1.526
0.00150	(0.0147)	4.280	3.356	2.287	1.698	1.332	1.654	1.616	1.590	1.557	1.536	1.519	1.509	1.496	1.489
0.00200	(0.0196)	3.706	2.907	1.981	1.471	*	1.612	1.577	1.554	1.524	1.505	1.490	1.481	1.470	1.464
0.00300	(0.0294)	3.026	2.373	1.617	*	*	1.556	1.526	1.505	1.479	1.464	1.451	1.443	1.435	1.431
0.00400	(0.0392)	2.621	2.055	1.400	*	1.562	1.519	1.492	1.473	1.450	1.436	1.425	1.418	1.412	1.409
0.00600	(0.0588)	2.140	1.678	*	1.536	1.507	1.470	1.446	1.430	1.410	1.399	1.390	1.385	1.380	1.379
0.00800	(0.0785)	1.853	1.453	*	1.497	1.471	1.437	1.416	1.401	1.384	1.374	1.366	1.362	1.359	1.359
0.01000	(0.0981)	1.658	1.300	*	1.469	1.444	1.412	1.393	1.380	1.364	1.355	1.349	1.346	1.343	1.344
0.01500	(0.147)	1.353	*	1.450	1.419	1.398	1.371	1.354	1.343	1.330	1.323	1.319	1.317	1.316	1.318
0.02000	(0.196)	1.172	*	1.415	1.387	1.367	1.343	1.328	1.319	1.307	1.302	1.298	1.297	1.298	1.301
0.03000	(0.294)	*	1.407	1.368	1.344	1.327	1.306	1.294	1.286	1.277	1.273	1.271	1.271	1.274	1.278
0.04000	(0.392)	*	1.372	1.337	1.315	1.300	1.281	1.271	1.264	1.257	1.254	1.253	1.254	1.258	1.263
0.06000	(0.588)	1.351	1.327	1.296	1.277	1.264	1.249	1.240	1.235	1.230	1.229	1.229	1.231	1.237	1.243
0.08000	(0.785)	1.318	1.297	1.269	1.252	1.241	1.227	1.220	1.216	1.212	1.212	1.213	1.216	1.222	1.229
0.10000	(0.981)	1.294	1.274	1.249	1.233	1.223	1.211	1.205	1.201	1.199	1.199	1.202	1.205	1.212	1.219
0.15000	(1.47)	1.254	1.237	1.215	1.202	1.193	1.184	1.179	1.177	1.176	1.178	1.182	1.186	1.194	1.203
0.20000	(1.96)	1.227	1.211	1.192	1.181	1.173	1.166	1.162	1.161	1.162	1.164	1.169	1.173	1.183	1.192
0.30000	(2.94)	1.192	1.179	1.162	1.153	1.147	1.142	1.140	1.140	1.142	1.146	1.152	1.157	1.168	1.178
0.40000	(3.92)	1.169	1.157	1.143	1.135	1.130	1.126	1.125	1.126	1.130	1.134	1.140	1.147	1.158	1.169
0.60000	(5.88)	1.138	1.129	1.117	1.111	1.108	1.106	1.106	1.108	1.113	1.119	1.126	1.133	1.146	1.158
0.80000	(7.85)	1.118	1.110	1.100	1.095	1.093	1.092	1.094	1.096	1.102	1.109	1.117	1.124	1.138	1.150
1.00000	(9.81)	1.104	1.096	1.088	1.084	1.082	1.083	1.085	1.088	1.095	1.102	1.110	1.118	1.132	1.145
1.50000	(14.7)	1.079	1.073	1.067	1.065	1.065	1.066	1.070	1.074	1.082	1.090	1.099	1.108	1.123	1.136
2.00000	(19.6)	1.063	1.058	1.054	1.053	1.053	1.056	1.060	1.065	1.074	1.082	1.092	1.101	1.117	1.131
3.00000	(29.4)	1.043	1.039	1.037	1.037	1.038	1.042	1.048	1.053	1.063	1.073	1.084	1.093	1.110	1.124
S	$(\Delta p/\rho l)$	125	150	200	250	300	400	500	600	800	1000	1250	1500	2000	2500

Grad'nt   k.p.g.   (Equivalent) Pipe diameters in mm   Roughness size, $k_s = 0.150$ mm

Kin. visc., $\nu = 80 \times 10^{-6}$ m^{2}s^{-1};
$S = 0.00010$ to 3.00000

i.e. kin. pr. grad., $\Delta p/\rho l =$
(0.00098) to (29.4) ms^{-2}

Roughness size, $k_s = 0.30$ mm
This table shows values of m, as follows

m_C for Colebrook-White solutions; or,
where $R \leq 2000$, m_P for laminar flow

E258

Grad'nt S	k.p.g. $(\Delta p/\rho l)$	125	150	200	250	300	400	500	600	800	1000	1250	1500	2000	2500
0.00010	(0.00098)	16.58	13.00	8.857	6.578	5.158	3.515	2.610	2.047	*	1.918	1.875	1.845	1.806	1.782
0.00015	(0.00147)	13.53	10.61	7.232	5.371	4.212	2.870	2.131	1.671	*	1.850	1.812	1.786	1.752	1.732
0.00020	(0.00196)	11.72	9.191	6.263	4.651	3.648	2.486	1.846	*	1.846	1.804	1.770	1.746	1.716	1.698
0.00030	(0.00294)	9.570	7.505	5.114	3.798	2.978	2.029	1.507	*	1.780	1.744	1.714	1.694	1.668	1.653
0.00040	(0.00392)	8.288	6.499	4.429	3.289	2.579	1.758	*	*	1.737	1.704	1.677	1.659	1.636	1.622
0.00060	(0.00588)	6.767	5.307	3.616	2.685	2.106	1.435	*	1.725	1.679	1.651	1.628	1.612	1.592	1.581
0.00080	(0.00785)	5.860	4.596	3.132	2.326	1.824	*	1.714	1.682	1.641	1.615	1.594	1.580	1.563	1.554
0.00100	(0.00981)	5.242	4.110	2.801	2.080	1.631	*	1.681	1.651	1.613	1.589	1.570	1.557	1.542	1.534
0.00150	(0.0147)	4.280	3.356	2.287	1.698	1.332	1.661	1.623	1.597	1.564	1.544	1.527	1.517	1.505	1.498
0.00200	(0.0196)	3.706	2.907	1.981	1.471	*	1.619	1.585	1.561	1.531	1.513	1.499	1.490	1.480	1.475
0.00300	(0.0294)	3.026	2.373	1.617	*	*	1.564	1.534	1.514	1.488	1.473	1.461	1.454	1.446	1.443
0.00400	(0.0392)	2.621	2.055	1.400	*	1.571	1.528	1.501	1.483	1.460	1.446	1.436	1.430	1.424	1.422
0.00600	(0.0588)	2.140	1.678	*	1.546	1.517	1.480	1.456	1.441	1.422	1.411	1.403	1.398	1.394	1.394
0.00800	(0.0785)	1.853	1.453	*	1.508	1.482	1.448	1.427	1.413	1.396	1.387	1.380	1.377	1.375	1.376
0.01000	(0.0981)	1.658	1.300	*	1.480	1.455	1.424	1.405	1.393	1.377	1.369	1.364	1.361	1.360	1.362
0.01500	(0.147)	1.353	*	1.463	1.432	1.411	1.384	1.368	1.358	1.345	1.339	1.335	1.334	1.335	1.339
0.02000	(0.196)	1.172	*	1.428	1.400	1.382	1.358	1.343	1.334	1.324	1.319	1.317	1.316	1.319	1.323
0.03000	(0.294)	*	1.422	1.383	1.359	1.343	1.323	1.311	1.304	1.296	1.293	1.292	1.293	1.297	1.303
0.04000	(0.392)	*	1.388	1.353	1.332	1.317	1.300	1.290	1.284	1.278	1.276	1.276	1.278	1.283	1.290
0.06000	(0.588)	1.368	1.345	1.314	1.296	1.284	1.269	1.262	1.257	1.253	1.253	1.255	1.258	1.265	1.273
0.08000	(0.785)	1.337	1.316	1.289	1.273	1.262	1.250	1.243	1.240	1.238	1.238	1.241	1.245	1.253	1.262
0.10000	(0.981)	1.315	1.295	1.270	1.255	1.246	1.235	1.230	1.227	1.226	1.228	1.231	1.236	1.245	1.254
0.15000	(1.47)	1.276	1.260	1.239	1.226	1.219	1.210	1.207	1.206	1.207	1.210	1.215	1.220	1.230	1.240
0.20000	(1.96)	1.251	1.236	1.218	1.207	1.201	1.194	1.192	1.192	1.194	1.198	1.204	1.210	1.221	1.232
0.30000	(2.94)	1.219	1.207	1.191	1.183	1.178	1.174	1.173	1.174	1.178	1.183	1.190	1.197	1.210	1.221
0.40000	(3.92)	1.198	1.187	1.174	1.167	1.163	1.161	1.161	1.163	1.168	1.174	1.182	1.189	1.203	1.215
0.60000	(5.88)	1.171	1.162	1.152	1.147	1.144	1.144	1.145	1.148	1.155	1.162	1.171	1.179	1.193	1.206
0.80000	(7.85)	1.154	1.146	1.137	1.134	1.132	1.133	1.135	1.139	1.147	1.154	1.164	1.172	1.188	1.201
1.00000	(9.81)	1.141	1.134	1.127	1.124	1.123	1.125	1.128	1.132	1.141	1.149	1.159	1.168	1.184	1.198
1.50000	(14.7)	1.120	1.115	1.110	1.109	1.109	1.112	1.117	1.122	1.131	1.140	1.151	1.161	1.177	1.192
2.00000	(19.6)	1.107	1.103	1.099	1.099	1.100	1.104	1.109	1.115	1.125	1.135	1.146	1.156	1.173	1.188
3.00000	(29.4)	1.090	1.087	1.085	1.086	1.089	1.094	1.100	1.106	1.118	1.128	1.140	1.151	1.169	1.184
S	$(\Delta p/\rho l)$	125	150	200	250	300	400	500	600	800	1000	1250	1500	2000	2500

Grad'nt   k.p.g.   (Equivalent) Pipe diameters in mm   Roughness size, $k_s = 0.30$ mm

E259

Kin. visc., $\nu = 100 \times 10^{-6}$ m^{2}s^{-1};
$S = 0.00010$ to 30.0000

i.e. kin. pr. grad., $\Delta p/\rho l =$
(0.00098) to (294) ms^{-2}

Roughness size, $k_s = 0.006$ mm
This table shows values of m, as follows

m_C for Colebrook-White solutions; or,
where $\mathbf{R} \leq 2000$, m_P for laminar flow

Grad'nt S	k.p.g. $(\Delta p/\rho l)$	6	8	10	12.5	15	20	25	30	40	50	60	80	100	125
0.00030	(0.00294)	685.7	467.3	347.0	257.7	202.1	137.7	102.3	80.20	54.65	40.59	31.83	21.69	16.11	11.96
0.00040	(0.00392)	593.9	404.7	300.5	223.2	175.0	119.3	88.57	69.46	47.33	35.15	27.56	18.78	13.95	10.36
0.00060	(0.00588)	484.9	330.4	245.4	182.2	142.9	97.38	72.32	56.71	38.65	28.70	22.51	15.34	11.39	8.459
0.00080	(0.00785)	419.9	286.1	212.5	157.8	123.8	84.33	62.63	49.12	33.47	24.86	19.49	13.28	9.864	7.325
0.00100	(0.00981)	375.6	255.9	190.1	141.2	110.7	75.43	56.02	43.93	29.93	22.23	17.43	11.88	8.822	6.552
0.00150	(0.0147)	306.7	209.0	155.2	115.3	90.38	61.59	45.74	35.87	24.44	18.15	14.23	9.700	7.204	5.350
0.00200	(0.0196)	265.6	181.0	134.4	99.81	78.27	53.34	39.61	31.06	21.17	15.72	12.33	8.400	6.238	4.633
0.00300	(0.0294)	216.9	147.8	109.7	81.50	63.91	43.55	32.34	25.36	17.28	12.84	10.07	6.859	5.094	3.783
0.00400	(0.0392)	187.8	128.0	95.04	70.58	55.35	37.72	28.01	21.96	14.97	11.12	8.717	5.940	4.411	3.276
0.00600	(0.0588)	153.3	104.5	77.60	57.63	45.19	30.79	22.87	17.93	12.22	9.076	7.117	4.850	3.602	2.675
0.00800	(0.0785)	132.8	90.49	67.20	49.91	39.14	26.67	19.81	15.53	10.58	7.860	6.164	4.200	3.119	2.316
0.01000	(0.0981)	118.8	80.93	60.11	44.64	35.01	23.85	17.71	13.89	9.466	7.030	5.513	3.757	2.790	2.072
0.01500	(0.147)	96.98	66.08	49.08	36.45	28.58	19.48	14.46	11.34	7.729	5.740	4.501	3.067	2.278	1.692
0.02000	(0.196)	83.99	57.23	42.50	31.56	24.75	16.87	12.53	9.823	6.694	4.971	3.898	2.656	1.973	1.465
0.03000	(0.294)	68.57	46.73	34.70	25.77	20.21	13.77	10.23	8.020	5.465	4.059	3.183	2.169	1.611	1.196
0.04000	(0.392)	59.39	40.47	30.05	22.32	17.50	11.93	8.857	6.946	4.733	3.515	2.756	1.878	1.395	*
0.06000	(0.588)	48.49	33.04	24.54	18.22	14.29	9.738	7.232	5.671	3.865	2.870	2.251	1.534	1.139	*
0.08000	(0.785)	41.99	28.61	21.25	15.78	12.38	8.433	6.263	4.912	3.347	2.486	1.949	1.328	*	1.353
0.10000	(0.981)	37.56	25.59	19.01	14.12	11.07	7.543	5.602	4.393	2.993	2.223	1.743	1.188	*	1.325
0.15000	(1.47)	30.67	20.90	15.52	11.53	9.038	6.159	4.574	3.587	2.444	1.815	1.423	*	1.308	1.278
0.20000	(1.96)	26.56	18.10	13.44	9.981	7.827	5.334	3.961	3.106	2.117	1.572	1.233	*	1.274	1.247
0.30000	(2.94)	21.69	14.78	10.97	8.150	6.391	4.355	3.234	2.536	1.728	1.284	1.007	1.258	1.229	1.205
0.40000	(3.92)	18.78	12.80	9.504	7.058	5.535	3.772	2.801	2.196	1.497	1.112	*	1.225	1.199	1.178
0.60000	(5.88)	15.33	10.45	7.760	5.763	4.519	3.079	2.287	1.793	1.222	*	*	1.182	1.159	1.141
0.80000	(7.85)	13.28	9.049	6.720	4.991	3.914	2.667	1.981	1.553	1.058	*	1.187	1.154	1.133	1.116
1.00000	(9.81)	11.88	8.093	6.011	4.464	3.501	2.385	1.771	1.389	0.947	*	1.163	1.132	1.113	1.097
1.50000	(14.7)	9.698	6.608	4.908	3.645	2.858	1.948	1.446	1.134	*	1.143	1.122	1.095	1.079	1.065
2.00000	(19.6)	8.399	5.723	4.250	3.156	2.475	1.687	1.253	0.982	*	1.114	1.095	1.071	1.056	1.043
3.00000	(29.4)	6.857	4.673	3.470	2.577	2.021	1.377	1.023	*	1.099	1.075	1.059	1.038	1.025	1.014
4.00000	(39.2)	5.939	4.047	3.005	2.232	1.750	1.193	0.886	*	1.071	1.050	1.035	1.016	1.004	0.995
6.00000	(58.8)	4.849	3.304	2.454	1.822	1.429	0.974	*	1.065	1.035	1.016	1.003	0.986	0.976	0.969
8.00000	(78.5)	4.199	2.861	2.125	1.578	1.238	0.843	*	1.037	1.010	0.993	0.981	0.967	0.958	0.951
10.0000	(98.1)	3.756	2.559	1.901	1.412	1.107	*	1.037	1.017	0.992	0.976	0.965	0.952	0.944	0.938
15.0000	(147)	3.067	2.090	1.552	1.153	0.904	*	0.999	0.982	0.960	0.947	0.937	0.926	0.920	0.915
20.0000	(196)	2.656	1.810	1.344	0.998	*	0.997	0.975	0.959	0.939	0.927	0.919	0.909	0.903	0.899
30.0000	(294)	2.169	1.478	1.097	0.815	*	0.962	0.942	0.929	0.911	0.901	0.894	0.886	0.881	0.878
S	$(\Delta p/\rho l)$	6	8	10	12.5	15	20	25	30	40	50	60	80	100	125

Grad'nt k.p.g. (Equivalent) Pipe diameters in mm

Grad'nt S	k.p.g. $(\Delta p/\rho l)$	125	150	200	250	300	400	500	600	800	1000	1250	1500	2000	2500
0.00010	(0.00098)	20.72	16.25	11.07	8.222	6.448	4.394	3.263	2.559	1.744	*	1.944	1.909	1.864	1.836
0.00015	(0.00147)	16.92	13.27	9.040	6.714	5.265	3.588	2.664	2.089	*	1.919	1.875	1.845	1.805	1.781
0.00020	(0.00196)	14.65	11.49	7.829	5.814	4.559	3.107	2.307	1.809	*	1.869	1.829	1.801	1.766	1.744
0.00030	(0.00294)	11.96	9.381	6.392	4.747	3.723	2.537	1.884	1.477	1.845	1.803	1.768	1.744	1.713	1.694
0.00040	(0.00392)	10.36	8.124	5.536	4.111	3.224	2.197	1.632	*	1.797	1.759	1.727	1.705	1.677	1.660
0.00060	(0.00588)	8.459	6.633	4.520	3.357	2.632	1.794	*	*	1.734	1.700	1.672	1.653	1.629	1.615
0.00080	(0.00785)	7.325	5.745	3.914	2.907	2.280	1.553	*	1.740	1.691	1.661	1.636	1.618	1.597	1.584
0.00100	(0.00981)	6.552	5.138	3.501	2.600	2.039	1.389	*	1.705	1.660	1.632	1.608	1.592	1.573	1.561
0.00150	(0.0147)	5.350	4.195	2.859	2.123	1.665	*	1.675	1.645	1.605	1.581	1.561	1.547	1.531	1.521
0.00200	(0.0196)	4.633	3.633	2.476	1.839	1.442	*	1.632	1.605	1.569	1.547	1.529	1.517	1.502	1.494
0.00300	(0.0294)	3.783	2.966	2.021	1.501	*	1.611	1.576	1.552	1.520	1.501	1.486	1.476	1.464	1.458
0.00400	(0.0392)	3.276	2.569	1.751	1.300	*	1.570	1.538	1.516	1.488	1.471	1.457	1.448	1.438	1.433
0.00600	(0.0588)	2.675	2.098	1.429	*	1.559	1.515	1.487	1.468	1.444	1.429	1.418	1.410	1.402	1.399
0.00800	(0.0785)	2.316	1.817	1.238	*	1.519	1.479	1.454	1.436	1.415	1.401	1.391	1.385	1.379	1.376
0.01000	(0.0981)	2.072	1.625	*	1.518	1.489	1.452	1.429	1.413	1.392	1.381	1.372	1.366	1.361	1.359
0.01500	(0.147)	1.692	1.327	*	1.463	1.438	1.405	1.385	1.371	1.354	1.344	1.337	1.333	1.329	1.328
0.02000	(0.196)	1.465	*	1.459	1.426	1.404	1.374	1.356	1.343	1.328	1.320	1.313	1.310	1.308	1.308
0.03000	(0.294)	1.196	*	1.407	1.378	1.358	1.332	1.316	1.306	1.293	1.286	1.282	1.279	1.278	1.280
0.04000	(0.392)	*	*	1.372	1.345	1.327	1.304	1.290	1.281	1.270	1.264	1.260	1.259	1.259	1.261
0.06000	(0.588)	*	1.362	1.325	1.302	1.286	1.266	1.254	1.247	1.238	1.233	1.231	1.230	1.232	1.235
0.08000	(0.785)	1.353	1.327	1.294	1.273	1.259	1.241	1.230	1.224	1.216	1.213	1.211	1.211	1.213	1.217
0.10000	(0.981)	1.325	1.302	1.271	1.251	1.238	1.222	1.212	1.206	1.200	1.197	1.196	1.197	1.200	1.203
0.15000	(1.47)	1.278	1.258	1.231	1.214	1.203	1.189	1.181	1.176	1.171	1.170	1.170	1.171	1.175	1.180
0.20000	(1.96)	1.247	1.228	1.204	1.189	1.179	1.167	1.160	1.156	1.152	1.151	1.152	1.154	1.159	1.164
0.30000	(2.94)	1.205	1.189	1.168	1.155	1.147	1.137	1.131	1.128	1.126	1.126	1.128	1.130	1.136	1.142
0.40000	(3.92)	1.178	1.163	1.144	1.132	1.125	1.116	1.112	1.109	1.108	1.109	1.111	1.114	1.121	1.127
0.60000	(5.88)	1.141	1.128	1.111	1.102	1.096	1.089	1.085	1.084	1.084	1.086	1.089	1.092	1.100	1.106
0.80000	(7.85)	1.116	1.104	1.090	1.081	1.076	1.070	1.068	1.067	1.068	1.070	1.074	1.077	1.085	1.092
1.00000	(9.81)	1.097	1.086	1.073	1.066	1.061	1.056	1.054	1.054	1.055	1.058	1.062	1.066	1.074	1.082
1.50000	(14.7)	1.065	1.056	1.045	1.039	1.035	1.032	1.031	1.031	1.034	1.037	1.042	1.047	1.055	1.063
2.00000	(19.6)	1.043	1.035	1.026	1.020	1.017	1.015	1.015	1.016	1.019	1.023	1.028	1.033	1.042	1.051
3.00000	(29.4)	1.014	1.008	1.000	0.996	0.994	0.993	0.993	0.995	0.999	1.004	1.009	1.015	1.025	1.034
S	$(\Delta p/\rho l)$	125	150	200	250	300	400	500	600	800	1000	1250	1500	2000	2500

Grad'nt k.p.g. (Equivalent) Pipe diameters in mm

Kinematic viscosity, $\nu = 100 \times 10^{-6}$ m^{2}s^{-1} ;

Roughness size, $k_s = 0.006$ mm

Kin. visc., $\nu = 100 \times 10^{-6}$ m^{2}s^{-1};
$S = 0{\cdot}00010$ to $30{\cdot}0000$
i.e. kin. pr. grad., $\Delta p/\rho l =$
$(0{\cdot}00098)$ to (294) ms^{-2}

Roughness size, $k_s = 0{\cdot}015$ mm
This table shows values of m, as follows
m_C for Colebrook-White solutions; or,
where $\mathbf{R} \leq 2000$, m_P for laminar flow

Grad'nt S	k.p.g. $(\Delta p/\rho l)$	6	8	10	12·5	15	20	25	30	40	50	60	80	100	125
												(Equivalent) Pipe diameters in mm			
0·00030	(0·00294)	685·7	467·3	347·0	257·7	202·1	137·7	102·3	80·20	54·65	40·59	31·83	21·69	16·11	11·96
0·00040	(0·00392)	593·9	404·7	300·5	223·2	175·0	119·3	88·57	69·46	47·33	35·15	27·56	18·78	13·95	10·36
0·00060	(0·00588)	484·9	330·4	245·4	182·2	142·9	97·38	72·32	56·71	38·65	28·70	22·51	15·34	11·39	8·459
0·00080	(0·00785)	419·9	286·1	212·5	157·8	123·8	84·33	62·63	49·12	33·47	24·86	19·49	13·28	9·864	7·325
0·00100	(0·00981)	375·6	255·9	190·1	141·2	110·7	75·43	56·02	43·93	29·93	22·23	17·43	11·88	8·822	6·552
0·00150	(0·0147)	306·7	209·0	155·2	115·3	90·38	61·59	45·74	35·87	24·44	18·15	14·23	9·700	7·204	5·350
0·00200	(0·0196)	265·6	181·0	134·4	99·81	78·27	53·34	39·61	31·06	21·17	15·72	12·33	8·400	6·238	4·633
0·00300	(0·0294)	216·9	147·8	109·7	81·50	63·91	43·55	32·34	25·36	17·28	12·84	10·07	6·859	5·094	3·783
0·00400	(0·0392)	187·8	128·0	95·04	70·58	55·35	37·72	28·01	21·96	14·97	11·12	8·717	5·940	4·411	3·276
0·00600	(0·0588)	153·3	104·5	77·60	57·63	45·19	30·79	22·87	17·93	12·22	9·076	7·117	4·850	3·602	2·675
0·00800	(0·0785)	132·8	90·49	67·20	49·91	39·14	26·67	19·81	15·53	10·58	7·860	6·164	4·200	3·119	2·316
0·01000	(0·0981)	118·8	80·93	60·11	44·64	35·01	23·85	17·71	13·89	9·466	7·030	5·513	3·757	2·790	2·072
0·01500	(0·147)	96·98	66·08	49·08	36·45	28·58	19·48	14·46	11·34	7·729	5·740	4·501	3·067	2·278	1·692
0·02000	(0·196)	83·99	57·23	42·50	31·56	24·75	16·87	12·53	9·823	6·694	4·971	3·898	2·656	1·973	1·465
0·03000	(0·294)	68·57	46·73	34·70	25·77	20·21	13·77	10·23	8·020	5·465	4·059	3·183	2·169	1·611	1·196
0·04000	(0·392)	59·39	40·47	30·05	22·32	17·50	11·93	8·857	6·946	4·733	3·515	2·756	1·878	1·395	*
0·06000	(0·588)	48·49	33·04	24·54	18·22	14·29	9·738	7·232	5·671	3·865	2·870	2·251	1·534	1·139	*
0·08000	(0·785)	41·99	28·61	21·25	15·78	12·38	8·433	6·263	4·912	3·347	2·486	1·949	1·328	*	1·354
0·10000	(0·981)	37·56	25·59	19·01	14·12	11·07	7·543	5·602	4·393	2·993	2·223	1·743	1·188	*	1·326
0·15000	(1·47)	30·67	20·90	15·52	11·53	9·038	6·159	4·574	3·587	2·444	1·815	1·423	*	1·310	1·280
0·20000	(1·96)	26·56	18·10	13·44	9·981	7·827	5·334	3·961	3·106	2·117	1·572	1·233	*	1·276	1·248
0·30000	(2·94)	21·69	14·78	10·97	8·150	6·391	4·355	3·234	2·536	1·728	1·284	1·007	1·260	1·231	1·207
0·40000	(3·92)	18·78	12·80	9·504	7·058	5·535	3·772	2·801	2·196	1·497	1·112	*	1·227	1·201	1·179
0·60000	(5·88)	15·33	10·45	7·760	5·763	4·519	3·079	2·287	1·793	1·222	*	*	1·184	1·161	1·143
0·80000	(7·85)	13·28	9·049	6·720	4·991	3·914	2·667	1·981	1·553	1·058	*	1·189	1·156	1·135	1·118
1·00000	(9·81)	11·88	8·093	6·011	4·464	3·501	2·385	1·771	1·389	0·947	*	1·165	1·134	1·115	1·100
1·50000	(14·7)	9·698	6·608	4·908	3·645	2·858	1·948	1·446	1·134	*	1·145	1·125	1·098	1·081	1·068
2·00000	(19·6)	8·399	5·723	4·250	3·156	2·475	1·687	1·253	0·982	*	1·116	1·098	1·074	1·059	1·047
3·00000	(29·4)	6·857	4·673	3·470	2·577	2·021	1·377	1·023	*	1·102	1·078	1·062	1·041	1·028	1·018
4·00000	(39·2)	5·939	4·047	3·005	2·232	1·750	1·193	0·886	*	1·075	1·053	1·038	1·019	1·008	0·999
6·00000	(58·8)	4·849	3·304	2·454	1·822	1·429	0·974	*	1·068	1·038	1·019	1·007	0·990	0·981	0·973
8·00000	(78·5)	4·199	2·861	2·125	1·578	1·238	0·843	*	1·041	1·014	0·997	0·986	0·971	0·963	0·956
10·0000	(98·1)	3·756	2·559	1·901	1·412	1·107	*	1·041	1·021	0·996	0·980	0·970	0·957	0·949	0·943
15·0000	(147)	3·067	2·090	1·552	1·153	0·904	*	1·004	0·987	0·965	0·952	0·943	0·932	0·925	0·921
20·0000	(196)	2·656	1·810	1·344	0·998	*	1·002	0·980	0·965	0·945	0·933	0·925	0·915	0·910	0·906
30·0000	(294)	2·169	1·478	1·097	0·815	*	0·967	0·948	0·935	0·918	0·907	0·901	0·893	0·889	0·886
S $(\Delta p/\rho l)$		6	8	10	12·5	15	20	25	30	40	50	60	80	100	125

Grad'nt S	k.p.g. $(\Delta p/\rho l)$	125	150	200	250	300	400	500	600	800	1000	1250	1500	2000	2500
								(Equivalent) Pipe diameters in mm						
0·00010	(0·00098)	20·72	16·25	11·07	8·222	6·448	4·394	3·263	2·559	1·744	*	1·944	1·910	1·864	1·836
0·00015	(0·00147)	16·92	13·27	9·040	6·714	5·265	3·588	2·664	2·089	*	1·919	1·875	1·845	1·805	1·781
0·00020	(0·00196)	14·65	11·49	7·829	5·814	4·559	3·107	2·307	1·809	*	1·869	1·829	1·802	1·766	1·744
0·00030	(0·00294)	11·96	9·381	6·392	4·747	3·723	2·537	1·884	1·477	1·845	1·803	1·768	1·744	1·713	1·694
0·00040	(0·00392)	10·36	8·124	5·536	4·111	3·224	2·197	1·632	*	1·797	1·759	1·727	1·705	1·677	1·660
0·00060	(0·00588)	8·459	6·633	4·520	3·357	2·632	1·794	*	*	1·734	1·700	1·673	1·654	1·630	1·615
0·00080	(0·00785)	7·325	5·745	3·914	2·907	2·280	1·553	*	1·740	1·692	1·661	1·636	1·619	1·597	1·585
0·00100	(0·00981)	6·552	5·138	3·501	2·600	2·039	1·389	*	1·705	1·660	1·632	1·609	1·593	1·573	1·562
0·00150	(0·0147)	5·350	4·195	2·859	2·123	1·665	*	1·676	1·645	1·606	1·581	1·561	1·548	1·531	1·522
0·00200	(0·0196)	4·633	3·633	2·476	1·839	1·442	*	1·633	1·605	1·569	1·547	1·529	1·517	1·503	1·495
0·00300	(0·0294)	3·783	2·966	2·021	1·501	*	1·611	1·576	1·552	1·521	1·502	1·486	1·476	1·464	1·458
0·00400	(0·0392)	3·276	2·569	1·751	1·300	*	1·570	1·538	1·516	1·488	1·471	1·457	1·449	1·438	1·433
0·00600	(0·0588)	2·675	2·098	1·429	*	1·560	1·516	1·488	1·469	1·445	1·430	1·418	1·411	1·403	1·400
0·00800	(0·0785)	2·316	1·817	1·238	*	1·520	1·479	1·454	1·437	1·415	1·402	1·392	1·386	1·379	1·377
0·01000	(0·0981)	2·072	1·625	*	1·518	1·490	1·453	1·429	1·413	1·393	1·381	1·372	1·367	1·362	1·360
0·01500	(0·147)	1·692	1·327	*	1·464	1·439	1·406	1·386	1·372	1·355	1·345	1·338	1·334	1·330	1·330
0·02000	(0·196)	1·465	*	1·460	1·427	1·404	1·375	1·357	1·344	1·329	1·321	1·315	1·311	1·309	1·309
0·03000	(0·294)	1·196	*	1·408	1·379	1·359	1·333	1·317	1·307	1·294	1·288	1·283	1·281	1·280	1·281
0·04000	(0·392)	*	*	1·372	1·346	1·328	1·305	1·291	1·282	1·271	1·265	1·262	1·260	1·260	1·262
0·06000	(0·588)	*	1·363	1·326	1·303	1·288	1·268	1·256	1·248	1·239	1·235	1·233	1·232	1·234	1·237
0·08000	(0·785)	1·354	1·328	1·295	1·274	1·260	1·242	1·232	1·225	1·218	1·214	1·213	1·213	1·216	1·219
0·10000	(0·981)	1·326	1·303	1·272	1·253	1·240	1·223	1·214	1·208	1·202	1·199	1·198	1·199	1·202	1·206
0·15000	(1·47)	1·280	1·259	1·232	1·216	1·204	1·191	1·183	1·178	1·173	1·172	1·172	1·174	1·178	1·183
0·20000	(1·96)	1·248	1·230	1·205	1·190	1·181	1·168	1·162	1·158	1·154	1·154	1·155	1·157	1·162	1·167
0·30000	(2·94)	1·207	1·191	1·170	1·157	1·149	1·139	1·133	1·131	1·129	1·129	1·131	1·134	1·140	1·146
0·40000	(3·92)	1·179	1·165	1·146	1·134	1·127	1·118	1·114	1·112	1·111	1·112	1·115	1·118	1·125	1·131
0·60000	(5·88)	1·143	1·130	1·114	1·104	1·098	1·091	1·088	1·087	1·088	1·090	1·093	1·097	1·104	1·111
0·80000	(7·85)	1·118	1·106	1·092	1·084	1·079	1·073	1·071	1·070	1·072	1·074	1·078	1·082	1·090	1·098
1·00000	(9·81)	1·100	1·089	1·076	1·068	1·064	1·059	1·058	1·058	1·060	1·063	1·067	1·071	1·080	1·088
1·50000	(14·7)	1·068	1·059	1·048	1·042	1·038	1·035	1·035	1·036	1·039	1·042	1·047	1·052	1·062	1·071
2·00000	(19·6)	1·047	1·039	1·029	1·024	1·021	1·019	1·019	1·021	1·024	1·029	1·034	1·040	1·050	1·059
3·00000	(29·4)	1·018	1·011	1·004	1·000	0·998	0·997	0·999	1·001	1·005	1·010	1·017	1·023	1·033	1·043
S $(\Delta p/\rho l)$		125	150	200	250	300	400	500	600	800	1000	1250	1500	2000	2500

Grad'nt k.p.g. (Equivalent) Pipe diameters in mm

Kinematic viscosity, $\nu = 100 \times 10^{-6}$ m^{2}s^{-1} ; **Roughness size, $k_s = 0{\cdot}015$ mm**

E261

Kin. visc., $\nu = 100 \times 10^{-6}\ \text{m}^2\text{s}^{-1}$; $S = 0.00010$ to 3.00000

i.e. kin. pr. grad., $\Delta p/\rho l =$ (0.00098) to (29.4) ms^{-2}

Roughness size, $k_s = 0.030$ mm

This table shows values of m, as follows

m_C for Colebrook-White solutions; or, where $\mathbf{R} \leq 2000$, m_P for laminar flow

Grad'nt S	k.p.g. $(\Delta p/\rho l)$	\(Equivalent) Pipe diameters in mm													
		125	150	200	250	300	400	500	600	800	1000	1250	1500	2000	2500
0.00010	(0.00098)	20.72	16.25	11.07	8.222	6.448	4.394	3.263	2.559	1.744	*	1.945	1.910	1.865	1.836
0.00015	(0.00147)	16.92	13.27	9.040	6.714	5.265	3.588	2.664	2.089	*	1.920	1.876	1.845	1.806	1.781
0.00020	(0.00196)	14.65	11.49	7.829	5.814	4.559	3.107	2.307	1.809	*	1.870	1.830	1.802	1.766	1.744
0.00030	(0.00294)	11.96	9.381	6.392	4.747	3.723	2.537	1.884	1.477	1.846	1.804	1.769	1.744	1.713	1.694
0.00040	(0.00392)	10.36	8.124	5.536	4.111	3.224	2.197	1.632	*	1.798	1.760	1.728	1.706	1.678	1.661
0.00060	(0.00588)	8.459	6.633	4.520	3.357	2.632	1.794	*	*	1.734	1.701	1.673	1.654	1.630	1.616
0.00080	(0.00785)	7.325	5.745	3.914	2.907	2.280	1.553	*	1.741	1.692	1.662	1.637	1.619	1.598	1.585
0.00100	(0.00981)	6.552	5.138	3.501	2.600	2.039	1.389	*	1.706	1.661	1.632	1.609	1.593	1.574	1.563
0.00150	(0.0147)	5.350	4.195	2.859	2.123	1.665	*	1.676	1.646	1.606	1.582	1.562	1.548	1.532	1.523
0.00200	(0.0196)	4.633	3.633	2.476	1.839	1.442	*	1.633	1.606	1.570	1.548	1.530	1.518	1.504	1.496
0.00300	(0.0294)	3.783	2.966	2.021	1.501	*	1.612	1.577	1.553	1.522	1.503	1.487	1.477	1.465	1.459
0.00400	(0.0392)	3.276	2.569	1.751	1.300	*	1.571	1.539	1.517	1.489	1.472	1.458	1.450	1.439	1.435
0.00600	(0.0588)	2.675	2.098	1.429	*	1.561	1.517	1.489	1.470	1.446	1.431	1.420	1.412	1.405	1.401
0.00800	(0.0785)	2.316	1.817	1.238	*	1.521	1.480	1.455	1.438	1.416	1.403	1.393	1.387	1.381	1.379
0.01000	(0.0981)	2.072	1.625	*	1.519	1.491	1.454	1.430	1.414	1.394	1.383	1.374	1.368	1.363	1.361
0.01500	(0.147)	1.692	1.327	*	1.465	1.440	1.407	1.387	1.373	1.356	1.347	1.339	1.335	1.332	1.332
0.02000	(0.196)	1.465	*	1.461	1.428	1.406	1.376	1.358	1.346	1.331	1.322	1.316	1.313	1.311	1.311
0.03000	(0.294)	1.196	*	1.409	1.380	1.360	1.335	1.319	1.309	1.296	1.290	1.285	1.283	1.282	1.284
0.04000	(0.392)	*	*	1.374	1.348	1.330	1.307	1.293	1.284	1.273	1.267	1.264	1.263	1.263	1.265
0.06000	(0.588)	*	1.364	1.328	1.305	1.289	1.270	1.258	1.250	1.242	1.237	1.235	1.235	1.237	1.240
0.08000	(0.785)	1.355	1.330	1.297	1.276	1.262	1.245	1.234	1.228	1.220	1.217	1.216	1.216	1.219	1.223
0.10000	(0.981)	1.328	1.305	1.274	1.255	1.242	1.226	1.216	1.211	1.204	1.202	1.201	1.202	1.206	1.210
0.15000	(1.47)	1.282	1.261	1.234	1.218	1.207	1.193	1.186	1.181	1.177	1.176	1.176	1.178	1.182	1.188
0.20000	(1.96)	1.251	1.232	1.208	1.193	1.183	1.171	1.165	1.161	1.158	1.158	1.159	1.161	1.167	1.172
0.30000	(2.94)	1.210	1.194	1.173	1.160	1.152	1.142	1.137	1.135	1.133	1.134	1.136	1.139	1.145	1.152
0.40000	(3.92)	1.182	1.168	1.149	1.138	1.131	1.122	1.118	1.116	1.116	1.117	1.120	1.124	1.131	1.138
0.60000	(5.88)	1.146	1.133	1.117	1.108	1.102	1.096	1.093	1.092	1.093	1.095	1.099	1.103	1.112	1.119
0.80000	(7.85)	1.122	1.110	1.096	1.088	1.083	1.078	1.076	1.076	1.078	1.081	1.085	1.090	1.099	1.107
1.00000	(9.81)	1.104	1.093	1.080	1.073	1.069	1.065	1.063	1.064	1.066	1.070	1.075	1.079	1.089	1.098
1.50000	(14.7)	1.073	1.064	1.053	1.047	1.044	1.042	1.042	1.043	1.046	1.051	1.056	1.062	1.072	1.082
2.00000	(19.6)	1.052	1.044	1.035	1.030	1.028	1.026	1.027	1.028	1.033	1.038	1.044	1.050	1.061	1.071
3.00000	(29.4)	1.024	1.018	1.010	1.007	1.006	1.005	1.007	1.009	1.015	1.021	1.028	1.034	1.046	1.057
S	$(\Delta p/\rho l)$	125	150	200	250	300	400	500	600	800	1000	1250	1500	2000	2500

Grad'nt k.p.g. (Equivalent) Pipe diameters in mm Roughness size, $k_s = 0.030$ mm

E262

Kin. visc., $\nu = 100 \times 10^{-6}\ \text{m}^2\text{s}^{-1}$; $S = 0.00010$ to 3.00000

i.e. kin. pr. grad., $\Delta p/\rho l =$ (0.00098) to (29.4) ms^{-2}

Roughness size, $k_s = 0.060$ mm

This table shows values of m, as follows

m_C for Colebrook-White solutions; or, where $\mathbf{R} \leq 2000$, m_P for laminar flow

Grad'nt S	k.p.g. $(\Delta p/\rho l)$	\(Equivalent) Pipe diameters in mm													
		125	150	200	250	300	400	500	600	800	1000	1250	1500	2000	2500
0.00010	(0.00098)	20.72	16.25	11.07	8.222	6.448	4.394	3.263	2.559	1.744	*	1.945	1.910	1.865	1.837
0.00015	(0.00147)	16.92	13.27	9.040	6.714	5.265	3.588	2.664	2.089	*	1.920	1.876	1.846	1.806	1.782
0.00020	(0.00196)	14.65	11.49	7.829	5.814	4.559	3.107	2.307	1.809	*	1.870	1.830	1.803	1.767	1.745
0.00030	(0.00294)	11.96	9.381	6.392	4.747	3.723	2.537	1.884	1.477	1.846	1.804	1.769	1.745	1.714	1.695
0.00040	(0.00392)	10.36	8.124	5.536	4.111	3.224	2.197	1.632	*	1.799	1.760	1.729	1.707	1.679	1.662
0.00060	(0.00588)	8.459	6.633	4.520	3.357	2.632	1.794	*	*	1.735	1.702	1.674	1.655	1.631	1.617
0.00080	(0.00785)	7.325	5.745	3.914	2.907	2.280	1.553	*	1.742	1.693	1.663	1.638	1.621	1.599	1.587
0.00100	(0.00981)	6.552	5.138	3.501	2.600	2.039	1.389	*	1.707	1.662	1.634	1.610	1.595	1.575	1.564
0.00150	(0.0147)	5.350	4.195	2.859	2.123	1.665	*	1.677	1.647	1.608	1.583	1.563	1.550	1.534	1.524
0.00200	(0.0196)	4.633	3.633	2.476	1.839	1.442	*	1.635	1.607	1.572	1.549	1.532	1.520	1.505	1.498
0.00300	(0.0294)	3.783	2.966	2.021	1.501	*	1.613	1.578	1.554	1.523	1.504	1.489	1.479	1.467	1.461
0.00400	(0.0392)	3.276	2.569	1.751	1.300	*	1.572	1.541	1.519	1.491	1.474	1.460	1.452	1.442	1.437
0.00600	(0.0588)	2.675	2.098	1.429	*	1.563	1.518	1.491	1.472	1.448	1.433	1.422	1.415	1.407	1.404
0.00800	(0.0785)	2.316	1.817	1.238	*	1.522	1.482	1.457	1.440	1.419	1.406	1.396	1.390	1.384	1.382
0.01000	(0.0981)	2.072	1.625	*	1.521	1.493	1.456	1.433	1.417	1.397	1.385	1.376	1.371	1.366	1.365
0.01500	(0.147)	1.692	1.327	*	1.467	1.442	1.410	1.390	1.376	1.359	1.350	1.343	1.339	1.336	1.336
0.02000	(0.196)	1.465	*	1.464	1.431	1.408	1.379	1.361	1.349	1.334	1.326	1.320	1.317	1.315	1.316
0.03000	(0.294)	1.196	*	1.412	1.383	1.363	1.338	1.322	1.312	1.300	1.293	1.289	1.287	1.287	1.289
0.04000	(0.392)	*	*	1.377	1.351	1.333	1.310	1.296	1.287	1.277	1.272	1.268	1.267	1.268	1.271
0.06000	(0.588)	*	1.368	1.331	1.309	1.293	1.273	1.262	1.255	1.246	1.242	1.241	1.241	1.243	1.247
0.08000	(0.785)	1.359	1.334	1.301	1.280	1.266	1.249	1.239	1.232	1.226	1.223	1.222	1.222	1.226	1.230
0.10000	(0.981)	1.332	1.309	1.278	1.259	1.246	1.230	1.221	1.216	1.210	1.208	1.208	1.209	1.213	1.218
0.15000	(1.47)	1.286	1.266	1.239	1.223	1.212	1.199	1.191	1.187	1.183	1.182	1.183	1.185	1.191	1.197
0.20000	(1.96)	1.256	1.237	1.213	1.198	1.189	1.177	1.171	1.168	1.165	1.165	1.167	1.170	1.176	1.182
0.30000	(2.94)	1.215	1.199	1.178	1.166	1.158	1.149	1.144	1.142	1.141	1.142	1.145	1.149	1.156	1.164
0.40000	(3.92)	1.188	1.174	1.155	1.145	1.138	1.130	1.126	1.125	1.125	1.127	1.131	1.135	1.143	1.151
0.60000	(5.88)	1.153	1.140	1.125	1.116	1.110	1.105	1.102	1.102	1.104	1.107	1.111	1.116	1.125	1.134
0.80000	(7.85)	1.129	1.118	1.104	1.097	1.092	1.088	1.086	1.087	1.089	1.093	1.098	1.103	1.114	1.123
1.00000	(9.81)	1.112	1.101	1.089	1.082	1.078	1.075	1.074	1.075	1.079	1.083	1.089	1.094	1.105	1.115
1.50000	(14.7)	1.081	1.073	1.063	1.058	1.055	1.053	1.054	1.056	1.060	1.066	1.072	1.079	1.090	1.101
2.00000	(19.6)	1.062	1.054	1.046	1.042	1.040	1.039	1.040	1.043	1.048	1.054	1.062	1.068	1.081	1.092
3.00000	(29.4)	1.035	1.029	1.023	1.020	1.019	1.020	1.023	1.026	1.033	1.039	1.047	1.055	1.069	1.080
S	$(\Delta p/\rho l)$	125	150	200	250	300	400	500	600	800	1000	1250	1500	2000	2500

Grad'nt k.p.g. (Equivalent) Pipe diameters in mm Roughness size, $k_s = 0.060$ mm

Kin. visc., $\nu = 100\times10^{-6}$ m^{2}s^{-1};
$S = 0\cdot00010$ to $3\cdot00000$
i.e. kin. pr. grad., $\Delta p/\rho l =$
$(0\cdot00098)$ to $(29\cdot4)$ ms^{-2}

Roughness size, $k_s = 0\cdot150$ mm
This table shows values of m, as follows
m_C for Colebrook-White solutions; or,
where $R \leq 2000$, m_P for laminar flow

Grad'nt S	k.p.g. $(\Delta p/\rho l)$	125	150	200	250	300	400	500	600	800	1000	1250	1500	2000	2500
								(Equivalent) Pipe diameters in mm							
0·00010	(0·00098)	20·72	16·25	11·07	8·222	6·448	4·394	3·263	2·559	1·744	*	1·947	1·912	1·867	1·839
0·00015	(0·00147)	16·92	13·27	9·040	6·714	5·265	3·588	2·664	2·089	*	1·922	1·878	1·848	1·808	1·784
0·00020	(0·00196)	14·65	11·49	7·829	5·814	4·559	3·107	2·307	1·809	*	1·873	1·833	1·805	1·769	1·747
0·00030	(0·00294)	11·96	9·381	6·392	4·747	3·723	2·537	1·884	1·477	1·849	1·807	1·772	1·748	1·717	1·698
0·00040	(0·00392)	10·36	8·124	5·536	4·111	3·224	2·197	1·632	*	1·801	1·763	1·731	1·709	1·682	1·665
0·00060	(0·00588)	8·459	6·633	4·520	3·357	2·632	1·794	*	*	1·738	1·705	1·677	1·658	1·635	1·621
0·00080	(0·00785)	7·325	5·745	3·914	2·907	2·280	1·553	*	1·745	1·696	1·666	1·641	1·624	1·603	1·591
0·00100	(0·00981)	6·552	5·138	3·501	2·600	2·039	1·389	*	1·710	1·665	1·637	1·614	1·598	1·579	1·568
0·00150	(0·0147)	5·350	4·195	2·859	2·123	1·665	*	1·681	1·651	1·612	1·587	1·567	1·554	1·538	1·529
0·00200	(0·0196)	4·633	3·633	2·476	1·839	1·442	*	1·639	1·611	1·576	1·554	1·536	1·524	1·510	1·503
0·00300	(0·0294)	3·783	2·966	2·021	1·501	*	1·618	1·583	1·559	1·528	1·509	1·494	1·485	1·473	1·468
0·00400	(0·0392)	3·276	2·569	1·751	1·300	*	1·577	1·546	1·524	1·496	1·479	1·466	1·458	1·448	1·444
0·00600	(0·0588)	2·675	2·098	1·429	*	1·568	1·524	1·496	1·478	1·454	1·440	1·429	1·422	1·415	1·412
0·00800	(0·0785)	2·316	1·817	1·238	*	1·528	1·488	1·464	1·447	1·425	1·413	1·403	1·398	1·392	1·391
0·01000	(0·0981)	2·072	1·625	*	1·527	1·499	1·462	1·439	1·424	1·404	1·393	1·385	1·380	1·375	1·375
0·01500	(0·147)	1·692	1·327	*	1·474	1·449	1·417	1·397	1·384	1·368	1·358	1·352	1·349	1·346	1·347
0·02000	(0·196)	1·465	*	1·471	1·438	1·416	1·387	1·369	1·357	1·343	1·335	1·330	1·328	1·327	1·328
0·03000	(0·294)	1·196	*	1·420	1·391	1·372	1·347	1·332	1·322	1·310	1·305	1·301	1·300	1·301	1·303
0·04000	(0·392)	*	*	1·386	1·360	1·342	1·320	1·307	1·298	1·289	1·284	1·282	1·281	1·283	1·287
0·06000	(0·588)	*	1·377	1·341	1·319	1·304	1·285	1·274	1·267	1·260	1·257	1·256	1·256	1·260	1·265
0·08000	(0·785)	1·369	1·344	1·312	1·292	1·278	1·261	1·252	1·246	1·240	1·238	1·238	1·240	1·245	1·250
0·10000	(0·981)	1·343	1·320	1·290	1·271	1·259	1·244	1·235	1·231	1·226	1·225	1·226	1·228	1·233	1·240
0·15000	(1·47)	1·299	1·278	1·252	1·237	1·226	1·214	1·208	1·204	1·201	1·202	1·204	1·207	1·214	1·221
0·20000	(1·96)	1·269	1·251	1·228	1·214	1·205	1·194	1·189	1·186	1·185	1·186	1·189	1·193	1·201	1·209
0·30000	(2·94)	1·231	1·215	1·195	1·184	1·176	1·168	1·165	1·163	1·164	1·166	1·170	1·175	1·185	1·194
0·40000	(3·92)	1·205	1·191	1·174	1·164	1·157	1·151	1·148	1·148	1·150	1·153	1·158	1·163	1·174	1·183
0·60000	(5·88)	1·172	1·160	1·145	1·137	1·133	1·128	1·127	1·128	1·131	1·136	1·142	1·148	1·160	1·170
0·80000	(7·85)	1·150	1·139	1·127	1·120	1·116	1·113	1·113	1·115	1·119	1·125	1·132	1·138	1·151	1·162
1·00000	(9·81)	1·134	1·124	1·113	1·107	1·104	1·103	1·103	1·105	1·111	1·117	1·124	1·131	1·144	1·156
1·50000	(14·7)	1·107	1·099	1·090	1·086	1·084	1·084	1·086	1·089	1·096	1·103	1·111	1·119	1·134	1·146
2·00000	(19·6)	1·089	1·082	1·075	1·072	1·071	1·073	1·075	1·079	1·087	1·094	1·103	1·112	1·127	1·140
3·00000	(29·4)	1·066	1·061	1·056	1·054	1·055	1·057	1·061	1·066	1·075	1·083	1·093	1·102	1·118	1·132
S	$(\Delta p/\rho l)$	125	150	200	250	300	400	500	600	800	1000	1250	1500	2000	2500

Grad'nt   k.p.g.   (Equivalent) Pipe diameters in mm   **Roughness size, $k_s = 0\cdot150$ mm**

Kin. visc., $\nu = 100\times10^{-6}$ m^{2}s^{-1};
$S = 0\cdot00010$ to $3\cdot00000$
i.e. kin. pr. grad., $\Delta p/\rho l =$
$(0\cdot00098)$ to $(29\cdot4)$ ms^{-2}

Roughness size, $k_s = 0\cdot30$ mm
This table shows values of m, as follows
m_C for Colebrook-White solutions; or,
where $R \leq 2000$, m_P for laminar flow

Grad'nt S	k.p.g. $(\Delta p/\rho l)$	125	150	200	250	300	400	500	600	800	1000	1250	1500	2000	2500
								(Equivalent) Pipe diameters in mm							
0·00010	(0·00098)	20·72	16·25	11·07	8·222	6·448	4·394	3·263	2·559	1·744	*	1·950	1·915	1·870	1·842
0·00015	(0·00147)	16·92	13·27	9·040	6·714	5·265	3·588	2·664	2·089	*	1·925	1·881	1·851	1·812	1·788
0·00020	(0·00196)	14·65	11·49	7·829	5·814	4·559	3·107	2·307	1·809	*	1·876	1·836	1·808	1·773	1·751
0·00030	(0·00294)	11·96	9·381	6·392	4·747	3·723	2·537	1·884	1·477	1·852	1·811	1·776	1·752	1·721	1·703
0·00040	(0·00392)	10·36	8·124	5·536	4·111	3·224	2·197	1·632	*	1·805	1·767	1·736	1·714	1·686	1·670
0·00060	(0·00588)	8·459	6·633	4·520	3·357	2·632	1·794	*	*	1·743	1·710	1·682	1·664	1·640	1·626
0·00080	(0·00785)	7·325	5·745	3·914	2·907	2·280	1·553	*	1·749	1·701	1·671	1·646	1·630	1·609	1·597
0·00100	(0·00981)	6·552	5·138	3·501	2·600	2·039	1·389	*	1·715	1·671	1·643	1·620	1·605	1·586	1·575
0·00150	(0·0147)	5·350	4·195	2·859	2·123	1·665	*	1·687	1·657	1·618	1·594	1·574	1·561	1·546	1·537
0·00200	(0·0196)	4·633	3·633	2·476	1·839	1·442	*	1·645	1·618	1·583	1·561	1·544	1·532	1·519	1·512
0·00300	(0·0294)	3·783	2·966	2·021	1·501	*	1·625	1·590	1·566	1·536	1·518	1·503	1·494	1·483	1·478
0·00400	(0·0392)	3·276	2·569	1·751	1·300	*	1·585	1·554	1·532	1·505	1·488	1·476	1·468	1·459	1·455
0·00600	(0·0588)	2·675	2·098	1·429	*	1·576	1·533	1·505	1·487	1·464	1·450	1·439	1·433	1·427	1·425
0·00800	(0·0785)	2·316	1·817	1·238	*	1·537	1·498	1·473	1·457	1·436	1·424	1·415	1·410	1·406	1·405
0·01000	(0·0981)	2·072	1·625	*	1·537	1·509	1·472	1·450	1·435	1·416	1·405	1·397	1·393	1·390	1·390
0·01500	(0·147)	1·692	1·327	*	1·485	1·460	1·429	1·409	1·396	1·381	1·372	1·367	1·364	1·363	1·364
0·02000	(0·196)	1·465	*	1·483	1·450	1·428	1·400	1·382	1·371	1·358	1·351	1·346	1·345	1·345	1·347
0·03000	(0·294)	1·196	*	1·433	1·405	1·386	1·361	1·347	1·338	1·327	1·322	1·319	1·319	1·321	1·325
0·04000	(0·392)	*	*	1·400	1·375	1·358	1·336	1·323	1·315	1·307	1·303	1·302	1·302	1·306	1·311
0·06000	(0·588)	*	1·393	1·357	1·336	1·321	1·303	1·293	1·286	1·280	1·278	1·278	1·280	1·285	1·292
0·08000	(0·785)	1·386	1·361	1·329	1·310	1·297	1·281	1·272	1·267	1·263	1·262	1·263	1·266	1·272	1·279
0·10000	(0·981)	1·361	1·338	1·309	1·291	1·279	1·265	1·257	1·253	1·250	1·250	1·252	1·255	1·262	1·270
0·15000	(1·47)	1·319	1·299	1·274	1·259	1·249	1·238	1·232	1·229	1·228	1·230	1·233	1·237	1·246	1·255
0·20000	(1·96)	1·291	1·273	1·251	1·237	1·229	1·220	1·216	1·214	1·214	1·217	1·221	1·226	1·236	1·245
0·30000	(2·94)	1·255	1·240	1·221	1·210	1·203	1·197	1·194	1·194	1·196	1·200	1·205	1·211	1·223	1·233
0·40000	(3·92)	1·231	1·218	1·201	1·192	1·187	1·182	1·180	1·181	1·184	1·189	1·195	1·202	1·214	1·225
0·60000	(5·88)	1·201	1·190	1·176	1·169	1·165	1·162	1·163	1·164	1·169	1·175	1·183	1·190	1·204	1·216
0·80000	(7·85)	1·181	1·171	1·160	1·154	1·151	1·150	1·151	1·154	1·160	1·166	1·175	1·183	1·197	1·210
1·00000	(9·81)	1·167	1·158	1·148	1·144	1·141	1·141	1·143	1·146	1·153	1·160	1·169	1·177	1·192	1·205
1·50000	(14·7)	1·143	1·136	1·129	1·126	1·125	1·126	1·130	1·134	1·142	1·150	1·160	1·169	1·185	1·198
2·00000	(19·6)	1·128	1·122	1·116	1·114	1·114	1·117	1·121	1·126	1·135	1·144	1·154	1·163	1·180	1·194
3·00000	(29·4)	1·109	1·104	1·100	1·100	1·101	1·105	1·110	1·116	1·126	1·136	1·147	1·157	1·174	1·189
S	$(\Delta p/\rho l)$	125	150	200	250	300	400	500	600	800	1000	1250	1500	2000	2500

Grad'nt   k.p.g.   (Equivalent) Pipe diameters in mm   **Roughness size, $k_s = 0\cdot30$ mm**

E265

Kin. visc., $v = 150 \times 10^{-6}$ m²s⁻¹;
S = 0·00010 to 30·0000

i.e. kin. pr. grad., $\Delta p/\rho l$ =
(0·00098) to (294) ms⁻²

Roughness size, k_s = 0·015 mm
This table shows values of m, as follows

m_C for Colebrook-White solutions; or,
where $R \leq 2000$, m_P for laminar flow

Grad'nt S	k.p.g. $(\Delta p/\rho l)$	6	8	10	12·5	15	20	25	30	40	50	60	80	100	125
0·00030	(0·00294)	1029	700·9	520·5	386·6	303·2	206·6	153·4	120·3	81·98	60·88	47·74	32·53	24·16	17·94
0·00040	(0·00392)	890·8	607·0	450·8	334·8	262·5	178·9	132·9	104·2	71·00	52·73	41·35	28·17	20·92	15·54
0·00060	(0·00588)	727·3	495·6	368·1	273·4	214·4	146·1	108·5	85·07	57·97	43·05	33·76	23·00	17·08	12·69
0·00080	(0·00785)	629·9	429·2	318·8	236·7	185·6	126·5	93·95	73·67	50·20	37·28	29·24	19·92	14·80	10·99
0·00100	(0·00981)	563·4	383·9	285·1	211·7	166·0	113·1	84·03	65·89	44·90	33·35	26·15	17·82	13·23	9·828
0·00150	(0·0147)	460·0	313·5	232·8	172·9	135·6	92·38	68·61	53·80	36·66	27·23	21·35	14·55	10·81	8·025
0·00200	(0·0196)	398·4	271·5	201·6	149·7	117·4	80·01	59·42	46·59	31·75	23·58	18·49	12·60	9·358	6·949
0·00300	(0·0294)	325·3	221·6	164·6	122·2	95·87	65·33	48·51	38·04	25·92	19·25	15·10	10·29	7·640	5·674
0·00400	(0·0392)	281·7	192·0	142·6	105·9	83·02	56·57	42·01	32·95	22·45	16·67	13·08	8·910	6·617	4·914
0·00600	(0·0588)	230·0	156·7	116·4	86·44	67·79	46·19	34·30	26·90	18·33	13·61	10·68	7·275	5·403	4·012
0·00800	(0·0785)	199·2	135·7	100·8	74·86	58·71	40·00	29·71	23·30	15·88	11·79	9·246	6·300	4·679	3·475
0·01000	(0·0981)	178·2	121·4	90·16	66·96	52·51	35·78	26·57	20·84	14·20	10·55	8·269	5·635	4·185	3·108
0·01500	(0·147)	145·5	99·12	73·62	54·67	42·87	29·21	21·70	17·01	11·59	8·610	6·752	4·601	3·417	2·538
0·02000	(0·196)	126·0	85·84	63·75	47·35	37·13	25·30	18·79	14·73	10·04	7·457	5·847	3·985	2·959	2·198
0·03000	(0·294)	102·9	70·09	52·05	38·66	30·32	20·66	15·34	12·03	8·198	6·088	4·774	3·253	2·416	1·794
0·04000	(0·392)	89·08	60·70	45·08	33·48	26·25	17·89	13·29	10·42	7·100	5·273	4·135	2·817	2·092	1·554
0·06000	(0·588)	72·73	49·56	36·81	27·34	21·44	14·61	10·85	8·507	5·797	4·305	3·376	2·300	1·708	1·269
0·08000	(0·785)	62·99	42·92	31·88	23·67	18·56	12·65	9·395	7·367	5·020	3·728	2·924	1·992	1·480	*
0·10000	(0·981)	56·34	38·39	28·51	21·17	16·60	11·31	8·403	6·589	4·490	3·335	2·615	1·782	1·323	*
0·15000	(1·47)	46·00	31·35	23·28	17·29	13·56	9·238	6·861	5·380	3·666	2·723	2·135	1·455	*	*
0·20000	(1·96)	39·84	27·15	20·16	14·97	11·74	8·001	5·942	4·659	3·175	2·358	1·849	1·260	*	1·341
0·30000	(2·94)	32·53	22·16	16·46	12·22	9·587	6·533	4·851	3·804	2·592	1·925	1·510	*	*	1·293
0·40000	(3·92)	28·17	19·20	14·26	10·59	8·302	5·657	4·201	3·295	2·245	1·667	1·308	*	1·289	1·261
0·60000	(5·88)	23·00	15·67	11·64	8·644	6·779	4·619	3·430	2·690	1·833	1·361	1·068	*	1·244	1·219
0·80000	(7·85)	19·92	13·57	10·08	7·486	5·871	4·000	2·971	2·330	1·588	1·179	*	1·240	1·213	1·191
1·00000	(9·81)	17·82	12·14	9·016	6·696	5·251	3·578	2·657	2·084	1·420	1·055	*	1·216	1·191	1·170
1·50000	(14·7)	14·55	9·912	7·362	5·467	4·287	2·921	2·170	1·701	1·159	*	1·209	1·174	1·152	1·133
2·00000	(19·6)	12·60	8·584	6·375	4·735	3·713	2·530	1·879	1·473	1·004	*	1·177	1·146	1·126	1·109
3·00000	(29·4)	10·29	7·009	5·205	3·866	3·032	2·066	1·534	1·203	*	1·157	1·136	1·108	1·091	1·077
4·00000	(39·2)	8·908	6·070	4·508	3·348	2·625	1·789	1·329	1·042	*	1·128	1·109	1·083	1·068	1·055
6·00000	(58·8)	7·273	4·956	3·681	2·734	2·144	1·461	1·085	*	1·114	1·089	1·072	1·050	1·037	1·026
8·00000	(78·5)	6·299	4·292	3·188	2·367	1·856	1·265	0·939	*	1·086	1·063	1·048	1·028	1·016	1·007
10·0000	(98·1)	5·634	3·839	2·851	2·117	1·660	1·131	*	*	1·065	1·044	1·030	1·012	1·001	0·992
15·0000	(147)	4·600	3·135	2·328	1·729	1·356	0·924	*	1·058	1·029	1·011	0·999	0·983	0·974	0·967
20·0000	(196)	3·984	2·715	2·016	1·497	1·174	*	1·052	1·032	1·005	0·989	0·978	0·964	0·956	0·950
30·0000	(294)	3·253	2·216	1·646	1·222	0·959	*	1·014	0·997	0·974	0·960	0·950	0·939	0·932	0·927
S	$(\Delta p/\rho l)$	6	8	10	12·5	15	20	25	30	40	50	60	80	100	125

Grad'nt **k.p.g.** (Equivalent) Pipe diameters in mm

Grad'nt S	k.p.g. $(\Delta p/\rho l)$	125	150	200	250	300	400	500	600	800	1000	1250	1500	2000	2500
0·00010	(0·00098)	31·08	24·37	16·61	12·33	9·672	6·591	4·895	3·838	2·616	1·942	*	2·054	1·994	1·957
0·00015	(0·00147)	25·38	19·90	13·56	10·07	7·897	5·381	3·996	3·134	2·136	*	*	1·979	1·927	1·895
0·00020	(0·00196)	21·98	17·23	11·74	8·721	6·839	4·660	3·461	2·714	1·849	*	1·965	1·929	1·882	1·853
0·00030	(0·00294)	17·94	14·07	9·588	7·121	5·584	3·805	2·826	2·216	*	*	1·895	1·863	1·822	1·797
0·00040	(0·00392)	15·54	12·19	8·304	6·167	4·836	3·295	2·447	1·919	*	1·890	1·848	1·819	1·782	1·759
0·00060	(0·00588)	12·69	9·950	6·780	5·035	3·949	2·691	1·998	1·567	*	1·822	1·786	1·760	1·728	1·708
0·00080	(0·00785)	10·99	8·617	5·872	4·361	3·420	2·330	1·731	*	1·817	1·777	1·744	1·721	1·692	1·674
0·00100	(0·00981)	9·828	7·707	5·252	3·900	3·059	2·084	1·548	*	1·780	1·743	1·713	1·692	1·665	1·648
0·00150	(0·0147)	8·025	6·293	4·288	3·185	2·497	1·702	*	1·770	1·718	1·686	1·659	1·641	1·618	1·604
0·00200	(0·0196)	6·949	5·450	3·714	2·758	2·163	1·474	*	1·723	1·677	1·647	1·623	1·606	1·586	1·574
0·00300	(0·0294)	5·674	4·450	3·032	2·252	1·766	*	1·694	1·662	1·621	1·596	1·575	1·561	1·543	1·533
0·00400	(0·0392)	4·914	3·854	2·626	1·950	1·529	*	1·650	1·621	1·584	1·561	1·542	1·530	1·514	1·506
0·00600	(0·0588)	4·012	3·146	2·144	1·592	*	1·628	1·592	1·567	1·535	1·515	1·499	1·488	1·475	1·469
0·00800	(0·0785)	3·475	2·725	1·857	1·379	*	1·587	1·554	1·531	1·502	1·483	1·469	1·460	1·449	1·443
0·01000	(0·0981)	3·108	2·437	1·661	*	*	1·556	1·525	1·504	1·477	1·460	1·447	1·439	1·429	1·425
0·01500	(0·147)	2·538	1·990	1·356	*	1·545	1·502	1·476	1·457	1·434	1·420	1·409	1·402	1·394	1·391
0·02000	(0·196)	2·198	1·723	*	1·535	1·505	1·467	1·442	1·426	1·405	1·392	1·383	1·377	1·371	1·369
0·03000	(0·294)	1·794	1·407	*	1·479	1·453	1·419	1·398	1·384	1·366	1·355	1·348	1·343	1·339	1·338
0·04000	(0·392)	1·554	1·219	1·476	1·442	1·418	1·387	1·368	1·356	1·340	1·331	1·324	1·320	1·317	1·317
0·06000	(0·588)	1·269	*	1·422	1·392	1·372	1·345	1·329	1·318	1·304	1·297	1·292	1·289	1·288	1·289
0·08000	(0·785)	*	*	1·387	1·359	1·341	1·316	1·302	1·292	1·280	1·274	1·270	1·268	1·268	1·270
0·10000	(0·981)	*	1·400	1·360	1·335	1·317	1·295	1·282	1·273	1·262	1·257	1·254	1·253	1·253	1·256
0·15000	(1·47)	*	1·350	1·315	1·292	1·277	1·258	1·247	1·240	1·231	1·227	1·225	1·225	1·227	1·230
0·20000	(1·96)	1·341	1·316	1·284	1·264	1·250	1·233	1·223	1·217	1·210	1·207	1·206	1·206	1·209	1·213
0·30000	(2·94)	1·293	1·271	1·243	1·226	1·214	1·200	1·192	1·187	1·182	1·180	1·180	1·181	1·185	1·189
0·40000	(3·92)	1·261	1·241	1·216	1·201	1·190	1·177	1·170	1·166	1·162	1·161	1·162	1·164	1·168	1·173
0·60000	(5·88)	1·219	1·202	1·180	1·166	1·158	1·147	1·141	1·138	1·136	1·136	1·138	1·140	1·146	1·152
0·80000	(7·85)	1·191	1·175	1·155	1·144	1·136	1·127	1·122	1·120	1·118	1·119	1·121	1·124	1·131	1·137
1·00000	(9·81)	1·170	1·155	1·137	1·126	1·119	1·111	1·107	1·106	1·105	1·106	1·109	1·112	1·119	1·126
1·50000	(14·7)	1·133	1·121	1·106	1·097	1·091	1·085	1·082	1·081	1·082	1·084	1·087	1·091	1·099	1·107
2·00000	(19·6)	1·109	1·098	1·084	1·076	1·072	1·067	1·065	1·064	1·066	1·069	1·073	1·077	1·085	1·093
3·00000	(29·4)	1·077	1·067	1·056	1·049	1·046	1·042	1·041	1·042	1·045	1·048	1·053	1·058	1·067	1·076
S	$(\Delta p/\rho l)$	125	150	200	250	300	400	500	600	800	1000	1250	1500	2000	2500

Grad'nt **k.p.g.** (Equivalent) Pipe diameters in mm

Kinematic viscosity, $v = 150 \times 10^{-6}$ m²s⁻¹ ;

Roughness size, k_s = 0·015 mm

Kin. visc., $\nu = 150\times10^{-6}\ \mathrm{m^2\,s^{-1}}$;   Roughness size, $k_\mathrm{s} = 0.030$ mm   # E266

$S = 0.00010$ to 3.00000

i.e. kin. pr. grad., $\Delta p/\rho l = (0.00098)$ to $(29.4)\ \mathrm{ms^{-2}}$

This table shows values of m, as follows m_C for Colebrook-White solutions; or, where $\mathbf{R} \le 2000$, m_P for laminar flow

Grad'nt S	k.p.g. $(\Delta p/\rho l)$	125	150	200	250	300	400	500	600	800	1000	1250	1500	2000	2500
0.00010	(0.00098)	31.08	24.37	16.61	12.33	9.672	6.591	4.895	3.838	2.616	1.942	*	2.054	1.995	1.957
0.00015	(0.00147)	25.38	19.90	13.56	10.07	7.897	5.381	3.996	3.134	2.136	*	*	1.979	1.927	1.895
0.00020	(0.00196)	21.98	17.23	11.74	8.721	6.839	4.660	3.461	2.714	1.849	*	1.966	1.929	1.882	1.853
0.00030	(0.00294)	17.94	14.07	9.588	7.121	5.584	3.805	2.826	2.216	*	*	1.895	1.864	1.822	1.797
0.00040	(0.00392)	15.54	12.19	8.304	6.167	4.836	3.295	2.447	1.919	*	1.890	1.848	1.819	1.782	1.759
0.00060	(0.00588)	12.69	9.950	6.780	5.035	3.949	2.691	1.998	1.567	*	1.822	1.786	1.761	1.728	1.709
0.00080	(0.00785)	10.99	8.617	5.872	4.361	3.420	2.330	1.731	*	1.817	1.777	1.744	1.721	1.692	1.675
0.00100	(0.00981)	9.828	7.707	5.252	3.900	3.059	2.084	1.548	*	1.781	1.744	1.713	1.692	1.665	1.649
0.00150	(0.0147)	8.025	6.293	4.288	3.185	2.497	1.702	*	1.770	1.719	1.686	1.660	1.641	1.618	1.605
0.00200	(0.0196)	6.949	5.450	3.714	2.758	2.163	1.474	*	1.724	1.677	1.648	1.623	1.607	1.586	1.575
0.00300	(0.0294)	5.674	4.450	3.032	2.252	1.766	*	1.694	1.663	1.622	1.596	1.575	1.561	1.544	1.534
0.00400	(0.0392)	4.914	3.854	2.626	1.950	1.529	*	1.651	1.622	1.585	1.562	1.543	1.530	1.515	1.507
0.00600	(0.0588)	4.012	3.146	2.144	1.592	*	1.629	1.593	1.568	1.536	1.516	1.499	1.489	1.476	1.470
0.00800	(0.0785)	3.475	2.725	1.857	1.379	*	1.587	1.554	1.532	1.502	1.484	1.470	1.461	1.450	1.445
0.01000	(0.0981)	3.108	2.437	1.661	*	*	1.556	1.526	1.505	1.478	1.461	1.448	1.440	1.430	1.426
0.01500	(0.147)	2.538	1.990	1.356	*	1.546	1.503	1.476	1.458	1.435	1.421	1.410	1.403	1.396	1.393
0.02000	(0.196)	2.198	1.723	*	1.536	1.506	1.468	1.443	1.427	1.406	1.394	1.384	1.378	1.372	1.370
0.03000	(0.294)	1.794	1.407	*	1.480	1.454	1.420	1.399	1.385	1.367	1.357	1.349	1.345	1.341	1.340
0.04000	(0.392)	1.554	1.219	1.477	1.443	1.419	1.389	1.370	1.357	1.341	1.332	1.326	1.322	1.319	1.320
0.06000	(0.588)	1.269	*	1.424	1.394	1.373	1.346	1.330	1.319	1.306	1.299	1.294	1.292	1.290	1.292
0.08000	(0.785)	*	*	1.388	1.361	1.342	1.318	1.303	1.294	1.282	1.276	1.272	1.271	1.271	1.273
0.10000	(0.981)	*	1.402	1.362	1.336	1.319	1.297	1.284	1.275	1.265	1.259	1.256	1.255	1.256	1.259
0.15000	(1.47)	*	1.351	1.316	1.294	1.279	1.260	1.249	1.242	1.234	1.230	1.228	1.228	1.230	1.234
0.20000	(1.96)	1.342	1.318	1.286	1.266	1.253	1.236	1.226	1.219	1.213	1.210	1.209	1.210	1.213	1.217
0.30000	(2.94)	1.295	1.273	1.246	1.228	1.217	1.203	1.194	1.190	1.185	1.183	1.183	1.185	1.189	1.194
0.40000	(3.92)	1.263	1.244	1.219	1.203	1.193	1.180	1.173	1.169	1.166	1.165	1.166	1.168	1.173	1.179
0.60000	(5.88)	1.221	1.204	1.183	1.169	1.161	1.150	1.145	1.142	1.140	1.140	1.142	1.145	1.151	1.158
0.80000	(7.85)	1.193	1.178	1.159	1.147	1.139	1.130	1.126	1.124	1.123	1.124	1.127	1.130	1.137	1.144
1.00000	(9.81)	1.173	1.159	1.141	1.130	1.123	1.115	1.112	1.110	1.110	1.111	1.115	1.118	1.126	1.133
1.50000	(14.7)	1.137	1.125	1.110	1.101	1.095	1.089	1.087	1.086	1.087	1.090	1.094	1.098	1.107	1.115
2.00000	(19.6)	1.113	1.102	1.089	1.081	1.076	1.072	1.070	1.070	1.072	1.075	1.080	1.085	1.094	1.103
3.00000	(29.4)	1.081	1.072	1.061	1.055	1.051	1.048	1.048	1.049	1.052	1.056	1.061	1.067	1.077	1.086
S	$(\Delta p/\rho l)$	125	150	200	250	300	400	500	600	800	1000	1250	1500	2000	2500

Grad'nt   k.p.g.   (Equivalent) Pipe diameters in mm   Roughness size, $k_\mathrm{s} = 0.030$ mm

---

Kin. visc., $\nu = 150\times10^{-6}\ \mathrm{m^2\,s^{-1}}$;   Roughness size, $k_\mathrm{s} = 0.060$ mm   # E267

$S = 0.00010$ to 3.00000

i.e. kin. pr. grad., $\Delta p/\rho l = (0.00098)$ to $(29.4)\ \mathrm{ms^{-2}}$

This table shows values of m, as follows m_C for Colebrook-White solutions; or, where $\mathbf{R} \le 2000$, m_P for laminar flow

Grad'nt S	k.p.g. $(\Delta p/\rho l)$	125	150	200	250	300	400	500	600	800	1000	1250	1500	2000	2500
0.00010	(0.00098)	31.08	24.37	16.61	12.33	9.672	6.591	4.895	3.838	2.616	1.942	*	2.054	1.995	1.958
0.00015	(0.00147)	25.38	19.90	13.56	10.07	7.897	5.381	3.996	3.134	2.136	*	*	1.980	1.928	1.895
0.00020	(0.00196)	21.98	17.23	11.74	8.721	6.839	4.660	3.461	2.714	1.849	*	1.966	1.930	1.883	1.854
0.00030	(0.00294)	17.94	14.07	9.588	7.121	5.584	3.805	2.826	2.216	*	*	1.896	1.864	1.823	1.798
0.00040	(0.00392)	15.54	12.19	8.304	6.167	4.836	3.295	2.447	1.919	*	1.891	1.849	1.820	1.783	1.760
0.00060	(0.00588)	12.69	9.950	6.780	5.035	3.949	2.691	1.998	1.567	*	1.823	1.787	1.762	1.729	1.710
0.00080	(0.00785)	10.99	8.617	5.872	4.361	3.420	2.330	1.731	*	1.818	1.778	1.745	1.722	1.693	1.676
0.00100	(0.00981)	9.828	7.707	5.252	3.900	3.059	2.084	1.548	*	1.782	1.745	1.714	1.693	1.666	1.650
0.00150	(0.0147)	8.025	6.293	4.288	3.185	2.497	1.702	*	1.771	1.720	1.687	1.661	1.642	1.619	1.606
0.00200	(0.0196)	6.949	5.450	3.714	2.758	2.163	1.474	*	1.725	1.678	1.649	1.625	1.608	1.588	1.576
0.00300	(0.0294)	5.674	4.450	3.032	2.252	1.766	*	1.695	1.664	1.623	1.598	1.577	1.563	1.545	1.536
0.00400	(0.0392)	4.914	3.854	2.626	1.950	1.529	*	1.652	1.623	1.586	1.563	1.544	1.532	1.517	1.508
0.00600	(0.0588)	4.012	3.146	2.144	1.592	*	1.631	1.594	1.569	1.537	1.517	1.501	1.491	1.478	1.472
0.00800	(0.0785)	3.475	2.725	1.857	1.379	*	1.589	1.556	1.533	1.504	1.486	1.472	1.463	1.452	1.447
0.01000	(0.0981)	3.108	2.437	1.661	*	*	1.558	1.528	1.506	1.479	1.463	1.450	1.442	1.433	1.428
0.01500	(0.147)	2.538	1.990	1.356	*	1.548	1.505	1.478	1.460	1.437	1.423	1.412	1.406	1.399	1.396
0.02000	(0.196)	2.198	1.723	*	1.538	1.508	1.470	1.446	1.429	1.408	1.396	1.387	1.381	1.375	1.374
0.03000	(0.294)	1.794	1.407	*	1.482	1.456	1.423	1.402	1.388	1.370	1.360	1.352	1.348	1.344	1.344
0.04000	(0.392)	1.554	1.219	1.480	1.445	1.422	1.391	1.372	1.360	1.344	1.335	1.329	1.326	1.323	1.324
0.06000	(0.588)	1.269	*	1.426	1.396	1.376	1.349	1.333	1.322	1.310	1.303	1.298	1.296	1.295	1.297
0.08000	(0.785)	*	*	1.391	1.364	1.345	1.321	1.307	1.297	1.286	1.280	1.277	1.275	1.276	1.278
0.10000	(0.981)	*	1.405	1.365	1.340	1.322	1.300	1.287	1.279	1.269	1.264	1.261	1.260	1.262	1.264
0.15000	(1.47)	*	1.355	1.320	1.298	1.283	1.264	1.253	1.246	1.239	1.235	1.234	1.234	1.237	1.241
0.20000	(1.96)	1.346	1.322	1.290	1.270	1.257	1.240	1.230	1.224	1.218	1.216	1.215	1.216	1.220	1.224
0.30000	(2.94)	1.299	1.278	1.250	1.233	1.222	1.208	1.200	1.195	1.191	1.190	1.190	1.192	1.197	1.203
0.40000	(3.92)	1.268	1.248	1.224	1.208	1.198	1.186	1.179	1.176	1.172	1.172	1.174	1.176	1.182	1.188
0.60000	(5.88)	1.227	1.210	1.188	1.175	1.167	1.157	1.152	1.149	1.148	1.149	1.151	1.155	1.162	1.169
0.80000	(7.85)	1.199	1.184	1.165	1.153	1.146	1.138	1.134	1.132	1.132	1.133	1.136	1.140	1.148	1.156
1.00000	(9.81)	1.179	1.165	1.147	1.137	1.130	1.123	1.120	1.119	1.119	1.122	1.125	1.130	1.138	1.146
1.50000	(14.7)	1.144	1.132	1.117	1.109	1.103	1.098	1.096	1.096	1.098	1.101	1.106	1.111	1.121	1.130
2.00000	(19.6)	1.121	1.110	1.097	1.090	1.086	1.082	1.081	1.081	1.084	1.088	1.094	1.099	1.110	1.119
3.00000	(29.4)	1.090	1.081	1.070	1.065	1.062	1.059	1.060	1.061	1.065	1.070	1.077	1.083	1.095	1.105
S	$(\Delta p/\rho l)$	125	150	200	250	300	400	500	600	800	1000	1250	1500	2000	2500

Grad'nt   k.p.g.   (Equivalent) Pipe diameters in mm   Roughness size, $k_\mathrm{s} = 0.060$ mm

E268

Kin. visc., $\nu = 200\times10^{-6}$ m^{2}s^{-1};
$S = 0.00010$ to 30.0000

i.e. kin. pr. grad., $\Delta p/\rho l =$ (0.00098) to (294) ms^{-2}

Roughness size, $k_s = 0.015$ mm
This table shows values of m, as follows

m_C for Colebrook-White solutions; or, where $R \le 2000$, m_P for laminar flow

Grad'nt S	k.p.g. $(\Delta p/\rho l)$	6	8	10	12.5	15	20	25	30	40	50	60	80	100	125
0.00030	(0.00294)	1371	934.6	694.1	515.4	404.2	275.4	204.6	160.4	109.3	81.18	63.66	43.38	32.22	23.92
0.00040	(0.00392)	1188	809.3	601.1	446.4	350.1	238.5	177.1	138.9	94.66	70.30	55.13	37.57	27.90	20.72
0.00060	(0.00588)	969.8	660.8	490.8	364.5	285.8	194.8	144.6	113.4	77.29	57.40	45.01	30.67	22.78	16.92
0.00080	(0.00785)	839.9	572.3	425.0	315.6	247.5	168.7	125.3	98.23	66.94	49.71	38.98	26.56	19.73	14.65
0.00100	(0.00981)	751.2	511.9	380.1	282.3	221.4	150.9	112.0	87.86	59.87	44.46	34.87	23.76	17.64	13.10
0.00150	(0.0147)	613.3	417.9	310.4	230.5	180.8	123.2	91.48	71.74	48.88	36.30	28.47	19.40	14.41	10.70
0.00200	(0.0196)	531.2	362.0	268.8	199.6	156.5	106.7	79.22	62.13	42.33	31.44	24.65	16.80	12.48	9.266
0.00300	(0.0294)	433.7	295.5	219.5	163.0	127.8	87.10	64.69	50.73	34.57	25.67	20.13	13.72	10.19	7.566
0.00400	(0.0392)	375.6	255.9	190.1	141.2	110.7	75.43	56.02	43.93	29.93	22.23	17.43	11.88	8.822	6.552
0.00600	(0.0588)	306.7	209.0	155.2	115.3	90.38	61.59	45.74	35.87	24.44	18.15	14.23	9.700	7.204	5.350
0.00800	(0.0785)	265.6	181.0	134.4	99.81	78.27	53.34	39.61	31.06	21.17	15.72	12.33	8.400	6.238	4.633
0.01000	(0.0981)	237.5	161.9	120.2	89.28	70.01	47.71	35.43	27.78	18.93	14.06	11.03	7.513	5.580	4.144
0.01500	(0.147)	194.0	132.2	98.15	72.89	57.16	38.95	28.93	22.69	15.46	11.48	9.003	6.135	4.556	3.383
0.02000	(0.196)	168.0	114.5	85.00	63.13	49.51	33.73	25.05	19.65	13.39	9.942	7.797	5.313	3.946	2.930
0.03000	(0.294)	137.1	93.46	69.41	51.54	40.42	27.54	20.46	16.04	10.93	8.118	6.366	4.338	3.222	2.392
0.04000	(0.392)	118.8	80.93	60.11	44.64	35.01	23.85	17.71	13.89	9.466	7.030	5.513	3.757	2.790	2.072
0.06000	(0.588)	96.98	66.08	49.08	36.45	28.58	19.48	14.46	11.34	7.729	5.740	4.501	3.067	2.278	1.692
0.08000	(0.785)	83.99	57.23	42.50	31.56	24.75	16.87	12.53	9.823	6.694	4.971	3.898	2.656	1.973	1.465
0.10000	(0.981)	75.12	51.19	38.01	28.23	22.14	15.09	11.20	8.786	5.987	4.446	3.487	2.376	1.764	1.310
0.15000	(1.47)	61.33	41.79	31.04	23.05	18.08	12.32	9.148	7.174	4.888	3.630	2.847	1.940	1.441	*
0.20000	(1.96)	53.12	36.20	26.88	19.96	15.65	10.67	7.922	6.213	4.233	3.144	2.465	1.680	1.248	*
0.30000	(2.94)	43.37	29.55	21.95	16.30	12.78	8.710	6.469	5.073	3.457	2.567	2.013	1.372	*	1.362
0.40000	(3.92)	37.56	25.59	19.01	14.12	11.07	7.543	5.602	4.393	2.993	2.223	1.743	1.188	*	1.326
0.60000	(5.88)	30.67	20.90	15.52	11.53	9.038	6.159	4.574	3.587	2.444	1.815	1.423	*	1.310	1.280
0.80000	(7.85)	26.56	18.10	13.44	9.981	7.827	5.334	3.961	3.106	2.117	1.572	1.233	*	1.276	1.248
1.00000	(9.81)	23.75	16.19	12.02	8.928	7.001	4.771	3.543	2.778	1.893	1.406	1.103	*	1.251	1.225
1.50000	(14.7)	19.40	13.22	9.815	7.289	5.716	3.895	2.893	2.269	1.546	1.148	*	1.234	1.208	1.186
2.00000	(19.6)	16.80	11.45	8.500	6.313	4.951	3.373	2.505	1.965	1.339	0.994	*	1.203	1.179	1.159
3.00000	(29.4)	13.71	9.346	6.941	5.154	4.042	2.754	2.046	1.604	1.093	*	1.196	1.162	1.141	1.123
4.00000	(39.2)	11.88	8.093	6.011	4.464	3.501	2.385	1.771	1.389	0.947	*	1.165	1.134	1.115	1.100
6.00000	(58.8)	9.698	6.608	4.908	3.645	2.858	1.948	1.446	1.134	*	1.145	1.125	1.098	1.081	1.068
8.00000	(78.5)	8.399	5.723	4.250	3.156	2.475	1.687	1.253	0.982	*	1.116	1.098	1.074	1.059	1.047
10.0000	(98.1)	7.512	5.119	3.801	2.823	2.214	1.509	1.120	*	1.120	1.095	1.078	1.055	1.042	1.031
15.0000	(147)	6.133	4.179	3.104	2.305	1.808	1.232	0.915	*	1.081	1.059	1.044	1.024	1.012	1.003
20.0000	(196)	5.312	3.620	2.688	1.996	1.565	1.067	*	*	1.054	1.034	1.021	1.003	0.993	0.985
30.0000	(294)	4.337	2.955	2.195	1.630	1.278	0.871	*	1.047	1.019	1.002	0.990	0.975	0.967	0.960
S	$(\Delta p/\rho l)$	6	8	10	12.5	15	20	25	30	40	50	60	80	100	125

Grad'nt S k.p.g. $(\Delta p/\rho l)$ (Equivalent) Pipe diameters in mm

Grad'nt S	k.p.g. $(\Delta p/\rho l)$	125	150	200	250	300	400	500	600	800	1000	1250	1500	2000	2500
0.00010	(0.00098)	41.44	32.50	22.14	16.44	12.90	8.788	6.526	5.118	3.487	2.590	1.923	*	2.098	2.053
0.00015	(0.00147)	33.83	26.53	18.08	13.43	10.53	7.175	5.329	4.179	2.847	2.115	*	*	2.024	1.985
0.00020	(0.00196)	29.30	22.98	15.66	11.63	9.119	6.214	4.615	3.619	2.466	1.831	*	2.031	1.974	1.938
0.00030	(0.00294)	23.92	18.76	12.78	9.494	7.446	5.074	3.768	2.955	2.013	*	1.996	1.958	1.908	1.877
0.00040	(0.00392)	20.72	16.25	11.07	8.222	6.448	4.394	3.263	2.559	1.744	*	1.944	1.910	1.864	1.836
0.00060	(0.00588)	16.92	13.27	9.040	6.714	5.265	3.588	2.664	2.089	*	1.919	1.875	1.845	1.805	1.781
0.00080	(0.00785)	14.65	11.49	7.829	5.814	4.559	3.107	2.307	1.809	*	1.869	1.829	1.802	1.766	1.744
0.00100	(0.00981)	13.10	10.28	7.002	5.200	4.078	2.779	2.064	1.618	*	1.832	1.795	1.769	1.736	1.716
0.00150	(0.0147)	10.70	8.390	5.717	4.246	3.330	2.269	1.685	*	1.808	1.769	1.736	1.714	1.685	1.668
0.00200	(0.0196)	9.266	7.266	4.951	3.677	2.884	1.965	1.459	*	1.762	1.726	1.697	1.676	1.651	1.635
0.00300	(0.0294)	7.566	5.933	4.043	3.002	2.354	1.604	*	1.750	1.701	1.670	1.644	1.627	1.604	1.592
0.00400	(0.0392)	6.552	5.138	3.501	2.600	2.039	1.389	*	1.705	1.660	1.632	1.609	1.593	1.573	1.562
0.00600	(0.0588)	5.350	4.195	2.859	2.123	1.665	*	1.676	1.645	1.606	1.581	1.561	1.548	1.531	1.522
0.00800	(0.0785)	4.633	3.633	2.476	1.839	1.442	*	1.633	1.605	1.569	1.547	1.529	1.517	1.503	1.495
0.01000	(0.0981)	4.144	3.250	2.214	1.644	*	1.638	1.601	1.576	1.542	1.522	1.505	1.494	1.481	1.474
0.01500	(0.147)	3.383	2.653	1.808	1.343	*	1.579	1.547	1.524	1.496	1.478	1.464	1.455	1.444	1.439
0.02000	(0.196)	2.930	2.298	1.566	*	*	1.540	1.510	1.490	1.464	1.448	1.436	1.428	1.419	1.415
0.03000	(0.294)	2.392	1.876	1.278	*	1.528	1.487	1.462	1.444	1.422	1.408	1.398	1.391	1.385	1.382
0.04000	(0.392)	2.072	1.625	*	1.518	1.490	1.453	1.429	1.413	1.393	1.381	1.372	1.367	1.362	1.360
0.06000	(0.588)	1.692	1.327	*	1.464	1.439	1.406	1.386	1.372	1.355	1.345	1.338	1.334	1.330	1.330
0.08000	(0.785)	1.465	*	1.460	1.427	1.404	1.375	1.357	1.344	1.329	1.321	1.315	1.311	1.309	1.309
0.10000	(0.981)	1.310	*	1.431	1.400	1.379	1.352	1.335	1.323	1.310	1.302	1.297	1.294	1.293	1.294
0.15000	(1.47)	*	*	1.380	1.353	1.335	1.311	1.297	1.287	1.276	1.270	1.266	1.265	1.265	1.267
0.20000	(1.96)	*	1.385	1.347	1.322	1.306	1.284	1.271	1.263	1.253	1.248	1.245	1.245	1.245	1.248
0.30000	(2.94)	1.362	1.336	1.302	1.281	1.266	1.248	1.237	1.230	1.222	1.219	1.217	1.217	1.220	1.223
0.40000	(3.92)	1.326	1.303	1.272	1.253	1.240	1.223	1.214	1.208	1.202	1.199	1.198	1.199	1.202	1.206
0.60000	(5.88)	1.280	1.259	1.232	1.216	1.204	1.191	1.183	1.178	1.173	1.172	1.172	1.174	1.178	1.183
0.80000	(7.85)	1.248	1.230	1.205	1.190	1.181	1.168	1.162	1.158	1.154	1.154	1.155	1.157	1.162	1.167
1.00000	(9.81)	1.225	1.208	1.185	1.172	1.163	1.152	1.146	1.143	1.140	1.140	1.141	1.144	1.149	1.155
1.50000	(14.7)	1.186	1.170	1.151	1.139	1.132	1.123	1.118	1.116	1.115	1.116	1.118	1.121	1.128	1.134
2.00000	(19.6)	1.159	1.145	1.128	1.118	1.111	1.103	1.100	1.098	1.098	1.100	1.103	1.106	1.113	1.120
3.00000	(29.4)	1.123	1.112	1.097	1.088	1.083	1.077	1.075	1.074	1.075	1.078	1.081	1.085	1.093	1.101
S	$(\Delta p/\rho l)$	125	150	200	250	300	400	500	600	800	1000	1250	1500	2000	2500

Grad'nt   k.p.g.   (Equivalent) Pipe diameters in mm

Kinematic viscosity, $\nu = 200\times10^{-6}$ m^{2}s^{-1} ;

Roughness size, $k_s = 0.015$ mm

Kin. visc., $\nu = 200 \times 10^{-6}$ m^{2}s^{-1};
S = 0·00010 to 3·00000

i.e. kin. pr. grad., $\Delta p/\rho l =$
(0·00098) to (29·4) ms^{-2}

Roughness size, k_s = 0·030 mm
This table shows values of m, as follows

m_C for Colebrook-White solutions; or,
where $R \leq 2000$, m_P for laminar flow

E269

Grad'nt S	k.p.g. $(\Delta p/\rho l)$	125	150	200	250	300	400	500	600	800	1000	1250	1500	2000	2500
0·00010	(0·00098)	41·44	32·50	22·14	16·44	12·90	8·788	6·526	5·118	3·487	2·590	1·923	*	2·098	2·053
0·00015	(0·00147)	33·83	26·53	18·08	13·43	10·53	7·175	5·329	4·179	2·847	2·115	*	*	2·024	1·985
0·00020	(0·00196)	29·30	22·98	15·66	11·63	9·119	6·214	4·615	3·619	2·466	1·831	*	2·031	1·975	1·939
0·00030	(0·00294)	23·92	18·76	12·78	9·494	7·446	5·074	3·768	2·955	2·013	*	1·997	1·959	1·909	1·877
0·00040	(0·00392)	20·72	16·25	11·07	8·222	6·448	4·394	3·263	2·559	1·744	*	1·945	1·910	1·865	1·836
0·00060	(0·00588)	16·92	13·27	9·040	6·714	5·265	3·588	2·664	2·089	*	1·920	1·876	1·845	1·806	1·781
0·00080	(0·00785)	14·65	11·49	7·829	5·814	4·559	3·107	2·307	1·809	*	1·870	1·830	1·802	1·766	1·744
0·00100	(0·00981)	13·10	10·28	7·002	5·200	4·078	2·779	2·064	1·618	*	1·833	1·796	1·770	1·737	1·716
0·00150	(0·0147)	10·70	8·390	5·717	4·246	3·330	2·269	1·685	*	1·808	1·769	1·737	1·714	1·686	1·668
0·00200	(0·0196)	9·266	7·266	4·951	3·677	2·884	1·965	1·459	*	1·762	1·727	1·697	1·677	1·651	1·636
0·00300	(0·0294)	7·566	5·933	4·043	3·002	2·354	1·604	*	1·751	1·701	1·670	1·645	1·627	1·605	1·592
0·00400	(0·0392)	6·552	5·138	3·501	2·600	2·039	1·389	*	1·706	1·661	1·632	1·609	1·593	1·574	1·563
0·00600	(0·0588)	5·350	4·195	2·859	2·123	1·665	*	1·676	1·646	1·606	1·582	1·562	1·548	1·532	1·523
0·00800	(0·0785)	4·633	3·633	2·476	1·839	1·442	*	1·633	1·606	1·570	1·548	1·530	1·518	1·504	1·496
0·01000	(0·0981)	4·144	3·250	2·214	1·644	*	1·639	1·602	1·576	1·543	1·523	1·506	1·495	1·482	1·475
0·01500	(0·147)	3·383	2·653	1·808	1·343	*	1·580	1·548	1·525	1·496	1·479	1·465	1·456	1·445	1·440
0·02000	(0·196)	2·930	2·298	1·566	*	*	1·541	1·511	1·491	1·465	1·449	1·437	1·429	1·420	1·416
0·03000	(0·294)	2·392	1·876	1·278	*	1·529	1·488	1·463	1·445	1·423	1·409	1·399	1·393	1·386	1·384
0·04000	(0·392)	2·072	1·625	*	1·519	1·491	1·454	1·430	1·414	1·394	1·383	1·374	1·368	1·363	1·361
0·06000	(0·588)	1·692	1·327	*	1·465	1·440	1·407	1·387	1·373	1·356	1·347	1·339	1·335	1·332	1·332
0·08000	(0·785)	1·465	*	1·461	1·428	1·406	1·376	1·358	1·346	1·331	1·322	1·316	1·313	1·311	1·311
0·10000	(0·981)	1·310	*	1·432	1·401	1·380	1·353	1·336	1·325	1·312	1·304	1·299	1·296	1·295	1·296
0·15000	(1·47)	*	*	1·382	1·355	1·337	1·313	1·299	1·289	1·278	1·272	1·269	1·267	1·267	1·269
0·20000	(1·96)	*	1·387	1·348	1·324	1·307	1·286	1·273	1·265	1·255	1·251	1·248	1·247	1·248	1·251
0·30000	(2·94)	1·364	1·338	1·304	1·283	1·268	1·250	1·239	1·233	1·225	1·222	1·220	1·220	1·223	1·227
0·40000	(3·92)	1·328	1·305	1·274	1·255	1·242	1·226	1·216	1·211	1·204	1·202	1·201	1·202	1·206	1·210
0·60000	(5·88)	1·282	1·261	1·234	1·218	1·207	1·193	1·186	1·181	1·177	1·176	1·176	1·178	1·182	1·188
0·80000	(7·85)	1·251	1·232	1·208	1·193	1·183	1·171	1·165	1·161	1·158	1·158	1·159	1·161	1·167	1·172
1·00000	(9·81)	1·228	1·211	1·188	1·175	1·166	1·155	1·149	1·146	1·144	1·144	1·146	1·149	1·155	1·161
1·50000	(14·7)	1·188	1·173	1·154	1·143	1·135	1·127	1·122	1·120	1·120	1·121	1·124	1·127	1·134	1·141
2·00000	(19·6)	1·162	1·149	1·131	1·121	1·115	1·108	1·104	1·103	1·103	1·105	1·109	1·112	1·120	1·128
3·00000	(29·4)	1·127	1·115	1·101	1·093	1·087	1·082	1·080	1·080	1·081	1·084	1·088	1·093	1·101	1·110
S	$(\Delta p/\rho l)$	125	150	200	250	300	400	500	600	800	1000	1250	1500	2000	2500

Grad'nt k.p.g. (Equivalent) Pipe diameters in mm Roughness size, k_s = 0·030 mm

---

Kin. visc., $\nu = 200 \times 10^{-6}$ m^{2}s^{-1};
S = 0·00010 to 3·00000

i.e. kin. pr. grad., $\Delta p/\rho l =$
(0·00098) to (29·4) ms^{-2}

Roughness size, k_s = 0·060 mm
This table shows values of m, as follows

m_C for Colebrook-White solutions; or,
where $R \leq 2000$, m_P for laminar flow

E270

Grad'nt S	k.p.g. $(\Delta p/\rho l)$	125	150	200	250	300	400	500	600	800	1000	1250	1500	2000	2500
0·00010	(0·00098)	41·44	32·50	22·14	16·44	12·90	8·788	6·526	5·118	3·487	2·590	1·923	*	2·099	2·054
0·00015	(0·00147)	33·83	26·53	18·08	13·43	10·53	7·175	5·329	4·179	2·847	2·115	*	*	2·025	1·985
0·00020	(0·00196)	29·30	22·98	15·66	11·63	9·119	6·214	4·615	3·619	2·466	1·831	*	2·032	1·975	1·939
0·00030	(0·00294)	23·92	18·76	12·78	9·494	7·446	5·074	3·768	2·955	2·013	*	1·997	1·959	1·909	1·878
0·00040	(0·00392)	20·72	16·25	11·07	8·222	6·448	4·394	3·263	2·559	1·744	*	1·945	1·910	1·865	1·837
0·00060	(0·00588)	16·92	13·27	9·040	6·714	5·265	3·588	2·664	2·089	*	1·920	1·876	1·846	1·806	1·782
0·00080	(0·00785)	14·65	11·49	7·829	5·814	4·559	3·107	2·307	1·809	*	1·870	1·830	1·803	1·767	1·745
0·00100	(0·00981)	13·10	10·28	7·002	5·200	4·078	2·779	2·064	1·618	*	1·834	1·796	1·771	1·738	1·717
0·00150	(0·0147)	10·70	8·390	5·717	4·246	3·330	2·269	1·685	*	1·809	1·770	1·738	1·715	1·687	1·669
0·00200	(0·0196)	9·266	7·266	4·951	3·677	2·884	1·965	1·459	*	1·763	1·728	1·698	1·678	1·652	1·637
0·00300	(0·0294)	7·566	5·933	4·043	3·002	2·354	1·604	*	1·752	1·702	1·671	1·646	1·628	1·606	1·593
0·00400	(0·0392)	6·552	5·138	3·501	2·600	2·039	1·389	*	1·707	1·662	1·634	1·610	1·595	1·575	1·564
0·00600	(0·0588)	5·350	4·195	2·859	2·123	1·665	*	1·677	1·647	1·608	1·583	1·563	1·550	1·534	1·524
0·00800	(0·0785)	4·633	3·633	2·476	1·839	1·442	*	1·635	1·607	1·572	1·549	1·532	1·520	1·505	1·498
0·01000	(0·0981)	4·144	3·250	2·214	1·644	*	1·640	1·603	1·578	1·545	1·524	1·508	1·497	1·484	1·477
0·01500	(0·147)	3·383	2·653	1·808	1·343	*	1·581	1·549	1·527	1·498	1·481	1·467	1·458	1·447	1·442
0·02000	(0·196)	2·930	2·298	1·566	*	*	1·542	1·513	1·493	1·467	1·451	1·439	1·431	1·423	1·419
0·03000	(0·294)	2·392	1·876	1·278	*	1·531	1·490	1·465	1·447	1·425	1·412	1·402	1·395	1·389	1·387
0·04000	(0·392)	2·072	1·625	*	1·521	1·493	1·456	1·433	1·417	1·397	1·385	1·376	1·371	1·366	1·365
0·06000	(0·588)	1·692	1·327	*	1·467	1·442	1·410	1·390	1·376	1·359	1·350	1·343	1·339	1·336	1·336
0·08000	(0·785)	1·465	*	1·464	1·431	1·408	1·379	1·361	1·349	1·334	1·326	1·320	1·317	1·315	1·316
0·10000	(0·981)	1·310	*	1·435	1·404	1·383	1·356	1·339	1·328	1·315	1·308	1·303	1·300	1·299	1·301
0·15000	(1·47)	*	*	1·385	1·358	1·340	1·316	1·302	1·293	1·282	1·276	1·273	1·272	1·272	1·275
0·20000	(1·96)	*	1·390	1·351	1·327	1·311	1·290	1·277	1·269	1·260	1·255	1·253	1·252	1·254	1·257
0·30000	(2·94)	1·367	1·341	1·307	1·286	1·272	1·254	1·244	1·237	1·230	1·227	1·226	1·226	1·230	1·234
0·40000	(3·92)	1·332	1·309	1·278	1·259	1·246	1·230	1·221	1·216	1·210	1·208	1·208	1·209	1·213	1·218
0·60000	(5·88)	1·286	1·266	1·239	1·223	1·212	1·199	1·191	1·187	1·183	1·182	1·183	1·185	1·191	1·197
0·80000	(7·85)	1·256	1·237	1·213	1·198	1·189	1·177	1·171	1·168	1·165	1·165	1·167	1·170	1·176	1·182
1·00000	(9·81)	1·233	1·216	1·194	1·180	1·172	1·162	1·156	1·153	1·152	1·152	1·155	1·158	1·165	1·172
1·50000	(14·7)	1·194	1·179	1·160	1·149	1·142	1·134	1·130	1·129	1·129	1·130	1·134	1·138	1·146	1·154
2·00000	(19·6)	1·168	1·155	1·138	1·128	1·122	1·116	1·113	1·112	1·113	1·116	1·120	1·124	1·133	1·141
3·00000	(29·4)	1·134	1·123	1·109	1·101	1·096	1·091	1·090	1·090	1·092	1·096	1·101	1·106	1·116	1·125
S	$(\Delta p/\rho l)$	125	150	200	250	300	400	500	600	800	1000	1250	1500	2000	2500

Grad'nt k.p.g. (Equivalent) Pipe diameters in mm Roughness size, k_s = 0·060 mm

E271

Kin. visc., $\nu = 300 \times 10^{-6}$ m^{2}s^{-1}; S = 0·00010 to 30·0000

i.e. kin. pr. grad., $\Delta p/\rho l$ = (0·00098) to (294) ms^{-2}

Roughness size, k_s = 0·015 mm

This table shows values of m, as follows

m_C for Colebrook-White solutions; or, where $\mathbf{R} \leq 2000$, m_P for laminar flow

Grad'nt S	k.p.g. $(\Delta p/\rho l)$	6	8	10	12·5	15	20	25	30	40	50	60	80	100	125
0·00030	(0·00294)	2057	1402	1041	773·2	606·3	413·2	306·8	240·6	164·0	121·8	95·49	65·07	48·32	35·89
0·00040	(0·00392)	1782	1214	901·6	669·6	525·1	357·8	265·7	208·4	142·0	105·5	82·69	56·35	41·85	31·08
0·00060	(0·00588)	1455	991·2	736·2	546·7	428·7	292·1	217·0	170·1	115·9	86·10	67·52	46·01	34·17	25·38
0·00080	(0·00785)	1260	858·4	637·5	473·5	371·3	253·0	187·9	147·3	100·4	74·57	58·47	39·85	29·59	21·98
0·00100	(0·00981)	1127	767·8	570·2	423·5	332·1	226·3	168·1	131·8	89·80	66·69	52·30	35·64	26·47	19·66
0·00150	(0·0147)	920·0	626·9	465·6	345·8	271·2	184·8	137·2	107·6	73·32	54·46	42·70	29·10	21·61	16·05
0·00200	(0·0196)	796·8	542·9	403·2	299·4	234·8	160·0	118·8	93·19	63·50	47·16	36·98	25·20	18·72	13·90
0·00300	(0·0294)	650·6	443·3	329·2	244·5	191·7	130·7	97·03	76·09	51·85	38·51	30·20	20·58	15·28	11·35
0·00400	(0·0392)	563·4	383·9	285·1	211·7	166·0	113·1	84·03	65·89	44·90	33·35	26·15	17·82	13·23	9·828
0·00600	(0·0588)	460·0	313·5	232·8	172·9	135·6	92·38	68·61	53·80	36·66	27·23	21·35	14·55	10·81	8·025
0·00800	(0·0785)	398·4	271·5	201·6	149·7	117·4	80·01	59·42	46·59	31·75	23·58	18·49	12·60	9·358	6·949
0·01000	(0·0981)	356·3	242·8	180·3	133·9	105·0	71·56	53·14	41·68	28·40	21·09	16·54	11·27	8·370	6·216
0·01500	(0·147)	290·9	198·2	147·2	109·3	85·75	58·43	43·39	34·03	23·19	17·22	13·50	9·202	6·834	5·075
0·02000	(0·196)	252·0	171·7	127·5	94·69	74·26	50·60	37·58	29·47	20·08	14·91	11·69	7·969	5·918	4·395
0·03000	(0·294)	205·7	140·2	104·1	77·32	60·63	41·32	30·68	24·06	16·40	12·18	9·549	6·507	4·832	3·589
0·04000	(0·392)	178·2	121·4	90·16	66·96	52·51	35·78	26·57	20·84	14·20	10·55	8·269	5·635	4·185	3·108
0·06000	(0·588)	145·5	99·12	73·62	54·67	42·87	29·21	21·70	17·01	11·59	8·610	6·752	4·601	3·417	2·538
0·08000	(0·785)	126·0	85·84	63·75	47·35	37·13	25·30	18·79	14·73	10·04	7·457	5·847	3·985	2·959	2·198
0·10000	(0·981)	112·7	76·78	57·02	42·35	33·21	22·63	16·81	13·18	8·980	6·669	5·230	3·564	2·647	1·966
0·15000	(1·47)	92·00	62·69	46·56	34·58	27·12	18·48	13·72	10·76	7·332	5·446	4·270	2·910	2·161	1·605
0·20000	(1·96)	79·68	54·29	40·32	29·94	23·48	16·00	11·88	9·319	6·350	4·716	3·698	2·520	1·872	1·390
0·30000	(2·94)	65·06	44·33	32·92	24·45	19·17	13·07	9·703	7·609	5·185	3·851	3·020	2·058	1·528	1·135
0·40000	(3·92)	56·34	38·39	28·51	21·17	16·60	11·31	8·403	6·589	4·490	3·335	2·615	1·782	1·323	*
0·60000	(5·88)	46·00	31·35	23·28	17·29	13·56	9·238	6·861	5·380	3·666	2·723	2·135	1·455	*	*
0·80000	(7·85)	39·84	27·15	20·16	14·97	11·74	8·001	5·942	4·659	3·175	2·358	1·849	1·260	*	1·341
1·00000	(9·81)	35·63	24·28	18·03	13·39	10·50	7·156	5·314	4·168	2·840	2·109	1·654	1·127	*	1·314
1·50000	(14·7)	29·09	19·82	14·72	10·93	8·575	5·843	4·339	3·403	2·319	1·722	1·350	*	1·297	1·268
2·00000	(19·6)	25·20	17·17	12·75	9·469	7·426	5·060	3·758	2·947	2·008	1·491	1·169	*	1·264	1·237
3·00000	(29·4)	20·57	14·02	10·41	7·732	6·063	4·132	3·068	2·406	1·640	1·218	*	1·247	1·220	1·197
4·00000	(39·2)	17·82	12·14	9·016	6·696	5·251	3·578	2·657	2·084	1·420	1·055	*	1·216	1·191	1·170
6·00000	(58·8)	14·55	9·912	7·362	5·467	4·287	2·921	2·170	1·701	1·159	*	1·209	1·174	1·152	1·133
8·00000	(78·5)	12·60	8·584	6·375	4·735	3·713	2·530	1·879	1·473	1·004	*	1·177	1·146	1·126	1·109
10·0000	(98·1)	11·27	7·678	5·702	4·235	3·321	2·263	1·681	1·318	*	1·177	1·154	1·125	1·106	1·091
15·0000	(147)	9·200	6·269	4·656	3·458	2·712	1·848	1·372	1·076	*	1·134	1·115	1·089	1·073	1·060
20·0000	(196)	7·968	5·429	4·032	2·994	2·348	1·600	1·188	0·932	1·132	1·106	1·088	1·065	1·051	1·039
30·0000	(294)	6·506	4·433	3·292	2·445	1·917	1·307	0·970	*	1·092	1·069	1·053	1·033	1·021	1·011
S	$(\Delta p/\rho l)$	6	8	10	12·5	15	20	25	30	40	50	60	80	100	125

Grad'nt k.p.g. (Equivalent) Pipe diameters in mm

Grad'nt S	k.p.g. $(\Delta p/\rho l)$	125	150	200	250	300	400	500	600	800	1000	1250	1500	2000	2500
0·00010	(0·00098)	62·16	48·74	33·22	24·67	19·34	13·18	9·789	7·677	5·231	3·885	2·885	2·263	*	2·206
0·00015	(0·00147)	50·75	39·80	27·12	20·14	15·79	10·76	7·993	6·268	4·271	3·172	2·356	1·847	*	2·127
0·00020	(0·00196)	43·95	34·47	23·49	17·44	13·68	9·321	6·922	5·428	3·699	2·747	2·040	*	2·121	2·074
0·00030	(0·00294)	35·89	28·14	19·18	14·24	11·17	7·610	5·652	4·432	3·020	2·243	1·666	*	2·045	2·004
0·00040	(0·00392)	31·08	24·37	16·61	12·33	9·672	6·591	4·895	3·838	2·616	1·942	*	2·054	1·994	1·957
0·00060	(0·00588)	25·38	19·90	13·56	10·07	7·897	5·381	3·996	3·134	2·136	*	*	1·979	1·927	1·895
0·00080	(0·00785)	21·98	17·23	11·74	8·721	6·839	4·660	3·461	2·714	1·849	*	1·965	1·929	1·882	1·853
0·00100	(0·00981)	19·66	15·41	10·50	7·801	6·117	4·168	3·096	2·428	1·654	*	1·926	1·892	1·849	1·821
0·00150	(0·0147)	16·05	12·59	8·576	6·369	4·995	3·403	2·528	1·982	*	1·901	1·858	1·829	1·791	1·767
0·00200	(0·0196)	13·90	10·90	7·427	5·516	4·325	2·947	2·189	1·717	*	1·852	1·813	1·786	1·752	1·731
0·00300	(0·0294)	11·35	8·899	6·064	4·504	3·532	2·407	1·787	*	1·828	1·787	1·753	1·730	1·700	1·682
0·00400	(0·0392)	9·828	7·707	5·252	3·900	3·059	2·084	1·548	*	1·780	1·743	1·713	1·692	1·665	1·648
0·00600	(0·0588)	8·025	6·293	4·288	3·185	2·497	1·702	*	1·770	1·718	1·686	1·659	1·641	1·618	1·604
0·00800	(0·0785)	6·949	5·450	3·714	2·758	2·163	1·474	*	1·723	1·677	1·647	1·623	1·606	1·586	1·574
0·01000	(0·0981)	6·216	4·874	3·322	2·467	1·934	*	1·723	1·689	1·646	1·618	1·596	1·581	1·562	1·551
0·01500	(0·147)	5·075	3·980	2·712	2·014	1·579	*	1·660	1·630	1·592	1·569	1·549	1·536	1·521	1·512
0·02000	(0·196)	4·395	3·447	2·349	1·744	1·368	1·656	1·618	1·591	1·557	1·535	1·518	1·506	1·493	1·485
0·03000	(0·294)	3·589	2·814	1·918	1·424	*	1·596	1·562	1·539	1·509	1·490	1·476	1·466	1·455	1·449
0·04000	(0·392)	3·108	2·437	1·661	*	*	1·556	1·525	1·504	1·477	1·460	1·447	1·439	1·429	1·425
0·06000	(0·588)	2·538	1·990	1·356	*	1·545	1·502	1·476	1·457	1·434	1·420	1·409	1·402	1·394	1·391
0·08000	(0·785)	2·198	1·723	*	1·535	1·505	1·467	1·442	1·426	1·405	1·392	1·383	1·377	1·371	1·369
0·10000	(0·981)	1·966	1·541	*	1·504	1·476	1·440	1·418	1·402	1·383	1·372	1·363	1·358	1·353	1·352
0·15000	(1·47)	1·605	1·259	*	1·450	1·426	1·394	1·375	1·362	1·345	1·336	1·329	1·325	1·322	1·322
0·20000	(1·96)	1·390	*	1·446	1·414	1·392	1·364	1·346	1·334	1·320	1·312	1·306	1·303	1·301	1·302
0·30000	(2·94)	1·135	*	1·394	1·367	1·347	1·323	1·308	1·298	1·286	1·279	1·275	1·273	1·273	1·274
0·40000	(3·92)	*	1·400	1·360	1·335	1·317	1·295	1·282	1·273	1·262	1·257	1·254	1·253	1·253	1·256
0·60000	(5·88)	*	1·350	1·315	1·292	1·277	1·258	1·247	1·240	1·231	1·227	1·225	1·225	1·227	1·230
0·80000	(7·85)	1·341	1·316	1·284	1·264	1·250	1·233	1·223	1·217	1·210	1·207	1·206	1·206	1·209	1·213
1·00000	(9·81)	1·314	1·291	1·261	1·243	1·230	1·215	1·206	1·200	1·194	1·192	1·191	1·192	1·196	1·200
1·50000	(14·7)	1·268	1·248	1·222	1·206	1·196	1·182	1·175	1·171	1·166	1·165	1·166	1·167	1·172	1·177
2·00000	(19·6)	1·237	1·219	1·196	1·182	1·172	1·161	1·154	1·151	1·148	1·147	1·148	1·151	1·156	1·161
3·00000	(29·4)	1·197	1·181	1·161	1·149	1·141	1·131	1·126	1·124	1·122	1·123	1·125	1·128	1·134	1·140
S	$(\Delta p/\rho l)$	125	150	200	250	300	400	500	600	800	1000	1250	1500	2000	2500

Grad'nt k.p.g. (Equivalent) Pipe diameters in mm

Kinematic viscosity, $\nu = 300 \times 10^{-6}$ m^{2}s^{-1} ; **Roughness size, k_s = 0·015 mm**

Kin. visc., $\nu = 300 \times 10^{-6}\ m^2 s^{-1}$;
$S = 0.00010$ to 3.00000

i.e. kin. pr. grad., $\Delta p / \rho l =$ (0.00098) to (29.4) ms^{-2}

Roughness size, $k_s = 0.030$ mm
This table shows values of m, as follows

m_C for Colebrook-White solutions; or, where $R \leq 2000$, m_P for laminar flow

Grad'nt S	k.p.g. $(\Delta p / \rho l)$	125	150	200	250	300	400	500	600	800	1000	1250	1500	2000	2500
0.00010	(0.00098)	62.16	48.74	33.22	24.67	19.34	13.18	9.789	7.677	5.231	3.885	2.885	2.263	*	2.206
0.00015	(0.00147)	50.75	39.80	27.12	20.14	15.79	10.76	7.993	6.268	4.271	3.172	2.356	1.847	*	2.127
0.00020	(0.00196)	43.95	34.47	23.49	17.44	13.68	9.321	6.922	5.428	3.699	2.747	2.040	*	2.121	2.074
0.00030	(0.00294)	35.89	28.14	19.18	14.24	11.17	7.610	5.652	4.432	3.020	2.243	1.666	*	2.045	2.004
0.00040	(0.00392)	31.08	24.37	16.61	12.33	9.672	6.591	4.895	3.838	2.616	1.942	*	2.054	1.995	1.957
0.00060	(0.00588)	25.38	19.90	13.56	10.07	7.897	5.381	3.996	3.134	2.136	*	*	1.979	1.927	1.895
0.00080	(0.00785)	21.98	17.23	11.74	8.721	6.839	4.660	3.461	2.714	1.849	*	1.966	1.929	1.882	1.853
0.00100	(0.00981)	19.66	15.41	10.50	7.801	6.117	4.168	3.096	2.428	1.654	*	1.926	1.893	1.849	1.822
0.00150	(0.0147)	16.05	12.59	8.576	6.369	4.995	3.403	2.528	1.982	*	1.901	1.859	1.829	1.791	1.767
0.00200	(0.0196)	13.90	10.90	7.427	5.516	4.325	2.947	2.189	1.717	*	1.852	1.813	1.787	1.752	1.731
0.00300	(0.0294)	11.35	8.899	6.064	4.504	3.532	2.407	1.787	*	1.828	1.787	1.753	1.730	1.700	1.682
0.00400	(0.0392)	9.828	7.707	5.252	3.900	3.059	2.084	1.548	*	1.781	1.744	1.713	1.692	1.665	1.649
0.00600	(0.0588)	8.025	6.293	4.288	3.185	2.497	1.702	*	1.770	1.719	1.686	1.660	1.641	1.618	1.605
0.00800	(0.0785)	6.949	5.450	3.714	2.758	2.163	1.474	*	1.724	1.677	1.648	1.623	1.607	1.586	1.575
0.01000	(0.0981)	6.216	4.874	3.322	2.467	1.934	*	1.723	1.690	1.646	1.619	1.597	1.582	1.563	1.552
0.01500	(0.147)	5.075	3.980	2.712	2.014	1.579	*	1.660	1.631	1.593	1.569	1.550	1.537	1.521	1.513
0.02000	(0.196)	4.395	3.447	2.349	1.744	1.368	1.657	1.618	1.592	1.557	1.536	1.519	1.507	1.493	1.486
0.03000	(0.294)	3.589	2.814	1.918	1.424	*	1.597	1.563	1.540	1.510	1.491	1.477	1.467	1.456	1.450
0.04000	(0.392)	3.108	2.437	1.661	*	*	1.556	1.526	1.505	1.478	1.461	1.448	1.440	1.430	1.426
0.06000	(0.588)	2.538	1.990	1.356	*	1.546	1.503	1.476	1.458	1.435	1.421	1.410	1.403	1.396	1.393
0.08000	(0.785)	2.198	1.723	*	1.536	1.506	1.468	1.443	1.427	1.406	1.394	1.384	1.378	1.372	1.370
0.10000	(0.981)	1.966	1.541	*	1.505	1.477	1.441	1.419	1.404	1.384	1.373	1.365	1.360	1.355	1.354
0.15000	(1.47)	1.605	1.259	*	1.451	1.427	1.396	1.376	1.363	1.347	1.338	1.331	1.327	1.324	1.324
0.20000	(1.96)	1.390	*	1.447	1.415	1.394	1.365	1.348	1.336	1.322	1.314	1.308	1.305	1.303	1.304
0.30000	(2.94)	1.135	*	1.396	1.368	1.349	1.324	1.309	1.299	1.288	1.281	1.277	1.275	1.275	1.277
0.40000	(3.92)	*	1.402	1.362	1.336	1.319	1.297	1.284	1.275	1.265	1.259	1.256	1.255	1.256	1.259
0.60000	(5.88)	*	1.351	1.316	1.294	1.279	1.260	1.249	1.242	1.234	1.230	1.228	1.228	1.230	1.234
0.80000	(7.85)	1.342	1.318	1.286	1.266	1.253	1.236	1.226	1.219	1.213	1.210	1.209	1.210	1.213	1.217
1.00000	(9.81)	1.316	1.293	1.263	1.245	1.233	1.217	1.208	1.203	1.197	1.195	1.195	1.196	1.199	1.204
1.50000	(14.7)	1.270	1.250	1.225	1.209	1.198	1.185	1.178	1.174	1.170	1.169	1.170	1.172	1.176	1.182
2.00000	(19.6)	1.240	1.222	1.199	1.184	1.175	1.164	1.158	1.154	1.151	1.151	1.153	1.155	1.161	1.167
3.00000	(29.4)	1.200	1.184	1.164	1.152	1.144	1.135	1.130	1.128	1.127	1.128	1.130	1.133	1.140	1.147
S	$(\Delta p / \rho l)$	125	150	200	250	300	400	500	600	800	1000	1250	1500	2000	2500
Grad'nt	k.p.g.														

(Equivalent) Pipe diameters in mm — Roughness size, $k_s = 0.030$ mm

Kin. visc., $\nu = 300 \times 10^{-6}\ m^2 s^{-1}$;
$S = 0.00010$ to 3.00000

i.e. kin. pr. grad., $\Delta p / \rho l =$ (0.00098) to (29.4) ms^{-2}

Roughness size, $k_s = 0.060$ mm
This table shows values of m, as follows

m_C for Colebrook-White solutions; or, where $R \leq 2000$, m_P for laminar flow

Grad'nt S	k.p.g. $(\Delta p / \rho l)$	125	150	200	250	300	400	500	600	800	1000	1250	1500	2000	2500
0.00010	(0.00098)	62.16	48.74	33.22	24.67	19.34	13.18	9.789	7.677	5.231	3.885	2.885	2.263	*	2.206
0.00015	(0.00147)	50.75	39.80	27.12	20.14	15.79	10.76	7.993	6.268	4.271	3.172	2.356	1.847	*	2.127
0.00020	(0.00196)	43.95	34.47	23.49	17.44	13.68	9.321	6.922	5.428	3.699	2.747	2.040	*	2.121	2.075
0.00030	(0.00294)	35.89	28.14	19.18	14.24	11.17	7.610	5.652	4.432	3.020	2.243	1.666	*	2.046	2.005
0.00040	(0.00392)	31.08	24.37	16.61	12.33	9.672	6.591	4.895	3.838	2.616	1.942	*	2.054	1.995	1.958
0.00060	(0.00588)	25.38	19.90	13.56	10.07	7.897	5.381	3.996	3.134	2.136	*	*	1.980	1.928	1.895
0.00080	(0.00785)	21.98	17.23	11.74	8.721	6.839	4.660	3.461	2.714	1.849	*	1.966	1.930	1.883	1.854
0.00100	(0.00981)	19.66	15.41	10.50	7.801	6.117	4.168	3.096	2.428	1.654	*	1.927	1.893	1.850	1.822
0.00150	(0.0147)	16.05	12.59	8.576	6.369	4.995	3.403	2.528	1.982	*	1.902	1.859	1.830	1.792	1.768
0.00200	(0.0196)	13.90	10.90	7.427	5.516	4.325	2.947	2.189	1.717	*	1.853	1.814	1.787	1.753	1.732
0.00300	(0.0294)	11.35	8.899	6.064	4.504	3.532	2.407	1.787	*	1.829	1.788	1.754	1.731	1.701	1.683
0.00400	(0.0392)	9.828	7.707	5.252	3.900	3.059	2.084	1.548	*	1.782	1.745	1.714	1.693	1.666	1.650
0.00600	(0.0588)	8.025	6.293	4.288	3.185	2.497	1.702	*	1.771	1.720	1.687	1.661	1.642	1.619	1.606
0.00800	(0.0785)	6.949	5.450	3.714	2.758	2.163	1.474	*	1.725	1.678	1.649	1.625	1.608	1.588	1.576
0.01000	(0.0981)	6.216	4.874	3.322	2.467	1.934	*	1.724	1.691	1.647	1.620	1.598	1.583	1.564	1.554
0.01500	(0.147)	5.075	3.980	2.712	2.014	1.579	*	1.662	1.632	1.594	1.571	1.551	1.539	1.523	1.514
0.02000	(0.196)	4.395	3.447	2.349	1.744	1.368	1.658	1.620	1.593	1.559	1.537	1.520	1.509	1.495	1.488
0.03000	(0.294)	3.589	2.814	1.918	1.424	*	1.598	1.564	1.541	1.511	1.493	1.478	1.469	1.458	1.452
0.04000	(0.392)	3.108	2.437	1.661	*	*	1.558	1.528	1.506	1.479	1.463	1.450	1.442	1.433	1.428
0.06000	(0.588)	2.538	1.990	1.356	*	1.548	1.505	1.478	1.460	1.437	1.423	1.412	1.406	1.399	1.396
0.08000	(0.785)	2.198	1.723	*	1.538	1.508	1.470	1.446	1.429	1.408	1.396	1.387	1.381	1.375	1.374
0.10000	(0.981)	1.966	1.541	*	1.507	1.479	1.443	1.421	1.406	1.387	1.376	1.368	1.363	1.358	1.357
0.15000	(1.47)	1.605	1.259	*	1.454	1.429	1.398	1.379	1.366	1.350	1.341	1.334	1.331	1.328	1.328
0.20000	(1.96)	1.390	*	1.450	1.418	1.396	1.368	1.351	1.339	1.325	1.317	1.312	1.309	1.308	1.309
0.30000	(2.94)	1.135	*	1.399	1.371	1.352	1.328	1.313	1.303	1.291	1.285	1.281	1.280	1.280	1.282
0.40000	(3.92)	*	1.405	1.365	1.340	1.322	1.300	1.287	1.279	1.269	1.264	1.261	1.260	1.262	1.264
0.60000	(5.88)	*	1.355	1.320	1.298	1.283	1.264	1.253	1.246	1.239	1.235	1.234	1.234	1.237	1.241
0.80000	(7.85)	1.346	1.322	1.290	1.270	1.257	1.240	1.230	1.224	1.218	1.216	1.215	1.216	1.220	1.224
1.00000	(9.81)	1.320	1.297	1.268	1.249	1.237	1.222	1.213	1.208	1.203	1.201	1.201	1.203	1.207	1.212
1.50000	(14.7)	1.275	1.255	1.229	1.214	1.203	1.191	1.184	1.180	1.177	1.176	1.177	1.180	1.185	1.191
2.00000	(19.6)	1.245	1.227	1.204	1.190	1.181	1.170	1.164	1.161	1.159	1.159	1.161	1.164	1.171	1.177
3.00000	(29.4)	1.205	1.190	1.170	1.158	1.151	1.142	1.138	1.136	1.135	1.137	1.140	1.143	1.151	1.159
S	$(\Delta p / \rho l)$	125	150	200	250	300	400	500	600	800	1000	1250	1500	2000	2500
Grad'nt	k.p.g.														

(Equivalent) Pipe diameters in mm — Roughness size, $k_s = 0.060$ mm

E274

Kin. visc., $\nu = 400 \times 10^{-6}$ m^{2}s^{-1};
$S = 0.00010$ to 30.0000

i.e. kin. pr. grad., $\Delta p/\rho l =$
(0.00098) to (294) ms^{-2}

Roughness size, $k_s = 0.015$ mm
This table shows values of m, as follows

m_C for Colebrook-White solutions; or,
where $R \leq 2000$, m_P for laminar flow

Grad'nt S	k.p.g. $(\Delta p/\rho l)$	6	8	10	12.5	15	20	25	30	40	50	60	80	100	125
0.00030	(0.00294)	2743	1869	1388	1031	808.4	550.9	409.1	320.8	218.6	162.4	127.3	86.76	64.43	47.85
0.00040	(0.00392)	2375	1619	1202	892.8	700.1	477.1	354.3	277.8	189.3	140.6	110.3	75.13	55.80	41.44
0.00060	(0.00588)	1940	1322	981.5	728.9	571.6	389.5	289.3	226.9	154.6	114.8	90.03	61.35	45.56	33.83
0.00080	(0.00785)	1680	1145	850.0	631.3	495.1	337.3	250.5	196.5	133.9	99.42	77.97	53.13	39.46	29.30
0.00100	(0.00981)	1502	1024	760.3	564.6	442.8	301.7	224.1	175.7	119.7	88.92	69.73	47.52	35.29	26.21
0.00150	(0.0147)	1227	835.9	620.8	461.0	361.5	246.4	183.0	143.5	97.77	72.61	56.94	38.80	28.81	21.40
0.00200	(0.0196)	1062	723.9	537.6	399.3	313.1	213.4	158.4	124.3	84.67	62.88	49.31	33.60	24.95	18.53
0.00300	(0.0294)	867.4	591.1	439.0	326.0	255.6	174.2	129.4	101.5	69.13	51.34	40.26	27.43	20.37	15.13
0.00400	(0.0392)	751.2	511.9	380.1	282.3	221.4	150.9	112.0	87.86	59.87	44.46	34.87	23.76	17.64	13.10
0.00600	(0.0588)	613.3	417.9	310.4	230.5	180.8	123.2	91.48	71.74	48.88	36.30	28.47	19.40	14.41	10.70
0.00800	(0.0785)	531.2	362.0	268.8	199.6	156.5	106.7	79.22	62.13	42.33	31.44	24.65	16.80	12.48	9.266
0.01000	(0.0981)	475.1	323.7	240.4	178.6	140.0	95.41	70.86	55.57	37.86	28.12	22.05	15.03	11.16	8.288
0.01500	(0.147)	387.9	264.3	196.3	145.8	114.3	77.90	57.86	45.37	30.92	22.96	18.01	12.27	9.112	6.767
0.02000	(0.196)	335.9	228.9	170.0	126.3	99.01	67.47	50.10	39.29	26.77	19.88	15.59	10.63	7.891	5.860
0.03000	(0.294)	274.3	186.9	138.8	103.1	80.84	55.09	40.91	32.08	21.86	16.24	12.73	8.676	6.443	4.785
0.04000	(0.392)	237.5	161.9	120.2	89.28	70.01	47.71	35.43	27.78	18.93	14.06	11.03	7.513	5.580	4.144
0.06000	(0.588)	194.0	132.2	98.15	72.89	57.16	38.95	28.93	22.69	15.46	11.48	9.003	6.135	4.556	3.383
0.08000	(0.785)	168.0	114.5	85.00	63.13	49.51	33.73	25.05	19.65	13.39	9.942	7.797	5.313	3.946	2.930
0.10000	(0.981)	150.2	102.4	76.03	56.46	44.28	30.17	22.41	17.57	11.97	8.892	6.973	4.752	3.529	2.621
0.15000	(1.47)	122.7	83.59	62.08	46.10	36.15	24.64	18.30	14.35	9.777	7.261	5.694	3.880	2.881	2.140
0.20000	(1.96)	106.2	72.39	53.76	39.93	31.31	21.34	15.84	12.43	8.467	6.288	4.931	3.360	2.495	1.853
0.30000	(2.94)	86.74	59.11	43.90	32.60	25.56	17.42	12.94	10.15	6.913	5.134	4.026	2.743	2.037	1.513
0.40000	(3.92)	75.12	51.19	38.01	28.23	22.14	15.09	11.20	8.786	5.987	4.446	3.487	2.376	1.764	1.310
0.60000	(5.88)	61.33	41.79	31.04	23.05	18.08	12.32	9.148	7.174	4.888	3.630	2.847	1.940	1.441	*
0.80000	(7.85)	53.12	36.20	26.88	19.96	15.65	10.67	7.922	6.213	4.233	3.144	2.465	1.680	1.248	*
1.00000	(9.81)	47.51	32.37	24.04	17.86	14.00	9.541	7.086	5.557	3.786	2.812	2.205	1.503	1.116	*
1.50000	(14.7)	38.79	26.43	19.63	14.58	11.43	7.790	5.786	4.537	3.092	2.296	1.801	1.227	*	1.334
2.00000	(19.6)	33.59	22.89	17.00	12.63	9.901	6.747	5.010	3.929	2.677	1.988	1.559	1.063	*	1.300
3.00000	(29.4)	27.43	18.69	13.88	10.31	8.084	5.509	4.091	3.208	2.186	1.624	1.273	*	1.283	1.255
4.00000	(39.2)	23.75	16.19	12.02	8.928	7.001	4.771	3.543	2.778	1.893	1.406	1.103	*	1.251	1.225
6.00000	(58.8)	19.40	13.22	9.815	7.289	5.716	3.895	2.893	2.269	1.546	1.148	*	1.234	1.208	1.186
8.00000	(78.5)	16.80	11.45	8.500	6.313	4.951	3.373	2.505	1.965	1.339	0.994	*	1.203	1.179	1.159
10.0000	(98.1)	15.02	10.24	7.603	5.646	4.428	3.017	2.241	1.757	1.197	*	*	1.180	1.158	1.139
15.0000	(147)	12.27	8.359	6.208	4.610	3.615	2.464	1.830	1.435	0.978	*	1.172	1.140	1.121	1.105
20.0000	(196)	10.62	7.239	5.376	3.993	3.131	2.134	1.584	1.243	*	1.164	1.142	1.114	1.096	1.082
30.0000	(294)	8.674	5.911	4.390	3.260	2.556	1.742	1.294	1.015	*	1.123	1.104	1.079	1.064	1.051
S	$(\Delta p/\rho l)$	6	8	10	12.5	15	20	25	30	40	50	60	80	100	125

Grad'nt k.p.g. (Equivalent) Pipe diameters in mm

Grad'nt S	k.p.g. $(\Delta p/\rho l)$	125	150	200	250	300	400	500	600	800	1000	1250	1500	2000	2500
0.00010	(0.00098)	82.88	64.99	44.29	32.89	25.79	17.58	13.05	10.24	6.975	5.180	3.847	3.017	2.056	*
0.00015	(0.00147)	67.67	53.07	36.16	26.85	21.06	14.35	10.66	8.357	5.695	4.229	3.141	2.463	*	2.241
0.00020	(0.00196)	58.60	45.96	31.32	23.26	18.24	12.43	9.229	7.238	4.932	3.663	2.720	2.133	*	2.182
0.00030	(0.00294)	47.85	37.52	25.57	18.99	14.89	10.15	7.536	5.910	4.027	2.991	2.221	1.742	2.154	2.105
0.00040	(0.00392)	41.44	32.50	22.14	16.44	12.90	8.788	6.526	5.118	3.487	2.590	1.923	*	2.098	2.053
0.00060	(0.00588)	33.83	26.53	18.08	13.43	10.53	7.175	5.329	4.179	2.847	2.115	*	*	2.024	1.985
0.00080	(0.00785)	29.30	22.98	15.66	11.63	9.119	6.214	4.615	3.619	2.466	1.831	*	2.031	1.974	1.938
0.00100	(0.00981)	26.21	20.55	14.00	10.40	8.156	5.558	4.128	3.237	2.206	1.638	*	1.990	1.937	1.904
0.00150	(0.0147)	21.40	16.78	11.43	8.492	6.660	4.538	3.370	2.643	1.801	*	1.956	1.920	1.874	1.845
0.00200	(0.0196)	18.53	14.53	9.903	7.354	5.767	3.930	2.919	2.289	1.560	*	1.906	1.873	1.831	1.805
0.00300	(0.0294)	15.13	11.87	8.086	6.005	4.709	3.209	2.383	1.869	*	1.880	1.839	1.811	1.775	1.752
0.00400	(0.0392)	13.10	10.28	7.002	5.200	4.078	2.779	2.064	1.618	*	1.832	1.795	1.769	1.736	1.716
0.00600	(0.0588)	10.70	8.390	5.717	4.246	3.330	2.269	1.685	*	1.808	1.769	1.736	1.714	1.685	1.668
0.00800	(0.0785)	9.266	7.266	4.951	3.677	2.884	1.965	1.459	*	1.762	1.726	1.697	1.676	1.651	1.635
0.01000	(0.0981)	8.288	6.499	4.429	3.289	2.579	1.758	*	*	1.728	1.695	1.667	1.649	1.625	1.611
0.01500	(0.147)	6.767	5.307	3.616	2.685	2.106	1.435	*	1.715	1.669	1.640	1.616	1.600	1.580	1.568
0.02000	(0.196)	5.860	4.596	3.132	2.326	1.824	*	1.704	1.672	1.630	1.604	1.582	1.568	1.550	1.540
0.03000	(0.294)	4.785	3.752	2.557	1.899	1.489	*	1.642	1.614	1.577	1.555	1.536	1.524	1.509	1.501
0.04000	(0.392)	4.144	3.250	2.214	1.644	*	1.638	1.601	1.576	1.542	1.522	1.505	1.494	1.481	1.474
0.06000	(0.588)	3.383	2.653	1.808	1.343	*	1.579	1.547	1.524	1.496	1.478	1.464	1.455	1.444	1.439
0.08000	(0.785)	2.930	2.298	1.566	*	*	1.540	1.510	1.490	1.464	1.448	1.436	1.428	1.419	1.415
0.10000	(0.981)	2.621	2.055	1.400	*	1.554	1.511	1.483	1.464	1.440	1.426	1.415	1.408	1.400	1.397
0.15000	(1.47)	2.140	1.678	*	1.528	1.498	1.460	1.436	1.420	1.400	1.387	1.378	1.372	1.367	1.365
0.20000	(1.96)	1.853	1.453	*	1.488	1.461	1.427	1.405	1.390	1.372	1.361	1.353	1.348	1.344	1.343
0.30000	(2.94)	1.513	1.187	1.469	1.435	1.412	1.382	1.363	1.350	1.335	1.326	1.320	1.316	1.314	1.314
0.40000	(3.92)	1.310	*	1.431	1.400	1.379	1.352	1.335	1.323	1.310	1.302	1.297	1.294	1.293	1.294
0.60000	(5.88)	*	*	1.380	1.353	1.335	1.311	1.297	1.287	1.276	1.270	1.266	1.265	1.265	1.267
0.80000	(7.85)	*	1.385	1.347	1.322	1.306	1.284	1.271	1.263	1.253	1.248	1.245	1.245	1.245	1.248
1.00000	(9.81)	*	1.358	1.322	1.299	1.284	1.264	1.252	1.245	1.236	1.232	1.230	1.229	1.231	1.234
1.50000	(14.7)	1.334	1.310	1.279	1.259	1.246	1.229	1.219	1.213	1.206	1.203	1.202	1.203	1.206	1.210
2.00000	(19.6)	1.300	1.278	1.250	1.232	1.220	1.205	1.197	1.191	1.186	1.184	1.184	1.184	1.185	1.193
3.00000	(29.4)	1.255	1.236	1.211	1.196	1.186	1.173	1.166	1.162	1.159	1.158	1.159	1.160	1.165	1.171
S	$(\Delta p/\rho l)$	125	150	200	250	300	400	500	600	800	1000	1250	1500	2000	2500

Grad'nt k.p.g. (Equivalent) Pipe diameters in mm

Kinematic viscosity, $\nu = 400 \times 10^{-6}$ m^{2}s^{-1} ;

Roughness size, $k_s = 0.015$ mm

Kin. visc., $\nu = 400 \times 10^{-6}$ m^{2}s^{-1}; $\quad$ Roughness size, $k_s = 0.030$ mm $\qquad$ **E275**
S = 0.00010 to 3.00000 $\qquad$ This table shows values of m, as follows

i.e. kin. pr. grad., $\Delta p/\rho l$ = (0.00098) to (29.4) ms^{-2} $\qquad$ m_C for Colebrook-White solutions; or, where $\mathbf{R} \leq 2000$, m_P for laminar flow

Grad'nt S	k.p.g. $(\Delta p/\rho l)$	125	150	200	250	300	400	500	600	800	1000	1250	1500	2000	2500
0.00010	(0.00098)	82.88	64.99	44.29	32.89	25.79	17.58	13.05	10.24	6.975	5.180	3.847	3.017	2.056	*
0.00015	(0.00147)	67.67	53.07	36.16	26.85	21.06	14.35	10.66	8.357	5.695	4.229	3.141	2.463	*	2.241
0.00020	(0.00196)	58.60	45.96	31.32	23.26	18.24	12.43	9.229	7.238	4.932	3.663	2.720	2.133	*	2.182
0.00030	(0.00294)	47.85	37.52	25.57	18.99	14.89	10.15	7.536	5.910	4.027	2.991	2.221	1.742	2.154	2.105
0.00040	(0.00392)	41.44	32.50	22.14	16.44	12.90	8.788	6.526	5.118	3.487	2.590	1.923	*	2.098	2.053
0.00060	(0.00588)	33.83	26.53	18.08	13.43	10.53	7.175	5.329	4.179	2.847	2.115	*	*	2.024	1.985
0.00080	(0.00785)	29.30	22.98	15.66	11.63	9.119	6.214	4.615	3.619	2.466	1.831	*	2.031	1.975	1.939
0.00100	(0.00981)	26.21	20.55	14.00	10.40	8.156	5.558	4.128	3.237	2.206	1.638	*	1.991	1.938	1.905
0.00150	(0.0147)	21.40	16.78	11.43	8.492	6.660	4.538	3.370	2.643	1.801	*	1.956	1.921	1.874	1.845
0.00200	(0.0196)	18.53	14.53	9.903	7.354	5.767	3.930	2.919	2.289	1.560	*	1.906	1.874	1.832	1.806
0.00300	(0.0294)	15.13	11.87	8.086	6.005	4.709	3.209	2.383	1.869	*	1.881	1.840	1.812	1.775	1.752
0.00400	(0.0392)	13.10	10.28	7.002	5.200	4.078	2.779	2.064	1.618	*	1.833	1.796	1.770	1.737	1.716
0.00600	(0.0588)	10.70	8.390	5.717	4.246	3.330	2.269	1.685	*	1.808	1.769	1.737	1.714	1.686	1.668
0.00800	(0.0785)	9.266	7.266	4.951	3.677	2.884	1.965	1.459	*	1.762	1.727	1.697	1.677	1.651	1.636
0.01000	(0.0981)	8.288	6.499	4.429	3.289	2.579	1.758	*	*	1.728	1.695	1.668	1.649	1.625	1.612
0.01500	(0.147)	6.767	5.307	3.616	2.685	2.106	1.435	*	1.716	1.670	1.641	1.617	1.601	1.581	1.569
0.02000	(0.196)	5.860	4.596	3.132	2.326	1.824	*	1.704	1.672	1.630	1.604	1.583	1.568	1.551	1.540
0.03000	(0.294)	4.785	3.752	2.557	1.899	1.489	*	1.643	1.615	1.578	1.555	1.537	1.525	1.510	1.502
0.04000	(0.392)	4.144	3.250	2.214	1.644	*	1.639	1.602	1.576	1.543	1.523	1.506	1.495	1.482	1.475
0.06000	(0.588)	3.383	2.653	1.808	1.343	*	1.580	1.548	1.525	1.496	1.479	1.465	1.456	1.445	1.440
0.08000	(0.785)	2.930	2.298	1.566	*	*	1.541	1.511	1.491	1.465	1.449	1.437	1.429	1.420	1.416
0.10000	(0.981)	2.621	2.055	1.400	*	1.555	1.511	1.484	1.465	1.441	1.427	1.416	1.409	1.401	1.398
0.15000	(1.47)	2.140	1.678	*	1.529	1.499	1.461	1.437	1.421	1.401	1.389	1.379	1.374	1.368	1.366
0.20000	(1.96)	1.853	1.453	*	1.489	1.462	1.428	1.406	1.392	1.373	1.363	1.355	1.350	1.346	1.345
0.30000	(2.94)	1.513	1.187	1.470	1.436	1.413	1.383	1.364	1.352	1.336	1.328	1.321	1.318	1.316	1.316
0.40000	(3.92)	1.310	*	1.432	1.401	1.380	1.353	1.336	1.325	1.312	1.304	1.299	1.296	1.295	1.296
0.60000	(5.88)	*	*	1.382	1.355	1.337	1.313	1.299	1.289	1.278	1.272	1.269	1.267	1.267	1.269
0.80000	(7.85)	*	1.387	1.348	1.324	1.307	1.286	1.273	1.265	1.255	1.251	1.248	1.247	1.248	1.251
1.00000	(9.81)	*	1.359	1.323	1.301	1.285	1.266	1.254	1.247	1.239	1.235	1.233	1.232	1.234	1.238
1.50000	(14.7)	1.336	1.312	1.281	1.261	1.248	1.231	1.221	1.215	1.209	1.206	1.206	1.206	1.209	1.214
2.00000	(19.6)	1.302	1.280	1.252	1.234	1.222	1.208	1.199	1.194	1.189	1.187	1.187	1.189	1.193	1.198
3.00000	(29.4)	1.258	1.239	1.214	1.199	1.189	1.176	1.170	1.166	1.162	1.162	1.163	1.165	1.170	1.176
S	$(\Delta p/\rho l)$	125	150	200	250	300	400	500	600	800	1000	1250	1500	2000	2500

Grad'nt $\quad$ k.p.g. $\qquad$ (Equivalent) Pipe diameters in mm $\qquad$ Roughness size, $k_s = 0.030$ mm

---

Kin. visc., $\nu = 400 \times 10^{-6}$ m^{2}s^{-1}; $\quad$ Roughness size, $k_s = 0.060$ mm $\qquad$ **E276**
S = 0.00010 to 3.00000 $\qquad$ This table shows values of m, as follows

i.e. kin. pr. grad., $\Delta p/\rho l$ = (0.00098) to (29.4) ms^{-2} $\qquad$ m_C for Colebrook-White solutions; or, where $\mathbf{R} \leq 2000$, m_P for laminar flow

Grad'nt S	k.p.g. $(\Delta p/\rho l)$	125	150	200	250	300	400	500	600	800	1000	1250	1500	2000	2500
0.00010	(0.00098)	82.88	64.99	44.29	32.89	25.79	17.58	13.05	10.24	6.975	5.180	3.847	3.017	2.056	*
0.00015	(0.00147)	67.67	53.07	36.16	26.85	21.06	14.35	10.66	8.357	5.695	4.229	3.141	2.463	*	2.241
0.00020	(0.00196)	58.60	45.96	31.32	23.26	18.24	12.43	9.229	7.238	4.932	3.663	2.720	2.133	*	2.183
0.00030	(0.00294)	47.85	37.52	25.57	18.99	14.89	10.15	7.536	5.910	4.027	2.991	2.221	1.742	2.155	2.105
0.00040	(0.00392)	41.44	32.50	22.14	16.44	12.90	8.788	6.526	5.118	3.487	2.590	1.923	*	2.099	2.054
0.00060	(0.00588)	33.83	26.53	18.08	13.43	10.53	7.175	5.329	4.179	2.847	2.115	*	*	2.025	1.985
0.00080	(0.00785)	29.30	22.98	15.66	11.63	9.119	6.214	4.615	3.619	2.466	1.831	*	2.032	1.975	1.939
0.00100	(0.00981)	26.21	20.55	14.00	10.40	8.156	5.558	4.128	3.237	2.206	1.638	*	1.991	1.938	1.905
0.00150	(0.0147)	21.40	16.78	11.43	8.492	6.660	4.538	3.370	2.643	1.801	*	1.957	1.921	1.875	1.846
0.00200	(0.0196)	18.53	14.53	9.903	7.354	5.767	3.930	2.919	2.289	1.560	*	1.907	1.874	1.832	1.806
0.00300	(0.0294)	15.13	11.87	8.086	6.005	4.709	3.209	2.383	1.869	*	1.881	1.841	1.812	1.776	1.753
0.00400	(0.0392)	13.10	10.28	7.002	5.200	4.078	2.779	2.064	1.618	*	1.834	1.796	1.771	1.738	1.717
0.00600	(0.0588)	10.70	8.390	5.717	4.246	3.330	2.269	1.685	*	1.809	1.770	1.738	1.715	1.687	1.669
0.00800	(0.0785)	9.266	7.266	4.951	3.677	2.884	1.965	1.459	*	1.763	1.728	1.698	1.678	1.652	1.637
0.01000	(0.0981)	8.288	6.499	4.429	3.289	2.579	1.758	*	*	1.729	1.696	1.669	1.650	1.627	1.613
0.01500	(0.147)	6.767	5.307	3.616	2.685	2.106	1.435	*	1.717	1.671	1.642	1.618	1.602	1.582	1.571
0.02000	(0.196)	5.860	4.596	3.132	2.326	1.824	*	1.706	1.673	1.632	1.605	1.584	1.570	1.552	1.542
0.03000	(0.294)	4.785	3.752	2.557	1.899	1.489	*	1.644	1.616	1.580	1.557	1.539	1.526	1.512	1.504
0.04000	(0.392)	4.144	3.250	2.214	1.644	*	1.640	1.603	1.578	1.545	1.524	1.508	1.497	1.484	1.477
0.06000	(0.588)	3.383	2.653	1.808	1.343	*	1.581	1.549	1.527	1.498	1.481	1.467	1.458	1.447	1.442
0.08000	(0.785)	2.930	2.298	1.566	*	*	1.542	1.513	1.493	1.467	1.451	1.439	1.431	1.423	1.419
0.10000	(0.981)	2.621	2.055	1.400	*	1.557	1.513	1.486	1.467	1.444	1.429	1.418	1.411	1.404	1.401
0.15000	(1.47)	2.140	1.678	*	1.530	1.501	1.463	1.440	1.423	1.403	1.391	1.382	1.377	1.371	1.370
0.20000	(1.96)	1.853	1.453	*	1.491	1.464	1.430	1.409	1.394	1.376	1.365	1.358	1.353	1.349	1.349
0.30000	(2.94)	1.513	1.187	1.472	1.439	1.416	1.386	1.367	1.355	1.339	1.331	1.325	1.322	1.320	1.320
0.40000	(3.92)	1.310	*	1.435	1.404	1.383	1.356	1.339	1.328	1.315	1.308	1.303	1.300	1.299	1.301
0.60000	(5.88)	*	*	1.385	1.358	1.340	1.316	1.302	1.293	1.282	1.276	1.273	1.272	1.272	1.275
0.80000	(7.85)	*	1.390	1.351	1.327	1.311	1.290	1.277	1.269	1.260	1.255	1.253	1.252	1.254	1.257
1.00000	(9.81)	*	1.363	1.327	1.304	1.289	1.270	1.259	1.251	1.243	1.240	1.238	1.238	1.240	1.244
1.50000	(14.7)	1.340	1.316	1.284	1.265	1.252	1.236	1.226	1.221	1.214	1.212	1.212	1.213	1.217	1.221
2.00000	(19.6)	1.306	1.285	1.256	1.239	1.227	1.213	1.205	1.200	1.195	1.194	1.194	1.196	1.201	1.206
3.00000	(29.4)	1.262	1.243	1.219	1.204	1.194	1.182	1.176	1.172	1.169	1.169	1.171	1.173	1.179	1.186
S	$(\Delta p/\rho l)$	125	150	200	250	300	400	500	600	800	1000	1250	1500	2000	2500

Grad'nt $\quad$ k.p.g. $\qquad$ (Equivalent) Pipe diameters in mm $\qquad$ Roughness size, $k_s = 0.060$ mm

Kin. visc., $\nu = 600 \times 10^{-6}$ m^{2}s^{-1};
$S = 0.00010$ to 30.0000

i.e. kin. pr. grad., $\Delta p/\rho l =$
(0.00098) to (294) ms^{-2}

Roughness size, $k_s = 0.015$ mm
This table shows values of m, as follows

m_C for Colebrook-White solutions; or,
where $R \le 2000$, m_P for laminar flow

Grad'nt S	k.p.g. $(\Delta p/\rho l)$	6	8	10	12·5	15	20	25	30	40	50	60	80	100	125
0·00030	(0·00294)	4114	2804	2082	1546	1213	826·3	613·7	481·2	327·9	243·5	191·0	130·1	96·65	71·77
0·00040	(0·00392)	3563	2428	1803	1339	1050	715·6	531·4	416·8	284·0	210·9	165·4	112·7	83·70	62·16
0·00060	(0·00588)	2909	1982	1472	1093	857·5	584·3	433·9	340·3	231·9	172·2	135·0	92·02	68·34	50·75
0·00080	(0·00785)	2520	1717	1275	946·9	742·6	506·0	375·8	294·7	200·8	149·1	116·9	79·69	59·18	43·95
0·00100	(0·00981)	2254	1536	1140	847·0	664·2	452·6	336·1	263·6	179·6	133·4	104·6	71·28	52·93	39·31
0·00150	(0·0147)	1840	1254	931·2	691·5	542·3	369·5	274·4	215·2	146·6	108·9	85·41	58·20	43·22	32·10
0·00200	(0·0196)	1594	1086	806·4	598·9	469·6	320·0	237·7	186·4	127·0	94·32	73·96	50·40	37·43	27·80
0·00300	(0·0294)	1301	886·6	658·4	489·0	383·5	261·3	194·1	152·2	103·7	77·01	60·39	41·15	30·56	22·70
0·00400	(0·0392)	1127	767·8	570·2	423·5	332·1	226·3	168·1	131·8	89·80	66·69	52·30	35·64	26·47	19·66
0·00600	(0·0588)	920·0	626·9	465·6	345·8	271·2	184·8	137·2	107·6	73·32	54·46	42·70	29·10	21·61	16·05
0·00800	(0·0785)	796·8	542·9	403·2	299·4	234·8	160·0	118·8	93·19	63·50	47·16	36·98	25·20	18·72	13·90
0·01000	(0·0981)	712·6	485·6	360·6	267·8	210·0	143·1	106·3	83·35	56·80	42·18	33·08	22·54	16·74	12·43
0·01500	(0·147)	581·9	396·5	294·5	218·7	171·5	116·9	86·78	68·06	46·37	34·44	27·01	18·40	13·67	10·15
0·02000	(0·196)	503·9	343·4	255·0	189·4	148·5	101·2	75·16	58·94	40·16	29·83	23·39	15·94	11·84	8·790
0·03000	(0·294)	411·4	280·4	208·2	154·6	121·3	82·63	61·37	48·12	32·79	24·35	19·10	13·01	9·665	7·177
0·04000	(0·392)	356·3	242·8	180·3	133·9	105·0	71·56	53·14	41·68	28·40	21·09	16·54	11·27	8·370	6·216
0·06000	(0·588)	290·9	198·2	147·2	109·3	85·75	58·43	43·39	34·03	23·19	17·22	13·50	9·202	6·834	5·075
0·08000	(0·785)	252·0	171·7	127·5	94·69	74·26	50·60	37·58	29·47	20·08	14·91	11·69	7·969	5·918	4·395
0·10000	(0·981)	225·4	153·6	114·0	84·70	66·42	45·26	33·61	26·36	17·96	13·34	10·46	7·128	5·293	3·931
0·15000	(1·47)	184·0	125·4	93·12	69·15	54·23	36·95	27·44	21·52	14·66	10·89	8·541	5·820	4·322	3·210
0·20000	(1·96)	159·4	108·6	80·64	59·89	46·96	32·00	23·77	18·64	12·70	9·432	7·396	5·040	3·743	2·780
0·30000	(2·94)	130·1	88·66	65·84	48·90	38·35	26·13	19·41	15·22	10·37	7·701	6·039	4·115	3·056	2·270
0·40000	(3·92)	112·7	76·78	57·02	42·35	33·21	22·63	16·81	13·18	8·980	6·669	5·230	3·564	2·647	1·966
0·60000	(5·88)	92·00	62·69	46·56	34·58	27·12	18·48	13·72	10·76	7·332	5·446	4·270	2·910	2·161	1·605
0·80000	(7·85)	79·68	54·29	40·32	29·94	23·48	16·00	11·88	9·319	6·350	4·716	3·698	2·520	1·872	1·390
1·00000	(9·81)	71·26	48·56	36·06	26·78	21·00	14·31	10·63	8·335	5·680	4·218	3·308	2·254	1·674	1·243
1·50000	(14·7)	58·19	39·65	29·45	21·87	17·15	11·69	8·678	6·806	4·637	3·444	2·701	1·840	1·367	*
2·00000	(19·6)	50·39	34·34	25·50	18·94	14·85	10·12	7·516	5·894	4·016	2·983	2·339	1·594	1·184	*
3·00000	(29·4)	41·14	28·04	20·82	15·46	12·13	8·263	6·137	4·812	3·279	2·435	1·910	1·301	*	1·349
4·00000	(39·2)	35·63	24·28	18·03	13·39	10·50	7·156	5·314	4·168	2·840	2·109	1·654	1·127	*	1·314
6·00000	(58·8)	29·09	19·82	14·72	10·93	8·575	5·843	4·339	3·403	2·319	1·722	1·350	*	1·297	1·268
8·00000	(78·5)	25·20	17·17	12·75	9·469	7·426	5·060	3·758	2·947	2·008	1·491	1·169	*	1·264	1·237
10·0000	(98·1)	22·54	15·36	11·40	8·470	6·642	4·526	3·361	2·636	1·796	1·334	1·046	1·269	1·239	1·215
15·0000	(147)	18·40	12·54	9·312	6·915	5·423	3·695	2·744	2·152	1·466	1·089	*	1·223	1·197	1·176
20·0000	(196)	15·94	10·86	8·064	5·989	4·696	3·200	2·377	1·864	1·270	*	*	1·192	1·169	1·149
30·0000	(294)	13·01	8·866	6·584	4·890	3·835	2·613	1·941	1·522	1·037	*	1·184	1·152	1·131	1·115
S	$(\Delta p/\rho l)$	6	8	10	12·5	15	20	25	30	40	50	60	80	100	125
Grad'nt	k.p.g.	(Equivalent) Pipe diameters in mm													

Grad'nt S	k.p.g. $(\Delta p/\rho l)$	125	150	200	250	300	400	500	600	800	1000	1250	1500	2000	2500
0·00010	(0·00098)	124·3	97·49	66·43	49·33	38·69	26·36	19·58	15·35	10·46	7·770	5·770	4·525	3·083	2·290
0·00015	(0·00147)	101·5	79·60	54·24	40·28	31·59	21·53	15·99	12·54	8·542	6·344	4·711	3·695	2·518	1·870
0·00020	(0·00196)	87·90	68·93	46·97	34·89	27·36	18·64	13·84	10·86	7·398	5·494	4·080	3·200	2·180	*
0·00030	(0·00294)	71·77	56·28	38·35	28·48	22·34	15·22	11·30	8·864	6·040	4·486	3·331	2·613	*	*
0·00040	(0·00392)	62·16	48·74	33·22	24·67	19·34	13·18	9·789	7·677	5·231	3·885	2·885	2·263	*	2·206
0·00060	(0·00588)	50·75	39·80	27·12	20·14	15·79	10·76	7·993	6·268	4·271	3·172	2·356	1·847	*	2·127
0·00080	(0·00785)	43·95	34·47	23·49	17·44	13·68	9·321	6·922	5·428	3·699	2·747	2·040	*	2·121	2·074
0·00100	(0·00981)	39·31	30·83	21·01	15·60	12·23	8·337	6·191	4·855	3·308	2·457	1·825	*	2·078	2·035
0·00150	(0·0147)	32·10	25·17	17·15	12·74	9·989	6·807	5·055	3·964	2·701	2·006	*	2·066	2·005	1·967
0·00200	(0·0196)	27·80	21·80	14·85	11·03	8·651	5·895	4·378	3·433	2·339	1·737	*	2·012	1·957	1·922
0·00300	(0·0294)	22·70	17·80	12·13	9·007	7·063	4·813	3·575	2·803	1·910	*	1·977	1·940	1·892	1·862
0·00400	(0·0392)	19·66	15·41	10·50	7·801	6·117	4·168	3·096	2·428	1·654	*	1·926	1·892	1·849	1·821
0·00600	(0·0588)	16·05	12·59	8·576	6·369	4·995	3·403	2·528	1·982	*	1·901	1·858	1·829	1·791	1·767
0·00800	(0·0785)	13·90	10·90	7·427	5·516	4·325	2·947	2·189	1·717	*	1·852	1·813	1·786	1·752	1·731
0·01000	(0·0981)	12·43	9·749	6·643	4·933	3·869	2·636	1·958	1·535	1·859	1·815	1·779	1·755	1·723	1·703
0·01500	(0·147)	10·15	7·960	5·424	4·028	3·159	2·153	1·599	*	1·791	1·753	1·722	1·700	1·672	1·656
0·02000	(0·196)	8·790	6·893	4·697	3·489	2·736	1·864	*	*	1·746	1·711	1·683	1·663	1·638	1·624
0·03000	(0·294)	7·177	5·628	3·835	2·848	2·234	1·522	*	1·734	1·686	1·656	1·631	1·614	1·593	1·581
0·04000	(0·392)	6·216	4·874	3·322	2·467	1·934	*	1·723	1·689	1·646	1·618	1·596	1·581	1·562	1·551
0·06000	(0·588)	5·075	3·980	2·712	2·014	1·579	*	1·660	1·630	1·592	1·569	1·549	1·536	1·521	1·512
0·08000	(0·785)	4·395	3·447	2·349	1·744	1·368	1·656	1·618	1·591	1·557	1·535	1·518	1·506	1·493	1·485
0·10000	(0·981)	3·931	3·083	2·101	1·560	*	1·622	1·587	1·562	1·530	1·510	1·494	1·484	1·471	1·465
0·15000	(1·47)	3·210	2·517	1·715	1·274	*	1·564	1·533	1·512	1·484	1·467	1·453	1·445	1·435	1·430
0·20000	(1·96)	2·780	2·180	1·485	*	1·571	1·526	1·497	1·478	1·453	1·438	1·426	1·418	1·410	1·406
0·30000	(2·94)	2·270	1·780	*	*	1·514	1·474	1·450	1·433	1·411	1·398	1·388	1·382	1·376	1·374
0·40000	(3·92)	1·966	1·541	*	1·504	1·476	1·440	1·418	1·402	1·383	1·372	1·363	1·358	1·353	1·352
0·60000	(5·88)	1·605	1·259	*	1·450	1·426	1·394	1·375	1·362	1·345	1·336	1·329	1·325	1·322	1·322
0·80000	(7·85)	1·390	*	1·446	1·414	1·392	1·364	1·346	1·334	1·320	1·312	1·306	1·303	1·301	1·302
1·00000	(9·81)	1·243	*	1·417	1·388	1·367	1·341	1·325	1·314	1·301	1·294	1·289	1·286	1·285	1·287
1·50000	(14·7)	*	1·409	1·368	1·342	1·324	1·301	1·287	1·278	1·268	1·262	1·259	1·257	1·258	1·260
2·00000	(19·6)	*	1·372	1·335	1·311	1·295	1·275	1·262	1·254	1·245	1·241	1·238	1·237	1·239	1·241
3·00000	(29·4)	1·349	1·323	1·291	1·270	1·256	1·239	1·228	1·222	1·215	1·212	1·210	1·210	1·213	1·217
S	$(\Delta p/\rho l)$	125	150	200	250	300	400	500	600	800	1000	1250	1500	2000	2500
Grad'nt	k.p.g.	(Equivalent) Pipe diameters in mm													

Kinematic viscosity, $\nu = 600 \times 10^{-6}$ m^{2}s^{-1} ;   **Roughness size, $k_s = 0.015$ mm**

E278

Kin. visc., $\nu = 600 \times 10^{-6}$ m^{2}s^{-1};
S = 0·00010 to 3·00000

i.e. kin. pr. grad., $\Delta p/\rho l$ =
(0·00098) to (29·4) ms^{-2}

Roughness size, k_s = 0·030 mm
This table shows values of m, as follows

m_C for Colebrook-White solutions; or,
where **R** $\leq$ 2000, m_P for laminar flow

Grad'nt S	k.p.g. $(\Delta p/\rho l)$	125	150	200	250	300	400	500	600	800	1000	1250	1500	2000	2500
0·00010	(0·00098)	124·3	97·49	66·43	49·33	38·69	26·36	19·58	15·35	10·46	7·770	5·770	4·525	3·083	2·290
0·00015	(0·00147)	101·5	79·60	54·24	40·28	31·59	21·53	15·99	12·54	8·542	6·344	4·711	3·695	2·518	1·870
0·00020	(0·00196)	87·90	68·93	46·97	34·89	27·36	18·64	13·84	10·86	7·398	5·494	4·080	3·200	2·180	*
0·00030	(0·00294)	71·77	56·28	38·35	28·48	22·34	15·22	11·30	8·864	6·040	4·486	3·331	2·613	*	*
0·00040	(0·00392)	62·16	48·74	33·22	24·67	19·34	13·18	9·789	7·677	5·231	3·885	2·885	2·263	*	2·206
0·00060	(0·00588)	50·75	39·80	27·12	20·14	15·79	10·76	7·993	6·268	4·271	3·172	2·356	1·847	*	2·127
0·00080	(0·00785)	43·95	34·47	23·49	17·44	13·68	9·321	6·922	5·428	3·699	2·747	2·040	*	2·121	2·074
0·00100	(0·00981)	39·31	30·83	21·01	15·60	12·23	8·337	6·191	4·855	3·308	2·457	1·825	*	2·079	2·035
0·00150	(0·0147)	32·10	25·17	17·15	12·74	9·989	6·807	5·055	3·964	2·701	2·006	*	2·066	2·006	1·968
0·00200	(0·0196)	27·80	21·80	14·85	11·03	8·651	5·895	4·378	3·433	2·339	1·737	*	2·012	1·957	1·922
0·00300	(0·0294)	22·70	17·80	12·13	9·007	7·063	4·813	3·575	2·803	1·910	*	1·977	1·940	1·892	1·862
0·00400	(0·0392)	19·66	15·41	10·50	7·801	6·117	4·168	3·096	2·428	1·654	*	1·926	1·893	1·849	1·822
0·00600	(0·0588)	16·05	12·59	8·576	6·369	4·995	3·403	2·528	1·982	*	1·901	1·859	1·829	1·791	1·767
0·00800	(0·0785)	13·90	10·90	7·427	5·516	4·325	2·947	2·189	1·717	*	1·852	1·813	1·787	1·752	1·731
0·01000	(0·0981)	12·43	9·749	6·643	4·933	3·869	2·636	1·958	1·535	1·859	1·816	1·780	1·755	1·723	1·704
0·01500	(0·147)	10·15	7·960	5·424	4·028	3·159	2·153	1·599	*	1·791	1·753	1·722	1·700	1·673	1·656
0·02000	(0·196)	8·790	6·893	4·697	3·489	2·736	1·864	*	*	1·746	1·712	1·683	1·664	1·639	1·624
0·03000	(0·294)	7·177	5·628	3·835	2·848	2·234	1·522	*	1·734	1·686	1·656	1·631	1·615	1·593	1·581
0·04000	(0·392)	6·216	4·874	3·322	2·467	1·934	*	1·723	1·690	1·646	1·619	1·597	1·582	1·563	1·552
0·06000	(0·588)	5·075	3·980	2·712	2·014	1·579	*	1·660	1·631	1·593	1·569	1·550	1·537	1·521	1·513
0·08000	(0·785)	4·395	3·447	2·349	1·744	1·368	1·657	1·618	1·592	1·557	1·536	1·519	1·507	1·493	1·486
0·10000	(0·981)	3·931	3·083	2·101	1·560	*	1·623	1·587	1·563	1·531	1·511	1·495	1·485	1·472	1·466
0·15000	(1·47)	3·210	2·517	1·715	1·274	*	1·565	1·534	1·512	1·485	1·468	1·454	1·446	1·436	1·431
0·20000	(1·96)	2·780	2·180	1·485	*	1·572	1·527	1·498	1·479	1·454	1·439	1·427	1·419	1·411	1·407
0·30000	(2·94)	2·270	1·780	*	*	1·515	1·475	1·451	1·434	1·412	1·400	1·390	1·384	1·378	1·375
0·40000	(3·92)	1·966	1·541	*	1·505	1·477	1·441	1·419	1·404	1·384	1·373	1·365	1·360	1·355	1·354
0·60000	(5·88)	1·605	1·259	*	1·451	1·427	1·396	1·376	1·363	1·347	1·338	1·331	1·327	1·324	1·324
0·80000	(7·85)	1·390	*	1·447	1·415	1·394	1·365	1·348	1·336	1·322	1·314	1·308	1·305	1·303	1·304
1·00000	(9·81)	1·243	*	1·419	1·389	1·369	1·342	1·326	1·316	1·303	1·296	1·291	1·289	1·288	1·289
1·50000	(14·7)	*	1·410	1·369	1·343	1·326	1·303	1·289	1·280	1·270	1·264	1·261	1·260	1·260	1·263
2·00000	(19·6)	*	1·374	1·336	1·313	1·297	1·276	1·264	1·256	1·247	1·243	1·241	1·240	1·242	1·245
3·00000	(29·4)	1·350	1·325	1·293	1·272	1·258	1·241	1·231	1·224	1·217	1·214	1·213	1·214	1·217	1·221
S $(\Delta p/\rho l)$		125	150	200	250	300	400	500	600	800	1000	1250	1500	2000	2500

Grad'nt   k.p.g.   (Equivalent) Pipe diameters in mm   Roughness size, k_s = 0·030 mm

E279

Kin. visc., $\nu = 600 \times 10^{-6}$ m^{2}s^{-1};
S = 0·00010 to 3·00000

i.e. kin. pr. grad., $\Delta p/\rho l$ =
(0·00098) to (29·4) ms^{-2}

Roughness size, k_s = 0·060 mm
This table shows values of m, as follows

m_C for Colebrook-White solutions; or,
where **R** $\leq$ 2000, m_P for laminar flow

Grad'nt S	k.p.g. $(\Delta p/\rho l)$	125	150	200	250	300	400	500	600	800	1000	1250	1500	2000	2500
0·00010	(0·00098)	124·3	97·49	66·43	49·33	38·69	26·36	19·58	15·35	10·46	7·770	5·770	4·525	3·083	2·290
0·00015	(0·00147)	101·5	79·60	54·24	40·28	31·59	21·53	15·99	12·54	8·542	6·344	4·711	3·695	2·518	1·870
0·00020	(0·00196)	87·90	68·93	46·97	34·89	27·36	18·64	13·84	10·86	7·398	5·494	4·080	3·200	2·180	*
0·00030	(0·00294)	71·77	56·28	38·35	28·48	22·34	15·22	11·30	8·864	6·040	4·486	3·331	2·613	*	*
0·00040	(0·00392)	62·16	48·74	33·22	24·67	19·34	13·18	9·789	7·677	5·231	3·885	2·885	2·263	*	2·206
0·00060	(0·00588)	50·75	39·80	27·12	20·14	15·79	10·76	7·993	6·268	4·271	3·172	2·356	1·847	*	2·127
0·00080	(0·00785)	43·95	34·47	23·49	17·44	13·68	9·321	6·922	5·428	3·699	2·747	2·040	*	2·121	2·075
0·00100	(0·00981)	39·31	30·83	21·01	15·60	12·23	8·337	6·191	4·855	3·308	2·457	1·825	*	2·079	2·035
0·00150	(0·0147)	32·10	25·17	17·15	12·74	9·989	6·807	5·055	3·964	2·701	2·006	*	2·067	2·006	1·968
0·00200	(0·0196)	27·80	21·80	14·85	11·03	8·651	5·895	4·378	3·433	2·339	1·737	*	2·012	1·957	1·923
0·00300	(0·0294)	22·70	17·80	12·13	9·007	7·063	4·813	3·575	2·803	1·910	*	1·978	1·941	1·893	1·863
0·00400	(0·0392)	19·66	15·41	10·50	7·801	6·117	4·168	3·096	2·428	1·654	*	1·927	1·893	1·850	1·822
0·00600	(0·0588)	16·05	12·59	8·576	6·369	4·995	3·403	2·528	1·982	*	1·902	1·859	1·830	1·792	1·768
0·00800	(0·0785)	13·90	10·90	7·427	5·516	4·325	2·947	2·189	1·717	*	1·853	1·814	1·787	1·753	1·732
0·01000	(0·0981)	12·43	9·749	6·643	4·933	3·869	2·636	1·958	1·535	1·860	1·817	1·781	1·756	1·724	1·705
0·01500	(0·147)	10·15	7·960	5·424	4·028	3·159	2·153	1·599	*	1·792	1·754	1·723	1·701	1·674	1·657
0·02000	(0·196)	8·790	6·893	4·697	3·489	2·736	1·864	*	*	1·747	1·713	1·684	1·665	1·640	1·625
0·03000	(0·294)	7·177	5·628	3·835	2·848	2·234	1·522	*	1·735	1·687	1·657	1·633	1·616	1·595	1·583
0·04000	(0·392)	6·216	4·874	3·322	2·467	1·934	*	1·724	1·691	1·647	1·620	1·598	1·583	1·564	1·554
0·06000	(0·588)	5·075	3·980	2·712	2·014	1·579	*	1·662	1·632	1·594	1·571	1·551	1·539	1·523	1·514
0·08000	(0·785)	4·395	3·447	2·349	1·744	1·368	1·658	1·620	1·593	1·559	1·537	1·520	1·509	1·495	1·488
0·10000	(0·981)	3·931	3·083	2·101	1·560	*	1·625	1·589	1·564	1·532	1·513	1·497	1·487	1·474	1·468
0·15000	(1·47)	3·210	2·517	1·715	1·274	*	1·567	1·536	1·514	1·487	1·470	1·456	1·448	1·438	1·434
0·20000	(1·96)	2·780	2·180	1·485	*	1·574	1·528	1·500	1·481	1·456	1·441	1·429	1·422	1·414	1·410
0·30000	(2·94)	2·270	1·780	*	*	1·517	1·477	1·453	1·436	1·415	1·402	1·392	1·386	1·381	1·379
0·40000	(3·92)	1·966	1·541	*	1·507	1·479	1·443	1·421	1·406	1·387	1·376	1·368	1·363	1·358	1·357
0·60000	(5·88)	1·605	1·259	*	1·454	1·429	1·398	1·379	1·366	1·350	1·341	1·334	1·331	1·328	1·328
0·80000	(7·85)	1·390	*	1·450	1·418	1·396	1·368	1·351	1·339	1·325	1·317	1·312	1·309	1·308	1·309
1·00000	(9·81)	1·243	*	1·421	1·392	1·371	1·345	1·329	1·319	1·306	1·299	1·295	1·293	1·292	1·294
1·50000	(14·7)	*	1·413	1·372	1·347	1·329	1·306	1·293	1·284	1·274	1·269	1·266	1·265	1·266	1·268
2·00000	(19·6)	*	1·377	1·340	1·316	1·300	1·280	1·268	1·261	1·252	1·248	1·246	1·246	1·248	1·251
3·00000	(29·4)	1·354	1·329	1·296	1·276	1·263	1·245	1·235	1·229	1·223	1·220	1·219	1·220	1·223	1·228
S $(\Delta p/\rho l)$		125	150	200	250	300	400	500	600	800	1000	1250	1500	2000	2500

Grad'nt   k.p.g.   (Equivalent) Pipe diameters in mm   Roughness size, k_s = 0·060 mm

E280

Kin. visc., $\nu = 800 \times 10^{-6}$ m²s⁻¹; $S = 0.00010$ to 30.0000

i.e. kin. pr. grad., $\Delta p/\rho l =$ (0.00098) to (294) ms⁻²

Roughness size, $k_s = 0.015$ mm
This table shows values of m, as follows
m_C for Colebrook-White solutions; or, where $\mathbf{R} \leq 2000$, m_P for laminar flow

Grad'nt S	k.p.g. $(\Delta p/\rho l)$	6	8	10	12.5	15	20	25	30	40	50	60	80	100	125
0.00030	(0.00294)	5486	3738	2776	2062	1617	1102	818.2	641.6	437.2	324.7	254.6	173.5	128.9	95.70
0.00040	(0.00392)	4751	3237	2404	1786	1400	954.1	708.6	555.7	378.6	281.2	220.5	150.3	111.6	82.88
0.00060	(0.00588)	3879	2643	1963	1458	1143	779.0	578.6	453.7	309.2	229.6	180.1	122.7	91.12	67.67
0.00080	(0.00785)	3359	2289	1700	1263	990.1	674.7	501.0	392.9	267.7	198.8	155.9	106.3	78.91	58.60
0.00100	(0.00981)	3005	2048	1521	1129	885.6	603.4	448.2	351.4	239.5	177.8	139.5	95.04	70.58	52.42
0.00150	(0.0147)	2453	1672	1242	922.0	723.1	492.7	365.9	286.9	195.5	145.2	113.9	77.60	57.63	42.80
0.00200	(0.0196)	2125	1448	1075	798.5	626.2	426.7	316.9	248.5	169.3	125.8	98.62	67.20	49.91	37.06
0.00300	(0.0294)	1735	1182	877.9	652.0	511.3	348.4	258.7	202.9	138.3	102.7	80.52	54.87	40.75	30.26
0.00400	(0.0392)	1502	1024	760.3	564.6	442.8	301.7	224.1	175.7	119.7	88.92	69.73	47.52	35.29	26.21
0.00600	(0.0588)	1227	835.9	620.8	461.0	361.5	246.4	183.0	143.5	97.77	72.61	56.94	38.80	28.81	21.40
0.00800	(0.0785)	1062	723.9	537.6	399.3	313.1	213.4	158.4	124.3	84.67	62.88	49.31	33.60	24.95	18.53
0.01000	(0.0981)	950.2	647.5	480.9	357.1	280.0	190.8	141.7	111.1	75.73	56.24	44.10	30.05	22.32	16.58
0.01500	(0.147)	775.8	528.7	392.6	291.6	228.7	155.8	115.7	90.74	61.83	45.92	36.01	24.54	18.22	13.53
0.02000	(0.196)	671.9	457.8	340.0	252.5	198.0	134.9	100.2	78.58	53.55	39.77	31.19	21.25	15.78	11.72
0.03000	(0.294)	548.6	373.8	277.6	206.2	161.7	110.2	81.82	64.16	43.72	32.47	25.46	17.35	12.89	9.570
0.04000	(0.392)	475.1	323.7	240.4	178.6	140.0	95.41	70.86	55.57	37.86	28.12	22.05	15.03	11.16	8.288
0.06000	(0.588)	387.9	264.3	196.3	145.8	114.3	77.90	57.86	45.37	30.92	22.96	18.01	12.27	9.112	6.767
0.08000	(0.785)	335.9	228.9	170.0	126.3	99.01	67.47	50.10	39.29	26.77	19.88	15.59	10.63	7.891	5.860
0.10000	(0.981)	300.5	204.8	152.1	112.9	88.56	60.34	44.82	35.14	23.95	17.78	13.95	9.504	7.058	5.242
0.15000	(1.47)	245.3	167.2	124.2	92.20	72.31	49.27	36.59	28.69	19.55	14.52	11.39	7.760	5.763	4.280
0.20000	(1.96)	212.5	144.8	107.5	79.85	62.62	42.67	31.69	24.85	16.93	12.58	9.862	6.720	4.991	3.706
0.30000	(2.94)	173.5	118.2	87.79	65.20	51.13	34.84	25.87	20.29	13.83	10.27	8.052	5.487	4.075	3.026
0.40000	(3.92)	150.2	102.4	76.03	56.46	44.28	30.17	22.41	17.57	11.97	8.892	6.973	4.752	3.529	2.621
0.60000	(5.88)	122.7	83.59	62.08	46.10	36.15	24.64	18.30	14.35	9.777	7.261	5.694	3.880	2.881	2.140
0.80000	(7.85)	106.2	72.39	53.76	39.93	31.31	21.34	15.84	12.43	8.467	6.288	4.931	3.360	2.495	1.853
1.00000	(9.81)	95.02	64.75	48.09	35.71	28.00	19.08	14.17	11.11	7.573	5.624	4.410	3.005	2.232	1.658
1.50000	(14.7)	77.58	52.87	39.26	29.16	22.87	15.58	11.57	9.074	6.183	4.592	3.601	2.454	1.822	1.353
2.00000	(19.6)	67.19	45.78	34.00	25.25	19.80	13.49	10.02	7.858	5.355	3.977	3.119	2.125	1.578	1.172
3.00000	(29.4)	54.86	37.38	27.76	20.62	16.17	11.02	8.182	6.416	4.372	3.247	2.546	1.735	1.289	*
4.00000	(39.2)	47.51	32.37	24.04	17.86	14.00	9.541	7.086	5.557	3.786	2.812	2.205	1.503	1.116	*
6.00000	(58.8)	38.79	26.43	19.63	14.58	11.43	7.790	5.786	4.537	3.092	2.296	1.801	1.227	*	1.334
8.00000	(78.5)	33.59	22.89	17.00	12.63	9.901	6.747	5.010	3.929	2.677	1.988	1.559	1.063	*	1.300
10.0000	(98.1)	30.05	20.48	15.21	11.29	8.856	6.034	4.482	3.514	2.395	1.778	1.395	*	1.305	1.275
15.0000	(147)	24.53	16.72	12.42	9.220	7.231	4.927	3.659	2.869	1.955	1.452	1.139	*	1.258	1.232
20.0000	(196)	21.25	14.48	10.75	7.985	6.262	4.267	3.169	2.485	1.693	1.258	*	1.255	1.227	1.203
30.0000	(294)	17.35	11.82	8.779	6.520	5.113	3.484	2.587	2.029	1.383	1.027	*	1.210	1.185	1.165
S	$(\Delta p/\rho l)$	6	8	10	12.5	15	20	25	30	40	50	60	80	100	125
Grad'nt	**k.p.g.**														

(Equivalent) Pipe diameters in mm

S	$(\Delta p/\rho l)$	125	150	200	250	300	400	500	600	800	1000	1250	1500	2000	2500
0.00010	(0.00098)	165.8	130.0	88.57	65.78	51.58	35.15	26.10	20.47	13.95	10.36	7.694	6.033	4.111	3.053
0.00015	(0.00147)	135.3	106.1	72.32	53.71	42.12	28.70	21.31	16.71	11.39	8.459	6.282	4.926	3.357	2.493
0.00020	(0.00196)	117.2	91.91	62.63	46.51	36.48	24.86	18.46	14.48	9.864	7.325	5.440	4.266	2.907	2.159
0.00030	(0.00294)	95.70	75.05	51.14	37.98	29.78	20.29	15.07	11.82	8.054	5.981	4.442	3.483	2.374	*
0.00040	(0.00392)	82.88	64.99	44.29	32.89	25.79	17.58	13.05	10.24	6.975	5.180	3.847	3.017	2.056	*
0.00060	(0.00588)	67.67	53.07	36.16	26.85	21.06	14.35	10.66	8.357	5.695	4.229	3.141	2.463	*	2.241
0.00080	(0.00785)	58.60	45.96	31.32	23.26	18.24	12.43	9.229	7.238	4.932	3.663	2.720	2.133	*	2.182
0.00100	(0.00981)	52.42	41.10	28.01	20.80	16.31	11.12	8.255	6.474	4.411	3.276	2.433	1.908	*	2.139
0.00150	(0.0147)	42.80	33.56	22.87	16.98	13.32	9.076	6.740	5.286	3.602	2.675	1.986	*	2.111	2.065
0.00200	(0.0196)	37.06	29.07	19.81	14.71	11.53	7.860	5.837	4.578	3.119	2.316	1.720	*	2.057	2.015
0.00300	(0.0294)	30.26	23.73	16.17	12.01	9.418	6.418	4.766	3.738	2.547	1.891	*	2.043	1.985	1.949
0.00400	(0.0392)	26.21	20.55	14.00	10.40	8.156	5.558	4.128	3.237	2.206	1.638	*	1.990	1.937	1.904
0.00600	(0.0588)	21.40	16.78	11.43	8.492	6.660	4.538	3.370	2.643	1.801	*	1.956	1.920	1.874	1.845
0.00800	(0.0785)	18.53	14.53	9.903	7.354	5.767	3.930	2.919	2.289	1.560	*	1.906	1.873	1.831	1.805
0.01000	(0.0981)	16.58	13.00	8.857	6.578	5.158	3.515	2.610	2.047	*	1.912	1.869	1.839	1.800	1.776
0.01500	(0.147)	13.53	10.61	7.232	5.371	4.212	2.870	2.131	1.671	*	1.843	1.805	1.779	1.745	1.724
0.02000	(0.196)	11.72	9.191	6.263	4.651	3.648	2.486	1.846	*	1.838	1.797	1.762	1.738	1.708	1.689
0.03000	(0.294)	9.570	7.505	5.114	3.798	2.978	2.029	1.507	*	1.772	1.736	1.705	1.685	1.658	1.642
0.04000	(0.392)	8.288	6.499	4.429	3.289	2.579	1.758	*	*	1.728	1.695	1.667	1.649	1.625	1.611
0.06000	(0.588)	6.767	5.307	3.616	2.685	2.106	1.435	*	1.715	1.669	1.640	1.616	1.600	1.580	1.568
0.08000	(0.785)	5.860	4.596	3.132	2.326	1.824	*	1.704	1.672	1.630	1.604	1.582	1.568	1.550	1.540
0.10000	(0.981)	5.242	4.110	2.801	2.080	1.631	*	1.669	1.639	1.601	1.576	1.557	1.543	1.527	1.518
0.15000	(1.47)	4.280	3.356	2.287	1.698	1.332	1.648	1.610	1.584	1.550	1.529	1.512	1.501	1.487	1.480
0.20000	(1.96)	3.706	2.907	1.981	1.471	*	1.605	1.571	1.547	1.516	1.497	1.482	1.472	1.461	1.455
0.30000	(2.94)	3.026	2.373	1.617	*	*	1.548	1.518	1.498	1.471	1.455	1.442	1.434	1.424	1.420
0.40000	(3.92)	2.621	2.055	1.400	*	1.554	1.511	1.483	1.464	1.440	1.426	1.415	1.408	1.400	1.397
0.60000	(5.88)	2.140	1.678	*	1.528	1.498	1.460	1.436	1.420	1.400	1.387	1.378	1.372	1.367	1.365
0.80000	(7.85)	1.853	1.453	*	1.488	1.461	1.427	1.405	1.390	1.372	1.361	1.353	1.348	1.344	1.343
1.00000	(9.81)	1.658	1.300	*	1.458	1.434	1.401	1.382	1.368	1.351	1.342	1.335	1.330	1.327	1.327
1.50000	(14.7)	1.353	*	1.439	1.408	1.386	1.358	1.341	1.329	1.315	1.307	1.302	1.299	1.297	1.298
2.00000	(19.6)	1.172	*	1.402	1.374	1.354	1.329	1.314	1.303	1.291	1.284	1.280	1.278	1.277	1.279
3.00000	(29.4)	*	1.394	1.354	1.329	1.312	1.290	1.277	1.268	1.258	1.253	1.250	1.249	1.250	1.252
S	$(\Delta p/\rho l)$	125	150	200	250	300	400	500	600	800	1000	1250	1500	2000	2500
Grad'nt	**k.p.g.**														

(Equivalent) Pipe diameters in mm

Kinematic viscosity, $\nu = 800 \times 10^{-6}$ m²s⁻¹ ;   **Roughness size, $k_s = 0.015$ mm**

Kin. visc., $\nu = 800 \times 10^{-6}$ m^{2}s^{-1};
$S = 0.00010$ to 3.00000

i.e. kin. pr. grad., $\Delta p / \rho l =$
(0.00098) to (29.4) ms^{-2}

Roughness size, $k_s = 0.030$ mm
This table shows values of m, as follows

m_C for Colebrook-White solutions; or,
where $\mathbf{R} \leq 2000$, m_P for laminar flow

Grad'nt S	k.p.g. $(\Delta p / \rho l)$	125	150	200	250	300	400	500	600	800	1000	1250	1500	2000	2500
0.00010	(0.00098)	165.8	130.0	88.57	65.78	51.58	35.15	26.10	20.47	13.95	10.36	7.694	6.033	4.111	3.053
0.00015	(0.00147)	135.3	106.1	72.32	53.71	42.12	28.70	21.31	16.71	11.39	8.459	6.282	4.926	3.357	2.493
0.00020	(0.00196)	117.2	91.91	62.63	46.51	36.48	24.86	18.46	14.48	9.864	7.325	5.440	4.266	2.907	2.159
0.00030	(0.00294)	95.70	75.05	51.14	37.98	29.78	20.29	15.07	11.82	8.054	5.981	4.442	3.483	2.374	*
0.00040	(0.00392)	82.88	64.99	44.29	32.89	25.79	17.58	13.05	10.24	6.975	5.180	3.847	3.017	2.056	*
0.00060	(0.00588)	67.67	53.07	36.16	26.85	21.06	14.35	10.66	8.357	5.695	4.229	3.141	2.463	*	2.241
0.00080	(0.00785)	58.60	45.96	31.32	23.26	18.24	12.43	9.229	7.238	4.932	3.663	2.720	2.133	*	2.182
0.00100	(0.00981)	52.42	41.10	28.01	20.80	16.31	11.12	8.255	6.474	4.411	3.276	2.433	1.908	*	2.139
0.00150	(0.0147)	42.80	33.56	22.87	16.98	13.32	9.076	6.740	5.286	3.602	2.675	1.986	*	2.111	2.065
0.00200	(0.0196)	37.06	29.07	19.81	14.71	11.53	7.860	5.837	4.578	3.119	2.316	1.720	*	2.057	2.015
0.00300	(0.0294)	30.26	23.73	16.17	12.01	9.418	6.418	4.766	3.738	2.547	1.891	*	2.044	1.985	1.949
0.00400	(0.0392)	26.21	20.55	14.00	10.40	8.156	5.558	4.128	3.237	2.206	1.638	*	1.991	1.938	1.905
0.00600	(0.0588)	21.40	16.78	11.43	8.492	6.660	4.538	3.370	2.643	1.801	*	1.956	1.921	1.874	1.845
0.00800	(0.0785)	18.53	14.53	9.903	7.354	5.767	3.930	2.919	2.289	1.560	*	1.906	1.874	1.832	1.806
0.01000	(0.0981)	16.58	13.00	8.857	6.578	5.158	3.515	2.610	2.047	*	1.912	1.869	1.839	1.800	1.776
0.01500	(0.147)	13.53	10.61	7.232	5.371	4.212	2.870	2.131	1.671	*	1.843	1.805	1.779	1.745	1.724
0.02000	(0.196)	11.72	9.191	6.263	4.651	3.648	2.486	1.846	*	1.839	1.797	1.763	1.739	1.708	1.690
0.03000	(0.294)	9.570	7.505	5.114	3.798	2.978	2.029	1.507	*	1.772	1.736	1.706	1.685	1.659	1.643
0.04000	(0.392)	8.288	6.499	4.429	3.289	2.579	1.758	*	*	1.728	1.695	1.668	1.649	1.625	1.612
0.06000	(0.588)	6.767	5.307	3.616	2.685	2.106	1.435	*	1.716	1.670	1.641	1.617	1.601	1.581	1.569
0.08000	(0.785)	5.860	4.596	3.132	2.326	1.824	*	1.704	1.672	1.630	1.604	1.583	1.568	1.551	1.540
0.10000	(0.981)	5.242	4.110	2.801	2.080	1.631	*	1.670	1.640	1.601	1.577	1.557	1.544	1.528	1.519
0.15000	(1.47)	4.280	3.356	2.287	1.698	1.332	1.649	1.611	1.585	1.551	1.530	1.513	1.502	1.488	1.481
0.20000	(1.96)	3.706	2.907	1.981	1.471	*	1.606	1.571	1.548	1.517	1.498	1.483	1.473	1.462	1.456
0.30000	(2.94)	3.026	2.373	1.617	*	*	1.549	1.519	1.498	1.472	1.456	1.443	1.435	1.426	1.421
0.40000	(3.92)	2.621	2.055	1.400	*	1.555	1.511	1.484	1.465	1.441	1.427	1.416	1.409	1.401	1.398
0.60000	(5.88)	2.140	1.678	*	1.529	1.499	1.461	1.437	1.421	1.401	1.389	1.379	1.374	1.368	1.366
0.80000	(7.85)	1.853	1.453	*	1.489	1.462	1.428	1.406	1.392	1.373	1.363	1.355	1.350	1.346	1.345
1.00000	(9.81)	1.658	1.300	*	1.459	1.435	1.403	1.383	1.369	1.353	1.343	1.336	1.332	1.329	1.329
1.50000	(14.7)	1.353	*	1.440	1.409	1.387	1.360	1.342	1.331	1.317	1.309	1.304	1.301	1.300	1.300
2.00000	(19.6)	1.172	*	1.404	1.375	1.356	1.331	1.315	1.305	1.293	1.286	1.282	1.280	1.280	1.281
3.00000	(29.4)	*	1.395	1.356	1.331	1.314	1.292	1.279	1.270	1.260	1.255	1.253	1.252	1.253	1.255
S	$(\Delta p / \rho l)$	125	150	200	250	300	400	500	600	800	1000	1250	1500	2000	2500

Grad'nt   k.p.g.   (Equivalent) Pipe diameters in mm   **Roughness size, $k_s = 0.030$ mm**

Kin. visc., $\nu = 800 \times 10^{-6}$ m^{2}s^{-1};
$S = 0.00010$ to 3.00000

i.e. kin. pr. grad., $\Delta p / \rho l =$
(0.00098) to (29.4) ms^{-2}

Roughness size, $k_s = 0.060$ mm
This table shows values of m, as follows

m_C for Colebrook-White solutions; or,
where $\mathbf{R} \leq 2000$, m_P for laminar flow

Grad'nt S	k.p.g. $(\Delta p / \rho l)$	125	150	200	250	300	400	500	600	800	1000	1250	1500	2000	2500
0.00010	(0.00098)	165.8	130.0	88.57	65.78	51.58	35.15	26.10	20.47	13.95	10.36	7.694	6.033	4.111	3.053
0.00015	(0.00147)	135.3	106.1	72.32	53.71	42.12	28.70	21.31	16.71	11.39	8.459	6.282	4.926	3.357	2.493
0.00020	(0.00196)	117.2	91.91	62.63	46.51	36.48	24.86	18.46	14.48	9.864	7.325	5.440	4.266	2.907	2.159
0.00030	(0.00294)	95.70	75.05	51.14	37.98	29.78	20.29	15.07	11.82	8.054	5.981	4.442	3.483	2.374	*
0.00040	(0.00392)	82.88	64.99	44.29	32.89	25.79	17.58	13.05	10.24	6.975	5.180	3.847	3.017	2.056	*
0.00060	(0.00588)	67.67	53.07	36.16	26.85	21.06	14.35	10.66	8.357	5.695	4.229	3.141	2.463	*	2.241
0.00080	(0.00785)	58.60	45.96	31.32	23.26	18.24	12.43	9.229	7.238	4.932	3.663	2.720	2.133	*	2.183
0.00100	(0.00981)	52.42	41.10	28.01	20.80	16.31	11.12	8.255	6.474	4.411	3.276	2.433	1.908	*	2.140
0.00150	(0.0147)	42.80	33.56	22.87	16.98	13.32	9.076	6.740	5.286	3.602	2.675	1.986	*	2.111	2.065
0.00200	(0.0196)	37.06	29.07	19.81	14.71	11.53	7.860	5.837	4.578	3.119	2.316	1.720	*	2.057	2.015
0.00300	(0.0294)	30.26	23.73	16.17	12.01	9.418	6.418	4.766	3.738	2.547	1.891	*	2.044	1.986	1.949
0.00400	(0.0392)	26.21	20.55	14.00	10.40	8.156	5.558	4.128	3.237	2.206	1.638	*	1.991	1.938	1.905
0.00600	(0.0588)	21.40	16.78	11.43	8.492	6.660	4.538	3.370	2.643	1.801	*	1.957	1.921	1.875	1.846
0.00800	(0.0785)	18.53	14.53	9.903	7.354	5.767	3.930	2.919	2.289	1.560	*	1.907	1.874	1.832	1.806
0.01000	(0.0981)	16.58	13.00	8.857	6.578	5.158	3.515	2.610	2.047	*	1.913	1.870	1.840	1.801	1.777
0.01500	(0.147)	13.53	10.61	7.232	5.371	4.212	2.870	2.131	1.671	*	1.844	1.806	1.780	1.746	1.725
0.02000	(0.196)	11.72	9.191	6.263	4.651	3.648	2.486	1.846	*	1.840	1.798	1.763	1.740	1.709	1.691
0.03000	(0.294)	9.570	7.505	5.114	3.798	2.978	2.029	1.507	*	1.773	1.737	1.707	1.686	1.660	1.644
0.04000	(0.392)	8.288	6.499	4.429	3.289	2.579	1.758	*	*	1.729	1.696	1.669	1.650	1.627	1.613
0.06000	(0.588)	6.767	5.307	3.616	2.685	2.106	1.435	*	1.717	1.671	1.642	1.618	1.602	1.582	1.571
0.08000	(0.785)	5.860	4.596	3.132	2.326	1.824	*	1.706	1.673	1.632	1.605	1.584	1.570	1.552	1.542
0.10000	(0.981)	5.242	4.110	2.801	2.080	1.631	*	1.671	1.641	1.603	1.578	1.559	1.546	1.529	1.521
0.15000	(1.47)	4.280	3.356	2.287	1.698	1.332	1.650	1.612	1.586	1.552	1.531	1.515	1.503	1.490	1.483
0.20000	(1.96)	3.706	2.907	1.981	1.471	*	1.607	1.573	1.549	1.519	1.500	1.485	1.475	1.464	1.458
0.30000	(2.94)	3.026	2.373	1.617	*	*	1.551	1.521	1.500	1.474	1.458	1.445	1.437	1.428	1.424
0.40000	(3.92)	2.621	2.055	1.400	*	1.557	1.513	1.486	1.467	1.444	1.429	1.418	1.411	1.404	1.401
0.60000	(5.88)	2.140	1.678	*	1.530	1.501	1.463	1.440	1.423	1.403	1.391	1.382	1.377	1.371	1.370
0.80000	(7.85)	1.853	1.453	*	1.491	1.464	1.430	1.409	1.394	1.376	1.365	1.358	1.353	1.349	1.349
1.00000	(9.81)	1.658	1.300	*	1.462	1.437	1.405	1.385	1.372	1.356	1.346	1.339	1.336	1.333	1.333
1.50000	(14.7)	1.353	*	1.443	1.412	1.390	1.362	1.345	1.334	1.320	1.313	1.308	1.305	1.304	1.305
2.00000	(19.6)	1.172	*	1.407	1.378	1.359	1.334	1.319	1.309	1.297	1.290	1.286	1.284	1.284	1.286
3.00000	(29.4)	*	1.398	1.359	1.334	1.317	1.296	1.283	1.274	1.265	1.260	1.257	1.257	1.258	1.261
S	$(\Delta p / \rho l)$	125	150	200	250	300	400	500	600	800	1000	1250	1500	2000	2500

Grad'nt   k.p.g.   (Equivalent) Pipe diameters in mm   **Roughness size, $k_s = 0.060$ mm**

E283

Kin. visc., $\nu = 1.00 \times 10^{-3}$ m²s⁻¹;
$S = 0.00010$ to 30.0000
i.e. kin. pr. grad., $\Delta p/\rho l =$ (0.00098) to (294) ms⁻²

Roughness size, $k_s = 0.015$ mm
This table shows values of m, as follows
m_C for Colebrook-White solutions; or,
where $\mathbf{R} \leq 2000$, m_P for laminar flow

Grad'nt S	k.p.g. $(\Delta p/\rho l)$	6	8	10	12.5	15	20	25	30	40	50	60	80	100	125
0.00030	(0.00294)	6857	4673	3470	2577	2021	1377	1023	802.0	546.5	405.9	318.3	216.9	161.1	119.6
0.00040	(0.00392)	5939	4047	3005	2232	1750	1193	885.7	694.6	473.3	351.5	275.6	187.8	139.5	103.6
0.00060	(0.00588)	4849	3304	2454	1822	1429	973.8	723.2	567.1	386.5	287.0	225.1	153.4	113.9	84.59
0.00080	(0.00785)	4199	2861	2125	1578	1238	843.3	626.3	491.2	334.7	248.6	194.9	132.8	98.64	73.25
0.00100	(0.00981)	3756	2559	1901	1412	1107	754.3	560.2	439.3	299.3	222.3	174.3	118.8	88.22	65.52
0.00150	(0.0147)	3067	2090	1552	1153	903.8	615.9	457.4	358.7	244.4	181.5	142.3	97.00	72.04	53.50
0.00200	(0.0196)	2656	1810	1344	998.1	782.7	533.4	396.1	310.6	211.7	157.2	123.3	84.00	62.38	46.33
0.00300	(0.0294)	2169	1478	1097	815.0	639.1	435.5	323.4	253.6	172.8	128.4	100.7	68.59	50.94	37.83
0.00400	(0.0392)	1878	1280	950.4	705.8	553.5	377.2	280.1	219.6	149.7	111.2	87.17	59.40	44.11	32.76
0.00600	(0.0588)	1533	1045	776.0	576.3	451.9	307.9	228.7	179.3	122.2	90.76	71.17	48.50	36.02	26.75
0.00800	(0.0785)	1328	904.9	672.0	499.1	391.4	266.7	198.1	155.3	105.8	78.60	61.64	42.00	31.19	23.16
0.01000	(0.0981)	1188	809.3	601.1	446.4	350.1	238.5	177.1	138.9	94.66	70.30	55.13	37.57	27.90	20.72
0.01500	(0.147)	969.8	660.8	490.8	364.5	285.8	194.8	144.6	113.4	77.29	57.40	45.01	30.67	22.78	16.92
0.02000	(0.196)	839.9	572.3	425.0	315.6	247.5	168.7	125.3	98.23	66.94	49.71	38.98	26.56	19.73	14.65
0.03000	(0.294)	685.7	467.3	347.0	257.7	202.1	137.7	102.3	80.20	54.65	40.59	31.83	21.69	16.11	11.96
0.04000	(0.392)	593.9	404.7	300.5	223.2	175.0	119.3	88.57	69.46	47.33	35.15	27.56	18.78	13.95	10.36
0.06000	(0.588)	484.9	330.4	245.4	182.2	142.9	97.38	72.32	56.71	38.65	28.70	22.51	15.34	11.39	8.459
0.08000	(0.785)	419.9	286.1	212.5	157.8	123.8	84.33	62.63	49.12	33.47	24.86	19.49	13.28	9.864	7.325
0.10000	(0.981)	375.6	255.9	190.1	141.2	110.7	75.43	56.02	43.93	29.93	22.23	17.43	11.88	8.822	6.552
0.15000	(1.47)	306.7	209.0	155.2	115.3	90.38	61.59	45.74	35.87	24.44	18.15	14.23	9.700	7.204	5.350
0.20000	(1.96)	265.6	181.0	134.4	99.81	78.27	53.34	39.61	31.06	21.17	15.72	12.33	8.400	6.238	4.633
0.30000	(2.94)	216.9	147.8	109.7	81.50	63.91	43.55	32.34	25.36	17.28	12.84	10.07	6.859	5.094	3.783
0.40000	(3.92)	187.8	128.0	95.04	70.58	55.35	37.72	28.01	21.96	14.97	11.12	8.717	5.940	4.411	3.276
0.60000	(5.88)	153.3	104.5	77.60	57.63	45.19	30.79	22.87	17.93	12.22	9.076	7.117	4.850	3.602	2.675
0.80000	(7.85)	132.8	90.49	67.20	49.91	39.14	26.67	19.81	15.53	10.58	7.860	6.164	4.200	3.119	2.316
1.00000	(9.81)	118.8	80.93	60.11	44.64	35.01	23.85	17.71	13.89	9.466	7.030	5.513	3.757	2.790	2.072
1.50000	(14.7)	96.98	66.08	49.08	36.45	28.58	19.48	14.46	11.34	7.729	5.740	4.501	3.067	2.278	1.692
2.00000	(19.6)	83.99	57.23	42.50	31.56	24.75	16.87	12.53	9.823	6.694	4.971	3.898	2.656	1.973	1.465
3.00000	(29.4)	68.57	46.73	34.70	25.77	20.21	13.77	10.23	8.020	5.465	4.059	3.183	2.169	1.611	1.196
4.00000	(39.2)	59.39	40.47	30.05	22.32	17.50	11.93	8.857	6.946	4.733	3.515	2.756	1.878	1.395	*
6.00000	(58.8)	48.49	33.04	24.54	18.22	14.29	9.738	7.232	5.671	3.865	2.870	2.251	1.534	1.139	*
8.00000	(78.5)	41.99	28.61	21.25	15.78	12.38	8.433	6.263	4.912	3.347	2.486	1.949	1.328	*	1.354
10.0000	(98.1)	37.56	25.59	19.01	14.12	11.07	7.543	5.602	4.393	2.993	2.223	1.743	1.188	*	1.326
15.0000	(147)	30.67	20.90	15.52	11.53	9.038	6.159	4.574	3.587	2.444	1.815	1.423	*	1.310	1.280
20.0000	(196)	26.56	18.10	13.44	9.981	7.827	5.334	3.961	3.106	2.117	1.572	1.233	*	1.276	1.248
30.0000	(294)	21.69	14.78	10.97	8.150	6.391	4.355	3.234	2.536	1.728	1.284	1.007	1.260	1.231	1.207
S	$(\Delta p/\rho l)$	6	8	10	12.5	15	20	25	30	40	50	60	80	100	125

Grad'nt k.p.g. (Equivalent) Pipe diameters in mm

Grad'nt S	k.p.g. $(\Delta p/\rho l)$	125	150	200	250	300	400	500	600	800	1000	1250	1500	2000	2500
0.00010	(0.00098)	207.2	162.5	110.7	82.22	64.48	43.94	32.63	25.59	17.44	12.95	9.617	7.542	5.139	3.817
0.00015	(0.00147)	169.2	132.7	90.40	67.14	52.65	35.88	26.64	20.89	14.24	10.57	7.852	6.158	4.196	3.116
0.00020	(0.00196)	146.5	114.9	78.29	58.14	45.59	31.07	23.07	18.09	12.33	9.157	6.800	5.333	3.634	2.699
0.00030	(0.00294)	119.6	93.81	63.92	47.47	37.23	25.37	18.84	14.77	10.07	7.476	5.552	4.354	2.967	2.203
0.00040	(0.00392)	103.6	81.24	55.36	41.11	32.24	21.97	16.32	12.79	8.718	6.475	4.809	3.771	2.570	1.908
0.00060	(0.00588)	84.59	66.33	45.20	33.57	26.32	17.94	13.32	10.45	7.119	5.287	3.926	3.079	2.098	*
0.00080	(0.00785)	73.25	57.45	39.14	29.07	22.80	15.53	11.54	9.047	6.165	4.578	3.400	2.666	1.817	*
0.00100	(0.00981)	65.52	51.38	35.01	26.00	20.39	13.89	10.32	8.092	5.514	4.095	3.041	2.385	*	2.227
0.00150	(0.0147)	53.50	41.95	28.59	21.23	16.65	11.34	8.425	6.607	4.502	3.344	2.483	1.947	*	2.147
0.00200	(0.0196)	46.33	36.33	24.76	18.39	14.42	9.825	7.296	5.722	3.899	2.896	2.150	*	2.141	2.093
0.00300	(0.0294)	37.83	29.66	20.21	15.01	11.77	8.022	5.958	4.672	3.184	2.364	1.756	*	2.064	2.022
0.00400	(0.0392)	32.76	25.69	17.51	13.00	10.20	6.947	5.159	4.046	2.757	2.048	*	*	2.013	1.974
0.00600	(0.0588)	26.75	20.98	14.29	10.62	8.324	5.672	4.213	3.304	2.251	1.672	*	1.998	1.944	1.910
0.00800	(0.0785)	23.16	18.17	12.38	9.193	7.209	4.912	3.648	2.861	1.950	*	1.985	1.947	1.898	1.868
0.01000	(0.0981)	20.72	16.25	11.07	8.222	6.448	4.394	3.263	2.559	1.744	*	1.944	1.910	1.864	1.836
0.01500	(0.147)	16.92	13.27	9.040	6.714	5.265	3.588	2.664	2.089	*	1.919	1.875	1.845	1.805	1.781
0.02000	(0.196)	14.65	11.49	7.829	5.814	4.559	3.107	2.307	1.809	*	1.869	1.829	1.802	1.766	1.744
0.03000	(0.294)	11.96	9.381	6.392	4.747	3.723	2.537	1.884	1.477	1.845	1.803	1.768	1.744	1.713	1.694
0.04000	(0.392)	10.36	8.124	5.536	4.111	3.224	2.197	1.632	*	1.797	1.759	1.727	1.705	1.677	1.660
0.06000	(0.588)	8.459	6.633	4.520	3.357	2.632	1.794	*	*	1.734	1.700	1.673	1.654	1.630	1.615
0.08000	(0.785)	7.325	5.745	3.914	2.907	2.280	1.553	*	1.740	1.692	1.661	1.636	1.619	1.597	1.585
0.10000	(0.981)	6.552	5.138	3.501	2.600	2.039	1.389	*	1.705	1.660	1.632	1.609	1.593	1.573	1.562
0.15000	(1.47)	5.350	4.195	2.859	2.123	1.665	*	1.676	1.645	1.606	1.581	1.561	1.548	1.531	1.522
0.20000	(1.96)	4.633	3.633	2.476	1.839	1.442	*	1.633	1.605	1.569	1.547	1.529	1.517	1.503	1.495
0.30000	(2.94)	3.783	2.966	2.021	1.501	*	1.611	1.576	1.552	1.521	1.502	1.486	1.476	1.464	1.458
0.40000	(3.92)	3.276	2.569	1.751	1.300	*	1.570	1.538	1.516	1.488	1.471	1.457	1.449	1.438	1.433
0.60000	(5.88)	2.675	2.098	1.429	*	1.560	1.516	1.488	1.469	1.445	1.430	1.418	1.411	1.403	1.400
0.80000	(7.85)	2.316	1.817	1.238	*	1.520	1.479	1.454	1.437	1.415	1.402	1.392	1.386	1.379	1.377
1.00000	(9.81)	2.072	1.625	*	1.518	1.490	1.453	1.429	1.413	1.393	1.381	1.372	1.367	1.362	1.360
1.50000	(14.7)	1.692	1.327	*	1.464	1.439	1.406	1.386	1.372	1.355	1.345	1.338	1.334	1.330	1.330
2.00000	(19.6)	1.465	*	1.460	1.427	1.404	1.375	1.357	1.344	1.329	1.321	1.315	1.311	1.309	1.309
3.00000	(29.4)	1.196	*	1.408	1.379	1.359	1.333	1.317	1.307	1.294	1.288	1.283	1.281	1.280	1.281
S	$(\Delta p/\rho l)$	125	150	200	250	300	400	500	600	800	1000	1250	1500	2000	2500

Grad'nt k.p.g. (Equivalent) Pipe diameters in mm

Kinematic viscosity, $\nu = 1.00 \times 10^{-3}$ m²s⁻¹ ; **Roughness size, $k_s = 0.015$ mm**

Kin. visc., $\nu = 1{\cdot}00 \times 10^{-3}$ m^{2}s^{-1};
S = 0·00010 to 3·00000

i.e. kin. pr. grad., $\Delta p/\rho l$ =
(0·00098) to (29·4) ms^{-2}

Roughness size, k_s = 0·030 mm
This table shows values of m, as follows

m_C for Colebrook-White solutions; or,
where $\mathbf{R} \leq 2000$, m_P for laminar flow

Grad'nt S	k.p.g. $(\Delta p/\rho l)$	125	150	200	250	300	400	500	600	800	1000	1250	1500	2000	2500
0·00010	(0·00098)	207·2	162·5	110·7	82·22	64·48	43·94	32·63	25·59	17·44	12·95	9·617	7·542	5·139	3·817
0·00015	(0·00147)	169·2	132·7	90·40	67·14	52·65	35·88	26·64	20·89	14·24	10·57	7·852	6·158	4·196	3·116
0·00020	(0·00196)	146·5	114·9	78·29	58·14	45·59	31·07	23·07	18·09	12·33	9·157	6·800	5·333	3·634	2·699
0·00030	(0·00294)	119·6	93·81	63·92	47·47	37·23	25·37	18·84	14·77	10·07	7·476	5·552	4·354	2·967	2·203
0·00040	(0·00392)	103·6	81·24	55·36	41·11	32·24	21·97	16·32	12·79	8·718	6·475	4·809	3·771	2·570	1·908
0·00060	(0·00588)	84·59	66·33	45·20	33·57	26·32	17·94	13·32	10·45	7·119	5·287	3·926	3·079	2·098	*
0·00080	(0·00785)	73·25	57·45	39·14	29·07	22·80	15·53	11·54	9·047	6·165	4·578	3·400	2·666	1·817	*
0·00100	(0·00981)	65·52	51·38	35·01	26·00	20·39	13·89	10·32	8·092	5·514	4·095	3·041	2·385	*	2·228
0·00150	(0·0147)	53·50	41·95	28·59	21·23	16·65	11·34	8·425	6·607	4·502	3·344	2·483	1·947	*	2·147
0·00200	(0·0196)	46·33	36·33	24·76	18·39	14·42	9·825	7·296	5·722	3·899	2·896	2·150	*	2·142	2·093
0·00300	(0·0294)	37·83	29·66	20·21	15·01	11·77	8·022	5·958	4·672	3·184	2·364	1·756	*	2·064	2·022
0·00400	(0·0392)	32·76	25·69	17·51	13·00	10·20	6·947	5·159	4·046	2·757	2·048	*	*	2·013	1·974
0·00600	(0·0588)	26·75	20·98	14·29	10·62	8·324	5·672	4·213	3·304	2·251	1·672	*	1·998	1·944	1·911
0·00800	(0·0785)	23·16	18·17	12·38	9·193	7·209	4·912	3·648	2·861	1·950	*	1·985	1·947	1·899	1·868
0·01000	(0·0981)	20·72	16·25	11·07	8·222	6·448	4·394	3·263	2·559	1·744	*	1·945	1·910	1·865	1·836
0·01500	(0·147)	16·92	13·27	9·040	6·714	5·265	3·588	2·664	2·089	*	1·920	1·876	1·845	1·806	1·781
0·02000	(0·196)	14·65	11·49	7·829	5·814	4·559	3·107	2·307	1·809	*	1·870	1·830	1·802	1·766	1·744
0·03000	(0·294)	11·96	9·381	6·392	4·747	3·723	2·537	1·884	1·477	1·846	1·804	1·769	1·744	1·713	1·694
0·04000	(0·392)	10·36	8·124	5·536	4·111	3·224	2·197	1·632	*	1·798	1·760	1·728	1·706	1·678	1·661
0·06000	(0·588)	8·459	6·633	4·520	3·357	2·632	1·794	*	*	1·734	1·701	1·673	1·654	1·630	1·616
0·08000	(0·785)	7·325	5·745	3·914	2·907	2·280	1·553	*	1·741	1·692	1·662	1·637	1·619	1·598	1·585
0·10000	(0·981)	6·552	5·138	3·501	2·600	2·039	1·389	*	1·706	1·661	1·632	1·609	1·593	1·574	1·563
0·15000	(1·47)	5·350	4·195	2·859	2·123	1·665	*	1·676	1·646	1·606	1·582	1·562	1·548	1·532	1·523
0·20000	(1·96)	4·633	3·633	2·476	1·839	1·442	*	1·633	1·606	1·570	1·548	1·530	1·518	1·504	1·496
0·30000	(2·94)	3·783	2·966	2·021	1·501	*	1·612	1·577	1·553	1·522	1·503	1·487	1·477	1·465	1·459
0·40000	(3·92)	3·276	2·569	1·751	1·300	*	1·571	1·539	1·517	1·489	1·472	1·458	1·450	1·439	1·435
0·60000	(5·88)	2·675	2·098	1·429	*	1·561	1·517	1·489	1·470	1·446	1·431	1·420	1·412	1·405	1·401
0·80000	(7·85)	2·316	1·817	1·238	*	1·521	1·480	1·455	1·438	1·416	1·403	1·393	1·387	1·381	1·379
1·00000	(9·81)	2·072	1·625	*	1·519	1·491	1·454	1·430	1·414	1·394	1·383	1·374	1·368	1·363	1·361
1·50000	(14·7)	1·692	1·327	*	1·465	1·440	1·407	1·387	1·373	1·356	1·347	1·339	1·335	1·332	1·332
2·00000	(19·6)	1·465	*	1·461	1·428	1·406	1·376	1·358	1·346	1·331	1·322	1·316	1·313	1·311	1·311
3·00000	(29·4)	1·196	*	1·409	1·380	1·360	1·335	1·319	1·309	1·296	1·290	1·285	1·283	1·282	1·284
S	$(\Delta p/\rho l)$	125	150	200	250	300	400	500	600	800	1000	1250	1500	2000	2500

Grad'nt   k.p.g.   (Equivalent) Pipe diameters in mm   **Roughness size, k_s = 0·030 mm**

Kin. visc., $\nu = 1{\cdot}00 \times 10^{-3}$ m^{2}s^{-1};
S = 0·00010 to 3·00000

i.e. kin. pr. grad., $\Delta p/\rho l$ =
(0·00098) to (29·4) ms^{-2}

Roughness size, k_s = 0·060 mm
This table shows values of m, as follows

m_C for Colebrook-White solutions; or,
where $\mathbf{R} \leq 2000$, m_P for laminar flow

Grad'nt S	k.p.g. $(\Delta p/\rho l)$	125	150	200	250	300	400	500	600	800	1000	1250	1500	2000	2500
0·00010	(0·00098)	207·2	162·5	110·7	82·22	64·48	43·94	32·63	25·59	17·44	12·95	9·617	7·542	5·139	3·817
0·00015	(0·00147)	169·2	132·7	90·40	67·14	52·65	35·88	26·64	20·89	14·24	10·57	7·852	6·158	4·196	3·116
0·00020	(0·00196)	146·5	114·9	78·29	58·14	45·59	31·07	23·07	18·09	12·33	9·157	6·800	5·333	3·634	2·699
0·00030	(0·00294)	119·6	93·81	63·92	47·47	37·23	25·37	18·84	14·77	10·07	7·476	5·552	4·354	2·967	2·203
0·00040	(0·00392)	103·6	81·24	55·36	41·11	32·24	21·97	16·32	12·79	8·718	6·475	4·809	3·771	2·570	1·908
0·00060	(0·00588)	84·59	66·33	45·20	33·57	26·32	17·94	13·32	10·45	7·119	5·287	3·926	3·079	2·098	*
0·00080	(0·00785)	73·25	57·45	39·14	29·07	22·80	15·53	11·54	9·047	6·165	4·578	3·400	2·666	1·817	*
0·00100	(0·00981)	65·52	51·38	35·01	26·00	20·39	13·89	10·32	8·092	5·514	4·095	3·041	2·385	*	2·228
0·00150	(0·0147)	53·50	41·95	28·59	21·23	16·65	11·34	8·425	6·607	4·502	3·344	2·483	1·947	*	2·147
0·00200	(0·0196)	46·33	36·33	24·76	18·39	14·42	9·825	7·296	5·722	3·899	2·896	2·150	*	2·142	2·094
0·00300	(0·0294)	37·83	29·66	20·21	15·01	11·77	8·022	5·958	4·672	3·184	2·364	1·756	*	2·065	2·022
0·00400	(0·0392)	32·76	25·69	17·51	13·00	10·20	6·947	5·159	4·046	2·757	2·048	*	*	2·013	1·975
0·00600	(0·0588)	26·75	20·98	14·29	10·62	8·324	5·672	4·213	3·304	2·251	1·672	*	1·999	1·945	1·911
0·00800	(0·0785)	23·16	18·17	12·38	9·193	7·209	4·912	3·648	2·861	1·950	*	1·985	1·948	1·899	1·869
0·01000	(0·0981)	20·72	16·25	11·07	8·222	6·448	4·394	3·263	2·559	1·744	*	1·945	1·910	1·865	1·837
0·01500	(0·147)	16·92	13·27	9·040	6·714	5·265	3·588	2·664	2·089	*	1·920	1·876	1·846	1·806	1·782
0·02000	(0·196)	14·65	11·49	7·829	5·814	4·559	3·107	2·307	1·809	*	1·870	1·830	1·803	1·767	1·745
0·03000	(0·294)	11·96	9·381	6·392	4·747	3·723	2·537	1·884	1·477	1·846	1·804	1·769	1·745	1·714	1·695
0·04000	(0·392)	10·36	8·124	5·536	4·111	3·224	2·197	1·632	*	1·799	1·760	1·729	1·707	1·679	1·662
0·06000	(0·588)	8·459	6·633	4·520	3·357	2·632	1·794	*	*	1·735	1·702	1·674	1·655	1·631	1·617
0·08000	(0·785)	7·325	5·745	3·914	2·907	2·280	1·553	*	1·742	1·693	1·663	1·638	1·621	1·599	1·587
0·10000	(0·981)	6·552	5·138	3·501	2·600	2·039	1·389	*	1·707	1·662	1·634	1·610	1·595	1·575	1·564
0·15000	(1·47)	5·350	4·195	2·859	2·123	1·665	*	1·677	1·647	1·608	1·583	1·563	1·550	1·534	1·524
0·20000	(1·96)	4·633	3·633	2·476	1·839	1·442	*	1·635	1·607	1·572	1·549	1·532	1·520	1·505	1·498
0·30000	(2·94)	3·783	2·966	2·021	1·501	*	1·613	1·578	1·554	1·523	1·504	1·489	1·479	1·467	1·461
0·40000	(3·92)	3·276	2·569	1·751	1·300	*	1·572	1·541	1·519	1·491	1·474	1·460	1·452	1·442	1·437
0·60000	(5·88)	2·675	2·098	1·429	*	1·563	1·518	1·491	1·472	1·448	1·433	1·422	1·415	1·407	1·404
0·80000	(7·85)	2·316	1·817	1·238	*	1·522	1·482	1·457	1·440	1·419	1·406	1·396	1·390	1·384	1·382
1·00000	(9·81)	2·072	1·625	*	1·521	1·493	1·456	1·433	1·417	1·397	1·385	1·376	1·371	1·366	1·365
1·50000	(14·7)	1·692	1·327	*	1·467	1·442	1·410	1·390	1·376	1·359	1·350	1·343	1·339	1·336	1·336
2·00000	(19·6)	1·465	*	1·464	1·431	1·408	1·379	1·361	1·349	1·334	1·326	1·320	1·317	1·315	1·316
3·00000	(29·4)	1·196	*	1·412	1·383	1·363	1·338	1·322	1·312	1·300	1·293	1·289	1·287	1·287	1·289
S	$(\Delta p/\rho l)$	125	150	200	250	300	400	500	600	800	1000	1250	1500	2000	2500

Grad'nt   k.p.g.   (Equivalent) Pipe diameters in mm   **Roughness size, k_s = 0·060 mm**

E286

Kin. visc., $\nu = 1.5 \times 10^{-3}$ m^{2}s^{-1};
$S = 0.00010$ to 30.0000

i.e. kin. pr. grad., $\Delta p/\rho l =$
(0.00098) to (294) ms^{-2}

Roughness size, $k_s = 0.015$ mm
This table shows values of m, as follows

m_C for Colebrook-White solutions; or,
where $\mathbf{R} \le 2000$, m_P for laminar flow

Grad'nt S	k.p.g. $(\Delta p/\rho l)$	6	8	10	12·5	15	20	25	30	40	50	60	80	100	125
0·00030	(0·00294)	10286	7009	5205	3866	3032	2066	1534	1203	819·8	608·8	477·4	325·3	241·6	179·4
0·00040	(0·00392)	8908	6070	4508	3348	2625	1789	1329	1042	710·0	527·3	413·5	281·7	209·2	155·4
0·00060	(0·00588)	7273	4956	3681	2734	2144	1461	1085	850·7	579·7	430·5	337·6	230·0	170·8	126·9
0·00080	(0·00785)	6299	4292	3188	2367	1856	1265	939·5	736·7	502·0	372·8	292·4	199·2	148·0	109·9
0·00100	(0·00981)	5634	3839	2851	2117	1660	1131	840·3	658·9	449·0	333·5	261·5	178·2	132·3	98·28
0·00150	(0·0147)	4600	3135	2328	1729	1356	923·8	686·1	538·0	366·6	272·3	213·5	145·5	108·1	80·25
0·00200	(0·0196)	3984	2715	2016	1497	1174	800·1	594·2	465·9	317·5	235·8	184·9	126·0	93·58	69·49
0·00300	(0·0294)	3253	2216	1646	1222	958·7	653·3	485·1	380·4	259·2	192·5	151·0	102·9	76·40	56·74
0·00400	(0·0392)	2817	1920	1426	1059	830·2	565·7	420·1	329·5	224·5	166·7	130·8	89·10	66·17	49·14
0·00600	(0·0588)	2300	1567	1164	864·4	677·9	461·9	343·0	269·0	183·3	136·1	106·8	72·75	54·03	40·12
0·00800	(0·0785)	1992	1357	1008	748·6	587·1	400·0	297·1	233·0	158·8	117·9	92·46	63·00	46·79	34·75
0·01000	(0·0981)	1782	1214	901·6	669·6	525·1	357·8	265·7	208·4	142·0	105·5	82·69	56·35	41·85	31·08
0·01500	(0·147)	1455	991·2	736·2	546·7	428·7	292·1	217·0	170·1	115·9	86·10	67·52	46·01	34·17	25·38
0·02000	(0·196)	1260	858·4	637·5	473·5	371·3	253·0	187·9	147·3	100·4	74·57	58·47	39·85	29·59	21·98
0·03000	(0·294)	1029	700·9	520·5	386·6	303·2	206·6	153·4	120·3	81·98	60·88	47·74	32·53	24·16	17·94
0·04000	(0·392)	890·8	607·0	450·8	334·8	262·5	178·9	132·9	104·2	71·00	52·73	41·35	28·17	20·92	15·54
0·06000	(0·588)	727·3	495·6	368·1	273·4	214·4	146·1	108·5	85·07	57·97	43·05	33·76	23·00	17·08	12·69
0·08000	(0·785)	629·9	429·2	318·8	236·7	185·6	126·5	93·95	73·67	50·20	37·28	29·24	19·92	14·80	10·99
0·10000	(0·981)	563·4	383·9	285·1	211·7	166·0	113·1	84·03	65·89	44·90	33·35	26·15	17·82	13·23	9·828
0·15000	(1·47)	460·0	313·5	232·8	172·9	135·6	92·38	68·61	53·80	36·66	27·23	21·35	14·55	10·81	8·025
0·20000	(1·96)	398·4	271·5	201·6	149·7	117·4	80·01	59·42	46·59	31·75	23·58	18·49	12·60	9·358	6·949
0·30000	(2·94)	325·3	221·6	164·6	122·2	95·87	65·33	48·51	38·04	25·92	19·25	15·10	10·29	7·640	5·674
0·40000	(3·92)	281·7	192·0	142·6	105·9	83·02	56·57	42·01	32·95	22·45	16·67	13·08	8·910	6·617	4·914
0·60000	(5·88)	230·0	156·7	116·4	86·44	67·79	46·19	34·30	26·90	18·33	13·61	10·68	7·275	5·403	4·012
0·80000	(7·85)	199·2	135·7	100·8	74·86	58·71	40·00	29·71	23·30	15·88	11·79	9·246	6·300	4·679	3·475
1·00000	(9·81)	178·2	121·4	90·16	66·96	52·51	35·78	26·57	20·84	14·20	10·55	8·269	5·635	4·185	3·108
1·50000	(14·7)	145·5	99·12	73·62	54·67	42·87	29·21	21·70	17·01	11·59	8·610	6·752	4·601	3·417	2·538
2·00000	(19·6)	126·0	85·84	63·75	47·35	37·13	25·30	18·79	14·73	10·04	7·457	5·847	3·985	2·959	2·198
3·00000	(29·4)	102·9	70·09	52·05	38·66	30·32	20·66	15·34	12·03	8·198	6·088	4·774	3·253	2·416	1·794
4·00000	(39·2)	89·08	60·70	45·08	33·48	26·25	17·89	13·29	10·42	7·100	5·273	4·135	2·817	2·092	1·554
6·00000	(58·8)	72·73	49·56	36·81	27·34	21·44	14·61	10·85	8·507	5·797	4·305	3·376	2·300	1·708	1·269
8·00000	(78·5)	62·99	42·92	31·88	23·67	18·56	12·65	9·395	7·367	5·020	3·728	2·924	1·992	1·480	*
10·0000	(98·1)	56·34	38·39	28·51	21·17	16·60	11·31	8·403	6·589	4·490	3·335	2·615	1·782	1·323	*
15·0000	(147)	46·00	31·35	23·28	17·29	13·56	9·238	6·861	5·380	3·666	2·723	2·135	1·455	*	*
20·0000	(196)	39·84	27·15	20·16	14·97	11·74	8·001	5·942	4·659	3·175	2·358	1·849	1·260	*	1·341
30·0000	(294)	32·53	22·16	16·46	12·22	9·587	6·533	4·851	3·804	2·592	1·925	1·510	*	*	1·293
S	$(\Delta p/\rho l)$	6	8	10	12·5	15	20	25	30	40	50	60	80	100	125

Grad'nt   k.p.g.   (Equivalent) Pipe diameters in mm

Grad'nt S	k.p.g. $(\Delta p/\rho l)$	125	150	200	250	300	400	500	600	800	1000	1250	1500	2000	2500
0·00010	(0·00098)	310·8	243·7	166·1	123·3	96·72	65·91	48·95	38·38	26·16	19·42	14·43	11·31	7·709	5·725
0·00015	(0·00147)	253·8	199·0	135·6	100·7	78·97	53·81	39·96	31·34	21·36	15·86	11·78	9·237	6·294	4·674
0·00020	(0·00196)	219·8	172·3	117·4	87·21	68·39	46·60	34·61	27·14	18·49	13·74	10·20	7·999	5·451	4·048
0·00030	(0·00294)	179·4	140·7	95·88	71·21	55·84	38·05	28·26	22·16	15·10	11·21	8·329	6·531	4·451	3·305
0·00040	(0·00392)	155·4	121·9	83·04	61·67	48·36	32·95	24·47	19·19	13·08	9·712	7·213	5·656	3·854	2·862
0·00060	(0·00588)	126·9	99·50	67·80	50·35	39·49	26·91	19·98	15·67	10·68	7·930	5·889	4·618	3·147	2·337
0·00080	(0·00785)	109·9	86·17	58·72	43·61	34·20	23·30	17·31	13·57	9·247	6·868	5·100	4·000	2·725	2·024
0·00100	(0·00981)	98·28	77·07	52·52	39·00	30·59	20·84	15·48	12·14	8·271	6·143	4·562	3·577	2·438	*
0·00150	(0·0147)	80·25	62·93	42·88	31·85	24·97	17·02	12·64	9·911	6·753	5·015	3·725	2·921	1·990	*
0·00200	(0·0196)	69·49	54·50	37·14	27·58	21·63	14·74	10·94	8·583	5·849	4·343	3·226	2·530	*	2·252
0·00300	(0·0294)	56·74	44·50	30·32	22·52	17·66	12·03	8·936	7·008	4·775	3·546	2·634	2·065	*	2·170
0·00400	(0·0392)	49·14	38·54	26·26	19·50	15·29	10·42	7·739	6·069	4·136	3·071	2·281	1·789	2·165	2·115
0·00600	(0·0588)	40·12	31·46	21·44	15·92	12·49	8·509	6·319	4·955	3·377	2·508	1·862	*	2·086	2·042
0·00800	(0·0785)	34·75	27·25	18·57	13·79	10·81	7·369	5·472	4·291	2·924	2·172	*	*	2·033	1·993
0·01000	(0·0981)	31·08	24·37	16·61	12·33	9·672	6·591	4·895	3·838	2·616	1·942	*	2·054	1·994	1·957
0·01500	(0·147)	25·38	19·90	13·56	10·07	7·897	5·381	3·996	3·134	2·136	*	*	1·979	1·927	1·895
0·02000	(0·196)	21·98	17·23	11·74	8·721	6·839	4·660	3·461	2·714	1·849	*	1·965	1·929	1·882	1·853
0·03000	(0·294)	17·94	14·07	9·588	7·121	5·584	3·805	2·826	2·216	*	*	1·895	1·863	1·822	1·797
0·04000	(0·392)	15·54	12·19	8·304	6·167	4·836	3·295	2·447	1·919	*	1·890	1·848	1·819	1·782	1·759
0·06000	(0·588)	12·69	9·950	6·780	5·035	3·949	2·691	1·998	1·567	*	1·822	1·786	1·760	1·728	1·708
0·08000	(0·785)	10·99	8·617	5·872	4·361	3·420	2·330	1·731	*	1·817	1·777	1·744	1·721	1·692	1·674
0·10000	(0·981)	9·828	7·707	5·252	3·900	3·059	2·084	1·548	*	1·780	1·743	1·713	1·692	1·665	1·648
0·15000	(1·47)	8·025	6·293	4·288	3·185	2·497	1·702	*	1·770	1·718	1·686	1·659	1·641	1·618	1·604
0·20000	(1·96)	6·949	5·450	3·714	2·758	2·163	1·474	*	1·723	1·677	1·647	1·623	1·606	1·586	1·574
0·30000	(2·94)	5·674	4·450	3·032	2·252	1·766	*	1·694	1·662	1·621	1·596	1·575	1·561	1·543	1·533
0·40000	(3·92)	4·914	3·854	2·626	1·950	1·529	*	1·650	1·621	1·584	1·561	1·542	1·530	1·514	1·506
0·60000	(5·88)	4·012	3·146	2·144	1·592	*	1·628	1·592	1·567	1·535	1·515	1·499	1·488	1·475	1·469
0·80000	(7·85)	3·475	2·725	1·857	1·379	*	1·587	1·554	1·531	1·502	1·483	1·469	1·460	1·449	1·443
1·00000	(9·81)	3·108	2·437	1·661	*	*	1·556	1·525	1·504	1·477	1·460	1·447	1·439	1·429	1·425
1·50000	(14·7)	2·538	1·990	1·356	*	1·545	1·502	1·476	1·457	1·434	1·420	1·409	1·402	1·394	1·391
2·00000	(19·6)	2·198	1·723	*	1·535	1·505	1·467	1·442	1·426	1·405	1·392	1·383	1·377	1·371	1·369
3·00000	(29·4)	1·794	1·407	*	1·479	1·453	1·419	1·398	1·384	1·366	1·355	1·348	1·343	1·339	1·338
S	$(\Delta p/\rho l)$	125	150	200	250	300	400	500	600	800	1000	1250	1500	2000	2500

Grad'nt   k.p.g.   (Equivalent) Pipe diameters in mm

Kinematic viscosity, $\nu = 1.5 \times 10^{-3}$ m^{2}s^{-1} ; **Roughness size, $k_s = 0.015$ mm**

Kin. visc., $\nu = 1{\cdot}5 \times 10^{-3}\ \mathrm{m^2 s^{-1}}$;
S = 0·00010 to 3·00000

i.e. kin. pr. grad., $\Delta p/\rho l =$
(0·00098) to (29·4) ms^{-2}

Roughness size, k_s = 0·030 mm
This table shows values of m, as follows

m_C for Colebrook-White solutions; or,
where $\mathbf{R} \le 2000$, m_P for laminar flow

Grad'nt S	k.p.g. $(\Delta p/\rho l)$	125	150	200	250	300	400	500	600	800	1000	1250	1500	2000	2500
0·00010	(0·00098)	310·8	243·7	166·1	123·3	96·72	65·91	48·95	38·38	26·16	19·42	14·43	11·31	7·709	5·725
0·00015	(0·00147)	253·8	199·0	135·6	100·7	78·97	53·81	39·96	31·34	21·36	15·86	11·78	9·237	6·294	4·674
0·00020	(0·00196)	219·8	172·3	117·4	87·21	68·39	46·60	34·61	27·14	18·49	13·74	10·20	7·999	5·451	4·048
0·00030	(0·00294)	179·4	140·7	95·88	71·21	55·84	38·05	28·26	22·16	15·10	11·21	8·329	6·531	4·451	3·305
0·00040	(0·00392)	155·4	121·9	83·04	61·67	48·36	32·95	24·47	19·19	13·08	9·712	7·213	5·656	3·854	2·862
0·00060	(0·00588)	126·9	99·50	67·80	50·35	39·49	26·91	19·98	15·67	10·68	7·930	5·889	4·618	3·147	2·337
0·00080	(0·00785)	109·9	86·17	58·72	43·61	34·20	23·30	17·31	13·57	9·247	6·868	5·100	4·000	2·725	2·024
0·00100	(0·00981)	98·28	77·07	52·52	39·00	30·59	20·84	15·48	12·14	8·271	6·143	4·562	3·577	2·438	*
0·00150	(0·0147)	80·25	62·93	42·88	31·85	24·97	17·02	12·64	9·911	6·753	5·015	3·725	2·921	1·990	*
0·00200	(0·0196)	69·49	54·50	37·14	27·58	21·63	14·74	10·94	8·583	5·849	4·343	3·226	2·530	*	2·252
0·00300	(0·0294)	56·74	44·50	30·32	22·52	17·66	12·03	8·936	7·008	4·775	3·546	2·634	2·065	*	2·170
0·00400	(0·0392)	49·14	38·54	26·26	19·50	15·29	10·42	7·739	6·069	4·136	3·071	2·281	1·789	2·165	2·115
0·00600	(0·0588)	40·12	31·46	21·44	15·92	12·49	8·509	6·319	4·955	3·377	2·508	1·862	*	2·086	2·042
0·00800	(0·0785)	34·75	27·25	18·57	13·79	10·81	7·369	5·472	4·291	2·924	2·172	*	*	2·034	1·993
0·01000	(0·0981)	31·08	24·37	16·61	12·33	9·672	6·591	4·895	3·838	2·616	1·942	*	2·054	1·995	1·957
0·01500	(0·147)	25·38	19·90	13·56	10·07	7·897	5·381	3·996	3·134	2·136	*	*	1·979	1·927	1·895
0·02000	(0·196)	21·98	17·23	11·74	8·721	6·839	4·660	3·461	2·714	1·849	*	1·966	1·929	1·882	1·853
0·03000	(0·294)	17·94	14·07	9·588	7·121	5·584	3·805	2·826	2·216	*	*	1·895	1·864	1·822	1·797
0·04000	(0·392)	15·54	12·19	8·304	6·167	4·836	3·295	2·447	1·919	*	1·890	1·848	1·819	1·782	1·759
0·06000	(0·588)	12·69	9·950	6·780	5·035	3·949	2·691	1·998	1·567	*	1·822	1·786	1·761	1·728	1·709
0·08000	(0·785)	10·99	8·617	5·872	4·361	3·420	2·330	1·731	*	1·817	1·777	1·744	1·721	1·692	1·675
0·10000	(0·981)	9·828	7·707	5·252	3·900	3·059	2·084	1·548	*	1·781	1·744	1·713	1·692	1·665	1·649
0·15000	(1·47)	8·025	6·293	4·288	3·185	2·497	1·702	*	1·770	1·719	1·686	1·660	1·641	1·618	1·605
0·20000	(1·96)	6·949	5·450	3·714	2·758	2·163	1·474	*	1·724	1·677	1·648	1·623	1·607	1·586	1·575
0·30000	(2·94)	5·674	4·450	3·032	2·252	1·766	*	1·694	1·663	1·622	1·596	1·575	1·561	1·544	1·534
0·40000	(3·92)	4·914	3·854	2·626	1·950	1·529	*	1·651	1·622	1·585	1·562	1·543	1·530	1·515	1·507
0·60000	(5·88)	4·012	3·146	2·144	1·592	*	1·629	1·593	1·568	1·536	1·516	1·499	1·489	1·476	1·470
0·80000	(7·85)	3·475	2·725	1·857	1·379	*	1·587	1·554	1·532	1·502	1·484	1·470	1·461	1·450	1·445
1·00000	(9·81)	3·108	2·437	1·661	*	*	1·556	1·526	1·505	1·478	1·461	1·448	1·440	1·430	1·426
1·50000	(14·7)	2·538	1·990	1·356	*	1·546	1·503	1·476	1·458	1·435	1·421	1·410	1·403	1·396	1·393
2·00000	(19·6)	2·198	1·723	*	1·536	1·506	1·468	1·443	1·427	1·406	1·394	1·384	1·378	1·372	1·370
3·00000	(29·4)	1·794	1·407	*	1·480	1·454	1·420	1·399	1·385	1·367	1·357	1·349	1·345	1·341	1·340
S	$(\Delta p/\rho l)$	125	150	200	250	300	400	500	600	800	1000	1250	1500	2000	2500

Grad'nt   k.p.g.   (Equivalent) Pipe diameters in mm   Roughness size, k_s = 0·030 mm

Kin. visc., $\nu = 1{\cdot}5 \times 10^{-3}\ \mathrm{m^2 s^{-1}}$;
S = 0·00010 to 3·00000

i.e. kin. pr. grad., $\Delta p/\rho l =$
(0·00098) to (29·4) ms^{-2}

Roughness size, k_s = 0·060 mm
This table shows values of m, as follows

m_C for Colebrook-White solutions; or,
where $\mathbf{R} \le 2000$, m_P for laminar flow

Grad'nt S	k.p.g. $(\Delta p/\rho l)$	125	150	200	250	300	400	500	600	800	1000	1250	1500	2000	2500
0·00010	(0·00098)	310·8	243·7	166·1	123·3	96·72	65·91	48·95	38·38	26·16	19·42	14·43	11·31	7·709	5·725
0·00015	(0·00147)	253·8	199·0	135·6	100·7	78·97	53·81	39·96	31·34	21·36	15·86	11·78	9·237	6·294	4·674
0·00020	(0·00196)	219·8	172·3	117·4	87·21	68·39	46·60	34·61	27·14	18·49	13·74	10·20	7·999	5·451	4·048
0·00030	(0·00294)	179·4	140·7	95·88	71·21	55·84	38·05	28·26	22·16	15·10	11·21	8·329	6·531	4·451	3·305
0·00040	(0·00392)	155·4	121·9	83·04	61·67	48·36	32·95	24·47	19·19	13·08	9·712	7·213	5·656	3·854	2·862
0·00060	(0·00588)	126·9	99·50	67·80	50·35	39·49	26·91	19·98	15·67	10·68	7·930	5·889	4·618	3·147	2·337
0·00080	(0·00785)	109·9	86·17	58·72	43·61	34·20	23·30	17·31	13·57	9·247	6·868	5·100	4·000	2·725	2·024
0·00100	(0·00981)	98·28	77·07	52·52	39·00	30·59	20·84	15·48	12·14	8·271	6·143	4·562	3·577	2·438	*
0·00150	(0·0147)	80·25	62·93	42·88	31·85	24·97	17·02	12·64	9·911	6·753	5·015	3·725	2·921	1·990	*
0·00200	(0·0196)	69·49	54·50	37·14	27·58	21·63	14·74	10·94	8·583	5·849	4·343	3·226	2·530	*	2·252
0·00300	(0·0294)	56·74	44·50	30·32	22·52	17·66	12·03	8·936	7·008	4·775	3·546	2·634	2·065	*	2·170
0·00400	(0·0392)	49·14	38·54	26·26	19·50	15·29	10·42	7·739	6·069	4·136	3·071	2·281	1·789	2·166	2·115
0·00600	(0·0588)	40·12	31·46	21·44	15·92	12·49	8·509	6·319	4·955	3·377	2·508	1·862	*	2·087	2·043
0·00800	(0·0785)	34·75	27·25	18·57	13·79	10·81	7·369	5·472	4·291	2·924	2·172	*	*	2·034	1·994
0·01000	(0·0981)	31·08	24·37	16·61	12·33	9·672	6·591	4·895	3·838	2·616	1·942	*	2·054	1·995	1·958
0·01500	(0·147)	25·38	19·90	13·56	10·07	7·897	5·381	3·996	3·134	2·136	*	*	1·980	1·928	1·895
0·02000	(0·196)	21·98	17·23	11·74	8·721	6·839	4·660	3·461	2·714	1·849	*	1·966	1·930	1·883	1·854
0·03000	(0·294)	17·94	14·07	9·588	7·121	5·584	3·805	2·826	2·216	*	*	1·896	1·864	1·823	1·798
0·04000	(0·392)	15·54	12·19	8·304	6·167	4·836	3·295	2·447	1·919	*	1·891	1·849	1·820	1·783	1·760
0·06000	(0·588)	12·69	9·950	6·780	5·035	3·949	2·691	1·998	1·567	*	1·823	1·787	1·762	1·729	1·710
0·08000	(0·785)	10·99	8·617	5·872	4·361	3·420	2·330	1·731	*	1·818	1·778	1·745	1·722	1·693	1·676
0·10000	(0·981)	9·828	7·707	5·252	3·900	3·059	2·084	1·548	*	1·782	1·745	1·714	1·693	1·666	1·650
0·15000	(1·47)	8·025	6·293	4·288	3·185	2·497	1·702	*	1·771	1·720	1·687	1·661	1·642	1·619	1·606
0·20000	(1·96)	6·949	5·450	3·714	2·758	2·163	1·474	*	1·725	1·678	1·649	1·625	1·608	1·588	1·576
0·30000	(2·94)	5·674	4·450	3·032	2·252	1·766	*	1·695	1·664	1·623	1·598	1·577	1·563	1·545	1·536
0·40000	(3·92)	4·914	3·854	2·626	1·950	1·529	*	1·652	1·623	1·586	1·563	1·544	1·532	1·517	1·508
0·60000	(5·88)	4·012	3·146	2·144	1·592	*	1·631	1·594	1·569	1·537	1·517	1·501	1·491	1·478	1·472
0·80000	(7·85)	3·475	2·725	1·857	1·379	*	1·589	1·556	1·533	1·504	1·486	1·472	1·463	1·452	1·447
1·00000	(9·81)	3·108	2·437	1·661	*	*	1·558	1·528	1·506	1·479	1·463	1·450	1·442	1·433	1·428
1·50000	(14·7)	2·538	1·990	1·356	*	1·548	1·505	1·478	1·460	1·437	1·423	1·412	1·406	1·399	1·396
2·00000	(19·6)	2·198	1·723	*	1·538	1·508	1·470	1·446	1·429	1·408	1·396	1·387	1·381	1·375	1·374
3·00000	(29·4)	1·794	1·407	*	1·482	1·456	1·423	1·402	1·388	1·370	1·360	1·352	1·348	1·344	1·344
S	$(\Delta p/\rho l)$	125	150	200	250	300	400	500	600	800	1000	1250	1500	2000	2500

Grad'nt   k.p.g.   (Equivalent) Pipe diameters in mm   Roughness size, k_s = 0·060 mm

E289

Kin. visc., $\nu = 2.0 \times 10^{-3}\ m^2 s^{-1}$;
$S = 0.00010$ to 30.0000

i.e. kin. pr. grad., $\Delta p / \rho l =$
(0.00098) to (294) ms^{-2}

Roughness size, $k_s = 0.015$ mm
This table shows values of m, as follows

m_C for Colebrook-White solutions; or,
where $\mathbf{R} \le 2000$, m_P for laminar flow

Grad'nt S	k.p.g. $(\Delta p/\rho l)$	6	8	10	12.5	15	20	25	30	40	50	60	80	100	125
0.00030	(0.00294)	13715	9346	6941	5154	4042	2754	2046	1604	1093	811.8	636.6	433.8	322.2	239.2
0.00040	(0.00392)	11877	8093	6011	4464	3501	2385	1771	1389	946.6	703.0	551.3	375.7	279.0	207.2
0.00060	(0.00588)	9698	6608	4908	3645	2858	1948	1446	1134	772.9	574.0	450.1	306.7	227.8	169.2
0.00080	(0.00785)	8399	5723	4250	3156	2475	1687	1253	982.3	669.4	497.1	389.8	265.6	197.3	146.5
0.00100	(0.00981)	7512	5119	3801	2823	2214	1509	1120	878.6	598.7	444.6	348.7	237.6	176.4	131.0
0.00150	(0.0147)	6133	4179	3104	2305	1808	1232	914.8	717.4	488.8	363.0	284.7	194.0	144.1	107.0
0.00200	(0.0196)	5312	3620	2688	1996	1565	1067	792.2	621.3	423.3	314.4	246.5	168.0	124.8	92.66
0.00300	(0.0294)	4337	2955	2195	1630	1278	871.0	646.9	507.3	345.7	256.7	201.3	137.2	101.9	75.66
0.00400	(0.0392)	3756	2559	1901	1412	1107	754.2	560.2	439.3	299.3	222.3	174.3	118.8	88.22	65.52
0.00600	(0.0588)	3067	2090	1552	1153	903.8	615.9	457.4	358.7	244.4	181.5	142.3	97.00	72.04	53.50
0.00800	(0.0785)	2656	1810	1344	998.1	782.7	533.4	396.1	310.6	211.7	157.2	123.3	84.00	62.38	46.33
0.01000	(0.0981)	2375	1619	1202	892.8	700.1	477.1	354.3	277.8	189.3	140.6	110.3	75.13	55.80	41.44
0.01500	(0.147)	1940	1322	981.5	728.9	571.6	389.5	289.3	226.9	154.6	114.8	90.03	61.35	45.56	33.83
0.02000	(0.196)	1680	1145	850.0	631.3	495.1	337.3	250.5	196.5	133.9	99.42	77.97	53.13	39.46	29.30
0.03000	(0.294)	1371	934.6	694.1	515.4	404.2	275.4	204.6	160.4	109.3	81.18	63.66	43.38	32.22	23.92
0.04000	(0.392)	1188	809.3	601.1	446.4	350.1	238.5	177.1	138.9	94.66	70.30	55.13	37.57	27.90	20.72
0.06000	(0.588)	969.8	660.8	490.8	364.5	285.8	194.8	144.6	113.4	77.29	57.40	45.01	30.67	22.78	16.92
0.08000	(0.785)	839.9	572.3	425.0	315.6	247.5	168.7	125.3	98.23	66.94	49.71	38.98	26.56	19.73	14.65
0.10000	(0.981)	751.2	511.9	380.1	282.3	221.4	150.9	112.0	87.86	59.87	44.46	34.87	23.76	17.64	13.10
0.15000	(1.47)	613.3	417.9	310.4	230.5	180.8	123.2	91.48	71.74	48.88	36.30	28.47	19.40	14.41	10.70
0.20000	(1.96)	531.2	362.0	268.8	199.6	156.5	106.7	79.22	62.13	42.33	31.44	24.65	16.80	12.48	9.266
0.30000	(2.94)	433.7	295.5	219.5	163.0	127.8	87.10	64.69	50.73	34.57	25.67	20.13	13.72	10.19	7.566
0.40000	(3.92)	375.6	255.9	190.1	141.2	110.7	75.43	56.02	43.93	29.93	22.23	17.43	11.88	8.822	6.552
0.60000	(5.88)	306.7	209.0	155.2	115.3	90.38	61.59	45.74	35.87	24.44	18.15	14.23	9.700	7.204	5.350
0.80000	(7.85)	265.6	181.0	134.4	99.81	78.27	53.34	39.61	31.06	21.17	15.72	12.33	8.400	6.238	4.633
1.00000	(9.81)	237.5	161.9	120.2	89.28	70.01	47.71	35.43	27.78	18.93	14.06	11.03	7.513	5.580	4.144
1.50000	(14.7)	194.0	132.2	98.15	72.89	57.16	38.95	28.93	22.69	15.46	11.48	9.003	6.135	4.556	3.383
2.00000	(19.6)	168.0	114.5	85.00	63.13	49.51	33.73	25.05	19.65	13.39	9.942	7.797	5.313	3.946	2.930
3.00000	(29.4)	137.1	93.46	69.41	51.54	40.42	27.54	20.46	16.04	10.93	8.118	6.366	4.338	3.222	2.392
4.00000	(39.2)	118.8	80.93	60.11	44.64	35.01	23.85	17.71	13.89	9.466	7.030	5.513	3.757	2.790	2.072
6.00000	(58.8)	96.98	66.08	49.08	36.45	28.58	19.48	14.46	11.34	7.729	5.740	4.501	3.067	2.278	1.692
8.00000	(78.5)	83.99	57.23	42.50	31.56	24.75	16.87	12.53	9.823	6.694	4.971	3.898	2.656	1.973	1.465
10.0000	(98.1)	75.12	51.19	38.01	28.23	22.14	15.09	11.20	8.786	5.987	4.446	3.487	2.376	1.764	1.310
15.0000	(147)	61.33	41.79	31.04	23.05	18.08	12.32	9.148	7.174	4.888	3.630	2.847	1.940	1.441	*
20.0000	(196)	53.12	36.20	26.88	19.96	15.65	10.67	7.922	6.213	4.233	3.144	2.465	1.680	1.248	*
30.0000	(294)	43.37	29.55	21.95	16.30	12.78	8.710	6.469	5.073	3.457	2.567	2.013	1.372	*	1.362
S	$(\Delta p/\rho l)$	6	8	10	12.5	15	20	25	30	40	50	60	80	100	125

Grad'nt k.p.g. **(Equivalent) Pipe diameters in mm**

Grad'nt S	k.p.g. $(\Delta p/\rho l)$	125	150	200	250	300	400	500	600	800	1000	1250	1500	2000	2500
0.00010	(0.00098)	414.4	325.0	221.4	164.4	129.0	87.88	65.26	51.18	34.87	25.90	19.23	15.08	10.28	7.633
0.00015	(0.00147)	338.3	265.3	180.8	134.3	105.3	71.75	53.29	41.79	28.47	21.15	15.70	12.32	8.392	6.232
0.00020	(0.00196)	293.0	229.8	156.6	116.3	91.19	62.14	46.15	36.19	24.66	18.31	13.60	10.67	7.268	5.397
0.00030	(0.00294)	239.2	187.6	127.8	94.94	74.46	50.74	37.68	29.55	20.13	14.95	11.10	8.708	5.934	4.407
0.00040	(0.00392)	207.2	162.5	110.7	82.22	64.48	43.94	32.63	25.59	17.44	12.95	9.617	7.542	5.139	3.817
0.00060	(0.00588)	169.2	132.7	90.40	67.14	52.65	35.88	26.64	20.89	14.24	10.57	7.852	6.158	4.196	3.116
0.00080	(0.00785)	146.5	114.9	78.29	58.14	45.59	31.07	23.07	18.09	12.33	9.157	6.800	5.333	3.634	2.699
0.00100	(0.00981)	131.0	102.8	70.02	52.00	40.78	27.79	20.64	16.18	11.03	8.190	6.082	4.770	3.250	2.414
0.00150	(0.0147)	107.0	83.90	57.17	42.46	33.30	22.69	16.85	13.21	9.004	6.687	4.966	3.895	2.654	1.971
0.00200	(0.0196)	92.66	72.66	49.51	36.77	28.84	19.65	14.59	11.44	7.798	5.791	4.301	3.373	2.298	*
0.00300	(0.0294)	75.66	59.33	40.43	30.02	23.54	16.04	11.92	9.344	6.367	4.729	3.512	2.754	1.877	*
0.00400	(0.0392)	65.52	51.38	35.01	26.00	20.39	13.89	10.32	8.092	5.514	4.095	3.041	2.385	*	2.227
0.00600	(0.0588)	53.50	41.95	28.59	21.23	16.65	11.34	8.425	6.607	4.502	3.344	2.483	1.947	*	2.147
0.00800	(0.0785)	46.33	36.33	24.76	18.39	14.42	9.825	7.296	5.722	3.899	2.896	2.150	*	2.141	2.093
0.01000	(0.0981)	41.44	32.50	22.14	16.44	12.90	8.788	6.526	5.118	3.487	2.590	1.923	*	2.098	2.053
0.01500	(0.147)	33.83	26.53	18.08	13.43	10.53	7.175	5.329	4.179	2.847	2.115	*	*	2.024	1.985
0.02000	(0.196)	29.30	22.98	15.66	11.63	9.119	6.214	4.615	3.619	2.466	1.831	*	2.031	1.974	1.938
0.03000	(0.294)	23.92	18.76	12.78	9.494	7.446	5.074	3.768	2.955	2.013	*	1.996	1.958	1.908	1.877
0.04000	(0.392)	20.72	16.25	11.07	8.222	6.448	4.394	3.263	2.559	1.744	*	1.944	1.910	1.864	1.836
0.06000	(0.588)	16.92	13.27	9.040	6.714	5.265	3.588	2.664	2.089	*	1.919	1.875	1.845	1.805	1.781
0.08000	(0.785)	14.65	11.49	7.829	5.814	4.559	3.107	2.307	1.809	*	1.869	1.829	1.802	1.766	1.744
0.10000	(0.981)	13.10	10.28	7.002	5.200	4.078	2.779	2.064	1.618	*	1.832	1.795	1.769	1.736	1.716
0.15000	(1.47)	10.70	8.390	5.717	4.246	3.330	2.269	1.685	*	1.808	1.769	1.736	1.714	1.685	1.668
0.20000	(1.96)	9.266	7.266	4.951	3.677	2.884	1.965	1.459	*	1.762	1.726	1.697	1.676	1.651	1.635
0.30000	(2.94)	7.566	5.933	4.043	3.002	2.354	1.604	*	1.750	1.701	1.670	1.644	1.627	1.604	1.592
0.40000	(3.92)	6.552	5.138	3.501	2.600	2.039	1.389	*	1.705	1.660	1.632	1.609	1.593	1.573	1.562
0.60000	(5.88)	5.350	4.195	2.859	2.123	1.665	*	1.676	1.645	1.606	1.581	1.561	1.548	1.531	1.522
0.80000	(7.85)	4.633	3.633	2.476	1.839	1.442	*	1.633	1.605	1.569	1.547	1.529	1.517	1.503	1.495
1.00000	(9.81)	4.144	3.250	2.214	1.644	*	1.638	1.601	1.576	1.542	1.522	1.505	1.494	1.481	1.474
1.50000	(14.7)	3.383	2.653	1.808	1.343	*	1.579	1.547	1.524	1.496	1.478	1.464	1.455	1.444	1.439
2.00000	(19.6)	2.930	2.298	1.566	*	*	1.540	1.510	1.490	1.464	1.448	1.436	1.428	1.419	1.415
3.00000	(29.4)	2.392	1.876	1.278	*	1.528	1.487	1.462	1.444	1.422	1.408	1.398	1.391	1.385	1.382
S	$(\Delta p/\rho l)$	125	150	200	250	300	400	500	600	800	1000	1250	1500	2000	2500

Grad'nt k.p.g. **(Equivalent) Pipe diameters in mm**

Kinematic viscosity, $\nu = 2.0 \times 10^{-3}\ m^2 s^{-1}$;   **Roughness size, $k_s = 0.015$ mm**

E290

Kin. visc., $\nu = 2{\cdot}0\times10^{-3}$ m^{2}s^{-1}; $S = 0{\cdot}00010$ to $3{\cdot}00000$

i.e. kin. pr. grad., $\Delta p/\rho l =$ ($0{\cdot}00098$) to ($29{\cdot}4$) ms^{-2}

Roughness size, $k_s = 0{\cdot}030$ mm
This table shows values of m, as follows
m_C for Colebrook-White solutions; or,
where $\mathbf{R} \le 2000$, m_P for laminar flow

Grad'nt S	k.p.g. $(\Delta p/\rho l)$	125	150	200	250	300	400	500	600	800	1000	1250	1500	2000	2500
0·00010	(0·00098)	414·4	325·0	221·4	164·4	129·0	87·88	65·26	51·18	34·87	25·90	19·23	15·08	10·28	7·633
0·00015	(0·00147)	338·3	265·3	180·8	134·3	105·3	71·75	53·29	41·79	28·47	21·15	15·70	12·32	8·392	6·232
0·00020	(0·00196)	293·0	229·8	156·6	116·3	91·19	62·14	46·15	36·19	24·66	18·31	13·60	10·67	7·268	5·397
0·00030	(0·00294)	239·2	187·6	127·8	94·94	74·46	50·74	37·68	29·55	20·13	14·95	11·10	8·708	5·934	4·407
0·00040	(0·00392)	207·2	162·5	110·7	82·22	64·48	43·94	32·63	25·59	17·44	12·95	9·617	7·542	5·139	3·817
0·00060	(0·00588)	169·2	132·7	90·40	67·14	52·65	35·88	26·64	20·89	14·24	10·57	7·852	6·158	4·196	3·116
0·00080	(0·00785)	146·5	114·9	78·29	58·14	45·59	31·07	23·07	18·09	12·33	9·157	6·800	5·333	3·634	2·699
0·00100	(0·00981)	131·0	102·8	70·02	52·00	40·78	27·79	20·64	16·18	11·03	8·190	6·082	4·770	3·250	2·414
0·00150	(0·0147)	107·0	83·90	57·17	42·46	33·30	22·69	16·85	13·21	9·004	6·687	4·966	3·895	2·654	1·971
0·00200	(0·0196)	92·66	72·66	49·51	36·77	28·84	19·65	14·59	11·44	7·798	5·791	4·301	3·373	2·298	*
0·00300	(0·0294)	75·66	59·33	40·43	30·02	23·54	16·04	11·92	9·344	6·367	4·729	3·512	2·754	1·877	*
0·00400	(0·0392)	65·52	51·38	35·01	26·00	20·39	13·89	10·32	8·092	5·514	4·095	3·041	2·385	*	2·228
0·00600	(0·0588)	53·50	41·95	28·59	21·23	16·65	11·34	8·425	6·607	4·502	3·344	2·483	1·947	*	2·147
0·00800	(0·0785)	46·33	36·33	24·76	18·39	14·42	9·825	7·296	5·722	3·899	2·896	2·150	*	2·142	2·093
0·01000	(0·0981)	41·44	32·50	22·14	16·44	12·90	8·788	6·526	5·118	3·487	2·590	1·923	*	2·098	2·053
0·01500	(0·147)	33·83	26·53	18·08	13·43	10·53	7·175	5·329	4·179	2·847	2·115	*	*	2·024	1·985
0·02000	(0·196)	29·30	22·98	15·66	11·63	9·119	6·214	4·615	3·619	2·466	1·831	*	2·031	1·975	1·939
0·03000	(0·294)	23·92	18·76	12·78	9·494	7·446	5·074	3·768	2·955	2·013	*	1·997	1·959	1·909	1·877
0·04000	(0·392)	20·72	16·25	11·07	8·222	6·448	4·394	3·263	2·559	1·744	*	1·945	1·910	1·865	1·836
0·06000	(0·588)	16·92	13·27	9·040	6·714	5·265	3·588	2·664	2·089	*	1·920	1·876	1·845	1·806	1·781
0·08000	(0·785)	14·65	11·49	7·829	5·814	4·559	3·107	2·307	1·809	*	1·870	1·830	1·802	1·766	1·744
0·10000	(0·981)	13·10	10·28	7·002	5·200	4·078	2·779	2·064	1·618	*	1·833	1·796	1·770	1·737	1·716
0·15000	(1·47)	10·70	8·390	5·717	4·246	3·330	2·269	1·685	*	1·808	1·769	1·737	1·714	1·686	1·668
0·20000	(1·96)	9·266	7·266	4·951	3·677	2·884	1·965	1·459	*	1·762	1·727	1·697	1·677	1·651	1·636
0·30000	(2·94)	7·566	5·933	4·043	3·002	2·354	1·604	*	1·751	1·701	1·670	1·645	1·627	1·605	1·592
0·40000	(3·92)	6·552	5·138	3·501	2·600	2·039	1·389	*	1·706	1·661	1·632	1·609	1·593	1·574	1·563
0·60000	(5·88)	5·350	4·195	2·859	2·123	1·665	*	1·676	1·646	1·606	1·582	1·562	1·548	1·532	1·523
0·80000	(7·85)	4·633	3·633	2·476	1·839	1·442	*	1·633	1·606	1·570	1·548	1·530	1·518	1·504	1·496
1·00000	(9·81)	4·144	3·250	2·214	1·644	*	1·639	1·602	1·576	1·543	1·523	1·506	1·495	1·482	1·475
1·50000	(14·7)	3·383	2·653	1·808	1·343	*	1·580	1·548	1·525	1·496	1·479	1·465	1·456	1·445	1·440
2·00000	(19·6)	2·930	2·298	1·566	*	*	1·541	1·511	1·491	1·465	1·449	1·437	1·429	1·420	1·416
3·00000	(29·4)	2·392	1·876	1·278	*	1·529	1·488	1·463	1·445	1·423	1·409	1·399	1·393	1·386	1·384
S	$(\Delta p/\rho l)$	125	150	200	250	300	400	500	600	800	1000	1250	1500	2000	2500

Grad'nt **k.p.g.** (Equivalent) Pipe diameters in mm — **Roughness size, $k_s = 0{\cdot}030$ mm**

E291

Kin. visc., $\nu = 2{\cdot}0\times10^{-3}$ m^{2}s^{-1}; $S = 0{\cdot}00010$ to $3{\cdot}00000$

i.e. kin. pr. grad., $\Delta p/\rho l =$ ($0{\cdot}00098$) to ($29{\cdot}4$) ms^{-2}

Roughness size, $k_s = 0{\cdot}060$ mm
This table shows values of m, as follows
m_C for Colebrook-White solutions; or,
where $\mathbf{R} \le 2000$, m_P for laminar flow

Grad'nt S	k.p.g. $(\Delta p/\rho l)$	125	150	200	250	300	400	500	600	800	1000	1250	1500	2000	2500
0·00010	(0·00098)	414·4	325·0	221·4	164·4	129·0	87·88	65·26	51·18	34·87	25·90	19·23	15·08	10·28	7·633
0·00015	(0·00147)	338·3	265·3	180·8	134·3	105·3	71·75	53·29	41·79	28·47	21·15	15·70	12·32	8·392	6·232
0·00020	(0·00196)	293·0	229·8	156·6	116·3	91·19	62·14	46·15	36·19	24·66	18·31	13·60	10·67	7·268	5·397
0·00030	(0·00294)	239·2	187·6	127·8	94·94	74·46	50·74	37·68	29·55	20·13	14·95	11·10	8·708	5·934	4·407
0·00040	(0·00392)	207·2	162·5	110·7	82·22	64·48	43·94	32·63	25·59	17·44	12·95	9·617	7·542	5·139	3·817
0·00060	(0·00588)	169·2	132·7	90·40	67·14	52·65	35·88	26·64	20·89	14·24	10·57	7·852	6·158	4·196	3·116
0·00080	(0·00785)	146·5	114·9	78·29	58·14	45·59	31·07	23·07	18·09	12·33	9·157	6·800	5·333	3·634	2·699
0·00100	(0·00981)	131·0	102·8	70·02	52·00	40·78	27·79	20·64	16·18	11·03	8·190	6·082	4·770	3·250	2·414
0·00150	(0·0147)	107·0	83·90	57·17	42·46	33·30	22·69	16·85	13·21	9·004	6·687	4·966	3·895	2·654	1·971
0·00200	(0·0196)	92·66	72·66	49·51	36·77	28·84	19·65	14·59	11·44	7·798	5·791	4·301	3·373	2·298	*
0·00300	(0·0294)	75·66	59·33	40·43	30·02	23·54	16·04	11·92	9·344	6·367	4·729	3·512	2·754	1·877	*
0·00400	(0·0392)	65·52	51·38	35·01	26·00	20·39	13·89	10·32	8·092	5·514	4·095	3·041	2·385	*	2·228
0·00600	(0·0588)	53·50	41·95	28·59	21·23	16·65	11·34	8·425	6·607	4·502	3·344	2·483	1·947	*	2·147
0·00800	(0·0785)	46·33	36·33	24·76	18·39	14·42	9·825	7·296	5·722	3·899	2·896	2·150	*	2·142	2·094
0·01000	(0·0981)	41·44	32·50	22·14	16·44	12·90	8·788	6·526	5·118	3·487	2·590	1·923	*	2·099	2·054
0·01500	(0·147)	33·83	26·53	18·08	13·43	10·53	7·175	5·329	4·179	2·847	2·115	*	*	2·025	1·985
0·02000	(0·196)	29·30	22·98	15·66	11·63	9·119	6·214	4·615	3·619	2·466	1·831	*	2·032	1·975	1·939
0·03000	(0·294)	23·92	18·76	12·78	9·494	7·446	5·074	3·768	2·955	2·013	*	1·997	1·959	1·909	1·878
0·04000	(0·392)	20·72	16·25	11·07	8·222	6·448	4·394	3·263	2·559	1·744	*	1·945	1·910	1·865	1·837
0·06000	(0·588)	16·92	13·27	9·040	6·714	5·265	3·588	2·664	2·089	*	1·920	1·876	1·846	1·806	1·782
0·08000	(0·785)	14·65	11·49	7·829	5·814	4·559	3·107	2·307	1·809	*	1·870	1·830	1·803	1·767	1·745
0·10000	(0·981)	13·10	10·28	7·002	5·200	4·078	2·779	2·064	1·618	*	1·834	1·796	1·771	1·738	1·717
0·15000	(1·47)	10·70	8·390	5·717	4·246	3·330	2·269	1·685	*	1·809	1·770	1·738	1·715	1·687	1·669
0·20000	(1·96)	9·266	7·266	4·951	3·677	2·884	1·965	1·459	*	1·763	1·728	1·698	1·678	1·652	1·637
0·30000	(2·94)	7·566	5·933	4·043	3·002	2·354	1·604	*	1·752	1·702	1·671	1·646	1·628	1·606	1·593
0·40000	(3·92)	6·552	5·138	3·501	2·600	2·039	1·389	*	1·707	1·662	1·634	1·610	1·595	1·575	1·564
0·60000	(5·88)	5·350	4·195	2·859	2·123	1·665	*	1·677	1·647	1·608	1·583	1·563	1·550	1·534	1·524
0·80000	(7·85)	4·633	3·633	2·476	1·839	1·442	*	1·635	1·607	1·572	1·549	1·532	1·520	1·505	1·498
1·00000	(9·81)	4·144	3·250	2·214	1·644	*	1·640	1·603	1·578	1·545	1·524	1·508	1·497	1·484	1·477
1·50000	(14·7)	3·383	2·653	1·808	1·343	*	1·581	1·549	1·527	1·498	1·481	1·467	1·458	1·447	1·442
2·00000	(19·6)	2·930	2·298	1·566	*	*	1·542	1·513	1·493	1·467	1·451	1·439	1·431	1·423	1·419
3·00000	(29·4)	2·392	1·876	1·278	*	1·531	1·490	1·465	1·447	1·425	1·412	1·402	1·395	1·389	1·387
S	$(\Delta p/\rho l)$	125	150	200	250	300	400	500	600	800	1000	1250	1500	2000	2500

Grad'nt **k.p.g.** (Equivalent) Pipe diameters in mm — **Roughness size, $k_s = 0{\cdot}060$ mm**

Tables F give properties of non-circular ducts, on a unit size basis where appropriate.

F1   Rectangle

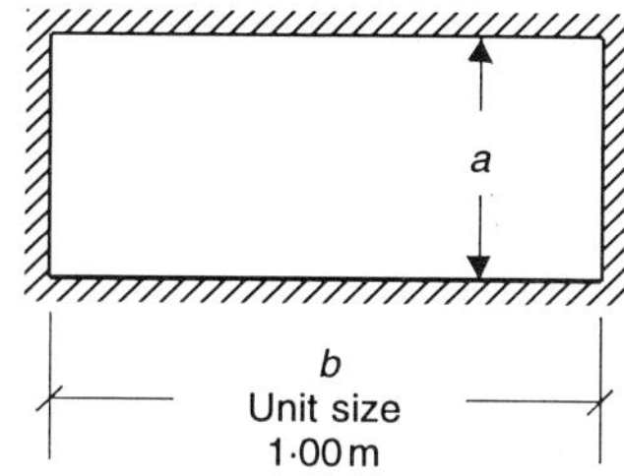

a/b	U.equ. dia. $D_{ep(u)}$ (m)	Equiv. disch. factor J	Unit sect. area A_u (m²)	Unit wetted perim. P_u (m)	Gradient factor C_L laminar	Gradient factor C_T turbulent
0·02	0·0392	0·0604	0·0200	2·0400	0·68	0·91
0·04	0·0769	0·1162	0·0400	2·0800	0·70	0·92
0·06	0·1132	0·1678	0·0600	2·1200	0·72	0·92
0·08	0·1481	0·2155	0·0800	2·1600	0·75	0·93
0·10	0·1818	0·2596	0·1000	2·2000	0·77	0·94
0·12	0·2143	0·3005	0·1200	2·2400	0·79	0·94
0·14	0·2456	0·3384	0·1400	2·2800	0·81	0·95
0·16	0·2759	0·3735	0·1600	2·3200	0·82	0·95
0·18	0·3051	0·4061	0·1800	2·3600	0·84	0·96
0·20	0·3333	0·4363	0·2000	2·4000	0·85	0·96
0·22	0·3607	0·4643	0·2200	2·4400	0·87	0·97
0·24	0·3871	0·4904	0·2400	2·4800	0·89	0·97
0·26	0·4127	0·5145	0·2600	2·5200	0·90	0·97
0·28	0·4375	0·5369	0·2800	2·5600	0·91	0·98
0·30	0·4615	0·5577	0·3000	2·6000	0·93	0·98
0·32	0·4848	0·5770	0·3200	2·6400	0·94	0·98
0·34	0·5075	0·5949	0·3400	2·6800	0·95	0·99
0·36	0·5294	0·6115	0·3600	2·7200	0·96	0·99
0·38	0·5507	0·6269	0·3800	2·7600	0·97	0·99
0·40	0·5714	0·6411	0·4000	2·8000	0·98	1·00
0·42	0·5915	0·6544	0·4200	2·8400	0·99	1·00
0·44	0·6111	0·6666	0·4400	2·8800	1·00	1·00
0·46	0·6301	0·6779	0·4600	2·9200	1·01	1·00
0·48	0·6486	0·6884	0·4800	2·9600	1·02	1·01
0·50	0·6667	0·6981	0·5000	3·0000	1·03	1·01
0·52	0·6842	0·7071	0·5200	3·0400	1·04	1·01
0·54	0·7013	0·7153	0·5400	3·0800	1·05	1·01
0·56	0·7179	0·7229	0·5600	3·1200	1·06	1·01
0·58	0·7342	0·7299	0·5800	3·1600	1·07	1·02
0·60	0·7500	0·7363	0·6000	3·2000	1·08	1·02
0·62	0·7654	0·7422	0·6200	3·2400	1·08	1·02
0·64	0·7805	0·7475	0·6400	3·2800	1·09	1·02
0·66	0·7952	0·7524	0·6600	3·3200	1·09	1·02
0·68	0·8095	0·7569	0·6800	3·3600	1·09	1·02
0·70	0·8235	0·7609	0·7000	3·4000	1·10	1·02
0·72	0·8372	0·7646	0·7200	3·4400	1·10	1·02
0·74	0·8506	0·7678	0·7400	3·4800	1·10	1·02
0·76	0·8636	0·7708	0·7600	3·5200	1·10	1·03
0·78	0·8764	0·7734	0·7800	3·5600	1·10	1·03
0·80	0·8889	0·7757	0·8000	3·6000	1·11	1·03
0·82	0·9011	0·7777	0·8200	3·6400	1·11	1·03
0·84	0·9130	0·7794	0·8400	3·6800	1·12	1·03
0·86	0·9247	0·7809	0·8600	3·7200	1·12	1·03
0·88	0·9362	0·7822	0·8800	3·7600	1·12	1·03
0·90	0·9474	0·7832	0·9000	3·8000	1·12	1·03
0·92	0·9583	0·7840	0·9200	3·8400	1·12	1·03
0·94	0·9691	0·7846	0·9400	3·8800	1·12	1·03
0·96	0·9796	0·7851	0·9600	3·9200	1·12	1·03
0·98	0·9899	0·7853	0·9800	3·9600	1·12	1·03
1·00	1·0000	0·7854	1·0000	4·0000	1·12	1·03

Isosceles triangle

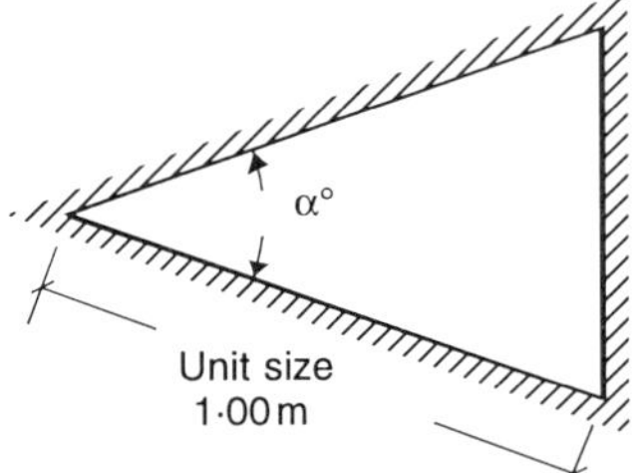

$\alpha°$	U.equ. dia. $D_{ep(u)}$ (m)	Equiv. disch. factor J	Unit sect. area A_u (m^2)	Unit wetted perim. P_u (m)	Gradient factor C_L laminar	C_T turbulent
4	0·0674	0·1023	0·0349	2·0698	1·54	1·11
6	0·0993	0·1483	0·0523	2·1047	1·48	1·10
8	0·1301	0·1910	0·0696	2·1395	1·45	1·10
10	0·1597	0·2308	0·0868	2·1743	1·41	1·09
12	0·1882	0·2677	0·1040	2·2091	1·37	1·08
14	0·2156	0·3019	0·1210	2·2437	1·35	1·08
16	0·2420	0·3336	0·1378	2·2783	1·33	1·07
18	0·2672	0·2630	0·1545	2·3129	1·31	1·07
20	0·2914	0·3900	0·1710	2·3473	1·29	1·07
22	0·3146	0·4150	0·1873	2·3816	1·28	1·06
24	0·3367	0·4379	0·2034	2·4158	1·26	1·06
30	0·3972	0·4956	0·2500	2·5176	1·23	1·05
35	0·4410	0·5325	0·2868	2·6014	1·21	1·05
40	0·4790	0·5606	0·3239	2·6840	1·18	1·04
45	0·5114	0·5810	0·3536	2·7654	1·17	1·04
50	0·5385	0·5946	0·3830	2·8452	1·16	1·04
55	0·5604	0·6022	0·4096	2·9235	1·15	1·04
60	0·5774	0·6046	0·4330	3·0000	1·13	1·03
65	0·5895	0·6024	0·4532	3·0746	-	-
70	0·5972	0·5961	0·4698	3·1472	-	-
75	0·6004	0·5862	0·4830	3·2175	-	-
80	0·5995	0·5732	0·4924	3·2856	-	-
85	0·5945	0·5574	0·4981	3·3512	-	-
90	0·5858	0·5390	0·5000	3·4142	-	-
100	0·5576	0·4960	0·4924	3·5321	-	-
110	0·5166	0·4460	0·4698	3·6383	-	-
120	0·4641	0·3907	0·4330	3·7321	-	-
130	0·4018	0·3311	0·3830	3·8126	-	-
140	0·3314	0·2687	0·3214	3·8794	-	-
150	0·2543	0·2032	0·2500	3·9319	-	-

Sector of circle

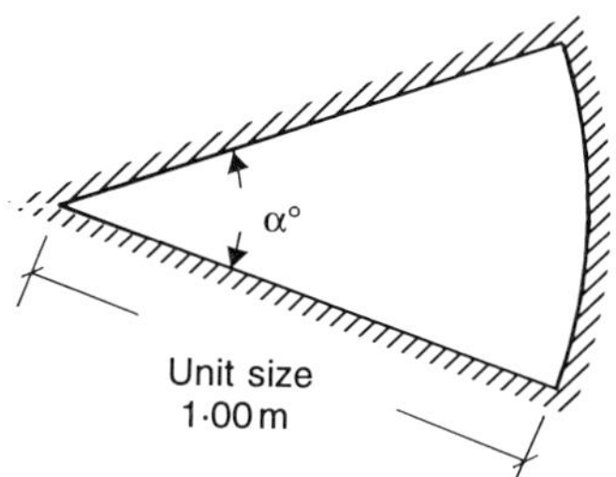

$\alpha°$	U.equ. dia. $D_{ep(u)}$ (m)	Equiv. disch. factor J	Unit sect. area A_u (m^2)	Unit wetted perim. P_u (m)	Gradient factor C_L laminar	C_T turbulent
14	0·2177	0·3048	0·1222	2·2443	-	-
16	0·2450	0·3377	0·1396	2·2793	-	-
18	0·2715	0·3686	0·1571	2·3142	-	-
20	0·2972	0·3975	0·1745	2·3491	-	-
24	0·3463	0·4498	0·2094	2·4189	-	-
30	0·4150	0·5166	0·2618	2·5236	-	-
35	0·4679	0·5631	0·3054	2·6109	-	-
40	0·5175	0·6025	0·3491	2·6981	-	-
45	0·5639	0·6361	0·3927	2·7854	-	-
50	0·6076	0·6644	0·4363	2·8727	-	-
60	0·6873	0·7086	0·5236	3·0472	-	-
70	0·7584	0·7396	0·6109	3·2217	-	-
80	0·8222	0·7606	0·6981	3·3963	-	-
90	0·8798	0·7741	0·7854	3·5708	-	-
100	0·9320	0·7818	0·8727	3·7453	-	-
110	0·9796	0·7851	0·9599	3·9199	-	-
120	1·0231	0·7850	1·0472	4·0944	-	-
140	1·0998	0·7776	1·2217	4·4435	-	-
160	1·1654	0·7639	1·3963	4·7925	-	-
180	1·2220	0·7467	1·5708	5·1416	-	-

Concentric annulus

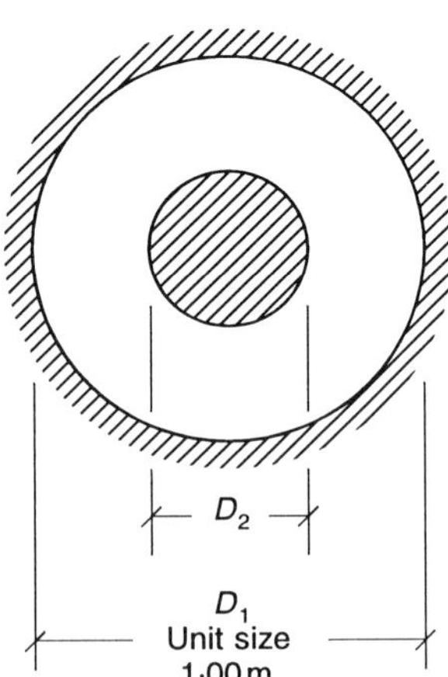

D_2/D_1	U.equ. dia. $D_{ep(u)}$ (m)	Equiv. disch. factor J	Unit sect. area A_u (m²)	Unit wetted perim. P_u (m)	Gradient factor C_L laminar	C_T turbulent
0·02	0·9800	0·9608	0·7851	3·2044	0·89	0·97
0·04	0·9600	0·9231	0·7841	3·2673	0·85	0·96
0·05	0·9500	0·9048	0·7834	3·2987	0·83	0·96
0·06	0·9400	0·8868	0·7826	3·3301	0·82	0·95
0·08	0·9200	0·8519	0·7804	3·3929	0·80	0·95
0·10	0·9000	0·8182	0·7775	3·4558	0·79	0·94
0·12	0·8800	0·7857	0·7741	3·5186	0·78	0·94
0·14	0·8600	0·7544	0·7700	3·5814	0·76	0·93
0·15	0·8500	0·7391	0·7677	3·6128	0·76	0·93
0·16	0·8400	0·7213	0·7653	3·6442	0·75	0·93
0·18	0·8200	0·6949	0·7600	3·7071	0·74	0·93
0·20	0·8000	0·6667	0·7540	3·7699	0·74	0·93
0·22	0·7800	0·6393	0·7474	3·8327	0·73	0·92
0·24	0·7600	0·6129	0·7402	3·8956	0·72	0·92
0·25	0·7500	0·6000	0·7363	3·9270	0·72	0·92
0·26	0·7400	0·5873	0·7323	3·9584	0·72	0·92
0·28	0·7200	0·5625	0·7238	4·0212	0·71	0·92
0·30	0·7000	0·5385	0·7147	4·0841	0·71	0·92
0·32	0·6800	0·5152	0·7050	4·1469	0·70	0·92
0·34	0·6600	0·4925	0·6946	4·2097	0·70	0·91
0·35	0·6500	0·4815	0·6892	4·2412	0·70	0·91
0·36	0·6400	0·4706	0·6836	4·2726	0·69	0·91
0·38	0·6200	0·4493	0·6720	4·3354	0·69	0·91
0·40	0·6000	0·4286	0·6597	4·3982	0·68	0·91
0·42	0·5800	0·4085	0·6469	4·4611	-	-
0·44	0·5600	0·3889	0·6333	4·5239	-	-
0·46	0·5400	0·3699	0·6192	4·5867	-	-
0·48	0·5200	0·3514	0·6044	4·6496	-	-
0·50	0·5000	0·3333	0·5890	4·7124	0·68	0·91
0·52	0·4800	0·3158	0·5730	4·7752	-	-
0·54	0·4600	0·2987	0·5564	4·8381	-	-
0·56	0·4400	0·2821	0·5391	4·9001	-	-
0·58	0·4200	0·2658	0·5212	4·9637	-	-
0·60	0·4000	0·2500	0·5027	5·0265	0·67	0·91
0·62	0·3800	0·2346	0·4835	5·0894	-	-
0·64	0·3600	0·2195	0·4637	5·1522	-	-
0·66	0·3400	0·2048	0·4433	5·2150	-	-
0·68	0·3200	0·1905	0·4222	5·2779	-	-
0·70	0·3000	0·1765	0·4006	5·3407	0·67	0·90
0·72	0·2800	0·1628	0·3782	5·4035	-	-
0·74	0·2600	0·1498	0·3553	5·4664	-	-
0·76	0·2400	0·1364	0·3318	5·5292	-	-
0·78	0·2200	0·1236	0·3076	5·5920	-	-
0·80	0·2000	0·1111	0·2827	5·6549	0·67	0·90
0·82	0·1800	0·0989	0·2573	5·7177	-	-
0·84	0·1600	0·0870	0·2312	5·7805	-	-
0·86	0·1400	0·0753	0·2045	5·8434	-	-
0·88	0·1200	0·0638	0·1772	5·9062	-	-
0·90	0·1000	0·0526	0·1492	5·9690	0·67	0·90
0·92	0·0800	0·0417	0·1206	6·0319	-	-
0·94	0·0600	0·0309	0·0914	6·0947	-	-
0·96	0·0400	0·0204	0·0616	6·1575	-	-
0·98	0·0200	0·0101	0·0311	6·2204	0·67	0·90

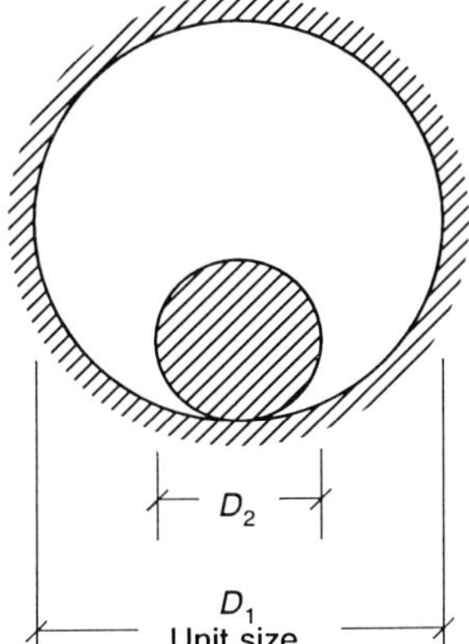

Fully eccentric annulus

F5

For Tables F5 and F6, the values of unit equivalent diameter, discharge factor, unit sectional area and unit wetted perimeter are as for concentric annulus of the same ratio of diameters, D_2/D_1. Indicative values of gradient factors, C_L and C_T are as follows.

D_2/D_1	C_L laminar	C_T turbulent	D_2/D_1	C_L laminar	C_T turbulent
0·05	1·09	1·02	0·40	1·47	1·10
0·10	1·16	1·04	0·50	1·53	1·11
0·15	1·22	1·05	0·60	1·56	1·12
0·20	1·28	1·06	0·70	1·58	1·12
0·25	1·35	1·08	0·80	1·60	1·13
0·30	1·41	1·09	0·90	1·61	1·13
0·35	1·43	1·09			

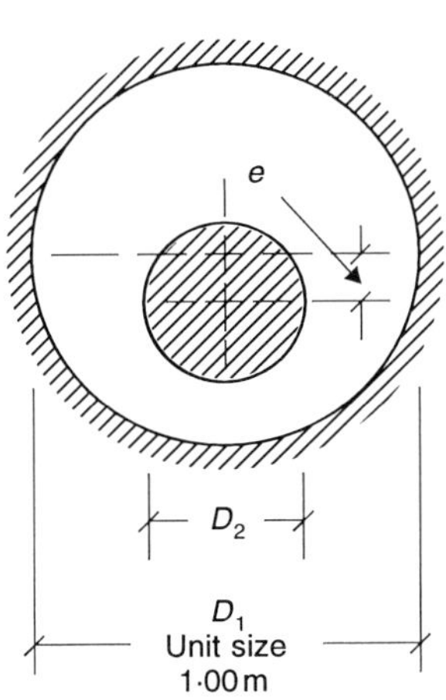

Eccentric annulus

F6

The degree of eccentricity is indicated by the ratio e/D_1.

D_2/D_1	e/D_1	C_L laminar	C_T turbulent	D_2/D_1	e/D_1	C_L laminar	C_T turbulent
0·05	0·05	0·84	0·96	0·25	0·04	0·72	0·92
0·05	0·15	0·85	0·96	0·25	0·11	0·78	0·94
0·05	0·25	0·91	0·98	0·25	0·19	0·89	0·97
0·05	0·34	0·95	0·99	0·25	0·26	1·06	1·02
0·05	0·43	1·05	1·01	0·25	0·34	1·25	1·06
0·10	0·05	0·79	0·94	0·35	0·03	0·70	0·92
0·10	0·14	0·83	0·95	0·35	0·10	0·77	0·94
0·10	0·23	0·89	0·97	0·35	0·16	0·90	0·97
0·10	0·32	0·98	1·00	0·35	0·23	1·08	1·02
0·10	0·40	1·11	1·03	0·35	0·29	1·30	1·07
0·15	0·04	0·76	0·93	0·50	0·03	0·69	0·91
0·15	0·13	0·79	0·94	0·50	0·08	0·76	0·93
0·15	0·21	0·88	0·97	0·50	0·13	0·91	0·98
0·15	0·30	1·02	1·01	0·50	0·18	1·14	1·03
0·15	0·38	1·15	1·04	0·50	0·23	1·41	1·09
0·20	0·04	0·74	0·93	0·90	0·005	0·68	0·91
0·20	0·12	0·79	0·94	0·90	0·015	0·75	0·93
0·20	0·20	0·89	0·97	0·90	0·025	0·91	0·98
0·20	0·28	1·04	1·01	0·90	0·035	1·15	1·04
0·20	0·36	1·20	1·05	0·90	0·045	1·47	1·10

F7

Tube interstices
(1·00 m dia. unit size throughout)

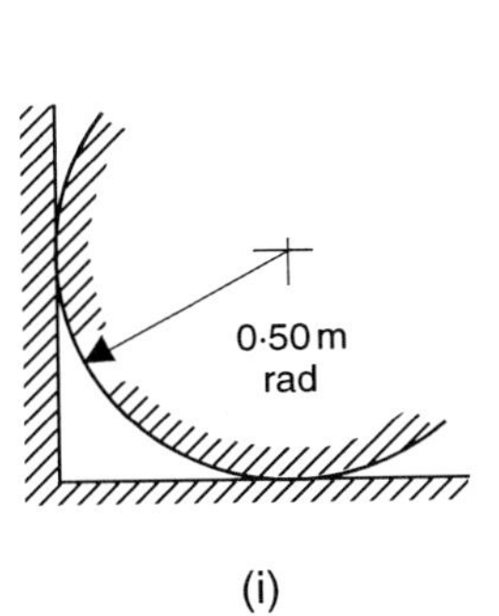

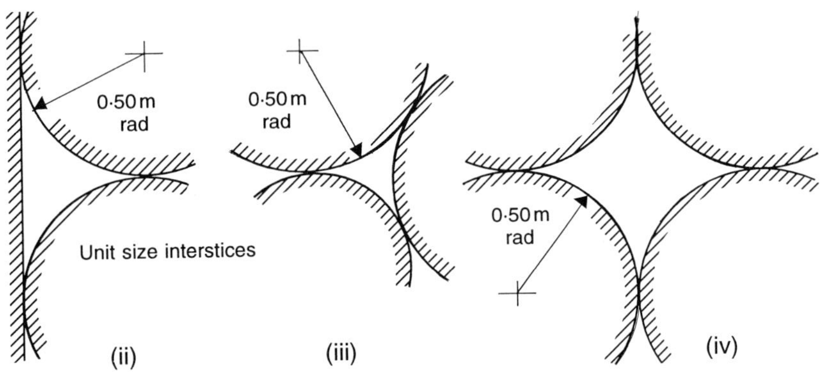

shape	U.equ. dia. $D_{ep(u)}$ (m)	Equiv. disch. factor J	Unit sect. area A_u (m^2)	Unit wetted perim. P_u (m)	Gradient factor C_L laminar	C_T turbulent
(i)	0·1202	0·2115	0·0537	1·7854	2·27	1·43
(ii)	0·1670	0·2040	0·1073	2·5708	2·46	1·52
(iii)	0·1027	0·2053	0·0403	1·5708	c. 2·4	c. 1·5
(iv)	0·2732	0·2732	0·2146	3·1416	2·46	1·43